SOLUTIONS MANUAL

Kathleen Thrush Shaginaw
Community College of Philadelphia
Particular Solutions, Inc.

CHEMISTRY

STRUCTURE AND PROPERTIES

NIVALDO J. TRO

PEARSON

Boston Columbus Indianapolis New York San Francisco Upper Saddle River
Amsterdam Cape Town Dubai London Madrid Milan Munich Paris Montréal Toronto
Delhi Mexico City São Paulo Sydney Hong Kong Seoul Singapore Taipei Tokyo

Editor in Chief: Adam Jaworski
Senior Acquisitions Editor: Terry Haugen
Executive Marketing Manager: Jonathan Cottrell
Associate Team Lead, Program Management, Chemistry and Geosciences: Jessica Moro
Team Lead, Project Management, Chemistry and Geosciences: Gina M. Cheselka
Project Manager: Beth Sweeten
Full-Service Project Management/Composition: CodeMantra
Operations Specialist: Christy Hall
Cover Art: Quade Paul

1 2 3 4 5 6 7 8 9 10—**B2V**—17 16 15 14

PEARSON

www.pearsonhighered.com

ISBN-10: 0-321-96529-9; ISBN-13: 978-0-321-96529-5

Contents

Student Guide to Using This Solutions Manual

The vision behind this solutions manual is to provide guidance that is useful for both the struggling student and the advanced student.

An important feature of this solutions manual is that answers for the review questions are given. This will help in the review of the major concepts in the chapter.

The format of the solutions follows the format in the textbook very closely. Each mathematical problem includes **Given, Find, Conceptual Plan, Solution**, and **Check** sections.

Given and Find: Many students struggle with taking the written problem, parsing the information into categories, and determining the goal of the problem. It is also important to know which pieces of information in the problem are not necessary to solve the problem and if additional information needs to be gathered from sources such as tables in the textbook.

Conceptual Plan: The conceptual plan shows a step-by-step method to solve the problem. In many cases, the given quantities need to be converted to a different unit. Under each of the arrows is the equation, constant, or conversion factor needed to complete this portion of the problem. In the "Problems by Topic" section of the end-of-chapter exercises, the odd-numbered and even-numbered problems are paired. This allows you to use a conceptual plan from an odd-numbered problem in this manual as a starting point to solve the following even-numbered problem. Students should keep in mind that the examples shown are one way to solve the problems. Other mathematically equivalent solutions may be possible.

11.45 **Given:** $m\,(CO_2) = 28.8\text{ g}$, $P = 742\text{ mmHg}$, and $T = 22\,°\text{C}$ **Find:** V

Conceptual Plan: **°C → K and mmHg → atm and g → mol then $n, P, T \rightarrow V$**

$$\text{K} = °\text{C} + 273.15 \qquad \frac{1\text{ atm}}{760\text{ mmHg}} \qquad \frac{1\text{ mol}}{44.01\text{ g}} \qquad PV = nRT$$

Solution: $T_1 = 22\,°\text{C} + 273.15 = 295\text{ K}$, $P = 742\,\cancel{\text{mmHg}} \times \dfrac{1\text{ atm}}{760\,\cancel{\text{mmHg}}} = 0.976316\text{ atm}$

$n = 28.8\,\cancel{\text{g}} \times \dfrac{1\text{ mol}}{44.01\,\cancel{\text{g}}} = 0.654397\text{ mol}$ $PV = nRT$ Rearrange to solve for V.

$$V = \frac{nRT}{P} = \frac{0.654397\,\cancel{\text{mol}} \times 0.08206\,\dfrac{\text{L}\cdot\cancel{\text{atm}}}{\cancel{\text{mol}}\cdot\cancel{\text{K}}} \times 295\,\cancel{\text{K}}}{0.976316\,\cancel{\text{atm}}} = 16.2\text{ L}$$

Check: The units (L) are correct. The magnitude of the answer (16 L) makes sense because one mole of an ideal gas under standard conditions (273 K and 1 atm) occupies 22.4 L. Although these are not standard conditions, they are close enough for a ballpark check of the answer. Because this gas sample contains 0.65 mole, a volume of 16 L is reasonable.

Solution: The Solution section walks you through solving the problem after the conceptual plan. Equations are rearranged to solve for the appropriate quantity. Intermediate results are shown with additional digits to minimize round-off error. The units are canceled in each appropriate step.

Check: The Check section confirms that the units in the answer are correct. This section also challenges the student to think about whether the magnitude of the answer makes sense. Thinking about what is a reasonable answer can help uncover errors such as calculation errors.

1 Atoms

Review Questions

1.1 "The properties of the substances around us depend on the atoms, ions, or molecules that compose them" means that the specific types of atoms and molecules that compose something tell us a great deal about which properties to expect from a substance. A material composed of only sodium and chloride ions will have the properties of table salt. A material composed of molecules with one carbon atom and two oxygen atoms will have the properties of the gas carbon dioxide. If the atoms and molecules change, so do the properties that we expect the material to have.

1.2 The main goal of chemistry is to seek to understand the behavior of matter by studying the behavior of atoms and molecules.

1.3 Matter can be classified according to its state—solid, liquid, or gas—and according to its composition.

1.4 In solid matter, atoms or molecules pack close to each other in fixed locations. Although the atoms and molecules in a solid vibrate, they do not move around or past each other. Consequently, a solid has a fixed volume and rigid shape.

In liquid matter, atoms or molecules pack about as closely as they do in solid matter, but they are free to move relative to each other, giving liquids a fixed volume but not a fixed shape. Liquids assume the shape of their container.

In gaseous matter, atoms or molecules have a lot of space between them and are free to move relative to one another, making gases compressible. Gases always assume the shape and volume of their container.

1.5 A pure substance is composed of only one type of atom or molecule. In contrast, a mixture is a substance composed of two or more different types of atoms or molecules that can be combined in variable proportions.

1.6 An element is a pure substance that cannot be decomposed into simpler substances. A compound is composed of two or more elements in fixed proportions.

1.7 A homogeneous mixture has the same composition throughout, while a heterogeneous mixture has different compositions in different regions.

1.8 The scientific approach to knowledge is based on observation and experiment. Scientists observe and perform experiments on the physical world to learn about it. Observations often lead scientists to formulate a hypothesis, a tentative interpretation or explanation of their observations. Hypotheses are tested by experiments, highly controlled procedures designed to generate such observations. The results of an experiment may support a hypothesis or prove it wrong—in which case the hypothesis must be modified or discarded. A series of similar observations can lead to the development of scientific law, a brief statement that summarizes past observations and predicts future ones. One or more well-established hypotheses may form the basis for a scientific theory. A scientific theory is a model for the way nature is and tries to explain not merely what nature does, but why.

The Greek philosopher Plato (427–347 B.C.) took an opposite approach. He thought that the best way to learn about reality was not through the senses, but through reason. He believed that the physical world was an imperfect representation of a perfect and transcendent world (a world beyond space and time). For him, true knowledge came, not through observing the real physical world, but through reasoning and thinking about the ideal one.

1.9 A hypothesis is a tentative interpretation or explanation of the observed phenomena. A law is a concise statement that summarizes observed behaviors and observations and predicts future observations. A theory attempts to explain why the observed behavior is happening.

1.10 Antoine Lavoisier studied combustion and made careful measurements of the mass of objects before and after burning them in closed containers. He noticed that there was no change in the total mass of material within the container during combustion. Lavoisier summarized his observations on combustion with the Law of Conservation of Mass, which states, "In a chemical reaction, matter is neither created nor destroyed."

1.11 John Dalton formulated the atomic theory of matter. Dalton explained the law of conservation of mass, as well as other laws and observations of the time, by proposing that matter was composed of small, indestructible particles called atoms. Because these particles were merely rearranged in chemical changes (and not created or destroyed), the total amount of mass would remain the same.

1.12 The statement "that is just a theory" is generally taken to mean that there is no scientific proof behind the statement. This statement is the opposite of the meaning in the context of the scientific theory, where theories are tested again and again.

1.13 The first people to propose that matter was composed of small, indestructible particles were Leucippus and Democritus. These Greek philosophers theorized that matter was ultimately composed of small, indivisible particles called *atomos*. In the sixteenth century, modern science began to emerge. A greater emphasis on observation brought rapid advancement as the scientific method became the established way to learn about the physical world. By the early 1800s, certain observations led the English chemist John Dalton to offer convincing evidence that supported the early atomic ideas of Leucippus and Democritus. The theory that all matter is composed of atoms grew out of observations and laws. The three most important laws that led to the development and acceptance of the atomic theory were the law of conservation of mass, the law of definite proportions, and the law of multiple proportions. John Dalton explained the laws with his atomic theory.

1.14 The law of conservation of mass states the following: In a chemical reaction, matter is neither created nor destroyed. In other words, when you carry out any chemical reaction, the total mass of the substances involved in the reaction does not change.

1.15 The law of definite proportions states the following: All samples of a given compound, regardless of their source or how they were prepared, have the same proportions of their constituent elements. This means that elements composing a given compound always occur in fixed (or definite) proportions in all samples of the compound.

1.16 The law of multiple proportions states the following: When two elements (call them A and B) form two different compounds, the masses of element B that combine with 1 g of element A can be expressed as a ratio of small whole numbers. This means that when two atoms (A and B) combine to form more than one compound, the ratio of B in one compound to B in the second compound will be a small whole number. The Law of Definite Proportions refers to the composition of a particular compound, not to a comparison of different compounds. The Law of Definite Proportions states that a particular compound is always made of the same elements in the same ratio.

1.17 The main ideas of John Dalton's atomic theory are as follows: (1) Each element is composed of tiny, indestructible particles called atoms. (2) All atoms of a given element have the same mass and other properties that distinguish them from the atoms of other elements. (3) Atoms combine in simple whole-number ratios to form compounds. (4) Atoms of one element cannot change into atoms of another element. In a chemical reaction, atoms change the way they are bound together with other atoms to form a new substance. The law of conservation of mass is explained by the fourth idea. Because the atoms cannot change into another element, and just change how they are bound together, the total mass will remain constant. The law of constant composition is supported by ideas 2 and 3. Because the atoms of a given element always have the same mass and other distinguishing properties, and they combine in simple whole-number ratios, different samples of the same compound will have the same properties and the same composition. The law of multiple proportions is also supported by ideas 2 and 3 because the atoms can combine in simple whole-number ratios; the ratio of the mass of B in one compound to the mass of B in a second compound will also be a small whole number.

1.18 In the late 1800s, an English physicist named J. J. Thomson performed experiments to probe the properties of cathode rays. Thomson found that these rays were actually streams of particles with the following properties: They traveled in straight lines, they were independent of the composition of the material from which they originated, and they carried a negative electrical charge. He measured the charge-to-mass ratio of the particles and found that the cathode ray particle was about 2000 times lighter than hydrogen.

1.19 In Millikan's oil drop experiment, oil was sprayed into fine droplets using an atomizer. The droplets were allowed to fall under the influence of gravity through a small hole into the lower portion of the apparatus where they could be viewed. During their fall, the drops would acquire electrons that had been produced by the interaction of high-energy radiation with air. These charged drops interacted with two electrically charged plates within the apparatus. The negatively charged plate at the bottom of the apparatus repelled the negatively charged drops. By varying the voltage on the plates, the fall of the charged drops could be slowed, stopped, or even reversed. From the voltage required to halt the free fall of the drops and from the masses of the drops themselves, Millikan calculated the charge of each drop. He then reasoned that because each drop must contain an integral number of electrons, the charge of each drop must be a whole-number multiple of the electron's charge. The magnitude of the charge of the electron is of tremendous importance because it determines how strongly an atom holds its electrons.

1.20 The plum-pudding model of the atom proposed by J. J. Thomson hypothesized that the negatively charged electrons were small particles electrostatically held within a positively charged sphere.

1.21 Rutherford's gold foil experiment directed positively charged α particles at an ultrathin sheet of gold foil. These particles were to act as probes of the gold atoms' structures. If the gold atoms were indeed like plum pudding—with their mass and charge spread throughout the entire volume of the atom—these speeding probes should pass right through the gold foil with minimum deflection. A majority of the particles did pass directly through the foil, but some particles were deflected, and some even bounced back. Rutherford realized that to account for the deflections, the mass and positive charge of an atom must all be concentrated in a space much smaller than the size of the atom itself.

1.22 Rutherford's nuclear model of the atom has three basic parts: (1) Most of the atom's mass and all of its positive charge are contained in a small core called the nucleus. (2) Most of the volume of the atom is empty space, throughout which tiny negatively charged electrons are dispersed. (3) There are as many negatively charged electrons outside the nucleus as there are positively charged particles within the nucleus, so that the atom is electrically neutral. The revolutionary part of this theory is the idea that matter, at its core, is much less uniform than it appears.

1.23 Matter appears solid because the variation in its density is on such a small scale that our eyes cannot see it.

1.24 The three subatomic particles that compose atoms are as follows:

Protons, which have a mass of 1.67262×10^{-27} kg or 1.00727 amu and a relative charge of $+1$;
Neutrons, which have a mass of 1.67493×10^{-27} kg or 1.00866 amu and a relative charge of 0;
Electrons, which have a mass of 0.00091×10^{-27} kg or 0.00055 amu and a relative charge of -1.

1.25 The number of protons in the nucleus defines the identity of an element.

1.26 The atomic number, Z, is the number of protons in an atom's nucleus. The atomic mass number A is the sum of the neutrons and protons in an atom.

1.27 The names of the elements were often given to describe their properties. For example, argon originates from the Greek word *argos*, meaning "inactive." Other elements were named after figures from Greek or Roman mythology or astronomical bodies. Still others were named for the places where they were discovered or where their discoverer was born. More recently, elements have been named after scientists.

1.28 Isotopes are atoms with the same number of protons but different numbers of neutrons. The percent natural abundance is the relative amount of each different isotope in a naturally occurring sample of a given element.

1.29 Isotopes can be symbolized as ${}^{A}_{Z}X$, where A is the mass number, Z is the atomic number, and X is the chemical symbol. A second notation is the chemical symbol (or chemical name) followed by a hyphen and the mass number of the isotope, such as X-A, where X is the chemical symbol or name and A is the mass number. The carbon isotope with a mass of 12 would have the symbol ${}^{12}_{6}C$ or C-12 or carbon-12.

1.30 An ion is a charged particle. Positively charged ions are called cations. Negatively charged ions are called anions.

1.31 Atomic mass represents the average mass of the isotopes that compose the element. The average calculated atomic mass is weighted according to the natural abundance of each isotope.
Atomic mass $= \sum_{n} (\text{fraction of isotope } n) \times (\text{mass of isotope } n)$

1.32 In a mass spectrometer, the sample is injected into the instrument and vaporized. The vaporized atoms are then ionized by an electron beam. The electrons in the beam collide with the vaporized atoms, removing electrons from the atoms and creating positively charged ions. Charged plates with slits in them accelerate the positively charged ions into a magnetic field, which deflects them. The amount of deflection depends on the mass of the ions—lighter ions are deflected more than heavier ones are. Finally, the ions strike a detector and produce an electrical signal that is recorded.

The result of the mass spectrometer is the separation of the atoms in the sample according to their mass, producing a mass spectrum. The position of each peak on the x-axis gives the mass of the isotope, and the intensity (indicated by the height of the peak) gives the relative abundance of that isotope.

Problems by Topic

The Classification of Matter

1.33 (a) The single type of blue cubes indicates that this is a pure substance.
(b) The single type of red cylinders indicates that this is a pure substance.
(c) The random and uniform distribution of the blue cubes and red cylinders indicates that this is a homogeneous mixture.
(d) The region of only red cylinders above a region of only blue cubes indicates that this is a heterogeneous mixture.

1.34 Mixture of two atoms:

Compound composed of two atoms:

1.35 (a) Sweat is a homogeneous mixture of water, sodium chloride, and other components.
(b) Carbon dioxide is a pure substance that is a compound (two or more elements bonded together).
(c) Aluminum is a pure substance that is an element (element 13 in the periodic table).
(d) Vegetable soup is a heterogeneous mixture of broth, chunks of vegetables, and extracts from the vegetables.

1.36 (a) Wine is a generally homogeneous mixture of water, ethyl alcohol, and other components from the grapes. In some cases, there may be sediment present; so it would be a heterogeneous mixture.
(b) Beef stew is a heterogeneous mixture of thick broth with chunks of beef and vegetables.
(c) Iron is a pure substance that is an element (element 26 in the periodic table).
(d) Carbon monoxide is a pure substance that is a compound (two or more elements bonded together).

1.37

Substance	Pure or Mixture	Type (element or compound)
aluminum	pure	element
apple juice	mixture	neither—mixture
hydrogen peroxide	pure	compound
chicken soup	mixture	neither—mixture

1.38

Substance	Pure or Mixture	Type (element or compound)
water	pure	compound
coffee	mixture	neither—mixture
ice	pure	compound
carbon	pure	element

1.39 (a) pure substance that is a compound (one type of molecule that contains two different elements)
(b) heterogeneous mixture (two different molecules that are segregated into regions)
(c) homogeneous mixture (two different molecules that are randomly mixed)
(d) pure substance that is an element (individual atoms of one type)

1.40 (a) pure substance that is an element (individual atoms of one type)
(b) homogeneous mixture (two different molecules that are randomly mixed)
(c) pure substance that is a compound (one type of molecule that contains two different elements)
(d) pure substance that is a compound (one type of molecule that contains two different elements)

The Scientific Approach to Knowledge

1.41 (a) This statement is a theory because it attempts to explain why. It is not possible to observe individual atoms.
(b) This statement is an observation.
(c) This statement is a law because it summarizes many observations and can explain future behavior.
(d) This statement is an observation.

1.42 (a) This statement is a law because it summarizes many observations and can explain future behavior.
(b) This statement is a law because it summarizes many observations and can explain future behavior.
(c) This statement is a law because it summarizes many observations and can explain future behavior.
(d) This statement is a theory because it attempts to explain why.

1.43 (a) If we divide the mass of the oxygen by the mass of the carbon, the result is always 4/3.
(b) If we divide the mass of the oxygen by the mass of the hydrogen, the result is always 16.
(c) These observations suggest that the masses of elements in molecules are ratios of whole numbers [4:3 and 16:1, respectively, for parts (a) and (b)].
(d) Atoms combine in small whole-number ratios and not as random weight ratios.

1.44 Many hypotheses may be developed. One hypothesis is that a large explosion generated galaxies with fragments that are still moving away from each other.

The Laws of Conservation of Mass, Definite Proportions, and Multiple Proportions

1.45 **Given:** 1.50 g hydrogen; 12.0 g oxygen **Find:** grams of water vapor
Conceptual Plan: total mass reactants = total mass products
Solution: Mass of reactants = 1.50 g hydrogen + 12.0 g oxygen = 13.5 grams
Mass of products = mass of reactants = 13.5 grams water vapor

Check: According to the law of conservation of mass, matter is neither created nor destroyed in a chemical reaction. Since water vapor is the only product, the masses of hydrogen and oxygen must combine to form the mass of water vapor.

1.46 **Given:** 21 kg gasoline; 84 kg oxygen **Find:** mass of carbon dioxide and water
Conceptual Plan: total mass reactants = total mass products

Solution: Mass of reactants = 21 kg gasoline + 84 kg oxygen = 105 kg
Mass of products = mass of reactants = 105 kg of carbon dioxide and water

Check: According to the law of conservation of mass, matter is neither created nor destroyed in a chemical reaction. Since carbon dioxide and water are the only products, the masses of gasoline and oxygen must combine to form the mass of carbon dioxide and water.

1.47 **Given:** sample 1: 38.9 g carbon, 448 g chlorine; sample 2: 14.8 g carbon, 134 g chlorine
Find: consistent with definite proportions
Conceptual Plan: Determine mass ratio of samples 1 and 2 and compare.

$$\frac{\text{mass of chlorine}}{\text{mass of carbon}}$$

Solution: sample 1: $\frac{448 \text{ g chlorine}}{38.9 \text{ g carbon}} = 11.5$ sample 2: $\frac{134 \text{ g chlorine}}{14.8 \text{ g carbon}} = 9.05$

Results are not consistent with the law of definite proportions because the ratio of chlorine to carbon is not the same.

Check: According to the law of definite proportions, the mass ratio of one element to another is the same for all samples of the compound.

1.48 **Given:** sample 1: 6.98 grams sodium, 10.7 grams chlorine; sample 2: 11.2 g sodium, 17.3 grams chlorine
Find: consistent with definite proportions
Conceptual Plan: Determine mass ratio of samples 1 and 2 and compare.

$$\frac{\text{mass of chlorine}}{\text{mass of sodium}}$$

Solution: sample 1: $\frac{10.7 \text{ g chlorine}}{6.98 \text{ g sodium}} = 1.53$ sample 2: $\frac{17.3 \text{ g chlorine}}{11.2 \text{ g sodium}} = 1.54$

Results are consistent with the law of definite proportions.

Check: According to the law of definite proportions, the mass ratio of one element to another is the same for all samples of the compound.

1.49 **Given:** mass ratio sodium to fluorine = 1.21:1; sample = 28.8 g sodium **Find:** g fluorine
Conceptual Plan: g sodium → g fluorine

$$\frac{\text{mass of fluorine}}{\text{mass of sodium}}$$

Solution: $28.8 \text{ g }\cancel{\text{sodium}} \times \frac{1 \text{ g fluorine}}{1.21 \text{ g }\cancel{\text{sodium}}} = 23.8 \text{ g fluorine}$

Check: The units of the answer (g fluorine) are correct. The magnitude of the answer is reasonable because it is less than the grams of sodium.

1.50 **Given:** sample 1: 1.65 kg magnesium, 2.57 kg fluorine; sample 2: 1.32 kg magnesium
Find: g fluorine in sample 2
Conceptual Plan: mass magnesium and mass fluorine → mass ratio → mass fluorine (kg) → mass fluorine (g)

$$\frac{\text{mass of fluorine}}{\text{mass of magnesium}} \qquad \frac{1000 \text{ g}}{\text{kg}}$$

Solution: mass ratio $= \frac{2.57 \text{ kg fluorine}}{1.65 \text{ kg magnesium}} = \frac{1.56 \text{ kg fluorine}}{1.00 \text{ kg magnesium}}$

$$1.32 \cancel{\text{kg magnesium}} \times \frac{1.56 \text{ kg fluorine}}{1.00 \cancel{\text{kg magnesium}}} \times \frac{1000 \text{ g}}{\text{kg}} = 2.06 \times 10^3 \text{ g fluorine}$$

Check: The units of the answer (g fluorine) are correct. The magnitude of the answer is reasonable because it is greater than the mass of magnesium and the ratio is greater than 1.

1.51 **Given:** 1 g osmium: sample 1 = 0.168 g oxygen; sample 2 = 0.3369 g oxygen
Find: consistent with multiple proportions
Conceptual Plan: Determine mass ratio of oxygen.

$$\frac{\text{mass of oxygen sample 2}}{\text{mass of oxygen sample 1}}$$

Solution: $\frac{0.3369 \text{ g oxygen}}{0.168 \text{ g oxygen}} = 2.00$ Ratio is a small whole number. Results are consistent with multiple proportions.

Check: According to the law of multiple proportions, when two elements form two different compounds, the masses of element B that combine with 1 g of element A can be expressed as a ratio of small whole numbers.

1.52 **Given:** 1 g palladium: compound A: 0.603 g S; compound B: 0.301 g S; compound C: 0.151 g S
Find: consistent with multiple proportions
Conceptual Plan: Determine mass ratio of sulfur in the three compounds.

$$\frac{\text{mass of sulfur sample A}}{\text{mass of sulfur sample B}} \quad \frac{\text{mass of sulfur sample A}}{\text{mass of sulfur sample C}} \quad \frac{\text{mass of sulfur sample B}}{\text{mass of sulfur sample C}}$$

Solution: $\frac{0.603 \text{ g S in compound A}}{0.301 \text{ g S in compound B}} = 2.00 \qquad \frac{0.603 \text{ g S in compound A}}{0.151 \text{ g S in compound C}} = 3.99 \sim 4$

$$\frac{0.301 \text{ g S in compound B}}{0.151 \text{ g S in compound C}} = 1.99 \sim 2$$

Ratio of each is a small whole number. Results are consistent with multiple proportions.

Check: According to the law of multiple proportions, when two elements form two different compounds, the masses of element B that combine with 1 g of element A can be expressed as a ratio of small whole numbers.

1.53 **Given:** sulfur dioxide = 3.49 g oxygen and 3.50 g sulfur; sulfur trioxide = 6.75 g oxygen and 4.50 g sulfur
Find: mass oxygen per g S for each compound and then determine the mass ratio of oxygen

$$\frac{\text{mass of oxygen in sulfur dioxide}}{\text{mass of sulfur in sulfur dioxide}} \quad \frac{\text{mass of oxygen in sulfur trioxide}}{\text{mass of sulfur in sulfur trioxide}} \quad \frac{\text{mass of oxygen in sulfur trioxide}}{\text{mass of oxygen in sulfur dioxide}}$$

Solution: sulfur dioxide $= \frac{3.49 \text{ g oxygen}}{3.50 \text{ g sulfur}} = \frac{0.997 \text{ g oxygen}}{1 \text{ g sulfur}}$ sulfur trioxide $= \frac{6.75 \text{ g oxygen}}{4.50 \text{ g sulfur}} = \frac{1.50 \text{ g oxygen}}{1 \text{ g sulfur}}$

$\frac{1.50 \text{ g oxygen in sulfur trioxide}}{0.997 \text{ g oxygen in sulfur dioxide}} = \frac{1.50}{1} = \frac{3}{2}.$ The ratio is converted from 1.50:1 to 3:2 because the law of multiple proportions states that the ratio is in small whole numbers. Ratio is in small whole numbers and is consistent with multiple proportions.

Check: According to the law of multiple proportions, when two elements form two different compounds, the masses of element B that combine with 1 g of element A can be expressed as a ratio of small whole numbers.

1.54 **Given:** sulfur hexafluoride = 4.45 g fluorine and 1.25 g sulfur; sulfur tetrafluoride = 4.43 g fluorine and 1.87 g sulfur
Find: mass fluorine per g S for each compound and then determine the mass ratio of fluorine

$$\frac{\text{mass of fluorine in sulfur hexafluoride}}{\text{mass of sulfur in sulfur hexafluoride}} \quad \frac{\text{mass of fluorine in sulfur tetrafluoride}}{\text{mass of sulfur in sulfur tetrafluoride}} \quad \frac{\text{mass of fluorine in sulfur hexafluoride}}{\text{mass of fluorine in sulfur tetrafluoride}}$$

Solution: sulfur hexafluoride $= \frac{4.45 \text{ g fluorine}}{1.25 \text{ g sulfur}} = \frac{3.56 \text{ g fluorine}}{1 \text{ g sulfur}}$

sulfur tetrafluoride $= \frac{4.43 \text{ g fluorine}}{1.87 \text{ g sulfur}} = \frac{2.3\underline{6}9 \text{ g fluorine}}{1 \text{ g sulfur}}$

$\frac{3.56 \text{ g fluorine in sulfur hexafluoride}}{2.3\underline{6}9 \text{ g fluorine in sulfur tetrafluoride}} = \frac{1.50}{1} = \frac{3}{2}$ The ratio is converted from 1.50:1 to 3:2 because the law of multiple proportions states that the ratio is in small whole numbers. Ratio is in small whole numbers and is consistent with multiple proportions.

Check: According to the law of multiple proportions, when two elements form two different compounds, the masses of element B that combine with 1 g of element A can be expressed as a ratio of small whole numbers.

Atomic Theory, Nuclear Theory, and Subatomic Particles

1.55
(a) Sulfur and oxygen atoms have the same mass. *Inconsistent* with Dalton's atomic theory because only atoms of the same element have the same mass.
(b) All cobalt atoms are identical. *Consistent* with Dalton's atomic theory because all atoms of a given element have the same mass and other properties that distinguish them from atoms of other elements.
(c) Potassium and chlorine atoms combine in a 1:1 ratio to form potassium chloride. *Consistent* with Dalton's atomic theory because atoms combine in simple whole-number ratios to form compounds.
(d) Lead atoms can be converted into gold. *Inconsistent* with Dalton's atomic theory because atoms of one element cannot change into atoms of another element.

1.56 (a) All carbon atoms are identical. Consistent with Dalton's atomic theory because all atoms of a given element have the same mass and other properties that distinguish them from atoms of other elements.

(b) An oxygen atom combines with 1.5 hydrogen atoms to form a water molecule. *Inconsistent* with Dalton's atomic theory because atoms combine in simple whole-number ratios to form compounds. An oxygen atom actually combines with two hydrogen atoms to form a water molecule.

(c) Two oxygen atoms combine with a carbon atom to form a carbon dioxide molecule. *Consistent* with Dalton's atomic theory because atoms combine in simple whole-number ratios to form compounds.

(d) The formation of a compound often involves the destruction of one or more atoms. *Inconsistent* with Dalton's atomic theory. Atoms change the way they are bound together with other atoms when they form a new substance, but they are neither created nor destroyed.

1.57 (a) The volume of an atom is mostly empty space. *Consistent* with Rutherford's nuclear theory because most of the volume of the atom is empty space, throughout which tiny, negatively charged electrons are dispersed.

(b) The nucleus of an atom is small compared to the size of the atom. *Consistent* with Rutherford's nuclear theory because most of the atom's mass and all of its positive charge are contained in a small core called the nucleus.

(c) Neutral lithium atoms contain more neutrons than protons. *Inconsistent* with Rutherford's nuclear theory because it did not distinguish where the mass of the nucleus came from other than from the protons.

(d) Neutral lithium atoms contain more protons than electrons. *Inconsistent* with Rutherford's nuclear theory because there are as many negatively charged particles outside the nucleus as there are positively charged particles in the nucleus.

1.58 (a) Since electrons are smaller than protons and since a hydrogen atom contains only one proton and one electron, it must follow that the volume of a hydrogen atom is mostly due to the proton. Inconsistent with Rutherford's nuclear theory because most of the volume of the atom is empty space, throughout which tiny, negatively charged electrons are dispersed.

(b) A nitrogen atom has seven protons in its nucleus and seven electrons outside its nucleus. *Consistent* with Rutherford's nuclear theory because there are as many negatively charged particles outside the nucleus as there are positively charged particles in the nucleus.

(c) A phosphorus atom has 15 protons in its nucleus and 150 electrons outside its nucleus. *Inconsistent* with Rutherford's nuclear theory because there are as many negatively charged particles outside the nucleus as there are positively charged particles in the nucleus.

(d) The majority of the mass of a fluorine atom is due to its nine electrons. *Inconsistent* with Rutherford's nuclear theory because most of the atom's mass and all of its positive charge are contained in a small core called the nucleus.

1.59 **Given:** drop A $= -6.9 \times 10^{-19}$ C; drop B $= -9.2 \times 10^{-19}$ C; drop C $= -11.5 \times 10^{-19}$ C; drop D $= -4.6 \times 10^{-19}$ C

Find: the charge on a single electron

Conceptual Plan: Determine the ratio of charge for each set of drops.

$$\frac{\text{charge on drop 1}}{\text{charge on drop 2}}$$

Solution: $\dfrac{-6.9 \times 10^{-19}\text{ C drop A}}{-4.6 \times 10^{-19}\text{ C drop D}} = 1.5$ $\quad \dfrac{-9.2 \times 10^{-19}\text{ C drop B}}{-4.6 \times 10^{-19}\text{ C drop D}} = 2$ $\quad \dfrac{-11.5 \times 10^{-19}\text{ C drop C}}{-4.6 \times 10^{-19}\text{ C drop D}} = 2.5$

The ratios obtained are not whole numbers, but they can be converted to whole numbers by multiplying by 2. Therefore, the charge on the electron has to be 1/2 the smallest value experimentally obtained. The charge on the electron $= -2.3 \times 10^{-19}$ C.

Check: The units of the answer (Coulombs) are correct. The magnitude of the answer is reasonable because all of the values experimentally obtained are integer multiples of -2.3×10^{-19}.

1.60 **Given:** drop A $= -4.8 \times 10^{-9}$ z; drop B $= -9.6 \times 10^{-9}$ z; drop C $= -6.4 \times 10^{-9}$ z; drop D $= -12.8 \times 10^{-9}$ z

Find: the charge on a single electron

Conceptual Plan: Determine the ratio of charge for each set of drops and determine the charge on an electron.

$$\frac{\text{charge on drop 1}}{\text{charge on drop 2}}$$

Then determine the number of electrons in each drop.

$$\frac{\text{charge on drop}}{\text{charge on one electron}}$$

Solution: $\frac{-9.6 \times 10^{-9}\text{ z drop B}}{-4.8 \times 10^{-9}\text{ z drop A}} = 2 \quad \frac{-6.4 \times 10^{-9}\text{ z drop C}}{-4.8 \times 10^{-9}\text{ z drop A}} = 1.33 \quad \frac{-12.8 \times 10^{-9}\text{ z drop D}}{-4.8 \times 10^{-9}\text{ z drop A}} = 2.66$

The ratios obtained are not all whole numbers, but they can be converted to whole numbers by multiplying by 3. Therefore, the charge on the electron has to be 1/3 the smallest value experimentally obtained. The charge on the electron $= \frac{1}{3} \times -4.8 \times 10^{-9}\text{ z} = -1.6 \times 10^{-9}\text{ z}$.

Number of electrons in

drop A: $\frac{-4.8 \times 10^{-9}\text{ z}}{-1.6 \times 10^{-9}\text{ z}} = 3$ electrons drop B: $\frac{-9.6 \times 10^{-9}\text{ z}}{-1.6 \times 10^{-9}\text{ z}} = 6$ electrons

drop C: $\frac{-6.4 \times 10^{-9}\text{ z}}{-1.6 \times 10^{-9}\text{ z}} = 4$ electrons drop D: $\frac{-12.8 \times 10^{-9}\text{ z}}{-1.6 \times 10^{-9}\text{ z}} = 8$ electrons

Check: The units of the answer (zorg) are correct. The magnitude of the answer is reasonable because all of the values experimentally obtained are integer multiples of -1.6×10^{-9}.

1.61 (a) True: Protons and electrons have equal and opposite charges.
(b) True: Protons and electrons have opposite charges, so they will attract each other.
(c) True: The mass of the electron is much less than the mass of the neutron.
(d) False: The mass of the proton and the mass of the neutron are about the same.

1.62 (a) True: Protons and electrons have equal and opposite charges.
(b) True: The mass of the proton and the mass of the neutron are about the same.
(c) False: All atoms contain protons. The lightest element, hydrogen, contains one proton.
(d) False: Protons have a positive charge, while neutrons are neutral.

Isotopes and Ions

1.63 For each of the isotopes determine Z (the number of protons) from the periodic table and determine A (protons + neutrons). Then write the symbol in the form X-A.
(a) The silver isotope with 60 neutrons: $Z = 47; A = 47 + 60 = 107;$ Ag-107
(b) The silver isotope with 62 neutrons: $Z = 47; A = 47 + 62 = 109;$ Ag-109
(c) The uranium isotope with 146 neutrons: $Z = 92; A = 92 + 146 = 238;$ U-238
(d) The hydrogen isotope with 1 neutron: $Z = 1; A = 1 + 1 = 2;$ H-2

1.64 For each of the isotopes determine Z (the number of protons) from the periodic table and determine A (protons + neutrons). Then write the symbol in the form ${}^{A}_{Z}\text{X}$.
(a) The copper isotope with 34 neutrons: $Z = 29; A = 29 + 34 = 63;$ ${}^{63}_{29}\text{Cu}$
(b) The copper isotope with 36 neutrons: $Z = 29; A = 29 + 36 = 65;$ ${}^{65}_{29}\text{Cu}$
(c) The potassium isotope with 21 neutrons: $Z = 19; A = 19 + 21 = 40;$ ${}^{40}_{19}\text{K}$
(d) The argon isotope with 22 neutrons: $Z = 18; A = 18 + 22 = 40;$ ${}^{40}_{18}\text{Ar}$

1.65 (a) ${}^{14}_{7}\text{N}$: $Z = 7; A = 14;$ protons $= Z = 7;$ neutrons $= A - Z = 14 - 7 = 7$
(b) ${}^{23}_{11}\text{Na}$: $Z = 11; A = 23;$ protons $= Z = 11;$ neutrons $= A - Z = 23 - 11 = 12$
(c) ${}^{222}_{86}\text{Rn}$: $Z = 86; A = 222;$ protons $= Z = 86;$ neutrons $= A - Z = 222 - 86 = 136$
(d) ${}^{208}_{82}\text{Pb}$: $Z = 82; A = 208;$ protons $= Z = 82;$ neutrons $= A - Z = 208 - 82 = 126$

1.66 (a) ${}^{40}_{19}K$: $Z = 19; A = 40$; protons $= Z = 19$; neutrons $= A - Z = 40 - 19 = 21$
(b) ${}^{226}_{88}Ra$: $Z = 88; A = 226$; protons $= Z = 88$; neutrons $= A - Z = 226 - 88 = 138$
(c) ${}^{99}_{43}Tc$: $Z = 43; A = 99$; protons $= Z = 43$; neutrons $= A - Z = 99 - 43 = 56$
(d) ${}^{33}_{15}P$: $Z = 15; A = 33$; protons $= Z = 15$; neutrons $= A - Z = 33 - 15 = 18$

1.67 Carbon-14: $A = 14, Z = 6; {}^{14}_{6}C$ # protons $= Z = 6$ # neutrons $= A - Z = 14 - 6 = 8$

1.68 Uranium-235: $A = 235, Z = 92; {}^{235}_{92}U$ # protons $= Z = 92$ # neutrons $= A - Z = 235 - 92 = 143$

1.69 In a neutral atom, the number of protons = the number of electrons = Z. For an ion, electrons are lost (cations) or gained (anions).
(a) Ni^{2+}: $Z = 28$ = protons; $Z - 2 = 26$ = electrons
(b) S^{2-}: $Z = 16$ = protons; $Z + 2 = 18$ = electrons
(c) Br^{-}: $Z = 35$ = protons; $Z + 1 = 36$ = electrons
(d) Cr^{3+}: $Z = 24$ = protons; $Z - 3 = 21$ = electrons

Check: Ions with a positive charge (Ni^{2+} and Cr^{3+}) have more protons than electrons. Ions with a negative charge (S^{2-} and Br^{-}) have more electrons than protons.

1.70 In a neutral atom, the number of protons = the number of electrons = Z. For an ion, electrons are lost (cations) or gained (anions).
(a) Al^{3+}: $Z = 13$ = protons; $Z - 3 = 10$ = electrons
(b) Se^{2-}: $Z = 34$ = protons; $Z + 2 = 36$ = electrons
(c) Ga^{3+}: $Z = 31$ = protons; $Z - 3 = 28$ = electrons
(d) Sr^{2+}: $Z = 38$ = protons; $Z - 2 = 36$ = electrons

Check: Ions with a positive charge (Al^{3+}, Ga^{3+}, and Sr^{2+}) have more protons than electrons. Ions with a negative charge (Se^{2-}) have more electrons than protons.

Atomic Mass and Mass Spectrometry

1.71 **Given:** Ga-69; mass = 68.92558 amu; 60.108%: Ga-71; mass = 70.92470 amu; 39.892% **Find:** sketch the mass spectrum

Conceptual Plan: → **show relative % abundance versus mass on the graph**

Solution:

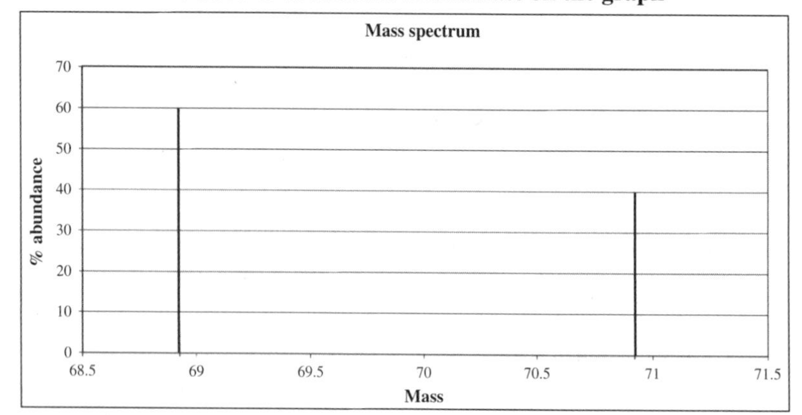

Check: The mass spectrum is reasonable because it has two mass lines corresponding to the two isotopes and the line at 68.92588 is about 1.5 times larger than the line at 70.92470.

1.72 **Given:** Mg-24: mass = 23.9850 amu; 78.99%: Mg-25; mass = 24.9858 amu; 10.00%: Mg-26: mass = 25.9826 amu; 11.01%
Find: sketch the mass spectrum
Conceptual Plan: → **show relative % abundance versus mass on the graph**
Solution:

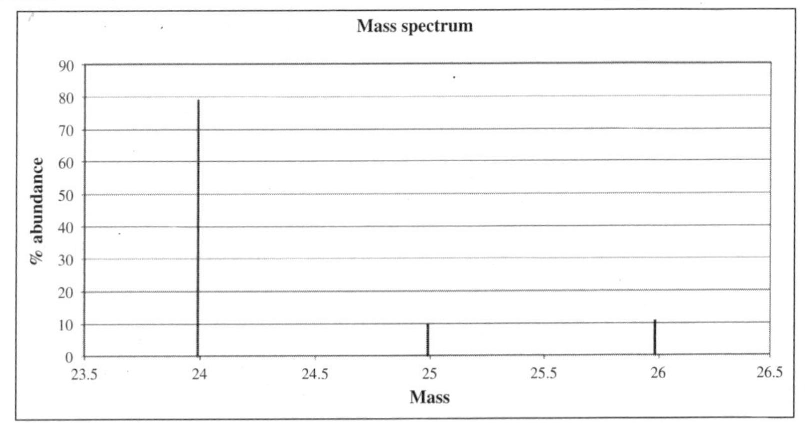

Check: The mass spectrum is reasonable because it has three mass lines corresponding to the three isotopes and the line at 23.9850 is about eight times larger than the other two lines.

1.73 Fluorine has an isotope F-19 with a very large abundance so that the mass of fluorine is very close to the mass of the isotope and the line in the mass spectrum reflects the abundance of F-19. Chlorine has two isotopes (Cl-35 and Cl-37), and the mass of 35.45 amu is the weighted average of these two isotopes; so there is no line at 35.45 amu.

1.74 Copper would have no isotope with a mass of 63.546 amu. Because the mass of the isotope comes primarily from the sum of the protons and neutrons, the mass of the isotope must have a value close to a whole number. Copper must be composed of two or more isotopes, at least one with a mass less than 63.546 amu and at least one with a mass greater than 63.546 amu.

1.75 **Given:** isotope 1 mass = 120.903 amu, 57.4%; isotope 2 mass = 122.9042 amu
Find: atomic mass of the element and identify the element
Conceptual Plan: **% abundance isotope 2 → and then % abundance → fraction and then find atomic mass**

$$100\% - \%\text{ abundance isotope 1} \qquad \frac{\%\text{ abundance}}{100} \qquad \text{Atomic mass} = \sum_n (\text{fraction of isotope } n) \times (\text{mass of isotope } n)$$

Solution: 100.0% − 57.4% isotope 1 = 42.6% isotope 2

$$\text{Fraction isotope 1} = \frac{57.4}{100} = 0.574 \qquad \text{Fraction isotope 2} = \frac{42.6}{100} = 0.426$$

$$\text{Atomic mass} = \sum_n (\text{fraction of isotope } n) \times (\text{mass of isotope } n)$$

$$= 0.574(120.9038\text{ amu}) + 0.426(122.9042\text{ amu}) = 121.8\text{ amu}$$

From the periodic table, Sb has a mass of 121.757 amu; so it is the closest mass, and the element is antimony.

Check: The units of the answer (amu) are correct. The magnitude of the answer is reasonable because it lies between 120.9038 and 122.9042 and is slightly less than halfway between the two values because the lower value has a slightly greater abundance.

1.76 **Given:** isotope = 1 mass = 135.90714 amu, 50.19%; isotope = 2 mass = 137.90599 amu, 0.25%; isotope = 3 mass = 139.90543 amu, 88.43%; isotope = 4, mass = 141.90924, 11.13%.
Find: atomic mass of the element and identify the element
Conceptual Plan: **% abundance → fraction and then find atomic mass**

$$\frac{\%\text{ abundance}}{100} \qquad \text{Atomic mass} = \sum_n (\text{fraction of isotope } n) \times (\text{mass of isotope } n)$$

Solution: $\text{Fraction isotope 1} = \frac{0.19}{100} = 0.0019 \qquad \text{Fraction isotope 2} = \frac{0.25}{100} = 0.0025$

$$\text{Fraction isotope 3} = \frac{88.43}{100} = 0.8843 \qquad \text{Fraction isotope 4} = \frac{11.13}{100} = 0.1113$$

$$\text{Atomic mass} = \sum_n (\text{fraction of isotope } n) \times (\text{mass of isotope } n)$$
$$= 0.0019(135.90714 \text{ amu}) + 0.0025(137.90599 \text{ amu}) + 0.8843(139.90543 \text{ amu}) + 0.1113(141.90924 \text{ amu})$$
$$= 140.1 \text{ amu}$$

From the periodic table, Ce has a mass of 140.12 amu; so it is the closest mass, and the element is cerium.

Check: The units of the answer (amu) are correct. The magnitude of the answer is reasonable because it lies between 135.90714 and 141.90924 and is closer to the higher value because isotopes 3 and 4 make up most of the mass of the element with a combined abundance of 99.56%.

1.77 **Given:** Br-81; mass = 80.9163 amu; 49.31%: atomic mass Br = 79.904 amu
Find: mass and abundance
Conceptual Plan: % abundance Br-79 → then % abundance → fraction → mass Br-79

$100\% - \%\text{ Br-81}$ $\qquad \frac{\%\text{ abundance}}{100}$ $\qquad \text{Atomic mass} = \sum_n (\text{fraction of isotope } n) \times (\text{mass of isotope } n)$

Solution: 100.00% − 49.31% = 50.69%

$$\text{Fraction Br-79} = \frac{50.69}{100} = 0.5069 \qquad \text{Fraction Br-81} = \frac{49.31}{100} = 0.4931$$

Let x be the mass of Br-79.

$$\text{Atomic mass} = \sum_n (\text{fraction of isotope } n) \times (\text{mass of isotope } n)$$
$$79.904 \text{ amu} = 0.5069(x \text{ amu}) + 0.4931(80.9163 \text{ amu})$$
$$x = 78.92 \text{ amu} = \text{mass Br-79}$$

Check: The units of the answer (amu) are correct. The magnitude of the answer is reasonable because it is less than the mass of the atom and the second isotope (Br-81) has a mass greater than the mass of the atom.

1.78 **Given:** Si-28; mass = 27.9769 amu; 92.2%: Si-29; mass = 28.9765 amu; 4.67%: atomic mass Si = 28.09 amu
Find: mass and abundance Si-30
Conceptual Plan: % abundance Si-30 → then % abundance → fraction → mass Si-30

$100\% - \%\text{Si-30}$ $\qquad \frac{\%\text{ abundance}}{100}$ $\qquad \text{Atomic mass} = \sum_n (\text{fraction of isotope } n) \times (\text{mass of isotope } n)$

Solution: 100.00% − 92.2% − 4.67% = 3.1% Si-30

$$\text{Fraction Si-28} = \frac{92.2}{100} = 0.922 \quad \text{Fraction Si-29} = \frac{4.67}{100} = 0.0467 \quad \text{Fraction Si-30} = \frac{3.1}{100} = 0.031$$

Let x be the mass of Si-30.

$$\text{Atomic mass} = \sum_n (\text{fraction of isotope } n) \times (\text{mass of isotope } n)$$
$$28.09 \text{ amu} = 0.922(27.9769 \text{ amu}) + 0.0467(28.9765 \text{ amu}) + 0.031(x)$$
$$x = 30 \text{ amu} = \text{mass Si-30}$$

Check: The units of the answer (amu) are correct. The magnitude of the answer is reasonable because it is greater than the mass of the second isotope (Si-29), which has a mass greater than the mass of the atom.

1.79 **Given:** graph of mass and % abundance of Eu **Find:** atomic mass of Eu
Conceptual Plan: % abundance → relative fraction abundance → mass Eu

$\frac{\%\text{ abundance of each isotope}}{\text{sum of all }\%\text{ abundance}}$ $\qquad \text{Atomic mass} = \sum_n (\text{fraction of isotope } n) \times (\text{mass of isotope } n)$

Solution: $\text{Fraction } ^{151}\text{Eu} = \frac{91.6}{191.6} = 0.478\underline{1} \qquad \text{Fraction } ^{153}\text{Eu} = \frac{100.0}{191.6} = 0.521\underline{9}$

$$\text{Atomic mass} = \sum_n (\text{fraction of isotope } n) \times (\text{mass of isotope } n)$$
$$= 0.47\underline{8}1(151 \text{ amu}) + 0.521\underline{9}(153 \text{ amu})$$
$$= 15\underline{2}.04 \text{ amu} = 152 \text{ amu}$$

Check: The units of the answer (amu) are correct. The magnitude of the answer is reasonable because the fraction of each isotope is about 0.5.

1.80 **Given:** graph of mass and % abundance of Rb **Find:** atomic mass of Rb
Conceptual Plan: % abundance → relative fraction abundance → mass Eu

$$\frac{\text{\% abundance of each isotope}}{\text{sum of all \% abundance}} \qquad \text{Atomic mass} = \sum_n (\text{fraction of isotope } n) \times (\text{mass of isotope } n)$$

Solution: Fraction ^{85}Rb $= \frac{100.0}{138.6} = 0.721\underline{5}$ Fraction ^{87}Rb $= \frac{38.6}{138.6} = 0.27\underline{8}5$

$$\begin{aligned}\text{Atomic mass} &= \sum_n (\text{fraction of isotope } n) \times (\text{mass of isotope } n)\\ &= 0.721\underline{5}(85.0 \text{ amu}) + 0.27\underline{8}5(87.0 \text{ amu})\\ &= 85.\underline{5}6 \text{ amu} = 85.6 \text{ amu}\end{aligned}$$

Check: The units of the answer (amu) are correct. The magnitude of the answer is reasonable because the fraction of each of isotope ^{85}Rb is greater than the fraction of isotope ^{87}Rb.

Cumulative Problems

1.81 **Given:** 7.83 g HCN sample 1: 0.290 g H; 4.06 g N; 3.37 g HCN sample 2 **Find:** g C in sample 2
Conceptual Plan: g HCN sample 1 → g C in HCN sample 1 → ratio g C to g HCN → g C in HCN sample 2

$$\text{g HCN} - \text{g H} - \text{g N} \qquad \frac{\text{g C}}{\text{g HCN}} \qquad \text{g HCN} \times \frac{\text{g C}}{\text{g HCN}}$$

Solution: 7.83 g HCN − 0.290 g H − 4.06 g N = 3.48 g C

$$3.37 \text{ g } \cancel{\text{HCN}} \times \frac{3.48 \text{ g C}}{7.83 \text{ g } \cancel{\text{HCN}}} = 1.50 \text{ g C}$$

Check: The units of the answer (g C) are correct. The magnitude of the answer is reasonable because the sample size is about half the original sample size and the g C are about half the original g C.

1.82 (a) **Given:** mass ratio S:O = 1.0:1.0 in SO_2 **Find:** mass ratio S:O in SO_3
Conceptual Plan: Determine the ratio of O:O in SO_3 and SO_2; then determine g O per g S in SO_3.

Solution: For a fixed amount of S, the ratio of O is $\frac{3\text{ O}}{2\text{ O}} = 1.5$. So for 1 gram S, SO_3 would have 1.5 g O. The mass ratio of S:O = 1.0:1.5, that is 2.0:3.0, in SO_3.

Check: The answer is reasonable because the ratio is smaller than the ratio for SO_2 and SO_3 has to contain more O per gram of S.

(b) **Given:** mass ratio S:O = 1.0:1.0 in SO_2 **Find:** mass ratio S:O in S_2O
Conceptual Plan: Determine the ratio of S:S in S_2O and SO_2; then determine g O per g S in S_2O.

Solution: For a fixed amount of O, the ratio of S is $\frac{2\text{ S}}{0.5\text{ S}} = 4.0$. So for 1 gram O, S_2O would have 4 grams S. The mass ratio of S:O = 4.0:1.0 in S_2O.

Check: The answer is reasonable because the ratio is larger than the ratio for SO_2 and S_2O has to contain more S per gram of O.

1.83 **Given:** graph of mass and % abundance of Pb **Find:** atomic mass of Pb
Conceptual Plan: % abundance → relative fraction abundance → mass Eu

$$\frac{\text{\% abundance of each isotope}}{\text{sum of all \% abundance}} \qquad \text{Atomic mass} = \sum_n (\text{fraction of isotope } n) \times (\text{mass of isotope } n)$$

Solution: Fraction ^{204}Pb $= \frac{2.7}{190.9} = 0.01\underline{4}1$ Fraction ^{206}Pb $= \frac{46.0}{190.9} = 0.24\underline{1}0$

Fraction ^{207}Pb $= \frac{42.2}{190.9} = 0.22\underline{1}1$ Fraction ^{208}Pb $= \frac{100.0}{190.9} = 0.523\underline{8}$

$$\begin{aligned}\text{Atomic mass} &= \sum_n (\text{fraction of isotope } n) \times (\text{mass of isotope } n)\\ &= 0.01\underline{4}1(204 \text{ amu}) + 0.24\underline{1}0(206 \text{ amu}) + 0.22\underline{1}1(207 \text{ amu}) + 0.523\underline{8}(208 \text{ amu})\\ &= 20\underline{7}.24 \text{ amu} = 207 \text{ amu}\end{aligned}$$

Check: The units of the answer (amu) are correct. The magnitude of the answer is reasonable because the fraction of the heavier isotopes is greater than that of the lighter isotopes.

1.84 **Given:** graph of mass and % abundance of Hg **Find:** atomic mass of Hg

Conceptual Plan: % abundance → relative fraction abundance → mass Eu

$\frac{\text{\% abundance of each isotope}}{\text{sum of all \% abundance}}$ Atomic mass $= \sum_n (\text{fraction of isotope } n) \times (\text{mass of isotope } n)$

Solution: Fraction $^{196}Hg = \frac{0.5}{334.9} = 0.001493$ Fraction $^{198}Hg = \frac{33.4}{334.9} = 0.09973$

Fraction $^{199}Hg = \frac{56.5}{334.9} = 0.1687$ Fraction $^{200}Hg = \frac{77.4}{334.9} = 0.2311$

Fraction $^{201}Hg = \frac{44.1}{334.9} = 0.1317$ Fraction $^{202}Hg = \frac{100.0}{334.9} = 0.2986$

Fraction $^{204}Hg = \frac{23.0}{334.9} = 0.06868$

$$\text{Atomic mass} = \sum_n (\text{fraction of isotope n}) \times (\text{mass of isotope } n)$$
$$= 0.00149(196\text{ amu}) + 0.09973(198\text{ amu}) + 0.1687(199\text{ amu}) + 0.2311(200\text{ amu}) + 0.1317(201\text{ amu}) + 0.2986(202\text{ amu}) + 0.06868(204\text{ amu})$$
$$= 200.6.29\text{ amu} = 200.6\text{ amu}$$

Check: The units of the answer (amu) are correct. The magnitude of the answer is reasonable because there are seven isotopes of Hg and the fraction of isotopes in the middle mass region is fairly close.

1.85 $^{236}_{90}Th$ A − Z = number of neutrons. 236 − 90 = 146 neutrons. So any nucleus with 146 neutrons is an isotone of $^{236}_{90}Th$.

Some would be $^{238}_{92}U$, $^{239}_{93}Np$, $^{241}_{95}Am$, $^{237}_{91}Pa$, $^{235}_{89}Ac$, and $^{244}_{98}Cf$.

1.86

Symbol	Z	A	Number of Protons	Number of Electrons	Number of Neutrons	Charge
Si	14	**28**	**14**	14	14	**0**
S^{2-}	**16**	32	**16**	**18**	**16**	2−
Cu^{2+}	**29**	**63**	**29**	**27**	34	2+
P	15	**31**	**15**	15	16	**0**

Check: The element is defined by the value of Z = number of protons. The mass number = A = (the number of protons) + (number of neutrons). The charge = (the number of protons) − (number of electrons).

Challenge Problems

1.87 **Given:** Ag-107, 51.839%, Ag-109, $\frac{\text{mass Ag-109}}{\text{mass Ag-107}} = 1.0187$ **Find:** mass Ag-107

Conceptual Plan: % abundance Ag-107 → % abundance Ag-109 → fraction → mass Ag-107

100% − (% Ag-107) $\frac{\text{\% abundance}}{100}$ Atomic mass $= \sum_n (\text{fraction of isotope } n) \times (\text{mass of isotope } n)$

Solution: 100.00% − 51.839% = 48.161% Ag − 109

Fraction Ag-107 $= \frac{51.839}{100.00} = 0.51839$ Fraction Ag-109 $= \frac{48.161}{100.00} = 0.48161$

Let x be the mass of Ag-107, then mass Ag-109 $= 1.0187x$

$$\text{Atomic mass} = \sum_n (\text{fraction of isotope } n) \times (\text{mass of isotope } n)$$

$$107.87\text{ amu} = 0.51839(x\text{ amu}) + 0.48161(1.0187x\text{ amu})$$
$$x = 106.907\text{ amu} = 106.91\text{ amu mass Ag-107}$$

Check: The units of the answer (amu) are correct. The answer is reasonable because it is close to the atomic mass number of Ag-107.

1.88 (a) **Given:** 36 atoms Wt-296; 2 atoms Wt-297; 12 atoms Wt-298 **Find:** % abundance of each

Conceptual Plan: total atoms → fraction of each isotope → % abundance

sum of atoms; $\frac{\text{number of each isotope}}{\text{total atoms}}$; fraction × 100%

Solution: Total atoms = 36 + 2 + 12 = 50

$\frac{36}{50} \times 100\% = 72\%$ Wt-296, $\frac{2}{50} \times 100\% = 4\%$ Wt-297, $\frac{12}{50} \times 100\% = 24\%$ Wt-298

Check: The units of the answers (% abundance) are correct. The values of the answers are reasonable because they add up to 100%.

(b)

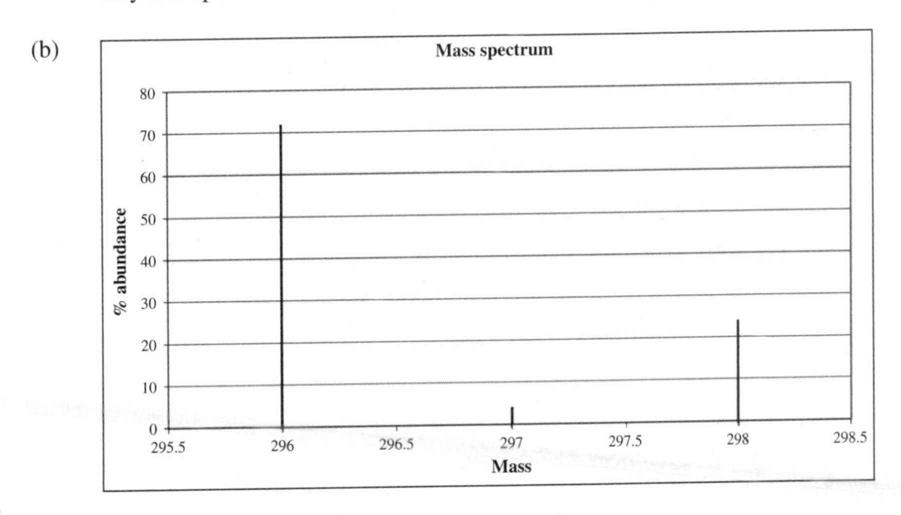

(c) **Given:** Wt-296 m = 24.6630 × mass ^{12}C, 72%; Wt-297 m = 24.7490 × mass ^{12}C, 4%; Wt-298 m = 24.8312 × mass ^{12}C, 24%

Find: atomic mass Wt

Conceptual Plan: mass of isotope relative to ^{12}C → mass of isotope and then % abundance → fraction abundance, then determine atomic mass

(Mass relative to ^{12}C)(12.00 amu); $\frac{\%\ \text{abundance}}{100}$

Atomic mass $= \sum_n (\text{fraction of isotope } n) \times (\text{mass of isotope } n)$

Solution:

Wt-296 = 24.6630 × 12.00 amu = 295.956 amu; Wt-297 = 24.7490 × 12.00 amu = 296.988 amu; W-298 = 24.8312 × 12.00 amu = 297.974 amu

fraction Wt-296 $= \frac{72}{100} = 0.72$ fraction Wt-297 $= \frac{4}{100} = 0.04$ fraction Wt-298 $= \frac{24}{100} = 0.24$

$$\begin{aligned}\text{Atomic mass} &= \sum_n (\text{fraction of isotope } n) \times (\text{mass of isotope } n)\\ &= (0.72)(295.956\ \text{amu}) + (0.04)(296.988\ \text{amu}) + (0.24)(297.974\ \text{amu})\\ &= 296.482\ \text{amu}\end{aligned}$$

Check: The units of the answer (amu) are correct. The magnitude of the answer is reasonable because it lies between 295.956 and 297.974 and is closer to 296, which has the highest abundance.

1.89 **Given:** $\frac{\text{mass 2 O}}{\text{mass 1 N}} = \frac{2.29}{1.00}$; $\frac{\text{mass 3 F}}{\text{mass 1 N}} = \frac{4.07}{1.00}$ **Find:** $\frac{\text{mass O}}{\text{mass 2 F}}$

Conceptual Plan: Mass O/N and mass F/N → mass O/F → mass O/2F

$\frac{\text{mass 2 O}}{\text{mass 1 N}}$ $\frac{\text{mass 3 F}}{\text{mass 1 N}}$; $\frac{\text{mass 2 O}}{\text{mass 3 F}}$

Solution: $\frac{\text{mass 2 O}}{\text{mass 1 N}} = \frac{2.29}{1.00}; \frac{\text{mass 3 F}}{\text{mass 1 N}} = \frac{4.07}{1.00}$ $\left(\frac{2.29 \text{ mass } \cancel{2O}}{4.07 \text{ mass } \cancel{3F}}\right)\left(\frac{1O}{\cancel{2O}}\right)\left(\frac{\cancel{3F}}{2\text{ F}}\right) = \frac{0.422 \text{ mass O}}{\text{mass 2 F}}$

Check: Mass ratio of O to F is reasonable because the mass of O is slightly less than the mass of fluorine.

1.90 **Given:** sample = 1.5886 g; ^{59}Co = 58.9332 amu; ^{60}Co = 59.9338 amu; apparent mass = 58.9901 amu
Find: mass of ^{60}Co in sample
Conceptual Plan: apparent mass → fraction ^{60}Co → mass ^{60}Co

Atomic mass $= \sum_n$(fraction of isotope n) × (mass of isotope n)

Solution: Let x = fraction of ^{60}Co; so $1.00 - x$ = fraction ^{59}Co.

$58.9901 \text{ amu} = (1.00 - x)(58.9332 \text{ amu}) + (x)(59.9338 \text{ amu})$

$x = 0.05686$

$1.5886 \text{ g sample} \times 0.05686 = 0.090328 \text{ g } ^{60}Co = 0.0903 \text{ g } ^{60}Co$

Check: The units of the answer (g ^{60}Co) are correct. The magnitude of the answer is reasonable because the apparent mass is very close to the mass ^{59}Co.

1.91 **Given:** 7.36 g Cu; 0.51 g Zn **Find:** atomic mass of sample
Conceptual Plan: fraction Cu and Zn → atomic mass

Atomic mass $= \sum_n$(fraction of atom n) × (mass of atom n)

Solution: 7.36 g Cu + 0.51 g Zn = 7.87 g sample

$$\left(\frac{7.36 \text{ g Cu}}{7.87 \text{ g sample}}\right)\left(\frac{63.55 \text{ g Cu}}{\text{mol Cu}}\right) + \left(\frac{0.51 \text{ g Zn}}{7.87 \text{ g sample}}\right)\left(\frac{65.41 \text{ g Zn}}{\text{mol Zn}}\right) = 63.67 \text{ g/mol}$$

Check: The units of the answer (g/mol) are correct. The magnitude of the answer is reasonable because it is between the mass of Cu (63.55 g/mol) and Zn (65.41 g/mol) and is closer to the mass of Cu.

1.92 **Given:** sample 1 = $N_2O_3 = \frac{\text{mass O}}{\text{mass N}} = \frac{12}{7}$; sample 2 = $\frac{\text{mass O}}{\text{mass N}} = \frac{16}{7}$ **Find:** formula of sample 2 and sample 3 (next in series)
Conceptual Plan: ratio O/N for N_2O_3 → x ratio O/N for sample 2 → ratio if O/O

Solution: $\frac{\text{mass O}}{\text{mass N}} = \frac{12}{7} = \frac{3\text{ O}}{2\text{ N}}$ $\frac{\text{mass O}}{\text{mass N}} = \frac{16}{7} = \frac{x\text{ O}}{2\text{ N}}$ $\frac{\text{mass O}}{\text{mass O}} = \frac{16}{12} = \frac{x\text{ O}}{3\text{ O}}$ so $x = 4$

Therefore, the formula is N_2O_4.

The next member of the series would be N_2O_5. Let y = ratio of O/N for sample 3.

$\frac{\text{mass O}}{\text{mass O}} = \frac{5\text{ O}}{3\text{ O}} = \frac{Y}{12}$ $y = 20$

So $\frac{\text{mass O}}{\text{mass N}} = \frac{20}{7}$

1.93 **Given:** Mg = 24.312 amu, ^{24}Mg = 23.98504, 78.99%, ^{26}Mg = 25.98259 amu, $\frac{\text{abundance } ^{25}Mg}{\text{abundance } ^{26}Mg} = \frac{0.9083}{1}$
Find: mass ^{25}Mg
Conceptual Plan: abundance of ^{24}Mg and ratio ^{25}Mg/^{26}Mg → abundance ^{25}Mg and ^{26}Mg → mass ^{25}Mg

Atomic mass $= \sum_n$(fraction of isotope n) × (mass of isotope n)

Solution: 100.00% − % abundance ^{24}Mg = % abundance ^{25}Mg and ^{26}Mg

100.00% − 78.99% = 21.01% ^{25}Mg and ^{26}Mg

fraction ^{25}Mg and ^{26}Mg $= \frac{21.01}{100.0} = 0.2101$

$\frac{\text{abundance } ^{25}Mg}{\text{abundance } ^{26}Mg} = \frac{0.9083}{1}$

Let x = fraction ^{26}Mg, $0.9083x$ = fraction ^{25}Mg

fraction ^{25}Mg and ^{26}Mg $= x + 0.9083x = 0.2101$

x = fraction ^{26}Mg = 0.1101, $0.9083x$ = fraction ^{25}Mg = 0.1000

$$\text{Atomic mass} = \sum_{n} (\text{fraction of isotope } n) \times (\text{mass of isotope } n)$$

$$24.312 = (0.7899)(23.98504 \text{ amu}) + (0.1000)(\text{mass } ^{25}\text{Mg}) + (0.1101)(25.98259 \text{ amu})$$

$$\text{mass } ^{25}\text{Mg} = 25.0\underline{5}6 \text{ amu} = 25.06 \text{ amu}$$

Check: The units of the answer (amu) are correct. The magnitude of the answer is reasonable because it is between the masses of ^{24}Mg and ^{26}Mg.

Conceptual Problems

1.94 No. Because the container is sealed, the atoms and molecules can move around but they cannot leave. If no atoms or molecules can leave, the mass must be constant.

1.95 (c) is the best representation. When solid carbon dioxide (dry ice) sublimes, it changes phase from a solid to a gas. Phase changes are physical changes, so no molecular bonds are broken. This diagram shows molecules with one carbon atom and two oxygen atoms bonded together in every molecule. The other diagrams have no carbon dioxide molecules.

1.96 Homogeneous mixture of elements A and B:

Compound AB in a liquid state:

Compound AB after evaporation:

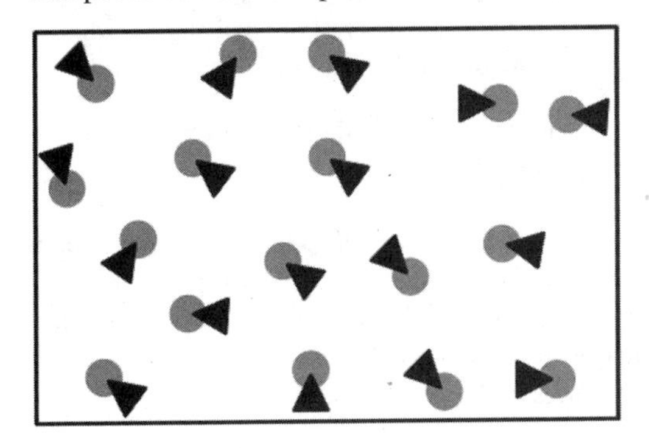

After decomposition of AB into elements A and B:

1.97 Remember that an observation is the information collected when studying phenomena. A law is a concise statement that summarizes observed behaviors and observations and predicts future observations. A theory attempts to explain why the observed behavior is happening.

(a) This statement is most like a law because it summarizes many observations and can explain future behavior—many places and many days.

(b) This statement is a theory because it attempts to explain why (gravitational forces).

(c) This statement is most like an observation because it is information collected to understand tidal behavior.

(d) This statement is most like a law because it summarizes many observations and can explain future behavior—many places and many days and months.

2 Measurement, Problem Solving, and the Mole Concept

Review Questions

2.1 The reliability of a measurement depends on the instrument used to make the measurement. The more precise the instrument, the more reliably we can differentiate between two measurements. We must consider the reliability of measurements when reporting and manipulating them.

2.2 Measured quantities are reported so that the number of digits reflects the uncertainty in the measurement. The non–place-holding digits in a reported number are called significant figures.

2.3 Accuracy refers to how close the measured value is to the actual value. Precision refers to how close a series of measurements are to one another or how reproducible they are. A series of measurements can be precise (close to one another in value and reproducible) but not accurate (not close to the true value).

2.4 Random error is error that has equal probability of being too high or too low. Almost all measurements have some degree of random error. Random error can, with enough trials, average itself out. Systematic error is error that tends toward being either too high or too low. Systematic error does not average out with repeated trials.

2.5 The density (d) of a substance is the ratio of its mass (m) to its volume (V):

$$\text{Density} = \frac{\text{Mass}}{\text{Volume}} \quad \text{or} \quad d = \frac{m}{V}$$

The density of a substance is an example of an intensive property, one that is independent of the amount of the substance. Mass is one of the properties used to calculate the density of a substance. Mass, in contrast, is an extensive property, one that depends on the amount of the substance.

2.6 An intensive property is a property that is independent of the amount of the substance. An extensive property is a property that depends on the amount of the substance.

2.7 Energy is the capacity to do work. Work is the result of a force acting through a distance. Kinetic energy is associated with the motion of an object. Potential energy is associated with the position or composition of an object. Examples of kinetic energy are a moving billiard ball, movement of gas molecules, and a raging river. Examples of potential energy are a billiard ball raised above the surface of a billiard table, a compressed spring, and bonds in molecules.

2.8 The law of conservation of energy states that energy can be neither created nor destroyed. Energy can be transferred from one object to another, and it can assume different forms. In an energy exchange, energy is transferred between the system and the surroundings. If the system loses energy, the surroundings gain energy, and vice versa.

In chemical and physical changes, matter often exchanges energy with its surroundings. In these exchanges, the total energy is always conserved; energy is neither created nor destroyed. Systems with high potential energy tend to change in the direction of lower potential energy, releasing energy into the surroundings.

2.9 Chemical energy is potential energy. It is the energy that is contained in the bonds that hold the molecules together. This energy arises primarily from electrostatic forces between the electrically charged particles (protons and electrons) that compose atoms and molecules. Some of these arrangements—such as the one within the molecules that compose gasoline—have a much higher potential energy than others. When gasoline undergoes combustion, the arrangement of these particles changes, creating molecules with much lower potential energy and transferring a great deal of energy (mostly in the form of heat) to the surroundings. A raised weight has a certain amount of potential energy (dependent on the height the weight is raised) that can be converted to kinetic energy when the weight is released.

2.10 A process in which a system transfers energy to the surroundings is exothermic, while a process in which a system gains energy from the surroundings is endothermic.

2.11 Using units as a guide to solving problems is often called dimensional analysis. This approach to solving many chemical problems involves four steps: sorting the information in the problem; strategizing about how to solve the problem; solving the problem; and checking the answer.

2.12 Units should always be included in calculations; they are multiplied, divided, and canceled like any other algebraic quantity.

2.13 A mole is an amount of material. It is defined as the amount of material containing 6.0221421×10^{23} particles (Avogadro's number). The numerical value of the mole is defined as being equal to the number of atoms in exactly 12 grams of pure carbon-12. It is useful for converting number of atoms to moles of atoms and moles of atoms to number of atoms.

2.14 The mass corresponding to a mole of one element is different from the mass corresponding to a mole of another element because the mass of an atom of each element is different. A mole is a specific number of atoms; so the heavier the mass of each atom, the heavier the mass of one mole of atoms.

The Reliability of a Measurement and Significant Figures

2.15 (c) is the correct measurement to report. The usual procedure is to divide the space between the finest markings into ten and make that estimation the last digit reported. In this case the markings are every 1 mm, so the measurement is to be reported to ± 0.1 mm.

2.16 (a) is the correct measurement to report. The usual procedure is to divide the space between the finest markings into ten and make that estimation the last digit reported. In this case the markings are every 0.1 kg, so the measurement is to be reported to ± 0.01 kg.

2.17 To obtain the readings, look to see where the bottom of the meniscus lies. Estimate the distance between two markings on the device.

(a) 73.2 mL—the bottom of the meniscus appears to be sitting just above the 73 mL mark.

(b) 88.2 °C—the mercury is between the 84 °C mark and the 85 °C mark, but it is closer to the lower number.

(c) 645 mL—the meniscus appears to be just above the 640 mL mark.

2.18 To obtain the readings, look to see where the bottom of the meniscus lies. Estimate the distance between two markings on the device. Use all digits on a digital device.

(a) 4.50 mL—the meniscus appears to be on the 4.5 mL mark.

(b) 27.43 °C—the mercury is just above the 27.4 °C mark. Note that the tens digit is only labeled every 10 °C.

(c) 0.873 g—read all of the places on the digital display.

Density

2.19 **Given:** $m = 2.49$ g; $V = 0.349\ \text{cm}^3$ **Find:** d in g/cm^3 and compare to pure copper.

Conceptual Plan: $m, V \rightarrow d$

$d = m/V$

Compare to the published value: d (pure copper) $= 8.96\ \text{g/cm}^3$. (This value is in Table 2.1.)

Solution: $d = \dfrac{2.49\ \text{g}}{0.349\ \text{cm}^3} = 7.13\ \dfrac{\text{g}}{\text{cm}^3}$

The density of the penny is much smaller than the density of pure copper ($7.13\ g/cm^3$; $8.96\ g/cm^3$), so the penny is not pure copper.

Check: The units (g/cm^3) are correct. The magnitude of the answer seems correct. Many coins are layers of metals, so it is not surprising that the penny is not pure copper.

2.20 **Given:** $m = 1.41$ kg; $V = 0.314$ L **Find:** d in g/cm^3 and compare to pure titanium.

Conceptual Plan: $m, V \rightarrow d$ **then kg** $\rightarrow$ **g then L** $\rightarrow$ cm^3

$$d = m/V \qquad \frac{1000\ g}{1\ kg} \qquad \frac{1000\ cm^3}{1\ L}$$

Compare to the published value: d (pure titanium) $= 4.51\ g/cm^3$. (This value is in Table 2.1.)

Solution: $d = \frac{1.41\ kg}{0.314\ L} \times \frac{1000\ g}{1\ kg} \times \frac{1\ L}{1000\ cm^3} = 4.49 \frac{g}{cm^3}$

Check: The units (g/cm^3) are correct. The magnitude of the answer seems correct. The density of the frame is almost exactly the density of pure titanium ($4.49\ g/cm^3$ versus $4.51\ g/cm^3$), so the frame could be titanium.

2.21 **Given:** $m = 4.10 \times 10^3$ g; $V = 3.25$ L **Find:** d in g/cm^3

Conceptual Plan: $m, V \rightarrow d$ **then L** $\rightarrow$ cm^3

$$d = m/V \qquad \frac{1000\ cm^3}{1\ L}$$

Solution: $d = \frac{4.10 \times 10^3\ g}{3.25\ L} \times \frac{1\ L}{1000\ cm^3} = 1.26 \frac{g}{cm^3}$

Check: The units (g/cm^3) are correct. The magnitude of the answer seems correct.

2.22 **Given:** $m = 371$ grams; $V = 19.3$ mL **Find:** d in g/cm^3 and compare to pure gold.

Conceptual Plan: $m, V \rightarrow d$

$$d = m/V$$

Compare to the published value: d (pure gold) $= 19.3$ g/mL. (This value is in Table 2.1.)

Solution: $d = \frac{371\ g}{19.3\ mL} = 19.2 \frac{g}{mL}$

The density of the nugget is essentially the same as the density of pure gold (19.2 g/mL versus 19.3 g/mL), so the nugget could be gold.

Check: The units (g/cm^3) are correct. The magnitude of the answer seems correct and is essentially the same as the density of pure gold.

2.23 (a) **Given:** $d = 1.11\ g/cm^3$; $V = 417$ mL **Find:** m

Conceptual Plan: $d, V \rightarrow m$ **then** $cm^3 \rightarrow$ **mL**

$$d = m/V \qquad \frac{1\ mL}{1\ cm^3}$$

Solution: $d = m/V$ Rearrange by multiplying both sides of the equation by V. $m = d \times V$

$$m = 1.11 \frac{g}{cm^3} \times \frac{1\ cm^3}{1\ mL} \times 417\ mL = 4.63 \times 10^2\ g$$

Check: The units (g) are correct. The magnitude of the answer seems correct considering that the value of the density is about $1\ g/cm^3$.

(b) **Given:** $d = 1.11\ g/cm^3$; $m = 4.1$ kg **Find:** V in L

Conceptual Plan: $d, V \rightarrow m$ **then kg** $\rightarrow$ **g and** $cm^3 \rightarrow$ **L**

$$d = m/V \qquad \frac{1000\ g}{1\ kg} \qquad \frac{1\ L}{1000\ cm^3}$$

Solution: $d = m/V$ Rearrange by multiplying both sides of the equation by V and dividing both sides of the equation by d.

$$V = \frac{m}{d} = \frac{4.1\ kg}{1.11 \frac{g}{cm^3}} \times \frac{1000\ g}{1\ kg} = 3.7 \times 10^3\ cm^3 \times \frac{1\ L}{1000\ cm^3} = 3.7\ L$$

Check: The units (L) are correct. The magnitude of the answer seems correct considering that the value of the density is about $1\ g/cm^3$.

2.24 (a) **Given:** $d = 0.7857\ \text{g/cm}^3$; $V = 28.56$ mL **Find:** m

Conceptual Plan: $d, V \rightarrow m$

$d = m/V$

Solution: $d = m/V$ Rearrange by multiplying both sides of the equation by V. $m = d \times V$

$$m = \left(0.7857\ \frac{\text{g}}{\text{cm}^3}\right) \times \frac{1\ \text{cm}^3}{1\ \text{mL}} \times (28.56\ \text{mL}) = 22.44\ \text{g}$$

Check: The units (g) are correct. The magnitude of the answer seems correct considering that the value of the density is less than $1\ \text{g/cm}^3$.

(b) **Given:** $d = 0.7857\ \text{g/cm}^3$; $m = 6.54$ g **Find:** V

Conceptual Plan: $d, m \rightarrow V$ **then** $\text{cm}^3 \rightarrow$ **mL**

$d = m/V$ $\quad \frac{1\ \text{mL}}{1\ \text{cm}^3}$

Solution: $d = m/V$ Rearrange by multiplying both sides of the equation by V and dividing both sides of the equation by d.

$$V = \frac{m}{d} = \frac{6.54\ \text{g}}{0.7857\ \frac{\text{g}}{\text{cm}^3}} = 8.32\ \text{cm}^3 \times \frac{1\ \text{mL}}{1\ \text{cm}^3} = 8.32\ \text{mL}$$

Check: The units (mL) are correct. The magnitude of the answer seems correct considering that the value of the density is less than $1\ \text{g/cm}^3$.

2.25 **Given:** $V = 245$ L, $d = 0.821$ g/mL **Find:** m

Conceptual Plan: g/mL $\rightarrow$ g/L then $d, V \rightarrow m$

$\frac{1000\ \text{mL}}{1\ \text{L}}$ $\quad d = m/V$

Solution: $d = m/V$ Rearrange by multiplying both sides of the equation by V. $m = d \times V$

$$m = 245\ \text{L} \times \frac{1000\ \text{mL}}{1\ \text{L}} \times \left(0.821\ \frac{\text{g}}{\text{mL}}\right) = 2.01 \times 10^5\ \text{g}$$

Check: The units (g) are correct. The magnitude of the answer seems correct considering the value of the density is less than 1 g/mL and the volume is very large.

2.26 **Given:** $d = 0.918\ \text{g/cm}^3$, $m = 10.0$ lbs **Find:** V

Conceptual Plan: lb $\rightarrow$ g then $d, m \rightarrow V$

$\frac{453.59\ \text{g}}{1\ \text{lb}}$ $\quad d = m/V$

Solution: $10.0\ \text{lbs} \times \frac{453.59\ \text{g}}{1\ \text{lbs}} = 4.5\underline{3}59 \times 10^3$ g; then $d = m/V$. Rearrange by multiplying both sides of the equation by V and dividing both sides of the equation by d.

$$V = \frac{m}{d} = \frac{4.5\underline{3}59 \times 10^3\ \text{g}}{0.918\ \frac{\text{g}}{\text{cm}^3}} = 4.94 \times 10^3\ \text{cm}^3$$

Check: The units (mL) are correct. The magnitude of the answer seems correct considering the value of the density is less than $1\ \text{g/cm}^3$.

Unit Conversions

2.27 (a) **Given:** 27.8 L **Find:** cm^3

Conceptual Plan: $\text{L} \rightarrow \text{cm}^3$

$\frac{1000\ \text{cm}^3}{1\ \text{L}}$

Solution: $27.8\ \text{L} \times \frac{1000\ \text{cm}^3}{1\ \text{L}} = 2.78 \times 10^4\ \text{cm}^3$

Check: The units (cm^3) are correct. The magnitude of the answer (10^4) makes physical sense because cm^3 is much smaller than a liter; so the answer should go up several orders of magnitude. Three significant figures are allowed because of the limitation of 27.8 L (three significant figures).

(b) **Given:** 1898 mg **Find:** kg

Conceptual Plan: $\mathbf{mg \rightarrow g \rightarrow kg}$

$$\frac{1\ \text{g}}{1000\ \text{mg}} \quad \frac{1\ \text{kg}}{1000\ \text{g}}$$

Solution: $1898\ \cancel{\text{mg}} \times \dfrac{1\ \cancel{\text{g}}}{1000\ \cancel{\text{mg}}} \times \dfrac{1\ \text{kg}}{1000\ \cancel{\text{g}}} = 1.898 \times 10^{-3}\ \text{kg}$

Check: The units (kg) are correct. The magnitude of the answer (10^{-3}) makes physical sense because a kilogram is a much larger unit than a milligram. Four significant figures are allowed because 1898 mg has four significant figures.

(c) **Given:** 198 km **Find:** cm

Conceptual Plan: $\mathbf{km \rightarrow m \rightarrow cm}$

$$\frac{1000\ \text{m}}{1\ \text{km}} \quad \frac{100\ \text{cm}}{1\ \text{m}}$$

Solution: $198\ \cancel{\text{km}} \times \dfrac{1000\ \cancel{\text{m}}}{1\ \cancel{\text{km}}} \times \dfrac{100\ \text{cm}}{1\ \cancel{\text{m}}} = 1.98 \times 10^{7}\ \text{cm}$

Check: The units (cm) are correct. The magnitude of the answer (10^7) makes physical sense because a kilometer is a much larger unit than a centimeter. Three significant figures are allowed because 198 km has three significant figures.

2.28 (a) **Given:** 28.9 nm **Find:** μm

Conceptual Plan: $\mathbf{nm \rightarrow m \rightarrow \mu m}$

$$\frac{10^{-9}\ \text{m}}{1\ \text{nm}} \quad \frac{10^{6}\ \mu\text{m}}{1\ \text{m}}$$

Solution: $28.9\ \cancel{\text{nm}} \times \dfrac{10^{-9}\ \cancel{\text{m}}}{1\ \cancel{\text{nm}}} \times \dfrac{10^{6}\ \mu\text{m}}{1\ \cancel{\text{m}}} = 2.89 \times 10^{-2}\ \mu\text{m} = 0.0289\ \mu\text{m}$

Check: The units (μm) are correct. The magnitude of the answer (10^{-2}) makes physical sense because a micrometer is a much larger unit than a nanometer. Three significant figures are allowed because 28.9 nm has three significant figures.

(b) **Given:** 1432 cm^3 **Find:** L

Conceptual Plan: $\mathbf{cm^3 \rightarrow L}$

$$\frac{1\ \text{L}}{1000\ \text{cm}^3}$$

Solution: $1432\ \cancel{\text{cm}^3} \times \dfrac{1\ \text{L}}{1000\ \cancel{\text{cm}^3}} = 1.432\ \text{L}$

Check: The units (L) are correct. The magnitude of the answer (1) makes physical sense because cm^3 is much smaller than a liter; so the answer should go down several orders of magnitude. Four significant figures are allowed because of the limitation of 1432 cm^3 (four significant figures).

(c) **Given:** 1211 Tm **Find:** Gm

Conceptual Plan: $\mathbf{Tm \rightarrow m \rightarrow Gm}$

$$\frac{10^{12}\ \text{m}}{1\ \text{Tm}} \quad \frac{10^{-9}\ \text{Gm}}{1\ \text{m}}$$

Solution: $1211\ \cancel{\text{Tm}} \times \dfrac{10^{12}\ \cancel{\text{m}}}{1\ \cancel{\text{Tm}}} \times \dfrac{10^{-9}\ \text{Gm}}{1\ \cancel{\text{m}}} = 1.211 \times 10^{6}\ \text{Gm}$

Check: The units (Gm) are correct. The magnitude of the answer (10^6) makes physical sense because a terameter is a much larger unit than a gigameter. Four significant figures are allowed because 1211 Tm has four significant figures.

2.29 (a) **Given:** 154 cm **Find:** in

Conceptual Plan: cm → in

$\frac{1 \text{ in}}{2.54 \text{ cm}}$

Solution: $154 \text{ cm} \times \frac{1 \text{ in}}{2.54 \text{ cm}} = 60.\underline{6}2992 \text{ in} = 60.6 \text{ in}$

Check: The units (in) are correct. The magnitude of the answer (60.6) makes physical sense because an inch is a larger unit than a cm. Three significant figures are allowed because 154 cm has three significant figures.

(b) **Given:** 3.14 kg **Find:** g

Conceptual Plan: kg → g

$\frac{1000 \text{ g}}{1 \text{ kg}}$

Solution: $3.14 \text{ kg} \times \frac{1000 \text{ g}}{1 \text{ kg}} = 3.14 \times 10^3 \text{ g}$

Check: The units (g) are correct. The magnitude of the answer (10^3) makes physical sense because a kilogram is a much larger unit than a gram. Three significant figures are allowed because 3.14 kg has three significant figures.

(c) **Given:** 3.5 L **Find:** qt

Conceptual Plan: L → qt

$\frac{1.057 \text{ qt}}{1 \text{ L}}$

Solution: $3.5 \text{ L} \times \frac{1.057 \text{ qt}}{1 \text{ L}} = 3.6995 \text{ qt} = 3.7 \text{ qt}$

Check: The units (qt) are correct. The magnitude of the answer (3.7) makes physical sense because a L is a smaller unit than a qt. Two significant figures are allowed because 3.5 L has two significant figures. Round the last digit up because the first nonsignificant digit is a 9.

(d) **Given:** 109 mm **Find:** in

Conceptual Plan: mm → m → in

$\frac{1 \text{ m}}{1000 \text{ mm}}$ $\frac{39.37 \text{ in}}{1 \text{ m}}$

Solution: $109 \text{ mm} \times \frac{1 \text{ m}}{1000 \text{ mm}} \times \frac{39.37 \text{ in}}{1 \text{ m}} = 4.2\underline{9}133 \text{ in} = 4.29 \text{ in}$

Check: The units (in) are correct. The magnitude of the answer (4) makes physical sense because a mm is a much smaller unit than an inch. Three significant figures are allowed because 109 mm has three significant figures.

2.30 (a) **Given:** 1.4 in **Find:** mm

Conceptual Plan: in → cm → m → mm

$\frac{2.54 \text{ cm}}{1 \text{ in}}$ $\frac{1 \text{ m}}{100 \text{ cm}}$ $\frac{1000 \text{ mm}}{1 \text{ m}}$

Solution: $1.4 \text{ in} \times \frac{2.54 \text{ cm}}{1 \text{ in}} \times \frac{1 \text{ m}}{100 \text{ cm}} \times \frac{1000 \text{ mm}}{1 \text{ m}} = 3\underline{5}.56 \text{ mm} = 36 \text{ mm}$

Check: The units (mm) are correct. The magnitude of the answer (36) makes physical sense because a mm is smaller than an inch. Two significant figures are allowed because 1.4 in has two significant figures. Round the last digit up because the first nonsignificant digit is a 5.

(b) **Given:** 116 ft **Find:** cm

Conceptual Plan: ft → in → cm

$\frac{12 \text{ in}}{1 \text{ ft}}$ $\frac{2.54 \text{ cm}}{1 \text{ in}}$

Solution: $116 \text{ ft} \times \frac{12 \text{ in}}{1 \text{ ft}} \times \frac{2.54 \text{ cm}}{1 \text{ in}} = 3.5\underline{3}57 \times 10^3 \text{ cm} = 3.54 \times 10^3 \text{ cm}$

Check: The units (cm) are correct. The magnitude of the answer (10^3) makes physical sense because a foot is a much larger unit than a cm. Three significant figures are allowed because 116 ft has three significant figures. Round the last digit up because the first nonsignificant digit is a 5.

(c) **Given:** 1845 kg **Find:** lb

Conceptual Plan: **kg → g → lb**

$$\frac{1000\ g}{1\ kg} \quad \frac{1\ lb}{453.6\ g}$$

Solution: $1845\ \cancel{kg} \times \frac{1000\ \cancel{g}}{1\ \cancel{kg}} \times \frac{1\ lb}{453.6\ \cancel{g}} = 4.06\underline{7}5 \times 10^3\ lb = 4.068 \times 10^3\ lb$

Check: The units (lb) are correct. The magnitude of the answer (10^3) makes physical sense because a lb is a smaller unit than a kg. Four significant figures are allowed because 1845 kg and 453.6 g/lb each have four significant figures. Round the last digit up because the first nonsignificant digit is a 5.

(d) **Given:** 815 yd **Find:** km

Conceptual Plan: **yd → m → km**

$$\frac{1\ m}{1.094\ yd} \quad \frac{1\ km}{1000\ m}$$

Solution: $815\ \cancel{yd} \times \frac{1\ \cancel{m}}{1.094\ \cancel{yd}} \times \frac{1\ km}{1000\ \cancel{m}} = 0.74\underline{4}9726\ km = 0.745\ km$

Check: The units (km) are correct. The magnitude of the answer (0.7) makes physical sense because a yard is a much smaller unit than a kilometer. Three significant figures are allowed because 815 yd has three significant figures. Round the last digit up because the first nonsignificant digit is a 9.

2.31 **Given:** 10.0 km **Find:** minutes **Other:** running pace = 7.5 miles per hour

Conceptual Plan: **km → mi → hr → min**

$$\frac{0.6214\ mi}{1\ km} \quad \frac{1\ hr}{7.5\ mi} \quad \frac{60\ min}{1\ hr}$$

Solution: $10.0\ \cancel{km} \times \frac{0.6214\ \cancel{mi}}{1\ \cancel{km}} \times \frac{1\ \cancel{hr}}{7.5\ \cancel{mi}} \times \frac{60\ min}{1\ \cancel{hr}} = 4\underline{9}.712\ min = 50.\ min = 5.0 \times 10^1\ min$

Check: The units (min) are correct. The magnitude of the answer (50) makes physical sense because she is running almost 7.5 miles (which would take her 60 min = 1 hr). Two significant figures are allowed because of the limitation of 7.5 mi/hr (two significant figures). Round the last digit up because the first nonsignificant digit is a 7.

2.32 **Given:** 212 km **Find:** hours **Other:** riding pace = 18 miles per hour

Conceptual Plan: **km → mi → hr**

$$\frac{0.6214\ mi}{1\ km} \quad \frac{1\ hr}{18\ mi}$$

Solution: $212\ \cancel{km} \times \frac{0.6214\ \cancel{mi}}{1\ \cancel{km}} \times \frac{1\ hr}{18\ \cancel{mi}} = 7.\underline{3}18711\ hr = 7.3\ hr$

Check: The units (hr) are correct. The magnitude of the answer (7) makes physical sense because she is riding over 100 miles (which would take her over 4 hr). Two significant figures are allowed because of the limitation of 24 mi/hr (two significant figures). Truncate the last digit because the first nonsignificant digit is a 1.

2.33 **Given:** 17 km/L **Find:** miles per gallon

Conceptual Plan: $\frac{\mathbf{km}}{\mathbf{L}} \rightarrow \frac{\mathbf{mi}}{\mathbf{L}} \rightarrow \frac{\mathbf{mi}}{\mathbf{gal}}$

$$\frac{0.6214\ mi}{1\ km} \quad \frac{3.785\ L}{1\ gal}$$

Solution: $\frac{17\ \cancel{km}}{1\ \cancel{L}} \times \frac{0.6214\ mi}{1\ \cancel{km}} \times \frac{3.785\ \cancel{L}}{1\ gal} = 3\underline{9}.98398\ \frac{mi}{gal} = 40.\ \frac{mi}{gal}$

Check: The units (mi/gal) are correct. The magnitude of the answer (40) makes physical sense because the dominating factor is that a liter is much smaller than a gallon; so the answer should go up. Two significant figures are allowed because of the limitation of 17 km/L (two significant figures). Round the last digit up because the first nonsignificant digit is a 9.

2.34 **Given:** 5.0 gallons **Find:** cm^3

Conceptual Plan: gal → L → cm^3

$\frac{3.785 \text{ L}}{1 \text{ gal}}$ $\frac{1000 \text{ cm}^3}{1 \text{ L}}$

Solution: $5.0 \text{ gal} \times \frac{3.785 \text{ L}}{1 \text{ gal}} \times \frac{1000 \text{ cm}^3}{1 \text{ L}} = 1.8925 \times 10^4 \text{ cm}^3 = 1.9 \times 10^4 \text{ cm}^3$

Check: The units (cm^3) are correct. The magnitude of the answer (10^4) makes physical sense because cm^3 is much smaller than a gallon; so the answer should go up several orders of magnitude. Two significant figures are allowed because of the limitation of 5.0 gallons (two significant figures). Round the last digit up because the first nonsignificant digit is a 9.

2.35 (a) **Given:** 195 m^2 **Find:** km^2

Conceptual Plan: $m^2 \rightarrow km^2$

$\frac{(1 \text{ km})^2}{(1000 \text{ m})^2}$

Notice that for squared units, the conversion factors must be squared.

Solution: $195 \text{ m}^2 \times \frac{(1 \text{ km})^2}{(1000 \text{ m})^2} = 1.95 \times 10^{-4} \text{ km}^2$

Check: The units (km^2) are correct. The magnitude of the answer (10^{-4}) makes physical sense because a kilometer is a much larger unit than a meter.

(b) **Given:** 195 m^2 **Find:** dm^2

Conceptual Plan: $m^2 \rightarrow dm^2$

$\frac{(10 \text{ dm})^2}{(1 \text{ m})^2}$

Notice that for squared units, the conversion factors must be squared.

Solution: $195 \text{ m}^2 \times \frac{(10 \text{ dm})^2}{(1 \text{ m})^2} = 1.95 \times 10^4 \text{ dm}^2$

Check: The units (dm^2) are correct. The magnitude of the answer (10^4) makes physical sense because a decimeter is a much smaller unit than a meter.

(c) **Given:** 195 m^2 **Find:** cm^2

Conceptual Plan: $m^2 \rightarrow cm^2$

$\frac{(100 \text{ cm})^2}{(1 \text{ m})^2}$

Notice that for squared units, the conversion factors must be squared.

Solution: $195 \text{ m}^2 \times \frac{(100 \text{ cm})^2}{(1 \text{ m})^2} = 1.95 \times 10^6 \text{ cm}^2$

Check: The units (cm^2) are correct. The magnitude of the answer (10^6) makes physical sense because a centimeter is a much smaller unit than a meter.

2.36 (a) **Given:** 115 m^3 **Find:** km^3

Conceptual Plan: $m^3 \rightarrow km^3$

$\frac{(1 \text{ km})^3}{(1000 \text{ m})^3}$

Notice that for cubed units, the conversion factors must be cubed.

Solution: $115 \text{ m}^3 \times \frac{(1 \text{ km})^3}{(1000 \text{ m})^3} = 1.15 \times 10^{-7} \text{ km}^3$

Check: The units (km^3) are correct. The magnitude of the answer (10^{-7}) makes physical sense because a kilometer is a much larger unit than a meter.

(b) **Given:** 115 m^3 **Find:** dm^3

Conceptual Plan: $m^3 \rightarrow dm^3$

$$\frac{(10\ dm)^3}{(1\ m)^3}$$

Notice that for cubed units, the conversion factors must be cubed.

Solution: $115\ m^3 \times \frac{(10\ dm)^3}{(1\ m)^3} = 1.15 \times 10^5\ dm^3$

Check: The units (dm^3) are correct. The magnitude of the answer (10^5) makes physical sense because a decimeter is a much smaller unit than a meter.

(c) **Given:** 115 m^3 **Find:** cm^3

Conceptual Plan: $m^3 \rightarrow cm^3$

$$\frac{(100\ cm)^3}{(1\ m)^3}$$

Notice that for cubed units, the conversion factors must be cubed.

Solution: $115\ m^3 \times \frac{(100\ cm)^3}{(1\ m)^3} = 1.15 \times 10^8\ cm^3$

Check: The units (cm^3) are correct. The magnitude of the answer (10^8) makes physical sense because a centimeter is a much smaller unit than a meter.

2.37 **Given:** 435 acres **Find:** square miles **Other:** 1 acre = 43,560 ft^2; 1 mile = 5280 ft

Conceptual Plan: acres $\rightarrow ft^2 \rightarrow mi^2$

$$\frac{43560\ ft^2}{1\ acre} \quad \frac{(1\ mi)^2}{(5280\ ft)^2}$$

Notice that for squared units, the conversion factors must be squared.

Solution: $435\ acres \times \frac{43560\ ft^2}{1\ acre} \times \frac{(1\ mi)^2}{(5280\ ft)^2} = 0.6796875\ mi^2 = 0.680\ mi^2$

Check: The units (mi^2) are correct. The magnitude of the answer (0.7) makes physical sense because an acre is much smaller than a mi^2; so the answer should go down several orders of magnitude. Three significant figures are allowed because of the limitation of 435 acres (three significant figures). Round the last digit up because the first nonsignificant digit is a 6.

2.38 (a) **Given:** 954 million acres **Find:** square miles **Other:** 1 acre = 43,560 ft^2; 1 mile = 5280 ft

Conceptual Plan: Substitute 10^6 for million then acres $\rightarrow ft^2 \rightarrow mi^2$

$$\frac{43560\ ft^2}{1\ acre} \quad \frac{(1\ mi)^2}{(5280\ ft)^2}$$

Notice that for squared units, the conversion must be squared.

Solution: 954 million acres = 954×10^6 acres

$$954 \times 10^6\ acres \times \frac{43560\ ft^2}{1\ acre} \times \frac{(1\ mi)^2}{(5280\ ft)^2} = 1.490625 \times 10^6\ mi^2 = 1.49 \times 10^6\ mi^2$$

Check: The units (mi^2) are correct. The magnitude of the answer (10^6) makes physical sense because an acre is much smaller than a mi^2; so the answer should go down several orders of magnitude. Three significant figures are allowed because of the limitation of 954 acres (three significant figures). Truncate the last digits because the first nonsignificant digit is a 0.

(b) **Given:** 3.537 million square miles **Find:** percentage of U.S. land is farmland

Conceptual Plan: Substitute 10^6 for million then $\text{farm} = \frac{\text{farmland}}{\text{total land}} \times 100\%$

Note: Units of farmland and total land must be the same.

Solution: 3.537 million mi^2 = $3.537 \times 10^6\ mi^2$

$$\% \text{ farmland} = \frac{1.49 \times \cancel{10^6}\,\cancel{\text{mi}^2}}{3.537 \times \cancel{10^6}\,\cancel{\text{mi}^2}} \times 100\% = 42.\underline{1}2609556\% \text{ farmland} = 42.1\% \text{ farmland}$$

Check: The units (%) are correct. The magnitude of the answer (42%) makes physical sense because less and less land is devoted to farmland. Three significant figures are allowed because of the limitation of 954 acres (three significant figures). Truncate the last digit because the first nonsignificant digit is a 2.

2.39 **Given:** 14 lb **Find:** mL **Other:** 80 mg/0.80 mL; 15 mg/kg body

Conceptual Plan: **lb → kg body → mg → mL**

$$\frac{1 \text{ kg body}}{2.205 \text{ lb}} \quad \frac{15 \text{ mg}}{1 \text{ kg body}} \quad \frac{0.80 \text{ mL}}{80 \text{ mg}}$$

Solution: $14\,\cancel{\text{lb}} \times \frac{1\,\cancel{\text{kg body}}}{2.205\,\cancel{\text{lb}}} \times \frac{15\,\cancel{\text{mg}}}{1\,\cancel{\text{kg body}}} \times \frac{0.80 \text{ mL}}{80\,\cancel{\text{mg}}} = 0.9\underline{5}23809524 \text{ mL} = 0.95 \text{ mL}$

Check: The units (mL) are correct. The magnitude of the answer (1 mL) makes physical sense because it is a reasonable amount of liquid to give to a baby. Two significant figures are allowed because of the statement in the problem. Truncate after the last significant digit because the first nonsignificant digit is a 2.

2.40 **Given:** 18 lb **Find:** mL **Other:** 100 mg/5.0 mL; 10 mg/kg body

Conceptual Plan: **lb → kg body → mg → mL**

$$\frac{1 \text{ kg body}}{2.205 \text{ lb}} \quad \frac{10 \text{ mg}}{1 \text{ kg body}} \quad \frac{5.0 \text{ mL}}{100 \text{ mg}}$$

Solution: $18\,\cancel{\text{lb}} \times \frac{1\,\cancel{\text{kg body}}}{2.205\,\cancel{\text{lb}}} \times \frac{10\,\cancel{\text{mg}}}{1\,\cancel{\text{kg body}}} \times \frac{5.0 \text{ mL}}{100\,\cancel{\text{mg}}} = 4.\underline{0}81632653 \text{ mL} = 4.1 \text{ mL}$

Check: The units (mL) are correct. The magnitude of the answer (4 mL) makes physical sense because it is a reasonable amount of liquid to give to a baby. Two significant figures are allowed because of the statement in the problem. Round the last digit up because the first nonsignificant digit is an 8.

2.41 (a) **Given:** 534 kWh **Find:** J

Conceptual Plan: **kWh → J**

$$\frac{3.60 \times 10^6 \text{ J}}{1 \text{ kWh}}$$

Solution: $534\,\cancel{\text{kWh}} \times \frac{3.60 \times 10^6 \text{ J}}{1\,\cancel{\text{kWh}}} = 1.92 \times 10^9 \text{ J}$

Check: The units (J) are correct. The magnitude of the answer (10^9) makes physical sense because a kWh is much larger than a Joule; so the answer increases.

(b) **Given:** 215 kJ **Find:** Cal

Conceptual Plan: **kJ → J → Cal**

$$\frac{1000 \text{ J}}{1 \text{ kJ}} \quad \frac{1 \text{ Cal}}{4184 \text{ J}}$$

Solution: $215\,\cancel{\text{kJ}} \times \frac{1000\,\cancel{\text{J}}}{1\,\cancel{\text{kJ}}} \times \frac{1 \text{ Cal}}{4184\,\cancel{\text{J}}} = 51.4 \text{ Cal}$

Check: The units (Cal) are correct. The magnitude of the answer (51) makes physical sense because a Calorie is about $\frac{1}{4}$ of a kJ; so the answer decreases by a factor of about four.

(c) **Given:** 567 Cal **Find:** J

Conceptual Plan: **Cal → J**

$$\frac{4184 \text{ J}}{1 \text{ Cal}}$$

Solution: $567\,\cancel{\text{Cal}} \times \frac{4184 \text{ J}}{1\,\cancel{\text{Cal}}} = 2.37 \times 10^6 \text{ J}$

Check: The units (J) are correct. The magnitude of the answer (10^6) makes physical sense because a Calorie is much larger than a Joule; so the answer increases.

(d) **Given:** 2.85×10^3 J **Find:** cal

Conceptual Plan: **J → cal**

$\frac{1 \text{ cal}}{4.184 \text{ J}}$

Solution: $2.85 \times 10^3 \text{ J} \times \frac{1 \text{ cal}}{4.184 \text{ J}} = 681 \text{ cal}$

Check: The units (cal) are correct. The magnitude of the answer (680) makes physical sense because a J is about $\frac{1}{4}$ the size of a calorie; so the answer decreases by a factor of about four.

2.42 (a) **Given:** 231 cal **Find:** kJ

Conceptual Plan: **cal → J → kJ**

$\frac{4.184 \text{ J}}{1 \text{ cal}}$ $\frac{1 \text{ kJ}}{1000 \text{ J}}$

Solution: $231 \text{ cal} \times \frac{4.184 \text{ J}}{1 \text{ cal}} \times \frac{1 \text{ kJ}}{1000 \text{ J}} = 0.967 \text{ kJ}$

Check: The units (kJ) are correct. The magnitude of the answer (1) makes physical sense because a kJ is much larger than a calorie; so the answer decreases.

(b) **Given:** 132×10^4 kJ **Find:** kcal

Conceptual Plan: **kJ → J → cal → kcal**

$\frac{1000 \text{ J}}{1 \text{k J}}$ $\frac{1 \text{ cal}}{4.184 \text{ J}}$ $\frac{1 \text{ kcal}}{1000 \text{ cal}}$

Solution: $132 \times 10^4 \text{ kJ} \times \frac{1000 \text{ J}}{1 \text{ kJ}} \times \frac{1 \text{ cal}}{4.184 \text{ J}} \times \frac{1 \text{ kcal}}{1000 \text{ cal}} = 3.15 \times 10^5 \text{ kcal}$

Check: The units (kcal) are correct. The magnitude of the answer (10^5) makes physical sense because a kcal is smaller than a kJ; so the answer increases.

(c) **Given:** 4.99×10^3 kJ **Find:** kWh

Conceptual Plan: **kJ → J → kWh**

$\frac{1000 \text{ J}}{1 \text{ kJ}}$ $\frac{1 \text{ kWh}}{3.60 \times 10^6 \text{ J}}$

Solution: $4.99 \times 10^3 \text{ kJ} \times \frac{1000 \text{ J}}{1 \text{ kJ}} \times \frac{1 \text{ kWh}}{3.60 \times 10^6 \text{ J}} = 1.39 \text{ kWh}$

Check: The units (kWh) are correct. The magnitude of the answer (1) makes physical sense because a kWh is much smaller than a Joule; so the answer increases.

(d) **Given:** 2.88×10^4 J **Find:** Cal

Conceptual Plan: **J → Cal**

$\frac{1 \text{ Cal}}{4184 \text{ J}}$

Solution: $2.88 \times 10^4 \text{ J} \times \frac{1 \text{ Cal}}{4184 \text{ J}} = 6.88 \text{ Cal}$

Check: The units (Cal) are correct. The magnitude of the answer (7) makes physical sense because a J is much smaller than a Calorie; so the answer decreases.

2.43 (a) **Given:** 2387 Cal **Find:** J

Conceptual Plan: **Cal → J**

$\frac{4184 \text{ J}}{1 \text{ Cal}}$

Solution: $2387 \text{ Cal} \times \frac{4184 \text{ J}}{1 \text{ Cal}} = 9.987 \times 10^6 \text{ J}$

Check: The units (J) are correct. The magnitude of the answer (10^7) makes physical sense because a Calorie is much larger than a Joule; so the answer increases.

(b) **Given:** 2387 Cal **Find:** kJ
Conceptual Plan: Cal → J → kWh

$\frac{4184 \text{ J}}{1 \text{ Cal}}$ $\frac{1 \text{ kJ}}{1000 \text{ J}}$

Solution: $2387 \text{ Cal} \times \frac{4184 \text{ J}}{1 \text{ Cal}} \times \frac{1 \text{ kJ}}{1000 \text{ J}} = 9.987 \times 10^3 \text{ kJ}$

Check: The units (kJ) are correct. The magnitude of the answer (10^4) makes physical sense because a Calorie is larger than a kJ; so the answer increases.

(c) **Given:** 2387 Cal **Find:** kWh
Conceptual Plan: Cal → J → kWh

$\frac{4184 \text{ J}}{1 \text{ Cal}}$ $\frac{1 \text{ kWh}}{3.60 \times 10^6 \text{ J}}$

Solution: $2387 \text{ Cal} \times \frac{4184 \text{ J}}{1 \text{ Cal}} \times \frac{1 \text{ kWh}}{3.60 \times 10^6 \text{ J}} = 2.774 \text{ kWh}$

Check: The units (kWh) are correct. The magnitude of the answer (3) makes physical sense because a Calorie is much smaller than a kWh; so the answer decreases.

2.44 (a) **Given:** 745 kWh **Find:** J
Conceptual Plan: kWh → J

$\frac{3.60 \times 10^6 \text{ J}}{1 \text{ kWh}}$

Solution: $745 \text{ kWh} \times \frac{3.60 \times 10^6 \text{ J}}{1 \text{ kWh}} = 2.68 \times 10^9 \text{ J}$

Check: The units (J) are correct. The magnitude of the answer (10^9) makes physical sense because a kWh is much larger than a Joule; so the answer increases.

(b) **Given:** 745 kWh **Find:** kJ
Conceptual Plan: kWh → J → kJ

$\frac{3.60 \times 10^6 \text{ J}}{1 \text{ kWh}}$ $\frac{1 \text{ kJ}}{1000 \text{ J}}$

Solution: $745 \text{ kWh} \times \frac{3.60 \times 10^6 \text{ J}}{1 \text{ kWh}} \times \frac{1 \text{ kJ}}{1000 \text{ J}} = 2.68 \times 10^6 \text{ kJ}$

Check: The units (J) are correct. The magnitude of the answer (10^6) makes physical sense because a kWh is much larger than a Joule; so the answer increases.

(c) **Given:** 745 kWh **Find:** Cal
Conceptual Plan: kWh → J → Cal

$\frac{3.60 \times 10^6 \text{ J}}{1 \text{ kWh}}$ $\frac{1 \text{ Cal}}{4184 \text{ J}}$

Solution: $745 \text{ kWh} \times \frac{3.60 \times 10^6 \text{ J}}{1 \text{ kWh}} \times \frac{1 \text{ Cal}}{4184 \text{ J}} = 6.41 \times 10^5 \text{ Cal}$

Check: The units (Cal) are correct. The magnitude of the answer (10^5) makes physical sense because a kWh is much larger than a Calorie; so the answer increases.

2.45 **Given:** \$145 electricity bill; electricity cost = \$0.120/kWh **Find:** J
Conceptual Plan: Cost (\$) → kWh → J

$\frac{1 \text{ kWh}}{\$0.120}$ $\frac{3.60 \times 10^6 \text{ J}}{1 \text{ kWh}}$

Solution: $\$145 \times \frac{1 \text{ kWh}}{\$0.120} \times \frac{3.60 \times 10^6 \text{J}}{1 \text{ kWh}} = 4.35 \times 10^{10} \text{ J}$

Check: The units (J) are correct. The magnitude of the answer (10^{10}) makes physical sense because a kWh is much larger than a Joule; so the answer increases.

2.46 **Given:** 2700 Cal **Find:** J
Conceptual Plan: Cal → J

$\frac{4184 \text{ J}}{1 \text{ Cal}}$

Solution: $2700 \text{ Cal} \times \frac{4184 \text{ J}}{1 \text{ Cal}} = 1.1 \times 10^7 \text{ J}$

Check: The units (J) are correct. The magnitude of the answer (10^7) makes physical sense because a Calorie is much larger than a Joule; so the answer increases.

The Mole Concept

2.47 **Given:** 5.52 mol sulfur **Find:** atoms of sulfur

Conceptual Plan: mol S → atoms S

$\frac{6.022 \times 10^{23} \text{ atoms}}{1 \text{ mol}}$

Solution: $5.52 \text{ mol S} \times \frac{6.022 \times 10^{23} \text{ atoms S}}{1 \text{ mol S}} = 3.32 \times 10^{24} \text{ atoms S}$

Check: The units of the answer (atoms S) are correct. The magnitude of the answer is reasonable because more than 1 mole of material is present.

2.48 **Given:** 3.7×10^{24} aluminum atoms **Find:** mol Al

Conceptual Plan: atoms Al → mol Al

$\frac{1 \text{ mol}}{6.022 \times 10^{23} \text{ atoms}}$

Solution: $3.7 \times 10^{24} \text{ atoms Al} \times \frac{1 \text{ mol Al}}{6.022 \times 10^{23} \text{ atoms Al}} = 6.1 \text{ mol Al}$

Check: The units of the answer (mol Al) are correct. The magnitude of the answer is reasonable because greater than Avogadro's number of atoms is present.

2.49 (a) **Given:** 11.8 g Ar **Find:** mol Ar

Conceptual Plan: g Ar → mol Ar

$\frac{1 \text{ mol Ar}}{39.95 \text{ g Ar}}$

Solution: $11.8 \text{ g Ar} \times \frac{1 \text{ mol Ar}}{39.95 \text{ g Ar}} = 0.295 \text{ mol Ar}$

Check: The units of the answer (mol Ar) are correct. The magnitude of the answer is reasonable because less than the mass of 1 mol is present.

(b) **Given:** 3.55 g Zn **Find:** mol Zn

Conceptual Plan: g Zn → mol Zn

$\frac{1 \text{ mol Zn}}{65.41 \text{ g Zn}}$

Solution: $3.55 \text{ g Zn} \times \frac{1 \text{ mol Zn}}{65.41 \text{ g Zn}} = 0.0543 \text{ mol Zn}$

Check: The units of the answer (mol Zn) are correct. The magnitude of the answer is reasonable because less than the mass of 1 mol is present.

(c) **Given:** 26.1 g Ta **Find:** mol Ta

Conceptual Plan: g Ta → mol Ta

$\frac{1 \text{ mol Ta}}{180.95 \text{ g Ta}}$

Solution: $26.1 \text{ g Ta} \times \frac{1 \text{ mol Ta}}{180.95 \text{ g Ta}} = 0.144 \text{ mol Ta}$

Check: The units of the answer (mol Ta) are correct. The magnitude of the answer is reasonable because less than the mass of 1 mol is present.

(d) **Given:** 0.211 g Li **Find:** mol Li

Conceptual Plan: g Li → mol Li

$\frac{1 \text{ mol Li}}{6.941 \text{ g Li}}$

Solution: $0.211 \text{ g Li} \times \frac{1 \text{ mol Li}}{6.941 \text{ g Li}} = 0.0304 \text{ mol Li}$

Check: The units of the answer (mol Li) are correct. The magnitude of the answer is reasonable because less than the mass of 1 mol is present.

2.50 (a) **Given:** 2.3×10^{-3} mol Sb **Find:** grams Sb

Conceptual Plan: mol Sb → g Sb

$\frac{121.76 \text{ g Sb}}{1 \text{ mol Sb}}$

Solution: $2.3 \times 10^{-3} \text{ mol Sb} \times \frac{121.76 \text{ g Sb}}{1 \text{ mol Sb}} = 0.28 \text{ grams Sb}$

Check: The units of the answer (grams Sb) are correct. The magnitude of the answer is reasonable because less than 1 mol of Sb is present.

(b) **Given:** 0.0355 mol Ba **Find:** grams Ba

Conceptual Plan: mol Ba → g Ba

$\frac{137.33 \text{ g Ba}}{1 \text{ mol Ba}}$

Solution: $0.0355 \text{ mol Ba} \times \frac{137.33 \text{ g Ba}}{1 \text{ mol Ba}} = 4.88 \text{ grams Ba}$

Check: The units of the answer (grams Ba) are correct. The magnitude of the answer is reasonable because less than 1 mol of Ba is present.

(c) **Given:** 43.9 mol Xe **Find:** grams Xe

Conceptual Plan: mol Xe → g Xe

$\frac{131.29 \text{ g Xe}}{1 \text{ mol Xe}}$

Solution: $43.9 \text{ mol Xe} \times \frac{131.29 \text{ g Xe}}{1 \text{ mol Xe}} = 5.76 \times 10^3 \text{ grams Xe}$

Check: The units of the answer (grams Xe) are correct. The magnitude of the answer is reasonable because much more than 1 mol of Xe is present.

(d) **Given:** 1.3 mol W **Find:** grams W

Conceptual Plan: mol W → g W

$\frac{183.84 \text{ g W}}{1 \text{ mol W}}$

Solution: $1.3 \text{ mol W} \times \frac{183.84 \text{ g W}}{1 \text{ mol W}} = 2.4 \times 10^2 \text{ grams W}$

Check: The units of the answer (grams W) are correct. The magnitude of the answer is reasonable because slightly more than 1 mol of W is present.

2.51 **Given:** 3.78 g silver **Find:** atoms Ag

Conceptual Plan: g Ag → mol Ag → atoms Ag

$\frac{1 \text{ mol Ag}}{107.87 \text{ g Ag}} \quad \frac{6.022 \times 10^{23} \text{ atoms}}{1 \text{ mol}}$

Solution: $3.78 \text{ g Ag} \times \frac{1 \text{ mol Ag}}{107.87 \text{ g Ag}} \times \frac{6.022 \times 10^{23} \text{ atoms Ag}}{1 \text{ mol Ag}} = 2.11 \times 10^{22} \text{ atoms Ag}$

Check: The units of the answer (atoms Ag) are correct. The magnitude of the answer is reasonable because less than the mass of 1 mol of Ag is present.

2.52 **Given:** 4.91×10^{21} Pt atoms **Find:** g Pt

Conceptual Plan: atoms Pt → mol Pt → g Pt

$\frac{1 \text{ mol}}{6.022 \times 10^{23} \text{ atoms}} \quad \frac{195.08 \text{ g Pt}}{1 \text{ mol Pt}}$

$$\textbf{Solution: } 4.91 \times 10^{21} \text{ atoms Pt} \times \frac{1 \text{ mol Pt}}{6.022 \times 10^{23} \text{ atoms Pt}} \times \frac{195.08 \text{ g Pt}}{1 \text{ mol Pt}} = 1.59 \text{ g Pt}$$

Check: The units of the answer (g Pt) are correct. The magnitude of the answer is reasonable because less than 1 mol of Pt atoms is present.

2.53 (a) **Given:** 5.18 g P **Find:** atoms P

Conceptual Plan: g P → mol P → atoms P

$$\frac{1 \text{ mol P}}{30.97 \text{ g P}} \quad \frac{6.022 \times 10^{23} \text{ atoms}}{1 \text{ mol}}$$

$$\textbf{Solution: } 5.18 \text{ g P} \times \frac{1 \text{ mol P}}{30.97 \text{ g P}} \times \frac{6.022 \times 10^{23} \text{ atoms P}}{1 \text{ mol P}} = 1.01 \times 10^{23} \text{ atoms P}$$

Check: The units of the answer (atoms P) are correct. The magnitude of the answer is reasonable because less than the mass of 1 mol of P is present.

(b) **Given:** 2.26 g Hg **Find:** atoms Hg

Conceptual Plan: g Hg → mol Hg → atoms Hg

$$\frac{1 \text{ mol Hg}}{200.59 \text{ g Hg}} \quad \frac{6.022 \times 10^{23} \text{ atoms}}{1 \text{ mol}}$$

$$\textbf{Solution: } 2.26 \text{ g Hg} \times \frac{1 \text{ mol Hg}}{200.59 \text{ g Hg}} \times \frac{6.022 \times 10^{23} \text{ atoms Hg}}{1 \text{ mol Hg}} = 6.78 \times 10^{21} \text{ atoms Hg}$$

Check: The units of the answer (atoms Hg) are correct. The magnitude of the answer is reasonable because much less than the mass of 1 mol of Hg is present.

(c) **Given:** 1.87 g Bi **Find:** atoms Bi

Conceptual Plan: g Bi → mol Bi → atoms Bi

$$\frac{1 \text{ mol Bi}}{208.98 \text{ g Bi}} \quad \frac{6.022 \times 10^{23} \text{ atoms}}{1 \text{ mol}}$$

$$\textbf{Solution: } 1.87 \text{ g Bi} \times \frac{1 \text{ mol Bi}}{208.98 \text{ g Bi}} \times \frac{6.022 \times 10^{23} \text{ atoms Bi}}{1 \text{ mol Bi}} = 5.39 \times 10^{21} \text{ atoms Bi}$$

Check: The units of the answer (atoms Bi) are correct. The magnitude of the answer is reasonable because less than the mass of 1 mol of Bi is present.

(d) **Given:** 0.082 g Sr **Find:** atoms Sr

Conceptual Plan: g Sr → mol Sr → atoms Sr

$$\frac{1 \text{ mol Sr}}{87.62 \text{ g Sr}} \quad \frac{6.022 \times 10^{23} \text{ atoms}}{1 \text{ mol}}$$

$$\textbf{Solution: } 0.082 \text{ g Sr} \times \frac{1 \text{ mol Sr}}{87.62 \text{ g Sr}} \times \frac{6.022 \times 10^{23} \text{ atoms Sr}}{1 \text{ mol Sr}} = 5.6 \times 10^{20} \text{ atoms Sr}$$

Check: The units of the answer (atoms Sr) are correct. The magnitude of the answer is reasonable because less than the mass of 1 mol of Sr is present.

2.54 (a) **Given:** 14.955 g Cr **Find:** atoms Cr

Conceptual Plan: g Cr → mol Cr → atoms Cr

$$\frac{1 \text{ mol Cr}}{52.00 \text{ g Cr}} \quad \frac{6.022 \times 10^{23} \text{ atoms}}{1 \text{ mol}}$$

$$\textbf{Solution: } 14.955 \text{ g Cr} \times \frac{1 \text{ mol Cr}}{52.00 \text{ g Cr}} \times \frac{6.022 \times 10^{23} \text{ atoms Cr}}{1 \text{ mol Cr}} = 1.732 \times 10^{23} \text{ atoms Cr}$$

Check: The units of the answer (atoms Cr) are correct. The magnitude of the answer is reasonable because less than the mass of 1 mol of Cr is present.

(b) **Given:** 39.733 g S **Find:** atoms S
Conceptual Plan: g S → mol S → atoms S

$$\frac{1 \text{ mol S}}{32.07 \text{ g S}} \quad \frac{6.0221 \times 10^{23} \text{ atoms}}{1 \text{ mol}}$$

Solution: $39.733 \cancel{\text{g S}} \times \frac{1 \cancel{\text{mol S}}}{32.07 \cancel{\text{g S}}} \times \frac{6.0221 \times 10^{23} \text{ atoms S}}{1 \cancel{\text{mol S}}} = 7.461 \times 10^{23} \text{ atoms S}$

Check: The units of the answer (atoms S) are correct. The magnitude of the answer is reasonable because slightly more than the mass of 1 mol of S is present.

(c) **Given:** 12.899 g Pt **Find:** atoms Pt
Conceptual Plan: g Pt → mol Pt → atoms Pt

$$\frac{1 \text{ mol Pt}}{195.08 \text{ g Pt}} \quad \frac{6.0221 \times 10^{23} \text{ atoms}}{1 \text{ mol}}$$

Solution: $12.899 \cancel{\text{g Pt}} \times \frac{1 \cancel{\text{mol Pt}}}{195.08 \cancel{\text{g Pt}}} \times \frac{6.0221 \times 10^{23} \text{ atoms Pt}}{1 \cancel{\text{mol Pt}}} = 3.9819 \times 10^{22} \text{ atoms Pt}$

Check: The units of the answer (atoms Pt) are correct. The magnitude of the answer is reasonable because less than the mass of 1 mol of Pt is present.

(d) **Given:** 97.552 g Sn **Find:** atoms Sn
Conceptual Plan: g Sn → mol Sn → atoms Sn

$$\frac{1 \text{ mol Sn}}{118.71 \text{ g Sn}} \quad \frac{6.0221 \times 10^{23} \text{ atoms}}{1 \text{ mol}}$$

Solution: $97.552 \cancel{\text{g Sn}} \times \frac{1 \cancel{\text{mol Sn}}}{118.71 \cancel{\text{g Sn}}} \times \frac{6.0221 \times 10^{23} \text{ atoms Sn}}{1 \cancel{\text{mol Sn}}} = 4.948 \times 10^{23} \text{ atoms Sn}$

Check: The units of the answer (atoms Sn) are correct. The magnitude of the answer is reasonable because slightly less than the mass of 1 mol of Sn is present.

2.55 (a) **Given:** 1.1×10^{23} gold atoms **Find:** grams Au
Conceptual Plan: atoms Au → mol Au → g Au

$$\frac{1 \text{ mol}}{6.022 \times 10^{23} \text{ atoms}} \quad \frac{196.97 \text{ g Au}}{1 \text{ mol Au}}$$

Solution: $1.1 \times 10^{23} \cancel{\text{atoms Au}} \times \frac{1 \cancel{\text{mol Au}}}{6.022 \times 10^{23} \cancel{\text{atoms Au}}} \times \frac{196.97 \text{ g Au}}{1 \cancel{\text{mol Au}}} = 36 \text{ g Au}$

Check: The units of the answer (g Au) are correct. The magnitude of the answer is reasonable because fewer than Avogadro's number of atoms are in the sample.

(b) **Given:** 2.82×10^{22} helium atoms **Find:** grams He
Conceptual Plan: atoms He → mol He → g He

$$\frac{1 \text{ mol}}{6.022 \times 10^{23} \text{ atoms}} \quad \frac{4.002 \text{ g He}}{1 \text{ mol He}}$$

Solution: $2.82 \times 10^{22} \cancel{\text{atoms He}} \times \frac{1 \cancel{\text{mol He}}}{6.022 \times 10^{23} \cancel{\text{atoms He}}} \times \frac{4.002 \text{ g He}}{1 \cancel{\text{mol He}}} = 0.187 \text{ g He}$

Check: The units of the answer (g He) are correct. The magnitude of the answer is reasonable because fewer than Avogadro's number of atoms are in the sample.

(c) **Given:** 1.8×10^{23} lead atoms **Find:** grams Pb
Conceptual Plan: atoms Pb → mol Pb → g Pb

$$\frac{1 \text{ mol}}{6.022 \times 10^{23} \text{ atoms}} \quad \frac{207.2 \text{ g Pb}}{1 \text{ mol Pb}}$$

Solution: $1.8 \times 10^{23} \cancel{\text{atoms Pb}} \times \frac{1 \cancel{\text{mol Pb}}}{6.022 \times 10^{23} \cancel{\text{atoms Pb}}} \times \frac{207.2 \text{ g Pb}}{1 \cancel{\text{mol Pb}}} = 62 \text{ g Pb}$

Check: The units of the answer (g Pb) are correct. The magnitude of the answer is reasonable because fewer than Avogadro's number of atoms are in the sample.

(d) **Given:** 7.9×10^{21} uranium atoms **Find:** grams U

Conceptual Plan: atoms U → mol U → g U

$$\frac{1 \text{ mol}}{6.022 \times 10^{23} \text{ atoms}} \quad \frac{238.029 \text{ g U}}{1 \text{ mol U}}$$

Solution: $7.9 \times 10^{21} \cancel{\text{atoms U}} \times \dfrac{1 \cancel{\text{mol U}}}{6.022 \times 10^{23} \cancel{\text{atoms U}}} \times \dfrac{238.029 \text{ g U}}{1 \cancel{\text{mol U}}} = 3.1 \text{ g U}$

Check: The units of the answer (g U) are correct. The magnitude of the answer is reasonable because fewer than Avogadro's number of atoms are in the sample.

2.56 (a) **Given:** 7.55×10^{26} cadmium atoms **Find:** kg Cd

Conceptual Plan: atoms Cd → mol Cd → g Cd → kg Cd

$$\frac{1 \text{ mol}}{6.022 \times 10^{23} \text{ atoms}} \quad \frac{112.41 \text{ g Cd}}{1 \text{ mol Cd}} \quad \frac{1 \text{ kg}}{1000 \text{ g}}$$

Solution: $7.55 \times 10^{26} \cancel{\text{atoms Cd}} \times \dfrac{1 \cancel{\text{mol Cd}}}{6.022 \times 10^{23} \cancel{\text{atoms Cd}}} \times \dfrac{112.41 \cancel{\text{g Cd}}}{1 \cancel{\text{mol Cd}}} \times \dfrac{1 \text{ kg Cd}}{1000 \cancel{\text{g Cd}}} = 141 \text{ g Cd}$

Check: The units of the answer (kg Cd) are correct. The magnitude of the answer is reasonable because many more than Avogadro's number of atoms are in the sample.

(b) **Given:** 8.15×10^{27} nickel atoms **Find:** kg Ni

Conceptual Plan: atoms Ni → mol Ni → g Ni → kg Ni

$$\frac{1 \text{ mol}}{6.022 \times 10^{23} \text{ atoms}} \quad \frac{58.69 \text{ g Ni}}{1 \text{ mol Ni}} \quad \frac{1 \text{ kg}}{1000 \text{ g}}$$

Solution: $8.15 \times 10^{27} \cancel{\text{atoms Ni}} \times \dfrac{1 \cancel{\text{mol Ni}}}{6.022 \times 10^{23} \cancel{\text{atoms Ni}}} \times \dfrac{58.69 \cancel{\text{g Ni}}}{1 \cancel{\text{mol Ni}}} \times \dfrac{1 \text{ kg Ni}}{1000 \cancel{\text{g Ni}}} = 794 \text{ kg Ni}$

Check: The units of the answer (kg Ni) are correct. The magnitude of the answer is reasonable because many more than Avogadro's number of atoms are in the sample.

(c) **Given:** 1.22×10^{27} manganese atoms **Find:** kg Mn

Conceptual Plan: atoms Mn → mol Mn → g Mn → kg Mn

$$\frac{1 \text{ mol}}{6.022 \times 10^{23} \text{ atoms}} \quad \frac{54.94 \text{ g Mn}}{1 \text{ mol Mn}} \quad \frac{1 \text{ kg}}{1000 \text{ g}}$$

Solution: $1.22 \times 10^{27} \cancel{\text{atoms Mn}} \times \dfrac{1 \cancel{\text{mol Mn}}}{6.022 \times 10^{23} \cancel{\text{atoms Mn}}} \times \dfrac{54.94 \cancel{\text{g Mn}}}{1 \cancel{\text{mol Mn}}} \times \dfrac{1 \text{ kg Mn}}{1000 \cancel{\text{g Mn}}} = 111 \text{ kg Mn}$

Check: The units of the answer (kg Mn) are correct. The magnitude of the answer is reasonable because many more than Avogadro's number of atoms are in the sample.

(d) **Given:** 5.48×10^{29} lithium atoms **Find:** kg Li

Conceptual Plan: atoms Li → mol Li → g Li → kg Li

$$\frac{1 \text{ mol}}{6.022 \times 10^{23} \text{ atoms}} \quad \frac{6.941 \text{ g Li}}{1 \text{ mol Li}} \quad \frac{1 \text{ kg}}{1000 \text{ g}}$$

Solution: $5.48 \times 10^{29} \cancel{\text{atoms Li}} \times \dfrac{1 \cancel{\text{mol Li}}}{6.022 \times 10^{23} \cancel{\text{atoms Li}}} \times \dfrac{6.941 \cancel{\text{g Li}}}{1 \cancel{\text{mol Li}}} \times \dfrac{1 \text{ kg Li}}{1000 \cancel{\text{g Li}}} = 6.32 \times 10^{3} \text{ kg Li}$

Check: The units of the answer (kg Li) are correct. The magnitude of the answer is reasonable because many more than Avogadro's number of atoms are in the sample.

2.57 **Given:** 52 mg diamond (carbon) **Find:** atoms C

Conceptual Plan: mg C → g C → mol C → atoms C

$$\frac{1 \text{ g C}}{1000 \text{ mg C}} \quad \frac{1 \text{ mol C}}{12.011 \text{ g C}} \quad \frac{6.022 \times 10^{23} \text{ atoms}}{1 \text{ mol}}$$

Solution: $52 \cancel{\text{mg C}} \times \dfrac{1 \cancel{\text{g C}}}{1000 \cancel{\text{mg C}}} \times \dfrac{1 \cancel{\text{mol C}}}{12.011 \cancel{\text{g C}}} \times \dfrac{6.022 \times 10^{23} \text{ atoms C}}{1 \cancel{\text{mol C}}} = 7.2 \times 10^{21} \text{ atoms C}$

Check: The units of the answer (atoms C) are correct. The magnitude of the answer is reasonable because less than the mass of 1 mol of C is present.

2.58 **Given:** 536 kg helium **Find:** atoms He

Conceptual Plan: kg He → g He → mol He → atoms He

$\frac{1000 \text{ g He}}{1 \text{ kg He}}$ $\frac{1 \text{ mol He}}{4.0026 \text{ g He}}$ $\frac{6.022 \times 10^{23} \text{ atoms}}{1 \text{ mol}}$

Solution: $536 \cancel{\text{kg He}} \times \frac{1000 \cancel{\text{g He}}}{1 \cancel{\text{kg He}}} \times \frac{1 \cancel{\text{mol He}}}{4.0026 \cancel{\text{g He}}} \times \frac{6.022 \times 10^{23} \text{ atoms He}}{1 \cancel{\text{mol He}}} = 1.29 \times 10^{30} \text{ atoms He}$

Check: The units of the answer (atoms He) are correct. The magnitude of the answer is reasonable because much more than the mass of 1 mol of He is present.

2.59 **Given:** 1 atom platinum **Find:** g Pt

Conceptual Plan: atoms Pt → mol Pt → g Pt

$\frac{1 \text{ mol}}{6.0221 \times 10^{23} \text{ atoms}}$ $\frac{195.08 \text{ g Pt}}{1 \text{ mol Pt}}$

Solution: $1 \cancel{\text{atom Pt}} \times \frac{1 \cancel{\text{mol Pt}}}{6.0221 \times 10^{23} \cancel{\text{atoms Pt}}} \times \frac{195.08 \text{ g Pt}}{1 \cancel{\text{mol Pt}}} = 3.2394 \times 10^{-22} \text{ g Pt}$

Check: The units of the answer (g Pt) are correct. The magnitude of the answer is reasonable because only one atom is in the sample.

2.60 **Given:** 35 atoms xenon **Find:** g Xe

Conceptual Plan: atoms Xe → mol Xe → g Xe

$\frac{1 \text{ mol}}{6.022 \times 10^{23} \text{ atoms}}$ $\frac{131.29 \text{ g Xe}}{1 \text{ mol Xe}}$

Solution: $35 \cancel{\text{atom Xe}} \times \frac{1 \cancel{\text{mol Xe}}}{6.022 \times 10^{23} \cancel{\text{atoms Xe}}} \times \frac{131.29 \text{ g Xe}}{1 \cancel{\text{mol Xe}}} = 7.6305 \times 10^{-21} \text{ g Xe}$

Check: The units of the answer (g Xe) are correct. The magnitude of the answer is reasonable because only 35 atoms are in the sample.

Cumulative Problems

2.61 (a) **Given:** cylinder dimensions: length = 22 cm; radius = 3.8 cm; $d(\text{gold}) = 19.3 \text{ g/cm}^3$; $d(\text{sand}) = 3.00 \text{ g/cm}^3$ **Find:** m(gold) and m(sand)

Conceptual Plan: $l, r \rightarrow V$ **then** $d, V \rightarrow m$

$V = l\pi r^2$ $d = m/V$

Solution: $V(\text{gold}) = V(\text{sand}) = (22 \text{ cm})(\pi)(3.8 \text{ cm})^2 = 9\underline{9}8.0212 \text{ cm}^3$ $d = m/V$

Rearrange by multiplying both sides of equation by V. $\rightarrow m = d \times V$

$$m(\text{gold}) = \left(19.3 \frac{\text{g}}{\cancel{\text{cm}^3}}\right) \times (9\underline{9}8.0212 \cancel{\text{cm}^3}) = 1.\underline{9}26181 \times 10^4 \text{ g} = 1.9 \times 10^4 \text{ g}$$

Check: The units (g) are correct. The magnitude of the answer seems correct considering that the value of the density is ~20 g/cm^3. Two significant figures are allowed to reflect the significant figures in 22 cm and 3.8 cm. Truncate the nonsignificant digits because the first nonsignificant digit is a 2.

$$m(\text{sand}) = \left(3.00 \frac{\text{g}}{\cancel{\text{cm}^3}}\right) \times (9\underline{9}8.0212 \cancel{\text{cm}^3}) = 2.\underline{9}9406 \times 10^3 \text{ g} = 3.0 \times 10^3 \text{ g}$$

Check: The units (g) are correct. The magnitude of the answer seems correct considering that the value of the density is 3 g/cm^3. This number is much lower than the gold mass. Two significant figures are allowed to reflect the significant figures in 22 cm and 3.8 cm. Round the last digit up because the first nonsignificant digit is a 9.

(b) Comparing the two values 1.9×10^4 g versus 3.0×10^3 g shows a difference in weight of almost a factor of 10. This difference should be enough to trip the alarm and alert the authorities to the presence of the thief.

2.62 **Given:** $r = 1.0 \times 10^{-13}$ cm; $m = 1.7 \times 10^{-24}$ g **Find:** density **Other:** $V = (4/3)\,\pi r^3$

Conceptual Plan: $r \rightarrow V$ **then** $m, V \rightarrow d$

$V = (4/3)\pi r^3$ $d = m/V$

Solution: $V = (4/3)\pi r^3 = (4/3)(\pi)(1.0 \times 10^{-13}\ \text{cm})^3 = 4.\underline{1}88790205 \times 10^{-39}\ \text{cm}^3$

$$d = \frac{m}{V} = \frac{1.7 \times 10^{-24}\ \text{g}}{4.\underline{1}88790205 \times 10^{-39}\ \text{cm}^3} = 4.\underline{0}58451049 \times 10^{14} \frac{\text{g}}{\text{cm}^3} = 4.1 \times 10^{14} \frac{\text{g}}{\text{cm}^3}$$

Check: The units (g/cm^3) are correct. The magnitude of the answer seems correct considering how small a nucleus is compared to an atom. Two significant figures are allowed to reflect the significant figures in 1.0×10^{-13} cm. Round the last digit up because the first nonsignificant digit is a 5.

2.63 **Given:** 3.5 lb of titanium **Find:** volume in in^3 **Other:** density of titanium is 4.51 g/cm^3

Conceptual Plan: lb → g then *m, d* → *V* then cm^3 → in^3

$$\frac{453.6\ \text{g}}{1\ \text{lb}} \qquad d = m/V \qquad \frac{(1\ \text{in})^3}{(2.54\ \text{cm})^3}$$

Solution: $3.5\ \cancel{\text{lb}} \times \frac{453.6\ \text{g}}{1\ \cancel{\text{lb}}} = 1.\underline{5}876 \times 10^3\ \text{g}$

$d = m/V$ Rearrange by multiplying both sides of the equation by V and dividing both sides of the equation by d.

$$V = \frac{m}{d} = \frac{1.\underline{5}876 \times 10^3\ \cancel{\text{g}}}{4.51 \frac{\cancel{\text{g}}}{\text{cm}^3}} = 3.\underline{5}20 \times 10^2\ \text{cm}^3 = 3.5 \times 10^2\ \cancel{\text{cm}^3} \times \frac{(1\ \text{in})^3}{(2.54\ \cancel{\text{cm}})^3} = 21.48\ \text{in}^3 = 21\ \text{in}^3$$

Check: The units (in^3) are correct. The magnitude of the answer seems correct considering the number of grams. Two significant figures are allowed to reflect the significant figures in 3.5 lb. Truncate the nonsignificant digits because the first nonsignificant digit is a 4.

2.64 **Given:** density (g/cm^3) **Find:** density (lb/in^3) **Other:** density of iron is 7.86 g/cm^3

Conceptual Plan: $\frac{\text{g}}{\text{cm}^3} \rightarrow \frac{\text{lb}}{\text{cm}^3} \rightarrow \frac{\text{lb}}{\text{in}^3}$

$$\frac{1\ \text{lb}}{453.6\ \text{g}} \qquad \frac{(2.54\ \text{cm})^3}{(1\ \text{in})^3}$$

Solution: $\frac{7.86\ \cancel{\text{g}}}{\cancel{\text{cm}^3}} \times \frac{1\ \text{lb}}{453.6\ \cancel{\text{g}}} \times \frac{(2.54\ \cancel{\text{cm}})^3}{(1\ \text{in})^3} = 0.28\underline{3}9557386 \frac{\text{lb}}{\text{in}^3} = 0.284 \frac{\text{lb}}{\text{in}^3}$

Check: The units (lb/in^3) are correct. The magnitude of the answer seems correct considering the dominating factor that a gram is smaller than a pound; so the answer should go down. Three significant figures are allowed to reflect the significant figures in 7.86 g/cm^3. Round the last digit up because the first nonsignificant digit is a 9.

2.65 **Given:** cylinder dimensions: length = 2.16 in; radius = 0.22 in; m = 41 g **Find:** density (g/m^3)

Conceptual Plan: in → cm then *l, r* → *V* then *m, V* → *d*

$$\frac{2.54\ \text{cm}}{1\ \text{in}} \qquad V = l\pi r^2 \qquad d = m/V$$

Solution: $2.16\ \cancel{\text{in}} \times \frac{2.54\ \text{cm}}{1\ \cancel{\text{in}}} = 5.4\underline{8}64\ \text{cm} = l$ $\quad 0.22\ \cancel{\text{in}} \times \frac{2.54\ \text{cm}}{1\ \cancel{\text{in}}} = 0.5\underline{5}88\ \text{cm} = r$

$V = l\pi r^2 = (5.4\underline{8}64\ \text{cm})(\pi)(0.5\underline{5}88\ \text{cm})^2 = 5.\underline{3}820798\ \text{cm}^3$

$$d = \frac{m}{V} = \frac{41\ \text{g}}{5.\underline{3}820798\ \text{cm}^3} = 7.\underline{6}178729 \frac{\text{g}}{\text{cm}^3} = 7.6 \frac{\text{g}}{\text{cm}^3}$$

Check: The units (g/cm^3) are correct. The magnitude of the answer seems correct considering that the value of the density of iron (a major component in steel) is 7.86 g/cm^3. Two significant figures are allowed to reflect the significant figures in 0.22 in and 41 g. Truncate the nonsignificant digits because the first nonsignificant digit is a 1.

2.66 **Given:** m = 85 g **Find:** radius of the sphere (inches)

Other: density (aluminum) = 2.7 g/cm^3

Conceptual Plan: *m, d* → *V* then *V* → *r* then cm → in

$$d = m/V \qquad V = (4/3)\pi r^3 \qquad \frac{1\ \text{in}}{2.54\ \text{cm}}$$

Solution: $d = m/V$ Rearrange by multiplying both sides of the equation by V and dividing both sides of the equation by d.

$$V = \frac{m}{d} = \frac{85\ \text{g}}{2.7\frac{\text{g}}{\text{cm}^3}} = 3\underline{1}.48148148\ \text{cm}^3$$

$V = (4/3)\pi r^3$ Rearrange by dividing both sides of the equation by $(4/3)\pi$. $r^3 = \frac{3V}{4\pi}$

Take the cube root of both sides of the equation.

$$r = \left(\frac{3V}{4\pi}\right)^{1/3} = \left(\frac{(3)(3\underline{1}.48148148\ \text{cm}^3)}{4\pi}\right)^{1/3} = (7.\underline{5}1565009\ \text{cm}^3)^{1/3} = 1.\underline{9}58794386\ \text{cm}$$

$$1.\underline{9}58794386\ \text{cm} \times \frac{1\ \text{in}}{2.54\ \text{cm}} = 0.7\underline{7}11788923\ \text{in} = 0.77\ \text{in}$$

Check: The units (in) are correct. The magnitude of the answer seems correct. The magnitude of the volume is about a third of the mass (density is about $3\ \text{g/cm}^3$). The radius (in cm) seems correct considering the geometry involved. The magnitude goes down when we convert from centimeters to inches because an inch is bigger than a centimeter. Two significant figures are allowed to reflect the significant figures in $2.7\ \text{g/cm}^3$ and 85 g. Truncate the nonsignificant digits because the first nonsignificant digit is a 1.

2.67 **Given:** 185 cubic yards (yd^3) of H_2O **Find:** mass of the H_2O (pounds)
Other: $d(H_2O) = 1.00\ \text{g/cm}^3$ at 4 °C
Conceptual Plan: $\text{yd}^3 \rightarrow \text{m}^3 \rightarrow \text{cm}^3 \rightarrow \text{g} \rightarrow \text{lb}$

$$\frac{(1\ \text{m})^3}{(1.094\ \text{yd})^3} \quad \frac{(100\ \text{cm})^3}{(1\ \text{m})^3} \quad \frac{1.00\ \text{g}}{1.00\ \text{cm}^3} \quad \frac{1\ \text{lb}}{453.59\ \text{g}}$$

Solution: $185\ \text{yd}^3 \times \frac{(1\ \text{m})^3}{(1.094\ \text{yd})^3} \times \frac{(100\ \text{cm})^3}{(1\ \text{m})^3} \times \frac{1.00\ \text{g}}{1.00\ \text{cm}^3} \times \frac{1\ \text{lb}}{453.59\ \text{g}} = 3.1\underline{1}4987377 \times 10^5\ \text{lb} = 3.11 \times 10^5\ \text{lb}$

Check: The units (lb) are correct. The magnitude of the answer (10^5) makes physical sense because a pool is not a small object. Three significant figures are allowed because the conversion factor with the least precision is the density [$1.00\ \text{g/cm}^3$ – (3 significant figures)] and the initial size has three significant figures. Truncate after the last digit because the first nonsignificant digit is a 4.

2.68 **Given:** 7655 cubic feet (ft^3) of ice **Find:** mass of the ice (kg) **Other:** $d(\text{ice}) = 0.917\ \text{g/cm}^3$ at 0 °C
Conceptual Plan: $\text{ft}^3 \rightarrow \text{cm}^3 \rightarrow \text{g} \rightarrow \text{kg}$

$$\frac{(30.48\ \text{cm})^3}{(1\ \text{ft})^3} \quad \frac{0.917\ \text{g}}{1.00\ \text{cm}^3} \quad \frac{1\ \text{kg}}{1000\ \text{g}}$$

Solution: $7655\ \text{ft}^3 \times \frac{(30.48\ \text{cm})^3}{(1\ \text{ft})^3} \times \frac{0.917\ \text{g}}{1.00\ \text{cm}^3} \times \frac{1\ \text{kg}}{1000\ \text{g}} = 1.9\underline{8}7739274 \times 10^5\ \text{kg} = 1.99 \times 10^5\ \text{kg}$

Check: The units (kg) are correct. The magnitude of the answer (10^5) makes physical sense because an iceberg is a large object. Three significant figures are allowed because the conversion factor with the least precision is the density [($0.917\ \text{g/cm}^3$ – (3 significant figures)]. Round the last digit up because the first nonsignificant digit is a 7.

2.69 **Given:** 15 liters of gasoline **Find:** kilometers **Other:** 52 mi/gal in the city
Conceptual Plan: $\text{L} \rightarrow \text{gal} \rightarrow \text{mi} \rightarrow \text{km}$

$$\frac{1\ \text{gal}}{3.785\ \text{L}} \quad \frac{52\ \text{mi}}{1\ \text{gal}} \quad \frac{1\ \text{km}}{0.6214\ \text{mi}}$$

Solution: $15\ \text{L} \times \frac{1\ \text{gal}}{3.785\ \text{L}} \times \frac{52\ \text{mi}}{1\ \text{gal}} \times \frac{1\ \text{km}}{0.6214\ \text{mi}} = 3.\underline{3}16327941 \times 10^2\ \text{km} = 3.3 \times 10^2\ \text{km}$

Check: The units (km) are correct. The magnitude of the answer (10^2) makes physical sense because the dominating conversion factor is the mileage, which increases the answer. Two significant figures are allowed because the conversion factor with the least precision is 52 mi/gal (two significant figures) and the initial volume (15 L) has two significant figures. Truncate the last digit because the first nonsignificant digit is a 1. It is best to put the answer in scientific notation so that it is clear how many significant figures are expressed.

2.70 **Given:** 355 mL of gasoline **Find:** kilometers **Other:** 57 mi/gal in the city
Conceptual Plan: $\text{mL} \rightarrow \text{L} \rightarrow \text{gal} \rightarrow \text{mi} \rightarrow \text{km}$

$$\frac{1\ \text{L}}{1000\ \text{mL}} \quad \frac{1\ \text{gal}}{3.785\ \text{L}} \quad \frac{57\ \text{mi}}{1\ \text{gal}} \quad \frac{1\ \text{km}}{0.6214\ \text{mi}}$$

Solution: $355\ \text{mL} \times \frac{1\ \text{L}}{1000\ \text{mL}} \times \frac{1\ \text{gal}}{3.785\ \text{L}} \times \frac{57\ \text{mi}}{1\ \text{gal}} \times \frac{1\ \text{km}}{0.6214\ \text{mi}} = 8.\underline{6}14700037\ \text{km} = 8.6\ \text{km}$

Check: The units (km) are correct. The magnitude of the answer (8.6) makes physical sense because the dominating conversion factor is the conversion from mL to L, which decreases the answer. Two significant figures are allowed because the conversion factor with the least precision is 57 mi/gal (two significant figures). Truncate the last digit because the first nonsignificant digit is a 1.

2.71 **Given:** radius of nucleus of the hydrogen atom $= 1.0 \times 10^{-13}$ cm; radius of the hydrogen atom $= 52.9$ pm
Find: percent of volume occupied by nucleus (%)
Conceptual Plan: cm → m then pm → m then r → V then V_{atom}, $V_{nucleus}$ → % $V_{nucleus}$

$$\frac{1\text{ m}}{100\text{ cm}} \qquad \frac{1\text{ m}}{10^{12}\text{ pm}} \qquad V = (4/3)\pi r^3 \qquad \% V_{nucleus} = \frac{V_{nucleus}}{V_{atom}} \times 100\%$$

Solution: $1.0 \times 10^{-13}\,\cancel{\text{cm}} \times \dfrac{1\text{ m}}{100\,\cancel{\text{cm}}} = 1.0 \times 10^{-15}$ m and $52.9\,\cancel{\text{pm}} \times \dfrac{1\text{ m}}{10^{12}\,\cancel{\text{pm}}} = 5.29 \times 10^{-11}$ m

$V = (4/3)\pi r^3$ Substitute into $\%V$ equation.

$$\% V_{nucleus} = \frac{V_{nucleus}}{V_{atom}} \times 100\% \quad \rightarrow \% V_{nucleus} = \frac{(4/3)\,\pi r^3_{nucleus}}{(4/3)\,\pi r^3_{atom}} \times 100\% \quad \text{Simplify equation.}$$

$$\% V_{nucleus} = \frac{r^3_{nucleus}}{r^3_{atom}} \times 100\% \quad \text{Substitute numbers and calculate result.}$$

$$\% V_{nucleus} = \frac{(1.0 \times 10^{-15}\text{ m})^3}{(5.29 \times 10^{-11}\text{ m})^3} \times 100\% = (1.\underline{8}90359168 \times 10^{-5})^3 \times 100\% = 6.\underline{7}55118686 \times 10^{-13}$$

$$= 6.8 \times 10^{-13}$$

Check: The units (none) are correct. The magnitude of the answer seems correct (10^{-13}) because a proton is so small. Two significant figures are allowed to reflect the significant figures in 1.0×10^{-13} cm. Round the last digits up because the first nonsignificant digit is a 5.

2.72 **Given:** radius of neon $= 69$ pm; 2.69×10^{22} atoms per liter **Find:** percent of volume occupied by neon (%)
Conceptual Plan: Assume 1L total volume.
pm → m → cm then r → V then cm³ → L then L/atom → L then V_{Ne}, V_{Total} → % V_{Ne}

$$\frac{1\text{ m}}{10^{12}\text{ pm}} \quad \frac{100\text{ cm}}{1\text{ m}} \qquad V = (4/3)\pi r^3 = 1\text{ atom} \qquad \frac{1\text{ L}}{1000\text{ cm}^3} \qquad 2.69 \times 10^{22}\text{ atoms} \qquad \% V_{Ne} = \frac{V_{Ne}}{V_{Total}} \times 100\%$$

Solution: $69\,\cancel{\text{pm}} \times \dfrac{1\,\cancel{\text{m}}}{10^{12}\,\cancel{\text{pm}}} \times \dfrac{100\text{ cm}}{1\,\cancel{\text{m}}} = 6.9 \times 10^{-9}$ cm

$$V = (4/3)\pi r^3 = (4/3)\pi(6.9 \times 10^{-9}\text{ cm})^3 = 1.\underline{3}7605528 \times 10^{-24}\text{ cm}^3$$

$$1.\underline{3}7605528 \times 10^{-24}\,\cancel{\text{cm}^3} \times \frac{1\text{ L}}{1000\,\cancel{\text{cm}^3}} = 1.\underline{3}7605528 \times 10^{-27}\text{ L}$$

$$\frac{1.\underline{3}7605528 \times 10^{-27}\text{ L}}{\text{atom}} \times 2.69 \times 10^{22}\text{ atoms} = 3.\underline{7}01588707 \times 10^{-5}\text{ L} \quad \text{Substitute into \% } V \text{ equation.}$$

$$\% V_{Ne} = \frac{V_{Ne}}{V_{Total}} \times 100\% = \frac{3.\underline{7}01588707 \times 10^{-5}\,\cancel{\text{L}}}{1\,\cancel{\text{L}}} \times 100\%s = 3.\underline{7}01588707 \times 10^{-3}\% = 3.7 \times 10^{-3}\%$$

Check: This says that the separation between atoms is very large in the gas phase. The units (%) are correct. The magnitude of the answer seems correct (10^{-3}); it is known that gases are primarily empty space. Two significant figures are allowed to reflect the significant figures in 69 pm. Truncate the nonsignificant digits because the first nonsignificant digit is a 0.

2.73 **Given:** radius of hydrogen $= 212$ pm; radius of Ping-Pong ball $= 4.0$ cm, 6.02×10^{23} atoms and balls in a row
Find: row length (km)
Conceptual Plan: atoms → pm → m → km and ball → cm → m → km

$$\frac{212\text{ pm}}{1\text{ atom}} \quad \frac{1\text{ m}}{10^{12}\text{ pm}} \quad \frac{1\text{ km}}{1000\text{ m}} \qquad \frac{4.0\text{ cm}}{1\text{ ball}} \quad \frac{1\text{ m}}{100\text{ cm}} \quad \frac{1\text{ km}}{1000\text{ m}}$$

Solution: $6.02 \times 10^{23}\,\cancel{\text{atoms}} \times \dfrac{212\,\cancel{\text{pm}}}{1\,\cancel{\text{atom}}} \times \dfrac{1\,\cancel{\text{m}}}{10^{12}\,\cancel{\text{pm}}} \times \dfrac{1\text{ km}}{1000\,\cancel{\text{m}}} = 1.28 \times 10^{11}$ km

$$6.02 \times 10^{23} \text{ balls} \times \frac{4.0 \text{ cm}}{1 \text{ ball}} \times \frac{1 \text{ m}}{100 \text{ cm}} \times \frac{1 \text{ km}}{1000 \text{ m}} = 2.4 \times 10^{19} \text{ km}$$

Check: The units (km) are correct. The magnitude of the answers seem correct (10^{11} and 10^{19}). The answers are driven by the large number of atoms or balls. The Ping-Pong ball row is 108 times longer. Three significant figures are allowed to reflect the significant figures in 212 pm. Two significant figures are allowed to reflect the significant figures in 4.0 cm.

2.74 **Given:** 100 m in 9.58 s; 100 yards in 9.07 s **Find:** miles/hr

Conceptual Plan: speed is distance/time; m/s → km/s → miles/s → miles/hr → miles/hr and

$$\frac{1 \text{ km}}{1000 \text{ m}} \quad \frac{0.62137 \text{ mi}}{1 \text{ km}} \quad \frac{60 \text{ s}}{1 \text{ min}} \quad \frac{60 \text{ min}}{1 \text{ hr}}$$

yd/s → m/s → km/s → miles/s → miles/hr → miles/hr and

$$\frac{1 \text{ m}}{1.0936 \text{ yd}} \quad \frac{1 \text{ km}}{1000 \text{ m}} \quad \frac{0.62137 \text{ mi}}{1 \text{ km}} \quad \frac{60 \text{ s}}{1 \text{ min}} \quad \frac{60 \text{ min}}{1 \text{ hr}}$$

Solution: $\frac{100 \text{ m}}{9.58 \text{ s}} \times \frac{1 \text{ km}}{1000 \text{ m}} \times \frac{0.62137 \text{mi}}{1 \text{km}} \times \frac{60 \text{ s}}{1 \text{ min}} \times \frac{60 \text{ min}}{1 \text{ hr}} = 23.\underline{3}5002088 \text{ mi/hr} = 23.4 \text{ mi/hr}$

$$\frac{100 \text{ yd}}{9.07 \text{ s}} \times \frac{1 \text{ m}}{1.0936 \text{ yd}} \times \frac{1 \text{ km}}{1000 \text{ m}} \times \frac{0.62137 \text{ mi}}{1 \text{km}} \times \frac{60 \text{ s}}{1 \text{ min}} \times \frac{60 \text{ min}}{1 \text{ hr}} = 22.\underline{5}5210026 \text{ mi/hr} = 22.6 \text{ mi/hr}$$

Check: The units (mi/hr) are correct. The magnitude of the answers seems correct (23 and 23), and they are close to each other. Assuming that 100 m and 100 yd have three significant figures, three significant figures are allowed to reflect the significant figures in the times.

2.75 **Given:** 39.33 g sodium/100 g salt; 1.25 g salt/100 g snack mix; FDA maximum 2.40 g sodium/day

Find: g snack mix

Conceptual Plan: g sodium → g salt → g snack mix

$$\frac{100 \text{ g salt}}{39.33 \text{ g sodium}} \quad \frac{100 \text{ g snack mix}}{1.25 \text{ g salt}}$$

Solution: $\frac{2.40 \text{ g sodium}}{1 \text{ day}} \times \frac{100 \text{ g salt}}{39.33 \text{ g sodium}} \times \frac{100 \text{ g snack mix}}{1.25 \text{ g salt}} = 48\underline{8}.1770 \text{ g snack mix/day}$

$= 488$ g snack mix/day

Check: The units (g) are correct. The magnitude of the answer seems correct (500) because salt is less than half sodium and there is a little over a gram of salt per 100 grams of snack mix. Three significant figures are allowed to reflect the significant figures in the FDA maximum and in the amount of salt in the snack mix.

2.76 **Given:** 86.6 g lead/100 g galena; 68.5 g galena/100 g ore; 92.5 g lead extracted/100 g lead available; 1.500 cm radius sphere **Other:** $d(\text{lead}) = 11.4 \text{ g/cm}^3$ **Find:** g ore

Conceptual Plan:

$r_{\text{sphere}} \rightarrow V_{\text{sphere}}$ then $V_{\text{sphere}}, d_{\text{sphere}} \rightarrow m_{\text{sphere (g lead)}} \rightarrow$ g lead available → g galena → g ore

$V = (4/3)\pi r^3 \quad d = m/V \quad \frac{100 \text{ g lead available}}{92.5 \text{ g lead extracted}} \quad \frac{100 \text{ g galena}}{86.6 \text{ g lead available}} \quad \frac{100 \text{ g ore}}{68.5 \text{ g galena}}$

Solution: Calculate m_{sphere} $V_{\text{sphere}} = (4/3)\,\pi\, r^3_{\text{sphere}} = (4/3)\,\pi\,(5.00 \text{ cm})^3 = 52\underline{3}.5988 \text{ cm}^3$ $d = m/V$ Solve for m by multiplying both sides of the equation by V. $m = V \times d$

$$m = 52\underline{3}.5988 \text{ cm}^3 \times \frac{11.4 \text{ g lead}}{1 \text{ cm}^3} = 59\underline{6}9.026 \text{ g lead}$$

$$59\underline{6}9.026 \text{ g lead} \times \frac{100 \text{ g lead available}}{92.5 \text{ g lead extracted}} \times \frac{100 \text{ g galena}}{86.6 \text{ g lead available}} \times \frac{100 \text{ g ore}}{68.5 \text{ g galena}}$$

$= 1.0\underline{8}7811 \times 10^4 \text{ g ore} = 1.09 \times 10^4 \text{ g ore}$

Check: The units (g) are correct. The magnitude of the answer seems correct (10^4) because lead is so dense and the sphere is not small. Three significant figures are allowed to reflect the significant figures in all of the information given.

2.77 **Given:** 24.0 kg copper wire; wire is a cylinder of radius = 1.63 mm **Find:** resistance (Ω)
Other: d(copper) = 8.96g/cm^3; resistance = 2.061 Ω/km
Conceptual Plan: mm → m → cm and kg → g then

$$\frac{10^{-3}\text{ m}}{1\text{ m}} \quad \frac{100\text{ cm}}{1\text{ m}} \qquad \frac{1000\text{ g}}{1\text{ kg}}$$

***d*, *m* → *V* then *V*, *r* → *l*(cm) → m → km → Ω**

$$d = m/V \quad V = l\,\pi\,r^2 \quad \frac{1\text{ m}}{100\text{ cm}} \quad \frac{1\text{ km}}{1000\text{ m}} \quad \frac{2.061\ \Omega}{1\text{ km}}$$

Solution: $1.63\ \text{mm} \times \frac{10^{-3}\text{ m}}{1\text{ m}} \times \frac{100\text{ cm}}{1\text{ m}} = 0.163\text{ cm}$

$24.0\text{ kg} \times \frac{1000\text{ g}}{1\text{ kg}} = 2.40 \times 10^4\text{ g}$; then $d = m/V$. Rearrange by multiplying both sides of the equation by V to get $m = d \times V$; then divide both sides by d. $V = \frac{m}{d} = \frac{2.40 \times 10^4\text{ g}}{\frac{8.96\text{ g}}{1\text{ cm}^3}} = 2.6\underline{7}85714 \times 10^3\text{ cm}^3$; then $V = l\pi r^2$

Rearrange by dividing both sides of the equation by πr^2 to get

$$l = \frac{V}{\pi r^2} = \frac{2.6\underline{7}85714 \times 10^3\text{ cm}^3}{\pi(0.163\text{ cm})^2} = 3.2\underline{0}90623 \times 10^4\text{ cm}$$

then

$$3.2\underline{0}90623 \times 10^4\text{ cm} \times \frac{1\text{ m}}{100\text{ cm}} \times \frac{1\text{ km}}{1000\text{ m}} \times \frac{2.061\ \Omega}{1\text{ km}} = 0.66\underline{1}3877\ \Omega = 0.661\ \Omega$$

Check: The units (Ω) are correct. The magnitude of the answer seems correct because we expect a small resistance for a material that is commonly used for electrical wiring. Three significant figures are allowed to reflect the significant figures in 1.63 mm and 24.0 kg. Truncate the nonsignificant digits because the first nonsignificant digit is a 3.

2.78 **Given:** aluminum foil; 304 mm wide and 0.016 mm thick; 1.10 kg **Find:** length of foil
Other: d(aluminum) = 2.70 g/cm^3
Conceptual Plan: for each dimension, mm → m → cm and kg → g then

$$\frac{10^{-3}\text{ m}}{1\text{ m}} \quad \frac{100\text{ cm}}{1\text{ m}} \qquad \frac{1000\text{ g}}{1\text{ kg}}$$

***d*, *m* → *V* then *V*, *w*, *h* → *l*(cm) → m**

$$d = m/V \quad V = l\,\pi\,r^2 \quad \frac{1\text{ m}}{100\text{ cm}}$$

Solution: $304\ \text{mm} \times \frac{10^{-3}\text{ m}}{1\text{ m}} \times \frac{100\text{ cm}}{1\text{ m}} = 30.4\text{ cm}$ and $0.016\ \text{mm} \times \frac{10^{-3}\text{ m}}{1\text{ m}} \times \frac{100\text{ cm}}{1\text{ m}} = 0.0016\text{ cm}$

$1.10\text{ kg} \times \frac{1000\text{ g}}{1\text{ kg}} = 1.10 \times 10^3\text{ g}$; then $d = m/V$. Rearrange by multiplying both sides of the equation by V to get $m = d \times V$; then divide both sides by d. $V = \frac{m}{d} = \frac{1.10 \times 10^3\text{ g}}{\frac{2.70\text{ g}}{1\text{ cm}^3}} = 40\underline{7}.40741\text{ cm}^3$ then $V = l\,w\,h$

Rearrange by dividing both sides of equation by $(w\,h)$ to get

$$l = \frac{V}{w\,h} = \frac{40\underline{7}.40741\text{ cm}^3}{(30.4\text{ cm})(0.0016\text{ cm})} = 8.\underline{3}75974712 \times 10^3\text{ cm}$$

then

$$8.\underline{3}75974712 \times 10^3\text{ cm} \times \frac{1\text{ m}}{100\text{ cm}} = 8\underline{3}.75974712\text{ m} = 84\text{ m}$$

Check: The units (m) are correct. The magnitude of the answer seems correct because we expect a long length of the foil. Two significant figures are allowed to reflect the significant figures in 0.016 mm. Round up the nonsignificant digits because the first nonsignificant digit is 7.

2.79 **Given:** d(liquid nitrogen) = 0.808 g/mL; d(gaseous nitrogen) = 1.15 g/L; 175 L liquid nitrogen; 10.00 m × 10.00 m × 2.50 m room **Find:** fraction of room displaced by nitrogen gas

Conceptual Plan: L → mL then $V_{liquid}, d_{liquid} \rightarrow m_{liquid}$ then set $m_{liquid} = m_{gas}$ then $m_{gas}, d_{gas} \rightarrow V_{gas}$ then

$\frac{1000 \text{ mL}}{1 \text{ L}}$ $\quad d = m/V \quad$ $d = m/V$

calculate the V_{room} → cm³ → L then calculate the fraction displaced

$V = l \times w \times h \quad \frac{(100 \text{ cm})^3}{(1 \text{ m})^3} \quad \frac{1 \text{ L}}{1000 \text{ cm}^3} \quad \frac{V_{gas}}{V_{room}}$

Solution: $175 \text{ L} \times \frac{1000 \text{ mL}}{1 \text{ L}} = 1.75 \times 10^5$ mL. Solve for m by multiplying both sides of the equation by V.

$$m = V \times d = 1.75 \times 10^5 \text{ mL} \times \frac{0.808 \text{ g}}{1 \text{ mL}} = 1.4\underline{1}4 \times 10^5 \text{ g nitrogen liquid} = 1.4\underline{1}4 \times 10^5 \text{ g nitrogen gas}$$

$d = m/V$ Rearrange by multiplying both sides of the equation by V and dividing both sides of the equation by d.

$$V = \frac{m}{d} = \frac{1.4\underline{1}4 \times 10^5 \text{ g}}{1.15 \frac{\text{g}}{\text{L}}} = 1.2\underline{2}9565 \times 10^5 \text{ L nitrogen gas}$$

$$V_{room} = l \times w \times h = 10.00 \text{ m} \times 10.00 \text{ m} \times 2.50 \text{ m} \times \frac{(100 \text{ cm})^3}{(1 \text{ m})^3} \times \frac{1 \text{ L}}{1000 \text{ cm}^3} = 2.50 \times 10^5 \text{ L}$$

$$\frac{V_{gas}}{V_{room}} = \frac{1.2\underline{2}9565 \times 10^5 \text{ L}}{2.50 \times 10^5 \text{ L}} = 0.49\underline{1}826 = 0.492$$

Check: The units (none) are correct. The magnitude of the answer seems correct (0.5) because there is a large volume of liquid and the density of the gas is about a factor of 1000 less than the density of the liquid. Three significant figures are allowed to reflect the significant figures in the densities and the volume of the liquid given.

2.80 **Given:** d(mercury at 0.0 °C) = 13.596 g/cm³; d(mercury at 25.0 °C) = 13.534 g/cm³; 3.380 g; 0.200 mm diameter capillary **Find:** distance mercury rises

Conceptual Plan: at each temperature $m, d \rightarrow V$ then mm → m → cm then $V, r \rightarrow h$

$d = m/V \quad \frac{1 \text{ m}}{1000 \text{ mm}} \quad \frac{100 \text{ cm}}{1 \text{ m}} \quad V = \pi r^2 h$

then calculate the difference between the two heights

Solution: $d = m/V$ Rearrange by multiplying both sides of the equation by V and dividing both sides of the equation by d. $V = \frac{m}{d}$

$$\text{at 0.0 °C: } V = \frac{m}{d} = \frac{3.380 \text{ g}}{13.596 \frac{\text{g}}{\text{cm}^3}} = 0.248\underline{6}0253 \text{ cm}^3$$

$$\text{and at 25.0 °C: } V = \frac{m}{d} = \frac{3.380 \text{ g}}{13.534 \frac{\text{g}}{\text{cm}^3}} = 0.249\underline{7}41392 \text{ cm}^3$$

$$r = 0.200 \text{ mm} \times \frac{1 \text{ m}}{1000 \text{ mm}} \times \frac{100 \text{ cm}}{1 \text{ m}} = 0.0200 \text{ cm; then } V = \pi r^2 h.$$

Rearrange by dividing both sides of the equation by πr^2. $h = \frac{V}{\pi r^2}$

$$\text{at 0.0 °C: } h = \frac{V}{\pi r^2} = \frac{0.248\underline{6}0253 \text{ cm}^3}{\pi (0.0200 \text{ cm})^2} = 19\underline{7}.8316077 \text{ cm}$$

$$\text{and at 25.0 °C: } h = \frac{V}{\pi r^2} = \frac{0.249\underline{7}41392 \text{ cm}^3}{\pi (0.0200 \text{ cm})^2} = 19\underline{8}.7378852 \text{ cm}$$

The difference in height is $19\underline{8}.7378852 \text{ cm} - 19\underline{7}.8316077 \text{ cm} = \underline{0}.9062775 \text{ cm} = 1 \text{ cm}$

Check: The units (cm) are correct. The magnitude of the answer seems correct (1 cm) because there is a relatively small change in temperature and the two densities are very close to each other. Only one significant figure is allowed because the heights have four significant figures; so the error is in the tenths place.

2.81 **Given:** $r_{nucleus} = 2.7$ fm; $r_{atom} = 70$ pm (assume two significant figures)

Find: $V_{nucleus}$; V_{atom}; % $V_{nucleus}$

Conceptual Plan:

$r_{nucleus}$**(fm)** $\rightarrow$ $r_{nucleus}$**(pm)** $\rightarrow$ $V_{nucleus}$ **and then** r_{atom} $\rightarrow$ V_{atom} **and then** **% *V***

$\frac{10^{-15}\text{ m}}{1\text{ fm}}\ \frac{1\text{ pm}}{10^{-12}\text{ m}}$ $\quad V = \frac{4}{3}\pi r^3$ $\quad V = \frac{4}{3}\pi r^3$ $\quad \frac{V_{nucleus}}{V_{atom}} \times 100\%$

Solution:

$$2.7\ \cancel{\text{fm}} \times \frac{10^{-15}\ \cancel{\text{m}}}{1\ \cancel{\text{fm}}} \times \frac{1\text{ pm}}{10^{-12}\ \cancel{\text{m}}} = 2.7 \times 10^{-3}\text{ pm} \qquad \ll_{nucleus} = \frac{4}{3}\pi\,(2.7 \times 10^{-3}\text{ pm})^3 = 8.2 \times 10^{-8}\text{ pm}^3$$

$$V_{atom} = \frac{4}{3}\pi(70\text{ pm})^3 = 1.4 \times 10^6\text{ pm}^3 \qquad \frac{8.2 \times 10^{-8}\ \cancel{\text{pm}^3}}{1.4 \times 10^6\ \cancel{\text{pm}^3}} \times 100\% = 5.9 \times 10^{-12}\%\ V$$

Check: The units of the answer (% *V*) are correct. The magnitude of the answer is reasonable because the nucleus occupies only a very small percentage of the *V* of the atom.

2.82 **Given:** 1 penny = 1.0 mm **Find:** height in km of Avogadro's number of pennies

Conceptual Plan: height of 1 penny $\rightarrow$ height of Avogadro's number of pennies

6.022×10^{23}

$$\textbf{Solution: } \frac{1.0\ \cancel{\text{mm}}}{\cancel{\text{penny}}} \times 6.022 \times 10^{23}\ \cancel{\text{pennies}} \times \frac{1\ \cancel{\text{m}}}{1000\ \cancel{\text{mm}}} \times \frac{1\text{ km}}{1000\ \cancel{\text{m}}} = 6.0 \times 10^{17}\text{ km}$$

Check: The units of the answer (km) are correct. The magnitude of the answer shows just how large Avogadro's number is.

2.83 **Given:** 6.022×10^{23} pennies **Find:** the amount in dollars; the dollars/person

Conceptual Plan: pennies $\rightarrow$ dollars $\rightarrow$ dollars/person

$\frac{1\text{ dollar}}{100\text{ pennies}}$ $\quad$ 6.5 billion people

Solution:

$$6.022 \times 10^{23}\ \cancel{\text{pennies}} \times \frac{1\text{ dollar}}{100\ \cancel{\text{pennies}}} = 6.022 \times 10^{21}\text{ dollars} \qquad \frac{6.022 \times 10^{21}\text{ dollars}}{6.5 \times 10^9\text{ people}} = 9.3 \times 10^{11}\text{ dollars/person}$$

They are billionaires.

2.84 **Given:** 1 mol blueberries, m = 0.75 g; m(automobile) = 2.0×10^3 kg **Find:** number of autos for 1 mol blueberries

Conceptual Plan:

mol blueberries $\rightarrow$ mass blueberries (g) $\rightarrow$ mass blueberries (kg) $\rightarrow$ number of automobiles

$\frac{0.75\text{ g}}{1\text{ blueberry}}$ $\quad \frac{1\text{ kg}}{1000\text{ g}}$ $\quad \frac{1\text{ automobile}}{2.0 \times 10^3\text{ kg}}$

Solution:

$$1\ \cancel{\text{mol blueberries}} \times \frac{6.022 \times 10^{23}\ \cancel{\text{blueberries}}}{1\ \cancel{\text{mol blueberries}}} \times \frac{0.75\ \cancel{\text{g}}}{1\ \cancel{\text{blueberry}}} \times \frac{1\ \cancel{\text{kg}}}{1000\ \cancel{\text{g}}} \times \frac{1\text{ automobile}}{2.0 \times 10^3\ \cancel{\text{kg}}} = 2.3 \times 10^{17}\text{ automobiles}$$

Check: The units of the answer (automobiles) are correct. The magnitude of the answer is reasonable because Avogadro's number is so large.

2.85 **Given:** Cu sphere: $r = 0.935$ in; $d = 8.96$ g/cm^3 **Find:** number of Cu atoms

Conceptual Plan: *r* (inch) $\rightarrow$ *r*(cm) $\rightarrow$ V_{sphere} $\rightarrow$ g Cu $\rightarrow$ mol Cu $\rightarrow$ atoms Cu

$\frac{2.54\text{ cm}}{1\text{ in}}$ $\quad V = \frac{4}{3}\pi r^3$ $\quad \frac{8.96\text{ g}}{\text{cm}^3}$ $\quad \frac{1\text{ mol Cu}}{63.546\text{ g}}$ $\quad \frac{6.022 \times 10^{23}\text{ atoms}}{1\text{ mol}}$

$$\textbf{Solution:} \qquad 0.935\ \cancel{\text{in}} \times \frac{2.54\text{ cm}}{1\ \cancel{\text{in}}} = 2.3\underline{7}49\text{ cm}$$

$$\frac{4}{3}\pi(2.3\underline{7}49\ \cancel{cm})^3 \times \frac{8.96\ \cancel{g}}{\cancel{cm^3}} \times \frac{1\ \cancel{mol\ Cu}}{63.546\ \cancel{g}} \times \frac{6.022 \times 10^{23}\ \text{atoms Cu}}{1\ \cancel{mol\ Cu}} = 4.76 \times 10^{24}\ \text{atoms Cu}$$

Check: The units of the answer (atoms Cu) are correct. The magnitude of the answer is reasonable because about 8 mol Cu are present.

2.86 **Given:** Ti cube: $d = 4.50\ \text{g/cm}^3$; $l = 2.78$ in **Find:** number of Ti atoms

Conceptual Plan: e in inch → e in cm → vol cube → g Ti → mol Ti → atoms Ti

$$\frac{2.54\ \text{cm}}{1\ \text{in}} \quad V = l^3 \quad \frac{4.50\ \text{g}}{1\ \text{cm}^3} \quad \frac{1\ \text{mol Ti}}{47.87\ \text{g}} \quad \frac{6.022 \times 10^{23}\ \text{atoms}}{1\ \text{mol}}$$

Solution:

$$2.78\ \cancel{in} \times \frac{2.54\ \text{cm}}{1\ \cancel{in}} = 7.0\underline{6}1\ \text{cm}$$

$$(7.0\underline{6}1\ \cancel{cm})^3 \times \frac{4.50\ \cancel{g}}{\cancel{cm^3}} \times \frac{1\ \cancel{mol\ Ti}}{47.87\ \cancel{g}} \times \frac{6.022 \times 10^{23}\ \text{atoms Ti}}{1\ \cancel{mol\ Ti}} = 1.99 \times 10^{25}\ \text{atoms Ti}$$

Check: The units of the answer (atoms Ti) are correct. The magnitude of the answer is reasonable because about 30 mol of Ti is in the cube.

2.87 **Given:** Alloy of Au and Pd = 67.2 g; 2.49×10^{23} atoms **Find:** % composition by mass

Conceptual Plan: atoms Au and Pd → mol Au and Pd → g Au and Pd → g Au

$$\frac{1\ \text{mol}}{6.022 \times 10^{23}\ \text{atoms}} \quad \frac{196.97\text{g Au}}{1\ \text{mol Au}} \quad \frac{106.42\ \text{g Pd}}{1\ \text{mol Pd}}$$

Solution: Let x = atoms Au and y = atoms Pd; develop expressions that will permit atoms to be related to moles and then to grams.

$$(x\ \cancel{\text{atoms Au}})\left(\frac{1\ \text{mol Au}}{6.022 \times 10^{23}\ \cancel{\text{atoms Au}}}\right) = \frac{x}{6.022 \times 10^{23}}\ \text{mol Au}$$

$$(y\ \cancel{\text{atoms Pd}})\left(\frac{1\ \text{mol Pd}}{6.022 \times 10^{23}\ \cancel{\text{atoms Pd}}}\right) = \frac{y}{6.022 \times 10^{23}}\ \text{mol Pd}$$

$x + y = 2.49 \times 10^{23}$atoms; $y = 2.49 \times 10^{23} - x$

$$\left(\frac{x}{6.022 \times 10^{23}}\ \cancel{\text{mol Au}}\right)\left(\frac{196.97\ \text{g Au}}{1\ \cancel{\text{mol Au}}}\right) = \frac{196.97x}{6.022 \times 10^{23}}\ \text{g Au}$$

$$\left(\frac{2.49 \times 10^{23} - x}{6.022 \times 10^{23}}\ \cancel{\text{mol Pd}}\right)\left(\frac{106.42\ \text{g Pd}}{1\ \cancel{\text{mol Pd}}}\right) = \frac{106.42(2.49 \times 10^{23} - x)}{6.022 \times 10^{23}}\ \text{g Pd}$$

g Au + g Pd = 67.2 g total

$$\frac{196.97x}{6.022 \times 10^{23}}\text{g Au} + \frac{106.42(2.49 \times 10^{23} - x)}{6.022 \times 10^{23}}\text{g Pd} = 67.2\ \text{g}$$

$X = 1.5\underline{4}26 \times 10^{23}$ atoms Au

$$(1.54 \times 10^{23}\ \cancel{\text{atoms Au}})\left(\frac{1\ \cancel{\text{mol Au}}}{6.022 \times 10^{23}\ \cancel{\text{atoms Au}}}\right)\left(\frac{196.97\ \text{g Au}}{1\ \cancel{\text{mol Au}}}\right) = 50.\underline{3}7\ \text{g Au}$$

$$\left(\frac{50.\underline{3}7\ \text{g Au}}{67.2\ \text{g sample}}\right) \times 100 = 74.\underline{9}6\%\ \text{Au} = 75.0\%\ \text{Au}$$

% Pd = 100.0% − 75.0% Au = 25.0% Pd

Check: Units of the answer (% composition) are correct.

2.88 **Given:** brass: 37.0% Zn; $d = 8.48\text{g/cm}^3$; volume = 112 cm^3 **Find:** atoms of Zn and Cu

Conceptual Plan: volume sample → g sample → g Zn → mole Zn → atoms Zn

$$\frac{8.48\ \text{g}}{\text{cm}^3} \quad \frac{37.0\ \text{g Zn}}{100.0\ \text{g sample}} \quad \frac{1\ \text{mol Zn}}{65.41\ \text{g Zn}} \quad \frac{6.022 \times 10^{23}\ \text{atoms}}{1\ \text{mol}}$$

→ g Cu → moles Cu → atoms Cu

$$\text{g sample} - \text{g Zn} \quad \frac{63.55\ \text{g Cu}}{1\ \text{mol Cu}} \quad \frac{6.022 \times 10^{23}\ \text{atoms}}{1\ \text{mol}}$$

Solution: $112\ \cancel{cm^3} \times \dfrac{8.48\ \text{g}}{\cancel{cm^3}} = 95\underline{4}.0\ \text{g sample}$ $\quad 95\underline{4}.0\ \cancel{\text{g sample}} \times \dfrac{37.0\ \text{g Zn}}{100.0\ \cancel{\text{g sample}}} = 35\underline{2}.98\ \text{g Zn}$

$$352.98\text{ g Zn} \times \frac{1\text{ mol Zn}}{65.41\text{ g Zn}} \times \frac{6.022 \times 10^{23}\text{ atoms Zn}}{1\text{ mol Zn}} = 3.2497 \times 10^{24}\text{ atoms Zn} = 3.25 \times 10^{24}\text{ atoms Zn}$$

$$954.0\text{ g sample} - 352.98\text{ g Zn} = 601.02\text{ g Cu}$$

$$601.02\text{ g Cu} \times \frac{1\text{ mol Cu}}{63.55\text{ g Cu}} \times \frac{6.022 \times 10^{23}\text{ atoms Cu}}{1\text{ mol Cu}} = 5.6953 \times 10^{24}\text{ atoms Cu} = 5.70 \times 10^{24}\text{ atoms Cu}$$

Check: The units of the answer (atoms of Zn and atoms of Cu) are correct. The magnitude is reasonable because more than 1 mole of each element is in the sample.

2.89 **Given:** air contains 1.5 μg Pb/m^3; lung volume = 5.50 L **Find:** atoms of Pb in lungs

Conceptual Plan: lung in L → mL → cm^3 → m^3 → μg Pb → g Pb → mol Pb → atoms Pb

$$\frac{1000\text{ mL}}{\text{L}} \quad \frac{1\text{ cm}^3}{\text{mL}} \quad \frac{1\text{ m}^3}{(100\text{ cm})^3} \quad \frac{1.5\ \mu\text{g Pb}}{\text{m}^3} \quad \frac{1\text{ g Pb}}{10^6\ \mu\text{g Pb}} \quad \frac{1\text{ mol Pb}}{207.2\text{ g Pb}} \quad \frac{6.022 \times 10^{23}\text{ atoms Pb}}{1\text{ mol Pb}}$$

Solution: $5.50\text{ L} \times \left(\frac{1000\text{ mL}}{\text{L}}\right) \times \left(\frac{1\text{ cm}^3}{1\text{ mL}}\right) \times \left(\frac{1\text{ m}^3}{(100\text{ cm})^3}\right) \times \left(\frac{1.5\ \mu\text{g Pb}}{\text{m}^3}\right) \times \left(\frac{1\text{ g Pb}}{10^6\ \mu\text{g Pb}}\right)$

$$\times \left(\frac{1\text{ mol Pb}}{207.2\text{ g Pb}}\right) \times \left(\frac{6.022 \times 10^{23}\text{ atoms Pb}}{1\text{ mol Pb}}\right) = 2.398 \times 10^{13}\text{ atoms Pb} = 2.4 \times 10^{13}\text{ atoms Pb}$$

Check: The units of the answer (atoms Pb) are correct. The magnitude of the answer is reasonable because about 400 nmol of Pb is present.

2.90 **Given:** 0.255 oz 18K Au **Find:** atoms Au

Conceptual Plan: oz 18K Au → oz pure Au → g Au → mol Au → atoms Au

$$\frac{75\text{ oz Au}}{100\text{ oz 18K Au}} \quad \frac{453.59\text{ g Au}}{16\text{ oz Au}} \quad \frac{1\text{ mol Au}}{196.97\text{ g Au}} \quad \frac{6.022 \times 10^{23}\text{ atoms Au}}{1\text{ mol Au}}$$

Solution: $0.255\text{ oz 18K Au} \times \left(\frac{75\text{ oz pure Au}}{100\text{ oz 18K Au}}\right) \times \left(\frac{453.59\text{ g Au}}{16\text{ oz Au}}\right) \times \left(\frac{1\text{ mol Au}}{196.97\text{ g Au}}\right) \times \left(\frac{6.022 \times 10^{23}\text{ atoms Au}}{1\text{ mol Au}}\right)$

$$= 1.658 \times 10^{22}\text{ atoms Au} = 1.7 \times 10^{22}\text{ atoms Au}$$

Check: The units of the answer (atoms Au) are correct. The magnitude of the answer is reasonable because less than 1 mol of Au is in the sample.

Challenge Problems

2.91 **Given:** mass of black hole (BH) = 1×10^3 suns; radius of black hole = one-half the radius of our moon

Find: density (g/cm^3) **Other:** radius of our sun = 7.0×10^5 km; average density of our sun = $1.4 \times 10^3\text{ kg/m}^3$; diameter of the moon = 2.16×10^3 miles

Conceptual Plan: $d_{BH} = m_{BH}/V_{BH}$

Calculate m_{BH}: $r_{sun} \rightarrow V_{sun}\text{ km}^3_{sun} \rightarrow \text{m}^3_{sun}$ then $V_{sun}, d_{sun} \rightarrow m_{sun}$ then $m_{sun} \rightarrow m_{BH}$ kg → g

$$V = (4/3)\pi r^3 \quad \frac{(1000\text{ m})^3}{(1\text{ km})^3} \quad d_{sun} = \frac{m_{sun}}{V_{sun}} \quad m_{BH} = (1 \times 10^3) \times m_{sun} \quad \frac{1000\text{ g}}{1\text{ kg}}$$

Calculate V_{BH} : $d_{moon} \rightarrow r_{moon} \rightarrow r_{BH}$ mi → km → m → cm then $r \rightarrow V$

$$r_{moon} = 1/2\, d_{moon} \quad r_{BH} = 1/2\, r_{moon} \quad \frac{1\text{ km}}{0.6214\text{ mi}} \quad \frac{1000\text{ m}}{1\text{ km}} \quad \frac{100\text{ cm}}{1\text{ m}} \quad V = (4/3)\pi r^3$$

Substitute into $d_{BH} = m_{BH}/V_{BH}$

Solution: Calculate m_{BH} : $V_{sun} = (4/3)\pi r^3_{sun} = (4/3)\pi(7.0 \times 10^5\text{ km})^3 = 1.43675504 \times 10^{18}\text{ km}^3$

$$1.43675504 \times 10^{18}\text{ km}^3 \times \frac{(1000\text{ m})^3}{(1\text{ km})^3} = 1.43675504 \times 10^{27}\text{ m}^3$$

$d_{sun} = m_{sun}/V_{sun}$ Solve for m by multiplying both sides of the equation by V_{sun}. $m_{sun} = V_{sun} \times d_{sun}$

$m_{sun} = (1.43675504 \times 10^{27}\text{ m}^3)(1.4 \times 10^3\text{ kg/m}^3) = 2.011457056 \times 10^{30}\text{ kg}$

$m_{BH} = (1 \times 10^3) \times m_{sun} = (1 \times 10^3) \times (2.011457056 \times 10^{30}\text{ kg}) = 2.011457056 \times 10^{33}\text{ kg}$

$$2.011457056 \times 10^{33}\text{ kg} \times \frac{1000\text{ g}}{1\text{ kg}} = 2.011457056 \times 10^{36}\text{ g}$$

Calculate V_{BH} : $r_{moon} = \frac{1}{2}d_{moon} = \frac{1}{2}(2.16 \times 10^3 \text{ miles}) = 1.08 \times 10^3 \text{ miles}$

$$r_{BH} = \frac{1}{2}r_{moon} = \frac{1}{2}(1.08 \times 10^3 \text{ miles}) = 540. \text{ miles}$$

$$540. \cancel{\text{miles}} \times \frac{1 \cancel{\text{km}}}{0.6214 \cancel{\text{mi}}} \times \frac{1000 \cancel{\text{m}}}{1 \cancel{\text{km}}} \times \frac{100 \text{ cm}}{1 \cancel{\text{m}}} = 8.6\underline{9}00547 \times 10^7 \text{ cm}$$

$$V = (4/3)\pi r^3 = (4/3)\pi(8.6\underline{9}00547 \times 10^7 \text{ cm})^3 = 2.7\underline{4}888227 \times 10^{24} \text{ cm}^3$$

Substitute into $d_{BH} = \frac{m_{BH}}{V_{BH}} = \frac{2.0\underline{1}1457056 \times 10^{36} \text{ g}}{2.7\underline{4}888227 \times 10^{24} \text{ cm}^3} = 7.\underline{3}1736342 \times 10^{11} \frac{\text{g}}{\text{cm}^3} = 7.3 \times 10^{11} \frac{\text{g}}{\text{cm}^3}$

Check: The units (g/cm^3) are correct. The magnitude of the answer seems correct (10^{12}) because we expect extremely high numbers for black holes. Two significant figures are allowed to reflect the significant figures in the radius of our sun $(7.0 \times 10^5 \text{ km})$ and the average density of the sun $(1.4 \times 10^3 \text{ kg/m}^3)$. Truncate the nonsignificant digits because the first nonsignificant digit is a 1.

2.92 **Given:** 15.0 ppm CO; 8-hour period **Find:** milligrams of carbon monoxide
Other: 0.50 L of air per breath; 20 breaths per minute; carbon monoxide has a density of 1.2 g/L; 15.0 ppm CO means 15.0 L CO per 10^6 L air

Conceptual plan: hr → min → breaths → L_{air} → L_{CO} → g_{CO} → mg_{CO}

$$\frac{60 \text{ min}}{1 \text{ hr}} \quad \frac{20 \text{ breaths}}{1 \text{ min}} \quad \frac{0.50 \text{ L}_{air}}{1 \text{ breath}} \quad \frac{15.0 \text{ L}_{CO}}{1 \times 10^6 \text{ L}_{air}} \quad \frac{1.2 \text{ g}_{CO}}{1 \text{ L}_{CO}} \quad \frac{1000 \text{ mg}_{CO}}{1 \text{ g}_{CO}}$$

Solution:

$$8 \cancel{\text{hr}} \times \frac{60 \cancel{\text{min}}}{1 \cancel{\text{hr}}} \times \frac{20 \cancel{\text{breaths}}}{1 \cancel{\text{min}}} \times \frac{0.50 \cancel{\text{L}_{air}}}{1 \cancel{\text{breath}}} \times \frac{15.0 \cancel{\text{L}_{CO}}}{1 \times 10^6 \cancel{\text{L}_{air}}} \times \frac{1.2 \cancel{\text{g}_{CO}}}{1 \cancel{\text{L}_{CO}}} \times \frac{1000 \text{ mg}_{CO}}{1 \cancel{\text{g}_{CO}}} = \underline{8}6.4 \text{ mg}_{CO} = 9 \times 10^1 \text{ mg}_{CO}$$

Check: The units (mg) are correct. The magnitude of the answer (10^2) makes physical sense because there are more than 6 powers of 10 visible in these conversion factors in the numerator and one factor of 10^6 in the denominator. This means that most of the conversions cancel each other out, but there is still some left in the numerator. One significant figure is allowed because the conversion factor with the least precision is 20 breaths/minute (one significant figure); the starting time (8 hours) also has one significant figure. Round the last digit up because the first nonsignificant digit is a 6.

2.93 **Given:** cubic nanocontainers with an edge length = 25 nanometers
Find: (a) volume of one nanocontainer; (b) grams of oxygen could be contained by each nanocontainer; (c) grams of oxygen inhaled per hour; (d) minimum number of nanocontainers per hour; (e) minimum volume of nanocontainers.
Other: (pressurized oxygen) = 85 g/L; 0.28 g of oxygen per liter; the average human inhales about 0.50 L of air per breath and takes about 20 breaths per minute; adult total blood volume = ~5 L
Conceptual Plan:

(a) **nm → m → cm then l → V then cm^3 → L**

$$\frac{1 \text{ m}}{10^9 \text{ nm}} \quad \frac{100 \text{ cm}}{1 \text{ m}} \quad V = l^3 \quad \frac{1 \text{ L}}{1000 \text{ cm}^3}$$

(b) **L → g pressurized oxygen**

$$\frac{85 \text{ g oxygen}}{1 \text{ L nanocontainers}}$$

(c) **hr → min → breaths → L_{air} → g_{O_2}**

$$\frac{60 \text{ min}}{1 \text{ hr}} \quad \frac{20 \text{ breaths}}{1 \text{ min}} \quad \frac{0.50 \text{ L}_{air}}{1 \text{ breaths}} \quad \frac{0.28 \text{ g}_{O_2}}{1 \text{ L}_{air}}$$

(d) **grams oxygen → number nanocontainers**

$$\frac{1 \text{ nanocontainer}}{\text{part (b) grams of oxygen}}$$

(e) **number nanocontainers → volume nanocontainers**

$$\frac{\text{part (a) volume}}{\text{of 1 nanocontainer}}$$

Solution:

(a) $25\ \text{nm} \times \frac{1\ \text{m}}{10^9\ \text{nm}} \times \frac{100\ \text{cm}}{1\ \text{m}} = 2.5 \times 10^{-6}\ \text{cm}$

$V = l^3 = (2.5 \times 10^{-6}\ \text{cm})^3 = 1.5625 \times 10^{-17}\ \text{cm}^3 \times \frac{1\ \text{L}}{1000\ \text{cm}^3} = 1.5625 \times 10^{-20}\ \text{L} = 1.6 \times 10^{-20}\ \text{L}$

(b) $1.5625 \times 10^{-20}\ \text{L} \times \frac{85\ \text{g oxygen}}{1\ \text{L nanocontainers}} = 1.328125 \times 10^{-18} \frac{\text{g pressurized O}_2}{\text{nanocontainer}}$

$= 1.3 \times 10^{-18} \frac{\text{g pressurized O}_2}{\text{nanocontainer}}$

(c) $1\ \text{hr} \times \frac{60\ \text{min}}{1\ \text{hr}} \times \frac{20\ \text{breaths}}{1\ \text{min}} \times \frac{0.50\ \text{L}_{\text{air}}}{1\ \text{breath}} \times \frac{0.28\ \text{gO}_2}{1\ \text{L}_{\text{air}}} = 1.68 \times 10^2\ \text{g oxygen} = 1.7 \times 10^2\ \text{g oxygen}$

(d) $1.68 \times 10^2\ \text{g oxygen} \times \frac{1\ \text{nanocontainer}}{1.3 \times 10^{-18}\ \text{g oxygen}} = 1.292307692 \times 10^{20}\ \text{nanocontainers}$

$= 1.3 \times 10^{20}\ \text{nanocontainers}$

(e) $1.292307692 \times 10^{20}\ \text{nanocontainers} \times \frac{1.5625 \times 10^{-20}\ \text{L}}{\text{nanocontainer}} = 2.019230769\ \text{L} = 2.0\ \text{L}$

This volume is much too large to be feasible because the volume of blood in the average human is 5 L.

Check:

(a) The units (L) are correct. The magnitude of the answer $(10-20)$ makes physical sense because these are very, very tiny containers. Two significant figures are allowed, reflecting the significant figures in the starting dimension [25 nm (2 significant figures)]. Round the last digit up because the first nonsignificant digit is a 6.

(b) The units (g) are correct. The magnitude of the answer (10^{-18}) makes physical sense because these are very, very tiny containers and very few molecules can fit inside. Two significant figures are allowed, reflecting the significant figures in the starting dimension (25 nm) and the given concentration (85 g/L)—two significant figures in each. Truncate the nonsignificant digits because the first nonsignificant digit is a 2.

(c) The units (g oxygen) are correct. The magnitude of the answer (10^2) makes physical sense because of the conversion factors involved and the fact that air is not very dense. Two significant figures are allowed because it is stated in the problem. Round the last digit up because the first nonsignificant digit is an 8.

(d) The units (nanocontainers) are correct. The magnitude of the answer (10^{20}) makes physical sense because these are very, very tiny containers and we need a macroscopic quantity of oxygen in them. Two significant figures are allowed, reflecting the significant figures in both of the quantities in the calculation—two significant figures. Round the last digit up because the first nonsignificant digit is a 9.

(e) The units (L) are correct. The magnitude of the answer (2) makes physical sense because of the magnitudes of the numbers in this step. Two significant figures are allowed, reflecting the significant figures in both of the quantities in the calculation—two significant figures. Truncate the nonsignificant digits because the first nonsignificant digit is a 1.

2.94 Because the person weighs 155 lb and has a density of $1.0\ \text{g/cm}^3$, the volume of the person can be calculated as

$155\ \text{lb} \times \frac{453.59\ \text{g}}{1\ \text{lb}} \times \frac{1\ \text{cm}^3}{1.0\ \text{g}} = 7.030645 \times 10^4\ \text{cm}^3$.

Approximating the volume of a person as a cylinder 4.0 feet tall, $V = l\pi r^2$. Rearranging the equation, solve for r.

$$r = \left(\frac{V}{l\pi}\right)^{\frac{1}{2}} = \sqrt{\frac{7.030645 \times 10^4\ \text{cm}^3}{4.0\ \text{ft} \times \frac{30.48\ \text{cm}}{1\ \text{ft}} \times \pi}} = 13.54831\ \text{cm}$$

The circumference is $2\pi r = 2\pi(13.54831\ \text{cm}) = 85.12655\ \text{cm}$. When the person gains 40.0 lb of fat, the volume increase is $40.0\ \text{lb} \times \frac{453.59\ \text{g}}{1\ \text{lb}} \times \frac{1\ \text{cm}^3}{0.918\ \text{g}} = 1.97643 \times 10^4\ \text{cm}^3$.

Thus, the new volume is $7.030645 \times 10^4\ cm^3 + 1.97643 \times 10^4\ cm^3 = 9.00708 \times 10^4\ cm^3$. So the new radius is $r = \left(\frac{V}{l\pi}\right)^{\frac{1}{2}} = \sqrt{\frac{9.00708 \times 10^4\ cm^3}{4.0\ ft \times \frac{30.48\ cm}{1\ ft} \times \pi}} = 15.33485$ cm, and the new circumference is

$2\pi r = 2\pi(15.33485\ cm) = 96.35170$ cm.

The percent increase in circumference $\frac{96.35170\ cm - 85.12655\ cm}{85.12655\ cm} \times 100\% = 13.1864\% = 13\%$.

2.95 Assume that all of the spheres are the same size. Let $x =$ the percentage of spheres that are copper (expressed as a fraction); so the volume of copper $= (427\ cm^3)x$, and the volume of lead $= (427\ cm^3)(1 - x)$.

Because the density of copper is $8.96\ g/cm^3$, the mass of copper $= (427\ cm^3)x \times \frac{8.96\ g}{cm^3} = 3825.92(x)$ g.

Because the density of lead is $11.4\ g/cm^3$, the mass of lead $= (427\ cm^3)(1 - x) \times \frac{11.46\ g}{cm^3} = 4893.42\ (1 - x)$ g.

Because the total mass is 4.36 kg = 4360 g, 4360 g = 3825.92(x) g + 4893.42 (1 − x) g. Solving for x, 1067.50(x) g = 533.42 g → x = 0.499691, or 50.% of the spheres are copper.

Check: This answer makes sense because the average density of the spheres $= 4360\ g/427\ cm^3 = 10.2\ g/cm^3$ and the average of the density of copper and the density of lead $= (8.96 + 11.4)/2\ g/cm^3 = 10.2\ g/cm^3$.

2.96 **Given:** 1 mol sand grains; cube edge (e) = 0.10 mm; area Texas = 268,601 sq mi **Find:** height of sand ft

Conceptual Plan:

$mol_{sand} \rightarrow grains_{sand} \rightarrow Vol_{sand}\ (mm^3) \rightarrow Vol_{sand}\ (ft^3)$ and then $A_{Texas}\ (mi^2) \rightarrow$ area (ft^2)

$\frac{6.022 \times 10^{23}\ grains}{1\ mol}$ $\quad V = e^3\ mm^3 \times \left(\frac{cm}{10\ mm}\right)^3\left(\frac{1\ in}{2.54\ cm}\right)^3\left(\frac{1\ ft}{12\ in}\right)^3$ $\quad \left(\frac{1\ mi}{5280\ ft}\right)^2$

and then $V, A \rightarrow$ height (ft)

$h = \frac{V}{A}$

Solution:

$$1\ mol\ sand\ grains \times \frac{6.022 \times 10^{23}\ grains}{1\ mol} \times \frac{(0.10\ mm)^3}{grain} \times \left(\frac{cm}{10\ mm}\right)^3 \times \left(\frac{1\ in}{2.54\ cm}\right)^3 \times \left(\frac{1\ ft}{12\ in}\right)^3$$

$= 2.1266 \times 10^{13}\ ft^3$ sand

$$\frac{2.1266 \times 10^{13}\ ft^3\ sand}{268{,}601\ mi^2} \times \left(\frac{1\ mi}{5280\ ft}\right)^2 = 2.8\ \text{ft of sand}$$

Check: The units of the answer (ft of sand) are correct. The magnitude of the answer seems reasonable.

2.97 **Given:** $d = 1.4\ g/cm^3$, $r = 7 \times 10^8$ m; 100 billion stars/galaxy; 10 billion galaxies/universe

Find: number of atoms in the universe

Conceptual Plan: $r_{(star)}\ (m) \rightarrow r_{(star)}\ (cm) \rightarrow V_{(star)} \rightarrow$ g H/star → mol H/star → atoms H/star

$\frac{100\ cm}{m}$ $\quad V = \frac{4}{3}\pi r^3$ $\quad \frac{1.4\ g\ H}{cm^3}$ $\quad \frac{1\ mol\ H}{1.008\ gH}$ $\quad \frac{6.022 \times 10^{23}\ atoms}{1\ mol}$

atoms H/star → atoms H/galaxy → atoms H/universe

$\frac{100 \times 10^9\ stars}{galaxy}$ $\quad \frac{10 \times 10^9\ galaxies}{universe}$

Solution: $7 \times 10^8\ m \times \frac{100\ cm}{m} = 7 \times 10^{10}$cm

$$\frac{4}{3}\pi\frac{(7 \times 10^{10}\ cm)^3}{star} \times \frac{1.4\ g\ H}{cm^3} \times \frac{1\ mol\ H}{1.008\ g\ H} \times \frac{6.022 \times 10^{23}\ atoms\ H}{1\ mol\ H} \times \frac{100 \times 10^9\ stars}{galaxy} \times \frac{10 \times 10^9\ galaxies}{universe}$$

$= 1 \times 10^{78}$ atoms/universe

Check: The units of the answer (atoms/universe) are correct.

Conceptual Problems

2.98 **Given:** 7 cm on each edge cube **Find:** cm^3

Conceptual plan: Read the information carefully. The cube is 7 cm on each side.

$l, w, h \rightarrow V$

$V = l\,w\,h$

in a cube $l = w = h$

Solution: $7\text{ cm} \times 7\text{ cm} \times 7\text{ cm} = (7\text{ cm})^3 = 343\text{ cm}^3$, or 343 cubes

2.99 To determine which number is large, the units need to be compared. There is a factor of 1000 between grams and kilograms in the numerator. There is a factor of $(100)^3$, or 1,000,000, between cm^3 and m^3. This second factor more than compensates for the first factor. Thus, Substance A with a density of 1.7 g/cm^3 is denser than Substance B with a density of 1.7 kg/m^3.

2.100 Remember that density = mass/volume.

(a) The darker-colored box has a heavier mass but a smaller volume, so it is denser than the lighter-colored box.

(b) The lighter-colored box is heavier than the darker-colored box, and both boxes have the same volume; so the lighter-colored box is denser.

(c) The larger box is the heavier box, so it cannot be determined with this information which box is denser.

2.101 If the amu and mole were not based on the same isotope, the numerical values obtained for an atom of material and a mole of material would not be the same. If, for example, the mole was based on the number of particles in C–12 but the amu was changed to a fraction of the mass of an atom of Ne–20 the number of particles and the number of amu that make up one mole of material would no longer be the same. We would no longer have the relationship where the mass of an atom in amu is numerically equal to the mass of a mole of those atoms in grams.

2.102 **Given:** (a) Cr: 55.0 g; atomic mass = 52 g/mol
(b) Ti: 45.0 g; atomic mass = 48 g/mol
(c) Zn: 60.0 g; atomic mass = 65 g/mol

Find: which has the greatest amount in moles and which has the greatest mass

Conceptual Plan: Without calculation, compare grams of material to g/mol for each.

Solution: Cr would have the greatest mole amount of the elements. It is the only one whose mass is greater than the molar mass. Zn would be the greatest mass amount because it is the largest mass value.

3 The Quantum-Mechanical Model of the Atom

Review Questions

3.1 The quantum-mechanical model of the atom is important because it explains how electrons exist in atoms and how those electrons determine the chemical and physical properties of elements.

3.2 Light is electromagnetic radiation, a type of energy embodied in oscillating electric and magnetic fields. Light in a vacuum travels at 3.00×10^8 m/s.

3.3 The wavelength (λ) of the wave is the distance in space between adjacent crests and is measured in units of distance. The amplitude of the wave is the vertical height of a crest. The more closely spaced the waves (i.e., the shorter the wavelength), the more energy there is. The amplitudes of the electric and magnetic field waves in light determine the intensity or brightness of the light. The higher the amplitude, the more energy the wave has.

3.4 The frequency, (ν), is the number of cycles (or wave crests) that pass through a stationary point in a given period of time. The units of frequency are cycles per second. The frequency is inversely proportional to the wavelength (λ). Frequency and wavelength are related by the equation $\nu = \frac{c}{\lambda}$.

3.5 For visible light, wavelength determines the color. Red light has a wavelength of 750 nm, the longest wavelength of visible light, and blue has a wavelength of 500 nm.

3.6 The presence of a variety of wavelengths in white light is responsible for the way we perceive colors in objects. When a substance absorbs some colors while reflecting others, it appears colored. Grass appears green because it reflects primarily the wavelength associated with green light and absorbs the others.

3.7 (a) Gamma rays (γ)—the wavelength range is 10^{-11} to 10^{-15} m. Gamma rays are produced by the sun, other stars, and certain unstable atomic nuclei on Earth. Human exposure to gamma rays is dangerous because the high energy of gamma rays can damage biological molecules.

(b) X-rays—the wavelength range is 10^{-8} to 10^{-11} m. X-rays are used in medicine. X-rays pass through many substances that block visible light and are therefore used to image bones and internal organs. X-rays are sufficiently energetic to damage biological molecules, so while several yearly exposures to X-rays are harmless, excessive exposure increases cancer risk.

(c) Ultraviolet radiation (UV)—the wavelength range is 0.4×10^{-6} to 10^{-8} m. Ultraviolet radiation is most familiar as the component of sunlight that produces a sunburn or suntan. While not as energetic as gamma rays or X-rays, ultraviolet light still carries enough energy to damage biological molecules. Excessive exposure to ultraviolet light increases the risk of skin cancer and cataracts and causes premature wrinkling of the skin.

(d) Visible light—the wavelength range is 0.75×10^{-6} to 0.4×10^{-6} m (750 nm to 400 nm). Visible light, as long as the intensity is not too high, does not carry enough energy to damage biological molecules. It does, however, cause certain molecules in our eyes to change their shape, sending a signal to the brain that results in vision.

(e) Infrared radiation (IR)—the wavelength range is 0.75×10^{-6} to 10^{-3} m. The heat you feel when you place your hand near a hot object is infrared radiation. All warm objects, including human bodies, emit infrared light. Although infrared light is invisible to our eyes, infrared sensors can detect it and are often used in night vision technology to "see" in the dark.

(f) Microwave radiation—the wavelength range is 10^{-3} to 10^{-1} m. Microwave radiation is used in radar and in microwave ovens. Microwave radiation is efficiently absorbed by water and can therefore heat substances that contain water.

(g) Radio waves—the wavelength range is 10^{-1} to 10^{5} m. Radio waves are used to transmit the signals responsible for AM and FM radio, cell telephones, television, and other forms of communication.

3.8 Waves interact with each other in a characteristic way called interference. They can cancel each other out or build each other up, depending on their alignment upon interaction. Constructive interference occurs when waves of equal amplitude from two sources are in phase (they align with overlapping crests), and a wave with twice the amplitude results. Destructive interference occurs when the waves are completely out of phase—they align so that the crest from one source overlaps the trough from the other source—and the waves cancel each other.

3.9 Diffraction occurs when a wave encounters an obstacle or a slit that is comparable in size to its wavelength. The wave bends around the slit. The diffraction of light through two slits separated by a distance comparable to the wavelength of the light results in an interference pattern. Each slit acts as a new wave source, and the two new waves interfere with each other. This results in a pattern of bright and dark lines.

Interference from Two Slits

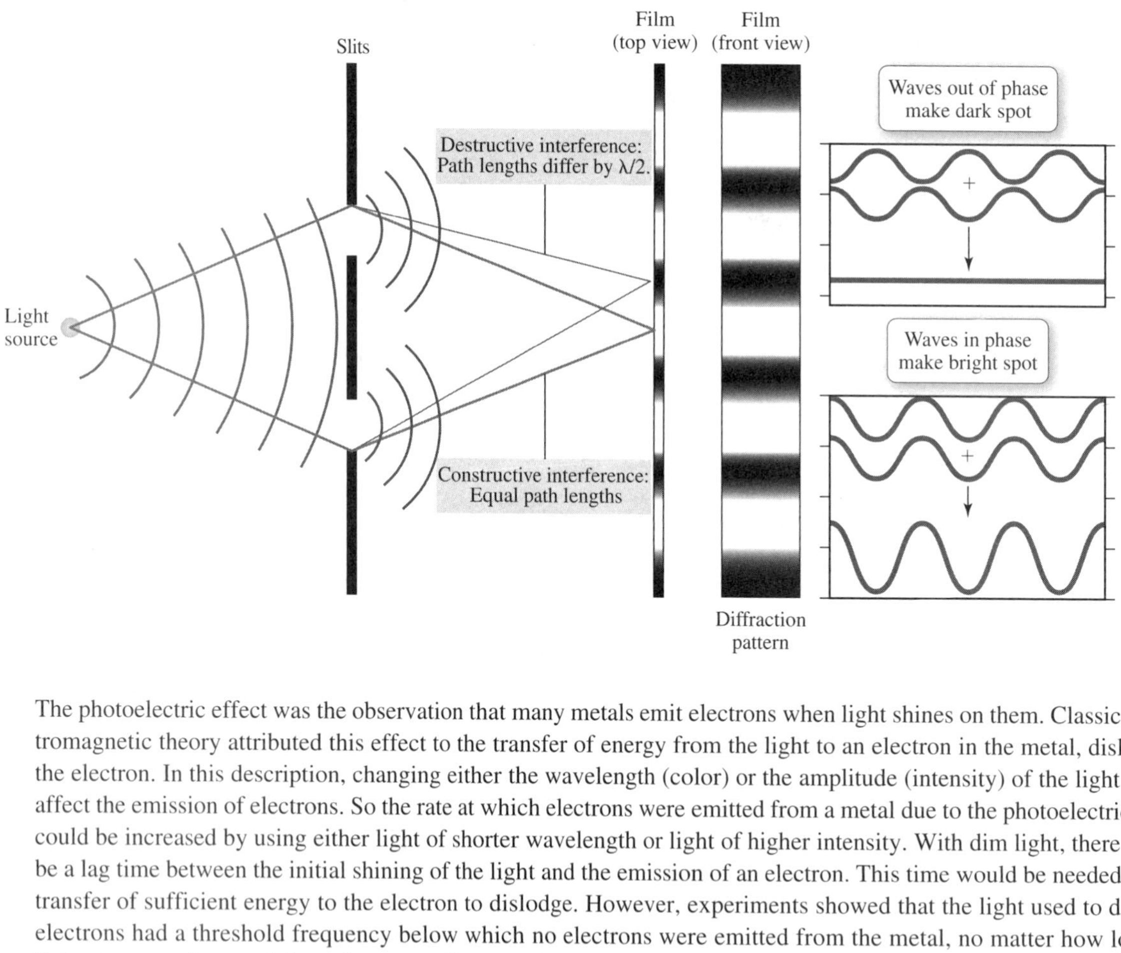

Slits
Film (top view)
Film (front view)
Waves out of phase make dark spot
Destructive interference: Path lengths differ by λ/2.
Light source
Waves in phase make bright spot
Constructive interference: Equal path lengths
Diffraction pattern

3.10 The photoelectric effect was the observation that many metals emit electrons when light shines on them. Classical electromagnetic theory attributed this effect to the transfer of energy from the light to an electron in the metal, dislodging the electron. In this description, changing either the wavelength (color) or the amplitude (intensity) of the light should affect the emission of electrons. So the rate at which electrons were emitted from a metal due to the photoelectric effect could be increased by using either light of shorter wavelength or light of higher intensity. With dim light, there should be a lag time between the initial shining of the light and the emission of an electron. This time would be needed for the transfer of sufficient energy to the electron to dislodge. However, experiments showed that the light used to dislodge electrons had a threshold frequency below which no electrons were emitted from the metal, no matter how long the light shone on the metal. Low-frequency light would not eject electrons from a metal regardless of intensity or duration. But high-frequency light would eject electrons even at low intensity without any lag time.

3.11 Because of the results of the experiments with the photoelectric effect, Einstein proposed that light energy must come in packets. The amount of energy in a light packet depends on its frequency (wavelength). The emission of electrons depends on whether a single photon has sufficient energy to dislodge a single electron.

3.12 A photon is a packet of light. The energy of the photon can be expressed in terms of wavelength as $E = \frac{hc}{\lambda}$ or in terms of frequency as $E = h\nu$.

3.13 An emission spectrum occurs when an atom absorbs energy and re-emits that energy as light. The light emitted contains distinct wavelengths for each element. The emission spectrum of a particular element is always the same and can be used to identify the element. A white light spectrum is continuous, meaning that there are no sudden interruptions in the intensity of the light as a function of wavelengths. It consists of all wavelengths. Emission spectra are not continuous. They consist of bright lines at specific wavelengths, with complete darkness in between.

3.14 In the Bohr model, electrons travel around the nucleus in circular orbits. Bohr's orbits could exist only at specific, fixed distances from the nucleus. The energy of each orbit was also fixed, or quantized. Bohr called these orbits stationary states and suggested that although they obeyed the laws of classical mechanics, they also possessed "a peculiar, mechanically unexplainable, stability." Bohr further proposed that in contradiction to classical electromagnetic theory, no radiation was emitted by an electron orbiting the nucleus in a stationary state. It was only when an electron jumped, or made a transition, from one stationary state to another that radiation was emitted or absorbed. The emission spectrum of an atom consisted of discrete lines because the stationary states existed only at specific, fixed energies. The energy of the photon created when an electron made a transition from one stationary state to another was simply the energy difference between the two stationary states.

3.15 Electron diffraction occurs when an electron beam is aimed at two closely spaced slits and a series of detectors is arranged to detect the electrons after they pass through the slits. An interference pattern similar to that observed for light is recorded behind the slits. Electron diffraction is evidence of the wave nature of electrons.

3.16 The de Broglie wavelength is associated with an electron traveling through space. It is related to its kinetic energy. The wavelength, λ, associated with an electron of mass, m, moving at velocity, v, is given by the de Broglie relation: $\lambda = \frac{h}{mv}$.

3.17 Complementary properties exclude one another. The more you know about one, the less you know about the other. Which of two complementary properties you observe depends on the experiment you perform. In electron diffraction, when you try to observe which hole the electron goes through (particle nature), you lose the interference pattern (wave nature). When you try to observe the interference pattern, you cannot determine which hole the electron goes through.

3.18 Heisenberg's uncertainty principle states that the product of Δx and $m\Delta v$ must be greater than or equal to a finite number. In other words, the more accurately you know the position of an electron (the smaller Δx), the less accurately you can know its velocity (the bigger Δv) and vice versa. The complementarity of the wave nature and particle nature of the electron results in the complementarity of velocity and position. Heisenberg solved the contradiction of an object as both a particle and a wave by introducing complementarity—an electron is observed as either a particle or a wave, but never both at once.

3.19 A trajectory is a path that is determined by the particle's velocity (the speed and direction of travel), its position, and the forces acting on it. Both position and velocity are required to predict a trajectory.

3.20 Because the uncertainty principle says that you cannot know both the position and velocity of the electron simultaneously, you cannot predict the trajectory.

3.21 Deterministic means that the present determines the future. That means that under the identical condition, identical results will occur.

3.22 The indeterminate behavior of an electron means that under identical conditions, the electron does not have the same trajectory and does not "land" in the same spot each time.

3.23 A probability distribution map is a statistical map that shows where an electron is likely to be found under a given set of conditions.

3.24 Using the Schrödinger equation we describe the probability distribution maps for electron states. In these states the electron has a well-defined energy, but not a well-defined position. In other words, for each state, we can specify the energy of the electron precisely, but not its location at a given instant. The electron's position is described in terms of an orbital.

3.25 An orbital is a probability distribution map showing where the electron is likely to be found.

3.26 The mathematical derivation of energies and orbitals for electrons in atoms comes from solving the Schrödinger equation. The general form of the Schrödinger equation is $H\psi = E\psi$. The symbol H stands for the Hamiltonian operator, a set of mathematical operations that represent the total energy (kinetic and potential) of the electron within the atom. The symbol E is the actual energy of the electron. The symbol ψ is the wave function, a mathematical function that describes the wave-like nature of the electron. A plot of the wave function squared (ψ^2) represents an orbital, a position probability distribution map of the electron.

3.27 The principal quantum number (n) is an integer that has possible values of 1, 2, 3, etc. The principal quantum number determines the overall size and energy of an orbital.

3.28 The angular momentum quantum number (l) is an integer that has possible values of 0, 1, 2, 3, etc. The angular momentum quantum number determines the shape of the orbital. It can take values from 0 to ($n - 1$) where n is the principal quantum number.

3.29 The magnetic quantum number (m_l) is an integer ranging from $-l$ to $+l$. For example, if $l = 1, m_l = -1, 0, +1$. The magnetic quantum number specifies the orientation of the orbital.

3.30

	n	Possible l values	Possible m_l values
(a)	1	0	0
(b)	2	0	0
		1	−1, 0, +1
(c)	3	0	0
		1	−1, 0, +1
		2	−2, −1, 0, +1, +2
(d)	4	0	0
		1	−1, 0, +1
		2	−2, −1, 0, +1, +2
		3	−3, −2, −1, 0, +1, +2, +3

3.31 The probability density is the probability per unit volume of finding the electron at a point in space. The radial distribution function represents the total probability of finding the electron within a thin spherical shell at a distance r from the nucleus. In contrast to probability density, which has a maximum at the nucleus for an s orbital, the radial distribution function has a value of zero at the nucleus. It increases to a maximum and then decreases again with increasing r.

3.32

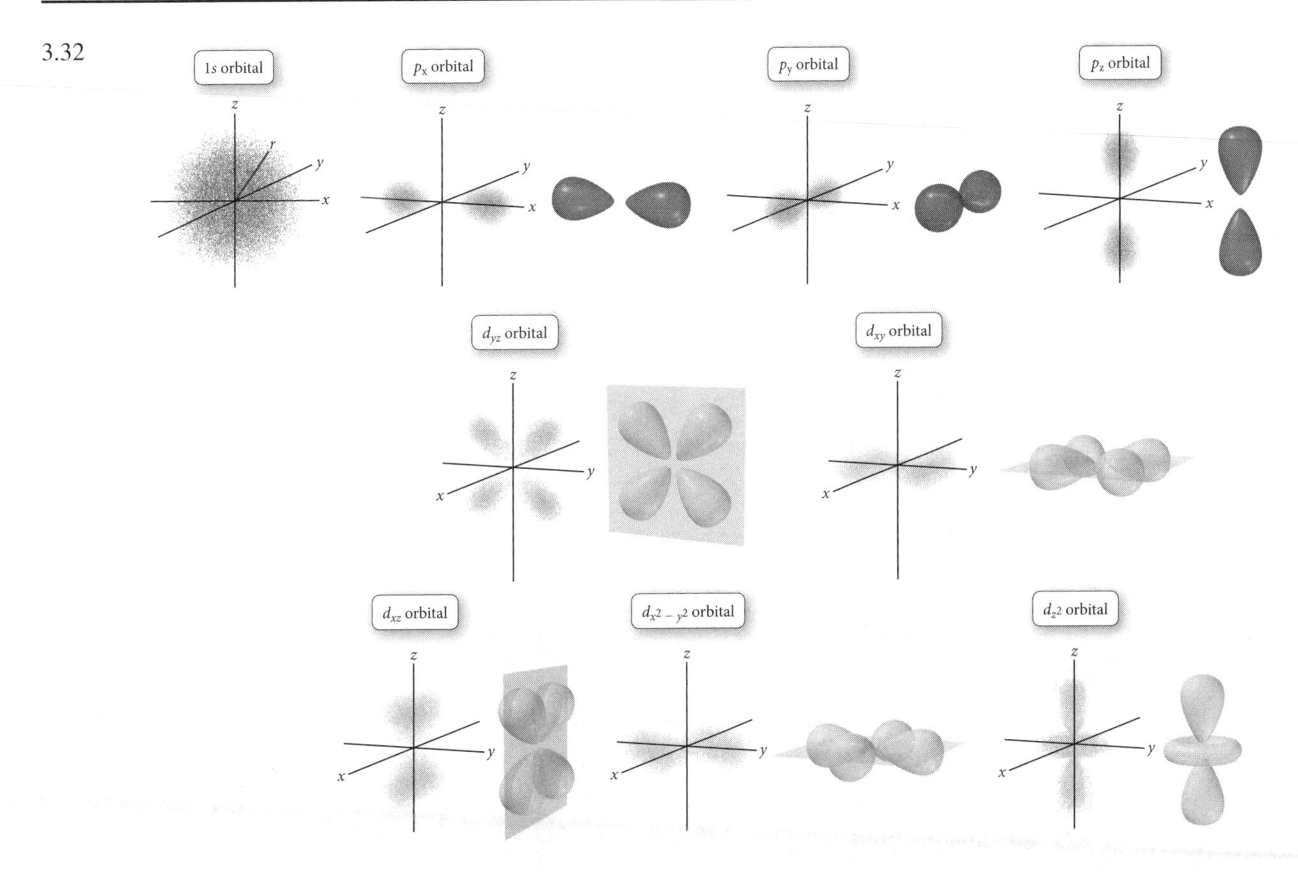

3.33 The sublevels are s ($l = 0$), which can hold a maximum of 2 electrons; p ($l = 1$), which can hold a maximum of 6 electrons; d ($l = 2$), which can hold a maximum of 10 electrons; and f ($l = 3$), which can hold a maximum of 14 electrons.

3.34 Atoms are usually drawn as spheres because most atoms contain many electrons occupying a number of different orbitals. Therefore, the shape of an atom is obtained by superimposing all of its orbitals. If the s, p, and d orbitals are superimposed, they have a spherical shape.

Problems by Topic

Electromagnetic Radiation

3.35 **Given:** distance to sun $= 1.496 \times 10^8$ km **Find:** time for light to travel from sun to Earth

Conceptual Plan: distance km → distance m → time

$$\frac{1000\text{ m}}{\text{km}} \qquad \text{time} = \frac{\text{distance}}{2.998 \times 10^8\text{ m/s}}$$

Solution: $1.496 \times 10^8\ \cancel{\text{km}} \times \frac{1000\ \cancel{\text{m}}}{\cancel{\text{km}}} \times \frac{\text{s}}{2.998 \times 10^8\ \cancel{\text{m}}} = 499.0\text{ s}$

Check: The units of the answer (s) are correct. The magnitude of the answer is reasonable because it corresponds to about 8 minutes.

3.36 **Given:** 4.3 light-years to star **Find:** distance in km

Conceptual Plan: light-years → days → hours → seconds → m → km

$$\frac{365\text{ days}}{\text{yr}} \quad \frac{24\text{ hr}}{\text{day}} \quad \frac{3600\text{ s}}{\text{hr}} \quad \frac{3.00 \times 10^8\text{ m}}{\text{s}} \quad \frac{\text{km}}{1000\text{ m}}$$

Solution: $4.3\text{ light }\cancel{\text{yr}} \times \frac{365\ \cancel{\text{days}}}{\cancel{\text{yr}}} \times \frac{24\ \cancel{\text{hr}}}{\cancel{\text{day}}} \times \frac{3600\ \cancel{\text{s}}}{\cancel{\text{hr}}} \times \frac{3.00 \times 10^8\ \cancel{\text{m}}}{\cancel{\text{s}}} \times \frac{\text{km}}{1000\ \cancel{\text{m}}} = 4.1 \times 10^{13}\text{ km}$

Check: The units of the answer (km) are correct. The magnitude of the answer is reasonable because it takes much longer for the light to reach Earth from Proxima Centauri than from the sun; so the distance should be much greater.

3.37 (i) By increasing wavelength, the order is (d) ultraviolet < (c) infrared < (b) microwave < (a) radio waves.
(ii) By increasing energy, the order is (a) radio waves < (b) microwaves < (c) infrared < (d) ultraviolet.

3.38 (i) By increasing frequency, the order is (b) radio waves < (c) microwaves < d) visible light < (a) gamma rays.
(ii) By decreasing energy, the order is (a) gamma rays > (d) visible light > (c) microwaves > (b) radio waves.

3.39 (a) **Given:** $\lambda = 632.8$ nm **Find:** frequency (ν)
Conceptual Plan: **nm → m → ν**

$\frac{1\text{ m}}{10^9\text{ nm}}$ $\quad \nu = \frac{c}{\lambda}$

Solution:

$$632.8\ \cancel{\text{nm}} \times \frac{1\text{ m}}{10^9\ \cancel{\text{nm}}} = 6.328 \times 10^{-7}\text{ m};\ \nu = \frac{2.998 \times 10^8\ \cancel{\text{m}}}{\text{s}} \times \frac{1}{6.328 \times 10^{-7}\ \cancel{\text{m}}} = 4.738 \times 10^{14}\text{ s}^{-1}$$

Check: The units of the answer (s^{-1}) are correct. The magnitude of the answer seems reasonable because wavelength and frequency are inversely proportional.

(b) **Given:** $\lambda = 503$ nm **Find:** frequency (ν)
Conceptual Plan: **nm → m → ν**

$\frac{1\text{ m}}{10^9\text{ nm}}$ $\quad \nu = \frac{c}{\lambda}$

Solution: $503\ \cancel{\text{nm}} \times \frac{1\text{ m}}{10^9\ \cancel{\text{nm}}} = 5.03 \times 10^{-7}\text{ m};\ \nu = \frac{3.00 \times 10^8\ \cancel{\text{m}}}{\text{s}} \times \frac{1}{5.03 \times 10^{-7}\ \cancel{\text{m}}} = 5.96 \times 10^{14}\text{ s}^{-1}$

Check: The units of the answer (s^{-1}) are correct. The magnitude of the answer seems reasonable because wavelength and frequency are inversely proportional.

(c) **Given:** $\lambda = 0.052$ nm **Find:** frequency (ν)
Conceptual Plan: **nm → m → ν**

$\frac{1\text{ m}}{10^9\text{ nm}}$ $\quad \nu = \frac{c}{\lambda}$

Solution: $0.052\ \cancel{\text{nm}} \times \frac{1\text{ m}}{10^9\ \cancel{\text{nm}}} = 5.2 \times 10^{-11}\text{ m};\ \nu = \frac{3.00 \times 10^8\ \cancel{\text{m}}}{\text{s}} \times \frac{1}{5.2 \times 10^{-11}\ \cancel{\text{m}}} = 5.8 \times 10^{18}\text{ s}^{-1}$

Check: The units of the answer (s^{-1}) are correct. The magnitude of the answer seems reasonable because wavelength and frequency are inversely proportional.

3.40 (a) **Given:** $\nu = 100.2$ MHz **Find:** wavelength (λ)
Conceptual Plan: **MHz → Hz → s^{-1} → λ**

$\frac{10^6\text{ Hz}}{1\text{ MHz}}$ $\quad 1\text{ Hz} = 1\text{ s}^{-1}$ $\quad \lambda = \frac{c}{\nu}$

Solution:

$$100.2\ \cancel{\text{MHz}} \times \frac{10^6\ \cancel{\text{Hz}}}{1\ \cancel{\text{MHz}}} \times \frac{1\text{ s}^{-1}}{1\ \cancel{\text{Hz}}} = 1.002 \times 10^8\text{ s}^{-1}; \qquad \lambda = \frac{2.998 \times 10^8\text{ m}}{\cancel{\text{s}}} \times \frac{\cancel{\text{s}}}{1.002 \times 10^8} = 2.992\text{ m}$$

Check: The units of the answer (m) are correct. The magnitude of the answer is reasonable because FM wavelengths are generally in the 3 to 8 m range.

(b) **Given:** $\nu = 1070$ kHz **Find:** wavelength (λ)
Conceptual Plan: **kHz → Hz → s^{-1} → λ**

$\frac{10^3\text{ Hz}}{1\text{ kHz}}$ $\quad 1\text{ Hz} = 1\text{ s}^{-1}$ $\quad \lambda = \frac{c}{\nu}$

Solution:

$$1070\ \cancel{\text{kHz}} \times \frac{10^3\ \cancel{\text{Hz}}}{1\ \cancel{\text{kHz}}} \times \frac{1\text{ s}^{-1}}{1\ \cancel{\text{Hz}}} = 1.070 \times 10^6\text{ s}^{-1}; \qquad \lambda = \frac{2.998 \times 10^8\text{ m}}{\cancel{\text{s}}} \times \frac{\cancel{\text{s}}}{1.070 \times 10^6} = 280.2\text{ m}$$

Check: The units of the answer (m) are correct. The magnitude of the answer is reasonable because AM wavelengths are generally in the 100 to 1000 m range.

(c) **Given:** $\nu = 835.6$ MHz **Find:** wavelength (λ)
Conceptual Plan: **MHz → Hz → s^{-1} → λ**

$\frac{10^6\text{ Hz}}{1\text{ MHz}}$ $\quad 1\text{ Hz} = 1\text{ s}^{-1}$ $\quad \lambda = \frac{c}{\nu}$

Solution:

$$835.6\ \cancel{\text{MHz}} \times \frac{10^6\ \cancel{\text{Hz}}}{1\ \cancel{\text{MHz}}} \times \frac{1\ \text{s}^{-1}}{1\ \cancel{\text{Hz}}} = 8.356 \times 10^8\ \text{s}^{-1};\ \lambda = \frac{2.998 \times 10^8\ \text{m}}{\cancel{\text{s}}} \times \frac{\cancel{\text{s}}}{8.356 \times 10^8} = 3.588 \times 10^{-1}\ \text{m}$$

Check: The units of the answer (m) are correct. The magnitude of the answer is reasonable because cell phone wavelengths are generally in the 0.1 to 0.5 m range.

3.41 (a) **Given:** frequency (ν) from Problem 3.39a $= 4.74 \times 10^{14}\ \text{s}^{-1}$ **Find:** Energy

Conceptual Plan: $\nu \rightarrow E$

$E = h\nu \quad h = 6.626 \times 10^{-34}\ \text{J} \cdot \text{s}$

Solution: $6.626 \times 10^{-34}\ \text{J} \cdot \cancel{\text{s}} \times \dfrac{4.74 \times 10^{14}}{\cancel{\text{s}}} = 3.14 \times 10^{-19}\ \text{J}$

Check: The units of the answer (J) are correct. The magnitude of the answer is reasonable because we are talking about the energy of one photon.

(b) **Given:** frequency (ν) from Problem 3.39b $= 5.96 \times 10^{14}\ \text{s}^{-1}$ **Find:** Energy

Conceptual Plan: $\nu \rightarrow E$

$E = h\nu \quad h = 6.626 \times 10^{-34}\ \text{J} \cdot \text{s}$

Solution: $6.626 \times 10^{-34}\ \text{J} \cdot \cancel{\text{s}} \times \dfrac{5.96 \times 10^{14}}{\cancel{\text{s}}} = 3.95 \times 10^{-19}\ \text{J}$

Check: The units of the answer (J) are correct. The magnitude of the answer is reasonable because we are talking about the energy of one photon.

(c) **Given:** frequency (ν) from Problem 3.39c $= 5.8 \times 10^{18}\ \text{s}^{-1}$ **Find:** Energy

Conceptual Plan: $\nu \rightarrow E$

$E = h\nu \quad h = 6.626 \times 10^{-34}\ \text{J} \cdot \text{s}$

Solution: $6.626 \times 10^{-34}\ \text{J} \cdot \cancel{\text{s}} \times \dfrac{5.8 \times 10^{18}}{\cancel{\text{s}}} = 3.8 \times 10^{-15}\ \text{J}$

Check: The units of the answer (J) are correct. The magnitude of the answer is reasonable because we are talking about the energy of one photon.

3.42 (a) **Given:** frequency (ν) from Problem 3.40a $= 100.2\ \text{MHz}$ **Find:** Energy

Conceptual Plan: $\text{MHz} \rightarrow \text{Hz} \rightarrow \text{s}^{-1} \rightarrow E$

$\dfrac{10^6\ \text{Hz}}{1\ \text{MHz}} \quad 1\ \text{Hz} = 1\ \text{s}^{-1} \quad E = h\nu, h = 6.626 \times 10^{-34}\ \text{J} \cdot \text{s}$

Solution:

$$100.2\ \cancel{\text{MHz}} \times \frac{10^6\ \cancel{\text{Hz}}}{1\ \cancel{\text{MHz}}} \times \frac{1\ \text{s}^{-1}}{1\ \cancel{\text{Hz}}} = 1.002 \times 10^8\ \text{s}^{-1};\quad 6.626 \times 10^{-34}\ \text{J} \cdot \cancel{\text{s}} \times \frac{1.002 \times 10^8}{\cancel{\text{s}}} = 6.639 \times 10^{-26}\ \text{J}$$

Check: The units of the answer (J) are correct. The magnitude of the answer is reasonable because we are talking about the energy of one photon and have a relatively long wavelength.

(b) **Given:** frequency (ν) from 3.40b $= 1070\ \text{kHz}$ **Find:** Energy

Conceptual Plan: $\text{kHz} \rightarrow \text{Hz} \rightarrow \text{s}^{-1} \rightarrow E$

$\dfrac{10^3\ \text{Hz}}{1\ \text{kHz}} \quad 1\ \text{Hz} = 1\ \text{s}^{-1} \quad E = h\nu, h = 6.626 \times 10^{-34}\ \text{J} \cdot \text{s}$

Solution:

$$1070\ \cancel{\text{kHz}} \times \frac{10^3\ \cancel{\text{Hz}}}{1\ \cancel{\text{kHz}}} \times \frac{1\ \text{s}^{-1}}{1\ \cancel{\text{Hz}}} = 1.070 \times 10^6\ \text{s}^{-1};\quad 6.626 \times 10^{-34}\ \text{J} \cdot \cancel{\text{s}} \times \frac{1.070 \times 10^6}{\cancel{\text{s}}} = 7.090 \times 10^{-28}\ \text{J}$$

Check: The units of the answer (J) are correct. The magnitude of the answer is reasonable because we are talking about the energy of one photon and have a relatively long wavelength.

(c) **Given:** frequency (ν) from 3.40c $= 835.6\ \text{MHz}$ **Find:** Energy

Conceptual Plan: $\text{MHz} \rightarrow \text{Hz} \rightarrow \text{s}^{-1} \rightarrow E$

$\dfrac{10^6\ \text{Hz}}{1\ \text{MHz}} \quad 1\ \text{Hz} = 1\ \text{s}^{-1} \quad E = h\nu, h = 6.626 \times 10^{-34}\ \text{J} \cdot \text{s}$

Solution:

$$835.6\ \text{MHz} \times \frac{10^6\ \text{Hz}}{1\ \text{MHz}} \times \frac{1\ \text{s}^{-1}}{1\ \text{Hz}} = 8.356 \times 10^8\ \text{s}^{-1} \quad 6.626 \times 10^{-34}\ \text{J} \cdot \text{s} \times \frac{8.356 \times 10^8}{\text{s}} = 5.537 \times 10^{-25}\ \text{J}$$

Check: The units of the answer (J) are correct. The magnitude of the answer is reasonable because we are talking about the energy of one photon and have a relatively long wavelength.

3.43 **Given:** $\lambda = 532$ nm and $E_{\text{pulse}} = 3.85$ mJ **Find:** number of photons

Conceptual Plan: nm → m → E_{photon} → number of photons

$$\frac{1\ \text{m}}{10^9\ \text{nm}} \quad E = \frac{hc}{\lambda};\ h = 6.626 \times 10^{-34}\ \text{J} \cdot \text{s} \quad \frac{E_{\text{pulse}}}{E_{\text{photon}}}$$

Solution:

$$532\ \text{nm} \times \frac{1\ \text{m}}{10^9\ \text{nm}} = 5.32 \times 10^{-7}\ \text{m}; \quad E = \frac{6.626 \times 10^{-34}\ \text{J} \cdot \text{s} \times \frac{3.00 \times 10^8\ \text{m}}{\text{s}}}{5.32 \times 10^{-7}\ \text{m}} = 3.7\underline{3}65 \times 10^{-19}\ \text{J/photon}$$

$$3.85\ \text{mJ} \times \frac{1\ \text{J}}{1000\ \text{mJ}} \times \frac{1\ \text{photon}}{3.7\underline{3}65 \times 10^{-19}\ \text{J}} = 1.03 \times 10^{16}\ \text{photons}$$

Check: The units of the answer (number of photons) are correct. The magnitude of the answer is reasonable for the amount of energy involved.

3.44 **Given:** $\lambda = 6.5\ \mu\text{m}$; power = 32.8 watts **Find:** photons/second

Conceptual Plan: μm → m → E_{photon} → and then watts → J/s → number of photons

$$\frac{1\ \text{m}}{10^6\ \mu\text{m}} \quad E = \frac{hc}{\lambda};\ h = 6.626 \times 10^{-34}\ \text{J} \cdot \text{s} \quad \frac{\text{J/s}}{1\ \text{watt}} \quad \frac{1\ \text{photon}}{3.0\underline{5}8 \times 10^{-20}\ \text{J}}$$

Solution:

$$6.5\ \mu\text{m} \times \frac{1\ \text{m}}{10^6\ \mu\text{m}} = 6.5 \times 10^{-6}\ \text{m}; \quad E = \frac{6.626 \times 10^{-34}\ \text{J} \cdot \text{s} \times \frac{3.00 \times 10^8\ \text{m}}{\text{s}}}{6.5 \times 10^{-6}\ \text{m}} = 3.0\underline{5}8 \times 10^{-20}\ \text{J/photon}$$

$$32.8\ \text{watts} \times \frac{1\ \text{J/s}}{1\ \text{watt}} \times \frac{1\ \text{photon}}{3.0\underline{5}8 \times 10^{-20}\ \text{J}} = 1.1 \times 10^{21}\ \text{photons/s}$$

Check: The units of the answer (photons/s) are correct. The magnitude of the answer is reasonable for the amount of energy involved.

3.45 (a) **Given:** $\lambda = 1500$ nm (assume 3 significant figures) **Find:** E for 1 mol photons

Conceptual Plan: nm → m → E_{photon} → $E(\text{J})_{\text{mol}}$ → $E(\text{kJ})_{\text{mol}}$

$$\frac{1\ \text{m}}{10^9\ \text{nm}} \quad E = \frac{hc}{\lambda};\ h = 6.626 \times 10^{-34}\ \text{J} \cdot \text{s} \quad \frac{6.022 \times 10^{23}\ \text{photons}}{\text{mol}} \quad \frac{1\ \text{kJ}}{1000\ \text{J}}$$

Solution: $1500\ \text{nm} \times \frac{1\ \text{m}}{10^9\ \text{nm}} = 1.50 \times 10^{-6}\ \text{m};$

$$E = \frac{6.626 \times 10^{-34}\ \text{J} \cdot \text{s} \times \frac{3.00 \times 10^8\ \text{m}}{\text{s}}}{1.500 \times 10^{-6}\ \text{m}} = 1.3\underline{2}52 \times 10^{-19}\ \text{J/photon}$$

$$\frac{1.3\underline{2}52 \times 10^{-19}\ \text{J}}{\text{photon}} \times \frac{6.022 \times 10^{23}\ \text{photons}}{\text{mol}} \times \frac{1\ \text{kJ}}{1000\ \text{J}} = 79.8\ \text{kJ/mol}$$

Check: The units of the answer (kJ/mol) are correct. The magnitude of the answer is reasonable for a wavelength in the infrared region.

(b) **Given:** $\lambda = 500.$ nm **Find:** E for 1 mol photons

Conceptual Plan: nm → m → E_{photon} → $E(\text{J})_{\text{mol}}$ → $E(\text{kJ})_{\text{mol}}$

$$\frac{1\ \text{m}}{10^9\ \text{nm}} \quad E = \frac{hc}{\lambda};\ h = 6.626 \times 10^{-34}\ \text{J} \cdot \text{s} \quad \frac{6.022 \times 10^{23}\ \text{photons}}{\text{mol}} \quad \frac{1\ \text{kJ}}{1000\ \text{J}}$$

Solution:

$$500.\ \cancel{\text{nm}} \times \frac{1\ \text{m}}{10^9\ \cancel{\text{nm}}} = 5.00 \times 10^{27}\ \text{m};\ E = \frac{6.626 \times 10^{-34}\ \text{J} \cdot \cancel{\text{s}} \times \dfrac{3.00 \times 10^8\ \cancel{\text{m}}}{\cancel{\text{s}}}}{5.00 \times 10^{27}\ \cancel{\text{m}}} = 3.9\underline{7}56 \times 10^{-19}\ \text{J/photon}$$

$$\frac{3.9\underline{7}56 \times 10^{-19}\ \cancel{\text{J}}}{\cancel{\text{photon}}} \times \frac{6.022 \times 10^{23}\ \cancel{\text{photons}}}{\text{mol}} \times \frac{1\ \text{kJ}}{1000\ \cancel{\text{J}}} = 239\ \text{kJ/mol}$$

Check: The units of the answer (kJ/mol) are correct. The magnitude of the answer is reasonable for a wavelength in the visible region.

(c) **Given:** $\lambda = 150.\ \text{nm}$ **Find:** E for 1 mol photons

Conceptual Plan: **nm** → **m** → $\boldsymbol{E_{\text{photon}}}$ → $\boldsymbol{E(\text{J})_{\text{mol}}}$ → $\boldsymbol{E(\text{kJ})_{\text{mol}}}$

$$\frac{1\ \text{m}}{10^9\ \text{nm}} \qquad E = \frac{hc}{\lambda};\ h = 6.626 \times 10^{-34}\ \text{J} \cdot \text{s} \qquad \frac{6.022 \times 10^{23}\ \text{photons}}{\text{mol}} \qquad \frac{1\ \text{kJ}}{1000\ \text{J}}$$

Solution:

$$150.\ \cancel{\text{nm}} \times \frac{1\ \text{m}}{10^9\ \cancel{\text{nm}}} = 1.50 \times 10^{-7}\ \text{m};\quad E = \frac{6.626 \times 10^{-34}\ \text{J} \cdot \cancel{\text{s}} \times \dfrac{3.00 \times 10^8\ \cancel{\text{m}}}{\cancel{\text{s}}}}{1.50 \times 10^{-7}\ \cancel{\text{m}}} = 1.3\underline{2}52 \times 10^{-18}\ \text{J/photon}$$

$$\frac{1.3\underline{2}52 \times 10^{-18}\ \cancel{\text{J}}}{\cancel{\text{photon}}} \times \frac{6.022 \times 10^{23}\ \cancel{\text{photons}}}{\text{mol}} \times \frac{1\ \text{kJ}}{1000\ \cancel{\text{J}}} = 798\ 1\ \text{kJ/mol}$$

Check: The units of the answer (kJ/mol) are correct. The magnitude of the answer is reasonable for a wavelength in the ultraviolet region. Note: The energy increases from the IR to the Vis to the UV as expected.

3.46 (a) **Given:** $\lambda = 0.135\ \text{nm}$ **Find:** E for 1 mol photons

Conceptual Plan: **nm** → **m** → $\boldsymbol{E_{\text{photon}}}$ → $\boldsymbol{E_{\text{mol}}}$ → $\boldsymbol{E(\text{kJ})_{\text{mol}}}$

$$\frac{1\ \text{m}}{10^9\ \text{nm}} \qquad E = \frac{hc}{\lambda};\ h = 6.626 \times 10^{-34}\ \text{J} \cdot \text{s} \qquad \frac{6.022 \times 10^{23}\ \text{photons}}{\text{mol}} \qquad \frac{1\ \text{kJ}}{1000\ \text{J}}$$

Solution:

$$0.135\ \cancel{\text{nm}} \times \frac{1\ \text{m}}{10^9\ \cancel{\text{nm}}} = 1.35 \times 10^{-10}\ \text{m};\quad E = \frac{6.626 \times 10^{-34}\ \text{J} \cdot \cancel{\text{s}} \times \dfrac{3.00 \times 10^8\ \cancel{\text{m}}}{\cancel{\text{s}}}}{1.35 \times 10^{-10}\ \cancel{\text{m}}} = 1.4\underline{7}2 \times 10^{-15}\ \text{J/photon}$$

$$\frac{1.4\underline{7}2 \times 10^{-15}\ \cancel{\text{J}}}{\cancel{\text{photon}}} \times \frac{6.022 \times 10^{23}\ \cancel{\text{photons}}}{\text{mol}} \times \frac{1\ \text{kJ}}{1000\ \cancel{\text{J}}} = 8.86 \times 10^5\ \text{kJ/mol}$$

Check: The units of the answer (kJ/mol) are correct. The magnitude of the answer is reasonable for a wavelength in the X-ray region.

(b) **Given:** $\lambda = 2.15 \times 10^{-5}\ \text{nm}$ **Find:** E for 1 mol photons

Conceptual Plan: **nm** → **m** → $\boldsymbol{E_{\text{photon}}}$ → $\boldsymbol{E_{\text{mol}}}$ → $\boldsymbol{E(\text{kJ})_{\text{mol}}}$

$$\frac{1\ \text{m}}{10^9\ \text{nm}} \qquad E = \frac{hc}{\lambda};\ h = 6.626 \times 10^{-34}\ \text{J} \cdot \text{s} \qquad \frac{6.022 \times 10^{23}\ \text{photons}}{\text{mol}} \qquad \frac{1\ \text{kJ}}{1000\ \text{J}}$$

Solution:

$$2.15 \times 10^{-5}\ \cancel{\text{nm}} \times \frac{1\ \text{m}}{10^9\ \cancel{\text{nm}}} = 2.15 \times 10^{-14}\ \text{m};\ E = \frac{6.626 \times 10^{-34}\ \text{J} \cdot \cancel{\text{s}} \times \dfrac{3.00 \times 10^8\ \cancel{\text{m}}}{\cancel{\text{s}}}}{2.15 \times 10^{-14}\ \cancel{\text{m}}} = 9.2\underline{4}6 \times 10^{-12}\ \text{J/photon}$$

$$\frac{9.2\underline{4}6 \times 10^{-12}\ \cancel{\text{J}}}{\cancel{\text{photon}}} \times \frac{6.022 \times 10^{23}\ \cancel{\text{photons}}}{\text{mol}} \times \frac{1\ \text{kJ}}{1000\ \cancel{\text{J}}} = 5.57 \times 10^9\ \text{kJ/mol}$$

Check: The units of the answer (kJ/mol) are correct. The magnitude of the answer is reasonable for a wavelength in the gamma ray region.

The Wave Nature of Matter and the Uncertainty Principle

3.47 The interference pattern would be a series of light and dark lines.

3.48 Because the interference pattern is caused by single electrons interfering with themselves, the pattern remains the same even when the rate of the electrons passing through the slits is one electron per hour. It will simply take longer for the full pattern to develop. When a laser is placed behind the slits to determine which hole the electron passes through, the laser flashes when a photon is scattered at the point of crossing, indicating the slit used, but the interference pattern is now absent. With the laser on, the electrons hit positions directly behind each slit, as if they were ordinary particles.

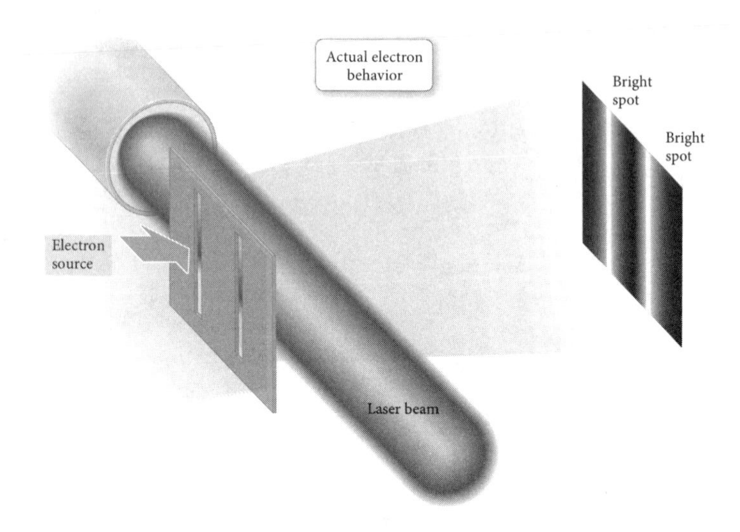

3.49 **Given:** $m = 9.109 \times 10^{-31}$ kg, $\lambda = 0.20$ nm **Find:** v

Conceptual Plan: $m, \lambda \rightarrow v$

$$v = \frac{h}{m\lambda}$$

Solution: $$\frac{6.626 \times 10^{-34} \frac{\text{kg} \cdot \text{m}^2}{\text{s}^2} \cdot \text{s}}{(9.109 \times 10^{-31} \text{ kg})(0.20 \text{ nm})\left(\frac{1 \text{ m}}{10^9 \text{ nm}}\right)} = 3.6 \times 10^6 \text{ m/s}$$

Check: The units of the answer (m/s) are correct. The magnitude of the answer is large, as would be expected for the speed of the electron.

3.50 **Given:** one proton; $v = 475$ m/s **Find:** λ

Conceptual Plan: Mass of 1 mol H atoms → mass of one H atom (g) → mass of one H atom (kg) $m, v \rightarrow \lambda$

$$\frac{1 \text{ mol H atoms}}{6.02214 \times 10^{23} \text{ H atoms}} \qquad \frac{1 \text{ kg}}{1000 \text{ g}} \qquad \lambda = \frac{h}{mv}$$

Solution: $$\frac{1.008 \text{ g H}}{1 \text{ mol H atoms}} \times \frac{1 \text{ mol H atoms}}{6.02214 \times 10^{23} \text{ H atoms}} \times \frac{1 \text{ kg H}}{1000 \text{ g H}} = 1.673 \times 10^{-27} \text{ kg H}$$

$$\frac{6.626 \times 10^{-34} \frac{\text{kg} \cdot \text{m}^2}{\text{s}^2} \cdot \text{s}\left(\frac{10^{12} \text{ pm}}{\text{m}}\right)}{(1.673 \times 10^{-27} \text{ kg})\left(\frac{475 \text{ m}}{\text{s}}\right)} = 833 \text{ pm}$$

Check: The units of the answer (pm) are correct. The magnitude of the wavelength is small, as expected.

3.51 **Given:** $m = 9.109 \times 10^{-31}$ kg; $v = 1.35 \times 10^5$ m/s **Find:** λ

Conceptual Plan: $\boldsymbol{m, v \rightarrow \lambda}$

$$\lambda = \frac{h}{mv}$$

Solution: $$\frac{6.626 \times 10^{-34} \frac{\text{kg} \cdot \text{m}^2}{\text{s}^2} \cdot \text{s}}{(9.109 \times 10^{-31}\ \text{kg})\left(\frac{1.35 \times 10^5\ \text{m}}{\text{s}}\right)} = 5.39 \times 10^{-9}\ \text{m} = 5.39\ \text{nm}$$

Check: The units of the answer (m) are correct. The magnitude is reasonable because we are looking at an electron.

3.52 **Given:** $m = 1.673 \times 10^{-27}$ kg; $\lambda = 122$ pm **Find:** v

Conceptual Plan: $\boldsymbol{m, v \rightarrow \lambda}$

$$v = \frac{h}{m\lambda}$$

Solution: $$\frac{6.626 \times 10^{-34} \frac{\text{kg} \cdot \text{m}^2}{\text{s}^2} \cdot \text{s}}{(1.673 \times 10^{-27}\ \text{kg})(122\ \text{pm})\left(\frac{1\ \text{m}}{10^{12}\ \text{pm}}\right)} = 3.25 \times 10^3\ \text{m/s}$$

Check: The units of the answer (m/s) are correct. The magnitude is reasonable because we are looking at an electron.

3.53 **Given:** $m = 143$ g; $v = 95$ mph **Find:** λ

Conceptual Plan: $\boldsymbol{m, v \rightarrow \lambda}$

$$\lambda = \frac{h}{mv}$$

Solution: $$\frac{6.626 \times 10^{-34} \frac{1\ \text{kg} \cdot \text{m}^2}{\text{s}^2} \cdot \text{s}}{(143\ \text{g})\left(\frac{1\ \text{kg}}{1000\ \text{g}}\right)\left(\frac{95\ \text{mi}}{\text{hr}}\right)\left(\frac{1.609\ \text{km}}{\text{mi}}\right)\left(\frac{1000\ \text{m}}{1\ \text{km}}\right)\left(\frac{1\ \text{hr}}{3600\ \text{s}}\right)} = 1.1 \times 10^{-34}\ \text{m}$$

The value of the wavelength (1.1×10^{-34} m) is so small that it will not have an effect on the trajectory of the baseball.

Check: The units of the answer (m) are correct. The magnitude of the answer is very small, as would be expected for the de Broglie wavelength of a baseball.

3.54 **Given:** $m = 27$ g; $v = 765$ m/s **Find:** λ

Conceptual Plan: $\boldsymbol{m, v \rightarrow \lambda}$

$$\lambda = \frac{h}{mv}$$

Solution: $$\frac{6.626 \times 10^{-34} \frac{\text{kg} \cdot \text{m}^2}{\text{s}^2} \cdot \text{s}}{(27\ \text{g})\left(\frac{1\ \text{kg}}{1000\ \text{g}}\right)\left(\frac{765\ \text{m}}{\text{s}}\right)} = 3.2 \times 10^{-35}\ \text{m}$$

The value of the wavelength (3.2×10^{-35} m) is so small that it will not have an effect on the trajectory of the bullet. The wave nature of matter is irrelevant to bullets.

Check: The units of the answer (m) are correct. The magnitude of the answer is very small, as would be expected for the de Broglie wavelength of a bullet.

3.55 **Given:** $\Delta x = 552$ pm, $m = 9.109 \times 10^{-31}$ kg **Find:** Δv

Conceptual Plan: $\boldsymbol{\Delta x, m \rightarrow \Delta v}$

$$\Delta x \times m\Delta v \geq \frac{h}{4\pi}$$

Solution: $\dfrac{6.626 \times 10^{-34} \dfrac{\text{kg} \cdot \text{m}^2}{\text{s}^2} \cdot \text{s}}{4(3.141)(9.109 \times 10^{-31}\ \text{kg})(552\ \text{pm})\left(\dfrac{1\ \text{m}}{10^{12}\ \text{pm}}\right)} = 1.05 \times 10^5\ \text{m/s}$

Check: The units of the answer (m/s) are correct. The magnitude is reasonable for the uncertainty in the speed of an electron.

3.56 **Given:** $m = 9.109 \times 10^{-31}$ kg, $v = 3.7 \times 10^5$m/s, $\Delta v = 1.88 \times 10^5$m/s **Find:** Δx

Conceptual Plan: $\Delta v, m \rightarrow \Delta x$

$$\Delta x \times m\Delta v \geq \frac{h}{4\pi}$$

Solution: $\dfrac{6.626 \times 10^{-34} \dfrac{\text{kg} \cdot \text{m}^2}{\text{s}^2} \cdot \text{s}\left(\dfrac{10^{12}\ \text{pm}}{1\ \text{m}}\right)}{4(3.141)(9.109 \times 10^{-31}\ \text{kg})\left(\dfrac{1.88 \times 10^5\ \text{m}}{\text{s}}\right)} = 308\ \text{pm}$

Check: The units of the answer (pm) are correct. The magnitude is reasonable when compared to the speed of the electron.

Orbitals and Quantum Numbers

3.57 Because the size of the orbital is determined by the n quantum, with the size increasing with increasing n, an electron in a $2s$ orbital is closer, on average, to the nucleus than is an electron in a $3s$ orbital.

3.58 Because the size of the orbital is determined by the n quantum, with the size increasing with increasing n, an electron in a $4p$ orbital is farther away, on average, from the nucleus than is an electron in a $3p$ orbital.

3.59 The value of l is an integer that lies between 0 and $n - 1$.

(a) When $n = 1$, l can only be $l = 0$.
(b) When $n = 2$, l can be $l = 0$ or $l = 1$.
(c) When $n = 3$, l can be $l = 0$, $l = 1$, or $l = 2$.
(d) When $n = 4$, l can be $l = 0$, $l = 1$, $l = 2$, or $l = 3$.

3.60 The value of m_l is an integer that lies between $-l$ and $+l$.

(a) When $l = 0$, m_l can only be $m_l = 0$.
(b) When $l = 1$, m_l can be $m_l = -1$, $m_l = 0$, or $m_l = +1$.
(c) When $l = 2$, m_l can be $m_l = -2$, $m_l = -1$, $m_l = 0$, $m_l = +1$, or $m_l = +2$.
(d) When $l = 3$, m_l can be $m_l = -3$, $m_l = -2$, $m_l = -1$, $m_l = 0$, $m_l = +1$, $m_l = +2$, or $m_l = +3$.

3.61 Set c cannot occur together as a set of quantum numbers to specify an orbital. l must lie between 0 and $n - 1$; so for $n = 3$, l can only be as high as 2.

3.62 (a) $1s$ is a real orbital, $n = 1$, $l = 0$.
(b) $2p$ is a real orbital, $n = 2$, $l = 1$.
(c) $4s$ is a real orbital, $n = 4$, $l = 0$.
(d) $2d$ is an impossible representation. $n = 2$, $l = 2$ is not allowed. l must lie between 0 and $(n - 1)$; so for $n = 2$, l can only be as high as 1. So it can be $n = 2$, $l = 1$ which is a $2p$ orbital.

3.63 The $2s$ orbital would be the same shape as the $1s$ orbital but would be larger in size, and the $3p$ orbitals would have the same shape as the $2p$ orbitals but would be larger in size. Also, the $2s$ and $3p$ orbitals would have more nodes.

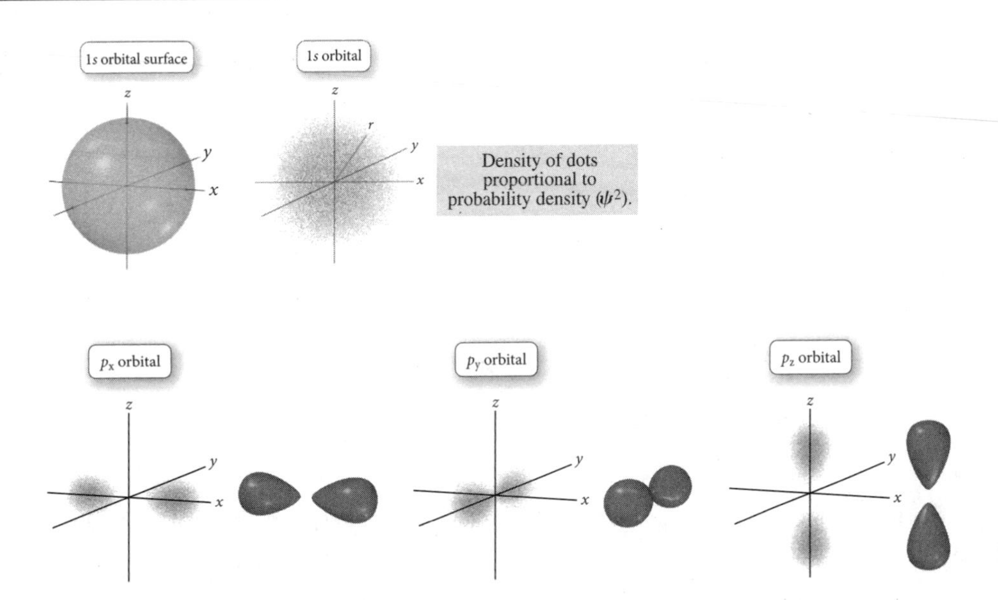

3.64 The 4*d* orbitals would be the same shape as the 3*d* orbitals but would be larger in size, and the 4*d* orbital would have more nodes.

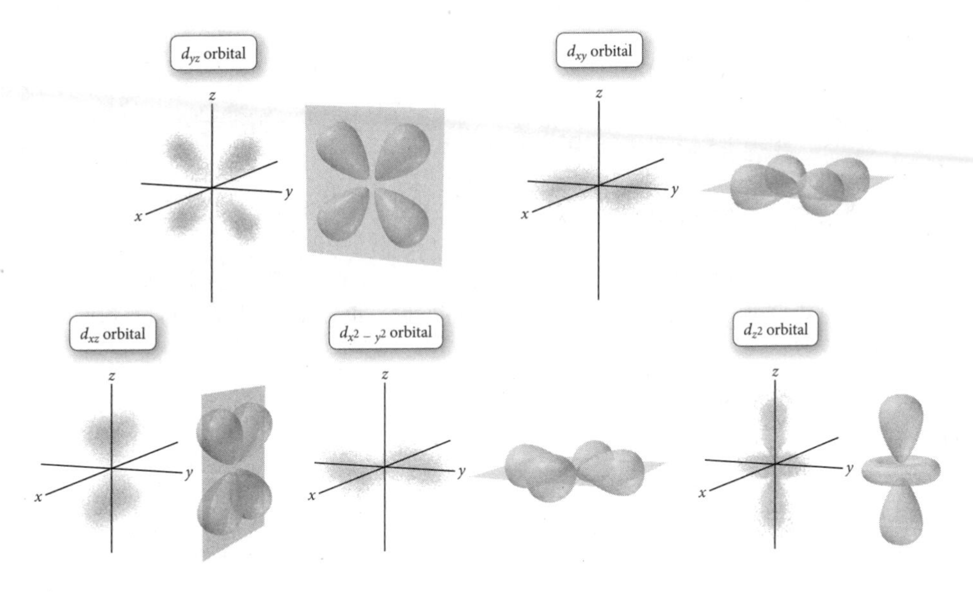

Atomic Spectroscopy

3.65 When the atom emits the photon of energy that was needed to raise the electron to the $n = 2$ level, the photon has the same energy as the energy absorbed to move the electron to the excited state. Therefore, the electron has to be in $n = 1$ (the ground state) following the emission of the photon.

3.66 (a) From $n = 3 \rightarrow n = 1$, the electron is moving to a lower energy level; therefore, there is an emission of energy.
(b) From $n = 2 \rightarrow n = 4$, the electron is moving to a higher energy level; therefore, there is an absorption of energy.
(c) From $n = 4 \rightarrow n = 3$, the electron is moving to a lower energy level; therefore, there is an emission of energy.

3.67 According to the quantum-mechanical model, the higher the *n* level, the higher the energy. So the transition from $3p \rightarrow 1s$ would be a greater energy difference than a transition from $2p \rightarrow 1s$. The lower energy transition would have the longer wavelength. Therefore, the $2p \rightarrow 1s$ transition would produce a longer wavelength.

3.68 According to the quantum-mechanical model, the higher the *n* level, the higher the energy and the higher in energy, the closer the levels are to each other. So the transition from $3p \rightarrow 2s$ would be a greater energy difference than the transition from $4p \rightarrow 3p$. The lower energy transition would have the longer wavelength. Therefore, the $4p \rightarrow 3p$ transition would produce a longer wavelength.

3.69 (a) **Given:** $n = 2 \rightarrow n = 1$ **Find:** λ

Conceptual Plan: $n = 1, n = 2 \rightarrow \Delta E_{atom} \rightarrow \Delta E_{photon} \rightarrow \lambda$

$\Delta E_{atom} = E_1 - E_2 \quad \Delta E_{atom} \rightarrow -\Delta E_{photon} \qquad E = \frac{hc}{\lambda}$

Solution:

$\Delta E = E_1 - E_2$

$$= -2.18 \times 10^{-18}\,\text{J}\left(\frac{1}{1^2}\right) - \left[-2.18 \times 10^{-18}\,\text{J}\left(\frac{1}{2^2}\right)\right] = -2.18 \times 10^{-18}\,\text{J}\left[\left(\frac{1}{1^2}\right) - \left(\frac{1}{2^2}\right)\right] = -1.6\underline{3}5 \times 10^{-18}\,\text{J}$$

$$\Delta E_{photon} = -\Delta E_{atom} = 1.6\underline{3}5 \times 10^{-18}\,\text{J};\ \lambda = \frac{hc}{E} = \frac{(6.626 \times 10^{-34}\,\cancel{\text{J}} \cdot \cancel{\text{s}})(3.00 \times 10^8\,\text{m}/\cancel{\text{s}})}{1.6\underline{3}5 \times 10^{-18}\,\cancel{\text{J}}} = 1.22 \times 10^{-7}\,\text{m}$$

This transition would produce a wavelength in the UV region.

Check: The units of the answer (m) are correct. The magnitude of the answer is reasonable because it is in the region of UV radiation.

(b) **Given:** $n = 3 \rightarrow n = 1$ **Find:** λ

Conceptual Plan: $n = 1, n = 3 \rightarrow \Delta E_{atom} \rightarrow \Delta E_{photon} \rightarrow \lambda$

$\Delta E_{atom} = E_1 - E_3 \quad \Delta E_{atom} \rightarrow -\Delta E_{photon} \qquad E = \frac{hc}{\lambda}$

Solution:

$\Delta E = E_1 - E_3$

$$= -2.18 \times 10^{-18}\,\text{J}\left(\frac{1}{1^2}\right) - \left[-2.18 \times 10^{-18}\,\text{J}\left(\frac{1}{3^2}\right)\right] = -2.18 \times 10^{-18}\,\text{J}\left[\left(\frac{1}{1^2}\right) - \left(\frac{1}{3^2}\right)\right] = -1.9\underline{3}8 \times 10^{-18}\,\text{J}$$

$$\Delta E_{photon} = -\Delta E_{atom} = 1.9\underline{3}8 \times 10^{-18}\,\text{J};\ \lambda = \frac{hc}{E} = \frac{(6.626 \times 10^{-34}\,\cancel{\text{J}} \cdot \cancel{\text{s}})(3.00 \times 10^8\,\text{m}/\cancel{\text{s}})}{1.9\underline{3}8 \times 10^{-18}\,\cancel{\text{J}}} = 1.03 \times 10^{-7}\,\text{m}$$

This transition would produce a wavelength in the UV region.

Check: The units of the answer (m) are correct. The magnitude of the answer is reasonable because it is in the region of UV radiation.

(c) **Given:** $n = 4 \rightarrow n = 2$ **Find:** λ

Conceptual Plan: $n = 2, n = 4 \rightarrow \Delta E_{atom} \rightarrow \Delta E_{photon} \rightarrow \lambda$

$\Delta E_{atom} = E_2 - E_4 \quad \Delta E_{atom} \rightarrow -\Delta E_{photon} \qquad E = \frac{hc}{\lambda}$

Solution:

$\Delta E = E_2 - E_4$

$$= 2.18 \times 10^{-18}\,\text{J}\left(\frac{1}{2^2}\right) - \left[-2.18 \times 10^{-18}\,\text{J}\left(\frac{1}{4^2}\right)\right] = -2.18 \times 10^{-18}\,\text{J}\left[\left(\frac{1}{2^2}\right) - \left(\frac{1}{4^2}\right)\right] = -4.0\underline{8}8 \times 10^{-19}\,\text{J}$$

$$\Delta E_{photon} = -\Delta E_{atom} = 4.0\underline{8}8 \times 10^{-19}\,\text{J} \quad \lambda = \frac{hc}{\text{E}} = \frac{(6.626 \times 10^{-34}\,\cancel{\text{J}} \cdot \cancel{\text{s}})(3.00 \times 10^8\,\text{m}/\cancel{\text{s}})}{4.0\underline{8}8 \times 10^{-19}\,\cancel{\text{J}}} = 4.86 \times 10^{-7}\,\text{m}$$

This transition would produce a wavelength in the visible region.

Check: The units of the answer (m) are correct. The magnitude of the answer is reasonable because it is in the region of visible light.

(d) **Given:** $n = 5 \rightarrow n = 2$ **Find:** λ

Conceptual Plan: $n = 2, n = 5 \rightarrow \Delta E_{atom} \rightarrow \Delta E_{photon} \rightarrow \lambda$

$\Delta E_{atom} = E_2 - E_5 \quad \Delta E_{atom} \rightarrow -\Delta E_{photon} \qquad E = \frac{hc}{\lambda}$

Solution:

$\Delta E = E_2 - E_5$

$$= -2.18 \times 10^{-18}\,\text{J}\left(\frac{1}{2^2}\right) - \left[-2.18 \times 10^{-18}\,\text{J}\left(\frac{1}{5^2}\right)\right] = -2.18 \times 10^{-18}\,\text{J}\left[\left(\frac{1}{2^2}\right) - \left(\frac{1}{5^2}\right)\right] = -4.5\underline{7}8 \times 10^{-19}\,\text{J}$$

$\Delta E_{photon} = -\Delta E_{atom} = 4.5\underline{7}8 \times 10^{-19}\text{ J} \quad \lambda = \frac{hc}{E} = \frac{(6.626 \times 10^{-34}\ \cancel{\text{J}}\cdot\cancel{\text{s}})(3.00 \times 10^{8}\text{ m}/\cancel{\text{s}})}{4.5\underline{7}8 \times 10^{-19}\ \cancel{\text{J}}} = 4.34 \times 10^{-7}\text{ m}$

This transition would produce a wavelength in the visible region.

Check: The units of the answer (m) are correct. The magnitude of the answer is reasonable because it is in the region of visible light.

3.70 (a) **Given:** $n = 4 \rightarrow n = 3$ **Find:** ν

Conceptual Plan: $n = 3, n = 4 \rightarrow \Delta E_{atom} \rightarrow \Delta E_{photon} \rightarrow \nu$

$\Delta E_{atom} = E_3 - E_4 \quad \Delta E_{atom} \rightarrow -\Delta E_{photon} \quad E = h\nu$

Solution:

$\Delta E = E_3 - E_4$

$= -2.18 \times 10^{-18}\text{ J}\left(\frac{1}{3^2}\right) - \left[-2.18 \times 10^{-18}\text{ J}\left(\frac{1}{4^2}\right)\right] = -2.18 \times 10^{-18}\text{ J}\left[\left(\frac{1}{3^2}\right) - \left(\frac{1}{4^2}\right)\right] = -1.0\underline{6}0 \times 10^{-19}\text{ J}$

$\Delta E_{photon} = -\Delta E_{atom} = 1.0\underline{6}0 \times 10^{-19}\text{ J} \quad \nu = \frac{E}{h} = \frac{1.0\underline{6}0 \times 10^{-19}\ \cancel{\text{J}}}{6.626 \times 10^{-34}\ \cancel{\text{J}}\cdot\text{s}} = 1.60 \times 10^{14}\text{ s}^{-1}$

Check: The units of the answer (s^{-1}) are correct. The magnitude of the answer is reasonable because it is a transition between two close levels and the levels become closer as the n value increases. Therefore, the energy difference is smaller and the frequency is smaller.

(b) **Given:** $n = 5 \rightarrow n = 1$ **Find:** ν

Conceptual Plan: $n = 1, n = 5 \rightarrow \Delta E_{atom} \rightarrow \Delta E_{photon} \rightarrow \nu$

$\Delta E_{atom} = E_1 - E_5 \quad \Delta E_{atom} \rightarrow -\Delta E_{photon} \quad E = h\nu$

Solution:

$\Delta E = E_1 - E_5$

$= -2.18 \times 10^{-18}\text{ J}\left(\frac{1}{1^2}\right) - \left[-2.18 \times 10^{-18}\text{ J}\left(\frac{1}{5^2}\right)\right] = -2.18 \times 10^{-18}\text{ J}\left[\left(\frac{1}{1^2}\right) - \left(\frac{1}{5^2}\right)\right] = -2.0\underline{9}3 \times 10^{-18}\text{ J}$

$\Delta E_{photon} = -\Delta E_{atom} = 2.0\underline{9}3 \times 10^{-18}\text{ J}; \nu = \frac{E}{h} = \frac{2.0\underline{9}3 \times 10^{-18}\ \cancel{\text{J}}}{6.626 \times 10^{-34}\ \cancel{\text{J}}\cdot\text{s}} = 3.16 \times 10^{15}\text{ s}^{-1}$

Check: The units of the answer (s^{-1}) are correct. The magnitude of the answer is reasonable because it is a transition that will produce a wavelength in the UV region and the frequency is correct for the UV region.

(c) **Given:** $n = 5 \rightarrow n = 4$ **Find:** ν

Conceptual Plan: $n = 4, n = 5 \rightarrow \Delta E_{atom} \rightarrow \Delta E_{photon} \rightarrow \nu$

$\Delta E_{atom} = E_4 - E_5 \quad \Delta E_{atom} \rightarrow -\Delta E_{photon} \quad E = h\nu$

Solution:

$\Delta E = E_4 - E_5$

$= -2.18 \times 10^{-18}\text{ J}\left(\frac{1}{4^2}\right) - \left[-2.18 \times 10^{-18}\text{ J}\left(\frac{1}{5^2}\right)\right] = -2.18 \times 10^{-18}\text{ J}\left[\left(\frac{1}{4^2}\right) - \left(\frac{1}{5^2}\right)\right] = -4.9\underline{0}5 \times 10^{-20}\text{ J}$

$\Delta E_{photon} = -\Delta E_{atom} = 4.9\underline{0}5 \times 10^{-20}\text{ J} \quad \nu = \frac{E}{h} = \frac{4.9\underline{0}5 \times 10^{-20}\ \cancel{\text{J}}}{6.626 \times 10^{-34}\ \cancel{\text{J}}\cdot\text{s}} = 7.40 \times 10^{13}\text{ s}^{-1}$

Check: The units of the answer (s^{-1}) are correct. The magnitude of the answer is reasonable because it is a transition between two close levels and the levels become closer as the n value increases. Therefore, the energy difference is smaller and the frequency is smaller.

(d) **Given:** $n = 6 \rightarrow n = 5$ **Find:** ν

Conceptual Plan: $n = 5, n = 6 \rightarrow \Delta E_{atom} \rightarrow \Delta E_{photon} \rightarrow \nu$

$\Delta E_{atom} = E_5 - E_6 \quad \Delta E_{atom} \rightarrow -\Delta E_{photon} \quad E = h\nu$

Solution:

$\Delta E = E_5 - E_6$

$= -2.18 \times 10^{-18}\text{ J}\left(\frac{1}{5^2}\right) - \left[-2.18 \times 10^{-18}\text{ J}\left(\frac{1}{6^2}\right)\right] = -2.18 \times 10^{-18}\text{ J}\left[\left(\frac{1}{5^2}\right) - \left(\frac{1}{6^2}\right)\right] = -2.6\underline{6}4 \times 10^{-20}\text{ J}$

$$\Delta E_{photon} = -\Delta E_{atom} = 2.6\underline{6}4 \times 10^{-20}\ \text{J} \quad \nu = \frac{E}{h} = \frac{2.6\underline{6}4 \times 10^{-20}\ \cancel{\text{J}}}{6.626 \times 10^{-34}\ \cancel{\text{J}} \cdot \text{s}} = 4.02 \times 10^{13}\ \text{s}^{-1}$$

Check: The units of the answer (s^{-1}) are correct. The magnitude of the answer is reasonable because it is a transition between two close levels and the levels become closer as the n value increases. Therefore, the energy difference is smaller and the frequency is smaller.

3.71 **Given:** $n(\text{initial}) = 7$; $\lambda = 397$ nm **Find:** $n(\text{final})$

Conceptual Plan: $\boldsymbol{\lambda \rightarrow \Delta E_{photon} \rightarrow \Delta E_{atom} \rightarrow n = x, n = 7}$

$$E = \frac{hc}{\lambda} \quad \Delta E_{photon} \rightarrow -\Delta E_{atom} \quad \Delta E_{atom} = E_x - E_7$$

$$\textbf{Solution: } E = \frac{hc}{\lambda} = \frac{(6.626 \times 10^{-34}\ \text{J} \cdot \cancel{\text{s}})(3.00 \times 10^8\ \cancel{\text{m}}/\cancel{\text{s}})}{(397\ \cancel{\text{nm}})\left(\dfrac{1\ \cancel{\text{m}}}{10^9\ \cancel{\text{nm}}}\right)} = 5.0\underline{0}7 \times 10^{-19}\ \text{J}$$

$$\Delta E_{atom} = -\Delta E_{photon} = -5.0\underline{0}7 \times 10^{-19}\ \text{J}$$

$$\Delta E = E_x - E_7 = -5.0\underline{0}7 \times 10^{-19} = -2.18 \times 10^{-18}\ \text{J}\left(\frac{1}{x^2}\right) - \left[-2.18 \times 10^{-18}\ \text{J}\left(\frac{1}{7^2}\right)\right] = -2.18 \times 10^{-18}\ \text{J}\left[\left(\frac{1}{x^2}\right) - \left(\frac{1}{7^2}\right)\right]$$

$$0.2297 = \left(\frac{1}{x^2}\right) - \left(\frac{1}{7^2}\right) \quad 0.25011 = \left(\frac{1}{x^2}\right) \quad x^2 = 3.998 \quad x = 2$$

Check: The answer is reasonable because it is an integer less than the initial value of 7.

3.72 **Given:** $n(\text{final}) = 4$; $\nu = 11.4$ THz **Find:** $n(\text{initial})$

Conceptual Plan: $\boldsymbol{\nu \rightarrow \Delta E_{photon} \rightarrow \Delta E_{atom} \rightarrow n = 4, n = x}$

$$E = h\nu \quad \Delta E_{photon} \rightarrow -\Delta E_{atom} \quad \Delta E_{atom} = E_4 - E_x$$

$$\textbf{Solution: } E = h\nu = (6.626 \times 10^{-34}\ \text{J} \cdot \cancel{\text{s}})(11.4\ \cancel{\text{THz}})\left(\frac{10^{12}\ \cancel{\text{Hz}}}{1\ \cancel{\text{THz}}}\right)\left(\frac{1\ \cancel{\text{s}^{-1}}}{1\ \cancel{\text{Hz}}}\right) = 7.5\underline{5}4 \times 10^{-20}\ \text{J}$$

$$\Delta E_{atom} = -\Delta E_{photon} = -7.5\underline{5}4 \times 10^{-20}\ \text{J}$$

$$\Delta E = E_4 - E_x = -7.5\underline{5}4 \times 10^{-20}\ \text{J} = -2.18 \times 10^{-18}\ \text{J}\left(\frac{1}{4^2}\right) - \left[-2.18 \times 10^{-18}\ \text{J}\left(\frac{1}{x^2}\right)\right] =$$

$$-2.18 \times 10^{-18}\ \text{J}\left[\left(\frac{1}{4^2}\right) - \left(\frac{1}{x^2}\right)\right]; \quad 0.03465 = \left(\frac{1}{4^2}\right) - \left(\frac{1}{x^2}\right); \quad 0.02785 = \left(\frac{1}{x^2}\right); \quad x^2 = 35.9 \quad x = 6$$

Check: The answer is reasonable because it is an integer greater than the final value of 4.

Cumulative Problems

3.73 **Given:** 348 kJ/mol **Find:** λ

Conceptual Plan: **kJ/mol → kJ/molec → kJ/molec → $\boldsymbol{\lambda}$**

$$\frac{\text{mol C} - \text{C bonds}}{6.022 \times 10^{23}\ \text{C} - \text{C bonds}} \quad \frac{1000\ \text{J}}{1\ \text{kJ}} \quad E = \frac{hc}{\lambda}$$

$$\textbf{Solution: } \frac{348\ \cancel{\text{kJ}}}{\cancel{\text{mol C} - \text{C bonds}}} \times \frac{\cancel{\text{mol C} - \text{C bonds}}}{6.022 \times 10^{23}\ \cancel{\text{C} - \text{C bonds}}} \times \frac{1000\ \text{J}}{1\ \cancel{\text{kJ}}} = 5.7\underline{7}9 \times 10^{-19}\ \text{J}$$

$$\lambda = \frac{(6.626 \times 10^{-34}\ \cancel{\text{J}} \cdot \cancel{\text{s}})(3.00 \times 10^8\ \text{m}/\cancel{\text{s}})}{5.7\underline{7}9 \times 10^{-19}\ \cancel{\text{J}}} = 3.44 \times 10^{-7}\ \text{m} = 344\ \text{nm}$$

Check: The units of the answer (m or nm) are correct. The magnitude of the answer is reasonable because this wavelength is in the UV region.

3.74 **Given:** 164 kJ/mol **Find:** λ

Conceptual Plan: **kJ/mol → kJ/molecule → J/molecule → $\boldsymbol{\lambda}$**

$$\frac{\text{mol}}{6.022 \times 10^{23}\ \text{molecules}} \quad \frac{1000\ \text{J}}{1\ \text{kJ}} \quad E = \frac{hc}{\lambda}$$

Solution: $\frac{164\ \cancel{kJ}}{\cancel{mol}} \times \frac{\cancel{mol}}{6.022 \times 10^{23}\ \text{molecules}} \times \frac{1000\ J}{1\ \cancel{kJ}} = 2.7\underline{2}3 \times 10^{-19}\ \text{J/molecule}$

$$\lambda = \frac{(6.626 \times 10^{-34}\ \cancel{J} \cdot \cancel{s})(3.00 \times 10^{8}\ \text{m}/\cancel{s})}{2.7\underline{2}3 \times 10^{-19}\ \cancel{J}} = 7.30 \times 10^{27}\ \text{m} = 730.\ \text{nm}$$

Check: The units of the answer (m or nm) are correct. The magnitude of the answer is reasonable because this wavelength is in the red region of visible light.

3.75 **Given:** $E_{pulse} = 5.0$ watts; d = 5.5 mm; hole = 1.2 mm; $\lambda = 532$ nm **Find:** photons/s

Conceptual Plan: fraction of beam through hole → fraction of power and then E_{photon} → number photons/s

$\frac{\text{area hole}}{\text{area beam}}$ fraction × power $E = \frac{hc}{\lambda}$ $\frac{\text{power/s}}{E/\text{photon}}$

Solution: $A = \pi r^2 \frac{\pi(0.60\ \cancel{mm})^2}{\pi(2.75\ \cancel{mm})^2} = 0.0476 \qquad 0.0476 \times 5.0\ \cancel{watts} \times \frac{1\ J/s}{1\ \cancel{watt}} = 0.2\underline{3}8\ J/s$

$$E_{photon} = \frac{(6.626 \times 10^{-34}\ J \cdot \cancel{s})(3.00 \times 10^{8}\ \cancel{m}/\cancel{s})}{(532\ \cancel{nm})\left(\frac{1\ \cancel{m}}{10^9\ \cancel{nm}}\right)} = 3.7\underline{3}6 \times 10^{-19}\ \text{J/photon}$$

$$\frac{0.2\underline{3}8\ \cancel{J}/s}{3.7\underline{3}6 \times 10^{-19}\ \cancel{J}/\text{photon}} = 6.4 \times 10^{17}\ \text{photons/s}$$

Check: The units of the answer (number of photons/s) are correct. The magnitude of the answer is reasonable.

3.76 **Given:** $A_{leaf} = 2.50\ \text{cm}^2$; $E_{rad} = 1000\ \text{W/m}^2$; $\lambda = 504$ nm **Find:** photons/s

Conceptual Plan: E_{rad}/s → E_{leaf}/s and then E_{photon} → number photons/s

$E_{rad} \times A_{leaf}$ $E = \frac{hc}{\lambda}$ $\frac{E_{leaf}/s}{E/\text{photon}}$

Solution: $E_{rad} = 2.50\ \cancel{cm^2} \times \frac{1000\ \cancel{W}}{1\ \cancel{m^2}} \times \frac{1\ \cancel{m^2}}{(100\ \cancel{cm})^2} \times \frac{1\ J/s}{1\ \cancel{W}} = 0.250\ J/s$

$$E_{photon} = \frac{hc}{\lambda} = \frac{(6.626 \times 10^{-34}\ J \cdot \cancel{s})(3.00 \times 10^{8}\ \cancel{m}/\cancel{s})}{(504\ \cancel{nm})\left(\frac{1\ \cancel{m}}{10^9\ \cancel{nm}}\right)} = 3.9\underline{4}4 \times 10^{-19}\ \text{J/photon}$$

$$\frac{E_{rad}}{E_{photon}} = \frac{0.250\ \cancel{J}/s}{3.9\underline{4}4 \times 10^{-19}\ \cancel{J}/\text{photon}} = 6.34 \times 10^{17}\ \text{photons/s}$$

Check: The units of the answer (photons/s) are correct. The magnitude of the answer is reasonable compared to the radiation from the sun.

3.77 **Given:** KE = 506 eV **Find:** λ

Conceptual Plan: KE_{eV} → KE_J → v → λ

$\frac{1.602 \times 10^{-19}\ J}{1\ eV}$ $KE = 1/2\,mv^2$ $\lambda = \frac{h}{mv}$

Solution:

$$506\ eV\left(\frac{1.602 \times 10^{-19}\ J}{1\ eV}\right)\left(\frac{\frac{kg \cdot m^2}{s^2}}{J}\right) = \frac{1}{2}(9.11 \times 10^{-31}\ kg)\, v^2$$

$$v^2 = \frac{506\ 1\ \cancel{eV}\left(\frac{1.602 \times 10^{-19}\ \cancel{J}}{1\ \cancel{eV}}\right)\left(\frac{\frac{\cancel{kg} \cdot m^2}{s^2}}{\cancel{J}}\right)}{\frac{1}{2}(9.11 \times 10^{-31}\ \cancel{kg})} = 1.7796 \times 10^{14}\ \frac{m^2}{s^2}$$

$$v = 1.33 \times 10^7 \text{ m/s} \qquad \lambda = \frac{h}{mv} = \frac{6.626 \times 10^{-34} \frac{\text{kg} \cdot \text{m}^2}{\text{s}^2} \cdot \text{s}}{(9.11 \times 10^{-31} \text{ kg})(1.33 \times 10^7 \text{ m/s})} = 5.47 \times 10^{-11} \text{ m} = 0.0547 \text{ nm}$$

Check: The units of the answer (m or nm) are correct. The magnitude of the answer is reasonable because a de Broglie wavelength is usually a very small number.

3.78 **Given:** $\lambda = 0.989$ nm; KE = 969 eV **Find:** *BE*/mol

Conceptual Plan: $\lambda \rightarrow E_{\text{photon}} \rightarrow BE_{\text{photon}} \rightarrow BE_{\text{mol}}$

$E = \frac{hc}{\lambda}$ $\quad (BE_{\text{photon}} = E_{\text{photon}} - \text{KE})$ $\quad \frac{6.022 \times 10^{23} \text{ photons}}{\text{mol}}$

Solution: $E_{\text{photon}} = \frac{hc}{\lambda} = \frac{(6.626 \times 10^{-34} \text{ J} \cdot \text{s})(3.00 \times 10^8 \text{ m/s})}{(0.989 \text{ nm})\left(\frac{1 \text{ m}}{10^9 \text{ nm}}\right)} = 2.0\underline{1}0 \times 10^{-16} \text{ J/photon}$

$$BE_{\text{photon}} = 2.0\underline{1}0 \times 10^{-16} \text{ J/photon} - \left[(969 \text{ eV})\left(\frac{1.602 \times 10^{-19} \text{ J}}{\text{eV}}\right)\right] = 4.5\underline{7}6 \times 10^{-17} \text{ J/photon}$$

$$\frac{4.5\underline{7}6 \times 10^{-17} \text{ J}}{\text{photon}} \times \frac{6.022 \times 10^{23} \text{ photons}}{\text{mol}} \times \frac{1 \text{ kJ}}{1000 \text{ J}} = 2.76 \times 10^4 \text{ kJ/mol}$$

Check: The units of the answer (kJ/mol) are correct. The magnitude of the answer is reasonable because it should require a large amount of energy to remove an electron from a metal surface.

3.79 **Given:** $n = 1 \rightarrow n = \infty$ **Find:** E; λ

Conceptual Plan: $n = \infty, n = 1 \rightarrow \Delta E_{\text{atom}} \rightarrow \Delta E_{\text{photon}} \rightarrow \lambda$

$\Delta E_{\text{atom}} = E_\infty - E_1$ $\quad \Delta E_{\text{atom}} = \Delta E_{\text{photon}}$ $\quad E = \frac{hc}{\lambda}$

Solution: $\Delta E = E_\infty - E_1 = 0 - \left[-2.18 \times 10^{-18} \text{ J}\left(\frac{1}{1^2}\right)\right] = +2.18 \times 10^{-18} \text{ J}$

$$\Delta E_{\text{photon}} = -\Delta E_{\text{atom}} = +2.18 \times 10^{-18} \text{ J}$$

$$\lambda = \frac{hc}{E} = \frac{(6.626 \times 10^{-34} \text{ J} \cdot \text{s})(3.00 \times 10^8 \text{ m/s})}{2.18 \times 10^{-18} \text{ J}} = 9.12 \times 10^{-8} \text{ m} = 91.2 \text{ nm}$$

Check: The units of the answers (J for E and m or nm for the first part) are correct. The magnitude of the answer is reasonable because it would require more energy to completely remove the electron than just moving it to a higher n level. This results in a shorter wavelength.

3.80 **Given:** $E = 496$ kJ/mol **Find:** ν

Conceptual Plan: kJ/mol $\rightarrow$ kJ/molecule $\rightarrow$ J/molecule $\rightarrow \nu$

$\frac{\text{mol}}{6.022 \times 10^{23} \text{ atoms}}$ $\quad \frac{1000 \text{ J}}{1 \text{ kJ}}$ $\quad E = h\nu$

Solution: $\nu = \frac{E}{h} = \frac{\left(\frac{496 \text{ kJ}}{\text{mol}}\right)\left(\frac{\text{mol}}{6.022 \times 10^{23} \text{ atoms}}\right)\left(\frac{1000 \text{ J}}{1 \text{ kJ}}\right)}{6.626 \times 10^{-34} \text{ J} \cdot \text{s}} = 1.24 \times 10^{15} \text{ s}^{-1}$

Check: The units of the answer (s^{-1}) are correct. The magnitude of the answer is reasonable because the frequency is slightly higher than the visible region of the spectrum. This is expected because the excitation of sodium produces a line in the visible region.

3.81 (a) **Given:** $n = 1$ **Find:** number of orbitals if $l = 0 \rightarrow n$

Conceptual Plan: value $n \rightarrow$ values $l \rightarrow$ values $m_l \rightarrow$ number of orbitals

$l = 0 \rightarrow n$ $\quad m_l = -1 \rightarrow +1$ total m_l

Solution:

$n =$	1	
$l =$	0	1
$m_l =$	0	$-1, 0, +1$

total 4 orbitals

Check: The total orbitals will be equal to the (number of l sublevels)^2.

(b) **Given:** $n = 2$ **Find:** number of orbitals if $l = 0 \rightarrow n$

Conceptual Plan: value $n \rightarrow$ values $l \rightarrow$ values $m_l \rightarrow$ number of orbitals

$l = 0 \rightarrow n$ $\quad m_l = -1 \rightarrow +1$ total m_l

Solution:

$n = 2$

$l = 0 \quad 1 \quad 2$

$m_l = 0 \quad -1, 0, +1 \quad -2, -1, 0, 1, 2$

total 9 orbitals

Check: The total orbitals will be equal to the (number of l sublevels)^2.

(c) **Given:** $n = 3$ **Find:** number of orbitals if $l = 0 \rightarrow n$

Conceptual Plan: value $n \rightarrow$ values $l \rightarrow$ values $m_l \rightarrow$ number of orbitals

$l = 0 \rightarrow n$ $\quad m_l = -1 \rightarrow +1$ total m_l

Solution:

$n = 3$

$l = 0 \quad 1 \quad 2 \quad 3$

$m_l = 0 \quad -1, 0, +1 \quad -2, -1, 0, 1, 2 \quad -3, -2, -1, 0, 1, 2, 3$

total 16 orbitals

Check: The total orbitals will be equal to the (number of l sublevels)^2.

3.82 (a) **Given:** s sublevel **Find:** number of orbitals if $m_l = -l - 1 \rightarrow l + 1$

Conceptual Plan: value $l \rightarrow$ values $m_l \rightarrow$ number of orbitals

$m_l = -l - 1 \rightarrow +l + 1$ $\quad$ total m_l

Solution: sublevel $s \rightarrow l = 0$

$m_l = -1, 0, +1$

total 3 orbitals

(b) **Given:** p sublevel **Find:** number of orbitals if $m_l = -l - 1 \rightarrow l + 1$

Conceptual Plan: value $l \rightarrow$ values $m_l \rightarrow$ number of orbitals

$m_l = -l - 1 \rightarrow +l + 1$ $\quad$ total m_l

Solution: sublevel $p \rightarrow l = 1$

$m_l = -2, -1, 0, +1, +2$

total 5 orbitals

(c) **Given:** d sublevel **Find:** number of orbitals if $m_l = -l - 1 \rightarrow l + 1$

Conceptual Plan: value $l \rightarrow$ values $m_l \rightarrow$ number of orbitals

$m_l = -l - 1 \rightarrow +l + 1$ $\quad$ total m_l

Solution: sublevel $d \rightarrow l = 2$

$m_1 = -3, -2, -1, 0, +1, +2, +3$

total 7 orbitals

3.83 **Given:** $\lambda = 1875$ nm; 1282 nm; 1093 nm **Find:** equivalent transitions

Conceptual Plan: $\lambda \rightarrow E_{\text{photon}} \rightarrow E_{\text{atom}} \rightarrow n$

$E = \frac{hc}{\lambda}$ $\quad E_{\text{photon}} \Rightarrow E_{\text{atom}}$ $\quad E = -2.18 \times 10^{-18}\,\text{J}\left(\frac{1}{n_f^2} - \frac{1}{n_i^2}\right)$

Solution: Because the wavelengths of the transitions are longer wavelengths than those obtained in the visual region, the electron must relax to a higher n level. Therefore, we can assume that the electron returns to the $n = 3$ level.

For $\lambda = 1875$ nm: $E = \dfrac{(6.626 \times 10^{-34}\,\text{J}\cdot\cancel{\text{s}})(3.00 \times 10^8\,\cancel{\text{m}}/\cancel{\text{s}})}{1875\,\cancel{\text{nm}}\left(\dfrac{1\,\cancel{\text{m}}}{10^9\,\cancel{\text{nm}}}\right)} = 1.060 \times 10^{-19}\,\text{J}$ $\quad 1.060 \times 10^{-19}\,\text{J} = -1.060 \times 10^{-19}\,\text{J}$

$$-1.060 \times 10^{-19}\,\text{J} = -2.18 \times 10^{-18}\,\text{J}\left(\frac{1}{3^2} - \frac{1}{n^2}\right);\ n = 4$$

For $\lambda = 1282$ nm: $E = \dfrac{(6.626 \times 10^{-34}\,\text{J}\cdot\text{s})(3.00 \times 10^{8}\,\text{m/s})}{1282\ 1\,\text{nm}\left(\dfrac{1\,\text{m}}{10^{9}\,\text{nm}}\right)} = 1.551 \times 10^{-19}\,\text{J} \quad 1.551 \times 10^{-19}\,\text{J} = -1.551 \times 10^{-19}\,\text{J}$

$$-1.551 \times 10^{-19}\,\text{J} = -2.18 \times 10^{-18}\,\text{J}\left(\frac{1}{3^2} - \frac{1}{n^2}\right); n = 5$$

For $\lambda = 1093$ nm: $E = \dfrac{(6.626 \times 10^{-34}\,\text{J}\cdot\text{s})(3.00 \times 10^{8}\,\text{m/s})}{1093\,\text{nm}\left(\dfrac{1\,\text{m}}{10^{9}\,\text{nm}}\right)} = 1.819 \times 10^{-19}\,\text{J}; \quad 1.819 \times 10^{-19}\,\text{J} = -1.819 \times 10^{-19}\,\text{J}$

$$-1.819 \times 10^{-19}\,\text{J} = -2.18 \times 10^{-18}\,\text{J}\left(\frac{1}{3^2} - \frac{1}{n^2}\right); n = 6$$

Check: The values obtained are all integers, which is correct. The values of n (4, 5, and 6) are reasonable. The values of n increase as the wavelength decreases because the two n levels involved are farther apart and more energy is released as the electron relaxes to the $n = 3$ level.

3.84 **Given:** $\lambda = 121.5$ nm; 102.6 nm; 97.23 nm **Find:** equivalent transitions
Conceptual Plan: $\lambda \rightarrow E_{\text{photon}} \rightarrow E_{\text{atom}} \rightarrow n$

$$E = \frac{hc}{\lambda} \qquad E_{\text{photon}} = -E_{\text{atom}} \qquad E = -2.18 \times 10^{-18}\,\text{J}\left(\frac{1}{n_f^2} - \frac{1}{n_i^2}\right)$$

Solution: Because the wavelengths of the transitions are shorter wavelengths than those obtained in the visual region, the electron must relax to a lower n level. Therefore, we can assume that the electron returns to the $n = 1$ level.

For $\lambda = 121.5$ nm: $E = \dfrac{(6.626 \times 10^{-34}\,\text{J}\cdot\text{s})(3.00 \times 10^{8}\,\text{m/s})}{121.5\,\text{nm}\left(\dfrac{1\,\text{m}}{10^{9}\,\text{nm}}\right)} = 1.636 \times 10^{-18}\,\text{J} \quad 1.636 \times 10^{-18}\,\text{J} = -1.636 \times 10^{-18}\,\text{J}$

$$-1.636 \times 10^{-18}\,\text{J} = -2.18 \times 10^{-18}\,\text{J}\left(\frac{1}{1^2} - \frac{1}{n^2}\right); n = 2$$

For $\lambda = 102.6$ nm: $E = \dfrac{(6.626 \times 10^{-34}\,\text{J}\cdot\text{s})(3.00 \times 10^{8}\,\text{m/s})}{102.6\,\text{nm}\left(\dfrac{1\,\text{m}}{10^{9}\,\text{nm}}\right)} = 1.937 \times 10^{-18}\,\text{J} \quad 1.937 \times 10^{-18}\,\text{J} = -1.937 \times 10^{-18}\,\text{J}$

$$-1.937 \times 10^{-18}\,\text{J} = -2.18 \times 10^{-18}\,\text{J}\left(\frac{1}{1^2} - \frac{1}{n^2}\right); n = 3$$

For $\lambda = 97.23$ nm: $E = \dfrac{(6.626 \times 10^{-34}\,\text{J}\cdot\text{s})(3.00 \times 10^{8}\,\text{m/s})}{97.23\,\text{nm}\left(\dfrac{1\,\text{m}}{10^{9}\,\text{nm}}\right)} = 2.044 \times 10^{-18}\,\text{J} \quad 2.044 \times 10^{-18}\,\text{J} = -2.044 \times 10^{-18}\,\text{J}$

$$-2.044 \times 10^{-18}\,\text{J} = -2.18 \times 10^{-18}\,\text{J}\left(\frac{1}{1^2} - \frac{1}{n^2}\right); n = 4$$

Check: The values obtained are all integers, which is correct. The values of n (2, 3, and 4) are reasonable. The values of n increase as the wavelength decreases because the two n levels involved are farther apart and more energy is released as the electron relaxes to the $n = 1$ level.

3.85 **Given:** $\Phi = 193$ kJ/mol **Find:** threshold frequency(ν)
Conceptual Plan: Φ **kJ/mol** $\rightarrow \Phi$ **kJ/atom** $\rightarrow \Phi$ **J/atom** $\rightarrow \nu$

$$\frac{\text{mol}}{6.022 \times 10^{23}\,\text{atoms}} \qquad \frac{1000\,\text{J}}{1\,\text{kJ}} \qquad \Phi = h\nu$$

Solution: $\nu = \dfrac{\Phi}{h} = \dfrac{\left(\dfrac{193\,\text{kJ}}{\text{mol}}\right)\left(\dfrac{\text{mol}}{6.022 \times 10^{23}\,\text{atoms}}\right)\left(\dfrac{1000\,\text{J}}{1\,\text{kJ}}\right)}{6.626 \times 10^{-34}\,\text{J}\cdot\text{s}} = 4.84 \times 10^{14}\,\text{s}^{-1}$

Check: The units of the answer (s^{-1}) are correct. The magnitude of the answer puts the frequency in the infrared range and is a reasonable answer.

3.86 **Given:** $m = 2$ amu; $v = 1 \times 10^6$ m/s **Find:** λ

Conceptual Plan: $\mathbf{m(amu) \rightarrow m(g) \rightarrow m(kg)}$ **and then** $\boldsymbol{m, v \rightarrow \lambda}$

$$\frac{1.661 \times 10^{-24}\text{ g}}{\text{amu}} \quad \frac{1\text{ kg}}{1000\text{ g}} \qquad \lambda = \frac{h}{mv}$$

Solution: $$\frac{6.626 \times 10^{-34} \frac{\text{kg} \cdot \text{m}^2}{\text{s}^2} \cdot \text{s}}{(2\text{ amu})\left(\frac{1.661 \times 10^{-24}\text{ g}}{\text{amu}}\right)\left(\frac{1\text{ kg}}{1000\text{ g}}\right)(1 \times 10^6\text{ m/s})} = 2 \times 10^{-13}\text{ m}$$

Check: The units of the answer (m) are correct. The magnitude of the answer is reasonable because it is a smaller wavelength than for an electron and a deuteron has a much larger mass than an electron.

3.87 **Given:** $\nu_{\text{low}} = 30\text{s}^{-1}$ $\nu_{\text{hi}} = 1.5 \times 10^4\text{ s}^{-1}$; speed $= 344$ m/s **Find:** $\lambda_{\text{low}} - \lambda_{\text{hi}}$

Conceptual Plan: $\boldsymbol{\nu_{\text{low}} \rightarrow \lambda_{\text{low}}}$ **and** $\boldsymbol{\nu_{\text{hi}} = \lambda_{\text{hi}}}$ **then** $\boldsymbol{\lambda_{\text{low}} - \lambda_{\text{hi}}}$

$$\lambda\nu = \text{speed}$$

Solution: $$\lambda = \frac{\text{speed}}{\nu}; \quad \lambda_{\text{low}} = \frac{344\text{ m/s}}{30\text{ s}^{-1}} = 11\text{ m}; \quad \lambda_{\text{hi}} = \frac{344\text{ m/s}}{1.5 \times 10^4\text{ s}^{-1}} = 0.023\text{ m}; \quad 11\text{ m} - 0.023\text{ m} = 11\text{ m}$$

Check: The units of the answer (m) are correct. The magnitude is reasonable because the value is only determined by the low-frequency value because of significant figures.

3.88 **Given:** d $= 1.5 \times 10^8$ km; $\nu = 1.0 \times 10^{14}\text{ s}^{-1}$ **Find:** number of wave crests

Conceptual Plan: $\boldsymbol{\nu \rightarrow \lambda}$ **and then d(km)** $\rightarrow$ **d(m)** $\rightarrow$ **number of waves** $\rightarrow$ **number of crests**

$$\nu = \frac{c}{\lambda} \qquad \frac{1000\text{ m}}{1\text{ km}} \qquad \frac{d}{\lambda}$$

Solution: $$\frac{3.00 \times 10^8\text{ m/s}}{1.0 \times 10^{14}\text{ s}^{-1}} = 3.0 \times 10^{-6}\text{ m}; \quad \frac{1.5 \times 10^8\text{ km} \times \frac{1000\text{ m}}{1\text{ km}}}{3.0 \times 10^{-6}\text{ m}} = 5.0 \times 10^{16}\text{ waves}$$

Because wavelength is measured crest to crest, the number of wave crests would be $5.0 \times 10^{16} + 1$.

Check: The answer is reasonable because the wavelength is small and the distance traveled is large.

3.89 **Given:** $\lambda = 792$ nm; $V = 100.0$ mL; $P = 55.7$ mtorr; $T = 25\,°\text{C}$ **Find:** E to dissociate 15.0%

Conceptual Plan: $\boldsymbol{\lambda \rightarrow E/}$ **molecule and then** $\boldsymbol{P,V,T \rightarrow n \rightarrow}$ **molecules**

$$E = \frac{hc}{\lambda} \qquad n = \frac{PV}{RT} \quad \frac{6.022 \times 10^{23}\text{ molecules}}{\text{mol}}$$

Solution: $$E = \frac{(6.626 \times 10^{-34}\text{ J} \cdot \text{s})(3.00 \times 10^8\text{ m/s})}{792\text{ nm}\left(\frac{1\text{ m}}{10^9\text{ 1 nm}}\right)} = 2.51 \times 10^{-19}\text{ J/molecule}$$

$PV = nRT$. Rearrange to solve for n, so $n = PV/RT =$

$$\frac{(55.7\text{ mtorr})\left(\frac{1\text{ torr}}{1000\text{ mtorr}}\right)\left(\frac{1\text{ atm}}{760\text{ torr}}\right)(100.0\text{ mL})\left(\frac{1\text{ L}}{1000\text{ mL}}\right)}{\left(\frac{0.0821\text{ L} \cdot \text{atm}}{\text{mol} \cdot \text{K}}\right)(298\text{ K})} = 2.7\underline{5}503 \times 10^{-4}\text{ mol}$$

$$2.7\underline{5}503 \times 10^{-4}\text{ mol} \times \left(\frac{6.022 \times 10^{23}\text{ molecules}}{\text{mol}}\right) = 1.80 \times 10^{17}\text{ molecules}$$

$(1.80 \times 10^{17}\text{ molecules})(0.15\%) = 2.70 \times 10^{16}$ molecules dissociated

$(2.51 \times 10^{-19}\text{ J/molecule})(2.70 \times 10^{16}\text{ molecules}) = 6.777 \times 10^{-3}\text{ J} = 6.78 \times 10^{-3}\text{ J}$

Check: The units of the answer (J) are correct. The magnitude is reasonable because it is for a part of a mole of molecules.

3.90 **Given:** 5.00 mL; 0.100 M; $E = 15.5$ J; $\lambda = 349$ nm **Find:** % molecules emitting a photon

Conceptual Plan: **mL,** $\boldsymbol{M}$ $\rightarrow$ **mol** $\rightarrow$ **molecules and then** $\boldsymbol{\lambda \rightarrow E/}$**molecule and then** $\boldsymbol{E \rightarrow}$ **% molecules**

$$M = \frac{\text{mol}}{1000\text{ mL}} \quad \frac{6.022 \times 10^{23}\text{ molecules}}{\text{mol}} \qquad E = \frac{hc}{\lambda} \qquad [E\text{ given}/(E \times \text{molecules})] \times 100\%$$

Solution: $(5.00\ \text{mL})\left(\frac{0.100\ \text{mol}}{1000\ \text{mL}}\right)\left(\frac{6.022 \times 10^{23}\ \text{molecules}}{\text{mol}}\right) = 3.0\underline{1}1 \times 10^{20}\ \text{molecules}$

$$E = \frac{(6.626 \times 10^{-34}\ \text{J}\cdot\text{s})(3.00 \times 10^{8}\ \text{m/s})}{349\ \text{nm}\left(\frac{1\ \text{m}}{10^{9}\ \text{nm}}\right)} = 5.6\underline{9}6 \times 10^{-19}\ \text{J/molecule}$$

$$(3.01 \times 10^{20}\ \text{molecules})(5.70 \times 10^{-19}\ \text{J/molecule}) = 17\underline{1}.57\ \text{J} = 172\ \text{J}$$

$$\frac{15.5\ \text{J}}{172\ \text{J}} \times 100\% = 9.01\%$$

Check: The units of the answer are correct; the magnitude is reasonable because it is less than 100%.

3.91 **Given:** 20.0 mW; 1.00 hr; 2.29×10^{20} photons **Find:** λ

Conceptual Plan: **mW → W → J → J/photon → λ**

$\frac{1\ \text{W}}{1000\ \text{mW}}$ $\quad E = \text{W} \times \text{s}$ $\quad \frac{E}{\text{number of photons}}$ $\quad \lambda = \frac{hc}{E}$

Solution: $(20.0\ \text{mW})\left(\frac{1\ \text{W}}{1000\ \text{mW}}\right)\left(\frac{\frac{1\ \text{J}}{1\ \text{s}}}{1\ \text{W}}\right)\left(\frac{3600\ \text{s}}{2.29 \times 10^{20}\ \text{photons}}\right) = 3.14 \times 10^{-19}\ \text{J/photon}$

$$\frac{(6.626 \times 10^{-34}\ \text{J}\cdot\text{s})(3.00 \times 10^{8}\ \text{m/s})\left(\frac{10^{9}\ \text{nm}}{1\ \text{m}}\right)}{3.14 \times 10^{-19}\ \text{J}} = 632\ \text{nm}$$

Check: The units of the answer (nm) are correct. The magnitude is reasonable because it is in the red range.

3.92 **Given:** 150.0 W; 1.33×10^{19} photons/s; $\lambda = 1064$ nm **Find:** % efficiency

Conceptual Plan: **λ → *E*/photon → *E* → W → %**

$E = \frac{hc}{\lambda}$ $\quad \times\ \text{photons}\ \frac{E}{\text{s}}$ $\quad (\text{W}/\text{W}_{\text{total}}) \times 100\%$

Solution: $$E = \frac{\left(6.626 \times 10^{-34}\ \frac{\text{J}}{\text{photon}}\cdot\text{s}\right)(3.00 \times 10^{8}\ \text{m/s})\left(1.33 \times 10^{19}\ \frac{\text{photon}}{\text{s}}\right)\left(\frac{1\ \text{W}}{1\ \text{J/s}}\right)}{1064\ \text{nm}\left(\frac{1\ \text{m}}{10^{9}\ \text{nm}}\right)} = 2.4\underline{8}4\ \text{W}$$

$$\frac{2.48\ \text{W}}{150.0\ \text{W}} \times 100\% = 1.65\%$$

Check: The units of the answer (%) are correct. The magnitude is reasonable because it is less than 100%.

Challenge Problems

3.93 (a) **Given:** $n = 1; n = 2; n = 3;$ L = 155 pm **Find:** E_1, E_2, E_3

Conceptual Plan: ***n* → *E***

$E_n = \frac{n^2h^2}{8\ \text{m}\ \text{L}^2}$

Solution:

$$E_1 = \frac{1^2(6.626 \times 10^{-34}\ \text{J}\cdot\text{s})^2}{8(9.11 \times 10^{-31}\ \text{kg})(155\ \text{pm})^2\left(\frac{1\ \text{m}}{10^{12}\ \text{pm}}\right)^2} = \frac{1(6.626 \times 10^{-34})^2\ \text{J}^2\text{s}^2}{8(9.11 \times 10^{-31}\ \text{kg})(155 \times 10^{-12})^2\ \text{m}^2}$$

$$= \frac{1(6.626 \times 10^{-34})^2\left(\frac{\text{kg}\cdot\text{m}^2}{\text{s}^2}\right)\text{J}\ \text{s}^2}{8(9.11 \times 10^{-31}\ \text{kg})(155 \times 10^{-12})^2\ \text{m}^2} = 2.51 \times 10^{-18}\ \text{J}$$

$$E_2 = \frac{2^2(6.626 \times 10^{-34}\ \text{J}\cdot\text{s})^2}{8(9.11 \times 10^{-31}\ \text{kg})(155\ \text{pm})^2\left(\frac{1\ \text{m}}{10^{12}\ \text{pm}}\right)^2} = \frac{4(6.626 \times 10^{-34})^2\ \text{J}^2\ \text{s}^2}{8(9.11 \times 10^{-31}\ \text{kg})(155 \times 10^{-12})^2\ \text{m}^2}$$

$$= \frac{4(6.626 \times 10^{-34})^2\left(\frac{\cancel{kg} \cdot \cancel{m^2}}{\cancel{s^2}}\right)\text{J } \cancel{s^2}}{8(9.11 \times 10^{-31}\ \cancel{kg})(155 \times 10^{-12})^2\ \cancel{m^2}} = 1.00 \times 10^{-17}\ \text{J}$$

$$E_3 = \frac{3^2(6.626 \times 10^{-34}\ \text{J} \cdot \text{s})^2}{8(9.11 \times 10^{-31}\ \text{kg})(155\ \cancel{pm})^2\left(\frac{1\ \text{m}}{10^{12}\ \cancel{pm}}\right)^2} = \frac{9(6.626 \times 10^{-34})^2\ \text{J}^2\,\text{s}^2}{8(9.11 \times 10^{-31}\ \text{kg})(155 \times 10^{-12})^2\ \text{m}^2}$$

$$= \frac{9(6.626 \times 10^{-34})^2\left(\frac{\cancel{kg} \cdot \cancel{m^2}}{\cancel{s^2}}\right)\text{J } \cancel{s^2}}{8(9.11 \times 10^{-31}\ \cancel{kg})(155 \times 10^{-12})^2\ \cancel{m^2}} = 2.26 \times 10^{-17}\ \text{J}$$

Check: The units of the answers (J) are correct. The answers seem reasonable because the energy is increasing with increasing n level.

(b) **Given:** $n = 1 \rightarrow n = 2$ and $n = 2 \rightarrow n = 3$ **Find:** λ

Conceptual Plan: $\boldsymbol{n = 1, n = 2 \rightarrow \Delta E_{\text{atom}} \rightarrow \Delta E_{\text{photon}} \rightarrow \lambda}$

$\Delta E_{\text{atom}} = E_2 - E_1 \qquad \Delta E_{\text{atom}} \rightarrow -\Delta E_{\text{photon}} \quad E = \frac{hc}{\lambda}$

Solution: Use the energies calculated in part a.

$$E_2 - E_1 = (1.00 \times 10^{-17}\ \text{J} - 2.51 \times 10^{-18}\ \text{J}) = 7.49 \times 10^{-18}\ \text{J}$$

$$\lambda = \frac{(6.626 \times 10^{-34}\ \cancel{J} \cdot \cancel{s})(3.00 \times 10^8\ \text{m}/\cancel{s})}{7.49 \times 10^{-18}\ \cancel{J}} = 2.65 \times 10^{-8}\ \text{m} = 26.5\ \text{nm}$$

$$E_3 - E_2 = (2.26 \times 10^{-17}\ \text{J} - 1.00 \times 10^{-17}\ \text{J}) = 1.26 \times 10^{-17}\ \text{J}$$

$$\lambda = \frac{(6.626 \times 10^{-34}\ \cancel{J} \cdot \cancel{s})(3.00 \times 10^8\ \text{m}/\cancel{s})}{1.26 \times 10^{-17}\ \cancel{J}} = 1.58 \times 10^{-8}\ \text{m} = 15.8\ \text{nm}$$

These wavelengths would lie in the UV region.

Check: The units of the answers (m) are correct. The magnitude of the answers is reasonable based on the energies obtained for the levels.

3.94 **Given:** $n = 1;\ \nu = 8.85 \times 10^{13}\ \text{s}^{-1}$ **Find:** E, λ

Conceptual Plan: $\boldsymbol{n, \nu \rightarrow E \rightarrow \lambda}$

$E = \left(n + \frac{1}{2}\right)h\nu \quad E = \frac{hc}{\lambda}$

Solution: $E = \left(1 + \frac{1}{2}\right)(6.626 \times 10^{-34}\ \text{J} \cdot \cancel{s})(8.85 \times 10^{13}\ \cancel{s^{-1}}) = 8.80 \times 10^{-20}\ \text{J}$

$$\lambda = \frac{hc}{E} = \frac{(6.626 \times 10^{-34}\ \cancel{J} \cdot \cancel{s})(3.00 \times 10^8\ \text{m}/\cancel{s})}{8.80 \times 10^{-20}\ \cancel{J}} = 2.26 \times 10^{-6}\ \text{m}$$

Check: The units of the answer (J and m) are correct. The magnitude of the answer puts the vibrational frequency in the infrared region, which is reasonable.

3.95 For the 1*s* orbital in the Excel® spreadsheet, label the columns as follows: column A as *r* and column B as ψ(1*s*). Make the values for *r* column A as follows: 0–200. In column B, write the equation for the wave function as follows: = (POWER(1/3.1415,1/2))*(1/POWER(53,3/2))*(EXP(−A2/53)). Go to make chart, choose *xy* scatter.

e.g., sample values	
r	ψ (1*s*)
0	7.000146224
1	7.000143491
2	7.000140809
3	7.000138177
4	7.000135594
5	7.000133060
6	7.000130573

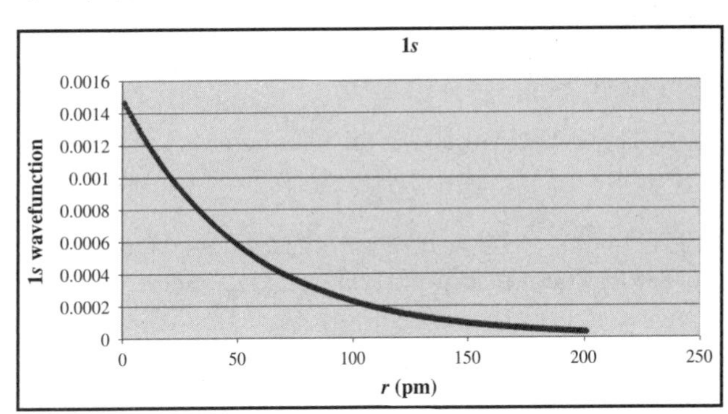

For the 2s orbital in the same Excel® spreadsheet, label the columns as follows: column A as r and column C as ψ (2s). Use the same values for r in column A: 0–200. In column C, write the equation for the wave function as follows: =(POWER(1/((32)*(3.1415)),1/2))*(1/POWER(53,3/2))*(2–(A2/53))*(EXP(–A2/53)). Go to make chart, choose xy scatter.

e.g., sample values	
r	ψ (2s)
0	7.0000516979
1	7.0000502530
2	7.0000488441
3	7.0000474702
4	7.0000461307
5	7.0000448247
6	7.0000435513

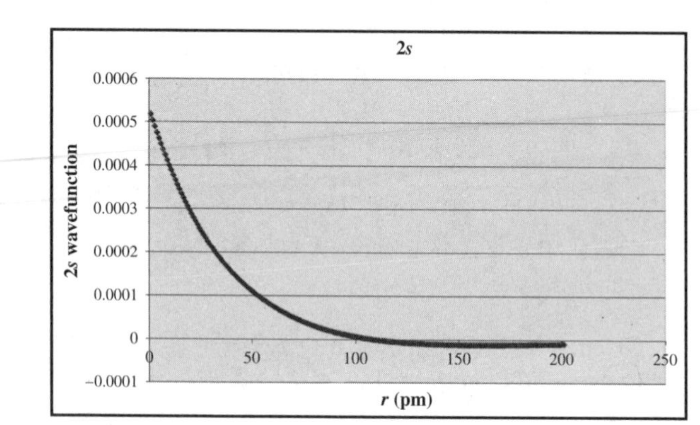

Note: The plot for the 2s orbital extends below the x-axis. The x-intercept represents the radial node of the orbital.

3.96 **Given:** $\Delta E = E_m - E_n = -2.18 \times 10^{-18}(1/\text{m}^2) - [-2.18 \times 10^{-18}(1/\text{n}^2)]$; $E = \text{hc}/\lambda$ **Find:** $1/\lambda = \text{R}(1/\text{m}^2 - 1/\text{n}^2)$

Conceptual Plan: $\Delta E_{\text{atom}} \rightarrow \Delta E_{\text{photon}} \rightarrow 1/\lambda$

$\Delta E_{\text{atom}} = E_m - E_n \Delta E_{\text{atom}} \rightarrow -\Delta E_{\text{photon}}\ E = \dfrac{hc}{\lambda}$

Solution:

$$\Delta E = E_m - E_n = -2.18 \times 10^{-18}\left(\frac{1}{m^2}\right) - \left[-2.18 \times 10^{-18}\left(\frac{1}{n^2}\right)\right] = -2.18 \times 10^{-18}\left(\frac{1}{m^2} - \frac{1}{n^2}\right)$$

$$\Delta E_{\text{atom}} = -\Delta E_{\text{photon}}$$

$$E_{\text{photon}} = -\left[-2.18 \times 10^{-18}\left(\frac{1}{m^2} - \frac{1}{n^2}\right)\right] = \frac{hc}{\lambda} \qquad \frac{1}{\lambda} = \frac{2.18 \times 10^{-18}}{hc}\left(\frac{1}{m^2} - \frac{1}{n^2}\right) = 1.1 \times 10^7\left(\frac{1}{m^2} - \frac{1}{n^2}\right)$$

$$\frac{1}{\lambda} = R\left(\frac{1}{m^2} - \frac{1}{n^2}\right)$$

3.97 **Given:** threshold frequency $= 2.25 \times 10^{14}\ \text{s}^{-1}$; $\lambda = 5.00 \times 10^{-7}$ m **Find:** v of electron

Conceptual Plan: $v \rightarrow \Phi$ **and then** $\lambda \rightarrow E$ **and then** $\rightarrow$ **KE** $\rightarrow v$

$\Phi = hv \qquad E = \dfrac{hc}{\lambda} \qquad \text{KE} = E - \Phi \quad \text{KE} = 1/2\, mv^2$

Solution:

$$\Phi = (6.626 \times 10^{-34}\ \text{J} \cdot \cancel{\text{s}})(2.25 \times 10^{14}\ \cancel{\text{s}^{-1}}) = 1.4\underline{9}1 \times 10^{-19}\ \text{J}; \quad E = \frac{(6.626 \times 10^{-34}\ \text{J} \cdot \cancel{\text{s}})(3.00 \times 10^{8}\ \cancel{\text{m}}/\cancel{\text{s}})}{5.00 \times 10^{-7}\ \cancel{\text{m}}} = 3.9\underline{7}6 \times 10^{-19}\ \text{J}$$

$\text{KE} = 3.9\underline{7}6 \times 10^{-19}\ \text{J} - 1.4\underline{9}1 \times 10^{-19}\ \text{J} = 2.485 \times 10^{-19}\ \text{J}$; $\text{KE} = 1/2\text{mv}^2$; Substituting and rearranging gives

$$v^2 = \frac{2.485 \times 10^{-19}\ \dfrac{\cancel{\text{kg}} \cdot \text{m}^2}{\text{s}^2}}{\dfrac{1}{2}(9.11 \times 10^{-31}\ \cancel{\text{kg}})} = 5.455 \times 10^{11}\ \frac{\text{m}^2}{\text{s}^2}$$

$v = 7.39 \times 10^5\ \text{m/s}$

Check: The units of the answer (m/s) are correct. The magnitude of the answer is reasonable for the speed of an electron.

3.98 **Given:** $\lambda = 2.8 \times 10^{-4}$ cm; $q = 16.72$ J; $\Delta T = 2.0$ K **Find:** number of photons

Conceptual Plan: $\lambda(\text{cm}) \rightarrow \lambda(\text{m}) \rightarrow E_{\text{photon}}$ **and** $q_{\text{water}} \rightarrow$ **number photons**

$\dfrac{m}{100\ \text{cm}} \qquad E = \dfrac{hc}{\lambda} \qquad \dfrac{q_{\text{water}}}{E_{\text{photon}}}$

Solution: $E_{photon} = \dfrac{(6.626 \times 10^{-34}\ \text{J}\cdot\text{s})(3.00 \times 10^{8}\ \text{m/s})}{(2.8 \times 10^{-4}\ \text{cm})\left(\dfrac{1\ \text{m}}{100\ \text{cm}}\right)} = 7.1 \times 10^{-20}\ \text{J/photon}$

$$\text{number of photons} = \frac{16.72\ \text{J}}{7.1 \times 10^{-20}\ \text{J/photon}} = 2.4 \times 10^{20}\ \text{photons}$$

Check: The units of the answer (photons) are correct. The magnitude of the answer seems reasonable because a large amount of heat energy is needed to raise the temperature of the water.

3.99 **Given:** $t = 5.0$ fs; $\lambda_{low} = 722$ nm **Find:** ΔE and λ_{high}

Conceptual Plan: $t \rightarrow \Delta E$ **and then** $\lambda_{low} \rightarrow E_{high} \rightarrow E_{low} \rightarrow \lambda_{high}$

$$\Delta t \times \Delta E \geq \frac{h}{4\pi} \qquad E = \frac{hc}{\lambda} \qquad E - \Delta E \qquad \lambda = \frac{hc}{E}$$

Solution: $\dfrac{6.626 \times 10^{-34}\ \text{J}\cdot\text{s}}{4(3.141)(5.0\ \text{fs})\left(\dfrac{\text{s}}{10^{15}\ \text{fs}}\right)} = 1.0\underline{5}5 \times 10^{-20}\ \text{J}$

$$E = \frac{(6.626 \times 10^{-34}\ \text{J}\cdot\text{s})(3.00 \times 10^{8}\ \text{m/s})}{722\ \text{nm}\left(\dfrac{1\ \text{m}}{10^{9}\ \text{nm}}\right)} = 2.7\underline{5} \times 10^{-19}\ \text{J}$$

$$2.75 \times 10^{-19}\ \text{J} - 1.06 \times 10^{-20}\ \text{J} = 2.\underline{6}4 \times 10^{-19}\ \text{J}$$

$$\frac{(6.626 \times 10^{-34}\ \text{J}\cdot\text{s})(3.00 \times 10^{8}\ \text{m/s})\left(\dfrac{10^{9}\ \text{nm}}{1\ \text{m}}\right)}{(2.64 \times 10^{-19}\ \text{J})} = 751.8\ \text{nm} = 7.5 \times 10^{2}\ \text{nm}$$

Check: The units of the answer (nm) are correct. The magnitude of the answer is reasonable because it is a longer wavelength, but it is close to the original wavelength.

3.100 **Given:** threshold $\nu = 6.71 \times 10^{14}\ \text{s}^{-1}$; $v = 6.95 \times 10^{5}$ m/s; $\nu = 1.01 \times 10^{15}\ \text{s}^{-1}$ **Find:** mass of electron

Conceptual Plan: $\nu \rightarrow \Phi$ **and then** $\nu \rightarrow E$ **and then** $\text{KE} \rightarrow m$

$$\Phi = h\nu \qquad E = h\nu \qquad \text{KE} = E - \Phi \quad m = \frac{2\text{KE}}{v^2}$$

Solution: $\Phi = (6.626 \times 10^{-34}\ \text{J}\cdot\text{s})(6.71 \times 10^{14}\ \text{s}^{-1}) = 4.45 \times 10^{-19}\ \text{J}$

$$E = (6.626 \times 10^{-34}\ \text{J}\cdot\text{s})(1.01 \times 10^{15}\ \text{s}^{-1}) = 6.69 \times 10^{-19}\ \text{J}$$

$$\text{KE} = E - \Phi = 2.24 \times 10^{-19}\ \text{J}$$

$\text{KE} = 1/2\ mv^2$ Rearrange to solve for m.

$$m = \frac{2\left(2.24 \times 10^{-19}\ \dfrac{\text{kg}\cdot\text{m}^2}{\text{s}^2}\right)}{\left(6.95 \times 10^{5}\ \dfrac{\text{m}}{\text{s}}\right)^2} = 9.27 \times 10^{-31}\ \text{kg}$$

Check: The units of the answer (kg) are correct. The magnitude of the answer is reasonable because it is very close to the accepted mass of an electron.

3.101 **Given:** $r = 1.8$ m **Find:** λ

Conceptual Plan: $r \rightarrow C$

$$C = 2\pi r$$

Solution: $(2)(3.141)(1.8\text{m}) = 1\underline{1}.3\ \text{m} =$ the circumference of the orbit. So the largest wavelength that would fit the orbit would be 11 m.

Check: The units of the answer (m) are correct. The magnitude of the wave is about the circumference of the orbit.

3.102 **Given:** $\Delta H_{fusion} = 0.333$ kJ/g; $\lambda = 6.42 \times 10^{-6}$ m; 5.55×10^{-2} mol water **Find:** number of photons

Conceptual Plan: mol → grams → kJ → J and then λ → *E*/photon → photons

$$\frac{18.02\text{ g}}{1\text{ mol}} \quad \frac{0.333\text{ kJ}}{1\text{g}} \quad \frac{1000\text{ J}}{1\text{ kJ}} \qquad E = \frac{hc}{\lambda}$$

Solution: $5.55 \times 10^{-2}\ \text{mol} \times \dfrac{18.02\ \text{g}}{1\ \text{mol}} \times \dfrac{0.333\ \text{kJ}}{1\ \text{g}} \times \dfrac{1000\ \text{J}}{1\ \text{kJ}} = 33\underline{3}.0366\ \text{J}$ then

$$E = \frac{hc}{\lambda} = \frac{(6.626 \times 10^{-34}\ \text{J} \cdot \text{s})(3.00 \times 10^{8}\ \text{m/s})}{6.42 \times 10^{-6}\ \text{m}} = 3.0\underline{9}6262 \times 10^{-20}\ \text{J/photon}$$

$$\frac{33\underline{3}.0366\ \text{J}}{3.0\underline{9}6262 \times 10^{-20}\ \text{J/photon}} = 1.08 \times 10^{22}\ \text{photons}$$

Check: The units of the answer (number of photons) are correct. The magnitude is reasonable because you will need a fairly large number of photons to melt 0.1 mol of ice.

Conceptual Problems

3.103 In the Bohr model of the atom, the electron travels in a circular orbit around the nucleus. It is a two-dimensional model. The electron is constrained to move only from one orbit to another orbit. But the electron is treated as a particle that behaves according to the laws of classical physics.

The quantum-mechanical model of the atom is three dimensional. In this model, we treat the electron, an absolutely small particle, differently than we treat particles with classical physics. The electron is in an orbital, which gives us the probability of finding the electron within a volume of space.

Because the electron in the Bohr model is constrained to a circular orbit, it would theoretically be possible to know both the position and the velocity of the electron simultaneously. This contradicts the Heisenberg uncertainty principle, which states that position and velocity are complementary terms that cannot both be known with precision. So in the quantum-mechanical model we can only know the position or the velocity of an electron at any given time, never both.

3.104 The transition from $n = 3 \rightarrow n = 2$ would cause the photoelectric effect, while the transition from $n = 4 \rightarrow n = 3$ would not. Because the n levels get closer together as n increases, the energy difference between the 4 and 3 levels would be less than the energy difference between the 3 and 2 levels. Therefore, the energy of the photon emitted when the electron moves from 4 to 3 would not be above the threshold energy for the metal. The energy of the photon emitted when the electron makes the transition from $n = 3$ to $n = 2$ is larger and surpasses the threshold energy, thus causing the photoelectric effect.

3.105 (a) Because the interference pattern is caused by single electrons interfering with themselves, the pattern remains the same even when the rate of the electrons passing through the slits is one electron per minute. It will simply take longer for the full pattern to develop.

(b) When a light is placed behind the slits, it flashes to indicate which hole the electron passed through, but the interference pattern is now absent. With the laser on, the electrons hit positions directly behind each slit, as if they were ordinary particles.

(c) Diffraction occurs when a wave encounters an obstacle of a slit that is comparable in size to its wavelength. The wave bends around the slit. The diffraction of light through two slits separated by a distance comparable to the wavelength of the light results in an interference pattern. Each slit acts as a new wave source, and the two new waves interfere with each other, which results in a pattern of bright and dark lines.

(d) Because the mass of the bullets and their particle size are not absolutely small, the bullets will not produce an interference pattern when they pass through the slits. The de Broglie wavelength produced by the bullets will not be sufficiently large enough to interfere with the bullet trajectory, and no interference pattern will be observed.

3.106 The transition from a, $n = 4 \rightarrow n = 3$, would result in emitted light with the longest wavelength. Because the n levels get closer together as n increases, the energy difference between the $n = 4$ and 3 levels would be less energy than the energy difference between the $n = 3$ and 2 levels and the $n = 2$ and 1 levels. Because energy and wavelength are inversely proportional, the smaller the energy, the longer the wavelength.

4 Periodic Properties of the Elements

Review Questions

4.1 A periodic property is predictable based on the element's position within the periodic table.

4.2 The densities of the elements generally follow a fairly well-defined, periodic trend of increasing as you move down a column in the periodic table. Aluminum is near the top of its column in the periodic table, and so it has a low density. As you move down a column in the periodic table, both the radius and the mass increase. The mass (due to additional protons and neutrons) increases more than its volume (due to an increasing radius) does, so the density increases.

4.3 The first attempt to organize the elements according to similarities in their properties was made by the German chemist Johann Döbereiner. He grouped elements into triads: three elements with similar properties. The English chemist John Newlands attempted a more complex approach. He organized elements into octaves, analogous to musical notes. When arranged this way, the properties of every eighth element were similar.

4.4 The modern periodic table is credited primarily to the Russian chemist Dmitri Mendeleev. Mendeleev's table is based on the periodic law, which states that when elements are arranged in order of increasing mass, their properties recur periodically. Mendeleev arranged the elements in a table in which mass increased from left to right and elements with similar properties fell in the same columns.

4.5 Julius Lothar Meyer proposed an organization of the known elements based on some periodic properties. Henry Moseley listed elements according to the atomic number rather than the atomic mass. This resolved the problems in Mendeleev's table where an increase in atomic mass did not correlate with similar properties.

4.6 The periodic law was based on the observations that the properties of elements recur and certain elements have similar properties. The theory that explains the existence of the periodic law is the quantum-mechanical theory.

4.7 An electron configuration shows the particular orbitals that are occupied by electrons in an atom. Some examples are H $= 1s^1$, He $= 1s^2$, and Li $= 1s^2 2s^1$.

4.8 Coulomb's law states that the potential energy (E) of two charged particles depends on their charges (q_1 and q_2) and on their separation, (r). $E = \frac{1}{4\pi\varepsilon_o}\frac{q_1 q_2}{r}$. The potential energy is positive for charges of the same sign and negative for charges of opposite sign. The magnitude of the potential energy depends inversely on the separation between the charged particles.

4.9 Shielding or screening occurs when one electron is blocked from the full effects of the nuclear charge so that the electron experiences only a part of the nuclear charge. It is the inner (core) electrons that shield the outer electrons from the full nuclear charge.

4.10 Penetration occurs when an electron penetrates the electron cloud of the $1s$ orbital and experiences the charge of the nucleus more fully because it is less shielded by the intervening electrons. As the outer electron undergoes penetration into the region occupied by the inner electrons, it experiences a greater nuclear charge and, therefore, according to Coulomb's law, a lower energy.

4.11 The sublevels within a principal level split in multielectron atoms because of the penetration of the outer electrons into the region of the core electrons. The sublevels in hydrogen are not split because they are empty in the ground state.

4.12 An orbital diagram is a different way to show the electron configuration of an atom. It symbolizes the electron as an arrow in a box that represents the orbital.

The orbital diagram for a hydrogen atom: [↿] H
$1s$

4.13 The Pauli exclusion principle states the following: No two electrons in an atom can have the same four quantum numbers. Because two electrons occupying the same orbital have three identical quantum numbers (n, l, m_l), they must have different spin quantum numbers. The Pauli exclusion principle implies that each orbital can have a maximum of only two electrons, with opposing spins.

4.14 Degenerate orbitals are orbitals of the same energy. In a multielectron atom, the orbitals in a sublevel are degenerate. Hund's rule states that when filling degenerate orbitals, electrons fill them singly first, with parallel spins. This is a result of an atom's tendency to find the lowest energy state possible.

4.15 In order of increasing energy, the orbitals are $1s < 2s < 2p < 3s < 3p < 4s < 3d < 4p < 5s$. The $4s$ orbital fills before the $3d$ and the $5s$ fills before the $4d$. They are lower in energy because of greater penetration of the $4s$ and $5s$ orbitals.

4.16 Valence electrons are important in chemical bonding. For main-group elements, the valence electrons are in the outermost principal energy level. For transition elements, we also count the outermost *d* electrons among the valence, even though they are not in the outermost principal energy level. The chemical properties of an element depend on its valence electrons, which are important in bonding because they are held most loosely. This is why the elements in a column of the periodic table have similar chemical properties: They have the same number of valence electrons.

4.17

s-block elements
d-block elements
p-block elements
f-block elements

Periods \ Groups	1 1A	2 2A	3 3B	4 4B	5 5B	6 6B	7 7B	8 8B	9 8B	10 8B	11 1B	12 2B	13 3A	14 4A	15 5A	16 6A	17 7A	18 8A
1	1 **H** $1s^1$																	2 **He** $1s^2$
2	3 **Li** $2s^1$	4 **Be** $2s^2$											5 **B** $2s^22p^1$	6 **C** $2s^22p^2$	7 **N** $2s^22p^3$	8 **O** $2s^22p^4$	9 **F** $2s^22p^5$	10 **Ne** $2s^22p^6$
3	11 **Na** $3s^1$	12 **Mg** $3s^2$											13 **Al** $3s^23p^1$	14 **Si** $3s^23p^2$	15 **P** $3s^23p^3$	16 **S** $3s^23p^4$	17 **Cl** $3s^23p^5$	18 **Ar** $3s^23p^6$
4	19 **K** $4s^1$	20 **Ca** $4s^2$	21 **Sc** $4s^23d^1$	22 **Ti** $4s^23d^2$	23 **V** $4s^23d^3$	24 **Cr** $4s^13d^5$	25 **Mn** $4s^23d^5$	26 **Fe** $4s^23d^6$	27 **Co** $4s^23d^7$	28 **Ni** $4s^23d^8$	29 **Cu** $4s^13d^{10}$	30 **Zn** $4s^23d^{10}$	31 **Ga** $4s^24p^1$	32 **Ge** $4s^24p^2$	33 **As** $4s^24p^3$	34 **Se** $4s^24p^4$	35 **Br** $4s^24p^5$	36 **Kr** $4s^24p^6$
5	37 **Rb** $5s^1$	38 **Sr** $5s^2$	39 **Y** $5s^24d^1$	40 **Zr** $5s^24d^2$	41 **Nb** $5s^14d^4$	42 **Mo** $5s^14d^5$	43 **Tc** $5s^24d^5$	44 **Ru** $5s^14d^7$	45 **Rh** $5s^14d^8$	46 **Pd** $4d^{10}$	47 **Ag** $5s^14d^{10}$	48 **Cd** $5s^24d^{10}$	49 **In** $5s^25p^1$	50 **Sn** $5s^25p^2$	51 **Sb** $5s^25p^3$	52 **Te** $5s^25p^4$	53 **I** $5s^25p^5$	54 **Xe** $5s^25p^6$
6	55 **Cs** $6s^1$	56 **Ba** $6s^2$	57 **La** $6s^25d^1$	72 **Hf** $6s^25d^2$	73 **Ta** $6s^25d^3$	74 **W** $6s^25d^4$	75 **Re** $6s^25d^5$	76 **Os** $6s^25d^6$	77 **Ir** $6s^25d^7$	78 **Pt** $6s^15d^9$	79 **Au** $6s^15d^{10}$	80 **Hg** $6s^25d^{10}$	81 **Tl** $6s^26p^1$	82 **Pb** $6s^26p^2$	83 **Bi** $6s^26p^3$	84 **Po** $6s^26p^4$	85 **At** $6s^26p^5$	86 **Rn** $6s^26p^6$
7	87 **Fr** $7s^1$	88 **Ra** $7s^2$	89 **Ac** $7s^26d^1$	104 **Rf** $7s^26d^2$	105 **Db** $7s^26d^3$	106 **Sg** $7s^26d^4$	107 **Bh**	108 **Hs**	109 **Mt**	110 **Ds**	111 **Rg**	112	113	114	115	116		

Lanthanides	58 **Ce** $6s^24f^15d^1$	59 **Pr** $6s^24f^3$	60 **Nd** $6s^24f^4$	61 **Pm** $6s^24f^5$	62 **Sm** $6s^24f^6$	63 **Eu** $6s^24f^7$	64 **Gd** $6s^24f^75d^1$	65 **Tb** $6s^24f^9$	66 **Dy** $6s^24f^{10}$	67 **Ho** $6s^24f^{11}$	68 **Er** $6s^24f^{12}$	69 **Tm** $6s^24f^{13}$	70 **Yb** $6s^24f^{14}$	71 **Lu** $6s^24f^{14}6d^1$
Actinides	90 **Th** $7s^26d^2$	91 **Pa** $7s^25f^26d^1$	92 **U** $7s^25f^36d^1$	93 **Np** $7s^25f^46d^1$	94 **Pu** $7s^25f^6$	95 **Am** $7s^25f^7$	96 **Cm** $7s^25f^76d^1$	97 **Bk** $7s^25f^9$	98 **Cf** $7s^25f^{10}$	99 **Es** $7s^25f^{11}$	100 **Fm** $7s^25f^{12}$	101 **Md** $7s^25f^{13}$	102 **No** $7s^25f^{14}$	103 **Lr** $7s^25f^{14}6d^1$

4.18 The number of columns in a block corresponds to the maximum number of electrons that can occupy the particular sublevel of that block. The *s* block has two columns corresponding to one *s* orbital holding a maximum of two electrons. The *p* block has six columns corresponding to the three *p* orbitals with two electrons each.

4.19 The rows in the periodic table grow progressively longer because you are adding sublevels as the *n* level increases.

4.20 The lettered group number of a main-group element is equal to the number of valence electrons for that element.

4.21 The row number of a main-group element is equal to the highest principal quantum number of that element. However, the principal quantum number of the *d* orbital being filled across each row in the transition series is equal to the row number minus one. For the inner transition elements, the principal quantum number of the *f* orbital being filled across each row is the row number minus two.

4.22 In the first transition series of the *d* block, Cr and Cu have anomalous electron configurations. Cr is expected to be $[Ar]4s^2 3d^4$ but is found to be $[Ar]4s^1 3d^5$, and Cu is expected to be $[Ar]4s^2 3d^9$ but is found to be $[Ar]4s^1 3d^{10}$.

4.23 To use the periodic table to write the electron configuration, find the noble gas that precedes the element. The element has the inner electron configuration of that noble gas. Place the symbol for the noble gas in [ ]. Obtain the outer electron configuration by tracing the element across the period and assigning electrons in the appropriate orbitals.

4.24 The chemical properties of elements are largely determined by the number of valence electrons they contain. Their properties are periodic because the number of valence electrons is periodic. Because elements within a column in the periodic table have the same number of valence electrons, they also have similar chemical properties.

4.25
(a) The alkali metals (group 1A) have one valence electron and are among the most reactive metals because their outer electron configuration (ns^1) is one electron beyond a noble gas configuration. They react to lose the ns^1 electron, obtaining a noble gas configuration. This is why the group 1A metals tend to form 1+ cations.
(b) The alkaline earth metals (group 2A) have two valence electrons and an outer electron configuration of ns^2, and also tend to be reactive metals. They lose their ns^2 electrons to form 2+ cations.
(c) The halogens (group 7A) have seven valence electrons and an outer electron configuration of ns^2np^5. They are among the most reactive nonmetals. They are only one electron short of a noble gas configuration and tend to react to gain that one electron, forming 1− anions.
(d) The oxygen family (group 6A) has six valence electrons and an outer electron configuration of ns^2np^4. They are two electrons short of a noble gas configuration and tend to react to gain those two electrons, forming 2− anions.

4.26 One way to define atomic radii is to consider the distance between nonbonding atoms in molecules or atoms that are touching each other but are not bonded together. An atomic radius determined this way is called the nonbonding atomic radius or the van der Waals radius. The van der Waals radius represents the radius of an atom when it is not bonded to another atom.

Another way to define the size of an atom, called bonding an atomic radius or covalent radius, is defined differently for nonmetals and metals as follows:

Nonmetals: one-half the distance between two of the atoms bonded together

Metal: one-half the distance between two of the atoms next to each other in a crystal of the metal

A more general term, the atomic radius, refers to a set of average bonding radii determined from measurements on a large number of elements and compounds. The atomic radius represents the radius of an atom when it is bonded to another atom and is always smaller than the van der Waals radius.
(a) As you move to the right across a period in the periodic table, atomic radius decreases.
(b) As you move down a column in the periodic table, atomic radius increases.

4.27 The effective nuclear charge (Z_{eff}) is the average or net charge from the nucleus experienced by the electrons in the outermost levels. Shielding is the blocking of nuclear charge from the outermost electrons. The shielding is primarily due to the inner (core) electrons, although there is some interaction and shielding from the electron repulsions of the outer electrons with each other.

4.28 When an alkali metal forms an ion, it loses its valence electron and forms an ion with a 1+ charge. When an alkaline earth metal forms an ion, it loses both of its valence electrons and forms an ion with a 2+.

4.29 When a halogen forms an ion, it gains one valence electron and forms an ion with a 1− charge. When an atom in the oxygen family forms an ion, it gains two valence electrons and forms an ion with a 2−. Nitrogen will gain three electrons, forming a 3− ion. Aluminum will lose three electrons, forming a 3+ ion.

4.30 As you move to the right across a row in the periodic table, the *n* level stays the same. However, the nuclear charge increases and the amount of shielding stays about the same because the number of inner electrons stays the same. So the effective nuclear charge experienced by the electrons in the outermost principal energy level increases, resulting in a stronger attraction between the outermost electrons and the nucleus and, therefore, smaller atomic radii.

4.31 (a) The radii of transition elements stay roughly constant across each row instead of decreasing in size as in the main-group elements. The difference is that across a row of transition elements, the number of electrons in the outermost principal energy level is nearly constant. As another proton is added to the nucleus with each successive element, another electron is added as well, but the electron goes into an $n_{\text{highest}} - 1$ orbital. The number of outermost electrons stays constant, and they experience a roughly constant effective nuclear charge, keeping the radius approximately constant.

(b) As you go down the first two rows of a column within the transition metals, the elements follow the same general trend in atomic radii and the main-group elements; that is, the radii get larger because you are adding outermost electrons into higher *n* levels.

4.32 The electron configuration of a main-group monatomic ion can be deduced from the electron configuration of the neutral atom and the charge of the ion. For anions, we simply add the number of electrons required by the magnitude of the charge of the anion. The electron configuration of cations is obtained by subtracting the number of electrons required by the magnitude of the charge.

4.33 An important exception to simply subtracting the number of electrons occurs for transition metal cations. When writing the electron configuration of a transition metal cation, remove the electrons in the highest *n*-value orbitals first, even if this does not correspond to the reverse order of filling. Normally, even though the *d* orbital electrons add after the *s* orbital electrons, the *s* orbital electrons are lost first. This is because (1) the *ns* and $(n - 1)d$ orbitals are extremely close in energy and, depending on the exact configuration, can vary in relative energy ordering and (2) as the $(n - 1)d$ orbitals begin to fill in the first transition series, the increasing nuclear charge stabilizes the $(n - 1)d$ orbitals relative to the *ns* orbitals. This happens because the $(n - 1)d$ orbitals are not outermost orbitals and, therefore, are not effectively shielded from the increasing nuclear charge by the *ns* orbitals.

4.34 (a) In general, cations are much smaller than their corresponding parent. This is because the outermost electrons are shielded from the nuclear charge in the atom and contribute greatly to the size of the atom. When these electrons are removed to form the cation, the same nuclear charge is now acting only on the core electrons.

(b) In general, anions are much larger than their corresponding atoms. This is because the extra electrons are added to the outermost electrons but no additional protons are added to increase the nuclear charge. The extra electrons increase the repulsions among the outermost electrons, resulting in an anion that is larger than the atom.

4.35 The ionization energy (IE) of an atom or ion is the energy required to remove an electron from the atom or ion in the gaseous state. The ionization energy is always positive because removing an electron takes energy. The energy required to remove the first electron is called the first ionization energy (IE_1). The energy required to remove the second electron is called the second ionization energy (IE_2). The second IE is always greater than the first IE.

4.36 Ionization energy generally decreases as you move down a column in the periodic table because electrons in the outermost principal level become farther from the positively charged nucleus and are, therefore, held less tightly.

Ionization energy generally increases as you move to the right across a period in the periodic table because electrons in the outermost principal energy level generally experience a greater effective nuclear charge; therefore, the electrons are closer to the nucleus.

4.37 Exceptions occur with elements Be, Mg, and Ca in group 2A having a higher first ionization energy than elements B, Al, and Ga in group 3A. This exception is caused by the change in going from the *s* block to the *p* block. The result is that the electrons in the *s* orbital shield the electron in the *p* orbital from nuclear charge, making it easier to remove.

Another exception occurs with N, P, and As in group 5A having a higher first ionization energy than O, S, and Se in group 6A. This exception is caused by the repulsion between electrons when they must occupy the same orbital. Group 5A has 3*p* electrons, while group 6A has 4*p* electrons. In the group 5A elements, the *p* orbitals are half-filled, which makes the configuration particularly stable. The fourth group 6A electron must pair with another electron, making it easier to remove.

4.38 The second ionization energy of Mg involves removing the second outermost electron, leading to an ion with a noble gas configuration for the core electrons. The third ionization energy requires removing a core electron from an ion with a noble gas configuration. This requires a tremendous amount of energy, making IE_3 very high.

For Al, IE_3 involves removing the third outermost electron for Al, leaving the ion with a noble gas configuration of the core electrons. IE_4 then requires removing a core electron from an ion with a noble gas configuration. This requires a tremendous amount of energy and makes IE_4 very high.

You can predict whether the IE energy is going to be very high by looking for the ionization that requires removing a core electron.

4.39 The electron affinity (EA) of an atom or ion is the energy change associated with the gaining of an electron by the atom in the gaseous state. The electron affinity is usually—though not always—negative because an atom or ion usually releases energy when it gains an electron. The trends in electron affinity are not as regular as trends in other properties. For main-group elements, electron affinity generally becomes more negative as you move to the right across a row in the periodic table. There is no corresponding trend in electron affinity going down a column, with the exception of group IA which becomes more positive as you go down the column.

4.40 Metals are good conductors of heat and electricity, they can be pounded into flat sheets (malleability), they can be drawn into wires (ductility), they are often shiny, and they tend to lose electrons in chemical reactions. As you move to the right across a period in the periodic table, metallic character decreases. As you move down a column in the periodic table, metallic character increases.

Problems by Topic

The Periodic Table

4.41
(a) K — Potassium is a metal.
(b) Ba — Barium is a metal.
(c) I — Iodine is a nonmetal.
(d) O — Oxygen is a nonmetal.
(e) Sb — Antimony is a metalloid.

4.42
(a) gold — Au is a metal.
(b) fluorine — F is a nonmetal.
(c) sodium — Na is a metal.
(d) tin — Sn is a metal.
(e) argon — Ar is a nonmetal.

4.43
(a) tellurium — Te is in group 6A and is a main-group element.
(b) potassium — K is in group 1A and is a main-group element.
(c) vanadium — V is in group 5B and is a transition element.
(d) manganese — Mn is in group 7B and is a transition element.

4.44
(a) Cr — Chromium is in group 6B and is a transition element.
(b) Br — Bromine is in group 7A and is a main-group element.
(c) Mo — Molybdenum is in group 6B and is a transition element.
(d) Cs — Cesium is in group 1A and is a main-group element.

Electron Configurations

4.45
(a) Si — Silicon has 14 electrons. Distribute two of these into the 1*s* orbital, two into the 2*s* orbital, six into the 2*p* orbital, two into the 3*s* orbital, and two into the 3*p* orbital. $1s^22s^22p^63s^23p^2$
(b) O — Oxygen has 8 electrons. Distribute two of these into the 1*s* orbital, two into the 2*s* orbital, and four into the 2*p* orbital. $1s^22s^22p^4$

(c) K Potassium has 19 electrons. Distribute two of these into the 1*s* orbital, two into the 2*s* orbital, six into the 2*p* orbital, two into the 3*s* orbital, six into the 3*p* orbital, and one into the 4*s* orbital. $1s^2 2s^2 2p^6 3s^2 3p^6 4s^1$

(d) Ne Neon has 10 electrons. Distribute two of these into the 1*s* orbital, two into the 2*s* orbital, and six into the 2*p* orbital. $1s^2 2s^2 2p^6$

4.46 (a) C Carbon has 6 electrons. Distribute two of these into the 1*s* orbital, two into the 2*s* orbital, and two into the 2*p* orbital. $1s^2 2s^2 2p^2$

(b) P Phosphorus has 15 electrons. Distribute two of these into the 1*s* orbital, two into the 2*s* orbital, six into the 2*p* orbital, two into the 3*s* orbital, and three into the 3*p* orbital. $1s^2 2s^2 2p^6 3s^2 3p^3$

(c) Ar Argon has 18 electrons. Distribute two of these into the 1*s* orbital, two into the 2*s* orbital, six into the 2*p* orbital, two into the 3*s* orbital, and six into the 3*p* orbital. $1s^2 2s^2 2p^6 3s^2 3p^6$

(d) Na Sodium has 11 electrons. Distribute two of these into the 1*s* orbital, two into the 2*s* orbital, six into the 2*p* orbital, and one into the 3*s* orbital. $1s^2 2s^2 2p^6 3s^1$

4.47 (a) N Nitrogen has seven electrons and has the electron configuration $1s^2 2s^2 2p^3$. Draw a box for each orbital, putting the lowest energy orbital (1*s*) on the far left and proceeding to orbitals of higher energy to the right. Distribute the seven electrons into the boxes representing the orbitals, allowing a maximum of two electrons per orbital and remembering Hund's rule. You can see from the diagram that nitrogen has three unpaired electrons.

1s	2s	2p
[↑↓]	[↑↓]	[↑][↑][↑]

(b) F Fluorine has nine electrons and has the electron configuration $1s^2 2s^2 2p^5$. Draw a box for each orbital, putting the lowest energy orbital (1*s*) on the far left and proceeding to orbitals of higher energy to the right. Distribute the nine electrons into the boxes representing the orbitals, allowing a maximum of two electrons per orbital and remembering Hund's rule. You can see from the diagram that fluorine has one unpaired electron.

1s	2s	2p
[↑↓]	[↑↓]	[↑↓][↑↓][↑]

(c) Mg Magnesium has 12 electrons and has the electron configuration $1s^2 2s^2 2p^6 3s^2$. Draw a box for each orbital, putting the lowest energy orbital (1*s*) on the far left and proceeding to orbitals of higher energy to the right. Distribute the 12 electrons into the boxes representing the orbitals, allowing a maximum of two electrons per orbital and remembering Hund's rule. You can see from the diagram that magnesium has no unpaired electrons.

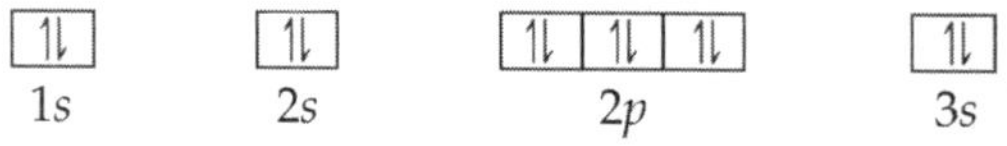

1s	2s	2p	3s
[↑↓]	[↑↓]	[↑↓][↑↓][↑↓]	[↑↓]

(d) Al Aluminum has 13 electrons and has the electron configuration $1s^2 2s^2 2p^6 3s^2 3p^1$. Draw a box for each orbital, putting the lowest energy orbital (1*s*) on the far left and proceeding to orbitals of higher energy to the right. Distribute the 13 electrons into the boxes representing the orbitals, allowing a maximum of two electrons per orbital and remembering Hund's rule. You can see from the diagram that aluminum has one unpaired electron.

1s	2s	2p	3s	3p
[↑↓]	[↑↓]	[↑↓][↑↓][↑↓]	[↑↓]	[↑][ ][ ]

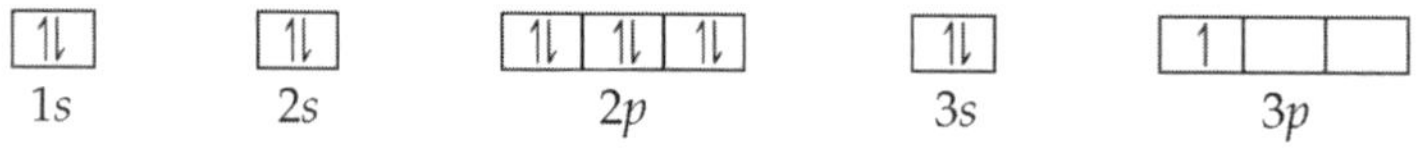

4.48 (a) S Sulfur has 16 electrons and has the electron configuration $1s^2 2s^2 2p^6 3s^2 3p^4$. Draw a box for each orbital, putting the lowest energy orbital (1*s*) on the far left and proceeding to orbitals of higher energy to the right. Distribute the 16 electrons into the boxes representing the orbitals, allowing a maximum of two electrons per orbital and remembering Hund's rule. You can see from the diagram that sulfur has two unpaired electrons.

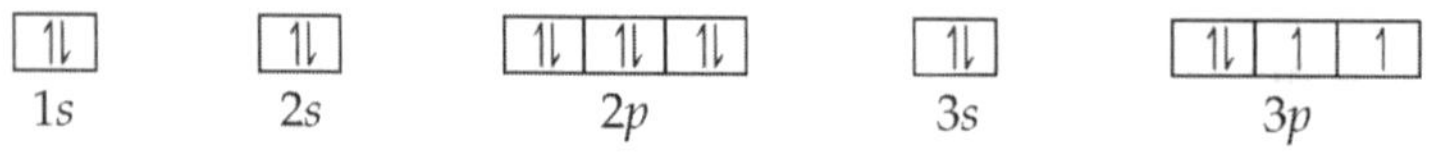

1s	2s	2p	3s	3p
[↑↓]	[↑↓]	[↑↓][↑↓][↑↓]	[↑↓]	[↑↓][↑][↑]

(b) Ca Calcium has 20 electrons and has the electron configuration $1s^2 2s^2 2p^6 3s^2 3p^6 4s^2$. Draw a box for each orbital, putting the lowest energy orbital (1*s*) on the far left and proceeding to orbitals of higher energy to the right. Distribute the 20 electrons into the boxes representing the orbitals, allowing a maximum of

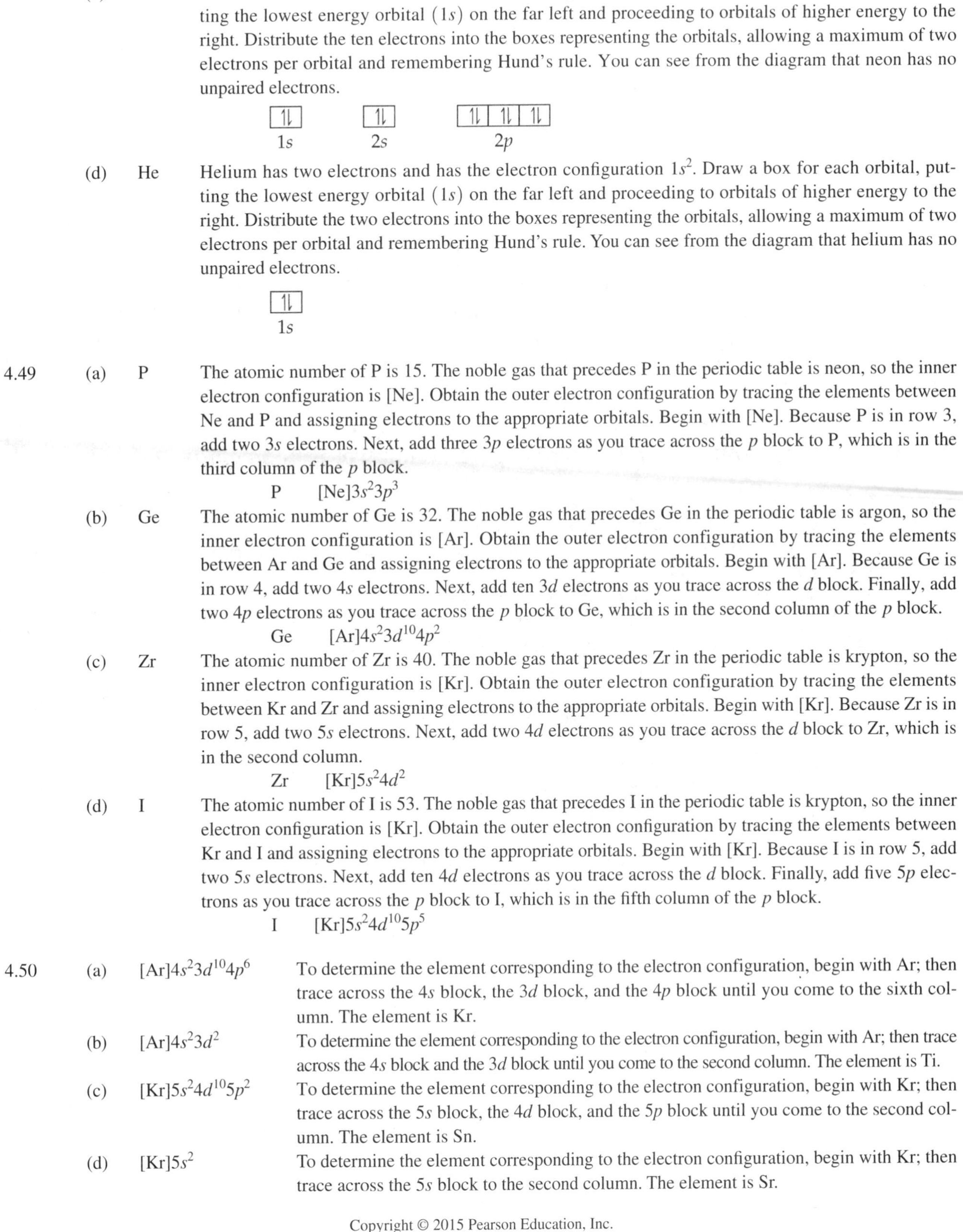

two electrons per orbital and remembering Hund's rule. You can see from the diagram that calcium has no unpaired electrons.

⥮	⥮	⥮ ⥮ ⥮	⥮	⥮ ⥮ ⥮	⥮
1*s*	2*s*	2*p*	3*s*	3*p*	4*s*

(c) Ne Neon has ten electrons and has the electron configuration $1s^2 2s^2 2p^6$. Draw a box for each orbital, putting the lowest energy orbital ($1s$) on the far left and proceeding to orbitals of higher energy to the right. Distribute the ten electrons into the boxes representing the orbitals, allowing a maximum of two electrons per orbital and remembering Hund's rule. You can see from the diagram that neon has no unpaired electrons.

⥮	⥮	⥮ ⥮ ⥮
1*s*	2*s*	2*p*

(d) He Helium has two electrons and has the electron configuration $1s^2$. Draw a box for each orbital, putting the lowest energy orbital ($1s$) on the far left and proceeding to orbitals of higher energy to the right. Distribute the two electrons into the boxes representing the orbitals, allowing a maximum of two electrons per orbital and remembering Hund's rule. You can see from the diagram that helium has no unpaired electrons.

⥮
1*s*

4.49 (a) P The atomic number of P is 15. The noble gas that precedes P in the periodic table is neon, so the inner electron configuration is [Ne]. Obtain the outer electron configuration by tracing the elements between Ne and P and assigning electrons to the appropriate orbitals. Begin with [Ne]. Because P is in row 3, add two 3*s* electrons. Next, add three 3*p* electrons as you trace across the *p* block to P, which is in the third column of the *p* block.

P $[Ne]3s^2 3p^3$

(b) Ge The atomic number of Ge is 32. The noble gas that precedes Ge in the periodic table is argon, so the inner electron configuration is [Ar]. Obtain the outer electron configuration by tracing the elements between Ar and Ge and assigning electrons to the appropriate orbitals. Begin with [Ar]. Because Ge is in row 4, add two 4*s* electrons. Next, add ten 3*d* electrons as you trace across the *d* block. Finally, add two 4*p* electrons as you trace across the *p* block to Ge, which is in the second column of the *p* block.

Ge $[Ar]4s^2 3d^{10} 4p^2$

(c) Zr The atomic number of Zr is 40. The noble gas that precedes Zr in the periodic table is krypton, so the inner electron configuration is [Kr]. Obtain the outer electron configuration by tracing the elements between Kr and Zr and assigning electrons to the appropriate orbitals. Begin with [Kr]. Because Zr is in row 5, add two 5*s* electrons. Next, add two 4*d* electrons as you trace across the *d* block to Zr, which is in the second column.

Zr $[Kr]5s^2 4d^2$

(d) I The atomic number of I is 53. The noble gas that precedes I in the periodic table is krypton, so the inner electron configuration is [Kr]. Obtain the outer electron configuration by tracing the elements between Kr and I and assigning electrons to the appropriate orbitals. Begin with [Kr]. Because I is in row 5, add two 5*s* electrons. Next, add ten 4*d* electrons as you trace across the *d* block. Finally, add five 5*p* electrons as you trace across the *p* block to I, which is in the fifth column of the *p* block.

I $[Kr]5s^2 4d^{10} 5p^5$

4.50 (a) $[Ar]4s^2 3d^{10} 4p^6$ To determine the element corresponding to the electron configuration, begin with Ar; then trace across the 4*s* block, the 3*d* block, and the 4*p* block until you come to the sixth column. The element is Kr.

(b) $[Ar]4s^2 3d^2$ To determine the element corresponding to the electron configuration, begin with Ar; then trace across the 4*s* block and the 3*d* block until you come to the second column. The element is Ti.

(c) $[Kr]5s^2 4d^{10} 5p^2$ To determine the element corresponding to the electron configuration, begin with Kr; then trace across the 5*s* block, the 4*d* block, and the 5*p* block until you come to the second column. The element is Sn.

(d) $[Kr]5s^2$ To determine the element corresponding to the electron configuration, begin with Kr; then trace across the 5*s* block to the second column. The element is Sr.

4.51 (a) Li is in period 2 and the first column in the *s* block, so Li has one 2*s* electron.
(b) Cu is in period 4 and the ninth column in the *d* block $(n - 1)$, so Cu should have nine 3*d* electrons. However, it is one of our exceptions; so it has ten 3*d* electrons.
(c) Br is in period 4 and the fifth column of the *p* block, so Br has five 4*p* electrons.
(d) Zr is in period 5 and the second column of the *d* block $(n - 1)$, so Zr has two 4*d* electrons.

4.52 (a) Mg is in period 3 and the second column of the *s* block, so Mg has two 3*s* electrons.
(b) Cr is in period 4 and the fourth column of the *d* block $(n - 1)$, so Cr should have four 3*d* electrons. However, Cr is one of our exceptions; so it has five 3*d* electrons.
(c) Y is in period 5 and the first column of the *d* block $(n - 1)$, so Y has one 4*d* electron.
(d) Pb is in period 6 and the second column of the *p* block, so Pb has two 6*p* electrons.

4.53 (a) In period 4, an element with five valence electrons could be V or As.
(b) In period 4, an element with four 4*p* electrons would be in the fourth column of the *p* block and is Se.
(c) In period 4, an element with three 3*d* electrons would be in the third column of the *d* block $(n - 1)$ and is V.
(d) In period 4, an element with a complete outer shell would be in the sixth column of the *p* block and is Kr.

4.54 (a) In period 3, an element with three valence electrons would be in the first column of the *p* block and is Al.
(b) In period 3, an element with four 3*p* electrons would be in the fourth column of the *p* block and is S.
(c) In period 3, an element with six 3*p* electrons would be in the sixth column of the *p* block and is Ar.
(d) In period 3, an element with two 3*s* electrons and no 3*p* electrons would be in the second column of the *s* block and is Mg.

Valence Electrons and Simple Chemical Behavior from the Periodic Table

4.55 (a) Ba is in column 2A, so it has two valence electrons.
(b) Cs is in column 1A, so it has one valence electron.
(c) Ni is in column 8 of the *d* block, so it has ten valence electrons (eight from the *d* block and two from the *s* block).
(d) S is in column 6A, so it has six valence electrons.

4.56 (a) Al is in column 3A, so it has three valence electrons. Al is a metal and will tend to lose the three valence electrons to achieve the noble gas configuration of Ne.
(b) Sn is in column 4A, so it has four valence electrons. Sn is a metal and will tend to lose the four valence electrons to achieve the noble gas configuration of Kr or lose only two valence electrons to have the configuration of Cd.
(c) Br is in column 7A, so it has seven valence electrons. Br is a nonmetal and will tend to gain an electron to achieve the noble gas configuration of Kr.
(d) Se is in column 6A, so it has six valence electrons. Se is a nonmetal and will tend to gain two electrons to achieve the noble gas configuration of Kr.

4.57 (a) The outer electron configuration ns^2 would belong to a reactive metal in the alkaline earth family.
(b) The outer electron configuration ns^2np^6 would belong to an unreactive nonmetal in the noble gas family.
(c) The outer electron configuration ns^2np^5 would belong to a reactive nonmetal in the halogen family.
(d) The outer electron configuration ns^2np^2 would belong to an element in the carbon family. If $n = 2$, the element is a nonmetal; if $n = 3$ or 4, the element is a metalloid; and if $n = 5$ or 6, the element is a metal.

4.58 (a) The outer electron configuration ns^2 would belong to a metal in the alkaline earth family for period $n = 2$ and greater. He is a noble gas with a $1s^2$ electron configuration.
(b) The outer electron configuration ns^2np^6 would belong to a nonmetal in the noble gas family.
(c) The outer electron configuration ns^2np^5 would belong to a nonmetal in the halogen family.
(d) The outer electron configuration ns^2np^2 would belong to an element in the carbon family. If $n = 2$, the element is a nonmetal; if $n = 3$ or 4, the element is a metalloid; and if $n = 5$ or 6, the element is a metal.

4.59
(a) sodium — Na is in group 1A, has 1 valence electron, and is an alkali metal.
(b) iodine — I is in group 7A, has 7 valence electrons, and is a halogen.
(c) calcium — Ca is in group 2A, has 2 valence electrons, and is an alkaline earth metal.
(d) barium — Ba is in group 2A, has 2 valence electrons, and is an alkaline earth metal.
(e) krypton — Kr is in group 8A, has 8 valence electrons, and is a noble gas.

4.60
(a) F — Fluorine is in group 7A, has 7 valence electrons, and is a halogen.
(b) Sr — Strontium is in group 2A, has 2 valence electrons, and is an alkaline earth metal.
(c) K — Potassium is in group 1A, has 1 valence electron, and is an alkali metal.
(d) Ne — Neon is in group 8A, has 8 valence electrons, and is a noble gas.
(e) At — Astatine is in group 7A, has 7 valence electrons, and is a halogen.

4.61
(a) N and Ni would not be most similar. Nitrogen is a nonmetal; nickel is a metal.
(b) Mo and Sn would not be most similar. Although both are metals, molybdenum is a transition metal and tin is a main-group metal.
(c) Na and Mg would not be most similar. Although both are main-group metals, sodium is in group 1A and magnesium is in group 2A.
(d) Cl and F would be most similar. Chlorine and fluorine are both in group 7A. Elements in the same group have similar chemical properties.
(e) Si and P would not be most similar. Silicon is a metalloid, and phosphorus is a nonmetal.

4.62
(a) Nitrogen and oxygen would not be most similar. Although both are nonmetals, N is in group 5A and O is in group 6A.
(b) Titanium and gallium would not be most similar. Although both are metals, Ti is a transition metal and Ga is a main-group metal.
(c) Lithium and sodium would be most similar. Li and Na are both in group 1A. Elements in the same group have similar chemical properties.
(d) Germanium and arsenic would not be most similar. Ge and As are both metalloids and would share some properties, but Ge is in group 4A, and As is in group 5A.
(e) Argon and bromine would not be most similar. Although both are nonmetals, Ar is in group 8A and Br is in group 7A.

4.63 Main-group metal atoms will lose electrons to form a cation with the same number of electrons as the nearest previous noble gas. Nonmetal atoms will gain electrons to form an anion with the same number of electrons as the nearest noble gas.
(a) O^{2-} — O is a nonmetal and has eight electrons. It will gain electrons to form an anion. The nearest noble gas is neon with ten electrons, so O will gain two electrons.
(b) K^{+} — K is a main-group metal and has 19 electrons. It will lose electrons to form a cation. The nearest noble gas is argon with 18 electrons, so K will lose 1 electron.
(c) Al^{3+} — Al is a main-group metal and has 13 electrons. It will lose electrons to form a cation. The nearest noble gas is neon with 10 electrons, so Al will lose 3 electrons.
(d) Rb^{+} — Rb is a main-group metal and has 37 electrons. It will lose electrons to form a cation. The nearest noble gas is krypton with 36 electrons, so Rb will lose 1 electron.

4.64 Main-group metal atoms will lose electrons to form a cation with the same number of electrons as the nearest previous noble gas. Nonmetal atoms will gain electrons to form an anion with the same number of electrons as the nearest noble gas.
(a) Mg^{2+} — Mg is a main-group metal and has 12 electrons. It will lose electrons to form a cation. The nearest noble gas is neon with 10 electrons, so Mg will lose 2 electrons.
(b) N^{3-} — N is a nonmetal and has seven electrons. It will gain electrons to form an anion. The nearest noble gas is neon with ten electrons, so N will gain three electrons.
(c) F^{-} — F is a nonmetal and has nine electrons. It will gain electrons to form an anion. The nearest noble gas is neon with ten electrons, so F will gain one electron.
(d) Na^{+} — Na is a main-group metal and has eleven electrons. It will lose electrons to form a cation. The nearest noble gas is neon with ten electrons, so Na will lose one electron.

Coulomb's Law and Effective Nuclear Charge

4.65 Coulomb's law states that the potential energy (E) of two charge particles depends on their charges (q_1 and q_2) and on their separation (r). $E = \frac{1}{4\pi\varepsilon_o}\frac{q_1 q_2}{r}$. The potential energy is positive for charges of the same sign and negative for charges of opposite signs. The magnitude of the potential energy depends inversely on the separation between the charged particles.

(a) **Given:** $q_1 = 1-$; $q_2 = 2+$; $r = 150.$ pm **Find:** E
Conceptual Plan: Magnitude of Potential Energy depends on the charge and the separation.

$$E = \frac{1}{4\pi\varepsilon_o}\frac{q_1 q_2}{r}$$

Solution: $E = \frac{1}{4\pi\varepsilon_o}\frac{(1-)(2+)}{150} = \frac{-0.013}{4\pi\varepsilon_o}$

(b) **Given:** $q_1 = 1-$; $q_2 = 1+$; $r = 150.$ pm **Find:** E
Conceptual Plan: Magnitude of Potential Energy depends on the charge and the separation.

$$E = \frac{1}{4\pi\varepsilon_o}\frac{q_1 q_2}{r}$$

Solution: $E = \frac{1}{4\pi\varepsilon_o}\frac{(1-)(1+)}{150} = \frac{-0.0067}{4\pi\varepsilon_o}$

(c) **Given:** $q_1 = 1-$; $q_2 = 3+$; $r = 150.$ pm **Find:** E
Conceptual Plan: Magnitude of Potential Energy depends on the charge and the separation.

$$E = \frac{1}{4\pi\varepsilon_o}\frac{q_1 q_2}{r}$$

Solution: $E = \frac{1}{4\pi\varepsilon_o}\frac{(1-)(3+)}{150.} = \frac{-0.030}{4\pi\varepsilon_o}$

(c) is most negative and will have the lowest potential energy.

4.66 According to Coulomb's law, for opposite charges, the potential energy increases (becomes less negative) with increasing distance; so (d) is greater than (a) or (c).

Because the charge of (b) is twice the charge of (a) at the same distance and is the opposite charge, it has the lowest (most negative) potential energy. Because both of the charges in (c) are the same (1+), it is the only one with a positive potential energy. Ranking from lowest to highest: (b) $<$ (a) $<$ (d) $<$ (c).

4.67 The valence electrons in nitrogen would experience a greater effective nuclear charge. Be has four protons, and N has seven protons. Both atoms have two core electrons that predominately contribute to the shielding, while the valence electrons will contribute a slight shielding effect. So Be has an effective nuclear charge of slightly more than 2+, and N has an effective nuclear charge of slightly more than 5+.

4.68 S(16) = [Ne]$3s^2 3p^4$ Mg(12) = [Ne]$3s^2$ Al(13) = [Ne]$3s^2 3p^1$ Si(14) = [Ne]$3s^2 3p^2$

All four atoms have the same number of core electrons that contribute to shielding. So the effective nuclear charge will decrease with a decreasing number of protons. S $>$ Si $>$ Al $>$ Mg

4.69
(a) K(19) = [Ar]$4s^1$ Z_{eff} = Z − core electrons = 19 − 18 = 1+
(b) Ca(20) = [Ar]$4s^2$ Z_{eff} = Z − core electrons = 20 − 18 = 2+
(c) O(8) = [He]$2s^2 2p^4$ Z_{eff} = Z − core electrons = 8 − 2 = 6+
(d) C(6) = [He]$2s^2 2p^2$ Z_{eff} = Z − core electrons = 6 − 2 = 4+

4.70 B has an electron configuration of $1s^2 2s^2 2p^1$. To estimate the effective nuclear charge experienced by the outer electrons, we need to distinguish between two types of shielding: (1) the shielding of the outermost electrons by the core electrons and (2) the shielding of the outermost electrons by each other. The three outermost electrons in boron experience the 5+ charge of the nucleus through the shield of the two $1s$ core electrons. We can estimate that the shielding experienced by any one of the outermost electrons due to the core electrons is nearly 2. For the 2*s* electrons, the shielding due to the other 2*s* electron is nearly zero. For the 2*p* electron, however, we would expect the 2*s* electrons to contribute some shielding because although the 2*p* orbital penetrates the 2*s* orbital to some degree, most of the 2*p* orbital lies outside the 2*s* orbital. So the effective nuclear charge would be slightly greater than 3+, and the effective nuclear charge felt by the 2*s* electrons would be greater than the effective nuclear charge felt by the 2*p* electrons.

Atomic Radius

4.71 (a) Al or In — In atoms are larger than Al atoms because as you trace the path between Al and In on the periodic table, you move down a column. Atomic size increases as you move down a column because the outermost electrons occupy orbitals with a higher principal quantum number that are larger, resulting in a larger atom.

(b) Si or N — Si atoms are larger than N atoms because as you trace the path between N and Si on the periodic table, you move down a column (atomic size increases) and then to the left across a period (atomic size increases). These effects add together for an overall increase.

(c) P or Pb — Pb atoms are larger than P atoms because as you trace the path between P and Pb on the periodic table, you move down a column (atomic size increases) and then to the left across a period (atomic size increases). These effects add together for an overall increase.

(d) C or F — C atoms are larger than F atoms because as you trace the path between C and F on the periodic table, you move to the right within the same period. As you move to the right across a period, the effective nuclear charge experienced by the outermost electrons increases, which results in a smaller size.

4.72 (a) Sn or Si — Sn atoms are larger than Si atoms because as you trace the path between Si and Sn on the periodic table, you move down a column. Atomic size increases as you move down a column because the outermost electrons occupy orbitals with a higher principal quantum number that is larger, resulting in a larger atom.

(b) Br or Ga — Ga atoms are larger than Br atoms because as you trace the path between Ga and Br on the periodic table, you move to the right within the same period. As you move to the right across a period, the effective nuclear charge experienced by the outermost electrons increases, which results in a smaller size.

(c) Sn or Bi — Based on periodic trends alone, you cannot tell which atom is larger because as you trace the path between Sn and Bi, you move to the right across a period (atomic size decreases) and then down a column (atomic size increases). These effects tend to oppose each other, and it is not easy to tell which will predominate.

(d) Se or Sn — Sn atoms are larger than Se atoms because as you trace the path between Se and Sn on the periodic table, you move down a column (atomic size increases) and then to the left across a period (atomic size increases). These effects add together for an overall increase.

4.73 Ca, Rb, S, Si, Ge, F — F is above and to the right of the other elements, so you start with F as the smallest atom. As you trace a path from F to S, you move to the left (size increases) and down (size increases). Next, you move left from S to Si (size increases), then down to Ge (size increases), then to the left to Ca (size increases), and then to the left and down to Rb (size increases). So in order of increasing atomic radii, F $<$ S $<$ Si $<$ Ge $<$ Ca $<$ Rb.

4.74 Cs, Sb, S, Pb, Se — Cs is below and to the left of the other elements, so you start with Cs as the largest atom. As you trace a path from Cs to Pb, you move to the right in the same period (size decreases). Next, going from Pb to Sb, you move up a column and then to the right (size decreases); from Sb to Se, you move up the column and then to the right (size decreases); and finally, from Se to S, you move up the column (size decreases). So in order of decreasing radii, Cs $>$ Pb $>$ Sb $>$ Se $>$ S.

Ionic Electron Configurations, Ionic Radii, Magnetic Properties, and Ionization Energy

4.75 (a) O^{2-} — Begin by writing the electron configuration of the neutral atom.

O $1s^2 2s^2 2p^4$

Because this ion has a 2− charge, add two electrons to write the electron configuration of the ion.
O^{2-} $1s^22s^22p^6$ This is isoelectronic with Ar.

(b) Br^- Begin by writing the electron configuration of the neutral atom.
Br $[Ar]4s^23d^{10}4p^5$
Because this ion has a 1− charge, add one electron to write the electron configuration of the ion.
Br^- $[Ar]4s^23d^{10}4p^6$ This is isoelectronic with Kr.

(c) Sr^{2+} Begin by writing the electron configuration of the neutral atom.
Sr $[Kr]5s^2$
Because this ion has a 2+ charge, remove two electrons to write the electron configuration of the ion.
Sr^{2+} [Kr]

(d) Co^{3+} Begin by writing the electron configuration of the neutral atom.
Co $[Ar]4s^23d^7$
Because this ion has a 3+ charge, remove three electrons to write the electron configuration of the ion. Because it is a transition metal, remove the electrons from the 4*s* orbital before removing electrons from the 3*d* orbitals.
Co^{3+} $[Ar]4s^03d^6$ or $[Ar]3d^6$

(e) Cu^{2+} Begin by writing the electron configuration of the neutral atom. Remember, Cu is one of our exceptions.
Cu $[Ar]4s^13d^{10}$
Because this ion has a 2+ charge, remove two electrons to write the electron configuration of the ion. Because it is a transition metal, remove the electrons from the 4*s* orbital before removing electrons from the 3*d* orbitals.
Cu^{2+} $[Ar]4s^03d^9$ or $[Ar]3d^9$

4.76 (a) Cl^- Begin by writing the electron configuration of the neutral atom.
Cl $[Ne]3s^23p^5$
Because this ion has a 1− charge, add one electron to write the electron configuration of the ion.
Cl^- $[Ne]3s^23p^6$ This is isoelectronic with Ar.

(b) P^{3-} Begin by writing the electron configuration of the neutral atom.
P $[Ne]3s^23p^3$
Because this ion has a 3− charge, add three electrons to write the electron configuration of the ion.
P^{3-} $[Ne]3s^23p^6$ This is isoelectronic with Ar.

(c) K^+ Begin by writing the electron configuration of the neutral atom.
K $[Ar]4s^1$
Because this ion has a 1+ charge, remove one electron to write the electron configuration of the ion.
K^+ [Ar]

(d) Mo^{3+} Begin by writing the electron configuration of the neutral atom. Remember, Mo is one of our exceptions.
Mo $[Kr]5s^14d^5$
Because this ion has a 3+ charge, remove three electrons to write the electron configuration of the ion. Because it is a transition metal, remove the electrons from the 5*s* orbital before removing electrons from the 4*d* orbitals.
Mo^{3+} $[Kr]5s^04d^3$ or $[Kr]4d^3$

(e) V^{3+} Begin by writing the electron configuration of the neutral atom.
V $[Ar]4s^23d^3$
Because this ion has a 3+ charge, remove three electrons to write the electron configuration of the ion. Because it is a transition metal, remove the electrons from the 4*s* orbital before removing electrons from the 3*d* orbitals.
V^{3+} $[Ar]4s^03d^2$ or $[Ar]3d^2$

4.77 (a) V^{5+} Begin by writing the electron configuration of the neutral atom.
V $[Ar]4s^23d^3$
Because this ion has a 5+ charge, remove five electrons to write the electron configuration of the ion. Because it is a transition metal, remove the electrons from the 4*s* orbital before removing electrons from the 3*d* orbitals.
V^{5+} $[Ar]4s^03d^0 = [Ne]3s^23p^6$

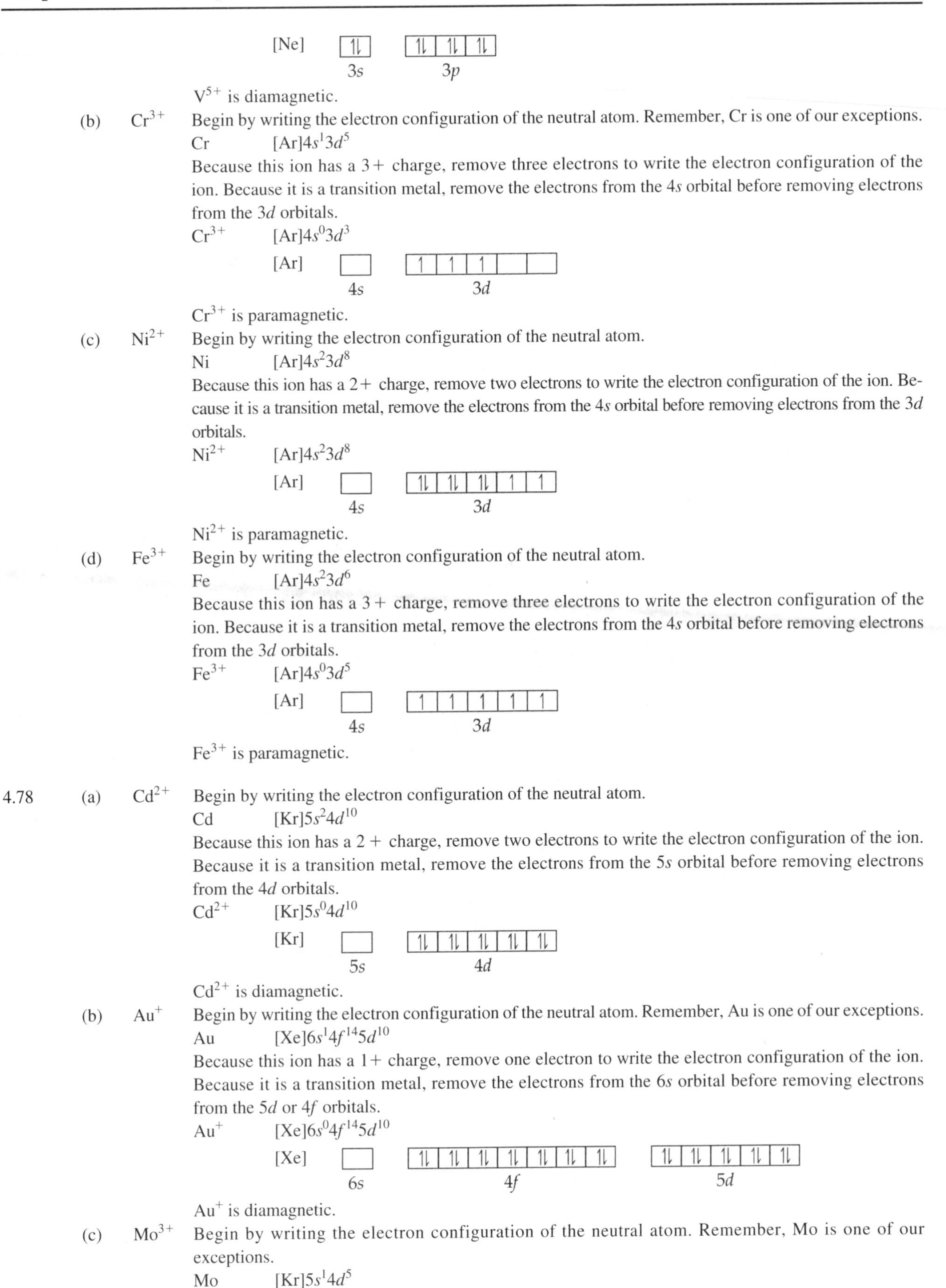

[Ne] [↿⇂] [↿⇂ | ↿⇂ | ↿⇂]
3s 3p

V^{5+} is diamagnetic.

(b) Cr^{3+} Begin by writing the electron configuration of the neutral atom. Remember, Cr is one of our exceptions.

Cr $[Ar]4s^1 3d^5$

Because this ion has a 3+ charge, remove three electrons to write the electron configuration of the ion. Because it is a transition metal, remove the electrons from the 4*s* orbital before removing electrons from the 3*d* orbitals.

Cr^{3+} $[Ar]4s^0 3d^3$

[Ar] [ ] [↿ | ↿ | ↿ | |]
4s 3d

Cr^{3+} is paramagnetic.

(c) Ni^{2+} Begin by writing the electron configuration of the neutral atom.

Ni $[Ar]4s^2 3d^8$

Because this ion has a 2+ charge, remove two electrons to write the electron configuration of the ion. Because it is a transition metal, remove the electrons from the 4*s* orbital before removing electrons from the 3*d* orbitals.

Ni^{2+} $[Ar]4s^2 3d^8$

[Ar] [ ] [↿⇂ | ↿⇂ | ↿⇂ | ↿ | ↿]
4s 3d

Ni^{2+} is paramagnetic.

(d) Fe^{3+} Begin by writing the electron configuration of the neutral atom.

Fe $[Ar]4s^2 3d^6$

Because this ion has a 3+ charge, remove three electrons to write the electron configuration of the ion. Because it is a transition metal, remove the electrons from the 4*s* orbital before removing electrons from the 3*d* orbitals.

Fe^{3+} $[Ar]4s^0 3d^5$

[Ar] [ ] [↿ | ↿ | ↿ | ↿ | ↿]
4s 3d

Fe^{3+} is paramagnetic.

4.78 (a) Cd^{2+} Begin by writing the electron configuration of the neutral atom.

Cd $[Kr]5s^2 4d^{10}$

Because this ion has a 2+ charge, remove two electrons to write the electron configuration of the ion. Because it is a transition metal, remove the electrons from the 5*s* orbital before removing electrons from the 4*d* orbitals.

Cd^{2+} $[Kr]5s^0 4d^{10}$

[Kr] [ ] [↿⇂ | ↿⇂ | ↿⇂ | ↿⇂ | ↿⇂]
5s 4d

Cd^{2+} is diamagnetic.

(b) Au^{+} Begin by writing the electron configuration of the neutral atom. Remember, Au is one of our exceptions.

Au $[Xe]6s^1 4f^{14} 5d^{10}$

Because this ion has a 1+ charge, remove one electron to write the electron configuration of the ion. Because it is a transition metal, remove the electrons from the 6*s* orbital before removing electrons from the 5*d* or 4*f* orbitals.

Au^{+} $[Xe]6s^0 4f^{14} 5d^{10}$

[Xe] [ ] [↿⇂ | ↿⇂ | ↿⇂ | ↿⇂ | ↿⇂ | ↿⇂ | ↿⇂] [↿⇂ | ↿⇂ | ↿⇂ | ↿⇂ | ↿⇂]
6s 4f 5d

Au^{+} is diamagnetic.

(c) Mo^{3+} Begin by writing the electron configuration of the neutral atom. Remember, Mo is one of our exceptions.

Mo $[Kr]5s^1 4d^5$

Because this ion has a 3+ charge, remove three electrons to write the electron configuration of the ion. Because it is a transition metal, remove the electrons from the 5*s* orbital before removing electrons from the 4*d* orbitals.

Mo^{3+} $[Kr]5s^04d^3$

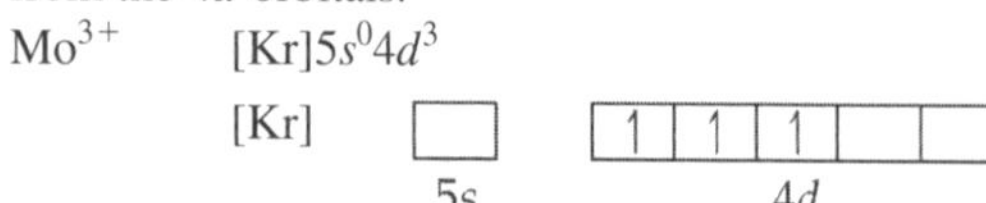

Mo^{3+} is paramagnetic.

(d) Zr^{2+} Begin by writing the electron configuration of the neutral atom.

Zr $[Kr]5s^24d^2$

Because this ion has a 2+ charge, remove two electrons to write the electron configuration of the ion. Because it is a transition metal, remove the electrons from the 5*s* orbital before removing electrons from the 4*d* orbitals.

Zr^{2+} $[Kr]5s^04d^2$

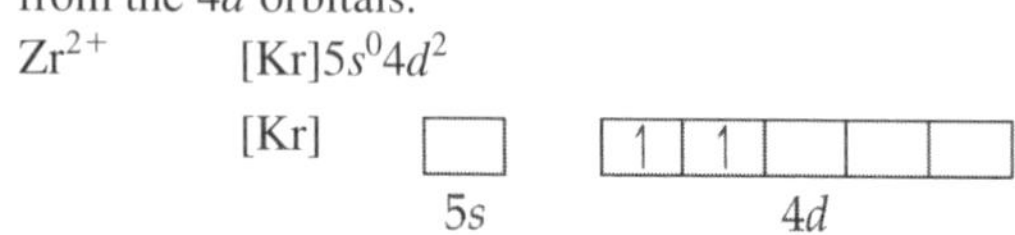

Zr^{2+} is paramagnetic.

4.79 (a) Li or Li^+ A Li atom is larger than Li^+ because cations are smaller than the atoms from which they are formed.

(b) I^- or Cs^+ An I^- ion is larger than a Cs^+ ion because, although they are isoelectronic, I^- has two fewer protons than Cs^+, resulting in a lesser pull on the electrons and, therefore, a larger radius.

(c) Cr or Cr^{3+} A Cr atom is larger than Cr^{3+} because cations are smaller than the atoms from which they are formed.

(d) O or O^{2-} An O^{2-} ion is larger than an O atom because anions are larger than the atoms from which they are formed.

4.80 (a) Sr or Sr^{2+} A Sr atom is larger than Sr^{2+} because cations are smaller than the atoms from which they are formed.

(b) N or N^{3-} A N^{3-} ion is larger than an N atom because anions are larger than the atoms from which they are formed.

(c) Ni or Ni^{2+} A Ni atom is larger than Ni^{2+} because cations are smaller than the atoms from which they are formed.

(d) S^{2-} or Ca^{2+} A S^{2-} ion is larger than a Ca^{2+} ion because, although they are isoelectronic, S^{2-} has four fewer protons than Ca^{2+}, resulting in a lesser pull on the electrons and, therefore, a larger radius.

4.81 Because all of the species are isoelectronic, the radius will depend on the number of protons in each species. The fewer the protons, the larger the radius.

F: $Z = 9$; Ne: $Z = 10$; O: $Z = 8$; Mg: $Z = 12$; Na: $Z = 11$

So the radii of the ions is: $O^{2-} > F^- > Ne > Na^+ > Mg^{2+}$

4.82 Because all of the species are isoelectronic, the radius will depend on the number of protons in each species. The fewer the protons, the larger the radius.

Se: $Z = 34$; Kr: $Z = 36$; Sr: $Z = 38$; Rb: $Z = 37$; Br: $Z = 35$

So the radii of the ions is: $Sr^{2+} < Rb^+ < Kr < Br^- < Se^{2-}$

4.83 (a) Br or Bi Br has a higher ionization energy than Bi because as you trace the path between Br and Bi on the periodic table, you move down a column (ionization energy decreases) and then to the left across a period (ionization energy decreases). These effects sum together for an overall decrease.

(b) Na or Rb Na has a higher ionization energy than Rb because as you trace a path between Na and Rb on the periodic table, you move down a column. Ionization energy decreases as you go down a column because of the increasing size of orbitals with increasing *n*.

(c) As or At Based on periodic trends alone, it is impossible to tell which one has a higher ionization energy because as you trace the path between As and At, you move to the right across a period (ionization energy increases) and then down a column (ionization energy decreases). These effects tend to oppose each other, and it is not obvious which one will dominate.

(d) P or Sn — P has a higher ionization energy than Sn because as you trace the path between P and Sn on the periodic table, you move down a column (ionization energy decreases) and then to the left across a period (ionization energy decreases). These effects sum together for an overall decrease.

4.84 (a) P or I — Based on periodic trends alone, it is impossible to tell which has a higher ionization energy because as you trace the path between P and I, you move to the right across a period (ionization energy increases) and then down a column (ionization energy decreases). These effects tend to oppose each other, and it is not obvious which one will dominate.

(b) Si or Cl — Cl has a higher ionization energy than Si because as you trace the path between Si and Cl on the periodic table, you move right across a period. Ionization energy increases as you go to the right because of increasing nuclear charge.

(c) P or Sb — P has a higher ionization energy than Sb because as you trace a path between P and Sb on the periodic table, you move down a column. Ionization energy decreases as you go down a column because of the increasing size of orbitals with increasing n.

(d) Ga or Ge — Ge has a higher ionization energy than Ga because as you trace a path between Ga and Ge on the periodic table, you move to the right within the same period. Ionization energy increases as you move to the right because of increasing effective nuclear charge.

4.85 Because ionization energy increases as you move to the right across a period and increases as you move up a column, the element with the smallest first ionization energy would be the element farthest to the left and lowest on the periodic table. So In has the smallest ionization energy. As you trace a path to the right and up on the periodic table, the next element you reach is Si; continuing up and to the right, you reach N; and continuing to the right, you reach F. So in the order of increasing first ionization energy, the elements are In $<$ Si $<$ N $<$ F.

4.86 Because ionization energy increases as you move to the right across a period and increases as you move up a column, the element with the largest first ionization energy would be the element farthest to the right and highest on the periodic table. So Cl has the largest ionization energy. As you trace a path to the left on the periodic table, you reach S; as you move down a column and to the left, you reach Sn; and as you move down the column, you reach Pb. So in the order of decreasing first ionization energy, the elements are Cl $>$ S $>$ Sn $>$ Pb.

4.87 The jump in ionization energy occurs when you change from removing a valence electron to removing a core electron. To determine where this jump occurs, you need to look at the electron configuration of the atom.

(a) Be $1s^2 2s^2$ — The first and second ionization energies involve removing $2s$ electrons, while the third ionization energy removes a core electron; so the jump will occur between the second and third ionization energies.

(b) N $1s^2 2s^2 2p^3$ — The first five ionization energies involve removing the $2p$ and $2s$ electrons, while the sixth ionization energy removes a core electron; so the jump will occur between the fifth and sixth ionization energies.

(c) O $1s^2 2s^2 2p^4$ — The first six ionization energies involve removing the $2p$ and $2s$ electrons, while the seventh ionization energy removes a core electron; so the jump will occur between the sixth and seventh ionization energies.

(d) Li $1s^2 2s^1$ — The first ionization energy involves removing a $2s$ electron, while the second ionization energy removes a core electron; so the jump will occur between the first and second ionization energies.

4.88 The jump occurs between IE_3 and IE_4. So removing the first three electrons involves removing valence electrons, and the fourth electron is a core electron; so the valence electron configuration would be ns^2np^1. This puts the element in column 3A and would be Al.

Electron Affinities and Metallic Character

4.89 (a) Na or Rb — Na has a more negative electron affinity than Rb. In column 1A, electron affinity becomes less negative as you go down the column.

(b) B or S — S has a more negative electron affinity than B. As you trace from B to S in the periodic table, you move to the right, which shows the value of the electron affinity becoming more negative. Also, as you move from period 2 to period 3, the value of the electron affinity becomes more negative. Both of these trends sum together for the value of the electron affinity to become more negative.

(c) C or N — C has the more negative electron affinity. As you trace from C to N across the periodic table, you normally expect N to have the more negative electron affinity. However, N has a half-filled *p* sublevel, which lends it extra stability; therefore, it is harder to add an electron.

(d) Li or F — F has the more negative electron affinity. As you trace from Li to F on the periodic table, you move to the right in the period. As you move to the right across a period, the value of the electron affinity generally becomes more negative.

4.90 (a) Mg or S — S has the more negative electron affinity. As you trace from Mg to S on the periodic table, you move to the right in the period. As you move to the right across a period, the value of the electron affinity generally becomes more negative.

(b) K or Cs — K has the more negative electron affinity. In column 1A, as you go down the column, the electron affinity becomes less negative.

(c) Si or P — Si has the more negative electron affinity. As you trace from Si to P across the periodic table, you normally expect P to have the more negative electron affinity. However, P has a half-filled *p* sublevel, which lends extra stability; therefore, it is harder to add an electron.

(d) Ga or Br — Br has the more negative electron affinity. As you trace from Ga to Br on the periodic table, you move to the right in the period. As you move to the right across a period, the value of the electron affinity generally becomes more negative.

4.91 (a) Sr or Sb — Sr is more metallic than Sb because as we trace the path between Sr and Sb on the periodic table, we move to the right within the same period. Metallic character decreases as we move to the right.

(b) As or Bi — Bi is more metallic because as we trace a path between As and Bi on the periodic table, we move down a column in the same family (metallic character increases).

(c) Cl or O — Based on periodic trends alone, we cannot tell which is more metallic because as we trace the path between O and Cl, we move to the right across a period (metallic character decreases) and then down a column (metallic character increases). These effects tend to oppose each other, and it is not easy to tell which will predominate.

(d) S or As — As is more metallic than S because as we trace the path between S and As on the periodic table, we move down a column (metallic character increases) and then to the left across a period (metallic character increases). These effects add together for an overall increase.

4.92 (a) Sb or Pb — Pb is more metallic than Sb because as we trace the path between Sb and Pb on the periodic table, we move down a column (metallic character increases) and then to the left across a period (metallic character increases). These effects add together for an overall increase.

(b) K or Ge — K is more metallic than Ge because as we trace the path between K and Ge on the periodic table, we move to the right within the same period. Metallic character decreases as we move to the right.

(c) Ge or Sb — Based on periodic trends alone, we cannot tell which is more metallic because as we trace the path between Ge and Sb, we move to the right across a period (metallic character decreases) and then down a column (metallic character increases). These effects tend to oppose each other, and it is not easy to tell which will predominate.

(d) As or Sn — Sn is more metallic than As because as we trace the path between As and Sn on the periodic table, we move down a column (metallic character increases) and then to the left across a period (metallic character increases). These effects add together for an overall increase.

4.93 The order of increasing metallic character is S $<$ Se $<$ Sb $<$ In $<$ Ba $<$ Fr.

Metallic character decreases as you move left to right across a period and decreases as you move up a column; therefore, the element with the least metallic character will be to the top right of the periodic table. So of these elements, S has the least metallic character. As you move down the column, the next element is Se. As you continue down and to the left, you reach Sb; continuing to the left, you reach In; moving down the column and to the left, you come to Ba; and down the column and to the left is Fr.

4.94 The order of decreasing metallic character is Sr $>$ Ga $>$ Al $>$ Si $>$ P $>$ N.

Metallic character decreases as you move left to right across a period and decreases as you move up a column; therefore, the element with the greatest metallic character will be at the bottom left of the periodic table. So of these elements, Sr has the most metallic character. As you trace up the column and then to the right across the period, the next element is Ga; trace up the column to Al, then to the right to Si and P; finally, trace up the column to N.

Cumulative Problems

4.95 Br: $1s^2 2s^2 2p^6 3s^2 3p^6 4s^2 3d^{10} 4p^5$
Kr: $1s^2 2s^2 2p^6 3s^2 3p^6 4s^2 3d^{10} 4p^6$
Krypton has a completely filled *p* sublevel, giving it chemical stability. Bromine needs one electron to achieve a completely filled *p* sublevel and thus has a highly negative electron affinity. Therefore, it easily takes on an electron and is reduced to the bromide ion, giving it the added stability of the filled *p* sublevel.

4.96 K: $1s^2 2s^2 2p^6 3s^2 3p^6 4s^1$
Ar: $1s^2 2s^2 2p^6 3s^2 3p^6$
Argon has a completely filled *p* sublevel, giving it chemical stability. Potassium has one electron in the 4*s* sublevel and can easily lose this electron, so it has a low first ionization energy. Therefore, it loses the 4*s* electron to achieve an argon electron configuration, giving it the added stability of the filled *p* sublevel.

4.97 Write the electron configuration of vanadium.
V: $[Ar]4s^2 3d^3$
Because this ion has a 3+ charge, remove three electrons to write the electron configuration of the ion. Because it is a transition metal, remove the electrons from the 4*s* orbital before removing electrons from the 3*d* orbitals.
V^{3+}: $[Ar]4s^0 3d^2$ or $[Ar]3d^2$
Both vanadium and the V^{3+} ion have unpaired electrons and are paramagnetic.

4.98 Begin by writing the electron configuration of the neutral atom. Remember, Cu is one of our exceptions.
Cu: $[Ar]4s^1 3d^{10}$
Because this ion has a 1+ charge, remove one electron to write the electron configuration of the ion. Because it is a transition metal, remove the electrons from the 4*s* orbital before removing electrons from the 3*d* orbitals.
Cu^+: $[Ar]4s^0 3d^{10}$ or $[Ar]3d^{10}$
Cu contains one unpaired electron in the 4*s* orbital and is paramagnetic; Cu^+ has all paired electrons in the 3*d* orbitals and is diamagnetic.

4.99 Because K^+ has a 1+ charge, we would need a cation with a similar size and a 1+ charge. Looking at the ions in the same family, we see that Na^+ would be too small and Rb^+ would be too large. If we consider Ar^+ and Ca^+, we would have ions of similar size and charge. Between these two, Ca^+ would be easier to achieve because the first ionization energy of Ca is similar to that of K, while the first ionization energy of Ar is much larger. However, the second ionization energy of Ca is relatively low, making it easy to lose the second electron.

4.100 Because Na^+ has a 1+ charge, we would need a cation with a similar size and a 1+ charge. Looking at the ions in the same family, Li^+ would be too small and K^+ would be too large. If we consider Ne^+ and Mg^+, we would have ions of similar size and charge. Between these two, Mg^+ would be easier to achieve because the first ionization energy of Mg is similar to that of Na, while the first ionization energy of Ne is much larger. However, the second ionization energy of Mg is relatively low, making it easy to lose the second electron.

4.101 C has an outer shell electron configuration of ns^2np^2; we would therefore expect Si and Ge, which are in the same family, to be most like carbon. Ionization energies for both Si and Ge are similar and tend to be slightly lower than that of C, but all are intermediate in the range of first ionization energies. The electron affinities of Si and Ge are close to that of C.

4.102 (a) Si and Ga — Ga would be larger than Si because as you trace from Si to Ga on the periodic table, you move down a column (radius increases) and then to the left across the period (radius increases). The sum of these two trends would give you a larger radius for Ga.

(b) Si and Ge — Ge would be larger than Si because as you trace from Si to Ge on the periodic table, you move down a column and the radius increases.

(c) Si and As — As would be most similar to Si in atomic radius because as you trace from Si to As on the periodic table, you move down the column (radius increase) and then to the right across the period (radius decreases). The sum of these two trends would make As smaller than Ga and Ge and thus closer to the radius of Si.

4.103 (a) N: [He]$2s^22p^3$ Mg: [Ne]$3s^2$ O: [He]$2s^22p^4$
F: [He]$2s^22p^5$ Al: [Ne]$3s^23p^1$

(b) Mg > Al > N > O > F

(c) Al < Mg < O < N < F (from the table)

(d) Mg and Al would have the largest radius because they are in period $n = 3$; Al is smaller than Mg because radius decreases as you move to the right across the period. F is smaller than O and O is smaller than N because as you move to the right across the period, radius decreases.

The first ionization energy of Al is smaller than the first ionization energy of Mg because Al loses the electron from the 3*p* orbital, which is shielded by the electrons in the 3*s* orbital. Mg loses the electron from the filled 3*s* orbital, which has added stability because it is a filled orbital. The first ionization energy of O is lower than the first ionization energy of N because N has a half-filled 2*p* orbital, which adds extra stability, thus making it harder to remove the electron. The fourth electron in the O 2*p* orbitals experiences added electron-electron repulsion because it must pair with another electron in the same 2*p* orbital, thus making it easier to remove.

4.104 (a) P: [Ne]$3s^23p^3$ Ca: [Ar]$4s^2$ Si: [Ne]$3s^23p^2$
S: [Ne]$3s^23p^4$ Ga: [Ar]$4s^23d^{10}4p^1$

(b) Ca > Ga > Si > P > S

(c) Ga < Ca < Si < S < P (from the table)

(d) Ca and Ga would have the largest radius because they are in period $n = 4$; Ga is smaller than Ca because radius increases as you move to the left across the period. S is smaller than P and P is smaller than Si because as you move to the right across the period, radius decreases.

The first ionization energy of Ga is smaller than the first ionization energy of Ca because Ga loses the electron from the 4*p* orbital, which is shielded by the electrons in the 4*s* orbital. Ca loses the electron from the filled 4*s* orbital, which has added stability because it is a filled orbital. The first ionization energy of S is lower than the first ionization energy of P because P has a half-filled 3*p* orbital, which adds extra stability, thus making it harder to remove the electron. The fourth electron in the S 3*p* orbitals experiences added electron-electron repulsion because it must pair with another electron in the same 3*p* orbital, thus making it easier to remove.

4.105 As you move to the right across a row in the periodic table for the main-group elements, the effective nuclear charge (Z_{eff}) experienced by the electrons in the outermost principal energy level increases, resulting in a stronger attraction between the outermost electrons and the nucleus and therefore a smaller atomic radii.

Across the row of transition elements, the number of electrons in the outermost principal energy level (highest *n* value) is nearly constant. As another proton is added to the nucleus with each successive element, another electron is added, but that electron goes into an $n_{highest} - 1$ orbital (a core level). The number of outermost electrons stays constant, and they experience a roughly constant effective nuclear charge, keeping the radius approximately constant after the first couple of elements in the series.

4.106 Across the row of transition elements, the number of electrons in the outermost principal energy level (highest *n* value) is nearly constant. As another proton is added to the nucleus with each successive element, another electron is added, but the electron goes into an $n_{highest} - 1$ orbital. So even though the atomic number of Cu is higher than that of V, the outermost electron experiences roughly the same effective nuclear charge; thus, the radii of the two elements are nearly the same. Because the radii of the two elements are nearly the same, the volume occupied by the element will be nearly the same. Because the mass increases as the atomic number increases, the mass of Cu is greater than the mass of V; density is mass/volume, so the density of Cu should be greater than the density of V. We find that the densities are Cu $= 8.96\ g/cm^3$ and V $= 5.49\ g/cm^3$, and our prediction was correct.

4.107 All of the noble gases have a filled outer quantum level, very high first ionization energies, and positive values for the electron affinity; thus, the noble gases are particularly unreactive. The lighter noble gases will not form any compounds because the ionization energies of both He and Ne are over 2000 kJ/mol. Because ionization energy decreases as you move down a column, you find that the heavier noble gases (Ar, Kr, and Xe) do form some compounds. They have ionization energies that are close to the ionization energy of H and thus can be forced to lose an electron.

4.108 All of the halogens will add an electron to achieve the stability of the noble gas configuration; thus, they are all powerful oxidizing agents (they are reduced). F would be the strongest because it adds the electron to the $n = 2$ level, achieving the electron configuration of Ne. Because the $n = 2$ level lies lower in energy than the outermost level of the other halogens, it is more energetically favorable for F to gain the noble gas configuration than for the other halogens. This combined with the high ionization energy and relatively exothermic electron affinity makes F very reactive. As you move down the column, the n level of the outermost electrons increases, making it less energetically favorable for each of the successive halogens to gain an electron.

4.109 Group 6A: ns^2np^4 Group 7A: ns^2np^5

The electron affinity of the group 7A elements is more negative than that of the group 6A elements in the same period because group 7A requires only one electron to achieve the noble gas configuration ns^2np^6, while the group 6A elements require two electrons. Adding one electron to the group 6A element will not give them any added stability and leads to extra electron-electron repulsions; so the value of the electron affinity is less negative than that for group 7A.

4.110 Group 5A: ns^2np^3 Group 4A: ns^2np^2

The electron affinity of the group 5A elements is more positive than that of the group 4A elements in the same period because group 5A has a half-filled p sublevel. Adding an electron to this group adds a fourth electron to the p sublevel and increases the electron-electron repulsions. It also eliminates the stability of the half-filled sublevel. Adding an electron to a group 4A element, however, adds a third electron to the p sublevel, giving it the added stability of the half-filled sublevel.

4.111 $Br(35) = [Ar]4s^23d^{10}4p^5$ $I(53) = [Kr]5s^24d^{10}5p^5$

Br and I are both halogens with an outermost electron configuration of ns^2np^5; the next element with the same outermost electron configuration is 85, At.

4.112 Begin by writing the electron configuration of the neutral atom. $S = 1s^22s^22p^63s^23p^4$

S^+ loses one electron: $1s^22s^22p^63s^23p^3$

1s	2s	2p	3s	3p
↑↓	↑↓	↑↓ ↑↓ ↑↓	↑↓	↑ ↑ ↑

S^{2+} loses two electrons: $1s^22s^22p^63s^23p^2$

1s	2s	2p	3s	3p
↑↓	↑↓	↑↓ ↑↓ ↑↓	↑↓	↑ ↑ _

S^{3+} loses three electrons: $1s^22s^22p^63s^23p^1$

1s	2s	2p	3s	3p
↑↓	↑↓	↑↓ ↑↓ ↑↓	↑↓	↑ _ _

S^{4+} loses four electrons: $1s^22s^22p^63s^23p^0$

1s	2s	2p	3s	3p
↑↓	↑↓	↑↓ ↑↓ ↑↓	↑↓	_ _ _

S^{5+} loses five electrons: $1s^22s^22p^63s^13p^0$

1s	2s	2p	3s	3p
↑↓	↑↓	↑↓ ↑↓ ↑↓	↑	_ _ _

S^{6+} loses six electrons: $1s^22s^22p^63s^03p^0$

1s	2s	2p	3s	3p
↑↓	↑↓	↑↓ ↑↓ ↑↓	_	_ _ _

4.113 (a) $10 - 2 = 8 \rightarrow$ O; $12 - 2 = 10 \rightarrow$ Ne; $58 - 5 = 53 \rightarrow$ I; $11 - 2 = 9 \rightarrow$ F; $7 - 2 = 5 \rightarrow$ B; $44 - 5 = 39 \rightarrow$ Y; $63 - 6 = 57 \rightarrow$ La; $66 - 6 = 60 \rightarrow$ Nd One If by Land

(b) $9 - 2 = 7 \rightarrow$ N; $99 - 7 = 92 \rightarrow$ U; $30 - 4 = 26 \rightarrow$ Fe; $95 - 7 = 88 \rightarrow$ Ra; $19 - 3 = 16 \rightarrow$ S; $47 - 5 = 42 \rightarrow$ Mo; $79 - 6 = 73 \rightarrow$ Ta (backwards) Atoms Are Fun

4.114 The electron affinity of sodium is lower than lithium. Going from lithium to sodium, the electron is entering an orbital with a larger principal quantum number and is farther from the nucleus. Because it is farther from the nucleus, the orbital is higher in energy and the addition of the electron is less exothermic.

With chlorine and fluorine, the electron affinity of chlorine is higher. Even though the electron is entering an orbital with a larger principal quantum number, if we look at the atomic radius of the two atoms, we see that fluorine has a very small radius of 72 pm when compared to chlorine with a radius of 99 pm. Because of the much smaller radius, the repulsive forces between the electrons entering the p orbitals will cause the electron affinity to be less exothermic.

4.115 **Given:** $r = 100.00$ pm, $q_{proton} = 1.60218 \times 10^{-19}$ C, and $q_{electron} = -1.60218 \times 10^{-19}$ C
Find: IE in kJ/mol and λ of ionization
Conceptual Plan: $r, q_{proton}, q_{electron}, \rightarrow E_{atom} \rightarrow E_{mol}$ **and then** $E_{atom} \rightarrow \lambda$

$$E = \frac{1}{4\pi_0}\frac{q_p q_e}{r} \quad \frac{1\text{ kJ}}{(1000\text{ J})} \quad \frac{6.022 \times 10^{23}\text{ atoms}}{\text{mol}} \qquad \lambda = \frac{hc}{E}$$

Solution:

$$E = \frac{1}{(4)(3.141)\left(8.85 \times 10^{-12}\dfrac{\text{C}^2}{\text{J}\cdot\text{m}}\right)} \times \frac{(1.60218 \times 10^{-19}\text{ C})(-1.60218 \times 10^{-19}\text{ C})}{(100.00\text{ pm})\left(\dfrac{1\text{ m}}{1 \times 10^{12}\text{ pm}}\right)} = -2.3\underline{0}89 \times 10^{-18}\text{ J/atom}$$

$$-2.3\underline{0}9 \times 10^{-18}\text{ J/atom} \times \frac{6.022 \times 10^{23}\text{ atoms}}{\text{mol}} \times \frac{1\text{ kJ}}{(1000\text{ J})} = -1.39 \times 10^3\text{ kJ/mol}$$

$$\text{IE} = 0 - (-1.39 \times 10^3\text{ kJ/mol}) = 1.39 \times 10^3\text{ kJ/mol}$$

$$\lambda = \frac{(6.626 \times 10^{-34}\text{ J}\cdot\text{s})(3.00 \times 10^8\text{ m}\cdot\text{s})\left(\dfrac{1 \times 10^9\text{ nm}}{1\text{ m}}\right)}{(2.3\underline{0}9 \times 10^{-18}\text{ J})} = 86.1\text{ nm}$$

Check: The units of the answer (kJ/mol) are correct. The magnitude of the answer is reasonable because the value is positive and energy must be added to the atom to remove the electron. The units of the wavelength (nm) are correct and the magnitude is reasonable based on the ionization energy.

4.116 **Given:** IE $= 496$ kJ/mol, $q_{proton} = 1.60218 \times 10^{-19}$ C, and $q_{electron} = -1.60218 \times 10^{-19}$ C **Find:** r
Conceptual Plan: kJ/mol → J/mol → J/atom → r

$$\frac{1000\text{ J}}{1\text{ kJ}} \quad \frac{1\text{ mol}}{6.022 \times 10^{23}\text{ atoms}} \quad r = \frac{1}{4\pi\varepsilon_o}\frac{q_p q_e}{E}$$

Solution: $E = -\text{IE} = -496$ kJ/mol

$$-496\frac{\text{kJ}}{\text{mol}} \times \frac{1000\text{ J}}{1\text{ kJ}} \times \frac{\text{mol}}{6.022 \times 10^{23}\text{ atoms}} = -8.2\underline{3}6 \times 10^{-19}\text{ J/atom}$$

$$r = \frac{1}{(4)(3.141)\left(8.85 \times 10^{-12}\dfrac{\text{C}^2}{\text{J}\cdot\text{m}}\right)\left(\dfrac{1\text{ m}}{1 \times 10^{12}\text{ pm}}\right)} \times \frac{(1.602 \times 10^{-19}\text{ C})(-1.602 \times 10^{219}\text{ C})}{(-8.236 \times 10^{-19}\text{ J})} = 280.\text{ pm}$$

The actual atomic radius of sodium is 186 pm. The $3s^1$ electron that is being removed is shielded from the nuclear charge by the inner shell electrons. Thus, the energy of the electron in the $3s$ orbital is less negative than what would be expected for an electron at a distance of 186 pm. Because the energy is less negative, the ionization energy is smaller.

Challenge Problems

4.117 (a) Using Excel, make a table of radius, atomic number, and density. Using xy scatter, make a chart of radius versus density. With an exponential trendline, estimate the density of argon and xenon. Also make a chart of atomic number versus density. With a linear trendline, estimate the density of argon and xenon.

Element	Atomic Radius (pm)	Atomic Number	Density (g/mL)
He	32	2	0.18
Ne	70	10	0.90
Ar	98	18	–
Kr	112	36	3.75
Xe	130	54	–
Rn	–	86	9.73

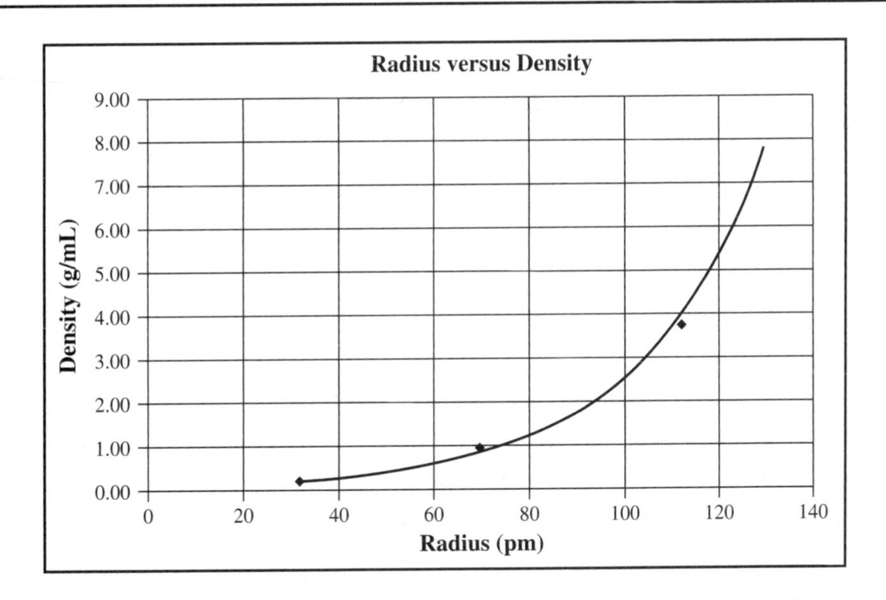

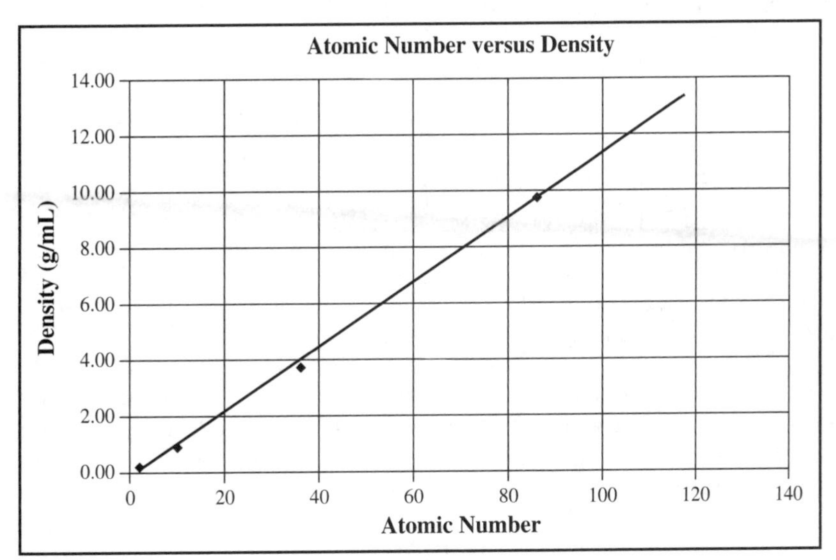

From the radius versus density chart, Ar has a density of ~2 g/L and Xe has a density of ~7.7 g/L. From the atomic number versus density chart, Ar has a density of ~1.8 g/L and Xe has a density of ~6 g/L.

(b) Using the chart of atomic number versus density, element 118 would be predicted to have a density of ~13 g/L.

(c) **Given:** Ne: $M = 20.18$ g/mol; $r = 70.$ pm **Find:** mass of neon; d neon

Conceptual Plan: $\boldsymbol{M \rightarrow m_{\text{atom}}}$ **and then** $\boldsymbol{r \rightarrow \text{vol}_{\text{atom}}}$ **and then** $\boldsymbol{\rightarrow d}$

$$\frac{\text{mol}}{6.022 \times 10^{23}\ \text{atoms}} \qquad V = \frac{4}{3}\pi r^3 \qquad d = \frac{m}{V}$$

Solution: $\dfrac{20.18\ \text{g}}{\cancel{\text{mol}}} \times \dfrac{\cancel{\text{mol}}}{6.022 \times 10^{23}\ \text{atoms}} = 3.35 \times 10^{-23}$ g/atom

$$V = \frac{4}{3} \times 3.14 \times (70\ \cancel{\text{pm}})^3 \times \left(\frac{1\ \cancel{\text{m}}}{1 \times 10^{12}\ \cancel{\text{pm}}}\right)^3 \times \frac{1\ \text{L}}{0.0010\ \cancel{\text{m}^3}} = 1.\underline{4}4 \times 10^{-27}\ \text{L}$$

$$d = \frac{3.35 \times 10^{-23}\ \text{g}}{1.\underline{4}4 \times 10^{-27}\ \text{L}} = 2.\underline{3}3 \times 10^4 = 2.3 \times 10^4\ \text{g/L}$$

Check: The units of the answer (g/L) are correct. This density is significantly larger than the actual density of neon gas. This suggests that a liter of neon is composed primarily of empty space.

(d) **Given:** Ne: $M = 20.18$ g/mol, $d = 0.90$ g/L; Kr: $M = 83.30$ g/mol, $d = 3.75$ g/L; Ar: $M = 39.95$ g/mol
Find: d of argon in g/L
Conceptual Plan: $d \rightarrow$ mol/L $\rightarrow$ atoms/L for Kr and Ne and then atoms/L $\rightarrow$ mol/L $\rightarrow d$ for Ar

$$\text{mol} = \frac{\text{mass}}{\text{molar mass}} \qquad \frac{6.022 \times 10^{23}\ \text{atoms}}{\text{mol}} \qquad \frac{\text{mol}}{6.022 \times 10^{23}\ \text{atoms}} \qquad \frac{39.95\ \text{g}}{\text{mol}}$$

Solution: for Ne: $\dfrac{0.90\ \cancel{g}}{\text{L}} \times \dfrac{\cancel{\text{mol}}}{20.18\ \cancel{g}} \times \dfrac{6.022 \times 10^{23}\ \text{atoms}}{\cancel{\text{mol}}} = 2.7 \times 10^{22}\ \text{atoms/L}$

for Kr: $\dfrac{3.75\ \cancel{g}}{\text{L}} \times \dfrac{\cancel{\text{mol}}}{83.80\ \cancel{g}} \times \dfrac{6.022 \times 10^{23}\ \text{atoms}}{\cancel{\text{mol}}} = 2.69 \times 10^{22}\ \text{atoms/L}$

for Ar: $\dfrac{2.69 \times 10^{22}\ \cancel{\text{atoms}}}{\text{L}} \times \dfrac{\cancel{\text{mol}}}{6.022 \times 10^{23}\ \cancel{\text{atoms}}} \times \dfrac{39.95\ \text{g}}{\cancel{\text{mol}}} = 1.78\ \text{g/L}$

This value is similar to the value calculated in part a. The value of the density calculated from the radius was 2 g/L, and the value of the density calculated from the atomic number was 1.8 g/L.

Check: The units of the answer (g/L) are correct. The value of the answer agrees with the published value.

4.118 If there were only two p orbitals, there would be only four p-block columns, and if there were only three d orbitals, there would be only six d-block columns. So the periodic table would have 12 columns.

H											He
Li	Be							B	C	N	O
F	Ne							Na	Mg	Al	Si
P	S	Cl	Ar	K	Ca	Sc	Ti	V	Cr	Mn	Fe

The noble gas equivalent elements would be He, O, Si, and Fe.
The halogen equivalent elements would be N, Al, and Mn.
The alkali metal equivalent elements would be Li, F, and P.

4.119 The density increases as you move to the right across the first transition series. For the first transition series, the mass increases as you move to the right across the periodic table. However, the radius of the transition series elements stays nearly constant as you move to the right across the periodic table; thus, the volume will remain nearly constant. Because density is mass/volume, the density of the elements increases.

4.120 If there are three possible spin quantum numbers, there will be 3s electrons, 9p electrons, and 15d electrons.

(a) Ne(10 e^-): $1s^3 2s^3 2p^4$

(b) Completed $n = 2$ level: $1s^3 2s^3 2p^9$ Atomic number $= 15$

(c) F(9 e^-): $1s^3 2s^3 2p^3$ There will be one unpaired electron in the 1s and 2s and three unpaired electrons in the 2p; therefore, there are five unpaired electrons

4.121 The longest wavelength would be associated with the lowest energy state next to the ground state of carbon, which has two unpaired electrons.
Ground state of carbon:

1s: [↑↓] 2s: [↑↓] 2p: [↑][↑][ ]

Longest wavelength: One of the p electrons flipped in its orbital, which requires the least amount of energy.

1s: [↑↓] 2s: [↑↓] 2p: [↑][↓][ ]

The next wavelength would be associated with the pairing of the two p electrons in the same orbital because this requires energy and raises the energy.

1s: [↑↓] 2s: [↑↓] 2p: [↑↓][ ][ ]

The next wavelength would be associated with the energy needed to promote one of the s electrons to a p orbital.

4.122 Element darmstadtium (110) would be in the column with Ni, Pd, and Pt, and so it might be expected to have an electron configuration similar to that of Ni, Pd, or Pt.
Similar to Ni: $[Rn]7s^25f^{14}6d^8$
Similar to Pd: $[Rn]7s^05f^{14}6d^{10}$
Similar to Pt: $[Rn]7s^15f^{14}6d^9$

4.123 The element that would fill the 8*s* and 8*p* orbitals would have atomic number 168. The element is in the noble gas family and would have the properties of noble gases. It would have the electron configuration of $[118]8s^25g^{18}6f^{14}7d^{10}8p^6$. The outer shell electron (highest *n* level) configuration would be $8s^28p^6$. The element would be relatively inert, have a first ionization energy less than 1037 kJ/mol (the first ionization energy of Rn), and have a positive electron affinity. It would be difficult to form compounds with most elements but would be able to form compounds with fluorine.

4.124 To determine the second ionization energies, look at the electron configuration of the 1+ ions.

First, write the electron configuration of the atom:		Then write the electron configuration of the 1+ ion:	
O	$1s^22s^1$	Li^+	$1s^22s^0$
Be	$1s^22s^2$	Be^+	$1s^22s^1$
B	$1s^22s^22p^1$	B^+	$1s^22s^2$
C	$1s^22s^22p^2$	C^+	$1s^22s^22p^1$
N	$1s^22s^22p^3$	N^+	$1s^22s^22p^2$
O	$1s^22s^22p^4$	O^+	$1s^22s^22p^3$
F	$1s^22s^22p^5$	F^+	$1s^22s^22p^4$

Based on the electron configuration of the ions, Li^+ should have the highest second ionization energy because the removal of the second electron involves removing a core electron. The lowest second ionization energy should be Be^+ because removing the second electron takes you to the $1s^2$ (stable) configuration.

O would have the highest second ionization energy because the electron configuration of O^+ has a half-filled *p* orbital, which is particularly stable; therefore, it would require more energy to remove the second electron. N^+ would have the lowest second ionization energy because the size of N^+ would be larger than the radius of the F^+; so the attraction between the outer electron and the nucleus would be less in N^+ than in F^+, making it easier to remove the electron.

4.125 When you move down the column from Al to Ga, the size of the atom actually decreases because the 3*d* electrons in the Ga atom do not contribute much shielding, while there is a large increase in the nuclear charge. Therefore, the effective nuclear charge is greater for Ga than for Al; so the ionization energy does not decrease. As you go from In to Tl, the ionization energy actually increases because the 4*f* electrons do not contribute to the shielding of the outermost electrons and there is a large increase in the effective nuclear charge.

4.126 ΔE for the reaction based on the ionization energy and the electron affinity = +147 kJ/mol

$Na(g) \rightarrow Na^+(g) + e^-$	IE = +496 kJ/mol
$Na(g) + e^- \rightarrow Cl^-(g)$	EA = −349 kJ/mol
$Na(g) + Cl(g) \rightarrow Na^+(g) + Cl^-(g)$	ΔE = +147 kJ/mol

4.127 The second electron is added to an ion with a 1− charge, so a large repulsive force must be overcome to add the second electron. Thus, it will require energy to add the second electron, and the second electron affinity will have a positive value.

4.128 The diagonal relationship between some elements could be explained because the atomic size of the atoms on the diagonal would be about the same. On the diagonal, the radius would increase as you go down the column but would decrease as you move to the right across the period. Also, the size of the ion formed would be about the same. Therefore, you might expect the elements to have similar behavior.

4.129 **Given:** Ra, $Z = 88$ **Find:** Z for next two alkaline earth metals
Solution: The next element would lie in period 8, column 2A. The largest currently known element is 116 in period 7, column 6A. To reach period 8 column 2A, you need to add four protons and would have $Z = 120$.

The alkaline earth metal following 120 would lie in period 9, column 2A. To reach this column, you need to add 18 g-block element protons, 10 d-block element protons, 14 f-block element protons, 6 p-block element protons, and 2 s-block element protons. This would give $Z = 170$.

4.130 **Given:** Pd = $[Kr]4d^{10}$ **Find:** electron configuration of first two excited states
Solution: The first excited state of Pd would move an electron from a $4d$ to a $5s$ orbital. The second excited state would move the electron from a $4d$ to the $5p$ orbital.
First excited state: $[Kr]4d^9 5s^1$
Second excited state: $[Kr]4d^9 5p^1$

Conceptual Problems

4.131 If six electrons rather than eight electrons led to a stable configuration, the electron configuration of the stable configuration would be ns^2np^4.

(a) A noble gas would have the electron configuration ns^2np^4. This could correspond to the O atom.

(b) A reactive nonmetal would have one less electron than the stable configuration. This would have the electron configuration ns^2np^3. This could correspond to the N atom.

(c) A reactive metal would have one more electron than the stable configuration. This would have the electron configuration of ns^1. This could correspond to the Li atom.

4.132 Atom B would have the higher first ionization energy. Even though the effective nuclear charge is less, the outermost electron is closer to the nucleus and the potential energy becomes more negative with decreasing distance, making it harder to remove and requiring a larger ionization energy.

4.133 (a) True: An electron in a $3s$ orbital is more shielded than an electron in a $2s$ orbital. This is true because there are more core electrons below a $3s$ orbital.

(b) True: An electron in a $3s$ orbital penetrates the region occupied by the core electrons more than electrons in a $3p$ orbital. Examine Figure 4.8, the radial distribution functions for the $3s$, $3p$, and $3d$ orbitals. You will see that the $3s$ electrons penetrate more deeply than the $3p$ electrons and more than the $3d$ electrons.

(c) False: An electron in an orbital that penetrates closer to the nucleus will experience *less* shielding than an electron in an orbital that does not penetrate as far.

(d) True: An electron in an orbital that penetrates close to the nucleus will tend to experience a higher effective nuclear charge than one that does not. Because the orbital penetrates closer to the nucleus, the electron will experience less shielding and therefore a higher effective nuclear charge.

4.134 An electron in a $5p$ orbital could have any one of the following combinations of quantum numbers:
$(5, 1, -1, +1/2)$ $(5, 1, -1, -1/2)$ $(5, 1, 0, +1/2)$ $(5, 1, 0, -1/2)$ $(5, 1, 1, +1/2)$ $(5, 1, 1, -1/2)$
An electron in a $6d$ orbital could have any one of the following combinations of quantum numbers:
$(6, 2, -2, +1/2)$ $(6, 2, -2, -1/2)$ $(6, 2, -1, +1/2)$ $(6, 2, -1, -1/2)$ $(6, 2, 0, +1/2)$ $(6, 2, 0, -1/2)$
$(6, 2, 1, +1/2)$ $(6, 2, 1, -1/2)$ $(6, 2, 2, +1/2)$ $(6, 2, 2, -1/2)$

4.135 The $4s$ electrons in calcium have relatively low ionization energies ($IE_1 = 590$ kJ/mol; $IE_2 = 1145$ kJ/mol) because they are valence electrons. The energetic cost for calcium to lose a third electron is extraordinarily high because the next electron to be lost is a core electron. Similarly, the electron affinity of fluorine to gain one electron (-328 kJ/mol) is highly exothermic because the added electron completes fluoride's valence shell. The gain of a second electron by the negatively charged fluoride anion would not be favorable. Therefore, we would expect calcium and fluoride to combine in a 1:2 ratio.

5 Molecules and Compounds

Review Questions

5.1 The properties of compounds are generally very different from the properties of the elements that compose them. When two elements combine to form a compound, an entirely new substance results.

5.2 Chemical bonds are the result of interactions between charged particles—electrons and protons—that compose atoms.

5.3 Ionic bonds, which occur between metals and nonmetals, involve the transfer of electrons from one atom to another. Covalent bonds, which occur between two or more nonmetals, involve the sharing of electrons between two atoms.

5.4 Chemical compounds can be represented by chemical formulas and molecular models. The type of formula or model you use depends on how much information you have about the compound and how much you want to communicate. An empirical formula gives the relative number of atoms of each element in the compound. It contains the smallest whole-number ratio of the elements in the compound. A molecular formula gives the actual number of atoms of each element in the compound. A structural formula shows how the atoms are connected. A ball-and-stick model shows the geometry of the compound. A space-filling model shows the relative sizes of the atoms and how they merge together.

5.5 An empirical formula gives the relative number of atoms of each element in a compound.

A molecular formula gives the actual number of atoms of each element in a molecule of a compound.

5.6 In a Lewis structure, the valence electrons of main-group elements are represented as dots surrounding the symbol for the element. The valence electrons can be determined from the group they are in on the periodic table.

5.7 Bonds are formed when atoms attain a stable electron configuration. Because the stable configuration usually has eight electrons in the outermost shell, this is known as the octet rule.

5.8 In Lewis theory, a chemical bond is the sharing or transferring of electrons to attain stable electron configurations for the bonding atoms. If electrons are transferred, the bond is an ionic bond. If the electrons are shared, the bond is a covalent bond.

5.9 To use Lewis theory to determine the formula of an ionic compound, determine the number of valence electrons lost by the metal to form an octet and the number of electrons gained by the nonmetal to form an octet. Consider the ionic compound formed between sodium and sulfur. The Lewis structures for sodium and sulfur are as follows:

Na· ·S̤̈:

Sodium must lose one electron to achieve an octet, while sulfur must gain two electrons to achieve an octet. So the compound requires two sodium atoms for each sulfur atom, giving the formula Na_2S.

5.10 Lattice energy is associated with forming a crystalline lattice of alternating cations and anions from the gaseous ions. Because the cations are positively charged and the anions are negatively charged, there is a lowering of potential—as described by Coulomb's law—when the ions come together to form a lattice. That energy is emitted as heat when the lattice forms.

5.11 The formation of the crystalline NaCl lattice from sodium cations and chloride anions is highly exothermic and more than compensates for the endothermicity of the electron transfer process. In other words, the formation of ionic compounds is not exothermic because sodium "wants" to lose electrons and chlorine "wants" to gain them; rather, it is exothermic because of the large amount of heat released when sodium and chlorine ions coalesce to form a crystal lattice.

5.12 To write a formula for an ionic compound, do the following: (1) Write the symbol for the metal cation and its charge followed by the symbol for the nonmetal or polyatomic anion and its charge. (2) Adjust the subscript on each cation and anion to balance the overall charge. (3) Check that the sum of the charges of the cations equals the sum of the charges of the anions.

5.13 Binary ionic compounds are named using the name of the cation (metal) and the base name of the anion (nonmetal) + the suffix *-ide*. Ionic compounds that contain a polyatomic anion are named using the name of the cation (metal) and the name of the polyatomic anion.

5.14 Ionic compounds formed from metals that can form more than one cation must include the charge of the cation in the name. The charge is indicated by putting the charge of the metal in Roman numerals in parentheses after the metal name. Metals that can form only one cation do not need the charge specified.

5.15 A pair of electrons that is shared between two atoms is called a bonding pair, while a pair of electrons that is associated with only one atom—and therefore, not involved in bonding—is called a lone pair.

5.16 A single bond occurs when one pair of electrons is shared between two atoms. A double bond results when two electron pairs are shared between the same two atoms. Double bonds are shorter and stronger than single bonds. A triple bond results when three electron pairs are shared between the same two atoms. Triple bonds are even shorter and stronger than double bonds.

5.17 Generally, combinations of atoms that can satisfy the octet rule on each atom are stable, while those combinations that do not satisfy the octet rule are not stable.

5.18 Lewis theory shows that covalent bonds are highly directional. The attraction between two covalently bonded atoms is due to the sharing of one or more electron pairs. Thus, each bond links just one specific pair of atoms—in contrast to ionic bonds, which are nondirectional and hold together the entire array of ions. The fundamental units of covalently bonded compounds are individual molecules. These molecules can interact with one another in a number of different ways; however, the interactions between molecules are generally much weaker than the bonding interactions within a molecule. When a molecular compound melts or boils, the molecules themselves remain intact. Only the weak interactions between the molecules must be overcome. Consequently, molecular compounds tend to have lower melting and boiling points than ionic compounds.

5.19 To name a binary molecular inorganic compound, list the name of the first element with a prefix to indicate the number of atoms in the compound if there is more than one, followed by the base name of the second element with a prefix to indicate the number of atoms in the compound if there is more than one, followed by the suffix *-ide*.

5.20 The prefix mono = 1; di = 2; tri = 3; tetra = 4; penta = 5; hexa = 6.

5.21 The formula mass is the average mass of the molecule (or formula unit) of a compound. The formula mass allows the conversion between the mass of molecules and the number of molecules present.

5.22 The chemical formula indicates the elements present in the compound and the relative number of atoms of each type. The chemical formula gives the conversion factor between the kind of element and the formula; it also allows the determination of mass percent composition.

5.23 Mass percent composition is the mass of an element of the compound divided by the total mass of the compound times 100. Mass percent composition is used as a conversion factor between the mass of the element and the mass of the compound.

5.24 Chemical formulas contain inherent relationships between atoms (or moles of atoms) and molecules (or moles of molecules). For example, the formula CCl_2F_2 tells us that one mole of CCl_2F_2 contains one mole of C atoms, two moles of Cl atoms, and two moles of F atoms.

5.25 The experimental data showing the relative masses of the elements in a compound can be used to obtain an empirical formula.

5.26 The molecular formula is a whole-number multiple of the empirical formula. To find the molecular formula, the molar mass of the compound must be known. The molecular molar mass divided by the empirical molar mass gives the whole-number multiple used to convert the empirical formula to the molecular formula.

5.27 In combustion analysis, the unknown compound undergoes combustion (burning) in the presence of pure oxygen. All of the carbon in the sample is converted to CO_2, and all of the hydrogen is converted to H_2O. Combustion analysis can be used to determine the empirical formula of a hydrocarbon.

5.28 Organic compounds are composed of carbon; hydrogen; and a few other elements, including nitrogen, oxygen, and sulfur.

Problems by Topic

Types of Compounds and Chemical Formulas

5.29
(a) CO_2 is a compound composed of a nonmetal and a nonmetal; therefore, it is a molecular compound.
(b) $NiCl_2$ is a compound composed of a metal and a nonmetal; therefore, it is an ionic compound.
(c) NaI is a compound composed of a metal and a nonmetal; therefore, it is an ionic compound.
(d) PCl_3 is a compound composed of a nonmetal and a nonmetal; therefore, it is a molecular compound.

5.30
(a) CF_2Cl_2 is a compound composed of a nonmetal and two other nonmetals; therefore, it is a molecular compound.
(b) CCl_4 is a compound composed of a nonmetal and a nonmetal; therefore, it is a molecular compound.
(c) PtO_2 is a compound composed of a metal and a nonmetal; therefore, it is an ionic compound.
(d) SO_3 is a compound composed of a nonmetal and a nonmetal; therefore, it is a molecular compound.

5.31 To determine the empirical formula from a molecular formula, divide the subscripts by the greatest common factor (the largest number that divides exactly into all of the subscripts).
(a) For N_2O_4, the greatest common factor is 2. The empirical formula is therefore NO_2.
(b) For C_5H_{12}, the only common factor is 1, so the empirical formula and the molecular formula are the same.
(c) For C_4H_{10}, the greatest common factor is 2. The empirical formula is therefore C_2H_5.

5.32 To determine the empirical formula from a molecular formula, divide the subscripts by the greatest common factor (the largest number that divides exactly into all of the subscripts).
(a) For C_2H_4, the greatest common factor is 2. The empirical formula is therefore CH_2.
(b) For $C_6H_{12}O_6$, the greatest common factor is 6. The empirical formula is therefore CH_2O.
(c) For NH_3, the only common factor is 1, so the empirical formula and the molecular formula are the same.

5.33 The chemical formula gives you the kind of atom and the number of each atom in the compound.
(a) $Mg_3(PO_4)_2$ contains: 3 magnesium atoms, 2 phosphorus atoms, and 8 oxygen atoms
(b) $BaCl_2$ contains: 1 barium atom and 2 chlorine atoms
(c) $Fe(NO_2)_2$ contains: 1 iron atom, 2 nitrogen atoms, and 4 oxygen atoms
(d) $Ca(OH)_2$ contains: 1 calcium atom, 2 oxygen atoms, and 2 hydrogen atoms

5.34 The chemical formula gives you the kind of atom and the number of each atom in the compound.
(a) $Ca(NO_2)_2$ contains: 1 calcium atom, 2 nitrogen atoms, and 4 oxygen atoms
(b) $CuSO_4$ contains: 1 copper atom, 1 sulfur atom, and 4 oxygen atoms
(c) $Al(NO_3)_3$ contains: 1 aluminum atom, 3 nitrogen atoms, and 9 oxygen atoms
(d) $Mg(HCO_3)_2$ contains: 1 magnesium atom, 2 hydrogen atoms, 2 carbon atoms, and 6 oxygen atoms

5.35 (a) 1 blue = nitrogen, 3 white = hydrogen: NH_3
(b) 2 black = carbon, 6 white = hydrogen: C_2H_6
(c) 1 yellow = sulfur, 3 red = oxygen: SO_3

5.36 (a) 1 blue = nitrogen, 2 red = oxygen: NO_2
(b) 1 yellow = sulfur, 2 white = hydrogen: H_2S
(c) 1 black = carbon, 4 white = hydrogen: CH_4

Valence Electrons and Lewis Dot Structures

5.37 N: $1s^22s^22p^3$ ·N: The electrons included in the Lewis structure are $2s^22p^3$.

5.38 Ne: $1s^22s^22p^6$:Ne: The electrons included in the Lewis structure are $2s^22p^6$.

5.39 (a) Al: $1s^22s^22p^63s^23p^1$
·Al·
(b) Na^+: $1s^22s^22p^6$
Na^+
(c) Cl: $1s^22s^22p^63s^23p^5$
:Cl·
(d) Cl^-: $1s^22s^22p^63s^23p^6$
[:Cl:]$^-$

5.40 (a) S^{2-}: $1s^22s^22p^63s^23p^6$
[:S:]$^{2-}$
(b) Mg: $1s^22s^22p^63s^2$
Mg:
(c) Mg^{2+}: $1s^22s^22p^6$
Mg^{2+}
(d) P: $1s^22s^22p^63s^23p^3$
·P·

Ionic Bonding and Lattice Energy

5.41 (a) NaF: Draw the Lewis structures for Na and F based on their valence electrons.
Na: $3s^1$ F: $2s^22p^5$
Na· :F·

Sodium must lose one electron and be left with the octet from the previous shell, while fluorine needs to gain one electron to get an octet.

Na^+ [:F:]$^-$

(b) CaO: Draw the Lewis structures for Ca and O based on their valence electrons. Ca: $4s^2$ O: $2s^22p^4$

Ca: :O·

Calcium must lose two electrons and be left with two $1s$ electrons from the previous shell, while oxygen needs to gain two electrons to get an octet.

Ca^{2+} [:O:]$^{2-}$

(c) $SrBr_2$: Draw the Lewis structures for Sr and Br based on their valence electrons. Sr: $5s^2$ Br: $4s^24p^5$

Sr: :Br·

Strontium must lose two electrons and be left with the octet from the previous shell, while bromine needs to gain one electron to get an octet.

Sr^{2+} 2[:B̤̈r:]$^-$

(d) K_2O: Draw the Lewis structures for K and O based on their valence electrons. K: $4s^1$ O: $2s^22p^4$

K· :Ö·

Potassium must lose one electron and be left with the octet from the previous shell, while oxygen needs to gain two electrons to get an octet.

$2K^+$ [:Ö̤:]$^{2-}$

5.42 (a) SrO: Draw the Lewis structures for Sr and O based on their valence electrons. Sr: $5s^2$ O: $2s^22p^4$

Sr: :Ö·

Strontium must lose two electrons and be left with the octet from the previous shell, while oxygen needs to gain two electrons to get an octet.

Sr^{2+} [:Ö̤:]$^{2-}$

(b) Li_2S: Draw the Lewis structures for Li and S based on their valence electrons. Li: $2s^1$ S: $3s^23p^4$

Li· :S̈·

Lithium must lose one electron and be left with two 1*s* electrons from the previous shell, while sulfur needs to gain two electrons to get an octet.

2 Li^+ [:S̤̈:]$^{2-}$

(c) CaI_2: Draw the Lewis structures for Ca and I based on their valence electrons. Ca: $4s^2$ I: $5s^25p^5$

Ca: :Ï̤·

Calcium must lose two electrons and be left with the octet from the previous shell, while iodine needs to gain one electron to get an octet.

Ca^{2+} 2[:Ï̤:]$^-$

(d) RbF: Draw the Lewis structures for Rb and F based on their valence electrons. Rb: $5s^1$ F: $2s^22p^5$

Rb· :F̤̈·

Rubidium must lose one electron and be left with the octet from the previous shell, while fluorine needs to gain one electron to get an octet.

Rb^+ [:F̤̈:]$^-$

5.43 (a) Sr and Se: Draw the Lewis structures for Sr and Se based on their valence electrons.

Sr: $5s^2$ Se: $4s^24p^4$

Sr: :S̈e·

Strontium must lose two electrons and be left with the octet from the previous shell, while selenium needs to gain two electrons to get an octet.

Sr^{2+} [:S̤̈e:]$^{2-}$

Thus, we need one Sr^{2+} and one Sr^{2-}. Write the formula with subscripts (if necessary) to indicate the number of atoms.

SrSe

(b) Ba and Cl: Draw the Lewis structures for Ba and Cl based on their valence electrons.

Ba: $6s^2$ Cl: $3s^23p^5$

Ba: :C̤̈l·

Barium must lose two electrons and be left with the octet from the previous shell, while chlorine needs to gain one electron to get an octet.

Ba^{2+} 2[:C̈l:]$^-$

Thus, we need one Ba^{2+} and two Cl^-. Write the formula with subscripts (if necessary) to indicate the number of atoms.

$BaCl_2$

(c) Na and S: Draw the Lewis structures for Na and S based on their valence electrons.

Na: $3s^1$ S: $3s^23p^4$

Na· :Ṡ·

Sodium must lose one electron and be left with the octet from the previous shell, while sulfur needs to gain two electrons to get an octet.

2 Na^+ [:S̈:]$^{2-}$

Thus, we need two Na^+ and one S^{2-}. Write the formula with subscripts (if necessary) to indicate the number of atoms.

Na_2S

(d) Al and O: Draw the Lewis structures for Al and O based on their valence electrons.

Al: $3s^23p^1$ O: $2s^22p^4$

Äl· :Ö·

Aluminum must lose three electrons and be left with the octet from the previous shell, while oxygen needs to gain two electrons to get an octet.

2 Al^{3+} 3[:Ö:]$^{2-}$

Thus, we need two Al^{3+} and three O^{2-} to lose and gain the same number of electrons. Write the formula with subscripts (if necessary) to indicate the number of atoms.

Al_2O_3

5.44 (a) Ca and N: Draw the Lewis structures for Ca and N based on their valence electrons.

Ca: $4s^2$ N: $2s^22p^3$

Ca: ·Ṅ·

Calcium must lose two electrons and be left with the octet from the previous shell, while nitrogen needs to gain three electrons to get an octet.

3 Ca^{2+} 2[:N̈:]$^{3-}$

Thus, we need three Ca^{2+} and two N^{3-} to lose and gain the same number of electrons. Write the formula with subscripts (if necessary) to indicate the number of atoms.

Ca_3N_2

(b) Mg and I: Draw the Lewis structures for Mg and I based on their valence electrons.

Mg: $3s^2$ I: $5s^25p^5$

Mg: :Ï·

Magnesium must lose two electrons and be left with the octet from the previous shell, while iodine needs to gain one electron to get an octet.

Mg^{2+} 2[:Ï:]$^-$

Thus, we need one Mg^{2+} and two I^-. Write the formula with subscripts (if necessary) to indicate the number of atoms.

MgI_2

(c) Ca and S: Draw the Lewis structures for Ca and S based on their valence electrons.

Ca: $4s^2$ S: $3s^2 3p^4$

Ca: :S̈·

Calcium must lose two electrons and be left with the octet from the previous shell, while sulfur needs to gain two electrons to get an octet.

Ca^{2+} $[:\ddot{S}:]^{2-}$

Thus, we need one Ca^{2+} and one S^{2-}. Write the formula with subscripts (if necessary) to indicate the number of atoms.

CaS

(d) Cs and F: Draw the Lewis structures for Cs and F based on their valence electrons.

Cs: $6s^1$ F: $2s^2 2p^5$

Cs· :F̤̈·

Cesium must lose one electron and be left with the octet from the previous shell, while fluorine needs to gain one electron to get an octet.

Cs^+ $[:\ddot{F}:]^-$

Thus, we need one Cs^+ and one F^-. Write the formula with subscripts (if necessary) to indicate the number of atoms.

CsF

5.45 Cesium is slightly larger than barium, but oxygen is slightly larger than fluorine; so we cannot use size to explain the difference in the lattice energy. However, the charge on the cesium ion is 1+ and the charge on the fluoride ion is 1−, while the charge on the barium ion is 2+ and the charge on the oxide ion is 2−. The coulombic equation states that the magnitude of the potential also depends on the product of the charges. Because the product of the charges for CsF = 1− and the product of the charges for BaO = 4−, the stabilization for BaO relative to CsF should be about four times greater, which is what we see in its more exothermic lattice energy.

5.46 Rubidium is below potassium on the periodic table, and iodine is below bromine on the periodic table. Therefore, both the rubidium ion and the iodide ion are larger than the potassium ion and the bromide ion. So the rubidium ion and the iodide ion cannot get as close to each other as the potassium ion and the bromide ion can; thus, the rubidium and iodide ions do not release as much energy, and the lattice energy of potassium bromide is more exothermic.

Formulas and Names for Ionic Compounds

5.47 To write the formula for an ionic compound, do the following: (1) Write the symbol for the metal cation and its charge and the symbol for the nonmetal anion and its charge. (2) Adjust the subscript on each cation and anion to balance the overall charge. (3) Check that the sum of the charges of the cations equals the sum of the charges of the anions.

(a)	calcium and oxygen:	Ca^{2-}	O^{2-}	CaO	cations 2+, anions 2−
(b)	zinc and sulfur:	Zn^{2+}	S^{2-}	ZnS	cations 2+, anions 2−
(c)	rubidium and bromine:	Rb^+	Br^-	RbBr	cation +, anions −
(d)	aluminum and oxygen:	Al^{3+}	O^{2-}	Al_2O_3	cation 2(3+) = 6+, anions 3(2−) = 6−

5.48 To write the formula for an ionic compound, do the following: (1) Write the symbol for the metal cation and its charge and the symbol for the nonmetal anion and its charge. (2) Adjust the subscript on each cation and anion to balance the overall charge. (3) Check that the sum of the charges of the cations equals the sum of the charges of the anions.

(a)	silver and chlorine:	Ag^+	Cl^-	AgCl	cation +, anions−
(b)	sodium and sulfur:	Na^+	S^{2-}	Na_2S	cation 2(1+) = 2+, anion 2−
(c)	aluminum and sulfur:	Al^{3+}	S^{2-}	Al_2S_3	cation 2(3+) = 6+, anions 3(2−) = 6−
(d)	potassium and chlorine:	K^+	Cl^-	KCl	cation +, anion−

5.49 To write the formula for an ionic compound, do the following: (1) Write the symbol for the metal cation and its charge and the symbol for the polyatomic anion and its charge. (2) Adjust the subscript on each cation and anion to balance the overall charge. (3) Check that the sum of the charges of the cations equals the sum of the charges of the anions.

Cation = calcium: Ca^{2+}

(a)	hydroxide:	OH^-	$Ca(OH)_2$	cation 2+, anion 2(1−) = 2−
(b)	chromate:	CrO_4^{2-}	$CaCrO_4$	cation 2+, anion 2−
(c)	phosphate:	PO_4^{3-}	$Ca_3(PO_4)_2$	cation 3(2+) = 6+, anion 2(3−) = 6−
(d)	cyanide:	CN^-	$Ca(CN)_2$	cation 2+, anion 2(1−) = 2−

5.50 To write the formula for an ionic compound, do the following: (1) Write the symbol for the metal cation and its charge and the symbol for the nonmetal anion and its charge. (2) Adjust the subscript on each cation and anion to balance the overall charge. (3) Check that the sum of the charges of the cations equals the sum of the charges of the anions.

Cation = potassium: K^+

(a)	carbonate:	CO_3^{2-}	K_2CO_3	cation 2(1+) = 2+, anion 2−
(b)	phosphate:	PO_4^{3-}	K_3PO_4	cation 3(1+) = 3+, anion 3−
(c)	hydrogen phosphate:	HPO_4^{2-}	K_2HPO_4	cation 2(1+) = 2+, anion 2−
(d)	acetate:	$C_2H_3O_2^-$	$KC_2H_3O_2$	cation 1+, anion 1−

5.51 To name a binary ionic compound, name the metal cation followed by the base name of the anion + *-ide*.

(a) Mg_3N_2: The cation is magnesium; the anion is from nitrogen, which becomes nitride: magnesium nitride.
(b) KF: The cation is potassium; the anion is from fluorine, which becomes fluoride: potassium fluoride.
(c) Na_2O: The cation is sodium; the anion is from oxygen, which becomes oxide: sodium oxide.
(d) Li_2S: The cation is lithium; the anion is from sulfur, which becomes sulfide: lithium sulfide.
(e) CsF: The cation is cesium; the anion is from fluorine, which becomes fluoride: cesium fluoride.
(f) KI: The cation is potassium; the anion is from iodine, which becomes iodide: potassium iodide.

5.52 To name an ionic compound with a metal cation that can have more than one charge, name the metal cation followed by parentheses with the charge in Roman numerals followed by the base name of the anion + *-ide*.

(a) $SnCl_4$: The charge on Sn must be 4+ for the compound to be charge neutral: The cation is tin(IV); the anion is from chlorine, which becomes chloride: tin(IV) chloride.
(b) PbI_2: The charge on Pb must be 2+ for the compound to be charge neutral: The cation is lead(II); the anion is from iodine, which becomes iodide: lead(II) iodide.
(c) Fe_2O_3: The charge on Fe must be 3+ for the compound to be charge neutral: The cation is iron(III); the anion is from oxygen, which becomes oxide: iron(III) oxide.
(d) CuI_2: The charge on Cu must be 2+ for the compound to be charge neutral: The cation is copper(II); the anion is from iodine, which becomes iodide: copper(II) iodide.
(e) $HgBr_2$: The charge of Hg must be 2+ for the compound to charge neutral: The cation is mercury(II); the anion is from bromine, which becomes bromide: mercury(II) bromide.
(f) $CrCl_2$: The charge on Cr must be 2+ for the compound to be charge neutral: The cation is chromium(II); the anion is from chlorine, which becomes chloride: chromium(II) chloride.

5.53 To name these compounds, you must first decide whether the metal cation is invariant or can have more than one charge. Then, name the metal cation followed by the base name of the anion + *-ide*.

(a) SnO: Sn can have more than one charge. The charge on Sn must be 2+ for the compound to be charge neutral: The cation is tin(II); the anion is from oxygen, which becomes oxide: tin(II) oxide.
(b) Cr_2S_3: Cr can have more than one charge. The charge on Cr must be 3+ for the compound to be charge neutral: The cation is chromium(III); the anion is from sulfur, which becomes sulfide: chromium(III) sulfide.
(c) RbI: Rb is invariant: The cation is rubidium; the anion is from iodine, which becomes iodide: rubidium iodide.
(d) $BaBr_2$: Ba is invariant: The cation is barium; the anion is from bromine, which becomes bromide: barium bromide.

5.54 To name these compounds, you must first decide whether the metal cation is invariant or can have more than one charge. Then, name the metal cation followed by the base name of the anion + *-ide*.

(a) BaS: Ba is invariant: The cation is barium; the anion is from sulfur, which becomes sulfide: barium sulfide.

(b) $FeCl_3$: Fe can have more than one charge. The charge on Fe must be 3+ for the compound to be charge neutral: The cation is iron(III); the anion is from chlorine, which becomes chloride: iron(III) chloride.

(c) PbI_4: Pb can have more than one charge. The charge on Pb must be 4+ for the compound to be charge neutral: The cation is lead(IV); the anion is from iodine, which becomes iodide: lead(IV) iodide.

(d) $SrBr_2$: Sr is invariant: The cation is strontium; the anion is from bromine, which becomes bromide: strontium bromide.

5.55 To name these compounds, you must first decide whether the metal cation is invariant or can have more than one charge. Then, name the metal cation followed by the name of the polyatomic anion.

(a) $CuNO_2$: Cu can have more than one charge. The charge on Cu must be 1+ for the compound to be charge neutral: The cation is copper(I); the anion is nitrite: copper(I) nitrite.

(b) $Mg(C_2H_3O_2)_2$: Mg is invariant: The cation is magnesium; the anion is acetate: magnesium acetate.

(c) $Ba(NO_3)_2$: Ba is invariant: The cation is barium; the anion is nitrate: barium nitrate.

(d) $Pb(C_2H_3O_2)_2$: Pb can have more than one charge. The charge on Pb must be 2+ for the compound to be charge neutral: The cation is lead(II); the anion is acetate: lead(II) acetate.

5.56 To name these compounds, you must first decide whether the metal cation is invariant or can have more than one charge. Then name the metal cation followed by the name of the polyatomic anion.

(a) $Ba(OH)_2$: Ba is invariant: The cation is barium; the anion is hydroxide: barium hydroxide.

(b) NH_4I: The cation is ammonium; the anion is from iodine, which becomes iodide: ammonium iodide.

(c) $NaBrO_4$: Na is invariant: The cation is sodium; the anion is perbromate: sodium perbromate.

(d) $Fe(OH)_3$: Fe can have more than one charge. The charge on Fe must be 3+ for the compound to be charge neutral: The cation is iron(III); the anion is hydroxide: iron(III) hydroxide.

5.57 To write the formula for an ionic compound, do the following: (1) Write the symbol for the metal cation and its charge and the symbol for the nonmetal anion or polyatomic anion and its charge. (2) Adjust the subscript on each cation and anion to balance the overall charge. (3) Check that the sum of the charges of the cations equals the sum of the charges of the anions.

(a)	sodium hydrogen sulfite:	Na^+	HSO^{3-}	$NaHSO_3$	cation 1+, anion 1−
(b)	lithium permanganate:	Li^+	MnO_4^-	$LiMnO_4$	cation 1+, anion 1−
(c)	silver nitrate:	Ag^+	NO_3^-	$AgNO_3$	cation 1+, anion 1−
(d)	potassium sulfate:	K^+	SO_4^{2-}	K_2SO_4	cation 2(1+) = 2+, anion 2−
(e)	rubidium hydrogen sulfate:	Rb^+	HSO_4^-	$RbHSO_4$	cation 1+, anion 1−
(f)	potassium hydrogen carbonate:	K^+	HCO_3^-	$KHCO_3$	cation 1+, anion 1−

5.58 To write the formula for an ionic compound, do the following: (1) Write the symbol for the metal cation and its charge and the symbol for the nonmetal anion or polyatomic anion and its charge. (2) Adjust the subscript on each cation and anion to balance the overall charge. (3) Check that the sum of the charges of the cations equals the sum of the charges of the anions.

(a)	copper(II) chloride:	Cu^{2+}	Cl^-	$CuCl_2$	cation 2+, anion 2(1−) = 2−
(b)	copper(I) iodate:	Cu^+	IO_3^-	$CuIO_3$	cation 1+, anion 1−
(c)	lead(II) chromate:	Pb^{2+}	CrO_4^{2-}	$PbCrO_4$	cation 2+, anion 2−
(d)	calcium fluoride:	Ca^{2+}	F^-	CaF_2	cation 2+, anion 2(1−) = 2−
(e)	potassium hydroxide:	K^+	OH^-	KOH	cation 1+, anion 1−
(f)	iron(II) phosphate:	Fe^{2+}	PO_4^{3-}	$Fe_3(PO_4)_2$	cation 3(2+) = 6+, anion 2(3−) = 6−

5.59 Hydrates are named the same way as other ionic compounds with the addition of the term *prefix*hydrate, where the prefix is the number of water molecules associated with each formula unit.

(a)	$CoSO_4 \cdot 7H_2O$	cobalt(II) sulfate heptahydrate
(b)	iridium(III) bromide tetrahydrate	$IrBr_3 \cdot 4H_2O$
(c)	$Mg(BrO_3)_2 \cdot 6H_2O$	magnesium bromate hexahydrate
(d)	potassium carbonate dihydrate	$K_2CO_3 \cdot 2H_2O$

5.60 Hydrates are named the same way as other ionic compounds with the addition of the term *prefix*hydrate, where the prefix is the number of water molecules associated with each formula unit.

(a) cobalt(II) phosphate octahydrate $Co_3(PO_4)_2 \cdot 8H_2O$

(b) $BeCl_2 \cdot 2H_2O$ beryllium chloride dihydrate

(c) chromium(III) phosphate trihydrate $CrPO_4 \cdot 3H_2O$

(d) $LiNO_2 \cdot H_2O$ lithium nitrite monohydrate

Simple Lewis Structures, Formulas, and Names for Molecular Compounds

5.61 (a) Hydrogen: Write the Lewis structure of each atom based on the number of valence electrons.

H· ·H

When the two hydrogen atoms share their electrons, they each get a duet, which is a stable configuration for hydrogen.

H—H

(b) The halogens: Write the Lewis structure of each atom based on the number of valence electrons.

If the two halogens pair together, each can achieve an octet, which is a stable configuration. So the halogens are predicted to exist as diatomic molecules.

(c) Oxygen: Write the Lewis structure of each atom based on the number of valence electrons.

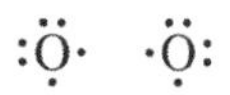

To achieve a stable octet on each oxygen, the oxygen atoms will need to share two electron pairs. So oxygen is predicted to exist as a diatomic molecule with a double bond.

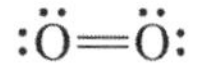

(d) Nitrogen: Write the Lewis structure of each atom based on the number of valence electrons.

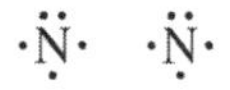

To achieve a stable octet on each nitrogen, the nitrogen atoms will need to share three electron pairs. So nitrogen is predicted to exist as a diatomic molecule with a triple bond.

5.62 Write the Lewis structure for N and H based on the number of valence electrons.

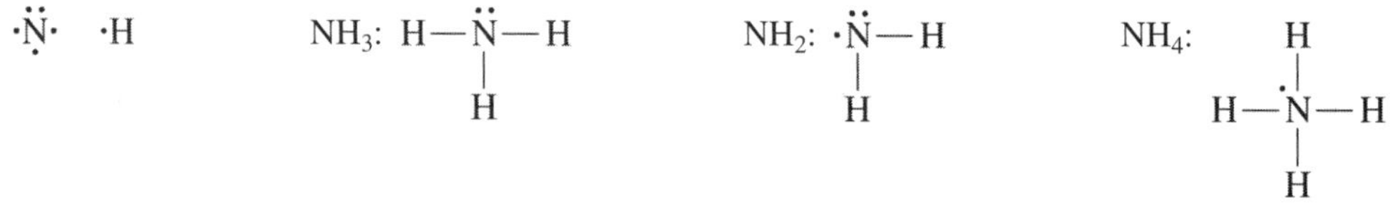

If nitrogen combines with three hydrogen atoms, the nitrogen will achieve a stable octet and each hydrogen will have a duet of electrons. This is a stable configuration. If the nitrogen were to combine with only two hydrogen atoms, the nitrogen could only achieve a seven-electron configuration, which is not stable. Also, if the nitrogen were to combine with four hydrogen atoms, the nitrogen would have a nine-electron configuration, which is not stable. So Lewis theory predicts that nitrogen will combine with three hydrogen atoms.

5.63 (a) CO The name of the compound is the name of the first element, *carbon*, followed by the base name of the second element, *ox*, prefixed by *mono-* to indicate 1 and given the suffix *-ide*. Because the prefix ends with "o" and the base name begins with "o," the first "o" is dropped: carbon monoxide.

(b) NI_3 The name of the compound is the name of the first element, *nitrogen*, followed by the base name of the second element, *iod*, prefixed by *tri-* to indicate 3 and given the suffix *-ide*: nitrogen triiodide.

(c) $SiCl_4$ The name of the compound is the name of the first element, *silicon*, followed by the base name of the second element, *chlor*, prefixed by *tetra-* to indicate 4 and given the suffix *-ide*: silicon tetrachloride.

(d) N_4Se_4 The name of the compound is the name of the first element, *nitrogen*, prefixed by *tetra-* to indicate 4 followed by the base name of the second element, *selen*, prefixed by *tetra-* to indicate 4 and given the suffix *-ide*: tetranitrogen tetraselenide.

5.64 (a) SO_3 The name of the compound is the name of the first element, *sulfur*, followed by the base name of the second element, *ox*, prefixed by *tri-* to indicate 3 and given the suffix *-ide*: sulfur trioxide.

(b) SO_2 The name of the compound is the name of the first element, *sulfur*, followed by the base name of the second element, *ox*, prefixed by *di-* to indicate 2 and given the suffix *-ide*: sulfur dioxide.

(c) BrF_5 The name of the compound is the name of the first element, *bromine*, followed by the base name of the second element, *fluor*, prefixed by *penta-* to indicate 5 and given the suffix *-ide*: bromine pentafluoride.

(d) NO The name of the compound is the name of the first element, *nitrogen*, followed by the base name of the second element, *ox*, prefixed by *mono-* to indicate 1 and given the suffix *-ide*. Because the prefix ends with "o" and the base name begins with "o," the first "o" is dropped: nitrogen monoxide.

5.65 (a) phosphorus trichloride: The name of the compound begins with the name of the first element, *phosphorous (P)*, followed by the base name of the second element, *chlor*, prefixed by *tri-* to indicate 3 chlorine (Cl) atoms, so the formula is PCl_3.

(b) chlorine monoxide: The name of the compound begins with the name of the first element, *chlorine (Cl)*, followed by the base name of the second element, *ox*, prefixed by *mono-* to indicate 1 oxygen (O) atom, so the formula is ClO.

(c) disulfur tetrafluoride: The name of the compound begins with the name of the first element, *sulfur (S)* prefixed by *di-* to indicate 2 S atoms, followed by the base name of the second element, *fluor*, prefixed by *tetra-* to indicate 4 fluorine (F) atoms, so the formula is S_2F_4.

(d) phosphorus pentafluoride: The name of the compound begins with the name of the first element, *phosphorous (P)*, followed by the base name of the second element, *fluor*, prefixed by *penta-* to indicate 5 fluorine (F) atoms, so the formula is PF_5.

5.66 (a) boron tribromide: The name of the compound begins with the name of the first element, *boron (B)*, followed by the base name of the second element, *brom*, prefixed by *tri-* to indicate 3 bromine (Br) atoms, so the formula is BBr_3.

(b) dichlorine monoxide: The name of the compound begins with the name of the first element, *chlorine (Cl)* prefixed by *di-* to indicate 2 Cl atoms, followed by the base name of the second element, *ox*, prefixed by *mono-* to indicate 1 oxygen (O) atom, so the formula is Cl_2O.

(c) xenon tetrafluoride: The name of the compound begins with the name of the first element, *xenon (Xe)*, followed by the base name of the second element, *fluor*, prefixed by *tetra-* to indicate 4 fluorine (F) atoms, so the formula is XeF_4.

(d) carbon tetrabromide: The name of the compound begins with the name of the first element, *carbon (C)*, followed by the base name of the second element, *brom*, prefixed by *tetra-* to indicate 4 bromine (Br) atoms, so the formula is CBr_4.

Naming Compounds (When the Type Is Not Specified)

5.67 To use the flowchart, begin by determining the type of compound you are trying to name—ionic, molecular, or acid.

(a) $SrCl_2$ is composed of a metal and a nonmetal, so it is ionic. Begin at the box labeled "IONIC" at the far left of the flowchart. The metal in the compound can only form one type of ion; therefore, take the left branch in the flowchart. Then name the compound according to the blocks at the end of the path in the flowchart. Write the name of the cation followed by the base name of the anion appended with the ending *-ide*. The full name is strontium chloride.

(b) SnO_2 is composed of a metal and a nonmetal, so it is ionic. Begin at the box labeled "IONIC" at the far left of the flowchart. The metal in the compound forms more than one type of ion; therefore, take the right branch in the flowchart. Then name the compound according to the blocks at the end of the path in the flowchart. Write the name of the cation followed by the charge of the cation in Roman numerals followed by the base name of the anion appended with the ending *-ide*. The full name is tin(IV) oxide.

(c) P_2S_5 is composed of two nonmetals and is a molecular compound. Begin at the box labeled "MOLECULAR" in the middle of the flowchart. Write the prefix to indicate the number of atoms of the first element, then the name of the first element followed by the prefix to indicate the number of atoms of the second element, then the base name of the second element followed by the ending *-ide*. The full name is diphorphorus pentasulfide.

5.68 To use the flowchart, begin by determining the type of compound you are trying to name—ionic, molecular, or acid.

(a) B_2Cl_2 is composed of two nonmetals and is a molecular compound. Begin at the box labeled "MOLECULAR" in the middle of the flowchart. Write the prefix to indicate the number of atoms of the first element, then the name of the first element followed by the prefix to indicate the number of atoms of the second element, then the base name of the second element followed by the ending *-ide*. The full name is diboron dichloride.

(b) $BaCl_2$ is composed of a metal and a nonmetal, so it is ionic. Begin at the box labeled "IONIC" at the far left of the flowchart. The metal in the compound can only form one type of ion; therefore, take the left branch in the flowchart. Then name the compound according to the blocks at the end of the path in the flowchart. Write the name of the cation followed by the base name of the anion appended with the ending *-ide*. The full name is barium chloride.

(c) $CrCl_3$ is composed of a metal and a nonmetal, so it is ionic. Begin at the box labeled "IONIC" at the far left of the flowchart. The metal in the compound forms more than one type of ion; therefore, take the right branch in the flowchart. Then name the compound according to the blocks at the end of the path in the flowchart. Write the name of the cation followed by the charge of the cation in Roman numerals followed by the base name of the anion appended with the ending *-ide*. The full name is chromium(III) chloride.

5.69 To use the flowchart, begin by determining the type of compound you are trying to name—ionic, molecular, or acid.

(a) $KClO_3$ is composed of a metal and an oxyanion, so it is ionic. Begin at the box labeled "IONIC" at the far left of the flowchart. The metal in the compound can only form one type of ion; therefore, take the left branch in the flowchart. Then name the compound according to the blocks at the end of the path in the flowchart. Write the name of the cation followed by the base name of the anion appended with the ending *-ate*. ClO_3^- is the polyatomic ion chlorate, so the name of the compound is potassium chlorate.

(b) I_2O_5 is composed of two nonmetals and is a molecular compound. Begin at the box labeled "MOLECULAR" in the middle of the flowchart. Write the prefix to indicate the number of atoms of the first element, then the name of the first element followed by the prefix to indicate the number of atoms of the second element, then the base name of the second element followed by the ending *-ide*. The full name is diiodine pentoxide.

(c) $PbSO_4$ is composed of a metal and an oxyanion, so it is ionic. Begin at the box labeled "IONIC" at the far left of the flowchart. The metal in the compound forms more than one type of ion; therefore, take the right branch in the flowchart. Then name the compound according to the blocks at the end of the path in the flowchart. Write the name of the cation followed by the charge of the cation in Roman numerals followed by the base name of the anion appended with the ending *-ate*. SO_4^{2-} is the polyatomic ion sulfate, so the name of the compound is lead (II) sulfate.

5.70 To use the flowchart, begin by determining the type of compound you are trying to name—ionic, molecular, or acid.

(a) XeO_3 is composed of two nonmetals and is a molecular compound. Begin at the box labeled "MOLECULAR" in the middle of the flowchart. Write the prefix to indicate the number of atoms of the first element, then the name of the first element followed by the prefix to indicate the number of atoms of the second element, then the base name of the second element followed by the ending *-ide*. The full name is xenon trioxide.

(b) $KClO$ is composed of a metal and an oxyanion, so it is ionic. Begin at the box labeled "IONIC" at the far left of the flowchart. The metal in the compound can only form one type of ion; therefore, take the left branch in the flowchart. Then name the compound according to the blocks at the end of the path in the flowchart. Write the name of the cation followed by the base name of the anion appended with the ending *-ite*. ClO^- is the polyatomic ion hypochlorite, so the name of the compound is potassium hypochlorite.

(c) $CoSO_4$ is composed of a metal and an oxyanion, so it is ionic. Begin at the box labeled "IONIC" at the far left of the flowchart. The metal in the compound forms more than one type of ion; therefore, take the right branch in the flowchart. Then name the compound according to the blocks at the end of the path in the flowchart. Write the name of the cation followed by the charge of the cation in Roman numerals followed by the base name of the anion appended with the ending *-ate*. SO_4^{2-} is the polyatomic ion sulfate. So the name of the compound is cobalt(II) sulfate.

Formula Mass and the Mole Concept for Compounds

5.71 To find the formula mass, sum the atomic masses of each atom in the chemical formula.

(a) NO_2

$$\begin{aligned}\text{formula mass} &= 1 \times (\text{atomic mass N}) + 2 \times (\text{atomic mass O})\\ &= 1 \times (14.01\ \text{amu}) + 2 \times (16.00\ \text{amu})\\ &= 46.01\ \text{amu}\end{aligned}$$

(b) C_4H_{10} formula mass $= 4 \times (\text{atomic mass C}) + 10 \times (\text{atomic mass H})$
$= 4 \times (12.01\ \text{amu}) + 10 \times (1.008\ \text{amu})$
$= 58.12\ \text{amu}$

(c) $C_6H_{12}O_6$ formula mass $= 6 \times (\text{atomic mass C}) + 12 \times (\text{atomic mass H}) + 6 \times (\text{atomic mass O})$
$= 6 \times (12.01\ \text{amu}) + 12 \times (1.008\ \text{amu}) + 6 \times (16.00\ \text{amu})$
$= 180.16\ \text{amu}$

(d) $Cr(NO_3)_3$ formula mass $= 1 \times (\text{atomic mass Cr}) + 3 \times (\text{atomic mass N}) + 9 \times (\text{atomic mass O})$
$= 1 \times (52.00\ \text{amu}) + 3 \times (14.01\ \text{amu}) + 9 \times (16.00\ \text{amu})$
$= 238.03\ \text{amu}$

Check: The units (amu) are correct. Each formula mass can be reported to two places to the right of the decimal point because each atom in the compound is reported to two or three places to the right of the decimal point.

5.72 To find the formula mass, sum the atomic masses of each atom in the chemical formula.

(a) $MgBr_2$ formula mass $= 1 \times (\text{atomic mass Mg}) + 2 \times (\text{atomic mass Br})$
$= 1 \times (24.31\ \text{amu}) + 2 \times (79.90\ \text{amu})$
$= 184.11\ \text{amu}$

(b) HNO_2 formula mass $= 1 \times (\text{atomic mass H}) + 1 \times (\text{atomic mass N}) + 2 \times (\text{atomic mass O})$
$= 1 \times (1.008\ \text{amu}) + 1 \times (14.01\ \text{amu}) + 2 \times (16.00\ \text{amu})$
$= 47.02\ \text{amu}$

(c) CBr_4 formula mass $= 1 \times (\text{atomic mass C}) + 4 \times (\text{atomic mass Br})$
$= 1 \times (12.01\ \text{amu}) + 4 \times (79.90\ \text{amu})$
$= 331.61\ \text{amu}$

(d) $Ca(NO_3)_2$ formula mass $= 1 \times (\text{atomic mass Ca}) + 2 \times (\text{atomic mass N}) + 6 \times (\text{atomic mass O})$
$= 1 \times (40.08\ \text{amu}) + 2 \times (14.01\ \text{amu}) + 6 \times (16.00\ \text{amu})$
$= 164.10\ \text{amu}$

Check: The units (amu) are correct. Each formula mass can be reported to two places to the right of the decimal point because each atom in the compound is reported to two or three places to the right of the decimal point.

5.73 (a) **Given:** 72.5 g CCl_4 **Find:** number of moles

Conceptual Plan: g CCl_4 → mole CCl_4

$$\frac{1\ \text{mol}}{153.81\ \text{g CCl}_4}$$

Solution: $72.5\ \cancel{\text{g CCl}_4} \times \frac{1\ \text{mol CCl}_4}{153.81\ \cancel{\text{g CCl}_4}} = 0.47\underline{1}4\ \text{mol CCl}_4 = 0.471\ \text{mol CCl}_4$

Check: Units of the answer (mole CCl_4) are correct. The magnitude is appropriate because it is less than 1 mole of NO_2.

(b) **Given:** 12.4 g $C_{12}H_{22}O_{11}$ **Find:** number of moles

Conceptual Plan: g $C_{12}H_{22}O_{11}$ → mole KNO_3

$$\frac{1\ \text{mol}}{342.296\ \text{g C}_{12}\text{H}_{22}\text{O}_{11}}$$

Solution: $12.4\ \cancel{\text{g C}_{12}\text{H}_{22}\text{O}_{11}} \times \frac{1\ \text{mol C}_{12}\text{H}_{22}\text{O}_{11}}{342.296\ \cancel{\text{g C}_{12}\text{H}_{22}\text{O}_{11}}} = 0.036\underline{2}3\ \text{mol C}_{12}\text{H}_{22}\text{O}_{11} = 0.0362\ \text{mol C}_{12}\text{H}_{22}\text{O}_{11}$

Check: Units of the answer (mole $C_{12}H_{22}O_{11}$) are correct. The magnitude is appropriate because there is less than 1 mole of $C_{12}H_{22}O_{11}$.

(c) **Given:** 25.2 kg C_2H_2 **Find:** number of moles

Conceptual Plan: kg C_2H_2 → g C_2H_2 → mole C_2H_2

$$\frac{1000\ \text{g C}_2\text{H}_2}{1\ \text{kg C}_2\text{H}_2} \qquad \frac{1\ \text{mol}}{26.036\ \text{g C}_2\text{H}_2}$$

Solution: $25.2\ \cancel{\text{kg C}_2\text{H}_2} \times \frac{1000\ \cancel{\text{g C}_2\text{H}_2}}{1\ \cancel{\text{kg C}_2\text{H}_2}} \times \frac{1\ \text{mol C}_2\text{H}_2}{26.036\ \cancel{\text{g C}_2\text{H}_2}} = 96\underline{7}.9\ \text{mol C}_2\text{H}_2 = 968\ \text{mol C}_2\text{H}_2$

Check: Units of the answer (mole C_2H_2) are correct. The magnitude is appropriate because more than a kilogram of C_2H_2 is present.

(d) **Given:** 12.3 g dinitrogen monoxide **Find:** number of moles

Conceptual Plan: Dinitrogen monoxide → formula and then g N_2O → mole N_2O

$$\frac{1 \text{ mol}}{44.02 \text{ g } N_2O}$$

Solution: Dinitrogen monoxide is N_2O.

$$12.3 \text{ g } N_2O \times \frac{1 \text{ mol } N_2O}{44.02 \text{ g } N_2O} = 0.27\underline{9}4 \text{ mol } N_2O = 0.279 \text{ mol } N_2O$$

Check: Units of the answer (mole N_2O) are correct. The magnitude is appropriate because less than a mole of N_2O is present.

5.74 (a) **Given:** 15.7 mol HNO_3 **Find:** number of grams

Conceptual Plan: Mole HNO_3 → g HNO_3

$$\frac{63.018 \text{ g } HNO_3}{1 \text{ mol}}$$

Solution: $15.7 \text{ mol } HNO_3 \times \frac{63.018 \text{ g } HNO_3}{1 \text{ mol } HNO_3} = 98\underline{9}.38 \text{ g } HNO_3 = 989 \text{ g } HNO_3$

Check: Units of the answer (g HNO_3) are correct. The magnitude is appropriate because there are more than 15 moles of HNO_3.

(b) **Given:** 1.04×10^{-3} mol H_2O_2 **Find:** number of grams

Conceptual Plan: Mole H_2O_2 → g H_2O_2

$$\frac{34.016 \text{ g } H_2O_2}{1 \text{ mol}}$$

Solution: $1.04 \times 10^{-3} \text{ mol } H_2O_2 \times \frac{34.016 \text{ g } H_2O_2}{1 \text{ mol } H_2O_2} = 0.035\underline{3}8 \text{ g } H_2O_2 = 0.0354 \text{ g } H_2O_2$

Check: Units of the answer (g H_2O_2) are correct. The magnitude is appropriate because significantly less than 1 mole of H_2O_2 is present.

(c) **Given:** 72.1 mol SO_2 **Find:** number of grams

Conceptual Plan: Mole SO_2 → g SO_2

$$\frac{64.07 \text{ g } SO_2}{1 \text{ mol}}$$

Solution: $72.1 \text{ mol } SO_2 \times \frac{64.07 \text{ g } SO_2}{1 \text{ mol } SO_2} = 46\underline{1}9.4 \text{ g } SO_2 = 4.62 \times 10^3 \text{ g } SO_2$

Check: Units of the answer (g SO_2) are correct. The magnitude is appropriate because there are more than 70 moles of SO_2.

(d) **Given:** 1.23 mol xenon difluoride **Find:** number of grams

Conceptual Plan: Name xenon difluoride → formula and then mole XeF_2 → g XeF_2

$$\frac{169.29 \text{ g } XeF_2}{1 \text{ mol}}$$

Solution: $1.23 \text{ mol } XeF_2 \times \frac{169.29 \text{ g } XeF_2}{1 \text{ mol } XeF_2} = 20\underline{8}.23 \text{ g } XeF_2 = 208 \text{ g } XeF_2$

Check: Units of the answer, g XeF_2, are correct. The magnitude is appropriate because there is more than a mole of XeF_2.

5.75 (a) **Given:** 25.5 g NO_2 **Find:** number of moles

Conceptual Plan: g NO_2 → mole NO_2

$$\frac{1 \text{ mol}}{46.01 \text{ g } NO_2}$$

Solution: $25.5 \text{ g } NO_2 \times \frac{1 \text{ mol } NO_2}{46.01 \text{ g } NO_2} = 0.554 \text{ mol } NO_2$

Check: The units of the answer (mole NO_2) are correct. The magnitude is appropriate because it is less than 1 mole of NO_2.

(b) **Given:** 1.25 kg CO_2 **Find:** number of moles

Conceptual Plan: kg CO_2 → g CO_2 → mole CO_2

$$\frac{1000\text{ g }CO_2}{1\text{ kg }CO_2} \quad \frac{1\text{ mol}}{44.01\text{ g }CO_2}$$

Solution: $1.25\text{ kg }CO_2 \times \frac{1000\text{ g }CO_2}{1\text{ kg }CO_2} \times \frac{1\text{ mol }CO_2}{44.01\text{ g }CO_2} = 28.4\text{ mol }CO_2$

Check: The units of the answer (mole CO_2) are correct. The magnitude is appropriate because over a kg of CO_2 is present.

(c) **Given:** 38.2 g KNO_3 **Find:** number of moles

Conceptual Plan: g KNO_3 → mole KNO_3

$$\frac{1\text{ mol}}{101.11\text{ g }KNO_3}$$

Solution: $38.2\text{ g }KNO_3 \times \frac{1\text{ mol }KNO_3}{101.11\text{ g }KNO_3} = 0.378\text{ mol }KNO_3$

Check: The units of the answer (mole KNO_3) are correct. The magnitude is appropriate because there is less than 1 mole of KNO_3.

(d) **Given:** 155.2 kg Na_2SO_4 **Find:** number of moles

Conceptual Plan: kg Na_2SO_4 → g Na_2SO_4 → mole Na_2SO_4

$$\frac{1000\text{ g }Na_2SO_4}{1\text{ kg }Na_2SO_4} \quad \frac{1\text{ mol}}{142.05\text{ g }Na_2SO_4}$$

Solution: $155.2\text{ kg }Na_2SO_4 \times \frac{1000\text{g }Na_2SO_4}{1\text{ kg }Na_2SO_4} \times \frac{1\text{ mol }Na_2SO_4}{142.05\text{ g }Na_2SO_4} = 1092\text{ mol }Na_2SO_4$

Check: The units of the answer (mole Na_2SO_4) are correct. The magnitude is appropriate because over 100 kg of Na_2SO_4 is present.

5.76 (a) **Given:** 55.98 g CF_2Cl_2 **Find:** number of moles

Conceptual Plan: g CF_2Cl_2 → mole CF_2Cl_2

$$\frac{1\text{ mol}}{120.91\text{ g }CF_2Cl_2}$$

Solution: $55.98\text{ g }CF_2Cl_2 \times \frac{1\text{ mol }CF_2Cl_2}{120.91\text{ g }CF_2Cl_2} = 0.462\underline{9}8\text{ mol }CF_2Cl_2 = 0.4630\text{ mol }CF_2Cl_2$

Check: The units of the answer (mole CF_2Cl_2) are correct. The magnitude is appropriate because it is less than 1 mole of CF_2Cl_2.

(b) **Given:** 23.6 kg $Fe(NO_3)_2$ **Find:** number of moles

Conceptual Plan: kg $Fe(NO_3)_2$ → g $Fe(NO_3)_2$ → mole $Fe(NO_3)_2$

$$\frac{1000\text{ g }Fe(NO_3)_2}{1\text{ kg }Fe(NO_3)_2} \quad \frac{1\text{ mol}}{179.87\text{ g }Fe(NO_3)_2}$$

Solution: $23.6\text{ kg }Fe(NO_3)_2 \times \frac{1000\text{ g }Fe(NO_3)_2}{1\text{ kg }Fe(NO_3)_2} \times \frac{1\text{ mol }Fe(NO_3)_2}{179.87\text{ g }Fe(NO_3)_2} = 131\text{ mol }Fe(NO_3)_2$

Check: The units of the answer (mole $Fe(NO_3)_2$) are correct. The magnitude is appropriate because over a kg of $Fe(NO_3)_2$ is present.

(c) **Given:** 0.1187 g C_8H_{18} **Find:** number of moles

Conceptual Plan: g C_8H_{18} → mole C_8H_{18}

$$\frac{1\text{ mol}}{114.22\text{ g }C_8H_{18}}$$

Solution: $0.1187\text{ g }C_8H_{18} \times \frac{1\text{ mol }C_8H_{18}}{114.22\text{ g }C_8H_{18}} = 1.039 \times 10^{-3}\text{ mol }C_8H_{18}$

Check: The units of the answer (mole C_8H_{18}) are correct. The magnitude is appropriate because it is much less than 1 mole of C_8H_{18}.

(d) **Given:** 195 kg CaO **Find:** number of moles

Conceptual Plan: kg CaO → g CaO → mole CaO

$$\frac{1000 \text{ g CaO}}{1 \text{ kg CaO}} \quad \frac{1 \text{ mol}}{56.08 \text{ g CaO}}$$

Solution: $195 \text{ kg CaO} \times \frac{1000 \text{ g CaO}}{1 \text{ kg CaO}} \times \frac{1 \text{ mol CaO}}{56.08 \text{ g CaO}} = 3477 \text{ mol CaO} = 3.48 \times 10^3 \text{ mol CaO}$

Check: The units of the answer (mole CaO) are correct. The magnitude is appropriate because over a kg of CaO is present.

5.77 (a) **Given:** 6.5 g H_2O **Find:** number of molecules

Conceptual Plan: g H_2O → mole H_2O → number H_2O molecules

$$\frac{1 \text{ mol}}{18.02 \text{ g } H_2O} \quad \frac{6.022 \times 10^{23} \text{ } H_2O \text{ molecules}}{\text{mol } H_2O}$$

Solution: $6.5 \text{ g } H_2O \times \frac{1 \text{ mol } H_2O}{18.02 \text{ g } H_2O} \times \frac{6.022 \times 10^{23} \text{ } H_2O \text{ molecules}}{\text{mol } H_2O} = 2.2 \times 10^{23} \text{ } H_2O \text{ molecules}$

Check: The units of the answer (H_2O molecules) are correct. The magnitude is appropriate; it is smaller than Avogadro's number, as expected, because we have less than 1 mole of H_2O.

(b) **Given:** 389 g CBr_4 **Find:** number of molecules

Conceptual Plan: g CBr_4 → mole CBr_4 → number CBr_4 molecules

$$\frac{1 \text{ mol}}{331.6 \text{ g } CBr_4} \quad \frac{6.022 \times 10^{23} \text{ } CBr_4 \text{ molecules}}{\text{mol } CBr_4}$$

Solution: $389 \text{ g } CBr_4 \times \frac{1 \text{ mol } CBr_4}{331.6 \text{ g } CBr_4} \times \frac{6.022 \times 10^{23} \text{ } CBr_4 \text{ molecules}}{\text{mol } CBr_4} = 7.06 \times 10^{23} \text{ } CBr_4 \text{ molecules}$

Check: The units of the answer (CBr_4 molecules) are correct. The magnitude is appropriate; it is larger than Avogadro's number, as expected, because we have more than 1 mole of CBr_4.

(c) **Given:** 22.1 g O_2 **Find:** number of molecules

Conceptual Plan: g O_2 → mole O_2 → number O_2 molecules

$$\frac{1 \text{ mol}}{32.00 \text{ g } O_2} \quad \frac{6.022 \times 10^{23} \text{ } O_2 \text{ molecules}}{\text{mol } O_2}$$

Solution: $22.1 \text{ g } O_2 \times \frac{1 \text{ mol } O_2}{32.00 \text{ g } O_2} \times \frac{6.022 \times 10^{23} \text{ } O_2 \text{ molecules}}{\text{mol } O_2} = 4.16 \times 10^{23} \text{ } O_2 \text{ molecules}$

Check: The units of the answer (O_2 molecules) are correct. The magnitude is appropriate; it is smaller than Avogadro's number, as expected, because we have less than 1 mole of O_2.

(d) **Given:** 19.3 g C_8H_{10} **Find:** number of molecules

Conceptual Plan: g C_8H_{10} → mole C_8H_{10} → number C_8H_{10} molecules

$$\frac{1 \text{ mol}}{106.16 \text{ g } C_8H_{10}} \quad \frac{6.022 \times 10^{23} \text{ } C_8H_{10} \text{ molecules}}{\text{mol } C_8H_{10}}$$

Solution:

$$19.3 \text{ g } C_8H_{10} \times \frac{1 \text{ mol } C_8H_{10}}{106.16 \text{ g } C_8H_{10}} \times \frac{6.022 \times 10^{23} \text{ } C_8H_{10} \text{ molecules}}{\text{mol } C_8H_{10}} = 1.09 \times 10^{23} \text{ } C_8H_{10} \text{ molecules}$$

Check: The units of the answer (C_8H_{10} molecules) are correct. The magnitude is appropriate; it is smaller than Avogadro's number, as expected, because we have less than 1 mole of C_8H_{10}.

5.78 (a) **Given:** 85.26 g CCl_4 **Find:** number of molecules

Conceptual Plan: g CCl_4 → mole CCl_4 → number CCl_4 molecules

$$\frac{1 \text{ mol}}{153.81 \text{ g } CCl_4} \quad \frac{6.022 \times 10^{23} \text{ } CCl_4 \text{ molecules}}{\text{mol } CCl_4}$$

Solution: $85.26 \text{ g } CCl_4 \times \frac{1 \text{ mol } CCl_4}{153.81 \text{ g } CCl_4} \times \frac{6.022 \times 10^{23} \text{ } CCl_4 \text{ molecules}}{\text{mol } CCl_4}$

$= 3.3381 \times 10^{23} \text{ } CCl_4 \text{ molecules} = 3.338 \times 10^{23} \text{ } CCl_4 \text{ molecules}$

Check: The units of the answer (CCl_4 molecules) are correct. The magnitude is appropriate; it is smaller than Avogadro's number, as expected, because we have less than 1 mole of CCl_4.

(b) **Given:** 55.93 kg $NaHCO_3$ **Find:** number of molecules

Conceptual Plan: kg $NaHCO_3$ → g $NaHCO_3$ → mole $NaHCO_3$ → number $NaHCO_3$ molecules

$$\frac{1000\text{ g}}{1\text{ kg}} \quad \frac{1\text{ mol}}{84.01\text{ g NaHCO}_3} \quad \frac{6.022\times10^{23}\text{ NaHCO}_3\text{ molecules}}{\text{mol NaHCO}_3}$$

Solution:

$$55.93\text{ kg NaHCO}_3 \times \frac{1000\text{ g NaHCO}_3}{1\text{ kg NaHCO}_3} \times \frac{1\text{ mol NaHCO}_3}{84.01\text{ g NaHCO}_3} \times \frac{6.022\times10^{23}\text{ NaHCO}_3\text{ molecules}}{\text{mol NaHCO}_3}$$

$= 4.009 \times 10^{26}$ $NaHCO_3$ molecules

Check: The units of the answer ($NaHCO_3$ molecules) are correct. The magnitude is appropriate; it is more than Avogadro's number, as expected, because we have many moles of $NaHCO_3$.

(c) **Given:** 119.78 g C_4H_{10} **Find:** number of molecules

Conceptual Plan: g C_4H_{10} → mole C_4H_{10} → number C_4H_{10} molecules

$$\frac{1\text{ mol}}{58.12\text{ g C}_4\text{H}_{10}} \quad \frac{6.022\times10^{23}\text{ C}_4\text{H}_{10}\text{ molecules}}{\text{mol C}_4\text{H}_{10}}$$

Solution:

$$119.78\text{ g C}_4\text{H}_{10} \times \frac{1\text{ mol C}_4\text{H}_{10}}{58.12\text{ g C}_4\text{H}_{10}} \times \frac{6.022\times10^{23}\text{ C}_4\text{H}_{10}\text{ molecules}}{\text{mol C}_4\text{H}_{10}} = 1.241\times10^{24}\text{ C}_4\text{H}_{10}\text{ molecules}$$

Check: The units of the answer (C_4H_{10} molecules) are correct. The magnitude is appropriate; it is larger than Avogadro's number, as expected, because we have more than 1 mole of C_4H_{10}.

(d) **Given:** 4.59×10^5 g Na_3PO_4 **Find:** number of molecules

Conceptual Plan: g Na_3PO_4 → mole Na_3PO_4 → number Na_3PO_4 molecules

$$\frac{1\text{ mol}}{163.94\text{ g Na}_3\text{PO}_4} \quad \frac{6.022\times10^{23}\text{ Na}_3\text{PO}_4\text{ molecules}}{\text{mol Na}_3\text{PO}_4}$$

Solution: $$4.59\times10^5\text{ g Na}_3\text{PO}_4 \times \frac{1\text{ mol Na}_3\text{PO}_4}{163.94\text{ g Na}_3\text{PO}_4} \times \frac{6.022\times10^{23}\text{ Na}_3\text{PO}_4\text{ molecules}}{\text{mol Na}_3\text{PO}_4}$$

$= 1.6\underline{8}6 \times 10^{27}$ Na_3PO_4 molecules $= 1.69 \times 10^{27}$ Na_3PO_4 molecules

Check: The units of the answer (Na_3PO_4 molecules) are correct. The magnitude is appropriate; it is larger than Avogadro's number, as expected, because we have more than 1 mole of Na_3PO_4.

5.79 (a) **Given:** 5.94×10^{20} SO_3 molecules **Find:** mass in g

Conceptual Plan: Number SO_3 molecules → mole SO_3 → g SO_3

$$\frac{1\text{ mol SO}_3}{6.022\times10^{23}\text{ SO}_3\text{ molecules}} \quad \frac{80.07\text{ g SO}_3}{1\text{ mol SO}_3}$$

Solution: $$5.94\times10^{20}\text{ SO}_3\text{ molecules} \times \frac{1\text{ mol SO}_3}{6.022\times10^{23}\text{ SO}_3\text{ molecules}} \times \frac{80.07\text{ g SO}_3}{1\text{ mol SO}_3} = 0.0790\text{ g SO}_3$$

Check: The units of the answer (grams SO_3) are correct. The magnitude is appropriate; there is less than Avogadro's number of molecules, so we have less than 1 mole of SO_3.

(b) **Given:** 2.8×10^{22} H_2O molecules **Find:** mass in g

Conceptual Plan: Number H_2O molecules → mole H_2O → g H_2O

$$\frac{1\text{ mol H}_2\text{O}}{6.022\times10^{23}\text{ H}_2\text{O molecules}} \quad \frac{18.02\text{ g H}_2\text{O}}{1\text{ mol H}_2\text{O}}$$

Solution: $$2.8\times10^{22}\text{ H}_2\text{O molecules} \times \frac{1\text{ mol H}_2\text{O}}{6.022\times10^{23}\text{ H}_2\text{O molecules}} \times \frac{18.02\text{ g H}_2\text{O}}{1\text{ mol H}_2\text{O}} = 0.84\text{ g H}_2\text{O}$$

Check: The units of the answer (grams H_2O) are correct. The magnitude is appropriate; there is less than Avogadro's number of molecules, so we have less than 1 mole of H_2O.

(c) **Given:** 1 $C_6H_{12}O_6$ molecule **Find:** mass in g
Conceptual Plan: Number $C_6H_{12}O_6$ molecules → mole $C_6H_{12}O_6$ → g $C_6H_{12}O_6$

$$\frac{1 \text{ mol } C_6H_{12}O_6}{6.022 \times 10^{23}\ C_6H_{12}O_6 \text{ molecules}} \qquad \frac{180.16 \text{ g } C_6H_{12}O_6}{1 \text{ mol } C_6H_{12}O_6}$$

Solution:

$$1\ \cancel{C_6H_{12}O_6 \text{ molecule}} \times \frac{1\ \cancel{\text{mol } C_6H_{12}O_6}}{6.022 \times 10^{23}\ \cancel{C_6H_{12}O_6 \text{ molecules}}} \times \frac{180.16 \text{ g } C_6H_{12}O_6}{1\ \cancel{\text{mol } C_6H_{12}O_6}} = 2.992 \times 10^{-22} \text{ g } C_6H_{12}O_6$$

Check: The units of the answer (grams $C_6H_{12}O_6$) are correct. The magnitude is appropriate; there is much less than Avogadro's number of molecules, so we have much less than 1 mole of $C_6H_{12}O_6$.

5.80 (a) **Given:** 4.5×10^{25} O_3 molecules **Find:** mass in g
Conceptual Plan: Number O_3 molecules → mole O_3 → g O_3

$$\frac{1 \text{ mol } O_3}{6.022 \times 10^{23}\ O_3 \text{ molecules}} \qquad \frac{48.00 \text{ g } O_3}{1 \text{ mol } O_3}$$

Solution: $4.5 \times 10^{25}\ \cancel{O_3 \text{ molecules}} \times \frac{1\ \cancel{\text{mol } O_3}}{6.022 \times 10^{23}\ \cancel{O_3 \text{ molecules}}} \times \frac{48.00 \text{ g } O_3}{1\ \cancel{\text{mol } O_3}} = 3.6 \times 10^{3} \text{ g } O_3$

Check: The units of the answer (grams O_3) are correct. The magnitude is appropriate; there is more than Avogadro's number of molecules, so we have more than 1 mole of O_3.

(b) **Given:** 9.85×10^{19} CCl_2F_2 molecules **Find:** mass in g
Conceptual Plan: Number CCl_2F_2 molecules → mole CCl_2F_2 → g CCl_2F_2

$$\frac{1 \text{ mol } CCl_2F_2}{6.022 \times 10^{23}\ CCl_2F_2 \text{ molecules}} \qquad \frac{120.91 \text{ g } CCl_2F_2}{1 \text{ mol } CCl_2F_2}$$

Solution:

$$9.85 \times 10^{19}\ \cancel{CCl_2F_2 \text{ molecules}} \times \frac{1\ \cancel{\text{mol } CCl_2F_2}}{6.022 \times 10^{23}\ \cancel{CCl_2F_2 \text{ molecules}}} \times \frac{120.91 \text{ g } CCl_2F_2}{1\ \cancel{\text{mol } CCl_2F_2}} = 1.98 \times 10^{-2} \text{ g } CCl_2F_2$$

Check: The units of the answer (grams CCl_2F_2) are correct. The magnitude is appropriate; there is less than Avogadro's number of molecules, so we have less than 1 mole of CCl_2F_2.

(c) **Given:** 1 H_2O molecule **Find:** mass in g
Conceptual Plan: Number H_2O molecules → mole H_2O → g H_2O

$$\frac{1 \text{ mol } H_2O}{6.022 \times 10^{23}\ H_2O \text{ molecules}} \qquad \frac{18.02 \text{ g } H_2O}{1 \text{ mol } H_2O}$$

Solution: $1\ \cancel{H_2O \text{ molecule}} \times \frac{1\ \cancel{\text{mol } H_2O}}{6.022 \times 10^{23}\ \cancel{H_2O \text{ molecules}}} \times \frac{18.02 \text{ g } H_2O}{1\ \cancel{\text{mol } H_2O}} = 2.992 \times 10^{-23} \text{ g } H_2O$

Check: The units of the answer (grams H_2O) are correct. The magnitude is appropriate; there is much less than Avogadro's number of molecules, so we have much less than 1 mole of H_2O.

5.81 **Given:** 1.8×10^{17} $C_{12}H_{22}O_{11}$ molecules **Find:** mass in mg
Conceptual Plan: Number $C_{12}H_{22}O_{11}$ molecules → mole $C_{12}H_{22}O_{11}$ → g $C_{12}H_{22}O_{11}$ → mg $C_{12}H_{22}O_{11}$

$$\frac{1 \text{ mol } C_{12}H_{22}O_{11}}{6.022 \times 10^{23}\ C_{12}H_{22}O_{11} \text{ molecules}} \qquad \frac{342.3 \text{ g } C_{12}H_{22}O_{11}}{1 \text{ mol } C_{12}H_{22}O_{11}} \qquad \frac{1 \times 10^{3} \text{ mg } C_{12}H_{22}O_{11}}{1 \text{ g } C_{12}H_{22}O_{11}}$$

Solution:

$$1.8 \times 10^{17}\ \cancel{C_{12}H_{22}O_{11} \text{ molecules}} \times \frac{1\ \cancel{\text{mol } C_{12}H_{22}O_{11}}}{6.022 \times 10^{23}\ \cancel{C_{12}H_{22}O_{11} \text{ molecules}}} \times \frac{342.3\ \cancel{\text{g } C_{12}H_{22}O_{11}}}{1\ \cancel{\text{mol } C_{12}H_{22}O_{11}}} \times \frac{1 \times 10^{3} \text{ mg } C_{12}H_{22}O_{11}}{1\ \cancel{\text{g } C_{12}H_{22}O_{11}}}$$

$= 0.10 \text{ mg } C_{12}H_{22}O_{11}$

Check: The units of the answer (mg $C_{12}H_{22}O_{11}$) are correct. The magnitude is appropriate; there is much less than Avogadro's number of molecules, so we have much less than 1 mole of $C_{12}H_{22}O_{11}$.

5.82 **Given:** 0.12 mg NaCl **Find:** number of formula units
Conceptual Plan: mg NaCl → g NaCl → mole NaCl → number of formula units NaCl

$$\frac{1 \text{ g NaCl}}{1 \times 10^{3} \text{ mg NaCl}} \qquad \frac{1 \text{ mol NaCl}}{58.44 \text{ g NaCl}} \qquad \frac{6.022 \times 10^{23} \text{ NaCl formula units}}{1 \text{ mol NaCl}}$$

Solution: $0.12\ \text{mg NaCl} \times \frac{1\ \text{g NaCl}}{1 \times 10^{3}\ \text{mg NaCl}} \times \frac{1\ \text{mol NaCl}}{58.44\ \text{g NaCl}} \times \frac{6.022 \times 10^{23}\ \text{formula units NaCl}}{1\ \text{mol NaCl}}$
$= 1.2 \times 10^{18}$ formula units NaCl

Check: The units of the answer (formula units NaCl) are correct. The magnitude is appropriate; there is less than 1 mole of NaCl, so we have less than Avogadro's number of formula units.

Composition of Compounds

5.83 (a) **Given:** CH_4 **Find:** mass percent C

Conceptual Plan: $\text{Mass \% C} = \frac{1 \times \text{molar mass C}}{\text{molar mass } CH_4} \times 100\%$

Solution:

$$1 \times \text{molar mass C} = 1(12.01\ \text{g/mol}) = 12.01\ \text{g C}$$
$$\text{molar mass } CH_4 = 1(12.01\ \text{g/mol}) + 4(1.008\ \text{g/mol}) = 16.04\ \text{g/mol}$$
$$\text{mass \% C} = \frac{1 \times \text{molar mass C}}{\text{molar mass } CH_4} \times 100\%$$
$$= \frac{12.01\ \text{g/mol}}{16.04\ \text{g/mol}} \times 100\%$$
$$= 74.87\%$$

Check: The units of the answer (%) are correct. The magnitude is reasonable because it is between 0 and 100% and carbon is the heaviest element.

(b) **Given:** C_2H_6 **Find:** mass percent C

Conceptual Plan: $\text{Mass \% C} = \frac{2 \times \text{molar mass C}}{\text{molar mass } C_2H_6} \times 100\%$

Solution:

$$2 \times \text{molar mass C} = 2(12.01\ \text{g/mol}) = 24.02\ \text{g C}$$
$$\text{molar mass } C_2H_6 = 2(12.01\ \text{g/mol}) + 6(1.008\ \text{g/mol}) = 30.07\ \text{g/mol}$$
$$\text{mass \% C} = \frac{2 \times \text{molar mass C}}{\text{molar mass } C_2H_6} \times 100\%$$
$$= \frac{24.02\ \text{g/mol}}{30.07\ \text{g/mol}} \times 100\%$$
$$= 79.88\%$$

Check: The units of the answer (%) are correct. The magnitude is reasonable because it is between 0 and 100% and carbon is the heaviest element.

(c) **Given:** C_2H_2 **Find:** mass percent C

Conceptual Plan: $\text{Mass \% C} = \frac{2 \times \text{molar mass C}}{\text{molar mass } C_2H_2} \times 100\%$

Solution:

$$2 \times \text{molar mass C} = 2(12.01\ \text{g/mol}) = 24.02\ \text{g C}$$
$$\text{molar mass } C_2H_2 = 2(12.01\ \text{g/mol}) + 2(1.008\ \text{g/mol}) = 26.04\ \text{g/mol}$$
$$\text{mass \% C} = \frac{2 \times \text{molar mass C}}{\text{molar mass } C_2H_2} \times 100\%$$
$$= \frac{24.02\ \text{g/mol}}{26.04\ \text{g/mol}} \times 100\%$$
$$= 92.24\%$$

Check: The units of the answer (%) are correct. The magnitude is reasonable because it is between 0 and 100% and carbon is the heaviest element.

(d) **Given:** C_2H_5Cl **Find:** mass percent C

Conceptual Plan: $\text{Mass \% C} = \dfrac{2 \times \text{molar mass C}}{\text{molar mass } C_2H_5Cl} \times 100\%$

Solution:

$$2 \times \text{molar mass C} = 2(12.01\text{ g/mol}) = 24.02\text{ g C}$$

$$\text{molar mass } C_2H_5Cl = 2(12.01\text{ g/mol}) + 5(1.008\text{ g/mol}) + 1(35.45\text{ g/mol}) = 64.51\text{ g/mol}$$

$$\begin{aligned}\text{mass \% C} &= \frac{2 \times \text{molar mass C}}{\text{molar mass } C_2H_5Cl} \times 100\% \\ &= \frac{24.02\ \cancel{\text{g/mol}}}{64.51\ \cancel{\text{g/mol}}} \times 100\% \\ &= 37.23\%\end{aligned}$$

Check: The units of the answer (%) are correct. The magnitude is reasonable because it is between 0 and 100% and chlorine is heavier than carbon.

5.84 (a) **Given:** N_2O **Find:** mass percent N

Conceptual Plan: $\text{Mass \% N} = \dfrac{2 \times \text{molar mass N}}{\text{molar mass } N_2O} \times 100\%$

Solution:

$$2 \times \text{molar mass N} = 2(14.01\text{ g/mol}) = 28.02\text{ g N}$$

$$\text{molar mass } N_2O = 2(14.01\text{ g/mol}) + (16.00\text{ g/mol}) = 44.02\text{ g/mol}$$

$$\begin{aligned}\text{mass \% N} &= \frac{2 \times \text{molar mass N}}{\text{molar mass } N_2O} \times 100\% \\ &= \frac{28.02\ \cancel{\text{g/mol}}}{44.02\ \cancel{\text{g/mol}}} \times 100\% \\ &= 63.65\%\end{aligned}$$

Check: The units of the answer (%) are correct. The magnitude is reasonable because it is between 0 and 100% and there are two nitrogens per molecule.

(b) **Given:** NO **Find:** mass percent N

Conceptual Plan: $\text{Mass \% N} = \dfrac{1 \times \text{molar mass N}}{\text{molar mass NO}} \times 100\%$

Solution:

$$1 \times \text{molar mass N} = 1(14.01\text{ g/mol}) = 14.01\text{ g N}$$

$$\text{molar mass NO} = (14.01\text{ g/mol}) + (16.00\text{ g/mol}) = 30.01\text{ g/mol}$$

$$\begin{aligned}\text{mass \% N} &= \frac{1 \times \text{molar mass N}}{\text{molar mass NO}} \times 100\% \\ &= \frac{14.01\ \cancel{\text{g/mol}}}{30.01\ \cancel{\text{g/mol}}} \times 100\% \\ &= 46.68\%\end{aligned}$$

Check: The units of the answer (%) are correct. The magnitude is reasonable because it is between 0 and 100% and the mass of nitrogen is less than the mass of oxygen.

(c) **Given:** NO_2 **Find:** mass percent N

Conceptual Plan: $\text{Mass \% N} = \dfrac{1 \times \text{molar mass N}}{\text{molar mass } NO_2} \times 100\%$

Solution:

$$1 \times \text{molar mass N} = 1(14.01\text{ g/mol}) = 14.01\text{ g N}$$

$$\text{molar mass } NO_2 = (14.01\text{ g/mol}) + 2(16.00\text{ g/mol}) = 46.01\text{ g/mol}$$

$$\text{mass \% N} = \frac{1 \times \text{molar mass N}}{\text{molar mass } NO_2} \times 100\%$$

$$= \frac{14.01\ \cancel{g/mol}}{46.01\ \cancel{g/mol}} \times 100\%$$

$$= 30.45\%$$

Check: The units of the answer (%) are correct. The magnitude is reasonable because it is between 0 and 100%. The mass of nitrogen is less than the mass of oxygen, and there are two oxygens per molecule.

(d) **Given:** HNO_3 **Find:** mass percent N

Conceptual Plan: Mass % N $= \frac{1 \times \text{molar mass N}}{\text{molar mass } HNO_3} \times 100\%$

Solution:

$$1 \times \text{molar mass N} = 1(14.01 \text{g/mol}) = 14.01 \text{ g N}$$

$$\text{molar mass } HNO_3 = (1.008 \text{ g/mol}) + (14.01 \text{ g/mol}) + 3(16.00 \text{ g/mol}) = 63.02 \text{ g/mol}$$

$$\text{mass \% N} = \frac{1 \times \text{molar mass N}}{\text{molar mass } HNO_3} \times 100\%$$

$$= \frac{14.01\ \cancel{g/mol}}{63.02\ \cancel{g/mol}} \times 100\%$$

$$= 22.23\%$$

Check: The units of the answer (%) are correct. The magnitude is reasonable because it is between 0 and 100%. The mass of nitrogen is less than the mass of oxygen, and there are three oxygens per molecule.

5.85 **Given:** NH_3 **Find:** mass percent N

Conceptual Plan: Mass % N $= \frac{1 \times \text{molar mass N}}{\text{molar mass } NH_3} \times 100\%$

Solution:

$$1 \times \text{molar mass N} = 1(14.01 \text{ g/mol}) = 14.01 \text{ g N}$$

$$\text{molar mass } NH_3 = 3(1.008 \text{ g/mol}) + (14.01 \text{ g/mol}) = 17.03 \text{ g/mol}$$

$$\text{mass \% N} = \frac{1 \times \text{molar mass N}}{\text{molar mass } NH_3} \times 100\%$$

$$= \frac{14.01\ \cancel{g/mol}}{17.03\ \cancel{g/mol}} \times 100\%$$

$$= 82.27\%$$

Check: The units of the answer (%) are correct. The magnitude is reasonable because it is between 0 and 100% and nitrogen is the heaviest atom present.

Given: $CO(NH_2)_2$ **Find:** mass percent N

Conceptual Plan: Mass % N $= \frac{2 \times \text{molar mass N}}{\text{molar mass } CO(NH_2)_2} \times 100\%$

Solution:

$$2 \times \text{molar mass N} = 1(14.01 \text{ g/mol}) = 28.02 \text{ g N}$$

$$\text{molar mass } CO(NH_2)_2 = (12.01 \text{ g/mol}) + (16.00 \text{ g/mol}) + 2(14.01 \text{ g/mol}) + 4(1.008 \text{ g/mol}) = 60.06 \text{ g/mol}$$

$$\text{mass \% N} = \frac{2 \times \text{molar mass N}}{\text{molar mass } CO(NH_2)_2} \times 100\%$$

$$= \frac{28.02\ \cancel{g/mol}}{60.06\ \cancel{g/mol}} \times 100\%$$

$$= 46.65\%$$

Check: The units of the answer (%) are correct. The magnitude is reasonable because it is between 0 and 100% and there are two nitrogens and only one carbon and one oxygen per molecule.

Given: NH_4NO_3 **Find:** mass percent N

Conceptual Plan: $\text{Mass \% N} = \dfrac{2 \times \text{molar mass N}}{\text{molar mass } NH_4NO_3} \times 100\%$

Solution:

$$2 \times \text{molar mass N} = 2(14.01\ \text{g/mol}) = 28.02\ \text{g N}$$

$$\text{molar mass } NH_4NO_3 = 2(14.01\ \text{g/mol}) + 4(1.008\ \text{g/mol}) + 3(16.00\ \text{g/mol}) = 80.05\ \text{g/mol}$$

$$\text{mass \% N} = \frac{2 \times \text{molar mass N}}{\text{molar mass } NH_4NO_3} \times 100\% = \frac{28.02\ \cancel{\text{g/mol}}}{80.05\ \cancel{\text{g/mol}}} \times 100\% = 35.00\%$$

Check: The units of the answer (%) are correct. The magnitude is reasonable because it is between 0 and 100%. The mass of nitrogen is less than the mass of oxygen, and there are two nitrogens and three oxygens per molecule.

Given: $(NH_4)_2SO_4$ **Find:** mass percent N

Conceptual Plan: $\text{Mass \% N} = \dfrac{2 \times \text{molar mass N}}{\text{molar mass } (NH_4)_2SO_4} \times 100\%$

Solution:

$$2 \times \text{molar mass N} = 2(14.01\ \text{g/mol}) = 28.02\ \text{g N}$$

$$\text{molar mass } (NH_4)_2SO_4 = 2(14.01\ \text{g/mol}) + 8(1.008\ \text{g/mol}) + (32.07\ \text{g/mol}) + 4(16.00\ \text{g/mol}) = 132.15\ \text{g/mol}$$

$$\text{mass \% N} = \frac{2 \times \text{molar mass N}}{\text{molar mass } (NH_4)_2SO_4} \times 100\% = \frac{28.02\ \cancel{\text{g/mol}}}{132.15\ \cancel{\text{g/mol}}} \times 100\% = 21.20\%$$

Check: The units of the answer (%) are correct. The magnitude is reasonable because it is between 0 and 100% and the mass of nitrogen is less than the mass of oxygen and sulfur.

The fertilizer with the highest nitrogen content is NH_3 with a N content of 82.27%.

5.86 **Given:** Fe_2O_3 **Find:** mass percent Fe

Conceptual Plan: $\text{Mass \% Fe} = \dfrac{2 \times \text{molar mass Fe}}{\text{molar mass } Fe_2O_3} \times 100\%$

Solution:

$$2 \times \text{molar mass Fe} = 2(55.85\ \text{g/mol}) = 111.7\ \text{g Fe}$$

$$\text{molar mass } Fe_2O_3 = 2(55.85\ \text{g/mol}) + 3(16.00\ \text{g/mol}) = 159.7\ \text{g/mol}$$

$$\text{mass \% Fe} = \frac{2 \times \text{molar mass Fe}}{\text{molar mass } Fe_2O_3} \times 100\% = \frac{111.7\ \cancel{\text{g/mol}}}{159.7\ \cancel{\text{g/mol}}} \times 100\% = 69.94\%$$

Check: The units of the answer (%) are correct. The magnitude is reasonable because it is between 0 and 100% and iron provides most of the formula mass.

Given: Fe_3O_4 **Find:** mass percent Fe

Conceptual Plan: $\text{Mass \% Fe} = \dfrac{3 \times \text{molar mass Fe}}{\text{molar mass } Fe_3O_4} \times 100\%$

Solution:

$$3 \times \text{molar mass Fe} = 3(55.85\ \text{g/mol}) = 167.6\ \text{g Fe}$$

$$\text{molar mass } Fe_3O_4 = 3(55.85\ \text{g/mol}) + 4(16.00\ \text{g/mol}) = 231.6\ \text{g/mol}$$

$$\text{mass \% Fe} = \frac{3 \times \text{molar mass Fe}}{\text{molar mass } Fe_3O_4} \times 100\% = \frac{167.6\ \cancel{\text{g/mol}}}{231.6\ \cancel{\text{g/mol}}} \times 100\% = 72.37\%$$

Check: The units of the answer (%) are correct. The magnitude is reasonable because it is between 0 and 100% and iron provides most of the formula mass.

Given: $FeCO_3$ **Find:** mass percent Fe

Conceptual Plan: $\text{Mass \% Fe} = \frac{1 \times \text{molar mass Fe}}{\text{molar mass FeCO}_3} \times 100\%$

Solution:

$$1 \times \text{molar mass Fe} = (55.85 \text{ g/mol}) = 55.85 \text{ g Fe}$$

$$\text{molar mass FeCO}_3 = 1(55.85 \text{ g/mol}) + 1(12.01 \text{ g/mol}) + 3(16.00 \text{ g/mol}) = 115.86 \text{ g/mol}$$

$$\text{mass \% Fe} = \frac{1 \times \text{molar mass Fe}}{\text{molar mass FeCO}_3} \times 100\% = \frac{55.85 \text{ g/mol}}{115.86 \text{ g/mol}} \times 100\% = 48.20\%$$

Check: The units of the answer (%) are correct. The magnitude is reasonable because it is between 0 and 100% and iron provides slightly less than half of the formula mass. The ore with the highest iron content is Fe_3O_4 with an Fe content of 72.37%.

5.87 **Given:** 55.5 g CuF_2; 37.42 **Find:** g F in CuF_2

Conceptual Plan: g CuF_2 → g F

$$\frac{37.42 \text{ g F}}{100.0 \text{ g CuF}_2}$$

Solution: $55.5 \text{ g CuF}_2 \times \frac{37.42 \text{ g F}}{100.0 \text{ g CuF}_2} = 20.\underline{7}7 = 20.8 \text{ g F}$

Check: The units of the answer (g F) are correct. The magnitude is reasonable because it is less than the original mass.

5.88 **Given:** 155 mg Ag; 75.27% Ag in AgCl **Find:** mg AgCl

Conceptual Plan: mg Ag → g Ag → g AgCl → mg AgCl

$$\frac{1 \text{ g Ag}}{1000 \text{ mg Ag}} \quad \frac{100.0 \text{ g AgCl}}{75.27 \text{ g Ag}} \quad \frac{1000 \text{ mg AgCl}}{1 \text{ g AgCl}}$$

Solution: $155 \text{ mg Ag} \times \frac{1 \text{ g Ag}}{1000 \text{ mg Ag}} \times \frac{100.0 \text{ g AgCl}}{75.27 \text{ g Ag}} \times \frac{1000 \text{ mg AgCl}}{1 \text{ g AgCl}} = 20\underline{5}.9 \text{ mg AgCl} = 206 \text{ mg AgCl}$

Check: The units of the answer (mg AgCl) are correct. The magnitude is reasonable because it is greater than the original mass.

5.89 **Given:** 150 μg I; 76.45% I in KI **Find:** μg KI

Conceptual Plan: μg I → g I → g KI → μg KI

$$\frac{1 \text{ g I}}{1 \times 10^6 \ \mu\text{g I}} \quad \frac{100.0 \text{ g KI}}{76.45 \text{ g I}} \quad \frac{1 \times 10^6 \ \mu\text{g KI}}{1 \text{ g KI}}$$

Solution: $150 \ \mu\text{g I} \times \frac{1 \text{ g I}}{1 \times 10^6 \ \mu\text{g I}} \times \frac{100.0 \text{ g KI}}{76.45 \text{ g I}} \times \frac{1 \times 10^6 \ \mu\text{g KI}}{1 \text{ g KI}} = 1\underline{9}6.207 \ \mu\text{g KI} = 2.0 \times 10^2 \ \mu\text{g KI}$

Check: The units of the answer (μg KI) are correct. The magnitude is reasonable because it is greater than the original mass.

5.90 **Given:** 3.0 mg F; 45.24% F in NaF **Find:** mg NaF

Conceptual Plan: mg F → g F → g NaF → mg NaF

$$\frac{1 \text{ g F}}{1000 \text{ mg F}} \quad \frac{100.0 \text{ g NaF}}{45.24 \text{ g F}} \quad \frac{1000 \text{ mg NaF}}{1 \text{ g NaF}}$$

Solution: $3.0 \text{ mg F} \times \frac{1 \text{ g F}}{1000 \text{ mg F}} \times \frac{100.0 \text{ g NaF}}{45.24 \text{ g NaF}} \times \frac{1000 \text{ mg NaF}}{1 \text{ g NaF}} = 6.6 \text{ mg NaF}$

Check: The units of the answer (mg NaF) are correct. The magnitude is reasonable because it is greater than the original mass.

5.91
(a) red = oxygen, white = hydrogen: 2H:O H_2O
(b) black = carbon, white = hydrogen: C:4H CH_4
(c) black = carbon, white = hydrogen, red = oxygen: 2C:6H:O CH_3CH_2OH or C_2H_6O

5.92 (a) black = carbon, red = oxygen: 2O:C CO_2
(b) red = oxygen, white = hydrogen: 2H:2O H_2O_2
(c) red = oxygen, white = hydrogen: 2H:O H_2O

5.93 (a) **Given:** 0.0885 mol C_4H_{10} **Find:** mol H atoms

Conceptual Plan: Mol C_4H_{10} → mole H atom

$$\frac{10 \text{ mol H}}{1 \text{ mol } C_4H_{10}}$$

Solution: $0.0885 \text{ mol } C_4H_{10} \times \frac{10 \text{ mol H}}{1 \text{ mol } C_4H_{10}} = 0.885 \text{ mol H atoms}$

Check: The units of the answer (mol H atoms) are correct. The magnitude is reasonable because it is greater than the original mol C_4H_{10}.

(b) **Given:** 1.3 mol CH_4 **Find:** mol H atoms

Conceptual Plan: Mol CH_4 → mole H atom

$$\frac{4 \text{ mol H}}{1 \text{ mol } CH_4}$$

Solution: $1.3 \text{ mol } CH_4 \times \frac{4 \text{ mol H}}{1 \text{ mol } CH_4} = 5.2 \text{ mol H atoms}$

Check: The units of the answer (mol H atoms) are correct. The magnitude is reasonable because it is greater than the original mol CH_4.

(c) **Given:** 2.4 mol C_6H_{12} **Find:** mol H atoms

Conceptual Plan: Mol C_6H_{12} → mole H atom

$$\frac{12 \text{ mol H}}{1 \text{ mol } C_6H_{12}}$$

Solution: $2.4 \text{ mol } C_6H_{12} \times \frac{12 \text{ mol H}}{1 \text{ mol } C_6H_{12}} = 29 \text{ mol H atoms}$

Check: The units of the answer (mol H atoms) are correct. The magnitude is reasonable because it is greater than the original mol C_6H_{12}.

(d) **Given:** 1.87 mol C_8H_{18} **Find:** mol H atoms

Conceptual Plan: Mol C_8H_{18} → mole H atom

$$\frac{18 \text{ mol H}}{1 \text{ mol } C_8H_{18}}$$

Solution: $1.87 \text{ mol } C_8H_{18} \times \frac{18 \text{ mol H}}{1 \text{ mol } C_8H_{18}} = 33.7 \text{ mol H atoms}$

Check: The units of the answer (mol H atoms) are correct. The magnitude is reasonable because it is greater than the original mol C_8H_{18}.

5.94 (a) **Given:** 4.88 mol H_2O_2 **Find:** mol O atoms

Conceptual Plan: Mol H_2O_2 → mole O atom

$$\frac{2 \text{ mol O}}{1 \text{ mol } H_2O_2}$$

Solution: $4.88 \text{ mol } H_2O_2 \times \frac{2 \text{ mol O}}{1 \text{ mol } H_2O_2} = 9.76 \text{ mol O atoms}$

Check: The units of the answer (mol O atoms) are correct. The magnitude is reasonable because it is greater than the original mol H_2O_2.

(b) **Given:** 2.15 mol N_2O **Find:** mol O atoms

Conceptual Plan: Mol N_2O → mole O atom

$$\frac{1 \text{ mol O}}{1 \text{ mol } N_2O}$$

Solution: $2.15 \text{ mol } N_2O \times \frac{1 \text{ mol O}}{1 \text{ mol } N_2O} = 2.15 \text{ mol O atoms}$

Check: The units of the answer (mol O atoms) are correct. The magnitude is reasonable because it is the same as the original mol N_2O.

(c) **Given:** 0.0237 mol H_2CO_3 **Find:** mol O atoms
Conceptual Plan: Mol H_2CO_3 → mole O atom

$$\frac{3 \text{ mol O}}{1 \text{ mol } H_2CO_3}$$

Solution: $0.0237 \text{ mol } H_2CO_3 \times \frac{3 \text{ mol O}}{1 \text{ mol } H_2CO_3} = 0.0711 \text{ mol O atoms}$

Check: The units of the answer (mol O atoms) are correct. The magnitude is reasonable because it is greater than the original mol H_2CO_3.

(d) **Given:** 24.1 mol CO_2 **Find:** mol O atoms
Conceptual Plan: Mol CO_2 → mole O atom

$$\frac{2 \text{ mol O}}{1 \text{ mol } CO_2}$$

Solution: $24.1 \text{ mol } CO_2 \times \frac{2 \text{ mol O}}{1 \text{ mol } CO_2} = 48.2 \text{ mol O atoms}$

Check: The units of the answer (mol O atoms) are correct. The magnitude is reasonable because it is greater than the original mol CO_2.

5.95 (a) **Given:** 8.5 g NaCl **Find:** g Na
Conceptual Plan: g NaCl → mole NaCl → mol Na → g Na

$$\frac{1 \text{ mol NaCl}}{58.44 \text{ g NaCl}} \quad \frac{1 \text{ mol Na}}{1 \text{ mol NaCl}} \quad \frac{22.99 \text{ g Na}}{1 \text{ mol Na}}$$

Solution: $8.5 \text{ g NaCl} \times \frac{1 \text{ mol NaCl}}{58.44 \text{ g NaCl}} \times \frac{1 \text{ mol Na}}{1 \text{ mol NaCl}} \times \frac{22.99 \text{ g Na}}{1 \text{ mol Na}} = 3.3 \text{ g Na}$

Check: The units of the answer (g Na) are correct. The magnitude is reasonable because it is less than the original g NaCl.

(b) **Given:** 8.5 g Na_3PO_4 **Find:** g Na
Conceptual Plan: g Na_3PO_4 → mole Na_3PO_4 → mol Na → g Na

$$\frac{1 \text{ mol } Na_3PO_4}{163.94 \text{ g } Na_3PO_4} \quad \frac{3 \text{ mol Na}}{1 \text{ mol } Na_3PO_4} \quad \frac{22.99 \text{ g Na}}{1 \text{ mol Na}}$$

Solution: $8.5 \text{ g } Na_3PO_4 \times \frac{1 \text{ mol } Na_3PO_4}{163.94 \text{ g } Na_3PO_4} \times \frac{3 \text{ mol Na}}{1 \text{ mol } Na_3PO_4} \times \frac{22.99 \text{ g Na}}{1 \text{ mol Na}} = 3.6 \text{ g Na}$

Check: The units of the answer (g Na) are correct. The magnitude is reasonable because it is less than the original g Na_3PO_4.

(c) **Given:** 8.5 g $NaC_7H_5O_2$ **Find:** g Na
Conceptual Plan: g $NaC_7H_5O_2$ → mole $NaC_7H_5O_2$ → mol Na → g Na

$$\frac{1 \text{ mol } NaC_7H_5O_2}{144.10 \text{ g } NaC_7H_5O_2} \quad \frac{1 \text{ mol Na}}{1 \text{ mol } NaC_7H_5O_2} \quad \frac{22.99 \text{ g Na}}{1 \text{ mol Na}}$$

Solution: $8.5 \text{ g } NaC_7H_5O_2 \times \frac{1 \text{ mol } NaC_7H_5O_2}{144.10 \text{ g } NaC_7H_5O_2} \times \frac{1 \text{ mol Na}}{1 \text{ mol } NaC_7H_5O_2} \times \frac{22.99 \text{ g Na}}{1 \text{ mol Na}} = 1.4 \text{ g Na}$

Check: The units of the answer (g Na) are correct. The magnitude is reasonable because it is less than the original g $NaC_7H_5O_2$.

(d) **Given:** 8.5 g $Na_2C_6H_6O_7$ **Find:** g Na
Conceptual Plan: g $Na_2C_6H_6O_7$ → mole $Na_2C_6H_6O_7$ → mol Na → g Na

$$\frac{1 \text{ mol } Na_2C_6H_6O_7}{236.1 \text{ g } Na_2C_6H_6O_7} \quad \frac{2 \text{ mol Na}}{1 \text{ mol } Na_2C_6H_6O_7} \quad \frac{22.99 \text{ g Na}}{1 \text{ mol Na}}$$

Solution: $8.5 \text{ g } Na_2C_6H_6O_7 \times \frac{1 \text{ mol } Na_2C_6H_6O_7}{236.1 \text{ g } Na_2C_6H_6O_7} \times \frac{2 \text{ mol Na}}{1 \text{ mol } Na_2C_6H_6O_7} \times \frac{22.99 \text{ g Na}}{1 \text{ mol Na}} = 1.7 \text{ g Na}$

Check: The units of the answer (g Na) are correct. The magnitude is reasonable because it is less than the original g $Na_2C_6H_6O_7$.

5.96 (a) **Given:** 25 kg CF_2Cl_2 **Find:** kg Cl

Conceptual Plan: kg CF_2Cl_2 → g CF_2Cl_2 → mole CF_2Cl_2 → mol Cl → g Cl → kg Cl

$$\frac{1000\text{ g }CF_2Cl_2}{1\text{ kg }CF_2Cl_2} \quad \frac{1\text{ mol }CF_2Cl_2}{120.91\text{ g }CF_2Cl_2} \quad \frac{2\text{ mol Cl}}{1\text{ mol }CF_2Cl_2} \quad \frac{35.45\text{ g Cl}}{1\text{ mol Cl}} \quad \frac{1\text{ kg Cl}}{1000\text{ g Cl}}$$

Solution: $25\text{ kg }CF_2Cl_2 \times \frac{1000\text{ g }CF_2Cl_2}{1\text{ kg }CF_2Cl_2} \times \frac{1\text{ mol }CF_2Cl_2}{120.91\text{ g }CF_2Cl_2} \times \frac{2\text{ mol Cl}}{1\text{ mol }CF_2Cl_2} \times \frac{35.45\text{ g Cl}}{1\text{ mol Cl}} \times \frac{1\text{ kg Cl}}{1000\text{ g Cl}}$

$= 15\text{ kg Cl}$

Check: The units of the answer (kg Cl) are correct. The magnitude is reasonable because it is less than the original kg CF_2Cl_2.

(b) **Given:** 25 kg $CFCl_3$ **Find:** kg Cl

Conceptual Plan: kg $CFCl_3$ → g $CFCl_3$ → mole $CFCl_3$ → mol Cl → g Cl → kg Cl

$$\frac{1000\text{ g }CFCl_3}{1\text{ kg }CFCl_3} \quad \frac{1\text{ mol }CFCl_3}{137.4\text{ g }CFCl_3} \quad \frac{3\text{ mol Cl}}{1\text{ mol }CFCl_3} \quad \frac{35.45\text{ g Cl}}{1\text{ mol Cl}} \quad \frac{1\text{ kg Cl}}{1000\text{ g Cl}}$$

Solution: $25\text{ kg }CFCl_3 \times \frac{1000\text{ g }CFCl_3}{1\text{ kg }CFCl_3} \times \frac{1\text{ mol }CFCl_3}{137.4\text{ g }CFCl_3} \times \frac{3\text{ mol Cl}}{1\text{ mol }CFCl_3} \times \frac{35.45\text{ g Cl}}{1\text{ mol Cl}} \times \frac{1\text{ kg Cl}}{1000\text{ g Cl}}$

$= 19\text{ kg Cl}$

Check: The units of the answer (kg Cl) are correct. The magnitude is reasonable because it is less than the original kg $CFCl_3$.

(c) **Given:** 25 kg $C_2F_3Cl_3$ **Find:** kg Cl

Conceptual Plan: kg $C_2F_3Cl_3$ → g $C_2F_3Cl_3$ → mole $C_2F_3Cl_3$ → mol Cl → g Cl → kg Cl

$$\frac{1000\text{ g }C_2F_3Cl_3}{1\text{ kg }C_2F_3Cl_3} \quad \frac{1\text{ mol }C_2F_3Cl_3}{187.4\text{ g }C_2F_3Cl_3} \quad \frac{3\text{ mol Cl}}{1\text{ mol }C_2F_3Cl_3} \quad \frac{35.45\text{ g Cl}}{1\text{ mol Cl}} \quad \frac{1\text{ kg Cl}}{1000\text{ g Cl}}$$

Solution:

$25\text{ kg }C_2F_3Cl_3 \times \frac{1000\text{ g }C_2F_3Cl_3}{1\text{ kg }C_2F_3Cl_3} \times \frac{1\text{ mol }C_2F_3Cl_3}{187.4\text{ g }C_2F_3Cl_3} \times \frac{3\text{ mol Cl}}{1\text{ mol }C_2F_3Cl_3} \times \frac{35.45\text{ g Cl}}{1\text{ mol Cl}} \times \frac{1\text{ kg Cl}}{1000\text{ g Cl}}$

$= 14\text{ kg Cl}$

Check: The units of the answer (kg Cl) are correct. The magnitude is reasonable because it is less than the original kg $C_2F_3Cl_3$.

(d) **Given:** 25 kg CF_3Cl **Find:** kg Cl

Conceptual Plan: kg CF_3Cl → g CF_3Cl → mole CF_3Cl → mol Cl → g Cl → kg Cl

$$\frac{1000\text{ g }CF_3Cl}{1\text{ kg }CF_3Cl} \quad \frac{1\text{ mol }CF_3Cl}{104.46\text{ g }CF_3Cl} \quad \frac{1\text{ mol Cl}}{1\text{ mol }CF_3Cl} \quad \frac{35.45\text{ g Cl}}{1\text{ mol Cl}} \quad \frac{1\text{ kg Cl}}{1000\text{ g Cl}}$$

Solution: $25\text{ kg }CF_3Cl \times \frac{1000\text{ g }CF_3Cl}{1\text{ kg }CF_3Cl} \times \frac{1\text{ mol }CF_3Cl}{104.46\text{ g }CF_3Cl} \times \frac{1\text{ mol Cl}}{1\text{ mol }CF_3Cl} \times \frac{35.45\text{ g Cl}}{1\text{ mol Cl}} \times \frac{1\text{ kg Cl}}{1000\text{ g Cl}}$

$= 8.5\text{ kg Cl}$

Check: The units of the answer (kg Cl) are correct. The magnitude is reasonable because it is less than the original kg CF_3Cl.

Chemical Formulas from Experimental Data

5.97 (a) **Given:** 1.651 g Ag; 0.1224 g O **Find:** empirical formula

Conceptual Plan:

Convert mass to mol of each element → write pseudoformula → write empirical formula

$$\frac{1\text{ mol Ag}}{107.9\text{ g Ag}} \quad \frac{1\text{ mol O}}{16.00\text{ g O}}$$ divide by smallest number

Solution: $1.651\text{ g Ag} \times \frac{1\text{ mol Ag}}{107.9\text{ g Ag}} = 0.01530\text{ mol Ag}$

$0.1224\text{ g O} \times \frac{1\text{ mol O}}{16.00\text{ g O}} = 0.007650\text{ mol O}$

$Ag_{0.01530}O_{0.007650}$

$Ag_{\frac{0.01530}{0.007650}}O_{\frac{0.007650}{0.007650}} \rightarrow Ag_2O$

The correct empirical formula is Ag_2O.

Check: The formula makes sense since silver forms a 1+ ion and oxygen typically forms a 2− ion.

(b) **Given:** 0.672 g Co; 0.569 g As; 0.486 g O **Find:** empirical formula

Conceptual Plan: Convert mass to mol of each element → write pseudoformula → write empirical formula

$\frac{1 \text{ mol Co}}{58.93 \text{ g Co}}$ $\frac{1 \text{ mol As}}{74.92 \text{ g As}}$ $\frac{1 \text{ mol O}}{16.00 \text{ g O}}$ divide by smallest number

Solution: $0.672 \text{ g Co} \times \frac{1 \text{ mol Co}}{58.93 \text{ g Co}} = 0.0114 \text{ mol Co}$

$0.569 \text{ g As} \times \frac{1 \text{ mol As}}{74.92 \text{ g As}} = 0.00759 \text{ mol As}$

$0.486 \text{ g O} \times \frac{1 \text{ mol O}}{16.00 \text{ g O}} = 0.0304 \text{ mol O}$

$Co_{0.0114}As_{0.00759}O_{0.0304}$

$Co_{\frac{0.0114}{0.00759}}As_{\frac{0.00759}{0.00759}}O_{\frac{0.0304}{0.00759}} \rightarrow Co_{1.5}As_1O_4$

$Co_{1.5}As_1O_4 \times 2 \rightarrow Co_3As_2O_8$

The correct empirical formula is $Co_3As_2O_8$.

(c) **Given:** 1.443 g Se; 5.841 g Br **Find:** empirical formula

Conceptual Plan:

Convert mass to mol of each element → write pseudoformula → write empirical formula

$\frac{1 \text{ mol Se}}{78.96 \text{ g Se}}$ $\frac{1 \text{ mol Br}}{79.90 \text{ g Br}}$ divide by smallest number

Solution: $1.443 \text{ g Se} \times \frac{1 \text{ mol Se}}{78.96 \text{ g Se}} = 0.01828 \text{ mol Se}$

$5.841 \text{ g Br} \times \frac{1 \text{ mol Br}}{79.90 \text{ g Br}} = 0.07310 \text{ mol Br}$

$Se_{0.01828}Br_{0.07310}$

$Se_{\frac{0.01828}{0.01828}}Br_{\frac{0.07310}{0.01828}} \rightarrow SeBr_4$

The correct empirical formula is $SeBr_4$.

5.98 (a) **Given:** 1.245 g Ni; 5.381 g I **Find:** empirical formula

Conceptual Plan:

Convert mass to mol of each element → write pseudoformula → write empirical formula

$\frac{1 \text{ mol Ni}}{58.69 \text{ g Ni}}$ $\frac{1 \text{ mol I}}{126.9 \text{ g I}}$ divide by smallest number

Solution: $1.245 \text{ g Ni} \times \frac{1 \text{ mol Ni}}{58.69 \text{ g Ni}} = 0.02121 \text{ mol Ni}$

$5.381 \text{ g I} \times \frac{1 \text{ mol I}}{126.9 \text{ g I}} = 0.04240 \text{ mol I}$

$Ni_{0.02121}I_{0.04240}$

$Ni_{\frac{0.02121}{0.02121}}I_{\frac{0.04240}{0.02121}} \rightarrow NiI_2$

The correct empirical formula is NiI_2.

(b) **Given:** 2.677 g Ba; 3.115 g Br **Find:** empirical formula

Conceptual Plan:

Convert mass to mol of each element → write pseudoformula → write empirical formula

$\frac{1 \text{ mol Ba}}{137.3 \text{ g Ba}}$ $\frac{1 \text{ mol Br}}{79.90 \text{ g Br}}$ divide by smallest number

Solution: $2.677\ \cancel{g\ Ba} \times \frac{1\ mol\ Ba}{137.3\ \cancel{g\ Ba}} = 0.01950\ mol\ Ba$

$3.115\ \cancel{g\ Br} \times \frac{1\ mol\ Br}{79.90\ \cancel{g\ Br}} = 0.03899\ mol\ Br$

$Ba_{0.01950}Br_{0.03899}$

$Ba_{\frac{0.01950}{0.01950}}Br_{\frac{0.03899}{0.01950}} \rightarrow BaBr_2$

The correct empirical formula is $BaBr_2$.

Check: The formula makes sense since barium forms a 2+ ion and bromine forms a 1− ion.

(c) **Given:** 2.128 g Be; 7.557 g S; 15.107 g O **Find:** empirical formula

Conceptual Plan:

Convert mass to mol of each element → write pseudoformula → write empirical formula

$\frac{1\ mol\ Be}{9.012\ g\ Be}$ $\frac{1\ mol\ S}{32.07\ g\ S}$ $\frac{1\ mol\ O}{16.00\ g\ O}$ divide by smallest number

Solution: $2.128\ \cancel{g\ Be} \times \frac{1\ mol\ Be}{9.012\ \cancel{g\ Be}} = 0.2361\ mol\ Be$

$7.557\ \cancel{g\ S} \times \frac{1\ mol\ S}{32.07\ \cancel{g\ S}} = 0.2356\ mol\ S$

$15.107\ \cancel{g\ O} \times \frac{1\ mol\ O}{16.00\ \cancel{g\ O}} = 0.9442\ mol\ O$

$Be_{0.2361}S_{0.2356}O_{0.9442}$

$Be_{\frac{0.2361}{0.2356}}S_{\frac{0.2356}{0.2356}}O_{\frac{0.9442}{0.2356}} \rightarrow BeSO_4$

The correct empirical formula is $BeSO_4$.

5.99 (a) **Given:** in a 100 g sample of nicotine: 74.03 g C; 8.70 g H; 17.27 g N **Find:** empirical formula

Conceptual Plan:

Convert mass to mol of each element → write pseudoformula → write empirical formula

$\frac{1\ mol\ C}{12.01\ g\ C}$ $\frac{1\ mol\ H}{1.008\ g\ H}$ $\frac{1\ mol\ N}{14.01\ g\ N}$ divide by smallest number

Solution: $74.03\ \cancel{g\ C} \times \frac{1\ mol\ C}{12.01\ \cancel{g\ C}} = 6.164\ mol\ C$

$8.70\ \cancel{g\ H} \times \frac{1\ mol\ H}{1.008\ \cancel{g\ H}} = 8.63\ mol\ H$

$17.27\ \cancel{g\ N} \times \frac{1\ mol\ N}{14.01\ \cancel{g\ N}} = 1.233\ mol\ N$

$C_{6.164}H_{8.63}N_{1.233}$

$C_{\frac{6.164}{1.233}}H_{\frac{8.63}{1.233}}N_{\frac{1.233}{1.233}} \rightarrow C_5H_7N$

The correct empirical formula is C_5H_7N.

(b) **Given:** in a 100 g sample of caffeine: 49.48 g C; 5.19 g H; 28.85 g N; 16.48 g O **Find:** empirical formula

Conceptual Plan:

Convert mass to mol of each element → write pseudoformula → write empirical formula

$\frac{1\ mol\ C}{12.01\ g\ C}$ $\frac{1\ mol\ H}{1.008\ g\ H}$ $\frac{1\ mol\ N}{14.01\ g\ N}$ $\frac{1\ mol\ O}{16.00\ g\ O}$ divide by smallest number

Solution $49.48\ \cancel{g\ C} \times \frac{1\ mol\ C}{12.01\ \cancel{g\ C}} = 4.120\ mol\ C$

$5.19\ \cancel{g\ H} \times \frac{1\ mol\ H}{1.008\ \cancel{g\ H}} = 5.15\ mol\ H$

$28.85\ \cancel{g\ N} \times \frac{1\ mol\ N}{14.01\ \cancel{g\ N}} = 2.059\ mol\ N$

$$16.48 \text{ g O} \times \frac{1 \text{ mol O}}{16.00 \text{ g O}} = 1.030 \text{ mol O}$$

$C_{4.120}H_{5.15}N_{2.059}O_{1.030}$

$C_{\frac{4.120}{1.030}}H_{\frac{5.15}{1.030}}N_{\frac{2.059}{1.030}}O_{\frac{1.030}{1.030}} \rightarrow C_4H_5N_2O$

The correct empirical formula is $C_4H_5N_2O$.

5.100 (a) **Given:** in a 100 g sample of methyl butyrate: 58.80 g C; 9.87 g H; 31.33 g O **Find:** empirical formula

Conceptual Plan:

Convert mass to mol of each element → write pseudoformula → write empirical formula

$\frac{1 \text{ mol C}}{12.01 \text{ g C}}$ $\frac{1 \text{ mol H}}{1.008 \text{ g H}}$ $\frac{1 \text{ mol O}}{16.00 \text{ g O}}$ divide by smallest number

Solution: $58.80 \text{ g C} \times \frac{1 \text{ mol C}}{12.01 \text{ g C}} = 4.896 \text{ mol C}$

$$9.87 \text{ g H} \times \frac{1 \text{ mol H}}{1.008 \text{ g H}} = 9.79 \text{ mol H}$$

$$31.33 \text{ g O} \times \frac{1 \text{ mol O}}{16.00 \text{ g O}} = 1.958 \text{ mol O}$$

$C_{4.896}H_{9.79}O_{1.958}$

$C_{\frac{4.896}{1.958}}H_{\frac{9.79}{1.958}}O_{\frac{1.958}{1.958}} \rightarrow C_{2.5}H_5O$

$C_{2.5}H_5O \times 2 = C_5H_{10}O_2$

The correct empirical formula is $C_5H_{10}O_2$.

(b) **Given:** in a 100 g sample of vanillin: 63.15 g C; 5.30 g H; 31.55 g O **Find:** empirical formula

Conceptual Plan:

Convert mass to mol of each element → write pseudoformula → write empirical formula

$\frac{1 \text{ mol C}}{12.01 \text{ g C}}$ $\frac{1 \text{ mol H}}{1.008 \text{ g H}}$ $\frac{1 \text{ mol O}}{16.00 \text{ g O}}$ divide by smallest number

Solution: $63.15 \text{ g C} \times \frac{1 \text{ mol C}}{12.01 \text{ g C}} = 5.258 \text{ mol C}$

$$5.30 \text{ g H} \times \frac{1 \text{ mol H}}{1.008 \text{ g H}} = 5.26 \text{ mol H}$$

$$31.55 \text{ g O} \times \frac{1 \text{ mol O}}{16.00 \text{ g O}} = 1.972 \text{ mol O}$$

$C_{5.258}H_{5.26}O_{1.972}$

$C_{\frac{5.258}{1.972}}H_{\frac{5.26}{1.972}}O_{\frac{1.972}{1.972}} \rightarrow C_{2.67}H_{2.67}O$

$C_{2.67}H_{2.67}O \times 3 = C_8H_8O_3$

The correct empirical formula is $C_8H_8O_3$.

5.101 **Given:** in a 100 g sample of ibuprofen 75.69 g C; 8.80 g H; 15.51 g O **Find:** empirical formula

Conceptual Plan: Convert mass to mol of each element → write pseudoformula → write empirical formula

$\frac{1 \text{ mol C}}{12.01 \text{ g C}}$ $\frac{1 \text{ mol H}}{1.008 \text{ g H}}$ $\frac{1 \text{ mol O}}{16.00 \text{ g O}}$ divide by smallest number

Solution: $75.69 \text{ g C} \times \frac{1 \text{ mol C}}{12.01 \text{ g C}} = 6.302 \text{ mol C}$

$$8.80 \text{ g H} \times \frac{1 \text{ mol H}}{1.008 \text{ g H}} = 8.73 \text{ mol H}$$

$$15.51 \text{ g O} \times \frac{1 \text{ mol O}}{16.00 \text{ g O}} = 0.9694 \text{ mol O}$$

$C_{6.302}H_{8.73}O_{0.9694}$

$C_{\frac{6.302}{0.9694}}H_{\frac{8.73}{0.9694}}O_{\frac{0.9694}{0.9694}} \rightarrow C_{6.50}H_{9.01}O$

$C_{6.50}H_{9.01}O \times 2 = C_{13}H_{18}O_2$

The correct empirical formula is $C_{13}H_{18}O_2$.

5.102 **Given:** in a 100 g sample of ascorbic acid 40.92 g C; 4.58 g H; 54.50 g O **Find:** empirical formula

Conceptual Plan: Convert mass to mol of each element → write pseudoformula → write empirical formula

$\frac{1\text{ mol C}}{12.01\text{ g C}}$ $\frac{1\text{ mol H}}{1.008\text{ g H}}$ $\frac{1\text{ mol O}}{16.00\text{ g O}}$ divide by smallest number

Solution: $40.92\text{ g C} \times \frac{1\text{ mol C}}{12.01\text{ g C}} = 3.407\text{ mol C}$

$4.58\text{ g H} \times \frac{1\text{ mol H}}{1.008\text{ g H}} = 4.54\text{ mol H}$

$54.50\text{ g O} \times \frac{1\text{ mol O}}{16.00\text{ g O}} = 3.406\text{ mol O}$

$C_{3.407}H_{4.54}O_{3.406}$

$C_{\frac{3.407}{3.406}}H_{\frac{4.54}{3.406}}O_{\frac{3.406}{3.406}} \rightarrow C_{1.00}H_{1.33}O_{1.00}$

$C_{1.00}H_{1.33}O_{1.00} \times 3 = C_3H_4O_3$

The correct empirical formula is $C_3H_4O_3$.

5.103 **Given:** 0.77 mg N; 6.61 mg N_xCl_y **Find:** empirical formula

Conceptual Plan: Find mg Cl → convert mg to g for each element → convert mass to mol of each element → write pseudoformula → write empirical formula

mg N_xCl_y − mg N; $\frac{1\text{ g}}{1000\text{ mg}}$; $\frac{1\text{ mol N}}{14.01\text{ g N}}$ $\frac{1\text{ mol Cl}}{35.45\text{ g Cl}}$; divide by smallest number

Solution: 6.61 mg N_xCl_y − 0.77 mg N = 5.84 mg Cl

$0.77\text{ mg N} \times \frac{1\text{ g N}}{1000\text{ mg N}} \times \frac{1\text{ mol N}}{14.01\text{ g N}} = 5.5 \times 10^{-5}\text{ mol N}$

$5.84\text{ mg Cl} \times \frac{1\text{ g Cl}}{1000\text{ mg Cl}} \times \frac{1\text{ mol Cl}}{35.45\text{ g Cl}} = 1.70 \times 10^{-4}\text{ mol Cl}$

$N_{5.5\times10^{-5}}Cl_{1.6\times10^{-4}}$

$N_{\frac{5.5\times10^{-5}}{5.5\times10^{-5}}}Cl_{\frac{1.6\times10^{-4}}{5.5\times10^{-5}}} \rightarrow NCl_3$

The correct empirical formula is NCl_3.

5.104 **Given:** 45.2 mg P; 131.6 mg P_xSe_y **Find:** empirical formula

Conceptual Plan: Find mg Se → convert mg to g for each element → convert mass to mol of each element → write pseudoformula → write empirical formula

mg P_xSe_y − mg P; $\frac{1\text{ g}}{1000\text{ mg}}$; $\frac{1\text{ mol P}}{30.97\text{ g P}}$ $\frac{1\text{ mol Se}}{78.96\text{ g Se}}$; divide by smallest number

Solution: 131.6 mg P_xSe_y − 45.2 mg P = 86.4 mg Se

$45.2\text{ mg P} \times \frac{1\text{ g P}}{1000\text{ mg P}} \times \frac{1\text{ mol P}}{30.97\text{ g P}} = 0.00146\text{ mol P}$

$86.4\text{ mg Se} \times \frac{1\text{ g Se}}{1000\text{ mg Se}} \times \frac{1\text{ mol Se}}{78.96\text{ g Se}} = 0.00109\text{ mol Se}$

$P_{0.00146}Se_{0.00109}$

$P_{\frac{0.00146}{0.00109}}Se_{\frac{0.00109}{0.00109}} \rightarrow P_{1.34}Se$

$P_{1.34}Se \times 3 = P_4Se_3$

The correct empirical formula is P_4Se_3.

5.105 (a) **Given:** empirical formula = C_6H_7N; molar mass = 186.24 g/mol **Find:** molecular formula

Conceptual Plan: Molecular formula = empirical formula × *n* $n = \frac{\text{molar mass}}{\text{empirical formula mass}}$

Solution: empirical formula mass = 6(12.01 g/mol) + 7(1.008 g/mol) + 1(14.01 g/mol) = 93.13 g/mol

$$n = \frac{\text{molar mass}}{\text{formula molar mass}} = \frac{186.24\ \text{g/mol}}{93.13\ \text{g/mol}} = 1.9998 = 2$$

$$\text{molecular formula} = C_6H_7N \times 2 = C_{12}H_{14}N_2$$

Check: The units (g/mol) are correct. Check the answer by calculating the molar mass of the formula as follows: 12(12.01 g/mol) + 14(1.008 g/mol) + 2(14.01 g/mol) = 186.24 g/mol. The calculated molar mass is in agreement with the given molar mass.

(b) **Given:** empirical formula = C_2HCl; molar mass = 181.44 g/mol **Find:** molecular formula

Conceptual Plan: Molecular formula = empirical formula × *n* $n = \frac{\text{molar mass}}{\text{empirical formula mass}}$

Solution: empirical formula mass = 2(12.01 g/mol) + 1(1.008 g/mol) + 1(35.45 g/mol) = 60.48 g/mol

$$n = \frac{\text{molar mass}}{\text{formula molar mass}} = \frac{181.44\ \text{g/mol}}{60.48\ \text{g/mol}} = 3$$

$$\text{molecular formula} = C_2HCl \times 3 = C_6H_3Cl_3$$

Check: The units (g/mol) are correct. Check the answer by calculating the molar mass of the formula as follows: 6(12.01 g/mol) + 3(1.008 g/mol) + 3(35.45 g/mol) = 181.44 g/mol. The calculated molar mass is in agreement with the given molar mass.

(c) **Given:** empirical formula = $C_5H_{10}NS_2$; molar mass = 296.54 g/mol **Find:** molecular formula

Conceptual Plan: Molecular formula = empirical formula × *n* $n = \frac{\text{molar mass}}{\text{empirical formula mass}}$

Solution: empirical formula mass = 5(12.01 g/mol) + 10(1.008 g/mol) + 1(14.01 g/mol) + 2(32.07) = 148.28 g/mol

$$n = \frac{\text{molar mass}}{\text{formula molar mass}} = \frac{296.54\ \text{g/mol}}{148.28\ \text{g/mol}} = 2$$

$$\text{molecular formula} = C_5H_{10}NS_2 \times 2 = C_{10}H_{20}N_2S_4$$

Check: The units (g/mol) are correct. Check the answer by calculating the molar mass of the formula as follows: 10(12.01 g/mol) + 20(1.008 g/mol) + 2(14.01 g/mol) + 1(32.07 g/mol) = 296.54 g/mol. The calculated molar mass is in agreement with the given molar mass.

5.106 (a) **Given:** empirical formula = C_4H_9; molar mass = 114.22 g/mol **Find:** molecular formula

Conceptual Plan: Molecular formula = empirical formula × *n* $n = \frac{\text{molar mass}}{\text{empirical formula mass}}$

Solution: empirical formula mass = 4(12.01 g/mol) + 9(1.008 g/mol) = 57.11 g/mol

$$n = \frac{\text{molar mass}}{\text{formula molar mass}} = \frac{114.22\ \text{g/mol}}{57.11\ \text{g/mol}} = 2$$

$$\text{molecular formula} = C_4H_9 \times 2 = C_8H_{18}$$

Check: The units (g/mol) are correct. Check the answer by calculating the molar mass of the formula as follows: 8(12.01 g/mol) + 18(1.008 g/mol) = 114.22 g/mol. The calculated molar mass is in agreement with the given molar mass.

(b) **Given:** empirical formula = CCl; molar mass = 284.77 g/mol **Find:** molecular formula

Conceptual Plan: Molecular formula = empirical formula × *n* $n = \frac{\text{molar mass}}{\text{empirical formula mass}}$

Solution: empirical formula mass = 1(12.01 g/mol) + 1(35.45 g/mol) = 47.46 g/mol

$$n = \frac{\text{molar mass}}{\text{formula molar mass}} = \frac{284.77\ \text{g/mol}}{47.46\ \text{g/mol}} = 6$$

$$\text{molecular formula} = \text{CCl} \times 6 = C_6Cl_6$$

Check: The units (g/mol) are correct. Check the answer by calculating the molar mass of the formula as follows: 6(12.01 g/mol) + 6(34.45 g/mol) = 284.77 g/mol. The calculated molar mass is in agreement with the given molar mass.

(c) **Given:** empirical formula = C_3H_2N; molar mass = 312.29 g/mol **Find:** molecular formula

Conceptual Plan: Molecular formula = empirical formula × *n* $n = \frac{\text{molar mass}}{\text{empirical formula mass}}$

Solution: empirical formula mass = 3(12.01 g/mol) + 2(1.008 g/mol) + 1(14.01 g/mol) = 52.06 g/mol

$$n = \frac{\text{molar mass}}{\text{formula molar mass}} = \frac{312.29\ \text{g/mol}}{52.06\ \text{g/mol}} = 6$$

$$\text{molecular formula} = C_3H_2N \times 6 = C_{18}H_{12}N_6$$

Check: The units (g/mol) are correct. Check the answer by calculating the molar mass of the formula as follows: 18(12.01 g/mol) + 12(1.008 g/mol) + 1(14.01 g/mol) = 52.06 g/mol. The calculated molar mass is in agreement with the given molar mass.

5.107 **Given:** 33.01 g CO_2; 13.51 g H_2O **Find:** empirical formula

Conceptual Plan: Mass CO_2, H_2O → mol CO_2, H_2O → mol C, mol H → pseudoformula → empirical formula

$$\frac{1\ \text{mol}\ CO_2}{44.01\ \text{g}\ CO_2}\quad \frac{1\ \text{mol}\ H_2O}{18.02\ \text{g}\ H_2O}\quad \frac{1\ \text{mol C}}{1\ \text{mol}\ CO_2}\quad \frac{2\ \text{mol H}}{1\ \text{mol}\ H_2O}\qquad \text{divide by smallest number}$$

Solution:

$$33.01\ \cancel{\text{g}\ CO_2} \times \frac{1\ \text{mol}\ CO_2}{44.01\ \cancel{\text{g}\ CO_2}} = 0.7501\ \text{mol}\ CO_2$$

$$13.51\ \cancel{\text{g}\ H_2O} \times \frac{1\ \text{mol}\ H_2O}{18.02\ \cancel{\text{g}\ H_2O}} = 0.7497\ \text{mol}\ H_2O$$

$$0.7501\ \cancel{\text{mol}\ CO_2} \times \frac{1\ \text{mol C}}{1\ \cancel{\text{mol}\ CO_2}} = 0.7501\ \text{mol C}$$

$$0.7497\ \cancel{\text{mol}\ H_2O} \times \frac{2\ \text{mol H}}{1\ \cancel{\text{mol}\ H_2O}} = 1.499\ \text{mol H}$$

$$C_{0.7501}H_{1.499}$$

$$C_{\frac{0.7501}{0.7501}}H_{\frac{1.499}{0.7501}} \rightarrow CH_2$$

The correct empirical formula is CH_2.

5.108 **Given:** 8.80 g CO_2; 1.44 g H_2O **Find:** empirical formula

Conceptual Plan: Mass CO_2, H_2O → mol CO_2, H_2O → mol C, mol H → pseudoformula → empirical formula

$$\frac{1\ \text{mol}\ CO_2}{44.01\ \text{g}\ CO_2}\quad \frac{1\ \text{mol}\ H_2O}{18.02\ \text{g}\ H_2O}\quad \frac{1\ \text{mol C}}{1\ \text{mol}\ CO_2}\quad \frac{2\ \text{mol H}}{1\ \text{mol}\ H_2O}\qquad \text{divide by smallest number}$$

Solution:

$$8.80\ \cancel{\text{g}\ CO_2} \times \frac{1\ \text{mol}\ CO_2}{44.01\ \cancel{\text{g}\ CO_2}} = 0.200\ \text{mol}\ CO_2$$

$$1.44\ \cancel{\text{g}\ H_2O} \times \frac{1\ \text{mol}\ H_2O}{18.02\ \cancel{\text{g}\ H_2O}} = 0.0799\ \text{mol}\ H_2O$$

$$0.200\ \cancel{\text{mol CO}_2} \times \frac{1\ \text{mol C}}{1\ \cancel{\text{mol CO}_2}} = 0.200\ \text{mol C}$$

$$0.0799\ \cancel{\text{mol H}_2\text{O}} \times \frac{2\ \text{mol H}}{1\ \cancel{\text{mol H}_2\text{O}}} = 0.160\ \text{mol H}$$

$$C_{0.200}H_{0.160}$$

$$C_{\frac{0.200}{0.160}}H_{\frac{0.160}{0.160}} \rightarrow C_{1.25}H_1$$

$$C_{1.25}H_1 \times 4 = C_5H_4$$

The correct empirical formula is C_5H_4.

5.109 **Given:** 4.30 g sample; 8.59 g CO_2; 3.52 g H_2O **Find:** empirical formula

Conceptual Plan: Mass CO_2, H_2O $\rightarrow$ mol CO_2, H_2O $\rightarrow$ mol C, mol H $\rightarrow$ mass C, mass H, mass O $\rightarrow$ mol O $\rightarrow$

$$\frac{1\ \text{mol CO}_2}{44.01\ \text{g CO}_2}\quad \frac{1\ \text{mol H}_2\text{O}}{18.02\ \text{g H}_2\text{O}}\quad \frac{1\ \text{mol C}}{1\ \text{mol CO}_2}\quad \frac{2\ \text{mol H}}{1\ \text{mol H}_2\text{O}}\quad \frac{12.01\ \text{g C}}{1\ \text{mol C}}\quad \frac{1.008\ \text{g H}}{1\ \text{mol H}}\quad \text{g sample} - \text{g C} - \text{g H}\quad \frac{1\ \text{mol O}}{16.00\ \text{g O}}$$

pseudoformula $\rightarrow$ empirical formula

divide by smallest number

Solution:

$$8.59\ \cancel{\text{g CO}_2} \times \frac{1\ \text{mol CO}_2}{44.01\ \cancel{\text{g CO}_2}} = 0.195\ \text{mol CO}_2$$

$$3.52\ \cancel{\text{g H}_2\text{O}} \times \frac{1\ \text{mol H}_2\text{O}}{18.02\ \cancel{\text{g H}_2\text{O}}} = 0.195\ \text{mol H}_2\text{O}$$

$$0.195\ \cancel{\text{mol CO}_2} \times \frac{1\ \text{mol C}}{1\ \cancel{\text{mol CO}_2}} = 0.195\ \text{mol C}$$

$$0.195\ \cancel{\text{mol H}_2\text{O}} \times \frac{2\ \text{mol H}}{1\ \cancel{\text{mol H}_2\text{O}}} = 0.390\ \text{mol H}$$

$$0.195\ \cancel{\text{mol C}} \times \frac{12.01\ \text{g C}}{1\ \cancel{\text{mol C}}} = 2.34\ \text{g C}$$

$$0.390\ \cancel{\text{mol H}} \times \frac{1.008\ \text{g H}}{1\ \cancel{\text{mol H}}} = 0.393\ \text{g H}$$

$$4.30\ \text{g} - 2.34\ \text{g} - 0.393\ \text{g} = 1.57\ \text{g O}$$

$$1.57\ \cancel{\text{g O}} \times \frac{1\ \text{mol O}}{16.00\ \cancel{\text{g O}}} = 0.0981\ \text{mol O}$$

$$C_{0.195}H_{0.390}O_{0.0981}$$

$$C_{\frac{0.195}{0.0981}}H_{\frac{0.390}{0.0981}}O_{\frac{0.0981}{0.0981}} \rightarrow C_2H_4O$$

The correct empirical formula is C_2H_4O.

5.110 **Given:** 12.01 g sample; 14.08 g CO_2; 4.32 g H_2O **Find:** empirical formula

Conceptual Plan: Mass CO_2, H_2O $\rightarrow$ mol CO_2, H_2O $\rightarrow$ mol C, mol H $\rightarrow$ mass C, mass H, mass O $\rightarrow$ mol O $\rightarrow$

$$\frac{1\ \text{mol CO}_2}{44.01\ \text{g CO}_2}\quad \frac{1\ \text{mol H}_2\text{O}}{18.02\ \text{g H}_2\text{O}}\quad \frac{1\ \text{mol C}}{1\ \text{mol CO}_2}\quad \frac{2\ \text{mol H}}{1\ \text{mol H}_2\text{O}}\quad \frac{12.01\ \text{g C}}{1\ \text{mol C}}\quad \frac{1.008\ \text{g H}}{1\ \text{mol H}}\quad \text{g sample} - \text{g C} - \text{g H}\quad \frac{1\ \text{mol O}}{16.00\ \text{g O}}$$

pseudoformula $\rightarrow$ empirical formula

divide by smallest number

Solution:

$$14.08\ \cancel{\text{g CO}_2} \times \frac{1\ \text{mol CO}_2}{44.01\ \cancel{\text{g CO}_2}} = 0.3199\ \text{mol CO}_2$$

$$4.32\ \cancel{\text{g H}_2\text{O}} \times \frac{1\ \text{mol H}_2\text{O}}{18.02\ \cancel{\text{g H}_2\text{O}}} = 0.23\underline{9}7\ \text{mol H}_2\text{O}$$

$$0.3199\ \cancel{\text{mol CO}_2} \times \frac{1\ \text{mol C}}{1\ \cancel{\text{mol CO}_2}} = 0.3199\ \text{mol C}$$

$$0.23\underline{9}7\ \cancel{\text{mol H}_2\text{O}} \times \frac{2\ \text{mol H}}{1\ \cancel{\text{mol H}_2\text{O}}} = 0.47\underline{9}5\ \text{mol H}$$

$$0.3199 \text{ mol C} \times \frac{12.01 \text{ g C}}{1 \text{ mol C}} = 3.842 \text{ g C}$$

$$0.47\underline{9}5 \text{ mol H} \times \frac{1.008 \text{ g H}}{1 \text{ mol H}} = 0.48\underline{3}3 \text{ g H}$$

$$12.01 \text{ g} - 3.842 \text{ g} - 0.48\underline{3}3 \text{ g} = 7.68 \text{ g O}$$

$$7.68 \text{ g O} \times \frac{1 \text{ mol O}}{16.00 \text{ g O}} = 0.480 \text{ mol O}$$

$$C_{0.3199}H_{0.4795}O_{0.480}$$

$$C_{\frac{0.3199}{0.3199}}H_{\frac{0.4795}{0.3199}}O_{\frac{0.480}{0.3199}} \rightarrow CH_{1.5}O_{1.5}$$

$$CH_{1.5}O_{1.5} \times 2 = C_2H_3O_3$$

The correct empirical formula is $C_2H_3O_3$.

Organic Compounds

5.111 (a) composed of metal cation and polyatomic anion—inorganic compound
(b) composed of carbon and hydrogen—organic compound
(c) composed of carbon, hydrogen, and oxygen—organic compound
(d) composed of metal cation and nonmetal anion—inorganic compound

5.112 (a) composed of carbon and hydrogen—organic compound
(b) composed of carbon, hydrogen, and nitrogen—organic compound
(c) composed of metal cation and nonmetal anion—inorganic compound
(d) composed of metal cation and polyatomic anion—inorganic compound

5.113 (a) This compound contains oxygen, in addition to carbon and hydrogen, so it is not a hydrocarbon.
(b) This compound contains only carbon and hydrogen, so it is a hydrocarbon.
(c) This compound contains oxygen, in addition to carbon and hydrogen, so it is not a hydrocarbon.
(d) This compound contains nitrogen, in addition to carbon and hydrogen, so it is not a hydrocarbon.

5.114 (a) This compound contains oxygen, in addition to carbon and hydrogen, so it is not a hydrocarbon.
(b) This compound contains oxygen, in addition to carbon and hydrogen, so it is not a hydrocarbon.
(c) This compound contains only carbon and hydrogen, so it is a hydrocarbon.
(d) This compound contains oxygen, in addition to carbon and hydrogen, so it is not a hydrocarbon.

Cumulative Problems

5.115 **Given:** 145 mL C_2H_5OH; $d = 0.789 \text{ g/cm}^3$ **Find:** number of molecules

Conceptual Plan: $\text{cm}^3 \rightarrow \text{mL} : \text{mL } C_2H_5OH \rightarrow \text{g } C_2H_5OH \rightarrow \text{mol } C_2H_5OH \rightarrow \text{molecules } C_2H_5OH$

$$\frac{1 \text{ cm}^3}{1 \text{ mL}} \quad \frac{0.789 \text{ g } C_2H_5OH}{\text{cm}^3} \quad \frac{1 \text{ mol } C_2H_5OH}{46.07 \text{ g } C_2H_5OH} \quad \frac{6.022 \times 10^{23} \text{ molecules } C_2H_5OH}{1 \text{ mol } C_2H_5OH}$$

Solution:

$$145 \text{ mL } C_2H_5OH \times \frac{1 \text{ cm}^3}{1 \text{ mL}} \times \frac{0.789 \text{ g } C_2H_5OH}{1 \text{ cm}^3} \times \frac{1 \text{ mol } C_2H_5OH}{46.07 \text{ g } C_2H_5OH} \times \frac{6.022 \times 10^{23} \text{ molecules } C_2H_5OH}{1 \text{ mol } C_2H_5OH}$$

$$= 1.50 \times 10^{24} \text{ molecules } C_2H_5OH$$

Check: The units of the answer (molecules C_2H_5OH) are correct. The magnitude is reasonable because we had more than two moles of C_2H_5OH and we have more than two times Avogadro's number of molecules.

5.116 **Given:** 0.05 mL H_2O; $d = 1.0 \text{ g/cm}^3$ **Find:** number of molecules

Conceptual Plan: $\text{cm}^3 \rightarrow \text{mL}: \text{mL } H_2O \rightarrow \text{g } H_2O \rightarrow \text{mol } H_2O \rightarrow \text{molecules } H_2O$

$$\frac{1 \text{ cm}^3}{1 \text{ mL}} \quad \frac{1.0 \text{ g } H_2O}{1 \text{ cm}^3 \text{ } H_2O} \quad \frac{1 \text{ mol } H_2O}{18.02 \text{ g } H_2O} \quad \frac{6.022 \times 10^{23} \text{ molecules } H_2O}{1 \text{ mol } H_2O}$$

Solution:

$$0.05 \text{ mL } H_2O \times \frac{1 \text{ cm}^3}{1 \text{ mL}} \times \frac{1.0 \text{ g } H_2O}{1 \text{ cm}^3 \text{ } H_2O} \times \frac{1 \text{ mol } H_2O}{18.02 \text{ g } H_2O} \times \frac{6.022 \times 10^{23} \text{ molecules } H_2O}{1 \text{ mol } H_2O} = 2 \times 10^{21} \text{ molecules } H_2O$$

Check: The units of the answer (molecules H_2O) are correct. The magnitude is reasonable because we have less than 1 mole of H_2O and we have less than Avogadro's number of molecules.

5.117 (a) To write the formula for an ionic compound, do the following: (1) Write the symbol for the metal cation and its charge and the symbol for the nonmetal anion or polyatomic anion and its charge. (2) Adjust the subscript on each cation and anion to balance the overall charge. (3) Check that the sum of the charges of the cations equals the sum of the charges of the anions.

potassium chromate: K^+ CrO_4^{2-}; K_2CrO_4 cation $2(1+) = 2+$; anion $2-$

Given: K_2CrO_4 **Find:** mass percent of each element

Conceptual Plan: % K, then % Cr, then % O

$$\text{mass \% K} = \frac{2 \times \text{molar mass K}}{\text{molar mass } K_2CrO_4} \times 100\% \quad \text{mass \% Cr} = \frac{1 \times \text{molar mass Cr}}{\text{molar mass } K_2CrO_4} \times 100\% \quad \text{mass \% O} = \frac{4 \times \text{molar mass O}}{\text{molar mass } K_2CrO_4} \times 100\%$$

molar mass of K = 39.10 g/mol; molar mass Cr = 52.00 g/mol; molar mass O = 16.00 g/mol

Solution: molar mass K_2CrO_4 = 2(39.10 g/mol) + 1(52.00 g/mol) + 4(16.00 g/mol) = 194.20 g/mol

2 × molar mass K = 2(39.10 g/mol) = 78.20 g K 1 × molar mass Cr = 1(52.00 g/mol) = 52.00 g Cr

$$\text{mass \% K} = \frac{2 \times \text{molar mass K}}{\text{molar mass } K_2CrO_4} \times 100\% = \frac{78.20 \text{ g/mol}}{194.20 \text{ g/mol}} \times 100\% = 40.27\%$$

$$\text{mass \% Cr} = \frac{1 \times \text{molar mass Cr}}{\text{molar mass } K_2CrO_4} \times 100\% = \frac{52.00 \text{ g/mol}}{194.20 \text{ g/mol}} \times 100\% = 26.78\%$$

4 × molar mass O = 4(16.00 g/mol) = 64.00 g O

$$\text{mass \% O} = \frac{4 \times \text{molar mass O}}{\text{molar mass } K_2CrO_4} \times 100\% = \frac{64.00 \text{ g/mol}}{194.20 \text{ g/mol}} \times 100\% = 32.96\%$$

Check: The units of the answer (%) are correct. The magnitude is reasonable because each is between 0 and 100% and the total is 100%.

(b) To write the formula for an ionic compound, do the following: (1) Write the symbol for the metal cation and its charge and the symbol for the nonmetal anion or polyatomic anion and its charge. (2) Adjust the subscript on each cation and anion to balance the overall charge. (3) Check that the sum of the charges of the cations equals the sum of the charges of the anions.

Lead(II)phosphate: Pb^{2+} PO_4^{3-}; $Pb_3(PO_4)_2$ cation $3(2+) = 6+$; anion $2(3-) = 6-$

Given: $Pb_3(PO_4)_2$ **Find:** mass percent of each element

Conceptual Plan: % Pb, then % P, then % O

$$\text{mass \% Pb} = \frac{3 \times \text{molar mass Pb}}{\text{molar mass } Pb_3(PO_4)_2} \times 100\% \quad \text{mass \% P} = \frac{2 \times \text{molar mass P}}{\text{molar mass } Pb_3(PO_4)_2} \times 100\% \quad \text{mass \% O} = \frac{8 \times \text{molar mass O}}{\text{molar mass } Pb_3(PO_4)_2} \times 100\%$$

Solution: molar mass $Pb_3(PO_4)_2$ = 3(207.2 g/mol) + 2(30.97 g/mol) + 8(16.00 g/mol) = 811.5 g/mol

3 × molar mass Pb = 3(207.2 g/mol) = 621.6 g Pb 2 × molar mass P = 2(30.97 g/mol) = 61.94 g P

$$\text{mass \% Pb} = \frac{3 \times \text{molar mass Pb}}{\text{molar mass } Pb_3(PO_4)_2} \times 100\% = \frac{621.6 \text{ g/mol}}{811.5 \text{ g/mol}} \times 100\% = 76.60\%$$

$$\text{mass \% P} = \frac{2 \times \text{molar mass P}}{\text{molar mass } Pb_3(PO_4)_2} \times 100\% = \frac{61.94 \text{ g/mol}}{811.5 \text{ g/mol}} \times 100\% = 7.633\%$$

4 × molar mass O = 8(16.00 g/mol) = 128.0 g O

$$\text{mass \% O} = \frac{8 \times \text{molar mass O}}{\text{molar mass } Pb_3(PO_4)_2} \times 100\% = \frac{128.0 \text{ g/mol}}{811.5 \text{ g/mol}} \times 100\% = 15.77\%$$

Check: The units of the answer (%) are correct. The magnitude is reasonable because each is between 0 and 100% and the total is 100%.

(c) To write the formula for an ionic compound, do the following: (1) Write the symbol for the metal cation and its charge and the symbol for the nonmetal anion or polyatomic anion and its charge. (2) Adjust the subscript on each cation and anion to balance the overall charge. (3) Check that the sum of the charges of the cations equals the sum of the charges of the anions.

cobalt(II)bromide: Co^{2+} Br^-; $CoBr_2$ cation 2+ = 2+; anion 2(1−) = 2−

Given: $CoBr_2$ **Find:** mass percent of each element

Conceptual Plan: % Co, then % Br

$\text{mass \% Co} = \frac{1 \times \text{molar mass Co}}{\text{molar mass CoBr}_2} \times 100\%$ $\text{mass \% Br} = \frac{2 \times \text{molar mass Br}}{\text{molar mass CoBr}_2} \times 100\%$

Solution: molar mass $CoBr_2$ = (58.93 g/mol) + 2(79.90 g/mol) = 218.73 g/mol

2 × molar mass Co = 1(58.93 g/mol) = 58.93 g Co 1 × molar mass Br = 2(79.90 g/mol) = 159.80 g Br

$$\text{mass \% Co} = \frac{1 \times \text{molar mass Co}}{\text{molar mass CoBr}_2} \times 100\% = \frac{58.93 \text{ g/mol}}{218.73 \text{ g/mol}} \times 100\% = 26.94\%$$

$$\text{mass \% Br} = \frac{2 \times \text{molar mass Br}}{\text{molar mass CoBr}_2} \times 100\% = \frac{159.80 \text{ g/mol}}{218.73 \text{ g/mol}} \times 100\% = 73.058\%$$

Check: The units of the answer (%) are correct. The magnitude is reasonable because each is between 0 and 100% and the total is 100%.

5.118 (a) phosphorus pentachloride: PCl_5

Given: PCl_5 **Find:** mass percent of each element

Conceptual Plan: % P, then % Cl

$\text{mass \% P} = \frac{1 \times \text{molar mass P}}{\text{molar mass PCl}_5} \times 100\%$ $\text{mass \% Cl} = \frac{5 \times \text{molar mass Cl}}{\text{molar mass PCl}_5} \times 100\%$

Solution: molar mass PCl_5 = 1(30.97 g/mol) + 5(35.45 g/mol) = 208.2 g/mol

1 × molar mass P = 1(30.97 g/mol) = 30.97 g P; 5 × molar mass Cl = 5(35.45 g/mol) = 177.25 g Cl

$$\text{mass \% P} = \frac{1 \times \text{molar mass P}}{\text{molar mass PCl}_5} \times 100\% = \frac{30.97 \text{ g/mol}}{208.2 \text{ g/mol}} \times 100\% = 14.88\%$$

$$\text{mass \% Cl} = \frac{5 \times \text{molar mass Cl}}{\text{molar mass PCl}_5} \times 100\% = \frac{177.25 \text{ g/mol}}{208.2 \text{ g/mol}} \times 100\% = 85.13\%$$

Check: The units of the answer (%) are correct. The magnitude is reasonable because each is between 0 and 100% and the total is 100%.

(b) nitrogen triiodide: NI_3

Given: NI_3 **Find:** mass percent of each element

Conceptual Plan: % N, then % I

$\text{mass \% N} = \frac{1 \times \text{molar mass N}}{\text{molar mass NI}_3} \times 100\%$ $\text{mass \% I} = \frac{3 \times \text{molar mass N}}{\text{molar mass NI}_3} \times 100\%$

Solution: molar mass NI_3 = 1(14.01 g/mol) + 3(126.9 g/mol) = 394.7 g/mol

1 × molar mass N = 1(14.01 g/mol) = 14.01 g N 1 × molar mass I = 3(126.9 g/mol) = 380.7 g I

$$\text{mass \% N} = \frac{1 \times \text{molar mass N}}{\text{molar mass NI}_3} \times 100\% = \frac{14.01 \text{ g/mol}}{394.7 \text{ g/mol}} \times 100\% = 3.549\%$$

$$\text{mass \% I} = \frac{3 \times \text{molar mass I}}{\text{molar mass NI}_3} \times 100\% = \frac{380.7 \text{ g/mol}}{394.7 \text{ g/mol}} \times 100\% = 96.45\%$$

Check: The units of the answer (%) are correct. The magnitude is reasonable because each is between 0 and 100% and the total is 100%.

(c) carbon dioxide: CO_2

Given: CO_2 **Find:** mass percent of each element

Conceptual Plan: % C, then % O

$\text{mass \% C} = \frac{1 \times \text{molar mass C}}{\text{molar mass CO}_2} \times 100\%$ $\text{mass \% O} = \frac{2 \times \text{molar mass O}}{\text{molar mass CO}_2} \times 100\%$

Solution: molar mass $CO_2 = 1(12.01 \text{ g/mol}) + 2(16.00 \text{ g/mol}) = 44.01 \text{ g/mol}$
$1 \times \text{molar mass C} = 1(12.01 \text{ g/mol}) = 12.01 \text{ g C}$ $2 \times \text{molar mass O} = 2(16.00 \text{ g/mol}) = 32.00 \text{ g O}$

$$\text{mass \% C} = \frac{1 \times \text{molar mass C}}{\text{molar mass CO}_2} \times 100\% = \frac{12.01 \text{ g/mol}}{44.01 \text{ g/mol}} \times 100\% = 27.29\%$$

$$\text{mass \% O} = \frac{2 \times \text{molar mass O}}{\text{molar mass CO}_2} \times 100\% = \frac{32.00 \text{ g/mol}}{44.01 \text{ g/mol}} \times 100\% = 72.71\%$$

Check: The units of the answer (%) are correct. The magnitude is reasonable because each is between 0 and 100% and the total is 100%.

5.119 **Given:** 25 g CF_2Cl_2/mo **Find:** g Cl_2/yr
Conceptual Plan: g CF_2Cl_2/mo → g Cl_2/mo → g Cl_2/yr

$\frac{70.90 \text{ g Cl}_2}{120.91 \text{ g CF}_2\text{Cl}_2}$ $\frac{12 \text{ mo}}{1 \text{ yr}}$

Solution: $$\frac{25 \text{ g CF}_2\text{Cl}_2}{\text{mo}} \times \frac{70.90 \text{ g Cl}_2}{120.91 \text{ g CF}_2\text{Cl}_2} \times \frac{12 \text{ mo}}{1 \text{ yr}} = 1.8 \times 10^2 \text{ g Cl}_2/\text{yr}$$

Check: The units of the answer (g Cl_2/yr) are correct. Magnitude is reasonable because it is less than the total CF_2Cl_2/yr.

5.120 **Given:** 12 kg CHF_2Cl/mo **Find:** kg Cl_2/yr
Conceptual Plan: kg CHF_2Cl/mo → kg Cl_2/mo → kg Cl_2/yr

$\frac{35.45 \text{ kg Cl}_2}{86.47 \text{ kg CHF}_2\text{Cl}}$ $\frac{12 \text{ mo}}{1 \text{ yr}}$

Solution: $$\frac{12 \text{ kg CHF}_2\text{Cl}}{\text{mo}} \times \frac{35.45 \text{ kg Cl}_2}{86.47 \text{ kg CHF}_2\text{Cl}} \times \frac{12 \text{ mo}}{1 \text{ yr}} = 59 \text{ kg Cl}_2/\text{yr}$$

Check: The units of the answer (kg Cl_2/yr) are correct. Magnitude is reasonable because it is less than the total CHF_2Cl/yr.

5.121 **Given:** MCl_3; 65.57% Cl **Find:** identify M
Conceptual Plan: g Cl → mol Cl → mol M → atomic mass M

$\frac{1 \text{ mol Cl}}{35.45 \text{ g Cl}}$ $\frac{1 \text{ mol M}}{3 \text{ mol Cl}}$ $\frac{\text{g M}}{\text{mol M}}$

Solution: in 100 g sample: 65.57 g Cl; 34.43 g M

$$65.57 \text{ g Cl} \times \frac{1 \text{ mol Cl}}{35.45 \text{ g Cl}} \times \frac{1 \text{ mol M}}{3 \text{ mol Cl}} = 0.6165 \text{ mol M} \quad \frac{34.43 \text{ g M}}{0.6165 \text{ mol M}} = 55.85 \text{ g/mol M}$$

molar mass of 55.85 = Fe
The identity of M = Fe.

5.122 **Given:** M_2O; 16.99% O **Find:** identify M
Conceptual Plan: g O → mol O → mol M → atomic mass M

$\frac{1 \text{ mol O}}{16.00 \text{ g O}}$ $\frac{2 \text{ mol M}}{1 \text{ mol O}}$ $\frac{\text{g M}}{\text{mol M}}$

Solution: in 100 g sample: 16.99 g O; 83.01 g M

$$16.99 \text{ g O} \times \frac{1 \text{ mol O}}{16.00 \text{ g O}} \times \frac{2 \text{ mol M}}{1 \text{ mol O}} = 2.124 \text{ mol M} \quad \frac{83.01 \text{ g M}}{2.124 \text{ mol M}} = 39.08 \text{ g/mol M}$$

molar mass of 39.08 = K
The identity of M = K.

5.123 **Given:** in a 100 g sample of estradiol: 79.37 g C; 8.88 g H; 11.75 g O; molar mass = 272.37g/mol **Find:** molecular formula

Conceptual Plan:
Convert mass to mol of each element → pseudoformula → empirical formula → molecular formula

$\frac{1 \text{ mol C}}{12.01 \text{ g C}}$ $\frac{1 \text{ mol H}}{1.008 \text{ g H}}$ $\frac{1 \text{ mol O}}{16.00 \text{ g O}}$ divide by smallest number empirical formula $\times n$

Solution: $79.37 \text{ g C} \times \frac{1 \text{ mol C}}{12.01 \text{ g C}} = 6.609 \text{ mol C}$

$8.88 \text{ g H} \times \frac{1 \text{ mol H}}{1.008 \text{ g H}} = 8.81 \text{ mol H}$

$11.75 \text{ g O} \times \frac{1 \text{ mol O}}{16.00 \text{ g O}} = 0.7344 \text{ mol O}$

$C_{6.609}H_{8.81}O_{0.7344}$

$C_{\frac{6.609}{0.7344}}H_{\frac{8.81}{0.7344}}O_{\frac{0.7344}{0.7344}} \rightarrow C_9H_{12}O$

The correct empirical formula is $C_9H_{12}O$.

empirical formula mass $= 9(12.01 \text{ g/mol}) + 12(1.008 \text{ g/mol}) + 1(16.00 \text{ g/mol}) = 136.19 \text{ g/mol}$

$$n = \frac{\text{molar mass}}{\text{formula molar mass}} = \frac{272.37 \text{ g/mol}}{136.19 \text{ g/mol}} = 2$$

molecular formula $= C_9H_{12}O \times 2 = C_{18}H_{24}O_2$

5.124 **Given:** in a 100 g sample of fructose: 40.00 g C; 6.72 g H; 53.28 g O; molar mass = 180.16 g/mol
Find: molecular formula

Conceptual Plan:
Convert mass to mol of each element → pseudoformula → empirical formula → molecular formula

$\frac{1 \text{ mol C}}{12.01 \text{ g C}}$ $\frac{1 \text{ mol H}}{1.008 \text{ g H}}$ $\frac{1 \text{ mol O}}{16.00 \text{ g O}}$ divide by smallest number empirical formula $\times n$

Solution: $40.00 \text{ g C} \times \frac{1 \text{ mol C}}{12.01 \text{ g C}} = 3.331 \text{ mol C}$

$6.72 \text{ g H} \times \frac{1 \text{ mol H}}{1.008 \text{ g H}} = 6.67 \text{ mol H}$

$53.28 \text{ g O} \times \frac{1 \text{ mol O}}{16.00 \text{ g O}} = 3.330 \text{ mol O}$

$C_{3.331}H_{6.67}O_{3.330}$

$C_{\frac{3.331}{3.331}}H_{\frac{6.67}{3.331}}O_{\frac{3.330}{3.331}} \rightarrow CH_2O$

The correct empirical formula is CH_2O.

empirical formula mass $= 1(12.01 \text{ g/mol}) + 2(1.008 \text{ g/mol}) + 1(16.00 \text{ g/mol}) = 30.03 \text{ g/mol}$

$$n = \frac{\text{molar mass}}{\text{formula molar mass}} = \frac{180.16 \text{ g/mol}}{30.03 \text{ g/mol}} = 6$$

molecular formula $= CH_2O \times 6 = C_6H_{12}O_6$

5.125 **Given:** 13.42 g sample of equilin; 39.61 g CO_2; 9.01 g H_2O; molar mass = 268.34 g/mol
Find: molecular formula

Conceptual Plan:
Mass CO_2, H_2O → mol CO_2, H_2O → mol C, mol H → mass C, mass H, mass O → mol O →

$\frac{1 \text{ mol } CO_2}{44.01 \text{ g } CO_2}$ $\frac{1 \text{ mol } H_2O}{18.02 \text{ g } H_2O}$ $\frac{1 \text{ mol C}}{1 \text{ mol } CO_2}$ $\frac{2 \text{ mol H}}{1 \text{ mol } H_2O}$ $\frac{12.01 \text{ g C}}{1 \text{ mol C}}$ $\frac{1.008 \text{ g H}}{1 \text{ mol H}}$ g sample − g C − g H $\frac{1 \text{ mol O}}{16.00 \text{ g O}}$

pseudoformula → empirical formula → molecular formula

divide by smallest number empirical formula $\times n$

$$39.61 \text{ g } \cancel{CO_2} \times \frac{1 \text{ mol } CO_2}{44.01 \text{ g } \cancel{CO_2}} = 0.9000 \text{ mol } CO_2$$

$$9.01 \text{ g } \cancel{H_2O} \times \frac{1 \text{ mol } H_2O}{18.02 \text{ g } \cancel{H_2O}} = 0.500 \text{ mol } H_2O$$

$$0.9000 \cancel{\text{mol } CO_2} \times \frac{1 \text{ mol C}}{1 \cancel{\text{mol } CO_2}} = 0.9000 \text{ mol C}$$

$$0.500 \cancel{\text{mol } H_2O} \times \frac{2 \text{ mol H}}{1 \cancel{\text{mol } H_2O}} = 1.00 \text{ mol H}$$

$$0.9000 \cancel{\text{mol C}} \times \frac{12.01 \text{ g C}}{1 \cancel{\text{mol C}}} = 10.81 \text{ g C}$$

$$1.00 \cancel{\text{mol H}} \times \frac{1.008 \text{ g H}}{1 \cancel{\text{mol H}}} = 1.01 \text{ g H}$$

$$13.42 \text{ g} - 10.81 \text{ g} - 1.01 \text{ g} = 1.60 \text{ g O}$$

$$1.60 \text{ g } \cancel{O} \times \frac{1 \text{ mol O}}{16.00 \text{ g } \cancel{O}} = 0.100 \text{ mol O}$$

$$C_{0.9000}H_{1.000}O_{0.100}$$

$$C_{\frac{0.9000}{0.100}}H_{\frac{1.00}{0.100}}O_{\frac{0.100}{0.100}} \rightarrow C_9H_{10}O$$

The correct empirical formula is $C_9H_{10}O$.

$$\text{empirical formula mass} = 9(12.01 \text{ g/mol}) + 10(1.008 \text{ g/mol}) + 1(16.00 \text{ g/mol}) = 134.2 \text{ g/mol}$$

$$n = \frac{\text{molar mass}}{\text{formula molar mass}} = \frac{268.34 \text{ g/mol}}{134.2 \text{ g/mol}} = 2$$

$$\text{molecular formula} = C_9H_{10}O \times 2 = C_{18}H_{20}O_2$$

5.126 **Given:** 1.893 g sample; 5.545 g CO_2; 1.388 g H_2O; molar mass = 270.36 g/mol
Find: molecular formula

Conceptual Plan:

Mass CO_2, H_2O → mol CO_2, H_2O → mol C, mol H → mass C, mass H, mass O → mol O →

$$\frac{1 \text{ mol } CO_2}{44.01 \text{ g } CO_2} \quad \frac{1 \text{ mol } H_2O}{18.02 \text{ g } H_2O} \quad \frac{1 \text{ mol C}}{1 \text{ mol } CO_2} \quad \frac{2 \text{ mol H}}{1 \text{ mol } H_2O} \quad \frac{12.01 \text{ g C}}{1 \text{ mol C}} \quad \frac{1.008 \text{ g H}}{1 \text{ mol H}} \quad \text{g sample} - \text{g C} - \text{g H} \quad \frac{1 \text{ mol O}}{16.00 \text{ g O}}$$

pseudoformula → empirical formula → molecular formula

divide by smallest number   empirical formula × n

Solution:

$$5.545 \text{ g } \cancel{CO_2} \times \frac{1 \text{ mol } CO_2}{44.01 \text{ g } \cancel{CO_2}} = 0.1260 \text{ mol } CO_2$$

$$1.388 \text{ g } \cancel{H_2O} \times \frac{1 \text{ mol } H_2O}{18.02 \text{ g } \cancel{H_2O}} = 0.07703 \text{ mol } H_2O$$

$$0.1260 \cancel{\text{mol } CO_2} \times \frac{1 \text{ mol C}}{1 \cancel{\text{mol } CO_2}} = 0.1260 \text{ mol C}$$

$$0.07703 \cancel{\text{mol } H_2O} \times \frac{2 \text{ mol H}}{1 \cancel{\text{mol } H_2O}} = 0.1541 \text{ mol H}$$

$$0.1260 \cancel{\text{mol C}} \times \frac{12.01 \text{ g C}}{1 \cancel{\text{mol C}}} = 1.513 \text{ g C}$$

$$0.1541 \cancel{\text{mol H}} \times \frac{1.008 \text{ g H}}{1 \cancel{\text{mol H}}} = 0.1553 \text{ g H}$$

$1.893\text{ g} - 1.513\text{ g} - 0.1553\text{ g} = 0.225\text{ g O}$

$$0.225\ \cancel{\text{g O}} \times \frac{1\text{ mol O}}{16.00\ \cancel{\text{g O}}} = 0.0141\text{ mol O}$$

$C_{0.1260}H_{0.1541}O_{0.0141}$

$$C_{\frac{0.1260}{0.0141}}H_{\frac{0.1541}{0.0141}}O_{\frac{0.0141}{0.0141}} \rightarrow C_9H_{11}O$$

The correct empirical formula is $C_9H_{11}O$.

empirical formula mass $= 9(12.01\text{ g/mol}) + 11(1.008\text{ g/mol}) + 1(16.00\text{ g/mol}) = 135.2\text{ g/mol}$

$$n = \frac{\text{molar mass}}{\text{formula molar mass}} = \frac{270.36\text{ g/mol}}{135.2\text{ g/mol}} = 2$$

molecular formula $= C_9H_{11}O \times 2$

$= C_{18}H_{22}O_2$

5.127 **Given:** 4.93 g $MgSO_4 \cdot xH_2O$; 2.41 g $MgSO_4$ **Find:** value of x

Conceptual Plan: g $MgSO_4$ → mol $MgSO_4$ g H_2O → mol H_2O Determine mole ratio.

$$\frac{1\text{ mol }MgSO_4}{120.38\text{ g }MgSO_4} \qquad \frac{1\text{ mol }H_2O}{18.02\text{ g }H_2O} \qquad \frac{\text{mol }HO_2}{\text{mol }MgSO_4}$$

Solution:

$$2.41\ \cancel{\text{g }MgSO_4} \times \frac{1\text{ mol }MgSO_4}{120.38\ \cancel{\text{g }MgSO_4}} = 0.0200\text{ mol }MgSO_4$$

Determine g H_2O: 4.93 g $MgSO_4 \cdot xH_2O - 2.41$ g $MgSO_4 = 2.52$ g H_2O

$$2.52\ \cancel{\text{g }H_2O} \times \frac{1\text{ mol }H_2O}{18.02\ \cancel{\text{g }H_2O}} = 0.140\text{ mol }H_2O$$

$$\frac{0.140\text{ mol }H_2O}{0.0200\text{ mol }MgSO_4} = 7$$

$x = 7$

5.128 **Given:** 3.41 g $CuCl_2 \cdot xH_2O$; 2.69 g $CuCl_2$ **Find:** value of x

Conceptual Plan: g $CuCl_2$ → mol $CuCl_2$ g H_2O → mol H_2O Determine mole ratio.

$$\frac{1\text{ mol }CuCl_2}{134.45\text{ g }CuCl_2} \qquad \frac{1\text{ mol }H_2O}{18.02\text{ g }H_2O} \qquad \frac{\text{mol }HO_2}{\text{mol }CuCl_2}$$

Solution:

$$2.69\ \cancel{\text{g }CuCl_2} \times \frac{1\text{ mol }CuCl_2}{134.45\ \cancel{\text{g }CuCl_2}} = 0.0200\text{ mol }CuCl_2$$

Determine g H_2O: 3.41 g $CuCl_2 \cdot x\,H_2O - 2.69$ g $CuCl_2 = 0.72$ g H_2O

$$0.72\ \cancel{\text{g }H_2O} \times \frac{1\text{ mol }H_2O}{18.02\ \cancel{\text{g }H_2O}} = 0.040\text{ mol }H_2O$$

$$\frac{0.040\text{ mol }H_2O}{0.0200\text{ mol }CuCl_2} = 2$$

$x = 2$

5.129 **Given:** molar mass $= 177$ g/mol; g C $= 8$(g H) **Find:** molecular formula

Conceptual Plan: C_xH_yBrO

Solution: in 1 mol compound, let $x =$ mol C and $y =$ mol H, assume mol Br $= 1$, assume mol O $= 1$

$177\ g/mol = x(12.01\ g/mol) + y(1.008\ g/mol) + 1(79.90\ g/mol) + 1(16.00\ g/mol)$

$x(12.01\ g/mol) = 8[y(1.008\ g/mol)]$

$177\ g/mol = 8y(1.008\ g/mol) + y(1.008\ g/mol) + 79.90\ g/mol + 16.00\ g/mol$

$81 = 9y(1.008)$

$y = 9 = \text{mol H}$

$x(12.01) = 8 \times 9(1.008)$

$x = 6 = \text{mol C}$

molecular formula $= C_6H_9BrO$

Check: molar mass $= 6(12.01\ g/mol) + 9(1.008\ g/mol) + 1(79.90\ g/mol) + 1(16.00\ g/mol) = 177.0\ g/mol$

5.130 **Given:** 3.54 g sample yields 8.49 g CO_2 and 2.14 g H_2O; 2.35 g sample yields 0.199 g N; molar mass = 165
Find: molecular formula

Conceptual Plan:

Mass N → mol N; then mass CO_2, H_2O → mol CO_2, H_2O → mol C, mol H → mass C, mass H;

$\frac{1\ mol\ N}{14.01\ g\ N}$ $\frac{1\ mol\ CO_2}{44.01\ g\ CO_2}$ $\frac{1\ mol\ H_2O}{18.02\ g\ H_2O}$ $\frac{1\ mol\ C}{1\ mol\ CO_2}$ $\frac{2\ mol\ H}{1\ mol\ H_2O}$ $\frac{12.01\ g\ C}{1\ mol\ C}$ $\frac{1.008\ g\ H}{1\ mol\ H}$

mass O → mol O → pseudoformula → empirical formula → molecular formula

g sample − g C − g H $\frac{1\ mol\ O}{16.00\ g\ O}$ divide by smallest number empirical formula × n

Solution:

$$\frac{0.199\ g\ N}{2.35\ g\ sample} = \frac{x\ g\ N}{3.54\ g\ sample};\ x = 0.300\ g\ N$$

$$0.300\ \cancel{g\ N} \times \frac{1\ mol\ N}{14.01\ \cancel{g\ N}} = 0.0214\ mol\ N$$

$$8.49\ \cancel{g\ CO_2} \times \frac{1\ mol\ CO_2}{44.01\ \cancel{g\ CO_2}} = 0.193\ mol\ CO_2$$

$$2.14\ \cancel{g\ H_2O} \times \frac{1\ mol\ H_2O}{18.02\ \cancel{g\ H_2O}} = 0.119\ mol\ H_2O$$

$$0.193\ \cancel{mol\ CO_2} \times \frac{1\ mol\ C}{1\ \cancel{mol\ CO_2}} = 0.193\ mol\ C$$

$$0.119\ \cancel{mol\ H_2O} \times \frac{2\ mol\ H}{1\ \cancel{mol\ H_2O}} = 0.238\ mol\ H$$

$$0.193\ \cancel{mol\ C} \times \frac{12.01\ g\ C}{1\ \cancel{mol\ C}} = 2.32\ g\ C$$

$$0.238\ \cancel{mol\ H} \times \frac{1.008\ g\ H}{1\ \cancel{mol\ H}} = 0.240\ g\ H$$

$$3.54\ g - 2.32\ g\ C - 0.240\ g\ H - 0.300\ g\ N = 0.680\ g\ O$$

$$0.680\ \cancel{g\ O} \times \frac{1\ mol\ O}{16.00\ \cancel{g\ O}} = 0.0425\ mol\ O$$

$C_{0.193}H_{0.238}N_{0.0214}O_{0.0425}$

$$C_{\frac{0.193}{0.0214}}H_{\frac{0.238}{0.0214}}N_{\frac{0.0214}{0.0214}}O_{\frac{0.0425}{0.0214}} \rightarrow C_9H_{11}NO_2$$

The correct empirical formula is $C_9H_{11}NO_2$.

empirical formula mass $= 9(12.01\ g/mol) + 11(1.008\ g/mol) + 1(14.01\ g/mol) + 2(16.00\ g/mol) = 165.19\ g/mol$

$$n = \frac{\text{molar mass}}{\text{formula molar mass}} = \frac{165\ g/mol}{165.19\ g/mol} = 1$$

molecular formula $= C_9H_{11}NO_2 \times 1$
$= C_9H_{11}NO_2$

5.131 **Given:** 23.5 mg $C_{17}H_{22}ClNO_4$ **Find:** total number of atoms

Conceptual Plan: mg compound → g compound → mol compound → mol atoms → number of atoms

$$\frac{1\ \text{g}}{1000\ \text{mg}} \quad \frac{1\ \text{mol compound}}{339.8\ \text{g}} \quad \frac{45\ \text{mol atoms}}{1\ \text{mol compound}} \quad \frac{6.022 \times 10^{23}\ \text{atoms}}{1\ \text{mol atoms}}$$

Solution: $23.5\ \cancel{\text{mg}} \times \frac{1\ \cancel{\text{g}}}{1000\ \cancel{\text{mg}}} \times \frac{1\ \cancel{\text{mol compound}}}{339.8\ \cancel{\text{g}}} \times \frac{45\ \cancel{\text{mol atoms}}}{1\ \cancel{\text{mol compound}}} \times \frac{6.022 \times 10^{23}\ \text{atoms}}{1\ \cancel{\text{mol atoms}}} = 1.87 \times 10^{21}\ \text{atoms}$

Check: The units of the answer (number of atoms) are correct. The magnitude of the answer is reasonable because the molecule is so complex.

5.132 (a) **Given:** in a 100 g sample: 76 g V; 24 g O **Find:** formula and name

Conceptual Plan:

Convert mass to mol of each element → write pseudoformula → write empirical formula

$$\frac{1\ \text{mol V}}{50.94\ \text{g V}} \quad \frac{1\ \text{mol O}}{16.00\ \text{g O}} \qquad \text{divide by smallest number}$$

Solution:

$$76\ \cancel{\text{g V}} \times \frac{1\ \text{mol V}}{50.94\ \cancel{\text{g V}}} = 1.5\ \text{mol V}$$

$$24\ \cancel{\text{g O}} \times \frac{1\ \text{mol O}}{16.00\ \cancel{\text{g O}}} = 1.5\ \text{mol O}$$

$V_{1.5}O_{1.5}$

$V_{\frac{1.5}{1.5}}O_{\frac{1.5}{1.5}} \rightarrow VO$

The correct formula is VO: vanadium(II) oxide.

(b) **Given:** in a 100 g sample: 68 g V; 32 g O **Find:** formula and name

Conceptual Plan:

Convert mass to mol of each element → write pseudoformula → write empirical formula

$$\frac{1\ \text{mol V}}{50.94\ \text{g V}} \quad \frac{1\ \text{mol O}}{16.00\ \text{g O}} \qquad \text{divide by smallest number}$$

Solution:

$$68\ \cancel{\text{g V}} \times \frac{1\ \text{mol V}}{50.94\ \cancel{\text{g V}}} = 1.33\ \text{mol V}$$

$$32\ \cancel{\text{g O}} \times \frac{1\ \text{mol O}}{16.00\ \cancel{\text{g O}}} = 2\ \text{mol O}$$

$V_{1.33}O_2$

$V_{\frac{1.33}{1.33}}O_{\frac{2}{1.33}} \rightarrow VO_{1.5} \rightarrow V_2O_3$

The correct formula is V_2O_3: vanadium(III) oxide.

(c) **Given:** in a 100 g sample: 61 g V; 39 g O **Find:** formula and name

Conceptual Plan:

Convert mass to mol of each element → write pseudoformula → write empirical formula

$$\frac{1\ \text{mol V}}{50.94\ \text{g V}} \quad \frac{1\ \text{mol O}}{16.00\ \text{g O}} \qquad \text{divide by smallest number}$$

Solution:

$$61\ \cancel{\text{g V}} \times \frac{1\ \text{mol V}}{50.94\ \cancel{\text{g V}}} = 1.2\ \text{mol V}$$

$$39\ \cancel{\text{g O}} \times \frac{1\ \text{mol O}}{16.00\ \cancel{\text{g O}}} = 2.4\ \text{mol O}$$

$V_{1.2}O_{2.4}$

$V_{\frac{1.2}{1.2}}O_{\frac{2.4}{1.2}} \rightarrow VO_2$

The correct formula is VO_2: vanadium(IV) oxide.

(d) **Given:** in a 100 g sample: 56 g V; 44 g O **Find:** formula and name
Conceptual Plan:
Convert mass to mol of each element → write pseudoformula → write empirical formula

$\frac{1 \text{ mol V}}{50.94 \text{ g V}}$ $\frac{1 \text{ mol O}}{16.00 \text{ g O}}$ divide by smallest number

Solution:

$$56 \text{ g V} \times \frac{1 \text{ mol V}}{50.94 \text{ g V}} = 1.1 \text{ mol V}$$

$$44 \text{ g O} \times \frac{1 \text{ mol O}}{16.00 \text{ g O}} = 2.75 \text{ mol O}$$

$V_{1.1}O_{2.75}$

$V_{\frac{1.1}{1.1}}O_{\frac{2.75}{1.1}} \rightarrow VO_{2.5} \rightarrow V_2O_5$

The correct formula is V_2O_5: vanadium(V) oxide.

5.133 **Given:** MCl_3; 2.395 g sample; 3.606×10^{-2} mol Cl **Find:** atomic mass M
Conceptual Plan: mol Cl → g Cl → g X

$\frac{35.45 \text{ g Cl}}{1 \text{ mol Cl}}$ g sample − g Cl = g M

mol Cl → mol M → atomic mass M

$\frac{1 \text{ mol M}}{3 \text{ mol Cl}}$ $\frac{\text{g M}}{\text{mol M}}$

Solution:

$$3.606 \times 10^{-2} \text{ mol Cl} \times \frac{35.45 \text{ g Cl}}{1 \text{ mol Cl}} = 1.278 \text{ g Cl}$$

$$2.395 \text{ g} - 1.278 \text{ g} = 1.117 \text{ g M}$$

$$3.606 \times 10^{-2} \text{ mol Cl} \times \frac{1 \text{ mol M}}{3 \text{ mol Cl}} = 1.202 \times 10^{-2} \text{ mol M}$$

$$\frac{1.117 \text{ g M}}{0.01202 \text{ mol M}} = 92.93 \text{ g/mol M}$$

molar mass of M = 92.93 g/mol

5.134

```
    H   H   H   H
    |   |   |   |
H — C — C — C — C — H
    |   |   |   |
    H   H   H   H
```

```
          H
          |
      H — C — H
    H     |     H
    |     C     |
 H— C  /  |  \  C —H
    |     H     |
    H           H
```

5.135 **Given:** $Fe_xCr_yO_4$; 28.59% O **Find:** x and y
Conceptual Plan: % O → molar mass $Fe_xCr_yO_4$ → mass Fe + Cr

$\frac{\text{mass O}}{\text{molar mass compound}} \times 100\% = \%\text{ O}$ mass cpd − mass O = mass Fe + Cr

Solution: $\frac{28.59 \text{ g O}}{100.0} = \frac{64.00 \text{ g O}}{\text{molar mass cpd}}$ molar mass = 223.9 g/mol

Mass Fe + Cr = molar mass − (4 × molar mass O) = 223.9 − 64.00 = 159.9 g
Molar mass Fe = 55.85; molar mass Cr = 52.00
Because the mass of the two metals is close, the average mass can be used to determine the total moles of Fe and Cr present in the compound. Average mass of Fe and Cr = 53.9. $\frac{159.9 \text{ g}}{53.9 \text{ g/mol}} = 2.97 = 3$ mol metal

Let x = mol Fe and y = mol Cr
x mol Fe + y mol Cr = 3 mol total
x mol Fe(55.85 g Fe/mol) + y mol Cr(52.00 gCr/mol) = 159.8
y mol Cr = 3 − x mol Fe
$x(55.85) + (3 - x)(52.00) = 159.8$
So $x = 1$ and $y = 2$.

Check: Formula = $FeCr_2O_4$ would have a molar mass of Fe + 2Cr + 4O = 55.85 + 2(52.00) + 4(16.00) = 223.85, and the molar mass of the compound is 223.8.

5.136 **Given:** X_3P_2; 34.00 P; 100 g sample contains 34.00 g P **Find:** X
Conceptual Plan: g P → mol P → mol X

$\frac{1 \text{ mol P}}{30.97 \text{ g P}} \quad \frac{3 \text{ mol X}}{2 \text{ mol P}}$

and then g P → g X → molar mass X

100.00 g sample − 34.00 g P $\quad \frac{\text{grams X}}{\text{mol X}}$

Solution: $34.00 \text{ g P} \times \frac{1 \text{ mol P}}{30.97 \text{ g P}} \times \frac{3 \text{ mol X}}{2 \text{ mol P}} = 1.647 \text{ mol X}$

100.00 g sample − 34.00 g P = 66.00 g X

$\frac{66.00 \text{ g X}}{1.647 \text{ mol X}} = 40.07 \text{ g/mol} = \text{Ca}$

Check: The units (g/mol) are correct. The answer, Ca, is reasonable because Ca_3P_2 is a molecule that exists.

5.137 **Given:** 0.0552% $NaNO_2$; 8.00 oz bag **Find:** mass Na in bag
Conceptual Plan: oz bag → g bag → g $NaNO_2$ → g Na → mg Na

$\frac{453.6 \text{ g}}{16.00 \text{ oz}} \quad \frac{0.0552 \text{ g } NaNO_2}{100.0 \text{ g bag}} \quad \frac{22.99 \text{ g Na}}{69.00 \text{ g } NaNO_2} \quad \frac{1000 \text{ mg Na}}{1 \text{ mg Na}}$

Solution: $8.00 \text{ oz bag} \times \frac{453.6 \text{ g bag}}{16.00 \text{ oz bag}} \times \frac{0.0552 \text{ g } NaNO_2}{100.0 \text{ g bag}} \times \frac{22.99 \text{ g Na}}{69.00 \text{ g } NaNO_2} \times \frac{1000 \text{ mg Na}}{1 \text{ mg Na}} = 41.7 \text{ mg Na}$

Check: The units of the answer (mg Na) are correct. The magnitude of the answer is reasonable because only a small percentage of the total mass is Na.

5.138 **Given:** ore is 57.8% $Ca_3(PO_4)_2$ **Find:** mass of ore to get 1.00 kg P
Conceptual Plan: Mass ore → mass $Ca_3(PO_4)_2$ → mass P then kg P → g P → g ore → kg ore

$\frac{57.8 \text{ g } Ca_3(PO_4)_2}{100.0 \text{ g ore}} \quad \frac{61.94 \text{ g P}}{310.18 \text{ g } Ca_3(PO_4)_2} \quad \frac{1000 \text{ g P}}{1 \text{ kg P}} \quad \frac{100.0 \text{ g ore}}{11.54 \text{ g P}} \quad \frac{1\text{k g ore}}{1000 \text{ g ore}}$

Solution: Assume a 100.0 gram sample of ore.

$$100.0 \text{ g ore} \times \frac{57.8 \text{ g } Ca_3(PO_4)_2}{100.0 \text{ g ore}} \times \frac{61.94 \text{ g P}}{310.18 \text{ g } Ca_3(PO_4)_2} = 11.54 \text{ g P}$$

$$1.00 \text{ kg P} \times \frac{1000 \text{ g P}}{1 \text{ kg P}} \times \frac{100.0 \text{ g ore}}{11.54 \text{ g P}} \times \frac{1\text{k g ore}}{1000 \text{ g ore}} = 8.6\underline{6}6 \text{ kg ore} = 8.67 \text{ kg ore}$$

Check: The units of the answer (kg ore) are correct. The magnitude of the answer is reasonable because the amount is greater than 1 kilogram.

Challenge Problems

5.139 **Given:** g NaCl + g NaBr = 2.00 g; g Na = 0.75 g **Find:** g NaBr
Conceptual Plan:
let x = mol NaCl, y = mol NaBr, then x(molar mass NaCl) = g NaCl, y(molar mass NaBr) = g NaBr
Solution: $x(58.4) + y(102.9) = 2.00$
$x(23.0) + y(23.0) = 0.75 \qquad y = 0.0326 - x$
$58.4x + 102.9(0.0326 - x) = 2.00$
$58.4x + 3.354 - 102.9x = 2.00$
$44.5x = 1.354$

$$x = 0.03043 \text{ mol NaCl}$$
$$y = 0.0326 - 0.03043 = 0.00217 \text{ mol NaBr}$$
$$\text{g NaBr} = (0.00217)(102.9\text{g/mol}) = 0.223 \text{ g NaBr}$$

Check: The units of the answer (g NaBr) are correct. The magnitude is reasonable because it is less than the total mass.

5.140 **Given:** sample 1:1.00 g X, 0.472 g Z, X_2Z_3; sample 2: 1.00 g X, 0.630 g Z; sample 3: 1.00 g X, 0.789 g Z
Find: empirical formula for samples 2 and 3
Conceptual Plan: Moles X remains constant; determine relative moles of Z for three samples.
Solution: Let X = atomic mass X, Z = atomic mass Z

$$n_X = \frac{1.00 \text{ g X}}{X} \qquad n_Z = \frac{0.472 \text{ g Z}}{Z}$$

for sample 1: $\frac{n_X}{n_Z} = \frac{2}{3}$

for sample 2: $\frac{0.630 \text{ g}}{0.472 \text{ g}} = 1.33$; mol $= 1.33n_Z$

mol ratio: $\frac{n_X}{1.33n_Z} = \frac{2}{(1.33)3} = \frac{2}{4} = \frac{1}{2}$

Empirical formula sample 2: XZ_2

for sample 3: $\frac{0.789 \text{ g}}{0.472 \text{ g}} = 1.67$; mol $= 1.67n_Z$

mol ratio: $\frac{n_X}{1.67n_Z} = \frac{2}{(1.67)3} = \frac{2}{5}$

Empirical formula sample 3: X_2Z_5

5.141 **Given:** sample of $CaCO_3$ and $(NH_4)_2CO_3$ is 61.9% CO_3^{2-} **Find:** % $CaCO_3$
Conceptual Plan: Let $x = CaCO_3$, $y = (NH_4)_2CO_3$, then x(molar mass $CaCO_3$) = g $CaCO_3$, y[molar mass $(NH_4)_2CO_3$] = g $(NH_4)_2CO_3$
then a 100.0 g sample contains $x(100.0)$ g $CaCO_3$; $y(96.1)$ g $(NH_4)_2CO_3$; 61.9 g CO_3^{2-}

Solution:
$$x(100.0) + y(96.1) = 100.0$$
$$x(60.0) + y(60.0) = 61.9 \qquad y = 1.03\underline{1}67 - x$$
$$100.0x + 96.1(1.03 - x) = 100$$
$$100.0x + 99.0 - 96.1x = 100$$
$$3.9x = 1.0$$
$$x = 0.26 \text{ mol } CaCO_3$$
$$y = 1.032 - 0.26 = 0.77 \text{ mol } (NH_4)_2CO_3$$
$$\text{g } CaCO_3 = (0.26 \text{ mol})(100.0\text{g/mol}) = 26 \text{ g } CaCO_3 \text{ in a 100 g sample mass \% } CaCO_3 = 26\%$$

Check: The units of the answer (mass % $CaCO_3$) are correct. The magnitude is reasonable because it is between 0 and 100%.

5.142 **Given:** 50.0 g S; 1.00×10^2g Cl_2; 150. g mixture S_2Cl_2 and SCl_2 **Find:** g S_2Cl_2
Conceptual Plan: Total mol S = 2(mol S_2Cl_2) + mol SCl_2; mol $S_2Cl_2 \rightarrow$ g S_2Cl_2

$$\frac{135.04 \text{ g}}{1 \text{ mol } S_2Cl_2} \times 100$$

then S_2Cl_2 = 135.04 g/mol, SCl_2 = 102.97 g/mol, let x = mol S_2Cl_2, y = mol SCl_2
$x(135.04)$ = g S in S_2Cl_2, $y(102.97)$ = g S in SCl_2

Solution:

$$\text{mol S} = 50.0 \text{ g S} \times \frac{1 \text{ mol S}}{32.1 \text{ g S}} = 1.56 \text{ mol}$$

$2x$ = mol S in S_2Cl_2, y = mol S in SCl_2
$$2x + y = 1.56$$
$$x(135.04) + y(102.97) = 150.0$$
$$135.04x + 102.97(1.56 - 2x) = 150.0$$

$$70.96x = 10.6y$$
$$x = 0.149$$
$$y = 1.26$$

$$0.149\ \cancel{mol\ S_2Cl_2} \times \frac{135.04\ g\ S_2Cl_2}{1\ \cancel{mol\ S_2Cl_2}} = 20.1\ g\ S_2Cl_2$$

Check: The units of the answer (g S_2Cl_2) are correct. The magnitude is reasonable because there would be fewer moles of S_2Cl_2 than SCl_2.

5.143 **Given:** 1.1 kg CF_2Cl_2/automobile; 25% leaked/year; 100×10^6 automobiles **Find:** kg Cl/yr

Conceptual Plan: kg CF_2Cl_2/auto → kg CF_2Cl_2 leaked/yr → kg Cl/yr/auto → kg Cl

$\frac{25\ kg\ CF_2Cl_2}{100\ kg\ CF_2Cl_2}$ $\frac{70.9\ kg\ Cl}{120.91\ kg\ CF_2Cl_2}$ 100×10^6 auto

Solution: $\frac{1.1\ \cancel{kg\ CF_2Cl_2}}{\cancel{auto}} \times \frac{25\ \cancel{kg\ CF_2Cl_2}}{100\ \cancel{kg\ CF_2Cl_2}} \times \frac{70.9\ kg\ Cl}{120.91\ \cancel{kg\ CF_2Cl_2}} \times 100 \times 10^6\ \cancel{auto} = 1.6 \times 10^7\ kg\ Cl/yr$

Check: The units of the answer (kg Cl) are correct. The magnitude is reasonable because it is less than the kilogram CF_2Cl_2 leaked per year.

5.144 **Given:** coal = 2.55% S; H_2SO_4; 1.0 metric ton coal **Find:** metric ton H_2SO_4 produced

Conceptual Plan: H_2SO_4 → % S

$\frac{32.07\ g\ S}{98.09\ g\ H_2SO_4} \times 100\%$

Solution: $\frac{32.07\ g\ S}{98.09\ g\ H_2SO_4} \times 100\% = 32.69\%\ S$

Conceptual Plan: Metric ton coal → kg coal → kg S → kg H_2SO_4 → metric ton H_2SO_4

$\frac{1000\ kg}{1\ metric\ ton}$ $\frac{2.55\ kg\ S}{100\ kg\ coal}$ $\frac{100\ kg\ H_2SO_4}{32.69\ kg\ S}$ $\frac{1\ metric\ ton}{1000\ kg}$

Solution: $1.0\ \cancel{metric\ ton\ coal} \times \frac{1000\ \cancel{kg\ coal}}{1\ \cancel{metric\ ton\ coal}} \times \frac{2.55\ \cancel{kg\ S}}{100\ \cancel{kg\ coal}} \times \frac{100\ \cancel{kg\ H_2SO_4}}{32.69\ \cancel{kg\ S}} \times \frac{1\ metric\ ton\ H_2SO_4}{1000\ \cancel{kg\ H_2SO_4}}$

$= 0.078$ metric ton H_2SO_4

Check: The units of the answer (metric ton H_2SO_4) are correct. Magnitude is reasonable because it is more than 2.55% of a metric ton and the mass of H_2SO_4 is greater than the mass of S.

5.145 **Given:** rock contains: 38.0% PbS; 25.0% $PbCO_3$; 17.4% $PbSO_4$ **Find:** kg rock needed for 5.0 metric ton Pb

Conceptual Plan: Determine kg Pb/100 kg rock then ton Pb → kg Pb → kg rock

$\frac{1000\ kg}{1\ metric\ ton}$ $\frac{100\ kg\ rock}{64.2\ kg\ rock}$

Solution: in 100 kg rock:

$$\left(38.0\ \cancel{kg\ PbS} \times \frac{207.2\ kg\ Pb}{239.3\ \cancel{kg\ PbS}}\right) + \left(25.0\ \cancel{kg\ PbCO_3} \times \frac{207.2\ kg\ Pb}{267.2\ \cancel{kg\ PbCO_3}}\right) + \left(17.4\ \cancel{kg\ PbSO_4} \times \frac{207.2\ kg\ Pb}{303.3\ \cancel{kg\ PbSO_4}}\right)$$

$= 64.2$ kg Pb

$$5.0\ \cancel{metric\ ton\ Pb} \times \frac{1000\ \cancel{kg\ Pb}}{1\ \cancel{metric\ ton\ Pb}} \times \frac{100\ kg\ rock}{64.2\ \cancel{kg\ Pb}} = 7.8 \times 10^3\ kg\ rock$$

Check: The units of the answer (kg rock) are correct. Magnitude is reasonable because it is greater than the amount of Pb needed.

5.146 **Given:** sample 1: 2.52 g sample, 4.23 g CO_2, 1.01 g H_2O; sample 2: 4.14 g, 2.11 g SO_3; sample 3: 5.66 g, 2.27 g HNO_3 **Find:** empirical formula of the compound

Conceptual Plan: g CO_2 → g C → % C; g H_2O → g H → % H; g SO_2 → g S → % S;

$\frac{12.01\ g\ C}{44.01\ g\ CO_2}$ $\frac{g\ C}{g\ sample} \times 100\%$ $\frac{2.02\ g\ H}{18.02\ g\ H_2O}$ $\frac{g\ H}{g\ sample} \times 100\%$ $\frac{32.07\ g\ S}{80.07\ g\ SO_3}$ $\frac{g\ S}{g\ sample} \times 100\%$

g HNO_3 → g N → % N and then → % O and then % composition → mol of each atom →

$\frac{14.01\ g\ N}{63.02\ g\ HNO_3}$ $\frac{g\ N}{g\ sample} \times 100\%$ $100\% - \%C - \%H - \%N - \%S = \%O$

pseudoformula → empirical formula

divide by smallest number

Solution: $4.23 \text{ g } \cancel{CO_2} \times \frac{12.01 \text{ g C}}{44.01 \text{ g } \cancel{CO_2}} = 1.15\underline{4} \text{ g C}$ $\qquad \frac{1.154 \text{ g C}}{2.52 \text{ g sample}} \times 100\% = 45.\underline{8}1\% \text{ C}$

$1.01 \text{ g } \cancel{H_2O} \times \frac{2.02 \text{ g H}}{18.02 \text{ g } \cancel{H_2O}} = 0.11\underline{3}2 \text{ g H}$ $\qquad \frac{0.1132 \text{ g H}}{2.52 \text{ g sample}} \times 100\% = 4.4\underline{9}\% \text{ H}$

$2.11 \text{ g } \cancel{SO_3} \times \frac{32.07 \text{ g S}}{80.07 \text{ g } \cancel{SO_3}} = 0.84\underline{5}1 \text{ g S}$ $\qquad \frac{0.8451 \text{ g S}}{4.14 \text{ g sample}} \times 100\% = 20.\underline{4}1\% \text{ S}$

$2.27 \text{ g } \cancel{HNO_3} \times \frac{14.01 \text{ g N}}{63.02 \text{ g } \cancel{HNO_3}} = 0.50\underline{4}6 \text{ g N}$ $\qquad \frac{0.5046 \text{ g N}}{5.66 \text{ g sample}} \times 100\% = 8.9\underline{2}\% \text{ N}$

$\% \text{ O} = 100 - 45.\underline{8}1 - 4.4\underline{9} - 20.\underline{4}1 - 8.9\underline{2} = 20.\underline{3}7\% \text{ O}$

Assume a 100 g sample:

$45.\underline{8}1 \text{ g } \cancel{C} \times \frac{1 \text{ mol C}}{12.01 \text{ g } \cancel{C}} = 3.8\underline{1}4 \text{ mol C}$ $\qquad 4.4\underline{9} \text{ g } \cancel{H} \times \frac{1 \text{ mol H}}{1.01 \text{ g } \cancel{H}} = 4.4\underline{4}5 \text{ mol H}$

$20.\underline{4}1 \text{ g } \cancel{S} \times \frac{1 \text{ mol S}}{32.07 \text{ g } \cancel{S}} = 0.63\underline{6}4 \text{ mol S}$ $\qquad 8.9\underline{2} \text{ g } \cancel{N} \times \frac{1 \text{ mol N}}{14.01 \text{ g } \cancel{N}} = 0.63\underline{6}7 \text{ mol N}$

$20.\underline{3}7 \text{ g } \cancel{O} \times \frac{1 \text{ mol O}}{16.00 \text{ g } \cancel{O}} = 1.2\underline{7}3 \text{ mol O}$

$C_{3.8\underline{1}4}\, H_{4.4\underline{4}5}\, S_{0.63\underline{6}4}\, N_{0.63\underline{6}7}\, O_{1.2\underline{7}3}$

$C_{\frac{3.8\underline{1}4}{0.63\underline{6}4}} H_{\frac{4.4\underline{4}5}{0.63\underline{6}4}} S_{\frac{0.63\underline{6}4}{0.63\underline{6}4}} N_{\frac{0.63\underline{6}7}{0.63\underline{6}4}} O_{\frac{1.2\underline{7}3}{0.63\underline{6}4}} \rightarrow C_6H_7SNO_2$

5.147 **Given:** molar mass $= 229$ g/mol, 6 times mass C as H **Find:** molecular formula

Conceptual Plan: Let x = mass of C, then $6x$ = mass of C

Solution: in 1 mol of the compound: g C + g H + g S + g I = 229 g

Because the molar mass of I = 127, there cannot be more than 1 mol of I in the compound; so

$x + 6x + \text{g S} + 127 = 229.$

$x + 6x + \text{g S} = 102$

If the compound contains 1 mol S, then $7x = 102 - 32 = 70$ and $x = 10$ g H and $6x = 60$ g C.

$10 \text{ g } \cancel{H} \times \frac{1 \text{ mol H}}{1.0 \text{ g } \cancel{H}} = 10 \text{ mol H}$

$60 \text{ g } \cancel{C} \times \frac{1 \text{ mol C}}{12 \text{ g } \cancel{C}} = 5 \text{ mol C}$

1 mol I and 1 mol S; so empirical formula is $C_5H_{10}SI$

Check: Molar mass of $C_5H_{10}SI = 5(12) + 10(1.0) + 32 + 127 = 229$ g/mol, which is the mass given.

5.148 **Given:** compound is 40% X and 60% Y; atomic mass X = 2 (atomic mass Y) **Find:** empirical formula

Conceptual Plan: Mass X and Y → mass ratio X : Y and then g X → mol X and g Y → mol Y and then mole ratio

$\frac{\text{g X}}{\text{atomic mass X}} \qquad \frac{\text{g Y}}{\text{atomic mass Y}}$

Solution: $\frac{\text{mass X}}{\text{mass Y}} = \frac{40}{60} = \frac{2}{3}.$ $\qquad \text{mol X} = \frac{2 \text{ g}}{\text{atomic mass X}}$ and $\text{mol Y} = \frac{3 \text{ g}}{\text{atomic mass Y}}$

But atomic mass X = 2(atomic mass Y)

$\text{mol X} = \frac{2 \text{ g}}{2(\text{atomic mass Y})}$ and $\text{mol Y} = \frac{3 \text{ g}}{\text{atomic mass Y}}$

$\frac{\text{mol X}}{\text{mol Y}} = \frac{\frac{\cancel{2} \cancel{g}}{\cancel{2}(\cancel{\text{atomic mass Y}})}}{\frac{3 \cancel{g}}{\cancel{\text{atomic mass Y}}}} = \frac{1}{3}$ $\qquad$ empirical formula: XY_3

5.149 **Given:** compound is 1/3 X by mass; atomic mass X is 1/3 atomic mass Y **Find:** empirical formula
Conceptual Plan: Mass X and Y → mass ratio X: Y and then g X → mol X and g Y → mol Y and then mole ratio

$$\frac{\text{g X}}{\text{atomic mass X}} \qquad \frac{\text{g Y}}{\text{atomic mass Y}}$$

Solution: $\dfrac{\text{mass X}}{\text{mass Y}} = \dfrac{\frac{1}{3}}{\frac{2}{3}} = \dfrac{1}{2}$. $\quad \text{mol X} = \dfrac{1\text{ g}}{\text{atomic mass X}}$ and $\text{mol Y} = \dfrac{2\text{ g}}{\text{atomic mass Y}}$

But atomic mass X = 1/3 (atomic mass Y)

$$\text{mol X} = \frac{1\text{ g}}{1/3(\text{atomic mass Y})} \text{ and mol Y} = \frac{2\text{ g}}{\text{atomic mass Y}}$$

$$\frac{\text{mol X}}{\text{mol Y}} = \frac{\dfrac{1\text{ g}}{1/3(\text{atomic mass Y})}}{\dfrac{2\text{ g}}{\text{atomic mass Y}}} = 1/6 \qquad \text{empirical formula} = XY_6$$

5.150 **Given:** 9.0 g sample of C and S; 23.3 g mixture of CO_2 and SO_2 **Find:** mass of S in sample
Conceptual Plan: Let x = g C and y = g S and then x g C → mol C → mol CO_2 → g CO_2 and then

$$\frac{x\text{ g C}}{12.01\text{ g C}} \qquad \frac{1\text{ mol }CO_2}{1\text{ mol C}} \qquad \frac{44.01\text{ g }CO_2}{1\text{ mol }CO_2}$$

y g S → mol S → mol SO_2 → g SO_2

$$\frac{y\text{ g S}}{32.07\text{ g S}} \qquad \frac{1\text{ mol }SO_2}{1\text{ mol S}} \qquad \frac{64.07\text{ g }SO_2}{1\text{ mol }SO_2}$$

Solution: $(x\text{ g C}) \times \dfrac{1\text{ mol C}}{12.01\text{ g C}} \times \dfrac{1\text{ mol }CO_2}{1\text{ mol C}} \times \dfrac{44.01\text{ g }CO_2}{1\text{ mol }CO_2} = 3.66\,x\text{ g }CO_2$

$$(y\text{ g S}) \times \frac{1\text{ mol S}}{32.07\text{ g S}} \times \frac{1\text{ mol }SO_2}{1\text{ mol S}} \times \frac{64.07\text{ g }SO_2}{1\text{ mol }SO_2} = 2.00\,y\text{ g }SO_2$$

$x + y = 9.0$
$3.66x + 2.00y = 23.3$
$3.66x + (9 - x)2.00 = 23.3$
$x = 3.2$ g C and $y = 5.8$ g S

Check: The units of the answer (g S) are correct. The magnitude is reasonable because it is less than 9.0 g.

Conceptual Problems

5.151 The sphere in the molecular models represents the electron cloud of the atom. On this scale, the nucleus would be too small to see.

5.152
(a) Atomic mass O > atomic mass C; % O would be higher.
(b) Atomic mass N and O close; molecule contains 2N to 1O; % N would be higher.
(c) Atomic mass O > atomic mass C and both are much greater than the atomic mass of H; same number of atoms of C and O; % O would be higher.
(d) Atomic mass N much greater than atomic mass H; % N would be higher.

5.153 The statement is incorrect because a chemical formula is based on the ratio of atoms combined, not the ratio of grams combined. The statement should read as follows: The chemical formula for ammonia (NH_3) indicates that ammonia contains three hydrogen atoms to each nitrogen atom.

5.154 H_2SO_4: Atomic mass S is approximately twice atomic mass O; both are much greater than atomic mass H. The order of % mass is % O >% S > % H.

6 Chemical Bonding I: Drawing Lewis Structures and Determining Molecular Shapes

Review Questions

6.1 Electronegativity is the ability of an atom to attract electrons to itself in a chemical bond. This results in a polar bond. Electronegativity generally increases across a period in the periodic table. And electronegativity generally decreases down a column (group) in the periodic table. The most electronegative element is fluorine.

6.2 If two elements with identical electronegativities form a covalent bond, they share the electrons equally, and the bond is purely covalent or nonpolar.

If there is an intermediate electronegativity difference between the two elements, such as between two different nonmetals, the bond is polar covalent.

If there is a large electronegativity difference between the two elements in a bond, such as normally occurs between a metal and a nonmetal, the electron from the metal is almost completely transferred to the nonmetal, and the bond is ionic.

6.3 Percent ionic character is defined as the ratio of a bond's actual dipole moment to the dipole moment it would have if the electron were completely transferred from one atom to the other, multiplied by 100.

A bond in which an electron is completely transferred from one atom to another would have 100% ionic character. However, no bond is 100% ionic. Percent ionic character generally increases as the electronegativity difference increases. In general, bonds with greater than 50% ionic character are referred to as ionic bonds.

6.4 A dipole moment (μ) occurs anytime there is a separation of positive and negative charge. It is used to quantify the polarity of a bond. The magnitude of the dipole moment created by separating two particles of equal but opposite charges of magnitude (q) by a distance (r) is given by $\mu = qr$.

6.5 To calculate the dipole moment, we use $\mu = qr$:

For 100 pm: $\mu = 1.6 \times 10^{-19}\,\cancel{\text{C}} \times 100\,\cancel{\text{pm}} \times \dfrac{\cancel{\text{m}}}{10^{12}\,\cancel{\text{pm}}} \times \dfrac{\text{D}}{3.34 \times 10^{-30}\,\cancel{\text{C}} \cdot \cancel{\text{m}}} = 4.8\ \text{D}$

For 200 pm: $\mu = 1.6 \times 10^{-19}\,\cancel{\text{C}} \times 200\,\cancel{\text{pm}} \times \dfrac{\cancel{\text{m}}}{10^{12}\,\cancel{\text{pm}}} \times \dfrac{\text{D}}{3.34 \times 10^{-30}\,\cancel{\text{C}} \cdot \cancel{\text{m}}} = 9.6\ \text{D}$

6.6 The steps involved in writing a Lewis structure are as follows:

1. Write the correct skeletal structure for the molecule.
2. Calculate the total number of electrons for the Lewis structure by summing the valence electrons of each atom in the molecule.
3. Distribute the electrons among the atoms, giving octets (or duets for hydrogen) to as many atoms as possible.
4. If any atoms lack an octet, form double or triple bonds as necessary to give them octets.

6.7 The total number of electrons for a Lewis structure of a molecule is the sum of the valence electrons of each atom in the molecule.

The total number of electrons for the Lewis structure of an ion is found by summing the number of valence electrons for each atom and then subtracting one electron for each positive charge or adding one electron for each negative charge.

6.8 Resonance structures result when you can write two or more Lewis structures for the same molecule. A resonance hybrid is then the weighted average of the resonance structures.

6.9 In some cases, we can write resonance structures that are not equivalent. One possible resonance structure may be somewhat better than another. In such cases, the true structure may still be represented as an average of the resonance structures, but with the better resonance structure contributing more to the true structure. Multiple nonequivalent resonance structures may be weighted differently in their contributions to the true overall structure of a molecule.

6.10 Formal charge is a fictitious charge assigned to each atom in a Lewis structure that helps us to distinguish among competing Lewis structures. The formal charge of an atom in a Lewis structure is the charge it would have if all bonding electrons were shared equally between the bonded atoms. Formal charge can be calculated simply by taking the number of valence electrons in the atom and subtracting the number of electrons that it "owns" in a Lewis structure. An atom in a Lewis structure "owns" all of its lone pair of electrons and one-half of its bonding electrons.

$$\text{Formal charge} = \text{number of valence electrons} - (\text{number of lone pair electrons} + 1/2 \text{ number of bonding electrons}).$$

The concept of formal charge is useful because it helps us distinguish between competing skeletal structures or competing resonance structures.

6.11 The octet rule has some exceptions because not all atoms have eight electrons surrounding them. The three major categories are (1) odd octets—electron species, molecules, or ions with an odd number of electron (e.g., NO); (2) incomplete octets—molecules or ions with fewer than eight electrons around an atom (e.g., BF_3); and (3) expanded octets—molecules or ions with more than eight electrons around an atom (e.g., AsF_5).

6.12 Elements in the third row of the periodic table and beyond often exhibit expanded octets. Elements in the first or second row of the periodic table never have expanded octets.

6.13 The bond energy of a chemical bond is the energy required to break 1 mole of the bond in the gas phase. Because breaking bonds is endothermic and forming bonds is exothermic, we can calculate the overall enthalpy change as a sum of the enthalpy changes associated with breaking the required bonds in the reactants and forming the required bonds in the products.

6.14 In general, smaller atoms have shorter bond lengths. Some examples are: H—H bond length = 74 pm; F—F bond length = 143 pm; Cl—Cl bond length = 199 pm; Br—Br bond length = 228 pm; and I—I bond length = 266 pm. In general, triple bonds are shorter and stronger than double bonds, which are in turn shorter and stronger than single bonds. Some examples are: C—C bond length = 154 pm; C═C bond length = 134 pm; and C≡C bond length = 120 pm.

6.15 The properties of molecules are directly related to their shape. The sensation of taste, immune response, the sense of smell, and many types of drug action all depend on shape-specific interactions between molecules and proteins.

6.16 According to VSEPR theory, the repulsion between electron groups on interior atoms of a molecule determines the geometry of the molecule.

6.17 The five basic electron geometries are as follows:

1. Linear, which has two electron groups
2. Trigonal planar, which has three electron groups
3. Tetrahedral, which has four electron groups
4. Trigonal bipyramidal, which has five electron groups
5. Octahedral, which has six electron groups

An electron group is defined as a lone pair of electrons, a single bond, a multiple bond, or even a single electron.

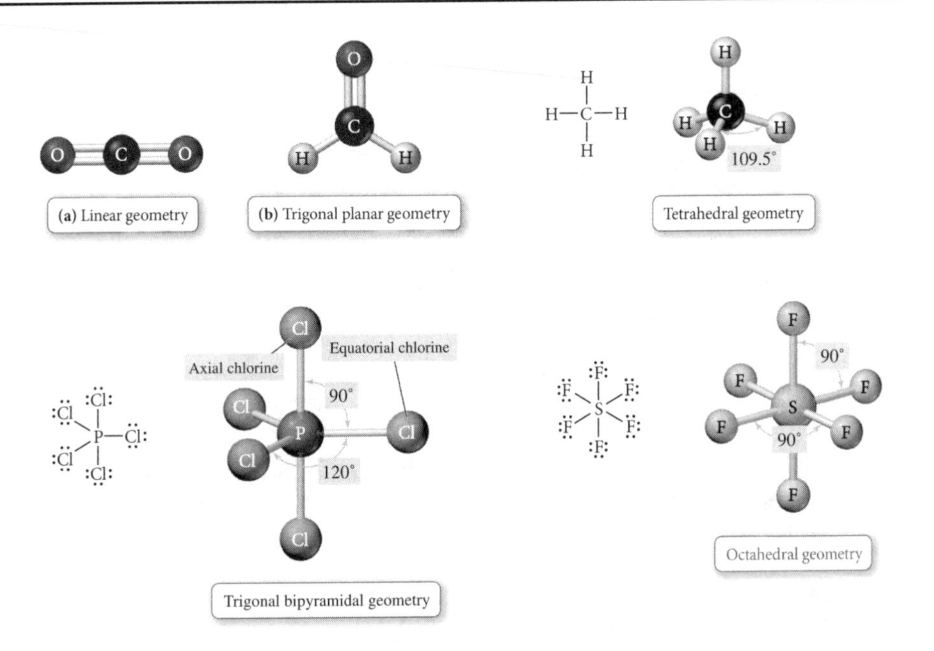

6.18 The electron geometry is the geometrical arrangement of the electron groups around the central atom.

The molecular geometry is the geometrical arrangement of the atoms around the central atom.

The electron geometry and the molecular geometry are the same when every electron group bonds two atoms together. The presence of an unbonded lone pair of electrons gives a different molecular geometry and electron geometry.

6.19
(a) Four electron groups give a tetrahedral electron geometry, while three bonding groups and one lone pair give a trigonal pyramidal molecular geometry.
(b) Four electron groups give a tetrahedral electron geometry, while two bonding groups and two lone pairs give a bent molecular geometry.
(c) Five electron groups give a trigonal bipyramidal electron geometry, while four bonding groups and one lone pair give a seesaw molecular geometry.
(d) Five electron groups give a trigonal bipyramidal electron geometry, while three bonding groups and two lone pairs give a T-shaped molecular geometry.
(e) Five electron groups give a trigonal bipyramidal electron geometry, while two bonding groups and three lone pairs give a linear geometry.
(f) Six electron groups give an octahedral electron geometry, while five bonding groups and one lone pair give a square pyramidal molecular geometry.
(g) Six electron groups give an octahedral electron geometry, while four bonding groups and two lone pairs give a square planar molecular geometry.

6.20 Larger molecules may have two or more interior atoms. When predicting the shapes of these molecules, determine the geometry about each interior atom and use these geometries to determine the entire three-dimensional shape of the molecules.

6.21 To determine whether a molecule is polar, do the following:
1. Draw the Lewis structure for the molecule and determine the molecular geometry.
2. Determine whether the molecule contains polar bonds.
3. Determine whether the polar bonds add together to form a net dipole moment.

6.22 Polarity is important because polar and nonpolar molecules have different properties. Polar molecules interact strongly with other polar molecules but do not interact with nonpolar molecules, and vice versa.

Problems by Topic

Electronegativity and Bond Polarity

6.23 (a) Br and Br: pure covalent From Figure 6.3, we find that the electronegativity of Br is 2.5. Because both atoms are the same, the electronegativity difference (Δ EN) = 0, and using Table 6.1, we classify this bond as pure covalent.

(b) C and Cl: polar covalent From Figure 6.3, we find that the electronegativity of C is 2.5 and Cl is 3.0. The electronegativity difference (Δ EN) is Δ EN = 3.0 − 2.5 = 0.5. Using Table 6.1, we classify this bond as polar covalent.

(c) C and S: pure covalent From Figure 6.3, we find that the electronegativity of C is 2.5 and S is 2.5. The electronegativity difference (Δ EN) is Δ EN = 2.5 − 2.5 = 0. Using Table 6.1, we classify this bond as pure covalent.

(d) Sr and O: ionic From Figure 6.3, we find that the electronegativity of Sr is 1.0 and O is 3.5. The electronegativity difference (Δ EN) is Δ EN = 3.5 − 1.0 = 2.5. Using Table 6.1, we classify this bond as ionic.

6.24 (a) C and N: polar covalent From Figure 6.3, we find that the electronegativity of C is 2.5 and N is 3.0. The electronegativity difference (Δ EN) is Δ EN = 3.0 − 2.5 = 0.5. Using Table 6.1, we classify this bond as polar covalent.

(b) N and S: polar covalent From Figure 6.3, we find that the electronegativity of S is 2.5 and N is 3.0. The electronegativity difference (Δ EN) is Δ EN = 3.0 − 2.5 = 0.5. Using Table 6.1, we classify this bond as polar covalent.

(c) K and F: ionic From Figure 6.3, we find that the electronegativity of K is 0.8 and F is 4.0. The electronegativity difference (Δ EN) is Δ EN = 4.0 − 0.8 = 3.2. Using Table 6.1, we classify this bond as ionic.

(d) N and N: pure covalent From Figure 6.3, we find that the electronegativity of N is 3.0. Because both atoms are the same, the electronegativity difference (Δ EN) = 0, and using Table 6.1, we classify this bond as pure covalent.

6.25 CO: Write the Lewis structure for each atom based on the number of valence electrons.

·C̈· :Ö·

The carbon will share three electron pairs with oxygen to achieve a stable octet.
The oxygen atom is more electronegative than the carbon atom; so the oxygen will have a partial negative charge, and the carbon will have a partial positive charge.

:C≡O: (dipole arrow pointing toward O)

To estimate the percent ionic character, determine the difference in electronegativity between carbon and oxygen. From Figure 6.3, we find that the electronegativity of C is 2.5 and O is 3.5. The electronegativity difference (Δ EN) is Δ EN = 3.5 − 2.5 = 1.0. From Figure 6.5, we can estimate a percent ionic character of 25%.

6.26 BrF: Write the Lewis structure for each atom based on the number of valence electrons.

:B̤̈r· ·F̤̈:

The bromine and fluorine will share an electron pair to achieve a stable octet.
The fluorine atom is more electronegative than the bromine atom; so the fluorine will have a partial negative charge, and the bromine will have a partial positive charge.

:B̤̈r—F̤̈: (dipole arrow pointing toward F)

To estimate the percent ionic character, determine the difference in electronegativity between bromine and fluorine. From Figure 6.3, we find that the electronegativity of Br = 2.8 and F = 4.0. The electronegativity difference (Δ EN) is Δ EN = 4.0 − 2.8 = 1.2. From Figure 6.5, we can estimate a percent ionic character of 30%.

Writing Lewis Structures, Resonance, and Formal Charge

6.27 (a) PH_3: Write the Lewis structure for each atom based on the number of valence electrons.

·P̤̈· ·H

Phosphorus will share an electron pair with each hydrogen to achieve a stable octet.

```
  ..
H—P—H
  |
  H
```

Check: The phosphorous atom has an octet, and the structure has 1(5) + 3(1) = 8 electrons.

(b) SCl_2: Write the Lewis structure for each atom based on the number of valence electrons.

:S̤̈· :C̤̈l·

The sulfur will share an electron pair with each chlorine to achieve a stable octet.

```
 ..  ..
:S—Cl:
 |   ..
:Cl:
 ..
```

Check: All of the atoms have octets, and the structure has 1(6) + 2(7) = 20 electrons.

(c) HI: Write the Lewis structure for each atom based on the number of valence electrons.

H· ·Ï̤:

The iodine will share an electron pair with hydrogen to achieve a stable octet.

H—Ï̤:

Check: The iodine atom has an octet, and the structure has 1(1) + 1(7) = 8 electrons.

(d) CH_4: Write the Lewis structure for each atom based on the number of valence electrons.

·Ċ̣· H·

The carbon will share an electron pair with each hydrogen to achieve a stable octet.

```
  H
  |
H—C—H
  |
  H
```

Check: The carbon atom has an octet, and the structure has 1(4) + 4(1) = 8 electrons.

6.28 (a) NF_3: Write the Lewis structure for each atom based on the number of valence electrons.

·Ṇ̈· ·F̤̈:

The nitrogen will share an electron pair with each fluorine to achieve a stable octet.

```
 ..  ..  ..
:F—N—F:
 ..  |  ..
    :F:
     ..
```

Check: All of the atoms have octets, and the structure has 1(5) + 3(7) = 26 electrons.

(b) HBr: Write the Lewis structure for each atom based on the number of valence electrons.

H· ·B̤̈r:

The bromine will share an electron pair with hydrogen to achieve a stable octet.

H—B̤̈r:

Check: The bromine atom has an octet, and the structure has 1(1) + 1(7) = 8 electrons.

(c) SBr_2: Write the Lewis structure for each atom based on the number of valence electrons.

:S̤̈· :B̤̈r·

The sulfur will share an electron pair with each bromine to achieve a stable octet.

:S̈—B̤̈r:
|
:B̤r:

Check: All of the atoms have octets, and the structure has 1(6) + 2(7) = 20 electrons.

(d) CCl_4: Write the Lewis structure for each atom based on the number of valence electrons.

·Ċ· :C̤̈l·

The carbon will share an electron pair with each chlorine to achieve a stable octet.

:C̈l:
|
:C̤̈l—C—C̤̈l:
|
:C̤l:

Check: All of the atoms have octets, and the structure has 1(4) + 4(7) = 32 electrons.

6.29 (a) SF_2: Write the Lewis structure for each atom based on the number of valence electrons.

:S̤̈· :F̤̈·

The sulfur will share an electron pair with each fluorine to achieve a stable octet.

:S̈—F̤̈:
|
:F̤:

Check: All of the atoms have octets, and the structure has 1(6) + 2(7) = 20 electrons.

(b) SiH_4: Write the Lewis structure for each atom based on the number of valence electrons.

H· ·Ṡ̤i·

The silicon will share an electron pair with each hydrogen to achieve a stable octet.

H
|
H—Si—H
|
H

Check: The silicon atom has an octet, and the structure has 1(4) + 4(1) = 8 electrons.

(c) HCOOH Write the Lewis structure for each atom based on the number of valence electrons.

H· ·Ċ· :Ö·

The carbon will share an electron pair with hydrogen, an electron pair with the interior oxygen, and two electron pairs with the terminal oxygen to achieve a stable octet. The terminal oxygen will share two electron pairs with carbon to achieve a stable octet. The interior oxygen will share an electron pair with carbon and an electron pair with hydrogen to achieve a stable octet.

:O:
‖
H—C—Ö̤—H

Check: The carbon and oxygen atoms have octets, and the structure has 1(4) + 2(1) + 2(6) = 18 electrons.

(d) CH_3SH Write the Lewis structure for each atom based on the number of valence electrons.

H· ·Ċ· :S̤̈·

The carbon will share an electron pair with each hydrogen and an electron pair with sulfur to achieve a stable octet. The sulfur will share an electron pair with carbon and an electron pair with hydrogen to achieve a stable octet.

```
        H
        |      ..
    H — C — S — H
        |      ..
        H
```

Check: The carbon and sulfur atoms have octets, and the structure has 1(4) + 1(6) + 4(1) = 14 electrons.

6.30 (a) CH_2O Write the Lewis structure for each atom based on the number of valence electrons.

H· ·Ċ· :Ö·

The carbon will share an electron pair with each hydrogen and two electron pairs with oxygen to achieve a stable octet.

```
       :O:
        ‖
    H — C — H
```

Check: The carbon and oxygen atoms have octets, and the structure has 1(4) + 2(1) + 1(6) = 12 electrons.

(b) C_2Cl_2 Write the Lewis structure for each atom based on the number of valence electrons.

·Ċ· :C̤̈l·

Each carbon will share three electron pairs with the other carbon and an electron pair with chlorine to achieve a stable octet.

:C̤̈l — C ≡ C — C̤̈l:

Check: All of the atoms have octets, and the structure has 2(4) + 2(7) = 22 electrons.

6.31 (a) CI_4: Write the correct skeletal structure for the molecule.

```
        I
        |
    I — C — I
        |
        I
```

Calculate the total number of electrons for the Lewis structure by summing the number of valence electrons of each atom in the molecule.

$$(\text{number of valence } e^- \text{ for C}) + 4(\text{number of valence } e^- \text{ for I}) = 4 + 4(7) = 32$$

Distribute the electrons among the atoms, giving octets to as many atoms as possible. Begin with the bonding electrons; then proceed to lone pairs on terminal atoms and finally to lone pairs on the central atom.

```
        ..
       :I:
   ..   |   ..
  :I — C — I:
   ..   |   ..
       :I:
        ..
```

All 32 valence electrons are used.

If any atom lacks an octet, form double or triple bonds as necessary to give them octets.

All atoms have octets; the structure is complete.

(b) N_2O: Write the correct skeletal structure for the molecule.

N is less electronegative, so it is central.

N — N — O

Calculate the total number of electrons for the Lewis structure by summing the number of valence electrons of each atom in the molecule.

$$2(\text{number of valence } e^- \text{ for N}) + (\text{number of valence } e^- \text{ for O}) = 2(5) + 6 = 16$$

Distribute the electrons among the atoms, giving octets to as many atoms as possible. Begin with the bonding electrons; then proceed to lone pairs on terminal atoms and finally to lone pairs on the central atom.

:N̤̈—N—Ö̤:

All 16 valence electrons are used.

If any atom lacks an octet, form double or triple bonds as necessary.

:N≡N—Ö̤:

All atoms have octets; the structure is complete. The formal charges are 0, +1, and −1 for the N, N, and O, respectively.

An alternate structure with two double bonds is not preferred because the formal charges would be −1, +1, and 0, and the most negative formal charges should be on the O atom.

(c) SiH_4: Write the correct skeletal structure for the molecule.

H is always terminal, so Si is the central atom.

```
     H
     |
H — Si — H
     |
     H
```

Calculate the total number of electrons for the Lewis structure by summing the number of valence electrons of each atom in the molecule.

$$(\text{number of valence e}^- \text{ for Si}) + 4(\text{number of valence e}^- \text{ for H}) = 4 + 4(1) = 8$$

Distribute the electrons among the atoms, giving octets (or duets for H) to as many atoms as possible. Begin with the bonding electrons; then proceed to lone pairs on terminal atoms and finally to lone pairs on the central atom.

```
     H
     |
H — Si — H
     |
     H
```

All eight valence electrons are used.

If any atom lacks an octet, form double or triple bonds as necessary to give them octets.

All atoms have octets; the structure is complete.

(d) Cl_2CO: Write the correct skeletal structure for the molecule.

C is least electronegative, so it is the central atom.

```
      O
      |
Cl — C — Cl
```

Calculate the total number of electrons for the Lewis structure by summing the number of valence electrons of each atom in the molecule.

$$(\text{number of valence e}^- \text{ for C}) + 2(\text{number of valence e}^- \text{ for Cl}) + (\text{number of valence e}^- \text{ for O}) = 4 + 2(7) + 6 = 24$$

Distribute the electrons among the atoms, giving octets to as many atoms as possible. Begin with the bonding electrons; then proceed to lone pairs on terminal atoms and finally to lone pairs on the central atom.

```
        :Ö:
         |
:C̤̈l — C — C̤̈l:
```

All 24 valence electrons are used.

If any atom lacks an octet, form double or triple bonds as necessary.

```
        :O:
         ||
:C̤̈l — C — C̤̈l:
```

All atoms have octets; the structure is complete.

6.32 (a) H_3COH: Write the correct skeletal structure for the molecule.

C is less electronegative, and H is terminal.

```
      O—H
      |
  H—C—H
      |
      H
```

Calculate the total number of electrons for the Lewis structure by summing the number of valence electrons of each atom in the molecule.

$(\text{number of valence e}^- \text{ for C}) + 4(\text{number of valence e}^- \text{ for H}) + (\text{number of valence e}^- \text{ for O}) = 4 + 4(1) + 6 = 14$

Distribute the electrons among the atoms, giving octets (or duets for H) to as many atoms as possible. Begin with the bonding electrons; then proceed to lone pairs on terminal atoms and finally to lone pairs on the central atoms.

```
      ..
     :O—H
      ..
      |
  H—C—H
      |
      H
```

All 14 valence electrons are used.

If any atom lacks an octet, form double or triple bonds as necessary to give them octets.

All atoms have octets (duets for H); the structure is complete.

(b) OH^-: Write the correct skeletal structure for the ion.

O—H

Calculate the total number of electrons for the Lewis structure by summing the number of valence electrons of each atom in the ion and adding 1 for the 1− charge.

$(\text{number of valence e}^- \text{ for O}) + (\text{number of valence e}^- \text{ for H}) + 1 = 6 + 1 + 1 = 8$

Distribute the electrons among the atoms, giving octets (or duets for H) to as many atoms as possible. Begin with the bonding electrons; then proceed to lone pairs on terminal atoms and finally to lone pairs on the central atom.

```
  ..
 :O—H
  ..
```

All 8 valence electrons are used.

If any atom lacks an octet, form double or triple bonds as necessary to give them octets.

Finally, write the Lewis structure in brackets with the charge of the ion in the upper right-hand corner.

```
   ..     -
 [:O—H]
   ..
```

(c) BrO^-: Write the correct skeletal structure for the ion.

Br—O

Calculate the total number of electrons for the Lewis structure by summing the number of valence electrons of each atom in the ion and adding 1 for the 1− charge.

$(\text{number of valence e}^- \text{ for O}) + (\text{number of valence e}^- \text{ for Br}) + 1 = 6 + 7 + 1 = 14$

Distribute the electrons among the atoms, giving octets to as many atoms as possible. Begin with the bonding electrons; then proceed to lone pairs on terminal atoms and finally to lone pairs on the central atom.

```
  ..   ..
 :Br—O:
  ..   ..
```

All 14 valence electrons are used.

If any atom lacks an octet, form double or triple bonds as necessary to give them octets.

Finally, write the Lewis structure in brackets with the charge of the ion in the upper right-hand corner.

```
   ..   ..   -
 [:Br—O:]
   ..   ..
```

(d) O_2^{2-}: Write the correct skeletal structure for the ion.

O—O

Calculate the total number of electrons for the Lewis structure by summing the number of valence electrons of each atom in the ion and adding 2 for the 2− charge.

$2(\text{number of valence e}^- \text{ for O}) + 2 = 2(6) + 2 = 14$

Distribute the electrons among the atoms, giving octets to as many atoms as possible. Begin with the bonding electrons; then proceed to lone pairs on terminal atoms and finally to lone pairs on the central atom.

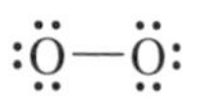

All 14 valence electrons are used.

If any atom lacks an octet, form double or triple bonds as necessary to give them octets.

Finally, write the Lewis structure in brackets with the charge of the ion in the upper right-hand corner.

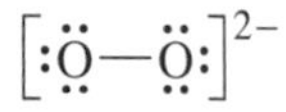

6.33 (a) N_2H_2: Write the correct skeletal structure for the molecule.

H—N—N—H

Calculate the total number of electrons for the Lewis structure by summing the number of valence electrons of each atom in the molecule.

$2(\text{number of valence e}^- \text{ for N}) + 2(\text{number of valence e}^- \text{ for H}) = 2(5) + 2(1) = 12$

Distribute the electrons among the atoms, giving octets (or duets for H) to as many atoms as possible. Begin with the bonding electrons; then proceed to lone pairs on terminal atoms and finally to lone pairs on the central atom.

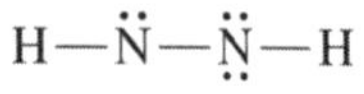

All 12 valence electrons are used.

If any atom lacks an octet, form double or triple bonds as necessary.

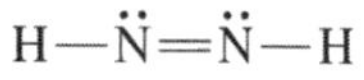

All atoms have octets (duets for H); the structure is complete.

(b) N_2H_4: Write the correct skeletal structure for the molecule.

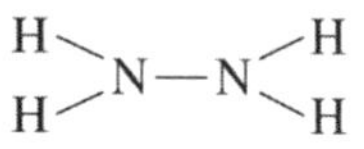

Calculate the total number of electrons for the Lewis structure by summing the valence electrons of each atom in the molecule.

$2(\text{number of valence e}^- \text{ for N}) + 4(\text{number of valence e}^- \text{ for H}) = 2(5) + 4(1) = 14$

Distribute the electrons among the atoms, giving octets (or duets for H) to as many atoms as possible. Begin with the bonding electrons; then proceed to lone pairs on terminal atoms and finally to lone pairs on the central atom.

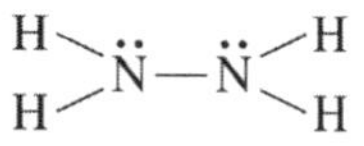

All 14 valence electrons are used.

If any atom lacks an octet, form double or triple bonds as necessary to give them octets.

All atoms have octets (duets for H); the structure is complete.

(c) C_2H_2: Write the correct skeletal structure for the molecule.

H—C—C—H

Calculate the total number of electrons for the Lewis structure by summing the number of valence electrons of each atom in the molecule.

$2(\text{number of valence e}^- \text{ for C}) + 2(\text{number of valence e}^- \text{ for H}) = 2(4) + 2(1) = 10$

Distribute the electrons among the atoms, giving octets (or duets for H) to as many atoms as possible. Begin with the bonding electrons; then proceed to lone pairs on terminal atoms and finally to lone pairs on the central atom.

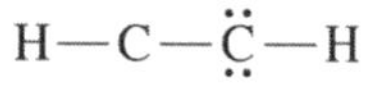

All 10 valence electrons are used.

If any atom lacks an octet, form double or triple bonds as necessary.

$$H{-}C{\equiv}C{-}H$$

All atoms have octets (duets for H); the structure is complete.

(d) C_2H_4: Write the correct skeletal structure for the molecule.

$$H_2C{-}CH_2$$

Calculate the total number of electrons for the Lewis structure by summing the number of valence electrons of each atom in the molecule.

$$2(\text{number of valence e}^- \text{ for C}) + 4(\text{number of valence e}^- \text{ for H}) = 2(4) + 4(1) = 12$$

Distribute the electrons among the atoms, giving octets (or duets for H) to as many atoms as possible. Begin with the bonding electrons; then proceed to lone pairs on terminal atoms and finally to lone pairs on the central atom.

$$H_2\ddot{C}{-}CH_2$$

All 12 valence electrons are used.

If any atom lacks an octet, form double or triple bonds as necessary.

$$H_2C{=}CH_2$$

All atoms have octets (duets for H); the structure is complete.

6.34 (a) H_3COCH_3: Write the correct skeletal structure for the molecule.

$$H_3C{-}O{-}CH_3$$

Calculate the total number of electrons for the Lewis structure by summing the number of valence electrons of each atom in the molecule.

$$2(\text{number of valence e}^- \text{ for C}) + (\text{number of valence e}^- \text{ for O}) + 6(\text{number of valence e}^- \text{ for H}) = 2(4) + 6 + 6(1) = 20$$

Distribute the electrons among the atoms, giving octets (or duets for H) to as many atoms as possible. Begin with the bonding electrons; then proceed to lone pairs on terminal atoms and finally to lone pairs on the central atom.

$$H_3C{-}\ddot{\underset{..}{O}}{-}CH_3$$

All 20 valence electrons are used.

If any atom lacks an octet, form double or triple bonds as necessary to give them octets. All atoms have octets (duets for H); the structure is complete.

(b) CN^-: Write the correct skeletal structure for the ion.

$$C{-}N$$

Calculate the total number of electrons for the Lewis structure by summing the number of valence electrons of each atom in the ion and adding 1 for the 1− charge.

$$(\text{number of valence e}^- \text{ for C}) + (\text{number of valence e}^- \text{ for N}) + 1 = 4 + 5 + 1 = 10$$

Distribute the electrons among the atoms, giving octets to as many atoms as possible. Begin with the bonding electrons; then proceed to lone pairs on terminal atoms and finally to lone pairs on the central atom.

$$\ddot{C}{-}\ddot{\underset{..}{N}}{:}$$

All 10 valence electrons are used.

If any atom lacks an octet, form double or triple bonds as necessary.

:C≡N:

Finally, write the Lewis structure in brackets with the charge of the ion in the upper right-hand corner.

$[:C{\equiv}N:]^-$

(c) NO_2^-: Write the correct skeletal structure for the ion.

O—N—O

Calculate the total number of electrons for the Lewis structure by summing the number of valence electrons of each atom in the ion and adding 1 for the 1− charge.

$$2(\text{number of valence e}^- \text{ for O}) + (\text{number of valence e}^- \text{ for N}) + 1 = 2(6) + 5 + 1 = 18$$

Distribute the electrons among the atoms, giving octets to as many atoms as possible. Begin with the bonding electrons; then proceed to lone pairs on terminal atoms and finally to lone pairs on the central atom.

:Ö—N̈—Ö:

All 18 valence electrons are used.

If any atom lacks an octet, form double or triple bonds as necessary.

:Ö—N̈=Ö:

Finally, write the Lewis structure in brackets with the charge of the ion in the upper right-hand corner. The molecule can have two resonance structures based on which oxygen forms the double bond.

[:Ö—N̈=Ö:]⁻ ⟷ [:Ö=N̈—Ö:]⁻

(d) ClO^-: Write the correct skeletal structure for the ion.

Cl—O

Calculate the total number of electrons for the Lewis structure by summing the number of valence electrons of each atom in the ion and adding 1 for the 1− charge.

$$(\text{number of valence e}^- \text{ for O}) + (\text{number of valence e}^- \text{ for Cl}) + 1 = 6 + 7 + 1 = 14$$

Distribute the electrons among the atoms, giving octets to as many atoms as possible. Begin with the bonding electrons; then proceed to lone pairs on terminal atoms and finally to lone pairs on the central atom.

:C̈l—Ö:

All 14 valence electrons are used.

All atoms have octets, so there is no need to form double or triple bonds.

Finally, write the Lewis structure in brackets with the charge of the ion in the upper right-hand corner.

[:C̈l—Ö:]⁻

6.35 (a) SeO_2: Write the correct skeletal structure for the molecule.

Se is less electronegative, so it is central.

O—Se—O

Calculate the total number of electrons for the Lewis structure by summing the number of valence electrons of each atom in the molecule.

$$(\text{number of valence e}^- \text{ for Se}) + 2(\text{number of valence e}^- \text{ for O}) = 6 + 2(6) = 18$$

Distribute the electrons among the atoms, giving octets to as many atoms as possible. Begin with the bonding electrons; then proceed to lone pairs on terminal atoms and finally to lone pairs on the central atom.

:Ö—S̈e—Ö:

All 18 valence electrons are used.

If any atom lacks an octet, form double or triple bonds as necessary.

:Ö—S̈e=Ö:

All atoms have octets; the structure is complete. However, the double bond can form from either oxygen atom, so there are two resonance forms.

:Ö—S̈e=Ö: ⟷ :Ö=S̈e—Ö:

Calculate the formal charge on each atom by finding the number of valence electrons and subtracting the number of lone pair electrons and one-half the number of bonding electrons.

	:Ö—S̈e=Ö:			⟷ :Ö=S̈e—Ö:		
number of valence electrons	6	6	6	6	6	6
− number of lone pair electrons	6	2	4	4	2	6
− 1/2(number of bonding electrons)	1	3	2	2	3	1
Formal charge	−1	+1	0	0	+1	−1

Check: All of the atoms have octets, and the structure has 18 electrons. Atoms with fewer bonds than is expected (oxygen typically has 2 bonds) will have a negative formal charge, and atoms with more bonds than is expected (selenium typically has 2 bonds) will have a positive formal charge.

(b) CO_3^{2-}: Write the correct skeletal structure for the ion.

```
     O
     |
 O — C — O
```

Calculate the total number of electrons for the Lewis structure by summing the number of valence electrons of each atom in the ion and adding 2 for the 2− charge.

$$3(\text{number of valence e}^- \text{ for O}) + (\text{number of valence e}^- \text{ for C}) + 2 = 3(6) + 4 + 2 = 24$$

Distribute the electrons among the atoms, giving octets to as many atoms as possible. Begin with the bonding electrons; then proceed to lone pairs on terminal atoms and finally to lone pairs on the central atom.

```
     :Ö:
      |
 :Ö — C — Ö:
```

All 24 valence electrons are used.

If any atom lacks an octet, form double or triple bonds as necessary.

```
     :Ö:
      |
 :Ö — C = Ö:
```

Finally, write the Lewis structure in brackets with the charge of the ion in the upper right-hand corner.

```
 [     :Ö:      ]2−
 [      |       ]
 [ :Ö — C = Ö:  ]
```

All atoms have octets; the structure is complete. However, the double bond can form from any oxygen atom, so there are three resonance forms.

```
 [     :Ö:      ]2−      [     :Ö:      ]2−      [     :O:      ]2−
 [      |       ]   ⟷   [      |       ]   ⟷   [      ‖       ]
 [ :Ö — C = Ö:  ]        [ :Ö = C — Ö:  ]        [ :Ö — C — Ö:  ]
```

Calculate the formal charge on each atom by finding the number of valence electrons and subtracting the number of lone pair electrons and one-half the number of bonding electrons.

```
 [     :Ö:      ]2−
 [      |       ]
 [ :Ö — C = Ö:  ]
```

	O_{left}	O_{top}	O_{right}	C
number of valence electrons	6	6	6	4
− number of lone pair electrons	6	6	4	0
− 1/2(number of bonding electrons)	1	1	2	4
Formal charge	−1	−1	0	0

The sum of the formal charges is −2, which is the overall charge of the ion. The other resonance forms would have the same values for the single- and double-bonded oxygen atoms.

Check: All of the atoms have octets, and the structure has 24 electrons. Atoms with fewer bonds than is expected (oxygen typically has 2 bonds) will have a negative formal charge, and atoms with more bonds than is expected will have a positive formal charge.

(c) ClO^-: Write the correct skeletal structure for the ion.

Cl—O

Calculate the total number of electrons for the Lewis structure by summing the number of valence electrons of each atom in the ion and adding 1 for the 1− charge.

$(\text{number of valence e}^- \text{ for O}) + (\text{number of valence e}^- \text{ for Cl}) + 1 = 6 + 7 + 1 = 14$

Distribute the electrons among the atoms, giving octets to as many atoms as possible. Begin with the bonding electrons; then proceed to lone pairs on terminal atoms and finally to lone pairs on the central atom.

:C̤̈l—Ö̤:

All 14 valence electrons are used.

If any atom lacks an octet, form double or triple bonds as necessary to give them octets.

Finally, write the Lewis structure in brackets with the charge of the ion in the upper right-hand corner.

[:C̤̈l—Ö̤:]⁻

All atoms have octets; the structure is complete.

Calculate the formal charge on each atom by finding the number of valence electrons and subtracting the number of lone pair electrons and one-half the number of bonding electrons.

	Cl	O
number of valence electrons	7	6
− number of lone pair electrons	6	6
− 1/2(number of bonding electrons)	1	1
Formal charge	0	−1

The sum of the formal charges is −1, which is the overall charge of the ion.

Check: All of the atoms have octets, and the structure has 14 electrons. Atoms with fewer bonds than is expected (oxygen typically has 2 bonds) will have a negative formal charge.

(d) NO_2^-: Write the correct skeletal structure for the ion.

O—N—O

Calculate the total number of electrons for the Lewis structure by summing the number of valence electrons of each atom in the ion and adding 1 for the 1− charge.

$2(\text{number of valence e}^- \text{ for O}) + (\text{number of valence e}^- \text{ for N}) + 1 = 2(6) + 5 + 1 = 18$

Distribute the electrons among the atoms, giving octets to as many atoms as possible. Begin with the bonding electrons; then proceed to lone pairs on terminal atoms and finally to lone pairs on the central atom.

:Ö̤—N̈—Ö̤:

All 18 valence electrons are used.

If any atom lacks an octet, form double or triple bonds as necessary.

$$:\underset{\cdot\cdot}{O}=\ddot{N}-\underset{\cdot\cdot}{\ddot{O}}:$$

Finally, write the Lewis structure in brackets with the charge of the ion in the upper right-hand corner.

$$\left[:\underset{\cdot\cdot}{O}=\ddot{N}-\underset{\cdot\cdot}{\ddot{O}}:\right]^-$$

All atoms have octets; the structure is complete. However, the double bond can form from either oxygen atom, so there are two resonance forms.

$$\left[:\underset{\cdot\cdot}{O}=\ddot{N}-\underset{\cdot\cdot}{\ddot{O}}:\right]^- \longleftrightarrow \left[:\underset{\cdot\cdot}{\ddot{O}}-\ddot{N}=\underset{\cdot\cdot}{O}:\right]^-$$

Calculate the formal charge on each atom by finding the number of valence electrons and subtracting the number of lone pair electrons and one-half the number of bonding electrons. Using the left-side structure:

	O	N	O
number of valence electrons	6	5	6
− number of lone pair electrons	4	2	6
− 1/2(number of bonding electrons)	2	3	1
Formal charge	0	0	−1

The sum of the formal charges is −1, which is the overall charge of the ion.

Check: All of the atoms have octets, and the structure has 18 electrons. Atoms with fewer bonds than is expected (oxygen typically has 2 bonds) will have a negative formal charge.

6.36 (a) ClO_3^-: Write the correct skeletal structure for the ion.

```
      O
      |
  O — Cl — O
```

Calculate the total number of electrons for the Lewis structure by summing the number of valence electrons of each atom in the ion and adding 1 for the 1− charge.

$$3(\text{number of valence e}^- \text{ for O}) + (\text{number of valence e}^- \text{ for Cl}) + 1 = 3(6) + 7 + 1 = 26$$

Distribute the electrons among the atoms, giving octets to as many atoms as possible. Begin with the bonding electrons; then proceed to lone pairs on terminal atoms and finally to lone pairs on the central atom.

```
       ..
      :O:
       |
  :O — Cl — O:
   ..   ..   ..
```

All 26 valence electrons are used.

If any atom lacks an octet, form double or triple bonds as necessary to give them octets.

Finally, write the Lewis structure in brackets with the charge of the ion in the upper right-hand corner.

```
 [      ..       ]-
 [     :O:       ]
 [      |        ]
 [ :O — Cl — O:  ]
 [  ..  ..   ..  ]
```

All atoms have octets; the structure is complete.

Calculate the formal charge on each atom by finding the number of valence electrons and subtracting the number of lone pair electrons and one-half the number of bonding electrons.

	O_{left}	O_{top}	O_{right}	Cl
number of valence electrons	6	6	6	7
− number of lone pair electrons	6	6	6	2
− 1/2(number of bonding electrons)	1	1	1	3
Formal charge	−1	−1	−1	+2

The sum of the formal charges is −1, which is the overall charge of the ion.

Check: All of the atoms have octets, and the structure has 26 electrons. Atoms with fewer bonds than is expected (oxygen typically has 2 bonds) will have a negative formal charge, and atoms with more bonds than is expected (chlorine typically has 1 bond) will have a positive formal charge.

(b) ClO_4^-: Write the correct skeletal structure for the ion.

```
       O
       |
   O—Cl—O
       |
       O
```

Calculate the total number of electrons for the Lewis structure by summing the number of valence electrons of each atom in the ion and adding 1 for the 1− charge.

$$4(\text{number of valence } e^- \text{ for O}) + (\text{number of valence } e^- \text{ for Cl}) + 1 = 4(6) + 7 + 1 = 32$$

Distribute the electrons among the atoms, giving octets to as many atoms as possible. Begin with the bonding electrons; then proceed to lone pairs on terminal atoms and finally to lone pairs on the central atom.

```
        :Ö:
         |
   :Ö—Cl—Ö:
         |
        :Ö:
```

All 32 valence electrons are used.

If any atom lacks an octet, form double or triple bonds as necessary to give them octets.

Finally, write the Lewis structure in brackets with the charge of the ion in the upper right-hand corner.

```
 ⎡      :Ö:      ⎤ −
 ⎢       |       ⎥
 ⎢ :Ö—Cl—Ö:      ⎥
 ⎢       |       ⎥
 ⎣      :Ö:      ⎦
```

All atoms have octets; the structure is complete.

Calculate the formal charge on each atom by finding the number of valence electrons and subtracting the number of lone pair electrons and one-half the number of bonding electrons.

Using the left-side structure:

	O_{left}	O_{top}	O_{right}	O_{bottom}	Cl
number of valence electrons	6	6	6	6	7
− number of lone pair electrons	6	6	6	6	0
− 1/2(number of bonding electrons)	1	1	1	1	4
Formal charge	−1	−1	−1	−1	+3

The sum of the formal charges is −1, which is the overall charge of the ion.

Check: All of the atoms have octets, and the structure has 32 electrons. Atoms with fewer bonds than is expected (oxygen typically has 2 bonds) will have a negative formal charge, and atoms with more bonds than is expected (chlorine typically has 1 bond) will have a positive formal charge.

(c) NO_3^-: Write the correct skeletal structure for the ion.

```
      O
      |
   O—N—O
```

Calculate the total number of electrons for the Lewis structure by summing the number of valence electrons of each atom in the ion and adding 1 for the 1− charge.

$$3(\text{number of valence } e^- \text{ for O}) + (\text{number of valence } e^- \text{ for N}) + 1 = 3(6) + 5 + 1 = 24$$

Distribute the electrons among the atoms, giving octets to as many atoms as possible. Begin with the bonding electrons; then proceed to lone pairs on terminal atoms and finally to lone pairs on the central atom.

```
     :Ö:
      |
  :Ö—N—Ö:
```

All 24 valence electrons are used.

If any atom lacks an octet, form double or triple bonds as necessary.

```
     :Ö:
      |
  :Ö—N=Ö:
```

Finally, write the Lewis structure in brackets with the charge of the ion in the upper right-hand corner.

```
 [    :Ö:    ]−
 [     |     ]
 [ :Ö—N=Ö:   ]
```

All atoms have octets; the structure is complete. However, the double bond can form from any oxygen atom, so there are three resonance forms.

```
 [    :Ö:   ]−       [    :Ö:   ]−       [    :O:   ]−
 [     |    ]  ←→   [     |    ]  ←→   [     ‖    ]
 [ :Ö—N=Ö:  ]       [ :Ö=N—Ö:  ]       [ :Ö—N—Ö:  ]
```

Calculate the formal charge on each atom by finding the number of valence electrons and subtracting the number of lone pair electrons and one-half the number of bonding electrons.

Using the left-hand structure:

```
 [    :Ö:    ]−
 [     |     ]
 [ :Ö—N=Ö:   ]
```

	O_{left}	O_{top}	O_{right}	N
number of valence electrons	6	6	6	5
− number of lone pair electrons	6	6	4	0
− 1/2(number of bonding electrons)	1	1	2	4
Formal charge	−1	−1	0	+1

The sum of the formal charges is −1, which is the overall charge of the ion. The other resonance forms would have the same values for the single- and double-bonded oxygen atoms.

Check: All of the atoms have octets, and the structure has 24 electrons. Atoms with fewer bonds than is expected (oxygen typically has 2 bonds) will have a negative formal charge, and atoms with more bonds than is expected (nitrogen typically has 3 bonds) will have a positive formal charge.

(d) NH_4^+: Write the correct skeletal structure for the ion.

```
     H
     |
  H—N—H
     |
     H
```

Calculate the total number of electrons for the Lewis structure by summing the valence electrons of each atom in the ion and subtracting 1 for the 1+ charge.

$$4(\text{number of valence } e^- \text{ for H}) + (\text{number of valence } e^- \text{ for N}) - 1 = 4(1) + 5 - 1 = 8$$

Distribute the electrons among the atoms, giving octets (or duets for H) to as many atoms as possible. Begin with the bonding electrons; then proceed to lone pairs on terminal atoms and finally to lone pairs on the central atom.

```
     H
     |
 H — N — H
     |
     H
```

All 8 valence electrons are used.

If any atom lacks an octet, form double or triple bonds as necessary to give them octets.

Finally, write the Lewis structure in brackets with the charge of the ion in the upper right-hand corner.

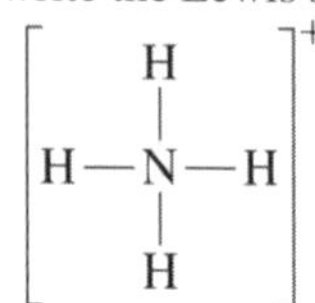

All atoms have octets (duets for H); the structure is complete.

Calculate the formal charge on each atom by finding the number of valence electrons and subtracting the number of lone pair electrons and one-half the number of bonding electrons.

	H_{left}	H_{top}	H_{right}	H_{bottom}	N
number of valence electrons	1	1	1	1	5
− number of lone pair electrons	0	0	0	0	0
− 1/2(number of bonding electrons)	1	1	1	1	4
Formal charge	0	0	0	0	+1

The sum of the formal charges is +1, which is the overall charge of the ion.

Check: All of the atoms have octets, and the structure has eight electrons. Atoms with more bonds than is expected (nitrogen typically has 3 bonds) will have a positive formal charge.

6.37

```
     H               H
     |               |
 H — C = S:      H — S = C:
        ..              ..
     I               II
```

Calculate the formal charge on each atom in structure I by finding the number of valence electrons and subtracting the number of lone pair electrons and one-half the number of bonding electrons.

	H_{left}	H_{top}	C	S
number of valence electrons	1	1	4	6
− number of lone pair electrons	0	0	0	4
− 1/2(number of bonding electrons)	1	1	4	2
Formal charge	0	0	0	0

The sum of the formal charges is 0, which is the overall charge of the molecule.

Calculate the formal charge on each atom in structure II by finding the number of valence electrons and subtracting the number of lone pair electrons and one-half the number of bonding electrons.

	H_{left}	H_{top}	S	C
number of valence electrons	1	1	6	4
− number of lone pair electrons	0	0	0	4
− 1/2(number of bonding electrons)	1	1	4	2
Formal charge	0	0	+2	−2

The sum of the formal charges is 0, which is the overall charge of the molecule.

Structure I is the better Lewis structure because it has the least amount of formal charge on each atom.

6.38

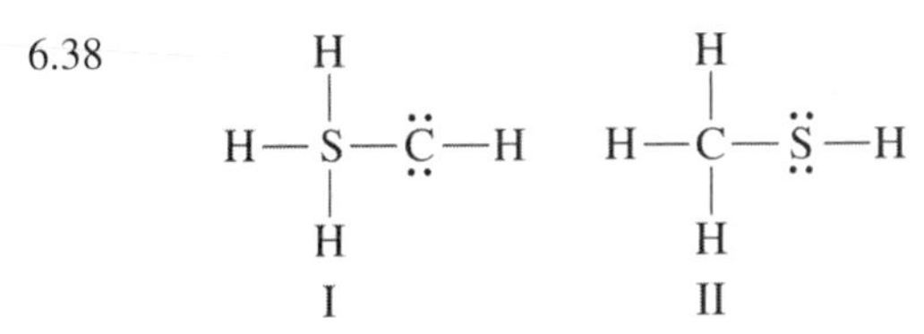

Calculate the formal charge on each atom in structure I by finding the number of valence electrons and subtracting the number of lone pair electrons and one-half the number of bonding electrons.

	H_{left}	H_{top}	H_{right}	H_{bottom}	S	C
number of valence electrons	1	1	1	1	6	4
− number of lone pair electrons	0	0	0	0	0	4
− 1/2(number of bonding electrons)	1	1	1	1	4	2
Formal charge	0	0	0	0	+2	−2

The sum of the formal charges is 0, which is the overall charge of the molecule.
Calculate the formal charge on each atom in structure II by finding the number of valence electrons and subtracting the number of lone pair electrons and one-half the number of bonding electrons.

	H_{left}	H_{top}	H_{right}	H_{bottom}	C	S
number of valence electrons	1	1	1	1	4	6
− number of lone pair electrons	0	0	0	0	0	4
− 1/2(number of bonding electrons)	1	1	1	1	4	2
Formal charge	0	0	0	0	0	0

The sum of the formal charges is 0, which is the overall charge of the molecule.
Structure II is the better Lewis structure because it has the least amount of formal charge on each atom.

6.39 :O≡C—Ö: (with three lone pairs on the right O)

This resonance structure does not provide a significant contribution to the resonance hybrid as it has a +1 formal charge on a very electronegative oxygen.

	O_{left}	O_{right}	C
number of valence electrons	6	6	4
− number of lone pair electrons	2	6	0
− 1/2(number of bonding electrons)	3	1	4
Formal charge	+1	−1	0

Check: Atoms with fewer bonds than is expected (oxygen typically has 2 bonds) will have a negative formal charge, and atoms with more bonds than is expected will have a positive formal charge.

6.40 Compare the two forms of each molecule with O as a central atom and a terminal atom. Determine the formal charge on the central atom for all of the structures.

Ö=N=N̈ (I) N̈=O=N̈ (II) :F̈—Ö—F̈: (III) :Ö—F̈—F̈: (IV)

	I	II	III	IV
	N	O	O	F
number of valence electrons	5	6	6	7
− number of lone pair electrons	0	0	4	4
− 1/2(number of bonding electrons)	4	4	2	2
Formal charge	+1	+2	0	+1

For the N_2O molecule, when O is the central atom, it has a +2 formal charge and is the more electronegative atom. So this would not be a good structure. For the OF_2 molecule, O has to be the central atom. When O is central, it has a formal charge of 0; when F is central, it has a formal charge of +1. This puts a positive formal charge on the most electronegative atom, which is not acceptable.

6.41 CH_3COO^-: Write the correct skeletal structure for the ion.

```
    H  O
    |  |
H — C — C
    |  |
    H  O
```

Calculate the total number of electrons for the Lewis structure by summing the valence electrons of each atom in the ion and adding 1 for the 1− charge.

3(number of valence e^- for H) + 2(number of valence e^- for C) + 2(number of valence e^- for O) + 1 = 3(1) + 2(4) + 2(6) + 1 = 24

Distribute the electrons among the atoms, giving octets (or duets for H) to as many atoms as possible. Begin with the bonding electrons; then proceed to lone pairs on terminal atoms and finally to lone pairs of the central atom.

```
    H  :Ö:
    |  |
H — C — C
    |  |
    H  :O:
        ..
```

All 24 valence electrons are used.

If any atom lacks an octet, form double or triple bonds as necessary to give them octets.

```
    H  :Ö:
    |  |
H — C — C
    |  ||
    H  :O:
```

Finally, write the Lewis structure in brackets with the charge of the ion in the upper right-hand corner.

```
[    H  :Ö: ]−
[    |  |   ]
[ H— C — C  ]
[    |  ||  ]
[    H  :O: ]
```

All atoms have octets (duets for H); the structure is complete. However, the double bond can form from any oxygen atom, so there are two resonance forms.

```
[    H  :Ö: ]−          [    H  :O: ]−
[    |  |   ]           [    |  ||  ]
[ H— C — C  ]  <——>     [ H— C — C  ]
[    |  ||  ]           [    |  |   ]
[    H  :O: ]           [    H  :O: ]
                                 ..
```

Calculate the formal charge on each atom by finding the number of valence electrons and subtracting the number of lone pair electrons and one-half the number of bonding electrons.

```
[    H  :Ö: ]−          [    H  :O: ]−
[    |  |   ]           [    |  ||  ]
[ H— C — C  ]  <——>     [ H— C — C  ]
[    |  ||  ]           [    |  |   ]
[    H  :O: ]           [    H  :O: ]
                                 ..
```

	C_{left}	C_{right}	O_{top}	O_{bottom}	C_{left}	C_{right}	O_{top}	O_{bottom}
number of valence electrons	4	4	6	6	4	4	6	6
− number of lone pair electrons	0	0	6	4	0	0	4	6
− 1/2(number of bonding electrons)	4	4	1	2	4	4	2	1
Formal charge	0	0	−1	0	0	0	0	−1

The sum of the formal charges is −1, which is the overall charge of the ion.

6.42 CH_3N_3 Write the correct skeletal structure for the ion.

```
      H
      |
  H—C—N—N—N
      |
      H
```

Calculate the total number of electrons for the Lewis structure by summing the valence electrons of each atom in the molecule.

3(number of valence e^- for H) + (number of valence e^- for C) + 3 (number of valence e^- for N) = 3(1) + 4 + 3(5) = 22

Distribute the electrons among the atoms, giving octets (or duets for H) to as many atoms as possible. Begin with the bonding electrons; then proceed to lone pairs on terminal atoms and finally to lone pairs on the central atom.

```
      H       ..    ..    ..
      |
  H—C—N—N—N
      |       ..    ..    ..
      H
```

All 22 valence electrons are used.

If any atom lacks an octet, form double or triple bonds as necessary to give them octets.

```
      H       ..          ..
      |
  H—C—N=N=N
      |                   ..
      H
```

All atoms have octets (duets for H); the structure is complete. However, the double bonds can also form as a triple bond, so there are three resonance forms.

```
      H       ..         ..              H       ..                     H                    ..
      |                                  |                              |
  H—C—N=N=N   ⟷   H—C—N—N≡N:   ⟷   H—C—N≡N—N:
      |                  ..              |       ..                     |                    ..
      H                                  H                              H
```

Calculate the formal charge on each atom by finding the number of valence electrons and subtracting the number of lone pair electrons and one-half the number of bonding electrons.

```
      H       ..         ..              H       ..                     H                    ..
      |                                  |                              |
  H—C—N=N=N   ⟷   H—C—N—N≡N:   ⟷   H—C—N≡N—N:
      |                  ..              |       ..                     |                    ..
      H                                  H                              H
```

	Left structure			Center structure			Right structure		
	N_{left}	N_{center}	N_{right}	N_{left}	N_{center}	N_{right}	N_{left}	N_{center}	N_{right}
number of valence electrons	5	5	5	5	5	5	5	5	5
– number of lone pair electrons	2	0	4	4	0	2	0	0	6
– 1/2(number of bonding electrons)	3	4	2	2	4	3	4	4	1
Formal charge	0	+1	−1	−1	+1	0	+1	+1	−2

The sum of the formal charges is 0, which is the overall charge of the molecule.

Check: All of the carbon and nitrogen atoms have octets, and the structure has 22 electrons. Atoms with fewer bonds than is expected (nitrogen typically has 3 bonds) will have a negative formal charge, and atoms with more bonds than is expected will have a positive formal charge.

6.43

```
         CH3
          |        ..
  CH3—N—O:
          |        ..
         CH3
```

	N	O
number of valence electrons	5	6
− number of lone pair electrons	0	6
− 1/2(number of bonding electrons)	4	1
Formal charge	+1	−1

Check: Atoms with fewer bonds than is expected (oxygen typically has 2 bonds) will have a negative formal charge, and atoms with more bonds than is expected (nitrogen typically has 3 bonds) will have a positive formal charge.

6.44

:Ö:
|
CH_3—S—CH_3

	S	O
number of valence electrons	6	6
− number of lone pair electrons	2	6
− 1/2(number of bonding electrons)	3	1
Formal charge	+1	−1

Check: Atoms with fewer bonds than is expected (oxygen typically has 2 bonds) will have a negative formal charge, and atoms with more bonds than is expected (sulfur typically has 2 bonds) will have a positive formal charge.

Odd-Electron Species, Incomplete Octets, and Expanded Octets

6.45 (a) BCl_3: Write the correct skeletal structure for the molecule.

B is less electronegative, so it is central.

Cl
|
Cl—B—Cl

Calculate the total number of electrons for the Lewis structure by summing the number of valence electrons of each atom in the molecule.

$$(\text{number of valence e}^- \text{ for B}) + 3(\text{number of valence e}^- \text{ for Cl}) = 3 + 3(7) = 24$$

Distribute the electrons among the atoms, giving octets to as many atoms as possible. Begin with the bonding electrons; then proceed to lone pairs on terminal atoms and finally to lone pairs on the central atom.

:C̈l:
|
:C̈l—B—C̈l:

All 24 valence electrons are used.

B has an incomplete octet. If we complete the octet, there is a formal charge of −1 on the B, which is less electronegative than Cl.

(b) NO_2: Write the correct skeletal structure for the molecule.

N is less electronegative, so it is central.

O—N—O

Calculate the total number of electrons for the Lewis structure by summing the number of valence electrons of each atom in the molecule.

$$(\text{number of valence e}^- \text{ for N}) + 2(\text{number of valence e}^- \text{ for O}) = 5 + 2(6) = 17$$

Distribute the electrons among the atoms, giving octets to as many atoms as possible. Begin with the bonding electrons; then proceed to lone pairs on terminal atoms and finally to lone pairs on the central atom.

Ö=Ṅ—Ö: ⟷ :Ö—Ṅ=Ö

All 17 valence electrons are used.

N has an incomplete octet. It has seven electrons because we have an odd number of valence electrons.

(c) BH_3: Write the correct skeletal structure for the molecule.
B is less electronegative, so it is central.

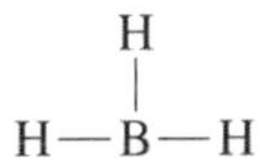

```
   H
   |
H—B—H
```

Calculate the total number of electrons for the Lewis structure by summing the number of valence electrons of each atom in the molecule.

$$(\text{number of valence } e^- \text{ for B}) + 3(\text{number of valence } e^- \text{ for H}) = 3 + 3(1) = 6$$

Distribute the electrons among the atoms, giving octets (or duets for H) to as many atoms as possible. Begin with the bonding electrons; then proceed to lone pairs on terminal atoms and finally to lone pairs on the central atom.

```
   H
   |
H—B—H
```

All 6 valence electrons are used.

B has an incomplete octet. H cannot double-bond, so it is not possible to complete the octet on B with a double-bond.

6.46 (a) BBr_3: Write the correct skeletal structure for the molecule.
B is less electronegative, so it is central.

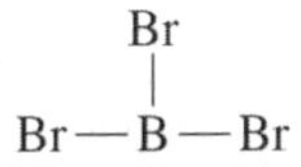

```
     Br
     |
Br—B—Br
```

Calculate the total number of electrons for the Lewis structure by summing the number of valence electrons of each atom in the molecule.

$$(\text{number of valence } e^- \text{ for B}) + 3(\text{number of valence } e^- \text{ for Br}) = 3 + 3(7) = 24$$

Distribute the electrons among the atoms, giving octets to as many atoms as possible. Begin with the bonding electrons; then proceed to lone pairs on terminal atoms and finally to lone pairs on the central atom.

```
       ..
      :Br:
  ..   |   ..
:Br — B — Br:
  ..       ..
```

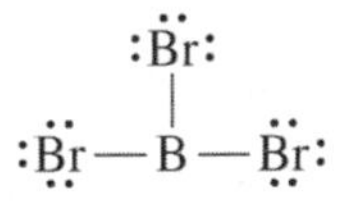

All 24 valence electrons are used.

B has an incomplete octet. If we complete the octet, there is a formal charge of −1 on the B, which is less electronegative than Br.

(b) NO: Write the correct skeletal structure for the molecule.

N—O

Calculate the total number of electrons for the Lewis structure by summing the number of valence electrons of each atom in the molecule.

$$(\text{number of valence } e^- \text{ for N}) + (\text{number of valence } e^- \text{ for O}) = 5 + 6 = 11$$

Distribute the electrons among the atoms, giving octets to as many atoms as possible. Begin with the bonding electrons; then proceed to lone pairs on terminal atoms and finally to lone pairs on the central atom.

```
.   ..
N = O:
..
```

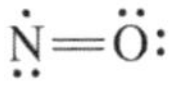

All 11 valence electrons are used.

N has an incomplete octet. It has seven electrons because we have an odd number of valence electrons.

(c) ClO_2: Write the correct skeletal structure for the molecule.
Cl is less electronegative, so it is central.

O—Cl—O

Calculate the total number of electrons for the Lewis structure by summing the number of valence electrons of each atom in the molecule.

$$(\text{number of valence } e^- \text{ for Cl}) + 2(\text{number of valence } e^- \text{ for O}) = 7 + 2(6) = 19$$

Distribute the electrons among the atoms, giving octets to as many atoms as possible. Begin with the bonding electrons; then proceed to lone pairs on terminal atoms and finally to lone pairs on the central atom.

:Ö—Ċl—Ö: or Ö=Ċl=Ö

All 19 valence electrons are used.

Cl will have either an incomplete octet or an expanded octet. Because Cl brings seven electrons, there is an odd number of electrons in either structure.

6.47 (a) PO_4^{3-}: Write the correct skeletal structure for the ion.

```
      O
      |
  O — P — O
      |
      O
```

Calculate the total number of electrons for the Lewis structure by summing the number of valence electrons of each atom in the ion and adding 3 for the 3− charge.

$4(\text{number of valence e}^- \text{ for O}) + (\text{number of valence e}^- \text{ for P}) + 3 = 4(6) + 5 + 3 = 32$

Distribute the electrons among the atoms, giving octets to as many atoms as possible. Begin with the bonding electrons; then proceed to lone pairs on terminal atoms and finally to lone pairs on the central atom.

```
      :Ö:
       |
  :Ö — P — Ö:
       |
      :Ö:
```

All 32 valence electrons are used.

Finally, write the Lewis structure in brackets with the charge of the ion in the upper right-hand corner.

```
  ┌    :Ö:     ┐3−
  |     |      |
  |:Ö — P — Ö: |
  |     |      |
  └    :Ö:     ┘
```

All atoms have octets (duets for H); the structure is complete.

Calculate the formal charge on each atom by finding the number of valence electrons and subtracting the number of lone pair electrons and one-half the number of bonding electrons.

```
  ┌    :Ö:     ┐3−
  |     |      |
  |:Ö — P — Ö: |
  |     |      |
  └    :Ö:     ┘
```

	O_{left}	O_{top}	O_{right}	O_{bottom}	P
number of valence electrons	6	6	6	6	5
− number of lone pair electrons	6	6	6	6	0
− 1/2(number of bonding electrons)	1	1	1	1	4
Formal charge	−1	−1	−1	−1	+1

The sum of the formal charges is −3, which is the overall charge of the ion. However, we can write a resonance structure with a double bond to an oxygen because P can expand its octet. This leads to lower formal charges on P and O.

$$\left[\begin{array}{c} :\ddot{O}: \\ | \\ \ddot{O}{=}P{-}\ddot{O}: \\ | \\ :\ddot{O}: \end{array}\right]^{3-} \longleftrightarrow \left[\begin{array}{c} :O: \\ \| \\ :\ddot{O}{-}P{-}\ddot{O}: \\ | \\ :\ddot{O}: \end{array}\right]^{3-} \longleftrightarrow \left[\begin{array}{c} :\ddot{O}: \\ | \\ :\ddot{O}{-}P{=}\ddot{O} \\ | \\ :\ddot{O}: \end{array}\right]^{3-} \longleftrightarrow \left[\begin{array}{c} :\ddot{O}: \\ | \\ :\ddot{O}{-}P{-}\ddot{O}: \\ \| \\ :O: \end{array}\right]^{3-}$$

Using the leftmost structure, calculate the formal charge on each atom by finding the number of valence electrons and subtracting the number of lone pair electrons and one-half the number of bonding electrons.

	O_{left}	O_{top}	O_{right}	O_{bottom}	P
number of valence electrons	6	6	6	6	5
− number of lone pair electrons	4	6	6	6	0
− 1/2(number of bonding electrons)	2	1	1	1	5
Formal charge	0	−1	−1	−1	0

The sum of the formal charges is −3, which is the overall charge of the ion. These resonance forms would all have the lower formal charges associated with the double-bonded O and P.

(b) CN^-: Write the correct skeletal structure for the ion.

C—N

Calculate the total number of electrons for the Lewis structure by summing the number of valence electrons of each atom in the ion and adding 1 for the 1− charge.

$$(\text{number of valence e}^- \text{ for C}) + (\text{number of valence e}^- \text{ for N}) + 1 = 4 + 5 + 1 = 10$$

Distribute the electrons among the atoms, giving octets to as many atoms as possible. Begin with the bonding electrons; then proceed to lone pairs on terminal atoms and finally to lone pairs on the central atom.

$:C{-}\ddot{\underset{\cdot\cdot}{N}}:$

All 10 valence electrons are used.

If any atom lacks an octet, form double or triple bonds as necessary.

$:C{\equiv}N:$

Finally, write the Lewis structure in brackets with the charge of the ion in the upper right-hand corner.

$[:C{\equiv}N:]^-$

All atoms have octets; the structure is complete.

Calculate the formal charge on each atom by finding the number of valence electrons and subtracting the number of lone pair electrons and one-half the number of bonding electrons.

$[:C{\equiv}N:]^-$

	C	N
number of valence electrons	4	5
− number of lone pair electrons	2	2
− 1/2(number of bonding electrons)	3	3
Formal charge	−1	0

The sum of the formal charges is −1, which is the overall charge of the ion.

(c) SO_3^{2-}: Write the correct skeletal structure for the ion.

$$\begin{array}{c} O \\ | \\ O{-}S{-}O \end{array}$$

Calculate the total number of electrons for the Lewis structure by summing the valence electrons of each atom in the ion and adding 2 for the 2− charge.

$$3(\text{number of valence e}^- \text{ for O}) + (\text{number of valence e}^- \text{ for S}) + 2 = 3(6) + 6 + 2 = 26$$

Distribute the electrons among the atoms, giving octets to as many atoms as possible. Begin with the bonding electrons; then proceed to lone pairs on terminal atoms and finally to lone pairs on the central atom.

```
     :Ö:
      |
:Ö — S — Ö:
```

All 26 valence electrons are used.

Finally, write the Lewis structure in brackets with the charge of the ion in the upper right-hand corner.

```
[     :Ö:      ]2−
[      |       ]
[:Ö — S — Ö:   ]
```

Calculate the formal charge on each atom by finding the number of valence electrons and subtracting the number of lone pair electrons and one-half the number of bonding electrons.

```
[     :Ö:      ]2−
[      |       ]
[:Ö — S — Ö:   ]
```

	O_{left}	O_{top}	O_{right}	S
number of valence electrons	6	6	6	6
− number of lone pair electrons	6	6	6	2
− 1/2(number of bonding electrons)	1	1	1	3
Formal charge	−1	−1	−1	+1

The sum of the formal charges is −2, which is the overall charge of the ion. However, we can write a resonance structure with a double bond to an oxygen because S can expand its octet. This leads to a lower formal charge.

```
[    :Ö:    ]2−      [    :O:    ]2−      [    :Ö:    ]2−
[     |     ]   ↔    [     ‖     ]   ↔    [     |     ]
[Ö = S — Ö: ]        [:Ö — S — Ö:]        [:Ö — S = Ö ]
```

Using the leftmost resonance form, calculate the formal charge on each atom by finding the number of valence electrons and subtracting the number of lone pair electrons and one-half the number of bonding electrons.

	O_{left}	O_{top}	O_{right}	S
number of valence electrons	6	6	6	6
− number of lone pair electrons	4	6	6	2
− 1/2(number of bonding electrons)	2	1	1	4
Formal charge	0	−1	−1	0

The sum of the formal charges is −2, which is the overall charge of the ion. These resonance forms would all have the lower formal charge on the double-bonded O and S.

(d) ClO_2^-: Write the correct skeletal structure for the ion.

O—Cl—O

Calculate the total number of electrons for the Lewis structure by summing the number of valence electrons of each atom in the ion and adding 1 for the 1− charge.

$$2(\text{number of valence e}^- \text{ for O}) + (\text{number of valence e}^- \text{ for Cl}) + 1 = 2(6) + 7 + 1 = 20$$

Distribute the electrons among the atoms, giving octets (or duets for H) to as many atoms as possible. Begin with the bonding electrons; then proceed to lone pairs on terminal atoms and finally to lone pairs on the central atom.

```
:Ö — C̈l — Ö:
```

All 20 valence electrons are used.

Finally, write the Lewis structure in brackets with the charge of the ion in the upper right-hand corner.

```
[:Ö — C̈l — Ö:]−
```

All atoms have octets; the structure is complete.

Calculate the formal charge on each atom by finding the number of valence electrons and subtracting the number of lone pair electrons and one-half the number of bonding electrons.

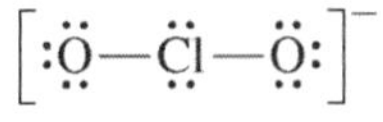

[:Ö—C̈l—Ö:]⁻

	O_{left}	O_{right}	Cl
number of valence electrons	6	6	7
− number of lone pair electrons	6	6	4
− 1/2(number of bonding electrons)	1	1	2
Formal charge	−1	−1	+1

The sum of the formal charges is −1, which is the overall charge of the ion. However, we can write a resonance structure with a double bond to an oxygen because Cl can expand its octet. This leads to a lower formal charge.

[Ö=C̈l—Ö:]⁻ ⟷ [:Ö—C̈l=Ö]⁻

Using the leftmost resonance form, calculate the formal charge on each atom by finding the number of valence electrons and subtracting the number of lone pair electrons and one-half the number of bonding electrons.

	O_{left}	O_{right}	Cl
number of valence electrons	6	6	7
− number of lone pair electrons	4	6	4
− 1/2(number of bonding electrons)	2	1	3
Formal charge	0	−1	0

The sum of the formal charges is −1, which is the overall charge of the ion. These resonance forms would all have the lower formal charge on the double-bonded O and Cl.

6.48 (a) SO_4^{2-}: Write the correct skeletal structure for the ion.

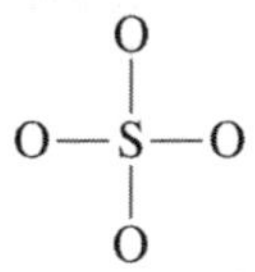

```
     O
     |
O—S—O
     |
     O
```

Calculate the total number of electrons for the Lewis structure by summing the valence electrons of each atom in the molecule and adding 2 for the 2− charge.

$$4(\text{number of valence e}^- \text{ for O}) + (\text{number of valence e}^- \text{ for S}) + 2 = 4(6) + 6 + 2 = 32$$

Distribute the electrons among the atoms, giving octets to as many atoms as possible. Begin with the bonding electrons; then proceed to lone pairs on terminal atoms and finally to lone pairs on the central atom.

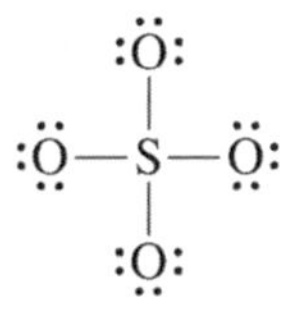

```
      :Ö:
       |
:Ö—S—Ö:
       |
      :Ö:
```

All 32 valence electrons are used.

Finally, write the Lewis structure in brackets with the charge of the ion in the upper right-hand corner.

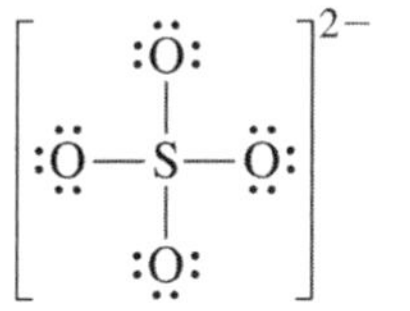

```
[      :Ö:      ]2−
[       |       ]
[ :Ö—S—Ö:  ]
[       |       ]
[      :Ö:      ]
```

All atoms have octets; the structure is complete.

Calculate the formal charge on each atom by finding the number of valence electrons and subtracting the number of lone pair electrons and one-half the number of bonding electrons.

	O_{left}	O_{top}	O_{right}	O_{bottom}	S
number of valence electrons	6	6	6	6	6
− number of lone pair electrons	6	6	6	6	0
− 1/2(number of bonding electrons)	1	1	1	1	4
Formal charge	−1	−1	−1	−1	+2

The sum of the formal charges is −2, which is the overall charge of the ion. However, we can write a resonance structure with double bonds to two oxygen atoms because S can expand its octet. This leads to lower formal charges.

Using the leftmost resonance form, calculate the formal charge on each atom by finding the number of valence electrons and subtracting the number of lone pair electrons and one-half the number of bonding electrons.

	O_{left}	O_{top}	O_{right}	O_{bottom}	S
number of valence electrons	6	6	6	6	6
− number of lone pair electrons	4	4	6	6	0
− 1/2(number of bonding electrons)	2	2	1	1	6
Formal charge	0	0	−1	−1	0

The sum of the formal charges is −2, which is the overall charge of the ion. These resonance forms would all have the lower formal charges on the double-bonded O and S.

(b) HSO_4^-: Write the correct skeletal structure for the ion.

Calculate the total number of electrons for the Lewis structure by summing the valence electrons of each atom in the ion and adding 1 for the 1− charge.

$$4(\text{number of valence } e^- \text{ for O}) + (\text{number of valence } e^- \text{ for S}) + (\text{number of valence } e^- \text{ for H}) + 1 = 4(6) + 6 + 1 + 1 = 32$$

Distribute the electrons among the atoms, giving octets (or duets for H) to as many atoms as possible. Begin with the bonding electrons; then proceed to lone pairs on terminal atoms and finally to lone pairs on the central atoms.

```
       :Ö:
        |
  :Ö — S — Ö — H
        |
       :Ö:
```

All 32 valence electrons are used.

Finally, write the Lewis structure in brackets with the charge of the ion in the upper right-hand corner.

```
 ┌      :Ö:          ┐−
 │       |           │
 │ :Ö — S — Ö — H    │
 │       |           │
 └      :Ö:          ┘
```

All atoms have octets (duets for H); the structure is complete.

Calculate the formal charge on each atom by finding the number of valence electrons and subtracting the number of lone pair electrons and one-half the number of bonding electrons.

```
 ┌      :Ö:          ┐−
 │       |           │
 │ :Ö — S — Ö — H    │
 │       |           │
 └      :Ö:          ┘
```

	O_{left}	O_{top}	O_{right}	O_{bottom}	S	H
number of valence electrons	6	6	6	6	6	1
− number of lone pair electrons	6	6	4	6	0	0
− 1/2(number of bonding electrons)	1	1	2	1	4	1
Formal charge	−1	−1	0	−1	+2	0

The sum of the formal charges is −1, which is the overall charge of the ion. However, we can write a resonance structure with double bonds to two oxygen atoms because S can expand its octet. This leads to lower formal charges.

```
 ┌    :O:        ┐−    ┌    :Ö:        ┐−    ┌    :O:         ┐−
 │     ||        │     │     |         │     │     ||         │
 │ Ö = S — Ö — H │ ⟷  │ Ö = S = Ö — H │ ⟷  │ :Ö — S — Ö — H │
 │     |         │     │     |         │     │      ||        │
 └    :Ö:        ┘     └    :Ö:        ┘     └     :O:        ┘
```

Using the leftmost resonance form, calculate the formal charge on each atom by finding the number of valence electrons and subtracting the number of lone pair electrons and one-half the number of bonding electrons.

	O_{left}	O_{top}	O_{right}	O_{bottom}	S	H
number of valence electrons	6	6	6	6	6	1
− number of lone pair electrons	4	4	4	6	0	0
− 1/2(number of bonding electrons)	2	2	2	1	6	1
Formal charge	0	0	0	−1	0	0

The sum of the formal charges is −1, which is the overall charge of the ion. Each of these resonance forms would have lower formal charges on O and S.

(c) SO_3: Write the correct skeletal structure for the molecule.

```
     O
     |
 O — S — O
```

Calculate the total number of electrons for the Lewis structure by summing the number of valence electrons of each atom in the molecule.

$$3(\text{number of valence e}^- \text{ for O}) + (\text{number of valence e}^- \text{ for S}) = 3(6) + 6 = 24$$

Distribute the electrons among the atoms, giving octets to as many atoms as possible. Begin with the bonding electrons; then proceed to lone pairs on terminal atoms and finally to lone pairs on the central atom.

```
      :Ö:
       |
   :Ö—S—Ö:
```

All 24 valence electrons are used.

If any atoms lack an octet, form double or triple bonds as necessary to give them octets.

```
      :Ö:
       |
   :Ö—S=Ö
```

All atoms have octets; the structure is complete.

Calculate the formal charge on each atom by finding the number of valence electrons and subtracting the number of lone pair electrons and one-half the number of bonding electrons.

This can have three resonance structures based on which oxygen has the double bond, and all three need to be shown.

```
    :Ö:              :Ö:              :Ö:
     |                ||               |
 :Ö—S=Ö   ⟷    :Ö—S—Ö:   ⟷    :Ö=S—Ö:
```

	O_{left}	O_{top}	O_{right}	S
number of valence electrons	6	6	6	6
− number of lone pair electrons	6	6	4	0
− 1/2(number of bonding electrons)	1	1	2	4
Formal charge	−1	−1	0	+2

The sum of the formal charges is 0, which is the overall charge of the molecule. However, we can write a resonance structure with a double bond to all oxygen atoms because S can expand its octet. This leads to lower formal charges.

```
      :O:
       ||
    Ö=S=Ö
```

Calculate the formal charge on each atom by finding the number of valence electrons and subtracting the number of lone pair electrons and one-half the number of bonding electrons.

	O_{left}	O_{top}	O_{right}	S
number of valence electrons	6	6	6	6
− number of lone pair electrons	4	4	4	0
− 1/2(number of bonding electrons)	2	2	2	6
Formal charge	0	0	0	0

The sum of the formal charges is 0, which is the overall charge of the molecule. This resonance form would have the lower formal charges on each atom.

(d) BrO_2^-: Write the correct skeletal structure for the ion.

O—Br—O

Calculate the total number of electrons for the Lewis structure by summing the number of valence electrons of each atom in the molecule and adding 1 for the 1− charge.

$$2(\text{number of valence e}^- \text{ for O}) + (\text{number of valence e}^- \text{ for Br}) + 1 = 2(6) + 7 + 1 = 20$$

Distribute the electrons among the atoms, giving octets to as many atoms as possible. Begin with the bonding electrons; then proceed to lone pairs on terminal atoms and finally to lone pairs on the central atom.

```
:Ö—B̈r—Ö:
```

All 20 valence electrons are used.

Finally, write the Lewis structure in brackets with the charge of the ion in the upper right-hand corner.

```
[:Ö—B̈r—Ö:]⁻
```

All atoms have octets; the structure is complete.
Calculate the formal charge on each atom by finding the number of valence electrons and subtracting the number of lone pair electrons and one-half the number of bonding electrons.

$$\left[:\ddot{\underset{..}{O}}-\ddot{\underset{..}{Br}}-\ddot{\underset{..}{O}}:\right]^-$$

	O_{left}	O_{right}	Br
number of valence electrons	6	6	7
− number of lone pair electrons	6	6	4
− 1/2(number of bonding electrons)	1	1	2
Formal charge	−1	−1	+1

The sum of the formal charges is −1, which is the overall charge of the ion. However, we can write a resonance structure with a double bond to an oxygen because Br can expand its octet. This leads to a lower formal charge.

$$\left[\ddot{\underset{..}{O}}=\ddot{\underset{..}{Br}}-\ddot{\underset{..}{O}}:\right]^- \longleftrightarrow \left[:\ddot{\underset{..}{O}}-\ddot{\underset{..}{Br}}=\ddot{\underset{..}{O}}\right]^-$$

Using the leftmost resonance form, calculate the formal charge on each atom by finding the number of valence electrons and subtracting the number of lone pair electrons and one-half the number of bonding electrons.

	O_{left}	O_{right}	Br
number of valence electrons	6	6	7
− number of lone pair electrons	4	6	4
− 1/2(number of bonding electrons)	2	1	3
Formal charge	0	−1	0

The sum of the formal charges is −1, which is the overall charge of the ion. These resonance forms would both have the lower formal charges on the double-bonded O and Br.

6.49 (a) PF_5: Write the correct skeletal structure for the molecule.

```
     F
     | /F
F — P
     | \F
     F
```

Calculate the total number of electrons for the Lewis structure by summing the number of valence electrons of each atom in the molecule.

$$(\text{number of valence e}^- \text{ for P}) + 5(\text{number of valence e}^- \text{ for F}) = 5 + 5(7) = 40$$

Distribute the electrons among the atoms, giving octets to as many atoms as possible. Begin with the bonding electrons; then proceed to lone pairs on terminal atoms and finally to lone pairs on the central atom. Arrange additional electrons around the central atom, giving it an expanded octet of up to 12 electrons.

```
      :F̤̈:
       | /F̤̈:
:F̤̈ — P
       | \F̤̈:
      :F̤̈:
```

(b) I_3^-: Write the correct skeletal structure for the ion.

I—I—I

Calculate the total number of electrons for the Lewis structure by summing the number of valence electrons of each atom in the ion and adding 1 for the 1− charge.

$$3(\text{number of valence e}^- \text{ for I}) + 1 = 3(7) + 1 = 22$$

Distribute the electrons among the atoms, giving octets to as many atoms as possible. Begin with the bonding electrons; then proceed to lone pairs on terminal atoms and finally to lone pairs on the central atom. Arrange additional electrons around the central atom, giving it an expanded octet of up to 12 electrons.

:Ï—Ï—Ï:

Finally, write the Lewis structure in brackets with the charge of the ion in the upper right-hand corner.

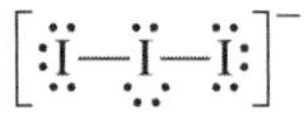

$[:Ï—Ï—Ï:]^-$

(c) SF_4: Write the correct skeletal structure for the molecule.

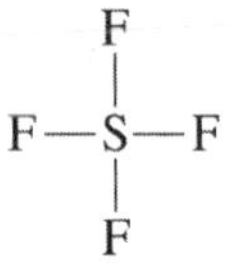

```
    F
    |
F — S — F
    |
    F
```

Calculate the total number of electrons for the Lewis structure by summing the number of valence electrons of each atom in the molecule.

(number of valence e^- for S) + 4(number of valence e^- for F) = 6 + 4(7) = 34

Distribute the electrons among the atoms, giving octets (or duets for H) to as many atoms as possible. Begin with the bonding electrons; then proceed to lone pairs on terminal atoms and finally to lone pairs on the central atom. Arrange additional electrons around the central atom, giving it an expanded octet of up to 12 electrons.

```
     :F:
      |
:F — S — F:
      |
     :F:
```

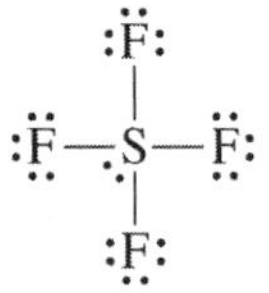

(d) GeF_4: Write the correct skeletal structure for the molecule.

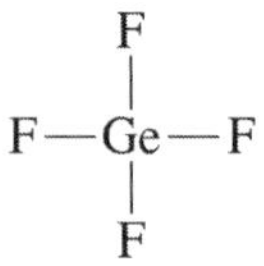

```
    F
    |
F — Ge — F
    |
    F
```

Calculate the total number of electrons for the Lewis structure by summing the number of valence electrons of each atom in the molecule.

(number of valence e^- for Ge) + 4(number of valence e^- for F) = 4 + 4(7) = 32

Distribute the electrons among the atoms, giving octets to as many atoms as possible. Begin with the bonding electrons; then proceed to lone pairs on terminal atoms and finally to lone pairs on the central atom.

```
     :F:
      |
:F — Ge — F:
      |
     :F:
```

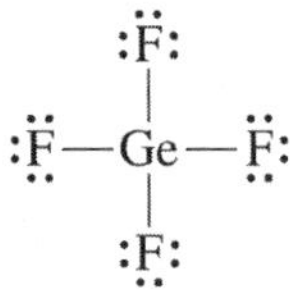

6.50 (a) ClF_5: Write the correct skeletal structure for the molecule.

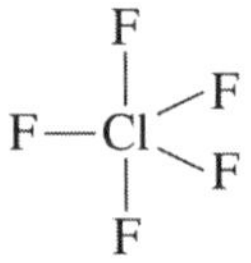

```
     F
     |  / F
F — Cl
     |  \ F
     F
```

Calculate the total number of electrons for the Lewis structure by summing the valence electrons of each atom in the molecule.

(number of valence e^- for Cl) + 5(number of valence e^- for F) = 7 + 5(7) = 42

Distribute the electrons among the atoms, giving octets to as many atoms as possible. Begin with the bonding electrons; then proceed to lone pairs on terminal atoms and finally to lone pairs on the central atom. Arrange additional electrons around the central atom, giving it an expanded octet of up to 12 electrons.

(b) AsF_6^-: Write the correct skeletal structure for the ion.

Calculate the total number of electrons for the Lewis structure by summing the valence electrons of each atom in the ion and adding 1 for the 1− charge.

(number of valence e^- for As) + 6(number of valence e^- for F) + 1 = 5 + 6(7) + 1 = 48

Distribute the electrons among the atoms, giving octets (or duets for H) to as many atoms as possible. Begin with the bonding electrons; then proceed to lone pairs on terminal atoms and finally to lone pairs on the central atom. Arrange additional electrons around the central atom, giving it an expanded octet of up to 12 electrons.

Finally, write the Lewis structure in brackets with the charge of the ion in the upper right-hand corner.

(c) Cl_3PO: Write the correct skeletal structure for the molecule.

Calculate the total number of electrons for the Lewis structure by summing the valence electrons of each atom in the molecule.

(number of valence e^- for P) + (number of valence e^- for O) + 3(number of valence e^- for Cl) = 5 + 6 + 3(7) = 32

Distribute the electrons among the atoms, giving octets to as many atoms as possible. Begin with the bonding electrons; then proceed to lone pairs on terminal atoms and finally to lone pairs on the central atom. Arrange additional electrons around the central atom, giving it an expanded octet of up to 12 electrons. This last structure, without double bonds, is a possibility. The formal charges are 0 except for P and O that are +1 and −1, respectively.

(d) IF_5: Write the correct skeletal structure for the molecule.

Calculate the total number of electrons for the Lewis structure by summing the valence electrons of each atom in the molecule.

$$(\text{number of valence e}^- \text{ for I}) + 5(\text{number of valence e}^- \text{ for F}) = 7 + 5(7) = 42$$

Distribute the electrons among the atoms, giving octets to as many atoms as possible. Begin with the bonding electrons; then proceed to lone pairs on terminal atoms and finally to lone pairs on the central atom. Arrange additional electrons around the central atom, giving it an expanded octet of up to 12 electrons.

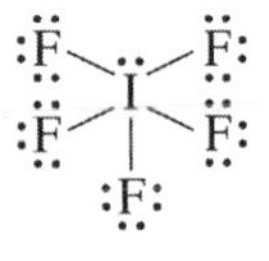

:F: :F: I :F: :F: :F:

Bond Energies and Bond Lengths

6.51 Bond strength: $H_3CCH_3 < H_2CCH_2 < HCCH$

Bond length: $H_3CCH_3 > H_2CCH_2 > HCCH$

Write the Lewis structures for the three compounds. Compare the C—C bonds. Triple bonds are stronger than double bonds; double bonds are stronger than single bonds. Also, single bonds are longer than double bonds, which are longer than triple bonds.

HCCH ($10\ e^-$) H_2CCH_2 ($12\ e^-$) H_3CCH_3($14\ e^-$)

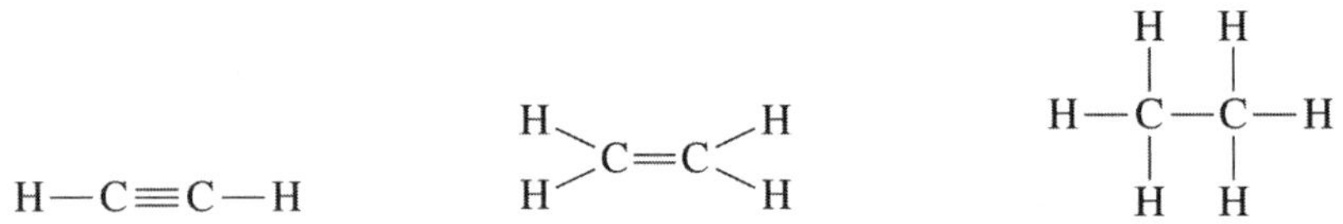

H—C≡C—H H₂C=CH₂ H—C(H)(H)—C(H)(H)—H

6.52 Stronger bond: HNNH

Shorter bond: HNNH

Write the Lewis structures for the two compounds. Compare the N—N bonds. Double bonds are stronger than single bonds. Also, single bonds are longer than double bonds. So HNNH will have the shorter and stronger N—N bond.

H_2NNH_2($14\ e^-$) HNNH($12\ e^-$)

H₂N̈—N̈H₂ H—N̈=N̈—H

VSEPR Theory and Molecular Geometry

6.53 Four electron groups: A trigonal pyramidal molecular geometry has three bonding groups and one lone pair of electrons, so there are four electron groups on atom A.

6.54 Three electron groups: A trigonal planar molecular geometry has three bonding groups and no lone pairs of electrons, so there are three electron groups on atom A.

6.55 (a) 4 total electron groups, 4 bonding groups, 0 lone pairs

A tetrahedral molecular geometry has four bonding groups and no lone pairs. So there are four total electron groups, four bonding groups, and no lone pairs.

(b) 5 total electron groups, 3 bonding groups, 2 lone pairs

A T-shaped molecular geometry has three bonding groups and two lone pairs. So there are five total electron groups, three bonding groups, and two lone pairs.

(c) 6 total electron groups, 5 bonding groups, 1 lone pair

A square pyramidal molecular geometry has five bonding groups and one lone pair. So there are six total electron groups, five bonding groups, and one lone pair.

6.56 (a) 6 total electron groups, 6 bonding groups, 0 lone pairs

An octahedral molecular geometry has six bonding groups and no lone pairs. So there are six total electron groups, six bonding groups, and no lone pairs.

(b) 6 electron groups, 4 bonding groups, 2 lone pairs
A square planar molecular geometry has four bonding groups and two lone pairs. So there are six total electron groups, four bonding groups, and two lone pairs.

(c) 5 electron groups, 4 bonding groups, 1 lone pair
A seesaw molecular geometry has four bonding groups and one lone pair. So there are five total electron groups, four bonding groups, and one lone pair.

6.57 (a) PF_3: Electron geometry—tetrahedral; molecular geometry—trigonal pyramidal; bond angle = 109.5°
Because of the lone pair, the bond angle will be less than 109.5°.
Draw a Lewis structure for the molecule:
PF_3 has 26 valence electrons.

```
      ..
     :F:
      |
 ..   ..   ..
:F — P — F:
 ..   ..   ..
```

Determine the total number of electron groups around the central atom:
There are four electron groups on P.
Determine the number of bonding groups and the number of lone pairs around the central atom:
There are three bonding groups and one lone pair.
Use Table 6.5 to determine the electron geometry, molecular geometry, and bond angles:
Four electron groups give a tetrahedral electron geometry; three bonding groups and one lone pair give a trigonal pyramidal molecular geometry; the idealized bond angles for tetrahedral geometry are 109.5°. The lone pair will make the bond angle less than idealized.

(b) SBr_2: Electron geometry—tetrahedral; molecular geometry—bent; bond angle = 109.5°
Because of the lone pairs, the bond angle will be less than 109.5°.
Draw a Lewis structure for the molecule:
SBr_2 has 20 valence electrons.

```
 ..   ..
:S — Br:
 |    ..
:Br:
 ..
```

Determine the total number of electron groups around the central atom:
There are four electron groups on S.
Determine the number of bonding groups and the number of lone pairs around the central atom:
There are two bonding groups and two lone pairs.
Use Table 6.5 to determine the electron geometry, molecular geometry, and bond angles:
Four electron groups give a tetrahedral electron geometry; two bonding groups and two lone pairs give a bent molecular geometry; the idealized bond angles for tetrahedral geometry are 109.5°. The lone pairs will make the bond angle less than idealized.

(c) $CHCl_3$: Electron geometry—tetrahedral; molecular geometry—tetrahedral; bond angle = 109.5°
Because there are no lone pairs, the bond angle will be 109.5°.
Draw a Lewis structure for the molecule:
$CHCl_3$ has 26 valence electrons.

```
         H
         |
 ..            ..
:Cl — C — Cl:
 ..      |     ..
        :Cl:
         ..
```

Determine the total number of electron groups around the central atom:
There are four electron groups on C.
Determine the number of bonding groups and the number of lone pairs around the central atom:
There are four bonding groups and no lone pairs.
Use Table 6.5 to determine the electron geometry, molecular geometry, and bond angles:
Four electron groups give a tetrahedral electron geometry; four bonding groups and no lone pairs give a tetrahedral molecular geometry; the idealized bond angles for tetrahedral geometry are 109.5°. However, because the attached atoms have different electronegativities, the bond angles are less than idealized.

(d) CS_2: Electron geometry—linear; molecular geometry—linear; bond angle = 180°
Because there are no lone pairs, the bond angle will be 180°.
Draw a Lewis structure for the molecule:
CS_2 has 16 valence electrons.

$\ddot{\underset{\cdot\cdot}{S}}=C=\ddot{\underset{\cdot\cdot}{S}}$

Determine the total number of electron groups around the central atom:
There are two electron groups on C.
Determine the number of bonding groups and the number of lone pairs around the central atom:
There are two bonding groups and no lone pairs.
Use Table 6.5 to determine the electron geometry, molecular geometry, and bond angles:
Two electron groups give a linear geometry; two bonding groups and no lone pairs give a linear molecular geometry; the idealized bond angle is 180°. The molecule will not deviate from this.

6.58 (a) CF_4: Electron geometry—tetrahedral; molecular geometry—tetrahedral; bond angle = 109.5°
Draw a Lewis structure for the molecule:
CF_4 has 32 valence electrons.

```
     ..
    :F:
     |
 ..  |  ..
:F—C—F:
 ..  |  ..
     |
    :F:
     ..
```

Determine the total number of electron groups around the central atom:
There are four electron groups on C.
Determine the number of bonding groups and the number of lone pairs around the central atom:
There are four bonding groups and no lone pairs.
Use Table 6.5 to determine the electron geometry, molecular geometry, and bond angles:
Four electron groups give a tetrahedral electron geometry; four bonding groups and no lone pairs give a tetrahedral molecular geometry; idealized tetrahedral bond angles for tetrahedral geometry are 109.5°.

(b) NF_3: Electron geometry—tetrahedral; molecular geometry—trigonal pyramidal; bond angle = 109.5°
Because of the lone pair, the bond angle will be less than 109.5°.
Draw a Lewis structure for the molecule:
NF_3 has 26 valence electrons.

```
 ..  ..  ..
:F—N—F:
 ..  |  ..
    :F:
     ..
```

Determine the total number of electron groups around the central atom:
There are four electron groups on N.
Determine the number of bonding groups and the number of lone pairs around the central atom:
There are three bonding groups and one lone pair.
Use Table 6.5 to determine the electron geometry, molecular geometry, and bond angles:
Four electron groups give a tetrahedral electron geometry; three bonding groups and one lone pair give a trigonal pyramidal molecular geometry; the idealized bond angles for tetrahedral geometry are 109.5°. The lone pair will make the bond angles less than idealized.

(c) OF_2: Electron geometry—tetrahedral; molecular geometry—bent; bond angle = 109.5°
Because of the lone pairs, the bond angle will be less than 109.5°.
Draw a Lewis structure for the molecule:
OF_2 has 20 valence electrons.

```
 ..  ..  ..
:F—O—F:
 ..  ..  ..
```

Determine the total number of electron groups around the central atom:
There are four electron groups on O.
Determine the number of bonding groups and the number of lone pairs around the central atom:
There are two bonding groups and two lone pairs.
Use Table 6.5 to determine the electron geometry, molecular geometry, and bond angles:

Four electron groups give a tetrahedral electron geometry; two bonding groups and two lone pairs give a bent molecular geometry; the idealized bond angles for tetrahedral geometry are 109.5°. The lone pairs will make the bond angles less than idealized.

(d) H_2S: Electron geometry—tetrahedral; molecular geometry—bent; bond angle = 109.5°

Because of the lone pair, the bond angle will be less than 109.5°.

Draw a Lewis structure for the molecule:

H_2S has eight valence electrons.

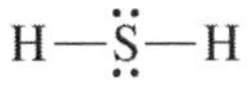

H—S̤̈—H

Determine the total number of electron groups around the central atom:

There are four electron groups on S.

Determine the number of bonding groups and the number of lone pairs around the central atom:

There are two bonding groups and two lone pairs.

Use Table 6.5 to determine the electron geometry, molecular geometry, and bond angles:

Four electron groups give a tetrahedral electron geometry; two bonding groups and two lone pairs give a bent molecular geometry; the idealized bond angles for tetrahedral geometry are 109.5°. However, the lone pairs will make the bond angle less than idealized.

6.59 H_2O will have the smaller bond angle because lone pair–lone pair repulsions are greater than lone pair–bonding pair repulsions.

Draw the Lewis structures for both structures:

H_3O^+ has 8 valence electrons.

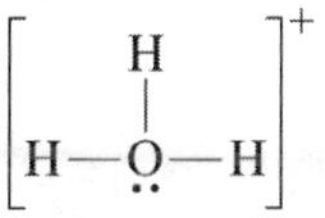

[H—O̤(—H)—H]$^+$

There are three bonding groups and one lone pair.

H_2O has 8 valence electrons.

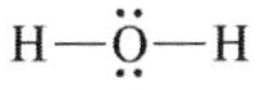

H—Ö̤—H

There are two bonding groups and two lone pairs.

Both have four electron groups, but the two lone pairs in H_2O will cause the bond angle to be smaller because of the lone pair–lone pair repulsions.

6.60 ClO_3^- will have the smaller bond angle because lone pair–bonding pair repulsions are greater than bonding pair–bonding pair repulsions.

Draw the Lewis structures for both structures:

ClO_3^- has 26 valence electrons.

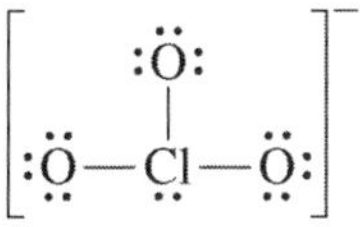

[:Ö̤—C̤l(—Ö̤:)—Ö̤:]$^-$

There are three bonding groups and one lone pair.

ClO_4^- has 32 valence electrons.

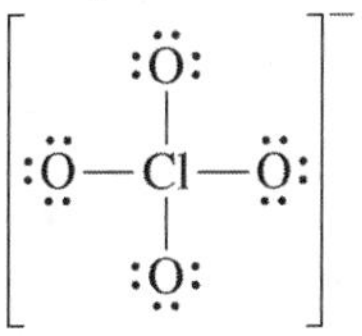

[:Ö̤—Cl(—Ö̤:)(—Ö̤:)—Ö̤:]$^-$

There are four bonding groups and no lone pairs.

Both have four electron groups, but the lone pair in ClO_3^- will cause the bond angle to be smaller because of the lone pair–bonding pair repulsions.

6.61 (a) SF_4 Draw a Lewis structure for the molecule:

SF_4 has 34 valence electrons.

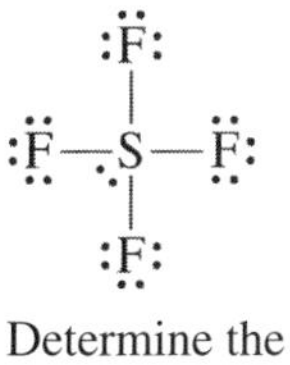

:F̤̈—S̤(—F̤̈:)(—F̤̈:)—F̤̈:

Determine the total number of electron groups around the central atom:

There are five electron groups on S.

Determine the number of bonding groups and the number of lone pairs around the central atom:

There are four bonding groups and one lone pair.

Use Table 6.5 to determine the electron geometry and molecular geometry:
The electron geometry is trigonal bipyramidal, so the molecular geometry is seesaw.
Sketch the molecule:

(b) ClF_3 Draw a Lewis structure for the molecule:
ClF_3 has 28 valence electrons.

Determine the total number of electron groups around the central atom:
There are five electron groups on Cl.
Determine the number of bonding groups and the number of lone pairs around the central atom:
There are three bonding groups and two lone pairs.
Use Table 6.5 to determine the electron geometry and molecular geometry:
The electron geometry is trigonal bipyramidal, so the molecular geometry is T-shaped.
Sketch the molecule:

(c) IF_2^- Draw a Lewis structure for the ion:
IF_2^- has 22 valence electrons.

Determine the total number of electron groups around the central atom:
There are five electron groups on I.
Determine the number of bonding groups and the number of lone pairs around the central atom:
There are two bonding groups and three lone pairs.
Use Table 6.5 to determine the electron geometry and molecular geometry:
The electron geometry is trigonal bipyramidal, so the molecular geometry is linear.
Sketch the ion:
[F—I—F]⁻

(d) IBr_4^- Draw a Lewis structure for the ion:
IBr_4^- has 36 valence electrons.

Determine the total number of electron groups around the central atom:
There are six electron groups on I.
Determine the number of bonding groups and the number of lone pairs around the central atom:
There are four bonding groups and two lone pairs.
Use Table 6.5 to determine the electron geometry and molecular geometry:
The electron geometry is octahedral, so the molecular geometry is square planar.
Sketch the ion:

6.62 (a) BrF_5 Draw a Lewis structure for the molecule:

BrF_5 has 42 valence electrons.

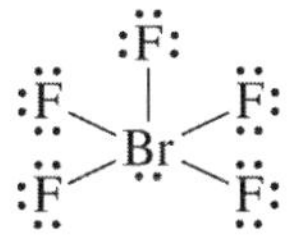

:F:
:F: Br :F:
:F: :F:

Determine the total number of electron groups around the central atom:
There are six electron groups on Br.
Determine the number of bonding groups and the number of lone pairs around the central atom:
There are five bonding groups and one lone pair.
Use Table 6.5 to determine the electron geometry and molecular geometry:
The electron geometry is octahedral, so the molecular geometry is square pyramidal.
Sketch the molecule:

F
F Br F
F F

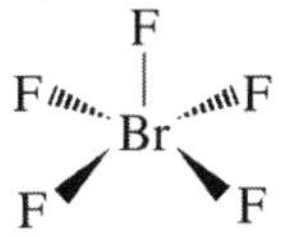

(b) SCl_6 Draw a Lewis structure for the molecule:

SCl_6 has 48 valence electrons.

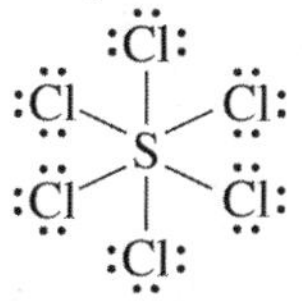

:Cl:
:Cl: S :Cl:
:Cl: :Cl:
:Cl:

Determine the total number of electron groups around the central atom:
There are six electron groups on S.
Determine the number of bonding groups and the number of lone pairs around the central atom:
There are six bonding groups and no lone pairs.
Use Table 6.5 to determine the electron geometry and molecular geometry:
The electron geometry is octahedral, so the molecular geometry is octahedral.
Sketch the molecule:

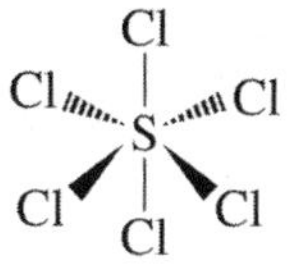

Cl
Cl S Cl
Cl Cl
Cl

(c) PF_5 Draw a Lewis structure for the molecule:

PF_5 has 40 valence electrons.

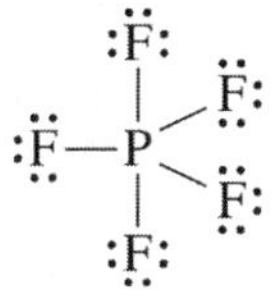

:F:
:F:—P :F:
:F:
:F:

Determine the total number of electron groups around the central atom:
There are five electron groups on P.
Determine the number of bonding groups and the number of lone pairs around the central atom:
There are five bonding groups and no lone pairs.
Use Table 6.5 to determine the electron geometry and molecular geometry:
The electron geometry is trigonal bipyramidal, so the molecular geometry is trigonal bipyramidal.
Sketch the molecule:

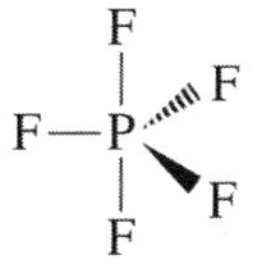

F
F—P F
F
F

(d) IF_4^+ Draw a Lewis structure for the ion:

IF_4^+ has 34 valence electrons.

$$\left[\begin{array}{c} :\ddot{F}: \\ | \\ :\ddot{F}-\dot{\underset{\cdot}{I}}-\ddot{F}: \\ | \\ :\ddot{F}: \end{array}\right]^+$$

Determine the total number of electron groups around the central atom:

There are five electron groups on I.

Determine the number of bonding groups and the number of lone pairs around the central atom:

There are four bonding groups and one lone pair.

Use Table 6.5 to determine the electron geometry and molecular geometry:

The electron geometry is trigonal bipyramidal, so the molecular geometry is seesaw.

Sketch the ion:

$$\left[\begin{array}{l} F \\ | \quad F \\ I \\ | \quad F \\ F \end{array}\right]^+$$

6.63 (a) C_2H_2 Draw the Lewis structure:

H—C≡C—H

Atom	Number of Electron Groups	Number of Lone Pairs	Molecular Geometry
Left C	2	0	Linear
Right C	2	0	Linear

Sketch the molecule:

H—C≡C—H

(b) C_2H_4 Draw the Lewis structure:

$H_2C=CH_2$

Atom	Number of Electron Groups	Number of Lone Pairs	Molecular Geometry
Left C	3	0	Trigonal planar
Right C	3	0	Trigonal planar

Sketch the molecule:

$H_2C=CH_2$

(c) C_2H_6 Draw the Lewis structure:

$$\begin{array}{c} H \quad H \\ | \quad\; | \\ H-C-C-H \\ | \quad\; | \\ H \quad H \end{array}$$

Atom	Number of Electron Groups	Number of Lone Pairs	Molecular Geometry
Left C	4	0	Tetrahedral
Right C	4	0	Tetrahedral

Sketch the molecule:

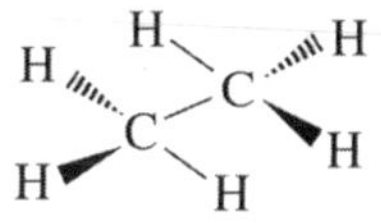

6.64 (a) N_2 Draw the Lewis structure:

:N≡N:

Atom	Number of Electron Groups	Number of Lone Pairs	Molecular Geometry
Left N	2	1	Linear
Right N	2	1	Linear

Sketch the molecule:

N≡N

(b) N_2H_2 Draw the Lewis structure:

H—N̈=N̈—H

Atom	Number of Electron Groups	Number of Lone Pairs	Molecular Geometry
Left N	3	1	Bent
Right N	3	1	Bent

Sketch the molecule:

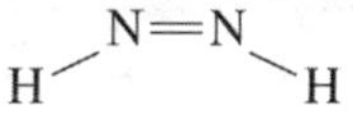

(c) N_2H_4 Draw the Lewis structure:

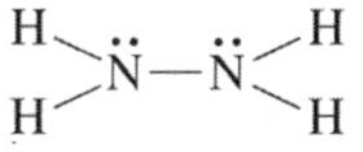

Atom	Number of Electron Groups	Number of Lone Pairs	Molecular Geometry
Left N	4	1	Trigonal pyramidal
Right N	4	1	Trigonal pyramidal

Sketch the molecule:

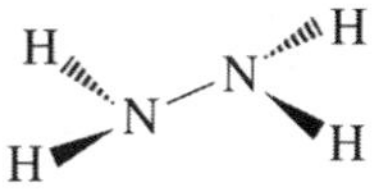

6.65 (a) Four pairs of electrons give a tetrahedral electron geometry. The lone pair would cause lone pair–bonded pair repulsions and would have a trigonal pyramidal molecular geometry.

(b) Five pairs of electrons give a trigonal bipyramidal electron geometry. The lone pair occupies an equatorial position to minimize lone pair–bonded pair repulsions, and the molecule would have a seesaw molecular geometry.

(c) Six pairs of electrons give an octahedral electron geometry. The two lone pairs would occupy opposite positions to minimize lone pair–lone pair repulsions. The molecular geometry would be square planar.

6.66 (a) Four pairs of electrons give a tetrahedral electron geometry. The two lone pairs would cause repulsions that would lead to a bent molecular geometry.

(b) Five pairs of electrons give a trigonal bipyramidal geometry. The three lone pairs would occupy equatorial positions to minimize the lone pair–lone pair repulsions. This would give a linear molecular geometry.

(c) Six pairs of electrons give an octahedral electron geometry. The lone pairs would occupy a position to minimize the lone pair–bonded pair repulsions. This would give a square pyramidal molecular geometry.

6.67 (a) CH_3OH Draw the Lewis structure and determine the geometry about each interior atom:

```
      ..
     :O—H
      |
   H—C—H
      |
      H
```

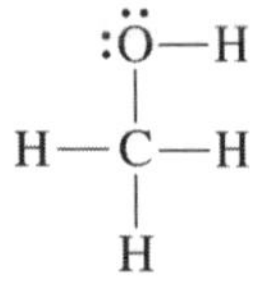

Atom	Number of Electron Groups	Number of Lone Pairs	Molecular Geometry
C	4	0	Tetrahedral
O	4	2	Bent

Sketch the molecule:

```
 H,,,   O
     C     ◀H
 H◀    H
```

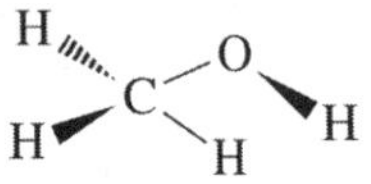

(b) CH_3OCH_3 Draw the Lewis structure and determine the geometry about each interior atom:

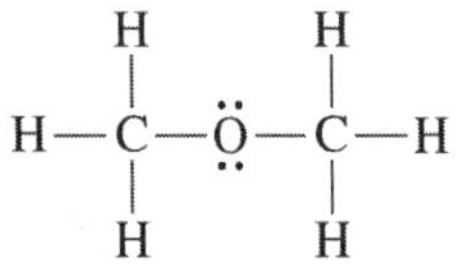

```
   H       H
   |  ..   |
H—C—O—C—H
   |  ..   |
   H       H
```

Atom	Number of Electron Groups	Number of Lone Pairs	Molecular Geometry
C	4	0	Tetrahedral
O	4	2	Bent
C	4	0	Tetrahedral

Sketch the molecule:

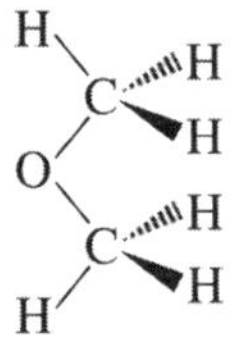

```
H
  C,,,H
    ◀H
O
  C,,,H
    ◀H
H
```

(c) H_2O_2 Draw the Lewis structure and determine the geometry about each interior atom:

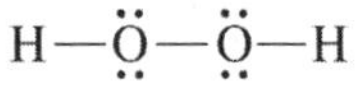

```
    ..  ..
H—O—O—H
    ..  ..
```

Atom	Number of Electron Groups	Number of Lone Pairs	Molecular Geometry
O	4	2	Bent
O	4	2	Bent

Sketch the molecule:

```
   O—O
H       H
```

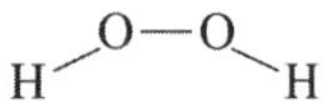

6.68 (a) CH_3NH_2 Draw the Lewis structure and determine the geometry about each interior atom:

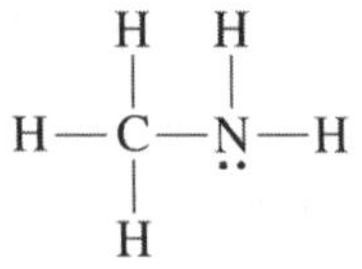

```
   H  H
   |  |
H—C—N—H
   |  ..
   H
```

Atom	Number of Electron Groups	Number of Lone Pairs	Molecular Geometry
C	4	0	Tetrahedral
N	4	1	Trigonal pyramidal

Sketch the molecule:

(b) $CH_3CO_2CH_3$ Draw the Lewis structure and determine the geometry about each interior atom:

Atom	Number of Electron Groups	Number of Lone Pairs	Molecular Geometry
Left C	4	0	Tetrahedral
Center C	3	0	Trigonal planar
O	4	2	Bent
Right C	4	0	Tetrahedral

Sketch the molecule:

(c) NH_2CO_2H Draw the Lewis structure and determine the geometry about each interior atom:

Atom	Number of Electron Groups	Number of Lone Pairs	Molecular Geometry
N	4	1	Trigonal pyramidal
C	3	0	Trigonal planar
O	4	2	Bent

Sketch the molecule:

Molecular Shape and Polarity

6.69 Draw the Lewis structure for CO_2 and CCl_4; determine the molecular geometry and the polarity.

	CO_2	CCl_4
Number of electron groups on C	2	4
Number of lone pairs	0	0
Molecular geometry	linear	tetrahedral

Even though each molecule contains polar bonds, the sum of the bond dipoles gives a net dipole of zero for each molecule.

The linear molecular geometry of CO_2 will have bond vectors that are equal and opposite. ←——→

The tetrahedral molecular geometry of CCl_4 will have bond vectors that are equal and have a net dipole of zero.

6.70 Draw the Lewis structure of CH_3F; determine the molecular geometry and then the polarity.

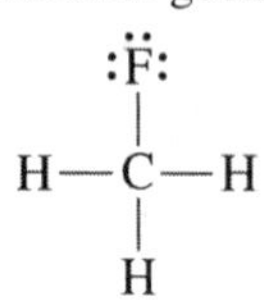

Number of electron groups on C	4
Number of lone pairs	0
Molecular geometry	tetrahedral

The molecule is tetrahedral but is polar because the C—H bond dipoles are different from the C—F bond dipoles because the electronegativities of C = 2.5, H = 2.1, and F = 4.0. Because the bond dipoles are different, the sum of the bond dipoles is *not* zero. Therefore, the molecule is polar.

The tetrahedral molecular geometry of CH_3F will have unequal bond vectors, so the molecule will have a net dipole.

6.71 (a) PF_3—polar

Use the Lewis structure and molecular geometry determined in Problem 6.57, which is trigonal pyramidal.
Determine whether the molecule contains polar bonds:
The electronegativity of P = 2.1 and F = 4. Therefore, the bonds are polar.
Determine whether the polar bonds add together to form a net dipole:
Because the molecule is trigonal pyramidal, the three dipole moments sum to a nonzero net dipole moment. The molecule is polar. Refer to Table 6.6 to see how dipole moments add to determine polarity.

(b) SBr_2—nonpolar

Use the Lewis structure and molecular geometry determined in Problem 6.57, which is bent.
Determine whether the molecule contains polar bonds:
The electronegativity of S = 2.5 and Br = 2.8. Therefore, the bonds are nonpolar.
Even though the molecule is bent, because the bonds are nonpolar, the molecule is nonpolar.

(c) $CHCl_3$—polar

Use the Lewis structure and molecular geometry determined in Problem 6.57, which is tetrahedral.
Determine whether the molecule contains polar bonds:
The electronegativity of C = 2.5, H = 2.1, and Cl = 3.0. Therefore, the bonds are polar.
Determine whether the polar bonds add together to form a net dipole:
Because the bonds have different dipole moments due to the different atoms involved, the four dipole moments sum to a nonzero net dipole moment. The molecule is polar. Refer to Table 6.6 to see how dipole moments add to determine polarity.

(d) CS_2—nonpolar

Use the Lewis structure and molecular geometry determined in Problem 6.57, which is linear.
Determine whether the molecule contains polar bonds:
The electronegativity of C = 2.5 and S = 2.5. Therefore, the bonds are nonpolar. Also, the molecule is linear, which would result in a zero net dipole even if the bonds were polar.
The molecule is nonpolar. Refer to Table 6.6 to see how dipole moments add to determine polarity.

6.72 (a) CF_4—nonpolar

Use the Lewis structure and molecular geometry determined in Problem 6.58, which is tetrahedral.
Determine whether the molecule contains polar bonds:
The electronegativity of C = 2.5 and F = 4.0. Therefore, the bonds are polar.
Determine whether the polar bonds add together to form a net dipole:
Because the molecular geometry is tetrahedral, the four equal dipole moments sum to a zero net dipole moment. The molecule is nonpolar. Refer to Table 6.6 to see how dipole moments add to determine polarity.

(b) NF_3—polar

Use the Lewis structure and molecular geometry determined in Problem 6.58, which is trigonal pyramidal.
Determine whether the molecule contains polar bonds:
The electronegativity of N = 3.0 and F = 4.0. Therefore, the bonds are polar.
Determine whether the polar bonds add together to form a net dipole:
Because the molecular geometry is trigonal pyramidal, the three dipole moments sum to a nonzero net dipole moment. The molecule is polar. Refer to Table 6.6 to see how dipole moments add to determine polarity.

(c) OF_2—polar

Use the Lewis structure and molecular geometry determined in Problem 6.58, which is bent.
Determine whether the molecule contains polar bonds:
The electronegativity of O = 3.5 and F = 4.0. Therefore, the bonds are polar.
Determine whether the polar bonds add together to form a net dipole:
Because the molecular geometry is bent, the two dipole moments sum to a nonzero net dipole moment. The molecule is polar. Refer to Table 6.6 to see how dipole moments add to determine polarity.

(d) H_2S—polar

Use the Lewis structure and molecular geometry determined in Problem 6.58, which is bent.
Determine whether the molecule contains polar bonds:
The electronegativity of H = 2.1 and S = 2.5. Therefore, the bonds are polar.
Determine whether the polar bonds add together to form a net dipole:
Because the molecular geometry is bent, the two dipole moments sum to a nonzero net dipole moment. The molecule is polar. Refer to Table 6.6 to see how dipole moments add to determine polarity.

6.73 (a) ClO_3^-—polar

Draw the Lewis structure and determine the molecular geometry:

```
 [      :Ö:      ]⁻
 [       |       ]
 [ :Ö—C̤l—Ö: ]
```

Four electron pairs with one lone pair give a trigonal pyramidal molecular geometry.
Determine whether the molecule contains polar bonds:
The electronegativities of Cl = 3.0 and O = 3.5. Therefore, the bonds are polar.
Determine whether the polar bonds add together to form a net dipole:
Because the molecular geometry is trigonal pyramidal, the three dipole moments sum to a nonzero net dipole moment. The molecule is polar. Refer to Table 6.6 to see how dipole moments add to determine polarity.

(b) SCl_2—polar

Draw the Lewis structure and determine the molecular geometry:

```
:S̈—C̤̈l:
 |
:C̤l:
```

Four electron pairs with two lone pairs give a bent molecular geometry.
Determine whether the molecule contains polar bonds:
The electronegativity of S = 2.5 and Cl = 3.0. Therefore, the bonds are polar.
Determine whether the polar bonds add together to form a net dipole:
Because the molecular geometry is bent, the two dipole moments sum to a nonzero net dipole moment. The molecule is polar. Refer to Table 6.6 to see how dipole moments add to determine polarity.

(c) SCl_4—polar

Draw the Lewis structure and determine the molecular geometry:

```
       :Cl:
        |
  :Cl — S — Cl:
        |
       :Cl:
```

Five electron pairs with one lone pair give a seesaw molecular geometry.

Determine whether the molecule contains polar bonds:

The electronegativity of S = 2.5 and Cl = 3.0. Therefore, the bonds are polar.

Determine whether the polar bonds add together to form a net dipole:

Because the molecular geometry is seesaw, the four equal dipole moments sum to a nonzero net dipole moment. The molecule is polar.

The seesaw molecular geometry will not have offsetting bond vectors.

(d) $BrCl_5$—nonpolar

Draw the Lewis structure and determine the molecular geometry.

```
        :Cl:
  :Cl \  |  / Cl:
        Br
  :Cl /  ··  \ Cl:
```

Six electron pairs with one lone pair give square pyramidal molecular geometry.

Determine whether the molecule contains polar bonds:

The electronegativity of Br = 2.8 and Cl = 3.0. The difference is only 0.2; therefore, the bonds are nonpolar. Even though the molecular geometry is square pyramidal, the five bonds are nonpolar; so there is no net dipole. The molecule is nonpolar.

6.74 (a) $SiCl_4$—nonpolar

Draw the Lewis structure and determine the molecular geometry:

```
        :Cl:
         |
  :Cl — Si — Cl:
         |
        :Cl:
```

Four electron pairs with no lone pairs give a tetrahedral molecular geometry.

Determine whether the molecule contains polar bonds:

The electronegativity of Cl = 3.0 and Si = 1.8. Therefore, the bonds are polar.

Determine whether the polar bonds add together to form a net dipole:

Because the molecular geometry is tetrahedral, the four equal dipole moments sum to a zero net dipole moment. The molecule is nonpolar. Refer to Table 6.6 to see how dipole moments add to determine polarity.

(b) CF_2Cl_2—polar

Draw the Lewis structure and determine the molecular geometry:

```
        :F:
         |
  :Cl — C — F:
         |
        :Cl:
```

Four electron pairs with no lone pairs give a tetrahedral molecular geometry.

Determine whether the molecule contains polar bonds:

The electronegativity of C = 2.5, F = 4.0, and Cl = 3.0. Therefore, the bonds are polar.

Determine whether the polar bonds add together to form a net dipole:

Even though the molecular geometry is tetrahedral, which normally yields a nonpolar molecule, the four dipole moments sum to a nonzero net dipole moment because of the different electronegativities of Cl and F. The molecule is polar. Refer to Table 6.6 to see how dipole moments add to determine polarity.

(c) SeF_6—nonpolar

Draw the Lewis structure and determine the molecular geometry:

```
       :F:
  :F\   |   /F:
       Se
  :F/   |   \F:
       :F:
```

Six electron pairs with no lone pairs give an octahedral molecular geometry.

Determine whether the molecule contains polar bonds:

The electronegativity of Se = 3.0 and F = 4.0. Therefore, the bonds are polar.

Determine whether the polar bonds add together to form a net dipole:

Because the molecular geometry is octahedral, the six equal dipole moments sum to a zero net dipole moment. The molecule is nonpolar. Refer to Table 6.6 to see how dipole moments add to determine polarity.

(d) IF_5—polar

Draw the Lewis structure and determine the molecular geometry:

```
       :F:
  :F\   |   /F:
        I
  :F/   ··  \F:
```

Six electron pairs with one lone pair give square pyramidal molecular geometry.

Determine whether the molecule contains polar bonds:

The electronegativity of I = 2.0 and F = 4.0. Therefore, the bonds are polar.

Determine whether the polar bonds add together to form a net dipole:

Because the molecular geometry is square pyramidal, the five dipole moments sum to a nonzero net dipole moment. The molecule is polar.

The square pyramid structure has offsetting bond vectors in the equatorial plane but not in the axial positions.

Cumulative Problems

6.75 (a) $BaCO_3$: Ba^{2+}

```
 [     :Ö:      ]2−
 [      |       ]
 [ Ö = C — Ö:   ]
```

Determine the cation and anion.

Ba^{2+} CO_3^{2-}

Write the Lewis structure for the barium cation based on the valence electrons.

Ba $5s^2$ Ba^{2+} $5s^0$

Ba: Ba^{2+}

Ba must lose two electrons and be left with the octet from the previous shell.

Write the Lewis structure for the covalent anion.

Write the correct skeletal structure for the ion.

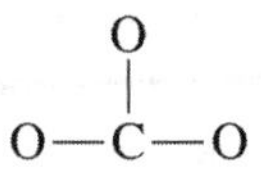

```
    O
    |
O—C—O
```

Calculate the total number of electrons for the Lewis structure by summing the number of valence electrons of each atom in the ion and adding two for the 2− charge.

$$(\text{number of valence } e^- \text{ for C}) + 3(\text{number of valence } e^- \text{ for O}) = 4 + 3(6) + 2 = 24$$

Distribute the electrons among the atoms, giving octets to as many atoms as possible. Begin with the bonding electrons; then proceed to lone pairs on terminal atoms and finally to lone pairs on the central atom.

```
     :Ö:
      |
:Ö—C—Ö:
```

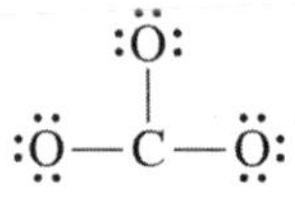

If any atom lacks an octet, form double or triple bonds as necessary.

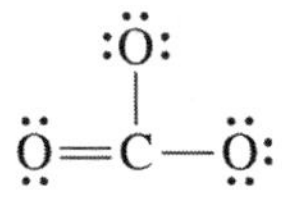

```
    :Ö:
     |
Ö=C—Ö:
```

Finally, write the Lewis structure in brackets with the charge of the ion in the upper right-hand corner.

```
        ⎡    :Ö:     ⎤2−
Ba2+    ⎢     |      ⎥
        ⎣ Ö=C—Ö:  ⎦
```

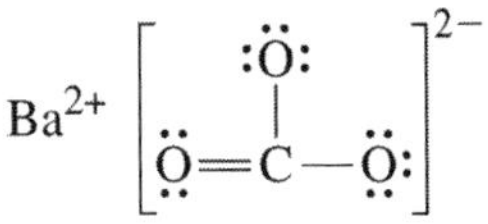

The double bond can be between the C and any of the oxygen atoms, so there are resonance structures.

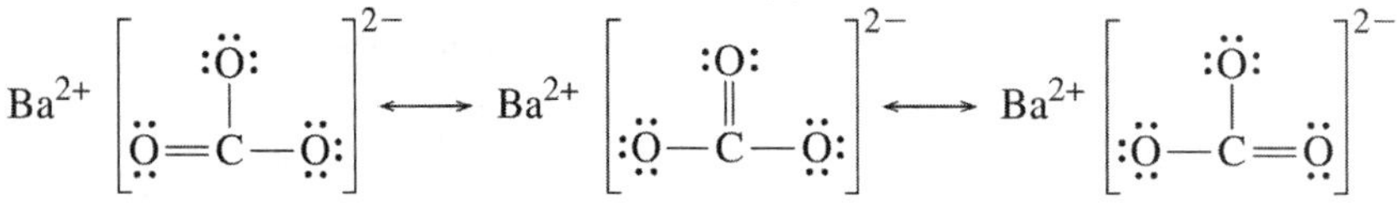

```
      ⎡   :Ö:    ⎤2−           ⎡    :O:     ⎤2−           ⎡    :Ö:    ⎤2−
Ba2+  ⎢    |     ⎥   ←→  Ba2+  ⎢    ‖       ⎥   ←→  Ba2+  ⎢     |     ⎥
      ⎣ Ö=C—Ö: ⎦              ⎣ :Ö—C—Ö:  ⎦              ⎣ :Ö—C=Ö  ⎦
```

(b) $Ca(OH)_2$: Ca^{2+}

$2[:\ddot{O}—H]^-$

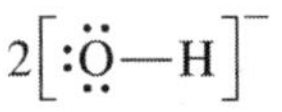

Determine the cation and anion.

Ca^{2+} OH^-

Write the Lewis structure for the calcium cation based on the valence electrons.

Ca $4s^2$	Ca^{2+} $4s^0$
Ca:	Ca^{2+}

Ca must lose two electrons and be left with the octet from the previous shell.

Write the Lewis structure for the covalent anion.

Write the correct skeletal structure for the ion.

O—H

Calculate the total number of electrons for the Lewis structure by summing the valence electrons of each atom in the ion and adding 1 for the 1− charge.

$$(\text{number of valence } e^- \text{ for H}) + (\text{number of valence } e^- \text{ for O}) + 1 = 1 + 6 + 1 = 8$$

Distribute the electrons among the atoms, giving octets (or duets for H) to as many atoms as possible. Begin with the bonding electrons; then proceed to lone pairs on terminal atoms and finally to lone pairs on the central atom.

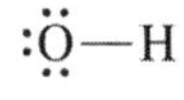

:Ö—H

Finally, write the Lewis structure in brackets with the charge of the ion in the upper right-hand corner.

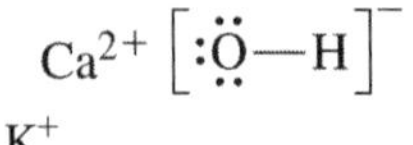

Ca^{2+} $[:\ddot{O}—H]^-$

(c) KNO_3: K^+

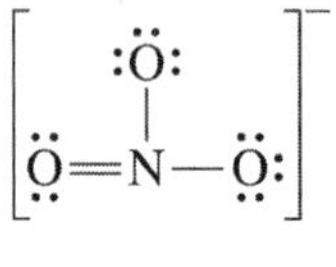

```
⎡    :Ö:    ⎤−
⎢     |     ⎥
⎣ Ö=N—Ö: ⎦
```

Determine the cation and anion.

K^+ NO_3^-

Write the Lewis structure for the potassium cation based on the valence electrons.

K $4s^1$ K^+ $4s^0$

K· K^+

K must lose one electron and be left with the octet from the previous shell.

Write the Lewis structure for the covalent anion.

Write the correct skeletal structure for the ion.

```
    O
    |
O — N — O
```

Calculate the total number of electrons for the Lewis structure by summing the valence electrons of each atom in the ion and adding 1 for the 1− charge.

$(\text{number of valence e}^- \text{ for N}) + (\text{number of valence e}^- \text{ for O}) = 5 + 3(6) + 1 = 24$

Distribute the electrons among the atoms, giving octets to as many atoms as possible. Begin with the bonding electrons; then proceed to lone pairs on terminal atoms and finally to lone pairs on the central atom.

```
     ..
    :O:
     |
 ..      ..
:O — N — O:
 ..      ..
```

If any atom lacks an octet, form double or triple bonds as necessary.

```
     ..
    :O:
     |
 ..      ..
 O = N — O:
 ..      ..
```

Finally, write the Lewis structure in brackets with the charge of the ion in the upper right-hand corner.

```
    [     ..    ]⁻
    [    :O:    ]
K+  [     |     ]
    [ ..      ..]
    [ O = N — O:]
    [ ..      ..]
```

The double bond can be between the N and any of the oxygen atoms, so there are resonance structures.

```
    [     ..    ]⁻        [     :O:    ]⁻        [     ..     ]⁻
    [    :O:    ]         [      ||    ]         [    :O:     ]
K+  [     |     ]  <--> K+[ ..       ..]  <--> K+[     |      ]
    [ ..      ..]         [:O — N — O: ]         [ ..       ..]
    [ O = N — O:]         [ ..       ..]         [:O — N = O  ]
    [ ..      ..]                                [ ..       ..]
```

(d) LiIO: Li^+ $[:\ddot{\underset{..}{I}}-\ddot{\underset{..}{O}}:]^-$

Determine the cation and anion.

Li^+ IO^-

Write the Lewis structure for the lithium cation based on the valence electrons.

Li $2s^1$ Li^+ $2s^0$

Li· Li^+

Li must lose one electron and be left with the octet from the previous shell.

Write the Lewis structure for the covalent anion.

Write the correct skeletal structure for the ion.

I—O

Calculate the total number of electrons for the Lewis structure by summing the number of valence electrons of each atom in the ion and adding 1 for the 1− charge.

$(\text{number of valence e}^- \text{ for I}) + (\text{number of valence e}^- \text{ for O}) = 7 + 6 + 1 = 14$

Distribute the electrons among the atoms, giving octets to as many atoms as possible. Begin with the bonding electrons; then proceed to lone pairs on terminal atoms and finally to lone pairs on the central atom.

$:\ddot{\underset{..}{I}}-\ddot{\underset{..}{O}}:$

Finally, write the Lewis structure in brackets with the charge of the ion in the upper right-hand corner.

$Li^+ [:\ddot{\underset{..}{I}}-\ddot{\underset{..}{O}}:]^-$

6.76 (a) $RbIO_2$: Rb^+

$$\left[:\ddot{\underset{..}{O}}-\ddot{\underset{..}{I}}-\ddot{\underset{..}{O}}:\right]^-$$

Determine the cation and anion.

Rb^+ $\quad IO_2^-$

Write the Lewis structure for the Rubidium cation based on the valence electrons.

Rb $5s^1$ $\quad Rb^+$ $5s^0$

Rb· $\quad Rb^+$

Rb must lose one electron and be left with the octet from the previous shell.

Write the Lewis structure for the covalent anion.

Write the correct skeletal structure for the ion.

$$O-I-O$$

Calculate the total number of electrons for the Lewis structure by summing the valence electrons of each atom in the ion and adding 1 for the 1− charge.

$$(\text{number of valence } e^- \text{ for I}) + 2(\text{number of valence } e^- \text{ for O}) = 7 + 2(6) + 1 = 20$$

Distribute the electrons among the atoms, giving octets to as many atoms as possible. Begin with the bonding electrons; then proceed to lone pairs on terminal atoms and finally to lone pairs on the central atom.

$$:\ddot{\underset{..}{O}}-\ddot{\underset{..}{I}}-\ddot{\underset{..}{O}}:$$

Then form double bonds to eliminate the formal charge on I.

$$\ddot{\underset{..}{O}}=\ddot{I}-\ddot{\underset{..}{O}}: \longleftrightarrow :\ddot{\underset{..}{O}}-\ddot{I}=\ddot{\underset{..}{O}}$$

Finally, write the Lewis structure in brackets with the charge of the ion in the upper right-hand corner.

$$Rb^+\left[\ddot{\underset{..}{O}}=\ddot{\underset{..}{I}}-\ddot{\underset{..}{O}}:\right]^- \longleftrightarrow Rb^+\left[:\ddot{\underset{..}{O}}-\ddot{\underset{..}{I}}=\ddot{\underset{..}{O}}\right]^-$$

(b) NH_4Cl:

$$\left[\begin{array}{c} \text{H} \\ | \\ \text{H}-\text{N}-\text{H} \\ | \\ \text{H} \end{array}\right]^+ \quad \left[:\ddot{\underset{..}{Cl}}:\right]^-$$

Determine the cation and anion.

NH_4^+ $\quad Cl^-$

Write the Lewis structure for the covalent cation.

Write the correct skeletal structure for the ion.

$$\begin{array}{c} \text{H} \\ | \\ \text{H}-\text{N}-\text{H} \\ | \\ \text{H} \end{array}$$

Calculate the total number of electrons for the Lewis structure by summing the number of valence electrons of each atom in the ion and subtracting 1 for the 1+ charge.

$$(\text{number of valence } e^- \text{ for N}) + 4(\text{number of valence } e^- \text{ for H}) - 1 = 5 + 4(1) - 1 = 8$$

Distribute the electrons among the atoms, giving octets (or duets for H) to as many atoms as possible. Begin with the bonding electrons; then proceed to lone pairs on terminal atoms and finally to lone pairs on the central atom.

$$\begin{array}{c} \text{H} \\ | \\ \text{H}-\text{N}-\text{H} \\ | \\ \text{H} \end{array}$$

Finally, write the Lewis structure in brackets with the charge of the ion in the upper right-hand corner.

$$\left[\begin{array}{c} \text{H} \\ | \\ \text{H}-\text{N}-\text{H} \\ | \\ \text{H} \end{array}\right]^+$$

Write the Lewis structure for the chlorine anion based on the valence electrons.

Cl $3s^2 3p^5$

$\cdot\ddot{\underset{\cdot\cdot}{Cl}}:$

Cl must gain one electron to complete its octet.

$[:\ddot{\underset{\cdot\cdot}{Cl}}:]^-$

$[:\ddot{\underset{\cdot\cdot}{Cl}}:]^-$ $\left[\begin{array}{c} H \\ | \\ H-N-H \\ | \\ H \end{array}\right]^+$

(c) KOH: K^+

$[:\ddot{\underset{\cdot\cdot}{O}}-H]^-$

Determine the cation and anion.

K^+ OH^-

Write the Lewis structure for the potassium cation based on the valence electrons.

K $4s^1$ K^+ $4s^0$

$K\cdot$ K^+

K must lose one electron and be left with the octet from the previous shell.

Write the Lewis structure for the covalent anion.

Write the correct skeletal structure for the ion.

O—H

Calculate the total number of electrons for the Lewis structure by summing the valence electrons of each atom in the ion and adding 1 for the 1− charge.

(number of valence e^- for H) + (number of valence e^- for O) + 1 = 1 + 6 + 1 = 8

Distribute the electrons among the atoms, giving octets (or duets for H) to as many atoms as possible. Begin with the bonding electrons; then proceed to lone pairs on terminal atoms and finally to lone pairs on the central atom.

$:\ddot{\underset{\cdot\cdot}{O}}-H$

Finally, write the Lewis structure in brackets with the charge of the ion in the upper right-hand corner.

$K^+ [:\ddot{\underset{\cdot\cdot}{O}}-H]^-$

(d) $Sr(CN)_2$: Sr^{2+}

$2[:C\equiv N:]^-$

Determine the cation and anion.

Sr^{2+} $2CN^-$

Write the Lewis structure for the strontium cation based on the valence electrons.

Sr $5s^2$ Sr^{2+} $5s^0$

$\cdot Sr \cdot$ Sr^{2+}

Sr must lose two electrons and be left with the octet from the previous shell.

Write the Lewis structure for the covalent anion.

Write the correct skeletal structure for the ion.

C—N

Calculate the total number of electrons for the Lewis structure by summing the valence electrons of each atom in the ion and adding 1 for the 1− charge.

(number of valence e^- for N) + (number of valence e^- for C) = 5 + 4 + 1 = 10

Distribute the electrons among the atoms, giving octets to as many atoms as possible. Begin with the bonding electrons; then proceed to lone pairs on terminal atoms and finally to lone pairs on the central atom.

$:C-\ddot{\underset{\cdot\cdot}{N}}:$

Complete octets on both atoms by forming a triple bond.

$:C\equiv N:$

Finally, write the Lewis structure in brackets with the charge of the ion in the upper right-hand corner.

$$Sr^{2+}\ 2\left[:C{\equiv}N:\right]^-$$

6.77 (a) C_4H_8: Write the correct skeletal structure for the molecule.

```
     H   H
     |   |
 H — C — C — H
     |   |
 H — C — C — H
     |   |
     H   H
```

Calculate the total number of electrons for the Lewis structure by summing the number of valence electrons of each atom in the molecule.

$$4(\text{number of valence e}^- \text{ for C}) + 8(\text{number of valence e}^- \text{ for H}) = 4(4) + 8(1) = 24$$

Distribute the electrons among the atoms, giving octets (or duets for H) to as many atoms as possible.

```
     H   H
     |   |
 H — C — C — H
     |   |
 H — C — C — H
     |   |
     H   H
```

All atoms have octets or duets for H.

(b) C_4H_4: Write the correct skeletal structure for the molecule.

```
 H — C — C — H
     |   |
 H — C — C — H
```

Calculate the total number of electrons for the Lewis structure by summing the number of valence electrons of each atom in the molecule.

$$4(\text{number of valence e}^- \text{ for C}) + 4(\text{number of valence e}^- \text{ for H}) = 4(4) + 4(1) = 20$$

Distribute the electrons among the atoms, giving octets (or duets for H) to as many atoms as possible.

```
     ..  ..
 H — C — C — H
     |   |
 H — C — C — H
```

Complete octets by forming double bonds on alternating carbons; draw resonance structures.

```
 H — C — C — H          H — C = C — H
     ‖   ‖      <——>        |   |
 H — C — C — H          H — C = C — H
```

(c) C_6H_{12}: Write the correct skeletal structure for the molecule.

```
        H H
         \/
 H       C       H
   \   /   \   /
     C       C
   /  |     |  \
 H    |     |    H
 H — C       C — H
    /  \   /  \
  H      C      H
        / \
        H H
```

Calculate the total number of electrons for the Lewis structure by summing the valence electrons of each atom in the molecule.

$$6(\text{number of valence e}^- \text{ for C}) + 12(\text{number of valence e}^- \text{ for H}) = 6(4) + 12(1) = 36$$

Distribute the electrons among the atoms, giving octets (or duets for H) to as many atoms as possible. Begin with the bonding.

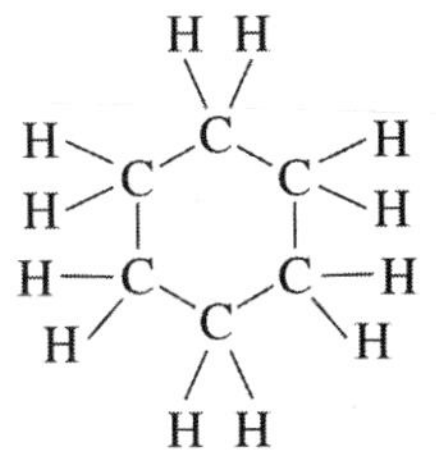

```
        H H
         \/
  H\    C     /H
  H/ C     C \H
     |     |
  H—C       C—H
  H/    C     \H
        /\
       H  H
```

All 36 electrons are used, and all atoms have octets or duets for H.

(d) C_6H_6: Write the correct skeletal structure for the molecule.

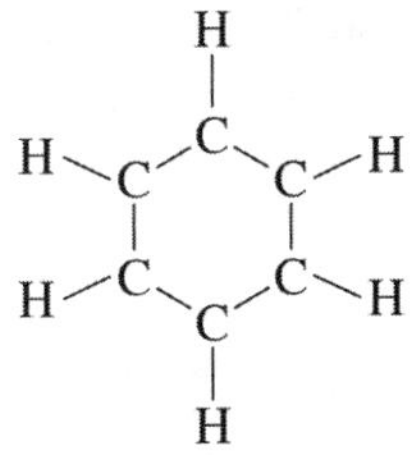

```
        H
        |
  H\    C    /H
     C     C
     |     |
     C     C
  H/    C    \H
        |
        H
```

Calculate the total number of electrons for the Lewis structure by summing the number of valence electrons of each atom in the molecule.

$$6(\text{number of valence e}^- \text{ for C}) + 6(\text{number of valence e}^- \text{ for H}) = 6(4) + 6(1) = 30$$

Distribute the electrons among the atoms, giving octets (or duets for H) to as many atoms as possible.

```
        H
        |
  H\    C    /H
     C  ..  C
     |      |
     C:    :C
  H/    C    \H
        |
        H
```

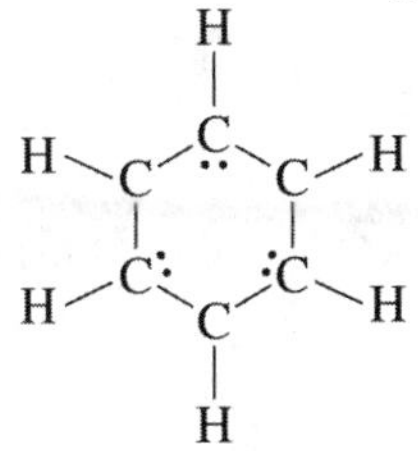

Complete octets by forming double bonds on alternating carbons; draw resonance structures.

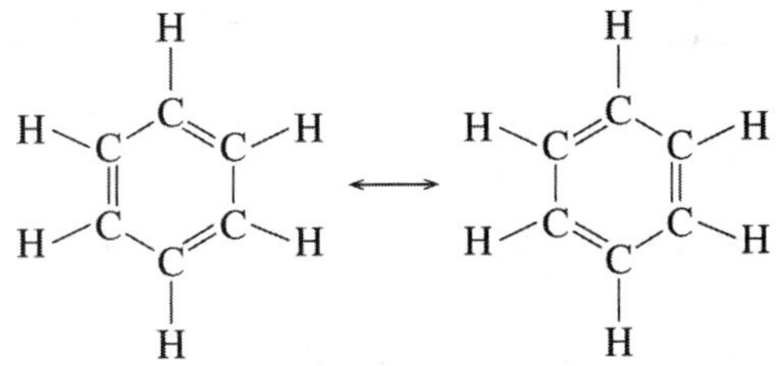

```
        H                      H
        |                      |
  H\    C    /H          H\    C    /H
     C     =C               C=     C
     ||     |     <—>       |      ||
     C      C               C      C
  H/    C=    \H         H/   =C     \H
        |                      |
        H                      H
```

6.78 H_2NCH_2COOH

Write the correct skeletal structure for the molecule.

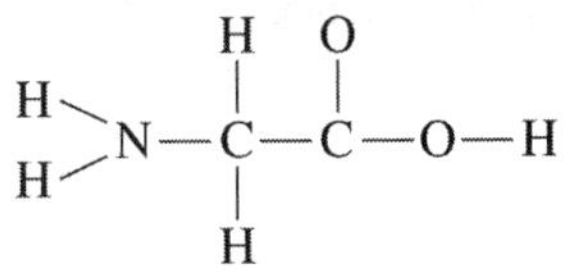

```
       H   O
       |   |
  H\
    N—C—C—O—H
  H/
       |
       H
```

Calculate the total number of electrons for the Lewis structure by summing the number of valence electrons of each atom in the molecule.

$$2(\text{number of valence e}^- \text{ for C}) + 2(\text{number of valence e}^- \text{ for O}) + (\text{number of valence e}^- \text{ for N}) +$$
$$5(\text{number of valence e}^- \text{ for H}) = 2(4) + 2(6) + 5 + 5(1) = 30$$

Distribute the electrons among the atoms, giving octets (or duets for H) to as many atoms as possible. Begin with the bonding electrons; then proceed to lone pairs on terminal atoms and finally to lone pairs on the central atoms.

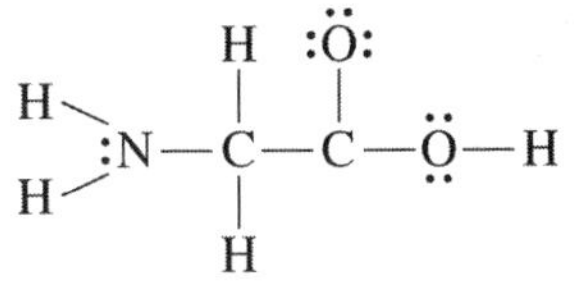

```
        H  :Ö:
        |   |
  H\
    :N—C—C—Ö—H
  H/         ..
        |
        H
```

All 30 electrons are used.

Draw a double bond to satisfy the octet on C.

```
       H  :O:
H\     |   ||   ..
  :N — C — C — O — H
H/     |        ..
       H
```

6.79 **Given:** 26.01% C; 4.38% H; 69.52% O; molar mass = 46.02 g/mol

Find: molecular formula and Lewis structure

Conceptual Plan: Convert mass to mol of each element → pseudoformula → empirical formula → molecular formula → Lewis structure

$\frac{1 \text{ mol C}}{12.01 \text{ g C}}$ $\frac{1 \text{ mol H}}{1.008 \text{ g H}}$ $\frac{1 \text{ mol O}}{16.00 \text{ g O}}$ divide by smallest number

empirical formula × n

Solution:

$$26.01 \text{ g C} \times \frac{1 \text{ mol C}}{12.01 \text{ g C}} = 2.166 \text{ mol C}$$

$$4.38 \text{ g H} \times \frac{1 \text{ mol H}}{1.008 \text{ g H}} = 4.3\underline{4}5 \text{ mol H}$$

$$69.52 \text{ g O} \times \frac{1 \text{ mol O}}{16.00 \text{ g O}} = 4.345 \text{ mol O}$$

$$C_{2.166}H_{4.345}O_{4.345}$$

$$C_{\frac{2.166}{2.166}}H_{\frac{4.345}{2.166}}O_{\frac{4.345}{2.166}} \rightarrow CH_2O_2$$

The correct empirical formula is CH_2O_2.

empirical formula mass = (12.01 g/mol) + 2(1.008 g/mol) + 2(16.00 g/mol) = 46.03 g/mol

$$n = \frac{\text{molar mass}}{\text{formula molar mass}} = \frac{46.02 \text{ g/mol}}{46.03 \text{ g/mol}} = 1$$

molecular formula $= CH_2O_2 \times 1$

$= CH_2O_2$

Write the correct skeletal structure for the molecule.

```
    O
    |
H — C — O — H
```

Calculate the total number of electrons for the Lewis structure by summing the number of valence electrons of each atom in the molecule.

(number of valence e^- for C) + 2(number of valence e^- for O) + 2(number of valence e^- for H)

= 4 + 2(6) + 2(1) = 18

Distribute the electrons among the atoms, giving octets (or duets for H) to as many atoms as possible. Begin with the bonding electrons; then proceed to lone pairs on terminal atoms and finally to lone pairs on the central atoms.

```
   ..
   :O:
    |     ..
H — C — O — H
          ..
```

Complete the octet on C by forming a double bond.

```
   :O:
    ||    ..
H — C — O — H
          ..
```

6.80 **Given:** 28.57% C; 4.80% H; 66.64% N; molar mass = 42.04 g/mol

Find: molecular formula and Lewis structure

Conceptual Plan: Convert mass to mol of each element → pseudoformula → empirical formula → molecular formula → Lewis structure

$\frac{1 \text{ mol C}}{12.01 \text{ g C}}$ $\frac{1 \text{ mol H}}{1.008 \text{ g H}}$ $\frac{1 \text{ mol N}}{14.00 \text{ g N}}$ divide by smallest number

empirical formula × n

Solution:

$$28.57 \text{ g C} \times \frac{1 \text{ mol C}}{12.01 \text{ g C}} = 2.380 \text{ mol C}$$

$$4.80 \text{ g H} \times \frac{1 \text{ mol H}}{1.008 \text{ g H}} = 4.7\underline{6}2 \text{ mol H}$$

$$66.64 \text{ g N} \times \frac{1 \text{ mol N}}{14.00 \text{ g N}} = 4.760 \text{ mol N}$$

$C_{2.380}H_{4.762}N_{4.760}$

$C_{\frac{2.380}{2.380}}H_{\frac{4.762}{2.380}}N_{\frac{4.760}{2.380}} \rightarrow CH_2N_2$

The correct empirical formula is CH_2N_2.

empirical formula mass $= (12.01 \text{ g/mol}) + 2(1.008 \text{ g/mol}) + 2(14.00 \text{ g/mol}) = 42.03 \text{ g/mol}$

$$n = \frac{\text{molar mass}}{\text{formula molar mass}} = \frac{42.04 \text{ g/mol}}{42.03 \text{ g/mol}} = 1$$

molecular formula $= CH_2N_2 \times 1 = CH_2N_2$

Write the correct skeletal structure for the molecule.

```
    H
    |
H—C—N—N
```

Calculate the total number of electrons for the Lewis structure by summing the number of valence electrons of each atom in the molecule.

(number of valence e^- for C) + 2(number of valence e^- for N) + 2(number of valence e^- for H)
$= 4 + 2(5) + 2(1) = 16$

Distribute the electrons among the atoms, giving octets (or duets for H) to as many atoms as possible. Begin with the bonding electrons; then proceed to lone pairs on terminal atoms and finally to lone pairs on the central atoms.

```
    H
    |      ..
H—C—N—N:
       ..  ..
```

Complete the octet on C and N by forming double bonds.

```
    H
    |       ..
H—C=N=N
            ..
```

Calculate the formal charge on each atom in the structure by finding the number of valence electrons and subtracting the number of lone pair electrons and one-half the number of bonding electrons.

	H_{left}	H_{top}	C	N_{left}	N_{right}
number of valence electrons	1	1	4	5	5
− number of lone pair electrons	0	0	0	0	4
− 1/2(number of bonding electrons)	1	1	4	4	2
Formal charge	0	0	0	+1	−1

The diazomethane molecule has nitrogen atoms next to each other with a +1 and a −1 charge. Nitrogen is more electronegative than C, which has a 0 formal charge. The nitrogen with the +1 charge is not a very stable configuration for nitrogen, particularly next to the 0 formal charge C atom.

6.81 HNO_3 Write the correct skeletal structure for the molecule.

```
         O
         |
H—O—N—O
```

Calculate the total number of electrons for the Lewis structure by summing the number of valence electrons of each atom in the molecule.

3(number of valence e^- for O) + (number of valence e^- for N) + (number of valence e^- for H) = 3(6) + 5 + 1 = 24

Distribute the electrons among the atoms, giving octets (or duets for H) to as many atoms as possible. Begin with the bonding electrons; then proceed to lone pairs on terminal atoms and finally to lone pairs on the central atom.

:Ö:
|
H—Ö—N—Ö:

All 24 valence electrons are used.

If any atoms lack an octet, form double or triple bonds as necessary. The double bond can be formed to any of the three oxygen atoms, so there are three resonance forms.

:O: (double bond to N) H—Ö—N—Ö: ⟷ :Ö: (single bond to N) H—Ö—N=Ö ⟷ :Ö: (single bond to N) H—Ö=N—Ö:

I   II   III

All atoms have octets (duets for H); the structure is complete.

To determine which resonance hybrid(s) is most important, calculate the formal charge on each atom in each structure by finding the number of valence electrons and subtracting the number of lone pair electrons and one-half the number of bonding electrons.

	Structure I					Structure II				
	O_{left}	O_{top}	O_{right}	N	H	O_{left}	O_{top}	O_{right}	N	H
number of valence electrons	6	6	6	5	1	6	6	6	5	1
− number of lone pair electrons	4	4	6	0	0	4	6	4	0	0
− 1/2(number of bonding electrons)	2	2	1	4	1	2	1	2	4	1
Formal charge	0	0	−1	+1	0	0	−1	0	+1	0

	Structure III				
	O_{left}	O_{top}	O_{right}	N	H
number of valence electrons	6	6	6	5	1
− number of lone pair electrons	2	6	6	0	0
− 1/2(number of bonding electrons)	3	1	1	4	1
Formal charge	+1	−1	−1	+1	0

The sum of the formal charges is 0 for each structure, which is the overall charge of the molecule. However, in structures I and II, the individual formal charges are lower. These two forms would contribute equally to the structure of HNO_3. Structure III would be less important because the individual formal charges are higher. In addition, oxygen is more electronegative than N, and in this structure N has the positive formal charge. So structure III will be unacceptable.

6.82 Cl_2CO Write the correct skeletal structure for the molecule.

O
|
Cl—C—Cl

Calculate the total number of electrons for the Lewis structure by summing the number of valence electrons of each atom in the molecule.

(number of valence e^- for O) + (number of valence e^- for C) + 2(number of valence e^- for Cl) = 6 + 4 + 2(7) = 24

Distribute the electrons among the atoms, giving octets to as many atoms as possible. Begin with the bonding electrons; then proceed to lone pairs on terminal atoms and finally to lone pairs on the central atom.

:Ö:
|
:C̈l—C—C̈l:

All 24 valence electrons are used.

If any atoms lack an octet, form double or triple bonds as necessary. The double bond can be formed to any of the three terminal atoms, so there are three resonance forms.

:O: ‖ :C̤l̤—C—C̤l̤: (I)   :Ö: | C̤l̤=C—C̤l̤: (II)   :Ö: | :C̤l̤—C=C̤l̤ (III)

All atoms have octets; the structures are complete.

To determine which resonance hybrid(s) is most important, calculate the formal charge on each atom in each structure by finding the number of valence electrons and subtracting the number of lone pair electrons and one-half the number of bonding electrons.

	Structure I				Structure II			
	Cl_{left}	Cl_{right}	O	C	Cl_{left}	Cl_{right}	O	C
number of valence electrons	7	7	6	4	7	7	6	4
− number of lone pair electrons	6	6	4	0	4	6	6	0
− 1/2(number of bonding electrons)	1	1	2	4	2	1	1	4
Formal charge	0	0	0	0	+1	0	−1	0

	Structure III			
	Cl_{left}	Cl_{right}	O	C
number of valence electrons	7	7	6	4
− number of lone pair electrons	6	4	6	0
− 1/2(number of bonding electrons)	1	2	1	4
Formal charge	0	+1	−1	0

The sum of the formal charges is 0 for each structure, which is the overall charge of the molecule. However, in structure I, the individual formal charges are lower, and this form would be more important to Cl_2CO than to structure II and structure III.

6.83 CNO^- Write the skeletal structure.

C—N—O

Calculate the total number of electrons for the Lewis structure by summing the number of valence electrons of each atom in the molecule.

(valence e^- from C) + (valence e^- from N) + (valence e^- from O) + 1 (from the negative charge) = 4 + 5 + 6 + 1 = 16

:C̤̈—N—Ö̤:

Distribute the electrons to complete octets if possible.

$[$C̤̈=N=Ö̤$]^-$ ⟷ $[$:C≡N—Ö̤:$]^-$ ⟷ $[$:C̤̈—N≡O:$]^-$

I   II   III

Determine the formal charge on each atom for each structure.

	Structure I			Structure II		
	C	N	O	C	N	O
number of valence electrons	4	5	6	4	5	6
− number of lone pair electrons	4	0	4	2	0	6
− 1/2(number of bonding electrons)	2	4	2	3	4	1
Formal charge	−2	+1	0	−1	+1	−1

	Structure III		
	C	N	O
number of valence electrons	4	5	6
− number of lone pair electrons	6	0	2
− 1/2(number of bonding electrons)	1	4	3
Formal charge	−3	+1	+1

Structures I, II, and III all follow the octet rule but have varying degrees of negative formal charge on carbon, which is the least electronegative atom. Also, the amount of formal charge is very high in all three resonance forms. Although structure II is the best of the resonance forms, it has a −1 charge on the least electronegative atom, C, and a +1 charge on the more electronegative atom, N. Therefore, none of these resonance forms contributes strongly to the stability of the fulminate ion and the ion is not very stable.

6.84 (a) C_3H_8: Write the correct skeletal structure for the molecule.

```
    H   H   H
    |   |   |
H — C — C — C — H
    |   |   |
    H   H   H
```

Calculate the total number of electrons for the Lewis structure by summing the valence electrons of each atom in the molecule.

$$3(\text{number of valence } e^- \text{ for C}) + 8(\text{number of valence } e^- \text{ for H}) = 3(4) + 8(1) = 20$$

Distribute the electrons among the atoms, giving octets (or duets for H) to as many atoms as possible. Begin with the bonding electrons; then proceed to lone pairs on terminal atoms and finally to lone pairs on the central atom.

```
    H   H   H
    |   |   |
H — C — C — C — H
    |   |   |
    H   H   H
```

All 20 valence electrons are used.

(b) CH_3OCH_3: Write the correct skeletal structure for the molecule.

```
    H       H
    |       |
H — C — O — C — H
    |       |
    H       H
```

Calculate the total number of electrons for the Lewis structure by summing the valence electrons of each atom in the molecule.

$$2(\text{number of valence } e^- \text{ for C}) + 6(\text{number of valence } e^- \text{ for H}) + (\text{number of valence } e^- \text{ for O}) = 2(4) + 6(1) + 6 = 20$$

Distribute the electrons among the atoms, giving octets (or duets for H) to as many atoms as possible. Begin with the bonding electrons; then proceed to lone pairs on terminal atoms and finally to lone pairs on the central atom.

```
    H   ..   H
    |        |
H — C — O — C — H
    |   ..   |
    H        H
```

All 20 valence electrons are used.

(c) CH_3COCH_3: Write the correct skeletal structure for the molecule.

```
    H   O   H
    |   |   |
H — C — C — C — H
    |       |
    H       H
```

Calculate the total number of electrons for the Lewis structure by summing the valence electrons of each atom in the molecule.

$$3(\text{number of valence } e^- \text{ for C}) + 6(\text{number of valence } e^- \text{ for H}) + (\text{number of valence } e^- \text{ for O}) = 3(4) + 6(1) + 6 = 24$$

Distribute the electrons among the atoms, giving octets (or duets for H) to as many atoms as possible. Begin with the bonding electrons; then proceed to lone pairs on terminal atoms and finally to lone pairs on the central atom.

```
     H  :Ö:  H
     |   |   |
 H — C — C — C — H
     |       |
     H       H
```

All 24 valence electrons are used.

If any atom lacks an octet, form double or triple bonds to give them octets.

```
     H  :O:  H
     |   ||  |
 H — C — C — C — H
     |       |
     H       H
```

All atoms have octets (duets for hydrogen); the structure is complete.

(d) CH_3COOH: Write the correct skeletal structure for the molecule.

```
     H   O
     |   |
 H — C — C — O — H
     |
     H
```

Calculate the total number of electrons for the Lewis structure by summing the valence electrons of each atom in the molecule.

2(number of valence e^- for C) + 4(number of valence e^- for H) + 2(number of valence e^- for O) $= 2(4) + 4(1) + 2(6) = 24$

Distribute the electrons among the atoms, giving octets (or duets for H) to as many atoms as possible. Begin with the bonding electrons; then proceed to lone pairs on terminal atoms and finally to lone pairs on the central atom.

```
     H  :Ö:
     |   |    ..
 H — C — C — O — H
     |        ..
     H
```

All 24 valence electrons are used.

If any atom lacks an octet, form double or triple bonds to give them octets.

```
     H  :O:
     |   ||   ..
 H — C — C — O — H
     |        ..
     H
```

All atoms have octets (duets for hydrogen); the structure is complete.

(e) CH_3CHO: Write the correct skeletal structure for the molecule.

```
     H   O
     |   |
 H — C — C — H
     |
     H
```

Calculate the total number of electrons for the Lewis structure by summing the valence electrons of each atom in the molecule.

2(number of valence e^- for C) + 4(number of valence e^- for H) + (number of valence e^- for O) $= 2(4) + 4(1) + 6 = 18$

Distribute the electrons among the atoms, giving octets (or duets for H) to as many atoms as possible. Begin with the bonding electrons; then proceed to lone pairs on terminal atoms and finally to lone pairs on the central atom.

```
     H  :Ö:
     |   |
 H — C — C — H
     |
     H
```

All 18 valence electrons are used.
If any atom lacks an octet, form double or triple bonds to give them octets.

```
    H  :O:
    |   ||
H — C — C — H
    |
    H
```

All atoms have octets (duets for hydrogen); the structure is complete.

6.85 (a) C_2H_4: Write the correct skeletal structure for the molecule.

```
H — C — C — H
    |   |
    H   H
```

Calculate the total number of electrons for the Lewis structure by summing the valence electrons of each atom in the molecule.

$$2(\text{number of valence } e^- \text{ for C}) + 4(\text{number of valence } e^- \text{ for H}) = 2(4) + 4(1) = 12$$

Distribute the electrons among the atoms, giving octets (or duets for H) to as many atoms as possible. Begin with the bonding electrons; then proceed to lone pairs on terminal atoms and finally to lone pairs on the central atom.

```
    ..
H — C — C — H
    |   |
    H   H
```

All 12 valence electrons are used.
If any atom lacks an octet, form double or triple bonds to give them octets.

```
H — C = C — H
    |   |
    H   H
```

All atoms have octets (duets for hydrogen); the structure is complete.

(b) CH_3NH_2: Write the correct skeletal structure for the molecule.

```
    H
    |
H — C — N — H
    |   |
    H   H
```

Calculate the total number of electrons for the Lewis structure by summing the valence electrons of each atom in the molecule.

$$(\text{number of valence } e^- \text{ for C}) + 5(\text{number of valence } e^- \text{ for H}) + (\text{number of valence } e^- \text{ for N}) = 4 + 5(1) + 5 = 14$$

Distribute the electrons among the atoms, giving octets (or duets for H) to as many atoms as possible. Begin with the bonding electrons; then proceed to lone pairs on terminal atoms and finally to lone pairs on the central atom.

```
    H
    |   ..
H — C — N — H
    |   |
    H   H
```

All 14 valence electrons are used.
All atoms have octets (duets for hydrogen); the structure is complete.

(c) HCHO: Write the correct skeletal structure for the molecule.

```
    O
    |
H — C — H
```

Calculate the total number of electrons for the Lewis structure by summing the valence electrons of each atom in the molecule.

$$(\text{number of valence } e^- \text{ for C}) + 2(\text{number of valence } e^- \text{ for H}) + (\text{number of valence } e^- \text{ for O}) = 4 + 2(1) + 6 = 12$$

Distribute the electrons among the atoms, giving octets (or duets for H) to as many atoms as possible. Begin with the bonding electrons; then proceed to lone pairs on terminal atoms and finally to lone pairs on the central atom.

```
   ..
  :O:
   |
H—C—H
```

All 12 valence electrons are used.

If any atom lacks an octet, form double or triple bonds to give them octets.

```
  :O:
   ||
H—C—H
```

All atoms have octets (duets for hydrogen); the structure is complete.

(d) CH_3CH_2OH: Write the correct skeletal structure for the molecule.

```
   H  H
   |  |
H—C—C—O—H
   |  |
   H  H
```

Calculate the total number of electrons for the Lewis structure by summing the valence electrons of each atom in the molecule.

2(number of valence e^- for C) + 6(number of valence e^- for H) + (number of valance e^- for O) $= 2(4) + 6(1) + 6 = 20$

Distribute the electrons among the atoms, giving octets (or duets for H) to as many atoms as possible. Begin with the bonding electrons; then proceed to lone pairs on terminal atoms and finally to lone pairs on the central atom.

```
   H  H
   |  |   ..
H—C—C—O—H
   |  |   ..
   H  H
```

All 20 valence electrons are used.

All atoms have octets (or duets for H); the structure is complete.

(e) HCOOH: Write the correct skeletal structure for the molecule.

```
   O
   |
H—C—O—H
```

Calculate the total number of electrons for the Lewis structure by summing the valence electrons of each atom in the molecule.

(number of valence e^- for C) + 2(number of valence e^- for H) + 2(number of valance e^- for O) $= 4 + 2(1) + 2(6) = 18$

Distribute the electrons among the atoms, giving octets (or duets for H) to as many atoms as possible. Begin with the bonding electrons; then proceed to lone pairs on terminal atoms and finally to lone pairs on the central atom.

```
   ..
  :O:
   |   ..
H—C—O—H
       ..
```

All 18 valence electrons are used.

If any atom lacks an octet, form double or triple bonds to give them octets.

```
  :O:
   ||  ..
H—C—O—H
       ..
```

All atoms have octets (or duets for H); the structure is complete.

6.86 The Lewis structures for the three ions will be similar. So we can write the Lewis structure for one ion and use it to determine the other two.

Br_3^-: Write the correct skeletal structure for the ion.

Br—Br—Br

Calculate the total number of electrons for the Lewis structure by summing the number of valence electrons of each atom in the ion and adding 1 for the 1− charge.

3(number of valence e$^-$ for Br) + 1 = 3(7) + 1 = 22

Distribute the electrons among the atoms, giving octets to as many atoms as possible. Begin with the bonding electrons; then proceed to lone pairs on terminal atoms and finally to lone pairs on the central atom. Assign electrons above 8 to the central atom.

:Br—Br—Br:

Finally, place the ion in brackets and place the charge in the upper right-hand corner.

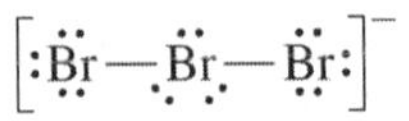

[:Br—Br—Br:]$^-$

Because Br, I, and F are all in the same family, each of these ions would have 22 electrons and should have the same Lewis structure.

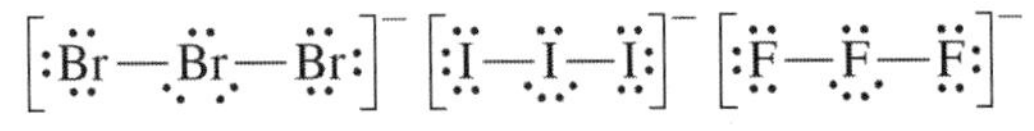

[:Br—Br—Br:]$^-$ [:I—I—I:]$^-$ [:F—F—F:]$^-$

All three ions are written with five electron groups around the central atom. Bromine and iodine can accommodate 10 electrons around the central atom; fluorine cannot. Fluorine is in period 2 and can accommodate, at most, eight electrons around the central atom because there are no orbitals low enough in energy to hybridize with the 2*s* and 2*p* orbitals. Therefore, F_3^- does not exist.

6.87 $HCSNH_2$: Write the correct skeletal structure for the molecule.

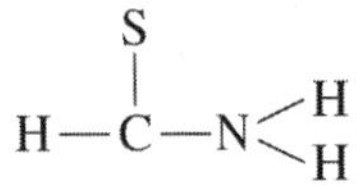

H—C(—S)—N(H)H

Calculate the total number of electrons for the Lewis structure by summing the number of valence electrons of each atom in the molecule.

(number of valence e$^-$ for N) + (number of valence e$^-$ for S) + (number of valence e$^-$ for C) + 3(number of valence e$^-$ for H) = 5 + 6 + 4 + 3(1) = 18

Distribute the electrons among the atoms, giving octets (or duets for H) to as many atoms as possible. Begin with the bonding electrons; then proceed to lone pairs on terminal atoms and finally to lone pairs on the central atoms.

H—C(—:S:)—N:(H)H

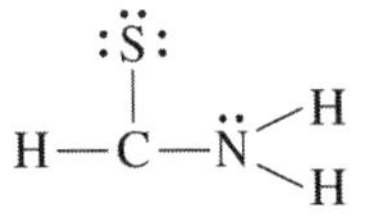

Complete the octet on C by forming a double bond.

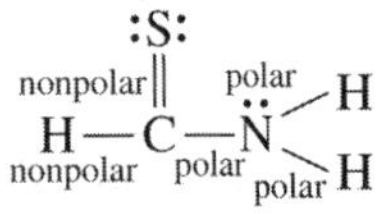

H—C(=:S:)—N:(H)H

6.88 H_2NCONH_2: Write the correct skeletal structure for the molecule.

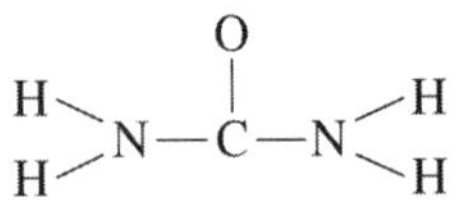

H(H)N—C(—O)—N(H)H

Calculate the total number of electrons for the Lewis structure by summing the number of valence electrons of each atom in the molecule.

2(number of valence e$^-$ for N) + (number of valence e$^-$ for O) + 4(number of valence e$^-$ for H) + (number of valence e$^-$ for C)
= 2(5) + 6 + 4(1) + 4 = 24

Distribute the electrons among the atoms, giving octets (or duets for H) to as many atoms as possible. Begin with the bonding electrons; then proceed to lone pairs on terminal atoms and finally to lone pairs on the central atoms.

```
           ..
          :O:
           |
 H\ ..     |    .. /H
   >N — C — N<
 H/              \H
```

Complete the octet on C with a double bond.

```
            :O:
  H polar    ||polar  polar H
   \  ..     ||      ..   /
    N ——  C  —— N
   / polar  polar  polar \
  H polar              polar H
```

The C—O bond would be most polar because it has the greatest difference in electronegativity. From Figure 6.3, Δ EN C—O = 1.0 Δ EN C—N = 0.5 Δ EN N—H = 0.9

6.89 (a) O_2^-: Write the correct skeletal structure for the radical.

O—O

Calculate the total number of electrons for the Lewis structure by summing the number of valence electrons of each atom in the radical and adding 1 for the 1− charge.

2(number of valence e^- for O) + 1 = 2(6) + 1 = 13

Distribute the electrons among the atoms, giving octets to as many atoms as possible. Begin with the bonding electrons; then proceed to lone pairs on terminal atoms and finally to lone pairs on the central atom.

$[\cdot\ddot{\underset{..}{O}}—\ddot{\underset{..}{O}}:]^-$

All 13 valence electrons are used.

O has an incomplete octet. It has seven electrons because we have an odd number of valence electrons.

(b) O^-: Write the Lewis structure based on the valence electrons $2s^22p^5$.

$[\cdot\ddot{\underset{..}{O}}:]^-$

(c) OH: Write the correct skeletal structure for the molecule.

H—O

Calculate the total number of electrons for the Lewis structure by summing the number of valence electrons of each atom in the molecule.

(number of valence e^- for O) + (number of valence e^- for H) = 6 + 1 = 7

Distribute the electrons among the atoms, giving octets (or duets for H) to as many atoms as possible. Begin with the bonding electrons; then proceed to lone pairs on terminal atoms and finally to lone pairs on the central atom.

$H—\ddot{\underset{..}{O}}\cdot$

All 7 valence electrons are used.

O has an incomplete octet. It has seven electrons because we have an odd number of valence electrons.

(d) CH_3OO: Write the correct skeletal structure for the radical.

C is the least electronegative atom, so it is central.

```
    H
    |
H — C — O — O
    |
    H
```

Calculate the total number of electrons for the Lewis structure by summing the number of valence electrons of each atom in the molecule.

3(number of valence e^- for H) + (number of valence e^- for C) + 2(number of valence e^- for O) = 3(1) + 4 + 2(6) = 19

Distribute the electrons among the atoms, giving octets (or duets for H) to as many atoms as possible. Begin with the bonding electrons; then proceed to lone pairs on terminal atoms and finally to lone pairs on the central atoms.

```
      H
      |
  H—C—Ö—Ö·
      |  ¨  ¨
      H
```

All 19 valence electrons are used.

O has an incomplete octet. It has seven electrons because we have an odd number of valence electrons.

6.90 $:\ddot{O}=\dot{N}-\ddot{O}: \rightarrow \dot{N}=\ddot{O}: + \cdot\ddot{O}:$

$\cdot\ddot{O}: + \ddot{O}=\ddot{O} \rightarrow \ddot{O}=\ddot{O}-\ddot{O}:$ Where NO_2, O, and NO are the free radicals.

6.91 **Given:** 7.743% H **Find:** Lewis structure

Conceptual Plan: **%H → %C → mass C, H → mol C, H → pseudoformula → empirical formula**

100% − %H; Assume 100 g sample; $\frac{1 \text{ mol C}}{12.01 \text{ g C}}$ $\frac{1 \text{ mol H}}{1.008 \text{ g H}}$; divide by smallest number

Solution: %C = 100% − 7.743% = 92.568% C

In a 100.00 g sample: 7.743 g H, 92.568 g C

$$7.743 \text{ g H} \times \frac{1 \text{ mol H}}{1.008 \text{ g H}} = 7.682 \text{ mol H}$$

$$92.568 \text{ g C} \times \frac{1 \text{ mol C}}{12.01 \text{ g C}} = 7.708 \text{ mol C}$$

$C_{7.708}H_{7.682}$

$C_{\frac{7.708}{7.682}}H_{\frac{7.682}{7.682}} \rightarrow CH$

The smallest molecular formula would be C_2H_2.

Write the correct skeletal structure for the molecule.

H—C—C—H

Calculate the total number of electrons for the Lewis structure by summing the valence electrons of each atom in the molecule.

$$2(\text{number of valence } e^- \text{ for C}) + 2(\text{number of valence } e^- \text{ for H}) = 2(4) + 2(1) = 10$$

Distribute the electrons among the atoms, giving octets (or duets for H) to as many atoms as possible. Begin with the bonding electrons; then proceed to lone pairs on terminal atoms and finally to lone pairs on the central atom.

$H-\underset{..}{C}-\underset{..}{C}-H$

Complete the octet on C by forming a triple bond.

H—C≡C—H

6.92 **Given:** 85.5% Cl **Find:** Lewis structure

Conceptual Plan: **%Cl → %C → mass Cl, C → mol Cl, C → pseudoformula → empirical formula**

100% − %Cl; Assume 100 g sample; $\frac{1 \text{ mol Cl}}{35.45 \text{ g Cl}}$ $\frac{1 \text{ mol C}}{12.01 \text{ g C}}$; divide by smallest number

Solution: %C = 100% − 85.5% = 14.5% C

In a 100.00 g sample: 85.5 g Cl, 14.5 g C

$$85.5 \text{ g Cl} \times \frac{1 \text{ mol Cl}}{35.45 \text{ g Cl}} = 2.41 \text{ mol Cl}$$

$$14.5 \text{ g C} \times \frac{1 \text{ mol C}}{12.01 \text{ g C}} = 1.21 \text{ mol C}$$

$C_{1.21}Cl_{2.41}$

$C_{\frac{1.21}{1.21}}Cl_{\frac{2.41}{1.21}} \rightarrow CCl_2$

The smallest molecular formula would be C_2Cl_4.

Write the correct skeletal structure for the molecule.

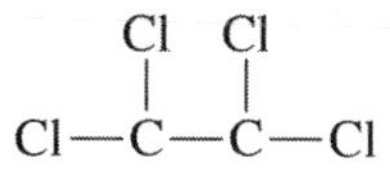

```
   Cl  Cl
   |   |
Cl—C—C—Cl
```

Calculate the total number of electrons for the Lewis structure by summing the valence electrons of each atom in the molecule.

$$2(\text{number of valence } e^- \text{ for C}) + 4(\text{number of valence } e^- \text{ for Cl}) = 2(4) + 4(7) = 36$$

Distribute the electrons among the atoms, giving octets to as many atoms as possible. Begin with the bonding electrons; then proceed to lone pairs on terminal atoms and finally to lone pairs on the central atoms.

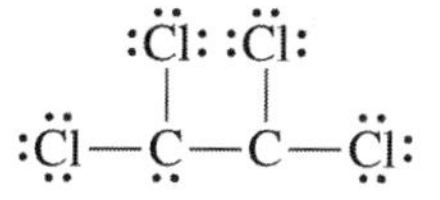

```
     :Cl: :Cl:
       |    |
:Cl—C—C—Cl:
```

Complete the octet on C by forming a double bond.

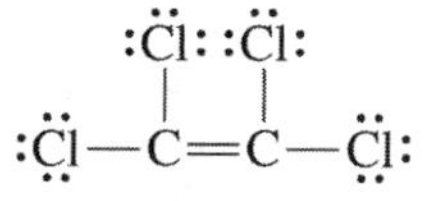

```
     :Cl: :Cl:
       |    |
:Cl—C=C—Cl:
```

6.93 (a) serine

```
     H   H  :O:
     |   |   ||
H—N—C₁—C₂—O₁—H
     ..   |        ..
       H—C₃—H
          |
         :O₂—H
```

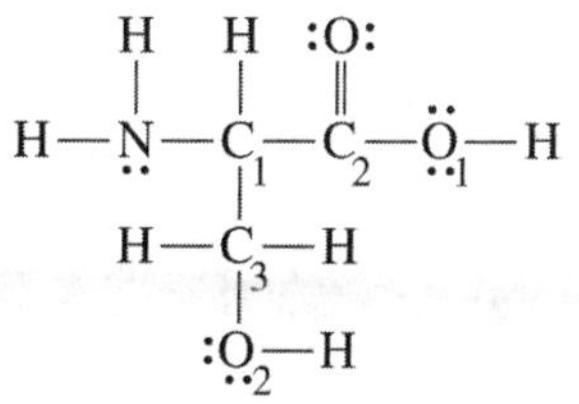

The Cs in positions 1 and 3 have four electron groups around the atom. Four electron groups give a tetrahedral electron geometry; tetrahedral electron geometry has sp^3 hybridization. Four bonding groups and zero lone pairs give a tetrahedral molecular geometry.

C in position 2 has three electron groups around the atom. Three electron groups give a trigonal planar geometry; trigonal planar geometry has sp^2 hybridization. Three bonding groups and zero lone pairs give a trigonal planar molecular geometry.

N has four electron groups around the atom. Four electron groups give a tetrahedral electron geometry; tetrahedral electron geometry has sp^3 hybridization. Three bonding groups and one lone pair give a trigonal pyramidal molecular geometry.

The Os in positions 1 and 2 each have four electron groups around the atom. Four electron groups give a tetrahedral electron geometry; tetrahedral electron geometry has sp^3 hybridization. Two bonding groups and two lone pairs give a bent molecular geometry.

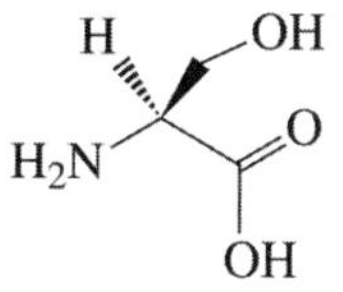

```
   H      OH
      \  /
H2N—C      O
         \ //
          C
          |
          OH
```

(b) asparagine

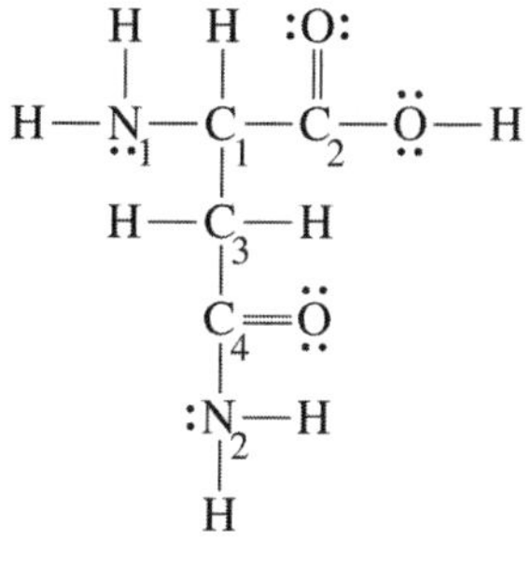

```
     H   H  :O:
     |   |   ||
H—N₁—C₁—C₂—O—H
     ..   |        ..
       H—C₃—H
          |
          C₄=O:
          |
         :N₂—H
          |
          H
```

The Cs in positions 1 and 3 each have four electron groups around the atom. Four electron groups give a tetrahedral electron geometry; tetrahedral electron geometry has sp^3 hybridization. Four bonding groups and zero lone pairs give a tetrahedral molecular geometry.
The Cs in positions 2 and 4 each have three electron groups around the atom. Three electron groups give a trigonal planar geometry; trigonal planar geometry has sp^2 hybridization. Three bonding groups and zero lone groups give a trigonal planar molecular geometry.
The Ns in positions 1 and 2 each have four electron groups around the atom. Four electron groups give a tetrahedral electron geometry; tetrahedral electron geometry has sp^3 hybridization. Three bonding groups and one lone pair give a trigonal pyramidal molecular geometry.
O has four electron groups around the atom. Four electron groups give a tetrahedral electron geometry; tetrahedral electron geometry has sp^3 hybridization. Two bonding groups and two lone pairs give a bent molecular geometry.

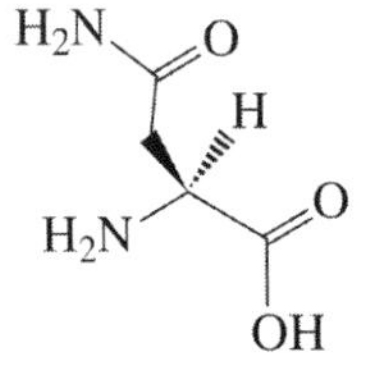

```
H2N   O
   H
H2N      O
   OH
```

(c) cysteine

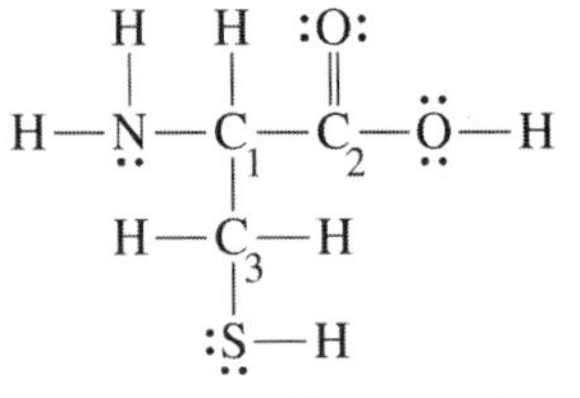

```
     H   H  :O:
     |   |   ||   ..
H — N — C1 — C2 — O — H
    ..   |        ..
    H — C3 — H
         |
        :S — H
         ..
```

The Cs in positions 1 and 3 each have four electron groups around the atom. Four electron groups give a tetrahedral electron geometry; tetrahedral electron geometry has sp^3 hybridization. Four bonding groups and zero lone pairs give a tetrahedral molecular geometry.
C in position 2 has three electron groups around the atom. Three electron groups give a trigonal planar geometry; trigonal planar geometry has sp^2 hybridization. Three bonding groups and zero lone pairs give a trigonal planar molecular geometry.

N has four electron groups around the atom. Four electron groups give a tetrahedral electron geometry; tetrahedral electron geometry has sp^3 hybridization. Three bonding groups and one lone pair give a trigonal pyramidal molecular geometry.
O and S have four electron groups around the atom. Four electron groups give a tetrahedral electron geometry; tetrahedral electron geometry has sp^3 hybridization. Two bonding groups and two lone pairs give bent molecular geometry.

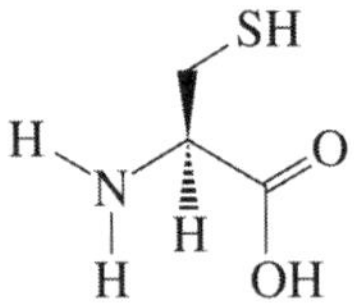

```
      SH
H
 N        O
 |  H
 H    OH
```

6.94 (a) cytosine

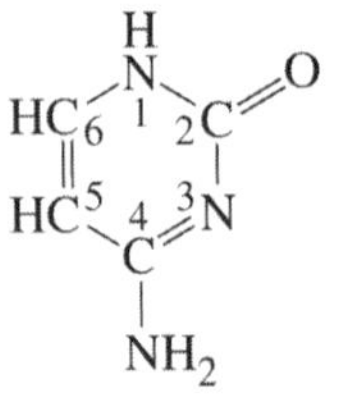

```
     H
     N
HC6  1  2C = O
 ||      |
HC5  4  3N
     C
     |
    NH2
```

N in position 1 has three bonding groups of electrons and one lone pair; four electron groups give a tetrahedral electron geometry and sp^3 hybridization. Three bonding groups of electrons and one lone pair give a trigonal pyramidal molecular geometry.

C in position 2 has three bonding groups of electrons and zero lone pairs; three electron groups give a trigonal planar geometry and sp^2 hybridization. Three bonding groups of electrons give a trigonal planar molecular geometry.
N in position 3 has two bonding groups of electrons and one lone pair; three electron groups give a trigonal planar electron geometry and sp^2 hybridization. Two bonding groups and one lone pair give a bent molecular geometry.
C in position 4 has three bonding groups of electrons and zero lone pairs; three electron groups give a trigonal planar geometry and sp^2 hybridization. Three bonding groups of electrons give a trigonal planar molecular geometry.
C in positions 5 and 6 have three bonding groups of electrons and zero lone pairs; three electron groups give a trigonal planar geometry and sp^2 hybridization. Three bonding groups of electrons give a trigonal planar molecular geometry.
N outside the ring has three bonding groups of electrons and one lone pair; four electron groups give a tetrahedral electron geometry and sp^3 hybridization. Three bonding groups of electrons give a trigonal pyramidal molecular geometry.

(b) adenine

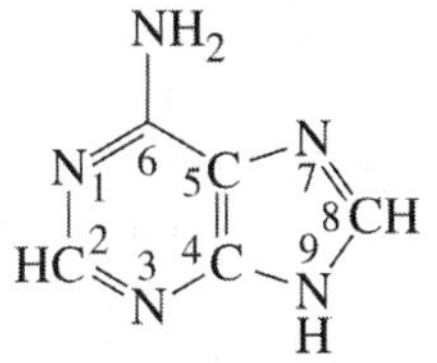

N in positions 1, 3, and 7 have two bonding groups of electrons and one lone pair; three electron groups give a trigonal planar electron geometry and sp^2 hybridization. Two bonding groups and one lone pair give a bent molecular geometry.
C in positions 2, 4, 5, 6, and 8 have three bonding groups of electrons and zero lone groups; three electron groups give a trigonal planar geometry and sp^2 hybridization. Three bonding groups of electrons give a trigonal planar molecular geometry.
N in position 9 and the N outside the ring have three bonding groups of electrons and one lone pair; four electron groups give a tetrahedral electron geometry and sp^3 hybridization. Three bonding groups of electrons give a trigonal pyramidal molecular geometry.

(c) thymine

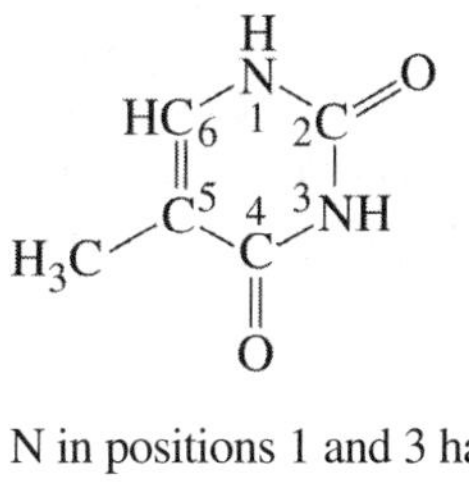

N in positions 1 and 3 have three bonding groups of electrons and one lone pair; four electron groups give a tetrahedral electron geometry and sp^3 hybridization. Three bonding groups of electrons give a trigonal pyramidal molecular geometry.
C in position 2 has three bonding groups of electrons and zero lone pairs; three electron groups give a trigonal planar geometry and sp^2 hybridization. Three bonding groups of electrons give a trigonal planar molecular geometry.
C in positions 4, 5, and 6 have three bonding groups of electrons and zero lone pairs; three electron groups give a trigonal planar geometry and sp^2 hybridization. Three bonding groups of electrons give a trigonal planar molecular geometry.
C outside the ring has four bonding groups of electrons and zero lone pairs; four electron groups give a tetrahedral electron geometry and sp^3 hybridization. Four bonding groups of electrons give a tetrahedral molecular geometry.

(d) guanine

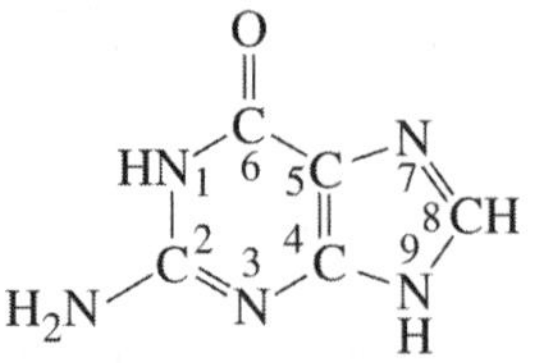

O
‖
C
HN_1 6 $_5C$ N 7 8 CH
2 3 4 9
H_2N C N C N
H

N in positions 1 and 9 and the N outside the ring have three bonding groups of electrons and one lone pair; four electron groups give a tetrahedral electron geometry and sp^3 hybridization. Three bonding groups of electrons give a trigonal pyramidal molecular geometry.

C in positions 2, 4, 5, 6, and 8 have three bonding groups of electrons and zero lone pairs; three electron groups give a trigonal planar geometry and sp^2 hybridization. Three bonding groups of electrons give a trigonal planar molecular geometry.

N in positions 3 and 7 have two bonding groups of electrons and one lone pair; three electron groups give a trigonal planar electron geometry and sp^2 hybridization. Two bonding groups and one lone pair give a bent molecular geometry.

6.95 (a) Water-soluble: The four C—OH bonds, the C═O bond, and the C—O bonds in the ring make the molecule polar. Because of the large electronegativity difference between the C and O, each of the bonds will have a dipole moment. The sum of the dipole moments does *not* give a net zero dipole moment, so the molecule is polar. Because it is polar, it will be water-soluble.

(b) Fat-soluble: There is only one C—O bond in the molecule. The dipole moment from this bond is not enough to make the molecule polar because of all of the nonpolar components of the molecule. The C—H bonds in the structure lead to a net dipole of zero for most of the sites in the molecule. Because the molecule is nonpolar, it is fat-soluble.

(c) Water-soluble: The carboxylic acid function (COOH group) along with the N atom in the ring make the molecule polar. Because of the electronegativity difference between the C and O and the C and N atoms, the bonds will have a dipole moment and the net dipole moment of the molecule is *not* zero; so the molecule is polar. Because the molecule is polar, it is water-soluble.

(d) Fat-soluble: The two O atoms in the structure contribute a very small amount to the net dipole moment of this molecule. The majority of the molecule is nonpolar because there is no net dipole moment around the interior C atoms. Because the molecule is nonpolar, it is fat-soluble.

6.96 The soap molecule has a nonpolar hydrocarbon end and an anionic end when it is dissolved in water. When it dissolves in water, the sodium stearate congregates to form small spheres (called micelles) with the nonpolar ends on the insides and the anionic ends on the surface. The anionic end interacts with the polar water molecules, while the nonpolar hydrocarbon end can attract and interact with the nonpolar grease. This allows the soapy water to remove the grease by trapping the grease inside the micelle.

Challenge Problems

6.97 N_3^-: Write the correct skeletal structure for the ion.

$$N—N—N$$

Calculate the total number of electrons for the Lewis structure by summing the valence electrons of each atom in the ion and adding 1 for the 1− charge.

$$3(\text{number of valence e}^- \text{ for N}) + 1 = 3(5) + 1 = 16$$

Distribute the electrons among the atoms, giving octets to as many atoms as possible. Begin with the bonding electrons; then proceed to lone pairs on terminal atoms and finally to lone pairs on the central atom.

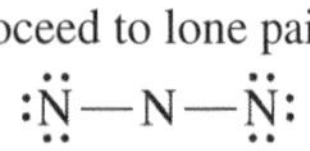

$$:\ddot{\underset{..}{N}}—N—\ddot{\underset{..}{N}}:$$

Complete octets with double or triple bonds.

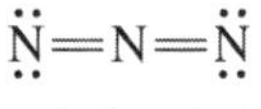

$$\ddot{\underset{..}{N}}=N=\ddot{\underset{..}{N}}$$

Finally, write the ion in brackets with the charge in the upper right-hand corner.

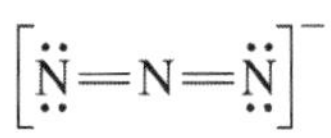

$$\left[\ddot{\underset{..}{N}}=N=\ddot{\underset{..}{N}}\right]^-$$

Write the resonance forms.

$$\left[\ddot{\underset{..}{N}}=N=\ddot{\underset{..}{N}}\right]^- \longleftrightarrow \left[:N\equiv N-\ddot{\underset{..}{N}}:\right]^- \longleftrightarrow \left[:\ddot{\underset{..}{N}}-N\equiv N:\right]^-$$

6.98 **Given:** 0.167 g acid; 27.8 mL 0.100 M NaOH; 40.00% C; 6.71% H; 53.29% O

Find: molar mass, molecular formula, Lewis structure

Conceptual Plan: mL → L → mol NaOH → mol acid → molar mass and then:

$$\frac{L}{1000\text{ mL}} \qquad \text{mol} = VM \qquad \text{mol acid} = \text{mol base} \qquad \frac{\text{mass}}{\text{mol}}$$

convert mass to mol of each element → pseudoformula → empirical formula → molecular formula

$$\frac{1\text{ mol C}}{12.01\text{ g C}}\ \frac{1\text{ mol H}}{1.008\text{ g H}}\ \frac{1\text{ mol O}}{16.00\text{ g O}} \qquad \text{divide by smallest number} \qquad \text{empirical formula} \times n$$

→ Lewis structure

Solution: $27.8\text{ mL} \times \frac{1\text{ L}}{1000\text{ mL}} \times \frac{0.100\text{ mol NaOH}}{\text{L}} \times \frac{1\text{ mol acid}}{1\text{ mol NaOH}} = 0.00278\text{ mol monoprotic acid}$

$$\frac{0.167\text{ g acid}}{0.00278\text{ mol acid}} = 60.1\text{ g/mol}$$

$$40.00\text{ g C} \times \frac{1\text{ mol C}}{12.01\text{ g C}} = 3.331\text{ mol C}$$

$$6.71\text{ g H} \times \frac{1\text{ mol H}}{1.008\text{ g H}} = 6.6\underline{5}7\text{ mol H}$$

$$53.29\text{ g O} \times \frac{1\text{ mol O}}{16.00\text{ g O}} = 3.331\text{ mol O}$$

$C_{3.331}H_{6.657}O_{3.331}$

$C_{\frac{3.331}{3.331}}H_{\frac{6.657}{3.331}}O_{\frac{3.331}{3.331}} \rightarrow CH_2O$

The correct empirical formula is CH_2O.

empirical formula mass $= (12.01\text{ g/mol}) + 2(1.008\text{ g/mol}) + (16.00\text{ g/mol}) = 30.03\text{ g/mol}$

$$n = \frac{\text{molar mass}}{\text{formula molar mass}} = \frac{60.1\text{ g/mol}}{30.03\text{ g/mol}} = 2$$

molecular formula $= CH_2O \times 2$
$= C_2H_4O_2$

Write the correct skeletal structure for the molecule.

```
   H   O
   |   |
H—C—C—O—H
   |
   H
```

Calculate the total number of electrons for the Lewis structure by summing the number of valence electrons of each atom in the molecule.

2(number of valence e^- for C) + 2(number of valence e^- for O) + 4(number of valence e^- for H)
$= 2(4) + 2(6) + 4(1) = 24$

Distribute the electrons among the atoms, giving octets (or duets for H) to as many atoms as possible. Begin with the bonding electrons; then proceed to lone pairs on terminal atoms and finally to lone pairs on the central atoms.

```
       ..
   H  :O:
   |   |      ..
H—C—C—O—H
   |          ..
   H
```

Complete the octet on C by forming a double bond.

```
   H  :O:
   |   ||     ..
H—C—C—O—H
   |          ..
   H
```

6.99 **Given:** μ = 1.08 D HCl, 20% ionic and μ = 1.82 D HF, 45% ionic **Find:** r

Conceptual Plan: $\mu \rightarrow \mu_{calc} \rightarrow r$

$$\% \text{ ionic character} = \frac{\mu}{\mu_{calc}} \times 100\%$$

Solution: For HCl $\mu_{calc} = \frac{1.08}{0.20} = 5.4 \text{ D}$ $\frac{5.4 \cancel{D} \times \frac{3.34 \times 10^{-30} \cancel{C} \cdot \cancel{m}}{\cancel{D}} \times \frac{10^{12} \text{ pm}}{\cancel{m}}}{1.6 \times 10^{-19} \cancel{C}} = 113 \text{ pm}$

For HF $\mu_{calc} = \frac{1.82}{0.45} = 4.0\underline{4} \text{ D}$ $\frac{4.0\underline{4} \cancel{D} \times \frac{3.34 \times 10^{-30} \cancel{C} \cdot \cancel{m}}{\cancel{D}} \times \frac{10^{12} \text{ pm}}{\cancel{m}}}{1.6 \times 10^{-19} \cancel{C}} = 84 \text{ pm}$

From Table 6.4, the bond length of HCl = 127 pm, and HF = 92 pm. Both of these values are slightly higher than the calculated values.

6.100 For the four P atoms to be equivalent, they must all be in the same electronic environment. That is, they must all see the same number of bonds and lone pair electrons. The only way to achieve this is with a tetrahedral configuration where the P atoms are at the four points of the tetrahedron.

6.101 **Given:** C_8H_8 all C's identical, all H's identical **Find:** Lewis structure

For the eight C atoms to be equivalent, they must all be in the same electronic environment; for the eight H atoms to be equivalent, they must all be in the same electronic environment as well. That is, they must all see the same number and kinds of bonds. One way to achieve this is with a cubic arrangement of the C atoms and then a H atom attached to each C.

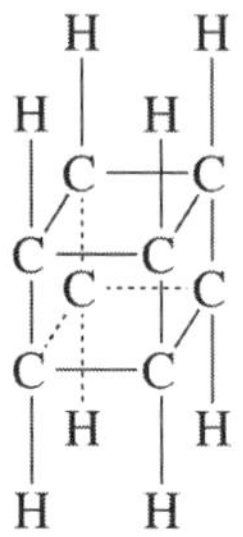

6.102 For each, write the Lewis structure:

Determine electron pair geometry around each central atom.

Determine the molecular geometry and idealized bond angles, and predict actual bond angles.

NO_2

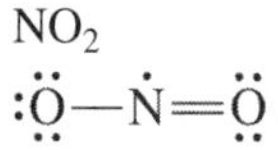

Two bonding groups and a lone electron give a trigonal planar electron geometry; the molecular geometry will be bent. Trigonal planar electron geometry has idealized bond angles of 120°. The bond angle is expected to be slightly less than 120° because of the lone electron occupying the third sp^2 orbital.

NO_2^+

Two bonding groups of electrons and no lone pairs give a linear electron geometry and molecular geometry. Linear electron geometry has a bond angle of 180°.

NO_2^-

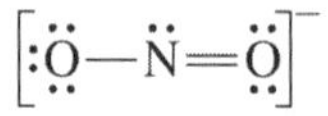

Two bonding groups of electrons and one lone pair give a trigonal planar electron geometry; the molecular geometry will be bent.

Trigonal planar electron geometry has idealized bond angles of 120°. The bond angle is expected to be less than 120° because of the lone pair of electrons occupying the third sp^2 orbital. Further, the bond angle should be less than the bond angle in NO_2 because the presence of lone pairs lowers the tendency for the central atom's orbitals to hybridize. As a result, as lone pairs are added, the bond angle moves further from the 120° hybrid angle to the 90° unhybridized angle, and the two electrons will increase this tendency.

6.103 As you move down the column from F to Cl to Br to I, the atomic radius of the atoms increases. As a result, the larger atoms cannot be accommodated with the smaller bond angle. The attached atoms themselves would begin to overlap their orbitals. So as the size of the attached atom increases, the bond angle becomes larger, approaching the hybridized 109.5° angle.

6.104 CH_3CONH_2 Draw the Lewis structure: (24 valence electrons)

```
      H   :O:
      |    ||       ..
  H — C₁ — C₂ — N — H
      |         |
      H         H
```

Structure I

In structure I: C_1 has 4σ bonds and no lone pairs. The electron geometry would be tetrahedral.
C_2 has 3σ bonds, 1π bond, and no lone pairs. The electron geometry would be trigonal planar.
N has 3σ bonds and 1 lone pair. The electron geometry would be tetrahedral.

This structure would have a trigonal pyramidal molecular geometry around the nitrogen and would not be planar. A second resonance form can be drawn.

```
           ..
      H   :O:
      |    |
  H — C₁ — C₂ = N — H
      |         |
      H         H
```

Structure II

In structure II: C_1 has 4σ bonds and no lone pairs. The electron geometry would be tetrahedral.
C_2 has 3σ bonds, 1π bond, and no lone pairs. The electron geometry would be trigonal planar.
N has 3σ bonds, 1π bond, and no lone pairs. The electron geometry would be trigonal planar.

This resonance form would account for a planar configuration around the N.

6.105 CH_3N_3 Predict geometry and sketch the molecule.

```
      H                              H                               H
      |    ..         ..             |    ..        ..               |                 ..
  H — C — N₁ = N₂ = N₃   ⟷   H — C — N₁ — N₂ ≡ N₃   ⟷   H — C — N₁ ≡ N₂ — N₃:
      |               ..             |    ..                         |                 ..
      H                              H                               H

              I                              II                              III
Formal charge 0   +1   −1             −1   +1   0                    +1   +1   −2
```

There are three resonance forms for the methyl azide molecule. Structures I and II would be most likely because they have the smaller formal charges. So we will look at structures I and II.

In structure I: C has four electron groups around the atom. Four electron groups give a tetrahedral electron geometry. Four bonding groups and zero lone pairs give a tetrahedral molecular geometry. N_1 has three electron groups around the atom. Three electron groups give a trigonal planar electron geometry. Two bonding groups and one lone pair give a bent molecular geometry. N_2 has two electron groups around the atom. Two electron groups give a linear electron geometry. Two bonding groups and zero lone pairs give a linear molecular geometry.

In structure II: C has four electron groups around the atom. Four electron groups give a tetrahedral electron geometry. Four bonding groups and zero lone pairs give a tetrahedral molecular geometry. N_1 has four electron groups around the atom. Four electron groups give a tetrahedral electron geometry. Two bonding groups and two lone pairs give a bent molecular geometry. N_2 has two electron groups around the atom. Two electron groups give a linear electron geometry. Two bonding groups and zero lone pairs give a linear molecular geometry.

I II

If structure I is the resonance, the C—N—N bond angle will be closer to 120° and the N—N—N bond angle will be 180°. The two N—N bond lengths would be expected to be the same. If structure II is the resonance, the C—N—N bond angle will be closer to 109° and the N—N—N bond angle will be 180°. The N_1—N_2 bond length would be expected to be longer than the N_2—N_3 bond length.

Conceptual Problems

6.106 Lewis theory is successful because it allows us to understand and predict many chemical observations. We can use it to determine the formulae of ionic compounds and to account for the low melting points and boiling points of molecular compounds compared to ionic compounds. Lewis theory allows us to predict what molecules or ions will be stable, which will be more reactive, and which will not exist. Lewis theory, however, does not tell us anything about how the bonds in the molecules and ions form. It does not give us a way to account for the paramagnetism of oxygen. And by itself, Lewis theory does not tell us anything about the shape of the molecule or ion.

6.107 Statement a is the best statement.
Statement b neglects the lowering of potential energy that arises from the interaction of the lone pair electrons with the bonding electrons.
Statement c neglects the interaction of the electrons altogether. The molecular geometries are determined by the number and types of electron groups around the central atom.

6.108 A molecule with four bond groups and one lone pair would need five equivalent positions around the central atom. In two dimensions, this could be accommodated with a pentagon shape around the central atom. The idealized bond angles would be 72°; however, because of the lone pair occupying one of the positions, the bond angles would be less than 72°.

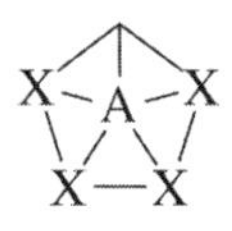

7 Chemical Bonding II: Valence Bond Theory and Molecular Orbital Theory

Review Questions

7.1 While the Lewis model is a simple model that can describe chemical bonding, it cannot predict all of the properties of a molecule or ion. More sophisticated models, such as valence bond theory and molecular orbital theory, are needed to adequately describe bonding and magnetism in certain species.

7.2 According to valence bond theory, a chemical bond results from the overlap of two half-filled orbitals with spin-pairing of the two valence electrons.

7.3 According to valence bond theory, the shape of the molecule is determined by the geometry of the overlapping orbitals.

7.4 In valence bond theory, the interaction energy is usually negative (or stabilizing) when the interacting atomic orbitals contain a total of two electrons that can spin-pair.

7.5 Hybridization is a mathematical procedure in which the standard atomic orbitals are combined to form new atomic orbitals called hybrid orbitals. Hybrid orbitals are still localized on individual atoms, but they have different shapes and energies from those of standard atomic orbitals. They are necessary in valence bond theory because they correspond more closely to the actual distribution of electrons in chemically bonded atoms.

7.6 Hybrid orbitals minimize the energy of the molecule by maximizing the orbital overlap in a bond.

7.7 The number of standard atomic orbitals added together always equals the number of hybrid orbitals formed. The total number of orbitals is conserved.

7.8

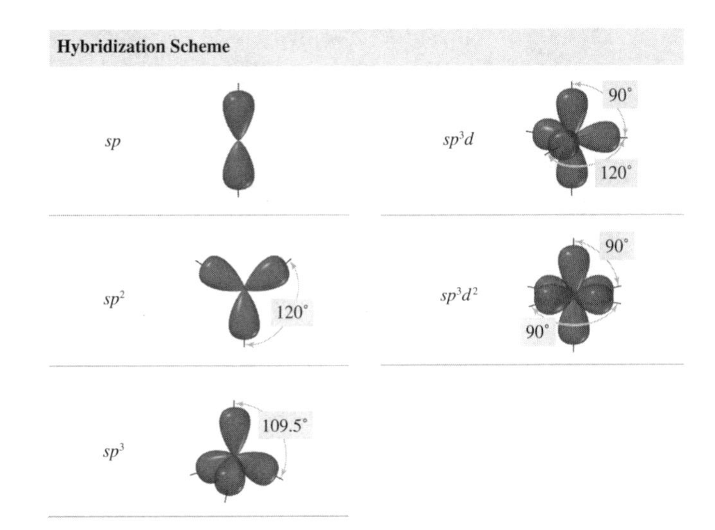

Each sketch indicates the number of hybrid orbitals formed.

7.9 The double bond in Lewis theory is simply two pairs of electrons that are shared between the same two atoms. However, in valence bond theory, we see that the double bond is made up of two different kinds of bonds. The double bond in valence bond theory consists of one σ bond and one π bond. Valence bond theory shows us that rotation about a double bond is severely restricted. Because of the side-by-side overlap of the *p* orbitals, the π bond must essentially break for rotation to occur. The σ bond consists of end-to-end overlap. Because the overlap is linear, rotation is not restricted.

7.10
(a) A linear electron geometry corresponds to *sp* hybridization.
(b) A trigonal planar electron geometry corresponds to sp^2 hybridization.
(c) A tetrahedral electron geometry corresponds to sp^3 hybridization.
(d) A trigonal bipyramidal electron geometry corresponds to sp^3d hybridization.
(e) An octahedral electron geometry corresponds to sp^3d^2 hybridization.

7.11 In molecular orbital theory, atoms will bond when the electrons in the atoms can lower their energy by occupying the molecular orbitals of the resultant molecule.

7.12 In valence bond theory, hybrid orbitals are weighted linear sums of the valence atomic orbitals of a particular atom, and the hybrid orbitals remain localized on that atom. In molecular orbital theory, the molecular orbitals are weighted linear sums of the valence atomic orbitals of all of the atoms in a molecule, and many of the molecular orbitals are delocalized over the entire molecule.

7.13 A bonding molecular orbital is lower in energy than the atomic orbitals from which it is formed. There is an increased electron density in the internuclear region.

7.14 An antibonding molecular orbital is higher in energy than the atomic orbitals from which it is formed. There is less electron density in the internuclear region, which results in a node.

7.15 The electrons in orbitals behave like waves. The bonding molecular orbital arises from the constructive interference between the atomic orbitals and is lower in energy than the atomic orbitals. The antibonding molecular orbital arises from the destructive interference between the atomic orbitals and is higher in energy than the atomic orbitals.

7.16 The bond order in a diatomic molecule is the number of electrons in bonding molecular orbitals (MOs) minus the number in antibonding MOs divided by two. The higher the bond order, the stronger the bond. A negative or zero bond order indicates that a bond will not form between the atoms.

7.17 Molecular orbitals can be approximated by a linear combination of atomic orbitals (AOs). The total number of MOs formed from a particular set of AOs will always equal the number of AOs used.

7.18

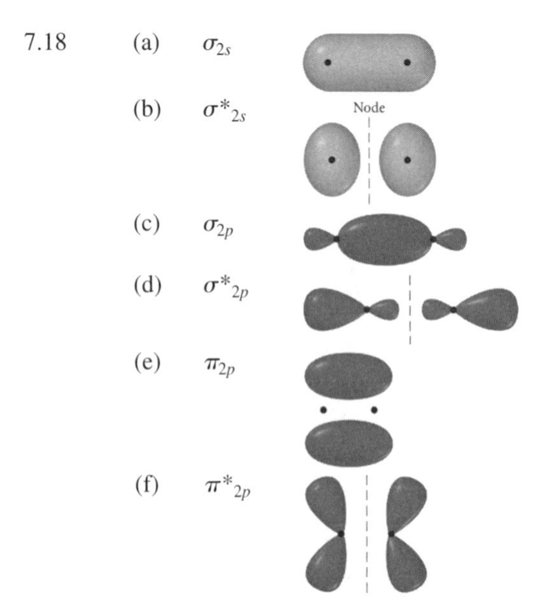

Dashed lines represent nodes.

7.19

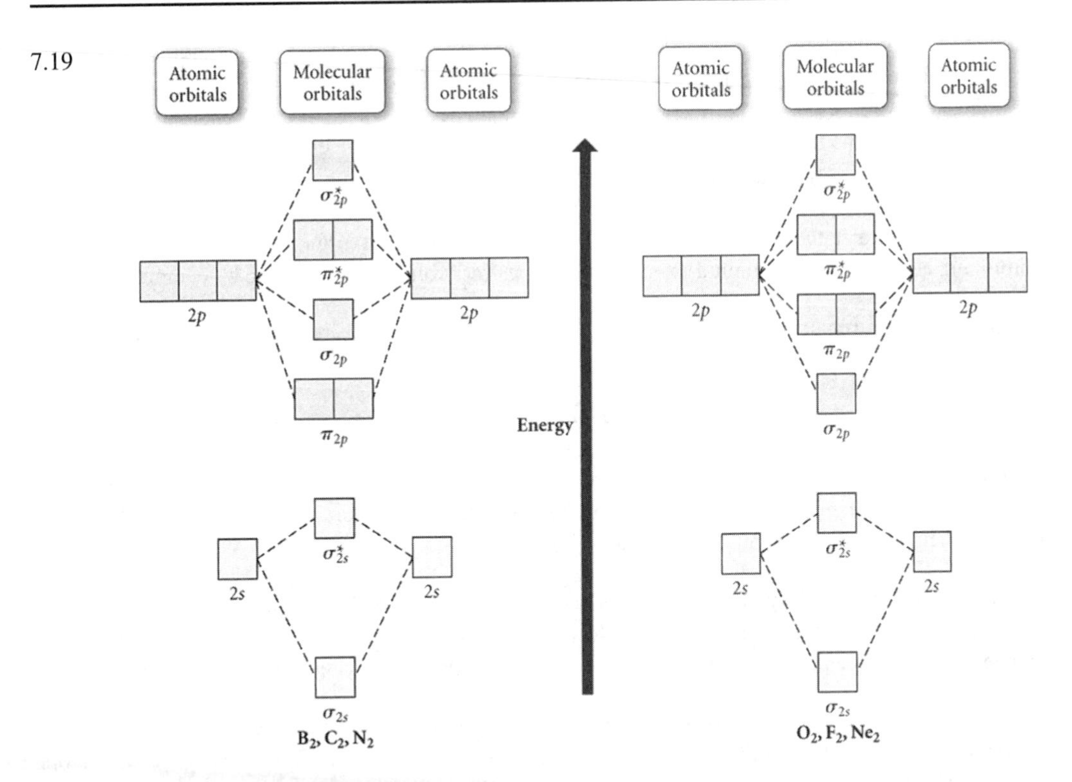

7.20 The degree of mixing between two orbitals decreases with increasing energy difference between them. Mixing of the 2*s* and $2p_x$ orbitals is greater in B_2, C_2, and N_2 than in O_2, F_2, and Ne_2 because in B, C, and N, the energy levels of the atomic orbitals are more closely spaced than in O, F, and Ne. This mixing produces a change in energy ordering for the π_{2p} and the σ_{2p} molecular orbitals.

7.21 A paramagnetic species has unpaired electrons in one or more molecular orbitals. A paramagnetic species is attracted to a magnetic field. The magnetic property is a direct result of the unpaired electron(s). The spin and angular momentum of the electrons generate tiny magnetic fields. A diamagnetic species has all of its electrons paired. The magnetic fields caused by the electron spin, and orbital angular momentum tend to cancel each other. A diamagnetic species is not attracted to a magnetic field, and is, in fact, slightly repelled.

7.22 When two atomic orbitals are different, the weighting of each orbital in forming a molecular orbital may be different. When a molecular orbital is approximated as a linear combination of atomic orbitals of different energies, the lower energy atomic orbital makes a greater contribution to the bonding molecular orbital and the higher energy atomic orbital makes a greater contribution to the antibonding molecular orbital. The shape of the molecular orbital shows a greater electron density at the atom that has the lower atomic orbital energy.

7.23 Nonbonding orbitals are atomic orbitals not involved in a bond that remain localized on the atom.

7.24 In Lewis theory, a chemical bond is the transfer or sharing of electrons represented as dots. Lewis theory allows us to predict the combination of atoms that form stable molecules and the general shape of a molecule.

Lewis theory is a quick way to predict the stability and shapes of molecules based on the number of valence electrons. However, it does not deal with how the bonds that we make are formed. Valence bond theory is a more advanced bonding theory that treats electrons in a quantum-mechanical manner. A quantitative approach is extremely complicated, but a qualitative approach allows an understanding of how the bonds are formed. In valence bond theory, electrons reside in quantum-mechanical orbitals localized on individual atoms. When two atoms approach each other, the electrons and nucleus of one atom interact with the electron and nucleus of the other atom. If the energy of the system is lowered, a chemical bond forms. So valence bond theory portrays a chemical bond as the overlap of two half-filled atomic orbitals. The shape of the molecule can be predicted from the geometry of the overlapping orbitals. Also, valence bond theory explains the rigidity of the double bond. However, valence bond theory falls short in explaining certain phenomena, such as magnetism, and certain bond properties. Valence bond theory treats the electrons as if they reside in the quantum-mechanical orbitals that we calculate for an atom. This is an oversimplification that is partially compensated for by introducing the concept of hybridization. An even more complex quantum-mechanical model is molecular orbital theory. In molecular orbital theory, a chemical bond occurs when the electrons in the atoms can lower their energy by occupying the molecular orbitals of the resultant molecule. The chemical bonds in MO theory are not localized between atoms, but are spread throughout the entire molecule. Molecular orbital theory uses trial functions to solve the Schrödinger equation for the molecules. To determine how well the trial function works, you calculate the energy, trying to minimize the energy. However, no matter how "good" your guess, you can never do better than nature at minimizing energy. These minimum-energy calculations for orbitals must be done by computer.

All three of these models have strengths and weaknesses; none is "correct." The information you need depends on which approach you use.

7.25 When metal atoms bond together to form a solid, each metal atom donates one or more electrons to an electron sea.

7.26 Metals conduct electricity because the electrons in a metal are free to move. The movement or flow of electrons in response to an electric potential is an electric current. Metals are also excellent conductors of heat because of the highly mobile electrons, which help to disperse thermal energy throughout the metal.

The malleability of metals and the ductility of metals are also accounted for by the electron sea model. Because there are no localized or specific "bonds" in a metal, it can be deformed relatively easily by forcing the metal ions to slide past one another. The electron sea can easily accommodate these deformations by flowing into the new shape.

7.27 In band theory, bands are the combination atomic orbitals of the atoms within a solid crystal that form orbitals that are not localized on individual atoms, but are delocalized over the entire crystal. In band theory, electrons become mobile when they make a transition from the highest occupied molecular orbital into higher-energy empty molecular orbitals. For this reason, the occupied molecular orbitals are often called the valence band and the unoccupied orbitals are called the conduction band.

7.28 The band gap is an energy gap that exists between the valence band and conduction band. In metals, the valence band and conduction band are energetically continuous—the energy difference between the top of the valence band and the bottom of the conduction band is infinitesimally small. In semiconductors, the band gap is small, allowing some electrons to be promoted at ordinary temperatures, resulting in limited conductivity. In insulators, the band gap is large, and electrons are not promoted into the conduction band at ordinary temperatures, resulting in no electrical conductivity.

7.29 Doped semiconductors contain minute amounts of impurities that result in additional electrons in the conduction band or electron "holes" in the valence band.

7.30 For example, silicon is a group 4A semiconductor. Its valence electrons just fill its valence band. When silicon is doped with phosphorus, a group 5A element with five valence electrons, its conductivity increases. The phosphorus atoms are incorporated into the silicon crystal structure, but each phosphorus atom brings with it one additional electron. Because the valence band is completely full, the additional electrons must go into the conduction band. These electrons are then mobile and can conduct electric current. This type of semiconductor is called an n-type semiconductor because the charge carriers are negatively charged electrons in the conduction band. Silicon can also be doped with a group 3A element, such as gallium, which has only three valence electrons. When gallium is incorporated into the silicon crystal structure, it results in electron "holes," empty molecular orbitals in the valence band. The presence of holes also allows for the movement of electric current because electrons in the valence band can move between holes. In this way, the holes move in the opposite direction of the electrons. This type of semiconductor is called a p-type semiconductor because the hole acts as a positive charge.

Problems by Topic

Valence Bond Theory

7.31 (a) Be $2s^2$ No bonds can form. Beryllium contains no unpaired electrons, so no bonds can form without hybridization.

(b) P $3s^23p^3$ Three bonds can form. Phosphorus contains three unpaired electrons, so three bonds can form without hybridization.

(c) F $2s^22p^5$ One bond can form. Fluorine contains one unpaired electron, so one bond can form without hybridization.

7.32 (a) B $2s^22p^1$ One bond can form. Boron contains one unpaired electron, so one bond can form without hybridization.

(b) N $2s^22p^3$ Three bonds can form. Nitrogen contains three unpaired electrons, so three bonds can form without hybridization.

(c) O $2s^22p^4$ Two bonds can form. Oxygen contains two unpaired electrons, so two bonds can form without hybridization.

7.33 PH_3

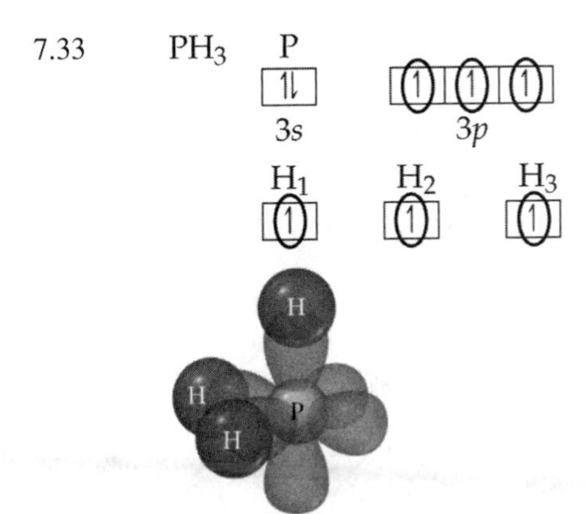

The unhybridized bond angles should be 90°. So without hybridization, there is good agreement between valence bond theory and the actual bond angle of 93.3°.

7.34 SF_2

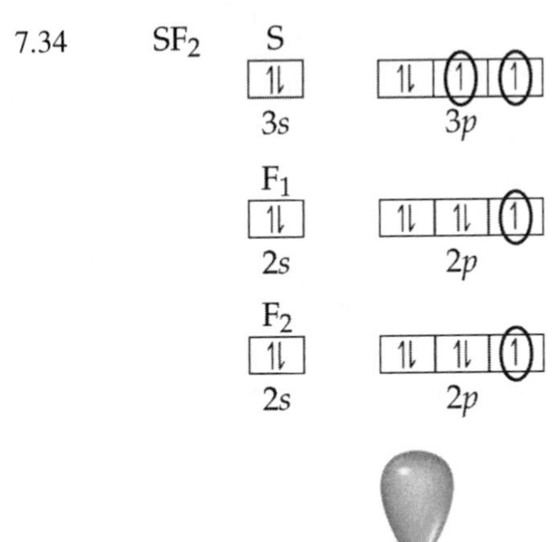

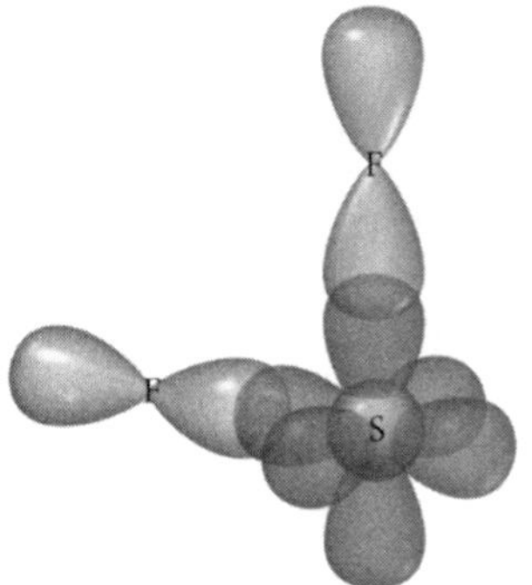

The unhybridized bond angles should be 90°. So without hybridization, there is not very good agreement between valence bond theory and the actual bond angle of 98.2°.

7.35 C $2s^22p^2$

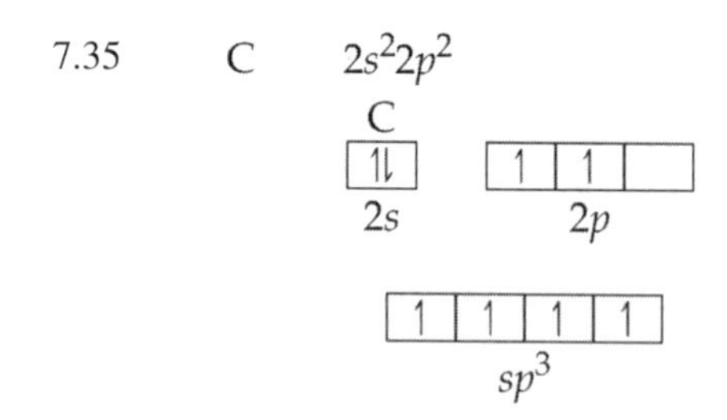

7.36 C $2s^22p^2$

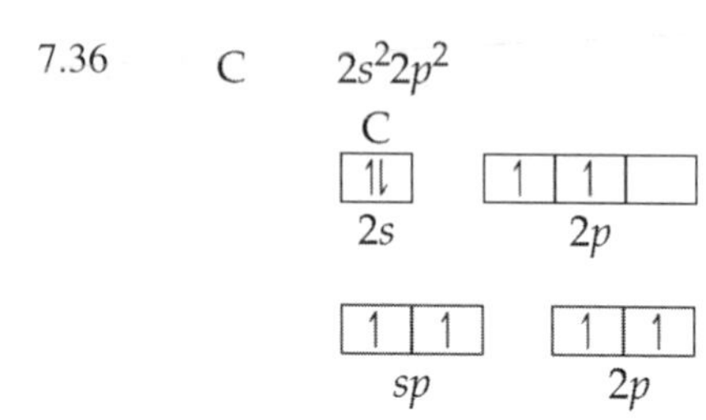

7.37 sp^2 Only sp^2 hybridization of this set of orbitals has a remaining *p* orbital to form a π bond.
sp^3 hybridization utilizes all 3*p* orbitals.
sp^3d^2 hybridization utilizes all 3*p* orbitals and 2*d* orbitals.

7.38 sp^3d sp^3d hybridization utilizes an *s* orbital, 3*p* orbitals, and a *d* orbital. Because five orbitals are used, five hybrid orbitals form and five bonds can form.
sp^3 Hybridization utilizes an *s* orbital and 3*p* orbitals. Four orbitals are used, so four hybrid orbitals form and four bonds can form.
sp^2 Hybridization utilizes an *s* orbital and 2*p* orbitals. Three orbitals are used, so three hybrid orbitals form. This allows three σ and one π bond to form, for a total of four bonds formed.

7.39 (a) CCl_4 Write the Lewis structure for the molecule:

```
         ..
        :Cl:
     ..  |  ..
    :Cl—C—Cl:
     ''  |  ''
        :Cl:
         ''
```

Use valence shell electron pair repulsion (VSEPR) to predict the electron geometry:
Four electron groups around the central atom give a tetrahedral electron geometry.
Select the correct hybridization for the central atom based on the electron geometry:
Tetrahedral electron geometry has sp^3 hybridization.
Sketch the molecule and label the bonds:

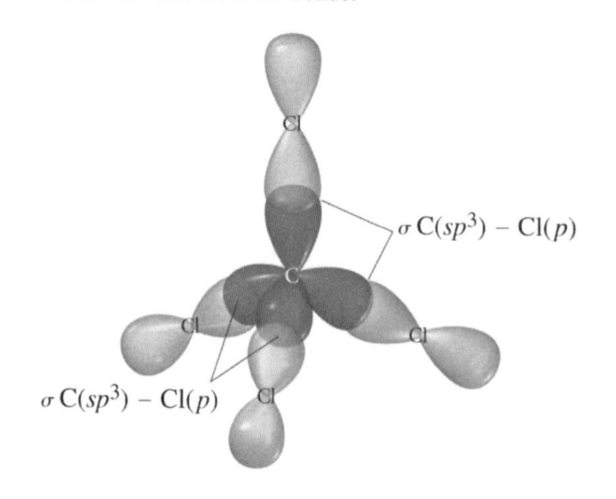

(b) NH_3 Write the Lewis structure for the molecule:

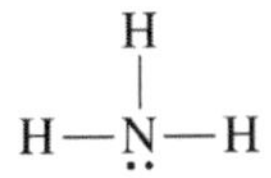

```
      H
      |
  H—N—H
     ··
```

Use VSEPR to predict the electron geometry:
Four electron groups around the central atom give a tetrahedral electron geometry.
Select the correct hybridization for the central atom based on the electron geometry:
Tetrahedral electron geometry has sp^3 hybridization.
Sketch the molecule and label the bonds:

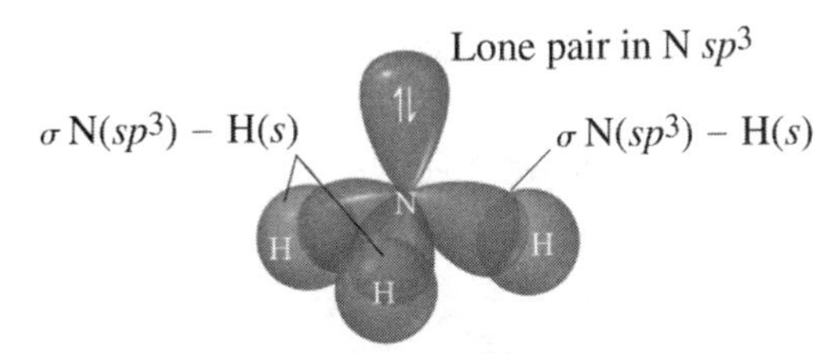

(c) OF_2 Write the Lewis structure for the molecule:

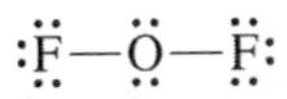

:F̤̈—Ö̤—F̤̈:

Use VSEPR to predict the electron geometry:
Four electron groups around the central atom give a tetrahedral electron geometry.
Select the correct hybridization for the central atom based on the electron geometry:
Tetrahedral electron geometry has sp^3 hybridization.
Sketch the molecule and label the bonds:

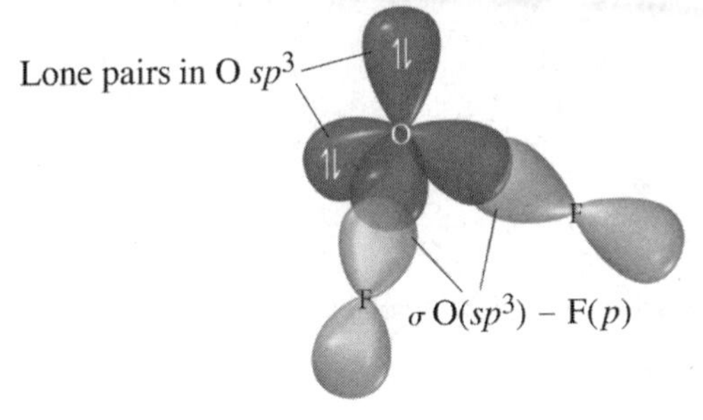

(d) CO_2 Write the Lewis structure for the molecule:

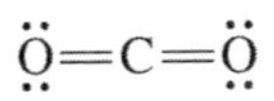

Ö̤=C=Ö̤

Use VSEPR to predict the electron geometry:
Two electron groups around the central atom give a linear electron geometry.
Select the correct hybridization for the central atom based on the electron geometry:
Linear electron geometry has sp hybridization.
Sketch the molecule and label the bonds:

π C(p_y) – O(p_y) π C(p_z) – O(p_z)

σ C(sp) – O(p)

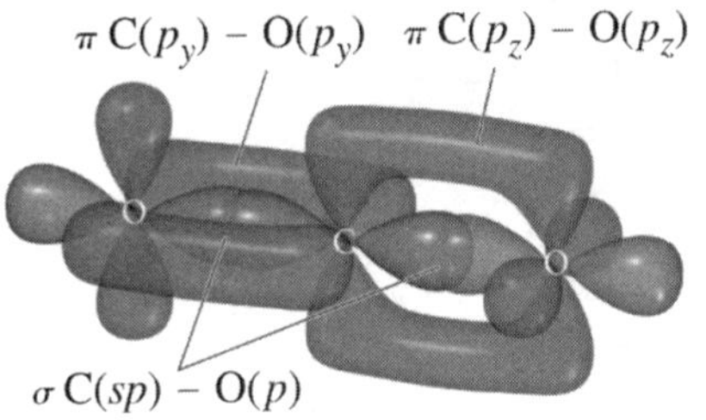

7.40 (a) CH_2Br_2 Write the Lewis structure for the molecule:

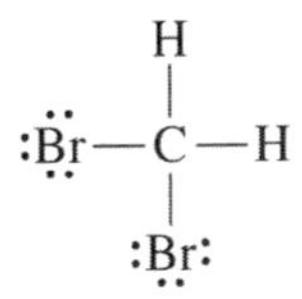

```
          H
          |
   :Br̤̈—C—H
          |
        :Br̤̈:
```

Use VSEPR to predict the electron geometry:
Four electron groups around the central atom give a tetrahedral electron geometry.
Select the correct hybridization for the central atom based on the electron geometry:
Tetrahedral electron geometry has sp^3 hybridization.
Sketch the molecule and label the bonds:

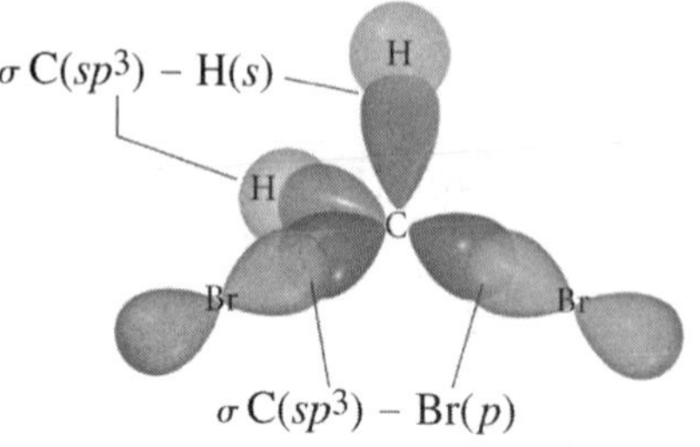

(b) SO_2 Write the Lewis structure for the molecule:

$$\ddot{\underset{\cdot\cdot}{O}}{=}\underset{\cdot\cdot}{S}{-}\ddot{\underset{\cdot\cdot}{O}}:$$

Use VSEPR to predict the electron geometry:
Three electron groups around the central atom give a trigonal planar electron geometry.
Select the correct hybridization for the central atom based on the electron geometry:
Trigonal planar electron geometry has sp^2 hybridization.
Sketch the molecule and label the bonds:

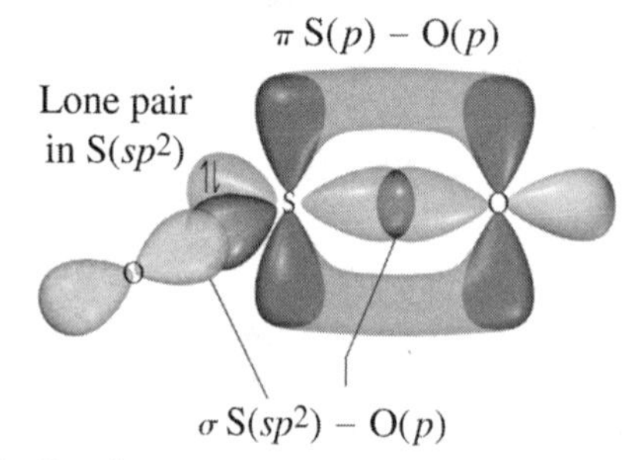

(c) NF_3 Write the Lewis structure for the molecule:

$$:\ddot{\underset{\cdot\cdot}{F}}{-}\ddot{N}{-}\ddot{\underset{\cdot\cdot}{F}}: \quad (\text{N also bonded to } :\ddot{\underset{\cdot\cdot}{F}}:)$$

Use VSEPR to predict the electron geometry:
Four electron groups around the central atom give a tetrahedral electron geometry.
Select the correct hybridization for the central atom based on the electron geometry:
Tetrahedral electron geometry has sp^3 hybridization.
Sketch the molecule and label the bonds:

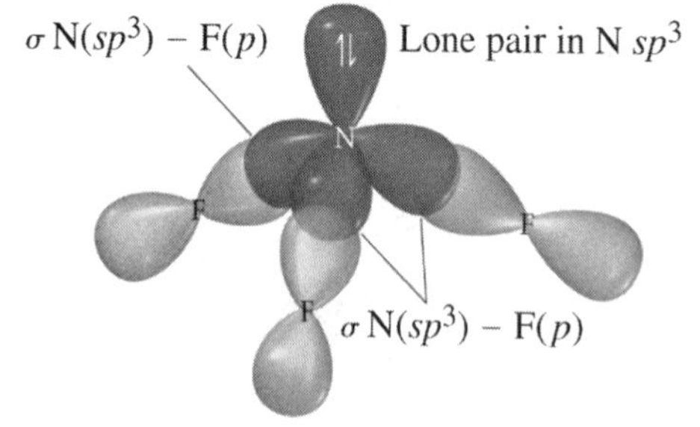

(d) BF_3 Write the Lewis structure for the molecule:

$$:\ddot{\underset{\cdot\cdot}{F}}{-}B{-}\ddot{\underset{\cdot\cdot}{F}}: \quad (\text{B also bonded to } :\ddot{F}:)$$

Use VSEPR to predict the electron geometry:
Three electron groups around the central atom give a trigonal planar electron geometry.

Select the correct hybridization for the central atom based on the electron geometry:
Trigonal planar electron geometry has sp^2 hybridization.
Sketch the molecule and label the bonds:

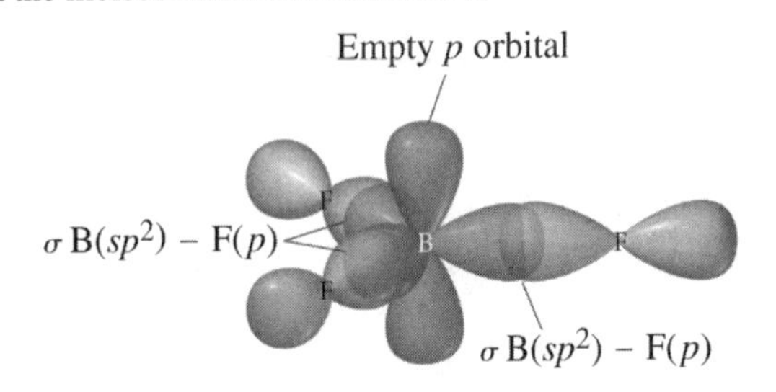

7.41 (a) $COCl_2$ Write the Lewis structure for the molecule:

```
       :O:
        ‖
:C̤l̤—C—C̤l̤:
```

Use VSEPR to predict the electron geometry:
Three electron groups around the central atom give a trigonal planar electron geometry.
Select the correct hybridization for the central atom based on the electron geometry:
Trigonal planar electron geometry has sp^2 hybridization.
Sketch the molecule and label the bonds:

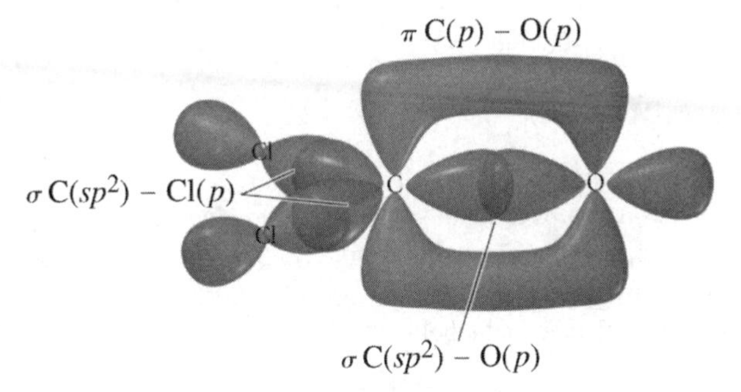

(b) BrF_5 Write the Lewis structure for the molecule:

```
       :F̤:
:F̤      |     F̤:
    \   Br  /
:F̤  /   ¨   \ F̤:
```

Use VSEPR to predict the electron geometry:
Six electron pairs around the central atoms give an octahedral electron geometry.
Select the correct hybridization for the central atom based on the electron geometry:
Octahedral electron geometry has sp^3d^2 hybridization.
Sketch the molecule and label the bonds:

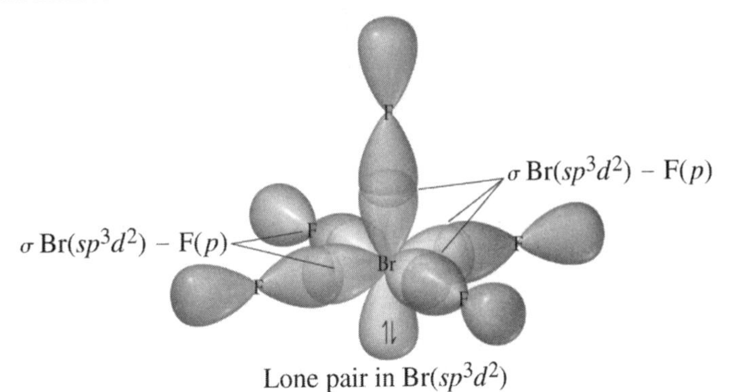

(c) XeF_2 Write the Lewis structure for the molecule:

:F̤̈—Xe—F̤̈:

Use VSEPR to predict the electron geometry:
Five electron groups around the central atom give a trigonal bipyramidal geometry.
Select the correct hybridization for the central atom based on the electron geometry:
Trigonal bipyramidal geometry has sp^3d hybridization.
Sketch the molecule and label the bonds:

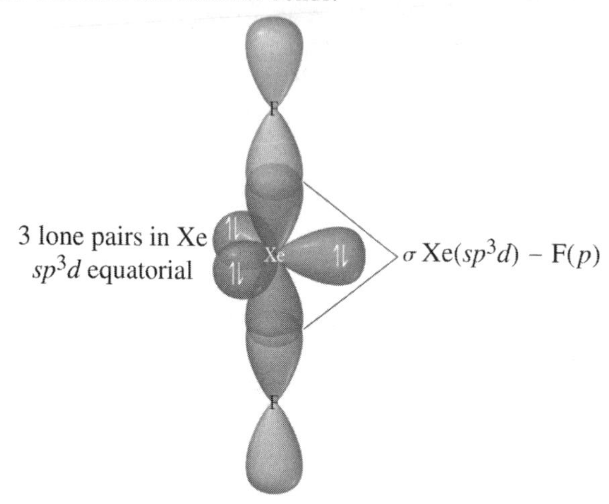

(d) I_3^- Write the Lewis structure for the molecule:

$[:\ddot{I}—\ddot{I}—\ddot{I}:]^-$

Use VSEPR to predict the electron geometry:
Five electron groups around the central atom give a trigonal bipyramidal geometry.
Select the correct hybridization for the central atom based on the electron geometry:
Trigonal bipyramidal geometry has sp^3d hybridization.
Sketch the molecule and label the bonds:

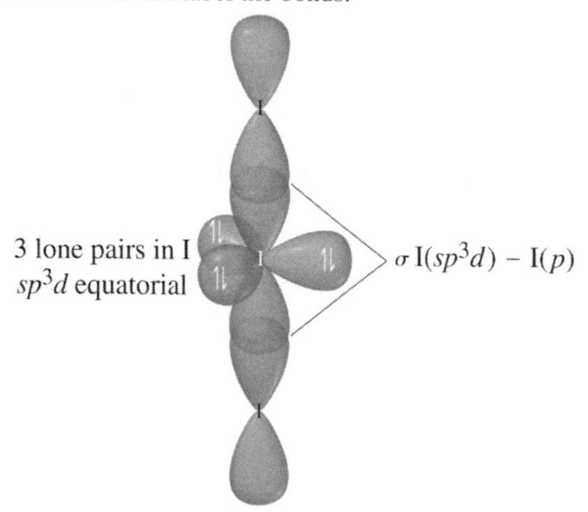

7.42 (a) SO_3^{2-} Write the Lewis structure for the ion:

$$\left[\begin{array}{c} :\ddot{O}: \\ | \\ :\ddot{O}—\ddot{S}—\ddot{O}: \end{array}\right]^{2-}$$

Use VSEPR to predict the electron geometry:
Four electron groups around the central atom give a tetrahedral electron geometry.
Select the correct hybridization for the central atom based on the electron geometry:
Tetrahedral electron geometry has sp^3 hybridization.

Sketch the molecule and label the bonds:

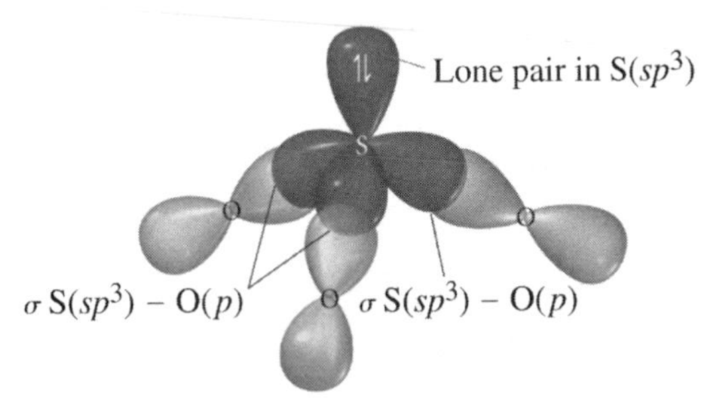

(b) PF_6^- Write the Lewis structure for the ion:

```
[      :F:       ]⁻
   :F:  |  :F:
       P
   :F:  |  :F:
       :F:
```

Use VSEPR to predict the electron geometry:
Six electron pairs around the central atom give an octahedral electron geometry.
Select the correct hybridization for the central atom based on the electron geometry:
Octahedral electron geometry has sp^3d^2 hybridization.
Sketch the molecule and label the bonds:

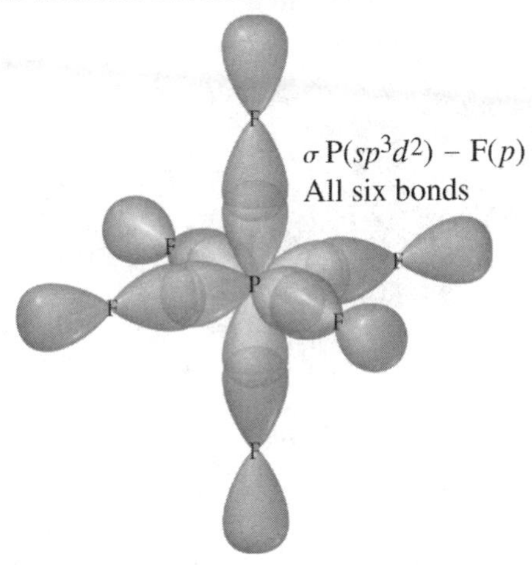

(c) BrF_3 Write the Lewis structure for the molecule:

```
 :F:
  |
:Br—F:
  |
 :F:
```

Use VSEPR to predict the electron geometry:
Five electron groups around the central atom give a trigonal bipyramidal geometry.
Select the correct hybridization for the central atom based on the electron geometry:
Trigonal bipyramidal geometry has sp^3d hybridization.

Sketch the molecule and label the bonds:

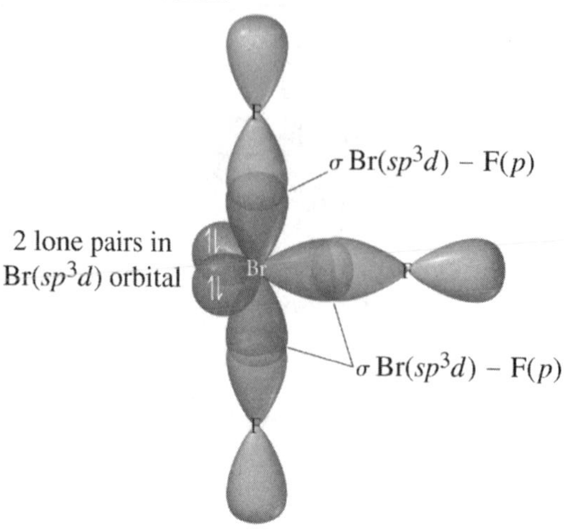

(d) HCN Write the Lewis structure for the molecule:

H—C≡N:

Use VSEPR to predict the electron geometry:
Two electron groups around the central atom give a linear electron geometry.
Select the correct hybridization for the central atom based on the electron geometry:
Linear electron geometry has *sp* hybridization.
Sketch the molecule and label the bonds:

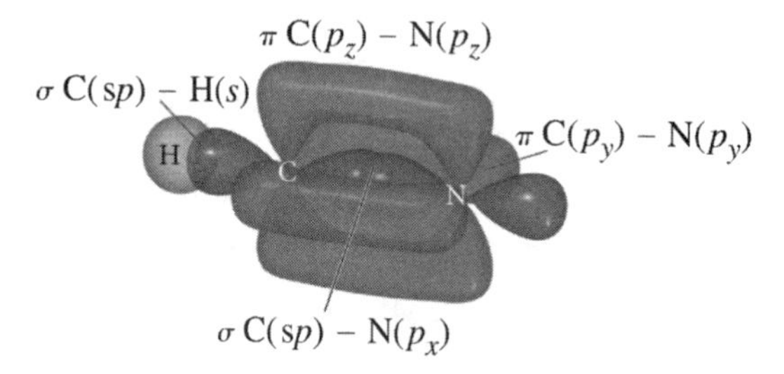

7.43 (a) N_2H_2 Write the Lewis structure for the molecule:

H—N̈=N̈—H

Use VSEPR to predict the electron geometry:
Three electron groups around each interior atom give a trigonal planar electron geometry.
Select the correct hybridization for the central atoms based on the electron geometry:
Trigonal planar electron geometry has sp^2 hybridization.
Sketch the molecule and label the bonds:

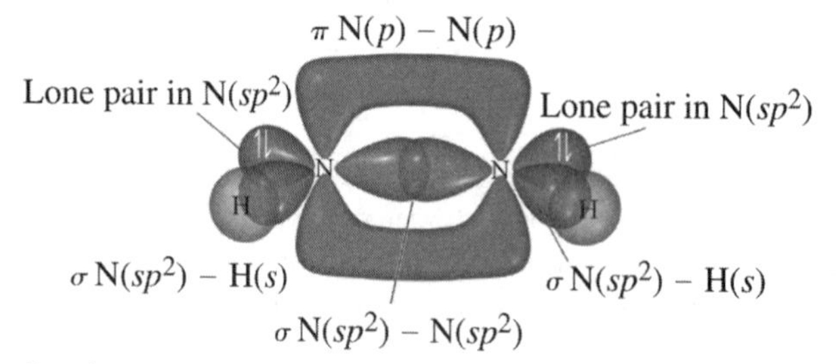

(b) N_2H_4 Write the Lewis structure for the molecule:

H_2N̈—N̈H_2 (each N bonded to two H)

Use VSEPR to predict the electron geometry:
Four electron groups around each interior atom give tetrahedral electron geometry.
Select the correct hybridization for the central atoms based on the electron geometry:
Tetrahedral electron geometry has sp^3 hybridization.

Sketch the molecule and label the bonds:

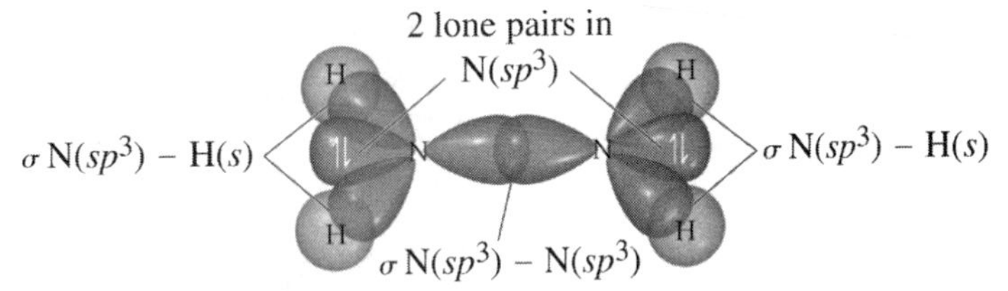

(c) CH_3NH_2 Write the Lewis structure for the molecule:

```
       H
       |    ..
   H — C — N — H
       |    |
       H    H
```

Use VSEPR to predict the electron geometry:
Four electron groups around the C give a tetrahedral electron geometry around the C atom, and four electron groups around the N give a tetrahedral geometry around the N atom.
Select the correct hybridization for the central atoms based on the electron geometry:
Tetrahedral electron geometry has sp^3 hybridization of both C and N.
Sketch the molecule and label the bonds:

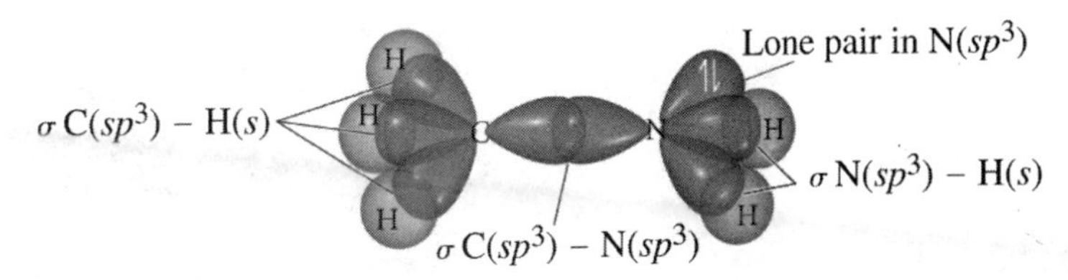

7.44 (a) C_2H_2 Write the Lewis structure for the molecule:

H—C≡C—H

Use VSEPR to predict the electron geometry:
Two electron groups around each interior atom give a linear electron geometry.
Select the correct hybridization for the central atoms based on the electron geometry:
Linear electron geometry has sp hybridization.
Sketch the molecule and label the bonds:

π C(p) – C(p)
2 bonds formed
σ C(sp) – H(s)
σ C(sp) – H(s)
σ C(sp) – C(sp)

(b) C_2H_4 Write the Lewis structure for the molecule:

```
H       H
  \   /
   C=C
  /   \
H       H
```

Use VSEPR to predict the electron geometry:
Three electron groups around each interior atom give a trigonal planar electron geometry.
Select the correct hybridization for the central atoms based on the electron geometry:
Trigonal planar electron geometry has sp^2 hybridization.

Sketch the molecule and label the bonds:

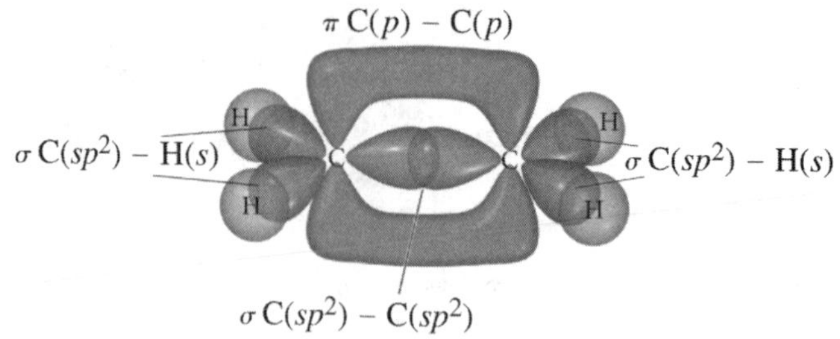

(c) C_2H_6 Write the Lewis structure for the molecule:

```
   H  H
   |  |
H—C—C—H
   |  |
   H  H
```

Use VSEPR to predict the electron geometry:
Four electron groups around each interior atom give a tetrahedral electron geometry.
Select the correct hybridization for the central atoms based on the electron geometry:
Tetrahedral electron geometry has sp^3 hybridization.
Sketch the molecule and label the bonds:

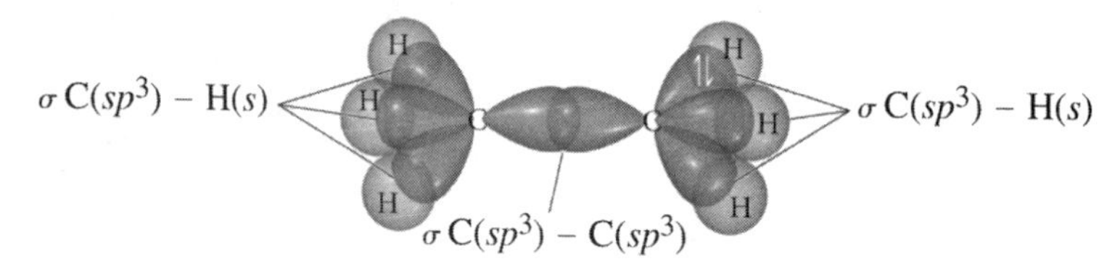

7.45

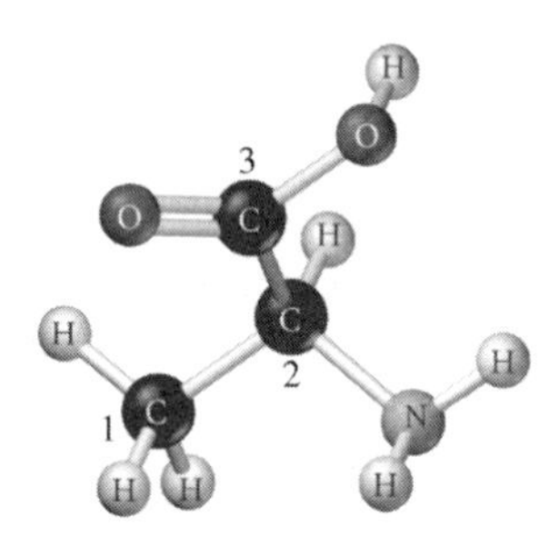

The Cs in positions 1 and 2 have four electron groups around the atom, which is tetrahedral electron geometry. Tetrahedral electron geometry is sp^3 hybridization.
C in position 3 has three electron groups around the atom, which is trigonal planar electron geometry. Trigonal planar electron geometry is sp^2 hybridization.
O (bonded to C and H) has four electron groups around the atom, which is tetrahedral electron geometry. Tetrahedral electron geometry is sp^3 hybridization.
N has four electron groups around the atom, which is tetrahedral electron geometry. Tetrahedral electron geometry is sp^3 hybridization.

7.46

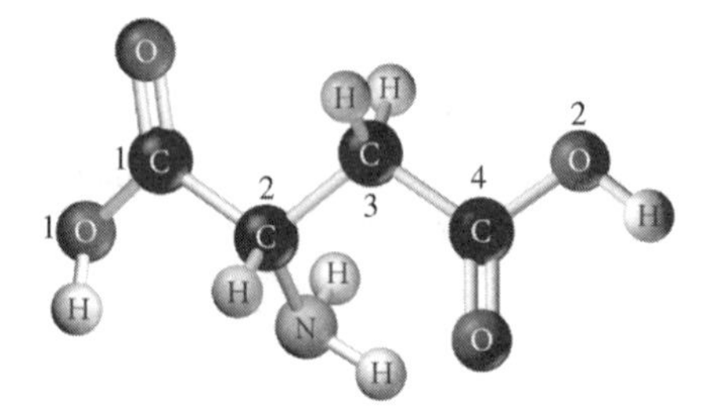

The Cs in positions 1 and 4 have three electron groups around the atom, which is trigonal planar electron geometry. Trigonal planar electron geometry is sp^2 hybridization.
The Cs in positions 2 and 3 have four electron groups around the atom, which is tetrahedral electron geometry. Tetrahedral electron geometry is sp^3 hybridization.
The Os in positions 1 and 2 have four electron groups around the atom, which is tetrahedral electron geometry. Tetrahedral electron geometry is sp^3 hybridization.
N has four electron groups around the atom, which is tetrahedral electron geometry. Tetrahedral electron geometry is sp^3 hybridization.

Molecular Orbital Theory

7.47 1*s* + 1*s* constructive interference results in a bonding orbital:

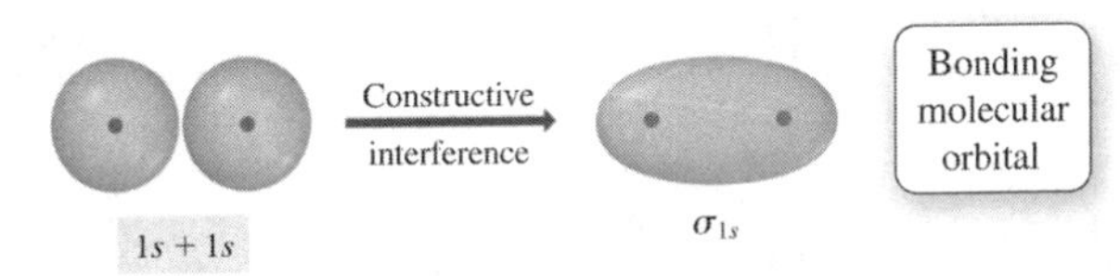

7.48 1*s* − 1*s* destructive interference results in an antibonding orbital:

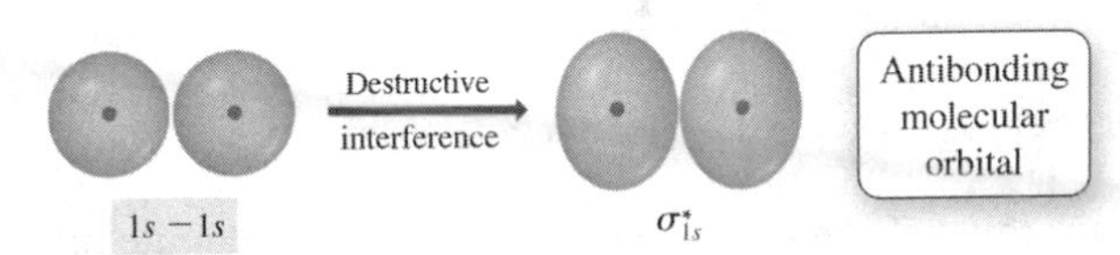

7.49 Be_2^+ has seven electrons. Be_2^- has nine electrons.

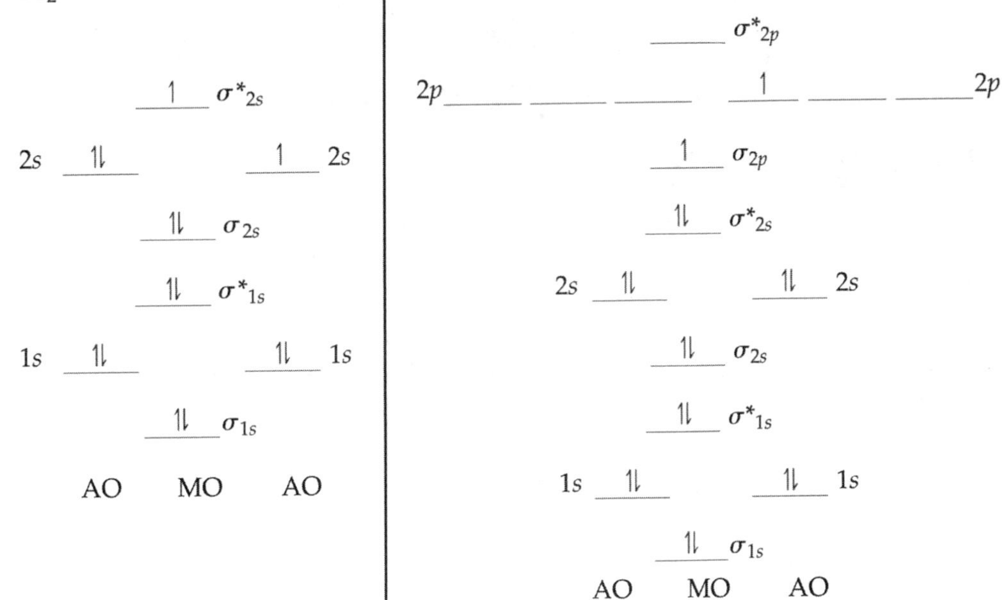

AO = Atomic Orbital; MO = Molecular Orbital

Bond order $= \frac{4-3}{2} = \frac{1}{2}$; stable Bond order $= \frac{5-4}{2} = \frac{1}{2}$; stable

Both of these ions are able to exist in the gas phase because the bond order is positive.

7.50 Li_2^+ has five electrons.

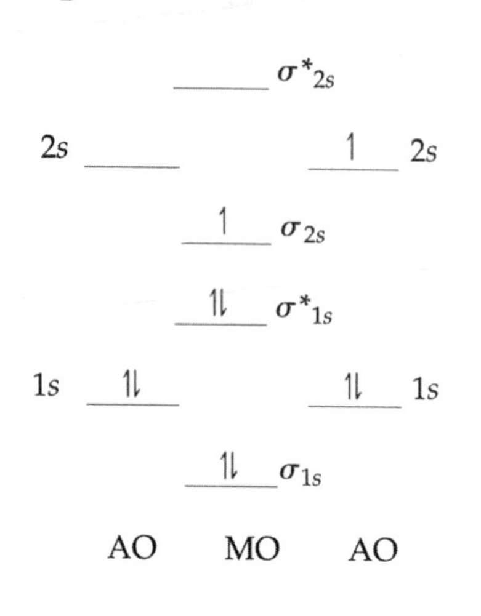

Li_2^- has seven electrons.

1 σ^*_{2s}

2s 1l   1 2s

1l σ_{2s}

1l σ^*_{1s}

1s 1l   1l 1s

1l σ_{1s}

AO   MO   AO

AO = Atomic Orbital; MO = Molecular Orbital

Bond order $= \dfrac{3-2}{2} = \dfrac{1}{2}$; stable   Bond order $= \dfrac{4-3}{2} = \dfrac{1}{2}$; stable

Both of these ions are able to exist in the gas phase because the bond order is positive.

7.51 The bonding and antibonding molecular orbitals from the combination of p_x and p_x atomic orbitals lie along the internuclear axis.

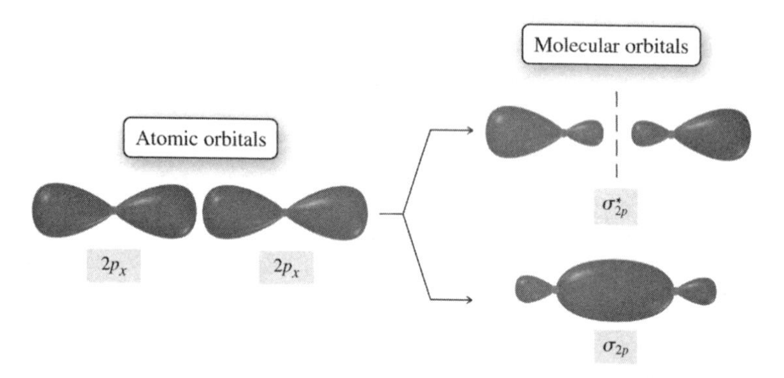

7.52 The bonding and antibonding molecular orbitals from the combination of p_y and p_y atomic orbitals lie above and below the internuclear axis.

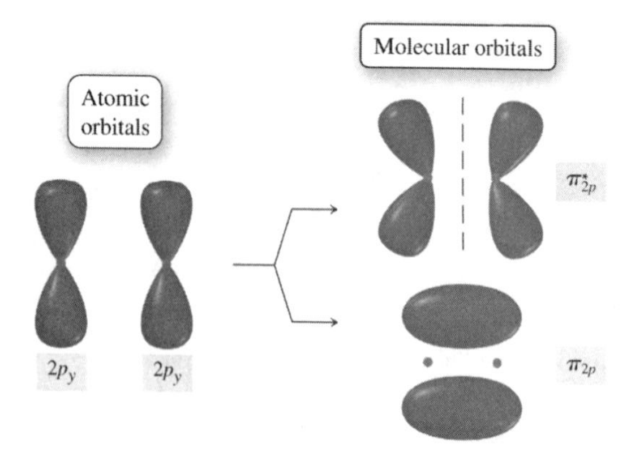

When the p_z and p_z orbitals combine, similar bonding and antibonding molecular orbitals form. The only difference between the resulting MOs is a rotation about the internuclear axis. The energies and the names of the bonding and

antibonding MOs obtained from the combination of the p_z atomic orbitals are identical to those obtained from the combination of the p_y atomic orbitals, which lie in front of and behind the internuclear axis.

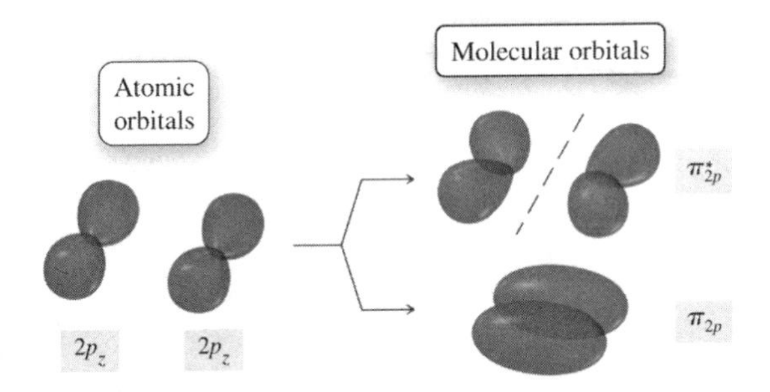

7.53

	(a) 4 valence electrons	(b) 6 valence electrons	(c) 8 valence electrons	(d) 9 valence electrons
σ^*_{2p}	___	___	___	___
π^*_{2p}	___ ___	___ ___	___ ___	___ ___
σ_{2p}	___	___	___	↑
π_{2p}	___ ___	↑ ↑	↑↓ ↑↓	↑↓ ↑↓
σ^*_{2s}	↑↓	↑↓	↑↓	↑↓
σ_{2s}	↑↓	↑↓	↑↓	↑↓
	Bond order $= \frac{2-2}{2} = 0$	Bond order $= \frac{4-2}{2} = 1$	Bond order $= \frac{6-2}{2} = 2$	Bond order $= \frac{7-2}{2} = 2.5$
	diamagnetic	paramagnetic	diamagnetic	paramagnetic

7.54

	(a) 10 valence electrons	(b) 12 valence electrons	(c) 13 valence electrons	(d) 14 valence electrons
σ^*_{2p}	___	___	___	___
π^*_{2p}	___ ___	↑ ↑	↑↓ ↑	↑↓ ↑↓
π_{2p}	↑↓ ↑↓	↑↓ ↑↓	↑↓ ↑↓	↑↓ ↑↓
σ_{2p}	↑↓	↑↓	↑↓	↑↓
σ^*_{2s}	↑↓	↑↓	↑↓	↑↓
σ_{2s}	↑↓	↑↓	↑↓	↑↓
	Bond order $= \frac{8-2}{2} = 3$	Bond order $= \frac{8-4}{2} = 2$	Bond order $= \frac{8-5}{2} = 1.5$	Bond order $= \frac{8-6}{2} = 1$
	diamagnetic	paramagnetic	paramagnetic	diamagnetic

7.55 (a) Write an energy level diagram for the molecular orbitals in H_2^{2-}. The ion has four valence electrons. Assign the electrons to the molecular orbitals beginning with the lowest energy orbitals and following Hund's rule.

σ^*_{1s} ↑↓

σ_{1s} ↑↓ Bond order $= \frac{2-2}{2} = 0$. With a bond order of 0, the ion will not exist.

(b) Write an energy level diagram for the molecular orbitals in Ne_2. The molecule has 16 valence electrons. Assign the electrons to the molecular orbitals beginning with the lowest energy orbitals and following Hund's rule.

σ^*_{2p} ↑↓

π^*_{2p} ↑↓ ↑↓

π_{2p} ↑↓ ↑↓

σ_{2p} ↑↓

σ^*_{2s} ↑↓

σ_{2s} ↑↓

Bond order $= \frac{8 - 8}{2} = 0$. With a bond order of 0, the molecule will exist.

(c) Write an energy level diagram for the molecular orbitals in He_2^{2+}. The ion has two valence electrons. Assign the electrons to the molecular orbitals beginning with the lowest energy orbitals and following Hund's rule.

σ^*_{1s} ____

σ_{1s} ↑↓

Bond order $= \frac{2 - 0}{2} = 1$. With a bond order of 1, the ion will exist.

(d) Write an energy level diagram for the molecular orbitals in F_2^{2-}. The molecule has 16 valence electrons. Assign the electrons to the molecular orbitals beginning with the lowest energy orbitals and following Hund's rule.

σ^*_{2p} ↑↓

π^*_{2p} ↑↓ ↑↓

π_{2p} ↑↓ ↑↓

σ_{2p} ↑↓

σ^*_{2s} ↑↓

σ_{2s} ↑↓

Bond order $= \frac{8 - 8}{2} = 0$. With a bond order of 0, the ion will not exist.

7.56 (a) Write an energy level diagram for the molecular orbitals in C_2^{2+}. The ion has six valence electrons. Assign the electrons to the molecular orbitals beginning with the lowest energy orbitals and following Hund's rule.

σ^*_{2p} ____

π^*_{2p} ____ ____

σ_{2p} ____

π_{2p} ↑ ↑

σ^*_{2s} ↑↓

σ_{2s} ↑↓

Bond order $= \frac{4 - 2}{2} = 1$. With a bond order of 1, the ion will exist.

(b) Write an energy level diagram for the molecular orbitals in Li_2. The molecule has two valence electrons. Assign the electrons to the molecular orbitals beginning with the lowest energy orbitals and following Hund's rule.

σ^*_{2s} ______

σ_{2s} ↑↓

$$\text{Bond order} = \frac{2-0}{2} = 1$$

With a bond order of 1, the molecule will exist.

(c) Write an energy level diagram for the molecular orbitals in Be_2^{2+}. The ion has two valence electrons. Assign the electrons to the molecular orbitals beginning with the lowest energy orbitals and following Hund's rule.

σ^*_{2s} ______

σ_{2s} ↑↓

$$\text{Bond order} = \frac{2-0}{2} = 1$$

With a bond order of 1, the ion will exist.

(d) Write an energy level diagram for the molecular orbitals in Li_2^{2-}. The ion has four valence electrons. Assign the electrons to the molecular orbitals beginning with the lowest energy orbitals and following Hund's rule.

σ^*_{2s} ↑↓

σ_{2s} ↑↓

$$\text{Bond order} = \frac{2-2}{2} = 0$$

With a bond order of 0, the ion will not exist.

7.57 C_2^- has the highest bond order, the highest bond energy, and the shortest bond.
Write an energy level diagram for the molecular orbitals in each of the C_2 species.
Assign the electrons to the molecular orbitals beginning with the lowest energy orbitals and following Hund's rule for each of the species.

C_2 (8 valence electrons) C_2^+ (7 valence electrons) C_2^- (9 valence electrons)

	C_2	C_2^+	C_2^-
σ^*_{2p}	____	____	____
π^*_{2p}	____ ____	____ ____	____ ____
σ_{2p}	____	____	↑
π_{2p}	↑↓ ↑↓	↑↓ ↑	↑↓ ↑↓
σ^*_{2s}	↑↓	↑↓	↑↓
σ_{2s}	↑↓	↑↓	↑↓

$$\text{Bond order} = \frac{6-2}{2} = 2 \qquad \text{Bond order} = \frac{5-2}{2} = 1.5 \qquad \text{Bond order} = \frac{7-2}{2} = 2.5$$

C_2^- has the highest bond order at 2.5. Bond order is directly related to bond energy, so C_2^- has the largest bond energy. Bond order is inversely related to bond length, so C_2^- has the shortest bond length.

7.58 O_2 has the highest bond order, the highest bond energy, and the shortest bond.
Write an energy level diagram for the molecular orbitals in each of the O_2 species.
Assign the electrons to the molecular orbitals beginning with the lowest energy orbitals and following Hund's rule for each of the species.

	O_2 (12 valence electrons)	O_2^- (13 valence electrons)	O_2^{2-} (14 valence electrons)
σ^*_{2p}	___	___	___
π^*_{2p}	1 1	1l 1	1l 1l
π_{2p}	1l 1l	1l 1l	1l 1l
σ_{2p}	1l	1l	1l
σ^*_{2s}	1l	1l	1l
σ_{2s}	1l	1l	1l

Bond order $= \dfrac{8-4}{2} = 2$ Bond order $= \dfrac{8-5}{2} = 1.5$ Bond order $= \dfrac{8-6}{2} = 1$

O_2 has the highest bond order at 2. Bond order is directly related to bond energy, so O_2 has the largest bond energy Bond order is inversely related to bond length, so O_2 has the shortest bond length.

7.59 Write an energy level diagram for the molecular orbitals in CO using O_2 energy ordering.
Assign the electrons to the molecular orbitals beginning with the lowest energy orbitals and following Hund's rule.
CO has 10 valence electrons.

	CO
σ^*_{2p}	___
π^*_{2p}	___ ___
π_{2p}	1l 1l
σ_{2p}	1l
σ^*_{2s}	1l
σ_{2s}	1l

Bond order $= \dfrac{8-2}{2} = 3$

The electron density is toward the O atom because it is more electronegative.

7.60 HCl (8 valence electrons)

		HCl	Cl
		σ^* ___	
		p 1l 1l (nonbonding)	1l 1l 1 2p Cl
		s 1l (nonbonding)	1l 2s Cl
H	1s 1		
		σ 1l	

Bond order $= \dfrac{2-0}{2} = 1$

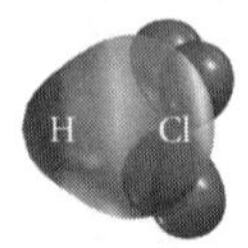

nonbonding
p orbitals on Cl

The electron density will be skewed toward the Cl atom because it is more electronegative.

Electron Sea Model and Band Theory

7.61 **Given:** 5.68 g magnesium metal **Find:** number of electrons in electron sea

Conceptual plan: Remember that only the valence electrons are in the electron sea

grams → mol Mg → moles electrons → electrons

$$\frac{1 \text{ mol Mg}}{24.31 \text{ g Mg}} \quad \frac{2 \text{ mol electrons}}{1 \text{mol Mg}} \quad \frac{6.022 \times 10^{23} \text{ electrons}}{1 \text{ mol electrons}}$$

Solution: $5.68 \text{ g Mg} \times \frac{1 \text{ mol Mg}}{24.31 \text{ g Mg}} \times \frac{2 \text{ mol electrons}}{1 \text{ mol Mg}} \times \frac{6.022 \times 10^{23} \text{ electrons}}{1 \text{ mol electrons}} = 1.41 \times 10^{23} \text{ electrons}$

Check: The units of the answer (number of electrons) are correct. The magnitude (10^{23}) is reasonable because there is about a quarter of a mole of magnesium.

7.62 **Given:** 28.5 g potassium metal **Find:** number of electrons in electron sea

Conceptual plan: Remember that only the valence electrons are in the electron sea

grams → mol K → moles electrons → electrons

$$\frac{1 \text{ mol K}}{39.10 \text{ g K}} \quad \frac{1 \text{ mol electrons}}{1 \text{mol K}} \quad \frac{6.022 \times 10^{23} \text{ electrons}}{1 \text{ mol electrons}}$$

Solution: $28.5 \text{ g K} \times \frac{1 \text{ mol K}}{39.10 \text{ g K}} \times \frac{1 \text{ mol electrons}}{1 \text{ mol K}} \times \frac{6.022 \times 10^{23} \text{ electrons}}{1 \text{ mol electrons}} = 4.39 \times 10^{23} \text{ electrons}$

Check: The units of the answer (number of electrons) are correct. The magnitude (10^{23}) is reasonable because there is almost a mole of potassium.

7.63 (a) Zn should have little or no band gap because it is the only metal in the group.

7.64 **Given:** 5.45 g sodium crystal **Find:** number of molecular orbitals in the valence band

Conceptual Plan: g → mol → Na_N → number of valence electrons → number of molecular orbitals

$$\frac{1 \text{ mol Na}}{22.99 \text{ g Na}} \quad \frac{6.022 \times 10^{23} \text{ atoms}}{1 \text{ mol}} \quad \frac{1\ 3s \text{ electron}}{1 \text{ Na atom}} \quad \frac{1 \text{ molecular orbital}}{1\ 3s \text{ electron}}$$

Solution: $5.45 \text{ g Na} \times \frac{1 \text{ mol Na}}{22.99 \text{ g Na}} \times \frac{6.022 \times 10^{23} \text{ Na atoms}}{1 \text{ mol Na}} \times \frac{1\ 3s \text{ electron}}{1 \text{ Na atom}} \times \frac{1 \text{ molecular orbital}}{1\ 3s \text{ electron}}$

$= 1.43 \times 10^{23}$ molecular orbitals

Check: The units (number of molecular orbitals) are correct. The magnitude (10^{23} molecular orbitals) is expected because there are so many orbitals since we have about $\frac{1}{4}$ mole of atoms. Metals can conduct electricity because of these large numbers of orbitals.

7.65 (a) p-type semiconductor: Ge is Group 4A, and Ga is Group 3A; so the Ga will generate electron "holes."

(b) n-type semiconductor: Si is Group 4A, and As is Group 5A; so the As will add electrons to the conduction band.

7.66 (a) p-type semiconductor: Si is Group 4A, and Ga is Group 3A; so the Ga will generate electron "holes."

(b) n-type semiconductor: Ge is Group 4A, and Sb is Group 5A; so the Sb will add electrons to the conduction band.

Cumulative Problems

7.67 (a) COF_2 Write the Lewis structure for the molecule:

```
     :O:
      ‖
 :F̤̈—C—F̤̈:
```

Use VSEPR to predict the electron geometry:

Three electron groups around the central atom give a trigonal planar electron geometry. Three bonding pairs of electrons give a trigonal planar molecular geometry.

Determine whether the molecule contains polar bonds:

The electronegativity of C = 2.5, O = 3.5, and F = 4.0. Therefore, the bonds are polar.

Determine whether the polar bonds add together to form a net dipole:
Even though a trigonal planar molecular geometry normally is nonpolar, because the bonds have different dipole moments, the sum of the dipole moments is not zero. The molecule is polar. See Table 10.2 to see how dipole moments add to determine polarity.
Select the correct hybridization for the central atom based on the electron geometry:
Trigonal planar geometry has sp^2 hybridization.
Sketch the molecule and label the bonds:

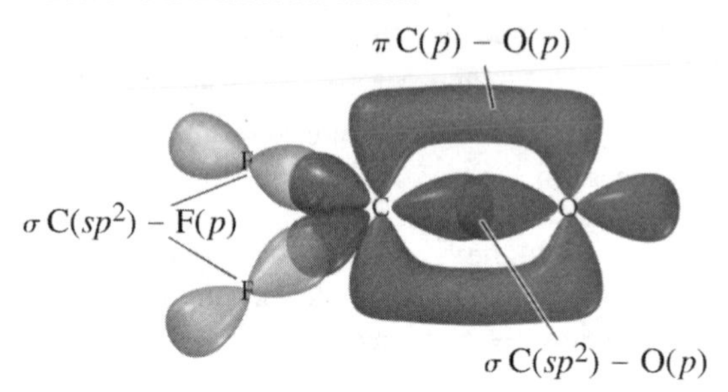

(b) S_2Cl_2 Write the Lewis structure for the molecule:

:C̤̈l—S̤̈—S̤̈—C̤̈l:

Use VSEPR to predict the electron geometry:
Four electron groups around a central atom give a tetrahedral electron geometry. Two bonding pairs and two lone pairs of electrons give a bent molecular geometry.
Determine whether the molecule contains polar bonds:
The electronegativity of S = 2.5 and Cl = 3.0. Therefore, the bonds are polar.
Determine whether the polar bonds add together to form a net dipole:
In a bent molecular geometry, the sum of the dipole moments is not zero. The molecule is polar. See Table 10.2 to see how dipole moments add to determine polarity.
Select the correct hybridization for the central atom based on the electron geometry:
Tetrahedral geometry has sp^3 hybridization.
Sketch the molecule and label the bonds:

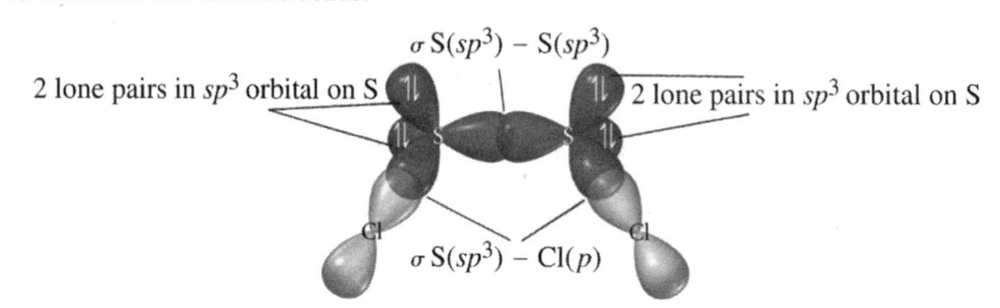

(c) SF_4 Write the Lewis structure for the molecule:

```
     :F̈:
      |
:F̤̈—S̤—F̤̈:
      |
     :F̤:
```

Use VSEPR to predict the electron geometry:
Five electron groups around the central atom give a trigonal bipyramidal electron geometry.
Four bonding pairs and one lone pair of electrons give a seesaw molecular geometry.
Determine whether the molecule contains polar bonds:
The electronegativity of S = 2.5 and F = 4.0. Therefore, the bonds are polar.
Determine whether the polar bonds add together to form a net dipole:
In a seesaw molecular geometry, the sum of the dipole moments is not zero. The molecule is polar.
See Table 10.2 to see how dipole moments add to determine polarity.
Select the correct hybridization for the central atom based on the electron geometry:

Trigonal bipyramidal electron geometry has sp^3d hybridization.
Sketch the molecule and label the bonds:

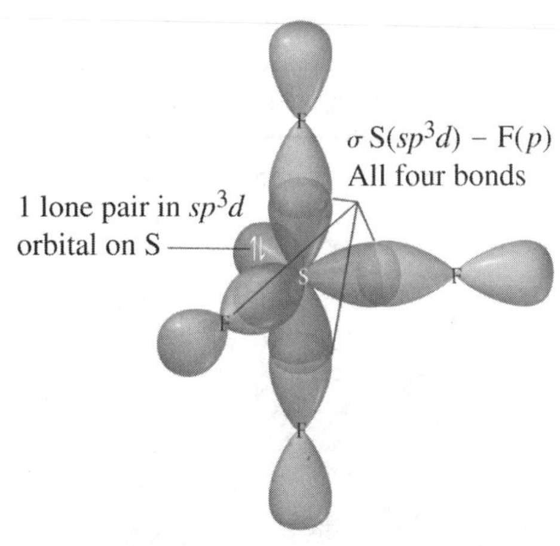

7.68 (a) IF_5 Write the Lewis structure for the molecule:

```
        ..
       :F:
  ..    |    ..
 :F \   |   / F:
  ..   \ | /  ..
        I
  ..   /..\   ..
 :F /       \ F:
  ..          ..
```

Use VSEPR to predict the electron geometry:
Six electron groups around the central atom give an octahedral electron geometry. Five bonding pairs and one lone pair of electrons give a square pyramidal molecular geometry.
Determine whether the molecule contains polar bonds:
The electronegativity of I = 2.5 and F = 4.0. Therefore, the bonds are polar.
Determine whether the polar bonds add together to form a net dipole:
In a square pyramidal molecular geometry, the sum of the dipole moments is not zero. The molecule is polar. See Table 10.2 to see how dipole moments add to determine polarity.
Select the correct hybridization for the interior atoms based on the electron geometry:
Octahedral electron geometry has sp^3d^2 hybridization.
Sketch the molecule and label the bonds:

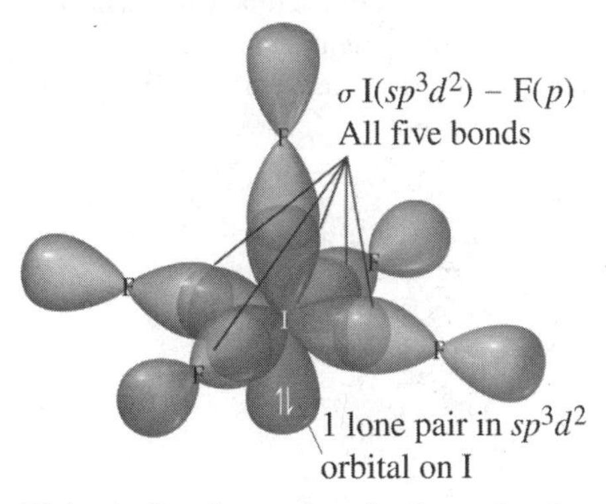

(b) CH_2CHCH_3 Write the Lewis structure for the molecule:

```
         H   H
 H       |   |
   \ C1 = C2 — C3 — H
 H /          |
              H
```

Use VSEPR to predict the electron geometry:
The Cs in positions 1 and 2 have three electron groups around the atom, which is trigonal planar electron geometry. Three bonding pairs of electrons give a trigonal planar molecular geometry.
C in position 3 has four electron groups around the C atom; four electron groups give a tetrahedral electron geometry. Four bonding groups give a tetrahedral molecular geometry.

Determine whether the molecule contains polar bonds:
The electronegativity of C = 2.5 and H = 2.1. Therefore, the bonds are slightly polar because the difference in electronegativity is less than 0.5.
Determine whether the polar bonds add together to form a net dipole:
The trigonal planar molecular geometry and the tetrahedral molecular geometry give a net dipole moment of zero. The molecule is nonpolar.
Select the correct hybridization for the interior atoms based on the electron geometry:
Trigonal planar electron geometry has sp^2 hybridization, and tetrahedral electron geometry has sp^3 hybridization.
Sketch the molecule and label the bonds:

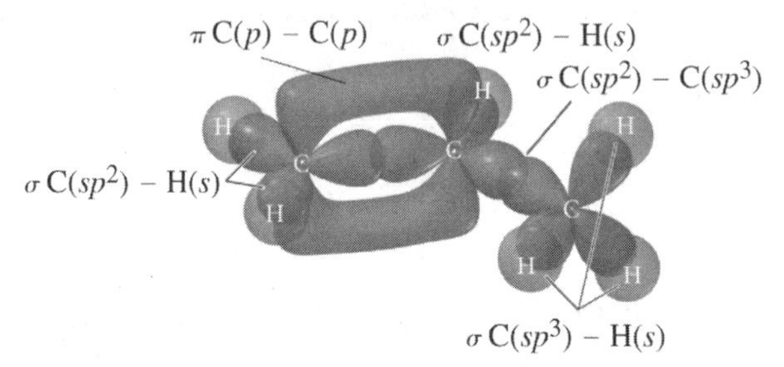

(c) CH_3SH Write the Lewis structure for the molecule:

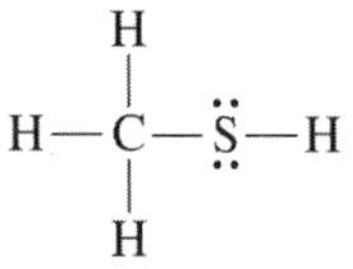

```
        H
        |    ..
    H—C—S—H
        |    ..
        H
```

Use VSEPR to predict the electron geometry:
Four electron groups around the C atom and the S atom give a tetrahedral electron geometry.
Four bonding pairs of electrons around the C give a tetrahedral molecular geometry, and two bonding groups and two lone pairs around the S give a bent molecular geometry.
Determine whether the molecule contains polar bonds:
The electronegativity of C = 2.5, S = 2.5, and H = 2.1. The C—H bonds and the S—H bond will be slightly polar, and the C—S bond will be nonpolar.
Determine whether the polar bonds add together to form a net dipole:
In both molecular geometries, the sum of the dipole moments is not zero. The molecule is polar.
Select the correct hybridization for the central atom based on the electron geometry:
Tetrahedral electron geometry has sp^3 hybridization on both C and S.
Sketch the molecule and label the bonds:

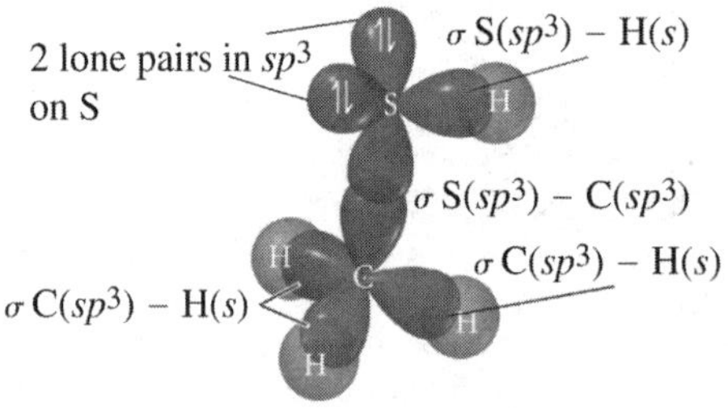

2 lone pairs in sp^3 on S
σ $S(sp^3) - H(s)$
σ $S(sp^3) - C(sp^3)$
σ $C(sp^3) - H(s)$
σ $C(sp^3) - H(s)$

7.69 (a) serine

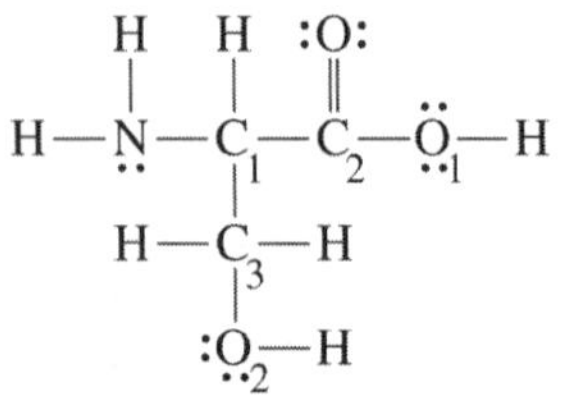

```
     H   H  :O:
     |   |   ||    ..
 H—N—C1—C2—O1—H
    ..   |         ..
     H—C3—H
         |
       :O2—H
        ..
```

The Cs in positions 1 and 3 have four electron groups around the atom. Four electron groups give a tetrahedral electron geometry; tetrahedral electron geometry has sp^3 hybridization. Four bonding groups and zero lone pairs give a tetrahedral molecular geometry.

C in position 2 has three electron groups around the atom. Three electron groups give a trigonal planar geometry; trigonal planar geometry has sp^2 hybridization. Three bonding groups and zero lone pairs give a trigonal planar molecular geometry.

N has four electron groups around the atom. Four electron groups give a tetrahedral electron geometry; tetrahedral electron geometry has sp^3 hybridization. Three bonding groups and one lone pair give a trigonal pyramidal molecular geometry.

The Os in positions 1 and 2 each have four electron groups around the atom. Four electron groups give a tetrahedral electron geometry; tetrahedral electron geometry has sp^3 hybridization. Two bonding groups and two lone pairs give a bent molecular geometry.

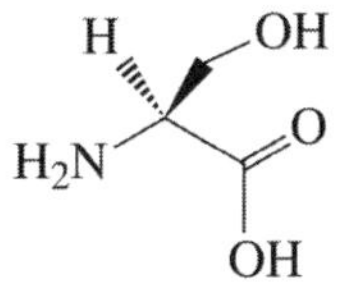

(b) asparagine

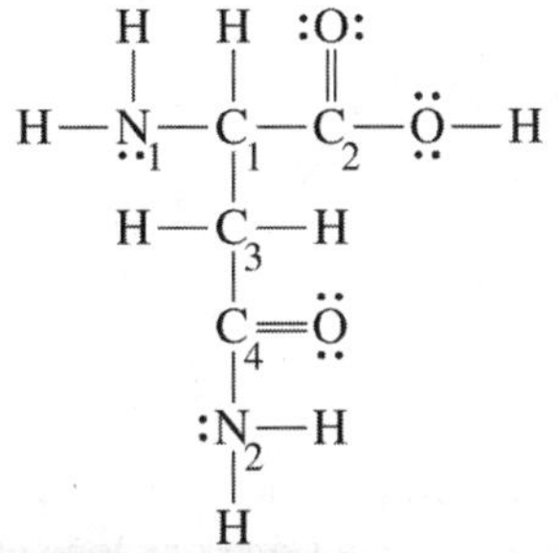

The Cs in positions 1 and 3 each have four electron groups around the atom. Four electron groups give a tetrahedral electron geometry; tetrahedral electron geometry has sp^3 hybridization. Four bonding groups and zero lone pairs give a tetrahedral molecular geometry.

The Cs in positions 2 and 4 each have three electron groups around the atom. Three electron groups give a trigonal planar geometry; trigonal planar geometry has sp^2 hybridization. Three bonding groups and zero lone groups give a trigonal planar molecular geometry.

The Ns in positions 1 and 2 each have four electron groups around the atom. Four electron groups give a tetrahedral electron geometry; tetrahedral electron geometry has sp^3 hybridization. Three bonding groups and one lone pair give a trigonal pyramidal molecular geometry.

O has four electron groups around the atom. Four electron groups give a tetrahedral electron geometry; tetrahedral electron geometry has sp^3 hybridization. Two bonding groups and two lone pairs give a bent molecular geometry.

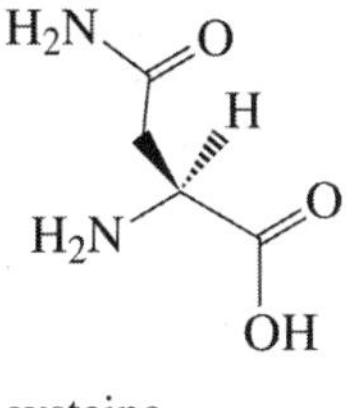

(c) cysteine

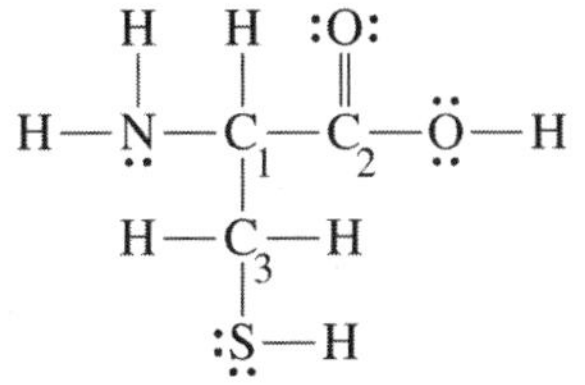

The Cs in positions 1 and 3 each have four electron groups around the atom. Four electron groups give a tetrahedral electron geometry; tetrahedral electron geometry has sp^3 hybridization. Four bonding groups and zero lone pairs give a tetrahedral molecular geometry.

C in position 2 has three electron groups around the atom. Three electron groups give a trigonal planar geometry; trigonal planar geometry has sp^2 hybridization. Three bonding groups and zero lone pairs give a trigonal planar molecular geometry.

N has four electron groups around the atom. Four electron groups give a tetrahedral electron geometry; tetrahedral electron geometry has sp^3 hybridization. Three bonding groups and one lone pair give a trigonal pyramidal molecular geometry.

O and S have four electron groups around the atom. Four electron groups give a tetrahedral electron geometry; tetrahedral electron geometry has sp^3 hybridization. Two bonding groups and two lone pairs give bent molecular geometry.

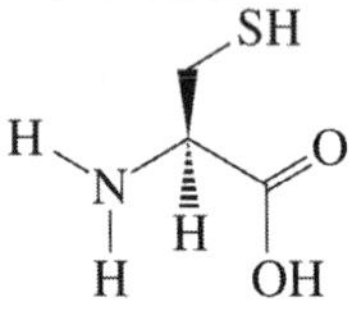

SH
H N H O
H OH

7.70 (a) cytosine

H
N
HC 6 1 2 C O
HC 5 4 3 N
C
NH_2

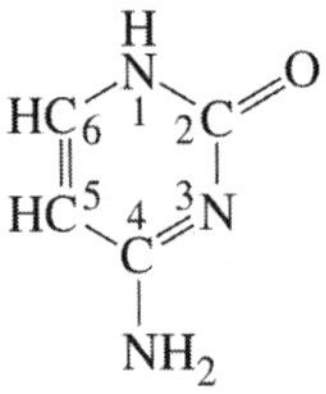

N in position 1 has three bonding groups of electrons and one lone pair; four electron groups give a tetrahedral electron geometry and sp^3 hybridization. Three bonding groups of electrons and one lone pair give a trigonal pyramidal molecular geometry.

C in position 2 has three bonding groups of electrons and zero lone pairs; three electron groups give a trigonal planar geometry and sp^2 hybridization. Three bonding groups of electrons give a trigonal planar molecular geometry.

N in position 3 has two bonding groups of electrons and one lone pair; three electron groups give a trigonal planar electron geometry and sp^2 hybridization. Two bonding groups and one lone pair give a bent molecular geometry.

C in position 4 has three bonding groups of electrons and zero lone pairs; three electron groups give a trigonal planar geometry and sp^2 hybridization. Three bonding groups of electrons give a trigonal planar molecular geometry.

The Cs in positions 5 and 6 have three bonding groups of electrons and zero lone pairs; three electron groups give a trigonal planar geometry and sp^2 hybridization. Three bonding groups of electrons give a trigonal planar molecular geometry.

N outside the ring has three bonding groups of electrons and one lone pair; four electron groups give a tetrahedral electron geometry and sp^3 hybridization. Three bonding groups of electrons give a trigonal pyramidal molecular geometry.

(b) adenine

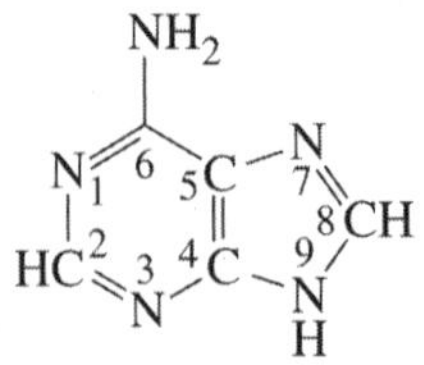

NH_2
N 1 6 5 C N 7
8 CH
HC 2 3 4 C 9 N
N H

N in positions 1, 3, and 7 have two bonding groups of electrons and one lone pair; three electron groups give a trigonal planar electron geometry and sp^2 hybridization. Two bonding groups and one lone pair give a bent molecular geometry.

The Cs in positions 2, 4, 5, 6, and 8 have three bonding groups of electrons and zero lone groups; three electron groups give a trigonal planar geometry and sp^2 hybridization. Three bonding groups of electrons give a trigonal planar molecular geometry.

N in position 9 and the N outside the ring have three bonding groups of electrons and one lone pair; four electron groups give a tetrahedral electron geometry and sp^3 hybridization. Three bonding groups of electrons give a trigonal pyramidal molecular geometry.

(c) thymine

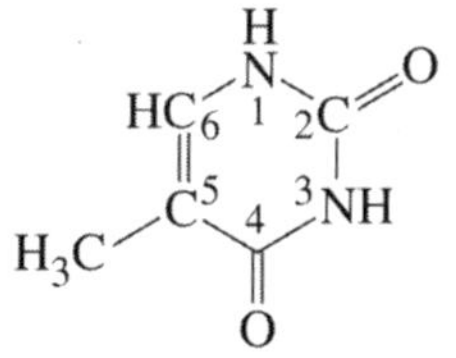

The Ns in positions 1 and 3 have three bonding groups of electrons and one lone pair; four electron groups give a tetrahedral electron geometry and sp^3 hybridization. Three bonding groups of electrons give a trigonal pyramidal molecular geometry.

C in position 2 has three bonding groups of electrons and zero lone pairs; three electron groups give a trigonal planar geometry and sp^2 hybridization. Three bonding groups of electrons give a trigonal planar molecular geometry.

The Cs in positions 4, 5, and 6 have three bonding groups of electrons and zero lone pairs; three electron groups give a trigonal planar geometry and sp^2 hybridization. Three bonding groups of electrons give a trigonal planar molecular geometry.

C outside the ring has four bonding groups of electrons and zero lone pairs; four electron groups give a tetrahedral electron geometry and sp^3 hybridization. Four bonding groups of electrons give a tetrahedral molecular geometry.

(d) guanine

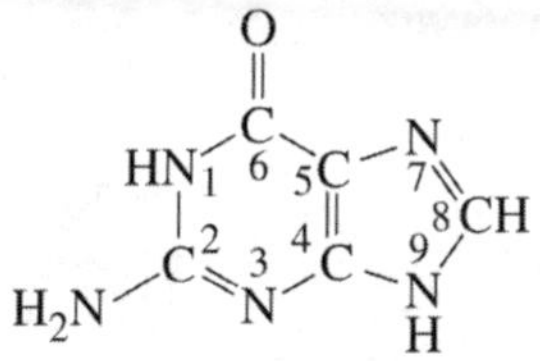

The Ns in positions 1 and 9 and the N outside the ring have three bonding groups of electrons and one lone pair; four electron groups give a tetrahedral electron geometry and sp^3 hybridization. Three bonding groups of electrons give a trigonal pyramidal molecular geometry.

The Cs in positions 2, 4, 5, 6, and 8 have three bonding groups of electrons and zero lone pairs; three electron groups give a trigonal planar geometry and sp^2 hybridization. Three bonding groups of electrons give a trigonal planar molecular geometry.

The Ns in positions 3 and 7 have two bonding groups of electrons and one lone pair; three electron groups give a trigonal planar electron geometry and sp^2 hybridization. Two bonding groups and one lone pair give a bent molecular geometry.

7.71 4π bonds, 25σ bonds; the lone pair on the Os and N in position 2 occupies sp^2 orbitals; the lone pairs on N in position 1, N in position 3, and N in position 4 occupy sp^3 orbitals.

caffeine

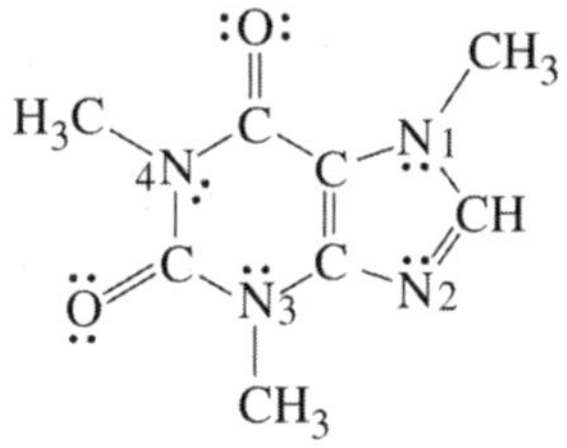

7.72 5π bonds, 21σ bonds

aspirin

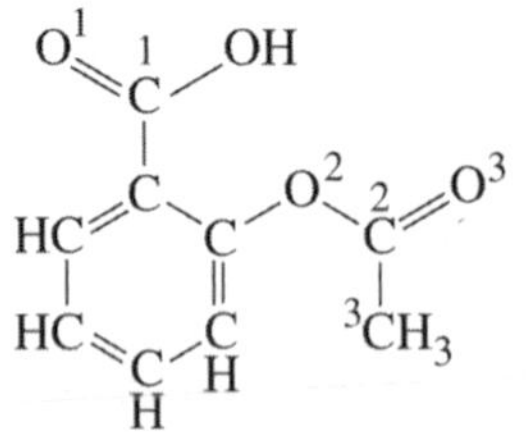

There is rotation around the bond from C in position 1 to the ring and from C in position 1 to the OH bond. There is rotation around the O in position 2 to the ring bond and around the O in position 2 to C in position 2 bond. There is rotation around the C in position 2 to C in position 3 bond. The C in position 1 to O in position 1 bond is rigid, the ring structure is rigid, and the C in position 2 to O in position 3 bond is rigid.

7.73 ClF has 14 valence electrons. Assign the electrons to the lowest energy MOs first and then follow Hund's rule. The MOs are formed from the 2*s* and 2*p* orbitals on F and the 3*s* and 3*p* orbitals on Cl.

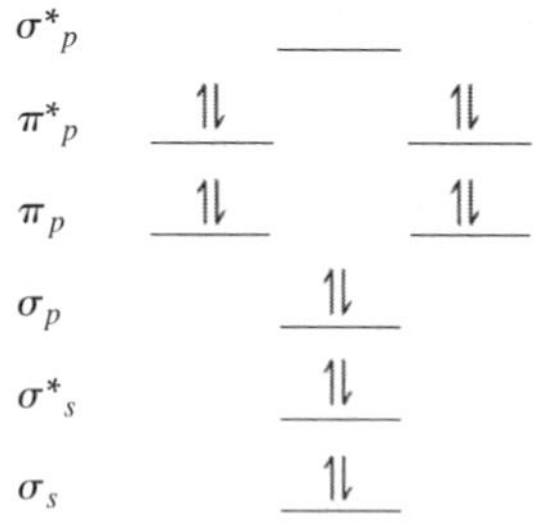

MO			
σ^*_p		___	
π^*_p	⇅		⇅
π_p	⇅		⇅
σ_p		⇅	
σ^*_s		⇅	
σ_s		⇅	

$$\text{Bond order} = \frac{8-6}{2} = 1$$

7.74 CN^+ (8 valence electrons) CN (9 valence electrons) CN^- (10 valence electrons)

$[C{\equiv}N:]^+$ $\cdot C{\equiv}N:$ $[:C{\equiv}N:]^-$

MO	CN^+			CN			CN^-		
σ^*_{2p}		___			___			___	
π^*_{2p}	___		___	___		___	___		___
σ_{2p}		⇅			⇅			⇅	
π_{2p}	1		1	⇅		1	⇅		⇅
σ^*_{2s}		⇅			⇅			⇅	
σ_{2s}		⇅			⇅			⇅	

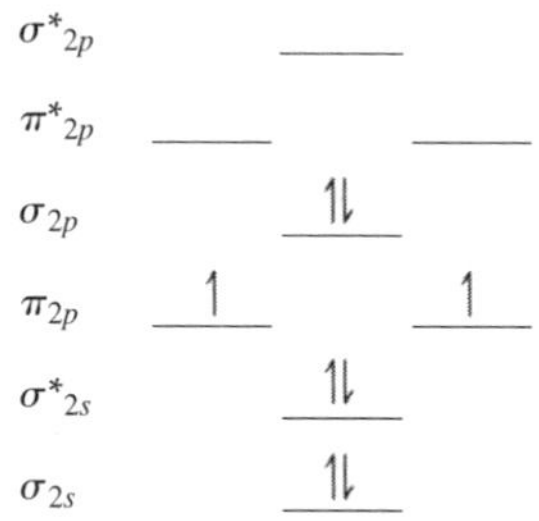

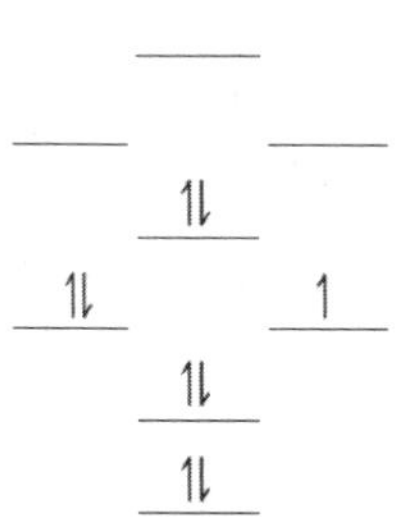

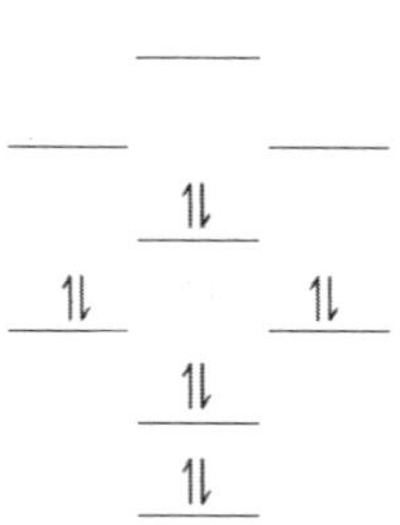

$$\text{Bond order} = \frac{6-2}{2} = 2 \quad \text{Bond order} = \frac{7-2}{2} = 2.5 \quad \text{Bond order} = \frac{8-2}{2} = 3$$

According to Lewis theory, the CN^- ion would be most stable because it completes the octet on both atoms. The other two (CN^+ and CN) have an incomplete octet on C.

According to molecular orbital theory, the CN^- ion would be most stable because it has the highest bond order, 3. The other two species would also exist but would be less stable. Both theories agree that the CN^- ion would be most stable.

7.75 BrF (14 valence electrons)

$:\ddot{\underset{..}{Br}}-\ddot{\underset{..}{F}}:$ no central atom, no hybridization, no electron structure.

BrF_2^- (22 valence electrons)

$[:\ddot{F}-\ddot{Br}-\ddot{F}:]^-$ There are five electron groups on the central atom, so the electron geometry is trigonal bipyramidal. The two bonding groups and three lone pairs give a linear molecular geometry. An electron geometry of trigonal bipyramidal has sp^3d hybridization.

BrF_3 (28 valence electrons)

There are five electron groups on the central atom, so the electron geometry is trigonal bipyramidal. The three bonding groups and two lone pairs give a T-shaped molecular geometry. An electron geometry of trigonal bipyramidal has sp^3d hybridization.

BrF_4^- (36 valence electrons)

There are six electron groups on the central atom, so the electron geometry is octahedral. The four bonding groups and two lone pairs give a square planar molecular geometry. An electron geometry of octahedral has sp^3d^2 hybridization.

BrF_5 (42 valence electrons)

There are six electron groups on the central atom, so the electron geometry is octahedral. The five bonding groups and one lone pair give a square pyramidal molecular geometry. An electron geometry of octahedral has sp^3d^2 hybridization.

7.76 Write the Lewis structure:

$H_2C_1=C_2=C_3H_2$

C in position 1 and C in position 3 both have three groups of electrons and a trigonal planar structure giving sp^2 hybridization on the C with a p orbital left for the π bond. C in position 2 has two groups of electrons and is linear, which shows sp hybridization and $2p$ orbitals left for the π bonds. According to valence bond theory, the π bonds are formed by the sideways overlap of p orbitals. Because the remaining p orbitals on C in position 2 are perpendicular to each other, the π bonds formed between C in position 1 and C in position 2 and between C in position 2 and C in position 3 also must be perpendicular to each other. Therefore, the two trigonal planar structures at C in position 1 and C in position 3 will be perpendicular to each other.

7.77 (a) N_2O_5 Draw the Lewis structure: (40 valence electrons)

Each N has a trigonal planar electron geometry, so there are 3 sp^2 hybrid orbitals on each N. The central O has tetrahedral electron geometry, so there are 4 sp^3 hybrid orbitals. There are a total of 10 hybrid orbitals.

(b) C_2H_5NO Draw the Lewis structure: (24 valence electrons)

$H_2C_A=\ddot{N}-C_BH_2-\ddot{O}-H$

C_A has a trigonal planar electron geometry, so there are 3 sp^2 hybrid orbitals.
N has a trigonal planar electron geometry, so there are 3 sp^2 hybrid orbitals.
C_B has a tetrahedral electron geometry, so there are 4 sp^3 hybrid orbitals.
O has a tetrahedral electron geometry, so there are 4 sp^3 hybrid orbitals.
There are a total of 14 hybrid orbitals.

(c) BrCN Draw the Lewis structure: (16 valence electrons)

:B̤̈r—C≡N:

The C has linear electron geometry, so there are 2 *sp* hybrid orbitals.

7.78 (a) $BeBr_2$ Draw the Lewis structure: (16 valence electrons)

:B̤̈r—Be—B̤̈r:

Sketch the molecule.

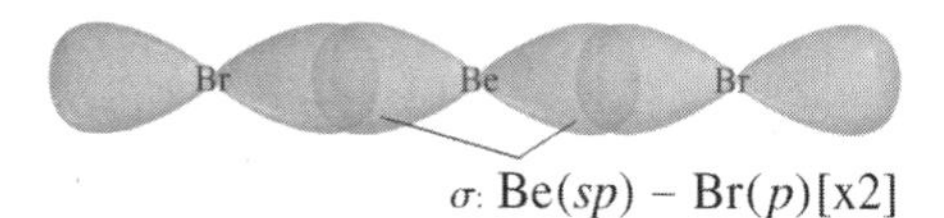

The 2σ bonds form between a hybrid *sp* orbital on Be and a *p* orbital on Br.

(b) $HgCl_2$ Draw the Lewis structure: (16 valence electrons)

:C̤̈l—Hg—C̤̈l:

Sketch the molecule.

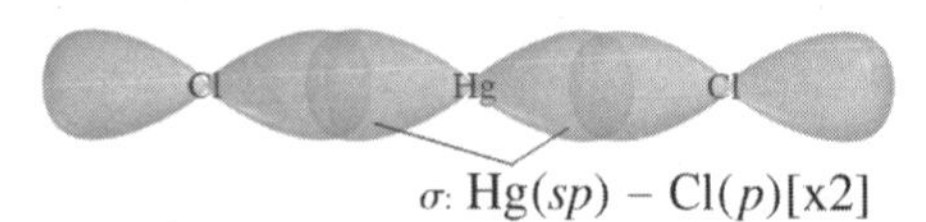

The 2σ bonds form between a hybrid *sp* orbital on Hg and a *p* orbital on Cl.

(c) ICN Draw the Lewis structure: (16 valence electrons)

:Ï̤—C≡N:

Sketch the molecule.

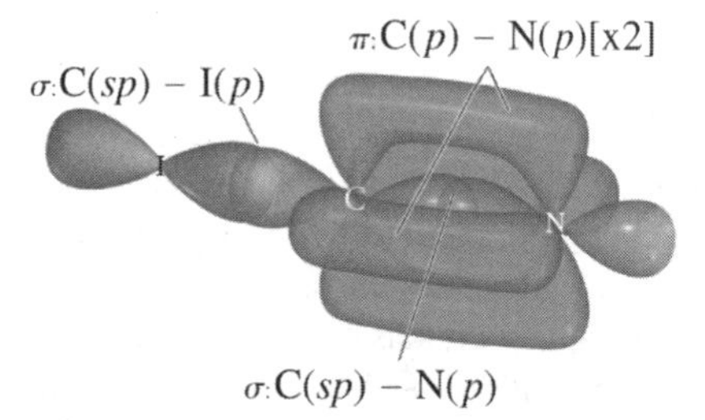

1σ bond forms between a hybrid *sp* orbital on C and a *p* orbital on I; 1σ bond forms between a hybrid *sp* orbital on C and a *p* orbital on N; 2π bonds form between unhybridized *p* orbitals on C and *p* orbitals on N.

Challenge Problems

7.79 According to valence bond theory, CH_4, NH_3, and H_2O are all sp^3 hybridized. This hybridization results in a tetrahedral electron group configuration with a 109.5° bond angle. NH_3 and H_2O deviate from this idealized bond angle because their lone electron pairs exist in their own sp^3 orbitals. The presence of lone pairs lowers the tendency for the central atom's orbitals to hybridize. As a result, as lone pairs are added, the bond angle moves from the 109.5° hybrid angle toward the 90° unhybridized angle.

7.80 Using the MO diagram for H_2O, assign the eight valence electrons to the molecular orbitals. Start with the lowest energy orbital first and follow Hund's rule.

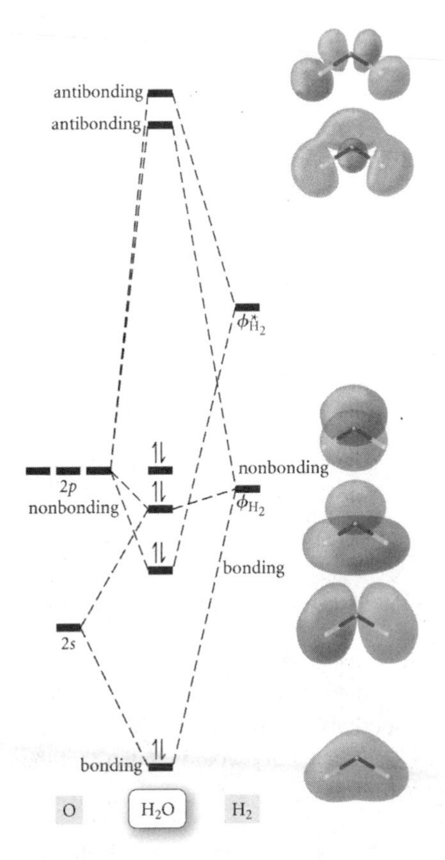

$$\text{Bond order} = \frac{4 - 0}{2} = 2$$

With a bond order of 2, the molecule is stable.

7.81 Using the MO diagram for NH_3, assign the eight valence electrons to the molecular orbitals. Start with the lowest energy orbital first and follow Hund's rule.

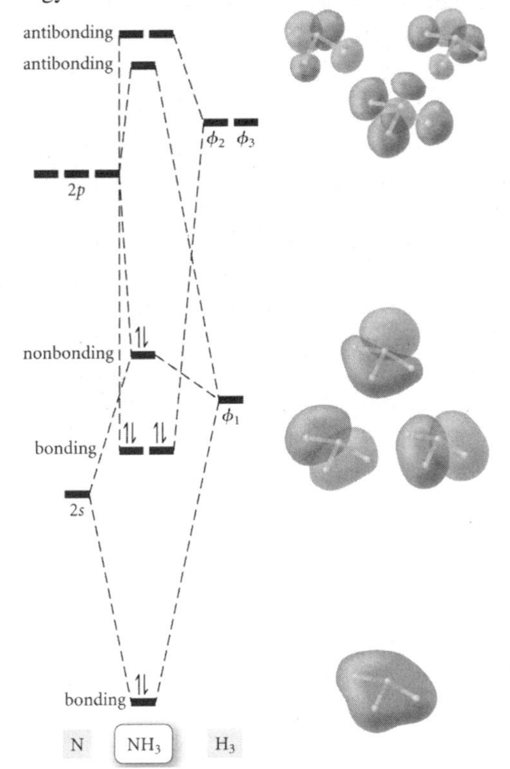

Bond order $= \frac{6 - 0}{2} = 3$

With a bond order of 3, the molecule is stable.

7.82 (a) In the isomerization, you need to break the C—C π bond but not the σ bond. So the energy needed would be the difference between the bond energy of a C=C bond and a C—C bond.

C=C 611 kJ/mol

C—C 347 kJ/mol

Therefore, the energy needed to break the π bond would be $264\ \text{kJ/mol} = 2.64 \times 10^5\ \text{J/mol}$

$$\frac{264\ \text{kJ}}{\text{mol}} \times \frac{1000\ \text{J}}{1\ \text{kJ}} \times \frac{1\ \text{mol}}{6.022 \times 10^{23}\ \text{molecules}} = 4.38 \times 10^{-19} \frac{\text{J}}{\text{molecule}}$$

(b) **Given:** 4.38×10^{-19} J/molecule **Find:** ν and part of the electromagnetic spectrum

Conceptual Plan: $E \rightarrow \nu$

$E = h\nu$

Solution: $\nu = \frac{E}{h} = \frac{4.38 \times 10^{-19}\ \text{J}}{6.626 \times 10^{-34}\ \text{J} \cdot \text{s}} = 6.61 \times 10^{14}\ \text{s}^{-1}$

This frequency is near the border of the ultraviolet-visible region of the electromagnetic spectrum.

7.83 To accommodate the five σ bonds, you need five equal energy hybrid orbitals. So you need to combine five atomic orbitals. The valence electrons on C are in the 2*s* and 2*p* orbitals. You can, therefore, combine the 2*s* and 2*p* orbitals, but this would only give you four hybrid orbitals. The next lowest energy orbital available on C is the 3*s*, so this would be the next atomic orbital added in. This gives a hybridization of s^2p^3. VSEPR theory would predict that the geometry is trigonal bipyramidal to accommodate the five bonds, as five identical substituents.

7.84 Both VSEPR theory and hybridization would predict a bond angle of 180°. However, gaseous BaF_2 has a bond angle of 108°. This may be because the 5*d* orbitals are very close in energy to the 6*s* orbitals on the Ba. Thus, these may contribute to the bonding scheme rather than the 6*p* orbitals.

Conceptual Problems

7.85 In Lewis theory, a covalent bond comes from the sharing of electrons.

A single bond shares two electrons (one pair).
A double bond shares four electrons (two pairs).
A triple bond shares six electrons (three pairs).

In valence bond theory, a covalent bond forms when orbitals overlap. The orbitals can be unhybridized or hybridized orbitals.

A single bond forms when a σ bond is formed from the overlap of an *s* orbital with an *s* orbital, an *s* orbital with a *p* orbital, or a *p* orbital and a *p* orbital overlapping end to end. A σ can also form from the overlap of a hybridized orbital on the central atom with an *s* orbital or with a *p* orbital overlapping end to end.
A double bond is a combination of a σ bond and a π bond. The π bond forms from the sideways overlap of a *p* orbital on each of the atoms involved in the bond. The *p* orbitals must have the same orientation.
A triple bond is a combination of a σ bond and 2π bonds. The π bonds form from the sideways overlap of a *p* orbital on each of the atoms involved in the bond. The *p* orbitals must have the same orientation, so each π bond is formed from a different set of *p* orbitals.

In molecular orbital theory, molecular orbitals form. These are combinations of the atomic orbitals of the atoms involved in the bond. The bonds form when the valence electrons occupy more bonding molecular orbitals than antibonding molecular orbitals. This is calculated by the bond order.

A single bond has a bond order of 1.
A double bond has a bond order of 2.
A triple bond has a bond order of 3.

All three models show the formation of bonds between two atoms. All three models show the formation of the same number of bonds between the atoms involved. Lewis theory tells us only about the number of bonds formed and, combined with VSEPR theory, allows us to predict the shape of the molecule. It does not, however, tell us anything about how the bonds are formed. Valence bond theory addresses the formation of the different types of bonds, sigma and pi. In valence bond theory, the bonds form from the overlap of atomic orbitals on the individual atoms involved in the bonds, and the atoms are localized between the two atoms involved in the bond. Molecular orbital theory approaches the formation of bonds by looking at the entire molecule. The electrons are not restricted to any two individual atoms, but are treated as belonging to the whole molecule. The electrons reside in molecular orbitals that are part of the entire molecule rather than being restricted to individual atoms. Each model gives us information about the molecule. The amount and type of information we need determine the model we choose to use.

7.86 In period 2, the atoms are smaller and do not have *d* orbitals available to hybridize; so they cannot accommodate as many atoms around the central atom. To complete the octet of electrons, multiple bonds must form. In the period 3 and higher atoms, more atoms can surround the central atom, which is larger, and the *d* orbitals can hybridize with the *s* and *p* orbitals. There are now more orbitals available to overlap, and there is space for them to do so. The central atom, therefore, can attain a stable configuration of eight or more electrons without having to multiple-bond.

7.87 Metals have a tendency to lose electrons, which means that they have relatively low ionization energies. When metal atoms bond together to form a solid, each metal atom donates one or more electrons to an *electron sea*. The metal cations are held together by their attraction to the sea of electrons. Metals are excellent conductors of heat, because of these highly mobile electrons, which help to disperse thermal energy throughout the metal.

7.88 According to valence bond theory, the bonds in each of these halide molecules result from overlap of atomic orbitals. Smaller atoms (Cl is the smallest atom and I is the largest atom) have smaller atomic orbitals, and hence, shorter bonds. The shorter the bond the higher the bond energy.

8 Chemical Reactions and Chemical Quantities

Review Questions

8.1 The greenhouse effect is the heating of the Earth's atmosphere that occurs when greenhouse gasses trap some of the sunlight that penetrates the Earth's atmosphere. Carbon dioxide is the most significant greenhouse gas.

8.2 Scientists are concerned about the increase in atmospheric carbon dioxide because carbon dioxide enhances the atmosphere's ability to heat. This trapped heat leads to global warming. The primary cause of rising carbon dioxide levels is the burning of fossil fuels (natural gas, petroleum, and coal). Other sources of carbon dioxide are volcanic eruptions, fermentation, and wildfires.

8.3 A physical property is one that a substance displays without changing its composition, whereas a chemical property is one that a substance displays only by changing its composition via a chemical change.

8.4 Changes that alter only state or appearance, but not composition, are called physical changes. The atoms or molecules that make up a substance *do not change* their identity during a physical change. For example, when water boils, it changes its state from a liquid to a gas, but the gas remains composed of water molecules; so this constitutes a physical change. When sugar dissolves in water, the sugar molecules are separated from each other, but the molecules of sugar and water remain intact.

In contrast, changes that alter the composition of matter are called chemical changes. During a chemical change, atoms rearrange themselves, transforming the original substances into different substances. For example, the rusting of iron, the combustion of natural gas to form carbon dioxide and water, and the denaturing of proteins when an egg is cooked are examples of chemical changes.

8.5 In chemistry, we represent chemical reactions with chemical equations where the substances on the left-hand side of a chemical equation are the reactants and the substances on the right-hand side are the products. Chemical equations are balanced when the number of each type of atom on the left side of the equation is equal to the number on the right side.

8.6 In a balanced chemical equation, the coefficients in front of each reactant and product allow us to predict, for example, the amount of product that can form from a given amount of reactant, or how much of one reactant is required to react with a given amount of another. These numerical relationships are referred to as reaction stoichiometry.

8.7 Reaction stoichiometry is the numerical relationship between chemical amounts in a balanced chemical equation. The coefficients in a chemical reaction specify the relative amounts in moles of each of the substances involved in the reaction.

8.8 The limiting reactant is the reactant that is completely consumed in a chemical reaction and limits the amount of product. The reactant in excess is any reactant that occurs in a quantity greater than that required to completely react with the limiting reactant. Some of this reactant will be left over when the reaction is complete.

8.9 The theoretical yield is the amount of product that can be made in a chemical reaction based on the amount of limiting reactant. The percent yield is calculated as $\frac{\text{actual yield}}{\text{theoretical yield}} \times 100\%$.

8.10 No, the percent yield would not be different if the actual yield and theoretical yield were calculated in moles. The relationship between grams and moles is the molar mass. This would be the same value for the actual yield and the theoretical yield.

8.11 Most U.S. energy comes from the combustion of fossil fuels, which include petroleum, natural gas, and coal.

8.12 Combustion reactions are characterized by the reaction of a substance with O_2 to form one or more oxygen-containing compounds, often including water. Combustion reactions emit heat. Combustion reactions are important because most of society's energy is derived from them. An example is $CH_4(g) + 2\,O_2(g) \rightarrow CO_2(g) + 2\,H_2O(g)$.

8.13 (a) The reactions of the alkali metals with halogens result in the formation of metal halides.
$2\,M(s) + X_2 \rightarrow 2\,MX(s)$
(b) Alkali metals react with water to form the dissolved alkali metal ion, the hydroxide ion, and hydrogen gas.
$2\,M(s) + 2\,H_2O(l) \rightarrow 2\,M^+(aq) + 2\,OH^-(aq) + H_2(g)$

8.14 All of the halogens are powerful oxidizing agents.
(a) The halogens react with metals to form metal halides.
$2\,M(s) + n\,X_2 \rightarrow 2\,MX_n(s)$
(b) The halogens react with hydrogen to form hydrogen halides.
$H_2(g) + X_2 \rightarrow 2\,HX(g)$
(c) The halogens react with each other to form interhalogen compounds.
For example, $Br_2(l) + F_2(g) \rightarrow 2\,BrF(g)$

Problems by Topic

Chemical and Physical Changes

8.15 (a) chemical change (new compounds are formed as methane and oxygen react to form carbon dioxide and water)
(b) physical change (vaporization is a phase change and does not involve the making or breaking of chemical bonds)
(c) chemical change (new compounds are formed as propane and oxygen react to form carbon dioxide and water)
(d) chemical change (new compounds are formed as the metal in the frame is converted to oxides)

8.16 (a) chemical change (new compounds are formed as the sugar burns)
(b) physical change (dissolution is a phase change and does not involve the making or breaking of chemical bonds)
(c) physical change (this is simply the rearrangement of the atoms)
(d) chemical change (new compounds are formed as the silver converts to an oxide)

8.17 (a) physical change (vaporization is a phase change and does not involve the making or breaking of chemical bonds)
(b) chemical change (new compounds are formed)
(c) physical change (vaporization is a phase change and does not involve the making or breaking of chemical bonds)

8.18 (a) physical change (vaporization of butane is a phase change and does not involve the making or breaking of chemical bonds)
(b) chemical change (new compounds are formed as the butane combusts)
(c) physical change (vaporization of water is a phase change and does not involve the making or breaking of chemical bonds)

8.19 (a) physical property (color can be observed without making or breaking chemical bonds)
(b) chemical property (must observe by making or breaking chemical bonds)

(c) physical property (the phase can be observed without making or breaking chemical bonds)
(d) physical property (density can be observed without making or breaking chemical bonds)
(e) physical property (mixing does not involve making or breaking chemical bonds, so this can be observed without making or breaking chemical bonds)

8.20
(a) physical property (color can be observed without making or breaking chemical bonds)
(b) physical property (odor can be observed without making or breaking chemical bonds)
(c) chemical property (must observe by making or breaking chemical bonds)
(d) chemical property (decomposition involves breaking bonds, so bonds must be broken to observe this property)
(e) physical property (the phase of a substance can be observed without making or breaking chemical bonds)

8.21
(a) chemical property (burning involves breaking and making bonds, so bonds must be broken and made to observe this property)
(b) physical property (shininess is a physical property and so can be observed without making or breaking chemical bonds)
(c) physical property (odor can be observed without making or breaking chemical bonds)
(d) chemical property (burning involves breaking and making bonds, so bonds must be broken and made to observe this property)

8.22
(a) physical property (vaporization is a phase change and so can be observed without making or breaking chemical bonds)
(b) physical property (sublimation is a phase change and so can be observed without making or breaking chemical bonds)
(c) chemical property (rusting involves the reaction of iron with oxygen to form iron oxide; observing this process involves making and breaking chemical bonds)
(d) physical property (color can be observed without making or breaking chemical bonds)

Writing and Balancing Chemical Equations

8.23 **Conceptual Plan: Write a skeletal reaction → balance atoms in more complex compounds → balance elements that occur as free elements → clear fractions**

Solution:

Skeletal reaction: $SO_2(g) + O_2(g) + H_2O(l) \rightarrow H_2SO_4(aq)$

Balance O: $SO_2(g) + 1/2\, O_2(g) + H_2O(l) \rightarrow H_2SO_4(aq)$

Clear fraction: $2\, SO_2(g) + O_2(g) + 2\, H_2O(l) \rightarrow 2\, H_2SO_4(aq)$

Check:

left side	right side
2 S atoms	2 S atoms
8 O atoms	8 O atoms
4 H atoms	4 H atoms

8.24 **Conceptual Plan: Write a skeletal reaction → balance atoms in more complex compounds → balance elements that occur as free elements → clear fractions**

Solution:

Skeletal reaction: $NO_2(g) + O_2(g) + H_2O(l) \rightarrow HNO_3(aq)$

Balance H: $NO_2(g) + O_2(g) + H_2O(l) \rightarrow 2\, HNO_3(aq)$

Balance N: $2\, NO_2(g) + O_2(g) + H_2O(l) \rightarrow 2\, HNO_3(aq)$

Balance O: $2\, NO_2(g) + 1/2\, O_2(g) + H_2O(l) \rightarrow 2\, HNO_3(aq)$

Clear fraction: $4\, NO_2(g) + O_2(g) + 2\, H_2O(l) \rightarrow 4\, HNO_3(aq)$

Check:

left side	right side
4 N atoms	4 N atoms
12 O atoms	12 O atoms
4 H atoms	4 H atoms

8.25 **Conceptual Plan: Write a skeletal reaction → balance atoms in more complex compounds → balance elements that occur as free elements → clear fractions**

Solution:

Skeletal reaction: $Na(s) + H_2O(l) \rightarrow H_2(g) + NaOH(aq)$

Balance H: $Na(s) + H_2O(l) \rightarrow 1/2\, H_2(g) + NaOH(aq)$

Clear fraction: $2\, Na(s) + 2\, H_2O(l) \rightarrow H_2(g) + 2\, NaOH(aq)$

Check:	left side	right side
	2 Na atoms	2 Na atoms
	4 H atoms	4 H atoms
	2 O atoms	2 O atoms

8.26 **Conceptual Plan: Write a skeletal reaction → balance atoms in more complex compounds → balance elements that occur as free elements → clear fractions**

Solution:
Skeletal reaction: $Fe(s) + O_2(g) \rightarrow Fe_2O_3(s)$
Balance O: $Fe(s) + 3\,O_2(g) \rightarrow 2\,Fe_2O_3(s)$
Balance Fe: $4\,Fe(s) + 3\,O_2(g) \rightarrow 2\,Fe_2O_3(s)$

Check:	left side	right side
	4 Fe atoms	4 Fe atoms
	6 O atoms	6 O atoms

8.27 **Conceptual Plan: Write a skeletal reaction → balance atoms in more complex compounds → balance elements that occur as free elements → clear fractions**

Solution:
Skeletal reaction: $C_{12}H_{22}O_{11}(aq) + H_2O(l) \rightarrow C_2H_5OH(aq) + CO_2(g)$
Balance H: $C_{12}H_{22}O_{11}(aq) + H_2O(l) \rightarrow 4\,C_2H_5OH(aq) + CO_2(g)$
Balance C: $C_{12}H_{22}O_{11}(aq) + H_2O(l) \rightarrow 4\,C_2H_5OH(aq) + 4\,CO_2(g)$

Check:	left side	right side
	12 C atoms	12 C atoms
	24 H atoms	24 H atoms
	12 O atoms	12 O atoms

8.28 **Conceptual Plan: Write a skeletal reaction → balance atoms in more complex compounds → balance elements that occur as free elements → clear fractions**

Solution:
Skeletal reaction: $CO_2(g) + H_2O(l) \rightarrow C_6H_{12}O_6(aq) + O_2(g)$
Balance C: $6\,CO_2(g) + H_2O(l) \rightarrow C_6H_{12}O_6(aq) + O_2(g)$
Balance H: $6\,CO_2(g) + 6\,H_2O(l) \rightarrow C_6H_{12}O_6(aq) + O_2(g)$
Balance O: $6\,CO_2(g) + 6\,H_2O(l) \rightarrow C_6H_{12}O_6(aq) + 6\,O_2(g)$

Check:	left side	right side
	6 C atoms	6 C atoms
	18 O atoms	18 O atoms
	12 H atoms	12 H atoms

8.29 (a) **Conceptual Plan: Write a skeletal reaction → balance atoms in more complex compounds → balance elements that occur as free elements → clear fractions**

Solution:
Skeletal reaction: $PbS(s) + HBr(aq) \rightarrow PbBr_2(s) + H_2S(g)$
Balance Br: $PbS(s) + 2\,HBr(aq) \rightarrow PbBr_2(s) + H_2S(g)$

Check:	left side	right side
	1 Pb atom	1 Pb atom
	1 S atom	1 S atom
	2 H atoms	2 H atoms
	2 Br atoms	2 Br atoms

(b) **Conceptual Plan: Write a skeletal reaction → balance atoms in more complex compounds → balance elements that occur as free elements → clear fractions**

Solution:
Skeletal reaction: $CO(g) + H_2(g) \rightarrow CH_4(g) + H_2O(l)$
Balance H: $CO(g) + 3\,H_2(g) \rightarrow CH_4(g) + H_2O(l)$

Check:	left side	right side
	1 C atom	1 C atom
	1 O atom	1 O atom
	6 H atoms	6 H atoms

(c) **Conceptual Plan: Write a skeletal reaction → balance atoms in more complex compounds → balance elements that occur as free elements → clear fractions**

Solution:
Skeletal reaction: $HCl(aq) + MnO_2(s) \rightarrow MnCl_2(aq) + H_2O(l) + Cl_2(g)$
Balance Cl: $4\ HCl(aq) + MnO_2(s) \rightarrow MnCl_2(aq) + H_2O(l) + Cl_2(g)$
Balance O: $4\ HCl(aq) + MnO_2(s) \rightarrow MnCl_2(aq) + 2\ H_2O(l) + Cl_2(g)$

Check:

left side	right side
4 H atoms	4 H atoms
4 Cl atoms	4 Cl atoms
1 Mn atom	1 Mn atom
2 O atoms	2 O atoms

(d) **Conceptual Plan: Write a skeletal reaction → balance atoms in more complex compounds → balance elements that occur as free elements → clear fractions**

Solution:
Skeletal reaction: $C_5H_{12}(l) + O_2(g) \rightarrow CO_2(g) + H_2O(l)$
Balance C: $C_5H_{12}(l) + O_2(g) \rightarrow 5\ CO_2(g) + H_2O(l)$
Balance H: $C_5H_{12}(l) + O_2(g) \rightarrow 5\ CO_2(g) + 6\ H_2O(l)$
Balance O: $C_5H_{12}(l) + 8\ O_2(g) \rightarrow 5\ CO_2(g) + 6\ H_2O(l)$

Check:

left side	right side
5 C atoms	5 C atoms
12 H atoms	12 H atoms
16 O atoms	16 O atoms

8.30 (a) **Conceptual Plan: Write a skeletal reaction → balance atoms in more complex compounds → balance elements that occur as free elements → clear fractions**

Solution:
Skeletal reaction: $Cu(s) + S(s) \rightarrow Cu_2S(s)$
Balance Cu: $2\ Cu(s) + S(s) \rightarrow Cu_2S(s)$

Check:

left side	right side
2 Cu atoms	2 Cu atoms
1 S atom	1 S atom

(b) **Conceptual Plan: Write a skeletal reaction → balance atoms in more complex compounds → balance elements that occur as free elements → clear fractions**

Solution:
Skeletal reaction: $Fe_2O_3(s) + H_2(g) \rightarrow Fe(s) + H_2O(l)$
Balance O: $Fe_2O_3(s) + H_2(g) \rightarrow Fe(s) + 3\ H_2O(l)$
Balance Fe: $Fe_2O_3(s) + H_2(g) \rightarrow 2\ Fe(s) + 3\ H_2O(l)$
Balance H: $Fe_2O_3(s) + 3\ H_2(g) \rightarrow 2\ Fe(s) + 3\ H_2O(l)$

Check:

left side	right side
2 Fe atoms	2 Fe atoms
3 O atoms	3 O atoms
6 H atoms	6 H atoms

(c) **Conceptual Plan: Write a skeletal reaction → balance atoms in more complex compounds → balance elements that occur as free elements → clear fractions**

Solution:
Skeletal reaction: $SO_2(g) + O_2(g) \rightarrow SO_3(g)$
Balance O: $SO_2(g) + 1/2\ O_2(g) \rightarrow SO_3(g)$
Clear fraction: $2\ SO_2(g) + O_2(g) \rightarrow 2\ SO_3(g)$

Check:

left side	right side
2 S atoms	2 S atoms
6 O atoms	6 O atoms

(d) **Conceptual Plan: Write a skeletal reaction → balance atoms in more complex compounds → balance elements that occur as free elements → clear fractions**

Solution:
Skeletal reaction: $NH_3(g) + O_2(g) \rightarrow NO(g) + H_2O(g)$
Balance H: $2\ NH_3(g) + O_2(g) \rightarrow NO(g) + 3\ H_2O(g)$
Balance N: $2\ NH_3(g) + O_2(g) \rightarrow 2\ NO(g) + 3\ H_2O(g)$
Balance O: $2\ NH_3(g) + 5/2\ O_2(g) \rightarrow 2\ NO(g) + 3\ H_2O(g)$
Clear fraction: $4\ NH_3(g) + 5\ O_2(g) \rightarrow 4\ NO(g) + 6\ H_2O(g)$

Check:

left side	right side
4 N atoms	4 N atoms
12 H atoms	12 H atoms
10 O atoms	10 O atoms

8.31 **Conceptual Plan: Write a skeletal reaction → balance atoms in more complex compounds → balance elements that occur as free elements → clear fractions**

Solution: Skeletal reaction: $Na_2CO_3(aq) + CuCl_2(aq) \rightarrow CuCO_3(s) + NaCl(aq)$

Balance Na: $Na_2CO_3(aq) + CuCl_2(aq) \rightarrow CuCO_3(s) + 2\ NaCl(aq)$

Check:

left side	right side
2 Na atoms	2 Na atoms
1 C atom	1 C atom
3 O atoms	3 O atoms
1 Cu atom	1 Cu atom
2 Cl atoms	2 Cl atoms

8.32 **Conceptual Plan: Write a skeletal reaction → balance atoms in more complex compounds → balance elements that occur as free elements → clear fractions**

Solution: Skeletal reaction: $KOH(aq) + FeCl_3(aq) \rightarrow Fe(OH)_3(s) + KCl(aq)$

Balance Cl: $KOH(aq) + FeCl_3(aq) \rightarrow Fe(OH)_3(s) + 3\ KCl(aq)$

Balance K: $3\ KOH(aq) + FeCl_3(aq) \rightarrow Fe(OH)_3(s) + 3\ KCl(aq)$

Check:

left side	right side
3 K atoms	3 K atoms
3 O atoms	3 O atoms
3 H atoms	3 H atoms
1 Fe atom	1 Fe atom
3 Cl atoms	3 Cl atoms

8.33 (a) **Conceptual Plan: Balance atoms in more complex compounds → balance elements that occur as free elements → clear fractions**

Solution: Skeletal reaction: $CO_2(g) + CaSiO_3(s) + H_2O(l) \rightarrow SiO_2(s) + Ca(HCO_3)_2(aq)$

Balance C: $2\ CO_2(g) + CaSiO_3(s) + H_2O(l) \rightarrow SiO_2(s) + Ca(HCO_3)_2(aq)$

Check:

left side	right side
2 C atoms	2 C atoms
8 O atoms	8 O atoms
1 Ca atom	1 Ca atom
1 Si atom	1 Si atom
2 H atoms	2 H atoms

(b) **Conceptual Plan: Balance atoms in more complex compounds → balance elements that occur as free elements → clear fractions**

Solution: Skeletal reaction: $Co(NO_3)_3(aq) + (NH_4)_2S(aq) \rightarrow Co_2S_3(s) + NH_4NO_3(aq)$

Balance S: $Co(NO_3)_3(aq) + 3\ (NH_4)_2S(aq) \rightarrow Co_2S_3(s) + NH_4NO_3(aq)$

Balance Co: $2\ Co(NO_3)_3(aq) + 3\ (NH_4)_2S(aq) \rightarrow Co_2S_3(s) + NH_4NO_3(aq)$

Balance N: $2\ Co(NO_3)_3(aq) + 3\ (NH_4)_2S(aq) \rightarrow Co_2S_3(s) + 6\ NH_4NO_3(aq)$

Check:

left side	right side
2 Co atoms	2 Co atoms
12 N atoms	12 N atoms
18 O atoms	18 O atoms
24 H atoms	24 H atoms
3 S atoms	3 S atoms

(c) **Conceptual Plan: Balance atoms in more complex compounds → balance elements that occur as free elements → clear fractions**

Solution: Skeletal reaction: $Cu_2O(s) + C(s) \rightarrow Cu(s) + CO(g)$

Balance Cu: $Cu_2O(s) + C(s) \rightarrow 2\ Cu(s) + CO(g)$

Check:

left side	right side
2 Cu atoms	2 Cu atoms
1 O atom	1 O atom
1 C atom	1 C atom

(d) **Conceptual Plan: Balance atoms in more complex compounds → balance elements that occur as free elements → clear fractions**

Solution:

Skeletal reaction: $H_2(g) + Cl_2(g) \rightarrow HCl(g)$

Balance Cl: $H_2(g) + Cl_2(g) \rightarrow 2\ HCl(g)$

Check:

left side	right side
2 H atoms	2 H atoms
2 Cl atoms	2 Cl atoms

8.34 (a) **Conceptual Plan: Balance atoms in more complex compounds → balance elements that occur as free elements → clear fractions**

Solution:

Skeletal reaction: $Na_2S(aq) + Cu(NO_3)_2(aq) \rightarrow NaNO_3(aq) + CuS(s)$

Balance Na: $Na_2S(aq) + Cu(NO_3)_2(aq) \rightarrow 2\ NaNO_3(aq) + CuS(s)$

Check:

left side	right side
2 Na atoms	2 Na atoms
1 S atom	1 S atom
1 Cu atom	1 Cu atom
2 N atoms	2 N atoms
6 O atoms	6 O atoms

(b) **Conceptual Plan: Balance atoms in more complex compounds → balance elements that occur as free elements → clear fractions**

Solution:

Skeletal reaction: $N_2H_4(l) \rightarrow NH_3(g) + N_2(g)$

Balance H: $3\ N_2H_4(l) \rightarrow 4\ NH_3(g) + N_2(g)$

Check:

left side	right side
6 N atoms	6 N atoms
12 H atoms	12 H atoms

(c) **Conceptual Plan: Balance atoms in more complex compounds → balance elements that occur as free elements → clear fractions**

Solution:

Skeletal reaction: $HCl(aq) + O_2(g) \rightarrow H_2O(l) + Cl_2(g)$

Balance Cl: $2\ HCl(aq) + O_2(g) \rightarrow H_2O(l) + Cl_2(g)$

Balance O: $2\ HCl(aq) + 1/2\ O_2(g) \rightarrow H_2O(l) + Cl_2(g)$

Clear fraction: $4\ HCl(aq) + O_2(g) \rightarrow 2\ H_2O(l) + 2\ Cl_2(g)$

Check:

left side	right side
4 H atoms	4 H atoms
4 Cl atoms	4 Cl atoms
2 O atoms	2 O atoms

(d) **Conceptual Plan: Balance atoms in more complex compounds → balance elements that occur as free elements → clear fractions**

Solution:

Skeletal reaction: $FeS(s) + HCl(aq) \rightarrow FeCl_2(aq) + H_2S(g)$

Balance Cl: $FeS(s) + 2\ HCl(aq) \rightarrow FeCl_2(aq) + H_2S(g)$

Check:

left side	right side
1 Fe atom	1 Fe atom
1 S atom	1 S atom
2 H atoms	2 H atoms
2 Cl atoms	2 Cl atoms

Reaction Stoichiometry

8.35 **Given:** 7.2 moles C_6H_{14} **Find:** balanced reaction, moles O_2 required

Conceptual Plan: Balance the equation then mol $C_6H_{14} \rightarrow$ mol O_2

$2\ C_6H_{14}(g) + 19\ O_2(g) \rightarrow 12\ CO_2(g) + 14\ H_2O(g)$ $\quad \frac{19\ \text{mol}\ O_2}{2\ \text{mol}\ C_6H_{14}}$

Solution: $7.2\ \cancel{\text{mol}\ C_6H_{14}} \times \frac{19\ \text{mol}\ O_2}{2\ \cancel{\text{mol}\ C_6H_{14}}} = 6\underline{8}.4\ \text{mol}\ O_2 = 68\ \text{mol}\ O_2$

Check: The units of the answer (mol O_2) are correct. The magnitude is reasonable because much more O_2 is needed than C_6H_{14}.

8.36 **Given:** 0.461 mole $HC_2H_3O_2$ **Find:** balanced reaction, moles $Ba(OH)_2$ required

Conceptual Plan: Balance the reaction then mol $HC_2H_3O_2$ → mol $Ba(OH)_2$

$2\ HC_2H_3O_2(aq) + Ba(OH)_2(aq) \rightarrow 2\ H_2O(l) + Ba(C_2H_3O_2)_2(aq)$ $\qquad \frac{1\text{ mol }Ba(OH)_2}{2\text{ mol }HC_2H_3O_2}$

Solution: $0.461\ \cancel{\text{mol }HC_2H_3O_2} \times \frac{1\text{ mol }Ba(OH)_2}{2\ \cancel{\text{mol }HC_2H_3O_2}} = 0.23\underline{0}5\text{ mol }Ba(OH)_2 = 0.231\text{ mol }Ba(OH)_2$

Check: The units of the answer (mol $Ba(OH)_2$) are correct. The magnitude is reasonable because much less $Ba(OH)_2$ is needed than $HC_2H_3O_2$.

8.37 (a) **Given:** 2.5 mol N_2O_5 **Find:** mol NO_2

Conceptual Plan: mol N_2O_5 → mol NO_2

$\frac{4\text{ mol }NO_2}{2\text{ mol }N_2O_5}$

Solution: $2.5\ \cancel{\text{mol }N_2O_5} \times \frac{4\text{ mol }NO_2}{2\ \cancel{\text{mol }N_2O_5}} = 5.0\text{ mol }NO_2$

Check: The units of the answer (mol NO_2) are correct. The magnitude is reasonable because it is greater than mol N_2O_5.

(b) **Given:** 6.8 mol N_2O_5 **Find:** mol NO_2

Conceptual Plan: mol N_2O_5 → mol NO_2

$\frac{4\text{ mol }NO_2}{2\text{ mol }N_2O_5}$

Solution: $6.8\ \cancel{\text{mol }N_2O_5} \times \frac{4\text{ mol }NO_2}{2\ \cancel{\text{mol }N_2O_5}} = 1\underline{3}.6\text{ mol }NO_2 = 14\text{ mol }NO_2$

Check: The units of the answer (mol NO_2) are correct. The magnitude is reasonable because it is greater than mol N_2O_5.

(c) **Given:** 15.2 g N_2O_5 **Find:** mol NO_2

Conceptual Plan: g N_2O_5 → mol N_2O_5 → mol NO_2

$\frac{1\text{ mol }N_2O_5}{108.02\text{ g }N_2O_5} \qquad \frac{4\text{ mol }NO_2}{2\text{ mol }N_2O_5}$

Solution: $15.2\ \cancel{\text{g }N_2O_5} \times \frac{1\ \cancel{\text{mol }N_2O_5}}{108.02\ \cancel{\text{g }N_2O_5}} \times \frac{4\text{ mol }NO_2}{2\ \cancel{\text{mol }N_2O_5}} = 0.28\underline{1}4\text{ mol }NO_2 = 0.281\text{ mol }NO_2$

Check: The units of the answer (mol NO_2) are correct. The magnitude is reasonable because 10 g is about 0.09 mol N_2O_5 and the answer is greater than mol N_2O_5.

(d) **Given:** 2.87 kg N_2O_5 **Find:** mol NO_2

Conceptual Plan: kg N_2O_5 → g N_2O_5 → mol N_2O_5 → mol NO_2

$\frac{1000\text{ g }N_2O_5}{1\text{ kg }N_2O_5} \qquad \frac{1\text{ mol }N_2O_5}{108.02\text{ g }N_2O_5} \qquad \frac{4\text{ mol }NO_2}{2\text{ mol }N_2O_5}$

Solution:

$$2.87\ \cancel{\text{kg }N_2O_5} \times \frac{1000\ \cancel{\text{g }N_2O_5}}{1\ \cancel{\text{kg }N_2O_5}} \times \frac{1\ \cancel{\text{mol }N_2O_5}}{108.02\ \cancel{\text{g }N_2O_5}} \times \frac{4\text{ mol }NO_2}{2\ \cancel{\text{mol }N_2O_5}} = 53.\underline{1}4\text{ mol }NO_2 = 53.1\text{ mol }NO_2$$

Check: The units of the answer (mol NO_2) are correct. The magnitude is reasonable because 1.55 kg is about 14 mol N_2O_5 and the answer is greater than mol N_2O_5.

8.38 (a) **Given:** 2.6 mol N_2H_4 **Find:** mol NH_3

Conceptual Plan: mol N_2H_4 → mol NH_3

$\frac{4\text{ mol }NH_3}{3\text{ mol }N_2H_4}$

Solution: $2.6\ \cancel{\text{mol }N_2H_4} \times \frac{4\text{ mol }NH_3}{3\ \cancel{\text{mol }N_2H_4}} = 3.\underline{4}6\text{ mol }NH_3 = 3.5\text{ mol }NH_3$

Check: The units of the answer (mol NH_3) are correct. The magnitude is reasonable because it is greater than mol N_2H_4.

(b) **Given:** 3.55 mol N_2H_4 **Find:** mol NH_3
Conceptual Plan: mol N_2H_4 → mol NH_3

$$\frac{4 \text{ mol } NH_3}{3 \text{ mol } N_2H_4}$$

Solution: $3.55 \text{ mol } N_2H_4 \times \frac{4 \text{ mol } NH_3}{3 \text{ mol } N_2H_4} = 4.7\underline{3}3 \text{ mol } NH_3 = 4.73 \text{ mol } NH_3$

Check: The units of the answer (mol NH_3) are correct. The magnitude is reasonable because it is greater than mol N_2H_4.

(c) **Given:** 65.3 g N_2H_4 **Find:** mol NH_3
Conceptual Plan: g N_2H_4 → mol N_2H_4 → mol NH_3

$$\frac{1 \text{ mol } N_2H_4}{32.05 \text{ g } N_2H_4} \qquad \frac{4 \text{ mol } NH_3}{3 \text{ mol } N_2H_4}$$

Solution: $65.3 \text{ g } N_2H_4 \times \frac{1 \text{ mol } N_2H_4}{32.05 \text{ g } N_2H_4} \times \frac{4 \text{ mol } NH_3}{3 \text{ mol } N_2H_4} = 2.7\underline{1}6 \text{ mol } NH_3 = 2.72 \text{ mol } NH_3$

Check: The units of the answer (mol NH_3) are correct. The magnitude is reasonable because there is about 2 mol N_2H_4 and the answer is greater than mol N_2H_4.

(d) **Given:** 4.88 kg N_2H_4 **Find:** mol NH_3
Conceptual Plan: kg N_2H_4 → g N_2H_4 → mol N_2H_4 → mol NH_3

$$\frac{1000 \text{ g } N_2H_4}{1 \text{ kg } N_2H_4} \qquad \frac{1 \text{ mol } N_2H_4}{32.05 \text{ g } N_2H_4} \qquad \frac{4 \text{ mol } NH_3}{3 \text{ mol } N_2H_4}$$

Solution:

$$4.88 \text{ kg } N_2H_4 \times \frac{1000 \text{ g } N_2H_4}{1 \text{ kg } N_2H_4} \times \frac{1 \text{ mol } N_2H_4}{32.05 \text{ g } N_2H_4} \times \frac{4 \text{ mol } NH_3}{3 \text{ mol } N_2H_4} = 20\underline{3}.0 \text{ mol } NH_3 = 203 \text{ mol } NH_3$$

Check: The units of the answer (mol NH_3) are correct. The magnitude is reasonable because 4.88 kg is about 152 mol N_2H_4 and the answer is greater than mol N_2H_4.

8.39 **Given:** 3 mol SiO_2 **Find:** mol C; mol SiC; mol CO
Conceptual Plan: mol SiO_2 → mol C → mol SiC → mol CO

$$\frac{3 \text{ mol C}}{1 \text{ mol } SiO_2} \qquad \frac{1 \text{ mol SiC}}{1 \text{ mol } SiO_2} \qquad \frac{2 \text{ mol CO}}{1 \text{ mol } SiO_2}$$

Solution: $3 \text{ mol } SiO_2 \times \frac{3 \text{ mol C}}{1 \text{ mol } SiO_2} = 9 \text{ mol C}$ $\quad 3 \text{ mol } SiO_2 \times \frac{1 \text{ mol SiC}}{1 \text{ mol } SiO_2} = 3 \text{ mol SiC}$

$3 \text{ mol } SiO_2 \times \frac{2 \text{ mol CO}}{1 \text{ mol } SiO_2} = 6 \text{ mol CO}$

Given: 6 mol C **Find:** mol SiO_2; mol SiC; mol CO
Conceptual Plan: mol C → mol SiO_2 → mol SiC → mol CO

$$\frac{1 \text{ mol } SiO_2}{3 \text{ mol C}} \qquad \frac{1 \text{ mol SiC}}{3 \text{ mol C}} \qquad \frac{2 \text{ mol CO}}{3 \text{ mol C}}$$

Solution: $6 \text{ mol C} \times \frac{1 \text{ mol } SiO_2}{3 \text{ mol C}} = 2 \text{ mol } SiO_2$ $\quad 6 \text{ mol C} \times \frac{1 \text{ mol SiC}}{3 \text{ mol C}} = 2 \text{ mol SiC}$

$6 \text{ mol C} \times \frac{2 \text{ mol CO}}{3 \text{ mol C}} = 4 \text{ mol CO}$

Given: 10 mol CO **Find:** mol SiO_2; mol C; mol SiC
Conceptual Plan: mol CO → mol SiO_2 → mol C → mol SiC

$$\frac{1 \text{ mol } SiO_2}{2 \text{ mol CO}} \qquad \frac{3 \text{ mol C}}{2 \text{ mol CO}} \qquad \frac{1 \text{ mol SiC}}{2 \text{ mol CO}}$$

Solution: $10 \text{ mol CO} \times \frac{1 \text{ mol } SiO_2}{2 \text{ mol CO}} = 5 \text{ mol } SiO_2$ $\quad 10 \text{ mol CO} \times \frac{3 \text{ mol C}}{2 \text{ mol CO}} = \underline{1}5 \text{ mol C} = 20 \text{ mol C}$

$10 \text{ mol CO} \times \frac{1 \text{ mol SiC}}{2 \text{ mol CO}} = 5 \text{ mol SiC}$

Given: 2.8 mol SiO_2 **Find:** mol C; mol SiC; mol CO

Conceptual Plan: mol SiO_2 → mol C → mol SiC → mol CO

$$\frac{3 \text{ mol C}}{1 \text{ mol SiO}_2} \quad \frac{1 \text{ mol SiC}}{1 \text{ mol SiO}_2} \quad \frac{2 \text{ mol CO}}{1 \text{ mol SiO}_2}$$

Solution: $2.8 \text{ mol SiO}_2 \times \frac{3 \text{ mol C}}{1 \text{ mol SiO}_2} = 8.4 \text{ mol C}$ $\quad 2.8 \text{ mol SiO}_2 \times \frac{1 \text{ mol SiC}}{1 \text{ mol SiO}_2} = 2.8 \text{ mol SiC}$

$2.8 \text{ mol SiO}_2 \times \frac{2 \text{ mol CO}}{1 \text{ mol SiO}_2} = 5.6 \text{ mol CO}$

Given: 1.55 mol C **Find:** mol SiO_2; mol SiC; mol CO

Conceptual Plan: mol C → mol SiO_2 → mol SiC → mol CO

$$\frac{1 \text{ mol SiO}_2}{3 \text{ mol C}} \quad \frac{1 \text{ mol SiC}}{3 \text{ mol C}} \quad \frac{2 \text{ mol CO}}{3 \text{ mol C}}$$

Solution: $1.55 \text{ mol C} \times \frac{1 \text{ mol SiO}_2}{3 \text{ mol C}} = 0.517 \text{ mol SiO}_2$ $\quad 1.55 \text{ mol C} \times \frac{1 \text{ mol SiC}}{3 \text{ mol C}} = 0.517 \text{ mol SiC}$

$1.55 \text{ mol C} \times \frac{2 \text{ mol CO}}{3 \text{ mol C}} = 1.03 \text{ mol CO}$

Mol SiO_2	Mol C	Mol SiC	Mol CO
3	9	3	6
2	**6**	2	4
5	20	5	**10**
2.8	8.4	2.8	5.6
0.517	**1.55**	0.517	1.03

8.40 **Given:** 2 mol N_2H_4 **Find:** mol N_2O_4; mol N_2; mol H_2O

Conceptual Plan: mol N_2H_4 → mol N_2O_4 → mol N_2 → mol H_2O

$$\frac{1 \text{ mol N}_2\text{O}_4}{2 \text{ mol N}_2\text{H}_4} \quad \frac{3 \text{ mol N}_2}{2 \text{ mol N}_2\text{H}_4} \quad \frac{4 \text{ mol H}_2\text{O}}{2 \text{ mol N}_2\text{H}_4}$$

Solution: $2 \text{ mol N}_2\text{H}_4 \times \frac{1 \text{ mol N}_2\text{O}_4}{2 \text{ mol N}_2\text{H}_4} = 1 \text{ mol N}_2\text{O}_4$ $\quad 2 \text{ mol N}_2\text{H}_4 \times \frac{3 \text{ mol N}_2}{2 \text{ mol N}_2\text{H}_4} = 3 \text{ mol N}_2$

$2 \text{ mol N}_2\text{H}_4 \times \frac{4 \text{ mol H}_2\text{O}}{2 \text{ mol N}_2\text{H}_4} = 4 \text{ mol H}_2\text{O}$

Given: 5 mol N_2O_4 **Find:** mol N_2H_4; mol N_2; mol H_2O

Conceptual Plan: mol N_2O_4 → mol N_2H_4 → mol N_2 → mol H_2O

$$\frac{2 \text{ mol N}_2\text{H}_4}{1 \text{ mol N}_2\text{O}_4} \quad \frac{3 \text{ mol N}_2}{1 \text{ mol N}_2\text{O}_4} \quad \frac{4 \text{ mol H}_2\text{O}}{1 \text{ mol N}_2\text{O}_4}$$

Solution: $5 \text{ mol N}_2\text{O}_4 \times \frac{2 \text{ mol N}_2\text{H}_4}{1 \text{ mol N}_2\text{O}_4} = 10 \text{ mol N}_2\text{H}_4$ $\quad 5 \text{ mol N}_2\text{O}_4 \times \frac{3 \text{ mol N}_2}{1 \text{ mol N}_2\text{O}_4} = 15 \text{ mol N}_2$

$5 \text{ mol N}_2\text{O}_4 \times \frac{4 \text{ mol H}_2\text{O}}{1 \text{ mol N}_2\text{O}_4} = 20 \text{ mol H}_2\text{O}$

Given: 10 mol H_2O **Find:** mol N_2H_4; mol N_2; mol N_2O_4

Conceptual Plan: mol H_2O → mol N_2H_4 → mol N_2 → mol N_2O_4

$$\frac{2 \text{ mol N}_2\text{H}_4}{4 \text{ mol H}_2\text{O}} \quad \frac{3 \text{ mol N}_2}{4 \text{ mol H}_2\text{O}} \quad \frac{1 \text{ mol N}_2\text{O}_4}{4 \text{ mol H}_2\text{O}}$$

Solution: $10 \text{ mol H}_2\text{O} \times \frac{2 \text{ mol N}_2\text{H}_4}{4 \text{ mol H}_2\text{O}} = 5 \text{ mol N}_2\text{H}_4$ $\quad 10 \text{ mol H}_2\text{O} \times \frac{3 \text{ mol N}_2}{4 \text{ mol H}_2\text{O}} = \underline{7}.5 \text{ mol N}_2 = 8 \text{ mol N}_2$

$10 \text{ mol H}_2\text{O} \times \frac{1 \text{ mol N}_2\text{O}_4}{4 \text{ mol H}_2\text{O}} = \underline{2}.5 \text{ mol N}_2\text{O}_4 = 3 \text{ mol N}_2\text{O}_4$

Given: 2.5 mol N_2H_4 **Find:** mol N_2O_4; mol N_2; mol H_2O

Conceptual Plan: mol N_2H_4 → mol N_2O_4 → mol N_2 → mol H_2O

$$\frac{1 \text{ mol N}_2\text{O}_4}{2 \text{ mol N}_2\text{H}_4} \quad \frac{3 \text{ mol N}_2}{2 \text{ mol N}_2\text{H}_4} \quad \frac{4 \text{ mol H}_2\text{O}}{2 \text{ mol N}_2\text{H}_4}$$

Solution: $2.5 \text{ mol N}_2\text{H}_4 \times \frac{1 \text{ mol N}_2\text{O}_4}{2 \text{ mol N}_2\text{H}_4} = 1.3 \text{ mol N}_2\text{O}_4$ $\quad 2.5 \text{ mol N}_2\text{H}_4 \times \frac{3 \text{ mol N}_2}{2 \text{ mol N}_2\text{H}_4} = 3.8 \text{ mol N}_2$

$$2.5 \text{ mol } N_2H_4 \times \frac{4 \text{ mol } H_2O}{2 \text{ mol } N_2H_4} = 5.0 \text{ mol } H_2O$$

Given: 4.2 mol N_2O_4 **Find:** mol N_2H_4; mol N_2; mol H_2O

Conceptual Plan: **mol N_2O_4 → mol N_2H_4 → mol N_2 → mol H_2O**

$$\frac{2 \text{ mol } N_2H_4}{1 \text{ mol } N_2O_4} \quad \frac{3 \text{ mol } N_2}{1 \text{ mol } N_2O_4} \quad \frac{4 \text{ mol } H_2O}{1 \text{ mol } N_2O_4}$$

Solution:

$$4.2 \text{ mol } N_2O_4 \times \frac{2 \text{ mol } N_2H_4}{1 \text{ mol } N_2O_4} = 8.4 \text{ mol } N_2H_4 \qquad 4.2 \text{ mol } N_2O_4 \times \frac{3 \text{ mol } N_2}{1 \text{ mol } N_2O_4} = 12.6 \text{ mol } N_2 = 13 \text{ mol } N_2$$

$$4.2 \text{ mol } N_2O_4 \times \frac{4 \text{ mol } H_2O}{1 \text{ mol } N_2O_4} = 16.8 \text{ mol } H_2O = 17 \text{ mol } H_2O$$

Given: 11.8 mol N_2 **Find:** mol N_2H_4; mol N_2O_4; mol H_2O

Conceptual Plan: **mol N_2 → mol N_2H_4 → mol N_2O_4 → mol H_2O**

$$\frac{2 \text{ mol } N_2H_4}{3 \text{ mol } N_2} \quad \frac{1 \text{ mol } N_2O_4}{3 \text{ mol } N_2} \quad \frac{4 \text{ mol } H_2O}{3 \text{ mol } N_2}$$

Solution: $11.8 \text{ mol } N_2 \times \frac{2 \text{ mol } N_2H_4}{3 \text{ mol } N_2} = 7.87 \text{ mol } N_2H_4 \qquad 11.8 \text{ mol } N_2 \times \frac{1 \text{ mol } N_2O_4}{3 \text{ mol } N_2} = 3.93 \text{ mol } N_2O_4$

$$11.8 \text{ mol } N_2 \times \frac{4 \text{ mol } H_2O}{3 \text{ mol } N_2} = 15.7 \text{ mol } H_2O$$

Mol N_2H_4	Mol N_2O_4	Mol N_2	Mol H_2O
2	1	3	**4**
10	**5**	15	20
5	2.5	7.5	**10**
2.5	1.3	3.8	5.0
8.4	**4.2**	13	17
7.87	3.93	**11.8**	15.7

8.41 **Given:** 3.2 g Fe **Find:** g HBr; gH_2

Conceptual Plan: **g Fe → mol Fe → mol HBr → g HBr**

$$\frac{1 \text{ mol Fe}}{55.8 \text{ g Fe}} \quad \frac{2 \text{ mol HBr}}{1 \text{ mol Fe}} \quad \frac{80.9 \text{ g HBr}}{1 \text{ mol HBr}}$$

g Fe → mol Fe → mol H_2 → g H_2

$$\frac{1 \text{ mol Fe}}{55.8 \text{ g Fe}} \quad \frac{1 \text{ mol } H_2}{1 \text{ mol Fe}} \quad \frac{2.02 \text{ g } H_2}{1 \text{ mol } H_2}$$

Solution: $3.2 \text{ g Fe} \times \frac{1 \text{ mol Fe}}{55.8 \text{ g Fe}} \times \frac{2 \text{ mol HBr}}{1 \text{ mol Fe}} \times \frac{80.9 \text{ g HBr}}{1 \text{ mol HBr}} = 9.\underline{2}8 \text{ g HBr} = 9.3 \text{ g HBr}$

$$3.2 \text{ g Fe} \times \frac{1 \text{ mol Fe}}{55.8 \text{ g Fe}} \times \frac{1 \text{ mol } H_2}{1 \text{ mol Fe}} \times \frac{2.02 \text{ g } H_2}{1 \text{ mol } H_2} = 0.1\underline{1}6 \text{ g } H_2 = 0.12 \text{ g } H_2$$

Check: The units of the answers (g HBr, g H_2) are correct. The magnitude of the answers is reasonable because molar mass HBr is greater than Fe and molar mass H_2 is much less than Fe.

8.42 **Given:** 15.2 g Al **Find:** g H_2SO_4; g H_2

Conceptual Plan: **g Al → mol Al → mol H_2SO_4 → g H_2SO_4**

$$\frac{1 \text{ mol Al}}{26.98 \text{ g Al}} \quad \frac{3 \text{ mol } H_2SO_4}{2 \text{ mol Al}} \quad \frac{98.09 \text{ g } H_2SO_4}{1 \text{ mol } H_2SO_4}$$

g Al → mol Al → mol H_2 → g H_2

$$\frac{1 \text{ mol Al}}{26.98 \text{ g Al}} \quad \frac{3 \text{ mol } H_2}{2 \text{ mol Al}} \quad \frac{2.016 \text{ g } H_2}{1 \text{ mol } H_2}$$

Solution: $15.2 \text{ g Al} \times \frac{1 \text{ mol Al}}{26.98 \text{ g Al}} \times \frac{3 \text{ mol } H_2SO_4}{2 \text{ mol Al}} \times \frac{98.09 \text{ g } H_2SO_4}{1 \text{ mol } H_2SO_4} = 82.9 \text{ g } H_2SO_4$

$$15.2 \text{ g Al} \times \frac{1 \text{ mol Al}}{26.98 \text{ g Al}} \times \frac{3 \text{ mol } H_2}{2 \text{ mol Al}} \times \frac{2.016 \text{ g } H_2}{1 \text{ mol } H_2} = 1.70 \text{ g } H_2$$

Check: The units of the answers (g H_2SO_4, g H_2) are correct. The magnitude of the answers is reasonable because molar mass H_2SO_4 is greater than Al and molar mass H_2 is much less than Al.

8.43 (a) **Given:** 3.67 g Ba **Find:** g $BaCl_2$

Conceptual Plan: g Ba → mol Ba → mol $BaCl_2$ → g $BaCl_2$

$$\frac{1 \text{ mol Ba}}{137.33 \text{ g Ba}} \quad \frac{1 \text{ mol BaCl}_2}{1 \text{ mol Ba}} \quad \frac{208.23 \text{ g BaCl}_2}{1 \text{ mol BaCl}_2}$$

Solution: $3.67 \text{ g Ba} \times \frac{1 \text{ mol Ba}}{137.33 \text{ g Ba}} \times \frac{1 \text{ mol BaCl}_2}{1 \text{ mol Ba}} \times \frac{208.23 \text{ g BaCl}_2}{1 \text{ mol BaCl}_2} = 5.5\underline{6}47 \text{ g BaCl}_2 = 5.56 \text{ g BaCl}_2$

Check: The units of the answer (g $BaCl_2$) are correct. The magnitude of the answer is reasonable because it is larger than grams Ba.

(b) **Given:** 3.67 g CaO **Find:** g $CaCO_3$

Conceptual Plan: g CaO → mol CaO → mol $CaCO_3$ → g $CaCO_3$

$$\frac{1 \text{ mol CaO}}{56.08 \text{ g CaO}} \quad \frac{1 \text{ mol CaCO}_3}{1 \text{ mol CaO}} \quad \frac{100.09 \text{ g CaCO}_3}{1 \text{ mol CaCO}_3}$$

Solution:

$$3.67 \text{ g CaO} \times \frac{1 \text{ mol CaO}}{56.08 \text{ g CaO}} \times \frac{1 \text{ mol CaCO}_3}{1 \text{ mol CaO}} \times \frac{100.09 \text{ g CaCO}_3}{1 \text{ mol CaCO}_3} = 6.5\underline{5}0 \text{ g CaCO}_3 = 6.55 \text{ g CaCO}_3$$

Check: Units of answer (g $CaCO_3$) are correct. The magnitude of the answer is reasonable because it is larger than grams CaO.

(c) **Given:** 3.67 g Mg **Find:** g MgO

Conceptual Plan: g Mg → mol Mg → mol MgO → g MgO

$$\frac{1 \text{ mol Mg}}{24.30 \text{ g Mg}} \quad \frac{1 \text{ mol MgO}}{1 \text{ mol Mg}} \quad \frac{40.30 \text{ g MgO}}{1 \text{ mol MgO}}$$

Solution: $3.67 \text{ g Mg} \times \frac{1 \text{ mol Mg}}{24.30 \text{ g Mg}} \times \frac{1 \text{ mol MgO}}{1 \text{ mol Mg}} \times \frac{40.30 \text{ g MgO}}{1 \text{ mol MgO}} = 6.0\underline{8}6 \text{ g MgO} = 6.09 \text{ g MgO}$

Check: The units of the answer (g MgO) are correct. The magnitude of the answer is reasonable because it is larger than grams Mg.

(d) **Given:** 3.67 g Al **Find:** g Al_2O_3

Conceptual Plan: g Al → mol Al → mol Al_2O_3 → g Al_2O_3

$$\frac{1 \text{ mol Al}}{26.98 \text{ g Al}} \quad \frac{2 \text{ mol Al}_2\text{O}_3}{4 \text{ mol Al}} \quad \frac{101.96 \text{ g Al}_2\text{O}_3}{1 \text{ mol Al}_2\text{O}_3}$$

Solution: $3.67 \text{ g Al} \times \frac{1 \text{ mol Al}}{26.98 \text{ g Al}} \times \frac{2 \text{ mol Al}_2\text{O}_3}{4 \text{ mol Al}} \times \frac{101.96 \text{ g Al}_2\text{O}_3}{1 \text{ mol Al}_2\text{O}_3} = 6.9\underline{3}4 \text{ g Al}_2\text{O}_3 = 6.93 \text{ g Al}_2\text{O}_3$

Check: The units of the answer (g Al_2O_3) are correct. The magnitude of the answer is reasonable because it is larger than grams Al.

8.44 (a) **Given:** 15.39 g Cl_2 **Find:** g KCl

Conceptual Plan: g Cl_2 → mol Cl_2 → mol KCl → g KCl

$$\frac{1 \text{ mol Cl}_2}{70.90 \text{ g Cl}_2} \quad \frac{2 \text{ mol KCl}}{1 \text{ mol Cl}_2} \quad \frac{74.55 \text{ g KCl}}{1 \text{ mol KCl}}$$

Solution: $15.39 \text{ g Cl}_2 \times \frac{1 \text{ mol Cl}_2}{70.90 \text{ g Cl}_2} \times \frac{2 \text{ mol KCl}}{1 \text{ mol Cl}_2} \times \frac{74.55 \text{ g KCl}}{1 \text{ mol KCl}} = 32.3\underline{6}4 \text{ g KCl} = 32.36 \text{ g KCl}$

Check: The units of the answer (g KCl) are correct. The magnitude of the answer is reasonable because it is larger than grams Cl_2.

(b) **Given:** 15.39 g Br_2 **Find:** g KBr

Conceptual Plan: g Br_2 → mol Br_2 → mol KBr → g KBr

$$\frac{1 \text{ mol Br}_2}{159.80 \text{ g Br}_2} \quad \frac{2 \text{ mol KBr}}{1 \text{ mol Br}_2} \quad \frac{119.00 \text{ g KBr}}{1 \text{ mol KBr}}$$

Solution: $15.39 \text{ g Br}_2 \times \frac{1 \text{ mol Br}_2}{159.80 \text{ g Br}_2} \times \frac{2 \text{ mol KBr}}{1 \text{ mol Br}_2} \times \frac{119.00 \text{ g KBr}}{1 \text{ mol KBr}} = 22.9\underline{2}1 \text{ g KBr} = 22.92 \text{ g KBr}$

Check: The units of the answer (g KBr) are correct. The magnitude of the answer is reasonable because it is larger than grams Br_2.

(c) **Given:** 15.39 g O_2 **Find:** g Cr_2O_3

Conceptual Plan: **g O_2 → mol O_2 → mol Cr_2O_3 → g Cr_2O_3**

$$\frac{1 \text{ mol } O_2}{32.00 \text{ g } O_2} \quad \frac{2 \text{ mol } Cr_2O_3}{3 \text{ mol } O_2} \quad \frac{152.00 \text{ g } Cr_2O_3}{1 \text{ mol } Cr_2O_3}$$

Solution:

$$15.39 \text{ g } O_2 \times \frac{1 \text{ mol } O_2}{32.00 \text{ g } O_2} \times \frac{2 \text{ mol } Cr_2O_3}{3 \text{ mol } O_2} \times \frac{152.00 \text{ g } Cr_2O_3}{1 \text{ mol } Cr_2O_3} = 48.7\underline{3}5 \text{ g } Cr_2O_3 = 48.74 \text{ g } Cr_2O_3$$

Check: The units of the answer (g Cr_2O_3) are correct. The magnitude of the answer is reasonable because it is larger than grams O_2.

(d) **Given:** 15.39 g Sr **Find:** g SrO

Conceptual Plan: **g Sr → mol Sr → mol SrO → g SrO**

$$\frac{1 \text{ mol Sr}}{87.62 \text{ g Sr}} \quad \frac{2 \text{ mol SrO}}{2 \text{ mol Sr}} \quad \frac{103.62 \text{ g SrO}}{1 \text{ mol SrO}}$$

Solution: $15.39 \text{ g Sr} \times \frac{1 \text{ mol Sr}}{87.62 \text{ g Sr}} \times \frac{2 \text{ mol SrO}}{2 \text{ mol Sr}} \times \frac{103.62 \text{ g SrO}}{1 \text{ mol SrO}} = 18.2\underline{0}0 \text{ g SrO} = 18.20 \text{ g SrO}$

Check: The units of the answer (g SrO) are correct. The magnitude of the answer is reasonable because it is larger than grams Sr.

Limiting Reactant, Theoretical Yield, and Percent Yield

8.45 (a) **Given:** 2 mol Na; 2 mol Br_2 **Find:** limiting reactant

Conceptual Plan: **mol Na → mol NaBr**

$$\frac{2 \text{ mol NaBr}}{2 \text{ mol Na}}$$ **→ smaller mol amount determines limiting reactant**

mol Br_2 → mol NaBr

$$\frac{2 \text{ mol NaBr}}{1 \text{ mol } Br_2}$$

Solution: $2 \text{ mol Na} \times \frac{2 \text{ mol NaBr}}{2 \text{ mol Na}} = 2 \text{ mol NaBr}$

$2 \text{ mol } Br_2 \times \frac{2 \text{ mol NaBr}}{1 \text{ mol } Br_2} = 4 \text{ mol NaBr}$

Na is the limiting reactant.

Check: The answer is reasonable because Na produced the smallest amount of product.

(b) **Given:** 1.8 mol Na; 1.4 mol Br_2 **Find:** limiting reactant

Conceptual Plan: **mol Na → mol NaBr**

$$\frac{2 \text{ mol NaBr}}{2 \text{ mol Na}}$$ **→ smaller mol amount determines limiting reactant**

mol Br_2 → mol NaBr

$$\frac{2 \text{ mol NaBr}}{1 \text{ mol } Br_2}$$

Solution: $1.8 \text{ mol Na} \times \frac{2 \text{ mol NaBr}}{2 \text{ mol Na}} = 1.8 \text{ mol NaBr}$

$1.4 \text{ mol } Br_2 \times \frac{2 \text{ mol NaBr}}{1 \text{ mol } Br_2} = 2.8 \text{ mol NaBr}$

Na is the limiting reactant.

Check: The answer is reasonable because Na produced the smallest amount of product.

(c) **Given:** 2.5 mol Na; 1 mol Br_2 **Find:** limiting reactant

Conceptual Plan: **mol Na → mol NaBr**

$$\frac{2 \text{ mol NaBr}}{2 \text{ mol Na}}$$ **→ smaller mol amount determines limiting reactant**

$$\textbf{mol Br}_2 \rightarrow \textbf{mol NaBr} \quad \frac{2 \text{ mol NaBr}}{1 \text{ mol Br}_2}$$

Solution: $2.5 \text{ mol Na} \times \dfrac{2 \text{ mol NaBr}}{2 \text{ mol Na}} = 2.5 \text{ mol NaBr}$

$1 \text{ mol Br}_2 \times \dfrac{2 \text{ mol NaBr}}{1 \text{ mol Br}_2} = 2 \text{ mol NaBr}$

Br_2 is the limiting reactant.

Check: The answer is reasonable because Br_2 produced the smallest amount of product.

(d) **Given:** 12.6 mol Na; 6.9 mol Br_2 **Find:** limiting reactant

Conceptual Plan: mol Na → mol NaBr $\dfrac{2 \text{ mol NaBr}}{2 \text{ mol Na}}$ **→ smaller mol amount determines limiting reactant**

$$\textbf{mol Br}_2 \rightarrow \textbf{mol NaBr} \quad \frac{2 \text{ mol NaBr}}{1 \text{ mol Br}_2}$$

Solution: $12.6 \text{ mol Na} \times \dfrac{2 \text{ mol NaBr}}{2 \text{ mol Na}} = 12.6 \text{ mol NaBr}$

$6.9 \text{ mol Br}_2 \times \dfrac{2 \text{ mol NaBr}}{1 \text{ mol Br}_2} = 13.8 \text{ mol NaBr} = 14 \text{ mol NaBr}$

Na is the limiting reactant.

Check: The answer is reasonable because Na produced the smallest amount of product.

8.46 (a) **Given:** 1 mol Al; 1 mol O_2 **Find:** limiting reactant

Conceptual Plan: mol Al → mol Al_2O_3 $\dfrac{2 \text{ mol Al}_2\text{O}_3}{4 \text{ mol Al}}$ **→ smaller mol amount determines limiting reactant**

$$\textbf{mol O}_2 \rightarrow \textbf{mol Al}_2\textbf{O}_3 \quad \frac{2 \text{ mol Al}_2\text{O}_3}{3 \text{ mol O}_2}$$

Solution: $1 \text{ mol Al} \times \dfrac{2 \text{ mol Al}_2\text{O}_3}{4 \text{ mol Al}} = 0.5 \text{ mol Al}_2\text{O}_3$

$1 \text{ mol O}_2 \times \dfrac{2 \text{ mol Al}_2\text{O}_3}{3 \text{ mol O}_2} = 0.7 \text{ mol Al}_2\text{O}_3$

Al is the limiting reactant.

Check: The answer is reasonable because Al produced the smallest amount of product.

(b) **Given:** 4 mol Al; 2.6 mol O_2 **Find:** limiting reactant

Conceptual Plan: mol Al → mol Al_2O_3 $\dfrac{2 \text{ mol Al}_2\text{O}_3}{4 \text{ mol Al}}$ **→ smaller mol amount determines limiting reactant**

$$\textbf{mol O}_2 \rightarrow \textbf{mol Al}_2\textbf{O}_3 \quad \frac{2 \text{ mol Al}_2\text{O}_3}{3 \text{ mol O}_2}$$

Solution: $4 \text{ mol Al} \times \dfrac{2 \text{ mol Al}_2\text{O}_3}{4 \text{ mol Al}} = 2 \text{ mol Al}_2\text{O}_3$

$2.6 \text{ mol O}_2 \times \dfrac{2 \text{ mol Al}_2\text{O}_3}{3 \text{ mol O}_2} = 1.7 \text{ mol Al}_2\text{O}_3$

O_2 is the limiting reactant.

Check: The answer is reasonable because O_2 produced the smallest amount of product.

(c) **Given:** 16 mol Al; 13 mol O_2 **Find:** limiting reactant

Conceptual Plan: mol Al → mol Al_2O_3 $\dfrac{2 \text{ mol Al}_2\text{O}_3}{4 \text{ mol Al}}$ **→ smaller mol amount determines limiting reactant**

mol O_2 → mol Al_2O_3

$\frac{2 \text{ mol } Al_2O_3}{3 \text{ mol } O_2}$

Solution: $16 \text{ mol Al} \times \frac{2 \text{ mol } Al_2O_3}{4 \text{ mol Al}} = 8.0 \text{ mol } Al_2O_3$

$13 \text{ mol } O_2 \times \frac{2 \text{ mol } Al_2O_3}{3 \text{ mol } O_2} = 8.7 \text{ mol } Al_2O_3$

Al is the limiting reactant.

Check: The answer is reasonable because Al produced the smallest amount of product.

(d) **Given:** 7.4 mol Al; 6.5 mol O_2 **Find:** limiting reactant

Conceptual Plan: mol Al → mol Al_2O_3

$\frac{2 \text{ mol } Al_2O_3}{4 \text{ mol Al}}$ → **smaller mol amount determines limiting reactant**

mol O_2 → mol Al_2O_3

$\frac{2 \text{ mol } Al_2O_3}{3 \text{ mol } O_2}$

Solution: $7.4 \text{ mol Al} \times \frac{2 \text{ mol } Al_2O_3}{4 \text{ mol Al}} = 3.7 \text{ mol } Al_2O_3$

$6.5 \text{ mol } O_2 \times \frac{2 \text{ mol } Al_2O_3}{3 \text{ mol } O_2} = 4.3 \text{ mol } Al_2O_3$

Al is the limiting reactant.

Check: The answer is reasonable because Al produced the smallest amount of product.

8.47 The greatest number of Cl_2 molecules will be formed from reaction mixture (b) and would be 3 molecules Cl_2.

(a) **Given:** 7 molecules HCl; 1 molecule O_2 **Find:** theoretical yield Cl_2

Conceptual Plan: molecules HCl → molecules Cl_2

$\frac{2 \text{ molecules } Cl_2}{4 \text{ molecules HCl}}$ → **smaller molecule amount determines limiting reactant**

molecules O_2 → molecules Cl_2

$\frac{2 \text{ molecules } Cl_2}{1 \text{ molecule } O_2}$

Solution: $7 \text{ molecules HCl} \times \frac{2 \text{ molecules } Cl_2}{4 \text{ molecules HCl}} = 3 \text{ molecules } Cl_2$

$1 \text{ molecule } O_2 \times \frac{2 \text{ molecules } Cl_2}{1 \text{ molecule } O_2} = 2 \text{ molecules } Cl_2$

theoretical yield = 2 molecules Cl_2

(b) **Given:** 6 molecules HCl; 3 molecules O_2 **Find:** theoretical yield Cl_2

Conceptual Plan: molecules HCl → molecules Cl_2

$\frac{2 \text{ molecules } Cl_2}{4 \text{ molecules HCl}}$ → **smaller molecule amount determines limiting reactant**

molecules O_2 → molecules Cl_2

$\frac{2 \text{ molecules } Cl_2}{1 \text{ molecule } O_2}$

Solution: $6 \text{ molecules HCl} \times \frac{2 \text{ molecules } Cl_2}{4 \text{ molecules HCl}} = 3 \text{ molecules } Cl_2$

$3 \text{ molecules } O_2 \times \frac{2 \text{ molecules } Cl_2}{1 \text{ molecule } O_2} = 6 \text{ molecules } Cl_2$

theoretical yield = 3 molecules

(c) **Given:** 4 molecules HCl; 5 molecules O_2 **Find:** theoretical yield Cl_2

Conceptual Plan: molecules HCl → molecules Cl_2

$\frac{2 \text{ molecules } Cl_2}{4 \text{ molecules HCl}}$ → **smaller molecule amount determines limiting reactant**

molecules O_2 → molecules Cl_2

$$\frac{2 \text{ molecules } Cl_2}{1 \text{ molecule } O_2}$$

Solution: $4 \text{ molecules HCl} \times \frac{2 \text{ molecules } Cl_2}{4 \text{ molecules HCl}} = 2 \text{ molecules } Cl_2$

$5 \text{ molecules } O_2 \times \frac{2 \text{ molecules } Cl_2}{1 \text{ molecule } O_2} = 10 \text{ molecules } Cl_2$

theoretical yield = 2 molecules Cl_2

Check: The units of the answer (molecules Cl_2) are correct. The answer is reasonable based on the limiting reactant in each mixture.

8.48 The greatest number of CO_2 molecules will be formed from reaction mixture (a) and would be 2 molecules CO_2.

(a) **Given:** 3 molecules CH_3OH; 3 molecules O_2 **Find:** theoretical yield CO_2

Conceptual Plan: molecules CH_3OH → molecules CO_2

$$\frac{2 \text{ molecules } CO_2}{2 \text{ molecules } CH_3OH}$$ → **smaller molecule amount determines limiting reactant**

molecules O_2 → molecules CO_2

$$\frac{2 \text{ molecules } CO_2}{3 \text{ molecules } O_2}$$

Solution: $3 \text{ molecules } CH_3OH \times \frac{2 \text{ molecules } CO_2}{2 \text{ molecules } CH_3OH} = 3 \text{ molecules } CO_2$

$3 \text{ molecules } O_2 \times \frac{2 \text{ molecules } CO_2}{3 \text{ molecules } O_2} = 2 \text{ molecules } CO_2$

theoretical yield = 2 molecules CO_2

(b) **Given:** 1 molecule CH_3OH; 6 molecules O_2 **Find:** theoretical yield CO_2

Conceptual Plan: molecules CH_3OH → molecules CO_2

$$\frac{2 \text{ molecules } CO_2}{2 \text{ molecules } CH_3OH}$$ → **smaller molecule amount determines limiting reactant**

molecules O_2 → molecules CO_2

$$\frac{2 \text{ molecules } CO_2}{3 \text{ molecules } O_2}$$

Solution: $1 \text{ molecule } CH_3OH \times \frac{2 \text{ molecules } CO_2}{2 \text{ molecules } CH_3OH} = 1 \text{ molecule } CO_2$

$6 \text{ molecules } O_2 \times \frac{2 \text{ molecules } CO_2}{3 \text{ molecules } O_2} = 4 \text{ molecules } CO_2$

theoretical yield = 1 molecule CO_2

(c) **Given:** 4 molecules CH_3OH; 2 molecules O_2 **Find:** theoretical yield CO_2

Conceptual Plan: molecules CH_3OH → molecules CO_2

$$\frac{2 \text{ molecules } CO_2}{2 \text{ molecules } CH_3OH}$$ → **smaller molecule amount determines limiting reactant**

molecules O_2 → molecules CO_2

$$\frac{2 \text{ molecules } CO_2}{3 \text{ molecules } O_2}$$

Solution: $4 \text{ molecules } CH_3OH \times \frac{2 \text{ molecules } CO_2}{2 \text{ molecules } CH_3OH} = 4 \text{ molecules } CO_2$

$2 \text{ molecules } O_2 \times \frac{2 \text{ molecules } CO_2}{3 \text{ molecules } O_2} = 1.3 \text{ molecules } CO_2 = 1 \text{ molecule } CO_2$ because you cannot have a fraction of a molecule

theoretical yield = 1 molecule CO_2

Check: The units of the answer (molecules CO_2) are correct. The answer is reasonable based on the limiting reactant in each mixture.

8.49 (a) **Given:** 4 mol Ti; 4 mol Cl_2 **Find:** theoretical yield $TiCl_4$

Conceptual Plan: mol Ti → mol $TiCl_4$

$$\frac{1 \text{ mol } TiCl_4}{1 \text{ mol Ti}}$$ **→ smaller mol amount determines limiting reactant**

mol Cl_2 → mol $TiCl_4$

$$\frac{1 \text{ mol } TiCl_4}{2 \text{ mol } Cl_2}$$

Solution: $4 \cancel{\text{mol Ti}} \times \frac{1 \text{ mol } TiCl_4}{1 \cancel{\text{mol Ti}}} = 4 \text{ mol } TiCl_4$

$$4 \cancel{\text{mol } Cl_2} \times \frac{1 \text{ mol } TiCl_4}{2 \cancel{\text{mol } Cl_2}} = 2 \text{ mol } TiCl_4$$

theoretical yield $= 2$ mol $TiCl_4$

Check: The units of the answer (mol $TiCl_4$) are correct. The answer is reasonable because Cl_2 produced the smallest amount of product and is the limiting reactant.

(b) **Given:** 7 mol Ti; 17 mol Cl_2 **Find:** theoretical yield $TiCl_4$

Conceptual Plan: mol Ti → mol $TiCl_4$

$$\frac{1 \text{ mol } TiCl_4}{1 \text{ mol Ti}}$$ **→ smaller mol amount determines limiting reactant**

mol Cl_2 → mol $TiCl_4$

$$\frac{1 \text{ mol } TiCl_4}{2 \text{ mol } Cl_2}$$

Solution: $7 \cancel{\text{mol Ti}} \times \frac{1 \text{ mol } TiCl_4}{1 \cancel{\text{mol Ti}}} = 7 \text{ mol } TiCl_4$

$$17 \cancel{\text{mol } Cl_2} \times \frac{1 \text{ mol } TiCl_4}{2 \cancel{\text{mol } Cl_2}} = 8.5 \text{ mol } TiCl_4$$

theoretical yield $= 7$ mol $TiCl_4$

Check: The units of the answer (mol $TiCl_4$) are correct. The answer is reasonable because Ti produced the smallest amount of product and is the limiting reactant.

(c) **Given:** 12.4 mol Ti; 18.8 mol Cl_2 **Find:** theoretical yield $TiCl_4$

Conceptual Plan: mol Ti → mol $TiCl_4$

$$\frac{1 \text{ mol } TiCl_4}{1 \text{ mol Ti}}$$ **→ smaller mol amount determines limiting reactant**

mol Cl_2 → mol $TiCl_4$

$$\frac{1 \text{ mol } TiCl_4}{2 \text{ mol } Cl_2}$$

Solution: $12.4 \cancel{\text{mol Ti}} \times \frac{1 \text{ mol } TiCl_4}{1 \cancel{\text{mol Ti}}} = 12.4 \text{ mol } TiCl_4$

$$18.8 \cancel{\text{mol } Cl_2} \times \frac{1 \text{ mol } TiCl_4}{2 \cancel{\text{mol } Cl_2}} = 9.40 \text{ mol } TiCl_4$$

theoretical yield $= 9.40$ mol $TiCl_4$

Check: The units of the answer (mol $TiCl_4$) are correct. The answer is reasonable because Cl_2 produced the smallest amount of product and is the limiting reactant.

8.50 (a) **Given:** 3 mol Mn; 3 mol O_2 **Find:** theoretical yield MnO_2

Conceptual Plan: mol Mn → mol MnO_2

$$\frac{2 \text{ mol } MnO_2}{2 \text{ mol Mn}}$$ **→ smaller mol amount determines limiting reactant**

mol O_2 → mol MnO_2

$$\frac{2 \text{ mol } MnO_2}{2 \text{ mol } O_2}$$

Solution: $3 \cancel{\text{mol Mn}} \times \frac{2 \text{ mol } MnO_2}{2 \cancel{\text{mol Mn}}} = 3 \text{ mol } MnO_2$

$$3 \cancel{\text{mol O}_2} \times \frac{2 \text{ mol MnO}_2}{2 \cancel{\text{mol O}_2}} = 3 \text{ mol MnO}_2$$

theoretical yield $= 3$ mol MnO_2

Check: The units of the answer (mol MnO_2) are correct. The answer is reasonable because equal mol are produced for both reactants.

(b) **Given:** 4 mol Mn; 7 mol O_2 **Find:** theoretical yield MnO_2

Conceptual Plan: mol Mn → mol MnO_2

$\frac{2 \text{ mol MnO}_2}{2 \text{ mol Mn}}$ → **smaller mol amount determines limiting reactant**

mol O_2 → mol MnO_2

$\frac{2 \text{ mol MnO}_2}{2 \text{ mol O}_2}$

Solution: $4 \cancel{\text{mol Mn}} \times \frac{2 \text{ mol MnO}_2}{2 \cancel{\text{mol Mn}}} = 4 \text{ mol MnO}_2$

$$7 \cancel{\text{mol O}_2} \times \frac{2 \text{ mol MnO}_2}{2 \cancel{\text{mol O}_2}} = 7 \text{ mol MnO}_2$$

theoretical yield $= 4$ mol MnO_2

Check: The units of the answer (mol MnO_2) are correct. The answer is reasonable because Mn produced the smallest amount of product and is the limiting reactant.

(c) **Given:** 27.5 mol Mn; 43.8 mol O_2 **Find:** theoretical yield MnO_2

Conceptual Plan: mol Mn → mol MnO_2

$\frac{2 \text{ mol MnO}_2}{2 \text{ mol Mn}}$ → **smaller mol amount determines limiting reactant**

mol O_2 → mol MnO_2

$\frac{2 \text{ mol MnO}_2}{2 \text{ mol O}_2}$

Solution: $27.5 \cancel{\text{mol Mn}} \times \frac{2 \text{ mol MnO}_2}{2 \cancel{\text{mol Mn}}} = 27.5 \text{ mol MnO}_2$

$$43.8 \cancel{\text{mol O}_2} \times \frac{2 \text{ mol MnO}_2}{2 \cancel{\text{mol O}_2}} = 43.8 \text{ mol MnO}_2$$

theoretical yield $= 27.5$ mol MnO_2

Check: The units of the answer (mol MnO_2) are correct. The answer is reasonable because Mn produced the smallest amount of product and is the limiting reactant.

8.51 **Given:** 4.2 mol ZnS; 6.8 mol O_2 **Find:** mole amount of excess reactant left

Conceptual Plan: mol ZnS → mol ZnO

$\frac{2 \text{ mol ZnO}}{2 \text{ mol ZnS}}$ → **smaller mol amount determines limiting reactant**

mol O_2 → mol ZnO

$\frac{2 \text{ mol ZnO}}{3 \text{ mol O}_2}$

mol limiting reactant → mol excess reactant required → mol excess reactant left

$\frac{3 \text{ mol O}_2}{2 \text{ mol ZnS}}$

Solution: $4.2 \cancel{\text{mol ZnS}} \times \frac{2 \text{ mol ZnO}}{2 \cancel{\text{mol ZnS}}} = 4.2 \text{ mol ZnO}$

$$6.8 \cancel{\text{mol O}_2} \times \frac{2 \text{ mol ZnO}}{3 \cancel{\text{mol O}_2}} = 4.5 \text{ mol ZnO}$$

ZnS is the limiting reactant; therefore, O_2 is the excess reactant.

$$4.2 \cancel{\text{mol ZnS}} \times \frac{3 \text{ mol O}_2}{2 \cancel{\text{mol ZnS}}} = 6.3 \text{ mol O}_2 \text{ required}$$

$$6.8 \text{ mol O}_2 - 6.3 \text{ mol O}_2 = 0.5 \text{ mol O}_2 \text{ left}$$

Check: The units of the answer (mol O_2) are correct. The magnitude is reasonable because it is less than the original amount of O_2.

8.52 **Given:** 0.223 mol FeS; 0.652 mol HCl **Find:** mole amount of excess reactant left

Conceptual Plan: mol FeS → mol $FeCl_2$

$\frac{1 \text{ mol FeCl}_2}{1 \text{ mol FeS}}$ → **smaller mol amount determines limiting reactant**

mol HCl → mol $FeCl_2$

$\frac{1 \text{ mol FeCl}_2}{2 \text{ mol HCl}}$

mol limiting reactant → mol excess reactant required → mol excess reactant left

$\frac{2 \text{ mol HCl}}{1 \text{ mol FeS}}$

Solution: $0.223 \text{ mol FeS} \times \frac{1 \text{ mol FeCl}_2}{1 \text{ mol FeS}} = 0.223 \text{ mol FeCl}_2$

$$0.652 \text{ mol HCl} \times \frac{1 \text{ mol FeCl}_2}{2 \text{ mol HCl}} = 0.326 \text{ mol FeCl}_2$$

FeS is the limiting reactant; therefore, HCl is the excess reactant.

$$0.223 \text{ mol FeS} \times \frac{2 \text{ mol HCl}}{1 \text{ mol FeS}} = 0.446 \text{ mol HCl required}$$

$$0.652 \text{ mol HCl} - 0.446 \text{ mol HCl} = 0.206 \text{ mol HCl left}$$

Check: The units of the answer (mol HCl) are correct. The magnitude is reasonable because it is less than the original amount of HCl.

8.53 (a) **Given:** 2.0 g Al; 2.0 g Cl_2 **Find:** theoretical yield in g $AlCl_3$

Conceptual Plan: g Al → mol Al → mol $AlCl_3$

$\frac{1 \text{ mol Al}}{26.98 \text{ g Al}}$ $\frac{2 \text{ mol AlCl}_3}{2 \text{ mol Al}}$ → **smaller mol amount determines limiting reactant**

g Cl_2 → mol Cl_2 → mol $AlCl_3$

$\frac{1 \text{ mol Cl}_2}{70.90 \text{ g Cl}_2}$ $\frac{2 \text{ mol AlCl}_3}{3 \text{ mol Cl}_2}$

then mol $AlCl_3$ → g $AlCl_3$

$\frac{133.3 \text{ g AlCl}_3}{\text{mol AlCl}_3}$

Solution: $2.0 \text{ g Al} \times \frac{1 \text{ mol Al}}{26.98 \text{ g Al}} \times \frac{2 \text{ mol AlCl}_3}{2 \text{ mol Al}} = 0.074 \text{ mol AlCl}_3$

$$2.0 \text{ g Cl}_2 \times \frac{1 \text{ mol Cl}_2}{70.90 \text{ g Cl}_2} \times \frac{2 \text{ mol AlCl}_3}{3 \text{ mol Cl}_2} = 0.01\underline{88} \text{ mol AlCl}_3$$

$$0.01\underline{88} \text{ mol AlCl}_3 \times \frac{133.3 \text{ g AlCl}_3}{\text{mol AlCl}_3} = 2.5 \text{ g AlCl}_3$$

Check: The units of the answer (g $AlCl_3$) are correct. The answer is reasonable because Cl_2 produced the smallest amount of product and is the limiting reactant.

(b) **Given:** 7.5 g Al; 24.8 g Cl_2 **Find:** theoretical yield in g $AlCl_3$

Conceptual Plan: g Al → mol Al → mol $AlCl_3$

$\frac{1 \text{ mol Al}}{26.98 \text{ g Al}}$ $\frac{2 \text{ mol AlCl}_3}{2 \text{ mol Al}}$ → **smaller mol amount determines limiting reactant**

g Cl_2 → mol Cl_2 → mol $AlCl_3$

$\frac{1 \text{ mol Cl}_2}{70.90 \text{ g Cl}_2}$ $\frac{2 \text{ mol AlCl}_3}{3 \text{ mol Cl}_2}$

then mol $AlCl_3$ → g $AlCl_3$

$\frac{133.3 \text{ g AlCl}_3}{\text{mol AlCl}_3}$

Solution: $7.5 \text{ g Al} \times \frac{1 \text{ mol Al}}{26.98 \text{ g Al}} \times \frac{2 \text{ mol AlCl}_3}{2 \text{ mol Al}} = 0.2\underline{7}80 \text{ mol AlCl}_3$

$24.8 \text{ g Cl}_2 \times \frac{1 \text{ mol Cl}_2}{70.90 \text{ g Cl}_2} \times \frac{2 \text{ mol AlCl}_3}{3 \text{ mol Cl}_2} = 0.23\underline{3}2 \text{ mol AlCl}_3$

$0.23\underline{3}2 \text{ mol AlCl}_3 \times \frac{133.3 \text{ g AlCl}_3}{\text{mol AlCl}_3} = 31.1 \text{ g AlCl}_3$

Check: The units of the answer (g AlCl_3) are correct. The answer is reasonable because Cl_2 produced the smallest amount of product and is the limiting reactant.

(c) **Given:** 0.235 g Al; 1.15 g Cl_2 **Find:** theoretical yield in g $AlCl_3$

Conceptual Plan: g A → mol Al → mol $AlCl_3$

$\frac{1 \text{ mol Al}}{26.98 \text{ g Al}} \quad \frac{2 \text{ mol AlCl}_3}{2 \text{ mol Al}}$ **→ smaller mol amount determines limiting reactant**

g Cl_2 → mol Cl_2 → mol $AlCl_3$

$\frac{1 \text{ mol Cl}_2}{70.90 \text{ g Cl}_2} \quad \frac{2 \text{ mol AlCl}_3}{3 \text{ mol Cl}_2}$

then mol $AlCl_3$ → g $AlCl_3$

$\frac{133.3 \text{ g AlCl}_3}{\text{mol AlCl}_3}$

Solution: $0.235 \text{ g Al} \times \frac{1 \text{ mol Al}}{26.98 \text{ g Al}} \times \frac{2 \text{ mol AlCl}_3}{2 \text{ mol Al}} = 0.0087\underline{1}0 \text{ mol AlCl}_3$

$1.15 \text{ g Cl}_2 \times \frac{1 \text{ mol Cl}_2}{70.90 \text{ g Cl}_2} \times \frac{2 \text{ mol AlCl}_3}{3 \text{ mol Cl}_2} = 0.01081 \text{ mol AlCl}_3$

$0.0087\underline{1}0 \text{ mol AlCl}_3 \times \frac{133.3 \text{ g AlCl}_3}{\text{mol AlCl}_3} = 1.16 \text{ g AlCl}_3$

Check: The units of the answer (g AlCl_3) are correct. The answer is reasonable because Al produced the smallest amount of product and is the limiting reactant.

8.54 (a) **Given:** 5.0 g Ti; 5.0 g F_2 **Find:** theoretical yield in g TiF_4

Conceptual Plan: g Ti → mol Ti → mol TiF_4

$\frac{1 \text{ mol Ti}}{47.87 \text{ g Ti}} \quad \frac{1 \text{ mol TiF}_4}{1 \text{ mol Ti}}$ **→ smaller mol amount determines limiting reactant**

g F_2 → mol F_2 → mol TiF_4

$\frac{1 \text{ mol F}_2}{38.00 \text{ g F}_2} \quad \frac{1 \text{ mol TiF}_4}{2 \text{ mol F}_2}$

then mol TiF_4 → g TiF_4

$\frac{123.87 \text{ g TiF}_4}{\text{mol TiF}_4}$

Solution: $5.0 \text{ g Ti} \times \frac{1 \text{ mol Ti}}{47.87 \text{ g Ti}} \times \frac{1 \text{ mol TiF}_4}{1 \text{ mol Ti}} = 0.1\underline{0}4 \text{ mol TiF}_4$

$5.0 \text{ g F}_2 \times \frac{1 \text{ mol F}_2}{38.00 \text{ g F}_2} \times \frac{1 \text{ mol TiF}_4}{2 \text{ mol F}_2} = 0.06\underline{5}8 \text{ mol TiF}_4$

$0.06\underline{5}8 \text{ mol TiF}_4 \times \frac{123.87 \text{ g TiF}_4}{1 \text{ mol TiF}_4} = 8.2 \text{ g TiF}_4$

Check: The units of the answer (g TiF_4) are correct. The answer is reasonable because F_2 produced the smallest amount of product and is the limiting reactant.

(b) **Given:** 2.4 g Ti; 1.6 g F_2 **Find:** theoretical yield in g TiF_4

Conceptual Plan: g Ti → mol Ti → mol TiF_4

$\frac{1 \text{ mol Ti}}{47.87 \text{ g Ti}} \quad \frac{1 \text{ mol TiF}_4}{1 \text{ mol Ti}}$ **→ smaller mol amount determines limiting reactant**

g F_2 → mol F_2 → mol TiF_4

$\frac{1 \text{ mol F}_2}{38.00 \text{ g F}_2} \quad \frac{1 \text{ mol TiF}_4}{2 \text{ mol F}_2}$

then mol $TiF_4 \rightarrow$ g TiF_4

$\frac{123.87 \text{ g TiF}_4}{\text{mol TiF}_4}$

Solution: $2.4 \text{ g Ti} \times \frac{1 \text{ mol Ti}}{47.87 \text{ g Ti}} \times \frac{1 \text{ mol TiF}_4}{1 \text{ mol Ti}} = 0.0501 \text{ mol TiF}_4$

$1.6 \text{ g F}_2 \times \frac{1 \text{ mol F}_2}{38.00 \text{ g F}_2} \times \frac{1 \text{ mol TiF}_4}{2 \text{ mol F}_2} = 0.02\underline{1}1 \text{ mol TiF}_4$

$0.02\underline{1}1 \text{ mol TiF}_4 \times \frac{123.87 \text{ g TiF}_4}{\text{mol TiF}_4} = 2.6 \text{ g TiF}_4$

Check: The units of the answer (g TiF_4) are correct. The answer is reasonable because F_2 produced the smallest amount of product and is the limiting reactant.

(c) **Given:** 0.233 g Ti; 0.288 g F_2 **Find:** theoretical yield in g TiF_4

Conceptual Plan: g Ti $\rightarrow$ mol Ti $\rightarrow$ mol TiF_4

$\frac{1 \text{ mol Ti}}{47.87 \text{ g Ti}} \quad \frac{1 \text{ mol TiF}_4}{1 \text{ mol Ti}}$ **$\rightarrow$ smaller mol amount determines limiting reactant**

g $F_2 \rightarrow$ mol $F_2 \rightarrow$ mol TiF_4

$\frac{1 \text{ mol F}_2}{38.00 \text{ g F}_2} \quad \frac{1 \text{ mol TiF}_4}{2 \text{ mol F}_2}$

then mol $TiF_4 \rightarrow$ g TiF_4

$\frac{123.87 \text{ g TiF}_4}{\text{mol TiF}_4}$

Solution: $0.233 \text{ g Ti} \times \frac{1 \text{ mol Ti}}{47.87 \text{ g Ti}} \times \frac{1 \text{ mol TiF}_4}{1 \text{ mol Ti}} = 0.0048\underline{6}7 \text{ mol TiF}_4$

$0.288 \text{ g F}_2 \times \frac{1 \text{ mol F}_2}{38.00 \text{ g F}_2} \times \frac{1 \text{ mol TiF}_4}{2 \text{ mol F}_2} = 0.0037\underline{8}9 \text{ mol TiF}_4$

$0.0037\underline{8}9 \text{ mol TiF}_4 \times \frac{123.87 \text{ g TiF}_4}{\text{mol TiF}_4} = 0.469 \text{ g TiF}_4$

Check: The units of the answer (g TiF_4) are correct. The answer is reasonable because F_2 produced the smallest amount of product and is the limiting reactant.

8.55 **Given:** 22.55 g Fe_2O_3; 14.78 g CO **Find:** mole amount of excess reactant left

Conceptual Plan: g $Fe_2O_3 \rightarrow$ mol $Fe_2O_3 \rightarrow$ mol Fe

$\frac{1 \text{ mol Fe}_2\text{O}_3}{159.7 \text{ g Fe}_2\text{O}_3} \quad \frac{2 \text{ mol Fe}}{1 \text{ mol Fe}_2\text{O}_3}$ **$\rightarrow$ smaller mol amount determines limiting reactant**

g CO $\rightarrow$ mol CO $\rightarrow$ mol Fe

$\frac{1 \text{ mol CO}}{28.01 \text{ g CO}} \quad \frac{2 \text{ mol Fe}}{3 \text{ mol CO}}$

then mol limiting reactant $\rightarrow$ mol excess reactant required $\rightarrow$ mol excess reactant left $\rightarrow$ g excess reactant left

$\frac{3 \text{ mol CO}}{1 \text{ mol Fe}_2\text{O}_3} \quad \frac{1 \text{ mol Fe}_2\text{O}_3}{159.7 \text{ g Fe}_2\text{O}_3}$ or $\frac{28.01 \text{ g CO}}{1 \text{ mol CO}}$

Solution: $22.55 \text{ g Fe}_2\text{O}_3 \times \frac{1 \text{ mol Fe}_2\text{O}_3}{159.7 \text{ g Fe}_2\text{O}_3} \times \frac{2 \text{ mol Fe}}{1 \text{ mol Fe}_2\text{O}_3} = 0.2824 \text{ mol Fe}$

$14.78 \text{ g CO} \times \frac{1 \text{ mol CO}}{28.01 \text{ g CO}} \times \frac{2 \text{ mol Fe}}{3 \text{ mol CO}} = 0.3518 \text{ mol Fe}$

Fe_2O_3 is the limiting reactant; therefore, CO is the excess reactant.

$22.55 \text{ g Fe}_2\text{O}_3 \times \frac{1 \text{ mol Fe}_2\text{O}_3}{159.7 \text{ g Fe}_2\text{O}_3} \times \frac{3 \text{ mol CO}}{1 \text{ mol Fe}_2\text{O}_3} \times \frac{28.01 \text{ g CO}}{1 \text{ mol CO}} = 11.8\underline{6}5 \text{ g CO required}$

$14.78 \text{ g CO} - 11.87 \text{ g CO} = 2.91 \text{ g CO left}$

Check: The units of the answer (g CO) are correct. The magnitude is reasonable because it is less than the original amount of CO.

8.56 **Given:** 45.69 g P_4; 131.3 g Cl_2 **Find:** mole amount of excess reactant left

Conceptual Plan: $\mathbf{g\ P_4 \rightarrow mol\ P_4 \rightarrow mol\ PCl_3}$

$\frac{1 \text{ mol } P_4}{123.88 \text{ g } P_4} \quad \frac{4 \text{ mol } PCl_3}{1 \text{ mol } P_4}$ $\rightarrow$ **smaller mol amount determines limiting reactant**

$\mathbf{g\ Cl_2 \rightarrow mol\ Cl_2 \rightarrow mol\ PCl_3}$

$\frac{1 \text{ mol } Cl_2}{70.90 \text{ g } Cl_2} \quad \frac{4 \text{ mol } PCl_3}{6 \text{ mol } Cl_2}$

then: mol limiting reactant → mol excess reactant required → mol excess reactant left → g excess reactant left

$\frac{1 \text{ mol } P_4}{6 \text{ mol } Cl_2} \quad \frac{123.88 \text{ g } P_4}{1 \text{ mol } P_4} \quad \text{or} \quad \frac{70.90 \text{ g } Cl_2}{1 \text{ mol } Cl_2}$

Solution: $45.69 \text{ g } P_4 \times \frac{1 \text{ mol } P_4}{123.88 \text{ g } P_4} \times \frac{4 \text{ mol } PCl_3}{1 \text{ mol } P_4} = 1.475 \text{ mol } PCl_3$

$131.3 \text{ g } Cl_2 \times \frac{1 \text{ mol } Cl_2}{70.90 \text{ g } Cl_2} \times \frac{4 \text{ mol } PCl_3}{6 \text{ mol } Cl_2} = 1.235 \text{ mol } PCl_3$

Cl_2 is the limiting reactant; therefore, P_4 is the excess reactant.

$131.3 \text{ g } Cl_2 \times \frac{1 \text{ mol } Cl_2}{70.90 \text{ g } Cl_2} \times \frac{1 \text{ mol } P_4}{6 \text{ mol } Cl_2} \times \frac{123.88 \text{ g } P_4}{1 \text{ mol } P_4} = 38.2\underline{3}6 \text{ g } P_4 \text{ required}$

$45.69 \text{ g } P_4 - 38.2\underline{3}6 \text{ g } P_4 = 7.45 \text{ g } P_4 \text{ left}$

Check: Units of the answer (g P_4) are correct. The magnitude is reasonable because it is less than the original amount of P_4.

8.57 **Given:** 28.5 g KCl; 25.7 g Pb^{2+}; 29.4 g $PbCl_2$ **Find:** limiting reactant; theoretical yield $PbCl_2$; % yield

Conceptual Plan: $\mathbf{g\ KCl \rightarrow mol\ KCl \rightarrow mol\ PbCl_2}$

$\frac{1 \text{ mol KCl}}{74.55 \text{ g KCl}} \quad \frac{1 \text{ mol } PbCl_2}{2 \text{ mol KCl}}$ $\rightarrow$ **smaller mol amount determines limiting reactant**

$\mathbf{g\ Pb^{2+} \rightarrow mol\ Pb^{2+} \rightarrow mol\ PbCl_2}$

$\frac{1 \text{ mol } Pb^{2+}}{207.2 \text{ g } Pb^{2+}} \quad \frac{1 \text{ mol } PbCl_2}{1 \text{ mol } Pb^{2+}}$

then $\mathbf{mol\ PbCl_2 \rightarrow g\ PbCl_2}$ **then determine % yield**

$\frac{278.1 \text{ g } PbCl_2}{1 \text{ mol } PbCl_2} \quad \frac{\text{actual yield g } PbCl_2}{\text{theoretical yield g } PbCl_2} \times 100\%$

Solution: $28.5 \text{ g KCl} \times \frac{1 \text{ mol KCl}}{74.55 \text{ g KCl}} \times \frac{1 \text{ mol } PbCl_2}{2 \text{ mol KCl}} = 0.19\underline{1}1 \text{ mol } PbCl_2$

$25.7 \text{ g } Pb^{2+} \times \frac{1 \text{ mol } Pb^{2+}}{207.2 \text{ g } Pb^{2+}} \times \frac{1 \text{ mol } PbCl_2}{1 \text{ mol } Pb^{2+}} = 0.12\underline{4}0 \text{ mol } PbCl_2$ Pb^{2+} is the limiting reactant.

$0.12\underline{4}0 \text{ mol } PbCl_2 \times \frac{278.1 \text{ g } PbCl_2}{1 \text{ mol } PbCl_2} = 34.\underline{5} \text{ g } PbCl_2$

$\frac{29.4 \text{ g } PbCl_2}{34.\underline{5} \text{ g } PbCl_2} \times 100\% = 85.2\%$

Check: The theoretical yield has the correct units (g $PbCl_2$) and has a reasonable magnitude compared to the mass of Pb^{2+}, the limiting reactant. The % yield is reasonable, under 100%.

8.58 **Given:** 10.1 g Mg; 10.5 g O_2; 11.9 g MgO **Find:** limiting reactant; theoretical yield MgO; % yield

Conceptual Plan: $\mathbf{g\ Mg \rightarrow mol\ Mg \rightarrow mol\ MgO}$

$\frac{1 \text{ mol Mg}}{24.31 \text{ g Mg}} \quad \frac{2 \text{ mol MgO}}{2 \text{ mol Mg}}$ $\rightarrow$ **smaller mol amount determines limiting reactant**

$\mathbf{g\ O_2 \rightarrow mol\ O_2 \rightarrow mol\ MgO}$

$\frac{1 \text{ mol } O_2}{32.00 \text{ g } O_2} \quad \frac{2 \text{ mol MgO}}{1 \text{ mol } O_2}$

then mol MgO → g MgO then determine % yield

$\frac{40.31 \text{ g MgO}}{1 \text{ mol MgO}} \quad \frac{\text{actual yield g MgO}}{\text{theoretical yield g MgO}} \times 100\%$

Solution: $10.1\ \cancel{g\ Mg} \times \frac{1\ \cancel{mol\ Mg}}{24.31\ \cancel{g\ Mg}} \times \frac{2\ mol\ MgO}{2\ \cancel{mol\ Mg}} = 0.41\underline{5}5\ mol\ MgO$ Mg is the limiting reactant.

$$10.5\ \cancel{g\ O_2} \times \frac{1\ \cancel{mol\ O_2}}{32.00\ \cancel{g\ O_2}} \times \frac{2\ mol\ MgO}{1\ \cancel{mol\ O_2}} = 0.65\underline{6}3\ mol\ MgO$$

$$0.41\underline{5}5\ \cancel{mol\ MgO} \times \frac{40.31\ g\ MgO}{1\ \cancel{mol\ MgO}} = 16.\underline{7}5\ g\ MgO$$

$$\frac{11.9\ \cancel{g\ MgO}}{16.\underline{7}5\ \cancel{g\ MgO}} \times 100\% = 71.0\%$$

Check: The theoretical yield has the correct units (g MgO) and has a reasonable magnitude compared to the mass of Mg, the limiting reactant. The % yield is reasonable, under 100%.

8.59 **Given:** 136.4 kg NH_3; 211.4 kg CO_2; 168.4 kg CH_4N_2O **Find:** limiting reactant; theoretical yield CH_4N_2O; % yield

Conceptual Plan: kg NH_3 → g NH_3 → mol NH_3 → mol CH_4N_2O

$\frac{1000\ g}{1\ kg}$ $\frac{1\ mol\ NH_3}{17.03\ g\ NH_3}$ $\frac{1\ mol\ CH_4N_2O}{2\ mol\ NH_3}$ **→ smaller amount determines limiting reactant**

kg CO_2 → g CO_2 → mol CO_2 → mol CH_4N_2O

$\frac{1000\ g}{1\ kg}$ $\frac{1\ mol\ CO_2}{44.01\ g\ CO_2}$ $\frac{1\ mol\ CH_4N_2O}{1\ mol\ CO_2}$

then mol CH_4N_2O → g CH_4N_2O → kg CH_4N_2O then determine % yield

$\frac{60.06\ g\ CH_4N_2O}{1\ mol\ CH_4N_2O}$ $\frac{1\ kg}{1000\ g}$ $\frac{actual\ yield\ kg\ CH_4N_2O}{theoretical\ yield\ kg\ CH_4N_2O} \times 100\%$

Solution: $136.4\ \cancel{kg\ NH_3} \times \frac{1000\ \cancel{g}}{1\ \cancel{kg}} \times \frac{1\ \cancel{mol\ NH_3}}{17.03\ \cancel{g\ NH_3}} \times \frac{1\ mol\ CH_4N_2O}{2\ \cancel{mol\ NH_3}} = 400\underline{4}.7\ mol\ CH_4N_2O$

$$211.4\ \cancel{kg\ CO_2} \times \frac{1000\ \cancel{g}}{1\ \cancel{kg}} \times \frac{1\ \cancel{mol\ CO_2}}{44.01\ \cancel{g\ CO_2}} \times \frac{1\ mol\ CH_4N_2O}{1\ \cancel{mol\ CO_2}} = 480\underline{3}.5\ mol\ CH_4N_2O$$

NH_3 is the limiting reactant.

$$400\underline{4}.7\ \cancel{mol\ CH_4N_2O} \times \frac{60.06\ \cancel{g\ CH_4N_2O}}{1\ \cancel{mol\ CH_4N_2O}} \times \frac{1\ kg}{1000\ \cancel{g}} = 240.\underline{5}2\ kg\ CH_4N_2O$$

$$\frac{168.4\ \cancel{kg\ CH_4N_2O}}{240.\underline{5}2\ \cancel{kg\ CH_4N_2O}} \times 100\% = 70.01\%$$

Check: The theoretical yield has the correct units (kg CH_4N_2O) and has a reasonable magnitude compared to the mass of NH_3, the limiting reactant. The % yield is reasonable, under 100%.

8.60 **Given:** 155.8 kg SiO_2; 78.3 kg C; 66.1 kg Si **Find:** limiting reactant; theoretical yield Si; % yield

Conceptual Plan: Write and balance the reaction, then

kg SiO_2 → g SiO_2 → mol SiO_2 → mol Si

$\frac{1000\ g}{1\ kg}$ $\frac{1\ mol\ SiO_2}{60.09\ g\ SiO_2}$ $\frac{1\ mol\ Si}{1\ mol\ SiO_2}$ **→ smaller amount determines limiting reactant**

kg C → g C → mol C → mol Si

$\frac{1000\ g}{1\ kg}$ $\frac{1\ mol\ C}{12.01\ g\ C}$ $\frac{1\ mol\ Si}{2\ mol\ C}$

then mol Si → g Si → kg Si **then determine % yield**

$\frac{28.09\ g\ Si}{1\ mol\ Si}$ $\frac{1\ kg}{1000\ g}$ $\frac{actual\ yield\ kg\ Si}{theoretical\ yield\ kg\ Si} \times 100\%$

Solution: $SiO_2(l) + 2\ C(s) \rightarrow Si(l) + 2\ CO(g)$

$$155.8\ \cancel{kg\ SiO_2} \times \frac{1000\ \cancel{g}}{1\ \cancel{kg}} \times \frac{1\ \cancel{mol\ SiO_2}}{60.09\ \cancel{g\ SiO_2}} \times \frac{1\ mol\ Si}{1\ \cancel{mol\ SiO_2}} = 259\underline{2}.8\ mol\ Si$$

$78.3\ \cancel{kg\ C} \times \frac{1000\ \cancel{g}}{1\ \cancel{kg}} \times \frac{1\ \cancel{mol\ C}}{12.01\ \cancel{g\ C}} \times \frac{1\ mol\ Si}{2\ \cancel{mol\ C}} = 32\underline{5}9.8\ mol\ Si$ SiO_2 is the limiting reactant.

$$259\underline{2}.8\ \cancel{mol\ Si} \times \frac{28.09\ \cancel{g\ Si}}{1\ \cancel{mol\ Si}} \times \frac{1\ kg}{1000\ \cancel{g}} = 72.8\underline{3}2\ kg\ Si$$

$$\frac{66.1 \text{ kg Si}}{72.8\underline{3} \text{ kg Si}} \times 100\% = 90.8\%$$

Check: The theoretical yield has the correct units (kg Si) and has a reasonable magnitude compared to the mass of SiO_2, the limiting reactant. The % yield is reasonable, under 100%.

Combustion, Alkali Metal, and Halogen Reactions

8.61 (a)

Skeletal reaction: $S(s) + O_2(g) \rightarrow SO_2(g)$
Balanced reaction: $S(s) + O_2(g) \rightarrow SO_2(g)$

Check:

left side	right side
1 S atom	1 S atom
2 O atoms	2 O atoms

(b)

Skeletal reaction: $C_3H_6(g) + O_2(g) \rightarrow CO_2(g) + H_2O(g)$
Balance C: $C_3H_6(g) + O_2(g) \rightarrow 3\,CO_2(g) + H_2O(g)$
Balance H: $C_3H_6(g) + O_2(g) \rightarrow 3\,CO_2(g) + 3\,H_2O(g)$
Balance O: $C_3H_6(g) + 9/2\,O_2(g) \rightarrow 3\,CO_2(g) + 3\,H_2O(g)$
Clear fraction: $2\,C_3H_6(g) + 9\,O_2(g) \rightarrow 6\,CO_2(g) + 6\,H_2O(g)$

Check:

left side	right side
6 C atoms	6 C atoms
12 H atoms	12 H atoms
18 O atoms	18 O atoms

(c)

Skeletal reaction: $Ca(s) + O_2(g) \rightarrow CaO(s)$
Balance O: $Ca(s) + O_2(g) \rightarrow 2\,CaO(s)$
Balance Ca: $2\,Ca(s) + O_2(g) \rightarrow 2\,CaO(s)$

Check:

left side	right side
2 Ca atoms	2 Ca atoms
2 O atoms	2 O atoms

(d)

Skeletal reaction: $C_5H_{12}S(l) + O_2(g) \rightarrow CO_2(g) + H_2O(g) + SO_2(g)$
Balance C: $C_5H_{12}S(l) + O_2(g) \rightarrow 5\,CO_2(g) + H_2O(g) + SO_2(g)$
Balance H: $C_5H_{12}S(l) + O_2(g) \rightarrow 5\,CO_2(g) + 6\,H_2O(g) + SO_2(g)$
Balance S: $C_5H_{12}S(l) + O_2(g) \rightarrow 5\,CO_2(g) + 6\,H_2O(g) + SO_2(g)$
Balance O: $C_5H_{12}S(l) + 9\,O_2(g) \rightarrow 5\,CO_2(g) + 6\,H_2O(g) + SO_2(g)$

Check:

left side	right side
5 C atoms	5 C atoms
12 H atoms	12 H atoms
18 O atoms	18 O atoms
1 S atom	1 S atom

8.62 (a)

Skeletal reaction: $C_4H_6(g) + O_2(g) \rightarrow CO_2(g) + H_2O(g)$
Balance C: $C_4H_6(g) + O_2(g) \rightarrow 4\,CO_2(g) + H_2O(g)$
Balance H: $C_4H_6(g) + O_2(g) \rightarrow 4\,CO_2(g) + 3\,H_2O(g)$
Balance O: $C_4H_6(g) + 11/2\,O_2(g) \rightarrow 4\,CO_2(g) + 3\,H_2O(g)$
Clear fraction: $2\,C_4H_6(g) + 11\,O_2(g) \rightarrow 8\,CO_2(g) + 6\,H_2O(g)$

Check:

left side	right side
8 C atoms	8 C atoms
12 H atoms	12 H atoms
22 O atoms	22 O atoms

(b)

Skeletal reaction: $C(s) + O_2(g) \rightarrow CO_2(g)$
Balanced reaction: $C(s) + O_2(g) \rightarrow CO_2(g)$

Check:

left side	right side
1 C atom	1 C atom
2 O atoms	2 O atoms

(c)

Skeletal reaction:	$CS_2(s) + O_2(g) \rightarrow CO_2(g) + SO_2(g)$	
Balance C:	$CS_2(s) + O_2(g) \rightarrow CO_2(g) + SO_2(g)$	
Balance S:	$CS_2(s) + O_2(g) \rightarrow CO_2(g) + 2\ SO_2(g)$	
Balance O:	$CS_2(s) + 3\ O_2(g) \rightarrow CO_2(g) + 2\ SO_2(g)$	

Check:

left side	right side
1 C atom	1 C atom
2 S atoms	2 S atoms
6 O atoms	6 O atoms

(d)

Skeletal reaction:	$C_3H_8O(l) + O_2(g) \rightarrow CO_2(g) + H_2O(g)$
Balance C:	$C_3H_8O(l) + O_2(g) \rightarrow 3\ CO_2(g) + H_2O(g)$
Balance H:	$C_3H_8O(l) + O_2(g) \rightarrow 3\ CO_2(g) + 4\ H_2O(g)$
Balance O:	$C_3H_8O(l) + 9/2\ O_2(g) \rightarrow 3\ CO_2(g) + 4\ H_2O(g)$
Clear fraction:	$2\ C_3H_8O(l) + 9\ O_2(g) \rightarrow 6\ CO_2(g) + 8\ H_2O(g)$

Check:

left side	right side
6 C atoms	6 C atoms
16 H atoms	16 H atoms
20 O atoms	20 O atoms

8.63 Alkaline earth metals react with halogens to form metal halides. Write the formulas for the reactants and the metal halide product.

$$Sr(s) + I_2(g) \rightarrow SrI_2(s)$$

8.64 Fr would be the alkali metal with the smallest ionization energy, so it should have the most exothermic reaction with chlorine gas.

$$2\ Fr(s) + Cl_2(g) \rightarrow 2\ FrCl(s)$$

8.65 Alkali metals react with water to form the dissolved metal ion, the hydroxide ion, and hydrogen gas. Write the skeletal equation including each of these and then balance it.

$$Li(s) + H_2O(l) \rightarrow Li^+(aq) + OH^-(aq) + H_2(g)$$
$$2\ Li(s) + 2\ H_2O(l) \rightarrow 2\ Li^+(aq) + 2\ OH^-(aq) + H_2(g)$$

8.66 Alkali metals react with water to form the dissolved metal ion, the hydroxide ion, and hydrogen gas. Write the skeletal equation, including each of these, and then balance it.

$$K(s) + H_2O(l) \rightarrow K^+(aq) + OH^-(aq) + H_2(g)$$
$$2\ K(s) + 2\ H_2O(l) \rightarrow 2\ K^+(aq) + 2\ OH^-(aq) + H_2(g)$$

8.67 The halogens react with hydrogen to form hydrogen halides. Write the skeletal reaction with each of the halogens and hydrogens as the reactants and the hydrogen halide compound as the product and balance the equation.

$$H_2(g) + Br_2(g) \rightarrow HBr(g)$$
$$H_2(g) + Br_2(g) \rightarrow 2\ HBr(g)$$

8.68 Halogens react with each other to form interhalogen compounds. Write the skeletal reaction with each of the halogens as the reactants and the interhalogen compound as the product and balance the equation.

$$Cl_2(g) + F_2(g) \rightarrow ClF(g)$$
$$Cl_2(g) + F_2(g) \rightarrow 2\ ClF(g)$$

Cumulative Problems

8.69 **Given:** 1.0 kg C_8H_{18} **Find:** kg CO_2

Conceptual Plan: **kg C_8H_{18} → g C_8H_{18} → mol C_8H_{18} → mol CO_2 → g CO_2 → kg CO_2**

$$\frac{1000\ g}{1\ kg} \quad \frac{1\ mol\ C_8H_{18}}{114.22\ g\ C_8H_{18}} \quad \frac{16\ mol\ CO_2}{2\ mol\ C_8H_{18}} \quad \frac{44.01\ g\ CO_2}{1\ mol\ CO_2} \quad \frac{1\ kg}{1000\ g}$$

Solution: $2\ C_8H_{18}(g) + 25\ O_2(g) \rightarrow 16\ CO_2(g) + 18\ H_2O(g)$

$$1.0\ \text{kg}\ C_8H_{18} \times \frac{1000\ g}{1\ kg} \times \frac{1\ mol\ C_8H_{18}}{114.22\ g\ C_8H_{18}} \times \frac{16\ mol\ CO_2}{2\ mol\ C_8H_{18}} \times \frac{44.01\ g\ CO_2}{1\ mol\ CO_2}\ \frac{1\ kg}{1000\ g} = 3.1\ kg\ CO_2$$

Check: The units of the answer (kg CO_2) are correct. The magnitude of the answer is reasonable because the mole ratio of CO_2 to C_8H_{18} is 8:1.

8.70 **Given:** 18.9 L C_3H_8; $d = 0.621$ g/mL **Find:** kg CO_2

Conceptual Plan: L $C_3H_8 \rightarrow$ mL $C_3H_8 \rightarrow$ g $C_3H_8 \rightarrow$ mol $C_3H_8 \rightarrow$ mol $CO_2 \rightarrow$ g $CO_2 \rightarrow$ kg CO_2

$$\frac{1000\text{ mL}}{1\text{ L}} \quad \frac{0.621\text{ g}}{\text{mL}} \quad \frac{1\text{ mol } C_3H_8}{44.09\text{ g } C_3H_8} \quad \frac{3\text{ mol } CO_2}{1\text{ mol } C_3H_8} \quad \frac{44.01\text{ g } CO_2}{1\text{ mol } CO_2} \quad \frac{1\text{ kg}}{1000\text{ g}}$$

Solution: $C_3H_8(g) + 5\,O_2(g) \rightarrow 3\,CO_2(g) + 4\,H_2O(g)$

$$18.9\text{ L } C_3H_8 \times \frac{1000\text{ mL}}{1\text{ L}} \times \frac{0.621\text{ g}}{\text{mL}} \times \frac{1\text{ mol } C_3H_8}{44.09\text{ g } C_3H_8} \times \frac{3\text{ mol } CO_2}{1\text{ mol } C_3H_8} \times \frac{44.01\text{ g } CO_2}{1\text{ mol } CO_2} \times \frac{1\text{ kg}}{1000\text{ g}}$$

$= 35.1$ kg CO_2

Check: The units of the answer (kg CO_2) are correct. The magnitude of the answer is reasonable because the molar mass of CO_2 and C_3H_8 are close and the mole ratio is 1:3.

8.71 **Given:** 3.00 mL $C_4H_6O_3$, $d = 1.08$ g/mL; 1.25 g $C_7H_6O_3$; 1.22 g $C_9H_8O_4$ **Find:** limiting reactant; theoretical yield $C_9H_8O_4$; % yield $C_9H_8O_4$

Conceptual Plan: mL $C_4H_6O_3 \rightarrow$ g $C_4H_6O_3 \rightarrow$ mol $C_4H_6O_3 \rightarrow$ mol $C_9H_8O_4$

$$\frac{1.08\text{ g } C_4H_6O_3}{1.00\text{ mL } C_4H_6O_3} \quad \frac{1\text{ mol } C_4H_6O_3}{102.09\text{ g } C_4H_6O_3} \quad \frac{1\text{ mol } C_9H_8O_4}{1\text{ mol } C_4H_6O_3}$$

g $C_7H_6O_3 \rightarrow$ mol $C_7H_6O_3 \rightarrow$ mol $C_9H_8O_4$ $\rightarrow$ smaller amount determines limiting reactant

$$\frac{1\text{ mol } C_7H_6O_3}{138.12\text{ g } C_7H_6O_3} \quad \frac{1\text{ mol } C_9H_8O_4}{1\text{ mol } C_7H_6O_3}$$

then mol $C_9H_8O_4 \rightarrow$ g $C_9H_8O_4$ then determine % yield

$$\frac{180.1\text{ g } C_9H_8O_4}{\text{mol } C_9H_8O_4} \quad \frac{\text{actual yield g } C_9H_8O_4}{\text{theoretical yield g } C_9H_8O_4} \times 100\%$$

Solution:

$$3.00\text{ mL } C_4H_6O_3 \times \frac{1.08\text{ g } C_4H_6O_3}{\text{mL } C_4H_6O_3} \times \frac{1\text{ mol } C_4H_6O_3}{102.09\text{ g } C_4H_6O_3} \times \frac{1\text{ mol } C_9H_8O_4}{1\text{ mol } C_4H_6O_3} = 0.031\underline{7}4\text{ mol } C_9H_8O_4$$

$$1.25\text{ g } C_7H_6O_3 \times \frac{1\text{ mol } C_7H_6O_3}{138.12\text{ g } C_7H_6O_3} \times \frac{1\text{ mol } C_9H_8O_4}{1\text{ mol } C_7H_6O_3} = 0.0090\underline{5}0\text{ mol } C_9H_8O_4$$

Salicylic acid is the limiting reactant.

$$0.0090\underline{5}0\text{ mol } C_9H_8O_4 \times \frac{180.1\text{ g } C_9H_8O_4}{1\text{ mol } C_9H_8O_4} = 1.6\underline{3}0\text{ g } C_9H_8O_4$$

$$\frac{1.22\text{ g } C_9H_8O_4}{1.6\underline{3}0\text{ g } C_9H_8O_4} \times 100\% = 74.8\%$$

Check: The theoretical yield has the correct units (g $C_9H_8O_4$) and has a reasonable magnitude compared to the mass of $C_7H_6O_3$, the limiting reactant. The % yield is reasonable, under 100%.

8.72 **Given:** 4.62 mL C_2H_5OH, $d = 0.789$ g/mL; 15.55 g O_2; 3.72 mL H_2O, $d = 1.00$g/mL

Find: limiting reactant; theoretical yield H_2O; % yield H_2O

Conceptual Plan: mL $C_2H_5OH \rightarrow$ g $C_2H_5OH \rightarrow$ mol $C_2H_5OH \rightarrow$ mol H_2O

$$\frac{0.789\text{ g } C_2H_5OH}{1.00\text{ mL } C_2H_5OH} \quad \frac{1\text{ mol } C_2H_5OH}{46.07\text{ g } C_2H_5OH} \quad \frac{3\text{ mol } H_2O}{1\text{ mol } C_2H_5OH}$$

g $O_2 \rightarrow$ mol $O_2 \rightarrow$ mol H_2O $\rightarrow$ smaller amount determines limiting reactant

$$\frac{1\text{ mol } O_2}{32.00\text{ g } O_2} \quad \frac{3\text{ mol } H_2O}{3\text{ mol } O_2}$$

then mol $H_2O \rightarrow$ g H_2O and mL $H_2O \rightarrow$ g H_2O then determine % yield

$$\frac{18.02\text{ g } H_2O}{\text{mol } H_2O} \quad \frac{1.0\text{ g } H_2O}{1\text{ mL } H_2O} \quad \frac{\text{actual yield g } H_2O}{\text{theoretical yield g } H_2O} \times 100\%$$

Solution:

$C_2H_5OH(l) + 3\,O_2(g) \rightarrow 2\,CO_2(g) + 3\,H_2O(l)$

$$4.62\text{ mL } C_2H_5OH \times \frac{0.789\text{ g } C_2H_5OH}{\text{mL } C_2H_5OH} \times \frac{1\text{ mol } C_2H_5OH}{46.07\text{ g } C_2H_5OH} \times \frac{3\text{ mol } H_2O}{1\text{ mol } C_2H_5OH} = 0.23\underline{7}4\text{ mol } H_2O$$

$$15.55 \text{ g } O_2 \times \frac{1 \text{ mol } O_2}{32.00 \text{ g } O_2} \times \frac{3 \text{ mol } H_2O}{3 \text{ mol } O_2} = 0.48594 \text{ mol } H_2O$$

C_2H_5OH is the limiting reactant.

$$0.2374 \text{ mol } H_2O \times \frac{18.02 \text{ g } H_2O}{1 \text{ mol } H_2O} = 4.278 \text{ g } H_2O;\ 3.72 \text{ mL } H_2O \times \frac{1.00 \text{ g } H_2O}{1 \text{ mL } H_2O} = 3.72 \text{ g } H_2O \ 1 \text{ mL } H_2O$$

$$\frac{3.72 \text{ g } H_2O}{4.278 \text{ g } H_2O} \times 100\% = 87.0\%$$

Check: The theoretical yield has the correct units (g H_2O) and has a reasonable magnitude compared to the mass of C_2H_5OH, the limiting reactant. The % yield is reasonable, under 100%.

8.73 **Given:** (a) 11 molecules H_2, 2 molecules O_2; (b) 8 molecules H_2, 4 molecules O_2; (c) 4 molecules H_2, 5 molecules O_2; (d) 3 molecules H_2, 6 molecules O_2 **Find:** loudest explosion based on equation

Conceptual Plan: Loudest explosion will occur in the balloon with the mol ratio closest to the balanced equation that contains the most H_2.

Solution: $2\ H_2(g) + O_2(g) \rightarrow H_2O(l)$

Balloon (a) has enough O_2 to react with 4 molecules H_2; balloon (b) has enough O_2 to react with 8 molecules H_2; balloon (c) has enough O_2 to react with 10 molecules H_2; balloon (d) has enough O_2 for 3 molecules of H_2 to react. Balloon (b) also has the proper stoichiometric ratio of 2 H_2:1 O_2 unlike the other three. Therefore, balloon (b) will have the loudest explosion because it has the most H_2 that will react.

Check: The answer seems correct because it has the most H_2 with enough O_2 in the balloon to react completely.

8.74 **Given:** 30.0% $NaNO_3$, \$9.00/100 lb; 20.0% $(NH_4)_2SO_4$, \$8.10/100 lb **Find:** cost/lb N

Conceptual Plan: mass fertilizer → mass $NaNO_3$ → mass N → cost/lb N

$$\frac{30.0 \text{ lb NaNO}_3}{100 \text{ lb fertilizer}} \quad \frac{16.48 \text{ lb N}}{100 \text{ lb NaNO}_3} \quad \frac{\$9.00}{100 \text{ lb fertilizer}}$$

and mass fertilizer → mass $(NH_4)_2SO_4$ → mass N → cost/lb N

$$\frac{20.0 \text{ lb } (NH_4)_2SO_4}{100 \text{ lb fertilizer}} \quad \frac{21.2 \text{ lb N}}{100 \text{ lb } (NH_4)_2SO_4} \quad \frac{\$8.10}{100 \text{ lb fertilizer}}$$

Solution:

$$100 \text{ lb fertilizer} \times \frac{30.0 \text{ lb NaNO}_3}{100 \text{ lb fertilizer}} \times \frac{16.48 \text{ lb N}}{100 \text{ lb NaNO}_3} = 4.944 \text{ lb N}$$

$$\frac{\$9.00}{100 \text{ lb fertilizer}} \times \frac{100 \text{ lb fertilizer}}{4.944 \text{ lb N}} = \$1.82/\text{lb N}$$

$$100 \text{ lb fertilizer} \times \frac{20.0 \text{ lb } (NH_4)_2SO_4}{100 \text{ lb fertilizer}} \times \frac{21.2 \text{ lb N}}{100 \text{ lb } (NH_4)_2SO_4} = 4.240 \text{ lb N}$$

$$\frac{\$8.10}{100 \text{ lb fertilizer}} \times \frac{100 \text{ lb fertilizer}}{4.24 \text{ lb N}} = \$1.91/\text{lb N}$$

The more economical fertilizer is the $NaNO_3$ because it costs less/lb N.

Check: The units of the cost (\$/lb N) are correct. The answer is reasonable because you compare the cost/lb N directly.

8.75 **Given:** 1.00 g NH_3 **Find:** g PH_3

Conceptual Plan: Determine reaction sequence, then g NH_3 → mol NH_3 → mol PH_3 → g PH_3

$$\frac{1 \text{ mol NH}_3}{17.04 \text{ g NH}_3} \quad \frac{34.00 \text{ g PH}_3}{1 \text{ mol PH}_3}$$

Solution: Balance the reaction sequence:

$6\ NH_3 + 7\ 1/2\ O_2 \rightarrow 6\ NO + 9\ H_2O$

$6\ NO + P_4 \rightarrow P_4O_6 + 3\ N_2$

$P_4O_6 + 6\ H_2O \rightarrow 4\ H_3PO_3$

$4\ H_3PO_3 \rightarrow PH_3 + 3\ H_3PO_4$

$6\ NH_3 + 7\ 1/2\ O_2 + P_4 \rightarrow 3\ H_2O + 3\ N_2 + PH_3 + 3\ H_3PO_4$

Therefore, 6 mol NH_3 produces 1 mol PH_3.

$$1.00 \text{ g NH}_3 \times \frac{1 \text{ mol NH}_3}{17.04 \text{ g NH}_3} \times \frac{1 \text{ mol PH}_3}{6 \text{ mol NH}_3} \times \frac{34.00 \text{ g PH}_3}{1 \text{ mol PH}_3} = 0.333 \text{ g PH}_3$$

Check: The answer is g PH_3, which is correct. The magnitude is reasonable. Even though the molar mass of PH_3 is greater than NH_3, 6 mol of NH_3 are required to produce 1 mol PH_3.

8.76 **Given:** 910. kg Fe **Find:** kg Fe_2O_3 and kg CO_2

Conceptual Plan: kg Fe → g Fe → mol Fe → mol Fe_2O_3 → g Fe_2O_3 → kg Fe_2O_3

$$\frac{1000 \text{ g Fe}}{1 \text{ kg Fe}} \quad \frac{1 \text{ mol Fe}}{55.85 \text{ g Fe}} \quad \frac{1 \text{ mol Fe}_2\text{O}_3}{2 \text{ mol Fe}} \quad \frac{159.70 \text{ g Fe}_2\text{O}_3}{1 \text{ mol Fe}_2\text{O}_3} \quad \frac{1 \text{ kg Fe}_2\text{O}_3}{1000 \text{ g Fe}_2\text{O}_3}$$

and then kg Fe → g Fe → mol Fe → mol CO_2 → g CO_2 → kg CO_2

$$\frac{1000 \text{ g Fe}}{1 \text{ kg Fe}} \quad \frac{1 \text{ mol Fe}}{55.85 \text{ g Fe}} \quad \frac{3 \text{ mol CO}_2}{2 \text{ mol Fe}} \quad \frac{44.01 \text{ g CO}_2}{1 \text{ mol CO}_2} \quad \frac{1 \text{ kg CO}_2}{1000 \text{ g CO}_2}$$

Solution: Balanced Reaction: $Fe_2O_3 + 3\,CO \rightarrow 2\,Fe + 3\,CO_2$

$$910 \text{ kg Fe} \times \frac{1000 \text{ g Fe}}{1 \text{ kg Fe}} \times \frac{1 \text{ mol Fe}}{55.85 \text{ g Fe}} \times \frac{1 \text{ mol Fe}_2\text{O}_3}{2 \text{ mol Fe}} \times \frac{159.70 \text{ g Fe}_2\text{O}_3}{1 \text{ mol Fe}_2\text{O}_3} \times \frac{1 \text{ kg Fe}_2\text{O}_3}{1000 \text{ g Fe}_2\text{O}_3}$$

$$= 1.3\underline{0}1 \times 10^3 \text{ kg Fe}_2\text{O}_3 = 1.3 \times 10^3 \text{ kg Fe}_2\text{O}_3$$

$$910 \text{ kg Fe} \times \frac{1000 \text{ g Fe}}{1 \text{ kg Fe}} \times \frac{1 \text{ mol Fe}}{55.85 \text{ g Fe}} \times \frac{3 \text{ mol CO}_2}{2 \text{ mol Fe}} \times \frac{44.01 \text{ g CO}_2}{1 \text{ mol CO}_2} \times \frac{1 \text{ kg CO}_2}{1000 \text{ g CO}_2}$$

$$= 1.\underline{0}75 \times 10^3 \text{ kg CO}_2 = 1.1 \times 10^3 \text{ kg CO}_2$$

Check: The answers (kg Fe_2O_3 and kg CO_2) are correct. The magnitudes are reasonable because the molar mass of Fe_2O_3 is greater than the molar mass of Fe and the mole of CO_2 is greater than the mole of Fe.

8.77 **Given:** 10.0 kg mixture; 30.35% hexane; 15.85% heptane; 53.80% octane **Find:** total mass CO_2

Conceptual Plan: kg hexane → kmol hexane → kmol CO_2 → kg CO_2

$$\frac{1 \text{ kmol C}_6\text{H}_{14}}{86.20 \text{ kg C}_6\text{H}_{14}} \quad \frac{12 \text{ kmol CO}_2}{2 \text{ kmol C}_6\text{H}_{14}} \quad \frac{44.01 \text{ kg CO}_2}{1 \text{ kmol CO}_2}$$

kg heptane → kmol heptane → kmol CO_2 → kg CO_2

$$\frac{1 \text{ kmol C}_7\text{H}_{16}}{100.23 \text{ kg C}_7\text{H}_{16}} \quad \frac{7 \text{ kmol CO}_2}{1 \text{ kmol C}_7\text{H}_{16}} \quad \frac{44.01 \text{ kg CO}_2}{1 \text{ kmol CO}_2}$$

kg octane → kmol octane → kmol CO_2 → kg CO_2

$$\frac{1 \text{ kmol C}_8\text{H}_{18}}{114.26 \text{ kg C}_8\text{H}_{18}} \quad \frac{16 \text{ kmol CO}_2}{2 \text{ kmol C}_8\text{H}_{18}} \quad \frac{44.01 \text{ kg CO}_2}{1 \text{ kmol CO}_2}$$

Solution: Balanced Reactions:

$$2\,C_6H_{14}(l) + 19\,O_2(g) \rightarrow 12\,CO_2(g) + 14\,H_2O(l)$$
$$C_7H_{16}(l) + 11\,O_2(g) \rightarrow 7\,CO_2(g) + 8\,H_2O(l)$$
$$2\,C_8H_{18}(l) + 25\,O_2(g) \rightarrow 16\,CO_2(g) + 18\,H_2O(l)$$

$$10.0 \text{ kg mix} \times \frac{30.35 \text{ kg C}_6\text{H}_{14}}{100.0 \text{ kg mix}} \times \frac{1 \text{ kmol C}_6\text{H}_{14}}{86.20 \text{ kg C}_6\text{H}_{14}} \times \frac{12 \text{ kmol CO}_2}{2 \text{ kmol C}_6\text{H}_{14}} \times \frac{44.01 \text{ kg CO}_2}{1 \text{ kmol CO}_2} = 9.2\underline{9}7 \text{ kg CO}_2$$

$$10.0 \text{ kg mix} \times \frac{15.85 \text{ kg C}_7\text{H}_{16}}{100.0 \text{ kg mix}} \times \frac{1 \text{ kmol C}_7\text{H}_{16}}{100.23 \text{ kg C}_7\text{H}_{16}} \times \frac{7 \text{ kmol CO}_2}{1 \text{ kmol C}_7\text{H}_{16}} \times \frac{44.01 \text{ kg CO}_2}{1 \text{ kmol CO}_2} = 4.8\underline{7}2 \text{ kg CO}_2$$

$$10.0 \text{ kg mix} \times \frac{53.80 \text{ kg C}_8\text{H}_{18}}{100.0 \text{ kg mix}} \times \frac{1 \text{ kmol C}_8\text{H}_{18}}{114.26 \text{ kg C}_8\text{H}_{18}} \times \frac{16 \text{ kmol CO}_2}{2 \text{ kmol C}_8\text{H}_{18}} \times \frac{44.01 \text{ kg CO}_2}{1 \text{ kmol CO}_2} = 16.\underline{5}78 \text{ kg CO}_2$$

Total CO_2 = 9.30 kg + 4.87 kg + 16.6 kg = 30.8 kg CO_2

Check: The units of the answer (kg CO_2) are correct. The magnitude of the answer is reasonable because a large amount of CO_2 is produced per mole of hydrocarbon.

8.78 **Given:** 1.00 kg sand; 22.8% ilemite($FeTiO_3$) **Find:** g Ti

Conceptual Plan: kg sand → g sand → g $FeTiO_3$ → mol $FeTiO_3$ → mol $TiCl_4$ → mol Ti → g Ti

$$\frac{1000 \text{ g sand}}{1 \text{ kg sand}} \quad \frac{22.8 \text{ g FeTiO}_3}{100 \text{ g sand}} \quad \frac{1 \text{ mol FeTiO}_3}{151.72 \text{ g FeTiO}_3} \quad \frac{0.908 \text{ mol TiCl}_4}{1 \text{ mol FeTiO}_3} \quad \frac{0.859 \text{ mol Ti}}{1 \text{ mol TiCl}_4} \quad \frac{47.87 \text{ g Ti}}{1 \text{ mol Ti}}$$

Solution: $$1.00 \text{ kg sand} \times \frac{1000 \text{ g sand}}{1 \text{ kg sand}} \times \frac{22.8 \text{ g FeTiO}_3}{100 \text{ g sand}} \times \frac{1 \text{ mol FeTiO}_3}{151.72 \text{ g FeTiO}_3} \times \frac{0.908 \text{ mol TiCl}_4}{1 \text{ mol FeTiO}_3} \times \frac{0.859 \text{ mol Ti}}{1 \text{ mol TiCl}_4}$$

$$\times \frac{47.87 \text{ g Ti}}{1 \text{ mol Ti}} = 56.\underline{1}09 \text{ g Ti} = 56.1 \text{ g Ti}$$

Check: The units of the answer (g Ti) are correct. The magnitude is reasonable because the percentage of ilemite in the sand is small.

Challenge Problems

8.79 **Given:** g C_3H_8 + C_2H_2 = 2.0 g; mol CO_2 = 1.5 mol H_2O **Find:** original g C_2H_2

Conceptual Plan: mol $C_3H_8 \rightarrow$ mol CO_2 and mol H_2O and mol $C_2H_2 \rightarrow$ mol CO_2 and mol H_2O

Solution: Let a = mol C_3H_8 and b = mol C_2H_2

$C_3H_8 + 5\,O_2 \rightarrow 3\,CO_2 + 4\,H_2O$ $C_2H_2 + 3/2\,O_2 \rightarrow 2\,CO_2 + H_2O$

a $3a$ $4a$ b $2b$ b

Total mol $CO_2 = 3a + 2b$ and total mol $H_2O = 4a + b$

mol CO_2 = 1.5(mol H_2O)

So $3a + 2b = 1.5(4a + b)$

$$\text{And}\left(a \text{ mol } C_3H_8 \times \frac{44.11\ C_3H_8}{1 \text{ mol } C_3H_8}\right) + \left(b \text{ mol } C_2H_2 \times \frac{26.01\ C_2H_2}{\text{mol } C_2H_2}\right) = 2.0 \text{ g}$$

Solve the simultaneous equations: $a = 9.98 \times 10^{-3}$ mol C_3H_8 and $b = 0.05\underline{9}9$ mol C_2H_2.

Substitute for b and solve for grams C_2H_2.

$$0.06\underline{0} \text{ mol } C_2H_2 \times \frac{26.01\ C_2H_2}{\text{mol } C_2H_2} = 1.\underline{5}6 \text{ g } C_2H_2 = 1.6 \text{ g } C_2H_2$$

Check: The units of the answer (g C_2H_2) are correct. The magnitude is reasonable because it is less than the total mass.

8.80 **Given:** 20.6 g P; 79.4 g Cl_2 **Find:** g PCl_3

Conceptual Plan: g P $\rightarrow$ mol P

$\frac{1 \text{ mol P}}{30.97 \text{ g P}}$

g $Cl_2 \rightarrow$ mol $Cl_2 \rightarrow$ smaller amount determines limiting reactant

$\frac{1 \text{ mol } Cl_2}{70.90 \text{ g } Cl_2}$

then: mol $PCl_3 \rightarrow$ g PCl_3

$\frac{137.32 \text{ g } PCl_3}{1 \text{ mol } PCl_3}$

Solution: $$20.6 \text{ g P} \times \frac{1 \text{ mol P}}{30.97 \text{ g P}} = 0.66\underline{5}2 \text{ mol P}$$

$$79.4 \text{ g } Cl_2 \times \frac{1 \text{ mol } Cl_2}{70.90 \text{ g } Cl_2} \times \frac{2 \text{ mol } Cl_2}{1 \text{ mol } Cl_2} = 2.2\underline{4}0 \text{ mol } Cl_2$$

mol P = mol PCl_3 + mol PCl_5 and mol Cl = 3(mol PCl_3) + 5(mol PCl_5)

let x = mol PCl_3; therefore, $0.6652 - x$ = mol PCl_5

mol Cl = $3x + 5(0.6652 - x) = 2.240$

$3x + 3.326 - 5x = 2.240$

$x = 0.543$ = mol PCl_3

$0.6652 - x = 0.122$ = mol PCl_5

$$0.543 \text{ mol } PCl_3 \times \frac{137.32 \text{ g } PCl_3}{1 \text{ mol } PCl_3} = 74.\underline{5}6 \text{ g } PCl_3 = 74.6 \text{ g } PCl_3$$

Check: The units of the answer (g PCl_3) are correct. The magnitude is reasonable because it is less than the total mass of P and Cl_2.

8.81 **Given:** 45 μg Pb/dL blood; V = 5.0 L; 1 mol succimer($C_4H_6O_4S_2$) = 1 mol Pb **Find:** mass $C_4H_6O_4S_2$ in mg

Conceptual Plan: volume blood L $\rightarrow$ volume blood dL $\rightarrow$ μg Pb $\rightarrow$ g Pb $\rightarrow$ mol Pb $\rightarrow$

$\frac{10 \text{ dL}}{1 \text{ L}}$ $\frac{45\ \mu\text{g}}{\text{dL}}$ $\frac{1 \text{ g}}{10^6\ \mu\text{g}}$ $\frac{1 \text{ mol Pb}}{207.2 \text{ g Pb}}$ $\frac{1 \text{ mol succimer}}{1 \text{ mol Pb}}$

mol succimer → g succimer → mg succimer

$\frac{182.23 \text{ g succimer}}{1 \text{ mol succimer}}$ $\frac{1000 \text{ mg succimer}}{1 \text{ g succimer}}$

Solution:

$$5.0 \text{ L blood} \times \frac{10 \text{ dL}}{1 \text{ L}} \times \frac{45 \text{ } \mu\text{g}}{\text{dL}} \times \frac{1 \text{ g}}{10^6 \text{ } \mu\text{g}} \times \frac{1 \text{ mol Pb}}{207.2 \text{ g}} \times \frac{1 \text{ mol succimer}}{1 \text{ mol Pb}} \times \frac{182.23 \text{ g succimer}}{1 \text{ mol succimer}} \times \frac{1000 \text{ mg}}{1 \text{ g}}$$

$= 2.0$ mg succimer

Check: The units of the answer (mg succimer) are correct. The magnitude is reasonable for the volume of blood and the concentration.

8.82 In designing the unit, you would need to consider the theoretical yield and % yield of the reaction, how changing the limiting reactant would affect the reaction, and the stoichiometry between KO_2 and O_2 to determine the mass of KO_2 required to produce enough O_2 for 10 minutes. You might also consider the speed of the reaction and whether the reaction produced heat. In addition, because your body does not use 100% of the oxygen taken in with each breath, the apparatus would need to replenish only the oxygen used. The percentage of oxygen in air is about 20%, and the percentage in exhaled air is about 16%; so we will assume that 4% of the air would need to be replenished with oxygen. (Note: The problem can also be solved by finding the amount of KO_2 that would be required to react with all of the exhaled CO_2.)

Given: air = 4% O_2; volume = 5 − 8 L/min; 1 mol gas = 22.4 L gas **Find:** O_2 for 10 min breathing time

Conceptual Plan: 10 min → vol air → vol O_2 → mol O_2 → mol KO_2 → g KO_2

$\frac{8 \text{ L air}}{1 \text{ min}}$ $\frac{4 \text{ L } O_2}{100 \text{ L air}}$ $\frac{1 \text{ mol } O_2}{22.4 \text{ L } O_2}$ $\frac{4 \text{ mol } KO_2}{3 \text{ mol } O_2}$ $\frac{71.10 \text{ g } KO_2}{1 \text{ mol } KO_2}$

Solution: $$10 \text{ min} \times \frac{8 \text{ L air}}{1 \text{ min}} \times \frac{4 \text{ L } O_2}{100 \text{ L air}} \times \frac{1 \text{ mol } O_2}{22.4 \text{ L } O_2} \times \frac{4 \text{ mol } KO_2}{3 \text{ mol } O_2} \times \frac{71.10 \text{ g } KO_2}{1 \text{ mol } KO_2} = 14 \text{ g } KO_2$$

Check: The units of the answer (g KO_2) are correct. The magnitude of the answer is reasonable because it is an amount that could be carried in a portable device.

8.83 **Given:** 250. g sample; 67.2 mol %Al **Find:** theoretical yield in g of Mn

Conceptual Plan: mol Al → g Al and mol MnO_2 → g MnO_2, then mass %Al

$\frac{26.98 \text{ g Al}}{\text{mol Al}}$ $\frac{86.94 \text{ g } MnO_2}{\text{mol } MnO_2}$ $\frac{\text{g Al}}{\text{total g}} \times 100\%$

then sample → g Al → mol Al → mol Mn

$\frac{38.86 \text{ g Al}}{100 \text{ g sample}}$ $\frac{\text{mol Al}}{26.98 \text{ g Al}}$ $\frac{3 \text{ mol Mn}}{4 \text{ mol Al}}$ **→ smaller mol amount determines limiting reactant**

sample → g MnO_2 → mol MnO_2 → mol Mn

$\frac{61.14 \text{ g } MnO_2}{100 \text{ g sample}}$ $\frac{\text{mol } MnO_2}{86.94 \text{ g } MnO_2}$ $\frac{1 \text{ mol Mn}}{\text{mol } MnO_2}$

then mol Mn → g Mn

$\frac{54.94 \text{ g Mn}}{\text{mol Mn}}$

Solution: $4 \text{ Al}(s) + 3 \text{ MnO}_2(s) \rightarrow 3 \text{ Mn} + 2 \text{ Al}_2\text{O}_3(s)$

Assume 1 mole: $0.672 \text{ mol Al} \times \frac{26.98 \text{ g Al}}{\text{mol Al}} = 18.13 \text{ g Al}$

$$0.328 \text{ mol } MnO_2 \times \frac{86.94 \text{ g } MnO_2}{\text{mol } MnO_2} = 28.52 \text{ g } MnO_2$$

$$\frac{18.13 \text{ g Al}}{(18.13 \text{ g Al} + 28.52 \text{ g } MnO_2)} \times 100\% = 38.86\% \text{ Al} \quad \text{So } 61.14\% \text{ } MnO_2$$

$$250 \text{ g sample} \times \frac{38.86 \text{ g Al}}{100 \text{ g sample}} \times \frac{\text{mol Al}}{26.98 \text{ g Al}} \times \frac{3 \text{ mol Mn}}{4 \text{ mol Al}} = 2.\underline{7}01 \text{ mol Mn}$$

$$250 \text{ g sample} \times \frac{61.14 \text{ g } MnO_2}{100 \text{ g sample}} \times \frac{\text{mol } MnO_2}{86.94 \text{ g } MnO_2} \times \frac{1 \text{ mol Mn}}{1 \text{ mol } MnO_2} = 1.\underline{7}58 \text{ mol Mn}$$

$$1.7\underline{5}8 \text{ mol Mn} \times \frac{54.94 \text{ g Mn}}{1 \text{ mol Mn}} = 96.6 \text{ g Mn}$$

Check: The units of the answer (g Mn) are correct. The magnitude of the answer is reasonable based on the amount of the limiting reactant, MnO_2.

8.84 **Given:** 151 g $Na_2B_4O_7$ **Find:** g B_5H_9

Conceptual Plan: g $Na_2B_4O_7$ → mol $Na_2B_4O_7$ → mol B_5H_9 → g B_5H_9

$$\frac{1 \text{ mol } Na_2B_4O_7}{201.22 \text{ g } Na_2B_4O_7} \quad \frac{4 \text{ mol } B_5H_9}{5 \text{ mol } Na_2B_4O_7} \quad \frac{63.13 \text{ g } B_5H_9}{1 \text{ mol } B_5H_9}$$

Solution: All of the B in B_5H_9 goes to the $Na_2B_4O_7$, so the mole ratio between the two can be used.

$$151 \text{ g } Na_2B_4O_7 \times \frac{1 \text{ mol } Na_2B_4O_7}{201.22 \text{ g } Na_2B_4O_7} \times \frac{4 \text{ mol } B_5H_9}{5 \text{ mol } Na_2B_4O_7} \times \frac{63.13 \text{ g } B_5H_9}{1 \text{ mol } B_5H_9} = 37.9 \text{ g } B_5H_9$$

Check: The units of the answer (g B_5H_9) are correct. The magnitude of the answer is reasonable because the molar mass of B_5H_9 is less than the molar mass of $Na_2B_4O_7$.

Conceptual Problems

8.85 The correct answer is (d). The molar masses of K and O_2 are comparable. Because the stoichiometry has a ratio of 4 mol K to 1 mol O_2, K will be the limiting reactant when mass of K is less than four times the mass of O_2.

8.86 **Given:** 5 mol NO; 10 mol H_2 **Find:** conditions of product mixture

Conceptual Plan: mol H_2 → mol NO and mol H_2 → mol NH_3 and mol H_2 → mol H_2O

Solution: The correct answer is (a). Because the mol ratio of H_2 to NO is 5:2, the 10 mol of H_2 will require 4 mol NO and H_2 is the limiting reactant. This eliminates answers (b) and (c). Because there is excess NO, this eliminates (d), leaving answer (a).

8.87 **Given:** 6 molecules N_2H_4; 4 molecules N_2O_4; (a) contains 9 molecules N_2, 12 molecules H_2O, and 1 molecule N_2H_4; solution (b) contains 12 molecules N_2, 16 molecules H_2O, and 2 molecules N_2H_4; solution (c) contains 9 molecules N_2 and 12 molecules H_2O **Find:** theoretical yield N_2, H_2O

Conceptual Plan: molecules N_2H_4 → molecules N_2

$$\frac{3 \text{ molecules } N_2}{2 \text{ molecules } N_2H_4}$$

→ smaller molecules amount determines limiting reactant

molecules N_2O_4 → molecules N_2

$$\frac{3 \text{ molecules } N_2}{1 \text{ molecule } N_2O_4}$$

molecules N_2H_4 → molecules H_2O

$$\frac{4 \text{ molecules } H_2O}{2 \text{ molecules } N_2H_4}$$

molecules N_2H_4 → molecules N_2O_4

$$\frac{1 \text{ molecule } N_2O_4}{2 \text{ molecules } N_2H_4}$$

Solution: $6 \text{ molecules } N_2H_4 \times \frac{3 \text{ molecules } N_2}{2 \text{ molecules } N_2H_4} = 9 \text{ molecules } N_2$

$$6 \text{ molecules } N_2O_4 \times \frac{3 \text{ molecules } N_2}{1 \text{ molecule } N_2O_4} = 18 \text{ molecules } N_2$$

Limiting reactant = N_2H_4 because it produced the least molecules of N_2.

$$6 \text{ molecules } N_2H_4 \times \frac{4 \text{ molecules } H_2O}{2 \text{ molecules } N_2H_4} = 12 \text{ molecules } H_2O$$

$$6 \text{ molecules } N_2H_4 \times \frac{1 \text{ molecule } N_2O_4}{2 \text{ molecules } N_2H_4} = 3 \text{ molecules } N_2O_4 \text{ used}$$

Reaction mixture should contain 9 molecules N_2, 12 molecules H_2O, and 1 molecule N_2O_4; this is best represented by (a).

9 Introduction to Solutions and Aqueous Reactions

Review Questions

9.1 An aqueous solution is a solution in which water acts as the solvent. The solvent is the majority component of the mixture, and the solute is the minority component of the mixture.

9.2 Molarity is a concentration term. It is the amount of solute (in moles) divided by the volume of solution (in liters). The molarity of a solution can be used as a conversion factor between moles of the solute and liters of the solution.

9.3 Substances that completely dissociate into ions when they dissolve in water are called strong electrolytes and conduct electricity easily. Substances that do not completely dissociate in water are called weak electrolytes and conduct electricity only weakly. Compounds that do not dissociate into ions when dissolved in water are called nonelectrolytes and do not conduct electricity.

9.4 Acids are molecular compounds that ionize—form ions—when they dissolve in water. An Arrhenius acid is a substance that produces H^+ ions in aqueous solutions. A strong acid completely ionizes in solution. A weak acid does not completely ionize in water. A solution of a weak acid is composed mostly of the nonionized acid.

9.5 A compound is soluble if it dissolves in water. A compound is insoluble if it does not dissolve in water.

9.6 The solubility rules are a set of empirical rules that have been inferred from observations on many ionic compounds. The solubility rules allow us to predict whether a compound is soluble or insoluble.

9.7 Cations that usually form soluble compounds are Li^+, Na^+, K^+, and NH_4^+. The anions that usually form soluble compounds are NO_3^- and $C_2H_3O_2^-$, which have no exceptions; Cl^-, Br^-, and I^- except when these ions pair with Ag^+, Hg_2^{2+}, or Pb^{2+}, which result in insoluble compounds; and SO_4^{2-} except with Sr^{2+}, Ba^{2+}, Pb^{2+}, Ag^+, or Ca^{2+}, which form insoluble compounds. The anions that usually form insoluble compounds are OH^- and S^{2-} except with Li^+, Na^+, K^+, and NH_4^+, which form soluble compounds; and when S^{2-} pairs with Ca^{2+}, Sr^{2+}, or Ba^{2+}, which form soluble compounds; CO_3^{2-} and PO_4^{3-} are insoluble except when paired with Li^+, Na^+, K^+, and NH_4^+.

9.8 A precipitation reaction is one in which a solid or precipitate forms upon mixing two solutions. An example is $2\ KI(aq) + Pb(NO_3)_2(aq) \rightarrow PbI_2(s) + 2\ KNO_3(aq)$.

9.9 The key to predicting precipitation reactions is to understand that only insoluble compounds form precipitates. In a precipitation reaction, two solutions containing soluble compounds combine and an insoluble compound precipitates.

9.10 A molecular equation shows the complete neutral formulas for each compound in the reaction as if they existed as molecules. Equations that list individually all of the ions present as either reactants or products in a chemical reaction are complete ionic equations. Equations that show only the species that actually change during the reaction are net ionic equations.

9.11 An Arrhenius base is a substance that produces OH^- ions in aqueous solutions.

9.12 Binary acids are composed of hydrogen and a nonmetal. The names for binary acids have the form: hydro plus the base name of the nonmetal + *-ic* acid. Oxyacids contain hydrogen and an oxyanion. The names of oxyacids depend on the ending of the oxyanion and have the following forms: oxyanions ending with *-ate*: base name of the oxyanion + *ic* acid; oxyanions ending with *-ite*: base name of the oxyanion + *ous* acid.

9.13 When an acid and a base are mixed, the $H^+(aq)$ from the acid combines with the $OH^-(aq)$ from the base to form $H_2O(l)$. An example is $HCl(aq) + NaOH(aq) \rightarrow H_2O(l) + NaCl(aq)$.

9.14 In a titration, a substance in a solution of known concentration is reacted with another substance in a solution of unknown concentration. The acid–base titration is continued until the neutralization is complete. At the equivalence point, the point when the number of moles of OH^- equals the number of moles of H^+, the titration is complete. An indicator is a dye whose color depends on the acidity or basicity of the solution.

9.15 Aqueous reactions that form a gas upon mixing two solutions are called gas-evolution reactions. An example is $H_2SO_4(aq) + Li_2S(aq) \rightarrow H_2S(g) + Li_2SO_4(aq)$.

9.16 The reactant types that give rise to gas-evolution reactions are sulfides, carbonates, bicarbonates, sulfites, bisulfites, and ammonium compounds.

9.17 Oxidation–reduction reactions, or redox reactions, are reactions in which electrons are transferred from one reactant to the other. An example is $4\,Fe(s) + 3\,O_2(g) \rightarrow 2\,Fe_2O_3(s)$.

9.18 The oxidation state or oxidation number is a number given to each atom based on the electron assignments. It is the charge an atom would have if all shared electrons were assigned to the atom with a greater attraction for those electrons.

To identify redox reactions using oxidation states, begin by assigning oxidation states to each atom in the reaction. A change in oxidation state for the atoms indicates a redox reaction.

9.19 When a substance is oxidized, it loses electrons and there is an increase in oxidation state. When a substance is reduced, it gains electrons and there is a reduction in oxidation state.

9.20 A substance that causes the oxidation of another substance is called an oxidizing agent. A substance that causes the reduction of another substance is called a reducing agent.

Problems by Topic

Solution Concentration and Solution Stoichiometry

9.21 (a) **Given:** 3.25 mol LiCl; 2.78 L solution **Find:** molarity LiCl

Conceptual Plan: mol LiCl, L solution → molarity

$$\text{molarity (M)} = \frac{\text{amount of solute (in moles)}}{\text{volume of solution (in L)}}$$

Solution: $\dfrac{3.25 \text{ mol LiCl}}{2.78 \text{ L solution}} = 1.1\underline{6}9 \text{ M} = 1.17 \text{ M}$

Check: The units of the answer (M) are correct. The magnitude of the answer is reasonable. Concentrations are usually between 0 M and 18 M.

(b) **Given:** 28.33 g $C_6H_{12}O_6$; 1.28 L solution **Find:** molarity $C_6H_{12}O_6$

Conceptual Plan: g $C_6H_{12}O_6$ → mol $C_6H_{12}O_6$, L solution → molarity

$$\frac{1 \text{ mol } C_6H_{12}O_6}{180.16 \text{ g } C_6H_{12}O_6} \qquad \text{molarity (M)} = \frac{\text{amount of solute (in moles)}}{\text{volume of solution (in L)}}$$

Solution: $28.33 \text{ g } \cancel{C_6H_{12}O_6} \times \dfrac{1 \text{ mol } C_6H_{12}O_6}{180.16 \text{ g } \cancel{C_6H_{12}O_6}} = 0.157\underline{2}4 \text{ mol } C_6H_{12}O_6$

$$\frac{0.157\underline{2}4 \text{ mol } C_6H_{12}O_6}{1.28 \text{ L solution}} = 0.12\underline{2}8 \text{ M} = 0.123 \text{ M}$$

Check: The units of the answer (M) are correct. The magnitude of the answer is reasonable. Concentrations are usually between 0 M and 18 M.

(c) **Given:** 32.4 mg NaCl; 122.4 mL solution **Find:** molarity NaCl

Conceptual Plan: mg NaCl → g NaCl → mol NaCl, and mL solution → L solution then molarity

$$\frac{1\text{ g NaCl}}{1000\text{ mg NaCl}} \quad \frac{\text{mol NaCl}}{58.45\text{ g NaCl}} \quad \frac{1\text{ L solution}}{1000\text{ mL solution}} \quad \text{molarity (M)} = \frac{\text{amount of solute (in moles)}}{\text{volume of solution (in L)}}$$

Solution: $32.4\ \cancel{\text{mg NaCl}} \times \frac{1\ \cancel{\text{g}}}{1000\ \cancel{\text{mg}}} \times \frac{1\text{ mol NaCl}}{58.45\ \cancel{\text{g NaCl}}} = 5.5\underline{4}3 \times 10^{-4}\text{ mol NaCl}$

$$122.4\ \cancel{\text{mL solution}} \times \frac{1\text{ L}}{1000\ \cancel{\text{mL}}} = 0.1224\text{ L}$$

$$\frac{5.5\underline{4}3 \times 10^{-4}\text{ mol NaCl}}{0.1224\text{ L}} = 0.0045\underline{2}8\text{ M NaCl} = 0.00453\text{ M NaCl}$$

Check: The units of the answer (M) are correct. The magnitude of the answer is reasonable. Concentrations are usually between 0 M and 18 M.

9.22 (a) **Given:** 0.38 mol $LiNO_3$; 6.14 L solution **Find:** molarity $LiNO_3$

Conceptual Plan: mol $LiNO_3$; L solution → molarity

$$\text{molarity (M)} = \frac{\text{amount of solute (in moles)}}{\text{volume of solution (in L)}}$$

Solution: $\frac{0.38\text{ mol LiNO}_3}{6.14\text{ L solution}} = 0.06\underline{1}89\text{ M} = 0.062\text{ M}$

Check: The units of the answer (M) are correct. The magnitude of the answer is reasonable. Concentrations are usually between 0 M and 18 M.

(b) **Given:** 72.8 g C_2H_6O; 2.34 L solution **Find:** molarity $C_6H_{12}O_6$

Conceptual Plan: g C_2H_6O → mol C_2H_6O, L solution → molarity

$$\frac{1\text{ mol C}_2\text{H}_6\text{O}}{46.068\text{ g C}_2\text{H}_6\text{O}} \quad \text{molarity (M)} = \frac{\text{amount of solute (in moles)}}{\text{volume of solution (in L)}}$$

Solution: $72.8\ \cancel{\text{g C}_2\text{H}_6\text{O}} \times \frac{1\text{ mol C}_2\text{H}_6\text{O}}{46.068\ \cancel{\text{g C}_2\text{H}_6\text{O}}} = 1.5\underline{8}0\text{ mol C}_2\text{H}_6\text{O}$

$$\frac{1.5\underline{8}0\text{ mol C}_2\text{H}_6\text{O}}{2.34\text{ L solution}} = 0.67\underline{5}3\text{ M} = 0.675\text{ M}$$

Check: The units of the answer (M) are correct. The magnitude of the answer is reasonable. Concentrations are usually between 0 M and 18 M.

(c) **Given:** 12.87 mg KI; 112.4 mL solution **Find:** molarity KI

Conceptual Plan: mg KI → g KI → mol KI, and mL solution → L solution then molarity

$$\frac{1\text{ g KI}}{1000\text{ mg KI}} \quad \frac{1\text{ mol KI}}{166.00\text{ g KI}} \quad \frac{1\text{ L solution}}{1000\text{ mL solution}} \quad \text{molarity (M)} = \frac{\text{amount of solute (in moles)}}{\text{volume of solution (in L)}}$$

Solution: $12.87\ \cancel{\text{mg KI}} \times \frac{1\ \cancel{\text{g}}}{1000\ \cancel{\text{mg}}} \times \frac{1\text{ mol KI}}{166.00\ \cancel{\text{g KI}}} = 7.75\underline{3}0 \times 10^{-5}\text{ mol KI}$

$$112.4\ \cancel{\text{mL solution}} \times \frac{1\text{ L}}{1000\ \cancel{\text{mL}}} = 0.1124\text{ L}$$

$$\frac{7.75\underline{3}0 \times 10^{-5}\text{ mol KI}}{0.1124\text{ L}} = 6.89\underline{7}7 \times 10^{-4}\text{ M KI} = 6.898 \times 10^{-4}\text{ M KI}$$

Check: The units of the answer (M) are correct. The magnitude of the answer is reasonable. Concentrations are usually between 0 M and 18 M.

9.23 (a) **Given:** 0.150 M KNO_3 **Find:** M NO_3^-

Conceptual Plan: M KNO_3 → M NO_3^-

$$\frac{1\text{ M NO}_3^-}{1\text{ M KNO}_3}$$

Solution: $0.150\ \cancel{\text{M KNO}_3} \times \dfrac{1\ \text{M NO}_3^-}{1\ \cancel{\text{M KNO}_3}} = 0.150\ \text{M NO}_3^-$

Check: The units of the answer, M, are correct. The answer is reasonable because it is greater than or equal to the concentration of KNO_3.

(b) **Given:** 0.150 M $Ca(NO_3)_2$ **Find:** M NO_3^-

Conceptual Plan: M $Ca(NO_3)_2$ → M NO_3^-

$$\frac{2\ \text{M NO}_3^-}{1\ \text{M Ca(NO}_3)_2}$$

Solution: $0.150\ \cancel{\text{M Ca(NO}_3)_2} \times \dfrac{2\ \text{M NO}_3^-}{1\ \cancel{\text{M Ca(NO}_3)_2}} = 0.300\ \text{M NO}_3^-$

Check: The units of the answer, M, are correct. The answer is reasonable because it is greater than or equal to the concentration of $Ca(NO_3)_2$.

(c) **Given:** 0.150 M $Al(NO_3)_3$ **Find:** M NO_3^-

Conceptual Plan: M $Al(NO_3)_3$ → M NO_3^-

$$\frac{3\ \text{M NO}_3^-}{1\ \text{M Al(NO}_3)_3}$$

Solution: $0.150\ \cancel{\text{M Al(NO}_3)_3} \times \dfrac{3\ \text{M NO}_3^-}{1\ \cancel{\text{M Al(NO}_3)_3}} = 0.450\ \text{M NO}_3^-$

Check: The units of the answer, M, are correct. The answer is reasonable because it is greater than or equal to the concentration of $Al(NO_3)_3$.

9.24 (a) **Given:** 0.200 M NaCl **Find:** M Cl^-

Conceptual Plan: M NaCl → M Cl^-

$$\frac{1\ \text{M Cl}^-}{1\ \text{M NaCl}}$$

Solution: $0.200\ \cancel{\text{M NaCl}} \times \dfrac{1\ \text{M Cl}^-}{1\ \cancel{\text{M NaCl}}} = 0.200\ \text{M Cl}^-$

Check: The units of the answer, M, are correct. The answer is reasonable because it is greater than or equal to the concentration of NaCl.

(b) **Given:** 0.150 M $SrCl_2$ **Find:** M Cl^-

Conceptual Plan: M $SrCl_2$ → M Cl^-

$$\frac{2\ \text{M Cl}^-}{1\ \text{M SrCl}_2}$$

Solution: $0.150\ \cancel{\text{M SrCl}_2} \times \dfrac{2\ \text{M Cl}^-}{1\ \cancel{\text{M SrCl}_2}} = 0.300\ \text{M Cl}^-$

Check: The units of the answer, M, are correct. The answer is reasonable because it is greater than or equal to the concentration of $SrCl_2$.

(c) **Given:** 0.100 M $AlCl_3$ **Find:** M Cl^-

Conceptual Plan: M $SrCl_2$ → M Cl^-

$$\frac{3\ \text{M Cl}^-}{1\ \text{M AlCl}_3}$$

Solution: $0.100\ \cancel{\text{M AlCl}_3} \times \dfrac{3\ \text{M Cl}^-}{1\ \cancel{\text{M AlCl}_3}} = 0.300\ \text{M Cl}^-$

Check: The units of the answer, M, are correct. The answer is reasonable because it is greater than or equal to the concentration of $AlCl_3$.

9.25 (a) **Given:** 0.556 L; 2.3 M KCl **Find:** mol KCl

Conceptual Plan: volume solution × M = mol

$$\text{volume solution (L)} \times \text{M} = \text{mol}$$

Solution: $0.556\ \cancel{\text{L solution}} \times \dfrac{2.3\ \text{mol KCl}}{\cancel{\text{L solution}}} = 1.3\ \text{mol KCl}$

Check: The units of the answer (mol KCl) are correct. The magnitude is reasonable because it is less than 1 L solution.

(b) **Given:** 1.8 L; 0.85 M KCl **Find:** mol KCl
Conceptual Plan: volume solution × M = mol

volume solution (L) × M = mol

Solution: $1.8\ \cancel{\text{L solution}} \times \dfrac{0.85\ \text{mol KCl}}{\cancel{\text{L solution}}} = 1.5\ \text{mol KCl}$

Check: The units of the answer (mol KCl) are correct. The magnitude is reasonable because it is less than 2 L solution.

(c) **Given:** 114 mL; 1.85 M KCl **Find:** mol KCl
Conceptual Plan: mL solution → L solution, then volume solution × M = mol

$\dfrac{1\ \text{L}}{1000\ \text{mL}}$ volume solution (L) × M = mol

Solution: $114\ \cancel{\text{mL solution}} \times \dfrac{1\ \cancel{\text{L}}}{1000\ \cancel{\text{mL}}} \times \dfrac{1.85\ \text{mol KCl}}{\cancel{\text{L solution}}} = 0.211\ \text{mol KCl}$

Check: The units of the answer (mol KCl) are correct. The magnitude is reasonable because it is less than 1 L solution.

9.26 (a) **Given:** 0.45 mol C_2H_5OH; 0.200 M C_2H_5OH **Find:** volume solution
Conceptual Plan: mol C_2H_5OH → volume solution

$\dfrac{1\ \text{mol}\ C_2H_5OH}{1\ \text{M}\ C_2H_5OH}$

Solution:

$$\frac{0.45\ \cancel{\text{mol}\ C_2H_5OH}}{0.200\ \dfrac{\cancel{\text{mol}\ C_2H_5OH}}{\text{L solution}}} = 2.3\ \text{L}\ C_2H_5OH$$

Check: The units of the answer (L C_2H_5OH) are correct. The magnitude is reasonable for the amount and volume of solution.

(b) **Given:** 1.22 mol C_2H_5OH; 0.200 M C_2H_5OH **Find:** volume solution
Conceptual Plan: mol C_2H_5OH → volume solution

$\dfrac{1\ \text{mol}\ C_2H_5OH}{1\ \text{M}\ C_2H_5OH}$

Solution:

$$\frac{1.22\ \cancel{\text{mol}\ C_2H_5OH}}{0.200\ \dfrac{\cancel{\text{mol}\ C_2H_5OH}}{\text{L solution}}} = 6.10\ \text{L}\ C_2H_5OH$$

Check: The units of the answer (L C_2H_5OH) are correct. The magnitude is reasonable for the amount and volume of solution.

(c) **Given:** 1.2×10^{-2} mol C_2H_5OH; 0.200 M C_2H_5OH **Find:** volume solution
Conceptual Plan: mol C_2H_5OH → volume solution

$\dfrac{1\ \text{mol}\ C_2H_5OH}{1\ \text{M}\ C_2H_5OH}$

Solution:

$$\frac{1.2 \times 10^{-2}\ \cancel{\text{mol}\ C_2H_5OH}}{0.200\ \dfrac{\cancel{\text{mol}\ C_2H_5OH}}{\text{L solution}}} = 0.060\ \text{L}\ C_2H_5OH$$

Check: The units of the answer (L C_2H_5OH) are correct. The magnitude is reasonable for the amount and volume of solution.

9.27 **Given:** 400.0 mL; 1.1 M $NaNO_3$ **Find:** g $NaNO_3$

Conceptual Plan: mL solution → L solution, then volume solution × M = mol $NaNO_3$

$\frac{1 \text{ L solution}}{1000 \text{ mL solution}}$ volume solution (L) × M = mol

then mol $NaNO_3$ → g $NaNO_3$

$\frac{85.00 \text{ g NaNO}_3}{1 \text{ mol NaNO}_3}$

Solution: $400.0 \text{ mL solution} \times \frac{1 \text{ L}}{1000 \text{ mL}} \times \frac{1.1 \text{ mol NaNO}_3}{1 \text{ L solution}} \times \frac{85.00 \text{ g NaNO}_3}{1 \text{ mol NaNO}_3} = 3\underline{7}.40 \text{ g NaNO}_3 = 37 \text{ g NaNO}_3$

Check: The units of the answer (g $NaNO_3$) are correct. The magnitude is reasonable for the concentration and volume of solution.

9.28 **Given:** 5.5 L; 0.300 M $CaCl_2$ **Find:** g $CaCl_2$

Conceptual Plan: volume solution × M = mol $CaCl_2$ then mol $CaCl_2$ → g $CaCl_2$

volume solution (L) × M = mol $\frac{110.98 \text{ g CaCl}_2}{1 \text{ mol CaCl}_2}$

Solution: $5.5 \text{ L solution} \times \frac{0.300 \text{ mol CaCl}_2}{\text{L solution}} \times \frac{110.98 \text{ g CaCl}_2}{1 \text{ mol CaCl}_2} = 1.8 \times 10^2 \text{ g CaCl}_2$

Check: The units of the answer (g $CaCl_2$) are correct. The magnitude is reasonable for the concentration and volume of solution.

9.29 **Given:** $V_1 = 123$ mL; $M_1 = 1.1$ M; $V_2 = 500.0$ mL **Find:** M_2

Conceptual Plan: mL → L then $V_1, M_1, V_2 \rightarrow M_2$

$\frac{1 \text{ L}}{1000 \text{ mL}}$ $V_1M_1 = V_2M_2$

Solution: $123 \text{ mL} \times \frac{1 \text{ L}}{1000 \text{ mL}} = 0.123 \text{ L}$ $500.0 \text{ mL} \times \frac{1 \text{ L}}{1000 \text{ mL}} = 0.5000 \text{ L}$

$$M_2 = \frac{V_1M_1}{V_2} = \frac{(0.123 \text{ L})(1.1 \text{ M})}{(0.5000 \text{ L})} = 0.27 \text{ M}$$

Check: The units of the answer (M) are correct. The magnitude of the answer is reasonable because it is less than the original concentration.

9.30 **Given:** $V_1 = 3.5$ L; $M_1 = 4.8$ M; $V_2 = 45$ L **Find:** M_2

Conceptual Plan: $V_1, M_1, V_2 \rightarrow M_2$

$V_1M_1 = V_2M_2$

Solution: $M_2 = \frac{V_1M_1}{V_2} = \frac{(3.5 \text{ L})(4.8 \text{ M})}{(45 \text{ L})} = 0.37 \text{ M}$

Check: The units of the answer (M) are correct. The magnitude of the answer is reasonable because it is less than the original concentration.

9.31 **Given:** $V_1 = 50.0$ mL; $M_1 = 12$ M; $M_2 = 0.100$ M **Find:** V_2

Conceptual Plan: mL → L then $V_1, M_1, M_2 \rightarrow V_2$

$\frac{1 \text{ L}}{1000 \text{ mL}}$ $V_1M_1 = V_2M_2$

Solution: $50.0 \text{ mL} \times \frac{1 \text{ L}}{1000 \text{ mL}} = 0.050 \text{ L}$

$$V_2 = \frac{V_1M_1}{M_2} = \frac{(0.0500 \text{ L})(12 \text{ M})}{(0.100 \text{ M})} = 6.0 \text{ L}$$

Check: The units of the answer (L) are correct. The magnitude of the answer is reasonable because the new concentration is much less than the original; the volume must be larger.

9.32 **Given:** $V_1 = 25$ mL; $M_1 = 10.0$ M; $M_2 = 0.150$ M **Find:** V_2

Conceptual Plan: mL → L then $V_1, M_1, M_2 \rightarrow V_2$

$\frac{1 \text{ L}}{1000 \text{ mL}}$ $V_1M_1 = V_2M_2$

Solution: $25\ \cancel{\text{mL}} \times \dfrac{1\ \text{L}}{1000\ \cancel{\text{mL}}} = 0.025\ \text{L}$

$$V_2 = \frac{V_1 M_1}{M_2} = \frac{(0.025\ \text{L})(10.0\ \cancel{M})}{(0.150\ \cancel{M})} = 1.7\ \text{L}$$

Check: The units of the answer (L) are correct. The magnitude of the answer is reasonable because the new concentration is much less than the original; the volume must be larger.

9.33 **Given:** 95.4 mL; 0.102 M $CuCl_2$; 0.175 M Na_3PO_4 **Find:** volume Na_3PO_4

Conceptual Plan: mL $CuCl_2$ → L $CuCl_2$ → mol $CuCl_2$ → mol Na_3PO_4 → L Na_3PO_4 → mL Na_3PO_4

$$\frac{1\ \text{L}}{1000\ \text{mL}} \quad \frac{0.102\ \text{mol CuCl}_2}{1\ \text{L}} \quad \frac{2\ \text{mol Na}_3\text{PO}_4}{3\ \text{mol CuCl}_2} \quad \frac{1\ \text{L}}{0.175\ \text{mol Na}_3\text{PO}_4} \quad \frac{1000\ \text{mL}}{1\ \text{L}}$$

Solution: $95.4\ \cancel{\text{mL CuCl}_2} \times \dfrac{1\ \cancel{\text{L}}}{1000\ \cancel{\text{mL}}} \times \dfrac{0.102\ \cancel{\text{mol CuCl}_2}}{1\ \cancel{\text{L}}} \times \dfrac{2\ \cancel{\text{mol Na}_3\text{PO}_4}}{3\ \cancel{\text{mol CuCl}_2}} \times \dfrac{1\ \cancel{\text{L}}}{0.175\ \cancel{\text{mol Na}_3\text{PO}_4}} \times \dfrac{1000\ \text{mL}}{1\ \cancel{\text{L}}}$

$= 37.1\ \text{mL Na}_3\text{PO}_4$

Check: The units of the answer (mL Na_3PO_4) are correct. The magnitude of the answer is reasonable because the concentration of Na_3PO_4 is greater.

9.34 **Given:** 125 mL; 0.150 M $Co(NO_3)_2$; 0.150 M Li_2S **Find:** volume Li_2S

Conceptual Plan: mL $Co(NO_3)_2$ → L $Co(NO_3)_2$ → mol $Co(NO_3)_2$ → mol Li_2S → L Li_2S → mL Li_2S

$$\frac{1\ \text{L}}{1000\ \text{mL}} \quad \frac{0.150\ \text{mol Co(NO}_3)_2}{1\ \text{L}} \quad \frac{1\ \text{mol Li}_2\text{S}}{1\ \text{mol Co(NO}_3)_2} \quad \frac{1\ \text{L}}{0.150\ \text{mol Li}_2\text{S}} \quad \frac{1000\ \text{mL}}{1\ \text{L}}$$

Solution:

$$125\ \cancel{\text{mL Co(NO}_3)_2} \times \frac{1\ \cancel{\text{L}}}{1000\ \cancel{\text{mL}}} \times \frac{0.150\ \cancel{\text{mol Co(NO}_3)_2}}{1\ \cancel{\text{L}}} \times \frac{1\ \cancel{\text{mol Li}_2\text{S}}}{1\ \cancel{\text{mol Co(NO}_3)_2}} \times \frac{1\ \cancel{\text{L}}}{0.150\ \cancel{\text{mol Li}_2\text{S}}} \times \frac{1000\ \text{mL}}{1\ \cancel{\text{L}}}$$

$= 125\ \text{mL Li}_2\text{S}$

Check: The units of the answer (mL Li_2S) are correct. The magnitude of the answer is reasonable because the concentrations are the same and the mole ratio is 1:1.

9.35 **Given:** 25.0 g H_2; 6.0 M H_2SO_4 **Find:** volume H_2SO_4

Conceptual Plan: g H_2 → mol H_2 → mol H_2SO_4 → L H_2SO_4

$$\frac{1\ \text{mol H}_2}{2.016\ \text{g H}_2} \quad \frac{3\ \text{mol H}_2\text{SO}_4}{3\ \text{mol H}_2} \quad \frac{1\ \text{L}}{6.0\ \text{mol H}_2\text{SO}_4}$$

Solution: $25.0\ \cancel{\text{g H}_2} \times \dfrac{1\ \cancel{\text{mol H}_2}}{2.016\ \cancel{\text{g H}_2}} \times \dfrac{3\ \cancel{\text{mol H}_2\text{SO}_4}}{3\ \cancel{\text{mol H}_2}} \times \dfrac{1\ \text{L}}{6.0\ \cancel{\text{mol H}_2\text{SO}_4}} = 2.1\ \text{L H}_2\text{SO}_4$

Check: The units of the answer (L H_2SO_4) are correct. The magnitude is reasonable because there are approximately 12 mol H_2 and the mole ratio is 1:1.

9.36 **Given:** 25.0 g Zn; 275 mL solution **Find:** M $ZnCl_2$

Conceptual Plan: g Zn → mol Zn → mol $ZnCl_2$ → M $ZnCl_2$

$$\frac{1\ \text{mol Zn}}{65.38\ \text{g Zn}} \quad \frac{1\ \text{mol ZnCl}_2}{1\ \text{mol Zn}} \quad \frac{\text{mol ZnCl}_2}{\text{volume solution}}$$

Solution:

$$25.0\ \cancel{\text{g Zn}} \times \frac{1\ \cancel{\text{mol Zn}}}{65.38\ \cancel{\text{g Zn}}} \times \frac{1\ \text{mol ZnCl}_2}{1\ \cancel{\text{mol Zn}}} = 0.38\underline{2}4\ \text{mol ZnCl}_2$$

$$\frac{0.38\underline{2}4\ \text{mol ZnCl}_2}{275\ \cancel{\text{mL}}} \times \frac{1000\ \cancel{\text{mL}}}{1\ \text{L}} = 1.39\ \text{M ZnCl}_2$$

Check: The units of the answer (M $ZnCl_2$) are correct. The magnitude is reasonable because the stoichiometry is 1:1 and the mol Zn is less than 0.5.

9.37 **Given:** 25.0 mL, 1.20 M KCl; 15.0 mL, 0.900 M $Ba(NO_3)_2$; 2.45 g $BaCl_2$

Find: limiting reactant, theoretical yield $BaCl_2$, % yield

Conceptual Plan: volume KCl solution × M → mol KCl → mol $BaCl_2$

volume solution (L) × M = mol $\quad \frac{1 \text{ mol } BaCl_2}{2 \text{ mol KCl}}$ → **smaller mol amount determines limiting reactant**

volume $Ba(NO_3)_2$ solution × M → mol $Ba(NO_3)_2$ → mol $BaCl_2$

volume solution (L) × M = mol $\quad \frac{1 \text{ mol } BaCl_2}{1 \text{ mol } Ba(NO_3)_2}$

then mol $BaCl_2$ → g $BaCl_2$ then determine % yield

$\frac{208.23 \text{ g } BaCl_2}{1 \text{ mol } BaCl_2} \quad \frac{\text{actual yield g } BaCl_2}{\text{theoretical yield g } BaCl_2} \times 100\%$

Solution: $25.0 \text{ mL solution} \times \frac{1 \text{ L solution}}{1000 \text{ mL solution}} \times \frac{1.20 \text{ mol KCl}}{1 \text{ L solution}} \times \frac{1 \text{ mol } BaCl_2}{2 \text{ mol KCl}} = 0.015\underline{0} \text{ mol } BaCl_2$

$15.0 \text{ mL solution} \times \frac{1 \text{ L solution}}{1000 \text{ mL solution}} \times \frac{0.900 \text{ mol } Ba(NO_3)_2}{1 \text{ L solution}} \times \frac{1 \text{ mol } BaCl_2}{1 \text{ mol } Ba(NO_3)_2} = 0.013\underline{5} \text{ mol } BaCl_2$

$0.013\underline{5} \text{ mol } BaCl_2 \times \frac{208.23 \text{ g } BaCl_2}{1 \text{ mol } BaCl_2} = 2.8\underline{1}1 \text{ g } BaCl_2$

$\frac{2.45 \text{ g } BaCl_2}{2.8\underline{1}1 \text{ g } BaCl_2} \times 100\% = 87.2\%$

Check: The units of the answer (%) are correct. The magnitude is reasonable since the value is between 0 and 100%. It is much closer to 100% because the theoretical yield is not much more than the actual yield.

9.38 **Given:** 55.0 mL, 0.102 M K_2SO_4; 35.0 mL, 0.114 M $Pb(C_2H_3O_2)_2$; 1.01 g $PbSO_4$
Find: limiting reactant; theoretical yield $PbSO_4$; % yield
Conceptual Plan: volume K_2SO_4 solution × M → mol K_2SO_4 → mol $PbSO_4$

volume solution (L) × M = mol $\quad \frac{1 \text{ mol } PbSO_4}{1 \text{ mol } K_2SO_4}$ → **smaller mol amount determines limiting reactant**

volume $Pb(C_2H_3O_2)_2$ solution × M → mol $Pb(C_2H_3O_2)_2$ → mol $PbSO_4$

volume solution (L) × M = mol $\quad \frac{1 \text{ mol } PbSO_4}{1 \text{ mol } Pb(C_2H_3O_2)_2}$

then mol $PbSO_4$ → g $PbSO_4$ then determine % yield

$\frac{263.3 \text{ g } PbSO_4}{1 \text{ mol } PbSO_4} \quad \frac{\text{actual yield g } PbSO_4}{\text{theoretical yield g } PbSO_4} \times 100\%$

Solution:

$55.0 \text{ mL solution} \times \frac{1 \text{ L solution}}{1000 \text{ mL solution}} \times \frac{0.102 \text{ mol } K_2SO_4}{1 \text{ L solution}} \times \frac{1 \text{ mol } PbSO_4}{1 \text{ mol } K_2SO_4} = 0.0056\underline{1} \text{ mol } PbSO_4$

$35.0 \text{ mL solution} \times \frac{1 \text{ L solution}}{1000 \text{ mL solution}} \times \frac{0.114 \text{ mol } Pb(C_2H_3O_2)_2}{1 \text{ L solution}} \times \frac{1 \text{ mol } PbSO_4}{1 \text{ mol } Pb(C_2H_3O_2)_2} = 0.0039\underline{9} \text{ mol } PbSO_4$

$0.0039\underline{9} \text{ mol } PbSO_4 \times \frac{263.3 \text{ g } PbSO_4}{1 \text{ mol } PbSO_4} = 1.0\underline{5}06 \text{ g } PbSO_4$

$\frac{1.01 \text{ g } PbSO_4}{1.0\underline{5}06 \text{ g } PbSO_4} \times 100\% = 96.1\%$

Check: The units of the answer (%) are correct. The magnitude is reasonable since the value is between 0 and 100%. It is much closer to 100% because the theoretical yield is not much more than the actual yield.

Types of Aqueous Solutions and Solubility

9.39
(a) CsCl is an ionic compound. An aqueous solution is an electrolyte solution, so it conducts electricity.
(b) CH_3OH is a molecular compound that does not dissociate. An aqueous solution is a nonelectrolyte solution, so it does not conduct electricity.
(c) $Ca(NO_2)_2$ is an ionic compound. An aqueous solution is an electrolyte solution, so it conducts electricity.
(d) $C_6H_{12}O_6$ is a molecular compound that does not dissociate. An aqueous solution is a nonelectrolyte solution, so it does not conduct electricity.

9.40 (a) $MgBr_2$ is an ionic compound. An aqueous solution is a strong electrolyte.
(b) $C_{12}H_{22}O_{11}$ is a molecular compound that does not dissociate. An aqueous solution is a nonelectrolyte.
(c) Na_2CO_3 is an ionic compound. An aqueous solution is a strong electrolyte.
(d) KOH is a strong base. An aqueous solution is a strong electrolyte.

9.41 (a) $AgNO_3$ is soluble. Compounds containing NO_3^- are always soluble with no exceptions. The ions in the solution are $Ag^+(aq)$ and $NO_3^-(aq)$.
(b) $Pb(C_2H_3O_2)_2$ is soluble. Compounds containing $C_2H_3O_2^-$ are always soluble with no exceptions. The ions in the solution are $Pb^{2+}(aq)$ and $C_2H_3O_2^-(aq)$.
(c) KNO_3 is soluble. Compounds containing K^+ or NO_3^- are always soluble with no exceptions. The ions in solution are $K^+(aq)$ and $NO_3^-(aq)$.
(d) $(NH_4)_2S$ is soluble. Compounds containing NH_4^+ are always soluble with no exceptions. The ions in solution are $NH_4^+(aq)$ and $S^{2-}(aq)$.

9.42 (a) AgI is insoluble. Compounds containing I^- are normally soluble, but Ag^+ is an exception.
(b) $Cu_3(PO_4)_2$ is insoluble. Compounds containing PO_4^{3-} are normally insoluble, and Cu^{2+} is not an exception.
(c) $CoCO_3$ is insoluble. Compounds containing CO_3^{2-} are normally insoluble, and Co^{2+} is not an exception.
(d) K_3PO_4 is soluble. Compounds containing PO_4^{3-} are normally insoluble, but K^+ is an exception. The ions in solution are $K^+(aq)$ and $PO_4^{3-}(aq)$.

Precipitation Reactions

9.43 (a) $LiI(aq) + BaS(aq) \rightarrow$ Possible products: Li_2S and BaI_2. Li_2S is soluble. Compounds containing S^{2-} are normally insoluble, but Li^+ is an exception. BaI_2 is soluble. Compounds containing I^- are normally soluble, and Ba^{2+} is not an exception. $LiI(aq) + BaS(aq) \rightarrow$ No Reaction
(b) $KCl(aq) + CaS(aq) \rightarrow$ Possible products: K_2S and $CaCl_2$. K_2S is soluble. Compounds containing S^{2-} are normally insoluble, but K^+ is an exception. $CaCl_2$ is soluble. Compounds containing Cl^- are normally soluble, and Ca^{2+} is not an exception. $KCl(aq) + CaS(aq) \rightarrow$ No Reaction
(c) $CrBr_2(aq) + Na_2CO_3(aq) \rightarrow$ Possible products: $CrCO_3$ and NaBr. $CrCO_3$ is insoluble. Compounds containing CO_3^{2-} are normally insoluble, and Cr^{2+} is not an exception. NaBr is soluble. Compounds containing Br^- are normally soluble, and Na^+ is not an exception.
$CrBr_2(aq) + Na_2CO_3(aq) \rightarrow CrCO_3(s) + 2\,NaBr(aq)$
(d) $NaOH(aq) + FeCl_3(aq) \rightarrow$ Possible products: NaCl and $Fe(OH)_3$. NaCl is soluble. Compounds containing Na^+ are normally soluble—no exceptions. $Fe(OH)_3$ is insoluble. Compounds containing OH^- are normally insoluble, and Fe^{3+} is not an exception.
$3\,NaOH(aq) + FeCl_3(aq) \rightarrow 3\,NaCl(aq) + Fe(OH)_3(s)$

9.44 (a) $NaNO_3(aq) + KCl(aq) \rightarrow$ Possible products: NaCl and KNO_3. NaCl is soluble. Compounds containing Na^+ are always soluble—no exceptions. KNO_3 is soluble. Compounds containing K^+ are always soluble—no exceptions. $NaNO_3(aq) + KCl(aq) \rightarrow$ No Reaction
(b) $NaCl(aq) + Hg_2(C_2H_3O_2)_2(aq) \rightarrow$ Possible products: $NaC_2H_3O_2$ and Hg_2Cl_2. $NaC_2H_3O_2$ is soluble. Compounds containing Na^+ are always soluble—no exceptions. Hg_2Cl_2 is insoluble. Compounds containing Cl^- are normally soluble, but Hg_2^{2+} is an exception.
$2\,NaCl(aq) + Hg_2(C_2H_3O_2)_2(aq) \rightarrow 2\,NaC_2H_3O_2(aq) + Hg_2Cl_2(s)$
(c) $(NH_4)_2SO_4(aq) + SrCl_2(aq) \rightarrow$ Possible products: NH_4Cl and $SrSO_4$. NH_4Cl is soluble. Compounds containing NH_4^+ are always soluble—no exceptions. $SrSO_4$ is insoluble. Compounds containing SO_4^{2-} are normally soluble, but Sr^{2+} is an exception.
$(NH_4)_2SO_4(aq) + SrCl_2(aq) \rightarrow 2\,NH_4Cl(aq) + SrSO_4(s)$
(d) $NH_4Cl(aq) + AgNO_3(aq) \rightarrow$ Possible products: NH_4NO_3 and AgCl. NH_4NO_3 is soluble. Compounds containing NH_4^+ are always soluble—no exceptions. AgCl is insoluble. Compounds containing Cl^- are normally soluble, but Ag^+ is an exception.
$NH_4Cl(aq) + AgNO_3(aq) \rightarrow NH_4NO_3(aq) + AgCl(s)$

9.45 (a) $K_2CO_3(aq) + Pb(NO_3)_2(aq) \rightarrow$ Possible products: KNO_3 and $PbCO_3$. KNO_3 is soluble. Compounds containing K^+ are always soluble—no exceptions. $PbCO_3$ is insoluble. Compounds containing CO_3^{2-} are normally insoluble, and Pb^{2+} is not an exception.
$K_2CO_3(aq) + Pb(NO_3)_2(aq) \rightarrow 2\,KNO_3(aq) + PbCO_3(s)$

(b) $Li_2SO_4(aq) + Pb(C_2H_3O_2)_2(aq) \rightarrow$ Possible products: $LiC_2H_3O_2$ and $PbSO_4$. $LiC_2H_3O_2$ is soluble. Compounds containing Li^+ are always soluble—no exceptions. $PbSO_4$ is insoluble. Compounds containing SO_4^{2-} are normally soluble, but Pb^{2+} is an exception.
$Li_2SO_4(aq) + Pb(C_2H_3O_2)_2(aq) \rightarrow 2\,LiC_2H_3O_2(aq) + PbSO_4(s)$

(c) $Cu(NO_3)_2(aq) + MgS(s) \rightarrow$ Possible products: CuS and $Mg(NO_3)_2$. CuS is insoluble. Compounds containing S^{2-} are normally insoluble, and Cu^{2+} is not an exception. $Mg(NO_3)_2$ is soluble. Compounds containing NO_3^- are always soluble—no exceptions.
$Cu(NO_3)_2(aq) + MgS(s) \rightarrow CuS(s) + Mg(NO_3)_2(aq)$

(d) $Sr(NO_3)_2(aq) + KI(aq) \rightarrow$ Possible products: SrI_2 and KNO_3. SrI_2 is soluble. Compounds containing I^- are normally soluble, and Sr^{2+} is not an exception. KNO_3 is soluble. Compounds containing K^+ are always soluble—no exceptions. $Sr(NO_3)_2(aq) + KI(aq) \rightarrow$ No Reaction

9.46 (a) $NaCl(aq) + Pb(C_2H_3O_2)_2(aq) \rightarrow$ Possible products: $NaC_2H_3O_2$ and $PbCl_2$. $NaC_2H_3O_2$ is soluble. Compounds containing Na^+ are always soluble—no exceptions. $PbCl_2$ is insoluble. Compounds containing Cl^- are normally soluble, but Pb^{2+} is an exception.
$2\,NaCl(aq) + Pb(C_2H_3O_2)_2(aq) \rightarrow 2\,NaC_2H_3O_2(aq) + PbCl_2(s)$

(b) $K_2SO_4(aq) + SrI_2(aq) \rightarrow$ Possible products: KI + $SrSO_4$. KI is soluble. Compounds containing K^+ are always soluble—no exceptions. $SrSO_4$ is insoluble. Compounds containing SO_4^{2-} are normally soluble, but Sr^{2+} is an exception. $K_2SO_4(aq) + SrI_2(aq) \rightarrow 2\,KI(aq) + SrSO_4(s)$

(c) $CsCl(aq) + CaS(aq) \rightarrow$ Possible products: Cs_2S and $CaCl_2$. Cs_2S is soluble. Compounds containing S^{2-} are normally insoluble, but Cs^+ is an exception. $CaCl_2$ is soluble. Compounds containing Cl^- are normally soluble, and Ca^{2+} is not an exception. $CsCl(aq) + CaS(aq) \rightarrow$ No Reaction

(d) $Cr(NO_3)_3(aq) + Na_3PO_4(aq) \rightarrow$ Possible products: $CrPO_4$ and $NaNO_3$. $CrPO_4$ is insoluble. Compounds containing PO_4^{3-} are normally insoluble, and Cr^{3+} is not an exception. $NaNO_3$ is soluble. Compounds containing Na^+ are always soluble—no exceptions.
$Cr(NO_3)_3(aq) + Na_3PO_4(aq) \rightarrow CrPO_4(s) + 3\,NaNO_3(aq)$

Ionic and Net Ionic Equations

9.47 (a) $H^+(aq) + \cancel{Cl^-}(aq) + \cancel{Li^+}(aq) + OH^-(aq) \rightarrow H_2O(l) + \cancel{Li^+}(aq) + \cancel{Cl^-}(aq)$
$H^+(aq) + OH^-(aq) \rightarrow H_2O(l)$

(b) $\cancel{Mg^{2+}}(aq) + S^{2-}(aq) + Cu^{2+}(aq) + \cancel{2\,Cl^-}(aq) \rightarrow CuS(s) + \cancel{Mg^{2+}}(aq) + \cancel{2\,Cl^-}(aq)$
$Cu^{2+}(aq) + S^{2-}(aq) \rightarrow CuS(s)$

(c) $\cancel{Na^+}(aq) + OH^-(aq) + H^+(aq) + \cancel{NO_3^-}(aq) \rightarrow H_2O(l) + \cancel{Na^+}(aq) + \cancel{NO_3^-}(aq)$
$H^+(aq) + OH^-(aq) \rightarrow H_2O(l)$

(d) $\cancel{6\,Na^+}(aq) + 2\,PO_4^{3-}(aq) + 3\,Ni^{2+}(aq) + \cancel{6\,Cl^-}(aq) \rightarrow Ni_3(PO_4)_2(s) + \cancel{6\,Na^+}(aq) + \cancel{6\,Cl^-}(aq)$
$3\,Ni^{2+}(aq) + 2\,PO_4^{3-}(aq) \rightarrow Ni_3(PO_4)_2(s)$

9.48 (a) $\cancel{2\,K^+}(aq) + SO_4^{2-}(aq) + Ca^{2+}(aq) + \cancel{2\,I^-}(aq) \rightarrow CaSO_4(s) + \cancel{2\,K^+}(aq) + \cancel{2\,I^-}(aq)$
$Ca^{2+}(aq) + SO_4^{2-}(aq) \rightarrow CaSO_4(s)$

(b) $NH_4^+(aq) + \cancel{Cl^-}(aq) + \cancel{Na^+}(aq) + OH^-(aq) \rightarrow H_2O(l) + NH_3(g) + \cancel{Na^+}(aq) + \cancel{Cl^-}(aq)$
$NH_4^+(aq) + OH^-(aq) \rightarrow H_2O(l) + NH_3(g)$

(c) $Ag^+(aq) + \cancel{NO_3^-}(aq) + \cancel{Na^+}(aq) + Cl^-(aq) \rightarrow AgCl(s) + \cancel{Na^+}(aq) + \cancel{NO_3^-}(aq)$
$Ag^+(aq) + Cl^-(aq) \rightarrow AgCl(s)$

(d) $2\,H\cancel{C_2H_3O_2}(aq) + \cancel{2\,K^+}(aq) + CO_3^{2-}(aq) \rightarrow H_2O(l) + CO_2(g) + \cancel{2\,K^+}(aq) + \cancel{2\,C_2H_3O_2^-}(aq)$
$2\,H^+(aq) + CO_3^{2-}(aq) \rightarrow H_2O(l) + CO_2(g)$

9.49 $Hg_2^{2+}(aq) + \cancel{2\,NO_3^-}(aq) + \cancel{2\,Na^+}(aq) + 2\,Cl^-(aq) \rightarrow Hg_2Cl_2(s) + \cancel{2\,Na^+}(aq) + \cancel{2\,NO_3^-}(aq)$
$Hg_2^{2+}(aq) + 2\,Cl^-(aq) \rightarrow Hg_2Cl_2(s)$

9.50 $Pb^{2+}(aq) + \cancel{2\,NO_3^-}(aq) + \cancel{2\,K^+}(aq) + SO_4^{2-}(aq) \rightarrow PbSO_4(s) + \cancel{2\,K^+}(aq) + \cancel{2\,NO_3^-}(aq)$
$Pb^{2+}(aq) + SO_4^{2-}(aq) \rightarrow PbSO_4(s)$

Naming Acids

9.51 (a) HI: The base name of I is *iod*, so the name is hydroiodic acid.
(b) HNO_3: The oxyanion is *nitrate*, which ends in *-ate*; therefore, the name of the acid is nitric acid.
(c) H_2CO_3: The oxyanion is *carbonate*, which ends in *-ate*; therefore, the name of the acid is carbonic acid.

9.52 (a) HCl: The base name of Cl is *chlor*, so the name is hydrochloric acid.
(b) $HClO_2$: The oxyanion is *chlorite*, which ends in *-ite*; therefore, the name of the acid is chlorous acid.
(c) H_2SO_4: The oxyanion is *sulfate*, which ends in *-ate*; therefore, the name of the acid is sulfuric acid.

9.53 (a) hydrofluoric acid: HF
(b) hydrobromic acid: HBr
(c) sulfurous acid: H_2SO_3

9.54 (a) phosphoric acid: H_3PO_4
(b) hydrocyanic acid: HCN
(c) chlorous acid: $HClO_2$

Acid–Base Reactions

9.55 Skeletal reaction: $\underset{\text{acid}}{HBr(aq)} + \underset{\text{base}}{KOH(aq)} \rightarrow \underset{\text{water}}{H_2O(l)} + \underset{\text{salt}}{KBr(aq)}$
Net ionic equation: $H^+(aq) + OH^-(aq) \rightarrow H_2O(l)$

9.56 Skeletal reaction: $\underset{\text{acid}}{HNO_3(aq)} + \underset{\text{base}}{Ca(OH)_2(aq)} \rightarrow \underset{\text{water}}{H_2O(l)} + \underset{\text{salt}}{Ca(NO_3)_2(aq)}$
Balanced reaction: $2\,HNO_3(aq) + Ca(OH)_2(aq) \rightarrow 2\,H_2O(l) + Ca(NO_3)_2(aq)$
Net ionic equation: $H^+(aq) + OH^-(aq) \rightarrow H_2O(l)$

9.57 (a) Skeletal reaction: $\underset{\text{acid}}{H_2SO_4(aq)} + \underset{\text{base}}{Ca(OH)_2(aq)} \rightarrow \underset{\text{water}}{H_2O(l)} + \underset{\text{salt}}{CaSO_4(s)}$
Balanced reaction: $H_2SO_4(aq) + Ca(OH)_2(aq) \rightarrow 2\,H_2O(l) + CaSO_4(s)$
(b) Skeletal reaction: $\underset{\text{acid}}{HClO_4(aq)} + \underset{\text{base}}{KOH(aq)} \rightarrow \underset{\text{water}}{H_2O(l)} + \underset{\text{salt}}{KClO_4(aq)}$
Balanced reaction: $HClO_4(aq) + KOH(aq) \rightarrow H_2O(l) + KClO_4(aq)$
(c) Skeletal reaction: $\underset{\text{acid}}{H_2SO_4(aq)} + \underset{\text{base}}{NaOH(aq)} \rightarrow \underset{\text{water}}{H_2O(l)} + \underset{\text{salt}}{Na_2SO_4(aq)}$
Balanced reaction: $H_2SO_4(aq) + 2\,NaOH(aq) \rightarrow 2\,H_2O(l) + Na_2SO_4(aq)$

9.58 (a) Skeletal reaction: $\underset{\text{acid}}{HI(aq)} + \underset{\text{base}}{LiOH(aq)} \rightarrow \underset{\text{water}}{H_2O(l)} + \underset{\text{salt}}{LiI(aq)}$
Balanced reaction: $HI(aq) + LiOH(aq) \rightarrow H_2O(l) + LiI(aq)$
(b) Skeletal reaction: $\underset{\text{acid}}{HC_2H_3O_2(aq)} + \underset{\text{base}}{Ca(OH)_2(aq)} \rightarrow \underset{\text{water}}{H_2O(l)} + \underset{\text{salt}}{Ca(C_2H_3O_2)_2(aq)}$
Balanced reaction: $2\,HC_2H_3O_2(aq) + Ca(OH)_2(aq) \rightarrow 2\,H_2O(l) + Ca(C_2H_3O_2)_2(aq)$
(c) Skeletal reaction: $\underset{\text{acid}}{HCl(aq)} + \underset{\text{base}}{Ba(OH)_2(aq)} \rightarrow \underset{\text{water}}{H_2O(l)} + \underset{\text{salt}}{BaCl_2(aq)}$
Balanced reaction: $2\,HCl(aq) + Ba(OH)_2(aq) \rightarrow 2\,H_2O(l) + BaCl_2(aq)$

9.59 **Given:** 22.62 mL, 0.2000 M NaOH solution; 25.00 mL $HClO_4$ solution **Find:** M $HClO_4$ solution
Conceptual Plan: mL NaOH → L NaOH → mol NaOH → mol $HClO_4$

$$\frac{1\text{ L}}{1000\text{ mL}} \qquad \frac{0.2000\text{ mol NaOH}}{1\text{ L NaOH}} \qquad \frac{1\text{ mol HClO}_4}{1\text{ mol NaOH}}$$

mol $HClO_4$, volume $HClO_4$ solution → M

$$M = \frac{1\text{ mol HClO}_4}{\text{L HClO}_4\text{ solution}}$$

Solution: $22.62\ \cancel{\text{mL NaOH}} \times \frac{1\ \cancel{\text{L}}}{1000\ \cancel{\text{mL}}} \times \frac{0.2000\ \cancel{\text{mol NaOH}}}{1\ \cancel{\text{L NaOH}}} \times \frac{1\ \text{mol HClO}_4}{1\ \cancel{\text{mol NaOH}}} = 0.00452\underline{4}\ \text{mol HClO}_4$

$$\frac{0.00452\underline{4}\ \text{mol HClO}_4}{25.00\ \cancel{\text{mL}}\ \text{HClO}_4} \times \frac{1000\ \cancel{\text{mL}}}{1\ \text{L}} = 0.180\underline{9}6\ \text{M HClO}_4 = 0.1810\ \text{M HClO}_4$$

Check: The units of the answer (M $HClO_4$) are correct. The magnitude of the answer is reasonable because it is less than the M of NaOH.

9.60 **Given:** 26.38 mL, 0.100 M NaOH solution; 30.00 mL H_3PO_4 solution **Find:** M H_3PO_4 solution

Conceptual Plan: mL NaOH → L NaOH → mol NaOH → mol H_3PO_4

$$\frac{1\ \text{L}}{1000\ \text{mL}} \qquad \frac{0.100\ \text{mol NaOH}}{1\ \text{L NaOH}} \qquad \frac{1\ \text{mol H}_3\text{PO}_4}{3\ \text{mol NaOH}}$$

mol H_3PO_4, volume H_3PO_4 solution → M

$$\text{M} = \frac{1\ \text{mol H}_3\text{PO}_4}{\text{L H}_3\text{PO}_4\ \text{solution}}$$

Solution: $26.38\ \cancel{\text{mL NaOH}} \times \frac{1\ \cancel{\text{L}}}{1000\ \cancel{\text{mL}}} \times \frac{0.100\ \cancel{\text{mol NaOH}}}{1\ \cancel{\text{L NaOH}}} \times \frac{1\ \text{mol H}_3\text{PO}_4}{3\ \cancel{\text{mol NaOH}}} = 0.00087\underline{9}33\ \text{mol H}_3\text{PO}_4$

$$\frac{0.00087\underline{9}33\ \text{mol H}_3\text{PO}_4}{30.00\ \cancel{\text{mL}}\ \text{H}_3\text{PO}_4} \times \frac{1000\ \cancel{\text{mL}}}{1\ \text{L}} = 0.029\underline{3}1\ \text{M H}_3\text{PO}_4 = 0.0293\ \text{M H}_3\text{PO}_4$$

Check: The units of the answer (M H_3PO_4) are correct. The magnitude of the answer is reasonable because it is less than the concentration of NaOH.

Gas-Evolution Reactions

9.61 (a) Skeletal reaction: $HBr(aq) + NiS(s) \rightarrow NiBr_2(aq) + \underset{\text{gas}}{H_2S(g)}$

Balanced reaction: $2\,HBr(aq) + NiS(s) \rightarrow NiBr_2(aq) + H_2S(g)$

(b) Skeletal reaction:

$NH_4I(aq) + NaOH(aq) \rightarrow \underset{\text{decomposes}}{NH_4OH(aq)} + NaI(aq) \rightarrow H_2O(l) + \underset{\text{gas}}{NH_3(g)} + NaI(aq)$

Balanced reaction: $NH_4I(aq) + NaOH(aq) \rightarrow H_2O(l) + NH_3(g) + NaI(aq)$

(c) Skeletal reaction: $HBr(aq) + Na_2S(aq) \rightarrow NaBr(aq) + \underset{\text{gas}}{H_2S(g)}$

Balanced reaction: $2\,HBr(aq) + Na_2S(aq) \rightarrow 2\,NaBr(aq) + H_2S(g)$

9.62 (a) Skeletal reaction: $HCl(aq) + KHCO_3(aq) \rightarrow \underset{\text{decomposes}}{H_2CO_3(aq)} + KCl(aq) \rightarrow H_2O(l) + \underset{\text{gas}}{CO_2(g)} + KCl(aq)$

Balanced reaction: $HCl(aq) + KHCO_3(aq) \rightarrow H_2O(l) + CO_2(g) + KCl(aq)$

(b) Skeletal reaction:

$HC_2H_3O_2(aq) + NaHSO_3(aq) \rightarrow NaC_2H_3O_2(aq) + \underset{\text{decomposes}}{H_2SO_3(aq)} \rightarrow H_2O(l) + \underset{\text{gas}}{SO_2(g)} + NaC_2H_3O_2(aq)$

Balanced reaction: $HC_2H_3O_2(aq) + NaHSO_3(aq) \rightarrow H_2O(l) + SO_2(g) + NaC_2H_3O_2(aq)$

(c) Skeletal reaction:

$(NH_4)_2SO_4(aq) + Ca(OH)_2(aq) \rightarrow \underset{\text{decomposes}}{NH_4OH(aq)} + CaSO_4(s) \rightarrow H_2O(l) + \underset{\text{gas}}{NH_3(g)} + CaSO_4(s)$

Balanced reaction: $(NH_4)_2SO_4(aq) + Ca(OH)_2(aq) \rightarrow 2H_2O(l) + 2NH_3(g) + CaSO_4(s)$

9.63 Skeletal reaction:

$HClO_4(aq) + Li_2CO_3(aq) \rightarrow \underset{\text{decomposes}}{H_2CO_3(aq)} + LiClO_4(aq) \rightarrow H_2O(l) + \underset{\text{gas}}{CO_2(g)} + LiClO_4(aq)$

Balanced reaction: $2\,HClO_4(aq) + Li_2CO_3(aq) \rightarrow H_2O(l) + CO_2(g) + 2\,LiClO_4(aq)$

9.64 Skeletal reaction:

$HNO_3(aq) + Na_2SO_3(aq) \rightarrow \underset{\text{decomposes}}{H_2SO_3(aq)} + NaNO_3(aq) \rightarrow H_2O(l) + \underset{\text{gas}}{SO_2(g)} + NaNO_3(aq)$

Balanced reaction: $2\,HNO_3(aq) + Na_2SO_3(aq) \rightarrow H_2O(l) + SO_2(g) + 2\,NaNO_3(aq)$

Oxidation and Reduction

9.65 (a) Ag. The oxidation state of Ag = 0. The oxidation state of an atom in a free element is 0.
(b) Ag^+. The oxidation state of $Ag^+ = +1$. The oxidation state of a monatomic ion is equal to its charge.
(c) CaF_2. The oxidation state of Ca = +2, and the oxidation state of F = −1. The oxidation state of a group 2A metal always has an oxidation state of +2, and the oxidation of F is −1 because the sum of the oxidation states in a neutral formula unit = 0.
(d) H_2S. The oxidation state of H = +1, and the oxidation state of S = −2. The oxidation state of H when listed first is +1, and the oxidation state of S is −2 because S is in group 6A and the sum of the oxidation states in a neutral molecular unit = 0.
(e) CO_3^{2-}. The oxidation state of C = +4, and the oxidation state of O = −2. The oxidation state of O is normally −2, and the oxidation state of C is deduced from the formula because the sum of the oxidation states must equal the charge on the ion. (C ox state) +3(O ox state) = −2; (C ox state) + 3(−2) = −2, so C ox state = +4.
(f) CrO_4^{2-}. The oxidation state of Cr = +6, and the oxidation state of O = −2. The oxidation state of O is normally −2, and the oxidation state of Cr is deduced from the formula because the sum of the oxidation states must equal the charge on the ion. (Cr ox state) + 4(O ox state) = −2; (Cr ox state) + 4(−2) = −2, so Cr ox state = +6.

9.66 (a) Cl_2. The oxidation state of both Cl atoms = 0. Because Cl_2 is a free element, the oxidation state of Cl = 0.
(b) Fe^{3+}. The oxidation state of Fe = +3. The oxidation state of a monatomic ion is equal to its charge.
(c) $CuCl_2$. The oxidation state of Cu = +2, and the oxidation state of each Cl = −1. The oxidation state of group 7A atoms is normally −1, and the oxidation state of Cu is deduced from the formula because the sum of the oxidation states in a neutral formula unit = 0.
(d) CH_4. The oxidation state of C = −4, and the oxidation state of H = +1. The oxidation state of H is normally +1, and the oxidation state of C is deduced from the formula because the sum of the oxidation states in a neutral molecular unit = 0. (C ox state) + 4(H ox state) = 0; (C ox state) + 4(+1) = 0, so C ox state = −4.
(e) $Cr_2O_7^{2-}$. The oxidation state of Cr = +6, and the oxidation state of O =−2. The oxidation state of O is normally −2, and the oxidation state of Cr is deduced from the formula because the sum of the oxidation states must equal the charge of the ion. 2(Cr ox state) + 7(O ox state) = −2; 2(Cr ox state) + 7(−2) = −2, so Cr ox state = +6.
(f) HSO_4^-. The oxidation state of H = +1, the oxidation state of S = +6, and the oxidation state of O = −2. The oxidation state of H is normally +1, the oxidation state of O is normally −2, and the oxidation state of S is deduced from the formula because the sum of the oxidation states must equal the charge of the ion. (H ox state) + (S ox state) + 4(O ox state) = −1;(+1) + (S ox state) + 4(−2) = −1, so S ox state = +6.

9.67 (a) CrO. The oxidation state of Cr = +2, and the oxidation state of O = −2. The oxidation state of O is normally −2, and the oxidation state of Cr is deduced from the formula because the sum of the oxidation states must = 0. (Cr ox state) + (O ox state) = 0; (Cr ox state) + (−2) = 0, so Cr = +2.
(b) CrO_3. The oxidation state of Cr = +6, and the oxidation state of O = −2. The oxidation state of O is normally −2, and the oxidation state of Cr is deduced from the formula because the sum of the oxidation states must = 0. (Cr ox state) + 3(O ox state) = 0; (Cr ox state) + 3 (−2) = 0, so Cr = +6.
(c) Cr_2O_3. The oxidation state of Cr = +3, and the oxidation state of O = −2. The oxidation state of O is normally −2, and the oxidation state of Cr is deduced from the formula because the sum of the oxidation states must = 0. 2(Cr ox state) + 3(O ox state) = 0; 2(Cr ox state) + 3(−2) = 0, so Cr = +3.

9.68 (a) ClO^-. The oxidation state of Cl = +1, and the oxidation state of O = −2. The oxidation state of O is normally −2, and the oxidation state of Cl is deduced from the formula because the sum of the oxidation states must equal the charge of the ion. (Cl ox state) + (O ox state) = −1; (Cl ox state) + (−2) = −1, so Cl = +1.
(b) ClO_2^-. The oxidation state of Cl = +3, and the oxidation state of O = −2. The oxidation state of O is normally −2, and the oxidation state of Cl is deduced from the formula because the sum of the oxidation states must equal the charge of the ion. (Cl ox state) + 2(O ox state) = −1; (Cl ox state) + 2(−2) = −1, so Cl = +3.
(c) ClO_3^-. The oxidation state of Cl = + 5, and the oxidation state of O = −2. The oxidation state of O is normally −2, and the oxidation state of Cl is deduced from the formula because the sum of the oxidation states must equal the charge of the ion. (Cl ox state) + 3(O ox state) = −1; (Cl ox state) + 3(−2) = −1, so Cl = +5.

(d) ClO_4^-. The oxidation state of Cl = +7, and the oxidation state of O = −2. The oxidation state of O is normally −2, and the oxidation state of Cl is deduced from the formula because the sum of the oxidation states must equal the charge of the ion. (Cl ox state) + 4(O ox state) = −1; (Cl ox state) + 4(−2) = −1, so Cl = +7.

9.69 (a)
$$4\,Li(s) + O_2(g) \rightarrow 2\,Li_2O(s)$$
Oxidation states; 0 0 +1 −2
This is a redox reaction because Li increases in oxidation number (oxidation) and O decreases in number (reduction). O_2 is the oxidizing agent, and Li is the reducing agent.

(b)
$$Mg(s) + Fe^{2+}(aq) \rightarrow Mg^{2+}(aq) + Fe(s)$$
Oxidation states; 0 +2 +2 0
This is a redox reaction because Mg increases in oxidation number (oxidation) and Fe decreases in number (reduction). Fe^{2+} is the oxidizing agent, and Mg is the reducing agent.

(c)
$$Pb(NO_3)_2(aq) + Na_2SO_4(aq) \rightarrow PbSO_4(s) + 2\,NaNO_3(aq)$$
Oxidation states; +2 +5−2 +1+6−2 +2+6−2 +1+5−2
This is not a redox reaction because none of the atoms undergoes a change in oxidation number.

(d)
$$HBr(aq) + KOH(aq) \rightarrow H_2O(l) + KBr(aq)$$
Oxidation states; +1−1 +1−2+1 +1−2 +1−1
This is not a redox reaction because none of the atoms undergoes a change in oxidation number.

9.70 (a)
$$Al(s) + 3\,Ag^+(aq) \rightarrow Al^{3+}(aq) + 3\,Ag(s)$$
Oxidation states; 0 +1 +3 0
This is a redox reaction because Al increases in oxidation number (oxidation) and Ag decreases in number (reduction). Ag^+ is the oxidizing agent, and Al is the reducing agent.

(b)
$$SO_3(g) + H_2O(l) \rightarrow H_2SO_4(aq)$$
Oxidation states; +6−2 +1−2 +1+6−2
This is not a redox reaction because none of the atoms undergoes a change in oxidation number.

(c)
$$Ba(s) + Cl_2(g) \rightarrow BaCl_2(s)$$
Oxidation states; 0 0 +2−1
This is a redox reaction because Ba increases in oxidation number (oxidation) and Cl decreases in number (reduction). Cl_2 is the oxidizing agent, and Ba is the reducing agent.

(d)
$$Mg(s) + Br_2(l) \rightarrow MgBr_2(s)$$
Oxidation states; 0 0 +2−1
This is a redox reaction because Mg increases in oxidation number (oxidation) and Br decreases in number (reduction). Br_2 is the oxidizing agent, and Mg is the reducing agent.

Cumulative Problems

9.71 **Given:** Assume 100.0 g solution, 20.0 g $C_2H_6O_2$; density of solution = 1.03 g/mL **Find:** M of solution

Conceptual Plan: g $C_2H_6O_2$ → mol $C_2H_6O_2$ and g solution → mL solution → L solution

$$\frac{1\ \text{mol}\ C_2H_6O_2}{62.07\ \text{g}\ C_2H_6O_2} \qquad \frac{1.00\ \text{mL}}{1.03\ \text{g}} \qquad \frac{1\ \text{L}}{1000\ \text{mL}}$$

then M $C_2H_6O_2$

$$M = \frac{\text{mol}\ C_2H_6O_2}{\text{L solution}}$$

Solution:

$$20.0\ \cancel{\text{g}\ C_2H_6O_2} \times \frac{1\ \text{mol}\ C_2H_6O_2}{62.07\ \cancel{\text{g}\ C_2H_6O_2}} = 0.32\underline{2}2\ \text{mol}\ C_2H_6O_2$$

$$100.0\ \cancel{\text{g solution}} \times \frac{1.00\ \cancel{\text{mL solution}}}{1.03\ \cancel{\text{g solution}}} \times \frac{1\ \text{L}}{1000\ \cancel{\text{mL}}} = 0.097\underline{0}9\ \text{L}\ H_2O$$

$$M = \frac{0.32\underline{2}2\ \text{mol}\ C_2H_6O_2}{0.097\underline{0}9\ \text{L}} = 3.32\ \text{M}\ C_2H_6O_2$$

Check: The units of the answer (M) are correct. The magnitude of the answer is reasonable because the concentration of solutions is usually between 0 M and 18 M.

9.72 **Given:** 1.35 M NaCl; density of solution = 1.05 g /mL **Find:** % NaCl by mass

Conceptual Plan: mol NaCl → g NaCl and L solution → mL solution → g solution then % NaCl

$$\frac{58.45 \text{ g NaCl}}{1 \text{ mol NaCl}} \quad \frac{1000 \text{ mL}}{1 \text{ L}} \quad \frac{1.05 \text{ g}}{1.00 \text{ mL}} \quad \frac{\text{g NaCl}}{\text{g solution}} \times 100\%$$

Solution: $1.35 \text{ M} = \frac{1.35 \text{ mol NaCl}}{1 \text{ L solution}}$

$$1.35 \text{ mol NaCl} \times \frac{58.45 \text{ g NaCl}}{1 \text{ mol NaCl}} = 78.\underline{9}1 \text{ g NaCl} \qquad 1 \text{ L solution} \times \frac{1000 \text{ mL}}{1 \text{ L}} \times \frac{1.05 \text{ g solution}}{\text{mL solution}} = 10\underline{5}0 \text{ g solution}$$

$$\frac{78.91 \text{ g NaCl}}{1050 \text{ g solution}} \times 100\% = 7.52\% \text{ NaCl}$$

Check: The units of the answer (% NaCl) are correct. The magnitude of the answer is reasonable for the concentration of the solution.

9.73 **Given:** 2.5 g $NaHCO_3$ **Find:** g HCl

Conceptual Plan: g $NaHCO_3$ → mol $NaHCO_3$ → mol HCl → g HCl

$$\frac{1 \text{ mol NaHCO}_3}{84.02 \text{ g NaHCO}_3} \quad \frac{1 \text{ mol HCl}}{1 \text{ mol NaHCO}_3} \quad \frac{36.46 \text{ g HCl}}{1 \text{ mol HCl}}$$

Solution: $HCl(aq) + NaHCO_3(aq) \rightarrow H_2O(l) + CO_2(g) + NaCl(aq)$

$$2.5 \text{ g NaHCO}_3 \times \frac{1 \text{ mol NaHCO}_3}{84.02 \text{ g NaHCO}_3} \times \frac{1 \text{ mol HCl}}{1 \text{ mol NaHCO}_3} \times \frac{36.46 \text{ g HCl}}{1 \text{ mol HCl}} = 1.1 \text{ g HCl}$$

Check: The units of the answer (g HCl) are correct. The magnitude of the answer is reasonable because the molar mass of HCl is less than the molar mass of $NaHCO_3$.

9.74 **Given:** 3.8 g HCl **Find:** g $CaCO_3$

Conceptual Plan: g HCl → mol HCl → mol $CaCO_3$ → g $CaCO_3$

$$\frac{1 \text{ mol HCl}}{36.46 \text{ g HCl}} \quad \frac{1 \text{ mol CaCO}_3}{2 \text{ mol HCl}} \quad \frac{100.09 \text{ g CaCO}_3}{1 \text{ mol CaCO}_3}$$

Solution: $2 \, HCl(aq) + CaCO_3(s) \rightarrow H_2O(l) + CO_2(g) + CaCl_2(aq)$

$$3.8 \text{ g HCl} \times \frac{1 \text{ mol HCl}}{36.46 \text{ g HCl}} \times \frac{1 \text{ mol CaCO}_3}{2 \text{ mol HCl}} \times \frac{100.09 \text{ g CaCO}_3}{1 \text{ mol CaCO}_3} = 5.2 \text{ g CaCO}_3$$

Check: The units of the answer (g $CaCO_3$) are correct. The magnitude of the answer is reasonable because the molar mass of $CaCO_3$ is greater than the molar mass of HCl.

9.75 **Given:** Beaker containing 4 ions H^+, 4 ions Cl^- **Find:** which NaOH beaker will just neutralize HCl beaker

Conceptual Plan: molecules H^+ → molecules OH^- then compare to four beakers

$$\frac{1 \text{ ion OH}^-}{1 \text{ ion H}^+}$$

Solution: $HCl(aq) + NaOH(aq) \rightarrow NaCl(aq) + H_2O(l)$

Net Ionic: $H^+(aq) + OH^-(aq) \rightarrow H_2O(l)$

$$4 \text{ ions H}^+ \times \frac{1 \text{ ion OH}^-}{1 \text{ ion H}^+} = 4 \text{ ions OH}^-$$

Beaker (a) contains 2 ions OH^-; beaker (b) contains 4 ions OH^-; beaker (c) contains 5 ions OH^-; beaker (d) contains 8 ions OH^-. Beaker (b) will completely neutralize the HCl beaker with no excess.

Check: The answer is correct because it will completely neutralize the H^+ with no excess OH^-.

9.76 **Given:** Beaker containing 4 ions H^+, 4 ions Cl^- and beaker containing 5 ions Na^+, 5 ions OH^- **Find:** draw a beaker after NaOH beaker reacts with HCl beaker

Conceptual Plan: molecules H^+ → molecules OH^- then compare to four beakers

$$\frac{1 \text{ ion OH}^-}{1 \text{ ion H}^+}$$

Solution: $HCl(aq) + NaOH(aq) \rightarrow NaCl(aq) + H_2O(l)$

Net Ionic: $H^+(aq) + OH^-(aq) \rightarrow H_2O(l)$

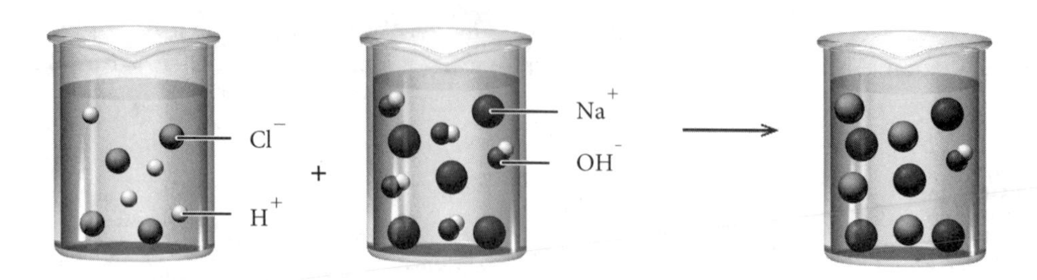

$$4 \text{ ions H}^+ \times \frac{1 \text{ ion OH}^-}{1 \text{ ion H}^+} = 4 \text{ ions OH}^-$$

NaOH beaker contains 5 ions OH^-, so 1 OH^- ion will remain. The 4 Cl^- ions and 5 Na^+ ions will also remain after the reaction.

Check: The answer is correct because there is 1 excess OH^- and NaCl will dissolve and form ions.

9.77

(a) Skeletal reaction: $HCl(aq) + Hg_2(NO_3)_2(aq) \rightarrow Hg_2Cl_2(s) + HNO_3(aq)$
Balance Cl: $2\,HCl(aq) + Hg_2(NO_3)_2(aq) \rightarrow Hg_2Cl_2(s) + 2\,HNO_3(aq)$

(b) Skeletal reaction: $KHSO_3(aq) + HNO_3(aq) \rightarrow H_2O(l) + SO_2(g) + KNO_3(aq)$
Balanced reaction: $KHSO_3(aq) + HNO_3(aq) \rightarrow H_2O(l) + SO_2(g) + KNO_3(aq)$

(c) Skeletal reaction: $NH_4Cl(aq) + Pb(NO_3)_2(aq) \rightarrow PbCl_2(s) + NH_4NO_3(aq)$
Balance Cl: $2\,NH_4Cl(aq) + Pb(NO_3)_2(aq) \rightarrow PbCl_2(s) + NH_4NO_3(aq)$
Balance N: $2\,NH_4Cl(aq) + Pb(NO_3)_2(aq) \rightarrow PbCl_2(s) + 2\,NH_4NO_3(aq)$

(d) Skeletal reaction: $NH_4Cl(aq) + Ca(OH)_2(aq) \rightarrow NH_3(g) + H_2O(l) + CaCl_2(aq)$
Balance Cl: $2\,NH_4Cl(aq) + Ca(OH)_2(aq) \rightarrow NH_3(g) + H_2O(l) + CaCl_2(aq)$
Balance N: $2\,NH_4Cl(aq) + Ca(OH)_2(aq) \rightarrow 2\,NH_3(g) + H_2O(l) + CaCl_2(aq)$
Balance H: $2\,NH_4Cl(aq) + Ca(OH)_2(aq) \rightarrow 2\,NH_3(g) + 2\,H_2O(l) + CaCl_2(aq)$

9.78

(a) Skeletal reaction: $H_2SO_4(aq) + HNO_3(aq) \rightarrow$ No Reaction

(b) Skeletal reaction: $Cr(NO_3)_3(aq) + LiOH(aq) \rightarrow Cr(OH)_3(s) + LiNO_3(aq)$
Balance OH: $Cr(NO_3)_3(aq) + 3\,LiOH(aq) \rightarrow Cr(OH)_3(s) + LiNO_3(aq)$
Balance Li: $Cr(NO_3)_3(aq) + 3\,LiOH(aq) \rightarrow Cr(OH)_3(s) + 3\,LiNO_3(aq)$

(c) Skeletal reaction: $C_5H_{12}O(l) + O_2(g) \rightarrow CO_2(g) + H_2O(g)$
Balance C: $C_5H_{12}O(l) + O_2(g) \rightarrow 5\,CO_2(g) + H_2O(g)$
Balance H: $C_5H_{12}O(l) + O_2(g) \rightarrow 5\,CO_2(g) + 6\,H_2O(g)$
Balance O: $C_5H_{12}O(l) + 15/2\,O_2(g) \rightarrow 5\,CO_2(g) + 6\,H_2O(g)$
Clear fraction: $2\,C_5H_{12}O(l) + 15\,O_2(g) \rightarrow 10\,CO_2(g) + 12\,H_2O(g)$

(d) Skeletal reaction: $SrS(aq) + CuSO_4(aq) \rightarrow SrSO_4(s) + CuS(s)$
Balanced reaction: $SrS(aq) + CuSO_4(aq) \rightarrow SrSO_4(s) + CuS(s)$

9.79 **Given:** 1.5 L solution; 0.050 M $CaCl_2$; 0.085 M $Mg(NO_3)_2$ **Find:** g Na_3PO_4

Conceptual Plan: $V, M\ CaCl_2 \rightarrow$ mol $CaCl_2$ and $V, M\ Mg(NO_3)_2 \rightarrow$ mol $Mg(NO_3)_2$

$V \times M = \text{mol}$ $\quad$ $V \times M = \text{mol}$

then (mol $CaCl_2$ + mol $Mg(NO_3)_2$) $\rightarrow Na_3PO_4 \rightarrow$ g Na_3PO_4

$$\frac{2 \text{ mol } Na_3PO_4}{3 \text{ mol } (CaCl_2 + Mg(NO_3)_2)} \quad \frac{163.97 \text{ g } Na_3PO_4}{1 \text{ mol } Na_3PO_4}$$

Solution: $3\,CaCl_2(aq) + 2\,Na_3PO_4(aq) \rightarrow Ca_3(PO_4)_2(s) + 6\,NaCl(aq)$
$3\,Mg(NO_3)_2(aq) + 2\,Na_3PO_4(aq) \rightarrow Mg_3(PO_4)_2(s) + 6\,NaCl(aq)$
$1.5\text{ L} \times 0.050\text{ M } CaCl_2 = 0.07\underline{5}\text{ mol } CaCl_2$
$1.5\text{ L} \times 0.085\text{ M } Mg(NO_3)_2 = 0.1\underline{2}75\text{ mol } Mg(NO_3)_2$

$$0.2\underline{0}25 \text{ mol } CaCl_2 \text{ and } Mg(NO_3)_2 \times \frac{2 \text{ mol } Na_3PO_4}{3 \text{ mol } CaCl_2 \text{ and } Mg(NO_3)_2} \times \frac{163.97 \text{ g } Na_3PO_4}{1 \text{ mol } Na_3PO_4} = 22 \text{ g mol } Na_3PO_4$$

Check: The units of the answer (g Na_3PO_4) are correct. The magnitude of the answer is reasonable because it is needed to remove both the Ca and Mg ions.

9.80 **Given:** 500.0 mL 0.100 M HCl and 0.200 M H_2SO_4; 0.150 M KOH **Find:** volume KOH to neutralize the acid

Conceptual Plan:

mL → L, then V, M(HCl) → mol HCl → mol H^+ and $V, M(H_2SO_4)$ → mol H_2SO_4 → mol H^+

$\frac{1\text{ L}}{1000\text{ mL}}$ $V \times M = \text{mol}$ $\frac{1\text{ mol H}^+}{1\text{ mol HCl}}$ $V \times M = \text{mol}$ $\frac{2\text{ mol H}^+}{1\text{ mol H}_2\text{SO}_4}$

Total mol H^+ → mol OH^- → mol KOH → volume KOH

$\frac{\text{mol OH}^-}{\text{mol H}^+}$ $\frac{\text{mol KOH}}{\text{mol OH}^-}$ $V = \frac{\text{mol KOH}}{\text{M KOH}}$

Solution: $500.0\ \cancel{\text{mL}} \times \frac{1\ \cancel{\text{L}}}{1000\ \cancel{\text{mL}}} \times \frac{0.100\ \cancel{\text{mol HCl}}}{1\ \cancel{\text{L}}} \times \frac{1\text{ mol H}^+}{1\ \cancel{\text{mol HCl}}} = 0.05\underline{0}00\text{ mol H}^+$

$$500.0\ \cancel{\text{mL}} \times \frac{1\ \cancel{\text{L}}}{1000\ \cancel{\text{mL}}} \times \frac{0.200\ \cancel{\text{mol H}_2\text{SO}_4}}{1\ \cancel{\text{L}}} \times \frac{2\text{ mol H}^+}{1\ \cancel{\text{mol H}_2\text{SO}_4}} = 0.20\underline{0}0\text{ mol H}^+$$

$$(0.20\underline{0}0 + 0.05\underline{0}00)\ \cancel{\text{mol H}^+} \times \frac{1\ \cancel{\text{mol OH}^-}}{1\ \cancel{\text{mol H}^+}} \times \frac{1\ \cancel{\text{mol KOH}}}{1\ \cancel{\text{mol OH}^-}} \times \frac{1\text{ L solution}}{0.150\ \cancel{\text{mol KOH}}} = 1.67\text{ L KOH solution}$$

Check: The units of the answer (L KOH) are correct. The magnitude of the answer is reasonable because the average concentration of the acid is greater than the concentration of the base.

9.81 **Given:** 1.0 L; 0.10 M OH^- **Find:** g Ba

Conceptual Plan: V, M → mol OH^- → mol $Ba(OH)_2$ → mol BaO → mol Ba → g Ba

$V \times M = \text{mol}$ $\frac{1\text{ mol Ba(OH)}_2}{2\text{ mol OH}^-}$ $\frac{1\text{ mol BaO}}{1\text{ mol Ba(OH)}_2}$ $\frac{1\text{ mol Ba}}{1\text{ mol BaO}}$ $\frac{137.3\text{ g Ba}}{1\text{ mol Ba}}$

Solution: $Ba(s) + H_2O(l) \rightarrow BaO(s)$; $BaO(s) + H_2O(l) \rightarrow Ba(OH)_2(aq)$

$$1.0\ \cancel{\text{L}} \times \frac{0.10\ \cancel{\text{mol OH}^-}}{1\ \cancel{\text{L}}} \times \frac{1\ \cancel{\text{mol Ba(OH)}_2}}{2\ \cancel{\text{mol OH}^-}} \times \frac{1\ \cancel{\text{mol BaO}}}{1\ \cancel{\text{mol Ba(OH)}_2}} \times \frac{1\ \cancel{\text{mol Ba}}}{1\ \cancel{\text{mol BaO}}} \times \frac{137.3\text{ g Ba}}{1\ \cancel{\text{mol Ba}}} = 6.9\text{ g Ba}$$

Check: The units of the answer (g Ba) are correct. The magnitude is reasonable because the molar mass of Ba is large and there are 2 moles hydroxide per mole Ba.

9.82 **Given:** 1.00 L, 1.51 M NaF; 49.6 g sample; mixture Cr^{3+} and Mg^{2+} **Find:** g Cr^{3+}

Conceptual Plan: V, M NaF → mol NaF → mol F^- and let x = mol CrF_3 and y = mol MgF_2 → mol F

$V \times M = \text{mol}$ $\frac{1\text{ mol F}^-}{1\text{ mol NaF}}$ $\frac{3\text{ mol F}}{\text{mol CrF}_3}$ $\frac{2\text{ mol F}}{\text{mol MgF}_2}$

and → g sample then solve for x = mol CrF_3 → mol Cr^{3+} → g Cr^{3+}

$x(\text{molar mass CrF}_3) = \text{g CrF}_3$ $y(\text{molar mass MgF}_2) = \text{g MgF}_2$ $\frac{1\text{ mol Cr}^{3+}}{1\text{ mol CrF}_3}$ $\frac{52.00\text{ g Cr}^{3+}}{1\text{ mol Cr}^{3+}}$

Solution: $1.00\ \cancel{\text{L}} \times \frac{1.51\ \cancel{\text{mol NaF}}}{\cancel{\text{L}}} \times \frac{1\text{ mol F}^-}{1\ \cancel{\text{mol NaF}}} = 1.51\text{ mol F}^-$

Let x = mol CrF_3 and y = mol MgF_2.

$3x$ = mol F^- from CrF_3; $2y$ = mol F^- from MgF_2

$3x + 2y = 1.51\text{ mol F}^-$

$2y = 1.51 - 3x$

$y = 0.755 - 3/2x$

$$x\ \cancel{\text{mol CrF}_3} \times \frac{109.00\text{ g CrF}_3}{\cancel{\text{mol CrF}_3}} = \text{g CrF}_3$$

$$y\ \cancel{\text{mol MgF}_2} \times \frac{62.30\text{ g MgF}_2}{\cancel{\text{mol MgF}_2}} = \text{g MgF}_2$$

$x(109.00) + y(62.30) = 49.6\text{ g sample}$

$$y = 0.755 - \frac{3}{2x}$$

Solve simultaneous equations by substituting for y: $x = 0.16\underline{4}8$ = mol CrF_3.

$$0.1648 \text{ mol CrF}_3 \times \frac{1 \text{ mol Cr}^{3+}}{1 \text{ mol CrF}_3} \times \frac{52.00 \text{ g Cr}^{3+}}{1 \text{ mol Cr}^{3+}} = 8.57 \text{ g Cr}^{3+}$$

Check: The units of the answer ($g\ Cr^{3+}$) are correct. The magnitude of the answer is reasonable because it is less than the mass of the sample.

9.83 **Given:** 0.110 M HCl; 1.52 g $Al(OH)_3$ **Find:** volume HCl needed to neutralize

Conceptual Plan: g $Al(OH)_3$ → mol $Al(OH)_3$ → mol HCl → vol HCl

$$\frac{\text{mol Al(OH)}_3}{78.00 \text{ g Al(OH)}_3} \quad \frac{3 \text{ mol HCl}}{1 \text{ mol Al(OH)}_3} \quad \frac{\text{mol HCl}}{\text{M HCl}}$$

Solution: $3\ HCl(aq) + Al(OH)_3(aq) \rightarrow 3\ H_2O(l) + AlCl_3(aq)$

$$1.52 \text{ g Al(OH)}_3 \times \frac{1 \text{ mol Al(OH)}_3}{78.00 \text{ g Al(OH)}_3} \times \frac{3 \text{ mol HCl}}{1 \text{ mol Al(OH)}_3} \times \frac{1 \text{ L HCl}}{0.110 \text{ mol HCl}} = 0.531 \text{ L HCl}$$

Check: The units of the answer (L HCl) are correct. The magnitude (0.5 L) of the answer is reasonable because we have much less than one mole of $Al(OH)_3$ and the mole ratio of HCl to $Al(OH)_3$ is 3:1.

9.84 **Given:** 0.150 M H_2SO_4; 75.3 g NaOH **Find:** volume H_2SO_4 needed to neutralize

Conceptual Plan: g NaOH → mol NaOH → mol H_2SO_4 → vol H_2SO_4

$$\frac{1 \text{ mol NaOH}}{40.00 \text{ g NaOH}} \quad \frac{1 \text{ mol H}_2\text{SO}_4}{2 \text{ mol NaOH}} \quad \frac{1 \text{ L H}_2\text{SO}_4}{0.150 \text{ mol H}_2\text{SO}_4}$$

Solution: $H_2SO_4(aq) + 2NaOH(aq) \rightarrow 2\ H_2O(l) + Na_2SO_4(aq)$

$$75.3 \text{ g NaOH} \times \frac{1 \text{ mol NaOH}}{40.00 \text{ g NaOH}} \times \frac{1 \text{ mol H}_2\text{SO}_4}{2 \text{ mol NaOH}} \times \frac{1 \text{ L H}_2\text{SO}_4}{0.150 \text{ mol H}_2\text{SO}_4} = 6.28 \text{ L H}_2\text{SO}_4$$

Check: The units of the answer (L H_2SO_4) are correct. The magnitude (6 L) of the answer is reasonable because we are starting with almost 2 moles of NaOH and the mole ratio of H_2SO_4 to NaOH is 2.

9.85 **Given:** 24.5 g Au; 24.5 g BrF_3; 24.5 g KF **Find:** g $KAuF_4$

Conceptual Plan: g Au → mol Au → mol $KAuF_4$

$$\frac{1 \text{ mol Au}}{196.97 \text{ g Au}} \quad \frac{2 \text{ mol KAuF}_4}{2 \text{ mol Au}}$$

g BrF_3 → mol BrF_3 → mol $KAuF_4$ **→ smaller mol amount determines limiting reactant**

$$\frac{1 \text{ mol BrF}_3}{136.90 \text{ g BrF}_3} \quad \frac{2 \text{ mol KAuF}_4}{2 \text{ mol BrF}_3}$$

g KF → mol KF → mol $KAuF_4$

$$\frac{1 \text{ mol KF}}{58.10 \text{ g KF}} \quad \frac{2 \text{ mol KAuF}_4}{2 \text{ mol KF}}$$

then mol $KAuF_4$ → g $KAuF_4$

$$\frac{312.07 \text{ g KAuF}_4}{1 \text{ mol KAuF}_4}$$

$$2\ Au(s) + 2\ BrF_3(l) + 2\ KF(s) \rightarrow Br_2(l) + 2\ KAuF_4(s)$$

Oxidation states; 0 +3−1 +1−1 0 +1 +3−1

This is a redox reaction because Au increases in oxidation number (oxidation) and Br decreases in number (reduction). BrF_3 is the oxidizing agent, and Au is the reducing agent.

Solution:

$$24.5 \text{ g Au} \times \frac{1 \text{ mol Au}}{196.97 \text{ g Au}} \times \frac{2 \text{ mol KAuF}_4}{2 \text{ mol Au}} = 0.1244 \text{ mol KAuF}_4$$

$$24.5 \text{ g BrF}_3 \times \frac{1 \text{ mol BrF}_3}{136.90 \text{ g BrF}_3} \times \frac{2 \text{ mol KAuF}_4}{2 \text{ mol BrF}_3} = 0.1790 \text{ mol KAuF}$$

$$24.5 \text{ g KF} \times \frac{1 \text{ mol KF}}{58.10 \text{ g KF}} \times \frac{2 \text{ mol KAuF}_4}{2 \text{ mol KF}} = 0.4217 \text{ mol KAuF}_4$$

$$0.1244 \text{ mol KAuF}_4 \times \frac{312.07 \text{ g KAuF}_4}{1 \text{ mol KAuF}_4} = 38.8 \text{ g KAuF}_4$$

Check: The units of the answer (g $KAuF_4$) are correct. The magnitude of the answer is reasonable compared to the mass of the limiting reactant Au.

9.86 **Given:** 0.10 L of 0.12 M NaCl; 0.23 L of 0.18 M $MgCl_2$; 0.20 M $AgNO_3$ **Find:** volume $AgNO_3$ to precipitate all of the Cl^-

Conceptual Plan : $V, M(\text{NaCl}) \rightarrow$ **mol NaCl** $\rightarrow$ **mol** Cl^- **and** $V, M(MgCl_2) \rightarrow$ **mol** $MgCl_2 \rightarrow$ **mol** Cl^-

$$V \times M = \text{mol} \qquad \frac{1 \text{ mol Cl}^-}{1 \text{ mol NaCl}} \qquad\qquad V \times M = \text{mol} \qquad \frac{2 \text{ mol Cl}^-}{1 \text{ mol MgCl}_2}$$

then total mol $Cl^- \rightarrow$ **mol** $Ag^+ \rightarrow$ **mol** $AgNO_3 \rightarrow$ **vol** $AgNO_3$

$$\frac{1 \text{ mol Ag}^+}{1 \text{ mol Cl}^-} \qquad \frac{1 \text{ mol AgNO}_3}{1 \text{ mol Ag}^+} \qquad \frac{1 \text{ L AgNO}_3}{0.20 \text{ mol AgNO}_3}$$

Solution:

$$0.10 \text{ L NaCl} \times \frac{0.12 \text{ mol NaCl}}{1 \text{ L NaCl}} \times \frac{1 \text{ mol Cl}^-}{1 \text{ mol NaCl}} = 0.01\underline{2} \text{ mol Cl}^-$$

$$0.23 \text{ L MgCl}_2 \times \frac{0.18 \text{ mol MgCl}_2}{1 \text{ L MgCl}_2} \times \frac{2 \text{ mol Cl}^-}{1 \text{ mol MgCl}_2} = 0.08\underline{2}8 \text{ mol Cl}^-$$

$$\text{Total mol Cl}^- = 0.01\underline{2} \text{ mol Cl}^- + 0.08\underline{2}8 \text{ mol Cl}^- = 0.09\underline{4}8 \text{ mol Cl}^-$$

$$0.09\underline{4}8 \text{ mol Cl}^- \times \frac{1 \text{ mol Ag}^+}{1 \text{ mol Cl}^-} \times \frac{1 \text{ mol AgNO}_3}{1 \text{ mol Ag}^+} \times \frac{1 \text{ L AgNO}_3}{0.20 \text{ mol AgNO}_3} = 0.47 \text{ L AgNO}_3$$

Check: The units of the answer (L $AgNO_3$) are correct. The magnitude of the answer is reasonable because the Cl^- comes from two sources.

Challenge Problems

9.87 **Given:** 0.100 L, 1.22 M NaI; total mass = 28.1 g **Find:** g AgI

Conceptual Plan: vol, M $\rightarrow$ **mol NaI** $\rightarrow$ **mol** I^-**; total mol** $I^- \rightarrow$ **mol AgI and** HgI_2

Solution: $0.100 \text{ L soln} \times \frac{1.22 \text{ mol NaI}}{1 \text{ L soln}} \times \frac{1 \text{ mol I}^-}{1 \text{ mol NaI}} = 0.122 \text{ mol I}^-$

Let x = mol AgI and y = mol HgI_2

$x + 2y = 0.122$ mol I^- so $y = 0.061 - 0.5x$

$$\left(x \text{ mol AgI} \times \frac{234.77 \text{ g AgI}}{1 \text{ mol AgI}}\right) + \left(y \text{ mol HgI}_2 \times \frac{454.39 \text{ g HgI}_2}{1 \text{ mol HgI}_2}\right) = 28.1 \text{ g}$$

Solve the simultaneous equations and $x = 0.0504$ mol AgI.

$$0.0504 \text{ mol AgI} \times \frac{234.77 \text{ g AgI}}{1 \text{ mol AgI}} = 11.8 \text{ g AgI}$$

Check: The units of the answer (g AgI) are correct. The magnitude is reasonable because it is less than the total mass.

9.88 **Given:** 15.2 billion L lake water; 1.8×10^{-5} M H_2SO_4; 8.7×10^{-6} M HNO_3 **Find:** kg $CaCO_3$ needed to neutralize

Conceptual Plan: vol lake $\rightarrow$ **mol** $H_2SO_4 \rightarrow$ **mol** H^+ **and vol lake** $\rightarrow$ **mol** $HNO_3 \rightarrow$ **mol** H^+

$$\text{vol} \times \text{M} = \text{mol} \qquad \frac{2 \text{ mol H}^+}{1 \text{ mol H}_2\text{SO}_4} \qquad\qquad \text{vol} \times \text{M} = \text{mol} \qquad \frac{1 \text{ mol H}^+}{1 \text{ mol HNO}_3}$$

then total mol $H^+ \rightarrow$ **mol** $CO_3^{2-} \rightarrow$ **mol** $CaCO_3 \rightarrow$ **g** $CaCO_3 \rightarrow$ **kg** $CaCO_3$

$$\frac{1 \text{ mol CO}_3^{2-}}{2 \text{ mol H}^+} \qquad \frac{1 \text{ mol CaCO}_3}{1 \text{ mol CO}_3^{2-}} \qquad \frac{100.09 \text{ g CaCO}_3}{1 \text{ mol CaCO}_3} \qquad \frac{1 \text{ kg}}{1000 \text{ g}}$$

Solution: $2\,H^+(aq) + CO_3^{2-}(aq) \rightarrow H_2O(l) + CO_2(g)$

$$15.2 \times 10^9 \text{ L} \times \frac{1.8 \times 10^{-5} \text{ mol H}_2\text{SO}_4}{1 \text{ L soln}} \times \frac{2 \text{ mol H}^+}{1 \text{ mol H}_2\text{SO}_4} = 5\underline{4}7200 \text{ mol H}^+$$

$$15.2 \times 10^9 \text{ L} \times \frac{8.7 \times 10^{-6} \text{ mol HNO}_3}{1 \text{ L soln}} \times \frac{1 \text{ mol H}^+}{1 \text{ mol HNO}_3} = 1\underline{3}2240 \text{ mol H}^+$$

$$\text{Total H}^+ = 5\underline{4}7200 \text{ mol H}^+ + 1\underline{3}2240 \text{ mol H}^+ = 6\underline{7}9440 \text{ mol H}^+$$

$$679440 \text{ mol H}^+ \times \frac{1 \text{ mol CO}_3^{2-}}{2 \text{ mol H}^+} \times \frac{1 \text{ mol CaCO}_3}{1 \text{ mol CO}_3^{2-}} \times \frac{100.09 \text{ g CaCO}_3}{1 \text{ mol CaCO}_3} \times \frac{1 \text{ kg}}{1000 \text{ g}} = 3.4 \times 10^4 \text{ kg CaCO}_3$$

Check: The units of the answer (kg $CaCO_3$) are correct. The magnitude of the answer is reasonable based on the size of the lake.

9.89 **Given:** 3.5×10^{-3} M Ca^{2+}; 1.1×10^{-3} M Mg^{2+}; 19.5 gal H_2O; 0.65 kg detergent/load **Find:** % by mass Na_2CO_3

Conceptual Plan: gal $H_2O \rightarrow$ L H_2O then $V, M \rightarrow$ mol Ca^{2+} and $V, M \rightarrow$ mol Mg^{2+}

$\frac{3.785 \text{ L}}{1 \text{ gal}}$ vol × M = mol vol × M = mol

then total moles ions → mol CO_3^{2-} → mol Na_2CO_3 → g Na_2CO_3 → kg Na_2CO_3 → % Na_2CO_3

$\frac{1 \text{ mol CO}_3^{2-}}{1 \text{ mol ion}}$ $\frac{1 \text{ mol Na}_2\text{CO}_3}{1 \text{ mol CO}_3^{2-}}$ $\frac{106.01 \text{ g Na}_2\text{CO}_3}{1 \text{ mol Na}_2\text{CO}_3}$ $\frac{1 \text{ kg}}{1000 \text{ g}}$ $\frac{\text{kg Na}_2\text{CO}_3}{\text{kg detergent}} \times 100\%$

Solution: $19.5 \text{ gal} \times \frac{3.785 \text{ L}}{1 \text{ gal}} \times \frac{3.5 \times 10^{-3} \text{ mol Ca}^{2+}}{1 \text{ L}} = 0.2\underline{5}8 \text{ mol Ca}^{2+}$

$$19.5 \text{ gal} \times \frac{3.785 \text{ L}}{1 \text{ gal}} \times \frac{1.1 \times 10^{-3} \text{ mol Mg}^{2+}}{1 \text{ L}} = 0.08\underline{1}19 \text{ mol Mg}^{2+}$$

$$0.3\underline{3}92 \text{ mol ions} \times \frac{1 \text{ mol CO}_3^{2-}}{1 \text{ mol ions}} \times \frac{1 \text{ mol Na}_2\text{CO}_3}{1 \text{ mol CO}_3^{2-}} \times \frac{106.01 \text{ g Na}_2\text{CO}_3}{1 \text{ mol Na}_2\text{CO}_3} \times \frac{1 \text{ kg Na}_2\text{CO}_3}{1000 \text{ g Na}_2\text{CO}_3} = 0.03\underline{5}96 \text{ kg Na}_2\text{CO}_3$$

$$\frac{0.03\underline{5}96 \text{ kg Na}_2\text{CO}_3}{0.65 \text{ kg detergent}} \times 100\% = 5.5\% \text{ Na}_2\text{CO}_3$$

Check: The units of the answer (% Na_2CO_3) are correct. The magnitude of the answer is reasonable. The percent is less than 100%.

9.90 **Given:** Solution may contain Ag^+, Ca^{2+}, and Cu^{2+} **Find:** which ions are present

Conceptual Plan: Test the solution sequentially with NaCl, Na_2SO_4, and Na_2CO_3 and see if precipitates form.

Solution: Original solution + NaCl yields no reaction: Ag^+ is not present because chlorides are normally soluble, but Ag^+ is an exception.

Original solution with Na_2SO_4 yields a precipitate and solution 2. The precipitate is $CaSO_4$, so Ca^{2+} is present. Sulfates are normally soluble, but Ca^{2+} is an exception.

Solution 2 with Na_2CO_3 yields a precipitate. The precipitate is $CuCO_3$, so Cu^{2+} is present. All carbonates are insoluble.

Net Ionic Equations:

$$Ca^{2+}(aq) + SO_4^{2-}(aq) \rightarrow CaSO_4(s)$$
$$Cu^{2+}(aq) + CO_3^{2-}(aq) \rightarrow CuCO_3(s)$$

Check: The answer is reasonable because two different precipitates formed and all of the Ca^{2+} was removed before the carbonate was added.

9.91 **Given:** Solution may contain Hg_2^{2+}, Ba^{2+}, and Fe^{2+} **Find:** Which ions are present

Conceptual Plan: Test the solution sequentially with KCl, K_2SO_4, and K_2CO_3 and see if precipitates form.

Solution: Original solution + KCl yields a precipitate and solution 2. The precipitate is Hg_2Cl_2, so Hg_2^{2+} is present. Chlorides are normally soluble, but Hg_2^{2+} is an exception.

Solution 2 with K_2SO_4 yields no precipitate, so Ba^{2+} is not present. Sulfates are normally soluble, but Ba^{2+} is an exception.

Solution 2 with K_2CO_3 yields a precipitate. The precipitate is $FeCO_3$, so Fe^{2+} is present. All carbonates are insoluble.

Net Ionic Equations:

$$Hg_2^{2+}(aq) + 2Cl^-(aq) \rightarrow Hg_2Cl_2(s)$$
$$Fe^{2+}(aq) + CO_3^{2-}(aq) \rightarrow FeCO_3(s)$$

Check: The answer is reasonable because two different precipitates formed and all of the Hg_2^{2+} was removed before the carbonate was added.

Conceptual Problems

9.92 **Given:** 1 M solution contains 8 particles **Find:** amount of solute or solvent needed to obtain new concentration

Conceptual Plan: Determine amount of solute particles in each new solution; then determine whether solute (if the number is greater) or solvent (if the number is less) needs to be added to obtain the new concentration.

Solution: Solution (a) contains 12 particles solute. Concentration is greater than the original, so solute needs to be added. $12 \text{ particles} \times \frac{1 \text{ mol}}{8 \text{ particles}} = 1.5 \text{ mol}$ $(1.5 \text{ mol} - 1.0 \text{ mol}) = 0.5 \text{ mol solute added}$

$0.5 \text{ mol solute} \times \frac{8 \text{ particles}}{1 \text{ mol solute}} = 4 \text{ solute particles added}$

Solution (a) is obtained by adding 4 particles solute to 1 L of original solution.
Solution (b) contains 4 particles. Concentration is less than the original, so solvent needs to be added.

$4 \text{ particles} \times \frac{1 \text{ mol}}{8 \text{ particles}} = 0.5 \text{ mol solute}$ So 1 L solution contains 0.5 mol = 0.5 M

$(1 \text{ M})(1 \text{ L}) = (0.5 \text{ M})(x) \quad x = 2 \text{ L}$

Solution (b) is obtained by diluting 1 L of the original solution to 2 L.
Solution (c) contains 6 particles. Concentration is less than the original, so solvent needs to be added.

$6 \text{ particles} \times \frac{1 \text{ mol}}{8 \text{ particles}} = 0.75 \text{ mol solute}$ So 1 L solution contains 0.75 mol = 0.75 M

$(1 \text{ M})(1 \text{ L}) = (0.75 \text{ M})(x) \quad x = 1.3 \text{ L}$

Solution (c) is obtained by diluting 1 L of the original solution to 1.3 L.

9.93 **Given:** A_2X soluble; BY_2 soluble; AY insoluble; BX soluble; 2 molecules A_2X mixed with 2 molecules BY_2
Find: a molecular representation of the mixture; an equation for the reaction
Solution:

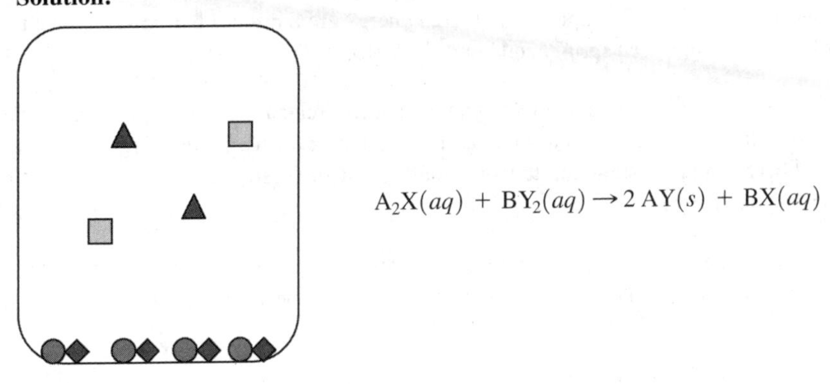

$A_2X(aq) + BY_2(aq) \rightarrow 2\,AY(s) + BX(aq)$

9.94 (c) Since the density of water is 1.0 g/mL, the mass of water is 155 g. The total mass will be 155 g + 27 g = 182 g. Since sugar will take up additional volume, the new volume is greater than the volume of the water (155 mL).

9.95 The charge of an atomic ion (O^{2-}) is associated with the individual atom (O) and is equal to its oxidation state. For a molecule, the oxidation states of the individual atoms must sum to 0 since molecules have no charge, but they are assigned to the individual atoms themselves. The oxidation states of atoms are not real physical properties, but an imposed electron bookkeeping scheme.

10 Thermochemistry

Review Questions

10.1 Thermochemistry is the study of the relationship between chemistry and energy. It is important because energy and its uses are critical to society. It is important to understand how much energy is required or released in a process.

10.2 Energy is the capacity to do work. Work is the result of a force acting through a distance. Examples of energy are kinetic energy, heat energy, electrical energy, chemical energy, and light or radiant energy. Examples of work are moving an object, expansion of a cylinder, and the running of a marathon.

10.3 Kinetic energy is associated with the motion of an object. Potential energy is associated with the position or composition of an object. Examples of kinetic energy are a moving billiard ball, movement of gas molecules, and a raging river. Examples of potential energy are a billiard ball raised above the surface of a billiard table, a compressed spring, and bonds in molecules.

10.4 The law of conservation of energy states that energy can be neither created nor destroyed. Energy can be transferred from one object to another, and it can assume different forms. In an energy exchange, energy is transferred between the system and the surroundings. If the system loses energy, the surroundings gain energy, and vice versa.

10.5 According to the first law, a device that would continually produce energy with no energy input, sometimes known as a perpetual motion machine, cannot exist because the best we can do with energy is break even.

10.6 A state function is a function whose value depends only on the state of the system, not on how the system arrived at that state. Examples are pressure, volume, and internal energy.

10.7 The internal energy (E) of a system is the sum of the kinetic and potential energies of all of the particles that compose the system. Internal energy is a state function.

10.8 Energy flowing out of the system is like a withdrawal from a checking account; therefore, it carries a negative sign.

10.9 If the reactants have a lower internal energy than the products, ΔE_{sys} is positive and energy flows into the system from the surroundings.

10.10 Heat is the flow of thermal energy caused by a temperature difference. Thermal energy is a type of kinetic energy because it arises from the motions of atoms or molecules within a substance. The higher the temperature, the greater the motion of atoms and molecules. Heat is measured in units of energy (e.g., joules, calories, and kilowatt-hours), while temperature is measured in units of Kelvins, degrees Celsius, and degrees Fahrenheit.

10.11 The internal energy (E) of a system is the sum of the kinetic and potential energies of all of the particles that compose the system. The change in the internal energy of the system (ΔE) must be the sum of the heat transferred (q) and the work done (w): $\Delta E = q + w$.

10.12 According to the first law of thermodynamics, the change in the internal energy of the system (ΔE) must be the sum of the heat transferred (q) and the work done (w): $\Delta E = q + w$. The total change in internal energy (ΔE) is the difference between its initial energy and its final energy. The amount of work done and the amount of heat transferred are dependent on the details of the path. In one path, more energy may be transferred through conversion to heat energy (if, for example, there is more friction). In another path, more energy may be transferred through work. Work and heat are not state functions, but their sum (ΔE) is constant.

10.13 The heat capacity of a system is usually defined as the quantity of heat required to change its temperature by 1 °C. Heat capacity (C) is a measure of the system's ability to hold thermal energy without undergoing a large change in temperature. The difference between heat capacity (C) and specific heat capacity (C_s) is that the specific heat capacity is the amount of heat required to raise the temperature of *1 gram* of the substance by 1 °C.

10.14 Because water has such a high heat capacity, it can moderate temperature changes. This keeps coastal temperatures more constant. Changing the temperature of water absorbs or releases large quantities of energy for a relatively small change in temperature. This serves to keep the air temperature of coastal areas more constant than the air temperature in inland areas.

10.15 When two objects of different temperatures come in direct contact, heat flows from the higher temperature object to the lower temperature object. The amount of heat lost by the warmer object is equal to the amount of heat gained by the cooler object. The warmer object's temperature will drop and the cooler object's temperature will rise until they reach the same temperature. The magnitude of these temperature changes depends on the mass and heat capacities of the two objects.

10.16 The work caused by an expansion of volume is simply the negative of the pressure that the volume expands against multiplied by the change in volume that occurs during the expansion: $w = -P\Delta V$.

10.17 In calorimetry, the thermal energy exchanged between the reaction (defined as the system) and the surroundings is measured by observing the change in temperature of the surroundings. A bomb calorimeter is used to measure the ΔE_{rxn} for combustion reactions. The calorimeter includes a tight-fitting, sealed container that forces the reaction to occur at constant volume. A coffee-cup calorimeter is used to measure ΔH_{rxn} for many aqueous reactions. The calorimeter consists of two Styrofoam® coffee cups, one inserted into the other, to provide insulation from the laboratory environment. Because the reaction happens under conditions of constant pressure (open to the atmosphere), $q_{rxn} = q_p = \Delta H_{rxn}$.

10.18 ΔH is the heat exchanged with the surroundings under conditions of constant pressure. ΔH is equal to q_p, the heat at constant pressure. Conceptually (and often numerically), ΔH and ΔE are similar: They both represent changes in a state function for the system. However, ΔE is a measure of all of the energy (heat and work) exchanged with the surroundings. $\Delta H = \Delta E + P\Delta V$.

10.19 An endothermic reaction has a positive ΔH and absorbs heat from the surroundings. An endothermic reaction feels cold to the touch. An exothermic reaction has a negative ΔH and gives off heat to the surroundings. An exothermic reaction feels warm to the touch.

10.20 The internal energy of a chemical system is the sum of its kinetic energy and its potential energy. It is this potential energy that is the energy source in an exothermic chemical reaction. Under normal circumstances, chemical potential energy (or simply chemical energy) arises primarily from the electrostatic forces between the protons and electrons that compose the atoms and molecules within the system. In an exothermic reaction, some bonds break and new ones form, and the protons and electrons go from an arrangement of higher potential energy to one of lower potential energy. As they rearrange, their potential energy is converted into kinetic energy, the heat emitted in the reaction. This increase in kinetic energy is detected as an increase in temperature.

10.21 The internal energy of a chemical system is the sum of its kinetic energy and its potential energy. It is this potential energy that absorbs the energy in an endothermic chemical reaction. In an endothermic reaction, as some bonds break and others form, the protons and electrons go from an arrangement of lower potential energy to one of higher potential energy, absorbing thermal energy in the process. This absorption of thermal energy reduces the kinetic energy of the system. This is detected as a drop in temperature.

10.22 ΔH_{rxn} is an extensive property; therefore, it depends on the quantity of reactants undergoing reaction. ΔH_{rxn} is usually reported for a reaction involving stoichiometric amounts of reactants and is dependent on the specific chemical reaction. For example, for a reaction A + 2B $\rightarrow$ C, ΔH_{rxn} is usually reported as the amount of heat emitted or absorbed when 1 mole of A reacts with 2 moles of B to form 1 mole of C.

10.23 (a) If a reaction is multiplied by a factor, ΔH is multiplied by the same factor.
(b) If a reaction is reversed, the sign of ΔH is reversed.
The relationships hold because H is a state function. Twice as much energy is contained in twice the quantity of reactants or products. If the reaction is reversed, the final and initial states have been switched and the direction of heat flow is reversed.

10.24 Hess's law states that if a chemical equation can be expressed as the sum of a series of steps, then ΔH_{rxn} for the overall equation is the sum of the heats of reactions for each step. This makes it possible to determine ΔH for a reaction without directly measuring it in the laboratory. If you can find related reactions (with known ΔH) that sum to the reaction of interest, you can find ΔH for the reaction of interest.

10.25 The standard state is defined as follows: for a gas, the pure gas at a pressure of exactly 1 atm; for a liquid or solid, the pure substance in its most stable form at a pressure of 1 atm and the temperature of interest (often taken to be 25 °C); and for a substance in solution, a concentration of exactly 1 M. The standard enthalpy change ($\Delta H°$) is the change in enthalpy for a process when all reactants and products are in their standard states. The superscript degree sign indicates standard states.

10.26 The bond energy of a chemical bond is the energy required to break 1 mole of the bond in the gas phase. Because breaking bonds is endothermic and forming bonds is exothermic, we can calculate the overall enthalpy change as a sum of the enthalpy changes associated with breaking the required bonds in the reactants and forming the required bonds in the products.

10.27 A reaction is exothermic when weak bonds break and strong bonds form. A reaction is endothermic when strong bonds break and weak bonds form.

10.28 The standard enthalpy of formation (ΔH_f°) for a pure compound is the change in enthalpy when 1 mole of the compound forms from its constituent elements in their standard states. For a pure element in its standard state, $\Delta H_f^\circ = 0$.

10.29 To calculate ΔH°_{rxn}, subtract the heats of formations of the reactants multiplied by their stoichiometric coefficients from the heats of formation of the products multiplied by their stoichiometric coefficients. In the form of an equation:

$$\Delta H^\circ_{rxn} = \sum n_p \Delta H_f^\circ(\text{products}) - \sum n_r \Delta H_f^\circ(\text{reactants})$$

10.30 Lattice energy is associated with forming a crystalline lattice of alternating cations and anions from the gaseous ions. Because the cations are positively charged and the anions are negatively charged, there is a lowering of potential—as described by Coulomb's law—when the ions come together to form a lattice. That energy is emitted as heat when the lattice forms.

As the ionic radii increase as you move down a group, the ions cannot get as close to each other and therefore do not release as much energy when the lattice forms. Thus, the lattice energy decreases (becomes less negative) as the radius increases.

Because the magnitude of the potential energy of two interacting charges depends not only on the distance between the charges but also on the product of the charges, the lattice energies become more exothermic (more negative) with increasing magnitude of ionic charge.

Problem by Topic

Internal Energy, Heat, and Work

10.31 (d) $\Delta E_{sys} = -\Delta E_{surr}$ If energy change of the system is negative, energy is being transferred from the system to the surroundings, decreasing the energy of the system and increasing the energy of the surroundings. The amount of energy lost by the system must go somewhere; so the amount gained by the surroundings is equal and opposite to that lost by the system.

10.32 The sign is positive because the energy is being taken in by or deposited into the system.

10.33 (a) The energy exchange is primarily heat because the skin (part of the surroundings) is cooled. There is a small expansion (work) because water is being converted from a liquid to a gas. The sign of ΔE_{sys} is positive because the surroundings cool.

(b) The energy exchange is primarily work. The sign of ΔE_{sys} is negative because the system is expanding (doing work on the surroundings).

(c) The energy exchange is primarily heat. The sign of ΔE_{sys} is positive because the system is being heated by the flame.

10.34 (a) The energy exchange is primarily work because there is a lot of motion. A small amount of heat is transferred because there is some friction as the balls roll. The sign of ΔE_{sys} is negative because the kinetic energy of the first ball is transferred to the second ball.

(b) The energy exchange is primarily work. The sign of ΔE_{sys} is negative because the potential energy of the book decreases as it falls.

(c) The energy exchange is primarily work. The sign of ΔE_{sys} is positive because the father is doing work to move the girl and the swing.

10.35 **Given:** 622 kJ heat released; 105 kJ work done on surroundings **Find:** ΔE_{sys}

Conceptual Plan: Interpret language to determine the sign of the two terms then $q, w \rightarrow \Delta E_{sys}$

$\Delta E = q + w$

Solution: Because heat is released from the system to the surroundings, $q = -622$ kJ; because the system is doing work on the surroundings, $w = -105$ kJ. $\Delta E = q + w = -622\text{ kJ} - 105\text{ kJ} = -727\text{ kJ} = -7.27 \times 10^2\text{ kJ}$.

Check: The units (kJ) are correct. The magnitude of the answer (-730) makes physical sense because both terms are negative.

10.36 **Given:** 196 kJ heat absorbed; surroundings do 117 kJ work Find: ΔE_{sys}

Conceptual Plan: Interpret language to determine the sign of the two terms then $q, w \rightarrow \Delta E_{sys}$

$\Delta E = q + w$

Solution: Because heat is absorbed by the system, $q = +196$ kJ; because the surroundings are doing work on the system, $w = +117$ kJ. $\Delta E = q + w = 196\text{ kJ} + 117\text{ J} = 313\text{ kJ} = 3.13 \times 10^2\text{ kJ}$.

Check: The units (kJ) are correct. The magnitude of the answer $(+300)$ makes physical sense because both terms are positive.

10.37 **Given:** 655 J heat absorbed; 344 J work done on surroundings **Find:** ΔE_{sys}

Conceptual Plan: Interpret language to determine the sign of the two terms then $q, w \rightarrow \Delta E_{sys}$

$\Delta E = q + w$

Solution: Because heat is absorbed by the system, $q = +655$ J; because the system is doing work on the surroundings, $w = -344$ J. $\Delta E = q + w = 655\text{ J} - 344\text{ J} = 311\text{ J}$.

Check: The units (J) are correct. The magnitude of the answer (+300) makes physical sense because the heat term dominates over the work term.

10.38 **Given:** 155 J heat absorbed; 77 kJ work done on surroundings **Find:** ΔE_{sys}

Conceptual Plan: Interpret language to determine the sign of the two terms J $\rightarrow$ kJ then $q, w \rightarrow \Delta E_{sys}$

$\frac{1000\text{ J}}{1\text{ kJ}}$ $\qquad \Delta E = q + w$

Solution: Because heat is absorbed by the system, $q = +155$ J; because the system is doing work on the surroundings, $w = -77$ kJ. Thus, $-77\text{ kJ} \times \frac{1000\text{ J}}{1\text{ kJ}} = -7\underline{7},000\text{ J}$

and $\Delta E = q + w = 155\text{ J} - 7\underline{7},000\text{ J} = -76,845\text{ J} = -77\text{ kJ}$.

Check: The units (kJ) are correct. The magnitude of the answer (-77 kJ) makes physical sense because the work term dominates over the heat term. In fact, the heat term is negligible compared to the work term.

Heat, Heat Capacity, and Work

10.39 Cooler A had more ice after 3 hours because most of the ice in cooler B was melted to cool the soft drinks that started at room temperature. In cooler A, the drinks were already cold; so the ice only needed to maintain this cool temperature.

10.40 Because the specific heat capacity of water is much larger than the specific heat capacity of aluminum, much more heat needs to be released by the water than the aluminum for each 1 °C of temperature drop. This means that more heat is stored in each kilogram of water than aluminum.

10.41 **Given:** 1.50 L water, $T_i = 25.0$ °C, $T_f = 100.0$ °C, $d = 1.0$ g/mL **Find:** q

Conceptual Plan: L → mL → g **and pull** C_s **from Table 10.2 and** $T_i, T_f \rightarrow \Delta T$ **then** $m, C_s, \Delta T \rightarrow q$

$\frac{1000\text{ mL}}{1\text{ L}}$ $\frac{1.0\text{ g}}{1.0\text{ mL}}$ $4.18\frac{\text{J}}{\text{g}\cdot{}^\circ\text{C}}$ $m\Delta T = T_f - T_i$ $q = mC_s\Delta T$

Solution: $1.50\text{ L} \times \frac{1000\text{ mL}}{1\text{ L}} \times \frac{1.0\text{ g}}{1.0\text{ mL}} = 1500\text{ g}$ and $\Delta T = T_f - T_i = 100.0\ ^\circ\text{C} - 25.0\ ^\circ\text{C} = 75.0\ ^\circ\text{C}$

then $q = mC_s\Delta T = 1500\text{ g} \times 4.18\frac{\text{J}}{\text{g}\cdot{}^\circ\text{C}} \times 75.0\ ^\circ\text{C} = 4.7 \times 10^5\text{ J}$

Check: The units (J) are correct. The magnitude of the answer (10^5) makes physical sense because there is such a large mass, a significant temperature change, and a high specific heat capacity material.

10.42 **Given:** 1.50 kg sand, $T_i = 25.0$°C, $T_f = 100.0$°C **Find:** q

Conceptual Plan: kg → g **and pull** C_s **from Table 10.2 and** $T_i, T_f \rightarrow \Delta T$ **then** $m, C_s, \Delta T \rightarrow q$

$\frac{1000\text{ g}}{1\text{ kg}}$ $0.84\frac{\text{J}}{\text{g}\cdot{}^\circ\text{C}}$ $\Delta T = T_f - T_i$ $q = mC_s\Delta T$

Solution: $1.50\text{ kg} \times \frac{1000\text{ g}}{1\text{ kg}} = 1500\text{ g}$ and $\Delta T = T_f - T_i = 100.0^\circ\text{C} - 25.0\ ^\circ\text{C} = 75.0\ ^\circ\text{C}$

then $q = mC_s\Delta T = 1500\text{ g} \times 0.84\frac{\text{J}}{\text{g}\cdot{}^\circ\text{C}} \times 75.0\ ^\circ\text{C} = 9.5 \times 10^4\text{ J}$

Check: The units (J) are correct. The magnitude of the answer ($\sim 10^5$) makes physical sense because there is such a large mass and a significant temperature change.

10.43 (a) **Given:** 25 g gold; $T_i = 27.0$ °C; $q = 2.35$ kJ **Find:** T_f

Conceptual Plan: kJ → J **and pull** C_s **from Table 10.2 then** $m, C_s, q \rightarrow \Delta T$ **then** $T_i, \Delta T \rightarrow T_f$

$\frac{1000\text{ J}}{1\text{ kJ}}$ $0.128\frac{\text{J}}{\text{g}\cdot{}^\circ\text{C}}$ $q = mC_s\Delta T$ $\Delta T = T_f - T_i$

Solution: $2.35\text{ kJ} \times \frac{1000\text{ J}}{1\text{ kJ}} = 2350\text{ J}$ then $q = mC_s\Delta T$. Rearrange to solve for ΔT.

$$\Delta T = \frac{q}{mC_s} = \frac{2350\text{ J}}{25\text{ g} \times 0.128\frac{\text{J}}{\text{g}\cdot{}^\circ\text{C}}} = 734.375\ ^\circ\text{C}$$ finally $\Delta T = T_f - T_i$. Rearrange to solve for T_f.

$T_f = \Delta T + T_i = 734.375\ ^\circ\text{C} + 27.0\ ^\circ\text{C} = 760\ ^\circ\text{C}$

Check: The units (°C) are correct. The magnitude of the answer (760) makes physical sense because such a large amount of heat is absorbed and there are a small mass and specific heat capacity. The temperature change should be very large.

(b) **Given:** 25 g silver; $T_i = 27.0$ °C; $q = 2.35$ kJ **Find:** T_f

Conceptual Plan: kJ → J **and pull** C_s **from Table 10.2 then** $m, C_s, q \rightarrow \Delta T$ **then** $T_i, \Delta T \rightarrow T_f$

$\frac{1000\text{ J}}{1\text{ kJ}}$ $0.235\frac{\text{J}}{\text{g}\cdot{}^\circ\text{C}}$ $q = mC_s\Delta T$ $\Delta T = T_f - T_i$

Solution: $2.35\text{ kJ} \times \frac{1000\text{ J}}{1\text{ kJ}} = 2350\text{ J}$ then $q = mC_s\Delta T$. Rearrange to solve for ΔT.

$$\Delta T = \frac{q}{mC_s} = \frac{2350\text{ J}}{25\text{ g} \times 0.235\frac{\text{J}}{\text{g}\cdot{}^\circ\text{C}}} = 400\ ^\circ\text{C}$$ finally $\Delta T = T_f - T_i$. Rearrange to solve for T_f.

$T_f = \Delta T + T_i = 4\underline{0}0\ ^\circ C + 27.0\ ^\circ C = 430\ ^\circ C$

Check: The units (°C) are correct. The magnitude of the answer (430) makes physical sense because such a large amount of heat is absorbed and there are a small mass and specific heat capacity. The temperature change should be very large. The temperature change should be less than that of the gold because the specific heat capacity is greater.

(c) **Given:** 25 g aluminum; $T_i = 27.0\ ^\circ C$; $q = 2.35$ kJ **Find:** T_f
Conceptual Plan: kJ → J and pull C_s from Table 10.2 then $m, C_s, q \rightarrow \Delta T$ then $T_i, \Delta T \rightarrow T_f$

$\frac{1000\ J}{1\ kJ}$ $0.903\frac{J}{g \cdot ^\circ C}$ $q = mC_s\Delta T$ $\Delta T = T_f - T_i$

Solution: $2.35\ kJ \times \frac{1000\ J}{1\ kJ} = 23\underline{5}0\ J$ then $q = mC_s\Delta T$. Rearrange to solve for ΔT.

$$\Delta T = \frac{q}{mC_s} = \frac{23\underline{5}0\ J}{25\ g \times 0.903\frac{J}{g \cdot ^\circ C}} = 1\underline{0}4.10\ ^\circ C \text{ finally } \Delta T = T_f - T_i. \text{ Rearrange to solve for } T_f.$$

$T_f = \Delta T + T_i = 1\underline{0}4.10\ ^\circ C + 27.0\ ^\circ C = 130\ ^\circ C$

Check: The units (°C) are correct. The magnitude of the answer (130) makes physical sense because such a large amount of heat is absorbed and there is such a small mass. The temperature change should be less than that of the silver because the specific heat capacity is greater.

(d) **Given:** 25 g water; $T_i = 27.0\ ^\circ C$; $q = 2.35$ kJ **Find:** T_f
Conceptual Plan: kJ → J and pull C_s from Table 10.2 then $m, C_s, q \rightarrow \Delta T$ then $T_i, \Delta T \rightarrow T_f$

$\frac{1000\ J}{1\ kJ}$ $4.18\frac{J}{g \cdot ^\circ C}$ $q = mC_s\Delta T$ $\Delta T = T_f - T_i$

Solution: $2.35\ kJ \times \frac{1000\ J}{1\ kJ} = 23\underline{5}0\ J$ then $q = mC_s\Delta T$. Rearrange to solve for ΔT.

$$\Delta T = \frac{q}{mC_s} = \frac{23\underline{5}0\ J}{25\ g \times 4.18\frac{J}{g \cdot ^\circ C}} = 2\underline{2}.488\ ^\circ C \text{ finally } \Delta T = T_f - T_i. \text{ Rearrange to solve for } T_f.$$

$T_f = \Delta T + T_i = 2\underline{2}.488\ ^\circ C + 27.0\ ^\circ C = 49\ ^\circ C$

Check: The units (°C) are correct. The magnitude of the answer (49) makes physical sense because such a large amount of heat is absorbed and there is such a small mass. The temperature change should be less than that of the aluminum because the specific heat capacity is greater.

10.44 (a) **Given:** Pyrex(R) glass; $q = 1.95 \times 10^3$ J; $T_i = 23.0\ ^\circ C$; $T_f = 55.4\ ^\circ C$ **Find:** m
Conceptual Plan: Pull C_s from Table 10.2 then $T_i, T_f \rightarrow \Delta T$ then $\Delta T, C_s, q \rightarrow m$

$0.75\frac{J}{g \cdot ^\circ C}$ $\Delta T = T_f - T_i$ $q = mC_s\Delta T$

Solution: $\Delta T = T_f - T_i = 55.4\ ^\circ C - 23.0\ ^\circ C = 32.4\ ^\circ C$ and $q = mC_s\Delta T$. Rearrange to solve for m.

$$m = \frac{q}{C_s\Delta T} = \frac{1.95 \times 10^3\ J}{0.75\frac{J}{g \cdot ^\circ C} \times 32.4\ ^\circ C} = 80.\ g, \text{ or } 8.0 \times 10^1\ g$$

Check: The units (g) are correct. The magnitude of the answer (80) makes physical sense because such a large amount of heat is absorbed and there are a moderate temperature rise and specific heat capacity.

(b) **Given:** sand; $q = 1.95 \times 10^3$ J; $T_i = 23.0\ ^\circ C$; $T_f = 62.1\ ^\circ C$ **Find:** m
Conceptual Plan: Pull C_s from Table 10.2 then $T_i, T_f \rightarrow \Delta T$ then $\Delta T, C_s, q \rightarrow m$

$0.84\frac{J}{g \cdot ^\circ C}$ $\Delta T = T_f - T_i$ $q = mC_s\Delta T$

Solution: $\Delta T = T_f - T_i = 62.1\ ^\circ C - 23.0\ ^\circ C = 39.1\ ^\circ C$ then $q = mC_s\Delta T$. Rearrange to solve for m.

$$m = \frac{q}{C_s\Delta T} = \frac{1.95 \times 10^3\ J}{0.84\frac{J}{g \cdot ^\circ C} \times 39.1\ ^\circ C} = 59\ g$$

Check: The units (g) are correct. The magnitude of the answer (60) makes physical sense because such a large amount of heat is absorbed and there are a moderate temperature rise and specific heat capacity.

(c) **Given:** ethanol; $q = 1.95 \times 10^3$ J; $T_i = 23.0$ °C; $T_f = 44.2$ °C **Find:** m

Conceptual Plan: Pull C_s from Table 10.2 then $T_i, T_f \rightarrow \Delta T$ then $\Delta T, C_s, q \rightarrow m$

$2.42 \frac{\text{J}}{\text{g} \cdot °\text{C}}$ $\quad \Delta T = T_f - T_i \quad q = mC_s\Delta T$

Solution: $\Delta T = T_f - T_i = 44.2\,°\text{C} - 23.0\,°\text{C} = 21.2\,°\text{C}$ then $q = mC_s\Delta T$. Rearrange to solve for m.

$$m = \frac{q}{C_s\Delta T} = \frac{1.95 \times 10^3 \text{ J}}{2.42 \frac{\text{J}}{\text{g} \cdot °\text{C}} \times 21.2\,°\text{C}} = 38.0 \text{ g}$$

Check: The units (g) are correct. The magnitude of the answer (40) makes physical sense because such a large amount of heat is absorbed and there are a small temperature rise and specific heat capacity.

(d) **Given:** water; $q = 1.95 \times 10^3$ J; $T_i = 23.0$ °C; $T_f = 32.4$ °C **Find:** m

Conceptual Plan: Pull C_s from Table 10.2 then $T_i, T_f \rightarrow \Delta T$ then $\Delta T, C_s, q \rightarrow m$

$4.18 \frac{\text{J}}{\text{g} \cdot °\text{C}}$ $\quad \Delta T = T_f - T_i \quad q = mC_s\Delta T$

Solution: $\Delta T = T_f - T_i = 32.4\,°\text{C} - 23.0\,°\text{C} = 9.4\,°\text{C}$ then $q = mC_s\Delta T$. Rearrange to solve for m.

$$m = \frac{q}{C_s\Delta T} = \frac{1.95 \times 10^3 \text{ J}}{4.18 \frac{\text{J}}{\text{g} \cdot °\text{C}} \times 9.4\,°\text{C}} = 50.\text{ g or } 5.0 \times 10^1 \text{ g}$$

Check: The units (g) are correct. The magnitude of the answer (50) makes physical sense because such a large amount of heat is absorbed and there are a small temperature rise and very large specific heat capacity.

10.45 **Given:** $V_i = 0.0$ L; $V_f = 2.5$ L; $P = 1.1$ atm **Find:** w (J)

Conceptual Plan: $V_i, V_f \rightarrow \Delta V$ then $P, \Delta V \rightarrow w$ (L atm) $\rightarrow w$ (J)

$\Delta V = V_f - V_i \quad w = -P\Delta V \quad \frac{101.3 \text{ J}}{1 \text{ L} \cdot \text{atm}}$

Solution: $\Delta V = V_f - V_i = 2.5 \text{ L} - 0.0 \text{ L} = 2.5 \text{ L}$ then

$$w = -P\Delta V = -1.1 \text{ atm} \times 2.5 \text{ L} \times \frac{101.3 \text{ J}}{1 \text{ L} \cdot \text{atm}} = -280 \text{ J or } -2.8 \times 10^2 \text{ J}$$

Check: The units (J) are correct. The magnitude of the answer (−280) makes physical sense because this is an expansion (negative work) and we have atmospheric pressure and a small volume of expansion.

10.46 **Given:** $\Delta V = 0.50$ L; $P = 1.0$ atm **Find:** w (J)

Conceptual Plan: $P, \Delta V \rightarrow w$ (L atm) $\rightarrow w$ (J)

$w = -P\Delta V \quad \frac{101.3 \text{ J}}{1 \text{ L} \cdot \text{atm}}$

Solution: $w = -P\Delta V = -1.0 \text{ atm} \times 0.50 \text{ L} \times \frac{101.3 \text{ J}}{1 \text{ L} \cdot \text{atm}} = -51 \text{ J}$

Check: The units (J) are correct. The magnitude of the answer (−51) makes physical sense because this is a small expansion (negative work) and we do not expect breathing to take much energy.

10.47 **Given:** $q = 565$ J absorbed; $V_i = 0.10$ L; $V_f = 0.85$ L; $P = 1.0$ atm **Find:** ΔE_{sys}

Conceptual Plan: $V_i, V_f \rightarrow \Delta V$ and interpret language to determine the sign of the heat

$\Delta V = V_f - V_i \quad q = +565 \text{ J}$

then $P, \Delta V \rightarrow w$ (L atm) $\rightarrow w$ (J) finally $q, w \rightarrow \Delta E_{sys}$

$w = -P\Delta V \quad \frac{101.3 \text{ J}}{1 \text{ L} \cdot \text{atm}} \quad \Delta E = q + w$

Solution: $\Delta V = V_f - V_i = 0.85 \text{ L} - 0.10 \text{ L} = 0.75 \text{ L}$ then

$$w = -P\Delta V = -1.0 \text{ atm} \times 0.75 \text{ L} \times \frac{101.3 \text{ J}}{1 \text{ L} \cdot \text{atm}} = -7\underline{5}.975 \text{ J} \qquad \Delta E = q + w = +565 \text{ J} - 7\underline{5}.975 \text{ J} = 489 \text{ J}$$

Check: The units (J) are correct. The magnitude of the answer (500) makes physical sense because the heat absorbed dominated the small expansion work (negative work).

10.48 **Given:** $q = 124$ J released; $V_i = 5.55$ L; $V_f = 1.22$ L; $P = 1.00$ atm **Find:** ΔE_{sys}

Conceptual Plan: $V_i, V_f \rightarrow \Delta V$ **and interpret language to determine the sign of the heat**

$\Delta V = V_f - V_i$ $\quad$ $q = -124$ J

then $P, \Delta V \rightarrow w$ **(L atm)** $\rightarrow w$ **(J) finally** $q, w \rightarrow \Delta E_{sys}$

$w = -P\Delta V$ $\quad$ $\frac{101.3\ \text{J}}{1\ \text{L}\cdot\text{atm}}$ $\quad$ $\Delta E = q + w$

Solution: $\Delta V = V_f - V_i = 1.22\ \text{L} - 5.55\ \text{L} = -4.33\ \text{L}$ then

$$w = -P\Delta V = -1.00\ \text{atm} \times (-4.33\ \text{L}) \times \frac{101.3\ \text{J}}{1\ \text{L}\cdot\text{atm}} = +438.629\ \text{J then}$$

$\Delta E = q + w = -124\ \text{J} + 438.629\ \text{J} = 315\ \text{J}$

Check: The units (J) are correct. The magnitude of the answer (300) makes physical sense because the compression work dominated the small amount of heat released.

Enthalpy and Thermochemical Stoichiometry

10.49 **Given:** 1 mol fuel, 3452 kJ heat produced; 11 kJ work done on surroundings **Find:** ΔE_{sys}, ΔH

Conceptual Plan: Interpret language to determine the sign of the two terms then $q \rightarrow \Delta H$ **and** $q, w \rightarrow \Delta E_{sys}$

$\Delta H = q_p$ $\quad$ $\Delta E = q + w$

Solution: Because heat is produced by the system to the surroundings, $q = -3452$ kJ; because the system is doing work on the surroundings, $w = -11$ kJ. $\Delta H = q_p = -3452$ kJ and $\Delta E = q + w = -3452\ \text{kJ} - 11\ \text{kJ} = -3463\ \text{kJ}$

Check: The units (kJ) are correct. The magnitude of the answer (-3500) makes physical sense because both terms are negative. We expect significant amounts of energy from fuels.

10.50 **Given:** 1 mol octane, $P = 1.0$ atm, $\Delta E_{sys} = 5084.3$ kJ; $\Delta H = 5074.1$ kJ **Find:** w

Conceptual Plan: Interpret language to determine the sign of the two terms then $q \rightarrow \Delta H$ **and** $q, \Delta E_{sys} \rightarrow w$

$\Delta E_{sys} = -5084.3$ kJ; $\Delta H = -5074.1$ kJ $\quad$ $\Delta H = q_p$ $\quad$ $\Delta E = q + w$

Solution: Because heat is produced by the system to the surroundings, $\Delta H = q_p = -5074.1$ kJ; $\Delta E_{sys} = -5084.3$ kJ; $\Delta E = q + w$. Rearrange to solve for w. $w = \Delta E - q = -5084.3\ \text{kJ} - (-5074.1\ \text{kJ}) = -10.2\ \text{kJ}$

Check: The units (kJ) are correct. The magnitude of the answer (-10) makes physical sense because the work should be negative in an expansion. We expect more heat than work in an engine.

10.51
(a) Combustion is an exothermic process; ΔH is negative.
(b) Evaporation requires an input of energy, so it is endothermic; ΔH is positive.
(c) Condensation is the reverse of evaporation, so it is exothermic; ΔH is negative.

10.52
(a) Sublimation requires an input of energy, so it is endothermic; ΔH is positive.
(b) Combustion is an exothermic process; ΔH is negative.
(c) Because the temperature drops, this is an endothermic process; ΔH is positive.

10.53 **Given:** 177 mL acetone (C_3H_6O), $\Delta H^\circ_{rxn} = -1790$ kJ; $d = 0.788$ g/mL **Find:** q

Conceptual Plan: mL acetone $\rightarrow$ **g acetone** $\rightarrow$ **mol acetone** $\rightarrow q$

$\frac{0.788\ \text{g}}{1\ \text{mL}}$ $\quad$ $\frac{1\ \text{mol}}{58.08\ \text{g}}$ $\quad$ $\frac{-1790\ \text{kJ}}{1\ \text{mol}}$

Solution: $177\ \text{mL} \times \frac{0.788\ \text{g}}{1\ \text{mL}} \times \frac{1\ \text{mol}}{58.08\ \text{g}} \times \frac{-1790\ \text{kJ}}{1\ \text{mol}} = -4.30 \times 10^3\ \text{kJ or } 4.30 \times 10^3\ \text{kJ released}$

Check: The units (kJ) are correct. The magnitude of the answer (-10^3) makes physical sense because the enthalpy change is negative and we have more than a mole of acetone. We expect more than 1790 kJ to be released.

10.54 **Given:** natural gas (CH_4), $\Delta H^\circ_{rxn} = -802.3$ kJ; $q = 267$ kJ **Find:** m (CH_4)

Conceptual Plan: $q \rightarrow$ **mol natural gas** $\rightarrow$ **g natural gas**

$$\frac{1 \text{ mol}}{-802.3 \text{ kJ}} \qquad \frac{16.04 \text{ g}}{1 \text{ mol}}$$

Solution: $-267 \text{ kJ} \times \frac{1 \text{ mol}}{-802.3 \text{ kJ}} \times \frac{16.04 \text{ g}}{1 \text{ mol}} = 5.34 \text{ g}$

Check: The units (g) are correct. The magnitude of the answer (5) makes physical sense because the enthalpy change per mole is so large and we need to burn less than a mole.

10.55 **Given:** 5.56 kg nitromethane (CH_3NO_2); $\Delta H^\circ_{rxn} = -1418 \text{ kJ}/2$ mol nitromethane **Find:** q

Conceptual Plan: kg nitromethane $\rightarrow$ **g nitromethane** $\rightarrow$ **mol nitromethane** $\rightarrow q$

$$\frac{1000 \text{ g}}{1 \text{ kg}} \qquad \frac{1 \text{ mol}}{61.04 \text{ g}} \qquad \frac{-1418 \text{ kJ}}{2 \text{ mol}}$$

Solution: $5.56 \text{ kg} \times \frac{1000 \text{ g}}{1 \text{ kg}} \times \frac{1 \text{ mol}}{61.04 \text{ g}} \times \frac{-1418 \text{ kJ}}{2 \text{ mol}} = -6.46 \times 10^4 \text{ kJ or } 6.46 \times 10^4 \text{ kJ released}$

Check: The units (kJ) are correct. The magnitude of the answer (-10^4) makes physical sense because the enthalpy change is negative and we have more than a mole of acetone. We expect more than 1418 kJ to be released.

10.56 **Given:** titanium and iodine, $\Delta H^\circ_{rxn} = -839$ kJ; $q = 1.55 \times 10^3$ kJ **Find:** m(Ti) and $m(I_2)$

Conceptual Plan: $q \rightarrow$ **mol Ti** $\rightarrow$ **g Ti and** $q \rightarrow$ **mol** $I_2 \rightarrow$ **g** I_2

$$\frac{2 \text{ mol}}{-839 \text{ kJ}} \qquad \frac{47.87 \text{ g}}{1 \text{ mol}} \qquad \frac{3 \text{ mol}}{-839 \text{ kJ}} \qquad \frac{253.80 \text{ g}}{1 \text{ mol}}$$

Solution: $-1.55 \times 10^3 \text{ kJ} \times \frac{2 \text{ mol Ti}}{-839 \text{ kJ}} \times \frac{47.87 \text{ g Ti}}{1 \text{ mol Ti}} = 1.77 \times 10^2 \text{ g Ti and}$

$-1.55 \times 10^3 \text{ kJ} \times \frac{3 \text{ mol } I_2}{-839 \text{ kJ}} \times \frac{253.80 \text{ g } I_2}{1 \text{ mol } I_2} = 1.41 \times 10^3 \text{ g } I_2$

Check: The units (g and g) are correct. The magnitude of the answers (10^2 and 10^3) makes physical sense because the heat emitted is much more than the enthalpy of the reaction; so we will generate much more than a mole of titanium and iodine.

10.57 **Given:** pork roast, $\Delta H^\circ_{rxn} = -2217 \text{ kJ}/3 \text{ mol } CO_2$; q needed $= 1.6 \times 10^3$ kJ, 10% efficiency **Find:** $m(CO_2)$

Conceptual Plan: q **used** $\rightarrow q$ **generated** $\rightarrow$ **mol** $CO_2 \rightarrow$ **g** CO_2

$$\frac{100 \text{ kJ generated}}{10 \text{ kJ used}} \qquad \frac{3 \text{ mol}}{2217 \text{ kJ}} \qquad \frac{44.01 \text{ g}}{1 \text{ mol}}$$

Solution: $1.6 \times 10^3 \text{ kJ} \times \frac{100 \text{ kJ generated}}{10 \text{ kJ used}} \times \frac{3 \text{ mol } CO_2}{2217 \text{ kJ}} \times \frac{44.01 \text{ g } CO_2}{1 \text{ mol } CO_2} = 950 \text{ g } CO_2$

Check: The units (g) are correct. The magnitude of the answer (~1000) makes physical sense because the process is not very efficient and a great deal of energy is needed.

10.58 **Given:** carbon, $\Delta H^\circ_{rxn} = -393.5$ kJ; q needed $= 5.00 \times 10^2$ kJ **Find:** $m(CO_2)$

Conceptual Plan: q **needed** $\rightarrow$ **mol** $CO_2 \rightarrow$ **g** CO_2

$$\frac{1 \text{ mol}}{393.5 \text{ kJ}} \qquad \frac{44.01 \text{ g}}{1 \text{ mol}}$$

Solution: $5.00 \times 10^2 \text{ kJ} \times \frac{1 \text{ mol } CO_2}{393.5 \text{ kJ}} \times \frac{44.01 \text{ g } CO_2}{1 \text{ mol } CO_2} = 55.9 \text{ g } CO_2$

Check: The units (g) are correct. The magnitude of the answer (~60) makes physical sense as there is a 1:1 mol ratio between the carbon burned and the carbon dioxide produced.

Thermal Energy Transfer

10.59 **Given:** silver block; $T_{Ag,\,i} = 58.5$ °C; 100.0 g water; $T_{H_2O,\,i} = 24.8$ °C; $T_f = 26.2$ °C **Find:** mass of silver block

Conceptual Plan: Pull C_s **values from table then** $H_2O: m, C_s, T_i, T_f \rightarrow q$ **and** $Ag: C_s, T_i, T_f \rightarrow m$

$$\text{Ag: } 0.235 \frac{\text{J}}{\text{g} \cdot {}^\circ\text{C}} \qquad H_2O\text{: } 4.18 \frac{\text{J}}{\text{g} \cdot {}^\circ\text{C}} \qquad q = mC_s(T_f - T_i) \text{ then set } q_{Ag} = -q_{H_2O}$$

Solution: $q = mC_s(T_f - T_i)$ substitute in values and set $q_{Ag} = -q_{H_2O}$.

$$q_{Ag} = m_{Ag}C_{s,\,Ag}(T_f - T_{Ag,\,i}) = m_{Ag} \times 0.235\,\frac{J}{g\cdot{}^\circ C} \times (26.2\,{}^\circ C - 58.5\,{}^\circ C) =$$

$$-q_{H_2O} = -m_{H_2O}C_{s,\,H_2O}(T_f - T_{H_2O,\,i}) = -100.0\,g \times 4.18\,\frac{J}{g\cdot{}^\circ C} \times (26.2\,{}^\circ C - 24.8\,{}^\circ C)$$

Rearrange to solve for m_{Ag}.

$$m_{Ag} \times \left(-7.5905\,\frac{J}{g}\right) = -585.2\,J \rightarrow m_{Ag} = \frac{-585.2\,J}{-7.5905\,\frac{J}{g}} = 77.0964\,g\,Ag = 77.1\,g\,Ag$$

Check: The units (g) are correct. The magnitude of the answer (77) makes physical sense because the specific heat capacity of water is much greater than the specific heat capacity of silver.

10.60 **Given:** 32.5 g iron rod; $T_{Fe,\,i} = 22.7\,{}^\circ C$; $T_{H_2O,\,i} = 63.2\,{}^\circ C$; $T_f = 59.5\,{}^\circ C$ **Find:** mass of water
Conceptual Plan: Pull C_s values from table then Fe : $m, C_s, T_i, T_f \rightarrow q$ and H_2O : $C_s, T_i, T_f \rightarrow m$

Fe: $0.449\,\frac{J}{g\cdot{}^\circ C}$ H_2O: $4.18\,\frac{J}{g\cdot{}^\circ C}$ $q = mC_s(T_f - T_i)$ then set $q_{Fe} = -q_{H_2O}$

Solution: $q = mC_{s,\,Ag}(T_f - T_i)$ substitute in values and set $q_{Fe} = -q_{H_2O}$.

$$q_{Fe} = m_{Fe}C_{s,\,Fe}(T_f - T_{Fe,\,i}) = 32.5\,g \times 0.449\,\frac{J}{g\cdot{}^\circ C} \times (59.5\,{}^\circ C - 22.7\,{}^\circ C) =$$

$$-q_{H_2O} = -m_{H_2O}C_{s,\,H_2O}(T_f - T_{H_2O,\,i}) = -m_{H_2O} \times 4.18\,\frac{J}{g\cdot{}^\circ C} \times (59.5\,{}^\circ C - 63.2\,{}^\circ C)$$

Rearrange to solve for m_{H_2O}.

$$537.004\,J = -m_{H_2O} \times \left(15.466\,\frac{J}{g}\right) \rightarrow m_{H_2O} = \frac{537.004\,J}{15.466\,\frac{J}{g}} = 34.72158\,g\,H_2O = 34.7\,g\,H_2O$$

Check: The units (g) are correct. The magnitude of the answer (35) makes physical sense because the specific heat capacity of water is much greater than the specific heat capacity of iron.

10.61 **Given:** 31.1 g gold; $T_{Au,\,i} = 69.3\,{}^\circ C$; 64.2 g water; $T_{H_2O,\,i} = 27.8\,{}^\circ C$ **Find:** T_f
Conceptual Plan: Pull C_s values from table then $m, C_s, T_i \rightarrow T_f$

Au: $0.128\,\frac{J}{g\cdot{}^\circ C}$ H_2O: $4.18\,\frac{J}{g\cdot{}^\circ C}$ $q = mC_s(T_f - T_i)$ then set $q_{Au} = -q_{H_2O}$

Solution: $q = mC_s(T_f - T_i)$ substitute in values and set $q_{Au} = -q_{H_2O}$.

$$q_{Au} = m_{Au}C_{s,\,Au}(T_f - T_{Au,\,i}) = 31.1\,g \times 0.128\,\frac{J}{g\cdot{}^\circ C} \times (T_f - 69.3\,{}^\circ C) =$$

$$-q_{H_2O} = -m_{H_2O}C_{s,\,H_2O}(T_f - T_{H_2O,\,i}) = -64.2\,g \times 4.18\,\frac{J}{g\cdot{}^\circ C} \times (T_f - 27.8\,{}^\circ C)$$

Rearrange to solve for T_f.

$$3.9808\,\frac{J}{{}^\circ C} \times (T_f - 69.3\,{}^\circ C) = -268.356\,\frac{J}{{}^\circ C} \times (T_f - 27.8\,{}^\circ C) \rightarrow$$

$$3.9808\,\frac{J}{{}^\circ C}\,T_f - 275.8694\,J = -268.356\,\frac{J}{{}^\circ C}\,T_f + 7460.2968\,J \rightarrow$$

$$268.356\,\frac{J}{{}^\circ C}\,T_f + 3.9808\,\frac{J}{{}^\circ C}\,T_f = 275.8694\,J + 7460.2968\,J \rightarrow 272.3368\,\frac{J}{{}^\circ C}\,T_f = 7736.1662\,J \rightarrow$$

$$T_f = \frac{7736.1662\,J}{272.3368\,\frac{J}{{}^\circ C}} = 28.4\,{}^\circ C$$

Check: The units (°C) are correct. The magnitude of the answer (28) makes physical sense because the heat transfer is dominated by the water (larger mass and larger specific heat capacity). The final temperature should be closer to the initial temperature of water than of gold.

10.62 **Given:** 2.85 g lead; $T_{Pb,\,i} = 10.3\ °C$; 7.55 g water; $T_{H_2O,\,i} = 52.3\ °C$ **Find:** T_f

Conceptual Plan: Pull C_s values from table then $m, C_s, T_i \rightarrow T_f$

Pb: $0.128 \frac{J}{g \cdot °C}$ H_2O: $4.18 \frac{J}{g \cdot °C}$ $q = mC_s(T_f - T_i)$ then set $q_{Pb} = -q_{H_2O}$

Solution: $q = mC_s(T_f - T_i)$ substitute in values and set $q_{Pb} = -q_{H_2O}$.

$$q_{Pb} = m_{Pb}C_{s,\,Pb}(T_f - T_{Pb,\,i}) = 2.85\ g \times 0.128 \frac{J}{g \cdot °C} \times (T_f - 10.3\ °C) =$$

$$-q_{H_2O} = -m_{H_2O}C_{s,\,H_2O}(T_f - T_{H_2O,\,i}) = -7.55\ g \times 4.18 \frac{J}{g \cdot °C} \times (T_f - 52.3\ °C)$$

Rearrange to solve for T_f.

$$0.3648 \frac{J}{°C} \times (T_f - 10.3\ °C) = -31.559 \frac{J}{°C} \times (T_f - 52.3\ °C) \rightarrow$$

$$0.3648 \frac{J}{°C} T_f - 3.75744\ J = -31.559 \frac{J}{°C} T_f + 1650.5357\ J \rightarrow$$

$$0.3648 \frac{J}{°C} T_f + 31.559 \frac{J}{°C} T_f = 3.75744\ J + 1650.5357\ J \rightarrow 31.9238 \frac{J}{°C} T_f = 1654.2931\ J \rightarrow$$

$$T_f = \frac{1654.2931\ J}{31.9238 \frac{J}{°C}} = 51.8\ °C$$

Check: The units (°C) are correct. The magnitude of the answer (52) makes physical sense because the heat transfer is dominated by the water (larger mass and larger specific heat capacity). The final temperature should be closer to the initial temperature of water than of lead.

10.63 **Given:** 6.15 g substance A; $T_{A,\,i} = 20.5\ °C$; 25.2 g substance B; $T_{B,\,i} = 52.7\ °C$; $C_s = 1.17\ J/g \cdot °C$; $T_f = 46.7\ °C$
Find: specific heat capacity of substance A

Conceptual Plan: A: $m, T_i, T_f \rightarrow q$ B: $m, C_s, T_i, T_f \rightarrow q$ and solve for C_s

$q = mC_s(T_f - T_i)$ then set $q_A = -q_B$

Solution: $q = mC_s(T_f - T_i)$ substitute in values and set $q_A = -q_B$.

$$q_A = m_A C_{s,\,A}(T_f - T_{A,\,i}) = 6.15\ g \times C_{s,\,A} \times (46.7\ °C - 20.5\ °C) =$$

$$-q_B = -m_B C_{s,\,B}(T_f - T_{B,\,i}) = -25.2\ g \times 1.17 \frac{J}{g \cdot °C} \times (46.7\ °C - 52.7\ °C)$$

Rearrange to solve for C_A.

$$C_{s,\,A} \times (161.13\ g \cdot °C) = 176.904\ J \rightarrow C_{s,\,A} = \frac{176.904\ J}{161.13\ g \cdot °C} = 1.097896 \frac{J}{g \cdot °C} = 1.10 \frac{J}{g \cdot °C}$$

Check: The units $(J/g \cdot °C)$ are correct. The magnitude of the answer (1) makes physical sense because the mass of substance B is greater than the mass of substance A by a factor of ~4.1 and the temperature change for substance A is greater than the temperature change of substance B by a factor of ~4.4; so the specific heat capacity of substance A will be slightly smaller.

10.64 **Given:** 2.74 g substance that may be gold; $T_{Au?,\,i} = 72.1\ °C$; 15.2 g water; $T_{H_2O,\,i} = 24.7\ °C$; $T_f = 26.3\ °C$
Find: specific heat capacity of substance that may be gold, and determine if it could be gold

Conceptual Plan: Pull C_s values from Table 10.2 substance : $m, T_i, T_f \rightarrow q$ H_2O: $m, C_s, T_i, T_f \rightarrow q$

Au: $0.128 \frac{J}{g \cdot °C}$ H_2O: $4.18 \frac{J}{g \cdot °C}$ $q = mC_s(T_f - T_i)$ then set $q_{Au?} = -q_{H_2O}$.

and solve for C_s

Solution: $q = mC_s(T_f - T_i)$ substitute in values and set $q_{Au?} = -q_{H_2O}$.

$$q_{Au?} = m_{Au?}C_{s,\,Au?}(T_f - T_{Au?,\,i}) = 2.74\ g \times C_{s,\,Au?} \times (26.3\ °C - 72.1\ °C) =$$

$$-q_{H_2O} = -m_{H_2O}C_{s,\,H_2O}(T_f - T_{H_2O,\,i}) = -15.2\ g \times 4.18 \frac{J}{g \cdot °C} \times (26.3\ °C - 24.7\ °C)$$

Rearrange to solve for $C_{s,\,Au?}$.

$$C_{s,\,Au?} \times (-125.492\ g \cdot °C) = -101.6576\ J \rightarrow C_{s,\,Au?} = \frac{-101.6576\ J}{-125.492\ g \cdot °C} = 0.8100724 \frac{J}{g \cdot °C} = 0.81 \frac{J}{g \cdot °C}$$

Because the specific heat capacity of gold is $0.128 \text{ J/g} \cdot °\text{C}$, this substance is not gold.

Check: The units $(\text{J/g} \cdot °\text{C})$ are correct. The magnitude of the answer (1) makes physical sense because the mass of the substance is much less than the mass of the water by a factor of ~5.5 and the temperature change for the substance is greater than the temperature change of the water by a factor of ~29; so the specific heat capacity of the unknown substance will be a factor of ~5.2 smaller than the water, but not as low as $0.128 \text{ J/g} \cdot °\text{C}$.

Calorimetry

10.65 $\Delta H_{rxn} = q_p$ and $\Delta E_{rxn} = q_V = \Delta H - P\Delta V$. Because combustion always involves expansions, expansions do work and therefore have a negative value. Combustions are always exothermic and therefore have a negative value. This means that ΔE_{rxn} is more negative than ΔH°_{rxn}; so A (−25.9 kJ) is the constant volume process, and B (−23.3 kJ) is the constant pressure process.

10.66 Constant volume conditions should be used. Because $\Delta E = q + w$ and $w = -P\Delta V$, at constant V, $w = 0$ and all of the energy is released as heat $(\Delta E_{rxn} = q_V)$. At constant P, $\Delta H_{rxn} = q_p$ and $\Delta E_{rxn} = q_p = \Delta H - P\Delta V$. Because combustions always involve expansions, expansions do work; so they have a negative value. Combustions are always exothermic, so they have a negative value. This means that ΔE_{rxn} is more negative than ΔH°_{rxn}, so more heat will be generated in a constant V process.

10.67 **Given:** 0.514 g biphenyl($C_{12}H_{10}$); bomb calorimeter; $T_i = 25.8\ °\text{C}$; $T_f = 29.4\ °\text{C}$; $C_{cal} = 5.86\ \text{kJ/°C}$ **Find:** ΔE_{rxn}

Conceptual Plan: $T_i, T_f \rightarrow \Delta T$ **then** $\Delta T, C_{cal} \rightarrow q_{cal} \rightarrow q_{rxn}$ **then g** $C_{12}H_{10} \rightarrow$ **mol** $C_{12}H_{10}$

$\Delta T = T_f - T_i$ $\quad q_{cal} = C_{cal}\Delta T \quad q_{cal} = -q_{rxn} \quad \frac{1 \text{ mol}}{154.20 \text{ g}}$

then q_{rxn}, **mol** $C_{12}H_{10} \rightarrow \Delta E_{rxn}$

$\Delta E_{rxn} = \frac{q_V}{\text{mol } C_{12}H_{10}}$

Solution: $\Delta T = T_f - T_i = 29.4\ °\text{C} - 25.8\ °\text{C} = 3.6\ °\text{C}$ then $q_{cal} = C_{cal}\Delta T = 5.86 \frac{\text{kJ}}{°\text{C}} \times 3.6\ °\text{C} = 21.096 \text{ kJ}$

then $q_{cal} = -q_{rxn} = -21.096$ kJ and $0.514 \text{ g } C_{12}H_{10} \times \frac{1 \text{ mol } C_{12}H_{10}}{154.20 \text{ g } C_{12}H_{10}} = 0.00333333 \text{ mol } C_{12}H_{10}$ then

$$\Delta E_{rxn} = \frac{q_V}{\text{mol } C_{12}H_{10}} = \frac{-21.096 \text{ kJ}}{0.00333333 \text{ mol } C_{12}H_{10}} = -6.3 \times 10^3 \text{ kJ/mol}$$

Check: The units (kJ/mol) are correct. The magnitude of the answer (−6000) makes physical sense because a large amount of heat is generated from a very small amount of biphenyl.

10.68 **Given:** 1.025 g naphthalene ($C_{10}H_8$); bomb calorimeter; $T_i = 24.25\ °\text{C}$; $T_f = 32.33\ °\text{C}$; $C_{cal} = 5.11\ \text{kJ/°C}$
Find: ΔE_{rxn}

Conceptual Plan: $T_i, T_f \rightarrow \Delta T$ **then** $\Delta T, C_{cal} \rightarrow q_{cal} \rightarrow q_{rxn}$ **then g** $C_{10}H_8 \rightarrow$ **mol** $C_{10}H_8$

$\Delta T = T_f - T_i$ $\quad q_{cal} = C_{cal}\Delta T \quad q_{cal} = -q_{rxn} \quad \frac{1 \text{ mol}}{128.16 \text{ g}}$

then q_{rxn}, **mol** $C_{10}H_8 \rightarrow \Delta E_{rxn}$

$\Delta E_{rxn} = \frac{q_V}{\text{mol } C_{10}H_8}$

Solution: $\Delta T = T_f - T_i = 32.33\ °\text{C} - 24.25\ °\text{C} = 8.08\ °\text{C}$ then

$q_{cal} = C_{cal}\Delta T = 5.11 \frac{\text{kJ}}{°\text{C}} \times 8.08\ °\text{C} = 41.2888$ kJ then $q_{cal} = -q_{rxn} = -41.2888$ kJ and

$1.025 \text{ g } C_{10}H_8 \times \frac{1 \text{ mol } C_{10}H_8}{128.16 \text{ g } C_{10}H_8} = 0.007997815 \text{ mol } C_{10}H_8$ then

$$\Delta E_{rxn} = \frac{q_V}{\text{mol } C_{10}H_8} = \frac{-41.2888 \text{ kJ}}{0.007997815 \text{ mol } C_{10}H_8} = -5.16 \times 10^3 \text{ kJ/mol}$$

Check: The units (kJ/mol) are correct. The magnitude of the answer (−5000) makes physical sense because such a large amount of heat is generated from a very small amount of naphthalene.

10.69 **Given:** 0.103 g zinc; coffee-cup calorimeter; $T_i = 22.5\ °C$; $T_f = 23.7\ °C$; 50.0 mL solution; $d(\text{solution}) = 1.0\ g/mL$; $C_{s,\,soln} = 4.18\ kJ/g \cdot °C$ **Find:** ΔH_{rxn}

Conceptual Plan: $T_i, T_f \rightarrow \Delta T$ **and mL soln** $\rightarrow$ **g soln then** $\Delta T, C_{s,\,soln} \rightarrow q_{cal} \rightarrow q_{rxn}$ **then**

$\Delta T = T_f - T_i$ $\quad \frac{1.0\ g}{1.0\ mL}$ $\quad q_{soln} = mC_{s,\,soln}\,\Delta T \quad q_{soln} = -q_{rxn}$

g Zn $\rightarrow$ **mol Zn then** q_{rxn}, **mol Zn** $\rightarrow \Delta H_{rxn}$

$\frac{1\ mol}{65.37\ g}$ $\quad \Delta H_{rxn} = \frac{q_p}{mol\ Zn}$

Solution: $\Delta T = T_f - T_i = 23.7\ °C - 22.5\ °C = 1.2\ °C$ and $50.0\ mL \times \frac{1.0\ g}{1.0\ mL} = 50.0\ g$

then $q_{soln} = mC_{s,\,soln}\,\Delta T = 50.0\ g \times 4.18\ \frac{J}{g \cdot °C} \times 1.2\ °C = 250.8\ J$ then

$q_{soln} = -q_{rxn} = -250.8\ J$ and $0.103\ g\ Zn \times \frac{1\ mol\ Zn}{65.37\ g\ Zn} = 0.00157565\ mol\ Zn$ then

$$\Delta H_{rxn} = \frac{q_p}{mol\ Zn} = \frac{-250.8\ J}{0.00157565\ mol\ Zn} = -1.6 \times 10^5\ J/mol = -1.6 \times 10^2\ kJ/mol$$

Check: The units (kJ/mol) are correct. The magnitude of the answer (−160) makes physical sense because a large amount of heat is generated from a very small amount of zinc.

10.70 **Given:** 1.25 g NH_4NO_3; coffee-cup calorimeter; $T_i = 25.8\ °C$; $T_f = 21.9\ °C$; 25.0 mL solution; $d(\text{solution}) = 1.0\ g/mL$; $C_{s,\,soln} = 4.18\ kJ/g\ °C$ **Find:** ΔH_{rxn}

Conceptual Plan: $T_i, T_f \rightarrow \Delta T$ **and mL soln** $\rightarrow$ **g soln then** $\Delta T, C_{s,\,soln} \rightarrow q_{cal} \rightarrow q_{rxn}$ **then**

$\Delta T = T_f - T_i$ $\quad \frac{1.0\ g}{1.0\ mL}$ $\quad q_{soln} = m\,C_{s,\,soln}\,\Delta T \quad q_{soln} = -q_{rxn}$

g $NH_4NO_3 \rightarrow$ **mol** NH_4NO_3 **then** q_{rxn}, **mol** $NH_4NO_3 \rightarrow \Delta H_{rxn}$

$\frac{1\ mol}{80.05\ g}$ $\quad \Delta H_{rxn} = \frac{q_p}{mol\ NH_4NO_3}$

Solution: $\Delta T = T_f - T_i = 21.9\ °C - 25.8\ °C = -3.9\ °C$ and $25.0\ mL \times \frac{1.0\ g}{1.0\ mL} = 25.0\ g$

then $q_{soln} = mC_{s,\,soln}\Delta T = 25.0\ g \times 4.18\ \frac{J}{g \cdot °C} \times (-3.9\ °C) = -407.55\ J$ then

$q_{soln} = -q_{rxn} = 407.55\ J$ and $1.25\ g\ NH_4NO_3 \times \frac{1\ mol\ NH_4NO_3}{80.05\ g\ NH_4NO_3} = 0.0156152\ mol\ NH_4NO_3$ then

$$\Delta H_{rxn} = \frac{q_P}{mol\ NH_4NO_3} = \frac{407.55\ J}{0.0156152\ mol\ NH_4NO_3} = 2.6 \times 10^4\ J/mol = 26\ kJ/mol$$

Check: The units (kJ/mol) are correct. The magnitude of the answer (26) makes physical sense because a small amount of heat is absorbed.

Quantitative Relationships Involving ΔH and Hess's Law

10.71 (a) Because $A + B \rightarrow 2C$ has ΔH_1, $2C \rightarrow A + B$ will have a $\Delta H_2 = -\Delta H_1$. When the reaction direction is reversed, it changes from exothermic to endothermic (or vice versa); so the sign of ΔH changes.

(b) Because $A + \frac{1}{2}B \rightarrow C$ has ΔH_1, $2A + B \rightarrow 2C$ will have a $\Delta H_2 = 2\,\Delta H_1$. When the reaction amount doubles, the amount of heat (or ΔH) doubles.

(c) Because $A \rightarrow B + 2C$ has ΔH_1, $\frac{1}{2}A \rightarrow \frac{1}{2}B + C$ will have a $\Delta H_1' = \frac{1}{2}\Delta H_1$. When the reaction amount is cut in half, the amount of heat (or ΔH) is cut in half. Then $\frac{1}{2}B + C \rightarrow \frac{1}{2}A$ will have a $\Delta H_2 = -H_1' = -\frac{1}{2}\Delta H_1$. When the reaction direction is reversed, it changes from exothermic to endothermic (or vice versa); so the sign of ΔH changes.

10.72 (a) Because $A + 2B \rightarrow C + 3D$ has $\Delta H = 155\ kJ$, $3A + 6B \rightarrow 3C + 9D$ will have a $\Delta H' = 3\,\Delta H = 3(155\ kJ) = 465\ kJ$. When the reaction amount triples, the amount of heat (or ΔH) triples.

(b) Because $A + 2B \rightarrow C + 3D$ has $\Delta H = 155\ kJ$, $C + 3D \rightarrow 3A + 6B$ will have a $\Delta H' = -\Delta H = -155\ kJ$. When the reaction direction is reversed, it changes from endothermic to exothermic; so the sign of ΔH changes.

(c) Because $A + 2B \rightarrow C + 3D$ has $\Delta H = 155$ kJ, $\frac{1}{2}A + B \rightarrow \frac{1}{2}C + \frac{3}{2}D$ will have a $\Delta H' = \frac{1}{2}\Delta H = \frac{1}{2}(155\text{ kJ}) = 77.5$ kJ. When the reaction amount is cut in half, the amount of heat (or ΔH) is cut in half. Then $\frac{1}{2}C + \frac{3}{2}D \rightarrow \frac{1}{2}A + B$ will have a $\Delta H'' = -H' = -77.5$ kJ. When the reaction direction is reversed, the sign of it changes from endothermic to exothermic; so the sign of ΔH changes.

10.73 Because the first reaction has Fe_2O_3 as a product and the reaction of interest has it as a reactant, we need to reverse the first reaction. When the reaction direction is reversed, ΔH changes.

$Fe_2O_3(s) \rightarrow 2\,Fe(s) + 3/2\,O_2(g)$ $\quad \Delta H = +824.2$ kJ

Because the second reaction has 1 mole CO as a reactant and the reaction of interest has 3 moles of CO as a reactant, we need to multiply the second reaction and the ΔH by 3.

$3[CO(g) + 1/2\,O_2(g) \rightarrow CO_2(g)]$ $\quad \Delta H = 3(-282.7\text{ kJ}) = -848.1$ kJ

Hess's law states that the ΔH of the net reaction is the sum of the ΔH of the steps.

The rewritten reactions are as follows:

$Fe_2O_3(s) \rightarrow 2\,Fe(s) + \cancel{3/2\,O_2(g)}$ $\quad \Delta H = +824.2$ kJ
$3\,CO(g) + \cancel{3/2\,O_2(g)} \rightarrow 3\,CO_2(g)$ $\quad \Delta H = -848.1$ kJ

$Fe_2O_3(s) + 3\,CO(g) \rightarrow 2\,Fe(s) + 3\,CO_2(g)$ $\quad \Delta H_{rxn} = -23.9$ kJ

10.74 Because the first reaction has $CaCO_3$ as a product and the reaction of interest has it as a product, we simply write the first reaction and the ΔH is unchanged.

$Ca(s) + CO_2(g) + 1/2\,O_2(g) \rightarrow CaCO_3(s)]$ $\quad \Delta H = -812.8$ kJ

Because the second reaction has 2 moles CaO as a product and the reaction of interest has 1 mole of CaO as a reactant, we need to reverse the direction of the reaction of the second reaction and multiply it by $\frac{1}{2}$. The sign of the ΔH in the second reaction is changed and is multiplied by $\frac{1}{2}$.

$1/2[2\,CaO(s) \rightarrow 2\,Ca(s) + O_2(g)]$ $\quad \Delta H = -1/2(-1269.8\text{ kJ}) = +634.9$ k

Hess's law states that the ΔH of the net reaction is the sum of the ΔH of the steps.

The rewritten reactions are as follows:

$\cancel{Ca(s)} + CO_2(g) + \cancel{1/2\,O_2(g)} \rightarrow CaCO_3(s)$ $\quad \Delta H = -812.8$ kJ
$CaO(s) \rightarrow \cancel{Ca(s)} + \cancel{1/2\,O_2(g)}$ $\quad \Delta H = +634.9$ kJ

$CaO(s) + CO_2(g) \rightarrow CaCO_3(s)$ $\quad \Delta H_{rxn} = -177.9$ kJ

10.75 Because the first reaction has C_5H_{12} as a reactant and the reaction of interest has it as a product, we need to reverse the first reaction. When the reaction direction is reversed, ΔH changes.

$5\,CO_2(g) + 6\,H_2O(g) \rightarrow C_5H_{12}(l) + 8\,O_2(g)$ $\quad \Delta H = +3244.8$ kJ

Because the second reaction has 1 mole C as a reactant and the reaction of interest has 5 moles of C as a reactant, we need to multiply the second reaction and the ΔH by 5.

$5[C(s) + O_2(g) \rightarrow CO_2(g)]$ $\quad \Delta H = 5(-393.5\text{ kJ}) = -1967.5$ kJ

Because the third reaction has 2 moles H_2 as a reactant and the reaction of interest has 6 moles of H_2 as a reactant, we need to multiply the third reaction and the ΔH by 3.

$3[2\,H_2(g) + O_2(g) \rightarrow 2\,H_2O(g)]$ $\quad \Delta H = 3(-483.5\text{ kJ}) = -1450.5$ kJ

Hess's law states that the ΔH of the net reaction is the sum of the ΔH of the steps. The rewritten reactions are as follows:

$\cancel{5\,CO_2(g)} + \cancel{6\,H_2O(g)} \rightarrow C_5H_{12}(l) + \cancel{8\,O_2(g)}$ $\quad \Delta H = +3244.8$ kJ
$5\,C(s) + \cancel{5\,O_2(g)} \rightarrow \cancel{5\,CO_2(g)}$ $\quad \Delta H = -1967.5$ kJ
$6\,H_2(g) + \cancel{3\,O_2(g)} \rightarrow \cancel{6\,H_2O(g)}$ $\quad \Delta H = -1450.5$ kJ

$5\,C(s) + 6\,H_2(g) \rightarrow C_5H_{12}(l)$ $\quad \Delta H_{rxn} = -173.2$ kJ

10.76 Because the first reaction has CH_4 as a product and the reaction of interest has it as a reactant, we need to reverse the first reaction. When the reaction direction is reversed, ΔH changes.

$CH_4(g) \rightarrow C(s) + 2\,H_2(g)$ $\quad \Delta H = +74.6$ kJ

Because the first reaction has CCl_4 as a product and the reaction of interest has it as a product, we simply write the first reaction and the ΔH is unchanged.

$C(s) + 2\,Cl_2(g) \rightarrow CCl_4(g)$ $\quad \Delta H = -95.7$ kJ

Because the third reaction has 2 moles HCl as a product and the reaction of interest has 4 moles of HCl as a product, we need to multiply the third reaction and the ΔH by 2.

$2[H_2(g) + Cl_2(g) \rightarrow 2\,HCl(g)]$ $\quad \Delta H = 2(-92.3\text{ kJ}) = -184.6$ kJ

Hess's law states that the ΔH of the net reaction is the sum of the ΔH of the steps.
The rewritten reactions are as follows:

$CH_4(g) \rightarrow \cancel{C(s)} + \cancel{2H_2(g)}$ $\Delta H = +74.6$ kJ
$\cancel{C(s)} + 2Cl_2(g) \rightarrow CCl_4(g)$ $\Delta H = -95.7$ kJ
$\cancel{2H_2(g)} + 2Cl_2(g) \rightarrow 4HCl(g)$ $\Delta H = -184.6$ kJ

$CH_4(g) + 4Cl_2(g) \rightarrow CCl_4(g) + 4HCl(g)$ $\Delta H_{rxn} = -205.7$ kJ

Using Bond Energies to Calculate ΔH_{rxn}

10.77 Rewrite the reaction using the Lewis structures of the molecules involved.

$$H_2C{=}CH_2 + H{-}H \longrightarrow H_3C{-}CH_3$$

Determine which bonds are broken in the reaction and sum the bond energies of the following:

$\Sigma(\Delta H\text{'s bonds broken})$
$= 4\text{ mol}(C{-}H) + 1\text{ mol}(C{=}C) + 1\text{ mol}(H{-}H)$
$= 4\text{ mol}(414\text{ kJ/mol}) + 1\text{ mol}(611\text{ kJ/mol}) + 1\text{ mol}(436\text{ kJ/mol})$
$= 2703\text{ kJ/mol}$

Determine which bonds are formed in the reaction and sum the negatives of the bond energies of the following:

$\Sigma(-\Delta H\text{'s bonds formed})$
$= -6\text{ mol}(C{-}H) - 1\text{ mol}(C{-}C)$
$= -6\text{ mol}(414\text{ kJ/mol}) - 1\text{ mol}(347\text{ kJ/mol})$
$= -2831\text{ kJ/mol}$

Find ΔH_{rxn} by summing the results of the two steps.

$\Delta H_{rxn} = \Sigma(\Delta H\text{'s bonds broken}) + \Sigma(\Delta H{-}\Delta H\text{'s bonds formed})$
$= 2703\text{ kJ/mol} - 2831\text{ kJ/mol}$
$= -128\text{ kJ/mol}$

10.78 Rewrite the reaction using the Lewis structures of the molecules involved.

$$H_3C{-}CH_2{-}\ddot{\underset{..}{O}}{-}H + 3\,\ddot{\underset{..}{O}}{=}\ddot{\underset{..}{O}} \longrightarrow 2\,\ddot{\underset{..}{O}}{=}C{=}\ddot{\underset{..}{O}} + 3\,H{-}\ddot{\underset{..}{O}}{-}H$$

Determine which bonds are broken in the reaction and sum the bond energies of the following:

$\Sigma(\Delta H\text{'s bonds broken})$
$= 5\text{ mol}(C{-}H) + 1\text{ mol}(C{-}C) + 1\text{ mol}(C{-}O) + 1\text{ mol}(O{-}H) + 3\text{ mol}(O{=}O)$
$= 5\text{ mol}(414\text{ kJ/mol}) + 1\text{ mol}(347\text{ kJ/mol}) + 1\text{ mol}(360.\text{ kJ/mol}) + 1\text{ mol}(464\text{ kJ/mol}) + 3\text{ mol}(498\text{kJ/mol})$
$= 4735\text{ kJ/mol}$

Determine which bonds are formed in the reaction and sum the negatives of the bond energies of the following:

$\Sigma(-\Delta H\text{'s bonds formed})$
$= -4\text{ mol}(C{=}O) - 6\text{ mol}(O{-}H)$
$= -4\text{ mol}(799\text{ kJ/mol}) - 6\text{ mol}(464\text{ kJ/mol})$
$= -5980\text{ kJ/mol}$

Find ΔH_{rxn} by summing the results of the two steps.

$\Delta H_{rxn} = \Sigma(\Delta H\text{'s bonds broken}) + \Sigma(-\Delta H\text{'s bonds formed})$
$= 4735\text{ kJ/mol} - 5980\text{ kJ/mol}$
$= -1245\text{ kJ/mol}$

10.79 Rewrite the reaction using the Lewis structures of the molecules involved.

$$C + 2H{-}\ddot{\underset{..}{O}}{-}H \longrightarrow 2H{-}H + \ddot{\underset{..}{O}}{=}C{=}\ddot{\underset{..}{O}}$$

Determine which bonds are broken in the reaction and sum the bond energies of the following:

$$\begin{aligned}&\Sigma(\Delta H\text{'s bonds broken})\\&= 4(\text{O—H})\\&= 4(464 \text{ kJ/mol})\\&= 1856 \text{ kJ/mol}\end{aligned}$$

Determine which bonds are formed in the reaction and sum the negatives of the bond energies of the following:

$$\begin{aligned}&\Sigma(-\Delta H\text{'s bonds formed})\\&= -2(\text{C=O}) - 2(\text{H—H})\\&= -2(799 \text{ kJ/mol}) - 2(436 \text{ kJ/mol})\\&= -2470 \text{ kJ/mol}\end{aligned}$$

Find ΔH_{rxn} by summing the results of the two steps.

$$\begin{aligned}\Delta H_{rxn} &= \Sigma(\Delta H\text{'s bonds broken}) + \Sigma(-\Delta H\text{'s bonds formed})\\&= 1856 \text{ kJ/mol} - 2470 \text{ kJ/mol}\\&= -614 \text{ kJ/mol}\end{aligned}$$

10.80 Rewrite the reaction using the Lewis structures of the molecules involved.

$$\cdot\ddot{\underset{\cdot\cdot}{\text{O}}}\text{—H} + \text{:}\ddot{\underset{\cdot\cdot}{\text{F}}}\text{—}\underset{\text{:Cl:}}{\overset{\text{:F:}}{\text{C}}}\text{—}\ddot{\underset{\cdot\cdot}{\text{Cl}}}\text{:} \longrightarrow \text{H—}\ddot{\underset{\cdot\cdot}{\text{O}}}\text{—}\ddot{\underset{\cdot\cdot}{\text{F}}}\text{:} + \cdot\underset{\text{:Cl:}}{\overset{\text{:F:}}{\text{C}}}\text{—}\ddot{\underset{\cdot\cdot}{\text{Cl}}}\text{:}$$

Because the C—F bond energy is not given, it is necessary to look it up (online). The C—F bond energy is 485 kJ/mol.

Determine which bonds are broken in the reaction and sum the bond energies of the following:

$$\begin{aligned}&\Sigma(\Delta H\text{'s bonds broken})\\&= (\text{O—H}) + 2(\text{C—F}) + 2(\text{C—Cl})\\&= (464 \text{ kJ/mol}) + 2(485 \text{ kJ/mol}) + 2(339 \text{ kJ/mol})\\&= 2112 \text{ kJ/mol}\end{aligned}$$

Determine which bonds are formed in the reaction and sum the negatives of the bond energies of the following:

$$\begin{aligned}&\Sigma(-\Delta H\text{'s bonds formed})\\&= -(\text{O—H}) - (\text{O—F}) - (\text{C—F}) - 2(\text{C—Cl})\\&= -(464 \text{ kJ/mol}) - (190 \text{ kJ/mol}) - (485 \text{ kJ/mol}) - 2(339 \text{ kJ/mol})\\&= -1817 \text{ kJ/mol}\end{aligned}$$

Find ΔH_{rxn} by summing the results of the two steps.

$$\begin{aligned}\Delta H_{rxn} &= \Sigma(\Delta H\text{'s bonds broken}) + \Sigma(-\Delta H\text{'s bonds formed})\\&= 2112 \text{ kJ/mol} - 1817 \text{ kJ/mol}\\&= 295 \text{ kJ/mol}\end{aligned}$$

From the bond energies, we find that the ΔH_{rxn} is positive. Therefore, you need to put more energy into the reaction than you get out of the reaction, and this makes the reaction less likely to happen.

Enthalpies of Formation and ΔH

10.81 (a) $\frac{1}{2}N_2(g) + \frac{3}{2}H_2(g) \rightarrow NH_3(g)$ $\quad \Delta H_f^\circ = -45.9 \text{ kJ/mol}$

(b) $C(s) + O_2(g) \rightarrow CO_2(g)$ $\quad \Delta H_f^\circ = -393.5 \text{ kJ/mol}$

(c) $2\,Fe(s) + \frac{3}{2}O_2(g) \rightarrow Fe_2O_3(s)$ $\quad \Delta H_f^\circ = -824.2 \text{ kJ/mol}$

(d) $C(s) + 2\,H_2(g) \rightarrow CH_4(g)$ $\quad \Delta H_f^\circ = -74.6 \text{ kJ/mol}$

10.82 (a) $\frac{1}{2}N_2(g) + O_2(g) \rightarrow NO_2(g)$ $\Delta H_f^\circ = 33.2$ kJ/mol

(b) $Mg(s) + C(s) + \frac{3}{2}O_2(g) \rightarrow MgCO_3(s)$ $\Delta H_f^\circ = -1095.8$ kJ/mol

(c) $2\,C(s) + 2\,H_2(g) \rightarrow C_2H_4(g)$ $\Delta H_f^\circ = 52.4$ kJ/mol

(d) $C(s) + 2\,H_2(g) + \frac{1}{2}O_2(g) \rightarrow CH_3OH(l)$ $\Delta H_f^\circ = -238.6$ kJ/mol

10.83 **Given:** $N_2H_4(l) + N_2O_4(g) \rightarrow 2\,N_2O(g) + 2\,H_2O(g)$ **Find:** ΔH°_{rxn}
Conceptual Plan: $\Delta H^\circ_{rxn} = \sum n_p \Delta H_f^\circ(products) - \sum n_r \Delta H_f^\circ(reactants)$
Solution:

Reactant/Product	ΔH_f° (kJ/mol from Appendix IVB)
$N_2H_4(l)$	50.6
$N_2O_4(g)$	9.16
$N_2O(g)$	81.6
$H_2O(g)$	−241.8

Be sure to pull data for the correct formula and phase.

$$\begin{aligned}\Delta H^\circ_{rxn} &= \sum n_p \Delta H_f^\circ(products) - \sum n_r \Delta H_f^\circ(reactants)\\ &= [2(\Delta H_f^\circ(N_2O(g))) + 2(\Delta H_f^\circ(H_2O(g)))] - [1(\Delta H_f^\circ(N_2H_4(l))) + 1(\Delta H_f^\circ(N_2O_4(g)))]\\ &= [2(81.6\text{ kJ}) + 2(-241.8\text{ kJ})] - [1(50.6\text{ kJ}) + 1(9.16\text{ kJ})]\\ &= [-320.4\text{ kJ}] - [59.8\text{ kJ}]\\ &= -380.2\text{ kJ}\end{aligned}$$

Check: The units (kJ) are correct. The answer is negative, which means that the reaction is exothermic. The answer is dominated by the negative heat of formation of water.

10.84 **Given:** $C_5H_{12}(l) + 8\,O_2(g) \rightarrow 5\,CO_2(g) + 6\,H_2O(g)$ **Find:** ΔH°_{rxn}
Conceptual Plan: $\Delta H^\circ_{rxn} = \sum n_p \Delta H_f^\circ(products) - \sum n_r \Delta H_f^\circ(reactants)$
Solution:

Reactant/Product	ΔH_f° (kJ/mol from Appendix IVB)
$C_5H_{12}(l)$	−146.8
$O_2(g)$	0.0
$CO_2(g)$	−393.5
$H_2O(g)$	−241.8

Be sure to pull data for the correct formula and phase.

$$\begin{aligned}\Delta H^\circ_{rxn} &= \sum n_p \Delta H_f^\circ(products) - \sum n_r \Delta H_f^\circ(reactants)\\ &= [5(\Delta H_f^\circ(CO_2(g))) + 6(\Delta H_f^\circ(H_2O(g)))] - [1(\Delta H_f^\circ(C_5H_{12}(l))) + 8(\Delta H_f^\circ(O_2(g)))]\\ &= [5(-393.5\text{ kJ}) + 6(-241.8\text{ kJ})] - [1(-146.8\text{ kJ}) + 8(0.0\text{ kJ})]\\ &= [-3418.3\text{ kJ}] - [-146.8\text{ kJ}]\\ &= -3271.5\text{ kJ}\end{aligned}$$

Check: The units (kJ) are correct. The answer is negative, which means that the reaction is exothermic, which is typical for combustion reactions.

10.85 (a) **Given:** $C_2H_4(g) + H_2(g) \rightarrow C_2H_6(g)$ **Find:** ΔH°_{rxn}
Conceptual Plan: $\Delta H^\circ_{rxn} = \sum n_p \Delta H_f^\circ(products) - \sum n_r \Delta H_f^\circ(reactants)$
Solution:

Reactant/Product	ΔH_f° (kJ/mol from Appendix IVB)
$C_2H_4(g)$	52.4
$H_2(g)$	0.0
$C_2H_6(g)$	−84.68

Be sure to pull data for the correct formula and phase.

$$\begin{aligned}\Delta H^\circ_{rxn} &= \sum n_p \Delta H^\circ_f(products) - \sum n_r \Delta H^\circ_f(reactants)\\ &= [1(\Delta H^\circ_f(C_2H_6(g)))] - [1(\Delta H^\circ_f(C_2H_4(g))) + 1(\Delta H^\circ_f(H_2(g)))]\\ &= [1(-84.68\text{ kJ})] - [1(52.4\text{ kJ}) + 1(0.0\text{ kJ})]\\ &= [-84.68\text{ kJ}] - [52.4\text{ kJ}]\\ &= -137.1\text{ kJ}\end{aligned}$$

Check: The units (kJ) are correct. The answer is negative, which means that the reaction is exothermic. Both hydrocarbon terms are negative, so the final answer is negative.

(b) **Given:** $CO(g) + H_2O(g) \rightarrow H_2(g) + CO_2(g)$ **Find:** ΔH°_{rxn}
Conceptual Plan: $\Delta H^\circ_{rxn} = \sum n_p \Delta H^\circ_f(products) - \sum n_r \Delta H^\circ_f(reactants)$
Solution:

Reactant/Product	ΔH°_f (kJ/mol from Appendix IVB)
$CO(g)$	−110.5
$H_2O(g)$	−241.8
$H_2(g)$	0.0
$CO_2(g)$	−393.5

Be sure to pull data for the correct formula and phase.

$$\begin{aligned}\Delta H^\circ_{rxn} &= \sum n_p \Delta H^\circ_f(products) - \sum n_r \Delta H^\circ_f(reactants)\\ &= [1(\Delta H^\circ_f(H_2(g))) + 1(\Delta H^\circ_f(CO_2(g)))] - [1(\Delta H^\circ_f(CO(g))) + 1(\Delta H^\circ_f(H_2O(g)))]\\ &= [1(0.0\text{ kJ}) + 1(-393.5\text{ kJ})] - [1(-110.5\text{ kJ}) + 1(-241.8\text{ kJ})]\\ &= [-393.5\text{ kJ}] - [-352.3\text{ kJ}]\\ &= -41.2\text{ kJ}\end{aligned}$$

Check: The units (kJ) are correct. The answer is negative, which means that the reaction is exothermic.

(c) **Given:** $3\ NO_2(g) + H_2O(l) \rightarrow 2\ HNO_3(aq) + NO(g)$ **Find:** ΔH°_{rxn}
Conceptual Plan: $\Delta H^\circ_{rxn} = \sum n_p \Delta H^\circ_f(products) - \sum n_r \Delta H^\circ_f(reactants)$
Solution:

Reactant/Product	ΔH°_f (kJ/mol from Appendix IVB)
$NO_2(g)$	33.2
$H_2O(l)$	−285.8
$HNO_3(aq)$	−207
$NO(g)$	91.3

Be sure to pull data for the correct formula and phase.

$$\begin{aligned}\Delta H^\circ_{rxn} &= \sum n_p \Delta H^\circ_f(products) - \sum n_r \Delta H^\circ_f(reactants)\\ &= [2(\Delta H^\circ_f(HNO_3(aq))) + 1(\Delta H^\circ_f(NO(g)))] - [3(\Delta H^\circ_f(NO_2(g))) + 1(\Delta H^\circ_f(H_2O(l)))]\\ &= [2(-207\text{ kJ}) + 1(91.3\text{ kJ})] - [3(33.2\text{ kJ}) + 1(-285.8\text{ kJ})]\\ &= [-32\underline{2}.7\text{ kJ}] - [-186.2\text{ kJ}]\\ &= -137\text{ kJ}\end{aligned}$$

Check: The units (kJ) are correct. The answer is negative, which means that the reaction is exothermic.

(d) **Given:** $Cr_2O_3(s) + 3\ CO(g) \rightarrow 2\ Cr(s) + 3\ CO_2(g)$ **Find:** ΔH°_{rxn}
Conceptual Plan: $\Delta H^\circ_{rxn} = \sum n_p \Delta H^\circ_f(products) - \sum n_r \Delta H^\circ_f(reactants)$
Solution:

Reactant/Product	ΔH°_f (kJ/mol from Appendix IVB)
$Cr_2O_3(s)$	−1139.7
$CO(g)$	−110.5
$Cr(s)$	0.0
$CO_2(g)$	−393.5

Be sure to pull data for the correct formula and phase.

$$\begin{aligned}\Delta H^\circ_{rxn} &= \sum n_p \Delta H^\circ_f(products) - \sum n_r \Delta H^\circ_f(reactants)\\ &= [2(\Delta H^\circ_f(Cr(s))) + 3(\Delta H^\circ_f(CO_2(g)))] - [1(\Delta H^\circ_f(Cr_2O_3(s))) + 3(\Delta H^\circ_f(CO(g)))]\\ &= [2(0.0\text{ kJ}) + 3(-393.5\text{ kJ})] - [1(-1139.7\text{ kJ}) + 3(-110.5\text{ kJ})]\\ &= [-1180.5\text{ kJ}] - [-1471.2\text{ kJ}]\\ &= 290.7\text{ kJ}\end{aligned}$$

Check: The units (kJ) are correct. The answer is positive, which means that the reaction is endothermic.

10.86 (a) **Given:** $2\ H_2S(g) + 3\ O_2(g) \rightarrow 2\ H_2O(l) + 2\ SO_2(g)$ **Find:** ΔH°_{rxn}
Conceptual Plan: $\Delta H^\circ_{rxn} = \sum n_p \Delta H^\circ_f(products) - \sum n_r \Delta H^\circ_f(reactants)$
Solution:

Reactant/Product	ΔH°_f (kJ/mol from Appendix IVB)
$H_2S(g)$	−20.6
$O_2(g)$	0.0
$H_2O(l)$	−285.8
$SO_2(g)$	−296.8

Be sure to pull data for the correct formula and phase.

$$\begin{aligned}\Delta H^\circ_{rxn} &= \sum n_p \Delta H^\circ_f(products) - \sum n_r \Delta H^\circ_f(reactants)\\ &= [2(\Delta H^\circ_f(H_2O(l))) + 2(\Delta H^\circ_f(SO_2(g)))] - [2(\Delta H^\circ_f(H_2S(g))) + 3(\Delta H^\circ_f(O_2(g)))]\\ &= [2(-285.8\text{ kJ}) + 2(-296.8\text{ kJ})] - [2(-20.6\text{ kJ}) + 3(0.0\text{ kJ})]\\ &= [-1165.2\text{ kJ}] - [-41.2\text{ kJ}]\\ &= -1124.0\text{ kJ}\end{aligned}$$

Check: The units (kJ) are correct. The answer is negative, which means that the reaction is exothermic.

(b) **Given:** $SO_2(g) + 1/2\ O_2(g) \rightarrow SO_3(g)$ **Find:** ΔH°_{rxn}
Conceptual Plan: $\Delta H^\circ_{rxn} = \sum n_p \Delta H^\circ_f(products) - \sum n_r \Delta H^\circ_f(reactants)$
Solution:

Reactant/Product	ΔH°_f (kJ/mol from Appendix IVB)
$SO_2(g)$	−296.8
$O_2(g)$	0.0
$SO_3(g)$	−395.7

Be sure to pull data for the correct formula and phase.

$$\begin{aligned}\Delta H^\circ_{rxn} &= \sum n_p \Delta H^\circ_f(products) - \sum n_r \Delta H^\circ_f(reactants)\\ &= [1(\Delta H^\circ_f(SO_3(g)))] - [1(\Delta H^\circ_f(SO_2(g))) + 1/2(\Delta H^\circ_f(O_2(g)))]\\ &= [1(-395.7\text{ kJ})] - [1(-296.8\text{ kJ}) + 1/2(0.0\text{ kJ})]\\ &= [-395.7\text{ kJ}] - [-296.8\text{ kJ}]\\ &= -98.9\text{ kJ}\end{aligned}$$

Check: The units (kJ) are correct. The answer is negative, which means that the reaction is exothermic. The SO_3 has a lower heat of formation than SO_2, so we expect an exothermic reaction.

(c) **Given:** $C(s) + H_2O(g) \rightarrow CO(g) + H_2(g)$ **Find:** ΔH°_{rxn}
Conceptual Plan: $\Delta H^\circ_{rxn} = \sum n_p \Delta H^\circ_f(products) - \sum n_r \Delta H^\circ_f(reactants)$
Solution:

Reactant/Product	ΔH°_f (kJ/mol from Appendix IVB)
$C(s)$	0.0
$H_2O(g)$	−241.8
$CO(g)$	−110.5
$H_2(g)$	0.0

Be sure to pull data for the correct formula and phase.

$\Delta H^\circ_{rxn} = \sum n_p \Delta H^\circ_f(products) - \sum n_r \Delta H^\circ_f(reactants)$

$= [1(\Delta H^\circ_f(CO(g))) + 1(\Delta H^\circ_f(H_2(g)))] - [1(\Delta H^\circ_f(C(s))) + 1(\Delta H^\circ_f(H_2O(g)))]$

$= [1(-110.5 \text{ kJ}) + 1(0.0 \text{ kJ})] - [1(0.0 \text{ kJ}) + 1(-241.8 \text{ kJ})]$

$= [-110.5 \text{ kJ}] - [-241.8 \text{ kJ}]$

$= 131.3 \text{ kJ}$

Check: The units (kJ) are correct. The answer is positive, which means that the reaction is endothermic. The CO has a smaller (less negative) heat of formation than H_2O, so we expect an endothermic reaction.

(d) **Given:** $N_2O_4(g) + 4\,H_2(g) \rightarrow N_2(g) + 4\,H_2O(g)$ **Find:** ΔH°_{rxn}
Conceptual Plan: $\Delta H^\circ_{rxn} = \sum n_p \Delta H^\circ_f(products) - \sum n_r \Delta H^\circ_f(reactants)$
Solution:

Reactant/Product	ΔH°_f (kJ/mol from Appendix IVB)
$N_2O_4(g)$	9.16
$H_2(g)$	0.0
$N_2(g)$	0.0
$H_2O(g)$	−241.8

Be sure to pull data for the correct formula and phase.

$\Delta H^\circ_{rxn} = \sum n_p \Delta H^\circ_f(products) - \sum n_r \Delta H^\circ_f(reactants)$

$= [1(\Delta H^\circ_f(N_2(g))) + 4(\Delta H^\circ_f(H_2O(g)))] - [1(\Delta H^\circ_f(N_2O_4(g))) + 4(\Delta H^\circ_f(H_2(g)))]$

$= [1(0.0 \text{ kJ}) + 4(-241.8 \text{ kJ})] - [1(9.16 \text{ kJ}) + 4(0.0 \text{ kJ})]$

$= [-967.2 \text{ kJ}] - [9.16 \text{ kJ}]$

$= -976.36 \text{ kJ}$

Check: The units (kJ) are correct. The answer is negative, which means that the reaction is exothermic. The H_2O has a lower heat of formation than N_2O_4, so we expect an exothermic reaction.

10.87 **Given:** form glucose ($C_6H_{12}O_6$) and oxygen from sunlight, carbon dioxide, and water **Find:** ΔH°_{rxn}
Conceptual Plan: Write balanced reaction then $\Delta H^\circ_{rxn} = \sum n_p \Delta H^\circ_f(products) - \sum n_r \Delta H^\circ_f(reactants)$
Solution: $6\,CO_2(g) + 6\,H_2O(l) \rightarrow C_6H_{12}O_6(s) + 6\,O_2(g)$

Reactant/Product	ΔH°_f (kJ/mol from Appendix IVB)
$CO_2(g)$	−393.5
$H_2O(l)$	−285.8
$C_6H_{12}O_6(s, glucose)$	−1273.3
$O_2(g)$	0.0

Be sure to pull data for the correct formula and phase.

$\Delta H^\circ_{rxn} = \sum n_p \Delta H^\circ_f(products) - \sum n_r \Delta H^\circ_f(reactants)$

$= [1(\Delta H^\circ_f(C_6H_{12}O_6(s))) + 6(\Delta H^\circ_f(O_2(g)))] - [6(\Delta H^\circ_f(CO_2(g))) + 6(\Delta H^\circ_f(H_2O(l)))]$

$= [1(-1273.3 \text{ kJ}) + 6(0.0 \text{ kJ})] - [6(-393.5 \text{ kJ}) + 6(-285.8 \text{ kJ})]$

$= [-1273.3 \text{ kJ}] - [-4075.8 \text{ kJ}]$

$= +2802.5 \text{ kJ}$

Check: The units (kJ) are correct. The answer is positive, which means that the reaction is endothermic. The reaction requires the input of light energy, so we expect that this will be an endothermic reaction.

10.88 **Given:** ethanol (C_2H_5OH) combustion **Find:** ΔH°_{rxn}
Conceptual Plan: Write balanced reaction then $\Delta H^\circ_{rxn} = \sum n_p \Delta H^\circ_f(products) - \sum n_r \Delta H^\circ_f(reactants)$
Solution: Combustion is the combination with oxygen to form carbon dioxide and water:
$C_2H_5OH(l) + 3\,O_2(g) \rightarrow 2\,CO_2(g) + 3\,H_2O(g)$

Reactant/Product	ΔH_f° (kJ/mol from Appendix IVB)
$C_2H_5OH(l)$	−277.6
$O_2(g)$	0.0
$CO_2(g)$	−393.5
$H_2O(g)$	−241.8

Be sure to pull data for the correct formula and phase.

$$\begin{aligned}\Delta H^\circ_{rxn} &= \sum n_p \Delta H_f^\circ(products) - \sum n_r \Delta H_f^\circ(reactants)\\ &= [2(\Delta H_f^\circ(CO_2(g))) + 3(\Delta H_f^\circ(H_2O(g)))] - [1(\Delta H_f^\circ(C_2H_5OH(l))) + 3(\Delta H_f^\circ(O_2(g)))]\\ &= [2(-393.5\ kJ) + 3(-241.8\ kJ)] - [1(-277.6\ kJ) + 3(0.0\ kJ)]\\ &= [-1512.4\ kJ] - [-277.6\ kJ]\\ &= -1234.8\ kJ\end{aligned}$$

Check: The units (kJ) are correct. The answer is negative, which means that the reaction is exothermic; this is typical for combustion reactions.

10.89 **Given:** $2\ CH_3NO_2(l) + 3/2\ O_2(g) \rightarrow 2\ CO_2(g) + 3\ H_2O(l) + N_2(g)$ and $\Delta H^\circ_{rxn} = -1418.4\ kJ$
Find: $\Delta H_f^\circ\ (CH_3NO_2(l))$
Conceptual Plan: Fill known values into $\Delta H^\circ_{rxn} = \sum n_p \Delta H_f^\circ(products) - \sum n_r \Delta H_f^\circ(reactants)$ and rearrange to solve for $\Delta H_f^\circ(CH_3NO_2(l))$.
Solution:

Reactant/Product	ΔH_f° (kJ/mol from Appendix IVB)
$O_2(g)$	0.0
$CO_2(g)$	−393.5
$H_2O(l)$	−285.8
$N_2(g)$	0.0

Be sure to pull data for the correct formula and phase.

$$\begin{aligned}\Delta H^\circ_{rxn} &= \sum n_p \Delta H_f^\circ(products) - \sum n_r \Delta H_f^\circ(reactants)\\ &= [2(\Delta H_f^\circ(CO_2(g))) + 3(\Delta H_f^\circ(H_2O(l))) + 1(\Delta H_f^\circ(N_2(g)))] - [2(\Delta H_f^\circ(CH_3NO_2(l))) + 3/2(\Delta H_f^\circ(O_2(g)))]\end{aligned}$$

$$(-1418.4\ kJ) = [2(-393.5\ kJ) + 3(-285.8\ kJ) + 1(0.0\ kJ)] - [2(\Delta H_f^\circ(CH_3NO_2(l)) + 3/2(0.0\ kJ)]$$

$$-1418.4\ kJ = [-1644.4\ kJ] - [2(\Delta H_f^\circ(CH_3NO_2(l)))]$$

$$\Delta H_f^\circ(CH_3NO_2(l)) = -113.0\ kJ/mol$$

Check: The units (kJ/mol) are correct. The answer is negative (but not as negative as water and carbon dioxide), which is consistent with an exothermic combustion reaction.

10.90 **Given:** $4\ C_3H_5N_3O_9(l) \rightarrow 12\ CO_2(g) + 10\ H_2O(g) + 6\ N_2(g) + O_2(g)$ and $\Delta H^\circ_{rxn} = -5678\ kJ$
Find: $\Delta H_f^\circ\ (C_3H_5N_3O_9(l))$
Conceptual Plan: Fill known values into $\Delta H^\circ_{rxn} = \sum n_p \Delta H_f^\circ(products) - \sum n_r \Delta H_f^\circ(reactants)$ and rearrange to solve for $\Delta H_f^\circ(C_3H_5N_3O_9(l))$.
Solution:

Reactant/Product	ΔH_f° (kJ/mol from Appendix IVB)
$CO_2(g)$	−393.5
$H_2O(g)$	−241.8
$N_2(g)$	0.0
$O_2(g)$	0.0

Be sure to pull data for the correct formula and phase.

$$\begin{aligned}\Delta H^\circ_{rxn} &= \sum n_p \Delta H_f^\circ(products) - \sum n_r \Delta H_f^\circ(reactants)\\ &= [12(\Delta H_f^\circ(CO_2(g))) + 10(\Delta H_f^\circ(H_2O(g))) + 6(\Delta H_f^\circ(N_2(g))) + 1(\Delta H_f^\circ(O_2(g)))] - [4(\Delta H_f^\circ(C_3H_5N_3O_9(l)))]\end{aligned}$$

$$(-5678\ kJ) = [12(-393.5\ kJ) + 10(-241.8\ kJ) + 6(0.0\ kJ) + 1(0.0\ kJ)] - [4(\Delta H_f^\circ(C_3H_5N_3O_9(l)))]$$

$$-5678\ kJ = [-7140.\ kJ] - [4(\Delta H_f^\circ(C_3H_5N_3O_9(l)))]$$

$$\Delta H_f^\circ(C_3H_5N_3O_9(l)) = -365.5\ kJ/mol$$

Check: The units (kJ/mol) are correct. The answer is positive, which is not surprising because this is a complex molecule with very high potential energy.

Lattice Energies

10.91 As the size of the alkaline metal ions increases down the column, so does the distance between the metal cation and the oxide anion. Therefore, the magnitude of the lattice energy of the oxides decreases, making the formation of the oxides less exothermic and the compounds less stable. Because the ions cannot get as close to each other, they do not release as much energy.

10.92 On the periodic table, rubidium is below potassium, and iodine is below bromine. Therefore, both the rubidium ion and the iodide ion are larger than the potassium ion and the bromide ion. So the rubidium ion and the iodide ion cannot get as close to each other as the potassium ion and the bromide ion can; thus, the rubidium and iodide ions do not release as much energy, and the lattice energy of potassium bromide is more exothermic.

10.93 Cesium is slightly larger than barium, but oxygen is slightly larger than fluorine; so we cannot use size to explain the difference in the lattice energy. However, the charge on the cesium ion is 1+ and the charge on the fluoride ion is 1−, while the charge on the barium ion is 2+ and the charge on the oxide ion is 2−. The coulombic equation states that the magnitude of the potential also depends on the product of the charges. Because the product of the charges for CsF = 1− and the product of the charges for BaO = 4−, the stabilization for BaO relative to CsF should be about four times greater, which is what we see in its more exothermic lattice energy.

10.94 RbBr < KCl < SrO < CaO. KCl and RbBr both have a product of the charges of 1−, while SrO and CaO have a product of the charges of 4−. So the lattice energies of KCl and RbBr are less than those of SrO and CaO. Within KCl and RbBr, the rubidium ion is larger than the potassium ion and the bromide ion is larger than the chloride ion. Therefore, the rubidium ion and the bromide ion will be farther apart, leading to a smaller lattice energy. Between SrO and CaO, the strontium ion is larger than the calcium ion; so the strontium oxide will have a smaller (less negative) lattice energy.

10.95 **Given:** $\Delta H_f^\circ(KCl) = -436.5\ kJ/mol$; $IE_1(K) = 419\ kJ/mol$; $\Delta H_{sub}(K) = 89.0\ kJ/mol$; $Cl_2(g)$ bond energy = 243 kJ/mol; $EA(Cl) = -349\ kJ/mol$ **Find:** lattice energy

Conceptual Plan:

$$K(s) + 1/2\,Cl_2(g) \xrightarrow{\Delta H_{sub}} K(g) + 1/2\,Cl_2(g) \xrightarrow{IE_1} K^+(g) + 1/2\,Cl_2(g) \xrightarrow{\text{bond energy}} K^+(g) + Cl(g) \xrightarrow{EA} K^+(g) + Cl^-(g) \xrightarrow{\text{lattice energy}} KCl(s)$$

ΔH_f° (from $K(s) + 1/2\,Cl_2(g)$ directly to $KCl(s)$)

Solution: $\Delta H_f^\circ = \Delta H_{sub} + IE_1 + 1/2\text{ bond energy} + EA + \text{lattice energy}$

$$-436.5\frac{kJ}{mol} = +89.0\frac{kJ}{mol} + 419\frac{kJ}{mol} + \frac{1}{2}(243)\frac{kJ}{mol} + (-349)\frac{kJ}{mol} + \text{lattice energy}$$

$$\text{lattice energy} = -717\ kJ/mol$$

10.96 **Given:** $\Delta H_f^\circ(CaO) = -634.9\ kJ/mol$; $IE_1(Ca) = 590\ kJ/mol$; $IE_2(Ca) = 1145\ kJ/mol$; $\Delta H_{sub}(Ca) = 178\ kJ/mol$; $O_2(g)$ bond energy = 498 kJ/mol; $EA_1(O) = -141\ kJ/mol$; $EA_2(O) = 744\ kJ/mol$ **Find:** lattice energy

Conceptual Plan:

$$Ca(s) + 1/2\,O_2(g) \xrightarrow{\Delta H_{sub}} Ca(g) + 1/2\,O_2(g) \xrightarrow{IE_1} Ca^+(g) + 1/2\,O_2(g) \xrightarrow{IE_2} Ca^{2+}(g) + 1/2\,O_2(g) \xrightarrow{\text{bond energy}} Ca^{2+}(g) + O(g) \xrightarrow{EA_1} Ca^{2+}(g) + O^-(g)$$

$$\xrightarrow{EA_2} Ca^{2+}(g) + O^{2-}(g) \xrightarrow{\text{lattice energy}} CaBr_2(s)$$

ΔH_f°

Solution: $\Delta H_f^\circ = \Delta H_{sub} + IE_1 + IE_2 + 1/2\text{ bond energy} + EA_1 + EA_2 + \text{lattice energy}$

$$-634.9\frac{kJ}{mol} = +178\frac{kJ}{mol} + 590\frac{kJ}{mol} + 1145\frac{kJ}{mol} + \frac{1}{2}(498)\frac{kJ}{mol} + (-141)\frac{kJ}{mol} + 744\frac{kJ}{mol} + \text{lattice energy}$$

$$\text{lattice energy} = -3400\ kJ/mol$$

Cumulative Problems

10.97 **Given:**

billiard $ball_A$ = system: $m_A = 0.17$ kg, $v_{A1} = 4.5$ m/s slows to $v_{A2} = 3.8$ m/s and $v_{A3} = 0$; $ball_B$: $m_B = 0.17$ kg, $v_{B1} = 0$ and $v_{B2} = 3.8$ m/s, and KE $= \frac{1}{2}mv^2$ **Find:** w, q, ΔE_{sys}

Conceptual Plan: $m, v \rightarrow$ **KE then** $KE_{A3}, KE_{A1} \rightarrow \Delta E_{sys}$ **and** $KE_{A2}, KE_{A1} \rightarrow q$ **and** $KE_{B2}, KE_{B1} \rightarrow w_B$

$KE = \frac{1}{2}mv^2$ $\quad \Delta E_{sys} = KE_{A3} - KE_{A1}$ $\quad q = KE_{A2} - KE_{A1}$ $\quad w_B = KE_{B2} - KE_{B1}$

$\Delta E_{sys}, q \rightarrow w_A$ **verify that** $w_A = -w_B$ **so that no heat is transferred to** $ball_B$

$\Delta E = q + w$

Solution: KE $= \frac{1}{2}mv^2$ because m is in kg and v is in m/s; KE will be in kg · m²/s², which is the definition of a joule.

$$KE_{A1} = \frac{1}{2}(0.17\text{ kg})\left(4.5\frac{\text{m}}{\text{s}}\right)^2 = 1.7213\frac{\text{kg}\cdot\text{m}^2}{\text{s}^2} = 1.7213\text{ J}$$

$$KE_{A2} = \frac{1}{2}(0.17\text{ kg})\left(3.8\frac{\text{m}}{\text{s}}\right)^2 = 1.2274\frac{\text{kg}\cdot\text{m}^2}{\text{s}^2} = 1.2274\text{ J}$$

$$KE_{A3} = \frac{1}{2}(0.17\text{ kg})\left(0\frac{\text{m}}{\text{s}}\right)^2 = 0\frac{\text{kg}\cdot\text{m}^2}{\text{s}^2} = 0\text{ J},$$

$$KE_{B1} = \frac{1}{2}(0.17\text{ kg})\left(0\frac{\text{m}}{\text{s}}\right)^2 = 0\frac{\text{kg}\cdot\text{m}^2}{\text{s}^2} = 0\text{ J}$$

$$KE_{B2} = \frac{1}{2}(0.17\text{ kg})\left(3.8\frac{\text{m}}{\text{s}}\right)^2 = 1.2274\frac{\text{kg}\cdot\text{m}^2}{\text{s}^2} = 1.2274\text{ J}$$

$\Delta E_{sys} = KE_{A3} - KE_{A1} = 0\text{ J} - 1.7213\text{ J} = -1.7213\text{ J} = -1.7\text{ J}$

$q = KE_{A2} - KE_{A1} = 1.2274\text{ J} - 1.7213\text{ J} = -0.4939\text{ J} = -0.5\text{ J}$

$w_B = KE_{B2} - KE_{B1} = 1.2274\text{ J} - 0\text{ J} = 1.2274\text{ J}$

$w = \Delta E - q = -1.7213\text{ J} - (-0.4939\text{ J}) = -1.2274\text{ J} = -1.2\text{ J}$

Because $w_A = -w_B$, no heat is transferred to $ball_B$.

Check: The units (J) are correct. Because the ball is initially moving and is stopped at the end, it has lost energy (negative ΔE_{sys}). As the ball slows due to friction, it is releasing heat (negative q). The kinetic energy is transferred to a second ball, so it does work (w negative).

10.98 **Given:** 100 W light bulb in a piston; bulb on for 0.015 hr, $V_i = 0.85$ L, $V_f = 5.88$ L, $P = 1.0$ atm **Find:** w, q, ΔE_{sys}

Conceptual Plan: bulb wattage, time $\rightarrow \Delta E_{sys}(\text{Wh}) \rightarrow \Delta E_{sys}(\text{kWh}) \rightarrow \Delta E_{sys}(\text{J})$ **and** $V_i, V_f \rightarrow \Delta V$ **then**

$\Delta E = (wattage)(time)$ $\quad \frac{1\text{ kW}}{1000\text{ W}}$ $\quad \frac{3.60 \times 10^6\text{J}}{1\text{ kWh}}$ $\quad \Delta V = V_f - V_i$

$P, \Delta V \rightarrow w(\text{L atm}) \rightarrow w(\text{J})$ **finally** $\Delta E_{sys}, w \rightarrow q$

$w = -P\Delta V$ $\quad \frac{101.3\text{ J}}{1\text{ L atm}}$ $\quad \Delta E = q + w$

Solution: $\Delta E = (wattage)(time) = (100\text{ W})(0.015\text{ hr}) = 1.5\text{ Wh} \times \frac{1\text{ kW}}{1000\text{ W}} \times \frac{3.60 \times 10^6\text{ J}}{1\text{ kWh}} = 5400\text{ J}$ and

$\Delta V = V_f - V_i = 5.88\text{ L} - 0.85\text{ L} = 5.03\text{ L}$ then

$$w = -P\Delta V = -1.0\text{ atm} \times 5.03\text{ L} \times \frac{101.3\text{ J}}{1\text{ L atm}} = -509.539\text{ J} = -5.1 \times 10^2\text{ J} \qquad \Delta E = q + w$$

Rearrange to solve for q. $q = \Delta E_{sys} - w = +5400\text{ J} - (-509.539\text{ J}) = 5900\text{ J}$

Check: The units (J) are correct. Electricity is added, so energy is added (positive ΔE_{sys}). The piston expands, so it does work (negative work). For the light bulb to generate light, it must be heated or it must absorb energy (positive q).

10.99 **Given:** $H_2O(l) \rightarrow H_2O(g)$ $\Delta H^\circ_{rxn} = +44.01$ kJ/mol; $\Delta T_{body} = -0.50$ °C, $m_{body} = 95$ kg, $C_{body} = 4.0$ J/g°C

Find: m_{H_2O}

Conceptual Plan: **kg → g then m_{body}, ΔT, $C_{s,\,body}$ → q_{body} → q_{rxn}(J) → q_{rxn}(kJ) → mol H_2O → g H_2O**

$$\frac{1000\text{ g}}{1\text{ kg}} \quad q_{body} = m_{body}C_{s,\,body}\Delta T_{body} \quad q_{rxn} = -q_{body} \quad \frac{1\text{ kJ}}{1000\text{ J}} \quad \frac{1\text{ mol}}{44.01\text{ kJ}} \quad \frac{18.02\text{ g}}{1\text{ mol}}$$

Solution: $95\text{ kg} \times \dfrac{1000\text{ g}}{1\text{ kg}} = 95000\text{ g}$ then

$$q_{body} = m_{body}C_{s,\,body}\Delta T_{body} = 95000\text{ g} \times 4.0\frac{\text{J}}{\text{g}\cdot{}^\circ\text{C}} \times (-0.50\,^\circ\text{C}) = -1\underline{9}0000\text{ J then}$$

$$q_{rxn} = -q_{body} = 1\underline{9}0000\text{ J} \times \frac{1\text{ kJ}}{1000\text{ J}} \times \frac{1\text{ mol}}{44.01\text{ kJ}} \times \frac{18.02\text{ g}}{1\text{ mol}} = 78\text{ g } H_2O$$

Check: The units (g) are correct. The magnitude of the answer (78) makes physical sense because a person can sweat this much on a hot day.

10.100 **Given:** LP gas combustion, $\Delta H^\circ_{rxn} = -2044$ kJ; 1.5 L water, $T_{H_2O,\,i} = 25.0\,^\circ$C, $T_{H_2O,\,f} = 100.0\,^\circ$C, 15% efficiency
Find: $m_{LP\text{ gas}}$
Conceptual Plan: **L → mL → g and T_i, T_f → ΔT then m_{H_2O}, $C_{s,\,H_2O}$ → q_{H_2O} → q_{rxn}**

$$\frac{1000\text{ mL}}{1\text{ L}} \quad \frac{1.0\text{ g}}{1.0\text{ mL}} \quad \Delta T = T_f - T_i \quad q_{H_2O} = m_{H_2O}C_{s,\,H_2O}\Delta T_{H_2O} \quad q_{rxn} = -q_{H_2O}$$

then q_{rxn} needed → q_{rxn} generated (J) → q_{rxn} (kJ) → mol LP gas → g LP gas

$$\frac{100\text{ J generated}}{15\text{ J needed}} \quad \frac{1\text{ kJ}}{1000\text{ J}} \quad \frac{1\text{ mol}}{-2044\text{ kJ}} \quad \frac{44.09\text{ g}}{1\text{ mol}}$$

Solution: $1.5\text{ L} \times \dfrac{1000\text{ mL}}{1\text{ L}} \times \dfrac{1.0\text{ g}}{1.0\text{ mL}} = 1500\text{ g}$ and $\Delta T = T_f - T_i = 100.0\,^\circ\text{C} - 25.0\,^\circ\text{C} = 75.0\,^\circ\text{C}$

then $q_{H_2O} = m_{H_2O}C_{s,\,H_2O}\Delta T_{H_2O} = 1500\text{ g} \times 4.184\dfrac{\text{J}}{\text{g}\cdot{}^\circ\text{C}} \times (75.0\,^\circ\text{C}) = 4\underline{7}0700$ J then

$$q_{rxn} = -q_{H_2O} = -4\underline{7}0700\text{ J needed} \times \frac{100\text{ J generated}}{15\text{ J needed}} \times \frac{1\text{ kJ}}{1000\text{ J}} \times \frac{1\text{ mol}}{-2044\text{ kJ}} \times \frac{44.09\text{ g}}{1\text{ mol}} = 68\text{ g LP gas}$$

Check: The units (g) are correct. The magnitude of the answer (68) makes physical sense because a tank of LP gas contains many orders of magnitude more than this amount.

10.101 **Given:** $H_2O(s) \rightarrow H_2O(l)$ $\Delta H^\circ_f(H_2O(s)) = -291.8$ kJ/mol; 355 mL beverage $T_{Bev,\,i} = 25.0\,^\circ$C, $T_{Bev,\,f} = 0.0\,^\circ$C, $C_{s,\,Bev} = 4.184$ J/g °C, $d_{Bev} = 1.0$ g/mL **Find:** ΔH°_{rxn} (ice melting) and m_{ice}
Conceptual Plan: $\Delta H^\circ_{rxn} = \sum n_p\Delta H^\circ_f(products) - \sum n_r\Delta H^\circ_f(reactants)$ **mL → g and T_i, T_f → ΔT then**

$$\frac{1.0\text{ g}}{1.0\text{ mL}} \quad \Delta T = T_f - T_i$$

m_{H_2O}, ΔT_{H_2O}, $C_{s,\,H_2O}$ → q_{H_2O} → q_{rxn}(J) → q_{rxn} (kJ) → mol ice → g ice

$$q_{Bev} = m_{Bev}C_{s,\,Bev}\Delta T_{Bev} \quad q_{rxn} = -q_{Bev} \quad \frac{1\text{ kJ}}{1000\text{ J}} \quad \frac{1\text{ mol}}{\Delta H^\circ_{rxn}} \quad \frac{18.02\text{ g}}{1\text{ mol}}$$

Solution:

Reactant/Product	ΔH°_f (kJ/mol from Appendix IVB)
$H_2O(s)$	−291.8
$H_2O(l)$	−285.8

Be sure to pull data for the correct formula and phase.

$$\begin{aligned}\Delta H^\circ_{rxn} &= \sum n_p\Delta H^\circ_f(products) - \sum n_r\Delta H^\circ_f(reactants)\\ &= [1(\Delta H^\circ_f(H_2O(l)))] - [1(\Delta H^\circ_f(H_2O(s)))]\\ &= [1(-285.8\text{ kJ})] - [1(-291.8\text{ kJ})]\\ &= +6.0\text{ kJ}\end{aligned}$$

$355\text{ mL} \times \dfrac{1.0\text{ g}}{1.0\text{ mL}} = 355\text{ g}$ and $\Delta T = T_f - T_i = 0.0\,^\circ\text{C} - 25.0\,^\circ\text{C} = -25.0\,^\circ\text{C}$ then

$$q_{Bev} = m_{Bev}C_{s,\,Bev}\Delta T_{Bev} = 355\text{ g} \times 4.184\frac{\text{J}}{\text{g}\cdot{}^\circ\text{C}} \times (-25.0\,^\circ\text{C}) = -37\underline{1}33\text{ J then}$$

$$q_{rxn} = -q_{Bev} = 37\underline{1}33\ \text{J} \times \frac{1\ \text{kJ}}{1000\ \text{J}} \times \frac{1\ \text{mol}}{6.0\ \text{kJ}} \times \frac{18.02\ \text{g}}{1\ \text{mol}} = 110\ \text{g ice}$$

Check: The units (J and g) are correct. The answer is positive, which means that the reaction is endothermic. We expect an endothermic reaction because we know that heat must be added to melt ice. The magnitude of the answer (110 g) makes physical sense because it is much smaller than the weight of the beverage and it would fit in a glass with the beverage.

10.102 **Given:** $CO_2(s) \rightarrow CO_2(g)$ $\Delta H^\circ_f(CO_2(s)) = -427.4$ kJ/mol; 15.0 L water $T_{H_2O,\,i} = 85\ °C$, $T_{H_2O,\,f} = 25\ °C$
Find: ΔH°_{rxn} (dry ice sublimation) and m_{dryice}
Conceptual Plan: $\Delta H^\circ_{rxn} = \sum n_p \Delta H^\circ_f(products) - \sum n_r \Delta H^\circ_f(reactants)$ **L → mL → g and T_i, T_f → ΔT then**

$$\frac{1000\ \text{mL}}{1\ \text{L}} \quad \frac{1.00\ \text{g}}{1.00\ \text{mL}} \quad \Delta T = T_f - T_i$$

m_{H_2O}, ΔT_{H_2O}, $C_{s,\,H_2O}$ → q_{H_2O} → q_{rxn}(J) → q_{rxn} (kJ) → mol ice → g ice

$$q_{H_2O} = m_{H_2O}C_{s,\,H_2O}\Delta T_{H_2O} \quad q_{rxn} = -q_{H_2O} \quad \frac{1\ \text{kJ}}{1000\ \text{J}} \quad \frac{1\ \text{mol}}{\Delta H^\circ_{rxn}} \quad \frac{44.01\ \text{g}}{1\ \text{mol}}$$

Solution:

Reactant/Product	ΔH°_f (kJ/mol from Appendix IVB)
$CO_2(s)$	−427.4
$CO_2(g)$	−393.5

Be sure to pull data for the correct formula and phase.

$$\begin{aligned}\Delta H^\circ_{rxn} &= \sum n_p \Delta H^\circ_f(products) - \sum n_r \Delta H^\circ_f(reactants)\\ &= [1(\Delta H^\circ_f(CO_2(g)))] - [1(\Delta H^\circ_f(CO_2(s)))]\\ &= [1(-393.5\ \text{kJ})] - [1(-427.4\ \text{kJ})]\\ &= +\ 33.9\ \text{kJ}\end{aligned}$$

$$15.0\ \text{L} \times \frac{1000\ \text{mL}}{1\ \text{L}} \times \frac{1.00\ \text{g}}{1.00\ \text{mL}} = 15\underline{0}00\ \text{g and } \Delta T = T_f - T_i = 25\ °C - 85\ °C = -60.\ °C$$

$$\text{then } q_{H_2O} = m_{H_2O}C_{s,\,H_2O}\Delta T_{H_2O} = 15\underline{0}00\ \text{g} \times 4.184\frac{\text{J}}{\text{g} \cdot °\text{C}} \times (-60.\ °\text{C}) = -3\underline{7}65600\ \text{J then}$$

$$q_{rxn} = -q_{H_2O} = 3\underline{7}65600\ \text{J} \times \frac{1\ \text{kJ}}{1000\ \text{J}} \times \frac{1\ \text{mol}}{33.9\ \text{kJ}} \times \frac{44.01\ \text{g}}{1\ \text{mol}} = 4900\ \text{g dry ice}$$

Check: The units (J and g) are correct. The answer is positive, which means that the reaction is endothermic. We expect an endothermic reaction because we know that heat must be added to sublime dry ice. The magnitude of the answer (4900) makes physical sense because the temperature change of the water is fairly large and the volume of water is large. It is a reasonable amount to put in a cooler.

10.103 **Given:** 25.5 g aluminum; $T_{Al,\,i} = 65.4\ °C$; 55.2 g water; $T_{H_2O,\,i} = 22.2\ °C$ **Find:** T_f
Conceptual Plan: Pull C_s values from table then m, C_s, T_i → T_f

$$\text{Al: } 0.903\frac{\text{J}}{\text{g} \cdot °\text{C}} \quad \text{H}_2\text{O: } 4.18\frac{\text{J}}{\text{g} \cdot °\text{C}} \quad q = mC_s(T_f - T) \text{ then set } q_{Al} = -q_{H_2O}$$

Solution: $q = mC_s(T_f - T_i)$ substitute in values and set $q_{Al} = -q_{H_2O}$.

$$q_{Al} = m_{Al}C_{s,\,Al}(T_f - T_{Al,\,i}) = 25.5\ \text{g} \times 0.903\frac{\text{J}}{\text{g} \cdot °\text{C}} \times (T_f - 65.4\ °\text{C}) =$$

$$-q_{H_2O} = -m_{H_2O}C_{s,\,H_2O}(T_f - T_{H_2O,\,i}) = -55.2\ \text{g} \times 4.18\frac{\text{J}}{\text{g} \cdot °\text{C}} \times (T_f - 22.2\ °\text{C})$$

Rearrange to solve for T_f.

$$23.\underline{0}265\frac{\text{J}}{°\text{C}} \times (T_f - 65.4\ °\text{C}) = -23\underline{0}.736\frac{\text{J}}{°\text{C}} \times (T_f - 22.2\ °\text{C}) \rightarrow$$

$$23.\underline{0}265\frac{\text{J}}{°\text{C}}T_f - 15\underline{0}5.93\ \text{J} = -23\underline{0}.736\frac{\text{J}}{°\text{C}}T_f + 51\underline{2}2.34\ \text{J} \rightarrow$$

$$+51\underline{2}2.34\ \text{J} + 15\underline{0}5.93\ \text{J} = +23\underline{0}.736\frac{\text{J}}{°\text{C}}T_f + 23.\underline{0}265\frac{\text{J}}{°\text{C}}T_f \rightarrow$$

$$66\underline{2}8.27\ \text{J} = 25\underline{3}.7625\frac{\text{J}}{°\text{C}}T_f \rightarrow$$

$$T_f = \frac{6628.27\ \text{J}}{253.7625\ \frac{\text{J}}{°\text{C}}} = 26.1\ °\text{C}$$

Check: The units (°C) are correct. The magnitude of the answer (26) makes physical sense because the heat transfer is dominated by the water (larger mass and larger specific heat capacity). The final temperature should be closer to the initial temperature of water than of aluminum.

10.104 **Given:** ethanol: 50.0 mL; $d = 0.789\ \text{g/mL}$, $T_{EtOH,\,i} = 7.0\ °\text{C}$, water: 50.0 mL; $d = 1.0\ \text{g/mL}$ $T_{H_2O,\,i} = 28.4\ °\text{C}$ **Find:** T_f

Conceptual Plan: pull C_s values from Table 10.2 mL → g then $m, C_s, T_i \rightarrow T_f$

EtOH: $2.42\ \frac{\text{J}}{\text{g} \cdot °\text{C}}$ H$_2$O: $4.18\ \frac{\text{J}}{\text{g} \cdot °\text{C}}$ EtOH: $\frac{0.789\ \text{g}}{1.0\ \text{mL}}$ H$_2$O: $\frac{1.0\ \text{g}}{1.0\ \text{mL}}$ $q = mC_s(T_f - T_i)$ then set $q_{EtOH} = -q_{H_2O}$

Solution: $50.0\ \text{mL} \times \frac{0.789\ \text{g}}{1.0\ \text{mL}} = 39.45\ \text{g EtOH}$ and $50.0\ \text{mL} \times \frac{1.0\ \text{g}}{1.0\ \text{mL}} = 50.0\ \text{g H}_2\text{O}$ then

$q = mC_s(T_f - T_i)$ substitute in values and set $q_{EtOH} = -q_{H_2O}$.

$$q_{EtOH} = m_{EtOH}C_{s,\,EtOH}(T_f - T_{EtOH,\,i}) = 39.45\ \text{g} \times 2.42\ \frac{\text{J}}{\text{g} \cdot °\text{C}} \times (T_f - 7.0\ °\text{C}) =$$

$$-q_{H_2O} = -m_{H_2O}C_{s,\,H_2O}(T_f - T_{H_2O,\,i}) = -50.0\ \text{g} \times 4.18\ \frac{\text{J}}{\text{g} \cdot °\text{C}} \times (T_f - 28.4\ °\text{C})$$

Rearrange to solve for T_f. $95.469\ \frac{\text{J}}{°\text{C}} \times (T_f - 7.0\ °\text{C}) = -209.0\ \frac{\text{J}}{°\text{C}} \times (T_f - 28.4\ °\text{C}) \rightarrow$

$95.469\ \frac{\text{J}}{°\text{C}}\ T_f - 668.283\ \text{J} = -209.0\ \frac{\text{J}}{°\text{C}}\ T_f + 5935.6\ \text{J} \rightarrow$

$+668.283\ \text{J} + 5935.6\ \text{J} = +209.2\ \frac{\text{J}}{°\text{C}}\ T_f + 95.469\ \frac{\text{J}}{°\text{C}}\ T_f \rightarrow 6603.883\ \text{J} = 304.669\ \frac{\text{J}}{°\text{C}}\ T_f \rightarrow$

$$T_f = \frac{6603.883\ \text{J}}{304.669\ \frac{\text{J}}{°\text{C}}} = 21.7\ °\text{C}$$

Check: The units (°C) are correct. The magnitude of the answer (22) makes physical sense because the heat transfer is dominated by the water (larger mass and larger specific heat capacity). The final temperature should be closer to the initial temperature of water than of ethanol.

10.105 **Given:** palmitic acid ($C_{16}H_{32}O_2$) combustion $\Delta H_f^\circ(C_{16}H_{32}O_2\,(s)) = -208\ \text{kJ/mol}$; sucrose ($C_{12}H_{22}O_{11}$) combustion $\Delta H_f^\circ(C_{12}H_{22}O_{11}(s)) = -2226.1\ \text{kJ/mol}$ **Find:** ΔH_{rxn}° in kJ/mol and Cal/g

Conceptual Plan: Write balanced reaction then $\Delta H_{rxn}^\circ = \sum n_p \Delta H_f^\circ(products) - \sum n_r \Delta H_f^\circ(reactants)$ then kJ/mol → J/mol → Cal/mol → Cal/g

$\frac{1000\ \text{J}}{1\ \text{kJ}}$ $\frac{1\ \text{Cal}}{4184\ \text{J}}$ PA: $\frac{1\ \text{mol}}{256.42\ \text{g}}$ S: $\frac{1\ \text{mol}}{342.30\ \text{g}}$

Solution: Combustion is the combination with oxygen to form carbon dioxide and water (l):

$C_{16}H_{32}O_2(s) + 23\ O_2(g) \rightarrow 16\ CO_2(g) + 16\ H_2O(l)$

Reactant/Product	ΔH_f° (kJ/mol from Appendix IVB)
$C_{16}H_{32}O_2(s)$	−208
$O_2(g)$	0.0
$CO_2(g)$	−393.5
$H_2O(l)$	−285.8

Be sure to pull data for the correct formula and phase.

$$\begin{aligned}\Delta H_{rxn}^\circ &= \sum n_p \Delta H_f^\circ(products) - \sum n_r \Delta H_f^\circ(reactants)\\ &= [16(\Delta H_f^\circ(CO_2(g))) + 16(\Delta H_f^\circ(H_2O(l)))] - [1(\Delta H_f^\circ(C_{16}H_{32}O_2(s))) + 23(\Delta H_f^\circ(O_2(g)))]\\ &= [16(-393.5\ \text{kJ}) + 16(-285.8\ \text{kJ})] - [1(-208\ \text{kJ}) + 23(0.0\ \text{kJ})]\\ &= [-10868.8\ \text{kJ}] - [-208\ \text{kJ}]\\ &= -10660.8\ \text{kJ/mol} = -10661\ \text{kJ/mol}\end{aligned}$$

$$-10660.8\ \frac{\text{kJ}}{\text{mol}} \times \frac{1000\ \text{J}}{1\ \text{kJ}} \times \frac{1\ \text{Cal}}{4184\ \text{J}} \times \frac{1\ \text{mol}}{256.42\ \text{g}} = -9.9378\ \text{Cal/g}$$

$C_{12}H_{22}O_{11}(s) + 12\,O_2(g) \rightarrow 12\,CO_2(g) + 11\,H_2O(l)$

Reactant/Product	ΔH°_f (kJ/mol from Appendix IVB)
$C_{12}H_{22}O_{11}(s)$	−2226.1
$O_2(g)$	0.0
$CO_2(g)$	−393.5
$H_2O(l)$	−285.8

Be sure to pull data for the correct formula and phase.

$$\begin{aligned}\Delta H^\circ_{rxn} &= \sum n_p \Delta H^\circ_f(products) - \sum n_r \Delta H^\circ_f(reactants)\\ &= [12(\Delta H^\circ_f(CO_2(g))) + 11(\Delta H^\circ_f(H_2O(l)))] - [1(\Delta H^\circ_f(C_{12}H_{22}O_{11}(s))) + 12(\Delta H^\circ_f(O_2(g)))]\\ &= [12(-393.5\text{ kJ}) + 11(-285.8\text{ kJ})] - [1(-2226.1\text{ kJ}) + 12(0.0\text{ kJ})]\\ &= [-7865.8\text{ kJ}] - [-2226.1\text{ kJ}]\\ &= -5639.7\text{ kJ/mol}\end{aligned}$$

$$-5639.7\frac{\text{kJ}}{\text{mol}} \times \frac{1000\text{ J}}{1\text{ kJ}} \times \frac{1\text{ Cal}}{4184\text{ J}} \times \frac{1\text{ mol}}{342.30\text{ g}} = -3.9378\text{ Cal/g}$$

Check: The units (kJ/mol and Cal/g) are correct. The magnitude of the answers are consistent with the food labels we see every day. Palmitic acid gives more Cal/g than does sucrose.

10.106 **Given:** hydrogen, methanol (CH_3OH), and octane combustion **Find:** q released in kJ/kg

Conceptual Plan: Write balanced reaction then $\Delta H^\circ_{rxn} = \sum n_p \Delta H^\circ_f(products) - \sum n_r \Delta H^\circ_f(reactants)$ then kJ/mol → kJ/g → kJ/kg

$H_2: \frac{1\text{ mol}}{2.016\text{ g}}$ $CH_3OH: \frac{1\text{ mol}}{32.04\text{ g}}$ O: $\frac{1\text{ mol}}{114.22\text{ g}}$ $\frac{1000\text{ g}}{1\text{ kg}}$

Solution: Combustion is the combination with oxygen to form carbon dioxide and water: $H_2(g) + 1/2\,O_2(g) \rightarrow H_2O(g)$. This reaction is the heat of formation of gaseous water, so $\Delta H^\circ_{rxn} = -241.8$ kJ/mol.

$$-241.8\frac{\text{kJ}}{\text{mol}} \times \frac{1\text{ mol}}{2.016\text{ g}} \times \frac{1000\text{ g}}{1\text{ kg}} = -1.199 \times 10^5\text{ kJ/kg } H_2 \text{ and}$$

$CH_3OH(l) + 3/2\,O_2(g) \rightarrow CO_2(g) + 2\,H_2O(g)$

Reactant/Product	ΔH°_f (kJ/mol from Appendix IVB)
$CH_3OH(l)$	−238.6
$O_2(g)$	0.0
$CO_2(g)$	−393.5
$H_2O(g)$	−241.8

Be sure to pull data for the correct formula and phase.

$$\begin{aligned}\Delta H^\circ_{rxn} &= \sum n_p \Delta H^\circ_f(products) - \sum n_r \Delta H^\circ_f(reactants)\\ &= [1(\Delta H^\circ_f(CO_2(g))) + 2(\Delta H^\circ_f(H_2O(g)))] - [1(\Delta H^\circ_f(CH_3OH(l))) + 3/2(\Delta H^\circ_f(O_2(g)))]\\ &= [1(-393.5\text{ kJ}) + 2(-241.8\text{ kJ})] - [1(-238.6\text{ kJ}) + 3/2(0.0\text{ kJ})]\\ &= [-877.1\text{ kJ}] - [-238.6\text{ kJ}]\\ &= -638.5\text{ kJ}\end{aligned}$$

$$-638.5\frac{\text{kJ}}{\text{mol}} \times \frac{1\text{ mol}}{32.04\text{ g}} \times \frac{1000\text{ g}}{1\text{ kg}} = -1.993 \times 10^4\text{ kJ/kg } CH_3OH \text{ and}$$

$$-5074.1\frac{\text{kJ}}{\text{mol}} \times \frac{1\text{ mol}}{114.22\text{ g}} \times \frac{1000\text{ g}}{1\text{ kg}} = -4.4424 \times 10^4\text{ kJ/kg } C_8H_{18}$$

Hydrogen delivers the most energy per weight of fuel. This is not surprising because hydrogen is so light. Octane delivers more energy per gram than does methanol.

Check: The units (kJ/kg fuel) are correct. The magnitude of the answers ($10^4 - 10^5$) makes physical sense because there are many moles of fuel in a kilogram and the heats of reactions are high.

10.107 **Given:** 16 g peanut butter; bomb calorimeter; $T_i = 22.2\text{ °C}$; $T_f = 25.4\text{ °C}$; $C_{cal} = 120.0\text{ kJ/°C}$ **Find:** calories in peanut butter

Conceptual Plan: $T_i, T_f \rightarrow \Delta T$ **then** $\Delta T, C_{cal} \rightarrow q_{cal} \rightarrow q_{rxn}$ **(kJ)** $\rightarrow q_{rxn}$ **(kJ)** $\rightarrow q_{rxn}$ **(Cal)**

$\Delta T = T_f - T_i \qquad q_{cal} = C_{cal}\Delta T \qquad q_{rxn} = -q_{cal} \qquad \frac{1000\text{ J}}{1\text{ kJ}} \qquad \frac{1\text{ Cal}}{4184\text{ J}}$

then q_{rxn} **(Cal)** $\rightarrow$ **Cal/g**

÷16 g peanut butter

Solution: $\Delta T = T_f - T_i = 25.4\ °\text{C} - 22.2\ °\text{C} = 3.2\ °\text{C}$ then $q_{cal} = C_{cal}\Delta T = 120.0\frac{\text{kJ}}{°\text{C}} \times 3.2\ °\text{C} = 384\text{ kJ}$

then $q_{rxn} = -q_{cal} = -384\text{ kJ} \times \frac{1000\text{ J}}{1\text{ kJ}} \times \frac{1\text{ Cal}}{4184\text{ J}} = 91.778\text{ Cal}$ then $\frac{91.778\text{ Cal}}{16\text{ g}} = 5.7\text{ Cal/g}$

Check: The units (Cal/g) are correct. The magnitude of the answer (6) makes physical sense because there is a significant percentage of fat and sugar in peanut butter. The answer is in line with the answers in Problem 10.103.

10.108 **Given:** 2.0 mol $H_2(g)$ + 1.0 mol $O_2(g)$ at 25 °C **Find:** temperature of water

Conceptual Plan: Write balanced reaction then $\Delta H°_{rxn} = \sum n_p \Delta H°_f(products) - \sum n_r \Delta H°_f(reactants)$ **then kJ/mol** $\rightarrow$ **q(kJ)** $\rightarrow$ **q(J) then 2.0 mol** H_2 **+ 1.0 mol** $O_2 \rightarrow$ **mol** $H_2O \rightarrow$ **g** H_2O **then**

x mol of limiting reagent $\frac{1000\text{ J}}{1\text{ kJ}} \qquad \frac{1\text{ mol }H_2O}{1\text{ mol }H_2} \qquad \frac{18.02\text{ g}}{1\text{ mol}}$

$q_{rxn} \rightarrow q_{H_2O}$ **then pull** C_s **for** $H_2O\ (l)$ **then** $q, m, C_s, T_i \rightarrow T_f$

$q_{rxn} = -q_{H_2O} \qquad 4.18\frac{\text{J}}{\text{g}\cdot°\text{C}} \qquad q = mC_s(T_f - T_i)$

Solution: Combustion is the combination with oxygen to form water; choose liquid water because $T = 25\ °\text{C}$.

$H_2(g) + \frac{1}{2}O_2(g) \rightarrow H_2O(l)$. This reaction is the heat of formation of gaseous water, so

$\Delta H°_{rxn} = -285.8\text{ kJ/mol}$. The two reactants are in the stoichiometric ratio, so either amount can be used.

$-285.8\frac{\text{kJ}}{1\text{ mol }H_2} \times 2.0\text{ mol }H_2 \times \frac{1000\text{ J}}{1\text{ kJ}} = -571600\text{ J}$ and $q_{H_2O} = -q_{rxn} = 571600\text{ J}$

$2.0\text{ mol }H_2 \times \frac{1\text{ mol }H_2O}{1\text{ mol }H_2} \times \frac{18.02\text{ g}}{1\text{ mol }H_2O} = 36.04\text{ g }H_2O$ then $q = mC_s(T_f - T_i)$. Rearrange to solve for T_f.

$$T_f = \frac{mC_sT_i + q}{mC_s} = \frac{\left(36.02\text{ g} \times 4.18\frac{\text{J}}{\text{g}\cdot°\text{C}} \times 25\ °\text{C}\right) + 571600\text{ J}}{36.02\text{ g} \times 4.18\frac{\text{J}}{\text{g}\cdot°\text{C}}} = 3821\ °\text{C}.$$ This is much higher than the

boiling point of water. The heat needed to raise the water to 100 °C is

$q = mC_s(T_f - T_i) = 36.02\text{ g} \times 4.18\frac{\text{J}}{\text{g}\cdot°\text{C}} \times (100\ °\text{C} - 25\ °\text{C}) = 11292\text{ J}$, so $571600\text{ J} - 11292\text{ J} = 560308\text{ J}$

is still available. 2.0 moles H_2O utilizes 81,400 J (=40.7 kJ/mol), so 478,908 J (=560,308 J − 81,400 J) is available to heat steam. Note: C_s (steam) = 2.04 J/g·°C.

Using equation from above $$T_f = \frac{mC_sT_i + q}{mC_s} = \frac{\left(36.04\text{ g} \times 2.04\frac{\text{J}}{\text{g}\cdot°\text{C}} \times 100.\ °\text{C}\right) + 478{,}908\text{ J}}{36.04\text{ g} \times 2.04\frac{\text{J}}{\text{g}\cdot°\text{C}}} = 6600\ °\text{C}$$

Check: The units (°C) are correct. The temperature is extremely high. A large amount of heat is liberated, and only a relatively small amount of mass absorbs it.

10.109 **Given:** $V_1 = 20.0\text{ L}$ at $P_1 = 3.0\text{ atm}$; $P_2 = 1.5\text{ atm}$; let it expand at constant T **Find:** $w, q, \Delta E_{sys}$

Conceptual Plan: $V_1, P_1, P_2 \rightarrow V_2$ **then** $V_1, V_2 \rightarrow \Delta V$ **then** $P, \Delta V \rightarrow w$ **(L atm)** $\rightarrow w$ **(J)**

$P_1V_1 = P_2V_2 \qquad \Delta V = V_2 - V_1 \qquad w = -P\Delta V \qquad \frac{101.3\text{ J}}{1\text{ L atm}}$

for an ideal gas $\Delta E_{sys} \alpha T$**; so because this is a constant temperature process,** $\Delta E_{sys} = 0$ **finally** $\Delta E_{sys}, w \rightarrow q$

$\Delta E = q + w$

Solution: $P_1V_1 = P_2V_2$. Rearrange to solve for V_2. $V_2 = V_1\dfrac{P_1}{P_2} = (20.0\text{ L}) \times \dfrac{3.0\text{ atm}}{1.5\text{ atm}} = 40.\text{ L}$ and

$\Delta V = V_2 - V_1 = 40.\text{ L} - 20.0\text{ L} = 20.\text{ L}$ then

$w = -P\Delta V = -1.5\text{ atm} \times 20.\text{ L} \times \dfrac{101.3\text{ J}}{1\text{ L}\cdot\text{atm}} = -3\underline{0}39\text{ J} = -3.0 \times 10^3\text{ J}$ $\quad\quad \Delta E = q + w$

Rearrange to solve for q. $q = \Delta E_{sys} - w = +0\text{ J} - (-3\underline{0}39\text{ J}) = 3.0 \times 10^3\text{ J}$

Check: The units (J) are correct. Because there is no temperature change, we expect no energy change ($\Delta E_{sys} = 0$). The piston expands as it does work (negative work), so heat is absorbed (positive q).

10.110 **Given:** 10.00 g $P_4(s) + O_2(g)$ to form $P_4O_{10}(s)$; q released heat 2950 g water from $T_i = 18.0\text{ °C}$ to $T_f = 38.0\text{ °C}$
Find: $\Delta H_f^\circ(P_4O_{10}(s))$
Conceptual Plan: Write balanced reaction then $\Delta H^\circ_{rxn} = \sum n_p\Delta H_f^\circ(products) - \sum n_r\Delta H_f^\circ(reactants)$ **then**
$m, C_s, T_i, T_f \rightarrow q_{H_2O} \rightarrow q_{rxn}(J) \rightarrow q(kJ)$ **then g** $(P_4) \rightarrow$ **mol** (P_4) **finally**

$q = mC_s(T_f - T_i) \quad q_{rxn} = -q_{H_2O} \quad \dfrac{1\text{ kJ}}{1000\text{ J}} \quad \dfrac{1\text{ mol }P_4}{123.90\text{ g }P_4}$

$q(kJ)$, **mol** $(P_4) \rightarrow \Delta H_f^\circ(P_4O_{10}(s))$

$\Delta H_f^\circ(P_4O_{10}(s)) = \dfrac{q}{\text{mol }P_4}$

Solution: $P_4(s) + 5\,O_2(g) \rightarrow P_4O_{10}(s)$. This reaction is the heat of formation of $P_4O_{10}(s)$, so $\Delta H^\circ_{rxn} = \Delta H_f^\circ(P_4O_{10}(s))$

then $q = mC_s(T_f - T_i) = 2950\text{ g} \times 4.18\dfrac{\text{J}}{\text{g}\cdot\text{°C}} \times (38.0\text{ °C} - 18.0\text{ °C}) = 24\underline{6}620\text{ J}$

then $q_{rxn} = -q_{H_2O} = -24\underline{6}620\text{ J} \times \dfrac{1\text{ kJ}}{1000\text{ J}} = -24\underline{6}.620\text{ kJ}$ then $10.00\text{ g }P_4 \times \dfrac{1\text{ mol }P_4}{123.90\text{ g }P_4} = 0.0807\underline{1}0\text{ mol }P_4$

then $\Delta H_f^\circ(P_4O_{10}(s)) = \dfrac{q}{\text{mol }P_4} = \dfrac{-24\underline{6}.620\text{ kJ}}{0.0807\underline{1}0\text{ mol}} = -3060\text{ kJ/mol}$

Check: The units (kJ/mol) are correct. The negative sign is consistent with the fact that heat was released to heat a large amount of water. The magnitude (3000) is not surprising because a small amount of phosphorus heated a great deal of water (a high specific heat-capacity material).

10.111 The oxidation of $S(g)$ to SO_3 can be written as follows:
$S(g) + 3/2\,O_2(g) \rightarrow SO_3(g) \quad\quad \Delta H = -204\text{ kJ}$
The oxidation of $SO_2(g)$ to SO_3 can be written as follows:
$SO_2(g) + 1/2\,O_2(g) \rightarrow SO_3(g) \quad\quad \Delta H = +89.5\text{ kJ}$
The enthalpy of formation reaction for $SO_2(g)$ under these conditions can be written as follows:
$S(g) + O_2(g) \rightarrow SO_2(g) \quad\quad \Delta H = ??$
Because the second reaction has 1 mole SO_2 as a reactant and the reaction of interest has 1 mole of SO_2 as a product, we need to reverse the second reaction. When the reaction direction is reversed, ΔH changes sign.
$SO_3(g) \rightarrow SO_2(g) + 1/2\,O_2(g) \quad\quad \Delta H = -89.5\text{ kJ}$
Hess's law states that the ΔH of the net reaction is the sum of the ΔH of the steps.
The rewritten reactions are as follows:
$S(g) + 3/2\,O_2(g) \rightarrow SO_3(g) \quad\quad \Delta H = -204\text{ kJ}$
$SO_3(g) \rightarrow SO_2(g) + 1/2\,O_2(g) \quad\quad \Delta H = -89.5\text{ kJ}$

$S(g) + O_2(g) \rightarrow SO_2(g) \quad\quad \Delta H = -294\text{ kJ} = \Delta H_f$
Note that this is not under standard conditions because S is not a solid.

10.112 The heat of formation of $TiI_3(s)$ can be written as follows:
$Ti(s) + 3/2\,I_2(s) \rightarrow TiI_3(s) \quad\quad \Delta H = -328\text{ kJ}$
The reaction of $Ti(s)$ with $I_2(g)$ can be written as follows:
$2\,Ti(s) + 3\,I_2(g) \rightarrow 2\,TiI_3(s) \quad\quad \Delta H = -839\text{ kJ}$
The sublimation of $I_2(s)$ can be written as follows:
$I_2(s) \rightarrow I_2(g) \quad\quad \Delta H = ??$

Because the first reaction has 3/2 moles $I_2(s)$ as a reactant and the reaction of interest has 1 mole of $I_2(s)$ as a reactant, we need to multiply it by 2/3.

$2/3\ Ti(s) + I_2(s) \rightarrow 2/3\ TiI_3(s)$ $\quad \Delta H = (2/3)(-328\ kJ) = -218.667\ kJ$

Because the second reaction has 3 moles $I_2(g)$ as a reactant and the reaction of interest has 1 mole of $I_2(g)$ as a product, we need to reverse the second reaction and divide it by 3. When the reaction direction is reversed, ΔH changes sign.

$2/3\ TiI_3(s) \rightarrow 2/3\ Ti(s) + I_2(g)$ $\quad \Delta H = (-1/3)(-839\ kJ) = +279.667\ kJ$

Hess's law states that the ΔH of the net reaction is the sum of the ΔH of the steps.

The rewritten reactions are as follows:

$\cancel{2/3\ Ti(s)} + I_2(s) \rightarrow \cancel{2/3\ TiI_3(s)}$ $\quad \Delta H = (2/3)(-328\ kJ) = -218.667\ kJ$

$\cancel{2/3\ TiI_3(s)} \rightarrow \cancel{2/3\ Ti(s)} + I_2(g)$ $\quad \Delta H = (-1/3)(-839\ kJ) = +279.667\ kJ$

$I_2(s) \rightarrow I_2(g)$ $\quad \Delta H = +61.0\ kJ = \Delta H_{sub}$

Note that this is endothermic, as expected.

10.113 **Given:** 1.55 cm copper cube and 1.62 cm aluminum cube, $T_{Metals,\,i} = 55.0\ °C$, 100.0 mL water, $T_{H_2O,\,i} = 22.2\ °C$ Other: density (water) = 0.998 g/mL **Find:** T_f

Conceptual Plan: Pull *d* values from Table 10.2 then edge length → *V* → *m* then

Cu: 8.96 g/mL Al: 2.70 g/mL $\quad V = l^3 d = m/V$

pull C_s values from Table 10.2 then $m, C_s, T_i \rightarrow T_f$

Cu: $0.385\ \frac{J}{g \cdot °C}$ Al: $0.903\ \frac{J}{g \cdot °C}$ H_2O: $4.18\ \frac{J}{g \cdot °C}$ $\quad q = mC_s(T_f - T_i)$ then set $q_{Cu} + q_{Al} = -q_{H_2O}$

Solution: $V_{Cu} = l^3 = (1.55\ cm)^3 = 3.723875\ cm^3 = 3.723875\ mL$ and $V_{Al} = l^3 = (1.62\ cm)^3 = 4.251528\ cm^3 = 4.251528\ mL$ then $d = m/V$. Rearrange to solve for m. $m = dV$

$$m_{Cu} = 8.96\ \frac{g}{\cancel{mL}} \times 3.723875\ \cancel{mL} = 33.36592\ g\ Cu$$

$$m_{Al} = 2.70\ \frac{g}{\cancel{mL}} \times 4.251528\ \cancel{mL} = 11.4791256\ g\ Al \text{ and } m_{H_2O} = 0.998\ \frac{g}{\cancel{mL}} \times 100.0\ \cancel{mL} = 99.8\ g\ H_2O$$

$q = mC_s(T_f - T_i)$ substitute in values and set $q_{Cu} + q_{Al} = -q_{H_2O}$

$$q_{Cu} + q_{Al} = m_{Cu}C_{s,\,Cu}(T_f - T_{Cu,\,i}) + m_{Al}C_{s,\,Al}(T_f - T_{Al,\,i}) =$$

$$33.36592\ \cancel{g} \times 0.385\ \frac{J}{\cancel{g} \cdot °C} \times (T_f - 55.0\ °C) + 11.4791256\ \cancel{g} \times 0.903\ \frac{J}{\cancel{g} \cdot °C} \times (T_f - 55.0\ °C) =$$

$$-q_{H_2O} = -m_{H_2O}C_{s,\,H_2O}(T_f - T_{H_2O,\,i}) = -99.8\ \cancel{g} \times 4.18\ \frac{J}{\cancel{g} \cdot °C} \times (T_f - 22.2\ °C)$$

Rearrange to solve for T_f.

$$12.84588\ \frac{J}{°C} \times (T_f - 55.0\ °C) + 10.36565\ \frac{J}{°C} \times (T_f - 55.0\ °C) = -417.164\ \frac{J}{°C} \times (T_f - 22.2\ °C) \rightarrow$$

$$12.84588\ \frac{J}{°C} T_f - 706.5234\ J + 10.36565\ \frac{J}{°C} T_f - 570.1108\ J = -417.164\ \frac{J}{°C} T_f + 9261.041\ J \rightarrow$$

$$12.84588\ \frac{J}{°C} T_f + 10.36565\ \frac{J}{°C} T_f + 417.164\ \frac{J}{°C} T_f = +706.5234\ J + 570.1108\ J + 9261.041\ J \rightarrow$$

$$440.3755\ \frac{J}{°C} T_f = 10537.675\ J \rightarrow T_f = \frac{10537.675\ \cancel{J}}{440.3755\ \frac{\cancel{J}}{°C}} = 23.92884\ °C = 23.9\ °C$$

Check: The units (°C) are correct. The magnitude of the answer (24) makes physical sense because the heat transfer is dominated by the water (larger mass and larger specific heat capacity). The final temperature should be closer to the initial temperature of water than of copper and aluminum.

10.114 **Given:** mass (gold + silver) = 14.9 g, $T_{Metals,\,i} = 62.0\ °C$, 15.0 mL water, $T_{H_2O,\,i} = 23.5\ °C$, $T_f = 25.0\ °C$ **Other:** density (water) = 0.998 g/mL **Find:** mass of each ring

Conceptual Plan: *V* → *m* then pull C_s values from Table 10.2 then

$d = m/V$

pull C_s values from Table 10.2, then let x = mass of gold ring, and then $m, C_s, T_i, T_f \rightarrow m$

Au: $0.128 \frac{J}{g \cdot °C}$ Ag: $0.235 \frac{J}{g \cdot °C}$ H_2O: $4.18 \frac{J}{g \cdot °C}$ $q = mC_s(T_f - T_i)$ then set $q_{Au} + q_{Ag} = -q_{H_2O}$

Solution: $d = m/V$. Rearrange to solve for m. $m = dV$ so $m_{H_2O} = 0.998 \frac{g}{mL} \times 15.0\ mL = 14.97 g\ H_2O$

Let x = mass of the gold ring. $q = mC_s(T_f - T_i)$ substitute in values and set $q_{Au} + q_{Ag} = -q_{H_2O}$.

$q_{Au} + q_{Ag} = m_{Au}C_{s,\,Au}(T_f - T_{Au,\,i}) + m_{Ag}C_{s,\,Ag}(T_f - T_{Ag,\,i}) =$

$$x \times 0.128 \frac{J}{g \cdot °C} \times (25.0\ °C - 62.0\ °C) + (14.9\ g - x) \times 0.235 \frac{J}{g \cdot °C} \times (25.0\ °C - 62.0\ °C) =$$

$$-q_{H_2O} = -m_{H_2O}C_{H_2O}(T_f - T_{H_2O,\,i}) = -14.97\ g \times 4.18 \frac{J}{g \cdot °C} \times (25.0\ °C - 23.5\ °C)$$

Rearrange to solve for mass of Au.

$$x\left(-4.736 \frac{J}{g}\right) - 129.5555\ J + x\left(-8.695 \frac{J}{g}\right) = -175.2089\ J \rightarrow$$

$$x\left(-4.736 \frac{J}{g}\right) + x\left(-8.695 \frac{J}{g}\right) = +129.5555\ J - 175.2089\ J \rightarrow x\left(-13.431 \frac{J}{g}\right) = -45.6534\ J \rightarrow$$

$$x = \frac{-45.6534\ J}{-13.431 \frac{J}{g}} = 3.39911\ g\ Au = 3.40\ g\ Au \text{ and } 14.9\ g - 11.5009\ g = 3.39911\ g\ Ag = 11.5\ g\ Ag$$

Check: The units (g) are correct. The magnitude of the answers (12 and 3) makes physical sense because the heat transfer is dominated by the larger specific heat capacity of water. The mass is mostly silver; the temperature change of the metals is large compared to the temperature change of the water, indicating a low specific heat metal.

10.115 To determine the values of the lattice energy, it is necessary to look them up online. The lattice energy of Al_2O_3 is $-15{,}916$ kJ/mol; the value for Fe_2O_3 is $-14{,}774$ kJ/mol. The thermite reaction is exothermic due to the energy released when the Al_2O_3 lattice forms. The lattice energy of Al_2O_3 is more negative than the lattice energy of Fe_2O_3.

10.116 For NaCl, E is proportional to $(1+)(1-) = -1$, while for XY, E is proportional to $(3+)(3-) = -9$. So the relative stabilization for XY relative to NaCl should be roughly nine times greater. $\Delta H_{lattice}$ (XY) $= 9 \times \Delta H_{lattice}$ (NaCl) $= 9(-787\ kJ/mol) = -7083\ kJ/mol$.

10.117 Rewrite the reaction using the Lewis structures of the molecules involved.

$$H{-}H(g) + 1/2\,\ddot{O}{=}\ddot{O}(g) \longrightarrow H{-}\ddot{O}{-}H$$

Determine which bonds are broken in the reaction and sum the bond energies of the following:

$$\begin{aligned} &\Sigma(\Delta H\text{'s bonds broken}) \\ &= 1\ mol(H{-}H) + 1/2\ mol(O{=}O) \\ &= 1\ mol(436\ kJ/mol) + 1/2\ mol(498) \\ &= 685\ kJ/mol \end{aligned}$$

Determine which bonds are formed in the reaction and sum the negatives of the bond energies of the following:

$$\begin{aligned} &\Sigma(-\Delta H\text{'s bonds formed}) \\ &= -2mol(O{-}H) \\ &= -2mol(464\ kJ/mol) \\ &= -928\ kJ/mol \end{aligned}$$

Find ΔH_{rxn} by summing the results of the two steps.

$$\begin{aligned} \Delta H_{rxn} &= \Sigma(\Delta H\text{'s bonds broken}) + \Sigma(-\Delta H\text{'s bonds formed}) \\ &= 685\ kJ/mol - 928\ kJ/mol \\ &= -243\ kJ/mol \end{aligned}$$

$CH_4(g) + 2\,O_2(g) \rightarrow CO_2(g) + 2\,H_2O(g)$

Rewrite the reaction using the Lewis structures of the molecules involved.

$$\mathrm{H{-}\underset{H}{\overset{H}{\underset{|}{\overset{|}{C}}}}{-}H} + 2\,\ddot{\underset{..}{O}}{=}\ddot{\underset{..}{O}} \longrightarrow \ddot{\underset{..}{O}}{=}C{=}\ddot{\underset{..}{O}} + 2\,H{-}\ddot{\underset{..}{O}}{-}H$$

Determine which bonds are broken in the reaction and sum the bond energies of the following:

$$\begin{aligned}&\Sigma(\Delta H\text{'s bonds broken})\\&= 4\text{ mol}(C{-}H) + 2\text{ mol}(O{=}O)\\&= 4\text{ mol}(414\text{ kJ/mol}) + 2\text{ mol}(498)\\&= 2652\text{ kJ/mol}\end{aligned}$$

Determine which bonds are formed in the reaction and sum the negatives of the bond energies of the following:

$$\begin{aligned}&\Sigma(-\Delta H\text{'s bonds formed})\\&= -2\text{ mol}(C{=}O) - 4\text{ mol}(O{-}H)\\&= -2\text{ mol}(799\text{ kJ/mol}) - 4\text{ mol}(464\text{ kJ/mol})\\&= -3454\text{ kJ/mol}\end{aligned}$$

Find ΔH_{rxn} by summing the results of the two steps.

$$\begin{aligned}\Delta H_{rxn} &= \Sigma(\Delta H\text{'s bonds broken}) + \Sigma(-\Delta H\text{'s bonds formed})\\&= 2652\text{ kJ/mol} - 3454\text{ kJ/mol}\\&= -802\text{ kJ/mol}\end{aligned}$$

Compare the following:

	kJ/mol	kJ/g
H_2	−243	−121 (using a molar mass of 2.016)
CH_4	−802	−50.0 (using a molar mass of 16.04)

$$\frac{-243\text{ kJ}}{\cancel{\text{mol }H_2}} \times \frac{\cancel{\text{mol }H_2}}{2.016\text{ g }H_2} = \frac{-121\text{ kJ}}{\text{g }H_2} \qquad \frac{-802\text{ kJ}}{\cancel{\text{mol }CH_4}} \times \frac{\cancel{\text{mol }CH_4}}{16.04\text{ g }CH_4} = \frac{-50.0\text{ kJ}}{\text{g }CH_4}$$

So methane yields more energy per mole, but hydrogen yields more energy per gram.

10.118 Octane = C_8H_{18}

$C_8H_{18}(l) + 25/2\,O_2(g) \rightarrow 8\,CO_2(g) + 9\,H_2O(g)$

Rewrite the reaction using the Lewis structures of the molecules involved.

$$\mathrm{H{-}\underset{H}{\overset{H}{C}}{-}\underset{H}{\overset{H}{C}}{-}\underset{H}{\overset{H}{C}}{-}\underset{H}{\overset{H}{C}}{-}\underset{H}{\overset{H}{C}}{-}\underset{H}{\overset{H}{C}}{-}\underset{H}{\overset{H}{C}}{-}\underset{H}{\overset{H}{C}}{-}H} + 25/2\,\ddot{\underset{..}{O}}{=}\ddot{\underset{..}{O}}$$

$$\longrightarrow 8\,\ddot{\underset{..}{O}}{=}C{=}\ddot{\underset{..}{O}} + 9\,H{-}\ddot{\underset{..}{O}}{-}H$$

Determine which bonds are broken in the reaction and sum the bond energies of the following:

$$\begin{aligned}&\Sigma(\Delta H\text{'s bonds broken})\\&= 18\text{ mol}(C{-}H) + 25/2\text{ mol}(O{=}O) + 7\text{ mol}(C{-}C)\\&= 18\text{ mol}(414\text{ kJ/mol}) + 25/2\text{ mol}(498) + 7\text{ mol}(347)\\&= 16106\text{ kJ/mol}\end{aligned}$$

Determine which bonds are formed in the reaction and sum the negatives of the bond energies of the following:

$$\begin{aligned}&\Sigma(-\Delta H\text{'s bonds formed})\\&= -16\text{ mol}(C{=}O) - 18\text{ mol}(O{-}H)\\&= -16\text{ mol}(799\text{ kJ/mol}) - 18\text{ mol}(464\text{ kJ/mol})\\&= -21136\text{ kJ/mol}\end{aligned}$$

Find ΔH_{rxn} by summing the results of the two steps.

$$\begin{aligned}\Delta H_{rxn} &= \Sigma(\Delta H\text{'s bonds broken}) + \Sigma(-\Delta H\text{'s bonds formed})\\&= 16106\text{ kJ/mol} - 21136\text{ kJ/mol}\\&= -5030\text{ kJ/mol}\end{aligned}$$

The ΔH_f for octane is given in Example 10.13. $\Delta H_f^\circ = -250.1 \text{ kJ/mol}$

$$\begin{aligned}\Delta H^\circ_{rxn} &= \Sigma[n_p \Delta H_f^\circ(\text{products})] - \Sigma[n_r \Delta H_f^\circ(\text{reactants})]\\ &= [8 \text{ mol } \Delta H_f^\circ(CO_2(g)) + 9 \text{ mol} \Delta H_f^\circ(H_2O(g))] - [1 \text{ mol } \Delta H_f^\circ(C_8H_{18}(l)) + 25/2 \text{ mol } \Delta H_f^\circ(O_2(g))]\\ &= [8 \text{ mol}(-393.5 \text{ kJ/mol}) + 9 \text{ mol}(-241.8 \text{ kJ/mol})] - [1 \text{ mol}(-250.1 \text{ kJ/mol}) + 25/2 \text{ mol}(0)]\\ &= -5074 \text{ kJ/mol}\end{aligned}$$

$$\% \text{ difference} = \frac{-5074 \text{ kJ} - (-5030 \text{ kJ})}{-5074 \text{ kJ}} \times 100\% = 0.8672\%$$

You would expect the value calculated from the heats of formation data to be more accurate. The bond energy values are average values, not values for a specific molecule. The heats of formation are for specific compounds.

10.119 **Given:** heat atomization CH_4 = 1660 kJ/mol, CH_2Cl_2 = 1495 kJ/mol **Find:** bond energy C—Cl Write the reaction using the Lewis structure.

$$H{-}\underset{\underset{H}{|}}{\overset{\overset{H}{|}}{C}}{-}H(g) \longrightarrow C(g) + 4\,H(g)$$

Determine the number and kinds of bonds broken and then ΔH atomization = Σ bonds broken.

ΔH atomization = $\Sigma\, 4\,(C{-}H)$ bonds broken

$$\frac{1660 \text{ kJ}}{1 \text{ mol } CH_4} \times \frac{1 \text{ mol } CH_4}{4 \text{ C—H bonds}} = 415 \text{ kJ/C—H bond}$$

Write the reaction using the Lewis structure.

$$:\ddot{\underset{\cdot\cdot}{Cl}}{-}\underset{\underset{:Cl:}{|}}{\overset{\overset{H}{|}}{C}}{-}H(g) \longrightarrow C(g) + 2\,H(g) + 2\,Cl(g)$$

Determine the number and kinds of bonds broken and ΔH atomization = Σ bonds broken.

ΔH atomization = $\Sigma\, 2\,(C{-}H)$ bonds broken + $2(C{-}Cl)$ bonds broken

$1495 \text{ kJ/mol} = 2(415 \text{ kJ/mol}) + 2(x)$ $\quad x = 333 \text{ kJ/mol}$ for the C—Cl bond energy

Check: The bond energy found (333 kJ/mol) is very close to the table value of 339 kJ/mol.

10.120 **Given:** bond energy: C—H = 414 kJ/mol; C—Cl = 339 kJ/mol; C=C = 611 kJ/ mol

Find: heat of atomization of C_2H_3Cl

Write the reaction using the Lewis structure.

$$:\ddot{\underset{\cdot\cdot}{Cl}}{-}\overset{\overset{H}{|}}{C}{=}\overset{\overset{H}{|}}{C}{-}H(g) \longrightarrow 2\,C(g) + 3\,H(g) + 2\,Cl(g)$$

Determine the number and kinds of bonds broken and then ΔH atomization = Σ bonds broken.

$$\begin{aligned}\Delta H \text{ atomization} &= \Sigma\, 3\,(C{-}H) \text{ bonds broken} + (C{-}Cl) \text{ bond broken} + (C{=}C) \text{ bond broken}\\ &= 3(414 \text{ kJ/mol}) + (339 \text{ kJ/mol}) + (611 \text{ kJ/mol})\\ &= 2192 \text{ kJ/mol}\end{aligned}$$

Challenge Problems

10.121 **Given:** 655 kWh/yr; coal is 3.2% S; remainder is C; S emitted as $SO_2(g)$ and gets converted to H_2SO_4 when reacting with water **Find:** $m(H_2SO_4)$/yr

Conceptual Plan: Write balanced reaction then $\Delta H^\circ_{rxn} = \sum n_p \Delta H_f^\circ(\textit{products}) - \sum n_r \Delta H_f^\circ(\textit{reactants})$

(because the form of sulfur is not given, assume that all heat is from combustion of only carbon) then

kWh → J → kJ → mol (C) → g (C) → g (S) → mol (H_2SO_4) → mol (H_2SO_4) → g (H_2SO_4)

$$\frac{3.60 \times 10^6 \text{J}}{1 \text{ kWh}} \quad \frac{1 \text{ kJ}}{1000 \text{ J}} \quad \frac{\text{mol C}}{\Delta H_f^\circ(CO_2(g))} \quad \frac{12.01 \text{ g}}{1 \text{ mol}} \quad \frac{3.2 \text{ g S}}{(100.0 - 3.2) \text{ g C}} \quad \frac{1 \text{ mol}}{32.07 \text{ g}} \quad \frac{1 \text{ mol } H_2SO_4}{1 \text{ mol S}} \quad \frac{98.09 \text{ g}}{1 \text{ mol}}$$

Solution: $C(s) + O_2(g) \rightarrow CO_2(g)$. This reaction is the heat of formation of $CO_2(g)$, so $\Delta H^\circ_{rxn} = \Delta H_f^\circ(CO_2(g)) = -393.5 \text{ kJ/mol}$ then

$$655\ \cancel{kWh} \times \frac{3.60 \times 10^6\ \cancel{J}}{1\ \cancel{kWh}} \times \frac{1\ \cancel{kJ}}{1000\ \cancel{J}} \times \frac{\cancel{mol\ C}}{393.5\ \cancel{kJ}} \times \frac{12.01\ \cancel{g\ C}}{1\ \cancel{mol\ C}} \times \frac{3.2\ \cancel{g\ S}}{(100.0 - 3.2)\ \cancel{g\ C}} \times \frac{1\ \cancel{mol\ S}}{32.07\ \cancel{g\ S}}$$

$$\times \frac{1\ \cancel{mol\ H_2SO_4}}{1\ \cancel{mol\ S}} \times \frac{98.09\ g\ H_2SO_4}{1\ \cancel{mol\ H_2SO_4}} = 7.3 \times 10^3\ g\ H_2SO_4$$

Check: The units (g) are correct. The magnitude (7300) is reasonable, considering this calculation is for only one home.

10.122 **Given:** 2.5×10^3 kg SUV; $v_1 = 0.0$ mph; $v_2 = 65.0$ mph; octane combustion, 30% efficiency **Find:** $m\ (CO_2)$
Conceptual Plan: **mi/hr → m/hr → m/min → m/s then m, v → KE then**

$$\frac{1000\ m}{0.6214\ mi} \quad \frac{1\ hr}{60\ min} \quad \frac{1\ min}{60\ sec} \quad KE = \frac{1}{2}mv^2$$

KE_1, KE_2 → ΔE used → ΔE generated

$$\Delta E_{sys} = KE_2 - KE_1 \quad \frac{100\ J\ generated}{30\ J\ used}$$

use reaction from Problem 10.116 $C_8H_{18}(l) + 25/2\ O_2(g) \rightarrow 8\ CO_2(g) + 9\ H_2O(g)$

with $\Delta H^\circ_{rxn} = -5074.1$ kJ

ΔE generated (J) → kJ → mol(C_8H_{18}) → mol CO_2 → g CO_2

$$\frac{1\ kJ}{1000\ J} \quad \frac{1\ mol\ C_8H_{18}}{5074.1\ kJ} \quad \frac{8\ mol\ CO_2}{1\ mol\ C_8H_{18}} \quad \frac{44.01\ g}{1\ mol}$$

Solution: $v_1 = 0.0$ m/s, $65.0\frac{\cancel{mi}}{\cancel{hr}} \times \frac{1000\ m}{0.6214\ \cancel{mi}} \times \frac{1\ \cancel{hr}}{60\ \cancel{min}} \times \frac{1\ \cancel{min}}{60\ sec} = 29.0563\frac{m}{s}$ then $KE = \frac{1}{2}mv^2$

$$KE_1 = \frac{1}{2}(2.5 \times 10^3\ kg)(0)^2 = 0$$

$$KE_2 = \frac{1}{2}(2.5 \times 10^3\ kg)\left(29.0563\frac{m}{s}\right)^2 = 1.05534 \times 10^6\frac{kg\ m^2}{s^2} = 1.05534 \times 10^6\ J$$

$$\Delta E_{sys} = KE_2 - KE_1 = 1.05534 \times 10^6\ J - 0\ J = 1.05534 \times 10^6\ \cancel{J\ used} \times \frac{100\ J\ generated}{30\ \cancel{J\ used}} =$$

3.51780×10^6 J generated

$$3.51780 \times 10^6\ \cancel{J\ generated} \times \frac{1\ \cancel{kJ}}{1000\ \cancel{J}} \times \frac{1\ \cancel{mol\ C_8H_{18}}}{5074.1\ \cancel{kJ}} \times \frac{8\ \cancel{mol\ CO_2}}{1\ \cancel{mol\ C_8H_{18}}} \times \frac{44.01\ g\ CO_2}{1\ \cancel{mol\ CO_2}} = 240\ g\ CO_2$$

Check: The units (g) are correct. The magnitude (240) is reasonable, considering the vehicle is so heavy and we generate 8 moles of CO_2 for each mole of octane.

10.123 **Given:** methane combustion, 100% efficiency; $\Delta T = 10.0\ °C$; house = 30.0 m × 30.0 m × 3.0 m; C_s (air) = 30 J/K · mol; 1.00 mol air = 22.4 L **Find:** $m\ (CH_4)$
Conceptual Plan: **$l, w, h \rightarrow V(m^3) \rightarrow V(cm^3) \rightarrow V(L) \rightarrow$ mol (air) then $m, C_s, \Delta T \rightarrow q_{air}$ (J)**

$$V = lwh \quad \frac{(100\ cm)^3}{(1\ m)^3} \quad \frac{1\ L}{1000\ cm^3} \quad \frac{1\ mol\ air}{22.4\ L} \quad q = mC_s\Delta T$$

then q_{air} (J) → q_{rxn} (J) → q (kJ) then write balanced reaction for methane combustion

$$q_{rxn} = -q_{air} \quad \frac{1\ kJ}{1000\ J}$$

then $\Delta H^\circ_{rxn} = \sum n_p \Delta H^\circ_f(products) - \sum n_r \Delta H^\circ_f(reactants)$, **and then q (kJ) → mol (CH_4) → g (CH_4)**

$$\Delta H^\circ_{rxn} \quad \frac{16.04\ g}{1\ mol}$$

Solution: $V = lwh = 30.0\ m \times 30.0\ m \times 3.0\ m = 2700\ m^3$, then

$$2700\ \cancel{m^3} \times \frac{(100\ \cancel{cm})^3}{(1\ \cancel{m^3})} \times \frac{1\ \cancel{L}}{1000\ \cancel{cm^3}} \times \frac{1\ mol\ air}{22.4\ \cancel{L}} = 1.20536 \times 10^5\ mol\ air, and then$$

$$q = mC_s\Delta T = 1.20536 \times 10^5\ \cancel{mol} \times 30\frac{J}{\cancel{mol} \cdot \cancel{°C}} \times 10.0\ \cancel{°C} = 3.6161 \times 10^7\ \cancel{J} \times \frac{1\ kJ}{1000\ \cancel{J}} = 3.6161 \times 10^4\ J\ needed$$

$CH_4(g) + 2\ O_2(g) \rightarrow CO_2(g) + 2\ H_2O(g)$

Reactant/Product	ΔH_f° (kJ/mol from Appendix IVB)
$CH_4(g)$	−74.6
$O_2(g)$	0.0
$CO_2(g)$	−393.5
$H_2O(g)$	−241.8

Be sure to pull data for the correct formula and phase.

$$\begin{aligned}\Delta H^\circ_{rxn} &= \sum n_p \Delta H_f^\circ(products) - \sum n_r \Delta H_f^\circ(reactants)\\ &= [1(\Delta H_f^\circ(CO_2(g))) + 2(\Delta H_f^\circ(H_2O(g)))] - [1(\Delta H_f^\circ(CH_4(g))) + 2(\Delta H_f^\circ(O_2(g)))]\\ &= [(-393.5\text{ kJ}) + 2(-241.8\text{ kJ})] - [1(-74.6\text{ kJ}) + 2(0.0\text{ kJ})]\\ &= [-877.1\text{ kJ}] - [-74.6\text{ kJ}]\\ &= -802.5\text{ kJ}\end{aligned}$$

$$q_{rxn} = -q_{air} = -3.6161 \times 10^4\ \text{kJ} \times \frac{1\ \text{mol CH}_4}{-802.5\ \text{kJ}} \times \frac{16.04\ \text{g CH}_4}{1\ \text{mol CH}_4} = 722.8\ \text{g CH}_4 = 700\ \text{g CH}_4$$

Check: The units (g) are correct. The magnitude (700) is not surprising because the volume of a house is large.

10.124 **Given:** water: $V = 35$ L, $T_i = 25.0$ °C, $T_f = 100.0$ °C; fuel = C_7H_{16}, 15% efficiency, $d = 0.78$ g/ml
Find: V (fuel)
Conceptual Plan: Write balanced reaction then $\Delta H^\circ_{rxn} = \sum n_p \Delta H_f^\circ(products) - \sum n_r \Delta H_f^\circ(reactants)$ then L → mL → g then $T_i, T_f \rightarrow \Delta T$ then $m, C_s, \Delta T \rightarrow q_{H_2O}$ (J) → q_{H_2O} (kJ) → q_{rxn} (kJ)

$\frac{1000\text{ mL}}{1\text{ L}}$ $\frac{1.0\text{ g}}{1.0\text{ mL}}$ $\Delta T = T_f - T_i$ $q_{H_2O} = m_{H_2O} C_{s,\,H_2O} \Delta T_{H_2O}$ $\frac{1\text{ kJ}}{1000\text{ J}}$ $q_{rxn} = -q_{H_2O}$

then q_{rxn} generated (J) → q_{rxn} used (kJ) → mol C_7H_{16} → g C_7H_{16} → mL C_7H_{16}

$\frac{100\text{ kJ generated}}{15\text{ kJ needed}}$ $\frac{1\text{ mol }C_7H_{16}}{\Delta H^\circ_{rxn}}$ $\frac{100.21\text{ g}}{1\text{ mol}}$ $\frac{1.0\text{ mL}}{0.78\text{ g}}$

Solution: Combustion is the combination with oxygen to form carbon dioxide and water:
$C_7H_{16}(l) + 11\,O_2(g) \rightarrow 7\,CO_2(g) + 8\,H_2O(g)$

Reactant/Product	ΔH_f° (kJ/mol from Appendix IVB)
$C_7H_{16}(l)$	−224.4
$O_2(g)$	0.0
$CO_2(g)$	−393.5
$H_2O(g)$	−241.8

Be sure to pull data for the correct formula and phase.

$$\begin{aligned}\Delta H^\circ_{rxn} &= \sum n_p \Delta H_f^\circ(products) - \sum n_r \Delta H_f^\circ(reactants)\\ &= [7(\Delta H_f^\circ(CO_2(g))) + 8(\Delta H_f^\circ(H_2O(g)))] - [1(\Delta H_f^\circ(C_7H_{16}(l))) + 11(\Delta H_f^\circ(O_2(g)))]\\ &= [7(-393.5\text{ kJ}) + 8(-241.8\text{ kJ})] - [1(-224.4\text{ kJ}) + 11(0.0\text{ kJ})]\\ &= [-4688.9\text{ kJ}] - [-224.4\text{ kJ}]\\ &= -4464.5\text{ kJ}\end{aligned}$$

$$35\ \text{L} \times \frac{1000\ \text{mL}}{1\ \text{L}} \times \frac{1.0\ \text{g}}{1.0\ \text{mL}} = 35000\ \text{g then } \Delta T = T_f - T_i = 100.0\ °\text{C} - 25.0\ °\text{C} = 75.0\ °\text{C then}$$

$$q_{H_2O} = m_{H_2O} C_{s,\,H_2O} \Delta T_{H_2O} = 35000\ \text{g} \times 4.18\ \frac{\text{J}}{\text{g} \cdot °\text{C}} \times 75.0\ °\text{C} = 1.09725 \times 10^7\ \text{J} \times \frac{1\ \text{kJ}}{1000\ \text{J}} = 1.09725 \times 10^4\ \text{kJ}$$

$$q_{rxn} = -q_{H_2O} = -1.09725 \times 10^4\ \text{kJ} \times \frac{100\ \text{kJ generated}}{15\ \text{kJ used}} \times \frac{1\text{mol C}_7\text{H}_{16}}{-4464.5\ \text{kJ}} \times \frac{100.21\ \text{g C}_7\text{H}_{16}}{1\ \text{mol C}_7\text{H}_{16}} \times \frac{1.0\ \text{mL C}_7\text{H}_{16}}{0.78\ \text{g C}_7\text{H}_{16}} =$$

2100 mL C_7H_{16} = 2.1 L C_7H_{16}

Check: The units (mL) are correct. The magnitude (2 L) is a reasonable volume to have to take on a back-packing trip.

10.125 **Given:** m(ice) = 9.0 g; coffee: $T_1 = 90.0$ °C, $m = 120.0$ g, $C_s = C_{s,\,H_2O}$, $\Delta H^\circ_{fus} = 6.0$ kJ/mol **Find:** T_f of coffee
Conceptual Plan: $q_{ice} = -q_{coffee}$ so g (ice) → mol (ice) → q_{fus} (kJ) → q_{fus} (J) → q_{coffee} (J) then

$\frac{1\text{ mol}}{18.01\text{ g}}$ $\frac{6.0\text{ kJ}}{1\text{ mol}}$ $\frac{1000\text{ J}}{1\text{ kJ}}$ $q_{coffee} = -q_{ice}$

$q, m, C_s \rightarrow \Delta T$ then $T_i, \Delta T \rightarrow T_2$; now we have slightly cooled coffee in contact with 0.0 °C water

$q = mC_s\Delta T$ $\quad\quad \Delta T = T_2 - T_i$

so $q_{ice} = -q_{coffee}$ with $m, C_s, T_i \rightarrow T_f$

$q = mC_s(T_f - T_i)$ then set $q_{H_2O} = -q_{coffee}$

Solution: $9.0\ \text{g} \times \frac{1\ \text{mol}}{18.01\ \text{g}} \times \frac{6.0\ \text{kJ}}{1\ \text{mol}} \times \frac{1000\ \text{J}}{1\ \text{kJ}} = 2.9983 \times 10^3\ \text{J}, q_{coffee} = -q_{ice} = -2.9983 \times 10^3\ \text{J}$

$q = mC_s\Delta T$. Rearrange to solve for ΔT. $\Delta T = \frac{q}{mC_s} = \frac{-2.9983 \times 10^3\ \text{J}}{120.0\ \text{g} \times 4.18 \frac{\text{J}}{\text{g} \cdot °\text{C}}} = -5.9775\ °\text{C}$ then

$\Delta T = T_2 - T_i$. Rearrange to solve for T_2. $T_2 = \Delta T + T_i = -5.9775\ °\text{C} + 90.0\ °\text{C} = 84.0225\ °\text{C}$
$q = mC_s(T_f - T_i)$ substitute in values and set $q_{H_2O} = -q_{coffee}$

$q_{H_2O} = m_{H_2O}C_{s,\ H_2O}(T_f - T_{H_2O,i}) = 9.0\ \text{g} \times 4.18 \frac{\text{J}}{\text{g} \cdot °\text{C}} \times (T_f - 0.0\ °\text{C}) =$

$-q_{coffee} = -m_{coffee}C_{s,\ coffee}(T_f - T_{coffee,\ i}) = -120.0\ \text{g} \times 4.18 \frac{\text{J}}{\text{g} \cdot °\text{C}} \times (T_f - 84.0225\ °\text{C})$

Rearrange to solve for T_f.
$9.0\ \text{g}\ T_f = -120.0\ \text{g}\ (T_f - 84.0225\ °\text{C}) \rightarrow 9.0\ \text{g}\ T_f = -120.0\ \text{g}\ T_f + 10082.7\ \text{g} \rightarrow$

$+10082.7\ \text{g} = +129.0 \frac{\text{g}}{°\text{C}} T_f \rightarrow T_f = \frac{+10082.7\ \text{g}}{+129.0 \frac{\text{g}}{°\text{C}}} = 78.2\ °\text{C}$

Check: The units (°C) are correct. The temperature is closer to the original coffee temperature because the mass of coffee is so much larger than the ice mass.

10.126 **Given:** liquid water at $-10.0\ °\text{C}$, C_s (ice) $= 2.04\ \text{J/g} \cdot °\text{C}$; $\Delta H°_{fus} = -332\ \text{J/g}$ (at 0.0 °C)
Find: ΔH, ΔE, q, and w for freezing at $-10.0\ °\text{C}$
Conceptual Plan: Assume exactly 1 g H_2O for all calculations (report answers as J/g) and constant $P = 1$ atm. Construct the following path: According to Hess's law, $\Delta H_1 + \Delta H_2 + \Delta H_3 = \Delta H_4 = \Delta H°_{fus}$ at $-10.0\ °\text{C}$

	step 2		
Liquid at 0.0 °C	→	**solid at 0.0 °C**	
↑ **step 1**		↓ **step 3**	**at constant P, $\Delta H = q$**
Liquid at −10.0 °C	→	**solid at −10.0 °C**	**For steps 1 and 3, $q = mC_s\Delta T$**
	step 4		

Look up the density of liquid and solid water at 0.0 °C. (Assume that the density of each phase does not change significantly at −10.0 °C.)

$d_L = 0.9998\ \text{g/mL}$ and $d_S = 0.917\ \text{g/mL}$

g → mL → L then $V_L, V_S \rightarrow \Delta V$ then $P, \Delta V \rightarrow w$(L atm) → w(J) then $q, w \rightarrow \Delta E$

L: $\frac{1\ \text{mL}}{0.9998\ \text{g}}$ S: $\frac{1\ \text{mL}}{0.917\ \text{g}}$ $\quad \frac{1\ \text{L}}{1000\ \text{mL}}$ $\quad \Delta V = V_S - V_L$ $\quad w = -P\Delta V$ $\quad \frac{101.3\ \text{J}}{1\ \text{L} \cdot \text{atm}}$ $\quad \Delta E = q + w$

Solution: $\Delta H_1 = q_1 = mC_S\Delta T = 1\ \text{g} \times 4.18 \frac{\text{J}}{\text{g} \cdot °\text{C}} \times (0.00\ °\text{C} - (-10.0\ °\text{C})) = +41.8\ \text{J}$

$\Delta H_2 = q_2 = m\Delta H = 1\ \text{g} \times -332 \frac{\text{J}}{\text{g}} = -332\ \text{J}$

$\Delta H_3 = q_3 = mC_s\Delta T = 1\ \text{g} \times 2.04 \frac{\text{J}}{\text{g} \cdot °\text{C}} \times (-10.00\ °\text{C} - 0.0\ °\text{C}) = -20.4\ \text{J}$

so $\Delta H_4 = q_4 = \Delta H_1 + \Delta H_2 + \Delta H_3 = +41.8\ \text{J} - 332\ \text{J} - 20.4\ \text{J} = -310.6\ \text{J} = -311\ \text{J/g}$

$V_L = 1\ \text{g} \times \frac{1\ \text{mL}}{0.9998\ \text{g}} \times \frac{1\ \text{L}}{1000\ \text{mL}} = 0.00100020004\ \text{L}$ and $V_S = 1\ \text{g} \times \frac{1\ \text{mL}}{0.917\ \text{g}} \times \frac{1\ \text{L}}{1000\ \text{mL}} = 0.0010905\ \text{L}$

then $\Delta V = V_S - V_L = 0.0010905\ \text{L} - 0.00100020004\ \text{L} = 9.029996 \times 10^{-5}\ \text{L}$

then $w = -P\Delta V = -1\ \cancel{\text{atm}} \times \underline{9}.029996 \times 10^{-5}\ \cancel{\text{L}} \times \dfrac{101.3\ \text{J}}{1\ \cancel{\text{L}}\cdot\cancel{\text{atm}}} = -0.00\underline{9}147\ \text{J} = -0.009\ \text{J/g}$

and $\Delta E = q + w = -31\underline{0}.6\ \text{J} - 0.00\underline{9}147\ \text{J} = -311\ \text{J/g}$

Check: The units (J/g) are correct. We expect freezing to release less energy at −10 °C because we are below the normal freezing point. The work is negligible because the volume change is so small.

10.127 $q = \Delta H = 454\ \cancel{\text{g}} \times \dfrac{1\ \cancel{\text{mol}}}{18.02\ \cancel{\text{g}}} \times \dfrac{40.7\ \text{kJ}}{1\ \cancel{\text{mol}}} = 10\underline{2}5.405\ \text{kJ} = 1030\ \text{kJ}$ and $w = -P\Delta V$. Assume that $P = 1$ atm (exactly) and $\Delta V = V_G - V_L$, where $V_L = 454\ \cancel{\text{g}} \times \dfrac{1\ \cancel{\text{mL}}}{0.9998\ \cancel{\text{g}}} \times \dfrac{1\ \text{L}}{1000\ \cancel{\text{mL}}} = 0.45\underline{4}0908\ \text{L}$ and $PV = nRT$.

Rearrange to solve for V_G. $V_G = \dfrac{nRT}{P} = \dfrac{454\ \cancel{\text{g}} \times \dfrac{1\ \cancel{\text{mol}}}{18.02\ \cancel{\text{g}}} \times 0.08206\ \dfrac{\text{L}\cdot\cancel{\text{atm}}}{\cancel{\text{mol}}\cdot\cancel{\text{K}}} \times 373\ \cancel{\text{K}}}{1\ \cancel{\text{atm}}} = 77\underline{1}.1545\ \text{L}$

$\Delta V = V_G - V_L = 77\underline{1}.1545\ \text{L} - 0.45\underline{4}0908\ \text{L} = 77\underline{0}.7004\ \text{L}$ and so

$w = -P\Delta V = -1.0\ \cancel{\text{atm}} \times 77\underline{0}.7004\ \cancel{\text{L}} \times \dfrac{101.3\ \text{J}}{1\ \cancel{\text{L}}\cdot\cancel{\text{atm}}} = -78\underline{0}71.9515\ \text{J} = -7.81 \times 10^4\ \text{J} = -78.1\ \text{kJ}$

Finally, $\Delta E = q + w = 10\underline{2}5.405\ \text{kJ} - 78.\underline{0}719515\ \text{kJ} = 9\underline{4}7.333\ \text{kJ} = 950\ \text{kJ}$.

10.128 $C_2H_6O(l) + 3\ O_2(g) \rightarrow 2\ CO_2(g) + 3\ H_2O(l)$

$q_{Cal} = C\Delta T = 34.65\ \dfrac{\text{kJ}}{\cancel{\text{K}}} \times (295.84\ \cancel{\text{K}} - 294.33\ \cancel{\text{K}}) = 52.\underline{3}215\ \text{kJ} = -q_{rxn} = -\Delta E_{rxn}$ and

$n = 1.765\ \cancel{\text{g}} \times \dfrac{1\ \text{mol}}{46.07\ \cancel{\text{g}}} = 0.0383\underline{1}2929\ \text{mol}$

There is no work because $w = -P\Delta V$, and there is no volume change in a bomb calorimeter. On a per-mole basis,

$q_{rxn} = \Delta E_{rxn} = \dfrac{-52.\underline{3}215\ \text{kJ}}{0.0383\underline{1}2929\ \text{mol}} = -13\underline{6}5.6356\ \text{kJ/mol} = -1370\ \text{kJ/mol}$. Because this is not a constant pressure problem, $\Delta E_{rxn} \neq \Delta H_{rxn}$. In fact, $\Delta H_{rxn} = \Delta E_{rxn} + \Delta nRT$. In this case, the change in the number of moles of gas $\Delta n_G = 2\ \text{mol} - 3\ \text{mol} = -1$ mol and so

$\Delta H_{rxn} = \Delta E_{rxn} + \Delta nRT = -13\underline{6}5.6356\ \text{kJ/mol} + \left(-1\ \cancel{\text{mol}} \times 0.08206\ \dfrac{\cancel{\text{L}}\cdot\cancel{\text{atm}}}{\cancel{\text{mol}}\cdot\cancel{\text{K}}} \times 295\ \cancel{\text{K}} \times \dfrac{101.3\ \cancel{\text{J}}}{1\ \cancel{\text{L}}\cdot\cancel{\text{atm}}} \times \dfrac{1\ \text{kJ}}{1000\ \cancel{\text{J}}}\right) =$

$-13\underline{6}8.0878\ \text{kJ/mol} = -1370\ \text{kJ/mol}$.

This answer is in the same ballpark as the answer in Problem 10.86, which had gaseous water as a product.

10.129

1 $\ddot{O}{=}\ddot{S}{=}\ddot{O} + H{-}\ddot{O}\cdot \longrightarrow\ :\!\ddot{O}{-}S(=O){-}\ddot{O}{-}H$

2 $:\!\ddot{O}{-}S(=O){-}\ddot{O}{-}H + \ddot{O}{=}\ddot{O} \longrightarrow\ :\!\ddot{O}{-}S(=O){-}\ddot{O}\!: + H{-}\ddot{O}{-}\ddot{O}\cdot$

3 $:\!\ddot{O}{-}S(=O){-}\ddot{O} + H{-}\ddot{O}{-}H \longrightarrow \ddot{O}{=}S(=O)(-O{-}H){-}\ddot{O}{-}H$

Step 1:

Bonds broken: $2\ mol(S{=}O) + 1\ mol(H{-}O) = 2\ mol(523\ kJ/mol) + 1\ mol(464\ kJ/mol) = 1510\ kJ/mol$

Bonds formed: $-2\ mol(S{-}O) - 1\ mol(S{=}O) - 1\ mol(O{-}H) =$
$-2\ mol(265\ kJ/mol) - 1\ mol(523\ kJ/mol) - 1\ mol(464\ kJ/mol) = -1517\ kJ/mol$

$\Delta H_{step} = -7\ kJ/mol$

Step 2:

Bonds broken: $2\ mol(S{-}O) + 1\ mol(S{=}O) + 1\ mol(O{-}H) + 1\ mol(O{=}O) =$
$2\ mol(265\ kJ/mol) + 1\ mol(523\ kJ/mol) + 1\ mol(464\ kJ/mol) + 1\ mol(498\ kJ/mol) =$
$2015\ kJ/mol$

Bonds formed: $-2\ mol(S{-}O) - 1\ mol(S{=}O) - 1\ mol(O{-}H) - 1\ mol(O{=}O) =$
$-2\ mol(265\ kJ/mol) - 1\ mol(523\ kJ/mol) - 1\ mol(464\ kJ/mol) - 1\ mol(142\ kJ/mol) =$
$-1659\ kJ/mol$

$\Delta H_{step} = +356\ kJ/mol$

Step 3:

Bonds broken: $2\ mol(S{-}O) + 1\ mol(S{=}O) + 2\ mol(O{-}H) =$
$2\ mol(265\ kJ/mol) + 1\ mol(523\ kJ/mol) + 2\ mol(464\ kJ/mol) = 1981\ kJ/mol$

Bonds formed: $-2\ mol(S{-}O) - 2\ mol(S{=}O) - 2\ mol(O{-}H) =$
$-2\ mol(265\ kJ/mol) + -2\ mol(523\ kJ/mol) + -2\ mol(464\ kJ/mol) = -2504\ kJ/mol$

$\Delta H_{step} = -523\ kJ/mol$

Hess's law states that ΔH for the reaction is the sum of ΔH of the steps:
$\Delta H_{rxn} = (-7\ kJ/mol) + (+356\ kJ/mol) + (-523\ kJ/mol) = -174\ kJ/mol$

10.130 Formation reaction: $6\ C(s) + 3\ H_2(g) \rightarrow C_6H_6(g)\ \Delta H_f^\circ = 82.9\ kJ/mol$
Using bond energies, we would have the reaction $6\ C(g) + 3\ H_2(g) \rightarrow C_6H_6(g)$; so we have to include in the bond energy calculation the energy needed to convert $C(s) \rightarrow C(g)(718.4\ kJ/mol)$.

$6\ molC(s) \rightarrow 6\ molC(g) \quad 6\ mol(718.4\ kJ/mol) = 4310.4\ kJ/mol$

Rewrite the reaction with the Lewis structures.

$6\ C(g) + 3\ H - H(g) \rightarrow$ [benzene Lewis structure: six C in a ring with alternating C=C double bonds, each C bonded to one H] (g)

Bonds broken: $3\ mol(H{-}H) = 3\ mol(436\ kJ/mol) = 1308\ kJ/mol$

Bonds formed: $-3\ mol(C{=}C)\ -3\ mol(C{-}C)\ -6\ mol(C{-}H) =$
$-3\ mol(611kJ/mol) - 3\ mol(347\ kJ/mol) - 6\ mol(414\ kJ/mol) = -5358\ kJ/mol$

ΔH from bond energies $= +(4310\ kJ) + (1308\ kJ) - 5358\ kJ = +260\ kJ/mol$

The difference between the value calculated from bond energies (260 kJ/mol) and $\Delta H_f^\circ = 82.9\ kJ/mol$ for benzene leads us to conclude that there is a great deal of stabilization from the two resonance forms and that they contribute much to the formation of benzene.

10.131 **Given:** $\Delta H_f^\circ\ CaBr_2 = -675\ kJ/mol$; $IE_1(Ca) = 590\ kJ/mol$; $IE_2(Ca) = 1145\ kJ/mol$; $\Delta H_{sub}(Ca) = 178\ kJ/mol$; $Br_2(g)$ bond energy $= 193\ kJ/mol$; $\Delta H_{vap}(Br_2(l)) = 31\ kJ/mol$; $EA(Br) = -325$ kJ/mol **Find:** lattice energy

Conceptual Plan:

$$\mathbf{Ca(s) + Br_2(l) \xrightarrow{\Delta H_{sub}} Ca(g) + Br_2(l) \xrightarrow{IE_1} Ca^+(g) + Br_2(l) \xrightarrow{IE_2} Ca^{2+}(g) + Br_2(l) \xrightarrow{\Delta H_{vap}} Ca^{2+}(g) + Br_2(g) \xrightarrow{\text{bond energy}} Ca^{2+}(g) + 2Br(g)}$$

$$\mathbf{\xrightarrow{EA} Ca^{2+}(g) + 2Br^-(g) \xrightarrow{\text{lattice energy}} CaBr_2(s)}$$

ΔH_f° (from $Ca(s) + Br_2(l)$ directly to $CaBr_2(s)$)

Solution: $\mathbf{\Delta H_f^\circ = \Delta H_{sub} + IE_1 + IE_2 + \Delta H_{vap} + bond\ energy + 2\,EA + lattice\ energy}$

$$-675\frac{kJ}{mol} = +178\frac{kJ}{mol} + 590\frac{kJ}{mol} + 1145\frac{kJ}{mol} + 31\frac{kJ}{mol} + 193\frac{kJ}{mol} + 2(-325)\frac{kJ}{mol} + \text{lattice energy}$$

lattice energy $= -2162\ kJ/mol$

10.132 **Given:** $\Delta H_f^\circ PI_3(s) = -24.7 kJ/mol$; $P—I = 184 kJ/mol$; $I—I = 151 kJ/mol$; $\Delta H_f^\circ P(g) = 334 kJ/mol$; $\Delta H_f^\circ I_2(g) = 62 kJ/mol$ **Find:** $\Delta H_{sub}(PI_3(s))$

Conceptual Plan: $PI_3(s) \rightarrow PI_3(g)$; use Hess's law

Solution:

Reaction	ΔH(kJ/mol)	
$PI_3(s) \rightarrow \cancel{P(s)} + \cancel{3/2 I_2(s)}$	+ 24.7	(this is the reverse of the formation reaction)
$\cancel{P(s)} \rightarrow \cancel{P(g)}$	+ 334	(formation of $P(g)$)
$\cancel{3/2 I_2(s)} \rightarrow \cancel{3/2 I_2(g)}$	3/2(62)	(formation of $I_2(g)$)
$\cancel{3/2 I_2(g)} \rightarrow \cancel{3 I(g)}$	3/2(151)	(breaking I—I bond)
$\cancel{P(g)} + \cancel{3 I(g)} \rightarrow PI_3(g)$	− 3(184)	(forming P—I bond)
$PI_3(s) \rightarrow PI_3(g)$	+126	(sublimation of $PI_3(s)$)

Conceptual Problems

10.133 (d) Only one answer is possible: $\Delta E_{sys} = -\Delta E_{surr}$.

10.134 (a) At constant P, $\Delta E_{sys} = q + w = q_P + w = \Delta H + w$; so $\Delta E_{sys} - w = \Delta H = q$.

10.135 Refrigerator A contains only air, which will cool quickly but will not stabilize the temperature. Refrigerator B is filled with containers of water, which require a great deal of energy to cool on day 1 but will remain stable at a cold temperature on day 2.

10.136 The aluminum cylinder will be cooler after 1 hour because it has a lower specific heat capacity than does water (less heat needs to be pulled out for every °C temperature change).

10.137 Because $q = mC_s\Delta T$, $m_A = 2\ m_B$, $C_{s,\ B} = 4\ C_{s,\ A}$, $q_A = -q_B$, we can substitute into the equation to get $q_A = m_A C_{s,\ A}\Delta T_A = -q_B = -m_B C_{s,\ B}\Delta T_B \rightarrow (2\ m_B)C_{s,\ A}\Delta T_A = -m_B(4\ C_{s,\ A})\Delta T_B$

$\rightarrow \Delta T_A = \dfrac{-\cancel{m_B}(4\ \cancel{C_{s,\ A}})}{(2\ \cancel{m_B})\cancel{C_{s,\ A}}}\Delta T_B = \dfrac{-4}{2}\Delta T_B = -2\Delta T_B$, or the temperature change for substance A is twice the magnitude of the temperature change for substance B.

10.138 **Given:** 2418 J heat produced; 5 J work done on surroundings at constant P **Find:** ΔE, ΔH, q, and w

Conceptual Plan: Interpret language to determine the sign of the two terms then $q, w \rightarrow \Delta E_{sys}$

$$\Delta E = q + w$$

Solution: Because heat is released from the system to the surroundings, $q = -2418$ J;
because the system is doing work on the surroundings, $w = -5$ kJ. At constant P, $\Delta H = q = -2.418$ kJ;
$\Delta E = q + w = -2418\ J - 5\ J = -2423\ J = -2$ kJ.

Check: The units (kJ) are correct. The magnitude of the answer (-2) makes physical sense because both terms are negative and the amount of work done is negligibly small.

10.139 The internal energy of a chemical system is the sum of its kinetic energy and its potential energy. This potential energy is the energy source in an exothermic chemical reaction. Under normal circumstances, chemical potential energy (or simply chemical energy) arises primarily from the electrostatic forces between the protons and electrons that compose the atoms and molecules within the system. In an exothermic reaction, some bonds break and new ones form, and the protons and electrons go from an arrangement of higher potential energy to one of lower potential energy. As they rearrange, their potential energy is converted into kinetic energy. Heat is emitted in the reaction, so it feels hot to the touch.

10.140 (b) If ΔV is positive, then $w = -P\Delta V < 0$. Because $\Delta E_{sys} = q + w = q_P + w = \Delta H + w$; if w is negative, then $\Delta H > \Delta E_{sys}$.

10.141 (a) is true: Strong bonds break, and weak bonds form. In an endothermic reaction, the energy required to break the bonds is greater than the energy given off when the bonds are formed $(\Delta H > 0)$; therefore, in an endothermic reaction, the bonds that are breaking are stronger than the bonds that are forming.

10.142 When we say that a compound is "energy rich," we mean that it gives off a great amount of energy when it reacts. It means that a lot of energy is stored in the compound. This energy is released when the weak bonds in the compound break and much stronger bonds are formed in the product, thereby releasing energy.

11 Gases

Review Questions

11.1 Pressure is the force exerted per unit area by gas molecules as they strike the surfaces around them. Pressure is caused by collisions of gas molecules with surfaces or other gas molecules.

11.2 The pressure drops as altitude increases because the gravitational forces are decreased at higher altitudes compared to sea level.

11.3 If a pressure drop occurs fast enough, a large pressure difference would quickly develop between the air in your lungs and the surrounding low pressure or vacuum. The pressure difference would cause your lungs to expand too much, resulting in severe lung damage.

11.4 $101{,}325 \text{ Pa} = 760 \text{ torr} = 760 \text{ mmHg} = 29.92 \text{ in Hg} = 14.7 \text{ psi} = 1 \text{ atm}$

11.5 A manometer is a U-shaped tube containing a dense liquid, usually mercury. In an open-ended manometer, one end of the tube is open to atmospheric pressure and the other is attached to a flask containing the gas sample. If the pressure of the gas sample is exactly equal to atmospheric pressure, the mercury levels on both sides of the tube are the same. If the pressure of the sample is greater than atmospheric pressure, the mercury level on the sample side of the tube is lower than on the side open to the atmosphere. If the pressure of the sample is less than atmospheric pressure, the mercury level on the sample side is higher than on the side open to the atmosphere. This type of manometer always measures the pressure of the gas sample relative to atmospheric pressure. The difference in height between the two levels is equal to the pressure difference from atmospheric pressure.

11.6 Boyle's law states that the volume of the gas varies inversely to the pressure on the gas, while temperature and number of moles are kept constant ($P_1 V_1 = P_2 V_2$). Charles's law states that the volume of a gas is directly proportional to the temperature of the gas, while pressure and number of moles are kept constant ($V_1/T_1 = V_2/T_2$). All temperatures must be in kelvins when used in gas law calculations. Avogadro's law states that the volume of a gas is directly proportional to the number of moles of the gas, while pressure and temperature are kept constant ($V_1/n_1 = V_2/n_2$).

11.7 This pain is caused by air-containing cavities within your ear. When you ascend a mountain, the external pressure (the pressure that surrounds you) drops, while the pressure within your ear cavities (the internal pressure) remains the same. This creates an imbalance—the greater internal pressure forces your eardrum to bulge outward, causing pain. With time and the help of a yawn or two, the excess air within your ear cavities escapes, equalizing the internal and external pressure and relieving the pain.

11.8 For every 10 m of depth a diver descends in water, she experiences an additional 1 atm of pressure due to the weight of the water above her. The pressure regulator used in scuba diving delivers air at a pressure that matches the external pressure; otherwise, the diver could not inhale the air. For example, when a diver is at a depth of 20 m below the surface, the regulator delivers air at a pressure of 3 atm to match the 3 atm of pressure around the diver (1 atm due to normal atmospheric pressure and 2 additional atmospheres due to the weight of the water at 20 m). Suppose a diver inhaled a lungful of air at a pressure of 3 atm and swam quickly to the surface (where the pressure drops to 1 atm) while holding his breath. What would happen to the volume of air in the diver's lungs? Because the pressure decreases by a factor of three, the volume of the air in the diver's lungs would increase by a factor of three, severely damaging his lungs and possibly killing him.

11.9 Charles's law explains why the second floor of a house is usually a bit warmer than the ground floor because when air is heated, its volume increases, resulting in a lower density. The warm, less dense air tends to rise in a room filled with colder, denser air.

11.10 Charles's law explains why a hot-air balloon can take flight. The gas that fills a hot-air balloon is warmed with a burner, increasing its volume and lowering its density and causing it to float in the colder, denser surrounding air.

11.11 The ideal gas law ($PV = nRT$) combines all of the relationships between the four variables relevant to gases [pressure, volume, number of moles, and temperature (in kelvins)] in one simple expression.

11.12 We know that $V \propto 1/P$ (Boyle's law), $V \propto T$ (Charles's law), $V \propto n$ (Avogadro's law).

Combining these three expressions, we get $V \propto nT/P$. Replace the proportional sign with an equal sign by incorporating R (the ideal gas constant): $V = RnT/P$. Rearranging, we get $PV = nRT$.

11.13 The molar volume of an ideal gas is the volume occupied by one mole of gas at STP, which is $T = 0\,°\text{C}$ (273 K) and $P = 1.00$ atm. Substituting these values into the ideal gas law, one can calculate this value as 22.414 L.

11.14 Because $d = \frac{P\mathcal{M}}{RT}$, the density will decrease as temperature increases. It will increase as pressure increases or as the molar mass of the gas increases.

11.15 The pressure due to any individual component in a gas mixture is called the partial pressure (P_n) of that component and can be calculated from the ideal gas law by assuming that each gas component acts independently. The sum of the partial pressures of the components in a gas mixture must equal the total pressure: $P_{total} = P_a + P_b + P_c + \ldots$ where P_{total} is the total pressure and $P_a, P_b, P_c \ldots$ are the partial pressures of the components.

11.16 Too much oxygen can also cause physiological problems. Scuba divers breathe pressurized air. At 30 m, a scuba diver breathes air at a total pressure of 4.0 atm, making P_{O_2} about 0.84 atm. This elevated partial pressure of oxygen raises the density of oxygen molecules in the lungs, resulting in a higher concentration of oxygen in body tissues. When P_{O_2} increases beyond 1.4 atm, the increased oxygen concentration in body tissues causes a condition called oxygen toxicity, which results in muscle twitching, tunnel vision, and convulsions.

11.17 No, when collecting a gas over water, it will contain some water molecules. The vapor pressure of water can be found in Table 11.3. Therefore, $P_{Gas} = P_{Total} - P_{H_2O}$.

11.18 The basic postulates of kinetic molecular theory are as follows: (1) The size of a particle is negligibly small, (2) the average kinetic energy of a particle is proportional to the temperature in kelvins, and (3) the collision of one particle with another (or with the walls) is completely elastic. Pressure is defined as force divided by area. According to kinetic molecular theory, a gas is a collection of particles in constant motion. The motion results in collisions between the particles and the surfaces around them. As each particle collides with a surface, it exerts a force upon that surface. The result of many particles in a gas sample exerting forces on the surfaces around them is constant pressure.

11.19 Boyle's law states that for a constant number of particles at constant temperature, the volume of a gas is inversely proportional to its pressure. If you decrease the volume of a gas, you force the gas particles to occupy a smaller space. It follows from kinetic molecular theory that as long as the temperature remains the same, the result is a greater number of collisions with the surrounding surfaces and therefore a greater pressure.

Charles's law states that for a constant number of particles at constant pressure, the volume of a gas is proportional to its temperature. According to kinetic molecular theory, when you increase the temperature of a gas, the average speed, and thus the average kinetic energy, of the particles increases. Because this greater kinetic energy results in more frequent collisions and more force per collision, the pressure of the gas would increase if its volume were held constant (Gay-Lussac's law). The only way for the pressure to remain constant is for the volume to increase. The greater volume spreads the collisions out over a greater area, so that the pressure (defined as force per unit area) is unchanged.

Avogadro's law states that at constant temperature and pressure, the volume of a gas is proportional to the number of particles. According to kinetic molecular theory, when you increase the number of particles in a gas sample, the number of collisions with the surrounding surfaces increases. Because the greater number of collisions results in a greater overall force on surrounding surfaces, the only way for the pressure to remain constant is for the volume to increase so that the number of particles per unit volume (and thus the number of collisions) remains constant.

Dalton's law states that the total pressure of a gas mixture is the sum of the partial pressures of its components. In other words, according to Dalton's law, the components in a gas mixture act identically to, and independently of, one another. According to kinetic molecular theory, the particles have negligible size and they do not interact. Consequently, the only property that would distinguish one type of particle from another is its mass. However, even particles of different masses have the same average kinetic energy at a given temperature; so they exert the same force upon a collision with a surface. Consequently, adding components to a gas mixture—even different *kinds* of gases—has the same effect as simply adding more particles. The partial pressures of all of the components sum to the overall pressure.

11.20 Postulate 2 of kinetic molecular theory states that the average kinetic energy is proportional to the temperature in kelvins. The root mean square velocity of a collection of gas particles is inversely proportional to the square root of the molar mass of the particles in kilograms per mole.

11.21 Gaseous particles travel at tremendous speeds along very haphazard paths. To a perfume molecule, the path from the perfume bottle in the bathroom to your nose 2 m away is much like the path through a busy shopping mall during a clearance sale. The molecule travels only a short distance before it collides with another molecule, changes direction, only to collide again, and so on. The average distance a molecule travels between collisions is called its mean free path.

11.22 The process by which gas molecules spread out in response to a concentration gradient is called diffusion. Effusion is the process by which a gas escapes from a container into a vacuum through a small hole. The rate of effusion is inversely proportional to the square root of the molar mass of the gas.

11.23 In Chapter 8, we learned how the coefficients in chemical equations can be used as conversion factors between number of moles of reactants and number of moles of products in a chemical reaction, and how the molar mass can be used to convert the number of moles to the mass. At STP, each mole of gas occupies 22.414 L. The mass of the product will be as follows:

$$\text{Volume of limiting reagent (L)} \times \frac{1 \text{ mol limiting reagent}}{22.414 \text{ L}} \times \frac{\text{c mol product}}{\text{a mol limiting reagent}} \times \frac{\text{g product}}{1 \text{ mol product}}$$

for the reaction:
$a\,A + b\,B \rightarrow c\,C + d\,D$ where A is the limiting reagent and C is the product of interest.

11.24 Gases behave ideally when both of the following are true: (1) The volume of the gas particles is small compared to the space between them. (2) The forces between the gas particles are not significant. At high pressures, the number of molecules increases, so the volume of the gas particles is much greater; and because the spacing between the particles is much smaller, the interactions become more significant. At low temperatures, the molecules are not moving as fast as at higher temperatures, so that when they collide, they have a greater opportunity to interact.

Problems by Topic

Converting between Pressure Units

11.25 (a) **Given:** 24.9 in Hg **Find:** atm

Conceptual Plan: in Hg → atm

$$\frac{1 \text{ atm}}{29.92 \text{ in Hg}}$$

Solution: $24.9 \cancel{\text{in Hg}} \times \frac{1 \text{ atm}}{29.92 \cancel{\text{in Hg}}} = 0.832 \text{ atm}$

Check: The units (atm) are correct. The magnitude of the answer (< 1) makes physical sense because we started with less than 29.92 in Hg.

(b) **Given:** 24.9 in Hg **Find:** mmHg
Conceptual Plan: Use answer from part (a), then convert atm → mmHg

$$\frac{760\ \text{mmHg}}{1\ \text{atm}}$$

Solution: $0.832\ \text{atm} \times \frac{760\ \text{mmHg}}{1\ \text{atm}} = 632\ \text{mmHg}$

Check: The units (mmHg) are correct. The magnitude of the answer (< 760) makes physical sense because we started with less than 1 atm.

(c) **Given:** 24.9 in Hg **Find:** psi
Conceptual Plan: Use answer from part (a), then convert atm → psi

$$\frac{14.7\ \text{psi}}{1\ \text{atm}}$$

Solution: $0.832\ \text{atm} \times \frac{14.7\ \text{psi}}{1\ \text{atm}} = 12.2\ \text{psi}$

Check: The units (psi) are correct. The magnitude of the answer (< 14.7) makes physical sense because we started with less than 1 atm.

(d) **Given:** 24.9 in Hg **Find:** Pa
Conceptual Plan: Use answer from part (a), then convert atm → Pa

$$\frac{101{,}325\ \text{Pa}}{1\ \text{atm}}$$

Solution: $0.832\ \text{atm} \times \frac{101{,}325\ \text{Pa}}{1\ \text{atm}} = 8.43 \times 10^4\ \text{Pa}$

Check: The units (Pa) are correct. The magnitude of the answer ($< 101, 325$) makes physical sense because we started with less than 1 atm.

11.26 (a) **Given:** 235 mmHg **Find:** torr
Conceptual Plan: mmHg → torr

$$\frac{1\ \text{torr}}{1\ \text{mm Hg}}$$

Solution: $235\ \text{mmHg} \times \frac{1\ \text{torr}}{1\ \text{mmHg}} = 235\ \text{torr}$

Check: The units (torr) are correct. The magnitude of the answer (235) makes physical sense because both units are of the same size.

(b) **Given:** 235 mmHg **Find:** psi
Conceptual Plan: mmHg → atm → psi

$$\frac{1\ \text{atm}}{760\ \text{mmHg}} \quad \frac{14.7\ \text{psi}}{1\ \text{atm}}$$

Solution: $235\ \text{mmHg} \times \frac{1\ \text{atm}}{760\ \text{mmHg}} \times \frac{14.7\ \text{psi}}{1\ \text{atm}} = 4.55\ \text{psi}$

Check: The units (psi) are correct. The magnitude of the answer (< 14.7) makes physical sense because we started with less than $760\ \text{mmHg} = 1\ \text{atm}$.

(c) **Given:** 235 mmHg **Find:** in Hg
Conceptual Plan: mmHg → in Hg

$$\frac{1\ \text{in Hg}}{25.4\ \text{mmHg}}$$

Solution: $235\ \text{mmHg} \times \frac{1\ \text{in Hg}}{25.4\ \text{mmHg}} = 9.25\ \text{in Hg}$

Check: The units (in Hg) are correct. The magnitude of the answer (9) makes physical sense because inches are larger than millimeters.

(d) **Given:** 235 mmHg **Find:** atm
Conceptual Plan: mmHg → atm

$$\frac{1\ \text{atm}}{760\ \text{mmHg}}$$

Solution: $235\ \cancel{\text{mmHg}} \times \frac{1\ \text{atm}}{760\ \cancel{\text{mmHg}}} = 0.309\ \text{atm}$

Check: The units (atm) are correct. The magnitude of the answer (< 1) makes physical sense because we started with less than 760 mmHg.

11.27 (a) **Given:** 31.85 in Hg **Find:** mmHg
Conceptual Plan: in Hg → mmHg

$$\frac{25.4\ \text{mmHg}}{1\ \text{in Hg}}$$

Solution: $31.85\ \cancel{\text{in Hg}} \times \frac{25.4\ \text{mmHg}}{1\ \cancel{\text{in Hg}}} = 809.0\ \text{mmHg}$

Check: The units (mmHg) are correct. The magnitude of the answer (809) makes physical sense because inches are larger than millimeters.

(b) **Given:** 31.85 in Hg **Find:** atm
Conceptual Plan: Use answer from part (a), then convert mmHg → atm

$$\frac{1\ \text{atm}}{760\ \text{mmHg}}$$

Solution: $809.0\ \cancel{\text{mmHg}} \times \frac{1\ \text{atm}}{760\ \cancel{\text{mmHg}}} = 1.064\ \text{atm}$

Check: The units (atm) are correct. The magnitude of the answer (> 1) makes physical sense because we started with more than 760 mmHg.

(c) **Given:** 31.85 in Hg **Find:** torr
Conceptual Plan: Use answer from part (a), then convert mmHg → torr

$$\frac{1\ \text{torr}}{1\ \text{mmHg}}$$

Solution: $809.0\ \cancel{\text{mmHg}} \times \frac{1\ \text{torr}}{1\ \cancel{\text{mmHg}}} = 809.0\ \text{torr}$

Check: The units (torr) are correct. The magnitude of the answer (809) makes physical sense because both units are of the same size.

(d) **Given:** 31.85 in Hg **Find:** kPa
Conceptual Plan: Use answer from part (b), then convert atm → Pa → kPa

$$\frac{101{,}325\ \text{Pa}}{1\ \text{atm}} \quad \frac{1\ \text{kPa}}{1000\ \text{Pa}}$$

Solution: $1.064\ \cancel{\text{atm}} \times \frac{101{,}325\ \cancel{\text{Pa}}}{1\ \cancel{\text{atm}}} \times \frac{1\ \text{kPa}}{1000\ \cancel{\text{Pa}}} = 107.8\ \text{kPa}$

Check: The units (kPa) are correct. The magnitude of the answer (108) makes physical sense because we started with more than 1 atm and there are ~101 kPa in an atm.

11.28 (a) **Given:** 652.5 mmHg **Find:** torr
Conceptual Plan: mmHg → torr

$$\frac{1\ \text{torr}}{1\ \text{mmHg}}$$

Solution: $652.5\ \cancel{\text{mmHg}} \times \frac{1\ \text{torr}}{1\ \cancel{\text{mmHg}}} = 652.5\ \text{torr}$

Check: The units (torr) are correct. The magnitude of the answer (653) makes physical sense because both units are of the same size.

(b) **Given:** 652.5 mmHg **Find:** atm
Conceptual Plan: mmHg → atm

$$\frac{1 \text{ atm}}{760 \text{ mmHg}}$$

Solution: $652.5 \cancel{\text{mmHg}} \times \frac{1 \text{ atm}}{760 \cancel{\text{mmHg}}} = 0.8586 \text{ atm}$

Check: The units (atm) are correct. The magnitude of the answer (< 14.7) makes physical sense because we started with less than 760 mmHg = 1 atm.

(c) **Given:** 652.5 mmHg **Find:** in Hg
Conceptual Plan: mmHg → in Hg

$$\frac{1 \text{ in Hg}}{25.4 \text{ mmHg}}$$

Solution: $652.5 \cancel{\text{mmHg}} \times \frac{1 \text{ in Hg}}{25.4 \cancel{\text{mmHg}}} = 25.69 \text{ in Hg}$

Check: The units (in Hg) are correct. The magnitude of the answer (26) makes physical sense because inches are larger than millimeters.

(d) **Given:** 652.5 mmHg **Find:** psi
Conceptual Plan: Use answer from part (b), then convert atm → psi

$$\frac{14.70 \text{ psi}}{1 \text{ atm}}$$

Solution: $0.8586 \cancel{\text{atm}} \times \frac{14.70 \text{ psi}}{1 \cancel{\text{atm}}} = 12.62 \text{ psi}$

Check: The units (psi) are correct. The magnitude of the answer (< 14.7) makes physical sense because we started with less than 1 atm.

11.29 (a) **Given:** $P_{\text{bar}} = 762.4$ mmHg and figure **Find:** P_{gas}
Conceptual Plan: Measure height difference, then convert cm Hg → mmHg → mmHg

$$\frac{10 \text{ mmHg}}{1 \text{ cm Hg}} \qquad P_{\text{gas}} = h + P_{\text{bar}}$$

Solution:

$$h = 7.0 \cancel{\text{cm Hg}} \times \frac{10 \text{ mmHg}}{1 \cancel{\text{cm Hg}}} = 70. \text{ mmHg} \qquad P_{\text{gas}} = 70. \text{ mmHg} + 762.4 \text{ mmHg} = 832 \text{ mmHg}$$

Check: The units (mmHg) are correct. The magnitude of the answer (832) makes physical sense because the mercury column is higher on the right, indicating that the pressure is above barometric pressure. No significant figures to the right of the decimal point can be reported because the mercury height is known only to the ones place. One digit beyond the markings can be estimated (and is included above) and it would be considered significant.

(b) **Given:** $P_{\text{bar}} = 762.4$ mmHg and figure **Find:** P_{gas}
Conceptual Plan: Measure height difference, then convert cm Hg → mmHg → mmHg

$$\frac{10 \text{ mmHg}}{1 \text{ cm Hg}} \qquad P_{\text{gas}} = h + P_{\text{bar}}$$

Solution:

$$h = -4.4 \cancel{\text{cm Hg}} \times \frac{10 \text{ mmHg}}{1 \cancel{\text{cm Hg}}} = -44 \text{ mmHg} \qquad P_{\text{gas}} = -44 \text{ mmHg} + 762.4 \text{ mmHg} = 718 \text{ mmHg}$$

Check: The units (mmHg) are correct. The magnitude of the answer (718) makes physical sense because the mercury column is higher on the left, indicating that the pressure is below barometric pressure. No significant figures to the right of the decimal point can be reported because the mercury height is known only to the ones place. One digit beyond the markings can be estimated (and is included above) and it would be considered significant.

11.30 (a) **Given:** $P_{bar} = 751.5$ mmHg and figure **Find:** P_{gas}

Conceptual Plan: Measure height difference, then convert cm Hg → mmHg → mmHg

$$\frac{10 \text{ mmHg}}{1 \text{ cm Hg}} \qquad P_{gas} = h + P_{bar}$$

Solution:

$$h = -2.2 \text{ cm Hg} \times \frac{10 \text{ mmHg}}{1 \text{ cm Hg}} = -22 \text{ mmHg} \quad P_{gas} = -22 \text{ mmHg} + 751.5 \text{ mmHg} = 730. \text{ mmHg}$$

Check: The units (mmHg) are correct. The magnitude of the answer (730.) makes physical sense because the mercury column is higher on the left, indicating that the pressure is below barometric pressure. No significant figures to the right of the decimal point can be reported because the mercury height is known only to the ones place. One digit beyond the markings can be estimated (and is included above) and it would be considered significant.

(b) **Given:** $P_{bar} = 751.5$ mmHg and figure **Find:** P_{gas}

Conceptual Plan: Measure height difference, then convert cm Hg → mmHg → mmHg

$$\frac{10 \text{ mmHg}}{1 \text{ cm Hg}} \qquad P_{gas} = h + P_{bar}$$

Solution:

$$h = 6.8 \text{ cm Hg} \times \frac{10 \text{ mmHg}}{1 \text{ cm Hg}} = 68 \text{ mmHg} \quad P_{gas} = 68 \text{ mmHg} + 751.5 \text{ mmHg} = 820. \text{ mmHg}$$

Check: The units (mmHg) are correct. The magnitude of the answer (820.) makes physical sense because the mercury column is higher on the right, indicating that the pressure is above barometric pressure. No significant figures to the right of the decimal point can be reported because the mercury height is known only to the ones place. One digit beyond the markings can be estimated (and is included above) and it would be considered significant.

Simple Gas Laws

11.31 **Given:** $V_1 = 5.6$ L, $P_1 = 735$ mmHg, and $V_2 = 9.4$ L **Find:** P_2

Conceptual Plan: $V_1, P_1, V_2 \rightarrow P_2$

$$P_1 V_1 = P_2 V_2$$

Solution:

$P_1 V_1 = P_2 V_2$ Rearrange to solve for P_2.

$$P_2 = P_1 \frac{V_1}{V_2} = 735 \text{ mmHg} \times \frac{5.6 \text{ L}}{9.4 \text{ L}} = 437.872 \text{ mmHg} = 4.4 \times 10^2 \text{ mmHg}$$

Check: The units (mmHg) are correct. The magnitude of the answer (440) makes physical sense because Boyle's law indicates that as the volume increases, the pressure decreases.

11.32 **Given:** $V_1 = 13.9$ L, $P_1 = 1.22$ atm, and $V_2 = 10.3$ L **Find:** P_2

Conceptual Plan: $V_1, P_1, V_2 \rightarrow P_2$

$$P_1 V_1 = P_2 V_2$$

Solution: $P_1 V_1 = P_2 V_2$ Rearrange to solve for P_2.

$$P_2 = P_1 \frac{V_1}{V_2} = 1.22 \text{ atm} \times \frac{13.9 \text{ L}}{10.3 \text{ L}} = 1.646408 \text{ atm} = 1.65 \text{ atm}$$

Check: The units (atm) are correct. The magnitude of the answer (2) makes physical sense because Boyle's law indicates that as the volume decreases, the pressure increases.

11.33 **Given:** $V_1 = 48.3$ mL, $T_1 = 22$ °C, and $T_2 = 87$ °C **Find:** V_2

Conceptual Plan: °C → K then $V_1, T_1, T_2 \rightarrow V_2$

$$\text{K} = °\text{C} + 273.15 \qquad \frac{V_1}{T_1} = \frac{V_2}{T_2}$$

Solution: $T_1 = 22\,°C + 273.15 = 295\,K$ and $T_2 = 87\,°C + 273.15 = 360.\,K$

$$\frac{V_1}{T_1} = \frac{V_2}{T_2} \quad \text{Rearrange to solve for } V_2.\ V_2 = V_1\frac{T_2}{T_1} = 48.3\,\text{mL} \times \frac{360\,\text{K}}{295\,\text{K}} = 58.9\,\text{mL}$$

Check: The units (mL) are correct. The magnitude of the answer (59) makes physical sense because Charles's law indicates that as the volume increases, the temperature increases.

11.34 **Given:** $V_1 = 1.55\,\text{mL}$, $T_1 = 95.3\,°C$, and $T_2 = 0.0\,°C$ **Find:** V_2

Conceptual Plan: **°C → K then $V_1, T_1, T_2 \rightarrow V_2$**

$K = °C + 273.15$ $\quad\quad$ $\frac{V_1}{T_1} = \frac{V_2}{T_2}$

Solution: $T_1 = 95.3\,°C + 273.15 = 368.5\,K$ and $T_2 = 0.0\,°C + 273.15 = 273.2\,K$

$$\frac{V_1}{T_1} = \frac{V_2}{T_2} \quad \text{Rearrange to solve for } V_2.\ V_2 = V_1\frac{T_2}{T_1} = 1.55\,\text{mL} \times \frac{273.2\,\text{K}}{368.5\,\text{K}} = 1.15\,\text{mL}$$

Check: The units (mL) are correct. The magnitude of the answer (1.15) makes physical sense because Charles's law indicates that as the volume decreases, the temperature decreases.

11.35 **Given:** $V_1 = 2.46\,\text{L}$, $n_1 = 0.158\,\text{mol}$, and $\Delta n = 0.113\,\text{mol}$ **Find:** V_2

Conceptual Plan: **$n_1 \rightarrow n_2$ then $V_1, n_1, n_2 \rightarrow V_2$**

$n_1 + \Delta n = n_2$ $\quad\quad$ $\frac{V_1}{n_1} = \frac{V_2}{n_2}$

Solution: $n_2 = 0.158\,\text{mol} + 0.113\,\text{mol} = 0.271\,\text{mol}$

$$\frac{V_1}{n_1} = \frac{V_2}{n_2} \quad \text{Rearrange to solve for } V_2.\ V_2 = V_1\frac{n_2}{n_1} = 2.46\,\text{L} \times \frac{0.271\,\text{mol}}{0.158\,\text{mol}} = 4.2\underline{1}937\,\text{L} = 4.22\,\text{L}$$

Check: The units (L) are correct. The magnitude of the answer (4) makes physical sense because Avogadro's law indicates that as the number of moles increases, the volume increases.

11.36 **Given:** $V_1 = 253\,\text{mL}$, $n_1 = 0.553\,\text{mol}$, and $\Delta n = 0.365\,\text{mol}$ **Find:** V_2

Conceptual Plan: **$n_1 \rightarrow n_2$ then $V_1, n_1, n_2 \rightarrow V_2$**

$n_1 + \Delta n = n_2$ $\quad\quad$ $\frac{V_1}{n_1} = \frac{V_2}{n_2}$

Solution: $n_2 = 0.553\,\text{mol} + 0.365\,\text{mol} = 0.918\,\text{mol}$

$$\frac{V_1}{n_1} = \frac{V_2}{n_2} \quad \text{Rearrange to solve for } V_2.\ V_2 = V_1\frac{n_2}{n_1} = 253\,\text{mL} \times \frac{0.918\,\text{mol}}{0.553\,\text{mol}} = 41\underline{9}.989\,\text{mL} = 4.20 \times 10^2\,\text{mL}$$

Check: The units (mL) are correct. The magnitude of the answer (420) makes physical sense because Avogadro's law indicates that as the number of moles increases, the volume increases.

Ideal Gas Law

11.37 **Given:** $n = 0.118\,\text{mol}$, $P = 0.97\,\text{atm}$, and $T = 305\,\text{K}$ **Find:** V

Conceptual Plan: **$n, P, T \rightarrow V$**

$PV = nRT$

Solution: $PV = nRT$ Rearrange to solve for V.

$$V = \frac{nRT}{P} = \frac{0.118\,\text{mol} \times 0.08206\,\frac{\text{L}\cdot\text{atm}}{\text{mol}\cdot\text{K}} \times 305\,\text{K}}{0.97\,\text{atm}} = 3.0\,\text{L}$$

The volume would be the same for argon gas because the ideal gas law does not care about the mass of the gas, only the number of moles of gas.

Check: The units (L) are correct. The magnitude of the answer (3) makes sense because, as you will see in the next section, one mole of an ideal gas under standard conditions (273 K and 1 atm) occupies 22.4 L. Although these are not standard conditions, they are close enough for a ballpark check of the answer. Because this gas sample contains 0.118 mole, a volume of 3 L is reasonable.

11.38 **Given:** 12.5 g argon or 12.5 g helium, $P = 1.05$ atm, and $T = 322$ K **Find:** V

Conceptual Plan: $m \rightarrow n$ **then** $n, P, T \rightarrow V$

$\frac{1 \text{ mol}}{39.95 \text{ g}}$ or $\frac{1 \text{ mol}}{4.003 \text{ g}}$ $\quad PV = nRT$

Solution: $12.5 \text{ g Ar} \times \dfrac{1 \text{ mol Ar}}{39.95 \text{ g Ar}} = 0.3128911 \text{ mol Ar}$ $PV = nRT$ Rearrange to solve for V.

$$V = \frac{nRT}{P} = \frac{0.3128911 \text{ mol Ar} \times 0.08206 \dfrac{\text{L} \cdot \text{atm}}{\text{mol} \cdot \text{K}} \times 322 \text{ K}}{1.05 \text{ atm}} = 7.87 \text{ L Ar}$$

$12.5 \text{ g He} \times \dfrac{1 \text{ mol He}}{4.003 \text{ g He}} = 3.122658 \text{ mol He}$ $PV = nRT$ Rearrange to solve for V.

$$V = \frac{nRT}{P} = \frac{3.122658 \text{ mol He} \times 0.08206 \dfrac{\text{L} \cdot \text{atm}}{\text{mol} \cdot \text{K}} \times 322 \text{ K}}{1.05 \text{ atm}} = 78.6 \text{ L He}$$

Check: The units (L) are correct. The magnitude of the answers (8 and 80) makes sense because, as you will see in the next section, one mole of an ideal gas under standard conditions (273 K and 1 atm) occupies 22.4 L. Although these are not standard conditions, they are close enough for a ballpark check of the answer. Because the molar mass of the two gases differ by a factor of ten, the resulting volumes will differ by a factor of ten.

11.39 **Given:** $V = 10.0$ L, $n = 0.448$ mol, and $T = 315$ K **Find:** P

Conceptual Plan: $n, V, T \rightarrow P$

$PV = nRT$

Solution:

$PV = nRT$ Rearrange to solve for P. $P = \dfrac{nRT}{V} = \dfrac{0.448 \text{ mol} \times 0.08206 \dfrac{\text{L} \cdot \text{atm}}{\text{mol} \cdot \text{K}} \times 315 \text{ K}}{10.0 \text{ L}} = 1.16 \text{ atm}$

Check: The units (atm) are correct. The magnitude of the answer (~1) makes sense because, as you will see in the next section, one mole of an ideal gas under standard conditions (273 K and 1 atm) occupies 22.4 L. Although these are not standard conditions, they are close enough for a ballpark check of the answer. Because this gas sample contains 0.448 mole in a volume of 10 L, a pressure of 1 atm is reasonable.

11.40 **Given:** $V = 15.0$ L, 32.7 g oxygen, and $T = 302$ K **Find:** P

Conceptual Plan: g $\rightarrow n$ **then** $n, P, T \rightarrow V$

$\frac{1 \text{ mol}}{32.00 \text{ g}}$ $\quad PV = nRT$

Solution: $32.7 \text{ g } O_2 \times \dfrac{1 \text{ mol } O_2}{32.00 \text{ g } O_2} = 1.021875 \text{ mol } O_2$ $PV = nRT$ Rearrange to solve for P.

$$P = \frac{nRT}{V} = \frac{1.021875 \text{ mol} \times 0.08206 \dfrac{\text{L} \cdot \text{atm}}{\text{mol} \cdot \text{K}} \times 302 \text{ K}}{15.0 \text{ L}} = 1.688282 \text{ atm} = 1.69 \text{ atm}$$

Check: The units (atm) are correct. The magnitude of the answer (~1.7) makes sense because, as you will see in the next section, one mole of an ideal gas under standard conditions (273 K and 1 atm) occupies 22.4 L. Although these are not standard conditions, they are close enough for a ballpark check of the answer. Because this gas sample contains ~1 mole in a volume of 15 L, a pressure of 1.7 atm is reasonable.

11.41 **Given:** $V = 28.5$ L, $P = 1.8$ atm, and $T = 298$ K **Find:** n

Conceptual Plan: $V, P, T \rightarrow n$

$PV = nRT$

Solution: $PV = nRT$ Rearrange to solve for n. $n = \dfrac{PV}{RT} = \dfrac{1.8 \text{ atm} \times 28.5 \text{ L}}{0.08206 \dfrac{\text{L} \cdot \text{atm}}{\text{mol} \cdot \text{K}} \times 298 \text{ K}} = 2.1 \text{ mol}$

Check: The units (mol) are correct. The magnitude of the answer (2) makes sense because, as you will see in the next section, one mole of an ideal gas under standard conditions (273 K and 1 atm) occupies 22.4 L. Although these are not standard conditions, they are close enough for a ballpark check of the answer. Because this gas sample has a volume of 28.5 L and a pressure of 1.8 atm, ~2 mol is reasonable.

11.42 **Given:** $V = 11.8$ L, $P = 1.3$ atm, and $n = 0.52$ mol **Find:** T

Conceptual Plan: $V, P, n \rightarrow T$

$PV = nRT$

Solution: $PV = nRT$ Rearrange to solve for T. $T = \frac{PV}{nR} = \frac{1.3 \text{ atm} \times 11.8 \text{ L}}{0.52 \text{ mol} \times 0.08206 \frac{\text{L} \cdot \text{atm}}{\text{mol} \cdot \text{K}}} = 360 \text{ K}$

Check: The units (K) are correct. The magnitude of the answer (360) makes sense because, as you will see in the next section, one mole of an ideal gas under standard conditions (273 K and 1 atm) occupies 22.4 L. Although these are not standard conditions, they are close enough for a ballpark check of the answer. Because this gas sample has 0.52 mol, a volume of 11.8 L, and a pressure of 1.3 atm, 360 K is reasonable.

11.43 **Given:** $P_1 = 36.0$ psi (gauge P), $V_1 = 11.8$ L, $T_1 = 12.0$ °C, $V_2 = 12.2$ L, and $T_2 = 65.0$ °C

Find: P_2 and compare to $P_{max} = 38.0$ psi (gauge P)

Conceptual Plan: **°C → K and gauge P → psi → atm then $P_1, V_1, T_1, V_2, T_2 \rightarrow P_2$**

$K = °C + 273.15$ $\quad$ $\text{psi} = \text{gauge P} + 14.7 \frac{1 \text{ atm}}{14.7 \text{ psi}}$ $\quad$ $\frac{P_1V_1}{T_1} = \frac{P_2V_2}{T_2}$

Solution: $T_1 = 12.0 \text{ °C} + 273.15 = 285.2$ K and $T_2 = 65.0 \text{ °C} + 273.15 = 338.2$ K

$P_1 = 36.0 \text{ psi (gauge P)} + 14.7 = 50.7 \text{ psi} \times \frac{1 \text{ atm}}{14.7 \text{ psi}} = 3.44898 \text{ atm}$

$P_{max} = 38.0 \text{ psi (gauge P)} + 14.7 = 52.7 \text{ psi} \times \frac{1 \text{ atm}}{14.7 \text{ psi}} = 3.59 \text{ atm}$

$\frac{P_1V_1}{T_1} = \frac{P_2V_2}{T_2}$ Rearrange to solve for P_2. $P_2 = P_1\frac{V_1}{V_2}\frac{T_2}{T_1} = 3.44898 \text{ atm} \times \frac{11.8 \text{ L}}{12.2 \text{ L}} \times \frac{338.2 \text{ K}}{285.2 \text{ K}} = 3.96 \text{ atm}$

This exceeds the maximum tire rating of 3.59 atm or 38.0 psi (gauge P).

Check: The units (atm) are correct. The magnitude of the answer (3.96) makes physical sense because the relative increase in T is greater than the relative increase in V; so P should increase.

11.44 **Given:** $P_1 = 748$ mmHg, $V_1 = 28.5$ L, $T_1 = 28.0$ °C, $P_2 = 385$ mmHg, and $T_2 = -15.0$ °C **Find:** V_2

Conceptual Plan: **°C → K then $P_1, V_1, T_1, V_2, T_2 \rightarrow P_2$**

$K = °C + 273.15$ $\quad$ $\frac{P_1V_1}{T_1} = \frac{P_2V_2}{T_2}$

Solution: $T_1 = 28.0 \text{ °C} + 273.15 = 301.2$ K and $T_2 = -15.0 \text{ °C} + 273.15 = 258.2$ K

$\frac{P_1V_1}{T_1} = \frac{P_2V_2}{T_2}$ Rearrange to solve for V_2. $V_2 = V_1\frac{P_1}{P_2}\frac{T_2}{T_1} = 28.5 \text{ L} \times \frac{748 \text{ mmHg}}{385 \text{ mmHg}} \times \frac{258.2 \text{ K}}{301.2 \text{ K}} = 47.5 \text{ L}$

Check: The units (L) are correct. The magnitude of the answer (48) makes physical sense because the relative decrease in P is greater than the relative decrease in T; so V should increase.

11.45 **Given:** $m_{CO_2} = 28.8$ g, $P = 742$ mmHg, and $T = 22$ °C **Find:** V

Conceptual Plan: **°C → K and mmHg → atm and g → mol then $n, P, T \rightarrow V$**

$K = °C + 273.15$ $\quad$ $\frac{1 \text{ atm}}{760 \text{ mmHg}}$ $\quad$ $\frac{1 \text{ mol}}{44.01 \text{ g}}$ $\quad$ $PV = nRT$

Solution: $T_1 = 22 \text{ °C} + 273.15 = 295 \text{ K}$, $P = 742 \text{ mmHg} \times \frac{1 \text{ atm}}{760 \text{ mmHg}} = 0.976316 \text{ atm}$

$n = 28.8 \text{ g} \times \frac{1 \text{ mol}}{44.01 \text{ g}} = 0.654397 \text{ mol}$ $PV = nRT$ Rearrange to solve for V.

$$V = \frac{nRT}{P} = \frac{0.654397 \text{ mol} \times 0.08206 \frac{\text{L} \cdot \text{atm}}{\text{mol} \cdot \text{K}} \times 295 \text{ K}}{0.976316 \text{ atm}} = 16.2 \text{ L}$$

Check: The units (L) are correct. The magnitude of the answer (16) makes sense because one mole of an ideal gas under standard conditions (273 K and 1 atm) occupies 22.4 L. Although these are not standard conditions, they are close enough for a ballpark check of the answer. Because this gas sample contains 0.65 mole, a volume of 16 L is reasonable.

11.46 **Given:** 1.0 L of liquid N_2 with $d = 0.807\ \text{g/mL}$, $T = 25.0\ °\text{C}$, $P = 1.0\ \text{atm}$, and closet is $1.0\ \text{m} \times 1.0\ \text{m} \times 2.0\ \text{m}$
Find: $V\%$ of closet displaced by evaporated liquid

Conceptual Plan: **°C → K and L → mL → g → mol then $n, P, T \rightarrow V_{evap}$**

$$\text{K} = °\text{C} + 273.15 \qquad \frac{1000\ \text{mL}}{1\ \text{L}} \qquad d = m/V \quad \frac{1\ \text{mol}}{28.02\ \text{g}} \qquad PV = nRT$$

then $l, w, h \rightarrow V_{closet}\text{m}^3 \rightarrow \text{cm}^3 \rightarrow$ L finally $V_{evap}, V_{closet} \rightarrow$ % *V displaced*

$$V = lwh \qquad \frac{(100\ \text{cm})^3}{(1\ \text{m})^3}\ \frac{1\ \text{L}}{1000\ \text{mL}} \qquad \%\ V\ displaced = \frac{V_{evap}}{V_{closet}} \times 100\%$$

Solution: $T_1 = 25.0\ °\text{C} + 273.15 = 298.2\ \text{K}$, $1.0\ \text{L} \times \dfrac{1000\ \text{mL}}{1\ \text{L}} = 1.0 \times 10^3\ \text{mL}$ $d = m/V$ Rearrange to solve for m.

$$m = d \times V = 0.807\frac{\text{g}}{\text{mL}} \times 1.0 \times 10^3\ \text{mL} = 8.07 \times 10^2\ \text{g} \times \frac{1\ \text{mol}}{28.02\ \text{g}} = 28.801\ \text{mol}$$

$PV = nRT$ Rearrange to solve for V.

$$V_{evap} = \frac{nRT}{P} = \frac{28.801\ \text{mol} \times 0.08206\dfrac{\text{L} \cdot \text{atm}}{\text{mol} \cdot \text{K}} \times 298.2\ \text{K}}{1.0\ \text{atm}} = 7.0477 \times 10^2\ \text{L}$$

$$V_{closet} = lwh = 1.0\ \text{m} \times 1.0\ \text{m} \times 2.0\ \text{m} = 2.0\ \text{m}^3$$

$$V_{closet} = 2.0\ \text{m}^3 \times \frac{(100\ \text{cm})^3}{(1\ \text{m})^3} \times \frac{1\ \text{L}}{1000\ \text{cm}^3} = 2.0 \times 10^3\ \text{L}$$

$$\%\ V\ displaced = \frac{V_{evap}}{V_{closet}} \times 100\% = \frac{7.0477 \times 10^2\ \text{L}}{2.0 \times 10^3\ \text{L}} \times 100\% = 35\%$$

Check: The units (%) are correct. The magnitude of the answer (35) makes sense because it should be between 0 and 100%. Looking at the two volumes, when a liquid evaporates, the volume increases by several orders of magnitude; when converting from cubic meters to liters, there is an increase of 3 orders of magnitude.

11.47 **Given:** 26.0 g argon, $V = 55.0\ \text{mL}$, and $T = 295\ \text{K}$ **Find:** P

Conceptual Plan: **$m \rightarrow n$ and mL → L then $n, V, T \rightarrow P$**

$$\frac{1\ \text{mol}}{39.95\ \text{g}} \qquad \frac{1\ \text{L}}{1000\ \text{mL}} \qquad \text{PV} = \text{nRT}$$

Solution: $26.0\ \text{g Ar} \times \dfrac{1\ \text{mol Ar}}{39.95\ \text{g Ar}} = 0.6508135\ \text{mol Ar}$ $55.0\ \text{mL} \times \dfrac{1\ \text{L}}{1000\ \text{mL}} = 0.0550\ \text{L}$

$PV = nRT$ Rearrange to solve for P.

$$P = \frac{nRT}{V} = \frac{0.6508135\ \text{mol} \times 0.08206\dfrac{\text{L} \cdot \text{atm}}{\text{mol} \cdot \text{K}} \times 295\ \text{K}}{0.0550\ \text{L}} = 286.44906\ \text{atm} = 286\ \text{atm}$$

Check: The units (atm) are correct. The magnitude of the answer (300) makes sense because, as you will see in the next section, one mole of an ideal gas under standard conditions (273 K and 1 atm) occupies 22.4 L. Although these are not standard conditions, they can be used for a ballpark check of the answer. Because the volume is ~ $1/400^{\text{th}}$ the molar volume and we have ~$2/3$ of a mole, the resulting pressure should be $(400)(2/3) = 270$ atm.

Given: $V_1 = 55.0\ \text{mL}$, $P_1 = 286\ \text{atm}$, and $P_2 = 1.20\ \text{atm}$ **Find:** V_2 (number of 750 mL bottles)

Conceptual Plan: **$V_1, P_1, P_2 \rightarrow V_2$**

$$P_1 V_1 = P_2 V_2$$

Solution: $P_1 V_1 = P_2 V_2$ Rearrange to solve for V_2.

$$V_2 = V_1\frac{P_1}{P_2} = 55.0\ \text{mL} \times \frac{286.44906\ \text{atm}}{1.20\ \text{atm}} = 1.31289 \times 10^4\ \text{mL} \times \frac{1\ \text{bottle}}{750.0\ \text{mL}} = 17.5\ \text{bottles}$$

Check: The units (bottles) are correct. The magnitude of the answer (18) makes physical sense because Boyle's law indicates that as the volume decreases, the pressure increases. The pressure is decreasing by a factor of ~250, so the volume should increase by this factor.

11.48 **Given:** 16.0 g CO_2, $V = 3.45$ L, and $T = 298$ K **Find:** P (psi)

Conceptual Plan: $m \rightarrow n$ **then** $n, V, T \rightarrow P$ **then atm → psi (absolute) → psi (gauge)**

$$\frac{1 \text{ mol}}{44.01 \text{ g}} \qquad PV = nRT \qquad \frac{14.7 \text{ psi}}{1 \text{ atm}} \qquad \text{subtract 14.7 psi (atmospheric pressure)}$$

Solution: $16.0 \text{ g} \times \dfrac{1 \text{ mol}}{44.01 \text{ g}} = 0.3635537 \text{ mol} \quad 55.0 \text{ mL} \times \dfrac{1 \text{ L}}{1000 \text{ mL}} = 0.0550 \text{ L}$

$PV = nRT$ Rearrange to solve for P.

$$P = \frac{nRT}{V} = \frac{0.3635537 \text{ mol} \times 0.08206 \dfrac{\text{L} \cdot \text{atm}}{\text{mol} \cdot \text{K}} \times 298 \text{ K}}{3.45 \text{ L}} = 2.576898 \text{ atm} \times \frac{14.7 \text{ psi}}{1 \text{ atm}} = 37.9 \text{ psi (absolute)}$$

or 23.2 psi (gauge) after subtracting 14.7 psi

Check: The units (psi) are correct. The magnitude of the answer (23) makes sense because, as you will see in the next section, one mole of an ideal gas under standard conditions (273 K and 1 atm) occupies 22.4 L. Although these are not standard conditions, they can be used for a ballpark check of the answer. Because the volume is almost 1/7th the molar volume and we have ~1/3 of a mole, the resulting pressure should be just under $(7)(1/3) = 3$ atm.

11.49 **Given:** sample a = 5 gas particles, sample b = 10 gas particles, and sample c = 8 gas particles, with all temperatures and volumes the same **Find:** sample with largest P

Conceptual Plan: $n, V, T \rightarrow P$

$$PV = nRT$$

Solution: $PV = nRT$ Because V and T are constant, $P \; \alpha \; n$. The sample with the largest number of gas particles will have the highest P. $P_b > P_c > P_a$

11.50 **Given:** $P_1 = 1$ atm, $V_1 = 1$ L, $T_1 = 25$ °C, $V_2 = 0.5$ L, and $T_2 = 250.$ °C

Find: Draw picture and P_2

Conceptual Plan: **°C → K then** $P_1, V_1, T_1, V_2, T_2 \rightarrow P_2$

$$\text{K} = {}^\circ\text{C} + 273.15 \qquad \frac{P_1 V_1}{T_1} = \frac{P_2 V_2}{T_2}$$

Solution: $T_1 = 25 \text{ °C} + 273.15 = 298 \text{ K}$ and $T_2 = 250. \text{ °C} + 273.15 = 523 \text{ K}$

$$\frac{P_1 V_1}{T_1} = \frac{P_2 V_2}{T_2} \quad \text{Rearrange to solve for } P_2.\ P_2 = P_1 \frac{V_1}{V_2} \frac{T_2}{T_1} = 1 \text{ atm} \times \frac{1 \text{ L}}{0.5 \text{ L}} \times \frac{523 \text{ K}}{298 \text{ K}} = 3.5101 \text{ atm} = 4 \text{ atm}$$

Check: The units (atm) are correct. The magnitude of the answer (4) makes physical sense because there is an increase in T and a decrease in V, both of which increase P.

11.51 **Given:** $P_1 = 755$ mmHg, $T_1 = 25$ °C, and $T_2 = 1155$ °C **Find:** P_2

Conceptual Plan: **°C → K and mmHg → atm then** $P_1, T_1, T_2 \rightarrow P_2$

$$\text{K} = {}^\circ\text{C} + 273.15 \qquad \frac{1 \text{ atm}}{760 \text{ mmHg}} \qquad \frac{P_1}{T_1} = \frac{P_2}{T_2}$$

Solution: $T_1 = 25 \text{ °C} + 273.15 = 298 \text{ K}$ and $T_2 = 1155 \text{ °C} + 273.15 = 1428 \text{ K}$

$$P = 755 \text{ mmHg} \times \frac{1 \text{ atm}}{760 \text{ mmHg}} = 0.993421 \text{ atm} \quad \frac{P_1}{T_1} = \frac{P_2}{T_2} \quad \text{Rearrange to solve for } P_2.$$

$$P_2 = P_1 \frac{T_2}{T_1} = 0.993421 \text{ atm} \times \frac{1428 \text{ K}}{298 \text{ K}} = 4.76 \text{ atm}$$

Check: The units (atm) are correct. The magnitude of the answer (5) makes physical sense because there is a significant increase in T, which will increase P significantly.

11.52 **Given:** $V_1 = 1.75$ L, $P_1 = 1.35$ atm, $T_1 = 25$ °C, $V_2 = 1.75$ L, and $T_2 = 355$ °C **Find:** P_2

Conceptual Plan: °C → K then $P_1, T_1, T_2 \rightarrow P_2$

$$K = °C + 273.15 \qquad \frac{P_1}{T_1} = \frac{P_2}{T_2}$$

Solution: $T_1 = 25\ °C + 273.15 = 298$ K and $T_2 = 355\ °C + 273.15 = 628$ K

$$\frac{P_1}{T_1} = \frac{P_2}{T_2} \quad \text{Rearrange to solve for } P_2. \quad P_2 = P_1 \frac{T_2}{T_1} = 1.35 \text{ atm} \times \frac{628 \cancel{\text{K}}}{298 \cancel{\text{K}}} = 2.84 \text{ atm}$$

Check: The units (atm) are correct. The magnitude of the answer (3) makes physical sense because there is a significant increase in T, which will increase P significantly.

Molar Volume, Density, and Molar Mass of a Gas

11.53 **Given:** STP and $m_{Ne} = 33.6$ g **Find:** V

Conceptual Plan: g → mol → V

$$\frac{1 \text{ mol}}{20.18 \text{ g}} \qquad \frac{22.414 \text{ L}}{1 \text{ mol}}$$

Solution: $33.6 \cancel{\text{g}} \times \frac{1 \cancel{\text{mol}}}{20.18 \cancel{\text{g}}} \times \frac{22.414 \text{ L}}{1 \cancel{\text{mol}}} = 37.3 \text{ L}$

Check: The units (L) are correct. The magnitude of the answer (37) makes sense because one mole of an ideal gas under standard conditions (273 K and 1 atm) occupies 22.4 L and we have about 1.7 mol.

11.54 **Given:** STP and N_2 **Find:** d

Conceptual Plan: mol → g then $m, V \rightarrow d$

$$\frac{28.02 \text{ g}}{1 \text{ mol}} \qquad d = \frac{m}{V}$$

Solution: $1 \cancel{\text{mol}} \times \frac{28.02 \text{ g}}{1 \cancel{\text{mol}}} = 28.02 \text{ g} = \text{m at STP } V = 22.414 \text{ L } d = \frac{m}{V} = \frac{28.02 \text{ g}}{22.414 \text{ L}} = 1.250 \text{ g/L}$

Check: The units (g/L) are correct. The magnitude of the answer (1) is reasonable for a gas density.

11.55 **Given:** H_2, $P = 1655$ psi, and $T = 20.0$ °C **Find:** d

Conceptual Plan: °C → K and psi → atm then $P, T, \mathcal{M} \rightarrow d$

$$K = °C + 273.15 \qquad \frac{1 \text{ atm}}{14.70 \text{ psi}} \qquad d = \frac{P\mathcal{M}}{RT}$$

Solution: $T = 20.0\ °C + 273.15 = 293.2$ K $\qquad P = 1655 \cancel{\text{psi}} \times \frac{1 \text{ atm}}{14.70 \cancel{\text{psi}}} = 112.\underline{5}85 \text{ atm}$

$$d = \frac{P\mathcal{M}}{RT} = \frac{112.\underline{5}85 \cancel{\text{atm}} \times 2.016 \frac{\text{g}}{\cancel{\text{mol}}}}{0.08206 \frac{\text{L} \cdot \cancel{\text{atm}}}{\cancel{\text{K}} \cdot \cancel{\text{mol}}} \times 293.2 \cancel{\text{K}}} = 9.434 \frac{\text{g}}{\text{L}}$$

Check: The units (g/L) are correct. The magnitude of the answer (9) makes physical sense because this is a high pressure; so the gas density will be on the high side.

11.56 **Given:** N_2O, $d = 2.85$ g/L, and $T = 298$ K **Find:** P (mmHg)

Conceptual Plan: $d, T, \mathcal{M} \rightarrow d$ then atm → mmHg

$$d = \frac{P\mathcal{M}}{RT} \qquad \frac{760 \text{ mmHg}}{1 \text{ atm}}$$

Solution: $d = \frac{P\mathcal{M}}{RT}$ Rearrange to solve for P.

$$P = \frac{dRT}{\mathcal{M}} = \frac{2.85 \frac{\cancel{\text{g}}}{\cancel{\text{L}}} \times 0.08206 \frac{\cancel{\text{L}} \cdot \text{atm}}{\cancel{\text{K}} \cdot \cancel{\text{mol}}} \times 298 \cancel{\text{K}}}{44.02 \frac{\cancel{\text{g}}}{\cancel{\text{mol}}}} = 1.5\underline{8}322 \text{ atm}$$

$$P = 1.58322\ \text{atm} \times \frac{760\ \text{mmHg}}{1\ \text{atm}} = 1.20 \times 10^3\ \text{mmHg}$$

Check: The units (mmHg) are correct. The magnitude of the answer (1200) makes physical sense because the gas density is reasonable; so we expect a $P \sim 1$ atm.

11.57 **Given:** $V = 248$ mL, $m = 0.433$ g, $P = 745$ mmHg, and $T = 28\ °\text{C}$ **Find:** $\mathcal{M}$
Conceptual Plan: °C → K mmHg → atm mL → L then $V, m \rightarrow d$ then $d, P, T \rightarrow \mathcal{M}$

$$\text{K} = °\text{C} + 273.15 \quad \frac{1\ \text{atm}}{760\ \text{mmHg}} \quad \frac{1\ \text{L}}{1000\ \text{mL}} \quad d = \frac{m}{V} \quad d = \frac{P\mathcal{M}}{RT}$$

Solution: $T = 28\ °\text{C} + 273.15 = 301\ \text{K}$ $P = 745\ \text{mmHg} \times \frac{1\ \text{atm}}{760\ \text{mmHg}} = 0.980263\ \text{atm}$

$V = 248\ \text{mL} \times \frac{1\ \text{L}}{1000\ \text{mL}} = 0.248\ \text{L}$ $d = \frac{m}{V} = \frac{0.433\ \text{g}}{0.248\ \text{L}} = 1.74597\ \text{g/L}$ $d = \frac{P\mathcal{M}}{RT}$ Rearrange to solve for $\mathcal{M}$.

$$\mathcal{M} = \frac{dRT}{P} = \frac{1.74597\ \frac{\text{g}}{\text{L}} \times 0.08206\ \frac{\text{L} \cdot \text{atm}}{\text{K} \cdot \text{mol}} \times 301\ \text{K}}{0.980263\ \text{atm}} = 44.0\ \text{g/mol}$$

Check: The units (g/mol) are correct. The magnitude of the answer (44) makes physical sense because this is a reasonable number for a molar mass of a gas.

11.58 **Given:** $V = 113$ mL, $m = 0.171$ g, $P = 721$ mmHg, and $T = 32\ °\text{C}$ **Find:** $\mathcal{M}$
Conceptual Plan: °C → K mmHg → atm mL → L then $V, m \rightarrow d$ then $d, P, T \rightarrow \mathcal{M}$

$$\text{K} = °\text{C} + 273.15 \quad \frac{1\ \text{atm}}{760\ \text{mmHg}} \quad \frac{1\ \text{L}}{1000\ \text{mL}} \quad d = \frac{m}{V} \quad d = \frac{P\mathcal{M}}{RT}$$

Solution: $T = 32\ °\text{C} + 273.15 = 305\ \text{K}$ $P = 721\ \text{mmHg} \times \frac{1\ \text{atm}}{760\ \text{mmHg}} = 0.948684\ \text{atm}$

$V = 113\ \text{mL} \times \frac{1\ \text{L}}{1000\ \text{mL}} = 0.113\ \text{L}$ $d = \frac{m}{V} = \frac{0.171\ \text{g}}{0.113\ \text{L}} = 1.51327\ \text{g/L}$ $d = \frac{P\mathcal{M}}{RT}$ Rearrange to solve for $\mathcal{M}$.

$$\mathcal{M} = \frac{dRT}{P} = \frac{1.51327\frac{\text{g}}{\text{L}} \times 0.08206\frac{\text{L} \cdot \text{atm}}{\text{K} \cdot \text{mol}} \times 305\ \text{K}}{0.948684\ \text{atm}} = 39.9\ \text{g/mol}$$

An alternative method to solve this problem would be to calculate the moles of gas using the ideal gas law (0.00428 mole) and then taking the mass (0.171 g) and dividing it by the number of moles.

Check: The units (g/mol) are correct. The magnitude of the answer (40) makes physical sense because this is a reasonable number for a molar mass of a gas.

11.59 **Given:** $m = 38.8$ mg, $V = 224$ mL, $T = 55\ °\text{C}$, and $P = 886$ torr **Find:** $\mathcal{M}$
Conceptual Plan: mg → g mL → L °C → K torr → atm then $V, m \rightarrow d$ then $d, P, T \rightarrow \mathcal{M}$

$$\frac{1\ \text{g}}{1000\ \text{mg}} \quad \frac{1\ \text{L}}{1000\ \text{mL}} \quad \text{K} = °\text{C} + 273.15 \quad \frac{1\ \text{atm}}{760\ \text{torr}} \quad d = \frac{m}{V} \quad d = \frac{P\mathcal{M}}{RT}$$

Solution:

$m = 38.8\ \text{mg} \times \frac{1\ \text{g}}{1000\ \text{mg}} = 0.0388\ \text{g}$ $V = 224\ \text{mL} \times \frac{1\ \text{L}}{1000\ \text{mL}} = 0.224\ \text{L}$ $T = 55\ °\text{C} + 273.15 = 328\ \text{K}$

$P = 886\ \text{torr} \times \frac{1\ \text{atm}}{760\ \text{torr}} = 1.165789\ \text{atm}$ $d = \frac{m}{V} = \frac{0.0388\ \text{g}}{0.224\ \text{L}} = 0.173214\ \text{g/L}$ $d = \frac{P\mathcal{M}}{RT}$

Rearrange to solve for $\mathcal{M}$. $\mathcal{M} = \frac{dRT}{P} = \frac{0.173214\ \frac{\text{g}}{\text{L}} \times 0.08206\frac{\text{L} \cdot \text{atm}}{\text{K} \cdot \text{mol}} \times 328\ \text{K}}{1.165789\ \text{atm}} = 4.00\ \text{g/mol}$

Check: The units (g/mol) are correct. The magnitude of the answer (4) makes physical sense because this is a reasonable number for a molar mass of a gas, especially because the density is on the low side.

11.60 **Given:** $m = 0.555$ g, $V = 117$ mL, $T = 85\ °\text{C}$, and $P = 753$ mmHg **Find:** $\mathcal{M}$

Conceptual Plan: mL → L °C → K mmHg → atm then $V, m \rightarrow d$ then $d, P, T \rightarrow \mathcal{M}$

$$\frac{1\text{ L}}{1000\text{ mL}} \quad K = °C + 273.15 \quad \frac{1\text{ atm}}{760\text{ mmHg}} \quad d = \frac{m}{V} \quad d = \frac{P\mathcal{M}}{RT}$$

Solution: $V = 117\text{ mL} \times \frac{1\text{ L}}{1000\text{ mL}} = 0.117\text{ L}$ $T = 85\text{ °C} + 273.15 = 358\text{ K}$

$$P = 753\text{ mmHg} \times \frac{1\text{ atm}}{760\text{ mmHg}} = 0.99\underline{0}7895\text{ atm} \qquad d = \frac{m}{V} = \frac{0.555\text{ g}}{0.117\text{ L}} = 4.7\underline{4}359\text{ g/L} \qquad d = \frac{P\mathcal{M}}{RT}$$

Rearrange to solve for $\mathcal{M}$. $\mathcal{M} = \frac{dRT}{P} = \frac{4.7\underline{4}359\frac{\text{g}}{\text{L}} \times 0.08206\frac{\text{L}\cdot\text{atm}}{\text{K}\cdot\text{mol}} \times 358\text{ K}}{0.99\underline{0}7895\text{ atm}} = 141\text{ g/mol}$

Check: The units (g/mol) are correct. The magnitude of the answer (141) makes physical sense because this is a reasonable number for a molar mass of a gas, especially because the density is on the high side.

Partial Pressure

11.61 **Given:** $P_{N_2} = 215$ torr, $P_{O_2} = 102$ torr, $P_{He} = 117$ torr, $V = 1.35$ L, and $T = 25.0$ °C **Find:** P_{Total}, m_{N_2}, m_{O_2}, m_{He}

Conceptual Plan: °C → K and torr → atm and $P, V, T \rightarrow n$ then $n \rightarrow$ g

$$K = °C + 273.15 \quad \frac{1\text{ atm}}{760\text{ torr}} \quad PV = nRT \quad \mathcal{M}$$

and $P_{N_2}, P_{O_2}, P_{He} \rightarrow P_{Total}$

$$P_{Total} = P_{N_2} + P_{O_2} + P_{He}$$

Solution: $T_1 = 25.0\text{ °C} + 273.15 = 298.2\text{ K}$, $PV = nRT$ Rearrange to solve for n.

$$n = \frac{PV}{RT}; P_{N_2} = 215\text{ torr} \times \frac{1\text{ atm}}{760\text{ torr}} = 0.28\underline{2}8947\text{ atm } n_{N_2} = \frac{0.28\underline{2}8947\text{ atm} \times 1.35\text{ L}}{0.08206\frac{\text{L}\cdot\text{atm}}{\text{mol}\cdot\text{K}} \times 298.2\text{ K}} = 0.015\underline{6}0700\text{ mol}$$

$$0.015\underline{6}0700\text{ mol} \times \frac{28.02\text{ g}}{1\text{ mol}} = 0.437\text{ g } N_2$$

$$P_{O_2} = 102\text{ torr} \times \frac{1\text{ atm}}{760\text{ torr}} = 0.13\underline{4}2105\text{ atm } n_{O_2} = \frac{0.13\underline{4}2105\text{ atm} \times 1.35\text{ L}}{0.08206\frac{\text{L}\cdot\text{atm}}{\text{mol}\cdot\text{K}} \times 298.2\text{ K}} = 0.0074\underline{0}4252\text{ mol}$$

$$0.0074\underline{0}4252\text{ mol} \times \frac{32.00\text{ g}}{1\text{ mol}} = 0.237\text{ g } O_2$$

$$P_{He} = 117\text{ torr} \times \frac{1\text{ atm}}{760\text{ torr}} = 0.15\underline{3}9474\text{ atm } n_{He} = \frac{0.15\underline{3}9474\text{ atm} \times 1.35\text{ L}}{0.08206\frac{\text{L}\cdot\text{atm}}{\text{mol}\cdot\text{K}} \times 298.2\text{ K}} = 0.0084\underline{9}3113\text{ mol}$$

$$0.0084\underline{9}3113\text{ mol} \times \frac{4.003\text{ g}}{1\text{ mol}} = 0.0340\text{ g He and}$$

$P_{Total} = P_{N_2} + P_{O_2} + P_{He} = 0.283\text{ atm} + 0.134\text{ atm} + 0.154\text{ atm} = 0.571\text{ atm}$ or
$P_{Total} = P_{N_2} + P_{O_2} + P_{He} = 215\text{ torr} + 102\text{ torr} + 117\text{ torr} = 434\text{ torr}$

Check: The units (g and atm) are correct. The magnitude of the answers (<1 and < 1) makes sense because gases are not very dense and these pressures are <1 atm. Because all of the pressures are small, the total is <1 atm.

11.62 **Given:** $P_{Total} = 745$ mmHg, $P_{CO_2} = 125$ mmHg, $P_{Ar} = 214$ mmHg, $P_{O_2} = 187$ mmHg, $V = 12.0$ L, and $T = 273$ K
Find: P_{He} and m_{He}
Conceptual Plan: $P_{Total}, P_{CO_2}, P_{Ar}, P_{O_2} \rightarrow P_{He}$ then mmHg → atm then $P, V, T \rightarrow n \rightarrow m$

$$P_{Total} = P_{CO_2} + P_{Ar} + P_{O_2} + P_{He} \quad \frac{1\text{ atm}}{760\text{ mmHg}} \quad V = nRT \quad \frac{4.003\text{ g}}{1\text{ mol}}$$

Solution: $P_{Total} = P_{CO_2} + P_{Ar} + P_{O_2} + P_{He}$ Rearrange to solve for P_{He}.

$$P_{He} = P_{Total} - P_{CO_2} + P_{Ar} + P_{O_2} = 745\text{ mmHg} - 125\text{ mmHg} - 214\text{ mmHg} - 187\text{ mmHg} = 219\text{ mmHg}$$

$P_{He} = 219\text{ mmHg} \times \frac{1\text{ atm}}{760\text{ mmHg}} = 0.288\text{ atm}$ $PV = nRT$ Rearrange to solve for n. $n = \frac{PV}{RT}$

$$n_{He} = \frac{0.288\ \cancel{atm} \times 12.0\ \cancel{L}}{0.08206 \frac{\cancel{L} \cdot \cancel{atm}}{mol \cdot \cancel{K}} \times 273\ \cancel{K}} = 0.15\underline{4}3539\ mol \qquad 0.15\underline{4}3539\ \cancel{mol} \times \frac{4.003\ g}{1\ \cancel{mol}} = 0.618\ g\ He$$

Check: The units (g) are correct. The magnitude of the answer (1) makes sense because gases are not very dense and these pressures are <1 atm.

11.63 **Given:** $m_{CO_2} = 1.20\ g$, $V = 755\ mL$, $P_{N_2} = 725\ mmHg$, and $T = 25.0\ °C$ **Find:** P_{Total}
Conceptual Plan: **mL → L and °C → K and $m \to n$ and $n, P, T \to V$ then atm → mmHg**

$$\frac{1\ L}{1000\ mL} \quad K = °C + 273.15 \quad \frac{1\ mol}{44.01\ g} \quad PV = nRT \quad \frac{760\ mmHg}{1\ atm}$$

finally $P_{CO_2}, P_{N_2} \to P_{Total}$

$$P_{Total} = P_{CO_2} + P_{N_2}$$

Solution: $V = 755\ \cancel{mL} \times \frac{1\ L}{1000\ \cancel{mL}} = 0.755\ L \qquad T = 25.0\ °C + 273.15 = 298.2\ K$

$n = 1.20\ \cancel{g} \times \frac{1\ mol}{44.01\ \cancel{g}} = 0.027\underline{2}665\ mol$, $PV = nRT$ Rearrange to solve for P.

$$P = \frac{nRT}{V} = \frac{0.027\underline{2}665\ \cancel{mol} \times 0.08206 \frac{\cancel{L} \cdot atm}{\cancel{mol} \cdot \cancel{K}} \times 298.2\ \cancel{K}}{0.755\ \cancel{L}} = 0.88\underline{3}734\ atm$$

$$P_{CO_2} = 0.88\underline{3}734\ \cancel{atm} \times \frac{760\ mmHg}{1\ \cancel{atm}} = 672\ mmHg$$

$$P_{Total} = P_{CO_2} + P_{N_2} = 672\ mmHg + 725\ mmHg = 1397\ mmHg \text{ or } 1397\ \cancel{torr} \times \frac{1\ atm}{760\ \cancel{torr}} = 1.84\ atm$$

Check: The units (mmHg) are correct. The magnitude of the answer (1400) makes sense because it must be greater than 725 mmHg.

11.64 **Given:** $V_{1,He} = 275\ mL$, $P_{1,He} = 752\ torr$, $V_{1,Ar} = 475\ mL$, and $P_{1,Ar} = 722\ torr$ **Find:** $P_{2,He}$, $P_{2,Ar}$, and P_{Total}
Conceptual Plan: **$V_{1,He}, V_{1,Ar} \to V_2$ $V_1, P_1, V_2 \to P_2$ then $P_{2,He}, P_{2,Ar} \to P_{Total}$**

$$V_{1,He} + V_{1,Ar} = V_2 \quad P_1V_1 = P_2V_2 \quad P_{Total} = P_{2,He} + P_{2,Ar}$$

Solution: $V_{1,He} + V_{1,Ar} = V_2 = 275\ mL + 475\ mL = 750.\ mL$ $P_1 V_1 = P_2 V_2$ Rearrange to solve for P_2.

$$P_2 = P_1 \frac{V_1}{V_2} \quad P_{2,He} = P_{1,He} \frac{V_{1,He}}{V_2} = 752\ torr \times \frac{275\ \cancel{mL}}{750.\ \cancel{mL}} = 27\underline{5}.733\ torr = 276\ torr\ He$$

$$P_{2,Ar} = P_{1,Ar} \frac{V_{1,Ar}}{V_2} = 722\ torr \times \frac{475\ \cancel{mL}}{750.\ \cancel{mL}} = 45\underline{7}.267\ torr = 457\ torr\ Ar$$

$P_{Total} = P_{2,He} + P_{2,Ar} = 27\underline{5}.733\ torr + 45\underline{7}.267\ torr = 733$ torr total pressure

Check: The units (torr) are correct. The magnitude of the answers (300, 450, and 700) makes physical sense because Boyle's law indicates that as the volume increases, the pressure decreases. Because both initial pressures are ~700 torr, the final total pressure should be about the same.

11.65 **Given:** $m_{N_2} = 1.25\ g$, $m_{O_2} = 0.85\ g$, $V = 1.55\ L$, and $T = 18\ °C$ **Find:** $\chi_{N_2}, \chi_{O_2}, P_{N_2}, P_{O_2}$
Conceptual Plan: **g → mol °C → K, then $n_{N_2}, n_{O_2} \to \chi_{N_2}$ and $n_{N_2}, n_{O_2} \to \chi_{O_2}$**

$$\mathcal{M} \quad K = °C + 273.15 \quad \chi_{N_2} = \frac{n_{N_2}}{n_{N_2} + n_{O_2}} \quad \chi_{O_2} = \frac{n_{O_2}}{n_{N_2} + n_{O_2}}$$

then $n, V, T \to P$

$$PV = nRT$$

Solution: $n_{N_2} = 1.25\ \cancel{g} \times \frac{1\ mol}{28.02\ \cancel{g}} = 0.044\underline{6}110\ mol$, $n_{O_2} = 0.85\ \cancel{g} \times \frac{1\ mol}{32.00\ \cancel{g}} = 0.02\underline{6}563\ mol$,

$$T = 18\ °C + 273.15 = 291\ K,\ \chi_{N_2} = \frac{n_{N_2}}{n_{N_2} + n_{O_2}} = \frac{0.044\underline{6}110\ \cancel{mol}}{0.044\underline{6}110\ \cancel{mol} + 0.02\underline{6}563\ \cancel{mol}} = 0.62\underline{6}792 = 0.627,$$

$$\chi_{O_2} = \frac{n_{O_2}}{n_{N_2} + n_{O_2}} = \frac{0.026563 \text{ mol}}{0.0446110 \text{ mol} + 0.026563 \text{ mol}} = 0.373212$$ We can also calculate this as

$\chi_{O_2} = 1 - \chi_{N_2} = 1 - 0.626792 = 0.373208 = 0.373$ $PV = nRT$ Rearrange to solve for P. $P = \frac{nRT}{V}$

$$P_{N_2} = \frac{0.0446110 \text{ mol} \times 0.08206 \frac{\text{L} \cdot \text{atm}}{\text{mol} \cdot \text{K}} \times 291 \text{ K}}{1.55 \text{ L}} = 0.687 \text{ atm}$$

$$P_{O_2} = \frac{0.026563 \text{ mol} \times 0.08206 \frac{\text{L} \cdot \text{atm}}{\text{mol} \cdot \text{K}} \times 291 \text{ K}}{1.55 \text{ L}} = 0.409 \text{ atm}$$

Check: The units (none and atm) are correct. The magnitude of the answers (<1 and <1) makes sense because the mole fractions should total 1, and because the weight of N_2 is greater than O_2, its mole fraction is larger. The number of moles is << 1, so we expect the pressures to be < 1 atm, given the V (1.55 L).

11.66 **Given:** Table 11.2, $m(O_2) = 10.0$ g, $T = 273$ K, and $P = 1.00$ atm **Find:** χ_{O_2} and V_{air}

Conceptual Plan: $\%V \rightarrow \chi_{O_2}$ **and** $m \rightarrow n \rightarrow V_{O_2} \rightarrow V_{air}$

$\frac{1}{100\%}$ $\frac{1 \text{ mol}}{32.00 \text{ g}}$ $\frac{22.414 \text{ L}}{1 \text{ mol}}$ $\frac{100 \text{ L air}}{21 \text{ L } O_2}$

Solution: from Table 11.2, $\%V_{O_2} = 21\%$ $21\% \times \frac{1}{100\%} = 0.21 = \chi_{O_2}$

$10.0 \text{ g} \times \frac{1 \text{ mol}}{32.00 \text{ g}} \times \frac{22.414 \text{ L}}{1 \text{ mol}} = 7.00 \text{ L} = V_{O_2}$ finally $7.00 \text{ L } O_2 \times \frac{100 \text{ L air}}{21 \text{ L } O_2} = 33 \text{ L air}$

Check: The units (none and L) are correct. The magnitude of the answer (0.21) makes sense because most of air is nitrogen. The magnitude of the answer (33 L) makes sense because one mole of an ideal gas under standard conditions (273 K and 1 atm) occupies 22.4 L and we have about 1/3 mol of O_2 (so, over a mole of air).

11.67 **Given:** $T = 30.0$ °C, $P_{Total} = 732$ mmHg, and $V = 722$ mL **Find:** P_{H_2} and m_{H_2}

Conceptual Plan: $T \rightarrow P_{H_2O}$ **then** $P_{Total}, P_{H_2O} \rightarrow P_{H_2}$ **then mmHg → atm and mL → L**

Table 11.3 $P_{Total} = P_{H_2O} + P_{H_2}$ $\frac{1 \text{ atm}}{760 \text{ mmHg}}$ $\frac{1 \text{ L}}{1000 \text{ mL}}$

and °C → K $P, V, T \rightarrow n$ **then** $n \rightarrow m$

K = °C + 273.15 $PV = nRT$ $\frac{2.016 \text{ g}}{1 \text{ mol}}$

Solution: Table 11.3 states that at 30 °C, $P_{H_2O} = 31.86$ mmHg $P_{Total} = P_{H_2O} + P_{H_2}$ Rearrange to solve for P_{H_2}. $P_{H_2} = P_{Total} - P_{H_2O} = 732 \text{ mmHg} - 31.86 \text{ mmHg} = 700. \text{ mmHg}$

$P_{H_2} = 700. \text{ mmHg} \times \frac{1 \text{ atm}}{760 \text{ mmHg}} = 0.921052 \text{ atm}$ $V = 722 \text{ mL} \times \frac{1 \text{ L}}{1000 \text{ mL}} = 0.722 \text{ L}$

$T = 30.0 \text{ °C} + 273.15 = 303.2 \text{ K}$, $PV = nRT$ Rearrange to solve for n. $n = \frac{PV}{RT}$

$$n_{H_2} = \frac{0.921052 \text{ atm} \times 0.722 \text{ L}}{0.08206 \frac{\text{L} \cdot \text{atm}}{\text{mol} \cdot \text{K}} \times 303.2 \text{ K}} = 0.0267276 \text{ mol then } 0.0267276 \text{ mol} \times \frac{2.016 \text{ g}}{1 \text{ mol}} = 0.0539 \text{ g } H_2$$

Check: The units (g) are correct. The magnitude of the answer (<< 1) makes sense because gases are not very dense, hydrogen is light, the volume is small, and the pressure is ~1 atm.

11.68 **Given:** $T = 25$ °C, $V = 5.45$ L, and $P_{Total} = 745$ mmHg **Find:** n

Conceptual Plan: $T \rightarrow P_{H_2O}$ **then** $P_{Total}, P_{H_2O} \rightarrow P_{air}$ **then mmHg → atm and °C → K**

Table 11.3 $P_{Total} = P_{H_2O} + P_{air}$ $\frac{1 \text{ atm}}{760 \text{ mmHg}}$ K = °C + 273.15

$P, V, T \rightarrow n$

$PV = nRT$

Solution: Table 11.3 states that at 25 °C, $P_{H_2O} = 23.78$ mmHg $P_{Total} = P_{H_2O} + P_{air}$ Rearrange to solve for P_{air}. $P_{air} = P_{Total} - P_{H_2O} = 745 \text{ mmHg} - 23.78 \text{ mmHg} = 721 \text{ mmHg}$

$P_{air} = 721 \text{ mmHg} \times \frac{1 \text{ atm}}{760 \text{ mmHg}} = 0.94\underline{8}684 \text{ atm}$ $T = 25\text{ °C} + 273.15 = 298 \text{ K}$, $PV = nRT$

Rearrange to solve for n. $n = \frac{PV}{RT} = \frac{0.94\underline{8}684 \text{ atm} \times 5.45 \text{ L}}{0.08206 \frac{\text{L} \cdot \text{atm}}{\text{mol} \cdot \text{K}} \times 298 \text{ K}} = 0.211 \text{ mol}$

Check: The units (mol) are correct. The magnitude of the answer (0.2) makes sense because 22.4 L of a gas at STP contains 1 mol. We have only 5.45 L, so the answer is reasonable.

11.69 **Given:** $T = 25$ °C, $P_{Total} = 748$ mmHg, and $V = 0.951$ L **Find:** P_{H_2} and m_{H_2}

Conceptual Plan: $T \rightarrow P_{H_2O}$ **then** $P_{Total}, P_{H_2O} \rightarrow P_{H_2}$ **then mmHg → atm and mL → L**

Table 11.3 $\quad P_{Total} = P_{H_2O} + P_{H_2} \quad \frac{1 \text{ atm}}{760 \text{ mmHg}} \quad \frac{1 \text{ L}}{1000 \text{ mL}}$

and °C → K $P, V, T \rightarrow n$ **then** $n \rightarrow m$

$K = °C + 273.15 \quad PV = nRT \quad \frac{2.016 \text{ g}}{1 \text{ mol}}$

Solution: Table 11.3 states that at 25 °C, $P_{H_2O} = 23.78$ mmHg $P_{Total} = P_{H_2O} + P_{H_2}$
Rearrange to solve for P_{H_2}. $P_{H_2} = P_{Total} - P_{H_2O} = 748 \text{ mmHg} - 23.78 \text{ mmHg} = 724 \text{ mmHg}$

$P_{H_2} = 724 \text{ mmHg} \times \frac{1 \text{ atm}}{760 \text{ mmHg}} = 0.95\underline{2}632 \text{ atm}$ $T = 25\text{ °C} + 273.15 = 298 \text{ K}$ $PV = nRT$

Rearrange to solve for n. $n_{H_2} = \frac{PV}{RT} = \frac{0.95\underline{2}632 \text{ atm} \times 0.951 \text{ L}}{0.08206 \frac{\text{L} \cdot \text{atm}}{\text{mol} \cdot \text{K}} \times 298 \text{ K}} = 0.037\underline{0}474 \text{ mol}$

$0.037\underline{0}474 \text{ mol} \times \frac{2.016 \text{ g}}{1 \text{ mol}} = 0.0747 \text{ g } H_2$

Check: The units (g) are correct. The magnitude of the answer ($<< 1$) makes sense because gases are not very dense, hydrogen is light, the volume is small, and the pressure is ~1 atm.

11.70 **Given:** $m_{O_2} = 2.0$ g, $m_{He} = 98.0$ g, $P_{Total} = 8.5$ atm **Find:** P_{O_2}

Conceptual Plan: $m \rightarrow n$ **then** $n_{O_2}, n_{He} \rightarrow \chi_{O_2}$ **then** $\chi_{O_2}, P_{Total} \rightarrow P_{O_2}$

$\mathcal{M} \quad \chi_{O_2} = \frac{n_{O_2}}{n_{O_2} + n_{He}} \quad P_{O_2} = \chi_{O_2} P_{Total}$

Solution: $n_{O_2} = 2.0 \text{ g} \times \frac{1 \text{ mol}}{32.00 \text{ g}} = 0.0625 \text{ mol}$, $n_{He} = 98.0 \text{ g} \times \frac{1 \text{ mol}}{4.003 \text{ g}} = 24.\underline{4}816 \text{ mol}$,

$\chi_{O_2} = \frac{n_{O_2}}{n_{O_2} + n_{He}} = \frac{0.0625 \text{ mol}}{0.0625 \text{ mol} + 24.\underline{4}816 \text{ mol}} = 0.002\underline{5}464$,

$P_{O_2} = \chi_{O_2} P_{Total} = 0.002\underline{5}464 \times 8.5 \text{ atm} = 0.022 \text{ atm}$

Check: The units (atm) are correct. The magnitude of the answer (0.022) makes sense because at these depths, high oxygen pressures can cause toxicity.

Kinetic Molecular Theory

11.71 (a) Yes, because the average kinetic energy of a particle is proportional to the temperature in kelvins and the two gases are at the same temperature, they have the same average kinetic energy.

(b) No, because the helium atoms are lighter, they must move faster to have the same kinetic energy as argon atoms.

(c) No, because the Ar atoms are moving slower to compensate for their larger mass, they will exert the same pressure on the walls of the container.

(d) Because He is lighter, it has the faster rate of effusion.

11.72 (a) Because both gases have a mole fraction of 0.5, they have the same partial pressure.

(b) The nitrogen molecules have a greater velocity because they are lighter than Xe atoms.

(c) Because the average kinetic energy of a particle is proportional to the temperature in kelvins and the two gases are at the same temperature, they have the same average kinetic energy.

(d) Because nitrogen is lighter, it has the faster rate of effusion.

11.73 **Given:** F_2, Cl_2, Br_2, and $T = 298$ K **Find:** u_{rms} and KE_{avg} for each gas and relative rates of effusion

Conceptual Plan: $\mathcal{M}, T \rightarrow u_{rms} \rightarrow KE_{avg}$

$$u_{rms} = \sqrt{\frac{3RT}{\mathcal{M}}} \qquad KE_{avg} = \frac{1}{2}N_A m u_{rms}^2 = \frac{3}{2}RT$$

Solution:

$$F_2: \mathcal{M} = \frac{38.00\ \text{g}}{1\ \text{mol}} \times \frac{1\ \text{kg}}{1000\ \text{g}} = 0.03800\ \text{kg/mol},\ u_{rms} = \sqrt{\frac{3RT}{\mathcal{M}}} = \sqrt{\frac{3 \times 8.314 \frac{\text{J}}{\text{K} \cdot \text{mol}} \times 298\ \text{K}}{0.03800 \frac{\text{kg}}{\text{mol}}}} = 442\ \text{m/s}$$

$$Cl_2: \mathcal{M} = \frac{70.90\ \text{g}}{1\ \text{mol}} \times \frac{1\ \text{kg}}{1000\ \text{g}} = 0.07090\ \text{kg/mol},\ u_{rms} = \sqrt{\frac{3RT}{\mathcal{M}}} = \sqrt{\frac{3 \times 8.314 \frac{\text{J}}{\text{K} \cdot \text{mol}} \times 298\ \text{K}}{0.07090 \frac{\text{kg}}{\text{mol}}}} = 324\ \text{m/s}$$

$$Br_2: \mathcal{M} = \frac{159.80\ \text{g}}{1\ \text{mol}} \times \frac{1\ \text{kg}}{1000\ \text{g}} = 0.15980\ \text{kg/mol},\ u_{rms} = \sqrt{\frac{3RT}{\mathcal{M}}} = \sqrt{\frac{3 \times 8.314 \frac{\text{J}}{\text{K} \cdot \text{mol}} \times 298\ \text{K}}{0.15980 \frac{\text{kg}}{\text{mol}}}} = 216\ \text{m/s}$$

All molecules have the same average kinetic energy:

$$KE_{avg} = \frac{3}{2}RT = \frac{3}{2} \times 8.314 \frac{\text{J}}{\text{K} \cdot \text{mol}} \times 298\ \text{K} = 3.72 \times 10^3\ \text{J/mol}$$

Because rate of effusion is proportional to $\sqrt{\frac{1}{\mathcal{M}}}$, F_2 will have the fastest rate and Br_2 will have the slowest rate.

Check: The units (m/s) are correct. The magnitude of the answers (200−450) makes sense because it is consistent with what was shown in the text and the heavier the molecule, the slower the molecule.

11.74 **Given:** CO, CO_2, SO_3, and $T = 298$ K

Find: u_{rms}, KE_{avg} for each gas, and greatest u_{rms}, KE_{avg}, and rate of effusion

Conceptual Plan: $\mathcal{M}, T \rightarrow u_{rms} \rightarrow KE_{avg}$

$$u_{rms} = \sqrt{\frac{3RT}{M}} \quad KE_{avg} = \frac{1}{2}N_A m u_{rms}^2 = \frac{3}{2}RT$$

Solution:

$$CO: \mathcal{M} = \frac{28.01\ \text{g}}{1\ \text{mol}} \times \frac{1\ \text{kg}}{1000\ \text{g}} = 0.02801\ \text{kg/mol}$$

$$u_{rms} = \sqrt{\frac{3RT}{\mathcal{M}}} = \sqrt{\frac{3 \times 8.314 \frac{\text{J}}{\text{K} \cdot \text{mol}} \times 298\ \text{K}}{0.02801 \frac{\text{kg}}{\text{mol}}}} = 515\ \text{m/s}$$

$$CO_2: \mathcal{M} = \frac{44.01\ \text{g}}{1\ \text{mol}} \times \frac{1\ \text{kg}}{1000\ \text{g}} = 0.04401\ \text{kg/mol}$$

$$u_{rms} = \sqrt{\frac{3RT}{\mathcal{M}}} = \sqrt{\frac{3 \times 8.314 \frac{\text{J}}{\text{K} \cdot \text{mol}} \times 298\ \text{K}}{0.04401 \frac{\text{kg}}{\text{mol}}}} = 411\ \text{m/s}$$

$$SO_3: \mathcal{M} = \frac{80.07\ \cancel{g}}{1\ mol} \times \frac{1\ kg}{1000\ \cancel{g}} = 0.08007\ kg/mol$$

$$u_{rms} = \sqrt{\frac{3RT}{\mathcal{M}}} = \sqrt{\frac{3 \times 8.314 \frac{J}{\cancel{K} \cdot \cancel{mol}} \times 298\ \cancel{K}}{0.08007 \frac{kg}{\cancel{mol}}}} = 305\ m/s$$

All molecules have the same average kinetic energy:

$KE_{avg} = \frac{3}{2}RT = \frac{3}{2} \times 8.314 \frac{J}{\cancel{K} \cdot mol} \times 298\ \cancel{K} = 3.72 \times 10^3\ J/mol$. CO has the fastest speed; all molecules have the same average kinetic energy; and because rate of effusion is proportional to $1/\sqrt{\mathcal{M}}$, CO has the fastest rate.

Check: The units (m/s) are correct. The magnitude of the answers $(300-520)$ makes sense because it is consistent with what was shown in the text and the heavier the molecule, the slower the molecule.

11.75 **Given:** $^{238}UF_6$ and $^{235}UF_6$ U-235 = 235.054 amu, U-238 = 238.051 amu
Find: ratio of effusion rates $^{238}UF_6/^{235}UF_6$
Conceptual Plan: $\mathcal{M}_{^{238}UF_6}, \mathcal{M}_{^{235}UF_6} \rightarrow Rate_{^{238}UF_6}/Rate_{^{235}UF_6}$

$$\frac{Rate_{^{238}UF_6}}{Rate_{^{235}UF_6}} = \sqrt{\frac{\mathcal{M}_{^{235}UF_6}}{\mathcal{M}_{^{238}UF_6}}}$$

Solution: $^{238}UF_6: \mathcal{M} = \frac{352.05\ \cancel{g}}{1\ mol} \times \frac{1\ kg}{1000\ \cancel{g}} = 0.35205\ kg/mol$

$$^{235}UF_6: \mathcal{M} = \frac{349.05\ \cancel{g}}{1\ mol} \times \frac{1\ kg}{1000\ \cancel{g}} = 0.34905\ kg/mol$$

$$\frac{Rate_{^{238}UF_6}}{Rate_{^{235}UF_6}} = \sqrt{\frac{\mathcal{M}_{^{235}UF_6}}{\mathcal{M}_{^{238}UF_6}}} = \sqrt{\frac{0.34905\ \cancel{kg/mol}}{0.35205\ \cancel{kg/mol}}} = 0.99573$$

Check: The units (none) are correct. The magnitude of the answer (< 1) makes sense because the heavier molecule has the lower effusion rate because it moves more slowly.

11.76 **Given:** Ar and Kr **Find:** ratio of effusion rates Ar/Kr
Conceptual Plan: $\mathcal{M}_{Ar}, \mathcal{M}_{Kr} \rightarrow Rate_{Ar}/Rate_{Kr}$

$$\frac{Rate_{Ar}}{Rate_{Kr}} = \sqrt{\frac{\mathcal{M}_{Kr}}{\mathcal{M}_{Ar}}}$$

Solution: Ar: $\mathcal{M} = \frac{39.95\ \cancel{g}}{1\ mol} \times \frac{1\ kg}{1000\ \cancel{g}} = 0.03995\ kg/mol$, Kr: $\mathcal{M} = \frac{83.80\ \cancel{g}}{1\ mol} \times \frac{1\ kg}{1000\ \cancel{g}} = 0.08380\ kg/mol$

$$\frac{Rate_{Ar}}{Rate_{Kr}} = \sqrt{\frac{\mathcal{M}_{Kr}}{\mathcal{M}_{Ar}}} = \sqrt{\frac{0.08380\ \cancel{kg/mol}}{0.03995\ \cancel{kg/mol}}} = 1.448$$

Check: The units (none) are correct. The magnitude of the answer (> 1) makes sense because the lighter molecule has the higher effusion rate because it moves faster.

11.77 **Given:** Ne and unknown gas; Ne effusion in 76 s and unknown in 155 s **Find:** identify unknown gas
Conceptual Plan: $\mathcal{M}_{Ne}, Rate_{Ne}, Rate_{Unk} \rightarrow \mathcal{M}_{Kr}$

$$\frac{Rate_{Ne}}{Rate_{Unk}} = \sqrt{\frac{\mathcal{M}_{Unk}}{\mathcal{M}_{Ne}}}$$

Solution: Ne: $\mathcal{M} = \frac{20.18\ \cancel{g}}{1\ mol} \times \frac{1\ kg}{1000\ \cancel{g}} = 0.02018\ kg/mol$, $\frac{Rate_{Ne}}{Rate_{Unk}} = \sqrt{\frac{\mathcal{M}_{unk}}{\mathcal{M}_{Ne}}}$

Rearrange to solve for $\mathcal{M}_{Unk}$. $\mathcal{M}_{Unk} = \mathcal{M}_{Ne}\left(\frac{Rate_{Ne}}{Rate_{Unk}}\right)^2$ Because Rate α 1/(effusion time),

$$\mathcal{M}_{Unk} = \mathcal{M}_{Ne}\left(\frac{Time_{Unk}}{Time_{Ne}}\right)^2 = 0.02018\frac{kg}{mol} \times \left(\frac{155\ \cancel{s}}{76\ \cancel{s}}\right)^2 = 0.084\frac{\cancel{kg}}{mol} \times \frac{1000\ g}{1\ \cancel{kg}} = 84\ g/mol \text{ or Kr}$$

Check: The units (g/mol) are correct. The magnitude of the answer (> Ne) makes sense because Ne effused faster; so it must be lighter.

11.78 **Given:** N_2O and I_2 gas; N_2O effusion in 42 s **Find:** effusion time for I_2 gas

Conceptual Plan: $\mathcal{M}_{N_2O}, \mathcal{M}_{Kr}, \text{Rate}_{N_2O}, \rightarrow \text{Rate}_{I_2}$

$$\frac{\text{Rate}_{N_2O}}{\text{Rate}_{I_2}} = \sqrt{\frac{\mathcal{M}_{I_2}}{\mathcal{M}_{N_2O}}}$$

Solution: N_2O: $\mathcal{M} = \frac{44.02\ \text{g}}{1\ \text{mol}} \times \frac{1\ \text{kg}}{1000\ \text{g}} = 0.04402\ \text{kg/mol}$, I_2: $\mathcal{M} = \frac{253.8\ \text{g}}{1\ \text{mol}} \times \frac{1\ \text{kg}}{1000\ \text{g}} = 0.2538\ \text{kg/mol}$

$\frac{\text{Rate}_{N_2O}}{\text{Rate}_{I_2}} = \sqrt{\frac{\mathcal{M}_{I_2}}{\mathcal{M}_{N_2O}}}$ Because Rate α 1/(effusion time), $\frac{Time_{I_2}}{Time_{N_2O}} = \sqrt{\frac{\mathcal{M}_{I_2}}{\mathcal{M}_{N_2O}}}$

Rearrange to solve for $Time_{I_2}$. $Time_{I_2} = Time_{N_2O} \times \sqrt{\frac{\mathcal{M}_{I_2}}{\mathcal{M}_{N_2O}}} = 42\ \text{s} \times \sqrt{\frac{0.2538\ \text{kg/mol}}{0.04402\ \text{kg/mol}}} = 1.0 \times 10^2\ \text{s}$

Check: The units (s) are correct. The magnitude of the answer (100) makes sense because the mass ratio of I_2 to N_2O is ~6; so the time should be over twice as long.

11.79 Gas A has the higher molar mass because it has slower average velocity. Gas B will have the higher effusion rate because it has the higher average velocity.

11.80 T_2 is the higher temperature because the molecules at T_2 have the higher average velocity.

Reaction Stoichiometry Involving Gases

11.81 **Given:** $m_C = 15.7$ g, $P = 1.0$ atm, and $T = 355$ K **Find:** V

Conceptual Plan: $m_C \rightarrow n_C \rightarrow n_{H_2}$ **then** $n_{H_2}, P, T \rightarrow V$

$$\frac{1\ \text{mol}}{12.01\ \text{g C}} \quad \frac{1\ \text{mol H}_2}{1\ \text{mol C}} \qquad PV = nRT$$

Solution: $15.7\ \text{g C} \times \frac{1\ \text{mol C}}{12.01\ \text{g C}} \times \frac{1\ \text{mol H}_2}{1\ \text{mol C}} = 1.30724\ \text{mol H}_2$ $PV = nRT$ Rearrange to solve for V.

$$V = \frac{nRT}{P} = \frac{1.30724\ \text{mol} \times 0.08206\ \frac{\text{L} \cdot \text{atm}}{\text{mol} \cdot \text{K}} \times 355\ \text{K}}{1.0\ \text{atm}} = 38\ \text{L}$$

Check: The units (L) are correct. The magnitude of the answer (38) makes sense because we have more than one mole of gas; so we expect more than 22 L.

11.82 **Given:** $V_{O_2} = 1.4$ L, $T = 315$ K, $P_{O_2} = 0.957$ atm **Find:** m_{H_2O}

Conceptual Plan: $P_{O_2}, V_{O_2}, T \rightarrow n_{O_2}$ **then** $n_{O_2} \rightarrow n_{H_2O} \rightarrow m_{H_2O}$

$$PV = nRT \qquad \frac{2\ \text{mol H}_2\text{O}}{1\ \text{mol O}_2} \quad \frac{18.02\ \text{g H}_2\text{O}}{1\ \text{mol H}_2\text{O}}$$

Solution: $PV = nRT$ Rearrange to solve for n. $n = \frac{PV}{RT} = \frac{0.957\ \text{atm} \times 1.4\ \text{L}}{0.08206\ \frac{\text{L} \cdot \text{atm}}{\text{mol} \cdot \text{K}} \times 315\ \text{K}} = 0.051832\ \text{mol O}_2$

$$0.051832\ \text{mol O}_2 \times \frac{2\ \text{mol H}_2\text{O}}{1\ \text{mol O}_2} \times \frac{18.02\ \text{g H}_2\text{O}}{1\ \text{mol H}_2\text{O}} = 1.9\ \text{g H}_2\text{O}$$

Check: The units (g) are correct. The magnitude of the answer (2) makes sense because we have much less than a mole of oxygen.

11.83 **Given:** $P = 748$ mmHg, $T = 86$ °C, and $m_{CH_3OH} = 25.8$ g **Find:** V_{H_2} and V_{CO}

Conceptual Plan: $m_{CH_3OH} \rightarrow n_{CH_3OH} \rightarrow n_{H_2}$ **and mmHg → atm and °C → K**

$$\frac{1\ \text{mol CH}_3\text{OH}}{32.04\ \text{g CH}_3\text{OH}} \quad \frac{2\ \text{mol H}_2}{1\ \text{mol CH}_3\text{OH}} \qquad \frac{1\ \text{atm}}{760\ \text{mmHg}} \qquad \text{K} = {^\circ}\text{C} + 273.15$$

then $n_{H_2}, P, T \rightarrow V$ and $n_{H_2} \rightarrow n_{CO}$ then $n_{CO}, P, T \rightarrow V$

$PV = nRT$ $\quad \frac{1 \text{ mol CO}}{2 \text{ mol } H_2}$ $\quad PV = nRT$

Solution: $25.8 \text{ g } CH_3OH \times \frac{1 \text{ mol } CH_3OH}{32.04 \text{ g } CH_3OH} \times \frac{2 \text{ mol } H_2}{1 \text{ mol } CH_3OH} = 1.61049 \text{ mol } H_2$

$P_{H_2} = 748 \text{ mmHg} \times \frac{1 \text{ atm}}{760 \text{ mmHg}} = 0.984211 \text{ atm}, T = 86\ °C + 273.15 = 359 \text{ K}, PV = nRT$

Rearrange to solve for V. $V = \frac{nRT}{P}$, $V_{H_2} = \frac{1.61049 \text{ mol} \times 0.08206 \frac{\text{L} \cdot \text{atm}}{\text{mol} \cdot \text{K}} \times 359 \text{ K}}{0.984211 \text{ atm}} = 48.2 \text{ L } H_2$

$1.61049 \text{ mol } H_2 \times \frac{1 \text{ mol CO}}{2 \text{ mol } H_2} = 0.80525 \text{ mol CO}, V_{CO} = \frac{0.80525 \text{ mol} \times 0.08206 \frac{\text{L} \cdot \text{atm}}{\text{mol} \cdot \text{K}} \times 359 \text{ K}}{0.984211 \text{ atm}} = 24.1 \text{ L CO}$

Check: The units (L) are correct. The magnitude of the answers (48 and 24) makes sense because we have more than one mole of hydrogen gas and half that of CO; so we expect significantly more than 22 L for hydrogen and half the volume of hydrogen for CO.

11.84 **Given:** $P = 782$ mmHg, $T = 25\ °C$, and $m_{Al} = 53.2$ g **Find:** V_{O_2}

Conceptual Plan: $m_{Al} \rightarrow n_{Al} \rightarrow n_{O_2}$ and mmHg $\rightarrow$ atm and °C $\rightarrow$ K then

$\frac{1 \text{ mol Al}}{26.98 \text{ g Al}}$ $\frac{3 \text{ mol } O_2}{4 \text{ mol Al}}$ $\quad \frac{1 \text{ atm}}{760 \text{ mmHg}}$ $\quad K = °C + 273.15$

$n_{O_2}, P, T \rightarrow V$

$PV = nRT$

Solution: $53.2 \text{ g Al} \times \frac{1 \text{ mol Al}}{26.98 \text{ g Al}} \times \frac{3 \text{ mol } O_2}{4 \text{ mol Al}} = 1.478873 \text{ mol } O_2$

$P_{O_2} = 782 \text{ mmHg} \times \frac{1 \text{ atm}}{760 \text{ mmHg}} = 1.028947 \text{ atm}, T = 25\ °C + 273.15 = 298 \text{ K}, PV = nRT$

Rearrange to solve for V. $V_{O_2} = \frac{nRT}{P} = \frac{1.478873 \text{ mol} \times 0.08206 \frac{\text{L} \cdot \text{atm}}{\text{mol} \cdot \text{K}} \times 298 \text{ K}}{1.028947 \text{ atm}} = 35.1 \text{ L } O_2$

Check: The units (L) are correct. The magnitude of the answer (35) makes sense because we have more than one mole of oxygen gas and more than 1 atm; so we expect significantly more than 22 L.

11.85 **Given:** $V = 11.8$ L at STP **Find:** m_{NaN_3}

Conceptual Plan: $V_{N_2} \rightarrow n_{N_2} \rightarrow n_{NaN_3} \rightarrow m_{NaN_3}$

$\frac{1 \text{ mol } N_2}{22.414 \text{ L } N_2}$ $\frac{2 \text{ mol } NaN_3}{3 \text{ mol } N_2}$ $\frac{65.02 \text{ g } NaN_3}{1 \text{ mol } NaN_3}$

Solution: $11.8 \text{ L } N_2 \times \frac{1 \text{ mol } N_2}{22.414 \text{ L } N_2} \times \frac{2 \text{ mol } NaN_3}{3 \text{ mol } N_2} \times \frac{65.02 \text{ g } NaN_3}{1 \text{ mol } NaN_3} = 22.8 \text{ g } NaN_3$

Check: The units (g) are correct. The magnitude of the answer (23 g) makes sense because we have about half a mole of nitrogen gas, which translates to even fewer moles of NaN_3; so we expect significantly less than 65 g.

11.86 **Given:** $V = 58.5$ mL, and STP **Find:** m_{Li}

Conceptual Plan: mL N_2 $\rightarrow$ L N_2 $\rightarrow n_{N_2} \rightarrow n_{Li} \rightarrow m_{Li}$

$\frac{1 \text{ L}}{1000 \text{ mL}}$ $\frac{1 \text{ mol } N_2}{22.414 \text{ L } N_2}$ $\frac{6 \text{ mol Li}}{1 \text{ mol } N_2}$ $\frac{6.941 \text{ g Li}}{1 \text{ mol Li}}$

Solution: $58.5 \text{ mL } N_2 \times \frac{1 \text{ L } N_2}{1000 \text{ mL } N_2} \times \frac{1 \text{ mol } N_2}{22.414 \text{ L } N_2} \times \frac{6 \text{ mol Li}}{1 \text{ mol } N_2} \times \frac{6.941 \text{ g Li}}{1 \text{ mol Li}} = 0.109 \text{ g Li}$

Check: The units (g) are correct. The magnitude of the answer (0.1) makes sense because we have such a small volume, which translates to a small fraction of a mole of Li; so we expect significantly less than 6.9 g.

11.87 **Given:** $V_{CH_4} = 25.5$ L, $P_{CH_4} = 732$ torr, and $T = 25$ °C; mixed with $V_{H_2O} = 22.8$ L, $P_{H_2O} = 702$ torr, and $T = 125$ °C; forms $P_{H_2} = 26.2$ L at STP **Find:** % yield

Conceptual Plan: CH_4 : torr → atm and °C → K and $P, V, T \rightarrow n_{CH_4} \rightarrow n_{H_2}$

$$\frac{1 \text{ atm}}{760 \text{ torr}} \quad K = °C + 273.15 \quad PV = nRT \quad \frac{3 \text{ mol } H_2}{1 \text{ mol } CH_4}$$

H_2O : torr → atm and °C → K and $P, V, T \rightarrow n_{H_2O} \rightarrow n_{H_2}$

$$\frac{1 \text{ atm}}{760 \text{ torr}} \quad K = °C + 273.15 \quad PV = nRT \quad \frac{3 \text{ mol } H_2}{1 \text{ mol } H_2O}$$

Select smaller n_{H_2} as theoretical yield,
then $V_{H_2} \rightarrow n_{H_2}$ (actual yield) then actual yield, theoretical yield → % yield

$$\frac{1 \text{ mol } H_2}{22.414 \text{ L } H_2} \qquad \% \text{ yield} = \frac{\text{actual yield}}{\text{theoretical yield}} \times 100\%$$

Solution: CH_4: $P_{CH_4} = 732 \text{ torr} \times \frac{1 \text{ atm}}{760 \text{ torr}} = 0.963158 \text{ atm}$, $T = 25 \text{ °C} + 273.15 = 298 \text{ K}$, $PV = nRT$

Rearrange to solve for n. $n = \frac{PV}{RT}$ $\quad n_{CH_4} = \frac{0.963158 \text{ atm} \times 25.5 \text{ L}}{0.08206 \frac{\text{L} \cdot \text{atm}}{\text{mol} \cdot \text{K}} \times 298 \text{ K}} = 1.00436 \text{ mol } CH_4$

$$1.00436 \text{ mol } CH_4 \times \frac{3 \text{ mol } H_2}{1 \text{ mol } CH_4} = 3.01308 \text{ mol } H_2$$

H_2O: $P_{H_2O} = 702 \text{ torr} \times \frac{1 \text{ atm}}{760 \text{ torr}} = 0.923684 \text{ atm}$, $T = 125 \text{ °C} + 273.15 = 398 \text{ K}$, $n = \frac{PV}{RT}$

$$n_{H_2O} = \frac{0.923684 \text{ atm} \times 22.8 \text{ L}}{0.08206 \frac{\text{L} \cdot \text{atm}}{\text{mol} \cdot \text{K}} \times 398 \text{ K}} = 0.644828 \text{ mol } H_2O \quad 0.644828 \text{ mol } H_2O \times \frac{3 \text{ mol } H_2}{1 \text{ mol } H_2O} = 1.93448 \text{ mol } H_2$$

Water is the limiting reagent because the moles of hydrogen generated are lower.
theoretical yield = 1.93448 mol H_2

$$26.2 \text{ L } H_2 \times \frac{1 \text{ mol } H_2}{22.414 \text{ L } H_2} = 1.16891 \text{ mol } H_2 = \text{actual yield}$$

$$\% \text{ yield} = \frac{\text{actual yield}}{\text{theoretical yield}} \times 100\% = \frac{1.16891 \text{ mol } H_2}{1.93448 \text{ mol } H_2} \times 100\% = 60.4\%$$

Check: The units (%) are correct. The magnitude of the answer (60) makes sense because it is between 0 and 100%.

11.88 **Given:** $P = 25.0$ mmHg, $T = 225$ K, and $m_{CF_3Cl} = 15.0$ g; 10 cycles **Find:** V_{O_3}

Conceptual Plan: $m_{CF_3Cl} \rightarrow n_{CF_3Cl} \rightarrow n_{O_3} \rightarrow n_{O_3}$ and mmHg → atm then

$$\frac{1 \text{ mol } CF_3Cl}{104.46 \text{ g } CF_3Cl} \quad \frac{2 \text{ mol } O_3/\text{cycle}}{1 \text{ mol } CF_3Cl} \quad 10 \text{ cycles} \qquad \frac{1 \text{ atm}}{760 \text{ mmHg}}$$

$n, P, T \rightarrow V$

$$PV = nRT$$

Solution: $15.0 \text{ g } CF_3Cl \times \frac{1 \text{ mol } CF_3Cl}{104.46 \text{ g } CF_3Cl} \times \frac{2 \text{ mol } O_3/\text{cycle}}{1 \text{ mol } CF_3Cl} \times 10 \text{ cycles} = 2.871913 \text{ mol } O_3$

$P_{O_3} = 25.0 \text{ mmHg} \times \frac{1 \text{ atm}}{760 \text{ mmHg}} = 0.0328947 \text{ atm}$, $PV = nRT$ Rearrange to solve for V. $V = \frac{nRT}{P}$

$$V_{O_3} = \frac{2.871913 \text{ mol} \times 0.08206 \frac{\text{L} \cdot \text{atm}}{\text{mol} \cdot \text{K}} \times 225 \text{ K}}{0.0328947 \text{ atm}} = 1.61 \times 10^3 \text{ L } O_3$$

Check: The units (L) are correct. The magnitude of the answer (1600) makes sense because we have ~3 moles of ozone gas and the pressure is so low (0.03 atm); so we expect a large volume.

11.89 **Given:** $V = 2.00$ L, $P_{Cl_2} = 337$ mmHg, $P_{F_2} = 729$ mmHg, $T = 298$ K **Find:** limiting reactant and m_{ClF_3}

Conceptual Plan: Determine limiting reactant by comparing the pressures of each reactant in the same way we would compare moles then

$$\frac{3 \text{ mmHg } F_2}{1 \text{ mmHg } Cl_2}$$

mmHg limiting reactant → mmHg ClF_3 and mmHg → atm then

$$\frac{2 \text{ mmHg ClF}_3}{1 \text{ mmHg Cl}_2} \text{ or } \frac{2 \text{ mmHg ClF}_3}{3 \text{ mmHg F}_2} \qquad \frac{1 \text{ atm}}{760 \text{ mmHg}}$$

P, V, T → n → m

$$PV = nRT \quad \frac{92.45 \text{ g ClF}_3}{1 \text{ mol ClF}_3}$$

Solution: To determine the limiting reactant, calculate the pressure of fluorine needed to react all of the chlorine and compare to the pressure of fluorine available.

$$337 \text{ mmHg Cl}_2 \times \frac{3 \text{ mmHg F}_2}{1 \text{ mmHg Cl}_2} = 1011 \text{ mmHg F}_2 \text{ needed.}$$

Because only 729 mmHg F_2 is available, F_2 is the limiting reactant; then

$$P_{ClF_3} = 729 \text{ mmHg F}_2 \times \frac{2 \text{ mmHg ClF}_3}{3 \text{ mmHg F}_2} \times \frac{1 \text{ atm ClF}_3}{760 \text{ mmHg ClF}_3} = 0.6394737 \text{ atm ClF}_3 \; PV = nRT \quad \text{Rearrange to}$$

$$\text{solve for } n.\; n = \frac{PV}{RT} \; n_{ClF_3} = \frac{0.6394737 \text{ atm ClF}_3 \times 2.00 \text{ L}}{0.08206 \frac{\text{L} \cdot \text{atm}}{\text{mol} \cdot \text{K}} \times 298 \text{ K}} = 0.052300387 \text{ mol ClF}_3$$

$$\text{Finally, } 0.052300387 \text{ mol ClF}_3 \times \frac{92.45 \text{ g ClF}_3}{1 \text{ mol ClF}_3} = 4.8351708 \text{ g ClF}_3 = 4.84 \text{ g ClF}_3$$

Check: The units (g) are correct. The magnitude of the answer (5) makes sense because we have much less than a mole of fluorine; so we expect a mass much less than the molar mass of ClF_3.

11.90 **Given:** $V = 1.50$ L, $P_{CO} = 232$ mmHg, $P_{H_2} = 397$ mmHg, $T = 305$ K **Find:** limiting reactant and m_{CH_3OH}

Conceptual Plan: Determine limiting reactant by comparing the pressures of each reactant, in the same way we would compare moles then

$$\frac{2 \text{ mmHg H}_2}{1 \text{ mmHg CO}}$$

mmHg limiting reactant → mmHg ClF_3 and mmHg → atm then

$$\frac{1 \text{ mmHg CH}_3\text{OH}}{1 \text{ mmHg CO}} \text{ or } \frac{1 \text{ mmHg CH}_3\text{OH}}{2 \text{ mmHg H}_2} \qquad \frac{1 \text{ atm}}{760 \text{ mmHg}}$$

P, V, T → n → m

$$PV = nRT \quad \frac{32.04 \text{ g CH}_3\text{OH}}{1 \text{ mol CH}_3\text{OH}}$$

Solution: To determine the limiting reactant, calculate the pressure of hydrogen needed to react all of the carbon monoxide and compare to the pressure of hydrogen available.

$$232 \text{ mmHg CO} \times \frac{2 \text{ mmHg H}_2}{1 \text{ mmHg CO}} = 464 \text{ mmHg H}_2 \text{ needed}$$

Because only 397 mmHg H_2 is available, H_2 is the limiting reactant; then

$$P_{CH_3OH} = 397 \text{ mmHg H}_2 \times \frac{1 \text{ mmHg CH}_3\text{OH}}{2 \text{ mmHg H}_2} \times \frac{1 \text{ atm CH}_3\text{OH}}{760 \text{ mmHg CH}_3\text{OH}} = 0.2611842 \text{ atm CH}_3\text{OH}. \; PV = nRT$$

$$\text{Rearrange to solve for } n.\; n = \frac{PV}{RT} \; n_{ClF_3} = \frac{0.2611842 \text{ atm CH}_3\text{OH} \times 1.50 \text{ L}}{0.08206 \frac{\text{L} \cdot \text{atm}}{\text{mol} \cdot \text{K}} \times 305 \text{ K}} = 0.015653332 \text{ mol CH}_3\text{OH}$$

$$\text{Finally, } 0.015653332 \text{ mol CH}_3\text{OH} \times \frac{32.04 \text{ g CH}_3\text{OH}}{1 \text{ mol CH}_3\text{OH}} = 0.50153279 \text{ g CH}_3\text{OH} = 0.502 \text{ g CH}_3\text{OH}$$

Check: The units (g) are correct. The magnitude of the answer (0.5) makes sense because we have much less than a mole of hydrogen; so we expect a mass much less than the molar mass of methanol.

Real Gases

11.91 The postulate that the volume of the gas particles is small compared to the space between them breaks down at high pressure. At high pressures, the number of molecules per unit volume increases; so the volume of the gas particles becomes more significant. Because the spacing between the particles is reduced, the molecules themselves occupy a significant portion of the volume.

11.92 The postulate that the forces between the gas particles are not significant breaks down at low temperatures. At low temperatures, the molecules are not moving as fast as at higher temperatures; so when they collide, they have a greater opportunity to interact.

11.93 **Given:** Ne, $n = 1.000$ mol, $P = 500.0$ atm, and $T = 355.0$ K **Find:** V(ideal) and V(van der Waals)

Conceptual Plan: $n, P, T \rightarrow V$ **and** $n, P, T \rightarrow V$

$$PV = nRT \qquad \left(P + \frac{an^2}{V^2}\right)(V - nb) = nRT$$

Solution: $PV = nRT$ Rearrange to solve for V.

$$V = \frac{nRT}{P} = \frac{1.000\ \text{mol} \times 0.08206\ \frac{\text{L} \cdot \text{atm}}{\text{mol} \cdot \text{K}} \times 355.0\ \text{K}}{500.0\ \text{atm}} = 0.05826\ \text{L}$$

$$\left(P + \frac{an^2}{V^2}\right)(V - nb) = nRT \quad \text{Rearrange to solve for } V = \frac{nRT}{\left(P + \frac{an^2}{V^2}\right)} + nb$$

Using a $= 0.211\ \text{L}^2\ \text{atm/mol}^2$ and b $= 0.0171$ L/mol from Table 11.4 and the V from the ideal gas law calculation above, solve for V by successive approximations.

$$V = \frac{1.000\ \text{mol} \times 0.08206\ \frac{\text{L} \cdot \text{atm}}{\text{mol} \cdot \text{K}} \times 355.0\ \text{K}}{500.0\ \text{atm} + \frac{0.211\ \frac{\text{L}^2 \cdot \text{atm}}{\text{mol}^2} \times (1.000\ \text{mol})^2}{(0.05826\ \text{L})^2}} + \left(1.000\ \text{mol} \times 0.0171\ \frac{\text{L}}{\text{mol}}\right) = 0.068\underline{9}20\ \text{L}$$

Plug in this new value.

$$V = \frac{1.000\ \text{mol} \times 0.08206\ \frac{\text{L} \cdot \text{atm}}{\text{mol} \cdot \text{K}} \times 355.0\ \text{K}}{500.0\ \text{atm} + \frac{0.211\ \frac{\text{L}^2 \cdot \text{atm}}{\text{mol}^2} \times (1.000\ \text{mol})^2}{(0.068920\ \text{L})^2}} + \left(1.000\ \text{mol} \times 0.0171\ \frac{\text{L}}{\text{mol}}\right) = 0.070\underline{6}09\ \text{L}$$

Plug in this new value.

$$V = \frac{1.000\ \text{mol} \times 0.08206\ \frac{\text{L} \cdot \text{atm}}{\text{mol} \cdot \text{K}} \times 355.0\ \text{K}}{500.0\ \text{atm} + \frac{0.211\ \frac{\text{L}^2 \cdot \text{atm}}{\text{mol}^2} \times (1.000\ \text{mol})^2}{(0.070609\ \text{L})^2}} + \left(1.000\ \text{mol} \times 0.0171\ \frac{\text{L}}{\text{mol}}\right) = 0.070\underline{8}16\ \text{L}$$

Plug in this new value.

$$V = \frac{1.000\ \text{mol} \times 0.08206\ \frac{\text{L} \cdot \text{atm}}{\text{mol} \cdot \text{K}} \times 355.0\ \text{K}}{500.0\ \text{atm} + \frac{0.211\ \frac{\text{L}^2 \cdot \text{atm}}{\text{mol}^2} \times (1.000\ \text{mol})^2}{(0.070816\ \text{L})^2}} + \left(1.000\ \text{mol} \times 0.0171\ \frac{\text{L}}{\text{mol}}\right) = 0.070\underline{8}40\ \text{L} = 0.0708\ \text{L}$$

The two values are different because we are at very high pressures. The pressure is corrected from 500.0 atm to 542.1 atm, and the final volume correction is 0.0171 L.

Check: The units (L) are correct. The magnitude of the answer (0.07) makes sense because we are at such a high pressure and have one mole of gas.

Cumulative Problems

11.94 **Given:** Cl_2, $n = 1.000$ mol, L = 5.000 L, and $T = 273.0$ K **Find:** P(ideal) and P(van der Waals)
Conceptual Plan: $n, V, T \rightarrow P$ **and** $n, V, T \rightarrow P$

$$PV = nRT \qquad \left(P + \frac{an^2}{V^2}\right)(V - nb) = nRT$$

Solution: $PV = nRT$ Rearrange to solve for P.

$$P = \frac{nRT}{V} = \frac{1.000 \text{ mol} \times 0.08206 \frac{\text{L} \cdot \text{atm}}{\text{mol} \cdot \text{K}} \times 273.0 \text{ K}}{5.000 \text{ L}} = 4.480 \text{ atm}$$

$\left(P + \frac{an^2}{V^2}\right)(V - nb) = nRT$ Rearrange to solve for P. $P = \frac{nRT}{(V - nb)} - \frac{an^2}{V^2}$

Using a = 6.49 L^2 atm/mol^2 and b = 0.0562 L/mol from Table 11.4,

$$P = \frac{1.000 \text{ mol} \times 0.08206 \frac{\text{L} \cdot \text{atm}}{\text{mol} \cdot \text{K}} \times 273.0 \text{ K}}{5.000 \text{ L} - \left(1.000 \text{ mol} \times 0.0562 \frac{\text{L}}{\text{mol}}\right)} - \frac{6.49 \frac{\text{L}^2 \cdot \text{atm}}{\text{mol}^2} \times (1.000 \text{ mol})^2}{(5.000 \text{ L})^2} = 4.272 \text{ atm}$$

The pressure values differ slightly because of the nonideal behavior of chlorine (a large molecule) at a slightly reduced temperature.

Check: The units (atm) are correct. The magnitude of the answers (4.5 and 4.3) makes sense because we are at such a low temperature and small volume with one mole of gas; we expect a difference in two pressures of less than 1 atm and a $P > 1$ atm.

11.95 **Given:** $m_{\text{penny}} = 2.482$ g, $T = 25$ °C, $V = 0.899$ L, and $P_{\text{Total}} = 791$ mmHg **Find:** % Zn in penny
Conceptual Plan: $T \rightarrow P_{H_2O}$ **then** $P_{\text{Total}}, P_{H_2O} \rightarrow P_{H_2}$ **then mmHg → atm and °C → K**

Table 11.3 $\qquad P_{\text{Total}} = P_{H_2O} + P_{H_2} \qquad \frac{1 \text{ atm}}{760 \text{ mmHg}} \qquad K = °C + 273.15$

and $P, V, T \rightarrow n_{H_2} \rightarrow n_{Zn} \rightarrow g_{Zn} \rightarrow$ **% Zn**

$PV = nRT \qquad \frac{1 \text{ mol Zn}}{1 \text{ mol } H_2} \qquad \frac{65.39 \text{ g Zn}}{1 \text{ mol Zn}} \qquad \% \text{ Zn} = \frac{g_{Zn}}{g_{\text{penny}}} \times 100\%$

Solution: Table 11.3 states that $P_{H_2O} = 23.78$ mmHg at 25 °C. $P_{\text{Total}} = P_{H_2O} + P_{H_2}$ Rearrange to solve for P_{H_2}.

$P_{H_2} = P_{\text{Total}} - P_{H_2O} = 791 \text{ mmHg} - 23.78 \text{ mmHg} = 767 \text{ mmHg}$ $P_{H_2} = 767 \text{ mmHg} \times \frac{1 \text{ atm}}{760 \text{ mmHg}} = 1.0\underline{0}92 \text{ atm}$

then $T = 25 \text{ °C} + 273.15 = 298$ K, $PV = nRT$

Rearrange to solve for n. $n_{H_2} = \frac{PV}{RT} = \frac{1.0\underline{0}92 \text{ atm} \times 0.899 \text{ L}}{0.08206 \frac{\text{L} \cdot \text{atm}}{\text{mol} \cdot \text{K}} \times 298 \text{ K}} = 0.037\underline{1}013 \text{ mol}$

$$0.037\underline{1}013 \text{ mol } H_2 \times \frac{1 \text{ mol Zn}}{1 \text{ mol } H_2} \times \frac{65.39 \text{ g Zn}}{1 \text{ mol Zn}} = 2.4\underline{2}605 \text{ g Zn}$$

$$\% \text{ Zn} = \frac{g_{Zn}}{g_{\text{penny}}} \times 100\% = \frac{2.4\underline{2}605 \text{ g}}{2.482 \text{ g}} \times 100\% = 97.7\% \text{ Zn}$$

Check: The units (% Zn) are correct. The magnitude of the answer (98) makes sense because it should be between 0 and 100%. We expect about 1/22 mole of gas because our conditions are close to STP and we have ~1 L of gas.

11.96 **Given:** $m_{\text{CFC}} = 2.85$ g, $V = 564$ mL, $P = 752$ mmHg, and $T = 298$ K **Find:** % Cl in CFC
Conceptual Plan: mmHg → atm and mL → L and $P, V, T \rightarrow n_{Cl_2} \rightarrow g_{Cl}$ **then** $g_{Cl}, g_{\text{CFC}} \rightarrow$ **% Cl**

$\frac{1 \text{ atm}}{760 \text{ mmHg}} \qquad \frac{1 \text{ L}}{1000 \text{ mL}} \qquad PV = nRT \quad \frac{70.90 \text{ g Cl}}{1 \text{ mol } Cl_2} \qquad \% \text{ Cl} = \frac{g_{Cl}}{g_{\text{CFC}}} \times 100\%$

Solution: $P_{Cl_2} = 752 \text{ mmHg} \times \frac{1 \text{ atm}}{760 \text{ mmHg}} = 0.98\underline{9}47 \text{ atm}$, $V_{Cl_2} = 564 \text{ mL} \times \frac{1 \text{ L}}{1000 \text{ mL}} = 0.564 \text{ L}$

$PV = nRT$ Rearrange to solve for n. $n_{Cl_2} = \frac{PV}{RT} = \frac{0.98\underline{9}47 \text{ atm} \times 0.564 \text{ L}}{0.08206 \frac{\text{L} \cdot \text{atm}}{\text{mol} \cdot \text{K}} \times 298 \text{ K}} = 0.022\underline{8}21 \text{ mol } Cl_2$

$0.022\underline{8}21 \text{ mol } Cl_2 \times \frac{70.90 \text{ g Cl}}{1 \text{ mol } Cl_2} = 1.6\underline{1}80 \text{ g Cl}$, $\% \text{ Cl} = \frac{g_{Cl}}{g_{CFC}} \times 100\ \% = \frac{1.6\underline{1}80 \text{ g}}{2.85 \text{ g}} \times 100\% = 56.8\% \text{ Cl}$

Check: The units (% Cl) are correct. The magnitude of the answer (57) makes sense because it should be between 0 and 100%. Because carbon and fluorine also will be in the compound, we do not expect it to be extremely close to 100%.

11.97 **Given:** $V = 255$ mL, $m_{flask} = 143.187$ g, $m_{flask+gas} = 143.289$ g, $P = 267$ torr, and $T = 25$ °C **Find:** $\mathcal{M}$

Conceptual Plan: °C → K torr → atm mL → L $m_{flask}, m_{flask+gas} \rightarrow m_{gas}$

$K = °C + 273.15$ $\frac{1 \text{ atm}}{760 \text{ torr}}$ $\frac{1 \text{ L}}{1000 \text{ mL}}$ $m_{gas} = m_{flask+gas} - m_{flask}$

then V, $m \rightarrow d$ then $d, P, T, \rightarrow \mathcal{M}$

$d = \frac{m}{V}$ $d = \frac{P\mathcal{M}}{RT}$

Solution: $T = 25\ °C + 273.15 = 298 \text{ K}$, $P = 267 \text{ torr} \times \frac{1 \text{ atm}}{760 \text{ torr}} = 0.35\underline{1}316 \text{ atm}$

$V = 255 \text{ mL} \times \frac{1 \text{ L}}{1000 \text{ mL}} = 0.255 \text{ L}$

$m_{gas} = m_{flask+gas} - m_{flask} = 143.289 \text{ g} - 143.187 \text{ g} = 0.102 \text{ g}$

$d = \frac{m}{V} = \frac{0.102 \text{ g}}{0.255 \text{ L}} = 0.400 \text{ g/L}$, $d = \frac{P\mathcal{M}}{RT}$ Rearrange to solve for $\mathcal{M}$.

$$\mathcal{M} = \frac{dRT}{P} = \frac{0.400 \frac{\text{g}}{\text{L}} \times 0.08206 \frac{\text{L} \cdot \text{atm}}{\text{K} \cdot \text{mol}} \times 298 \text{ K}}{0.35\underline{1}316 \text{ atm}} = 27.8 \text{ g/mol}$$

An alternative method to solve this problem would be to calculate the moles of gas using the ideal gas law (0.00366 mole) and then taking the mass (0.102 g) and dividing it by the number of moles.

Check: The units (g/mol) are correct. The magnitude of the answer (28) makes physical sense because this is a reasonable number for a molar mass of a gas.

11.98 **Given:** $V = 118$ mL, $m_{flask} = 97.129$ g, $m_{flask+gas} = 97.171$ g, $P = 768$ torr, and $T = 35$ °C

Find: whether gas is pure

Conceptual Plan: °C → K torr → atm mL → L $m_{flask}, m_{flask+gas} \rightarrow m_{gas}$

$K = °C + 273.15$ $\frac{1 \text{ atm}}{760 \text{ torr}}$ $\frac{1 \text{ L}}{1000 \text{ mL}}$ $m_{gas} = m_{flask+gas} - m_{flask}$

then $V, m \rightarrow d$ then $d, P, T, \rightarrow \mathcal{M}$

$d = \frac{m}{V}$ $d = \frac{P\mathcal{M}}{RT}$

Solution: $T = 35\ °C + 273.15 = 308 \text{ K}$, $P = 768 \text{ torr} \times \frac{1 \text{ atm}}{760 \text{ torr}} = 1.0\underline{1}053 \text{ atm}$

$V = 118 \text{ mL} \times \frac{1 \text{ L}}{1000 \text{ mL}} = 0.118 \text{ L}$

$m_{gas} = m_{flask+gas} - m_{flask} = 97.171 \text{ g} - 97.129 \text{ g} = 0.042 \text{ g}$,

$d = \frac{m}{V} = \frac{0.042 \text{ g}}{0.118 \text{ L}} = 0.3\underline{5}593 \text{ g/L}$, $d = \frac{P\mathcal{M}}{RT}$ Rearrange to solve for $\mathcal{M}$.

$$\mathcal{M} = \frac{dRT}{P} = \frac{0.3\underline{5}593 \frac{\text{g}}{\text{L}} \times 0.08206 \frac{\text{L} \cdot \text{atm}}{\text{K} \cdot \text{mol}} \times 308 \text{ K}}{1.0\underline{1}053 \text{ atm}} = 8.9 \text{ g/mol}$$

The gas is not pure because the molar mass is not 4.003 g/ml.

Check: The units (g/mol) are correct. The magnitude of the answer (9) makes physical sense because this is a reasonable number for a molar mass of a gas.

11.99 **Given:** $V = 158$ mL, $m_{gas} = 0.275$ g, $P = 556$ mmHg, $T = 25$ °C, gas = 82.66% C and 17.34% H
Find: molecular formula
Conceptual Plan: °C → K mmHg → atm mL → L then $V, m \rightarrow d$

$$K = °C + 273.15 \qquad \frac{1 \text{ atm}}{760 \text{ mmHg}} \qquad \frac{1 \text{ L}}{1000 \text{ mL}} \qquad d = \frac{m}{V}$$

then $d, P, T, \rightarrow \mathcal{M}$ then % C, % H, $\mathcal{M} \rightarrow$ formula

$$d = \frac{P\mathcal{M}}{RT} \qquad \#C = \frac{\mathcal{M} 0.8266 \text{ g C}}{12.01 \frac{\text{g C}}{\text{mol C}}} \qquad \#H = \frac{\mathcal{M} 0.1734 \text{ g H}}{1.008 \frac{\text{g H}}{\text{mol H}}}$$

Solution: $T = 25\text{ °C} + 273.15 = 298 \text{ K}, P = 556 \text{ mmHg} \times \frac{1 \text{ atm}}{760 \text{ mmHg}} = 0.731579 \text{ atm}$

$V = 158 \text{ mL} \times \frac{1 \text{ L}}{1000 \text{ mL}} = 0.158 \text{ L}, d = \frac{m}{V} = \frac{0.275 \text{ g}}{0.158 \text{ L}} = 1.74051 \text{ g/L}, d = \frac{P\mathcal{M}}{RT}$ Rearrange to solve for $\mathcal{M}$.

$$\mathcal{M} = \frac{dRT}{P} = \frac{1.74051 \frac{\text{g}}{\text{L}} \times 0.08206 \frac{\text{L} \cdot \text{atm}}{\text{K} \cdot \text{mol}} \times 298 \text{ K}}{0.731579 \text{ atm}} = 58.2 \text{ g/mol}$$

$$\#C = \frac{\mathcal{M} \times 0.8266 \text{ g C}}{12.01 \frac{\text{g C}}{\text{mol C}}} = \frac{58.2 \frac{\text{g HC}}{\text{mol HC}} \times \frac{0.8266 \text{ g C}}{1 \text{ g HC}}}{12.01 \frac{\text{g C}}{\text{mol C}}} = 4.00 \frac{\text{mol C}}{\text{mol HC}}$$

$$\#H = \frac{\mathcal{M} \times 0.1734 \text{ g H}}{1.0008 \frac{\text{g H}}{\text{mol H}}} = \frac{58.2 \frac{\text{g HC}}{\text{mol HC}} \times \frac{0.1734 \text{ g H}}{1 \text{ g HC}}}{1.008 \frac{\text{g H}}{\text{mol H}}} = 10.0 \frac{\text{mol H}}{\text{mol HC}}$$ Formula is C_4H_{10} or butane.

An alternative method would be to calculate the empirical formula (C_2H_5) and then use the ideal gas law to calculate the number of moles of gas (0.00473 mole). The molar mass can be calculated by dividing the mass (0.275 g) by the number of moles. Divide the molar mass by the molar mass (58.2 g/mol) of the empirical formula (29.07 g/mol), giving a multiplier of 2 for the empirical formula.

Check: The answer came up with integer numbers of C and H atoms in the formula and a molecular weight (58) that is reasonable for a gas.

11.100 **Given:** STP, $V = 258$ mL, $m_{gas} = 0.646$ g, gas = 85.63% C and 14.37% H. **Find:** $\mathcal{M}$
Conceptual Plan: mL → L then $V, m \rightarrow d$ then $d, P, T, \rightarrow \mathcal{M}$

$$\frac{1 \text{ L}}{1000 \text{ mL}} \qquad d = \frac{m}{V} \qquad d = \frac{P\mathcal{M}}{RT}$$

then % C, % H, $\mathcal{M} \rightarrow$ formula

$$\#C = \frac{\mathcal{M}\, 0.8563 \text{ g C}}{12.01 \frac{\text{g C}}{\text{mol C}}} \qquad \#H = \frac{\mathcal{M}\, 0.1437 \text{ g H}}{1.008 \frac{\text{g H}}{\text{mol H}}}$$

Solution: $V = 258 \text{ mL} \times \frac{1 \text{ L}}{1000 \text{ mL}} = 0.258 \text{ L}, d = \frac{m}{V} = \frac{0.646 \text{ g}}{0.258 \text{ L}} = 2.50388 \text{ g/L}, d = \frac{P\mathcal{M}}{RT}$ Rearrange to solve

for $\mathcal{M}$. $\mathcal{M} = \frac{dRT}{P} = \frac{2.50388 \frac{\text{g}}{\text{L}} \times 0.08206 \frac{\text{L} \cdot \text{atm}}{\text{K} \cdot \text{mol}} \times 273.15 \text{ K}}{1 \text{ atm}} = 56.12 \text{ g/mol},$

$$\#C = \frac{\mathcal{M} \times 0.8563 \text{ g C}}{12.01 \frac{\text{g C}}{\text{mol C}}} = \frac{56.12 \frac{\text{g HC}}{\text{mol HC}} \times \frac{0.8563 \text{ g C}}{1 \text{ g HC}}}{12.01 \frac{\text{g C}}{\text{mol C}}} = 4.00 \frac{\text{mol C}}{\text{mol HC}}$$

$$\#\text{H} = \frac{\mathcal{M} \times 0.1437\ \text{g H}}{1.008 \frac{\text{g H}}{\text{mol H}}} = \frac{56.12 \frac{\text{g HC}}{\text{mol HC}} \times \frac{0.1437\ \text{g H}}{1\ \text{g HC}}}{1.008 \frac{\text{g H}}{\text{mol H}}} = 8.00 \frac{\text{mol H}}{\text{mol HC}}$$ Formula is C_4H_8 or butene.

An alternative method would be to calculate the empirical formula (CH_2) and then use the ideal gas law to calculate the number of moles of gas (0.0115 mole). The molar mass can be calculated by dividing the mass (0.646 g) by the number of moles. Divide the molar mass by the molar mass (56.1 g/mol) of the empirical formula (14.03 g/mol), giving a multiplier of 4 for the empirical formula.

Check: The answer came up with integer numbers of C and H atoms in the formula and a molecular weight (56 g/mol) that is reasonable for a gas.

11.101 **Given:** $m_{NiO} = 24.78$ g, $T = 40.0$ °C, and $P_{Total} = 745$ mmHg **Find:** V_{O_2}

Conceptual Plan: $T \rightarrow P_{H_2O}$ **then** $P_{Total}, P_{H_2O} \rightarrow P_{O_2}$ **then mmHg → atm and °C → K**

Table 11.3 $\quad P_{Total} = P_{H_2O} + P_{O_2} \quad \frac{1\ \text{atm}}{760\ \text{mmHg}} \quad K = °C + 273.15$

and $m_{NiO} \rightarrow n_{NiO} \rightarrow n_{O_2}$ **then** $P, V, T \rightarrow n_{O_2}$

$\frac{1\ \text{mol NiO}}{74.69\ \text{g NiO}} \quad \frac{1\ \text{mol } O_2}{2\ \text{mol NiO}} \quad PV = nRT$

Solution: Table 11.3 states that $P_{H_2O} = 55.40$ mmHg at 40 °C $P_{Total} = P_{H_2O} + P_{O_2}$ Rearrange to solve for P_{O_2}.

$P_{O_2} = P_{Total} - P_{H_2O} = 745\ \text{mmHg} - 55.40\ \text{mmHg} = 689.6\ \text{mmHg}$

$$P_{O_2} = 689.6\ \text{mmHg} \times \frac{1\ \text{atm}}{760\ \text{mmHg}} = 0.907368\ \text{atm} \quad T = 40.0\ °C + 273.15 = 313.2\ K$$

$$24.78\ \text{g NiO} \times \frac{1\ \text{mol NiO}}{74.69\ \text{g NiO}} \times \frac{1\ \text{mol } O_2}{2\ \text{mol NiO}} = 0.1658857\ \text{mol } O_2 \qquad PV = nRT$$

Rearrange to solve for V. $V_{O_2} = \frac{nRT}{P} = \frac{0.1658857\ \text{mol} \times 0.08206 \frac{\text{L} \cdot \text{atm}}{\text{mol} \cdot \text{K}} \times 313.2\ \text{K}}{0.907368\ \text{atm}} = 4.70\ \text{L}$

Check: The units (L) are correct. The magnitude of the answer (5) makes sense because we have much less than 0.5 mole of NiO; so we get less than a mole of oxygen. Thus, we expect a volume much less than 22 L.

11.102 **Given:** $m_{Ag} = 15.8$ g, $T = 25$ °C, and $P_{Total} = 752$ mmHg **Find:** V_{O_2}

Conceptual Plan: $T \rightarrow P_{H_2O}$ **then** $P_{Total}, P_{H_2O} \rightarrow P_{O_2}$ **then mmHg → atm and °C → K**

Table 11.3 $\quad P_{Total} = P_{H_2O} + P_{O_2} \quad \frac{1\ \text{atm}}{760\ \text{mmHg}} \quad K = °C + 273.15$

and $m_{Ag} \rightarrow n_{Ag} \rightarrow n_{O_2}$ **then** $P, V, T \rightarrow n_{O_2}$

$\frac{1\ \text{mol Ag}}{107.9\ \text{g Ag}} \quad \frac{1\ \text{mol } O_2}{4\ \text{mol Ag}} \quad PV = nRT$

Solution: Table 11.3 states that $P_{H_2O} = 23.78$ mmHg at 25 °C. $P_{Total} = P_{H_2O} + P_{O_2}$ Rearrange to solve for P_{O_2}.

$P_{O_2} = P_{Total} - P_{H_2O} = 752\ \text{mmHg} - 23.78\ \text{mmHg} = 728.22\ \text{mmHg}$

$$P_{O_2} = 728.22\ \text{mmHg} \times \frac{1\ \text{atm}}{760\ \text{mmHg}} = 0.958184\ \text{atm} \quad T = 25\ °C + 273.15 = 298\ K$$

$$15.8\ \text{g Ag} \times \frac{1\ \text{mol Ag}}{107.9\ \text{g Ag}} \times \frac{1\ \text{mol } O_2}{4\ \text{mol Ag}} = 0.03660797\ \text{mol } O_2 \quad PV = nRT$$ Rearrange to solve for V.

$$V_{O_2} = \frac{nRT}{P} = \frac{0.03660797\ \text{mol} \times 0.08206 \frac{\text{L} \cdot \text{atm}}{\text{mol} \cdot \text{K}} \times 298\ \text{K}}{0.958184\ \text{atm}} = 0.934\ \text{L}$$

Check: The units (L) are correct. The magnitude of the answer (1) makes sense because we have ~0.1 mole of Ag; so we get less than 0.5 mol of oxygen. Thus, we expect a volume much, much less than 22 L.

11.103 **Given:** HCl, K_2S to H_2S, $V_{H_2S} = 42.9$ mL, $P_{H_2S} = 752$ mmHg, and $T = 25.8$ °C **Find:** m_{K_2S}

Conceptual Plan: Read description of reaction and convert words to equation, then °C → K

$K = °C + 273.15$

and mmHg → atm and mL → L then $P, V, T \rightarrow n_{H_2S} \rightarrow n_{K_2S} \rightarrow m_{K_2S}$

$\frac{1 \text{ atm}}{760 \text{ mmHg}}$ $\frac{1 \text{ L}}{1000 \text{ mL}}$ $PV = nRT$ $\frac{1 \text{ mol } K_2S}{1 \text{ mol } H_2S}$ $\frac{1 \text{ mol } K_2S}{110.27 \text{ g } K_2S}$

Solution: $2\ HCl(aq) + K_2S(s) \rightarrow H_2S(g) + 2\ KCl(aq)$

$T = 25.8\ ^\circ C + 273.15 = 299.0\ K$, $P_{H_2S} = 752\ \cancel{mmHg} \times \frac{1 \text{ atm}}{760\ \cancel{mmHg}} = 0.98\underline{9}474 \text{ atm}$

$V_{H_2S} = 42.9\ \cancel{mL} \times \frac{1 \text{ L}}{1000\ \cancel{mL}} = 0.0429 \text{ L}$ $PV = nRT$ Rearrange to solve for n_{H_2S}.

$$n_{H_2S} = \frac{PV}{RT} = \frac{0.98\underline{9}474\ \cancel{atm} \times 0.0429\ \cancel{L}}{0.08206 \frac{\cancel{L} \cdot \cancel{atm}}{mol \cdot \cancel{K}} \times 299.0\ \cancel{K}} = 0.0017\underline{3}005 \text{ mol}$$

$$0.0017\underline{3}005\ \cancel{mol\ H_2S} \times \frac{1\ \cancel{mol\ K_2S}}{1\ \cancel{mol\ H_2S}} \times \frac{110.27 \text{ g } K_2S}{1\ \cancel{mol\ K_2S}} = 0.191 \text{ g } K_2S$$

Check: The units (g) are correct. The magnitude of the answer (0.2) makes sense because such a small volume of gas is generated.

11.104 (a) **Given:** $T = 315$ K, $P = 50.0$ mmHg, $V_{SO_2} = 285.5$ mL, and $V_{O_2} = 158.9$ mL
Find: limiting reagent and theoretical yield
Conceptual Plan: mmHg → atm and $mL_{SO_2} \rightarrow L_{SO_2}$ then $P, V_{SO_2}, T \rightarrow n_{SO_2}, \rightarrow n_{SO_3}$

$\frac{1 \text{ atm}}{760 \text{ mmHg}}$ $\frac{1 \text{ L}}{1000 \text{ mL}}$ $PV = nRT$ $\frac{2 \text{ mol } SO_3}{2 \text{ mol } SO_2}$

and $mL_{O_2} \rightarrow L_{SO_2}$ then $P, V_{SO_2}, T \rightarrow n_{O_2} \rightarrow n_{SO_3}$

$\frac{1 \text{ L}}{1000 \text{ mL}}$ $PV = nRT$ $\frac{2 \text{ mol } SO_3}{1 \text{ mol } O_2}$

Solution: $50.0\ \cancel{mmHg} \times \frac{1 \text{ atm}}{760\ \cancel{mmHg}} = 0.065\underline{7}895 \text{ atm}$ and $285.5\ \cancel{mL\ SO_2} \times \frac{1 \text{ L } SO_2}{1000\ \cancel{mL\ SO_2}} = 0.2855 \text{ L}$

then $PV = nRT$ Rearrange to solve for n.

$$n_{SO_2} = \frac{PV}{RT} = \frac{0.065\underline{7}895\ \cancel{atm} \times 0.2855\ \cancel{L}}{0.08206 \frac{\cancel{L} \cdot \cancel{atm}}{mol \cdot \cancel{K}} \times 315\ \cancel{K}} = 7.2\underline{6}642 \times 10^{-4} \text{ mol } SO_2$$

$$7.2\underline{6}642 \times 10^{-4}\ \cancel{mol\ SO_2} \times \frac{2 \text{ mol } SO_3}{2\ \cancel{mol\ SO_2}} = 7.2\underline{6}642 \times 10^{-4} \text{ mol } SO_3$$

then $158.9\ \cancel{mL\ O_2} \times \frac{1 \text{ L } O_2}{1000\ \cancel{mL\ O_2}} = 0.1589 \text{ L } O_2$ then $PV = nRT$ Rearrange to solve for n.

$$n_{O_2} = \frac{PV}{RT} = \frac{0.065\underline{7}895\ \cancel{atm} \times 0.1589\ \cancel{L}}{0.08206 \frac{\cancel{L} \cdot \cancel{atm}}{mol \cdot \cancel{K}} \times 315\ \cancel{K}} = 4.0\underline{4}425 \times 10^{-4} \text{ mol } O_2 \text{ then}$$

$4.0\underline{4}425 \times 10^{-4}\ \cancel{mol\ O_2} \times \frac{2 \text{ mol } SO_3}{1\ \cancel{mol\ O_2}} = 8.0\underline{8}851 \times 10^{-4} \text{ mol } SO_3$ Because the amount generated from the SO_2 is less, it is the limiting reagent, and the theoretical yield is 7.27×10^{-4} mol SO_3.

Check: The units (mol) are correct. The magnitude of the answer (0.0007) makes sense because small volumes of gas are involved (compared to 22 L).

(b) **Given:** preceding info and $V_{SO_3} = 187.2$ mL, $T = 315$ K, and $P = 50.0$ mmHg **Find:** % yield
Conceptual Plan: mmHg → atm and mL SO_3 → L SO_3 then $P, V_{SO_2}, T \rightarrow n_{SO_2} \rightarrow n_{SO_3}$

$\frac{1 \text{ atm}}{760 \text{ mmHg}}$ $\frac{1 \text{ L}}{1000 \text{ mL}}$ $PV = nRT$

then actual yield, theoretical yield → % yield

$$\% \text{ yield} = \frac{\text{actual yield}}{\text{theoretical yield}} \times 100\%$$

Solution: $50.0\ \text{mmHg} \times \dfrac{1\ \text{atm}}{760\ \text{mmHg}} = 0.0657895\ \text{atm}$ and

$187.2\ \text{mL SO}_3 \times \dfrac{1\ \text{L SO}_3}{1000\ \text{mL SO}_3} = 0.1872\ \text{L SO}_3$ then $PV = nRT$ Rearrange to solve for n.

$$n_{SO_3} = \frac{PV}{RT} = \frac{0.0657895\ \text{atm} \times 0.1872\ \text{L}}{0.08206\ \dfrac{\text{L} \cdot \text{atm}}{\text{mol} \cdot \text{K}} \times 315\ \text{K}} = 4.76453 \times 10^{-4}\ \text{mol SO}_3$$

$$\text{then \% yield} = \frac{\text{actual yield}}{\text{theoretical yield}} \times 100\% = \frac{4.76453 \times 10^{-4}\ \text{mol SO}_3}{7.26642 \times 10^{-4}\ \text{mol SO}_3} \times 100\% = 65.6\%$$

Check: The units (%) are correct. The magnitude of the answer (66) makes sense because it should be between 0 and 100%. Because the volume of product is a bit over half the volume of the limiting reagent and there is a 2:2 mole ratio of the reactant and product, we expect a number a little over 50%.

11.105 **Given:** $T = 22\ °\text{C}, P = 1.02\ \text{atm}$, and $m = 11.83\ \text{g}$ **Find:** V_{Total}

Conceptual Plan: **°C → K and $m_{(NH_4)_2CO_3} \rightarrow n_{(NH_4)_2CO_3} \rightarrow n_{\text{Gas}}$ then $P, n, T \rightarrow V$**

$K = °C + 273.15$ $\qquad \dfrac{1\ \text{mol (NH}_4)_2\text{CO}_3}{96.09\ \text{g (NH}_4)_2\text{CO}_3}\ \dfrac{(2 + 1 + 1 = 4)\ \text{mol gas}}{1\ \text{mol (NH}_4)_2\text{CO}_3}$ $\qquad PV = nRT$

Solution: $T = 22\ °\text{C} + 273.15 = 295\ \text{K}$

$$11.83\ \text{g (NH}_4)_2\text{CO}_3 \times \frac{1\ \text{mol (NH}_4)_2\text{CO}_3}{96.09\ \text{g (NH}_4)_2\text{CO}_3} \times \frac{4\ \text{mol gas}}{1\ \text{mol (NH}_4)_2\text{CO}_3} = 0.492455\ \text{mol gas}$$

$PV = nRT$ Rearrange to solve for V_{Gas}.

$$V_{\text{Gas}} = \frac{nRT}{P} = \frac{0.492455\ \text{mol gas} \times 0.08206\ \dfrac{\text{L} \cdot \text{atm}}{\text{mol} \cdot \text{K}} \times 295\ \text{K}}{1.02\ \text{atm}} = 11.7\ \text{L}$$

Check: The units (L) are correct. The magnitude of the answer (12) makes sense because about half a mole of gas is generated.

11.106 **Given:** $T = 125\ °\text{C}, P = 748\ \text{mmHg}$, and $m = 1.55\ \text{kg}$ **Find:** V_{Total}

Conceptual Plan: **°C → K and mmHg → atm and kg $_{NH_4NO_3} \rightarrow m_{NH_4NO_3} \rightarrow n_{NH_4NO_3} \rightarrow n_{\text{Gas}}$**

$K = °C + 273.15$ $\qquad \dfrac{1\ \text{atm}}{760\ \text{mmHg}}$ $\qquad \dfrac{1000\ \text{g}}{1\ \text{kg}}\ \dfrac{1\ \text{mol NH}_4\text{NO}_3}{80.05\ \text{g NH}_4\text{NO}_3}\ \dfrac{(2 + 1 + 4 = 7)\ \text{mol gas}}{2\ \text{mol NH}_4\text{NO}_3}$

then $P, n, T \rightarrow V$

$PV = nRT$

Solution: $T = 125\ °\text{C} + 273.15 = 398\ \text{K}$, $748\ \text{mmHg} \times \dfrac{1\ \text{atm}}{760\ \text{mmHg}} = 0.984211\ \text{atm}$

$$1.55\ \text{kg NH}_4\text{NO}_3 \times \frac{1000\ \text{g NH}_4\text{NO}_3}{1\ \text{kg NH}_4\text{NO}_3} \times \frac{1\ \text{mol NH}_4\text{NO}_3}{80.05\ \text{g NH}_4\text{NO}_3} \times \frac{7\ \text{mol gas}}{2\ \text{mol NH}_4\text{NO}_3} = 67.7701\ \text{mol gas}$$

$PV = nRT$ Rearrange to solve for V_{Gas}.

$$V_{\text{Gas}} = \frac{nRT}{P} = \frac{67.7701\ \text{mol gas} \times 0.08206\ \dfrac{\text{L} \cdot \text{atm}}{\text{mol} \cdot \text{K}} \times 398\ \text{K}}{0.984211\ \text{atm}} = 2250\ \text{L}$$

Check: The units (L) are correct. The magnitude of the answer (2250) makes sense because about 67 moles of gas are generated; so we expect a volume a bit above 67 × 22 L.

11.107 **Given:** He and air; $V = 855\ \text{mL}, P = 125\ \text{psi}, T = 25\ °\text{C}, \mathcal{M}_{\text{air}} = 28.8\ \text{g/mol}$ **Find:** $\Delta m = m_{\text{air}} - m_{\text{He}}$

Conceptual Plan: **mL → L and psi → atm and °C → K then $P, T, \mathcal{M} \rightarrow d$**

$\dfrac{1\ \text{L}}{1000\ \text{mL}}$ $\qquad \dfrac{1\ \text{atm}}{14.7\ \text{psi}}$ $\qquad K = °C + 273.15$ $\qquad d = \dfrac{P\mathcal{M}}{RT}$

then $d, V \rightarrow m$ then $m_{\text{air}}, m_{\text{He}} \rightarrow \Delta m$

$d = \dfrac{m}{V}$ $\qquad \Delta = m_{\text{air}} - m_{\text{He}}$

Solution: $V = 855\ \text{mL} \times \frac{1\ \text{L}}{1000\ \text{mL}} = 0.855\ \text{L}$, $P = 125\ \text{psi} \times \frac{1\ \text{atm}}{14.7\ \text{psi}} = 8.50\underline{3}40\ \text{atm}$

$$T = 25\ °\text{C} + 273.15 = 298\ \text{K},\ d_{\text{air}} = \frac{P\mathcal{M}}{RT} = \frac{8.50\underline{3}40\ \text{atm} \times 28.8\ \frac{\text{g air}}{\text{mol air}}}{0.08206\ \frac{\text{L} \cdot \text{atm}}{\text{K} \cdot \text{mol}} \times 298\ \text{K}} = 10.\underline{0}147\frac{\text{g air}}{\text{L}},\ d = \frac{m}{V}$$

Rearrange to solve for m. $m = dV$

$$m_{\text{air}} = 10.\underline{0}147\frac{\text{g air}}{\text{L}} \times 0.855\ \text{L} = 8.5\underline{6}657\ \text{g air}$$

$$d_{\text{He}} = \frac{P\mathcal{M}}{RT} = \frac{8.50\underline{3}40\ \text{atm} \times 4.03\ \frac{\text{g He}}{\text{mol He}}}{0.08206\ \frac{\text{L} \cdot \text{atm}}{\text{K} \cdot \text{mol}} \times 298\ \text{K}} = 1.\underline{4}0136\frac{\text{g He}}{\text{L}}$$

$$m_{\text{He}} = 1.\underline{4}0136\frac{\text{g He}}{\text{L}} \times 0.855\ \text{L} = 1.1\underline{9}816\ \text{g He}$$

$$\Delta m = m_{\text{air}} - m_{\text{He}} = 8.5\underline{6}657\ \text{g air} - 1.1\underline{9}816\ \text{g He} = 7.37\ \text{g}$$

Check: The units (g) are correct. We expect the difference to be less than the difference in the molecular weights because we have less than a mole of gas.

11.108 **Given:** $V_1 = 2.95\ \text{L}$, $P = 0.998\ \text{atm}$, $T_1 = 25.0\ °\text{C}$, and $T_2 = -196\ °\text{C}$ **Find:** V_2 and compare to 0.61 L

Conceptual Plan: °C → K then $V_1, T_1, T_2 \rightarrow V_2$ and then compare to 0.61 L

$\text{K} = °\text{C} + 273.15$ $\quad \frac{V_1}{T_1} = \frac{V_2}{T_2}$

Solution: $T_1 = 25\ °\text{C} + 273.15 = 298\text{K}$, $T_2 = -196\ °\text{C} + 273.15 = 77\ \text{K}$, $\frac{V_1}{T_1} = \frac{V_2}{T_2}$ Rearrange to solve for V_2.

$V_2 = V_1 \times \frac{T_2}{T_1} = 2.95\ \text{L} \times \frac{77\ \text{K}}{298\ \text{K}} = 0.76\ \text{L}$ This is 25% larger than the measured volume. We expect gases to behave nonideally as the temperature drops. We are at the boiling point of the material; so the velocity dramatically decreases, and some nitrogen will be condensing.

Check: The units (L) are correct. We expect the volume to decrease dramatically because the temperature has dropped significantly.

11.109 **Given:** $\text{flow}_{\text{Total}} = 335\ \text{L/s}$, $P_{\text{NO}} = 22.4\ \text{torr}$, $T_{\text{NO}} = 955\ \text{K}$, $P_{\text{NH}_3} = 755\ \text{torr}$, $T_{\text{NO}} = 298\ \text{K}$, and NH_3 purity = 65.2% **Find:** $\text{flow}_{\text{NH}_3}$

Conceptual Plan: torr → atm then $P_{\text{NO}}, V_{\text{NO}}/\text{s}, T_{\text{NO}} \rightarrow n_{\text{NO}}/\text{s} \rightarrow n_{\text{NH}_3}/\text{s}$ (pure)

$\frac{1\ \text{atm}}{760\ \text{torr}}$ $\quad PV = nRT \quad \frac{4\ \text{mol NH}_3}{4\ \text{mol NO}}$

then n_{NH_3}/s (pure) → n_{NH_3}/s (impure) then n_{NH_3}/s (impure), $P_{\text{NH}_3}, T_{\text{NH}_3} \rightarrow V_{\text{NH}_3}/\text{s}$

$\frac{100\ \text{mol NH}_3\ \text{impure}}{65.2\ \text{mol NH}_3\ \text{pure}}$ $\quad PV = nRT$

Solution: $P_{\text{NO}} = 22.4\ \text{torr} \times \frac{1\ \text{atm}}{760\ \text{torr}} = 0.029\underline{4}737\ \text{atm}$, $P_{\text{NH}_3} = 755\ \text{torr} \times \frac{1\ \text{atm}}{760\ \text{torr}} = 0.99\underline{3}421\ \text{atm}$

$PV = nRT$ Rearrange to solve for n_{NO}. Note that we can substitute V/s for V and get n/s as a result.

$$\frac{n_{\text{NO}}}{\text{s}} = \frac{PV}{RT} = \frac{0.029\underline{4}737\ \text{atm} \times 335\ \text{L/s}}{0.08206\ \frac{\text{L} \cdot \text{atm}}{\text{mol} \cdot \text{K}} \times 955\ \text{K}} = 0.12\underline{5}992\ \frac{\text{mol NO}}{\text{s}}$$

$$0.12\underline{5}992\ \frac{\text{mol NO}}{\text{s}} \times \frac{4\ \text{mol NH}_3}{4\ \text{mol NO}} \times \frac{100\ \text{mol NH}_3\ \text{impure}}{65.2\ \text{mol NH}_3\ \text{pure}} = 0.19\underline{3}240\ \frac{\text{mol NH}_3\ \text{impure}}{\text{s}} \qquad PV = nRT$$

Rearrange to solve for V_{NH_3}. Note that we can substitute n/s for n and get V/s as a result.

$$\frac{V_{NH_3}}{s} = \frac{nRT}{P} = \frac{0.19\underline{3}240 \frac{\cancel{mol\ NH_3\ impure}}{s} \times 0.08206 \times \frac{L \cdot \cancel{atm}}{\cancel{mol} \cdot \cancel{K}} \times 298\cancel{K}}{0.99\underline{3}421\ \cancel{atm}} = 4.76 \frac{L}{s} \text{ impure } NH_3$$

Check: The units (L/s) are correct. The magnitude of the answer (5) makes sense because we expect it to be less than that for the NO. The NO is at a very low concentration and a high temperature. When this converts to a much higher pressure and lower temperature, it goes down significantly even though the ammonia is impure. From a practical standpoint, you would like a low flow rate to make it economical.

11.110 **Given:** flow$_{NO}$ = 2.55 L/s, P_{NO} = 12.4 torr, T_{NO} = 655 K, and 8.0 hours **Find:** m_{urea}

Conceptual Plan: torr → atm and hr → min → s then P_{NO}, V_{NO}/s, T_{NO} → n_{NO}/s

$\frac{1\ atm}{760\ torr}$ $\frac{60\ min}{1\ hr}$ $\frac{60\ s}{1\ min}$ $PV = nRT$

n_{NO}/s → n_{urea}/s then s → n_{urea} → g_{urea}

$\frac{2\ mol\ urea}{4\ mol\ NO}$ $n_{urea} = (n_{urea}/s)(s)$ $\frac{60.06\ g\ urea}{1\ mol\ urea}$

Solution: $P_{NO} = 12.4\ \cancel{torr} \times \frac{1\ atm}{760\ \cancel{torr}} = 0.016\underline{3}1579\ atm$, $8.0\ \cancel{hr} \times \frac{60\ \cancel{min}}{1\ \cancel{hr}} \times \frac{60\ s}{1\ \cancel{min}} = 2\underline{8}800\ s$

$PV = nRT$ Rearrange to solve for n_{NO}. Note that we can substitute V/s for V and get n/s as a result.

$$\frac{n_{NO}}{s} = \frac{PV}{RT} = \frac{0.016\underline{3}1579\ \cancel{atm} \times 2.55\ \cancel{L}/s}{0.08206 \frac{\cancel{L} \cdot \cancel{atm}}{mol \cdot \cancel{K}} \times 655\ \cancel{K}} = 0.000774\underline{4}062 \frac{mol\ NO}{s}$$

$$0.00077\underline{4}062 \frac{\cancel{mol\ NO}}{s} \times \frac{2\ mol\ urea}{4\ \cancel{mol\ NO}} = 0.00038\underline{7}031 \frac{mol\ urea}{s}$$

$$2\underline{8}800\ \cancel{s} \times 0.00038\underline{7}031 \frac{\cancel{mol\ urea}}{\cancel{s}} \times \frac{60.06\ g\ urea}{1\ \cancel{mol\ urea}} = 670\ g\ urea$$

Check: The units (g) are correct. The magnitude of the answer (670) is not an unreasonable mass to add to a car because many more grams of gasoline are burned in 8 hours of driving.

11.111 **Given:** l = 30.0 cm, w = 20.0 cm, h = 15.0 cm, 14.7 psi **Find:** force (lb)

Conceptual Plan: l, w, h → Surface Area, SA(cm^2) → Surface Area(in^2) → Force

$SA = 2(lh) + 2(wh) + 2(lw)$ $\frac{(1\ in)^2}{(2.54\ cm)^2}$ $\frac{14.7\ lb}{1\ in^2}$

Solution: $SA = 2(lh) + 2(wh) + 2(lw) = 2(30.0\ cm \times 15.0\ cm) + 2(20.0\ cm \times 15.0\ cm) + 2(30.0\ cm \times 20.0\ cm) = 27\underline{0}0\ cm^2$

$$27\underline{0}0\ \cancel{cm^2} \times \frac{(1\ in)^2}{(2.54\ \cancel{cm})^2} = 41\underline{8}.50\ in^2,\ 41\underline{8}.50\ \cancel{in^2} \times \frac{14.7\ lb}{1\ \cancel{in^2}} = 6150\ lb. \text{ The can would be crushed.}$$

Check: The units (lb) are correct. The magnitude of the answer (6150) is not unreasonable because there is a large surface area.

11.112 **Given:** l = 20.0 cm, r = 10.0 cm, 25 mL with d = 0.807 g/mL, and P_{air} = 760.0 mmHg = 1.000 atm

Find: force (lb)

Conceptual Plan: mL → m → n then l, r → V(cm^3) → V(L) then V, n, T → P_{N_2} then

$d = \frac{m}{V}$ $\frac{1\ mol}{28.02\ g}$ $V = \pi r^2 l$ $\frac{1\ L}{1000\ cm^3}$ $PV = nRT$

P_{N_2}, P_{air} → P_{Total} then atm → psi then l, r → Surface Area(cm^2) → Surface Area(in^2) → Force

$P_{Total} = P_{air} + P_{N_2}$ $\frac{14.7\ lbs}{1\ atm}$ $SA = 2\pi rl + 2\pi r^2$ $\frac{(1\ in)^2}{(2.54\ cm)^2}$ $p = \frac{F}{A}$

Solution: $d = \frac{m}{V}$ Rearrange to solve for m. $m = dV = 0.807 \frac{g}{\cancel{mL}} \times 25\ \cancel{mL} = 2\underline{0}.175\ g$

$20.175\ \text{g} \times \dfrac{1\ \text{mol}}{28.02\ \text{g}} = 0.72002\ \text{mol}, V = \pi r^2 l = \pi \times (10.0\ \text{cm})^2 \times 20.0\ \text{cm} = 6283.19\ \text{cm}^3$

$6283.19\ \text{cm}^3 \times \dfrac{1\ \text{L}}{1000\ \text{cm}^3} = 6.28319\ \text{L}, PV = nRT$ Rearrange to solve for P.

$$P_{N_2} = \frac{nRT}{V} = \frac{0.72002\ \text{mol} \times 0.08206 \dfrac{\text{L} \cdot \text{atm}}{\text{mol} \cdot \text{K}} \times 298\ \text{K}}{6.28319\ \text{L}} = 2.80228\ \text{atm}$$

$P_{\text{Total}} = P_{\text{air}} + P_{N_2} = 1.000\ \text{atm} + 2.80228\ \text{atm} = 3.80228\ \text{atm}, 3.80228\ \text{atm} \times \dfrac{14.7\ \text{psi}}{1\ \text{atm}} = 55.894\ \text{psi}$

$SA = 2\pi rl + 2\pi r^2 = (2 \times \pi \times 10.0\ \text{cm} \times 20.0\ \text{cm}) + [2 \times \pi \times (10.0\ \text{cm})^2] = 1884.956\ \text{cm}^2$

$1884.956\ \text{cm}^2 \times \dfrac{(1\ \text{in})^2}{(2.54\ \text{cm})^2} = 292.169\ \text{in}^2, P = \dfrac{F}{A}$ Rearrange to solve for F.

$292.169\ \text{in}^2 \times \dfrac{55.894\ \text{lbs}}{1\ \text{in}^2} = 1.6 \times 10^4\ \text{lb}$

Check: The units (lb) are correct. The magnitude of the answer (16,000) is not unreasonable because there is a large surface area and this is a high pressure.

11.113 **Given:** $V_1 = 160.0\ \text{L}, P_1 = 1855\ \text{psi}, 3.5\ \text{L/balloon}, P_2 = 1.0\ \text{atm} = 14.7\ \text{psi}$, and $T = 298\ \text{K}$ **Find:** # balloons

Conceptual Plan: $V_1, P_1, P_2 \rightarrow V_2$ **then L** $\rightarrow$ **# balloons**

$P_1V_1 = P_2V_2$ $\qquad \dfrac{1\ \text{balloon}}{3.5\ \text{L}}$

Solution: $P_1V_1 = P_2V_2$ Rearrange to solve for V_2. $V_2 = \dfrac{P_1}{P_2} V_1 = \dfrac{1855\ \text{psi}}{14.7\ \text{psi}} \times 160.0\ \text{L} = 20190.5\ \text{L}$

$20190.5\ \text{L} \times \dfrac{1\ \text{balloon}}{3.5\ \text{L}} = 5768\ \text{balloons} = 5800\ \text{balloons}$

Check: The units (balloons) are correct. The magnitude of the answer (5800) is reasonable because a store does not have to buy a new helium tank very often.

11.114 **Given:** 11.5 mL with $d = 0.573\ \text{g/ml}, T = 28.5\ °\text{C}$, and $P = 892$ torr **Find:** V

Conceptual Plan: mL $\rightarrow$ m $\rightarrow$ n **and °C,** $\rightarrow$ **K and torr** $\rightarrow$ **atm then** $P, n, T \rightarrow V$

$d = \dfrac{m}{V}$ $\quad \dfrac{1\ \text{mol}}{58.12\ \text{g}}$ $\quad \text{K} = °\text{C} + 273.15$ $\quad \dfrac{1\ \text{atm}}{760\ \text{torr}}$ $\quad PV = nRT$

Solution: $d = \dfrac{m}{V}$ Rearrange to solve for m. $\quad m = dV = 0.573 \dfrac{\text{g}}{\text{mL}} \times 11.5\ \text{mL} = 6.5895\ \text{g}$

$6.5895\ \text{g} \times \dfrac{1\ \text{mol}}{58.12\ \text{g}} = 0.113377\ \text{mol}, T = 28.5°\text{C} + 273.15 = 301.7\ \text{K}, 892\ \text{torr} \times \dfrac{1\ \text{atm}}{760\ \text{torr}} = 1.17368\ \text{atm}$

$PV = nRT$ Rearrange to solve for V.

$$V = \frac{nRT}{P} = \frac{0.113377\ \text{mol} \times 0.08206 \dfrac{\text{L} \cdot \text{atm}}{\text{mol} \cdot \text{K}} \times 301.7\ \text{K}}{1.17368\ \text{atm}} = 2.39\ \text{L}$$

Check: The units (L) are correct. The magnitude of the answer (2) is reasonable because there is a lot less than one mole of butane.

11.115 **Given:** $r_1 = 2.5\ \text{cm}, P_1 = 4.00\ \text{atm}, T = 298\ \text{K}$, and $P_2 = 1.00\ \text{atm}$ **Find:** r_2

Conceptual Plan: $r_1 \rightarrow V_1 V_1, P_1, P_2 \rightarrow V_2$ **then** $V_2 \rightarrow r_2$

$V = \dfrac{4}{3}\pi r^3$ $\quad P_1V_1 = P_2V_2$ $\quad V = \dfrac{4}{3}\pi r^3$

Solution: $V = \dfrac{4}{3}\pi r^3 = \dfrac{4}{3} \times \pi \times (2.5\ \text{cm})^3 = 65.450\ \text{cm}^3\ P_1V_1 = P_2V_2$ Rearrange to solve for V_2.

$V_2 = \dfrac{P_1}{P_2} V_1 = \dfrac{4.00\ \text{atm}}{1.00\ \text{atm}} \times 65.450\ \text{cm}^3 = 261.80\ \text{cm}^3, V = \dfrac{4}{3}\pi r^3$

Rearrange to solve for r. $r = \sqrt[3]{\frac{3V}{4\pi}} = \sqrt[3]{\frac{3 \times 261.80 \text{ cm}^3}{4 \times \pi}} = 4.0 \text{ cm}$

Check: The units (cm) are correct. The magnitude of the answer (4) is reasonable because the bubble will expand as the pressure is decreased.

11.116 **Given:** max $SA = 1257 \text{ cm}^2$, $V_1 = 3.0 \text{ L}$, $P_1 = 755 \text{ torr}$, $T_1 = 298 \text{ K}$, and $T_2 = 273 \text{ K}$ **Find:** P_2 to burst balloon
Conceptual Plan: torr → atm and $A \rightarrow r \rightarrow V(\text{cm}^3) \rightarrow V(\text{L})$ then

$\frac{1 \text{ atm}}{760 \text{ torr}}$ $\quad SA = 4\pi r^2 \; V = \frac{4}{3}\pi r^3 \quad$ $\frac{1 \text{ L}}{1000 \text{ cm}^3}$

$P_1, V_1, P_2, T_1, V_2, T_2 \rightarrow P_2$

$\frac{P_1 V_1}{T_1} = \frac{P_2 V_2}{T_2}$

Solution: $P_1 = 755 \text{ torr} \times \frac{1 \text{ atm}}{760 \text{ torr}} = 0.993421 \text{ atm}$, $SA = 4\pi r^2$ Rearrange to solve for r.

$r = \sqrt{\frac{SA}{4\pi}} = \sqrt{\frac{1257 \text{ cm}^2}{4\pi}} = 10.00144 \text{ cm}$, $V = \frac{4}{3}\pi r^3 = \frac{4}{3} \times \pi \times (10.00144 \text{ cm})^3 = 4190.600 \text{ cm}^3$

$4190.600 \text{ cm}^3 \times \frac{1 \text{ L}}{1000 \text{ cm}^3} = 4.190600 \text{ L}$ $\quad \frac{P_1 V_1}{T_1} = \frac{P_2 V_2}{T_2}$ Rearrange to solve for P_2.

$P_2 = P_1 \frac{V_1 T_2}{V_2 T_1} = 0.993421 \text{ atm} \times \frac{3.0 \text{ L}}{4.190600 \text{ L}} \times \frac{273 \text{ K}}{298 \text{ K}} = 0.65 \text{ atm}$

Check: The units (atm) are correct. The magnitude of the answer (0.65) is reasonable because the pressure must decrease for the balloon to expand.

11.117 **Given:** 2.0 mol CO : 1.0 mol O_2, $V = 2.45 \text{ L}$, $P_1 = 745 \text{ torr}$, $P_2 = 552 \text{ torr}$, and $T = 552 \text{ °C}$
Find: % reacted
Conceptual Plan: From $PV = nRT$, we know that $P \alpha n$; looking at the chemical reaction, we see that $2 + 1 = 3$ moles of gas gets converted to 2 moles of gas. If all of the gas reacts, $P_2 = 2/3\,P_1$. Calculate $-\Delta P$ for 100% reacted and for actual case. Then calculate % reacted.

$-\Delta P_{100\% \text{ reacted}} = P_1 - \frac{2}{3}P_1 \quad -\Delta P_{\text{actual}} = P_1 - P_2 \quad \% \text{ reacted} = \frac{\Delta P_{\text{actual}}}{\Delta P_{100\% \text{ reacted}}} \times 100\%$

Solution: $-\Delta P_{100\% \text{ reacted}} = P_1 - \frac{2}{3}P_1 = 745 \text{ torr} - \frac{2}{3}745 \text{ torr} = 248.333 \text{ torr}$

$-\Delta P_{\text{actual}} = P_1 - P_2 = 745 \text{ torr} - 552 \text{ torr} = 193 \text{ torr}$

$\% \text{ reacted} = \frac{\Delta P_{\text{actual}}}{\Delta P_{100\% \text{ reacted}}} \times 100\% = \frac{193 \text{ torr}}{248.333 \text{ torr}} \times 100\% = 77.7\%$

Check: The units (%) are correct. The magnitude of the answer (78) makes sense because the pressure dropped most of the way to the pressure if all of the reactants had reacted $(2/3(745 \text{ torr}) = 497 \text{ torr})$. **Note: This problem can be solved in many ways, including calculating the moles of reactants and products using $PV = nRT$.**

11.118 **Given:** N_2, $V_1 = 1.0 \text{ L}$, $P_1 = 1.0 \text{ atm}$, $T_1 = 300. \text{ K}$, and $V_2 = 3.0 \text{ L}$ **Find:** d_2
Conceptual Plan: $\mathcal{M}, V_1, P_1, T_1 \rightarrow d_1 \rightarrow d_2$

$d = \frac{P\mathcal{M}}{RT} \quad d = \frac{m}{V}$

Solution: $d_1 = \frac{P\mathcal{M}}{RT} = \frac{1.0 \text{ atm} \times 28.02 \frac{\text{g}}{\text{mol}}}{0.08206 \frac{\text{L} \cdot \text{atm}}{\text{K} \cdot \text{mol}} \times 300. \text{ K}} = 1.13819 \frac{\text{g}}{\text{L}}, d = \frac{m}{V}$

Because we have a sealed container, $m_1 = m_2$. Rearrange to solve for m. $m = dV$ or $m = d_1 V_1 = d_2 V_2$

Rearrange to solve for d_2. $d_2 = d_1 \frac{V_1}{V_2} = 1.13819 \frac{\text{g}}{\text{L}} \times \frac{1.0 \text{ L}}{3.0 \text{ L}} = 0.38 \frac{\text{g}}{\text{L}}$

An alternative method is to use the ideal gas law to calculate the number of moles of gas (0.04062 mol), convert this to the mass of nitrogen using the molar mass (28.02 g/mol) and finally calculating the density by dividing the mass (1.14 g) by the volume (3.00 L).

Check: The units (g/L) are correct. The magnitude of the answer (0.4) is a typical gas density. The density dropped as the volume went up.

11.119 **Given:** $P_{1,\text{ total}} = 2.2$ atm $=$ CO $+ O_2$, $P_{2,\text{ total}} = 1.9$ atm $=$ CO $+ O_2 + CO_2$, $V = 1.0$ L, and $T = 1.0 \times 10^3$ K **Find:** mass CO_2 made

Conceptual Plan:

$\mathbf{P_{1,\text{ total}} = 2.2\text{ atm} = P_{1,\text{ CO}} + P_{1,\text{ O}_2}, P_{2,\text{ total}} = 1.9\text{ atm} = P_{2,\text{ CO}} + P_{2,\text{ O}_2} + P_{2,\text{ CO}_2}}$

Let x = amount of P_{O_2} reacted. From stoichiometry: $P_{2,\text{ CO}} = P_{1,\text{ CO}} - 2x, P_{2,\text{ O}_2} = P_{1,\text{ O}_2} - x$, $P_{2,\text{ CO}_2} = 2x$. Thus, $P_{2,\text{ total}} = 1.9\text{ atm} = P_{1,\text{ CO}_2} - 2x + P_{1,\text{ O}_2} - x + 2x = P_{1,\text{ total}} - x$. Using the initial conditions: $1.9\text{ atm} = 2.2\text{ atm} - x$. So $x = 0.3$ atm, and because $2x = P_{2,\text{ CO}_2} = 0.6$ atm, $P, V, T \rightarrow n \rightarrow$ g.

$$PV = nRT \qquad \frac{44.01\text{ g}}{1\text{ mol}}$$

Solution: $PV = nRT$ Rearrange to solve for n.

$$n = \frac{PV}{RT} = \frac{0.6\text{ atm} \times 1.0\text{ L}}{0.08206 \frac{\text{L} \cdot \text{atm}}{\text{mol} \cdot \text{K}} \times 1000\text{ K}} = 0.0073117\text{ mol}$$

$$0.0073117\text{ mol} \times \frac{44.01\text{ g}}{1\text{ mol}} = 0.321788\text{ g } CO_2 = 0.3\text{ g } CO_2$$

Check: The units (g) are correct. The magnitude of the answer (0.3) makes sense because we have such a small volume at a very high temperature and such a small pressure. This leads us to expect a very small number of moles.

11.120 **Given:** $r = 1.3 \times 10^{-8}$ cm, $V = 100.$ mL, $P = 1.0$ atm, and $T_1 = 273$ K **Find:** V fraction occupied by Xe atoms

Conceptual Plan: mL $\rightarrow$ L then $V \rightarrow n \rightarrow$ atoms then $r \rightarrow V(\text{cm}^3)/\text{atom}$

$$\frac{1\text{ L}}{1000\text{ mL}} \qquad \frac{1\text{ mol}}{22.414\text{ L}}\text{ at STP} \qquad 6.022 \times 10^{23}\text{ atoms/mol} \qquad V = \frac{4}{3}\pi r^3$$

then atoms, $V(\text{cm}^3)/\text{atom} \rightarrow V_{\text{Xe}}$ then $V_{\text{Xe}}, V_{\text{container}} \rightarrow$ Fraction Xe

$$V_{\text{Xe}} = (V/atom)(atoms) \qquad \%V_{\text{Xe}} = \frac{V_{\text{Xe}}}{V_{\text{container}}} \times 100\%$$

Solution: $100\text{ mL} \times \frac{1\text{ L}}{1000\text{ mL}} = 0.100\text{ L}$

$$0.100\text{ L} \times \frac{1\text{ mol}}{22.414\text{ L}} = 0.00446150\text{ mol} \times 6.022 \times 10^{23}\frac{\text{atoms}}{\text{mol}} = 2.68672 \times 10^{21}\text{ atoms}$$

$$V = \frac{4}{3}\pi r^3 = \frac{4}{3} \times \pi \times (1.3 \times 10^{-8}\text{ cm})^3 = 9.2028 \times 10^{-24}\text{ cm}^3/\text{atom}$$

$$V_{\text{Xe}} = (V/atom)(atoms) = \frac{9.2028 \times 10^{-24}\text{ cm}^3}{\text{atom}} \times 2.68672 \times 10^{21}\text{ atoms} = 0.024725\text{ cm}^3$$

$$\%\,V_{\text{Xe}} = \frac{V_{\text{Xe}}}{V_{\text{container}}} \times 100\% = \frac{0.024725\text{ cm}^3}{100\text{ cm}^3} \times 100\% = 0.025\%\text{V}$$

Check: The units (%V) are correct. The magnitude of the answer (0.025) is reasonable because we expect the molecules to take up very little volume of a container of a gas.

11.121 **Given:** $h_1 = 22.6$ m, $T_1 = 22$ °C, and $h_2 = 23.8$ m **Find:** T_2

Conceptual Plan: °C $\rightarrow$ K because $V_{\text{cylinder}}\ \alpha\ h$ we do not need to know r to use $V_1, T_1, T_2 \rightarrow V_2$

$$\text{K} = {}^\circ\text{C} + 273.15 \qquad V = \pi r^2 h \qquad \frac{V_1}{T_1} = \frac{V_2}{T_2}$$

Solution: $T_1 = 22\text{ °C} + 273.15 = 295\text{K}, \frac{V_1}{T_1} = \frac{V_2}{T_2}$ Rearrange to solve for T_2.

$$T_2 = T_1 \times \frac{V_2}{V_1} = T_1 \times \frac{\pi r^2 h_2}{\pi r^2 h_1} = 295\text{ K} \times \frac{23.8\text{ m}}{22.6\text{ m}} = 311\text{ K}$$

Check: The units (K) are correct. We expect the temperature to increase because the volume increased.

11.122 **Given:** $m_{CH_4} = 8.0$ g, $m_{Xe} = 8.0$ g, and $P_{Total} = 0.44$ atm **Find:** P_{CH_4}

Conceptual Plan: $m \rightarrow n$ **then** $n_{CH_4}, n_{Xe} \rightarrow \chi_{CH_4}$ **then** $\chi_{CH_4}, P_{Total} \rightarrow P_{CH_4}$

$\mathcal{M}$ $\quad \chi_{CH_4} = \frac{n_{CH_4}}{n_{CH_4} + n_{Xe}}$ $\quad P_{CH_4} = \chi_{CH_4} P_{Total}$

Solution: $n_{CH_4} = 8.0 \text{ g} \times \frac{1 \text{ mol}}{16.04 \text{ g}} = 0.49875 \text{ mol}, n_{Xe} = 8.0 \text{ g} \times \frac{1 \text{ mol}}{131.3 \text{ g}} = 0.060929 \text{ mol}$

$$\chi_{CH_4} = \frac{n_{CH_4}}{n_{CH_4} + n_{Xe}} = \frac{0.49875 \text{ mol}}{0.49875 \text{ mol} + 0.060929 \text{ mol}} = 0.89114$$

$$P_{CH_4} = \chi_{CH_4} P_{Total} = 0.89114 \times 0.44 \text{ atm} = 0.39 \text{ atm}$$

Check: The units (atm) are correct. The magnitude of the answer (0.4) makes sense because the molecular weight of methane is so much lower than that of xenon; so we have many more moles of methane. The partial pressure of methane is almost as large as the total pressure.

11.123 **Given:** He, $V = 0.35$ L, $P_{max} = 88$ atm, and $T = 299$ K **Find:** m_{He}

Conceptual Plan: $P, V, T \rightarrow n$ **then mol** $\rightarrow m$

$PV = nRT$ $\quad \mathcal{M}$

Solution: $PV = nRT$ Rearrange to solve for n.

$$n_{He} = \frac{PV}{RT} = \frac{88 \text{ atm} \times 0.35 \text{ L}}{0.08206 \frac{\text{L} \cdot \text{atm}}{\text{mol} \cdot \text{K}} \times 299 \text{ K}} = 1.2553 \text{ mol}, 1.2553 \text{ mol} \times \frac{4.003 \text{ g}}{1 \text{ mol}} = 5.0 \text{ g He}$$

Check: The units (g) are correct. The magnitude of the answer (5) makes sense because the high pressure and the low volume cancel out (remember 22 L/mol at STP). So we expect ~1 mol; so ~4 g.

11.124 **Given:** NaH + water, $V = 0.490$ L, $P_{Total} = 758$ mmHg, and $T = 35$ °C **Find:** m_{H2} and m_{NaH}

Other: $P_{H_2O} = 42.23$ mmHg at 35 °C

Conceptual Plan: °C $\rightarrow$ K **and** P_{Total}(mmHg) $\rightarrow P_{H_2}$(mmHg) $\rightarrow P_{H_2}$(atm) **then**

K = °C + 273.15 $\quad P_{Total} = P_{H_2O} + P_{H_2}$ $\quad \frac{1 \text{ atm}}{760 \text{ mmHg}}$

Write balanced reaction. Show reaction as part of the solution and display in larger font

$NaH(s) + H_2O(l) \rightarrow NaOH(aq) + H_2(g)$

$P, V, T \rightarrow n_{H_2}$ **then** $n_{H_2} \rightarrow m_{H_2}$ **then** $n_{H_2} \rightarrow n_{NaH} \rightarrow m_{NaH}$

$PV = nRT$ $\quad \frac{2.016 \text{ g}}{1 \text{ mol}}$ $\quad \frac{1 \text{ mol NaH}}{1 \text{ mol } H_2}$ $\quad \frac{24.0 \text{ g}}{1 \text{ mol}}$

Solution: $T = 35 \text{ °C} + 273.15 = 308 \text{ K}, P_{Total} = P_{H_2O} + P_{H_2}$ Rearrange to solve for P_{H_2}.

$$P_{H_2} = 758 \text{ mmHg} - 42.23 \text{ mmHg} = 715.77 \text{ mmHg}, 715.77 \text{ mmHg} \times \frac{1 \text{ atm}}{760 \text{ mmHg}} = 0.941803 \text{ atm } PV = nRT$$

Rearrange to solve for n. $n_{H_2} = \frac{PV}{RT} = \frac{0.941803 \text{ atm} \times 0.490 \text{ L}}{0.08206 \frac{\text{L} \cdot \text{atm}}{\text{mol} \cdot \text{K}} \times 308 \text{ K}} = 0.0182589 \text{ mol}$

$$NaH(s) + H_2O(l) \rightarrow NaOH(aq) + H_2(g)$$

$$0.0182589 \text{ mol} \times \frac{2.016 \text{ g}}{1 \text{ mol}} = 0.0368 \text{ g } H_2 \text{ and } 0.0182589 \text{ mol } H_2 \times \frac{1 \text{ mol NaH}}{1 \text{ mol } H_2} = 0.0182589 \text{ mol NaH}$$

$$0.0182589 \text{ mol NaH} \times \frac{24.0 \text{ g}}{1 \text{ mol}} = 0.438 \text{ g NaH}$$

Check: The units (g) are correct. The magnitude of the answers (0.04 and 0.4) makes sense because we have much less than a mole of each material (remember 22 L/mol at STP); so we expect < 2 g gas and < 24 g solid.

11.125 **Given:** 15.0 mL HBr in 1.0 minute and 20.3 mL unknown hydrocarbon gas in 1.0 minute **Find:** formula of unknown gas

Conceptual Plan: Because these gases are under the same conditions, $V \alpha n, V$, **time** $\rightarrow$ **Rate then**

$\text{Rate} = \frac{V}{time}$

$\mathcal{M}_{HBr}, \text{Rate}_{HBr}, \text{Rate}_{Unk} \rightarrow \mathcal{M}_{Unk}$

$$\frac{\text{Rate}_{HBr}}{\text{Rate}_{Unk}} = \sqrt{\frac{\mathcal{M}_{Unk}}{\mathcal{M}_{HBr}}}$$

Solution: $\text{Rate}_{HBr} = \frac{V}{time} = \frac{15.0\text{ mL}}{1.0\text{ min}} = 15.0\frac{\text{mL}}{\text{min}}$, $\text{Rate}_{Unk} = \frac{V}{time} = \frac{20.3\text{ mL}}{1.0\text{ min}} = 20.3\frac{\text{mL}}{\text{min}}$,

$\frac{\text{Rate}_{HBr}}{\text{Rate}_{Unk}} = \sqrt{\frac{\mathcal{M}_{Unk}}{\mathcal{M}_{HBr}}}$ Rearrange to solve for $\mathcal{M}_{Unk}$.

$\mathcal{M}_{Unk} = \mathcal{M}_{HBr}\left(\frac{\text{Rate}_{HBr}}{\text{Rate}_{Unk}}\right)^2 = 80.91\frac{\text{g}}{\text{mol}} \times \left(\frac{15.0\frac{\text{mL}}{\text{min}}}{20.3\frac{\text{mL}}{\text{min}}}\right)^2 = 44.2\frac{\text{g}}{\text{mol}}$ The formula is C_3H_8, or propane, because we know the molecule contains only H and C and this combination of atoms leads to a molar mass of 44.2 g/mol

Check: The units (g/mol) are correct. The magnitude of the answer (< HBr) makes sense because the unknown diffused faster and so must be lighter.

11.126 Because $N_2O_3(g) \rightarrow NO_2(g) + NO(g)$ undergoes a complete reaction, according to Avogadro's law, the pressure will double because one mole of gas decomposes to two moles of gas. According to Charles's law, when the temperature in kelvins doubles, the pressure doubles. Thus, the pressure will increase by a factor of four or $4 \times 0.017\text{ atm} = 0.068\text{ atm}$.

11.127 **Given:** 0.583 g neon, $V = 8.00 \times 10^2\text{ cm}^3$, $P_{Total} = 1.17$ atm, and $T = 295$ K **Find:** m_{Ar}
Conceptual Plan: $m_{Ne} \rightarrow n_{Ne}$ **and mL** $\rightarrow$ **L then** $n_{Ne}, V, T \rightarrow P_{Ne}$ **then** $P_{Ne}, P_{Total} \rightarrow P_{Ar}$ **then**

$\frac{1\text{ mol}}{20.18\text{ g}}$ $\quad \frac{1\text{ L}}{1000\text{ mL}}$ $\quad PV = nRT$ $\quad P_{Total} = P_{Ne} + P_{Ar}$

$P_{Ar}, V, T \rightarrow n_{Ar} \rightarrow m_{Ar}$

$PV = nRT$ $\quad \frac{39.95\text{ g}}{1\text{ mol}}$

Solution: $0.583\text{ g Ne} \times \frac{1\text{ mol Ne}}{20.18\text{ g Ne}} = 0.02888999\text{ mol Ne}$ $\quad 8.00 \times 10^2\text{ mL} \times \frac{1\text{ L}}{1000\text{ mL}} = 0.800\text{ L}$

$PV = nRT$ Rearrange to solve for P.

$$P = \frac{nRT}{V} = \frac{0.02888999\text{ mol} \times 0.08206\frac{\text{L}\cdot\text{atm}}{\text{mol}\cdot\text{K}} \times 295\text{ K}}{0.800\text{ L}} = 0.874200\text{ atm Ne}$$

$P_{Total} = P_{Ne} + P_{Ar}$ Rearrange to solve for P_{Ar}. $P_{Ar} = P_{Total} - P_{Ne} = 1.17\text{ atm} - 0.874200\text{ atm} = 0.295800\text{ atm}$

$PV = nRT$ Rearrange to solve for n. $n_{Ar} = \frac{PV}{RT} = \frac{0.295800\text{ atm} \times 0.800\text{ L}}{0.08206\frac{\text{L}\cdot\text{atm}}{\text{mol}\cdot\text{K}} \times 295\text{ K}} = 0.00977539\text{ mol Ar}$

$$0.00977539\text{ mol Ar} \times \frac{39.95\text{ g Ar}}{1\text{ mol Ar}} = 0.390527\text{ g Ar} = 0.39\text{ g Ar}$$

Check: The units (g) are correct. The magnitude of the answer (0.4) makes sense because the pressure and volume are small.

11.128 **Given:** helium + argon density 0.670 g/L, $P = 755$ mmHg, and $T = 298$ K **Find:** composition
Solution: Assume 22.414 L/mol. Although not at STP, the percent composition will not change so we can assume a molar volume; so number of moles of gas = 755 mmHg/760 mmHg = 0.9934211 moles of gas

and the mass of the gas $= \frac{0.670\text{ g}}{1\text{ L}} \times 22.414\text{ L} = 15.01738$ g total. Let $x = n_{He}$,
$n_{Total} = 0.09934211 = x + y$. Solve for y. $y = 0.9934211 - x$ and $y = n_{Ar}$, so

$m_{Total} = 15.01738\text{ g total} = x\frac{4.003\text{ g}}{1\text{ mol}} + (0.9934211 - x)\frac{39.95\text{ g}}{1\text{ mol}}$ Solve for x.

$$x\text{ mol}\left(\frac{39.95\text{ g}}{1\text{ mol}} - \frac{4.003\text{ g}}{1\text{ mol}}\right) = (39.68717 - 15.01738)\text{ g} \rightarrow x\text{ mol} = \frac{(24.66979)\text{ g}}{\left(\frac{35.947\text{ g}}{1\text{ mol}}\right)} = 0.6862823\text{ mol He}$$

and $(0.9934211 - 0.6862823)$ mol Ar = 0.3071388 mol Ar. The composition on a volume basis is the same as the composition on a molar basis.

$$\frac{0.6862823 \text{ mol He}}{0.9934211 \text{ mol total}} \times 100\% = 69.1\% \text{ He and } 100\% - 69.1\% \text{ Ar} = 30.9\% \text{ Ar}$$

Check: The units (%) are correct. The magnitude of the answer (70) makes sense because the average molar mass is ~15 g/mol, which is closer to the molar mass of He than to Ar.

11.129 **Given:** 75.2% by mass nitrogen + 24.8% by mass krypton and $P_{Total} = 745$ mmHg **Find:** P_{Kr}
Solution: Assume that 100 g total, so we have 75.2 g N_2 and 24.8 g Kr. Converting these masses to moles,

$$75.2 \text{ g } N_2 \times \frac{1 \text{ mol } N_2}{28.02 \text{ g } N_2} = 2.683797 \text{ mol } N_2 \text{ and } 24.8 \text{ g Kr} \times \frac{1 \text{ mol Kr}}{83.80 \text{ g Kr}} = 0.2959427 \text{ mol Kr}$$

$$P_{Kr} = \chi_{Kr} P_{Total} = \frac{0.2959427 \text{ mol Kr}}{2.683797 \text{ mol } N_2 + 0.2959427 \text{ mol Kr}} 745 \text{ mmHg Kr} = 74.0 \text{ mmHg Kr}$$

Check: The units (mmHg) are correct. The magnitude of the answer (74) makes sense because the mixture is mostly nitrogen by mass and this dominance is magnified because the molar mass of krypton is larger than the molar mass of nitrogen.

Challenge Problems

11.130 **Given:** $V = 10$ L, 0.10 mol H_2 initially, $T = 3000$ K, $P_{Final} = 3.0$ atm **Find:** P_H
Conceptual Plan: Write balanced reaction to determine change in moles of gas.

$H_2(g) \rightarrow 2\,H(g)$; thus, $\frac{2 \text{ mol H}}{1 \text{ mol } H_2 \text{ reacted}}$. Because $P_{H_2} \alpha\, n_{H_2}$, the pressure will increase 1 atm for every 1 atm of H_2 that reacts.

$n, T, V \rightarrow P_{initial}$ $\quad P_{initial}, P_{final} \rightarrow \Delta P$ $\quad$ write expression for P_H

$PV = nRT \qquad \Delta P = P_{final} - P_{initial} \qquad P_H = \Delta P \frac{2 \text{ mol H}}{1 \text{ atm reacted}}$

Solution: $PV = nRT$ Rearrange to solve for P.

$$P = \frac{nRT}{V} = \frac{0.10 \text{ mol} \times 0.08206 \frac{\text{L} \cdot \text{atm}}{\text{mol} \cdot \text{K}} \times 3000 \text{ K}}{10 \text{ L}} = 2.4618 \text{ atm } H_2$$

Thus, if all of the gas remains as molecular hydrogen, the pressure would be 2.4618 atm. The increase in pressure is due to the decomposition of the molecules.
$\Delta P = P_{final} - P_{initial} = 3.0 \text{ atm} - 2.4618 \text{ atm} = 0.5382$ atm and

$$P_H = \Delta P \frac{2 \text{ mol H}}{1 \text{ atm reacted}} = 0.5382 \text{ atm} \times \frac{2 \text{ mol H}}{1 \text{ atm reacted}} = 1.0764 \text{ atm} = 1.1 \text{ atm H}$$

Check: The units (atm) are correct. The magnitude of the answer (1) makes sense because if all of the hydrogen dissociated, the final pressure would have been 5 atm. Because we are closer to the initial pressure than this maximum pressure, less than half of the hydrogen has dissociated.

11.131 **Given:** $2\,NH_3(g) \rightarrow N_2(g) + 3\,H_2(g)$; $N_2H_4(g) \rightarrow N_2(g) + 2\,H_2(g)$; initially $P = 0.50$ atm, $T = 300$ K; finally $P = 4.5$ atm, $T = 1200$ K **Find:** percent N_2H_4 initially
Conceptual Plan: $P_{initial}, T_{initial}, T_{final} \rightarrow P_{final}$ then determine change in moles of gas

$\frac{P_{initial}}{T_{initial}} = \frac{P_{final}}{T_{final}} \qquad \frac{2 \text{ atm added gas}}{2 \text{ atm } NH_3 \text{ reacted}}$ and $\frac{2 \text{ atm added gas}}{1 \text{ atm } N_2H_4 \text{ reacted}}$

$P_1, P_2 \rightarrow \Delta P$ write expression for ΔP, then solve for P_{1,N_2H_4} and P_{1,NH_3} finally $P_{1,N_2H_4}, P_{1,NH_3} \rightarrow \% N_2H_4$

$\Delta P = P_2 - P_1 \qquad \Delta P = P_{1,NH_3}\frac{3 \text{ atm added gas}}{2 \text{ atm reacted}} + P_{1,N_2H_4}\frac{2 \text{ atm added gas}}{1 \text{ atm reacted}}$ where $P_{1,1200\,K} = P_{1,NH_3} + P_{1,N_2H_4} \qquad \% N_2H_4 = \frac{P_{N_2H_4}}{P_{N_2H_4} + P_{NH_3}} \times 100\%$

Solution: $\frac{P_{initial}}{T_{initial}} = \frac{P_{final}}{T_{final}}$ Rearrange to solve for P_{final}. $P_2 = P_1 \times \frac{T_2}{T_1} = 0.50 \text{ atm} \times \frac{1200 \text{ K}}{300 \text{ K}} = 2.0$ atm if no reaction occurred. The increase in pressure from 2.0 atm to 4.5 atm (after the reaction occurs) is due to the decomposition reactions.
$\Delta P = P_{final} - P_{initial} = 4.5 \text{ atm} - 2.0 \text{ atm} = 2.5$ atm, and $P_{1,1200\,K} = 2.0 \text{ atm} = P_{1,NH_3} + P_{1,N_2H_4}$ or $P_{1,NH_3} = 2.0 \text{ atm} - P_{1,N_2H_4}$

Substitute this into $\Delta P = P_{1,NH_3}\frac{2 \text{ atm added gas}}{2 \text{ atm reacted}} + P_{1,N_2H_4}\frac{2 \text{ atm added gas}}{1 \text{ atm reacted}}$ and solve for P_{1,N_2H_4}.

$$\Delta P = 2.5 \text{ atm} = (2.0 \text{ atm} - P_{1,N_2H_4})\frac{2 \text{ atm added gas}}{2 \text{ atm reacted}} + P_{1,N_2H_4}\frac{2 \text{ atm added gas}}{1 \text{ atm reacted}} \rightarrow$$

$P_{1,N_2H_4} = 2.5 \text{ atm} - 2.0 \text{ atm} = 0.5 \text{ atm}$ and $P_{NH_3} = 2.0 \text{ atm} - 0.5 \text{ atm} = 1.5 \text{ atm}$; finally

$$\% \, N_2H_4 = \frac{P_{N_2H_4}}{P_{N_2H_4} + P_{NH_3}} \times 100\% = \frac{0.5 \text{ atm}}{0.5 \text{ atm} + 1.5 \text{ atm}} \times 100\% = 25\% \, N_2H_4 = 30\% \, N_2H_4$$

Check: The units (%) are correct. The magnitude of the answer (30) makes sense because if it were all N_2H_4, the final pressure would have been 6 atm. Because we are closer to the initial pressure than this maximum pressure, less than half of the gas is N_2H_4.

11.132 **Given:** CO gas; initial: $V = 0.48$ L, $P = 1.0$ atm, and $T = 275$ K; final: $V = 1.3$ L **Find:** final gas density

Conceptual Plan: $P, V, T \rightarrow n \rightarrow m$ **then** $m, V \rightarrow d$

$$PV = nRT \qquad \frac{28.01 \text{ g}}{1 \text{ mol}} \qquad d = m/V$$

Solution: $PV = nRT$ Rearrange to solve for n. $n = \dfrac{PV}{RT} = \dfrac{1.0 \text{ atm} \times 0.48 \text{ L}}{0.08206 \dfrac{\text{L} \cdot \text{atm}}{\text{mol} \cdot \text{K}} \times 275 \text{ K}} = 0.02127047 \text{ mol}$

$$0.02127047 \text{ mol} \times \frac{28.01 \text{ g}}{1 \text{ mol}} = 0.5957859 \text{ g then } d = \frac{m}{V} = \frac{0.5957859 \text{ g}}{1.3 \text{ L}} = 0.4582968 \text{ g/L} = 0.46 \text{ g/L}$$

Check: The units (g/L) are correct. The magnitude of the answer (0.5) makes sense because this is typical for a gas density.

11.133 **Given:** $2 \, CO_2(g) \rightarrow 2 \, CO(g) + O_2(g)$; initially $P = 10.0$ atm, $T = 701$ K; finally $P = 22.5$ atm, $T = 1401$ K
Find: mole percent decomposed

Conceptual Plan: $P_1, T_1, T_2 \rightarrow P_2$ **then determine change in moles of gas**

$$\frac{P_1}{T_1} = \frac{P_2}{T_2} \qquad \frac{1 \text{ atm added gas}}{2 \text{ atm } CO_2 \text{ reacted}}$$

$P_1, P_2 \rightarrow \Delta P$ **write expression for** ΔP, **then solve for** $P_{CO_2 \text{reacted}}$ **finally**

$$\Delta P = P_2 - P_1 \qquad \Delta P = P_{CO_2 \text{ reacted}} \frac{1 \text{ atm added gas}}{2 \text{ atm } CO_2 \text{ reacted}}$$

$P_2, P_{CO_2 \text{ reacted}} \rightarrow \% \, CO_2$ **decomposed**

$$\% CO_2 \text{ decomposed} = \frac{P_{CO_2 \text{ reacted}}}{P_2} \times 100\%$$

Solution: $\dfrac{P_1}{T_1} = \dfrac{P_2}{T_2}$ Rearrange to solve for P_1. $P_2 = P_1 \times \dfrac{T_2}{T_1} = 10.0 \text{ atm} \times \dfrac{1401 \text{ K}}{701 \text{ K}} = 19.985735 \text{ atm}$

$\Delta P = P_2 - P_1 = 22.5 \text{ atm} - 19.985735 \text{ atm} = 2.514265 \text{ atm}$

$\Delta P = P_{CO_2 \text{ reacted}} \dfrac{1 \text{ atm added gas}}{2 \text{ atm } CO_2 \text{ reacted}}$ or the pressure increases 1 atm for each 2 atm of gas decomposed, so 5.02853 atm decomposes and then

$$\% \, CO_2 \text{ decomposed} = \frac{P_{CO_2 \text{ reacted}}}{P_2} \times 100\% = \frac{5.02853 \text{ atm}}{19.985735 \text{ atm}} \times 100\% = 25.1606\% CO_2 \text{ decomposed} =$$

$25\% \, CO_2$ decomposed

Check: The units (%) are correct. The magnitude of the answer (25) makes sense because if all of the gas had decomposed, the final pressure would have been 40 atm. Because we are much closer to the initial pressure than this maximum pressure, much less than half of the gas had decomposed.

11.134 **Given:** 9.0×10^{12} kg/yr octane; atm = 387 ppm CO_2 by volume; atm thickness = 15 km; $r_{Earth} = 6371$ km; $P_{atm} = 381$ torr; $T_{atm} = 275$ K **Find:** m_{CO_2} and % increase in CO_2

Conceptual Plan: Write a balanced chemical reaction $\text{kg}_{C_3H_8} \rightarrow \text{g}_{C_3H_8} \rightarrow n_{C_3H_8} \rightarrow n_{CO_2}$

$$2 \, C_8H_{18}(g) + 25 \, O_2(g) \rightarrow 16 \, CO_2(g) + 18 \, H_2O(l) \qquad \frac{1000 \text{ g}}{1 \text{ kg}} \qquad \frac{1 \text{ mol}}{114.22 \text{ g}} \qquad \frac{16 \text{ mol } CO_2}{2 \text{ mol } C_8H_{18}}$$

then n_{CO_2} → m_{CO_2} and ppm_{CO_2} → χ_{CO_2} → P_{CO_2} torr → atm then r_{Earth} → V_{Earth} and

$\frac{44.01 \text{ g}}{1 \text{ mol}}$ $\frac{1 \text{ part}}{10^6 \text{ parts}}$ $P_{CO_2} = \chi_{CO_2} P_{atm}$ $\frac{1 \text{ atm}}{760 \text{ torr}}$ $V = \frac{4}{3}\pi r^3$

r_{Earth}, atm thickness → $r_{Earth+atm}$ then $r_{Earth+atm}$ → $V_{Earth+atm}$ $V_{Earth+atm}$, V_{Earth} → V_{atm} then

$r_{Earth+atm} = r_{Earth} + r_{atm}$ $V = \frac{4}{3}\pi r^3$ $V_{atm} = V_{Earth+atm} - V_{Earth}$

km^3 → m^3 → cm^3 → L then V_{atm}, P_{CO_2}, T_{atm} → n_{CO_2} → m_{CO_2} and

$\left(\frac{1000 \text{ m}}{1 \text{ km}}\right)^3 \left(\frac{100 \text{ cm}}{1 \text{ m}}\right)^3$ $\frac{1 \text{ L}}{1000 \text{ cm}^3}$ $PV = nRT$ $\frac{44.01 \text{ g}}{1 \text{ mol}}$

$m_{CO_2 added}$, $m_{CO_2 initially}$, → % $increase_{CO_2}$

$\% \text{ increase} = \frac{\text{added}}{\text{initial}} \times 100\%$

Solution: $9.0 \times 10^{12} \text{ kg} \times \frac{1000 \text{ g}}{1 \text{ kg}} \times \frac{1 \text{ mol}}{114.22 \text{ g}} \times \frac{16 \text{ mol } CO_2}{2 \text{ mol } C_8H_{18}} \times \frac{44.01 \text{ g}}{1 \text{ mol}} = 2.7742 \times 10^{16} \text{g } CO_2 \text{ added}$

$387 \text{ parts } CO_2 \frac{1 \text{ part}}{10^6 \text{ parts}} = 3.87 \times 10^{-4} = \chi_{CO_2}, P_{CO_2} = \chi_{CO_2} P_{atm} = 3.87 \times 10^{-4} \times 381 \text{ torr} = 0.147447 \text{ torr}$

$0.147447 \text{ torr} \times \frac{1 \text{ atm}}{760 \text{ torr}} = 0.00019401 \text{ atm}, V_{Earth} = \frac{4}{3}\pi r^3 = \frac{4}{3} \times \pi \times (6371 \text{ km})^3 = 1.08321 \times 10^{12} \text{ km}^3$

$r_{Earth+atm} = r_{Earth} + r_{atm} = 6371 \text{ km} + 15 \text{ km} = 6386 \text{ km}$

$V_{Earth+atm} = \frac{4}{3}\pi r^3 = \frac{4}{3} \times \pi \times (6386 \text{ km})^3 = 1.09088 \times 10^{12} \text{ km}^3$

$V_{atm} = V_{Earth+atm} - V_{Earth} = 1.09088 \times 10^{12} \text{ km}^3 - 1.08321 \times 10^{12} \text{ km}^3 = 7.670 \times 10^9 \text{ km}^3$

$7.670 \times 10^9 \text{ km}^3 \times \left(\frac{1000 \text{ m}}{1 \text{ km}}\right)^3 \times \left(\frac{100 \text{ cm}}{1 \text{ m}}\right)^3 \times \frac{1 \text{ L}}{1000 \text{ cm}^3} = 7.670 \times 10^{21} \text{ L } PV = nRT$

Rearrange to solve for n. $n_{CO_2 \text{ initial}} = \frac{PV}{RT} = \frac{0.00019401 \text{ atm} \times 7.670 \times 10^{21} \text{ L}}{0.08206 \frac{\text{L} \cdot \text{atm}}{\text{mol} \cdot \text{K}} \times 275 \text{ K}} = 6.594 \times 10^{16} \text{ mol}$

$6.594 \times 10^{16} \text{ mol} \times \frac{44.01 \text{ g}}{1 \text{ mol}} = 2.9020 \times 10^{18} \text{ g } CO_2$

$\% \text{ increase} = \frac{\text{added}}{\text{initial}} \times 100\% = \frac{2.7742 \times 10^{16} \text{ g } CO_2}{2.9020 \times 10^{18} \text{ g } CO_2} \times 100 = 1\% \text{ increase}$

Check: The units (g and %) are correct. The magnitude of the answer (10^{16}) is reasonable because we started with so much octane and the mass of CO_2 will be larger than the original octane weight because there is so much added oxygen. The % increase is reasonable because the volume of the atmosphere is so large.

11.135 **Given:** CH_4: $V = 155$ mL at STP; O_2: $V = 885$ mL at STP; NO: $V = 55.5$ mL at STP; mixed in a flask: $V = 2.0$ L, $T = 275$ K, and 90.0% of limiting reagent used. **Find:** Ps of all components and P_{Total}

Conceptual Plan: CH_4: mL → L → n_{CO_4} → n_{CO_2} and

$\frac{1 \text{ L}}{1000 \text{ mL}}$ $\frac{1 \text{ mol}}{22.414 \text{ L}}$ $\frac{1 \text{ mol } CO_2}{5 \text{ mol NO}}$

O_2: mL → L → n_{O_2} → n_{CO_2} and NO: mL → L → n_{NO} → n_{CO_2}

$\frac{1 \text{ L}}{1000 \text{ mL}}$ $\frac{1 \text{ mol}}{22.414 \text{ L}}$ $\frac{1 \text{ mol } CO_2}{5 \text{ mol } O_2}$ $\frac{1 \text{ L}}{1000 \text{ mL}}$ $\frac{1 \text{ mol}}{22.414 \text{ L}}$ $\frac{1 \text{ mol } CO_2}{5 \text{ mol NO}}$

the smallest yield determines the limiting reagent; then initial n_{NO} → reacted n_{NO} → final n_{NO}

NO is the limiting reagent 90.0% $0.100 \times \text{initial mol}_{NO}$

reacted n_{NO} → reacted n_{CH_4} then initial n_{CH_4}, reacted n_{CH_4} → final n_{CH_4} then

$\frac{1 \text{ mol } CH_4}{5 \text{ mol NO}}$ $\text{initial mol}_{CH_4} - \text{reacted mol}_{CH_4} = \text{final mol}_{CH_4}$

final n_{CH_4}, V, T → final P_{CH_4} and reacted n_{NO} → reacted n_{O_2} then

$PV = nRT$ $\frac{5 \text{ mol } O_2}{5 \text{ mol NO}}$

initial n_{O_2}, reacted $n_{O_2} \rightarrow$ final n_{O_2} then final n_{O_2} $V, T \rightarrow$ final P_{O_2} and

initial mol_{O_2} − reacted mol_{O_2} = final mol_{O_2} $\qquad PV = nRT$

final n_{NO}, $V, T \rightarrow$ final P_{NO} and theoretical n_{CO_2} from NO $\rightarrow$ final n_{CO_2}

$PV = nRT$ $\qquad$ 90.0%

final n_{CO_2}, $V, T \rightarrow P_{CO_2}$ then final $n_{CO_2} \rightarrow n_{H_2O}$, $V, T \rightarrow P_{H_2O}$ and

$PV = nRT$ $\qquad \dfrac{1 \text{ mol } H_2O}{1 \text{ mol } CO_2}$ $\qquad PV = nRT$

final $n_{CO_2} \rightarrow n_{NO_2}$ then n_{NO_2}, $V, T \rightarrow P_{NO_2}$ and final $n_{CO_2} \rightarrow n_{OH}$ then

$\dfrac{1 \text{ mol } NO_2}{1 \text{ mol } CO_2}$ $\qquad PV = nRT$ $\qquad \dfrac{2 \text{ mol OH}}{1 \text{ mol } CO_2}$

n_{OH}, $V, T \rightarrow P_{OH}$ finally $P_{CH_4}, P_{O_2}, P_{NO}, P_{CO_2}, P_{H_2O}, P_{NO_2}, P_{OH} \rightarrow P_{T\ total}$

$PV = nRT$ $\qquad P_{Total} = \Sigma P$

Solution: CH_4: $155 \text{ mL} \times \dfrac{1 \text{ L}}{1000 \text{ mL}} \times \dfrac{1 \text{ mol } CH_4}{22.414 \text{ L}} \times \dfrac{1 \text{ mol } CO_2}{1 \text{ mol } CH_4} = 0.00691532 \text{ mol } CO_2$

O_2: $885 \text{ mL} \times \dfrac{1 \text{ L}}{1000 \text{ mL}} \times \dfrac{1 \text{ mol } O_2}{22.414 \text{ L}} = 0.0394843 \text{ mol } O_2 \times \dfrac{1 \text{ mol } CO_2}{5 \text{ mol } O_2} = 0.00789686 \text{ mol } CO_2$

NO: $55.5 \text{ mL} \times \dfrac{1 \text{ L}}{1000 \text{ mL}} \times \dfrac{1 \text{ mol NO}}{22.414 \text{ L}} \times \dfrac{1 \text{ mol } CO_2}{5 \text{ mol NO}} = 0.000495226 \text{ mol } CO_2$

$0.000495226 \text{ mol } CO_2$ is the smallest yield, so NO is the limiting reagent.

$55.5 \text{ mL} \times \dfrac{1 \text{ L}}{1000 \text{ mL}} \times \dfrac{1 \text{ mol NO}}{22.414 \text{ L}} = 0.00247613 \text{ mol NO}$

reacted mol NO = 0.900 × mol NO = 0.900 × 0.00247613 mol NO = 0.00222852 mol NO
unreacted mol NO = 0.100 × mol NO = 0.100 × 0.00247613 mol NO = 0.000247613 mol NO

$$P = \frac{nRT}{V} = \frac{0.000247613 \text{ mol} \times 0.08206 \dfrac{\text{L} \cdot \text{atm}}{\text{mol} \cdot \text{K}} \times 275 \text{ K}}{2.0 \text{ L}} = 0.00279 \text{ atm NO remaining}$$

$0.00222852 \text{ mol NO} \times \dfrac{1 \text{ mol } CH_4}{5 \text{ mol NO}} = 0.000445704 \text{ mol } CH_4 \text{ reacted}$

$0.00691532 \text{ mol } CH_4 - 0.000445704 \text{ mol } CH_4 \text{ reacted} = 0.00646962 \text{ mol } CH_4$ unreacted then $PV = nRT$

Rearrange to solve for P. $P = \dfrac{nRT}{V} = \dfrac{0.00646962 \text{ mol} \times 0.08206 \dfrac{\text{L} \cdot \text{atm}}{\text{mol} \cdot \text{K}} \times 275 \text{ K}}{2.0 \text{ L}} = 0.0730 \text{ atm } CH_4 \text{ remaining}$

$0.00222852 \text{ mol NO} \times \dfrac{5 \text{ mol } O_2}{5 \text{ mol NO}} = 0.00222852 \text{ mol } O_2 \text{ reacted}$

$0.0394843 \text{ mol } O_2 - 0.00222852 \text{ mol } O_2 \text{ reacted} = 0.0372558 \text{ mol } O_2 \text{ unreacted}$

$$P = \frac{nRT}{V} = \frac{0.0372558 \text{ mol} \times 0.08206 \dfrac{\text{L} \cdot \text{atm}}{\text{mol} \cdot \text{K}} \times 275 \text{ K}}{2.0 \text{ L}} = 0.420 \text{ atm } O_2 \text{ remaining}$$

$0.00222852 \text{ mol NO} \times \dfrac{1 \text{ mol } CO_2}{5 \text{ mol NO}} = 0.000445704 \text{ mol } CO_2$

$$P = \frac{nRT}{V} = \frac{0.000445704 \text{ mol} \times 0.08206 \dfrac{\text{L} \cdot \text{atm}}{\text{mol} \cdot \text{K}} \times 275 \text{ K}}{2.0 \text{ L}} = 0.00503 \text{ atm } CO_2 \text{ produced}$$

$0.00222852 \text{ mol NO} \times \dfrac{1 \text{ mol } H_2O}{5 \text{ mol NO}} = 0.000445704 \text{ mol } H_2O$

$$P = \frac{nRT}{V} = \frac{0.000445704 \text{ mol} \times 0.08206 \dfrac{\text{L} \cdot \text{atm}}{\text{mol} \cdot \text{K}} \times 275 \text{ K}}{2.0 \text{ L}} = 0.00503 \text{ atm } H_2O \text{ produced}$$

$$0.00222852 \text{ mol NO} \times \frac{5 \text{ mol NO}_2}{5 \text{ mol NO}} = 0.00222852 \text{ mol NO}_2$$

$$P = \frac{nRT}{V} = \frac{0.00222852 \text{ mol} \times 0.08206 \frac{\text{L} \cdot \text{atm}}{\text{mol} \cdot \text{K}} \times 275 \text{ K}}{2.0 \text{ L}} = 0.0251 \text{ atm NO}_2 \text{ produced}$$

$$0.00222852 \text{ mol NO} \times \frac{2 \text{ mol OH}}{5 \text{ mol NO}} = 0.000891408 \text{ mol OH}$$

$$P = \frac{nRT}{V} = \frac{0.000891408 \text{ mol} \times 0.08206 \frac{\text{L} \cdot \text{atm}}{\text{mol} \cdot \text{K}} \times 275 \text{ K}}{2.0 \text{ L}} = 0.0101 \text{ atm OH produced}$$

$$P_{\text{Total}} = \sum P$$
$$= 0.0730 \text{ atm} + 0.420 \text{ atm} + 0.00279 \text{ atm} + 0.00503 \text{ atm} + 0.00503 \text{ atm} + 0.0251 \text{ atm} + 0.0101 \text{ atm}$$
$$= 0.541 \text{ atm}$$

Check: The units (atm) are correct. The magnitude of the answers is reasonable. The limiting reagent has the lowest pressure. The product pressures are in line with the ratios of the stoichiometric coefficients.

11.136 **Given:** He and air **Find:** % He diffused through balloon wall

Conceptual Plan: $\mathcal{M}_{N_2}, \mathcal{M}_{O_2} \rightarrow \mathcal{M}_{\text{air}}$ **then** $\mathcal{M}_{\text{air}}, \mathcal{M}_{\text{He}}$, **% air diffused** $\rightarrow$ **% He diffused**

$$\mathcal{M}_{\text{air}} = \chi_{N_2} \mathcal{M}_{N_2} + \chi_{O_2} \mathcal{M}_{O_2} \qquad \frac{\text{Rate}_{\text{He}}}{\text{Rate}_{\text{air}}} = \sqrt{\frac{\mathcal{M}_{\text{air}}}{\mathcal{M}_{\text{He}}}}$$

Solution: $\mathcal{M}_{\text{air}} = \chi_{N_2} \mathcal{M}_{N_2} + \chi_{O_2} \mathcal{M}_{O_2} = \left(\frac{4}{5} \times 28.02 \text{ g/mol}\right) + \left(\frac{1}{5} \times 32.00 \text{ g/mol}\right) = 28.82 \text{ g/mol}$

$\frac{\text{Rate}_{\text{He}}}{\text{Rate}_{\text{air}}} = \sqrt{\frac{\mathcal{M}_{\text{air}}}{\mathcal{M}_{\text{He}}}}$ Because rate α % diffused, substitute % diffused for rate and rearrange to solve for % He diffused. % He diffused = % air diffused$\sqrt{\frac{\mathcal{M}_{\text{air}}}{\mathcal{M}_{\text{He}}}} = 5.0\%\sqrt{\frac{28.82 \text{ g/mol}}{4.003 \text{ g/mol}}} = 13\%$

Check: The units (%) are correct. The magnitude of the answer (> 5) makes sense because He is lighter, so it has the higher diffusion rate.

11.137 **Given:** $P_{CH_4} + P_{C_2H_4} = 0.53$ atm and $P_{CO_2} + P_{H_2O} = 2.2$ atm **Find:** χ_{CH_4}

Conceptual Plan: Write balanced reactions to determine change in moles of gas for CH_4 and C_2H_6.

$2\ CH_4(g) + 4\ O_2(g) \rightarrow 4\ H_2O(g) + 2\ CO_2(g)$ and $2\ C_2H_6(g) + 7\ O_2(g) \rightarrow 6\ H_2O(g) + 4\ CO_2(g)$; thus, $\frac{6 \text{ mol gases}}{2 \text{ mol CH}_4}$ $\frac{10 \text{ mol gases}}{2 \text{ mol C}_2\text{H}_6}$

Write expression for final pressure, substituting in data given $\rightarrow$ χ_{CH_4}.

$$\chi_{CH_4} = \frac{n_{CH_4}}{n_{CH_4} + n_{C_2H_6}} \text{ and } \chi_{C_2H_6} = 1 - \chi_{CH_4}$$

$$P_{CH_4} = \chi_{CH_4} P_{\text{Total}} \qquad P_{C_2H_6} = \chi_{C_2H_6} P_{\text{Total}} \qquad P_{\text{Final}} = \left(\chi_{CH_4} P_{\text{Total}} \times \frac{6 \text{ mol gases}}{2 \text{ mol CH}_4}\right) + \left((1 - \chi_{CH_4}) P_{\text{Total}} \times \frac{10 \text{ mol gases}}{2 \text{ mol C}_2\text{H}_6}\right)$$

Solution:

$$P_{\text{Final}} = \left(\chi_{CH_4} \times 0.53 \text{ atm} \times \frac{6 \text{ mol gases}}{2 \text{ mol CH}_4}\right) + \left((1 - \chi_{CH_4}) \times 0.53 \text{ atm} \times \frac{10 \text{ mol gases}}{2 \text{ mol C}_2\text{H}_6}\right) = 2.2 \text{ atm}$$

Substitute as above for $\chi_{C_2H_6}$, then solve for $\chi_{CH_4} = 0.42$.

Check: The units (none) are correct. The magnitude of the answer (0.42) makes sense because if it had been all methane, the final pressure would have been 1.59 atm and if it had been all ethane, the final pressure would have been 2.65 atm. Because we are closer to the latter pressure, we expect the mole fraction of methane to be less than 0.5.

11.138 **Given:** $P_{C_2H_2} = 7.8$ kPa initially, $P_{C_2H_2} + P_{C_6H_6} = 3.9$ kPa **Find:** fraction of C_2H_2 reacted

Conceptual Plan: Write balanced reaction to determine change in moles of gas.

$3\ C_2H_2(g) \rightarrow C_6H_6(g)$; thus, $\frac{1 \text{ mol C}_6\text{H}_6}{3 \text{ mol C}_2\text{H}_2 \text{ reacted}}$ Because $P_{C_2H_2} \alpha\ n_{C_2H_2}$, the pressure will drop 2 kPa for every 3 kPa of ethylene that reacts.

$P_{initial}$, $P_{final} \rightarrow P_{drop}$ write expression for reacted $P_{C_2H_2}$, then

$$P_{drop} = P_{initial} - P_{final} \qquad \text{reacted } P_{C_2H_2} = \Delta P \frac{3 \text{ kPa } C_2H_2 \text{ reacted}}{2 \text{ kPa pressure drop}}$$

reacted $P_{C_2H_2}$, initial $P_{C_2H_2} \rightarrow$ % C_2H_2 reacted

$$\% \ C_2H_2 \text{ reacted} = \frac{\text{reacted } P_{C_2H_2}}{\text{initial } P_{C_2H_2}} \times 100\%$$

Solution: $P_{drop} = P_{initial} - P_{final} = 7.8 \text{ kPa} - 3.9 \text{ kPa} = 3.9 \text{ kPa}$

$$\text{reacted } P_{C_2H_2} = \Delta P \frac{3 \text{ kPa } C_2H_2 \text{ reacted}}{2 \text{ kPa pressure drop}} = 3.9 \text{ kPa} \times \frac{3 \text{ kPa } C_2H_2 \text{ reacted}}{2 \text{ kPa pressure drop}} = 5.85 \text{ kPa}$$

$$\% \ C_2H_2 \text{ reacted} = \frac{\text{reacted } P_{C_2H_2}}{\text{initial } P_{C_2H_2}} \times 100\% = \frac{5.85 \text{ kPa}}{7.8 \text{ kPa}} \times 100\% = 75\%$$

Check: The units (%) are correct. The magnitude of the answer (75) makes sense because if all of the ethylene had reacted, the final pressure would have been 2.6 kPa. Because we are most of the way to that, we expect the amount reacted to be higher than 50%.

Conceptual Problems

11.139 Because the passengers have more mass than the balloon, they have more momentum than the balloon. The passengers will continue to travel in their original direction longer. The car is slowing, so the relative position of the passengers is to move forward and the balloon to move backward. The opposite happens upon acceleration.

11.140 If a liquid is ten times denser than water, the force needed to move it will be ten times greater per unit volume. This means that the straw will be 1/10 the maximum length of a straw that can be used for water.

11.141 B is the limiting reactant (2.0 L of B requires 1.0 L A to completely react). The final container will have 0.5 L A and 2.0 L C, so the final volume will be 2.5 L. The change will be $[(2.5 \text{ L}/3.5 \text{ L}) \times 100\%] - 100\% = -29\%$.

11.142 Because each gas will occupy 22.414 L/mole at STP and we have 2 moles of gas, we will have a volume of 44.828 L.

11.143
(a) False—All gases have the same average kinetic energy at the same temperature.
(b) False—The gases will have the same partial pressures because we have the same number of moles of each.
(c) False—The average velocity of the B molecules will be less than that of the A molecules because the Bs are heavier.
(d) True—Because B molecules are heavier, they will contribute more to the density $(d = m/V)$.

11.144 Br_2 would deviate the most from ideal behavior because it is the largest of the three.

11.145 When the volume of a gas is cut in half, the pressure doubles. When the temperature of a gas in kelvins doubles, the pressure doubles. The next effect is that the pressure increases by a factor of four.

11.146 Because He has the lowest molar mass, it will have the most number of moles and the greatest volume.

11.147 Because the velocity is inversely proportional to the molar mass, the tails on the helium are $\sqrt{20/4} = {\sim}2.2$ times as long as those for neon and $\sqrt{84/4} = {\sim}4.6$ times as long as those for krypton.

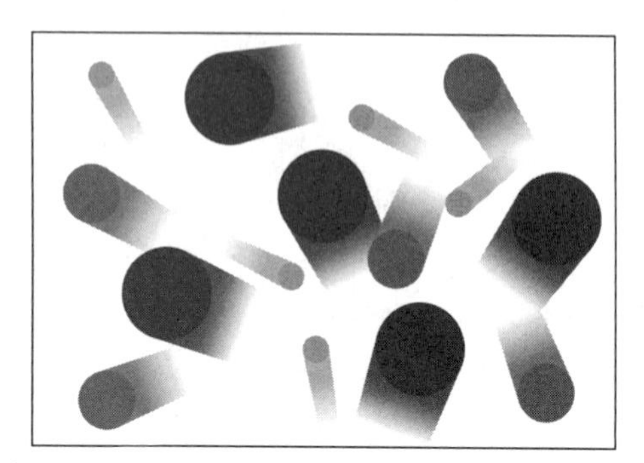

smallest particles = He

medium-sized particles = Ne

largest = Kr

12 Liquids, Solids, and Intermolecular Forces

Review Questions

12.1 Ethanol and dimethyl ether are isomers—they have the same chemical formula, C_2H_6O, but are different compounds. In ethanol, the oxygen atom is bonded to a carbon atom and a hydrogen atom. In dimethyl ether, the oxygen atom is bonded to two carbon atoms. The structures of these two molecules are different, and structure determines properties.

12.2 Intermolecular forces are important because they hold many liquids and solids—water and ice, for example—together. These intermolecular forces determine many of the physical properties of a substance. All living organisms depend on intermolecular forces for many physiological processes. Intermolecular forces are responsible for the very existence of the condensed phases.

12.3 The main properties of liquids are that liquids have much higher densities in comparison to gases and generally have lower densities in comparison to solids; liquids have an indefinite shape and assume the shape of their container; liquids have a definite volume; and liquids are not easily compressed.

12.4 The main properties of solids are that solids have much higher densities in comparison to gases (and usually higher densities than liquids); solids have a definite shape; they do not assume the shape of their container; solids have a definite volume, they are not easily compressed; and they may be crystalline (ordered) or amorphous (disordered).

12.5 Solids may be crystalline, in which case the atoms or molecules that compose them are arranged in a well-ordered three-dimensional array, or they may be amorphous, in which case the atoms or molecules that compose them have no long-range order.

12.6 One phase of matter can be transformed to another by changing the temperature, pressure, or both. A liquid can be converted to a gas by heating, and a gas can be condensed into a liquid by cooling. In general, increases in pressure favor the denser phase; so increasing the pressure of a gas sample can result in a transition to the liquid phase. A solid can be converted to a liquid by heating, and a liquid can be converted to a solid by cooling. In general, increases in pressure favor the denser phase (because atoms are pushed closer together); so increasing the pressure of most liquids can result in a transition to the solid phase.

12.7 Because the most molecular motion occurs in the gas phase and the least molecular motion occurs in the solid phase (atoms are pushed closer together), a substance will be converted from a solid to a liquid and then to a gas as the temperature increases. The strength of the intermolecular interactions is least in the gas phase because there are large distances between particles and they are moving very fast. Intermolecular forces are stronger in liquids and solids, where molecules are "touching" one another. The strength of the interactions in the condensed phases determines at what temperature the substance will melt and boil.

12.8 Intermolecular forces originate from the interactions between charges, partial charges, and temporary charges on molecules (or atoms and ions), much like bonding forces originate from interactions between charged particles in atoms.

12.9 Intermolecular forces, even the strongest ones, are generally much weaker than bonding forces. The reason for the relative weakness of intermolecular forces compared to bonding forces is related to Coulomb's law $\left(E = \frac{1}{4\pi\varepsilon_0}\frac{q_1 q_2}{r}\right)$. Bonding forces are the result of large charges (the charges on protons and electrons, q_1 and q_2) interacting at very close distances (r). Intermolecular forces are the result of smaller charges (as we will see in the following discussion) interacting at greater distances.

12.10 Dispersion forces (also called London forces) are the result of fluctuations in the electron distribution within molecules or atoms. Because all atoms and molecules have electrons, they all exhibit dispersion forces. The electrons in an atom or a molecule may, at any one instant, be unevenly distributed.

The magnitude of the dispersion force depends on how easily the electrons in the atom or molecule can move or polarize in response to an instantaneous dipole (a temporary change in charge distribution), which in turn depends on the size (or volume) of the electron cloud. A larger electron cloud results in a greater dispersion force because the electrons are held less tightly by the nucleus and can therefore polarize more easily. If all other variables are constant, the dispersion force increases with increasing molar mass because molecules or atoms of higher molar mass generally have more electrons dispersed over a greater volume. The shape of the molecules can also affect the magnitude of the dispersion forces. The larger the area of interaction between two molecules, the larger the dispersion forces.

12.11 The dipole–dipole force exists in all molecules that are polar. Polar molecules have permanent dipoles that interact with the permanent dipoles of neighboring molecules. The positive end of one permanent dipole is attracted to the negative end of another; this attraction is the dipole–dipole force.

12.12 Miscibility is the ability to mix without separating into two phases. The "rule of thumb" is like dissolves like. In general, polar liquids are miscible with other polar liquids but are not miscible with nonpolar liquids. Nonpolar liquids are miscible with other nonpolar liquids.

12.13 The hydrogen bond is a sort of super dipole–dipole force. Polar molecules containing hydrogen atoms bonded directly to fluorine, oxygen, or nitrogen exhibit an intermolecular force called hydrogen bonding. The large electronegativity difference between hydrogen and these electronegative elements means that the H atoms will have fairly large partial positive charges ($\delta+$), while the F, O, or N atoms will have fairly large partial negative charges ($\delta-$). In addition, because these atoms are all quite small, they can approach one another very closely. The result is a strong attraction between the hydrogen in each of these molecules and the F, O, or N on its neighbors, an attraction called a hydrogen bond.

12.14 The ion–dipole force occurs when an ionic compound is mixed with a polar compound, and it is especially important in aqueous solutions of ionic compounds. For example, when sodium chloride is mixed with water, the sodium and chloride ions interact with water molecules via ion–dipole forces. The positive sodium ions interact with the negative poles of water molecules, while the negative chloride ions interact with the positive poles. Ion–dipole forces are the strongest types of intermolecular forces discussed here and are responsible for the ability of ionic substances to form solutions with water.

12.15 Surface tension is the tendency of liquids to minimize their surface area. Molecules at the surface have relatively fewer neighbors with which to interact because there are no molecules above the surface. Consequently, molecules at the surface are inherently less stable—they have higher potential energy—than those in the interior. To increase the surface area of the liquid, some molecules from the interior must be moved to the surface, a process requiring energy. The surface tension of a liquid is the energy required to increase the surface area by a unit amount. Surface tension decreases with decreasing intermolecular forces.

12.16 Viscosity is the resistance of a liquid to flow. Viscosity is measured in a unit called the poise (P), defined as 1 g/cm • s. The centipoise (cP) is a convenient unit because the viscosity of water at room temperature is approximately one centipoise. Viscosity is greater in substances with stronger intermolecular forces because molecules are more strongly attracted to each other, preventing them from flowing around each other as freely. Viscosity also depends on molecular shape, increasing in longer molecules that can interact over a greater area and possibly become entangled. Viscosity increases with increasing molar mass (and therefore increasing magnitude of dispersion forces) and with increasing length (and therefore increasing potential for molecular entanglement). Viscosity also depends on temperature because thermal energy partially overcomes the intermolecular forces, allowing molecules to flow past each other more easily.

12.17 Capillary action is the ability of a liquid to flow against gravity up a narrow tube. Capillary action results from a combination of two forces: the attraction between molecules in a liquid, called cohesive forces, and the attraction between these molecules and the surface of the tube, called adhesive forces. The adhesive forces cause the liquid to spread out over the surface of the tube, while the cohesive forces cause the liquid to stay together. If the adhesive forces are greater than the cohesive forces (as is the case for water in a glass tube), the attraction to the surface draws the liquid up the tube while the cohesive forces pull along those molecules that are not in direct contact with the tube walls. The liquid rises up the tube until the force of gravity balances the capillary action—the thinner the tube, the higher the rise. If the adhesive forces are smaller than the cohesive forces (as is the case for liquid mercury), the liquid does not rise up the tube at all (and, in fact, will drop to a level below the level of the surrounding liquid).

12.18 Molecules are in constant motion. The higher the temperature, the greater the average energy of the collection of molecules. However, at any one time, some molecules will have more thermal energy than the average and some will have less. The molecules with the highest thermal energy have enough energy to break free from the surface—where molecules are held less tightly than in the interior due to fewer neighbor–neighbor interactions—and into the gas phase. This process is called vaporization, the phase transition from liquid to gas. The greater the temperature, the greater the rate of vaporization. Some of the water molecules in the gas phase, at the low end of the energy distribution curve for the gaseous molecules, can plunge back into the liquid and be captured by intermolecular forces. This process—the opposite of vaporization—is called condensation, the phase transition from gas to liquid.

12.19 The molecules that leave the liquid are at the high end of the energy curve—the most energetic. If no additional heat enters the liquid, the average energy of the entire collection of molecules goes down—much as the class average on an exam goes down when the highest-scoring students are eliminated. So vaporization is an endothermic process; it takes energy to vaporize the molecules in a liquid. Also, vaporization requires overcoming the intermolecular forces that hold liquids together. Because energy must be absorbed to pull the molecules apart, the process is endothermic. Condensation is the opposite process, so it must be exothermic. Also, gas particles have more energy than those in the liquid. It is the least energetic of these that condense, adding energy to the liquid.

12.20 The weaker the intermolecular forces are the more likely molecules will evaporate at a given temperature, making the liquid more volatile.

12.21 The heat of vaporization (ΔH_{vap}) is the amount of heat required to vaporize 1 mole of a liquid to a gas. The heat of vaporization of a liquid can be used to calculate the amount of heat energy required to vaporize a given mass of the liquid (or the amount of heat given off by the condensation of a given mass of liquid) and can be used to compare the volatility of two substances.

12.22 Molecules are in constant motion. Molecules leave the liquid for the gas phase, and gas-phase molecules condense to become a liquid. Dynamic equilibrium has been reached when the rate of condensation and the rate of vaporization become equal. Although condensation and vaporization continue, at equal rates, the concentration of water vapor above the liquid is constant. The pressure of a gas in dynamic equilibrium with its liquid is called its vapor pressure.

12.23 When a system in dynamic equilibrium is disturbed, the system responds so as to minimize the disturbance and return to a state of equilibrium.

12.24 The vapor pressure of a liquid increases with increasing temperature. However, the relationship is not linear; it is exponential. As the temperature of a liquid increases, the vapor increases more and more quickly. As the temperature is decreased, the vapor pressure decreases following this same relationship.

12.25 The boiling point of a liquid is the temperature at which its vapor pressure equals the external pressure. The normal boiling point of a liquid is the temperature at which its vapor pressure equals 1 atm.

12.26 The Clausius–Clapeyron equation is the relationship between vapor pressure and temperature. It can be expressed as $\ln P_{vap} = \frac{-\Delta H_{vap}}{R}\left(\frac{1}{T}\right) + \ln \beta$. In this expression, P_{vap} is the vapor pressure, β is a constant that depends on the gas, ΔH_{vap} is the heat of vaporization, R is the gas constant (8.314 J/mol K), and T is the temperature in kelvin. The Clausius–Clapeyron equation gives a linear relationship—not between the vapor pressure and the temperature (which have an exponential relationship), but between the natural log of the vapor pressure and the inverse of temperature.

The Clausius–Clapeyron equation leads to a convenient way to measure the heat of vaporization in the laboratory or to calculate the vapor pressure of a liquid at a temperature (if the heat of vaporization and a vapor pressure at one temperature are known).

12.27 As the temperature rises, more liquid vaporizes and the pressure within the container increases. As more and more gas is forced into the same amount of space, the density of the gas becomes higher and higher. At the same time, the increasing temperature causes the density of the liquid to become lower and lower. At the critical temperature, the meniscus between the liquid and gas disappears, and the gas and liquid phases commingle to form a supercritical fluid.

12.28 Sublimation is the phase transition from solid to gas without going through a liquid phase. A common example of sublimation is the carbon dioxide where "dry ice" converts from a solid to a gas without going through a "wet" (or liquid) phase.

12.29 Fusion, or melting, is the phase transition from solid to liquid. The term fusion is used for melting because if you heat several crystals of a solid, they will fuse into a continuous liquid upon melting. Fusion is endothermic because solids have less kinetic energy than liquids, so energy must be added to a solid to get it to melt.

12.30 The heat of fusion (ΔH_{fus}) is the amount of heat required to melt 1 mole of a solid. The heat of fusion of a solid, which is related to the strength of the intermolecular forces, can be used to calculate the amount of heat energy required to melt a given mass of the solid (or the amount of heat given off by the freezing of a given mass of liquid).

12.31 There are two horizontal lines (i.e., heat is added, but the temperature stays constant) in the heating curve because there are two endothermic phase changes. The heat that is added is used to change the phase from solid to liquid or from liquid to gas, and therefore there is no rise in temperature.

12.32 The slopes indicate how much heat is necessary to increase the temperature of the sample. The slope is proportional to $1/C_s$ (the specific heat capacity of the phase). The amount of molecular motion and the amount of intermolecular interactions are different in each phase, so the specific heat capacity and the slope are different for each phase and each substance.

Problem by Topic

Intermolecular Forces

12.33 ✓ (a) dispersion forces
✓ (b) dispersion forces, dipole–dipole forces, and hydrogen bonding
✓ (c) dispersion forces and dipole–dipole forces
✗ (d) dispersion forces

12.34 ✓ (a) dispersion forces
✓ (b) dispersion forces and dipole–dipole forces
✗ (c) dispersion forces
✓ (d) dispersion forces, dipole–dipole forces, and hydrogen bonding

12.35 ✓ (a) dispersion forces and dipole–dipole forces
✓ (b) dispersion forces, dipole–dipole forces, and hydrogen bonding
✓ (c) dispersion forces
✓ (d) dispersion forces

12.36 ✓ (a) dispersion forces and dipole–dipole forces
✓ (b) dispersion forces and dipole–dipole forces
✓ (c) dispersion forces, dipole–dipole forces, and hydrogen bonding
✓ (d) dispersion forces

12.37 (a) $CH_4 <$ (b) $CH_3CH_3 <$ (c) $CH_3CH_2Cl <$ (d) CH_3CH_2OH. The first two molecules only exhibit dispersion forces, so the boiling point increases with increasing molar mass. The third molecule also exhibits dipole–dipole forces, which are stronger than dispersion forces. The last molecule exhibits hydrogen bonding. Because these are by far the strongest intermolecular forces in this group, the last molecule has the highest boiling point.

12.38 (a) H_2S < (b) H_2Se < (c) H_2O. The first two molecules only exhibit dispersion forces and dipole–dipo so the boiling point increases with increasing molar mass. The third molecule also exhibits hydrogen Because these are by far the strongest intermolecular forces in this group, the last molecule has the high ing point.

12.39 (a) CH_3OH has the higher boiling point because it exhibits hydrogen bonding.
(b) CH_3CH_2OH has the higher boiling point because it exhibits hydrogen bonding.
(c) CH_3CH_3 has the higher boiling point because it has the larger molar mass.

12.40 (a) NH_3 has the higher boiling point because it exhibits hydrogen bonding.
(b) CS_2 has the higher boiling point because it has the larger molar mass.
(c) NO_2 has the higher boiling point because it exhibits dipole–dipole forces.

12.41 (a) Br_2 has the higher vapor pressure because it has the smaller molar mass.
(b) H_2S has the higher vapor pressure because it does not exhibit hydrogen bonding.
(c) PH_3 has the higher vapor pressure because it does not exhibit hydrogen bonding.

12.42 (a) CH_4 has the higher vapor pressure because it has the smaller molar mass and it does not exhibit dipole–dipole forces.
(b) CH_3OH has the higher vapor pressure because it has the smaller molar mass and both compounds exhibit hydrogen bonding.
(c) H_2CO has the higher vapor pressure because it has the smaller molar mass and it does not exhibit hydrogen bonding.

12.43 (a) This will not form a homogeneous solution because one is polar and one is nonpolar.
(b) This will form a homogeneous solution. There will be ion–dipole interactions between the K^+ and Cl^- ions and the water molecules. There will also be dispersion forces, dipole–dipole forces, and hydrogen bonding between the water molecules.
(c) This will form a homogeneous solution. Dispersion forces will be present among all of the molecules.
(d) This will form a homogeneous solution. There will be dispersion forces, dipole–dipole forces, and hydrogen bonding among all of the molecules.

12.44 (a) This will form a homogeneous solution. Only dispersion forces will be present.
(b) This will not form a homogeneous solution because one is polar and one is nonpolar.
(c) This will form a homogeneous solution. There will be ion–dipole interactions between the Li^+ and NO_3^- ions and the water molecules. There will also be dispersion forces, dipole–dipole forces, and hydrogen bonding between the water molecules.
(d) This will not form a homogeneous solution because one is polar and one is nonpolar.

Surface Tension, Viscosity, and Capillary Action

12.45 Water will have the higher surface tension because it exhibits hydrogen bonding, a strong intermolecular force. Acetone cannot form hydrogen bonds.

12.46 (a) Water "wets" surfaces that are capable of dipole–dipole interactions. The water will form strong adhesive forces with the surface when these dipole–dipole forces are present; so the water will spread to cover as much of the surface as possible. Water does not experience strong intermolecular forces with oil and other nonpolar surfaces. The water will bead up, maximizing the cohesive interactions, which involve strong hydrogen bonds. So water will bead up on surfaces that can exhibit only dispersion forces.
(b) Mercury will bead up on surfaces because it is not capable of forming strong intermolecular interactions (only dispersion forces).

12.47 Compound A will have the higher viscosity because it can interact with other molecules along the entire molecule. The more branched isomer has a smaller surface area, allowing for fewer interactions. Also, the molecule is very flexible, and the molecules can get tangled with each other.

ain polymers (long molecules made up of repeating structural units) that coil at low temperatures temperatures. At low temperatures, the coiled polymers—because of their compact shape—do not to the viscosity of the oil. As the temperature increases, however, the molecules unwind, and their ults in intermolecular forces and molecular entanglements that prevent the viscosity from decreasing as ould normally. The result is an oil whose viscosity is less temperature-dependent than it would be otherwing the same oil to be used over a wider range of temperatures.

ean glass tube, the water can generate strong adhesive interactions with the glass (due to the dipoles at the sur- of the glass). Water experiences adhesive forces with glass that are stronger than its cohesive forces, causing it to mb the surface of a glass tube. When grease or oil coats the glass, this interferes with the formation of these adhesive interactions with the glass because oils are nonpolar and cannot interact strongly with the dipoles in the water. Without experiencing these strong intermolecular forces with oil, the water's cohesive forces will be greater and water will be drawn away from the surface of the tube.

.50 Water can generate strong adhesive interactions with the glass (due to the dipoles at the surface of the glass), but hexane is nonpolar and cannot interact strongly with the glass surface.

Vaporization and Vapor Pressure

12.51 The water in the 12-cm-diameter beaker will evaporate more quickly because there is more surface area for the molecules to evaporate from. The vapor pressure will be the same in the two containers because the vapor pressure is the pressure of the gas when it is in dynamic equilibrium with the liquid (evaporation rate = condensation rate). The vapor pressure is dependent only on the substance and the temperature. The 12-cm-diameter container will reach this dynamic equilibrium faster.

12.52 The acetone will evaporate more quickly because it is not capable of forming hydrogen bonds, so the intermolecular forces are much weaker. This will result in a larger vapor pressure at the same temperature as the water.

12.53 The boiling point and higher heat of vaporization of oil are much higher than those of water, so it will not vaporize as quickly as the water. The evaporation of water cools your skin because evaporation is an endothermic process.

12.54 Water molecules have a lower kinetic energy at room temperature than at 100 °C. The heat of vaporization is the energy difference between the molecules in the liquid phase and the gas phase. Because the energy of the liquid is lower at room temperature, the energy difference that must be overcome to become steam is greater; so the heat of vaporization is greater.

12.55 **Given:** 915 kJ from candy bar, water $d = 1.00$ g/mL **Find:** L(H_2O) vaporized at 100.0 °C
Other: $\Delta H^\circ_{vap} = 40.7$ kJ/mol
Conceptual Plan: $q \rightarrow \text{mol } H_2O \rightarrow \text{g } H_2O \rightarrow \text{mL } H_2O \rightarrow \text{L } H_2O$

$$\frac{1 \text{ mol}}{40.7 \text{ kJ}} \quad \frac{18.02 \text{ g}}{1 \text{ mol}} \quad \frac{1.00 \text{ mL}}{1.00 \text{ g}} \quad \frac{1 \text{ L}}{1000 \text{ mL}}$$

Solution: $915 \text{ kJ} \times \frac{1 \text{ mol}}{40.7 \text{ kJ}} \times \frac{18.02 \text{ g}}{1 \text{ mol}} \times \frac{1.00 \text{ mL}}{1 \text{ g}} \times \frac{1 \text{ L}}{1000 \text{ mL}} = 0.405 \text{ L } H_2O$

Check: The units (L) are correct. The magnitude of the answer (< 1 L) makes physical sense because we are vaporizing about 22 moles of water.

12.56 **Given:** 100.0 mL water, $d = 1.00$ g/mL, heated to 100.0 °C **Find:** heat (kJ) to vaporize at 100.0 °C
Other: $\Delta H^\circ_{vap} = 40.7$ kJ/mol
Conceptual Plan: $\text{mL } H_2O \rightarrow \text{g } H_2O \rightarrow \text{mol } H_2O \rightarrow q$

$$\frac{1.00 \text{ g}}{1.00 \text{ mL}} \quad \frac{1 \text{ mol}}{18.02 \text{ g}} \quad \frac{40.7 \text{ kJ}}{1 \text{ mol}}$$

Solution: $100.0 \text{ mL} \times \frac{1.00 \text{ g}}{1.00 \text{ mL}} \times \frac{1 \text{ mol}}{18.02 \text{ g}} \times \frac{40.7 \text{ kJ}}{1 \text{ mol}} = 226 \text{ kJ}$

Check: The units (kJ) are correct. The magnitude of the answer (226 kJ) makes physical sense because we are vaporizing about 6 moles of water.

12.57 **Given:** 0.95 g water condenses on iron block 75.0 g at $T_i = 22\ °C$ **Find:** T_f (iron block)
Other: $\Delta H^\circ_{vap} = 44.0$ kJ/mol; $C_{Fe} = 0.449$ J/g·°C from text
Conceptual Plan: $g\ H_2O \rightarrow mol\ H_2O \rightarrow q_{H_2O}\ (kJ) \rightarrow q_{H_2O}\ (J) \rightarrow q_{Fe}$ **then** $q_{Fe}, m_{Fe}, T_i \rightarrow T_f$

$$\frac{1\ mol}{18.02\ g} \quad \frac{-44.0\ kJ}{1\ mol} \quad \frac{1000\ J}{1\ kJ} \quad -q_{H_2O} = q_{Fe} \quad q = mC_s(T_f - T_i)$$

Solution: $0.95\ g \times \frac{1\ mol}{18.02\ g} \times \frac{-44.0\ kJ}{1\ mol} \times \frac{1000\ J}{1\ kJ} = -2\underline{3}19.64\ J$ then $-q_{H_2O} = q_{Fe} = 2\underline{3}19.64\ J$ then

$q = mC_s(T_f - T_i)$. Rearrange to solve for T_f.

$$T_f = \frac{mC_sT_i + q}{mC_s} = \frac{\left(75.0\ g \times 0.449\ \frac{J}{g \cdot °C} \times 22\ °C\right) + 2\underline{3}19.64\ J}{75.0\ g \times 0.449\ \frac{J}{g \cdot °C}} = 91\ °C$$

Check: The units (°C) are correct. The temperature rose, which is consistent with heat being added to the block. The magnitude of the answer (91 °C) makes physical sense because even though we have $\sim\frac{1}{20}$ of a mole, the energy involved in condensation is very large.

12.58 **Given:** 1.15 g rubbing alcohol (C_3H_8O) evaporated from aluminum block 65.0 g at $T_i = 25\ °C$
Find: T_f (aluminum block) **Other:** $\Delta H^\circ_{vap} = 45.4$ kJ/mol; $C_{Al} = 0.903$ J/g·°C from text
Conceptual Plan: $g\ C_3H_8O \rightarrow mol\ C_3H_8O \rightarrow q_{C_3H_8O}\ (kJ) \rightarrow q_{C_3H_8O}\ (J) \rightarrow q_{Al}$ **then** $q_{Al}, m_{Fe}, T_i \rightarrow T_f$

$$\frac{1\ mol}{60.09\ g} \quad \frac{45.4\ kJ}{1\ mol} \quad \frac{1000\ J}{1\ kJ} \quad -q_{H_2O} = q_{Al} \quad q = mC_s(T_f - T_i)$$

Solution: $1.15\ g \times \frac{1\ mol}{60.09\ g} \times \frac{45.4\ kJ}{1\ mol} \times \frac{1000\ J}{1\ kJ} = 86\underline{8}.8634\ J$ then $-q_{H_2O} = q_{Al} = -86\underline{8}.8634\ J$ then

$q = mC_s(T_f - T_i)$. Rearrange to solve for T_f.

$$T_f = \frac{mC_sT_i + q}{mC_s} = \frac{\left(65.0\ g \times 0.903\ \frac{J}{g \cdot °C} \times 25\ °C\right) - 86\underline{8}.8634\ J}{65.0\ g \times 0.903\ \frac{J}{g \cdot °C}} = 1\underline{0}.19698\ °C = 10.\ °C$$

Check: The units (°C) are correct. The temperature dropped, which is consistent with heat being removed from the block. The magnitude of the answer (10 °C) makes physical sense because even though we have only a fraction of a mole, the energy involved in vaporization is very large.

12.59 **Given:** **Find:** $\Delta H^\circ_{vap}(NH_3)$ and normal boiling point

Temperature (K)	Vapor Pressure (torr)
200	65.3
210	134.3
220	255.7
230	456.0
235	597.0

Conceptual Plan: To find the heat of vaporization, use Excel or similar software to make a plot of the natural log of vapor pressure (ln *P*) as a function of the inverse of the temperature in K (1/*T*). Then fit the points to a line and determine the slope of the line. Because the *slope* = $-\Delta H_{vap}/R$, we find the heat of vaporization as follows:
$slope = -\Delta H_{vap}/R \rightarrow \Delta H_{vap} = -slope \times R$ **then J → kJ**

$$\frac{1\ kJ}{1000\ J}$$

For the normal boiling point, use the equation of the best-fit line, substitute 760 torr for the pressure, and calculate the temperature.
Solution: Data was plotted in Excel.
The slope of the best-fitting line is $-29\underline{6}9.9$ K.

$$\Delta H_{vap} = -slope \times R = -(-29\underline{6}9.9\ K) \times \frac{8.314\ J}{K\ mol} = \frac{2.4\underline{6}917 \times 10^4\ J}{mol} \times \frac{1\ kJ}{1000\ J} = 24.7\ \frac{kJ}{mol}$$

$\ln P = -2969.9 \text{ K}\left(\frac{1}{T}\right) + 19.036 \rightarrow$

$\ln 760 = -2969.9 \text{ K}\left(\frac{1}{T}\right) + 19.036 \rightarrow$

$2969.9 \text{ K}\left(\frac{1}{T}\right) = 19.036 - 6.63332 \rightarrow$

$T = \frac{2969.9 \text{ K}}{12.40268} = 239 \text{ K}$

Check: The units (kJ/mol) are correct. The magnitude of the answer (25 kJ/mol) is consistent with other values in the text.

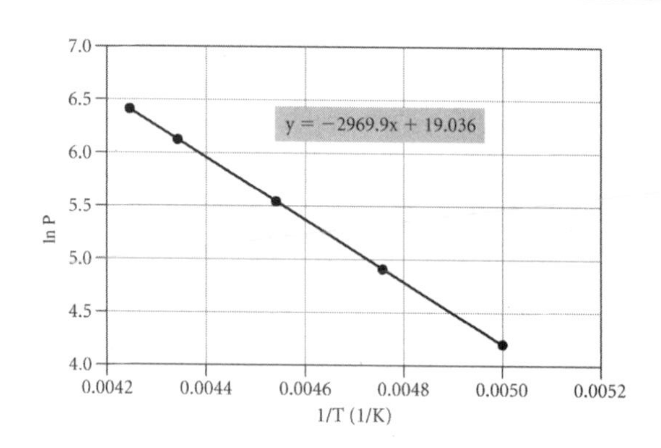

12.60 **Given:**

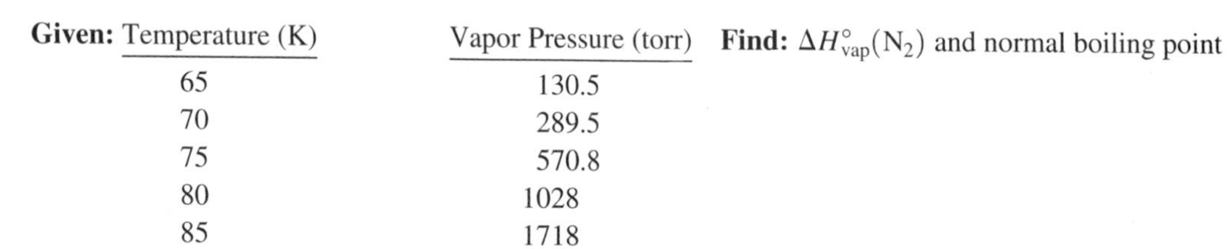

Temperature (K)	Vapor Pressure (torr)
65	130.5
70	289.5
75	570.8
80	1028
85	1718

Find: $\Delta H^\circ_{vap}(N_2)$ and normal boiling point

Conceptual Plan: To find the heat of vaporization, use Excel or similar software to make a plot of the natural log of vapor pressure (ln *P*) as a function of the inverse of the temperature in K (1/*T*). Then fit the points to a line and determine the slope of the line. Because the *slope* = $-\Delta H_{vap}/R$, we find the heat of vaporization as follows:

slope $= -\Delta H_{vap}/R \rightarrow \Delta H_{vap} = -slope \times R$ **then J → kJ**

$\frac{1 \text{ kJ}}{1000 \text{ J}}$

For the normal boiling point, use the equation of the best-fit line, substitute 760 torr for the pressure, and calculate the temperature.

Solution: Data was plotted in Excel.

The slope of the best-fitting line is −711.98 K.

$\Delta H_{vap} = -slope \times R = -(-711.98 \text{ K}) \times \frac{8.314 \text{ J}}{\text{K mol}} =$

$\frac{5.91940 \times 10^3 \text{ J}}{\text{mol}} \times \frac{1 \text{ kJ}}{1000 \text{ J}} = 5.92 \frac{\text{kJ}}{\text{mol}}$

$\ln P = -711.98 \text{ K}\left(\frac{1}{T}\right) + 15.833 \rightarrow$

$\ln 760 = -711.98 \text{ K}\left(\frac{1}{T}\right) + 15.833 \rightarrow$

$711.98 \text{ K}\left(\frac{1}{T}\right) = 15.833 - 6.63332 \rightarrow$

$T = \frac{711.98 \text{ K}}{9.19968} = 77.4 \text{ K}$

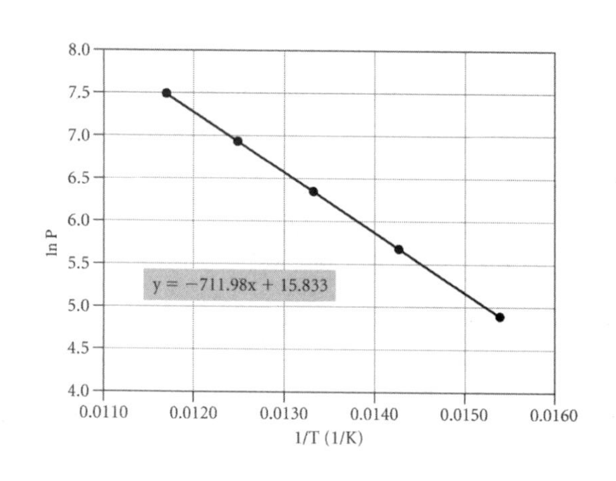

Check: The units (kJ/mol) are correct. The magnitude of the answer is lower than other values in the text. This is consistent with the fact that nitrogen boils at such a low temperature.

12.61 **Given:** ethanol, $\Delta H^\circ_{vap} = 38.56$ kJ/mol; normal boiling point = 78.4 °C **Find:** $P_{Ethanol}$ at 15 °C

Conceptual Plan: °C → K and kJ → J then $\Delta H^\circ_{vap}, T_1, P_1, T_2 \rightarrow P_2$

$K = °C + 273.15 \quad \frac{1000 \text{ J}}{1 \text{ kJ}} \qquad \ln\frac{P_2}{P_1} = \frac{-\Delta H_{vap}}{R}\left(\frac{1}{T_2} - \frac{1}{T_1}\right)$

Solution: $T_1 = 78.4 \text{ °C} + 273.15 = 351.6 \text{ K} \quad T_2 = 15 \text{ °C} + 273.15 = 288 \text{ K}$

$\frac{38.56\ \text{kJ}}{\text{mol}} \times \frac{1000\ \text{J}}{1\ \text{kJ}} = 3.856 \times 10^4 \frac{\text{J}}{\text{mol}}$; $P_1 = 760$ torr; $\ln\frac{P_2}{P_1} = \frac{-\Delta H_{vap}}{R}\left(\frac{1}{T_2} - \frac{1}{T_1}\right)$

Substitute values in equation. $\ln\frac{P_2}{760\ \text{torr}} = \frac{-3.856 \times 10^4 \frac{\text{J}}{\text{mol}}}{8.314 \frac{\text{J}}{\text{K}\cdot\text{mol}}}\left(\frac{1}{288\ \text{K}} - \frac{1}{351.6\ \text{K}}\right) = -2.91302 \rightarrow$

$\frac{P_2}{760\ \text{torr}} = e^{-2.91302} = 0.054311 \rightarrow P_2 = 0.054311 \times 760\ \text{torr} = 41\ \text{torr}$

Check: The units (torr) are correct. Because 15 °C is significantly below the boiling point, we expect the answer to be much less than 760 torr.

12.62 **Given:** benzene, $\Delta H^\circ_{vap} = 30.72$ kJ/mol; normal boiling point = 80.1 °C; $P_2 = 445$ torr **Find:** T_2
Conceptual Plan: °C → K and kJ → J then $\Delta H^\circ_{vap}, T_1, P_1, T_2 \rightarrow P_2$

$K = °C + 273.15$ $\frac{1000\ \text{J}}{1\ \text{kJ}}$ $\ln\frac{P_2}{P_1} = \frac{-\Delta H_{vap}}{R}\left(\frac{1}{T_2} - \frac{1}{T_1}\right)$

Solution: $T_1 = 80.1\ °\text{C} + 273.15 = 353.3\ \text{K}$ $\frac{30.72\ \text{kJ}}{\text{mol}} \times \frac{1000\ \text{J}}{1\ \text{kJ}} = 3.072 \times 10^4 \frac{\text{J}}{\text{mol}}$ $P_1 = 760$ torr

$P_2 = 445$ torr $\ln\frac{P_2}{P_1} = \frac{-\Delta H_{vap}}{R}\left(\frac{1}{T_2} - \frac{1}{T_1}\right)$ Substitute values in equation.

$\ln\frac{445\ \text{torr}}{760\ \text{torr}} = \frac{-3.072 \times 10^4 \frac{\text{J}}{\text{mol}}}{8.314 \frac{\text{J}}{\text{K}\cdot\text{mol}}}\left(\frac{1}{T_2} - \frac{1}{353.3\ \text{K}}\right) \rightarrow -0.535244 = -3.694972 \times 10^3\ \text{K}\left(\frac{1}{T_2} - 0.002830456\right)$

$\rightarrow \frac{1.44857 \times 10^{-4}}{\text{K}} = \left(\frac{1}{T_2} - \frac{0.002830456}{\text{K}}\right) \rightarrow \frac{1}{T_2} = \frac{2.97531 \times 10^{-3}}{\text{K}} \rightarrow T_2 = 336.0990\ \text{K} = 63\ °\text{C}$

Check: The units (°C) are correct. Because the pressure is over half of 760 torr, we expect a temperature a little lower than the boiling point.

Sublimation and Fusion

12.63 **Given:** 65.8 g water freezes **Find:** energy released **Other:** $\Delta H^\circ_{fus} = 6.02$ kJ/mol from text
Conceptual Plan: g H_2O → mol H_2O → q_{H_2O} (kJ) → q_{H_2O} (J)

$\frac{1\ \text{mol}}{18.02\ \text{g}}$ $\frac{-6.02\ \text{kJ}}{1\ \text{mol}}$ $\frac{1000\ \text{J}}{1\ \text{kJ}}$

Solution: $65.8\ \text{g} \times \frac{1\ \text{mol}}{18.02\ \text{g}} \times \frac{-6.02\ \text{kJ}}{1\ \text{mol}} \times \frac{1000\ \text{J}}{1\ \text{kJ}} = -21{,}982\ \text{J} = 2.20 \times 10^4\ \text{J}$ so 2.20×10^4 J released or 22.0 kJ released

Check: The units (J) are correct. The magnitude (22,000 J) makes sense because we are freezing about 3 moles of water. Freezing is exothermic, so heat is released.

12.64 **Given:** 50.0 g dry ice (CO_2) sublimation **Find:** heat required **Other:** $\Delta H^\circ_{sub} = 32.3$ kJ/mol
Conceptual Plan: g dry ice → mol dry ice → $q_{\text{dry ice}}$ (kJ) → $q_{\text{dry ice}}$ (J)

$\frac{1\ \text{mol}}{44.01\ \text{g}}$ $\frac{32.3\ \text{kJ}}{1\ \text{mol}}$ $\frac{1000\ \text{J}}{1\ \text{kJ}}$

Solution: $50.0\ \text{g} \times \frac{1\ \text{mol}}{44.01\ \text{g}} \times \frac{32.3\ \text{kJ}}{1\ \text{mol}} \times \frac{1000\ \text{J}}{1\ \text{kJ}} = 36{,}696$ J required = 36,700 J required or 36.7 kJ absorbed

Check: The units (J) are correct. The magnitude (36,700 J) makes sense because we are subliming just over 1 mole of dry ice. Sublimation is endothermic, so heat is required.

12.65 **Given:** 8.5 g ice; 255 g water **Find:** ΔT of water **Other:** $\Delta H^\circ_{fus} = 6.0$ kJ/mol; $C_{s,\ \text{water}} = 4.18$ J/g·°C from text
Conceptual Plan: The first step is to calculate how much heat is removed from the water to melt the ice.
$q_{ice} = -q_{water}$ so g ice → mol ice → q_{fus}(kJ) → q_{fus}(J) → q_{water}(J) then $q, m, C_s \rightarrow \Delta T_1$

$\frac{1\ \text{mol}}{18.02\ \text{g}}$ $\frac{6.0\ \text{kJ}}{1\ \text{mol}}$ $\frac{1000\ \text{J}}{1\ \text{kJ}}$ $q_{water} = -q_{ice}$ $q = mC_s\Delta T_1$

Now we have slightly cooled water (at a temperature of T_1) in contact with 0.0 °C water, and we can calculate a second temperature drop of the water due to mixing of the water that was ice with the water that was initially room temperature; so $q_{ice} = -q_{water}$ with $m, C_s \rightarrow \Delta T_2$ with $\Delta T_1, \Delta T_2 \rightarrow \Delta T_{Total}$.

$q = mC_s\Delta T_2$ *then set* $q_{ice} = -q_{water}$ $\Delta T_{Total} = \Delta T_1 + \Delta T_2$

Solution: $8.5\ g \times \frac{1\ mol}{18.02\ g} \times \frac{6.0\ kJ}{1\ mol} \times \frac{1000\ J}{1\ kJ} = 2.83176 \times 10^3\ J$, $q_{water} = -q_{ice} = -2.83176 \times 10^3\ J$

$q_{water} = mC_s\Delta T$ Rearrange to solve for ΔT. $\Delta T_1 = \frac{q}{mC_s} = \frac{-2.83176 \times 10^3\ J}{255\ g \times 4.18 \frac{J}{g \cdot °C}} = -2.6567\ °C$

$q = mC_s\Delta T$ Substitute values and set $q_{ice} = -q_{H_2O}$.

$q_{ice} = m_{ice}C_{s,\ ice}(T_f - T_{ice,\ i}) = 8.5\ g \times 2.09 \frac{J}{g \cdot °C} \times (T_f - 0.0\ °C) =$

$-q_{water} = -m_{water}C_{s,\ water}\Delta T_{water,\ 2} = -255\ g \times 4.18 \frac{J}{g \cdot °C} \times \Delta T_{water,\ 2} \rightarrow$

$17.765\ T_f = -1065.9\Delta T_{water,\ 2} = -1065.9(T_f - T_{f,\ 1})$ Rearrange to solve for T_f. $17.765\ T_f + 1065.9\ T_f = 1065.9\ T_{f,\ 1} \rightarrow$ $1083.665\ T_f = 1065.9\ T_{f,\ 1} \rightarrow T_f = 0.983607\ T_{f,\ 1}$ but $\Delta T_1 = (T_{f,\ 1} - T_{i,\ 1}) = -2.6567\ °C$, which says that $T_{f,\ 1} = T_{i,\ 1} - 2.6567\ °C$ and $\Delta T_{Total} = (T_f - T_{i,\ 1})$; so $\Delta T_{Total} = 0.983607\ T_{f,\ 1} - T_{i,\ 1} = 0.983607(T_{i,\ 1} - 2.6567\ °C) - T_{i,\ 1} = -2.6131\ °C - 0.016393\ T_{i,\ 1}$.

This implies that the larger the initial temperature of the water, the larger the temperature drop. If the initial temperature was 90 °C, the temperature drop would be 4.1 °C. If the initial temperature was 25 °C, the temperature drop would be 3.0 °C. If the initial temperature was 5 °C, the temperature drop would be 2.7 °C. This makes physical sense because the lower the initial temperature of the water, the less kinetic energy it initially has and the smaller the heat transfer from the water to the melted ice will be.

Check: The units (°C) are correct. The temperature drop from the melting of the ice is only 2.7 °C because the mass of the water is so much larger than that of the ice.

12.66 **Given:** 352 mL water, $T_i = 25\ °C$, $T_f = 5\ °C$, $d = 1.0\ g/mL$; ice $T_i = 0\ °C$, $T_f = 5\ °C$ **Find:** g (ice)
Other: $\Delta H°_{fus} = 6.02\ kJ/mol$; $C_{H_2O} = 4.18\ J/g \cdot °C$ from text
Conceptual Plan: mL $\rightarrow$ g then $m, C_s, T_i, T_f \rightarrow q_{water}(J) \rightarrow q_{ice}(J)$ then $q_{ice}, \Delta H°_{fus}, C_{s,\ H_2O}, T_i, T_f \rightarrow$ g(ice)

$\frac{1.0\ g}{1.0\ mL}$ $q = mC_s(T_f - T_i)$ $q_{ice} = -q_{water}$ $q_{ice} = mC_s(T_f - T_i) + m \times \frac{1\ mol}{18.02\ g} \times \frac{6.02\ kJ}{1\ mol} \times \frac{1000\ J}{1\ kJ}$

Solution: $352\ mL \times \frac{1.0\ g}{1.0\ mL} = 352\ g$

$q_{water} = m_{water}C_{s,\ water}(T_f - T_i) = 352\ g \times 4.18 \frac{J}{g \cdot °C} \times (5\ °C - 25\ °C) = -2.94272 \times 10^4\ J$

$q_{ice} = -q_{water} = 2.94272 \times 10^4\ J$ then $q_{ice} = mC_s(T_f - T_i) + m \times \frac{1\ mol}{18.02\ g} \times \frac{6.02\ kJ}{1\ mol} \times \frac{1000\ J}{1\ kJ}$

Substitute values. $2.94272 \times 10^4\ J = m \times 4.18 \frac{J}{g \cdot °C} \times (5\ °C - 0\ °C) + m \times \frac{334.07\ J}{1\ g}$. Rearrange to solve for m.

$$m = \frac{2.94272 \times 10^4\ J}{\frac{354.97\ J}{1\ g}} = 83\ g$$

Check: The units (g) are correct. Because the temperature drop and the amount of water we want to cool are large, the mass seems reasonable.

12.67 **Given:** 10.0 g ice $T_i = -10.0\ °C$ to steam at $T_f = 110.0\ °C$ **Find:** heat required (kJ)
Other: $\Delta H°_{fus} = 6.02\ kJ/mol$; $\Delta H°_{vap} = 40.7\ kJ/mol$; $C_{s,\ ice} = 2.09\ J/g \cdot °C$; $C_{s,\ water} = 4.18\ J/g \cdot °C$; $C_{s,\ steam} = 2.01\ J/g \cdot °C$
Conceptual Plan: Follow the heating curve in Figure 12.23. $q_{Total} = q_1 + q_2 + q_3 + q_4 + q_5$ where q_1, q_3, and q_5 are heating of a single phase then J $\rightarrow$ kJ and q_2 and q_4 are phase transitions.

$q = mC_s(T_f - T_i)$ $\frac{1\ kJ}{1000\ J}$ $q = m \times \frac{1\ mol}{18.02\ g} \times \frac{\Delta H}{1\ mol}$

Solution:

Heating ice from −10.0 °C to 0 °C:

$q_1 = m_{ice}C_{s,\,ice}(T_{ice,\,f} - T_{ice,\,i}) = 10.0\,\cancel{g} \times 2.09\,\frac{J}{\cancel{g}\cdot\cancel{°C}} \times (0.0\,\cancel{°C} - (-10.0\,\cancel{°C}) = 209\,\cancel{J} \times \frac{1\,kJ}{1000\,\cancel{J}} = 0.209\,kJ$

Melting the ice—phase change:

$q_2 = m \times \frac{1\,\cancel{mol}}{18.02\,g} \times \frac{\Delta H_{fus}}{1\,\cancel{mol}} = 10.0\,\cancel{g} \times \frac{1\,\cancel{mol}}{18.02\,\cancel{g}} \times \frac{6.02\,kJ}{1\,\cancel{mol}} = 3.3\underline{4}1\,kJ$

Heating water to100 °C:

$q_3 = m_{water}C_{s,\,water}(T_{water,\,f} - T_{water,\,i}) = 10.0\,\cancel{g} \times 4.18\,\frac{J}{\cancel{g}\cdot\cancel{°C}} \times (100.0\,\cancel{°C} - 0.0\,\cancel{°C}) = 4180\,\cancel{J} \times \frac{1\,kJ}{1000\,\cancel{J}} = 4.18\,kJ$

Converting water to steam—phase change:

$q_4 = m \times \frac{1\,\cancel{mol}}{18.02\,g} \times \frac{\Delta H_{vap}}{1\,\cancel{mol}} = 10.0\,\cancel{g} \times \frac{1\,\cancel{mol}}{18.02\,\cancel{g}} \times \frac{40.7\,kJ}{1\,\cancel{mol}} = 22.\underline{5}86\,kJ$

Heating steam to 110 °C:

$q_5 = m_{steam}C_{s,\,steam}(T_{steam,\,f} - T_{steam,\,i}) = 10.0\,\cancel{g} \times 2.01\,\frac{J}{\cancel{g}\cdot\cancel{°C}} \times (110.0\,\cancel{°C} - 100.0\,\cancel{°C})$

$= 201\,\cancel{J} \times \frac{1\,kJ}{1000\,\cancel{J}} = 0.201\,kJ$

$q_{Total} = q_1 + q_2 + q_3 + q_4 + q_5 = 0.209\,kJ + 3.3\underline{4}1\,kJ + 4.18\,kJ + 22.\underline{5}86\,kJ + 0.201\,kJ = 30.5\,kJ$

Check: The units (kJ) are correct. The total amount of heat is dominated by the vaporization step. Because we have less than 1 mole, we expect less than 41 kJ.

12.68 **Given:** 1.00 mole steam $T_i = 145.0\,°C$ to ice at $T_f = -50.0\,°C$ **Find:** heat evolved (kJ) **Other:** $\Delta H°_{fus} = 6.02\,kJ/mol$; $\Delta H°_{vap} = 40.7\,kJ/mol$; $C_{s,\,ice} = 2.09\,J/g\cdot°C$; $C_{s,\,water} = 4.18\,J/g\cdot°C$; $C_{s,\,steam} = 2.09\,J/g\cdot°C$

Conceptual Plan: mol → g Follow the heating curve in Figure 12.33, but in reverse.

$\frac{18.02\,g}{1\,mol}$

$q_{Total} = q_1 + q_2 + q_3 + q_4 + q_5$ **where q_1, q_3, and q_5 are heating of a single phase then J → kJ**

$q = mC_s(T_f - T_i)$ $\frac{1\,kJ}{1000\,J}$

and q_2 and q_4 are phase transitions.

$q = n \times \frac{\Delta H}{1\,mol}$

Solution: $1.00\,\cancel{mol} \times \frac{18.02\,g}{1\,\cancel{mol}} = 18.\underline{0}2\,g$

$q_1 = m_{steam}C_{s,\,steam}(T_{steam,\,f} - T_{steam,\,i}) = 18.\underline{0}2\,\cancel{g} \times 2.09\,\frac{J}{\cancel{g}\cdot\cancel{°C}} \times (100.0\,\cancel{°C} - 145.0\,\cancel{°C}) = -16\underline{9}3.84\,\cancel{J} \times \frac{1\,kJ}{1000\,\cancel{J}}$

$= -1.6\underline{9}4\,kJ$

Conversion of steam to water—phase change:

$q_2 = n \times \frac{-\Delta H_{vap}}{1\,mol} = 1.00\,\cancel{mol} \times \frac{-40.7\,kJ}{1\,\cancel{mol}} = -40.7\,kJ$

Cooling water to 0 °C:

$q_3 = m_{water}C_{s,\,water}(T_{water,\,f} - T_{water,\,i}) = 18.\underline{0}2\,\cancel{g} \times 4.18\,\frac{J}{\cancel{g}\cdot\cancel{°C}} \times (0.0\,\cancel{°C} - 100.0\,\cancel{°C}) = -75\underline{3}2.36\,\cancel{J} \times \frac{1\,kJ}{1000\,\cancel{J}}$

$= -7.5\underline{3}236\,kJ$

Conversion of water to ice—phase change:

$q_4 = n \times \frac{-\Delta H_{fus}}{1\,mol} = 1.00\,\cancel{mol} \times \frac{-6.02\,kJ}{1\,\cancel{mol}} = -6.02\,kJ$

Cooling of ice to −50 °C:

$q_5 = m_{ice}C_{s,\,ice}(T_{ice,\,f} - T_{ice,\,i}) = 18.\underline{0}2\,\cancel{g} \times 2.09\,\frac{J}{\cancel{g}\cdot\cancel{°C}} \times (-50.0\,\cancel{°C} - 0.0\,\cancel{°C}) = -18\underline{8}3.09\,\cancel{J} \times \frac{1\,kJ}{1000\,\cancel{J}}$

$= -1.8\underline{8}20\,kJ$

$q_{Total} = q_1 + q_2 + q_3 + q_4 + q_5 = -1.694 \text{ kJ} - 40.7 \text{ kJ} - 7.53236 \text{ kJ} - 6.02 \text{ kJ} - 1.88309 \text{ kJ}$
$= -57.8 \text{ kJ}$ or 57.8 kJ released

Check: The units (kJ) are correct. The amount of heat is dominated by the vaporization step. Because we have exactly 1 mole, we expect more than 41 kJ.

The Uniqueness of Water

12.69 Water has a low molar mass (18.02 g/mol), yet it is a liquid at room temperature. Water's high boiling point for its molar mass can be understood by examining the structure of the water molecule. The bent geometry of the water molecule and the highly polar nature of the O—H bonds result in a molecule with a significant dipole moment. Water's two O—H bonds (hydrogen directly bonded to oxygen) allow a water molecule to form very strong hydrogen bonds with four other water molecules, resulting in a relatively high boiling point.

12.70 Water's high polarity also allows it to dissolve many other polar and ionic compounds—and even a number of nonpolar gases such as oxygen and carbon dioxide (by inducing a dipole moment in their molecules). Consequently, water is the main solvent in living organisms, transporting nutrients and other important compounds throughout the body. Water is also the main solvent of the environment, allowing aquatic animals, for example, to survive by breathing dissolved oxygen and allowing aquatic plants to survive by using dissolved carbon dioxide for photosynthesis.

12.71 Water has an exceptionally high specific heat capacity, which has a moderating effect on the climate of coastal cities. Also, its high ΔH_{vap} causes water evaporation and condensation to have a strong effect on temperature. A tremendous amount of heat can be stored in large bodies of water. Heat will be absorbed or released from large bodies of water preferentially over the land around it. In some cities, such as San Francisco, the daily fluctuation in temperature can be less than 10 °C. This same moderating effect occurs over the entire planet, two-thirds of which is covered by water. In other words, without water, the daily temperature fluctuations on our planet might be more like those on Mars, where temperature fluctuations of 63 °C (113 °F) have been measured between early morning and midday.

12.72 One significant difference between the phase diagram of water and that of other substances is the fusion curve for water, which has a negative slope. The fusion curve in the phase diagrams for most substances has a positive slope because increasing pressure favors the denser phase, which for most substances is the solid phase. This negative slope means that ice is less dense than liquid water, so ice floats. The solids sink in the liquids of most other substances. The frozen layer of ice at the surface of a winter lake insulates the water in the lake from further freezing. If this ice layer sank, it would kill bottom-dwelling aquatic life and possibly allow the lake to freeze solid, eliminating virtually all life in the lake.

Cumulative Problems

12.73 The general trend is that melting point increases with increasing molar mass. This is because the electrons of the larger molecules are held more loosely and a stronger dipole moment can be induced more easily. HF is the exception to the rule. It has a relatively high melting point due to strong intermolecular forces due to hydrogen bonding.

12.74 The general trend is that boiling point increases with increasing molar mass. This is because the electrons of the larger molecules are held more loosely and a stronger dipole moment can be induced more easily. H_2O is the exception to the rule. It has a relatively high boiling point due to strong intermolecular forces due to hydrogen bonding.

12.75 **Given:** $P_{H_2O} = 23.76$ torr at 25 °C; 1.25 g water in 1.5 L container **Find:** m (H_2O) as liquid

Conceptual Plan: **°C → K and torr → atm then P, V, T → mol (g) → g (g) then g (g), g (l)$_i$ → g (l)$_f$**

$K = °C + 273.15$ $\qquad \frac{1 \text{ atm}}{760 \text{ torr}}$ $\qquad PV = nRT$ $\qquad \frac{18.02 \text{ g}}{1 \text{ mol}}$ $\qquad g(l)_f = g(l)_i - g(g)$

Solution: $T = 25 °C + 273.15 = 298 \text{ K} \quad 23.76 \text{ torr} \times \frac{1 \text{ atm}}{760 \text{ torr}} = 0.0312632 \text{ atm}$ then $PV = nRT$

Rearrange to solve for n. $n = \frac{PV}{RT} = \frac{0.0312632 \text{ atm} \times 1.5 \text{ L}}{0.08206 \frac{\text{L} \cdot \text{atm}}{\text{K} \cdot \text{mol}} \times 298 \text{ K}} = 0.00191768$ mol in the gas phase then

$0.00191768 \text{ mol} \times \frac{18.02 \text{ g}}{1 \text{ mol}} = 0.0345566$ g in the gas phase then

$g(l)_f = g(l)_i - g(g) = 1.25 \text{ g} - 0.0345566 \text{ g} = 1.22$ g remaining as liquid. Yes, there is 1.22 g of liquid.

Check: The units (g) are correct. The magnitude (1.2 g) is expected because very little material is expected to be in the gas phase.

12.76 **Given:** $P_{CCl_3F} = 856$ torr at 300 K; 11.5 g CCl_3F in 1.0 L container **Find:** m (CCl_3F) as liquid

Conceptual Plan: torr → atm then P, V, T → mol (g) → g (g) then g (g), g $(l)_i$ → g $(l)_f$

$\frac{1 \text{ atm}}{760 \text{ torr}}$ $\quad PV = nRT \quad$ $\frac{137.36 \text{ g}}{1 \text{ mol}}$ $\quad$ g $(l)_f$ = g$(l)_i$ − g(g)

Solution: $856 \text{ torr} \times \frac{1 \text{ atm}}{760 \text{ torr}} = 1.12\underline{6}32$ atm then $PV = nRT$ Rearrange to solve for n.

$$n = \frac{PV}{RT} = \frac{1.12\underline{6}32 \text{ atm} \times 1.0 \text{ L}}{0.08206 \frac{\text{L} \cdot \text{atm}}{\text{K} \cdot \text{mol}} \times 300 \text{ K}} = 0.04\underline{5}7519 \text{ mol in the gas phase then}$$

$$0.04\underline{5}7519 \text{ mol} \times \frac{137.36 \text{ g}}{1 \text{ mol}} = 6.\underline{2}845 \text{ g in the gas phase then}$$

$g(l)_f = g(l)_i - g(g) = 11.5 \text{ g} - 6.\underline{2}845 \text{ g} = 5.2$ g remaining as liquid. Yes, there is 5.2 g of liquid.

Check: The units (g) are correct. The magnitude (5 g) is expected because even at moderate pressures, little material is expected to be in the gas phase.

12.77 **Given:** ice: $T_1 = 0$ °C exactly, $m = 53.5$ g; water: $T_1 = 75$ °C, $m = 115$ g **Find:** T_f

Other: $\Delta H^\circ_{fus} = 6.0$ kJ/mol; $C_{water} = 4.18$ J/g·°C

Conceptual Plan: $q_{ice} = -q_{water}$ so g ice → mol ice → q_{fus}(kJ) → q_{fus}(J) → q_{water}(J) then

$\frac{1 \text{ mol}}{18.02 \text{ g}}$ $\quad$ $\frac{6.02 \text{ kJ}}{1 \text{ mol}}$ $\quad$ $\frac{1000 \text{ J}}{1 \text{ kJ}}$ $\quad$ $q_{water} = -q_{ice}$

q, m, C_s → ΔT then T_i, ΔT → T_2 now we have slightly cooled water in contact with 0.0 °C water

$q = mC_s\Delta T \quad \Delta T = T_2 - T_i$

so $q_{ice} = -q_{water}$ with m, C_s, T_i → T_f

$q = mC_s(T_f - T_i)$ *then set* $q_{ice} = -q_{water}$

Solution: $53.5 \text{ g} \times \frac{1 \text{ mol}}{18.02 \text{ g}} \times \frac{6.02 \text{ kJ}}{1 \text{ mol}} \times \frac{1000 \text{ J}}{1 \text{ kJ}} = 1.7\underline{8}828 \times 10^4 \text{ J}$ $q_{water} = -q_{ice} = -1.7\underline{8}828 \times 10^4$ J

$$q_{water} = mC_s\Delta T \quad \text{Rearrange to solve for } \Delta T.\ \Delta T = \frac{q}{mC_s} = \frac{-1.7\underline{8}828 \times 10^4 \text{ J}}{115 \text{ g} \times 4.18 \frac{\text{J}}{\text{g} \cdot ^\circ\text{C}}} = -37.\underline{2}016 \text{ °C then}$$

$\Delta T = T_2 - T_i$ Rearrange to solve for T_2. $T_2 = \Delta T + T_i = -37.\underline{2}016 \text{ °C} + 75 \text{ °C} = 3\underline{7}.798$ °C

$q = mC_s(T_f - T_i)$ Substitute values and set $q_{ice} = -q_{water}$.

$$q_{ice} = m_{ice}C_{s,\,ice}(T_f - T_{ice,\,i}) = 53.5 \text{ g} \times 4.18 \frac{\text{J}}{\text{g} \cdot ^\circ\text{C}} \times (T_f - 0.0 \text{ °C}) =$$

$$-q_{water} = -m_{water}C_{s,\,water}(T_f - T_{water,\,2}) = -115 \text{ g} \times 4.18 \frac{\text{J}}{\text{g} \cdot ^\circ\text{C}} \times (T_f - 3\underline{7}.798 \text{ °C})$$

Rearrange to solve for T_f.

$53.5T_f = -115(T_f - 3\underline{7}.798 \text{ °C}) \rightarrow 53.5\,T_f = -115\,T_f + 4\underline{3}46.8 \text{ °C} \rightarrow -4\underline{3}46.8 \text{ °C} = -16\underline{8}.5\,T_f$

$$\rightarrow T_f = \frac{-4\underline{3}46.8 \text{ °C}}{-16\underline{8}.5} = 2\underline{5}.8 \text{ °C} = 26 \text{ °C}$$

Check: The units (°C) are correct. The temperature is between the two initial temperatures. Because the ice mass is about half the water mass, we are not surprised that the temperature is closer to the original ice temperature.

12.78 **Given:** steam: $T_i = 100$ °C, $m = 0.552$ g; water : $T_i = 5.0$ °C, $m = 4.25$ g **Find:** T_f

Other: $\Delta H^\circ_{vap} = 40.7$ kJ/mol; $C_{water} = 4.18$ J/g·°C

Conceptual Plan: $q_{steam} = -q_{water}$ so g steam → mol steam → q_{vap}(kJ) → q_{fvap}(J) → q_{water}(J) then

$\frac{1 \text{ mol}}{18.02 \text{ g}}$ $\quad$ $\frac{-40.7 \text{ kJ}}{1 \text{ mol}}$ $\quad$ $\frac{1000 \text{ J}}{1 \text{ kJ}}$ $\quad$ $q_{steam} = -q_{ice}$

q, m, C_s → ΔT then T_i, ΔT → T_2 Now we have slightly warmed water in contact with 100.0 °C water

$q = mC_s\Delta T \quad \Delta T = T_2 - T_i$

so $q_{steam} = -q_{water}$ with $m, C_s, T_i \rightarrow T_f$

$q = mC_s(T_f - T_i)$ *then set* $q_{ice} = -q_{water}$

Solution: $0.552\ \text{g} \times \frac{1\ \text{mol}}{18.02\ \text{g}} \times \frac{-40.7\ \text{kJ}}{1\ \text{mol}} \times \frac{1000\ \text{J}}{1\ \text{kJ}} = -1.24744 \times 10^3\ \text{J}$

$q_{water} = -q_{steam} = -(-1.24744 \times 10^3\ \text{J})$ $q = mC_s\Delta T$ Rearrange to solve for ΔT.

$$\Delta T = \frac{q}{mC_s} = \frac{1.24744 \times 10^3\ \text{J}}{4.25\ \text{g} \times 4.18\ \frac{\text{J}}{\text{g} \cdot °\text{C}}} = 70.2190\ °\text{C} \text{ then } \Delta T = T_2 - T_i$$

Rearrange to solve for T_2. $T_2 = \Delta T + T_i = 70.2190\ °\text{C} + 5.0\ °\text{C} = 75.2190\ °\text{C}$
Recall that $q = mC_s(T_f - T_i)$ and substitute values and set $q_{steam} = -q_{water}$.

$$q_{steam} = m_{steam}C_{steam}(T_f - T_{steam,\,i}) = 0.552\ \text{g} \times 4.18\ \frac{\text{J}}{\text{g} \cdot °\text{C}} \times (T_f - 100.0\ °\text{C}) =$$

$$-q_{water} = -m_{water}C_{water}(T_f - T_{water,\,2}) = -4.25\ \text{g} \times 4.18\ \frac{\text{J}}{\text{g} \cdot °\text{C}} \times (T_f - 75.2190\ °\text{C})$$

Rearrange to solve for T_f.

$$0.552(T_f - 100.0\ °\text{C}) = -4.25(T_f - 75.2190\ °\text{C}) \rightarrow 0.552\ T_f - 55.2\ °\text{C} = -4.25\ T_f + 319.681\ °\text{C} \rightarrow$$

$$-374.881\ °\text{C} = -4.802\ T_f \rightarrow T_f = \frac{-374.881\ °\text{C}}{-4.802} = 78.1\ °\text{C}$$

Check: The units (°C) are correct. The temperature is between the two initial temperatures. Because so much heat is involved in the vaporization process, we are not surprised that the temperature is closer to the original steam temperature.

12.79 **Given:** 1 mole of methanol **Find:** Draw a heating curve beginning at 170 K and ending at 350 K
Other: melting point = 176 K, boiling point = 338 K, $\Delta H_{fus} = 2.2$ kJ/mol, $\Delta H_{vap} = 35.2$ kJ/mol, $C_{s,solid} = 105$ J/mol · K, $C_{s,liquid} = 81.3$ J/mol · K, $C_{s,gas} = 48$ J/mol · K
Conceptual Plan:
Calculate the temperature range of each phase:
$T_{Starting}, T_{Melting} \rightarrow \Delta T_{Solid}$ **and** $T_{Melting}, T_{Boiling} \rightarrow \Delta T_{Liquid}$ **and** $T_{Boiling}, T_{Ending} \rightarrow \Delta T_{Gas}$

$\Delta T_{Solid} = T_{Melting} - T_{Starting}$ $\Delta T_{Liquid} = T_{Boiling} - T_{Melting}$ $\Delta T_{Gas} = T_{Ending} - T_{Boiling}$

Because there is 1 mole of methanol, the heat for each phase change is simply the ΔH for that phase change.
Calculate the heat required for each step:
$\Delta T_{Solid}, C_{s,solid} \rightarrow q_{Solid}$ **and** $\Delta T_{Liquid}, C_{s,liquid} \rightarrow q_{Liquid}$ **and** $\Delta T_{Gas}, C_{s,gas} \rightarrow q_{Gas}$

$q = C_s \times \Delta T$ $q = C_s \times \Delta T$ $q = C_s \times \Delta T$

Finally, plot each of the segments.
Solution:
$\Delta T_{Solid} = T_{Melting} - T_{Starting} = 176\ \text{K} - 170\ \text{K} = 6\ \text{K}$ $\Delta T_{Liquid} = T_{Boiling} - T_{Melting} = 338\ \text{K} - 176\ \text{K} = 162\ \text{K}$ and $\Delta T_{Gas} = T_{Ending} - T_{Boiling} = 350\ \text{K} - 338\ \text{K} = 12\ \text{K}$

$q_{fus} = \Delta H_{fus} = 2.2\ \text{kJ/mol}$ and $q_{vap} = \Delta H_{vap} = 35.2\ \text{kJ/mol}$

$$q_{Solid} = C_{s,Solid} \times \Delta T_{Solid} = 105\frac{\text{J}}{\text{mol} \cdot \text{K}} \times 6\ \text{K} = 630\ \text{J/mol} = 0.63\ \text{kJ/mol}$$

$$q_{Liquid} = C_{s,Liquid} \times \Delta T_{Liquid} = 81.3\frac{\text{J}}{\text{mol} \cdot \text{K}} \times 162\ \text{K} = 13170.6\ \text{J/mol} = 13.1706\ \text{kJ/mol and}$$

$$q_{Gas} = C_{s,Gas} \times \Delta T_{Gas} = 48\frac{\text{J}}{\text{mol} \cdot \text{K}} \times 12\ \text{K} = 576\ \text{J/mol} = 0.576\ \text{kJ/mol}$$

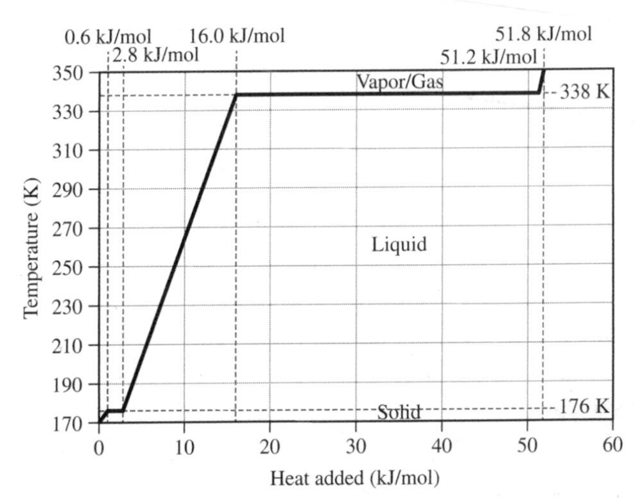

Check: The heating curve has a shape that is similar to that of water.

12.80 **Given:** 1 mole of benzene **Find:** Draw a heating curve beginning at 0 °C and ending at 100 °C.

Other: melting point = 5.4 °C, boiling point = 90.1 °C, ΔH_{fus} = 9.9 kJ/mol, ΔH_{vap} = 30.7 kJ/mol, $C_{s,solid}$ = 118 J/mol · °C, $C_{s,liquid}$ = 135 J/mol · °C, $C_{s,gas}$ = 104 J/mol · °C

Conceptual Plan:

Calculate the temperature range of each phase:

$T_{Starting}, T_{Melting} \rightarrow \Delta T_{Solid}$ **and** $T_{Melting}, T_{Boiling} \rightarrow \Delta T_{Liquid}$ **and** $T_{Boiling}, T_{Ending} \rightarrow \Delta T_{Gas}$

$\Delta T_{Solid} = T_{Melting} - T_{Starting}$ $\quad$ $\Delta T_{Liquid} = T_{Boiling} - T_{Melting}$ $\quad$ $\Delta T_{Gas} = T_{Ending} - T_{Boiling}$

Because there is 1 mole of benzene, the heat for each phase change is simply the ΔH for that phase change.

Calculate the heat required for each step:

$\Delta T_{Solid}, C_{s,solid} \rightarrow q_{Solid}$ **and** $\Delta T_{Liquid}, C_{s,liquid} \rightarrow q_{Liquid}$ **and** $\Delta T_{Gas}, C_{s,gas} \rightarrow q_{Gas}$

$q = C_s \times \Delta T$ $\quad$ $q = C_s \times \Delta T$ $\quad$ $q = C_s \times \Delta T$

Finally, plot each of the segments.

Solution:

$\Delta T_{Solid} = T_{Melting} - T_{Starting} = 5.4\,°C - 0.0\,°C = 5.4\,°C$

$\Delta T_{Liquid} = T_{Boiling} - T_{Melting} = 90.1\,°C - 5.4\,°C = 84.7\,°C$

$\Delta T_{Gas} = T_{Ending} - T_{Boiling} = 100.0\,°C - 90.1\,°C = 9.9\,°C$

$q_{fus} = \Delta H_{fus} = 9.9\,\text{kJ/mol}$ and $q_{vap} = \Delta H_{vap} = 30.7\,\text{kJ/mol}$

$$q_{Solid} = C_{s,Solid} \times \Delta T_{Solid} = 118\frac{\text{J}}{\text{mol}\cdot\cancel{°C}} \times 5.4\,\cancel{°C} = 6\underline{3}7.2\,\text{J/mol} = 0.6\underline{3}72\,\text{kJ/mol}$$

$$q_{Liquid} = C_{s,Liquid} \times \Delta T_{Liquid} = 135\frac{\text{J}}{\text{mol}\cdot\cancel{°C}} \times 84.7\,\cancel{°C} = 11\underline{4}34.5\,\text{J/mol} = 11.\underline{4}35\,\text{kJ/mol}$$

$$q_{Gas} = C_{s,Gas} \times \Delta T_{Gas} = 104\frac{\text{J}}{\text{mol}\cdot\cancel{°C}} \times 9.9\,\cancel{°C} = 1\underline{0}29.6\,\text{J/mol} = 1.\underline{0}296\,\text{kJ/mol}$$

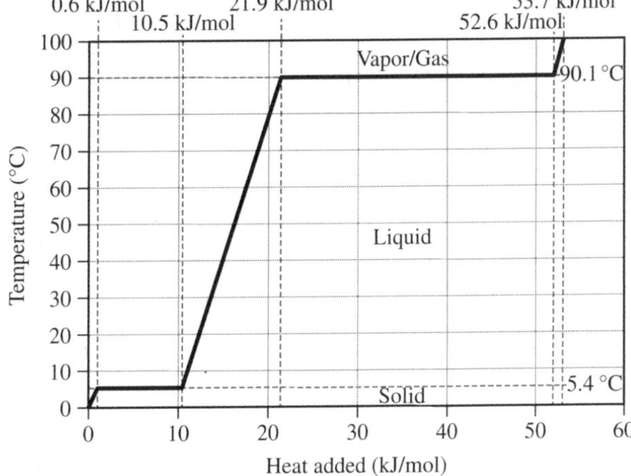

Check: The heating curve has a shape that is similar to that of water.

12.81 **Given:** home: 6.0 m × 10.0 m × 2.2 m; $T = 30\,°C$, $P_{H_2O} = 85\%$ of $P°_{H_2O}$ **Find:** $m(H_2O)$ removed
Other: $P°_{H_2O} = 31.8$ mmHg from text
Conceptual Plan: $l, w, h \rightarrow V(m^3) \rightarrow V(cm^3) \rightarrow V(L)$ **and** $P°_{H_2O} \rightarrow P_{H_2O}(mmHg) \rightarrow P_{H_2O}(atm)$ **and**

$$V = l\,w\,h \qquad \frac{(100\ cm)^3}{(1\ m)^3} \qquad \frac{1\ L}{1000\ cm^3} \qquad P_{H_2O} = 0.85\,P°_{H_2O} \qquad \frac{1\ atm}{760\ mmHg}$$

°C → K then $P, V, T \rightarrow$ **mol** $H_2O \rightarrow$ **g** H_2O

$$K = °C + 273.15 \qquad PV = nRT \qquad \frac{18.02\ g}{1\ mol}$$

Solution: $V = l\,w\,h = 6.0\ m \times 10.0\ m \times 2.2\ m = 132\ m^3 \times \dfrac{(100\ cm)^3}{(1\ m)^3} \times \dfrac{1\ L}{1000\ cm^3} = 1.32 \times 10^5\ L$

$P_{H_2O} = 0.85\,P°_{H_2O} = 0.85 \times 31.8\ mmHg \times \dfrac{1\ atm}{760\ mmHg} = 0.035566\ atm \quad T = 30\,°C + 273.15 = 303\ K$

then $PV = nRT$ Rearrange to solve for n.

$$n = \frac{PV}{RT} = \frac{0.035566\ atm \times 1.32 \times 10^5\ L}{0.08206\,\dfrac{L \cdot atm}{K \cdot mol} \times 303\ K} = 188.81\ mol \text{ then } 188.81\ mol \times \frac{18.02\ g}{1\ mol} = 3400\ g \text{ to remove}$$

Check: The units (g) are correct. The magnitude of the answer (3400 g) makes sense because the volume of the house is so large. We are removing almost 200 moles of water.

12.82 **Given:** flask with 0.55 g water at $T = 28\,°C$, $P°_{H_2O} = 28.36$ mmHg **Find:** minimum V of flask for all vapor
Conceptual Plan: g $H_2O \rightarrow$ **mol** H_2O **and** $P°_{H_2O}(mmHg) \rightarrow P_{H_2O}(atm)$ **and °C → K then** $P, n, T \rightarrow V$

$$\frac{1\ mol}{18.02\ g} \qquad \frac{1\ atm}{760\ mmHg} \qquad K = °C + 273.15 \qquad PV = nRT$$

Solution: $0.55\ g \times \dfrac{1\ mol}{18.02\ g} = 0.030539\ mol$, $P_{H_2O} = 28.63\ mmHg \times \dfrac{1\ atm}{760\ mmHg} = 0.03767105\ atm$

$T = 28\,°C + 273.15 = 301\ K$, then $PV = nRT$ Rearrange to solve for V.

$$V = \frac{nRT}{P} = \frac{0.030539\ mol \times 0.08206\,\dfrac{L \cdot atm}{K \cdot mol} \times 301\ K}{0.03767105\ atm} = 20.02\ L = 2.0 \times 10^1\ L$$

Check: The units (L) are correct. The magnitude of the answer (20 L) makes sense because we have about 1/30 mole and a pressure of about 1/30 atm and because at STP, 1 mole of a gas occupies 22 L.

Challenge Problems

12.83 $P_{Total} = P_A + P_B$. Looking at the data, we see that as the relative amount of A increases, the pressure decreases, because the vapor pressure of A is less than the vapor pressure of B. Because the solution is almost all A, we expect a vapor pressure just above 24 mmHg. In Chapter 14, we will more fully discuss this relationship and learn that

$P_A = \chi_A P°_A$, where $\chi_A = \dfrac{\text{mol A}}{\text{total mol} = \text{mol A} + \text{mol B}}$, $\chi_B = 1 - \chi_A$, and $P_B = \chi_B P°_B$. So

$$P_{Total} = P_A + P_B = \chi_A P°_A + (1 - \chi_A)P°_B = \frac{5\ mol}{5\ mol + 1\ mol} \times 24\ mmHg + \left(1 - \frac{5\ mol}{5\ mol + 1\ mol}\right) \times 36\ mmHg$$

$= 26$ mmHg

12.84 **Given:** butane (C_4H_{10}), $\Delta H°_{vap} = 22.44$ kJ/mol; normal boiling point $= -0.4\,°C$, 0.55 g; 250 mL flask
Find: amount of butane present as a liquid at $-22\,°C$ and at 25 °C
Conceptual Plan: At each temperature: °C → K and kJ → J then $\Delta H°_{vap}, T_1, P_1, T_2 \rightarrow P_2$

$$K = °C + 273.15 \qquad \frac{1000\ J}{1\ kJ}$$

$$\ln\frac{P_2}{P_1} = \frac{-\Delta H_{vap}}{R}\left(\frac{1}{T_2} - \frac{1}{T_1}\right)$$

and mL → L then $P_2, V, T_2 \rightarrow$ **mol** $\rightarrow g(g) \rightarrow g(l)$

$$\frac{1\ L}{1000\ mL} \qquad PV = nRT \qquad \frac{58.12\ g}{1\ mol} \qquad g(l) = g_{Total} - g(g)$$

Solution: $T_1 = -0.4\ °C + 273.15 = 272.8\ K$ $T_2 = -22\ °C + 273.15 = 251\ K$

$\frac{22.44\ kJ}{mol} \times \frac{1000\ J}{1\ kJ} = 2.244 \times 10^4 \frac{J}{mol}$ $P_1 = 1$ atm $\ln\frac{P_2}{P_1} = \frac{-\Delta H_{vap}}{R}\left(\frac{1}{T_2} - \frac{1}{T_1}\right)$

Substitute values in equation.

$$\ln\frac{P_2}{1\ atm} = \frac{-2.244 \times 10^4 \frac{J}{mol}}{8.314 \frac{J}{K \cdot mol}}\left(\frac{1}{251\ K} - \frac{1}{272.8\ K}\right) = -0.859313 \rightarrow \frac{P_2}{1\ atm} = e^{-0.859312} = 0.42345 \rightarrow$$

$P_2 = 0.42345 \times 1\ atm = 0.42345$ atm and $250\ mL \times \frac{1\ L}{1000\ mL} = 0.25$ L then $PV = nRT$

Rearrange to solve for n. $n = \frac{PV}{RT} = \frac{0.42345\ atm \times 0.25\ L}{0.08206 \frac{L \cdot atm}{K \cdot mol} \times 251\ K} = 0.0051397$ mol then

$0.0051397\ mol \times \frac{58.12\ g}{1\ mol} = 0.29872$ g in the gas phase then

$g(l) = g_{Total} - g(g) = 0.55\ g - 0.29872\ g = 0.25$ gas liquid at $-22\ °C$

$T_1 = -0.4\ °C + 273.15 = 272.8\ K$ $T_2 = 25\ °C + 273.15 = 298\ K$ $\frac{22.44\ kJ}{mol} \times \frac{1000\ J}{1\ kJ} = 2.244 \times 10^4 \frac{J}{mol}$

$P_1 = 1$ atm $\ln\frac{P_2}{P_1} = \frac{-\Delta H_{vap}}{R}\left(\frac{1}{T_2} - \frac{1}{T_1}\right)$ Substitute values in equation.

$$\ln\frac{P_2}{1\ atm} = \frac{-2.244 \times 10^4 \frac{J}{mol}}{8.314 \frac{J}{K \cdot mol}}\left(\frac{1}{298\ K} - \frac{1}{272.8\ K}\right) = 0.836667 \rightarrow \frac{P_2}{1\ atm} = e^{0.836667} = 2.30866 \rightarrow$$

$P_2 = 2.30866 \times 1\ atm = 2.30866$ atm and $250\ mL \times \frac{1\ L}{1000\ mL} = 0.25$ L then $PV = nRT$

Rearrange to solve for n. $n = \frac{PV}{RT} = \frac{2.30866\ atm \times 0.25\ L}{0.08206 \frac{L \cdot atm}{K \cdot mol} \times 298\ K} = 0.023602$ mol then

$0.023602\ mol \times \frac{58.12\ g}{1\ mol} = 1.3718$ g in the gas phase because the amount that can be put in the gas phase is greater than the available amount; no liquid is present at 25 °C.

Check: The units (g) are correct. The magnitude of the answers (0.25 g and 0.02 g) makes sense because we expect less butane to be in the liquid phase at a higher temperature. The second temperature is over the boiling point, so we expect more butane the gas phase.

12.85 Decreasing the pressure will decrease the temperature of liquid nitrogen. Because the nitrogen is boiling, its temperature must be constant at a given pressure. As the pressure decreases, the boiling point decreases, as does the temperature. Remember that vaporization is an endothermic process; so as the nitrogen vaporizes, it removes heat from the liquid, dropping its temperature. If the pressure drops below the pressure of the triple point, the phase change will shift from vaporization to sublimation and the liquid nitrogen will become solid.

12.86 **Given:** $\Delta H°_{fus} = -6.02$ kJ/mol @ 0.0 °C; $C_{s,\ liquid} = 7.52$ J/mol·K; $C_{s,\ ice} = 37.7$ J/mol·K
Find: ΔH_{fus} at -10.0 °C
Conceptual Plan: Assume exactly 1 mol H_2O for all calculations (report answer as kJ/mol). Because K = °C + 273.13, $\Delta T(°C) = \Delta T(K)$. Construct the following path:

step 2
Liquid @ 0.0 °C → solid @ 0.0 °C
↑ step 1 ↓ step 3
Liquid @ –10.0 °C → solid @ –10.0 °C
step 4

According to Hess's law, $\Delta H_1 + \Delta H_2 + \Delta H_3 = \Delta H_4 = \Delta H°_{fus}$ @ -10.0 °C
at constant P, $\Delta H = q$
for steps 1 and 3 $q = nC_s\Delta T$

for steps 1 and 3 J → kJ

$\frac{1\text{ kJ}}{1000\text{ J}}$

Solution:

$\Delta H_1 = q_1 = nC_s\Delta T = 1\text{ mol} \times 75.2\frac{\text{J}}{\text{mol}\cdot{}^\circ\text{C}} \times (0.00\,^\circ\text{C} - 10.0\,^\circ\text{C}) = +752\text{ J} \times \frac{1\text{ kJ}}{1000\text{ J}} = +0.752\text{ kJ}$

$\Delta H_2 = q_2 = n\Delta H = 1\text{ mol} \times -6.02\frac{\text{kJ}}{\text{mol}} = -6.02\text{ kJ}$

$\Delta H_3 = q_3 = nC_s\Delta T = 1\text{ mol} \times 37.7\frac{\text{J}}{\text{mol}\cdot{}^\circ\text{C}} \times (-10.00\,^\circ\text{C} - 0.0\,^\circ\text{C}) = -377\text{ J} \times \frac{1\text{ kJ}}{1000\text{ J}} = -0.377\text{ kJ}$ so

$\Delta H_4 = \Delta H_1 + \Delta H_2 + \Delta H_3 = 0.752\text{ kJ} - 6.02\text{ kJ} - 0.377\text{ kJ} = -5.65\text{ kJ/mol}$

Check: The units (kJ) are correct. We expect freezing to release less energy at –10 °C because we are below the normal freezing point.

12.87 **Given:** 1.00 L water, $T_i = 298$ K, $T_f = 373$ K-vapor; $P_{CH_4} = 1.00$ atm **Find:** $V(CH_4)$
Other: $\Delta H^\circ_{comb}(CH_4) = 890.4$ kJ/mol; $C_{water} = 75.2$ J/mol · K; $\Delta H^\circ_{vap}(H_2O) = 40.7$ kJ/mol, $d = 1.00$ g/mL
Conceptual Plan: L → mL → g → mol then heat liquid water: $n, C_s, T_i, T_f \rightarrow q_{1,\,water}$(J)

$\frac{1000\text{ mL}}{1\text{ L}}$ $\frac{1.00\text{ g}}{1.00\text{ mL}}$ $\frac{1\text{ mol}}{18.02\text{ g}}$ $q = mC_s(T_f - T_i)$

vaporize water: $n_{water}, \Delta H^\circ_{vap} \rightarrow q_{2,\,water}$(J) then calculate total heat $q_{1,\,water}, q_{2,\,water} \rightarrow q_{water}$(J) then

$q = n\Delta H$ $q_{1,\,water} + q_{2,\,water} = q_{water}$

q_{water}(J) → $-q_{CH_4,\,comb}$(J) → n_{CH_4} finally $n_{CH_4}, P, T \rightarrow V$

$q_{water} = -q_{CH_4,\,comb}$ $q = n\Delta H$ $PV = nRT$

Solution: $1.00\text{ L} \times \frac{1000\text{ mL}}{1\text{ L}} \times \frac{1.00\text{ g}}{1.00\text{ mL}} \times \frac{1\text{ mol}}{18.02\text{ g}} = 55.\underline{5}247\text{ mol H}_2\text{O}$

$q_{1,\,water} = n_{water}C_{water}(T_f - T_i) = 55.\underline{5}247\text{ mol} \times 75.2\frac{\text{J}}{\text{mol}\cdot\text{K}} \times (373\text{ K} - 298\text{ K}) = 3.1\underline{3}15931 \times 10^5\text{ J}$
$= 31\underline{3}.15931\text{ kJ}$

$q_{2,\,water} = n\Delta H = 55.\underline{5}247\text{ mol} \times 40.7\frac{\text{kJ}}{\text{mol}} = 2.2\underline{5}9855 \times 10^3\text{ kJ}$

$q_{1,\,water} + q_{2,\,water} = q_{water} = 31\underline{3}.15931\text{ kJ} + 2.2\underline{5}9855 \times 10^3\text{ kJ} = 2.5\underline{7}301431 \times 10^3\text{ kJ}$
$q_{water} = -q_{CH_4,\,comb} = 2.5\underline{7}301431 \times 10^3\text{ kJ}$ then $q_{CH_4,\,comb} = n\Delta H$ Rearrange to solve for n.

$n_{CH_4} = \frac{q_{CH_4}}{\Delta H_{CH_4,\,comb}} = \frac{-2.5\underline{7}301431 \times 10^3\text{ kJ}}{-890.4\frac{\text{kJ}}{\text{mol}}} = 2.8\underline{8}97286\text{ mol}$ then $PV = nRT$

Rearrange to solve for V. $V = \frac{nRT}{P} = \frac{2.8\underline{8}97286\text{ mol} \times 0.08206\frac{\text{L}\cdot\text{atm}}{\text{mol}\cdot\text{K}} \times 298\text{ K}}{1.00\text{ atm}} = 70.\underline{6}65076\text{ L} = 70.7\text{ L}$

Check: The units (L) are correct. The volume (71 L) is reasonable because we are using about 3 moles of methane.

12.88 $P_{Total} = P_{N_2} + P_{H_2O} + P_{ethanol}$
P_{N_2}: Use Boyle's law to calculate $P_1V_1 = P_2V_2$ Rearrange to solve for P_2.
$P_2 = P_1\frac{V_1}{V_2} = 1.0\text{ atm} \times \frac{1.0\text{ L}}{3.0\text{ L}} \times \frac{760\text{ mmHg}}{1\text{ atm}} = 2\underline{5}3.333\text{ mmHg}$

For water and ethanol, we need to calculate the pressure if all of the liquid were to vaporize in the 3.0 L apparatus. $PV = nRT$. Rearrange to solve for P.

$P = \frac{nRT}{V} = \frac{2.0\text{ g} \times \frac{1\text{ mol}}{18.02\text{ g}} \times 0.08206\frac{\text{L}\cdot\text{atm}}{\text{mol}\cdot\text{K}} \times \frac{760\text{ mmHg}}{1\text{ atm}} \times 308\text{ K}}{3.00\text{ L}} = 7\underline{1}0.640\text{ mmHg H}_2\text{O}$. Because this pressure is greater than the vapor pressure of water at this temperature, $P_{H_2O} = 42$ mmHg.

$$P = \frac{nRT}{V} = \frac{0.50\ \text{g} \times \frac{1\ \text{mol}}{46.07\ \text{g}} \times 0.08206 \frac{\text{L} \cdot \text{atm}}{\text{mol} \cdot \text{K}} \times \frac{760\ \text{mmHg}}{1\ \text{atm}} \times 308\ \text{K}}{3.00\ \text{L}} = 6\underline{9}.4937\ \text{mmHg ethanol}$$

Because this pressure is less than the vapor pressure of ethanol at this temperature (102 mmHg), all of the liquid will vaporize and $P_{\text{ethanol}} = 6\underline{9}.4937$ mmHg.

Finally, the total pressure is

$P_{\text{Total}} = P_{N_2} + P_{H_2O} + P_{\text{ethanol}} = 2\underline{5}3.333\ \text{mmHg} + 42\ \text{mmHg} + 6\underline{9}.4937\ \text{mmHg} = 3\underline{6}4.827\ \text{mmHg} = 360\ \text{mmHg}.$

Conceptual Problems

12.89 The melting of an ice cube in a glass of water will not raise or lower the level of the liquid in the glass as long as the ice is floating in the liquid. This is because the ice will displace a volume of water based on its mass. By the same logic, melting floating icebergs will not raise the ocean levels (assuming that the dissolved solids content, and thus the density, will not change when the icebergs melt). Dissolving ice formations that are supported by land will raise the ocean levels, just as pouring more water into a glass will raise the liquid level in the glass.

12.90 The water in a container with a larger surface area will evaporate more quickly because there is more surface area for the molecules to evaporate from. Vapor pressure is the pressure of the gas when it is in dynamic equilibrium with the liquid (evaporation rate = condensation rate). The vapor pressure is dependent only on the substance and the temperature. The larger the surface area, the more quickly it will reach this equilibrium state.

12.91 Substance A will have the larger change in vapor pressure with the same temperature change. To understand this, consider the Clausius–Clapeyron equation: $\ln \frac{P_2}{P_1} = \frac{-\Delta H_{\text{vap}}}{R}\left(\frac{1}{T_2} - \frac{1}{T_1}\right)$. If we use the same temperatures, we see that $\frac{P_2}{P_1} \alpha\, e^{-\Delta H_{\text{vap}}}$. So the smaller the heat of vaporization, the larger the final vapor pressure. We can also consider that the lower the heat of vaporization is, the easier the substance is converted from a liquid to a gas. This again leads to substance A having the larger change in vapor pressure.

12.92 $\Delta H_{\text{sub}} = \Delta H_{\text{fus}} + \Delta H_{\text{vap}}$ as long as the heats of fusion and vaporization are measured at the same temperatures.

12.93 The liquid segment will have the least steep slope because it takes the most kJ/mol to raise the temperature of the phase.

12.94 Water has an exceptionally high specific heat capacity, which has a moderating effect on the temperature of the root cellar. A large amount of heat can be stored in a large vat of water. The heat will be absorbed or released from the large bodies of water preferentially over the area around it. As the temperature of the air drops, the water will release heat, keeping the temperature more constant. If the temperature of the cellar falls enough to begin to freeze the water, the heat given off during the freezing will further protect the food in the cellar.

12.95 The heat of fusion of a substance is always smaller than the heat of vaporization because the number of interactions between particles that are broken is less in fusion than in vaporization. When we melt a solid, the particles have increased mobility but are still strongly interacting with other liquid particles. In vaporization, all of the interactions between particles must be broken (gas particles have essentially no intermolecular interactions), and the particles must absorb enough energy to move much more rapidly.

12.96
(a) In moving from line segment 5 to line segment 4, the water is condensing. Heat is being released, and q is negative.
(b) This line represents the heat absorbed when melting ice. The heat is converted to increase the kinetic energy of the water molecules as they become more mobile in the liquid phase.
(c) For other substances, the following things would change: the melting point, the boiling point, the length of segments, and the slopes of line segments 1, 3, and 5.

12.97 The oxygen atoms would be attracted to the $-CH_2$ groups through dipole–dipole forces.

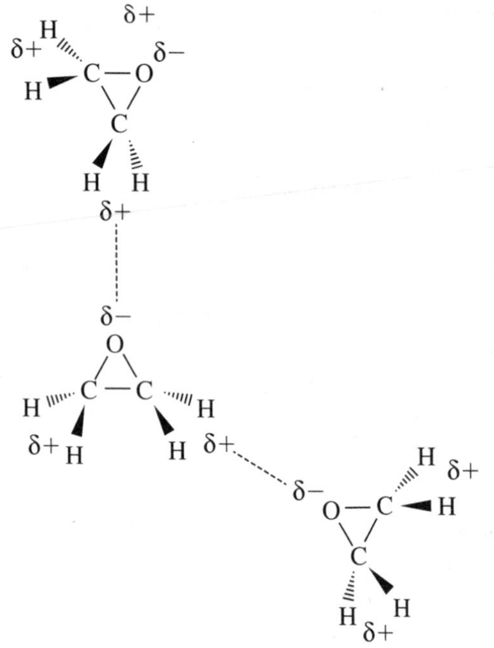

13 Phase Diagrams and Crystalline Solids

Review Questions

13.1 A phase diagram is a map of the phase of a substance as a function of pressure (on the y-axis) and temperature (on the x-axis).

13.2

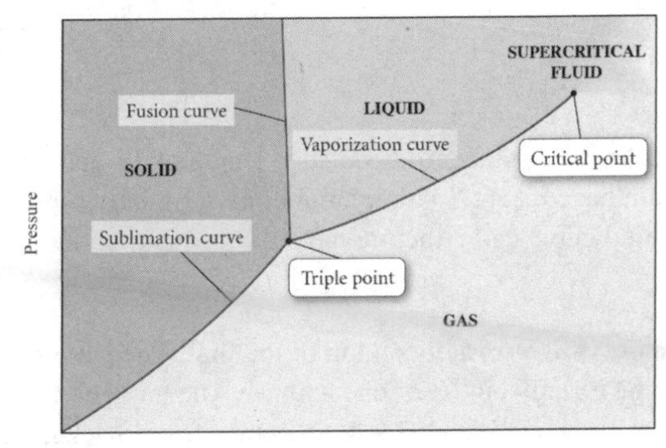

13.3 When a line is crossed in a phase diagram, it means that a phase transition has occurred.

13.4 X-ray crystallography is used to examine the structure of crystalline materials (materials that have an ordered array of atoms). The technique is based on the wave nature of X-rays. Atoms within crystal structures have spacings between them on the order of 10^2 pm. When light of similar wavelength (which happens to fall in the X-ray region of the electromagnetic spectrum) interacts with a sample, the waves are diffracted by the atoms. The waves interact with each other to constructively (they reinforce each other) or destructively (they cancel each other) interfere to form interference patterns or diffraction patterns.

13.5 The exact pattern of diffraction reveals the spacings between planes of atoms. How the waves interact depends on the path the waves take. Consider two planes of atoms within a crystalline lattice separated by a distance d, as shown in Figure 13.4. If two rays of light with wavelength λ that are initially in phase (i.e., the crests of one wave are aligned with the crests of the other) diffract from the two layers, the diffracted rays may interfere with each other constructively or destructively, depending on the difference between the path lengths traveled by each ray. If the difference between the two path lengths ($2a$) is an integral number (n) of wavelengths, the interference will be constructive, or $n\lambda = 2a$. Using trigonometry, we can see that the angle of reflection (θ) is related to the distance a and the separation between layers (d) by the following relation: $\sin\theta = a/d$. Rearranging and substituting into $n\lambda = 2a$, we get $n\lambda = 2\,d\sin\theta$, known as Bragg's law. For a given wavelength of light incident on atoms arranged in layers, we can measure the angle that produces constructive interference (which appears as a bright spot on the X-ray diffraction pattern) and then compute d, the distance between the atomic layers.

13.6 A crystalline lattice is the regular arrangements of atoms within a crystalline solid. The crystalline lattice can be represented by a small collection of atoms, ions, or molecules—a fundamental building block called the unit cell. When the unit cell is repeated over and over—like tiles in a floor or the pattern in a wallpaper design, but in three dimensions—the entire lattice can be reproduced.

13.7

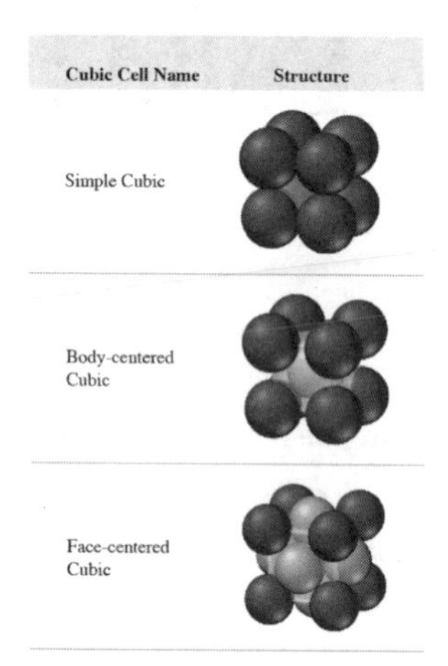

13.8 Atoms in a simple cubic cell structure have a coordination number of 6, an edge length of $2r$, and one atom in the unit cell. Atoms in a body-centered cubic cell structure have a coordination number of 8, an edge length of $4r/\sqrt{3}$, and two atoms in the unit cell. Atoms in a face-centered cubic cell structure have a coordination number of 12, an edge length of $2\sqrt{2}r$, and four atoms in the unit cell.

13.9 In hexagonal closest packing, the third layer of atoms aligns exactly on top of the first. The pattern from one layer to the next is ABAB... with alternating layers aligning exactly on top of one another. The unit cell for this crystal structure is not a cubic unit cell, but a hexagonal one, as shown in Figure 13.10. In cubic closest packing, the third layer of atoms is offset from the first. The pattern from one layer to the next is ABCABC... with every fourth layer aligning with the first. The unit cell for this crystal structure is the face-centered cubic unit cell.

13.10 The three types of solids are molecular solids, ionic solids, and atomic solids. Molecular solids are those solids whose composite units are molecules. The lattice sites in a crystalline molecular solid are therefore occupied by molecules. Ice (solid H_2O) and dry ice (solid CO_2) are examples of molecular solids. Molecular solids are held together by three kinds of intermolecular forces—dispersion forces, dipole–dipole forces, and hydrogen bonding. Ionic solids are those solids whose composite units are ions. Table salt (NaCl) and calcium fluoride (CaF_2) are good examples of ionic solids. Ionic solids are held together by the coulombic interactions that occur between the cations and anions occupying the lattice sites in the crystal, which are ionic bonds. Atomic solids are those solids whose composite units are individual atoms. Atomic solids can themselves be divided into three categories—nonbonding atomic solids, metallic atomic solids, and network covalent atomic solids—each held together by a different kind of force. Nonbonding atomic solids, which include only the noble gases in their solid form, are held together by relatively weak dispersion forces. Metallic atomic solids, such as iron and gold, are held together by metallic bonds, which in the simplest model are represented by the interaction of metal cations with a sea of electrons that surround them. Network covalent atomic solids, such as diamond, graphite, and silicon dioxide, are held together by covalent bonds.

13.11 The coordination number of the unit cell for an ionic compound represents the number of close cation–anion interactions. Because these interactions lower potential energy, the crystal structure of a particular ionic compound will be the one that maximizes the coordination number while accommodating both charge neutrality (each unit cell must be neutral) and the different sizes of the cations and anions that compose the particular compound. In general, the more similar the radii of the cation and the anion, the higher the possible coordination number.

13.12 Cesium chloride (CsCl) is a good example of an ionic compound containing cations and anions of similar size (Cs^+ radius $= 167$ pm; Cl^- radius $= 181$ pm). In the cesium chloride structure, the chloride ions occupy the lattice sites of a simple cubic cell, and one cesium ion lies in the very center of the cell, as shown in Figure 13.15. Notice that the cesium chloride unit cell contains one chloride anion ($8 \times 1/8 = 1$) and one cesium cation (the cesium ion in the middle belongs entirely to the unit cell) for a ratio of Cs to Cl of 1:1, as in the formula for the compound.

The crystal structure of sodium chloride must accommodate the more disproportionate sizes of Na^+ (radius $= 97$ pm) and Cl^- (radius $= 181$ pm). The larger chloride anion could theoretically fit many of the smaller sodium cations around it, but charge neutrality requires that each sodium cation be surrounded by an equal number of chloride anions. The structure that minimizes the energy is shown in Figure 13.16 and has a coordination number of 6 (each chloride anion is surrounded by six sodium cations, and vice versa). You can visualize this structure, called the rock salt structure, as chloride anions occupying the lattice sites of a face-centered cubic structure, with the smaller sodium cations occupying the holes between the anions. (Alternatively, you can visualize this structure as the sodium cations occupying the lattice sites of a face-centered cubic structure, with the larger chloride anions occupying the spaces between the cations.) Each unit cell contains four chloride anions $[(8 \times 1/8) + (6 \times \frac{1}{2}) = 4]$ and four sodium cations $(12 \times \frac{1}{4})$, resulting in a ratio of 1:1, as in the formula of the compound.

You can visualize this structure (shown in Figure 13.17), called the zinc blende structure, as sulfide anions occupying the lattice sites of a face-centered cubic structure, with the smaller zinc cations occupying four of the eight tetrahedral holes located directly beneath each corner atom. A tetrahedral hole is the empty space that lies in the center of a tetrahedral arrangement of four atoms. Each unit cell contains four sulfide anions $[(8 \times 1/8) + (6 \times \frac{1}{2}) = 4]$ and four zinc cations (each of the four zinc cations is completely contained within the unit cell), resulting in a ratio of 1:1, as in the formula of the compound.

13.13 The fluorite (CaF_2) structure is shown in Figure 13.18. You can visualize this structure as calcium cations occupying the lattice sites of a face-centered cubic structure with the larger fluoride anions occupying all eight of the tetrahedral holes located directly beneath each corner atom. Each unit cell contains four calcium cations $[(8 \times 1/8) + (6 \times \frac{1}{2}) = 4]$ and eight fluoride anions (each of the eight fluoride anions is completely contained within the unit cell), resulting in a cation to anion ratio of 1:2, as in the formula of the compound.

13.14 Atomic solids can be divided into three categories—nonbonding atomic solids, metallic atomic solids, and network covalent atomic solids. Nonbonding atomic solids, which include only the noble gases in their solid form, are held together by relatively weak dispersion forces. Metallic atomic solids, such as iron and gold, are held together by metallic bonds, which in the simplest model are represented by the interaction of metal cations with a sea of electrons around them. Network covalent atomic solids, such as diamond, graphite, and silicon dioxide, are held together by covalent bonds.

13.15 Graphite's structure consists of flat sheets of carbon atoms covalently bonded together as interconnected hexagonal rings. The forces *between* sheets are much different. There are no covalent bonds between sheets, only relatively weak dispersion forces. Consequently, the sheets slide past each other relatively easily, which explains the slippery feel of graphite and its extensive use as a lubricant. The electrons in the extended pi bonding network within a sheet make graphite a good electrical conductor in the direction of the plane of the sheets.

The diamond structure consists of carbon atoms covalently bonded to four other carbon atoms at the corners of a tetrahedron. This structure extends throughout the entire crystal, so that a diamond crystal can be thought of as a giant molecule, held together by these covalent bonds. Because covalent bonds are very strong, diamond has a very high melting point (it is estimated to melt at about 3800 °C). The electrons in diamond are confined to the covalent bonds and are not free to flow. Therefore, diamond does not conduct electricity. Diamond is very hard and is an excellent conductor of heat.

13.16 Fullerenes, nanotubes, and nanoribbons are new forms of carbon based on soccer-ball-shaped clusters of 60 carbon atoms (C_{60}). The atoms form five- and six-membered carbon rings wrapped into a 20-sided icosahedral structure. Carbon clusters similar to C_{60} containing from 36 to over 100 carbon atoms are called fullerenes and nicknamed buckyballs. Nanotubes are long carbon structures that consist of sheets of interconnected C_6 rings that assume the shape of a cylinder (like a roll of chicken wire). Today, two general types of nanotubes can be produced: (1) single-walled nanotubes (SWNT) that have one layer of interconnected C6 rings forming the walls, and (2) multiwalled nanotubes (MWNT) that have concentric layers of interconnected C6 rings forming the walls. Nanoribbons are nanotubes that have been sliced open.

13.17 Silicon does not form a double bond (one sigma and one pi) like carbon does because the silicon atom is too large for there to be substantial overlap between the *p* orbitals on the two atoms.

13.18 Silica melts when heated above 1500 °C. After melting, if cooled quickly, silica does not crystallize back into the quartz structure. Instead, the Si atoms and O atoms form a randomly ordered or amorphous structure called a glass. A slow cooling will result in the crystalline quartz structure.

Problem by Topic

Phase Diagrams

13.19
(a) solid
(b) liquid
(c) gas
(d) supercritical fluid
(e) solid/liquid equilibrium or fusion curve
(f) liquid/gas equilibrium or vaporization curve
(g) solid/liquid/gas equilibrium or triple point

13.20
(a) 184.4 °C
(b) 113.6 °C
(c) solid
(d) gas

13.21 **Given:** nitrogen; normal boiling point $= 77.3$ K; normal melting point $= 63.1$ K; critical temperature $= 126.2$ K; critical pressure $= 2.55 \times 10^4$ torr; triple point at 63.1 K and 94.0 torr
Find: Sketch phase diagram. Does nitrogen have a stable liquid phase at 1 atm?

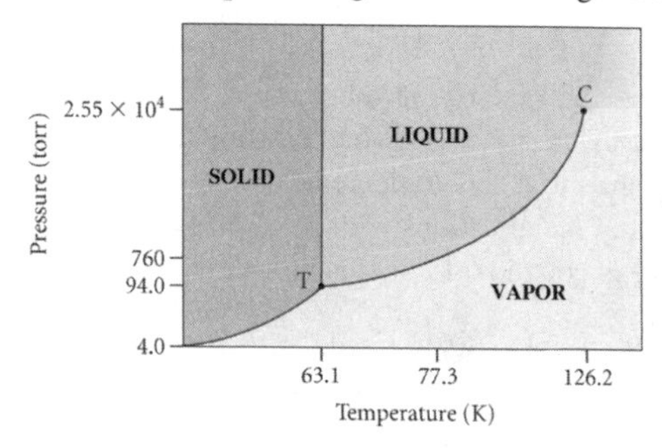

Nitrogen has a stable liquid phase at 1 atm.
Note that the axes are not to scale.

13.22 **Given:** argon; normal boiling point $= 87.2$ K; normal melting point $= 84.1$ K; critical temperature $= 150.8$ K; critical pressure $= 48.3$ atm; triple point at 83.7 K and 0.68 atm
Find: Sketch phase diagram. Which has the greater density, solid or liquid argon?

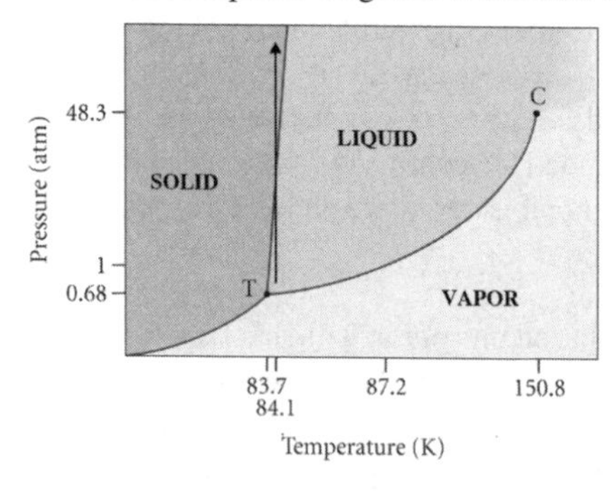

The solid has the higher density because the slope of the solid/liquid equilibrium line is positive. If we start in the liquid and increase the pressure, we will cross into the solid phase, the dense phase.
Note that the axes are not to scale.

13.23
(a) 0.027 mmHg, the higher of the two triple points
(b) The rhombic phase is denser because if we start in the monoclinic phase at 100 °C and increase the pressure, we will cross into the rhombic phase.

13.24 The triple point marked "O" shows the equilibrium of Ice II, Ice III, and Ice V. Ice II is denser than Ice I because you can generate Ice II from Ice I by increasing the pressure (pushing the molecules closer together). Ice III would sink in liquid water. Note that the Ice III/liquid line has the typical positive slope.

X-Ray Crystallography

13.25 **Given:** X-ray with $\lambda = 154$ pm, maximum reflection angle of $\theta = 28.3°$; assume $n = 1$
Find: distance between layers

Conceptual Plan: $\lambda, \theta, n \rightarrow d$

$n\lambda = 2\,d\sin\theta$

Solution: $n\lambda = 2\,d\sin\theta$. Rearrange to solve for d. $d = \dfrac{n\,\lambda}{2\sin\theta} = \dfrac{1 \times 154\text{ pm}}{2\sin 28.3°} = 162\text{ pm}$

Check: The units (pm) are correct. The magnitude (162 pm) makes sense because $n = 1$ and the sin is always <1. The number is consistent with interatomic distances.

13.26 **Given:** distance between layers $= 286$ pm, maximum reflection angle of $\theta = 7.23°$; assume $n = 1$ **Find:** λ (X-ray)
Conceptual Plan: $d, \theta, n \rightarrow \lambda$

$n\lambda = 2\,d\sin\theta$

Solution: $n\,\lambda = 2\,d\sin\theta$. Rearrange to solve for λ. $\lambda = \dfrac{2\,d\sin\theta}{n} = \dfrac{2 \times 286\text{ pm} \times \sin 7.23°}{1} = 72.0\text{ pm}$

Check: The units (pm) are correct. The magnitude (72 pm) makes sense because $n = 1$ and the sin is always <1. The number is consistent with X-ray wavelengths.

Crystalline Structures and Unit Cells

13.27 (a) 8 corner atoms $\times$ (1/8 atom/unit cell) = 1 atom/unit cell
(b) 8 corner atoms $\times$ (1/8 atom/unit cell) + 1 atom in center = (1 + 1) atoms/unit cell = 2 atoms/unit cell
(c) 8 corner atoms $\times$ (1/8 atom/unit cell) + 6 face-centered atoms $\times$ (1/2 atom/unit cell) = (1 + 3) atoms/unit cell = 4 atoms/unit cell

13.28 (a) coordination number of 12 because this is a face-centered cubic structure
(b) coordination number of 12 because this is a hexagonal closest-packed structure
(c) coordination number of 8 because this is a body-centered cubic structure

13.29 **Given:** platinum; face-centered cubic structure; $r = 139$ pm **Find:** edge length of unit cell and density (g/cm^3)
Conceptual Plan: $r \rightarrow l$ **and** $l \rightarrow V(\text{pm}^3) \rightarrow V(\text{cm}^3)$ **and** $\mathcal{M}$**, FCC structure** $\rightarrow m$ **then** $m, V \rightarrow d$

$l = 2\sqrt{2}\,r$ $\quad V = l^3 \quad \dfrac{(1\text{ cm})^3}{(10^{10}\text{ pm})^3} \quad m = \dfrac{4\text{ atoms}}{\text{unit cell}} \times \dfrac{\mathcal{M}}{N_A} \quad d = m/V$

Solution: $l = 2\sqrt{2}\,r = 2\sqrt{2} \times 139\text{ pm} = 39\underline{3}.151\text{ pm} = 393\text{ pm}$ and

$$V = l^3 = (39\underline{3}.151\text{ pm})^3 \times \frac{(1\text{ cm})^3}{(10^{10}\text{ pm})^3} = 6.0\underline{7}684 \times 10^{-23}\text{ cm}^3 \text{ and}$$

$$m = \frac{4\text{ atoms}}{\text{unit cell}} \times \frac{\mathcal{M}}{N_A} = \frac{4\text{ atoms}}{\text{unit cell}} \times \frac{195.09\text{ g}}{1\text{ mol}} \times \frac{1\text{ mol}}{6.022 \times 10^{23}\text{ atoms}} = 1.29\underline{5}849 \times 10^{-21}\frac{\text{g}}{\text{unit cell}} \text{ then}$$

$$d = \frac{m}{V} = \frac{1.29\underline{5}849 \times 10^{-21}\dfrac{\text{g}}{\text{unit cell}}}{6.0\underline{7}684 \times 10^{-23}\dfrac{\text{cm}^3}{\text{unit cell}}} = 21.3\frac{\text{g}}{\text{cm}^3}$$

Check: The units (pm and g/cm^3) are correct. The magnitude (393 pm) makes sense because it must be larger than the radius of an atom. The magnitude (21 g/cm^3) is consistent for Pt from Chapter 2.

13.30 **Given:** molybdenum; body-centered cubic structure; $r = 136$ pm **Find:** edge length of unit cell and density (g/cm^3)
Conceptual Plan: $r \rightarrow l$ **and** $l \rightarrow V(\text{pm}^3) \rightarrow V(\text{cm}^3)$ **and** $\mathcal{M}$**, BCC structure** $\rightarrow m$ **then** $m, V \rightarrow d$

$l = \dfrac{4r}{\sqrt{3}} \quad V = l^3 \quad \dfrac{(1\text{ cm})^3}{(10^{10}\text{ pm})^3} \quad m = \dfrac{2\text{ atoms}}{\text{unit cell}} \times \dfrac{\mathcal{M}}{N_A} \quad d = m/V$

Solution: $l = \dfrac{4\,r}{\sqrt{3}} = \dfrac{4 \times 136\text{ pm}}{\sqrt{3}} = 31\underline{4}.079\text{ pm} = 314\text{ pm}$ and

$$V = l^3 = (31\underline{4}.079\text{ pm})^3 \times \frac{(1\text{ cm})^3}{(10^{10}\text{ pm})^3} = 3.0\underline{9}825 \times 10^{-23}\text{ cm}^3 \text{ and}$$

$$m = \frac{2\text{ atoms}}{\text{unit cell}} \times \frac{\mathcal{M}}{N_A} = \frac{2\text{ atoms}}{\text{unit cell}} \times \frac{95.94\text{ g}}{1\text{ mol}} \times \frac{1\text{ mol}}{6.022 \times 10^{23}\text{ atoms}} = 3.18\underline{6}317 \times 10^{-22}\frac{\text{g}}{\text{unit cell}} \text{ then}$$

$$d = \frac{m}{V} = \frac{3.18\underline{6}317 \times 10^{-22}\,\frac{\text{g}}{\cancel{\text{unit cell}}}}{3.0\underline{9}825 \times 10^{-23}\,\frac{\text{cm}^3}{\cancel{\text{unit cell}}}} = 10.3\,\frac{\text{g}}{\text{cm}^3}$$

Check: The units (pm and g/cm^3) are correct. The magnitude (314 pm) makes sense because it must be larger than the radius of an atom. The magnitude ($10\ \text{g/cm}^3$) is reasonable for a metal density.

13.31 **Given:** body-centered cubic structure, $V = 2.62 \times 10^{-23}\ \text{cm}^3$ **Find:** r (pm)

Conceptual plan: $V(\text{cm}^3) \rightarrow l\ (\text{cm}) \rightarrow l\ (\text{pm})$ **then** $l \rightarrow r$

$V = l^3$ $\quad \frac{10^{10}\ \text{pm}}{1\ \text{cm}}$ $\quad l = \frac{4r}{\sqrt{3}}$

Solution: $V = l^3$ Rearrange to solve for l.

$$l = \sqrt[3]{V} = \sqrt[3]{2.62 \times 10^{-23}\ \text{cm}^3} = 2.97\underline{0}5876 \times 10^{-8}\ \cancel{\text{cm}} \times \frac{10^{10}\ \text{pm}}{1\ \cancel{\text{cm}}} = 2.97\underline{0}5876\ \text{pm then } l = \frac{4r}{\sqrt{3}}$$

Rearrange to solve for r. $r = \frac{l\sqrt{3}}{4} = \frac{2.97\underline{0}5876\ \text{pm} \times \sqrt{3}}{4} = 129\ \text{pm}$

Check: The units (pm) are correct. The magnitude (129 pm) is consistent with atomic diameters.

13.32 **Given:** face-centered cubic structure, r = 142 pm **Find:** volume of unit cell

Conceptual plan: $r \rightarrow l$ **then** $l \rightarrow V(\text{pm}^3) \rightarrow V(\text{cm}^3)$

$l = 2\sqrt{2}\,r$ $\quad V = l^3$ $\quad \frac{(1\ \text{cm})^3}{(10^{10}\ \text{pm})^3}$

Solution: $l = 2\sqrt{2}\,r = 2\sqrt{2} \times 142\ \text{pm} = 40\underline{1}.63665\ \text{pm}$ then

$$V = l^3 = (40\underline{1}.63665\ \cancel{\text{pm}})^3 \times \frac{(1\ \text{cm})^3}{(10^{10}\ \cancel{\text{pm}})^3} = 6.4\underline{7}8881 \times 10^{-23}\ \text{cm}^3 = 6.48 \times 10^{-23}\ \text{cm}^3$$

Check: The units (cm^3) are correct. The magnitude ($6 \times 10^{-23}\ \text{cm}^3$) is the same as that of the other unit cells.

13.33 **Given:** rhodium; face-centered cubic structure; $d = 12.41\ \text{g/cm}^3$ **Find:** $r(\text{Rh})$

Conceptual Plan: $\mathcal{M}$**, FCC structure** $\rightarrow m$ **then** $m, V \rightarrow d$ **then** $V(\text{cm}^3) \rightarrow l(\text{cm}) \rightarrow l(\text{pm})$ **then** $l \rightarrow r$

$m = \frac{4\ \text{atoms}}{\text{unit cell}} \times \frac{\mathcal{M}}{N_A}$ $\quad d = m/V$ $\quad V = l^3$ $\quad \frac{10^{10}\ \text{pm}}{1\ \text{cm}}$ $\quad l = 2\sqrt{2}\,r$

Solution: $m = \frac{4\ \text{atoms}}{\text{unit cell}} \times \frac{\mathcal{M}}{N_A} = \frac{4\ \cancel{\text{atoms}}}{\text{unit cell}} \times \frac{102.905\ \text{g}}{1\ \cancel{\text{mol}}} \times \frac{1\ \cancel{\text{mol}}}{6.022 \times 10^{23}\ \cancel{\text{atoms}}} = 6.83\underline{5}271 \times 10^{-22}\ \frac{\text{g}}{\text{unit cell}}$

then $d = \frac{m}{V}$ Rearrange to solve for V. $V = \frac{m}{d} = \frac{6.83\underline{5}271 \times 10^{-22}\ \frac{\cancel{\text{g}}}{\text{unit cell}}}{12.41\frac{\cancel{\text{g}}}{\text{cm}^3}} = 5.50\underline{7}873 \times 10^{-23}\ \frac{\text{cm}^3}{\text{unit cell}}$

then $V = l^3$ Rearrange to solve for l.

$$l = \sqrt[3]{V} = \sqrt[3]{5.50\underline{7}873 \times 10^{-23}\ \text{cm}^3} = 3.80\underline{4}831 \times 10^{-8}\ \cancel{\text{cm}} \times \frac{10^{10}\ \text{pm}}{1\ \cancel{\text{cm}}} = 380.\underline{4}831\ \text{pm then } l = 2\sqrt{2}r$$

Rearrange to solve for r. $r = \frac{1}{2\sqrt{2}} = \frac{380.\underline{4}831\ \text{pm}}{2\sqrt{2}} = 134.5\ \text{pm}$

Check: The units (pm) are correct. The magnitude (135 pm) is consistent with an atomic diameter.

13.34 **Given:** barium; body-centered cubic structure; $d = 3.59\ \text{g/cm}^3$ **Find:** $r\ (\text{Ba})$

Conceptual Plan: $\mathcal{M}$**, BCC structure** $\rightarrow m$ **then** $m, V \rightarrow d$ **then** $V(\text{cm}^3) \rightarrow l(\text{cm}) \rightarrow l(\text{pm})$ **then** $l \rightarrow r$

$m = \frac{2\ \text{atoms}}{\text{unit cell}} \times \frac{\mathcal{M}}{N_A}$ $\quad d = m/V$ $\quad V = l^3$ $\quad \frac{10^{10}\ \text{pm}}{1\ \text{cm}}$ $\quad l = \frac{4r}{\sqrt{3}}$

Solution: $m = \frac{2 \text{ atoms}}{\text{unit cell}} \times \frac{\mathcal{M}}{N_A} = \frac{2 \text{ atoms}}{\text{unit cell}} \times \frac{137.34 \text{ g}}{1 \text{ mol}} \times \frac{1 \text{ mol}}{6.022 \times 10^{23} \text{ atoms}} = 4.56\underline{1}275 \times 10^{-22} \frac{\text{g}}{\text{unit cell}}$ then

$d = \frac{m}{V}$ Rearrange to solve for V. $V = \frac{m}{d} = \frac{4.56\underline{1}275 \times 10^{-22} \frac{\text{g}}{\text{unit cell}}}{3.59 \frac{\text{g}}{\text{cm}^3}} = 1.27\underline{0}550 \times 10^{-22} \frac{\text{cm}^3}{\text{unit cell}}$

then $V = l^3$ Rearrange to solve for l.

$l = \sqrt[3]{V} = \sqrt[3]{1.27\underline{0}550 \times 10^{-22} \text{ cm}^3} = 5.02\underline{7}336 \times 10^{-8} \text{ cm} \times \frac{10^{10} \text{ pm}}{1 \text{ cm}} = 502.\underline{7}336 \text{ pm}$

then $l = \frac{4r}{\sqrt{3}}$ Rearrange to solve for r. $r = \frac{l\sqrt{3}}{4} = \frac{502.\underline{7}336 \text{ pm} \times \sqrt{3}}{4} = 217.7 \text{ pm}$

Check: The units (pm) are correct. The magnitude (218 pm) is consistent with an atomic diameter.

13.35 **Given:** polonium, simple cubic structure $d = 9.3 \text{ g/cm}^3$; $r = 167$ pm; $\mathcal{M} = 209$ g/mol **Find:** estimate N_A

Conceptual Plan: $r \rightarrow l$ **and** $l \rightarrow V(\text{pm}^3) \rightarrow V(\text{cm}^3)$ **then** $d, V \rightarrow m$ **then** $\mathcal{M}$**, SC structure** $\rightarrow N_A$

$l = 2r$ $\quad V = l^3 \quad \frac{(1 \text{ cm})^3}{(10^{10} \text{ pm})^3} \quad d = m/V \quad m = \frac{1 \text{ atom}}{\text{unit cell}} \times \frac{\mathcal{M}}{N_A}$

Solution: $l = 2r = 2 \times 167 \text{ pm} = 334 \text{ pm}$ and $V = l^3 = (334 \text{ pm})^3 \times \frac{(1 \text{ cm})^3}{(10^{10} \text{ pm})^3} = 3.7\underline{2}597 \times 10^{-23} \text{ cm}^3$ then

$d = \frac{m}{V}$ Rearrange to solve for m. $m = dV = 9.3 \frac{\text{g}}{\text{cm}^3} \times \frac{3.7\underline{2}597 \times 10^{-23} \text{ cm}^3}{\text{unit cell}} = 3.\underline{4}6515 \times 10^{-22} \frac{\text{g}}{\text{unit cell}}$

then $m = \frac{1 \text{ atom}}{\text{unit cell}} \times \frac{\mathcal{M}}{N_A}$ Rearrange to solve for N_A.

$N_A = \frac{1 \text{ atom}}{\text{unit cell}} \times \frac{\mathcal{M}}{m} = \frac{1 \text{ atom}}{\text{unit cell}} \times \frac{209 \text{ g}}{1 \text{ mol}} \times \frac{1 \text{ unit cell}}{3.\underline{4}6515 \times 10^{-22} \text{ g}} = 6.0 \times 10^{23} \frac{\text{atoms}}{\text{mol}}$

Check: The units (atoms/mol) are correct. The magnitude (6×10^{23} atoms/mol) is consistent with Avogadro's number.

13.36 **Given:** palladium, face-centered cubic structure, $d = 12.0 \text{ g/cm}^3$; $r = 138$ pm $\mathcal{M} = 106.42$ g/mol

Find: estimate N_A

Conceptual Plan: $r \rightarrow l$ **and** $l \rightarrow V(\text{pm}^3) \rightarrow V(\text{cm}^3)$ **then** $d, V \rightarrow m$ **then** $\mathcal{M}$**, FCC structure** $\rightarrow N_A$

$l = 2\sqrt{2}r \quad V = l^3 \quad \frac{(1 \text{ cm})^3}{(10^{10} \text{ pm})^3} \quad d = m/V \quad m = \frac{4 \text{ atoms}}{\text{unit cell}} \times \frac{\mathcal{M}}{N_A}$

Solution: $l = 2\sqrt{2}r = 2\sqrt{2} \times 138 \text{ pm} = 39\underline{0}.323 \text{ pm}$ and

$V = l^3 = (39\underline{0}.323 \text{ pm})^3 \times \frac{(1 \text{ cm})^3}{(10^{10} \text{ pm})^3} = 5.9\underline{4}665 \times 10^{-23} \text{ cm}^3$ then $d = \frac{m}{V}$ Rearrange to solve for m.

$m = dV = 12.0 \frac{\text{g}}{\text{cm}^3} \times \frac{5.9\underline{4}665 \times 10^{-23} \text{ cm}^3}{\text{unit cell}} = 7.1\underline{3}598 \times 10^{-22} \frac{\text{g}}{\text{unit cell}}$ then $m = \frac{4 \text{ atoms}}{\text{unit cell}} \times \frac{\mathcal{M}}{N_A}$

Rearrange to solve for N_A. $N_A = \frac{4 \text{ atoms}}{\text{unit cell}} \times \frac{\mathcal{M}}{m} = \frac{4 \text{ atoms}}{\text{unit cell}} \times \frac{106.42 \text{ g}}{1 \text{ mol}} \times \frac{1 \text{ unit cell}}{7.1\underline{3}598 \times 10^{-22} \text{ g}} = 5.97 \times 10^{23} \frac{\text{atoms}}{\text{mol}}$

Check: The units (atoms/mol) are correct. The magnitude (6×10^{23} atoms/mol) is consistent with Avogadro's number.

Types of Crystalline Solids

13.37
(a) atomic because argon (Ar) is an atom
(b) molecular because water (H_2O) is a molecule
(c) ionic because potassium oxide (K_2O) is an ionic solid
(d) atomic because iron (Fe) is an atom

13.38 (a) ionic because calcium chloride $(CaCl_2)$ is an ionic solid
(b) molecular because carbon dioxide (CO_2) is a molecule
(c) atomic because nickel (Ni) is an atom
(d) molecular because iodine (I_2) is a molecule

13.39 LiCl has the highest melting point because it is the only ionic solid in the group. The other three solids are held together by intermolecular forces, while LiCl is held together by stronger coulombic interactions between the cations and anions of the crystal lattice.

13.40 C (diamond) has the highest melting point (3800 °C). Both covalent network solids and ionic solids have high melting points. NaCl has a melting point of 801 °C. In diamond (Figure 13.19b), each carbon atom forms four covalent bonds to four other carbon atoms in a tetrahedral geometry. This structure extends throughout the entire crystal; so a diamond crystal can be thought of as a giant molecule held together by these covalent bonds. Because covalent bonds are very strong, covalent atomic solids have high melting points.

13.41 (a) TiO_2 because it is an ionic solid
(b) $SiCl_4$ because it has a higher molar mass and therefore has stronger dispersion forces
(c) Xe because it has a higher molar mass and therefore has stronger dispersion forces
(d) CaO because the ions have greater charge and therefore stronger dipole–dipole interactions

13.42 (a) Fe because it is an atomic solid held together by metallic bonding
(b) KCl because it is an ionic solid
(c) Ti because it is an atomic solid held together by metallic bonding
(d) H_2O because it is capable of hydrogen bonding

Ionic Solids

13.43 CuI is the most likely compound since the zinc blend structure has an equal number of the anions and cations and the size of the ions are dramatically different.

13.44 BaF_2 is the most likely to adopt the fluorite structure since it has a 1:2 cation to anion ratio.

13.45 The Ti atoms occupy the corner positions and the center of the unit cell: 8 corner atoms $\times$ (1/8 atom/unit cell) +1 atom in center = (1 + 1) Ti atoms/unit cell = 2 Ti atoms/unit cell. The O atoms occupy four positions on the top and bottom faces and two positions inside the unit cell: 4 face-centered atoms $\times$ (1/2 atom/unit cell) +2 atoms in the interior = (2 + 2) O atoms/unit cell = 4 O atoms/unit cell. Therefore, there are 2 Ti atoms/ unit cell and 4 O atoms/unit cell; so the ratio Ti:O is 2:4, or 1:2. The formula for the compound is TiO_2.

13.46 The Re atoms occupy the corner positions and the center of the unit cell: 8 corner atoms $\times$ (1/8 atom/unit cell) = 1 Re atom/unit cell. The O atoms occupy 12 edge positions: 12 edge atoms $\times$ (1/4 atom/unit cell) = 3 O atoms/ unit cell. Therefore, there are 1 Re atom/unit cell and 3 O atoms/unit cell, so the ratio Re:O is 1:3. The formula for the compound is ReO_3.

13.47 In CsCl: The Cs atoms occupy the center of the unit cell: 1 atom in center = 1 Cs atom/unit cell. The Cl atoms occupy corner positions of the unit cell: 8 corner atoms $\times$ (1/8 atom/unit cell) = 1 Cl atom/unit cell. Therefore, there are 1 Cs atom/unit cell and 1 Cl atom/unit cell, so the ratio Cs:Cl is 1:1. The formula for the compound is CsCl, as expected.

In $BaCl_2$: The Ba atoms occupy the corner positions and the face-centered positions of the unit cell: 8 corner atoms $\times$ (1/8 atom/unit cell) + 6 face-centered atoms $\times$ (1/2 atom/unit cell) = (1 + 3) Ba atoms/unit cell = 4 Ba atoms/unit cell. The Cl atoms occupy eight positions inside the unit cell: 8 Cl atoms/unit cell. Therefore, there are 4 Ba atoms/unit cell and 8 Cl atoms/unit cell, so the ratio Ba:Cl is 4:8, or 1:2. The formula for the compound is $BaCl_2$, as expected.

13.48 In Li_2O: The Li atoms occupy eight positions inside the unit cell: 8 Li atoms/unit cell. The O atoms occupy the corner positions and the face-centered positions of the unit cell: 8 corner atoms $\times$ (1/8 atom/unit cell) + 6 face-centered atoms $\times$ (1/2 atom/unit cell) = (1 + 3) O atoms/unit cell = 4 O atoms/unit cell. Therefore, there are 4 O atoms/ unit cell and 8 Li atoms/unit cell, so the ratio Li:O is 8:4, or 2:1. The formula for the compound is Li_2O, as expected.

In AgI : The Ag atoms occupy four positions inside the unit cell: 4 Ag atoms/unit cell. The I atoms occupy the corner positions and the face-centered positions of the unit cell: 8 corner atoms $\times$ (1/8 atom/unit cell) + 6 face-centered atoms $\times$ (1/2 atom/unit cell) = (1 + 3) I atoms/unit cell = 4 I atoms/unit cell. Therefore, there are 4 I atoms/unit cell and 4 Ag atoms/unit cell; so the ratio Ag:I is 4:4, or 1:1. The formula for the compound is AgI, as expected.

Network Covalent Atomic Solids

13.49 Graphite is a good lubricant because the structure has strong covalent bonds within the sheets of the structure, but weak interactions between the sheets that allow the sheets to easily slide over one another. The diamond structure consists of carbon atoms connected to four other carbon atoms at the corners of a tetrahedron. This bonding extends throughout three dimensions, making giant molecules described as network covalent solids. There are no weak interactions in the diamond structure.

13.50 The electrons in the extended pi bonding network within a sheet make graphite a good electric conductor in the direction of the plane of the sheets. Because the diamond structure is a three-dimensional network, this overlap of orbitals is not possible.

Cumulative Problems

13.51 Because we are starting at a temperature that is higher and a pressure that is lower than the triple point, the phase transitions will be gas $\rightarrow$ liquid $\rightarrow$ solid, or condensation followed by freezing.

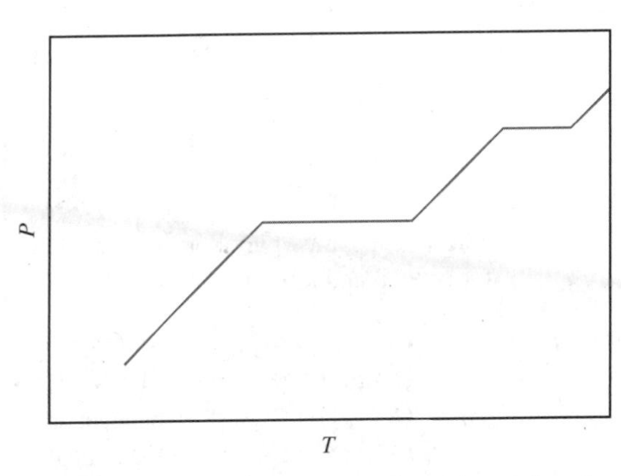

13.52 The solid is denser than the liquid. Because the triple point temperature is lower than the normal melting point, the slope of the fusion curve must be positive. This means that as you start in the liquid phase and increase the pressure, you will eventually cross into the solid phase. As pressure increases, the phases get denser and the atoms, molecules, or ions are pushed closer and closer together.

13.53 CsCl has a higher melting point than AgI because of its higher coordination number. In CsCl, one anion bonds to eight cations (and vice versa), while in AgI, one anion bonds only to four cations.

13.54 KCl has a higher melting point than copper iodide because of its higher coordination number. In KCl, one anion bonds to six cations (and vice versa) while in copper iodide, one anion bonds only to four cations.

13.55 (a) Atoms are connected across the face diagonal (c), so $c = 4r$.
(b) From the Pythagorean Theorem, $c^2 = a^2 + b^2$; from part (a), $c = 4r$; and for a cubic structure, $a = l, b = l$. So $(4r)^2 = l^2 + l^2 \rightarrow 16r^2 = 2l^2 \rightarrow 8r^2 = l^2 \rightarrow l = \sqrt{8r^2} \rightarrow l = 2\sqrt{2}r$.

13.56 (a) Atoms are connected across the cube diagonal (c), so $c = 4r$.
(b) Because b forms the diagonal of the face, where each edge length $a = l$, and combining this with the Pythagorean Theorem, $b^2 = l^2 + l^2 \rightarrow b^2 = 2l^2 \rightarrow b = \sqrt{2}l$.
(c) From the Pythagorean Theorem, $c^2 = a^2 + b^2$; from part (a), $c = 4r$; from part (b), $b = \sqrt{2}l$; and for a cubic structure, $a = l$. So $(4r)^2 = l^2 + (\sqrt{2}l)^2 \rightarrow 16r^2 = l^2 + 2l^2 \rightarrow 16r^2 = 3l^2 \rightarrow 4r = \sqrt{3}l \rightarrow l = \dfrac{4r}{\sqrt{3}}$.

13.57 **Given:** diamond, $V(\text{unit cell}) = 0.0454\ \text{nm}^3$; $d = 3.52\ \text{g/cm}^3$ **Find:** number of carbon atoms/unit cell
Conceptual Plan: $V(\text{nm}^3) \rightarrow V(\text{cm}^3)$ **then** $d, V \rightarrow m \rightarrow \text{mol} \rightarrow \text{atoms}$

$$\frac{(1\ \text{cm})^3}{(10^7\ \text{nm})^3} \qquad d = m/V \quad \frac{1\ \text{mol}}{12.01\ \text{g}} \quad \frac{6.022 \times 10^{23}\ \text{atoms}}{1\ \text{mol}}$$

Solution: $0.0454\ \cancel{nm^3} \times \frac{(1\ cm)^3}{(10^7\ \cancel{nm})^3} = 4.54 \times 10^{-23}\ cm^3$ then $d = \frac{m}{V}$ Rearrange to solve for m.

$m = dV = 3.52\frac{g}{\cancel{cm^3}} \times 4.54 \times 10^{-23}\ \cancel{cm^3} = 1.59808 \times 10^{-22}$ g then

$$\frac{1.59808 \times 10^{-22}\ \cancel{g}}{\text{unit cell}} \times \frac{1\ \cancel{mol}}{12.01\ \cancel{g}} \times \frac{6.022 \times 10^{23}\ \text{atoms}}{1\ \cancel{mol}} = 8.01\frac{\text{C atoms}}{\text{unit cell}} = 8\frac{\text{C atoms}}{\text{unit cell}}$$

Check: The units (atoms) are correct. The magnitude (8 atoms) makes sense because it is a fairly small number and our answer is within calculation error of an integer.

13.58 **Given:** metal, $d = 12.3\ g/cm^3$; $r = 0.134$ nm, face-centered cubic lattice **Find:** $\mathcal{M}$

Conceptual Plan: $r \rightarrow l \rightarrow V(nm^3) \rightarrow V(cm^3)$ **then** $d, V \rightarrow m$ **then** m, **FCC structure** $\rightarrow \mathcal{M}$

$l = 2\sqrt{2}r \quad V = l^3 \quad \frac{(1\ cm)^3}{(10^7\ nm)^3} \quad d = m/V \quad m = \frac{4\ \text{atoms}}{\text{unit cell}} \times \frac{\mathcal{M}}{N_A}$

Solution: $l = 2\sqrt{2}r = 2\sqrt{2} \times 0.134\ nm = 0.379009$ nm

$V = l^3 = (0.379009\ nm)^3 = 0.0544438\ \cancel{nm^3} \times \frac{(1\ cm)^3}{(10^7\ \cancel{nm})^3} = 5.44438 \times 10^{-23}\ cm^3$ then $d = \frac{m}{V}$ Rearrange to solve for m. $m = dV = 12.3\frac{g}{\cancel{cm^3}} \times 5.44438 \times 10^{-23}\ \cancel{cm^3} = 6.69659 \times 10^{-22}$ g then $m = \frac{4\ \text{atoms}}{\text{unit cell}} \times \frac{\mathcal{M}}{N_A}$

Rearrange to solve for $\mathcal{M}$.

$$\mathcal{M} = \frac{\text{unit cell}}{4\ \text{atoms}} \times N_A \times m = \frac{\cancel{\text{unit cell}}}{4\ \cancel{\text{atoms}}} \times \frac{6.022 \times 10^{23}\ \cancel{\text{atoms}}}{1\ \text{mol}} \times \frac{6.69659 \times 10^{-22}\ g}{\cancel{\text{unit cell}}} = 101\frac{g}{mol}\ \text{ruthenium}$$

Check: The units (g/mol) are correct. The magnitude (101 g/mol) makes sense because it is a reasonable atomic mass for a metal and it is close to the value of the molar mass of ruthenium.

13.59 (a) $CO_2(s) \rightarrow CO_2(g)$ at 194.7 K

(b) $CO_2(s) \rightarrow$ triple point at 216.5 K $\rightarrow CO_2(g)$ just above 216.5 K

(c) $CO_2(s) \rightarrow CO_2(l)$ at somewhat above 216 K $\rightarrow CO_2(g)$ at around 250 K

(d) $CO_2(s) \rightarrow CO_2$ above the critical point where there is no distinction between liquid and gas. This change occurs at about 300 K.

13.60 If atmospheric pressure was 2500 mmHg, water would still be a liquid. At a higher atmospheric pressure, water would remain a liquid to a lower temperature than 0 °C; this could reduce the damage done to organisms that are exposed to cold temperatures. At a higher atmospheric pressure, there would be more molecules in the gas phase and the atmosphere would not behave as ideally; water might condense more readily, lowering its vapor pressure. This could have an adverse effect on living organisms. At higher atmospheric pressures, cell walls would need to be stronger to withstand higher pressures. This would most likely make the cell walls less permeable and affect many biological systems.

13.61 **Given:** metal, $d = 7.8748\ g/cm^3$; $l = 0.28664$ nm, body-centered cubic lattice **Find:** $\mathcal{M}$

Conceptual Plan: $l \rightarrow V(nm^3) \rightarrow V(cm^3)$ **then** $d, V \rightarrow m$ **then** m, **FCC structure** $\rightarrow \mathcal{M}$

$V = l^3 \quad \frac{(1\ cm)^3}{(10^7\ nm)^3} \quad d = m/V \quad m = \frac{2\ \text{atoms}}{\text{unit cell}} \times \frac{\mathcal{M}}{N_A}$

Solution: $V = l^3 = (0.28664\ nm)^3 = 0.02355105602\ \cancel{nm^3} \times \frac{(1\ cm)^3}{(10^7\ \cancel{nm})^3} = 2.355105602 \times 10^{-23}\ cm^3$ then $d = \frac{m}{V}$

Rearrange to solve for m. $m = dV = 7.8748\frac{g}{\cancel{cm^3}} \times 2.355105602 \times 10^{-23}\ \cancel{cm^3} = 1.854598559 \times 10^{-22}$ g

then $m = \frac{2\ \text{atoms}}{\text{unit cell}} \times \frac{\mathcal{M}}{N_A}$ Rearrange to solve for $\mathcal{M}$.

$$\mathcal{M} = \frac{\text{unit cell}}{2\ \text{atoms}} \times N_A \times m = \frac{\cancel{\text{unit cell}}}{2\ \cancel{\text{atoms}}} \times \frac{6.022 \times 10^{23}\ \cancel{\text{atoms}}}{1\ \text{mol}} \times \frac{1.854598559 \times 10^{-22}\ g}{\cancel{\text{unit cell}}} = 55.842\frac{g}{mol}$$

$= 55.84\frac{g}{mol}$ iron

Check: The units (g/mol) are correct. The magnitude (55.8 g/mol) makes sense because it is a reasonable atomic mass for a metal and it is close to the value of the molar mass of iron.

13.62 There are two spheres in each unit cell. The volume of the unit cell $V = a^3$. Because the spheres occupy 68.0% of the available volume and the volume of a sphere $= \frac{4}{3}\pi r^3$, $0.680V = 0.680a^3 = 2\left(\frac{4}{3}\pi r^3\right)$. Rearranging this to solve for a, we have $a^3 = \frac{2\left(\frac{4}{3}\pi r^3\right)}{0.680} \rightarrow a = \sqrt[3]{\frac{2\left(\frac{4}{3}\pi\right)}{0.680}}\; r \rightarrow a = 1.58\sqrt[3]{\pi}\, r \rightarrow a = 2.31\, r$, which agrees with the solution in Problem 13.56.

Challenge Problems

13.63 **Given:** KCl, rock salt structure **Find:** density (g/cm^3) **Other:** $r(\text{K}^+) = 133$ pm; $r(\text{Cl}^-) = 181$ pm from Chapter 4

Conceptual Plan: Rock salt structure is a face-centered cubic structure with anions at the lattice points and cations in the holes between lattice sites → assume $r = r(\text{Cl}^-)$, but $\mathcal{M} = \mathcal{M}(\text{KCl})$
$r(\text{K}^+), r(\text{Cl}^-) \rightarrow l$ and $l \rightarrow V(\text{pm}^3) \rightarrow V(\text{cm}^3)$, and FCC structure → m then $m, V \rightarrow d$

from Figure 13.16 $l = 2r(\text{Cl}^-) + 2r(\text{K}^+)$ $V = l^3$ $\frac{(1\text{ cm})^3}{(10^{10}\text{ pm})^3}$ $m = \frac{4\text{ formula units}}{\text{unit cell}} \times \frac{\mathcal{M}}{N_A}$ $d = m/V$

Solution: $l = 2r(\text{Cl}^-) + 2r(\text{K}^+) = 2(181\text{ pm}) + 2(133\text{ pm}) = 628\text{ pm}$ and

$$V = l^3 = (628\text{ pm})^3 \times \frac{(1\text{ cm})^3}{(10^{10}\text{ pm})^3} = 2.4\underline{7}673 \times 10^{-22}\text{ cm}^3 \text{ and}$$

$$m = \frac{4\text{ formula units}}{\text{unit cell}} \times \frac{\mathcal{M}}{N_A} = \frac{4\text{ formula units}}{\text{unit cell}} \times \frac{74.55\text{ g}}{1\text{ mol}} \times \frac{1\text{ mol}}{6.022 \times 10^{23}\text{ formula units}}$$

$$= 4.95\underline{1}843 \times 10^{-22}\frac{\text{g}}{\text{unit cell}}$$

$$\text{then } d = \frac{m}{V} = \frac{4.95\underline{1}843 \times 10^{-22}\frac{\text{g}}{\text{unit cell}}}{2.4\underline{7}673 \times 10^{-22}\frac{\text{cm}^3}{\text{unit cell}}} = 1.9\underline{9}935\frac{\text{g}}{\text{cm}^3} = 2.00\frac{\text{g}}{\text{cm}^3}$$

Check: The units (g/cm^3) are correct. The magnitude (2 g/cm^3) is reasonable for a salt density. The published value is 1.98 g/cm^3. This method of estimating the density gives a value that is close to the experimentally measured density.

13.64 **Given:** cubic closest packing structure **Find:** fraction of empty space to five significant figures

Conceptual Plan: Cubic closest packing is the same as face-centered cubic structure, so calculate the volume of the atoms and the volume of the unit cell then $V_{\text{atoms}}, V_{\text{unit cell}} \rightarrow \%V_{\text{empty}}$

$\frac{4\text{ atoms}}{\text{unit cell}}$ and $V_{\text{atom}} = \frac{\frac{4}{3}\pi r^3}{\text{atom}}$ $V_{\text{unit cell}} = l^3$ and $l = 2\sqrt{2}r$ $\%V_{\text{empty}} = \frac{V_{\text{unit cell}} - V_{\text{atoms}}}{V_{\text{unit cell}}} \times 100\%$

Solution: $V_{\text{atoms}} = \frac{4\text{ atoms}}{\text{unit cell}} \times \frac{\frac{4}{3}\pi r^3}{\text{atom}} = \frac{\frac{16}{3}\pi r^3}{\text{unit cell}}$ and $V_{\text{unit cell}} = l^3 = (2\sqrt{2}r)^3$ so

$$\%V_{\text{empty}} = \frac{V_{\text{unit cell}} - V_{\text{atoms}}}{V_{\text{unit cell}}} \times 100\% \text{ or}$$

$$\%V_{\text{empty}} = \frac{(2\sqrt{2}r)^3 - \frac{16}{3}\pi r^3}{(2\sqrt{2}r)^3} \times 100\% = \frac{2^3 2^{3/2} r^3 - \frac{2^4}{3}\pi r^3}{2^3 2^{3/2} r^3} \times 100\% = \frac{\sqrt{2} - \frac{\pi}{3}}{\sqrt{2}} \times 100\% = 25.952\%$$

Check: The units (%) are correct. The magnitude (26%) is consistent with what is stated in the text.

13.65 **Given:** cubic closest packing structure = cube with touching spheres of radius = r on alternating corners of a cube
Find: body diagonal of cube and radius of tetrahedral hole
Solution: The cell edge length = l and $l^2 + l^2 = (2r)^2 \rightarrow 2l^2 = 4r^2 \rightarrow l^2 = 2r^2$. So, $l = \sqrt{2}r$. Because body diagonal = BD is the hypotenuse of the right triangle formed by the face diagonal and the cell edge, we have $(BD)^2 = l^2 + (2r)^2 = 2r^2 + 4r^2 = 6r^2 \rightarrow BD = \sqrt{6}r$. The radius of the tetrahedral hole = r_T is half the body diagonal minus the radius of the sphere, or

$$r_T = \frac{BD}{2} - r = \frac{\sqrt{6}r}{2} - r = \left(\frac{\sqrt{6}}{2} - 1\right)r = \left(\frac{\sqrt{6} - 2}{2}\right)r = \left(\frac{\sqrt{3}\sqrt{2} - \sqrt{2}\sqrt{2}}{\sqrt{2}\sqrt{2}}\right)r$$

$$= \left(\frac{\sqrt{3} - \sqrt{2}}{\sqrt{2}}\right)r \approx 0.22474r$$

Conceptual Problems

13.66 The triple point will be at a lower temperature because the fusion equilibrium line has a positive slope. This means that we will be increasing both temperature and pressure as we travel from the triple point to the normal melting point.

13.67 The substance can exist as a solid or a gas. It cannot exist as a liquid because a liquid cannot exist below 3.2 atm.

14 Solutions

Review Questions

14.1 A solution is a homogeneous mixture of two or more substances. A solution has at least two components. The majority component is usually called the solvent, and the minority component is usually called the solute.

14.2 A substance is soluble in another substance if they can form a homogeneous mixture. The solubility of a substance is the amount of the substance that will dissolve in a given amount of solvent. Many different units can be used to express solubility, including grams of solute per 100 grams of solvent, grams of solute per liter of solvent, moles of solute per liter of solution, and moles of solute per kilogram of solvent.

14.3 Ideal gases do not interact with each other in any way (i.e., there are no significant forces between their constituent particles). When the two gases mix, their potential energy remains unchanged; so this does not drive the mixing. The tendency to mix is related, rather, to a concept called entropy. Entropy is a measure of energy randomization or energy dispersal in a system. Recall that a gas at any temperature above 0 K has kinetic energy due to the motion of its atoms. When the gases are separated, their kinetic energies are also confined to those regions. However, when the gases mix, the kinetic energy of each gas becomes spread out, or dispersed, over a larger volume. Therefore, the mixture of the two gases has greater energy dispersal, or greater entropy, than the separated components. The pervasive tendency for all kinds of energy to spread out, or disperse, whenever they are not restrained from doing so is the reason two ideal gases mix.

14.4 Entropy is a measure of energy randomization or energy dispersal in a system. When two substances mix to form a solution, there is an increase in randomness because the components are no longer segregated to separate regions. This makes the formation of a solution energetically favorable, even when it is endothermic.

14.5 Whether two substances will spontaneously mix to form a homogeneous solution is dependent upon a number of different types of intermolecular forces, including dispersion forces, dipole–dipole forces, hydrogen bonding, and ion–dipole forces.

14.6 A solution always forms if the solvent–solute interactions are comparable to, or stronger than, the solvent–solvent interactions and the solute–solute interactions.

14.7 The expression "like dissolves like" means that similar kinds of solvents dissolve similar kinds of solutes. Polar solvents, such as water, dissolve many polar or ionic solutes, and nonpolar solvents, such as hexane, dissolve many nonpolar solutes.

14.8 Step 1: Separate the solute into its constituent particles. This step is always endothermic (positive ΔH) because energy is required to overcome the forces that hold the solute together.

Step 2: Separate the solvent particles from each other to make room for the solute particles. This step is also endothermic because energy is required to overcome the intermolecular forces among the solvent particles.

Step 3: Mix the solute particles with the solvent particles. This step is exothermic because energy is released as the solute particles interact with the solvent particles through the various types of intermolecular forces.

14.9 The heat of hydration is the enthalpy change that occurs when 1 mol of gaseous solute ions are dissolved in water. In aqueous solutions, $\Delta H_{solvent}$ and ΔH_{mix} can be combined into a single term called the heat of hydration ($\Delta H_{hydration}$). Because the ion–dipole interactions that occur between a dissolved ion and the surrounding water molecules are much stronger than the hydrogen bonds in water, $\Delta H_{hydration}$ is always largely negative (exothermic) for ionic compounds. Using the heat of hydration, we can write the enthalpy of solution as a sum of just two terms, one endothermic and one exothermic: $\Delta H_{soln} = \Delta H_{solute} + \Delta H_{solvent} + \Delta H_{mix} = \Delta H_{solute} + \Delta H_{hydration} = \text{endothermic}(+)\text{ term} + \text{exothermic}(-)\text{ term}$. For ionic compounds, ΔH_{solute}, the energy required to separate the solute into its constituent particles, is simply the negative of the solute's lattice energy ($\Delta H_{solute} = -\Delta H_{lattice}$), discussed in Section 9.4. For ionic aqueous solutions then, the overall enthalpy of solution depends on the relative magnitudes of ΔH_{solute} and $\Delta H_{hydration}$, with three possible scenarios (in each case, we refer to the magnitude or absolute value of ΔH): (1) If $\Delta H_{solute} < \Delta H_{hydration}$, the amount of energy required to separate the solute into its constituent ions is less than the energy given off when the ions are hydrated. ΔH_{soln} is therefore negative, the solution process is exothermic, and the solution feels warm to the touch. (2) If $\Delta H_{solute} > \Delta H_{hydration}$, the amount of energy required to separate the solute into its constituent ions is greater than the energy given off when the ions are hydrated. ΔH_{soln} is therefore positive, the solution process is endothermic (if a solution forms at all), and the resulting solution feels cool to the touch. (3) If $\Delta H_{solute} \approx \Delta H_{hydration}$, the amount of energy required to separate the solute into its constituent ions is about equal to the energy given off when the ions are hydrated. ΔH_{soln} is therefore approximately zero, the solution process is neither appreciably exothermic nor appreciably endothermic, and there is no noticeable change in temperature.

14.10 In any solution formation, the initial rate of dissolution far exceeds the rate of deposition. But as the concentration of dissolved solute increases, the rate of deposition also increases. Eventually, the rates of dissolution and deposition become equal—dynamic equilibrium has been reached.

A saturated solution is a solution in which the dissolved solute is in dynamic equilibrium with the solid (or undissolved) solute. If you add more solute to a saturated solution, it will not dissolve.

An unsaturated solution is a solution containing less than the equilibrium amount of solute. If you add more solute to an unsaturated solution, it will dissolve.

A supersaturated solution is a solution containing more than the equilibrium amount of solute. Such solutions are unstable, and the excess solute normally precipitates out of the solution. However, in some cases, if left undisturbed, a supersaturated solution can exist for an extended period of time.

14.11 Although there are exceptions, the solubility of most solids in water increases with increasing temperature. Recrystallization is a common technique to purify a solid. In this technique, the solid is put into water (or some other solvent) at an elevated temperature. Enough solid is added to the solvent to create a saturated solution at the elevated temperature. As the solution cools, it becomes supersaturated and the excess solid begins to come out of solution. If the solution cools slowly, the solid forms crystals as it comes out of solution. The crystalline structure tends to reject impurities, resulting in a purer solid. The solvent is chosen so that the solubility of the impurities is high at the lower temperature, reducing the tendency to co-precipitate.

14.12 The solubility of gases in liquids decreases with increasing temperature. The decreasing solubility of gases with increasing temperature results in a lower oxygen concentration available for fish and other aquatic life in warm waters.

14.13 The higher the pressure of a gas above a liquid, the more soluble the gas in the liquid. In a sealed can of soda pop, for example, the carbon dioxide is maintained in solution by a high pressure of carbon dioxide within the can. When the can is opened, this pressure is released and the solubility of carbon dioxide decreases, resulting in bubbling.

14.14 Henry's law quantifies the solubility of gases with increasing pressure as follows: $S_{gas} = k_H P_{gas}$, where S_{gas} is the solubility of the gas; k_H is a constant of proportionality (called the Henry's law constant) that depends on the specific solute, solvent, and temperature; and P_{gas} is the partial pressure of the gas. The equation simply shows that the solubility of a gas in a liquid is directly proportional to the pressure of the gas above the liquid. If the solubility of a gas is known at a certain temperature, the solubility at another pressure at this temperature can be calculated.

14.15 The common ways of reporting solution concentration include the following:

(a) Molarity: $M = \frac{\text{amount solute (moles)}}{\text{volume solution (L)}}$

(b) molality: $m = \frac{\text{amount solute (moles)}}{\text{mass solvent (kg)}}$

(c) parts by mass: $\frac{\text{mass solute}}{\text{mass solution}} \times$ multiplication factor, where percent by mass (%) factor = 100, parts per million by mass (ppm) factor = 10^6, and parts per billion by mass (ppb) factor = 10^9

(d) parts by volume (%, ppm, ppb): $\frac{\text{volume solute}}{\text{volume solution}} \times$ multiplication factor, where the same factors as in parts by mass are used

(e) mole fraction: $\chi = \frac{\text{amount solute (in moles)}}{\text{total amount of solute and solvent (in moles)}}$

(f) mole percent $\chi \times 100\%$

14.16 Parts by mass and parts by volume are ratios of masses and volume, respectively. A parts by mass concentration is the ratio of the mass of the solute to the mass of the solution, all multiplied by a multiplication factor, where percent by mass (%) is the desired unit, the factor = 100; where parts per million by mass (ppm) is the desired unit, the factor = 10^6; and for parts per billion by mass (ppb), the factor = 10^9. The size of the multiplication factor depends on the concentration of the solution. For example, in percent by mass, the multiplication factor is 100%; so percent by mass = $\frac{\text{mass solute}}{\text{mass solution}} \times 100\%$. A solution with a concentration of 28% by mass contains 28 g of solute per 100 g of solution.

14.17 The vapor pressure of the solution is lower than the vapor pressure of the pure solvent. The simplest explanation for the lowering of the vapor pressure of a solution relative to that of the pure solvent is related to the concept of dynamic equilibrium. In dynamic equilibrium, the rate of vaporization is equal to the rate of condensation. When a nonvolatile solute is added, however, the solute particles interfere with the ability of the solvent particles to vaporize because they occupy some of the surface area formerly occupied by the solvent. The rate of vaporization is therefore diminished compared to that of the pure solvent. The change in the rate of vaporization creates an imbalance in the rates; the rate of condensation is now greater than the rate of vaporization. The net effect is that some of the molecules that were in the gas phase condense into the liquid. As they condense, the reduced number of molecules in the gas phase causes the rate of condensation to decrease. Eventually, the two rates become equal again, but only after the concentration of molecules in the gas phase has decreased, which means a lower vapor pressure for the solution compared to the pure solvent.

14.18 Raoult's law quantifies the relationship between the vapor pressure of a solution and its concentration as $P_{\text{solution}} = \chi_{\text{solvent}} P^\circ_{\text{solvent}}$, where P_{solution} is the vapor pressure of the solution, χ_{solvent} is the mole fraction of the solvent, and P°_{solvent} is the vapor pressure of the pure solvent. This equation allows you to calculate the vapor pressure of a solution or to calculate the concentration of a solution, given the vapor pressure of the solution.

14.19 An ideal solution follows Raoult's law at all concentrations for both the solute and the solvent. A nonideal solution exhibits deviations from Raoult's law in the vapor pressure of a component as the mole fraction of this component decreases from 1 (i.e., the pure component).

14.20 If the solute–solvent interactions are particularly strong (stronger than solvent–solvent interactions), the solute tends to prevent the solvent from vaporizing as easily as it would otherwise and the vapor pressure of the solution will be less than that predicted by Raoult's law. If the solute–solvent interactions are particularly weak (weaker than solvent–solvent interactions), the solute tends to allow more vaporization than would occur with just the solvent and the vapor pressure of the solution will be greater than predicted by Raoult's law.

14.21 A nonvolatile solute lowers the vapor pressure of a solution relative to that of the pure solvent. The vapor pressure lowering occurs at all temperatures, which shifts the vaporization curve in the phase diagram. This means that the temperature must be raised above the pure solvent normal boiling point for the solution vapor pressures to be raised to

1 atm. This shift also results in a lowering of where the vapor pressure curve intersects the solid–gas curve. The net effect is that the solution has a lower melting point and a higher boiling point than the pure solvent.

14.22 Colligative properties depend on the amount of solute and not the type of solute. Examples of colligative properties are vapor pressure lowering, freezing point depression, boiling point elevation, and osmotic pressure.

14.23 Osmosis is defined as the flow of solvent from a solution of lower solute concentration to one of higher solute concentration through a semipermeable membrane—a membrane that selectively allows some substances, but not others, to pass through. The osmotic pressure is the pressure required to stop the osmotic flow and is given by the following equation: $\Pi = MRT$.

14.24 The van't Hoff factor (i) is the ratio of moles of particles in solution to moles of formula units dissolved: $i = \dfrac{\text{moles of particles}}{\text{moles of formula units dissolved}}$.

The van't Hoff factor often does not match its theoretical value because the ionic solute is not completely dissolved into the expected number of ions, leaving ion pairs in solution. The result is that the number of particles in the solution is not as high as theoretically expected.

Problems by Topic

Solubility

14.25
(a) hexane, toluene, or CCl_4; dispersion forces
(b) water, methanol, acetone; dispersion, dipole–dipole, hydrogen bonding (except for acetone)
(c) hexane, toluene, or CCl_4; dispersion forces
(d) water, acetone, methanol, ethanol; dispersion, ion–dipole

14.26
(a) water, methanol, ethanol; dispersion, dipole–dipole, hydrogen bonding
(b) water, acetone, methanol, ethanol; dispersion, ion–dipole
(c) hexane, toluene, or CCl_4; dispersion forces
(d) water, acetone, methanol, ethanol; dispersion, ion–dipole

14.27 $HOCH_2CH_2CH_2OH$ would be more soluble in water because it has —OH groups on both ends of the molecule; so it can hydrogen bond on both ends.

14.28 CH_2Cl_2 would be more soluble in water because it is a polar molecule and can exhibit dipole–dipole interactions with the water molecules. CCl_4 is a nonpolar molecule.

14.29
(a) water; dispersion, dipole–dipole, hydrogen bonding
(b) hexane; dispersion forces
(c) water; dispersion, dipole–dipole
(d) water; dispersion, dipole–dipole, hydrogen bonding

14.30
(a) hexane; dispersion forces
(b) water; dispersion, dipole–dipole, hydrogen bonding
(c) hexane; dispersion forces
(d) water; dispersion, dipole–dipole, hydrogen bonding

Energetics of Solution Formation

14.31
(a) endothermic
(b) The lattice energy is greater in magnitude than is the heat of hydration.

(c)

E

$NH_4^+(g) + Cl^-(g)$

$\Delta H_{hydration}$

$\Delta H = -$lattice energy

$NH_4^+(aq) + Cl^-(aq)$

ΔH_{sol}

$NH_4Cl(s)$

(d) The solution forms because chemical systems tend toward greater entropy.

14.32 (a) exothermic

(b) The lattice energy is smaller in magnitude than is the heat of hydration.

(c)

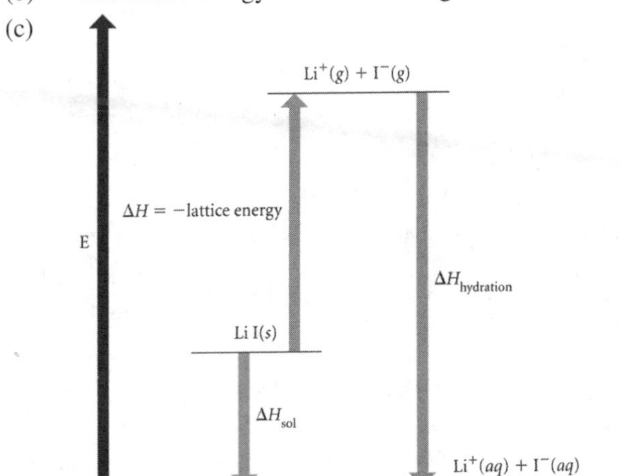

(d) The solution forms because chemical systems tend toward lower energy and greater entropy.

14.33 **Given:** $AgNO_3$: lattice energy $= -820.$ kJ/mol, $\Delta H_{soln} = +22.6$ kJ/mol **Find:** $\Delta H_{hydration}$

Conceptual Plan: lattice energy, $\Delta H_{soln} \rightarrow \Delta H_{hydration}$

$$\Delta H_{soln} = \Delta H_{solute} + \Delta H_{hydration} \text{ where } \Delta H_{solute} = -\Delta H_{lattice}$$

Solution: $\Delta H_{soln} = \Delta H_{solute} + \Delta H_{hydration}$ where $\Delta H_{solute} = -\Delta H_{lattice}$ so $\Delta H_{hydration} = \Delta H_{soln} + \Delta H_{lattice}$

$\Delta H_{hydration} = 22.6 \text{ kJ/mol} - 820. \text{ kJ/mol} = -797 \text{ kJ/mol}$

Check: The units (kJ/mol) are correct. The magnitude of the answer (−800 kJ/mol) makes physical sense because the lattice energy is so negative; thus, it dominates the calculation.

14.34 **Given:** LiCl: lattice energy $= -834$ kJ/mol, $\Delta H_{soln} = -37.0$ kJ/mol; NaCl: lattice energy $= -769$ kJ/mol, $\Delta H_{soln} = +3.88$ kJ/mol **Find:** $\Delta H_{hydration}$ and which has stronger ion–dipole interactions

Conceptual Plan: lattice energy, $\Delta H_{soln} \rightarrow \Delta H_{hydration}$ then compare values

$$\Delta H_{soln} = \Delta H_{solute} + \Delta H_{hydration} \text{ where } \Delta H_{solute} = -\Delta H_{lattice}$$

Solution: $\Delta H_{soln} = \Delta H_{solute} + \Delta H_{hydration}$ where $\Delta H_{solute} = -\Delta H_{lattice}$ so $\Delta H_{hydration} = \Delta H_{soln} + \Delta H_{lattice}$

LiCl: $\Delta H_{hydration} = -37.0 \text{ kJ/mol} - 834 \text{ kJ/mol} = -871 \text{ kJ/mol}$

NaCl: $\Delta H_{hydration} = +3.88 \text{ kJ/mol} - 769 \text{ kJ/mol} = -765 \text{ kJ/mol}$

Because $\Delta H_{hydration}$ of LiCl is more negative than for NaCl, LiCl has the stronger ion–dipole interactions.

Check: The units (kJ/mol) are correct. The magnitude of the answers (−900 and −800 kJ/mol) makes physical sense because the lattice energies are so negative; thus, they dominate the calculation. We expect stronger interactions with lithium because the Li^+ ion is smaller than the Na^+ ion. Its charge density is higher, and it will interact more strongly with the dipoles of the water molecules.

14.35 **Given:** LiI: lattice energy $= -7.3 \times 10^2$ kJ/mol, $\Delta H_{hydration} = -793$ kJ/mol; 15.0 g LiI
Find: ΔH_{soln} and heat evolved
Conceptual Plan: lattice energy, $\Delta H_{hydration} \rightarrow \Delta H_{soln}$ and g → mol then mol, $\Delta H_{soln} \rightarrow q$

$$\Delta H_{soln} = \Delta H_{solute} + \Delta H_{hydration} \text{ where } \Delta H_{solute} = -\Delta H_{lattice} \qquad \frac{1 \text{ mol}}{133.84 \text{ g}} \qquad q = n\Delta H_{soln}$$

Solution: $\Delta H_{soln} = \Delta H_{solute} + \Delta H_{hydration}$ where $\Delta H_{solute} = -\Delta H_{lattice}$ so $\Delta H_{soln} = \Delta H_{hydration} - \Delta H_{lattice}$

$\Delta H_{soln} = -793 \text{ kJ/mol} - (-7\underline{3}0 \text{ kJ/mol}) = -\underline{6}3 \text{ kJ/mol} = -6.3 \times 10^1 \text{ kJ/mol}$ and

$$15.0 \text{ g} \times \frac{1 \text{ mol}}{133.84 \text{ g}} = 0.11\underline{2}074 \text{ mol then}$$

$$q = n\Delta H_{soln} = 0.11\underline{2}074 \text{ mol} \times -6.3 \times 10^1 \frac{\text{kJ}}{\text{mol}} = -7.1 \text{ kJ or } 7.0 \text{ kJ released}$$

Check: The units (kJ/mol and kJ) are correct. The magnitude of the answer (−60 kJ/mol) makes physical sense because the lattice energy and the heat of hydration are about the same. The magnitude of the heat (7 kJ) makes physical sense because 15 g is much less than a mole; thus, the amount of heat released is going to be small.

14.36 **Given:** KNO_3: lattice energy $= -163.8$ kcal/mol, $\Delta H_{hydration} = -155.5$ kcal/mol; 1.00×10^2 kJ absorbed
Find: ΔH_{soln} and m (KNO_3)
Conceptual Plan: lattice energy, $\Delta H_{hydration} \rightarrow \Delta H_{soln}$ (kcal) → ΔH_{soln} (kcal) then

$$\Delta H_{soln} = \Delta H_{solute} + \Delta H_{hydration} \text{ where } \Delta H_{solute} = -\Delta H_{lattice}$$

mol, $q \rightarrow \Delta H_{soln}$ then mol → g

$$q = n\Delta H_{soln} \qquad \frac{101.11 \text{ g}}{1 \text{ mol}}$$

Solution: $\Delta H_{soln} = \Delta H_{solute} + \Delta H_{hydration}$ where $\Delta H_{solute} = -\Delta H_{lattice}$ so $\Delta H_{soln} = \Delta H_{hydration} - \Delta H_{lattice}$
$\Delta H_{soln} = [-155.5 \text{ kcal/mol} - (-163.8 \text{ kcal/mol})](4.184 \text{ kJ/kcal}) = 3\underline{4}.7 \text{ kJ/mol}$ then $q = n\Delta H_{soln}$

Rearrange to solve for n. $n = \dfrac{q}{\Delta H_{soln}} = \dfrac{1.00 \times 10^2 \text{ kJ}}{34.7 \dfrac{\text{kJ}}{\text{mol}}} = 2.\underline{8}818 \text{ mol then } 2.\underline{8}818 \text{ mol} \times \dfrac{101.11 \text{ g}}{1 \text{ mol}} = 2.9 \times 10^2 \text{ g}$

Check: The units (kJ/mol and g) are correct. The magnitude of the answer (+35 kJ/mol) makes physical sense because the lattice energy and the heat of hydration are about the same, with the lattice energy dominating; therefore, the answer is positive. The problem hints at a positive heat of solution by saying that heat is absorbed. The magnitude of the mass (290) makes physical sense because 100 kJ will require over 2 moles of salt.

Solution Equilibrium and Factors Affecting Solubility

14.37 The solution is unsaturated because we are dissolving 25 g of NaCl per 100 g of water and the solubility from the figure is ~35 g NaCl per 100 g of water at 25 °C.

14.38 The solution is almost saturated because we are dissolving 32 g of KNO_3 per 100 g of water and the solubility from the figure is ~36 g KNO_3 per 100 g of water at 25 °C.

14.39 At 40 °C, the solution has 45 g of KNO_3 per 100 g of water, and it can contain up to 63 g of KNO_3 per 100 g of water. At 0 °C, the solubility from the figure is ~14 g KNO_3 per 100 g of water; so ~31 g KNO_3 per 100 g of water will precipitate out of solution.

14.40 At 60 °C, the solution has 42 g of KCl per 100 g of water, and it can contain up to 45 g of KCl per 100 g of water. At 0 °C, the solubility from the figure is ~26 g KCl per 100 g of water; so ~16 g KCl per 100 g of water will precipitate out of solution.

14.41 Because the solubility of gases decreases as the temperature increases, boiling will cause dissolved oxygen to be removed from the solution.

14.42 Because the solubility of gases decreases as the temperature increases, dissolved oxygen was removed from the solution and there was no oxygen in the water for the fish to breathe.

14.43 Henry's law states that as pressure increases, nitrogen will more easily dissolve in blood. To reverse this process, divers should ascend to lower pressures.

14.44 Henry's law states that as pressure increases, oxygen will more easily dissolve in blood. To reverse this process, divers should ascend to lower pressures or breathe special gas mixtures with lower oxygen levels.

14.45 **Given:** room temperature, 80.0 L aquarium, $P_{Total} = 1.0$ atm; $\chi_{N_2} = 0.78$ **Find:** $m\ (N_2)$
Other: $k_H(N_2) = 6.1 \times 10^{-4}$ M/L at 25 °C
Conceptual Plan: $P_{Total}, \chi_{N_2} \rightarrow P_{N_2}$ **then** $P_{N_2}, k_H(N_2) \rightarrow S_{N_2}$ **then L → mol → g**

$P_{N_2} = \chi_{N_2} P_{Total}$ $\quad S_{N_2} = k_H(N_2)P_{N_2}$ $\quad \mathrm{M} = \dfrac{\text{amount solute (moles)}}{\text{volume solution (L)}}\ \dfrac{28.01\ \text{g}\ N_2}{1\ \text{mol}\ N_2}$

Solution: $P_{N_2} = \chi_{N_2} P_{Total} = 0.78 \times 1.0\ \text{atm} = 0.78$ atm then

$S_{N_2} = k_H(N_2)P_{N_2} = 6.1 \times 10^{-4} \dfrac{\text{M}}{\text{atm}} \times 0.78\ \text{atm} = 4.\underline{7}58 \times 10^{-4}$ M then

$80.0\ \text{L} \times 4.\underline{7}58 \times 10^{-4} \dfrac{\text{mol}}{\text{L}} \times \dfrac{28.01\ \text{g}}{1\ \text{mol}} = 1.1\ \text{g}$

Check: The units (g) are correct. The magnitude of the answer (1 g) seems reasonable because we have 80 L of water and expect much less than a mole of nitrogen.

14.46 **Given:** helium, 25 °C, $P_{He} = 1.0$ atm **Find:** S_{He} (M) **Other:** $k_H(He) = 3.7 \times 10^{-4}$ M/L at 25 °C
Conceptual Plan: $P_{He}, k_H(He) \rightarrow S_{He}$

$S_{He} = k_H(He)P_{He}$

Solution: $S_{He} = k_H(He)P_{He} = 3.7 \times 10^{-4} \dfrac{\text{M}}{\text{atm}} \times 1.0\ \text{atm} = 3.7 \times 10^{-4}$ M

Check: The units (M) are correct. The magnitude of the answer (10^{-4} M) seems reasonable because this is the value of k_H.

Concentrations of Solutions

14.47 **Given:** NaCl and water; 112 g NaCl in 1.00 L solution **Find:** M, m, and mass percent **Other:** $d = 1.08$ g/mL
Conceptual Plan: g_{NaCl} **→ mol and L → mL →** g_{soln} **and** $g_{soln}, g_{NaCl} \rightarrow g_{H_2O} \rightarrow kg_{H_2O}$ **then**

$\dfrac{1\ \text{mol NaCl}}{58.44\ \text{g NaCl}}$ $\quad \dfrac{1000\ \text{mL}}{1\ \text{L}}$ $\quad \dfrac{1.08\ \text{g}}{1\ \text{mL}}$ $\quad g_{H_2O} = g_{soln} - g_{NaCl}$ $\quad \dfrac{1\ \text{kg}}{1000\ \text{g}}$

mol, V **→ M and mol,** kg_{H_2O} **→** m **and** g_{soln}, g_{NaCl} **→ mass percent**

$\mathrm{M} = \dfrac{\text{amount solute (moles)}}{\text{volume solution (L)}}$ $\quad m = \dfrac{\text{amount solute (moles)}}{\text{mass solvent (kg)}}$ $\quad \text{mass percent} = \dfrac{\text{mass solute}}{\text{mass solution}} \times 100\%$

Solution: $112\ \text{g NaCl} \times \dfrac{1\ \text{mol NaCl}}{58.44\ \text{g NaCl}} = 1.9\underline{1}64956$ mol NaCl and

$1.00\ \text{L} \times \dfrac{1000\ \text{mL}}{1\ \text{L}} \times \dfrac{1.08\ \text{g}}{1\ \text{mL}} = 10\underline{8}0$ g soln and

$g_{H_2O} = g_{soln} - g_{NaCl} = 10\underline{8}0\ \text{g} - 112\ \text{g} = 9\underline{6}8\ \text{g}\ H_2O \times \dfrac{1\ \text{kg}}{1000\ \text{g}} = 0.9\underline{6}8\ \text{kg}\ H_2O$ then

$\mathrm{M} = \dfrac{\text{amount solute (moles)}}{\text{volume solution (L)}} = \dfrac{1.9\underline{1}64956\ \text{mol NaCl}}{1.00\ \text{L soln}} = 1.92\ \text{M}$ and

$m = \dfrac{\text{amount solute (moles)}}{\text{mass solvent (kg)}} = \dfrac{1.9\underline{1}64956\ \text{mol NaCl}}{0.9\underline{6}8\ \text{kg}\ H_2O} = 2.0\ m$ and

$\text{mass percent} = \dfrac{\text{mass solute}}{\text{mass solution}} \times 100\% = \dfrac{112\ \text{g NaCl}}{10\underline{8}0\ \text{g soln}} \times 100\% = 10.4\%$ by mass

Check: The units (M, m, and percent by mass) are correct. The magnitude of the answer (2 M) seems reasonable because we have 112 g NaCl, which is a couple of moles and we have 1 L. The magnitude of the answer (2 m) seems reasonable because it is a little higher than the molarity, which we expect because we only use the solvent weight in the denominator. The magnitude of the answer (10%) seems reasonable because we have 112 g NaCl and just over 1000 g of solution.

14.48 **Given:** KNO_3 and water; 72.5 g KNO_3 in 2.00 L solution **Find:** M, m, and mass percent **Other:** d = 1.05 g/mL
Conceptual Plan: $g_{KNO_3} \rightarrow$ **mol and L** $\rightarrow$ **mL** $\rightarrow g_{soln}$ **and** $g_{soln}, g_{KNO_3} \rightarrow g_{H_2O} \rightarrow kg_{H_2O}$ **then**

$$\frac{1 \text{ mol KNO}_3}{101.11 \text{ g KNO}_3} \qquad \frac{1000 \text{ mL}}{1 \text{ L}} \quad \frac{1.05 \text{ g}}{1 \text{ mL}} \qquad g_{H_2O} = g_{soln} - g_{KNO_3} \quad \frac{1 \text{ kg}}{1000 \text{ g}}$$

mol, $V \rightarrow$ **M and mol,** $kg_{H_2O} \rightarrow m$ **and** $g_{soln}, g_{KNO_3} \rightarrow$ **mass percent**

$$\text{M} = \frac{\text{amount solute (moles)}}{\text{volume solution (L)}} \quad m = \frac{\text{amount solute (moles)}}{\text{mass solvent (kg)}} \quad \text{mass percent} = \frac{\text{mass solute}}{\text{mass solution}} \times 100\%$$

Solution: $72.5 \text{ g KNO}_3 \times \frac{1 \text{ mol KNO}_3}{101.11 \text{ g KNO}_3} = 0.71\underline{7}04085 \text{ mol KNO}_3$ and

$$2.00 \text{ L} \times \frac{1000 \text{ mL}}{1 \text{ L}} \times \frac{1.05 \text{ g}}{1 \text{ mL}} = 21\underline{0}0 \text{ g soln and}$$

$$g_{H_2O} = g_{soln} - g_{KNO_3} = 21\underline{0}0 \text{ g} - 72.5 \text{ g} = 20\underline{2}7.5 \text{ g H}_2\text{O} \times \frac{1 \text{ kg H}_2\text{O}}{1000 \text{ g H}_2\text{O}} = 2.0\underline{2}75 \text{ kg H}_2\text{O then}$$

$$\text{M} = \frac{\text{amount solute (moles)}}{\text{volume solution (L)}} = \frac{0.71\underline{7}04085 \text{ mol KNO}_3}{2.00 \text{ L soln}} = 0.358 \text{ M and}$$

$$m = \frac{\text{amount solute (moles)}}{\text{mass solvent (kg)}} = \frac{0.71\underline{7}04085 \text{ mol KNO}_3}{2.0\underline{2}75 \text{ kg H}_2\text{O}} = 0.354\ m \text{ and}$$

$$\text{mass percent} = \frac{\text{mass solute}}{\text{mass solution}} \times 100\% = \frac{72.5 \text{ g KNO}_3}{21\underline{0}0 \text{ g soln}} \times 100\% = 3.45\% \text{ by mass}$$

Check: The units (M, m, and percent by mass) are correct. The magnitude of the answer (0.358 M) seems reasonable because we have 72.5 g KNO_3, which is less than a mole and we have 2 L. The magnitude of the answer (0.354 m) seems reasonable because it is a little higher than the molarity, which we expect because we only use the solvent weight in the denominator. The magnitude of the answer (3.45%) seems reasonable because we have 72.5 g KNO_3 and ~2010 g of solution.

14.49 **Given:** initial solution: 50.0 mL of 5.00 M KI; final solution contains 3.05 g KI in 25.0 mL
Find: final volume of diluted solution
Conceptual Plan: final solution: $g_{KI} \rightarrow$ **mol and mL** $\rightarrow$ **L then mol,** $V \rightarrow M_2$ **then** $M_1, V_1, M_2 \rightarrow V_2$

$$\frac{1 \text{ mol KI}}{166.006 \text{ g KI}} \qquad \frac{1 \text{ L}}{1000 \text{ mL}} \qquad \text{M} = \frac{\text{amount solute (moles)}}{\text{volume solution (L)}} \qquad M_1V_1 = M_2V_2$$

Solution: $3.05 \text{ g KI} \times \frac{1 \text{ mol KI}}{166.006 \text{ g KI}} = 0.018\underline{3}7283 \text{ mol KI}$ and $25.0 \text{ mL} \times \frac{1 \text{ L}}{1000 \text{ mL}} = 0.0250 \text{ L}$

then $\text{M} = \frac{\text{amount solute (moles)}}{\text{volume solution (L)}} = \frac{0.018\underline{3}7283 \text{ mol KI}}{0.0250 \text{ L soln}} = 0.73\underline{4}9132 \text{ M}$ then $M_1V_1 = M_2V_2$

Rearrange to solve for V_2. $V_2 = \frac{M_1}{M_2} \times V_1 = \frac{5.00 \text{ M}}{0.73\underline{4}9132 \text{ M}} \times 50.0 \text{ mL} = 340. \text{ mL diluted volume}$

Check: The units (mL) are correct. The magnitude of the answer (340 mL) seems reasonable because we are starting with a concentration of 5 M and ending with a concentration of less than 1 M.

14.50 **Given:** initial solution: 125 mL of 8.00 M $CuCl_2$; final solution contains 4.67 g $CuCl_2$ in 50.0 mL
Find: final volume of diluted solution
Conceptual Plan: final solution: $g_{CuCl_2} \rightarrow$ **mol and mL** $\rightarrow$ **L then mol,** $V \rightarrow M_2$ **then** $M_1, V_1, M_2 \rightarrow V_2$

$$\frac{1 \text{ mol CuCl}_2}{134.45 \text{ g CuCl}_2} \qquad \frac{1 \text{ L}}{1000 \text{ mL}} \qquad \text{M} = \frac{\text{amount solute (moles)}}{\text{volume solution (L)}} \qquad M_1V_1 = M_2V_2$$

Solution: $4.67 \text{ g CuCl}_2 \times \frac{1 \text{ mol CuCl}_2}{134.45 \text{ g CuCl}_2} = 0.034\underline{7}3410 \text{ mol CuCl}_2$ and $50.0 \text{ mL} \times \frac{1 \text{ L}}{1000 \text{ mL}} = 0.0500 \text{ L}$

then $\text{M} = \frac{\text{amount solute (moles)}}{\text{volume solution (L)}} = \frac{0.034\underline{7}3410 \text{ mol CuCl}_2}{0.0500 \text{ L soln}} = 0.69\underline{4}6820 \text{ M}$ then $M_1V_1 = M_2V_2$

Rearrange to solve for V_2.

$$V_2 = \frac{M_1}{M_2} \times V_1 = \frac{8.00 \text{ M}}{0.69\underline{4}6820 \text{ M}} \times 125 \text{ mL} = 1440 \text{ mL diluted volume}$$

Check: The units (mL) are correct. The magnitude of the answer (1440 mL) seems reasonable because we are starting with a concentration of 8 M and ending with a concentration of less than 1 M.

14.51 **Given:** $AgNO_3$ and water; 3.4% Ag by mass, 4.8 L solution **Find:** m (Ag) **Other:** $d = 1.01\ g/mL$

Conceptual Plan: $\mathbf{L \rightarrow mL \rightarrow g_{soln} \rightarrow g_{Ag}}$

$$\frac{1000\ mL}{1\ L} \quad \frac{1.01\ g}{1\ mL} \quad \frac{3.4\ g\ Ag}{100\ g\ soln}$$

Solution: $4.8\ \cancel{L} \times \dfrac{1000\ \cancel{mL}}{1\ \cancel{L}} \times \dfrac{1.01\ g}{1\ \cancel{mL}} = 4\underline{8}48$ g soln then

$$4\underline{8}48\ \cancel{g\ soln} \times \frac{3.4\ g\ Ag}{100\ \cancel{g\ soln}} = 160\ g\ Ag = 1.6 \times 10^2\ g\ Ag$$

Check: The units (g) are correct. The magnitude of the answer (160 g) seems reasonable because we have almost 5000 g of solution.

14.52 **Given:** dioxin and water; 0.085% dioxin by mass, 2.5 L solution **Find:** m (dioxin) **Other:** $d = 1.00\ g/mL$

Conceptual Plan: $\mathbf{L \rightarrow mL \rightarrow g_{soln} \rightarrow g_{dioxin}}$

$$\frac{1000\ mL}{1\ L} \quad \frac{1.00\ g}{1\ mL} \quad \frac{0.085\ g\ dioxin}{100\ g\ soln}$$

Solution: $2.5\ \cancel{L} \times \dfrac{1000\ \cancel{mL}}{1\ \cancel{L}} \times \dfrac{1.00\ g}{1\ \cancel{mL}} = 2\underline{5}00$ g soln then $2\underline{5}00\ \cancel{g\ soln} \times \dfrac{0.085\ g\ dioxin}{100\ \cancel{g\ soln}} = 2.1$ g dioxin

Check: The units (g) are correct. The magnitude of the answer (2 g) seems reasonable because we have 2500 g of solution and a low concentration.

14.53 **Given:** Ca^{2+} and water; 0.0085% Ca^{2+} by mass, 1.2 g Ca **Find:** m (water)

Conceptual Plan: $\mathbf{g_{Ca} \rightarrow g_{soln} \rightarrow g_{H_2O}}$

$$\frac{100\ g\ soln}{0.0085\ g\ Ca} \qquad g_{H_2O} = g_{soln} - g_{Ca}$$

Solution: $1.2\ \cancel{g\ Ca} \times \dfrac{100\ g\ soln}{0.0085\ \cancel{g\ Ca}} = 1\underline{4}118$ g soln then

$$g_{H_2O} = g_{soln} - g_{Ca} = 1\underline{4}118\ g - 1.2\ g = 1.4 \times 10^4\ g\ water$$

Check: The units (g) are correct. The magnitude of the answer (10^4 g) seems reasonable because we have such a low concentration of Ca.

14.54 **Given:** Pb and water; 0.0011% Pb by mass, 150 mg Pb **Find:** V (mL) **Other:** $d = 1.0\ g/mL$

Conceptual Plan: $\mathbf{mg_{Pb} \rightarrow g_{Pb} \rightarrow g_{soln} \rightarrow mL}$

$$\frac{1\ g\ Pb}{1000\ mg\ Pb} \quad \frac{100\ g\ soln}{0.0011\ g\ Pb} \quad \frac{1\ mL}{1.0\ g\ soln}$$

Solution: $150\ \cancel{mg\ Pb} \times \dfrac{1\ \cancel{g\ Pb}}{1000\ \cancel{mg\ Pb}} \times \dfrac{100\ \cancel{g\ soln}}{0.0011\ \cancel{g\ Pb}} \times \dfrac{1\ mL}{1.0\ \cancel{g\ soln}} = 1.4 \times 10^4\ mL$

Check: The units (mL) are correct. The magnitude of the answer (10^4 mL) seems reasonable because we have such a low concentration of Pb.

14.55 **Given:** concentrated HNO_3: 70.3% HNO_3 by mass, $d = 1.41\ g/mL$; final solution: 1.15 L of 0.100 M HNO_3

Find: describe final solution preparation

Conceptual Plan: $\mathbf{M_2, V_2 \rightarrow mol_{HNO_3} \rightarrow g_{HNO_3} \rightarrow g_{conc\ acid} \rightarrow mL_{conc\ acid}}$ **then describe method**

$$mol = MV \quad \frac{63.02\ g\ HNO_3}{1\ mol\ HNO_3} \quad \frac{100\ g\ conc\ acid}{70.3\ g\ HNO_3} \quad \frac{1\ mL}{1.41\ g}$$

Solution: $mol = MV = 0.100 \dfrac{mol\ HNO_3}{1\ \cancel{L\ soln}} \times 1.15\ \cancel{L\ soln} = 0.115\ mol\ HNO_3$ then

$$0.115\ \cancel{mol\ HNO_3} \times \frac{63.02\ \cancel{g\ HNO_3}}{1\ \cancel{mol\ HNO_3}} \times \frac{100\ \cancel{g\ conc\ acid}}{70.3\ \cancel{g\ HNO_3}} \times \frac{1\ mL\ conc\ acid}{1.41\ \cancel{g\ conc\ acid}} = 7.31\ mL\ conc\ acid$$

Prepare the solution by putting about 1.00 L of distilled water in a container. Carefully pour in the 7.31 mL of the concentrated acid, mix the solution, and allow it to cool. Finally, add enough water to generate the total volume of solution (1.15 L). It is important to add acid to water, not the reverse, because such a large amount of heat is released upon mixing.

Check: The units (mL) are correct. The magnitude of the answer (7 mL) seems reasonable because we are starting with such a concentrated solution and diluting it to a low concentration.

14.56 **Given:** concentrated HCl: 37.0% HCl by mass, $d = 1.20 \text{ g/mL}$; final solution: 2.85 L of 0.500 M HCl
Find: describe final solution preparation
Conceptual Plan: $M_2, V_2 \rightarrow \text{mol}_{HCl} \rightarrow \text{g}_{HCl} \rightarrow \text{g}_{\text{conc acid}} \rightarrow \text{mL}_{\text{conc acid}}$ **then describe method**

$$mol = MV \qquad \frac{36.46 \text{ g HCl}}{1 \text{ mol HCl}} \qquad \frac{100 \text{ g conc acid}}{37.0 \text{ g HCl}} \qquad \frac{1 \text{ mL}}{1.20 \text{ g}}$$

Solution: $mol = MV = 0.500 \dfrac{\text{mol HCl}}{1 \text{ L soln}} \times 2.85 \text{ L soln} = 1.4\underline{2}5 \text{ mol HCl}$ then

$$1.4\underline{2}5 \text{ mol HCl} \times \frac{36.46 \text{ g HCl}}{1 \text{ mol HCl}} \times \frac{100 \text{ g conc acid}}{37.0 \text{ g HCl}} \times \frac{1 \text{ mL conc acid}}{1.20 \text{ g conc acid}} = 117 \text{ mL conc acid}$$

Prepare the solution by putting about 2.5 L of distilled water in a container. Carefully pour in the 117 mL of the concentrated acid, mix the solution, and allow it to cool. Finally, add enough water to generate the total volume of solution required (2.85 L). It is important to add acid to water, not the reverse, because such a large amount of heat is released upon mixing.

Check: The units (mL) are correct. The magnitude of the answer (117 mL) seems reasonable because we are starting with such a concentrated solution and diluting it to a low concentration.

14.57 (a) **Given:** 1.00×10^2 mL of 0.500 M KCl **Find:** describe final solution preparation
Conceptual Plan: **mL → L then** $M, V \rightarrow \text{mol}_{KCl} \rightarrow \text{g}_{KCl}$ **then describe method**

$$\frac{1 \text{ L}}{1000 \text{ mL}} \qquad mol = MV \qquad \frac{74.55 \text{ g KCl}}{1 \text{ mol KCl}}$$

Solution: $1.00 \times 10^2 \text{ mL} \times \dfrac{1 \text{ L}}{1000 \text{ mL}} = 0.100 \text{ L}$

$$mol = MV = 0.500 \frac{\text{mol KCl}}{1 \text{ L soln}} \times 0.100 \text{ L soln} = 0.0500 \text{ mol KCl}$$

then $0.0500 \text{ mol KCl} \times \dfrac{74.55 \text{ g KCl}}{1 \text{ mol KCl}} = 3.73 \text{ g KCl}$

Prepare the solution by carefully adding 3.73 g KCl to a 100 mL volumetric flask. Add ~75 mL of distilled water and agitate the solution until the salt dissolves completely. Finally, add enough water to generate a total volume of solution (add water to the mark on the flask).

Check: The units (g) are correct. The magnitude of the answer (4 g) seems reasonable because we are making a small volume of solution and the formula weight of KCl is ~75 g/mol.

(b) **Given:** 1.00×10^2 g of 0.500 m KCl **Find:** describe final solution preparation
Conceptual Plan:
$m \rightarrow \text{mol}_{KCl}/1 \text{ kg solvent} \rightarrow \text{g}_{KCl}/1 \text{ kg solvent}$ **then** $\text{g}_{KCl}/1 \text{ kg solvent}, \text{g}_{soln} \rightarrow \text{g}_{KCl}, \text{g}_{H_2O}$

$$m = \frac{\text{amount solute (moles)}}{\text{mass solvent (kg)}} \qquad \frac{74.55 \text{ g KCl}}{1 \text{ mol KCl}} \qquad \text{g}_{soln} = \text{g}_{KCl} + \text{g}_{H_2O}$$

then describe method

Solution: $m = \dfrac{\text{amount solute (moles)}}{\text{mass solvent (kg)}}$ so $0.500\ m = \dfrac{0.500 \text{ mol KCl}}{1 \text{ kg } H_2O}$ so

$$\frac{0.500 \text{ mol KCl}}{1 \text{ kg } H_2O} \times \frac{74.55 \text{ g KCl}}{1 \text{ mol KCl}} = \frac{37.\underline{2}8 \text{ g KCl}}{1000 \text{ g } H_2O} \quad \text{g}_{soln} = \text{g}_{KCl} + \text{g}_{H_2O} \text{ so } \text{g}_{soln} - \text{g}_{KCl} = \text{g}_{H_2O}$$

Substitute into ratio. $\dfrac{37.\underline{2}8 \text{ g KCl}}{1037.28 \text{ g solution}} = \dfrac{x \text{ g KCl}}{100 \text{ g solution}}$. Multiply both sides by 100 g solution.

$$\frac{37.\underline{2}8 \text{ g KCl (100 g soln)}}{1037.28 \text{ g soln}} = 3.59 \text{ g}$$

$g_{H_2O} = g_{soln} - g_{KCl} = 100.\ g - 3.59\ g = 96.41\ g\ H_2O$

Prepare the solution by carefully adding 3.59 g KCl to a container with 96.41 g of distilled water and agitate the solution until the salt dissolves completely.

Check: The units (g) are correct. The magnitude of the answer (3.6 g) seems reasonable because we are making a small volume of solution and the formula weight of KCl is ~75 g/mol.

(c) **Given:** 1.00×10^2 g of 5.0% KCl by mass **Find:** describe final solution preparation

Conceptual Plan: $g_{soln} \rightarrow g_{KCl}$ **then** $g_{KCl}, g_{soln} \rightarrow g_{H_2O}$

$$\frac{5.0\ g\ KCl}{100\ g\ soln} \qquad g_{soln} = g_{KCl} + g_{H_2O}$$

then describe method

Solution: $1.00 \times 10^2\ \cancel{g\ soln} \times \frac{5.0\ g\ KCl}{100\ \cancel{g\ soln}} = 5.0\ g\ KCl$ then $g_{soln} = g_{KCl} + g_{H_2O}$

So $g_{H_2O} = g_{soln} - g_{KCl} = 100.\ g - 5.0\ g = 95\ g\ H_2O$

Prepare the solution by carefully adding 5.0 g KCl to a container with 95 g of distilled water and agitate the solution until the salt dissolves completely.

Check: The units (g) are correct. The magnitude of the answer (5 g) seems reasonable because we are making a small volume of solution and the solution is 5% by mass KCl.

14.58 (a) **Given:** 125 mL of 0.100 M $NaNO_3$ **Find:** describe final solution preparation

Conceptual Plan: **mL → L then** $M, V \rightarrow mol_{NaNO_3} \rightarrow g_{NaNO_3}$ **then describe method**

$$\frac{1\ L}{1000\ mL} \qquad mol = MV \quad \frac{85.00\ g\ NaNO_3}{1\ mol\ NaNO_3}$$

Solution: $125\ \cancel{mL} \times \frac{1\ L}{1000\ \cancel{mL}} = 0.125\ L$

$mol = MV = 0.100 \frac{mol\ NaNO_3}{1\ \cancel{L\ soln}} \times 0.125\ \cancel{L\ soln} = 0.0125\ mol\ NaNO_3$ then

$0.0125\ \cancel{mol\ NaNO_3} \times \frac{85.00\ g\ NaNO_3}{1\ \cancel{mol\ NaNO_3}} = 1.06\ g\ NaNO_3$

Prepare the solution by carefully adding 1.06 g $NaNO_3$ to a container. Add ~100 mL of distilled water and agitate the solution until the salt dissolves completely. Finally, add enough water to generate a total volume of solution (125 mL).

Check: The units (g) are correct. The magnitude of the answer (1 g) seems reasonable because we are making a small volume of solution and the formula weight of $NaNO_3$ is ~85 g/mol.

(b) **Given:** 125 g of 0.100 *m* $NaNO_3$ **Find:** describe final solution preparation

Conceptual Plan: $m \rightarrow mol_{NaNO_3}/1$ **kg solvent** $\rightarrow g_{NaNO_3}/1$ **kg solvent then**

$$m = \frac{\text{amount solute (moles)}}{\text{mass solvent (kg)}} \qquad \frac{85.00\ g\ NaNO_3}{1\ mol\ NaNO_3}$$

$g_{NaNO_3}/1$ **kg solvent,** $g_{soln} \rightarrow g_{NaNO_3}, g_{H_2O}$ **then describe method**

$$g_{soln} = g_{NaNO_3} + g_{H_2O}$$

Solution: $m = \frac{\text{amount solute (moles)}}{\text{mass solvent (kg)}}$ so $0.100\ m = \frac{0.500\ mol\ NaNO_3}{1\ kg\ H_2O}$ so

$\frac{0.100\ \cancel{mol\ NaNO_3}}{1\ kg\ H_2O} \times \frac{85.00\ g\ NaNO_3}{1\ \cancel{mol\ NaNO_3}} = \frac{8.5\underline{0}\ g\ NaNO_3}{1000\ g\ H_2O}$ then $g_{soln} = g_{NaNO_3} + g_{H_2O}$ so

$g_{soln} - g_{NaNO_3} = g_{H_2O}$ Substitute into ratio. $\frac{0.0085\underline{0}\ g\ NaNO_3}{1\ g\ H_2O} = \frac{x\ g\ NaNO_3}{125\ g\ soln - x\ g\ NaNO_3}$

Rearrange and solve for x g $NaNO_3$. $0.00850 \frac{g\ NaNO_3}{1\ g\ H_2O}(125\ g\ soln - x\ g\ NaNO_3) = x\ g\ NaNO_3 \rightarrow$

$1.0\underline{6}25 - 0.00850\,(x\ g\ NaNO_3) = x\ g\ NaNO_3 \rightarrow 1.0\underline{6}25 = 1.00850\,(x\ g\ NaNO_3) \rightarrow$

$\frac{1.0\underline{6}25}{1.00850} = x\ g\ NaNO_3 = 1.05\ g\ NaNO_3$ then $g_{H_2O} = g_{soln} - g_{NaNO_3} = 125\ g - 1.05\ g = 124\ g\ H_2O$

Prepare the solution by carefully adding 1.05 g $NaNO_3$ to a container with 124 g of distilled water and agitate the solution until the salt dissolves completely.

Check: The units (g) are correct. The magnitude of the answer (1 g) seems reasonable because we are making a small volume of solution and the formula weight of $NaNO_3$ is 85 g/mol.

(c) **Given:** 125 g of 1.0 % $NaNO_3$ by mass **Find:** describe final solution preparation

Conceptual Plan: $g_{soln} \rightarrow g_{NaNO_3}$ **then** $g_{NaNO_3}, g_{soln} \rightarrow g_{H_2O}$

$\frac{1.0 \text{ g NaNO}_3}{100 \text{ g soln}}$ $\quad g_{soln} = g_{NaNO_3} + g_{H_2O}$

then describe method

Solution: $125 \text{ g soln} \times \frac{1.0 \text{ g NaNO}_3}{100 \text{ g soln}} = 1.\underline{2}5 \text{ g NaNO}_3$ then $g_{soln} = g_{NaNO_3} + g_{H_2O}$

So $g_{H_2O} = g_{soln} - g_{NaNO_3} = 125 \text{ g} - 1.25 \text{ g} = 124 \text{ g H}_2\text{O}$

Prepare the solution by carefully adding 1.3 g $NaNO_3$ to a container with 124 g of distilled water and agitate the solution until the salt dissolves completely.

Check: The units (g) are correct. The magnitude of the answer (1 g) seems reasonable because we are making a small volume of solution and the solution is 1% by mass $NaNO_3$.

14.59 (a) **Given:** 28.4 g of glucose ($C_6H_{12}O_6$) in 355 g water; final volume = 378 mL **Find:** molarity

Conceptual Plan: $\text{mL} \rightarrow \text{L}$ **and** $g_{C_6H_{12}O_6} \rightarrow mol_{C_6H_{12}O_6}$ **then** $mol_{C_6H_{12}O_6}, V \rightarrow M$

$\frac{1 \text{ L}}{1000 \text{ mL}}$ $\quad \frac{1 \text{ mol C}_6\text{H}_{12}\text{O}_6}{180.16 \text{ g C}_6\text{H}_{12}\text{O}_6}$ $\quad M = \frac{\text{amount solute (moles)}}{\text{volume solution (L)}}$

Solution: $378 \text{ mL} \times \frac{1 \text{ L}}{1000 \text{ mL}} = 0.378 \text{ L}$ and

$$28.4 \text{ g C}_6\text{H}_{12}\text{O}_6 \times \frac{1 \text{ mol C}_6\text{H}_{12}\text{O}_6}{180.16 \text{ g C}_6\text{H}_{12}\text{O}_6} = 0.15\underline{7}638 \text{ mol C}_6\text{H}_{12}\text{O}_6$$

$$M = \frac{\text{amount solute (moles)}}{\text{volume solution (L)}} = \frac{0.15\underline{7}638 \text{ mol C}_6\text{H}_{12}\text{O}_6}{0.378 \text{ L}} = 0.417 \text{ M}$$

Check: The units (M) are correct. The magnitude of the answer (0.4 M) seems reasonable because we have 1/8 mole in about 1/3 L.

(b) **Given:** 28.4 g of glucose ($C_6H_{12}O_6$) in 355 g water; final volume = 378 mL **Find:** molality

Conceptual Plan: $g_{H_2O} \rightarrow kg_{H_2O}$ **and** $g_{C_6H_{12}O_6} \rightarrow mol_{C_6H_{12}O_6}$ **then** $mol_{C_6H_{12}O_6}, kg_{H_2O} \rightarrow m$

$\frac{1 \text{ kg}}{1000 \text{ g}}$ $\quad \frac{1 \text{ mol C}_6\text{H}_{12}\text{O}_6}{180.16 \text{ g C}_6\text{H}_{12}\text{O}_6}$ $\quad m = \frac{\text{amount solute (moles)}}{\text{mass solvent (kg)}}$

Solution: $355 \text{ g} \times \frac{1 \text{ kg}}{1000 \text{ g}} = 0.355 \text{ kg}$ and

$$28.4 \text{ g C}_6\text{H}_{12}\text{O}_6 \times \frac{1 \text{ mol C}_6\text{H}_{12}\text{O}_6}{180.16 \text{ g C}_6\text{H}_{12}\text{O}_6} = 0.15\underline{7}638 \text{ mol C}_6\text{H}_{12}\text{O}_6$$

$$m = \frac{\text{amount solute (moles)}}{\text{mass solvent (kg)}} = \frac{0.15\underline{7}638 \text{ mol C}_6\text{H}_{12}\text{O}_6}{0.355 \text{ kg}} = 0.444\ m$$

Check: The units (m) are correct. The magnitude of the answer (0.4 m) seems reasonable because we have 1/8 mole in about 1/3 kg.

(c) **Given:** 28.4 g of glucose ($C_6H_{12}O_6$) in 355 g water; final volume = 378 mL **Find:** percent by mass

Conceptual Plan: $g_{C_6H_{12}O_6}, g_{H_2O} \rightarrow g_{soln}$ **then** $g_{C_6H_{12}O_6}, g_{soln} \rightarrow$ **percent by mass**

$g_{soln} = g_{C_6H_{12}O_6} + g_{H_2O}$ $\quad \text{mass percent} = \frac{\text{mass solute}}{\text{mass solution}} \times 100\%$

Solution: $g_{soln} = g_{C_6H_{12}O_6} + g_{H_2O} = 28.4 \text{ g} + 355 \text{ g} = 38\underline{3}.4 \text{ g soln}$ then

$$\text{mass percent} = \frac{\text{mass solute}}{\text{mass solution}} \times 100\% = \frac{28.4 \text{ g C}_6\text{H}_{12}\text{O}_6}{38\underline{3}.4 \text{ g soln}} \times 100\% = 7.41\% \text{ by mass}$$

Check: The units (percent by mass) are correct. The magnitude of the answer (7%) seems reasonable because we are dissolving 28 g in 355 g.

(d) **Given:** 28.4 g of glucose ($C_6H_{12}O_6$) in 355 g water; final volume = 378 mL **Find:** mole fraction

Conceptual Plan: $g_{C_6H_{12}O_6} \rightarrow mol_{C_6H_{12}O_6}$ **and** $g_{H_2O} \rightarrow mol_{H_2O}$ **then** $mol_{C_6H_{12}O_6}, mol_{H_2O} \rightarrow \chi_{C_6H_{12}O_6}$

$$\frac{1 \text{ mol } C_6H_{12}O_6}{180.16 \text{ g } C_6H_{12}O_6} \qquad \frac{1 \text{ mol } H_2O}{18.02 \text{ g } H_2O} \qquad \chi = \frac{\text{amount solute (in moles)}}{\text{total amount of solute and solvent (in moles)}}$$

Solution: $28.4 \text{ g } C_6H_{12}O_6 \times \frac{1 \text{ mol } C_6H_{12}O_6}{180.16 \text{ g } C_6H_{12}O_6} = 0.15\underline{7}638 \text{ mol } C_6H_{12}O_6$ and

$$355 \text{ g } H_2O \times \frac{1 \text{ mol } H_2O}{18.02 \text{ g } H_2O} = 19.\underline{7}003 \text{ mol } H_2O \text{ then}$$

$$\chi = \frac{\text{amount solute (in moles)}}{\text{total amount of solute and solvent (in moles)}} = \frac{0.15\underline{7}638 \text{ mol}}{0.15\underline{7}638 \text{ mol} + 19.\underline{7}003 \text{ mol}} = 0.00794$$

Check: The units (none) are correct. The magnitude of the answer (0.008) seems reasonable because we have many more grams of water and water has a much lower molar mass.

(e) **Given:** 28.4 g of glucose ($C_6H_{12}O_6$) in 355 g water; final volume = 378 mL **Find:** mole percent

Conceptual Plan: use answer from part (d) then $\chi_{C_6H_{12}O_6} \rightarrow$ **mole percent**

$$\chi \times 100\%$$

Solution: mole percent $= \chi \times 100\% = 0.00794 \times 100\% = 0.794$ mole percent

Check: The units (mol %) are correct. The magnitude of the answer (0.8) seems reasonable because we have many more grams of water, water has a much lower molar mass than glucose, and we are increasing the answer from part (d) by a factor of 100.

14.60 (a) **Given:** 20.2 mL of methanol (CH_3OH) in 100.0 mL water; final volume = 118 mL **Find:** molarity

Other: $d\,(CH_3OH) = 0.782$ g/mL; $d\,(H_2O) = 1.00$ g/mL

Conceptual Plan: mL → L and $mL_{CH_3OH} \rightarrow g_{CH_3OH} \rightarrow mol_{CH_3OH}$ **then** $mol_{CH_3OH}, V \rightarrow M$

$$\frac{1 \text{ L}}{1000 \text{ mL}} \qquad \frac{0.782 \text{ g}}{1 \text{ mL}} \quad \frac{1 \text{ mol } CH_3OH}{32.04 \text{ g } CH_3OH} \qquad M = \frac{\text{amount solute (moles)}}{\text{volume solution (L)}}$$

Solution: $118 \text{ mL} \times \frac{1 \text{ L}}{1000 \text{ mL}} = 0.118 \text{ L}$ and

$$20.2 \text{ mL } CH_3OH \times \frac{0.782 \text{ g } CH_3OH}{1 \text{ mL } CH_3OH} \times \frac{1 \text{ mol } CH_3OH}{32.04 \text{ g } CH_3OH} = 0.49\underline{3}021 \text{ mol } CH_3OH$$

$$M = \frac{\text{amount solute (moles)}}{\text{volume solution (L)}} = \frac{0.49\underline{3}021 \text{ mol } CH_3OH}{0.118 \text{ L}} = 4.18 \text{ M}$$

Check: The units (M) are correct. The magnitude of the answer (4 M) seems reasonable because we have 1/2 mole in about 1/8 L.

(b) **Given:** 20.2 mL of methanol (CH_3OH) in 100.0 mL water; final volume = 118 mL **Find:** molality

Other: $d(CH_3OH) = 0.782$ g/mL; $d(H_2O) = 1.00$ g/mL

Conceptual Plan:

$mL_{CH_3OH} \rightarrow g_{CH_3OH} \rightarrow mol_{CH_3OH}$ **and** $mL_{H_2O} \rightarrow g_{H_2O} \rightarrow kg_{H_2O}$ **then** $mol_{CH_3OH}, kg_{H_2O} \rightarrow m$

$$\frac{0.782 \text{ g}}{1 \text{ mL}} \quad \frac{1 \text{ mol } CH_3OH}{32.04 \text{ g } CH_3OH} \qquad \frac{1.00 \text{ g}}{1 \text{ mL}} \quad \frac{1 \text{ kg}}{1000 \text{ g}} \qquad m = \frac{\text{amount solute (moles)}}{\text{mass solvent (kg)}}$$

Solution: $20.2 \text{ mL } CH_3OH \times \frac{0.782 \text{ g } CH_3OH}{1 \text{ mL } CH_3OH} \times \frac{1 \text{ mol } CH_3OH}{32.04 \text{ g } CH_3OH} = 0.49\underline{3}021 \text{ mol } CH_3OH$ and

$$100.0 \text{ mL} \times \frac{1.00 \text{ g}}{1 \text{ mL}} \times \frac{1 \text{ kg}}{1000 \text{ g}} = 0.1000 \text{ kg then}$$

$$m = \frac{\text{amount solute (moles)}}{\text{mass solvent (kg)}} = \frac{0.49\underline{3}021 \text{ mol } CH_3OH}{0.1000 \text{ kg}} = 4.93\ m$$

Check: The units (m) are correct. The magnitude of the answer (5 m) seems reasonable because we have 1/2 mole in 1/10 kg.

(c) **Given:** 20.2 mL of methanol (CH_3OH) in 100.0 mL water; final volume = 118 mL **Find:** percent by mass
Other: d (CH_3OH) = 0.782 g/mL; d (H_2O) = 1.00 g/mL
Conceptual Plan: $mL_{CH_3OH} \rightarrow g_{CH_3OH}$ **and** $mL_{H_2O} \rightarrow g_{H_2O}$ **then** $g_{CH_3OH}, g_{H_2O} \rightarrow g_{soln}$ **then**

$\frac{0.782\ g}{1\ mL}$ $\frac{1.00\ g}{1\ mL}$ $g_{soln} = g_{CH_3OH} + g_{H_2O}$

$g_{CH_3OH}, g_{soln} \rightarrow$ **percent by mass**

$\text{mass percent} = \frac{\text{mass solute}}{\text{mass solution}} \times 100\%$

Solution: $20.2\ \cancel{mL\ CH_3OH} \times \frac{0.782\ g\ CH_3OH}{1\ \cancel{mL\ CH_3OH}} = 15.\underline{7}964\ g\ CH_3OH$

$100.0\ \cancel{mL} \times \frac{1.00\ g}{1\ \cancel{mL}} = 100.0\ g\ H_2O$ $g_{soln} = g_{CH_3OH} + g_{H_2O} = 15.\underline{7}964\ g + 100.0\ g = 115.\underline{7}964\ g$ soln then

$\text{mass percent} = \frac{\text{mass solute}}{\text{mass solution}} \times 100\% = \frac{15.\underline{7}964\ \cancel{g\ CH_3OH}}{115.\underline{7}964\ \cancel{g\ soln}} \times 100\% = 13.6\%$ by mass

Check: The units (percent by mass) are correct. The magnitude of the answer (14%) seems reasonable because we are dissolving 16 g in 100 g.

(d) **Given:** 20.2 mL of methanol (CH_3OH) in 100.0 mL water; final volume = 118 mL **Find:** mole fraction
Other: d (CH_3OH) = 0.782 g/mL; d (H_2O) = 1.00 g/mL
Conceptual Plan: $mL_{CH_3OH} \rightarrow g_{CH_3OH} \rightarrow mol_{CH_3OH}$ **and** $mL_{H_2O} \rightarrow g_{H_2O} \rightarrow mol_{H_2O}$ **then**

$\frac{0.782\ g}{1\ mL}$ $\frac{1\ mol\ CH_3OH}{32.04\ g\ CH_3OH}$ $\frac{1.00\ g}{1\ mL}$ $\frac{1\ mol\ H_2O}{18.01\ g\ H_2O}$

$mol_{CH_3OH}, mol_{H_2O} \rightarrow \chi_{CH_3OH}$

$\chi = \frac{\text{amount solute (in moles)}}{\text{total amount of solute and solvent (in moles)}}$

Solution: $20.2\ \cancel{mL\ CH_3OH} \times \frac{0.782\ \cancel{g\ CH_3OH}}{1\ \cancel{mL\ CH_3OH}} \times \frac{1\ mol\ CH_3OH}{32.04\ \cancel{g\ CH_3OH}} = 0.49\underline{3}021\ mol\ CH_3OH$ and

$100.0\ \cancel{mL} \times \frac{1.00\ \cancel{g}}{1\ \cancel{mL}} \times \frac{1\ mol\ H_2O}{18.01\ \cancel{g\ H_2O}} = 5.55\underline{2}471\ mol\ H_2O$ then

$\chi = \frac{\text{amount solute (in moles)}}{\text{total amount of solute and solvent (in moles)}} = \frac{0.49\underline{3}021\ \cancel{mol}}{0.49\underline{3}021\ \cancel{mol} + 5.55\underline{2}471\ \cancel{mol}} = 0.0816$

Check: The units (none) are correct. The magnitude of the answer (0.08) seems reasonable because we have many more grams of water and water has a lower molar mass than methanol.

(e) **Given:** 20.2 mL of methanol (CH_3OH) in 100.0 mL water; final volume = 118 mL **Find:** mole percent
Other: d (CH_3OH) = 0.782 g/mL; d (H_2O) = 1.00 g/mL
Conceptual Plan: use answer from part (d) then $\chi_{C_6H_{12}O_6} \rightarrow$ **mole percent**

$\chi \times 100\%$

Solution: mole percent = $\chi \times 100\% = 0.0816 \times 100\% = 8.16$ mole percent

Check: The units (mol %) are correct. The magnitude of the answer (8) seems reasonable because we have many more grams of water, water has a lower molar mass than methanol, and we are increasing the answer from part (d) by a factor of 100.

14.61 **Given:** 3.0% H_2O_2 by mass, d = 1.01 g/mL **Find:** molarity
Conceptual Plan:
Assume exactly 100 g of solution; $g_{solution} \rightarrow g_{H_2O_2} \rightarrow mol_{H_2O_2}$ **and** $g_{solution} \rightarrow mL_{solution} \rightarrow L_{solution}$

$\frac{3.0\ g\ H_2O_2}{100\ g\ solution}$ $\frac{1\ mol\ H_2O_2}{34.02\ g\ H_2O_2}$ $\frac{1\ mL}{1.01\ g}$ $\frac{1\ L}{1000\ mL}$

then $mol_{H_2O_2}, L_{solution} \rightarrow M$

$M = \frac{\text{amount solute (moles)}}{\text{volume solution (L)}}$

Solution: $100\ \cancel{g\ solution} \times \frac{3.0\ \cancel{g\ H_2O_2}}{100\ \cancel{g\ solution}} \times \frac{1\ mol\ H_2O_2}{34.02\ \cancel{g\ H_2O_2}} = 0.08\underline{8}1834\ mol\ H_2O_2$ and

$$100 \text{ g solution} \times \frac{1 \text{ mL solution}}{1.01 \text{ g solution}} \times \frac{1 \text{ L solution}}{1000 \text{ mL solution}} = 0.0990\underline{0}99 \text{ L solution then}$$

$$M = \frac{\text{amount solute (moles)}}{\text{volume solution (L)}} = \frac{0.08\underline{8}1834 \text{ mol } H_2O_2}{0.099\underline{0}099 \text{ L solution}} = 0.89 \text{ M } H_2O_2$$

Check: The units (M) are correct. The magnitude of the answer (1) seems reasonable because we are starting with a low concentration solution and pure water is ~55.5 M.

14.62 **Given:** 4.55% NaOCl by mass, $d = 1.02$ g/mL **Find:** molarity

Conceptual Plan: Assume exactly 100 g of solution; $g_{solution} \rightarrow g_{NaOCl} \rightarrow mol_{NaOCl}$ **and** $g_{solution} \rightarrow mL_{solution} \rightarrow L_{solution}$ **then** $mol_{NaOCl}, L_{solution} \rightarrow M$

$$\frac{4.55 \text{ g NaOCl}}{100 \text{ g solution}} \quad \frac{1 \text{ mol NaOCl}}{74.44 \text{ g NaOCl}} \quad \frac{1 \text{ mL}}{1.02 \text{ g}} \quad \frac{1 \text{ L}}{1000 \text{ mL}} \quad M = \frac{\text{amount solute (moles)}}{\text{volume solution (L)}}$$

Solution: $$100 \text{ g solution} \times \frac{4.55 \text{ g NaOCl}}{100 \text{ g solution}} \times \frac{1 \text{ mol NaOCl}}{74.44 \text{ g NaOCl}} = 0.061\underline{1}2305 \text{ mol NaOCl and}$$

$$100 \text{ g solution} \times \frac{1 \text{ mL solution}}{1.02 \text{ g solution}} \times \frac{1 \text{ L solution}}{1000 \text{ mL solution}} = 0.0980\underline{3}92 \text{ L solution then}$$

$$M = \frac{\text{amount solute (moles)}}{\text{volume solution (L)}} = \frac{0.061\underline{1}2305 \text{ mol NaOCl}}{0.0980\underline{3}92 \text{ L solution}} = 0.623 \text{ M NaOCl}$$

Check: The units (M) are correct. The magnitude of the answer (1) seems reasonable because we are starting with a low concentration solution and pure water is ~55.5 M.

14.63 **Given:** 36% HCl by mass **Find:** molality and mole fraction

Conceptual Plan: Assume exactly 100 g of solution; $g_{solution} \rightarrow g_{HCl} \rightarrow mol_{HCl}$ **and** $g_{HCl}, g_{solution} \rightarrow g_{solvent} \rightarrow kg_{solvent}$ **then** $mol_{HCl}, kg_{solvent} \rightarrow m$ **and** $g_{solvent} \rightarrow mol_{solvent}$ **then** $mol_{HCl}, mol_{solvent} \rightarrow \chi_{HCl}$

$$\frac{36 \text{ g HCl}}{100 \text{ g solution}} \quad \frac{1 \text{ mol HCl}}{36.46 \text{ g HCl}} \quad g_{soln} = g_{HCl} + g_{H_2O} \quad \frac{1 \text{ kg}}{1000 \text{ g}} \quad m = \frac{\text{amount solute (moles)}}{\text{mass solvent (kg)}} \quad \frac{1 \text{ mol } H_2O}{18.02 \text{ g } H_2O} \quad \chi = \frac{\text{amount solute (in moles)}}{\text{total amount of solute and solvent (in moles)}}$$

Solution: $$100 \text{ g solution} \times \frac{36 \text{ g HCl}}{100 \text{ g solution}} = 36 \text{ g HCl} \times \frac{1 \text{ mol HCl}}{36.46 \text{ g HCl}} = 0.9\underline{8}7383 \text{ mol HCl and}$$

$g_{soln} = g_{HCl} + g_{H_2O}$ Rearrange to solve for $g_{solvent}$. $g_{H_2O} = g_{soln} - g_{HCl} = 100 \text{ g} - 36 \text{ g} = 64 \text{ g } H_2O$

$$64 \text{ g } H_2O \times \frac{1 \text{ kg } H_2O}{1000 \text{ g } H_2O} = 0.064 \text{ kg } H_2O \text{ then}$$

$$m = \frac{\text{amount solute (moles)}}{\text{mass solvent (kg)}} = \frac{0.9\underline{8}7383 \text{ mol HCl}}{0.064 \text{ kg}} = 15 \, m \text{ HCl and}$$

$$64 \text{ g } H_2O \times \frac{1 \text{ mol } H_2O}{18.02 \text{ g } H_2O} = 3.\underline{5}5161 \text{ mol } H_2O \text{ then}$$

$$\chi = \frac{\text{amount solute (in moles)}}{\text{total amount of solute and solvent (in moles)}} = \frac{0.9\underline{8}7383 \text{ mol}}{0.9\underline{8}7383 \text{ mol} + 3.\underline{5}5161 \text{ mol}} = 0.22$$

Check: The units (m and unitless) are correct. The magnitude of the answers (15 and 0.2) seems reasonable because we are starting with a high concentration solution and the molar mass of water is much less than that of HCl.

14.64 **Given:** 5.0% NaCl by mass **Find:** molality and mole fraction

Conceptual Plan: Assume exactly 100 g of solution; $g_{solution} \rightarrow g_{HCl} \rightarrow mol_{NaCl}$ **and** $g_{NaCl}, g_{solution} \rightarrow g_{solvent} \rightarrow kg_{solvent}$ **then** $mol_{NaCl}, kg_{solvent} \rightarrow m$ **and** $g_{solvent} \rightarrow mol_{solvent}$ **then** $mol_{NaCl}, mol_{solvent} \rightarrow \chi_{NaCl}$

$$\frac{5.0 \text{ gNaCl}}{100 \text{ g solution}} \quad \frac{1 \text{ mol NaCl}}{58.44 \text{ g NaCl}} \quad g_{soln} = g_{NaCl} + g_{H_2O} \quad \frac{1 \text{ kg}}{1000 \text{ g}} \quad m = \frac{\text{amount solute (moles)}}{\text{mass solvent (kg)}} \quad \frac{1 \text{ mol } H_2O}{18.02 \text{ g } H_2O} \quad \chi = \frac{\text{amount solute (in moles)}}{\text{total amount of solute and solvent (in moles)}}$$

Solution: $100 \text{ g solution} \times \dfrac{5.0 \text{ g NaCl}}{100 \text{ g solution}} = 5.0 \text{ g HCl}$

$5.0 \text{ g HCl} \times \dfrac{1 \text{ mol NaCl}}{58.44 \text{ g NaCl}} = 0.0855578 \text{ mol NaCl}$ and $g_{soln} = g_{NaCl} + g_{H_2O}$

Rearrange to solve for $g_{solvent}$. $g_{H_2O} = g_{soln} - g_{NaCl} = 100.0 \text{ g} - 5.0 \text{ g} = 95.0 \text{ g } H_2O$

$95.0 \text{ g } H_2O \times \dfrac{1 \text{ kg } H_2O}{1000 \text{ g } H_2O} = 0.0950 \text{ kg } H_2O$ then

$$m = \frac{\text{amount solute (moles)}}{\text{mass solvent (kg)}} = \frac{0.0855578 \text{ mol NaCl}}{0.0950 \text{ kg}} = 0.901\ m \text{ NaCl and}$$

$95.0 \text{ g } H_2O \times \dfrac{1 \text{ mol } H_2O}{18.02 \text{ g } H_2O} = 5.27192 \text{ mol } H_2O$ then

$$\chi = \frac{\text{amount solute (in moles)}}{\text{total amount of solute and solvent (in moles)}} = \frac{0.0855578 \text{ mol}}{0.0855578 \text{ mol} + 5.27192 \text{ mol}} = 0.016$$

Check: The units (m and unitless) are correct. The magnitude of the answers (1 and 0.02) seems reasonable because we are starting with a low concentration solution and the molar mass of water is much less than that of NaCl.

Vapor Pressure of Solutions

14.65 The level has decreased more in the beaker filled with pure water. The dissolved salt in the seawater decreases the vapor pressure and subsequently lowers the rate of vaporization.

14.66 Assume that the solutions obey Raoult's law ($P_{solution} = \chi_{solvent} P^{\circ}_{solvent}$). Each solution has the same amount of solvent, so we need to compare the number of moles of particles in the solvent. Without doing any calculations, we can see that (b) will have a lower number of particles than (a) because (b) has the higher molar mass. Potassium acetate will generate ~2 moles of particles per mole of the salt, so it will generate more than the other two. So solution (b) will have the highest vapor pressure.

14.67 **Given:** 24.5 g of glycerin ($C_3H_8O_3$) in 135 mL water at 30 °C; $P^{\circ}_{H_2O} = 31.8$ torr **Find:** P_{H_2O}
Other: $d\ (H_2O) = 1.00$ g/mL; glycerin is not an ionic solid
Conceptual Plan: $g_{C_3H_8O_3} \rightarrow mol_{C_3H_8O_3}$ **and** $mL\ H_2O \rightarrow gH_2O \rightarrow mol_{H_2O}$ **then** $mol_{C_3H_8O_3}, mol_{H_2O} \rightarrow \chi_{H_2O}$

$\dfrac{1 \text{ mol } C_3H_8O_3}{92.09 \text{ g } C_3H_8O_3}$ $\quad \dfrac{1.00 \text{ g}}{1 \text{ mL}}$ $\quad \dfrac{1 \text{ mol } H_2O}{18.01 \text{ g } H_2O}$ $\quad \chi = \dfrac{\text{amount solvent (in moles)}}{\text{total amount of solute and solvent (in moles)}}$

then $\chi_{H_2O}, P^{\circ}_{H_2O} \rightarrow P_{H_2O}$

$P_{solution} = \chi_{solvent} P^{\circ}_{solvent}$

Solution: $24.5 \text{ g } C_3H_8O_3 \times \dfrac{1 \text{ mol } C_3H_8O_3}{92.09 \text{ g } C_3H_8O_3} = 0.2660441 \text{ mol } C_3H_8O_3$ and

$135 \text{ mL} \times \dfrac{1.00 \text{ g}}{1 \text{ mL}} \times \dfrac{1 \text{ mol } H_2O}{18.01 \text{ g } H_2O} = 7.495836 \text{ mol } H_2O$ then

$$\chi = \frac{\text{amount solvent (in moles)}}{\text{total amount of solute and solvent (in moles)}} = \frac{7.495836 \text{ mol}}{0.2660441 \text{ mol} + 7.495836 \text{ mol}} = 0.9657243 \text{ then}$$

$P_{solution} = \chi_{solvent} P^{\circ}_{solvent} = 0.9657243 \times 31.8 \text{ torr} = 30.7 \text{ torr}$

Check: The units (torr) are correct. The magnitude of the answer (31 torr) seems reasonable because it is a drop from the pure vapor pressure. Very few moles of glycerin are added, so the pressure will not drop much.

14.68 **Given:** 12.35 % naphthalene ($C_{10}H_8$) by mass in hexane (C_6H_{14}) at 25 °C; $P^{\circ}_{C_6H_{14}} = 151$ torr **Find:** $P_{C_6H_{14}}$
Conceptual Plan: **% naphthalene ($C_{10}H_8$) by mass** $\rightarrow g_{C_{10}H_8}, g_{C_6H_{14}}$ **then** $g_{C_{10}H_8} \rightarrow mol_{C_{10}H_8}$ **and**

$\dfrac{10.85 \text{ g } C_{10}H_8}{100 \text{ g } (C_{10}H_8 + C_6H_{14})}$ $\quad \dfrac{1 \text{ mol } C_{10}H_8}{128.16 \text{ g } C_{10}H_8}$

$g_{C_6H_{14}} \rightarrow mol_{C_6H_{14}}$ **then** $mol_{C_{10}H_8}, mol_{C_6H_{14}} \rightarrow \chi_{C_6H_{14}}$ **then** $\chi_{C_6H_{14}}, P^{\circ}_{C_6H_{14}} \rightarrow P_{C_6H_{14}}$

$\dfrac{1 \text{ mol } C_6H_{14}}{86.17 \text{ g } C_6H_{14}}$ $\quad \chi = \dfrac{\text{amount solvent (in moles)}}{\text{total amount of solute and solvent (in moles)}}$ $\quad P_{solution} = \chi_{solvent} P^{\circ}_{solvent}$

Solution: $\frac{12.35 \text{ g } C_{10}H_8}{100 \text{ g } (C_{10}H_8 + C_6H_{14})}$ means 12.35 g $C_{10}H_8$ and $(100 \text{ g} - 12.35 \text{ g}) = 87.65 \text{ g } C_6H_{14}$ then

$$12.35 \text{ g } C_{10}H_8 \times \frac{1 \text{ mol } C_{10}H_8}{128.16 \text{ g } C_{10}H_8} = 0.096363920 \text{ mol } C_{10}H_8 \text{ and}$$

$$87.65 \text{ g } C_6H_{14} \times \frac{1 \text{ mol } C_6H_{14}}{86.17 \text{ g } C_6H_{14}} = 1.0171754 \text{ mol } C_6H_{14} \text{ then}$$

$$\chi = \frac{\text{amount solvent (in moles)}}{\text{total amount of solute and solvent (in moles)}} = \frac{1.0171754 \text{ mol}}{0.096363920 \text{ mol} + 1.0171754 \text{ mol}} = 0.91346159 \text{ then}$$

$$P_{\text{solution}} = \chi_{\text{solvent}} P^\circ_{\text{solvent}} = 0.91346159 \times 151 \text{ torr} = 138 \text{ torr } C_6H_{14}$$

Check: The units (torr) are correct. The magnitude of the answer (140 torr) seems reasonable because it is a drop from the pure vapor pressure. Only a fraction of a mole of naphthalene is added, so the pressure will not drop much.

14.69 **Given:** 50.0 g of heptane (C_7H_{16}) and 50.0 g of octane (C_8H_{18}) at 25 °C; $P^\circ_{C_7H_{16}} = 45.8$ torr; $P^\circ_{C_8H_{18}} = 10.9$ torr

(a) **Find:** $P_{C_7H_{16}}, P_{C_8H_{18}}$

Conceptual Plan: $g_{C_7H_{16}} \rightarrow mol_{C_7H_{16}}$ **and** $g_{C_8H_{18}} \rightarrow mol_{C_8H_{18}}$ **then** $mol_{C_7H_{16}}$, $mol_{C_8H_{18}} \rightarrow \chi_{C_7H_{16}}, \chi_{C_8H_{18}}$ **then** $\chi_{C_7H_{16}}, P^\circ_{C_7H_{16}} \rightarrow P_{C_7H_{16}}$ **and** $\chi_{C_8H_{18}}, P^\circ_{C_8H_{18}} \rightarrow P_{C_8H_{18}}$

$\frac{1 \text{ mol } C_7H_{16}}{100.20 \text{ g } C_7H_{16}}$; $\frac{1 \text{ mol } C_8H_{18}}{114.22 \text{ g } C_8H_{18}}$; $\chi_{C_7H_{16}} = \frac{\text{amount } C_7H_{16} \text{ (in moles)}}{\text{total amount (in moles)}}$; $\chi_{C_8H_{18}} = 1 - \chi_{C_7H_{16}}$; $P_{C_7H_{16}} = \chi_{C_7H_{16}} P^\circ_{C_7H_{16}}$; $P_{C_8H_{18}} = \chi_{C_8H_{18}} P^\circ_{C_8H_{18}}$

Solution: $50.0 \text{ g } C_7H_{16} \times \frac{1 \text{ mol } C_7H_{16}}{100.20 \text{ g } C_7H_{16}} = 0.499002 \text{ mol } C_7H_{16}$ and

$$50.0 \text{ g } C_8H_{18} \times \frac{1 \text{ mol } C_8H_{18}}{114.22 \text{ g } C_8H_{18}} = 0.437752 \text{ mol } C_8H_{18} \text{ then}$$

$$\chi_{C_7H_{16}} = \frac{\text{amount } C_7H_{16} \text{ (in moles)}}{\text{total amount (in moles)}} = \frac{0.499002 \text{ mol}}{0.499002 \text{ mol} + 0.437752 \text{ mol}} = 0.532693 \text{ and}$$

$$\chi_{C_8H_{18}} = 1 - \chi_{C_7H_{16}} = 1 - 0.532693 = 0.467307 \text{ then}$$

$$P_{C_7H_{16}} = \chi_{C_7H_{16}} P^\circ_{C_7H_{16}} = 0.532693 \times 45.8 \text{ torr} = 24.4 \text{ torr and}$$

$$P_{C_8H_{18}} = \chi_{C_8H_{18}} P^\circ_{C_8H_{18}} = 0.467307 \times 10.9 \text{ torr} = 5.09 \text{ torr}$$

Check: The units (torr) are correct. The magnitude of the answers (24 and 5 torr) seems reasonable because we expect a drop in half from the pure vapor pressures because we have roughly a 50:50 mole ratio of the two components.

(b) **Find:** P_{Total}

Conceptual Plan: $P_{C_7H_{16}}, P_{C_8H_{18}} \rightarrow P_{\text{Total}}$

$P_{\text{Total}} = P_{C_7H_{16}} + P_{C_8H_{18}}$

Solution: $P_{\text{Total}} = P_{C_7H_{16}} + P_{C_8H_{18}} = 24.4 \text{ torr} + 5.09 \text{ torr} = 29.5 \text{ torr}$

Check: The units (torr) are correct. The magnitude of the answer (30 torr) seems reasonable considering the two pressures.

(c) **Find:** mass percent composition of the gas phase

Conceptual Plan: because $n \ \alpha \ P$ and we are calculating a mass percent, which is a ratio of masses, we can simply convert 1 torr to 1 mole so

$P_{C_7H_{16}}, P_{C_8H_{18}} \rightarrow n_{C_7H_{16}}, n_{C_8H_{18}}$ **then** $mol_{C_7H_{16}} \rightarrow g_{C_7H_{16}}$ **and** $mol_{C_8H_{18}} \rightarrow g_{C_8H_{18}}$

$\frac{100.20 \text{ g } C_7H_{16}}{1 \text{ mol } C_7H_{16}}$; $\frac{114.22 \text{ g } C_8H_{18}}{1 \text{ mol } C_8H_{18}}$

then $g_{C_7H_{16}}, g_{C_8H_{18}} \rightarrow$ **mass percents**

$$\text{mass percent} = \frac{\text{mass solute}}{\text{mass solution}} \times 100\%$$

Solution: so $n_{C_7H_{16}} = 24.4$ mol and $n_{C_8H_{18}} = 5.09$ mol then

$$24.4 \text{ mol } C_7H_{16} \times \frac{100.20 \text{ g } C_7H_{16}}{1 \text{ mol } C_7H_{16}} = 2444.88 \text{ g } C_7H_{16} \text{ and}$$

$5.09 \text{ mol } C_8H_{18} \times \frac{114.22 \text{ g } C_8H_{18}}{1 \text{ mol } C_8H_{18}} = 58\underline{1}.380 \text{ g } C_8H_{18}$ then

$\text{mass percent} = \frac{\text{mass solute}}{\text{mass solution}} \times 100\% = \frac{24\underline{4}4.88 \text{ g } C_7H_{16}}{24\underline{4}4.88 \text{ g } C_7H_{16} + 58\underline{1}.380 \text{ g } C_8H_{18}} \times 100\% =$

80.8% by mass C_7H_{16}

then 100% − 80.8% = 19.2% by mass C_8H_{18}

Check: The units (mol %) are correct. The magnitude of the answers (81% and 19%) seems reasonable considering the two pressures.

(d) The two mass percents are different because the vapor is richer in the more volatile component (the lighter molecule).

14.70 **Given:** pentane (C_5H_{12}) and hexane (C_6H_{14}) $P_{Total} = 258$ torr; $P^\circ_{C_5H_{12}} = 425$ torr; $P^\circ_{C_6H_{14}} = 151$ torr at 25 °C
Find: $\chi_{C_5H_{12}}, \chi_{C_6H_{14}}$

Conceptual Plan: $P_{Total} = P_{C_5H_{12}} + P_{C_6H_{14}}$ **where** $P_{C_5H_{12}} = \chi_{C_5H_{12}} P^\circ_{C_5H_{12}}$ **and** $P_{C_6H_{14}} = \chi_{C_6H_{14}} P^\circ_{C_6H_{14}}$ **but** $\chi_{C_6H_{14}} = 1 - \chi_{C_5H_{12}}$ **so** $P_{Total} = \chi_{C_5H_{12}} P^\circ_{C_5H_{12}} + (1 - \chi_{C_5H_{12}}) P^\circ_{C_6H_{14}}$ **substitute values and solve for** $\chi_{C_5H_{12}}, \chi_{C_6H_{14}}$

Solution: $P_{Total} = \chi_{C_5H_{12}} P^\circ_{C_5H_{12}} + (1 - \chi_{C_5H_{12}}) P^\circ_{C_6H_{14}}$ so

$258 \text{ torr} = \chi_{C_5H_{12}} 425 \text{ torr} + (1 - \chi_{C_5H_{12}})\, 151 \text{ torr} \rightarrow 258 - 151 = \chi_{C_5H_{12}}(425 - 151) \rightarrow$

$\chi_{C_5H_{12}} = \frac{107}{274} = 0.391$ and $\chi_{C_6H_{14}} = 1 - \chi_{C_5H_{12}} = 1 - 0.391 = 0.609$

Check: The units (none) are correct. The magnitude of the answers (0.4 and 0.6) seems reasonable because the total vapor pressure is closer to the vapor pressure of hexane; so we expect more hexane in the liquid.

14.71 **Given:** 4.08 g of chloroform ($CHCl_3$) and 9.29 g of acetone (CH_3COCH_3); at 35 °C, $P^\circ_{CHCl_3} = 295$ torr; $P^\circ_{CH_3COCH_3} = 332$ torr; assume ideal behavior; $P_{Total\ measured} = 312$ torr
Find: $P_{CHCl_3}, P_{CH_3COCH_3}, P_{Total}$, and if the solution is ideal.

Conceptual Plan: $g_{CHCl_3} \rightarrow mol_{CHCl_3}$ **and** $g_{CH_3COCH_3} \rightarrow mol_{CH_3COCH_3}$ **then**

$\frac{1 \text{ mol } CHCl_3}{119.38 \text{ g } CHCl_3}$ $\frac{1 \text{ mol } CH_3COCH_3}{58.08 \text{ g } CH_3COCH_3}$

$mol_{CHCl_3}, mol_{CH_3COCH_3} \rightarrow \chi_{CHCl_3}, \chi_{CH_3COCH_3}$ **then** $\chi_{CHCl_3}, P^\circ_{CHCl_3} \rightarrow P_{CHCl_3}$ **and**

$\chi_{CHCl_3} = \frac{\text{amount } CHCl_3 \text{ (in moles)}}{\text{total amount (in moles)}}$ $\chi_{CH_3COCH_3} = 1 - \chi_{CHCl_3}$ $P_{CHCl_3} = \chi_{CHCl_3} P^\circ_{CHCl_3}$

$\chi_{CH_3COCH_3}, P^\circ_{CH_3COCH_3} \rightarrow P_{CH_3COCH_3}$ **then** $P_{CHCl_3}, P_{CH_3COCH_3} \rightarrow P_{Total}$ **then compare values**

$P_{CH_3COCH_3} = \chi_{CH_3COCH_3} P^\circ_{CH_3COCH_3}$ $P_{Total} = P_{CHCl_3} + P_{CH_3COCH_3}$

Solution:

$4.08 \text{ g } CHCl_3 \times \frac{1 \text{ mol } CHCl_3}{119.38 \text{ g } CHCl_3} = 0.034\underline{1}766 \text{ mol } CHCl_3$ and

$9.29 \text{ g } CH_3COCH_3 \times \frac{1 \text{ mol } CH_3COCH_3}{58.08 \text{ g } CH_3COCH_3} = 0.15\underline{9}952 \text{ mol } CH_3COCH_3$ then

$\chi_{CHCl_3} = \frac{\text{amount } CHCl_3 \text{ (in moles)}}{\text{total amount (in moles)}} = \frac{0.034\underline{1}766 \text{ mol}}{0.034\underline{1}766 \text{ mol} + 0.15\underline{9}952 \text{ mol}} = 0.17\underline{6}052$ and

$\chi_{CH_3COCH_3} = 1 - \chi_{CHCl_3} = 1 - 0.17\underline{6}052 = 0.82\underline{3}948$ then
$P_{CHCl_3} = \chi_{CHCl_3} P^\circ_{CHCl_3} = 0.17\underline{6}052 \times 295 \text{ torr} = 51.9 \text{ torr}$ and
$P_{CH_3COCH_3} = \chi_{CH_3COCH_3} P^\circ_{CH_3COCH_3} = 0.82\underline{3}948 \times 332 \text{ torr} = 274 \text{ torr}$ then
$P_{Total} = P_{CHCl_3} + P_{CH_3COCH_3} = 51.9 \text{ torr} + 274 \text{ torr} = 326 \text{ torr}$
Because 326 torr ≠ 312 torr, the solution is not behaving ideally. The chloroform–acetone interactions are stronger than the chloroform–chloroform and acetone–acetone interactions.

Check: The units (torr) are correct. The magnitude of the answer (326) seems reasonable because each is a fraction of the pure vapor pressure. We are not surprised that the solution is not ideal because the types of bonds in the two molecules are very different.

14.72 **Given:** methanol (CH_3OH) and water; $\chi_{H_2O} = 0.312$; at 39.9 °C, $P_{Total\ measured} = 211$ torr; $P^{\circ}_{CH_3OH} = 256$ torr; $P^{\circ}_{H_2O} = 55.3$ torr **Find:** whether solution is ideal

Conceptual Plan: $\chi_{H_2O} \rightarrow \chi_{CH_3OH}$ **then** $\chi_{H_2O}, P^{\circ}_{H_2O} \rightarrow P_{H_2O}$ **and** $\chi_{CH_3OH}, P^{\circ}_{CH_3OH} \rightarrow P_{CH_3OH}$

$\chi_{CH_3OH} = 1 - \chi_{H_2O}$ $\quad P_{H_2O} = \chi_{H_2O} P^{\circ}_{H_2O}$ $\quad P_{CH_3OH} = \chi_{CH_3OH} P^{\circ}_{CH_3OH}$

then $P_{H_2O}, P_{CH_3OH} \rightarrow P_{Total}$ **then compare values**

$P_{Total} = P_{H_2O} + P_{CH_3OH}$

Solution:

$\chi_{CH_3OH} = 1 - \chi_{H_2O} = 1 - 0.312 = 0.688$ then $P_{H_2O} = \chi_{H_2O} P^{\circ}_{H_2O} = 0.312 \times 55.3 \text{ torr} = 17.\underline{2}536$ torr and

$P_{CH_3OH} = \chi_{CH_3OH} P^{\circ}_{CH_3OH} = 0.688 \times 256 \text{ torr} = 17\underline{6}.128$ torr then

$P_{Total} = P_{H_2O} + P_{CH_3OH} = 17.\underline{2}536 \text{ torr} + 17\underline{6}.128 \text{ torr} = 193$ torr

Because 193 torr ≠ 211 torr, the solution is not behaving ideally. The methanol–water interactions are weaker than the methanol–methanol and water–water interactions.

Check: The units (torr) are correct. The magnitude of the answer (193) seems reasonable because each is a fraction of the pure vapor pressures. We are not surprised that the solution is not ideal because the types of bonds in the two molecules are very different.

Freezing Point Depression, Boiling Point Elevation, and Osmosis

14.73 **Given:** 55.8 g of glucose ($C_6H_{12}O_6$) in 455 g water **Find:** T_f and T_b **Other:** $K_f = 1.86$ °C/m; $K_b = 0.512$ °C/m;

Conceptual Plan: $g_{H_2O} \rightarrow kg_{H_2O}$ **and** $g_{C_6H_{12}O_6} \rightarrow mol_{C_6H_{12}O_6}$ **then** $mol_{C_6H_{12}O_6}, kg_{H_2O} \rightarrow m$

$\frac{1 \text{ kg}}{1000 \text{ g}}$ $\quad \frac{1 \text{ mol } C_6H_{12}O_6}{180.16 \text{ g } C_6H_{12}O_6}$ $\quad m = \frac{\text{amount solute (moles)}}{\text{mass solvent (kg)}}$

$m, K_f \rightarrow \Delta T_f \rightarrow T_f$ **and** $m, K_b \rightarrow \Delta T_b \rightarrow T_b$

$\Delta T_f = K_f \times m \quad T_f = T^{\circ}_f - \Delta T_f \quad \Delta T_b = K_b \times m \quad \Delta T_b = T_b - T^{\circ}_b$

Solution: $455 \cancel{g} \times \frac{1 \text{ kg}}{1000 \cancel{g}} = 0.455$ kg and $55.8 \cancel{g\ C_6H_{12}O_6} \times \frac{1 \text{ mol } C_6H_{12}O_6}{180.16 \cancel{g\ C_6H_{12}O_6}} = 0.30\underline{9}725 \text{ mol } C_6H_{12}O_6$ then

$$m = \frac{\text{amount solute (moles)}}{\text{mass solvent (kg)}} = \frac{0.30\underline{9}725 \text{ mol } C_6H_{12}O_6}{0.455 \text{ kg}} = 0.68\underline{0}714\ m \text{ then}$$

$$\Delta T_f = K_f \times m = 1.86 \frac{°C}{\cancel{m}} \times 0.68\underline{0}714 \cancel{m} = 1.27\ °C \text{ then } T_f = T^{\circ}_f - \Delta T_f = 0.00\ °C - 1.27\ °C = -1.27\ °C \text{ and}$$

$$\Delta T_b = K_b \times m = 0.512 \frac{°C}{\cancel{m}} \times 0.68\underline{0}714 \cancel{m} = 0.349\ °C \text{ and } \Delta T_b = T_b - T^{\circ}_b \text{ so}$$

$$T_b = T^{\circ}_b + \Delta T_b = 100.000\ °C + 0.349\ °C = 100.349\ °C$$

Check: The units (°C) are correct. The magnitude of the answers (−1.3 and 100.3) seems reasonable because the molality is ~2/3. The shift in boiling point is less than the shift in freezing point because the constant for boiling is smaller than the constant for freezing.

14.74 **Given:** 21.2 g of ethylene glycol ($C_2H_6O_2$) in 85.4 g water **Find:** T_f and T_b

Other: $K_f = 1.86$ °C/m; $K_b = 0.512$ °C/m

Conceptual Plan: $g_{H_2O} \rightarrow kg_{H_2O}$ **and** $g_{C_2H_6O_2} \rightarrow mol_{C_2H_6O_2}$ **then** $mol_{C_2H_6O_2}, kg_{H_2O} \rightarrow m$

$\frac{1 \text{ kg}}{1000 \text{ g}}$ $\quad \frac{1 \text{ mol } C_2H_6O_2}{62.07 \text{ g } C_2H_6O_2}$ $\quad m = \frac{\text{amount solute (moles)}}{\text{mass solvent (kg)}}$

$m, K_f \rightarrow \Delta T_f \rightarrow T_f$ **and** $m, K_b \rightarrow \Delta T_b \rightarrow T_b$

$\Delta T_f = K_f \times m \quad T_f = T^{\circ}_f - \Delta T_f \quad \Delta T_b = K_b \times m \quad \Delta T_b = T_b - T^{\circ}_b$

Solution: $85.4 \cancel{g} \times \frac{1 \text{ kg}}{1000 \cancel{g}} = 0.0854$ kg and $21.2 \cancel{g\ C_2H_6O_2} \times \frac{1 \text{ mol } C_2H_6O_2}{62.07 \cancel{g\ C_2H_6O_2}} = 0.34\underline{1}550 \text{ mol } C_2H_6O_2$ then

$$m = \frac{\text{amount solute (moles)}}{\text{mass solvent (kg)}} = \frac{0.34\underline{1}550 \text{ mol } C_2H_6O_2}{0.0854 \text{ kg}} = 3.9\underline{9}941\ m \text{ then}$$

$$\Delta T_f = K_f \times m = 1.86 \frac{°C}{\cancel{m}} \times 3.9\underline{9}941 \cancel{m} = 7.44\ °C \text{ then } T_f = T^{\circ}_f - \Delta T_f = 0.00\ °C - 7.44\ °C = -7.44\ °C \text{ and}$$

$\Delta T_b = K_b \times m = 0.512 \frac{°C}{m} \times 3.99941\ m = 2.05\ °C$ and $\Delta T_b = T_b - T_b^\circ$ so

$T_b = T_b^\circ + \Delta T_b = 100.00\ °C + 2.05\ °C = 102.05\ °C$

Check: The units (°C) are correct. The magnitude of the answers (−7.4 and 102.05) seems reasonable because the molality is ~4. The shift in boiling point is less than the shift in freezing point because the constant for boiling is smaller than the constant for freezing.

14.75 **Given:** 10.0 g of naphthalene ($C_{10}H_8$) in 100.0 mL benzene (C_6H_6) **Find:** T_f and T_b

Other: d(benzene) $= 0.877\ g/cm^3$; $K_f = 5.12\ °C/m$; $K_b = 2.53\ °C/m$; $T_f^\circ = 5.5\ °C$; $T_b^\circ = 80.1\ °C$

Conceptual Plan: $\mathbf{mL_{C_6H_6} \rightarrow g_{C_6H_6} \rightarrow kg_{C_6H_6}}$ **and** $\mathbf{g_{C_{10}H_8} \rightarrow mol_{C_{10}H_8}}$ **then** $\mathbf{mol_{C_{10}H_8}, kg_{C_6H_6} \rightarrow m}$

$\frac{0.877\ g}{1\ cm^3}$ $\frac{1\ kg}{1000\ g}$ $\frac{1\ mol\ C_{10}H_8}{128.2\ g\ C_{10}H_8}$ $m = \frac{\text{amount solute (moles)}}{\text{mass solvent (kg)}}$

$\mathbf{m, K_f \rightarrow \Delta T_f \rightarrow T_f}$ **and** $\mathbf{m, K_b \rightarrow \Delta T_b \rightarrow T_b}$

$\Delta T_f = K_f \times m$ $T_f = T_f^\circ - \Delta T_f$ $\Delta T_b = K_b \times m$ $\Delta T_b = T_b - T_b^\circ$

Solution: $100.0\ cm^3 \times \frac{0.877\ g}{1\ cm^3} \times \frac{1\ kg}{1000\ g} = 0.0877\ kg$ and

$10.0\ g\ C_{10}H_8 \times \frac{1\ mol\ C_{10}H_8}{128.2\ g\ C_{10}H_8} = 0.07802503\ mol\ C_{10}H_8$ then

$m = \frac{\text{amount solute (moles)}}{\text{mass solvent (kg)}} = \frac{0.07802503\ mol\ C_{10}H_8}{0.0877\ kg} = 0.8896811\ m$ then

$\Delta T_f = K_f \times m = 5.12 \frac{°C}{m} \times 0.8896811\ m = 4.56\ °C$ then $T_f = T_f^\circ - \Delta T_f = 5.5\ °C - 4.56\ °C = 0.94\ °C = 0.9\ °C$

and $\Delta T_b = K_b \times m = 2.53 \frac{°C}{m} \times 0.8896811\ m = 2.25\ °C$ and $\Delta T_b = T_b - T_b^\circ$ so

$T_b = T_b^\circ + \Delta T_b = 80.1\ °C + 2.25\ °C = 82.35\ °C = 82.4\ °C$

Check: The units (°C) are correct. The magnitude of the answers (0.9 and 82.4) seems reasonable because the molality is almost 1. Because the constants are larger for benzene than for water, we expect larger temperature shifts. The shift in boiling point is less than the shift in freezing point because the constant is smaller for freezing.

14.76 **Given:** 7.55 g of ethylene glycol ($C_2H_6O_2$) in 85.7 mL ethanol (C_2H_6O) **Find:** T_f and T_b

Other: d(ethanol) $= 0.789\ g/cm^3$; $K_f = 1.99\ °C/m$; $K_b = 1.22\ °C/m$; $T_f^\circ = -114.1\ °C$; $T_b^\circ = 78.3\ °C$

Conceptual Plan: $\mathbf{mL_{C_2H_6O} \rightarrow g_{C_2H_6O} \rightarrow kg_{C_2H_6O}}$ **and** $\mathbf{g_{C_2H_6O_2} \rightarrow mol_{C_2H_6O_2}}$ **then**

$\frac{0.789\ g}{1\ cm^3}$ $\frac{1\ kg}{1000\ g}$ $\frac{1\ mol\ C_2H_6O_2}{62.07\ g\ C_2H_6O_2}$

$\mathbf{mol_{C_2H_6O_2}, kg_{C_2H_6O} \rightarrow m}$ **then** $\mathbf{m, K_f \rightarrow \Delta T_f \rightarrow T_f}$ **and** $\mathbf{m, K_b \rightarrow \Delta T_b \rightarrow T_b}$

$m = \frac{\text{amount solute (moles)}}{\text{mass solvent (kg)}}$ $\Delta T_f = K_f \times m$ $T_f = T_f^\circ - \Delta T_f$ $\Delta T_b = K_b \times m$ $\Delta T_b = T_b - T_b^\circ$

Solution: $85.7\ cm^3\ C_2H_6O \times \frac{0.789\ g\ C_2H_6O}{1\ cm^3\ C_2H_6O} \times \frac{1\ kg}{1000\ g} = 0.0676173\ kg\ C_2H_6O$ and

$7.55\ g\ C_2H_6O_2 \times \frac{1\ mol\ C_2H_6O_2}{62.07\ g\ C_2H_6O_2} = 0.1216369\ mol\ C_2H_6O_2$ then

$m = \frac{\text{amount solute (moles)}}{\text{mass solvent (kg)}} = \frac{0.1216369\ mol\ C_2H_6O_2}{0.0676173\ kg\ C_2H_6O} = 1.798902\ m$ then

$\Delta T_f = K_f \times m = 1.99 \frac{°C}{m} \times 1.798902\ m = 3.58\ °C$ then

$T_f = T_f^\circ - \Delta T_f = -114.1\ °C - 3.58\ °C = -117.68\ °C = -117.7\ °C$ and

$\Delta T_b = K_b \times m = 1.22 \frac{°C}{m} \times 1.798902\ m = 2.19\ °C$ and $\Delta T_b = T_b - T_b^\circ$ so

$T_b = T_b^\circ + \Delta T_b = 78.3\ °C + 2.19\ °C = 79.49\ °C = 79.5\ °C$

Check: The units (°C) are correct. The magnitude of the answers (−117.7 and 79.5) seems reasonable because the molality is almost 2. Because the constants are larger for ethanol than for water, we expect larger temperature shifts. The shift in boiling point is less than the shift in freezing point because the constant is smaller for freezing.

14.77 **Given:** 17.5 g of unknown nonelectrolyte in 100.0 g water, $T_f = -1.8\ °C$ **Find:** $\mathcal{M}$
Other: $K_f = 1.86\ °C/m$
Conceptual Plan: $g_{H_2O} \rightarrow kg_{H_2O}$ **and** $T_f \rightarrow \Delta T_f$ **then** $\Delta T_f, K_f \rightarrow m$ **then** $m, kg_{H_2O} \rightarrow mol_{Unk}$

$$\frac{1\ kg}{1000\ g} \qquad T_f = T_f^\circ - \Delta T_f \qquad \Delta T_f = K_f \times m \qquad m = \frac{\text{amount solute (moles)}}{\text{mass solvent (kg)}}$$

then $g_{Unk}, mol_{Unk} \rightarrow \mathcal{M}$

$$\mathcal{M} = \frac{g_{Unk}}{mol_{Unk}}$$

Solution: $100.0\ \cancel{g}\ H_2O \times \frac{1\ kg}{1000\ \cancel{g}} = 0.1000\ kg\ H_2O$ and $T_f = T_f^\circ - \Delta T_f$ so

$\Delta T_f = T_f^\circ - T_f = 0.00\ °C - (-1.8\ °C) = +1.8\ °C$; $\Delta T_f = K_f \times m$ Rearrange to solve for m.

$$m = \frac{\Delta T_f}{K_f} = \frac{1.8\ \cancel{°C}}{1.86 \frac{\cancel{°C}}{m}} = 0.9\underline{6}774\ m \text{ then } m = \frac{\text{amount solute (moles)}}{\text{mass solvent (kg)}} \text{ so}$$

$$mol_{Unk} = m_{Unk} \times kg_{H_2O} = 0.9\underline{6}774 \frac{mol\ Unk}{\cancel{kg}} \times 0.1000\ \cancel{kg} = 0.09\underline{6}774\ mol\ Unk \text{ then}$$

$$\mathcal{M} = \frac{g_{Unk}}{mol_{Unk}} = \frac{17.5\ g}{0.09\underline{6}774\ mol} = 180 \frac{g}{mol} = 1.8 \times 10^2 \frac{g}{mol}$$

Check: The units (g/mol) are correct. The magnitude of the answer (180 g/mol) seems reasonable because the molality is ~0.1 and we have ~18 g. It is a reasonable molar mass for a solid or liquid.

14.78 **Given:** 35.9 g of unknown nonelectrolyte in 150.0 g water, $T_f = -1.3\ °C$ **Find:** $\mathcal{M}$
Other: $K_f = 1.86\ °C/m$
Conceptual Plan: $g_{H_2O} \rightarrow kg_{H_2O}$ **and** $T_f \rightarrow \Delta T_f$ **then** $\Delta T_f, K_f \rightarrow m$ **then** $m, kg_{H_2O} \rightarrow mol_{Unk}$

$$\frac{1\ kg}{1000\ g} \qquad T_f = T_f^\circ - \Delta T_f \qquad \Delta T_f = K_f \times m \qquad m = \frac{\text{amount solute (moles)}}{\text{mass solvent (kg)}}$$

then $g_{Unk}, mol_{Unk} \rightarrow \mathcal{M}$

$$\mathcal{M} = \frac{g_{Unk}}{mol_{Unk}}$$

Solution: $150.0\ \cancel{g}\ H_2O \times \frac{1\ kg}{1000\ \cancel{g}} = 0.1500\ kg\ H_2O$ and $T_f = T_f^\circ - \Delta T_f$ so

$\Delta T_f = T_f^\circ - T_f = 0.00\ °C - (-1.3\ °C) = +1.3\ °C$ $\Delta T_f = K_f \times m$ Rearrange to solve for m.

$$m = \frac{\Delta T_f}{K_f} = \frac{1.3\ \cancel{°C}}{1.86 \frac{\cancel{°C}}{m}} = 0.6\underline{9}892\ m \text{ then } m = \frac{\text{amount solute (moles)}}{\text{mass solvent (kg)}} \text{ so}$$

$$mol_{Unk} = m_{Unk} \times kg_{H_2O} = 0.6\underline{9}892 \frac{mol\ Unk}{\cancel{kg}} \times 0.1500\ \cancel{kg} = 0.1\underline{0}4838\ mol\ Unk \text{ then}$$

$$\mathcal{M} = \frac{g_{Unk}}{mol_{Unk}} = \frac{35.9\ g}{0.1\underline{0}4838\ mol} = 340 \frac{g}{mol}$$

Check: The units (g/mol) are correct. The magnitude of the answer (340 g/mol) seems reasonable because the molality is ~0.7 and we have ~36 g. It is a reasonable molar mass for a solid or liquid.

14.79 **Given:** 24.6 g of glycerin ($C_3H_8O_3$) in 250.0 mL of solution at 298 K **Find:** Π
Conceptual Plan: $mL_{soln} \rightarrow L_{soln}$ **and** $g_{C_3H_8O_3} \rightarrow mol_{C_3H_8O_3}$ **then** $mol_{C_3H_8O_3}, L_{soln} \rightarrow$ **M then**

$$\frac{1\ L}{1000\ mL} \qquad \frac{1\ mol\ C_3H_8O_3}{92.09\ g\ C_3H_8O_3} \qquad M = \frac{\text{amount solute (moles)}}{\text{volume solution (L)}}$$

M, $T \rightarrow \Pi$

$$\Pi = MRT$$

Solution:

$$250.0\ \text{mL} \times \frac{1\ \text{L}}{1000\ \text{mL}} = 0.2500\ \text{L and } 24.6\ \text{g}\ C_3H_8O_3 \times \frac{1\ \text{mol}\ C_3H_8O_3}{92.09\ \text{g}\ C_3H_8O_3} = 0.26\underline{7}130\ \text{mol}\ C_3H_8O_3 \text{ then}$$

$$\text{M} = \frac{\text{amount solute (moles)}}{\text{volume solution (L)}} = \frac{0.26\underline{7}130\ \text{mol}\ C_3H_8O_3}{0.2500\ \text{L}} = 1.0\underline{6}852\ \text{M then}$$

$$\Pi = MRT = 1.0\underline{6}852\frac{\text{mol}}{\text{L}} \times 0.08206\frac{\text{L} \cdot \text{atm}}{\text{K} \cdot \text{mol}} \times 298\ \text{K} = 26.1\ \text{atm}$$

Check: The units (atm) are correct. The magnitude of the answer (26 atm) seems reasonable because the molarity is ~1.

14.80 **Given:** sucrose ($C_{12}H_{22}O_{11}$) in 5.00×10^2 g water; $\Pi = 8.55$ atm at 298 K **Find:** m ($C_{12}H_{22}O_{11}$)
Other: $d = 1.0$ g/mL
Conceptual Plan: $\Pi, T \rightarrow \text{M}$ **then** $g_{H_2O}, d, \mathcal{M} \rightarrow g_{C_{12}H_{22}O_{11}}$

$$\Pi = MRT \quad \text{M} = \frac{\text{amount solute (moles)}}{\text{volume solution (L)}} \text{ with } \frac{1\ \text{L}}{1000\ \text{mL}}, \frac{1\ \text{mol}\ C_{12}H_{22}O_{11}}{342.30\ \text{g}\ C_{12}H_{22}O_{11}}, \text{ and } \frac{1.0\ \text{mL}}{1.0\ \text{g}}$$

Solution: $\Pi = MRT$ Rearrange to solve for M. $\text{M} = \frac{\Pi}{RT} = \frac{8.55\ \text{atm}}{0.08206\frac{\text{L} \cdot \text{atm}}{\text{K} \cdot \text{mol}} \times 298\ \text{K}} = 0.34\underline{9}638\frac{\text{mol}}{\text{L}}$

Substitute quantities into the definition of M.

$$\text{M} = \frac{\text{amount solute (moles)}}{\text{volume solution (L)}} = \frac{(x\ \text{g}\ C_{12}H_{22}O_{11})\left(\frac{1\ \text{mol}\ C_{12}H_{22}O_{11}}{342.30\ \text{g}\ C_{12}H_{22}O_{11}}\right)}{(5.00 \times 10^2\ \text{g}\ H_2O + x\ \text{g}\ C_{12}H_{22}O_{11}) \times \frac{1.0\ \text{mL}}{1.0\ \text{g}} \times \frac{1\ \text{L}}{1000\ \text{mL}}} = 0.34\underline{9}638\frac{\text{mol}}{\text{L}}$$

Rearrange to solve for x g $C_{12}H_{22}O_{11}$. $x\ \text{g}\ C_{12}H_{22}O_{11} = 0.11\underline{9}681 \times (5.00 \times 10^2\ \text{g}\ H_2O + x\ \text{g}\ C_{12}H_{22}O_{11}) \rightarrow$

$$0.88\underline{0}312 \times (x\ \text{g}\ C_{12}H_{22}O_{11}) = 59.\underline{8}405 \rightarrow x\ \text{g}\ C_{12}H_{22}O_{11} = \frac{59.\underline{8}405}{0.88\underline{0}312} = 68.0\ \text{g}\ C_{12}H_{22}O_{11}$$

Check: The units (g) are correct. The magnitude of the answer (68 g) seems reasonable because the molarity is ~1/3 and we have 0.5 L of water.

14.81 **Given:** 27.55 mg unknown protein in 25.0 mL solution; $\Pi = 3.22$ torr at 25 °C **Find:** $\mathcal{M}_{\text{unknown protein}}$
Conceptual Plan: **°C → K and torr → atm then** $\Pi, T \rightarrow \text{M}$ **then** $\text{mL}_{\text{soln}} \rightarrow \text{L}_{\text{soln}}$ **then**

$$\text{K} = {}^\circ\text{C} + 273.15 \qquad \frac{1\ \text{atm}}{760\ \text{torr}} \qquad \Pi = MRT \qquad \frac{1\ \text{L}}{1000\ \text{mL}}$$

$\text{L}_{\text{soln}}, \text{M} \rightarrow \text{mol}_{\text{unknown protein}}$ **and mg → g then** $g_{\text{unknown protein}}, \text{mol}_{\text{unknown protein}} \rightarrow \mathcal{M}_{\text{unknown protein}}$

$$\text{M} = \frac{\text{amount solute (moles)}}{\text{volume solution (L)}} \qquad \frac{1\ \text{g}}{1000\ \text{mg}} \qquad \mathcal{M} = \frac{g_{\text{unknown protein}}}{\text{mol}_{\text{unknown protein}}}$$

Solution: $25\ {}^\circ\text{C} + 273.15 = 298\ \text{K}$ and $3.22\ \text{torr} \times \frac{1\ \text{atm}}{760\ \text{torr}} = 0.0042\underline{3}684\ \text{atm}$ then solve $\Pi = MRT$ for M.

$$\text{M} = \frac{\Pi}{RT} = \frac{0.0042\underline{3}684\ \text{atm}}{0.08206\frac{\text{L} \cdot \text{atm}}{\text{K} \cdot \text{mol}} \times 298\ \text{K}} = 1.7\underline{3}258 \times 10^{-4}\frac{\text{mol}}{\text{L}} \text{ then } 25.0\ \text{mL} \times \frac{1\ \text{L}}{1000\ \text{mL}} = 0.0250\ \text{L then}$$

$$\text{M} = \frac{\text{amount solute (moles)}}{\text{volume solution (L)}} \quad \text{Rearrange to solve for } \text{mol}_{\text{unknown protein}}.$$

$$\text{mol}_{\text{unknown protein}} = \text{M} \times \text{L} = 1.7\underline{3}258 \times 10^{-4}\frac{\text{mol}}{\text{L}} \times 0.0250\ \text{L} = 4.3\underline{3}146 \times 10^{-6}\ \text{mol and}$$

$$27.55\ \text{mg} \times \frac{1\ \text{g}}{1000\ \text{mg}} = 0.02755\ \text{g then } \mathcal{M} = \frac{g_{\text{unknown protein}}}{\text{mol}_{\text{unknown protein}}} = \frac{0.02755\ \text{g}}{4.3\underline{3}146 \times 10^{-6}\ \text{mol}} = 6.36 \times 10^3\frac{\text{g}}{\text{mol}}$$

Check: The units (g/mol) are correct. The magnitude of the answer (6400 g/mol) seems reasonable for a large biological molecule. A small amount of material is put into 0.025 L; so the concentration is very small, and the molar mass is large.

14.82 **Given:** 18.75 mg of hemoglobin in 15.0 mL of solution at 25 °C, $\mathcal{M}_{\text{hemoglobin}} = 6.5 \times 10^4 \text{g/mol}$ **Find:** Π

Conceptual Plan: $\textbf{mL}_{\textbf{soln}} \rightarrow \textbf{L}_{\textbf{soln}}$ **and** $\textbf{mg}_{\textbf{Hgb}} \rightarrow \textbf{g}_{\textbf{Hgb}} \rightarrow \textbf{mol}_{\textbf{Hgb}}$ **then** $\textbf{mol}_{\textbf{Hgb}}, \textbf{L}_{\textbf{soln}} \rightarrow \textbf{M}$ **then**

$\frac{1 \text{ L}}{1000 \text{ mL}}$ $\frac{1 \text{ g}}{1000 \text{ mg}}$ $\frac{1 \text{ mol Hgb}}{6.5 \times 10^4 \text{ g Hgb}}$ $\text{M} = \frac{\text{amount solute (moles)}}{\text{volume solution (L)}}$

$\textbf{M}, \boldsymbol{T} \rightarrow \boldsymbol{\Pi}$

$\Pi = MRT$

Solution:

$$15.0 \text{ mL} \times \frac{1 \text{ L}}{1000 \text{ mL}} = 0.0150 \text{ L and } 18.75 \text{ mg Hgb} \times \frac{1 \text{ g}}{1000 \text{ mg}} \times \frac{1 \text{ mol Hgb}}{6.5 \times 10^4 \text{ g Hgb}} = 2.\underline{8}8462 \times 10^{-7} \text{ mol Hgb then}$$

$$\text{M} = \frac{\text{amount solute (moles)}}{\text{volume solution (L)}} = \frac{2.\underline{8}8462 \times 10^{-7} \text{ mol Hgb}}{0.0150 \text{ L}} = 1.\underline{9}231 \times 10^{-5} \text{ M then}$$

$$\Pi = MRT = 1.\underline{9}231 \times 10^{-5} \frac{\text{mol}}{\text{L}} \times 0.08206 \frac{\text{L} \cdot \text{atm}}{\text{K} \cdot \text{mol}} \times 298 \text{ K} = 4.7 \times 10^{-4} \text{ atm} = 0.36 \text{ torr}$$

Check: The units (atm) are correct. The magnitude of the answer (10^{-4} atm) seems reasonable because the molarity is so small.

14.83 (a) **Given:** 0.100 m of K_2S, completely dissociated **Find:** T_f, T_b

Other: $K_f = 1.86 \text{ °C}/m$; $K_b = 0.512 \text{ °C}/m$

Conceptual Plan: $\boldsymbol{m, i, K_f} \rightarrow \boldsymbol{\Delta T_f}$ **then** $\boldsymbol{\Delta T_f} \rightarrow \boldsymbol{T_f}$ **and** $\boldsymbol{m, i, K_b} \rightarrow \boldsymbol{\Delta T_b}$ **then** $\boldsymbol{\Delta T_b} \rightarrow \boldsymbol{T_b}$

$\Delta T_f = K_f \times i \times m_{K_2S}$ $T_f = T_f^\circ - \Delta T_f$ $\Delta T_b = K_b \times i \times m_{K_2S}$ $T_b = T_b^\circ + \Delta T_b$

Solution: $\Delta T_f = K_f \times i \times m = 1.86 \frac{\text{°C}}{m} \times 3 \times 0.100\, m = 0.558 \text{ °C then}$

$T_f = T_f^\circ - \Delta T_f = 0.000 \text{ °C} - 0.558 \text{ °C} = -0.558 \text{ °C and}$

$\Delta T_b = K_b \times i \times m = 0.512 \frac{\text{°C}}{m} \times 3 \times 0.100\, m = 0.154 \text{ °C then}$

$T_b = T_b^\circ + \Delta T_b = 100.000 \text{ °C} + 0.154 \text{ °C} = 100.154 \text{ °C}$

Check: The units (°C) are correct. The magnitude of the answers (−0.6 °C and 100.2 °C) seems reasonable because the molality of the particles is 0.3. The shift in boiling point is less than the shift in freezing point because the constant for boiling is larger than the constant for freezing.

(b) **Given:** 21.5 g $CuCl_2$ in 4.50×10^2 g water, completely dissociated **Find:** T_f, T_b

Other: $K_f = 1.86 \text{ °C}/m$; $K_b = 0.512 \text{ °C}/m$

Conceptual Plan: $\textbf{g}_{\textbf{H}_2\textbf{O}} \rightarrow \textbf{kg}_{\textbf{H}_2\textbf{O}}$ **and** $\textbf{g}_{\textbf{CuCl}_2} \rightarrow \textbf{mol}_{\textbf{CuCl}_2}$ **then** $\textbf{mol}_{\textbf{CuCl}_2}, \textbf{kg}_{\textbf{H}_2\textbf{O}} \rightarrow \boldsymbol{m}$

$\frac{1 \text{ kg}}{1000 \text{ g}}$ $\frac{1 \text{ mol CuCl}_2}{134.45 \text{ g CuCl}_2}$ $m = \frac{\text{amount solute (moles)}}{\text{mass solvent (kg)}}$

$\boldsymbol{m, i, K_f} \rightarrow \boldsymbol{\Delta T_f} \rightarrow \boldsymbol{T_f}$ **and** $\boldsymbol{m, i, K_b} \rightarrow \boldsymbol{\Delta T_b} \rightarrow \boldsymbol{T_b}$

$\Delta T_f = K_f \times i \times m_{CuCl_2}$ $T_f = T_f^\circ - \Delta T_f$ $\Delta T_b = K_b \times i \times m_{CuCl_2}$ $T_b = T_b^\circ + \Delta T_b$

Solution:

$$4.5 \times 10^2 \text{ g H}_2\text{O} \times \frac{1 \text{ kg}}{1000 \text{ g}} = 0.450 \text{ kg H}_2\text{O and } 21.5 \text{ g CuCl}_2 \times \frac{1 \text{ mol CuCl}_2}{134.45 \text{ g CuCl}_2} = 0.15\underline{9}911 \text{ mol CuCl}_2 \text{ then}$$

$$m = \frac{\text{amount solute (moles)}}{\text{mass solvent (kg)}} = \frac{0.15\underline{9}911 \text{ mol CuCl}_2}{0.450 \text{ kg H}_2\text{O}} = 0.35\underline{5}358\, m \text{ then}$$

$\Delta T_f = K_f \times i \times m = 1.86 \frac{\text{°C}}{m} \times 3 \times 0.35\underline{5}358\, m = 1.98 \text{ °C then}$

$T_f = T_f^\circ - \Delta T_f = 0.000 \text{ °C} - 1.98 \text{ °C} = -1.98 \text{ °C and}$

$\Delta T_b = K_b \times i \times m = 0.512 \frac{\text{°C}}{m} \times 3 \times 0.35\underline{5}358\, m = 0.546 \text{ °C then}$

$T_b = T_b^\circ + \Delta T_b = 100.000 \text{ °C} + 0.546 \text{ °C} = 100.546 \text{ °C}$

Check: The units (°C) are correct. The magnitude of the answers (−2 °C and 100.5 °C) seems reasonable because the molality of the particles is ~1. The shift in boiling point is less than the shift in freezing point because the constant for boiling is larger than the constant for freezing.

(c) **Given:** 5.5% by mass $NaNO_3$, completely dissociated **Find:** T_f, T_b
Other: $K_f = 1.86\ °C/m$; $K_b = 0.512\ °C/m$
Conceptual Plan: Percent by mass → g_{NaNO_3}, g_{H_2O} then $g_{H_2O} \rightarrow kg_{H_2O}$ and $g_{NaNO_3} \rightarrow mol_{NaNO_3}$

$$\text{mass percent} = \frac{\text{mass solute}}{\text{mass solution}} \times 100\% \qquad \frac{1\ \text{kg}}{1000\ \text{g}} \qquad \frac{1\ \text{mol NaNO}_3}{85.00\ \text{g NaNO}_3}$$

then mol_{NaNO_3}, kg_{H_2O} → m then $m, i, K_f \rightarrow \Delta T_f \rightarrow T_f$ and $m, i, K_b \rightarrow \Delta T_b \rightarrow T_b$

$$m = \frac{\text{amount solute (moles)}}{\text{mass solvent (kg)}} \qquad \Delta T_f = K_f \times i \times m_{NaNO_3}\quad T_f = T_f^\circ - \Delta T_f \qquad \Delta T_b = K_b \times i \times m_{NaNO_3}\ T_b = T_b^\circ + \Delta T_b$$

Solution: mass percent $= \dfrac{\text{mass solute}}{\text{mass solution}} \times 100\%$, so 5.5% by mass $NaNO_3$ means 5.5 g $NaNO_3$ and

$100.0\ \text{g} - 5.5\ \text{g} = 94.5$ g water. Then $94.5\ \text{g H}_2\text{O} \times \dfrac{1\ \text{kg}}{1000\ \text{g}} = 0.0945$ kg H_2O and

$$5.5\ \text{g NaNO}_3 \times \frac{1\ \text{mol NaNO}_3}{85.00\ \text{g NaNO}_3} = 0.06\underline{4}706\ \text{mol NaNO}_3 \text{ then}$$

$$m = \frac{\text{amount solute (moles)}}{\text{mass solvent (kg)}} = \frac{0.06\underline{4}706\ \text{mol NaNO}_3}{0.0945\ \text{kg H}_2\text{O}} = 0.6\underline{8}472\ m \text{ then}$$

$$\Delta T_f = K_f \times i \times m = 1.86\frac{°C}{m} \times 2 \times 0.6\underline{8}472\ m = 2.5\ °C \text{ then}$$

$$T_f = T_f^\circ - \Delta T_f = 0.000\ °C - 2.5\ °C = -2.5\ °C \text{ and}$$

$$\Delta T_b = K_b \times i \times m = 0.512\frac{°C}{m} \times 2 \times 0.6\underline{8}472\ m = 0.70\ °C \text{ then}$$

$$T_b = T_b^\circ + \Delta T_b = 100.000\ °C + 0.70\ °C = 100.70\ °C$$

Check: The units (°C) are correct. The magnitude of the answers (−2.5 °C and 100.7 °C) seems reasonable because the molality of the particles is ~1. The shift in boiling point is less than the shift in freezing point because the constant for boiling is larger than the constant for freezing.

14.84 (a) **Given:** 10.5 g $FeCl_3$ in 1.50×10^2 g water, completely dissociated **Find:** T_f, T_b
Other: $K_f = 1.86\ °C/m$; $K_b = 0.512\ °C/m$
Conceptual Plan: $g_{H_2O} \rightarrow kg_{H_2O}$ and $g_{FeCl_3} \rightarrow mol_{FeCl_3}$ then mol_{FeCl_3}, $kg_{H_2O} \rightarrow m$

$$\frac{1\ \text{kg}}{1000\ \text{g}} \qquad \frac{1\ \text{mol FeCl}_3}{162.20\ \text{g FeCl}_3} \qquad m = \frac{\text{amount solute (moles)}}{\text{mass solvent (kg)}}$$

$m, i, K_f \rightarrow \Delta T_f \rightarrow T_f$ and $m, i, K_b \rightarrow \Delta T_b \rightarrow T_b$

$$\Delta T_f = K_f \times i \times m_{FeCl_3}\quad T_f = T_f^\circ - \Delta T_f \quad \Delta T_b = K_b \times i \times m_{FeCl_3}\ T_b = T^\circ{}_b + \Delta T_b$$

Solution: $4.5 \times 10^2\ \text{g H}_2\text{O} \times \dfrac{1\ \text{kg}}{1000\ \text{g}} = 0.450$ kg H_2O and

$$10.5\ \text{g FeCl}_3 \times \frac{1\ \text{mol FeCl}_3}{162.20\ \text{g FeCl}_3} = 0.064\underline{7}349\ \text{mol FeCl}_3 \text{ then}$$

$$m = \frac{\text{amount solute (moles)}}{\text{mass solvent (kg)}} = \frac{0.064\underline{7}349\ \text{mol FeCl}_3}{0.150\ \text{kg H}_2\text{O}} = 0.431\underline{5}66\ m \text{ then}$$

$$\Delta T_f = K_f \times i \times m = 1.86\frac{°C}{m} \times 4 \times 0.431\underline{5}66\ m = 3.21\ °C \text{ then}$$

$$T_f = T_f^\circ - \Delta T_f = 0.000\ °C - 3.21\ °C = -3.21\ °C \text{ and}$$

$$\Delta T_b = K_b \times i \times m = 0.512\frac{°C}{m} \times 4 \times 0.431\underline{5}66\ m = 0.884\ °C \text{ then}$$

$$T_b = T_b^\circ + \Delta T_b = 100.000\ °C + 0.884\ °C = 100.884\ °C$$

Check: The units (°C) are correct. The magnitude of the answers (−3 °C and 100.9 °C) seems reasonable because the molality of the particles is ~2. The shift in boiling point is less than the shift in freezing point because the constant for boiling is smaller than the constant for freezing.

(b) **Given:** 3.5% by mass KCl, completely dissociated **Find:** T_f, T_b
Other: $K_f = 1.86\ °C/m$; $K_b = 0.512\ °C/m$

Conceptual Plan: percent by mass $\rightarrow g_{KCl}, g_{H_2O}$ then $g_{H_2O} \rightarrow kg_{H_2O}$ and $g_{KCl} \rightarrow mol_{KCl}$ then

$$\text{mass percent} = \frac{\text{mass solute}}{\text{mass solution}} \times 100\% \qquad \frac{1\ \text{kg}}{1000\ \text{g}} \qquad \frac{1\ \text{mol KCl}}{74.55\ \text{g KCl}}$$

$mol_{NaNO_3}, kg_{H_2O} \rightarrow m$ then $m, i, K_f \rightarrow \Delta T_f \rightarrow T_f$ and $m, i, K_b \rightarrow \Delta T_b \rightarrow T_b$

$$m = \frac{\text{amount solute (moles)}}{\text{mass solvent (kg)}} \quad \Delta T_f = K_f \times i \times m_{KCl} \quad T_f = T_f^\circ - \Delta T_f \quad \Delta T_b = K_b \times i \times m_{KCl} \quad T_b = T_b^\circ + \Delta T_b$$

Solution: mass percent $= \dfrac{\text{mass solute}}{\text{mass solution}} \times 100\%$ so 3.5% by mass KCl means

3.5 g KCl and 100.0 g $-$ 3.5 g $=$ 96.5 g water. Then 96.5 g $H_2O \times \dfrac{1\ \text{kg}}{1000\ \text{g}} = 0.0965$ kg H_2O and

$3.5\ \text{g KCl} \times \dfrac{1\ \text{mol KCl}}{74.55\ \text{g KCl}} = 0.04\underline{6}948$ mol KCl

then $m = \dfrac{\text{amount solute (moles)}}{\text{mass solvent (kg)}} = \dfrac{0.04\underline{6}948\ \text{mol KCl}}{0.0965\ \text{kg H}_2\text{O}} = 0.4\underline{8}651\ m$ then

$\Delta T_f = K_f \times i \times m = 1.86 \dfrac{°\text{C}}{m} \times 2 \times 0.4\underline{8}651\ m = 1.8\ °\text{C}$ then

$T_f = T_f^\circ - \Delta T_f = 0.000\ °\text{C} - 1.8\ °\text{C} = -1.8\ °\text{C}$

and $\Delta T_b = K_b \times i \times m = 0.512 \dfrac{°\text{C}}{m} \times 2 \times 0.4\underline{8}651\ m = 0.50\ °\text{C}$ then

$T_b = T_b^\circ + \Delta T_b = 100.000\ °\text{C} + 0.50\ °\text{C} = 100.50\ °\text{C}$

Check: The units (°C) are correct. The magnitude of the answers (-2 °C and 100.5 °C) seems reasonable because the molality of the particles is ~1. The shift in boiling point is less than the shift in freezing point because the constant for boiling is smaller than the constant for freezing.

(c) **Given:** 0.150 m of MgF_2, completely dissociated **Find:** T_f, T_b
Other: $K_f = 1.86\ °\text{C}/m$; $K_b = 0.512\ °\text{C}/m$
Conceptual Plan: $m, i, K_f \rightarrow \Delta T_f$ then $\Delta T_f \rightarrow T_f$ and $m, i, K_b \rightarrow \Delta T_b$ then $\Delta T_b \rightarrow T_b$

$$\Delta T_f = K_f \times i \times m_{MgF_2} \quad T_f = T_f^\circ - \Delta T_f \quad \Delta T_b = K_b \times i \times m_{MgF_2} \quad T_b = T_b^\circ + \Delta T_b$$

Solution: $\Delta T_f = K_f \times i \times m = 1.86 \dfrac{°\text{C}}{m} \times 3 \times 0.150\ m = 0.837\ °\text{C}$ then

$T_f = T_f^\circ - \Delta T_f = 0.000\ °\text{C} - 0.837\ °\text{C} = -0.837\ °\text{C}$ and

$\Delta T_b = K_b \times i \times m = 0.512 \dfrac{°\text{C}}{m} \times 3 \times 0.150\ m = 0.230\ °\text{C}$ then

$T_b = T_b^\circ + \Delta T_b = 100.000\ °\text{C} + 0.230\ °\text{C} = 100.230\ °\text{C}$

Check: The units (°C) are correct. The magnitude of the answers (-0.8 °C and 100.2 °C) seems reasonable because the molality of the particles is 0.5. The shift in boiling point is less than the shift in freezing point because the constant for boiling is smaller than the constant for freezing.

14.85 **Given:** NaCl complete dissociation; 1.0 L water and $T_f = -10.0\ °\text{C}$ **Find:** mass of NaCl
Other: $K_f = 1.86\ °\text{C}/m$; $d(\text{water}) = 1.0\ \text{g/mL}$
Conceptual Plan:
$T_f \rightarrow \Delta T_f$ then $\Delta T_f, i, K_f \rightarrow m$ then $L_{H_2O} \rightarrow mL_{H_2O} \rightarrow g_{H_2O} \rightarrow kg_{H_2O}$ then $m, kg_{H_2O} \rightarrow mol_{NaCl}$

$$T_f = T_f^\circ - \Delta T_f \qquad \Delta T_f = K_f \times i \times m \qquad \frac{1000\ \text{mL}}{1\ \text{L}} \quad \frac{1.0\ \text{g}}{1\ \text{mL}} \quad \frac{1\ \text{kg}}{1000\ \text{g}} \qquad m = \frac{\text{amount solute (moles)}}{\text{mass solvent (kg)}}$$

then $mol_{NaCl} \rightarrow g_{NaCl}$

$$\frac{58.44\ \text{g NaCl}}{1\ \text{mol NaCl}}$$

Solution: $T_f = T_f^\circ - \Delta T_f$ so $\Delta T_f = T_f^\circ - T_f = 0.0\ °\text{C} - 10.0\ °\text{C} = -10.0\ °\text{C}$ then $\Delta T_f = K_f \times i \times m$

Rearrange to solve for m. Because the salt completely dissolves, $i = 2$.

$m = \dfrac{\Delta T_f}{K_f \times i} = \dfrac{10.0\ °\text{C}}{1.86 \dfrac{°\text{C}}{m} \times 2} = 2.\underline{6}88172\ m$ NaCl then $1.0\ \text{L} \times \dfrac{1000\ \text{mL}}{1\ \text{L}} \times \dfrac{1.0\ \text{g}}{1\ \text{mL}} \times \dfrac{1\ \text{kg}}{1000\ \text{g}} = 1.0$ kg then

$m = \frac{\text{amount solute (moles)}}{\text{mass solvent (kg)}}$ so $\text{mol}_{NaCl} = m \times \text{kg}_{H_2O} = 2.688172 \frac{\text{mol NaCl}}{\cancel{\text{kg}_{H_2O}}} \times 1.0\ \cancel{\text{kg}_{H_2O}} = 2.688172\ \text{mol NaCl}$

then $2.688172\ \cancel{\text{mol NaCl}} \times \frac{58.44\ \text{g NaCl}}{1\ \cancel{\text{mol NaCl}}} = 157.097\ \text{g NaCl} = 160\ \text{g NaCl}$

Check: The units (g) are correct. The magnitude of the answer (160 g) seems reasonable because the temperature change is moderate and NaCl has a low formula mass.

14.86 **Given:** ethylene glycol ($C_2H_6O_2$) in water and $T_b = 104.0\ °C$ **Find:** ethylene glycol % by mass
Other: $K_b = 0.512\ °C/m$; $d(\text{water}) = 1.0\ \text{g/mL}$
Conceptual Plan: $T_b \rightarrow \Delta T_b$ **then** $\Delta T_b, K_b \rightarrow m$ **then assume exactly 1 kg of water**

$\Delta T_b = T_b - T_b^\circ$ $\quad \Delta T_b = K_b \times m$

then $m, \text{kg}_{H_2O} \rightarrow \text{mol}_{C_2H_6O_2} \rightarrow \text{g}_{C_2H_6O_2}$ **and** $\text{kg}_{H_2O} \rightarrow \text{g}_{H_2O}$ **then** $\text{g}_{C_2H_6O_2}, \text{g}_{H_2O} \rightarrow C_2H_6O_2$ **% by mass**

$m = \frac{\text{amount solute (moles)}}{\text{mass solvent (kg)}}$ $\quad \frac{62.07\ \text{g}\ C_2H_6O_2}{1\ \text{mol}\ C_2H_6O_2}$ $\quad \frac{1000\ \text{g}}{1\ \text{kg}}$ $\quad \%\ \text{by mass} = \frac{\text{grams solute}}{\text{grams solute + grams solvent}} \times 100\%$

Solution: $\Delta T_b = T_b - T_b^\circ = 104.0\ °C - 100.0\ °C = 4.0\ °C$ then $\Delta T_b = K_b \times m$

Rearrange to solve for m. $m = \frac{\Delta T_b}{K_b} = \frac{4.0\ \cancel{°C}}{0.512 \frac{\cancel{°C}}{m}} = 7.8125\ m\ C_2H_6O_2$ then assume exactly 1 kg of water

$m = \frac{\text{amount solute (moles)}}{\text{mass solvent (kg)}}$ so $\text{mol}_{C_2H_6O_2} = m \times \text{kg}_{H_2O} = 7.8125 \frac{\text{mol}\ C_2H_6O_2}{\cancel{\text{kg}_{H_2O}}} \times 1\ \cancel{\text{kg}_{H_2O}} = 7.8125\ \text{mol}\ C_2H_6O_2$

then $7.8125\ \cancel{\text{mol}\ C_2H_6O_2} \times \frac{62.07\ \text{g}\ C_2H_6O_2}{1\ \cancel{\text{mol}\ C_2H_6O_2}} = 484.922\ \text{g}\ C_2H_6O_2$ and $1\ \cancel{\text{kg}} \times \frac{1000\ \text{g}}{1\ \cancel{\text{kg}}} = 1000\ \text{g (exactly)}$ finally

$\%\ \text{by mass} = \frac{\text{grams solute}}{\text{grams solute + grams solvent}} \times 100\% = \frac{484.922\ \cancel{\text{g}}}{484.922\ \cancel{\text{g}} + 1000\ \cancel{\text{g}}} \times 100\% = 32.65640\%$

$= 33\%\ C_2H_6O_2$ by mass

Check: The units (% by mass) are correct. The magnitude of the answer (33) seems reasonable because the temperature change is moderate and ethylene glycol has a low molar mass.

14.87 (a) **Given:** $0.100\ m$ of $FeCl_3$ **Find:** T_f **Other:** $K_f = 1.86\ °C/m$; $i_{\text{measured}} = 3.4$
Conceptual Plan: $m, i, K_f \rightarrow \Delta T_f$ **then** $\Delta T_f \rightarrow T_f$

$\Delta T_f = K_f \times i \times m \quad T_f = T_f^\circ - \Delta T_f$

Solution: $\Delta T_f = K_f \times i \times m = 1.86 \frac{°C}{\cancel{m}} \times 3.4 \times 0.100\ \cancel{m} = 0.632\ °C$ then

$T_f = T_f^\circ - \Delta T_f = 0.000\ °C - 0.632\ °C = -0.632\ °C$

Check: The units (°C) are correct. The magnitude of the answer (−0.6 °C) seems reasonable because the theoretical molality of the particles is 0.4.

(b) **Given:** 0.085 M of K_2SO_4 at 298 K **Find:** Π **Other:** $i_{\text{measured}} = 2.6$
Conceptual Plan: $M, i, T \rightarrow \Pi$

$\Pi = i \times MRT$

Solution: $\Pi = i \times MRT = 2.6 \times 0.085 \frac{\cancel{\text{mol}}}{\cancel{\text{L}}} \times 0.08206 \frac{\cancel{\text{L}} \cdot \text{atm}}{\cancel{\text{K}} \cdot \cancel{\text{mol}}} \times 298\ \cancel{\text{K}} = 5.4\ \text{atm}$

Check: The units (atm) are correct. The magnitude of the answer (5 atm) seems reasonable because the molarity of particles is ~0.2 m.

(c) **Given:** 1.22% by mass $MgCl_2$ **Find:** T_b **Other:** $K_b = 0.512\ °C/m$; $i_{\text{measured}} = 2.7$
Conceptual Plan: percent by mass $\rightarrow \text{g}_{MgCl_2}, \text{g}_{H_2O}$ **then** $\text{g}_{H_2O} \rightarrow \text{kg}_{H_2O}$ **and** $\text{g}_{MgCl_2} \rightarrow \text{mol}_{MgCl_2}$ **then**

$\text{mass percent} = \frac{\text{mass solute}}{\text{mass solution}} \times 100\%$ $\quad \frac{1\ \text{kg}}{1000\ \text{g}}$ $\quad \frac{1\ \text{mol}\ MgCl_2}{95.21\ \text{g}\ MgCl_2}$

$\text{mol}_{MgCl_2}, \text{kg}_{H_2O} \rightarrow m$ **then** m, **i**, $K_b \rightarrow \Delta T_b \rightarrow T_b$

$m = \frac{\text{amount solute (moles)}}{\text{mass solvent (kg)}}$ $\quad \Delta T_b = K_b \times i \times m\ T_b = T_b^\circ + \Delta T_b$

Solution: mass percent $= \frac{\text{mass solute}}{\text{mass solution}} \times 100\%$ so 1.22% by mass $MgCl_2$ means 1.22 g $MgCl_2$ and

100.00 g − 1.22 g = 98.78 g water then $98.78 \text{ g } H_2O \times \frac{1 \text{ kg}}{1000 \text{ g}} = 0.09878 \text{ kg } H_2O$ and

$1.22 \text{ g } MgCl_2 \times \frac{1 \text{ mol } MgCl_2}{95.21 \text{ g } MgCl_2} = 0.01\underline{2}8138 \text{ mol } MgCl_2$ then

$m = \frac{\text{amount solute (moles)}}{\text{mass solvent (kg)}} = \frac{0.01\underline{2}8138 \text{ mol } MgCl_2}{0.09878 \text{ kg } H_2O} = 0.12\underline{9}721\ m$ then

$\Delta T_b = K_b \times i \times m = 0.512 \frac{°C}{m} \times 2.7 \times 0.12\underline{9}721\ m = 0.18\ °C$ then

$T_b = T_b^\circ - \Delta T_b = 100.000\ °C + 0.18\ °C = 100.18\ °C$

Check: The units (°C) are correct. The magnitude of the answer (100.2 °C) seems reasonable because the molality of the particles is ~1/3.

14.88 (a) **Given:** NaCl; 1.50×10^2 g water and $T_f = -1.0\ °C$ **Find:** g_{NaCl}

Other: $K_f = 1.86\ °C/m$; $i_{measured} = 1.9$

Conceptual Plan: $T_f \rightarrow \Delta T_f$ **then** $\Delta T_f, i, K_f \rightarrow m$ **then** $g_{H_2O} \rightarrow kg_{H_2O}$ **then** $m, kg_{H_2O} \rightarrow mol_{NaCl}$

$T_f = T_f^\circ - \Delta T_f$ $\qquad \Delta T_f = K_f \times i \times m$ $\qquad \frac{1 \text{ kg}}{1000 \text{ g}}$ $\qquad m = \frac{\text{amount solute (moles)}}{\text{mass solvent (kg)}}$

then $mol_{NaCl} \rightarrow g_{NaCl}$

$\frac{58.44 \text{ gNaCl}}{1 \text{ mol NaCl}}$

Solution: $T_f = T_f^\circ - \Delta T_f$ so $\Delta T_f = T_f^\circ - T_f = 0.0\ °C - 1.0\ °C = -1.0\ °C$ then $\Delta T_f = K_f \times i \times m$

Rearrange to solve for m. $m = \frac{\Delta T_f}{K_f \times i} = \frac{1.0\ °C}{1.86 \frac{°C}{m} \times 1.9} = 0.2\underline{8}297\ m$ NaCl then

$1.50 \times 10^2 \text{ g } H_2O \times \frac{1 \text{ kg}}{1000 \text{ g}} = 0.150 \text{ kg } H_2O$ then $m = \frac{\text{amount solute (moles)}}{\text{mass solvent (kg)}}$ so

$mol_{NaCl} = m \times kg_{H_2O} = 0.2\underline{8}297 \frac{\text{mol NaCl}}{kg_{H_2O}} \times 0.150\ kg_{H_2O} = 0.04\underline{2}446$ mol NaCl then

$0.04\underline{2}446 \text{ mol NaCl} \times \frac{58.44 \text{ g NaCl}}{1 \text{ mol NaCl}} = 2.5 \text{ g NaCl}$

Check: The units (g) are correct. The magnitude of the answer (2.5 g) seems reasonable because the temperature change is moderate and NaCl has a low formula mass.

(b) **Given:** $MgSO_4$; 2.50×10^2 mL solution and $\Pi = 3.82$ atm at 298 K **Find:** $m(MgSO_4)$

Other: $i_{measured} = 1.3$

Conceptual Plan: $i, \Pi, T \rightarrow M$ **then** $mL_{soln} \rightarrow L_{soln}$ **then** $L_{soln}, M \rightarrow mol_{MgSO_4} \rightarrow g_{MgSO_4}$

$\Pi = i \times MRT$ $\qquad \frac{1 \text{ L}}{1000 \text{ mL}}$ $\qquad M = \frac{\text{amount solute (moles)}}{\text{volume solution (L)}}$ $\qquad \frac{120.38 \text{ g } MgSO_4}{1 \text{ mol } MgSO_4}$

Solution: $\Pi = i \times MRT$. Rearrange to solve for M.

$M = \frac{\Pi}{i \times RT} = \frac{3.82 \text{ atm}}{1.3 \times 0.08206 \frac{L \cdot atm}{K \cdot mol} \times 298 \text{ K}} = 0.1\underline{2}016 \frac{\text{mol } MgSO_4}{L}$ then

$2.50 \times 10^2 \text{ mL} \times \frac{1 \text{ L}}{1000 \text{ mL}} = 0.250 \text{ L}$ then $M = \frac{\text{amount solute (moles)}}{\text{volume solution (L)}}$ so

$mol_{MgSO_4} = M \times L_{soln} = 0.1\underline{2}016 \frac{\text{mol } MgSO_4}{L} \times 0.250 \text{ L} = 0.03\underline{0}040 \text{ mol } MgSO_4$ finally

$0.03\underline{0}040 \text{ mol } MgSO_4 \times \frac{120.38 \text{ g } MgSO_4}{1 \text{ mol } MgSO_4} = 3.6 \text{ g } MgSO_4$

Check: The units (g) are correct. The magnitude of the answer (4 g) seems reasonable because the pressure is moderate and $MgSO_4$ has a low formula mass.

(c) **Given:** $FeCl_3$; 2.50×10^2 g water and $T_b = 102\ °C$ **Find:** $m(FeCl_3)$
Other: $K_b = 0.512\ °C/m$; $i_{measured} = 3.4$
Conceptual Plan: $T_b \rightarrow \Delta T_b$ **then** $\Delta T_b, i, K_b \rightarrow m$ **then** $g_{H_2O} \rightarrow kg_{H_2O}$ **then** $m, kg_{H_2O} \rightarrow mol_{FeCl_3}$

$T_b = T_b^\circ + \Delta T_b$ $\quad \Delta T_b = K_b \times i \times m$ $\quad \frac{1\ kg}{1000\ g}$ $\quad m = \frac{\text{amount solute (moles)}}{\text{mass solvent (kg)}}$

then $mol_{FeCl_3} \rightarrow g_{FeCl_3}$

$\frac{162.20\ g\ FeCl_3}{1\ mol\ FeCl_3}$

Solution: $T_b = T_b^\circ + \Delta T_b$ so $\Delta T_b = T_b - T_b^\circ = 102\ °C - 100\ °C = 2\ °C$ then $\Delta T_b = K_b \times i \times m$

Rearrange to solve for m. $m = \frac{\Delta T_b}{K_b \times i} = \frac{2\ °C}{0.512 \frac{°C}{m} \times 3.4} = 1.149\ m\ FeCl_3$ then

$2.50 \times 10^2\ g\ H_2O \times \frac{1\ kg}{1000\ g} = 0.250\ kg\ H_2O$ then $m = \frac{\text{amount solute (moles)}}{\text{mass solvent (kg)}}$ so

$mol_{FeCl_3} = m \times kg_{H_2O} = 1.149 \frac{mol\ FeCl_3}{kg_{H_2O}} \times 0.250\ kg_{H_2O} = 0.2873\ mol\ FeCl_3$ then

$0.2873\ mol\ FeCl_3 \times \frac{162.20\ g\ FeCl_3}{1\ mol\ FeCl_3} = 47\ g\ FeCl_3 = 50\ g\ FeCl_3$

Check: The units (g) are correct. The magnitude of the answer (50 g) seems reasonable because the temperature change is significant and we are making 0.25 L of solution.

14.89 **Given:** 1.2 m MX_2 in water and $T_b = 101.4\ °C$ **Find:** i
Other: $K_b = 0.512\ °C/m$
Conceptual Plan: $T_b \rightarrow \Delta T_b$ **then** $m, \Delta T_b, K_b \rightarrow i$

$\Delta T_b = T_b - T_b^\circ$ $\quad \Delta T_b = K_b \times i \times m$

Solution: $\Delta T_b = T_b - T_b^\circ = 101.4\ °C - 100.0\ °C = 1.4\ °C$ then $\Delta T_b = K_b \times i \times m$

Rearrange to solve for i. $i = \frac{\Delta T_b}{K_b \times m} = \frac{1.4\ °C}{0.512 \frac{°C}{m} \times 1.2\ m} = 2.27865 = 2.3$

Check: The units (none) are correct. The magnitude of the answer (2.3) seems reasonable because the formula of the salt is MX_2, where $i = 3$ if it completely dissociated.

14.90 **Given:** 0.95 m MX in water and $T_f = -3.0\ °C$ **Find:** i
Other: $K_f = 1.86\ °C/m$
Conceptual Plan: $T_f \rightarrow \Delta T_f$ **then** $m, \Delta T_f, K_f \rightarrow i$

$T_f = T_f^\circ - \Delta T_f$ $\quad \Delta T_f = K_f \times i \times m.$

Solution: $T_f = T_f^\circ - \Delta T_f$ so $\Delta T_f = T_f^\circ - T_f = 0.0\ °C - 3.0\ °C = -3.0\ °C$ then $\Delta T_f = K_f \times i \times m$

Rearrange to solve for i. $i = \frac{\Delta T_f}{K_f \times m} = \frac{3.0\ °C}{1.86 \frac{°C}{m} \times 0.95\ m} = 1.69779 = 1.7$

Check: The units (none) are correct. The magnitude of the answer (1.7) seems reasonable because the formula of the salt is MX, where $i = 2$ if it completely dissociated.

14.91 **Given:** 0.100 M of ionic solution, $\Pi = 8.3$ atm at 25 °C **Find:** $i_{measured}$
Conceptual Plan: **°C → K then** Π, **M**, $T \rightarrow i$

$K = °C + 273.15$ $\quad \Pi = i \times MRT$

Solution: 25 °C + 273.15 = 298 K then $\Pi = i \times MRT$. Rearrange to solve for i.

$$i_{measured} = \frac{\Pi}{MRT} = \frac{8.3\ \text{atm}}{0.100\frac{\text{mol}}{\text{L}} \times 0.08206\frac{\text{L} \cdot \text{atm}}{\text{K} \cdot \text{mol}} \times 298\ \text{K}} = 3.4$$

Check: The units (none) are correct. The magnitude of the answer (3) seems reasonable for an ionic solution with a high osmotic pressure.

14.92 **Given:** 8.92 g of KBr in 500.0 mL solution, Π = 6.97 atm at 25 °C **Find:** $i_{measured}$

Conceptual Plan: **°C → K and $mL_{soln} \to L_{soln}$ and $g_{KBr} \to mol_{KBr}$ then mol_{KBr}, $L_{soln} \to$ M then**

$K = °C + 273.15$ $\quad \frac{1\ \text{L}}{1000\ \text{mL}}$ $\quad \frac{1\ \text{mol KBr}}{119.00\ \text{g KBr}}$ $\quad M = \frac{\text{amount solute (moles)}}{\text{volume solution (L)}}$

Π, M, $T \to i$

$\Pi = i \times MRT$

Solution: 25 °C + 273.15 = 298 K and $500.0\ \text{mL} \times \frac{1\ \text{L}}{1000\ \text{mL}} = 0.5000\ \text{L}$ and

$$8.92\ \text{g KBr} \times \frac{1\ \text{mol KBr}}{119.00\ \text{g KBr}} = 0.0749580\ \text{mol KBr then}$$

$$M = \frac{\text{amount solute (moles)}}{\text{volume solution (L)}} = \frac{0.0749580\ \text{mol KBr}}{0.5000\ \text{L}} = 0.149916\frac{\text{mol KBr}}{\text{L}} \text{ then } \Pi = i \times MRT$$

Rearrange to solve for i. $$i_{measured} = \frac{\Pi}{MRT} = \frac{6.97\ \text{atm}}{0.149916\frac{\text{mol}}{\text{L}} \times 0.08206\frac{\text{L} \cdot \text{atm}}{\text{K} \cdot \text{mol}} \times 298\ \text{K}} = 1.90$$

Check: The units (none) are correct. The magnitude of the answer (1.9) seems reasonable for KBr because we expect i to be 2 if it completely dissociates. Because both ions are large and have only one charge each, we expect i to be close to the theoretical value.

14.93 **Given:** 5.50 % NaCl by mass in water at 25 °C **Find:** P_{H_2O} **Other:** $P^{\circ}_{H_2O}$ = 23.78 torr

Conceptual Plan: **% NaCl by mass → g_{NaCl}, g_{H_2O} then $g_{NaCl} \to mol_{NaCl}$ and**

$\frac{5.50\ \text{g NaCl}}{100\ \text{g (NaCl + H}_2\text{O)}}$ $\quad \frac{1\ \text{mol NaCl}}{58.44\ \text{g NaCl}}$

$g_{H_2O} \to mol_{H_2O}$ then mol_{NaCl}, $mol_{H_2O} \to \chi_{H_2O}$ then χ_{H_2O}, $P^{\circ}_{H_2O} \to P_{H_2O}$

$\frac{1\ \text{mol H}_2\text{O}}{18.01\ \text{g H}_2\text{O}}$ $\quad \chi = \frac{\text{amount solvent (in moles)}}{\text{total amount of solute and solvent particles (in moles)}}$ $\quad P_{solution} = \chi_{solvent} P^{\circ}_{solvent}$

Solution: $\frac{5.50\ \text{g NaCl}}{100\ \text{g (NaCl + H}_2\text{O)}}$ means 5.50 g NaCl and (100 g − 5.50 g) = 94.5 g H_2O then

$$5.50\ \text{g NaCl} \times \frac{1\ \text{mol NaCl}}{58.44\ \text{g NaCl}} = 0.0941136\ \text{mol NaCl and } 94.5\ \text{g H}_2\text{O} \times \frac{1\ \text{mol H}_2\text{O}}{18.01\ \text{g H}_2\text{O}} = 5.24708\ \text{mol H}_2\text{O then}$$

number of moles of solute = $i_{NaCl} \times n_{NaCl}$ so $\chi_{solv} = \frac{\text{amount solvent (in moles)}}{\text{total amount solute and solvent particles (in moles)}} =$

$$\frac{5.24708\ \text{mol}}{5.24708\ \text{mol} + 2(0.0941136\ \text{mol})} = 0.96536 \text{ then } P_{soln} = \chi_{solv}P^{\circ}_{solv} = 0.96536 \times 23.78\ \text{torr} = 23.0\ \text{torr}$$

Check: The units (torr) are correct. The magnitude of the answer (23 torr) seems reasonable because it is a drop from the pure vapor pressure. Only a fraction of a mole of NaCl is added, so the pressure will not drop much.

14.94 **Given:** $CaCl_2$ and water; $P^{\circ}_{H_2O}$ = 92.6 mmHg, $P^{\circ}_{H_2O}$ = 81.6 mmHg at 50 °C **Find:** mass percent $CaCl_2$

Conceptual Plan: **P_{H_2O}, $P^{\circ}_{H_2O} \to \chi_{H_2O}$ then assume 1 mol water, $\chi_{H_2O} \to mol_{CaCl_2}$ then**

$P_{solution} = \chi_{solvent} P^{\circ}_{solvent}$ $\quad \chi_{H_2O} = \frac{\text{mol H}_2\text{O}}{\text{mol H}_2\text{O} + 3 \times \text{mol CaCl}_2}$

$mol_{CaCl_2} \to g_{CaCl_2}$ and $mol_{H_2O} \to g_{H_2O}$ then g_{CaCl_2}, $g_{H_2O} \to$ percent by mass $CaCl_2$

$\frac{110.99\ \text{g CaCl}_2}{1\ \text{mol CaCl}_2}$ $\quad \frac{18.01\ \text{g H}_2\text{O}}{1\ \text{mol H}_2\text{O}}$ $\quad \text{mass percent} = \frac{\text{mass solute}}{\text{mass solution}} \times 100\%$

Solution: $P_{solution} = \chi_{solvent} P^\circ_{solvent}$. Rearrange to solve for χ_{H_2O}.

$$\chi_{solvent} = \frac{P_{solution}}{P^\circ_{solvent}} = \frac{81.6\ \cancel{mm\ Hg}}{92.6\ \cancel{mm\ Hg}} = 0.88\underline{1}210 \text{ then assume 1 mol water}$$

$$\chi_{H_2O} = \frac{\text{mol } H_2O}{\text{mol } H_2O + 3 \times \text{mol } CaCl_2} \text{ so } 0.88\underline{1}210 = \frac{1 \text{ mol}}{1 \text{ mol} + 3 \times \text{mol } CaCl_2}$$

Rearrange to solve for mol_{CaCl_2}.

$$0.88\underline{1}210(1 \text{ mol} + 3 \times \text{mol } CaCl_2) = 1 \text{ mol} \rightarrow \text{mol } CaCl_2 = \frac{1 \text{ mol} - 0.88\underline{1}210 \text{ mol}}{2.6\underline{4}363} = 0.044\underline{9}345 \text{ mol } CaCl_2$$

$$\text{then } 0.044\underline{9}345\ \cancel{\text{mol } CaCl_2} \times \frac{110.99 \text{ g } CaCl_2}{1\ \cancel{\text{mol } CaCl_2}} = 4.9\underline{8}727 \text{ g } CaCl_2 \text{ and } 1 \text{ mol } H_2O = 18.01 \text{ g } H_2O \text{ then}$$

$$\text{mass percent} = \frac{\text{mass solute}}{\text{mass solution}} \times 100\% = \frac{4.9\underline{8}727\ \cancel{\text{g } CaCl_2}}{4.9\underline{8}727\ \cancel{g} + 18.01\ \cancel{g}} \times 100\% = 21.69 \text{ percent by mass } CaCl_2$$

Check: The units (mol %) are correct. The magnitude of the answer (22%) seems reasonable because there is a significant drop of the pure vapor pressure and the formula mass of the salt is much larger than water's molar mass.

Cumulative Problems

14.95 Chloroform is polar and has stronger solute–solvent interactions than does nonpolar carbon tetrachloride.

14.96 Each molecule has one —OH group that is capable of hydrogen bonding. Because phenol is a smaller molecule than naphthol, the —OH group has a bigger impact on the overall polarity of the phenol molecule.

14.97 **Given:** $KClO_4$: lattice energy $= -599$ kJ/mol, $\Delta H_{hydration} = -548$ kJ/mol; 10.0 g $KClO_4$ in 100.00 mL solution
Find: ΔH_{soln} and ΔT **Other:** $C_s = 4.05$ J/g °C; $d = 1.05$ g/mL
Conceptual Plan: Lattice energy, $\Delta H_{hydration} \rightarrow \Delta H_{soln}$ and g → mol then mol, $\Delta H_{soln} \rightarrow q(\text{kJ}) \rightarrow q(\text{J})$

$\Delta H_{soln} = \Delta H_{solute} + \Delta H_{hydration}$ *where* $\Delta H_{solute} = -\Delta H_{lattice}$ $\quad \frac{1 \text{ mol}}{138.55 \text{ g}} \quad q = n\,\Delta H_{soln} \quad \frac{1000 \text{ J}}{1 \text{ kJ}}$

then $\text{mL}_{soln} \rightarrow \text{g}_{soln}$ then q, g_{soln}, $C_s \rightarrow \Delta T$

$\frac{1.05 \text{ g}}{1 \text{ mL}} \quad q = mC_s\Delta T$

Solution: $\Delta H_{soln} = \Delta H_{solute} + \Delta H_{hydration}$ where $\Delta H_{solute} = -\Delta H_{lattice}$ so $\Delta H_{soln} = \Delta H_{hydration} - \Delta H_{lattice}$

$$\Delta H_{soln} = -548 \text{ kJ/mol} - (-599 \text{ kJ/mol}) = +51 \text{ kJ/mol and } 10.0\ \cancel{g} \times \frac{1 \text{ mol}}{138.55\ \cancel{g}} = 0.072\underline{1}761 \text{ mol then}$$

$$q = n\,\Delta H_{soln} = 0.072\underline{1}761\ \cancel{mol} \times 51 \frac{\text{kJ}}{\cancel{mol}} = +3.\underline{6}810\ \cancel{kJ} \times \frac{1000 \text{ J}}{1\ \cancel{kJ}} = +3\underline{6}81.0 \text{ J absorbed then}$$

$$100.0\ \cancel{mL} \times \frac{1.05 \text{ g}}{1\ \cancel{mL}} = 105 \text{ g. Because heat is absorbed when } KClO_4 \text{ dissolves, the temperature will drop or}$$

$q = -3\underline{6}81.0$ J and $q = mC_s\Delta T$. Rearrange to solve for ΔT.

$$\Delta T = \frac{q}{mC_s} = \frac{-3\underline{6}81.0\ \cancel{J}}{105\ \cancel{g} \times 4.05 \frac{\cancel{J}}{\cancel{g} \cdot °\text{C}}} = -8.7\ °\text{C}$$

Check: The units (kJ/mol and °C) are correct. The magnitude of the answer (51 kJ/mol) makes physical sense because the lattice energy is larger than the heat of hydration. The magnitude of the temperature change (−9 °C) makes physical sense because heat is absorbed and the heat of solution is fairly small.

14.98 **Given:** NaOH: lattice energy $= -887$ kJ/mol, $\Delta H_{hydration} = -932$ kJ/mol; 25.0 g NaOH in solution, $T_i = 25.0$ °C; $T_f = 100.0$ °C **Find:** ΔH_{soln} and m (solution) **Other:** $C_s = 4.01$ J/g °C; $d = 1.05$ g/mL
Conceptual Plan: Lattice energy, $\Delta H_{hydration} \rightarrow \Delta H_{soln}$ and g → mol then mol, $\Delta H_{soln} \rightarrow q(\text{kJ}) \rightarrow q(\text{J})$

$\Delta H_{soln} = \Delta H_{solute} + \Delta H_{hydration}$ *where* $\Delta H_{solute} = -\Delta H_{lattice}$ $\quad \frac{1 \text{ mol}}{40.00 \text{ g}} \quad q = n\,\Delta H_{soln} \quad \frac{1000 \text{ J}}{1 \text{ kJ}}$

and T_i, $T_f \rightarrow \Delta T$ then q, ΔT, $C_s \rightarrow \text{g}_{soln}$ then $\text{g}_{soln} \rightarrow \text{mL}_{soln}$

$\Delta T = T_f - T_i \quad q = mC_s\Delta T \quad \frac{1 \text{ mL}}{1.05 \text{ g}}$

Solution: $\Delta H_{soln} = \Delta H_{solute} + \Delta H_{hydration}$ where $\Delta H_{solute} = -\Delta H_{lattice}$ so $\Delta H_{soln} = \Delta H_{hydration} - \Delta H_{lattice}$

$\Delta H_{soln} = -932\ \text{kJ/mol} - (-887\ \text{kJ/mol}) = -45\ \text{kJ/mol}$ and $25.0\ \text{g} \times \dfrac{1\ \text{mol}}{40.00\ \text{g}} = 0.625\ \text{mol}$ then

$$q = n\,\Delta H_{soln} = 0.625\ \text{mol} \times \left(-45\frac{\text{kJ}}{\text{mol}}\right) = -2\underline{8}.125\ \text{kJ} \times \frac{1000\ \text{J}}{1\ \text{kJ}} = -2\underline{8}125\ \text{J released then}$$

$\Delta T = T_f - T_i = 100.0\ °\text{C} - 25.0\ °\text{C} = 75.0\ °\text{C}$ Because heat is released when NaOH dissolves, the temperature will rise or $q = +2\underline{8}125\ \text{J}$ and $q = mC_s\Delta T$. Rearrange to solve for m.

$$m = \frac{q}{C_s\Delta T} = \frac{+2\underline{8}125\ \text{J}}{4.01\dfrac{\text{J}}{\text{g}\cdot °\text{C}} \times 75.0\ °\text{C}} = 9\underline{3}.516\ \text{g soln then } 9\underline{3}.516\ \text{g} \times \frac{1\ \text{mL}}{1.05\ \text{g}} = 89\ \text{mL soln}$$

Check: The units (kJ/mol and mL) are correct. The magnitude of the answer (−45 kJ/mol) makes physical sense because the lattice energy is smaller than the heat of hydration. The magnitude of the solution volume (90 mL) makes physical sense because we have 2/3 mole of NaOH and a large temperature change. NaOH is a strong base, so we expect heat to be released and need to take precautions in the lab.

14.99 **Given:** argon, 0.0537 L; 25 °C, $P_{Ar} = 1.0\ \text{atm}$ to make 1.0 L saturated solution **Find:** $k_H(\text{Ar})$

Conceptual Plan: °C → K and P_{Ar}, V, T → mol_{Ar} then mol_{Ar}, V_{soln}, P_{Ar} → $k_H(\text{Ar})$

$\text{K} = °\text{C} + 273.15$ $\quad PV = nRT \quad S_{Ar} = k_H(\text{Ar})\,P_{Ar}$ with $S_{Ar} = \dfrac{\text{mol}_{Ar}}{\text{L}_{soln}}$

Solution: $25\ °\text{C} + 273.15 = 298\ \text{K}$ and $PV = nRT$ Rearrange to solve for n.

$$n = \frac{PV}{RT} = \frac{1.0\ \text{atm} \times 0.0537\ \text{L}}{0.08206\dfrac{\text{L}\cdot\text{atm}}{\text{K}\cdot\text{mol}} \times 298\ \text{K}} = 0.002\underline{1}9597\ \text{mol then } S_{Ar} = k_H(\text{Ar})P_{Ar} \text{ with } S_{Ar} = \frac{\text{mol}_{Ar}}{\text{L}_{soln}}$$

Substitute in values and rearrange to solve for k_H.

$$k_H(\text{Ar}) = \frac{mol_{Ar}}{\text{L}_{soln}\,P_{Ar}} = \frac{0.002\underline{1}9597\ \text{mol}}{1.0\ \text{L}_{soln} \times 1.0\ \text{atm}} = 2.2 \times 10^{-3}\frac{\text{M}}{\text{atm}}$$

Check: The units (M/atm) are correct. The magnitude of the answer (10^{-3}) seems reasonable because it is consistent with other values in the text.

14.100 **Given:** gas: 1.65 L; 25 °C, $P = 725$ torr; and $k_H = 0.112\ \text{M/atm}$ **Find:** volume of saturated solution

Conceptual Plan: °C → K and torr → atm P, V, T → mol_{gas} then mol_{gas}, k_H, P → V_{soln}

$\text{K} = °\text{C} + 273.15 \quad \dfrac{1\ \text{atm}}{760\ \text{torr}} \quad PV = nRT \quad S_{gas} = k_H(\text{gas})P_{gas}$ with $S_{gas} = \dfrac{\text{mol}_{gas}}{\text{L}_{soln}}$

Solution: $25\ °\text{C} + 273.15 = 298\ \text{K}$ and $725\ \text{torr} \times \dfrac{1\ \text{atm}}{760\ \text{torr}} = 0.95\underline{3}947\ \text{atm}$ then $PV = nRT$ Rearrange to solve for n. $n = \dfrac{PV}{RT} = \dfrac{0.95\underline{3}947\ \text{atm} \times 1.65\ \text{L}}{0.08206\dfrac{\text{L}\cdot\text{atm}}{\text{K}\cdot\text{mol}} \times 298\ \text{K}} = 0.064\underline{3}666\ \text{mol}$ then $S_{gas} = k_H(\text{gas})P_{gas}$ with $S_{gas} = \dfrac{mol_{gas}}{\text{L}_{soln}}$. Substitute values and rearrange to solve for V_{soln}.

$$\text{L}_{soln} = \frac{mol_{gas}}{k_H(\text{gas})P_{gas}} = \frac{0.064\underline{3}666\ \text{mol}}{0.112\dfrac{\text{mol}}{\text{L}\cdot\text{atm}} \times 0.95\underline{3}947\ \text{atm}} = 0.602\ \text{L soln}$$

Check: The units (L) are correct. The magnitude of the answer (0.6 L) seems reasonable because the Henry's law constant is so large.

14.101 **Given:** 0.0020 ppm by mass Hg = legal limit; 0.0040 ppm by mass Hg = contaminated water; 50.0 mg Hg ingested **Find:** volume of contaminated water

Conceptual Plan: mg_{Hg} → g_{Hg} → g_{H_2O} → mL_{H_2O} → L_{H_2O}

$\dfrac{1\ \text{g}}{1000\ \text{mg}} \quad \dfrac{10^6\ \text{g water}}{0.0040\ \text{g Hg}} \quad \dfrac{1\ \text{mL}}{1.00\ \text{g}} \quad \dfrac{1\ \text{L}}{1000\ \text{mL}}$

Solution: $50.0\ \text{mg Hg} \times \frac{1\ \text{g Hg}}{1000\ \text{mg Hg}} \times \frac{10^6\ \text{g water}}{0.0040\ \text{g Hg}} \times \frac{1\ \text{mL water}}{1.00\ \text{g water}} \times \frac{1\ \text{L water}}{1000\ \text{mL water}} = 1.3 \times 10^4\ \text{L water}$

Check: The units (L) are correct. The magnitude of the answer (10^4 L) seems reasonable because the concentration is so low.

14.102 **Given:** 2.4 g Na ingested/day; 0.050% Na by mass in water; $d = 1.0$ g/mL **Find:** volume of water
Conceptual Plan: $g_{Na} \rightarrow g_{H_2O} \rightarrow mL_{H_2O} \rightarrow L_{H_2O}$

$\frac{100.000\ \text{g water}}{0.050\ \text{g Na}}$ $\frac{1\ \text{mL}}{1.0\ \text{g}}$ $\frac{1\ \text{L}}{1000\ \text{mL}}$

Solution: $2.4\ \text{g Na} \times \frac{100.000\ \text{g solution}}{0.050\ \text{g Na}} \times \frac{1\ \text{mL water}}{1.0\ \text{g solution}} \times \frac{1\ \text{L water}}{1000\ \text{mL water}} = 4.8\ \text{L water}$

Check: The units (L) are correct. The magnitude of the answer (5 L) seems reasonable because the concentration is low but not extremely low.

14.103 **Given:** 12.5% NaCl by mass in water at 55 °C; 2.5 L vapor **Find:** g_{H_2O} in vapor
Other: $P^\circ_{H_2O} = 118$ torr, $i_{NaCl} = 2.0$ (complete dissociation)
Conceptual Plan: % NaCl by mass $\rightarrow g_{NaCl}, g_{H_2O}$ then $g_{NaCl} \rightarrow mol_{NaCl}$ and $g_{H_2O} \rightarrow mol_{H_2O}$

$\frac{12.5\ \text{g NaCl}}{100\ \text{g (NaCl + H}_2\text{O)}}$ $\frac{1\ \text{mol NaCl}}{58.44\ \text{g NaCl}}$ $\frac{1\ \text{mol H}_2\text{O}}{18.02\ \text{g H}_2\text{O}}$

then $mol_{NaCl}, mol_{H_2O} \rightarrow \chi_{NaCl} \rightarrow \chi_{H_2O}$ then $\chi_{H_2O}, P^\circ_{H_2O} \rightarrow P_{H_2O}$

$\chi = \frac{\text{amount solute (in moles)}}{\text{total amount of solute and solvent (in moles)}}$ $\chi_{H_2O} = 1 - i_{NaCl}\,\chi_{NaCl}$ $P_{solution} = \chi_{solvent}P^\circ_{solvent}$

then torr $\rightarrow$ atm and °C $\rightarrow$ K $P, V, T \rightarrow mol_{H_2O} \rightarrow g_{H_2O}$

$\frac{1\ \text{atm}}{760\ \text{torr}}$ $K = °C + 273.15$ $PV = nRT$ $\frac{18.02\ \text{g H}_2\text{O}}{1\ \text{mol H}_2\text{O}}$

Solution: $\frac{12.5\ \text{g NaCl}}{100\ \text{g (NaCl + H}_2\text{O)}}$ means 12.5 g NaCl and (100 g − 12.5 g) = 87.5 g H_2O then

$12.5\ \text{g NaCl} \times \frac{1\ \text{mol NaCl}}{58.44\ \text{g NaCl}} = 0.21\underline{3}895\ \text{mol NaCl}$ and $87.5\ \text{g H}_2\text{O} \times \frac{1\ \text{mol H}_2\text{O}}{18.02\ \text{g H}_2\text{O}} = 4.8\underline{5}841\ \text{mol H}_2\text{O}$

then $\chi = \frac{\text{amount solute (in moles)}}{\text{total amount of solute and solvent (in moles)}} = \frac{0.21\underline{3}895\ \text{mol}}{0.21\underline{3}895\ \text{mol} + 4.8\underline{5}841\ \text{mol}} = 0.042\underline{1}692$

then $\chi_{H_2O} = 1 - i_{NaCl}\chi_{NaCl} = 1 - (2.0 \times 0.042\underline{1}492) = 0.91\underline{5}662$ then

$P_{solution} = \chi_{solvent}P^\circ_{solvent} = 0.91\underline{5}662 \times 118\ \text{torr} = 10\underline{8}.048\ \text{torr H}_2\text{O}$ then

$10\underline{8}.048\ \text{torr H}_2\text{O} \times \frac{1\ \text{atm}}{760\ \text{torr}} = 0.14\underline{2}168\ \text{atm}$

and 55 °C + 273.15 = 328 K then $PV = nRT$. Rearrange to solve for n.

$n = \frac{PV}{RT} = \frac{0.14\underline{2}168\ \text{atm} \times 2.5\ \text{L}}{0.08206 \frac{\text{L} \cdot \text{atm}}{\text{K} \cdot \text{mol}} \times 328\ \text{K}} = 0.01\underline{3}205\ \text{mol}$ then

$0.01\underline{3}205\ \text{mol H}_2\text{O} \times \frac{18.02\ \text{g H}_2\text{O}}{1\ \text{mol H}_2\text{O}} = 0.24\ \text{g H}_2\text{O}$

Check: The units (g) are correct. The magnitude of the answer (0.2 g) seems reasonable because there is very little mass in a vapor.

14.104 **Given:** 19.5 mg water in 1 L vapor at 25 °C **Find:** mole percent solute in solution **Other:** $P^\circ_{H_2O} = 23.78$ torr
Conceptual Plan: $mg_{H_2O} \rightarrow g_{H_2O} \rightarrow mol_{H_2O}$ and °C $\rightarrow$ K then $V, mol_{H_2O}, T \rightarrow P_{H_2O}$

$\frac{1\ \text{g}}{1000\ \text{mg}}$ $\frac{1\ \text{mol H}_2\text{O}}{18.02\ \text{g H}_2\text{O}}$ $K = °C + 273.15$ $PV = nRT$

then atm $\rightarrow$ torr then $P_{H_2O}, P^\circ_{H_2O} \rightarrow \chi_{H_2O} \rightarrow \chi_{solute} \rightarrow$ mole percent solute

$\frac{760\ \text{torr}}{1\ \text{atm}}$ $P_{H_2O} = \chi_{H_2O}P^\circ_{H_2O}$ $\chi_{solute} = 1 - \chi_{H_2O}$ mole percent solute $= \chi_{solute} \times 100\%$

Solution: $19.5\ \text{mg H}_2\text{O} \times \frac{1\ \text{g H}_2\text{O}}{1000\ \text{mg H}_2\text{O}} \times \frac{1\ \text{mol H}_2\text{O}}{18.02\ \text{g H}_2\text{O}} = 0.00108213\ \text{mol H}_2\text{O}$ and

$25\ °\text{C} + 273.15 = 298\ \text{K}$ then $PV = nRT$ Rearrange to solve for P.

$$P = \frac{nRT}{V} = \frac{0.00108213\ \text{mol} \times 0.08206 \frac{\text{L} \cdot \text{atm}}{\text{K} \cdot \text{mol}} \times 298\ \text{K}}{1.00\ \text{L}} = 0.02621774\ \text{atm}$$

$0.02621774\ \text{atm} \times \frac{760\ \text{torr}}{1\ \text{atm}} = 19.92548\ \text{torr}$ and $P_{H_2O} = \chi_{H_2O} P^{\circ}_{H_2O}$ Rearrange to solve for χ_{H_2O}.

$\chi_{H_2O} = \frac{P_{H_2O}}{P^{\circ}_{H_2O}} = \frac{19.92548\ \text{torr}}{23.78\ \text{torr}} = 0.837909$ then $\chi_{\text{solute}} = 1 - \chi_{H_2O} = 1 - 0.837909 = 0.16209$ then

mole percent solute $= \chi_{\text{solute}} \times 100\% = 0.16209 \times 100\% = 16.2$ mole percent

Check: The units (mole percent) are correct. The magnitude of the answer (16 mole percent) seems reasonable because we expect more water than solute.

14.105 **Given:** $T_b = 106.5\ °\text{C}$ aqueous solution **Find:** T_f **Other:** $K_f = 1.86\ °\text{C}/m$; $K_b = 0.512\ °\text{C}/m$

Conceptual Plan: $T_b \rightarrow \Delta T_b$ **then** $\Delta T_b, K_b \rightarrow m$ **then** $m, K_f \rightarrow \Delta T_f \rightarrow T_f$

$T_b = T^{\circ}_b + \Delta T_b$ $\quad \Delta T_b = K_b \times m$ $\quad \Delta T_f = K_f \times m$ $\quad T_f = T^{\circ}_f - \Delta T_f$

Solution: $T_b = T^{\circ}_b + \Delta T_b$ so $\Delta T_b = T_b - T^{\circ}_b = 106.5\ °\text{C} - 100.0\ °\text{C} = 6.5\ °\text{C}$ then $\Delta T_b = K_b \times m$

Rearrange to solve for m. $m = \frac{\Delta T_b}{K_b} = \frac{6.5\ °\text{C}}{0.512 \frac{°\text{C}}{m}} = 12.695\ m$ then

$\Delta T_f = K_f \times m = 1.86 \frac{°\text{C}}{m} \times 12.695\ m = 23.6\ °\text{C}$ then $T_f = T^{\circ}_f - \Delta T_f = 0.000\ °\text{C} - 23.6\ °\text{C} = -24\ °\text{C}$

Check: The units (°C) are correct. The magnitude of the answer (−24 °C) seems reasonable because the shift in boiling point is less than the shift in freezing point because the constant for boiling is smaller than the constant for freezing.

14.106 **Given:** $P_{H_2O} = 20.5$ torr at 25 °C aqueous solution **Find:** T_b **Other:** $P^{\circ}_{H_2O} = 23.78$ torr; $K_b = 0.512\ °\text{C}/m$

Conceptual Plan: $P_{H_2O}, P^{\circ}_{H_2O} \rightarrow \chi_{H_2O}$ **assume 1 kg water** $\text{kg}_{H_2O} \rightarrow \text{mol}_{H_2O}$ **then**

$P_{H_2O} = \chi_{H_2O} P^{\circ}_{H_2O}$ $\quad \frac{1\ \text{mol H}_2\text{O}}{18.02\ \text{g H}_2\text{O}}$

$\text{mol}_{H_2O}, \chi_{H_2O} \rightarrow \text{mol}_{\text{solute}}$ **then** $\text{mol}_{\text{solute}}, \text{kg}_{H_2O} \rightarrow m_{\text{solute}}$ **then** $m, K_b \rightarrow \Delta T_b \rightarrow T_b$

$\chi_{H_2O} = \frac{\text{moles H}_2\text{O}}{\text{moles H}_2\text{O} + \text{moles solute}}$ $\quad m = \frac{\text{amount solute (moles)}}{\text{mass solvent (kg)}}$ $\quad \Delta T_b = K_b \times m$ $\quad T_b = T^{\circ}_b + \Delta T_b$

Solution: $P_{H_2O} = \chi_{H_2O} P^{\circ}_{H_2O}$ Rearrange to solve for χ_{H_2O}. $\chi_{H_2O} = \frac{P_{H_2O}}{P^{\circ}_{H_2O}} = \frac{20.5\ \text{torr}}{23.78\ \text{torr}} = 0.862069$ then

$1000\ \text{g H}_2\text{O} \times \frac{1\ \text{mol H}_2\text{O}}{18.02\ \text{g H}_2\text{O}} = 55.49390\ \text{mol H}_2\text{O}$ then

$\chi_{H_2O} = \frac{\text{moles H}_2\text{O}}{\text{moles H}_2\text{O} + \text{moles solute}} = \frac{55.49390\ \text{mol}}{55.49390\ \text{mol} + x\ \text{mol}} = 0.862069$ Solve for x moles of solute.

$55.49390\ \text{mol} = 0.862069\ (55.49390\ \text{mol} + x\ \text{mol}) \rightarrow x = \frac{(55.49390 - 47.8396)\ \text{mol}}{0.862069} = 8.87899\ \text{mol}$ then

$m = \frac{\text{amount solute (moles)}}{\text{mass solvent (kg)}} = \frac{8.87899\ \text{mol}}{1\ \text{kg}} = 8.87899\ m$ then $\Delta T_b = K_b \times m = 0.512 \frac{°\text{C}}{m} \times 8.87899\ m = 4.5\ °\text{C}$

then $T_b = T^{\circ}_b + \Delta T_b = 100.0\ °\text{C} + 4.5\ °\text{C} = 104.5\ °\text{C}$

Check: The units (°C) are correct. The magnitude of the answer (4.5 °C) seems reasonable because there is a significant lowering of the vapor pressure.

14.107 (a) **Given:** 0.90% NaCl by mass per volume; isotonic aqueous solution at 25 °C; KCl; $i = 1.9$
Find: % KCl by mass per volume
Conceptual Plan: Isotonic solutions will have the same number of particles. Because *i* is the same,

$$\frac{1 \text{ mol KCl}}{1 \text{ mol NaCl}}$$

the new % mass per volume will be the mass ratio of the two salts.

$$\text{percent by mass per volume} = \frac{\text{mass solute}}{V} \times 100\% \quad \frac{1 \text{ mol NaCl}}{58.44 \text{ g NaCl}} \text{ and } \frac{74.55 \text{ g KCl}}{1 \text{ mol KCl}}$$

Solution: percent by mass per volume $= \dfrac{\text{mass solute}}{V} \times 100\%$

$$= \frac{0.0090 \text{ g NaCl}}{V} \times \frac{1 \text{ mol NaCl}}{58.44 \text{ g NaCl}} \times \frac{1 \text{ mol KCl}}{1 \text{ mol NaCl}} \times \frac{74.55 \text{ g KCl}}{1 \text{ mol KCl}} \times 100\%$$

$= 1.1\%$ KCl by mass per volume

Check: The units (% KCl by mass per volume) are correct. The magnitude of the answer (1.1%) seems reasonable because the molar mass of KCl is larger than the molar mass of NaCl.

(b) **Given:** 0.90% NaCl by mass per volume; isotonic aqueous solution at 25 °C; NaBr; $i = 1.9$
Find: % NaBr by mass per volume
Conceptual Plan: Isotonic solutions will have the same number of particles. Because *i* is the same,

$$\frac{1 \text{ mol NaBr}}{1 \text{ mol NaCl}}$$

the new % mass per volume will be the mass ratio of the two salts.

$$\text{percent by mass per volume} = \frac{\text{mass solute}}{V} \times 100\% \quad \frac{1 \text{ mol NaCl}}{58.44 \text{ g NaCl}} \text{ and } \frac{102.89 \text{ g NaBr}}{1 \text{ mol NaBr}}$$

Solution: percent by mass per volume $= \dfrac{\text{mass solute}}{V} \times 100\%$

$$= \frac{0.0090 \text{ g NaCl}}{V} \times \frac{1 \text{ mol NaCl}}{58.44 \text{ g NaCl}} \times \frac{1 \text{ mol NaBr}}{1 \text{ mol NaCl}} \times \frac{102.89 \text{ g NaBr}}{1 \text{ mol NaBr}} \times 100\%$$

$= 1.6\%$ NaBr by mass per volume

Check: The units (% NaBr by mass per volume) are correct. The magnitude of the answer (1.6%) seems reasonable because the molar mass of NaBr is larger than the molar mass of NaCl.

(c) **Given:** 0.90% NaCl by mass per volume; isotonic aqueous solution at 25 °C; glucose ($C_6H_{12}O_6$); $i = 1.9$
Find: % glucose by mass per volume
Conceptual Plan: Isotonic solutions will have the same number of particles. Because glucose is a nonelectrolyte, the *i* is not the same; then use the mass ratio of the two compounds.

$$\frac{1.9 \text{ mol } C_6H_{12}O_6}{1 \text{ mol NaCl}} \quad \text{percent by mass per volume} = \frac{\text{mass solute}}{V} \times 100\% \quad \frac{1 \text{ mol NaCl}}{58.44 \text{ g NaCl}} \text{ and } \frac{180.16 \text{ g } C_6H_{12}O_6}{1 \text{ mol } C_6H_{12}O_6}$$

Solution: percent by mass per volume $= \dfrac{\text{mass solute}}{V} \times 100\% =$

$$\frac{0.0090 \text{ g NaCl}}{V} \times \frac{1 \text{ mol NaCl}}{58.44 \text{ g NaCl}} \times \frac{1.9 \text{ mol } C_6H_{12}O_6}{1 \text{ mol NaCl}} \times \frac{180.16 \text{ g } C_6H_{12}O_6}{1 \text{ mol } C_6H_{12}O_6} \times 100\% =$$

5.3 $C_6H_{12}O_6$ by mass per volume

Check: The units (% $C_6H_{12}O_6$ by mass per volume) are correct. The magnitude of the answer (5.3%) seems reasonable because the molar mass of $C_6H_{12}O_6$ is larger than the molar mass of NaCl and we need more moles of $C_6H_{12}O_6$ because it is a nonelectrolyte.

14.108 **Given:** 28.5 g of magnesium citrate ($Mg_3(C_6H_5O_3)_2$) in 235 mL of solution at 37 °C, complete dissociation
Find: Π
Conceptual Plan: $\textbf{mL}_{\textbf{soln}} \rightarrow \textbf{L}_{\textbf{soln}}$ **and** $\textbf{g}_{Mg_3(C_6H_5O_3)_2} \rightarrow \textbf{mol}_{Mg_3(C_6H_5O_3)_2}$ **then** $\textbf{mol}_{Mg_3(C_6H_5O_3)_2}, \textbf{L}_{\textbf{soln}} \rightarrow \textbf{M}$

$$\frac{1 \text{ L}}{1000 \text{ mL}} \quad \frac{1 \text{ mol } Mg_3(C_6H_5O_3)_2}{323.13 \text{ g } Mg_3(C_6H_5O_3)_2} \quad M = \frac{\text{amount solute (moles)}}{\text{volume solution (L)}}$$

then M, i, $T \rightarrow \Pi$

$\Pi = i \times MRT$ where $i = 5$

Solution: $235 \text{ mL} \times \frac{1 \text{ L}}{1000 \text{ mL}} = 0.235 \text{ L}$ and

$28.5 \text{ g } Mg_3(C_6H_5O_3)_2 \times \frac{1 \text{ mol } Mg_3(C_6H_5O_3)_2}{323.13 \text{ g } Mg_3(C_6H_5O_3)_2} = 0.08\underline{8}1998 \text{ mol } Mg_3(C_6H_5O_3)_2$ then

$M = \frac{\text{amount solute (moles)}}{\text{volume solution (L)}} = \frac{0.08\underline{8}1998 \text{ mol } Mg_3(C_6H_5O_3)_2}{0.235 \text{ L}} = 0.37\underline{5}318 \text{ M}$ then

$\Pi = i \times MRT = 5 \times 0.37\underline{5}318 \frac{\text{mol}}{\text{L}} \times 0.08206 \frac{\text{L} \cdot \text{atm}}{\text{K} \cdot \text{mol}} \times 310. \text{ K} = 47.7 \text{ atm}$

Check: The units (atm) are correct. The magnitude of the answer (48 atm) seems reasonable because the molarity is ~ 1.5.

14.109 **Given:** 4.5701 g of $MgCl_2$ and 43.238 g water, $P_{soln} = 0.3624$ atm, $P^{\circ}_{soln} = 0.3804$ atm at 348.0 K
Find: $i_{measured}$
Conceptual Plan: $g_{MgCl_2} \rightarrow mol_{MgCl_2}$ **and** $g_{H_2O} \rightarrow mol_{H_2O}$ **then** $P_{soln}, P^{\circ}_{soln}, \rightarrow \chi_{MgCl_2}$

$\frac{1 \text{ mol } MgCl_2}{95.21 \text{ g } MgCl_2}$ $\quad$ $\frac{1 \text{ mol } H_2O}{18.02 \text{ g } H_2O}$ $\quad$ $P_{Soln} = (1 - \chi_{MgCl_2})P^{\circ}_{H_2O}$

then $mol_{MgCl_2}, mol_{H_2O}, \chi_{MgCl_2} \rightarrow i_{measured}$

$\chi_{MgCl_2} = \frac{i(\text{moles } MgCl_2)}{\text{moles } H_2O + i(\text{moles } MgCl_2)}$

Solution: $4.5701 \text{ g } MgCl_2 \times \frac{1 \text{ mol } MgCl_2}{95.21 \text{ g } MgCl_2} = 0.04800\underline{0}21 \text{ mol } MgCl_2$ and

$43.238 \text{ g } H_2O \times \frac{1 \text{ mol } H_2O}{18.015 \text{ g } H_2O} = 2.400\underline{1}110 \text{ mol } H_2O$ then $P_{soln} = (1 - \chi_{MgCl_2})P^{\circ}_{H_2O}$ so

$\chi_{MgCl_2} = 1 - \frac{P_{soln}}{P^{\circ}_{H_2O}} = 1 - \frac{0.3624 \text{ atm}}{0.3804 \text{ atm}} = 0.0473\underline{1}861$. Solve for i.

$i_{measured}(0.0480\underline{0}021) = 0.0473\underline{1}861(2.400\underline{1}110 + i_{measured}(0.0480\underline{0}021)) \rightarrow$

$i_{measured}(0.0480\underline{0}021 - 0.00227\underline{1}303) = 0.113\underline{5}699 \rightarrow i_{measured} = \frac{0.113\underline{5}699}{0.0457\underline{2}891} = 2.484$

Check: The units (none) are correct. The magnitude of the answer (2.5) seems reasonable for $MgCl_2$ because we expect i to be 3 if it completely dissociates. Because Mg is small and doubly charged, we expect a significant drop from 3.

14.110 **Given:** 7.050 g of HNO_2 and 1.000 kg of water, $T_f = -0.2929$ °C **Find:** fraction dissociated
Other: $K_f = 1.86$ °C/m
Conceptual Plan: $g_{HNO_2} \rightarrow mol_{HNO_2}$ **then** $mol_{HNO_2}, kg_{H_2O} \rightarrow m$ **then** $m, \Delta T_{f,\,actual}, K_f \rightarrow i_{actual}$ **then**

$\frac{1 \text{ mol } HNO_2}{47.02 \text{ g } HNO_2}$ $\quad$ $m = \frac{\text{amount solute (moles)}}{\text{mass solvent (kg)}}$ $\quad$ $\Delta T_{f,\,actual} = i_{actual} \times m \times K_f$

$i_{actual} \rightarrow$ **fraction dissociated**

fraction dissociated $= i_{actual} - 1$

Solution: $7.050 \text{ g } HNO_2 \times \frac{1 \text{ mol } HNO_2}{47.02 \text{ g } HNO_2} = 0.149\underline{9}362 \text{ mol } HNO_2$ then

$m = \frac{\text{amount solute (moles)}}{\text{mass solvent (kg)}} = \frac{0.149\underline{9}362 \text{ mol } HNO_2}{1.000 \text{ kg}} = 0.149\underline{9}362\, m$ then $\Delta T_{f,\,actual} = i_{actual} \times m \times K_f \rightarrow$

$0.2929 = i(0.149\underline{9}362\, m)\left(\frac{1.86 \text{ °C}}{m}\right) \rightarrow i = 1.05\underline{0}3$ then fraction dissociated $= i_{actual} - 1 = 1.05\underline{0}3 - 1 = 0.050$

Check: The units (none) are correct. The magnitude of the answer (0.05) seems reasonable because weak acids do not fully dissociate.

14.111 **Given:** $T_b = 375.3$ K aqueous solution **Find:** P_{H_2O} **Other:** $P^\circ_{H_2O} = 0.2467$ atm; $K_b = 0.512\ °C/m$

Conceptual Plan: $T_b \rightarrow \Delta T_b$ **then** $\Delta T_b, K_b \rightarrow m$ **assume 1 kg water** $kg_{H_2O} \rightarrow mol_{H_2O}$ **then**

$T_b = T^\circ_b + \Delta T_b$ $\quad \Delta T_b = K_b \times m$ $\quad \dfrac{1 \text{ mol } H_2O}{18.02 \text{ g } H_2O}$

$m \rightarrow mol_{solute}$ **then** $mol_{H_2O}, mol_{solute} \rightarrow \chi_{H_2O}$ **then** $\chi_{H_2O}, P^\circ_{H_2O} \rightarrow P_{H_2O}$

$m = \dfrac{\text{amount solute (moles)}}{\text{mass solvent (kg)}}$ $\quad \chi_{H_2O} = \dfrac{\text{moles } H_2O}{\text{moles } H_2O + \text{moles solute}}$ $\quad P_{H_2O} = \chi_{H_2O}P^\circ_{H_2O}$

Solution: $T_b = T^\circ_b + \Delta T_b$ so $\Delta T_b = T_b - T^\circ_b = 375.3\text{ K} - 373.15\text{ K} = 2.2\text{ K} = 2.2\ °C$ then

$\Delta T_b = K_b \times m$ Rearrange to solve for m. $m = \dfrac{\Delta T_b}{K_b} = \dfrac{2.2\ °C}{0.512\ \dfrac{°C}{m}} = 4.\underline{2}96875\ m$ then

$1000 \text{ g } H_2O \times \dfrac{1 \text{ mol } H_2O}{18.02 \text{ g } H_2O} = 55.4\underline{9}390 \text{ mol } H_2O$ then

$m = \dfrac{\text{amount solute (moles)}}{\text{mass solvent (kg)}} = \dfrac{x \text{ mol}}{1 \text{ kg}} = 4.\underline{2}96875\ m$ so $x = 4.\underline{2}96875$ mol then

$\chi_{H_2O} = \dfrac{\text{moles } H_2O}{\text{moles } H_2O + \text{ moles solute}} = \dfrac{55.4\underline{9}390 \text{ mol}}{55.4\underline{9}390 \text{ mol} + 4.\underline{2}96875 \text{ mol}} = 0.92\underline{8}1348$

then $P_{H_2O} = \chi_{H_2O}P^\circ_{H_2O} = 0.92\underline{8}1348 \times 0.2467 \text{ atm} = 0.229$ atm

Check: The units (atm) are correct. The magnitude of the answer (0.229 atm) seems reasonable because the mole fraction is lowered by ~7%.

14.112 **Given:** 0.438 M K_2CrO_4 aqueous solution; $d = 1.063$ g/mL at 298 K; complete dissociation **Find:** P_{soln}
Other: $P^\circ_{H_2O} = 0.0313$ atm

Conceptual Plan: Assume 1 L solution; so we have 0.438 $mol_{K_2CrO_4} \rightarrow g_{K_2CrO_4}$ **then** $mL_{soln} \rightarrow g_{soln}$ **then**

$\dfrac{194.20 \text{ g } K_2CrO_4}{1 \text{ mol } K_2CrO_4}$ $\quad \dfrac{1.063 \text{ g}}{1 \text{ mL}}$

$g_{K_2CrO_4}, g_{soln} \rightarrow g_{H_2O} \rightarrow mol_{H_2O}$ **then** $mol_{H_2O}, mol_{solute} \rightarrow \chi_{H_2O}$ **then** $\chi_{H_2O}, P^\circ_{H_2O} \rightarrow P_{H_2O}$

$gH_2O = g \text{ soln} - g\ K_2CrO_4$ $\quad \dfrac{1 \text{ mol } H_2O}{18.02 \text{ g } H_2O}$ $\quad \chi_{H_2O} = \dfrac{\text{moles } H_2O}{\text{moles } H_2O + \text{moles solute}}$ $\quad P_{soln} = (1 - i\chi_{K_2CrO_4})P^\circ_{H_2O};\ i = 3$

Solution: $0.438 \text{ mol } K_2CrO_4 \times \dfrac{194.20 \text{ g } K_2CrO_4}{1 \text{ mol } K_2CrO_4} = 85.\underline{0}596 \text{ g } K_2CrO_4$ then

$1000 \text{ mL} \times \dfrac{1.063 \text{ g}}{1 \text{ mL}} = 1063$ g soln then

$g\ H_2O = g \text{ soln} - g\ K_2CrO_4 = 1063 \text{ g soln} - 85.\underline{0}596 \text{ g } K_2CrO_4 = 97\underline{7}.9404 \text{ g } H_2O$ then

$97\underline{7}.9404 \text{ g } H_2O \times \dfrac{1 \text{ mol } H_2O}{18.02 \text{ g } H_2O} = 54.\underline{2}697 \text{ mol } H_2O$ then

$\chi_{K_2CrO_4} = \dfrac{\text{moles } K_2CrO_4}{\text{moles } H_2O + \text{moles } K_2CrO_4} = \dfrac{0.438 \text{ mol}}{54.\underline{2}697 \text{ mol} + 0.438 \text{ mol}} = 0.00800\underline{6}19$ then

$P_{soln} = (1 - i\chi_{K_2CrO_4})P^\circ_{H_2O} = (1 - 3 \times 0.00800\underline{6}19)0.0313 \text{ atm} = 0.0305$ atm

Check: The units (atm) are correct. The magnitude of the answer (0.03 atm) seems reasonable because the mole fraction is lowered by <1%.

14.113 **Given:** equal masses of carbon tetrachloride (CCl_4) and chloroform ($CHCl_3$) at 316 K; $P^\circ_{CCl_4} = 0.354$ atm; $P^\circ_{CHCl_3} = 0.526$ atm **Find:** $\chi_{CCl_4}, \chi_{CHCl_3}$ in vapor and P_{CHCl_3} in flask of condensed vapor

Conceptual Plan: Assume 100 grams of each $g_{CCl_4} \rightarrow mol_{CCl_4}$ **and** $g_{CHCl_3} \rightarrow mol_{CHCl_3}$ **then**

$\dfrac{1 \text{ mol } CCl_4}{153.81 \text{ g } CCl_4}$ $\quad \dfrac{1 \text{ mol } CHCl_3}{119.37 \text{ g } CHCl_3}$

$mol_{CCl_4}, mol_{CHCl_3} \rightarrow \chi_{CCl_4}, \chi_{CHCl_3}$ **then** $\chi_{CCl_4}, P^\circ_{CCl_4} \rightarrow P_{CCl_4}$ **and** $\chi_{CHCl_3}, P^\circ_{CHCl_3} \rightarrow P_{CHCl_3}$ **then**

$\chi_{CCl_4} = \dfrac{\text{amount } CCl_4 \text{ (in moles)}}{\text{total amount (in moles)}}$ $\quad \chi_{CHCl_3} = 1 - \chi_{CCl_4}$ $\quad P_{CCl_4} = \chi_{CCl_4}P^\circ_{CCl_4}$ $\quad P_{CHCl_3} = \chi_{CHCl_3}P^\circ_{CHCl_3}$

$P^\circ_{CCl_4}$, $P_{CHCl_3} \rightarrow P_{Total}$ then because $n \propto P$ and we are calculating a mass percent, which is a ratio of masses,

$P_{Total} = P_{CCl_4} + P_{CHCl_3}$

we can simply convert 1 atm to 1 mole so P_{CCl_4}, $P_{CHCl_3} \rightarrow n_{CHCl_4}$, n_{CHCl_3} then

$$\chi_{CCl_4} = \frac{\text{amount CCl}_4\text{ (in moles)}}{\text{total amount (in moles)}}$$

mol_{CCl_4}, $mol_{CHCl_3} \rightarrow \chi_{CCl_4}$, χ_{CHCl_3} then for the second vapor χ_{CHCl_3}, $P^\circ_{CHCl_3} \rightarrow P_{CHCl_3}$

$\chi_{CHCl_3} = 1 - \chi_{CCl_4}$ $\qquad$ $P_{CHCl_3} = \chi_{CHCl_3} P^\circ_{CHCl_3}$

Solution: $100.00 \text{ g CCl}_4 \times \frac{1 \text{ mol CCl}_4}{153.81 \text{ g CCl}_4} = 0.65015279 \text{ mol CCl}_4$ and

$100.00 \text{ g CHCl}_3 \times \frac{1 \text{ mol CHCl}_3}{119.37 \text{ g CHCl}_3} = 0.83773142 \text{ mol CHCl}_3$ then

$$\chi_{CCl_4} = \frac{\text{amount CCl}_4\text{ (in moles)}}{\text{total amount (in moles)}} = \frac{0.65015279 \text{ mol}}{0.65015279 \text{ mol} + 0.83773142 \text{ mol}} = 0.43696464 \text{ and}$$

$\chi_{CHCl_3} = 1 - \chi_{CCl_4} = 1 - 0.43696464 = 0.56303536$ then

$P_{CCl_4} = \chi_{CCl_4} P^\circ_{CCl_4} = 0.43696464 \times 0.354 \text{ atm} = 0.154685 \text{ atm}$ and

$P_{CHCl_3} = \chi_{CHCl_3} P^\circ_{CHCl_3} = 0.56303536 \times 0.526 \text{ atm} = 0.296157 \text{ atm}$ then

$P_{Total} = P_{CCl_4} + P_{CHCl_3} = 0.154685 \text{ atm} + 0.296157 \text{ atm} = 0.450841 \text{ atm}$ then

$mol_{CCl_4} = 0.154687$ mol and $mol_{CHCl_3} = 0.296157$ mol then

$$\chi_{CCl_4} = \frac{\text{amount CCl}_4\text{ (in moles)}}{\text{total amount (in moles)}} = \frac{0.154685 \text{ mol}}{0.154685 \text{ mol} + 0.296157 \text{ mol}} = 0.343102 = 0.343 \text{ in the first vapor and}$$

$\chi_{CHCl_3} = 1 - \chi_{CCl_4} = 1 - 0.343102 = 0.656898 = 0.657$ in the first vapor; then in the second vapor,

$P_{CHCl_3} = \chi_{CHCl_3} P^\circ_{CHCl_3} = 0.656898 \times 0.526 \text{ atm} = 0.345528 \text{ atm} = 0.346 \text{ atm}$

Check: The units (none and atm) are correct. The magnitude of the answers seems reasonable because we expect the lighter component to be found preferentially in the vapor phase. This effect is magnified in the second vapor.

14.114 In the previous problem, we saw that the original liquid has the $\chi_{CHCl_3} = 0.563$, and when the second vapor is condensed, it rose to 0.657. Continue the preceding calculation scheme from Problem 14.113.

$P_{CCl_4} = \chi_{CCl_4} P^\circ_{CCl_4} = 0.343102 \times 0.354 \text{ atm} = 0.121458 \text{ atm}$ converting to moles, $mol_{CCl_4} = 0.121458$ mol and $mol_{CHCl_3} = 0.345528$ mol then

$$\chi_{CCl_4} = \frac{\text{amount CCl}_4\text{ (in moles)}}{\text{total amount (in moles)}} = \frac{0.121458 \text{ mol}}{0.121458 \text{ mol} + 0.345528 \text{ mol}} = 0.260089 = 0.260 \text{ in the second vapor and}$$

$\chi_{CHCl_3} = 1 - \chi_{CCl_4} = 1 - 0.260089 = 0.739911 = 0.740$ in the second vapor then in the third vapor,

$P_{CHCl_3} = \chi_{CHCl_3} P^\circ_{CHCl_3} = 0.739911 \times 0.526 \text{ atm} = 0.389193 \text{ atm}$ and

$P_{CCl_4} = \chi_{CCl_4} P^\circ_{CCl_4} = 0.260089 \times 0.354 \text{ atm} = 0.0920719 \text{ atm}$ converting to moles

$mol_{CCl_4} = 0.0920732$ mol and $mol_{CHCl_3} = 0.389193$ mol then

$$\chi_{CCl_4} = \frac{\text{amount CCl}_4\text{ (in moles)}}{\text{total amount (in moles)}} = \frac{0.0920719 \text{ mol}}{0.0920719 \text{ mol} + 0.389193 \text{ mol}} = 0.191312 = 0.191 \text{ in the third vapor and}$$

$\chi_{CHCl_3} = 1 - \chi_{CCl_4} = 1 - 0.191312 = 0.808688 = 0.809$ in the third vapor then in the fourth vapor,

$P_{CHCl_3} = \chi_{CHCl_3} P^\circ_{CHCl_3} = 0.808688 \times 0.526 \text{ atm} = 0.425370 \text{ atm}$ and

$P_{CCl_4} = \chi_{CCl_4} P^\circ_{CCl_4} = 0.191312 \times 0.354 \text{ atm} = 0.0677244 \text{ atm}$ converting to moles,

$mol_{CCl_4} = 0.0677244$ mol and $mol_{CHCl_3} = 0.425370$ mol then

$$\chi_{CCl_4} = \frac{\text{amount CCl}_4\text{ (in moles)}}{\text{total amount (in moles)}} = \frac{0.0677244 \text{ mol}}{0.0677244 \text{ mol} + 0.425370 \text{ mol}} = 0.137346 = 0.137 \text{ in the fourth vapor}$$

and $\chi_{CHCl_3} = 1 - \chi_{CCl_4} = 1 - 0.137346 = 0.862654 = 0.863$ in the fourth vapor. The concentration of the lighter component (chloroform) in the gas phase increases with each step.

14.115 **Given:** 49.0% H_2SO_4 by mass, $d = 1.39 \text{ g/cm}^3$, 25.0 mL diluted to 99.8 cm^3 **Find:** molarity

Conceptual Plan: Initial $mL_{solution} \rightarrow g_{solution} \rightarrow g_{H_2SO_4} \rightarrow mol_{H_2SO_4}$ and final $mL_{solution} \rightarrow L_{solution}$

$\frac{1.39 \text{ g}}{1 \text{ mL}}$ $\quad$ $\frac{49.0 \text{ g H}_2\text{SO}_4}{100 \text{ g solution}}$ $\quad$ $\frac{1 \text{ mol H}_2\text{SO}_4}{98.09 \text{ g H}_2\text{SO}_4}$ $\qquad$ $\frac{1 \text{ L}}{1000 \text{ mL}}$

then $mol_{H_2SO_4}$, $L_{solution} \rightarrow M$

$$M = \frac{\text{amount solute (moles)}}{\text{volume solution (L)}}$$

Solution:

$$25.0\ \text{mL solution} \times \frac{1.39\ \text{g solution}}{1\ \text{mL solution}} \times \frac{49.0\ \text{g H}_2\text{SO}_4}{100\ \text{g solution}} \times \frac{1\ \text{mol H}_2\text{SO}_4}{98.09\ \text{g H}_2\text{SO}_4} = 0.1735906\ \text{mol H}_2\text{SO}_4 \text{ and}$$

$$99.8\ \text{mL solution} \times \frac{1\ \text{L solution}}{1000\ \text{mL solution}} = 0.0998\ \text{L solution then}$$

$$\text{M} = \frac{\text{amount solute (moles)}}{\text{volume solution (L)}} = \frac{0.1735906\ \text{mol H}_2\text{SO}_4}{0.0998\ \text{L solution}} = 1.74\ \text{M H}_2\text{SO}_4$$

Check: The units (M) are correct. The magnitude of the answer (1.74 M) seems reasonable because the solution is ~1/6 sulfuric acid.

14.116 **Given:** 50.0 g of solution in water, $\chi_{CH_4N_2O} = 0.0770$ **Find:** mass CH_4N_2O

Conceptual Plan: Set up equations for mass and moles.

$$g_{CH_4N_2O} + g_{H_2O} = 50.0\ \text{g} \quad \chi_{CH_4N_2O} = \frac{\text{mol CH}_4\text{N}_2\text{O}}{\text{mol CH}_4\text{N}_2\text{O} + \text{mol H}_2\text{O}} \quad \frac{1\ \text{mol H}_2\text{O}}{18.02\ \text{g H}_2\text{O}} \text{ and } \frac{1\ \text{mol CH}_4\text{N}_2\text{O}}{60.06\ \text{g CH}_4\text{N}_2\text{O}}$$

Combine relationships and solve for $g_{CH_4N_2O}$.

Solution: $g_{CH_4N_2O} + g_{H_2O} = 50.0\ \text{g}$, $\chi_{CH_4N_2O} = \dfrac{\text{mol CH}_4\text{N}_2\text{O}}{\text{mol CH}_4\text{N}_2\text{O} + \text{mol H}_2\text{O}}$, $\dfrac{1\ \text{mol H}_2\text{O}}{18.02\ \text{g H}_2\text{O}}$, and $\dfrac{1\ \text{mol CH}_4\text{N}_2\text{O}}{60.06\ \text{g CH}_4\text{N}_2\text{O}}$. Combine relationships and solve for $g_{CH_4N_2O}$.

Start with $g_{H_2O} = 50.0\ \text{g} - g_{CH_4N_2O}$ and substitute into $\chi_{CH_4N_2O}$.

$$\chi_{CH_4N_2O} = 0.0770 = \frac{\text{g CH}_4\text{N}_2\text{O} \times \dfrac{1\ \text{mol CH}_4\text{N}_2\text{O}}{60.06\ \text{g CH}_4\text{N}_2\text{O}}}{\text{g CH}_4\text{N}_2\text{O} \times \dfrac{1\ \text{mol CH}_4\text{N}_2\text{O}}{60.06\ \text{g CH}_4\text{N}_2\text{O}} + (50.0\ \text{g} - \text{g CH}_4\text{N}_2\text{O}) \times \dfrac{1\ \text{mol H}_2\text{O}}{18.02\ \text{g H}_2\text{O}}} \text{ and}$$

solve for $g_{CH_4N_2O}$. $0.0770\left(\text{g CH}_4\text{N}_2\text{O} \times \dfrac{1\ \text{mol CH}_4\text{N}_2\text{O}}{60.06\ \text{g CH}_4\text{N}_2\text{O}} + (50.0\ \text{g} - \text{g CH}_4\text{N}_2\text{O}) \times \dfrac{1\ \text{mol H}_2\text{O}}{18.02\ \text{g H}_2\text{O}}\right)$

$$= \text{g CH}_4\text{N}_2\text{O} \times \frac{1\ \text{mol CH}_4\text{N}_2\text{O}}{60.06\ \text{g CH}_4\text{N}_2\text{O}}$$

$\rightarrow \text{g CH}_4\text{N}_2\text{O}\ (0.001282051) + 0.21365150\ \text{g} - \text{g CH}_4\text{N}_2\text{O}\ (0.0042730300) = \text{g CH}_4\text{N}_2\text{O}(0.01665002)$

$\rightarrow 0.21365150\ \text{g} = \text{g CH}_4\text{N}_2\text{O}(0.01964100) \rightarrow$

$$\text{g CH}_4\text{N}_2\text{O} = \frac{0.21365150\ \text{g}}{0.01964100} = 10.877832\ \text{g CH}_4\text{N}_2\text{O} = 10.9\ \text{g CH}_4\text{N}_2\text{O}$$

Check: The units (g) are correct. The magnitude of the answer (11 g) seems reasonable because the molar mass of urea is over three times the molar mass of water and the mole fraction is almost 0.1.

14.117 **Given:** 10.05 g of unknown compound in 50.0 g water, $T_f = -3.16\ °\text{C}$, mass percent composition of the compound is 60.97% C and 11.94% H; the rest is O **Find:** molecular formula

Other: $K_f = 1.86\ °\text{C}/m$; $d = 1.00\ \text{g/mL}$

Conceptual Plan: $g_{H_2O} \rightarrow kg_{H_2O}$ **and** $T_f \rightarrow \Delta T_f$ **then** $\Delta T_f, K_f \rightarrow m$ **then** $m, kg_{H_2O} \rightarrow mol_{Unk}$

$$\frac{1\ \text{kg}}{1000\ \text{g}} \qquad T_f = T_f^\circ - \Delta T_f \qquad \Delta T_f = K_f \times m \qquad m = \frac{\text{amount solute (moles)}}{\text{mass solvent (kg)}}$$

then $g_{Unk}, mol_{Unk} \rightarrow \mathcal{M} \rightarrow g_C, g_H, g_O \rightarrow mol_C, mol_H, mol_O \rightarrow$ **molecular formula**

$$\mathcal{M} = \frac{g_{Unk}}{mol_{Unk}} \quad \text{mass percents} \quad \frac{1\ \text{mol C}}{12.01\ \text{g C}} \quad \frac{1\ \text{mol H}}{1.008\ \text{g H}} \quad \frac{1\ \text{mol O}}{16.00\ \text{g O}}$$

Solution: $50.0\ \text{g} \times \dfrac{1\ \text{kg}}{1000\ \text{g}} = 0.0500\ \text{kg}$ and $T_f = T_f^\circ - \Delta T_f$ so

$\Delta T_f = T_f^\circ - T_f = 0.00\ °\text{C} - (-3.16\ °\text{C}) = +3.16\ °\text{C}$ $\Delta T_f = K_f \times m$. Rearrange to solve for m.

$$m = \frac{\Delta T_f}{K_f} = \frac{3.16\ °\text{C}}{1.86\ \frac{°\text{C}}{m}} = 1.69892\ m \text{ then } m = \frac{\text{amount solute (moles)}}{\text{mass solvent (kg)}} \text{ so}$$

$$mol_{Unk} = m_{Unk} \times kg_{H_2O} = 1.6\underline{9}892 \frac{mol\ Unk}{\cancel{kg}} \times 0.0500\ \cancel{kg} = 0.084\underline{9}4600\ mol\ Unk\ then$$

$$\mathcal{M} = \frac{g_{Unk}}{mol_{Unk}} = \frac{10.05\ g}{0.084\underline{9}4600\ mol} = 11\underline{8}.3105 \frac{g}{mol}\ then$$

$$\frac{11\underline{8}.3105\ g\ \cancel{Unk}}{1\ mol\ Unk} \times \frac{60.97\ \cancel{g\ C}}{100\ \cancel{g\ Unk}} \times \frac{1\ mol\ C}{12.01\ \cancel{g\ C}} = \frac{6.01\ mol\ C}{1\ mol\ Unk}$$

$$\frac{11\underline{8}.3105\ g\ \cancel{Unk}}{1\ mol\ Unk} \times \frac{11.94\ \cancel{g\ H}}{100\ \cancel{g\ Unk}} \times \frac{1\ mol\ H}{1.008\ \cancel{g\ H}} = \frac{14.0\ mol\ H}{1\ mol\ Unk}\ and$$

$$\frac{11\underline{8}.3105\ g\ \cancel{Unk}}{1\ mol\ Unk} \times \frac{(100 - (60.97 + 11.94))\ \cancel{g\ O}}{100\ \cancel{g\ Unk}} \times \frac{1\ mol\ O}{16.00\ \cancel{g\ O}} = \frac{2.00\ mol\ O}{1\ mol\ Unk}$$

So the molecular formula is $C_6H_{14}O_2$.

Check: The units (formula) are correct. The magnitude of the answer (formula with ~118 g/mol) seems reasonable because the molality is ~1.7 and we have ~10 g. It is a reasonable molecular weight for a solid or liquid. The formula does have the correct molar mass.

14.118 **Given:** 2.10 g unknown compound in 175.0 mL solution; $\Pi = 1.93$ atm at 25 °C; combustion of 24.02 g of the unknown compound produced 28.16 g CO_2 and 8.64 g H_2O **Find:** molecular formula

Conceptual Plan: °C → K then Π, $T \rightarrow M$ then $mL_{soln} \rightarrow L_{soln}$ then L_{soln}, M → $mol_{unknown}$

$K = °C + 273.15$ $\quad \Pi = MRT \quad \frac{1\ L}{1000\ mL} \quad M = \frac{amount\ solute\ (moles)}{volume\ solution\ (L)}$

then $g_{unknown}$, $mol_{unknown} \rightarrow \mathcal{M}_{unknown}$ then $g_{CO_2} \rightarrow mol_C$ and $g_{H_2O} \rightarrow mol_H$

$\mathcal{M} = \frac{g_{unknown\ protein}}{mol_{unknown\ protein}} \quad \frac{1\ mol\ CO_2}{44.01\ g\ CO_2} \cdot \quad \frac{1\ mol\ H_2O}{18.02\ g\ H_2O}$ and $\frac{2\ mol\ H}{1\ mol\ H_2O}$

then $g_{unknown}$, mol_C, $mol_H \rightarrow mol_O$ then mol_C, mol_H, mol_O, $\mathcal{M}_{unknown} \rightarrow$ molecular formula

$\frac{12.01\ g\ C}{1\ mol\ C}\ \frac{1.008\ g\ H}{1\ mol\ H} \quad g\ O = g\ Total - g\ C - g\ H \frac{1\ mol\ O}{16.00\ g\ O}$

Solution: 25 °C + 273.15 = 298 K and $\Pi = MRT$ for M.

$$M = \frac{\Pi}{RT} = \frac{1.93\ \cancel{atm}}{0.08206 \frac{L \cdot \cancel{atm}}{\cancel{K} \cdot mol} \times 298\ \cancel{K}} = 0.078\underline{9}2408 \frac{mol}{L}\ then\ 175.0\ \cancel{mL} \times \frac{1\ L}{1000\ \cancel{mL}} = 0.1750\ L\ then$$

$$M = \frac{amount\ solute\ (moles)}{volume\ solution\ (L)}.$$ Rearrange to solve for $mol_{unknown}$.

$$mol_{unknown} = M \times L = 0.078\underline{9}2408 \frac{mol}{\cancel{L}} \times 0.1750\ \cancel{L} = 0.013\underline{8}11714\ mol\ then$$

$$\mathcal{M} = \frac{g_{unknown}}{mol_{unknown}} = \frac{2.10\ g}{0.013\underline{8}11714\ mol} = 15\underline{2}.04485 \frac{g}{mol}$$ then using the combustion data

$$28.16\ \cancel{g\ CO_2} \times \frac{1\ \cancel{mol\ CO_2}}{44.01\ \cancel{g\ CO_2}} \times \frac{1\ mol\ C}{1\ \cancel{mol\ CO_2}} = 0.639\underline{8}54579\ mol\ C$$

$$8.64\ \cancel{g\ H_2O} \times \frac{1\ \cancel{mol\ H_2O}}{18.02\ \cancel{g\ H_2O}} \times \frac{2\ mol\ H}{1\ \cancel{mol\ H_2O}} = 0.95\underline{8}93452\ mol\ H\ then\ g\ O = g\ Total - g\ C - g\ H =$$

$$24.02\ g\ Total - \left(0.639\underline{8}54579\ \cancel{mol\ C} \times \frac{12.01\ g\ C}{1\ \cancel{mol\ C}} + 0.95\underline{8}93452\ \cancel{mol\ H} \times \frac{1.008\ g\ H}{1\ \cancel{mol\ H}}\right) = 15.3\underline{6}87405\ g\ O\ then$$

$$15.3\underline{6}87405\ \cancel{g\ O} \times \frac{1\ mol\ O}{16.00\ \cancel{g\ O}} = 0.960\underline{5}4628\ mol\ O.$$ Finally, use the molar mass and the moles of each element to get the molecular formula.

$$\frac{15\underline{2}.04485\ g\ \cancel{Unk}}{1\ mol\ Unk} \times \frac{0.639\underline{8}54579\ mol\ C}{24.03\ \cancel{g\ Unk}} = \frac{4.05\ mol\ C}{1\ mol\ Unk}$$

$$\frac{15\underline{2}.04485\ g\ \cancel{Unk}}{1\ mol\ Unk} \times \frac{0.95\underline{8}93452\ mol\ H}{24.03\ \cancel{g\ Unk}} = \frac{6.07\ mol\ H}{1\ mol\ Unk}$$

and $$\frac{15\underline{2}.04485\ g\ \cancel{Unk}}{1\ mol\ Unk} \times \frac{0.960\underline{5}4628\ mol\ O}{24.03\ \cancel{g\ Unk}} = \frac{6.08\ mol\ O}{1\ mol\ Unk}$$

So the molecular formula is $C_4H_6O_6$.

Check: The units (formula) are correct. The magnitude of the answer (formula with ~152 g/mol) seems reasonable because the molarity is ~0.08 and we have ~2 g. It is a reasonable molecular weight for a solid or liquid. The formula does have the correct molar mass.

14.119 **Given:** 100.0 mL solution 13.5% by mass NaCl, $d = 1.12$ g/mL; $T_b = 104.4\ °C$ **Find:** g NaCl or water to add
Other: $K_b = 0.512\ °C/m$; $i_{measured} = 1.8$
Conceptual Plan: $T_b \rightarrow \Delta T_b$ **then** $\Delta T_b, i, K_b \rightarrow m$ **then** $mL_{solution} \rightarrow g_{solution} \rightarrow g_{NaCl} \rightarrow mol_{NaCl}$ **then**

$\Delta T_b = T_b - T_b^\circ$ $\quad \Delta T_b = K_b \times i \times m \quad \frac{1.12 \text{ g solution}}{1 \text{ mL solution}} \quad \frac{13.5 \text{ g NaCl}}{100 \text{ g solution}} \quad \frac{1 \text{ mol NaCl}}{58.44 \text{ g NaCl}}$

$m, mol_{NaCl} \rightarrow kg_{H_2O} \rightarrow g_{H_2O}$ **and** $g_{solution}, g_{NaCl} \rightarrow g_{H_2O}$ **then compare the initial and final** g_{H_2O} **then**

$m = \frac{\text{amount solute (moles)}}{\text{mass solvent (kg)}} \quad \frac{1000 \text{ g}}{1 \text{ kg}} \quad g_{solution} = g_{NaCl} + g_{H_2O}$

calculate the total NaCl in final solution by scaling up the amount from the initial solution. Then calculate the difference between the needed and starting amounts of NaCl.

Solution: $\Delta T_b = T_b - T_b^\circ = 104.4\ °C - 100.0\ °C = 4.4\ °C$ then $\Delta T_b = K_b \times i \times m$

Rearrange to solve for m. $m = \frac{\Delta T_b}{K_b\, i} = \frac{4.4\ \cancel{°C}}{0.512 \frac{\cancel{°C}}{m} \times 1.8} = 4.\underline{7}74306\ m$ NaCl then

$$100.0 \text{ }\cancel{\text{mL solution}} \times \frac{1.12 \text{ g solution}}{1 \text{ }\cancel{\text{mL solution}}} = 112 \text{ g solution} \qquad 112 \text{ }\cancel{\text{g solution}} \times \frac{13.5 \text{ g NaCl}}{100 \text{ }\cancel{\text{g solution}}} = 15.\underline{1}2 \text{ g NaCl}$$

$$15.\underline{1}2 \text{ }\cancel{\text{g NaCl}} \times \frac{1 \text{ mol NaCl}}{58.44 \text{ }\cancel{\text{g NaCl}}} = 0.25\underline{8}7269 \text{ mol NaCl}$$

then $m = \frac{\text{amount solute (moles)}}{\text{mass solvent (kg)}}$ Rearrange to solve for kg_{H_2O}.

$$kg_{H_2O} = \frac{mol_{NaCl}}{m} = \frac{0.25\underline{8}7269 \text{ }\cancel{\text{mol NaCl}}}{\frac{4.\underline{7}74306 \text{ }\cancel{\text{mol NaCl}}}{1 \text{ kg}_{H_2O}}} = 0.05\underline{4}1915 \text{ }\cancel{kg_{H_2O}} \times \frac{1000 \text{ g}_{H_2O}}{1 \text{ }\cancel{kg_{H_2O}}} = 5\underline{4}.1915 \text{ g}_{H_2O} \text{ in final solution}$$

then $g_{solution} = g_{NaCl} + g_{H_2O} = 112$ g solution $- 15.\underline{1}2$ g NaCl $= 9\underline{6}.88$ g H_2O in initial solution. Comparing the initial and final solutions, there is a lot more water in the initial solution; so NaCl needs to be added.

In the solution with a boiling point of 104.4 °C, $\frac{15.\underline{1}2 \text{ g NaCl}}{5\underline{4}.1915 \text{ g } H_2O} = \frac{x \text{ g NaCl}}{9\underline{6}.88 \text{ g } H_2O}$. Solve for x g NaCl.

$$x \text{ g NaCl} = \frac{15.\underline{1}2 \text{ g NaCl}}{5\underline{4}.1915 \text{ }\cancel{\text{g } H_2O}} \times 9\underline{6}.88 \text{ }\cancel{\text{g } H_2O} = 2\underline{7}.031 \text{ g NaCl; so the amount to be added is}$$

$2\underline{7}.031$ g NaCl $- 15.\underline{1}2$ g NaCl $= 1\underline{1}.911$ g NaCl $= 12$ g NaCl

Check: The units (g) are correct. The magnitude of the answer (12 g) seems reasonable because there is approximately twice as much water in the initial solution as is desired; so the NaCl amount needs to be approximately doubled.

14.120 **Given:** 50.0 mL solution 1.55% by mass $MgCl_2$, $d = 1.05$ g/mL; add 1.35 g $MgCl_2$ **Find:** T_f
Other: $K_b = 1.86\ °C/m$; $i_{measured} = 2.5$
Conceptual Plan: $mL_{solution} \rightarrow g_{solution} \rightarrow g_{MgCl_2}$ **then calculate final mass of** $MgCl_2$ $g_{MgCl_2} \rightarrow mol_{MgCl_2}$ **and**

$\frac{1.05 \text{ g solution}}{1 \text{ mL solution}} \quad \frac{1.55 \text{ g } MgCl_2}{100 \text{ g solution}} \quad \text{add 1.35 g } MgCl_2 \quad \frac{1 \text{ mol } MgCl_2}{95.21 \text{ g } MgCl_2}$

$g_{solution}, g_{MgCl_2} \rightarrow g_{H_2O} \rightarrow kg_{H_2O}$ **then** $g_{H_2O}, mol_{MgCl_2} \rightarrow m$ **then** $m, i, K_f \rightarrow \Delta T_f \rightarrow T_f$

$g_{solution} = g_{MgCl_2} + g_{H_2O} \quad \frac{1 \text{ kg}}{1000 \text{ g}} \quad m = \frac{\text{amount solute (moles)}}{\text{mass solvent (kg)}} \quad \Delta T_f = K_f \times i \times m \quad T_f = T_f^\circ - \Delta T_f$

Solution:

$$50.0 \text{ }\cancel{\text{mL solution}} \times \frac{1.05 \text{ g solution}}{1 \text{ }\cancel{\text{mL solution}}} = 52.5 \text{ g solution} \qquad 52.5 \text{ }\cancel{\text{g solution}} \times \frac{1.55 \text{ g } MgCl_2}{100 \text{ }\cancel{\text{g solution}}} = 0.\underline{8}1375 \text{ g } MgCl_2$$

$(0.\underline{8}1375 + 1.35)$ $\cancel{\text{g } MgCl_2} \times \frac{1 \text{ mol } MgCl_2}{95.21 \text{ }\cancel{\text{g } MgCl_2}} = 0.022\underline{7}261$ mol $MgCl_2$ and $g_{solution} = g_{MgCl_2} + g_{H_2O}$ so

$g_{H_2O} = g_{solution} - g_{MgCl_2} = 52.5 \text{ g solution} - 0.81375 \text{ g } MgCl_2 = 51.6863 \text{ g } H_2O$ and

$$51.6863 \text{ g } H_2O \times \frac{1 \text{ kg } H_2O}{1000 \text{ g } H_2O} = 0.0516863 \text{ kg } H_2O \text{ then } m = \frac{\text{amount solute (moles)}}{\text{mass solvent (kg)}}$$

$$m = \frac{0.0227261 \text{ mol } MgCl_2}{0.0516863 \text{ kg } H_2O} = 0.439693\ m \text{ then } \Delta T_f = K_f \times i \times m = 1.86 \frac{°C}{m} \times 2.5 \times 0.439693\ m = 2.0446\ °C$$

then $T_f = T_f^\circ - \Delta T_f = 0.00\ °C - 2.0446\ °C = -2.0\ °C$

Check: The units (°C) are correct. The magnitude of the answer (−2 °C) seems reasonable because the freezing point drops and almost three particles are generated for each $MgCl_2$.

Challenge Problems

14.121 **Given:** N_2: $k_H(N_2) = 6.1 \times 10^{-4}$ M/L at 25 °C; 14.6 mg/L at 50 °C and 1.00 atm $P_{N_2} = 0.78$ atm
O_2: $k_H(O_2) = 1.3 \times 10^{-3}$ M/L at 25 °C; 27.8 mg/L at 50 °C and 1.00 atm; $P_{O_2} = 0.21$ atm; and 1.5 L water
Find: $V(N_2)$ and $V(O_2)$
Conceptual Plan: at 25 °C : $P_{Total}, \chi_{N_2} \rightarrow P_{N_2}$ **then** $P_{N_2}, k_H(N_2) \rightarrow S_{N_2}$ **then L → mol**

$P_{N_2} = \chi_{N_2} P_{Total}$ $S_{N_2} = k_H(N_2) P_{N_2}$ S_{N_2}

at 50 °C : L → mL → mg → g → mol then $mol_{25\ °C}, mol_{50\ °C} \rightarrow mol_{removed}$ **then °C → K**

$\frac{1000 \text{ mL}}{1 \text{ L}}$ $\frac{14.6 \text{ mg}}{1 \text{ L}}$ $\frac{1 \text{ g}}{1000 \text{ mg}}$ $\frac{1 \text{ mol}}{28.01 \text{ g}}$ $mol_{removed} = mol_{25\ °C} - mol_{50\ °C}$ $K = °C + 273.15$

then $P, n, T \rightarrow V$

$PV = nRT$

at 25 °C : $P_{Total}, \chi_{O_2} \rightarrow P_{O_2}$ **then** $P_{O_2}, k_H(O_2) \rightarrow S_{O_2}$ **then L → mol**

$P_{O_2} = \chi_{O_2} P_{Total}$ $S_{O_2} = k_H(O_2) P_{O_2}$ S_{O_2}

at 50 °C : L → mL → mg → g → mol then $mol_{25\ °C}, mol_{50\ °C} \rightarrow mol_{removed}$ **then °C → K**

$\frac{1000 \text{ mL}}{1 \text{ L}}$ $\frac{27.8 \text{ mg}}{1 \text{ L}}$ $\frac{1 \text{ g}}{1000 \text{ mg}}$ $\frac{1 \text{ mol}}{32.00 \text{ g}}$ $mol_{removed} = mol_{25\ °C} - mol_{50\ °C}$ $K = °C + 273.15$

then $P, n, T \rightarrow V$

$PV = nRT$

Solution: at 25 °C: $P_{N_2} = \chi_{N_2} P_{Total} = 0.78 \times 1.0 \text{ atm} = 0.78 \text{ atm}$ then

$$S_{N_2} = k_H(N_2) P_{N_2} = 6.1 \times 10^{-4} \frac{M}{atm} \times 0.78 \text{ atm} = 4.758 \times 10^{-4} \text{ M then}$$

$$1.5 \text{ L} \times 4.758 \times 10^{-4} \frac{mol}{L} = 0.00071370 \text{ mol}$$

$$\text{at 50 °C: } 1.5 \text{ L} \times \frac{14.6 \text{ mg}}{1 \text{ L} \cdot \text{atm}} \times 0.78 \text{ atm} \times \frac{1 \text{ g}}{1000 \text{ mg}} \times \frac{1 \text{ mol}}{28.01 \text{ g}} = 0.00060985 \text{ mol then}$$

$mol_{removed} = mol_{25\ °C} - mol_{50\ °C} = 0.00071370 \text{ mol} - 0.00060985 \text{ mol} = 1.039 \times 10^{-4} \text{ mol } N_2$
then 50 °C + 273.15 = 323 K then $PV = nRT$. Rearrange to solve for V.

$$V = \frac{nRT}{P} = \frac{1.039 \times 10^{-4} \text{ mol} \times 0.08206 \frac{L \cdot atm}{K \cdot mol} \times 323 \text{ K}}{1.00 \text{ atm}} = 0.0027539 \text{ L } N_2$$

at 25 °C: $P_{O_2} = \chi_{O_2} P_{Total} = 0.21 \times 1.0 \text{ atm} = 0.21 \text{ atm}$ then

$$S_{O_2} = k_H(O_2) P_{O_2} = 1.3 \times 10^{-3} \frac{M}{atm} \times 0.21 \text{ atm} = 2.73 \times 10^{-4} \text{ M then}$$

$$1.5 \text{ L} \times 2.73 \times 10^{-4} \frac{mol}{L} = 0.0004095 \text{ mol}$$

$$\text{at 50 °C: } 1.5 \text{ L} \times \frac{27.8 \text{ mg}}{1 \text{ L} \cdot \text{atm}} \times 0.21 \text{ atm} \times \frac{1 \text{ g}}{1000 \text{ mg}} \times \frac{1 \text{ mol}}{32.00 \text{ g}} = 0.00027366 \text{ mol then}$$

$mol_{removed} = mol_{25\ °C} - mol_{50\ °C} = 0.0004095 \text{ mol} - 0.00027366 \text{ mol} = 1.358 \times 10^{-4} \text{ mol } O_2$
then 50 °C + 273.15 = 323 K then $PV = nRT$. Rearrange to solve for V.

$$V = \frac{nRT}{P} = \frac{1.3\underline{5}8 \times 10^{-4}\ \cancel{\text{mol}} \times 0.08206 \frac{\text{L} \cdot \cancel{\text{atm}}}{\cancel{\text{K}} \cdot \cancel{\text{mol}}} \times 323\ \cancel{\text{K}}}{1.00\ \cancel{\text{atm}}} = 0.003\underline{5}994\ \text{L O}_2 \text{ finally}$$

$V_{Total} = V_{N_2} + V_{O_2} = 0.002\underline{7}526\ \text{L} + 0.003\underline{5}994\ \text{L} = 0.0064\ \text{L}$

Check: The units (L) are correct. The magnitude of the answer (0.006 L) seems reasonable because we have so little dissolved gas at room temperature and most is still soluble at 50 °C.

14.122 **Given:** pentane (C_5H_{12}) and hexane (C_6H_{14}): 35.5% by mass C_5H_{12} in vapor at 25 °C; $P^{\circ}_{C_5H_{12}} = 425$ torr; $P^{\circ}_{C_6H_{14}} = 151$ torr **Find:** percent by mass C_5H_{12} and percent by mass C_6H_{14} in solution

Conceptual Plan: mass percents → $g_{C_5H_{12}}, g_{C_6H_{14}}$, then $g_{C_5H_{12}} \rightarrow mol_{C_5H_{12}}$ and $g_{C_6H_{14}} \rightarrow mol_{C_6H_{14}}$ then

$$\text{mass percent} = \frac{\text{mass solute}}{\text{mass solution}} \times 100\% \qquad \frac{1\ \text{mol } C_5H_{12}}{72.15\ \text{g } C_5H_{12}} \qquad \frac{1\ \text{mol } C_6H_{14}}{86.17\ \text{g } C_6H_{14}}$$

$mol_{C_5H_{12}}, mol_{C_6H_{14}} \rightarrow \chi_{C_5H_{12}vapor}$ then $\chi_{C_5H_{12}vapor}, P^{\circ}_{C_5H_{12}}, P^{\circ}_{C_6H_{14}} \rightarrow \chi_{C_5H_{12}soln}$ then assume

$$\chi_{C_5H_{12}} = \frac{\text{amount } C_5H_{12} \text{ (in moles)}}{\text{total amount (in moles)}} \qquad \chi_{C_5H_{12},\ vapor} = \frac{P_{C_5H_{12}}}{P_{Total}} = \frac{\chi_{C_5H_{12},\ soln}P^{\circ}_{C_5H_{12}}}{\chi_{C_5H_{12},\ soln}P^{\circ}_{C_5H_{12}} + (1 - \chi_{C_5H_{12},\ soln})P^{\circ}_{C_6H_{14}}}$$

1 total mole of solution → $mol_{C_5H_{12}}, mol_{C_6H_{14}}$ then $mol_{C_5H_{12}} \rightarrow g_{C_5H_{12}}$ and $mol_{C_6H_{14}} \rightarrow g_{C_6H_{14}}$

$$\chi_{C_6H_{14},\ soln} = 1 - \chi_{C_5H_{12},\ soln} \qquad \frac{72.15\ \text{g } C_5H_{12}}{1\ \text{mol } C_5H_{12}} \qquad \frac{86.17\ \text{g } C_6H_{14}}{1\ \text{mol } C_6H_{14}}$$

finally $g_{C_5H_{12}}, g_{C_6H_{14}} \rightarrow$ mass percents

$$\text{mass percent} = \frac{\text{mass solute}}{\text{mass solution}} \times 100\%$$

Solution: mass percent $= \dfrac{\text{mass solute}}{\text{mass solution}} \times 100\%$ means that 35.5 g C_5H_{12} and 100.0 g − 35.5 g = 64.5 g C_6H_{14}

then $35.5\ \cancel{\text{g } C_5H_{12}} \times \dfrac{1\ \text{mol } C_5H_{12}}{72.15\ \cancel{\text{g } C_5H_{12}}} = 0.49\underline{2}030\ \text{mol } C_5H_{12}$ and

$64.5\ \cancel{\text{g } C_6H_{14}} \times \dfrac{1\ \text{mol } C_6H_{14}}{86.17\ \cancel{\text{g } C_6H_{14}}} = 0.74\underline{8}520\ \text{mol } C_6H_{14}$ then

$$\chi_{C_5H_{12}} = \frac{\text{amount } C_5H_{12} \text{ (in moles)}}{\text{total amount (in moles)}} = \frac{0.49\underline{2}030\ \cancel{\text{mol}}}{0.49\underline{2}030\ \cancel{\text{mol}} + 0.74\underline{8}520\ \cancel{\text{mol}}} = 0.39\underline{6}622 \text{ then}$$

$$\chi_{C_5H_{12},\ vapor} = \frac{P_{C_5H_{12}}}{P_{Total}} = \frac{\chi_{C_5H_{12},\ soln}P^{\circ}_{C_5H_{12}}}{\chi_{C_5H_{12},\ soln}P^{\circ}_{C_5H_{12}} + (1 - \chi_{C_5H_{12},\ soln})P^{\circ}_{C_6H_{14}}}$$

Substitute in values and solve for $\chi_{C_5H_{12},\ soln}$.

$$0.39\underline{6}622 = \frac{\chi_{C_5H_{12},\ soln} \times 425\ \cancel{\text{torr}}}{\chi_{C_5H_{12},\ soln} \times 425\ \cancel{\text{torr}} + (1 - \chi_{C_5H_{12},\ soln}) \times 151\ \cancel{\text{torr}}} \rightarrow$$

$0.39\underline{6}622(425\chi_{C_5H_{12},\ soln} + 151(1 - \chi_{C_5H_{12},\ soln})) = 425\chi_{C_5H_{12},\ soln} \rightarrow$

$16\underline{8}.564\chi_{C_5H_{12},\ soln} + 59.\underline{8}900 - 59.\underline{8}900\chi_{C_5H_{12},\ soln} = 425\chi_{C_5H_{12},\ soln} \rightarrow 31\underline{6}.326\chi_{C_5H_{12},\ soln} = 59.\underline{8}900 \rightarrow$

$\chi_{C_5H_{12},\ soln} = \dfrac{59.\underline{8}900}{31\underline{6}.326} = 0.18\underline{9}330$ then $\chi_{C_6H_{14},\ soln} = 1 - \chi_{C_5H_{12},\ soln} = 1 - 0.18\underline{9}330 = 0.81\underline{0}670$ so

$mol_{C_5H_{12}} = 0.18\underline{9}330$ mol and $mol_{C_6H_{14}} = 0.81\underline{0}670$ mol then

$0.18\underline{9}330\ \cancel{\text{mol } C_5H_{12}} \times \dfrac{72.15\ \text{g } C_5H_{12}}{1\ \cancel{\text{mol } C_5H_{12}}} = 13.\underline{6}602\ \text{g } C_5H_{12}$ and

$0.81\underline{0}670\ \cancel{\text{mol } C_6H_{14}} \times \dfrac{86.17\ \text{g } C_6H_{14}}{1\ \cancel{\text{mol } C_6H_{14}}} = 69.\underline{8}554\ \text{g } C_6H_{14}$ finally

$$\text{mass percent} = \frac{\text{mass solute}}{\text{mass solution}} \times 100\% = \frac{13.\underline{6}602\ \cancel{\text{g } C_5H_{12}}}{13.\underline{6}602\ \cancel{\text{g } C_5H_{12}} + 69.\underline{8}554\ \cancel{\text{g } C_6H_{14}}} \times 100\%$$

= 16.4% by mass C_5H_{12}

and 100.0 − 16.4 = 83.6 mass percent C_6H_{14}

Check: The units (mass percent) are correct. We expect the mass percent of C_6H_{14} to be much higher than the mass percent of C_5H_{12} because the vapor is richer in hexane and it is the less volatile phase.

14.123 **Given:** 1.10 g glucose ($C_6H_{12}O_6$) and sucrose ($C_{12}H_{22}O_{11}$) mixture in 25.0 mL solution and Π = 3.78 atm at 298 K
Find: percent composition of mixture
Conceptual Plan: $\Pi, T \rightarrow M$ **then** $mL_{soln} \rightarrow L_{soln}$ **then** $L_{soln}, M \rightarrow mol_{mixture}$ **then**

$$\Pi = MRT \qquad \frac{1\ L}{1000\ mL} \qquad M = \frac{\text{amount solute (moles)}}{\text{volume solution (L)}}$$

$mol_{mixture}, g_{mixture} \rightarrow mol_{C_6H_{12}O_6}, mol_{C_{12}H_{22}O_{11}}$ **then**

$$g_{mixture} = \text{mol } C_6H_{12}O_6 \times \frac{180.16\ g\ C_6H_{12}O_6}{1\ mol\ C_6H_{12}O_6} + \text{mol } C_{12}H_{22}O_{11} \times \frac{342.30\ g\ C_{12}H_{22}O_{11}}{1\ mol\ C_{12}H_{22}O_{11}} \text{ with } mol_{mixture} = mol_{C_6H_{12}O_6} + mol_{C_{12}H_{22}O_{11}}$$

$mol_{C_6H_{12}O_6} \rightarrow g_{C_6H_{12}O_6}$ **and** $mol_{C_{12}H_{22}O_{11}} \rightarrow g_{C_{12}H_{22}O_{11}}$ **and** $g_{C_6H_{12}O_6}, g_{C_{12}H_{22}O_{11}} \rightarrow$ **mass percents**

$$\frac{180.16\ g\ C_6H_{12}O_6}{1\ mol\ C_6H_{12}O_6} \qquad \frac{342.30\ g\ C_{12}H_{22}O_{11}}{1\ mol\ C_{12}H_{22}O_{11}} \qquad \text{mass percent} = \frac{\text{mass solute}}{\text{mass solution}} \times 100\%$$

Solution: $\Pi = MRT$. Rearrange to solve for M.

$$M = \frac{\Pi}{RT} = \frac{3.78\ \cancel{atm}}{0.08206\ \frac{L \cdot \cancel{atm}}{\cancel{K} \cdot mol} \times 298\ \cancel{K}} = 0.15\underline{4}577 \frac{\text{mol mixture}}{L} \text{ then } 25.0\ \cancel{mL} \times \frac{1\ L}{1000\ \cancel{mL}} = 0.0250\ L$$

$$\text{then } M = \frac{\text{amount solute (moles)}}{\text{volume solution (L)}} \text{ so}$$

$$mol_{mixture} = M \times L_{soln} = 0.15\underline{4}577 \frac{\text{mol mixture}}{\cancel{L}} \times 0.0250\ \cancel{L} = 0.0038\underline{6}442 \text{ mol mixture then}$$

$$g_{mixture} = \cancel{mol\ C_6H_{12}O_6} \times \frac{180.16\ g\ C_6H_{12}O_6}{1\ \cancel{mol\ C_6H_{12}O_6}} + \cancel{mol\ C_{12}H_{22}O_{11}} \times \frac{342.30\ g\ C_{12}H_{22}O_{11}}{1\ \cancel{mol\ C_{12}H_{22}O_{11}}} \text{ with}$$

$$mol_{mixture} = mol_{C_6H_{12}O_6} + mol_{C_{12}H_{22}O_{11}} \text{ so}$$

$$1.10\ g = \text{mol } C_6H_{12}O_6 \times \frac{180.16\ g\ C_6H_{12}O_6}{1\ mol\ C_6H_{12}O_6} + (0.0038\underline{6}442\ mol - mol\ C_6H_{12}O_6) \times \frac{342.30\ g\ C_{12}H_{22}O_{11}}{1\ mol\ C_{12}H_{22}O_{11}} \rightarrow$$

$$1.10 = 180.16 \times mol\ C_6H_{12}O_6 + 1.3\underline{2}228 - 342.30 \times mol\ C_6H_{12}O_6 \rightarrow 162.14\,x\ mol\ C_6H_{12}O_6 = 0.2\underline{2}228 \rightarrow$$

$$x\ mol\ C_6H_{12}O_6 = \frac{0.2\underline{2}228}{162.14} = 0.001\underline{3}7091\ mol\ C_6H_{12}O_6 \text{ then}$$

$$mol_{C_{12}H_{22}O_{11}} = mol_{mixture} - mol_{C_6H_{12}O_6} = 0.0038\underline{6}442\ mol - 0.001\underline{3}7091\ mol$$
$$= 0.002\underline{4}935\ mol\ C_{12}H_{22}O_{11} \text{ then}$$

$$0.001\underline{3}7091\ \cancel{mol\ C_6H_{12}O_6} \times \frac{180.16\ g\ C_6H_{12}O_6}{1\ \cancel{mol\ C_6H_{12}O_6}} = 0.2\underline{4}698\ g\ C_6H_{12}O_6 \text{ and}$$

$$0.002\underline{4}935\ \cancel{mol\ C_{12}H_{22}O_{11}} \times \frac{342.30\ g\ C_{12}H_{22}O_{11}}{1\ \cancel{mol\ C_{12}H_{22}O_{11}}} = 0.8\underline{5}353\ g\ C_{12}H_{22}O_{11} \text{ finally}$$

$$\text{mass percent} = \frac{\text{mass solute}}{\text{mass solution}} \times 100\% = \frac{0.2\underline{4}698\ \cancel{g\ C_6H_{12}O_6}}{0.2\underline{4}698\ \cancel{g\ C_6H_{12}O_6} + 0.8\underline{5}353\ \cancel{g\ C_{12}H_{22}O_{11}}} \times 100\%$$

$= 2\underline{2}.44\%$ $C_6H_{12}O_6$ by mass and $100.00\% - 2\underline{2}.44\% = 7\underline{7}.56\%$ $C_{12}H_{22}O_{11}$ by mass

Check: The units (% by mass) are correct. We expect the percent by $C_6H_{12}O_6$ to be larger than that for $C_{12}H_{22}O_{11}$ because the $g_{mixture}/mol_{mixture}$ = 285 g/mol, which is closer to $C_{12}H_{22}O_{11}$ than $C_6H_{12}O_6$ and the molar mass of $C_{12}H_{22}O_{11}$ is larger than the molar mass of $C_6H_{12}O_6$. In addition, and most definitively, the masses obtained for sucrose and glucose sum to 1.1 g, the initial amount of solid dissolved.

14.124 **Given:** 631 mL methanol (CH_3OH) and 501 mL water; solution = 14.29 M CH_3OH, $d(CH_3OH)$ = 0.792 g/mL
Find: volume change on mixing
Conceptual Plan: $V_{CH_3OH}, V_{H_2O} \rightarrow V_{before\ mixing}$ **then** $mL_{CH_3OH} \rightarrow g_{CH_3OH} \rightarrow mol_{CH_3OH}$ **then**

$$V_{before\ mixing} = V_{CH_3OH} + V_{H_2O} \qquad \frac{0.792\ g}{1\ mL} \qquad \frac{1\ mol\ CH_3OH}{32.04\ g\ CH_3OH}$$

$mol_{CH_3OH}, M \rightarrow L_{soln} \rightarrow mL_{soln}$ **then** $V_{before\ mixing}, L_{soln} \rightarrow \Delta V_{mixing}$

$$M = \frac{\text{amount solute (moles)}}{\text{volume solution (L)}} \qquad \frac{1000\ mL}{1\ L} \qquad \Delta V_{mixing} = V_{before\ mixing} - V_{soln}$$

Solution: $V_{\text{before mixing}} = V_{CH_3OH} + V_{H_2O} = 631 \text{ mL} + 501 \text{ mL} = 1132 \text{ mL}$ then

$$631 \text{ mL} \times \frac{0.792 \text{ g}}{1 \text{ mL}} \times \frac{1 \text{ mol } CH_3OH}{32.04 \text{ g } CH_3OH} = 15.6174 \text{ mol } CH_3OH \text{ then } M = \frac{\text{amount solute (moles)}}{\text{volume solution (L)}} \text{ so}$$

$$L_{\text{soln}} = \frac{\text{mol}_{CH_3OH}}{M} = \frac{15.6174 \text{ mol } CH_3OH}{14.29 \frac{\text{mol } CH_3OH}{\text{L}}} = 1.09289 \text{ L then } 1.09289 \text{ L} \times \frac{1000 \text{ mL}}{1 \text{ L}} = 1092.89 \text{ mL then}$$

$$\Delta V_{\text{mixing}} = V_{\text{before mixing}} - V_{\text{soln}} = 1132 \text{ mL} - 1092.89 \text{ mL} = 39.11 \text{ mL} = 4 \times 10^1 \text{ mL}$$

Check: The units (mL) are correct. Because the intermolecular forces between a methanol molecule and a water molecule are different from those between two water molecules or between two methanol molecules, the spacing between molecules changes and thus the volume changes. The amount of the change ($\sim$40) is reasonable.

14.125 **Given:** isopropyl alcohol ($(CH_3)_2CHOH$) and propyl alcohol ($CH_3CH_2CH_2OH$) at 313 K; solution 2/3 by mass isopropyl alcohol $P_{2/3} = 0.110$ atm; solution 1/3 by mass isopropyl alcohol $P_{1/3} = 0.089$ atm
Find: P°_{iso} and P°_{pro} and explain why they are different
Conceptual Plan: Because these are isomers, they have the same molar mass and so the fraction by mass is the same as the mole fraction χs, P_{soln}s $\rightarrow P^\circ$s

$$\chi_{iso} = \frac{\text{amount iso (in moles)}}{\text{total amount (in moles)}} \quad \chi_{pro} = 1 - \chi_{iso} \quad P_{iso} = \chi_{iso}P^\circ_{iso} \quad P_{pro} = \chi_{pro}P^\circ_{pro} \text{ and } P_{soln} = P_{iso} + P_{pro}$$

Solution:
Solution 1: $\chi_{iso} = 2/3$ and $\chi_{iso} = 1/3$ $P_{soln} = P_{iso} + P_{pro}$ so $0.110 \text{ atm} = 2/3P^\circ_{iso} + 1/3P^\circ_{pro}$
Solution 2: $\chi_{iso} = 1/3$ and $\chi_{iso} = 2/3$ $P_{soln} = P_{iso} + P_{pro}$ so $0.089 \text{ atm} = 1/3P^\circ_{iso} + 2/3P^\circ_{pro}$ We now have two equations and two unknowns and a number of ways to solve this. One way is to rearrange the first equation for P°_{iso} and substitute into the other equation. Thus, $P^\circ_{iso} = 3/2(0.110 \text{ atm} - 1/3P^\circ_{pro})$ and

$$0.089 \text{ atm} = \frac{1\cancel{3}}{\cancel{3}2}(0.110 \text{ atm} - 1/3\, P^\circ_{pro}) + \frac{2}{3}P^\circ_{pro} \rightarrow 0.089 \text{ atm} = 0.0550 \text{ atm} - \frac{1}{6}P^\circ_{pro} + \frac{2}{3}P^\circ_{pro} \rightarrow$$

$$\frac{1}{2}P^\circ_{pro} = 0.0340 \text{ atm} \rightarrow P^\circ_{pro} = 0.0680 \text{ atm} = 0.068 \text{ atm and then}$$

$$P^\circ_{iso} = 3/2(0.110 \text{ atm} - 1/3P^\circ_{pro}) = 3/2(0.110 \text{ atm} - 1/3(0.0680 \text{ atm})) = 0.131 \text{ atm}$$

The major intermolecular attractions are between the OH groups. The OH group at the end of the chain in propyl alcohol is more accessible than the one in the middle of the chain in isopropyl alcohol. In addition, the molecular shape of propyl alcohol is a straight chain of carbon atoms, while that of isopropyl alcohol has a branched chain and is more like a ball. The contact area between two ball-like objects is smaller than that of two chain-like objects. The smaller contact area in isopropyl alcohol means that the molecules do not attract each other as strongly as do those of propyl alcohol. As a result of both of these factors, the vapor pressure of isopropyl alcohol is higher.

Check: The units (atm) are correct. The magnitude of the answers seems reasonable because both solution partial pressures are $\sim$0.1 atm.

14.126 **Given:** metal, M, of atomic weight 96 amu forms MF_x salt; 9.18 g of MF_x completely dissociates in 100.0 g water, $T_b = 374.38$ K
Find: x and formula unit **Other:** $K_b = 0.512\ °C/m$
Conceptual Plan: $g_{H_2O} \rightarrow kg_{H_2O}$ and $T_b \rightarrow \Delta T_b$ then $\Delta T_f, i, K_f \rightarrow m$ then $m, kg_{H_2O} \rightarrow mol_{Unk}$

$$\frac{1 \text{ kg}}{1000 \text{ g}} \quad T_b = T^\circ_b + \Delta T_b \quad \Delta T_b = K_b \times i \times m \text{ where } i = 1 + x \quad m = \frac{\text{amount solute (moles)}}{\text{mass solvent (kg)}}$$

then g_{MF_x}, $mol_{MF_x} \rightarrow \mathcal{M} \rightarrow x$

$$\mathcal{M} = \frac{g_{MF_x}}{mol_{MF_x}} \quad \frac{1 \text{ mol } MF_x}{(96 + 19x)\text{g } MF_x}$$

Solution: $100.0 \text{ g} \times \frac{1 \text{ kg}}{1000 \text{ g}} = 0.1000 \text{ kg}$ and $T_b = T^\circ_b + \Delta T_b$ so

$\Delta T_b = T_b - T^\circ_b = 374.38 \text{ K} - 373.15 \text{ K} = +1.23 \text{ K} = +1.23\ °C$ then $\Delta T_b = K_b \times i \times m$ where $i = 1 + x$ so

$$\Delta T_b = K_b(1 + x)m \quad \text{Rearrange to solve for } m.\ m = \frac{\Delta T_b}{K_b(1 + x)} = \frac{1.23\ °C}{0.512 \frac{°C}{m}(1 + x)} = \frac{2.40234}{(1 + x)} m \text{ then}$$

$m = \frac{\text{amount solute (moles)}}{\text{mass solvent (kg)}}$ so

$\text{mol}_{MF_x} = m_{MF_x} \times \text{kg}_{H_2O} = \frac{2.40\underline{2}34 \text{ mol MF}_x}{(1 + x) \cancel{\text{kg}}} \times 0.1000 \cancel{\text{kg}} = \frac{0.240\underline{2}34}{(1 + x)} \text{mol MF}_x$

then $\mathcal{M} = \frac{g_{MF_x}}{\text{mol}_{MF_x}} = \frac{9.18 \text{ g}}{\frac{0.240\underline{2}34}{(1 + x)} \text{mol MF}_x} = \frac{(96 + 19x) \text{ g MF}_x}{1 \text{ mol MF}_x}$ Rearrrange and solve for x.

$(9.18 \cancel{g})(1 \cancel{\text{mol MF}_x}) = (96 + 19x)(\cancel{g \text{ MF}_x})\left(\frac{0.240\underline{2}34}{(1 + x)} \cancel{\text{mol MFx}}\right) \rightarrow 9.18(1 + x) = 0.240\underline{2}34(96 + 19x)$

$\rightarrow 9.18x - 4.\underline{5}645x = 2\underline{3}.062 - 9.18 \rightarrow 4.\underline{6}155x = 1\underline{3}.882 \rightarrow x = \frac{1\underline{3}.882}{4.\underline{6}155} = 3.0$; so the salt is MF_3.

Because molybdenum has an atomic mass ~96 amu, the salt is MoF_3.

Check: The units (none) are correct. The answer (3.0) seems reasonable because it is a small integer.

14.127 **Given:** 0.1000 m H_2SO_4 solution; complete dissociation to H^+ and HSO_4^-; limited dissociation to SO_4^{2-}; $T_f = 272.76$ K **Find:** $m(SO_4^{2-})$ **Other:** $K_f = 1.86$ K/m

Conceptual Plan: $T_f \rightarrow \Delta T_f$ **then** $m, \Delta T_f, K_f \rightarrow i \rightarrow m(SO_4^{2-})$

$T_f = T_f^\circ - \Delta T_f$ $\quad \Delta T_f = K_f i m \quad m(SO_4^{2-}) = m\, H_2SO_4\, (i - 2.0)/2$

Solution: $T_f = T_f^\circ - \Delta T_f$ Rearrange to solve for ΔT_f. So
$\Delta T_f = T_f^\circ - T_f = 273.15 \text{ K} - 272.76 \text{ K} = -0.39$ K then $\Delta T_f = K_f i m$. Rearrange to solve for i.

$i = \frac{\Delta T_f}{K_f m} = \frac{0.39 \cancel{K}}{1.86 \frac{\cancel{K}}{\cancel{m}} \times 0.1000 \cancel{m}} = 2.\underline{0}968.$

Remember that when H_2SO_4 completely dissociates, two particles are formed (H^+ and HSO_4^-); so $i = 2$. When HSO_4^- dissociates, two particles are generated (SO_4^{2-} and H^+).

$m(SO_4^{2-}) = m\, H_2SO_4 \frac{i - 2}{2} = (0.1000\, m) \frac{2.\underline{0}968 - 2}{2} = 0.0\underline{4}839\, m = 0.4\, m\, SO_4^{2-}$

Check: The units (m) are correct. The magnitude of the answer (0.4 m) seems reasonable because not much of the HSO_4^- dissociates.

14.128 **Given:** 75.0 g of benzene (C_6H_6) and 75.0 g of toluene (C_7H_8) at 303 K $P^\circ_{Total} = 80.9$ mmHg; 100.0 g of benzene (C_6H_6) and 50.0 g of toluene (C_7H_8) at 303 K $P^\circ_{Total} = 93.9$ mmHg **Find:** $P^\circ_{C_6H_6}$, $P^\circ_{C_8H_8}$

Conceptual Plan: For each solution,

$g_{C_6H_6} \rightarrow mol_{C_7H_8}$ **and** $g_{C_7H_8} \rightarrow mol_{C_7H_8}$ **then** $mol_{C_6H_6}, mol_{C_7H_8} \rightarrow \chi_{C_6H_6}, \chi_{C_7H_8}$ **then**

$\frac{1 \text{ mol } C_6H_6}{78.11 \text{ g } C_6H_6} \quad \frac{1 \text{ mol } C_7H_8}{92.13 \text{ g } C_7H_8} \quad \chi_{C_6H_6} = \frac{\text{amount } C_6H_6 \text{ (in moles)}}{\text{total amount (in moles)}} \quad \chi_{C_7H_8} = 1 - \chi_{C_6H_6}$

write expressions relating $\chi_{C_6H_6}, P^\circ_{C_6H_6} \rightarrow P^\circ_{C_6H_6}$ and $\chi_{C_7H_8}, P^\circ_{C_7H_8} \rightarrow P_{C_8H_{18}}$ and $P_{C_6H_6}, P_{C_7H_8} \rightarrow P_{Total}$

$P_{C_6H_6} = \chi_{C_6H_6} P^\circ_{C_6H_6} \quad P_{C_7H_8} = \chi_{C_7H_8} P^\circ_{C_7H_8} \quad P_{Total} = P_{C_6H_6} + P_{C_7H_8}$

then solve the two simultaneous equations for $P^\circ_{C_6H_6}$ **and** $P^\circ_{C_7H_8}$

Solution: For the first solution: $75.0 \text{ g } \cancel{C_6H_6} \times \frac{1 \text{ mol } C_6H_6}{78.11 \text{ g } \cancel{C_6H_6}} = 0.96\underline{0}18436 \text{ mol } C_6H_6$ and

$75.0 \text{ g } \cancel{C_7H_8} \times \frac{1 \text{ mol } C_7H_8}{92.13 \text{ g } \cancel{C_7H_8}} = 0.81\underline{4}06708 \text{ mol } C_7H_8$ then

$\chi_{C_6H_6} = \frac{\text{amount } C_6H_6 \text{ (in moles)}}{\text{total amount (in moles)}} = \frac{0.96\underline{0}18436 \cancel{\text{mol}}}{0.96\underline{0}18436 \cancel{\text{mol}} + 0.81\underline{4}06708 \cancel{\text{mol}}} = 0.54\underline{1}17716$ and

$\chi_{C_7H_8} = 1 - \chi_{C_6H_6} = 1 - 0.54\underline{1}17716 = 0.45\underline{8}82284$ then
$P_{Total} = P_{C_6H_6} + P_{C_7H_8} = \chi_{C_6H_6} P^\circ_{C_6H_6} + \chi_{C_7H_8} P^\circ_{C_7H_8}$
$= 0.54\underline{1}17716 P^\circ_{C_6H_6} + 0.45\underline{8}82284 P^\circ_{C_7H_8} = 80.9$ mmHg

For the second solution: $100.0\ \text{g}\ C_6H_6 \times \frac{1\ \text{mol}\ C_6H_6}{78.11\ \text{g}\ C_6H_6} = 1.28\underline{0}24581\ \text{mol}\ C_6H_6$ and

$50.0\ \text{g}\ C_7H_8 \times \frac{1\ \text{mol}\ C_7H_8}{92.13\ \text{g}\ C_7H_8} = 0.54\underline{2}71139\ \text{mol}\ C_7H_8$ then

$$\chi_{C_6H_6} = \frac{\text{amount } C_6H_6 \text{ (in moles)}}{\text{total amount (in moles)}} = \frac{1.28\underline{0}24581\ \text{mol}}{1.28\underline{0}24581\ \text{mol} + 0.54\underline{2}71139\ \text{mol}} = 0.70\underline{2}290657 \text{ and}$$

$\chi_{C_7H_8} = 1 - \chi_{C_6H_6} = 1 - 0.70\underline{2}290657 = 0.29\underline{7}709343$ then

$P_{\text{Total}} = P_{C_6H_6} + P_{C_7H_8} = \chi_{C_6H_6}P^{\circ}_{C_6H_6} + \chi_{C_7H_8}P_{C_7H_8}$

$= 0.70\underline{2}290657P^{\circ}_{C_6H_6} + 0.29\underline{7}709343P^{\circ}_{C_7H_8} = 93.9\ \text{mmHg}$

Solve the two simultaneous equations for $P^{\circ}_{C_6H_6}$ and $P^{\circ}_{C_7H_8}$.

$$P^{\circ}_{C_6H_6} = \frac{80.9\ \text{mmHg} - 0.45\underline{8}82284P^{\circ}_{C_7H_8}}{0.54\underline{1}17716} = 14\underline{9}.48894\ \text{mmHg} - 0.84\underline{7}82373\ P^{\circ}_{C_7H_8} \text{ then}$$

$0.70\underline{2}290657(14\underline{9}.48894\ \text{mmHg} - 0.84\underline{7}82373P^{\circ}_{C_7H_8}) + 0.29\underline{7}709343P^{\circ}_{C_7H_8} = 93.9\ \text{mmHg} \rightarrow$
$10\underline{4}.98468\ \text{mmHg} - 0.59\underline{5}41868P^{\circ}_{C_7H_8} + 0.29\underline{7}709343P^{\circ}_{C_7H_8} = 93.9\ \text{mmHg} \rightarrow$
$11.\underline{0}8469\ \text{mmHg} = 0.29\underline{7}70935P^{\circ}_{C_7H_8} \rightarrow$

$$P^{\circ}_{C_7H_8} = \frac{11.\underline{0}8469\ \text{mmHg}}{0.29\underline{7}70935} = 37.\underline{2}33261\ \text{mmHg} = 37.2\ \text{mmHg then}$$

$P^{\circ}_{C_6H_6} = 14\underline{9}.48894\ \text{mmHg} - 0.84\underline{7}82373P^{\circ}_{C_7H_8}$
$= 14\underline{9}.48894\ \text{mmHg} - 0.84\underline{7}82373 \times 37.\underline{2}33261\ \text{mmHg} = 11\underline{7}.9217\ \text{mmHg} = 118\ \text{mmHg}$

Check: The units (mmHg and mmHg) are correct. The magnitude of the answers (37.2 mmHg and 118 mmHg) seems reasonable because we expect toluene to have a lower vapor pressure than benzene (based on molar mass).

14.129 **Given:** $Na_2CO_3 + NaHCO_3 = 11.60$ g in 1.00 L; treat 300.0 cm^3 of solution with HNO_3 and collect 0.940 L CO_2 at 298 K and 0.972 atm **Find:** $M(Na_2CO_3)$ and $M(NaHCO_3)$

Conceptual Plan: $P, V, T \rightarrow n$ in 300.0 cm^3 then n in 300.0 cm$^3 \rightarrow n$ in 1.00 L then

$PV = nRT$ take ratio of volumes

set up equations for the total mass and the total moles and solve. Then calculate concentrations.

$$g_{Na_2CO_3} + g_{NaHCO_3} = 11.60\ \text{g} \quad \frac{1\ \text{mol}\ Na_2CO_3}{105.99\ \text{g}\ Na_2CO_3} \quad \frac{1\ \text{mol}\ NaHCO_3}{84.01\ \text{g}\ NaHCO_3} \quad n_{Na_2CO_3} + n_{NaHCO_3} = n$$

Solution: $PV = nRT$. Rearrange to solve for n. $n = \frac{PV}{RT}$

$$n_{CO_2} = \frac{0.972\ \text{atm} \times 0.940\ \text{L}}{0.08206\frac{\text{L}\cdot\text{atm}}{\text{mol}\cdot\text{K}} \times 298\ \text{K}} = 0.03\underline{7}36340\ \text{mol}\ CO_2 \text{ in } 300.0\ \text{cm}^3 \text{ then take ratio of moles to volume to get the}$$

moles in 1.00 L

$\frac{0.03\underline{7}36340\ \text{mol}\ CO_2}{300.0\ \text{cm}^3} \times 1000\ \text{cm}^3 = 0.12\underline{4}5447\ \text{mol}\ CO_2$ in 1.00 L because 1 mole of CO_2 is generated for each mole of carbonate. So $n = n_{Na_2CO_3} + n_{NaHCO_3} = 0.12\underline{4}5447$ mol and $g_{Na_2CO_3} + g_{NaHCO_3} = 11.60$ g or $g_{Na_2CO_3} = 11.60\ \text{g} - g_{NaHCO_3}$ using molar masses and substituting

$$n_{Na_2CO_3} + n_{NaHCO_3} = 0.12\underline{4}5447\ \text{mol} = g_{Na_2CO_3} \times \frac{1\ \text{mol}\ Na_2CO_3}{105.99\ \text{g}\ Na_2CO_3} + g_{NaHCO_3} \times \frac{1\ \text{mol}\ NaHCO_3}{84.01\ \text{g}\ NaHCO_3} =$$

$$(11.60\ \text{g} - g_{NaHCO_3}) \times \frac{1\ \text{mol}\ Na_2CO_3}{105.99\ \text{g}\ Na_2CO_3} + g_{NaHCO_3} \times \frac{1\ \text{mol}\ NaHCO_3}{84.01\ \text{g}\ NaHCO_3}$$

$$\rightarrow 0.12\underline{4}5447\ \text{mol} = 0.10\underline{9}44429\ \text{mol} - g_{NaHCO_3} \times 0.009\underline{4}34852\frac{\text{mol}}{\text{g}} + g_{NaHCO_3}0.011\underline{9}03345\frac{\text{mol}}{\text{g}}$$

$$\rightarrow 0.01\underline{5}10041\ \text{mol} = g_{NaHCO_3} \times 0.002\underline{4}68493\frac{\text{mol}}{\text{g}} \rightarrow$$

$$g_{NaHCO_3} = \frac{0.01\underline{5}10041 \cancel{mol}}{0.002\underline{4}68493 \frac{\cancel{mol}}{g}} = 6.\underline{1}17259 \text{ g NaHCO}_3 = 6.1 \text{ g NaHCO}_3 \text{ and then } g_{Na_2CO_3} =$$

$$11.60 \text{ g} - g_{NaHCO_3} = 11.60 \text{ g} - 6.\underline{1}17259 \text{ g NaHCO}_3 = 5.\underline{4}8274 \text{ g Na}_2\text{CO}_3 = 5.5 \text{ g Na}_2\text{CO}_3 \text{ then M} = \text{mol/L}$$

$$\frac{6.\underline{1}17259 \text{ g } \cancel{NaHCO_3} \times \frac{1 \text{ mol NaHCO}_3}{84.01 \text{ g } \cancel{NaHCO_3}}}{1.00 \text{ L}} = 0.073 \text{ M NaHCO}_3 \text{ and}$$

$$\frac{5.\underline{4}8274 \text{ g } \cancel{Na_2CO_3} \times \frac{1 \text{ mol Na}_2\text{CO}_3}{105.99 \text{ g } \cancel{Na_2CO_3}}}{1.00 \text{ L}} = 0.052 \text{ M Na}_2\text{CO}_3$$

Check: The units (M and M) are correct. The magnitude of the answers (0.073 M and 0.052 M) makes sense because the number of moles of CO_2 is small (0.12 M). The balance of the two components makes sense because if it were all Na_2CO_3, the number of moles would have been 0.109 mole (11.6/105.99) and if it were all $NaHCO_3$, the number of moles would have been 0.138 mole (11.6/84.01)—the actual number of moles is roughly in the middle of these two values.

Conceptual Problems

14.130 (a) The two substances mix because the intermolecular forces among all of the species are roughly equal and there is a pervasive tendency to increase randomness, which happens when the two substances mix.

(b) $\Delta H_{soln} \approx 0$ because the intermolecular forces between themselves are roughly equal to the forces between each other.

(c) ΔH_{solute} and $\Delta H_{solvent}$ are positive, and ΔH_{mix} is negative and equals the sum of ΔH_{solute} and $\Delta H_{solvent}$.

14.131 The warm coolant water should not be put directly in the river without being cooled because it will raise the temperature of the water. When water is warmed, there is less dissolved oxygen in the water; this will be detrimental to aquatic life that depends on the dissolved oxygen.

14.132 (d) More solute particles are found in an ionic solution because the solute breaks apart into its ions. The vapor pressure is lowered due to fewer solute particles in the vapor phase as more solute particles are present.

14.133 (e) NaCl. If all of the substances have the same cost per kilogram, we need to determine which substance will generate the largest number of particles per kilogram (or gram). $HOCH_2CH_2OH$ generates 1 mol particle/62.07 g, NaCl generates 2 mol particles/58.44 g, KCl generates 2 mol particles/74.56 g, $MgCl_2$ generates 3 mol particles/95.22 g, and $SrCl_2$ generates 3 particles/158.53 g. So NaCl will generate 1 mole of particles for each 29 g.

14.134 The balloon not only loses He, but it also takes in N_2 and O_2 from the air surrounding the balloon (due to the tendency for mixing), increasing the density of the gas inside the balloon and thus increasing the density of the balloon.

15 Chemical Kinetics

Review Questions

15.1 Unlike mammals, which actively regulate their body temperature through metabolic activity, lizards are ectotherms—their body temperature depends on their surroundings. When splashed with cold water, a lizard's body gets colder. The drop in body temperature immobilizes the lizard because its movement depends on chemical reactions that occur within its muscles, and the rates of those reactions—how fast they occur—are highly sensitive to temperature. In other words, when the temperature drops, the reactions that produce movement occur more slowly; therefore, the movement itself slows down. Cold reptiles are lethargic, unable to move very quickly. For this reason, reptiles try to maintain their body temperature in a narrow range by moving between sun and shade.

15.2 The rates of chemical reactions, and especially the ability to control those rates, are important phenomena in our everyday lives. For example, the human body's ability to switch a specific reaction on or off at a specific time is achieved largely by controlling the rate of that reaction through the use of enzymes. Chemical kinetics is an important subject to chemists and engineers. The launching of a rocket depends on controlling the rate at which fuel burns—too quickly and the rocket can explode, too slowly and it will not leave the ground. The rate of nuclear decay in a nuclear power plant must be carefully controlled to provide electricity safely and efficiently. Chemists must consider reaction rates when synthesizing compounds. No matter how stable a compound might be, its synthesis is impossible if the rate at which it forms is too slow. As we have seen with reptiles, reaction rates are important to life.

15.3 Since reactions occur as a result of collisions between particles, the more particles are in a reaction vessel, the higher the concentration of particles, the more collisions are possible. The more collisions that occur, the faster the reaction rate.

15.4 The higher the temperature, the faster the particles are moving. When particles are moving faster, they will be in more collisions and the particles will collide with a greater force. Since reactions occur as a result of collisions between particles, there will be more collisions with enough energy to cause a reaction and, thus, a faster reaction rate.

15.5 The rate of a chemical reaction is measured as a change in the amounts of reactants or products (usually in terms of concentration) divided by the change in time. Typical units are molarity per second (M/s), molarity per minute (M/min), and molarity per year (M/yr), depending on how fast the reaction proceeds.

15.6 The reaction rate is defined as the negative of the change in concentration of a reactant divided by the change in time because reactant concentrations decrease as a reaction proceeds; therefore, the change in the concentration of a reactant is negative. Thus, the negative sign in the definition makes the overall rate positive. In other words, the negative sign is the result of the convention that reaction rates are usually reported as positive quantities. Because the product concentrations are increasing, the concentration of a product divided by the change in time is positive.

15.7 The average rate of the reaction can be calculated for any time interval as $\text{Rate} = -\frac{1}{a}\frac{[\text{A}]_{t_2} - [\text{A}]_{t_1}}{t_2 - t_1} = -\frac{1}{b}\frac{[\text{B}]_{t_2} - [\text{B}]_{t_1}}{t_2 - t_1} = \frac{1}{c}\frac{[\text{C}]_{t_2} - [\text{C}]_{t_1}}{t_2 - t_1} = \frac{1}{d}\frac{[\text{D}]_{t_2} - [\text{D}]_{t_1}}{t_2 - t_1}$ for the chemical reaction $aA + bB \rightarrow cC + dD$. The instantaneous rate of the reaction is the rate at any one point in time, represented by the instantaneous slope of the plot of concentration versus time at that point. We can obtain the instantaneous rate from the slope of the tangent to this curve at the point of interest.

15.8 For a zero-order reaction, Rate $= k[\text{A}]^0 = k$; so doubling the concentration of A does nothing to the reaction rate. For a first-order reaction, Rate $= k[\text{A}]^1 = k[\text{A}]$; so doubling the concentration of A doubles the reaction rate. For a second-order reaction, Rate $= k[\text{A}]^2$; so doubling the concentration of A quadruples the reaction rate.

15.9 The reaction order cannot be determined by the stoichiometry of the reaction. It can only be determined by running controlled experiments where the concentrations of the reactants are varied and the reaction rates are measured and analyzed.

15.10 When multiple reactants are present, Rate $= k[\text{A}]^m[\text{B}]^n$, where m is the reaction order with respect to A and n is the reaction order with respect to B. The overall order is simply the sum of the exponents $(m + n)$.

15.11 The rate law shows the relationship between the rate of a reaction and the concentrations of the reactants. The integrated rate law for a chemical reaction is a relationship between the concentration of a reactant and time.

15.12 For a zero-order reaction, $[\text{A}]_t = -kt + [\text{A}]_0$. For a first-order reaction, $\ln[\text{A}]_t = -kt + \ln[\text{A}]_0$. For a second-order reaction, $\frac{1}{[\text{A}]_t} = kt + \frac{1}{[\text{A}]_0}$.

15.13 The half-life $(t_{1/2})$ of a reaction is the time required for the concentration of a reactant to fall to one-half of its initial value. For a zero-order reaction, $t_{1/2} = \frac{[\text{A}]_0}{2k}$. For a first-order reaction, $t_{1/2} = \frac{0.693}{k}$. For a second-order reaction, $t_{1/2} = \frac{1}{k[\text{A}]_0}$.

15.14 The rates of chemical reactions are, in general, highly sensitive to temperature. Reaction rates increase as the temperature increases. The temperature dependence of the reaction rate is contained in the rate constant (k), which is actually a constant only when the temperature remains constant.

15.15 The modern form of the Arrhenius equation, which relates the rate constant (k) and the temperature in kelvin (T), is as follows: $k = A\,e^{-E_a/RT}$, where R is the gas constant $(8.314\ \text{J/mol}\cdot\text{K})$, A is a constant called the frequency factor (or the pre-exponential factor), and E_a is called the activation energy (or activation barrier). The frequency factor is the number of times the reactants approach the activation barrier per unit time. The exponential factor $(-E_a/RT)$ is the fraction of approaches that are successful in surmounting the activation barrier and forming products. The exponential factor increases with increasing temperature, but decreases with an increasing value for the activation energy. As the temperature increases, the number of collisions increases and the number of molecules having enough thermal energy to surmount the activation barrier increases. At any given temperature, a sample of molecules will have a distribution of energies, as shown in Figure 15.14. Under common circumstances, only a small fraction of the molecules have enough energy to make it over the activation barrier. Because of the shape of the energy distribution curve, however, a small change in temperature results in a large difference in the number of molecules having enough energy to surmount the activation barrier.

15.16 An Arrhenius plot is a plot of the natural log of the rate constant ($\ln k$) versus the inverse of the temperature in kelvin ($1/T$). It yields a straight line with a slope of $-E_a/R$ and a y-intercept of $\ln A$.

15.17 The orientation factor says that if two molecules are to react with each other, they must collide in such a way that allows the necessary bonds to break and form. The small orientation factor indicates that the orientational requirements for this reaction are fairly stringent—the molecules must be aligned in a very specific way for the reaction to occur. When two molecules with sufficient energy and the correct orientation collide, something unique happens: The electrons on one of the atoms or molecules are attracted to the nuclei of the other; some bonds begin to weaken, and other bonds begin to form. If all goes well, the reactants go through the transition state and are transformed into the products and a chemical reaction occurs.

15.18 When we write a chemical equation to represent a chemical reaction, we usually represent the overall reaction, not the series of individual steps by which the reaction occurs. The overall equation shows the substances present at the beginning of the reaction and the substances formed by the reaction—it does not show the intermediate steps that may be involved. A reaction mechanism is a series of individual chemical steps by which an overall chemical reaction occurs.

15.19 An elementary step is a single step in a reaction mechanism. Elementary steps cannot be broken down into simpler steps—they occur as they are written. Elementary steps are characterized by their molecularity, the number of reactant particles involved in the step. The molecularity of the three most common types of elementary steps are as follows: unimolecular—$A \rightarrow$ products and Rate $= k[A]$; bimolecular—$A + A \rightarrow$ products and Rate $= k[A]^2$; and bimolecular—$A + B \rightarrow$ products and Rate $= k[A][B]$. Elementary steps in which three reactant particles collide, called termolecular steps, are very rare because the probability of three particles simultaneously colliding is small.

15.20 For a proposed reaction mechanism to be valid—mechanisms can only be validated, not proven—two conditions must be met: (1) the elementary steps in the mechanism must sum to the overall reaction, and (2) the rate law predicted by the mechanism must be consistent with the experimentally observed rate law.

15.21 Reaction intermediates are species that are formed in one step of a mechanism and consumed in another step. An intermediate is not found in the balanced equation for the overall reaction, but plays a key role in the mechanism.

15.22 A catalyst is a substance that increases the rate of a chemical reaction but is not consumed by the reaction. A catalyst works by providing an alternative mechanism for the reaction—one in which the rate-determining step has a lower activation energy.

15.23 In homogeneous catalysis, the catalyst exists in the same phase as the reactants. In heterogeneous catalysis, the catalyst exists in a phase different from that of the reactants. The most common type of heterogeneous catalyst is a solid catalyst.

15.24 Heterogeneous catalysis (involving solid catalysts) occurs by the following four-step process: (1) adsorption—the reactants are adsorbed onto the solid surface, (2) diffusion—the reactants diffuse on the surface until they approach each other, (3) reaction—the reactants react to form the products, and (4) desorption—the products desorb from the surface into the gas phase.

15.25 Enzymes are biological catalysts that increase the rates of biochemical reactions. Enzymes are large protein molecules with complex three-dimensional structures. Within that structure is a specific area called the active site. The properties and shape of the active site are just right to bind the reactant molecule, usually called the substrate. The substrate fits into the active site in a manner that is analogous to a key fitting into a lock.

15.26 The general mechanism by which an enzyme (E) binds a substrate (S) and then reacts to form the products (P) is as follows: (1) $E + S \rightleftharpoons ES$ (fast) and (2) $ES \rightarrow E + P$ (slow, rate-limiting).

Reaction Rates

15.27 (a) $$\text{Rate} = -\frac{1}{2}\frac{\Delta[HBr]}{\Delta t} = \frac{\Delta[H_2]}{\Delta t} = \frac{\Delta[Br_2]}{\Delta t}$$

(b) **Given:** first 25.0 s; 0.600 M to 0.512 M **Find:** average rate

Conceptual Plan: $t_1, t_2, [HBr]_1, [HBr]_2 \rightarrow$ **average rate**

$$\text{Rate} = -\frac{1}{2}\frac{\Delta[HBr]}{\Delta t}$$

Solution:

$$\text{Rate} = -\frac{1}{2}\frac{[HBr]_{t_2} - [HBr]_{t_1}}{t_2 - t_1} = -\frac{1}{2}\frac{0.512\text{ M} - 0.600\text{ M}}{25.0\text{ s} - 0.0\text{ s}} = 1.\underline{7}6 \times 10^{-3}\text{ M}\cdot\text{s}^{-1} = 1.8 \times 10^{-3}\text{ M}\cdot\text{s}^{-1}$$

Check: The units $(\text{M}\cdot\text{s}^{-1})$ are correct. The magnitude of the answer $(10^{-3}\text{ M}\cdot\text{s}^{-1})$ makes physical sense because rates are always positive and we are not changing the concentration much in 25 s.

(c) **Given:** 1.50 L vessel, first 15.0 s of reaction, and part (b) data **Find:** mol_{Br_2} formed

Conceptual Plan: Average rate, $t_1, t_2, \rightarrow \Delta[Br_2]$ then $\Delta[Br_2]$, L $\rightarrow$ mol_{Br_2} formed

$$\text{Rate} = \frac{\Delta[Br_2]}{\Delta t} \qquad\qquad M = \frac{mol_{Br_2}}{L}$$

Solution: $\text{Rate} = 1.7\underline{6} \times 10^{-3}\ M \cdot s^{-1} = \frac{\Delta[Br_2]}{\Delta t} = \frac{\Delta[Br_2]}{15.0\ s - 0.0\ s}$ Rearrange to solve for $\Delta[Br_2]$.

$\Delta[Br_2] = 1.7\underline{6} \times 10^{-3} \frac{M}{\cancel{s}} \times 15.0\ \cancel{s} = 0.02\underline{6}4\ M$ then $M = \frac{mol_{Br_2}}{L}$ Rearrange to solve for mol_{Br_2}.

$0.02\underline{6}4 \frac{mol\ Br_2}{L} \times 1.50\ L = 0.040\ mol\ Br_2$

Check: The units (mol) are correct. The magnitude of the answer (0.04 mol) makes physical sense because the time is shorter than in part (b) and we need to divide by 2 and multiply by 1.5 because of the stoichiometric coefficient difference and the volume of the vessel, respectively.

15.28 (a) $\text{Rate} = -\frac{1}{2}\frac{\Delta[N_2O]}{\Delta t} = \frac{1}{2}\frac{\Delta[N_2]}{\Delta t} = \frac{\Delta[O_2]}{\Delta t}$

(b) **Given:** first 15.0 s; 0.015 mol O_2 in 0.500 L **Find:** average rate

Conceptual Plan: mol_{O_2}, L $\rightarrow$ M then $t_1, t_2, [O_2]_1, [O_2]_2 \rightarrow$ average rate

$$M = \frac{mol_{O_2}}{L} \qquad\qquad \text{Rate} = \frac{\Delta[O_2]}{\Delta t}$$

Solution: At $t_1 = 0$ s, we have no O_2. $M = \frac{mol_{O_2}}{L} = \frac{0.015\ mol\ O_2}{0.500\ L} = 0.030\ M$ then

$$\text{Rate} = \frac{[O_2]_{t_2} - [O_2]_{t_1}}{t_2 - t_1} = \frac{0.030\ M - 0.000\ M}{15.0\ s - 0.0\ s} = 2.0 \times 10^{-3}\ M \cdot s^{-1}$$

Check: The units $(M \cdot s^{-1})$ are correct. The magnitude of the answer $(10^{-3}\ M \cdot s^{-1})$ makes physical sense because rates are always positive and we are not changing the concentration much in 15 s.

(c) **Given:** part (b) data **Find:** $\frac{\Delta[N_2O]}{\Delta t}$

Conceptual Plan: Average rate $\rightarrow \frac{\Delta[N_2O]}{\Delta t}$

$$\text{Rate} = -\frac{1}{2}\frac{\Delta[N_2O]}{\Delta t} = \frac{\Delta[O_2]}{\Delta t}$$

Solution: $\text{Rate} = -\frac{1}{2}\frac{\Delta[N_2O]}{\Delta t} = \frac{\Delta[O_2]}{\Delta t}$ Rearrange to solve for $\frac{\Delta[N_2O]}{\Delta t}$.

$$\frac{\Delta[N_2O]}{\Delta t} = -2\frac{\Delta[O_2]}{\Delta t} = -2 \times 2.0 \times 10^{-3}\ M \cdot s^{-1} = -0.0040\ M \cdot s^{-1}$$

Check: The units $(M \cdot s^{-1})$ are correct. The magnitude of the answer $(-0.004\ M \cdot s^{-1})$ makes physical sense because we multiply by 2 because of the different stoichiometric coefficients. The change in concentration with time is negative because this is a reactant.

15.29 (a) $\text{Rate} = -\frac{1}{2}\frac{\Delta[A]}{\Delta t} = -\frac{\Delta[B]}{\Delta t} = \frac{1}{3}\frac{\Delta[C]}{\Delta t}$

(b) **Given:** $\frac{\Delta[A]}{\Delta t} = -0.100\ M/s$ **Find:** $\frac{\Delta[B]}{\Delta t}$ and $\frac{\Delta[C]}{\Delta t}$

Conceptual Plan: $\frac{\Delta[A]}{\Delta t} \rightarrow \frac{\Delta[B]}{\Delta t}$ and $\frac{\Delta[C]}{\Delta t}$

$$\text{Rate} = -\frac{1}{2}\frac{\Delta[A]}{\Delta t} = -\frac{\Delta[B]}{\Delta t} = \frac{1}{3}\frac{\Delta[C]}{\Delta t}$$

Solution: $\text{Rate} = -\frac{1}{2}\frac{\Delta[A]}{\Delta t} = -\frac{\Delta[B]}{\Delta t} = \frac{1}{3}\frac{\Delta[C]}{\Delta t}$. Substitute value and solve for the two desired values. $-\frac{1}{2}\frac{-0.100\ M}{s} = -\frac{\Delta[B]}{\Delta t}$ so $\frac{\Delta[B]}{\Delta t} = -0.0500\ M \cdot s^{-1}$ and $-\frac{1}{2}\frac{-0.100\ M}{s} = \frac{1}{3}\frac{\Delta[C]}{\Delta t}$

so $\frac{\Delta[C]}{\Delta t} = 0.150\ M \cdot s^{-1}$

Check: The units $(M \cdot s^{-1})$ are correct. The magnitude of the answer $(-0.05\ M \cdot s^{-1})$ makes physical sense because fewer moles of B are reacting for every mole of A and the change in concentration with time is negative because this is a reactant. The magnitude of the answer $(0.15\ M \cdot s^{-1})$ makes physical sense because more moles of C are being formed for every mole of A reacting and the change in concentration with time is positive because this is a product.

15.30 (a) Rate $= -\frac{\Delta[A]}{\Delta t} = -2\frac{\Delta[B]}{\Delta t} = \frac{1}{2}\frac{\Delta[C]}{\Delta t}$

(b) **Given:** $\frac{\Delta[C]}{\Delta t} = 0.025\ M/s$ **Find:** $\frac{\Delta[B]}{\Delta t}$ and $\frac{\Delta[A]}{\Delta t}$

Conceptual Plan: $\frac{\Delta[C]}{\Delta t} \rightarrow \frac{\Delta[B]}{\Delta t}$ **and** $\frac{\Delta[A]}{\Delta t}$

Rate $= -\frac{\Delta[A]}{\Delta t} = -2\frac{\Delta[B]}{\Delta t} = \frac{1}{2}\frac{\Delta[C]}{\Delta t}$

Solution: Rate $= -\frac{\Delta[A]}{\Delta t} = -2\frac{\Delta[B]}{\Delta t} = \frac{1}{2}\frac{\Delta[C]}{\Delta t}$. Substitute value and solve for the two desired values.

$-2\frac{\Delta[B]}{\Delta t} = \frac{1}{2}\frac{0.025\ M}{s}$ so $\frac{\Delta[B]}{\Delta t} = -0.0063\ M \cdot s^{-1}$ and $-\frac{\Delta[A]}{\Delta t} = \frac{1}{2}\frac{0.025\ M}{s}$ so $\frac{\Delta[A]}{\Delta t} = -0.013\ M \cdot s^{-1}$

Check: The units $(M \cdot s^{-1})$ are correct. The magnitude of the answer $(-0.006\ M \cdot s^{-1})$ makes physical sense because fewer moles of B are reacting for every mole of C being formed and the change in concentration with time is negative because this is a reactant. The magnitude of the answer $(-0.01\ M \cdot s^{-1})$ makes physical sense because fewer moles of A are reacting for every mole of C being formed and the change in concentration with time is negative because this is a reactant. B has the smallest stoichiometric coefficient, so its rate of change has the smallest magnitude.

15.31 **Given:** $Cl_2(g) + 3\ F_2(g) \rightarrow 2\ ClF_3(g)$; $\Delta[Cl_2]/\Delta t = -0.012$ M/s **Find:** $\Delta[F_2]/\Delta t$; $\Delta[ClF_3]/\Delta t$; and Rate

Conceptual Plan: Write the expression for the rate with respect to each species then

Rate $= -\frac{\Delta[Cl_2]}{\Delta t} = -\frac{1}{3}\frac{\Delta[F_2]}{\Delta t} = \frac{1}{2}\frac{\Delta[ClF_3]}{\Delta t}$

rate expression, $\frac{\Delta[Cl_2]}{\Delta t} \rightarrow \frac{\Delta[F_2]}{\Delta t}; \frac{\Delta[ClF_3]}{\Delta t}$**; Rate**

Rate $= -\frac{\Delta[Cl_2]}{\Delta t} = -\frac{1}{3}\frac{\Delta[F_2]}{\Delta t} = \frac{1}{2}\frac{\Delta[ClF_3]}{\Delta t}$

Solution: Rate $= -\frac{\Delta[Cl_2]}{\Delta t} = -\frac{1}{3}\frac{\Delta[F_2]}{\Delta t} = \frac{1}{2}\frac{\Delta[ClF_3]}{\Delta t}$ so

$-\frac{\Delta[Cl_2]}{\Delta t} = -\frac{1}{3}\frac{\Delta[F_2]}{\Delta t}$ Rearrange to solve for $\frac{\Delta[F_2]}{\Delta t}$. $\frac{\Delta[F_2]}{\Delta t} = 3\frac{\Delta[Cl_2]}{\Delta t} = 3(-0.012\ M \cdot s^{-1}) = -0.036\ M \cdot s^{-1}$

and $-\frac{\Delta[Cl_2]}{\Delta t} = \frac{1}{2}\frac{\Delta[ClF_3]}{\Delta t}$ Rearrange to solve for $\frac{\Delta[ClF_3]}{\Delta t}$.

$\frac{\Delta[ClF_3]}{\Delta t} = -2\frac{\Delta[Cl_2]}{\Delta t} = -2(-0.012\ M \cdot s^{-1}) = 0.024\ M \cdot s^{-1}$

Check: The units $(M \cdot s^{-1})$ are correct. The magnitude of the answers $(-0.036\ M \cdot s^{-1}$ and $0.024\ M \cdot s^{-1})$ makes physical sense because F_2 is being used at three times the rate of Cl_2, ClF_3 is being formed at two times the rate of Cl_2 disappearance, and Cl_2 has a stoichiometric coefficient of 1.

15.32 **Given:** $8\ H_2S(g) + 4\ O_2(g) \rightarrow 8\ H_2O(g) + S_8(g)$; $\Delta[H_2S]/\Delta t = -0.080$ M/s

Find: $\Delta[O_2]/\Delta t$; $\Delta[H_2O]/\Delta t$; $\Delta[S_8]/\Delta t$; and Rate

Conceptual Plan: Write the expression for the rate with respect to each species then

Rate $= -\frac{1}{8}\frac{\Delta[H_2S]}{\Delta t} = -\frac{1}{4}\frac{\Delta[O_2]}{\Delta t} = \frac{1}{8}\frac{\Delta[H_2O]}{\Delta t} = \frac{\Delta[S_8]}{\Delta t}$

rate expression, $\frac{\Delta[H_2S]}{\Delta t} \rightarrow \frac{\Delta[O_2]}{\Delta t}; \frac{\Delta[H_2O]}{\Delta t}; \frac{\Delta[S_8]}{\Delta t}$**; Rate**

$$\text{Rate} = -\frac{1}{8}\frac{\Delta[H_2S]}{\Delta t} = -\frac{1}{4}\frac{\Delta[O_2]}{\Delta t} = \frac{1}{8}\frac{\Delta[H_2O]}{\Delta t} = \frac{\Delta[S_8]}{\Delta t}$$

Solution: Rate $= -\frac{1}{8}\frac{\Delta[H_2S]}{\Delta t} = -\frac{1}{4}\frac{\Delta[O_2]}{\Delta t} = \frac{1}{8}\frac{\Delta[H_2O]}{\Delta t} = \frac{\Delta[S_8]}{\Delta t}$ so

$-\frac{1}{8}\frac{\Delta[H_2S]}{\Delta t} = -\frac{1}{4}\frac{\Delta[O_2]}{\Delta t}$ Rearrange to solve for $\frac{\Delta[O_2]}{\Delta t}$.

$\frac{\Delta[O_2]}{\Delta t} = \frac{4}{8}\frac{\Delta[H_2S]}{\Delta t} = \frac{1}{2}(-0.080\ \text{M}\cdot\text{s}^{-1}) = -0.040\ \text{M}\cdot\text{s}^{-1}$

and $-\frac{1}{8}\frac{\Delta[H_2S]}{\Delta t} = \frac{1}{8}\frac{\Delta[H_2O]}{\Delta t}$ Rearrange to solve for $\frac{\Delta[H_2O]}{\Delta t}$.

$\frac{\Delta[H_2O]}{\Delta t} = -\frac{8}{8}\frac{\Delta[H_2S]}{\Delta t} = -(-0.080\ \text{M}\cdot\text{s}^{-1}) = 0.080\ \text{M}\cdot\text{s}^{-1}$

and $-\frac{1}{8}\frac{\Delta[H_2S]}{\Delta t} = \frac{\Delta[S_8]}{\Delta t}\ -\frac{1}{8}\frac{\Delta[H_2S]}{\Delta t} = -\frac{1}{8}(-0.080\ \text{M}\cdot\text{s}^{-1}) = 0.010\ \text{M}\cdot\text{s}^{-1}$

Check: The units ($\text{M}\cdot\text{s}^{-1}$) are correct. The magnitude of the answers ($-0.040\ \text{M}\cdot\text{s}^{-1}$, $0.080\ \text{M}\cdot\text{s}^{-1}$, and $0.010\ \text{M}\cdot\text{s}^{-1}$) makes physical sense because O_2 is being used at one-half times the rate of H_2S, H_2O is being formed at two times the rate of H_2S disappearance, S_8 is being formed at one-eighth times the rate of H_2S disappearance, and S_8 has a stoichiometric coefficient of 1.

15.33 (a) **Given:** $[C_4H_8]$ versus time data **Find:** average rate between 0 and 10 s and between 40 and 50 s
Conceptual Plan: $t_1, t_2, [C_4H_8]_1, [C_4H_8]_2 \rightarrow$ **average rate**

$$\text{Rate} = -\frac{\Delta[C_4H_8]}{\Delta t}$$

Solution: For 0 to 10 s, Rate $= -\frac{[C_4H_8]_{t_2} - [C_4H_8]_{t_1}}{t_2 - t_1} = -\frac{0.913\ \text{M} - 1.000\ \text{M}}{10.\ \text{s} - 0.\ \text{s}} = 8.7 \times 10^{-3}\ \text{M}\cdot\text{s}^{-1}$ and

for 40 to 50 s, Rate $= -\frac{[C_4H_8]_{t_2} - [C_4H_8]_{t_1}}{t_2 - t_1} = -\frac{0.637\ \text{M} - 0.697\ \text{M}}{50.\ \text{s} - 40.\ \text{s}} = 6.0 \times 10^{-3}\ \text{M}\cdot\text{s}^{-1}$

Check: The units ($\text{M}\cdot\text{s}^{-1}$) are correct. The magnitude of the answers ($10^{-3}\ \text{M}\cdot\text{s}^{-1}$) makes physical sense because rates are always positive and we are not changing the concentration much in 10 s. Also, reactions slow as they proceed because the concentration of the reactants is decreasing.

(b) **Given:** $[C_4H_8]$ versus time data **Find:** $\frac{\Delta[C_2H_4]}{\Delta t}$ between 20 and 30 s

Conceptual Plan: $t_1, t_2, [C_4H_8]_1, [C_4H_8]_2 \rightarrow \frac{\Delta[C_2H_4]}{\Delta t}$

$$\text{Rate} = -\frac{\Delta[C_4H_8]}{\Delta t} = \frac{1}{2}\frac{\Delta[C_2H_4]}{\Delta t}$$

Solution: Rate $= -\frac{[C_4H_8]_{t_2} - [C_4H_8]_{t_1}}{t_2 - t_1} = -\frac{0.763\ \text{M} - 0.835\ \text{M}}{30.\ \text{s} - 20.\ \text{s}} = 7.2 \times 10^{-3}\ \text{M}\cdot\text{s}^{-1} = \frac{1}{2}\frac{\Delta[C_2H_4]}{\Delta t}$

Rearrange to solve for $\frac{\Delta[C_2H_4]}{\Delta t}$. So $\frac{\Delta[C_2H_4]}{\Delta t} = 2(7.2 \times 10^{-3}\ \text{M}\cdot\text{s}^{-1}) = 1.4 \times 10^{-2}\ \text{M}\cdot\text{s}^{-1}$

Check: The units ($\text{M}\cdot\text{s}^{-1}$) are correct. The magnitude of the answer ($10^{-2}\ \text{M}\cdot\text{s}^{-1}$) makes physical sense because the rate of product formation is always positive and we are not changing the concentration much in 10 s. The rate of change of the product is faster than the decline of the reactant because of the stoichiometric coefficients.

15.34 (a) **Given:** $[NO_2]$ versus time data **Find:** average rate between 10 and 20 s and between 50 and 60 s
Conceptual Plan: $t_1, t_2, [NO_2]_1, [NO_2]_2 \rightarrow$ **average rate**

$$\text{Rate} = -\frac{\Delta[NO_2]}{\Delta t}$$

Solution: For 10 to 20 s, Rate $= -\dfrac{[NO_2]_{t_2} - [NO_2]_{t_1}}{t_2 - t_1} = -\dfrac{0.904\ \text{M} - 0.951\ \text{M}}{20.\ \text{s} - 10.\ \text{s}} = 4.7 \times 10^{-3}\ \text{M}\cdot\text{s}^{-1}$ and

for 50 to 60 s, Rate $= -\dfrac{[NO_2]_{t_2} - [NO_2]_{t_1}}{t_2 - t_1} = -\dfrac{0.740\ \text{M} - 0.778\ \text{M}}{60.\ \text{s} - 50.\ \text{s}} = 3.8 \times 10^{-3}\ \text{M}\cdot\text{s}^{-1}$

Check: The units $(\text{M}\cdot\text{s}^{-1})$ are correct. The magnitude of the answer $(10^{-3}\ \text{M}\cdot\text{s}^{-1})$ makes physical sense because rates are always positive and we are not changing the concentration much in 10 s. Also, reactions slow as they proceed because the concentration of the reactants is decreasing.

(b) **Given:** $[NO_2]$ versus time data **Find:** $\dfrac{\Delta[O_2]}{\Delta t}$ between 50 and 60 s

Conceptual Plan: Average rate from part (a) $\rightarrow \dfrac{\Delta[O_2]}{\Delta t}$

$$\text{Rate} = -\frac{\Delta[NO_2]}{\Delta t} = 2\frac{\Delta[O_2]}{\Delta t}$$

Solution: Rate $= -\dfrac{\Delta[NO_2]}{\Delta t} = 2\dfrac{\Delta[O_2]}{\Delta t}$. Substitute value and solve for the desired value.

$3.8 \times 10^{-3}\ \text{M}\cdot\text{s}^{-1} = 2\dfrac{\Delta[O_2]}{\Delta t}$ so $\dfrac{\Delta[O_2]}{\Delta t} = 1.9 \times 10^{-3}\ \text{M}\cdot\text{s}^{-1}$

Check: The units $(\text{M}\cdot\text{s}^{-1})$ are correct. The magnitude of the answer $(10^{-3}\ \text{M}\cdot\text{s}^{-1})$ makes physical sense because rate of product formation is always positive and we are not changing the concentration much in 10 s. The rate of change of the product is slower than the decline of the reactant because of the stoichiometric coefficients.

15.35 (a) **Given:** $[Br_2]$ versus time plot

Find: (i) average rate between 0 and 25 s; (ii) instantaneous rate at 25 s; (iii) instantaneous rate of HBr formation at 50 s

Conceptual Plan: (i) $t_1, t_2, [Br_2]_1, [Br_2]_2 \rightarrow$ average rate then

$$\text{Rate} = -\frac{\Delta[Br_2]}{\Delta t}$$

(ii) draw tangent at 25 s and determine slope $\rightarrow$ instantaneous rate then

$$\text{Rate} = -\frac{\Delta[Br_2]}{\Delta t}$$

(iii) draw tangent at 50 s and determine slope $\rightarrow$ instantaneous rate $\rightarrow \dfrac{\Delta[HBr]}{\Delta t}$

$$\text{Rate} = -\frac{\Delta[Br_2]}{\Delta t} \qquad \text{Rate} = \frac{1}{2}\frac{\Delta[HBr]}{\Delta t}$$

Solution:

(i) Rate $= -\dfrac{[Br_2]_{t_2} - [Br_2]_{t_1}}{t_2 - t_1} = -\dfrac{0.75\ \text{M} - 1.00\ \text{M}}{25\ \text{s} - 0.\ \text{s}} = 1.0 \times 10^{-2}\ \text{M}\cdot\text{s}^{-1}$ and (ii) at 25 s:

Slope $= \dfrac{\Delta y}{\Delta x} = \dfrac{0.68\ \text{M} - 0.85\ \text{M}}{35\ \text{s} - 15\ \text{s}} = -8.5 \times 10^{-3}\ \text{M}\cdot\text{s}^{-1}$

because the slope $= \dfrac{\Delta[Br_2]}{\Delta t}$ and Rate $= -\dfrac{\Delta[Br_2]}{\Delta t}$,

then Rate $= -(-8.5 \times 10^{-3}\ \text{M}\cdot\text{s}^{-1}) = 8.5 \times 10^{-3}\ \text{M}\cdot\text{s}^{-1}$

(iii) at 50 s:

Slope $= \dfrac{\Delta y}{\Delta x} = \dfrac{0.53\ \text{M} - 0.66\ \text{M}}{60.\ \text{s} - 40.\ \text{s}} = -6.5 \times 10^{-3}\ \text{M}\cdot\text{s}^{-1}$

because the slope $= \dfrac{\Delta[Br_2]}{\Delta t}$ and

Rate $= -\dfrac{\Delta[Br_2]}{\Delta t} = \dfrac{1}{2}\dfrac{\Delta[HBr]}{\Delta t}$ then

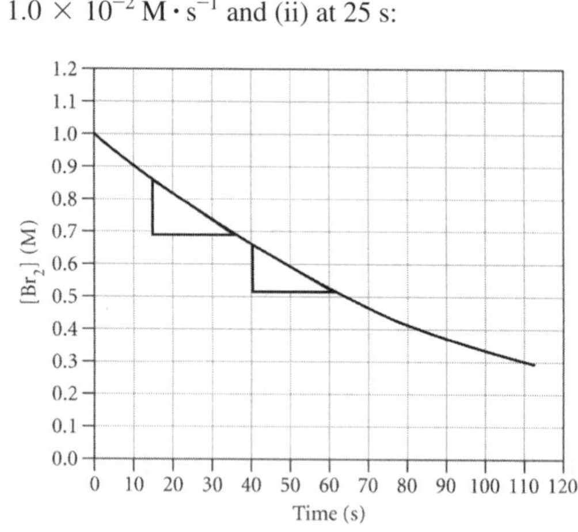

$\dfrac{\Delta[HBr]}{\Delta t} = -2\dfrac{\Delta[Br_2]}{\Delta t} = -2(-6.5 \times 10^{-3}\ \text{M}\cdot\text{s}^{-1}) = 1.3 \times 10^{-2}\ \text{M}\cdot\text{s}^{-1}$

Check: The units $(M \cdot s^{-1})$ are correct. The magnitude of the first answer is larger than that of the second answer because the rate is slowing down and the first answer includes the initial portion of the data. The magnitude of the answers $(10^{-3}\ M \cdot s^{-1})$ makes physical sense because rates are always positive and we are not changing the concentration much.

(b) **Given:** $[Br_2]$ versus time data and $[HBr]_0 = 0\ M$ **Find:** plot [HBr] with time

Conceptual Plan: Because Rate $= -\frac{\Delta[Br_2]}{\Delta t} = \frac{1}{2}\frac{\Delta[HBr]}{\Delta t}$**. The rate of change of [HBr] will be twice that of $[Br_2]$. The plot will start at the origin.**

Solution:

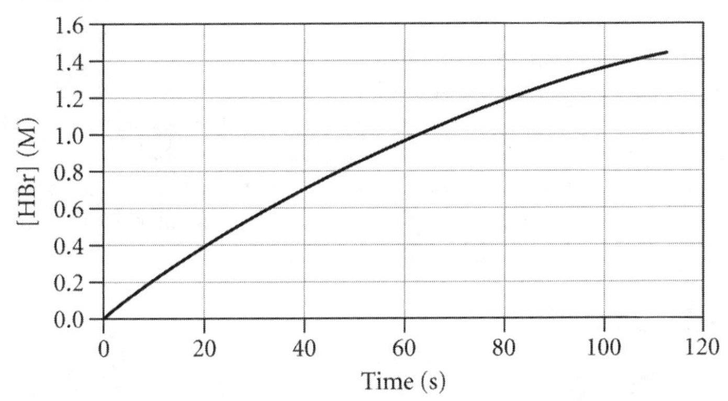

Check: The units (M versus s) are correct. The plot makes sense because it has the same general shape of the original plot except that we are increasing instead of decreasing our concentration axis by a factor of two (to account for the difference in stoichiometric coefficients).

15.36 **Given:** $[H_2O_2]$ versus time plot and 1.5 L H_2O_2 initially

Find: (a) average rate between 10 and 20 s; (b) instantaneous rate at 30 s; (c) instantaneous rate of O_2 formation at 50 s; (d) mol_{O_2} formed in first 50 s

Conceptual Plan: (a) $t_1, t_2, [H_2O_2]_1, [H_2O_2]_2 \rightarrow$ **average rate then**

$$\text{Rate} = -\frac{1}{2}\frac{\Delta[H_2O_2]}{\Delta t}$$

(b) draw tangent at 30 s and determine slope $\rightarrow$ instantaneous rate then

$$\text{Rate} = -\frac{1}{2}\frac{\Delta[H_2O_2]}{\Delta t}$$

(c) draw tangent at 50 s and determine slope $\rightarrow$ instantaneous rate $\rightarrow \frac{\Delta[O_2]}{\Delta t}$

$$\text{Rate} = -\frac{1}{2}\frac{\Delta[H_2O_2]}{\Delta t} \qquad \text{Rate} = \frac{\Delta[O_2]}{\Delta t}$$

(d) $[H_2O_2]_{0\,s}, [H_2O_2]_{50\,s} \rightarrow \Delta[H_2O_2] \rightarrow \Delta[O_2]$ **then** $\Delta[O_2], V \rightarrow mol_{O_2}$

$$\Delta[H_2O_2] = [H_2O_2]_{50\,s} - [H_2O_2]_{0\,s} \quad \text{Rate} = -\frac{1}{2}\frac{\Delta[H_2O_2]}{\Delta t} = \frac{\Delta[O_2]}{\Delta t} \quad M = \frac{mol_{O_2}}{L}$$

Solution:

(a) $\text{Rate} = -\frac{[H_2O_2]_{t_2} - [H_2O_2]_{t_1}}{t_2 - t_1} = -\frac{0.55M - 0.75\ M}{20.\ s - 10.\ s} = 2.0 \times 10^{-2}\ M \cdot s^{-1}$ and

(b) at 30 s, Slope $= \frac{\Delta y}{\Delta x} = \frac{0.28\ M - 0.52\ M}{40.\ s - 20.\ s} = -1.2 \times 10^{-2}\ M \cdot s^{-1}$ because the slope $= \frac{\Delta[H_2O_2]}{\Delta t}$ and

$\text{Rate} = -\frac{1}{2}\frac{\Delta[H_2O_2]}{\Delta t}$ then

$\text{Rate} = (-0.5)(-1.2 \times 10^{-2}\ M \cdot s^{-1}) = 6.0 \times 10^{-3}\ M \cdot s^{-1}$

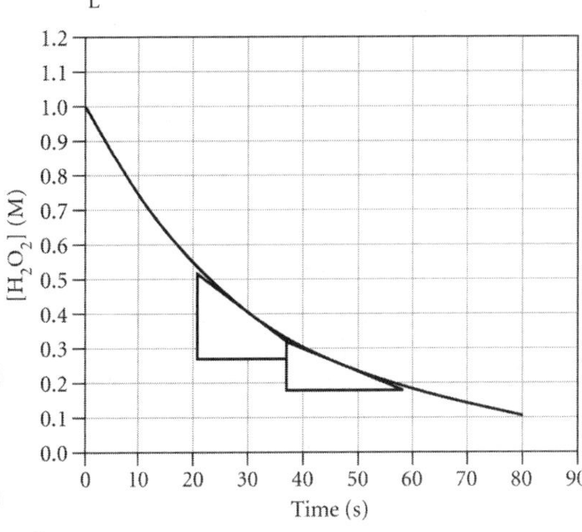

(c) at 50 s,

$$\text{Slope} = \frac{\Delta y}{\Delta x} = \frac{0.15\text{ M} - 0.28\text{ M}}{60.\text{ s} - 40.\text{ s}} = -6.5 \times 10^{-3}\text{ M}\cdot\text{s}^{-1} \text{ because the slope } = \frac{\Delta[\text{H}_2\text{O}_2]}{\Delta t} \text{ and}$$

$$\text{Rate} = -\frac{1}{2}\frac{\Delta[\text{H}_2\text{O}_2]}{\Delta t} = \frac{\Delta[\text{O}_2]}{\Delta t} \text{ then } \frac{\Delta[\text{O}_2]}{\Delta t} = (-0.5)(-6.5 \times 10^{-3}\text{ M}\cdot\text{s}^{-1}) = 3.3 \times 10^{-3}\text{ M}\cdot\text{s}^{-1}$$

(d) $\Delta[\text{H}_2\text{O}_2] = [\text{H}_2\text{O}_2]_{50\text{ s}} - [\text{H}_2\text{O}_2]_{0\text{ s}} = 0.23\text{ M} - 1.00\text{ M} = 0.77\text{ M}$ because

$$\Delta[\text{O}_2] = -\frac{1}{2}\Delta[\text{H}_2\text{O}_2] = (-0.5)(0.77\text{ M}) = 0.3\underline{8}5\text{ M then M} = \frac{\text{mol}_{\text{O}_2}}{\text{L}} \text{ so}$$

$$\text{mol}_{\text{O}_2} = \text{M}\cdot\text{L} = 0.3\underline{8}5\frac{\text{mol}_{\text{O}_2}}{\cancel{\text{L}}} \times 1.5\,\cancel{\text{L}} = 0.58\text{ mol}_{\text{O}_2}$$

Check: (a) The units $(\text{M}\cdot\text{s}^{-1})$ are correct. The magnitude of the first answer is reasonable considering the concentrations and times involved $(1\text{ M}/100\text{ s})$. (b) The units $(\text{M}\cdot\text{s}^{-1})$ are correct. We expect the answer in this part to be less than in the first part because the rate is decreasing as the reaction proceeds. (c) The units $(\text{M}\cdot\text{s}^{-1})$ are correct. We expect the answer in this part to be less than in the first part because the rate is decreasing as the reaction proceeds. (d) The units (mol) are correct. We expect an answer less than 1 mol because the drop in reactant concentration is less than 1 M, we have 1.5 L, and only half as much O_2 is generated as hydrogen peroxide is consumed.

The Rate Law and Reaction Orders

15.37 (a) **Given:** Rate versus [A] plot **Find:** reaction order

Conceptual Plan: Look at shape of plot and match to possibilities.

Solution: The plot is a linear plot, so Rate α [A] or the reaction is first order.

Check: The order of the reaction is a common reaction order.

(b) **Given:** part (a) **Find:** sketch plot of [A] versus time

Conceptual Plan: Using the result from part (a), shape plot of [A] versus time should be curved with [A] decreasing. Use 1.0 M as initial concentration.

Solution:

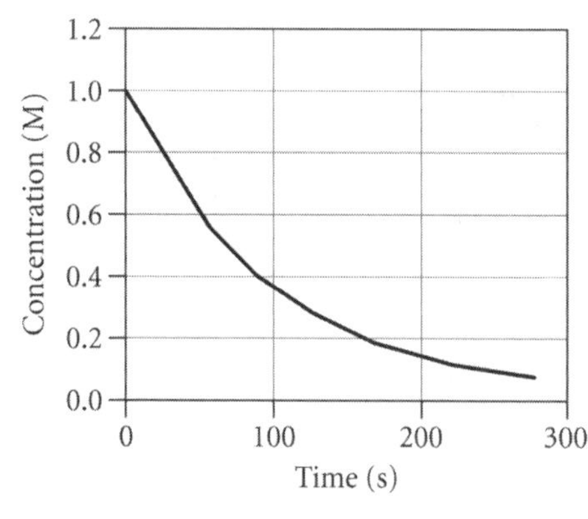

Rate = k [A]

plugging in any of the values from the graph

$0.002\text{ M}\cdot\text{s}^{-1} = k\,[0.2\text{ M}]$

$$k = \frac{0.002\,\cancel{\text{M}}\cdot\text{s}^{-1}}{0.2\,\cancel{\text{M}}}$$

$k = 0.01\text{ s}^{-1}$

Check: The plot has a shape that matches the one in the text for first-order plots.

(c) **Given:** part (a) **Find:** write a rate law and estimate k

Conceptual Plan: Using result from part (a), the slope of the plot is the rate constant.

$$\textbf{Solution: } \text{Slope} = \frac{\Delta y}{\Delta x} = \frac{0.010\frac{\cancel{\text{M}}}{\text{s}} - 0.00\frac{\cancel{\text{M}}}{\text{s}}}{1.0\,\cancel{\text{M}} - 0.0\,\cancel{\text{M}}} = 0.010\text{ s}^{-1} \text{ so Rate } = k[\text{A}]^1 \text{ or Rate } = k[\text{A}] \text{ or}$$

Rate $= 0.010\text{ s}^{-1}\,[\text{A}]$

Check: The units (s^{-1}) are correct. The magnitude of the answer (10^{-2} s^{-1}) makes physical sense because of the rate and concentration data. Remember that concentration is in units of M; so plugging the rate constant into the equation has the units of the rate as $\text{M}\cdot\text{s}^{-1}$, which is correct.

15.38 (a) **Given:** Rate versus [A] plot **Find:** reaction order
Conceptual Plan: Look at shape of plot and match to possibilities.
Solution: The plot is a linear plot that is horizontal, so rate is independent of [A] or the reaction is zero-order with respect to A.

Check: The order of the reaction is a common reaction order.

(b) **Given:** part (a) **Find:** sketch plot of [A] versus time
Conceptual Plan: Using the result from part (a), shape plot of [A] versus time should be a straight line with [A] decreasing. Use 1.0 M as initial concentration.
Solution:

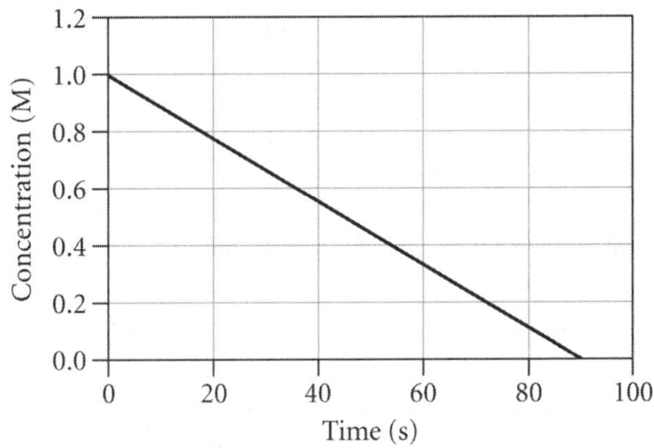

Check: The plot has a shape that matches the one in the text for zero-order plots.

(c) **Given:** part (a) **Find:** write a rate law and estimate k
Conceptual Plan: Using result from part (a), the rate is equal to the rate constant.
Solution: Rate $= k[\text{A}]^0$ or Rate $= k$ or Rate $= 0.011\ \text{M} \cdot \text{s}^{-1}$

Check: The units $(\text{M} \cdot \text{s}^{-1})$ are correct. The magnitude of the answer $(10^{-2}\ \text{M} \cdot \text{s}^{-1})$ makes physical sense because of the rate and concentration data. Plugging the rate constant into the equation, the rate has the units of $\text{M} \cdot \text{s}^{-1}$, which is correct.

15.39 **Given:** reaction order: (a) first order, (b) second order, and (c) zero order **Find:** units of k
Conceptual Plan: Using rate law, rearrange to solve for k.

$$\text{Rate} = k[\text{A}]^n, \text{ where } n = \text{reaction order}$$

Solution: For all cases, rate has units of $\text{M} \cdot \text{s}^{-1}$ and [A] has units of M.

(a) $$\text{Rate} = k[\text{A}]^1 = k[\text{A}] \text{ so } k = \frac{\text{Rate}}{[\text{A}]} = \frac{\frac{\cancel{\text{M}}}{\text{s}}}{\cancel{\text{M}}} = \text{s}^{-1}$$

(b) $$\text{Rate} = k[\text{A}]^2 \text{ so } k = \frac{\text{Rate}}{[\text{A}]^2} = \frac{\frac{\cancel{\text{M}}}{\text{s}}}{\cancel{\text{M}} \cdot \text{M}} = \text{M}^{-1} \cdot \text{s}^{-1}$$

(c) $$\text{Rate} = k[\text{A}]^0 = k = \text{M} \cdot \text{s}^{-1}$$

Check: The units $(\text{s}^{-1}, \text{M}^{-1} \cdot \text{s}^{-1}, \text{and M} \cdot \text{s}^{-1})$ are correct. The units for k change with the reaction order so that the units on the rate remain as $\text{M} \cdot \text{s}^{-1}$.

15.40 **Given:** $k = 0.053/\text{s}$ and $[N_2O_5] = 0.055$ M; reaction order: (a) first order, (b) second order, and zero order (change units on k as necessary) **Find:** rate
Conceptual Plan: Using rate law, substitute values to solve for Rate.

$$\text{Rate} = k[N_2O_5]^n, \text{ where } n = \text{reaction order}$$

Solution: For all cases, Rate has units of $\text{M} \cdot \text{s}^{-1}$ and [A] has units of M. Use the results from Problem 15.8 to choose the appropriate units for k.

(a) $$\text{Rate} = k[N_2O_5]^1 = k[N_2O_5] = \frac{0.053}{\text{s}} \times 0.055\ \text{M} = 2.9 \times 10^{-3} \frac{\text{M}}{\text{s}}$$

(b) $\text{Rate} = k[N_2O_5]^2 = \frac{0.053}{\text{M s}} \times (0.055\ \text{M})^2 = 1.6 \times 10^{-4} \frac{\text{M}}{\text{s}}$

$\text{Rate} = k[N_2O_5]^0 = k = 5.3 \times 10^{-2} \frac{\text{M}}{\text{s}}$

Check: The units $(\text{M} \cdot \text{s}^{-1})$ are correct. The magnitude of the rate changes as the order of the reaction changes because we are multiplying by the concentration a different number of times in each case. The higher the order, the lower the rate because the concentration is less than 1 M.

15.41 **Given:** A, B, and C react to form products. Reaction is first order in A, second order in B, and zero order in C. **Find:** (a) rate law; (b) overall order of reaction; (c) factor change in rate if [A] doubled; (d) factor change in rate if [B] doubled; (e) factor change in rate if [C] doubled; and (f) factor change in rate if [A], [B], and [C] doubled

Conceptual Plan:

(a) Using general rate law form, substitute values for orders.

$\text{Rate} = k[\text{A}]^m[\text{B}]^n[\text{C}]^p$, where m, n, and p = reaction orders

(b) Using rate law in part (a), add up all reaction orders.

overall reaction order $= m + n + p$

(c) Through (f), using rate law from part (a), substitute concentration changes.

$\frac{\text{Rate 2}}{\text{Rate 1}} = \frac{k[\text{A}]_2^1[\text{B}]_2^2}{k[\text{A}]_1^1[\text{B}]_1^2}$

Solution:

(a) $m = 1, n = 2$, and $p = 0$ so $\text{Rate} = k[\text{A}]^1[\text{B}]^2[\text{C}]^0$ or $\text{Rate} = k[\text{A}][\text{B}]^2$

(b) *overall reaction order* $= m + n + p = 1 + 2 + 0 = 3$, so it is a third-order reaction overall

(c) $\frac{\text{Rate 2}}{\text{Rate 1}} = \frac{k[\text{A}]_2^1[\text{B}]_2^2}{k[\text{A}]_1^1[\text{B}]_1^2}$ and $[\text{A}]_2 = 2[\text{A}]_1, [\text{B}]_2 = [\text{B}]_1, [\text{C}]_2 = [\text{C}]_1$, so $\frac{\text{Rate 2}}{\text{Rate 1}} = \frac{\cancel{k}(2\cancel{[\text{A}]_1})^1 \cancel{[\text{B}]_1^2}}{\cancel{k}\cancel{[\text{A}]_1^1}\cancel{[\text{B}]_1^2}} = 2$

so the reaction rate doubles (factor of 2)

(d) $\frac{\text{Rate 2}}{\text{Rate 1}} = \frac{k[\text{A}]_2^1[\text{B}]_2^2}{k[\text{A}]_1^1[\text{B}]_1^2}$ and $[\text{A}]_2 = [\text{A}]_1, [\text{B}]_2 = 2[\text{B}]_1, [\text{C}]_2 = [\text{C}]_1$, so $\frac{\text{Rate 2}}{\text{Rate 1}} = \frac{\cancel{k}\cancel{[\text{A}]_1}(2\cancel{[\text{B}]_1})^2}{\cancel{k}\cancel{[\text{A}]_1^1}\cancel{[\text{B}]_1^2}} = 2^2 = 4$

so the reaction rate quadruples (factor of 4)

(e) $\frac{\text{Rate 2}}{\text{Rate 1}} = \frac{k[\text{A}]_2^1[\text{B}]_2^2}{k[\text{A}]_1^1[\text{B}]_1^2}$ and $[\text{A}]_2 = [\text{A}]_1, [\text{B}]_2 = [\text{B}]_1, [\text{C}]_2 = 2[\text{C}]_1$, so $\frac{\text{Rate 2}}{\text{Rate 1}} = \frac{\cancel{k}\cancel{[\text{A}]_1^1}\cancel{[\text{B}]_1^2}}{\cancel{k}\cancel{[\text{A}]_1^1}\cancel{[\text{B}]_1^2}} = 1$

so the reaction rate is unchanged (factor of 1)

(f) $\frac{\text{Rate 2}}{\text{Rate 1}} = \frac{k[\text{A}]_2^1[\text{B}]_2^2}{k[\text{A}]_1^1[\text{B}]_1^2}$ and $[\text{A}]_2 = 2[\text{A}]_1, [\text{B}]_2 = 2[\text{B}]_1, [\text{C}]_2 = 2[\text{C}]_1$, so

$\frac{\text{Rate 2}}{\text{Rate 1}} = \frac{\cancel{k}(2\cancel{[\text{A}]_1})^1(2\cancel{[\text{B}]_1})^2}{\cancel{k}\cancel{[\text{A}]_1^1}\cancel{[\text{B}]_1^2}} = 2 \times 2^2 = 8$ so the reaction rate goes up by a factor of 8

Check: The units (none) are correct. The rate law is consistent with the orders given, and the overall order is larger than any of the individual orders. The factors are consistent with the reaction orders. The larger the order, the larger the factor. When all concentrations are changed, the rate changes the most. If a reactant is not in the rate law, then changing its concentration has no effect on the reaction rate.

15.42 **Given:** A, B, and C react to form products. Reaction is zero order in A, one-half order in B, and second order in C. **Find:** (a) rate law; (b) overall order of reaction; (c) factor change in rate if [A] doubled; (d) factor change in rate if [B] doubled; (e) factor change in rate if [C] doubled; and (f) factor change in rate if [A], [B], and [C] doubled

Conceptual Plan:

(a) Using general rate law form, substitute values for orders.

$\text{Rate} = k[\text{A}]^m[\text{B}]^n[\text{C}]^p$, where m, n, and p = reaction orders

(b) Using rate law in part (a), add up all reaction orders.

overall reaction order $= m + n + p$

(c) Through (f), using rate law from part (a), substitute concentration changes.

$\frac{\text{Rate 2}}{\text{Rate 1}} = \frac{k[\text{B}]_2^{1/2}[\text{C}]_2^2}{k[\text{B}]_1^{1/2}[\text{C}]_1^2}$

Solution:

(a) $m = 0, n = 1/2$, and $p = 2$ so Rate $= k[A]^0[B]^{1/2}[C]^2$ or Rate $= k[B]^{1/2}[C]^2$

(b) *overall reaction order* $= m + n + p = 0 + 1/2 + 2 = 5/2 = 2.5$, so it is a two-and-a-half-order reaction overall

(c) $\dfrac{\text{Rate 2}}{\text{Rate 1}} = \dfrac{k[B]_2^{1/2}[C]_2^2}{k[B]_1^{1/2}[C]_1^2}$ and $[A]_2 = 2[A]_1, [B]_2 = [B]_1, [C]_2 = [C]_1$, so $\dfrac{\text{Rate 2}}{\text{Rate 1}} = \dfrac{\cancel{k}\cancel{[B]_1^{1/2}}\cancel{[C]_1^2}}{\cancel{k}\cancel{[B]_1^{1/2}}\cancel{[C]_1^2}} = 1$

so the reaction rate is unchanged (factor of 1).

(d) $\dfrac{\text{Rate 2}}{\text{Rate 1}} = \dfrac{k[B]_2^{1/2}[C]_2^2}{k[B]_1^{1/2}[C]_1^2}$ and $[A]_2 = [A]_1, [B]_2 = 2[B]_1, [C]_2 = [C]_1$, so $\dfrac{\text{Rate 2}}{\text{Rate 1}} = \dfrac{\cancel{k}(2\cancel{[B]_1})^{1/2}\cancel{[C]_1^2}}{\cancel{k}\cancel{[B]_1^{1/2}}\cancel{[C]_1^2}} = 2^{1/2}$

so the reaction rate increases by a factor of $2^{1/2}$ or $\sqrt{2}$ or 1.414

(e) $\dfrac{\text{Rate 2}}{\text{Rate 1}} = \dfrac{k[B]_2^{1/2}[C]_2^2}{k[B]_1^{1/2}[C]_1^2}$ and $[A]_2 = [A]_1, [B]_2 = [B]_1, [C]_2 = 2[C]_1$, so

$\dfrac{\text{Rate 2}}{\text{Rate 1}} = \dfrac{\cancel{k}\cancel{[B]_1^{1/2}}(2\cancel{[C]_1})^2}{\cancel{k}\cancel{[B]_1^{1/2}}\cancel{[C]_1^2}} = 2^2 = 4$, so the reaction rate quadruples (factor of 4)

(f) $\dfrac{\text{Rate 2}}{\text{Rate 1}} = \dfrac{k[B]_2^{1/2}[C]_2^2}{k[B]_1^{1/2}[C]_1^2}$ and $[A]_2 = 2[A]_1, [B]_2 = 2[B]_1, [C]_2 = 2[C]_1$, so

$\dfrac{\text{Rate 2}}{\text{Rate 1}} = \dfrac{\cancel{k}(2\cancel{[B]_1})^{1/2}(2\cancel{[C]_1})^2}{\cancel{k}\cancel{[B]_1^{1/2}}\cancel{[C]_1^2}} = 2^{1/2} \times 2^2 = 2^{5/2}$, so the reaction rate goes up by a factor of $2^{5/2}$ or $4\sqrt{2}$ or 5.66

Check: The units (none) are correct. The rate law is consistent with the orders given, and the overall order is larger than any of the individual orders. The factors are consistent with the reaction orders. The larger the order, the larger the factor. When all concentrations are changed, the rate changes the most. If a reactant is not in the rate law, then changing its concentration has no effect on the reaction rate.

15.43 **Given:** table of [A] versus initial rate **Find:** rate law and k

Conceptual Plan: Using general rate law form, compare rate ratios to determine reaction order.

$$\frac{\text{Rate 2}}{\text{Rate 1}} = \frac{k[A]_2^n}{k[A]_1^n}$$

Then use one of the concentration/initial rate pairs to determine k.

$$\text{Rate} = k[A]^n$$

Solution: $\dfrac{\text{Rate 2}}{\text{Rate 1}} = \dfrac{k[A]_2^n}{k[A]_1^n}$ Comparing the first two sets of data, $\dfrac{0.210\ \cancel{M/s}}{0.053\ \cancel{M/s}} = \dfrac{\cancel{k}(0.200\ \cancel{M})^n}{\cancel{k}(0.100\ \cancel{M})^n}$ and $3.\underline{9}623 = 2^n$

so $n = 2$ If we compare the first and the last data sets, $\dfrac{0.473\ \cancel{M/s}}{0.053\ \cancel{M/s}} = \dfrac{\cancel{k}(0.300\ \cancel{M})^n}{\cancel{k}(0.100\ \cancel{M})^n}$ and $8.\underline{9}245 = 3^n$ so $n = 2$

This second comparison is not necessary, but it increases our confidence in the reaction order. So Rate $= k[A]^2$. Selecting the second data set and rearranging the rate equation,

$$k = \frac{\text{Rate}}{[A]^2} = \frac{0.210\dfrac{M}{s}}{(0.200\ M)^2} = 5.25\ M^{-1} \cdot s^{-1}, \text{ so Rate} = 5.25\ M^{-1} \cdot s^{-1}[A]^2$$

Check: The units (none and $M^{-1} \cdot s^{-1}$) are correct. The rate law is a common form. The rate is changing more rapidly than the concentration, so second order is consistent. The rate constant is consistent with the units necessary to get rate as M/s, and the magnitude is reasonable because we have a second-order reaction.

15.44 **Given:** table of [A] versus initial rate **Find:** rate law and k

Conceptual Plan: Using general rate law form, compare rate ratios to determine reaction order.

$$\frac{\text{Rate 2}}{\text{Rate 1}} = \frac{k[A]_2^n}{k[A]_1^n}$$

Then use one of the concentration/initial rate pairs to determine k.

$$\text{Rate} = k[A]^n$$

Solution: $\frac{\text{Rate 2}}{\text{Rate 1}} = \frac{k[\text{A}]_2^n}{k[\text{A}]_1^n}$ Comparing the first two sets of data, $\frac{0.016\ \cancel{\text{M/s}}}{0.008\ \cancel{\text{M/s}}} = \frac{\cancel{k}(0.30\ \cancel{\text{M}})^n}{\cancel{k}(0.15\ \cancel{\text{M}})^n}$ and $2 = 2^n$; so $n = 1$. If we compare the first and last data sets, $\frac{0.032\ \cancel{\text{M/s}}}{0.008\ \cancel{\text{M/s}}} = \frac{\cancel{k}(0.032\ \cancel{\text{M}})^n}{\cancel{k}(0.008\ \cancel{\text{M}})^n}$ and $4 = 4^n$, $n = 1$.

This second comparison is not necessary, but it increases our confidence in the reaction order. So Rate $= k[\text{A}]$.

Selecting the second data set and rearranging the rate equation, $k = \frac{\text{Rate}}{[\text{A}]} = \frac{0.016\frac{\cancel{\text{M}}}{\text{s}}}{0.30\ \cancel{\text{M}}} = 5.3 \times 10^{-2}\ \text{s}^{-1}$;

so Rate $= 5.3 \times 10^{-2}\ \text{s}^{-1}[\text{A}]$.

Check: The units (none and s^{-1}) are correct. The rate law is a common form. The rate is changing as rapidly as the concentration is consistent with first order. The rate constant is consistent with the units necessary to get rate as M/s, and the magnitude is reasonable because we have a first-order reaction.

15.45 **Given:** table of $[NO_2]$ and $[F_2]$ versus initial rate **Find:** rate law, *k*, and overall order

Conceptual Plan: Using general rate law form, compare rate ratios to determine reaction order of each reactant. Be sure to choose data that changes only one concentration at a time.

$$\frac{\text{Rate 2}}{\text{Rate 1}} = \frac{k[NO_2]_2^m[F_2]_2^n}{k[NO_2]_1^m[F_2]_1^n}$$

Then use one of the concentration/initial rate pairs to determine *k*.

$$\text{Rate} = k[NO_2]^m[F_2]^n$$

Solution: $\frac{\text{Rate 2}}{\text{Rate 1}} = \frac{k[NO_2]_2^m[F_2]_2^n}{k[NO_2]_1^m[F_2]_1^n}$ Comparing the first two sets of data,

$\frac{0.051\ \cancel{\text{M/s}}}{0.026\ \cancel{\text{M/s}}} = \frac{\cancel{k}(0.200\ \cancel{\text{M}})^m(\cancel{0.100\ \text{M}})^n}{\cancel{k}(0.100\ \cancel{\text{M}})^m(\cancel{0.100\ \text{M}})^n}$ and $1.\underline{9}615 = 2^m$; so $m = 1$. If we compare the second and third data sets, $\frac{0.103\ \cancel{\text{M/s}}}{0.051\ \cancel{\text{M/s}}} = \frac{\cancel{k}(\cancel{0.200\ \text{M}})^m(0.200\ \cancel{\text{M}})^n}{\cancel{k}(\cancel{0.200\ \text{M}})^m(0.100\ \cancel{\text{M}})^n}$ and $2 = 2^n$, $n = 1$. Other comparisons can be made, but they are not necessary. They should reinforce these values of the reaction orders. So Rate $= k[NO_2][F_2]$. Selecting the last data set and rearranging the rate equation, $k = \frac{\text{Rate}}{[NO_2][F_2]} = \frac{0.411\frac{\cancel{\text{M}}}{\text{s}}}{(0.400\ \text{M})(0.400\ \text{M})} = 2.57\ \text{M}^{-1}\cdot\text{s}^{-1}$; so Rate $= 2.57\ \text{M}^{-1}\cdot\text{s}^{-1}[NO_2][F_2]$, and the reaction is second order overall.

Check: The units (none and $\text{M}^{-1}\cdot\text{s}^{-1}$) are correct. The rate law is a common form. The rate is changing as rapidly as each concentration is changing, which is consistent with first order in each reactant. The rate constant is consistent with the units necessary to get rate as M/s, and the magnitude is reasonable because we have a second-order reaction.

15.46 **Given:** table of $[CH_3Cl]$ and $[Cl_2]$ versus initial rate **Find:** rate law, *k*, and overall order

Conceptual Plan: Using general rate law form, compare rate ratios to determine reaction order of each reactant. Be sure to choose data that changes only one concentration at a time.

$$\frac{\text{Rate 2}}{\text{Rate 1}} = \frac{k[CH_3Cl]_2^m[Cl_2]_2^n}{k[CH_3Cl]_1^m[Cl_2]_1^n}$$

Then use one of the concentration/initial rate pairs to determine *k*.

$$\text{Rate} = k[CH_3Cl]^m[Cl_2]^n$$

Solution: $\frac{\text{Rate 2}}{\text{Rate 1}} = \frac{k[CH_3Cl]_2^m[Cl_2]_2^n}{k[CH_3Cl]_1^m[Cl_2]_1^n}$. Comparing the first two sets of data,

$\frac{0.029\ \cancel{\text{M/s}}}{0.014\ \cancel{\text{M/s}}} = \frac{\cancel{k}(0.100\ \cancel{\text{M}})^m(\cancel{0.050\ \text{M}})^n}{\cancel{k}(0.050\ \cancel{\text{M}})^m(\cancel{0.050\ \text{M}})^n}$ and $2.\underline{0}714 = 2^m$; so $m = 1$. If we compare the second and third data sets, $\frac{0.041\ \cancel{\text{M/s}}}{0.029\ \cancel{\text{M/s}}} = \frac{\cancel{k}(\cancel{0.100\ \text{M}})^m(0.100\ \cancel{\text{M}})^n}{\cancel{k}(\cancel{0.100\ \text{M}})^m(0.050\ \cancel{\text{M}})^n}$ and $1.\underline{4}14 = 2^n$; so $n = 1/2$. Other comparisons can be made, but they are not necessary. They should reinforce these values of the reaction orders. So Rate $= k[CH_3Cl][Cl_2]^{1/2}$. Selecting the last data set and rearranging the rate equation,

$$k = \frac{\text{Rate}}{[\text{CH}_3\text{Cl}][\text{Cl}_2]^{1/2}} = \frac{0.115\frac{\cancel{\text{M}}}{\text{s}}}{(0.200\,\cancel{\text{M}})(0.200\,\text{M})^{1/2}} = 1.29\,\text{M}^{-1/2}\cdot\text{s}^{-1};\ \text{so Rate} = 1.29\,\text{M}^{-1/2}\cdot\text{s}^{-1}[\text{CH}_3\text{Cl}][\text{Cl}_2]^{1/2},$$

and the reaction is one-and-a-half order overall.

Check: The units (none and $\text{M}^{-1/2}\cdot\text{s}^{-1}$) are correct. The rate law is not as common as others, but it is reasonable. The rate is changing as rapidly as the CH_3Cl concentration is changing, which is consistent with first order in this reactant. The rate is changing a bit more slowly than the Cl_2 concentration, which is consistent with half order in this reactant. The rate constant is consistent with the units necessary to get rate as M/s, and the magnitude is reasonable because we have a one-and-a-half order reaction.

The Integrated Rate Law and Half-Life

15.47 (a) The reaction is zero order. Because the slope of the plot is independent of the concentration, there is no dependence of the concentration of the reactant in the rate law.

(b) The reaction is first order. The expression for the half-life of a first-order reaction is $t_{1/2} = \frac{0.693}{k}$, which is independent of the reactant concentration.

(c) The reaction is second order. The integrated rate expression for a second-order reaction is $\frac{1}{[\text{A}]_t} = kt + \frac{1}{[\text{A}]_0}$, which is linear when the inverse of the concentration is plotted versus time.

15.48 (a) The reaction is second order. The expression for the half-life of a second-order reaction, $t_{1/2} = \frac{1}{k[\text{A}]_0}$, shows that the half-life decreases as concentration increases.

(b) The reaction is first order. The integrated rate expression for a first-order reaction is $\ln[\text{A}]_t = -kt + \ln[\text{A}]_0$, which is linear when the natural log of the concentration is plotted versus time.

(c) The reaction is zero order. The expression for the half-life of a zero-order reaction, $t_{1/2} = \frac{[\text{A}]_0}{2k}$, shows that the half-life increases as concentration increases.

15.49 **Given:** table of [AB] versus time **Find:** reaction order, k, and [AB] at 25 s

Conceptual Plan: Look at the data and see if any common reaction orders can be eliminated. If the data does not show an equal concentration drop with time, zero order can be eliminated. Look for changes in the half-life (compare time for concentration to drop to one-half of any value). If the half-life is not constant, the first order can be eliminated. If the half-life is getting longer as the concentration drops, this might suggest second order. Plot the data as indicated by the appropriate rate law. Determine k from the slope of the plot. Finally, calculate the [AB] at 25 s by using the appropriate integrated rate expression.

Solution: By the preceding logic, we can eliminate both the zero-order and first-order reactions. (Alternatively, you could make all three plots and only one should be linear.) This suggests that we should have a second-order reaction. Plot 1/[AB] versus time.

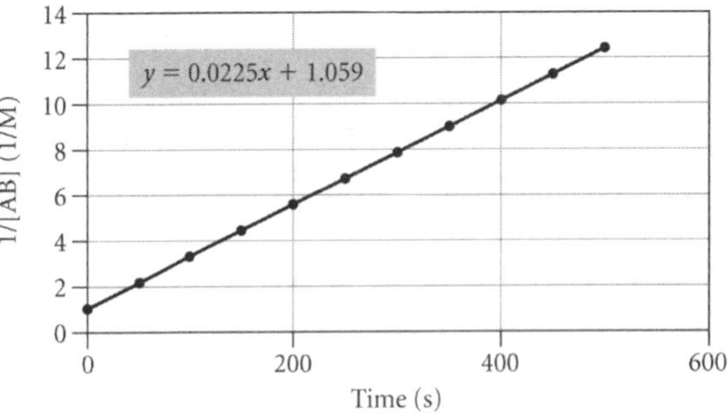

Because $\frac{1}{[\text{AB}]_t} = kt + \frac{1}{[\text{AB}]_0}$, the slope will be the rate constant. The slope can be determined by measuring $\Delta y/\Delta x$ on the plot or by using functions such as "add trendline" in Excel. Thus, the rate constant is $0.0225\,\text{M}^{-1}\cdot\text{s}^{-1}$, and the rate law is Rate $= 0.0225\,\text{M}^{-1}\cdot\text{s}^{-1}[\text{AB}]^2$.

Finally, use $\frac{1}{[\text{AB}]_t} = kt + \frac{1}{[\text{AB}]_0}$; substitute the values of $[\text{AB}]_0$, 25 s, and k; and rearrange to solve for [AB] at 25 s. $[\text{AB}]_t = \frac{1}{kt + \frac{1}{[\text{AB}]_0}} = \frac{1}{(0.0225\,\text{M}^{-1}\cdot\cancel{\text{s}^{-1}})(25\,\cancel{\text{s}}) + \left(\frac{1}{0.950\,\text{M}}\right)} = 0.619\,\text{M}$

Check: The units (none, $\text{M}^{-1}\cdot\text{s}^{-1}$, and M) are correct. The rate law is a common form. The plot was linear, confirming second-order kinetics. The rate constant is consistent with the units necessary to get the rate as M/s, and the magnitude is reasonable because we have a second-order reaction. The [AB] at 25 s is between the values at 0 s and 50 s.

15.50 **Given:** table of $[N_2O_5]$ versus time **Find:** reaction order, k, and $[N_2O_5]$ at 250 s

Conceptual Plan: Look at the data and see if any common reaction orders can be eliminated. If the data does not show an equal concentration drop with time, zero order can be eliminated. Look for changes in the half-life (compare time for concentration to drop to one-half of any value). If the half-life is not constant, the first order can be eliminated. If the half-life is getting longer as the concentration drops, this might suggest second order. Plot the data as indicated by the appropriate rate law. Determine k from the slope of the plot. Finally, calculate the $[N_2O_5]$ at 250 s by using the appropriate integrated rate expression.

Solution: By the preceding logic, we can see that the reaction is most likely first order. It takes just under 75 s for the concentration to be cut in half for any concentration. Plot $\ln[N_2O_5]$ versus time. Because $\ln[A]_t = -kt + \ln[A]_0$, the negative of the slope will be the rate constant. The slope can be determined by measuring $\Delta y/\Delta x$ on the plot or by using functions such as "add trendline" in Excel. Thus, the rate constant is $0.00780\ s^{-1}$, and the rate law is Rate $= 0.00780\ s^{-1}\ [N_2O_5]$. Finally, use $\ln[N_2O_5]_t = -kt + \ln[N_2O_5]_0$; substitute the values of $[N_2O_5]_0$, 250 s, and k; and rearrange to solve for $[N_2O_5]$ at 250 s. $\ln[N_2O_5]_{250\ s} = -(0.00780\ \cancel{s^{-1}})(250\ \cancel{s}) + \ln[\cancel{1.000}]_0$ then $[N_2O_5]_{250\ s} = e^{-1.95} = 0.142\ M$

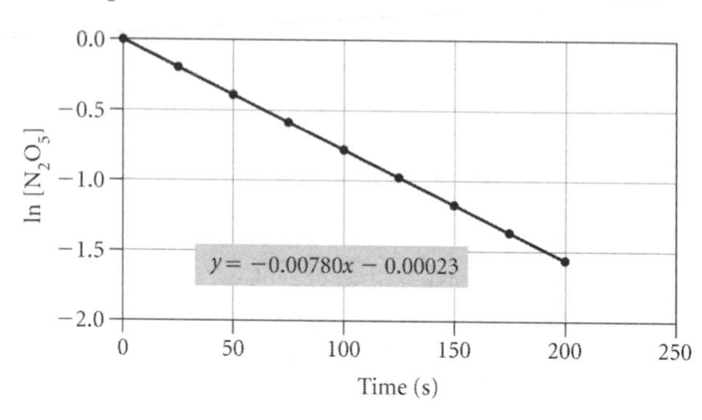

Check: The units (none, s^{-1}, and M) are correct. The rate law is a common form. The plot was linear, confirming first-order kinetics. The rate constant is consistent with the units necessary to get the rate as M/s, and the magnitude is reasonable because we have a first-order reaction. The $[N_2O_5]$ at 250 s is less than the value at 200 s.

15.51 **Given:** table of $[C_4H_8]$ versus time **Find:** reaction order, k, and reaction rate when $[C_4H_8] = 0.25\ M$

Conceptual Plan: Look at the data and see if any common reaction orders can be eliminated. If the data does not show an equal concentration drop with time, zero order can be eliminated. Look for changes in half-life (compare time for concentration to drop to one-half of any value). If the half-life is not constant, the first order can be eliminated. If the half-life is getting longer as the concentration drops, this might suggest second order. Plot the data as indicated by the appropriate rate law. Determine k from the slope of the plot. Finally, calculate the reaction rate when $[C_4H_8] = 0.25\ M$ by using the rate law.

Solution: By the preceding logic, we can see that the reaction is most likely first order. It takes about 60 s for any concentration to be cut to half its value. Plot $\ln[C_4H_8]$ versus time. Because $\ln[A]_t = -kt + \ln[A]_0$, the negative of the slope will be the rate constant. The slope can be determined by measuring $\Delta y/\Delta x$ on the plot or by using functions such as "add trendline" in Excel. Thus, the rate constant is $0.0112\ s^{-1}$, and the rate law is Rate $= 0.0112\ s^{-1}\ [C_4H_8]$. Finally, use Rate $= 0.0112\ s^{-1}[C_4H_8]$ and substitute the values of $[C_4H_8]$: Rate $= 0.0112\ s^{-1}[0.25\ M] = 2.8 \times 10^{-3}\ M \cdot s^{-1}$

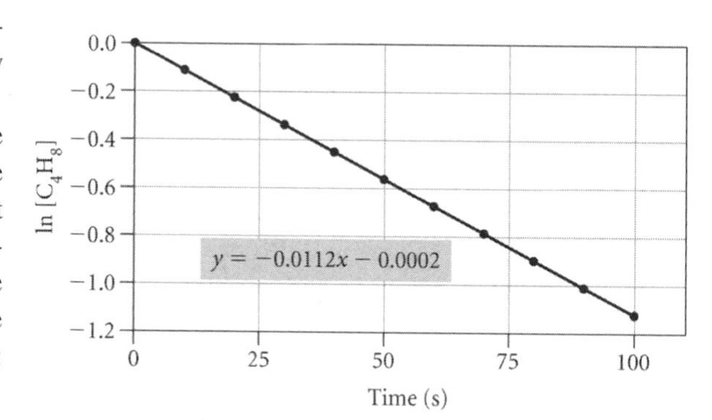

Check: The units (none, s^{-1}, and $M \cdot s^{-1}$) are correct. The rate law is a common form. The plot was linear, confirming first-order kinetics. The rate constant is consistent with the units necessary to get rate as M/s, and the magnitude is reasonable because we have a first-order reaction. The rate when $[C_4H_8] = 0.25\ M$ is consistent with the average rate using 90 s and 100 s.

15.52 **Given:** table of [A] versus time **Find:** reaction order, k, and reaction rate when $[A] = 0.10\ M$

Conceptual Plan: Look at the data and see if any common reaction orders can be eliminated. If the data does not show an equal concentration drop with time, zero order can be eliminated. Look for changes in half-life (compare time for concentration to drop to one-half of any value). If the half-life is not constant, the first order can be eliminated. If the half-life is getting longer as the concentration drops, this might suggest second order. Plot the data as indicated by the appropriate rate law. Determine k from the slope of the plot. Finally, calculate the reaction rate when $[A] = 0.10\ M$ by using the rate law.

Solution: By the preceding logic, we can see that the reaction is most likely zero order. There is a difference of about 0.085 M between each data point, so the rate is independent of the [A]. Plot [A] versus time. Because $[A]_t = -kt + [A]_0$, the negative of the slope will be the rate constant. The slope can be determined by measuring $\Delta y/\Delta x$ on the plot or by using functions such as "add trendline" in Excel. Thus, the rate constant is $3.41 \times 10^{-3}\ M \cdot s^{-1}$, and the rate law is Rate $= 3.41 \times 10^{-3}\ M \cdot s^{-1}$. Finally, because the rate is independent of concentration, Rate $= 3.41 \times 10^{-3}\ M \cdot s^{-1}$ at 0.10 M and all other concentrations. Note: A plot is not necessary because the kinetics are so simple.

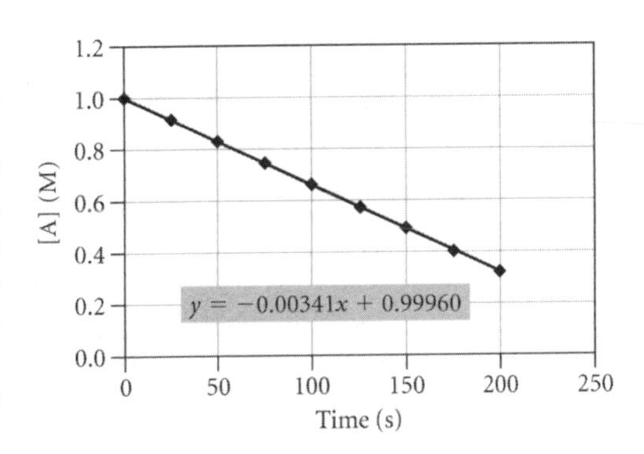

Check: The units (none, $M \cdot s^{-1}$, and $M \cdot s^{-1}$) are correct. The rate law is a common form. The plot was linear, confirming zero-order kinetics. The rate constant is consistent with the units necessary to get rate as M/s, and the magnitude is reasonable because we have a zero-order reaction. The rate is the same as any average rate that can be calculated using the data.

15.53 **Given:** plot of ln[A] versus time has slope $= -0.0045/s$; $[A]_0 = 0.250$ M **Find:** (a) k, (b) rate law, (c) $t_{1/2}$, and (d) [A] after 225 s

Conceptual Plan:

(a) **A plot of ln[A] versus time is linear for a first-order reaction. Using $\ln [A]_t = -kt + \ln [A]_0$, the rate constant is the negative of the slope.**

(b) **Rate law is first order. Add rate constant from part (a).**

(c) **For a first-order reaction, $t_{1/2} = \frac{0.693}{k}$. Substitute k from part (a).**

(d) **Use the integrated rate law, $\ln [A]_t = -kt + \ln [A]_0$, and substitute k and the initial concentration.**

Solution:

(a) Because the rate constant is the negative of the slope, $k = 4.5 \times 10^{-3}\ s^{-1}$.

(b) Because the reaction is first order, Rate $= k[A] = 4.5 \times 10^{-3}\ s^{-1}\ [A]$.

(c) $t_{1/2} = \frac{0.693}{k} = \frac{0.693}{0.0045/s} = 1.5 \times 10^2\ s$

(d) $\ln[A]_t = -kt + \ln[A]_0$ and substitute k and the initial concentration. So $\ln[A]_t = -(0.0045/\cancel{s})(225\ \cancel{s}) + \ln 0.250\ M = -2.39879$ and $[A]_{250\ s} = e^{-2.39879} = 0.0908$ M

Check: The units (s^{-1}, none, s, and M) are correct. The rate law is a common form. The rate constant is consistent with value of the slope. The half-life is consistent with a small value of k. The concentration at 225 s is consistent with being between one and two half-lives.

15.54 **Given:** plot of 1/[AB] versus time has slope $= 0.055/M\ s$; $[A]_0 = 0.250$ M
Find: (a) k, (b) rate law, (c) $t_{1/2}$ when $[AB]_0 = 0.55$ M, and (d) [A] and [B] after 75 s

Conceptual Plan:

(a) **A plot of 1/[AB] versus time is linear for a second-order reaction. Using $\frac{1}{[AB]_t} = kt + \frac{1}{[AB]_0}$, the rate constant is the slope.**

(b) **Rate law is second order. Add rate constant from part (a).**

(c) **For a second-order reaction, $t_{1/2} = \frac{1}{k[AB]_0}$. Substitute k from part (a).**

(d) **Use the integrated rate law, $\frac{1}{[AB]_t} = kt + \frac{1}{[AB]_0}$, and substitute k, t, and the initial concentration to get the [AB]. Then $[AB]_0, [AB]_{75} \rightarrow [A], [B]$.**

$$\Delta[AB] = [AB]_{0\,s} - [AB]_{75\,s} \text{ with } \frac{1\ \text{mol A}}{1\ \text{mol AB}} \text{ and } \frac{1\ \text{mol B}}{1\ \text{mol AB}}$$

Solution:

(a) Because the rate constant is the slope, $k = 5.5 \times 10^{-2}\ M^{-1} \cdot s^{-1}$.

(b) Because the reaction is second order, Rate $= k[A]^2 = 5.5 \times 10^{-2}\ M^{-1} \cdot s^{-1}\ [AB]^2$.

(c) $t_{1/2} = \frac{1}{k[AB]_0}$ so $t_{1/2} = \frac{1}{(5.5 \times 10^{-2}\ \cancel{M^{-1}} \cdot s^{-1})(0.550\ \cancel{M})} = 33\ s$

(d) $\frac{1}{[\text{AB}]_t} = kt + \frac{1}{[\text{AB}]_0}$ so $[\text{AB}]_t = \frac{1}{kt + \frac{1}{[\text{AB}]_0}} = \frac{1}{(5.5 \times 10^{-2}\ \text{M}^{-1} \cdot \cancel{\text{s}^{-1}})(75\ \cancel{\text{s}}) + \left(\frac{1}{0.250\ \text{M}}\right)} = 0.12\underline{3}08\ \text{M}$

then $\Delta[\text{AB}] = [\text{AB}]_{0\,\text{s}} - [\text{AB}]_{75\,\text{s}} = 0.250\ \text{M} - 0.12\underline{3}08\ \text{M} = 0.12\underline{6}92\ \text{M AB}$

so $0.12\underline{6}92 \frac{\cancel{\text{mol AB}}}{\text{L}} \times \frac{1\ \text{mol A}}{1\ \cancel{\text{mol AB}}} = 0.13\ \text{M A}$ and $0.12\underline{6}92 \frac{\cancel{\text{mol AB}}}{\text{L}} \times \frac{1\ \text{mol B}}{1\ \cancel{\text{mol AB}}} = 0.13\ \text{M B}$

Check: The units ($\text{M}^{-1} \cdot \text{s}^{-1}$, none, s, and M) are correct. The rate law is a common form. The rate constant is consistent with value of the slope. The half-life is consistent with a small value of k. The concentration at 75 s is consistent with being about one half-life.

15.55 **Given:** decomposition of SO_2Cl_2, first order; $k = 1.42 \times 10^{-4}\ \text{s}^{-1}$
Find: (a) $t_{1/2}$, (b) t to decrease to 25% of $[SO_2Cl_2]_0$, (c) t to 0.78 M when $[SO_2Cl_2]_0 = 1.00$ M, and (d) $[SO_2Cl_2]$ after 2.00×10^2 s and 5.00×10^2 s when $[SO_2Cl_2]_0 = 0.150$ M
Conceptual Plan:

(a) $k \rightarrow t_{1/2}$

$t_{1/2} = \frac{0.693}{k}$

(b) $[SO_2Cl_2]_0$, 25% of $[SO_2Cl_2]_0, k \rightarrow t$

$\ln[\text{A}]_t = -kt + \ln[\text{A}]_0$

(c) $[SO_2Cl_2]_0, [SO_2Cl_2]_t, k \rightarrow t$

$\ln[\text{A}]_t = -kt + \ln[\text{A}]_0$

(d) $[SO_2Cl_2]_0, t, k \rightarrow [SO_2Cl_2]_t$

$\ln[\text{A}]_t = -kt + \ln[\text{A}]_0$

Solution:

(a) $t_{1/2} = \frac{0.693}{k} = \frac{0.693}{1.42 \times 10^{-4}\ \text{s}^{-1}} = 4.88 \times 10^3\ \text{s}$

(b) $[SO_2Cl_2]_t = 0.25\,[SO_2Cl_2]_0$ Because $\ln[SO_2Cl_2]_t = -kt + \ln[SO_2Cl_2]_0$, rearrange to solve for t.

$t = -\frac{1}{k}\ln\frac{[SO_2Cl_2]_t}{[SO_2Cl_2]_0} = -\frac{1}{1.42 \times 10^{-4}\ \text{s}^{-1}}\ln\frac{0.25\,\cancel{[SO_2Cl_2]_0}}{\cancel{[SO_2Cl_2]_0}} = 9.8 \times 10^3\ \text{s}$

(c) $[SO_2Cl_2]_t = 0.78$ M; $[SO_2Cl_2]_0 = 1.00$ M Because $\ln[SO_2Cl_2]_t = -kt + \ln[SO_2Cl_2]_0$, rearrange to solve for t. $t = -\frac{1}{k}\ln\frac{[SO_2Cl_2]_t}{[SO_2Cl_2]_0} = -\frac{1}{1.42 \times 10^{-4}\ \text{s}^{-1}}\ln\frac{0.78\ \cancel{\text{M}}}{1.00\ \cancel{\text{M}}} = 1.7 \times 10^3\ \text{s}$

(d) $[SO_2Cl_2]_0 = 0.150$ M and $t = 2.00 \times 10^2$ s

$\ln[SO_2Cl_2]_t = -(1.42 \times 10^{-4}\ \cancel{\text{s}^{-1}})(2.00 \times 10^2\ \cancel{\text{s}}) + \ln 0.150\ \text{M} = -1.9\underline{2}552 \rightarrow$

$[SO_2Cl_2]_t = e^{-1.9\underline{2}552} = 0.146\ \text{M}$

$[SO_2Cl_2]_0 = 0.150$ M and $t = 5.00 \times 10^2$ s

$\ln[SO_2Cl_2]_t = -(1.42 \times 10^{-4}\ \cancel{\text{s}^{-1}})(5.00 \times 10^2\ \cancel{\text{s}}) + \ln 0.150\ \text{M} = -1.9\underline{6}812 \rightarrow$

$[SO_2Cl_2]_t = e^{-1.9\underline{6}812} = 0.140\ \text{M}$

Check: The units (s, s, s, and M) are correct. The rate law is a common form. The half-life is consistent with a small value of k. The time to 25% is consistent with two half-lives. The time to 0.78 M is consistent with being less than one half-life. The final concentrations are consistent with the time being less than one half-life.

15.56 **Given:** decomposition of XY, second order in XY; $k = 7.02 \times 10^{-3}\ \text{M}^{-1} \cdot \text{s}^{-1}$
Find: (a) $t_{1/2}$ when $[\text{XY}]_0 = 0.100$ M, (b) t to decrease to 12.5% of $[\text{XY}]_0 = 0.100$ M and 0.200 M, (c) t to 0.062 M when $[\text{XY}]_0 = 0.150$ M, and (d) [XY] after 5.0×10^1 s and 5.50×10^2 s when $[\text{XY}]_0 = 0.050$ M
Conceptual Plan:

(a) $[\text{XY}]_0, k \rightarrow t_{1/2}$

$t_{1/2} = \frac{1}{k[\text{A}]_0}$

(b) $[XY]_0$, **12.5% of** $[XY]_0, k \rightarrow t$

$$\frac{1}{[A]_t} = kt + \frac{1}{[A]_0}$$

(c) $[XY]_0, [XY]_t, k \rightarrow t$

$$\frac{1}{[A]_t} = kt + \frac{1}{[A]_0}$$

(d) $[XY]_0, t, k \rightarrow [XY]_t$

$$\frac{1}{[A]_t} = kt + \frac{1}{[A]_0}$$

Solution:

(a) $$t_{1/2} = \frac{1}{k[XY]_0} = \frac{1}{(7.02 \times 10^{-3}\ \cancel{M^{-1}}\ s^{-1})(0.100\ \cancel{M})} = 1.42 \times 10^3\ s$$

(b) $[XY]_t = 0.125\,[XY]_0 = 0.125 \times 0.100\ M = 0.0125\ M$ Because $\frac{1}{[XY]_t} = kt + \frac{1}{[XY]_0}$, rearrange to solve for t.

$$t = \frac{1}{k}\left(\frac{1}{[XY]_t} - \frac{1}{[XY]_0}\right) = \frac{1}{(7.02 \times 10^{-3}\ \cancel{M^{-1}} \cdot s^{-1})}\left(\frac{1}{0.0125\ \cancel{M}} - \frac{1}{0.100\ \cancel{M}}\right) = 9.97 \times 10^3\ s \text{ and}$$

$[XY]_t = 0.125\,[XY]_0 = 0.125 \times 0.200\ M = 0.0250\ M$ Because $\frac{1}{[XY]_t} = kt + \frac{1}{[XY]_0}$, rearrange to solve for t.

$$t = \frac{1}{k}\left(\frac{1}{[XY]_t} - \frac{1}{[XY]_0}\right) = \frac{1}{(7.02 \times 10^{-3}\ \cancel{M^{-1}} \cdot s^{-1})}\left(\frac{1}{0.0250\ \cancel{M}} - \frac{1}{0.200\ \cancel{M}}\right) = 4.99 \times 10^3\ s$$

(c) $[XY]_t = 0.062$; $[XY]_0 = 0.150$ M. Because $\frac{1}{[XY]_t} = kt + \frac{1}{[XY]_0}$ rearrange to solve for t.

$$t = \frac{1}{k}\left(\frac{1}{[XY]_t} - \frac{1}{[XY]_0}\right) = \frac{1}{(7.02 \times 10^{-3}\ \cancel{M^{-1}} \cdot s^{-1})}\left(\frac{1}{0.062\ \cancel{M}} - \frac{1}{0.150\ \cancel{M}}\right) = 1.3 \times 10^3\ s$$

(d) $[XY]_0 = 0.050$ M and 5.0×10^1 s in $\frac{1}{[XY]_t} = kt + \frac{1}{[XY]_0} \rightarrow$

$$\frac{1}{[XY]_t} = (7.02 \times 10^{-3}\ M^{-1} \cdot \cancel{s^{-1}})(5.0 \times 10^1\ \cancel{s}) + \frac{1}{0.050\ M} = \frac{20.\underline{3}51}{M} \text{ so } [XY] = 0.049\ M \text{ and}$$

$[XY]_0 = 0.050$ M and 5.50×10^2 s in $\frac{1}{[XY]_t} = (7.02 \times 10^{-3}\ M^{-1} \cdot \cancel{s^{-1}})(5.50 \times 10^2\ \cancel{s}) + \frac{1}{0.050\ M} = \frac{23.\underline{8}61}{M}$

so $[XY] = 0.042$ M

Check: The units (s, s, s, s, M, and M) are correct. The rate law is a common form. The half-life is consistent with a small value of k. The time to 12.5% is consistent with three half-lives, where the half-life time is increasing. The next time (5000 s) is shorter because the initial concentration is higher. The last time is the shortest because it is less than a half-life with an intermediate concentration. The final concentrations are consistent with the time being much less than one half-life.

15.57 **Given:** $t_{1/2}$ for radioactive decay of U-238 = 4.5 billion years and independent of $[\text{U-238}]_0$
Find: t to decrease by 10%; number U-238 atoms today, when 1.5×10^8 atoms formed 13.8 billion years ago
Conceptual Plan: $t_{1/2}$ **independent of concentration implies first-order kinetics,** $t_{1/2} \rightarrow k$ **then**

$$t_{1/2} = \frac{0.693}{k}$$

90% of $[\text{U-238}]_0, k \rightarrow t$ **and** $[\text{U-238}]_0, t, k \rightarrow [\text{U-238}]_t$

$$\ln[A]_t = -kt + \ln[A]_0 \qquad \ln[A]_t = -kt + \ln[A]_0$$

Solution: $t_{1/2} = \frac{0.693}{k}$ Rearrange to solve for k. $k = \frac{0.693}{t_{1/2}} = \frac{0.693}{4.5 \times 10^9\ yr} = 1.\underline{5}4 \times 10^{-10}\ yr^{-1}$ then

$[\text{U-238}]_t = 0.10\,[\text{U-238}]_0$ Because $\ln[\text{U-238}]_t = -kt + \ln[\text{U-238}]_0$, rearrange to solve for t.

$$t = -\frac{1}{k}\ln\frac{[\text{U-238}]_t}{[\text{U-238}]_0} = -\frac{1}{1.\underline{5}4 \times 10^{-10}\ yr^{-1}}\ln\frac{0.90\ \cancel{[\text{U-238}]_0}}{\cancel{[\text{U-238}]_0}} = 6.8 \times 10^8\ yr$$

and $[\text{U-238}]_0 = 1.15 \times 10^{18}$ atoms; $t = 13.8 \times 10^9$ yr

$\ln[\text{U-238}]_t = -kt + \ln[\text{U-238}]_0 = -(1.5\underline{4} \times 10^{-10}\ \cancel{\text{yr}^{-1}})(13.8 \times 10^9\ \cancel{\text{yr}})$
$+\ln(1.5 \times 10^{18}\ \text{atoms}) = 39.7\underline{2}6797 \rightarrow$
$[\text{U-238}]_t = e^{39.7\underline{2}6797} = 1.8 \times 10^{17}\ \text{atoms}$

Check: The units (yr and atoms) are correct. The time to 10% decay is consistent with less than one half-life. The final concentration is consistent with the time being about three half-lives.

15.58 **Given:** $t_{1/2}$ for radioactive decay of C-14 = 5730 years
Find: t to decrease by 25%; mmol C-14 atoms left, after 2255 yr in sample initially contains 1.5 mmol C-14
Conceptual Plan: Radioactive decay implies first-order kinetics, $t_{1/2} \rightarrow k$ then

$$t_{1/2} = \frac{0.693}{k}$$

75% of $[\text{C-14}]_0$, $k \rightarrow t$ and $[\text{C-14}]_0, t, k \rightarrow [\text{C-14}]_t$

$\ln[\text{A}]_t = -kt + \ln[\text{A}]_0 \quad \ln[\text{A}]_t = -kt + \ln[\text{A}]_0$

Solution: $t_{1/2} = \dfrac{0.693}{k}$ Rearrange to solve for k. $k = \dfrac{0.693}{t_{1/2}} = \dfrac{0.693}{5730\ \text{yr}} = 1.2\underline{0}942 \times 10^{-4}\ \text{yr}^{-1}$ then

$[\text{C-14}]_t = 0.75\,[\text{C-14}]_0$ Because $\ln[\text{C-14}]_t = -kt + \ln[\text{C-14}]_0$, rearrange to solve for t.

$$t = -\frac{1}{k}\ln\frac{[\text{C-14}]_t}{[\text{C-14}]_0} = -\frac{1}{1.2\underline{0}942 \times 10^{-4}\ \text{yr}^{-1}}\ln\frac{0.75\cancel{[\text{C-14}]_0}}{\cancel{[\text{C-14}]_0}} = 2.4 \times 10^3\ \text{yr and } [\text{C-14}]_0 = 1.5\ \text{mmol}$$

$t = 2255$ yr
$\ln[\text{C-14}]_t = -kt + \ln[\text{C-14}]_0 = -(1.2\underline{0}942 \times 10^{-4}\ \cancel{\text{yr}^{-1}})(2255\ \cancel{\text{yr}}) + \ln(1.5\ \text{mmol}) = 0.13\underline{2}741 \rightarrow$
$[\text{C-14}]_t = e^{0.13\underline{2}741} = 1.1\ \text{mmol}$

Check: The units (yr and mmol) are correct. The time to 25% decay is consistent with less than one half-life. The final concentration is consistent with the time being less than one half-life.

The Effect of Temperature and the Collision Model

15.59

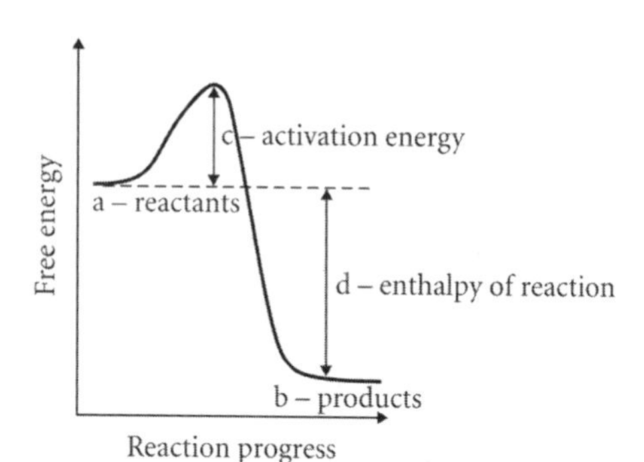

15.60

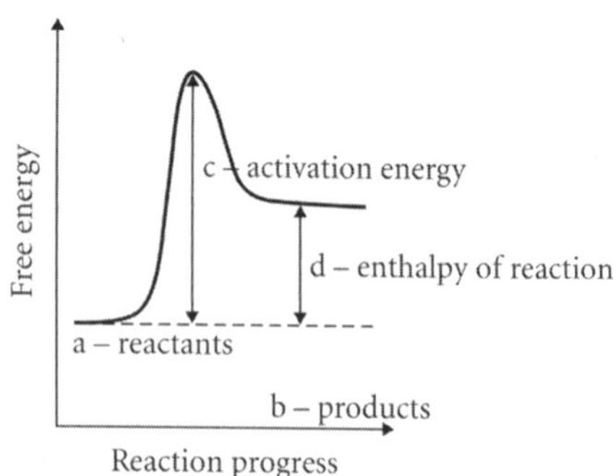

15.61 **Given:** activation energy = 56.8 kJ/mol, frequency factor = 1.5×10^{11}/s, 25 °C **Find:** rate constant
Conceptual Plan: °C → K and kJ/mol → J/mol then $E_a, T, A \rightarrow k$

$$\text{K} = {}^\circ\text{C} + 273.15 \qquad \frac{1000\ \text{J}}{1\ \text{kJ}} \qquad k = Ae^{-E_a/RT}$$

Solution: $T = 25\,°\text{C} + 273.15 = 298\text{ K}$ and $\frac{56.8\text{ kJ}}{\text{mol}} \times \frac{1000\text{ J}}{1\text{ kJ}} = 5.68 \times 10^4 \frac{\text{J}}{\text{mol}}$ then

$$k = Ae^{-E_a/RT} = (1.5 \times 10^{11}\text{ s}^{-1})e^{\frac{-5.68 \times 10^4 \frac{\text{J}}{\text{mol}}}{\left(8.314 \frac{\text{J}}{\text{K}\cdot\text{mol}}\right)298\text{ K}}} = 17\text{ s}^{-1}$$

Check: The units (s^{-1}) are correct. The rate constant is consistent with a large activation energy and a large frequency factor.

15.62 **Given:** 32°C, rate constant = 0.055/s, and frequency factor = 1.5×10^{13}/s **Find:** activation energy

Conceptual Plan: °C → K then $E_a, T, A \rightarrow k$ then J/mol → kJ/mol

$\text{K} = °\text{C} + 273.15$ $\quad k = Ae^{-E_a/RT}$ $\quad \frac{1\text{ kJ}}{1000\text{ J}}$

Solution: $T = 32\,°\text{C} + 273.15 = 305\text{ K}$ then $k = Ae^{-E_a/RT}$ Rearrange to solve for E_a.

$$E_a = -RT\ln\left(\frac{k}{A}\right) = -8.314 \frac{\text{J}}{\text{K mol}} \times 305\text{ K} \times \ln\left(\frac{0.055\text{ s}^{-1}}{1.2 \times 10^{13}\text{ s}^{-1}}\right) = 8.37 \times 10^4 \frac{\text{J}}{\text{mol}} \times \frac{1\text{ kJ}}{1000\text{ J}} = 83.7 \frac{\text{kJ}}{\text{mol}}$$

Check: The units (kJ/mol) are correct. The activation energy is consistent with a modest rate constant and a large frequency factor.

15.63 **Given:** plot of ln k versus $1/T$ (in K) is linear with a slope of −7445 K **Find:** E_a

Conceptual Plan: Because $\ln k = \frac{-E_a}{R}\left(\frac{1}{T}\right) + \ln A$ **plot of ln k versus $1/T$ will have a *slope* = $-E_a/R$.**

Solution: Because the *slope* = −7445 K = $-E_a/R$,

$$E_a = -(slope)R = -(-7445\text{ K})\left(8.314 \frac{\text{J}}{\text{K}\cdot\text{mol}}\right)\left(\frac{1\text{ kJ}}{1000\text{ J}}\right) = 61.90 \frac{\text{kJ}}{\text{mol}}$$

Check: The units (kJ/mol) are correct. The activation energy is typical for many reactions.

15.64 **Given:** Plot of ln k versus $1/T$ (in K) is linear with a slope of -1.01×10^4 K. **Find:** E_a

Conceptual Plan: Because $\ln k = \frac{-E_a}{R}\left(\frac{1}{T}\right) + \ln A$, **a plot of ln k versus $1/T$ will have a *slope* = $-E_a/R$.**

Solution: Because the *slope* = -1.01×10^4 K = $-E_a/R$,

$$E_a = -(slope)R = -(-1.01 \times 10^4\text{ K})\left(8.314 \frac{\text{J}}{\text{K}\cdot\text{mol}}\right)\left(\frac{1\text{ kJ}}{1000\text{ J}}\right) = 84.0 \frac{\text{kJ}}{\text{mol}}.$$

Check: The units (kJ/mol) are correct. The activation energy is typical for many reactions.

15.65 **Given:** table of rate constant versus T **Find:** E_a and A

Conceptual Plan: Because $\ln k = \frac{-E_a}{R}\left(\frac{1}{T}\right) + \ln A$, **a plot of ln k versus $1/T$ will have a *slope* = $-E_a/R$ and an *intercept* = ln A.**

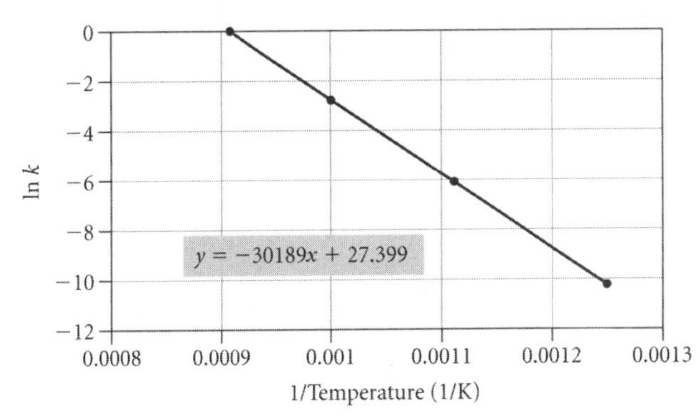

Solution: The slope can be determined by measuring $\Delta y/\Delta x$ on the plot or by using functions such as "add trendline" in Excel. Because the *slope* = $-30\underline{1}89$ K = $-E_a/R$,

$E_a = -(slope)R =$

$$-(-30\underline{1}89\text{ K})\left(8.314 \frac{\text{J}}{\text{K}\cdot\text{mol}}\right)\left(\frac{1\text{ kJ}}{1000\text{ J}}\right)$$

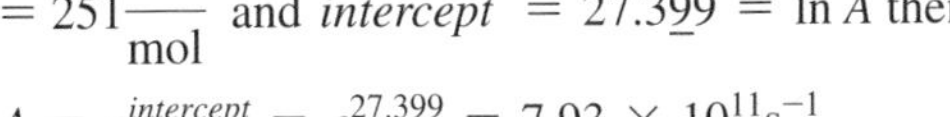

$= 251 \frac{\text{kJ}}{\text{mol}}$ and *intercept* = $27.3\underline{9}9$ = ln A then

$A = e^{intercept} = e^{27.3\underline{9}9} = 7.93 \times 10^{11}\text{s}^{-1}$.

Check: The units (kJ/mol and s^{-1}) are correct. The plot was linear, confirming Arrhenius behavior. The activation and frequency factor are typical for many reactions.

15.66 **Given:** table of rate constant versus T **Find:** E_a and A

Conceptual Plan: Because $\ln k = \frac{-E_a}{R}\left(\frac{1}{T}\right) + \ln A$**, a plot of** $\ln k$ **versus** $1/T$ **will have a** ***slope*** $= -E_a/R$ **and an** ***intercept*** $= \ln A$**.**

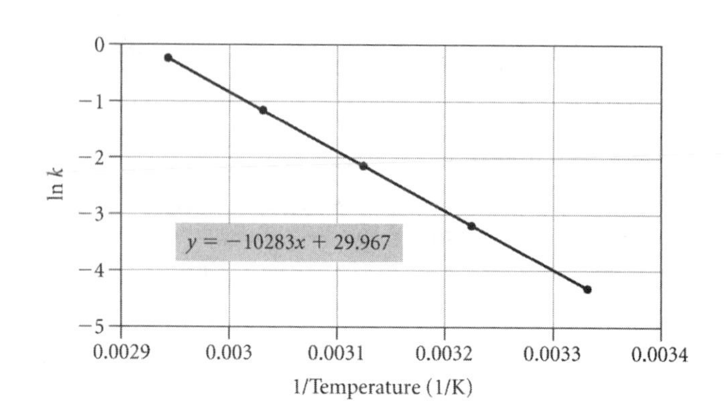

Solution: The slope can be determined by measuring $\Delta y/\Delta x$ on the plot or by using functions such as "add trendline" in Excel. Because the *slope* $= -10\underline{2}83\text{ K} = -E_a/R$, $E_a = -(slope)R =$

$$-(-10\underline{2}83\text{ K})\left(8.314\frac{\text{J}}{\text{K}\cdot\text{mol}}\right)\left(\frac{1\text{ kJ}}{1000\text{ J}}\right)$$ and

$= 85.5\frac{\text{kJ}}{\text{mol}}$ and *intercept* $= 29.9\underline{6}7 = \ln A$ then

$A = e^{intercept} = e^{29.9\underline{6}7} = 1.03 \times 10^{13}\text{ s}^{-1}$.

Check: The units (kJ/mol and s^{-1}) are correct. The plot was linear, confirming Arrhenius behavior. The activation and frequency factor are typical for many reactions.

15.67 **Given:** table of rate constant versus T **Find:** E_a and A

Conceptual Plan: Because $\ln k = \frac{-E_a}{R}\left(\frac{1}{T}\right) + \ln A$**, a plot of** $\ln k$ **versus** $1/T$ **will have a** ***slope*** $= -E_a/R$ **and an** ***intercept*** $= \ln A$**.**

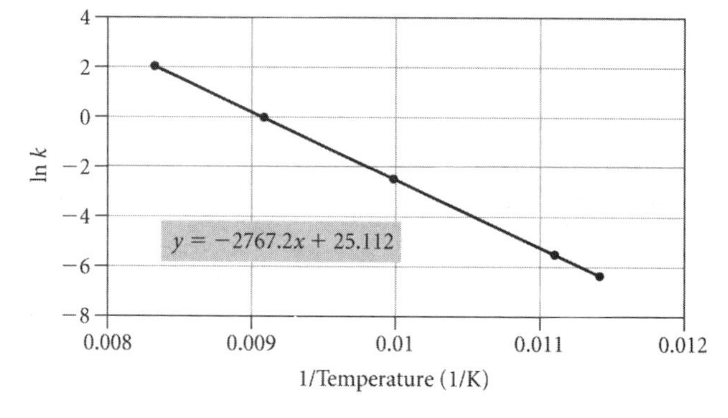

Solution: The slope can be determined by measuring $\Delta y/\Delta x$ on the plot or by using functions such as "add trendline" in Excel. Because the *slope* $= -27\underline{6}7.2\text{ K} = -E_a/R$, $E_a = -(slope)R$

$$= -(-27\underline{6}7.2\text{ K})\left(8.314\frac{\text{J}}{\text{K}\cdot\text{mol}}\right)\left(\frac{1\text{ kJ}}{1000\text{ J}}\right)$$

$= 23.0\frac{\text{kJ}}{\text{mol}}$ and *intercept* $= 25.1\underline{1}2 = \ln A$ then

$A = e^{intercept} = e^{25.1\underline{1}2} = 8.05 \times 10^{10}\text{ s}^{-1}$.

Check: The units (kJ/mol and s^{-1}) are correct. The plot was linear, confirming Arrhenius behavior. The activation and frequency factor are typical for many reactions.

15.68 **Given:** table of rate constant versus T **Find:** E_a and A

Conceptual Plan: Because $\ln k = \frac{-E_a}{R}\left(\frac{1}{T}\right) + \ln A$**, a plot of** $\ln k$ **versus** $1/T$ **will have a** ***slope*** $= -E_a/R$ **and an** ***intercept*** $= \ln A$**.**

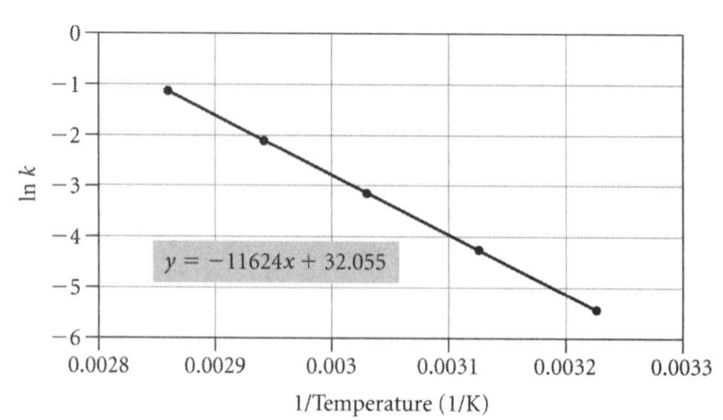

Solution: The slope can be determined by measuring $\Delta y/\Delta x$ on the plot or by using functions such as "add trendline" in Excel. Because the *slope* $= -11\underline{6}24\text{ K} = -E_a/R$, $E_a = -(slope)R$

$$= -(-11\underline{6}24\text{ K})\left(8.314\frac{\text{J}}{\text{K}\cdot\text{mol}}\right)\left(\frac{1\text{ kJ}}{1000\text{ J}}\right)$$

$= 96.6\frac{\text{kJ}}{\text{mol}}$ and *intercept* $= 32.0\underline{5}5 = \ln A$ then

$A = e^{intercept} = e^{32.0\underline{5}5} = 8.34 \times 10^{13}\text{ s}^{-1}$.

Check: The units (kJ/mol and s^{-1}) are correct. The plot was linear, confirming Arrhenius behavior. The activation and frequency factor are typical for many reactions.

15.69 **Given:** rate constant $= 0.0117/s$ at 400.0 K and 0.689/s at 450.0 K **Find:** (a) E_a and (b) rate constant at 425 K

Conceptual Plan:

(a) $k_1, T_1, k_2, T_2, \rightarrow E_a$ then J/mol $\rightarrow$ kJ/mol

$$\ln\left(\frac{k_2}{k_1}\right) = \frac{E_a}{R}\left(\frac{1}{T_1} - \frac{1}{T_2}\right) \qquad \frac{1\text{ kJ}}{1000\text{ J}}$$

(b) $E_a, k_1, T_1, T_2 \rightarrow k_2$

$$\ln\left(\frac{k_2}{k_1}\right) = \frac{E_a}{R}\left(\frac{1}{T_1} - \frac{1}{T_2}\right)$$

Solution:

(a) $\ln\left(\frac{k_2}{k_1}\right) = \frac{E_a}{R}\left(\frac{1}{T_1} - \frac{1}{T_2}\right)$ Rearrange to solve for E_a.

$$E_a = \frac{R\ln\left(\frac{k_2}{k_1}\right)}{\left(\frac{1}{T_1} - \frac{1}{T_2}\right)} = \frac{8.314\frac{\text{J}}{\text{K}\cdot\text{mol}}\ln\left(\frac{0.689\text{ s}^{-1}}{0.0117\text{ s}^{-1}}\right)}{\left(\frac{1}{400.0\text{ K}} - \frac{1}{450.0\text{ K}}\right)} = 1.22 \times 10^5\frac{\text{J}}{\text{mol}} \times \frac{1\text{ kJ}}{1000\text{ J}} = 122\frac{\text{kJ}}{\text{mol}}$$

(b) $\ln\left(\frac{k_2}{k_1}\right) = \frac{E_a}{R}\left(\frac{1}{T_1} - \frac{1}{T_2}\right)$ with $k_1, = 0.0117/s$, $T_1 = 400.0$ K, $T_2 = 425$ K Rearrange to solve for k_2.

$$\ln k_2 = \frac{E_a}{R}\left(\frac{1}{T_1} - \frac{1}{T_2}\right) + \ln k_1 = \frac{1.22 \times 10^5\frac{\text{J}}{\text{mol}}}{8.314\frac{\text{J}}{\text{K}\cdot\text{mol}}}\left(\frac{1}{400.0\text{ K}} - \frac{1}{425\text{ K}}\right) + \ln 0.0117\text{ s}^{-1} = -2.29\underline{0}2 \rightarrow$$

$$k_2 = e^{-2.29\underline{0}2} = 0.101\text{ s}^{-1}$$

Check: The units (kJ/mol and s^{-1}) are correct. The activation energy is typical for a reaction. The rate constant at 425 K is between the values given at 400 K and 450 K.

15.70 **Given:** rate constant $= 0.000122/s$ at 27 °C and 0.228/s at 77 °C **Find:** (a) E_a and (b) rate constant at 17 °C

Conceptual Plan:

(a) °C $\rightarrow$ K then $k_1, T_1, k_2, T_2 \rightarrow E_a$ then J/mol $\rightarrow$ kJ/mol

$$\text{K} = {}^\circ\text{C} + 273.15 \qquad \ln\left(\frac{k_2}{k_1}\right) = \frac{E_a}{R}\left(\frac{1}{T_1} - \frac{1}{T_2}\right) \qquad \frac{1\text{ kJ}}{1000\text{ J}}$$

(b) °C $\rightarrow$ K then $E_a, k_1, T_1, T_2 \rightarrow k_2$

$$\text{K} = {}^\circ\text{C} + 273.15 \qquad \ln\left(\frac{k_2}{k_1}\right) = \frac{E_a}{R}\left(\frac{1}{T_1} - \frac{1}{T_2}\right)$$

Solution: $T_1 = 27\,^\circ\text{C} + 273.15 = 300.$ K and $T_2 = 77\,^\circ\text{C} + 273.15 = 350.$ K then $\ln\left(\frac{k_2}{k_1}\right) = \frac{E_a}{R}\left(\frac{1}{T_1} - \frac{1}{T_2}\right)$

(a) Rearrange to solve for E_a.

$$E_a = \frac{R\ln\left(\frac{k_2}{k_1}\right)}{\left(\frac{1}{T_1} - \frac{1}{T_2}\right)} = \frac{8.314\frac{\text{J}}{\text{K}\cdot\text{mol}}\ln\left(\frac{0.228\text{ s}^{-1}}{0.000122\text{ s}^{-1}}\right)}{\left(\frac{1}{300.\text{ K}} - \frac{1}{350.\text{ K}}\right)} = 1.32 \times 10^5\frac{\text{J}}{\text{mol}} \times \frac{1\text{ kJ}}{1000\text{ J}} = 132\frac{\text{kJ}}{\text{mol}}$$

(b) $\ln\left(\frac{k_2}{k_1}\right) = \frac{E_a}{R}\left(\frac{1}{T_1} - \frac{1}{T_2}\right)$ with $k_1 = 0.000122/s$, $T_1 = 300.$ K, $T_2 = 17\,^\circ\text{C} + 273.15 = 290$ K

Rearrange to solve for k_2.

$$\ln k_2 = \frac{E_a}{R}\left(\frac{1}{T_1} - \frac{1}{T_2}\right) + \ln k_1 = \frac{1.32 \times 10^5\frac{\text{J}}{\text{mol}}}{8.314\frac{\text{J}}{\text{K}\cdot\text{mol}}}\left(\frac{1}{300.\text{ K}} - \frac{1}{290.\text{ K}}\right) + \ln 0.000122\text{ s}^{-1} = -10.8\underline{3}64 \rightarrow$$

$k_2 = e^{-10.8364} = 0.0000197\ \text{s}^{-1} = 1.97 \times 10^{-5}\ \text{s}^{-1}$

Check: The units (kJ/mol and s^{-1}) are correct. The activation energy is typical for a reaction. The rate constant at 17 °C is smaller than the values given at 27 °C.

15.71 **Given:** rate constant doubles from 10.0 °C to 20.0 °C **Find:** E_a

Conceptual Plan: °C → K then k_1, T_1, k_2, T_2 → E_a then J/mol → kJ/mol

$$\text{K} = {^\circ}\text{C} + 273.15 \qquad \ln\left(\frac{k_2}{k_1}\right) = \frac{E_a}{R}\left(\frac{1}{T_1} - \frac{1}{T_2}\right) \qquad \frac{1\ \text{kJ}}{1000\ \text{J}}$$

Solution: $T_1 = 10.0\ {^\circ}\text{C} + 273.15 = 283.2\ \text{K}$ and $T_2 = 20.0\ {^\circ}\text{C} + 273.15 = 293.2\ \text{K}$ and $k_2 = 2\,k_1$ then

$\ln\left(\frac{k_2}{k_1}\right) = \frac{E_a}{R}\left(\frac{1}{T_1} - \frac{1}{T_2}\right)$. Rearrange to solve for E_a.

$$E_a = \frac{R \ln\left(\frac{k_2}{k_1}\right)}{\left(\frac{1}{T_1} - \frac{1}{T_2}\right)} = \frac{8.314 \frac{\text{J}}{\text{K} \cdot \text{mol}} \ln\left(\frac{2\,k_1}{k_1}\right)}{\left(\frac{1}{283.2\ \text{K}} - \frac{1}{293.2\ \text{K}}\right)} = 4.7\underline{8}51 \times 10^4 \frac{\text{J}}{\text{mol}} \times \frac{1\ \text{kJ}}{1000\ \text{J}} = 47.85\ \frac{\text{kJ}}{\text{mol}}$$

Check: The units (kJ/mol) are correct. The activation energy is typical for a reaction.

15.72 **Given:** rate constant triples from 20.0 °C to 35.0 °C **Find:** E_a

Conceptual Plan: °C → K then k_1, T_1, k_2, T_2 → E_a then J/mol → kJ/mol

$$\text{K} = {^\circ}\text{C} + 273.15 \qquad \ln\left(\frac{k_2}{k_1}\right) = \frac{E_a}{R}\left(\frac{1}{T_1} - \frac{1}{T_2}\right) \qquad \frac{1\ \text{kJ}}{1000\ \text{J}}$$

Solution: $T_1 = 20.0\ {^\circ}\text{C} + 273.15 = 293.2\ \text{K}$ and $T_2 = 35.0\ {^\circ}\text{C} + 273.15 = 308.2\ \text{K}$ and $k_2 = 3\,k_1$ then

$\ln\left(\frac{k_2}{k_1}\right) = \frac{E_a}{R}\left(\frac{1}{T_1} - \frac{1}{T_2}\right)$ Rearrange to solve for E_a.

$$E_a = \frac{R \ln\left(\frac{k_2}{k_1}\right)}{\left(\frac{1}{T_1} - \frac{1}{T_2}\right)} = \frac{8.314 \frac{\text{J}}{\text{K} \cdot \text{mol}} \ln\left(\frac{3\,k_1}{k_1}\right)}{\left(\frac{1}{293.2\ \text{K}} - \frac{1}{308.2\ \text{K}}\right)} = 5.502 \times 10^4 \frac{\text{J}}{\text{mol}} \times \frac{1\ \text{kJ}}{1000\ \text{J}} = 55.02\ \frac{\text{kJ}}{\text{mol}}$$

Check: The units (kJ/mol) are correct. The activation energy is typical for a reaction.

15.73 Reaction (a) would have the faster rate because the orientation factor, p, would be larger for this reaction because the reactants are symmetrical.

15.74 Reaction (b) would have the smaller orientation factor because we are reacting an asymmetric molecule with a homonuclear diatomic molecule (symmetrical); so the orientation is important. In reaction (a), both reacting species are symmetrical; so orientation is unimportant.

Reaction Mechanisms

15.75 The overall reaction is the sum of the elementary steps in the proposed mechanism. Because the first reaction is the slow step, it is the rate-determining step. Using this first step to determine the rate law, Rate $= k_1[\text{AB}]^2$. Because this is the observed rate law, this mechanism is consistent with the experimental data.

15.76 (a) The reaction cannot occur in a single step in which X and Y collide because the rate law would be Rate $= k[\text{X}][\text{Y}]$. This is not consistent with the stated rate law of Rate $= k[\text{X}]^2[\text{Y}]$.

(b) The overall reaction is the sum of the elementary steps in the proposed mechanism. Because the second step is the rate-determining step, Rate $= k_3[\text{X}_2][\text{Y}]$. X_2 is an intermediate; so its concentration cannot appear in the rate law. Using the fast equilibrium in the first step, we see that $k_1[\text{X}]^2 = k_2[\text{X}_2]$ or $[\text{X}_2] = \frac{k_1}{k_2}[\text{X}]^2$. Substituting this into the first rate expression, we get Rate $= \frac{k_3 k_1}{k_2}[\text{X}]^2[\text{Y}]$. Simplifying this expression, we see that Rate $= k[\text{X}]^2[\text{Y}]$, which is consistent with the experimentally derived rate law.

15.77 (a) The overall reaction is the sum of the steps in the mechanism:

$$Cl_2(g) \underset{k_2}{\overset{k_1}{\rightleftharpoons}} 2\,\cancel{Cl(g)}$$

$$\cancel{Cl(g)} + CHCl_3(g) \xrightarrow[k_3]{} HCl(g) + \cancel{CCl_3(g)}$$

$$\cancel{Cl(g)} + \cancel{CCl_3(g)} \xrightarrow[k_4]{} CCl_4(g)$$

$$Cl_2(g) + CHCl_3(g) \longrightarrow HCl(g) + CCl_4(g)$$

(b) The intermediates are the species that are generated by one step and consumed by other steps. These are a $Cl(g)$ and $CCl_3(g)$.

(c) Because the second step is the rate-determining step, Rate $= k_3\,[Cl][CHCl_3]$. Because Cl is an intermediate, its concentration cannot appear in the rate law. Using the fast equilibrium in the first step, we see that $k_1[Cl_2] = k_2[Cl]^2$ or $[Cl] = \sqrt{\frac{k_1}{k_2}[Cl_2]}$. Substituting this into the first rate expression, we get Rate $= k_3\sqrt{\frac{k_1}{k_2}}[Cl_2]^{1/2}[CHCl_3]$. Simplifying this expression, we see that Rate $= k[Cl_2]^{1/2}\,[CHCl_3]$.

15.78 (a) The overall reaction is the sum of the steps in the mechanism:

$$NO_2(g) + Cl_2(g) \xrightarrow[k_1]{} ClNO_2(g) + \cancel{Cl(g)}$$

$$NO_2(g) + \cancel{Cl(g)} \xrightarrow[k_2]{} ClNO_2(g)$$

$$2\,NO_2(g) + Cl_2(g) \longrightarrow 2\,ClNO_2(g)$$

(b) The intermediates are the species that are generated by one step and consumed by other steps. This is $Cl(g)$.

(c) Because the first step is the rate-determining step, Rate $= k_1[NO_2][Cl_2]$. Because both of these species are reactants, this is the predicted rate law.

Catalysis

15.79 Heterogeneous catalysts require a large surface area because catalysis can only happen at the active sites on the surface. A greater surface area means greater opportunity for the substrate to react, which results in a speedier reaction.

15.80 The initial and final energies (reactants and products) remain the same. The activation energy drops from 75 kJ/mol to a smaller value (e.g., 30 kJ/mol). There are usually more steps in the reaction progress diagram.

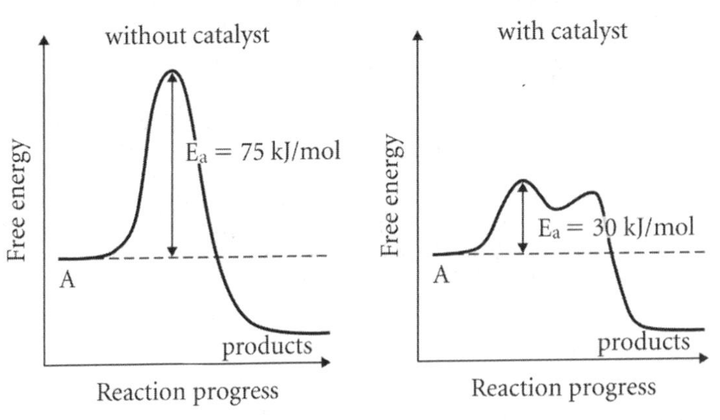

15.81 Assume rate ratio $\propto k$ ratio (because concentration terms will cancel each other) and $k = Ae^{-E_a/RT}$. $T = 25\,°C + 273.15 = 298$ K, $E_{a_1} = 1.25 \times 10^5$ J/mol, and $E_{a_2} = 5.5 \times 10^4$ J/mol. Ratio of rates will be

$$\frac{k_2}{k_1} = \frac{\cancel{A}e^{-E_{a_2}/RT}}{\cancel{A}e^{-E_{a_1}/RT}} = \frac{e^{\frac{-5.5\times10^4\frac{\cancel{J}}{\cancel{mol}}}{\left(8.314\frac{\cancel{J}}{\cancel{K}\cdot\cancel{mol}}\right)298\,\cancel{K}}}}{e^{\frac{-1.25\times10^5\frac{\cancel{J}}{\cancel{mol}}}{\left(8.314\frac{\cancel{J}}{\cancel{K}\cdot\cancel{mol}}\right)298\,\cancel{K}}}} = \frac{e^{-2\underline{2}.199}}{e^{-5\underline{0}.453}} = 10^{12}$$

Note that no additional significant figures are quoted since the exponent has its last significant digit in the 1's place.

15.82 Assume rate ratio $\propto$ k ratio (because concentration terms will cancel each other) and $k = Ae^{-E_a/RT}$. $T = 25\ °C + 273.15 = 298\ K$ and $E_{a_1} = 1.08 \times 10^5\ J/mol$. Ratio of rates will be

$$\frac{k_2}{k_1} = 10^6 = \frac{\cancel{A}e^{-E_{a_2}/RT}}{\cancel{A}e^{-E_{a_1}/RT}} = \frac{e^{\frac{-E_{a_2}}{\left(8.314\frac{J}{K\cdot mol}\right)298\ K}}}{e^{\frac{-1.08\times10^5\frac{J}{mol}}{\left(8.314\frac{J}{K\cdot mol}\right)298\ K}}} = \frac{e^{\frac{-E_{a_2}}{2.47756\times10^3\frac{J}{mol}}}}{1.17\times10^{-19}} \rightarrow e^{\frac{-E_{a_2}}{2.47756\times10^3\frac{J}{mol}}} = 1.17\times10^{-13} \rightarrow$$

$$\frac{-E_{a_2}}{2.47756\times10^3\frac{J}{mol}} = \ln(1.17\times10^{-13}) = -29.7766 \rightarrow E_{a_2} = 7.38\times10^4\frac{J}{mol} = 73.8\frac{kJ}{mol}$$

Cumulative Problems

15.83 **Given:** table of $[CH_3CN]$ versus time **Find:** (a) reaction order, k; (b) $t_{1/2}$; and (c) t for 90% conversion

Conceptual Plan: (a) and (b) Look at the data and see if any common reaction orders can be eliminated. If the data does not show an equal concentration drop with time, zero order can be eliminated. Look for changes in the half-life (compare time for concentration to drop to one-half of any value). If the half-life is not constant, the first order can be eliminated. If the half-life is getting longer as the concentration drops, this might suggest second order. Plot the data as indicated by the appropriate rate law, or if it is first order and there is an obvious half-life in the data, a plot is not necessary. Determine k from the slope of the plot (or using the half-life equation for first order). (c) Finally, calculate the time to 90% conversion using the appropriate integrated rate equation.

Solution: (a) and (b) By the preceding logic, we can see that the reaction is first order. It takes 15.0 h for the concentration to be cut in half for any concentration (1.000 M to 0.501 M, 0.794 M to 0.398 M, and 0.631 M to 0.316 M), so $t_{1/2} = 15.0$ h. Then use $t_{1/2} = \frac{0.693}{k}$ and rearrange to solve for k.

$$k = \frac{0.693}{t_{1/2}} = \frac{0.693}{15.0\ h} = 0.0462\ h^{-1}$$

(c) $[CH_3CN]_t = 0.10\,[CH_3CN]_0$ Because $\ln[CH_3CN]_t = -kt + \ln[CH_3CN]_0$, rearrange to solve for t.

$$t = -\frac{1}{k}\ln\frac{[CH_3CN]_t}{[CH_3CN]_0} = -\frac{1}{0.0462\ h^{-1}}\ln\frac{0.10\cancel{[CH_3CN]_0}}{\cancel{[CH_3CN]_0}} = 49.8\ h$$

Check: The units (none, h^{-1}, h, and h) are correct. The rate law is a common form. The data showed a constant half-life very clearly. The rate constant is consistent with the units necessary to get rate as M/s, and the magnitude is reasonable because we have a first-order reaction. The time to 90% conversion is consistent with a time between three and four half-lives.

15.84 **Given:** table of $[X_2Y]$ versus time

Find: (a) reaction order, k; (b) $t_{1/2}$ at initial concentration; and (c) [X] at 10.0 h

Conceptual Plan: (a) Look at the data and see if any common reaction orders can be eliminated. If the data does not show an equal concentration drop with time, zero order can be eliminated. Look for changes in the half-life (compare time for concentration to drop to one-half of any value). If the half-life is not constant, the first order can be eliminated. If the half-life is getting longer as the concentration drops, this might suggest second order. Plot the data as indicated by the appropriate rate law. Determine k from the slope of the plot. (b) Calculate the half-life with the appropriate equation. (c) Finally, calculate the $[X_2Y]$ at 10.0 h using the appropriate integrated rate expression and convert this to a change in $[X_2Y]$ and then to $[X]$ using the reaction stoichiometry.

Solution:

(a) By the preceding logic, we can eliminate both the zero-order and first-order reactions. (Alternatively, you could make all three plots, and only one should be linear.) This suggests that we should have a second-order reaction. Plot $1/[X_2Y]$ versus time. Because $\frac{1}{[X_2Y]_t} = kt + \frac{1}{[X_2Y]_0}$, the slope will be the rate constant. The slope can be determined by measuring $\Delta y/\Delta x$ on the plot or by using functions such as "add trendline" in Excel. Thus, the rate constant is $1.6827\ M^{-1} \cdot h^{-1}$, and the rate law is Rate $= 1.68\ M^{-1} \cdot h^{-1}[X_2Y]^2$.

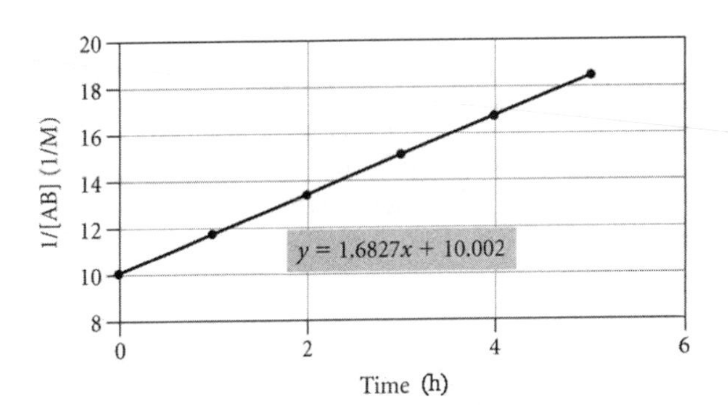

(b) $t_{1/2} = \frac{1}{k[X_2Y]_0}$ so

$$t_{1/2} = \frac{1}{(1.6827\ M^{-1} \cdot h^{-1})(0.100\ M)} = 5.94\ h$$

(c) Finally, use $\frac{1}{[X_2Y]_t} = kt + \frac{1}{[X_2Y]_0}$; substitute the values of $[X_2Y]_0$, 10.0 h, and k; and rearrange to solve for $[X_2Y]$ at 10.0 h.

$$[X_2Y]_t = \frac{1}{kt + \frac{1}{[X_2Y]_0}} = \frac{1}{(1.6827\ M^{-1} \cdot h^{-1})(10.0\ h) + \left(\frac{1}{0.100\ M}\right)} = 0.0372759\ M \text{ then}$$

$$\Delta[X_2Y] = [X_2Y]_0 - [X_2Y]_{10.0\,h} = 0.100\ M - 0.0372759\ M = 0.062724\ M \text{ then}$$

$$\frac{0.062724\ \text{mol } X_2Y}{L} \times \frac{2\ \text{mol X}}{1\ \text{mol } X_2Y} = 0.13\ M\ X$$

Check: The units (none, $M^{-1} \cdot h^{-1}$, h, and M) are correct. The rate law is a common form. The plot was linear, confirming second-order kinetics. The rate constant is consistent with the units necessary to get rate as M/s and the magnitude is reasonable because we have a second-order reaction. The half-life is consistent with the data table, which indicates that the half-life is a little over 5 h. The [X] at 10 h s is consistent with the changes that we see in the data table through 5 h.

15.85 **Given:** Rate $= k\frac{[A][C]^2}{[B]^{1/2}} = 0.0115$ M/s at certain initial concentrations of A, B, and C; double A and C concentration and triple B concentration **Find:** reaction rate

Conceptual Plan: $[A]_1, [B]_1, [C]_1$, Rate 1, $[A]_2, [B]_2, [C]_2 \rightarrow$ Rate 2

$$\frac{\text{Rate 2}}{\text{Rate 1}} = \frac{k\frac{[A]_2[C]_2^2}{[B]_2^{1/2}}}{k\frac{[A]_1[C]_1^2}{[B]_1^{1/2}}}$$

Solution: $\frac{\text{Rate 2}}{\text{Rate 1}} = \frac{k\frac{[A]_2[C]_2^2}{[B]_2^{1/2}}}{k\frac{[A]_1[C]_1^2}{[B]_1^{1/2}}}$ Rearrange to solve for Rate 2. Rate 2 $= \frac{k\frac{[A]_2[C]_2^2}{[B]_2^{1/2}}}{k\frac{[A]_1[C]_1^2}{[B]_1^{1/2}}}$ Rate 1

$[A]_2 = 2[A]_1$, $[B]_2 = 3[B]_1$, $[C]_2 = 2[C]_1$, and Rate 1 $= 0.0115$ M/s so

$$\text{Rate 2} = \frac{k\frac{2[A]_1(2[C]_1)^2}{(3[B]_1)^{1/2}}}{k\frac{[A]_1[C]_1^2}{[B]_1^{1/2}}} 0.0115\frac{M}{s} = \frac{2^3}{3^{1/2}} 0.0115\frac{M}{s} = 0.0531\frac{M}{s}$$

Check: The units $(M \cdot s^{-1})$ are correct. They should increase because we have a factor of eight (2^3) divided by the square root of three (1.73).

15.86 **Given:** Rate $= k\dfrac{[O_3]^2}{[O_2]}$; initially, 1.0 mol O_3 and 1.0 mol O_2 in 1.0 L
Find: fraction O_3 reacted when reaction rate is cut in half
Conceptual Plan:
mol, L → M then $[O_3]_1$, $[O_2]_1$, Rate 1, Rate 2 → $[O_3]_2$ then $[O_3]_1$, $[O_3]_2$ → O_3 fraction reacted

$$M = \frac{mol}{L} \qquad \frac{\text{Rate 2}}{\text{Rate 1}} = \frac{k\dfrac{[O_3]_2^2}{[O_2]_2}}{k\dfrac{[O_3]_1^2}{[O_2]_1}} \qquad O_{3\text{ fraction reacted}} = \frac{[O_3]_1 - [O_3]_2}{[O_3]_1}$$

Solution: $M = \dfrac{mol}{L}$ so $[O_3]_1 = \dfrac{1.0\text{ mol}}{1.0\text{ L}} = 1.0\text{ M}$ and $[O_2]_1 = \dfrac{1.0\text{ mol}}{1.0\text{ L}} = 1.0\text{ M}$. Rate 1 = 2 Rate 2

Let $x = \Delta[O_3]$ so $[O_3]_2 = [O_3]_{1-x}$ and $[O_2]_2 = [O_2]_{1+3/2x}$. Substitute values into $\dfrac{\text{Rate 2}}{\text{Rate 1}} = \dfrac{k\dfrac{[O_3]_2^2}{[O_2]_2}}{k\dfrac{[O_3]_1^2}{[O_2]_1}}$ and

rearrange to solve for x. $\dfrac{\cancel{\text{Rate 2}}}{2\,\cancel{\text{Rate 2}}} = \dfrac{\cancel{k}\dfrac{(1.0\text{ M} - x)^2}{(1.0\text{ M} + 3/2x)}}{\cancel{k}\dfrac{(1.0\text{ M})^2}{\cancel{(1.0\text{ M})}}} \rightarrow 0.50\text{ M}\,(1.0\text{ M} + 3/2x) = (1.0\text{ M} - x)^2 \rightarrow$

$0.50 + 0.75x = 1.0 - 2.0x + x^2 \rightarrow 0 = x^2 - 2.\underline{7}5x + 0.\underline{5}0$ Solve with quadratic equation

$\left(x = \dfrac{-b \pm \sqrt{b^2 - 4ac}}{2a}\right)$ So

$$x = \frac{2.\underline{7}5 \pm \sqrt{(-2.\underline{7}5)^2 - (4)(1.0)(0.50)}}{2(1.0)} = \frac{2.\underline{7}5 \pm \sqrt{5.\underline{5}625}}{2.0} = \frac{2.\underline{7}5 \pm 2.\underline{3}585}{2.0} = 0.\underline{1}9575\text{ M or } 2.\underline{5}5\text{ M}$$

The answer must be $0.\underline{1}9575$ M because the other answer is larger than our initial concentration (and therefore is impossible).

$$O_{3\text{ fraction reacted}} = \frac{[O_3]_1 - [O_3]_2}{[O_3]_1} = \frac{x}{[O_3]_1} = \frac{0.\underline{1}9575\text{ M}}{1.0\text{ M}} = 0.2$$

Check: The units (unitless) are correct. The concentration is reasonable because two forces are slowing down the reaction: (1) the decrease in the reactant and (2) the increase of the product (which appears in the rate law). The calculation can be double-checked by substituting the value of X; the resulting rate $= 0.5\,k$.

15.87 **Given:** table of P_{Total} versus time **Find:** rate law, k, and P_{Total} at 2.00×10^4 s
Conceptual Plan: Because two moles of gas are generated for each mole of CH_3CHO decomposed, this $P_{CH_3CHO} = P^\circ_{Total} - (P_{Total} - P^\circ_{Total})$. Look at the data and see if any common reaction orders can be eliminated. If the data does not show an equal P_{Total} rise (or P_{CH_3CHO} drop) with time, zero order can be eliminated. There is not enough data to look for changes in the half-life (compare time for P_{CH_3CHO} to drop to one-half of any value). It does appear that the half-life is getting longer, so the first order can be eliminated. Plot the data as indicated by the appropriate rate law. Determine k from the slope of the plot. Finally, calculate the P_{CH_3CHO} at 2.00×10^4 s using the appropriate integrated rate expression and convert this to P_{Total} using the reaction stoichiometry.
Solution: Calculate $P_{CH_3CHO} = P^\circ_{Total} - (P_{Total} - P^\circ_{Total})$.

Time(s)	P_{Total}(atm)	P_{CH_3CHO}(atm)
0	0.22	0.22
1000	0.24	0.20
3000	0.27	0.17
7000	0.31	0.13

By the preceding logic, we can eliminate both the zero-order and the first-order reactions. (Alternatively, you could make all three plots, and only one should be linear.) This suggests that we should have a second-order reaction. Plot $1/P_{CH_3CHO}$ versus time. Because $\dfrac{1}{P_{CH_3CHO}} = kt + \dfrac{1}{P^\circ_{CH_3CHO}}$, the slope will be the rate constant. The slope can be determined by measuring $\Delta y/\Delta x$ on the plot or by using functions such as "add trendline" in Excel. Thus, the rate constant is $4.5 \times 10^{-4}\ \text{atm}^{-1} \cdot \text{s}^{-1}$, and the rate law is Rate $= 4.5 \times 10^{-4}\ \text{atm}^{-1} \cdot \text{s}^{-1} P_{CH_3CHO}$.

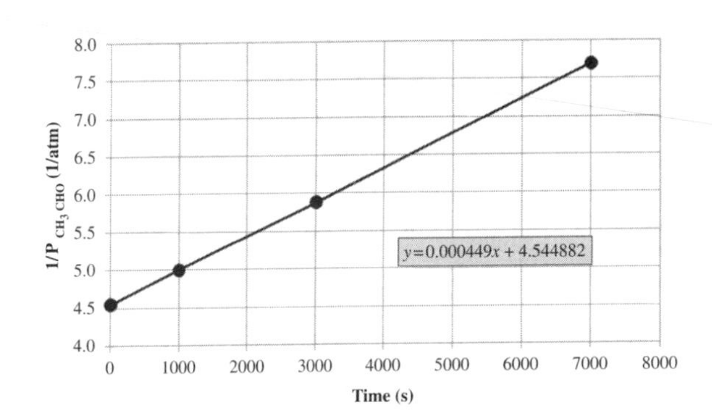

Finally, use $\dfrac{1}{P_{CH_3CHO}} = kt + \dfrac{1}{P^\circ_{CH_3CHO}}$; substitute the values of $P^\circ_{CH_3CHO}$, 2.00×10^4 s, and k and rearrange to solve for $P^\circ_{CH_3CHO}$ at 2.00×10^4 s.

$$P_{CH_3CHO} = \frac{1}{kt + \dfrac{1}{P^\circ_{CH_3CHO}}} =$$

$$\frac{1}{(4.5 \times 10^{-4}\ \text{atm}^{-1} \cdot \cancel{\text{s}^{-1}})(2.00 \times 10^4\ \cancel{\text{s}}) + \left(\dfrac{1}{0.22\ \text{atm}}\right)} = 0.07\underline{3}8255\ \text{atm} = 0.074\ \text{atm}$$

Finally, from the first equation in the solution, $P_{\text{Total}} = 2P^\circ_{\text{Total}} - P_{CH_3CHO} = 2(0.22\ \text{atm}) - 0.07\underline{3}8255\ \text{atm} = 0.3\underline{6}6175\ \text{atm} = 0.37\ \text{atm}$.

Check: The units (none, $\text{atm}^{-1} \cdot \text{s}^{-1}$, and atm) are correct. The rate law is a common form. The plot was linear, confirming second-order kinetics. The rate constant is consistent with the units necessary to get rate as atm/s, and the magnitude is reasonable because we have a second-order reaction. The P_{Total} at 2.00×10^4 s is consistent with the changes we see in the data table through 7000 s.

15.88 **Given:** table of $P^\circ_{H_2C_2O_4}$ versus P_{Total} at 20,000 s **Find:** rate law and k

Conceptual Plan: Because two moles of gas are generated for each mole of $H_2C_2O_4$ decomposed, this $P^\circ_{H_2C_2O_4} = P_{H_2C_2O_4} - (P_{\text{Total}} - P^\circ_{H_2C_2O_4})$ and the Rate $= (P^\circ_{H_2C_2O_4} - P_{\text{Total}})/20{,}000$ s. Using general rate law form, compare rate ratios to determine reaction order.

$$\frac{\text{Rate 2}}{\text{Rate 1}} = \frac{k[\text{A}]_2^n}{k[\text{A}]_1^n}$$

Then use one of the concentration/initial rate pairs to determine k.

$$\text{Rate} = k[\text{A}]^n$$

Solution: Calculate Rate $= (P^\circ_{H_2C_2O_4} - P_{\text{Total}})/20{,}000$ s.

	Experiment 1	Experiment 2	Experiment 3
$P^\circ_{H_2C_2O_4}$ (mmHg)	65.8	92.1	111
P_{Total} at 20,000 s (mmHg)	94.6	132	160
Rate (mmHg/s)	0.0014$\underline{4}$	0.001$\underline{9}$95	0.0024$\underline{5}$

$\dfrac{\text{Rate 2}}{\text{Rate 1}} = \dfrac{k[\text{A}]_2^n}{k[\text{A}]_1^n}$ Comparing the first two sets of data,

$\dfrac{0.001\underline{9}95\ \cancel{\text{mmHg/s}}}{0.0014\underline{4}\ \cancel{\text{mmHg/s}}} = \dfrac{\cancel{k}(92.1\ \cancel{\text{mmHg}})^n}{\cancel{k}(65.8\ \cancel{\text{mmHg}})^n}$ and $1.\underline{3}854 = 1.3\underline{9}97^n$; so $n = 1$. If we compare the first and third data sets,

$\dfrac{0.0024\underline{5}\ \cancel{\text{mmHg/s}}}{0.0014\underline{4}\ \cancel{\text{mmHg/s}}} = \dfrac{\cancel{k}(111\ \cancel{\text{mmHg}})^n}{\cancel{k}(65.8\ \cancel{\text{mmHg}})^n}$ and $1.\underline{7}0139 = 1.\underline{6}8693^n$; so $n = 1$. This second comparison is not necessary, but it increases our confidence in the reaction order. So Rate $= k\,P_{H_2C_2O_4}$. Selecting the first data set and rearranging the rate equation,

$$k = \frac{\text{Rate}}{P_{H_2C_2O_4}} = \frac{0.00144\dfrac{\cancel{\text{mmHg}}}{\text{s}}}{65.8\ \cancel{\text{mmHg}}} = 2.19 \times 10^{-5}\ \text{s}^{-1}\text{; so Rate} = 2.19 \times 10^{-5}\ \text{s}^{-1} P_{H_2C_2O_4}.$$

Check: The units (none and s^{-1}) are correct. The rate law is a common form. The rate is changing proportionately with the initial pressure, so first order is consistent. The rate constant is consistent with the units necessary to get rate as mmHg/s, and the magnitude is reasonable because we have a first-order reaction.

15.89 **Given:** N_2O_5 decomposes to NO_2 and O_2, first order in $[N_2O_5]$; $t_{1/2} = 2.81$ h at 25 °C; $V = 1.5$ L, $P^{\circ}_{N_2O_5} = 745$ torr
Find: P_{O_2} after 215 minutes
Conceptual Plan: Write a balanced reaction. Then $t_{1/2} \rightarrow k$ then °C → K and torr → atm then

$$N_2O_5 \rightarrow 2\,NO_2 + \tfrac{1}{2}O_2 \qquad t_{1/2} = \frac{0.693}{k} \qquad K = {}^{\circ}C + 273.15 \qquad \frac{1\text{ atm}}{760\text{ torr}}$$

$P^{\circ}_{N_2O_5}, V, T \rightarrow n/V$ then min → h then $[N_2O_5]_0$, $t, k \rightarrow [N_2O_5]_t$ then $[N_2O_5]_0, [N_2O_5]_t \rightarrow [O_2]_t$

$$PV = nRT \qquad \frac{1\text{ h}}{60\text{ min}} \qquad \ln[A]_t = -kt + \ln[A]_0 \qquad [O_2]_t = ([N_2O_5]_0 - [N_2O_5]_t) \times \frac{1/2\text{ mol }O_2}{1\text{ mol }N_2O_5}$$

then $[O_2]_t, V, T \rightarrow P^{\circ}_{O_2}$ and finally atm → torr

$$PV = nRT \qquad \frac{760\text{ torr}}{1\text{ atm}}$$

Solution: $t_{1/2} = \frac{0.693}{k}$ and rearrange to solve for k. $k = \frac{0.693}{t_{1/2}} = \frac{0.693}{2.81\text{ h}} = 0.24\underline{6}619\text{ h}^{-1}$ then

$T = 25\text{ °C} + 273.15 = 298$ K. $745\text{ torr} \times \frac{1\text{ atm}}{760\text{ torr}} = 0.98\underline{0}263$ atm then $PV = nRT$ Rearrange to solve for n/V.

$$\frac{n}{V} = \frac{P}{RT} = \frac{0.98\underline{0}263\text{ atm}}{0.08206\frac{\text{L}\cdot\text{atm}}{\text{K}\cdot\text{mol}} \times 298\text{ K}} = 0.040\underline{0}862\text{ M then } 215\text{ min} \times \frac{1\text{ h}}{60\text{ min}} = 3.5\underline{8}333\text{ h}$$

Because $\ln[N_2O_5]_t = -kt + \ln[N_2O_5]_0 = -(0.24\underline{6}619\text{ h}^{-1})(3.5\underline{8}333\text{ h}) + \ln(0.040\underline{0}862\text{ M}) = -4.1\underline{0}044 \rightarrow$
$[N_2O_5]_t = e^{-4.1\underline{0}044} = 0.016\underline{5}654$ M then

$$[O_2]_t = ([N_2O_5]_0 - [N_2O_5]_t) \times \frac{1/2\text{ mol }O_2}{1\text{ mol }N_2O_5} = \left(0.040\underline{0}862\frac{\text{mol }N_2O_5}{\text{L}} - 0.016\underline{5}654\frac{\text{mol }N_2O_5}{\text{L}}\right) \times \frac{1/2\text{ mol }O_2}{1\text{ mol }N_2O_5}$$
$= 0.011\underline{7}604$ M O_2

then finally $PV = nRT$ and rearrange to solve for P

$$P = \frac{n}{V}RT = 0.011\underline{7}604\frac{\text{mol}}{\text{L}} \times 0.08206\frac{\text{L atm}}{\text{K}\cdot\text{mol}} \times 298\text{ K} = 0.287587\text{ atm} \times \frac{760\text{ torr}}{1\text{ atm}} = 219\text{ torr}$$

Check: The units (torr) are correct. The pressure is reasonable because it must be less than one-half of the original pressure.

15.90 **Given:** cyclopropane (C_3H_6) reacts, first order in $[C_3H_6]$; $k = 5.87 \times 10^{-4}$/s at 485 °C; $V = 2.5$ L, $P^{\circ}_{C_3H_6} = 722$ torr
Find: t to $P_{C_3H_6} = 100.0$ torr
Conceptual Plan: Because $P \propto$ M, we do not need to convert P to M. $P^{\circ}_{C_3H_6}, P_{C_3H_6}, k \rightarrow t$

$$\ln[A]_t = -kt + \ln[A]_0$$

Solution: $\ln[C_3H_6]_t = -kt + \ln[C_3H_6]_0$ Rearrange to solve for t.

$$t = -\frac{1}{k}\ln\frac{[C_3H_6]_t}{[C_3H_6]_0} = -\frac{1}{k}\ln\frac{P_{C_3H_6}}{P^{\circ}_{C_3H_6}} = -\frac{1}{5.87 \times 10^{-4}\text{ s}^{-1}}\ln\frac{100.\text{ torr}}{722\text{ torr}} = 3.37 \times 10^3\text{ s} = 56.1\text{ min}$$

Check: The units (s or min) are correct. The time is reasonable because it is about three half-lives (pressure dropped to 14% of original pressure).

15.91 **Given:** I_2 formation from I atoms, second order in I; $k = 1.5 \times 10^{10}\text{ M}^{-1}\cdot\text{s}^{-1}$, $[I]_0 = 0.0100$ M
Find: t to decrease by 95%
Conceptual Plan: $[I]_0, [I]_t, k \rightarrow t$

$$\frac{1}{[A]_t} = kt + \frac{1}{[A]_0}$$

Solution: $[I]_t = 0.05\,[I]_0 = 0.05 \times 0.0100\text{ M} = 0.0005\text{ M}$ $\frac{1}{[I]_t} = kt + \frac{1}{[I]_0}$ Rearrange to solve for t.

$$t = \frac{1}{k}\left(\frac{1}{[I]_t} - \frac{1}{[I]_0}\right) = \frac{1}{(1.5 \times 10^{10}\text{ M}^{-1}\cdot\text{s}^{-1})}\left(\frac{1}{0.0005\text{ M}} - \frac{1}{0.0100\text{ M}}\right) = \underline{1}.267 \times 10^{-7}\text{ s} = 1 \times 10^{-7}\text{ s}$$

Check: The units (s) are correct. We expect the time to be extremely small because the rate constant is so large.

15.92 **Given:** sucrose hydrolysis, first order in $[C_{12}H_{22}O_{11}]$; $k = 1.8 \times 10^{-4}\text{ s}^{-1}$ at 25 °C; $V = 2.55$ L, $[C_{12}H_{22}O_{11}]_0 = 0.150$ M, and 195 min **Find:** $m(C_{12}H_{22}O_{11})$ hydrolyzed

Conceptual Plan: min → s then $[C_{12}H_{22}O_{11}]_0, t, k \rightarrow [C_{12}H_{22}O_{11}]_t$ then

$$\frac{60\text{ s}}{1\text{ min}} \qquad \ln[A]_t = -kt + \ln[A]_0$$

V, $[C_{12}H_{22}O_{11}]_0, [C_{12}H_{22}O_{11}]_t \rightarrow$ mol $C_{12}H_{22}O_{11}$ hydrolyzed → g $C_{12}H_{22}O_{11}$ hydrolyzed

$$\text{mol } C_{12}H_{22}O_{11} = ([C_{12}H_{22}O_{11}]_0 - [C_{12}H_{22}O_{11}]_t) \times V \qquad \frac{342.30\text{ g } C_{12}H_{22}O_{11}}{1\text{ mol } C_{12}H_{22}O_{11}}$$

Solution: $195\ \cancel{\text{min}} \times \dfrac{60\text{ sec}}{1\ \cancel{\text{min}}} = 11\underline{7}00\text{ s}$ Because

$\ln[C_{12}H_{22}O_{11}]_t = -kt + \ln[C_{12}H_{22}O_{11}]_0 = -(1.8 \times 10^{-4}\ \cancel{\text{s}^{-1}})(11\underline{7}00\ \cancel{\text{s}}) + \ln(0.150\text{ M}) = -4.0\underline{0}312 \rightarrow$

$[C_{12}H_{22}O_{11}]_t = e^{-4.00312} = 0.018\underline{2}586\text{ M}$ then

$\text{mol } C_{12}H_{22}O_{11} = ([C_{12}H_{22}O_{11}]_0 - [C_{12}H_{22}O_{11}]_t) \times V =$

$$\left(0.150\frac{\text{mol } C_{12}H_{22}O_{11}}{\cancel{\text{L}}} - 0.018\underline{2}586\frac{\text{mol } C_{12}H_{22}O_{11}}{\cancel{\text{L}}}\right) \times 2.55\ \cancel{\text{L}} = 0.33\underline{5}941\text{ mol } C_{12}H_{22}O_{11}$$

$$0.33\underline{5}941\ \cancel{\text{mol } C_{12}H_{22}O_{11}} \times \frac{342.30\text{ g } C_{12}H_{22}O_{11}}{1\ \cancel{\text{mol } C_{12}H_{22}O_{11}}} = 115\text{ g } C_{12}H_{22}O_{11}$$

Check: The units (g) are correct. The mass is reasonable because it must be less than the original amount in solution (131 g). The amount is close to the original amount in solution because the final sucrose concentration is so low inasmuch as we have gone over three half-lives.

15.93 **Given:** $AB(aq) \rightarrow A(g) + B(g)$; $k = 0.0118\text{ M}^{-1} \cdot \text{s}^{-1}$; 250.0 mL of 0.100 M AB; collect gas over water $T = 25.0\text{ °C}$, $P_{\text{Total}} = 755.1$ mmHg, and $V = 200.0$ mL; $P^{\circ}_{H_2O} = 23.8$ mmHg **Find:** t

Conceptual Plan: $P_{\text{Total}}, P_{H_2O} \rightarrow P_A + P_B$ then mmHg → atm and mL → L

$$P_{\text{Total}} = P_{H_2O} + P_A + P_B \qquad \frac{1\text{ atm}}{760\text{ mmHg}} \qquad \frac{1\text{ L}}{1000\text{ mL}}$$

and °C → K $P, V, T \rightarrow n_{A+B} \rightarrow \Delta n_{AB}$ then $[AB]_0, V_{AB}, \Delta n_{AB} \rightarrow [AB]$ then $k, [AB] \rightarrow t$

$$K = °C + 273.15 \quad PV = nRT \quad \Delta n_{AB} = \tfrac{1}{2}n_{A+B} \qquad [AB] = [AB]_0 - \frac{\Delta n_{AB}}{V_{AB} \times \dfrac{1\text{ L}}{1000\text{ mL}}} \qquad \frac{1}{[AB]_t} = kt + \frac{1}{[AB]_0}$$

Solution: $P_{\text{Total}} = P_{H_2O} + P_A + P_B$ Rearrange to solve for $P_A + P_B$. $P_A + P_B = P_{\text{Total}} - P_{H_2O} = 755.1$ mmHg -23.8 mmHg $= 731.3$ mmHg

$$P_A + P_B = 731.3\ \cancel{\text{mmHg}} \times \frac{1\text{ atm}}{760\ \cancel{\text{mmHg}}} = 0.962\underline{2}3684\text{ atm} \quad V = 200.0\ \cancel{\text{mL}} \times \frac{1\text{ L}}{1000\ \cancel{\text{mL}}} = 0.2000\text{ L}$$

$T = 25.0\text{ °C} + 273.15 = 298.2\text{ K}$ $PV = nRT$ Rearrange to solve for n. $n = \dfrac{PV}{RT}$

$$n_{A+B} = \frac{0.962\underline{2}3684\ \cancel{\text{atm}} \times 0.2000\ \cancel{\text{L}}}{0.08206\dfrac{\cancel{\text{L}} \cdot \cancel{\text{atm}}}{\text{mol} \cdot \cancel{\text{K}}} \times 298.2\ \cancel{\text{K}}} = 0.00786\underline{4}5309\text{ mol A + B}$$ Because one mole each of A and B is generated for each mole of AB reacting, $\Delta n_{AB} = \frac{1}{2}n_{A+B} = \frac{1}{2}(0.00786\underline{4}5309\text{ mol A + B}) = 0.00393\underline{2}2655$ mol AB then

$$[AB] = [AB]_0 - \frac{\Delta n_{AB}}{V_{AB} \times \dfrac{1\text{ L}}{1000\text{ mL}}} = 0.100\text{ M} - \frac{0.00393\underline{2}2655\text{ mol AB}}{250.0\ \cancel{\text{mL}} \times \dfrac{1\text{ L}}{1000\ \cancel{\text{mL}}}} = 0.100\text{ M} - 0.0157\underline{2}9062\text{ M}$$

$= 0.08\underline{4}2709$ M

Because $\dfrac{1}{[AB]_t} = kt + \dfrac{1}{[AB]_0}$, rearrange to solve for t.

$$t = \frac{1}{k}\left(\frac{1}{[XY]_t} - \frac{1}{[XY]_0}\right) = \frac{1}{(0.0118\ \cancel{\text{M}^{-1}} \cdot \text{s}^{-1})}\left(\frac{1}{0.08\underline{4}2709\ \cancel{\text{M}}} - \frac{1}{0.100\ \cancel{\text{M}}}\right) = 15\underline{8}.1773\text{ s} = 160\text{ s}$$

Check: The units (s) are correct. The magnitude of the answer (160 s) makes sense because the rate constant is $0.0118\text{ M}^{-1}\text{ s}^{-1}$ and a small volume of gas is generated.

15.94 **Given:** $2\,H_2O_2(aq) \rightarrow 2\,H_2O(l) + O_2(g)$; $k = 0.00752\text{ s}^{-1}$; 150.0 mL of 30.0 % H_2O_2 by mass, $d = 1.11$ g/mL; collect gas over water $T = 20.0\text{ °C}$, $P_{\text{Total}} = 742.5$ mmHg, and $t = 85.0$ s; $P^{\circ}_{H_2O} = 17.5$ mmHg **Find:** V_{O_2}

Conceptual Plan: $mL_{solution} \rightarrow g_{solution} \rightarrow g_{H_2O_2} \rightarrow mol_{H_2O_2}$ **and** $mL_{solution} \rightarrow L_{solution}$

$$\frac{1.11\text{ g}}{1\text{ mL}} \quad \frac{30.0\text{ g }H_2O_2}{100\text{ g solution}} \quad \frac{1\text{ mol }H_2O_2}{34.02\text{ g }H_2O_2} \qquad \frac{1\text{ L}}{1000\text{ mL}}$$

then $mol_{H_2O_2}, L_{solution} \rightarrow M\ H_2O_2$ **then** $k, [H_2O_2]_0, t \rightarrow [H_2O_2]_t$ **then**

$$M = \frac{\text{amount solute (moles)}}{\text{volume solution (L)}} \qquad \ln[A]_t = -kt + \ln[A]_0$$

$[H_2O_2]_0, [H_2O_2]_t, L_{solution} \rightarrow \Delta n_{H_2O_2} \rightarrow n_{O_2}$ **then** $P_{Total}, P_{H_2O} \rightarrow P_{O_2}$ **then mmHg → atm**

$$\Delta n_{H_2O_2} = ([H_2O_2]_0 - [H_2O_2]_t) \times V_{H_2O_2} \quad \Delta n_{H_2O_2} = n_{O_2} \qquad P_{Total} = P_{H_2O} + P_{O_2} \qquad \frac{1\text{ atm}}{760\text{ mmHg}}$$

and mL → L and °C → K then $P, n, T \rightarrow V_{O_2}$

$$\frac{1\text{ L}}{1000\text{ mL}} \qquad K = °C + 273.15 \qquad PV = nRT$$

Solution: $150.0\ \cancel{\text{mL solution}} \times \frac{1.11\ \cancel{\text{g solution}}}{1\ \cancel{\text{mL solution}}} \times \frac{30.0\ \cancel{\text{g }H_2O_2}}{100\ \cancel{\text{g solution}}} \times \frac{1\text{ mol }H_2O_2}{34.02\ \cancel{\text{g }H_2O_2}} = 1.4\underline{6}8254\text{ mol }H_2O_2$ and

$150.0\ \cancel{\text{mL solution}} \times \frac{1\text{ L solution}}{1000\ \cancel{\text{mL solution}}} = 0.1500\text{ L solution then}$

$M = \frac{\text{amount solute (moles)}}{\text{volume solution (L)}} = \frac{1.4\underline{6}8254\text{ mol }H_2O_2}{0.1500\text{ L solution}} = 9.7\underline{8}8360\text{ M }H_2O_2$

Because $\ln[H_2O_2]_t = -kt + \ln[H_2O_2]_0 = -(0.00752\ \cancel{s^{-1}})(85.0\ \cancel{s}) + \ln(9.7\underline{8}8360\text{ M}) = 1.6\underline{4}1994 \rightarrow$

$[H_2O_2]_t = e^{1.6\underline{4}1994} = 5.1\underline{6}5459\text{ M}, \Delta n_{H_2O_2} = ([H_2O_2]_0 - [H_2O_2]_t) \times V_{H_2O_2} \times \frac{1\text{ L}}{1000\text{ mL}}$

$\Delta n_{H_2O_2} = (9.7\underline{8}8360\text{ M} - 5.1\underline{6}5459\text{ M}) \times (0.1500\text{ L}) = 0.69\underline{3}4352\text{ mol }H_2O_2$

then $\Delta n_{H_2O_2} = 0.5 n_{O_2} = 0.34672\text{ mol }O_2$ then $P_{Total} = P_{H_2O} + P_{O_2}$ Rearrange to solve for P_{O_2}.

$P_{O_2} = P_{Total} - P_{H_2O} = 742.5\text{ mmHg} - 17.5\text{ mmHg} = 725.0\text{ mmHg}$

$P_{O_2} = 725.0\ \cancel{\text{mmHg}} \times \frac{1\text{ atm}}{760\ \cancel{\text{mmHg}}} = 0.95\underline{3}94737\text{ atm}, T = 20.0\text{ °C} + 273.15 = 293.2\text{ K}, PV = nRT$

Rearrange to solve for V. $V = \frac{nRT}{P}$ and

$$V_{O_2} = \frac{0.34672\ \cancel{\text{mol }O_2} \times 0.08206\frac{\text{L} \cdot \cancel{\text{atm}}}{\cancel{\text{mol}} \cdot \cancel{\text{K}}} \times 293.2\ \cancel{\text{K}}}{0.95\underline{3}94737\ \cancel{\text{atm}}} = 8.74480\text{ L }O_2 = 8.75\text{ L }O_2$$

Check: The units (L) are correct. The magnitude of the answer (9 L) makes sense because about 0.4 mole of gas is generated (22.4 L = 1 mole gas at STP).

15.95 (a) There are two elementary steps in the reaction mechanism because there are two peaks in the reaction progress diagram.

(b)

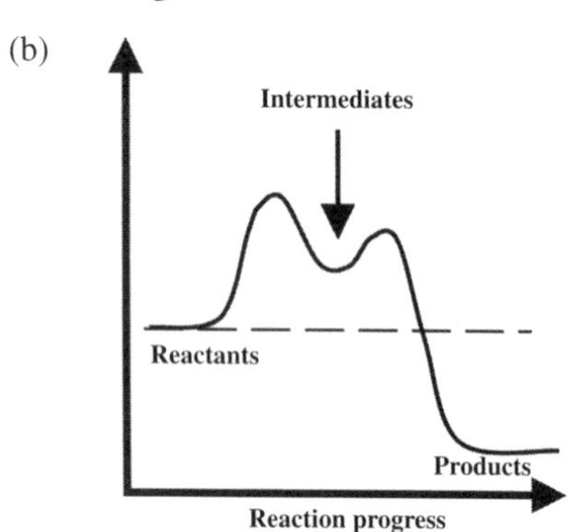

(c) The first step is the rate limiting step because it has the higher activation energy.

(d) The overall reaction is exothermic because the products are at a lower energy than the reactants.

15.96 (a) The first step is the rate-limiting step because it has the higher activation energy.

(b) Because the first step is the rate-determining step, Rate $= k_1[\text{HCl}][\text{H}_2\text{C} = \text{CH}_2]$. The reaction will be second-order overall.

(c) The overall reaction is exothermic because the products are at a lower energy than the reactants.

15.97 **Given:** *n*-butane desorption from single crystal aluminum oxide, first order; $k = 0.128\ \text{s}^{-1}$ at 150 K; initially completely covered

Find: (a) $t_{1/2}$; (b) t for 25% and for 50% to desorb; (c) fraction remaining after 10 s and 20 s

Conceptual Plan: (a) $k \rightarrow t_{1/2}$ **(b)** $[\text{C}_4\text{H}_{10}]_0, [\text{C}_4\text{H}_{10}]_t, k \rightarrow t$ **(c)** $[\text{C}_4\text{H}_{10}]_0, t, k \rightarrow [\text{C}_4\text{H}_{10}]_t$

$t_{1/2} = \frac{0.693}{k}$ $\qquad \ln[\text{A}]_t = -kt + \ln[\text{A}]_0 \qquad \ln[\text{A}]_t = -kt + \ln[\text{A}]_0$

Solution:

(a) $t_{1/2} = \frac{0.693}{k} = \frac{0.693}{0.128\ \text{s}^{-1}} = 5.41\ \text{s}$

(b) $\ln[\text{C}_4\text{H}_{10}]_t = -kt + \ln[\text{C}_4\text{H}_{10}]_0$ Rearrange to solve for t. For 25% desorbed, $[\text{C}_4\text{H}_{10}]_t = 0.75\,[\text{C}_4\text{H}_{10}]_0$

and $t = -\frac{1}{k}\ln\frac{[\text{C}_4\text{H}_{10}]_t}{[\text{C}_4\text{H}_{10}]_0} = -\frac{1}{0.128\ \text{s}^{-1}}\ln\frac{0.75\cancel{[\text{C}_4\text{H}_{10}]_0}}{\cancel{[\text{C}_4\text{H}_{10}]_0}} = 2.2\ \text{s}$. For 50% desorbed,

$[\text{C}_4\text{H}_{10}]_t = 0.50\,[\text{C}_4\text{H}_{10}]_0$ and $t = -\frac{1}{k}\ln\frac{[\text{C}_4\text{H}_{10}]_t}{[\text{C}_4\text{H}_{10}]_0} = -\frac{1}{0.128\ \text{s}^{-1}}\ln\frac{0.50\,\cancel{[\text{C}_4\text{H}_{10}]_0}}{\cancel{[\text{C}_4\text{H}_{10}]_0}} = 5.4\ \text{s}$

(c) For 10 s, $\ln[\text{C}_4\text{H}_{10}]_t = -kt + \ln[\text{C}_4\text{H}_{10}]_0 = -(0.128\,\cancel{\text{s}^{-1}})(10\,\cancel{\text{s}}) + \cancel{\ln(1.00)} = -1.\underline{2}8 \rightarrow$

$[\text{C}_4\text{H}_{10}]_t = e^{-1.\underline{2}8} = 0.28 =$ fraction covered

For 20 s, $\ln[\text{C}_4\text{H}_{10}]_t = -kt + \ln[\text{C}_4\text{H}_{10}]_0 = -(0.128\,\cancel{\text{s}^{-1}})(20\,\cancel{\text{s}}) + \cancel{\ln(1.00)} = -2.\underline{5}6 \rightarrow$

$[\text{C}_4\text{H}_{10}]_t = e^{-2.\underline{5}6} = 0.077 =$ fraction covered

Check: The units (s, s, s, none, and none) are correct. The half-life is reasonable considering the size of the rate constant. The time to 25% desorbed is less than one half-life. The time to 50% desorbed is the half-life. The fraction at 10 s is consistent with about two half-lives. The fraction covered at 20 s is consistent with about four half-lives.

15.98 **Given:** 120 nm film *n*-pentane evaporation from single crystal aluminum oxide, zero order; $k = 1.92 \times 10^{13}$ molecules/cm$^2 \cdot$s at 120 K; initially, coverage $= 8.9 \times 10^{16}$ molecules/cm^2

Find: (a) $t_{1/2}$; (b) fraction remaining after 10 s

Conceptual Plan:

(a) $[\text{C}_5\text{H}_{12}]_0, k \rightarrow t_{1/2}$

$t_{1/2} = \frac{[\text{A}]_0}{2k}$

(b) $[\text{C}_5\text{H}_{12}]_0, t, k \rightarrow [\text{C}_5\text{H}_{12}]_t$ **then** $[\text{C}_5\text{H}_{12}]_0, [\text{C}_5\text{H}_{12}]_t \rightarrow$ **fraction remaining**

$[\text{A}]_t = -kt + [\text{A}]_0 \qquad \text{fraction remaining} = \frac{[\text{C}_5\text{H}_{12}]_t}{[\text{C}_5\text{H}_{12}]_0}$

Solution:

(a) $t_{1/2} = \frac{[\text{C}_5\text{H}_{12}]_0}{2k} = \frac{8.9 \times 10^{16}\,\frac{\cancel{\text{molecules}}}{\cancel{\text{cm}^2}}}{2 \times 1.92 \times 10^{13}\,\frac{\cancel{\text{molecules}}}{\cancel{\text{cm}^2}\cdot\text{s}}} = 2.3 \times 10^3$ s for half of the film to evaporate

(b) $[\text{C}_5\text{H}_{12}]_t = -kt + [\text{C}_5\text{H}_{12}]_0 = -\left(1.92 \times 10^{13}\,\frac{\text{molecules}}{\text{cm}^2\,\cancel{\text{s}}}\right)(10 \cdot \cancel{\text{s}}) + 8.9 \times 10^{16}\,\frac{\text{molecules}}{\text{cm}^2}$

$= 8.\underline{8}808 \times 10^{16}\,\frac{\text{molecules}}{\text{cm}^2}$

$$\text{fraction remaining} = \frac{[C_5H_{12}]_t}{[C_5H_{12}]_0} = \frac{8.8808 \times 10^{16} \frac{\text{molecules}}{\text{cm}^2}}{8.9 \times 10^{16} \frac{\text{molecules}}{\text{cm}^2}} = 0.99784 = 1.0; \text{ so within experimental error,}$$

all are remaining on the surface.

Check: The units (s and none) are correct. The half-life is reasonable considering the size of the rate constant. The fraction at 10 s is reasonable given the fact that the time is very, very small compared to the half-life.

15.99 (a) **Given:** table of rate constant versus T **Find:** E_a and A

Conceptual Plan: First, convert temperature data into kelvin (°C + 273.15 = K). Because $\ln k = \frac{-E_a}{R}\left(\frac{1}{T}\right) + \ln A$**, a plot of ln k versus $1/T$ will have a *slope* = $-E_a/R$ and an *intercept* = ln A.**

Solution: The slope can be determined by measuring $\Delta y/\Delta x$ on the plot or by using functions such as "add trendline" in Excel. Because the *slope* $= -10759\ K = -E_a/R$,

$E_a = -(slope)R$

$$= -(-10759\ \text{K})\left(8.314 \frac{\text{J}}{\text{K} \cdot \text{mol}}\right)\left(\frac{1\ \text{kJ}}{1000\ \text{J}}\right)$$

$$= 89.5 \frac{\text{kJ}}{\text{mol}}$$

and *intercept* $= 26.769 = \ln A$ then

$A = e^{intercept} = e^{26.769} = 4.22 \times 10^{11}\ \text{s}^{-1}$

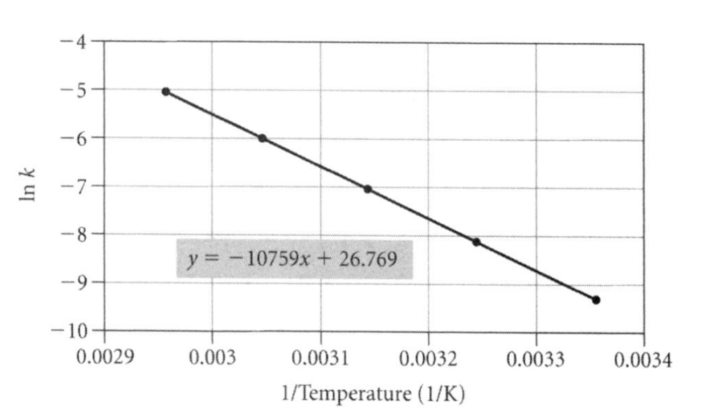

Check: The units (kJ/mol and s^{-1}) are correct. The plot was linear, confirming Arrhenius behavior. The activation and frequency factor are typical for many reactions.

(b) **Given:** part (a) results **Find:** k at 15 °C

Conceptual Plan: °C → K then T, E_a, A → k

$$°\text{C} + 273.15 = \text{K} \qquad \ln k = \frac{-E_a}{R}\left(\frac{1}{T}\right) + \ln A$$

Solution: 15 °C + 273.15 = 288 K then

$$\ln k = \frac{-E_a}{R}\left(\frac{1}{T}\right) + \ln A = \frac{-89.5 \frac{\text{kJ}}{\text{mol}} \times \frac{1000\ \text{J}}{1\ \text{kJ}}}{8.314 \frac{\text{J}}{\text{K} \cdot \text{mol}}}\left(\frac{1}{288\ \text{K}}\right) + \ln(4.22 \times 10^{11}\ \text{s}^{-1}) = -10.610 \rightarrow$$

$k = e^{-10.610} = 2.5 \times 10^{-5}\ \text{M}^{-1} \cdot \text{s}^{-1}$

Check: The units ($\text{M}^{-1} \cdot \text{s}^{-1}$) are correct. The value of the rate constant is less than the value at 25 °C.

(c) **Given:** part (a) results, 0.155 M C_2H_5Br and 0.250 M OH^- at 75 °C **Find:** initial reaction rate

Conceptual Plan: °C → K then T, E_a, A → k then k, $[C_2H_5Br]$, $[OH^-]$ → initial reaction rate

$$°\text{C} + 273.15 = \text{K} \qquad \ln k = \frac{-E_a}{R}\left(\frac{1}{T}\right) + A \qquad \text{Rate} = k[C_2H_5Br][OH^-]$$

Solution: 75 °C + 273.15 = 348 K then

$$\ln k = \frac{-E_a}{R}\left(\frac{1}{T}\right) + \ln A = \frac{-89.5 \frac{\text{kJ}}{\text{mol}} \times \frac{1000\ \text{J}}{1\ \text{kJ}}}{8.314 \frac{\text{J}}{\text{K} \cdot \text{mol}}}\left(\frac{1}{348\ \text{K}}\right) + \ln(4.22 \times 10^{11}\ \text{s}^{-1}) = -4.1656 \rightarrow$$

$k = e^{-4.1656} = 1.5521 \times 10^{-2}\ \text{M}^{-1} \cdot \text{s}^{-1}$

$\text{Rate} = k[C_2H_5Br][OH^-] = (1.5521 \times 10^{-2}\ \text{M}^{-1} \cdot \text{s}^{-1})(0.155\ \text{M})(0.250\ \text{M}) = 6.0 \times 10^{-4}\ \text{M} \cdot \text{s}^{-1}$

Check: The units $(M \cdot s^{-1})$ are correct. The value of the rate is reasonable considering the value of the rate constant (larger than in the table) and the fact that the concentrations are less than 1 M.

15.100 **Given:** $k = 2.35 \times 10^{-4}\ s^{-1}$ at 293 K and $k = 9.15 \times 10^{-4}\ s^{-1}$ at 303 K **Find:** A

Conceptual Plan: °C → K then k_1, T_1, k_2, T_2 → E_a then k_2, T_2, E_a → A

$$K = °C + 273.15 \qquad \ln k = \frac{-E_a}{R}\left(\frac{1}{T}\right) + \ln A \qquad k = Ae^{-E_a/RT}$$

Solution: $T_1 = 293$ K and $k_1 = 2.35 \times 10^{-4}\ s^{-1}$; $T_2 = 303$ K and $k_2 = 9.15 \times 10^{-4}\ s^{-1}$ then

$\ln\left(\frac{k_2}{k_1}\right) = \frac{E_a}{R}\left(\frac{1}{T_1} - \frac{1}{T_2}\right)$ Rearrange to solve for E_a.

$$E_a = \frac{R\ln\left(\frac{k_2}{k_1}\right)}{\left(\frac{1}{T_1} - \frac{1}{T_2}\right)} = \frac{8.314\frac{J}{K \cdot mol}\ln\left(\frac{9.15 \times 10^{-4}\ s^{-1}}{2.35 \times 10^{-4}\ s^{-1}}\right)}{\left(\frac{1}{293\ K} - \frac{1}{303\ K}\right)} = 1.0\underline{0}334 \times 10^5 \frac{J}{mol}$$

Because $k = Ae^{-E_a/RT}$, rearrange to solve for A. $A = ke^{E_a/RT} = 9.15 \times 10^{-4}\ s^{-1} e^{\frac{1.0\underline{0}334 \times 10^5 \frac{J}{mol}}{\left(8.314\frac{J}{K \cdot mol}\right)303\ K}} = 1.8 \times 10^{14}\ s^{-1}$

Check: The units (s^{-1}) are correct. The frequency factor is typical for a reaction.

15.101 (a) No, because the activation energy is zero. This means that the rate constant $(k = Ae^{-E_a/RT})$ will be independent of temperature.

(b) No bond is broken, and the two radicals (CH_3) attract each other.

(c) Formation of diatomic gases from atomic gases

15.102 (a) Nitrogen has a triple bond, so it will take more energy to break the N≡N bond than the H—H bond.

(b) **Given:** $E_a = 315$ kJ/mol for reaction 1 and $E_a = 23$ kJ/mol for reaction 2, frequency factor similar, and 25 °C **Find:** ratio of rate constants

Conceptual Plan: °C → K then T, E_{a1}, E_{a2}, A → k_1/k_2

$$°C + 273.15 = K \qquad \frac{k_1}{k_2} = \frac{Ae^{-E_{a_1}/RT}}{Ae^{-E_{a_2}/RT}}$$

Solution: $T = 25\ °C + 273.15 = 298$ K. $E_{a_1} = 315$ kJ/mol and $E_{a_2} = 23$ kJ/mol. Ratio of rate constants

will be $$\frac{k_1}{k_2} = \frac{Ae^{-E_{a_1}/RT}}{Ae^{-E_{a_2}/RT}} = \frac{e^{\frac{-315\frac{kJ}{mol} \times \frac{1000\ J}{1\ kJ}}{\left(8.314\frac{J}{K \cdot mol}\right)298\ K}}}{e^{\frac{-23\frac{kJ}{mol} \times \frac{1000\ J}{1\ kJ}}{\left(8.314\frac{J}{K \cdot mol}\right)298\ K}}} = 6.5 \times 10^{-52}.$$

Check: The units (none) are correct. Because there is a large difference between the activation energies, we expect a large difference in the rate constants.

15.103 **Given:** $t_{1/2}$ for radioactive decay of C-14 = 5730 years; bone has 19.5% C-14 in living bone

Find: age of bone

Conceptual Plan: Radioactive decay implies first-order kinetics, $t_{1/2}$ → k then 19.5% of $[C\text{-}14]_0$, k → t

$$t_{1/2} = \frac{0.693}{k} \qquad \ln[A]_t = -kt + \ln[A]_0$$

Solution: $t_{1/2} = \frac{0.693}{k}$ Rearrange to solve for k. $k = \frac{0.693}{t_{1/2}} = \frac{0.693}{5730\ yr} = 1.2\underline{0}942 \times 10^{-4}\ yr^{-1}$ then

$[C\text{-}14]_t = 0.195\ [C\text{-}14]_0$ Because $\ln[C\text{-}14]_t = -kt + \ln[C\text{-}14]_0$, rearrange to solve for t.

$$t = -\frac{1}{k}\ln\frac{[\text{C-14}]_t}{[\text{C-14}]_0} = -\frac{1}{1.2\underline{0}942 \times 10^{-4}\ \text{yr}^{-1}}\ln\frac{0.195\cancel{[\text{C-14}]_0}}{\cancel{[\text{C-14}]_0}} = 1.35 \times 10^4\ \text{yr}$$

Check: The units (yr) are correct. The time to 19.5% decay is consistent with the time being between two and three half-lives.

15.104 **Given:** $t_{1/2}$ for radioactive decay of U-238 = 4.5 billion years; rock has 83.2% of original U-238
Find: age of rock
Conceptual Plan: Radioactive decay implies first-order kinetics, $t_{1/2} \rightarrow k$ then 82.3 of $[\text{U-238}]_0$, $k \rightarrow t$

$$t_{1/2} = \frac{0.693}{k} \qquad \ln[\text{A}]_t = -kt + \ln[\text{A}]_0$$

Solution: $t_{1/2} = \dfrac{0.693}{k}$ Rearrange to solve for k. $k = \dfrac{0.693}{t_{1/2}} = \dfrac{0.693}{4.5 \times 10^9\ \text{yr}} = 1.54 \times 10^{-10}\ \text{yr}^{-1}$ then $[\text{U-238}]_t = 0.823[\text{U-238}]_0$ Because $\ln[\text{U-238}]_t = -kt + \ln[\text{U-238}]_0$, rearrange to solve for t.

$$t = -\frac{1}{k}\ln\frac{[\text{U-238}]_t}{[\text{U-238}]_0} = -\frac{1}{1.54 \times 10^{-10}\ \text{yr}^{-1}}\ln\frac{0.832\cancel{[\text{U-238}]_0}}{\cancel{[\text{U-238}]_0}} = 1.19 \times 10^9\ \text{yr}$$

Check: The units (yr) are correct. The time to 82.3% decay is consistent with the time being less than one half-life.

15.105 (a) For each, check that all steps sum to overall reaction and that the predicted rate law is consistent with experimental data (Rate = $k[\text{H}_2][\text{I}_2]$).
For the first mechanism, the single step is the overall reaction. The rate law is determined by the stoichiometry; so Rate = $k[\text{H}_2][\text{I}_2]$, and the mechanism is valid.
For the second mechanism, the overall reaction is the sum of the steps in the mechanism:

$$\text{I}_2(g) \underset{k_2}{\overset{k_1}{\rightleftharpoons}} \cancel{2\,\text{I}(g)}$$
$$\text{H}_2(g) + \cancel{2\,\text{I}(g)} \xrightarrow[k_3]{} 2\,\text{HI}(g) \qquad \text{So the sum matches the overall reaction.}$$
$$\text{H}_2(g) + \text{I}_2(g) \longrightarrow 2\,\text{HI}(g)$$

Because the second step is the rate-determining step, Rate = $k_3[\text{H}_2][\text{I}]^2$. Because I is an intermediate, its concentration cannot appear in the rate law. Using the fast equilibrium in the first step, we see that $k_1[\text{I}_2] = k_2[\text{I}]^2$ or $[\text{I}]^2 = \dfrac{k_1}{k_2}[\text{I}_2]$. Substituting this into the first rate expression, we get

Rate = $k_3\dfrac{k_1}{k_2}[\text{H}_2][\text{I}_2]$, which can be simplified to rate = $k[\text{H}_2][\text{I}_2]$ and the mechanism is valid.

(b) To distinguish between mechanisms, you could look for the buildup of I(g), the intermediate in the second mechanism.

15.106 (a) The overall reaction is the sum of the steps in the mechanism:

$$\text{NH}_3(aq) + \text{OCl}^-(aq) \underset{k_2}{\overset{k_1}{\rightleftharpoons}} \cancel{\text{NH}_2\text{Cl}(aq)} + \cancel{\text{OH}^-(aq)}$$
$$\cancel{\text{NH}_2\text{Cl}(aq)} + \text{NH}_3(aq) \xrightarrow[k_3]{} \cancel{\text{N}_2\text{H}_5^+(aq)} + \text{Cl}^-(aq)$$
$$\cancel{\text{N}_2\text{H}_5^+(aq)} + \cancel{\text{OH}^-(aq)} \xrightarrow[k_4]{} \text{N}_2\text{H}_4(aq) + \text{H}_2\text{O}(l)$$
$$2\,\text{NH}_3(aq) + \text{OCl}^-(aq) \longrightarrow \text{N}_2\text{H}_4(aq) + \text{H}_2\text{O}(l) + \text{Cl}^-(aq)$$

So the sum matches the overall reaction.

(b) Because the second step is the rate-determining step, Rate = $k_3[\text{NH}_2\text{Cl}][\text{NH}_3]$. Because NH_2Cl is an intermediate, its concentration cannot appear in the rate law. Using the fast equilibrium in the first step, we see that $k_1[\text{NH}_3][\text{OCl}^-] = k_2[\text{NH}_2\text{Cl}][\text{OH}^-]$ or $[\text{NH}_2\text{Cl}] = \dfrac{k_1}{k_2}\dfrac{[\text{NH}_3][\text{OCl}^-]}{[\text{OH}^-]}$. Substituting this into the first rate expression, we get Rate = $k_3\dfrac{k_1}{k_2}\dfrac{[\text{NH}_3][\text{OCl}^-]}{[\text{OH}^-]}[\text{NH}_3]$ or Rate = $k_3\dfrac{k_1}{k_2}\dfrac{[\text{NH}_3]^2[\text{OCl}^-]}{[\text{OH}^-]}$. In a pH-neutral solution, $[\text{OH}^-] = 10^{-7}$ (see Chapter 17), so that the rate law can be approximated as Rate = $k[\text{NH}_3]^2[\text{OCl}^-]$.

15.107 The steps in the mechanism are as follows:

$$\text{Br}_2(g) \underset{k_{-1}}{\overset{k_1}{\rightleftharpoons}} 2\,\text{Br}(g)$$

$\cancel{Br(g)} + H_2(g) \underset{k_2}{\rightarrow} HBr(g) + \cancel{H(g)}$

$\cancel{H(g)} + Br_2(g) \underset{k_3}{\rightarrow} HBr(g) + \cancel{Br(g)}$

Because the second step is the rate-determining step, Rate $= k_2[H_2][Br]$. Because Br is an intermediate, its concentration cannot appear in the rate law. Using the fast equilibrium in the first step, we see that

$k_1[Br_2] = k_{-1}[Br]^2$ or $[Br] = \sqrt{\frac{k_1}{k_{-1}}}[Br_2]^{1/2}$. Substituting this into the first rate expression, we get

Rate $= k_2\sqrt{\frac{k_1}{k_{-1}}}[H_2][Br_2]^{1/2}$. The rate law is 3/2 order overall.

15.108 The sum of the steps in the mechanism is as follows:

$I_2(g) \underset{k_{-1}}{\overset{k_1}{\rightleftharpoons}} 2\,\cancel{I(g)}$

$\cancel{I(g)} + H_2(g) \underset{k_{-2}}{\overset{k_2}{\rightleftharpoons}} \cancel{H_2I(g)}$

$\cancel{H_2I(g)} + \cancel{I(g)} \underset{k_3}{\rightarrow} 2\,HI(g)$

$H_2(g) + I_2(g) \rightarrow 2\,HI(g)$

Because the third step is the rate-determining step, Rate $= k_3[H_2I][I]$. Because H_2I and I are intermediates, their concentrations cannot appear in the rate law.

Using the fast equilibrium in the first step, we see that $k_1[I_2] = k_{-1}[I]^2$ or $[I] = \sqrt{\frac{k_1}{k_{-1}}}[I_2]^{1/2}$.

Using the fast equilibrium in the second step, we see that $k_2[H_2][I] = k_{-2}[H_2I]$ or $[H_2I] = \frac{k_2}{k_{-2}}[H_2][I]$.

Substituting these into the first rate expression, we get Rate $= k_3\frac{k_2}{k_{-2}}[H_2][I]\sqrt{\frac{k_1}{k_{-1}}}[I_2]^{1/2}$. This expression contains I, so another substitution of the first equilibrium expression needs to be done to give

Rate $= k_3\frac{k_2}{k_{-2}}[H_2]\sqrt{\frac{k_1}{k_{-1}}}[I_2]^{1/2}\sqrt{\frac{k_1}{k_{-1}}}[I_2]^{1/2}$. Simplifying, we get Rate $= k_3\frac{k_1}{k_{-1}}\frac{k_2}{k_{-2}}[H_2][I_2]$ or

Rate $= k_{obs}[H_2][I_2]$.

15.109 (a) For a zero-order reaction, the rate is independent of the concentration. If the first half goes in the first 100 minutes, the second half will go in the second 100 minutes. This means that none, or 0%, will be left at 200 minutes.

(b) For a first-order reaction, the half-life is independent of concentration. This means that if half of the reactant decomposes in the first 100 minutes, half of this (or another 25% of the original amount) will decompose in the second 100 minutes. This means that at 200 minutes, 50% + 25% = 75% has decomposed or 25% remains.

(c) For a second-order reaction, $t_{1/2} = \frac{1}{k[A]_0} = 100$ min and the integrated rate expression is $\frac{1}{[A]_t} = kt + \frac{1}{[A]_0}$.

We can rearrange the first expression to solve for k as $k = \frac{1}{100\text{ min }[A]_0}$. Substituting this and 200 minutes into the integrated rate expression, we get $\frac{1}{[A]_t} = \frac{200\,\cancel{\text{min}}}{100\,\cancel{\text{min}}\,[A]_0} + \frac{1}{[A]_0} \rightarrow \frac{1}{[A]_t} = \frac{3}{[A]_0} \rightarrow \frac{[A]_t}{[A]_0} = \frac{1}{3}$ or 33% remains.

15.110 **Given:** $t_{1/2}$ for radioactive decay of Pu-239 = 24,000 years; 1 mole initially to 1 atom **Find:** t

Conceptual Plan: Radioactive decay implies first-order kinetics, $t_{1/2} \rightarrow k$ then $[\text{Pu-239}]_0, [\text{Pu-239}]_t, k \rightarrow t$

$t_{1/2} = \frac{0.693}{k}$ $\quad\quad$ $\ln[A]_t = -kt + \ln[A]_0$

Solution: $t_{1/2} = \frac{0.693}{k}$ Rearrange to solve for k. $k = \frac{0.693}{t_{1/2}} = \frac{0.693}{24000\text{ yr}} = 2.8875 \times 10^{-5}\text{ yr}^{-1}$ then

$[\text{Pu-239}]_t = 1$ and $[\text{Pu-239}]_0 = 6.022 \times 10^{23}$ Because $\ln[\text{Pu-239}]_t = -kt + \ln[\text{Pu-239}]_0$, rearrange to solve

for t. $t = -\frac{1}{k}\ln\frac{[\text{Pu-239}]_t}{[\text{Pu-239}]_0} = -\frac{1}{2.8875 \times 10^{-5}\text{ yr}^{-1}}\ln\frac{1\ \cancel{\text{atom Pu-239}}}{6.022 \times 10^{23}\ \cancel{\text{atom Pu-239}}} = 1.9 \times 10^6\text{ yr}$

Check: The units (yr) are correct. The time to decay to 1 atom is consistent with the time being 79 half-lives, which makes sense because $2^{79} = 6.09 \times 10^{23}$.

15.111 Using the energy diagram shown and using Hess's law, we can see that the activation energy for the decomposition is equal to the activation energy for the formation reaction plus the heat of formation of 2 moles of HI, or $E_{a,\text{ formation}} = E_{a,\text{ decomposition}} + 2\Delta H_f^\circ(\text{HI})$. So $E_{a,\text{ formation}} = 185 \text{ kJ} + 2 \text{ mol}(-5.65 \text{ kJ/mol}) = 174 \text{ kJ}$.

Check: Because the reaction is endothermic, we expect the activation energy in the reverse direction to be less than that in the forward direction.

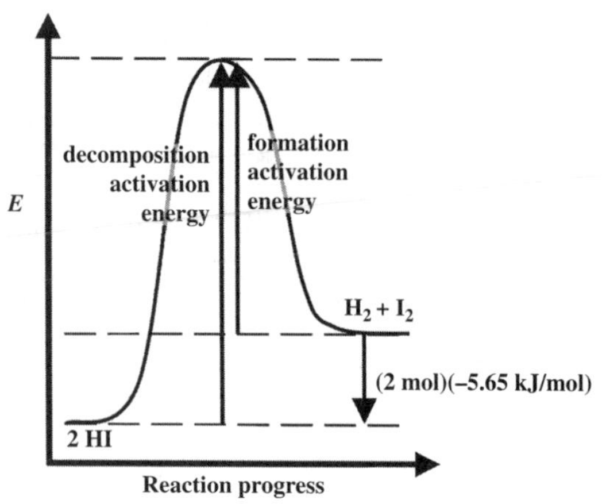

Note: Energy axis is not to scale.

15.112 **Given:** first-order reaction, $E_a = 249 \text{ kJ/mol}$, $A = 1.6 \times 10^{14} \text{ s}^{-1}$, and 710 K
Find: k, fraction decomposed in 15 min and T for double the reaction rate
Conceptual Plan: $T, A, E_a \rightarrow k$ **then min → s then** $k, t \rightarrow$ **fraction decomposed then** $k_2/k_1, T_1, E_a \rightarrow T_2$

$$k = Ae^{-E_a/RT} \qquad \frac{60 \text{ s}}{1 \text{ min}} \qquad \ln[\text{A}]_t = -kt + \ln[\text{A}]_0 \qquad \ln\left(\frac{k_2}{k_1}\right) = \frac{E_a}{R}\left(\frac{1}{T_1} - \frac{1}{T_2}\right)$$

Solution: Because $k = Ae^{-E_a/RT} = 1.6 \times 10^{14} \text{ s}^{-1} e^{\frac{-249 \frac{\text{kJ}}{\text{mol}} \times \frac{1000 \text{ J}}{1 \text{ kJ}}}{\left(8.314 \frac{\text{J}}{\text{K} \cdot \text{mol}}\right) 710 \text{ K}}} = 7.\underline{6}657 \times 10^{-5} \text{ s}^{-1} = 7.7 \times 10^{-5} \text{ s}^{-1}$,

then $15 \text{ min} \times \frac{60 \text{ s}}{1 \text{ min}} = 9\underline{0}0 \text{ s}$ in $\ln[\text{C}_2\text{H}_5\text{Cl}]_t = -kt + \ln[\text{C}_2\text{H}_5\text{Cl}]_0$. Rearrange to solve for

fraction remaining $\rightarrow \frac{[\text{C}_2\text{H}_5\text{Cl}]_t}{[\text{C}_2\text{H}_5\text{Cl}]_0} = e^{-kt} = e^{-(7.\underline{6}657 \times 10^{-5} \text{ s}^{-1})(9\underline{0}0 \text{ s})} = 0.9\underline{3}333$; thus, fraction decomposed $=$

$1 - 0.9\underline{3}333 = 0.0\underline{6}667 = 0.07$. $T_1 = 710 \text{ K}$, $k_2/k_1 = 2$, and $\ln\left(\frac{k_2}{k_1}\right) = \frac{E_a}{R}\left(\frac{1}{T_1} - \frac{1}{T_2}\right)$. Rearrange to solve for T_2.

$$T_2 = \frac{\frac{E_a}{R}}{\frac{E_a}{RT_1} - \ln\left(\frac{k_2}{k_1}\right)} = \frac{\frac{249 \frac{\text{kJ}}{\text{mol}} \times \frac{1000 \text{ J}}{1 \text{ kJ}}}{8.314 \frac{\text{J}}{\text{K} \cdot \text{mol}}}}{\left(\frac{249 \frac{\text{kJ}}{\text{mol}} \times \frac{1000 \text{ J}}{1 \text{ kJ}}}{8.314 \frac{\text{J}}{\text{K} \cdot \text{mol}} \times 710 \text{ K}}\right) - \ln(2)} = 7\underline{2}1.86 \text{ K} = 720 \text{ K}$$

Check: The units (s^{-1}, none, and K) are correct. The reaction rate is reasonable considering the frequency factor, the activation energy, and T. The fraction decomposed is reasonable because 900 s is a small fraction of the half-life. The temperature is reasonable because many reactions double their rate with an increase in temperature of 10 K.

Challenge Problems

15.113 (a) Because the rate-determining step involves the collision of two molecules, the expected reaction order would be second order.

(b) The proposed mechanism is

$\text{CH}_3\text{NC} + \cancel{\text{CH}_3\text{NC}} \underset{k_2}{\overset{k_1}{\rightleftharpoons}} \cancel{\text{CH}_3\text{NC*}} + \cancel{\text{CH}_3\text{NC}}$ (fast)

$\cancel{\text{CH}_3\text{NC*}} \xrightarrow{k_3} \text{CH}_3\text{CN}$ (slow)

$\text{CH}_3\text{NC} \longrightarrow \text{CH}_3\text{CN}$ So the sum matches the overall reaction.

$\text{CH}_3\text{NC*}$ is the activated molecule. Because the second step is the rate-determining step, Rate $= k_3[\text{CH}_3\text{NC*}]$. Because $\text{CH}_3\text{NC*}$ is an intermediate, its concentration cannot appear in the rate law. Using the fast equilibrium in the first step, we see that $k_1[\text{CH}_3\text{NC}]^2 = k_2[\text{CH}_3\text{NC*}][\text{CH}_3\text{NC}]$ or $[\text{CH}_3\text{NC*}] = \frac{k_1}{k_2}[\text{CH}_3\text{NC}]$. Substituting this into the first rate expression, we get Rate $= k_3\frac{k_1}{k_2}[\text{CH}_3\text{NC}]$, which simplifies to Rate $= k[\text{CH}_3\text{NC}]$. This matches the experimental observation of first order, and the mechanism is valid.

15.114 (a) Rate $= k[\text{A}]^{1/2}$ and Rate $= -\frac{d[\text{A}]}{dt}$, so $\frac{d[\text{A}]}{dt} = -k[\text{A}]^{1/2}$. Moving the A terms to the left and the t and constants to the right, we have $\frac{d[\text{A}]}{[\text{A}]^{1/2}} = -kdt$. Integrating, we get $\int_{[\text{A}]_0}^{[\text{A}]} \frac{d[\text{A}]}{[\text{A}]^{1/2}} = -\int_0^t kdt$. When we evaluate this integral, $2[\text{A}]^{1/2}\Big|_{[\text{A}]_0}^{[\text{A}]} = -kt\Big|_0^t \rightarrow 2[\text{A}]_t^{1/2} - 2[\text{A}]_0^{1/2} = -kt \rightarrow 2[\text{A}]_t^{1/2} = -kt + 2[\text{A}]_0^{1/2}$, which is the desired integrated rate law.

(b) To derive the half-life, set $[\text{A}]_t = 1/2[\text{A}]_0$. Substituting this into $2[\text{A}]_t^{1/2} = -kt + 2[\text{A}]_0^{1/2}$, we get $2(1/2[\text{A}]_0)^{1/2} = -kt_{1/2} + 2[\text{A}]_0^{1/2} \rightarrow t_{1/2} = \frac{2[\text{A}]_0^{1/2} - 2(1/2[\text{A}]_0)^{1/2}}{k} \rightarrow t_{1/2} = \frac{(2 - 2\sqrt{2})[\text{A}]_0^{1/2}}{k}$.

15.115 Rate $= k[\text{A}]^2$ and Rate $= -\frac{d[\text{A}]}{dt}$, so $\frac{d[\text{A}]}{dt} = -k[\text{A}]^2$. Moving the A terms to the left and the t and constants to the right, we have $\frac{d[\text{A}]}{[\text{A}]^2} = -kdt$. Integrating, we get $\int_{[\text{A}]_0}^{[\text{A}]} \frac{d[\text{A}]}{[\text{A}]^2} = -\int_0^t kdt$. When we evaluate this integral, $-[\text{A}]^{-1}\Big|_{[\text{A}]_0}^{[\text{A}]} = -kt\Big|_0^t \rightarrow -[\text{A}]_t^{-1} - (-[\text{A}]_0^{-1}) = -kt \rightarrow [\text{A}]_t^{-1} = kt + [\text{A}]_0^{-1}$ or $\frac{1}{[\text{A}]_t} = kt + \frac{1}{[\text{A}]_0}$, which is the desired integrated rate law.

15.116 (a) **Given:** N_2O_5 decomposes to NO_2 and O_2, first order in $[N_2O_5]$; $k = 7.48 \times 10^{-3}\ \text{s}^{-1}$; $P^\circ_{N_2O_5} = 0.100$ atm
Find: t to $P_{\text{Total}} = 0.145$ atm
Conceptual Plan: Write a balanced reaction. Assume that x is the amount of N_2O_5 that has reacted.

$$N_2O_5 \rightarrow 2\ NO_2 + \frac{1}{2}O_2$$

Then write the expression for P_{Total} in terms of amount reacted then x, $P^\circ_{N_2O_5}$, $k \rightarrow t$

let $x = P_{N_2O_5\ \text{reacted}}$ $P_{\text{Total}} = P_{N_2O_5} + P_{NO_2} + P_{O_2}$ $\ln[\text{A}]_t = -kt + \ln[\text{A}]_0$

Solution: $P_{\text{Total}} = P_{N_2O_5} + P_{NO_2} + P_{O_2} = (0.100\ \text{atm} - x) + (2x) + (1/2x) = 0.100\ \text{atm} + 1.5\,x$ Set $P_{\text{Total}} = 0.145\ \text{atm} = 0.100\ \text{atm} + 1.5\,x$ and solve for x. Note that for every 1 mole of reactant decomposing, 2.5 moles of product are generated; so the pressure increases by a factor of 1.5.

$$x = \frac{0.145\ \text{atm} - 0.100\ \text{atm}}{1.5} = 0.030\ \text{atm}$$

then $P_{N_2O_5} = 0.100\ \text{atm} - x = 0.100\ \text{atm} - 0.030\ \text{atm} = 0.070$ atm. Because $P \propto n/V$ or M, $\ln[N_2O_5]_t = -kt + \ln[N_2O_5]_0$. Rearrange to solve for t.

$$t = -\frac{\ln\frac{[N_2O_5]_t}{[N_2O_5]_0}}{k} = -\frac{\ln\left(\frac{0.070\ \cancel{\text{atm}}}{0.100\ \cancel{\text{atm}}}\right)}{7.48 \times 10^{-3}\ \text{s}^{-1}} = 4\underline{7}.684\ \text{s} = 48\ \text{s}$$

Check: The units (s) are correct. The time is reasonable because it is less than one half-life and the amount decomposing is less than 50%.

(b) **Given:** N_2O_5 decomposes to NO_2 and O_2, first order $[N_2O_5]$; $k = 7.48 \times 10^{-3}\ \text{s}^{-1}$; $P^\circ_{N_2O_5} = 0.100$ atm
Find: t to $P_{\text{Total}} = 0.200$ atm
Conceptual Plan: Write a balanced reaction.

$$N_2O_5 \rightarrow 2\ NO_2 + \frac{1}{2}O_2$$

Then write the expression for P_{Total} in terms of amount reacted then x, $P^\circ_{N_2O_5}$, $k \rightarrow t$

let $x = P_{N_2O_5\ \text{reacted}}$ $P_{\text{Total}} = P_{N_2O_5} + P_{NO_2} + P_{O_2}$ $\ln[\text{A}]_t = -kt + \ln[\text{A}]_0$

Solution: $P_{Total} = P_{N_2O_5} + P_{NO_2} + P_{O_2} = (0.100\text{ atm} - x) + (2x) + (1/2x) = 0.100\text{ atm} + 1.5x$ $P_{Total} = 0.200\text{ atm} = 0.100\text{ atm} + 1.5x$. Solve for x. $x = \frac{0.200\text{ atm} - 0.100\text{ atm}}{1.5} = 0.06\underline{6}667\text{ atm}$ then $P_{N_2O_5} = 0.100\text{ atm} - x = 0.100\text{ atm} - 0.06\underline{6}667\text{ atm} = 0.03\underline{3}333\text{ atm}$. Because $P \propto n/V$ or M,

$\ln[N_2O_5]_t = -kt + \ln[N_2O_5]_0$. Rearrange to solve for t.

$$t = -\frac{\ln\frac{[N_2O_5]_t}{[N_2O_5]_0}}{k} = -\frac{\ln\left(\frac{0.03\underline{3}333\text{ atm}}{0.100\text{ atm}}\right)}{7.48 \times 10^{-3}\text{ s}^{-1}} = 1\underline{4}6.875\text{ s} = 150\text{ s} = 1.5 \times 10^2\text{ s}$$

Check: The units (s) are correct. The time is reasonable because it is between one and two half-lives and the amount decomposing is 67%.

(c) **Given:** N_2O_5 decomposes to NO_2 and O_2, first order in $[N_2O_5]$; $k = 7.48 \times 10^{-3}\text{ s}^{-1}$; $P^\circ_{N_2O_5} = 0.100\text{ atm}$
Find: P_{Total} after 100 s
Conceptual Plan: $P^\circ_{N_2O_5}, k, t \rightarrow P_{N_2O_5}$ then write a balanced reaction. Then $P^\circ_{N_2O_5}, P_{N_2O_5} \rightarrow x$ then

$\ln[A]_t = -kt + \ln[A]_0$ $\quad N_2O_5 \rightarrow 2\,NO_2 + \frac{1}{2}O_2$ $\quad x = P^\circ_{N_2O_5} - P_{N_2O_5}$

write the expression for P_{Total} in terms of amount reacted

let $x = P_{N_2O_5\text{ reacted}}$ $\quad P_{Total} = P_{N_2O_5} + P_{NO_2} + P_{O_2}$

Solution: Because $P \propto n/V$ or M,
$\ln[N_2O_5]_t = -kt + \ln[N_2O_5]_0 = -(7.48 \times 10^{-3}\text{ s}^{-1})(100\text{ s}) + \ln(0.100\text{ atm}) = -3.0\underline{5}059$
$P_{N_2O_5} = e^{-3.0\underline{5}059} = 0.047\underline{3}310\text{ atm}$ so
$x = P^\circ_{N_2O_5} - P_{N_2O_5} = 0.100\text{ atm} - 0.047\underline{3}310\text{ atm} = 0.05\underline{2}669\text{ atm}$
finally, $P_{Total} = P_{N_2O_5} + P_{NO_2} + P_{O_2} = (0.100\text{ atm} - x) + (2x) + (1/2x) = 0.100\text{ atm} + 1.5x = 0.100\text{ atm} + 1.5(0.05\underline{2}669\text{ atm}) = 0.179\text{ atm}$

Check: The units (atm) are correct. The pressure is reasonable because the time is between those for parts (a) and (b).

15.117 For this mechanism, the overall reaction is the sum of the steps in the mechanism:

$$Cl_2(g) \underset{k_2}{\overset{k_1}{\rightleftharpoons}} 2\,Cl(g)$$
$$\cancel{Cl(g)} + CO(g) \underset{k_4}{\overset{k_3}{\rightleftharpoons}} \cancel{ClCO(g)}$$
$$\cancel{ClCO(g)} + Cl_2(g) \xrightarrow{k_5} \cancel{Cl_2CO(g)} + Cl(g)$$
$$\overline{CO(g) + 2\,Cl_2(g) \rightarrow Cl_2CO(g) + 2\,Cl(g)}$$

No overall reaction is given. Because the third step is the rate-determining step, Rate $= k_5[ClCO][Cl_2]$. Because ClCO is an intermediate, its concentration cannot appear in the rate law. Using the fast equilibrium in the second step, we see that $k_3[Cl][CO] = k_4[ClCO]$ or $[ClCO] = \frac{k_3}{k_4}[Cl][CO]$. Substituting this expression into the first rate expression, we get Rate $= k_5\frac{k_3}{k_4}[Cl][CO][Cl_2]$. Because Cl is an intermediate, its concentration cannot appear in the rate law. Using the fast equilibrium in the first step, we see that $k_1[Cl_2] = k_2[Cl]^2$ or $[Cl] = \sqrt{\frac{k_1}{k_2}[Cl_2]}$. Substituting this expression into last rate expression, we get Rate $= k_5\frac{k_3}{k_4}\sqrt{\frac{k_1}{k_2}[Cl_2]}[CO][Cl_2] = k_5\frac{k_3}{k_4}\sqrt{\frac{k_1}{k_2}}[CO][Cl_2]^{3/2}$.

Simplifying this expression, we see that Rate $= k\,[CO][Cl_2]^{3/2}$.

15.118 **Given:** N_2O_3 decomposes to NO_2 and NO, first order in $[N_2O_3]$, table of $[NO_2]$ versus time, at 50,000 $s = [N_2O_3] = 0$ **Find:** k
Conceptual Plan: Write a balanced reaction. Then write the expression for $[NO_2]$ in terms $[N_2O_3]$, then

$N_2O_3 \rightarrow NO_2 + NO$ $\quad [N_2O_3] = 0.784\text{ M} - [NO_2]$

plot $\ln[N_2O_3]$ versus time. Because $\ln[A]_t = -kt + \ln[A]_0$, the negative of the slope will be the rate constant.

Solution: $[N_2O_3] = 0.784\text{ M} - [NO_2]$
Plot $\ln[N_2O_3]$ versus time. Because $\ln[A]_t = -kt + \ln[A]_0$, the negative of the slope will be the rate constant. The slope can be determined by measuring $\Delta y/\Delta x$ on the plot or by using functions such as "add trendline" in Excel. The last point (50,000 s) cannot be plotted because the concentration is 0 and the ln0 is undefined. Thus, the rate constant is $3.20 \times 10^{-4}\text{ s}^{-1}$, and the rate law is Rate $= 3.20 \times 10^{-4}\text{ s}^{-1}[N_2O_3]$.

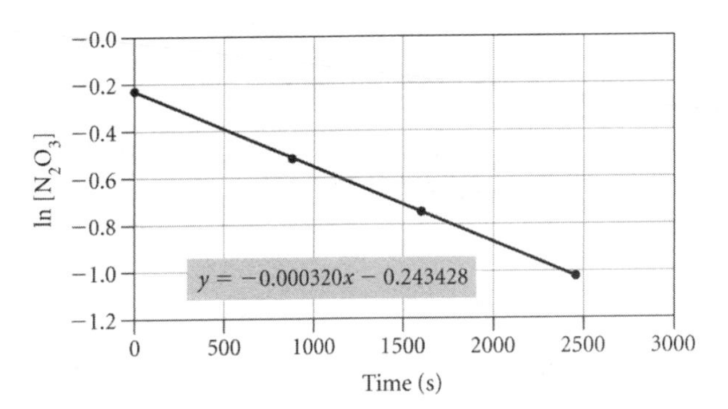

Check: The units (s^{-1}) are correct. The rate constant is typical for a reaction.

15.119 For the elementary reaction $2\,NOCl(g) \underset{k_{-1}}{\overset{k_1}{\rightleftharpoons}} 2\,NO(g) + Cl_2(g)$, we see that $k_1[NOCl]^2 = k_{-1}[NO]^2[Cl_2]$. For each mole of NOCl that reacts, 1 mole of NO and 0.5 mole of Cl_2 are generated. Because before any reaction only NOCl is present, $[NO] = 2[Cl_2]$. Substituting this into the first expression, we get $k_1[NOCl]^2 = k_{-1}(2[Cl_2])^2[Cl_2] \rightarrow k_1[NOCl]^2 = 4k_{-1}[Cl_2]^3$. Rearranging and substituting the specific values into this expression, we get

$$[Cl_2] = \sqrt[3]{\frac{k_1}{4k_{-1}}}[NOCl]^{2/3} = \sqrt[3]{\frac{7.8 \times 10^{-2}\,\frac{\cancel{L}^2}{\text{mol}^{\cancel{2}} \cdot \cancel{s}}}{4\left(4.7 \times 10^2 \frac{\cancel{L}^2}{\cancel{\text{mol}} \cdot \cancel{s}}\right)}}\left(0.12\,\frac{\text{mol}}{\text{L}}\right)^{2/3} = 0.00\underline{8}42235\,\frac{\text{mol}}{\text{L}}\,Cl_2$$

$$= 0.0084\text{ M }Cl_2 \text{ and } [Cl_2] = 2[NO] = 2\left(0.00\underline{8}42235\frac{\text{mol}}{\text{L}}\right) = 0.01\underline{6}8445\frac{\text{mol}}{\text{L}}NO = 0.017\text{ M NO}$$

Conceptual Problems

15.120 Because Rate $= k\,[CHCl_3][Cl_2]^{1/2}$, Rate a $= k(3)(3)^{1/2} = 5.2\,k$, Rate b $= k(4)(2)^{1/2} = 5.7\,k$, and Rate c $= k(2)(4)^{1/2} = 4k$. So b has the fastest rate.

15.121 Reactant concentrations drop more quickly for first-order reactions than for second-order reactions, so reaction A must be second order. A plot of 1/[A] versus time will be linear $\left(\frac{1}{[A]_t} = kt + \frac{1}{[A]_0}\right)$. Reaction B is first order. A plot of ln[A] versus time will be linear $(\ln[A]_t = -kt + \ln[A]_0)$.

15.122 A reaction that slows down as the reaction proceeds and has a half-life that is dependent on the concentration is a second-order reaction. Statement (a) is false because it describes a first-order reaction. Statement (b) is true because it describes a second-order reaction. Statements (c) and (d) are false because they describe a zero-order reaction.

16 Chemical Equilibrium

Review Questions

16.1 Like adult hemoglobin, fetal hemoglobin is in equilibrium with oxygen. However, the equilibrium constant for fetal hemoglobin is larger than the equilibrium constant for adult hemoglobin, meaning that the reaction tends to go farther in the direction of the product. Consequently, fetal hemoglobin loads oxygen at a lower oxygen concentration than does adult hemoglobin. In the placenta, fetal blood flows in close proximity to maternal blood, without the two ever mixing. Because of the different equilibrium constants, the maternal hemoglobin unloads oxygen, which the fetal hemoglobin then binds and carries into its own circulatory system. Nature has thus evolved a chemical system through which the mother's hemoglobin can in effect hand off oxygen to the hemoglobin of the fetus.

16.2 Dynamic equilibrium in a chemical reaction is the condition in which the rate of the forward reaction equals the rate of the reverse reaction. Dynamic equilibrium is called dynamic because the forward and reverse reactions are still occurring; however, they are occurring at the same rate.

16.3 The general expression of the equilibrium constant is $K = \dfrac{[\mathrm{C}]^c[\mathrm{D}]^d}{[\mathrm{A}]^a[\mathrm{B}]^b}$.

16.4 The equilibrium constant for a reaction is defined as the ratio—at equilibrium—of the concentrations of the products raised to their stoichiometric coefficients divided by the concentrations of the reactants raised to their stoichiometric coefficients. If the equilibrium constant for a reaction is large, the equilibrium point of the reaction lies far to the right—the concentration of products is large, and the concentration of reactants is small. If the equilibrium constant for a reaction is small, the equilibrium point of the reaction lies far to the left—the concentration of products is small, and the concentration of reactants is large.

16.5 If you reverse the equation, invert the equilibrium constant. $K_{\text{reverse}} = \dfrac{1}{K_{\text{forward}}}$

If you multiply the coefficients in the equation by a factor, raise the equilibrium constant to the same factor. That is, if you multiply the reaction by n, raise the equilibrium constant K to the n. $K' = K^n$

16.6 If you add two or more individual chemical equations to obtain an overall equation, multiply the corresponding equilibrium constants by each other to obtain the overall equilibrium constant. For two reactions with equilibrium constants of K_1 and K_2, the equilibrium constant for the combined reaction $K_3 = K_1K_2$.

16.7 K_c is the equilibrium constant with respect to concentration in molarity. K_p is the equilibrium constant with respect to partial pressures in atmospheres. The two can be related as follows: $K_p = K_c(RT)^{\Delta n}$

16.8 For the equilibrium constant, concentrations are expressed in terms of molarity and partial pressures are expressed in terms of atmospheres. The equilibrium constant is usually written with no units. Formally, the values of concentration or partial pressures that we substitute into the equilibrium constant expressions are ratios of the concentration or pressure to a reference concentration or reference pressure. As long as concentration units are expressed in molarity or atmospheres, we simply enter the quantities directly into the equilibrium expression, dropping their corresponding units.

16.9 The concentration of a solid does not change because a solid does not expand to fill its container. Its concentration, therefore, depends only on its density, which is constant as long as some solid is present. Consequently, solids are not included in the equilibrium expression. Similarly, the concentration of a pure liquid does not change, so liquids are also excluded from the equilibrium expression.

16.10 For any reaction, the equilibrium concentrations of the reactants and products will depend on the initial concentration. However, the equilibrium constant will always be the same at a given temperature, regardless of the initial concentrations.

16.11 When we know the initial concentrations of the reactants and products and the equilibrium concentration of one reactant or product, the other equilibrium concentrations can be deduced from the stoichiometry of the reaction. From the initial and equilibrium concentration of the one reactant, we can find the change in concentration for that reactant. Using the stoichiometry of the reaction, we can determine the equilibrium concentrations of the other reactants and products. We generally use an ICE table (I = Initial C = Change E = Equilibrium) to keep track of the changes.

16.12 The definition of the reaction quotient takes the same form as the definition of the equilibrium constant, except that the reaction need not be at equilibrium. So the reaction quotient (Q_c) is the ratio—at any point in the reaction—of the concentrations of the products raised to their stoichiometric coefficients divided by the concentration of the reactants raised to their stoichiometric coefficients. The value of Q relative to K is a measure of the progress of the reaction toward equilibrium. At equilibrium, the reaction quotient is equal to the equilibrium constant.

16.13 Standard states are defined as 1 M concentration, or 1 atm pressure. So the value of K_c and K_p when the reactants and products are in their standard states is 1.

16.14
(a) When $Q < K$, the reaction goes to the right (toward products).
(b) When $Q > K$, the reaction goes to the left (toward reactants).
(c) When $Q = K$, the reaction is at equilibrium.

16.15 To solve a problem given initial concentrations and the equilibrium constant, you would prepare an ICE table, calculate Q, compare Q and K_c, predict the direction of the reaction, represent the change with x, sum the table, determine the equilibrium values, put the equilibrium values in the equilibrium expression, and solve for x. Determine the reactant and product concentration.

16.16 When adding or subtracting a small value of x, we can often ignore x because of significant figures. In addition and subtraction, the number of significant figures depends on the number with the fewest decimal places. Therefore, a small value of x would not be significant. In multiplication and division, the number of significant figures depends on the number with the fewest significant figures; therefore, we cannot ignore x in this case.

16.17 According to Le Châtelier's principle, when a chemical system at equilibrium is disturbed, the system shifts in a direction that minimizes the disturbance.

16.18 If a system is at equilibrium:

- increasing the concentration of one or more of the reactants causes the reaction to shift to the right.
- increasing the concentration of one or more of the products causes the reaction to shift to the left.
- decreasing the concentration of one or more of the reactants causes the reaction to shift to the left.
- decreasing the concentration of one or more of the products causes the reaction to shift to the right.

16.19 If a chemical system is at equilibrium:

- decreasing the volume causes the reaction to shift in the direction that has the fewer moles of gas particles.
- increasing the volume causes the reaction to shift in the direction that has the greater number of moles of gas particles.
- if a reaction has an equal number of moles of gas on both sides of the chemical equation, then a change in volume produces no effect on the equilibrium.
- adding an inert gas to the mixture at a fixed volume has no effect on the equilibrium.

16.20 If the temperature of a system at equilibrium is changed, the system shifts in a direction to counter that change. So if the temperature is increased, the reaction shifts in the direction that tends to decrease the temperature and vice versa. In an exothermic chemical reaction:

- increasing the temperature causes the reaction to shift left and the value of the equilibrium constant decreases.
- decreasing the temperature causes the reaction to shift right and the value of the equilibrium constant increases.

In an endothermic chemical reaction:

- increasing the temperature causes the reaction to shift right and the equilibrium constant increases.
- decreasing the temperature causes the reaction to shift left and the equilibrium constant decreases.

Problems by Topic

Equilibrium and the Equilibrium Constant

16.21 The equilibrium constant is defined as the concentrations of the products raised to their stoichiometric coefficients divided by the concentrations of the reactants raised to their stoichiometric coefficients.

(a) $K = \dfrac{[SbCl_3][Cl_2]}{[SbCl_5]}$

(b) $K = \dfrac{[NO]^2[Br_2]}{[BrNO]^2}$

(c) $K = \dfrac{[CS_2][H_2]^4}{[CH_4][H_2S]^2}$

(d) $K = \dfrac{[CO_2]^2}{[CO]^2[O_2]}$

16.22 (a) The equilibrium constant is defined as the concentrations of the products raised to their stoichiometric coefficients divided by the concentrations of the reactants raised to their stoichiometric coefficients.

$K = \dfrac{[H_2]^2[S_2]}{[H_2S]^2}$

(b) The equilibrium constant is defined as the concentrations of the products raised to their stoichiometric coefficients divided by the concentrations of the reactants raised to their stoichiometric coefficients.

$K = \dfrac{[COCl_2]}{[CO][Cl_2]}$

16.23 With an equilibrium constant of 1.4×10^{-5}, the value of the equilibrium constant is small; therefore, the concentration of reactants will be greater than the concentration of products. This is independent of the initial concentration of the reactants and products.

16.24 Figure a at equilibrium has 8 $C_2H_4Cl_2$, 2 Cl_2, and 2 C_2H_4.
Figure b at equilibrium has 6 $C_2H_4Br_2$, 4 Br_2, and 4 C_2H_4.
Figure c at equilibrium has 3 $C_2H_4I_2$, 7 I_2, and 7 C_2H_4.
Because the equilibrium constant is concentration of products/concentration of reactants, the equilibrium situation that has the largest concentration of products will have the largest equilibrium constant. Therefore, $K_{Cl_2} > K_{Br_2} > K_{I_2}$.

16.25 (i) has 10 H_2 and 10 I_2
(ii) has 7 H_2 and 7 I_2 and 6 HI
(iii) has 5 H_2 and 5 I_2 and 10 HI
(iv) has 4 H_2 and 4 I_2 and 12 HI
(v) has 3 H_2 and 3 I_2 and 14 HI
(vi) has 3 H_2 and 3 I_2 and 14 HI

(a) Concentration of (v) and (vi) are the same, so the system reached equilibrium at (v).

(b) If a catalyst was added to the system, the system would reach the conditions at (v) sooner because a catalyst speeds up the reaction but does not change the equilibrium conditions.

(c) The final figure (vi) would have the same amount of reactants and products because a catalyst speeds up the reaction but does not change the equilibrium concentrations.

16.26 The equilibrium constant gives us the ratio of products to reactants at equilibrium; it does not say how long it takes to reach equilibrium. So after 15 minutes, if the smaller equilibrium constant has more products, the kinetics of that reaction are faster.

16.27 (a) If you reverse the reaction, invert the equilibrium constant. So $K' = \frac{1}{K_p} = \frac{1}{2.26 \times 10^4} = 4.42 \times 10^{-5}$.

The reactants will be favored.

(b) If you multiply the coefficients in the equation by a factor, raise the equilibrium constant to the same factor. So $K' = (K_p)^{1/2} = (2.26 \times 10^4)^{1/2} = 1.50 \times 10^2$. The products will be favored.

(c) Begin with the reverse of the reaction and invert the equilibrium constant.

$$K_{reverse} = \frac{1}{K_p} = \frac{1}{2.26 \times 10^4} = 4.42 \times 10^{-5}$$

Then multiply the reaction by 2 and raise the value of $K_{reverse}$ to the second power.

$K' = (K_{reverse})^2 = (4.42 \times 10^{-5})^2 = 1.95 \times 10^{-9}$. The reactants will be favored.

16.28 (a) The reaction is multiplied by 1/2, so raise the value of the equilibrium constant to 1/2.

$K' = (K_p)^{1/2} = (2.2 \times 10^6)^{1/2} = 1.5 \times 10^3$. The products will be favored.

(b) The reaction is multiplied by 3, so raise the value of the equilibrium constant to 3.

$K' = (K_p)^3 = (2.2 \times 10^6)^3 = 1.1 \times 10^{19}$. The products will be favored.

(c) Begin with the reverse of the reaction and invert the equilibrium constant.

$$K_{reverse} = \frac{1}{K_p} = \frac{1}{2.2 \times 10^6} = 4.5 \times 10^{-7}$$

Then multiply the reaction by 2 and raise the value of $K_{reverse}$ to the second power.

$K' = (K_{reverse})^2 = (4.5 \times 10^{-7})^2 = 2.0 \times 10^{-13}$. The reactants will be favored.

16.29 To find the equilibrium constant for reaction 3, you need to combine reactions 1 and 2 to get reaction 3. Begin by reversing reaction 2; then multiply reaction 1 by 2 and add the two new reactions. When you add reactions, you multiply the values of K.

$$N_2(g) + O_2(g) \rightleftharpoons \cancel{2\,NO}(g) \qquad K_1 = \frac{1}{K_p} = \frac{1}{2.1 \times 10^{30}} = 4.\underline{7}6 \times 10^{-31}$$

$$\cancel{2\,NO}(g) + Br_2(g) \rightleftharpoons 2\,NOBr(g) \qquad K_2 = (K_p)^2 = (5.3)^2 = 2\underline{8}.09$$

$$N_2(g) + O_2(g) + Br_2(g) \rightleftharpoons 2\,NOBr(g) \qquad K_3 = K_1K_2 = (4.\underline{7}6 \times 10^{-31})(2\underline{8}.09) = 1.3 \times 10^{-29}$$

16.30 To find the equilibrium constant for reaction 3, you need to combine reactions 1 and 2 to get reaction 3. Begin by multiplying reaction 1 by 2; then reverse reaction 2 and add the two new reactions. When you add reactions, you multiply the values of K.

$$2\,A(s) \rightleftharpoons \cancel{B}(g) + \cancel{2C}(g) \qquad K_1 = (K_p)^2 = (0.0334)^2 = 1.1 \times 10^{-3}$$

$$\cancel{B}(g) + \cancel{2C}(g) \rightleftharpoons 3\,D(g) \qquad K_2 = \frac{1}{K_p} = \frac{1}{2.35} = 0.426$$

$$2\,A(s) \rightleftharpoons 3\,D(g) \qquad K' = K_1K_2 = (1.1 \times 10^{-3})(0.426) = 4.69 \times 10^{-4}$$

K_p, K_c, and Heterogeneous Equilibria

16.31 (a) **Given:** $K_p = 6.26 \times 10^{-22}$ $T = 298\,K$ **Find:** K_c

Conceptual Plan: $K_p \rightarrow K_c$

$K_p = K_c(RT)^{\Delta n}$

Solution: Δn = mol product gas − mol reactant gas = 2 − 1 = 1

$$K_c = \frac{K_p}{(RT)^{\Delta n}} = \frac{6.26 \times 10^{-22}}{\left(0.08206\frac{L \cdot atm}{mol \cdot \cancel{K}} \times 298\,\cancel{K}\right)^1} = 2.56 \times 10^{-23}$$

Check: Substitute into the equation and confirm that you get the original value of K_p.

$$K_p = K_c(RT)^{\Delta n} = (2.56 \times 10^{-23})\left(0.08206\frac{\text{L} \cdot \text{atm}}{\text{mol} \cdot \text{K}} \times 298\right)^1 = 6.26 \times 10^{-22}$$

(b) **Given:** $K_p = 7.7 \times 10^{24}$ $T = 298\,\text{K}$ **Find:** K_c
Conceptual Plan: $K_p \rightarrow K_c$

$K_p = K_c(RT)^{\Delta n}$

Solution: Δn = mol product gas − mol reactant gas = 4 − 2 = 2

$$K_c = \frac{K_p}{(RT)^{\Delta n}} = \frac{7.7 \times 10^{24}}{\left(0.08206\frac{\text{L} \cdot \text{atm}}{\text{mol} \cdot \text{K}} \times 298\,\text{K}\right)^2} = 1.3 \times 10^{22}$$

Check: Substitute into the equation and confirm that you get the original value of K_p.

$$K_p = K_c(RT)^{\Delta n} = (1.3 \times 10^{22})\left(0.08206\frac{\text{L} \cdot \text{atm}}{\text{mol} \cdot \text{K}} \times 298\right)^2 = 7.7 \times 10^{24}$$

(c) **Given:** $K_p = 81.9$ $T = 298\,\text{K}$ **Find:** K_c
Conceptual Plan: $K_p \rightarrow K_c$

$K_p = K_c(RT)^{\Delta n}$

Solution: Δn = mol product gas − mol reactant gas = 2 − 2 = 0

$$K_c = \frac{K_p}{(RT)^{\Delta n}} = \frac{81.9}{\left(0.08206\frac{\text{L} \cdot \text{atm}}{\text{mol} \cdot \text{K}} \times 298\,\text{K}\right)^0} = 81.9$$

Check: Substitute into the equation and confirm that you get the original value of K_p.

$$K_p = K_c(RT)^{\Delta n} = (81.9)\left(0.08206\frac{\text{L} \cdot \text{atm}}{\text{mol} \cdot \text{K}} \times 298\right)^0 = 81.9$$

16.32 (a) **Given:** $K_c = 5.9 \times 10^{-3}$ $T = 298\,\text{K}$ **Find:** K_p
Conceptual Plan: $K_p \rightarrow K_c$

$K_p = K_c(RT)^{\Delta n}$

Solution: Δn = mol product gas − mol reactant gas = 2 − 1 = 1

$$K_p = K_c(RT)^{\Delta n} = 5.9 \times 10^{-3}\left(0.08206\frac{\text{L} \cdot \text{atm}}{\text{mol} \cdot \text{K}} \times 298\,\text{K}\right)^1 = 1.4 \times 10^{-1}$$

Check: Substitute into the equation and confirm that you get the original value of K_p.

$$K_c = \frac{K_p}{(RT)^{\Delta n}} = \frac{1.4 \times 10^{-1}}{\left(0.08206\frac{\text{L} \cdot \text{atm}}{\text{mol} \cdot \text{K}} \times 298\,\text{K}\right)^1} = 0.0059$$

(b) **Given:** $K_c = 3.7 \times 10^8$ $T = 298\,\text{K}$ **Find:** K_p
Conceptual Plan: $K_p \rightarrow K_c$

$K_p = K_c(RT)^{\Delta n}$

Solution: Δn = mol product gas − mol reactant gas = 2 − 4 = −2

$$K_p = K_c(RT)^{\Delta n} = 3.7 \times 10^8\left(0.08206\frac{\text{L} \cdot \text{atm}}{\text{mol} \cdot \text{K}} \times 298\,\text{K}\right)^{-2} = 6.2 \times 10^5$$

Check: Substitute into the equation and confirm that you get the original value of K_p.

$$K_c = \frac{K_p}{(RT)^{\Delta n}} = \frac{6.2 \times 10^5}{\left(0.08206\frac{\text{L} \cdot \text{atm}}{\text{mol} \cdot \text{K}} \times 298\,\text{K}\right)^{-2}} = 3.7 \times 10^8$$

(c) **Given:** $K_c = 4.10 \times 10^{-31}$ $T = 298\,\text{K}$ **Find:** K_p
Conceptual Plan: $K_p \rightarrow K_c$

$K_p = K_c(RT)^{\Delta n}$

Solution: Δn = mol product gas − mol reactant gas = 2 − 2 = 0

$$K_p = K_c(RT)^{\Delta n} = 4.10 \times 10^{-31}\left(0.08206\frac{\text{L}\cdot\text{atm}}{\text{mol}\cdot\text{K}} \times 298\,\text{K}\right)^0 = 4.10 \times 10^{-31}$$

Check: Substitute into the equation and confirm that you get the original value of K_p.

$$K_c = \frac{K_p}{(RT)^{\Delta n}} = \frac{4.10 \times 10^{-31}}{\left(0.08206\frac{\text{L}\cdot\text{atm}}{\text{mol}\cdot\text{K}} \times 298\,\text{K}\right)^0} = 4.10 \times 10^{-31}$$

16.33 (a) Because H_2O is a liquid, it is omitted from the equilibrium expression. $K_{eq} = \frac{[HCO_3^-][OH^-]}{[CO_3^{2-}]}$

(b) Because $KClO_3$ and KCl are both solids, they are omitted from the equilibrium expression. $K_{eq} = [O_2]^3$

(c) Because H_2O is a liquid, it is omitted from the equilibrium expression. $K_{eq} = \frac{[H_3O^+][F^-]}{[HF]}$

(d) Because H_2O is a liquid, it is omitted from the equilibrium expression. $K_{eq} = \frac{[NH_4^+][OH^-]}{[NH_3]}$

16.34 Because PCl_3 is a liquid, it is omitted from the equilibrium expression. $K_{eq} = \frac{[Cl_2]}{[PCl_5]}$

Relating the Equilibrium Constant to Equilibrium Concentrations and Equilibrium Partial Pressures

16.35 **Given:** at equilibrium: $[CO] = 0.105\,\text{M}$, $[H_2] = 0.114\,\text{M}$, $[CH_3OH] = 0.185\,\text{M}$ **Find:** K_c
Conceptual Plan: Balanced reaction → equilibrium expression → K_c

Solution: $K_c = \frac{[CH_3OH]}{[CO][H_2]^2} = \frac{(0.185)}{(0.105)(0.114)^2} = 136$

Check: The answer is reasonable because the concentration of products is greater than the concentration of reactants and the equilibrium constant should be greater than 1.

16.36 **Given:** at equilibrium: $[NH_3] = 0.278\,\text{M}$, $[H_2S] = 0.355\,\text{M}$ **Find:** K_c
Conceptual Plan: Balanced reaction → equilibrium expression → K_c
Solution: Because NH_4HS is a solid, it is omitted from the equilibrium expression.

$$K_c = [NH_3][H_2S] = (0.278)(0.355) = 0.0987$$

Check: The answer is reasonable because the concentration of products is less than 1 M and the equilibrium constant is less than 1.

16.37 At 500 K: **Given:** at equilibrium: $[N_2] = 0.115\,\text{M}$, $[H_2] = 0.105\,\text{M}$, and $[NH_3] = 0.439\,\text{M}$ **Find:** K_c
Conceptual Plan: Balanced reaction → equilibrium expression → K_c

Solution: $K_c = \frac{[NH_3]^2}{[N_2][H_2]^3} = \frac{(0.439)^2}{(0.115)(0.105)^3} = 1.45 \times 10^3$

Check: The value is reasonable because the concentration of products is greater than the concentration of reactants.

At 575 K: **Given:** at equilibrium: $[N_2] = 0.110\,\text{M}$, $[NH_3] = 0.128\,\text{M}$, $K_c = 9.6$ **Find:** $[H_2]$
Conceptual Plan: Balanced reaction → equilibrium expression → $[H_2]$

Solution: $K_c = \frac{[NH_3]^2}{[N_2][H_2]^3}$ $\quad 9.6 = \frac{(0.128)^2}{(0.110)(x)^3}$ $\quad x = 0.249$

Check: Plug the value for x back into the equilibrium expression and check the value.

$$9.6 = \frac{(0.128)^2}{(0.110)(0.249)^3}$$

At 775 K: **Given:** at equilibrium: $[N_2] = 0.120$ M, $[H_2] = 0.140$ M, $K_c = 0.0584$ **Find:** $[NH_3]$
Conceptual Plan: Balanced reaction → equilibrium expression → $[NH_3]$

Solution: $K_c = \dfrac{[NH_3]^2}{[N_2][H_2]^3}$ $\quad 0.0584 = \dfrac{(x)^2}{(0.120)(0.140)^3}$ $\quad x = 0.00439$

Check: Plug the value for x back into the equilibrium expression and check the value.

$$0.0584 = \frac{(0.00439)^2}{(0.120)(0.140)^3}$$

16.38 At 25 °C: **Given:** at equilibrium: $[H_2] = 0.0355$ M, $[I_2] = 0.0388$ M, and $[HI] = 0.922$ M **Find:** K_c
Conceptual Plan: Balanced reaction → equilibrium expression → K_c

Solution: $K_c = \dfrac{[HI]^2}{[H_2][I_2]} = \dfrac{(0.922)^2}{(0.0355)(0.0388)} = 617$

Check: The value is reasonable because the concentration of products is greater than the concentration of reactants.

At 340 °C: **Given:** at equilibrium: $[I_2] = 0.0455$, $[HI] = 0.387$, $K_c = 9.6$ **Find:** $[H_2]$
Conceptual Plan: Balanced reaction → equilibrium expression → $[H_2]$

Solution: $K_c = \dfrac{[HI]^2}{[H_2][I_2]}$ $\quad 9.6 = \dfrac{(0.387)^2}{(x)(0.0455)}$ $\quad x = 0.343$

Check: Plug the value for x back into the equilibrium expression and check the value.

$$9.6 = \frac{(0.387)^2}{(0.343)(0.0455)}$$

At 445 °C: **Given:** at equilibrium: $[H_2] = 0.0485$ M, $[I_2] = 0.0468$, $K_c = 50.2$ **Find:** $[HI]$
Conceptual Plan: Balanced reaction → equilibrium expression → $[HI]$

Solution: $K_c = \dfrac{[HI]^2}{[H_2][I_2]}$ $\quad 50.2 = \dfrac{(x)^2}{(0.0485)(0.0468)}$ $\quad x = 0.338$

Check: Plug the value for x back into the equilibrium expression and check the value.

$$50.2 = \frac{(0.338)^2}{(0.0485)(0.0468)}$$

16.39 **Given:** $P_{NO} = 108$ torr; $P_{Br_2} = 126$ torr, $K_p = 28.4$ **Find:** P_{NOBr}
Conceptual Plan: torr → atm and then balanced reaction → equilibrium expression → P_{NOBr}

$$\frac{1 \text{ atm}}{760 \text{ torr}}$$

Solution: $P_{NO} = 108 \cancel{\text{torr}} \times \dfrac{1 \text{ atm}}{760 \cancel{\text{torr}}} = 0.14\underline{2}1$ atm $\quad P_{Br_2} = 126 \cancel{\text{torr}} \times \dfrac{1 \text{ atm}}{760 \cancel{\text{torr}}} = 0.16\underline{5}8$ atm

$K_p = \dfrac{P_{NOBr}^2}{P_{NO}^2 P_{Br_2}}$ $\quad 28.4 = \dfrac{x^2}{(0.14\underline{2}1)^2(0.16\underline{5}8)}$ $\quad x = 0.308 \text{ atm} = 234 \text{ torr}$

Check: Plug the value for x back into the equilibrium expression and check the value.

$$28.3 = \frac{(0.308)^2}{(0.14\underline{2}1)^2(0.16\underline{5}8)}$$

16.40 **Given:** $P_{SO_2} = 137$ torr; $P_{Cl_2} = 285$ torr, $K_p = 2.91 \times 10^3$ **Find:** $P_{SO_2Cl_2}$
Conceptual Plan: torr → atm and then balanced reaction → equilibrium expression → $P_{SO_2Cl_2}$

$$\frac{1 \text{ atm}}{760 \text{ torr}}$$

Solution: $P_{SO_2} = 137 \text{ torr} \times \frac{1 \text{ atm}}{760 \text{ torr}} = 0.18\underline{0}3 \text{ atm}$ $P_{Cl_2} = 285 \text{ torr} \times \frac{1 \text{ atm}}{760 \text{ torr}} = 0.37\underline{5}0 \text{ atm}$

$$K_p = \frac{P_{SO_2}P_{Cl_2}}{P_{SO_2Cl_2}} \quad 2.91 \times 10^3 = \frac{(0.18\underline{0}3)(0.37\underline{5}0)}{(x)} \quad x = 2.32 \times 10^{-5} \text{ atm} = 0.0177 \text{ torr}$$

Check: Plug the value for x back into the equilibrium expression and check the value.

$$2.92 \times 10^3 = \frac{(0.18\underline{0}3)(0.37\underline{5}0)}{(2.32 \times 10^{-5})}$$

16.41 **Given:** $P_{A,\text{ initial}} = 1.32$ atm; $P_{A,\text{ eq}} = 0.25$ atm **Find:** K_p

Conceptual Plan:
1. **Prepare ICE table.**
2. **Calculate pressure change for known value.**
3. **Calculate pressure changes for other reactants/products.**
4. **Determine equilibrium pressures.**
5. **Write the equilibrium expression and determine K_p.**

Solution: $A(g) \rightleftharpoons 2B(g)$

	P_A	P_B
Initial	1.32 atm	0.00
Change	$-x$	$+2x$
Equil	0.25 atm	$2x$

$x = 1.32 - 0.25 = 1.07$ atm

$2x = 2(1.07 \text{ atm}) = 2.14 \text{ atm} = P_B$

$$K_p = \frac{P_B^2}{P_A} = \frac{(2.14)^2}{(0.25)} = 1\underline{8}.32 = 18$$

Check: The units (none) are correct. The magnitude is reasonable (18) since there are significant concentrations of both reactants and products.

16.42 **Given:** $P_{A,\text{ initial}} = 255$ mmHg; $P_{A,\text{ eq}} = 55$ mmHg **Find:** K_p

Conceptual Plan:
1. **Convert given pressures to units of atm.**
2. **Prepare ICE table.**
3. **Calculate pressure change for known value.**
4. **Calculate pressure changes for other reactants/products.**
5. **Determine equilibrium pressures.**
6. **Write the equilibrium expression and determine K_p.**

Solution: $P_{A,\text{ initial}} = 255 \text{ mmHg} \frac{(1 \text{ atm})}{760 \text{ mmHg}} = 0.336 \text{ atm}$

$P_{A,\text{ eq}} = 55 \text{ mmHg} \frac{(1 \text{ atm})}{760 \text{ mmHg}} = 0.00658 \text{ atm}$

$2\,A(g) \rightleftharpoons B(g) + 2\,C(g)$

	P_A	P_B	P_C
Initial	0.336 atm	0.00	0.00
Change	$-2x$	$+x$	$+2x$
Equil	0.0724 atm	x	$2x$

$2x = 0.336 - 0.0724 = 0.264 \text{ atm} = P_C$; $x = 0.132 \text{ atm} = P_B$

$$K_p = \frac{P_B P_C^2}{P_A^2} = \frac{(0.132)(0.264)^2}{(0.0724)^2} = 1.76$$

Check: The units (none) are correct. The magnitude is reasonable (1.8) since there are higher concentrations of the products than the reactants.

16.43 **Given:** $[Fe^{3+}]_{\text{initial}} = 1.0 \times 10^{-3}$ M; $[SCN^-]_{\text{initial}} = 8.0 \times 10^{-4}$ M; $[FeSCN^{2+}]_{\text{eq}} = 1.7 \times 10^{-4}$ M **Find:** K_c

Conceptual Plan:
1. **Prepare ICE table.**
2. **Calculate concentration change for known value.**
3. **Calculate concentration changes for other reactants/products.**

4. **Determine equilibrium concentration.**
5. **Write the equilibrium expression and determine K_c.**

Solution: $Fe^{3+}(aq) + SCN^-(aq) \rightleftharpoons FeSCN^{3+}(aq)$

	$[Fe^{3+}]$	$[SCN^-]$	$[FeSCN^{3+}]$
Initial	1.0×10^{-3}	8.0×10^{-4}	0.00
Change	-1.7×10^{-4}	-1.7×10^{-4}	$+1.7 \times 10^{-4}$
Equil	8.3×10^{-4}	6.3×10^{-4}	1.7×10^{-4}

$$K_c = \frac{[FeSCN^{2+}]}{[Fe^{3+}][SCN^-]} = \frac{(1.7 \times 10^{-4})}{(8.3 \times 10^{-4})(6.3 \times 10^{-4})} = 3.3 \times 10^2$$

Check: The units (none) are correct. The magnitude is reasonable (330) since there are significant concentrations of both reactants and products.

16.44 **Given:** $[SO_2Cl_2]_{initial} = 0.020$ M; $[Cl_2]_{eq} = 1.2 \times 10^{-2}$ M **Find:** K_c

Conceptual Plan:
1. **Prepare ICE table.**
2. **Calculate concentration change for known value.**
3. **Calculate concentration changes for other reactants/products.**
4. **Determine equilibrium concentration.**
5. **Write the equilibrium expression and determine K_c.**

Solution: $SO_2Cl_2(g) \rightleftharpoons SO_2(g) + Cl_2(g)$

	$[SO_2Cl_2]$	$[SO_2]$	$[Cl_2]$
Initial	0.020	0.00	0.00
Change	-1.2×10^{-2}	$+1.2 \times 10^{-2}$	$+1.2 \times 10^{-2}$
Equil	0.0080	1.2×10^{-2}	1.2×10^{-2}

$$K_c = \frac{[SO_2][Cl_2]}{[SO_2Cl_2]} = \frac{(1.2 \times 10^{-2})(1.2 \times 10^{-2})}{(0.0080)} = 0.018$$

Check: The units (none) are correct. The magnitude is reasonable since there are significant concentrations of both reactants and products.

16.45 **Given:** 3.67 L flask, 0.763 g H_2 initial, 96.9 g I_2 initial, 90.4 g HI equilibrium **Find:** K_c

Conceptual Plan: g → mol → M and then

$n = \frac{g}{\text{molar mass}}$ $M = \frac{n}{V}$

1. **Prepare ICE table.**
2. **Calculate concentration change for known value.**
3. **Calculate concentration changes for other reactants/products.**
4. **Determine equilibrium concentration.**
5. **Write the equilibrium expression and determine K_c.**

Solution: $0.763 \text{ g } H_2 \times \frac{1 \text{ mol } H_2}{2.016 \text{ g } H_2} = 0.37\underline{8}5 \text{ mol } H_2$ then $\frac{0.37\underline{8}5 \text{ mol } H_2}{3.67 \text{ L}} = 0.103$ M This is an initial concentration.

$96.9 \text{ g } I_2 \times \frac{1 \text{ mol } I_2}{253.8 \text{ g } I_2} = 0.38\underline{1}8 \text{ mol } I_2$ then $\frac{0.38\underline{1}8 \text{ mol } I_2}{3.67 \text{ L}} = 0.104$ M This is an initial concentration.

$90.4 \text{ g HI} \times \frac{1 \text{ mol HI}}{127.9 \text{ g } I_2} = 0.70\underline{6}8 \text{ mol HI}$ then $\frac{0.70\underline{6}8 \text{ mol HI}}{3.67 \text{ L}} = 0.193$ M This is an equilibrium concentration.

$H_2(g) + I_2(g) \rightleftharpoons 2\,HI(g)$

	$[H_2]$	$[I_2]$	$[HI]$
Initial	0.103	0.104	0.00
Change	−0.0965	−0.0965	+0.193
Equil	0.0065	0.0075	0.193

Because HI gained 0.193 M, H_2 and I_2 had to lose $0.193/2 = 0.0965$ M from the stoichiometry of the balanced reaction.

$$K_c = \frac{[HI]^2}{[H_2][I_2]} = \frac{(0.193)^2}{(0.0065)(0.0075)} = 764$$

Check: The units (none) are correct. The magnitude is reasonable (764) since there are higher concentrations of the products than the reactants.

16.46 **Given:** 5.19 L flask, 26.9 g CO initial, 2.34 g H_2 initial, 8.65 g CH_3OH equilibrium **Find:** K_c

Conceptual Plan: g → mol → M and then

$n = \frac{g}{\text{molar mass}}$ $M = \frac{n}{V}$

1. **Prepare ICE table.**
2. **Calculate concentration change for known value.**
3. **Calculate concentration changes for other reactants/products.**
4. **Determine equilibrium concentration.**
5. **Write the equilibrium expression and determine K_c.**

Solution: $26.9 \text{ g CO} \times \frac{1 \text{ mol CO}}{28.01 \text{ g CO}} = 0.9604 \text{ mol CO}$ then $\frac{0.9604 \text{ mol CO}}{5.19 \text{ L}} = 0.185 \text{ M}$

$2.34 \text{ g } H_2 \times \frac{1 \text{ mol } H_2}{2.016 \text{ g } H_2} = 1.161 \text{ mol } H_2$ then $\frac{1.161 \text{ mol } H_2}{5.19 \text{ L}} = 0.224 \text{ M}$

$8.65 \text{ g } CH_3OH \times \frac{1 \text{ mol } CH_3OH}{32.04 \text{ g } CH_3OH} = 0.2700 \text{ mol } CH_3OH$ then $\frac{0.2700 \text{ mol } CH_3OH}{5.19 \text{ L}} = 0.0520 \text{ M}$

$$CO(g) + 2\,H_2(g) \rightleftharpoons CH_3OH(g)$$

	[CO]	[H_2]	[CH_3OH]
Initial	0.185	0.224	0.00
Change	−0.0520	−0.104	+0.0520
Equil	0.133	0.120	0.0520

Because CH_3OH gained 0.5210 M, CO and H_2 had to lose 0.520 M and 0.104 M, respectively, because of the stoichiometry of the balanced reaction.

$$K_c = \frac{[CH_3OH]}{[CO][H_2]^2} = \frac{(0.0520)}{(0.133)(0.120)^2} = 27.2$$

Check: The units (none) are correct. The magnitude is reasonable (27) since there are higher concentrations of the products than the reactants.

The Reaction Quotient and Reaction Direction

16.47 **Given:** $K_c = 8.5 \times 10^{-3}$; $[NH_3] = 0.166$ M; $[H_2S] = 0.166$ M **Find:** whether solid will form or decompose

Conceptual Plan: Calculate Q → compare Q → and K_c

Solution: $Q = [NH_3][H_2S] = (0.166)(0.166) = 0.0276$

$Q = 0.0276$ and $K_c = 8.5 \times 10^{-3}$ so $Q > K_c$ and the reaction will shift to the left; so more solid will form.

16.48 **Given:** $K_p = 2.4 \times 10^{-4}$; $P_{H_2} = 0.112$ atm; $P_{S_2} = 0.055$ atm; $P_{H_2S} = 0.445$ atm **Find:** whether the reaction is at equilibrium

Conceptual Plan: Calculate Q → compare Q and K_p

Solution: $Q = \frac{P_{H_2}^2 P_{S_2}}{P_{H_2S}^2} = \frac{(0.112)^2(0.055)}{(0.445)^2} = 3.48 \times 10^{-3}$ $Q = 3.48 \times 10^{-3}$ and $K_p = 2.4 \times 10^{-4}$

$Q > K_p$, so the system is not at equilibrium, and the reaction will shift to the left.

16.49 **Given:** 6.55 g Ag_2SO_4, 1.5 L solution, $K_c = 1.1 \times 10^{-5}$ **Find:** whether more solid will dissolve

Conceptual Plan: g Ag_2SO_4 → mol Ag_2SO_4 → [Ag_2SO_4] → [Ag^+], [SO_4^{-2}] → calculate Q and compare to K_c

$\frac{1 \text{ mol } Ag_2SO_4}{311.81 \text{ g}}$ $[\,] = \frac{\text{mol } Ag_2SO_4}{\text{vol solution}}$ $Q = [Ag^+]^2[SO_4^{-2}]$

Solution: $6.55 \text{ g } Ag_2SO_4 \left(\frac{1 \text{ mol } Ag_2SO_4}{311.81 \text{ g } Ag_2SO_4}\right) = 0.0210 \text{ mol } Ag_2SO_4$

$$\frac{0.0210 \text{ mol } Ag_2SO_4}{1.5 \text{ L solution}} = 0.0140 \text{ M } Ag_2SO_4$$

$[Ag^+] = 2[Ag_2SO_4] = 2(0.0140\ M) = 0.0280\ M$; $[SO_4^{2-}] = [Ag_2SO_4] = 0.0140\ M$
$Q = [Ag^+]^2[SO_4^{2-}] = (0.0280)^2(0.0140) = 1.1 \times 10^{-5}$
$Q = K_c$, so the system is at equilibrium and is a saturated solution. Therefore, if more solid is added, it will not dissolve.

16.50 **Given:** $K_p = 6.7$ at 298 K; 2.25 L flask; $NO_2 = 0.055$ mol; $N_2O_4 = 0.082$ mol **Find:** whether the reaction is at equilibrium

Conceptual Plan: $K_p \rightarrow K_c$ and mol $\rightarrow$ M and then calculate Q and compare to K_c

$K_p = K_c(RT)^{\Delta n}$ $\quad$ $M = \frac{mol}{V}$

Solution: $K_c = \frac{K_p}{(RT)^{\Delta n}} = \frac{(6.7)}{\left[\left(0.08206\frac{L \cdot atm}{mol \cdot K}\right)(298\ K)\right]^{-1}} = 164$

$$Q = \frac{[N_2O_4]}{[NO]^2} = \frac{\left(\frac{0.082\ mol}{2.25\ L}\right)}{\left(\frac{0.055\ mol}{2.25\ L}\right)^2} = 61$$

$Q < K_c$, so the reaction is not at equilibrium and will shift to the right.

Finding Equilibrium Concentrations from Initial Concentrations and the Equilibrium Constant

16.51 (a) **Given:** $[A] = 1.0\ M$, $[B] = 0.0$, $K_c = 4.0$; $a = 1, b = 1$ **Find:** $[A]$, $[B]$ at equilibrium

Conceptual Plan: Prepare an ICE table, calculate Q, compare Q and K_c, predict the direction of the reaction, represent the change with x, sum the table, determine the equilibrium values, put the equilibrium values in the equilibrium expression, and solve for x. Determine [A] and [B].

Solution: $A(g) \rightleftharpoons B(g)$

	[A]	[B]
Initial	1.0	0.00
Change	$-x$	$+x$
Equil	$1.0 - x$	x

$Q = \frac{[B]}{[A]} = \frac{0}{1.0} = 0 \quad Q < K$ Therefore, the reaction will proceed to the right by x.

$K_c = \frac{[B]}{[A]} = \frac{(x)}{(1.0 - x)} = 4.0; \quad x = 0.80$

$[A] = 1 - 0.80 = 0.20\ M$; $\quad [B] = 0.80\ M$

Check: Plug the values into the equilibrium expression: $K_c = \frac{0.80}{0.20} = 4.0$

(b) **Given:** $[A] = 1.0\ M$, $[B] = 0.0$, $K_c = 4.0$; $a = 2, b = 2$ **Find:** $[A]$, $[B]$ at equilibrium

Conceptual Plan: Prepare an ICE table, calculate Q, compare Q and K_c, predict the direction of the reaction, represent the change with x, sum the table, determine the equilibrium values, put the equilibrium values in the equilibrium expression, and solve for x. Determine [A] and [B].

Solution: $2\ A(g) \rightleftharpoons 2\ B(g)$

	[A]	[B]
Initial	1.0	0.00
Change	$-2x$	$+2x$
Equil	$1.0 - 2x$	$2x$

$Q = \frac{[B]^2}{[A]^2} = \frac{0}{1.0} = 0 \quad Q < K$ Therefore, the reaction will proceed to the right by x.

$$K_c = \frac{[B]^2}{[A]^2} = \frac{(2x)^2}{(1.0 - 2x)^2} = 4.0$$

$\sqrt{\frac{(2x)^2}{(1.0 - 2x)^2}} = \sqrt{4.0}$ Take the square root of this expression. $x = 0.33$

$[A] = 1 - 2(0.33) = 0.34$ M; $[B] = 2(0.33) = 0.66$ M

Check: Plug the values into the equilibrium expression: $K_c = \frac{(0.66)^2}{(0.34)^2} = 3.8$.

(c) **Given:** $[A] = 1.0$ M, $[B] = 0.0$, $K_c = 4.0$; $a = 1, b = 2$ **Find:** $[A]$, $[B]$ at equilibrium

Conceptual Plan: Prepare an ICE table, calculate Q, compare Q and K_c, predict the direction of the reaction, represent the change with x, sum the table, determine the equilibrium values, put the equilibrium values in the equilibrium expression, and solve for x. Determine [A] and [B].

Solution: $A(g) \rightleftharpoons 2\,B(g)$

	[A]	[B]
Initial	1.0	0.00
Change	$-x$	$+2x$
Equil	$1.0 - x$	$2x$

$Q = \frac{[B]^2}{[A]} = \frac{0}{1.0} = 0$ $Q < K$ Therefore, the reaction will proceed to the right by x.

$K_c = \frac{[B]^2}{[A]} = \frac{(2x)^2}{(1.0 - x)} = 4.0$ $4x^2 + 4x - 4 = 0$ Solve using the quadratic equation, found in Appendix I.

$x = -1.6$ or $x = 0.62$; therefore, $x = 0.62$.

$[A] = 1.0 - 0.62 = 0.38$ M $[B] = 2x = 2(0.62) = 1.\underline{2}$ M

Check: Plug the values into the equilibrium expression: $K_c = \frac{(1.2)^2}{0.38} = 3.8$.

16.52 (a) **Given:** $[A] = 1.0$ M, $[B] = 1.0$ M, $[C] = 0.0$, $K_c = 5.0$; $a = 1, b = 1, c = 2$ **Find:** $[A]$, $[B]$, $[C]$ at equilibrium

Conceptual Plan: Prepare an ICE table, calculate Q, compare Q and K_c, predict the direction of the reaction, represent the change with x, sum the table, determine the equilibrium values, put the equilibrium values in the equilibrium expression, and solve for x. Determine [A], [B], and [C].

Solution: $A(g) + B(g) \rightleftharpoons 2\,C(g)$

	[A]	[B]	[C]
Initial	1.0	1.0	0.0
Change	$-x$	$-x$	$+2x$
Equil	$1.0 - x$	$1.0 - x$	$2x$

$Q = \frac{[C]^2}{[A][B]} = \frac{0}{(1.0)(1.0)} = 0$ $Q < K$ Therefore, the reaction will proceed to the right by x.

$K_c = \frac{[C]^2}{[A][B]} = \frac{(2x)^2}{(1.0 - x)(1.0 - x)} = 5.0$; $x = 0.53$ Solve by taking the square root of both sides.

$[A] = [B] = 1.0 - 0.53 = 0.47$ M $[C] = 2x = 2(0.53)$ M $= 1.1$ M

Check: Plug the values into the equilibrium expression: $K_c = \frac{(1.1)^2}{(0.47)(0.47)} = 5.4$.

(b) **Given:** $[A] = 1.0$ M, $[B] = 1.0$ M, $[C] = 0.0$, $K_c = 5.0$; $a = 1, b = 1, c = 1$ **Find:** $[A]$, $[B]$, $[C]$ at equilibrium

Conceptual Plan: Prepare an ICE table, calculate Q, compare Q and K_c, predict the direction of the reaction, represent the change with x, sum the table, determine the equilibrium values, put the equilibrium values in the equilibrium expression, and solve for x. Determine [A], [B], and [C].

Solution: $A(g) + B(g) \rightleftharpoons C(g)$

	[A]	[B]	[C]
Initial	1.0	1.0	0.0
Change	$-x$	$-x$	$+x$
Equil	$1.0 - x$	$1.0 - x$	x

$Q = \dfrac{[C]}{[A][B]} = \dfrac{0}{(1.0)(1.0)} = 0$ $\quad Q < K$ Therefore, the reaction will proceed to the right by x.

$K_c = \dfrac{[C]}{[A][B]} = \dfrac{(x)}{(1.0 - x)(1.0 - x)} = 5.0; \quad 5x^2 - 11x + 5 = 0$

Solve by using the quadratic equation. $x = 1.6$ or $x = 0.64$ Therefore, $x = 0.64$.
$[A] = [B] = 1.0 - 0.64 = 0.36$ M $\quad [C] = x = 0.64$ M

Check: Plug the values into the equilibrium expression: $K_c = \dfrac{(0.64)}{(0.36)(0.36)} = 4.9.$

(c) **Given:** $[A] = 1.0$ M, $[B] = 1.0$ M, $[C] = 0.0$, $K_c = 5.0$; $a = 2, b = 1, c = 1$ **Find:** [A], [B], [C] at equilibrium

Conceptual Plan: Prepare an ICE table, calculate Q, compare Q and K_c, predict the direction of the reaction, represent the change with x, sum the table, determine the equilibrium values, put the equilibrium values in the equilibrium expression, and solve for x. Determine [A], [B], and [C].

Solution: $2\,A(g) + B(g) \rightleftharpoons C(g)$

	[A]	[B]	[C]
Initial	1.0	1.0	0.0
Change	$-2x$	$-x$	$+x$
Equil	$1.0 - 2x$	$1.0 - x$	x

$Q = \dfrac{[C]}{[A]^2[B]} = \dfrac{0}{(1.0)(1.0)} = 0$ $\quad Q < K$ Therefore, the reaction will proceed to the right by x.

$K_c = \dfrac{[C]}{[A]^2[B]} = \dfrac{(x)}{(1.0 - 2x)^2(1.0 - x)} = 5.0$

$-20.x^3 + 40.x^2 - 26x + 5 = 0$ Solve by using successive approximations or a cubic equation calculator found on the Internet.

$x = 0.34$

$[A] = 1 - 2x = 0.32$ M, $[B] = 1 - x = 0.66$ M, $[C] = x = 0.34$ M

Check: Plug the values into the equilibrium expression: $K_c = \dfrac{(0.34)}{(0.32)^2(0.66)} = 5.0.$

16.53 **Given:** $[N_2O_4] = 0.0500$ M, $[NO_2] = 0.0$, $K_c = 0.513$ **Find:** $[N_2O_4]$, $[NO_2]$ at equilibrium

Conceptual Plan: Prepare an ICE table, calculate Q, compare Q and K_c, predict the direction of the reaction, represent the change with x, sum the table, determine the equilibrium values, put the equilibrium values in the equilibrium expression, and solve for x. Determine $[N_2O_4]$ and $[NO_2]$.

Solution:

$N_2O_4(g) \rightleftharpoons 2\,NO_2(g)$

	$[N_2O_4]$	$[NO_2]$
Initial	0.0500	0.00
Change	$-x$	$+2x$
Equil	$0.0500 - x$	$2x$

$Q = \frac{[NO_2]^2}{[N_2O_4]} = \frac{0}{0.0500} = 0$ $\quad Q < K$ Therefore, the reaction will proceed to the right by x.

$K_c = \frac{[NO_2]^2}{[N_2O_4]} = \frac{(2x)^2}{(0.0500 - x)} = 0.513 \quad 4x^2 + 0.513x - 0.02565 = 0$

$\frac{-b \pm \sqrt{b^2 - 4ac}}{2a} = \frac{-0.513 \pm \sqrt{(0.513)^2 - 4(4)(-0.02565)}}{2(4)} = \frac{-0.513 \pm \sqrt{0.6735}}{2(4)}$

$x = -0.1667$ or $x = 0.0385$ Therefore, $x = 0.0385$.
$[N_2O_4] = 0.0500 - 0.0385 = 0.0115$ M $[NO_2] = 2x = 2(0.0385) = 0.0770$ M

Check: Plug the values into the equilibrium expression: $K_c = \frac{(0.0770)^2}{0.0115} = 0.516.$

16.54 **Given:** $[CO] = 0.1500$ M, $[Cl_2] = 0.175$ M, $[COCl_2] = 0.0$, $K_c = 255$ **Find:** $[CO]$, $[Cl_2]$, $[COCl_2]$ at equilibrium
Conceptual Plan: Prepare an ICE table, calculate Q, compare Q and K_c, predict the direction of the reaction, represent the change with x, sum the table, determine the equilibrium values, put the equilibrium values in the equilibrium expression, and solve for x. Determine $[CO]$, $[Cl_2]$, and $[COCl_2]$.
Solution:

$CO(g) + Cl_2(g) \rightleftharpoons COCl_2(g)$

	$[CO]$	$[Cl_2]$	$[COCl_2]$
Initial	0.1500	0.175	0.0
Change	$-x$	$-x$	$+x$
Equil	$0.1500 - x$	$0.175 - x$	x

$Q = \frac{[COCl_2]}{[CO][Cl_2]} = \frac{0}{(0.1500)(0.175)} = 0$ $\quad Q < K$ Therefore, the reaction will proceed to the right by x.

$K_c = \frac{[COCl_2]}{[CO][Cl_2]} = \frac{x}{(0.1500 - x)(0.175 - x)} = 255 \quad 255x^2 - 83.875x + 6.69375 = 0$

$\frac{-b \pm \sqrt{b^2 - 4ac}}{2a} = \frac{-(-83.875) \pm \sqrt{(-83.875)^2 - 4(255)(6.69375)}}{2(255)}$

$x = 0.19\underline{2}7$ or $x = 0.13\underline{6}22$ Therefore, $x = 0.13\underline{6}22$.
$[CO] = 0.1500 - 0.13\underline{6}22 = 0.01\underline{3}78$ M
$[Cl_2] = 0.175 - 0.13\underline{6}22 = 0.03\underline{8}78$ M
$[COCl_2] = x = 0.13\underline{6}22$ M

Check: Plug the values into the equilibrium expression: $K_c = \frac{(0.13622)}{(0.01378)(0.03878)} = 255.$

16.55 **Given:** $[CO] = 0.20$ M, $[CO_2] = 0.0$, $K_c = 4.0 \times 10^3$ **Find:** $[CO_2]$ at equilibrium
Conceptual Plan: Prepare an ICE table, calculate Q, compare Q and K_c, predict the direction of the reaction, represent the change with x, sum the table, determine the equilibrium values, put the equilibrium values in the equilibrium expression, and solve for x. Determine $[CO_2]$.
Solution:

$NiO(s) + CO(g) \rightleftharpoons Ni(s) + CO_2(g)$

	$[CO]$	$[CO_2]$
Initial	0.20	0.0
Change	$-x$	$+x$
Equil	$0.20 - x$	x

$Q = \frac{[CO_2]}{[CO]} = \frac{0}{(0.20)} = 0$ $\quad Q < K$ Therefore, the reaction will proceed to the right by x.

$$K_c = \frac{[CO_2]}{[CO]} = \frac{x}{(0.20 - x)} = 4.0 \times 10^3 \qquad 4.0 \times 10^3(0.20 - x) = x$$

$x = 0.1\underline{9}9$

$[CO_2] = 0.1\underline{9}9\text{ M}$

Check: Because the equilibrium constant is so large, the reaction goes essentially to completion; therefore, it is reasonable that the concentration of the product is $0.1\underline{9}9$ M.

16.56 **Given:** $[CO] = 0.110\text{ M}$, $[H_2O] = 0.110\text{ M}$, $[CO_2] = 0.0$, $[H_2] = 0.0$, $K_c = 102$
Find: $[CO]$, $[H_2O]$, $[CO_2]$, $[H_2]$ at equilibrium
Conceptual Plan: Prepare an ICE table, calculate Q, compare Q and K_c, predict the direction of the reaction, represent the change with x, sum the table, determine the equilibrium values, put the equilibrium values in the equilibrium expression, and solve for x. Determine $[CO]$, $[H_2O]$, $[CO_2]$, and $[H_2]$.
Solution:

$$CO(g) + H_2O(g) \rightleftharpoons CO_2(g) + H_2(g)$$

	$[CO]$	$[H_2O]$	$[CO_2]$	$[H_2]$
Initial	0.110	0.110	0.0	0.0
Change	$-x$	$-x$	$+x$	$+x$
Equil	$0.110 - x$	$0.110 - x$	x	x

$$Q = \frac{[CO_2][H_2]}{[CO][H_2O]} = \frac{0}{(0.110)(0.110)} = 0 \qquad Q < K \quad \text{Therefore, the reaction will proceed to the right by } x.$$

$$K_c = \frac{[CO_2][H_2]}{[CO][H_2O]} = \frac{(x)(x)}{(0.110 - x)(0.110 - x)} = 102$$

$$\sqrt{\frac{(x)(x)}{(0.110 - x)(0.110 - x)}} = \sqrt{102}$$

$$\frac{x}{0.110 - x} = \pm 10.\underline{0}995$$

$$\pm 10.\underline{0}995(0.110 - x) = x$$

$x = 0.10\underline{0}09$ or $x = 0.12\underline{2}09$ Therefore, $x = 0.10\underline{0}09 = 0.10\underline{0}$.

$[CO] = [H_2O] = 0.110 - 0.10\underline{0} = 0.010 = 0.010\text{M}$
$[H_2] = [CO_2] = x = 0.100 = 0.10\underline{0}\text{M}$

Check: Plug the values into the equilibrium expression: $K_c = \frac{(0.100)(0.100)}{(0.010)(0.010)} = 101.2 = 1.0 \times 10^2$, which is close to the true value; so the answers are valid.

16.57 **Given:** $[HC_2H_3O_2] = 0.210\text{ M}$, $[H_3O^+] = 0.0$, $[C_2H_3O_2^-] = 0.0$, $K_c = 1.8 \times 10^{-5}$
Find: $[HC_2H_3O_2]$, $[H_3O^+]$, $[C_2H_3O_2^-]$ at equilibrium
Conceptual Plan: Prepare an ICE table, calculate Q, compare Q and K_c, predict the direction of the reaction, represent the change with x, sum the table, determine the equilibrium values, put the equilibrium values in the equilibrium expression, and solve for x. Determine $[H_2C_2H_3O_2]$, $[H_3O^+]$, and $[C_2H_3O_2^-]$.
Solution:

$$HC_2H_3O_2(aq) + H_2O(l) \rightleftharpoons H_3O^+(aq) + C_2H_3O_2^-(aq)$$

	$HC_2H_3O_2$	$[H_2O]$	$[H_3O^+]$	$[C_2H_3O_2^-]$
Initial	0.210		0.0	0.0
Change	$-x$		$+x$	$+x$
Equil	$0.210 - x$		x	x

$$Q = \frac{[H_3O^+][C_2H_3O_2^-]}{[HC_2H_3O_2]} = \frac{0}{(0.210)} = 0 \qquad Q < K \quad \text{Therefore, the reaction will proceed to the right by } x.$$

$$K_c = \frac{[H_3O^+][C_2H_3O_2^-]}{[HC_2H_3O_2]} = \frac{(x)(x)}{(0.210 - x)} = 1.8 \times 10^{-5}$$

Assume that x is small compared to 0.210.

$x^2 = 0.210(1.8 \times 10^{-5})$

$x = 0.00194$ Check assumption: $\frac{0.00194}{0.210} \times 100 = 0.92\%$; assumption is valid.

$[H_3O^+] = [C_2H_3O_2^-] = 0.00194$ M

$[HC_2H_3O_2] = 0.210 - 0.00194 = 0.20\underline{8}1 = 0.208$ M

Check: Plug the values into the equilibrium expression: $K_c = \frac{(0.00194)(0.00194)}{(0.208)} = 1.81 \times 10^{-5}$.

The answer is the same to two significant figures with the true value, so the answers are valid.

16.58 **Given:** $[SO_2Cl_2] = 0.175$ M, $[SO_2] = 0.0$, $[Cl_2] = 0.0$, $K_c = 2.99 \times 10^{-7}$
Find: $[SO_2Cl_2]$, $[SO_2]$, $[Cl_2]$ at equilibrium
Conceptual Plan: Prepare an ICE table, calculate Q, compare Q and K_c, predict the direction of the reaction, represent the change with x, sum the table, determine the equilibrium values, put the equilibrium values in the equilibrium expression, and solve for x. Determine $[SO_2Cl_2]$, $[SO_2]$, and $[Cl_2]$.
Solution:

$$SO_2Cl_2(g) \rightleftharpoons SO_2(g) + Cl_2(g)$$

	$[SO_2Cl_2]$	$[SO_2]$	$[Cl_2]$
Initial	0.175	0.0	0.0
Change	$-x$	$+x$	$+x$
Equil	$0.175 - x$	x	x

$$Q = \frac{[SO_2][Cl_2]}{[SO_2Cl_2]} = \frac{0}{(0.175)} = 0$$

$Q < K$ Therefore, the reaction will proceed to the right by x.

$$K_c = \frac{[SO_2][Cl_2]}{[SO_2Cl_2]} = \frac{(x)(x)}{(0.175 - x)} = 2.99 \times 10^{-7}$$

Assume that x is small compared to 0.175.

$x^2 = 2.99 \times 10^{-7}(0.175)$

$x = 2.2\underline{8}7 \times 10^{-4}$ Check assumption: $\frac{2.2\underline{8}7 \times 10^{-4}}{0.175} \times 100 = 0.13\%$; assumption is valid.

$[SO_2] = [Cl_2] = x = 2.2\underline{8}7 \times 10^{-4} = 2.29 \times 10^{-4}$ M

$[SO_2Cl_2] = 0.175 - 2.2\underline{8}7 \times 10^{-4} = 0.17\underline{4}8 = 0.175$ M

Check: Plug the values into the equilibrium expression:

$$K_c = \frac{(2.2\underline{8}7 \times 10^{-4})(2.2\underline{8}7 \times 10^{-4})}{(0.175)} = 2.989 \times 10^{-7} = 3.00 \times 10^{-7}$$

This is within 0.011×10^{-7} of the true value; therefore, the answers are valid.

16.59 **Given:** $P_{Br_2} = 755$ torr, $P_{Cl_2} = 735$ torr, $P_{BrCl} = 0.0$, $K_p = 1.11 \times 10^{-4}$ **Find:** P_{BrCl} at equilibrium
Conceptual Plan: Torr $\rightarrow$ atm and then prepare an ICE table, calculate Q, compare Q and K_c, predict the direction of the reaction, represent the change with x, sum the table, determine the equilibrium values, put the equilibrium values in the equilibrium expression, and solve for x. Determine P_{BrCl}.

Solution: $P_{Br_2} = 755 \text{ torr} \times \frac{1 \text{ atm}}{760 \text{ torr}} = 0.99\underline{3}4$ atm $P_{Cl_2} = 735 \text{ torr} \times \frac{1 \text{ atm}}{760 \text{ torr}} = 0.96\underline{7}1$ atm

$$Br_2(g) + Cl_2(g) \rightleftharpoons 2\ BrCl(g)$$

	P_{Br_2}	P_{Cl_2}	P_{BrCl}
Initial	0.9934	0.9671	0.0
Change	$-x$	$-x$	$+2x$
Equil	$0.9934 - x$	$0.9671 - x$	$2x$

$$Q = \frac{P^2_{BrCl}}{P_{Br_2}P_{Cl_2}} = \frac{0}{(0.9934)(0.9671)} = 0 \qquad Q < K \quad \text{Therefore, the reaction will proceed to the right by } x.$$

$$K_p = \frac{P^2_{BrCl}}{P_{Br_2}P_{Cl_2}} = \frac{(2x)^2}{(0.9934 - x)(0.9671 - x)} = 1.11 \times 10^{-4}$$

Assume that x is small compared to 0.9934 and 0.9671.

$$\frac{(2x)^2}{(0.9934)(0.9671)} = 1.11 \times 10^{-4} \quad 4x^2 = 1.066 \times 10^{-4}$$

$x = 0.00516 \text{ atm} = 3.92 \text{ torr}$

$P_{BrCl} = 2x = 2(3.92 \text{ torr}) = 7.84 \text{ torr}$

Check: Plug the values into the equilibrium expression:

$$K_c = \frac{(2(0.00516))^2}{(0.9934 - 0.00516)(0.9671 - 0.00516)} = 1.120 \times 10^{-4} = 1.12 \times 10^{-4}.$$

This is within 0.01×10^{-4} of the true value; therefore, the answers are valid.

16.60 **Given:** $P_{CO} = 1344$ torr, $P_{H_2O} = 1766$ torr, $P_{CO_2} = 0.0$, $P_{H_2} = 0.0$, $K_p = 0.0611$ **Find:** P_{CO_2}, P_{H_2} at equilibrium
Conceptual Plan: Torr → atm and then prepare an ICE table, calculate Q, compare Q and K_c, predict the direction of the reaction, represent the change with x, sum the table, determine the equilibrium values, put the equilibrium values in the equilibrium expression, and solve for x. Determine P_{CO_2} and P_{H_2}.

Solution: $P_{CO} = 1344 \text{ torr} \times \frac{1 \text{ atm}}{760 \text{ torr}} = 1.7684 \text{ atm} \quad P_{H_2O} = 1766 \text{ torr} \times \frac{1 \text{ atm}}{760 \text{ torr}} = 2.3237 \text{ atm}$

$$CO(g) + H_2O(g) \rightleftharpoons CO_2(g) + H_2(g)$$

	P_{CO}	P_{H_2O}	P_{CO_2}	P_{H_2}
Initial	1.7684	2.3237	0.0	0.0
Change	$-x$	$-x$	$+x$	$+x$
Equil	$1.7684 - x$	$2.3237 - x$	x	x

$$Q = \frac{P_{CO_2}P_{H_2}}{P_{CO}P_{H_2O}} = \frac{0}{(1.7684)(2.3237)} = 0 \qquad Q < K \quad \text{Therefore, the reaction will proceed to the right by } x.$$

$$K_p = \frac{P_{CO_2}P_{H_2}}{P_{CO}P_{H_2O}} = \frac{(x)(x)}{(1.7684 - x)(2.3237 - x)} = 0.0611$$

$x^2 = (0.0611)(4.1092 - 4.0921x + x^2)$

$x^2 = (0.25107 - 0.2500\,x + 0.0611\,x^2)$

$0.9389x^2 + 0.2500x - 0.25107 = 0$

$$\frac{-b \pm \sqrt{b^2 - 4ac}}{2a} = \frac{-0.2500 \pm \sqrt{(0.2500)^2 - 4(0.9389)(-0.25107)}}{2(0.9389)}$$

$x = 0.4008 \text{ or } -0.6671 \text{ so } x = 0.4008$

$P_{CO_2} = P_{H_2} = x = 0.4008 \text{ atm} = 305 \text{ torr}$

Check: Plug the values into the equilibrium expression:

$$K_p = \frac{(0.4008)^2}{(1.7684 - 0.4008)(2.3237 - 0.4008)} = 0.06109 = 0.061.$$

The answers are the same to two significant figures with the true value; therefore, the answers are valid.

16.61 (a) **Given:** $[A] = 1.0$ M, $[B] = [C] = 0.0$, $K_c = 1.0$ **Find:** $[A]$, $[B]$, $[C]$ at equilibrium
Conceptual Plan: Prepare an ICE table, calculate Q, compare Q and K_c, predict the direction of the reaction, represent the change with x, sum the table, determine the equilibrium values, put the equilibrium values in the equilibrium expression, and solve for x. Determine $[A]$, $[B]$, and $[C]$.

Solution:

$$A(g) \rightleftharpoons B(g) + C(g)$$

	[A]	[B]	[C]
Initial	1.0	0.0	0.0
Change	$-x$	$+x$	$+x$
Equil	$1.0 - x$	x	x

$Q = \dfrac{[B][C]}{[A]} = \dfrac{0}{(1.0)} = 0$ $\quad Q < K$ Therefore, the reaction will proceed to the right by x.

$$K_c = \frac{[B][C]}{[A]} = \frac{(x)(x)}{(1.0 - x)} = 1.0$$

$$x^2 = 1.0(1.0 - x)$$

$$x^2 + x - 1 = 0$$

$$\frac{-b \pm \sqrt{b^2 - 4ac}}{2a} = \frac{-1 \pm \sqrt{1^2 - 4(1)(-1)}}{2(1)}$$

$x = 0.6\underline{1}8$ or $x = -1.618$ Therefore, $x = 0.6\underline{1}8$.

$[B] = [C] = x = 0.6\underline{1}8 = 0.62$ M

$[A] = 1.0 - 0.6\underline{1}8 = 0.382 = 0.38$ M

Check: Plug the values into the equilibrium expression:

$K_c = \dfrac{(0.62)(0.62)}{(0.38)} = 1.01 = 1.0$, which is the equilibrium constant; so the values are correct.

(b) **Given:** $[A] = 1.0$ M, $[B] = [C] = 0.0$, $K_c = 0.010$ **Find:** $[A]$, $[B]$, $[C]$ at equilibrium
Conceptual Plan: Prepare an ICE table, calculate Q, compare Q and K_c, predict the direction of the reaction, represent the change with x, sum the table, determine the equilibrium values, put the equilibrium values in the equilibrium expression, and solve for x. Determine [A], [B], and [C].
Solution:

$$A(g) \rightleftharpoons B(g) + C(g)$$

	[A]	[B]	[C]
Initial	1.0	0.0	0.0
Change	$-x$	$+x$	$+x$
Equil	$1.0 - x$	x	x

$Q = \dfrac{[B][C]}{[A]} = \dfrac{0}{(1.0)} = 0$ $\quad Q < K$ Therefore, the reaction will proceed to the right by x.

$$K_c = \frac{[B][C]}{[A]} = \frac{(x)(x)}{(1.0 - x)} = 0.010$$

$$x^2 = 0.010(1.0 - x)$$

$$x^2 + 0.010x - 0.010 = 0$$

$$\frac{-b \pm \sqrt{b^2 - 4ac}}{2a} = \frac{-(0.010) \pm \sqrt{(0.010)^2 - 4(1)(-0.010)}}{2(1)}$$

$x = 0.09\underline{5}12$ or $x = -0.1051$ Therefore, $x = 0.09\underline{5}12$.

$[B] = [C] = x = 0.09\underline{5}12 = 0.095$ M

$[A] = 1.0 - 0.09\underline{5}12 = 0.90488 = 0.90$ M

Check: Plug the values into the equilibrium expression:

$K_c = \dfrac{(0.095)(0.095)}{(0.90)} = 0.01003 = 0.010$, which is the equilibrium constant; so the values are correct.

(c) **Given:** $[A] = 1.0\ M$, $[B] = [C] = 0.0$, $K_c = 1.0 \times 10^{-5}$ **Find:** $[A]$, $[B]$, $[C]$ at equilibrium
Conceptual Plan: Prepare an ICE table, calculate Q, compare Q and K_c, predict the direction of the reaction, represent the change with x, sum the table, determine the equilibrium values, put the equilibrium values in the equilibrium expression, and solve for x. Determine $[A]$, $[B]$, and $[C]$.
Solution:

$$A(g) \rightleftharpoons B(g) + C(g)$$

	$[A]$	$[B]$	$[C]$
Initial	1.0	0.0	0.0
Change	$-x$	$+x$	$+x$
Equil	$1.0 - x$	x	x

$Q = \dfrac{[B][C]}{[A]} = \dfrac{0}{(1.0)} = 0$ $\quad Q < K$ Therefore, the reaction will proceed to the right by x.

$K_c = \dfrac{[B][C]}{[A]} = \dfrac{(x)(x)}{(1.0 - x)} = 1.0 \times 10^{-5}$

Assume that x is small compared to 1.0.

$x^2 = 1.0(1.0 \times 10^{-5})$

$x = 0.00316$ Check assumption: $\dfrac{0.00316}{1.0} \times 100\% = 0.32\%$; assumption is valid.

$[B] = [C] = x = 0.00316 = 0.0032\ M$

$[A] = 1.0 - 0.00316 = 0.9968 = 1.0\ M$

Check: Plug the values into the equilibrium expression:

$K_c = \dfrac{(0.0032)(0.0032)}{(1.0)} = 1.024 \times 10^{-5} = 1.0 \times 10^{-5}$,

which is the equilibrium constant; so the values are correct

16.62 (a) **Given:** $P_B = 1.0$ atm, $P_A = 0.0$ atm, $K_p = 1.0$ **Find:** P_B, P_A at equilibrium
Conceptual Plan: Prepare an ICE table, calculate Q, compare Q and K_p, predict the direction of the reaction, represent the change with x, sum the table, determine the equilibrium values, put the equilibrium values in the equilibrium expression, and solve for x. Determine P_A and P_B.
Solution:

$$A(g) \rightleftharpoons 2\,B(g)$$

	P_A	P_B
Initial	0.0	1.0
Change	$+x$	$-2x$
Equil	x	$1.0 - 2x$

$Q = \dfrac{P_B^2}{P_A} = \dfrac{1.0^2}{0}$ Because there is no A, the reaction shifts to the left.

$K_p = \dfrac{P_B^2}{P_A} = \dfrac{(1.0 - 2x)^2}{(x)} = 1.0$

$1.0 - 4x + 4x^2 = 1.0x$

$4x^2 - 5x + 1.0 = 0$

$$\frac{-b \pm \sqrt{b^2 - 4ac}}{2a} = \frac{-(-5) \pm \sqrt{(-5)^2 - 4(4)(1)}}{2(4)}$$

$x = 0.25$ or $x = 1.0x$ Therefore, $x = 0.25$.

$P_A = x = 0.25$ atm; $P_B = 1.0 - 2x = 1.0 - 0.50 = 0.50$ atm

Check: Plug the values into the equilibrium expression:

$K_p = \dfrac{(0.50)^2}{(0.25)} = 1.0$, which is the equilibrium constant.

(b) **Given:** $P_B = 1.0$ atm, $P_A = 0.0$ atm, $K_p = 1.0 \times 10^{-4}$ **Find:** P_B, P_A at equilibrium
Conceptual Plan: Prepare an ICE table, calculate Q, compare Q and K_p, predict the direction of the reaction, represent the change with x, sum the table, determine the equilibrium values, put the equilibrium values in the equilibrium expression, and solve for x. Determine P_A and P_B.
Solution:

$$A(g) \rightleftharpoons 2\,B(g)$$

	P_A	P_B
Initial	0.0	1.0
Change	$+x$	$-2x$
Equil	x	$1.0 - 2x$

$Q = \dfrac{P_B^2}{P_A} = \dfrac{(1.0)^2}{0}$ Because there is no A, the reaction shifts to the left.

$$K_p = \frac{P_B^2}{P_A} = \frac{(1.0 - 2x)^2}{(x)} = 1.0 \times 10^{-4}$$

$$1.0 - 4x + 4x^2 = (1.0 \times 10^{-4})x$$

$$4x^2 - 4.0001x + 1.0 = 0$$

$$\frac{-b \pm \sqrt{b^2 - 4ac}}{2a} = \frac{-(-4.0001) \pm \sqrt{(-4.0001)^2 - 4(4)(1)}}{2(4)} = \frac{-(-4.0001) \pm \sqrt{8.0 \times 10^{-4}}}{2(4)}$$

$x = 0.5\underline{0}35$ or $x = 0.4\underline{9}65$ Therefore, $x = 0.4\underline{9}65$.

$P_A = x = 0.4\underline{9}65 = 0.50$ atm; $P_B = 1.0 - 2x = 1.0 - 2(0.4\underline{9}65) = 0.01$ atm

Check: Plug the values into the equilibrium expression:

$K_p = \dfrac{(0.007)^2}{(0.4\underline{9}65)} = \underline{9}.869 \times 10^{-5} = 1 \times 10^{-4}$, which is the equilibrium constant.

(c) **Given:** $P_B = 1.0$ atm, $P_A = 0.0$ atm, $K_p = 1.0 \times 10^{5}$ **Find:** P_B, P_A at equilibrium
Conceptual Plan: Prepare an ICE table, calculate Q, compare Q and K_p, predict the direction of the reaction, represent the change with x, sum the table, determine the equilibrium values, put the equilibrium values in the equilibrium expression, and solve for x. Determine P_A and P_B.
Solution:

$$A(g) \rightleftharpoons 2\,B(g)$$

	P_A	P_B
Initial	0.0	1.0
Change	$+x$	$-2x$
Equil	x	$1.0 - 2x$

$Q = \dfrac{P_B^2}{P_A} = \dfrac{1.0^2}{0}$ Because there is no A, the reaction shifts to the left.

$$K_p = \frac{P_B^2}{P_A} = \frac{(1.0 - 2x)^2}{(x)} = 1.0 \times 10^{5}$$

Assume that $2x$ is small compared to 1.0.

$$\frac{1.0}{x} = 1.0 \times 10^{5}$$

$x = 1.0 \times 10^{-5}$ Check assumption: $\dfrac{2(1.0 \times 10^{-5})}{1.0} \times 100\% = 0.002\%$; assumption is valid.

$P_A = x = 1.0 \times 10^{-5}$ atm; $P_B = 1.0 - 2x = 1.0 - 2(1.0 \times 10^{-5}) = 0.99998$ atm

Check: Plug the values into the equilibrium expression:

$K_p = \dfrac{(0.99998)^2}{(1.0 \times 10^{-5})} = 9.9996 \times 10^{4} = 1.0 \times 10^{5}$, which is the equilibrium constant.

Le Châtelier's Principle

16.63 **Given:** $CO(g) + Cl_2(g) \rightleftharpoons COCl_2(g)$ at equilibrium **Find:** What is the effect of each of the following?

(a) $COCl_2$ is added to the reaction mixture. Adding $COCl_2$ increases the concentration of $COCl_2$ and causes the reaction to shift to the left.

(b) Cl_2 is added to the reaction mixture. Adding Cl_2 increases the concentration of Cl_2 and causes the reaction to shift to the right.

(c) $COCl_2$ is removed from the reaction mixture. Removing the $COCl_2$ decreases the concentration of $COCl_2$ and causes the reaction to shift to the right.

16.64 **Given:** $2\ BrNO(g) \rightleftharpoons 2\ NO(g) + Br_2(g)$ at equilibrium **Find:** What is the effect of each of the following?

(a) NO is added to the reaction mixture. Adding NO increases the concentration of NO and causes the reaction to shift to the left.

(b) BrNO is added to the reaction mixture. Adding BrNO increases the concentration of BrNO and causes the reaction to shift to the right.

(c) Br_2 is removed from the reaction mixture. Removing Br_2 decreases the concentration of Br_2 and causes the reaction to shift to the right.

16.65 **Given:** $2\ KClO_3(s) \rightleftharpoons 2\ KCl(s) + 3\ O_2(g)$ at equilibrium **Find:** What is the effect of each of the following?

(a) O_2 is removed from the reaction mixture. Removing the O_2 decreases the concentration of O_2 and causes the reaction to shift to the right.

(b) KCl is added to the reaction mixture. Adding KCl does not cause any change in the reaction. KCl is a solid, and the concentration remains constant; so the addition of more solid does not change the equilibrium concentration.

(c) $KClO_3$ is added to the reaction mixture. Adding $KClO_3$ does not cause any change in the reaction. $KClO_3$ is a solid, and the concentration remains constant; so the addition of more solid does not change the equilibrium concentration.

(d) O_2 is added to the reaction mixture. Adding O_2 increases the concentration of O_2 and causes the reaction to shift to the left.

16.66 **Given:** $C(s) + H_2O(g) \rightleftharpoons CO(g) + H_2(g)$ **Find:** What is the effect of each of the following?

(a) C is added to the reaction mixture. Adding C does not cause any change in the reaction. C is a solid, and the concentration remains constant; so the addition of more solid does not change the equilibrium concentration.

(b) H_2O is condensed and removed from the reaction mixture. Removing the H_2O decreases the concentration of H_2O and causes the reaction to shift to the left.

(c) CO is added to the reaction mixture. Adding CO increases the concentration of CO and causes the reaction to shift to the left.

(d) H_2 is removed from the reaction mixture. Removing the H_2 decreases the concentration of H_2 and causes the reaction to shift to the right.

16.67 (a) **Given:** $I_2(g) \rightleftharpoons 2\ I(g)$ at equilibrium **Find:** the effect of increasing the volume

The chemical equation has 2 moles of gas on the right and 1 mole of gas on the left. Increasing the volume of the reaction mixture decreases the pressure and causes the reaction to shift to the right (toward the side with more moles of gas particles).

(b) **Given:** $2\ H_2S(g) \rightleftharpoons 2\ H_2(g) + S_2(g)$ **Find:** the effect of decreasing the volume

The chemical equation has 3 moles of gas on the right and 2 moles of gas on the left. Decreasing the volume of the reaction mixture increases the pressure and causes the reaction to shift to the left (toward the side with fewer moles of gas particles).

(c) **Given:** $I_2(g) + Cl_2(g) \rightleftharpoons 2\ ICl(g)$ **Find:** the effect of decreasing the volume

The chemical equation has 2 moles of gas on the right and 2 moles of gas on the left. Decreasing the volume of the reaction mixture increases the pressure but causes no shift in the reaction because the moles are equal on both sides.

16.68 (a) **Given:** $CO(g) + H_2O(g) \rightleftharpoons CO_2(g) + H_2(g)$ **Find:** the effect of decreasing the volume

The chemical equation has 2 moles of gas on the right and 2 moles of gas on the left. Decreasing the volume of the reaction mixture increases the pressure but causes no shift in the reaction because the moles are equal on both sides.

(b) **Given:** $PCl_3(g) + Cl_2(g) \rightleftharpoons PCl_5(g)$ **Find:** the effect of increasing the volume
The chemical equation has 2 moles of gas on the left and 1 mole of gas on the right. Increasing the volume of the reaction mixture decreases the pressure and causes the reaction to shift to the left (toward the side with more moles of gas particles).

(c) **Given:** $CaCO_3(s) \rightleftharpoons CaO(s) + CO_2(g)$ **Find:** the effect of increasing the volume
The chemical equation has 1 mole of gas on the right and 0 mole of gas on the left. Increasing the volume of the reaction mixture decreases the pressure and causes the reaction to shift to the right (toward the side with more moles of gas particles).

16.69 **Given:** $C(s) + CO_2(g) \rightleftharpoons 2\,CO(g)$ is endothermic. **Find:** the effect of increasing the temperature
Because the reaction is endothermic, we can think of the heat as a reactant. Increasing the temperature is equivalent to adding a reactant, causing the reaction to shift to the right. This will cause an increase in the concentration of products and a decrease in the concentration of reactants; therefore, the value of K will increase.
Find: the effect of decreasing the temperature
Because the reaction is endothermic, we can think of the heat as a reactant. Decreasing the temperature is equivalent to removing a reactant, causing the reaction to shift to the left. This will cause a decrease in the concentration of products and an increase in the concentration of reactants; therefore, the value of K will decrease.

16.70 **Given:** $C_6H_{12}O_6(s) + 6\,O_2(g) \rightleftharpoons 6\,CO_2(g) + 6\,H_2O(g)$ is exothermic.
Find: the effect of increasing the temperature
Because the reaction is exothermic, we can think of the heat as a product. Increasing the temperature is equivalent to adding a product, causing the reaction to shift to the left. This will cause a decrease in the concentration of products and an increase in the concentration of reactants; therefore, the value of K will decrease.
Find: the effect of decreasing the temperature
Because the reaction is exothermic, we can think of the heat as a product. Decreasing the temperature is equivalent to removing a product, causing the reaction to shift to the right. This will cause an increase in the concentration of products and a decrease in the concentration of reactants; therefore, the value of K will increase.

16.71 **Given:** $C(s) + 2\,H_2(g) \rightleftharpoons CH_4(g)$ is exothermic. **Find:** Determine which will favor CH_4.

(a) Adding more C to the reaction mixture does *not* favor CH_4. Adding C does not cause any change in the reaction. C is a solid, and the concentration remains constant; so the addition of more solid does not change the equilibrium concentration.

(b) Adding more H_2 to the reaction mixture favors CH_4. Adding H_2 increases the concentration of H_2, causing the reaction to shift to the right.

(c) Raising the temperature of the reaction mixture does *not* favor CH_4. Because the reaction is exothermic, we can think of heat as a product. Raising the temperature is equivalent to adding a product, causing the reaction to shift to the left.

(d) Lowering the volume of the reaction mixture favors CH_4. The chemical equation has 1 mole of gas on the right and 2 moles of gas on the left. Decreasing the volume of the reaction mixture increases the pressure and causes the reaction to shift to the right (toward the side with fewer moles of gas particles).

(e) Adding a catalyst to the reaction mixture does *not* favor CH_4. A catalyst added to the reaction mixture only speeds up the reaction; it does not change the equilibrium concentration.

(f) Adding neon gas to the reaction mixture does *not* favor CH_4. Adding an inert gas to a reaction mixture at a fixed volume has no effect on the equilibrium.

16.72 **Given:** $C(s) + H_2O(g) \rightleftharpoons CO(g) + H_2(g)$ is endothermic. **Find:** Determine which will favor H_2.

(a) Adding more C to the reaction mixture has no effect on the quantity of H_2. Adding C does not cause any change in the reaction. C is a solid, and the concentration remains constant; so the addition of more solid does not change the equilibrium concentration.

(b) Adding more H_2O to the reaction mixture results in the formation of additional H_2. Adding H_2O increases the concentration of H_2O and causes the reaction to shift to the right.

(c) Raising the temperature of the reaction mixture results in the formation of additional H_2. Because the reaction is endothermic, we can think of heat as a reactant. Raising the temperature is equivalent to adding a reactant, causing the reaction to shift to the right.

(d) Increasing the volume of the reaction mixture results in the formation of additional H_2. The chemical equation has 2 moles of gas on the right and 1 mole of gas on the left. Increasing the volume of the reaction mixture decreases the pressure and causes the reaction to shift to the right (toward the side with more moles of gas particles).

(e) Adding a catalyst to the reaction mixture has no effect on the quantity of H_2. A catalyst added to the reaction mixture only speeds up the reaction; it does not change the equilibrium concentration.

(f) Adding an inert gas to the reaction mixture has no effect on the quantity of H_2. Adding an inert gas to a reaction mixture at a fixed volume has no effect on the equilibrium.

Cumulative Problems

16.73 (a) To find the value of K for the new equation, combine the two given equations to yield the new equation. Reverse equation 1 and use $1/K_1$, then add to equation 2. To find K for equation 3, use $(1/K_1)(K_2)$.

$$\begin{array}{llll} HbO_2(aq) & \rightleftharpoons & \cancel{Hb(aq)} + O_2(aq) & K_1 = 1/1.8 \\ \cancel{Hb(aq)} + CO(aq) & \rightleftharpoons & HbCO(aq) & K_2 = 306 \\ \hline HbO_2(aq) + CO(aq) & \rightleftharpoons & HbCO(aq) + O_2(aq) & K_3 = K_1K_2 = (1/1.8)(306) = 170 \end{array}$$

(b) **Given:** $O_2 = 20\%, CO = 0.10\%$ **Find:** the ratio $\dfrac{[HbCO]}{[HbO_2]}$

Conceptual Plan: Determine the equilibrium expression and then determine $\dfrac{[HbCO]}{[HbO_2]}$.

Solution: $K = \dfrac{[HbCO][O_2]}{[HbO_2][CO]}$ $170 = \dfrac{[HbCO](20.0)}{[HbO_2](0.10)}$ then $\dfrac{[HbCO]}{[HbO_2]} = 170\left(\dfrac{0.10}{20.0}\right) = \dfrac{0.85}{1.0}$

Because the ratio is almost 1:1, 0.10% CO will replace about 50% of the O_2 in the blood. The CO blocks the uptake of O_2 by the blood and is therefore highly toxic.

16.74 **Given:** $P = 1$ atm, $T = 298$ K, $N_2 = 78\%$, $O_2 = 21\%$, $K_p = 4.1 \times 10^{-31}$ **Find:** $[NO]$ in molecules/cm^3

Conceptual Plan:

% vol → n → M and then $K_p \rightarrow K_c$ and then prepare an ICE table, represent the change with

$PV = nRT$ $\quad \dfrac{n}{1\text{ L air}}$ $\quad K_p = K_c(RT)^{\Delta n}$

x, sum the table, determine the equilibrium values, put the equilibrium values in the equilibrium expression, and solve for x. Determine [NO] in mol/L → molecules/cm^3.

$\dfrac{6.022 \times 10^{23}\text{ molecules}}{\text{mole}}$ $\dfrac{1\text{ L}}{1000\text{ mL}}$ $\dfrac{\text{mL}}{\text{cm}^3}$

Solution: Assume 1 L of air.

$$n_{N_2} = \frac{(1\ \cancel{\text{atm}})(0.78\ \cancel{\text{L}})}{\left(0.08206\dfrac{\cancel{\text{L}}\cdot\cancel{\text{atm}}}{\text{mol}\cdot\cancel{\text{K}}}\right)(298\ \cancel{\text{K}})} = 0.03\underline{1}90\text{ mol} \qquad n_{O_2} = \frac{(1\ \cancel{\text{atm}})(0.21\ \cancel{\text{L}})}{\left(0.08206\dfrac{\cancel{\text{L}}\cdot\cancel{\text{atm}}}{\text{mol}\cdot\cancel{\text{K}}}\right)(298\ \cancel{\text{K}})} = 0.008\underline{5}9\text{ mol}$$

$[N_2] = 0.03\underline{1}90\text{ mol/L}$ $\quad [O_2] = 0.008\underline{5}9\text{ mol/L}$

$$K_p = K_c(RT)^{\Delta n} \quad K_c = \frac{K_p}{(RT)^{\Delta n}} = \frac{4.1 \times 10^{-31}}{\left(\left(0.08206\dfrac{\text{L}\cdot\text{atm}}{\text{mol}\cdot\cancel{\text{K}}}\right)(298\ \cancel{\text{K}})\right)^0} = 4.1 \times 10^{-31}$$

	$N_2(g)$ +	$O_2(g) \rightleftharpoons$	$2\,NO(g)$
	$[N_2]$	$[O_2]$	$[NO]$
Initial	0.03$\underline{1}$9	0.008$\underline{5}$9	0.0
Change	$-x$	$-x$	$+2x$
Equil	0.03$\underline{1}$9 $- x$	0.008$\underline{5}$9 $- x$	$2x$

The reaction will proceed to the right by x.

$$K_c = \frac{[NO]^2}{[N_2][O_2]} = \frac{(2x)^2}{(0.0319 - x)(0.00859 - x)} = 4.1 \times 10^{-31}$$

Assume that x is small compared to 0.00859 and 0.0319.

$x = 5.\underline{3}0 \times 10^{-18}$ Check assumption: $\frac{5.\underline{3}0 \times 10^{-18}}{0.00858} \times 100\% = 6.2 \times 10^{-14}\%$

$[NO] = 2x = 2(5.\underline{3}0 \times 10^{-18}) = 1.\underline{0}6 \times 10^{-17}$ M

$$1.\underline{0}6 \times 10^{-17} \frac{\text{mol}}{\text{L}} \times \frac{6.022 \times 10^{23} \text{ molecules}}{\text{mol}} \times \frac{\text{L}}{1000 \text{ mL}} \times \frac{\text{mL}}{\text{cm}^3} = 6.38 \times 10^3 \frac{\text{molecules}}{\text{cm}^3}$$
$$= 6.4 \times 10^3 \frac{\text{molecules}}{\text{cm}^3}$$

Check: The answer is reasonable because the reaction has a small equilibrium constant; so you would not expect to produce much product.
The reaction to produce NO is endothermic, so we can think of heat as a reactant. Therefore, raising the temperature (as in an automobile engine) shifts the reaction to the right, producing more NO.

16.75 (a) **Given:** 4.45 g CO_2, 10.0 L, 1200 K, 2.00 g C, $K_p = 5.78$ **Find:** total pressure

Conceptual Plan: g $CO_2 \rightarrow$ mol CO_2 and g C $\rightarrow$ mol C and then determine limiting reactant

$\frac{1 \text{ mol } CO_2}{44.01 \text{ g } CO_2}$ $\frac{1 \text{ mol C}}{12.01 \text{ g C}}$

and then mol $CO_2 \rightarrow P$ CO_2. Prepare an ICE table, represent the change with x, sum the table,

$PV = nRT$

determine the equilibrium values, put the equilibrium values in the equilibrium expression, and solve for x.

Solution: $CO_2(g) + C(s) \rightleftharpoons 2\,CO(g)$

$$4.45 \text{ g } CO_2 \times \frac{1 \text{ mol}}{44.01 \text{ g } CO_2} = 0.10\underline{1}1 \text{ mol } CO_2 \quad 2.00 \text{ g C} \times \frac{1 \text{ mol}}{12.01 \text{ g C}} = 0.16\underline{6}5 \text{ mol C}$$

Because the stoichiometry is 1:1, the CO_2 is the limiting reactant.

$$P_{CO_2} = \frac{(0.10\underline{1}1 \text{ mol})\left(\frac{0.08206 \text{ L} \cdot \text{atm}}{\text{mol} \cdot \text{K}}\right)(1200 \text{ K})}{10.0 \text{ L}} = 0.99\underline{5}6 \text{ atm}$$

$CO_2(g) + C(s) \rightleftharpoons 2\,CO(g)$

	P_{CO_2}	P_{CO}
Initial	0.996	0.0
Change	$-x$	$+2x$
Equil	$0.996 - x$	$2x$

The reaction will proceed to the right by x.

$$K_p = \frac{P^2_{CO}}{P_{CO_2}} = \frac{(2x)^2}{(0.996 - x)} = 5.78.$$ Solve using the quadratic equation, found in Appendix I.

$x = 0.678$ atm

$P_{CO_2} = 0.996 \text{ atm} - 0.678 \text{ atm} = 0.318 \text{ atm}$ $P_{CO} = 2(0.678 \text{ atm}) = 1.3\underline{5}6 \text{ atm}$

$P_{total} = 1.67$ atm

Check: Plug the values for the partial pressure into the equilibrium expression. $\frac{(1.356)^2}{0.318} = 5.78$, which is the value of the equilibrium constant.

(b) **Given:** 4.45 g CO_2, 10.0 L, 1200 K, 0.50 g C, $K_p = 5.78$ **Find:** total pressure

Conceptual Plan: g $CO_2 \rightarrow$ mol CO_2 and g C $\rightarrow$ mol C and then determine limiting reactant

$\frac{1 \text{ mol } CO_2}{44.01 \text{ g } CO_2}$ $\frac{1 \text{ mol C}}{12.01 \text{ g C}}$

and then mol $CO_2 \rightarrow P_{CO_2}$ and mol C $\rightarrow$ mol CO $\rightarrow P_{CO}$

$PV = nRT$ $PV = nRT$

Solution: $CO_2(g) + C(s) \rightleftharpoons 2\,CO(g)$

$$4.45 \text{ g } CO_2 \times \frac{1 \text{ mol}}{44.01 \text{ g } CO_2} = 0.10\underline{1}1 \text{ mol } CO_2 \qquad 0.50 \text{ g C} \times \frac{1 \text{ mol}}{12.01 \text{ g C}} = 0.04\underline{1}6 \text{ mol C}$$

Because the stoichiometry is 1:1, the C is the limiting reactant; therefore, the moles of CO formed will be determined from the reaction, not the equilibrium.

$$0.0416 \text{ mol C} \times \frac{2 \text{ mol CO}}{1 \text{ mol C}} = 0.0832 \text{ mol CO}$$

	$CO_2(g)$	+ $C(s)$	$\rightleftharpoons$ $2\,CO(g)$
Initial	0.1011	0.0416	0.0
Change	−0.0416	−0.0416	+2(0.0416)
Equil	0.0595	0	0.0832

$$P_{CO_2} = \frac{(0.0595 \text{ mol})\left(\frac{0.08206 \text{ L} \cdot \text{atm}}{\text{mol} \cdot \text{K}}\right)(1200 \text{ K})}{10.0 \text{ L}} = 0.586 \text{ atm}$$

$$P_{CO} = \frac{(0.0832 \text{ mol})\left(\frac{0.08206 \text{ L} \cdot \text{atm}}{\text{mol} \cdot \text{K}}\right)(1200 \text{ K})}{10.0 \text{ L}} = 0.819 \text{ atm}$$

$P_{total} = 0.586 + 0.819 = 1.405 = 1.41 \text{ atm}$

Check: The pressure is less than the equilibrium pressure, which is reasonable because the C was the limiting reactant.

16.76 **Given:** at equilibrium: 0.13 mol H_2, 0.13 mol CO, 0.43 mol H_2O, then react all H_2
Find: CO at new equilibrium
Conceptual Plan: Assume 1 L. Equilibrium values → K_c and then new value of H_2O. Prepare an ICE table, represent the change with x, sum the table, determine the equilibrium values, put the equilibrium values in the equilibrium expression, and solve for x.
Solution: $H_2O(g) + C(s) \rightleftharpoons H_2(g) + CO(g)$

	$H_2O(g)$	$C(s)$	$H_2(g)$	$CO(g)$
Equil	0.43		0.13	0.13

$$K_c = \frac{[H_2][CO]}{[H_2O]} = \frac{(0.13)(0.13)}{(0.43)} = 0.0393$$

$2\,H_2(g) + O_2(g) \rightarrow 2\,H_2O(g)$ So 0.13 mol H_2 produces 0.13 mol H_2O, giving new initial conditions of 0.56 mol H_2O (0.13 + 0.43), 0 mol H_2 (reacts completely), and 0.13 mol CO (unaffected).

	$H_2O(g)$ + $C(s)$	$\rightleftharpoons$ $H_2(g)$	+ $CO(g)$
Initial	0.56	0	0.13
Change	$-x$	$+x$	$+x$
Equil	$0.56 - x$	x	$0.13 + x$

$$K_c = \frac{[H_2][CO]}{[H_2O]} = \frac{(x)(0.13 + x)}{(0.56 - x)} = 0.0393 \quad \text{so } x^2 + 0.1693x - 0.0220 = 0$$

Solve using the quadratic equation, found in Appendix I. $x = 0.086$

$[CO] = 0.13 + 0.086 = 0.216 \text{ M} = 0.22 \text{ M}$ therefore, 0.22 mol CO.

Check: Determine equilibrium concentration and plug into the equilibrium expression.

$$K_c = \frac{[H_2][CO]}{[H_2O]} = \frac{(0.086)(0.22)}{(0.47)} = 0.040.$$ This is within 0.001 of the equilibrium value, so the answer is reasonable.

16.77 **Given:** $V = 10.0 \text{ L}$, $T = 650 \text{ K}$, 1.0 g MgO, $P_{CO_2} = 0.026 \text{ atm}$, $K_p = 0.0260$
Find: mass $MgCO_3$ when volume is 0.100 L
Conceptual Plan: $P(10.0 \text{ L}) \rightarrow P(0.100 \text{ L})$. Prepare an ICE table, represent the change with x, sum the table, determine the equilibrium value, put the equilibrium values in the equilibrium expression, and solve for x. Then determine moles CO_2, the limiting reactant, and the mass of $MgCO_3$ formed.

$P_1V_1 = P_2V_2 \quad PV = nRT$

Solution:

$P_1V_1 = P_2V_2$ $\quad (0.0260 \text{ atm})(10.0 \text{ L}) = (x)(0.100 \text{ L})$ $\quad x = 2.60 \text{ atm}$

	$MgCO_3(s) \rightleftharpoons MgO(s) + CO_2(g)$
Initial	2.60
Change	$-x$
Equil	$2.60 - x$

$K_p = P_{CO_2} = 0.026 = 2.60 - x \quad x = 2.5\underline{7}9 \text{ atm}$

$$n_{CO_2} = \frac{(2.5\underline{7}9 \text{ atm})10.0 \text{ L}}{\left(\frac{0.08206 \text{ L} \cdot \text{atm}}{\text{mol} \cdot \text{K}}\right)(650 \text{ K})} = 0.48\underline{3}5 \text{ mol } CO_2 \quad 1.0 \text{ g MgO} \times \frac{1 \text{ mol}}{40.30 \text{ g MgO}} = 0.02\underline{4}8 \text{ mol MgO}$$

Therefore, MgO is the limiting reactant and produces 0.0248 mol $MgCO_3$.

$$0.0248 \text{ mol } MgCO_3 \times \frac{84.31 \text{ g } MgCO_3}{1 \text{ mol } MgCO_3} = 2.09 \text{ g } MgCO_3$$

Check: The units (g) are correct. The magnitude (2) is reasonable since the pressure and volume are small. Much less than one mole is anticipated.

16.78 **Given:** at equilibrium: $P_{I_2} = 0.21$ atm, $P_I = 0.23$ atm
Find: P of each gas when the volume is compressed to half the initial volume
Conceptual Plan: P at equilibrium to equilibrium constant, determine P at new volume, prepare an ICE table, represent the change with x, sum the table, determine the equilibrium value, put the equilibrium values in the equilibrium expression, and solve for x.

$P_1V_1 = P_2V_2$

Solution: $I_2(g) \rightleftharpoons 2\,I(g) \quad K_p = \frac{P_I^2}{P_{I_2}} = \frac{(0.23)^2}{(0.21)} = 0.2\underline{5}2$

$P_1V_1 = P_2V_2$ $\quad$ For I_2: $(0.21 \text{ atm})(1) = (x \text{ atm})(0.5)\ x = 0.42 \text{ atm}$
For I: $(0.23 \text{ atm})(1) = (y \text{ atm})(0.5)\ y = 0.46 \text{ atm}$

	$I_2(g)$	$\rightleftharpoons 2\,I(g)$
Initial	0.42	0.46
Change	$+x$	$-2x$
Equil	$0.42 + x$	$0.46 - 2x$

$Q = \frac{(0.46)^2}{(0.42)} = 0.50$

$Q > K$, so the reaction shifts to the left.

$K_p = \frac{(0.46 - 2x)^2}{(0.42 + x)} = 0.2\underline{5}2 \quad x = 0.05\underline{6}7$ Solve using the quadratic equation, found in Appendix I.

$P_{(I_2)} = 0.42 + 0.057 = 0.477 = 0.48 \text{ atm}$

$P_{(I)} = 0.46 - 2(0.057) = 0.346 = 0.35 \text{ atm}$

Check: Plug the equilibrium values into the equilibrium expression: $K_p = \frac{(0.35)^2}{(0.48)} = 0.2\underline{5}5 = 0.26$. The value is within one significant figure of the original K_p, so the answer is reasonable.

16.79 **Given:** $C_2H_4(g) + Cl_2(g) \rightleftharpoons C_2H_4Cl_2(g)$ is exothermic **Find:** Which of the following will maximize $C_2H_4Cl_2$?

(a) Increasing the reaction volume will not maximize $C_2H_4Cl_2$. The chemical equation has 1 mole of gas on the right and 2 moles of gas on the left. Increasing the volume of the reaction mixture decreases the pressure and causes the reaction to shift to the left (toward the side with more moles of gas particles).

(b) Removing $C_2H_4Cl_2$ as it forms will maximize $C_2H_4Cl_2$. Removing the $C_2H_4Cl_2$ will decrease the concentration of $C_2H_4Cl_2$ and will cause the reaction to shift to the right, producing more $C_2H_4Cl_2$.

(c) Lowering the reaction temperature will maximize $C_2H_4Cl_2$. The reaction is exothermic, so we can think of heat as a product. Lowering the temperature will cause the reaction to shift to the right, producing more $C_2H_4Cl_2$.

(d) Adding Cl_2 will maximize $C_2H_4Cl_2$. Adding Cl_2 increases the concentration of Cl_2, so the reaction shifts to the right, which will produce more $C_2H_4Cl_2$.

16.80 **Given:** $C_2H_4(g) + I_2(g) \rightleftharpoons C_2H_4I_2(g)$ is endothermic **Find:** Which of the following will maximize $C_2H_4I_2$?

(a) Decreasing the reaction volume will maximize $C_2H_4I_2$. The chemical equation has 1 mole of gas on the right and 2 moles of gas on the left. Decreasing the volume of the reaction mixture increases the pressure and causes the reaction to shift to the right (toward the side with fewer moles of gas particles).

(b) Removing I_2 from the reaction mixture will not maximize $C_2H_4I_2$. Removing I_2 decreases the concentration of I_2, and the reaction will shift to the left to produce more I_2.

(c) Raising the temperature of the reaction will maximize $C_2H_4I_2$. The reaction is endothermic, so we can think of heat as a reactant. Raising the temperature will cause the reaction to shift to the right, producing more $C_2H_4I_2$.

(d) Adding C_2H_4 to the reaction mixture will maximize $C_2H_4I_2$. Adding C_2H_4 will increase the concentration of C_2H_4 and cause the reaction to shift to the right to produce more $C_2H_4I_2$.

16.81 **Given:** reaction 1 at equilibrium: $P_{H_2} = 0.958$ atm, $P_{I_2} = 0.877$ atm, $P_{HI} = 0.020$ atm; reaction 2: $P_{H_2} = P_{I_2} = 0.621$ atm, $P_{HI} = 0.101$ atm **Find:** Is reaction 2 at equilibrium? If not, what is the P_{HI} at equilibrium?
Conceptual Plan: Use equilibrium partial pressures to determine K_p. Use K_p to determine whether reaction 2 is at equilibrium. Prepare an ICE table, calculate Q, compare Q and K_p, predict the direction of the reaction, represent the change with x, sum the table, determine the equilibrium values, put the equilibrium values in the equilibrium expression, and solve for x. Determine P_{HI}.

Solution: $H_2(g) + I_2(g) \rightleftharpoons 2\,HI(g)$

Reaction 1:	P_{H_2}	P_{I_2}	P_{HI}
Equil	0.958	0.877	0.020

$$K_p = \frac{P_{HI}^2}{P_{H_2}P_{I_2}} = \frac{(0.020)^2}{(0.958)(0.877)} = 4.7\underline{6}10 \times 10^{-4}$$

$$Q = \frac{P_{HI}^2}{P_{H_2}P_{I_2}} = \frac{(0.101)^2}{(0.621)(0.621)} = 0.0264$$

$Q > K$, so the reaction shifts to the left.

$H_2(g) + I_2(g) \rightleftharpoons 2\,HI(g)$

Reaction 2:	P_{H_2}	P_{I_2}	P_{HI}
Initial	0.621	0.621	0.101
Change	$+x$	$+x$	$-2x$
Equil	$0.621 + x$	$0.621 + x$	$0.101 - 2x$

$$K_p = \frac{P_{HI}^2}{P_{H_2}P_{I_2}} = \frac{(0.101 - 2x)^2}{(0.621 + x)(0.621 + x)} = 4.7\underline{6}10 \times 10^{-4}$$

$$\sqrt{\frac{(0.101 - 2x)^2}{(0.621 + x)(0.621 + x)}} = \sqrt{4.7\underline{6}10 \times 10^{-4}}$$

$$\frac{(0.101 - 2x)}{(0.621 + x)} = 2.1\underline{8}2 \times 10^{-2}$$

$x = 0.04325 = 0.0433$

$P_{H_2} = P_{I_2} = 0.621 + x = 0.621 + 0.0433 = 0.664$ atm

$P_{HI} = 0.101 - 2x = 0.101 - 2(0.0433) = 0.0144$ atm

Check: Plug the values into the equilibrium expression:

$$K_p = \frac{(0.0144)^2}{(0.664)^2} = 4.703 \times 10^{-4} = 4.70 \times 10^{-4}$$; this value is close to the original equilibrium constant.

16.82 **Given:** reaction 1 initial: $H_2S = 0.500$ M, $SO_2 = 0.500$ M, H_2O at equilibrium $= 0.0011$ M; reaction 2: $H_2S = 0.250$ M, $SO_2 = 0.325$ M **Find:** $[H_2O]$ at equilibrium in reaction 2
Conceptual Plan: Prepare an ICE table. Determine the equilibrium concentrations in reaction 1 and determine K_c. Use K_c to determine the equilibrium concentrations for reaction 2. Prepare an ICE table, calculate Q, compare Q and K_c, predict the direction of the reaction, represent the change with x, sum the table, determine the equilibrium values, put the equilibrium values in the equilibrium expression, and solve for x. Determine $[H_2O]$.

Solution: $2\,H_2S(g) + SO_2(g) \rightleftharpoons 3\,S(s) + 2\,H_2O(g)$

Reaction 1:	$[H_2S]$	$[SO_2]$	[S]	$[H_2O]$
Initial	0.500	0.500	constant	0
Change	$-2x$	$-x$		$+2x$
Equil	$0.500 - 2x$	$0.500 - x$		0.0011

$2x = 0.0011, x = 5.5 \times 10^{-4}$

$[H_2S] = 0.500 - 0.0011 = 0.4989$ $\qquad [SO_2] = 0.500 - 5.5 \times 10^{-4} = 0.49945$

$$K_c = \frac{[H_2O]^2}{[H_2S]^2[SO_2]} = \frac{(0.0011)^2}{(0.4989)^2(0.49945)} = 9.7\underline{3}3 \times 10^{-6}$$

$2\,H_2S(g) + SO_2(g) \rightleftharpoons 3\,S(s) + 2\,H_2O(g)$

Reaction 2:	$[H_2S]$	$[SO_2]$	[S]	$[H_2O]$
Initial	0.250	0.325	constant	0
Change	$-2x$	$-x$		$+2x$
Equil	$0.250 - 2x$	$0.325 - x$		$2x$

Reaction shifts to the right.

$$K_c = \frac{[H_2O]^2}{[H_2S]^2[SO_2]} = \frac{(2x)^2}{(0.250 - 2x)^2(0.325 - x)} = 9.7\underline{3}3 \times 10^{-6}$$

Assume that $2x$ and x are small, respectively, compared to 0.250 and 0.325.

$4x^2 = 1.9\underline{7}70 \times 10^{-7}$

$x^2 = 4.9\underline{4}25 \times 10^{-8}$

$x = 2.2\underline{2}3 \times 10^{-4}$ Check assumption: $\frac{2.22 \times 10^{-4}}{0.250} = 8.88 \times 10^{-4} \times 100\% = 0.089\%$; assumption is valid.

$[H_2O] = 2x = 2(2.22 \times 10^{-4}) = 4.44 \times 10^{-4}\,M$

$[H_2S] = 0.250 - 2(2.22 \times 10^{-4}) = 0.250\,M$

$[SO_2] = 0.325 - 2.22 \times 10^{-4} = 0.325\,M$

Check: Plug the values into the equilibrium expression:

$K_c = \frac{(4.44 \times 10^{-4})^2}{(0.250)^2(0.325)} = 9.71 \times 10^{-6}$; this value is close to the original equilibrium constant.

16.83 **Given:** 200.0 L container; 1.27 kg N_2; 0.310 kg H_2; 725 K; $K_p = 5.3 \times 10^{-5}$ **Find:** mass in g of NH_3 and % yield
Conceptual Plan: $K_p \rightarrow K_c$ and then kg $\rightarrow$ g $\rightarrow$ mol $\rightarrow$ M and then prepare an ICE table. Represent

$K_p = K_c(RT)^{\Delta n}$ $\quad \frac{1000\text{ g}}{\text{kg}} \quad \frac{\text{g}}{\text{molar mass}} \quad \frac{\text{mol}}{\text{vol}}$

the change with x, sum the table, determine the equilibrium values, put the equilibrium values in the equilibrium expression, and solve for x. Determine $[NH_3]$. Then M $\rightarrow$ mol $\rightarrow$ g and then determine

M × vol $\quad$ mol × molar mass

theoretical yield $NH_3 \rightarrow$ % yield.

determine limiting reactant $\quad \%\text{ yield} = \frac{\text{actual yield}}{\text{theoretical yield}} \times 100\%$

Solution: $K_p = K_c(RT)^{\Delta n}$ $\quad K_c = \frac{K_p}{(RT)^{\Delta n}} = \frac{5.3 \times 10^{-5}}{\left(\left(0.08206 \frac{\text{L} \cdot \text{atm}}{\text{mol} \cdot \text{K}}\right)(725\text{ K})\right)^{-2}} = 0.1\underline{8}76$

$$n_{N_2} = 1.27\text{ kg } N_2 \times \frac{1000\text{ g}}{\text{kg}} \times \frac{1\text{ mol } N_2}{28.02\text{ g } N_2} = 45.\underline{3}25\text{ mol } N_2 \quad [N_2] = \frac{45.\underline{3}25\text{ mol}}{200.0\text{ L}} = 0.22\underline{6}63\text{ M}$$

$$n_{H_2} = 0.310\text{ kg } H_2 \times \frac{1000\text{ g}}{\text{kg}} \times \frac{1\text{ mol } H_2}{2.016\text{ g } H_2} = 15\underline{3}.77\text{ mol } H_2 \quad [H_2] = \frac{15\underline{3}.77\text{ mol}}{200.0\text{ L}} = 0.76\underline{8}85\text{ M}$$

$$N_2(g) + 3\,H_2(g) \rightleftharpoons 2\,NH_3(g)$$

Reaction 1:	$[N_2]$	$[H_2]$	$[NH_3]$
Initial	0.2266	0.7689	0.0
Change	$-x$	$-3x$	$+2x$
Equil	$0.2266 - x$	$0.7689 - 3x$	$2x$

Reaction shifts to the right.

$$K_c = \frac{[NH_3]^2}{[N_2][H_2]^3} = \frac{(2x)^2}{(0.2266 - x)(0.7689 - 3x)^3} = 0.1876$$

Assume that x is small compared to 0.2268 and $3x$ is small compared to 0.7689.

$$\frac{(2x)^2}{(0.2266)(0.7689)^3} = 0.1877 \qquad x = 0.06951$$

Check assumptions: $\frac{0.06951}{0.2268} \times 100\% = 30.6$, which is not valid, and $\frac{3(0.06951)}{0.7689} \times 100\% = 27.1\%$.

Use method of successive substitution to solve for x. This yields $x = 0.0460$.
$[NH_3] = 2x = 2(0.0460) = 0.0920\ M$

Check: Plug the values into the equilibrium expression:

$$K_c = \frac{(0.0920)^2}{(0.2266 - 0.0460)((0.7689 - 3(0.0460))^3} = 0.1866;$$

this value is close to the original equilibrium constant.

Determine grams NH_3: $\frac{0.0920\ \cancel{mol\ NH_3}}{\cancel{L}} \times 200.0\ \cancel{L} \times \frac{17.02\ g\ NH_3}{\cancel{mol\ NH_3}} = 31\underline{3}.2\ g = 3.1 \times 10^2\ g$

Determine the theoretical yield and the limiting reactant:

$$1.27\ \cancel{kg\ N_2} \times \frac{1000\ \cancel{g}}{\cancel{kg}} \times \frac{1\ \cancel{mol\ N_2}}{28.0\ \cancel{g\ N_2}} \times \frac{2\ \cancel{mol\ NH_3}}{1\ \cancel{mol\ N_2}} \times \frac{17.02\ g\ NH_3}{\cancel{mol\ NH_3}} = 15\underline{4}4\ g\ NH_3$$

$$0.310\ \cancel{kg\ H_2} \times \frac{1000\ \cancel{g}}{\cancel{kg}} \times \frac{1\ \cancel{mol\ H_2}}{2.016\ \cancel{g\ H_2}} \times \frac{2\ \cancel{mol\ NH_3}}{3\ \cancel{mol\ H_2}} \times \frac{17.02\ g\ NH_3}{\cancel{mol\ NH_3}} = 1745\ g\ NH_3$$

N_2 produces the least amount of NH_3; therefore, it is the limiting reactant, and the theoretical yield is 1.54×10^3 g NH_3.

$$\%\ \text{yield} = \frac{3.1 \times 10^2\ \cancel{g}}{1.54 \times 10^3\ \cancel{g}} \times 100\% = 20.\%$$

16.84 **Given:** $V = 85.0$ L, 22.3 kg CH_4, 55.4 kg CO_2, $T = 825$ K, $K_p = 4.5 \times 10^2$ **Find:** g H_2 at equilibrium

Conceptual Plan: kg → g → mol → P and then prepare an ICE table. Represent the change

$\frac{1000\ g}{kg}$ $\frac{g}{\text{molar mass}}$ $PV = nRT$

with x, sum the table, determine the equilibrium values, put the equilibrium values in the equilibrium expression, and solve for x. Determine P_{H_2}. Then P → mol → g and then determine

$PV = nRT$ mol × molar mass

theoretical yield H_2 → % yield.

determine limiting reactant % yield $= \frac{\text{actual yield}}{\text{theoretical yield}} \times 100\%$

Solution:

$$n_{CH_4} = 22.3\ \cancel{kg} \times \frac{1000\ \cancel{g}}{\cancel{kg}} \times \frac{1\ mol}{16.0\ \cancel{g}} = 139\underline{3}.8\ mol$$

$$P_{CH_4} = \frac{nRT}{V} = \frac{(139\underline{4}\ \cancel{mol})\left(0.08206 \frac{\cancel{L} \cdot atm}{\cancel{mol} \cdot \cancel{K}}\right)(825\ \cancel{K})}{85.0\ \cancel{L}} = 11\underline{1}0\ atm$$

$$n_{CO_2} = 55.4\ \text{kg} \times \frac{1000\ \text{g}}{\text{kg}} \times \frac{1\ \text{mol}}{44.0\ \text{g}} = 12\underline{5}9\ \text{mol}$$

$$P_{CO_2} = \frac{nRT}{V} = \frac{(12\underline{5}9\ \text{mol})\left(0.08206\frac{\text{L}\cdot\text{atm}}{\text{mol}\cdot\text{K}}\right)(825\ \text{K})}{85.0\ \text{L}} = 10\underline{0}3\ \text{atm}$$

$$CH_4(g) + CO_2(g) \rightleftharpoons 2\,CO(g) + 2\,H_2(g)$$

	P_{CH_4}	P_{CO_2}	P_{CO}	P_{H_2}
Initial	1110	1003	0.00	0.00
Change	$-x$	$-x$	$+2x$	$+2x$
Equil	$1110 - x$	$1003 - x$	$2x$	$2x$

$$K_p = \frac{P_{CO}^2 P_{H_2}^2}{P_{CH_4} P_{CO_2}} = \frac{(2x)^2(2x)^2}{(1110 - x)(1003 - x)} = 4.5 \times 10^2$$

Assume that x is small compared to 1003 and 1110.
$x^4 = 3.13 \times 10^7$; $x = 74.8$

Check assumptions: $\frac{74.8}{1003} \times 100\% = 7.46\%$ and $\frac{74.8}{1160} \times 100\% = 6.45\%$; assumptions are not valid.

Use method of successive substitutions, which yields
$x = 72.0$.

Check: Plug into equilibrium expression:

$$K_p = \frac{P_{CO}^2 P_{H_2}^2}{P_{CH_4} P_{CO_2}} = \frac{(2(72.0))^2(2(72.0))^2}{(1160 - 72.0)(1003 - 72.0)} = 4.49 \times 10^2 = 4.5 \times 10^2$$, which is the equilibrium constant.

Determine grams of H_2.

$$P_{H_2} = 2(72.0) = 144\ \text{atm}; \quad n_{H_2} = \frac{PV}{RT} = \frac{(144\ \text{atm})(85.0\ \text{L})}{\left(0.08206\frac{\text{L}\cdot\text{atm}}{\text{mol}\cdot\text{K}}\right)(825\ \text{K})} = 18\underline{0}.8\ \text{mol}$$

$$18\underline{0}.8\ \text{mol} \times \frac{2.016\ \text{g}\ H_2}{\text{mol}} = 364.5\ \text{g} = 365\ \text{g}\ H_2$$

Determine % yield.

$$1394\ \text{mol}\ CH_4 \times \frac{2\ \text{mol}\ H_2}{\text{mol}\ CH_4} = 2788\ \text{mol}\ H_2$$

$$1259\ \text{mol}\ CO_2 \times \frac{2\ \text{mol}\ H_2}{\text{mol}\ CO_2} = 2518\ \text{mol}\ H_2$$

CO_2 is the limiting reactant.

$$\frac{18\underline{0}.8\ \text{mol}\ H_2}{2518\ \text{mol}\ H_2} \times 100\% = 7.18\%\ \text{yield}$$

16.85 **Given:** at equilibrium: $P_{CO} = 0.30$ atm; $P_{Cl_2} = 0.10$ atm; $P_{COCl_2} = 0.60$ atm, add 0.40 atm Cl_2
Find: P_{CO} when system returns to equilibrium
Conceptual Plan: Use equilibrium partial pressures to determine K_p. For the new conditions, prepare an ICE table, represent the change with x, sum the table, determine the equilibrium values, put the equilibrium values in the equilibrium expression, and solve for x. Determine P_{CO}.
Solution: $CO(g) + Cl_2(g) \rightleftharpoons COCl_2(g)$

Condition 1:	P_{CO}	P_{Cl_2}	P_{COCl_2}
	0.30	0.10	0.60

$$K_p = \frac{P_{COCl_2}}{P_{CO} P_{Cl_2}} = \frac{(0.60)}{(0.30)(0.10)} = 20$$

Condition 2:	$CO(g)$ + P_{CO}	$Cl_2(g)$ $\rightleftharpoons$ P_{Cl_2}	$COCl_2(g)$ P_{COCl_2}
Initial	0.30	0.10 + 0.40	0.60
Change	$-x$	$-x$	$+x$
Equil	$0.30 - x$	$0.50 - x$	$0.60 + x$

Reaction shifts to the right because the concentration of Cl_2 was increased.

$$K_p = \frac{P_{COCl_2}}{P_{CO}P_{Cl_2}} = \frac{(0.60 + x)}{(0.30 - x)(0.50 - x)} = 20$$

$20x^2 - 17x + 2.4 = 0$

$$\frac{-b \pm \sqrt{b^2 - 4ac}}{2a} = \frac{-(-17) \pm \sqrt{(-17)^2 - 4(20)(2.4)}}{2(20)}$$

$x = 0.67$ or 0.18 so $x = 0.18$

$P_{CO} = 0.30 - 0.18 = 0.12$ atm; $P_{Cl_2} = 0.50 - 0.18 = 0.32$; $P_{COCl_2} = 0.60 + 0.18 = 0.78$ atm

Check: Plug the values into the equilibrium expression:

$$K_p = \frac{(0.78)}{(0.12)(0.32)} = 2\underline{0}.3 = 20.$$; this is the same as the original equilibrium constant.

16.86 **Given:** $P_{SO_2} = 3.00$ atm; $P_{O_2} = 1.00$ atm; at equilibrium, $P_{total} = 3.75$ atm, $T = 27\ °C$ **Find:** K_c

Conceptual Plan: Prepare an ICE table, represent the change with x, sum the table, determine the equilibrium values, use the total pressure, and solve for x. Determine partial pressure of each at equilibrium. Determine $K_p \rightarrow K_c$.

$K_p = K_c(RT)^{\Delta n}$

Solution:	$2\ SO_2(g)$ + P_{SO_2}	$O_2(g)$ $\rightleftharpoons$ P_{O_2}	$2\ SO_3(g)$ P_{SO_3}
Initial	3.00	1.00	0.00
Change	$-2x$	$-x$	$+2x$
Equil	$3.00 - 2x$	$1.00 - x$	$2x$

$P_{total} = P_{SO_2} + P_{O_2} + P_{SO_3}$ $\quad 3.75 = 3.00 - 2x + (1.00 - x) + 2x$

$x = 0.25$ $\quad P_{SO_2} = (3.00 - 2(0.25)) = 2.50$ atm; $P_{O_2} = (1.00 - 0.25) = 0.75$ atm; $P_{SO_3} = 2(0.25) = 0.50$ atm

$$K_p = \frac{P^2_{SO_3}}{P^2_{SO_2}P_{O_2}} = \frac{(0.50)^2}{(2.50)^2(0.75)} = 0.05\underline{3}3 = 0.053$$

Check: The value of the pressure of SO_3 is small compared to the pressures of SO_2 and O_2; therefore, you would expect K_p to be less than 1.

$$K_p = K_c(RT)^{\Delta n} \quad 0.05\underline{3}3 = K_c\left(\left(0.08206\frac{\text{L} \cdot \text{atm}}{\text{mol} \cdot \text{K}}\right)(27 + 273\ \text{K})\right)^{-1}$$

$K_c = 1.\underline{3}12 = 1.3$

16.87 **Given:** $K_p = 0.76$; P_{total} at equilibrium $= 1.0$ atm **Find:** $P_{initial}$ CCl_4

Conceptual Plan: Prepare an ICE table, represent the P_{CCl_4} with A and the change with x, sum the table, determine the equilibrium values, use the total pressure, and solve for A in terms of x. Determine partial pressure of each at equilibrium, use the equilibrium expression to determine x, and determine A.

Solution: $CCl_4(g) \rightleftharpoons C(s) + 2\ Cl_2(g)$

	P_{CCl_4}	P_C	P_{Cl_2}
Initial	A	constant	0.00
Change	$-x$		$+2x$
Equil	$A - x$		$2x$

$P_{total} = P_{CCl_4} + P_{Cl_2}$ $1.0 = A - x + 2x$ $A = 1.0 - x$
$P_{CCl_4} = (A - x) = (1.0 - x) - x = 1.0 - 2x; P_{Cl_2} = 2x$

$$K_p = \frac{P^2_{Cl_2}}{P_{CCl_4}} = \frac{(2x)^2}{(1.0 - 2x)} = 0.76$$

$4x^2 + 1.52x - 0.76 = 0$ Solve quadratic equation. $x = 0.285$ or -0.665 so $x = 0.2\underline{8}5$
$A = 1.0 - x = 1.0 - 0.285 = 0.715 = 0.72$ atm

Check: Plug the values into the equilibrium expression:

$$K_p = \frac{P^2_{Cl_2}}{P_{CCl_4}} = \frac{(2x)^2}{(A - x)} = \frac{(2(0.285))^2}{(0.715 - 0.285)} = 0.756 = 0.76$$; this is the same as the original equilibrium constant.

16.88 **Given:** $K = 3.0$; $SO_2 = 2.4$ mol initial; $SO_3 = 1.2$ mol equilibrium **Find:** mol NO_2 initial
Conceptual Plan: Assume 1.0 L, prepare an ICE table, represent the change with x, sum the table, determine the equilibrium values, and determine the initial values.
Solution: $SO_2(g) + NO_2(g) \rightleftharpoons SO_3(g) + NO(g)$

	$[SO_2]$	$[NO_2]$	$[SO_3]$	$[NO]$
Initial	2.4	y	0.0	0.0
Change	$-x$	$-x$	$+x$	$+x$
Equil	1.2	$y - 1.2$	1.2	1.2

$x = 1.2$ Fill in the table.

$$K = \frac{[SO_3][NO]}{[SO_2][NO_2]} \quad 3.0 = \frac{(1.2)(1.2)}{(1.2)(y - 1.2)} \quad y = 1.6$$

mol NO_2 initial $= 1.6$ mol

Check: The initial amount 1.6 mol is greater than the amount lost.

16.89 **Given:** $V = 0.654$ L, $T = 1000$ K, $K_p = 3.9 \times 10^{-2}$ **Find:** mass CaO as equilibrium
Conceptual Plan: $K_p \rightarrow P_{CO_2} \rightarrow n_{(CO_2)} \rightarrow n_{(CaO)} \rightarrow$ **g**

$PV = nRT$ stoichiometry $g = n(\text{molar mass})$

Solution: Because $CaCO_3$ and CaO are solids, they are not included in the equilibrium expression.

$$K_p = P_{CO_2} = 3.9 \times 10^{-2}; \quad n = \frac{PV}{RT} = \frac{(3.9 \times 10^{-2}\ \cancel{\text{atm}})(0.654\ \cancel{\text{L}})}{\left(0.08206 \frac{\cancel{\text{L}} \cdot \cancel{\text{atm}}}{\text{mol} \cdot \cancel{\text{K}}}\right)(1000\ \cancel{\text{K}})} = 3.\underline{1}08 \times 10^{-4}\ \text{mol } CO_2$$

$$3.\underline{1}08 \times 10^{-4}\ \cancel{\text{mol } CO_2} \times \frac{1\ \cancel{\text{mol CaO}}}{1\ \cancel{\text{mol } CO_2}} \times \frac{56.1\ \text{g CaO}}{1\ \cancel{\text{mol CaO}}} = 0.0174\ \text{g} = 0.0174\ \text{g CaO}$$

Check: The small value of K would give a small amount of products, so we would not expect a large mass of CaO to form.

16.90 **Given:** at equilibrium: N_2O_4, $P = 0.28$ atm, NO_2, $P = 1.1$ atm; $T = 350$ K
Find: equilibrium pressures when volume doubles
Conceptual Plan: $P_{N_2O_4}, P_{NO_2} \rightarrow K_p$ **and then P when volume doubles. Then prepare an ICE table, represent the change with x, sum the table, and determine the equilibrium values.**

Solution: $$K_p = \frac{P^2_{NO_2}}{P_{N_2O_4}} = \frac{(1.1)^2}{0.28} = 4.\underline{3}21$$

When the volume is doubled, the partial pressure of each gas decreases by half.

$N_2O_2(g) \rightleftharpoons 2\ NO_2(g)$

	$P_{N_2O_4}$	P_{NO_2}
Initial	0.28/2	1.1/2
Change	$-x$	$+2x$
Equil	$0.14 - x$	$0.55 + 2x$

The reaction shifts to the side with more moles—to the right.

$K_p = \frac{P^2_{NO_2}}{P_{N_2O_4}} = \frac{(0.55 + 2x)^2}{(0.14 - x)} = 4.3\underline{2}1$

$4x^2 + 6.521x - 0.3024 = 0$

$\frac{-b \pm \sqrt{b^2 - 4ac}}{2a} = \frac{-(6.521) \pm \sqrt{(6.521)^2 - 4(4)(0.3024)}}{2(4)}$

$x = 0.0451$ or -1.675 so $x = 0.0451$

$P_{N_2O_4} = 0.14 - 0.0451 = 0.0949$ atm $= 0.095$ atm; $P_{NO_2} = 0.55 + 2(0.0451) = 0.6402 = 0.64$ atm

Check: Plug the values into the equilibrium expression:

$K_p = \frac{(0.6402)^2}{(0.0949)} = 4.32 = 4.3$; this is the same as the original equilibrium constant.

16.91 **Given:** $K_p = 3.10$, initial $P_{CO} = 215$ torr, $P_{Cl_2} = 245$ torr **Find:** mole fraction $COCl_2$

Conceptual Plan: P in torr → P in atm. Prepare an ICE table, represent the change with x, sum the table,

$\frac{1 \text{ atm}}{760 \text{ torr}}$

determine the equilibrium values, use the total pressure, and solve for mole fraction.

$\frac{P_{COCl_2}}{P_{total}}$

Solution: $P_{CO} = (215 \text{ torr})\left(\frac{1 \text{ atm}}{760 \text{ torr}}\right) = 0.28\underline{2}9$ atm $\quad P_{Cl_2} = (245 \text{ torr})\left(\frac{1 \text{ atm}}{760 \text{ torr}}\right) = 0.32\underline{2}4$ atm

	$CO(g)$ +	$Cl_2(g)$ ⇌	$COCl_2(g)$
Initial	$0.28\underline{2}9$	$0.32\underline{2}4$	0
Change	$-x$	$-x$	$+x$
Equil	$0.28\underline{2}9 - x$	$0.32\underline{2}4 - x$	x

$Q < K$, so the reaction shifts to the right.

$K_p = \frac{P_{COCl_2}}{P_{CO}P_{Cl_2}} = \frac{x}{(0.28\underline{2}9 - x)(0.32\underline{2}4 - x)} = 3.10$ so $3.10x^2 - 2.87634x + 0.2827 = 0$ Solve using the quadratic equation, found in Appendix I. $x = 0.81\underline{6}1$ or $0.11\underline{1}7$ so $x = 0.1117$ P is proportional to n under like conditions, so the mole fraction can be determined using the pressure values.

$P_{CO} = 0.28\underline{2}9 - 0.11\underline{1}7 = 0.17\underline{1}2 \quad P_{Cl_2} = 0.32\underline{2}4 - 0.11\underline{1}7 = 0.21\underline{0}7 \quad P_{COCl_2} = 0.11\underline{1}7$

mole fraction $COCl_2 = \frac{P_{COCl_2}}{P_{CO} + P_{Cl_2} + P_{COCl_2}} = \frac{0.11\underline{1}7}{0.17\underline{1}2 + 0.21\underline{0}7 + 0.11\underline{1}7} = 0.22\underline{6}3 = 0.226$

Check: Plug the equilibrium pressures into the equilibrium expression:

$K_p = \frac{0.11\underline{1}7}{(0.17\underline{1}2)(0.21\underline{0}7)} = 3.0966 = 3.10$, which is the equilibrium constant; so the answer is reasonable.

16.92 **Given:** $K_p = 1.60 \times 10^{-3}$, $T = 700$ K, 1.55 L, $P_{H_2O} = 145$ torr **Find:** % mass H_2 at equilibrium

Conceptual Plan: P(torr) → P(atm). Prepare an ICE table, represent the change with x, sum the table,

$\frac{1 \text{ atm}}{760 \text{ torr}}$

and determine the equilibrium values. P → n → mass for each → mass %

$PV = nRT$ $\quad \frac{28.01 \text{ g CO}}{1 \text{ mol CO}} \quad \frac{2.016 \text{ g } H_2}{1 \text{ mol } H_2} \quad \frac{18.02 \text{ gl } H_2O}{1 \text{ mol } H_2O} \quad \frac{\text{mass } H_2}{\text{total mass}} \times 100$

Solution: $P_{H_2O} = (145 \text{ torr})\left(\frac{1 \text{ atm}}{760 \text{ torr}}\right) = 0.1908$ atm

	$H_2O(g)$ +	$C(s)$ ⇌	$CO(g)$ +	$H_2(g)$
Initial	$0.19\underline{0}8$	constant	0	0
Change	$-x$		$+x$	$+x$
Equil	$0.19\underline{0}8 - x$		x	x

$Q < K$ reaction shifts to the right.

$$K_p = \frac{P_{CO}P_{H_2}}{P_{H_2O}} = \frac{(x)(x)}{(0.19\underline{0}8 - x)} = 1.60x10^{-3} \quad x = 0.016\underline{6}8 \text{ or } -0.01828 \text{ so } x = 0.016\underline{6}8$$

$P_{CO} = P_{H_2} = 0.016\underline{6}8$ atm $\quad P_{H_2O} = 0.19\underline{0}8 - 0.016\underline{6}8 = 0.17\underline{4}1$ atm

$$n_{CO} = n_{H_2} = \frac{(0.016\underline{6}8 \text{ atm})(1.55 \text{ L})}{\left(\frac{0.08206 \text{ L} \cdot \text{atm}}{\text{mol} \cdot \text{K}}\right)(700.0 \text{ K})} = 4.5\underline{0}1 \times 10^{-4} \text{ mol}$$

$$n_{H_2O} = \frac{(0.17\underline{4}1 \text{ atm})(1.55 \text{ L})}{\left(\frac{0.08206 \text{ L} \cdot \text{atm}}{\text{mol} \cdot \text{K}}\right)(700 \text{ K})} = 4.\underline{6}98 \times 10^{-3} \text{ mol}$$

$$(4.5\underline{0}1 \times 10^{-4} \text{ mol CO})\frac{28.01 \text{ g CO}}{1 \text{ mol CO}} = 0.012\underline{6}1 \text{ g CO}; \qquad (4.5\underline{0}1 \times 10^{-4} \text{ mol } H_2)\frac{2.016 \text{ g } H_2}{1 \text{ mol } H_2} = 9.0\underline{7}4 \times 10^{-4} \text{ g } H_2$$

$$(4.6\underline{9}8 \times 10^{-3} \text{ mol } H_2O)\frac{18.02 \text{ g } H_2O}{1 \text{ mol } H_2O} = 0.084\underline{6}6 \text{ g } H_2O$$

$$\frac{9.0\underline{7}4 \times 10^{-4} \text{ g } H_2}{(0.012\underline{6}1 \text{ g} + 9.0\underline{7}4 \times 10^{-4} \text{ g} + 0.084\underline{6}6 \text{ g})} \times 100\% = \frac{9.0\underline{7}4 \times 10^{-4} \text{ g } H_2}{(0.098\underline{1}774 \text{ g})} \times 100\% = 0.92\underline{4}2\% = 0.924\% \; H_2$$

Check: Plug the equilibrium values into the equilibrium expression:

$$K_p = \frac{(0.016\underline{6}8)(0.016\underline{6}8)}{(0.17\underline{4}1)} = 1.60 \times 10^{-3}$$, which is the equilibrium constant. Thus, the answer is reasonable.

Challenge Problems

16.93 (a) **Given:** $P_{NO} = 522$ torr, $P_{O_2} = 421$ torr; at equilibrium, $P_{total} = 748$ torr **Find:** K_p

Conceptual Plan: Prepare an ICE table, represent the change with x, sum the table, determine the equilibrium values, use the total pressure, and solve for x. torr $\rightarrow$ atm $\rightarrow K_p$

Solution: $2\,NO(g) + O_2(g) \rightleftharpoons 2\,NO_2(g)$

	P_{NO}	P_{O_2}	P_{NO_2}
Initial	522 torr	421 torr	0.00
Change	$-2x$	$-x$	$+2x$
Equil	$522 - 2x$	$421 - x$	$2x$

$P_{total} = P_{NO} + P_{O_2} + P_{NO_2} \quad 748 = 522 - 2x + (421 - x) + 2x$

$x = 195$ torr $\quad P_{NO} = (522 - 2(195)) = 132$ torr

$P_{O_2} = (421 - 195) = 226$ torr $\quad P_{NO_2} = 2(195) = 390$ torr

$$P_{NO} = 132 \text{ torr} \times \frac{1 \text{ atm}}{760 \text{ torr}} = 0.17\underline{3}7 \text{ atm}; \qquad P_{O_2} = 226 \text{ torr} \times \frac{1 \text{ atm}}{760 \text{ torr}} = 0.29\underline{7}4 \text{ atm}$$

$$P_{NO_2} = 390 \text{ torr} \times \frac{1 \text{ atm}}{760 \text{ torr}} = 0.51\underline{3}2 \text{ atm}$$

$$K_p = \frac{P^2_{NO_2}}{P^2_{NO}P_{O_2}} = \frac{(0.51\underline{3}2)^2}{(0.17\underline{3}7)^2(0.29\underline{7}4)} = 29.35 = 29.3$$

(b) **Given:** $P_{NO} = 255$ torr, $P_{O_2} = 185$ torr, $K_p = 29.3$ **Find:** equilibrium P_{NO_2}

Conceptual Plan:

torr $\rightarrow$ atm and then prepare an ICE table. Represent the change with x, sum the table,

$\frac{\text{atm}}{760 \text{ torr}}$

determine the equilibrium values, put the equilibrium values in the equilibrium expression, and solve for x. Determine P_{NO_2}.

Solution: $P_{NO} = 255 \text{ torr} \times \frac{1 \text{ atm}}{760 \text{ torr}} = 0.3355 \text{ atm}$ $\quad P_{O_2} = 185 \text{ torr} \times \frac{1 \text{ atm}}{760 \text{ torr}} = 0.2434 \text{ atm}$

	$2\,NO(g)$ +	$O_2(g) \rightleftharpoons$	$2\,NO_2(g)$
	P_{NO}	P_{O_2}	P_{NO_2}
Initial	0.3355	0.2434	0.00
Change	$-2x$	$-x$	$+2x$
Equil	$0.3355 - 2x$	$0.2434 - x$	$2x$

$$K_p = \frac{P^2_{NO_2}}{P^2_{NO}P_{O_2}} = \frac{(2x)^2}{(0.3355 - 2x)^2(0.2434 - x)} = 29.3$$

$-117.2x^3 + 63.847x^2 - 12.867x + 0.80282 = 0$. Solve using successive approximations or a cubic equation calculator found on the Internet.

$x = 0.1113 \quad P_{NO_2} = 2x = 2(0.1113) = 0.2226 \text{ atm}$

$P_{NO} = (0.3355 - 2(0.1113)) = 0.1129 \quad P_{O_2} = (0.2434 - 0.1113) = 0.1321$

$$0.2226 \text{ atm} \times \frac{760 \text{ torr}}{1 \text{ atm}} = 169.2 \text{ torr} = 169 \text{ torr}$$

Check: Plug the values into the equilibrium expression:

$K_p = \frac{(0.2226)^2}{(0.1129)^2(0.1321)} = 29.428 = 29.4$; this is within 0.1 of the original equilibrium constant.

16.94 **Given:** 2.75 L, 950 K, 0.100 mol SO_2, 0.100 mol O_2; $K_p = 0.355$ **Find:** P_{total} at equilibrium

Conceptual Plan: $n \rightarrow P$ ($PV = nRT$) and then prepare an ICE table. Represent the change with x, sum the table, determine the equilibrium values, put the equilibrium values in the equilibrium expression, solve for x, and then determine P for each reactant and product $\rightarrow P_{total}$.

$$P_{total} = P_{SO_2} + P_{O_2} + P_{SO_3}$$

Solution: $P = \frac{nRT}{V} \quad P_{O_2} = P_{SO_2} = \frac{(0.100 \text{ mol})\left(0.08206 \frac{\text{L} \cdot \text{atm}}{\text{mol} \cdot \text{K}}\right)(950 \text{ K})}{2.75 \text{ L}} = 2.835 \text{ atm}$

	$2\,SO_2(g)$ +	$O_2(g) \rightleftharpoons$	$2\,SO_3(g)$
	P_{SO_2}	P_{O_2}	P_{SO_3}
Initial	2.835	2.835	0.00
Change	$-2x$	$-x$	$+2x$
Equil	$2.835 - 2x$	$2.835 - x$	$2x$

$$K_p = \frac{P^2_{SO_3}}{P^2_{SO_2}P_{O_2}} = \frac{(2x)^2}{(2.835 - 2x)^2(2.835 - x)} = 0.355$$

$-1.42x^3 + 4.051x^2 - 14.266x + 8.089 = 0$. Solve using successive approximations or a cubic equation calculator found on the Internet.

$x = 0.6628 = 0.663$ and the other two roots are imaginary numbers.

$P_{SO_3} = 2x = 2(0.663) = 1.326 = 1.33 \text{ atm}$

$P_{SO_2} = (2.835 - 2x) = (2.835 - 2(0.663)) = 1.510 = 1.51 \text{ atm}$

$P_{O_2} = (2.835 - x) = (2.835 - 0.663) = 2.173 = 2.17 \text{ atm}$

$P_{total} = 1.33 + 1.51 + 2.17 = 5.01 \text{ atm}$

Check: Plug the values into the equilibrium expression:

$$K_p = \frac{P^2_{SO_3}}{P^2_{SO_2}P_{O_2}} = \frac{(1.33)^2}{(1.51)^2(2.17)} = 0.3575$$

This is within 0.7% of the original equilibrium constant.

16.95 **Given:** P_{NOCl} at equilibrium = 115 torr; $K_p = 0.27$, $T = 700$ K **Find:** initial pressure NO, Cl_2

Conceptual Plan: torr → atm and then prepare an ICE table. Represent the change with x, sum the table,

$\frac{\text{atm}}{760 \text{ torr}}$

determine the equilibrium values, put the equilibrium values in the equilibrium expression, and determine initial pressure.

Solution: $115 \text{ torr} \times \frac{1 \text{ atm}}{760 \text{ torr}} = 0.15\underline{1}3 \text{ atm}$

$2\,NO(g) + Cl_2(g) \rightleftharpoons 2\,NOCl(g)$

	P_{NO}	P_{Cl_2}	P_{NOCl}
Initial	A	A	0.00
Change	$-2x$	$-x$	$+2x$
Equil	$A - 2x$	$A - x$	0.151
	$A - 0.151$	$A - 0.0756$	

Let A = initial pressure of NO and Cl_2.

$2x = 0.151;$ $x = 0.0756$

$$K_p = \frac{P^2_{NOCl}}{P^2_{NO}P_{Cl_2}} = \frac{(0.151)^2}{(A - 0.151)^2(A - 0.0756)} = 0.27$$

$0.27A^3 - 0.1019A^2 + 0.01231A - 0.023266 = 0$. Solve using successive approximations or a cubic equation calculator found on the Internet.

$A = 0.566$

$P_{NO} = P_{Cl_2} = A = 0.566 \text{ atm} = 430 \text{ torr}$

Check: Plug the values into the equilibrium expression:

$$K_p = \frac{P^2_{NOCl}}{P^2_{NO}P_{Cl_2}} = \frac{(0.151)^2}{(0.566 - 0.151)^2(0.566 - 0.0756)} = 0.2699 = 0.27;$$

this is the same as the original equilibrium constant.

16.96 **Given:** $P_{N_2O_4} = 1$ atm, K_p reaction 1 $= 1 \times 10^4$, K_p reaction 2 $= 0.10$ **Find:** Which component will have $P > 0.2$ atm?

Conceptual Plan: Use reaction 2 to determine P_{NO_2}. Then prepare an ICE table, represent the change with x, sum the table, determine the equilibrium values, put the equilibrium values in the equilibrium expression, and solve for x.

Solution: $2\,NO(g) \rightleftharpoons N_2O_4(g)$

	P_{NO}	$P_{N_2O_4}$
Initial	0.00	1.00
Change	$+2x$	$-x$
Equil	$+2x$	$1.00 - x$

$$K_p = \frac{P_{N_2O_4}}{P^2_{NO}} = \frac{(1.00 - x)}{(2x)^2} = 0.10$$

$0.40x^2 + x - 1.00 = 0$

$$\frac{-b \pm \sqrt{b^2 - 4ac}}{2a} = \frac{-1 \pm \sqrt{1^2 - 4(0.40)(-1.00)}}{2(0.40)}$$

$x = 0.7655 = 0.77$

$P_{NO_2} = 2x = 1.54$

Because K_p for reaction 1 is so large, essentially all of the materials are products; so the P_{NO} and P_{O_2} in reaction 1 will be negligible.

16.97 **Given:** $P = 0.750$ atm, density $= 0.520$ g/L, $T = 337\,°C$ **Find:** K_c

Conceptual Plan: Prepare an ICE table, represent the P_{CCl_4} with A and the change with x, sum the table, determine the equilibrium values, use the total pressure, and solve for A in terms of x. Determine partial pressure of each at equilibrium in terms of x, use the density to determine the apparent molar mass, and use

$d = \frac{P\mathcal{M}}{RT}$

the mole fraction (in terms of P) and the molar mass of each gas to determine x.

$$\chi_A = \frac{P_A}{P_{\text{total}}}$$

Solution: $2\ NO_2(g) \rightleftharpoons 2\ NO(g) + O_2(g)$

	P_{NO_2}	P_{NO}	P_{O_2}
Initial	A	0.00	0.00
Change	$-2x$	$+2x$	$+x$
Equil	$A - 2x$	$2x$	x

$P_{\text{total}} = P_{NO_2} + P_{NO} + P_{O_2}$ $\quad 0.750 = A - 2x + 2x + x \quad A = 0.750 - x$

$P_{NO_2} = (A - 2x) = (0.750 - x) - 2x = (0.750 - 3x);\ P_{NO} = (2x);\ P_{O_2} = x$

$$d = \frac{P\mathcal{M}}{RT} \qquad \mathcal{M} = \frac{dRT}{P} = \frac{\left(0.520\frac{\text{g}}{\cancel{\text{L}}}\right)\left(0.08206\frac{\cancel{\text{L}} \cdot \cancel{\text{atm}}}{\text{mol} \cdot \cancel{\text{K}}}\right)(610\ \cancel{\text{K}})}{0.750\ \cancel{\text{atm}}} = 34.71\ \text{g/mol}$$

$$\mathcal{M} = \chi_{NO_2}\mathcal{M}_{NO_2} + \chi_{NO}\mathcal{M}_{NO} + \chi_{O_2}\mathcal{M}_{O_2} = \frac{P_{NO_2}}{P_{\text{total}}}\mathcal{M}_{NO_2} + \frac{P_{NO}}{P_{\text{total}}}\mathcal{M}_{NO} + \frac{P_{O_2}}{P_{\text{total}}}\mathcal{M}_{O_2}$$

$$P_{\text{total}}\mathcal{M} = P_{NO_2}\mathcal{M}_{NO_2} + P_{NO}\mathcal{M}_{NO} + P_{O_2}\mathcal{M}_{O_2}$$

$(0.750)(34.7) = (0.750 - 3x)(46.0) + 2x(30.0) + x(32.0)$

$x = 0.184$

$$K_p = \frac{P_{NO}^2 P_{O_2}}{P_{NO_2}^2} = \frac{(2x)^2(x)}{(0.750 - 3x)} = \frac{(2(0.184))^2(0.184)}{(0.750 - 3(0.184))^2} = 0.63\underline{5}6$$

$$K_p = K_c(RT)^{\Delta n} \quad 0.63\underline{5}6 = K_c\left(\left(0.08206\frac{\text{L} \cdot \text{atm}}{\text{mol} \cdot \cancel{\text{K}}}\right)(610\ \cancel{\text{K}})\right)^1$$

$K_c = 1.27 \times 10^{-2}$

Check: The units (none) are correct. The magnitude (0.01) is reasonable since there are higher concentrations of the products than the reactants.

16.98 **Given:** reaction 1: $K_c = 7.75$; reaction 2: $K_c = 4.00$, $[N_2O_5]_{\text{initial}} = 4.00\ \text{M}$, $[O_2]_{\text{equil}} = 4.50\ \text{M}$

Find: concentration of other species at equilibrium

Conceptual Plan: Combine reaction 1 and reaction 2 to get the overall reaction; then prepare an ICE table. Represent the change with x, sum the table, determine the equilibrium values, use the total pressure, and solve for x. Use x to determine equilibrium concentrations.

Solution:

$N_2O_5(g) \rightleftharpoons \cancel{N_2O_3(g)} + O_2(g) \qquad K_1 = 7.75$

$\cancel{N_2O_3(g)} \rightleftharpoons N_2O(g) + O_2(g) \qquad K_2 = 4.00$

$N_2O_5(g) \rightleftharpoons N_2O(g) + 2\ O_2(g) \qquad K = K_1K_2 = 31.0$

Initial	4.00	0.00	0.00
Change	$-x$	$+x$	$+2x$
Equil	$4.00 - x$	x	4.50

$$K_c = \frac{[N_2O][O_2]^2}{[N_2O_5]} = 31.0 = \frac{(x)(4.50)^2}{(4.00 - x)} \qquad 31.0\ (4.00 - x) = 20.\underline{2}5x \qquad 6.1\underline{2}346 = 2.5\underline{3}068x$$

$x = 2.42$

$[N_2O_5] = 4.00 - x = 4.00 - 2.42 = 1.58\ \text{M}$

$[N_2O] = x = 2.42\ \text{M}$

$[O_2] = 4.50\ \text{M}$

Use reaction 1 or reaction 2 and solve for $[N_2O_3]$ represented as y.

$$K_c = \frac{[N_2O_3][O_2]}{[N_2O_5]} = 7.75 = \frac{(y)(4.50)}{(1.58)}$$

$y = 2.72 \quad [N_2O_3] = y = 2.72\ \text{M}$

Check: Plug the values into any of the equilibrium expressions. For example, for the overall reaction:

$$K_c = \frac{[N_2O][O_2]^2}{[N_2O_5]} = \frac{(2.42)(4.50)^2}{(1.58)} = 31.0, \text{ the equilibrium constant.}$$

For example, for reaction 2:

$$K_c = \frac{[N_2O][O_2]}{[N_2O_3]} = \frac{(2.42)(4.50)}{(2.72)} = 4.00$$

16.99 **Given:** $P_{total} = 3.0$ atm, mole fraction $O_2 = 0.12$, $T = 600$ K **Find:** K_p

Conceptual Plan: mole fraction $\rightarrow P_{O_2} \rightarrow P_{SO_2} \rightarrow P_{SO_3} \rightarrow K_p$

$$\text{mole fraction} = \frac{P_{O_2}}{P_{total}} \qquad \frac{2P_{SO_2}}{P_{O_2}} \qquad P_{SO_3} = P_{total} - P_{SO_2} - P_{O_2} \qquad K_p = \frac{P^2_{SO_2}P_{O_2}}{P^2_{SO_3}}$$

Solution: $P_{O_2} = (0.12)(3.0) = 0.36$ atm $\quad P_{SO_2} = 2P_{O_2} = 2(0.36 \text{ atm}) = 0.72$ atm

$$P_{SO_3} = 3.0 - 0.36 - 0.72 = 1.\underline{9}2 \quad K_p = \frac{P^2_{SO_2}P_{O_2}}{P^2_{SO_3}} = \frac{(0.72)^2(0.36)}{(1.\underline{9}2)^2} = 0.05\underline{0}6 = 5.1 \times 10^{-2}$$

Check: The units (none) are correct. The magnitude (0.05) is reasonable since there are higher concentrations of the products than the reactants.

Conceptual Problems

16.100 The equilibrium constant is very small; therefore, at equilibrium, the concentration of products will be small compared to the concentration of reactants. In reaction mixture d, the initial concentration of reactants is large and there are no products. Because only a small amount of products will be formed, the x is small approximation will most likely apply. In reaction mixture b, there is initially no reactant and only products. To reach equilibrium, most of the product would have to go to reactants and the x is small approximation would not apply. In reaction mixture a, there is a smaller amount of reactants than in reaction mixture d. Because there are fewer initial reactants, the percentage needed to reach equilibrium is higher and the x is small approximation is less likely to apply. In reaction mixture c, the initial concentrations of A and B are equal. The reaction will have to lose a large amount of product to reach equilibrium and the x is small approximation will not apply.

16.101 Yes, the direction will depend on the volume. If the initial moles of A and B are equal, the initial concentrations of A and B are equal regardless of the volume. Because $K_c = \frac{[B]^2}{[A]} = 1$, if the volume is such that the $[A] = [B] < 1.0$, then $Q < K$ and the reaction goes to the right to reach equilibrium. However, if the volume is such that the $[A] = [B] > 1.0$, then $Q > K$ and the reaction goes to the left to reach equilibrium.

16.102 $K_p = 0.50$ means that $P_{products} < P_{reactants}$. If the reactants and products are in their standard states, then P of each reactant and product $= 1.0$; so $Q > K$. To reach equilibrium, the reaction will have to shift to the left.

16.103 An examination of the data shows that when $P_A = 1.0$, $P_B = 1.0$; therefore, $K_p = \frac{P^b_B}{P^a_A} = \frac{(1.0)^b}{(1.0)^a} = 1.0$. Therefore, the value of the numerator and denominator must be equal. We see from the data that $P_B = \sqrt{P_A}$, so $P^2_B = P_A$. Because the stoichiometric coefficients become exponents in the equilibrium expression, $a = 1$ and $b = 2$.

16.104 When the concentration of A is increased, the rate of the reaction in the forward direction is increased. This will increase the amount of product, B. The rate of reaction in the reverse direction will increase slightly as the concentration of product increases, but not to the same extent as the forward reaction. According to Le Châtelier's principle, increasing the amount of reactant will increase both the reactant and product at equilibrium. This is consistent with the rate of reaction increasing to a greater extent in the forward direction than in the reverse direction.

17 Acids and Bases

Review Questions

17.1 There are two problems with this treatment. First, the neutralization reaction between a strong acid and a strong base is highly exothermic, so Batman's treatment is likely to have produced a significant amount of heat, leading to potential burns for Black Mask. Second, strong bases are themselves caustic to the skin, so Batman's base has the potential to cause additional burns as well. The standard treatment for spilling concentrated acid on the skin is to rinse with large amounts of water for an extended period of time. The water dilutes the acid and washes it away.

17.2 Acids have the following general properties: a sour taste, the ability to dissolve many metals, the ability to turn blue litmus paper red, and the ability to neutralize bases.

Bases have the following general properties: a bitter taste, a slippery feel, the ability to turn red litmus paper blue, and the ability to neutralize acids.

17.3 A carboxylic acid is an organic acid that contains the following group of atoms: $H{-}O{-}\overset{\overset{\displaystyle O}{\|}}{C}{-}$. Carboxylic acids are often found in substances derived from living organisms. Examples include citric acid, malic acid, and acetic acid.

17.4 An Arrhenius acid is a substance that produces H^+ in aqueous solution.

An Arrhenius base is a substance that produces OH^- in aqueous solution.

17.5 The hydronium ion is H_3O^+. In water, H^+ ions always associate with H_2O molecules to form hydronium ions and other associated species with the general formula $H(H_2O)_n{}^+$.

17.6 A Brønsted–Lowry acid is a proton (H^+) donor.

A Brønsted–Lowry base is a proton (H^+) acceptor.

17.7 According to Huheey, "The differences between the various acid–base concepts are not concerned with which is right, but which is most convenient to use in a particular situation." There is no single correct definition; we use the definition that is best for a particular situation.

17.8 Amphoteric substances can act as acids or bases depending on the circumstances. Some amphoteric substances are H_2O and NH_3.

17.9 A conjugate acid–base pair consists of two substances related to each other by the transfer of a proton. In the reaction:

$NH_3(aq) + H_2O(l) \rightleftharpoons NH_4{}^+(aq) + OH^-(aq)$, the conjugate pairs are $NH_3/NH_4{}^+$ and H_2O/OH^-.

17.10 A strong acid completely ionizes in solution. An example is HCl.

A weak acid only partially ionizes in solution. An example is HF.

17.11 For an H—Y binary acid, the factors affecting the ease with which this hydrogen will be donated are the polarity of the bond and the strength of the bond.

17.12 The factors affecting the ease with which the hydrogen will be donated (and therefore acidic) are the electronegativity of the element Y and the number of oxygen atoms attached to the element Y.

17.13 A diprotic acid contains two ionizable protons. An example is H_2SO_4.

A triprotic acid contains three ionizable protons. An example is H_3PO_4.

17.14 An acid ionization constant (K_a) is the equilibrium constant for the ionization reaction of a weak acid. The smaller the acid ionization constant, the weaker the acid.

17.15 The autoionization of water is $H_2O(l) + H_2O(l) \rightleftharpoons H_3O^+(aq) + OH^-(aq)$ and has the ion product of water $K_w = [H_3O^+][OH^-]$.

It can also be written $H_2O(l) \rightleftharpoons H^+(aq) + OH^-(aq)$, and the ion product of water is $K_w = [H^+][OH^-]$. At 25 °C, $K_w = 1.0 \times 10^{-14}$.

17.16 Because $K_w = 1.0 \times 10^{-14}$, the concentration of H_3O^+ times the concentration of OH^- is always 1.0×10^{-14}; therefore, if $[H_3O^+]$ increases, $[OH^-]$ decreases and if the $[H_3O^+]$ decreases, $[OH^-]$ increases.

17.17 We define pH as $pH = -\log[H_3O^+]$. An acidic solution has a $pH < 7$, a basic solution has a $pH > 7$, and a neutral solution has a $pH = 7$.

17.18 We define pH as $pOH = -\log[OH^-]$. An acidic solution has $pOH > 7$, a basic solution has $pOH < 7$, and a neutral solution has $pOH = 7$.

17.19 We can neglect the contribution of the autoionization of water to the H_3O^+ concentration in a solution of a strong or weak acid because it is negligible compared to the concentration of H_3O^+ ions generated from the dissociation of the acid. Even the weakest of the weak acids has an ionization constant that is four orders of magnitude larger than that of water.

17.20 The x is small approximation is used when the equilibrium constant is relatively small and the initial concentration is relatively large. We make the assumption that x is small compared to the initial concentration. If the assumption is valid, x is $<5\%$ of the initial concentration.

17.21 The percent ionization is defined as $\text{percent ionization} = \dfrac{\text{concentration of ionized acid}}{\text{initial concentration of acid}} \times 100$. The percent ionization of a weak acid decreases with the increasing concentration of the acid.

17.22 The complete ionization of the strong acid produces a significant concentration of H_3O^+ and suppresses the formation of additional H_3O^+ by the ionization of the weak acid. Because of Le Châtelier's principle, the formation of H_3O^+ by the strong acid causes the weak acid to ionize even less than it would in the absence of the strong acid.

17.23 $B(aq) + H_2O(l) \rightleftharpoons BH^+(aq) + OH^-(aq)$

17.24 An anion that is the conjugate base of a weak acid is itself a weak base. $A^-(aq) + H_2O(l) \rightleftharpoons HA(aq) + OH^-(aq)$ An anion that is the conjugate base of a strong acid is pH-neutral.

17.25 At 25 °C, the product of K_a for an acid and K_b for its conjugate base is $K_a \times K_b = K_w = 1.0 \times 10^{-14}$.

17.26 A cation that is the conjugate acid of a weak base such as NH_4^+ is a weak acid. Small highly charged metal cations such as Al^{3+} and Fe^{3+} form weakly acidic solutions.

17.27 For most polyprotic acids, K_{a_1} is much larger than K_{a_2}. Therefore, the amount of H_3O^+ contributed by the first ionization step is much larger than that contributed by the second or third ionization step. In addition, the production of H_3O^+ by the first step inhibits additional production of H_3O^+ by the second step because of Le Châtelier's principle.

17.28 For a weak diprotic acid, $[X^{2-}] = K_{a_2}$. This result is general for all diprotic acids in which the x is small approximation is valid.

17.29 A Lewis acid is an electron pair acceptor. A Lewis base is an electron pair donor.

17.30 A Lewis acid has an empty orbital (or can rearrange electrons to create an empty orbital) that can accept an electron pair. A Lewis base has a lone pair of electrons that it can donate to the Lewis acid.

Problems by Topic

The Nature and Definitions of Acids and Bases

17.31
(a) acid $HNO_3(aq) \rightarrow H^+(aq) + NO_3^-(aq)$
(b) acid $NH_4^+(aq) \rightarrow H^+(aq) + NH_3(aq)$
(c) base $KOH(aq) \rightarrow K^+(aq) + OH^-(aq)$
(d) acid $HC_2H_3O_2(aq) \rightarrow H^+(aq) + C_2H_3O_2^-(aq)$

17.32
(a) base $NaOH(aq) \rightarrow Na^+(aq) + OH^-(aq)$
(b) acid $H_2SO_4(aq) \rightarrow 2H^+(aq) + SO_4^{2-}(aq)$
(c) acid $HBr(aq) \rightarrow H^+(aq) + Br^-(aq)$
(d) base $Sr(OH)_2(aq) \rightarrow Sr^{2+}(aq) + 2OH^-(aq)$

17.33
(a) Because H_2CO_3 donates a proton to H_2O, it is the acid. After H_2CO_3 donates the proton, it becomes HCO_3^-, the conjugate base. Because H_2O accepts a proton, it is the base. After H_2O accepts the proton, it becomes H_3O^+, the conjugate acid.
(b) Because H_2O donates a proton to NH_3, it is the acid. After H_2O donates the proton, it becomes OH^-, the conjugate base. Because NH_3 accepts a proton, it is the base. After NH_3 accepts the proton, it becomes NH_4^+, the conjugate acid.
(c) Because HNO_3 donates a proton to H_2O, it is the acid. After HNO_3 donates the proton, it becomes NO_3^-, the conjugate base. Because H_2O accepts a proton, it is the base. After H_2O accepts the proton, it becomes H_3O^+, the conjugate acid.
(d) Because H_2O donates a proton to C_5H_5N, it is the acid. After H_2O donates the proton, it becomes OH^-, the conjugate base. Because C_5H_5N accepts a proton, it is the base. After C_5H_5N accepts the proton, it becomes $C_5H_5NH^+$, the conjugate acid.

17.34
(a) Because HI donates a proton to H_2O, it is the acid. After HI donates the proton, it becomes I^-, the conjugate base. Because H_2O accepts a proton, it is the base. After H_2O accepts the proton, it becomes H_3O^+, the conjugate acid.
(b) Because H_2O donates a proton to CH_3NH_2, it is the acid. After H_2O donates the proton, it becomes OH^-, the conjugate base. Because CH_3NH_2 accepts a proton, it is the base. After CH_3NH_2 accepts the proton, it becomes $CH_3NH_3^+$, the conjugate acid.
(c) Because H_2O donates a proton to CO_3^{2-}, it is the acid. After H_2O donates the proton, it becomes OH^-, the conjugate base. Because CO_3^{2-} accepts a proton, it is the base. After CO_3^{2-} accepts the proton, it becomes HCO_3^-, the conjugate acid.
(d) Because HBr donates a proton to H_2O, it is the acid. After HBr donates the proton, it becomes Br^-, the conjugate base. Because H_2O accepts a proton, it is the base. After H_2O accepts the proton, it becomes H_3O^+, the conjugate acid.

17.35
(a) Cl^- $HCl(aq) + H_2O(l) \rightarrow H_3O^+(aq) + Cl^-(aq)$
(b) HSO_3^- $H_2SO_3(aq) + H_2O(l) \rightleftharpoons H_3O^+(aq) + HSO_3^-(aq)$
(c) CHO_2^- $HCHO_2(aq) + H_2O(l) \rightleftharpoons H_3O^+(aq) + CHO_2^-(aq)$
(d) F^- $HF(aq) + H_2O(l) \rightleftharpoons H_3O^+(aq) + F^-(aq)$

17.36
(a) NH_4^+ $NH_3(aq) + H_2O(l) \rightleftharpoons NH_4^+(aq) + OH^-(aq)$
(b) ClO_4^- $ClO_4^-(aq) + H_2O(l) \rightarrow HClO_4(aq) + OH^-(aq)$
(c) HSO_4^- $HSO_4^-(aq) + H_2O(l) \rightarrow H_2SO_4(aq) + OH^-(aq)$
(d) CO_3^{2-} $CO_3^{2-}(aq) + H_2O(l) \rightleftharpoons HCO_3^-(aq) + OH^-(aq)$

17.37 $H_2PO_4^-(aq) + H_2O(l) \rightleftharpoons H_3O^+(aq) + HPO_4^{2-}(aq)$
$H_2PO_4^-(aq) + H_2O(l) \rightleftharpoons H_3PO_4(aq) + OH^-(aq)$

17.38 $HCO_3^-(aq) + HS^-(aq) \rightleftharpoons H_2S(aq) + CO_3^{2-}(aq)$
$HCO_3^-(aq) + HS^-(aq) \rightleftharpoons H_2CO_3(aq) + S^{2-}(aq)$

Molecular Structure and Acid Strength

17.39
(a) HCl is the stronger acid. HCl is the weaker bond; therefore, it is more acidic.
(b) HF is the stronger acid. F is more electronegative than is O, so the bond is more polar and more acidic.
(c) H_2Se is the stronger acid. The H—Se bond is weaker; therefore, it is more acidic.

17.40 Increasing acid strength: $NaH < H_2S < H_2Te < HI$
H_2Te is a stronger acid than is H_2S because the H—Te bond is weaker. HI is a stronger acid than is H_2Te because I is more electronegative than is Te. NaH is not acidic because H is more electronegative than is Na.

17.41
(a) H_2SO_4 is the stronger acid because it has more oxygen atoms.
(b) $HClO_2$ is the stronger acid because it has more oxygen atoms.
(c) HClO is the stronger acid because Cl is more electronegative than is Br.
(d) CCl_3COOH is the stronger acid because Cl is more electronegative than is H.

17.42 Increasing acid strength: $HIO_3 < HBrO_3 < HClO_3$
Cl is more electronegative than is Br, which is more electronegative than is I.

17.43 S^{2-} is the stronger base. Base strength is determined from the corresponding acid. The weaker the acid, the stronger the base. H_2S is the weaker acid because it has a stronger bond.

17.44 AsO_4^{3-} is the stronger base. Base strength is determined from the corresponding acid. The weaker the acid, the stronger the base. H_3AsO_4 is the weaker acid because P is more electronegative than is As.

Acid Strength and K_a

17.45
(a) HNO_3 is a strong acid.
(b) HCl is a strong acid.
(c) HBr is a strong acid.
(d) H_2SO_3 is a weak acid. $H_2SO_3(aq) + H_2O(l) \rightleftharpoons H_3O^+(aq) + HSO_3^-(aq)$

$$K_a = \frac{[H_3O^+][HSO_3^-]}{[H_2SO_3]}$$

17.46
(a) HF is a weak acid. $HF(aq) + H_2O(l) \rightleftharpoons H_3O^+(aq) + F^-(aq)$

$$K_a = \frac{[H_3O^+][F^-]}{[HF]}$$

(b) $HCHO_2$ is a weak acid. $HCHO_2(aq) + H_2O(l) \rightleftharpoons H_3O^+(aq) + CHO_2^-(aq)$

$$K_a = \frac{[H_3O^+][CHO_2^-]}{[HCHO_2]}$$

(c) H_2SO_4 is a strong acid.
(d) H_2CO_3 is a weak acid. $H_2CO_3(aq) + H_2O(l) \rightleftharpoons H_3O^+(aq) + HCO_3^-(aq)$

$$K_{a_1} = \frac{[H_3O^+][HCO_3^-]}{[H_2CO_3]}$$

17.47 The strength of an acid is determined by how much it ionizes and how much undissociated acid is left over. The larger the fraction of the acid that ionized, the stronger the acid.
(a) contains no HA, 10 H^+, and 10 A^-
(b) contains 3 HA, 3 H^+, and 7 A^-
(c) contains 9 HA, 1 H^+, and 1 A^-
So solution a > solution b > solution c.

17.48 HCl is a strong acid, $K_a(HF) = 3.5 \times 10^{-4}$, $K_a(HClO) = 2.9 \times 10^{-8}$, $K_a(HC_6H_5O) = 1.3 \times 10^{-10}$.
The larger the value of K_a, the stronger the acid and the greater the $[H_3O^+]$.
The order of decreasing $[H_3O^+]$ is HCl > HF > HClO > HC_6H_5O.

Autoionization of Water and pH

17.49 (a) **Given:** $K_w = 1.0 \times 10^{-14}$, $[H_3O^+] = 1.2 \times 10^{-8}$ M **Find:** $[OH^-]$
Conceptual Plan: $[H_3O^+] \rightarrow [OH^-]$

$K_w = 1.0 \times 10^{-14} = [H_3O^+][OH^-]$

Solution:
$K_w = 1.0 \times 10^{-14} = (1.2 \times 10^{-8})[OH^-]$
$[OH^-] = 8.3 \times 10^{-7}$ M
$[OH^-] > [H_3O^+]$, so the solution is basic.

Check: The units (M) are correct. The magnitude (8×10^{-7} M) makes sense because $[H^+] < 10^{-7}$ M.

(b) **Given:** $K_w = 1.0 \times 10^{-14}$, $[H_3O^+] = 8.5 \times 10^{-5}$ M **Find:** $[OH^-]$
Conceptual Plan: $[H_3O^+] \rightarrow [OH^-]$

$K_w = 1.0 \times 10^{-14} = [H_3O^+][OH^-]$

Solution:
$K_w = 1.0 \times 10^{-14} = (8.5 \times 10^{-5})[OH^-]$
$[OH^-] = 1.2 \times 10^{-10}$ M
$[H_3O^+] > [OH^-]$, so the solution is acidic.

Check: The units (M) are correct. The magnitude (1×10^{-10} M) makes sense because $[H^+] > 10^{-7}$ M.

(c) **Given:** $K_w = 1.0 \times 10^{-14}$, $[H_3O^+] = 3.5 \times 10^{-2}$ M **Find:** $[OH^-]$
Conceptual Plan: $[H_3O^+] \rightarrow [OH^-]$

$K_w = 1.0 \times 10^{-14} = [H_3O^+][OH^-]$

Solution:
$K_w = 1.0 \times 10^{-14} = (3.5 \times 10^{-2})[OH^-]$
$[OH^-] = 2.9 \times 10^{-13}$ M
$[H_3O^+] > [OH^-]$, so the solution is acidic.

Check: The units (M) are correct. The magnitude (3×10^{-13} M) makes sense because $[H^+] > 10^{-7}$ M.

17.50 (a) **Given:** $K_w = 1.0 \times 10^{-14}$, $[OH^-] = 1.1 \times 10^{-9}$ M **Find:** $[H_3O^+]$
Conceptual Plan: $[OH^-] \rightarrow [H_3O^+]$

$K_w = 1.0 \times 10^{-14} = [H_3O^+][OH^-]$

Solution:
$K_w = 1.0 \times 10^{-14} = [H_3O^+](1.1 \times 10^{-9})$
$[H_3O^+] = 9.1 \times 10^{-6}$ M
$[H_3O^+] > [OH^-]$, so the solution is acidic.

Check: The units (M) are correct. The magnitude (9×10^{-6} M) makes sense because $[OH^-] < 10^{-7}$ M.

(b) **Given:** $K_w = 1.0 \times 10^{-14}$, $[OH^-] = 2.9 \times 10^{-2}$ M **Find:** $[H_3O^+]$
Conceptual Plan: $[OH^-] \rightarrow [H_3O^+]$

$K_w = 1.0 \times 10^{-14} = [H_3O^+][OH^-]$

Solution:
$K_w = 1.0 \times 10^{-14} = [H_3O^+](2.9 \times 10^{-2})$
$[H_3O^+] = 3.4 \times 10^{-13}$ M
$[OH^-] > [H_3O^+]$, so the solution is basic.

Check: The units (M) are correct. The magnitude (3×10^{-13} M) makes sense because $[OH^-] > 10^{-7}$ M.

(c) **Given:** $K_w = 1.0 \times 10^{-14}$, $[OH^-] = 6.9 \times 10^{-12}$ M **Find:** $[H_3O^+]$
Conceptual Plan: $[OH^-] \rightarrow [H_3O^+]$

$K_w = 1.0 \times 10^{-14} = [H_3O^+][OH^-]$

Solution:
$K_w = 1.0 \times 10^{-14} = [H_3O^+](6.9 \times 10^{-12})$
$[H_3O^+] = 1.4 \times 10^{-3}$ M
$[H_3O^+] > [OH^-]$, so the solution is acidic.
Check: The units (M) are correct. The magnitude (1×10^{-3} M) makes sense because $[OH^-] < 10^{-7}$ M.

17.51 (a) **Given:** $[H_3O^+] = 1.7 \times 10^{-8}$ M **Find:** pH and pOH
Conceptual Plan: $[H_3O^+] \rightarrow pH \rightarrow pOH$

$pH = -\log[H_3O^+]$ $pH + pOH = 14$

Solution: $pH = -\log(1.7 \times 10^{-8}) = 7.77$ $pOH = 14.00 - 7.77 = 6.23$
pH > 7, so the solution is basic.
Check: The units (none and none) are correct. When $[H^+] < 10^{-7}$ M the pH > 7, pOH < 7.

(b) **Given:** $[H_3O^+] = 1. \times 10^{-7}$ M **Find:** pH and pOH
Conceptual Plan: $[H_3O^+] \rightarrow pH \rightarrow pOH$

$pH = -\log[H_3O^+]$ $pH + pOH = 14$

Solution: $pH = -\log(1.0 \times 10^{-7}) = 7.00$ $pOH = 14.00 - 7.00 = 7.00$
pH = 7, so the solution is neutral.
Check: The units (none and none) are correct. When $[H^+] = 1 \times 10^{-7}$ M the pH = 7 = pOH.

(c) **Given:** $[H_3O^+] = 2.2 \times 10^{-6}$ M **Find:** pH and pOH
Conceptual Plan: $[H_3O^+] \rightarrow pH \rightarrow pOH$

$pH = -\log[H_3O^+]$ $pH + pOH = 14$

Solution: $pH = -\log(2.2 \times 10^{-6}) = 5.66$ $pOH = 14.00 - 5.66 = 8.34$
pH < 7, so the solution is acidic.
Check: The units (none and none) are correct. When $[H^+] > 10^{-7}$ M the pH < 7, pOH > 7.

17.52 (a) **Given:** pH = 8.55 **Find:** $[H_3O^+]$, $[OH^-]$
Conceptual Plan: $pH \rightarrow [H_3O^+] \rightarrow [OH^-]$

$pH = -\log[H_3O^+]$ $K_w = 1.0 \times 10^{-14} = [H_3O^+][OH^-]$

Solution: $pH = -\log[H_3O^+]$ $8.55 = -\log[H_3O^+]$
$-8.55 = \log[H_3O^+]$ $10^{-8.55} = 10^{\log[H_3O^+]}$
$10^{-8.55} = [H_3O^+]$ $[H_3O^+] = 2.8 \times 10^{-9}$ M
$K_w = 1 \times 10^{-14} = (2.8 \times 10^{-9})[OH^-]$
$[OH^-] = 3.6 \times 10^{-6}$ M
Check: The units (M and M) are correct. When pH > 7, $[H^+] < 10^{-7}$ M and $[OH^-] > 10^{-7}$ M.

(b) **Given:** pH = 11.23 **Find:** $[H_3O^+]$, $[OH^-]$
Conceptual Plan: $pH \rightarrow [H_3O^+] \rightarrow [OH^-]$

$pH = -\log[H_3O^+]$ $K_w = 1.0 \times 10^{-14} = [H_3O^+][OH^-]$

Solution: $pH = -\log[H_3O^+]$ $11.23 = -\log[H_3O^+]$
$-11.23 = \log[H_3O^+]$ $10^{-11.23} = 10^{\log[H_3O^+]}$
$10^{-11.23} = [H_3O^+]$ $[H_3O^+] = 5.9 \times 10^{-12}$ M
$K_w = 1 \times 10^{-14} = (5.9 \times 10^{-12})[OH^-]$
$[OH^-] = 1.7 \times 10^{-3}$ M
Check: The units (M and M) are correct. When pH > 7, $[H^+] < 10^{-7}$ M and $[OH^-] > 10^{-7}$ M.

(c) **Given:** pH = 2.87 **Find:** $[H_3O^+]$, $[OH^-]$
Conceptual Plan: $pH \rightarrow [H_3O^+] \rightarrow [OH^-]$

$pH = -\log[H_3O^+]$ $K_w = 1.0 \times 10^{-14} = [H_3O^+][OH^-]$

Solution: $pH = -\log[H_3O^+]$ $2.87 = -\log[H_3O^+]$
$-2.87 = \log[H_3O^+]$ $10^{-2.87} = 10^{\log[H_3O^+]}$

$10^{-2.87} = [H_3O^+]$ $\qquad [H_3O^+] = 1.3 \times 10^{-3}$
$K_w = 1 \times 10^{-14} = (1.3 \times 10^{-3})[OH^-]$
$[OH^-] = 7.7 \times 10^{-12}$ M

Check: The units (M and M) are correct. When pH < 7, $[H^+] > 10^{-7}$ M and $[OH^-] < 10^{-7}$ M.

17.53 $pH = -\log[H_3O^+]$ $K_w = 1.0 \times 10^{-14} = [H_3O^+][OH^-]$

$[H_3O^+]$	$[OH^-]$	pH	Acidic or basic
7.1×10^{-4} M	1.4×10^{-11} M	**3.15**	acidic
3.7×10^{-9} M	2.7×10^{-6} M	8.43	basic
8×10^{-12} M	1×10^{-3} M	**11.1**	basic
6.3×10^{-4} M	**1.6×10^{-11} M**	3.21	acidic

$[H_3O^+] = 10^{-3.15} = 7.1 \times 10^{-4}$ M $\qquad [OH^-] = \dfrac{1.0 \times 10^{-14}}{7.1 \times 10^{-4}} = 1.4 \times 10^{-11}$ M

$[OH^-] = \dfrac{1.0 \times 10^{-14}}{3.7 \times 10^{-9}} = 2.7 \times 10^{-6}$ M $\qquad pH = -\log(3.7 \times 10^{-9}) = 8.43$

$[H_3O^+] = 10^{-11.1} = 8 \times 10^{-12}$ M $\qquad [OH^-] = \dfrac{1.0 \times 10^{-14}}{8 \times 10^{-12}} = 1 \times 10^{-3}$ M

$[H_3O^+] = \dfrac{1.0 \times 10^{-14}}{1.6 \times 10^{-11}} = 6.3 \times 10^{-4}$ M $\qquad pH = -\log(6.2 \times 10^{-4}) = 3.21$

Check: When pH < 7, $[H^+] > 10^{-7}$ M and $[OH^-] < 10^{-7}$ M. When pH > 7, $[H^+] < 10^{-7}$ M and $[OH^-] > 10^{-7}$ M.

17.54 $pH = -\log[H_3O^+]$ $K_w = 1.0 \times 10^{-14} = [H_3O^+][OH^-]$

$[H_3O^+]$	$[OH^-]$	pH	Acidic or basic
3.5×10^{-3} M	2.9×10^{-12} M	2.46	acidic
2.6×10^{-8} M	**3.8×10^{-7} M**	7.58	basic
1.8×10^{-9} M	5.6×10^{-6} M	8.74	basic
7.1×10^{-8} M	1.4×10^{-7} M	**7.15**	basic

$[OH^-] = \dfrac{1.0 \times 10^{-14}}{3.5 \times 10^{-3}} = 2.9 \times 10^{-12}$ M $\qquad pH = -\log(3.5 \times 10^{-3}) = 2.46$

$[H_3O^+] = \dfrac{1.0 \times 10^{-14}}{3.8 \times 10^{-7}} = 2.6 \times 10^{-8}$ M $\qquad pH = -\log(2.6 \times 10^{-8}) = 7.58$

$[OH^-] = \dfrac{1.0 \times 10^{-14}}{1.8 \times 10^{-9}} = 5.6 \times 10^{-6}$ M $\qquad pH = -\log(1.8 \times 10^{-9}) = 8.74$

$[H_3O^+] = 10^{-7.15} = 7.1 \times 10^{-8}$ M $\qquad [OH^-] = \dfrac{1.0 \times 10^{-14}}{7.1 \times 10^{-8}} = 1.4 \times 10^{-7}$ M

Check: When pH < 7, $[H^+] > 10^{-7}$ M and $[OH^-] < 10^{-7}$ M. When pH > 7, $[H^+] < 10^{-7}$ M and $[OH^-] > 10^{-7}$ M.

17.55 **Given:** $K_w = 2.4 \times 10^{-14}$ at 37 °C **Find:** $[H_3O^+]$, pH

Conceptual Plan: $K_w \rightarrow [H_3O^+] \rightarrow$ **pH**

$K_w = [H_3O^+][OH^-]$ $\quad pH = -\log[H_3O^+]$

Solution: $H_2O(l) + H_2O(l) \rightleftharpoons H_3O^+(aq) + OH^-(aq)$

$K_w = [H_3O^+][OH^-]$

$[H_3O^+] = [OH^-] = \sqrt{K_w} = \sqrt{2.4 \times 10^{-14}} = 1.5 \times 10^{-7}$ M

$pH = -\log[H_3O^+] = -\log(1.5 \times 10^{-7}) = 6.82$

Check: The value of K_w increased, indicating more products formed; so the H_3O^+ increases and the pH decreases from the values at 25 °C.

17.56 The increasing value of K_w indicates that more products are formed as the temperature increases. According to Le Châtelier, this means that the heat is a reactant. Therefore, the autoionization of water is endothermic.

17.57 (a) **Given:** $[H_3O^+] = 0.044$ M **Find:** pH

Conceptual Plan: $[H_3O^+] \rightarrow pH$

$pH = -\log[H_3O^+]$

Solution: $pH = -\log(0.044) = 1.3\underline{5}7 = 1.36$

(b) **Given:** $[H_3O^+] = 0.045$ M **Find:** pH

Conceptual Plan: $[H_3O^+] \rightarrow pH$

$pH = -\log[H_3O^+]$

Solution: $pH = -\log(0.045) = 1.3\underline{4}7 = 1.35$

(c) **Given:** $[H_3O^+] = 0.046$ M **Find:** pH

Conceptual Plan: $[H_3O^+] \rightarrow pH$

$pH = -\log[H_3O^+]$

Solution: $pH = -\log(0.046) = 1.3\underline{3}7 = 1.34$

If the pH of the solution did not carry as many decimal places as the significant digits to the right of the decimal point, you would not be able to distinguish a difference in the pH of solutions (b) and (c). Also, solution (a) would have a pH that was too high.

17.58 (a) **Given:** pH = 2.50 **Find:** $[H_3O^+]$

Conceptual Plan: $pH \rightarrow [H_3O^+]$

$pH = -\log[H_3O^+]$

Solution: $pH = -\log[H_3O^+]$ $\quad 2.50 = -\log[H_3O^+]$

$-2.50 = \log[H_3O^+]$ $\quad 10^{-2.50} = 10^{\log[H_3O^+]}$

$10^{-2.50} = [H_3O^+]$ $\quad [H_3O^+] = 3.\underline{1}6 \times 10^{-3} = 3.2 \times 10^{-3}$ M

(b) **Given:** pH = 2.51 **Find:** $[H_3O^+]$

Conceptual Plan: $pH \rightarrow [H_3O^+]$

$pH = -\log[H_3O^+]$

Solution: $pH = -\log[H_3O^+]$ $\quad 2.51 = -\log[H_3O^+]$

$-2.51 = \log[H_3O^+]$ $\quad 10^{-2.51} = 10^{\log[H_3O^+]}$

$10^{-2.51} = [H_3O^+]$ $\quad [H_3O^+] = 3.\underline{0}9 \times 10^{-3} = 3.1 \times 10^{-3}$ M

(c) **Given:** pH = 2.52 **Find:** $[H_3O^+]$

Conceptual Plan: $pH \rightarrow [H_3O^+]$

$pH = -\log[H_3O^+]$

Solution: $pH = -\log[H_3O^+]$ $\quad 2.52 = -\log[H_3O^+]$

$-2.52 = \log[H_3O^+]$ $\quad 10^{-2.52} = 10^{\log[H_3O^+]}$

$10^{-2.52} = [H_3O^+]$ $\quad [H_3O^+] = 3.\underline{0}2 \times 10^{-3} = 3.0 \times 10^{-3}$ M

The pH values have two digits to the right of the decimal point. This gives two significant figures in the concentration of H_3O^+. If you do not carry out the concentration to two significant figures, all of the concentrations will appear to be the same.

Acid Solutions

17.59 (a) **Given:** 0.25 M HCl (strong acid) **Find:** $[H_3O^+]$, $[OH^-]$, pH

Conceptual Plan: $[HCl] \rightarrow [H_3O^+] \rightarrow$ **pH and then** $[H_3O^+] \rightarrow [OH^-]$

$[HCl] = [H_3O^+]$ $\quad pH = -\log[H_3O^+]$ $\quad [H_3O^+][OH^-] = 1.0 \times 10^{-14}$

Solution: 0.25 M HCl $= 0.25$ M H_3O^+ pH $= -\log(0.25) = 0.60$

$[OH^-] = 1.0 \times 10^{-14}/0.25 \text{ M} = 4.0 \times 10^{-14}$

Check: HCl is a strong acid with a relatively high concentration, so we expect the pH to be low and the $[OH^-]$ to be small.

(b) **Given:** 0.015 M HNO_3 (strong acid) **Find:** $[H_3O^+]$, $[OH^-]$, pH

Conceptual Plan: $[HNO_3] \rightarrow [H_3O^+] \rightarrow$ **pH and then** $[H_3O^+] \rightarrow [OH^-]$

$[HNO_3] = [H_3O^+]$ $\quad pH = -\log[H_3O^+]$ $\quad [H_3O^+][OH^-] = 1.0 \times 10^{-14}$

Solution: 0.015 M HNO_3 $= 0.015$ M H_3O^+ pH $= -\log(0.015) = 1.82$

$[OH^-] = 1.0 \times 10^{-14}/0.015 \text{ M} = 6.7 \times 10^{-13}$

Check: HNO_3 is a strong acid, so we expect the pH to be low and the $[OH^-]$ to be small.

(c) **Given:** 0.052 M HBr and 0.015 M HNO_3 (strong acids) **Find:** $[H_3O^+]$, $[OH^-]$, pH

Conceptual Plan: $[HBr] + [HNO_3] \rightarrow [H_3O^+] \rightarrow$ **pH and then** $[H_3O^+] \rightarrow [OH^-]$

$[HBr]+[HNO_3] = [H_3O^+]$ $\quad pH = -\log[H_3O^+]$ $\quad [H_3O^+][OH^-] = 1.0 \times 10^{-14}$

Solution: 0.052 M HBr $= 0.052$ M H_3O^+ and 0.020 M HNO_3 $= 0.020$ M H_3O^+

Total $H_3O^+ = 0.072 \text{ M} + 0.020 \text{ M} = 0.072$ M; pH $= -\log(0.072) = 1.14$

$[OH^-] = 1.0 \times 10^{-14}/0.072 \text{ M} = 1.4 \times 10^{-13}$ M

Check: HBr and HNO_3 are both strong acids that completely dissociate. This gives a relatively high concentration, so we expect the pH to be low and the $[OH^-]$ to be small.

(d) **Given:** $HNO_3 = 0.655\%$ by mass, $d_{solution} = 1.01$ g/mL **Find:** $[H_3O^+]$, $[OH^-]$, pH

Conceptual Plan:

% mass HNO_3 → g HNO_3 → mol HNO_3 and then g soln → mL soln → L soln → M HNO_3

$\frac{\%}{100}$ $\quad \frac{1 \text{ mol } HNO_3}{63.018 \text{ g } HNO_3}$ $\quad \frac{1.01 \text{ g soln}}{1 \text{ mL soln}}$ $\quad \frac{1000 \text{ mL soln}}{1 \text{ L soln}}$ $\quad \frac{\text{mol } HNO_3}{1 \text{ L soln}}$

→ M H_3O^+ → pH and then $[H_3O^+] \rightarrow [OH^-]$

$[HNO_3] = [H_3O^+]$ $\quad pH = -\log[H_3O^+]$ $\quad [H_3O^+][OH^-] = 1.0 \times 10^{-14}$

Solution: $$\frac{0.655 \text{ g } HNO_3}{100 \text{ g soln}} \times \frac{1 \text{ mol } HNO_3}{63.018 \text{ g } HNO_3} \times \frac{1.01 \text{ g soln}}{\text{mL soln}} \times \frac{1000 \text{ mL soln}}{1 \text{ L soln}} = 0.105 \text{ M } HNO_3$$

0.105 M HNO_3 $= 0.105$ M H_3O^+ pH $= -\log(0.105) = 0.979$

$[OH^-] = 1.00 \times 10^{-14}/0.105 \text{ M} = 9.52 \times 10^{-14}$ M

Check: HNO_3 is a strong acid that completely dissociates. This gives a relatively high concentration, so we expect the pH to be low and the $[OH^-]$ to be small.

17.60 (a) **Given:** 0.048 M HI (strong acid) **Find:** pH

Conceptual Plan: $[HI] \rightarrow [H_3O^+] \rightarrow$ **pH and then** $[H_3O^+] \rightarrow [OH^-]$

$[HI] = [H_3O^+]$ $\quad pH = -\log[H_3O^+]$ $\quad [H_3O^+][OH^-] = 1.0 \times 10^{-14}$

Solution: 0.048 M HI $= 0.048$ M H_3O^+ pH $= -\log(0.048) = 1.32$

Check: HI is a strong acid with a relatively high concentration, so we expect the pH to be low.

(b) **Given:** 0.0895 M $HClO_4$ (strong acid) **Find:** pH

Conceptual Plan: $[HClO_4] \rightarrow [H_3O^+] \rightarrow$ **pH**

$[HClO_4] = [H_3O^+]$ $\quad pH = -\log[H_3O^+]$

Solution: 0.0895 M $HClO_4$ $= 0.0895$ M H_3O^+ pH $= -\log(0.0895) = 1.048$

Check: $HClO_4$ is a strong acid, so we expect the pH to be low.

(c) **Given:** 0.045 M $HClO_4$ and 0.048 M HCl (strong acids) **Find:** pH

Conceptual Plan: $[HCl] + [HClO_4] \rightarrow [H_3O^+] \rightarrow$ **pH**

$[HCl]+[HClO_4] = [H_3O^+]$ $\quad pH = -\log[H_3O^+]$

Solution: 0.045 M $HClO_4$ $= 0.045$ M H_3O^+ and 0.048 M HCl $= 0.048$ M H_3O^+

Total $H_3O^+ = 0.045 \text{ M} + 0.048 \text{ M} = 0.093$ M pH $= -\log(0.093) = 1.03$

Check: $HClO_4$ and HCl are both strong acids that completely dissociate. This gives a relatively high concentration, so we expect the pH to be low.

(d) **Given:** HCl = 1.09% by mass, $d_{solution} = 1.01$ g/mL **Find:** pH

Conceptual Plan:

% mass HCl → g HCl → mol HCl and then g soln → mL soln → L soln → M HCl

$$\frac{\%}{100} \quad \frac{1 \text{ mol HCl}}{36.46 \text{ g HCl}} \quad \frac{1.01 \text{ g soln}}{1 \text{ mL soln}} \quad \frac{1000 \text{ mL soln}}{1 \text{ L soln}} \quad \frac{\text{mol HCl}}{1 \text{ L soln}}$$

→ M H_3O^+ → pH

$[HCl] = [H_3O^+]$ $pH = -\log[H_3O^+]$

Solution: $\frac{1.09 \text{ g HCl}}{100 \text{ g soln}} \times \frac{1 \text{ mol HCl}}{36.46 \text{ g HCl}} \times \frac{1.01 \text{ g soln}}{\text{mL soln}} \times \frac{1000 \text{ mL soln}}{1 \text{ L soln}} = 0.30\underline{1}9 \text{ M} = 0.302 \text{ M HCl}$

$0.302 \text{ M HCl} = 0.302 \text{ M } H_3O^+$ $\quad pH = -\log(0.302) = 0.520$

Check: HCl is a strong acid that completely dissociates. This gives a relatively high concentration, so we expect the pH to be low.

17.61 (a) **Given:** pH = 1.25, 0.250 L **Find:** g HI

Conceptual Plan: pH → $[H_3O^+]$ → [HI] → mol HI → g HI

$pH = -\log[H_3O^+]$ $\quad [H_3O^+] = [HI]$ $\quad mol = MV$ $\quad g = \frac{127.9 \text{ g HI}}{\text{mol HI}}$

Solution: $[H_3O^+] = 10^{-1.25} = 0.056 \text{ M} = [HI]$ $\frac{0.056 \text{ mol HI}}{1 \text{ L}} \times 0.250 \text{ L} \times \frac{127.9 \text{ g HI}}{1 \text{ mol HI}} = 1.8 \text{ g HI}$

(b) **Given:** pH = 1.75, 0.250 L **Find:** g HI

Conceptual Plan: pH → $[H_3O^+]$ → [HI] → mol HI → g HI

$pH = -\log[H_3O^+]$ $\quad [H_3O^+] = [HI]$ $\quad mol = MV$ $\quad g = \frac{127.9 \text{ g HI}}{\text{mol HI}}$

Solution: $[H_3O^+] = 10^{-1.75} = 0.01\underline{7}8 \text{ M} = [HI]$ $\frac{0.01\underline{7}8 \text{ mol HI}}{1 \text{ L}} \times 0.250 \text{ L} \times \frac{127.9 \text{ g HI}}{1 \text{ mol HI}} = 0.57 \text{ g HI}$

(c) **Given:** pH = 2.85, 0.250 L **Find:** g HI

Conceptual Plan: pH → $[H_3O^+]$ → [HI] → mol HI → g HI

$pH = -\log[H_3O^+]$ $\quad [H_3O^+] = [HI]$ $\quad mol = MV$ $\quad g = \frac{127.9 \text{ g HI}}{\text{mol HI}}$

Solution: $[H_3O^+] = 10^{-2.85} = 0.0014 \text{ M} = [HI]$ $\frac{0.0014 \text{ mol HI}}{1 \text{ L}} \times 0.250 \text{ L} \times \frac{127.9 \text{ g HI}}{1 \text{ mol HI}} = 0.045 \text{ g HI}$

Check: The units (g) are correct. The magnitudes of the answers make sense, since the volume is small (<1 L). The higher the pH the lower the amount of HI.

17.62 (a) **Given:** pH = 2.50, 0.500 L **Find:** g $HClO_4$

Conceptual Plan: pH → $[H_3O^+]$ → $[HClO_4]$ → mol $HClO_4$ → g $HClO_4$

$pH = -\log[H_3O^+]$ $\quad [H_3O^+] = [HClO_4]$ $\quad mol = MV$ $\quad g = \frac{100.46 \text{ g } HClO_4}{1 \text{ mol } HClO_4}$

Solution:

$[H_3O^+] = 10^{-2.50} = 0.003\underline{1}6 \text{ M} = [HClO_4]$

$\frac{0.003\underline{1}6 \text{ mol } HClO_4}{1 \text{ L}} \times 0.500 \text{ L} \times \frac{100.46 \text{ g } HClO_4}{1 \text{ mol } HClO_4} = 0.16 \text{ g } HClO_4$

(b) **Given:** pH = 1.50, 0.500 L **Find:** g $HClO_4$

Conceptual Plan: pH → $[H_3O^+]$ → $[HClO_4]$ → mol $HClO_4$ → g $HClO_4$

$pH = -\log[H_3O^+]$ $\quad [H_3O^+] = [HClO_4]$ $\quad mol = MV$ $\quad g = \frac{100.46 \text{ g } HClO_4}{1 \text{ mol } HClO_4}$

Solution:

$[H_3O^+] = 10^{-1.50} = 0.03\underline{1}6 \text{ M} = [HClO_4]$

$$\frac{0.0316\ \cancel{\text{mol HClO}_4}}{1\ \cancel{\text{L}}} \times 0.500\ \cancel{\text{L}} \times \frac{100.46\ \text{g HClO}_4}{1\ \cancel{\text{mol HClO}_4}} = 1.6\ \text{g HClO}_4$$

(c) **Given:** pH = 0.50, 0.500 L **Find:** g $HClO_4$

Conceptual Plan: $\text{pH} \rightarrow [\text{H}_3\text{O}^+] \rightarrow [\text{HClO}_4] \rightarrow \text{mol HClO}_4 \rightarrow \text{g HClO}_4$

$$\text{pH} = -\log[\text{H}_3\text{O}^+] \quad [\text{H}_3\text{O}^+] = [\text{HClO}_4] \quad \text{mol} = MV \quad \text{g} = \frac{100.46\ \text{g HClO}_4}{1\ \text{mol HClO}_4}$$

Solution:

$$[\text{H}_3\text{O}^+] = 10^{-0.50} = 0.316\ \text{M} = [\text{HClO}_4]$$

$$\frac{0.316\ \cancel{\text{mol HClO}_4}}{1\ \cancel{\text{L}}} \times 0.500\ \cancel{\text{L}} \times \frac{100.46\ \text{g HClO}_4}{1\ \cancel{\text{mol HClO}_4}} = 16\ \text{g HClO}_4$$

Check: The units (g) are correct. The magnitudes of the answers make sense, since the volume is small (<1 L). The higher the pH the lower the amount of $HClO_4$.

17.63 **Given:** 224 mL HCl, 27.2 °C, 1.02 atm, 1.5 L solution **Find:** pH

Conceptual Plan: $\text{vol HCl} \rightarrow \text{mol HCl} \rightarrow [\text{HCl}] \rightarrow [\text{H}_3\text{O}^+] \rightarrow \text{pH}$

$$PV = nRT \qquad \text{M} = \frac{\text{mol HCl}}{\text{L soln}} \quad [\text{HCl}] = [\text{H}_3\text{O}^+] \quad \text{pH} = -\log[\text{H}_3\text{O}^+]$$

Solution:
$$n = \frac{(1.02\ \cancel{\text{atm}})(224\ \cancel{\text{mL}})\left(\frac{1\ \cancel{\text{L}}}{1000\ \cancel{\text{mL}}}\right)}{\left(\frac{0.08206\ \cancel{\text{L}}\cdot\cancel{\text{atm}}}{\text{mol}\cdot\cancel{\text{K}}}\right)((27.2 + 273.15)\ \cancel{\text{K}})} = 0.00927\ \text{mol}$$

$$[\text{HCl}] = \frac{0.00927\ \text{mol}}{1.5\ \text{L}} = 0.00618\ \text{M} = [\text{H}_3\text{O}^+];\ \text{pH} = -\log(0.00618) = 2.21$$

Check: The units (none) are correct. The magnitude of the answer makes sense, since HCl is an acid and we will have a low pH.

17.64 **Given:** 36.0% HCl, $d_{soln} = 1.179\ \text{g/mL}$; 5.00 L, pH 1.8 **Find:** vol soln

Conceptual Plan: % mass HCl → g HCl → mol HCl and then g soln → mL soln → L soln → M HCl

$$\frac{\%}{100} \quad \frac{1\ \text{mol HCl}}{36.46\ \text{g HCl}} \qquad \frac{1.179\ \text{g soln}}{1\ \text{mL soln}} \quad \frac{1000\ \text{mL soln}}{1\ \text{L soln}} \quad \frac{\text{mol HCl}}{\text{L soln}}$$

→ $\text{M H}_3\text{O}^+$ **and then** $\text{pH} \rightarrow [\text{H}_3\text{O}^+]$ **and the** $V_1M_1 \rightarrow V_2M_2$

$$[\text{HCl}] = [\text{H}_3\text{O}^+] \qquad [\text{H}_3\text{O}^+] = 10^{-\text{pH}} \qquad V_1M_1 = V_2M_2$$

Solution:
$$\frac{36.0\ \cancel{\text{g HCl}}}{100\ \cancel{\text{g soln}}} \times \frac{1\ \text{mol HCl}}{36.46\ \cancel{\text{g HCl}}} \times \frac{1.179\ \cancel{\text{g soln}}}{1\ \cancel{\text{mL soln}}} \times \frac{1000\ \cancel{\text{mL soln}}}{1\ \text{L soln}} = 11.64\ \text{M HCl}$$

$$[\text{H}_3\text{O}^+] = 10^{-1.8} = 0.0158\ \text{M}$$

$$V_1M_1 = V_2M_2 \quad V_1(11.64\ \text{M}) = (5.00\ \text{L})(0.0158\ \text{M}) \quad V_1 = 0.006787\ \text{L} = 6.79\ \text{mL}$$

Check: The units (mL) are correct. The magnitude of the answer (7 mL) makes sense, since we are starting with a very high concentration of HCl (a strong acid).

17.65 **Given:** 0.100 M benzoic acid, $K_a = 6.5 \times 10^{-5}$ **Find:** $[\text{H}_3\text{O}^+]$, pH

Conceptual Plan: Write a balanced reaction. Prepare an ICE table, represent the change with *x*, sum the table, determine the equilibrium values, put the equilibrium values in the equilibrium expression, and solve for *x*. Determine $[\text{H}_3\text{O}^+]$ and pH.

Solution:	$HC_7H_5O_2(aq)$	+ $H_2O(l)$ ⇌	$H_3O^+(aq)$	+ $C_7H_5O_2^-(aq)$
Initial	0.100		0.0	0.0
Change	$-x$		$+x$	$+x$
Equil	$0.100 - x$		x	x

$$K_a = \frac{[\text{H}_3\text{O}^+][\text{C}_7\text{H}_5\text{O}_2^-]}{[\text{HC}_7\text{H}_5\text{O}_2]} = \frac{(x)(x)}{(0.100 - x)} = 6.5 \times 10^{-5}$$

Assume that x is small compared to 0.100.

$x^2 = (6.5 \times 10^{-5})(0.100); \quad x = 2.5 \times 10^{-3}\,\text{M} = [H_3O^+]$

Check assumption: $\dfrac{2.5 \times 10^{-3}}{0.100} \times 100\% = 2.5\%$; assumption is valid.

$\text{pH} = -\log(2.5 \times 10^{-3}) = 2.60$

Check: The units (M and none) are correct. The magnitude of the pH (3) makes sense, since we expect a pH $> -\log(0.100) = 1.00$ when we have a weak acid.

17.66 **Given:** 0.200 M formic acid, $K_a = 1.8 \times 10^{-4}$ **Find:** $[H_3O^+]$, pH

Conceptual Plan: Write a balanced reaction. Prepare an ICE table, represent the change with x, sum the table, determine the equilibrium values, put the equilibrium values in the equilibrium expression, and solve for x. Determine $[H_3O^+]$ and pH.

Solution: $HCH_2O(aq) + H_2O(l) \rightleftharpoons H_3O^+(aq) + CH_2O^-(aq)$

	HCH_2O	H_3O^+	CH_2O^-
Initial	0.200	0.0	0.0
Change	$-x$	$+x$	$+x$
Equil	$0.200 - x$	x	x

$$K_a = \frac{[H_3O^+][CH_2O^-]}{[HCH_2O]} = \frac{(x)(x)}{(0.200 - x)} = 1.8 \times 10^{-4}$$

Assume that x is small compared to 0.200.

$x^2 = (1.8 \times 10^{-4})(0.200); \quad x = 6.0 \times 10^{-3}\,\text{M} = [H_3O^+]$

Check assumption: $\dfrac{6.0 \times 10^{-3}}{0.200} \times 100\% = 3.0\%$; assumption is valid.

$\text{pH} = -\log(6.0 \times 10^{-3}) = 2.22$

Check: The units (M and none) are correct. The magnitude of the pH (2) makes sense, since we expect a pH $> -\log(0.200) = 0.700$ when we have a weak acid.

17.67 (a) **Given:** 0.500 M HNO_2, $K_a = 4.6 \times 10^{-4}$ **Find:** pH

Conceptual Plan: Write a balanced reaction. Prepare an ICE table, represent the change with x, sum the table, determine the equilibrium values, put the equilibrium values in the equilibrium expression, and solve for x. Determine $[H_3O^+]$ and pH.

Solution: $HNO_2(aq) + H_2O(l) \rightleftharpoons H_3O^+(aq) + NO_2^-(aq)$

	HNO_2	H_3O^+	NO_2^-
Initial	0.500	0.0	0.0
Change	$-x$	$+x$	$+x$
Equil	$0.500 - x$	x	x

$$K_a = \frac{[H_3O^+][NO_2^-]}{[HNO_2]} = \frac{(x)(x)}{(0.500 - x)} = 4.6 \times 10^{-4}$$

Assume that x is small compared to 0.500.

$x^2 = (4.6 \times 10^{-4})(0.500); \quad x = 0.015\,\text{M} = [H_3O^+]$

Check assumption: $\dfrac{0.015}{0.500} \times 100\% = 3.0\%$; assumption is valid.

$\text{pH} = -\log(0.015) = 1.82$

Check: The units (none) are correct. The magnitude of the answer (2) makes sense, since we expect a pH $> -\log(0.500) = 0.301$ when we have a weak acid.

(b) **Given:** 0.100 M HNO_2, $K_a = 4.6 \times 10^{-4}$ **Find:** pH

Conceptual Plan: Write a balanced reaction. Prepare an ICE table, represent the change with x, sum the table, determine the equilibrium values, put the equilibrium values in the equilibrium expression, and solve for x. Determine $[H_3O^+]$ and pH.

Solution: $HNO_2(aq) + H_2O(l) \rightleftharpoons H_3O^+(aq) + NO_2^-(aq)$

	$HNO_2(aq)$	$H_3O^+(aq)$	$NO_2^-(aq)$
Initial	0.100	0.0	0.0
Change	$-x$	$+x$	$+x$
Equil	$0.100 - x$	x	x

$$K_a = \frac{[H_3O^+][NO_2^-]}{[HNO_2]} = \frac{(x)(x)}{(0.100 - x)} = 4.6 \times 10^{-4}$$

Assume that x is small compared to 0.100.

$x^2 = (4.6 \times 10^{-4})(0.100)$ $x = 0.0068 \text{ M} = [H_3O^+]$

Check assumption: $\frac{0.0068}{0.100} \times 100\% = 6.8\%$; assumption is not valid; solve using the quadratic equation.

$x^2 = (4.6 \times 10^{-4})(0.100 - x)$; $x^2 + 4.6 \times 10^{-4}x - 4.6 \times 10^{-5} = 0$

$x = 0.00656$

$\text{pH} = -\log(0.00656) = 2.18$

Check: The units (none) are correct. The magnitude of the answer (2) makes sense, since we expect a $\text{pH} > -\log(0.100) = 1.00$ when we have a weak acid.

(c) **Given:** 0.0100 M HNO_2, $K_a = 4.6 \times 10^{-4}$ **Find:** pH

Conceptual Plan: Write a balanced reaction. Prepare an ICE table, represent the change with x, sum the table, determine the equilibrium values, put the equilibrium values in the equilibrium expression, and solve for x. Determine $[H_3O^+]$ and pH.

Solution: $HNO_2(aq) + H_2O(l) \rightleftharpoons H_3O^+(aq) + NO_2^-(aq)$

	$HNO_2(aq)$	$H_3O^+(aq)$	$NO_2^-(aq)$
Initial	0.0100	0.0	0.0
Change	$-x$	$+x$	$+x$
Equil	$0.0100 - x$	x	x

$$K_a = \frac{[H_3O^+][NO_2^-]}{[HNO_2]} = \frac{(x)(x)}{(0.0100 - x)} = 4.6 \times 10^{-4}$$

Assume that x is small compared to 0.100.

$x^2 = (4.6 \times 10^{-4})(0.0100)$; $x = 0.0021 \text{ M} = [H_3O^+]$

Check assumption: $\frac{0.0021}{0.0100} \times 100\% = 21\%$; assumption is not valid; solve using the quadratic equation.

$x^2 = (4.6 \times 10^{-4})(0.0100 - x)$ $x^2 + 4.6 \times 10^{-4}x - 4.6 \times 10^{-6} = 0$

$x = 0.0019$

$\text{pH} = -\log(0.0019) = 2.72$

Check: The units (none) are correct. The magnitude of the answer (3) makes sense, since we expect a $\text{pH} > -\log(0.0100) = 2.00$ when we have a weak acid.

17.68 (a) **Given:** 0.250 M HF, $K_a = 3.5 \times 10^{-4}$ **Find:** pH

Conceptual Plan: Write a balanced reaction. Prepare an ICE table, represent the change with x, sum the table, determine the equilibrium values, put the equilibrium values in the equilibrium expression, and solve for x. Determine $[H_3O^+]$ and pH.

Solution: $HF(aq) + H_2O(l) \rightleftharpoons H_3O^+(aq) + F^-(aq)$

	$HF(aq)$	$H_3O^+(aq)$	$F^-(aq)$
Initial	0.250	0.0	0.0
Change	$-x$	$+x$	$+x$
Equil	$0.250 - x$	x	x

$$K_a = \frac{[H_3O^+][F^-]}{[HF]} = \frac{(x)(x)}{(0.250 - x)} = 3.5 \times 10^{-4}$$

Assume that x is small compared to 0.250.

$x^2 = (3.5 \times 10^{-4})(0.250) \quad x = 0.00935 \text{ M} = [H_3O^+]$

Check assumption: $\frac{0.00935}{0.250} \times 100\% = 3.7\%$; assumption is valid.

$pH = -\log(0.00935) = 2.03$

Check: The units (none) are correct. The magnitude of the answer (2) makes sense, since we expect a $pH > -\log(0.250) = 0.602$ when we have a weak acid.

(b) **Given:** 0.0500 M HF, $K_a = 3.5 \times 10^{-4}$ **Find:** pH

Conceptual Plan: Write a balanced reaction. Prepare an ICE table, represent the change with x, sum the table, determine the equilibrium values, put the equilibrium values in the equilibrium expression, and solve for x. Determine $[H_3O^+]$ and pH.

Solution:	$HF(aq) + H_2O(l) \rightleftharpoons$	$H_3O^+(aq)$	$+ F^-(aq)$
Initial	0.0500	0.0	0.0
Change	$-x$	$+x$	$+x$
Equil	$0.0500 - x$	x	x

$$K_a = \frac{[H_3O^+][F^-]}{[HF]} = \frac{(x)(x)}{(0.0500 - x)} = 3.5 \times 10^{-4}$$

Assume that x is small compared to 0.0500.

$x^2 = (3.5 \times 10^{-4})(0.0500); \quad x = 0.00418 \text{ M} = [H_3O^+]$

Check assumption: $\frac{0.00418}{0.050} \times 100\% = 8.4\%$; assumption is not valid; solve using the quadratic equation.

$x^2 + 3.5 \times 10^{-4}x - 1.75 \times 10^{-5} = 0; \quad x = 0.0040 \text{ M} = [H_3O^+]$

$pH = -\log(0.0040) = 2.40$

Check: The units (none) are correct. The magnitude of the answer (2) makes sense, since we expect a $pH > -\log(0.500) = 1.301$ when we have a weak acid.

(c) **Given:** 0.0250 M HF, $K_a = 3.5 \times 10^{-4}$ **Find:** pH

Conceptual Plan: Write a balanced reaction. Prepare an ICE table, represent the change with x, sum the table, determine the equilibrium values, put the equilibrium values in the equilibrium expression, and solve for x. Determine $[H_3O^+]$ and pH.

Solution:	$HF(aq) + H_2O(l) \rightleftharpoons$	$H_3O^+(aq)$	$+ F^-(aq)$
Initial	0.0250	0.0	0.0
Change	$-x$	$+x$	$+x$
Equil	$0.0250 - x$	x	x

$$K_a = \frac{[H_3O^+][F^-]}{[HF]} = \frac{(x)(x)}{(0.0250 - x)} = 3.5 \times 10^{-4}$$

Assume that x is small compared to 0.0250.

$x^2 = (3.5 \times 10^{-4})(0.0250); \quad x = 0.00296 \text{ M} = [H_3O^+]$

Check assumption: $\frac{0.00296}{0.0250} \times 100\% = 11.8\%$; assumption is not valid; solve using the quadratic equation.

$x^2 + 3.5 \times 10^{-4}x - 8.75 \times 10^{-6} = 0 \quad x = 0.00279 \text{ M} = [H_3O^+]$

$pH = -\log(0.00279) = 2.55$

Check: The units (none) are correct. The magnitude of the answer (3) makes sense, since we expect a $pH > -\log(0.0250) = 1.602$ when we have a weak acid.

17.69 **Given:** 15.0 mL glacial acetic, $d = 1.05\ g/mL$, dilute to 1.50 L, $K_a = 1.8 \times 10^{-5}$ **Find:** pH

Conceptual Plan: mL acetic acid → g acetic acid → mol acetic acid → M and then write a balanced reaction.

$$\frac{1.05\ g}{mL} \qquad \frac{\text{mol acetic acid}}{60.05\ g} \qquad M = \frac{mol}{L}$$

Prepare an ICE table, represent the change with x, sum the table, determine the equilibrium values, put the equilibrium values in the equilibrium expression, and solve for x. Determine $[H_3O^+]$ and pH.

Solution: $15.0\ \text{mL} \times \frac{1.05\ g}{\text{mL}} \times \frac{1\ mol}{60.05\ g} \times \frac{1}{1.50\ L} = 0.1749\ M$

	$HC_2H_3O_2(aq) + H_2O(l) \rightleftharpoons$	$H_3O^+(aq) +$	$C_2H_3O_2^-(aq)$
Initial	0.1749	0.0	0.0
Change	$-x$	$+x$	$+x$
Equil	$0.1749 - x$	x	x

$$K_a = \frac{[H_3O^+][C_2H_3O_2^-]}{[HC_2H_3O_2]} = \frac{(x)(x)}{(0.1749 - x)} = 1.8 \times 10^{-5}$$

Assume that x is small compared to 0.1749.

$x^2 = (1.8 \times 10^{-5})(0.1749);\quad x = 0.00177\ M = [H_3O^+]$

Check assumption: $\frac{0.00177}{0.1749} \times 100\% = 1.0\%$; assumption is valid.

$pH = -\log(0.00177) = 2.75$

Check: The units (none) are correct. The magnitude of the answer (3) makes sense, since we have an acid concentration <1 M and we have a weak acid.

17.70 **Given:** 1.35% formic acid, $d = 1.01\ g/mL$, $K_a = 1.8 \times 10^{-4}$ **Find:** pH

Conceptual Plan: % formic acid → g formic acid → mol and g soln → mL soln → L soln and then M

$$\frac{1\ mol}{46.03\ g} \qquad \frac{1.01\ g\ soln}{1\ mL\ soln} \qquad \frac{1000\ mL}{1\ L\ soln}$$

Write a balanced reaction. Prepare an ICE table, represent the change with x, sum the table, determine the equilibrium values, put the equilibrium values in the equilibrium expression, and solve for x. Determine $[H_3O^+]$ and pH.

Solution: $\frac{1.35\ g\ HCHO_2}{100\ g\ soln} \times \frac{1\ mol\ HCHO_2}{46.03\ g} \times \frac{1.01\ g\ soln}{1\ mL\ soln} \times \frac{1000\ mL\ soln}{1\ L\ soln} = 0.2962\ M$

	$HCHO_2(aq) + H_2O(l) \rightleftharpoons$	$H_3O^+(aq) +$	$CHO_2^-(aq)$
Initial	0.2962	0.0	0.0
Change	$-x$	$+x$	$+x$
Equil	$0.2962 - x$	x	x

$$K_a = \frac{[H_3O^+][CHO_2^-]}{[HCHO_2]} = \frac{(x)(x)}{(0.2962 - x)} = 1.8 \times 10^{-4}$$

Assume that x is small compared to 0.2962.

$x^2 = (1.8 \times 10^{-4})(0.2962);\quad x = 0.00730\ M = [H_3O^+]$

Check assumption: $\frac{0.00730}{0.2962} \times 100\% = 2.5\%$; assumption is valid.

$pH = -\log(0.00730) = 2.14$

Check: The units (none) are correct. The magnitude of the answer (2) makes sense, since we have a low concentration of a weak acid.

17.71 **Given:** 0.185 M HA, pH = 2.95 **Find:** K_a

Conceptual Plan: pH → $[H_3O^+]$ and then write a balanced reaction. Prepare an ICE table, calculate equilibrium concentrations, and plug into the equilibrium expression to solve for K_a.

Solution: $[H_3O^+] = 10^{-2.95} = 0.00112\ M = [A^-]$

	$HA(aq) + H_2O(l) \rightleftharpoons$	$H_3O^+(aq)$	$+ A^-(aq)$
Initial	0.185	0.0	0.0
Change	$-x$	$+x$	$+x$
Equil	$0.185 - 0.00112$	0.00112	0.00112

$$K_a = \frac{[H_3O^+][A^-]}{[HA]} = \frac{(0.00112)(0.00112)}{(0.185 - 0.00112)} = 6.82 \times 10^{-6}$$

Check: The units (none) are correct. The magnitude of the answer makes sense, since we expect a pH $> -\log(0.185) = 0.733$. Thus, we have a weak acid with a small K.

17.72 **Given:** 0.115 M HA, pH = 3.29 **Find:** K_a

Conceptual Plan: pH $\rightarrow [H_3O^+]$ and then write a balanced reaction. Prepare an ICE table, calculate equilibrium concentrations, and plug into the equilibrium expression to solve for K_a.

Solution: $[H_3O^+] = 10^{-3.29} = 5.\underline{1}3 \times 10^{-4}\ M = [A^-]$

	$HA(aq) + H_2O(l) \rightleftharpoons$	$H_3O^+(aq)$	$+ A^-(aq)$
Initial	0.115	0.0	0.0
Change	$-x$	$+x$	$+x$
Equil	$(0.115 - 5.\underline{1}3 \times 10^{-4})$	$(5.\underline{1}3 \times 10^{-4})$	$(5.\underline{1}3 \times 10^{-4})$

$$K_a = \frac{[H_3O^+][A^-]}{[HA]} = \frac{(5.\underline{1}3 \times 10^{-4})(5.\underline{1}3 \times 10^{-4})}{(0.115 - 5.\underline{1}3 \times 10^{-4})} = 2.30 \times 10^{-6}$$

Check: The units (none) are correct. The magnitude of the answer makes sense, since we expect a pH $> -\log(0.115) = 0.939$. Thus, we have a weak acid with a small K.

17.73 **Given:** 0.125 M HCN, $K_a = 4.9 \times 10^{-10}$ **Find:** % ionization

Conceptual Plan: Write a balanced reaction. Prepare an ICE table, represent the change with x, sum the table, determine the equilibrium values, put the equilibrium values in the equilibrium expression, solve for x, and then $x \rightarrow$ % ionization.

$$\%\ \text{ionization} = \frac{x}{[HCN]_{original}} \times 100\%$$

Solution:

	$HCN(aq) + H_2O(l) \rightleftharpoons$	$H_3O^+(aq)$	$+ CN^-(aq)$
Initial	0.125	0.0	0.0
Change	$-x$	$+x$	$+x$
Equil	$0.125 - x$	x	x

$$K_a = \frac{[H_3O^+][CN^-]}{[HCN]} = \frac{(x)(x)}{(0.125 - x)} = 4.9 \times 10^{-10}$$

Assume that x is small compared to 0.125.

$x^2 = (4.9 \times 10^{-10})(0.125)$; $x = 7.\underline{8}3 \times 10^{-6}$

$$\%\ \text{ionization} = \frac{7.\underline{8}3 \times 10^{-6}}{0.125} \times 100\% = 0.0063\%\ \text{ionized}$$

Check: The units (%) are correct. The magnitude of the answer makes sense, since we have a very weak acid (very small K).

17.74 **Given:** 0.225 M $HC_7H_5O_2$, $K_a = 6.5 \times 10^{-5}$ **Find:** % ionization

Conceptual Plan: Write a balanced reaction. Prepare an ICE table, represent the change with x, sum the table, determine the equilibrium values, put the equilibrium values in the equilibrium expression, solve for x, and then $x \rightarrow$ % ionization.

$$\%\ \text{ionization} = \frac{x}{[HC_7H_5O_2]_{original}} \times 100\%$$

Solution: $HC_7H_5O_2(aq) + H_2O(l) \rightleftharpoons H_3O^+(aq) + C_7H_5O_2^-(aq)$

	$HC_7H_5O_2$	H_3O^+	$C_7H_5O_2^-$
Initial	0.225	0.0	0.0
Change	$-x$	$+x$	$+x$
Equil	$0.225 - x$	x	x

$$K_a = \frac{[H_3O^+][C_7H_5O_2^-]}{[HC_7H_5O_2]} = \frac{(x)(x)}{(0.225 - x)} = 6.5 \times 10^{-5}$$

Assume that x is small compared to 0.225.

$x^2 = (6.5 \times 10^{-5})(0.225) \quad x = 0.003\underline{8}2$

$$\% \text{ ionization} = \frac{0.003\underline{8}2}{0.225} \times 100\% = 1.7\% \text{ ionized}$$

Check: The units (%) are correct. The magnitude of the answer makes sense, since we have a weak acid (small K).

17.75 (a) **Given:** 1.00 M $HC_2H_3O_2$, $K_a = 1.8 \times 10^{-5}$ **Find:** % ionization

Conceptual Plan: Write a balanced reaction. Prepare an ICE table, represent the change with x, sum the table, determine the equilibrium values, put the equilibrium values in the equilibrium expression, solve for x, and then $x \rightarrow$ % ionization.

$$\% \text{ ionization} = \frac{x}{[HC_2H_3O_2]_{\text{original}}} \times 100\%$$

Solution: $HC_2H_3O_2(aq) + H_2O(l) \rightleftharpoons H_3O^+(aq) + C_2H_3O_2^-(aq)$

	$HC_2H_3O_2$	H_3O^+	$C_2H_3O_2^-$
Initial	1.00	0.0	0.0
Change	$-x$	$+x$	$+x$
Equil	$1.00 - x$	x	x

$$K_a = \frac{[H_3O^+][C_2H_3O_2^-]}{[HC_2H_3O_2]} = \frac{(x)(x)}{(1.00 - x)} = 1.8 \times 10^{-5}$$

Assume that x is small compared to 1.00.

$x^2 = (1.8 \times 10^{-5})(1.00); \quad x = 0.004\underline{2}4$

$$\% \text{ ionization} = \frac{0.004\underline{2}4}{1.00} \times 100\% = 0.42\% \text{ ionized}$$

Check: The units (%) are correct. The magnitude of the answer makes sense, since we have a weak acid (small K).

(b) **Given:** 0.500 M $HC_2H_3O_2$, $K_a = 1.8 \times 10^{-5}$ **Find:** % ionization

Conceptual Plan: Write a balanced reaction. Prepare an ICE table, represent the change with x, sum the table, determine the equilibrium values, put the equilibrium values in the equilibrium expression, solve for x, and then $x \rightarrow$ % ionization.

$$\% \text{ ionization} = \frac{x}{[HC_2H_3O_2]_{\text{original}}} \times 100\%$$

Solution: $HC_2H_3O_2(aq) + H_2O(l) \rightleftharpoons H_3O^+(aq) + C_2H_3O_2^-(aq)$

	$HC_2H_3O_2$	H_3O^+	$C_2H_3O_2^-$
Initial	0.500	0.0	0.0
Change	$-x$	$+x$	$+x$
Equil	$0.500 - x$	x	x

$$K_a = \frac{[H_3O^+][C_2H_3O_2^-]}{[HC_2H_3O_2]} = \frac{(x)(x)}{(0.500 - x)} = 1.8 \times 10^{-5}$$

Assume that x is small compared to 0.500.

$x^2 = (1.8 \times 10^{-5})(0.500) \quad x = 0.003\underline{0}0$

$$\% \text{ ionization} = \frac{0.003\underline{0}0}{0.500} \times 100\% = 0.60\% \text{ ionized}$$

Check: The units (%) are correct. The magnitude of the answer makes sense, since we have a weak acid (small K).

(c) **Given:** 0.100 M $HC_2H_3O_2$, $K_a = 1.8 \times 10^{-5}$ **Find:** % ionization
Conceptual Plan: Write a balanced reaction. Prepare an ICE table, represent the change with x, sum the table, determine the equilibrium values, put the equilibrium values in the equilibrium expression, solve for x, and then $x \rightarrow$ % ionization.

$$\% \text{ ionization} = \frac{x}{[HC_2H_3O_2]_{\text{original}}} \times 100\%$$

Solution:	$HC_2H_3O_2(aq) + H_2O(l) \rightleftharpoons$	$H_3O^+(aq)$	$+ C_2H_3O_2^-(aq)$
Initial	0.100	0.0	0.0
Change	$-x$	$+x$	$+x$
Equil	$0.100 - x$	x	x

$$K_a = \frac{[H_3O^+][C_2H_3O_2^-]}{[HC_2H_3O_2]} = \frac{(x)(x)}{(0.100 - x)} = 1.8 \times 10^{-5}$$

Assume that x is small compared to 0.100.

$x^2 = (1.8 \times 10^{-5})(0.100) \quad x = 0.00134$

$$\% \text{ ionization} = \frac{0.00134}{0.100} \times 100\% = 1.3\% \text{ ionized}$$

Check: The units (%) are correct. The magnitude of the answer makes sense, since we have a weak acid (small K).

(d) **Given:** 0.0500 M $HC_2H_3O_2$, $K_a = 1.8 \times 10^{-5}$ **Find:** % ionization
Conceptual Plan: Write a balanced reaction. Prepare an ICE table, represent the change with x, sum the table, determine the equilibrium values, put the equilibrium values in the equilibrium expression, solve for x, and then $x \rightarrow$ % ionization.

$$\% \text{ ionization} = \frac{x}{[HC_2H_3O_2]_{\text{original}}} \times 100\%$$

Solution:	$HC_2H_3O_2(aq) + H_2O(l) \rightleftharpoons$	$H_3O^+(aq)$	$+ C_2H_3O_2^-(aq)$
Initial	0.0500	0.0	0.0
Change	$-x$	$+x$	$+x$
Equil	$0.0500 - x$	x	x

$$K_a = \frac{[H_3O^+][C_2H_3O_2^-]}{[HC_2H_3O_2]} = \frac{(x)(x)}{(0.0500 - x)} = 1.8 \times 10^{-5}$$

Assume that x is small compared to 0.0500.

$x^2 = (1.8 \times 10^{-5})(0.0500); \quad x = 9.49 \times 10^{-4}$

$$\% \text{ ionization} = \frac{9.49 \times 10^{-4}}{0.0500} \times 100\% = 1.9\% \text{ ionized}$$

Check: The units (%) are correct. The magnitude of the answer makes sense, since we have a weak acid (small K).

17.76 (a) **Given:** 1.00 M $HCHO_2$, $K_a = 1.8 \times 10^{-4}$ **Find:** % ionization
Conceptual Plan: Write a balanced reaction. Prepare an ICE table, represent the change with x, sum the table, determine the equilibrium values, put the equilibrium values in the equilibrium expression, solve for x, and then $x \rightarrow$ % ionization.

$$\% \text{ ionization} = \frac{x}{[HCHO_2]_{\text{original}}} \times 100\%$$

Solution: $HCHO_2(aq) + H_2O(l) \rightleftharpoons H_3O^+(aq) + CHO_2^-(aq)$

	$HCHO_2(aq)$	$H_3O^+(aq)$	$CHO_2^-(aq)$
Initial	1.00	0.0	0.0
Change	$-x$	$+x$	$+x$
Equil	$1.00 - x$	x	x

$$K_a = \frac{[H_3O^+][CHO_2^-]}{[HCHO_2]} = \frac{(x)(x)}{(1.00 - x)} = 1.8 \times 10^{-4}$$

Assume that x is small compared to 1.00.

$x^2 = (1.8 \times 10^{-4})(1.00) \quad x = 0.01\underline{3}4$

$$\% \text{ ionization} = \frac{0.01\underline{3}4}{1.00} \times 100\% = 1.3\% \text{ ionized}$$

Check: The units (%) are correct. The magnitude of the answer makes sense, since we have a weak acid (small K).

(b) **Given:** 0.500 M $HCHO_2$, $K_a = 1.8 \times 10^{-4}$ **Find:** % ionization

Conceptual Plan: Write a balanced reaction. Prepare an ICE table, represent the change with x, sum the table, determine the equilibrium values, put the equilibrium values in the equilibrium expression, solve for x, and then $x \rightarrow$ % ionization.

$$\% \text{ ionization} = \frac{x}{[HCHO_2]_{\text{original}}} \times 100\%$$

Solution: $HCHO_2(aq) + H_2O(l) \rightleftharpoons H_3O^+(aq) + CHO_2^-(aq)$

	$HCHO_2(aq)$	$H_3O^+(aq)$	$CHO_2^-(aq)$
Initial	0.500	0.0	0.0
Change	$-x$	$+x$	$+x$
Equil	$0.500 - x$	x	x

$$K_a = \frac{[H_3O^+][CHO_2^-]}{[HCHO_2]} = \frac{(x)(x)}{(0.500 - x)} = 1.8 \times 10^{-4}$$

Assume that x is small compared to 1.00.

$x^2 = (1.8 \times 10^{-4})(0.500) \quad x = 0.009\underline{4}9$

$$\% \text{ ionization} = \frac{0.009\underline{4}9}{0.500} \times 100\% = 1.9\% \text{ ionized}$$

Check: The units (%) are correct. The magnitude of the answer makes sense, since we have a weak acid (small K).

(c) **Given:** 0.100 M $HCHO_2$, $K_a = 1.8 \times 10^{-4}$ **Find:** % ionization

Conceptual Plan: Write a balanced reaction. Prepare an ICE table, represent the change with x, sum the table, determine the equilibrium values, put the equilibrium values in the equilibrium expression, solve for x, and then $x \rightarrow$ % ionization.

$$\% \text{ ionization} = \frac{x}{[HCHO_2]_{\text{original}}} \times 100\%$$

Solution: $HCHO_2(aq) + H_2O(l) \rightleftharpoons H_3O^+(aq) + CHO_2^-(aq)$

	$HCHO_2(aq)$	$H_3O^+(aq)$	$CHO_2^-(aq)$
Initial	0.100	0.0	0.0
Change	$-x$	$+x$	$+x$
Equil	$0.100 - x$	x	x

$$K_a = \frac{[H_3O^+][CHO_2^-]}{[HCHO_2]} = \frac{(x)(x)}{(0.100 - x)} = 1.8 \times 10^{-4}$$

Assume that x is small compared to 0.100.

$x^2 = (1.8 \times 10^{-4})(0.100) \quad x = 0.004\underline{2}4$

$$\% \text{ ionization} = \frac{0.004\underline{2}4}{0.100} \times 100\% = 4.2\% \text{ ionized}$$

Check: The units (%) are correct. The magnitude of the answer makes sense, since we have a weak acid (small K).

(d) **Given:** 0.0500 M $HCHO_2$, $K_a = 1.8 \times 10^{-4}$ **Find:** % ionization

Conceptual Plan: Write a balanced reaction. Prepare an ICE table, represent the change with x, sum the table, determine the equilibrium values, put the equilibrium values in the equilibrium expression, solve for x, and then $x \rightarrow$ % ionization.

$$\% \text{ ionization} = \frac{x}{[HCHO_2]_{original}} \times 100\%$$

Solution:

	$HCHO_2(aq)$ + $H_2O(l)$ $\rightleftharpoons$	$H_3O^+(aq)$ +	$CHO_2^-(aq)$
Initial	0.0500	0.0	0.0
Change	$-x$	$+x$	$+x$
Equil	$0.0500 - x$	x	x

$$K_a = \frac{[H_3O^+][CHO_2^-]}{[HCHO_2]} = \frac{(x)(x)}{(0.0500 - x)} = 1.8 \times 10^{-4}$$

Assume that x is small compared to 0.0500.

$x^2 = (1.8 \times 10^{-4})(0.0500) \quad x = 0.00300$

$$\% \text{ ionization} = \frac{0.00300}{0.0500} \times 100\% = 6.0\% \text{ ionized}$$

x is small assumption is invalid because 6.0% is greater than the 5.0% limit.

$x^2 + 1.8 \times 10^{-4}x - 9.0 \times 10^{-6} = 0$

Solve for x using the quadratic equation. $x = 0.00291$

$$\% \text{ ionization} = \frac{0.00291}{0.0500} \times 100\% = 5.8\%$$

Check: The units (%) are correct. The magnitude of the answer makes sense, since we have a moderately weak acid (moderately small K).

17.77 Given: 0.148 M HA, 1.55% dissociation **Find:** K_a

Conceptual Plan: M $\rightarrow [H_3O^+] \rightarrow K_a$ and then write a balanced reaction, determine equilibrium concentration, and plug into the equilibrium expression.

Solution: (0.148 M HA)(0.0155) = 0.002294; $[H_3O^+] = [A^-]$

	$HA(aq)$ + $H_2O(l)$ $\rightleftharpoons$	$H_3O^+(aq)$ +	$A^-(aq)$
Initial	0.148	0.0	0.0
Change	$-x$	$+x$	$+x$
Equil	0.148 − 0.002294	0.002294	0.002294

$$K_a = \frac{[H_3O^+][A^-]}{[HA]} = \frac{(0.002294)(0.002294)}{(0.148 - 0.002294)} = 3.61 \times 10^{-5}$$

Check: The units (none) are correct. The magnitude of the answer makes sense, since we have a low amount of the acid that dissociates. Thus, we have a weak acid with a small K.

17.78 **Given:** 0.085 M HA, 0.59% dissociation **Find:** K_a

Conceptual Plan: M $\rightarrow [H_3O^+] \rightarrow K_a$ and then write a balanced reaction, determine equilibrium concentration, and plug into the equilibrium expression.

Solution: (0.085 M HA)(0.0059) = 5.02×10^{-4}; $[H_3O^+] = [A^-]$

	$HA(aq)$ + $H_2O(l)$ $\rightleftharpoons$	$H_3O^+(aq)$ +	$A^-(aq)$
Initial	0.085	0.0	0.0
Change	$-x$	$+x$	$+x$
Equil	$0.085 - 5.02 \times 10^{-4}$	5.02×10^{-4}	5.02×10^{-4}

$$K_a = \frac{[H_3O^+][A^-]}{[HA]} = \frac{(5.02 \times 10^{-4})(5.02 \times 10^{-4})}{(0.085 - 5.02 \times 10^{-4})} = 3.0 \times 10^{-6}$$

Check: The units (none) are correct. The magnitude of the answer makes sense, since we have a low amount of the acid that dissociates. Thus, we have a weak acid with a small K.

17.79 (a) **Given:** 0.250 M HF, $K_a = 3.5 \times 10^{-4}$ **Find:** pH, % dissociation

Conceptual Plan: Write a balanced reaction. Prepare an ICE table, represent the change with x, sum the table, determine the equilibrium values, put the equilibrium values in the equilibrium expression, solve for x, and then $x \rightarrow$ % ionization.

$$\% \text{ ionization} = \frac{x}{[\text{HF}]_{\text{original}}} \times 100\%$$

Solution:	$HF(aq)$ +	$H_2O(l) \rightleftharpoons$	$H_3O^+(aq)$ +	$F^-(aq)$
Initial	0.250		0.0	0.0
Change	$-x$		$+x$	$+x$
Equil	$0.250 - x$		x	x

$$K_a = \frac{[H_3O^+][F^-]}{[HF]} = \frac{(x)(x)}{(0.250 - x)} = 3.5 \times 10^{-4}$$

Assume that x is small compared to 0.250.

$x^2 = (3.5 \times 10^{-4})(0.250)$ $x = 0.00935 \text{ M} = [H_3O^+]$

$$\frac{0.00935}{0.250} \times 100\% = 3.7\%$$

$\text{pH} = -\log(0.00935) = 2.03$

Check: The units (% and none) are correct. The magnitude of the answers makes sense. We have a weak acid with a small K so we expect a pH > $-\log(0.250) = 0.602$ and a small percent dissociation.

(b) **Given:** 0.100 M HF, $K_a = 3.5 \times 10^{-4}$ **Find:** pH, % dissociation

Conceptual Plan: Write a balanced reaction. Prepare an ICE table, represent the change with x, sum the table, determine the equilibrium values, put the equilibrium values in the equilibrium expression, solve for x, and then $x \rightarrow$ % ionization.

$$\% \text{ ionization} = \frac{x}{[\text{HF}]_{\text{original}}} \times 100\%$$

Solution:	$HF(aq)$ +	$H_2O(l) \rightleftharpoons$	$H_3O^+(aq)$ +	$F^-(aq)$
Initial	0.100		0.0	0.0
Change	$-x$		$+x$	$+x$
Equil	$0.100 - x$		x	x

$$K_a = \frac{[H_3O^+][F^-]}{[HF]} = \frac{(x)(x)}{(0.100 - x)} = 3.5 \times 10^{-4}$$

Assume that x is small compared to 0.100.

$x^2 = (3.5 \times 10^{-4})(0.100);$ $x = 0.00592 \text{ M} = [H_3O^+]$

$$\frac{0.00592}{0.100} \times 100\% = 5.9\%$$

$x > 5.0\%$ Therefore, assumption is invalid; solve using the quadratic equation.

$x^2 + 3.5 \times 10^{-4}x - 3.5 \times 10^{-5} = 0$

$x = 0.00574$ or $-$ 0.00609

$\text{pH} = -\log(0.00574) = 2.24$

$$\% \text{ dissociation} = \frac{0.00574}{0.100} \times 100\% = 5.7\%$$

Check: The units (% and none) are correct. The magnitude of the answers makes sense. We have a weak acid with a small K so we expect a pH > $-\log(0.100) = 1.000$ and a small percent dissociation.

(c) **Given:** 0.050 M HF, $K_a = 3.5 \times 10^{-4}$ **Find:** pH, % dissociation

Conceptual Plan: Write a balanced reaction. Prepare an ICE table, represent the change with x, sum the table, determine the equilibrium values, put the equilibrium values in the equilibrium expression, solve for x, and then $x \rightarrow$ % ionization.

$$\% \text{ ionization} = \frac{x}{[\text{HF}]_{\text{original}}} \times 100\%$$

Solution: $HF(aq) + H_2O(l) \rightleftharpoons H_3O^+(aq) + F^-(aq)$

	HF	H_3O^+	F^-
Initial	0.050	0.0	0.0
Change	$-x$	$+x$	$+x$
Equil	$0.050 - x$	x	x

$$K_a = \frac{[H_3O^+][F^-]}{[HF]} = \frac{(x)(x)}{(0.050 - x)} = 3.5 \times 10^{-4}$$

Assume that x is small compared to 0.050.

$x^2 = (3.5 \times 10^{-4})(0.050)$; $\quad x = 0.00418 \text{ M} = [H_3O^+]$

$$\frac{0.00418}{0.050} \times 100\% = 8.4\%$$

Assumption is invalid; solve using the quadratic equation.

$x^2 + 3.5 \times 10^{-4}x - 1.75 \times 10^{-5} = 0$

$x = 0.00401$ or -0.00436

$\text{pH} = -\log(0.00401) = 2.40$

$$\% \text{ dissociation} = \frac{0.00401}{0.050} \times 100\% = 8.0\%$$

Check: The units (% and none) are correct. The magnitude of the answers makes sense. We have a weak acid with a small K so we expect a pH $> -\log(0.050) = 1.30$ and a small percent dissociation.

17.80 (a) **Given:** 0.100 M HA, $K_a = 1.0 \times 10^{-5}$ **Find:** pH, % dissociation

Conceptual Plan: Write a balanced reaction. Prepare an ICE table, represent the change with x, sum the table, determine the equilibrium values, put the equilibrium values in the equilibrium expression, solve for x, and then $x \rightarrow$ % ionization.

$$\% \text{ ionization} = \frac{x}{[\text{HA}]_{\text{original}}} \times 100\%$$

Solution: $HA(aq) + H_2O(l) \rightleftharpoons H_3O^+(aq) + A^-(aq)$

	HA	H_3O^+	A^-
Initial	0.100	0.0	0.0
Change	$-x$	$+x$	$+x$
Equil	$0.100 - x$	x	x

$$K_a = \frac{[H_3O^+][A^-]}{[HA]} = \frac{(x)(x)}{(0.100 - x)} = 1.0 \times 10^{-5}$$

Assume that x is small compared to 0.100.

$x^2 = (1.0 \times 10^{-5})(0.100) \quad x = 0.00100 \text{ M} = [H_3O^+]$

$$\frac{0.00100}{0.100} \times 100\% = 1.0\%$$

$\text{pH} = -\log(0.00100) = 3.00$

Check: The units (none and %) are correct. The magnitude of the answers makes sense. We have a weak acid with a small K so we expect a pH $> -\log(0.100) = 1.000$ and a small percent dissociation.

(b) **Given:** 0.100 M HA, $K_a = 1.0 \times 10^{-3}$ **Find:** pH, % dissociation

Conceptual Plan: Write a balanced reaction. Prepare an ICE table, represent the change with x, sum the table, determine the equilibrium values, put the equilibrium values in the equilibrium expression, solve for x, and then $x \rightarrow$ % ionization.

$$\% \text{ ionization} = \frac{x}{[\text{HA}]_{\text{original}}} \times 100\%$$

Solution:	$HA(aq) + H_2O(l) \rightleftharpoons$	$H_3O^+(aq)$	$+ A^-(aq)$
Initial	0.100	0.0	0.0
Change	$-x$	$+x$	$+x$
Equil	$0.100 - x$	x	x

$K_a = \frac{[H_3O^+][A^-]}{[HA]} = \frac{(x)(x)}{(0.100 - x)} = 1.0 \times 10^{-3}$ Solve using the quadratic equation.

$x^2 = (1.0 \times 10^{-3} - x)(0.100)$; $x^2 + 1 \times 10^{-3}x - 1 \times 10^{-4} = 0$
$x = 0.009\underline{5}\text{ M} = [H_3O^+]$

$\frac{0.009\underline{5}}{0.100} \times 100\% = 9.5\%$

$pH = -\log(0.009\underline{5}) = 2.02$

Check: The units (none and %) are correct. The magnitude of the answers makes sense. We have a weak acid with a small K so we expect a pH $> -\log(0.100) = 1.000$ and a small percent dissociation.

(c) **Given:** 0.100 M HA, $K_a = 1.0 \times 10^{-1}$ **Find:** pH, % dissociation
Conceptual Plan: Write a balanced reaction. Prepare an ICE table, represent the change with x, sum the table, determine the equilibrium values, put the equilibrium values in the equilibrium expression, solve for x, and then $x \rightarrow$ % ionization.

$\% \text{ ionization} = \frac{x}{[HA]_{original}} \times 100\%$

Solution:	$HA(aq) + H_2O(l) \rightleftharpoons$	$H_3O^+(aq)$	$+ A^-(aq)$
Initial	0.100	0.0	0.0
Change	$-x$	$+x$	$+x$
Equil	$0.100 - x$	x	x

$K_a = \frac{[H_3O^+][A^-]}{[HA]} = \frac{(x)(x)}{(0.100 - x)} = 1.0 \times 10^{-1}$

$x^2 = (1.0 \times 10^{-1} - x)(0.100)$ $x^2 + 0.10x - 0.01 = 0$ Solve using the quadratic equation.
$x = 0.06\underline{1}8\text{ M} = [H_3O^+]$

$\frac{0.06\underline{1}8}{0.100} \times 100\% = 61.8\%$

$pH = -\log(0.06\underline{1}8) = 1.21$

Check: The units (none and %) are correct. The magnitude of the answers makes sense. We have a fairly strong weak acid with a K so we expect a pH just above $-\log(0.100) = 1.000$ and a significant percent dissociation.

17.81 (a) **Given:** 0.115 M HBr (strong acid), 0.125 M $HCHO_2$ (weak acid) **Find:** pH
Conceptual Plan: Because the mixture is a strong acid and a weak acid, the strong acid will dominate. Use the concentration of the strong acid to determine $[H_3O^+]$ and then pH.
Solution: 0.115 M HBr = 0.115 M $[H_3O^+]$ $pH = -\log(0.115) = 0.939$

(b) **Given:** 0.150 M HNO_2 (weak acid), 0.085 M HNO_3 (strong acid) **Find:** pH
Conceptual Plan: Because the mixture is a strong acid and a weak acid, the strong acid will dominate. Use the concentration of the strong acid to determine $[H_3O^+]$ and then pH.
Solution: 0.085 M HNO_3 = 0.085 M $[H_3O^+]$ $pH = -\log(0.085) = 1.07$

(c) **Given:** 0.185 M $HCHO_2$, $K_a = 1.8 \times 10^{-4}$; 0.225 M $HC_2H_3O_2$, $K_a = 1.8 \times 10^{-5}$ **Find:** pH
Conceptual Plan: Because the mixture is a weak acid and a weak acid and the K values are only 10^1 apart, you need to find $[H_3O^+]$ from each reaction. Write a balanced reaction. Prepare an ICE table, represent the change with x, sum the table, determine the equilibrium values, put the equilibrium values in the equilibrium expression, and solve for x.

Solution: $HCHO_2(aq) + H_2O(l) \rightleftharpoons H_3O^+(aq) + CHO_2^-(aq)$

	$HCHO_2$	H_3O^+	CHO_2^-
Initial	0.185	0.0	0.0
Change	$-x$	$+x$	$+x$
Equil	$0.185 - x$	x	x

$$K_a = \frac{[H_3O^+][CHO_2^-]}{[HCHO_2]} = \frac{(x)(x)}{(0.185 - x)} = 1.8 \times 10^{-4}$$

Assume that x is small compared to 0.100.

$x^2 = (1.8 \times 10^{-4})(0.185)$; $x = 0.0057706$

$HC_2H_3O_2(aq) + H_2O(l) \rightleftharpoons H_3O^+(aq) + C_2H_3O_2^-(aq)$

	$HC_2H_3O_2$	H_3O^+	$C_2H_3O_2^-$
Initial	0.225	0.0057706	0.0
Change	$-x$	$+x$	$+x$
Equil	$0.225 - x$	$0.0057706 + x$	x

$$K_a = \frac{[H_3O^+][C_2H_3O_2^-]}{[HC_2H_3O_2]} = \frac{(0.0057706 + x)(x)}{(0.225 - x)} = 1.8 \times 10^{-5}$$

Assume that x is small compared to 0.225.

$x^2 + 0.0057706x - 4.05 \times 10^{-6} = 0$; $x = 0.0063250$

$[H_3O^+] = 0.0057706 + x = 0.0057706 + 0.0063250 = 0.0064031$ M

$pH = -\log(0.0064031) = 2.19$

Check: The units (none) are correct. The magnitude of the answer makes sense. We have weak acids so we expect a pH $> -\log(0.185 + 0.225) = 0.387$.

(d) **Given:** 0.050 M $HC_2H_3O_2$, $K_a = 1.8 \times 10^{-5}$; 0.050 M HCN, $K_a = 4.9 \times 10^{-10}$ **Find:** pH

Conceptual Plan: Because the values of K are more than 10^1 apart, the acid with the larger K will dominate the reaction. Write a balanced reaction. Prepare an ICE table, represent the change with x, sum the table, determine the equilibrium values, put the equilibrium values in the equilibrium expression, and solve for x.

Solution: $HC_2H_3O_2(aq) + H_2O(l) \rightleftharpoons H_3O^+(aq) + C_2H_3O_2^-(aq)$

	$HC_2H_3O_2$	H_3O^+	$C_2H_3O_2^-$
Initial	0.0500	0.0	0.0
Change	$-x$	$+x$	$+x$
Equil	$0.0500 - x$	x	x

$$K_a = \frac{[H_3O^+][C_2H_3O_2^-]}{[HC_2H_3O_2]} = \frac{(x)(x)}{(0.0500 - x)} = 1.8 \times 10^{-5}$$

Assume that x is small compared to 0.0500.

$x^2 = (1.8 \times 10^{-5})(0.0500)$ $x = 9.49 \times 10^{-4}$

$pH = -\log(9.49 \times 10^{-4}) = 3.02$

Check: The units (none) are correct. The magnitude of the answer makes sense. We have weak acids so we expect a pH $> -\log(0.050 + 0.050) = 1.000$.

17.82 (a) **Given:** 0.075 M HNO_3 (strong acid), 0.175 M $HC_7H_5O_2$ (weak acid) **Find:** pH

Conceptual Plan: Because the mixture is a strong acid and a weak acid, the strong acid will dominate. Use the concentration of the strong acid to determine $[H_3O^+]$ and then pH.

Solution: 0.075 M HNO_3 = 0.075 M $[H_3O^+]$; $pH = -\log(0.075) = 1.12$

(b) **Given**: 0.020 M HBr (strong acid), 0.015 M $HClO_4$ (strong acid) **Find:** pH

Conceptual Plan: Because the mixture is a strong acid and another strong acid, the $[H_3O^+]$ is the sum of the concentration of both acids. Determine pH.

Solution: 0.020 M HBr = 0.020 $[H_3O^+]$, 0.015 M $HClO_4$ = 0.015 M $[H_3O^+]$

$[H_3O^+] = 0.020 + 0.015 = 0.035$ M; $pH = -\log(0.035) = 1.46$

(c) **Given:** 0.095M HF, $K_a = 3.5 \times 10^{-4}$; 0.225 M $HC_6H_5O_2$, $K_a = 1.3 \times 10^{-10}$ **Find:** pH

Conceptual Plan: Because the values of *K* are more than 10^1 apart, the acid with the larger *K* will dominate the reaction. Write a balanced reaction. Prepare an ICE table, represent the change with *x*, sum the table, determine the equilibrium values, put the equilibrium values in the equilibrium expression, and solve for *x*.

Solution: $HF(aq) + H_2O(l) \rightleftharpoons H_3O^+(aq) + F^-(aq)$

Initial	0.095	0.0	0.0
Change	$-x$	$+x$	$+x$
Equil	$0.095 - x$	x	x

$$K_a = \frac{[H_3O^+][F^-]}{[HF]} = \frac{(x)(x)}{(0.095 - x)} = 3.5 \times 10^{-4}$$

$x^2 = (3.5 \times 10^{-4})(0.095) \quad x = 0.005\underline{7}7 \text{ M} = [H_3O^+]$

$(0.005\underline{7}7/0.095) \times 100 = 6.1\%$ Assumption is invalid because 6.1% is greater than the 5.0% limit.

Solve for x using the quadratic formula.

$x^2 + 3.5 \times 10^{-4}x - 3.325 \times 10^{-5} = 0$

$x = 0.005\underline{5}9 \text{ M} = [H_3O^+]$

$\text{pH} = -\log(0.005\underline{5}9) = 2.25$

Check: The units (none) are correct. The magnitude of the answer makes sense. We have weak acids so we expect a pH > $-\log(0.095 + 0.225) = 0.495$.

(d) **Given:** 0.100 M $HCHO_2$, $K_a = 1.8 \times 10^{-4}$; 0.050 M HClO, $K_a = 2.9 \times 10^{-8}$ **Find:** pH

Conceptual Plan: Because the values of *K* are more than 10^1 apart, the acid with the larger *K* will dominate the reaction. Write a balanced reaction. Prepare an ICE table, represent the change with *x*, sum the table, determine the equilibrium values, put the equilibrium values in the equilibrium expression, and solve for *x*.

Solution: $HCHO_2(aq) + H_2O(l) \rightleftharpoons H_3O^+(aq) + C_2H_3O_2^-(aq)$

Initial	0.100	0.0	0.0
Change	$-x$	$+x$	$+x$
Equil	$0.100 - x$	x	x

$$K_a = \frac{[H_3O^+][CHO_2^-]}{[HCHO_2]} = \frac{(x)(x)}{(0.100 - x)} = 1.8 \times 10^{-4}$$

Assume that x is small compared to 0.0500.

$x^2 = (1.8 \times 10^{-4})(0.100) \; x = 0.004\underline{2}4$

$\text{pH} = -\log(0.004\underline{2}4) = 2.37$

Check: The units (none) are correct. The magnitude of the answer makes sense. We have weak acids so we expect a pH > $-\log(0.100 + 0.050) = 0.824$.

Base Solutions

17.83 (a) **Given:** 0.15 M NaOH **Find:** $[OH^-]$, $[H_3O^+]$, pH, pOH

Conceptual Plan: $[NaOH] \rightarrow [OH^-] \rightarrow [H_3O^+] \rightarrow pH \rightarrow pOH$

$K_w = [H_3O^+][OH^-]$ $\quad \text{pH} = -\log[H_3O^+]$ $\quad \text{pH} + \text{pOH} = 14$

Solution: $[OH^-] = [NaOH] = 0.15 \text{ M}$

$$[H_3O^+] = \frac{K_w}{[OH^-]} = \frac{1.0 \times 10^{-14}}{0.15 \text{ M}} = 6.7 \times 10^{-14} \text{ M}$$

$\text{pH} = -\log(6.7 \times 10^{-14}) = 13.17$

$\text{pOH} = 14.00 - 13.17 = 0.83$

(b) **Given:** 1.5×10^{-3} M $Ca(OH)_2$ **Find:** $[OH^-]$, $[H_3O^+]$, pH, pOH

Conceptual Plan: $[Ca(OH)_2] \rightarrow [OH^-] \rightarrow [H_3O^+] \rightarrow pH \rightarrow pOH$

$K_w = [H_3O^+][OH^-]$ $\quad \text{pH} = -\log[H_3O^+]$ $\quad \text{pH} + \text{pOH} = 14$

Solution: $[OH^-] = 2[Ca(OH)_2] = 2(1.5 \times 10^{-3}) = 0.0030 \text{ M}$

$$[H_3O^+] = \frac{K_w}{[OH^-]} = \frac{1.0 \times 10^{-14}}{0.0030\ M} = 3.\underline{3}3 \times 10^{-12}\ M$$

$pH = -\log(3.\underline{3}3 \times 10^{-12}) = 11.48$

$pOH = 14.00 - 11.48 = 2.52$

(c) **Given:** 4.8×10^{-4} M $Sr(OH)_2$ **Find:** $[OH^-]$, $[H_3O^+]$, pH, pOH

Conceptual Plan: $[Sr(OH)_2] \rightarrow [OH^-] \rightarrow [H_3O^+] \rightarrow pH \rightarrow pOH$

$K_w = [H_3O^+][OH^-]$ $pH = -\log[H_3O^+]$ $pH + pOH = 14$

Solution: $[OH^+] = [Sr(OH)_2] = 2(4.8 \times 10^{-4}) = 9.6 \times 10^{-4}\ M$

$$[H_3O^-] = \frac{K_w}{[OH^-]} = \frac{1.0 \times 10^{-14}}{9.6 \times 10^{-4}\ M} = 1.\underline{0}4 \times 10^{-11}\ M$$

$pH = -\log(1.\underline{0}4 \times 10^{-11}) = 10.98$

$pOH = 14.00 - 10.98 = 3.02$

Check: The units (M, M, none, and none) are correct. All of these solutions are bases so $[OH^-] > 10^{-7}$ M, $[H^+] < 10^{-7}$ M, pH > 7, and pOH < 7. The higher the base concentration, the lower the $[H^+]$, the higher the pH, and the lower the pOH.

(d) **Given:** 8.7×10^{-5} M KOH **Find:** $[OH^-]$, $[H_3O^+]$, pH, pOH

Conceptual Plan: $[KOH] \rightarrow [OH^-] \rightarrow [H_3O^+] \rightarrow pH \rightarrow pOH$

$K_w = [H_3O^+][OH^-]$ $pH = -\log[H_3O^+]$ $pH + pOH = 14$

Solution: $[OH^-] = [KOH] = 8.7 \times 10^{-5}\ M$

$$[H_3O^+] = \frac{K_w}{[OH^-]} = \frac{1.0 \times 10^{-14}}{8.7 \times 10^{-5}\ M} = 1.\underline{1} \times 10^{-10}\ M$$

$pH = -\log(1.\underline{1} \times 10^{-10}) = 9.96$

$pOH = 14.00 - 9.96 = 4.04$

Check: The units (M, M, none, and none) are correct. All of these solutions are bases so $[OH^-] > 10^{-7}$ M, $[H^+] < 10^{-7}$ M, pH > 7, and pOH < 7. The higher the base concentration, the lower the $[H^+]$, the higher the pH, and the lower the pOH.

17.84 (a) **Given:** 8.77×10^{-3} M LiOH **Find:** $[OH^-]$, $[H_3O^+]$, pH, pOH

Conceptual Plan: $[LiOH] \rightarrow [OH^-] \rightarrow [H_3O^+] \rightarrow pH \rightarrow pOH$

$K_w = [H_3O^+][OH^-]$ $pH = -\log[H_3O^+]$ $pH + pOH = 14$

Solution: $[OH^-] = [LiOH] = 8.77 \times 10^{-3}\ M$

$$[H_3O^+] = \frac{K_w}{[OH^-]} = \frac{1.0 \times 10^{-14}}{8.77 \times 10^{-3}\ M} = 1.1\underline{4}0 \times 10^{-12}\ M$$

$pH = -\log(1.1\underline{4}0 \times 10^{-12}) = 11.943$

$pOH = 14.00 - 11.943 = 2.057$

(b) **Given:** 0.0112 M $Ba(OH)_2$ **Find:** $[OH^-]$, $[H_3O^+]$, pH, pOH

Conceptual Plan: $[Ba(OH)_2] \rightarrow [OH^-] \rightarrow [H_3O^+] \rightarrow pH \rightarrow pOH$

$K_w = [H_3O^+][OH^-]$ $pH = -\log[H_3O^+]$ $pH + pOH = 14$

Solution: $[OH^-] = 2[Ba(OH)_2] = 2(0.0112) = 0.0224\ M$

$$[H_3O^+] = \frac{K_w}{[OH^-]} = \frac{1.0 \times 10^{-14}}{0.0224\ M} = 4.4\underline{6}4 \times 10^{-13}\ M$$

$pH = -\log(4.4\underline{6}4 \times 10^{-13}) = 12.350$

$pOH = 14.000 - 12.350 = 1.650$

(c) **Given:** 1.9×10^{-4} M KOH **Find:** $[OH^-]$, $[H_3O^+]$, pH, pOH

Conceptual Plan: $[KOH] \rightarrow [OH^-] \rightarrow [H_3O^+] \rightarrow pH \rightarrow pOH$

$K_w = [H_3O^+][OH^-]$ $pH = -\log[H_3O^+]$ $pH + pOH = 14$

Solution: $[OH^-] = [KOH] = 1.9 \times 10^{-4}\ M$

$[H_3O^+] = \frac{K_w}{[OH^-]} = \frac{1.0 \times 10^{-14}}{1.9 \times 10^{-4}\ M} = 5.\underline{2}6 \times 10^{-11}\ M$

$pH = -\log(5.\underline{2}6 \times 10^{-11}) = 10.28$

$pOH = 14.00 - 10.28 = 3.72$

(d) **Given:** 5.0×10^{-4} M $Ca(OH)_2$ **Find:** $[OH^-]$, $[H_3O^+]$, pH, pOH

Conceptual Plan: $[Ca(OH)_2] \rightarrow [OH^-] \rightarrow [H_3O^+] \rightarrow pH \rightarrow pOH$

$K_w = [H_3O^+][OH^-]$ $pH = -\log[H_3O^+]$ $pH + pOH = 14$

Solution: $[OH^-] = [Ca(OH)_2] = 2(5.0 \times 10^{-4}) = 0.0010\ M$

$[H_3O^+] = \frac{K_w}{[OH^-]} = \frac{1.0 \times 10^{-14}}{0.0010\ M} = 1.\underline{0}0 \times 10^{-11}\ M$

$pH = -\log(1.\underline{0}0 \times 10^{-11}) = 11.00$

$pOH = 14.00 - 11.00 = 3.00$

Check: The units (M, M, none, and none) are correct. All of these solutions are bases so $[OH^-] > 10^{-7}$ M, $[H^+] < 10^{-7}$ M, pH > 7, and pOH < 7. The higher the base concentration, the lower the $[H^+]$, the higher the pH, and the lower the pOH.

17.85 **Given:** 3.85% KOH by mass, $d = 1.01$ g/mL **Find:** pH

Conceptual Plan:

% mass → g KOH → mol KOH and mass soln → mL soln → L soln → M KOH → $[OH^+]$

$\frac{1\ mol\ KOH}{56.11\ g\ KOH}$ $\frac{1.01\ g\ soln}{1\ mL\ soln}$ $\frac{1000\ mL\ soln}{1\ L\ soln}$ $\frac{mol\ KOH}{L\ soln}$

→ pOH → pH

$pOH = -\log[OH^-]$ $pH + pOH = 14$

Solution: $\frac{3.85\ g\ KOH}{100.0\ g\ soln} \times \frac{1\ mol\ KOH}{56.11\ g\ KOH} \times \frac{1.01\ g\ soln}{1\ mL\ soln} \times \frac{1000\ mL\ soln}{1\ L\ soln} = 0.69\underline{3}0\ M\ KOH$

$[OH^-] = [KOH] = 0.69\underline{3}0\ M$ $pOH = -\log(0.69\underline{3}0) = 0.159$; $pH = 14.000 - 0.159 = 13.841$

Check: The units (none) are correct. This is a solution so $[OH^-] > 10^{-7}$ M, pH > 7, and pOH < 7. The concentration is fairly high, so we expect a high pH.

17.86 **Given:** 1.55% NaOH by mass, $d = 1.01$ g/mL **Find:** pH

Conceptual Plan:

% mass → g NaOH → mol NaOH and mass soln → mL soln → L soln → M NaOH → $[OH^-]$

$\frac{1\ mol\ NaOH}{40.00\ g\ NaOH}$ $\frac{1.01\ g\ soln}{1\ mL\ soln}$ $\frac{1000\ mL\ soln}{1\ L\ soln}$ $\frac{mol\ NaOH}{L\ soln}$

→ pOH → pH

$pOH = -\log[OH^-]$ $pH + pOH = 14$

Solution: $\frac{1.55\ g\ NaOH}{100.0\ g\ soln} \times \frac{1\ mol\ NaOH}{40.00\ g\ NaOH} \times \frac{1.01\ g\ soln}{1\ mL\ soln} \times \frac{1000\ mL\ soln}{1\ L\ soln} = 0.39\underline{1}4\ M\ NaOH$

$[OH^-] = [NaOH] = 0.39\underline{1}4\ M$ $pOH = -\log(0.39\underline{1}4) = 0.407$ $pH = 14.000 - 0.407 = 13.593$

Check: The units (none) are correct. This is a solution so $[OH^-] > 10^{-7}$ M, pH > 7, and pOH < 7. The concentration is fairly high, so we expect a high pH.

17.87 **Given:** 3.55 L, pH = 12.4; 0.855 M KOH **Find:** Vol

Conceptual Plan: $pH \rightarrow [H_3O^+] \rightarrow [OH^-]$ **and then** $V_1M_1 = V_2M_2$

$[H_3O^+] = 10^{-pH}$ $1.0 \times 10^{-14} = [H_3O^+][OH^-]$ $V_1M_1 = V_2M_2$

Solution: $[H_3O^+] = 10^{-12.4} = 3.98 \times 10^{-13}$ $1.0 \times 10^{-14} = 3.98 \times 10^{-13}[OH^-]$

$[OH^-] = 0.025\underline{1}3\ M$

$V_1M_1 = V_2M_2$ $V_1(0.855\ M) = (3.55\ L)(0.0251\ M)$ $V_1 = 0.104\ L$

Check: The units (L) are correct. The magnitude of the answer (0.1 L) makes sense, since the concentration of the original solution is fairly high.

17.88 **Given:** 5.00 L, pH = 10.8; 15.0% NaOH, $d = 1.116$ g/mL **Find:** Vol

Conceptual Plan:

% mass → g NaOH → mol NaOH and mass soln → mL soln → L soln → M NaOH → $[OH^-]$

$\frac{1 \text{ mol NaOH}}{40.00 \text{ g NaOH}}$ $\frac{1.116 \text{ g soln}}{\text{mL soln}}$ $\frac{1000 \text{ mL soln}}{1 \text{ L soln}}$ $\frac{\text{mol NaOH}}{\text{L soln}}$

and then $V_1M_1 \rightarrow V_2M_2$

$V_1M_1 = V_2M_2$

Solution: $\frac{15.0 \text{ g NaOH}}{100.0 \text{ g soln}} \times \frac{1 \text{ mol NaOH}}{40.00 \text{ g NaOH}} \times \frac{1.116 \text{ g soln}}{1 \text{ mL soln}} \times \frac{1000 \text{ mL soln}}{1 \text{ L soln}} = 4.185 \text{ M NaOH}$

$[OH^-] = [NaOH] = 4.185$ M

$[H_3O^+] = 10^{-10.8} = 1.58 \times 10^{-11}$ M; $1.0 \times 10^{-14} = (1.58 \times 10^{-11})[OH^-]$

$[OH^-] = 6.33 \times 10^{-4}$ M

$V_1(4.19 \text{ M}) = (5.00 \text{ L})(6.33 \times 10^{-4} \text{ M})$ $V_1 = 7.55 \times 10^{-4} \text{ L} = 0.8 \text{ mL}$

Check: The units (mL) are correct. The magnitude of the answer (0.8 mL) makes sense since the concentration of the original solution is high.

17.89 (a) $NH_3(aq) + H_2O(l) \rightleftharpoons NH_4^+(aq) + OH^-(aq)$ $K_b = \frac{[NH_4^+][OH^-]}{[NH_3]}$

(b) $HCO_3^-(aq) + H_2O(l) \rightleftharpoons H_2CO_3(aq) + OH^-(aq)$ $K_b = \frac{[H_2CO_3][OH^-]}{[HCO_3^-]}$

(c) $CH_3NH_2(aq) + H_2O(l) \rightleftharpoons CH_3NH_3^+(aq) + OH^-(aq)$ $K_b = \frac{[CH_3NH_3^+][OH^-]}{[CH_3NH_2]}$

17.90 (a) $CO_3^{2-}(aq) + H_2O(l) \rightleftharpoons HCO_3^-(aq) + OH^-(aq)$ $K_b = \frac{[HCO_3^-][OH^-]}{[CO_3^{2-}]}$

(b) $C_6H_5NH_2(aq) + H_2O(l) \rightleftharpoons C_6H_5NH_3^+(aq) + OH^-(aq)$ $K_b = \frac{[C_6H_5NH_3^+][OH^-]}{[C_6H_5NH_2]}$

(c) $C_2H_5NH_2(aq) + H_2O(l) \rightleftharpoons C_2H_5NH_3^+(aq) + OH^-(aq)$ $K_b = \frac{[C_2H_5NH_3^+][OH^-]}{[C_2H_5NH_2]}$

17.91 **Given:** 0.15 M NH_3, $K_b = 1.76 \times 10^{-5}$ **Find:** $[OH^-]$, pH, pOH

Conceptual Plan: Write a balanced reaction. Prepare an ICE table, represent the change with x, sum the table, determine the equilibrium values, put the equilibrium values in the equilibrium expression, and solve for x.

$x = [OH^-] \rightarrow [pOH] \rightarrow pH$

$pOH = -\log[OH^-]$ $pH + pOH = 14$

Solution:	$NH_3(aq)$ + $H_2O(l)$	$\rightleftharpoons NH_4^+(aq)$	+ $OH^-(aq)$
Initial	0.15	0.0	0.0
Change	$-x$	$+x$	$+x$
Equil	$0.15 - x$	x	x

$K_b = \frac{[NH_4^+][OH^-]}{[NH_3]} = \frac{(x)(x)}{(0.15 - x)} = 1.76 \times 10^{-5}$

Assume that x is small.

$x^2 = (1.76 \times 10^{-5})(0.15)$ $x = [OH^-] = 0.00162$ M

pOH = $-\log(0.00162) = 2.79$

pH = $14.00 - 2.79 = 11.21$

Check: The units (M, none, and none) are correct. The magnitude of the pH makes sense. We have a weak base so we expect a pH $< 14 - \log(0.15) = 13.18$; $[OH^-] > 10^{-7}$ M, and pOH > 0.82.

17.92 **Given:** 0.125 M CO_3^{2-}, $K_b = 1.8 \times 10^{-4}$ **Find:** $[OH^-]$, pH, pOH

Conceptual Plan: Write a balanced reaction. Prepare an ICE table, represent the change with x, sum the table, determine the equilibrium values, put the equilibrium values in the equilibrium expression, and solve for x.

$x = [OH^-] \rightarrow [pOH] \rightarrow pH$

$pOH = -\log[OH^-]$ $pH + pOH = 14$

Solution: $CO_3^{2-}(aq) + H_2O(l) \rightleftharpoons HCO_3^-(aq) + OH^-(aq)$

	CO_3^{2-}	HCO_3^-	OH^-
Initial	0.125	0.0	0.0
Change	$-x$	$+x$	$+x$
Equil	$0.125 - x$	x	x

$$K_b = \frac{[HCO_3^-][OH^-]}{[CO_3^{2-}]} = \frac{(x)(x)}{(0.125 - x)} = 1.8 \times 10^{-4}$$

Assume that x is small.

$x^2 = (1.8 \times 10^{-4})(0.125)$ $x = [OH^-] = 0.004\underline{7}4$ M

$\frac{0.004\underline{7}4}{0.125} \times 100\% = 3.8\%$; assumption is valid.

$pOH = -\log(0.004\underline{7}4) = 2.32$

$pH = 14.00 - 2.32 = 11.68$

Check: The units (M, none, and none) are correct. The magnitude of the pH makes sense. We have a weak base so we expect a pH $< 14 - \log(0.125) = 13.10$; $[OH^-] > 10^{-7}$ M, and pOH > 0.90.

17.93 **Given:** $pK_b = 10.4$, 455 mg/L caffeine **Find:** pH

Conceptual Plan: $pK_b \rightarrow K_b$ and then mg/L $\rightarrow$ g/L $\rightarrow$ mol/L and then write a balanced reaction. Prepare an ICE table, represent the change with x, sum the table, determine the equilibrium values, put the equilibrium values in the equilibrium expression, and solve for x.

$x = [OH^-] \rightarrow [pOH] \rightarrow pH$

$pOH = -\log[OH^-]$ $pH + pOH = 14$

Solution: $K_b = 10^{-10.4} = \underline{3}.98 \times 10^{-11}$

$$\frac{455 \text{ mg caffeine}}{\text{L soln}} \times \frac{1 \text{ g caffeine}}{1000 \text{ mg caffeine}} \times \frac{1 \text{ mol caffeine}}{194.20 \text{ g}} = 0.0023\underline{4}3 \text{ M caffeine}$$

$C_8H_{10}N_4O_2(aq) + H_2O(l) \rightleftharpoons HC_8H_{10}N_4O_2^+(aq) + OH^-(aq)$

	$C_8H_{10}N_4O_2$	$HC_8H_{10}N_4O_2^+$	OH^-
Initial	0.0023\underline{4}3	0.0	0.0
Change	$-x$	$+x$	$+x$
Equil	$0.0023\underline{4}3 - x$	x	x

$$K_b = \frac{[HC_8H_{10}N_4O_2^+][OH^-]}{[C_8H_{10}N_4O_2]} = \frac{(x)(x)}{(0.0023\underline{4}3 - x)} = \underline{3}.98 \times 10^{-11}$$

Assume that x is small.

$x^2 = (\underline{3}.98 \times 10^{-11})(0.0023\underline{4}3)$ $x = [OH^-] = \underline{3}.05 \times 10^{-7}$ M

$\frac{\underline{3}.05 \times 10^{-7} \text{ M}}{0.0023\underline{4}3} \times 100\% = 0.013\%$; assumption is valid.

$pOH = -\log(\underline{3}.05 \times 10^{-7}) = 6.5$

$pH = 14.00 - 6.5 = 7.5$

Check: The units (none) are correct. We expect a pH just above neutral because we have a very low concentration of a weak base.

17.94 **Given:** $pK_b = 4.2$, 225 mg/L amphetamine **Find:** pH

Conceptual Plan: $pK_b \rightarrow K_b$ and then mg/L $\rightarrow$ g/L $\rightarrow$ mol/L and then write a balanced reaction. Prepare an ICE table, represent the change with x, sum the table, determine the equilibrium values, put the equilibrium values in the equilibrium expression, and solve for x.

$x = [OH^-] \rightarrow [pOH] \rightarrow pH$

$pOH = -\log[OH^-]$ $pH + pOH = 14$

Solution: $K_b = 10^{-4.2} = \underline{6}.31 \times 10^{-5}$

$$\frac{225 \text{ mg amphetamine}}{\text{L soln}} \times \frac{1 \text{ g amphetamine}}{1000 \text{ mg amphetamine}} \times \frac{1 \text{ mol amphetamine}}{135.20 \text{ g amphetamine}} = 0.001664 \text{ M amphetamine}$$

	$C_9H_{13}N(aq) + H_2O(l) \rightleftharpoons$	$C_9H_{13}NH^+(aq)$	$+ OH^-(aq)$
Initial	0.001664	0.0	0.0
Change	$-x$	$+x$	$+x$
Equil	$0.001664 - x$	x	x

$$K_b = \frac{[C_9H_{13}NH^+][OH^-]}{[C_9H_{13}N]} = \frac{(x)(x)}{(0.001664 - x)} = 6.31 \times 10^{-5}$$

Assume that x is small.
$x^2 = (6.31 \times 10^{-5})(0.001664)$ $\quad x = [OH^-] = 3.24 \times 10^{-4}$ M

$\frac{3.24 \times 10^{-4}}{0.001664} \times 100\% = 19.5\%$; assumption is not valid; solve using the quadratic equation.

$x^2 + 6.31 \times 10^{-5}x - 1.05 \times 10^{-7} = 0$
$x = [OH^-] = 2.94 \times 10^{-4}$ M
$pOH = -\log(2.94 \times 10^{-4}) = 3.5$
$pH = 14.00 - 3.5 = 10.5$

Check: The units (none) are correct. The magnitude of the pH makes sense. We have a weak base so we expect a $pH < 14 - \log(0.00166) = 11.220$.

17.95 **Given:** 0.150 M morphine, pH = 10.5 **Find:** K_b
Conceptual Plan:
pH → pOH → [OH⁻] and then write a balanced equation, prepare an ICE table, and determine
$pH + pOH = 14 \quad pOH = -\log[OH^-]$
equilibrium concentrations → K_b.
Solution: $pOH = 14.0 - 10.5 = 3.5$; $\quad [OH^-] = 10^{-3.5} = 3.16 \times 10^{-4} = [\text{Hmorphine}^+]$

	morphine$(aq) + H_2O(l) \rightleftharpoons$	Hmorphine$^+(aq)$	$+ OH^-(aq)$
Initial	0.150	0.0	0.0
Change	$-x$	$+x$	$+x$
Equil	$0.150 - x$	3.16×10^{-4}	3.16×10^{-4}

$$K_b = \frac{[\text{Hmorphine}^+][OH^-]}{[\text{morphine}]} = \frac{(3.16 \times 10^{-4})(3.16 \times 10^{-4})}{(0.150 - 3.16 \times 10^{-4})} = 6.67 \times 10^{-7} = 7 \times 10^{-7}$$

Check: The units (none) are correct. The magnitude of the K makes sense, since the pH is much less than $14 - \log(0.150) = 13.176$.

17.96 **Given:** 0.135 M base, pH = 11.23 **Find:** K_b
Conceptual Plan:
pH → pOH → [OH⁻] and then write a balanced equation, prepare an ICE table, and determine
$pH = pOH = 14 \; pOH = -\log[OH^-]$
equilibrium concentrations → K_b.
Solution: $pOH = 14.00 - 11.23 = 2.77$; $\quad [OH^-] = 10^{-2.77} = 1.698 \times 10^{-3} = [HB^+]$

	$B(aq) + H_2O(l) \rightleftharpoons$	$HB^+(aq)$	$+ OH^-(aq)$
Initial	0.135	0.0	0.0
Change	$-x$	$+x$	$+x$
Equil	$0.135 - x$	1.698×10^{-3}	1.698×10^{-3}

$$K_b = \frac{[HB^+][OH^-]}{[B]} = \frac{(1.698 \times 10^{-3})(1.698 \times 10^{-3})}{(0.135 - 1.698 \times 10^{-3})} = 2.16 \times 10^{-5} = 2.2 \times 10^{-5}$$

Check: The units (none) are correct. The magnitude of the K makes sense, since the pH is much less than $14 - \log(0.135) = 13.130$.

Acid–Base Properties of Ions and Salts

17.97
(a) pH-neutral: Br^- is the conjugate base of a strong acid; therefore, it is pH-neutral.
(b) weak base: ClO^- is the conjugate base of a weak acid; therefore, it is a weak base.
$ClO^-(aq) + H_2O(l) \rightleftharpoons HClO(aq) + OH^-(aq)$
(c) weak base: CN^- is the conjugate base of a weak acid; therefore, it is a weak base.
$CN^-(aq) + H_2O(l) \rightleftharpoons HCN(aq) + OH^-(aq)$
(d) pH-neutral: Cl^- is the conjugate base of a strong acid; therefore, it is pH-neutral.

17.98
(a) weak base: $C_7H_5O_2^-$ is the conjugate base of a weak acid; therefore, it is a weak base.
$C_7H_5O_2^-(aq) + H_2O(l) \rightleftharpoons HC_7H_5O_2(aq) + OH^-(aq)$
(b) pH-neutral: I^- is the conjugate base of a strong acid; therefore, it is pH-neutral.
(c) pH-neutral: NO_3^- is the conjugate base of a strong acid; therefore, it is pH-neutral.
(d) weak base: F^- is the conjugate base of a weak acid; therefore, it is a weak base.
$F^-(aq) + H_2O(l) \rightleftharpoons HF(aq) + OH^-(aq)$

17.99 **Given:** $[F^-] = 0.140$ M, $K_a(HF) = 3.5 \times 10^{-4}$ **Find:** $[OH^-]$, pH
Conceptual Plan: Determine K_b. Write a balanced reaction. Prepare an ICE table, represent the change

$$K_b = \frac{K_w}{K_a}$$

with x, sum the table, determine the equilibrium values, put the equilibrium values in the equilibrium expression, and solve for x. Determine $[OH^-] \rightarrow pOH \rightarrow pH$.

$$pOH = -\log[OH^-] \quad pH + pOH = 14$$

Solution:	$F^-(aq)$ + $H_2O(l)$ ⇌	$HF(aq)$ +	$OH^-(aq)$
Initial	0.140	0.0	0.0
Change	$-x$	$+x$	$+x$
Equil	$0.140 - x$	x	x

$$K_b = \frac{K_w}{K_a} = \frac{1 \times 10^{-14}}{3.5 \times 10^{-4}} = 2.\underline{8}6 \times 10^{-11} = \frac{(x)(x)}{(0.140 - x)}$$

Assume that x is small.
$x = 2.0 \times 10^{-6}\text{ M} = [OH^-]$ $\quad pOH = -\log(2.0 \times 10^{-6}) = 5.70$
$pH = 14.00 - 5.70 = 8.30$

Check: The units (M and none) are correct. We expect $[OH^-] > (1 \times 10^{-14}/0.140) = 7 \times 10^{-14}$ M and $7 > pH > -\log(0.140) = 0.854$.

17.100 **Given:** $[HCO_3^-] = 0.250$ M, $K_a(H_2CO_3) = 4.3 \times 10^{-7}$ **Find:** $[OH^-]$, pH
Conceptual Plan: Determine K_b. Write a balanced reaction. Prepare an ICE table, represent the change

$$K_b = \frac{K_w}{K_a}$$

with x, sum the table, determine the equilibrium values, put the equilibrium values in the equilibrium expression, and solve for x. Determine $[OH^-] \rightarrow pOH \rightarrow pH$.

$$pOH = -\log[OH^-] \quad pH + pOH = 14$$

Solution:	$HCO_3^-(aq)$ + $H_2O(l)$ ⇌	$H_2CO_3(aq)$ +	$OH^-(aq)$
Initial	0.250	0.0	0.0
Change	$-x$	$+x$	$+x$
Equil	$0.250 - x$	x	x

$$K_b = \frac{K_w}{K_a} = \frac{1 \times 10^{-14}}{4.3 \times 10^{-7}} = 2.\underline{3}3 \times 10^{-8} = \frac{(x)(x)}{(0.250 - x)}$$

Assume that x is small.
$x = 7.6 \times 10^{-5}\text{ M} = [OH^-]$ $\quad pOH = -\log(7.6 \times 10^{-5}) = 4.12$
$pH = 14.00 - 4.12 = 9.88$

Check: The units (M and none) are correct. We expect $[OH^-] > (1 \times 10^{-14}/0.250) = 4 \times 10^{-14}$ M and $7 < pH < 14 - \log(0.250) = 13.398$.

17.101 (a) weak acid: NH_4^+ is the conjugate acid of a weak base; therefore, it is a weak acid.
$NH_4^+(aq) + H_2O(l) \rightleftharpoons H_3O^+(aq) + NH_3(aq)$

(b) pH-neutral: Na^+ is the counterion of a strong base; therefore, it is pH-neutral.

(c) weak acid: The Co^{3+} cation is a small, highly charged metal cation; therefore, it is a weak acid.
$Co(H_2O)_6^{3+}(aq) + H_2O(l) \rightleftharpoons Co(H_2O)_5(OH)^{2+}(aq) + H_3O^+(aq)$

(d) weak acid: $CH_2NH_3^+$ is the conjugate acid of a weak base; therefore, it is a weak acid.
$CH_2NH_3^+(aq) + H_2O(l) \rightleftharpoons H_3O^+(aq) + CH_2NH_2(aq)$

17.102 (a) pH-neutral: Sr^{2+} is the counterion of a strong base; therefore, it is pH-neutral.

(b) weak acid: The Mn^{3+} cation is a small, highly charged metal cation; therefore, it is a weak acid.
$Mn(H_2O)_6^{3+}(aq) + H_2O(l) \rightleftharpoons Mn(H_2O)_5(OH)^{2+}(aq) + H_3O^+(aq)$

(c) weak acid: $C_5H_5NH^+$ is the conjugate acid of a weak base; therefore, it is a weak acid.
$C_5H_5NH^+(aq) + H_2O(l) \rightleftharpoons H_3O^+(aq) + C_5H_5N(aq)$

(d) pH-neutral: Li^+ is the counterion of a strong base; therefore, it is pH-neutral.

17.103 (a) acidic: $FeCl_3$ Fe^{3+} is a small, highly charged metal cation; therefore, it is acidic. Cl^- is the conjugate base of a strong acid; therefore, it is pH-neutral.

(b) basic: NaF Na^+ is the counterion of a strong base; therefore, it is pH-neutral. F^- is the conjugate base of a weak acid; therefore, it is basic.

(c) pH-neutral: $CaBr_2$ Ca^{2+} is the counterion of a strong base; therefore, it is pH-neutral. Br^- is the conjugate base of a strong acid; therefore, it is pH-neutral.

(d) acidic: NH_4Br NH_4^+ is the conjugate acid of a weak base; therefore, it is acidic. Br^- is the conjugate base of a strong acid; therefore, it is pH-neutral.

(e) acidic: $C_6H_5NH_3NO_2$ $C_6H_5NH_3^+$ is the conjugate acid of a weak base; therefore, it is acidic. NO_2^- is the conjugate base of a weak acid; therefore, it is basic. To determine pH, compare K values.

$$K_a(C_6H_5NH_3^+) = \frac{1.0 \times 10^{-14}}{3.9 \times 10^{-10}} = 2.6 \times 10^{-5} \quad K_b(NO_2^-) = \frac{1.0 \times 10^{-14}}{4.6 \times 10^{-4}} = 2.2 \times 10^{-11}$$

$K_a > K_b$; therefore, the solution is acidic.

17.104 (a) acidic: $Al(NO_3)_3$ Al^{3+} is a small, highly charged metal cation; therefore, it is acidic. NO_3^- is the conjugate base of a strong acid; therefore, it is pH-neutral.

(b) acidic: $C_2H_5NH_3NO_3$ $C_2H_5NH_3^+$ is the conjugate acid of a weak base; therefore, it is acidic. NO_3^- is the conjugate base of a strong acid; therefore, it is pH-neutral.

(c) basic: K_2CO_3 K^+ is the counterion of a strong base; therefore, it is pH-neutral. CO_3^{2-} is the conjugate base of a weak acid; therefore, it is basic.

(d) pH-neutral: RbI Rb^+ is the counterion of a strong base; therefore, it is pH-neutral. I^- is the conjugate base of a strong acid; therefore, it is pH-neutral.

(e) basic NH_4ClO NH_4^+ is the conjugate acid of a weak base; therefore, it is acidic. ClO^- is the conjugate base of a weak acid; therefore, it is basic. To determine pH, compare K values.

$$K_a(NH_4^+) = \frac{1.0 \times 10^{-14}}{1.8 \times 10^{-5}} = 5.6 \times 10^{-10} \qquad K_b(ClO^-) = \frac{1.0 \times 10^{-14}}{2.9 \times 10^{-8}} = 3.4 \times 10^{-7}$$

$K_b > K_a$; therefore, the solution is basic.

17.105 **Conceptual Plan:** Identify each species and determine whether it is acidic, basic, or neutral.

NaCl pH-neutral: Na^+ is the counterion of a strong base; therefore, it is pH-neutral. Cl^- is the conjugate base of a strong acid; therefore, it is pH-neutral.

NH_4Cl acidic: NH_4^+ is the conjugate acid of a weak base; therefore, it is acidic. Cl^- is the conjugate base of a strong acid; therefore, it is pH-neutral.

$NaHCO_3$ basic: Na^+ is the counterion of a strong base; therefore, it is pH-neutral. HCO_3^- is the conjugate base of a weak acid; therefore, it is basic.

NH_4ClO_2 acidic: NH_4^+ is the conjugate acid of a weak base; therefore, it is acidic. ClO_2^- is the conjugate base of a weak acid; therefore, it is basic. $K_a(NH_4^+) = 5.6 \times 10^{-10}$ $K_b(ClO_2^-) = 9.1 \times 10^{-13}$

$K_a > K_b$; therefore, the solution is acidic.

NaOH strong base

Increasing acidity: $NaOH < NaHCO_3 < NaCl < NH_4ClO_2 < NH_4Cl$

17.106 **Conceptual Plan:** Identify each species and determine whether it is acidic, basic, or neutral.
CH_3NH_3Br acidic: $CH_3NH_3^+$ is the conjugate acid of a weak base; therefore, it is acidic. Br^- is the conjugate base of a strong acid; therefore, it is pH-neutral.
KOH strong base
KBr pH-neutral: K^+ is the counterion of a strong base; therefore, it is pH-neutral. Br^- is the conjugate base of a strong acid; therefore, it is pH-neutral.
KCN basic: K^+ is the counterion of a strong base; therefore, it is pH-neutral. CN^- is the conjugate base of a weak acid; therefore, it is basic.
$C_5H_5NHNO_2$ acidic: $C_5H_5NH^+$ is the conjugate acid of a weak base; therefore, it is acidic. NO_2^- is the conjugate base of a weak acid; therefore, it is basic. $K_a(C_5H_5NH^+) = 5.9 \times 10^{-6}$ $K_b(NO_2^-) = 2.2 \times 10^{-11}$
$K_a > K_b$; therefore, the solution is acidic.
Increasing basicity: $CH_3NH_3Br < C_5H_5NHNO_2 < KBr < KCN < KOH$

17.107 (a) **Given:** 0.10 M NH_4Cl **Find:** pH
Conceptual Plan: Identify each species and determine which will contribute to pH. Write a balanced reaction. Prepare an ICE table, represent the change with x, sum the table, determine the equilibrium values, put the equilibrium values in the equilibrium expression, and solve for x. Determine $[H_3O^+] \rightarrow$ pH.
Solution: NH_4^+ is the conjugate acid of a weak base; therefore, it is acidic. Cl^- is the conjugate base of a strong acid; therefore, it is pH-neutral.

	$NH_4^+(aq)$ + $H_2O(l)$ $\rightleftharpoons$	$NH_3(aq)$ +	$H_3O^+(aq)$
Initial	0.10	0.0	0.0
Change	$-x$	$+x$	$+x$
Equil	$0.10 - x$	x	x

$$K_a = \frac{K_w}{K_b} = \frac{1.0 \times 10^{-14}}{1.76 \times 10^{-5}} = 5.\underline{6}8 \times 10^{-10} = \frac{(x)(x)}{(0.10 - x)}$$

Assume that x is small.
$x = 7.\underline{5}4 \times 10^{-6}\ M = [H_3O^+]$ $pH = -\log(7.\underline{5}4 \times 10^{-6}) = 5.12$

(b) **Given:** 0.10 M $NaC_2H_3O_2$ **Find:** pH
Conceptual Plan: Identify each species and determine which will contribute to pH. Write a balanced reaction. Prepare an ICE table, represent the change with x, sum the table, determine the equilibrium values, put the equilibrium values in the equilibrium expression, and solve for x. Determine $[OH^-] \rightarrow$ pOH $\rightarrow$ pH.
$pOH = -\log[OH^-]$ $pH + pOH = 14$
Solution: Na^+ is the counterion of a strong base; therefore, it is pH-neutral. $C_2H_3O_2^-$ is the conjugate base of a weak acid; therefore, it is basic.

	$C_2H_3O_2^-(aq)$ + $H_2O(l)$ $\rightleftharpoons$	$HC_2H_3O_2(aq)$ +	$OH^-(aq)$
Initial	0.10	0.0	0.0
Change	$-x$	$+x$	$+x$
Equil	$0.10 - x$	x	x

$$K_b = \frac{K_w}{K_a} = \frac{1.0 \times 10^{-14}}{1.8 \times 10^{-5}} = 5.\underline{5}6 \times 10^{-10} = \frac{(x)(x)}{(0.10 - x)}$$

Assume that x is small.
$x = 7.\underline{4}6 \times 10^{-6}\ M = [OH^-]$ $pOH = -\log(7.\underline{4}6 \times 10^{-6}) = 5.13$
$pH = 14.00 - 5.13 = 8.87$

(c) **Given:** 0.10 M NaCl **Find:** pH
Conceptual Plan: Identify each species and determine which will contribute to pH.
Solution: Na^+ is the counterion of a strong base; therefore, it is pH-neutral. Cl^- is the conjugate base of a strong acid; therefore, it is pH-neutral.
pH = 7.0

17.108 (a) **Given:** 0.20 M $NaCHO_2$ **Find:** pH

Conceptual Plan: Identify each species and determine which will contribute to pH. Write a balanced reaction. Prepare an ICE table, represent the change with x, sum the table, determine the equilibrium values, put the equilibrium values in the equilibrium expression, and solve for x. Determine $[OH^-] \rightarrow pOH \rightarrow pH$.

$pOH = -\log[OH^-]$ $pH + pOH = 14$

Solution: Na^+ is the counterion of a strong base; therefore, it is pH-neutral. CHO_2^- is the conjugate base of a weak acid; therefore, it is basic.

	$CHO_2^-(aq)$ + $H_2O(l)$ ⇌	$HCHO_2(aq)$ +	$OH^-(aq)$
Initial	0.20	0.0	0.0
Change	$-x$	$+x$	$+x$
Equil	$0.20 - x$	x	x

$$K_b = \frac{K_w}{K_a} = \frac{1.0 \times 10^{-14}}{1.8 \times 10^{-4}} = 5.\underline{5}6 \times 10^{-11} = \frac{(x)(x)}{(0.20 - x)}$$

Assume that x is small.

$x = 3.\underline{3}3 \times 10^{-6}\ M = [OH^-]$ $pOH = -\log(3.\underline{3}3 \times 10^{-6}) = 5.48$

$pH = 14.00 - 5.48 = 8.52$

(b) **Given:** 0.20 M CH_3NH_3I **Find:** pH

Conceptual Plan: Identify each species and determine which will contribute to pH. Write a balanced reaction. Prepare an ICE table, represent the change with x, sum the table, determine the equilibrium values, put the equilibrium values in the equilibrium expression, and solve for x. Determine $[H_3O^+] \rightarrow pH$.

Solution: $CH_3NH_3^+$ is the conjugate acid of a weak base; therefore, it is acidic. Cl^- is the conjugate base of a strong acid; therefore, it is pH-neutral.

	$CH_3NH_3^+(aq)$ + $H_2O(l)$ ⇌	$CH_3NH_2(aq)$ +	$H_3O^+(aq)$
Initial	0.20	0.0	0.0
Change	$-x$	$+x$	$+x$
Equil	$0.20 - x$	x	x

$$K_a = \frac{K_w}{K_b} = \frac{1.0 \times 10^{-14}}{4.4 \times 10^{-4}} = 2.\underline{2}7 \times 10^{-11} = \frac{(x)(x)}{(0.20 - x)}$$

Assume that x is small.

$x = 2.\underline{1}3 \times 10^{-6}\ M = [H_3O^+]$ $pH = -\log(2.\underline{1}3 \times 10^{-6}) = 5.67$

(c) **Given:** 0.20 M KI **Find:** pH

Conceptual Plan: Identify each species and determine which will contribute to pH.

Solution: K^+ is the counterion of a strong base; therefore, it is pH-neutral. I^- is the conjugate base of a strong acid; therefore, it is pH-neutral.

$pH = 7.0$

17.109 **Given:** 0.15 M KF **Find:** concentration of all species

Conceptual Plan: Identify each species and determine which will contribute to pH. Write a balanced reaction. Prepare an ICE table, represent the change with x, sum the table, determine the equilibrium values, put the equilibrium values in the equilibrium expression, and solve for x. Then $[OH^-] \rightarrow [H_3O^+]$.

$K_w = [H_3O^+][OH^-]$

Solution: K^+ is the counterion of a strong base; therefore, it is pH-neutral. F^- is the conjugate base of a weak acid; therefore, it is basic.

	$F^-(aq)$ + $H_2O(l)$ ⇌	$HF(aq)$ +	$OH^-(aq)$
Initial	0.15	0.0	0.0
Change	$-x$	$+x$	$+x$
Equil	$0.15 - x$	x	x

$$K_b = \frac{K_w}{K_a} = \frac{1.0 \times 10^{-14}}{3.5 \times 10^{-4}} = \frac{(x)(x)}{(0.15 - x)}$$

Assume that x is small.

$x = 2.1 \times 10^{-6}\text{ M} = [OH^-] = [HF]$; $\quad [H_3O^+] = \dfrac{K_w}{[OH^-]} = \dfrac{1 \times 10^{-14}}{2.1 \times 10^{-6}} = 4.8 \times 10^{-9}\text{ M}$

$[K^+] = 0.15\text{ M}$
$[F^-] = (0.15 - 2.1 \times 10^{-6}) = 0.15\text{ M}$
$[HF] = 2.1 \times 10^{-6}\text{ M}$
$[OH^-] = 2.1 \times 10^{-6}\text{ M}$
$[H_3O^+] = 4.8 \times 10^{-9}\text{ M}$

Check: The units (Ms) are correct. We expect $[K^+]$ and $[F^-]$ to be at or near 0.15 M and all other concentrations to be very low.

17.110 **Given:** 0.225 M $C_6H_5NH_3Cl$ **Find:** concentration of all species

Conceptual Plan: Identify each species and determine which will contribute to pH. Write a balanced reaction. Prepare an ICE table, represent the change with x, sum the table, determine the equilibrium values, put the equilibrium values in the equilibrium expression, and solve for x. Then determine $[H_3O^+] \rightarrow [OH^-]$.

$K_w = [H_3O^+][OH^-]$

Solution: $C_6H_5NH_3^+$ is the conjugate acid of a weak base; therefore, it is a weak acid. Cl^- is the conjugate base of a strong acid; therefore, it is pH-neutral.

	$C_6H_5NH_3^+(aq)$ + $H_2O(l)$ ⇌	$C_6H_5NH_2(aq)$ +	$H_3O^+(aq)$
Initial	0.225	0.0	0.0
Change	$-x$	$+x$	$+x$
Equil	$0.225 - x$	x	x

$$K_a = \frac{K_w}{K_b} = \frac{1.0 \times 10^{-14}}{3.9 \times 10^{-10}} = 2.\underline{5}6 \times 10^{-5} = \frac{(x)(x)}{(0.225 - x)}$$

Assume that x is small.

$x = 0.0024\text{ M} = [H_3O^+] = [CH_5NH_2]$ $\quad [OH^-] = \dfrac{K_w}{[H_3O^+]} = \dfrac{1.0 \times 10^{-14}}{0.0024} = 4.2 \times 10^{-12}\text{ M}$

$[C_6H_5NH_3^+] = 0.225 - 0.0024 = 0.223\text{ M}$
$[Cl^-] = 0.225\text{ M}$
$[C_6H_5NH_2] = 0.0024\text{ M}$
$[H_3O^+] = 0.0024\text{ M}$
$[OH^-] = 4.2 \times 10^{-12}\text{ M}$

Check: The units (Ms) are correct. We expect $[C_6H_5NH_3^+]$ and $[Cl^-]$ to be at or near 0.225 M and all other concentrations to be very low.

17.111 (a) F^- is a stronger base than is Cl^-.
F^- is the conjugate base of HF (a weak acid); Cl^- is the conjugate base of HCl (a strong acid); the weaker the acid, the stronger the conjugate base.

(b) NO_2^- is a stronger base than is NO_3^-.
NO_2^- is the conjugate base of HNO_2 (a weak acid); NO_3^- is the conjugate base of HNO_3 (a strong acid); the weaker the acid, the stronger the conjugate base.

(c) ClO^- is a stronger base than is F^-.
F^- is the conjugate base of HF ($K_a = 3.5 \times 10^{-4}$); ClO^- is the conjugate base of HClO ($K_a = 2.9 \times 10^{-8}$); HClO is the weaker acid so ClO^- will be the stronger base—the weaker the acid, the stronger the conjugate base.

17.112 (a) ClO_2^- is a stronger base than is ClO_4^-.
ClO_2^- is the conjugate base of $HClO_2$ (a weak acid); ClO_4^- is the conjugate base of $HClO_4$ (a strong acid); the weaker the acid, the stronger the conjugate base.

(b) H_2O is a stronger base than is Cl^-.
H_2O is the conjugate base of H_3O^+; Cl^- is the conjugate base of HCl (a strong acid); the weaker the acid, the stronger the conjugate base.

(c) CN^- is a stronger base than is ClO^-.
CN^- is the conjugate base of HCN ($K_a = 4.9 \times 10^{-10}$); ClO^- is the conjugate base of HClO ($K_a = 2.9 \times 10^{-8}$); the weaker the acid, the stronger the conjugate base.

Polyprotic Acids

17.113 $H_3PO_4(aq) + H_2O(l) \rightleftharpoons H_3O^+(aq) + H_2PO_4^-(aq)$ $K_{a_1} = \frac{[H_3O^+][H_2PO_4^-]}{[H_3PO_4]}$

$H_2PO_4^-(aq) + H_2O(l) \rightleftharpoons H_3O^+(aq) + HPO_4^{2-}(aq)$ $K_{a_2} = \frac{[H_3O^+][HPO_4^{2-}]}{[H_2PO_4^-]}$

$HPO_4^{2-}(aq) + H_2O(l) \rightleftharpoons H_3O^+(aq) + PO_4^{3-}(aq)$ $K_{a_3} = \frac{[H_3O^+][PO_4^{3-}]}{[HPO_4^{2-}]}$

17.114 $H_2CO_3(aq) + H_2O(l) \rightleftharpoons H_3O^+(aq) + HCO_3^-(aq)$ $K_{a_1} = \frac{[H_3O^+][HCO_3^-]}{[H_2CO_3]}$

$HCO_3^-(aq) + H_2O(l) \rightleftharpoons H_3O^+(aq) + CO_3^{2-}(aq)$ $K_{a_2} = \frac{[H_3O^+][CO_3^{2-}]}{[HCO_3^-]}$

17.115 (a) **Given:** 0.350 M H_3PO_4 $K_{a_1} = 7.5 \times 10^{-3}$, $K_{a_2} = 6.2 \times 10^{-8}$ **Find:** $[H_3O^+]$, pH

Conceptual Plan: K_{a_1} is much larger than K_{a_2}, so use K_{a_1} to calculate $[H_3O^+]$. Write a balanced reaction. Prepare an ICE table, represent the change with x, sum the table, determine the equilibrium values, put the equilibrium values in the equilibrium expression, and solve for x.

Solution: $H_3PO_4(aq) + H_2O(l) \rightleftharpoons H_3O^+(aq) + H_2PO_4^-(aq)$

	H_3PO_4	H_3O^+	$H_2PO_4^-$
Initial	0.350	0.0	0.0
Change	$-x$	$+x$	$+x$
Equil	$0.350 - x$	x	x

$$K_{a_1} = \frac{[H_3O^+][H_2PO_4^-]}{[H_3PO_4]} = \frac{(x)(x)}{(0.350 - x)} = 7.5 \times 10^{-3}$$

Assume that x is small compared to 0.350.

$x^2 = (7.5 \times 10^{-3})(0.350)$; $x = 0.0505 \text{ M} = [H_3O^+]$

Check assumption: $\frac{0.0505}{0.350} \times 100\% = 14.4\%$ assumption is not valid, solve using the quadratic equation.

$x^2 + 7.5 \times 10^{-3}x - 0.002625 = 0$; $x = 0.04\underline{7}62 \text{ M} = [H_3O^+]$

$\text{pH} = -\log(0.04\underline{7}62) = 1.32$

Check: The units (none) are correct. We expect pH slightly above $-\log((0.350 \times 0.0075)^{1/2}) = 1.29$.

(b) **Given:** 0.350 M $H_2C_2O_4$ $K_{a_1} = 6.0 \times 10^{-2}$, $K_{a_2} = 6.0 \times 10^{-5}$ **Find:** $[H_3O^+]$, pH

Conceptual Plan: K_{a_1} is much larger than K_{a_2}, so use K_{a_1} to calculate $[H_3O^+]$. Write a balanced reaction. Prepare an ICE table, represent the change with x, sum the table, determine the equilibrium values, put the equilibrium values in the equilibrium expression, and solve for x.

Solution: $H_2C_2O_4(aq) + H_2O(l) \rightleftharpoons H_3O^+(aq) + HC_2O_4^-(aq)$

	$H_2C_2O_4$	H_3O^+	$HC_2O_4^-$
Initial	0.350	0.0	0.0
Change	$-x$	$+x$	$+x$
Equil	$0.350 - x$	x	x

$$K_{a_1} = \frac{[H_3O^+][HC_2O_4^-]}{[H_2C_2O_4]} = \frac{(x)(x)}{(0.350 - x)} = 6.0 \times 10^{-2}$$

$x^2 + 6.0 \times 10^{-2}x - 0.021 = 0$ $x = 0.1\underline{1}79 = 0.12 \text{ M } [H_3O^+]$

$\text{pH} = -\log(0.1\underline{1}79) = 0.93$

Check: The units (none) are correct. We expect pH slightly above $-\log((0.350 \times 0.060)^{1/2}) = 0.84$.

17.116 (a) **Given:** 0.125 M H_2CO_3 $K_{a_1} = 4.3 \times 10^{-7}$, $K_{a_2} = 5.6 \times 10^{-11}$ **Find:** $[H_3O^+]$, pH

Conceptual Plan: K_{a_1} is much larger than K_{a_2}, so use K_{a_1} to calculate $[H_3O^+]$. Write a balanced reaction. Prepare an ICE table, represent the change with x, sum the table, determine the equilibrium values, put the equilibrium values in the equilibrium expression, and solve for x.

Solution:	$H_2CO_3(aq)$ + $H_2O(l)$ ⇌	$H_3O^+(aq)$ +	$HCO_3^-(aq)$
Initial	0.125	0.0	0.0
Change	$-x$	$+x$	$+x$
Equil	$0.125 - x$	x	x

$$K_{a_1} = \frac{[H_3O^+][HCO_3^-]}{[H_2CO_3]} = \frac{(x)(x)}{(0.125 - x)} = 4.3 \times 10^{-7}$$

Assume that x is small.

$x^2 = (4.3 \times 10^{-7})(0.125)\ x = 2.32 \times 10^{-4}\ M = [H_3O^+]$

$\frac{2.32 \times 10^{-4}}{0.125} \times 100\% = 0.19\%$; assumption is valid.

$pH = -\log(2.32 \times 10^{-4}) = 3.63$

Check: The units (none) are correct. We expect pH near $-\log((0.125 \times 4.3 \times 10^{-7})^{1/2}) = 3.63$.

(b) **Given:** 0.125 M $H_3C_6H_5O_3$ $K_{a_1} = 7.4 \times 10^{-4}$, $K_{a_2} = 1.7 \times 10^{-5}$, $K_{a_3} = 4.0 \times 10^{-7}$ **Find:** $[H_3O^+]$, pH

Conceptual Plan: K_{a_1} and K_{a_2} are only 10^{-1} apart, so use both to calculate $[H_3O^+]$. Write a balanced reaction. Prepare an ICE table, represent the change with x, sum the table, determine the equilibrium values, put the equilibrium values in the equilibrium expression, and solve for x. Repeat this process by preparing a second ICE table using these values and representing the change with y.

Solution:	$H_3C_6H_5O_3(aq)$ + $H_2O(l)$ ⇌	$H_3O^+(aq)$ +	$H_2C_6H_5O_3^-(aq)$
Initial	0.125	0.0	0.0
Change	$-x$	$+x$	$+x$
Equil	$0.125 - x$	x	x

$$K_{a_1} = \frac{[H_3O^+][H_2C_6H_5O_3^-]}{[H_3C_6H_5O_3]} = \frac{(x)(x)}{(0.125 - x)} = 7.4 \times 10^{-4}$$

Assume that x is small.

$x^2 = (7.4 \times 10^{-4})(0.125)\ x = 9.62 \times 10^{-3}\ M = [H_3O^+]$

$\frac{9.62 \times 10^{-3}}{0.125} \times 100\% = 7.8\%$; assumption is not valid; solve using the quadratic equation.

$x^2 + 7.4 \times 10^{-4}x - 9.25 \times 10^{-5} = 0 \quad x = 0.009255\ M = [H_3O^+] = [H_2C_6H_5O_3^-]$

and then

	$H_2C_6H_5O_3^-(aq)$ + $H_2O(l)$ ⇌	$H_3O^+(aq)$ +	$HC_6H_5O_3^{2-}(aq)$
Initial	0.009255	0.009255	0.0
Change	$-y$	$+y$	$+y$
Equil	$0.009255 - y$	$0.009255 + y$	y

$$K_{a_1} = \frac{[H_3O^+][HC_6H_5O_3^{2-}]}{[H_2C_6H_5O_3^-]} = \frac{(0.009255 + y)(y)}{(0.009255 - y)} = 1.7 \times 10^{-5}$$

Assume that y is small. $y = 1.7 \times 10^{-5}$

$\frac{1.7 \times 10^{-5}}{0.009255} \times 100\% = 1.8\%$; assumption is valid.

$[H_3O^+] = 1.7 \times 10^{-5}\ M$ (from second ionization)

$[H_3O^+] = 0.009255 + 1.7 \times 10^{-5} = 0.00927\ M \quad pH = -\log(0.00927) = 2.03$

Check: The units (none) are correct. We expect pH near $-\log((0.125 \times 0.00074)^{1/2}) = 2.02$.

17.117 **Given:** 0.500 M H_2SO_3 $K_{a_1} = 1.6 \times 10^{-2}$, $K_{a_2} = 6.4 \times 10^{-8}$ **Find:** concentration of all species

Conceptual Plan: K_{a_1} is much larger than K_{a_2}, so use K_{a_1} to calculate $[H_3O^+]$. Write a balanced reaction. Prepare an ICE table, represent the change with x, sum the table, determine the equilibrium values, put the equilibrium values in the equilibrium expression, and solve for x. Repeat this process by preparing a second ICE table using these values and representing the change with y.

Solution: $H_2SO_3(aq) + H_2O(l) \rightleftharpoons H_3O^+(aq) + HSO_3^-(aq)$

	H_2SO_3	H_3O^+	HSO_3^-
Initial	0.500	0.0	0.0
Change	$-x$	$+x$	$+x$
Equil	$0.500 - x$	x	x

$$K_{a_1} = \frac{[H_3O^+][HSO_3^-]}{[H_2SO_3]} = \frac{(x)(x)}{(0.500 - x)} = 1.6 \times 10^{-2}$$

$x^2 + 1.6 \times 10^{-2}x - 0.0080 = 0;\quad x = 0.08\underline{1}8 = 0.082\text{ M} = [H_3O^+] = [HSO_3^-]$

Use the values from reaction 1 in reaction 2.

$HSO_3^-(aq) + H_2O(l) \rightleftharpoons H_3O^+(aq) + SO_3^{2-}(aq)$

	HSO_3^-	H_3O^+	SO_3^{2-}
Initial	0.0818	0.0818	0.0
Change	$-y$	$+y$	$+y$
Equil	$0.0818 - y$	$0.0818 + y$	y

$$K_{a_1} = \frac{[H_3O^+][SO_3^{2-}]}{[HSO_3^-]} = \frac{(0.0818 + y)(y)}{(0.0818 - y)} = 6.4 \times 10^{-8}$$

Assume that y is small. $y = 6.4 \times 10^{-8}$

$[H_2SO_3] = 0.500 - 0.0818 = 0.418\text{ M}$

$[HSO_3^-] = x = 0.0818 = 0.082\text{ M}$

$[SO_3^{2-}] = y = 6.4 \times 10^{-8}\text{ M}$

$[H_3O^+] = x + y = 0.0818\text{ M} + 6.4 \times 10^{-8}\text{ M} = 0.082\text{ M}$

$$[OH^-] = \frac{K_w}{[H_3O^+]} = \frac{1.0 \times 10^{-14}}{0.0818} = 1.2 \times 10^{-13}\text{ M}$$

Check: The units (Ms) are correct. We expect $[H_2SO_3]$ to be the highest at just under 0.500 M and $[SO_3^{2-}]$ to be much lower than $[HSO_3^-]$ and near K_{a2}. We expect $[H_3O^+]$ near $(0.500 \times 0.016)^{1/2}) = 0.089$ M.

17.118 **Given:** 0.155 M H_2CO_3 $K_{a_1} = 4.3 \times 10^{-7}$, $K_{a_2} = 5.6 \times 10^{-11}$ **Find:** concentration of all species

Conceptual Plan: K_{a_1} is much larger than K_{a_2}, so use K_{a_1} to calculate $[H_3O^+]$ and $[HCO_3^-]$. Use K_{a_2} to find $[CO_3^-]$. Write a balanced reaction. Prepare an ICE table, represent the change with x, sum the table, determine the equilibrium values, put the equilibrium values in the equilibrium expression, and solve for x. Repeat this process by preparing a second ICE table using these values and representing the change with y.

Solution: $H_2CO_3(aq) + H_2O(l) \rightleftharpoons H_3O^+(aq) + HCO_3^-(aq)$

	H_2CO_3	H_3O^+	HCO_3^-
Initial	0.125	0.0	0.0
Change	$-x$	$+x$	$+x$
Equil	$0.125 - x$	x	x

$$K_{a_1} = \frac{[H_3O^+][HCO_3^-]}{[H_2CO_3]} = \frac{(x)(x)}{(0.155 - x)} = 4.3 \times 10^{-7}$$

Assume that x is small.

$x^2 = (4.3 \times 10^{-7})(0.155);\quad x = 2.58 \times 10^{-4}\text{ M} = [H_3O^+] = [HCO_3^-]$

Reaction 2

$HCO_3^-(aq) + H_2O(l) \rightleftharpoons H_3O^+(aq) + CO_3^{2-}(aq)$

Because K_{a_2} is small, $y = [CO_3^{2-}] = K_{a_2}$.

$[H_2CO_3] = 0.155 - 2.\underline{5}8 \times 10^{-4} = 0.1547 = 0.155\text{ M}$

$[HCO_3^-] = x = 2.58 \times 10^{-4} = 2.6 \times 10^{-4}\text{ M}$

$[CO_3^{2-}] = y = 5.6 \times 10^{-11}\text{ M}$

$[H_3O^+] = x = 2.58 \times 10^{-4} = 2.6 \times 10^{-4}\text{ M}$

$$[OH^-] = \frac{K_w}{[H_3O^+]} = \frac{1.0 \times 10^{-14}}{2.\underline{5}8 \times 10^{-4}} = 3.88 \times 10^{-11}\text{ M} = 3.9 \times 10^{-11}\text{ M}$$

Check: The units (Ms) are correct. We expect $[H_2CO_3]$ near 0.125 M and $[CO_3^{2-}]$ to be much lower than $[HCO_3^-]$ and near K_{a2}. We expect $[H_3O^+]$ near $(0.155 \times 4.3 \times 10^{-7})^{1/2}) = 2.3 \times 10^{-4}$ M.

17.119 (a) **Given:** $[H_2SO_4] = 0.50$ M $K_{a_2} = 0.012$ **Find:** $[H_3O^+]$, pH

Conceptual Plan: The first ionization step is strong. Use K_{a_2} and reaction 2. Write a balanced reaction. Prepare an ICE table, represent the change with x, sum the table, determine the equilibrium values, put the equilibrium values in the equilibrium expression, and solve for x.

Solution: $H_2SO_4(aq) + H_2O(l) \rightarrow H_3O^+(aq) + HSO_4^-(aq)$ strong

0.50 M

$[H_3O^+] = [HSO_4^-] = 0.50$ M

	$HSO_4^-(aq)$ + $H_2O(l)$ ⇌	$H_3O^+(aq)$ +	$SO_4^{2-}(aq)$
Initial	0.50	0.50	0.0
Change	$-x$	$+x$	$+x$
Equil	$0.500 - x$	$0.50 + x$	x

$$K_{a_2} = \frac{[H_3O^+][SO_4^{2-}]}{[HSO_4^-]} = \frac{(0.50 + x)(x)}{(0.50 - x)} = 0.012$$

$x^2 + 0.512\,x - 0.006 = 0;\quad x = 0.01\underline{1}5$ M $= [H_3O^+]$ from second ionization step

$[H_3O^+] = 0.50 + 0.012 = 0.51$ M

pH $= -\log(0.51) = 0.29$

(b) **Given:** $[H_2SO_4] = 0.10$ M $K_{a_2} = 0.012$ **Find:** $[H_3O^+]$, pH

Conceptual Plan: The first ionization step is strong. Use K_{a_2} and reaction 2. Write a balanced reaction. Prepare an ICE table, represent the change with x, sum the table, determine the equilibrium values, put the equilibrium values in the equilibrium expression, and solve for x.

Solution: $H_2SO_4(aq) + H_2O(l) \rightarrow H_3O^+(aq) + HSO_4^-(aq)$ strong

0.10 M

$[H_3O^+] = [HSO_4^-] = 0.10$ M

	$HSO_4^-(aq)$ + $H_2O(l)$ ⇌	$H_3O^+(aq)$ +	$SO_4^{2-}(aq)$
Initial	0.10	0.10	0.0
Change	$-x$	$+x$	$+x$
Equil	$0.10 - x$	$0.10 + x$	x

$$K_{a_2} = \frac{[H_3O^+][SO_4^{2-}]}{[HSO_4^-]} = \frac{(0.10 + x)(x)}{(0.10 - x)} = 0.012$$

$x^2 + 0.112x - 0.0012 = 0;\quad x = 0.009\underline{8}48$ M

$\frac{0.009\underline{8}48}{0.10} \times 100\% = 9.8\%$ contribution of second ionization step is not negligible.

$[H_3O^+] = 0.10 + 0.009\underline{8}48 = 0.1\underline{0}9848$ M $= 0.11$ M

pH $= -\log(0.11) = 0.96$

(c) **Given:** $[H_2SO_4] = 0.050$ M $K_{a_2} = 0.012$ **Find:** $[H_3O^+]$, pH

Conceptual Plan: The first ionization step is strong. Use K_{a_2} and reaction 2. Write a balanced reaction. Prepare an ICE table, represent the change with x, sum the table, determine the equilibrium values, put the equilibrium values in the equilibrium expression, and solve for x.

Solution: $H_2SO_4(aq) + H_2O(l) \rightarrow H_3O^+(aq) + HSO_4^-(aq)$ strong

0.050 M

$[H_3O^+] = [HSO_4^-] = 0.050$ M

	$HSO_4^-(aq)$ + $H_2O(l)$ ⇌	$H_3O^+(aq)$ +	$SO_4^{2-}(aq)$
Initial	0.050	0.050	0.0
Change	$-x$	$+x$	$+x$
Equil	$0.050 - x$	$0.050 + x$	x

$$K_{a_2} = \frac{[H_3O^+][SO_4^{2-}]}{[HSO_4^-]} = \frac{(0.050 + x)(x)}{(0.050 - x)} = 0.012$$

$x^2 + 0.062\,x - 0.0006 = 0;\quad x = 0.008\underline{5}09$

$\frac{0.008\underline{5}09}{0.05} \times 100\% = 17\%$ contribution of second ionization step is not negligible.

$[H_3O^+] = 0.050 + 0.0085 = 0.0585 = 0.059\text{ M} \qquad pH = -\log(0.059) = 1.23$

Check: The units (M and none) are correct. We expect pH to be just below $-\log[H_2SO_4]$ and $[H^+]$ to be just above $[H_2SO_4]$. The lower the initial acid concentration, the bigger the differences from these values.

17.120 (a) **Given:** 0.10 M H_2A $K_{a_1} = 1.0 \times 10^{-4}, K_{a_2} = 5.0 \times 10^{-5}$ **Find:** whether the second ionization is negligible

Conceptual Plan: Write a balanced reaction for both ionization steps. Prepare an ICE table, represent the change with *x*, sum the table, determine the equilibrium values, put the equilibrium values in the equilibrium expression, and solve for *x*. Repeat this process by preparing a second ICE table using these values and representing the change with *y*.

	$H_2A(aq)$ + $H_2O(l) \rightarrow$	$H_3O^+(aq)$ +	$HA^-(aq)$
Initial	0.10	0.0	0.0
Change	$-x$	$+x$	$+x$
Equil	$0.10 - x$	x	x

$$K_{a_1} = \frac{[H_3O^+][HA^-]}{[H_2A]} = \frac{(x)(x)}{(0.10 - x)} = 1.0 \times 10^{-4}$$

Assume $x << 0.10$; $\frac{x^2}{0.10} = 1.0 \times 10^{-4};\quad x^2 = 1.0 \times 10^{-5};\quad x = 3.16 \times 10^{-3}$

$\left(\frac{3.16 \times 10^{-3}}{0.1}\right) \times 100\% = 3.16\%$ The assumption is valid. $x = [H_3O^+] = 3.16 \times 10^{-3}\text{ M}$

	$HA^-(aq)$ + $H_2O(l) \rightleftharpoons$	$H_3O^+(aq)$ +	$A^{2-}(aq)$
Initial	0.003$\underline{1}$6	0.003$\underline{1}$6	0.0
Change	$-y$	$+y$	$+y$
Equil	$0.003\underline{1}6 - y$	$0.003\underline{1}6 + y$	y

$$K_{a_2} = \frac{[H_3O^+][A^{2-}]}{[HA^-]} = \frac{(0.003\underline{1}6 + y)(y)}{(0.003\underline{1}6 - y)} = 5.0 \times 10^{-5}$$

$y^2 + 0.003\underline{2}1\,y - 1.58 \times 10^{-7} = 0$; Solve the quadratic equation. $y = 4.\underline{8}5 \times 10^{-5}$

$\frac{4.\underline{8}3 \times 10^{-5}}{0.00316} \times 100\% = 1.5\%$ contribution of second ionization step is negligible.

$[H_3O^+] = 0.00316\ pH = -\log(0.00311) = 2.5$

(b) **Given:** 0.10 M H_2A $K_{a_1} = 1.0 \times 10^{-4}, K_{a_2} = 1.0 \times 10^{-5}$ **Find:** Whether the second ionization is negligible

Conceptual Plan: Write a balanced reaction for both ionization steps. Prepare an ICE table, represent the change with *x*, sum the table, determine the equilibrium values, put the equilibrium values in the equilibrium expression, and solve for *x*. Repeat this process by preparing a second ICE table using these values and representing the change with *y*.

	$H_2A(aq)$ + $H_2O(l) \rightarrow$	$H_3O^+(aq)$ +	$HA^-(aq)$
Initial	0.10	0.0	0.0
Change	$-x$	$+x$	$+x$
Equil	$0.10 - x$	x	x

$$K_{a_1} = \frac{[H_3O^+][HA^-]}{[H_2A]} = \frac{(x)(x)}{(0.10 - x)} = 1.0 \times 10^{-4}$$

Assume $x << 0.10$; $\frac{x^2}{0.10} = 1.0 \times 10^{-4};\quad x^2 = 1.0 \times 10^{-5};\quad x = 3.16 \times 10^{-3}$

$\left(\frac{3.16 \times 10^{-3}}{0.10}\right) \times 100\% = 3.16\%$ The assumption is valid. $x = [H_3O^+] = 3.16 \times 10^{-3}\text{M}$

	$HA^-(aq)$ + $H_2O(l)$ ⇌	$H_3O^+(aq)$ +	$A^{2-}(aq)$
Initial	0.00316	0.00316	0.0
Change	$-y$	$+y$	$+y$
Equil	$0.003\underline{1}6 - y$	$0.003\underline{1}6 + y$	y

$$K_{a_2} = \frac{[H_3O^+][A^{2-}]}{[HA^-]} = \frac{(0.003\underline{1}6 + y)(y)}{(0.003\underline{1}6 - y)} = 1.0 \times 10^{-5}$$

$y^2 + 0.00317\,y - 3.16 \times 10^{-8} = 0$; Solve the quadratic equation. $y = 9.\underline{9}4 \times 10^{-6}$

$\frac{9.\underline{9}4 \times 10^{-6}}{0.00316} \times 100\% = 0.32\%$ contribution of second ionization step is negligible.

$[H_3O^+] = 0.003\underline{1}1$ pH $= -\log(0.00311) = 2.5$

(c) **Given:** 0.10 M H_2A $K_{a_1} = 1.0 \times 10^{-4}$, $K_{a_2} = 1.0 \times 10^{-6}$ **Find:** whether the second ionization is negligible
Conceptual Plan: Write a balanced reaction for both ionization steps. Prepare an ICE table, represent the change with x, sum the table, determine the equilibrium values, put the equilibrium values in the equilibrium expression, and solve for x. Repeat this process by preparing a second ICE table using these values and representing the change with y.

	$H_2A(aq)$ + $H_2O(l)$ →	$H_3O^+(aq)$ +	$HA^-(aq)$
Initial	0.10	0.0	0.0
Change	$-x$	$+x$	$+x$
Equil	$0.10 - x$	x	x

$$K_{a_1} = \frac{[H_3O^+][HA^-]}{[H_2A]} = \frac{(x)(x)}{(0.10 - x)} = 1.0 \times 10^{-4}$$

Assume $x << 0.10$; $\frac{x^2}{0.10} = 1.0 \times 10^{-4}$; $x^2 = 1.0 \times 10^{-5}$; $x = 3.16 \times 10^{-3}$

$\left(\frac{3.16 \times 10^{-3}}{0.10}\right) \times 100\% = 3.16\%$ The assumption is valid. $x = [H_3O^+] = 3.16 \times 10^{-3}$ M

	$HA^-(aq)$ + $H_2O(l)$ ⇌	$H_3O^+(aq)$ +	$A^{2-}(aq)$
Initial	$0.003\underline{1}6$	$0.003\underline{1}6$	0.0
Change	$-y$	$+y$	$+y$
Equil	$0.003\underline{1}6 - y$	$(0.003\underline{1}6 + y)$	y

$$K_{a_2} = \frac{[H_3O^+][A^{2-}]}{[HA^-]} = \frac{(0.003\underline{1}6 + y)(y)}{(0.003\underline{1}6 - y)} = 1.0 \times 10^{-6}$$

$y^2 + 0.003\underline{1}6\,y - 3.16 \times 10^{-9} = 0$; Solve the quadratic equation. $y = 9.\underline{8}4 \times 10^{-7}$

$\frac{9.\underline{8}4 \times 10^{-7}}{0.003\underline{1}6} \times 100\% = 0.032\%$ contribution of second ionization step is negligible.

$[H_3O^+] = 0.003\underline{1}6$ M; pH $= -\log(0.003\underline{1}6) = 2.5$

Check: The units (M and none) are correct. We expect pH to be significantly higher than $-\log[H_2CO_3]$ and $[H^+]$ to be significantly below $[H_2CO_3]$. The lower the initial acid concentration, the smaller the differences from these values.

Lewis Acids and Bases

17.121 (a) Lewis acid: Fe^{3+} has an empty *d* orbital and can accept a lone pair of electrons.
(b) Lewis acid: BH_3 has an empty *p* orbital to accept a lone pair of electrons.
(c) Lewis base: NH_3 has a lone pair of electrons to donate.
(d) Lewis base: F^- has a lone pair of electrons to donate.

17.122 (a) Lewis acid: $BeCl_2$ has empty *p* orbitals to accept a lone pair of electrons.
(b) Lewis base: OH^- has a lone pair of electrons to donate.

(c) Lewis acid: $B(OH)_3$ has an empty *p* orbital to accept a lone pair of electrons.
(d) Lewis base: CN^- has a lone of pair electrons to donate.

17.123 (a) Fe^{3+} accepts an electron pair from H_2O; so Fe^{3+} is the Lewis acid, and H_2O is the Lewis base.
(b) Zn^{2+} accepts an electron pair from NH_3; so Zn^{2+} is the Lewis acid, and NH_3 is the Lewis base.
(c) The empty *p* orbital on B accepts an electron pair from $(CH_3)_3N$; so BF_3 is the Lewis acid, and $(CH_3)_3N$ is the Lewis base.

17.124 (a) Ag^+ accepts an electron pair from NH_3; so Ag^+ is the Lewis acid, and NH_3 is the Lewis base.
(b) The empty *p* orbital on Al accepts an electron pair from NH_3; so $AlBr_3$ is the Lewis acid, and NH_3 is the Lewis base.
(c) The empty *p* orbital on B accepts an electron pair from F^-; so BF_3 is the Lewis acid, and F^- is the Lewis base.

Cumulative Problems

17.125 (a) weak acid: The beaker contains 10 HF molecules, 2 H_3O^+ ions, and 2 F^- ions. Because both the molecules and the ions exist in solution, the acid is a weak acid.
(b) strong acid: The beaker contains 12 H_3O^+ ions and 12 I^- ions. Because the molecules are completely ionized in solution, the acid is a strong acid.
(c) weak acid: The beaker contains 10 $HCHO_2$ molecules, 2 H_3O^+ ions, and 2 CHO_2^- ions. Because both the molecules and the ions exist in solution, the acid is a weak acid.
(d) strong acid: The beaker contains 12 H_3O^+ ions and 12 NO_3^- ions. Because the molecules are completely ionized in solution, the acid is a strong acid.

17.126 (a) weak base: The beaker contains 11 NH_3 molecules, 2 NH_4^+ ions, and 2 OH^- ions. Because both the molecules and the ions exist in solution, this is a weak base.
(b) strong base: The beaker contains 12 Na^+ ions and 12 OH^- ions. Because the molecules are completely ionized in solution, it is a strong base.
(c) weak base: The beaker contains 12 Na^+, 10 H_2CO_3 molecules, 2 HCO_3^- ions, and 2 OH^- ions. The beaker contains the HCO_3^- ion from the $NaHCO_3$ and H_2CO_3 molecules from the reaction of the salt.
(d) strong base: The beaker contains 12 Sr^{2+} ions and 24 OH^- ions. Because the molecules are completely ionized in solution, it is a strong base.

17.127 $HbH^+(aq) + O_2(aq) \rightleftharpoons HbO_2(aq) + H^+(aq)$
Using Le Châtelier's principle, if the $[H^+]$ increases, the reaction will shift left, and if the $[H^+]$ decreases, the reaction will shift right. So if the pH of blood is too acidic (low pH; $[H^+]$ increased), the reaction will shift to the left. This will cause less of the HbO_2 in the blood and decrease the oxygen-carrying capacity of the hemoglobin in the blood.

17.128 $CO_2(g) + H_2O(l) \rightleftharpoons H_2CO_3(aq)$
$H_2CO_3(aq) + H_2O(l) \rightleftharpoons HCO_3^-(aq) + H_3O^+(aq)$
As the concentration of CO_2 in the atmosphere increases, more will dissolve in H_2O and form H_2CO_3. The H_2CO_3 will then act as a weak acid with H_2O and form HCO_3^- and H_3O^+, causing the water (oceans) to be more acidic. If the pH of the oceans decreases, the added H_3O^+ will react with the CO_3^{2-} ion in the $CaCO_3$ of the limestone structures and decompose the $CaCO_3$.
$2\,H_3O^+(aq) + CaCO_3(s) \rightarrow Ca^{2+}(aq) + H_2CO_3(aq) + 2\,H_2O(l)$; $H_2CO_3(aq) \rightarrow H_2O(l) + CO_2(g)$

17.129 **Given:** 4.00×10^2 mg $Mg(OH)_2$, 2.00×10^2 mL HCl solution, pH $= 1.3$ **Find:** volume neutralized, % neutralized
Conceptual Plan:
mg $Mg(OH)_2 \rightarrow$ g $Mg(OH)_2 \rightarrow$ mol $Mg(OH)_2$ and then pH $\rightarrow [H_3O^+]$ and then mol $Mg(OH)_2$

$$\frac{1\text{ g Mg(OH)}_2}{1000\text{ mg}} \quad \frac{1\text{ mol Mg(OH)}_2}{58.326\text{ g}} \qquad \text{pH} = -\log[\text{H}_3\text{O}^+]$$

mol $OH^- \rightarrow$ mol $H_3O^+ \rightarrow$ vol $H_3O^+ \rightarrow$ % neutralized

$$\frac{2\text{ OH}^-}{\text{Mg(OH)}_2}\ \frac{\text{H}_3\text{O}^+}{\text{OH}^-} \quad \frac{\text{mol H}_3\text{O}^+}{\text{M(H}_3\text{O}^+)} \quad \frac{\text{vol HCl neutralized}}{\text{total vol HCl}} \times 100$$

Solution: $[H_3O^+] = 10^{-1.3} = 0.05\underline{0}12\ M = 0.05\ M$

$$4.00 \times 10^2\ \text{mg Mg(OH)}_2 \times \frac{1\ \text{g Mg(OH)}_2}{1000\ \text{mg Mg(OH)}_2} \times \frac{1\ \text{mol Mg(OH)}_2}{58.3\underline{2}6\ \text{g Mg(OH)}_2} \times \frac{2\ \text{mol OH}^-}{1\ \text{mol Mg(OH)}_2}$$

$$\times \frac{1\ \text{mol H}_3\text{O}^+}{1\ \text{mol OH}^-} \times \frac{1\ \text{L}}{0.05\underline{0}12\ \text{mol H}_3\text{O}^+} \times \frac{1000\ \text{mL}}{1\ \text{L}} = 27\underline{3}.8\ \text{mL} = 274\ \text{mL neutralized}$$

The stomach contains 2.00×10^2 mL HCl at pH = 1.3, and 4.00×10^2 mg will neutralize 274 mL of pH 1.3 HCl; so all of the stomach acid will be neutralized.

17.130 **Given:** 4.3 billion L, pH = 5.5 **Find:** mass in kg $CaCO_3$

Conceptual Plan:

pH → $[H_3O^+]$ and then vol lake → mol $[H_3O^+]$ → mol $CaCO_3$ → g $CaCO_3$ → kg $CaCO_3$

$pH = -\log[H_3O^+]$ $\quad$ mol = vol × M $\quad \frac{1\ \text{mol CaCO}_3}{2\ \text{mol H}_3\text{O}^+} \quad \frac{100.09\ \text{g CaCO}_3}{1\ \text{mol CaCO}_3} \quad \frac{1\ \text{kg CaCO}_3}{1000\ \text{g CaCO}_3}$

Solution: $[H_3O^+] = 10^{-5.5} = 3.\underline{1}6 \times 10^{-6}\ M$

$$4.3 \times 10^9\ \text{L} \times \frac{3.\underline{1}6 \times 10^{-6}\ \text{mol H}_3\text{O}^+}{1\ \text{L}} \times \frac{1\ \text{mol CaCO}_3}{2\ \text{mol H}_3\text{O}^+} \times \frac{100.09\ \text{g CaCO}_3}{1\ \text{mol CaCO}_3} \times \frac{1\ \text{kg}}{1000\ \text{g}}$$

$= 6\underline{8}0.01\ \text{kg} = 6.8 \times 10^2\ \text{kg CaCO}_3$

Check: The units (kg) are correct. The magnitude 680 kg seems reasonable since the volume is so large (10^9 L). The volume more than compensates for the $[H^+]$ (10^{-6} M).

17.131 **Given:** pH of Great Lakes acid rain = 4.5, West Coast = 5.4 **Find:** $[H_3O^+]$ and ratio of Great Lakes/West Coast

Conceptual Plan: pH → $[H_3O^+]$ and then ratio of $[H_3O^+]$ Great Lakes to West Coast

$pH = -\log[H_3O^+]$

Solution: Great Lakes: $[H_3O^+] = 10^{-4.5} = \underline{3}.16 \times 10^{-5}\ M$ West Coast: $[H_3O^+] = 10^{-5.4} = \underline{3}.98 \times 10^{-6}\ M$

$$\frac{\text{Great Lakes}}{\text{West Coast}} = \frac{\underline{3}.16 \times 10^{-5}\ M}{\underline{3}.98 \times 10^{-6}\ M} = \underline{7}.94 = 8 \text{ times more acidic}$$

Check: The answers seem reasonable in light of the fact that pH is a log scale and the difference in pH is almost <1 pH unit.

17.132 **Given:** pH of Sauvignon Blanc = 3.23, Cabernet Sauvignon = 3.64

Find: $[H_3O^+]$ and ratio of Sauvignon Blanc to Cabernet Sauvignon

Conceptual Plan: pH → $[H_3O^+]$ and then ratio of $[H_3O^+]$ Sauvignon Blanc to Cabernet Sauvignon

$pH = -\log[H_3O^+]$

Solution:

Sauvignon Blanc: $[H_3O^+] = 10^{-3.23} = 5.8\underline{8}8 \times 10^{-4}\ M$

Cabernet Sauvignon: $[H_3O^+] = 10^{-3.64} = 2.2\underline{9}1 \times 10^{-4}\ M$

$$\frac{\text{Sauvignon Blanc}}{\text{Cabernet Sauvignon}} = \frac{5.8\underline{8}8 \times 10^{-4}\ M}{2.2\underline{9}1 \times 10^{-4}\ M} = 2.570 = 2.57 \text{ times more acidic}$$

Check: The answers seem reasonable in light of the fact that pH is a log scale and the difference in pH is almost <1 pH unit.

17.133 **Given:** 6.5×10^2 mg aspirin, 8 oz water, $pK_a = 3.5$ **Find:** pH of solution

Conceptual Plan:

mg aspirin → g aspirin → mol aspirin and ounces → quart → L and then [aspirin] and pK_a

$\frac{1\ \text{g aspirin}}{1000\ \text{mg}} \quad \frac{1\ \text{mol aspirin}}{180.15\ \text{g}} \quad \frac{32\ \text{oz}}{1\ \text{qt}} \quad \frac{1.0567\ \text{qt}}{1\ \text{L}} \quad \frac{\text{mol aspirin}}{\text{L soln}} \quad pK_a = -\log K_a$

→ K_a. Write a balanced reaction. Prepare an ICE table, represent the change with x, sum the table, determine the equilibrium values, put the equilibrium values in the equilibrium expression, and solve for x. Determine $[H_3O^+]$ → pH.

$pH = -\log[H_3O^+]$

Solution: $\frac{6.5 \times 10^2 \text{ mg aspirin}}{8 \text{ oz}} \times \frac{1 \text{ g aspirin}}{1000 \text{ mg aspirin}} \times \frac{1 \text{ mol aspirin}}{180.15 \text{ g}} \times \frac{32 \text{ oz}}{1 \text{ qt}} \times \frac{1.0567 \text{ qt}}{1 \text{ L}} = 0.0153 \text{ M}$

$K_a = 10^{-3.5} = 3.16 \times 10^{-4}$

	$\text{aspirin}(aq)$ + $H_2O(l)$ ⇌	$H_3O^+(aq)$ +	$\text{aspirin}^-(aq)$
Initial	0.0153	0.0	0.0
Change	$-x$	$+x$	$+x$
Equil	$0.0153 - x$	x	x

$$K_a = \frac{[H_3O^+][\text{aspirin}^-]}{[\text{aspirin}]} = \frac{(x)(x)}{(0.0153 - x)} = 3.16 \times 10^{-4}$$

$x^2 + 3.16 \times 10^{-4}x - 4.83 \times 10^{-6} = 0$; Solve the quadratic equation. $x = 2.04 \times 10^{-3} \text{ M} = [H_3O^+]$

$\text{pH} = -\log(2.04 \times 10^{-3}) = 2.69 = 2.7$

Check: The units (none) are correct. The answer seems reasonable since aspirin is an acid and patients are warned of potential stomach problems issuing from taking aspirin.

17.134 **Given:** 565 mg/L ddC, $pK_b = 9.8$ **Find:** % protonated

Conceptual Plan: mg → g → mol → M and then $pK_b \rightarrow K_b$ then write a balanced reaction. Prepare an ICE table, represent the change with x, sum the table, determine the equilibrium values, put the the equilibrium values in the equilibrium expression, and solve for x.

Solution: $\frac{565 \text{ mg ddC}}{1 \text{ L}} \times \frac{1 \text{ g ddC}}{1000 \text{ mg ddC}} \times \frac{1 \text{ mol ddC}}{224.22 \text{ g ddC}} = 0.0025198 \text{ M}$

$K_b = 10^{-9.8} = 1.585 \times 10^{-10}$

	$\text{ddC}(aq)$ + $H_2O(l)$ ⇌	$\text{HddC}^+(aq)$ +	$OH^-(aq)$
Initial	0.0025198	0.0	0.0
Change	$-x$	$+x$	$+x$
Equil	$0.0025198 - x$	x	x

$$K_b = \frac{[\text{HddC}^+][OH^-]}{[\text{ddC}]} = \frac{(x)(x)}{(0.0025198 - x)} = 1.585 \times 10^{-10}$$

Assume that x is small.

$x^2 = (1.585 \times 10^{-10})(0.0025198)$; $x = [OH^-] = [\text{HddC}^+] = 6.32 \times 10^{-7} \text{ M}$

$$\% \text{ protonated} = \frac{[\text{HddC}^+]_{\text{equilibrium}}}{[\text{ddC}]_{\text{original}}} \times 100\% = \frac{(6.32 \times 10^{-7})}{(0.0025198)} \times 100\% = 0.025\%$$

Check: The units (%) are correct. The answer seems reasonable since the K is so small.

17.135 (a) **Given:** 0.0100 M $HClO_4$ **Find:** pH

Conceptual Plan: $[HClO_4] \rightarrow [H_3O^+] \rightarrow \text{pH}$

$[HClO_4] = [H_3O^+]$ $\quad \text{pH} = -\log[H_3O^+]$

Solution: $HClO_4$ is a strong acid, so 0.0100 M $HClO_4$ = 0.0100 M H_3O^+ $\quad \text{pH} = -\log(0.0100) = 2.000$

Check: The units (none) are correct. The answer is reasonable since perchloric acid is a strong acid.

(b) **Given:** 0.115 M $HClO_2$, $K_a = 1.1 \times 10^{-2}$ **Find:** pH

Conceptual Plan: Write a balanced reaction. Prepare an ICE table, represent the change with x, sum the table, determine the equilibrium values, put the equilibrium values in the equilibrium expression, and solve for x. Determine $[H_3O^+]$ and pH.

Solution:

	$HClO_2(aq)$ + $H_2O(l)$ ⇌	$H_3O^+(aq)$ +	$ClO_2^-(aq)$
Initial	0.115	0.0	0.0
Change	$-x$	$+x$	$+x$
Equil	$0.115 - x$	x	x

$$K_a = \frac{[H_3O^+][ClO_2^-]}{[HClO_2]} = \frac{(x)(x)}{(0.115 - x)} = 1.1 \times 10^{-2}$$

Assume that x is small compared to 0.115.

$x^2 = (1.1 \times 10^{-2})(0.115)$; $x = 0.03\underline{5}6\text{ M} = [H_3O^+]$

Check assumption: $\frac{0.03\underline{5}6}{0.115} \times 100\% = 31.0\%$; assumption is not valid; solve using the quadratic equation.

$x^2 + 1.1 \times 10^{-2}x - 0.001265 = 0$; Solve the quadratic equation.

$x = 0.03\underline{0}49$ or -0.0415

$\text{pH} = -\log(0.03\underline{0}49) = 1.52$

Check: The units (none) are correct. We expect pH near $-\log((0.115 \times 0.011)^{1/2}) = 1.49$.

(c) **Given:** 0.045 M $Sr(OH)_2$ **Find:** pH

Conceptual Plan: $[Sr(OH)_2] \rightarrow [OH^-] \rightarrow [H_3O^+] \rightarrow \text{pH}$

$K_w = [H_3O^+][OH^-]$ $\text{pH} = -\log[H_3O^+]$ $\text{pH} + \text{pOH} = 14$

Solution: $Sr(OH)_2$ is a strong base, so $[OH^-] = 2[Sr(OH)_2] = 2(0.045) = 0.090\text{ M}$

$$[H_3O^+] = \frac{K_w}{[OH^-]} = \frac{1.0 \times 10^{-14}}{0.090\text{ M}} = 1.\underline{1}1 \times 10^{-13}\text{ M}$$

$\text{pH} = -\log(1.\underline{1}1 \times 10^{-13}) = 12.95$

Check: The units (none) are correct. The answer is reasonable since strontium hydroxide is a strong base.

(d) **Given:** 0.0852 KCN, $K_a(HCN) = 4.9 \times 10^{-10}$ **Find:** pH

Conceptual Plan: Identify each species and determine which will contribute to pH. Write a balanced reaction. Prepare an ICE table, represent the change with *x*, sum the table, determine the equilibrium values, put the equilibrium values in the equilibrium expression, and solve for *x*. Determine $[OH^-] \rightarrow \text{pOH} \rightarrow \text{pH}$.

$\text{pOH} = -\log[OH^-]$ $\text{pH} + \text{pOH} = 14$

Solution: K^+ is the counterion of a strong base; therefore, it is pH-neutral. CN^- is the conjugate base of a weak acid; therefore, it is basic.

	$CN^-(aq)$ + $H_2O(l)$ ⇌	$HCN(aq)$ +	$OH^-(aq)$
Initial	0.0852	0.0	0.0
Change	$-x$	$+x$	$+x$
Equil	$0.0852 - x$	x	x

$$K_b = \frac{K_w}{K_a} = \frac{1.0 \times 10^{-14}}{4.9 \times 10^{-10}} = 2.\underline{0}4 \times 10^{-5} = \frac{(x)(x)}{(0.0852 - x)}$$

Assume that x is small.

$x = 1.\underline{3}2 \times 10^{-3}\text{ M} = [OH^-]$; $\text{pOH} = -\log(1.\underline{3}2 \times 10^{-3}) = 2.88$

$\text{pH} = 14.00 - 2.88 = 11.12$

Check: The units (none) are correct. We expect pH significantly less than $14 - \log(0.0852) = 12.930$.

(e) **Given:** 0.155 NH_4Cl, $K_b(NH_3) = 1.76 \times 10^{-5}$ **Find:** pH

Conceptual Plan: Identify each species and determine which will contribute to pH. Write a balanced reaction. Prepare an ICE table, represent the change with *x*, sum the table, determine the equilibrium values, put the equilibrium values in the equilibrium expression, and solve for *x*. Determine $[H_3O^+] \rightarrow \text{pH}$.

Solution: NH_4^+ is the conjugate acid of a weak base; therefore, it is acidic. Cl^- is the conjugate base of a strong acid; therefore, it is pH-neutral.

	$NH_4^+(aq)$ + $H_2O(l)$ ⇌	$NH_3(aq)$ +	$H_3O^+(aq)$
Initial	0.155	0.0	0.0
Change	$-x$	$+x$	$+x$
Equil	$0.155 - x$	x	x

$$K_a = \frac{K_w}{K_b} = \frac{1 \times 10^{-14}}{1.76 \times 10^{-5}} = 5.6\underline{8}2 \times 10^{-10} = \frac{(x)(x)}{(0.155 - x)}$$

Assume that x is small.

$x = 9.3\underline{8}5 \times 10^{-6}$ M $= [H_3O^+]$; $\quad$ pH $= -\log(9.3\underline{8}5 \times 10^{-6}) = 5.028$

Check: The units (none) are correct. We expect pH significantly above log(0.155) = 0.810.

17.136 (a) **Given:** 0.0650 M HNO_3 **Find:** pH
Conceptual Plan: $[HNO_3] \rightarrow [H_3O^+] \rightarrow$ **pH**

$[HNO_3] = [H_3O] \quad pH = -\log[H_3O^+]$

Solution: HNO_3 is a strong acid, so 0.0650 M HNO_3 = 0.0650 M H_3O^+ $\quad$ pH $= -\log(0.0650) = 1.187$

Check: The units (none) are correct. The answer is reasonable since nitric acid is a strong acid.

(b) **Given:** 0.150 M HNO_2, $K_a = 4.6 \times 10^{-4}$ **Find:** pH
Conceptual Plan: Write a balanced reaction. Prepare an ICE table, represent the change with x, sum the table, determine the equilibrium values, put the equilibrium values in the equilibrium expression, and solve for x. Determine $[H_3O^+]$ and pH.

Solution: $HNO_2(aq) + H_2O(l) \rightleftharpoons H_3O^+(aq) + NO_2^-(aq)$

	$HNO_2(aq)$	$H_2O(l)$	$H_3O^+(aq)$	$NO_2^-(aq)$
Initial	0.150		0.0	0.0
Change	$-x$		$+x$	$+x$
Equil	$0.150 - x$		x	x

$$K_a = \frac{[H_3O^+][NO_2^-]}{[HNO_2]} = \frac{(x)(x)}{(0.150 - x)} = 4.6 \times 10^{-4}$$

Assume that x is small compared to 0.150.

$x^2 = (4.6 \times 10^{-4})(0.150)$; $\quad x = 0.008\underline{3}1$ M $= [H_3O^+]$

Check assumption: $\frac{0.008\underline{3}1}{0.150} \times 100\% = 5.54\%$; assumption is not valid; solve using the quadratic equation.

$x^2 + 4.6 \times 10^{-4}x - 6.9 \times 10^{-5} = 0$; $\quad$ Solve the quadratic equation.

$x = 0.008\underline{0}8$ M $= [H_3O^+]$

pH $= -\log(0.008\underline{0}8) = 2.09$

Check: The units (none) are correct. We expect pH near $-\log((0.150 \times 0.00046)^{1/2}) = 2.08$.

(c) **Given:** 0.0195 M KOH **Find:** pH
Conceptual Plan: $[KOH] \rightarrow [OH^-] \rightarrow [H_3O^+] \rightarrow$ **pH**

$K_w = [H_3O^+][OH^-] \quad pH = -\log[H_3O^+] \quad pH + pOH = 14$

Solution: KOH is a strong base, so $[OH^-] = [KOH] = (0.0195) = 0.0195$ M

$$[H_3O^+] = \frac{K_w}{[OH^-]} = \frac{1.0 \times 10^{-14}}{0.0195 \text{ M}} = 5.1\underline{2}8 \times 10^{-13} \text{ M}$$

pH $= -\log(5.1\underline{2}8 \times 10^{-13}) = 12.290$

Check: The units (none) are correct. The answer is reasonable since potassium hydroxide is a strong base.

(d) **Given:** 0.245 CH_3NH_3I, K_b $(CH_3NH_2) = 4.4 \times 10^{-4}$ **Find:** pH
Conceptual Plan: Identify each species and determine which will contribute to pH. Write a balanced reaction. Prepare an ICE table, represent the change with x, sum the table, determine the equilibrium values, put the equilibrium values in the equilibrium expression, and solve for x. Determine $[H_3O^+] \rightarrow$ pH.

Solution: $CH_3NH_3^+$ is the conjugate acid of a weak base; therefore, it is acidic. I^- is the conjugate base of a strong acid; therefore, it is pH-neutral.

$CH_3NH_4^+(aq) + H_2O(l) \rightleftharpoons CH_3NH_2(aq) + H_3O^+(aq)$

	$CH_3NH_4^+(aq)$	$H_2O(l)$	$CH_3NH_2(aq)$	$H_3O^+(aq)$
Initial	0.245		0.0	0.0
Change	$-x$		$+x$	$+x$
Equil	$0.245 - x$		x	x

$$K_a = \frac{K_w}{K_b} = \frac{1.0 \times 10^{-14}}{4.4 \times 10^{-4}} = 2.\underline{2}73 \times 10^{-11} = \frac{(x)(x)}{(0.245 - x)}$$

Assume that x is small.

$x = 2.\underline{3}6 \times 10^{-6}$ M $= [H_3O^+]$; pH $= -\log(2.\underline{3}6 \times 10^{-6}) = 5.63$

Check: The units (none) are correct. We expect pH significantly above $-\log(0.245) = 0.611$.

(e) **Given:** 0.318 KC_6H_5O, K_a (HC_6H_5O) $= 1.3 \times 10^{-10}$ **Find:** pH

Conceptual Plan: Identify each species and determine which will contribute to pH. Write a balanced reaction. Prepare an ICE table, represent the change with x, sum the table, determine the equilibrium values, put the equilibrium values in the equilibrium expression, and solve for x. Determine $[OH^-] \rightarrow$ pOH $\rightarrow$ pH.

pOH $= -\log[OH^-]$ pH + pOH = 14

Solution: K^+ is the counterion of a strong base; therefore, it is pH-neutral. $C_6H_5O^-$ is the conjugate base of a weak acid; therefore, it is basic.

	$C_6H_5O^-(aq) + H_2O(l) \rightleftharpoons$	$HC_6H_5O(aq)$ +	$OH^-(aq)$
Initial	0.318	0.0	0.0
Change	$-x$	$+x$	$+x$
Equil	$0.318 - x$	x	x

$$K_b = \frac{K_w}{K_a} = \frac{1.0 \times 10^{-14}}{1.3 \times 10^{-10}} = 7.\underline{6}9 \times 10^{-5} = \frac{(x)(x)}{(0.318 - x)}$$

Assume that x is small.

$$x = 4.\underline{9}5 \times 10^{-3}\text{ M} = [OH^-];\ \%\text{ dissociation} = \left(\frac{4.95 \times 10^{-3}}{0.318}\right) \times 100\% = 1.55\%$$

which confirms that the assumption is valid

pOH $= -\log(4.\underline{9}5 \times 10^{-3}) = 2.31$

pH $= 14.00 - 2.31 = 11.69$

Check: The units (none) are correct. We expect pH significantly less than $14 - \log(0.318) = 13.502$.

17.137 (a) **Given:** 0.0550M HI (strong acid), 0.00850M HF (weak acid) **Find:** pH

Conceptual Plan: Because the mixture is a strong acid and a weak acid, the strong acid will dominate. Use the concentration of the strong acid to determine $[H_3O^+]$ and then pH.

Solution: 0.0550M HI = 0.0550 M $[H_3O^+]$ pH $= -\log(0.0550) = 1.260$

Check: The units (none) are correct. The answer is reasonable since hydroiodic acid is a strong acid and will dominate the calculation of pH.

(b) **Given:** 0.112 M NaCl (salt), 0.0953 M KF (salt) **Find:** pH

Conceptual Plan: Identify each species and determine which will contribute to pH. Write a balanced reaction. Prepare an ICE table, represent the change with x, sum the table, determine the equilibrium values, put the equilibrium values in the equilibrium expression, and solve for x. Determine $[H_3O^+] \rightarrow$ pH.

Solution: Na^+ is the counterion of a strong base; therefore, it is pH-neutral. Cl^- is the conjugate base of a strong acid; therefore, it is pH-neutral. K^+ is the counterion of a strong base; therefore, it is pH-neutral. F^- is the conjugate base of a weak acid. Therefore, it will produce a basic solution.

	$F^-(aq) + H_2O(l) \rightleftharpoons$	$HF(aq)$ +	$OH^-(aq)$
Initial	0.0953	0.0	0.0
Change	$-x$	$+x$	$+x$
Equil	$0.0953 - x$	x	x

$$K_b = \frac{K_w}{K_a} = \frac{1.0 \times 10^{-14}}{3.5 \times 10^{-4}} = \frac{(x)(x)}{(0.0953 - x)}$$

Assume that x is small.

$x = 1.\underline{6}50 \times 10^{-6}$ M $= [OH^-]$; pOH $= -\log(1.\underline{6}50 \times 10^{-6}) = 5.78$

pH $= 14.00 - 5.78 = 8.22$

Check: The units (none) are correct. The answer is reasonable since NaCl will not affect the pH and KF will make the solution slightly basic.

(c) **Given:** 0.132 M NH_4Cl (salt), 0.150 M HNO_3 (strong acid) **Find:** pH
Conceptual Plan: Because the mixture is a strong acid and a salt, the strong acid will dominate. Use the concentration of the strong acid to determine $[H_3O^+]$ and then pH.
Solution: 0.150 M HNO_3 = 0.150 M $[H_3O^+]$ pH = $-\log(0.150)$ = 0.824
Check: The units (none) are correct. The answer is reasonable since nitric acid is a strong acid and will dominate the calculation of pH.

(d) **Given:** 0.0887 M $NaC_7H_5O_2$ (salt) 0.225 M KBr (salt)
Conceptual Plan: Identify each species and determine which will contribute to pH. Write a balanced reaction. Prepare an ICE table, represent the change with *x*, sum the table, determine the equilibrium values, put the equilibrium values in the equilibrium expression, and solve for *x*. Determine $[H_3O^+] \rightarrow$ pH.
Solution: Na^+ is the counterion of a strong base; therefore, it is pH-neutral. $C_7H_5O_2^-$ is the conjugate base of a weak acid. Therefore, it will produce a basic solution. K^+ is the counterion of a strong base; therefore, it is pH-neutral. Cl^- is the conjugate base of a strong acid; therefore, it is pH-neutral.

	$C_7H_5O_2^-(aq)$ + $H_2O(l)$ ⇌	$HC_7H_5O_2(aq)$ +	$OH^-(aq)$
Initial	0.0887	0.0	0.0
Change	$-x$	$+x$	$+x$
Equil	$0.0887 - x$	x	x

$$K_b = \frac{K_w}{K_a} = \frac{1.0 \times 10^{-14}}{6.5 \times 10^{-5}} = \frac{(x)(x)}{(0.0887 - x)}$$

Assume that *x* is small.

$x = 3.\underline{6}94 \times 10^{-6}\ M = [OH^-]$; pOH $= -\log(3.\underline{6}94 \times 10^{-6}) = 5.43$
pH = 14.00 − 5.43 = 8.57

Check: The units (none) are correct. The answer is reasonable since KBr will not affect the pH and $NaC_7H_5O_2$ will make the solution slightly basic.

(e) **Given:** 0.0450 M HCl (strong acid), 0.0225 M HNO_3 (strong acid) **Find:** pH
Conceptual Plan: Because the mixture is a strong acid and a strong acid, $[H_3O^+]$ is the sum of the concentration of both acids, and then determine pH.

Solution: 0.0450 M HCl = 0.0450 $[H_3O^+]$, 0.0225 M HNO_3 = 0.0225 M $[H_3O^+]$
$[H_3O^+]$ = 0.0450 + 0.0225 = 0.0675 M; pH = $-\log(0.0675)$ = 1.171

Check: The units (none) are correct. The answer is reasonable since hydrochloric acid and nitric acid are strong acids. The two acid concentrations determine the pH.

17.138 (a) **Given:** 0.050 M KOH (strong base), 0.015 M $Ba(OH)_2$ (strong base) **Find:** pH
Conceptual Plan: Because the mixture is a strong base and a strong base, $[OH^-]$ is the sum of the concentration of both acids. Determine pOH and then pH.
Solution: 0.050 M KOH = 0.050 $[OH^-]$, 0.015 M $Ba(OH)_2$ = 2(0.015 M) $[OH^-]$
$[OH^-]$ = 0.050 + 0.030 = 0.080 M pOH = $-\log(0.080)$ = 1.10, pH = 14.00 − 1.10 = 12.90

Check: The units (none) are correct. The answer is reasonable since potassium hydroxide and barium hydroxide are strong bases. The two base concentrations determine the pH.

(b) **Given:** 0.265 M NH_4NO_3 (salt), 0.102 M HCN (weak acid) **Find:** pH
Conceptual Plan: Identify each species and determine which will contribute to pH. Write a balanced reaction. Prepare an ICE table, represent the change with *x*, sum the table, determine the equilibrium values, put the equilibrium values in the equilibrium expression, and solve for *x*. Determine $[H_3O^+] \rightarrow$ pH.
Solution: NH_4^+ is the conjugate acid of a weak base; therefore, it is acidic. NO_3^- is the conjugate base of a strong acid; therefore, it is pH-neutral. HCN is a weak acid and will produce an acidic solution.

$$K_a(NH_4^+) = \frac{1 \times 10^{-14}}{1.8 \times 10^{-5}} = 5.56 \times 10^{-10}; \quad K_a(HCN) = 4.9 \times 10^{-10}$$

Solution: $NH_4^+(aq) + H_2O(l) \rightleftharpoons NH_3(aq) + H_3O^+(aq)$

	NH_4^+	NH_3	H_3O^+
Initial	0.265	0.0	0.0
Change	$-x$	$+x$	$+x$
Equil	$0.265 - x$	x	x

$$K_a = \frac{K_w}{K_b} = \frac{1.0 \times 10^{-14}}{1.8 \times 10^{-5}} = 5.\underline{5}6 \times 10^{-10} = \frac{(x)(x)}{(0.265 - x)}$$

Assume that x is small.

So $x = 1.\underline{2}1 \times 10^{-5}\ \text{M} = [H_3O^+]$

$$\%\ \text{ionization} = \left(\frac{1.\underline{2}1 \times 10^{-5}}{0.265}\right) \times 100\% = 0.005\%$$

The assumption is valid

	$HCN(aq) + H_2O(l)$	$\rightleftharpoons H_3O^+(aq)$	$+ CN^-(aq)$
Initial	0.102	$1.\underline{2}1 \times 10^{-5}$	0.0
Change	$-x$	$+x$	$+x$
Equil	$0.102 - x$	$(1.\underline{2}1 \times 10^{-5} + x)$	x

$$K_a = \frac{[H_3O^+][CN^-]}{[HCN]} = \frac{(1.\underline{2}1 \times 10^{-5} + x)(x)}{(0.102 - x)} = 4.9 \times 10^{-10}$$

Assume that x is small compared to 0.102.

$$\frac{(1.\underline{2}1 \times 10^{-5})x + x^2}{0.102} = 4.9 \times 10^{-10}$$

$(1.\underline{2}1 \times 10^{-5})x + x^2 = 4.\underline{9}98 \times 10^{-11}$

$x^2 + (1.\underline{2}1 \times 10^{-5})x - 4.\underline{9}98 \times 10^{-11} = 0$

$$x = \frac{-1.\underline{2}1 \times 10^{-5} \pm \sqrt{(1.\underline{2}1 \times 10^{-5})^2 - 4(1)(-4.\underline{9}98 \times 10^{-11})}}{2(1)}$$

$x = 3.\underline{2}55 \times 10^{-6}$ or $-1.\underline{5}36 \times 10^{-5}$ Concentration cannot be negative.

$[H_3O^+] = 1.\underline{2}1 \times 10^{-5}\ \text{M} + 3.\underline{2}6 \times 10^{-6}\ \text{M} = 1.\underline{5}36 \times 10^{-5}\ \text{M}$

$\text{pH} = -\log(1.\underline{5}36 \times 10^{-5}) = 4.81$

Check: The units (none) are correct. The answer is reasonable since we have a weak acid and a slightly acidic salt.

(c) **Given:** 0.075 RbOH (strong base), 0.100 M $NaHCO_3$ (salt)
Conceptual Plan: Because the mixture is a strong base and a salt, the strong base will dominate. Use the concentration of the strong base to determine $[OH^-]$ and then pOH → pH.
Solution: 0.075 M RbOH = 0.075 $[OH^-]$ pOH = $-\log(0.075)$ = 1.12 pH = 14.00 − 1.12 = 12.88

Check: The units (none) are correct. The answer is reasonable since RbOH is a strong base and will dominate the calculation of pH.

(d) **Given:** 0.088 $HClO_4$ (strong acid), 0.022 KOH (strong base) **Find:** pH
Conceptual Plan: A strong acid and a strong base will neutralize each other. Strong acid is in excess and is used to determine the pH.

Solution: $HClO_4(aq) + KOH(aq) \rightarrow H_2O(l) + KClO_4(aq)$

	$HClO_4$	KOH
Initial	0.088	0.022
Change	−0.022	−0.022
Equil	0.066	

$[H_3O^+] = 0.066\ \text{M}$; $\text{pH} = -\log(0.066) = 1.18$

Check: The units (none) are correct. The answer is reasonable since we have more of the strong acid than we have of the strong base. The excess acid concentration will dominate the calculation of pH.

(e) **Given:** 0.115 M NaClO (salt), 0.0500 KI (salt) **Find:** pH

Conceptual Plan: Identify each species and determine which will contribute to pH. Write a balanced reaction. Prepare an ICE table, represent the change with x, sum the table, determine the equilibrium values, put the equilibrium values in the equilibrium expression, and solve for x. Determine $[H_3O^+] \rightarrow$ pH.

Solution: Na^+ is the counterion of a strong base; therefore, it is pH-neutral. ClO^- is the conjugate base of a weak acid. Therefore, it will produce a basic solution. K^+ is the counterion of a strong base; therefore, it is pH-neutral. I^- is the conjugate base of a strong acid; therefore, it is pH-neutral.

	$ClO^-(aq) + H_2O(l) \rightleftharpoons$	$HClO(aq)$ +	$OH^-(aq)$
Initial	0.115	0.0	0.0
Change	$-x$	$+x$	$+x$
Equil	$0.115 - x$	x	x

$$K_b = \frac{K_w}{K_a} = \frac{1.0 \times 10^{-14}}{2.9 \times 10^{-8}} = \frac{(x)(x)}{(0.115 - x)}$$

Assume that x is small.

$$x = 1.\underline{9}9 \times 10^{-4}\ \text{M} = [OH^-];\ \%\ \text{ionization} = \left(\frac{1.99 \times 10^{-4}}{0.115}\right) \times 100\% = 0.173\%$$

$$\text{pOH} = -\log(1.\underline{9}9 \times 10^{-4}) = 3.70$$
$$\text{pH} = 14.00 - 3.70 = 10.30$$

Check: The units (none) are correct. The answer is reasonable since KI will not affect the pH and NaClO will make the solution basic.

17.139 (a) sodium cyanide = NaCN; nitric acid = HNO_3
$H^+(aq) + CN^-(aq) \rightleftharpoons HCN(aq)$

(b) ammonium chloride = NH_4Cl; sodium hydroxide = NaOH
$NH_4^+(aq) + OH^-(aq) \rightleftharpoons NH_3(aq) + H_2O(l)$

(c) sodium cyanide = NaCN; ammonium bromide = NH_4Br
$NH_4^+(aq) + CN^-(aq) \rightleftharpoons NH_3(aq) + HCN(aq)$

(d) potassium hydrogen sulfate = $KHSO_4$; lithium acetate = $LiC_2H_3O_2$
$HSO_4^-(aq) + C_2H_3O_2^-(aq) \rightleftharpoons SO_4^{2-}(aq) + HC_2H_3O_2(aq)$

(e) sodium hypochlorite = NaClO; ammonia = NH_3
No reaction; both are bases.

17.140 **Given:** 0.682 g opium, 8.92 mL of 0.0116 M H_2SO_4 **Find:** % morphine

Conceptual Plan: vol $H_2SO_4 \rightarrow$ mol $H_2SO_4 \rightarrow$ mol $H_3O^+ \rightarrow$ mol morphine $\rightarrow$ g morphine $\rightarrow$ % morphine

$$\text{mol} = \text{VM} \qquad \frac{2\ \text{mol}\ H_3O^+}{1\ \text{mol}\ H_2SO_4} \qquad \frac{1\ \text{mol morphine}}{1\ \text{mol}\ H_3O^+} \qquad \frac{285.3\ \text{g morphine}}{1\ \text{mol morphine}} \qquad \frac{\text{g morphine}}{\text{g opium}} \times 100\%$$

Solution:

$$8.92\ \text{mL}\ H_2SO_4 \times \frac{1\ \text{L}}{1000\ \text{mL}} \times \frac{0.0116\ \text{mol}\ H_2SO_4}{1\ \text{L}} \times \frac{2\ \text{mol}\ H_3O^+}{1\ \text{mol}\ H_2SO_4} \times \frac{1\ \text{mol morphine}}{1\ \text{mol}\ H_3O^+} \times \frac{285.3\ \text{g morphine}}{1\ \text{mol morphine}}$$

$= 0.059\underline{0}4$ g morphine

$$\frac{0.059\underline{0}4\ \text{g morphine}}{0.682\ \text{g opium}} \times 100\% = 8.66\%$$

Check: The units (%) are correct. The answer is reasonable since sulfuric acid is a diprotic acid and morphine accepts one proton.

17.141 **Given:** 1.0 M urea, pH = 7.050 **Find:** K_a Hurea$^+$

Conceptual Plan: pH $\rightarrow$ pOH $\rightarrow [OH^-] \rightarrow K_b$(urea) $\rightarrow K_a$(Hurea$^+$). Write a balanced reaction.

$$\text{pH} + \text{pOH} = 14 \qquad \text{pOH} = -\log[OH^-]$$

Prepare an ICE table, represent the change with x, sum the table, determine the equilibrium values, put the equilibrium values in the equilibrium expression, and determine K_b.

Solution: pOH = 14.000 − 7.050 = 6.950; $[OH^-] = 10^{-6.950} = 1.1\underline{2}2 \times 10^{-7}$ M

	$\text{urea}(aq) + H_2O(l) \rightleftharpoons$	$\text{Hurea}^+(aq)$ +	$OH^-(aq)$
Initial	1.0	0.0	0.0
Change	$-x$	$+x$	$+x$
Equil	$1.0 - 1.1\underline{2}2 \times 10^{-7}$	$1.1\underline{2}2 \times 10^{-7}$	$1.1\underline{2}2 \times 10^{-7}$

$$K_b = \frac{[\text{Hurea}^+][OH^-]}{[\text{urea}]} = \frac{(1.1\underline{2}2 \times 10^{-7})(1.1\underline{2}2 \times 10^{-7})}{(1.0 - 1.1\underline{2}2 \times 10^{-7})} = 1.2\underline{5}89 \times 10^{-14}$$

$$K_a = \frac{K_w}{K_b} = \frac{1.00 \times 10^{-14}}{1.2\underline{5}89 \times 10^{-14}} = 0.79\underline{4}3 = 0.794$$

Check: The units (none) are correct. The answer is reasonable since the pH of a 1 M solution is close to neutral.

17.142 **Given:** 0.10 M $HC_2H_3O_2$, $K_a = 1.8 \times 10^{-5}$; 0.10 M NH_4Cl, $K_b = 1.76 \times 10^{-5}$ **Find:** $[NH_3]$

Conceptual Plan: Use $HC_2H_3O_2$ to determine $[H_3O^+]$; then use NH_4Cl to determine $[NH_3]$. Write a balanced reaction. Prepare an ICE table, represent the change with x, sum the table, and determine the equilibrium values.

Solution: $HC_2H_3O_2(aq) + H_2O(l) \rightleftharpoons H_3O^+(aq) + C_2H_3O_2^-(aq)$

	$HC_2H_3O_2$	H_3O^+	$C_2H_3O_2^-$
Initial	0.10	0.0	0.0
Change	$-x$	$+x$	$+x$
Equil	$0.10 - x$	x	x

$$K_a = \frac{[H_3O^+][C_2H_3O_2^-]}{[HC_2H_3O_2]} = \frac{(x)(x)}{(0.10 - x)} = 1.76 \times 10^{-5}$$

Assume that x is small compared to 0.10.

$x^2 = (1.76 \times 10^{-5})(0.10)$

$x = [H_3O^+] = 0.00133$ M

	$NH_4^+(aq) + H_2O(l) \rightleftharpoons$	$NH_3(aq)$ +	$H_3O^+(aq)$
Initial	0.10	0.0	0.00133
Change	$-y$	$+y$	$+y$
Equil	$0.10 - y$	y	$0.00133 + y$

$$K_a = \frac{K_w}{K_b} = \frac{1.00 \times 10^{-14}}{1.76 \times 10^{-5}} = 5.6\underline{8}2 \times 10^{-10} = \frac{(y)(0.00133 + y)}{(0.10 - y)}$$

Assume that y is small compared to 0.10.

$y^2 + 0.00133y - 5.6\underline{8}2 \times 10^{-11} = 0$ Solve using the quadratic equation.

$y = [NH_3] = 4.\underline{2}7 \times 10^{-8} = 4.3 \times 10^{-8}$ M

Check: The units (M) are correct. The answer is reasonable since ammonia is a weak base.

17.143 **Given:** $Ca(Lact)_2$, $[Ca^{2+}] = 0.26$ M, pH = 8.40 **Find:** K_a lactic acid

Conceptual Plan: $[Ca^{2+}] \rightarrow [Lact^-]$; determine K_b lactate ion. Prepare an ICE table, represent the

$\frac{2 \text{ mol lactate ion}}{1 \text{ mol } Ca^{2+}}$

change with x, sum the table, and determine the equilibrium constant. $K_b \rightarrow K_a$

$K_a = \frac{K_w}{K_b}$

Solution: $0.26 \text{ M } Ca^{2+} \left(\frac{2 \text{ mol Lact}^-}{1 \text{ mol } Ca^{2+}}\right) = 0.52 \text{ M Lact}^-$

	$Lact^-(aq) + H_2O(l) \rightleftharpoons$	$H\,Lact(aq)$ +	$OH^-(aq)$
Initial	0.52	0	0
Change	$-x$	$+x$	$+x$
Equil	$0.52 - x$	x	x

$$\text{pH} = 8.40\ [H_3O^+] = 10^{-8.40} = 4.0 \times 10^{-9};\quad [OH^-] = \frac{1.0 \times 10^{-14}}{4.0 \times 10^{-9}} = 2.5 \times 10^{-6}\ \text{M} = x$$

$$K_b = \frac{[\text{H Lact}][OH^-]}{[\text{Lact}^-]} = \frac{(2.5 \times 10^{-6})(2.5 \times 10^{-6})}{(0.52 - 2.5 \times 10^{-6})} = 1.2 \times 10^{-11}$$

$$K_a = \frac{K_w}{K_b} = \frac{1.0 \times 10^{-14}}{1.2 \times 10^{-11}} = 8.3 \times 10^{-4}$$

Check: The units (none) are correct. The answer is reasonable since lactic acid is a weak acid.

17.144 **Given:** 0.23 mol QHCl, 1.0 L, pH = 4.58 **Find:** K_b quinine

Conceptual Plan: mol QHCl → [QHCl] → [QH$^+$]; determine K_a QH$^+$. Prepare an ICE table, represent the

$$M = \frac{\text{mol}}{\text{L}} \qquad \frac{1\ \text{mol QH}^+}{1\ \text{mol QHCl}}$$

change with x, sum the table, and determine the equilibrium constant. $K_a \rightarrow K_b$

$$K_b = \frac{K_w}{K_a}$$

Solution: $\dfrac{0.23\ \text{mol QHCl}}{1\ \text{L}}\left(\dfrac{1\ \text{mol QH}^+}{1\ \text{mol QHCl}}\right) = 0.23\ \text{M QH}^+$

	$QH^+(aq)$ + $H_2O(l)$	⇌ $Q(aq)$	+ $H_3O^+(aq)$
Initial	0.23	0	0
Change	$-x$	$+x$	$+x$
Equil	$0.23 - x$	x	x

$$\text{pH} = 4.58\ [H_3O^+] = 10^{-4.58} = 2.6 \times 10^{-5} = x$$

$$K_a = \frac{[Q][H_3O^+]}{[QH^+]} = \frac{(2.6 \times 10^{-5})(2.6 \times 10^{-5})}{(0.23 - 2.6 \times 10^{-5})} = 2.9 \times 10^{-9}$$

$$K_b = \frac{K_w}{K_a} = \frac{1.0 \times 10^{-14}}{2.9 \times 10^{-9}} = 3.4 \times 10^{-6}$$

Check: The units (none) are correct. The answer is reasonable since quinine is a weak base.

Challenge Problems

17.145 The calculation is incorrect because it neglects the contribution from the autoionization of water. HI is a strong acid, so $[H_3O^+]$ from HI = 1.0×10^{-7}.

	$H_2O(l)$ + $H_2O(l)$	⇌ $H_3O^+(aq)$	+ $OH^-(aq)$
Initial		1×10^{-7}	0.0
Change	$-x$	$+x$	$+x$
Equil		$1 \times 10^{-7} + x$	x

$$K_w = [H_3O^+][OH^-] = 1.0 \times 10^{-14}$$

$$(1 \times 10^{-7} + x)(x) = 1.0 \times 10^{-14} \quad x^2 + 1 \times 10^{-7}x - 1.0 \times 10^{-14} = 0$$

$$x = 6.18 \times 10^{-8}$$

$$[H_3O^+] = (1 \times 10^{-7} + x) = (1 \times 10^{-7} + 6.18 \times 10^{-8}) = 1.618 \times 10^{-7}\ \text{M}$$

$$\text{pH} = -\log(1.618 \times 10^{-7}) = 6.79$$

17.146 **Given:** 2.55 g HA, molar mass = 85.0 g/mol, 250.0 g H_2O, FP = −0.257 °C **Find:** K_a

Conceptual Plan:

g HA → mol HA → m HA → i → % ionization and then mol HA → M HA. Then write a

$$\text{mol HA} = \frac{\text{g HA}}{\text{molar mass}} \qquad \frac{\text{mol HA}}{\text{kg } H_2O} \qquad \Delta T = iK_f m$$

balanced reaction, prepare an ICE table, calculate equilibrium concentrations, and plug into the equilibrium expression to solve for K_a.

Solution: $2.55 \text{ g HA} \times \frac{1 \text{ mol}}{85.0 \text{ g HA}} = 0.0300 \text{ mol HA} \quad \frac{0.0300 \text{ mol HA}}{\left(250.0 \text{ g H}_2\text{O} \times \frac{1 \text{ kg}}{1000 \text{ g}}\right)} = 0.120\ m$

$\Delta T = iK_f m \quad \Delta T = 0.000\ °C - (-0.257\ °C) = 0.257\ °C$

$(0.257\ °C) = i\left(1.86 \frac{°C}{m}\right)(0.120\ m) \quad i = 1.151 = \text{moles of particles in solution/mol HA}$

$HA(aq) + H_2O(l) \rightleftharpoons H_3O^+(aq) + A^-(aq)$

$1 - y \qquad y \qquad y$

So $(1 - y) + y + y = 1.151 \qquad y = 0.151 = \text{fraction of HA dissociated}$

$$[HA] = \frac{0.0300 \text{ mol HA}}{(250.0 \text{ g} + 2.55 \text{ g})\left(\frac{1.00 \text{ mL}}{1.00 \text{ g}}\right)\left(\frac{1 \text{ L}}{1000 \text{ mL}}\right)} = 0.1188 \text{ M}$$

$(0.1188 \text{ M})(0.151) = 0.01794 \text{ M} = [H_3O^+] = [A^-]$ at equilibrium

	$HA(aq)$ + $H_2O(l)$ ⇌	$H_3O^+(aq)$ +	$A^-(aq)$
Initial	0.1188	0.0	0.0
Change	$-x$	$+x$	$+x$
Equil	$0.1188 - 0.01794$	0.01794	0.01794

$$K_a = \frac{[H_3O^+][A^-]}{[HA]} = \frac{(0.01794)(0.01794)}{(0.1188 - 0.01794)} = 0.003191 = 3.2 \times 10^{-3}$$

Check: The units (none) are correct. The answer is reasonable since we have a weak acid.

17.147 **Given:** 0.00115 M HCl, 0.0100 M $HClO_2$, $K_a = 1.1 \times 10^{-2}$ **Find:** pH

Conceptual Plan: Use HCl to determine $[H_3O^+]$. Use $[H_3O^+]$ and $HClO_2$ to determine dissociation of $HClO_2$. Write a balanced reaction, prepare an ICE table, calculate equilibrium concentrations, and plug into the equilibrium expression.

Solution: $0.00115 \text{ M HCl} = 0.00115 \text{ M } H_3O^+$

	$HClO_2(aq)$ + $H_2O(l)$ ⇌	$H_3O^+(aq)$ +	$ClO_2^-(aq)$
Initial	0.0100	0.00115	0.0
Change	$-x$	$+x$	$+x$
Equil	$0.01000 - x$	$0.00115 + x$	x

$$K_a = \frac{[H_3O^+][ClO_2^-]}{[HClO_2]} = \frac{(0.00115 + x)(x)}{(0.0100 - x)} = 1.1 \times 10^{-2}$$

$x^2 + 0.01215x - 1.1 \times 10^{-4} = 0; \qquad x = 0.006045$

$[H_3O^+] = 0.00115 + 0.006045 = 0.0071\underline{95}$

$pH = -\log(0.0071\underline{95}) = 2.14$

Check: The units (none) are correct. The answer is reasonable since hydrochloric acid is a strong acid and chlorous acid is a fairly strong weak acid.

17.148 Volume should be increased to 4 L.

$HA(aq) + H_2O(l) \rightleftharpoons H_3O^+(aq) + A^-(aq)$

Initial equilibrium conditions: $[HA] = A$, $[H_3O^+] = [A^-] = x$

$$K_a = \frac{[H_3O^+][A^-]}{[HA]} = \frac{(x)(x)}{(A)}$$

Second equilibrium conditions: $[HA] = A/y$, $[H_3O^+] = [A^-] = x/2$

$$K_a = \frac{[H_3O^+][A^-]}{[HA]} = \frac{(x)(x)}{(A)} = \frac{\left(\frac{x}{2}\right)\left(\frac{x}{2}\right)}{\left(\frac{A}{y}\right)}$$

$\frac{x^2A}{y} = \frac{x^2A}{4}$; $y = 4\%$; so for the concentration of H^+ to be halved, the [A] needs to decrease to one-fourth the original concentration. So the volume has to increase to 4 L.

17.149 **Given:** 1.0 M HA, $K_a = 1.0 \times 10^{-8}$, $K = 4.0$ for reaction 2 **Find:** $[H^+]$, $[A^-]$, $[HA_2^-]$

Conceptual Plan: Combine reaction 1 and reaction 2; then determine the equilibrium expression and the value of *K*. Prepare an ICE table, calculate equilibrium concentrations, and plug into the equilibrium expression.

Solution:

$HA(aq) \rightleftharpoons H^+(aq) + A^-(aq) \qquad K = 1.0 \times 10^{-8}$

$HA(aq) + A^-(aq) \rightleftharpoons HA_2^-(aq) \qquad K = 4.0$

$2\,HA(aq) \rightleftharpoons H^+(aq) + HA_2^-(aq) \qquad K = 4.0 \times 10^{-8}$

	$2\,HA(aq)$	$\rightleftharpoons H^+(aq)$	$+\ HA_2^-(aq)$
Initial	1.0	0.0	0.0
Change	$-2x$	$+x$	$+x$
Equil	$1.0 - 2x$	x	x

$$K = \frac{[H^+][HA_2^-]}{[HA]^2} = 4.0 \times 10^{-8} = \frac{(x)(x)}{(1.0 - 2x)^2}$$

Take the square root of both sides of the equation. $x = 2.0 \times 10^{-4}\,M = [H^+]$

	$HA(aq)$	$\rightleftharpoons H^+(aq)$	$+\ A^-(aq)$
Initial	1.0	2.0×10^{-4}	0.0
Change	$-y$	$+y$	$+y$
Equil	$1.0 - y$	$2.0 \times 10^{-4} + y$	y

$K = 1.0 \times 10^{-8}$

$$K = \frac{[H^+][A^-]}{[HA]} = 1.0 \times 10^{-8} = \frac{(2.0 \times 10^{-4} + y)(y)}{(1.0 - y)}$$

Assume that y is small compared to 1.0.

$y^2 + 2.0 \times 10^{-4}y - 1.0 \times 10^{-8} = 0; \quad y = 4.\underline{1}4 \times 10^{-5}$

$[H^+] = x + y = 2.0 \times 10^{-4}\,M + 4.\underline{1}4 \times 10^{-5}\,M = 2.1 \times 10^{-4}\,M$

$[A^-] = y = 4.1 \times 10^{-5}\,M$

$[HA_2^-] = x = 2.0 \times 10^{-4}\,M$

Check: The units (Ms) are correct. The values are reasonable since the value of *K* for the second reaction is greater than 1.

17.150 In the gas phase, $(CH_3)_3N$ is a stronger Lewis base than is CH_3NH_2 because the N—H bond is more polar than the N—C bond, making the lone pair on the N less accessible to the H^+ that needs to be added. In the liquid phase, the steric hindrance from the size of the CH_3 groups becomes more pronounced; therefore, it is harder to add the H^+ ion.

17.151 **Given:** 0.200 mol NH_4CN, 1.00 L, $K_b(NH_3) = 1.76 \times 10^{-5}$, $K_a(HCN) = 4.9 \times 10^{-10}$ **Find:** pH

Conceptual Plan: $K_b\ NH_3 \rightarrow K_a\ NH_4^+$; $K_a\ HCN \rightarrow K_b\ CN^-$ **Prepare an ICE table, represent the**

$K_aK_b = K_w$

change with *x*, sum the table, and determine the equilibrium conditions.

Solution: $K_a(NH_4^+) = \frac{1.0 \times 10^{-14}}{1.76 \times 10^{-5}} = 5.68 \times 10^{-10}$ $K_b(CN^-) = \frac{1.0 \times 10^{-14}}{4.9 \times 10^{-10}} = 2.0 \times 10^{-5}$

0.200 M NH_4CN = 0.200 M NH_4^+ and 0.200 M CN^-

Because the value of *K* for CN^- is greater than *K* for NH_4^+, the CN^- reaction will be larger and the solution will be basic.

	$CN^-(aq)$ +	$H_2O(l) \rightleftharpoons$	$HCN(aq)$ +	$OH^-(aq)$
Initial	0.200		0	0
Change	$-x$		$+x$	$+x$
Equil	$0.200 - x$		x	x

$K_b = \frac{[HCN][OH^-]}{[CN^-]}$; $2.0 \times 10^{-5} = \frac{(x)(x)}{(0.200 - x)}$ Solve using the quadratic equation.

$x = 0.0020\ M = [OH^-]$

$[H_3O^+] = \frac{1.0 \times 10^{-14}}{0.0020} = 5.0 \times 10^{-12}$; $pH = -\log(5.0 \times 10^{-12}) = 11.30$

Check: The units (none) are correct. The value is reasonable since the value of K for CN^- is greater than the K for NH_4^+.

17.152 **Given:** 1.0 L, 0.30 M $HClO_2$, 0.20 mol NaF, $K_a\ (HClO_2) = 1.1 \times 10^{-2}$, $K_a\ (HF) = 3.5 \times 10^{-4}$ **Find:** $[HClO_2]$
Conceptual Plan: Determine the products of the reaction of $HClO_2$ and NaF. Prepare an ICE table, represent the change with x, sum the table, and determine the equilibrium conditions.

Solution:

	$HClO_2(aq)$ +	$NaF(aq) \rightarrow$	$HF(aq)$ +	$NaClO_2(aq)$
initial	0.30	0.20		
	-0.20	-0.20	$+0.20$	$+0.20$
final	0.10	0	0.20	0.20

Because HF is the weaker acid, the reaction will proceed to the right. Because NaF is the smaller amount, NaF is the limiting reactant.

The $HClO_2$ has the largest equilibrium constant, so it will be the species that contributes to the pH to the largest extent.

	$HClO_2(aq)$ +	$H_2O(l) \rightleftharpoons$	$ClO_2^-(aq)$ +	$H_3O^+(aq)$
Initial	0.10		0.20	0
Change	$-x$		$+x$	$+x$
Equil	$0.10 - x$		$0.20 + x$	x

$K_a = \frac{[ClO_2^-][H_3O^+]}{[HClO_2]}$ $\quad 1.1 \times 10^{-2} = \frac{(0.20 + x)(x)}{(0.10 - x)}$

Solve the quadratic, $x = 0.0051$; $[HClO_2] = 0.10 - 0.0051 = 0.095\ M$

17.153 **Given:** mixture Na_2CO_3 and $NaHCO_3$ = 82.2 g, 1.0 L, pH = 9.95, $K_a\ (H_2CO_3) = K_{a_1} = 4.3 \times 10^{-7}$, $K_{a_2} = 5.6 \times 10^{-11}$ **Find:** mass $NaHCO_3$
Conceptual Plan: Let x = g $NaHCO_3$, y = g Na_2CO_3 → mol $NaHCO_3$, Na_2CO_3 → $[NaHCO_3]$, $[Na_2CO_3]$

$\frac{1\ mol\ NaHCO_3}{84.01\ g}$ $\quad \frac{1\ mol\ Na_2CO_3}{105.99\ g}$ $\quad M = \frac{mol}{L}$

→ $[HCO_3^-]$, $[CO_3^{2-}]$, $K_a\ (HCO_3^-)$ → $K_b(CO_3^{2-})$. Prepare an ICE table, represent the change with z,

$\frac{HCO_3^-}{NaHCO_3}$ $\quad \frac{CO_3^{-2}}{Na_2CO_3}$ $\quad K_b = \frac{K_w}{K_a}$

sum the table, and determine the equilibrium conditions.

Solution: Let x = g $NaHCO_3$ and y = g Na_2CO_3 $x + y = 82.2$, so $y = 82.2 - x$

$$\text{mol } NaHCO_3 = x \text{ g } \cancel{NaHCO_3}\left(\frac{1\ mol}{84.01\ g\ \cancel{NaHCO_3}}\right) = \frac{x}{84.01}$$

$$[HCO_3^-] = \left(\frac{\frac{x}{84.01}\ \cancel{mol\ NaHCO_3}}{1\ L}\right)\frac{1\ mol\ HCO_3^-}{1\ \cancel{mol\ NaHCO_3}} = \frac{x}{84.01} M\ HCO_3^-$$

$$\text{mol } Na_2CO_3 = 82.2 - x \text{ g } \cancel{Na_2CO_3}\left(\frac{1\ mol}{105.99\ g\ \cancel{Na_2CO_3}}\right) = \frac{82.2 - x}{105.99} Na_2CO_3$$

$$[CO_3^{-2}] = \frac{\left(\frac{82.2 - x}{105.99}\ \cancel{\text{mol Na}_2\text{CO}_3}\right)}{1\ \text{L}} \frac{1\ \text{mol CO}_3^{2-}}{1\ \cancel{\text{mol Na}_2\text{CO}_3}} = \frac{82.2 - x}{105.99}\ \text{M CO}_3^{2-}$$

Because the pH of the solution is basic, it is the hydrolysis of CO_3^{2-} that dominates in the solution.

$K_b(CO_3^{2-}) = \frac{1.0 \times 10^{-14}}{5.6 \times 10^{-11}} = 1.79 \times 10^{-4}$ and $[H_3O^+] = 10^{-9.95} = 1.12 \times 10^{-10}$ M; $[OH^-] = \frac{1.0 \times 10^{-14}}{1.12 \times 10^{-10}}$ $= 8.93 \times 10^{-5}$ M

	$CO^{2-}(aq)$ + $H_2O(l)$ $\rightleftharpoons$	$HCO_3^-(aq)$	+ $OH^-(aq)$
Initial	$\frac{82.2 - x}{105.99}$	$\frac{x}{84.01}$	
Change	$-z$	$+z$	
Equil	$\left(\frac{82.2 - x}{105.99}\right) - z$	$\left(\frac{x}{84.01}\right) + z$	8.93×10^{-5}

$$K_b(CO_3^{2-}) = 1.79 \times 10^{-4} = \frac{[HCO_3^-][OH^-]}{[CO_3^{2-}]} = \frac{\left(\left(\frac{x}{84.01}\right) - z\right)(8.93 \times 10^{-5})}{\left(\frac{82.2 - x}{105.99}\right) - z}$$ Assume that z is small.

$$\frac{1.79 \times 10^{-4}}{8.93 \times 10^{-5}} = \frac{\left(\frac{x}{84.01}\right)}{\left(\frac{82.2 - x}{105.99}\right)};\quad x = 50.\underline{3}8\ \text{g} = 50.4\ \text{g NaHCO}_3$$

17.154 **Given:** mixture NaCN and $NaHSO_4$ = 0.60 mol, 1.0 L, pH = 9.9; $K_a(HCN) = 4.9 \times 10^{-10}$, $K_a(HSO_4^-) = 1.1 \times 10^{-2}$ **Find:** amount NaCN

Conceptual Plan: Determine $[CN^-]$ in solution after the reaction of NaCN and $NaHSO_4$. Then prepare an ICE table, represent the change by x, and determine the equilibrium conditions.

Solution: Because HCN is the weaker acid, we can assume complete reaction between NaCN and $NaHSO_4$. Also, because the solution is basic, the CN^- must be in excess. So let x = moles of NaCN. So $NaHSO_4 = 0.60 - x$.

$NaHSO_4(aq)$ +	$NaCN(aq)$ →	$HCN(aq)$ +	$SO_4^{2-}(aq)$
$0.60 - x$	x	0	0
$-(0.60 - x)$	$-(0.60 - x)$	$0.60 - x$	$0.60 - x$
0	$2x - 0.60$	$0.60 - x$	$0.60 - x$

$[H_3O^+] = 10^{-9.9} = 1.26 \times 10^{-10}$ $[OH^-] = \frac{1.0 \times 10^{-14}}{1.\underline{26} \times 10^{-10}} = 7.9 \times 10^{-5}$

	$CN^-(aq) + H_2O(l)$ $\rightleftharpoons$	$HCN(aq)$ +	$OH^-(aq)$
Initial	$2x - 0.60$	$0.60 - x$	0
Change	-7.9×10^{-5}	$+7.9 \times 10^{-5}$	$+7.9 \times 10^{-5}$
Equil	$2x - 0.60 - 7.9 \times 10^{-5}$	$0.60 - x + 7.9 \times 10^{-5}$	7.9×10^{-5}

$$K_b(CN^-) = \frac{K_w}{K_a(HCN)} = \frac{1.0 \times 10^{-14}}{4.9 \times 10^{-10}} = 2.0 \times 10^{-5} = \frac{[HCN][OH^-]}{[CN^-]} = \frac{(0.60 - x)(7.9 \times 10^{-5})}{(2x - 0.60)}$$

$x = 0.50$ = mol NaCN

Conceptual Problems

17.155 Solution (b) would be most acidic.

(a) 0.0100 M HCl (strong acid) and 0.0100 M KOH (strong base) because the concentrations are equal, the acid and base will completely neutralize each other, and the resulting solution will be pH-neutral.

(b) 0.0100 M HF (weak acid) and 0.0100 M KBr (salt). K_a (HF) $= 3.5 \times 10^{-4}$. The weak acid will produce an acidic solution. K^+ is the counterion of a strong base and is pH-neutral. Br^- is the conjugate base of a strong acid and is pH-neutral.

(c) 0.0100 M NH_4Cl (salt) and 0.100 M CH_3NH_3Br. $K_b(NH_3) = 1.8 \times 10^{-5}$, $K_b(CH_3NH_2) = 4.4 \times 10^{-4}$. NH_4^+ is the conjugate acid of a weak base, and $CH_3NH_3^+$ is the conjugate acid of a weak base. Cl^- and Br^- are the conjugate bases of strong acids and will be pH-neutral. K_a of $NH_4^+ = 5.68 \times 10^{-10}$, K_a of $CH_3NH_3^+ = 2.27 \times 10^{-11}$. Since NH_4^+ has a higher K_a, it will have a higher concentration of $[H_3O^+]$ ions and therefore will be more acidic. The solution will be acidic. However, because the K_a for the conjugate acids in this solution is smaller K_a for HF, the solution will be acidic, but not as acidic as HF.

(d) 0.100 M NaCN (salt) and 0.100 M $CaCl_2$. Na^+ and Ca^{2+} ion are the counterions of a strong base; therefore, they are pH-neutral. Cl^- is the conjugate base of a strong acid and is pH-neutral. CN^- is the conjugate base of a weak acid and will produce a basic solution.

17.156 (a) 0.100 M NaClO (salt) and 0.100 M NaF (salt). Na^+ is the counterion of a strong base and is pH-neutral. ClO^- is the conjugate base of a weak acid (HClO, $K_a = 2.9 \times 10^{-8}$). F^- is the conjugate base of a weak acid (HF, $K_a = 3.5 \times 10^{-4}$). The solution will be basic, and ClO^- is a stronger base than is F^-. K_b of $ClO^- = 3.45 \times 10^{-7}$ and K_b of $F^- = 2.86 \times 10^{-11}$. Since ClO^- has a higher K_b, it will have a higher concentration of $[OH^-]$ ions and therefore, NaClO will be more basic.

(b) 0.0100 M KCl (salt) and 0.0100 M $KClO_2$ (salt). K^+ is the counterion of a strong base and is pH-neutral. Cl^- is the conjugate base of a strong acid and is pH-neutral. ClO_2^- is the conjugate base of a weak acid ($HClO_2$, $K_a = 1.1 \times 10^{-2}$) and will produce a basic solution. However, because the ClO_2^- is a weaker conjugate base than is ClO^-, solution (a) will be more basic. Therefore, $KClO_2$ will be the more basic of the two solutions.

(c) 0.0100 M HNO_3 (strong acid) and 0.0100 M NaOH (strong base) because the concentrations are equal, the acid and base will completely neutralize each other, and the resulting solution will be pH-neutral. NaOH is a strong base and hence is more basic.

(d) 0.0100 M NH_4Cl (salt) and 0.0100 M HCN (weak acid). NH_4^+ is the conjugate acid of a weak base and will be acidic. HCN is a weak acid and will be acidic. Cl^- is the conjugate base of a strong acid and is pH-neutral. K_a of $NH_4^+ = 5.68 \times 10^{-10}$, K_a of HCN $= 4.9 \times 10^{-10}$. They have almost the same K_a and hence almost the same acidity. Since NH_4^+ has a slightly higher K_a, it will have a slightly higher concentration of $[H_3O^+]$ ions; therefore, NH_4Cl will be slightly more acidic and HCN will be more basic.

17.157 $CH_3COOH < CH_2ClCOOH < CHCl_2COOH < CCl_3COOH$

Because Cl is more electronegative than is H, as you add Cl, you increase the number of electronegative atoms, which pulls the electron density away from the O—H group, polarizing the O—H bond, making it more acidic.

18 Aqueous Ionic Equilibrium

Review Questions

18.1 The pH range of human blood is between 7.36 and 7.42. This nearly constant blood pH is maintained by buffers that are chemical systems that resist pH changes, neutralizing an added acid or base. An important buffer system in blood is a mixture of carbonic acid (H_2CO_3) and bicarbonate ion (HCO_3^-).

18.2 A buffer is a chemical system that resists pH changes. The buffer works by neutralizing an added acid or base. Most buffers contain significant amounts of both a weak acid and its conjugate base (or a weak base and its conjugate acid). When additional base is added to a buffer, the weak acid reacts with the base, neutralizing it (generating more of the buffer system conjugate base). When additional acid is added to a buffer, the conjugate base reacts with the acid, neutralizing it (generating more of the buffer system conjugate acid). In this way, a buffer can maintain a nearly constant pH.

18.3 The common ion effect occurs when a solution contains two substances ($HC_2H_3O_2$ and $NaC_2H_3O_2$) that share a common ion ($C_2H_3O_2^-$). The presence of the $C_2H_3O_2^-(aq)$ ion causes the acid to ionize even less than it normally would, resulting in a less acidic solution (higher pH). This effect is an example of Le Châtelier's principle shifting an equilibrium because of the addition (or removal) of the common ion from the solution.

18.4 The Henderson–Hasselbalch equation $\left(\text{pH} = \text{p}K_\text{a} + \log\frac{[\text{base}]}{[\text{acid}]}\right)$ allows easy calculation of the pH of a buffer solution from the initial concentrations of the buffer components as long as the "x is small" approximation is valid.

18.5 When the concentration of the conjugate acid and base components of a buffer system are equal, the pH is equal to the pK_a of the weak acid of the buffer system. When more of the acid component is present, the pH becomes more acidic (pH drops). When more of the base component is present, the pH becomes more basic (pH rises). The pH in both cases can be calculated using the Henderson–Hasselbalch equation.

18.6 When the concentration of the conjugate acid and base components of a buffer system are equal, the pH is equal to the pK_a of the weak acid of the buffer system. When a small amount of a strong acid is added, it reacts with the conjugate base of the buffer system, converting it to the weak acid of the buffer system. Thus, the weak acid concentration increases, the conjugate base concentration decreases, and the pH drops slightly at most. When a small amount of a strong base is added, it reacts with the weak acid of the buffer system, converting it to the conjugate base of the buffer system. Thus, the weak acid concentration decreases, the conjugate base concentration increases, and the pH rises slightly at most.

18.7 To find the pH of this solution, determine which component is the acid, determine which component is the base, and substitute their concentrations into the Henderson–Hasselbalch equation. The pK_a is the negative of the log of the equilibrium constant of the acid dissociation reaction where the weak acid component and water are the reactants and the conjugate base and H_3O^+ are the products. At the end, confirm that the "x is small" approximation is valid by calculating the $[H_3O^+]$ from the pH. Because H_3O^+ is formed by ionization of the acid, the calculated $[H_3O^+]$ has to be less than 0.05 (or 5%) of the initial concentration of the acid for the "x is small" approximation to be valid.

18.8 The factors that influence the effectiveness of a buffer are the relative amounts of the acid and conjugate base (the closer they are to each other, the more effective the buffer) and the absolute concentrations of the acid and conjugate base (the higher the absolute concentrations, the more effective the buffer).

18.9 The relative concentrations of the acid and conjugate base should not differ by more than a factor of 10 for a buffer to be reasonably effective. Using the Henderson–Hasselbalch equation, this means that the pH should be within one pH unit of the weak acid's pK_a.

18.10 In an acid–base titration, a basic (or acidic) solution of unknown concentration is reacted with an acidic (or basic) solution of known concentration. The known solution is slowly added to the unknown solution while the pH is monitored with either a pH meter or an indicator (a substance whose color depends on the pH). As the acid and base combine, they neutralize each other. At the equivalence point—the point in the titration when the number of moles of base is stoichiometrically equal to the number of moles of acid—the titration is complete. When this point is reached, neither reactant is in excess and the number of moles of the reactants are related by the reaction stoichiometry.

18.11 The titration of weak acid by a strong base always has a basic equivalence point because at the equivalence point, all of the acid has been converted into its conjugate base, resulting in a weakly basic solution.

18.12 The volume required to get to the equivalence point is only dependent on the concentration and volume of acid or base to be titrated and the base or acid used to do the titration because the equivalence point is dependent on the stoichiometry of the balanced reaction of the acid and base. The stoichiometry only considers the number of moles involved, not the strength of the reactants involved.

18.13
(a) The initial pH of the solution is simply the pH of the strong acid. Because strong acids completely dissociate, the concentration of H_3O^+ is the concentration of the strong acid and $pH = -\log[H_3O^+]$.
(b) Before the equivalence point, H_3O^+ is in excess. Calculate the $[H_3O^+]$ by subtracting the number of moles of added OH^- from the initial number of moles of H_3O^+ and dividing by the *total* volume. Then convert to pH using $-\log[H_3O^+]$.
(c) At the equivalence point, neither reactant is in excess and the $pH = 7.00$.
(d) Beyond the equivalence point, OH^- is in excess. Calculate the $[OH^-]$ by subtracting the initial number of moles of H_3O^+ from the number of moles of added OH^- and dividing by the *total* volume. Then convert to pH using $-\log[H_3O^+]$.

18.14
(a) The initial pH is that of the weak acid solution to be titrated. Calculate the pH by working an equilibrium problem (similar to Examples 17.5 and 17.6) using the concentration of the weak acid as the initial concentration.
(b) Between the initial pH and the equivalence point, the solution becomes a buffer. Use the reaction stoichiometry to compute the amounts of each buffer component and then use the Henderson–Hasselbalch equation to compute the pH (as in Example 18.3).
(c) Halfway to the equivalence point, the buffer components are exactly equal and $pH = pK_a$.
(d) At the equivalence point, the acid has all been converted into its conjugate base. Calculate the pH by working an equilibrium problem for the ionization of water by the ion acting as a weak base (similar to Example 17.14). (Compute the concentration of the ion acting as a weak base by dividing the number of moles of the ion by the total volume at the equivalence point.)
(e) Beyond the equivalence point, OH^- is in excess. You can ignore the weak base and calculate the $[OH^-]$ by subtracting the initial number of moles of H_3O^+ from the number of moles of added OH^- and dividing by the total volume, then converting to pH using $-\log[H_3O^+]$.

18.15 When a polyprotic acid is titrated with a strong base and if K_{a_1} and K_{a_2} are sufficiently different, the pH curve will have two equivalence points because the two acidic protons will be titrated sequentially. The titration of the first acidic proton will be completed before the titration of the second acidic proton.

18.16 The volume required to get to the first equivalence point is identical to the volume between the first and second equivalence points because the equivalence point is dependent on the stoichiometry of the balanced reaction of the acid and base. The stoichiometry only considers the number of moles involved, not the strength of the reactants involved. There are the same number of moles of the first acidic proton and the second acidic proton.

18.17 The endpoint is the point when the indicator changes color in an acid–base titration. The equivalence point is when stoichiometrically equivalent amounts of acid and base have reacted. With the correct indicator, the endpoint of the titration will occur at the equivalence point.

18.18 An indicator (HIn) is itself a weak organic acid that has a different color than its conjugate base (In^-). The color of a solution containing an indicator depends on the relative concentrations of HIn and In^-. As the $[H_3O^+]$ changes during the titration, the above relative concentrations of HIn and In^- change accordingly. At low pH, the $[H_3O^+]$ is high, the equilibrium favors the acid species, and the color is that of the acid species. As the titration proceeds, the $[H_3O^+]$ decreases, shifting toward higher concentrations of the conjugate base, and the color changes. Because the color of an indicator is intense, only a small amount is required—an amount that will not affect the pH of the solution or the equivalence point of the neutralization reaction.

18.19 The solubility product constant (K_{sp}) is the equilibrium expression for a chemical equation representing the dissolution of an ionic compound. The expression of the solubility product constant of A_mX_n is $K_{sp} = [A^{n+}]^m [X^{m-}]^n$.

18.20 The molar solubility, S, is simply the solubility in units of moles per liter (mol/L). The molar solubility of a compound, A_mX_n, can be computed directly from K_{sp} by solving for S in the expression $K_{sp} = (mS)^m (nS)^n = m^m n^n S^{m+n}$.

18.21 In accordance with Le Châtelier's principle, the presence of a common ion in solution causes the equilibrium to shift to the left (compared to its position with pure water as the solvent), which means that less of the ionic compound dissolves. Thus, the solubility of an ionic compound is lower in a solution containing a common ion than in pure water. The exact value of the solubility can be calculated by working an equilibrium problem in which the concentration of the common ion is accounted for in the initial conditions. The molar solubility of a compound, A_mX_n, in a solution with an initial concentration of $A^{n+} = [A^{n+}]_0$ and an initial concentration of $X^{m-} = [X^{m-}]_0$ can be computed directly from K_{sp} by solving for S in the expression $K_{sp} = ([A^{n+}]_0 + mS)^m ([X^{m-}]_0 + nS)^n$.

18.22 In general, the solubility of an ionic compound with a strongly basic or weakly basic anion increases with increasing acidity (decreasing pH). If the anion is neutralized by reaction with H^+, to form the conjugate acid of the basic anion, Le Châtelier's principle says that the solubility equilibrium will shift to the right, allowing more solid to dissolve. Remember that the K_{sp} expression only has the terms of the cation and the anion concentrations. If the anion is converted to a different species (such as the conjugate acid), the concentration of the anion drops, allowing more solid to dissolve.

18.23 Q is the reaction quotient, the product of the concentrations of the ionic components raised to their stoichiometric coefficients, and K is the product of the concentrations of the ionic components raised to their stoichiometric coefficients at equilibrium. A solution contains an ionic compound: If $Q < K_{sp}$, the solution is unsaturated. More of the solid ionic compound can dissolve in the solution: If $Q = K_{sp}$, the solution is saturated, the solution is holding the equilibrium amount of the dissolved ions, and additional solid will not dissolve in the solution. If $Q > K_{sp}$, the solution is supersaturated and under most circumstances, the excess solid will precipitate out.

18.24 Selective precipitation is a process for separating metal cations from a solution containing several different dissolved metal cations. In the process, a metal cation can often be separated by the addition of a reagent that forms a precipitate with one of the dissolved cations but not the others. The appropriate reagent must form compounds with both metal cations with sufficiently different K_{sp} values (a difference of a factor of at least 10^3), so that one compound remains dissolved while the other compound significantly precipitates.

The Common Ion Effect and Buffers

18.25 The only solution in which HNO_2 will ionize less is (d) 0.10 M $NaNO_2$. It is the only solution that generates a common ion NO_2^- with nitrous acid.

18.26 Formic acid is $HCHO_2$, which dissociates to H^+ and CHO_2^-. The only solution that generates a common ion (CHO_2^-) with formic acid is (c) $NaCHO_2$.

18.27 (a) **Given:** 0.20 M $HCHO_2$ and 0.15 M $NaCHO_2$ **Find:** pH **Other:** $K_a(HCHO_2) = 1.8 \times 10^{-4}$

Conceptual Plan: M $NaCHO_2$ → M CHO_2^- then M $HCHO_2$, M CHO_2^- → $[H_3O^+]$ → pH

$NaCHO_2(aq) \rightarrow Na^+(aq) + CHO_2^-(aq)$ ICE table $pH = -\log[H_3O^+]$

Solution: Because 1 CHO_2^- ion is generated for each $NaCHO_2$, $[CHO_2^-] = 0.15$ M CHO_2^-.

$$HCHO_2(aq) + H_2O(l) \rightleftharpoons H_3O^+(aq) + CHO_2^-(aq)$$

	$[HCHO_2]$	$[H_3O^+]$	$[CHO_2^-]$
Initial	0.20	≈ 0.00	0.15
Change	$-x$	$+x$	$+x$
Equil	$0.20 - x$	$+x$	$0.15 + x$

$K_a = \dfrac{[H_3O^+][CHO_2^-]}{[HCHO_2]} = 1.8 \times 10^{-4} = \dfrac{x(0.15 + x)}{0.20 - x}$ Assume that x is small ($x << 0.15 < 0.20$), so $\dfrac{x(0.15 + \cancel{x})}{0.20 - \cancel{x}} = 1.8 \times 10^{-4} = \dfrac{x(0.15)}{0.20}$ and $x = 2.4 \times 10^{-4}$ M $= [H_3O^+]$. Confirm that the more stringent assumption is valid.

$\dfrac{2.4 \times 10^{-4}}{0.15} \times 100\% = 0.16\%$, so the assumption is valid. Finally,

$pH = -\log[H_3O^+] = -\log(2.4 \times 10^{-4}) = 3.62.$

Check: The units (none) are correct. The magnitude of the answer makes physical sense because pH should be greater than $-\log(0.20) = 0.70$ because this is a weak acid and there is a common ion effect.

(b) **Given:** 0.16 M NH_3 and 0.22 M NH_4Cl **Find:** pH **Other:** $K_b(NH_3) = 1.79 \times 10^{-5}$

Conceptual Plan: M $NH_4Cl \rightarrow$ M NH_4^+ then M NH_3, M $NH_4^+ \rightarrow [OH^-] \rightarrow [H_3O^+] \rightarrow$ pH

$NH_4Cl(aq) \rightarrow NH_4^+(aq) + Cl^-(aq)$ ICE table $K_w = [H_3O^+][OH^-]$ $pH = -\log[H_3O^+]$

Solution: Because 1 NH_4^+ ion is generated for each NH_4Cl, $[NH_4^+] = 0.22$ M NH_4^+.

$$NH_3(aq) + H_2O(l) \rightleftharpoons NH_4^+(aq) + OH^-(aq)$$

	$[NH_3]$	$[NH_4^+]$	$[OH^-]$
Initial	0.16	0.22	≈ 0.00
Change	$-x$	$+x$	$+x$
Equil	$0.16 - x$	$0.22 + x$	$+x$

$K_b = \dfrac{[NH_4^+][OH^-]}{[NH_3]} = 1.79 \times 10^{-5} = \dfrac{(0.22 + x)x}{0.16 - x}$

Assume that x is small ($x << 0.16 < 0.22$), so $\dfrac{(0.22 + \cancel{x})x}{0.16 - \cancel{x}} = 1.79 \times 10^{-5} = \dfrac{(0.22)x}{0.16}$ and $x = 1.\underline{3}0182 \times 10^{-5}$ M $= [OH^-]$. Confirm that the more stringent assumption is valid.

$\dfrac{1.\underline{3}0182 \times 10^{-5}}{0.16} \times 100\% = 8.1 \times 10^{-3}\%$, so the assumption is valid.

$K_w = [H_3O^+][OH^-]$ so $[H_3O^+] = \dfrac{K_w}{[OH^-]} = \dfrac{1.0 \times 10^{-14}}{1.\underline{3}0182 \times 10^{-5}} = 7.\underline{6}816 \times 10^{-10}$ M

Finally, $pH = -\log[H_3O^+] = -\log(7.\underline{6}816 \times 10^{-10}) = 9.11.$

Check: The units (none) are correct. The magnitude of the answer makes physical sense because pH should be less than $14 + \log(0.16) = 13.2$ because this is a weak base and there is a common ion effect.

18.28 (a) **Given:** 0.195 M $HC_2H_3O_2$ and 0.125 M $KC_2H_3O_2$ **Find:** pH

Other: $K_a(HC_2H_3O_2) = 1.8 \times 10^{-5}$

Conceptual Plan: M $KC_2H_3O_2 \rightarrow$ M $C_2H_3O_2^-$ then M $HC_2H_3O_2$, M $C_2H_3O_2^- \rightarrow [H_3O^+] \rightarrow$ pH

$KC_2H_3O_2(aq) \rightarrow K^+(aq) + C_2H_3O_2^-(aq)$ ICE table $pH = -\log[H_3O^+]$

Solution: Because 1 $C_2H_3O_2^-$ ion is generated for each $KC_2H_3O_2$, $[C_2H_3O_2^-] = 0.125$ M $C_2H_3O_2^-$.

$$HC_2H_3O_2(aq) + H_2O(l) \rightleftharpoons H_3O^+(aq) + C_2H_3O_2^-(aq)$$

	$[HC_2H_3O_2]$	$[H_3O^+]$	$[C_2H_3O_2^-]$
Initial	0.195	≈0.00	0.125
Change	$-x$	$+x$	$+x$
Equil	$0.195 - x$	$+x$	$0.125 + x$

$K_a = \dfrac{[H_3O^+][HC_2H_3O_2^-]}{[HC_2H_3O_2]} = 1.8 \times 10^{-5} = \dfrac{x(0.125 + x)}{0.195 - x}$ Assume that x is small ($x << 0.125 < 0.195$), so

$\dfrac{x(0.125 + \cancel{x})}{0.195 - \cancel{x}} = 1.8 \times 10^{-5} = \dfrac{x(0.125)}{0.195}$ and $x = 2.\underline{8}08 \times 10^{-5}\text{ M} = [H_3O^+]$. Confirm that the more stringent assumption is valid.

$\dfrac{2.\underline{8}08 \times 10^{-5}}{0.195} \times 100\% = 0.022\%$, so the assumption is valid. Finally,

$pH = -\log[H_3O^+] = -\log(2.\underline{8}08 \times 10^{-5}) = 4.55.$

Check: The units (none) are correct. The magnitude of the answer makes physical sense because pH should be greater than $-\log(0.125) = 0.90$ because this is a weak acid and there is a common ion effect.

(b) **Given:** 0.255 M CH_3NH_2 and 0.135 M CH_3NH_3Br **Find:** pH
Other: $K_b\,(CH_3NH_2) = 4.4 \times 10^{-4}$
Conceptual Plan: M $CH_3NH_3Br \rightarrow$ M $CH_3NH_3^+$ then

$CH_3NH_3Br(aq) \rightarrow CH_3NH_3^+(aq) + Br^-(aq)$

M CH_3NH_2, M $CH_3NH_3^+ \rightarrow [OH^-] \rightarrow [H_3O^+] \rightarrow$ pH

ICE table $\quad K_w = [H_3O^+][OH^-] \quad pH = -\log[H_3O^+]$

Solution: Because 1 $CH_3NH_3^+$ ion is generated for each CH_3NH_3Br, $[CH_3NH_3^+] = 0.135\text{ M } CH_3NH_3^+$.

$$CH_3NH_2(aq) + H_2O(l) \rightleftharpoons CH_3NH_3^+(aq) + OH^-(aq)$$

	$[CH_3NH_2]$	$[CH_3NH_3^+]$	$[OH^-]$
Initial	0.255	0.135	≈0.00
Change	$-x$	$+x$	$+x$
Equil	$0.255 - x$	$0.135 + x$	$+x$

$K_b = \dfrac{[CH_3NH_3^+][OH^-]}{[CH_3NH_2]} = 4.4 \times 10^{-4} = \dfrac{(0.135 + x)x}{0.255 - x}$

Assume that x is small ($x << 0.135 < 0.255$), so $\dfrac{(0.135 + \cancel{x})x}{0.255 - \cancel{x}} = 4.4 \times 10^{-4} = \dfrac{(0.135)x}{0.255}$ and

$x = 8.\underline{3}111 \times 10^{-4}\text{ M} = [OH^-]$. Confirm that the more stringent assumption is valid

$\dfrac{8.\underline{3}111 \times 10^{-4}}{0.135} \times 100\% = 0.62\%$, so the assumption is valid.

$K_w = [H_3O^+][OH^-]$ so $[H_3O^+] = \dfrac{K_w}{[OH^-]} = \dfrac{1.0 \times 10^{-14}}{8.\underline{3}111 \times 10^{-4}} = 1.\underline{2}0321 \times 10^{-11}\text{ M}$

Finally, $pH = -\log[H_3O^+] = -\log(1.\underline{2}0321 \times 10^{-11}) = 10.92.$

Check: The units (none) are correct. The magnitude of the answer makes physical sense because pH should be less than $14 + \log(0.255) = 13.4$ because this is a weak base and there is a common ion effect.

18.29 **Given:** 0.15 M $HC_7H_5O_2$ in pure water and in 0.10 M $NaC_7H_5O_2$
Find: % ionization in both solutions **Other:** $K_a(HC_7H_5O_2) = 6.5 \times 10^{-5}$
Conceptual Plan: Pure water: M $HC_7H_5O_2 \rightarrow [H_3O^+] \rightarrow$ % ionization then in $NaC_7H_5O_2$ solution:

ICE table $\quad \%\ ionization = \dfrac{[H_3O^+]_{equil}}{[HC_7H_5O_2]_0} \times 100\%$

M $NaC_7H_5O_2 \rightarrow$ M $C_7H_5O_2^-$ then M $HC_7H_5O_2$, M $C_7H_5O_2^- \rightarrow [H_3O^+] \rightarrow$ % ionization

$NaC_7H_5O_2(aq) \rightarrow Na^+(aq) + C_7H_5O_2^-(aq)$ $\quad$ ICE table $\quad \%\ ionization = \dfrac{[H_3O^-]_{equil}}{[HC_7H_5O_2]_0} \times 100\%$

Solution: in pure water:

$$HC_7H_5O_2(aq) + H_2O(l) \rightleftharpoons H_3O^+(aq) + C_7H_5O_2^-(aq)$$

	$[HC_7H_5O_2]$	$[H_3O^+]$	$[C_7H_5O_2^-]$
Initial	0.15	≈ 0.00	0.00
Change	$-x$	$+x$	$+x$
Equil	$0.15 - x$	$+x$	$+x$

$$K_a = \frac{[H_3O^+][C_7H_5O_2^-]}{[HC_7H_5O_2]} = 6.5 \times 10^{-5} = \frac{x^2}{0.15 - x}$$

Assume that x is small ($x << 0.10$), so $\frac{x^2}{0.15 - \cancel{x}} = 6.5 \times 10^{-5} = \frac{x^2}{0.15}$ and $x = 3.\underline{1}225 \times 10^{-3}\ M = [H_3O^+]$. Then

$\%\ \text{ionization} = \frac{[H_3O^+]_{equil}}{[HC_7H_5O_2]_0} \times 100\% = \frac{3.\underline{1}225 \times 10^{-3}}{0.15} \times 100\% = 2.1\%$, which also confirms that the assumption is valid (because it is less than 5%). In $NaC_7H_5O_2$ solution: because one $C_7H_5O_2^-$ ion is generated for each $NaC_7H_5O_2$, $[C_7H_5O_2^-] = 0.10\ M\ C_7H_5O_2^-$.

$$HC_7H_5O_2(aq) + H_2O(l) \rightleftharpoons H_3O^+(aq) + C_7H_5O_2^-(aq)$$

	$[HC_7H_5O_2]$	$[H_3O^+]$	$[C_7H_5O_2^-]$
Initial	0.15	≈ 0.00	0.10
Change	$-x$	$+x$	$+x$
Equil	$0.15 - x$	$+x$	$0.10 + x$

$K_a = \frac{[H_3O^+][C_7H_5O_2^-]}{[HC_7H_5O_2]} = 6.5 \times 10^{-5} = \frac{x(0.10 + x)}{0.15 - x}$ Assume that x is small ($x << 0.10 < 0.15$), so

$\frac{x(0.10 + \cancel{x})}{0.15 - \cancel{x}} = 6.5 \times 10^{-5} = \frac{x(0.10)}{0.15}$ and $x = 9.\underline{7}5 \times 10^{-5}\ M = [H_3O^+]$. Then

$\%\ \text{ionization} = \frac{[H_3O^+]_{equil}}{[HC_7H_5O_2]_0} \times 100\% = \frac{9.\underline{7}5 \times 10^{-5}}{0.15} \times 100\% = 0.065\%$, which also confirms that the assumption is valid (because it is less than 5%). The percent ionization in the sodium benzoate solution is less than in pure water because of the common ion effect. An increase in one of the products (benzoate ion) shifts the equilibrium to the left, so less acid dissociates.

Check: The units (%) are correct. The magnitude of the answer makes physical sense because the acid is weak; so the percent ionization is low. With a common ion present, the percent ionization decreases.

18.30 **Given:** 0.13 M $HCHO_2$ in pure water and in 0.11 M $KCHO_2$ **Find:** % ionization in both solutions
Other: $K_a(HCHO_2) = 1.8 \times 10^{-4}$

Conceptual Plan: Pure water: M $HCHO_2 \rightarrow [H_3O^+] \rightarrow$ % ionization then in $KCHO_2$ solution:

ICE table $\quad \%\ ionization = \frac{[H_3O^+]_{equil}}{[HCHO_2]_0} \times 100\%$

M $KCHO_2 \rightarrow$ M CHO_2^- then M $HCHO_2$, M $CHO_2^- \rightarrow [H_3O^+] \rightarrow$ % ionization

$KCHO_2(aq) \rightarrow K^+(aq) + CHO_2^-(aq)$ $\quad$ ICE table $\quad \%\ ionization = \frac{[H_3O^+]_{equil}}{[HCHO_2]_0} \times 100\%$

Solution: in pure water:

$$HCHO_2(aq) + H_2O(l) \rightleftharpoons H_3O^+(aq) + CHO_2^-(aq)$$

	$[HCHO_2]$	$[H_3O^+]$	$[CHO_2^-]$
Initial	0.13	≈ 0.00	0.00
Change	$-x$	$+x$	$+x$
Equil	$0.13 - x$	$+x$	$+x$

$$K_a = \frac{[H_3O^+][CHO_2^-]}{[HCHO_2]} = 1.8 \times 10^{-4} = \frac{x^2}{0.13 - x}$$

Assume that x is small ($x << 0.10$), so $\frac{x^2}{0.13 - x} = 1.8 \times 10^{-4} = \frac{x^2}{0.13}$ and $x = 4.\underline{8}374 \times 10^{-3}$ M $= [H_3O^+]$.

Then % ionization $= \frac{[H_3O^+]_{equil}}{[HCHO_2]_0} \times 100\% = \frac{4.\underline{8}374 \times 10^{-3}}{0.13} \times 100\% = 3.7\%$, which also confirms that the assumption is valid (because it is less than 5%).

In $KCHO_2$ solution: Because one CHO_2^- ion is generated for each $KCHO_2$, $[CHO_2^-] = 0.11$ M CHO_2^-.

$$HCHO_2(aq) + H_2O(l) \rightleftharpoons H_3O^+(aq) + CHO_2^-(aq)$$

	$[HCHO_2]$	$[H_3O^+]$	$[CHO_2^-]$
Initial	0.13	≈0.00	0.11
Change	$-x$	$+x$	$+x$
Equil	$0.13 - x$	$+x$	$0.11 + x$

$$K_a = \frac{[H_3O^+][CHO_2^-]}{[HCHO_2]} = 1.8 \times 10^{-4} = \frac{x(0.11 + x)}{0.13 - x}$$

Assume that x is small ($x << 0.11 < 0.13$), so $\frac{x(0.11 + x)}{0.13 - x} = 1.8 \times 10^{-4} = \frac{x(0.11)}{0.13}$ and $x = 2.\underline{1}273 \times 10^{-4}$ M $=$ $[H_3O^+]$. Then % ionization $= \frac{[H_3O^+]_{equil}}{[HCHO_2]_0} \times 100\% = \frac{2.\underline{1}273 \times 10^{-4}}{0.13} \times 100\% = 0.16\%$, which also confirms that the assumption is valid (because it is less than 5%). The percent ionization in the potassium formate solution is less than in pure water because of the common ion effect. An increase in one of the products (formate ion) shifts the equilibrium to the left, so less acid dissociates.

Check: The units (%) are correct. The magnitude of the answer makes physical sense because the acid is weak; so the percent ionization is low. With a common ion present, the percent ionization decreases.

18.31 (a) **Given:** 0.15 M HF **Find:** pH **Other:** $K_a(HF) = 3.5 \times 10^{-4}$

Conceptual Plan: $\textbf{M HF} \rightarrow [\mathbf{H_3O^+}] \rightarrow \textbf{pH}$

ICE table; $pH = -\log[H_3O^+]$

Solution:

$$HF(aq) + H_2O(l) \rightleftharpoons H_3O^+(aq) + F^-(aq)$$

	$[HF]$	$[H_3O^+]$	$[F^-]$
Initial	0.15	≈0.00	0.00
Change	$-x$	$+x$	$+x$
Equil	$0.15 - x$	$+x$	$+x$

$K_a = \frac{[H_3O^+][F^-]}{[HF]} = 3.5 \times 10^{-4} = \frac{x^2}{0.15 - x}$ Assume that x is small ($x << 0.15$), so $\frac{x^2}{0.15 - x} = 3.5 \times 10^{-4} = \frac{x^2}{0.15}$ and $x = 7.\underline{2}457 \times 10^{-3}$ M $= [H_3O^+]$. Confirm that the assumption is valid. $\frac{7.\underline{2}457 \times 10^{-3}}{0.15} \times 100\% = 4.8\% < 5\%$, so the assumption is valid.

Finally, $pH = -\log[H_3O^+] = -\log(7.\underline{2}457 \times 10^{-3}) = 2.14$.

Check: The units (none) are correct. The magnitude of the answer makes physical sense because the pH should be greater than $-\log(0.15) = 0.82$ because this is a weak acid.

(b) **Given:** 0.15 M NaF **Find:** pH **Other:** $K_a(HF) = 3.5 \times 10^{-4}$

Conceptual Plan: $\textbf{M NaF} \rightarrow \textbf{M F}^-$ **and** $K_a \rightarrow K_b$ **then** $\textbf{M F}^- \rightarrow [\mathbf{OH^-}] \rightarrow [\mathbf{H_3O^+}] \rightarrow \textbf{pH}$

$NaF(aq) \rightarrow Na^+(aq) + F^-(aq)$; $K_w = K_aK_b$; ICE table; $K_w = [H_3O^+][OH^-]$; $pH = -\log[H_3O^+]$

Solution:

Because one F^- ion is generated for each NaF, $[F^-] = 0.15$ M F^-. Because $K_w = K_aK_b$, rearrange to solve for K_b.

$$K_b = \frac{K_w}{K_a} = \frac{1.0 \times 10^{-14}}{3.5 \times 10^{-4}} = 2.\underline{8}571 \times 10^{-11}$$

$$F^-(aq) + H_2O(l) \rightleftharpoons HF(aq) + OH^-(aq)$$

	$[F^-]$	$[HF]$	$[OH^-]$
Initial	0.15	0.00	≈ 0.00
Change	$-x$	$+x$	$+x$
Equil	$0.15 - x$	$+x$	$+x$

$$K_b = \frac{[HF][OH^-]}{[F^-]} = 2.\underline{8}571 \times 10^{-11} = \frac{x^2}{0.15 - x}$$

Assume that x is small ($x << 0.15$), so $\frac{x^2}{0.15 - \cancel{x}} = 2.\underline{8}571 \times 10^{-11} = \frac{x^2}{0.15}$ and

$x = 2.\underline{0}702 \times 10^{-6}\,M = [OH^-]$.

Confirm that the assumption is valid. $\frac{2.\underline{0}702 \times 10^{-6}}{0.15} \times 100\% = 0.0014\% < 5\%$, so the assumption is valid.

$K_w = [H_3O^+][OH^-]$, so $[H_3O^+] = \frac{K_w}{[OH^-]} = \frac{1.0 \times 10^{-14}}{2.\underline{0}702 \times 10^{-6}} = 4.\underline{8}305 \times 10^{-9}\,M$.

Finally, $pH = -\log[H_3O^+] = -\log(4.\underline{8}305 \times 10^{-9}) = 8.32$.

Check: The units (none) are correct. The magnitude of the answer makes physical sense because the pH should be slightly basic because the fluoride ion is a very weak base.

(c) **Given:** 0.15 M HF and 0.15 M NaF **Find:** pH **Other:** $K_a(HF) = 3.5 \times 10^{-4}$

Conceptual Plan: M NaF → M F^- then M HF, M F^- → $[H_3O^+]$ → pH

$NaF(aq) \rightarrow Na^+(aq) + F^-(aq)$ ICE table $pH = -\log[H_3O^+]$

Solution: Because one F^- ion is generated for each NaF, $[F^-] = 0.15\,M\,F^-$.

$$HF(aq) + H_2O(l) \rightleftharpoons H_3O^+(aq) + F^-(aq)$$

	$[HF]$	$[H_3O^+]$	$[F^-]$
Initial	0.15	≈ 0.00	0.15
Change	$-x$	$+x$	$+x$
Equil	$0.15 - x$	$+x$	$0.15 + x$

$$K_a = \frac{[H_3O^+][F^-]}{[HF]} = 3.5 \times 10^{-4} = \frac{x(0.15 + x)}{0.15 - x}$$

Assume that x is small ($x << 0.15$), so $\frac{x(0.15 + \cancel{x})}{0.15 - \cancel{x}} = 3.5 \times 10^{-4} = \frac{x(\cancel{0.15})}{\cancel{0.15}}$ and

$x = 3.5 \times 10^{-4}\,M = [H_3O^+]$.

Confirm that the assumption is valid. $\frac{3.5 \times 10^{-4}}{0.15} \times 100\% = 0.23\% < 5\%$, so the assumption is valid.

Finally, $pH = -\log[H_3O^+] = -\log(3.5 \times 10^{-4}) = 3.46$.

Check: The units (none) are correct. The magnitude of the answer makes physical sense because the pH should be greater than that in part (a) (2.14) because of the common ion effect suppressing the dissociation of the weak acid.

18.32 (a) **Given:** 0.18 M CH_3NH_2 **Find:** pH **Other:** $K_b(CH_3NH_2) = 4.4 \times 10^{-4}$

Conceptual Plan: M CH_3NH_2 → $[OH^-]$ → $[H_3O^+]$ → pH

ICE table $K_w = [H_3O^+][OH^-]$ $pH = -\log[H_3O^+]$

Solution:

$$CH_3NH_2(aq) + H_2O(l) \rightleftharpoons CH_3NH_3^+(aq) + OH^-(aq)$$

	$[CH_3NH_2]$	$[CH_3NH_3^+]$	$[OH^-]$
Initial	0.18	0.00	≈ 0.00
Change	$-x$	$+x$	$+x$
Equil	$0.18 - x$	$+x$	$+x$

$$K_b = \frac{[CH_3NH_3^+][OH^-]}{[CH_3NH_2]} = 4.4 \times 10^{-4} = \frac{x^2}{0.18 - x}$$

Assume that x is small ($x << 0.18$), so $\frac{x^2}{0.18 - \cancel{x}} = 4.4 \times 10^{-4} = \frac{x^2}{0.18}$ and $x = 8.\underline{8}994 \times 10^{-3}$ M $= [OH^-]$.

Confirm that the assumption is valid. $\frac{8.\underline{8}994 \times 10^{-3}}{0.18} \times 100\% = 4.9\% < 5\%$, so the assumption is valid.

$K_w = [H_3O^+][OH^-]$, so $[H_3O^+] = \frac{K_w}{[OH^-]} = \frac{1.0 \times 10^{-14}}{8.\underline{8}994 \times 10^{-3}} = 1.\underline{1}237 \times 10^{-12}$ M.

Finally, pH $= -\log[H_3O^+] = -\log(1.\underline{1}237 \times 10^{-12}) = 11.95$.

Check: The units (none) are correct. The magnitude of the answer makes physical sense because the pH should be less than $14 + \log(0.18) = 13.3$ because this is a weak base.

(b) **Given:** 0.18 M CH_3NH_3Cl **Find:** pH **Other:** $K_b(CH_3NH_2) = 4.4 \times 10^{-4}$

Conceptual Plan: M $CH_3NH_3Cl \rightarrow$ M $CH_3NH_3^+$ and $K_b \rightarrow K_a$ then M $CH_3NH_3^+ \rightarrow [H_3O^+] \rightarrow$ pH

$CH_3NH_3Cl(aq) \rightarrow CH_3NH_3^+(aq) + Cl^-(aq)$ $K_w = K_aK_b$ ICE table $pH = -\log[H_3O^+]$

Solution: Because one $CH_3NH_3^+$ ion is generated for each CH_3NH_3Cl, $[CH_3NH_3^+] = 0.18$ M $CH_3NH_3^+$.

Because $K_w = K_aK_b$, rearrange to solve for K_a. $K_a = \frac{K_w}{K_b} = \frac{1.0 \times 10^{-14}}{4.4 \times 10^{-4}} = 2.\underline{2}727 \times 10^{-11}$

$$CH_3NH_3^+(aq) + H_2O(l) \rightleftharpoons H_3O^+(aq) + CH_3NH_2(aq)$$

	$[CH_3NH_3^+]$	$[H_3O^+]$	$[CH_3NH_2]$
Initial	0.18	≈0.00	0.00
Change	$-x$	$+x$	$+x$
Equil	$0.18 - x$	$+x$	$+x$

$K_a = \frac{[H_3O^+][CH_3NH_2]}{[CH_3NH_3^+]} = 2.\underline{2}727 \times 10^{-11} = \frac{x^2}{0.18 - x}$ Assume that x is small ($x << 0.18$), so

$\frac{x^2}{0.18 - \cancel{x}} = 2.\underline{2}727 \times 10^{-11} = \frac{x^2}{0.18}$ and $x = 2.\underline{0}226 \times 10^{-6}$ M $= [H_3O^+]$.

Confirm that the assumption is valid. $\frac{2.\underline{0}226 \times 10^{-6}}{0.18} \times 100\% = 0.0012\% < 5\%$, so the assumption is valid.

Finally, pH $= -\log[H_3O^+] = -\log(2.\underline{0}226 \times 10^{-6}) = 5.69$.

Check: The units (none) are correct. The magnitude of the answer makes physical sense because the pH should be slightly acidic because the methylammonium cation is a very weak acid.

(c) **Given:** 0.18 M CH_3NH_2 and 0.18 M CH_3NH_3Cl **Find:** pH **Other:** $K_b(CH_3NH_2) = 4.4 \times 10^{-4}$

Conceptual Plan: M $CH_3NH_3Cl \rightarrow$ M $CH_3NH_3^+$ then

$CH_3NH_3Cl(aq)CH_3NH_3^+(aq) + Cl^-(aq)$

M CH_3NH_2, M $CH_3NH_3^+ \rightarrow [OH^-] \rightarrow [H_3O^+] \rightarrow$ pH

ICE table $K_w = [H_3O^+][OH^-]$ $pH = -\log[H_3O^+]$

Solution: Because one $CH_3NH_3^+$ ion is generated for each CH_3NH_3Cl, $[CH_3NH_3^+] = 0.105$ M $CH_3NH_3^+$.

$$CH_3NH_2(aq) + H_2O(l) \rightleftharpoons CH_3NH_3^+(aq) + OH^-(aq)$$

	$[CH_3NH_2]$	$[CH_3NH_3^+]$	$[OH^-]$
Initial	0.18	0.18	≈0.00
Change	$-x$	$+x$	$+x$
Equil	$0.18 - x$	$0.18 + x$	$+x$

$$K_b = \frac{[CH_3NH_3^+][OH^-]}{[CH_3NH_2]} = 4.4 \times 10^{-4} = \frac{(0.18 + x)x}{0.18 - x}$$

Assume that x is small ($x << 0.18$), so $\frac{(0.18 + \cancel{x})x}{0.18 - \cancel{x}} = 4.4 \times 10^{-4} = \frac{(\cancel{0.18})x}{\cancel{0.18}}$ and $x = 4.4 \times 10^{-4}\,M = [OH^-]$.

Confirm that the assumption is valid. $\frac{4.4 \times 10^{-4}}{0.18} \times 100\% = 0.24\% < 5\%$, so the assumption is valid.

$K_w = [H_3O^+][OH^-]$, so $[H_3O^+] = \frac{K_w}{[OH^-]} = \frac{1.0 \times 10^{-14}}{4.4 \times 10^{-4}} = 2.\underline{2}727 \times 10^{-11}\,M$.

Finally, $pH = -\log[H_3O^+] = -\log(2.\underline{2}727 \times 10^{-11}) = 10.64$.

Check: The units (none) are correct. The magnitude of the answer makes physical sense because the pH should be less than $14 + \log(0.18) = 13.3$ because this is a weak base and there is a common ion effect.

18.33 When an acid (such as HCl) is added, it reacts with the conjugate base of the buffer system as follows: $HCl + NaC_2H_3O_2 \rightarrow HC_2H_3O_2 + NaCl$. When a base (such as NaOH) is added, it reacts with the weak acid of the buffer system as follows: $NaOH + HC_2H_3O_2 \rightarrow H_2O + NaC_2H_3O_2$. The reaction generates the other buffer system component.

18.34 When an acid (such as HCl) is added, it reacts with the conjugate base of the buffer system as follows: $HCl + NH_3 \rightarrow NH_4Cl$. When a base (such as NaOH) is added, it reacts with the weak acid of the buffer system as follows: $NaOH + NH_4Cl \rightarrow H_2O + NH_3 + NaCl$. The reaction generates the other buffer system component.

18.35 (a) **Given:** 0.15 M $HCHO_2$ and 0.10 M $NaCHO_2$ **Find:** pH **Other:** $K_a(HCHO_2) = 1.8 \times 10^{-4}$

Conceptual Plan: Identify acid and base components then M $NaCHO_2 \rightarrow$ M CHO_2^- then

acid = $HCHO_2$ base = CHO_2^- $NaCHO_2(aq) \rightarrow Na^+(aq) + CHO_2^-(aq)$

K_a, M $HCHO_2$, M $CHO_2^- \rightarrow$ pH

$$pH = pK_a + \log\frac{[base]}{[acid]}$$

Solution: Acid = $HCHO_2$, so $[acid] = [HCHO_2] = 0.15\,M$. Base = CHO_2^-. Because one CHO_2^- ion is generated for each $NaCHO_2$, $[CHO_2^-] = 0.10\,M\,CHO_2^- = [base]$. Then

$$pH = pK_a + \log\frac{[base]}{[acid]} = -\log(1.8 \times 10^{-4}) + \log\frac{0.10\,\cancel{M}}{0.15\,\cancel{M}} = 3.57.$$

Note that to use the Henderson–Hasselbalch equation, the assumption that x is small must be valid. This was confirmed in Problem 18.27.

Check: The units (none) are correct. The magnitude of the answer makes physical sense because the pH should be less than the pK_a of the acid because there is more acid than base. The answer agrees with Problem 18.27.

(b) **Given:** 0.12 M NH_3 and 0.18 M NH_4Cl **Find:** pH **Other:** $K_b(NH_3) = 1.76 \times 10^{-5}$

Conceptual Plan: Identify acid and base components then M $NH_4Cl \rightarrow$ M NH_4^+ and $K_b \rightarrow pK_b \rightarrow pK_a$

acid = NH_4^+ base = NH_3 $NH_4Cl(aq) \rightarrow NH_4^+(aq) + Cl^-(aq)$ $pK_b = -\log K_b$ $14 = pK_a + pK_b$

then pK_a, M NH_3, M $NH_4^+ \rightarrow$ pH

$$pH = pK_a + \log\frac{[base]}{[acid]}$$

Solution: Base = NH_3, $[base] = [NH_3] = 0.12\,M$. Acid = NH_4^+. Because one NH_4^+ ion is generated for each NH_4Cl, $[NH_4^+] = 0.18\,M\,NH_4^+ = [acid]$.

Because $K_b(NH_3) = 1.76 \times 10^{-5}$, $pK_b = -\log K_b = -\log(1.76 \times 10^{-5}) = 4.75$. Because $14 = pK_a + pK_b$, $pK_a = 14 - pK_b = 14 - 4.75 = 9.25$. Then $pH = pK_a + \log\frac{[base]}{[acid]} = 9.25 + \log\frac{0.12\,\cancel{M}}{0.18\,\cancel{M}} = 9.07$.

Note that to use the Henderson–Hasselbalch equation, the assumption that x is small must be valid. This was confirmed in Problem 18.27.

Check: The units (none) are correct. The magnitude of the answer makes physical sense because the pH should be less than the pK_a of the acid because there is more acid than base. The answer agrees with Problem 18.27 within the error of the value.

18.36 (a) **Given:** 0.175 M $HC_2H_3O_2$ and 0.110 M $KC_2H_3O_2$ **Find:** pH **Other:** $K_a(HC_2H_3O_2) = 1.8 \times 10^{-5}$
Conceptual Plan: Identify acid and base components then M $KC_2H_3O_2 \rightarrow$ M $C_2H_3O_2^-$ then

acid = $HC_2H_3O_2$ base = $C_2H_3O_2^-$ $KC_2H_3O_2(aq) \rightarrow K^+(aq) + C_2H_3O_2^-(aq)$

M $HC_2H_3O_2$, M $C_2H_3O_2^- \rightarrow$ pH

$$pH = pK_a + \log\frac{[base]}{[acid]}$$

Solution: Acid = $HC_2H_3O_2$, so [acid] = [$HC_2H_3O_2$] = 0.175 M. Base = $C_2H_3O_2^-$. Because one $C_2H_3O_2^-$ ion is generated for each $KC_2H_3O_2$, [$C_2H_3O_2^-$] = 0.110 M $C_2H_3O_2^-$ = [base].

Then $pH = pK_a + \log\frac{[base]}{[acid]} = -\log(1.8 \times 10^{-5}) + \log\frac{0.110\ \cancel{M}}{0.175\ \cancel{M}} = 4.54.$

Note that to use the Henderson–Hasselbalch equation, the assumption that x is small must be valid. This was confirmed in Problem 18.28.

Check: The units (none) are correct. The magnitude of the answer makes physical sense because the pH should be less than the pK_a of the acid because there is more acid than base. The answer agrees with Problem 18.28.

(b) **Given:** 0.195 M CH_3NH_2 and 0.105 M CH_3NH_3Br **Find:** pH **Other:** $K_b(CH_3NH_2) = 4.4 \times 10^{-4}$
Conceptual Plan: Identify acid and base components then M $CH_3NH_3Br \rightarrow$ M $CH_3NH_3^+$ and

acid = $CH_3NH_3^+$ base = CH_3NH_2 $CH_3NH_3Br(aq) \rightarrow CH_3NH_3^+(aq) + Br^-(aq)$

$K_b \rightarrow pK_b \rightarrow pK_a$ then pK_a, M CH_3NH_2, M $CH_3NH_3^+ \rightarrow$ pH

$pK_b = -\log K_b$ $14 = pK_a + pK_b$ $pH = pK_a + \log\frac{[base]}{[acid]}$

Solution: Base = CH_3NH_2, so [base] = [CH_3NH_2] = 0.195 M. Acid = $CH_3NH_3^+$. Because one $CH_3NH_3^+$ ion is generated for each CH_3NH_3Br, [$CH_3NH_3^+$] = 0.105 M $CH_3NH_3^+$ = [acid].
Because $K_b(CH_3NH_2) = 4.4 \times 10^{-4}$, $pK_b = -\log K_b = -\log(4.4 \times 10^{-4}) = 3.36$.
Because $14 = pK_a + pK_b$, $pK_a = 14 - pK_b = 14 - 3.36 = 10.64$. Then

$$pH = pK_a + \log\frac{[base]}{[acid]} = 10.64 + \log\frac{0.195\ \cancel{M}}{0.105\ \cancel{M}} = 10.91.$$

Note that to use the Henderson–Hasselbalch equation, the assumption that x is small must be valid. This was confirmed in Problem 18.28.

Check: The units (none) are correct. The magnitude of the answer makes physical sense because the pH should be greater than the pK_a of the acid because there is more base than acid. The answer agrees with Problem 18.28.

18.37 (a) **Given:** 0.135 M HClO and 0.155 M KClO **Find:** pH **Other:** $K_a(HClO) = 2.9 \times 10^{-8}$
Conceptual Plan: Identify acid and base components then M KClO $\rightarrow$ M ClO^- then

acid = HClO, base = ClO^- $KClO(aq) \rightarrow K^+(aq) + ClO^-(aq)$

M HClO, M $ClO^- \rightarrow$ pH

$$pH = pK_a + \log\frac{[base]}{[acid]}$$

Solution: Acid = HClO, so [acid] = [HClO] = 0.135 M. Base = ClO^-. Because 1 ClO^- ion is generated for each KClO, [ClO^-] = 0.155 M ClO^- = [base]. Then

$$pH = pK_a + \log\frac{[base]}{[acid]} = -\log(2.9 \times 10^{-8}) + \log\frac{0.155\ \cancel{M}}{0.135\ \cancel{M}} = 7.60.$$

Check: The units (none) are correct. The magnitude of the answer makes physical sense because pH should be greater than the pK_a of the acid because there is more base than acid.

(b) **Given:** 1.05% by mass $C_2H_5NH_2$ and 1.10% by mass $C_2H_5NH_3Br$ **Find:** pH
Other: $K_b(C_2H_5NH_2) = 5.6 \times 10^{-4}$
Conceptual Plan: Assume exactly 100 g of solution. Because both components are in the same solution (i.e., the same final volume of solution), the ratio of the moles of each component is the same as the ratio of the molarity of these components and only the relative number of moles needs to be calculated.

$\mathbf{g_{solution} \rightarrow g_{C_2H_5NH_2} \rightarrow mol_{C_2H_5NH_2}}$ **and** $\mathbf{g_{solution} \rightarrow g_{C_2H_5NH_3Br} \rightarrow mol_{C_2H_5NH_3Br}}$

$$\frac{1.05\text{ g }C_2H_5NH_2}{100\text{ g solution}} \quad \frac{1\text{ mol }C_2H_5NH_2}{45.09\text{ g }C_2H_5NH_2} \qquad \frac{1.10\text{ g }C_2H_5NH_3Br}{100\text{ g solution}} \quad \frac{1\text{ mol }C_2H_5NH_3Br}{125.99\text{ g }C_2H_5NH_3Br}$$

identify acid and base components then M $C_2H_5NH_3Br \rightarrow$ M $C_2H_5NH_3^+$ and

acid = $C_2H_5NH_3^+$, base = $C_2H_5NH_2$ $\quad C_2H_5NH_3Br(aq) \rightarrow C_2H_5NH_3^+(aq) + Br^-(aq)$

$K_b \rightarrow pK_b \rightarrow pK_a$ then pK_a, M $C_2H_5NH_2$, M $C_2H_5NH_3^+ \rightarrow$ pH

$pK_b = -\log K_b \quad 14 = pK_a + pK_b \qquad pH = pK_a + \log \frac{[\text{base}]}{[\text{acid}]}$

Solution:

$$100\text{ g solution} \times \frac{1.05\text{ g }C_2H_5NH_2}{100\text{ g solution}} \times \frac{1\text{ mol }C_2H_5NH_2}{45.09\text{ g }C_2H_5NH_2} = 0.02328676\text{ mol }C_2H_5NH_2 \text{ and}$$

$$100\text{ g solution} \times \frac{1.10\text{ g }C_2H_5NH_3Br}{100\text{ g solution}} \times \frac{1\text{ mol }C_2H_5NH_3Br}{125.99\text{ g }C_2H_5NH_3Br} = 0.008730852\text{ mol }C_2H_5NH_3Br$$

Then base = $C_2H_5NH_2$, moles of base = mol $C_2H_5NH_2$ = 0.02328676 mol; acid = $C_2H_5NH_3^+$. Because 1 $C_2H_5NH_3^+$ ion is generated for each $C_2H_5NH_3Br$, mol $C_2H_5NH_3^+$ = 0.008730852 mol. Because $K_b\,(C_2H_5NH_2) = 5.6 \times 10^{-4}$, $pK_b = -\log K_b = -\log(5.6 \times 10^{-4}) = 3.25$. Because $14 = pK_a + pK_b$, $pK_a = 14 - pK_b = 14 - 3.25 = 10.75$, then

$$pH = pK_a + \log \frac{\text{mol base}}{\text{mol acid}} = 10.75 + \log \frac{0.02328676\text{ mol}}{0.008730852\text{ mol}} = 11.18.$$

Check: The units (none) are correct. The magnitude of the answer makes physical sense because pH should be greater than the pK_a of the acid because there is more base than acid.

(c) **Given:** 10.0 g $HC_2H_3O_2$ and 10.0 g $NaC_2H_3O_2$ in 150.0 mL solution **Find:** pH
Other: $K_a\,(HC_2H_3O_2) = 1.8 \times 10^{-5}$
Conceptual Plan: Identify acid and base components then mL $\rightarrow$ L and g $HC_2H_3O_2 \rightarrow$ mol $HC_2H_3O_2$

acid = $HC_2H_3O_2$, base = $C_2H_3O_2^-$ $\qquad \frac{1\text{ L}}{1000\text{ mL}} \qquad \frac{1\text{ mol }HC_2H_3O_2}{60.05\text{ g }HC_2H_3O_2}$

then mol $HC_2H_3O_2$, L $\rightarrow$ M $HC_2H_3O_2$ and g $NaC_2H_3O_2 \rightarrow$ mol $NaC_2H_3O_2$ then

$M = \frac{\text{mol}}{\text{L}} \qquad \frac{1\text{ mol }NaC_2H_3O_2}{82.04\text{ g }NaC_2H_3O_2}$

mol $NaC_2H_3O_2$, L $\rightarrow$ M $NaC_2H_3O_2 \rightarrow$ M $C_2H_3O_2^-$ then M $HC_2H_3O_2$, M $C_2H_3O_2^- \rightarrow$ pH

$M = \frac{\text{mol}}{\text{L}} \qquad NaC_2H_3O_2(aq) \rightarrow Na^+(aq) + C_2H_3O_2^-(aq) \qquad pH = pK_a + \log \frac{[\text{base}]}{[\text{acid}]}$

Solution: $150.0\text{ mL} \times \frac{1\text{ L}}{1000\text{ mL}} = 0.1500\text{ L}$ and

$$10.0\text{ g }HC_2H_3O_2 \times \frac{1\text{ mol }HC_2H_3O_2}{60.05\text{ g }HC_2H_3O_2} = 0.166528\text{ mol }HC_2H_3O_2$$

$$\text{then } M = \frac{\text{mol}}{\text{L}} = \frac{0.166528\text{ mol }HC_2H_3O_2}{0.1500\text{ L}} = 1.11019\text{ M }HC_2H_3O_2 \text{ and}$$

$$10.0\text{ g }NaC_2H_3O_2 \times \frac{1\text{ mol }NaC_2H_3O_2}{82.04\text{ g }NaC_2H_3O_2} = 0.121892\text{ mol }NaC_2H_3O_2 \text{ then}$$

$$M = \frac{\text{mol}}{\text{L}} = \frac{0.121892\text{ mol }NaC_2H_3O_2}{0.1500\text{ L}} = 0.812612\text{ M }NaC_2H_3O_2$$

Acid = $HC_2H_3O_2$, so [acid] = $[HC_2H_3O_2]$ = 1.11019 M and base = $C_2H_3O_2^-$. Because 1 $C_2H_3O_2^-$ ion is generated for each $NaC_2H_3O_2$, $[C_2H_3O_2^-]$ = 0.812612 M $C_2H_3O_2^-$ = [base]. Then

$$pH = pK_a + \log \frac{[\text{base}]}{[\text{acid}]} = -\log(1.8 \times 10^{-5}) + \log \frac{0.812612\text{ M}}{1.11019\text{ M}} = 4.61.$$

Check: The units (none) are correct. The magnitude of the answer makes physical sense because pH should be less than the pK_a of the acid because there is more acid than base.

18.38 (a) **Given:** 0.145 M $HC_3H_5O_2$ (propanoic acid) and 0.115 M $KC_3H_5O_2$ (potassium propanoate)
Find: pH **Other:** $K_a\ (HC_3H_5O_2) = 1.3 \times 10^{-5}$
Conceptual Plan: Identify acid and base components then M $KC_3H_5O_2 \rightarrow$ M $C_3H_5O_2^-$ then

acid = $HC_3H_5O_2$, base = $C_3H_5O_2^-$ $\qquad$ $KC_3H_5O_2(aq) \rightarrow K^+(aq) + C_3H_5O_2^-(aq)$

M $HC_3H_5O_2$, M $C_3H_5O_2^- \rightarrow$ pH

$$pH = pK_a + \log\frac{[base]}{[acid]}$$

Solution: Acid = $HC_3H_5O_2$, so [acid] = $[HC_3H_5O_2]$ = 0.145 M. Base = $C_3H_5O_2^-$. Because 1 $C_3H_5O_2^-$ ion is generated for each $KC_3H_5O_2$, $[C_3H_5O_2^-]$ = 0.115 M $C_3H_5O_2^-$ = [base]. Then

$$pH = pK_a + \log\frac{[base]}{[acid]} = -\log(1.3 \times 10^{-5}) + \log\frac{0.115\ \cancel{M}}{0.145\ \cancel{M}} = 4.79.$$

Check: The units (none) are correct. The magnitude of the answer makes physical sense because pH should be less than the pK_a of the acid because there is more acid than base.

(b) **Given:** 0.785% by mass C_2H_5N and 0.985% by mass C_5H_5NHCl **Find:** pH **Other:** $K_b\ (C_5H_5N) = 1.9 \times 10^{-9}$
Conceptual Plan: Assume exactly 100 g of solution. Because both components are in the same solution (i.e., the same final volume of solution), the ratio of the moles of each component is the same as the ratio of the molarity of these components and only the relative number of moles needs to be calculated.
$g_{solution} \rightarrow g_{C_5H_5N} \rightarrow mol_{C_5H_5N}$ **and** $g_{solution} \rightarrow g_{C_5H_5NHCl} \rightarrow mol_{C_5H_5NHCl}$

$\frac{0.785\ g\ C_5H_5N}{100\ g\ solution}$ $\frac{1\ mol\ C_5H_5N}{79.10\ g\ C_5H_5N}$ $\qquad$ $\frac{0.985\ g\ C_5H_5NHCl}{100\ g\ solution}$ $\frac{1\ mol\ C_5H_5NHCl}{115.56\ g\ C_5H_5NHCl}$

identify acid and base components then M $CH_3NH_3Cl \rightarrow$ M $CH_3NH_3^+$ and

acid = $C_5H_5NH^+$, base = C_5H_5N $\qquad$ $C_5H_5NHCl(aq) \rightarrow C_5H_5NH^+(aq) + Cl^-(aq)$

$K_b \rightarrow pK_b \rightarrow pK_a$ then pK_a, M C_5H_5N, M $C_5H_5NH^+ \rightarrow$ pH

$pK_b = -\log K_b \quad 14 = pK_a + pK_b$ $\qquad$ $pH = pK_a + \log\frac{[base]}{[acid]}$

Solution:

$$100\ \cancel{g\ solution} \times \frac{0.785\ \cancel{g\ C_5H_5N}}{100\ \cancel{g\ solution}} \times \frac{1\ mol\ C_5H_5N}{79.10\ \cancel{g\ C_5H_5N}} = 0.009\underline{9}24147\ mol\ C_5H_5N \text{ and}$$

$$100\ \cancel{g\ solution} \times \frac{0.985\ \cancel{g\ C_5H_5NHCl}}{100\ \cancel{g\ solution}} \times \frac{1\ mol\ C_5H_5NHCl}{115.56\ \cancel{g\ C_5H_5NHCl}} = 0.008\underline{5}23711\ mol\ C_5H_5NHCl$$

Then base = C_5H_5N, moles of base = mol C_5H_5N = 0.009$\underline{9}$24147 mol; acid = $C_5H_5NH^+$.
Because 1 $C_5H_5NH^+$ ion is generated for each C_5H_5NHCl, mol $C_5H_5NH^+$ = 0.008$\underline{5}$23711 mol.
Because $K_b\ (C_5H_5N) = 1.7 \times 10^{-9}$, $pK_b = -\log K_b = -\log(1.7 \times 10^{-9}) = 8.77$.
Because $14 = pK_a + pK_b$, $pK_a = 14 - pK_b = 14 - 8.77 = 5.23$, then

$$pH = pK_a + \log\frac{mol\ base}{mol\ acid} = 5.23 + \log\frac{0.009\underline{9}24147\ \cancel{mol}}{0.008\underline{5}23711\ \cancel{mol}} = 5.30.$$

Check: The units (none) are correct. The magnitude of the answer makes physical sense because pH should be greater than the pK_a of the acid because there is more base than acid.

(c) **Given:** 15.0 g HF and 25.0 g NaF in 125 mL solution **Find:** pH **Other:** $K_a\ (HF) = 3.5 \times 10^{-4}$
Conceptual Plan: Identify acid and base components then mL $\rightarrow$ L and g HF $\rightarrow$ mol HF

acid = HF base = F^- $\qquad$ $\frac{1\ L}{1000\ mL}$ $\qquad$ $\frac{1\ mol\ HF}{20.01\ g\ HF}$

then mol HF, L $\rightarrow$ M HF and g NaF $\rightarrow$ mol NaF then

$M = \frac{mol}{L}$ $\qquad$ $\frac{1\ mol\ NaF}{41.99\ g\ NaF}$

mol NaF, L $\rightarrow$ M NaF $\rightarrow$ M F^- then M HF, M $F^- \rightarrow$ pH

$M = \frac{mol}{L}$ $\qquad$ $NaF(aq) \rightarrow Na^+(aq) + F^-(aq)$ $\qquad$ $pH = pK_a + \log\frac{[base]}{[acid]}$

Solution: $125\ \cancel{mL} \times \frac{1\ L}{1000\ \cancel{mL}} = 0.125\ L$ and $15.0\ \cancel{g\ HF} \times \frac{1\ mol\ HF}{20.01\ \cancel{g\ HF}} = 0.74\underline{9}625\ mol\ HF$ then

$M = \frac{mol}{L} = \frac{0.749625 \text{ mol HF}}{0.125 \text{ L}} = 5.997 \text{ M HF}$ and $25.0 \text{ g NaF} \times \frac{1 \text{ mol NaF}}{41.99 \text{ g NaF}} = 0.595380 \text{ mol NaF}$

then $M = \frac{mol}{L} = \frac{0.595380 \text{ mol NaF}}{0.125 \text{ L}} = 4.76304 \text{ M NaF}$. Acid = HF, so [acid] = [HF] = 5.997 M and

base = F^-. Because one F^- ion is generated for each NaF, $[F^-] = 4.76304 \text{ M } F^- = [\text{base}]$. Then

$pH = pK_a + \log\frac{[\text{base}]}{[\text{acid}]} = -\log(3.5 \times 10^{-4}) + \log\frac{4.76304 \text{ M}}{5.997 \text{ M}} = 3.36.$

Check: The units (none) are correct. The magnitude of the answer makes physical sense because pH should be less than the pK_a of the acid because there is more acid than base.

18.39 (a) **Given:** 50.0 mL of 0.15 M $HCHO_2$ and 75.0 mL of 0.13 M $NaCHO_2$ **Find:** pH

Other: $K_a(HCHO_2) = 1.8 \times 10^{-4}$

Conceptual Plan: Identify acid and base components then mL $HCHO_2$, mL $NaCHO_2$ → total mL then

acid = $HCHO_2$, base = CHO_2^- total mL = mL $HCHO_2$ + mL $NaCHO_2$

mL $HCHO_2$, M $HCHO_2$, total mL → buffer M $HCHO_2$ and

$M_1V_1 = M_2V_2$

mL $NaCHO_2$, M $NaCHO_2$, total mL → buffer M $NaCHO_2$ → buffer M CHO_2^- then

$M_1V_1 = M_2V_2$ $NaCHO_2(aq) \rightarrow Na^+(aq) + CHO_2^-(aq)$

K_a, M $HCHO_2$, M CHO_2^- → pH

$pH = pK_a + \log\frac{[\text{base}]}{[\text{acid}]}$

Solution: total mL = mL $HCHO_2$ + mL $NaCHO_2$ = 50.0 mL + 75.0 mL = 125.0 mL

Then because $M_1V_1 = M_2V_2$, rearrange to solve for M_2.

$M_2 = \frac{M_1V_1}{V_2} = \frac{(0.15 \text{ M})(50.0 \text{ mL})}{125.0 \text{ mL}} = 0.060 \text{ M } HCHO_2$ and

$M_2 = \frac{M_1V_1}{V_2} = \frac{(0.13 \text{ M})(75.0 \text{ mL})}{125.0 \text{ mL}} = 0.078 \text{ M } NaCHO_2$. Acid = $HCHO_2$, so [acid] = $[HCHO_2]$ =

0.060 M. Base = CHO_2^-. Because one CHO_2^- ion is generated for each $NaCHO_2$, $[CHO_2^-]$ =

$0.078 \text{ M } CHO_2^- = [\text{base}]$. Then $pH = pK_a + \log\frac{[\text{base}]}{[\text{acid}]} = -\log(1.8 \times 10^{-4}) + \log\frac{0.078 \text{ M}}{0.060 \text{ M}} = 3.86.$

Check: The units (none) are correct. The magnitude of the answer makes physical sense because the pH should be greater than the pK_a of the acid because there is more base than acid.

(b) **Given:** 125.0 mL of 0.10 M NH_3 and 250.0 mL of 0.10 M NH_4Cl **Find:** pH

Other: $K_b(NH_3) = 1.76 \times 10^{-5}$

Conceptual Plan: Identify acid and base components then mL NH_3, mL NH_4Cl → total mL then

acid = NH_4^+ base = NH_3 total mL = mL NH_3 + mL NH_4Cl

mL NH_3, M NH_3, total mL → buffer M NH_3 and

$M_1V_1 = M_2V_2$

mL NH_4Cl, M NH_4Cl, total mL → buffer M NH_4Cl → buffer M NH_4^+ and K_b → pK_b → pK_a then

$M_1V_1 = M_2V_2$ $NH_4Cl(aq) \rightarrow NH_4^+(aq) + Cl^-(aq)$ $pK_b = -\log K_b$ $14 = pK_a + pK_b$

pK_a, M NH_3, M NH_4^+ → pH

$pH = pK_a + \log\frac{[\text{base}]}{[\text{acid}]}$

Solution: total mL = mL NH_3 + mL NH_4Cl = 125.0 mL + 250.0 mL = 375.0 mL

Then because $M_1V_1 = M_2V_2$, rearrange to solve for M_2.

$M_2 = \frac{M_1V_1}{V_2} = \frac{(0.10 \text{ M})(125.0 \text{ mL})}{375.0 \text{ mL}} = 0.033333 \text{ M } NH_3$ and

$M_2 = \frac{M_1V_1}{V_2} = \frac{(0.10\ M)(250.0\ \cancel{mL})}{375.0\ \cancel{mL}} = 0.06\underline{6}667\ M\ NH_4Cl$. Base $= NH_3$, $[base] = [NH_3] =$

$0.03\underline{3}333$ M acid $= NH_4^+$. Because one NH_4^+ ion is generated for each NH_4Cl, $[NH_4^+] =$
$0.06\underline{6}6667\ M\ NH_4^+ = [acid]$. Because $K_b(NH_3) = 1.79 \times 10^{-5}$,
$pK_b = -\log K_b = -\log(1.76 \times 10^{-5}) = 4.75$. Because $14 = pK_a + pK_b$,

$pK_a = 14 - pK_b = 14 - 4.75 = 9.25$. Then $pH = pK_a + \log\frac{[base]}{[acid]} = 9.25 + \log\frac{0.03\underline{3}333\ \cancel{M}}{0.06\underline{6}667\ \cancel{M}} = 8.95$.

Check: The units (none) are correct. The magnitude of the answer makes physical sense because the pH should be less than the pK_a of the acid because there is more acid than base.

18.40 (a) **Given:** 150.0 mL of 0.25 M HF and 225.0 mL of 0.30 M NaF **Find:** pH
Other: $K_a(HF) = 3.5 \times 10^{-4}$
Conceptual Plan: Identify acid and base components then mL $HCHO_2$, mL $NaCHO_2 \rightarrow$ total mL then

acid = HF base = F^- total mL = mL HF + mL NaF

mL HF, M HF, total mL $\rightarrow$ buffer M HF and

$M_1V_1 = M_2V_2$

mL NaF, M NaF, total mL $\rightarrow$ buffer M NaF $\rightarrow$ buffer M F^-

$M_1V_1 = M_2V_2$ $NaF(aq) \rightarrow Na^+(aq) + F^-(aq)$

then K_a, M CHO_2, M $CHO_2^- \rightarrow$ pH

$pH = pK_a + \log\frac{[base]}{[acid]}$

Solution: total mL = mL HF + mL NaF = 150.0 mL + 225.0 mL = 375.0 mL

Then because $M_1V_1 = M_2V_2$, rearrange to solve for M_2. $M_2 = \frac{M_1V_1}{V_2} = \frac{(0.25\ M)(150.0\ \cancel{mL})}{375.0\ \cancel{mL}} = 0.10\ M\ HF$

and $M_2 = \frac{M_1V_1}{V_2} = \frac{(0.30\ M)(225.0\ \cancel{mL})}{375.0\ \cancel{mL}} = 0.18\ M\ NaF$. Acid = HF, so $[acid] = [HF] = 0.10\ M$.
Base $= F^-$. Because one F^- ion is generated for each NaF, $[F^-] = 0.18\ M\ F^- = [base]$. Then

$pH = pK_a + \log\frac{[base]}{[acid]} = -\log(3.5 \times 10^{-4}) + \log\frac{0.18\ \cancel{M}}{0.10\ \cancel{M}} = 3.71$.

Check: The units (none) are correct. The magnitude of the answer makes physical sense because the pH should be greater than the pK_a of the acid because there is more base than acid.

(b) **Given:** 175.0 mL of 0.10 M $C_2H_5NH_2$ and 275.0 mL of 0.20 M $C_2H_5NH_3Cl$ **Find:** pH
Other: $K_b(C_2H_5NH_2) = 5.6 \times 10^{-4}$
Conceptual Plan: Identify acid and base components then mL $C_2H_5NH_2$, mL $C_2H_5NH_3Cl \rightarrow$ total mL

acid = $C_2H_5NH_3^+$ base = $C_2H_5NH_2$ total mL = mL $C_2H_5NH_2$ + mL $C_2H_5NH_3Cl$

then mL NH_3, M $C_2H_5NH_2$, total mL $\rightarrow$ buffer M $C_2H_5NH_2$ and

$M_1V_1 = M_2V_2$

mL NH_4Cl, M $C_2H_5NH_3Cl$, total mL $\rightarrow$ buffer M $C_2H_5NH_3Cl \rightarrow$ buffer M $C_2H_5NH_3^+$ and

$M_1V_1 = M_2V_2$ $C_2H_5NH_3Cl(aq) \rightarrow C_2H_5NH_3^+(aq) + Cl^-(aq)$

$K_b \rightarrow pK_b \rightarrow pK_a$ then pK_a, M $C_2H_5NH_2$, M $C_2H_5NH_3^+ \rightarrow$ pH

$pK_b = -\log K_b$ $14 = pK_a + pK_b$ $pH = pK_a + \log\frac{[base]}{[acid]}$

Solution: total mL = mL $C_2H_5NH_2$ + mL $C_2H_5NH_3Cl$ = 175.0 mL + 275.0 mL = 450.0 mL
Then because $M_1V_1 = M_2V_2$, rearrange to solve for M_2.

$M_2 = \frac{M_1V_1}{V_2} = \frac{(0.10\ M)(175.0\ \cancel{mL})}{450.0\ \cancel{mL}} = 0.03\underline{8}889\ M\ C_2H_5NH_2$ and

$M_2 = \frac{M_1V_1}{V_2} = \frac{(0.20\ M)(275.0\ \cancel{mL})}{450.0\ \cancel{mL}} = 0.1\underline{2}222\ M\ C_2H_5NH_3Cl$. Base $= C_2H_5NH_2$, so

$[\text{base}] = [C_2H_5NH_2] = 0.03\underline{8}889$ M. Acid $= C_2H_5NH_3^+$. Because one $C_2H_5NH_3^+$ ion is generated for each $C_2H_5NH_3Cl$, $[C_2H_5NH_3^+] = 0.1\underline{2}222$ M $C_2H_5NH_3^+ = [\text{acid}]$. Because $K_b(C_2H_5NH_2) = 5.6 \times 10^{-4}$, $pK_b = -\log K_b = -\log(5.6 \times 10^{-4}) = 3.25$. Because $14 = pK_a + pK_b$, $pK_a = 14 - pK_b = 14 - 3.25 = 10.75$. Then

$$pH = pK_a + \log\frac{[\text{base}]}{[\text{acid}]} = 10.75 + \log\frac{0.03\underline{8}889 \text{ M}}{0.1\underline{2}222 \text{ M}} = 10.25.$$

Check: The units (none) are correct. The magnitude of the answer makes physical sense because the pH should be less than the pK_a of the acid because there is more acid than base.

18.41 **Given:** NaF/HF buffer at pH = 4.00 **Find:** [NaF]/[HF] **Other:** $K_a(HF) = 3.5 \times 10^{-4}$
Conceptual Plan: Identify acid and base components then pH, $K_a \rightarrow$ [NaF]/[HF]

acid = HF base = F^- $\quad\quad$ $pH = pK_a + \log\frac{[\text{base}]}{[\text{acid}]}$

Solution: $pH = pK_a + \log\frac{[\text{base}]}{[\text{acid}]} = -\log(3.5 \times 10^{-4}) + \log\frac{[\text{NaF}]}{[\text{HF}]} = 4.00$. Solve for [NaF]/[HF].

$$\log\frac{[\text{NaF}]}{[\text{HF}]} = 4.00 - 3.46 = 0.54 \rightarrow \frac{[\text{NaF}]}{[\text{HF}]} = 10^{0.54} = 3.5.$$

Check: The units (none) are correct. The magnitude of the answer makes physical sense because the pH is greater than the pK_a of the acid; so there needs to be more base than acid.

18.42 **Given:** CH_3NH_2/CH_3NH_3Cl buffer at pH = 10.24 **Find:** $[CH_3NH_2]/[CH_3NH_3Cl]$
Other: $K_b(CH_3NH_2) = 4.4 \times 10^{-4}$
Conceptual Plan: Identify acid and base components and $K_b \rightarrow pK_b \rightarrow pK_a$ then

acid = $CH_3NH_3^+$ base = CH_3N $\quad\quad$ $pK_b = -\log K_b \quad 14 = pK_a + pK_b$

pH, $K_a \rightarrow [CH_3NH_2]/[CH_3NH_3Cl]$

$pH = pK_a + \log\frac{[\text{base}]}{[\text{acid}]}$

Solution: Because $K_b(CH_3NH_2) = 4.4 \times 10^{-4}$, $pK_b = -\log K_b = -\log(4.4 \times 10^{-4}) = 3.36$. Because $14 = pK_a + pK_b$, $pK_a = 14 - pK_b = 14 - 3.36 = 10.64$. Then

$$pH = pK_a + \log\frac{[\text{base}]}{[\text{acid}]} = 10.64 + \log\frac{[CH_3NH_2]}{[CH_3NH_3Cl]} = 10.24. \text{ Solve for } [CH_3NH_2]/[CH_3NH_3Cl].$$

$$\log\frac{[CH_3NH_2]}{[CH_3NH_3Cl]} = 10.24 - 10.64 = -0.40 \rightarrow \frac{[CH_3NH_2]}{[CH_3NH_3Cl]} = 10^{-0.40} = 0.40$$

Check: The units (none) are correct. The magnitude of the answer makes physical sense because the pH is less than the pK_a of the acid; so there needs to be less base than acid.

18.43 **Given:** 150.0 mL buffer of 0.15 M benzoic acid at pH = 4.25 **Find:** mass sodium benzoate
Other: $K_a(HC_7H_5O_2) = 6.5 \times 10^{-5}$
Conceptual Plan: Identify acid and base components then pH, K_a, $[HC_7H_5O_2] \rightarrow [NaC_7H_5O_2]$

acid = $HC_7H_5O_2$ base = $C_7H_5O_2^-$ $\quad\quad$ $pH = pK_a + \log\frac{[\text{base}]}{[\text{acid}]}$

mL $\rightarrow$ L then $[NaC_7H_5O_2]$, L $\rightarrow$ mol $NaC_7H_5O_2 \rightarrow$ g $NaC_7H_5O_2$

$\frac{1\text{ L}}{1000\text{ mL}}$ $\quad\quad$ $M = \frac{\text{mol}}{\text{L}}$ $\quad\quad$ $\frac{144.11\text{ g } NaC_7H_5O_2}{1\text{ mol } NaC_7H_5O_2}$

Solution: $pH = pK_a + \log\frac{[\text{base}]}{[\text{acid}]} = -\log(6.5 \times 10^{-5}) + \log\frac{[NaC_7H_5O_2]}{0.15\text{ M}} = 4.25$. Solve for $[NaC_7H_5O_2]$.

$$\log\frac{[NaC_7H_5O_2]}{0.15\text{ M}} = 4.25 - 4.1\underline{8}70866 = 0.0\underline{6}291 \rightarrow \frac{[NaC_7H_5O_2]}{0.15\text{ M}} = 10^{0.0\underline{6}291} = 1.\underline{1}559 \rightarrow$$

$[NaC_7H_5O_2] = 0.1\underline{7}338$ M

Convert to moles using $M = \frac{\text{mol}}{\text{L}}$.

$$\frac{0.17338 \text{ mol NaC}_7\text{H}_5\text{O}_2}{1 \text{ L}} \times 0.150 \text{ L} = 0.026007 \text{ mol NaC}_7\text{H}_5\text{O}_2$$

$$0.026007 \text{ mol NaC}_7\text{H}_5\text{O}_2 \times \frac{144.11 \text{ g NaC}_7\text{H}_5\text{O}_2}{1 \text{ mol NaC}_7\text{H}_5\text{O}_2} = 3.7 \text{ g NaC}_7\text{H}_5\text{O}_2$$

Check: The units (g) are correct. The magnitude of the answer makes physical sense because the volume of solution is small and the concentration is low; so much less than a mole is needed.

18.44 **Given:** 2.55 L buffer of 0.155 M NH_3 at pH = 9.55 **Find:** mass ammonium chloride
Other: $K_b(NH_3) = 1.76 \times 10^{-5}$
Conceptual Plan: Identify acid and base components then $K_b \rightarrow pK_b \rightarrow pK_a$ then

acid = NH_4^+ base = NH_3 $pK_b = -\log K_b$ $14 = pK_a + pK_b$

pH, K_a, $[NH_3] \rightarrow [NH_4Cl]$ then $[NH_4Cl]$, L $\rightarrow$ mol $NH_4Cl \rightarrow$ g NH_4Cl

$\text{pH} = pK_a + \log\frac{[\text{base}]}{[\text{acid}]}$ $M = \frac{\text{mol}}{\text{L}}$ $\frac{53.49 \text{ g NH}_4\text{Cl}}{1 \text{ mol NH}_4\text{Cl}}$

Solution: Because $K_b(NH_3) = 1.76 \times 10^{-5}$, $pK_b = -\log K_b = -\log(1.76 \times 10^{-5}) = 4.75$. Because $14 = pK_a + pK_b$,

$pK_a = 14 - pK_b = 14 - 4.75 = 9.25$. Then $\text{pH} = pK_a + \log\frac{[\text{base}]}{[\text{acid}]} = 9.25 + \log\frac{0.155 \text{ M}}{[NH_4Cl]} = 9.55.$

Solve for $[NH_4Cl]$. $\log\frac{0.155 \text{ M}}{[NH_4Cl]} = 9.55 - 9.25 = 0.30 \rightarrow \frac{0.155 \text{ M}}{[NH_4Cl]} = 10^{0.30} = 1.99526 \rightarrow$

$[NH_4Cl] = 0.0776841$ M. Convert to moles using $M = \frac{\text{mol}}{\text{L}}$.

$$\frac{0.0776841 \text{ mol NH}_4\text{Cl}}{1 \text{ L}} \times 2.55 \text{ L} = 0.198094 \text{ mol NH}_4\text{Cl};\ 0.198094 \text{ mol NH}_4\text{Cl} \times \frac{53.49 \text{ g NH}_4\text{Cl}}{1 \text{ mol NH}_4\text{Cl}} = 10.6 \text{ g NH}_4\text{Cl}$$

Check: The units (g) are correct. The magnitude of the answer makes physical sense because the volume of solution is large and the concentration is low; so less than a mole is needed.

18.45 (a) **Given:** 250.0 mL buffer of 0.250 M $HC_2H_3O_2$ and 0.250 M $NaC_2H_3O_2$ **Find:** initial pH
Other: $K_a(HC_2H_3O_2) = 1.8 \times 10^{-5}$
Conceptual Plan: Identify acid and base components then M $NaC_2H_3O_2 \rightarrow$ M $C_2H_3O_2^-$ then

acid = $HC_2H_3O_2$ base = $C_2H_3O_2^-$ $NaC_2H_3O_2(aq) \rightarrow Na^+(aq) + C_2H_3O_2^-(aq)$

M $HC_2H_3O_2$, M $C_2H_3O_2^- \rightarrow$ pH

$\text{pH} = pK_a + \log\frac{[\text{base}]}{[\text{acid}]}$

Solution: Acid = $HC_2H_3O_2$, so $[\text{acid}] = [HC_2H_3O_2] = 0.250$ M. Base = $C_2H_3O_2^-$. Because one $C_2H_3O_2^-$ ion is generated for each $NaC_2H_3O_2$, $[C_2H_3O_2^-] = 0.250 \text{ M } C_2H_3O_2^- = [\text{base}]$. Then

$$\text{pH} = pK_a + \log\frac{[\text{base}]}{[\text{acid}]} = -\log(1.8 \times 10^{-5}) + \log\frac{0.250 \text{ M}}{0.250 \text{ M}} = 4.74.$$

Check: The units (none) are correct. The magnitude of the answer makes physical sense because the pH is equal to the pK_a of the acid because there are equal amounts of acid and base.

(b) **Given:** 250.0 mL buffer of 0.250 M $HC_2H_3O_2$ and 0.250 M $NaC_2H_3O_2$, add 0.0050 mol HCl **Find:** pH
Other: $K_a(HC_2H_3O_2) = 1.8 \times 10^{-5}$
Conceptual Plan: Part I: Stoichiometry:
mL $\rightarrow$ L then $[NaC_2H_3O_2]$, L $\rightarrow$ mol $NaC_2H_3O_2$ and $[HC_2H_3O_2]$, L $\rightarrow$ mol $HC_2H_3O_2$

$\frac{1 \text{ L}}{1000 \text{ mL}}$ $M = \frac{\text{mol}}{\text{L}}$ $M = \frac{\text{mol}}{\text{L}}$

write balanced equation then

$HCl + NaC_2H_3O_2 \rightarrow HC_2H_3O_2 + NaCl$

mol $NaC_2H_3O_2$, mol $HC_2H_3O_2$, mol HCl $\rightarrow$ mol $NaC_2H_3O_2$, mol $HC_2H_3O_2$ then

set up stoichiometry table

Part II: Equilibrium:
mol $NaC_2H_3O_2$, mol $HC_2H_3O_2$, L, $K_a \rightarrow$ pH

$$pH = pK_a + \log\frac{[base]}{[acid]}$$

Solution: $250.0\ \text{mL} \times \frac{1\ L}{1000\ \text{mL}} = 0.2500\ L$ then

$\frac{0.250\ \text{mol}\ HC_2H_3O_2}{1\ L} \times 0.250\ L = 0.0625\ \text{mol}\ HC_2H_3O_2$ and

$\frac{0.250\ \text{mol}\ NaC_2H_3O_2}{1\ L} \times 0.250\ L = 0.0625\ \text{mol}\ NaC_2H_3O_2$. Set up a table to track changes:

	$HCl(aq)$ +	$NaC_2H_3O_2(aq)$ →	$HC_2H_3O_2(aq)$ +	$NaCl(aq)$
Before addition	≈0.00 mol	0.0625 mol	0.0625 mol	0.00 mol
Addition	0.0050 mol	—	—	—
After addition	≈0.00 mol	0.0575 mol	0.0675 mol	0.0050 mol

Because the amount of HCl is small, there are still significant amounts of both buffer components; so the Henderson–Hasselbalch equation can be used to calculate the new pH.

$$pH = pK_a + \log\frac{[base]}{[acid]} = -\log(1.8 \times 10^{-5}) + \log\frac{\frac{0.0575\ \text{mol}}{0.250\ L}}{\frac{0.0675\ \text{mol}}{0.250\ L}} = 4.68$$

Check: The units (none) are correct. The magnitude of the answer makes physical sense because the pH dropped slightly when acid was added.

(c) **Given:** 250.0 mL buffer of 0.250 M $HC_2H_3O_2$ and 0.250 M $NaC_2H_3O_2$, add 0.0050 mol NaOH
Find: pH **Other:** $K_a(HC_2H_3O_2) = 1.8 \times 10^{-5}$
Conceptual Plan: Part I: Stoichiometry:
mL → L then $[NaC_2H_3O_2]$, L → mol $NaC_2H_3O_2$ and $[HC_2H_3O_2]$, L → mol $HC_2H_3O_2$

$\frac{1\ L}{1000\ mL}$ $M = \frac{mol}{L}$ $M = \frac{mol}{L}$

write balanced equation then

$NaOH + HC_2H_3O_2 \rightarrow H_2O + NaC_2H_3O_2$

mol $NaC_2H_3O_2$, mol $HC_2H_3O_2$, mol NaOH → mol $NaC_2H_3O_2$, mol $HC_2H_3O_2$ then

set up stoichiometry table

Part II: Equilibrium:
mol $NaC_2H_3O_2$, mol $HC_2H_3O_2$, L, $K_a \rightarrow$ pH

$$pH = pK_a + \log\frac{[base]}{[acid]}$$

Solution: $250.0\ \text{mL} \times \frac{1\ L}{1000\ \text{mL}} = 0.2500\ L$ then

$\frac{0.250\ \text{mol}\ HC_2H_3O_2}{1\ L} \times 0.2500\ L = 0.0625\ \text{mol}\ HC_2H_3O_2$ and

$\frac{0.250\ \text{mol}\ NaC_2H_3O_2}{1\ L} \times 0.2500\ L = 0.0625\ \text{mol}\ NaC_2H_3O_2$. Set up a table to track changes:

	$NaOH(aq)$ +	$HC_2H_3O_2(aq)$ →	$NaC_2H_3O_2(aq)$ +	$H_2O(l)$
Before addition	≈0.00 mol	0.0625 mol	0.0625 mol	—
Addition	0.0050 mol	—	—	—
After addition	≈0.00 mol	0.0575 mol	0.0675 mol	—

Because the amount of NaOH is small, there are still significant amounts of both buffer components; so the Henderson–Hasselbalch equation can be used to calculate the new pH.

$$pH = pK_a + \log \frac{[\text{base}]}{[\text{acid}]} = -\log(1.8 \times 10^{-5}) + \log \frac{\frac{0.0675 \cancel{\text{mol}}}{0.2500 \cancel{\text{L}}}}{\frac{0.0575 \cancel{\text{mol}}}{0.2500 \cancel{\text{L}}}} = 4.81$$

Check: The units (none) are correct. The magnitude of the answer makes physical sense because the pH rose slightly when base was added.

18.46 (a) **Given:** 100.0 mL buffer of 0.175 M HClO and 0.150 M NaClO **Find:** initial pH
Other: $K_a(\text{HClO}) = 2.9 \times 10^{-8}$
Conceptual Plan: Identify acid and base components then M NaClO → M ClO^- then

acid = HClO base = ClO^- $\text{NaClO}(aq) \rightarrow \text{Na}^+(aq) + \text{ClO}^-(aq)$

M HClO, M ClO^- → pH

$$pH = pK_a + \log \frac{[\text{base}]}{[\text{acid}]}$$

Solution: Acid = HClO, so [acid] = [HClO] = 0.175 M. Base = ClO^-. Because one ClO^- ion is generated for each NaClO, $[ClO^-]$ = 0.150 M ClO^- = [base].

$$\text{Then } pH = pK_a + \log \frac{[\text{base}]}{[\text{acid}]} = -\log(2.9 \times 10^{-8}) + \log \frac{0.150 \cancel{\text{M}}}{0.175 \cancel{\text{M}}} = 7.47.$$

Check: The units (none) are correct. The magnitude of the answer makes physical sense because the pH is less than the pK_a of the acid because there is more acid than base.

(b) **Given:** 100.0 mL buffer of 0.175 M HClO and 0.150 M NaClO, add 150.0 mg HBr **Find:** pH
Other: $K_a(\text{HClO}) = 2.9 \times 10^{-8}$
Conceptual Plan: Part I: Stoichiometry:
mL → L then [NaClO], L → mol NaClO and [HClO], L → mol HClO and

$\frac{1 \text{ L}}{1000 \text{ mL}}$ $M = \frac{\text{mol}}{\text{L}}$ $M = \frac{\text{mol}}{\text{L}}$

mg HBr → g HBr → mol HBr write balanced equation then

$\frac{1 \text{ g HBr}}{1000 \text{ mg HBr}}$ $\frac{1 \text{ mol HBr}}{80.91 \text{ g HBr}}$ HBr + NaClO → HClO + NaBr

mol NaClO, mol HClO, mol HBr → mol NaClO, mol HClO then

set up stoichiometry table

Part II: Equilibrium:
mol NaClO, mol HClO, L, K_a → pH

$$pH = pK_a + \log \frac{[\text{base}]}{[\text{acid}]}$$

Solution: $100.0 \cancel{\text{mL}} \times \frac{1 \text{ L}}{1000 \cancel{\text{mL}}} = 0.1000 \text{ L}$ then $\frac{0.175 \text{ mol HClO}}{1 \cancel{\text{L}}} \times 0.1000 \cancel{\text{L}} = 0.0175 \text{ mol HClO}$

and $\frac{0.150 \text{ mol NaClO}}{1 \cancel{\text{L}}} \times 0.1000 \cancel{\text{L}} = 0.0150 \text{ mol NaClO}$ and

$150.0 \cancel{\text{mg HBr}} \times \frac{1 \cancel{\text{g HBr}}}{1000 \cancel{\text{mg HBr}}} \times \frac{1 \text{ mol HBr}}{80.91 \cancel{\text{g HBr}}} = 0.00185\underline{3}912 \text{ mol HBr}$. Set up a table to track changes:

	HBr(*aq*)	+	NaClO(*aq*)	→	HClO(*aq*)	+	NaBr(*aq*)
Before addition	≈0.00 mol		0.0150 mol		0.0175 mol		0.00 mol
Addition	0.00185$\underline{3}$912 mol		—		—		—
After addition	≈0.00 mol		0.0131$\underline{4}$6 mol		0.0156$\underline{4}$6 mol		0.00185$\underline{3}$912 mol

Because the amount of HBr is small, there are still significant amounts of both buffer components; so the Henderson–Hasselbalch equation can be used to calculate the new pH.

$$pH = pK_a + \log\frac{[\text{base}]}{[\text{acid}]} = -\log(2.9 \times 10^{-8}) + \log\frac{\frac{0.013\underline{1}46\ \text{mol}}{0.1000\ \text{L}}}{\frac{0.015\underline{6}46\ \text{mol}}{0.1000\ \text{L}}} = 7.46$$

Check: The units (none) are correct. The magnitude of the answer makes physical sense because the pH dropped slightly when acid was added. The pH is closer to the pK_a of the acid than at the start.

(c) **Given:** 100.0 mL buffer of 0.175 M HClO and 0.150 M NaClO, add 85.0 mg NaOH **Find:** pH
Other: $K_a(\text{HClO}) = 2.9 \times 10^{-8}$
Conceptual Plan: Part I: Stoichiometry:
mL → L then [NaClO], L → mol NaClO and [HClO], L → mol HClO and

$\frac{1\ \text{L}}{1000\ \text{mL}}$ $\quad M = \frac{\text{mol}}{\text{L}}$ $\quad M = \frac{\text{mol}}{\text{L}}$

mg NaOH → g NaOH → mol NaOH write balanced equation then

$\frac{1\ \text{g NaOH}}{1000\ \text{mg NaOH}}$ $\quad \frac{1\ \text{mol NaOH}}{40.00\ \text{g NaOH}}$ $\quad NaOH + HClO \rightarrow H_2O + NaClO$

mol NaClO, mol HClO, mol NaOH → mol NaClO, mol HClO then

set up stoichiometry table

Part II: Equilibrium:
mol NaClO, mol HClO, L, K_a → pH

$$pH = pK_a + \log\frac{[\text{base}]}{[\text{acid}]}$$

Solution: $100.0\ \text{mL} \times \frac{1\ \text{L}}{1000\ \text{mL}} = 0.1000\ \text{L}$ then $\frac{0.175\ \text{mol HClO}}{1\ \text{L}} \times 0.1000\ \text{L} = 0.0175$ mol HClO and

$\frac{0.150\ \text{mol NaClO}}{1\ \text{L}} \times 0.1000\ \text{L} = 0.0150$ mol NaClO and

$$85.0\ \text{mg NaOH} \times \frac{1\ \text{g NaOH}}{1000\ \text{mg NaOH}} \times \frac{1\ \text{mol NaOH}}{40.00\ \text{g NaOH}} = 0.00213\ \text{mol NaOH}$$

Set up a table to track changes:

	$NaOH(aq)$ +	$HClO(aq)$ →	$NaClO(aq)$ +	$H_2O(l)$
Before addition	≈0.00 mol	0.0175 mol	0.0150 mol	—
Addition	0.00213 mol	—	—	—
After addition	≈0.00 mol	0.0154 mol	0.0171 mol	—

Because the amount of NaOH is small, there are still significant amounts of both buffer components; so the Henderson–Hasselbalch equation can be used to calculate the new pH.

$$pH = pK_a + \log\frac{[\text{base}]}{[\text{acid}]} = -\log(2.9 \times 10^{-8}) + \log\frac{\frac{0.0171\ \text{mol}}{0.1000\ \text{L}}}{\frac{0.0154\ \text{mol}}{0.1000\ \text{L}}} = 7.58$$

Check: The units (none) are correct. The magnitude of the answer makes physical sense because the pH rose slightly when base was added.

18.47 (a) **Given:** 500.0 mL pure water **Find:** initial pH and pH after adding 0.010 mol HCl
Conceptual Plan: Pure water has a pH of 7.00 then mL → L then mol HCl, L → $[H_3O^+]$ → pH

$\frac{1\ \text{L}}{1000\ \text{mL}}$ $\quad M = \frac{\text{mol}}{\text{L}}$ $\quad pH = -\log[H_3O^+]$

Solution: Pure water has a pH of 7.00, so initial pH = 7.00, $500.0\ \text{mL} \times \frac{1\ \text{L}}{1000\ \text{mL}} = 0.5000\ \text{L}$,

then $M = \frac{\text{mol}}{\text{L}} = \frac{0.010\ \text{mol HCl}}{0.5000\ \text{L}} = 0.020$ M HCl. Because HCl is a strong acid, it dissociates completely; so $pH = -\log[H_3O^+] = -\log(0.020) = 1.70$.

Check: The units (none) are correct. The magnitude of the answers makes physical sense because the pH starts neutral and then drops significantly when acid is added and no buffer is present.

(b) **Given:** 500.0 mL buffer of 0.125 M $HC_2H_3O_2$ and 0.115 M $NaC_2H_3O_2$
Find: initial pH and pH after adding 0.010 mol HCl **Other:** $K_a(HC_2H_3O_2) = 1.8 \times 10^{-5}$
Conceptual Plan: Initial pH:
Identify acid and base components then M $NaC_2H_3O_2 \rightarrow$ M $C_2H_3O_2^-$ then

acid = $HC_2H_3O_2$, base = $C_2H_3O_2^-$ $NaC_2H_3O_2(aq) \rightarrow Na^+(aq) + C_2H_3O_2^-(aq)$

M $HC_2H_3O_2$, M $C_2H_3O_2^- \rightarrow$ pH

$$pH = pK_a + \log\frac{[base]}{[acid]}$$

pH after HCl addition: Part I: Stoichiometry:
mL → L then $[NaC_2H_3O_2]$, L → mol $NaC_2H_3O_2$ and $[HC_2H_3O_2]$, L → mol $HC_2H_3O_2$

$\frac{1\text{ L}}{1000\text{ mL}}$ $\quad M = \frac{mol}{L}$ $\quad M = \frac{mol}{L}$

write balanced equation then

$HCl + NaC_2H_3O_2 \rightarrow HC_2H_3O_2 + NaCl$

mol $NaC_2H_3O_2$, mol $HC_2H_3O_2$, mol HCl → mol $NaC_2H_3O_2$, mol $HC_2H_3O_2$ then

set up stoichiometry table

Part II: Equilibrium:
mol $NaC_2H_3O_2$, mol $HC_2H_3O_2$, L, K_a → pH

$$pH = pK_a + \log\frac{[base]}{[acid]}$$

Solution: Initial pH: Acid = $HC_2H_3O_2$, so $[acid] = [HC_2H_3O_2] = 0.125$ M. Base = $C_2H_3O_2^-$. Because one $C_2H_3O_2^-$ ion is generated for each $NaC_2H_3O_2$, $[C_2H_3O_2^-] = 0.115$ M $C_2H_3O_2^- = [base]$. Then

$$pH = pK_a + \log\frac{[base]}{[acid]} = -\log(1.8 \times 10^{-5}) + \log\frac{0.115\ \cancel{M}}{0.125\ \cancel{M}} = 4.71.$$

pH after HCl addition:

$$500.0\ \cancel{mL} \times \frac{1\text{ L}}{1000\ \cancel{mL}} = 0.5000\text{ L then } \frac{0.125\text{ mol } HC_2H_3O_2}{1\ \cancel{L}} \times 0.5000\ \cancel{L} = 0.0625\text{ mol } HC_2H_3O_2 \text{ and}$$

$$\frac{0.115\text{ mol } NaC_2H_3O_2}{1\ \cancel{L}} \times 0.5000\ \cancel{L} = 0.0575\text{ mol } NaC_2H_3O_2.$$ Set up a table to track changes:

	$HCl(aq)$	+	$NaC_2H_3O_2(aq)$	→	$HC_2H_3O_2(aq)$	+	$NaCl(aq)$
Before addition	≈0.00 mol		0.0575 mol		0.0625 mol		0.00 mol
Addition	0.010 mol		—		—		—
After addition	≈0.00 mol		$0.04\underline{7}5$ mol		$0.07\underline{2}5$ mol		0.10 mol

Because the amount of HCl is small, there are still significant amounts of both buffer components; so the Henderson–Hasselbalch equation can be used to calculate the new pH.

$$pH = pK_a + \log\frac{[base]}{[acid]} = -\log(1.8 \times 10^{-5}) + \log\frac{\frac{0.04\underline{7}5\ \cancel{mol}}{\cancel{0.5000\ L}}}{\frac{0.07\underline{2}5\ \cancel{mol}}{\cancel{0.5000\ L}}} = 4.56$$

Check: The units (none) are correct. The magnitude of the answers makes physical sense because the pH started below the pK_a of the acid and it dropped slightly when acid was added.

(c) **Given:** 500.0 mL buffer of 0.155 M $C_2H_5NH_2$ and 0.145 M $C_2H_5NH_3Cl$
Find: initial pH and pH after adding 0.010 mol HCl **Other:** $K_b(C_2H_5NH_2) = 5.6 \times 10^{-4}$
Conceptual Plan: Initial pH:
Identify acid and base components then M $C_2H_5NH_3Cl \rightarrow$ M $C_2H_5NH_3^+$

acid = $C_2H_5NH_3^+$ base = $C_2H_5NH_2$ $\quad C_2H_5NH_3Cl(aq) \rightarrow C_2H_5NH_3^+(aq) + Cl^-(aq)$

and $K_b \rightarrow pK_b \rightarrow pK_a$ then pK_a, M $C_2H_5NH_2$, M $C_2H_5NH_3^+ \rightarrow$ pH

$pK_b = -\log K_b$ $14 = pK_a + pK_b$ $pH = pK_a + \log\frac{[base]}{[acid]}$

pH after HCl addition: Part I: Stoichiometry:
mL → L then $[C_2H_5NH_2]$, L → mol $C_2H_5NH_2$ and

$\frac{1\ L}{1000\ mL}$ $M = \frac{mol}{L}$

$[C_2H_5NH_3Cl]$, L → mol $C_2H_5NH_3Cl$ write balanced equation then

$M = \frac{mol}{L}$ $HCl + C_2H_5NH_2 \rightarrow C_2H_5NH_3Cl$

mol $C_2H_5NH_2$, mol $C_2H_5NH_3Cl$, mol HCl → mol $C_2H_5NH_2$, mol $C_2H_5NH_3Cl$ then

set up stoichiometry table

Part II: Equilibrium:
mol $C_2H_5NH_2$, mol $C_2H_5NH_3Cl$, L, K_a → pH

$pH = pK_a + \log\frac{[base]}{[acid]}$

Solution: Base = $C_2H_5NH_2$, so [base] = $[C_2H_5NH_2]$ = 0.155 M. Acid = $C_2H_5NH_3^+$. Because one $C_2H_5NH_3^+$ ion is generated for each $C_2H_5NH_3Cl$, $[C_2H_5NH_3^+]$ = 0.145 M $C_2H_5NH_3^+$ = [acid]. Because $K_b(C_2H_5NH_2) = 5.6 \times 10^{-4}$, $pK_b = -\log K_b = -\log(5.6 \times 10^{-4}) = 3.25$. Because $14 = pK_a + pK_b$, $pK_a = 14 - pK_b = 14 - 3.25 = 10.75$ then

$$pH = pK_a + \log\frac{[base]}{[acid]} = 10.75 + \log\frac{0.155\ M}{0.145\ M} = 10.78.$$

pH after HCl addition: $500.0\ mL \times \frac{1\ L}{1000\ mL} = 0.5000\ L$ then

$\frac{0.155\ mol\ C_2H_5NH_2}{1\ L} \times 0.5000\ L = 0.0775\ mol\ C_2H_5NH_2$ and

$\frac{0.145\ mol\ C_2H_5NH_3Cl}{1\ L} \times 0.5000\ L = 0.0725\ mol\ C_2H_5NH_3Cl$. Set up a table to track changes:

	$HCl(aq)$	+	$C_2H_5NH_2(aq)$	→	$C_2H_5NH_3Cl(aq)$
Before addition	≈0.00 mol		0.0775 mol		0.0725 mol
Addition	0.010 mol		—		—
After addition	≈0.00 mol		0.0675 mol		0.0825 mol

Because the amount of HCl is small, there are still significant amounts of both buffer components; so the Henderson–Hasselbalch equation can be used to calculate the new pH.

$$pH = pK_a + \log\frac{[base]}{[acid]} = 10.75 + \log\frac{\frac{0.0675\ mol}{0.5000\ L}}{\frac{0.0825\ mol}{0.5000\ L}} = 10.66$$

Check: The units (none) are correct. The magnitude of the answers makes physical sense because the initial pH should be greater than the pK_a of the acid because there is more base than acid and the pH drops slightly when acid is added.

18.48 (a) **Given:** 250.0 mL pure water **Find:** initial pH and pH after adding 0.010 mol NaOH
Conceptual Plan:
Pure water has a pH of 7.00 then mL → L then mol NaOH, L → $[OH^-] \rightarrow [H_3O^+] \rightarrow$ pH

$\frac{1\ L}{1000\ mL}$ $M = \frac{mol}{L}$ $K_w = [H_3O^+][OH^-]$ $pH = -\log[H_3O^+]$

Solution: Pure water has a pH of 7.00, so initial pH = 7.00 then $250.0\ mL \times \frac{1\ L}{1000\ mL} = 0.2500\ L$ then

$M = \frac{mol}{L} = \frac{0.010\ mol\ NaOH}{0.2500\ L} = 0.040\ M\ NaOH$. Because NaOH is a strong base, it dissociates completely; so $[OH^-] = 0.040$ M. $K_w = [H_3O^+][OH^-]$ so

$$[H_3O^-] = \frac{K_w}{[OH^-]} = \frac{1.0 \times 10^{-14}}{0.040} = 2.5 \times 10^{-13} \text{ M and}$$

$pH = -\log[H_3O^+] = -\log(2.5 \times 10^{-13}) = 12.60$

Check: The units (none) are correct. The magnitude of the answers make physical sense because the pH started neutral and then rose significantly when base was added and no buffer is present.

(b) **Given:** 250.0 mL buffer of 0.195 M $HCHO_2$ and 0.275 M $KCHO_2$
Find: initial pH and pH after adding 0.010 mol NaOH **Other:** $K_a(HCHO_2) = 1.8 \times 10^{-4}$
Conceptual Plan: Initial pH:
Identify acid and base components then M $KCHO_2 \rightarrow$ M CHO_2^- then

acid = $HCHO_2$ base = CHO_2^- $\quad$ $KCHO_2(aq) \rightarrow K^+(aq) + CHO_2^-(aq)$

M $HCHO_2$, M $CHO_2^- \rightarrow$ pH

$$pH = pK_a + \log\frac{[base]}{[acid]}$$

pH after NaOH addition: Part I: Stoichiometry:
mL → L then $[KCHO_2]$, L → mol $KCHO_2$ and $[HCHO_2]$, L → mol $HCHO_2$

$\frac{1\ L}{1000\ mL}$ $\quad$ $M = \frac{mol}{L}$ $\quad$ $M = \frac{mol}{L}$

write balanced equation then

$NaOH + HCHO_2 \rightarrow NaCHO_2 + H_2O$

mol CHO_2^-, mol $HCHO_2$, mol NaOH → mol $NaCHO_2$, mol $HCHO_2$ then

set up stoichiometry table

Part II: Equilibrium:
mol $NaC_2H_3O_2$, mol $HC_2H_3O_2$, L, K_a → pH

$$pH = pK_a + \log\frac{[base]}{[acid]}$$

Solution: Initial pH: Acid = $HCHO_2$, so $[acid] = [HCHO_2] = 0.195$ M. Base = CHO_2^-.
Because one CHO_2^- ion is generated for each $KCHO_2$, $[CHO_2^-] = 0.275$ M $CHO_2^- = [base]$. Then

$$pH = pK_a + \log\frac{[base]}{[acid]} = -\log(1.8 \times 10^{-4}) + \log\frac{0.275\ \cancel{M}}{0.195\ \cancel{M}} = 3.89.$$

pH after NaOH addition: $250.0\ \cancel{mL} \times \frac{1\ L}{1000\ \cancel{mL}} = 0.2500$ L then

$$\frac{0.195 \text{ mol } HCHO_2}{1\ \cancel{L}} \times 0.2500\ \cancel{L} = 0.048\underline{7}5 \text{ mol } HCHO_2 \text{ and}$$

$$\frac{0.275 \text{ mol } KCHO_2}{1\ \cancel{L}} \times 0.2500\ \cancel{L} = 0.068\underline{7}5 \text{ mol } KCHO_2$$

Set up a table to track changes:

	$NaOH(aq)$	+	$HCHO_2(aq)$	→	$NaCHO_2(aq)$	+	$H_2O(l)$
Before addition	≈ 0.00 mol		0.04875 mol		0.06875 mol		—
Addition	0.010 mol		—		—		—
After addition	≈ 0.00 mol		$0.03\underline{8}75$ mol		$0.07\underline{8}75$ mol		—

Because the amount of NaOH is small, there are still significant amounts of both buffer components; so the Henderson–Hasselbalch equation can be used to calculate the new pH.

$$pH = pK_a + \log\frac{[base]}{[acid]} = -\log(1.8 \times 10^{-4}) + \log\frac{\frac{0.07\underline{8}75\ \cancel{mol}}{0.2500\ \cancel{L}}}{\frac{0.03\underline{8}75\ \cancel{mol}}{0.2500\ \cancel{L}}} = 4.05$$

Check: The units (none) are correct. The magnitude of the answers makes physical sense because the pH started above the pK_a of the acid and it rose slightly when base was added.

(c) **Given:** 250.0 mL buffer of 0.255 M $C_2H_5NH_2$ and 0.235 M $C_2H_5NH_3Cl$
Find: initial pH and pH after adding 0.010 mol NaOH **Other:** $K_b(C_2H_5NH_2) = 5.6 \times 10^{-4}$
Conceptual Plan: Initial pH:
Identify acid and base components then M $C_2H_5NH_3Cl \rightarrow$ M $C_2H_5NH_3^+$

acid = $C_2H_5NH_3^+$ base = $C_2H_5NH_2$ $C_2H_5NH_3Cl(aq) \rightarrow C_2H_5NH_3^+(aq) + Cl^-(aq)$

and $K_b \rightarrow pK_b \rightarrow pK_a$ then pK_a, M $C_2H_5NH_2$, M $C_2H_5NH_3^+ \rightarrow$ pH

$pK_b = -\log K_b$ $14 = pK_a + pK_b$ $\text{pH} = pK_a + \log\frac{[\text{base}]}{[\text{acid}]}$

pH after NaOH addition: Part I: Stoichiometry:
mL $\rightarrow$ L then $[C_2H_5NH_2]$, L $\rightarrow$ mol $C_2H_5NH_2$ and

$\frac{1\text{ L}}{1000\text{ mL}}$ $\text{M} = \frac{\text{mol}}{\text{L}}$

$[C_2H_5NH_3Cl]$, L $\rightarrow$ mol $C_2H_5NH_3Cl$

$\text{M} = \frac{\text{mol}}{\text{L}}$

write balanced equation then

$NaOH + C_2H_5NH_3Cl \rightarrow C_2H_5NH_2 + NaCl + H_2O$

mol $C_2H_5NH_2$, mol $C_2H_5NH_3Cl$, mol NaOH $\rightarrow$ mol $C_2H_5NH_2$, mol $C_2H_5NH_3Cl$

set up stoichiometry table

Part II: Equilibrium:
mol $C_2H_5NH_2$, mol $C_2H_5NH_3Cl$, L, $K_a \rightarrow$ pH

$\text{pH} = pK_a + \log\frac{[\text{base}]}{[\text{acid}]}$

Solution: Base = $C_2H_5NH_2$, so $[\text{base}] = [C_2H_5NH_2] = 0.255$ M. Acid = $C_2H_5NH_3^+$. Because one $C_2H_5NH_3^+$ ion is generated for each $C_2H_5NH_3Cl$, $[C_2H_5NH_3^+] = 0.235$ M $C_2H_5NH_3^+ = [\text{acid}]$. Because $K_b(C_2H_5NH_2) = 5.6 \times 10^{-4}$, $pK_b = -\log K_b = -\log(5.6 \times 10^{-4}) = 3.25$. Because $14 = pK_a + pK_b$, $pK_a = 14 - pK_b = 14 - 3.25 = 10.75$. Then

$$\text{pH} = pK_a + \log\frac{[\text{base}]}{[\text{acid}]} = 10.75 + \log\frac{0.255\ \cancel{M}}{0.235\ \cancel{M}} = 10.78.$$

pH after HCl addition: $250.0\ \cancel{\text{mL}} \times \frac{1\text{ L}}{1000\ \cancel{\text{mL}}} = 0.2500\text{ L}$ then

$$\frac{0.255\text{ mol } C_2H_5NH_2}{1\ \cancel{\text{L}}} \times 0.2500\ \cancel{\text{L}} = 0.06\underline{3}75\text{ mol } C_2H_5NH_2 \text{ and}$$

$$\frac{0.235\text{ mol } C_2H_5NH_3Cl}{1\ \cancel{\text{L}}} \times 0.2500\ \cancel{\text{L}} = 0.05\underline{8}75\text{ mol } C_2H_5NH_3Cl.$$ Set up a table to track changes:

	$NaOH(aq)$	+ $C_2H_5NH_3Cl(aq)$	$\rightarrow$ $C_2H_5NH_2(aq)$	+ $NaCl(aq)$	+ $H_2O(aq)$
Before addition	$\approx$0.00 mol	0.05875 mol	0.06375 mol	0.00 mol	—
Addition	0.010 mol	—	—	—	—
After addition	$\approx$0.00 mol	0.04$\underline{8}$75 mol	0.07$\underline{3}$75 mol		

Because the amount of NaOH is small, there are still significant amounts of both buffer components; so the Henderson–Hasselbalch equation can be used to calculate the new pH.

$$\text{pH} = pK_a + \log\frac{[\text{base}]}{[\text{acid}]} = 10.75 + \log\frac{\frac{0.07\underline{3}75\ \cancel{\text{mol}}}{\cancel{0.2500\text{ L}}}}{\frac{0.04\underline{8}75\ \cancel{\text{mol}}}{\cancel{0.2500\text{ L}}}} = 10.93$$

Check: The units (none) are correct. The magnitude of the answers makes physical sense because the initial pH should be greater than the pK_a of the acid because there is more base than acid and the pH rises slightly when base is added.

18.49 **Given:** 350.00 mL 0.150 M HF and 0.150 M NaF buffer
Find: mass NaOH to raise pH to 4.00 and mass NaOH to raise pH to 4.00 with buffer concentrations raised to 0.350 M
Other: $K_a(\text{HF}) = 3.5 \times 10^{-4}$

Conceptual Plan: Identify acid and base components. Because [NaF] = [HF], then initial pH = pK_a

acid = HF, base = F^- — pH = pK_a

final pH, pK_a → [NaF]/[HF] and mL → L then [HF], L → mol HF and [NaF], L → mol NaF

$pH = pK_a + \log\frac{[base]}{[acid]}$ — $\frac{1\ L}{1000\ mL}$ — $M = \frac{mol}{L}$ — $M = \frac{mol}{L}$

then write balanced equation then

$NaOH + HF \rightarrow NaF + H_2O$

mol HF, mol NaF, [NaF]/[HF] → mol NaOH → g NaOH.

set up stoichiometry table — $\frac{40.00\ g\ NaOH}{1\ mol\ NaOH}$

Finally, when the buffer concentrations are raised to 0.350 M, simply multiply the g NaOH by the ratio of concentrations (0.350 M/0.150 M).

Solution: initial pH = pK_a = $-\log(3.5 \times 10^{-4}) = 3.46$ then

$pH = pK_a + \log\frac{[base]}{[acid]} = -\log(3.5 \times 10^{-4}) + \log\frac{[NaF]}{[HF]} = 4.00$. Solve for [NaF]/[HF].

$\log\frac{[NaF]}{[HF]} = 4.00 - 3.46 = 0.54 \rightarrow \frac{[NaF]}{[HF]} = 10^{0.54} = 3.5$. $350.0\ \cancel{mL} \times \frac{1\ L}{1000\ \cancel{mL}} = 0.3500\ L$ then

$\frac{0.150\ mol\ HF}{1\ \cancel{L}} \times 0.3500\ \cancel{L} = 0.0525\ mol\ HF$ and $\frac{0.150\ mol\ NaF}{1\ \cancel{L}} \times 0.3500\ \cancel{L} = 0.0525\ mol\ NaF$

Set up a table to track changes:

	$NaOH(aq)$	+	$HF(aq)$	→	$NaF(aq)$	+	$H_2O(aq)$
Before addition	≈0.00 mol		0.0525 mol		0.0525 mol		—
Addition	x		—		—		—
After addition	≈0.00 mol		$(0.0525 - x)$ mol		$(0.0525 + x)$ mol		—

Because $\frac{[NaF]}{[HF]} = 3.5 = \frac{(0.0525 + x)\ \cancel{mol}}{(0.0525 - x)\ \cancel{mol}}$, solve for x. Note that the ratio of moles is the same as the ratio of concentrations because the volume for both terms is the same. $3.5(0.0525 - x) = (0.0525 + x) \rightarrow$ $0.1\underline{8}375 - 3.5x = 0.0525 + x \rightarrow 0.1\underline{3}125 = 4.5x \rightarrow x = 0.02\underline{9}167$ mol NaOH then

$0.02\underline{9}167\ \cancel{mol\ NaOH} \times \frac{40.00\ g\ NaOH}{1\ \cancel{mol\ NaOH}} = 1.\underline{1}667\ g\ NaOH = 1.2\ g\ NaOH$

To scale the amount of NaOH up to a 0.350 M HF and NaF solution, multiply the NaOH mass by the ratio of concentrations. $1.1667\ g\ NaOH \times \frac{0.350\ \cancel{M}}{0.150\ \cancel{M}} = 2.7\ g\ NaOH$

Check: The units (g) are correct. The magnitude of the answers makes physical sense because there is much less than a mole of each of the buffer components; so there must be much less than a mole of NaOH. The higher the buffer concentrations, the higher the buffer capacity and the mass of NaOH it can neutralize.

18.50 **Given:** 100.00 mL of 0.100 M NH_3 and 0.125 M NH_4Br buffer

Find: mass HCl to lower pH to 9.00 and mass HCl to lower pH to 9.00 with buffer concentrations raised to 0.250 M NH_3 and 0.400 M NH_4Br

Other: $K_b(NH_3) = 1.76 \times 10^{-5}$

Conceptual Plan: Identify acid and base components then $K_b \rightarrow pK_b \rightarrow pK_a$ then

acid = NH_4^+ base = NH_3 — $pK_b = -\log K_b$ — $14 = pK_a + pK_b$

final pH, pK_a → $[NH_3]/[NH_4^+]$ and mL → L then $[NH_3]$, L → mol NH_3 and

$pH = pK_a + \log\frac{[base]}{[acid]}$ — $\frac{1\ L}{1000\ mL}$ — $M = \frac{mol}{L}$

$[NH_4^+]$, L → mol NH_4^+ then write balanced equation then

$M = \frac{mol}{L}$ — $H^+ + NH_3 \rightarrow NH_4^+$

mol NH_3, mol NH_4^+, $[NH_3]/[NH_4^+]$ → mol HCl → g HCl

set up stoichiometry table — $\frac{36.46\ g\ HCl}{1\ mol\ HCl}$

Solution: Because $K_b(NH_3) = 1.76 \times 10^{-5}$, $pK_b = -\log K_b = -\log(1.76 \times 10^{-5}) = 4.75$. Because $14 = pK_a + pK_b$, $pK_a = 14 - pK_b = 14 - 4.75 = 9.25$ then $pH = pK_a + \log\dfrac{[\text{base}]}{[\text{acid}]} = 9.25 + \log\dfrac{[NH_3]}{[NH_4^+]} = 9.00$

Solve for $[NH_4Br]$. $\log\dfrac{[NH_3]}{[NH_4^+]} = 9.00 - 9.25 = -0.25 \rightarrow \dfrac{[NH_3]}{[NH_4^+]} = 10^{-0.25} = 0.5\underline{6}2341$

$100.0\ \cancel{mL} \times \dfrac{1\ L}{1000\ \cancel{mL}} = 0.1000\ L$ then $\dfrac{0.100\ mol\ NH_3}{1\ \cancel{L}} \times 0.1000\ \cancel{L} = 0.0100\ mol\ NH_3$ and

$\dfrac{0.125\ mol\ NH_4Br}{1\ \cancel{L}} \times 0.1000\ \cancel{L} = 0.0125\ mol\ NH_4Br = 0.0125\ mol\ NH_4^+$. Because HCl is a strong acid,

$[HCl] = [H^+]$. Set up a table to track changes:

	$H^+(aq)$	+	$NH_3(aq)$	$\rightarrow$	$NH_4^+(aq)$
Before addition	≈ 0.00 mol		0.0100 mol		0.0125 mol
Addition	x		—		—
After addition	≈ 0.00 mol		$(0.0100 - x)$ mol		$(0.0125 + x)$ mol

Because $\dfrac{[NH_3]}{[NH_4^+]} = 0.5\underline{6}2341 = \dfrac{(0.0100 - x)\ \cancel{mol}}{(0.0125 + x)\ \cancel{mol}}$, solve for x. Note that the ratio of moles is the same as the ratio of concentrations because the volume for both terms is the same. $0.5\underline{6}2341(0.0125 + x) = (0.0100 - x) \rightarrow 0.007\underline{0}2926 + 0.5\underline{6}2341x = 0.0100 - x \rightarrow 1.5\underline{6}2341x = 0.00\underline{2}9707 \rightarrow x = 0.00\underline{1}9015$ mol HCl, then

$0.00\underline{1}9015\ \cancel{mol\ HCl} \times \dfrac{36.46\ g\ HCl}{1\ \cancel{mol\ HCl}} = 0.0\underline{6}933\ g\ HCl = 0.07\ g\ HCl$. For the higher concentration buffer,

repeat part of the previous calculations. $\dfrac{0.250\ mol\ NH_3}{1\ \cancel{L}} \times 0.1000\ \cancel{L} = 0.0250\ mol\ NH_3$ and

$\dfrac{0.400\ mol\ NH_4Br}{1\ \cancel{L}} \times 0.1000\ \cancel{L} = 0.0400\ mol\ NH_4Br = 0.0400\ mol\ NH_4^+$

Because HCl is a strong acid, $[HCl] = [H^+]$. Set up a table to track changes:

	$H^+(aq)$	+	$NH_3(aq)$	$\rightarrow$	$NH_4^+(aq)$
Before addition	≈ 0.00 mol		0.0250 mol		0.0400 mol
Addition	x		—		—
After addition	≈ 0.00 mol		$(0.0250 - x)$ mol		$(0.0400 + x)$ mol

Because $\dfrac{[NH_3]}{[NH_4^+]} = 0.5\underline{6}2341 = \dfrac{(0.0250 - x)\ \cancel{mol}}{(0.0400 + x)\ \cancel{mol}}$, solve for x. Note that the ratio of moles is the same as the ratio of concentrations because the volume for both terms is the same. $0.5\underline{6}2341(0.0400 + x) = (0.0250 - x) \rightarrow 0.02\underline{2}4936 + 0.5\underline{6}2341x = 0.0250 - x \rightarrow 1.5\underline{6}2341x = 0.00\underline{2}50636 \rightarrow x = 0.00\underline{1}6042$ mol HCl then

$0.00\underline{1}6042\ \cancel{mol\ HCl} \times \dfrac{36.46\ g\ HCl}{1\ \cancel{mol\ HCl}} = 0.0\underline{5}8489\ g\ HCl = 0.06\ g\ HCl$

Check: The units (g) are correct. The magnitude of the answers makes physical sense because there is much less than a mole of each of the buffer components; so there must be much less than a mole of HCl. Also, the pH of the initial buffer is less than the pK_a of the acid, so even less acid is necessary to drop the pH below 9.00. The higher concentration buffer solution requires about the same amount of acid because the initial pH of this buffer is closer to 9.00 than is the low concentration buffer.

18.51 (a) Yes, this will be a buffer because NH_3 is a weak base and NH_4^+ is its conjugate acid. The ratio of base to acid is $0.10/0.15 = 0.67$, so the pH will be within 1 pH unit of the pK_a.

(b) No, this will not be a buffer solution because HCl is a strong acid and NaOH is a strong base.

(c) Yes, this will be a buffer because HF is a weak acid and the NaOH will convert $(20.0/50.0) \times 100\% = 40\%$ of the acid to its conjugate base.

(d) No, this will not be a buffer solution because both components are bases.

(e) No, this will not be a buffer solution because both components are bases.

18.52 (a) Yes, this will be a buffer because HF is a weak acid and F^- is its conjugate base. The ratio of base to acid is $(55.0 \times 0.15)/(75.0 \times 0.10) = 1.1$, so the pH will be within 1 pH unit of the pK_a.

(b) No, this will not be a buffer solution because both components are acids.

(c) Yes, this will be a buffer because HF is a weak acid and the KOH will convert $[(135.0 \times 0.050)/(165.0 \times 0.10)] \times 100\% = 41\%$ of the acid to its conjugate base.

(d) Yes, this will be a buffer because CH_3NH_2 is a weak base and $CH_3NH_3^+$ is its conjugate acid. The ratio of base to acid is $(125.0 \times 0.15)/(120.0 \times 0.25) = 0.63$, so the pH will be within 1 pH unit of the pK_a.

(e) Yes, this will be a buffer because CH_3NH_2 is a weak base and the HCl will convert $[(95.0 \times 0.10)/(105.0 \times 0.15)] \times 100\% = 60\%$ of the base to its conjugate acid.

18.53 (a) **Given:** blood buffer 0.024 M HCO_3^- and 0.0012 M H_2CO_3, pK_a = 6.1 **Find:** initial pH

Conceptual Plan: Identify acid and base components then M HCO_3^-, M $H_2CO_3 \rightarrow$ pH

acid = H_2CO_3 base = HCO_3^- $\quad\quad$ $\text{pH} = \text{p}K_a + \log\frac{[\text{base}]}{[\text{acid}]}$

Solution: Acid = H_2CO_3, so [acid] = $[H_2CO_3]$ = 0.0012 M. Base = HCO_3^-, so [base] = $[HCO_3^-]$ = 0.024 M HCO_3^-. Then $\text{pH} = \text{p}K_a + \log\frac{[\text{base}]}{[\text{acid}]} = 6.1 + \log\frac{0.024\ \cancel{M}}{0.0012\ \cancel{M}} = 7.4$.

Check: The units (none) are correct. The magnitude of the answer makes physical sense because the pH is greater than the pK_a of the acid because there is more base than acid.

(b) **Given:** 5.0 L of blood buffer **Find:** mass HCl to lower pH to 7.0

Conceptual Plan: Final pH, p$K_a \rightarrow [HCO_3^-]/[H_2CO_3]$ then $[HCO_3^-]$, L $\rightarrow$ mol HCO_3^- and

$\text{pH} = \text{p}K_a + \log\frac{[\text{base}]}{[\text{acid}]}$ $\quad\quad$ $M = \frac{\text{mol}}{\text{L}}$

$[H_2CO_3]$, L $\rightarrow$ mol H_2CO_3 then write balanced equation then

$M = \frac{\text{mol}}{\text{L}}$ $\quad$ $H^+ + HCO_3^- \rightarrow H_2CO_3$

mol HCO_3^-, mol H_2CO_3, $[HCO_3^-]/[H_2CO_3] \rightarrow$ mol HCl $\rightarrow$ g HCl

set up stoichiometry table $\quad$ $\frac{36.46\text{ g HCl}}{1\text{ mol HCl}}$

Solution: $\text{pH} = \text{p}K_a + \log\frac{[\text{base}]}{[\text{acid}]} = 6.1 + \log\frac{[HCO_3^-]}{[H_2CO_3]} = 7.0$. Solve for $[HCO_3^-]/[H_2CO_3]$.

$\log\frac{[HCO_3^-]}{[H_2CO_3]} = 7.0 - 6.1 = 0.9 \rightarrow \frac{[HCO_3^-]}{[H_2CO_3]} = 10^{0.9} = \underline{7}.9433$. Then

$\frac{0.024\text{ mol }HCO_3^-}{1\ \cancel{L}} \times 5.0\ \cancel{L} = 0.12\text{ mol }HCO_3^-$ and $\frac{0.0012\text{ mol }H_2CO_3}{1\ \cancel{L}} \times 5.0\ \cancel{L} = 0.0060\text{ mol }H_2CO_3$.

Because HCl is a strong acid, [HCl] = $[H^+]$. Set up a table to track changes:

	$H^+(aq)$	+	$HCO_3^-(aq)$	$\rightarrow$	$H_2CO_3(aq)$
Before addition	$\approx$ 0.00 mol		0.12 mol		0.0060 mol
Addition	x		—		—
After addition	$\approx$ 0.00 mol		$(0.12 - x)$ mol		$(0.0060 + x)$ mol

Because $\frac{[HCO_3^-]}{[H_2CO_3]} = \underline{7}.9433 = \frac{(0.12 - x)\ \cancel{\text{mol}}}{(0.0060 + x)\ \cancel{\text{mol}}}$, solve for x. Note that the ratio of moles is the same as the ratio of concentrations because the volume for both terms is the same.

$\underline{7}.9433(0.0060 + x) = (0.12 - x) \rightarrow 0.04\underline{7}6598 + \underline{7}.9433x = 0.12 - x \rightarrow \underline{8}.9433x = 0.0\underline{7}234 \rightarrow$

$x = 0.00\underline{8}0888$ mol HCl then $0.00\underline{8}0888\ \cancel{\text{mol HCl}} \times \frac{36.46\text{ g HCl}}{1\ \cancel{\text{mol HCl}}} = 0.\underline{2}9492\text{ g HCl} = 0.3\text{ g HCl}$

Check: The units (g) are correct. The amount of acid needed is small because the concentrations of the buffer components are very low and the buffer starts only 0.4 pH unit above the final pH.

(c) **Given:** 5.0 L of blood buffer **Find:** mass NaOH to raise pH to 7.8

Conceptual Plan: Final pH, pK_a → $[HCO_3^-]/[H_2CO_3]$ then $[HCO_3^-]$, L → mol HCO_3^- and

$$pH = pK_a + \log\frac{[base]}{[acid]} \qquad M = \frac{mol}{L}$$

$[H_2CO_3]$, L → mol H_2CO_3 then write balanced equation then

$$M = \frac{mol}{L} \qquad OH^- + H_2CO_3 \rightarrow HCO_3^- + H_2O$$

mol HCO_3^-, mol H_2CO_3, $[HCO_3^-]/[H_2CO_3]$ → mol NaOH → g NaOH

$$\text{set up stoichiometry table} \qquad \frac{40.00\text{ g NaOH}}{1\text{ mol NaOH}}$$

Solution: $pH = pK_a + \log\frac{[base]}{[acid]} = 6.1 + \log\frac{[HCO_3^-]}{[H_2CO_3]} = 7.8$

Solve for $[HCO_3^-]/[H_2CO_3]$. $\log\frac{[HCO_3^-]}{[H_2CO_3]} = 7.8 - 6.1 = 1.7 \rightarrow \frac{[HCO_3^-]}{[H_2CO_3]} = 10^{1.7} = 50.\underline{1}1872$. Then

$\frac{0.024\text{ mol }HCO_3^-}{1\text{ L}} \times 5.0\text{ L} = 0.12\text{ mol }HCO_3^-$ and $\frac{0.0012\text{ mol }H_2CO_3}{1\text{ L}} \times 5.0\text{ L} = 0.0060\text{ mol }H_2CO_3$.

Because NaOH is a strong base, $[NaOH] = [OH^-]$. Set up a table to track changes:

	$OH^-(aq)$	+	$H_2CO_3(aq)$	→	$HCO_3^-(aq)$	+	$H_2O(l)$
Before addition	≈ 0.00 mol		0.0060 mol		0.12 mol		—
Addition	x		—		—		
After addition	≈ 0.00 mol		$(0.0060 - x)$ mol		$(0.12 + x)$ mol		

Because $\frac{[HCO_3^-]}{[H_2CO_3]} = 50.\underline{1}1872 = \frac{(0.12 + x)\text{ mol}}{(0.0060 - x)\text{ mol}}$, solve for x. Note that the ratio of moles is the same as the ratio of concentrations because the volume for both terms is the same.

$50.\underline{1}1872(0.0060 - x) = (0.12 + x) \rightarrow 0.3\underline{0}071 - 50.\underline{1}1872x = 0.12 + x \rightarrow 51.\underline{1}1872x = 0.1\underline{8}071$
$\rightarrow x = 0.003\underline{5}351$ mol NaOH then

$0.003\underline{5}351\text{ mol NaOH} \times \frac{40.00\text{ g NaOH}}{1\text{ mol NaOH}} = 0.1\underline{4}140\text{ g NaOH} = 0.14\text{ g NaOH}$

Check: The units (g) are correct. The amount of base needed is small because the concentrations of the buffer components are very low.

18.54 (a) **Given:** $HPO_4^{2-}/H_2PO_4^-$ buffer at pH = 7.1 **Find:** $[HPO_4^{2-}]/[H_2PO_4^-]$

Other: $K_{a_2}(H_3PO_4) = 6.2 \times 10^{-8}$

Conceptual Plan: Identify acid and base components then pH, K_{a_2} → $[HPO_4^{2-}]/[H_2PO^{4-}]$

$$\text{acid} = H_2PO_4^-, \text{base} = HPO_4^{2-} \qquad pH = pK_a + \log\frac{[base]}{[acid]}$$

Solution: $pH = pK_a + \log\frac{[base]}{[acid]} = -\log(6.2 \times 10^{-8}) + \log\frac{[HPO_4^{2-}]}{[H_2PO_4^-]} = 7.1$

Solve for $[HPO_4^{2-}]/[H_2PO_4^-]$.

$\log\frac{[HPO_4^{2-}]}{[H_2PO_4^-]} = 7.1 - 7.2 = -0.1 \rightarrow \frac{[HPO_4^{2-}]}{[H_2PO_4^-]} = 10^{-0.1} = 0.\underline{7}9433 = 0.8$

Check: The units (none) are correct. The magnitude of the answer makes physical sense because the pH is very close but less than the pK_a of the acid.

(b) No, H_3PO_4 and $H_2PO_4^-$ cannot be used as a buffer in the cell because the $K_{a_1}(H_3PO_4) = 7.5 \times 10^{-3}$; so the $pK_{a_1} = 2.1$. To have an effective buffer, the pK_a should be within 1 pH unit of the desired pH (not 5.0 pH units).

18.55 **Given:** $HC_2H_3O_2/KC_2H_3O_2$, $HClO_2/KClO_2$, NH_3/NH_4Cl, and $HClO/KClO$ potential buffer systems to create buffer at pH = 7.20 **Find:** best buffer system and ratio of component masses

Other: $K_a(HC_2H_3O_2) = 1.8 \times 10^{-5}$, $K_a(HClO_2) = 1.8 \times 10^{-4}$, $K_b(NH_3) = 1.76 \times 10^{-5}$, $K_a(HClO) = 2.9 \times 10^{-8}$

Conceptual Plan: Calculate pK_a of all potential buffer acids for the base $K_b \rightarrow pK_b \rightarrow pK_a$ and

$pK_a = -\log K_a$ $\quad pK_b = -\log K_b \quad 14 = pK_a + pK_b$

choose the pK_a that is closest to 7.20 then pH, $K_a \rightarrow [\text{base}]/[\text{acid}] \rightarrow$ mass base/mass acid

$pH = pK_a + \log\frac{[\text{base}]}{[\text{acid}]} \quad \frac{\mathcal{M}(\text{base})}{\mathcal{M}(\text{acid})}$

Solution: For $HC_2H_3O_2/KC_2H_3O_2$: $pK_a = -\log K_a = -\log(1.8 \times 10^{-5}) = 4.74$
For $HClO_2/KClO_2$: $pK_a = -\log K_a = -\log(1.8 \times 10^{-4}) = 3.74$
For NH_3/NH_4Cl: $pK_b = -\log K_b = -\log(1.76 \times 10^{-5}) = 4.75$
Because $14 = pK_a + pK_b$, $pK_a = 14 - pK_b = 14 - 4.75 = 9.25$.
And for HClO/KClO: $pK_a = -\log K_a = -\log(2.9 \times 10^{-8}) = 7.54$ So the HClO/KClO buffer system has the pK_a that is closest to 7.20. So $pH = pK_a + \log\frac{[\text{base}]}{[\text{acid}]} = 7.54 + \log\frac{[KClO]}{[HClO]} = 7.20$. Solve for $[KClO]/[HClO]$.

$$\log\frac{[KClO]}{[HClO]} = 7.20 - 7.54 = -0.34 \rightarrow \frac{[KClO]}{[HClO]} = 10^{-0.34} = 0.4\underline{5}7088$$ Then convert to mass ratio using

$$\frac{\mathcal{M}(\text{base})}{\mathcal{M}(\text{acid})}, 0.4\underline{5}7088\frac{\frac{\cancel{KClO\ mol}}{\cancel{L}}}{\frac{\cancel{HClO\ mol}}{\cancel{L}}} \times \frac{\frac{90.55\text{ g KClO}}{\cancel{mol\ KClO}}}{\frac{52.46\text{ g HClO}}{\cancel{mol\ HClO}}} = 0.79\frac{\text{g KClO}}{\text{g HClO}}.$$

Check: The units (none and g base/g acid) are correct. The buffer system with the K_a closest to 10^{-7} is the best choice. The magnitude of the answer makes physical sense because the buffer needs more acid than base (and this fact is not overcome by the heavier molar mass of the base).

18.56 **Given:** HF/KF, HNO_2/KNO_2, NH_3/NH_4Cl, and HClO/KClO potential buffer systems to create buffer at pH = 9.00
Find: best buffer system and ratio of component masses
Other: $K_a(HF) = 3.5 \times 10^{-4}$, $K_a(HNO_2) = 4.6 \times 10^{-4}$, $K_b(NH_3) = 1.76 \times 10^{-5}$, $K_a(HClO) = 2.9 \times 10^{-8}$
Conceptual Plan: Calculate pK_a of all potential buffer acids for the base $K_b \rightarrow pK_b \rightarrow pK_a$ and

$pK_a = -\log K_a$ $\quad pK_b = -\log K_b \quad 14 = pK_a + pK_b$

choose the pK_a that is closest to 9.00 then pH, $K_a \rightarrow [\text{base}]/[\text{acid}] \rightarrow$ mass base/mass acid

$pH = pK_a + \log\frac{[\text{base}]}{[\text{acid}]} \quad \frac{\mathcal{M}(\text{base})}{\mathcal{M}(\text{acid})}$

Solution: For HF/KF: $pK_a = -\log K_a = -\log(3.5 \times 10^{-4}) = 3.46$
For HNO_2/KNO_2: $pK_a = -\log K_a = -\log(4.6 \times 10^{-4}) = 3.34$
For NH_3/NH_4Cl: $pK_b = -\log K_b = -\log(1.76 \times 10^{-5}) = 4.75$
Because $14 = pK_a + pK_b$, $pK_a = 14 - pK_b = 14 - 4.75 = 9.25$.
And for HClO/KClO: $pK_a = -\log K_a = -\log(2.9 \times 10^{-8}) = 7.54$
So the NH_3/NH_4Cl buffer system has the pK_a that is closest to 9.00.
So $pH = pK_a + \log\frac{[\text{base}]}{[\text{acid}]} = 9.25 + \log\frac{[NH_3]}{[NH_4Cl]} = 9.00$ Solve for $[NH_3]/[NH_4Cl]$.

$$\log\frac{[NH_3]}{[NH_4Cl]} = 9.00 - 9.25 = -0.25 \rightarrow \frac{[NH_3]}{[NH_4Cl]} = 10^{-0.25} = 0.5\underline{6}2341$$ Then convert to mass ratio using

$$\frac{\mathcal{M}(\text{base})}{\mathcal{M}(\text{acid})}, 0.5\underline{6}2341\frac{\frac{\cancel{NH_3\ mol}}{\cancel{L}}}{\frac{\cancel{NH_4Cl\ mol}}{\cancel{L}}} \times \frac{\frac{17.03\text{ g } NH_3}{\cancel{mol\ NH_3}}}{\frac{53.49\text{ g } NH_4Cl}{\cancel{mol\ NH_4Cl}}} = 0.18\frac{\text{g } NH_3}{\text{g } NH_4Cl}.$$

Check: The units (none and g base/g acid) are correct. The buffer system with the K_a closest to 10^{-9} is the best choice. The magnitude of the answer makes physical sense because the buffer needs more acid than base and the acid component has a heavier molar mass; so the mass ratio is small.

18.57 **Given:** 500.0 mL of 0.100 M HNO_2/0.150 M KNO_2 buffer and (a) 250 mg NaOH, (b) 350 mg KOH, (c) 1.25 g HBr, and (d) 1.35 g HI **Find:** whether buffer capacity is exceeded

Conceptual Plan: mL → L then [HNO_2], L → mol HNO_2 and [KNO_2], L → mol KNO_2

$\frac{1\text{ L}}{1000\text{ mL}}$ $\quad$ $M = \frac{mol}{L}$ $\quad$ $M = \frac{mol}{L}$

then calculate moles of acid or base to be added to the buffer mg → g → mol then

$\frac{1\text{ g}}{1000\text{ mg}}$ $\quad \mathcal{M}$

compare the added amount to the buffer amount of the opposite component. Ratio of base/acid must be between 0.1 and 10 to maintain the buffer integrity.

Solution: $500.00\text{ mL} \times \frac{1\text{ L}}{1000\text{ mL}} = 0.5000\text{ L}$ then $\frac{0.100\text{ mol } HNO_2}{1\text{ L}} \times 0.5000\text{ L} = 0.0500\text{ mol } HNO_2$ and

$\frac{0.150\text{ mol } KNO_2}{1\text{ L}} \times 0.5000\text{ L} = 0.0750\text{ mol } KNO_2$

(a) For NaOH: $250\text{ mg NaOH} \times \frac{1\text{ g NaOH}}{1000\text{ mg NaOH}} \times \frac{1\text{ mol NaOH}}{40.00\text{ g NaOH}} = 0.00625\text{ mol NaOH}$. Because the buffer contains 0.0500 mol acid, the amount of acid is reduced by $0.00625/0.0500 = 0.125$ and the ratio of base/acid is still between 0.1 and 10. The buffer capacity is not exceeded.

(b) For KOH: $350\text{ mg KOH} \times \frac{1\text{ g KOH}}{1000\text{ mg KOH}} \times \frac{1\text{ mol KOH}}{56.11\text{ g KOH}} = 0.00624\text{ mol KOH}$. Because the buffer contains 0.0500 mol acid, the amount of acid is reduced by $0.00624/0.0500 = 0.125$ and the ratio of base/acid is still between 0.1 and 10. The buffer capacity is not exceeded.

(c) For HBr: $1.25\text{ g HBr} \times \frac{1\text{ mol HBr}}{80.91\text{ g HBr}} = 0.0154493\text{ mol HBr}$. Because the buffer contains 0.0750 mol base, the amount of acid is reduced by $0.0154/0.0750 = 0.206$ and the ratio of base/acid is still between 0.1 and 10. The buffer capacity is not exceeded.

(d) For HI: $1.35\text{ g HI} \times \frac{1\text{ mol HI}}{127.91\text{ g HI}} = 0.0105543\text{ mol HI}$. Because the buffer contains 0.0750 mol base, the amount of acid is reduced by $0.0106/0.0750 = 0.141$ and the ratio of base/acid is still between 0.1 and 10. The buffer capacity is not exceeded.

18.58 **Given:** 1.0 L of 0.125 M HNO_2/0.145 M $NaNO_2$ buffer and (a) 1.5 g HCl, (b) 1.5 g NaOH, and (c) 1.5 g HI
Find: [HNO_2] and [$NaNO_2$] after addition **Other:** $K_a(HNO_2) = 4.6 \times 10^{-4}$
Conceptual Plan: [HNO_2], L → mol HNO_2 and [$NaNO_2$], L → mol $NaNO_2$(= mol NO_2^-)

$M = \frac{mol}{L}$ $\quad$ $M = \frac{mol}{L}$

then calculate moles of acid or base to be added to the buffer g → mol

$\mathcal{M}$

then write balanced equation then mol HNO_2, mol NO_2^-, mol added species → mol HNO_2, mol NO_2^-

$H^+ + NO_2^- \rightarrow HNO_2$ $\quad$ set up stoichiometry table

$OH^- + HNO_2 \rightarrow NO_2^- + H_2O$

mol HNO_2, L → [HNO_2] and mol NO_2^-(= mol $NaNO_2$), L → [$NaNO_2$]

$M = \frac{mol}{L}$ $\quad$ $M = \frac{mol}{L}$

Solution: $\frac{0.125\text{ mol } HNO_2}{1\text{ L}} \times 1.0\text{ L} = 0.125\text{ mol } HNO_2$ and $\frac{0.145\text{ mol } KNO_2}{1\text{ L}} \times 1.0\text{ L} = 0.145\text{ mol } KNO_2$

(a) For HCl: $1.5\text{ g HCl} \times \frac{1\text{ mol HCl}}{36.46\text{ g HCl}} = 0.041141\text{ mol HCl}$. Because HCl is a strong acid, [HCl] = [H^+].

Set up a table to track changes:

	$H^1(aq)$	+	$NO_2^-(aq)$	→	$HNO_2(aq)$
Before addition	≈0.00 mol		0.145 mol		0.125 mol
Addition	0.041141 mol		—		—
After addition	≈0.00 mol		0.104 mol		0.166 mol

Because the concentrations of the acid and base components have not changed much, the buffer is still able to do its job. Finally, because there is 1.0 L of solution, $[HNO_2] = 0.17$ M and $[NaNO_2] = 0.10$ M.

(b) For NaOH: $1.5 \text{ g NaOH} \times \frac{1 \text{ mol NaOH}}{40.00 \text{ g NaOH}} = 0.0375 \text{ mol NaOH}$. Because NaOH is a strong base, $[NaOH] = [OH^-]$. Set up a table to track changes:

	$OH^-(aq)$	+	$HNO_2(aq)$	→	$NO_2^-(aq)$	+	$H_2O(l)$
Before addition	≈0.00 mol		0.125 mol		0.145 mol		—
Addition	0.0375 mol		—		—		—
After addition	≈0.00 mol		0.0875 mol		0.1825 mol		—

Because the concentrations of the acid and base components have not changed much, the buffer is still able to do its job. Finally, because there is 1.0 L of solution, $[HNO_2] = 0.09$ M and $[NaNO_2] = 0.18$ M.

(c) For HI: $1.5 \text{ g HI} \times \frac{1 \text{ mol HI}}{127.91 \text{ g HI}} = 0.011727 \text{ mol HI}$. Because HI is a strong acid, $[HI] = [H^+]$.

Set up a table to track changes:

	$H^1(aq)$	+	$NO_2^-(aq)$	→	$HNO_2(aq)$
Before addition	≈0.00 mol		0.145 mol		0.125 mol
Addition	0.011727 mol		—		—
After addition	≈0.00 mol		0.133 mol		0.137 mol

Because the concentrations of the acid and base components have not changed much, the buffer is still able to do its job. Finally, because there is 1.0 L of solution, $[HNO_2] = 0.14$ M and $[NaNO_2] = 0.13$ M.

Check: The units (M) are correct. Because the number of moles added is small compared to the buffer components, the buffer still remains active. Adding acid increases the amount of the conjugate base. Adding base increases the amount of the weak acid.

Titrations, pH Curves, and Indicators

18.59 (i) The equivalence point of a titration is where the pH rises sharply as base is added. The pH at the equivalence point is the midpoint of the sharp rise at ~50 mL added base. For (a), the pH = ~8, and for (b), the pH = ~7.

(ii) Graph (a) represents a weak acid, and graph (b) represents a strong acid. A strong acid titration starts at a lower pH, has a flatter initial region, and has a sharper rise at the equivalence point than does a weak acid. The pH at the equivalence point of a strong acid is neutral, while the pH at the equivalence point of a weak acid is basic.

18.60 **Given:** 25.0 mL 0.100 M HCl and 0.100 M HF titrated with 0.200 M KOH

(a) **Find:** volume of base to reach equivalence point

Conceptual Plan: The answer for both titrations will be the same because the initial concentration and volumes of the acids are the same and both acids are monoprotic. Write balanced equation

$HCl + KOH \rightarrow KCl + H_2O$ and $HF + KOH \rightarrow KF + H_2O$

then mL → L then [acid], L → mol acid then set mol acid = mol base and

$\frac{1 \text{ L}}{1000 \text{ mL}}$ $\quad M = \frac{\text{mol}}{\text{L}}$ $\quad$ balanced equation has 1:1 stoichiometry

[KOH], mol KOH → L KOH → mL KOH

$M = \frac{\text{mol}}{\text{L}}$ $\quad \frac{1000 \text{ mL}}{1 \text{ L}}$

Solution: $25.0 \text{ mL acid} \times \frac{1 \text{ L}}{1000 \text{ mL}} = 0.0250 \text{ L acid}$ then

$\frac{0.100 \text{ mol acid}}{1 \text{ L}} \times 0.0250 \text{ L} = 0.00250 \text{ mol acid}$. So mol acid $= 0.00250$ mol $=$ mol KOH then

$0.00250 \text{ mol KOH} \times \frac{1 \text{ L KOH}}{0.200 \text{ mol KOH}} = 0.0125 \text{ L KOH} \times \frac{1000 \text{ mL}}{1 \text{ L}} = 12.5 \text{ mL KOH}$ for both titrations

Check: The units (mL) are correct. The volume of base is half the volume of acids because the concentration of the base is twice that of the acids. The answer for both titrations is the same because the stoichiometry is the same for both titration reactions.

(b) The pH at the equivalence point will be neutral for HCl (because it is a strong acid), and it will be basic for HF (because it is a weak acid and will produce a conjugate base when titrated).

(c) The initial pH will be lower for HCl (because it is a strong acid), so it dissociates completely. The HF (because it is a weak acid) will only partially dissociate and not drop the pH as low as HCl will at the same acid concentration.

(d) The titration curves will look like the following:

HCl:

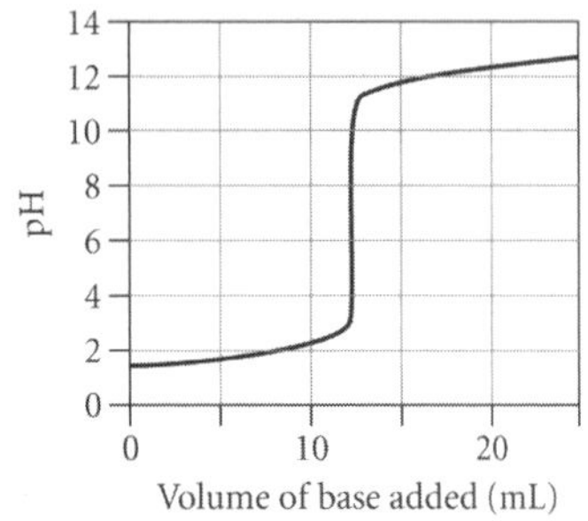

HF:

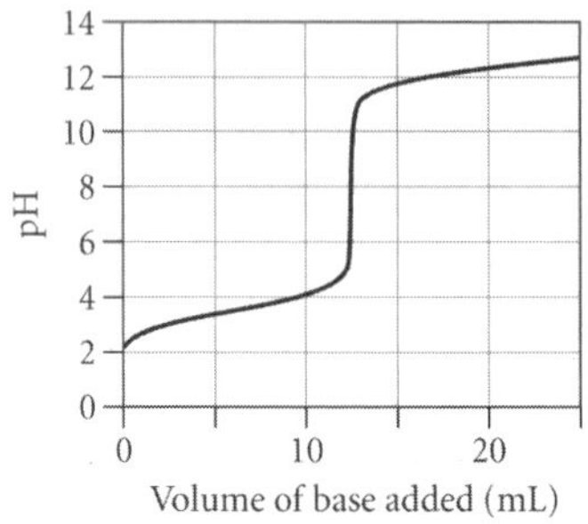

Important features to include are a low initial pH (pH is 1 for a strong acid and is higher for a weak acid), flat initial region (very flat for strong acid, not as flat for weak acid where pH halfway to equivalence point is the pK_a of the acid), sharp rise at equivalence point, pH at equivalence point (neutral for strong acid and higher for weak acid), and then flattening out at high pH.

18.61 **Given:** 20.0 mL 0.200 M KOH and 0.200 M CH_3NH_2 titrated with 0.100 M HI

(a) **Find:** volume of base to reach equivalence point

Conceptual Plan: The answer for both titrations will be the same because the initial concentration and volumes of the bases are the same. Write balanced equation then mL → L then

$$HI + KOH \rightarrow KI + H_2O \text{ and } HI + CH_3NH_2 \rightarrow CH_3NH_3I \qquad \frac{1\text{ L}}{1000\text{ mL}}$$

[base], L → mol base then set mol base = mol acid and [HI], mol HI → L HI → mL HI

$$M = \frac{\text{mol}}{\text{L}} \qquad \text{balanced equation has 1:1 stoichiometry} \qquad M = \frac{\text{mol}}{\text{L}} \quad \frac{1000\text{ mL}}{1\text{ L}}$$

Solution: $20.0\ \cancel{\text{mL base}} \times \frac{1\text{ L}}{1000\ \cancel{\text{mL}}} = 0.0200\text{ L base then}$

$$\frac{0.200\text{ mol base}}{1\ \cancel{\text{L}}} \times 0.0200\ \cancel{\text{L}} = 0.00400\text{ mol base. So mol base} = 0.00400\text{ mol} = \text{mol HI then}$$

$$0.00400\ \cancel{\text{mol HI}} \times \frac{1\text{ L HI}}{0.100\ \cancel{\text{mol HI}}} = 0.0400\ \cancel{\text{L HI}} \times \frac{1000\text{ mL}}{1\ \cancel{\text{L}}} = 40.0\text{ mL HI for both titrations}$$

Check: The units (mL) are correct. The volume of acid is twice the volume of bases because the concentration of the base is twice that of the acid in each case. The answer for both titrations is the same because the stoichiometry is the same for both titration reactions.

(b) The pH at the equivalence point will be neutral for KOH (because it is a strong base), and it will be acidic for CH_3NH_2 (because it is a weak base and will produce a conjugate acid when titrated).

(c) The initial pH will be lower for CH_3NH_2 (because it is a weak base and will only partially dissociate) and not raise the pH as high as KOH will (because it is a strong base and so dissociates completely) at the same base concentration.

(d) The titration curves will look like the following:

KOH:

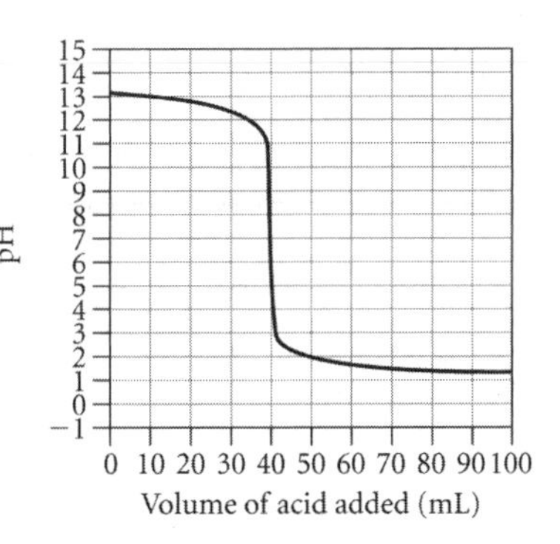

CH_3NH_2:

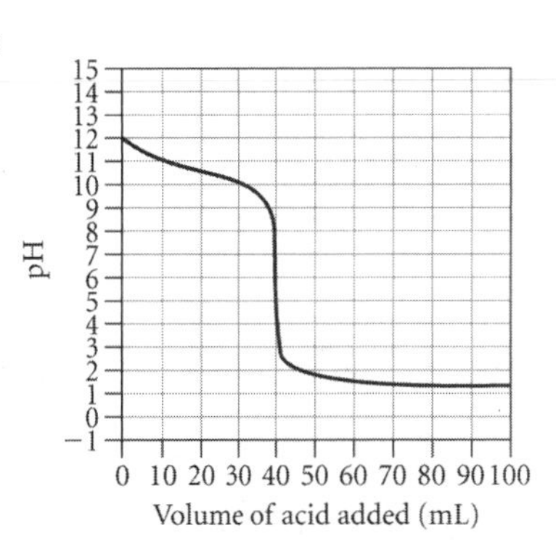

Important features to include are a high initial pH (pH is over 13 for a strong base and is lower for a weak base), flat initial region (very flat for strong base, not as flat for weak base where pH halfway to equivalence point is the pK_b of the base), sharp drop at equivalence point, pH at equivalence point (neutral for strong base and lower for weak base), and then flattening out at low pH.

18.62 (i) The equivalence point of a titration is where the pH drops sharply as acid is added. The pH at the equivalence point is the midpoint of the sharp drop at ~25 mL added acid. For (a), the pH = ~7, and for (b), the pH = ~5.

(ii) Graph (a) represents a strong base, and graph (b) represents a weak base. A strong base titration starts at a higher pH, has a flatter initial region, and has a sharper drop at the equivalence point than does a weak base. The pH at the equivalence point of a strong base is neutral, while the pH at the equivalence point of a weak base is acidic.

18.63 (a) The equivalence point of a titration is where the pH rises sharply as base is added. The volume at the equivalence point is ~30 mL. The pH at the equivalence point is the midpoint of the sharp rise at ~30 mL added base, which is a pH = ~9.

(b) At 0 mL, the pH is calculated by doing an equilibrium calculation of a weak acid in water (as done in Chapter 17).

(c) The pH halfway to the equivalence point is equal to the pK_a of the acid, or ~15 mL.

(d) The pH at the equivalence point, or ~30 mL, is calculated by doing an equilibrium problem with the K_b of the acid. At the equivalence point, all of the acid has been converted to its conjugate base.

(e) Beyond the equivalence point (30 mL), there is excess base. All of the acid has been converted to its conjugate base, so the pH is calculated by focusing on this excess base concentration.

18.64 (a) The equivalence point of a titration is where the pH drops sharply as acid is added. The volume at the equivalence point is ~25 mL. The pH at the equivalence point is the midpoint of the sharp drop at ~25 mL added acid, which is a pH = ~6.

(b) At 0 mL, the pH is calculated by doing an equilibrium calculation of a weak base in water (as done in Chapter 17).

(c) The pH halfway to the equivalence point is equal to the $14 - pK_b = pK_a$ of the base, or ~12 mL.

(d) The pH at the equivalence point, or ~25 mL, is calculated by doing an equilibrium problem with the K_a of the base. At the equivalence point, all of the base has been converted to its conjugate acid.

(e) Beyond the equivalence point (25 mL), there is excess acid. All of the base has been converted to its conjugate acid, so the pH is calculated by focusing on this excess acid concentration.

18.65 **Given:** 35.0 mL of 0.175 M HBr titrated with 0.200 M KOH

(a) **Find:** initial pH

Conceptual Plan: Because HBr is a strong acid, it will dissociate completely; so initial $pH = -\log[H_3O^+] = -\log[HBr]$.

Solution: $pH = -\log[HBr] = -\log 0.175 = 0.757$

Check: The units (none) are correct. The pH is reasonable because the concentration is greater than 0.1 M, and when the acid dissociates completely, the pH becomes less than 1.

(b) **Find:** volume of base to reach equivalence point

Conceptual Plan: Write balanced equation then mL → L then [HBr], L → mol HBr then

$HBr + KOH \rightarrow KBr + H_2O$ $\quad \frac{1\ \text{L}}{1000\ \text{mL}}$ $\quad M = \frac{\text{mol}}{\text{L}}$

set mol acid (HBr) = mol base (KOH) and [KOH], mol KOH → L KOH → mL KOH

balanced equation has 1:1 stoichiometry $\quad M = \frac{\text{mol}}{\text{L}}$ $\quad \frac{1000\ \text{mL}}{1\ \text{L}}$

Solution: $35.0\ \text{mL HBr} \times \frac{1\ \text{L}}{1000\ \text{mL}} = 0.0350\ \text{L HBr}$ then

$\frac{0.175\ \text{mol HBr}}{1\ \text{L}} \times 0.0350\ \text{L} = 0.006125\ \text{mol HBr}$

So mol acid = mol HBr = 0.006125 mol = mol KOH then

$0.006125\ \text{mol KOH} \times \frac{1\ \text{L}}{0.200\ \text{mol KOH}} = 0.030625\ \text{L KOH} \times \frac{1000\ \text{mL}}{1\ \text{L}} = 30.6\ \text{mL KOH}$

Check: The units (mL) are correct. The volume of base is a little less than the volume of acid because the concentration of the base is a little greater than that of the acid.

(c) **Find:** pH after adding 10.0 mL of base

Conceptual Plan: Use calculations from part (b). Then mL → L then [KOH], L → mol KOH then

$\frac{1\ \text{L}}{1000\ \text{mL}}$ $\quad M = \frac{\text{mol}}{\text{L}}$

mol HBr, mol KOH → mol excess HBr and L HBr, L KOH → total L then

set up stoichiometry table $\quad$ L HBr + L KOH = total L

mol excess HBr, L → [HBr] → pH.

$M = \frac{\text{mol}}{\text{L}}$ $\quad pH = -\log [HBr]$

Solution: $10.0\ \text{mL KOH} \times \frac{1\ \text{L}}{1000\ \text{mL}} = 0.0100\ \text{L KOH}$ then

$\frac{0.200\ \text{mol KOH}}{1\ \text{L}} \times 0.0100\ \text{L} = 0.00200\ \text{mol KOH}$

Because KOH is a strong base, $[KOH] = [OH^-]$. Set up a table to track changes:

	$KOH(aq)$	+	$HBr(aq)$	→	$KBr(aq)$	+	$H_2O(l)$
Before addition	≈0.00 mol		0.006125 mol		0.00 mol		—
Addition	0.00200 mol		—		—		—
After addition	≈0.00 mol		0.004125 mol		0.00200 mol		—

Then 0.0350 L HBr + 0.0100 L KOH = 0.0450 L total volume.

So mol excess acid = mol HBr = 0.004125 mol in 0.0450 L so

$[HBr] = \frac{0.004125\ \text{mol HBr}}{0.0450\ \text{L}} = 0.0916667\ \text{M}$ and

$pH = -\log [HBr] = -\log 0.0916667 = 1.038$

Check: The units (none) are correct. The pH is a little higher than the initial pH, which is expected because this is a strong acid.

(d) **Find:** pH at equivalence point

Solution: Because this is a strong acid–strong base titration, the pH at the equivalence point is neutral, or 7.

(e) **Find:** pH after adding 5.0 mL of base beyond the equivalence point

Conceptual Plan: Use calculations from parts (b) and (c). Then the pH is only dependent on the amount of excess base and the total solution volumes.

mL excess → L excess then [KOH], L excess → mol KOH excess

$\frac{1\ \text{L}}{1000\ \text{mL}}$ $\quad M = \frac{\text{mol}}{\text{L}}$

then L HBr, L KOH to equivalence point, L KOH excess → total L then

L HBr + L KOH to equivalence point + L KOH excess = total L

mol excess KOH, total L → [KOH] = [OH^-] → [H_3O^+] → pH

$M = \frac{mol}{L}$ $K_w = [H_3O^+][OH^-]$ $pH = -\log[H_3O^+]$

Solution: $5.0\ \text{mL KOH} \times \frac{1\ L}{1000\ mL} = 0.0050\ L$ KOH excess then

$\frac{0.200\ \text{mol KOH}}{1\ L} \times 0.0050\ L = 0.0010$ mol KOH excess. Then 0.0350 L HBr + 0.0306 L KOH + 0.0050 L KOH = 0.0706 L total volume. $[\text{KOH excess}] = \frac{0.0010\ \text{mol KOH excess}}{0.0706\ L} = 0.01\underline{4}164$ M KOH excess.

Because KOH is a strong base, [KOH] excess = [OH^-]. $K_w = [H_3O^+][OH^-]$, so

$$[H_3O^+] = \frac{K_w}{[OH^-]} = \frac{1.0 \times 10^{-14}}{0.01\underline{4}164} = 7.\underline{0}6 \times 10^{-13}\ M.$$

Finally, $pH = -\log[H_3O^+] = -\log(7.\underline{0}6 \times 10^{-13}) = 12.15$.

Check: The units (none) are correct. The pH is rising sharply at the equivalence point, so the pH after 5 mL past the equivalence point should be quite basic.

18.66 **Given:** 20.0 mL of 0.125 M HNO_3 titrated with 0.150 M NaOH
Find: pH at five different points and plot titration curve
Conceptual Plan: Choose points to calculate (a) initial pH, (b) pH after 5.0 mL, (c) pH after 10.0 mL, (d) pH at equivalence point, and (e) pH at 25.0 mL. Points should be on both sides of the equivalence point.

(a) **Because HNO_3 is a strong acid, it will dissociate completely; so initial $pH = -\log[H_3O^+] = -\log[HNO_3]$.**

Solution: $pH = -\log[HNO_3] = -\log 0.125 = 0.903$

Check: The units (none) are correct. The pH is reasonable because the concentration is greater than 0.1 M, and when a 0.1 M acid dissociates completely, the pH becomes less than 1.

(b) **Find:** pH after adding 5.0 mL of base
Conceptual Plan: Write balanced equation then mL → L then [HNO_3], L → mol HNO_3 then

$HNO_3 + NaOH \rightarrow NaNO_3 + H_2O$ $\frac{1\ L}{1000\ mL}$ $M = \frac{mol}{L}$

mL → L then [NaOH], L → mol NaOH then mol HNO_3, mol NaOH → mol excess HNO_3

$\frac{1\ L}{1000\ mL}$ $M = \frac{mol}{L}$ set up stoichiometry table

and L HNO_3, L NaOH → total L then mol excess HNO_3, L → [HNO_3] → pH

L HNO_3 + L NaOH = total L $M = \frac{mol}{L}$ $pH = -\log[HNO_3]$

Solution: $20.0\ \text{mL HNO}_3 \times \frac{1\ L}{1000\ mL} = 0.0200\ L\ HNO_3$ then

$\frac{0.125\ \text{mol HNO}_3}{1\ L} \times 0.0200\ L = 0.00250$ mol HNO_3 and $5.0\ \text{mL NaOH} \times \frac{1\ L}{1000\ mL} = 0.0050$ L NaOH

then $\frac{0.150\ \text{mol NaOH}}{1\ L} \times 0.0050\ L = 0.00075$ mol NaOH

This is a strong acid–strong base titration. Set up a table to track changes:

	NaOH(*aq*)	+	HNO_3(*aq*)	→	$NaNO_3$(*aq*)	+	H_2O(*l*)
Before addition	0.00 mol		0.00250 mol		0.00 mol		—
Addition	0.00075 mol		—		—		—
After addition	≈0.00 mol		0.00175 mol		0.00075 mol		—

Then 0.0200 L HNO_3 + 0.0050 L NaOH = 0.0250 L total volume. So mol excess acid = mol HNO_3 = 0.00175 mol in 0.0250 L so $[HNO_3] = \frac{0.00175\ \text{mol HNO}_3}{0.0250\ L} = 0.0700$ M and

$pH = -\log[HNO_3] = -\log 0.0700 = 1.155$

Check: The units (none) are correct. The pH remains very low in a strong acid–strong base titration before the equivalence point.

(c) **Find:** pH after adding 10.0 mL of base

Conceptual Plan: Use calculations from point (b) then mL → L then [NaOH], L → mol NaOH then

$\frac{1\ L}{1000\ mL}$ $M = \frac{mol}{L}$

mol HNO_3, mol NaOH → mol excess HNO_3 and L HNO_3, L NaOH → total L then

set up stoichiometry table $L\ HNO_3 + L\ NaOH = total\ L$

mol excess HNO_3, L → $[HNO_3]$ → pH

$M = \frac{mol}{L}$ $pH = -\log[HNO_3]$

Solution: $10.0\ \cancel{mL\ NaOH} \times \frac{1\ L}{1000\ \cancel{mL}} = 0.0100\ L\ NaOH$ then

$\frac{0.150\ mol\ NaOH}{1\ \cancel{L}} \times 0.0100\ \cancel{L} = 0.00150\ mol\ NaOH$. Set up a table to track changes:

	$NaOH(aq)$	+	$HNO_3(aq)$	→	$NaNO_3(aq)$	+	$H_2O(l)$
Before addition	0.00 mol		0.00250 mol		0.00 mol		—
Addition	0.00150 mol		—		—		—
After addition	≈0.00 mol		0.00100 mol		0.00150 mol		—

Then $0.0200\ L\ HNO_3 + 0.0100\ L\ NaOH = 0.0300\ L$ total volume. So mol excess acid = mol HNO_3 = 0.00100 mol in 0.0300 L so $[HNO_3] = \frac{0.00100\ mol\ HNO_3}{0.0300\ L} = 0.03\underline{3}3333\ M$ and

$pH = -\log[HNO_3] = -\log 0.03\underline{3}3333 = 1.477$

Check: The units (none) are correct. The pH remains very low in a strong acid–strong base titration before the equivalence point.

(d) **Find:** pH at equivalence point and volume of base to reach equivalence point

Conceptual Plan: Because this is a strong acid–strong base titration, the pH at the equivalence point is neutral, or 7. Use calculations from point (b) then set mol acid (HNO_3) = mol base (NaOH) and

balanced equation has 1:1 stoichiometry

[NaOH], mol NaOH → L NaOH → mL NaOH

$M = \frac{mol}{L}$ $\frac{1000\ mL}{1\ L}$

Solution: Because this is a strong acid–strong base titration, the pH at the equivalence point is neutral, or 7. So mol acid = mol HNO_3 = 0.00250 mol = mol NaOH then

$$0.00250\ \cancel{mol\ NaOH} \times \frac{1\ L}{0.150\ \cancel{mol\ NaOH}} = 0.016\underline{6}667\ \cancel{L\ NaOH} \times \frac{1000\ mL}{1\ \cancel{L}} = 16.7\ mL\ NaOH$$

Check: The units (none and mL) are correct. The equivalence point pH of a strong acid–strong base titration is neutral. The volume of base is a little less than the volume of acid because the concentration of the base is a little greater than that of the acid.

(e) **Find:** pH after adding 25.0 mL of base

Conceptual Plan: Use calculations from point (b) then mL → L then [NaOH], L → mol NaOH then

$\frac{1\ L}{1000\ mL}$ $M = \frac{mol}{L}$

mol HNO_3, mol NaOH → mol excess NaOH and L HNO_3, L NaOH → total L then

set up stoichiometry table $L\ HNO_3 + L\ NaOH = total\ L$

mol excess NaOH, total L → [NaOH] = $[OH^-]$ → $[H_3O^+]$ → pH

$M = \frac{mol}{L}$ $K_w = [H_3O^+][OH^-]$ $pH = -\log[H_3O^+]$

Solution: $25.0\ \cancel{\text{mL NaOH}} \times \frac{1\ \text{L}}{1000\ \cancel{\text{mL}}} = 0.0250\ \text{L NaOH}$ then

$\frac{0.150\ \text{mol NaOH}}{1\ \cancel{\text{L}}} \times 0.0250\ \cancel{\text{L}} = 0.00375\ \text{mol NaOH}$. Set up a table to track changes:

	$NaOH(aq)$	+	$HNO_3(aq)$	→	$NaNO_3(aq)$	+	$H_2O(l)$
Before addition	0.00 mol		0.00250 mol		0.00 mol		—
Addition	0.00375 mol		—		—		—
After addition	0.00125 mol		≈0.00 mol		0.00275 mol		—

Then $0.0200\ \text{L HNO}_3 + 0.0250\ \text{L NaOH} = 0.0450\ \text{L}$ total volume. So mol excess acid = mol NaOH = 0.00 125 mol in 0.0450 L, so

$[\text{NaOH excess}] = \frac{0.00125\ \text{mol NaOH excess}}{0.0450\ \text{L}} = 0.027\underline{7}778\ \text{M}$ NaOH excess. Because NaOH is a strong base, $[\text{NaOH}]\ \text{excess} = [\text{OH}^-]$. $K_w = [\text{H}_3\text{O}^+][\text{OH}^-]$, so

$$[\text{H}_3\text{O}^+] = \frac{K_w}{[\text{OH}^-]} = \frac{1.0 \times 10^{-14}}{0.027\underline{7}778} = 3.6 \times 10^{-13}\ \text{M}.$$

Finally, $\text{pH} = -\log[\text{H}_3\text{O}^+] = -\log(3.6 \times 10^{-13}) = 12.44$.

Check: The units (none) are correct. The pH is rising sharply at the equivalence point, so the pH over 5 mL past the equivalence point should be quite basic. Finally, after these five points are plotted, the titration curve looks like this:

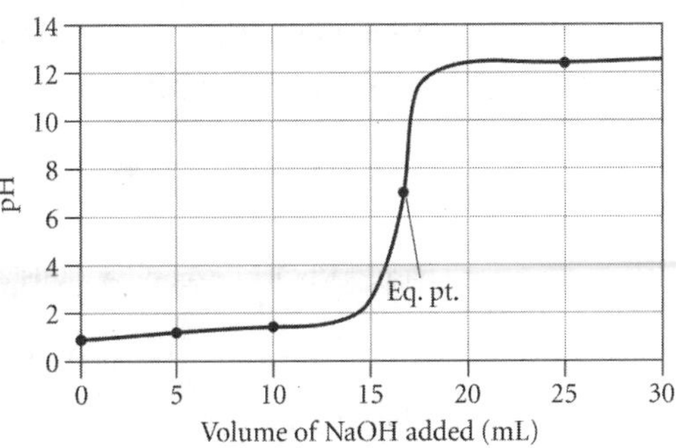

18.67 **Given:** 25.0 mL of 0.115 M RbOH titrated with 0.100 M HCl

(a) **Find:** initial pH

Conceptual Plan: Because RbOH is a strong base, it will dissociate completely, so $[\text{RbOH}] = [\text{OH}^-] \rightarrow [\text{H}_3\text{O}^+] \rightarrow \text{pH}$

$K_w = [\text{H}_3\text{O}^+][\text{OH}^-]$ $\quad$ $\text{pH} = -\log[\text{H}_3\text{O}^+]$

Solution: Because RbOH is a strong base, $[\text{RbOH}]\ \text{excess} = [\text{OH}^-]$. $K_w = [\text{H}_3\text{O}^+][\text{OH}^-]$, so

$$[\text{H}_3\text{O}^+] = \frac{K_w}{[\text{OH}^-]} = \frac{1.0 \times 10^{-14}}{0.115} = 8.\underline{6}9565 \times 10^{-14}\ \text{M}\ \text{and}$$

$\text{pH} = -\log[\text{H}_3\text{O}^+] = -\log(8.\underline{6}9565 \times 10^{-14}) = 13.06$.

Check: The units (none) are correct. The pH is reasonable because the concentration is greater than 0.1 M, and when the base dissociates completely, the pH becomes greater than 13.

(b) **Find:** volume of acid to reach equivalence point

Conceptual Plan: Write balanced equation then mL → L then [RbOH], L → mol RbOH then

$\text{HCl} + \text{RbOH} \rightarrow \text{RbCl} + \text{H}_2\text{O}$ $\quad$ $\frac{1\ \text{L}}{1000\ \text{mL}}$ $\quad$ $\text{M} = \frac{\text{mol}}{\text{L}}$

set mol base (RbOH) = mol acid (HCl) and [HCl], mol HCl → L HCl → mL HCl

balanced equation has 1:1 stoichiometry $\quad$ $\text{M} = \frac{\text{mol}}{\text{L}}$ $\quad$ $\frac{1000\ \text{mL}}{1\ \text{L}}$

Solution: $25.0\ \cancel{\text{mL RbOH}} \times \frac{1\ \text{L}}{1000\ \cancel{\text{mL}}} = 0.0250\ \text{L RbOH}$ then

$\frac{0.115\ \text{mol RbOH}}{1\ \cancel{\text{L}}} \times 0.0250\ \cancel{\text{L}} = 0.0028\underline{7}5\ \text{mol RbOH}$. So mol base = mol RbOH = $0.0028\underline{7}5$ mol = mol HCl then $0.0028\underline{7}5\ \cancel{\text{mol HCl}} \times \frac{1\ \text{L}}{0.100\ \cancel{\text{mol HCl}}} = 0.028\underline{7}5\ \cancel{\text{L HCl}} \times \frac{1000\ \text{mL}}{1\ \cancel{\text{L}}} = 28.8\ \text{mL HCl}$.

Check: The units (mL) are correct. The volume of acid is greater than the volume of base because the concentration of the base is a little greater than that of the acid.

(c) **Find:** pH after adding 5.0 mL of acid

Conceptual Plan: Use calculations from part (b). Then mL → L then [HCl], L → mol HCl then

$$\frac{1\ \text{L}}{1000\ \text{mL}} \qquad M = \frac{\text{mol}}{\text{L}}$$

mol RbOH, mol HCl → mol excess RbOH and L RbOH, L HCl → total L then

set up stoichiometry table $\qquad$ L RbOH + L HCl = total L

mol excess RbOH, L → [RbOH] = [OH⁻] → [H₃O⁺] → pH.

$$M = \frac{\text{mol}}{\text{L}} \qquad K_w = [H_3O^+][OH^-] \qquad pH = -\log[H_3O^+]$$

Solution: $5.0\ \cancel{\text{mL HCl}} \times \frac{1\ \text{L}}{1000\ \cancel{\text{mL}}} = 0.0050\ \text{L HCl}$ then $\frac{0.100\ \text{mol HCl}}{1\ \cancel{\text{L}}} \times 0.0050\ \cancel{\text{L}} = 0.00050\ \text{mol HCl}$

Because HCl is a strong acid, $[HCl] = [H_3O^+]$. Set up a table to track changes:

	$HCl(aq)$	+	$RbOH(aq)$	→	$RbCl(aq)$	+	$H_2O(l)$
Before addition	0.00 mol		0.002875 mol		0.00 mol		—
Addition	0.00050 mol		—		—		—
After addition	≈0.00 mol		0.002375 mol		0.00050 mol		—

Then 0.0250 L RbOH + 0.0050 L HCl = 0.0300 L total volume. So mol excess base = mol RbOH = 0.002375 mol in 0.0300 L, so $[RbOH] = \frac{0.002375\ \text{mol RbOH}}{0.0300\ \text{L}} = 0.0791667\ M$. Because RbOH is a strong base, $[RbOH]$ excess $= [OH^-]$. $K_w = [H_3O^+][OH^-]$ so $[H_3O^+] = \frac{K_w}{[OH^-]} = \frac{1.0 \times 10^{-14}}{0.0791667}$ $= 1.26316 \times 10^{-13}\ M$ and $pH = -\log[H_3O^+] = -\log(1.26316 \times 10^{-13}) = 12.90$

Check: The units (none) are correct. The pH is a little lower than the initial pH, which is expected because this is a strong base.

(d) **Find:** pH at equivalence point

Solution: Because this is a strong acid–strong base titration, the pH at the equivalence point is neutral, or 7.

(e) **Find:** pH after adding 5.0 mL of acid beyond the equivalence point

Conceptual Plan: Use calculations from parts (b) and (c). Then the pH is only dependent on the amount of excess acid and the total solution volumes. Then mL excess → L excess then [HCl], L excess → mol HCl excess

$$\frac{1\ \text{L}}{1000\ \text{mL}} \qquad M = \frac{\text{mol}}{\text{L}}$$

then L RbOH, L HCl to equivalence point, L HCl excess → total L then

L RbOH + L HCl to equivalence point + L HCl excess = total L

mol excess HCl, total L → [HCl] = [H₃O⁺] → pH.

$$M = \frac{\text{mol}}{\text{L}} \qquad pH = -\log[H_3O^+]$$

Solution: $5.0\ \cancel{\text{mL HCl}} \times \frac{1\ \text{L}}{1000\ \cancel{\text{mL}}} = 0.0050\ \text{L HCl excess}$ then

$\frac{0.100\ \text{mol HCl}}{1\ \cancel{\text{L}}} \times 0.0050\ \cancel{\text{L}} = 0.00050\ \text{mol HCl excess}$. Then 0.0250 L RbOH + 0.0288 L HCl + 0.0050 L HCl = 0.0588 L total volume. $[\text{HCl excess}] = \frac{0.00050\ \text{mol HCl excess}}{0.0588\ \text{L}} = 0.0085034\ M$ HCl excess

Because HCl is a strong acid, $[HCl]$ excess $= [H_3O^+]$.

Finally, $pH = -\log[H_3O^+] = -\log(0.0085034) = 2.07$.

Check: The units (none) are correct. The pH is dropping sharply at the equivalence point, so the pH after 5 mL past the equivalence point should be quite acidic.

18.68 **Given:** 15.0 mL of 0.100 M $Ba(OH)_2$ titrated with 0.125 M HCl
Find: pH at five different points and plot titration curve
Conceptual Plan: Choose points to calculate (a) initial pH, (b) pH after 10.0 mL, (c) pH after 20.0 mL, (d) pH at equivalence point, and (e) pH at 30.0 mL. Points should be on both sides of the equivalence point.

(a) **Find:** initial pH
Conceptual Plan: Because $Ba(OH)_2$ is a strong base, it will dissociate completely. Keep in mind that $Ba(OH)_2 \rightarrow Ba^{2+} + 2\,OH^-$, so two hydroxide ions are generated for each barium hydroxide and $2[Ba(OH)_2] = [OH^-] \rightarrow [H_3O^+] \rightarrow$ pH.

$K_w = [H_3O^+][OH^-]$ $\quad$ $pH = -\log[H_3O^+]$

Solution: Because $Ba(OH)_2$ is a strong base, $2[Ba(OH)_2] = [OH^-] = 2 \times 0.100\text{ M} = 0.200\text{ M}$.

$$K_w = [H_3O^+][OH^-], \text{ so } [H_3O^+] = \frac{K_w}{[OH^-]} = \frac{1.0 \times 10^{-14}}{0.200} = 5.0 \times 10^{-14}\text{ M and}$$

$$pH = -\log[H_3O^+] = -\log(5.0 \times 10^{-14}) = 13.30$$

Check: The units (none) are correct. The pH is reasonable because at concentrations greater than 0.1 M and when the base dissociates completely, the pH is greater than 13.

(b) **Find:** pH after adding 10.0 mL of acid
Conceptual Plan: Write a balanced equation then mL → L then $[Ba(OH)_2]$, L → mol $Ba(OH)_2$ then

$2\,HCl + Ba(OH)_2 \rightarrow BaCl_2 + 2\,H_2O$ $\quad$ $\frac{1\text{ L}}{1000\text{ mL}}$ $\quad$ $M = \frac{mol}{L}$

mL → L then [HCl], L → mol HCl then mol $Ba(OH)_2$, mol HCl → mol excess $Ba(OH)_2$ and

$\frac{1\text{ L}}{1000\text{ mL}}$ $\quad$ $M = \frac{mol}{L}$ $\quad$ set up stoichiometry table

L $Ba(OH)_2$, L HCl → total L then

$M = \frac{mol}{L}$

mol excess $Ba(OH)_2$, L → $2[Ba(OH)_2] = [OH^-] \rightarrow [H_3O^+] \rightarrow$ pH

$L\ Ba(OH)_2 + L\ HCl = \text{total L}$ $\quad$ $K_w = [H_3O^+][OH^-]$ $\quad$ $pH = -\log[H_3O^+]$

Solution: $15.0\text{ mL } Ba(OH)_2 \times \frac{1\text{ L}}{1000\text{ mL}} = 0.0150\text{ L } Ba(OH)_2$ then

$$\frac{0.100\text{ mol } Ba(OH)_2}{1\text{ L}} \times 0.0150\text{ L} = 0.00150\text{ mol } Ba(OH)_2 \text{ and } 10.0\text{ mL HCl} \times \frac{1\text{ L}}{1000\text{ mL}} = 0.0100\text{ L HCl}$$

then $\frac{0.125\text{ mol HCl}}{1\text{ L}} \times 0.0100\text{ L} = 0.00125$ mol HCl. Because HCl is a strong acid, $[HCl] = [H_3O^+]$.

Set up a table to track changes:

	$2\,HCl(aq)$	+	$Ba(OH)_2(aq)$	→	$BaCl_2(aq)$	+	$2\,H_2O(l)$
Before addition	0.00 mol		0.00150 mol		0.00 mol		—
Addition	0.00125 mol		—		—		—
After addition	≈0.00 mol		0.000875 mol		0.000625 mol		—

Then 0.0150 L $Ba(OH)_2$ + 0.0100 L HCl = 0.0250 L total volume. So mol excess base = mol $Ba(OH)_2$ =

$$0.000875\text{ mol in } 0.0250\text{ L, so } [Ba(OH)_2] = \frac{0.000875\text{ mol } Ba(OH)_2}{0.0250\text{ L}} = 0.035\text{ M.}$$ Because $Ba(OH)_2$ is a strong base, $2[Ba(OH)_2] = [OH^-] = 2 \times 0.035\text{ M} = 0.070\text{ M}$. $K_w = [H_3O^+][OH^-]$, so

$$[H_3O^+] = \frac{K_w}{[OH^-]} = \frac{1.0 \times 10^{-14}}{0.070} = 1.4286 \times 10^{-13}\text{ M and}$$

$$pH = -\log[H_3O^+] = -\log(1.4286 \times 10^{-13}) = 12.85.$$

Check: The units (none) are correct. The pH is a little lower than the initial pH, which is expected because this is a strong base.

(c) **Find:** pH after adding 20.0 mL of acid

Conceptual Plan: Use calculations from part (b) then mL → L then [HCl], L → mol HCl then

$\frac{1\text{ L}}{1000\text{ mL}}$ $\quad M = \frac{\text{mol}}{\text{L}}$

mol $Ba(OH)_2$, mol HCl → mol excess $Ba(OH)_2$ and L $Ba(OH)_2$, L HCl → total L then

set up stoichiometry table $\quad$ L $Ba(OH)_2$ + L HCl = total L

mol excess $Ba(OH)_2$, L → 2 [$Ba(OH)_2$] = [OH^-] → [H_3O^+] → pH

$M = \frac{\text{mol}}{\text{L}}$ $\quad K_w = [H_3O^+][OH^-]$ $\quad pH = -\log[H_3O^+]$

Solution: $20.0\ \cancel{\text{mL HCl}} \times \frac{1\text{ L}}{1000\ \cancel{\text{mL}}} = 0.0200\text{ L HCl}$ then $\frac{0.125\text{ mol HCl}}{1\ \cancel{\text{L}}} \times 0.0200\ \cancel{\text{L}} = 0.00250\text{ mol HCl}$

Because HCl is a strong acid, [HCl] = [H_3O^+]. Set up a table to track changes:

	$2\ HCl(aq)$ +	$Ba(OH)_2(aq)$ →	$BaCl_2(aq)$ +	$2\ H_2O(l)$
Before addition	0.00 mol	0.00150 mol	0.00 mol	—
Addition	0.00250 mol	—	—	—
After addition	≈0.00 mol	0.00025 mol	0.00125 mol	—

Then 0.0150 L $Ba(OH)_2$ + 0.0200 L HCl = 0.0350 L total volume. So mol excess base = mol $Ba(OH)_2$ = 0.00025 mol in 0.0350 L so $[Ba(OH)_2] = \frac{0.00025\text{ mol }Ba(OH)_2}{0.0350\text{ L}} = 0.007\underline{1}429\text{ M}$. Because $Ba(OH)_2$ is a strong base, $2[Ba(OH)_2] = [OH^-] = 2 \times 0.007\underline{1}429\text{ M} = 0.01\underline{4}286\text{ M}$. $K_w = [H_3O^+][OH^-]$, so

$[H_3O^+] = \frac{K_w}{[OH^-]} = \frac{1.0 \times 10^{-14}}{0.01\underline{4}286} = 6.\underline{9}9986 \times 10^{-13}\text{ M}$ and

$pH = -\log[H_3O^+] = -\log(6.\underline{9}9986 \times 10^{-13}) = 12.15$.

Check: The units (none) are correct. The pH is a little lower than the initial pH, which is expected because this is a strong base.

(d) **Find:** pH at equivalence point and volume of acid to reach equivalence point

Solution: Because this is a strong acid–strong base titration, the pH at the equivalence point is neutral, or 7.

Conceptual Plan: Use calculations from part (b) then set 2 mol base ($Ba(OH)_2$) = mol acid (HCl) and

balanced equation has 1:2 stoichiometry

[HCl], mol HCl → L HCl → mL HCl

$M = \frac{\text{mol}}{\text{L}}$ $\quad \frac{1000\text{ mL}}{1\text{ L}}$

Solution: $0.00150\ \cancel{\text{mol Ba(OH)}_2} \times \frac{2\text{ mol HCl}}{1\ \cancel{\text{mol Ba(OH)}_2}} = 0.00300\text{ mol HCl}$ then

$0.00300\ \cancel{\text{mol HCl}} \times \frac{1\text{ L}}{0.125\ \cancel{\text{mol HCl}}} = 0.0240\ \cancel{\text{L HCl}} \times \frac{1000\text{ mL}}{1\ \cancel{\text{L}}} = 24.0\text{ mL HCl}$

Check: The units (mL) are correct. The volume of acid is greater than the volume of base because two moles of acid are needed for each mole of base.

(e) **Find:** pH after adding 30.0 mL of acid

Conceptual Plan: Use calculations from earlier parts. Then the pH is only dependent on the amount of excess acid and the total solution volume.

mL added, mL at equiv. pt. → mL excess → L excess then

mL excess = mL added − mL at equiv. pt. $\quad \frac{1\text{ L}}{1000\text{ mL}}$

[HCl], L excess → mol HCl excess then L $Ba(OH)_2$, L HCl → total L then

$M = \frac{\text{mol}}{\text{L}}$ $\quad$ L $Ba(OH)_2$ + L HCl = total L

mol excess HCl, total L → [HCl] = [H_3O^+] → pH

$M = \frac{\text{mol}}{\text{L}}$ $\quad pH = -\log[H_3O^+]$

Solution: mL HCl excess = mL added − mL to equiv. pt. = 30.0 mL − 24.0 mL = 6.0 mL

$6.0\ \text{mL HCl} \times \frac{1\ \text{L}}{1000\ \text{mL}} = 0.0060\ \text{L HCl excess then}$

$\frac{0.125\ \text{mol HCl}}{1\ \text{L}} \times 0.0060\ \text{L} = 0.00075\ \text{mol HCl excess}$. Then $0.0150\ \text{L}\ Ba(OH)_2 + 0.0300\ \text{L HCl} =$ $0.0450\ \text{L}$ total volume. $[\text{HCl excess}] = \frac{0.00075\ \text{mol HCl excess}}{0.0450\ \text{L}} = 0.016667\ \text{M HCl excess}$. Because HCl is a strong acid, $[HCl]$ excess $= [H_3O^+]$. Finally, $pH = -\log[H_3O^+] = -\log(0.016667) = 1.78$.

Check: The units (none) are correct. The pH is dropping sharply at the equivalence point, so the pH after 6 mL past the equivalence point should be quite acidic. Finally, after these five points are plotted the titration curve looks like this:

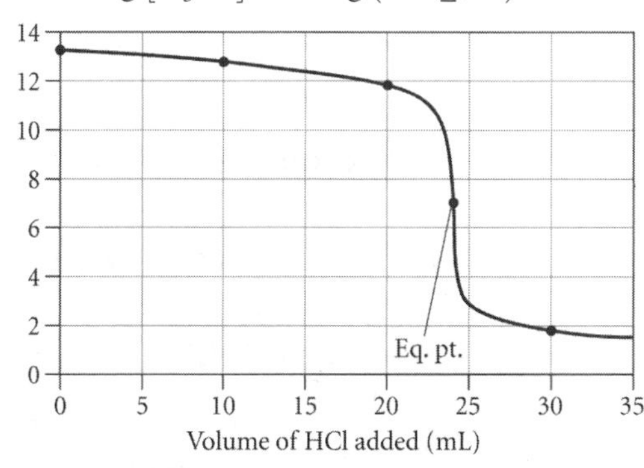

18.69 **Given:** 20.0 mL of 0.105 M $HC_2H_3O_2$ titrated with 0.125 M NaOH **Other:** $K_a(HC_2H_3O_2) = 1.8 \times 10^{-5}$

(a) **Find:** initial pH

Conceptual Plan: Because $HC_2H_3O_2$ is a weak acid, set up an equilibrium problem using the initial concentration.

So M $HC_2H_3O_2 \rightarrow [H_3O^+] \rightarrow$ pH

ICE table $\quad pH = -\log[H_3O^+]$

Solution:

$$HC_2H_3O_2(aq) + H_2O(l) \rightleftharpoons H_3O^+(aq) + C_2H_3O_2^-(aq)$$

	$[HC_2H_3O_2]$	$[H_3O^+]$	$[C_2H_3O_2^-]$
Initial	0.105	≈ 0.00	0.00
Change	$-x$	$+x$	$+x$
Equil	$0.105 - x$	$+x$	$+x$

$K_a = \frac{[H_3O^+][C_2H_3O_2^-]}{[HC_2H_3O_2]} = 1.8 \times 10^{-5} = \frac{x^2}{0.105 - x}$ Assume that x is small ($x << 0.105$), so

$\frac{x^2}{0.105 - x} = 1.8 \times 10^{-5} = \frac{x^2}{0.105}$ and $x = 1.3748 \times 10^{-3}\ \text{M} = [H_3O^+]$. Confirm that the assumption is valid.

$\frac{1.3748 \times 10^{-3}}{0.105} \times 100\% = 1.3\% < 5\%$, so the assumption is valid. Finally,

$pH = -\log[H_3O^+] = -\log(1.3748 \times 10^{-3}) = 2.86$.

Check: The units (none) are correct. The magnitude of the answer makes physical sense because the pH should be greater than $-\log(0.105) = 0.98$ because this is a weak acid.

(b) **Find:** volume of base to reach equivalence point

Conceptual Plan: Write a balanced equation then mL $\rightarrow$ L then $[HC_2H_3O_2]$, L $\rightarrow$ mol $HC_2H_3O_2$ then

$HC_2H_3O_2 + NaOH \rightarrow NaC_2H_3O_2 + H_2O \quad \frac{1\ \text{L}}{1000\ \text{mL}} \quad M = \frac{\text{mol}}{\text{L}}$

set mol acid($HC_2H_3O_2$) = mol base(NaOH) and [NaOH], mol NaOH $\rightarrow$ L NaOH $\rightarrow$ mL NaOH

balanced equation has 1:1 stoichiometry $\quad M = \frac{\text{mol}}{\text{L}} \quad \frac{1000\ \text{mL}}{1\ \text{L}}$

Solution: $20.0\ \text{mL}\ HC_2H_3O_2 \times \frac{1\ \text{L}}{1000\ \text{mL}} = 0.0200\ \text{L}\ HC_2H_3O_2$ then

$\frac{0.105\ \text{mol}\ HC_2H_3O_2}{1\ \text{L}} \times 0.0200\ \text{L} = 0.00210\ \text{mol}\ HC_2H_3O_2$

So mol acid = mol $HC_2H_3O_2$ = 0.00210 mol = mol NaOH then

$$0.00210\ \cancel{\text{mol NaOH}} \times \frac{1\ \text{L}}{0.125\ \cancel{\text{mol NaOH}}} = 0.0168\ \cancel{\text{L NaOH}} \times \frac{1000\ \text{mL}}{1\ \cancel{\text{L}}} = 16.8\ \text{mL NaOH}$$

Check: The units (mL) are correct. The volume of base is a little less than the volume of acid because the concentration of the base is a little greater than that of the acid.

(c) **Find:** pH after adding 5.0 mL of base

Conceptual Plan: Use calculations from part (b). Then mL → L then [NaOH], L → mol NaOH then

$\frac{1\ \text{L}}{1000\ \text{mL}}$ $\quad$ $M = \frac{\text{mol}}{\text{L}}$

mol $HC_2H_3O_2$, mol NaOH → mol excess $HC_2H_3O_2$, mol $C_2H_3O_2^-$ and

set up stoichiometry table

L $HC_2H_3O_2$, L NaOH → total L then

L $HC_2H_3O_2$ + L NaOH = total L

mol excess $HC_2H_3O_2$, L → $[HC_2H_3O_2]$ and mol excess $C_2H_3O_2^-$, L → $[C_2H_3O_2^-]$ then

$M = \frac{\text{mol}}{\text{L}}$ $\quad$ $M = \frac{\text{mol}}{\text{L}}$

M $HC_2H_3O_2$, M $C_2H_3O_2^-$ → $[H_3O^+]$ → pH.

ICE table $\quad$ $pH = -\log[H_3O^+]$

Solution: $5.0\ \cancel{\text{mL NaOH}} \times \frac{1\ \text{L}}{1000\ \cancel{\text{mL}}} = 0.0050\ \text{L NaOH}$ then

$\frac{0.125\ \text{mol NaOH}}{1\ \cancel{\text{L}}} \times 0.0050\ \cancel{\text{L}} = 0.0006\underline{2}5$ mol NaOH. Set up a table to track changes:

	$NaOH(aq)$	+	$HC_2H_3O_2(aq)$	→	$NaC_2H_3O_2(aq)$	+	$H_2O(l)$
Before addition	0.00 mol		0.00210 mol		0.00 mol		—
Addition	0.0006$\underline{2}$5 mol		—		—		—
After addition	≈0.00 mol		0.0014$\underline{7}$5 mol		0.000625 mol		—

Then 0.0200 L $HC_2H_3O_2$ + 0.0050 L NaOH = 0.0250 L total volume. Then

$$[HC_2H_3O_2] = \frac{0.0014\underline{7}5\ \text{mol}\ HC_2H_3O_2}{0.0250\ \text{L}} = 0.0590\ \text{M and}$$

$$[NaC_2H_3O_2] = \frac{0.0006\underline{2}5\ \text{mol}\ C_2H_3O_2^-}{0.0250\ \text{L}} = 0.025\ \text{M.}$$

Because one $C_2H_3O_2^-$ ion is generated for each $NaC_2H_3O_2$, $[C_2H_3O_2^-] = 0.025$ M $C_2H_3O_2^-$.

$$HC_2H_3O_2(aq) + H_2O(l) \rightleftharpoons H_3O^+(aq) + C_2H_3O_2^-(aq)$$

	$[HC_2H_3O_2]$	$[H_3O^+]$	$[C_2H_3O_2^-]$
Initial	0.0590	≈0.00	0.025
Change	$-x$	$+x$	$+x$
Equil	$0.0590 - x$	$+x$	$0.025 + x$

$K_a = \frac{[H_3O^+][C_2H_3O_2^-]}{[HC_2H_3O_2]} = 1.8 \times 10^{-5} = \frac{x(0.025 + x)}{0.0590 - x}$ Assume that x is small ($x << 0.025 < 0.0590$), so

$\frac{x(0.025 + \cancel{x})}{0.0590 - \cancel{x}} = 1.8 \times 10^{-5} = \frac{x(0.025)}{0.0590}$ and $x = 4.\underline{2}48 \times 10^{-5}\ \text{M} = [H_3O^+]$. Confirm that the assumption is valid.

$\frac{4.\underline{2}48 \times 10^{-5}}{0.025} \times 100\% = 0.17\% < 5\%$, so the assumption is valid.

Finally, $pH = -\log[H_3O^+] = -\log(4.\underline{2}48 \times 10^{-5}) = 4.37$.

Check: The units (none) are correct. The pH is a little higher than the initial pH, which is expected because some of the acid has been neutralized.

(d) **Find:** pH at one-half the equivalence point

Conceptual Plan: Because this is a weak acid–strong base titration, the pH at one-half the equivalence point is the pK_a of the weak acid.

Solution: $pH = pK_a = -\log K_a = -\log(1.8 \times 10^{-5}) = 4.74$

Check: The units (none) are correct. Because this is a weak acid–strong base titration, the pH at one-half the equivalence point is the pK_a of the weak acid; so it should be a little below 5.

(e) **Find:** pH at equivalence point

Conceptual Plan: Use calculations from part (b). Then because all of the weak acid has been converted to its conjugate base, the pH is only dependent on the hydrolysis reaction of the conjugate base. The mol $C_2H_3O_2^-$ = initial mol $HC_2H_3O_2$ and L $HC_2H_3O_2$, L NaOH to equivalence point → total L then

$L\ HC_2H_3O_2 + L\ NaOH = \text{total L}$

mol excess $C_2H_3O_2^-$, L → $[C_2H_3O_2^-]$ and $K_a \rightarrow K_b$ then do an equilibrium calculation:

$M = \frac{mol}{L}$ $K_w = K_aK_b$

$[C_2H_3O_2^-], K_b \rightarrow [OH^-] \rightarrow [H_3O^+] \rightarrow pH$

set up ICE table $K_w = [H_3O^+][OH^-]$ $pH = -\log[H_3O^+]$

Solution: mol $C_2H_3O_2^-$ = initial mol $HC_2H_3O_2$ = 0.00210 mol and total volume = L $HC_2H_3O_2$ + L NaOH = 0.020 L + 0.0168 L = 0.0368 L then

$[C_2H_3O_2^-] = \frac{0.00210 \text{ mol } C_2H_3O_2^-}{0.0368 \text{ L}} = 0.05\underline{7}0\underline{6}52 \text{ M}$ and $K_w = K_aK_b$. Rearrange to solve for K_b.

$K_b = \frac{K_w}{K_a} = \frac{1.0 \times 10^{-14}}{1.8 \times 10^{-5}} = 5.\underline{5}556 \times 10^{-10}$. Set up an ICE table:

$$C_2H_3O_2^-(aq) + H_2O(l) \rightleftharpoons HC_2H_3O_2(aq) + OH^-(aq)$$

	$[C_2H_3O_2^-]$	$[HC_2H_3O_2]$	$[OH^-]$
Initial	0.0570652	≈0.00	≈0.00
Change	$-x$	$+x$	$+x$
Equil	$0.0570652 - x$	$+x$	$+x$

$K_b = \frac{[HC_2H_3O_2][OH^-]}{[C_2H_3O_2^-]} = 5.\underline{5}556 \times 10^{-10} = \frac{x^2}{0.0570652 - x}$. Assume that x is small ($x << 0.057$), so

$\frac{x^2}{0.0570652 - \cancel{x}} = 5.\underline{5}556 \times 10^{-10} = \frac{x^2}{0.0570652}$ and $x = 5.\underline{6}306 \times 10^{-6} \text{ M} = [OH^-]$.

Confirm that the assumption is valid. $\frac{5.\underline{6}306 \times 10^{-6}}{0.0570652} \times 100\% = 0.0099\% < 5\%$, so the assumption is valid.

$K_w = [H_3O^+][OH^-]$, so $[H_3O^+] = \frac{K_w}{[OH^-]} = \frac{1.0 \times 10^{-14}}{5.\underline{6}306 \times 10^{-6}} = 1.\underline{7}760 \times 10^{-9} \text{ M}$.

Finally, $pH = -\log[H_3O^+] = -\log(1.\underline{7}760 \times 10^{-9}) = 8.75$.

Check: The units (none) are correct. Because this is a weak acid–strong base titration, the pH at the equivalence point is basic.

(f) **Find:** pH after adding 5.0 mL of base beyond the equivalence point

Conceptual Plan: Use calculations from parts (b) and (c). Then the pH is only dependent on the amount of excess base and the total solution volumes.

mL excess → L excess then [NaOH], L excess → mol NaOH excess

$\frac{1 \text{ L}}{1000 \text{ mL}}$ $M = \frac{mol}{L}$

then L $HC_2H_3O_2$, L NaOH to equivalence point, L NaOH excess → total L then

L $HC_2H_3O_2$ + L NaOH to equivalence point + L NaOH excess = total L

mol excess NaOH, total L → [NaOH] = $[OH^-] \rightarrow [H_3O^+] \rightarrow pH$

$M = \frac{mol}{L}$ $K_w = [H_3O^+][OH^-]$ $pH = -\log[H_3O^+]$

Solution: $5.0\ \cancel{\text{mL NaOH}} \times \frac{1\ \text{L}}{1000\ \cancel{\text{mL}}} = 0.0050\ \text{L NaOH excess then}$

$\frac{0.125\ \text{mol NaOH}}{1\ \cancel{\text{L}}} \times 0.0050\ \cancel{\text{L}} = 0.000625\ \text{mol NaOH excess}$. Then $0.0200\ \text{L}\ HC_2H_3O_2 + 0.0168\ \text{L NaOH} + 0.0050\ \text{L NaOH} = 0.0418\ \text{L}$ total volume.

$[\text{NaOH excess}] = \frac{0.000625\ \text{mol NaOH excess}}{0.0418\ \text{L}} = 0.0149\underline{5}22\ \text{M NaOH excess}$. Because NaOH is a strong base, $[\text{NaOH}]$ excess $= [OH^-]$. The strong base overwhelms the weak base, which is insignificant in the calculation. $K_w = [H_3O^+][OH^-]$, so $[H_3O^+] = \frac{K_w}{[OH^-]} = \frac{1.0 \times 10^{-14}}{0.0149\underline{5}22} = 6.\underline{6}88 \times 10^{-13}\ \text{M}$.

Finally, $\text{pH} = -\log[H_3O^+] = -\log(6.\underline{6}88 \times 10^{-13}) = 12.17$.

Check: The units (none) are correct. The pH is rising sharply at the equivalence point, so the pH after 5 mL past the equivalence point should be quite basic.

18.70 **Given:** 30.0 mL of 0.165 M $HC_3H_5O_2$ titrated with 0.300 M KOH **Other:** $K_a(HC_3H_5O_2) = 1.3 \times 10^{-5}$
Find: initial pH
Conceptual Plan: Because $HC_3H_5O_2$ is a weak acid, set up an equilibrium problem using the initial concentration. So M $HC_3H_5O_2 \rightarrow [H_3O^+] \rightarrow$ pH.

ICE table $\quad \text{pH} = -\log[H_3O^+]$

Solution:

$$HC_3H_5O_2(aq) + H_2O(l) \rightleftharpoons H_3O^+(aq) + C_3H_5O_2^-(aq)$$

	$[HC_3H_5O_2]$	$[H_3O^+]$	$[C_3H_5O_2^-]$
Initial	0.165	≈ 0.00	≈ 0.00
Change	$-x$	$+x$	$+x$
Equil	$0.165 - x$	$+x$	$+x$

$K_a = \frac{[H_3O^+][C_3H_5O_2^-]}{[HC_3H_5O_2]} = 1.3 \times 10^{-5} = \frac{x^2}{0.165 - x}$ Assume that x is small ($x << 0.165$), so

$\frac{x^2}{0.165 - \cancel{x}} = 1.3 \times 10^{-5} = \frac{x^2}{0.165}$ and $x = 1.\underline{4}646 \times 10^{-3}\ \text{M} = [H_3O^+]$. Confirm that the assumption is valid.

$\frac{1.\underline{4}646 \times 10^{-3}}{0.165} \times 100\% = 0.89\% < 5\%$, so the assumption is valid. Finally,

$\text{pH} = -\log[H_3O^+] = -\log(1.\underline{4}646 \times 10^{-3}) = 2.83$.

Check: The units (none) are correct. The magnitude of the answer makes physical sense because the pH should be greater than $-\log(0.165) = 0.78$ because this is a weak acid.
Find: pH after adding 5.0 mL of base
Conceptual Plan: Write a balanced equation then mL $\rightarrow$ L then $[HC_3H_5O_2]$, L $\rightarrow$ mol $HC_3H_5O_2$ then

$HC_3H_5O_2 + KOH \rightarrow KC_3H_5O_2 + H_2O \quad \frac{1\ \text{L}}{1000\ \text{mL}} \quad M = \frac{\text{mol}}{\text{L}}$

mL $\rightarrow$ L then [KOH], L $\rightarrow$ mol KOH then

$\frac{1\ \text{L}}{1000\ \text{mL}} \quad M = \frac{\text{mol}}{\text{L}}$

mol $HC_3H_5O_2$, mol KOH $\rightarrow$ mol excess $HC_3H_5O_2$, mol $C_3H_5O_2^-$

set up stoichiometry table

Because there are significant concentrations of both the acid and the conjugate base species, this is a buffer solution; so the Henderson–Hasselbalch equation $\left(\text{pH} = \text{p}K_a + \log\frac{[\text{base}]}{[\text{acid}]}\right)$ can be used. Also note that the ratio of concentrations is the same as the ratio of moles because the volume is the same for both species.

Solution: $30.0\ \cancel{\text{mL}\ HC_3H_5O_2} \times \frac{1\ \text{L}}{1000\ \cancel{\text{mL}}} = 0.0300\ \text{L}\ HC_3H_5O_2$ then

$\frac{0.165 \text{ mol } HC_3H_5O_2}{1 \text{ L}} \times 0.0300 \text{ L} = 0.00495 \text{ mol } HC_3H_5O_2$ and $5.0 \text{ mL KOH} \times \frac{1 \text{ L}}{1000 \text{ mL}} = 0.0050 \text{ L KOH}$ then

$\frac{0.300 \text{ mol KOH}}{1 \text{ L}} \times 0.0050 \text{ L} = 0.0015 \text{ mol KOH}$. Set up a table to track changes:

	$KOH(aq)$ +	$HC_3H_5O_2(aq)$ →	$KC_3H_5O_2(aq)$ +	$H_2O(l)$
Before addition	0.00 mol	0.00495 mol	0.00 mol	—
Addition	0.0015 mol	—	—	—
After addition	≈0.00 mol	0.00345 mol	0.0015 mol	—

Then use the Henderson–Hasselbalch equation because the solution is a buffer.

$$pH = pK_a + \log \frac{[\text{base}]}{[\text{acid}]} = -\log(1.3 \times 10^{-5}) + \log \frac{0.0015}{0.00345} = 4.52$$

Check: The units (none) are correct. The pH is a little higher than the initial pH, which is expected because some of the acid has been neutralized.

Find: pH after adding 10.0 mL of base

Conceptual Plan: Use previous calculations, then mL → L then [KOH], L → mol KOH then

$\frac{1 \text{ L}}{1000 \text{ mL}}$ $M = \frac{\text{mol}}{\text{L}}$

mol $HC_3H_5O_2$, mol KOH → mol excess $HC_3H_5O_2$, mol $C_3H_5O_2^-$

set up stoichiometry table

Because there are significant concentrations of both the acid and the conjugate base species, this is a buffer solution; so the Henderson–Hasselbalch equation $\left(pH = pK_a + \log \frac{[\text{base}]}{[\text{acid}]}\right)$ can be used. Also note that the ratio of concentrations is the same as the ratio of moles because the volume is the same for both species.

Solution: $10.0 \text{ mL KOH} \times \frac{1 \text{ L}}{1000 \text{ mL}} = 0.0100 \text{ L KOH}$ then $\frac{0.300 \text{ mol KOH}}{1 \text{ L}} \times 0.0100 \text{ L} = 0.0030 \text{ mol KOH}$

Set up a table to track changes:

	$KOH(aq)$ +	$HC_3H_5O_2(aq)$ →	$KC_3H_5O_2(aq)$ +	$H_2O(l)$
Before addition	0.00 mol	0.00495 mol	0.00 mol	—
Addition	0.0030 mol	—	—	—
After addition	≈0.00 mol	0.00195 mol	0.0030 mol	—

Then use the Henderson–Hasselbalch equation because the solution is a buffer.

$$pH = pK_a + \log \frac{[\text{base}]}{[\text{acid}]} = -\log(1.3 \times 10^{-5}) + \log \frac{0.0030}{0.00195} = 5.07$$

Check: The units (none) are correct. The pH is a little higher than the last pH, which is expected because some of the acid has been neutralized.

Find: pH at equivalence point

Conceptual Plan: Use previous calculations, then set mol acid $(HC_3H_5O_2)$ = mol base KOH

balanced equation has 1:1 stoichiometry

and [KOH], mol KOH → L KOH. Because all of the weak acid has been converted to its

$M = \frac{\text{mol}}{\text{L}}$

conjugate base, the pH is only dependent on the hydrolysis reaction of the conjugate base. The mol $C_3H_5O_2^-$ = initial mol $HC_3H_5O_2$ and L $HC_3H_5O_2$, L KOH to equivalence point → total L

$L\ HC_3H_5O_2 + L\ KOH = \text{total L}$

then mol $C_2H_3O_2^-$, L → $[C_2H_3O_2^-]$ and K_a → K_b

$M = \frac{\text{mol}}{\text{L}}$ $K_w = K_a K_b$

Then do an equilibrium calculation: $[C_2H_3O_2^-]$, K_b → $[OH^-]$ → $[H_3O^+]$ → pH.

set up ICE table $K_w = [H_3O^+][OH^-]$ $pH = -\log[H_3O^+]$

Solution: mol acid = mol $HC_3H_5O_2$ = 0.00495 mol = mol KOH then

$0.00495 \text{ mol KOH} \times \dfrac{1 \text{ L}}{0.300 \text{ mol KOH}} = 0.0165 \text{ L KOH}$. Then total volume $= \text{L } HC_3H_5O_2 + \text{L KOH} =$

$0.0300 \text{ L} + 0.0165 \text{ L} = 0.0465$ then $[C_3H_5O_2^-] = \dfrac{0.00495 \text{ mol } C_3H_5O_2^-}{0.0465 \text{ L}} = 0.10\underline{6}452 \text{ M}$ and $K_w = K_aK_b$

Rearrange to solve for K_b. $K_b = \dfrac{K_w}{K_a} = \dfrac{1.0 \times 10^{-14}}{1.3 \times 10^{-5}} = 7.\underline{6}923 \times 10^{-10}$ M. Set up an ICE table:

$$C_3H_5O_2^-(aq) + H_2O(l) \rightleftharpoons HC_3H_5O_2(aq) + OH^-(aq)$$

	$[C_3H_5O_2^-]$	$[HC_3H_5O_2]$	$[OH^-]$
Initial	$0.10\underline{6}452$	≈ 0.00	0.00
Change	$-x$	$+x$	$+x$
Equil	$0.10\underline{6}452 - x$	$+x$	$+x$

$K_b = \dfrac{[HC_3H_5O_2][OH^-]}{[C_3H_5O_2^-]} = 7.\underline{6}923 \times 10^{-10} = \dfrac{x^2}{0.10\underline{6}452 - x}$ Assume that x is small ($x << 0.106$), so

$\dfrac{x^2}{0.10\underline{6}452 - \cancel{x}} = 7.\underline{6}923 \times 10^{-10} = \dfrac{x^2}{0.10\underline{6}452}$ and $x = 9.\underline{0}491 \times 10^{-6} \text{ M} = [OH^-]$.

Confirm that the assumption is valid. $\dfrac{9.\underline{0}491 \times 10^{-6}}{0.10\underline{6}452} \times 100\% = 0.0085\% < 5\%$, so the assumption is valid.

$K_w = [H_3O^+][OH^-]$, so $[H_3O^+] = \dfrac{K_w}{[OH^-]} = \dfrac{1.0 \times 10^{-14}}{9.\underline{0}491 \times 10^{-6}} = 1.\underline{1}051 \times 10^{-9}$ M.

Finally, $\text{pH} = -\log[H_3O^+] = -\log(1.\underline{1}051 \times 10^{-9}) = 8.96$.

Check: The units (none) are correct. Because this is a weak acid–strong base titration, the pH at the equivalence point is basic.

Find: pH at one-half the equivalence point

Conceptual Plan: Because this is a weak acid–strong base titration, the pH at one-half the equivalence point is the pK_a of the weak acid.

Solution: $\text{pH} = \text{p}K_a = -\log K_a = -\log(1.3 \times 10^{-5}) = 4.89$, and the volume of added base is $0.5 \times 16.5 \text{ mL} = 8.3$ mL.

Check: The units (none) are correct. Because this is a weak acid–strong base titration, the pH at one-half the equivalence point is the pK_a of the weak acid; so it should be a little below 5.

Find: pH after adding 20.0 mL of base

Conceptual Plan: Use previous calculations. Then the pH is only dependent on the amount of excess base and the total solution volume. mL added, mL at equivalence point → mL excess → L excess

mL excess = mL added − mL at equiv. pt. $\qquad \dfrac{1 \text{ L}}{1000 \text{ mL}}$

then [KOH], L excess → mol KOH excess

$M = \dfrac{\text{mol}}{\text{L}}$

then L $HC_3H_5O_2$, L KOH to equivalence point, L KOH excess → total L then

L $HC_2H_3O_2$ + L KOH to equivalence point + L KOH excess = total L

mol excess KOH, total L → [KOH] = $[OH^-]$ → $[H_3O^+]$ → pH

$M = \dfrac{\text{mol}}{\text{L}} \qquad K_w = [H_3O^+][OH^-] \qquad \text{pH} = -\log[H_3O^+]$

Solution: mL KOH excess = mL added − mL to equiv. pt. = 20.0 mL − 16.5 mL = 3.5 mL

$3.5 \text{ mL KOH} \times \dfrac{1 \text{ L}}{1000 \text{ mL}} = 0.0035 \text{ L KOH}$ excess then

$\dfrac{0.300 \text{ mol KOH}}{1 \text{ L}} \times 0.0035 \text{ L} = 0.001\underline{0}5$ mol KOH excess. Then $0.0300 \text{ L } HC_3H_5O_2 + 0.0165 \text{ L KOH} +$

$0.0035 \text{ L KOH} = 0.0500$ L total volume. $[\text{KOH excess}] = \dfrac{0.001\underline{0}5 \text{ mol KOH excess}}{0.0500 \text{ L}} = 0.021$ M KOH excess.

Because KOH is a strong base, [KOH] excess $= [OH^-]$. The strong base overwhelms the weak base which is insignificant in the calculation. $K_w = [H_3O^+][OH^-]$, so $[H_3O^+] = \frac{K_w}{[OH^-]} = \frac{1.0 \times 10^{-14}}{0.021} = 4.\underline{7}619 \times 10^{-13}$ M.

Finally, pH $= -\log[H_3O^+] = -\log(4.\underline{7}619 \times 10^{-13}) = 12.32$.

Check: The units (none) are correct. The pH is rising sharply at the equivalence point, so the pH after 5 mL past the equivalence point should be quite basic.

Find: pH after adding 25.0 mL of base

Conceptual Plan: Use previous calculations. Then the pH is only dependent on the amount of excess base and the total solution volumes. mL added, mL at equivalence point → mL excess → L excess

mL excess = mL added − mL at equiv. pt. $\qquad \frac{1\text{ L}}{1000\text{ mL}}$

then [KOH], L excess → mol KOH excess

$M = \frac{\text{mol}}{\text{L}}$

then L $HC_3H_5O_2$, L KOH to equivalence point, L KOH excess → total L then

L $HC_2H_3O_2$ + L KOH to equivalence point + L KOH excess = total L

mol excess KOH, total L → [KOH] = $[OH^-] \rightarrow [H_3O^+] \rightarrow$ pH

$M = \frac{\text{mol}}{\text{L}} \qquad K_w = [H_3O^+][OH^-] \qquad \text{pH} = -\log[H_3O^+]$

Solution: mL KOH excess = mL added − mL to equiv. pt. = 25.0 mL − 16.5 mL = 8.5 mL

$8.5\ \text{mL KOH} \times \frac{1\text{ L}}{1000\text{ mL}} = 0.0085$ L KOH excess then

$\frac{0.300\text{ mol KOH}}{1\text{ L}} \times 0.0085\text{ L} = 0.002\underline{5}5$ mol KOH excess. Then 0.0300 L $HC_3H_5O_2$ + 0.0165 L KOH +

0.0085 L KOH = 0.0550 L total volume. $[\text{KOH excess}] = \frac{0.002\underline{5}5\text{ mol KOH excess}}{0.0550\text{ L}} = 0.04\underline{6}364$ M KOH excess.

Because KOH is a strong base, [KOH] excess = $[OH^-]$. The strong base overwhelms the weak base, which is insignificant in the calculation. $K_w = [H_3O^+][OH^-]$, so $[H_3O^+] = \frac{K_w}{[OH^-]} = \frac{1.0 \times 10^{-14}}{0.04\underline{6}364} = 2.\underline{1}568 \times 10^{-13}$ M.

Finally, pH $= -\log[H_3O^+] = -\log(2.\underline{1}568 \times 10^{-13}) = 12.67$.

Check: The units (none) are correct. The pH is rising sharply at the equivalence point, so the pH after 5 mL past the equivalence point should be quite basic. This pH is higher than the last pH. Plotting these data points is shown as follows:

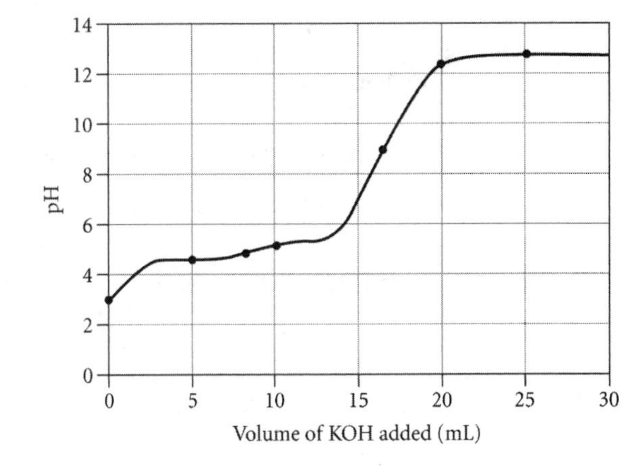

18.71 **Given:** 25.0 mL of 0.175 M CH_3NH_2 titrated with 0.150 M HBr **Other:** $K_b(CH_3NH_2) = 4.4 \times 10^{-4}$

(a) **Find:** initial pH

Conceptual Plan: Because CH_3NH_2 is a weak base, set up an equilibrium problem using the initial concentration, so M $CH_3NH_2 \rightarrow [OH^-] \rightarrow [H_3O^+] \rightarrow$ pH

ICE table $\quad K_w = [H_3O^+][OH^-] \quad \text{pH} = -\log[H_3O^+]$

Solution:

$$CH_3NH_2(aq) + H_2O(l) \rightleftharpoons CH_3NH_3^+(aq) + OH^-(aq)$$

	$[CH_3NH_2]$	$[CH_3NH_3^+]$	$[OH^-]$
Initial	0.175	0.00	≈0.00
Change	$-x$	$+x$	$+x$
Equil	$0.175 - x$	$+x$	$+x$

$$K_b = \frac{[CH_3NH_3^+][OH^-]}{[CH_3NH_2]} = 4.4 \times 10^{-4} = \frac{x^2}{0.175 - x}$$

Assume that x is small $(x << 0.175)$, so $\frac{x^2}{0.175 - \cancel{x}} = 4.4 \times 10^{-4} = \frac{x^2}{0.175}$ and $x = 8.\underline{7}750 \times 10^{-3}\ \text{M} = [\text{OH}^-]$.

Confirm that the assumption is valid. $\frac{8.\underline{7}750 \times 10^{-3}}{0.175} \times 100\% = 5.0\%$, so the assumption is valid.

$K_w = [\text{H}_3\text{O}^+][\text{OH}^-]$, so $[\text{H}_3\text{O}^+] = \frac{K_w}{[\text{OH}^-]} = \frac{1.0 \times 10^{-14}}{8.\underline{7}750 \times 10^{-3}} = 1.\underline{1}396 \times 10^{-12}\ \text{M}$.

Finally, $\text{pH} = -\log[\text{H}_3\text{O}^+] = -\log(1.\underline{1}396 \times 10^{-12}) = 11.94$.

Check: The units (none) are correct. The magnitude of the answer makes physical sense because the pH should be less than $14 + \log(0.175) = 13.2$ because this is a weak base.

(b) **Find:** volume of acid to reach equivalence point

Conceptual Plan: Write a balanced equation, then mL → L then $[CH_3NH_2]$, L → mol CH_3NH_2

$HBr + CH_3NH_2 \rightarrow CH_3NH_3Br + H_2O$ $\quad \frac{1\ \text{L}}{1000\ \text{mL}} \quad M = \frac{\text{mol}}{\text{L}}$

then set mol base $[CH_3NH_2]$ = mol acid (HBr) and [HBr], mol HBr → L HBr → mL HBr

balanced equation has 1:1 stoichiometry $\quad M = \frac{\text{mol}}{\text{L}} \quad \frac{1000\ \text{mL}}{1\ \text{L}}$

Solution: $25.0\ \cancel{\text{mL CH}_3\text{NH}_2} \times \frac{1\ \text{L}}{1000\ \cancel{\text{mL}}} = 0.0250\ \text{L CH}_3\text{NH}_2$ then

$\frac{0.175\ \text{mol CH}_3\text{NH}_2}{1\ \cancel{\text{L}}} \times 0.0250\ \cancel{\text{L}} = 0.0043\underline{7}5\ \text{mol CH}_3\text{NH}_2$. So mol base = mol CH_3NH_2 = 0.004375 mol

= mol HBr then $0.0043\underline{7}5\ \cancel{\text{mol HBr}} \times \frac{1\ \text{L}}{0.150\ \cancel{\text{mol HBr}}} = 0.029\underline{1}667\ \cancel{\text{L HBr}} \times \frac{1000\ \text{mL}}{1\ \cancel{\text{L}}} = 29.2\ \text{mL HBr}$.

Check: The units (mL) are correct. The volume of acid is greater than the volume of base because the concentration of the base is a little greater than that of the acid.

(c) **Find:** pH after adding 5.0 mL of acid

Conceptual Plan: Use calculations from part (b). Then mL → L then [HBr], L → mol HBr

$\frac{1\ \text{L}}{1000\ \text{mL}} \quad M = \frac{\text{mol}}{\text{L}}$

then mol CH_3NH_2, mol HBr → mol excess CH_3NH_2 and L CH_3NH_2, L HBr → total L.

set up stoichiometry table $\quad$ L CH_3NH_2 + L HBr = total L

Because there are significant concentrations of both the acid and the conjugate base species, this is a buffer solution; so the Henderson–Hasselbalch equation $\left(\text{pH} = \text{p}K_a + \log\frac{[\text{base}]}{[\text{acid}]}\right)$ can be used.

Convert K_b to K_a using $K_w = K_aK_b$. Also note that the ratio of concentrations is the same as the ratio of moles because the volume is the same for both species.

Solution: $5.0\ \cancel{\text{mL HBr}} \times \frac{1\ \text{L}}{1000\ \cancel{\text{mL}}} = 0.0050\ \text{L HBr}$ then $\frac{0.150\ \text{mol HBr}}{1\ \cancel{\text{L}}} \times 0.0050\ \cancel{\text{L}} = 0.00075\ \text{mol HBr}$

Set up a table to track changes:

	$HBr(aq)$	+	$CH_3NH_2(aq)$	→	$CH_3NH_3Br(aq)$
Before addition	0.00 mol		0.0043$\underline{7}$5 mol		0.00 mol
Addition	0.00075 mol		—		—
After addition	≈0.00 mol		0.0036$\underline{2}$5 mol		0.000750 mol

then $K_w = K_aK_b$ so

$K_a = \frac{K_w}{K_b} = \frac{1.0 \times 10^{-14}}{4.4 \times 10^{-4}} = 2.\underline{2}727 \times 10^{-11}\ \text{M}$. Then use the Henderson–Hasselbalch equation because the solution is a buffer. $\text{pH} = \text{p}K_a + \log\frac{[\text{base}]}{[\text{acid}]} = -\log(2.\underline{2}727 \times 10^{-11}) + \log\frac{0.0036\underline{2}5}{0.000750} = 11.33$

Check: The units (none) are correct. The pH is a little lower than the last pH, which is expected because some of the base has been neutralized.

(d) **Find:** pH at one-half the equivalence point

Conceptual Plan: Because this is a weak base–strong acid titration, the pH at one-half the equivalence point is the pK_a of the conjugate acid of a weak base.

Solution: $pH = pK_a = -\log K_a = -\log(2.\underline{2}727 \times 10^{-11}) = 10.64$

Check: The units (none) are correct. Because this is a weak acid–strong base titration, the pH at one-half the equivalence point is the pK_a of the conjugate acid of the weak base; so it should be a little below 11.

(e) **Find:** pH at equivalence point

Conceptual Plan: Use previous calculations. Because all of the weak base has been converted to its conjugate acid, the pH is only dependent on the hydrolysis reaction of the conjugate acid. The mol $CH_3NH_3^+$ = initial mol CH_3NH_2 and

L CH_3NH_2, L HBr to equivalence point → total L then mol $CH_3NH_3^+$, L → $[CH_3NH_3^+]$

$L\ CH_3NH_2 + L\ HBr = total\ L$ $\qquad M = \frac{mol}{L}$

then do an equilibrium calculation: $[CH_3NH_3^+], K_a \rightarrow [H_3O^+] \rightarrow pH$

set up ICE table $\qquad pH = -\log[H_3O^+]$

Solution: mol base = mol acid = mol $CH_3NH_3^+$ = 0.00437̲5 mol. Then total volume = L CH_3NH2 + L HBr = 0.0250 L + 0.0292 L = 0.0542 then $[CH_3NH_3^+] = \frac{0.0043\underline{7}5\ mol\ CH_3NH_3^+}{0.0542\ L} = 0.080\underline{7}196$ M. Set up an ICE table:

$$CH_3NH_3^+(aq) + H_2O(l) \rightleftharpoons CH_3NH_2(aq) + H_3O^+(aq)$$

	$[CH_3NH_3^+]$	$[CH_3NH_2]$	$[H_3O^+]$
Initial	$0.080\underline{7}196$	≈ 0.00	≈ 0.00
Change	$-x$	$+x$	$+x$
Equil	$0.080\underline{7}196 - x$	$+x$	$+x$

$$K_a = \frac{[CH_3NH_2][H_3O^+]}{[CH_3NH_3^+]} = 2.\underline{2}727 \times 10^{-11} = \frac{x^2}{0.080\underline{7}196 - x}$$ Assume that x is small ($x << 0.0807$), so

$$\frac{x^2}{0.080\underline{7}196 - \cancel{x}} = 2.\underline{2}727 \times 10^{-11} = \frac{x^2}{0.080\underline{7}196}$$ and $x = 1.\underline{3}544 \times 10^{-6} = [H_3O^+]$.

Confirm that the assumption is valid. $\frac{1.\underline{3}544 \times 10^{-6}}{0.080\underline{7}106} \times 100\% = 0.0017\% < 5\%$, so the assumption is valid.

Finally, $pH = -\log[H_3O^+] = -\log(1.\underline{3}544 \times 10^{-6}) = 5.87$.

Check: The units (none) are correct. Because this is a weak base–strong acid titration, the pH at the equivalence point is acidic.

(f) **Find:** pH after adding 5.0 mL of acid beyond the equivalence point

Conceptual Plan: Use calculations from parts (b) and (c). Then the pH is only dependent on the amount of excess acid and the total solution volumes.

mL excess → L excess then [HBr], L excess → mol HBr excess

$\frac{1\ L}{1000\ mL}$ $\qquad M = \frac{mol}{L}$

then L CH_3NH_2, L HBr to equivalence point, L HBr excess → total L then

$L\ CH_3NH_2 + L\ HBr\ to\ equivalence\ point + L\ HBr\ excess = total\ L$

mol excess HBr, total L → [HBr] = $[H_3O^+] \rightarrow pH$

$M = \frac{mol}{L}$ $\qquad pH = -\log[H_3O^+]$

Solution: $5.0\ \cancel{mL\ HBr} \times \frac{1\ L}{1000\ \cancel{mL}} = 0.0050$ L HBr excess then

$\frac{0.150\ mol\ HBr}{1\ \cancel{L}} \times 0.0050\ \cancel{L} = 0.00075$ mol HBr excess. Then 0.0250 L CH_3NH_2 + 0.0292 L HBr + 0.0050 L HBr = 0.0592 L total volume.

$$[\text{HBr excess}] = \frac{0.00075 \text{ mol HBr excess}}{0.0592 \text{ L}} = 0.01\underline{2}669 \text{ M HBr excess}$$

Because HBr is a strong acid, $[HBr]$ excess $= [H_3O^+]$. The strong acid overwhelms the weak acid which is insignificant in the calculation. Finally, $pH = -\log[H_3O^+] = -\log(0.01\underline{2}669) = 1.90$.

Check: The units (none) are correct. The pH is dropping sharply at the equivalence point, so the pH after 5 mL past the equivalence point should be quite acidic.

18.72 **Given:** 25.0 mL of 0.125 M pyridine (C_5H_5N) titrated with 0.100 M HCl **Other:** $K_b(C_5H_5N) = 1.7 \times 10^{-9}$
Find: initial pH

Conceptual Plan: Because C_5H_5N is a weak base, set up an equilibrium problem using the initial concentration, so M $C_5H_5N \rightarrow [OH^-] \rightarrow [H_3O^+] \rightarrow$ pH.

ICE table $K_w = [H_3O^+][OH^-]$ $pH = -\log[H_3O^+]$

Solution:

$$C_5H_5N(aq) + H_2O(l) \rightleftharpoons C_5H_5NH^+(aq) + OH^-(aq)$$

	$[C_5H_5N]$	$[C_5H_5NH^+]$	$[OH^-]$
Initial	0.125	0.00	≈0.00
Change	$-x$	$+x$	$+x$
Equil	$0.125 - x$	$+x$	$+x$

$$K_b = \frac{[C_5H_5NH^+][OH^-]}{[C_5H_5N]} = 1.7 \times 10^{-9} = \frac{x^2}{0.125 - x}$$

Assume that x is small ($x << 0.125$), so $\frac{x^2}{0.125 - x} = 1.7 \times 10^{-9} = \frac{x^2}{0.125}$ and $x = 1.\underline{4}577 \times 10^{-5} = [OH^-]$.

Confirm that the assumption is valid. $\frac{1.\underline{4}577 \times 10^{-5}}{0.125} \times 100\% = 0.012\% < 5.0\%$, so the assumption is valid.

$$K_w = [H_3O^+][OH^-], \text{ so } [H_3O^+] = \frac{K_w}{[OH^-]} = \frac{1.0 \times 10^{-14}}{1.\underline{4}577 \times 10^{-5}} = 6.\underline{8}601 \times 10^{-10} \text{ M}.$$

Finally, $pH = -\log[H_3O^+] = -\log(6.\underline{8}601 \times 10^{-10}) = 9.16$.

Check: The units (none) are correct. The magnitude of the answer makes physical sense because the pH should be less than $14 + \log(0.125) = 13.1$ because this is a weak base.

Find: pH after adding 10.0 mL of acid

Conceptual Plan: Write balanced equation then mL → L then $[C_5H_5N]$, L → mol C_5H_5N then

$HCl + C_5H_5N \rightarrow C_5H_5NHCl$ $\frac{1 \text{ L}}{1000 \text{ mL}}$ $M = \frac{\text{mol}}{\text{L}}$

mL → L then [HCl], L → mol HCl then mol C_5H_5N, mol HCl → mol excess C_5H_5N and

$\frac{1 \text{ L}}{1000 \text{ mL}}$ $M = \frac{\text{mol}}{\text{L}}$ set up stoichiometry table

L C_5H_5N, L HCl → total L. Because there are significant concentrations of both the acid and

$L\ C_5H_5N + L\ HCl = \text{total L}$

the conjugate base species, this is a buffer solution; so the Henderson–Hasselbalch equation $\left(pH = pK_a + \log\frac{[\text{base}]}{[\text{acid}]}\right)$ can be used. Convert K_b to K_a using $K_w = K_aK_b$. Also note that the ratio of concentrations is the same as the ratio of moles because the volume is the same for both species.

Solution: $25.0 \text{ mL } C_5H_5N \times \frac{1 \text{ L}}{1000 \text{ mL}} = 0.0250 \text{ L } C_5H_5N$ then

$\frac{0.125 \text{ mol } C_5H_5N}{1 \text{ L}} \times 0.0250 \text{ L} = 0.0031\underline{2}5 \text{ mol } C_5H_5N$ and $10.0 \text{ mL HCl} \times \frac{1 \text{ L}}{1000 \text{ mL}} = 0.0100 \text{ L HCl}$ then

$10.0 \text{ mL HCl} \times \frac{1 \text{ L}}{1000 \text{ mL}} = 0.0100 \text{ L HCl}$ then

$\frac{0.100 \text{ mol HCl}}{1 \text{ L}} \times 0.0100 \text{ L} = 0.00100 \text{ mol HCl}$. Set up a table to track changes:

	$HCl(aq)$	+	$C_5H_5N(aq)$	→	$C_5H_5NHCl(aq)$
Before addition	0.00 mol		0.003125 mol		0.00 mol
Addition	0.00100 mol		—		—
After addition	≈0.00 mol		0.002125 mol		0.00100 mol

Then $K_w = K_aK_b$, so $K_a = \frac{K_w}{K_b} = \frac{1.0 \times 10^{-14}}{1.7 \times 10^{-9}} = 5.8824 \times 10^{-6}$ M. Then use the Henderson–Hasselbalch equation because the solution is a buffer.

$$pH = pK_a + \log\frac{[\text{base}]}{[\text{acid}]} = -\log(5.8824 \times 10^{-6}) + \log\frac{0.002125}{0.00100} = 5.56$$

Check: The units (none) are correct. The pH is lower than the last pH, which is expected because some of the base has been neutralized.

Find: pH after adding 20.0 mL of acid

Conceptual Plan: Use previous calculations. Then mL → L then [HCl], L → mol HCl

$\frac{1\text{ L}}{1000\text{ mL}}$ $M = \frac{\text{mol}}{\text{L}}$

then mol C_5H_5N, mol HCl → mol excess C_5H_5N and L C_5H_5N, L HCl → total L.

set up stoichiometry table L C_5H_5N + L HCl = total L

Because there are significant concentrations of both the acid and the conjugate base species, this is a buffer solution; so the Henderson–Hasselbalch equation $\left(pH = pK_a + \log\frac{[\text{base}]}{[\text{acid}]}\right)$ can be used. Convert K_b to K_a using $K_w = K_aK_b$. Also note that the ratio of concentrations is the same as the ratio of moles because the volume is the same for both species.

Solution: $20.0 \text{ mL HCl} \times \frac{1\text{ L}}{1000\text{ mL}} = 0.0200$ L HCl then $\frac{0.100\text{ mol HCl}}{1\text{ L}} \times 0.0200\text{ L} = 0.00200$ mol HCl

Set up a table to track changes:

	$HCl(aq)$	+	$C_5H_5N(aq)$	→	$C_5H_5NHCl(aq)$
Before addition	0.00 mol		0.003125 mol		0.00 mol
Addition	0.00200 mol		—		—
After addition	≈0.00 mol		0.001125 mol		0.00200 mol

Then use the Henderson–Hasselbalch equation because the solution is a buffer.

$$pH = pK_a + \log\frac{[\text{base}]}{[\text{acid}]} = -\log(5.8824 \times 10^{-6}) + \log\frac{0.001125}{0.00200} = 4.98$$

Check: The units (none) are correct. The pH is lower than the last pH, which is expected because some of the base has been neutralized.

Find: pH at equivalence point

Conceptual Plan: Use previous calculations. Because all of the weak base has been converted to its conjugate acid, the pH is only dependent on the hydrolysis reaction of the conjugate acid. The mol $C_5H_5NH^+$ = initial mol C_5H_5N then set mol base (C_5H_5N) = mol acid (HCl) and [HCl], mol HCl → L HCl then

balanced equation has 1:1 stoichiometry $M = \frac{\text{mol}}{\text{L}}$

L C_5H_5N, L HCl to equivalence point → total L then mol $C_5H_5NH^+$, L → $[C_5H_5NH^+]$

L C_5H_5N + L HCl = total L $M = \frac{\text{mol}}{\text{L}}$

then do an equilibrium calculation: $[C_5H_5NH^+]$, K_a → $[H_3O^+]$ → pH

set up ICE table $pH = -\log[H_3O^+]$

Solution: mol base = mol acid = mol $C_5H_5NH^+$ = 0.003125 mol then

$$0.003125 \text{ mol HCl} \times \frac{1\text{ L}}{0.100\text{ mol HCl}} = 0.03125\text{ L HCl} \times \frac{1000\text{ mL}}{1\text{ L}} = 31.3\text{ mL HCl then}$$

total volume $= \text{L } C_5H_5N + \text{L HCl} = 0.0250 \text{ L} + 0.0313 \text{ L} = 0.0563 \text{ L}$ then

$[C_5H_5NH^+] = \frac{0.003125 \text{ mol } C_5H_5NH^+}{0.0563 \text{ L}} = 0.0555062 \text{ M}$. Set up an ICE table:

$$C_5H_5NH^+(aq) + H_2O(l) \rightleftharpoons C_5H_5N(aq) + H_3O^+(aq)$$

	$[C_5H_5NH^+]$	$[C_5H_5N]$	$[H_3O^+]$
Initial	0.0555062	≈0.00	≈0.00
Change	$-x$	$+x$	$+x$
Equil	$0.0555062 - x$	$+x$	$+x$

$K_a = \frac{[C_5H_5N][H_3O^+]}{[C_5H_5NH^+]} = 5.8824 \times 10^{-6} = \frac{x^2}{0.0555062 - x}$. Assume that x is small ($x << 0.0556$), so

$\frac{x^2}{0.0555062 - x} = 5.8824 \times 10^{-6} = \frac{x^2}{0.0555062}$ and $x = 5.7141 \times 10^{-4} \text{ M} = [H_3O^+]$. Confirm that the

assumption is valid. $\frac{5.7141 \times 10^{-4}}{0.0555062} \times 100\% = 1.0\% < 5\%$, so the assumption is valid.

Finally, $\text{pH} = -\log[H_3O^+] = -\log(5.7141 \times 10^{-4}) = 3.24$.

Check: The units (none) are correct. Because this is a weak base–strong acid titration, the pH at the equivalence point is acidic.

Find: pH at one-half the equivalence point

Conceptual Plan: Because this is a weak base–strong acid titration, the pH at one-half the equivalence point is the pK_a of the conjugate acid of weak base.

Solution: $\text{pH} = \text{p}K_a = -\log K_a = -\log(5.8824 \times 10^{-6}) = 5.23$, and the volume is $0.5 \times 31.3 \text{ mL} = 15.7 \text{ mL}$.

Check: The units (none) are correct. Because this is a weak base–strong acid titration, the pH at one-half the equivalence point is the pK_a of the conjugate acid of the weak base; so it should be a little below 6.

Find: pH after adding 40.0 mL of acid

Conceptual Plan: Use previous calculations. Then the pH is only dependent on the amount of excess acid and the total solution volumes.

mL added, mL at equiv. pt. → mL excess → L excess then

$$\text{mL excess} = \text{mL added} - \text{mL at equiv. pt.} \qquad \frac{1 \text{ L}}{1000 \text{ mL}}$$

mL excess → L excess then [HCl], L excess → mol HCl excess

$$\frac{1 \text{ L}}{1000 \text{ mL}} \qquad \text{M} = \frac{\text{mol}}{\text{L}}$$

then L C_5H_5N, L HCl to equivalence point, L HCl excess → total L then

$$\text{L } C_5H_5N + \text{L HCl to equivalence point} + \text{L HCl excess} = \text{total L}$$

mol excess HCl, total L → [HCl] = $[H_3O^+]$ → pH

$$\text{M} = \frac{\text{mol}}{\text{L}} \qquad \text{pH} = -\log[H_3O^+]$$

Solution: mL excess HCl = mL added HCl − mL at equiv. pt. HCl = 40.0 mL − 31.3 mL = 8.7 mL HCl

$8.7 \text{ mL HCl} \times \frac{1 \text{ L}}{1000 \text{ mL}} = 0.0087 \text{ L HCl excess}$ then $\frac{0.100 \text{ mol HCl}}{1 \text{ L}} \times 0.0087 \text{ L} = 0.00087 \text{ mol HCl excess}$

Then $0.0250 \text{ L } C_5H_5N + 0.0313 \text{ L HCl} + 0.0087 \text{ L HCl} = 0.0650 \text{ L}$ total volume

$[\text{HCl excess}] = \frac{0.00087 \text{ mol HCl excess}}{0.0650 \text{ L}} = 0.013385 \text{ M HCl excess}$

Because HCl is a strong acid, [HCl] excess $= [H_3O^+]$. The strong acid overwhelms the weak acid and is insignificant in the calculation. Finally, $\text{pH} = -\log[H_3O^+] = -\log(0.013385) = 1.87$.

Check: The units (none) are correct. The pH is dropping sharply at the equivalence point, so the pH after 5 mL past the equivalence point should be quite acidic.

Find: pH after adding 50.0 mL of acid

Conceptual Plan: Use previous calculations. Then the pH is only dependent on the amount of excess acid and the total solution volumes. mL added, mL at equiv. pt. → mL excess → L excess then

$$\text{mL excess} = \text{mL added} - \text{mL at equiv. pt.} \qquad \frac{1\ \text{L}}{1000\ \text{mL}}$$

mL excess → L excess then [HCl], L excess → mol HCl excess

$$\frac{1\ \text{L}}{1000\ \text{mL}} \qquad M = \frac{\text{mol}}{\text{L}}$$

then L C_5H_5N, L HCl to equivalence point, L HCl excess → total L then

$$\text{L } C_5H_5N + \text{L HCl to equivalence point} + \text{L HCl excess} = \text{total L}$$

mol excess HCl, total L → [HCl] = $[H_3O^+]$ → pH

$$M = \frac{\text{mol}}{\text{L}} \qquad \text{pH} = -\log[H_3O^+]$$

Solution: mL excess HCl = mL added HCl − mL at equiv. pt. HCl = 50.0 mL − 31.3 mL = 18.7 mL HCl

$$18.7\ \text{mL HCl} \times \frac{1\ \text{L}}{1000\ \text{mL}} = 0.0187\ \text{L HCl excess then}$$

$\frac{0.100\ \text{mol HCl}}{1\ \text{L}} \times 0.0187\ \text{L} = 0.00187$ mol HCl excess. Then 0.0250 L C_5H_5N + 0.0313 L HCl + 0.0187 L HBr = 0.0750 L total volume. $[\text{HCl excess}] = \frac{0.00187\ \text{mol HCl excess}}{0.0750\ \text{L}} = 0.024933$ M HCl excess

Because HCl is a strong acid, [HCl] excess = $[H_3O^+]$. The strong acid overwhelms the weak acid and is insignificant in the calculation. Finally, pH = $-\log[H_3O^+] = -\log(0.024933) = 1.60$.

Check: The units (none) are correct. The pH is dropping sharply at the equivalence point, so the pH after 5 mL past the equivalence point should be quite acidic. The pH is lower than at the last point. Plotting these points gives the following:

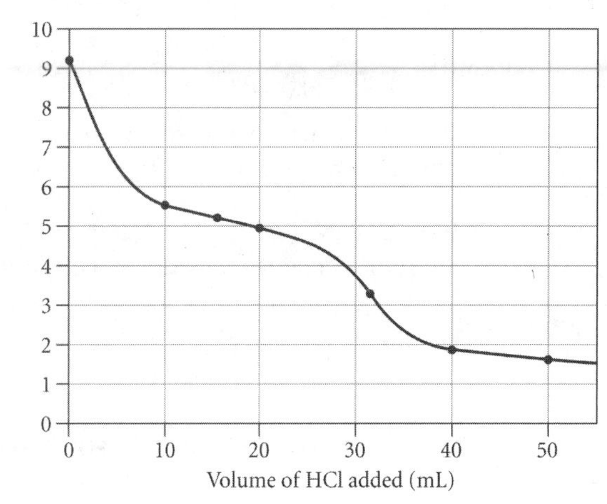

18.73 (i) Acid a is more concentrated because the equivalence point (where sharp pH rise occurs) is at a higher volume of added base.

(ii) Acid b has the larger K_a because the pH at a volume of added base equal to half the equivalence point volume is lower.

18.74 (i) Acid b is more concentrated because the equivalence point (where a sharp pH drop occurs) is at a higher volume of added base.

(ii) Acid b has the larger K_b because the pH at a volume of added base equal to half the equivalence point volume is higher.

18.75 **Given:** 0.229 g unknown monoprotic acid titrated with 0.112 M NaOH and curve

Find: molar mass and pK_a of acid

Conceptual Plan: The equivalence point is where a sharp pH rise occurs. The pK_a is the pH at a volume of added base equal to half the equivalence point volume. Then mL NaOH → L NaOH

$$\frac{1\ \text{L}}{1000\ \text{mL}}$$

then [NaOH], L NaOH → mol NaOH = mol acid then mol acid, g acid → molar mass.

$$M = \frac{\text{mol}}{\text{L}} \qquad \frac{\text{g acid}}{\text{mol acid}}$$

Solution: The equivalence point is at 25 mL NaOH. The pH at 0.5 × 25 mL = 13 mL is ~3 = pK_a. Then

$$25\ \cancel{\text{mL NaOH}} \times \frac{1\ \text{L}}{1000\ \cancel{\text{mL}}} = 0.025\ \text{L NaOH then}$$

$$\frac{0.112\ \text{mol NaOH}}{1\ \cancel{\text{L}}} \times 0.025\ \cancel{\text{L}} = 0.0028\ \text{mol NaOH} = 0.0028\ \text{mol acid then}$$

$$\text{molar mass} = \frac{0.229\ \text{g acid}}{0.0028\ \text{mol acid}} = 82\ \text{g/mol}$$

Check: The units (none and g/mol) are correct. The pK_a is consistent with a weak acid. The molar mass is reasonable for an acid (> 1 g/mol).

18.76 **Given:** 0.446 g unknown monoprotic acid titrated with 0.105 M KOH and curve
Find: molar mass and pK_a of acid
Conceptual Plan: The equivalence point is where a sharp pH rise occurs. The pK_a is the pH at a volume of added base equal to half the equivalence point volume. Then mL KOH → L KOH then

$$\frac{1\ \text{L}}{1000\ \text{mL}}$$

[KOH], L KOH → mol KOH = mol acid then mol acid, g acid → molar mass.

$$M = \frac{\text{mol}}{\text{L}} \qquad \frac{\text{g acid}}{\text{mol acid}}$$

Solution: The equivalence point is at 35 mL NaOH. The pH at 0.5 × 35 mL = 18 mL is ~4.5 = pK_a. Then

$$35\ \cancel{\text{mL KOH}} \times \frac{1\ \text{L}}{1000\ \cancel{\text{mL}}} = 0.035\ \text{L KOH then}$$

$$\frac{0.105\ \text{mol KOH}}{1\ \cancel{\text{L}}} \times 0.035\ \cancel{\text{L}} = 0.003\underline{6}75\ \text{mol KOH} = 0.003\underline{6}75\ \text{mol acid then}$$

$$\text{molar mass} = \frac{0.446\ \text{g acid}}{0.003\underline{6}75\ \text{mol acid}} = 120\ \text{g/mol}$$

Check: The units (none and g/mol) are correct. The pK_a is consistent with a weak acid. The molar mass is reasonable for an acid (> 1 g/mol).

18.77 **Given:** 20.0 mL of 0.115 M sulfurous acid (H_2SO_3) titrated with 0.1014 M KOH **Find:** volume of base added
Conceptual Plan: Because this is a diprotic acid, each proton is titrated sequentially. Write balanced equations.

$$H_2SO_3 + OH^- \rightarrow HSO_3^- + H_2O \text{ and } HSO_3^- + OH^- \rightarrow SO_3^{2-} + H_2O$$

Then mL → L then [H_2SO_3], L → mol H_2SO_3 then set mol base (H_2SO_3) = mol acid (KOH) and

$$\frac{1\ \text{L}}{1000\ \text{mL}} \qquad M = \frac{\text{mol}}{\text{L}} \qquad \textit{balanced equation has 1:1 stoichiometry}\ \text{(1st equivalence point)}$$

[KOH], mol KOH → L KOH → mL KOH the volume to the second equivalence point will be

$$M = \frac{\text{mol}}{\text{L}} \qquad \frac{1000\ \text{mL}}{1\ \text{L}}$$

double the volume to the first equivalence point.

Solution: $20.0\ \cancel{\text{mL}\ H_2SO_3} \times \frac{1\ \text{L}}{1000\ \cancel{\text{mL}}} = 0.0200\ \text{L}\ H_2SO_3$ then

$$\frac{0.115\ \text{mol}\ H_2SO_3}{1\ \cancel{\text{L}}} \times 0.0200\ \cancel{\text{L}} = 0.00230\ \text{mol}\ H_2SO_3.\ \text{So mol base} = \text{mol}\ H_2SO_3 = 0.00230\ \text{mol} = \text{mol KOH}$$

then $0.00230\ \cancel{\text{mol KOH}} \times \frac{1\ \text{L}}{0.1014\ \cancel{\text{mol KOH}}} = 0.022\underline{6}8245\ \cancel{\text{L KOH}} \times \frac{1000\ \text{mL}}{1\ \cancel{\text{L}}} = 22.7\ \text{mL KOH}$ to first equivalence point. The volume to the second equivalence point is simply twice this amount, or 45.4 mL, to the second equivalence point.

Check: The units (mL) are correct. The volume of base is greater than the volume of acid because the concentration of the acid is a little greater than that of the base. The volume to the second equivalence point is twice the volume to the first equivalence point.

18.78 **Given:** 20.0 mL of 0.125 M diprotic acid (H_2A) titrated with 0.1019 M KOH **Find:** volume of base added
Conceptual Plan: Because this is a diprotic acid, each proton is titrated sequentially. Write balanced equations.

$$H_2A + OH^- \rightarrow HA^- + H_2O \text{ and } HA^- + OH^- \rightarrow A^{2-} + H_2O$$

Then mL → L then [H_2A], L → mol H_2A then set mol base (H_2A) = mol acid (KOH) and

$\frac{1\ L}{1000\ mL}$ $M = \frac{mol}{L}$ *balanced equation has 1:1 stoichiometry*

[KOH], mol KOH → L KOH → mL KOH the volume to the second equivalence point will be

$M = \frac{mol}{L}$ $\frac{1000\ mL}{1\ L}$

double the volume to the first equivalence point.

Solution: $20.0\ \text{mL H}_2\text{A} \times \frac{1\ L}{1000\ mL} = 0.0200\ L\ H_2A$ then $\frac{0.125\ mol\ H_2A}{1\ L} \times 0.0200\ L = 0.00250\ mol\ H_2A$.

So mol base = mol H_2A = 0.00250 mol = mol KOH then

$0.00250\ \text{mol KOH} \times \frac{1\ L}{0.1019\ \text{mol KOH}} = 0.0245339\ \text{L KOH} \times \frac{1000\ mL}{1\ L} = 24.5\ mL$ KOH to first equivalence point. The volume to the second equivalence point is simply twice this amount, or 49.0 mL, to the second equivalence point.

Check: The units (mL) are correct. The volume of base is less than the volume of acid because the concentration of the acid is a little greater than that of the base. The volume to the second equivalence point is twice the volume to the first equivalence point.

18.79 The indicator will be in its acid form at an acidic pH, so the color in the HCl sample will be red. The color change will occur over the pH range from pH = $pK_a - 1.0$ to pH = $pK_a + 1.0$, so the color will start to change at pH = 5.0 − 1.0 = 4.0 and finish changing by pH = 5.0 + 1.0 = 6.0.

18.80 **Given:** Phenolphthalein $pK_a = 9.7$, acid form = colorless, base form = pink **Find:** $[In^-]/[HIn]$ and color
Conceptual Plan: Calculate [In]/[HIn] using the equation in chapter. If pH < (pK_a − 1.0), then the solution

$\frac{[In^-]}{[HIn]} = 10^{(pH - pK_a)}$

is colorless. If pH > pK_a + 1.0, then the solution is pink.
Solution:

(a) pH = 2.0 so $\frac{[In^-]}{[HIn]} = 10^{(pH - pK_a)} = 10^{(2.0-9.7)} = 10^{-7.7} = 2 \times 10^{-8}$ and the solution will be colorless.

(b) pH = 5.0 so $\frac{[In^-]}{[HIn]} = 10^{(pH - pK_a)} = 10^{(5.0-9.7)} = 10^{-4.7} = 2 \times 10^{-5}$ and the solution will be colorless.

(c) pH = 8.0 so $\frac{[In^-]}{[HIn]} = 10^{(pH - pK_a)} = 10^{(8.0-9.7)} = 10^{-1.7} = 2 \times 10^{-2}$ and the solution will be colorless.

(d) pH = 11.0 so $\frac{[In^-]}{[HIn]} = 10^{(pH - pK_a)} = 10^{(11.0-9.7)} = 10^{+1.3} = 2 \times 10^{1}$ and the solution will be pink.

Check: The units (none) are correct. The ratio is less than one, and the solution is colorless when the solution is more acidic than the pK_a. The ratio is greater than one, and the solution is pink when the solution is more basic than the pK_a.

18.81 Because the exact conditions of the titration are not given, a rough calculation will suffice. Recall that at the equivalence point, the moles of acid and base are equal. If it is assumed that the concentrations of the acid and the base are equal, the total volume of the solution will have doubled. Assuming an acid and base concentration of 0.1 M, the conjugate based formed must have a concentration of ~0.05 M. From earlier calculations, it can be seen that

the $K_b = \frac{K_w}{K_a} = \frac{[OH^-]^2}{0.05}$; thus,

$[OH^-] = \sqrt{\frac{0.05\ K_w}{K_a}} = \sqrt{\frac{5 \times 10^{-16}}{K_a}}$ and the pH $= 14 + \log \sqrt{\frac{5 \times 10^{-16}}{K_a}}$.

(a) For HF, the $K_a = 3.5 \times 10^{-4}$; so the above equation approximates the pH at the equivalence point of ~8.0. Looking at Table 18.1, phenol red or *m*-nitrophenol will change at the appropriate pH range.

(b) For HCl, the pH at the equivalence point is 7 because HCl is a strong acid. Looking at Table 18.1, alizarin, bromthymol blue, *m*-nitrophenol, or phenol red will change at the appropriate pH range.

(c) For HCN, the $K_a = 4.9 \times 10^{-10}$; so the preceding equation approximates the pH at the equivalence point of ~11.0. Looking at Table 18.1, alizarin yellow R will change at the appropriate pH range.

18.82 Because the exact conditions of the titration are not given, a rough calculation will suffice. Looking at the pattern of earlier problems, the pH at the equivalence point of a titration of a weak base and a strong acid is the hydrolysis of the conjugate acid of the weak base that has been diluted by a factor of roughly 2 with acid. If it is assumed that the initial concentration of the weak base is ~0.1 M, the conjugate acid concentration will be ~0.05 M. From earlier calculations, it can be seen that the $K_a = \frac{K_w}{K_b} = \frac{[H_3O^+]^2}{0.05}$; thus, $[H_3O^+] = \sqrt{\frac{0.05\,K_w}{K_b}} = \sqrt{\frac{5 \times 10^{-16}}{K_b}}$ and the $pH = -\log\sqrt{\frac{5 \times 10^{-16}}{K_b}}$.

(a) For CH_3NH_2, the $K_b = 4.4 \times 10^{-4}$; so the above equation approximates the pH at the equivalence point of ~6.0. Looking at Table 18.1, methyl red, Eriochrome Black T, bromocresol purple, or alizarin or bromthymol blue will change at the appropriate pH range.

(b) For NaOH, the pH at the equivalence point is 7 because NaOH is a strong base. Looking at Table 18.1, alizarin, bromthymol blue, or phenol red will change at the appropriate pH range.

(c) For $C_6H_5NH_2$, the $K_b = 3.9 \times 10^{-10}$; so the above equation approximates the pH at the equivalence point of ~2.9. Looking at Table 18.1, erythrosin B or 2,4-dinitrophenol will change at the appropriate pH range.

Solubility Equilibria

18.83 For the dissolution reaction, start with the ionic compound as a solid and put it in equilibrium with the appropriate cation and anion, making sure to include the appropriate stoichiometric coefficients. The K_{sp} expression is the product of the concentrations of the cation and anion concentrations raised to their stoichiometric coefficients.

(a) $BaSO_4(s) \rightleftharpoons Ba^{2+}(aq) + SO_4^{2-}(aq)$ and $K_{sp} = [Ba^{2+}][SO_4^{2-}]$

(b) $PbBr_2(s) \rightleftharpoons Pb^{2+}(aq) + 2\,Br^-(aq)$ and $K_{sp} = [Pb^{2+}][Br^-]^2$

(c) $Ag_2CrO_4(s) \rightleftharpoons 2\,Ag^+(aq) + CrO_4^{2-}(aq)$ and $K_{sp} = [Ag^+]^2\,[CrO_4^{2-}]$

18.84 For the dissolution reaction, start with the ionic compound as a solid and put it in equilibrium with the appropriate cation and anion, making sure to include the appropriate stoichiometric coefficients. The K_{sp} expression is the product of the concentrations of the cation and anion concentrations raised to their stoichiometric coefficients.

(a) $CaCO_3(s) \rightleftharpoons Ca^{2+}(aq) + CO_3^{2-}(aq)$ and $K_{sp} = [Ca^{2+}][CO_3^{2-}]$

(b) $PbCl_2(s) \rightleftharpoons Pb^{2+}(aq) + 2\,Cl^-(aq)$ and $K_{sp} = [Pb^{2+}][Cl^-]^2$

(c) $AgI(s) \rightleftharpoons Ag^+(aq) + I^-(aq)$ and $K_{sp} = [Ag^+][I^-]$

18.85 **Given:** ionic compound formula and Table 18.2 of K_{sp} values **Find:** molar solubility (S)

Conceptual Plan: Because the balanced chemical equation for the dissolution of A_mX_n is $A_mX_n(s) \rightleftharpoons mA^{n+}(aq) + nX^{m-}(aq)$, the expression of the solubility product constant of A_mX_n is $K_{sp} = [A^{n+}]^m[X^{m-}]^n$. The molar solubility of a compound, A_mX_n, can be computed directly from K_{sp} by solving for S in the expression $K_{sp} = (mS)^m(nS)^n = m^m n^n S^{m+n}$.

Solution:

(a) For AgBr, $K_{sp} = 5.35 \times 10^{-13}$, $A = Ag^+$, $m = 1$, $X = Br^-$, and $n = 1$; so $K_{sp} = 5.35 \times 10^{-13} = S^2$.

Rearrange to solve for S. $S = \sqrt{5.35 \times 10^{-13}} = 7.31 \times 10^{-7}$ M

(b) For $Mg(OH)_2$, $K_{sp} = 2.06 \times 10^{-13}$, $A = Mg^{2+}$, $m = 1$, $X = OH^-$, and $n = 2$; so $K_{sp} = 2.06 \times 10^{-13} = 2^2S^3$.

Rearrange to solve for S. $S = \sqrt[3]{\frac{2.06 \times 10^{-13}}{4}} = 3.72 \times 10^{-5}$ M

(c) For CaF_2, $K_{sp} = 1.46 \times 10^{-10}$, $A = Ca^{2+}$, $m = 1$, $X = F^-$, and $n = 2$; so $K_{sp} = 1.46 \times 10^{-10} = 2^2S^3$.

Rearrange to solve for S. $S = \sqrt[3]{\frac{1.46 \times 10^{-10}}{4}} = 3.32 \times 10^{-4}$ M

Check: The units (M) are correct. The molar solubilities are much less than 1 and dependent not only on the value of the K_{sp}, but also on the stoichiometry of the ionic compound. The more ions generated, the greater the molar solubility for the same value of the K_{sp}.

18.86 **Given:** ionic compound formula and Table 18.2 of K_{sp} values **Find:** molar solubility (S)
Conceptual Plan: Because the balanced chemical equation for the dissolution of A_mX_n is $A_mX_n(s) \rightleftharpoons mA^{n+}(aq) + nX^{m-}(aq)$, the expression of the solubility product constant of A_mX_n is $K_{sp} + [A^{n+}]^m[X^{m-}]^n$. The molar solubility of a compound, A_mX_n, can be computed directly from K_{sp} by solving for S in the expression $K_{sp} = (mS)^m(nS)^n = m^mn^nS^{m+n}$.

Solution:

(a) For MX, $K_{sp} = 1.27 \times 10^{-36}$, $A = M^+$, $m = 1$, $X = X^-$, and $n = 1$; so $K_{sp} = 1.27 \times 10^{-36} = S^2$.

Rearrange to solve for S. $S = \sqrt{1.27 \times 10^{-36}} = 1.13 \times 10^{-18}$ M

(b) For Ag_2CrO_4, $K_{sp} = 1.12 \times 10^{-12}$, $A = Ag^+$, $m = 2$, $X = CrO_4^{2-}$, and $n = 1$; so $K_{sp} = 1.12 \times 10^{-12} = 2^2S^3$.

Rearrange to solve for S. $S = \sqrt[3]{\frac{1.12 \times 10^{-12}}{4}} = 6.54 \times 10^{-5}$ M

(c) For $Ca(OH)_2$, $K_{sp} = 4.68 \times 10^{-6}$, $A = Ca^{2+}$, $m = 1$, $X = OH^-$, and $n = 2$; so $K_{sp} = 4.68 \times 10^{-6} = 2^2S^3$.

Rearrange to solve for S. $S = \sqrt[3]{\frac{4.68 \times 10^{-6}}{4}} = 1.05 \times 10^{-2}$ M

Check: The units (M) are correct. The molar solubilities are much less than 1 and dependent not only on the value of the K_{sp}, but also on the stoichiometry of the ionic compound. The more ions generated, the greater the molar solubility for the same value of the K_{sp}.

18.87 **Given:** ionic compound formula and molar solubility (S) **Find:** K_{sp}
Conceptual Plan: The expression of the solubility product constant of A_mX_n is $K_{sp} = [A^{n+}]^m[X^{m-}]^n$. The molar solubility of a compound, A_mX_n, can be computed directly from K_{sp} by solving for S in the expression $K_{sp} = (mS)^m(nS)^n = m^mn^nS^{m+n}$.

Solution:

(a) For MX, $S = 3.27 \times 10^{-11}$ M, $A = M^+$, $m = 1$, $X = X^-$, and $n = 1$; so $K_{sp} = S^2 = (3.27 \times 10^{-11})^2 = 1.07 \times 10^{-21}$.

(b) For PbF_2, $S = 5.63 \times 10^{-3}$ M, $A = Pb^{2+}$, $m = 1$, $X = F^-$, and $n = 2$; so $K_{sp} = 2^2S^3 = 2^2(5.63 \times 10^{-3})^3 = 7.14 \times 10^{-7}$.

(d) For MgF_2, $S = 2.65 \times 10^{-4}$ M, $A = Mg^{2+}$, $m = 1$, $X = F^-$, and $n = 2$; so $K_{sp} = 2^2S^3 = 2^2(2.65 \times 10^{-4})^3 = 7.44 \times 10^{-11}$.

Check: The units (none) are correct. The K_{sp} values are much less than 1 and dependent not only on the value of the solubility, but also on the stoichiometry of the ionic compound. The more ions generated, the smaller the K_{sp} for the same value of the S.

18.88 **Given:** ionic compound formula and molar solubility (S) **Find:** K_{sp}
Conceptual Plan: The expression of the solubility product constant of A_mX_n is $K_{sp} = [A^{n+}]^m[X^{m-}]^n$. The molar solubility of a compound, A_mX_n, can be computed directly from K_{sp} by solving for S in the expression $K_{sp} = (mS)^m(nS)^n = m^mn^nS^{m+n}$.

Solution:

(a) For $BaCrO_4$, $S = 1.08 \times 10^{-5}$ M, $A = Ba^{2+}$, $m = 1$, $X = CrO_4^{2-}$, and $n = 1$; so $K_{sp} = S^2 = (1.08 \times 10^{-5})^2 = 1.17 \times 10^{-10}$.

(b) For Ag_2SO_3, $S = 1.55 \times 10^{-5}$ M, A = Ag^+, $m = 2$, X = SO_3^{2-}, and $n = 1$; so $K_{sp} = 2^2S^3 = 2^2(1.55 \times 10^{-5})^3 = 1.49 \times 10^{-14}$.

(c) For $Pd(SCN)_2$, $S = 2.22 \times 10^{-8}$ M, A = Pd^{2+}, $m = 1$, X = SCN^-, and $n = 2$; so $K_{sp} = 2^2S^3 = 2^2(2.22 \times 10^{-8})^3 = 4.38 \times 10^{-23}$.

Check: The units (none) are correct. The K_{sp} values are much less than 1 and dependent not only on the value of the solubility, but also on the stoichiometry of the ionic compound. The more ions generated, the smaller the K_{sp} for the same value of the S.

18.89 **Given:** ionic compound formulas AX and AX_2 and $K_{sp} = 1.5 \times 10^{-5}$ **Find:** higher molar solubility (S)
Conceptual Plan: The expression of the solubility product constant of A_mX_n is $K_{sp} = [A^{n+}]^m[X^{m-}]^n$. The molar solubility of a compound, A_mX_n, can be computed directly from K_{sp} by solving for S in the expression $K_{sp} = (mS)^m(nS)^n = m^mn^nS^{m+n}$.
Solution: For AX, $K_{sp} = 1.5 \times 10^{-5}$, $m = 1$, and $n = 1$; so $K_{sp} = 1.5 \times 10^{-5} = S^2$. Rearrange to solve for S.
$S = \sqrt{1.5 \times 10^{-5}} = 3.9 \times 10^{-3}$ M. For AX_2, $K_{sp} = 1.5 \times 10^{-5}$, $m = 1$, and $n = 2$; so $K_{sp} = 1.5 \times 10^{-5} = 2^2S^3$. .

Rearrange to solve for S. $S = \sqrt[3]{\frac{1.5 \times 10^{-5}}{4}} = 1.6 \times 10^{-2}$ M. Because 10^{-2} M $> 10^{-3}$ M, AX_2 has a higher molar solubility.

Check: The units (M) are correct. The more ions generated, the greater the molar solubility for the same value of the K_{sp}.

18.90 **Given:** ionic compound formula and molar solubility (S) **Find:** K_{sp}
Conceptual Plan: The expression of the solubility product constant of A_mX_n is $K_{sp} = [A^{n+}]^m[X^{m-}]^n$. The molar solubility of a compound, A_mX_n, can be computed directly from K_{sp} by solving for S in the expression $K_{sp} = (mS)^m(nS)^n = m^mn^nS^{m+n}$.
Solution: For AX, $S = 1.35 \times 10^{-4}$ M, $m = 1$, and $n = 1$; so $K_{sp} = S^2 = (1.35 \times 10^{-4})^2 = 1.82 \times 10^{-8}$.
For AX_2, $S = 2.25 \times 10^{-4}$ M, $m = 1$, and $n = 2$; so $K_{sp} = 2^2S^3 = 2^2(2.25 \times 10^{-4})^3 = 4.56 \times 10^{-11}$.
For A_2X, $S = 1.75 \times 10^{-4}$ M, $m = 2$, and $n = 1$; so $K_{sp} = 2^2S^3 = 2^2(1.75 \times 10^{-4})^3 = 2.14 \times 10^{-11}$. So A_2X has the lowest K_{sp} because it has a lower S than for AX_2.

Check: The units (none) are correct. The K_{sp} values are much less than 1 and dependent not only on the value of the solubility, but also on the stoichiometry of the ionic compound. The more ions generated, the smaller the K_{sp} for the same value of the S.

18.91 **Given:** $Fe(OH)_2$ in 100.0 mL solution **Find:** grams of $Fe(OH)_2$ **Other:** $K_{sp} = 4.87 \times 10^{-17}$
Conceptual Plan: The expression of the solubility product constant of A_mX_n is $K_{sp} = [A^{n+}]^m[X^{m-}]^n$. The molar solubility of a compound, A_mX_n, can be computed directly from K_{sp} by solving for S in the expression $K_{sp} = (mS)^m(nS)^n = m^mn^nS^{m+n}$. Then solve for S, then mL $\rightarrow$ L then

$$\frac{1\ \text{L}}{1000\ \text{mL}}$$

S, L $\rightarrow$ mol $Fe(OH)_2 \rightarrow$ g $Fe(OH)_2$.

$$\text{M} = \frac{\text{mol}}{\text{L}} \qquad \frac{89.87\ \text{g Fe(OH)}_2}{1\ \text{mol Fe(OH)}_2}$$

Solution: For $Fe(OH)_2$, $K_{sp} = 4.87 \times 10^{-17}$, A = Fe^{2+}, $m = 1$, X = OH^-, and $n = 2$; so $K_{sp} = 4.87 \times 10^{-17} = 2^2S^3$.

Rearrange to solve for S. $S = \sqrt[3]{\frac{4.87 \times 10^{-17}}{4}} = 2.3\underline{0}050 \times 10^{-6}$ M. Then $100.0\ \cancel{\text{mL}} \times \frac{1\ \text{L}}{1000\ \cancel{\text{mL}}} = 0.1000$ L

then $\frac{2.3\underline{0}050 \times 10^{-6}\ \text{mol Fe(OH)}_2}{1\ \cancel{\text{L}}} \times 0.1000\ \cancel{\text{L}} = 2.3\underline{0}050 \times 10^{-7}\ \cancel{\text{mol Fe(OH)}_2} \times \frac{89.87\ \text{g Fe(OH)}_2}{1\ \cancel{\text{mol Fe(OH)}_2}}$

$= 2.07 \times 10^{-5}$g $Fe(OH)_2$.

Check: The units (g) are correct. The solubility rules from Chapter 9 (most hydroxides are insoluble) suggest that very little $Fe(OH)_2$ will dissolve; so the magnitude of the answer is not surprising.

18.92 **Given:** 3.91 mg CuCl in 100.0 mL solution **Find:** K_{sp}

Conceptual Plan: mL → L then mg CuCl → g CuCl → mol CuCl then L, mol CuCl → S

$\frac{1\ \text{L}}{1000\ \text{mL}}$ $\frac{1\ \text{g CuCl}}{1000\ \text{mg CuCl}}$ $\frac{1\ \text{mol CuCl}}{99.00\ \text{g CuCl}}$ $M = \frac{\text{mol}}{\text{L}}$

The expression of the solubility product constant of A_mX_n is $K_{sp} = [A^{n+}]^m[X^{m-}]^n$. The molar solubility of a compound, A_mX_n, can be computed directly from K_{sp} by solving for S in the expression $K_{sp} = (mS)^m(nS)^n = m^m n^n S^{m+n}$.

Solution: $100.0\ \text{mL} \times \frac{1\ \text{L}}{1000\ \text{mL}} = 0.1000\ \text{L}$ then

$3.91\ \text{mg CuCl} \times \frac{1\ \text{g CuCl}}{1000\ \text{mg CuCl}} \times \frac{1\ \text{mol CuCl}}{99.00\ \text{g CuCl}} = 3.94950 \times 10^{-5}\ \text{mol CuCl}$

then $\frac{3.94950 \times 10^{-5}\ \text{mol CuCl}}{0.1000\ \text{L}} = 3.94950 \times 10^{-4}\ \text{M CuCl} = S$ then for CuCl, $A = Cu^+$, $m = 1$, $X = Cl^-$, and $n = 1$

so $K_{sp} = S^2 = (3.94950 \times 10^{-4})^2 = 1.56 \times 10^{-7}$

Check: The units (none) are correct. The value of $K_{sp} << 1$ because only 3.91 mg dissolve in a liter of solution. The K_{sp} is not too low because CuCl dissociates into only two ions and S is 10^{-4}.

18.93 (a) **Given:** BaF_2 **Find:** molar solubility (S) in pure water **Other:** $K_{sp}(BaF_2) = 2.45 \times 10^{-5}$

Conceptual Plan: The expression of the solubility product constant of A_mX_n is $K_{sp} = [A^{n+}]^m[X^{m-}]^n$. The molar solubility of a compound, A_mX_n, can be computed directly from K_{sp} by solving for S in the expression $K_{sp} = (mS)^m(nS)^n = m^m n^n S^{m+n}$.

Solution: For BaF_2, $K_{sp} = 2.45 \times 10^{-5}$, $A = Ba^{2+}$, $m = 1$, $X = F^-$, and $n = 2$; so $K_{sp} = 2.45 \times 10^{-5} = 2^2S^3$.

Rearrange to solve for S. $S = \sqrt[3]{\frac{2.45 \times 10^{-5}}{4}} = 1.83 \times 10^{-2}\ \text{M}$

(b) **Given:** BaF_2 **Find:** molar solubility (S) in 0.10 M $Ba(NO_3)_2$ **Other:** $K_{sp}(BaF_2) = 2.45 \times 10^{-5}$

Conceptual Plan: M $Ba(NO_3)_2$ → M Ba^{2+} then M Ba^{2+}, K_{sp} → S

$Ba(NO_3)_2(s) \rightarrow Ba^{2+}(aq) + 2\ NO_3^-(aq)$ ICE table

Solution: Because one Ba^{2+} ion is generated for each $Ba(NO_3)_2$, $[Ba^{2+}] = 0.10$ M.

$BaF_2(s) \rightleftharpoons$	$Ba^{2+}(aq)$ +	$2\ F^-(aq)$
Initial	0.10	0.00
Change	S	$2S$
Equil	$0.10 + S$	$2S$

$K_{sp}(BaF_2) = [Ba^{2+}][F^-]^2 = 2.45 \times 10^{-5} = (0.10 + S)(2S)^2$

Assume that $S << 0.10$, $2.45 \times 10^{-5} = (0.10)(2S)^2$, and $S = 7.83 \times 10^{-3}$ M. Confirm that the assumption is valid. $\frac{7.83 \times 10^{-3}}{0.10} \times 100\% = 7.8\% > 5\%$, so the assumption is not valid. Because expanding the expression will give a third-order polynomial, that is not easily solved directly. Solve by successive approximations. Substitute $S = 7.83 \times 10^{-3}$ M for the S term that is part of a sum [i.e., the one in $(0.10 + S)$]. Thus, $2.45 \times 10^{-5} = (0.10 + 7.83 \times 10^{-3})(2S)^2$ and $S = 7.53 \times 10^{-3}$ M. Substitute this new S value again. Thus, $2.45 \times 10^{-5} = (0.10 + 7.53 \times 10^{-3})(2S)^2$ and $S = 7.55 \times 10^{-3}$ M. Substitute this new S value again. Thus, $2.45 \times 10^{-5} = (0.10 + 7.55 \times 10^{-3})(2S)^2$ and $S = 7.55 \times 10^{-3}$ M. So the solution has converged and $S = 7.55 \times 10^{-3}$ M.

(c) **Given:** BaF_2 **Find:** molar solubility (S) in 0.15 M NaF **Other:** $K_{sp}(BaF_2) = 2.45 \times 10^{-5}$

Conceptual Plan: M NaF → M F^- then M F^-, K_{sp} → S

$NaF(s) \rightarrow Na^+(aq) + F^-(aq)$ ICE table

Solution: Because one F^- ion is generated for each NaF, $[F^-] = 0.15$ M.

$BaF_2(s) \rightleftharpoons$	$Ba^{2+}(aq)$ +	$2\,F^-(aq)$
Initial	0.00	0.15
Change	S	$2S$
Equil	S	$0.15 + 2S$

$K_{sp}(BaF_2) = [Ba^{2+}][F^-]^2 = 2.45 \times 10^{-5} = (S)(0.15 + 2S)^2$

Because $2S << 0.15$, $2.45 \times 10^{-5} = (S)(0.15)^2$ and $S = 1.09 \times 10^{-3}$ M. Confirm that the assumption is valid. $\frac{2\,(1.09 \times 10^{-3})}{0.15} \times 100\% = 1.5\% < 5\%$, so the assumption is valid.

Check: The units (M) are correct. The solubility of the BaF_2 decreases in the presence of a common ion. The effect of the anion is greater because the K_{sp} expression has the anion concentration squared.

18.94 (a) **Given:** MX **Find:** molar solubility (S) in pure water **Other:** $K_{sp}(MX) = 1.27 \times 10^{-36}$
Conceptual Plan: The expression of the solubility product constant of A_mX_n is $K_{sp} = [A^{n+}]^m[X^{m-}]^n$. The molar solubility of a compound, A_mX_n, can be computed directly from K_{sp} by solving for S in the expression $K_{sp} = (mS)^m(nS)^n = m^m n^n S^{m+n}$.

Solution: For MX, $K_{sp} = 1.27 \times 10^{-36}$, $A = M^{2+}$, $m = 1$, $X = X^{2-}$, and $n = 1$. There is a 1:1 ratio of the cation:anion, so $K_{sp} = 1.27 \times 10^{-36} = S^2$. Rearrange to solve for S. $S = 1.13 \times 10^{-18}$ M

(b) **Given:** MX **Find:** molar solubility (S) in 0.25 M MCl_2 **Other:** K_{sp} (MX) $= 1.27 \times 10^{-36}$
Conceptual Plan: M $MCl_2 \rightarrow$ M M^{2+} then M M^{2+}, $K_{sp} \rightarrow S$

$MCl_2(s) \rightarrow M^{2+}(aq) + 2\,Cl^-(aq)$ ICE table

Solution: Because one M^{2+} ion is generated for each MCl_2, $[M^{2+}] = 0.25$ M.

$MX(s) \rightleftharpoons$	$M^{2+}(aq)$ +	$X^{2-}(aq)$
Initial	0.25	0.00
Change	S	S
Equil	$0.25 + S$	S

$K_{sp}(MX) = [M^{2+}][X^{2-}] = 1.27 \times 10^{-36} = (0.25 + S)S$

Assume that $S << 0.25$, $1.27 \times 10^{-36} = (0.25)S$, and $S = 5.08 \times 10^{-36}$ M. Confirm that the assumption is valid. $\frac{5.08 \times 10^{-36}}{0.25} \times 100\% = 2.0 \times 10^{-33}\% << 5\%$, so the assumption is valid.

(c) **Given:** MX **Find:** molar solubility (S) in 0.20 M Na_2X **Other:** $K_{sp}(MX) = 1.27 \times 10^{-36}$
Conceptual Plan: M $Na_2X \rightarrow$ M X^{2-} then M X^{2-}, $K_{sp} \rightarrow S$

$Na_2X(s) \rightarrow 2\,Na^+(aq) + X^{2-}(aq)$ ICE table

Solution: Because one X^{2-} ion is generated for each Na_2X, $[X^{2-}] = 0.20$ M.

$MX(s) \rightleftharpoons$	$M^{2+}(aq)$ +	$X^{2-}(aq)$
Initial	0.00	0.20
Change	S	S
Equil	S	$0.20 + S$

$K_{sp}(MX) = [M^{2+}][X^{2-}] = 1.27 \times 10^{-36} = (S)(0.20 + S)$.

Assume that $S << 0.20$, $1.27 \times 10^{-36} = (S)(0.20)$, and $S = 6.35 \times 10^{-36}$ M. Confirm that the assumption is valid.

$\frac{6.35 \times 10^{-36}}{0.20} \times 100\% = 3.2 \times 10^{-33}\% << 5$, so the assumption is valid.

Check: The units (M) are correct. The solubility of the MX decreases in the presence of a common ion.

18.95 **Given:** $Ca(OH)_2$ **Find:** molar solubility (S) in buffers at (a) pH = 4, (b) pH = 7, and (c) pH = 9
Other: K_{sp} $(Ca(OH)_2) = 4.68 \times 10^{-6}$
Conceptual Plan: pH $\rightarrow$ $[H_3O^+]$ $\rightarrow$ $[OH^-]$ then M OH^-, K_{sp} $\rightarrow$ S

$[H_3O^+] = 10^{-pH}$ $K_w = [H_3O^+][OH^-]$ set up ICE table

Solution:

(a) pH = 4, so $[H_3O^+] = 10^{-pH} = 10^{-4} = 1 \times 10^{-4}$ M then $K_w = [H_3O^+][OH^-]$ so

$$[OH^-] = \frac{K_w}{[H_3O^+]} = \frac{1.0 \times 10^{-14}}{1 \times 10^{-4}} = 1 \times 10^{-10} \text{ M then}$$

$Ca(OH)_2(s) \rightleftharpoons$	$Ca^{2+}(aq)$ +	$2\,OH^-(aq)$
Initial	0.00	1×10^{-10}
Change	S	—
Equil	S	1×10^{-10}

$K_{sp}(Ca(OH)_2) = [Ca^{2+}][OH^-]^2 = 4.68 \times 10^{-6} = S(1 \times 10^{-10})^2$ and $S = 5 \times 10^{14}$ M

(b) pH = 7, so $[H_3O^+] = 10^{-pH} = 10^{-7} = 1 \times 10^{-7}$ M then $K_w = [H_3O^+][OH^-]$ so

$$[OH^-] = \frac{K_w}{[H_3O^+]} = \frac{1.0 \times 10^{-14}}{1 \times 10^{-7}} = 1 \times 10^{-7} \text{ M then}$$

$Ca(OH)_2(s) \rightleftharpoons$	$Ca^{2+}(aq)$ +	$2\,OH^-(aq)$
Initial	0.00	1×10^{-7}
Change	S	—
Equil	S	1×10^{-7}

$K_{sp}(Ca(OH)_2) = [Ca^{2+}][OH^-]^2 = 4.68 \times 10^{-6} = S(1 \times 10^{-7})^2$ and $S = 5 \times 10^{8}$ M

(c) pH = 9, so $[H_3O^+] = 10^{-pH} = 10^{-9} = 1 \times 10^{-9}$ M then $K_w = [H_3O^+][OH^-]$ so

$$[OH^-] = \frac{K_w}{[H_3O^+]} = \frac{1.0 \times 10^{-14}}{1 \times 10^{-9}} = 1 \times 10^{-5} \text{ M then}$$

$Ca(OH)_2(s) \rightleftharpoons$	$Ca^{2+}(aq)$ +	$2\,OH^-(aq)$
Initial	0.00	1×10^{-5}
Change	S	—
Equil	S	1×10^{-5}

$K_{sp}(Ca(OH)_2) = [Ca^{2+}][OH^-]^2 = 4.68 \times 10^{-6} = S(1 \times 10^{-5})^2$ and $S = 5 \times 10^{4}$ M

Check: The units (M) are correct. The solubility of the $Ca(OH)_2$ decreases as the pH increases (and the hydroxide ion concentration increases). These molar solubilities are not achievable because the saturation point of pure $Ca(OH)_2$ is ~30 M. The bottom line is that as long as the hydroxide concentration can be controlled with a buffer, the $Ca(OH)_2$ will be very soluble.

18.96 **Given:** $Mg(OH)_2$ in 1.00×10^2 mL solution **Find:** grams of $Mg(OH)_2$ in pure water and buffer at pH = 10
Other: $K_{sp}(Mg(OH)_2) = 2.06 \times 10^{-13}$

Conceptual Plan: For pure water:

The expression of the solubility product constant of A_mX_n is $K_{sp} = [A^{n+}]^m[X^{m-}]^n$. The molar solubility of a compound, A_mX_n, can be computed directly from K_{sp} by solving for S in the expression $K_{sp} = (mS)^m(nS)^n = m^m n^n S^{m+n}$. Then mL → L then S, L → mol $Mg(OH)_2$ → g $Mg(OH)_2$.

$$\frac{1\text{ L}}{1000\text{ mL}} \qquad M = \frac{\text{mol}}{\text{L}} \qquad \frac{58.33\text{ g }Mg(OH)_2}{1\text{ mol }Mg(OH)_2}$$

For buffer solution: pH → $[H_3O^+]$ → $[OH^-]$ then

$$[H_3O^+] = 10^{-pH} \qquad K_w = [H_3O^+][OH^-]$$

M OH^-, K_{sp} → S then S, L → mol $Mg(OH)_2$ → g $Mg(OH)_2$

$$\text{set up ICE table} \qquad M = \frac{\text{mol}}{\text{L}} \qquad \frac{58.33\text{ g }Mg(OH)_2}{1\text{ mol }Mg(OH)_2}$$

Solution: For pure water, $K_{sp} = 2.06 \times 10^{-13}$, A = Mg^{2+}, $m = 1$, X = OH^-, and $n = 2$; so $K_{sp} = 2.06 \times 10^{-13} = 2^2S^3$.

Rearrange to solve for S. $S = \sqrt[3]{\dfrac{2.06 \times 10^{-13}}{4}} = 3.7\underline{2}051 \times 10^{-5}$ M. Then

$1.00 \times 10^2 \text{ mL} \times \frac{1 \text{ L}}{1000 \text{ mL}} = 0.100 \text{ L}$ then

$\frac{3.72051 \times 10^{-5} \text{ mol Mg(OH)}_2}{1 \text{ L}} \times 0.100 \text{ L} = 3.72051 \times 10^{-6} \text{ mol Mg(OH)}_2$

$3.72051 \times 10^{-6} \text{ mol Mg(OH)}_2 \times \frac{58.33 \text{ g Mg(OH)}_2}{1 \text{ mol Mg(OH)}_2} = 2.17 \times 10^{-4} \text{ g Mg(OH)}_2$

$\text{pH} = 10$, so $[H_3O^+] = 10^{-\text{pH}} = 10^{-10} = 1 \times 10^{-10}$ M then $K_w = [H_3O^+][OH^-]$ so

$[OH^-] = \frac{K_w}{[H_3O^+]} = \frac{1.0 \times 10^{-14}}{1 \times 10^{-10}} = 1 \times 10^{-4}$ M then

$Mg(OH)_2(s) \rightleftharpoons$	$Mg^{2+}(aq)$ +	$OH^-(aq)$
Initial	0.00	1×10^{-4}
Change	S	—
Equil	S	1×10^{-4}

$K_{sp}(Mg(OH)_2) = [Mg^{2+}][OH^-]^2 = 2.06 \times 10^{-13} = S(1 \times 10^{-4})^2$ and 2.06×10^{-5} M. Then

$\frac{2.06 \times 10^{-5} \text{ mol Mg(OH)}_2}{1 \text{ L}} \times 0.100 \text{ L} = 2.06 \times 10^{-6} \text{ mol Mg(OH)}_2 \times \frac{58.33 \text{ g Mg(OH)}_2}{1 \text{ mol Mg(OH)}_2} = 1 \times 10^{-4} \text{ g Mg(OH)}_2$

Check: The units (M) are correct. The solubility of the $Mg(OH)_2$ decreases as the pH increases (and the hydroxide ion concentration increases).

18.97 (a) $BaCO_3$ will be more soluble in acidic solutions because CO_3^{2-} is basic. In acidic solutions, it can be converted to HCO_3^- and H_2CO_3. These species are not CO_3^{2-}, so they do not appear in the K_{sp} expression.

(b) CuS will be more soluble in acidic solutions because S^{2-} is basic. In acidic solutions, it can be converted to HS^- and H_2S. These species are not S^{2-}, so they do not appear in the K_{sp} expression.

(c) AgCl will not be more soluble in acidic solutions because Cl^- will not react with acidic solutions because HCl is a strong acid.

(d) PbI_2 will not be more soluble in acidic solutions because I^- will not react with acidic solutions because HI is a strong acid.

18.98 (a) Hg_2Br_2 will not be more soluble in acidic solutions because Br^- will not react with acidic solutions because HBr is a strong acid.

(b) $Mg(OH)_2$ will be more soluble in acidic solutions because OH^- is basic. In acidic solutions, it can be converted to H_2O. This species is not OH^-, so it does not appear in the K_{sp} expression.

(c) $CaCO_3$ will be more soluble in acidic solutions because CO_3^{2-} is basic. In acidic solutions, it can be converted to HCO_3^- and H_2CO_3. These species are not CO_3^{2-}, so they do not appear in the K_{sp} expression.

(d) AgI will not be more soluble in acidic solutions because I^- will not react with acidic solutions because HI is a strong acid.

Precipitation

18.99 **Given:** 0.015 M NaF and 0.010 M $Ca(NO_3)_2$ **Find:** Will a precipitate form? If so, identify it.
Other: $K_{sp}(CaF_2) = 1.46 \times 10^{-10}$

Conceptual Plan: Look at all possible combinations and consider the solubility rules from Chapter 9. Salts of alkali metals (Na) are very soluble, so NaF and $NaNO_3$ will be very soluble. Nitrate compounds are very soluble, so $NaNO_3$ will be very soluble. The only possibility for a precipitate is CaF_2. Determine whether a precipitate will form by determining the concentration of the Ca^{2+} and F^- in solution. Then compute the reaction quotient, Q. If $Q > K_{sp}$, a precipitate will form.

Solution: Because the only possible precipitate is CaF_2, calculate the concentrations of Ca^{2+} and F^-. $NaF(s) \rightarrow Na^+(aq) + F^-(aq)$. Because one F^- ion is generated for each NaF, $[F^-] = 0.015$ M.
$Ca(NO_3)_2(s) \rightarrow Ca^{2+}(aq) + 2\,NO_3^-(aq)$. Because one Ca^{2+} ion is generated for each $Ca(NO_3)_2$, $[Ca^{2+}] = 0.010$ M.
Then calculate $Q\,(CaF_2)$, $A = Ca^{2+}$, $m = 1$, $X = F^-$, and $n = 2$. Because $Q = [A^{n+}]^m[X^{m-}]^n$,
$Q\,(CaF_2) = [Ca^{2+}][F^-]^2 = (0.010)(0.015)^2 = 2.3 \times 10^{-6} > 1.46 \times 10^{-10} = K_{sp}(CaF_2)$; so a precipitate will form.

Check: The units (none) are correct. The solubility of the CaF_2 is low, and the concentration of ions is extremely large compared to the K_{sp}; so a precipitate will form.

18.100 **Given:** 0.013 M KBr and 0.0035 M $Pb(C_2H_3O_2)_2$ **Find:** $Pb(C_2H_3O_2)_2$. Will a precipitate form? If so, identify it. **Other:** $K_{sp}(PbBr_2) = 4.67 \times 10^{-6}$

Conceptual Plan: Look at all possible combinations and consider the solubility rules from Chapter 9. Salts of alkali metals (K) are very soluble, so KBr and $KC_2H_3O_2$ will be very soluble. Acetate compounds are very soluble, so $Pb(C_2H_3O_2)_2$ and $KC_2H_3O_2$ will be very soluble. The only possibility for a precipitate is $PbBr_2$. Determine whether a precipitate will form by determining the concentration of the Pb^{2+} and Br^- in solution. Then compute the reaction quotient, Q. If $Q > K_{sp}$, a precipitate will form.

Solution: Because the only possible precipitate is $PbBr_2$, calculate the concentrations of Pb^{2+} and Br^-. $KBr(s) \rightarrow K^+(aq) + Br^-(aq)$. Because one Br^- ion is generated for each KBr, $[Br^-] = 0.013$ M. $Pb(C_2H_3O_2)_2(s) \rightarrow Pb^{2+}(aq) + 2\,C_2H_3O_2^-(aq)$. Because one Pb^{2+} ion is generated for each $Pb(C_2H_3O_2)_2$, $[Pb^+] = 0.0035$ M. Then calculate $Q(PbBr_2)$, $A = Pb^{2+}$, $m = 1$, $X = Br^-$, and $n = 2$. Because $Q = [A^{n+}]^m[X^{m-}]^n$, $Q(PbBr_2) = [Pb^{2+}][Br^-]^2 = (0.0035)(0.013)^2 = 6.0 \times 10^{-7} < 4.67 \times 10^{-6} = K_{sp}(PbBr_2)$; so a precipitate will not form.

Check: The units (none) are correct. The K_{sp} of the $PbBr_2$ is not too low compared to the solution ion concentrations, so a precipitate will not form.

18.101 **Given:** 75.0 mL of NaOH with pOH = 2.58 and 125.0 mL of 0.0018 M $MgCl_2$

Find: Will a precipitate form? If so, identify it. **Other:** $K_{sp}(Mg(OH)_2) = 2.06 \times 10^{-13}$

Conceptual Plan: Look at all possible combinations and consider the solubility rules from Chapter 9. Salts of alkali metals (Na) are very soluble, so NaOH and NaCl will be very soluble. Chloride compounds are generally very soluble, so $MgCl_2$ and NaCl will be very soluble. The only possibility for a precipitate is $Mg(OH)_2$. Determine whether a precipitate will form by determining the concentration of the Mg^{2+} and OH^- in solution. Because pOH, not NaOH concentration, is given, pOH $\rightarrow [OH^-]$ then

$$[OH^-] = 10^{-pOH}$$

mix solutions and calculate diluted concentrations mL NaOH, mL $MgCl_2 \rightarrow$ mL total then

$$\text{mL NaOH} + \text{mL MgCl}_2 = \text{total mL}$$

mL, initial M $\rightarrow$ final M then compute the reaction quotient, Q.

$$M_1V_1 = M_2V_2$$

If $Q > K_{sp}$, a precipitate will form.

Solution: Because the only possible precipitate is $Mg(OH)_2$, calculate the concentrations of Mg^{2+} and OH^-. For NaOH at pOH = 2.58, so $[OH^-] = 10^{-pOH} = 10^{-2.58} = 2.\underline{6}3027 \times 10^{-3}$ M and $MgCl_2(s) \rightarrow Mg^{2+}(aq) + 2\,Cl^-(aq)$. Because one Mg^{2+} ion is generated for each $MgCl_2$, $[Mg^{2+}] = 0.0018$ M. Then total mL = mL NaOH + mL $MgCl_2$ = 75.0 mL + 125.0 mL = 200.0 mL. Then $M_1V_1 = M_2V_2$;

rearrange to solve for M_2. $M_2 = M_1\dfrac{V_1}{V_2} = 2.\underline{6}3027 \times 10^{-3}\text{ M OH}^- \times \dfrac{75.0\text{ mL}}{200.0\text{ mL}} = 9.\underline{8}635 \times 10^{-4}\text{ M OH}^-$ and

$M_2 = M_1\dfrac{V_1}{V_2} = 0.0018\text{ M Mg}^+ \times \dfrac{125.0\text{ mL}}{200.0\text{ mL}} = 1.\underline{1}25 \times 10^{-3}\text{ M Mg}^{2+}$. Calculate $Q(Mg(OH)_2)$, $A = Mg^{2+}$,

$m = 1$, $X = OH^-$, and $n = 2$. Because $Q = [A^{n+}]^m[X^{m-}]^n$, $Q(Mg(OH)_2) = [Mg^{2+}][OH^-]^2 =$

$(1.\underline{1}25 \times 10^{-3})(9.\underline{8}635 \times 10^{-4})^2 = 1.1 \times 10^{-9} > 2.06 \times 10^{-13} = K_{sp}(Mg(OH)_2)$; so a precipitate will form.

Check: The units (none) are correct. The solubility of the $Mg(OH)_2$ is low, and the NaOH (a base) is high enough that the product of the concentration of ions is large compared to the K_{sp}; so a precipitate will form.

18.102 **Given:** 175.0 mL of 0.0055 M KCl and 145.0 mL of 0.0015 M $AgNO_3$

Find: Will a precipitate form? If so, identify it. **Other:** $K_{sp}(AgCl) = 1.77 \times 10^{-10}$

Conceptual Plan: Look at all possible combinations and consider the solubility rules from Chapter 9. Salts of alkali metals (K) are very soluble, so KCl and KNO_3 will be very soluble. Nitrate compounds are very soluble, so KNO_3 and $AgNO_3$ will be very soluble. The only possibility for a precipitate is AgCl. Determine whether a precipitate will form by determining the concentration of the Ag^+ and Cl^- in solution. Mix solutions and calculate diluted concentrations mL KCl, mL $AgNO_3 \rightarrow$ mL total then

$$\text{mL KCl} + \text{mL AgNO}_3 = \text{total mL}$$

mL, initial M → final M then compute the reaction quotient, Q. If $Q > K_{sp}$, a precipitate will form.

$M_1V_1 = M_2V_2$

Solution: Because the only possible precipitate is AgCl, calculate the concentrations of Ag^+ and Cl^-. $KCl(s) \rightarrow K^+(aq) + Cl^-(aq)$. Because one Cl^- ion is generated for each AgCl, $[Cl^-] = 0.0055$ M and $AgNO_3(s) \rightarrow Ag^+(aq) + NO_3^-(aq)$. Because one Ag^+ ion is generated for each $AgNO_3$, $[Ag^+] = 0.0015$ M. Then total mL = mL KCl + mL $AgNO_3$ = 175.0 mL + 145.0 mL = 320.0. Then $M_1V_1 = M_2V_2$; rearrange to solve for M_2.

$M_2 = M_1\dfrac{V_1}{V_2} = 0.0055 \text{ M Cl}^- \times \dfrac{175.0 \text{ mL}}{320.0 \text{ mL}} = 0.00300781 \text{ M Cl}^-$ and

$M_2 = M_1\dfrac{V_1}{V_2} = 0.0015 \text{ M Ag}^+ \times \dfrac{145.0 \text{ mL}}{320.0 \text{ mL}} = 0.00067969 \text{ M Ag}^+$. Calculate $Q(AgCl)$, $A = Ag^+$, $m = 1$, $X = Cl^-$, and $n = 1$. Because $Q = [A^{n+}]^m[X^{m-}]^n$, $Q(AgCl) = [Ag^+][Cl^-] = (0.00067969)(0.00300781) = 2.0 \times 10^{-6} > 1.77 \times 10^{-10} = K_{sp}(AgCl)$; so a precipitate will form.

Check: The units (none) are correct. The solubility of the AgCl is low, and the concentrations of the ions are high enough that the product of the concentration of ions is very large compared to the K_{sp}; so a precipitate will form.

18.103 **Given:** KOH as precipitation agent in (a) 0.015 M $CaCl_2$, (b) 0.0025 M $Fe(NO_3)_2$, and (c) 0.0018 M $MgBr_2$
Find: concentration of KOH necessary to form a precipitate
Other: $K_{sp}(Ca(OH)_2) = 4.68 \times 10^{-6}$, $K_{sp}(Fe(OH)_2) = 4.87 \times 10^{-17}$, $K_{sp}(Mg(OH)_2) = 2.06 \times 10^{-13}$
Conceptual Plan: The solubility rules from Chapter 9 state that most hydroxides are insoluble, so all precipitates will be hydroxides. Determine the concentration of the cation in solution. Because all metals have an oxidation state of +2 and $[OH^-] = [KOH]$, all of the $K_{sp} = [\text{cation}][KOH]^2$; so $[KOH] = \sqrt{\dfrac{K_{sp}}{[\text{cation}]}}$.

Solution:

(a) $CaCl_2(s) \rightarrow Ca^{2+}(aq) + 2\,Cl^-(aq)$. Because one Ca^{2+} ion is generated for each $CaCl_2$, $[Ca^{2+}] = 0.015$ M.

Then $[KOH] = \sqrt{\dfrac{K_{sp}}{[\text{cation}]}} = \sqrt{\dfrac{4.68 \times 10^{-6}}{0.015}} = 0.018$ M KOH.

(b) $Fe(NO_3)_2(s) \rightarrow Fe^{2+}(aq) + 2\,NO_3^-(aq)$. Because one Fe^{2+} ion is generated for each $Fe(NO_3)_2$, $[Fe^{2+}] = 0.0025$ M.

Then $[KOH] = \sqrt{\dfrac{K_{sp}}{[\text{cation}]}} = \sqrt{\dfrac{4.87 \times 10^{-17}}{0.0025}} = 1.4 \times 10^{-7}$ M KOH.

(c) $MgBr_2(s) \rightarrow Mg^{2+}(aq) + 2\,Br^-(aq)$. Because one Mg^{2+} ion is generated for each $MgBr_2$, $[Mg^{2+}] = 0.0018$ M.

Then $[KOH] = \sqrt{\dfrac{K_{sp}}{[\text{cation}]}} = \sqrt{\dfrac{2.06 \times 10^{-13}}{0.0018}} = 1.1 \times 10^{-5}$ M KOH.

Check: The units (none) are correct. Because all cations have an oxidation state of +2, it can be seen that the [KOH] needed to precipitate the hydroxide is lower the smaller the K_{sp}.

18.104 **Given:** solution and precipitation agent pairs (a) 0.035 M $Ba(NO_3)_2$: NaF, (b) 0.085 M CaI_2: K_2SO_4, and (c) 0.0018 M $AgNO_3$: RbCl **Find:** concentration of precipitation agent necessary to form a precipitate
Other: $K_{sp}(BaF_2) = 2.45 \times 10^{-5}$, $K_{sp}(CaSO_4) = 7.10 \times 10^{-5}$, $K_{sp}(AgCl) = 1.77 \times 10^{-10}$
Conceptual Plan: Determine the concentration of the cation in solution. The solubility product constant (K_{sp}) is the equilibrium expression for a chemical equation representing the dissolution of an ionic compound. The expression of the solubility product constant of A_mX_n is $K_{sp} = [A^{n+}]^m[X^{m-}]^n$. Substitute concentration of cation and solve for concentration of anion.
Solution:

(a) The precipitate is BaF_2. $Ba(NO_3)_2(s) \rightarrow Ba^{2+}(aq) + 2\,NO_3^-(aq)$. Because one Ba^{2+} ion is generated for each $Ba(NO_3)_2$, $[Ba^{2+}] = 0.035$ M. Then derive expression for $K_{sp}(BaF_2)$, $A = Ba^{2+}$, $m = 1$, $X = F^-$, and $n = 2$. Because $K_{sp} = [Ba^{2+}][F^-]^2$, $K_{sp}(BaF_2) = 2.45 \times 10^{-5} = 0.035\,[F^-]^2$. Solve for $[F^-]$. $[F^-] = 0.026$ M F^-. Because $NaF(s) \rightarrow Na^+(aq) + F^-(aq)$, one F^- ion is generated for each NaF, $[NaF] = 0.026$ M NaF.

(b) The precipitate is $CaSO_4$. $CaI_2(s) \rightarrow Ca^{2+}(aq) + 2\,I^-(aq)$. Because one Ca^{2+} ion is generated for each CaI_2, $[Ca^{2+}] = 0.085$ M. Then derive expression for $K_{sp}(CaSO_4)$, $A = Ca^{2+}$, $m = 1$, $X = SO_4^{2-}$, and $n = 1$.

Because $K_{sp} = [Ca^{2+}][SO_4^{2-}]$, $K_{sp}(CaSO_4) = 7.10 \times 10^{-5} = 0.085\,[SO_4^{2-}]$. Solve for $[SO_4^{2-}]$. $[SO_4^{2-}] = 0.00084$ M SO_4^{2-}. Because $K_2SO_4(s) \rightarrow 2\,K^+(aq) + SO_4^{2-}(aq)$, one SO_4^{2-} ion is generated for each K_2SO_4, $[K_2SO_4] = 0.00084$ M K_2SO_4.

(c) The precipitate is AgCl. Then $AgNO_3(s) \rightarrow Ag^+(aq) + NO_3^-(aq)$. Because one NO_3^- ion is generated for each $AgNO_3$, $[Ag^+] = 0.0018$ M. Then derive expression for K_{sp} (AgCl), A = Ag^+, $m = 1$, X = Cl^-, and $n = 1$. Because $K_{sp} = [Ag^+][Cl^-]$, $K_{sp}(AgCl) = 1.77 \times 10^{-10} = 0.0018\,[Cl^-]$. Solve for $[Cl^-]$. $[Cl^-] = 9.8 \times 10^{-8}$ M Cl^-. Because $RbCl(s) \rightarrow Rb^+(aq) + Cl^-(aq)$, one Cl^- ion is generated for each RbCl, $[RbCl] = 9.8 \times 10^{-8}$ M RbCl.

Check: The units (M) are correct. Comparing part (a) and part (b), the effect of the stoichiometry of the precipitate is seen and the concentration of the precipitation agent is much lower. Looking at part (c), the concentration of the precipitation agent is so low because the K_{sp} is so small.

18.105 **Given:** solution with 0.010 M Ba^{2+} and 0.020 M Ca^{2+}, add Na_2SO_4 to form precipitates
Find: (a) which ion precipitates first and minimum $[Na_2SO_4]$ needed and (b) [first cation] when second cation precipitates **Other:** $K_{sp}(BaSO_4) = 1.07 \times 10^{-10}$, $K_{sp}(CaSO_4) = 7.10 \times 10^{-5}$
Conceptual Plan: (a) The precipitates that will form are $BaSO_4$ and $CaSO_4$.
Use the equation derived in Problem 18.19 to define K_{sp}. Substitute concentration of cation and solve for

for ionic compound, A_mX_n, $K_{sp} = [A^{n+}]^m[X^{m-}]^n$

concentration of anion to form precipitate. The cation with the lower anion concentration will precipitate first.

(b) **Substitute the higher anion concentration into the K_{sp} expression for the first cation to precipitate and calculate the amount of this first cation to remain in solution.**

Solution:

(a) Derive expression for $K_{sp}(BaSO_4)$, $A = Ba^{2+}$, $m = 1$, $X = SO_4^{2-}$, and $n = 1$ Because $K_{sp} = [Ba^{2+}][SO_4^{2-}]$, $K_{sp}(BaSO_4) = 1.07 \times 10^{-10} = 0.010\,[SO_4^{2-}]$. Solve for $[SO_4^{2-}]$. $[SO_4^{2-}] = 1.07 \times 10^{-8}$ M SO_4^{2-}. Because $Na_2SO_4(s) \rightarrow 2\,Na^+(aq) + SO_4^{2-}(aq)$. Because one SO_4^{2-} ion is generated for each Na_2SO_4, $[Na_2SO_4] = 1.1 \times 10^{-8}$ M Na_2SO_4 to precipitate $BaSO_4$. Derive expression for $K_{sp}(CaSO_4)$, A = Ca^{2+}, $m = 1$, X = SO_4^{2-}, and $n = 1$ Because $K_{sp} = [Ca^{2+}][SO_4^{2-}]$, $K_{sp}(CaSO_4) = 7.10 \times 10^{-5} = 0.020\,[SO_4^{2-}]$. Solve for $[SO_4^{2-}]$. $[SO_4^{2-}] = 0.0036$ M SO_4^{2-} = 0.0036 M Na_2SO_4 = $[Na_2SO_4]$ to precipitate $CaSO_4$. Because 1.1×10^{-8} M Na_2SO_4 << 0.0036 M Na_2SO_4, the Ba^{2+} will precipitate first.

(b) Because Ca^{2+} will not precipitate until $[Na_2SO_4] = 0.00355$ M Na_2SO_4, substitute this value into the K_{sp} expression for $BaSO_4$. So $K_{sp}(BaSO_4) = [Ba^{2+}][SO_4^{2-}] = 1.07 \times 10^{-10} = [Ba^{2+}]\,0.0035$. Solve for $[Ba^{2+}] = 3.0 \times 10^{-8}$ M Ba^{2+}.

Check: The units (none, M, and M) are correct. Comparing the two K_{sp} values, it can be seen that the Ba^{2+} will precipitate first because the solubility product is so much lower. Because the K_{sp} value is so low, the concentration of precipitating agent is very low. Because the $CaSO_4$ K_{sp} value is so much higher, the higher $[SO_4^{2-}]$ to precipitate Ca will force the concentration of Ba^{2+} to very low levels.

18.106 **Given:** solution with 0.022 M Fe^{2+} and 0.014 M Mg^{2+} add K_2CO_3 to form precipitates
Find: (a) which ion precipitates first and minimum $[K_2CO_3]$ needed; and (b) [first cation] when second cation precipitates **Other:** $K_{sp}(FeCO_3) = 3.07 \times 10^{-11}$, $K_{sp}(MgCO_3) = 6.82 \times 10^{-6}$
Conceptual Plan: (a) The precipitates that will form are $FeCO_3$ and $MgCO_3$.
Use the equation derived in Problem 18.19 to define K_{sp}. Substitute concentration of cation and solve for

for ionic compound, A_mX_n, $K_{sp} = [A^{n+}]^m[X^{m-}]^n$

concentration of anion to form precipitate. The cation with the lower anion concentration will precipitate first.

(b) **Substitute the higher anion concentration into the K_{sp} expression for the first cation to precipitate and calculate the amount of this first cation to remain in solution.**

Solution:

(a) Derive expression for $K_{sp}(FeCO_3)$, A = Fe^{2+}, $m = 1$, X = CO_3^{2-}, and $n = 1$. Because $K_{sp} = [Fe^{2+}][CO_3^{2-}]$, $K_{sp}(FeCO_3) = 3.07 \times 10^{-11} = 0.022\,[CO_3^{2-}]$. Solve for $[CO_3^{2-}]$. $[CO_3^{2-}] = 1.4 \times 10^{-9}$ M CO_3^{2-}. Because $K_2CO_3(s) \rightarrow 2\,K^+(aq) + CO_3^{2-}(aq)$. Because one CO_3^{2-} ion is generated for each K_2CO_3, $[K_2CO_3] = 1.4 \times 10^{-9}$ M K_2CO_3 to precipitate $FeCO_3$. Derive expression for $K_{sp}(MgCO_3)$, A = Mg^{2+}, $m = 1$, X = CO_3^{2-}, and $n = 1$. Because $K_{sp} = [Mg^{2+}][CO_3^{2-}]$, $K_{sp}(MgCO_3) = 6.82 \times 10^{-6}$ =

$0.014\,[CO_3^{2-}]$. Solve for $[CO_3^{2-}]$. $[CO_3^{2-}] = 4.9 \times 10^{-4}$ M $CO_3^{2-} = 4.9 \times 10^{-4}$ M $K_2CO_3 = [K_2CO_3]$ to precipitate $MgCO_3$. Because 1.4×10^{-9} M $K_2CO_3 < 4.9 \times 10^{-4}$ M K_2CO_3, the Fe^{2+} will precipitate first.

(b) Because Mg^{2+} will not precipitate until $[K_2CO_3] = 4.9 \times 10^{-4}$ M K_2CO_3, substitute this value into the K_{sp} expression for $FeCO_3$. So $K_{sp}(FeCO_3) = [Fe^{2+}][CO_3^{2-}] = 3.07 \times 10^{-11} = [Fe^{2+}]\,4.9 \times 10^{-4}$. Solve for $[Fe^{2+}] = 6.3 \times 10^{-8}$ M Fe^{2+}.

Check: The units (none, M, and M) are correct. Comparing the two K_{sp} values, it can be seen that the Fe^{2+} will precipitate first because the solubility product is so much lower. Because the K_{sp} value is so low, the concentration of precipitating agent is very low. Because the $MgCO_3$ K_{sp} value is so much higher, the higher $[CO_3^{2-}]$ to precipitate Mg will force the concentration of Fe^{2+} to very low levels.

Complex Ion Equilibria

18.107 **Given:** solution with 1.1×10^{-3} M $Zn(NO_3)_2$ and 0.150 M NH_3 **Find:** $[Zn^{2+}]$ at equilibrium
Other: $K_f(Zn(NH_3)_4^{2+}) = 2.8 \times 10^9$

Conceptual Plan: Write a balanced equation and expression for K_f. Use initial concentrations to set up an ICE table. Because the K_f is so large, assume that reaction essentially goes to completion. Solve for $[Zn^{2+}]$ at equilibrium.

Solution: $Zn(NO_3)_2(s) \rightarrow Zn^{2+}(aq) + 2\,NO_3^-(aq)$. Because one Zn^{2+} ion is generated for each $Zn(NO_3)_2$, $[Zn^{2+}] = 1.1 \times 10^{-3}$ M. Balanced equation is:

$$Zn^{2+}(aq) \quad + \quad 4\,NH_3(aq) \rightleftharpoons Zn(NH_3)_4^{2+}(aq)$$

	$[Zn^{2+}]$	$[NH_3]$	$[Zn(NH_3)_4^{2+}]$
Initial	1.1×10^{-3}	0.150	0.00
Change	$\approx 1.1 \times 10^{-3}$	$\approx -4(1.1 \times 10^{-3})$	$\approx 1.1 \times 10^{-3}$
Equil	x	$0.14\underline{5}6$	1.1×10^{-3}

Set up an ICE table with initial concentrations. Because K_f is so large and because initially $[NH_3] > 4\,[Zn^{2+}]$, the reaction essentially goes to completion.

Write equilibrium expression and solve for x.

$$K_f = \frac{[Zn(NH_3)_4^{2+}]}{[Zn^{2+}][NH_3]^4} = 2.8 \times 10^9 = \frac{1.1 \times 10^{-3}}{x\,(0.14\underline{5}6)^4}.$$ So $x = 8.7 \times 10^{-10}$ M Zn^{2+}. Because x is insignificant compared to the initial concentration of Zn^{2+} the assumption is valid.

Check: The units (M) are correct. Because K_f is so large, the reaction essentially goes to completion and $[Zn^{2+}]$ is extremely small.

18.108 **Given:** 120.0 mL of 2.8×10^{-3} M $AgNO_3$ mixed with 225.0 mL of 0.10 M NaCN **Find:** $[Ag^+]$ at equilibrium
Other: $K_f(Ag(CN)_2^-) = 1 \times 10^{21}$

Conceptual Plan: Mix solutions and calculate diluted concentrations mL $AgNO_3$, mL NaCN $\rightarrow$ mL total

mL $AgNO_3$ + mL NaCN = total mL

then mL, initial M $\rightarrow$ final M then write balanced equation and expression for K_f.

$M_1V_1 = M_2V_2$

Use initial concentrations to set up an ICE table. Because the K_f is so large, assume that reaction essentially goes to completion. Solve for $[Ag^+]$ at equilibrium.

Solution: $AgNO_3(s) \rightarrow Ag^+(aq) + NO_3^-(aq)$. Because one Ag^+ ion is generated for each $AgNO_3$, $[Ag^+] = 2.8 \times 10^{-3}$ M and $NaCN(s) \rightarrow Na^+(aq) + CN^-(aq)$. Because one CN^- ion is generated for each NaCN, $[CN^-] = 0.10$ M. Then total mL = mL $AgNO_3$ + mL NaCN = 120.0 mL + 225.0 mL = 345.0 mL. Then $M_1V_1 = M_2V_2$.

Rearrange to solve for M_2. $M_2 = M_1\dfrac{V_1}{V_2} = 2.8 \times 10^{-3}\text{ M Ag}^+ \times \dfrac{120.0\text{ mL}}{345.0\text{ mL}} = 0.0009\underline{7}391\text{ M Ag}^+$ and

$M_2 = M_1\dfrac{V_1}{V_2} = 0.10\text{ M CN}^- \times \dfrac{225.0\text{ mL}}{345.0\text{ mL}} = 0.06\underline{5}217\text{ M CN}^-$. The balanced equation is as follows:

	$Ag^{+}(aq)$ +	$2\,CN^{-}(aq)$ $\rightleftharpoons$	$Ag(CN)_2^{-}(aq)$	
	$[Ag^{+}]$	$[CN^{-}]$	$[Ag(CN)_2^{-}]$	
Initial	0.00097391	0.065217	0.00	Set up an ICE table with initial concentrations.
Change	$\approx$0.00097391	$\approx -2(0.00097391)$	$\approx$0.00097391	Because K_f is so large and because initially $[CN^{-}] > 2\,[Ag^{+}]$, the reaction essentially goes to completion.
Equil	x	0.063269	0.00097391	

Write equilibrium expression and solve for x.

$K_f = \dfrac{[Ag(CN)_2^{-}]}{[Ag^{+}][CN^{-}]^2} = 1 \times 10^{21} = \dfrac{0.00097391}{x\,(0.063269)^2}$. So $x = 2 \times 10^{-22}$ M Ag^{+}. Because x is insignificant compared to the initial concentration, the assumption is valid.

Check: The units (M) are correct. Because K_f is so large, the reaction essentially goes to completion and $[Ag^{+}]$ is extremely small.

18.109 **Given:** $FeS(s) + 6\,CN^{-}(aq) \rightleftharpoons Fe(CN)_6^{4-}(aq) + S^{2-}(aq)$ use K_{sp} and K_f values **Find:** K
Other: $K_f\,(Fe(CN)_6^{4-}) = 1.5 \times 10^{35}$, $K_{sp}\,(FeS) = 3.72 \times 10^{-19}$
Conceptual Plan: Identify the appropriate solid and complex ion. Write balanced equations for dissolving the solid and forming the complex ion. Add these two reactions to get the desired overall reaction. Using the rules from Chapter 16, multiply the individual reaction *K*s to get the overall *K* for the sum of these reactions.
Solution: Identify the solid as FeS and the complex ion as $Fe(CN)_6^{4-}$. Write the individual reactions and add them together.

$FeS(s) \rightleftharpoons Fe^{2+}(aq) + S^{2-}(aq)$ $\quad K_{sp} = 1.5 \times 10^{-19}$

$Fe^{2+}(aq) + 6\,CN^{-}(aq) \rightleftharpoons Fe(CN)_6^{4-}(aq)$ $\quad K_f = 3.72 \times 10^{35}$

$FeS(s) + 6\,CN^{-}(aq) \rightleftharpoons Fe(CN)_6^{4-}(aq) + S^{2-}(aq)$

Because the overall reaction is the simple sum of the two reactions, the overall reaction $K = K_f \times K_{sp} = (1.5 \times 10^{35}) \times (3.72 \times 10^{-19}) = 5.6 \times 10^{16}$.

Check: The units (none) are correct. Because K_f is so large, it overwhelms the K_{sp} and the overall reaction is very spontaneous.

18.110 **Given:** $PbCl_2(s) + 3\,OH^{-}(aq) \rightleftharpoons Pb(OH)_3^{-}(aq) + 2\,Cl^{-}(aq)$ use K_{sp} and K_f values **Find:** K
Other: $K_f\,(Pb(OH)_3^{-}) = 8 \times 10^{13}$, $K_{sp}\,(PbCl_2) = 1.17 \times 10^{-5}$
Conceptual Plan: Identify the appropriate solid and complex ion. Write balanced equations for dissolving the solid and forming the complex ion. Add these two reactions to get the desired overall reaction. Using the rules from Chapter 16, multiply the individual reaction *K*s to get the overall *K* for the sum of these reactions.
Solution: Identify the solid as $PbCl_2$ and the complex ion as $Pb(OH)_3^{-}$. Write the individual reactions and add them together.

$PbCl_2(s) \rightleftharpoons Pb^{2+}(aq) + 2\,Cl^{-}(aq)$ $\quad K_{sp} = 1.17 \times 10^{-5}$

$Pb^{2+}(aq) + 3\,OH^{-}(aq) \rightleftharpoons Pb(OH)_3^{-}(aq)$ $\quad K_f = 8 \times 10^{13}$

$PbCl_2(s) + 3\,OH^{-}(aq) \rightleftharpoons Pb(OH)_3^{-}(aq) + 2\,Cl^{-}(aq)$

Because the overall reaction is the simple sum of the two reactions, the overall reaction $K = K_f \times K_{sp} = (8 \times 10^{13}) \times (1.17 \times 10^{-5}) = 9 \times 10^{8}$.

Check: The units (none) are correct. Because K_f is so large, it overwhelms the K_{sp} and the overall reaction is very spontaneous.

Cumulative Problems

18.111 **Given:** 150.0 mL solution of 2.05 g sodium benzoate and 2.47 g benzoic acid **Find:** pH
Other: $K_a(HC_7H_5O_2) = 6.5 \times 10^{-5}$
Conceptual Plan: g $NaC_7H_5O_2 \rightarrow$ mol $NaC_7H_5O_2$ and g $HC_7H_5O_2 \rightarrow$ mol $HC_7H_5O_2$

$\dfrac{1 \text{ mol } NaC_7H_5O_2}{144.10 \text{ g } NaC_7H_5O_2}$ $\qquad$ $\dfrac{1 \text{ mol } HC_7H_5O_2}{122.12 \text{ g } HC_7H_5O_2}$

Because the two components are in the same solution, the ratio of [base]/[acid] = (mol base)/(mol acid).

Then K_a, mol $NaC_7H_5O_2$, mol $HC_7H_5O_2 \rightarrow$ pH.

$$pH = pK_a + \log\frac{[base]}{[acid]}$$

Solution: $2.05\ \cancel{g\ NaC_7H_5O_2} \times \frac{1\ mol\ NaC_7H_5O_2}{144.10\ \cancel{g\ NaC_7H_5O_2}} = 0.01422262\ mol\ NaC_7H_5O_2$ and

$$2.47\ \cancel{g\ HC_7H_5O_2} \times \frac{1\ mol\ HC_7H_5O_2}{122.12\ \cancel{g\ HC_7H_5O_2}} = 0.02022260\ mol\ HC_7H_5O_2 \text{ then}$$

$$pH = pK_a + \log\frac{[base]}{[acid]} = pK_a + \log\frac{mol\ base}{mol\ acid} = -\log(6.5 \times 10^{-5}) + \log\frac{0.01422262\ \cancel{mol}}{0.02022260\ \cancel{mol}} = 4.03$$

Check: The units (none) are correct. The magnitude of the answer makes physical sense because the pH is a little lower than the pK_a of the acid because there is more acid than base in the buffer solution.

18.112 **Given:** 10.0 mL of 17.5 M acetic acid and 5.54 g sodium acetate diluted to 1.50 L **Find:** pH
Other: $K_a(HC_2H_3O_2) = 1.8 \times 10^{-5}$
Conceptual Plan: mL $\rightarrow$ L then L, initial $HC_2H_3O_2$ M $\rightarrow$ mol $HC_2H_3O_2$ then

$$\frac{1\ L}{1000\ mL} \qquad M = \frac{mol}{L}$$

g $NaC_2H_3O_2 \rightarrow$ mol $NaC_2H_3O_2$ then because the two components are in the same solution,

$$\frac{1\ mol\ NaC_2H_3O_2}{82.03\ g\ NaC_2H_3O_2}$$

the ratio of [base]/[acid] = (mol base)/(mol acid). Then K_a, mol $NaC_2H_3O_2$ mol $HC_2H_3O_2 \rightarrow$ pH.

$$pH = pK_a + \log\frac{[base]}{[acid]}$$

Solution: $10.0\ \cancel{mL} \times \frac{1\ L}{1000\ \cancel{mL}} = 0.0100\ L$ then

$$0.0100\ \cancel{L\ HC_2H_3O_2} \times \frac{17.5\ mol\ HC_2H_3O_2}{1\ \cancel{L\ HC_2H_3O_2}} = 0.175\ mol\ HC_2H_3O_2 \text{ then}$$

$$5.54\ \cancel{g\ NaC_2H_3O_2} \times \frac{1\ mol\ NaC_2H_3O_2}{82.03\ \cancel{g\ NaC_2H_3O_2}} = 0.06753627\ mol\ NaC_2H_3O_2 \text{ then}$$

$$pH = pK_a + \log\frac{[base]}{[acid]} = pK_a + \log\frac{mol\ base}{mol\ acid} = -\log(1.8 \times 10^{-5}) + \log\frac{0.06753627\ \cancel{mol}}{0.175\ \cancel{mol}} = 4.33$$

Check: The units (none) are correct. The magnitude of the answer makes physical sense because the pH is a little lower than the pK_a of the acid because there is more acid than base in the buffer solution.

18.113 **Given:** 150.0 mL of 0.25 M $HCHO_2$ and 75.0 ml of 0.20 M NaOH **Find:** pH
Other: $K_a(HCHO_2) = 1.8 \times 10^{-4}$
Conceptual Plan: In this buffer, the base is generated by converting some of the formic acid to the formate ion.
Part I: Stoichiometry:
mL $\rightarrow$ L then L, initial $HCHO_2$ M $\rightarrow$ mol $HCHO_2$ then mL $\rightarrow$ L then

$$\frac{1\ L}{1000\ mL} \qquad M = \frac{mol}{L} \qquad \frac{1\ L}{1000\ mL}$$

L, initial NaOH M $\rightarrow$ mol NaOH then write a balanced equation then

$$M = \frac{mol}{L} \qquad NaOH + HCHO_2 \rightarrow H_2O + NaCHO_2$$

mol $HCHO_2$, mol NaOH $\rightarrow$ mol $NaCHO_2$, mol $HCHO_2$ then

set up stoichiometry table

Part II: Equilibrium:
Because the two components are in the same solution, the ratio of [base]/[acid] = (mol base)/(mol acid). Then K_a, mol $NaCHO_2$, mol $HCHO_2 \rightarrow$ pH.

$$pH = pK_a + \log\frac{[base]}{[acid]}$$

Solution: $150.0\ \cancel{mL} \times \frac{1\ L}{1000\ \cancel{mL}} = 0.1500\ L$ then

$0.1500 \text{ L HCHO}_2 \times \frac{0.25 \text{ mol HCHO}_2}{1 \text{ L HCHO}_2} = 0.03\underline{7}5 \text{ mol HCHO}_2$

Then $75.0 \text{ mL} \times \frac{1 \text{ L}}{1000 \text{ mL}} = 0.0750 \text{ L}$ then $0.0750 \text{ L NaOH} \times \frac{0.20 \text{ mol NaOH}}{1 \text{ L NaOH}} = 0.015 \text{ mol NaOH}$

Set up a table to track changes:

	$NaOH(aq)$	+	$HCHO_2(aq)$	→	$NaCHO_2(aq)$	+	$H_2O(l)$
Before addition	0.00 mol		$0.03\underline{7}5$ mol		0.00 mol		—
Addition	0.015 mol		—		—		—
After addition	≈0.00 mol		$0.02\underline{2}5$ mol		0.015 mol		—

Because the amount of NaOH is small, there are significant amounts of both buffer components; so the Henderson–Hasselbalch equation can be used to calculate the pH.

$$pH = pK_a + \log\frac{[\text{base}]}{[\text{acid}]} = pK_a + \log\frac{\text{mol base}}{\text{mol acid}} = -\log(1.8 \times 10^{-4}) + \log\frac{0.015 \text{ mol}}{0.02\underline{2}5 \text{ mol}} = 3.57$$

Check: The units (none) are correct. The magnitude of the answer makes physical sense because the pH is a little lower than the pK_a of the acid because there is more acid than base in the buffer solution.

18.114 **Given:** 750.0 mL solution of 3.55 g NH_3 and 4.78 g HCl **Find:** pH
Other: $K_b(NH_3) = 1.76 \times 10^{-5}$
Conceptual Plan: In this buffer, the acid is generated by converting some of the ammonia to the ammonium ion. Part I: Stoichiometry:
g NH_3 → mol NH_3 and g HCl → mol HCl write a balanced equation then

$\frac{1 \text{ mol NH}_3}{17.03 \text{ g NH}_3}$ $\frac{1 \text{ mol HCl}}{36.46 \text{ g HCl}}$ $NH_3 + HCl \rightarrow NH_4Cl$

mol NH_3, mol HCl → mol NH_3, mol NH_4Cl then

set up stoichiometry table

Part II: Equilibrium:
K_b → pK_b → pK_a then because the two components are in the same solution,

$pK_b = -\log K_b$ $14 = pK_a + pK_b$

the ratio of [base]/[acid] = (mol base)/(mol acid). Then pK_a, mol NH_3, mol NH_4Cl → pH.

$pH = pK_a + \log\frac{[\text{base}]}{[\text{acid}]}$

Solution: $3.55 \text{ g NH}_3 \times \frac{1 \text{ mol NH}_3}{17.03 \text{ g NH}_3} = 0.20\underline{8}456 \text{ mol NH}_3$ and

$4.78 \text{ g HCl} \times \frac{1 \text{ mol HCl}}{36.46 \text{ g HCl}} = 0.13\underline{1}103 \text{ mol HCl}$. Set up a table to track changes:

	$HCl(aq)$	+	$NH_3(aq)$	→	$NH_4Cl(aq)$
Before addition	0.00 mol		$0.20\underline{8}456$ mol		0.00 mol
Addition	$0.13\underline{1}103$ mol		—		—
After addition	≈0.00 mol		$0.07\underline{7}353$ mol		$0.13\underline{1}103$ mol

Because the amount of HCl is small, there are significant amounts of both buffer components; so the Henderson–Hasselbalch equation can be used to calculate the pH.
Because $K_b(NH_3) = 1.79 \times 10^{-5}$, $pK_b = -\log K_b = -\log(1.76 \times 10^{-5}) = 4.75$. Because $14 = pK_a + pK_b$, $pK_a = 14 - pK_b = 14 - 4.75 = 9.25$ then

$$pH = pK_a + \log\frac{[\text{base}]}{[\text{acid}]} = pK_a + \log\frac{\text{mol base}}{\text{mol acid}} = 9.25 + \log\frac{0.07\underline{7}353 \text{ mol}}{0.13\underline{1}103 \text{ mol}} = 9.02$$

Check: The units (none) are correct. The magnitude of the answer makes physical sense because the pH is a little lower than the pK_a of the acid because there is more acid than base in the buffer solution.

18.115 **Given:** 1.0 L of buffer of 0.25 mol NH_3 and 0.25 mol NH_4Cl; adjust to pH = 8.75
Find: mass NaOH or HCl **Other:** $K_b(NH_3) = 1.76 \times 10^{-5}$

Conceptual Plan: To decide which reagent needs to be added to adjust pH, calculate the initial pH. Because the mol NH_3 = mol NH_4Cl, the pH = pK_a so $K_b \rightarrow pK_b \rightarrow pK_a$ then

acid = NH_4^+ base = NH_3 $\quad$ $pK_b = -\log K_b$ $\quad$ $14 = pK_a + pK_b$

final pH, pK_a $\rightarrow$ $[NH_3]/[NH_4^+]$ then $[NH_3]$, L $\rightarrow$ mol $[NH_3]$ and $[NH_4^+]$, L $\rightarrow$ mol $[NH_4^+]$

$pH = pK_a + \log\frac{[base]}{[acid]}$ $\quad$ $M = \frac{mol}{L}$ $\quad$ $M = \frac{mol}{L}$

then write a balanced equation then

$H^+ + NH_3 \rightarrow NH_4^+$

mol NH_3, mol NH_4^+, $[NH_3]/[NH_4^+]$ $\rightarrow$ mol HCl $\rightarrow$ g HCl.

set up stoichiometry table $\quad$ $\frac{36.46 \text{ g HCl}}{1 \text{ mol HCl}}$

Solution: Because $K_b(NH_3) = 1.76 \times 10^{-5}$, $pK_b = -\log K_b = -\log(1.76 \times 10^{-5}) = 4.75$. Because $14 = pK_a + pK_b$, $pK_a = 14 - pK_b = 14 - 4.75 = 9.25$. Because the desired pH is lower (8.75), HCl (a strong acid) needs to be added. Then $pH = pK_a + \log\frac{[base]}{[acid]} = 9.25 + \log\frac{[NH_3]}{[NH_4^+]} = 8.75$. Solve for $\frac{[NH_3]}{[NH_4^+]}$.

$\log\frac{[NH_3]}{[NH_4^+]} = 8.75 - 9.25 = -0.50 \rightarrow \frac{[NH_3]}{[NH_4^+]} = 10^{-0.50} = 0.3\underline{1}623$ then

$\frac{0.25 \text{ mol } NH_3}{1 \cancel{L}} \times 1.0 \cancel{L} = 0.25 \text{ mol } NH_3$ and

$\frac{0.25 \text{ mol } NH_4Cl}{1 \cancel{L}} \times 1.0 \cancel{L} = 0.25 \text{ mol } NH_4Cl = 0.25 \text{ mol } NH_4^+$. Because HCl is a strong acid, $[HCl] = [H^+]$.

Set up a table to track changes:

	$H^+(aq)$	+	$NH_3(aq)$	$\rightarrow$	$NH_4^+(aq)$
Before addition	$\approx$0.00 mol		0.25 mol		0.25 mol
Addition	x		—		—
After addition	$\approx$0.00 mol		$(0.25 - x)$ mol		$(0.25 + x)$ mol

Because $\frac{[NH_3]}{[NH_4^+]} = 0.3\underline{1}623 = \frac{(0.25 - x)\cancel{mol}}{(0.25 + x)\cancel{mol}}$, solve for x. Note that the ratio of moles is the same as the ratio of concentrations because the volume for both terms is the same. $0.3\underline{1}623(0.25 + x) = (0.25 - x) \rightarrow 0.07\underline{9}0575 + 0.3\underline{1}623x = 0.25 - x \rightarrow 1.3\underline{1}623x = 0.1\underline{7}094 \rightarrow x = 0.1\underline{2}987$ mol HCl then

$0.1\underline{2}987 \cancel{\text{mol HCl}} \times \frac{36.46 \text{ g HCl}}{1 \cancel{\text{mol HCl}}} = 4.7 \text{ g HCl}$

Check: The units (g) are correct. The magnitude of the answer makes physical sense because there is much less than a mole of each of the buffer components; so there must be much less than a mole of HCl.

18.116 **Given:** 250.0 mL of buffer of 0.025 mol $HCHO_2$ and 0.025 mol $NaCHO_2$; adjust to pH = 4.10
Find: mass NaOH or HCl **Other:** $K_a(HCHO_2) = 1.8 \times 10^{-4}$
Conceptual Plan: To decide which reagent needs to be added to adjust pH, calculate the initial pH. Because the mol $HCHO_2$ = mol $NaCHO_2$, the pH = pK_a then final pH, pK_a $\rightarrow$ $[NaCHO_2]/[HCHO_2]$

acid = $HCHO_2$ base = $HCHO_2^-$ $\quad$ $pK_a = -\log K_a$ $\quad$ $pH = pK_a + \log\frac{[base]}{[acid]}$

then mL $\rightarrow$ L then write a balanced equation then

$\frac{1 \text{ L}}{1000 \text{ mL}}$ $\quad$ $NaOH + HCHO_2 \rightarrow NaCHO_2 + H_2O$

mol $NaCHO_2$, mol $HCHO_2$, $[NaCHO_2]/[HCHO_2]$ $\rightarrow$ mol NaOH $\rightarrow$ g NaOH.

set up stoichiometry table $\quad$ $\frac{40.00 \text{ g NaOH}}{1 \text{ mol NaOH}}$

Solution: Because $K_a(HCHO_2) = 1.8 \times 10^{-4}$, $pK_a = -\log K_a = -\log(1.8 \times 10^{-4}) = 3.74$. Because the desired pH (4.10) is higher, NaOH (a strong base) needs to be added. Then

$pH = pK_a + \log\frac{[base]}{[acid]} = 3.74 + \log\frac{[NaCHO_2]}{[HCHO_2]} = 4.10$. Solve for $\frac{[NaCHO_2]}{[HCHO_2]}$.

$\log \frac{[NaCHO_2]}{[HCHO_2]} = 4.10 - 3.74 = 0.36 \rightarrow \frac{[NaCHO_2]}{[HCHO_2]} = 10^{+0.36} = 2.2\underline{9}087$. Because NaOH is a strong base, $[NaOH] = [OH^-]$. Set up a table to track changes:

	$NaOH(aq)$	+	$HCHO_2(aq)$	$\rightarrow$	$NaCHO_2(aq)$	+	$H_2O(l)$
Before addition	≈ 0.00 mol		0.025 mol		0.025 mol		—
Addition	$-x$		$-x$		$+x$		—
After addition	≈ 0.00 mol		$(0.025 - x)$ mol		$(0.025 + x)$ mol		—

Because $pH = pK_a + \log \frac{[A^-]}{[HA]}$, we get $4.10 = 3.74 + \log \frac{0.025 \text{ mol} + x}{0.025 \text{ mol} - x}$. Solve for x. Note that the ratio of moles is the same as the ratio of concentrations because the volume for both terms is the same.

$0.360 = \log \frac{0.025 \text{ mol} + x}{0.025 \text{ mol} - x} \rightarrow 2.29 = \frac{0.025 \text{ mol} + x}{0.025 \text{ mol} - x} \rightarrow 0.05\underline{7}3 - 2.29x = 0.025 + x \rightarrow 3.29x = 0.03\underline{2}3 \rightarrow$

$x = 0.00\underline{9}81$ mol then $0.00981 \text{ mol NaOH} \times \frac{40.00 \text{ g NaOH}}{1 \text{ mol NaOH}} = 0.39$ g NaOH

Check: The units (g) are correct. The magnitude of the answer makes physical sense because there is much less than a mole of each of the buffer components; so there must be much less than a mole of NaOH.

18.117 (a) **Given:** potassium hydrogen phthalate = KHP = $KHC_8H_4O_4$ titration with NaOH **Find:** balanced equation
Conceptual Plan: The reaction will be a titration of the acid proton, leaving the phthalate ion intact. The K will not be titrated because it is basic.
Solution: $NaOH(aq) + KHC_8H_4O_4(aq) \rightarrow Na^+(aq) + K^+(aq) + C_8H_4O_4^{2-}(aq) + H_2O(l)$

Check: An acid–base reaction generates a salt (soluble here) and water. There is only one acidic proton in KHP.

(b) **Given:** 0.5527 g KHP titrated with 25.87 mL of NaOH solution **Find:** [NaOH]
Conceptual Plan:
g KHP → mol KHP → mol NaOH and mL → L then mol NaOH and mL → M NaOH

$\frac{1 \text{ mol KHP}}{204.22 \text{ g KHP}}$ 1:1 from balanced equation $\frac{1 \text{ L}}{1000 \text{ mL}}$ $M = \frac{\text{mol}}{\text{L}}$

Solution: $0.5527 \text{ g KHP} \times \frac{1 \text{ mol KHP}}{204.22 \text{ g KHP}} = 0.00270\underline{6}395$ mol KHP; mol KHP = mol acid = mol base = $0.00270\underline{6}395$ mol NaOH then $25.87 \text{ mL} \times \frac{1 \text{ L}}{1000 \text{ mL}} = 0.02587$ L then

$[NaOH] = \frac{0.00270\underline{6}395 \text{ mol NaOH}}{0.02587 \text{ L}} = 0.1046$ M NaOH

Check: The units (M) are correct. The magnitude of the answer makes physical sense because there is much less than a mole of acid. The magnitudes of the moles of acid and base are smaller than the volume of base in liters.

18.118 **Given:** 0.5224 g monoprotic acid titrated with 23.82 mL of 0.0998 M NaOH solution **Find:** molar mass of acid
Conceptual Plan: mL → L then M NaOH, L → mol NaOH → mol acid then mol acid, g acid → $\mathcal{M}$

$\frac{1 \text{ L}}{1000 \text{ mL}}$ $M = \frac{\text{mol}}{\text{L}}$ 1:1 for monoprotic acid $\mathcal{M} = \frac{\text{g acid}}{\text{mol acid}}$

Solution: $23.82 \text{ mL} \times \frac{1 \text{ L}}{1000 \text{ mL}} = 0.02382$ L then

$0.02382 \text{ L NaOH} \times \frac{0.0998 \text{ mol NaOH}}{1 \text{ L NaOH}} = 0.0023\underline{7}724$ mol NaOH; $0.0023\underline{7}724$ mol NaOH = mol base =

mol acid = $0.0023\underline{7}724$ mol acid then $\mathcal{M} = \frac{\text{g acid}}{\text{mol acid}} = \frac{0.5224 \text{ g acid}}{0.0023\underline{7}724 \text{ mol acid}} = 220.$ g/mol

Check: The units (g/mol) are correct. The magnitude of the answer makes physical sense because there is much less than a mole of acid and about half a gram of acid; so the molar mass will be high. The number is reasonable for an acid (must be >20 g/mol—lightest acid is HF).

18.119 **Given:** 0.25 mol weak acid with 10.0 mL of 3.00 M KOH diluted to 1.5000 L has pH = 3.85 **Find:** pK_a of acid

Conceptual Plan: mL → L then M KOH, L → mol KOH then write a balanced reaction

$\frac{1\ L}{1000\ mL}$ $\quad M = \frac{mol}{L}$ $\quad KOH + HA \rightarrow NaA + H_2O$

added mol KOH, initial mol acid → equil. mol KOH, equil. mol acid then

set up stoichiometry table

equil. mol KOH, equil. mol acid, pH → pK_a

$$pH = pK_a + \log\frac{[base]}{[acid]}$$

Solution: $10.00\ \cancel{mL} \times \frac{1\ L}{1000\ \cancel{mL}} = 0.01000\ L$ then $0.01000\ \cancel{L\ KOH} \times \frac{3.00\ mol\ KOH}{1\ \cancel{L\ KOH}} = 0.0300\ mol\ KOH$

Because KOH is a strong base, $[KOH] = [OH^-]$. Set up a table to track changes:

	$KOH(aq)$ +	$HA(aq)$ →	$KA(aq)$ +	$H_2O(l)$
Before addition	≈0.00 mol	0.25 mol	0.00 mol	—
Addition	0.0300 mol	—	—	—
After addition	≈0.00 mol	0.22 mol	0.0300 mol	—

Because the ratio of base to acid is between 0.1 and 10, it is a buffer solution. Note that the ratio of moles is the same as the ratio of concentrations because the volume for both terms is the same.

$$pH = pK_a + \log\frac{[base]}{[acid]} = pK_a + \log\frac{0.0300\ mol}{0.22\ mol} = 3.85 \text{ Solve for } pK_a.$$

$$pK_a = 3.85 - \log\frac{0.0300\ mol}{0.22\ mol} = 4.72$$

Check: The units (none) are correct. The magnitude of the answer makes physical sense because there is more acid than base at equilibrium; so the pK_a is higher than the pH of the solution.

18.120 **Given:** 5.55 g weak acid with $K_a = 1.3 \times 10^{-4}$ with 5.00 mL of 6.00 M NaOH diluted to 750 mL has pH = 4.25 **Find:** molar mass of acid

Conceptual Plan: mL → L then M NaOH, L → mol NaOH then write a balanced reaction

$\frac{1\ L}{1000\ mL}$ $\quad M = \frac{mol}{L}$ $\quad NaOH + HA \rightarrow NaA + H_2O$

added mol NaOH, initial mol acid → equil. mol NaOH, equil. mol acid then

set up stoichiometry table

added mol NaOH, equil. mol acid, pH, pK_a → equil. mol NaOH, equil. mol acid then

$$pH = pK_a + \log\frac{[base]}{[acid]}$$

mol acid, g acid → $\mathcal{M}$

$$\mathcal{M} = \frac{g\ acid}{mol\ acid}$$

Solution: $5.00\ \cancel{mL} \times \frac{1\ L}{1000\ \cancel{mL}} = 0.00500\ L$ then $0.00500\ \cancel{L\ NaOH} \times \frac{6.00\ mol\ NaOH}{1\ \cancel{L\ NaOH}} = 0.0300\ mol\ NaOH$

Because NaOH is a strong base, $[NaOH] = [OH^-]$. Set up a table to track changes:

	$NaOH(aq)$ +	$HA(aq)$ →	$NaA(aq)$ +	$H_2O(l)$
Before addition	≈0.00 mol	x mol	0.00 mol	—
Addition	0.0300 mol	—	—	—
After addition	≈0.00 mol	$x - 0.0300$ mol	0.0300 mol	—

Because the pH is within 1 unit of the pK_a, it is a buffer solution. Note that the ratio of moles is the same as the ratio of concentrations because the volume for both terms is the same.

$$pH = pK_a + \log\frac{[base]}{[acid]} = -\log(1.3 \times 10^{-4}) + \log\frac{0.0300\ mol}{(x - 0.0300)\ mol} = 4.25$$

Solve for x. $\log 0.0300 - \log(x - 0.0300) = 4.25 - 3.89 \rightarrow -\log(x - 0.0300) = 1.8\underline{8}288 \rightarrow$ $x - 0.0300 = 10^{-1.8\underline{8}288} = 0.013\underline{0}954 \rightarrow x = 0.043\underline{0}854$ mol. Finally,

$$\mathcal{M} = \frac{\text{g acid}}{\text{mol acid}} = \frac{5.55 \text{ g acid}}{0.0430854 \text{ mol acid}} = 129 \text{ g/mol}.$$

Check: The units (g/mol) are correct. The magnitude of the answer makes physical sense because there is much less than a mole of acid and about 6 grams of acid; so the molar mass will be high. The number is reasonable for an acid (must be > 20 g/mol—lightest acid is HF).

18.121 **Given:** 0.552 g ascorbic acid dissolved in 20.00 mL and titrated with 28.42 mL of 0.1103 M KOH solution; pH = 3.72 when 10.0 mL KOH added **Find:** molar mass and K_a of acid

Conceptual Plan: mL → L then M KOH, L → mol KOH → mol acid then mol acid, g acid → $\mathcal{M}$

$\frac{1 \text{ L}}{1000 \text{ mL}}$ $\quad$ $\text{M} = \frac{\text{mol}}{\text{L}}$ $\quad$ 1:1 for monoprotic acid $\quad$ $\mathcal{M} = \frac{\text{g acid}}{\text{mol acid}}$

For the second part of the problem:

mL → L then M KOH, L → mol KOH then write a balanced reaction

$\frac{1 \text{ L}}{1000 \text{ mL}}$ $\quad$ $\text{M} = \frac{\text{mol}}{\text{L}}$ $\quad$ $\text{KOH} + \text{HA} \rightarrow \text{NaA} + \text{H}_2\text{O}$

added mol KOH, initial mol acid → equil. mol KOH, equil. mol acid then

set up stoichiometry table

equil. mol KOH, equil. mol acid, pH → pK_a → K_a.

$\text{pH} = \text{p}K_a + \log\frac{[\text{base}]}{[\text{acid}]}$ $\quad$ $\text{p}K_a = -\log K_a$

Solution: $28.42 \text{ mL} \times \frac{1 \text{ L}}{1000 \text{ mL}} = 0.02842 \text{ L}$ then

$$0.02842 \text{ L KOH} \times \frac{0.1103 \text{ mol KOH}}{1 \text{ L KOH}} = 0.003134726 \text{ mol KOH} = \text{mol base} = \text{mol acid} = 0.003134726 \text{ mol}$$

ascorbic acid then $\mathcal{M} = \frac{\text{g acid}}{\text{mol acid}} = \frac{0.552 \text{ g acid}}{0.003134726 \text{ mol acid}} = 176 \text{ g/mol}$. For the second part of the problem,

$$10.00 \text{ mL} \times \frac{1 \text{ L}}{1000 \text{ mL}} = 0.01000 \text{ L then } 0.01000 \text{ L KOH} \times \frac{0.1103 \text{ mol KOH}}{1 \text{ L KOH}} = 0.001103 \text{ mol KOH}$$

Because KOH is a strong base, $[\text{KOH}] = [\text{OH}^-]$. Set up a table to track changes:

	$\text{KOH}(aq)$ +	$\text{HA}(aq)$ →	$\text{KA}(aq)$ +	$\text{H}_2\text{O}(l)$
Before addition	≈0.00 mol	0.003134726 mol	0.00 mol	—
Addition	0.001103 mol	—	—	—
After addition	≈0.00 mol	0.002031726 mol	0.001103 mol	—

Because the ratio of base to acid is between 0.1 and 10, it is a buffer solution. Note that the ratio of moles is the same as the ratio of concentrations because the volume for both terms is the same.

$$\text{pH} = \text{p}K_a + \log\frac{[\text{base}]}{[\text{acid}]} = \text{p}K_a + \log\frac{0.001103 \text{ mol}}{0.002031726 \text{ mol}} = 3.72. \text{ Solve for p}K_a.$$

$$\text{p}K_a = 3.72 - \log\frac{0.001103}{0.002031726} = 3.98529 \text{ and so p}K_a = -\log K_a$$

or $K_a = 10^{-\text{p}K_a} = 10^{-3.98529} = 1.0 \times 10^{-4}$.

Check: The units (g/mol and none) are correct. The magnitude of the answer makes physical sense because there is much less than a mole of acid and about half a gram of acid; so the molar mass will be high. The number is reasonable for an acid (must be >20 g/mol—lightest acid is HF). The K_a is reasonable because the pK_a is within 1 unit of the pH when the titration solution is behaving as a buffer.

18.122 **Given:** titration data in Problem 18.121 **Find:** initial pH

Conceptual Plan: Because ascorbic acid is a weak acid, set up an equilibrium problem using the initial concentration so mL → L then mol HA, L → M HA → $[\text{H}_3\text{O}^+]$ → pH.

$\frac{1 \text{ L}}{1000 \text{ mL}}$ $\quad$ $\text{M} = \frac{\text{mol}}{\text{L}}$ $\quad$ ICE table $\quad$ $\text{pH} = -\log[\text{H}_3\text{O}^+]$

Solution: $20.00\ \cancel{mL} \times \frac{1\ L}{1000\ \cancel{mL}} = 0.02000\ L$ then $\frac{0.003134726\ mol\ HA}{0.02000\ L\ HA} = 0.1567363\ M\ HA$ then

$$HA(aq) + H_2O(l) \rightleftharpoons H_3O^+(aq) + A^-(aq)$$

	$[HA]$	$[H_3O^+]$	$[A^-]$
Initial	0.1567363	≈0.00	0.00
Change	$-x$	$+x$	$+x$
Equil	$0.1567363 - x$	$+x$	$+x$

and $K_a = \frac{[H_3O^+][A^-]}{[HA]} = 1.0 \times 10^{-4} = \frac{x^2}{0.1567363 - x}$

Assume that x is small ($x << 0.1567$), so $\frac{x^2}{0.1567363 - \cancel{x}} = 1.0 \times 10^{-4} = \frac{x^2}{0.1567363}$ and $x = 3.9590 \times 10^{-3} =$ $[H_3O^+]$. Confirm that the assumption is valid. $\frac{3.9590 \times 10^{-3}}{0.1567363} \times 100\% = 2.5\% < 5\%$, so the assumption is valid.

Finally, $pH = -\log[H_3O^+] = -\log(3.9590 \times 10^{-3}) = 2.40$

Check: The units (none) are correct. The magnitude of the answer makes physical sense because pH should be greater than $-\log(0.1567) = 0.80$ because this is a weak acid.

Find: pH at one-half the equivalence point

Conceptual Plan: Because this is a weak acid–strong base titration, the pH at one-half the equivalence point is the pK_a of the weak acid.

Solution: $pH = pK_a = -\log K_a = -\log(1.0 \times 10^{-4}) = 4.00$ and the volume of added base is $0.5 \times 28.42\ mL = 14.21\ mL$.

Check: The units (none) are correct. Because this is a weak acid–strong base titration, the pH at one-half the equivalence point is the pK_a of the weak acid; so it should be 4.

Find: pH at equivalence point

Conceptual Plan: Use the preceding calculations. Then since all of the weak acid has been converted to its conjugate base, the pH is only dependent on the hydrolysis reaction of the conjugate base. The mol A^- = initial mol HA and L HA, L KOH to equiv. pt. → total L then mol A^-, L → $[A^-]$

$L\ HA + L\ KOH = total\ L$ $\quad\quad$ $M = \frac{mol}{L}$

and $K_a \rightarrow K_b$ then do equilibrium calculation: $[A], K_b \rightarrow [OH^-] \rightarrow [H_3O^+] \rightarrow pH$.

$K_w = K_a K_b$ $\quad\quad$ set up ICE table $\quad$ $K_w = [H_3O^+][OH^-]$ $\quad$ $pH = -\log[H_3O^+]$

Solution: total volume $= L\ HA + L\ KOH = 0.0200\ L + 0.02842\ L = 0.04842\ L$ then

$[A^-] = \frac{0.003134726\ mol\ A^-}{0.04842\ L} = 0.0647403\ M$ and $K_w = K_a K_b$. Rearrange to solve for K_b.

$K_b = \frac{K_w}{K_a} = \frac{1.0 \times 10^{-14}}{1.0 \times 10^{-4}} = 1.0 \times 10^{-10}$. Set up an ICE table:

$$A^-(aq) + H_2O(l) \rightleftharpoons HA(aq) + OH^-(aq)$$

	$[A^-]$	$[HA]$	$[OH^-]$
Initial	0.0647403	≈0.00	≈0.00
Change	$-x$	$+x$	$+x$
Equil	$0.0647403 - x$	$+x$	$+x$

$K_b = \frac{[HA][OH^-]}{[A^-]} = 1.0 \times 10^{-10} = \frac{x^2}{0.0647403 - x}$ Assume that x is small ($x << 0.0647$), so

$\frac{x^2}{0.0647403 - \cancel{x}} = 1.0 \times 10^{-10} = \frac{x^2}{0.0647403}$ and $x = 2.5444 \times 10^{-6} = [OH^-]$. Confirm that the assumption is valid.

$\frac{2.5444 \times 10^{-6}}{0.0647403} \times 100\% = 0.0039\% < 5\%$, so the assumption is valid. $K_w = [H_3O^+][OH^-]$ so

$[H_3O^+] = \frac{K_w}{[OH^-]} = \frac{1.0 \times 10^{-14}}{2.5444 \times 10^{-6}} = 3.9302 \times 10^{-9}$ M. Finally,

$pH = -\log[H_3O^+] = -\log(3.9302 \times 10^{-9}) = 8.41.$

Check: The units (none) are correct. Because this is a weak acid–strong base titration, the pH at the equivalence point is basic.

Find: pH after adding 5.0 mL of excess base

Conceptual Plan: Use the preceding calculations. Then the pH is only dependent on the amount of excess base and the total solution volumes. mL excess → L excess then [KOH], L excess → mol KOH excess

$\frac{1\ L}{1000\ mL}$ $\qquad M = \frac{mol}{L}$

then L HA, L KOH to equiv. pt. L KOH excess → total L then

L HA + L NaOH to equiv. pt. + L NaOH excess = total L

mol excess KOH, total L → [KOH] = [OH⁻] → [H₃O⁺] → pH

$M = \frac{mol}{L}$ $\qquad K_w = [H_3O^+][OH^-]$ $\qquad pH = -\log[H_3O^+]$

Solution: $5.0\ \text{mL KOH} \times \frac{1\ L}{1000\ \text{mL}} = 0.0050$ L KOH excess then

$\frac{0.1103\ \text{mol KOH}}{1\ \text{L}} \times 0.0050\ \text{L} = 0.0005515$ mol KOH excess then

$0.020\ L\ HA + 0.02842\ L\ KOH + 0.0050\ L\ KOH = 0.05342$ L total volume

$[\text{KOH excess}] = \frac{0.0005515\ \text{mol KOH excess}}{0.05342\ L} = 0.010324$ M KOH excess. Because KOH is a strong base,

$[\text{KOH}]$ excess $= [OH^-]$. The strong base overwhelms the weak base and is insignificant in the calculation.

$K_w = [H_3O^+][OH^-]$ so $[H_3O^+] = \frac{K_w}{[OH^-]} = \frac{1.0 \times 10^{-14}}{0.010324} = 9.6863 \times 10^{-13}$ M. Finally,

$pH = -\log[H_3O^+] = -\log(9.6863 \times 10^{-13}) = 12.01.$

Check: The units (none) are correct. The pH is rising sharply at the equivalence point, so the pH after 5 mL past the equivalence point should be quite basic.

Given: above data

Find: titration curve and suitable indicator

Solution: Plotting these data points is shown in the accompanying graph.

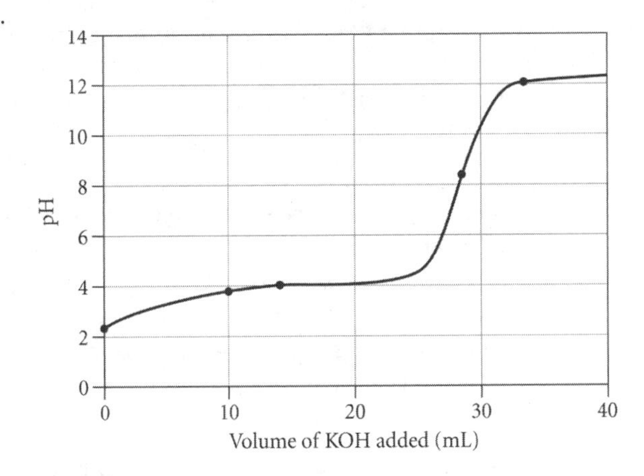

Because the pH at the equivalence point is 8.41, appropriate indicators are *m*-nitrophenol or thymol blue.

18.123 **Given:** saturated $CaCO_3$ solution; precipitate 1.00×10^2 mg $CaCO_3$ **Find:** volume of solution evaporated

Other: $K_{sp}(CaCO_3) = 4.96 \times 10^{-9}$

Conceptual Plan: mg $CaCO_3$ → g $CaCO_3$ → mol $CaCO_3$

$\frac{1\ g\ CaCO_3}{1000\ mg\ CaCO_3}$ $\qquad \frac{1\ mol\ CaCO_3}{100.09\ g\ CaCO_3}$

The expression of the solubility product constant of A_mX_n is $K_{sp} = [A^{n+}]^m[X^{m-}]^n$. The molar solubility of a compound, A_mX_n, can be computed directly from K_{sp} by solving for S in the expression $K_{sp} = (mS)^m (nS)^n = m^n n^n S^{m+n}$. Then mol $CaCO_3$, S → L.

$M = \frac{mol}{L}$

Solution: $1.00 \times 10^2\ \cancel{\text{mg CaCO}_3} \times \frac{1\ \cancel{\text{g CaCO}_3}}{1000\ \cancel{\text{mg CaCO}_3}} \times \frac{1\ \text{mol CaCO}_3}{100.09\ \cancel{\text{g CaCO}_3}} = 9.9\underline{9}101 \times 10^{-4}$ mol $CaCO_3$ then

$K_{sp} = 4.96 \times 10^{-9}$, A = Ca^{2+}, $m = 1$, X = CO_3^{2-}, and $n = 1$; so $K_{sp} = 4.96 \times 10^{-9} = S^2$. Rearrange to solve for S. $S = \sqrt{4.96 \times 10^{-9}} = 7.0\underline{4}273 \times 10^{-5}$ M. Finally,

$$9.9\underline{9}101 \times 10^{-4}\ \cancel{\text{mol CaCO}_3} \times \frac{1\ \text{L}}{7.0\underline{4}273 \times 10^{-5}\ \cancel{\text{mol CaCO}_3}} = 14.2\ \text{L}.$$

Check: The units (L) are correct. The volume should be large because the solubility is low.

18.124 **Given:** $[Na^+] = 0.140$ M and $K_{sp}(NaC_5H_3N_4) = 5.76 \times 10^{-8}$ **Find:** $[C_5H_3N_4^-]$ to form precipitate
Conceptual Plan: Write a balanced equation and expression for K_{sp}. Then $[Na^+], K_{sp} \rightarrow [C_5H_3N_4^-]$.
Solution: $NaC_5H_3N_4(s) \rightarrow Na^+(aq) + C_5H_3N_4^-(aq)$. So $K_{sp} = [Na^+][C_5H_3N_4^-] = 5.76 \times 10^{-8} = (0.140)[C_5H_3N_4^-]$. Solve for $[C_5H_3N_4^-]$ then $[C_5H_3N_4^-] = 4.11 \times 10^{-7}$ M.

Check: The units (M) are correct. Because K_{sp} is so small and the sodium concentration is fairly high, the urate concentration is driven to a very low level.

18.125 **Given:** $[Ca^{2+}] = 9.2$ mg/dL and $K_{sp}(Ca_2P_2O_7) = 8.64 \times 10^{-13}$ **Find:** $[P_2O_7^{4-}]$ to form precipitate
Conceptual Plan: mg Ca^{2+}/dL $\rightarrow$ g Ca^{2+}/dL $\rightarrow$ mol Ca^{2+}/dL $\rightarrow$ mol Ca^{2+}/L then

$$\frac{1\ \text{g Ca}^{2+}}{1000\ \text{mg Ca}^{2+}} \qquad \frac{1\ \text{mol Ca}^{2+}}{40.08\ \text{g Ca}^{2+}} \qquad \frac{10\ \text{dL}}{1\ \text{L}}$$

write a balanced equation and expression for K_{sp}. Then $[Ca^{2+}], K_{sp} \rightarrow [P_2O_7^{4-}]$.

Solution: $9.2\frac{\cancel{\text{mg Ca}^{2+}}}{\cancel{\text{dL}}} \times \frac{1\ \cancel{\text{g Ca}^{2+}}}{1000\ \cancel{\text{mg Ca}^{2+}}} \times \frac{1\ \text{mol Ca}^{2+}}{40.08\ \cancel{\text{g Ca}^{2+}}} \times \frac{10\ \cancel{\text{dL}}}{1\ \text{L}} = 2.\underline{2}9541 \times 10^{-3}$ M Ca^{2+} then write

equation $Ca_2P_2O_7(s) \rightarrow 2\,Ca^{2+}(aq) + P_2O_7^{4-}(aq)$. So $K_{sp} = [Ca^{2+}]^2[P_2O_7^{4-}] = 8.64 \times 10^{-13} = (2.\underline{2}9541 \times 10^{-3})^2[P_2O_7^{4-}]$. Solve for $[P_2O_7^{4-}]$ then $[P_2O_7^{4-}] = 1.6 \times 10^{-7}$ M.

Check: The units (M) are correct. Because K_{sp} is so small and the calcium concentration is relatively high, the diphosphate concentration required is at a very low level.

18.126 **Given:** AgCl in 0.100 M NH_3 **Find:** molar solubility (S)
Other: $K_f(Ag(NH_3)_2^+) = 1.7 \times 10^7$, $K_{sp}(AgCl) = 1.77 \times 10^{-10}$
Conceptual Plan: Identify the appropriate solid and complex ion. Write balanced equations for dissolving the solid and forming the complex ion. Add these two reactions to get the desired overall reaction. Using the rules from Chapter 16, multiply the individual reaction Ks to get the overall K for the sum of these reactions. Then M NH_3, $K \rightarrow S$.

ICE table

Solution: Identify the solid as AgCl and the complex ion as $Ag(NH_3)_2^+$. Write the individual reactions and add them together.

$AgCl(s) \rightleftharpoons \cancel{Ag^+(aq)} + Cl^-(aq)$ $\qquad K_{sp} = 1.77 \times 10^{-10}$
$\cancel{Ag^+(aq)} + 2\,NH_3(aq) \rightleftharpoons Ag(NH_3)_2^+(aq)$ $\qquad K_f = 1.7 \times 10^7$

$AgCl(s) + 2\,NH_3(aq) \rightleftharpoons Ag(NH_3)_2^+(aq) + Cl^-(aq)$
Because the overall reaction is the simple sum of the two reactions, the overall reaction $K = K_fK_{sp} = (1.7 \times 10^7) \times (1.77 \times 10^{-10}) = 3.\underline{0}09 \times 10^{-3}$. Set up an ICE table:

$AgCl(s) + 2\,NH_3(aq) \rightleftharpoons Ag(NH_3)_2^+(aq) + Cl^-(aq)$

	$[NH_3]$	$[Ag(NH_3)_2^+]$	$[Cl^-]$
Initial	0.100	0.00	0.00
Change	$-2S$	$+S$	$+S$
Equil	$0.100 - 2S$	$+S$	$+S$

$K = \frac{[Ag(NH_3)_2^+][Cl^-]}{[NH_3]^2} = 3.\underline{0}09 \times 10^{-3} = \frac{S^2}{(0.100 - 2S)^2}$. Simplify by taking the square root of the expression.

$\sqrt{3.\underline{0}09 \times 10^{-3}} = 5.\underline{4}854 \times 10^{-2} = \frac{S}{(0.100 - 2S)}$. Solve for S.

$(5.\underline{4}854 \times 10^{-2})(0.100 - 2S) = S \rightarrow 5.\underline{4}854 \times 10^{-3} = (1.1\underline{0}971)S \rightarrow S = 4.\underline{9}431 \times 10^{-3} = 4.9 \times 10^{-3}$ M

Check: The units (M) are correct. Because K_f is large, the overall K is larger than the original K_{sp} and the solubility of AgCl increases over that of pure water ($\sqrt{1.77 \times 10^{-10}} = 1.33 \times 10^{-5}$ M).

18.127 **Given:** CuX in 0.150 M NaCN **Find:** molar solubility (S)

Other: $K_f(Cu(CN)_4^{2-}) = 1.0 \times 10^{25}$, $K_{sp}(CuX) = [Cu^{2+}][X^{2-}] = 1.27 \times 10^{-36}$

Conceptual Plan: Identify the appropriate solid and complex ion. Write balanced equations for dissolving the solid and forming the complex ion. Add these two reactions to get the desired overall reaction. Using the rules from Chapter 16, multiply the individual reaction *K*s to get the overall *K* for the sum of these reactions. Then M NaCN, $K \rightarrow S$.

ICE table

Solution: Identify the solid as MX and the complex ion as $Cu(CN)_4^{2-}$. Write the individual reactions and add them together.

$CuX(s) \rightleftharpoons \cancel{Cu^{+}(aq)} + X^{2-}(aq)$ $K_{sp} = 1.27 \times 10^{-36}$

$\cancel{Cu^{2+}(aq)} + 4\,CN^-(aq) \rightleftharpoons Cu(CN)_4^{2-}(aq)$ $K_f = 1.0 \times 10^{25}$

$CuX(s) + 4\,CN^-(aq) \rightleftharpoons Cu(CN)_4^{2-}(aq) + X^{2-}(aq)$

Because the overall reaction is the simple sum of the two reactions, the overall reaction $K = K_f K_{sp} = (1.0 \times 10^{25}) \times (1.27 \times 10^{-36}) = 1.\underline{2}7 \times 10^{-11}$. $NaCN(s) \rightarrow Na^+(aq) + CN^-(aq)$. Because one CN^- ion is generated for each NaCN, $[CN^-] = 0.150$ M. Set up an ICE table:

$Cu(s) + 4\,CN^-(aq) \rightleftharpoons Cu(CN)_4^{2-}(aq) + X^{2-}(aq)$

	$[CN^-]$	$[Cu(CN)_4^{2-}]$	$[X^{2-}]$
Initial	0.150	0.00	0.00
Change	$-4S$	$+S$	$+S$
Equil	$0.150 - 4S$	$+S$	$+S$

$$K = \frac{[Cu(CN)_4^{2-}][X^{2-}]}{[CN^-]^4} = 1.\underline{2}7 \times 10^{-11} = \frac{S^2}{(0.150 - 4S)^4}$$

Assume that S is small ($4S << 0.150$), so $\frac{S^2}{(0.150 - \cancel{4S})^4} = 1.\underline{2}7 \times 10^{-11} = \frac{S^2}{(0.150)^4}$ and $S = 8.\underline{0}183 \times 10^{-8} = 8.0 \times 10^{-8}$ M. Confirm that the assumption is valid. $\frac{4(8.\underline{0}183 \times 10^{-8})}{0.150} \times 100\% = 0.00021\% << 5\%$, so the assumption is valid.

Check: The units (M) are correct. Because K_f is large, the overall K is larger than the original K_{sp} and the solubility of MX increases over that of pure water ($\sqrt{1.27 \times 10^{-36}} = 1.13 \times 10^{-18}$ M).

18.128 **Given:** 0.10 M $C_6H_5NH_2$, keep $[C_6H_5NH_3^+] < 1.0 \times 10^{-9}$ and $K_b(C_6H_5NH_2) = 4.3 \times 10^{-10}$ **Find:** [NaOH]

Conceptual Plan: M $C_6H_5NH_2$, maximum M $C_6H_5NH_3^+ \rightarrow [OH^-] = [NaOH]$

ICE table $K_w = [H_3O^+][OH^-]$ $pH = -\log[H_3O^+]$

Solution: Set up an ICE table. Because the amount of the conjugate acid is set so small, the concentration of the weak base does not significantly change.

$C_6H_5NH_2(aq) + H_2O(l) \rightleftharpoons C_6H_5NH_3^+(aq) + OH^-(aq)$

	$[C_6H_5NH_2]$	$[C_6H_5NH_3^+]$	$[OH^-]$
Initial	0.10	0.00	≈ 0.00
Change	—	—	$+x$
Equil	≈ 0.10	1.0×10^{-9}	$+x$

$$K_b = \frac{[C_6H_5NH_3^+][OH^-]}{[C_6H_5NH_2]} = 4.3 \times 10^{-10} = \frac{(1.0 \times 10^{-9})x}{0.10}$$

Solve for x. So $x = 0.043$ M $= [OH^-]$. $NaOH(aq) \rightarrow Na^+(aq) + OH^-(aq)$. Because one OH^- ion is generated for each NaOH, $[NaOH] = 0.043$ M NaOH.

Check: The units (M) are correct. The magnitude of the answer makes physical sense because $[OH^-]$ needs to be about one order of magnitude lower than the aniline concentration (comparing K_b with maximum conjugate acid concentrations).

18.129 **Given:** 100.0 mL of 0.36 M NH_2OH and 50.0 mL of 0.26 M HCl and $K_b(NH_2OH) = 1.10 \times 10^{-8}$ **Find:** pH

Conceptual Plan: Identify acid and base components mL → L then $[NH_2OH]$, L → mol NH_2OH

acid = NH_3OH^+ base = NH_2OH $\qquad \frac{1\ L}{1000\ mL} \qquad M = \frac{mol}{L}$

then mL → L then [HCl], L → mol HCl then write balanced equation then

$\frac{1\ L}{1000\ mL} \qquad M = \frac{mol}{L} \qquad HCl + NH_2OH \rightarrow NH_3OHCl$

mol NH_2OH, mol HCL → mol excess NH_2OH, mol NH_3OH^+.

set up stoichiometry table

Because there are significant amounts of both the acid and the conjugate base species, this is a buffer solution; so the Henderson–Hasselbalch equation $\left(pH = pK_a + \log\frac{[base]}{[acid]}\right)$ can be used. Convert K_b to K_a using $K_w = K_aK_b$. Also note that the ratio of concentrations is the same as the ratio of moles because the volume is the same for both species.

Solution: $100\ mL\ NH_2OH \times \frac{1\ L}{1000\ mL} = 0.1\ L\ NH_2OH$ then

$\frac{0.36\ mol\ NH_2OH}{1\ L} \times 0.1000\ L = 0.036\ mol\ NH_2OH$. $50.0\ mL\ HCl \times \frac{1\ L}{1000\ mL} = 0.0500\ L\ HCl$ then

$\frac{0.26\ mol\ HCl}{1\ L} \times 0.0500\ L = 0.013\ mol\ HCl$. Set up a table to track changes:

	$HCl(aq)$	+	$NH_2OH(aq)$	→	$NH_3OHCl(aq)$
Before addition	0.00 mol		0.036 mol		0.00 mol
Addition	0.013 mol		—		—
After addition	≈0.00 mol		0.023 mol		0.013 mol

Then $K_w = K_aK_b$, so $K_a = \frac{K_w}{K_b} = \frac{1.0 \times 10^{-14}}{1.10 \times 10^{-8}} = 9.\underline{0}909 \times 10^{-7}$ M. Then use the Henderson–Hasselbalch equation because the solution is a buffer. Note that the ratio of moles is the same as the ratio of concentrations because the volume for both terms is the same.

$$pH = pK_a + \log\frac{[base]}{[acid]} = -\log(9.\underline{0}909 \times 10^{-7}) + \log\frac{0.023\ mol}{0.013\ mol} = 6.2\underline{8}918 = 6.29$$

Check: The units (none) are correct. The magnitude of the answer makes physical sense because the pH should be more than the pK_a of the acid because there is more base than acid.

18.130 **Given:** 0.867 g diprotic acid titrated with 32.2 mL of 0.182 M $Ba(OH)_2$ solution **Find:** molar mass of acid

Conceptual Plan: Write a balanced reaction then mL → L then M $Ba(OH)_2$, L → mol $Ba(OH)_2$ → mol acid

$H_2A + Ba(OH)_2 \rightarrow BaA + 2\ H_2O \qquad \frac{1\ L}{1000\ mL} \qquad M = \frac{mol}{L} \qquad 1:1$

then mol acid, g acid → $\mathcal{M}$.

$\mathcal{M} = \frac{g\ acid}{mol\ acid}$

Solution: $32.2\ mL \times \frac{1\ L}{1000\ mL} = 0.0322\ L$ then

$$0.0322\ L\ Ba(OH)_2 \times \frac{0.182\ mol\ Ba(OH)_2}{1\ L\ Ba(OH)_2} = 0.0058\underline{6}04\ mol\ Ba(OH)_2 \times \frac{1\ mol\ H_2A}{1\ mol\ Ba(OH)_2} = 0.0058\underline{6}04\ mol\ H_2A$$

then $\mathcal{M} = \frac{g\ acid}{mol\ acid} = \frac{0.867\ g\ H_2A}{0.0058\underline{6}04\ mol\ H_2A} = 148\ g/mol$

Check: The units (g/mol) are correct. The magnitude of the answer makes physical sense because there is much less than a mole of acid and about half a gram of acid; so the molar mass will be high. The number is reasonable for an acid (must be >20 g/mol—lightest acid is HF).

18.131 **Given:** 25.0 mL of NaOH titrated with 19.6 mL of 0.189 M HCl solution; 10.0 mL of H_3PO_4 titrated with 34.9 mL NaOH **Find:** concentration of H_3PO_4 solution

Conceptual Plan: Write the first balanced reaction then mL → L then M HCl, L → mol HCl → mol NaOH

$HCl + NaOH \rightarrow NaCl + H_2O$ $\quad \frac{1\ L}{1000\ mL}$ $\quad M = \frac{mol}{L}$ $\quad$ 1:1

then mL → L then mol NaOH, L → M NaOH then write 2nd balanced reaction then mL → L

$\frac{1\ L}{1000\ mL}$ $\quad H_3PO_4 + 3\ NaOH \rightarrow Na_3PO_4 + 3\ H_2O$ $\quad \frac{1\ L}{1000\ mL}$

then M NaOH, L → mol NaOH → mol H_3PO_4 then mL → L then mol H_3PO_4, L → M H_3PO_4.

$M = \frac{mol}{L}$ $\quad$ 3:1 $\quad \frac{1\ L}{1000\ mL}$ $\quad M = \frac{mol}{L}$

Solution: In the first titration, $19.6\ \cancel{mL} \times \frac{1\ L}{1000\ \cancel{mL}} = 0.0196\ L$ then

$$0.0196\ \cancel{L\ HCl} \times \frac{0.189\ mol\ HCl}{1\ \cancel{L\ HCl}} = 0.0037\underline{0}44\ \cancel{mol\ HCl} \times \frac{1\ mol\ NaOH}{1\ \cancel{mol\ HCl}} = 0.0037\underline{0}44\ mol\ NaOH \text{ then}$$

$$25.0\ \cancel{mL} \times \frac{1\ L}{1000\ \cancel{mL}} = 0.0250\ L \text{ then } \frac{0.0037\underline{0}44\ mol\ NaOH}{0.0250\ L\ NaOH} = 0.14\underline{8}176\ M\ NaOH$$

In the second titration, $34.9\ \cancel{mL} \times \frac{1\ L}{1000\ \cancel{mL}} = 0.0349\ L$ then

$$0.0349\ \cancel{L\ NaOH} \times \frac{0.14\underline{8}176\ mol\ NaOH}{1\ \cancel{L\ NaOH}} = 0.0051\underline{7}134\ \cancel{mol\ NaOH} \times \frac{1\ mol\ H_3PO_4}{3\ \cancel{mol\ NaOH}} = 0.0017\underline{2}378\ mol\ H_3PO_4$$

$$\text{then } 10.0\ \cancel{mL} \times \frac{1\ L}{1000\ \cancel{mL}} = 0.0100\ L \text{ then } \frac{0.0017\underline{2}378\ mol\ H_3PO_4}{0.0100\ L\ H_3PO_4} = 0.172\ M\ H_3PO_4$$

Check: The units (M) are correct. The magnitude of the answer makes physical sense because the concentration of NaOH is a little lower than the HCl (because the volume of NaOH is greater than HCl) and the concentration of H_3PO_4 is more than the NaOH (because the ratio of the volume of NaOH to volume of H_3PO_4 is just over 3 and H_3PO_4 is a triprotic acid).

18.132 **Given:** 250.0 cm^3 of 1.4 M HCOOH; adjust to pH = 3.36 **Find:** mass NaCOOH

Other: $K_a(HCOOH) = 1.8 \times 10^{-4}$

Conceptual Plan: cm^3 → L and K_a → pK_a then [HCOOH], pH, pK_a → [NaCOOH]

$\frac{1\ L}{1000\ mL}$ $\quad pK_a = -\log K_a$ $\quad$ acid = HCOOH base = NaCOOH $\quad pH = pK_a + \log\frac{[base]}{[acid]}$

then [NaCOOH], L → mol NaCOOH → g NaCOOH

$M = \frac{mol}{L}$ $\quad \frac{68.01\ g\ NaCOOH}{1\ mol\ NaCOOH}$

Solution: $250.0\ \cancel{mL} \times \frac{1\ L}{1000\ \cancel{mL}} = 0.2500\ L$ because $K_a(HCOOH) = 1.8 \times 10^{-4}$,

$pK_a = -\log K_a = -\log(1.8 \times 10^{-4}) = 3.74$. Because acid = HCOOH and base = NaCOOH,

$$pH = pK_a + \log\frac{[base]}{[acid]} = 3.74 + \log\frac{[NaCOOH]}{1.4\ M} = 3.36. \text{ Solve for [NaCOOH].}$$

$$\log\frac{[NaCOOH]}{1.4\ M} = 3.36 - 3.74 = -0.38 \rightarrow \frac{[NaCOOH]}{1.4\ M} = 10^{-0.38} = 0.4\underline{1}6869 \rightarrow$$

$[NaCOOH] = 0.4\underline{1}6869 \times 1.4\ M = 0.5\underline{8}3617\ M\ NaCOOH$. Then

$$\frac{0.5\underline{8}3617\ mol\ NaCOOH}{1\ \cancel{L}} \times 0.2500\ \cancel{L} = 0.14\underline{5}90\ \cancel{mol\ NaCOOH} \times \frac{68.01\ g\ NaCOOH}{1\ \cancel{mol\ NaCOOH}} = 9.\underline{9}227\ g\ NaCOOH$$

$= 9.9\ g\ NaCOOH$

Check: The units (g) are correct. The magnitude of the answer makes physical sense because there is more acid than base in the buffer ($pH < pK_a$) and there is less than a mole of acid in the buffer; so there must be less than a mole of NaCOOH.

18.133 **Given:** $(CH_3)_2NH/(CH_3)_2NH_2Cl$ buffer at pH = 10.43 **Find:** relative masses of $(CH_3)_2NH$ and $(CH_3)_2NH_2Cl$

Other: $K_b((CH_3)_2NH) = 5.4 \times 10^{-4}$

Conceptual Plan: $K_b \rightarrow pK_b \rightarrow pK_a$ **then pH,** $pK_a \rightarrow [(CH_3)_2NH]/[(CH_3)_2NH_2^+]$ **then**

$pK_b = -\log K_b \quad 14 = pK_a + pK_b \quad \text{acid} = (CH_3)_2NH_2^+ \ \text{base} = (CH_3)_2NH \quad pH = pK_a + \log\frac{[\text{base}]}{[\text{acid}]}$

$[(CH_3)_2NH]/[(CH_3)_2NH_2^+] \rightarrow g(CH_3)_2NH/g(CH_3)_2NH_2^+$

$\frac{45.09\ g(CH_3)_2NH}{1\ mol\ (CH_3)_2NH} \quad \frac{1\ mol\ (CH_3)_2NH_2Cl}{81.54\ g\ (CH_3)_2NH_2Cl}$

Solution: Because $K_b((CH_3)_2NH) = 5.4 \times 10^{-4}$, $pK_b = -\log K_b = -\log(5.4 \times 10^{-4}) = 3.27$. Because $14 = pK_a + pK_b$, $pK_a = 14 - pK_b = 14 - 3.27 = 10.73$. Because $[\text{acid}] = [(CH_3)_2NH_2^+] = [(CH_3)_2NH_2Cl]$ and $[\text{base}] = [(CH_3)_2NH]$, $pH = pK_a + \log\frac{[\text{base}]}{[\text{acid}]} = 10.73 + \log\frac{[(CH_3)_2NH]}{[(CH_3)_2NH_2Cl]} = 10.43$. Solve for $\frac{[(CH_3)_2NH]}{[(CH_3)_2NH_2Cl]}$. $\log\frac{[(CH_3)_2NH]}{[(CH_3)_2NH_2Cl]} = 10.43 - 10.73 = -0.30 \rightarrow \frac{[(CH_3)_2NH]}{[(CH_3)_2NH_2Cl]} = 10^{-0.30} = 0.5\underline{0}1187$.

Then $\frac{0.5\underline{0}1187\ \cancel{mol(CH_3)_2NH}}{1\ \cancel{mol(CH_3)_2NH_2Cl}} \times \frac{45.09\ g(CH_3)_2NH}{1\ \cancel{mol(CH_3)_2NH}} \times \frac{1\ \cancel{mol(CH_3)_2NH_2Cl}}{81.54\ g(CH_3)_2NH_2Cl} = \frac{0.2\underline{7}7146\ g(CH_3)_2NH}{g(CH_3)_2NH_2Cl}$

$= \frac{0.28\ g\ (CH_3)_2NH}{g\ (CH_3)_2NH_2Cl}$ or $\frac{3.6\ g\ (CH_3)_2NH_2Cl}{g\ (CH_3)_2NH}$

Check: The units (g/g) are correct. The magnitude of the answer makes physical sense because there are more moles of acid than base in the buffer ($pH < pK_a$) and the molar mass of the acid is greater than the molar mass of the base. Thus, the ratio of the mass of the base to the mass of the acid is expected to be less than 1.

18.134 **Given:** 2.0 L HCN/NaCN buffer at pH = 9.8, complete dissociation of NaCN; osmotic pressure = 1.35 atm at 298 K
Find: masses of HCN and NaCN **Other:** $K_a(HCN) = 4.9 \times 10^{-10}$

Conceptual Plan: $K_a \rightarrow pK_a$ **then pH,** $pK_a \rightarrow [CN^-]/[HCN]$ **then** $\Pi, T \rightarrow M$ **then**

$pK_a = -\log K_a \quad \text{acid} = HCN\ \text{base} = CN^- \quad pH = pK_a + \log\frac{[\text{base}]}{[\text{acid}]} \quad \Pi = MRT$

assume that HCN does not dissociate and NaCN completely dissociates, then M = [HCN] + 2[NaCN]. Use [NaCN]/[HCN] and M = [HCN] + 2[NaCN] to solve for [HCN] and [NaCN]. Then [HCN], L → g HCN and [HCN], L → g HCN.

$M = \frac{\text{amount solute (moles)}}{\text{volume solution (L)}} \quad \frac{27.03\ g\ HCN}{1\ mol\ HCN} \quad \frac{49.01\ g\ NaCN}{1\ mol\ NaCN}$

Solution: Because $K_a(HCN) = 4.9 \times 10^{-10}$, $pK_a = -\log K_a = -\log(4.9 \times 10^{-10}) = 9.31$. Because $[\text{acid}] = [HCN]$ and $[\text{base}] = [CN^-] = [NaCN]$, $pH = pK_a + \log\frac{[\text{base}]}{[\text{acid}]} = 9.31 + \log\frac{[NaCN]}{[HCN]} = 9.8$. Solve for $\frac{[NaCN]}{[HCN]}$. $\log\frac{[NaCN]}{[HCN]} = 9.8 - 9.31 = 0.\underline{4}9 \rightarrow \frac{[NaCN]}{[HCN]} = 10^{0.49} = 3.\underline{0}9030$.

Then $\Pi = MRT$. Rearrange to solve for M.

$$M = \frac{\Pi}{RT} = \frac{1.35\ \cancel{atm}}{0.08206\frac{L \cdot \cancel{atm}}{\cancel{K} \cdot mol} \times 298\ \cancel{K}} = 0.055\underline{2}0596\frac{\text{mol particles}}{L}$$

Assume that HCN does not dissociate and that NaCN completely dissociates; then $M = [HCN] + 2[NaCN]$. Use $[NaCN]/[HCN] = 3.\underline{0}9030$ and $M = [HCN] + 2[NaCN]$ to solve for $[HCN]$ and $[NaCN]$. So $[NaCN] = 3.\underline{0}9030\ [HCN] \rightarrow M = 0.055\underline{2}0596\frac{\text{mol particles}}{L} = [HCN] + 2(3.\underline{0}9030[HCN]) \rightarrow$

$0.055\underline{2}0596\frac{\text{mol particles}}{L} = 7.\underline{1}8060\ [HCN] \rightarrow [HCN] = 0.007\underline{6}8821$ M and

$[NaCN] = 3.\underline{0}9030\ [HCN] = 3.\underline{0}9030 \times 0.007\underline{6}8821\ M = 0.02\underline{3}7589$ M. Finally,

$\frac{0.007\underline{6}8821\ \cancel{mol\ HCN}}{\cancel{L\ solution}} \times 2.0\ \cancel{L\ solution} \times \frac{27.03\ g\ HCN}{1\ \cancel{mol\ HCN}} = 0.4\underline{1}562\ g\ HCN = 0.42\ g\ HCN$ and

$\frac{0.02\underline{3}7589\ \cancel{mol\ NaCN}}{\cancel{L\ solution}} \times 2.0\ \cancel{L\ solution} \times \frac{49.01\ g\ NaCN}{1\ \cancel{mol\ NaCN}} = 2.\underline{3}288\ g\ NaCN = 2.3\ g\ NaCN.$

Check: The units (g and g) are correct. The magnitude of the answer makes physical sense because there are more moles of base than acid in the buffer ($pH > pK_a$) and the molar mass of the base is greater than the molar mass of the acid.

18.135 **Given:** $HC_7H_5O_2/C_7H_5O_2Na$ buffer at pH = 4.55, complete dissociation of $C_7H_5O_2Na$, $d = 1.01$ g/mL; $T_f = -2.0\ °C$
Find: $[HC_7H_5O_2]$ and $[C_7H_5O_2Na]$ **Other:** $K_a(HC_7H_5O_2) = 6.5 \times 10^{-5}$, $K_f = 1.86\ °C/m$
Conceptual Plan: $K_a \rightarrow pK_a$ **then pH,** $pK_a \rightarrow [C_7H_5O_2Na]/[HC_7H_5O_2]$ **and** $T_f \rightarrow \Delta T_f$ **then**

$$pK_a = -\log K_a \quad \text{acid} = HC_7H_5O_2 \text{ base} = C_7H_5O_2^- \quad pH = pK_a + \log\frac{[\text{base}]}{[\text{acid}]} \quad T_f = T_f^\circ - \Delta T_f$$

$\Delta T_f, i, K_f \rightarrow m$ **then assume 1 kg water (or 1000 g water)**

$$\Delta T_f = K_f \times m \quad m = \frac{\text{amount solute (moles)}}{\text{mass solvent (kg)}}$$

Assume that $HC_7H_5O_2$ **does not dissociate and that** $C_7H_5O_2Na$ **completely dissociates, then mol particles = mol** $HC_7H_5O_2$ **+ 2(mol** $C_7H_5O_2Na$**). Use** $[C_7H_5O_2Na]/[HC_7H_5O_2]$ **and total mol particles = mol** $[HC_7H_5O_2]$ **+ 2(mol** $C_7H_5O_2Na$**) to solve for mol** $HC_7H_5O_2$ **and mol** $C_7H_5O_2Na$**. Then mol** $HC_7H_5O_2 \rightarrow$ **g** $HC_7H_5O_2$ **and mol** $C_7H_5O_2Na \rightarrow$ **g** $C_7H_5O_2Na$ **then**

$$\frac{122.12\text{ g }HC_7H_5O_2}{1\text{ mol }HC_7H_5O_2} \qquad \frac{144.10\text{ g }C_7H_5O_2Na}{1\text{ mol }HC_7H_5O_2Na}$$

g $HC_7H_5O_2$**, g** $C_7H_5O_2Na$**, g water → g solution → mL solution → L solution then**

$$\text{g }HC_7H_5O_2 + \text{g }C_7H_5O_2Na + \text{g water} = \text{g solution} \qquad d = \frac{1\text{ mL}}{1.01\text{ g}} \qquad \frac{1\text{ L}}{1000\text{ mL}}$$

mol $HC_7H_5O_2$**, L → M** $HC_7H_5O_2$ **and mol** $C_7H_5O_2Na$**, L → M** $C_7H_5O_2Na$**.**

$$M = \frac{\text{mol}}{\text{L}} \qquad M = \frac{\text{mol}}{\text{L}}$$

Solution: Because $K_a(HC_7H_5O_2) = 6.5 \times 10^{-5}$, $pK_a = -\log K_a = -\log(6.5 \times 10^{-5}) = 4.19$. Because $[\text{acid}] = [HC_7H_5O_2]$ and $[\text{base}] = [C_7H_5O_2^-] = [C_7H_5O_2Na]$,

$pH = pK_a + \log\dfrac{[\text{base}]}{[\text{acid}]} = 4.19 + \log\dfrac{[C_7H_5O_2Na]}{[HC_7H_5O_2]} = 4.55$. Solve for $\dfrac{[C_7H_5O_2Na]}{[HC_7H_5O_2]}$.

$\log\dfrac{[C_7H_5O_2Na]}{[HC_7H_5O_2]} = 4.55 - 4.19 = 0.36 \rightarrow \dfrac{[C_7H_5O_2Na]}{[HC_7H_5O_2]} = 10^{0.36} = 2.2\underline{9}087$. Then $T_f = T_f^\circ - \Delta T_f$, so

$\Delta T_f = T_f^\circ - T_f = 0.0\ °C - (-2.0\ °C) = 2.0\ °C$. Then $\Delta T_f = K_f \times m$. Rearrange to solve for m.

$m = \dfrac{\Delta T_f}{K_f} = \dfrac{2.0\ \cancel{°C}}{1.86\dfrac{\cancel{°C}}{m}} = 1.\underline{0}7527\ \dfrac{\text{mol particles}}{\text{kg solvent}}$. Assume 1 kg water, so we have $1.\underline{0}7527$ mol particles.

Assume that $HC_7H_5O_2$ does not dissociate and that $C_7H_5O_2Na$ completely dissociates; then $1.\underline{0}7527$ mol particles = mol $HC_7H_5O_2$ + 2(mol $C_7H_5O_2Na$). Use $[C_7H_5O_2Na]/[HC_7H_5O_2] = 2.2\underline{9}087$ and $1.\underline{0}7527$ mol particles = mol $[HC_7H_5O_2]$ + 2(mol $C_7H_5O_2Na$) to solve for mol $HC_7H_5O_2$ and mol $C_7H_5O_2Na$. So mol $C_7H_5O_2Na$ = $2.2\underline{9}087$(mol $HC_7H_5O_2$) → $1.\underline{0}7527$ mol particles = mol $HC_7H_5O_2$ + 2($2.2\underline{9}087$ mol $HC_7H_5O_2$) → $1.\underline{0}7527$ mol particles = $5.\underline{5}8174$ mol $HC_7H_5O_2$ → mol $HC_7H_5O_2$ = $0.1\underline{9}2641$ mol and mol $C_7H_5O_2Na$ = 2.29087 × mol $HC_7H_5O_2$ = $2.2\underline{9}087 \times 0.1\underline{9}2641$ mol $HC_7H_5O_2$ = $0.4\underline{4}1315$ mol $C_7H_5O_2Na$.

$$0.1\underline{9}2641\ \cancel{\text{mol }HC_7H_5O_2} \times \frac{122.12\text{ g }HC_7H_5O_2}{1\ \cancel{\text{mol }HC_7H_5O_2}} = 2\underline{3}.5253\text{ g }HC_7H_5O_2 \text{ and}$$

$$0.4\underline{4}1315\ \cancel{\text{mol }C_7H_5O_2Na} \times \frac{144.10\text{ g }C_7H_5O_2Na}{1\ \cancel{\text{mol }C_7H_5O_2Na}} = 6\underline{3}.5935\text{ g }C_7H_5O_2Na \text{ then}$$

$2\underline{3}.5253$ g $HC_7H_5O_2$ + $6\underline{3}.5935$ g $C_7H_5O_2Na$ + 1000 g water = $108\underline{7}.119$ g solution then

$$108\underline{7}.119\ \cancel{\text{g solution}} \times \frac{1\ \cancel{\text{mL}}}{1.01\ \cancel{\text{g}}} \times \frac{1\text{ L}}{1000\ \cancel{\text{mL}}} = 1.0\underline{7}6355\text{ L. Finally,}$$

$$\frac{0.1\underline{9}2641\text{ mol }HC_7H_5O_2}{1.0\underline{7}6355\text{ L}} = 0.1\underline{7}8975\text{ M }HC_7H_5O_2 = 0.18\text{ M }HC_7H_5O_2 \text{ and}$$

$$\frac{0.4\underline{4}1315\text{ mol }C_7H_5O_2Na}{1.0\underline{7}6355\text{ L}} = 0.4\underline{1}0009\text{ M }C_7H_5O_2Na = 0.41\text{ M }C_7H_5O_2Na.$$

Check: The units (M and M) are correct. The magnitude of the answer makes physical sense because there are more moles of base than acid in the buffer ($pH > pK_a$).

Challenge Problems

18.136 The Henderson–Hasselbalch equation is $pH = pK_a + \log \frac{[base]}{[acid]}$. Remember that $14 = pH + pOH$ and that $14 = pK_a + pK_b$. Substituting these into the Henderson–Hasselbalch equation:

$14 - pOH = 14 - pK_b + \log \frac{[base]}{[acid]}$. Simplifying the expression gives $pOH = pK_b - \log \frac{[base]}{[acid]}$.

18.137 **Given:** 10.0 L of 75 ppm $CaCO_3$ and 55 ppm $MgCO_3$(by mass)
Find: mass Na_2CO_3 to precipitate 90.0% of ions

Other: $K_{sp}(CaCO_3) = 4.96 \times 10^{-9}$ and $K_{sp}(MgCO_3) = 6.82 \times 10^{-6}$

Conceptual Plan: Assume that the density of water is 1.00 g/mL L water → mL water → g water then

$\frac{1000\text{ mL}}{1\text{ L}}$ $\frac{1.00\text{ g water}}{1\text{ mL}}$

g water → g $CaCO_3$ → mol $CaCO_3$ → mol Ca^{2+} and g water → g $MgCO_3$ → mol $MgCO_3$ then

$\frac{75\text{ g }CaCO_3}{10^6\text{ g water}}$ $\frac{1\text{ mol }CaCO_3}{100.09\text{ g }CaCO_3}$ $\frac{1\text{ mol }Ca^{2+}}{1\text{ mol }CaCO_3}$ $\frac{55\text{ g }MgCO_3}{10^6\text{ g water}}$ $\frac{1\text{ mol }MgCO_3}{84.32\text{ g }MgCO_3}$

mol $MgCO_3$ → mol Mg^{2+} then comparing the two K_{sp} values, essentially all of the Ca^{2+} will

$\frac{1\text{ mol }Mg^{2+}}{1\text{ mol }MgCO_3}$

precipitate before the Mg^{2+} will begin to precipitate. Because 90.0% of the ions are to be precipitates, 10.0% of the ions will be left in solution (all will be Mg^{2+}).

$(0.100)(\text{mol }Ca^{2+} + \text{mol }Mg^{2+})$

Calculate the moles of ions remaining in solution. Then mol Mg^{2+}, L → M Mg^{2+}.

$M = \frac{mol}{L}$

The solubility product constant (K_{sp}) is the equilibrium expression for a chemical equation representing the dissolution of an ionic compound. The expression of the solubility product constant of A_mX_n is $K_{sp} = [A^{n+}]^m[X^{m-}]^n$. Use this equation to find M Mg^{2+}, K_{sp} → M CO_3^{2-} then M CO_3^{2-}, L → mol CO_3^{2-}

for ionic compound, $A_mX_n, K_{sp} = [A^{n+}]^m[X^{m-}]^n$ $M = \frac{mol}{L}$

then mol CO_3^{2-} → mol Na_2CO_3 → g Na_2CO_3.

$\frac{1\text{ mol }CO_3^{2-}}{1\text{ mol }Na_2CO_3}$ $\frac{105.99\text{ g }Na_2CO_3}{1\text{ mol }Na_2CO_3}$

Solution: $10.0\text{ L} \times \frac{1000\text{ mL}}{1\text{ L}} \times \frac{1.00\text{ g water}}{1\text{ mL}} = 1.00 \times 10^4$ g water then

$1.00 \times 10^4\text{ g water} \times \frac{75\text{ g }CaCO_3}{10^6\text{ g water}} \times \frac{1\text{ mol }CaCO_3}{100.09\text{ g }CaCO_3} \times \frac{1\text{ mol }Ca^{2+}}{1\text{ mol }CaCO_3} = 0.0074933\text{ mol }Ca^{2+}$ and

$1.00 \times 10^4\text{ g water} \times \frac{55\text{ g }MgCO_3}{10^6\text{ g water}} \times \frac{1\text{ mol }MgCO_3}{84.32\text{ g }MgCO_3} \times \frac{1\text{ mol }Mg^{2+}}{1\text{ mol }MgCO_3} = 0.0065228\text{ mol }Mg^{2+}$ so the ions remaining in solution after 90.0% precipitate out $= (0.100)(\text{mol }Ca^{2+} + \text{mol }Mg^{2+})$

$= (0.100)(0.0074933\text{ mol }Ca^{2+} + 0.0065228\text{ mol }Mg^{2+}) = 0.00140161$ mol ions

so $\frac{0.00140161\text{ mol }Mg^{2+}}{10.0\text{ L}} = 0.000140161\text{ M }Mg^{2+}$. Then $K_{sp} = 6.82 \times 10^{-6}$, $A = Mg^{2+}$, $m = 1$, $X = CO_3^{2-}$, and $n = 1$, so $K_{sp} = 6.82 \times 10^{-6} = [Mg^{2+}][CO_3^{2-}] = (0.00140161)[CO_3^{2-}]$. Rearrange to solve for $[CO_3^{2-}]$. So $[CO_3^{2-}] = 0.0486583$ M. Then

$\frac{0.0486583\text{ mol }CO_3^{2-}}{1\text{ L}} \times 10.0\text{ L} \times \frac{1\text{ mol }Na_2CO_3}{1\text{ mol }CO_3^{2-}} \times \frac{105.99\text{ g }Na_2CO_3}{1\text{ mol }Na_2CO_3} = 51.6\text{ g }Na_2CO_3$.

Check: The units (g) are correct. The mass is reasonable to put in a washing machine load.

18.138 **Given:** 0.558 g diprotic acid with molar mass = 255.8 g/mol dissolved in 25.00 mL water and titrated with a saturated $Ca(OH)_2$ solution **Other:** $K_{sp}\ (Ca(OH)_2) = 4.68 \times 10^{-6}$

(a) **Find:** volume of base to the first and second equivalence points

Conceptual Plan:

Calculate the $[Ca(OH)_2]$ by using the equation derived in Problem 18.20 and solve for S.

for ionic compound, A_mX_n, $K_{sp} = m^m n^n S^{m+n}$

Then write a balanced reaction for titration of both protons. (Volume to the first equivalence point will

$H_2A + Ca(OH)_2 \rightarrow CaA + 2\ H_2O$

be half of this.) Then g H_2A, $\mathcal{M}$ → mol H_2A → mol $Ca(OH)_2$ then

$\mathcal{M} = \frac{\text{g acid}}{\text{mol acid}}$ 1:1

M $Ca(OH)_2$, mol $Ca(OH)_2$ → L $Ca(OH)_2$ → mL $Ca(OH)_2$. Finally, divide the volume to

$M = \frac{\text{mol}}{\text{L}}$ $\frac{1000\ \text{mL}}{1\ \text{L}}$

the second equivalence point by 2 to get the volume to the first equivalence point.

Solution: for $Ca(OH)_2$, $K_{sp} = 4.68 \times 10^{-6}$, $A = Ca^{2+}$, $m = 1$, $X = OH^-$, and $n = 2$; so $K_{sp} = 4.68 \times 10^{-6} = 2^2S^3$.

Rearrange to solve for S. $S = \sqrt[3]{\frac{4.68 \times 10^{-6}}{4}} = 1.05 \times 10^{-2}\ M$

$0.558\ \cancel{\text{g H}_2\text{A}} \times \frac{1\ \text{mol H}_2\text{A}}{255.8\ \cancel{\text{g H}_2\text{A}}} = 0.00218139\ \text{mol H}_2\text{A}$. Because the stoichiometry of the overall titration reaction is 1:1, $0.00218139\ \text{mol H}_2\text{A} = 0.00218139\ \text{mol Ca(OH)}_2$.

Then $0.00218139\ \cancel{\text{mol Ca(OH)}_2} \times \frac{1\ \cancel{\text{L Ca(OH)}_2}}{0.0105\ \cancel{\text{mol Ca(OH)}_2}} \times \frac{1000\ \text{mL Ca(OH)}_2}{1\ \cancel{\text{L Ca(OH)}_2}} = 208\ \text{mL Ca(OH)}_2$ to reach the second equivalence point. Then $\frac{1}{2}(208\ \text{mL Ca(OH)}_2) = 104\ \text{mL Ca(OH)}_2$ to reach the first equivalence point.

Check: The units (mL and mL) are correct. The magnitude of the answer makes physical sense because the concentration of the $Ca(OH)_2$ solution is so low.

(b) **Given:** after adding 25.0 mL of $Ca(OH)_2$ solution pH = 3.82 **Find:** pK_{a_1} of acid

Conceptual Plan:

mL → L then M $Ca(OH)_2$, L → mol $Ca(OH)_2$ → mol OH^- then write a balanced reaction

$\frac{1\ \text{L}}{1000\ \text{mL}}$ $M = \frac{\text{mol}}{\text{L}}$ $Ca(OH)_2 \rightarrow Ca^{2+} + 2\ OH^-$ $OH^- + H_2A \rightarrow HA^- + H_2O$

added mol OH^-, initial mol acid → equil. mol HA^-, equil. mol H_2A then

set up stoichiometry table

equil. mol HA^-, equil. mol H_2A, pH → pK_{a_1}

$pH = pK_a + \log \frac{[\text{base}]}{[\text{acid}]}$

Solution: $25.0\ \cancel{\text{mL}} \times \frac{1\ \text{L}}{1000\ \cancel{\text{mL}}} = 0.0250\ \text{L}$ then

$0.0250\ \cancel{\text{L Ca(OH)}_2} \times \frac{0.0105\ \text{mol Ca(OH)}_2}{1\ \cancel{\text{L Ca(OH)}_2}} = 0.0002625\ \text{mol Ca(OH)}_2$

$0.0002625\ \cancel{\text{mol Ca(OH)}_2} \times \frac{2\ \text{mol OH}^-}{1\ \cancel{\text{mol Ca(OH)}_2}} = 0.000525\ \text{mol OH}^-$

Set up a table to track changes:

	$OH^-(aq)$	+	$H_2A(aq)$	→	$HA^-(aq)$	+	$H_2O(l)$
Before addition	≈0.00 mol		0.00218139 mol		0.00 mol		—
Addition	0.000525 mol		—		—		—
After addition	≈0.00 mol		0.00165639 mol		0.000525 mol		—

Because the ratio of base to acid is between 0.1 and 10, it is a buffer solution. Note that the ratio of moles is the same as the ratio of concentrations because the volume for both terms is the same.

$pH = pK_a + \log \frac{[\text{base}]}{[\text{acid}]} = pK_{a_1} + \log \frac{0.000525\ \cancel{\text{mol}}}{0.00165639\ \cancel{\text{mol}}} = 3.82$ Solve for pK_{a_1}.

$pK_{a_1} = 3.82 - \log \frac{0.000525}{0.00165639} = 4.32$

Check: The units (none) are correct. The magnitude of the answer makes physical sense because there is more acid than base at equilibrium; so the pK_{a_1} is higher than the pH of the solution.

(c) **Given:** after adding 20.0 mL of $Ca(OH)_2$ solution after first equivalence point pH = 8.25
Find: pK_{a_2} of acid
Conceptual Plan:
mL → L then M $Ca(OH)_2$, L → mol $Ca(OH)_2$ → mol OH^- then write balanced reaction

$\frac{1\text{ L}}{1000\text{ mL}}$ $M = \frac{\text{mol}}{\text{L}}$ $Ca(OH)_2 \rightarrow Ca^{2+} + 2\,OH^-$ and $OH^- + HA^- \rightarrow A^{2-} + H_2O$

added mol OH^+, initial mol acid → equil. mol HA^-, equil. mol A^{2-} then

set up stoichiometry table

equil. mol HA^-, equil. mol A^{2-}, pH → pK_{a_2}

$pH = pK_a + \log \frac{[\text{base}]}{[\text{acid}]}$

Solution: $20.0\text{ mL} \times \frac{1\text{ L}}{1000\text{ mL}} = 0.0200\text{ L}$ then

$0.0200\text{ L }Ca(OH)_2 \times \frac{0.0105\text{ mol }Ca(OH)_2}{1\text{ L }Ca(OH)_2} = 0.000210\text{ mol }Ca(OH)_2 \times \frac{2\text{ mol }OH^-}{1\text{ mol }Ca(OH)_2}$

$= 0.000420$ mol OH^- Then set up a table to track changes:

	$OH^-(aq)$	+	$HA^-(aq)$	→	$A^{2-}(aq)$	+	$H_2O(l)$
Before addition	≈0.00 mol		0.00218139 mol		0.00 mol		—
Addition	0.000420 mol		—		—		—
After addition	≈0.00 mol		0.00176139 mol		0.000420 mol		—

Because the ratio of base to acid is between 0.1 and 10, it is a buffer solution. Note that the ratio of moles is the same as the ratio of concentrations because the volume for both terms is the same.

$pH = pK_a + \log \frac{[\text{base}]}{[\text{acid}]} = pK_{a_2} + \log \frac{0.000420\text{ mol}}{0.00176139\text{ mol}} = 8.25$ Solve for pK_{a_2}.

$pK_{a_2} = 8.25 - \log \frac{0.000420}{0.00176139} = 8.87$

Check: The units (none) are correct. The magnitude of the answer makes physical sense because there is more acid than base at equilibrium; so the pK_{a_2} is higher than the pH of the solution.

18.139 **Given:** excess $Mg(OH)_2$ in 1.00 L of 1.0 M NH_4Cl has pH = 9.00 **Find:** $K_{sp}(Mg(OH)_2)$
Other: $K_b(NH_3) = 1.76 \times 10^{-5}$
Conceptual Plan: M NH_4Cl → M NH_4^+ and $K_b \rightarrow K_a$ then final pH → $[H_3O^+]$ then

$NH_4Cl(aq) \rightarrow NH_4^+(aq) + Cl^-(aq)$ $K_w = K_a K_b$ $[H_3O^+] = 10^{-pH}$

M NH_4^+, M H_3O^+, K_a → x. Because x is significant compared to initial M NH_4^+, this is a buffer solution.

ICE table

The NH_4^+ is neutralized with $Mg(OH)_2$. Because $Mg(OH)_2(s) \rightleftharpoons Mg^{2+}(aq) + 2\,OH^-(aq)$, two moles of OH^- are generated for each mole of $Mg(OH)_2$ dissolved. Thus, $\frac{1}{2}(x$ mol $OH^-)$ = mol $Mg(OH)_2$ dissolved in 1.00 L of solution. Because there is 1.00 L solution, mol $Mg(OH)_2 = [Mg^{2+}]$ and $[H_3O^+] \rightarrow [OH^-]$.

$K_w = [H_3O^+][OH^-]$

Finally, write an expression for $K_{sp}(Mg(OH)_2)$ and substitute values for $[Mg^{2+}]$ and $[OH^-]$.
Solution: Because one NH_4^+ ion is generated for each NH_4Cl, $[NH_4^+] = 1.0$ M NH_4^+. Because

$K_w = K_a K_b$, rearrange to solve for K_a. $K_a = \frac{K_w}{K_b} = \frac{1.0 \times 10^{-14}}{1.76 \times 10^{-5}} = 5.6818 \times 10^{-10}$. Final pH = 9.00, so

$[H_3O^+] = 10^{-pH} = 10^{-9.00} = 1.0 \times 10^{-9}$ M. Set up an ICE table:

$NH_4^+(aq) + H_2O(l) \rightleftharpoons H_3O^+(aq) + NH_3(aq)$

	$[NH_4^+]$	$[H_3O^+]$	$[NH_3]$
Initial	1.0	≈0.00	0.00
Change	$-x$	$+x$	$+x$
Equil	$1.0 - x$	1×10^{-9}	x

$$K_a = \frac{[H_3O^+][NH_3]}{[NH_4^+]} = 5.\underline{6}818 \times 10^{-10} = \frac{(1.0 \times 10^{-9})x}{1.0 - x}$$

Solve for x. $5.\underline{6}818 \times 10^{-10}(1.0 - x) = (1.0 \times 10^{-9})x \rightarrow 5.\underline{6}818 \times 10^{-10} = (1.0 \times 10^{-9} + 5.\underline{6}818 \times 10^{-10})x \rightarrow x =$ $0.3\underline{6}2318$, so this is a buffer solution. Because there is 1.00 L of solution, 0.3$\underline{6}$2318 mol of NH_4^+ is neutralized with $Mg(OH)_2$. Because $Mg(OH)_2(s) \rightleftharpoons Mg^{2+}(aq) + 2\,OH^-(aq)$, two moles of OH^- are generated for each mole of $Mg(OH)_2$ dissolved. Thus, $\frac{1}{2}(0.3\underline{6}2318 \text{ mol } OH^-) = 0.1\underline{8}1159 \text{ mol } Mg(OH)_2$ was dissolved in 1.00 L of solution. Thus, the $[Mg^{2+}] = 0.1\underline{8}1159$ M. Because $K_w = [H_3O^+][OH^-]$,

$$[OH^-] = \frac{K_w}{[H_3O^+]} = \frac{1.0 \times 10^{-14}}{1.0 \times 10^{-9}} = 1.0 \times 10^{-5} \text{ M then}$$

$K_{sp}(Mg(OH)_2) = [Mg^{2+}][OH^-]^2 = (0.1\underline{8}1159)(1.0 \times 10^{-5})^2 = 1.8 \times 10^{-11}$.

Check: The units (none) are correct. The magnitude of the answer makes physical sense because the concentration of NH_4Cl is high; so it took a significant amount of $Mg(OH)_2$ to raise the pH to 9.00. Note that this number disagrees with the accepted value for the $K_{sp}(Mg(OH)_2)$. This is most likely due to errors in the measurements in this experiment.

18.140 **Given:** 1.0 L of 0.10 M H_2CO_3 titrated with NaOH to $[H^+] = 3.2 \times 10^{-11}$ M (assume no volume change)
Find: mass NaOH **Other:** $K_{a_1}(H_2CO_3) = 4.3 \times 10^{-7}$, $K_{a_2}(H_2CO_3) = 5.6 \times 10^{-11}$
Conceptual Plan: Comparing the $[H^+]$ to the K_as, it can be seen that the pH is just below the pK_{a_2} so the final solution is a buffer solution where the acid is HCO_3^- and the base is CO_3^{2-}. The total NaOH will be the amount needed to get to the first equivalence point and then what is needed to get to final pH. To first equivalence point, L, M $CO_3^{2-} \rightarrow$ mol $CO_3^{2-} \rightarrow$ mol NaOH then use the Henderson–Hasselbalch equation to calculate

$M = \frac{mol}{L}$ 1:1 *ratio*

the $[CO_3^{2-}]/[HCO_3^-]$. Then solve for $[CO_3^{2-}]$ knowing that $[CO_3^{2-}] + [HCO_3^-] =$ initial $[H_2CO_3]$.

$pH = pK_a + \log\frac{[base]}{[acid]}$

Then L, M $CO_3^{2-} \rightarrow$ mol $CO_3^{2-} \rightarrow$ mol NaOH beyond first equivalence point then add two NaOH

$M = \frac{mol}{L}$ 1:1 *ratio*

moles and finally total mol NaOH $\rightarrow$ g NaOH.

$\frac{40.00 \text{ g NaOH}}{1 \text{ mol NaOH}}$

Solution: To the first equivalence point,

$$1.0 \text{ L } H_2CO_3 \times \frac{0.10 \text{ mol } H_2CO_3}{1 \text{ L } H_2CO_3} \times \frac{1 \text{ mol } HCO_3^-}{1 \text{ mol } H_2CO_3} \times \frac{1 \text{ mol NaOH}}{1 \text{ mol } HCO_3^-} = 0.10 \text{ mol NaOH}$$

To the final pH, $pH = pK_a + \log\frac{[base]}{[acid]}$ so $-\log[H^+] = -\log K_{a_2} + \log\frac{[CO_3^{2-}]}{[HCO_3^-]} \rightarrow$

$$-\log(3.2 \times 10^{-11}) = -\log(5.6 \times 10^{-11}) + \log\frac{[CO_3^{2-}]}{[HCO_3^-]} \rightarrow 10.49 = 10.25 + \log\frac{[CO_3^{2-}]}{[HCO_3^-]}$$

$$\rightarrow 0.24 = \log\frac{[CO_3^{2-}]}{[HCO_3^-]} \rightarrow \frac{[CO_3^{2-}]}{[HCO_3^-]} = 10^{+0.24} = 1.7\underline{3}780 \text{ and}$$

$[CO_3^{2-}] + [HCO_3^-] = \text{initial } [H_2CO_3] = 0.10 \text{ M}$

Thus, $[HCO_3^-] = 0.10 \text{ M} - [CO_3^{2-}]$ and $\frac{[CO_3^{2-}]}{0.10 \text{ M} - [CO_3^{2-}]} = 1.7\underline{3}780 \rightarrow [CO_3^{2-}] = 1.7\underline{3}780\,(0.10 \text{ M} - [CO_3^{2-}])$

$\rightarrow [CO_3^{2-}] = 0.17\underline{3}780 \text{ M} - 1.7\underline{3}780\,[CO_3^{2-}] \rightarrow 2.7\underline{3}780\,[CO_3^{2-}] = 0.17\underline{3}780 \text{ M} \rightarrow$

$$[CO_3^{2-}] = 0.0634743 \text{ M then } 1.0 \text{ L } CO_3^{2-} \times \frac{0.0634743 \text{ mol } CO_3^{2-}}{1 \text{ L } CO_3^{2-}} \times \frac{1 \text{ mol NaOH}}{1 \text{ mol } CO_3^{2-}} = 0.0634743 \text{ mol NaOH}.$$

So the total mole NaOH = 0.10 mol + 0.0634743 mol = 0.1634743 mol NaOH. Finally,

$$0.1634743 \text{ mol NaOH} \times \frac{40.00 \text{ g NaOH}}{1 \text{ mol NaOH}} = 6.5 \text{ g NaOH}.$$

Check: The units (g) are correct. The magnitude of the answer makes physical sense because if all of the acid were fully titrated, 0.20 mol of NaOH (or 8 g) would be required. The pH indicates that the titration is most of the way there.

18.141 (a) **Given:** $Au(OH)_3$ in pure water **Find:** molar solubility (S) **Other:** $K_{sp} = 5.5 \times 10^{-46}$

Conceptual Plan: Use equations derived in Problems 18.19 and 18.20 and solve for S. Then

for ionic compound, A_mX_n, $K_{sp} = [A^{n+}]^m [X^{m-}]^n = m^m n^n S^{m+n}$

check answer for validity.

Solution: For $Au(OH)_3$, $K_{sp} = 5.5 \times 10^{-46}$, $A = Au^{3+}$, $m = 1$, $X = OH^-$, and $n = 3$; so $K_{sp} = [Au^{3+}][OH^-]^3 = 5.5 \times 10^{-46} = 3^3S^4$. Rearrange to solve for S. $S \sqrt[4]{\frac{5.5 \times 10^{-46}}{27}} = 2.1 \times 10^{-12}$ M. This answer suggests that the $[OH^-] = 3(2.1 \times 10^{-12} \text{ M}) = 6.3 \times 10^{-12}$ M. This result is lower than what is found in pure water (1.0×10^{-7} M), so substitute this value for $[OH^-]$ and solve for $S = [Au^{3+}]$. So $K_{sp} = [Au^{3+}][OH^-]^3 = 5.5 \times 10^{-46} = S(1.0 \times 10^{-7} \text{ M})^3$ and solving for S gives $S = 5.5 \times 10^{-25}$ M.

Check: The units (M) are correct. Because K_{sp} is so small, the autoionization of water must be considered and the solubility is smaller than what is normally anticipated.

(b) **Given:** $Au(OH)_3$ in 1.0 M HNO_3 **Find:** molar solubility (S) **Other:** $K_{sp} = 5.5 \times 10^{-46}$

Conceptual Plan: Because HNO_3 is a strong acid, it will neutralize the gold(III) hydroxide (through the reaction of H^+ with OH^+ to form water (the reverse of the autoionization of water equilibrium). Write balanced equations for dissolving the solid and for the neutralization reaction. Add these two reactions to get the desired overall reaction. Using the rules from Chapter 16, multiply the individual reaction Ks to get the overall K for the sum of these reactions. Then M HNO_3, $K \rightarrow S$.

ICE table

Solution: Identify the solid as $Au(OH)_3$. Write the individual reactions and add them together.

$$Au(OH)_3(s) \rightleftharpoons Au^{3+}(aq) + \cancel{3\,OH^-(aq)} \qquad K_{sp} = 5.5 \times 10^{-46}$$

$$3\,H^+(aq) + \cancel{3\,OH^-(aq)} \rightleftharpoons 3\,H_2O(l) \qquad \left(\frac{1}{K_w}\right)^3 = \left(\frac{1}{1.0 \times 10^{-14}}\right)^3$$

$$Au(OH)_3(s) + 3\,H^+(aq) \rightleftharpoons Au^{3+}(aq) + 3\,H_2O(l)$$

Because the overall reaction is the sum of the dissolution reaction and three times the reverse of the autoionization of water reaction, the overall reaction

$K = K_{sp}\left(\frac{1}{K_w}\right)^3 = (5.5 \times 10^{-46})\left(\frac{1}{1.0 \times 10^{-14}}\right)^3 = 5.5 \times 10^{-4} = \frac{[Au^{3+}]}{[H^+]^3}$; then because HNO_3 is a strong acid, it will completely dissociate to H^+ and NO_3^-. Set up an ICE table:

$$Au(OH)_3(s) + 3\,H^+(aq) \rightleftharpoons Au^{3+}(aq) + 3\,H_2O(l)$$

	$[H^+]$	$[Au^{3+}]$
Initial	1.0	0.00
Change	$-3S$	$+S$
Equil	$1.0 - 3S$	$+S$

$$K = \frac{[Au^{3+}]}{[H^+]^3} = 5.5 \times 10^{-4} = \frac{S}{(1.0 - 3S)^3}.$$

Assume that S is small ($3S << 1.0$), so $\frac{S}{(1.0 - \cancel{3S})^3} = 5.5 \times 10^{-4}$ M $= \frac{S}{(1.0)^3} = S$. Confirm that the assumption is valid. $\frac{3(5.5 \times 10^{-4})}{1.0} \times 100\% = 0.017\% << 5\%$, so the assumption is valid.

Check: The units (M) are correct. K is much larger than the original K_{sp}, so the solubility of $Au(OH)_3$ increases over that of pure water.

18.142 **Given:** excess AgCl in 0.10 M KI **Find:** $[I^-]$ **Other:** $K_{sp}(AgCl) = 1.77 \times 10^{-10}$, $K_{sp}(AgI) = 8.51 \times 10^{-17}$

Conceptual Plan:

Because the KI does not generate a common ion with AgCl, use equations derived in Problems 18.19 and 18.20 and solve for *S*. So $S(AgCl) = [Ag^+]$ then use the equation derived in Problem 18.19 to get K_{sp} (AgI)

for ionic compound, A_mX_n, $K_{sp} = [A^{n+}]^m [X^{m-}]^n = m^m n^n S^{m+n}$

for ionic compound, A_mX_n, $K_{sp} = [A^{n+}]^m [X^{m-}]^n$

expression. Substitute value for $[Ag^+]$ into K_{sp} (AgI) expression and solve for $[I^-]$.

Solution: For AgCl, $K_{sp}(AgCl) = 1.77 \times 10^{-10}$, $A = Ag^+$, $m = 1$, $X = Cl^-$, and $n = 1$; so $K_{sp} = [Ag^+][Cl^-] = 1.77 \times 10^{-10} = S^2$. Rearrange to solve for *S*. $S = 1.33 \times 10^{-5}$ M. Then $K_{sp}(AgI) = 8.51 \times 10^{-17}$, $A = Ag^+$, $m = 1$, $X = I^-$, and $n = 1$; so $K_{sp} = [Ag^+][I^-] = 8.51 \times 10^{-17}$. Substitute value for $[Ag^+]$ into K_{sp} (AgI) expression and solve for $[I^-]$. $8.51 \times 10^{-17} = (1.33 \times 10^{-5})[I^-]$, so $[I^-] = 6.40 \times 10^{-12}$ M

Check: The units (M) are correct. Because silver ions are generated from the excess AgCl in solution, the iodide ions are converted to AgI and the concentration of remaining iodide ions is controlled by the solubility of the AgCl.

18.143 **Given:** 1.00 L of 0.100 M $MgCO_3$ **Find:** volume of 0.100 M Na_2CO_3 to precipitate 99% of Mg^{2+} ions

Other: $K_{sp}(MgCO_3) = 6.82 \times 10^{-6}$

Conceptual Plan: Because 99% of the Mg^{2+} ions are to be precipitated, 1% of the ions will be left in solution.

$(0.01)(0.100\ M\ Mg^{2+})$

Let x = required volume (in L). Calculate the amount of CO_3^{2-} added and the amount of Mg^{2+} that does not precipitate and remains in solution. Use these to calculate the $[Mg^{2+}]$ and $[CO_3^{2-}]$. The solubility product constant (K_{sp}) is the equilibrium expression for a chemical equation representing the dissolution of an ionic compound. The expression of the solubility product constant of A_mX_n is $K_{sp} = [A^{n+}]^m[X^{m-}]^n$. Substitute these expressions in this equation to $[Mg^{2+}], [CO_3^{2-}], K_{sp} \rightarrow x$.

for ionic compound, A_mX_n, $K_{sp} = [A^{n+}]^m[X^{m-}]^n$

Solution: Because 99% of the Mg^{2+} ions are to be precipitated, 1% of the ions will be left in solution, or $(0.01)(0.100\ M\ Mg^{2+}) = 0.001\ M\ Mg^{2+}$. Let x = required volume (in L). The volume of the solution after precipitation is $(1.00 + x)$. The amount of CO_3^{2-} added $= (0.100M)(x\ L) = 0.100x$ mol CO_3^{2-}. The amount of Mg^{2+} that does not precipitate and remains in solution is $(0.100\ M)(1.00\ L)(0.01) = 1.00 \times 10^{-3}$ mol, and the amount that precipitates = 0.099 mol, which is also equal to the amount of CO_3^{2-} used. The amount of CO_3^{2-} remaining in solution is $(0.10x - 0.099)$. Thus, $[Mg^{2+}] = 1.00 \times 10^{-3}\text{mol}/(1.00 + x)$ L and $[CO_3^{2-}] = (0.10x - 0.099)\ \text{mol}/(1.00 + x)$ L. Then $K_{sp} = 6.82 \times 10^{-6}$, $A = Mg^{2+}$, $m = 1$, $X = CO_3^{2-}$, and

$n = 1$; so $K_{sp} = 6.82 \times 10^{-6} = [Mg^{2+}][CO_3^{2-}] = \dfrac{(1.00 \times 10^{-3})(0.10x - 0.099)}{(1.00 + x)^2}$. Rearrange to solve for x.

$1.00 + 2.00x + x^2 = \dfrac{1.0 \times 10^{-4}x - 9.9 \times 10^{-5}}{6.82 \times 10^{-6}} \rightarrow 0 = x^2 - 12.6628x + 15.5161$. Using quadratic equation,

$x = 1.375\ L = 1.4\ L$.

Check: The units (L) are correct. The necessary concentration is very low, so the volume is fairly large.

18.144 **Given:** solution with 0.40 M HCN **Find:** solubility of CuI

Other: $K_{sp}(CuI) = 1.1 \times 10^{-12}$, $K_f(Cu(CN)_2^-) = 1 \times 10^{24}$

Conceptual Plan: Write balanced equations for the solubility of CuI and reaction with CN^- and expressions for K_{sp} and K_f. Use initial concentrations to set up an ICE table. Because the *K* is so large, assume that reaction essentially goes to completion. Solve for $[I^-]$ at equilibrium.

Solution: Write two reactions and combine.

$CuI(s) \rightleftharpoons \cancel{Cu^+(aq)} + I^-(aq)$ with $K_{sp} = [Cu^+][I^-] = 1.1 \times 10^{-12}$

$\cancel{Cu^+(aq)} + 2\,CN^-(aq) \rightleftharpoons Cu(CN)_2^-(aq)$ with $K_f = \dfrac{[Cu(CN)_2^-]}{[Cu^+][CN^-]^2} = 1 \times 10^{24}$

$CuI(s) + 2\,CN^-(aq) \rightleftharpoons Cu(CN)_2^-(aq) + I^-(aq)$

with $K = K_{sp}K_f = \cancel{[Cu^+]}[I^-]\dfrac{[Cu(CN)_2^-]}{\cancel{[Cu^+]}[CN^-]^2} = (1.1 \times 10^{-12})(1 \times 10^{24})$

$K = \dfrac{[Cu(CN)_2^-][I^-]}{[CN^-]^2} = 1.1 \times 10^{12}$

Because $K_{sp} = [Cu^-][I^-] = 1.1 \times 10^{-12}$ without HCN present,
$[Cu^+] = [I^-] = \sqrt{1.1 \times 10^{-12}} = 1.\underline{0}49 \times 10^{-6}$ M. Set up an ICE table with initial concentrations. Because K is so large and because initially $[CN^-] > [I^-]$, the reaction essentially goes to completion; then write the equilibrium expression and solve for x.

$CuI(s) + 2\,CN^-(aq) \rightleftharpoons Cu(CN)_2^-(aq) + I^-(aq)$

	$[CN^-]$	$[Cu(CN)_2^-]$	$[I^-]$
Initial	0.40	0.00	$1.\underline{0}49 \times 10^{-6}$
Change	≈ -0.40	$\approx +1/2\,(0.40)$	$\approx +1/2\,(0.40)$
Equil	$0.40 - 2x$	x	$1.\underline{0}49 \times 10^{-6} + x$

$K = \dfrac{[Cu(CN)_2^-][I^-]}{[CN^-]^2} = \dfrac{(x)(1.\underline{0}49 \times 10^{-6} + x)}{(0.40 - 2x)^2} = \underline{1}.1 \times 10^{12}$. Assume that $x >> 1.\underline{0}49 \times 10^{-6}$ M, so

$\dfrac{(x)(x)}{(0.40 - 2x)^2} = \underline{1}.1 \times 10^{12} = \dfrac{(x)^2}{(0.40 - 2x)^2} \rightarrow \sqrt{\underline{1}.1 \times 10^{12}} = \dfrac{x}{0.40 - 2x} = \underline{1}.049 \times 10^6 \rightarrow$

$x = 0.\underline{1}9995$. So $[I^-] = x = 0.\underline{1}9995\text{ M} = 0.2\text{ M } I^-$. Because $1.\underline{0}48 \times 10^{-6}$ M is insignificant compared to x, the assumption is valid. The solubility of CuI is 0.2 M, or one-half the initial concentration of HCN.

Check: The units (M) are correct. Because K is so large, the reaction essentially goes to completion and the solubility of CuI is dramatically increased.

18.145 **Given:** 1.0 L solution with 0.10 M $Ba(OH)_2$ and excess $Zn(OH)_2$ **Find:** pH
Other: $K_{sp}(Zn(OH)_2) = 3 \times 10^{-15}$, $K_f(Zn(OH)_4^{2-}) = 2 \times 10^{15}$
Conceptual Plan: Because $[Ba(OH)_2] = 0.10$ M, $[OH^-] = 0.20$ M. Write balanced equations for the solubility of $Zn(OH)_2$ and reaction with excess OH^- and expressions for K_{sp} and K_f. Use initial concentrations to set up an ICE table. Solve for $[OH^-]$ at equilibrium. Then $[OH^-] \rightarrow [H_3O^+] \rightarrow$ pH.

$K_w = [H_3O^+][OH^-]$ $\quad$ $pH = -\log[H_3O^+]$

Solution: Write two reactions and combine.

$Zn(OH)_2(s) \rightleftharpoons \cancel{Zn^{2+}(aq)} + \cancel{2}\,OH^-(aq)$ $\quad$ with $K_{sp} = [Zn^{2+}][OH^-]^2 = 3 \times 10^{-15}$

$\cancel{Zn^{2+}(aq)} + \cancel{4}^{2}\,OH^-(aq) \rightleftharpoons Zn(OH)_4^{2-}(aq)$ $\quad$ with $K_f = \dfrac{[Zn(OH)_4^{2-}]}{[Zn^{2+}][OH^-]^4} = 2 \times 10^{15}$

$Zn(OH)_2(s) + 2\,OH^-(aq) \rightleftharpoons Zn(OH)_4^{2-}(aq)$

with $K = K_{sp}K_f = \cancel{[Zn^{2+}][OH^-]^2}\dfrac{[Zn(OH)_4^{2-}]}{\cancel{[Zn^{2+}]}[OH^-]^{\cancel{4}\,2}} = (3 \times 10^{-15})(2 \times 10^{15})$

$K = \dfrac{[Zn(OH)_4^{2-}]}{[OH^-]^2} = 6$. Set up an ICE table with initial concentration and solve for x.

$Zn(OH)_2(s) + 2\,OH^-(aq) \rightleftharpoons Zn(OH)_4^{2-}(aq)$

	$[OH^-]$	$[Zn(OH)_4^{2-}]$
Initial	0.20	0.00
Change	$-2x$	$+x$
Equil	$0.20 - 2x$	x

$K = \dfrac{[Zn(OH)_4^{2-}]}{[OH^-]^2} = \dfrac{x}{(0.20 - 2x)^2} = 6$

$x = 6(0.20 - 2x)^2 \rightarrow x = 6(4x^2 - 0.80x + 0.040) \rightarrow x = 24x^2 - 4.8x + 0.24 \rightarrow 0 = 24x^2 - 5.8x + 0.24.$
Using quadratic equation, $x = 0.0530049 \rightarrow [OH^-] = 0.20 - 2x = 0.20 - 2(0.05\underline{3}0049) = 0.0\underline{9}3990$ M OH^-.
Then $K_w = [H_3O^+][OH^-]$, so

$[H_3O^+] = \dfrac{K_w}{[OH^-]} = \dfrac{1.0 \times 10^{-14}}{0.0\underline{9}3990} = 1.\underline{0}6394 \times 10^{-13}$ M. Finally,

$pH = -\log[H_3O^+] = -\log(1.\underline{0}6394 \times 10^{-13}) = 12.97.$

Check: The units (none) are correct. Because the pH of the solution before the addition of the $Zn(OH)_2$ is 13.30, the reaction decreases the $[OH^-]$ and the pH drops.

18.146 **Given:** 1.00 L of 2.0 M $HC_2H_3O_2$ and 1.0 M $C_2H_3O_2^-$ **Find:** amount of HCl to pH = 4.00
Other: $K_a(HC_2H_3O_2) = 1.8 \times 10^{-5}$
Conceptual Plan: Identify acid and base components then final pH, pK_a → $[C_2H_3O_2^-]/[HC_2H_3O_2]$ then

acid = $HC_2H_3O_2$ base = $C_2H_3O_2^-$ $\qquad$ $pH = pK_a + \log\frac{[\text{base}]}{[\text{acid}]}$

$[C_2H_3O_2^-]$, L → mol $C_2H_3O_2^-$ and $[HC_2H_3O_2]$, L → mol $HC_2H_3O_2$ then write balanced equation then

$M = \frac{\text{mol}}{\text{L}}$ $\qquad$ $M = \frac{\text{mol}}{\text{L}}$ $\qquad$ $HCl + C_2H_3O_2^- \rightarrow HC_2H_3O_2 + Cl^-$

initial mol $NaC_2H_3O_2$, initial mol $HC_2H_3O_2$, final $[C_2H_3O_2^-]/[HC_2H_3O_2]$ → mol HCl → g HCl.

set up stoichiometry table $\qquad$ $\frac{36.46 \text{ g HCl}}{1 \text{ mol HCl}}$

Solution: Because $K_a(HC_2H_3O_2) = 1.8 \times 10^{-5}$, $pK_a = -\log K_a = -\log(1.8 \times 10^{-5}) = 4.74$.

Then $pH = pK_a + \log\frac{[\text{base}]}{[\text{acid}]} = 4.74 + \log\frac{[C_2H_3O_2^-]}{[HC_2H_3O_2]} = 4.00$. Solve for $\frac{[C_2H_3O_2^-]}{[HC_2H_3O_2]}$.

$$\log\frac{[C_2H_3O_2^-]}{[HC_2H_3O_2]} = 4.00 - 4.74 = -0.74 \rightarrow \frac{[C_2H_3O_2^-]}{[HC_2H_3O_2]} = 10^{-0.74} = 0.1\underline{8}1970$$

Then $\frac{2.0 \text{ mol } HC_2H_3O_2}{1 \cancel{L}} \times 1.00 \cancel{L} = 2.0 \text{ mol } HC_2H_3O_2$ and

$\frac{1.0 \text{ mol } C_2H_3O_2^-}{1 \cancel{L}} \times 1.00 \cancel{L} = 1.0 \text{ mol } C_2H_3O_2^-$. Set up a table to track changes:

	$HCl(aq)$ +	$C_2H_3O_2^-(aq)$ →	$HC_2H_3O_2(aq)$ +	$Cl^-(aq)$
Before addition	0.0 mol	1.0 mol	2.0 mol	0.0 mol
Addition	$+x$ mol	$-x$ mol	$+x$ mol	$+x$ mol
After addition	≈0.00 mol	$(1.0 - x)$ mol	$(2.0 + x)$ mol	x mol

Combine the final line of table with $\frac{[C_2H_3O_2^-]}{[HC_2H_3O_2]} = 0.1\underline{8}1970$.

So $\frac{[C_2H_3O_2^-]}{[HC_2H_3O_2]} = 0.1\underline{8}1970 = \frac{(1.0 - x)}{(2.0 + x)}$. Solve for x. $0.1\underline{8}1970(2.0 + x) = 1.0 - x \rightarrow x + 0.1\underline{8}1970x =$
$1.0 - 0.3\underline{6}394 \rightarrow 1.1\underline{8}1970x = 0.6\underline{3}606 \rightarrow x = 0.5\underline{3}8135$ mol HCl = 0.54 mol HCl then

$$0.5\underline{3}8135 \cancel{\text{mol HCl}} \times \frac{36.46 \text{ g HCl}}{1 \cancel{\text{mol HCl}}} = 1\underline{9}.6204 \text{ g HCl} = 20. \text{ g HCl}$$

Check: The units (mol and g) are correct. The magnitude of the answer makes physical sense because the initial $[C_2H_3O_2^-]/[HC_2H_3O_2] = 0.50$ and is reduced to ~0.18 and the starting number of moles of base = 1.0 mol; so much less than a mole of HCl must be added.

Conceptual Problems

18.147 If the concentration of the acid is greater than the concentration of the base, the pH will be less than the pK_a. If the concentration of the acid is equal to the concentration of the base, the pH will be equal to the pK_a. If the concentration of the acid is less than the concentration of the base, the pH will be greater than the pK_a.

(a) $pH < pK_a$
(b) $pH > pK_a$
(c) $pH = pK_a$; the OH^- will convert half of the acid to base.
(d) $pH > pK_a$; the OH^- will convert more than half of the acid to base.

18.148 As long as the [base]/[acid] is between 0.1 and 10, the buffer will be active and the buffer capacity will not have been exceeded.

(a) No, the buffer capacity is not exceeded because [base]/[acid] = 0.22/0.08.
(b) No, the buffer capacity is not exceeded because [base]/[acid] = 0.18/0.12.
(c) Yes, the buffer capacity will be exceeded because all of the acid is converted to base.
(d) No, the buffer capacity is not exceeded because [base]/[acid] = 0.19/0.11.

18.149 Only (a) is the same for both solutions. The volume to the first equivalence point will be the same because the number of moles of acid is the same. The pH profiles of the two titrations will be different.

18.150 Only (c) is correct. If the volume of base is twice as high, the acid concentration is twice as high and the weaker acid has the higher pH at the equivalence point.

18.151 (a) The solubility will be unchanged because the pH is constant and no common ions are added.

(b) The solubility will be less because extra fluoride ions are added, suppressing the solubility of the fluoride ionic compound.

(c) The solubility will increase because some of the fluoride ion will be converted to HF; so more of the ionic compound can be dissolved.

19 Free Energy and Thermodynamics

Review Questions

19.1 The second law of thermodynamics can be summarized as energy spreads out (or disperses) from a more concentrated to a less concentrated form. An example is a hot slice of pizza is removed from the oven and cools, spreading its thermal energy to the plate and the surrounding air. A second example is burning gasoline in an internal combustion engine. Only a small amount of the energy (about 20%) goes toward driving the car forward. The rest of the energy generated by the combustion reaction is dissipated as heat to the car engine and then other car components.

19.2 A spontaneous process occurs without ongoing outside intervention (such as the performance of work by some external force). For example, when you drop a book in a gravitational field, the book spontaneously drops to the floor.

19.3 A nonspontaneous process will not occur without ongoing outside intervention (such as the performance of work by some external force)—for example, when you pick up a book from the floor and set it on a table, increasing its potential energy in a gravitational field.

19.4 In thermodynamics, the spontaneity of a reaction is the direction in which and the extent to which a chemical reaction proceeds. Kinetics is the speed of the reaction—how fast a reaction takes place. A reaction may be thermodynamically spontaneous but kinetically slow at a given temperature. For example, the conversion of diamond to graphite is thermodynamically spontaneous. But your diamonds will not become worthless anytime soon because the process is extremely slow kinetically. Although the rate of a spontaneous process can be increased by the use of a catalyst, a nonspontaneous process cannot be made spontaneous by the use of a catalyst. Catalysts affect only the rate of a reaction, not the spontaneity.

19.5 Entropy (S) is a thermodynamic function that is proportional to the number of energetically equivalent ways to arrange the components of a system to achieve a particular state. Entropy, like enthalpy, is a state function—its value depends only on the state of the system, not on how the system got to that state. Therefore, for any process, the change in entropy is just the entropy of the final state minus the entropy of the initial state, or $\Delta S = S_{\text{final}} - S_{\text{initial}}$.

19.6 The entropy of a gas increases when it expands into a vacuum because the number of possible states increases (such as where the particle is located).

19.7 Microstates are the number of internal arrangements. These microstates give rise to the same external arrangement, or macrostate. If there are three gas particles (A, B, and C) and two containers, the fact that the first container has two particles and the other container has one particle is a macrostate. The fact that particles A and C are in the first container and particle B is in the other container is a microstate. There are at least as many microstates as there are macrostates.

19.8 The entropy of a state is proportional to the number of energetically equivalent ways to arrange the components of the system to achieve a particular state. This implies that the state with the highest entropy also has the greatest dispersal of energy.

19.9 The second law of thermodynamics states that for any spontaneous process, the entropy of the universe increases ($\Delta S_{\text{univ}} > 0$). The criterion for spontaneity is the entropy of the universe. Processes that increase the entropy of the universe—those that result in greater dispersal or randomization of energy—occur spontaneously. Processes that decrease the entropy of the universe do not occur spontaneously. Heat travels from a substance at higher temperature to one at lower temperature because the process disperses thermal energy. The cooler object has less thermal energy, so transferring heat from the warmer object results in greater energy randomization—the energy that was concentrated in the hot substance becomes dispersed between the two substances.

19.10 The entropy of a sample of matter increases as it changes state from a solid to a liquid and then to a gas because the amount of molecular motion and the amount of thermal energy are greatest in the gas phase and least in the solid phase.

19.11 The third law of thermodynamics states that the entropy of a perfect crystal at absolute zero (0 K) is zero. For enthalpy, we defined a standard state so that we could define a "zero" for the scale. This is not necessary for entropy because there is an absolute zero.

19.12 Gases have much greater standard entropy because they have more energetically equivalent ways to arrange their components, which in turn results in greater energy dispersal at 25 °C.

19.13 The larger the molar mass, the greater the entropy at 25 °C. For a given state of matter, entropy generally increases with increasing molecular complexity.

19.14 To calculate ΔS°_{rxn}, subtract the standard entropies of the reactants multiplied by their stoichiometric coefficients, from the standard entropies of the products multiplied by their stoichiometric coefficients, or $\Delta S^\circ_{rxn} = \sum n_p S^\circ(\text{products}) - \sum n_r S^\circ(\text{reactants})$, where n_p represents the stoichiometric coefficients of the products, n_r represents the stoichiometric coefficients of the reactants, and S° represents the standard entropies. Keep in mind when using this equation that, unlike enthalpies of formation, which are equal to zero for elements in their standard states, standard entropies are always nonzero at 25 °C.

19.15 When water freezes at temperatures below 0 °C, the entropy of the water decreases, yet the process is spontaneous because the entropy of the universe increases ($\Delta S_{univ} > 0$). The entropy of the system can decrease ($\Delta S_{sys} < 0$) as long as the entropy of the surroundings increases by a greater amount ($\Delta S_{surr} > -\Delta S_{sys}$) so that the overall entropy of the universe undergoes a net increase. For liquid water freezing, the change in entropy for the system (ΔS_{sys}) is negative because the water becomes more ordered. For ΔS_{univ} to be positive, ΔS_{surr} must be positive and greater in absolute value (or magnitude) than ΔS_{sys}. We learned in Chapter 10 that freezing is an exothermic process: it gives off heat to the surroundings. If we think of entropy as the dispersal or randomization of energy, the release of heat energy by the system disperses that energy into the surroundings, increasing the entropy of the surroundings. The freezing of water below 0 °C increases the entropy of the universe because the heat given off to the surroundings increases the entropy of the surroundings to a sufficient degree to overcome the entropy decrease in the water. The freezing of water becomes nonspontaneous above 0 °C because the magnitude of the increase in the entropy of the surroundings due to the dispersal of energy into the surroundings is temperature-dependent. The higher the temperature, the smaller the percent increase in entropy for a given rise in temperature. Therefore, the impact of the heat released to the surroundings by the freezing of water depends on the temperature of the surroundings—the higher the temperature, the smaller the impact.

19.16 Exothermic processes ($-\Delta H_{sys}$) tend to be spontaneous at low temperatures because they increase the entropy of the surroundings ($+\Delta S_{surr}$). Because $\Delta S_{surr} = \dfrac{-\Delta H_{sys}}{T}$, as temperature increases, a given negative ΔH produces a smaller positive ΔS_{surr}; therefore, exothermicity becomes less of a determining factor for spontaneity as temperature increases.

19.17 The change in Gibbs free energy for a process is proportional to the negative of ΔS_{univ}. Because ΔS_{univ} is a criterion for spontaneity, ΔG is also a criterion for spontaneity (although opposite in sign).

19.18 (a) ΔH negative, ΔS positive. If a reaction is exothermic ($\Delta H < 0$) and the change in entropy for the reaction is positive ($\Delta S > 0$), the change in free energy will be negative at all temperatures and the reaction will therefore be spontaneous at all temperatures.

(b) ΔH positive, ΔS negative. If a reaction is endothermic ($\Delta H > 0$) and the change in entropy for the reaction is negative ($\Delta S < 0$), the change in free energy will be positive at all temperatures and the reaction will therefore be nonspontaneous at all temperatures.

(c) ΔH negative, ΔS negative. If a reaction is exothermic ($\Delta H < 0$) and the change in entropy for the reaction is negative ($\Delta S < 0$), the change in free energy will depend on temperature. The reaction will be spontaneous at low temperatures but nonspontaneous at high temperatures.

(d) ΔH positive, ΔS positive. If a reaction is endothermic ($\Delta H > 0$) and the change in entropy for the reaction is positive ($\Delta S > 0$), the change in free energy will also depend on temperature. The reaction will be nonspontaneous at low temperatures, but spontaneous at high temperatures.

19.19 The three ways of calculating the ΔG°_{rxn} are:

- Use tabulated values of standard enthalpies of formation to calculate ΔH°_{rxn} and use tabulated values of standard entropies to calculate ΔS°_{rxn}; then use the values of ΔH°_{rxn} and ΔS°_{rxn} calculated in these ways to calculate the standard free energy change for a reaction by using the equation $\Delta G^\circ_{rxn} = \Delta H^\circ_{rxn} - T\Delta S^\circ_{rxn}$.
- Use tabulated values of the standard free energies of formation to calculate ΔG°_{rxn} using an equation similar to that used for standard enthalpy of a reaction $\Delta G^\circ_{rxn} = \sum n_p \Delta G^\circ_f(\text{products}) - \sum n_r \Delta G^\circ_f(\text{reactants})$.
- Use a reaction pathway or stepwise reaction to sum the changes in free energy for each of the steps in a manner similar to Hess's law, used in Chapter 10 for the enthalpy of stepwise reactions.

The method to calculate the free energy of a reaction at temperatures other than at 25 °C is the first method. The second method is only applicable at 25 °C. The third method is only applicable at the temperature of the individual reactions, generally 25 °C.

19.20 The change in free energy of a chemical reaction represents the maximum amount of energy available, or free, to do work (if ΔG°_{rxn} is negative).

19.21 The standard free energy change for a reaction (ΔG°_{rxn}) applies only to standard conditions. For a gas, standard conditions are those in which the pure gas is present at a partial pressure of 1 atm. For nonstandard conditions, we need to calculate ΔG_{rxn} (not ΔG°_{rxn}) to predict spontaneity.

19.22 Even though ΔG°_{vap} is positive at room temperature, liquid water is in equilibrium with water vapor at a pressure of 0.0313 atm. The vapor pressure of water is just not as high as 1 atm, the standard conditions of a gas. This means that some water will evaporate.

19.23 The free energy of reaction under nonstandard conditions (ΔG_{rxn}) can be calculated from ΔG°_{rxn} using the relationship $\Delta G_{rxn} = \Delta G^\circ_{rxn} + RT\ln Q$, where Q is the reaction quotient (defined in Section 16.7), T is the temperature in K, and R is the gas constant in the appropriate units (8.314 J/mol K).

19.24 The relationship between ΔG°_{rxn} and K is $\Delta G^\circ_{rxn} = -RT\ln K$. When ΔG°_{rxn} is positive, the log of K is negative and $K < 1$. Under standard conditions (when $Q = 1$), the reaction is spontaneous in the reverse direction. When ΔG°_{rxn} is zero, the log of K is zero and $K = 1$. The reaction happens to be at equilibrium under standard conditions. When ΔG°_{rxn} is negative, the log of K is positive and $K > 1$. Under standard conditions (when $Q = 1$), the reaction is spontaneous in the forward direction.

Entropy, the Second Law of Thermodynamics, and the Direction of Spontaneous Change

19.25 (a) and (c) are spontaneous processes.

19.26 (a) and (c) are nonspontaneous processes. Nonspontaneous processes are not impossible. Work of some form must be added to make these processes proceed.

19.27 System B has the greatest entropy. There is only one energetically equivalent arrangement for System A. However, the particles of System B may exchange positions for a second energetically equivalent arrangement.

19.28 There is only one energetically equivalent arrangement for System A. There are 3!, or 6, energetically equivalent arrangements for System B. System B has the greatest entropy because there are more energetically equivalent arrangements for System B.

19.29 (a) $\Delta S > 0$ because a gas is being generated.
(b) $\Delta S < 0$ because 2 moles of gas are being converted to 1 mole of gas.
(c) $\Delta S < 0$ because a gas is being converted to a solid.
(d) $\Delta S < 0$ because 4 moles of gas are being converted to 2 moles of gas.

19.30 (a) $\Delta S < 0$ because a gas is being converted to a solid.
(b) $\Delta S < 0$ because 5 moles of gas are being converted to 4 moles of gas.
(c) $\Delta S > 0$ because 2 moles of gas are being converted to 3 moles of gas.
(d) $\Delta S < 0$ because 2 moles of gas are being converted to a solid.

19.31 The molar entropy of a substance increases with increasing temperatures. The kinetic energy and the molecular motion increase. The substance will have access to an increased number of energy levels.

19.32 The third law of thermodynamics states that the entropy of a perfect crystal at absolute zero (0 K) is zero. For enthalpy, we defined a standard state so that we could define a "zero" for the scale. This is not necessary for entropy because there is an absolute zero for the entropy scale.

19.33 (a) $CO_2(g)$ because it has greater molar mass/complexity
(b) $CH_3OH(g)$ because it is in the gas phase
(c) $CO_2(g)$ because it has greater molar mass/complexity
(d) $SiH_4(g)$ because it has greater molar mass
(e) $CH_3CH_2CH_3(g)$ because it has greater molar mass/complexity
(f) $NaBr(aq)$ because a solution has more entropy than does a solid crystal

19.34 (a) $NaNO_3(aq)$ because a solution has more entropy than does a solid crystal
(b) $CH_3CH_3(g)$ because it has greater molar mass/complexity
(c) $Br_2(g)$ because it is in the gas phase
(d) $Br_2(g)$ because it has greater molar mass
(e) $PCl_5(g)$ because it has greater molar mass/complexity
(f) $CH_3CH_2CH_2CH_3(g)$ because it has greater complexity

19.35 (a) $He(g) < Ne(g) < SO_2(g) < NH_3(g) < CH_3CH_2OH(g)$. All are in the gas phase. From He to Ne, there is an increase in molar mass; beyond that, the molecules increase in complexity.
(b) $H_2O(s) < H_2O(l) < H_2O(g)$. Entropy increases as we go from a solid to a liquid to a gas.
(c) $CH_4(g) < CF_4(g) < CCl_4(g)$. Entropy increases as the molar mass increases.

19.36 (a) $F_2(g) < Cl_2(g) < Br_2(g) < I_2(g)$. All are in the gas phase. Entropy increases as the molar mass increases.
(b) $H_2O(g) < H_2S(g) < H_2O_2(g)$. Entropy increases as the molar mass and the complexity of the molecules increase.
(c) $C(s, \text{diamond}) < C(s, \text{graphite}) < C(s, \text{amorphous})$. Entropy increases as the complexity increases. The diamond structure is ordered in all three dimensions. Graphite has ordered sheets that can slide with respect to each other. The amorphous carbon has no long-range order and therefore would have the highest entropy of the three.

19.37 (a) **Given:** $C_2H_4(g) + H_2(g) \rightarrow C_2H_6(g)$ **Find:** ΔS°_{rxn}
Conceptual Plan: $\Delta S^\circ_{rxn} = \sum n_p S^\circ(\text{products}) - \sum n_r S^\circ(\text{reactants})$
Solution:

Reactant/Product	S°(J/mol K from Appendix IVB)
$C_2H_4(g)$	219.3
$H_2(g)$	130.7
$C_2H_6(g)$	229.2

Be sure to pull data for the correct formula and phase.

$$\begin{aligned}\Delta S^\circ_{rxn} &= \sum n_p S^\circ(\text{products}) - \sum n_r S^\circ(\text{reactants})\\ &= [1(S^\circ(C_2H_6(g)))] - [1(S^\circ(C_2H_4(g))) + 1(S^\circ(H_2(g)))]\\ &= [1(229.2\ \text{J/K})] - [1(219.3\ \text{J/K}) + 1(130.7\ \text{J/K})]\\ &= [229.2\ \text{J/K}] - [350.0\ \text{J/K}]\\ &= -120.8\ \text{J/K} \quad \text{The moles of gas are decreasing.}\end{aligned}$$

Check: The units (J/K) are correct. The answer is negative, which is consistent with 2 moles of gas going to 1 mole of gas.

(b) **Given:** $C(s) + H_2O(g) \rightarrow CO(g) + H_2(g)$ **Find:** ΔS°_{rxn}
Conceptual Plan: $\boldsymbol{\Delta S^\circ_{rxn} = \sum n_p S^\circ(\text{products}) - \sum n_r S^\circ(\text{reactants})}$
Solution:

Reactant/Product	S°(J/mol K from Appendix IVB)
$C(s)$	5.7
$H_2O(g)$	188.8
$CO(g)$	197.7
$H_2(g)$	130.7

Be sure to pull data for the correct formula and phase.

$$\begin{aligned}\Delta S^\circ_{rxn} &= \sum n_p S^\circ(\text{products}) - \sum n_r S^\circ(\text{reactants}) \\ &= [1(S^\circ(CO(g))) + 1(S^\circ(H_2(g)))] - [1(S^\circ(C(s))) + 1(S^\circ(H_2O(g)))] \\ &= [1(197.7\ \text{J/K}) + 1(130.7\ \text{J/K})] - [1(5.7\ \text{J/K}) + 1(188.8\ \text{J/K})] \\ &= [328.4\ \text{J/K}] - [194.5\ \text{J/K}] \\ &= +133.9\ \text{J/K} \quad \text{The moles of gas are increasing.}\end{aligned}$$

Check: The units (J/K) are correct. The answer is positive, which is consistent with 1 mole of gas going to 2 moles of gas.

(c) **Given:** $CO(g) + H_2O(g) \rightarrow H_2(g) + CO_2(g)$ **Find:** ΔS°_{rxn}
Conceptual Plan: $\boldsymbol{\Delta S^\circ_{rxn} = \sum n_p S^\circ(\text{products}) - \sum n_r S^\circ(\text{reactants})}$
Solution:

Reactant/Product	S°(J/mol K from Appendix IVB)
$CO(g)$	197.7
$H_2O(g)$	188.8
$H_2(g)$	130.7
$CO_2(g)$	213.8

Be sure to pull data for the correct formula and phase.

$$\begin{aligned}\Delta S^\circ_{rxn} &= \sum n_p S^\circ(\text{products}) - \sum n_r S^\circ(\text{reactants}) \\ &= [1(S^\circ(H_2(g))) + 1(S^\circ(CO_2(g)))] - [1(S^\circ(CO(g))) + 1(S^\circ(H_2O(g)))] \\ &= [1(130.7\ \text{J/K}) + 1(213.8\ \text{J/K})] - [1(197.7\ \text{J/K}) + 1(188.8\ \text{J/K})] \\ &= [344.5\ \text{J/K}] - [386.5\ \text{J/K}] \\ &= -42.0\ \text{J/K}\end{aligned}$$

The change is small because the number of moles of gas is constant.

Check: The units (J/K) are correct. The answer is small and negative, which is consistent with a constant number of moles of gas. Water molecules are bent, and carbon dioxide molecules are linear; so the water has more complexity. Also, carbon monoxide is more complex than is hydrogen gas.

(d) **Given:** $2\,H_2S(g) + 3\,O_2(g) \rightarrow 2\,H_2O(l) + 2\,SO_2(g)$ **Find:** ΔS°_{rxn}
Conceptual Plan: $\boldsymbol{\Delta S^\circ_{rxn} = \sum n_p S^\circ(\text{products}) - \sum n_r S^\circ(\text{reactants})}$
Solution:

Reactant/Product	S°(J/mol K from Appendix IVB)
$H_2S(g)$	205.8
$O_2(g)$	205.2
$H_2O(l)$	70.0
$SO_2(g)$	248.2

Be sure to pull data for the correct formula and phase.

$$\begin{aligned}\Delta S^\circ_{rxn} &= \sum n_p S^\circ(\text{products}) - \sum n_r S^\circ(\text{reactants}) \\ &= [2(S^\circ(H_2O(l))) + 2(S^\circ(SO_2(g)))] - [2(S^\circ(H_2S(g))) + 3(S^\circ(O_2(g)))] \\ &= [2(70.0\ \text{J/K}) + 2(248.2\ \text{J/K})] - [2(205.8\ \text{J/K}) + 3(205.2\ \text{J/K})] \\ &= [636.4\ \text{J/K}] - [1027.2\ \text{J/K}] \\ &= -390.8\ \text{J/K}\end{aligned}$$

The number of moles of gas is decreasing.

Check: The units (J/K) are correct. The answer is negative, which is consistent with a decrease in the number of moles of gas.

19.38 (a) **Given:** $3\ NO_2(g) + H_2O(l) \rightarrow 2\ HNO_3(aq) + NO(g)$ **Find:** ΔS°_{rxn}

Conceptual Plan: $\Delta S^\circ_{rxn} = \sum n_p S^\circ(\text{products}) - \sum n_r S^\circ(\text{reactants})$

Solution:

Reactant/Product	S°(J/mol K from Appendix IVB)
$NO_2(g)$	240.1
$H_2O(l)$	70.0
$HNO_3(aq)$	146
$NO(g)$	210.8

Be sure to pull data for the correct formula and phase.

$$\begin{aligned}\Delta S^\circ_{rxn} &= \sum n_p S^\circ(\text{products}) - \sum n_r S^\circ(\text{reactants})\\ &= [2(S^\circ(HNO_3(aq))) + 1(S^\circ(NO(g)))] - [3(S^\circ(NO_2(g))) + 1(S^\circ(H_2O(l)))]\\ &= [2(146\ J/K) + 1(210.8\ J/K)] - [3(240.1\ J/K) + 1(70.0\ J/K)]\\ &= [50\underline{2}.8\ J/K] - [790.3\ J/K]\\ &= -288\ J/K\end{aligned}$$

The number of moles of gas is decreasing.

Check: The units (J/K) are correct. The answer is negative, which is consistent with a decrease in the number of moles of gas.

(b) **Given:** $Cr_2O_3(s) + 3\ CO(g) \rightarrow 2\ Cr(s) + 3\ CO_2(g)$ **Find:** ΔS°_{rxn}

Conceptual Plan: $\Delta S^\circ_{rxn} = \sum n_p S^\circ(\text{products}) - \sum n_r S^\circ(\text{reactants})$

Solution:

Reactant/Product	S°(J/mol K from Appendix IVB)
$Cr_2O_3(s)$	81.2
$CO(g)$	197.7
$Cr(s)$	23.8
$CO_2(g)$	213.8

Be sure to pull data for the correct formula and phase.

$$\begin{aligned}\Delta S^\circ_{rxn} &= \sum n_p S^\circ(\text{products}) - \sum n_r S^\circ(\text{reactants})\\ &= [2(S^\circ(Cr(s))) + 3(S^\circ(CO_2(g)))] - [1(S^\circ(Cr_2O_3(s))) + 3(S^\circ(CO(g)))]\\ &= [2(23.8\ J/K) + 3(213.8\ J/K)] - [1(81.2\ J/K) + 3(197.7\ J/K)]\\ &= [689.0\ J/K] - [674.3\ J/K]\\ &= +14.7\ J/K\end{aligned}$$

The change is small because the number of moles of gas is constant.

Check: The units (J/K) are correct. The answer is small and positive, which is consistent with a constant number of moles of gas. Carbon dioxide molecules have more complexity than do carbon monoxide molecules, but the chromium oxide is more complex than is chromium metal.

(c) **Given:** $SO_2(g) + \frac{1}{2}O_2(g) \rightarrow SO_3(g)$ **Find:** ΔS°_{rxn}

Conceptual Plan: $\Delta S^\circ_{rxn} = \sum n_p S^\circ(\text{products}) - \sum n_r S^\circ(\text{reactants})$

Solution:

Reactant/Product	S°(J/mol K from Appendix IVB)
$SO_2(g)$	248.2
$O_2(g)$	205.2
$SO_3(g)$	256.8

Be sure to pull data for the correct formula and phase.

$$\begin{aligned}\Delta S^\circ_{rxn} &= \sum n_p S^\circ(\text{products}) - \sum n_r S^\circ(\text{reactants})\\ &= [1(S^\circ(SO_3(g)))] - [1(S^\circ(SO_2(g))) + 1/2(S^\circ(O_2(g)))]\\ &= [1(256.8\ J/K)] - [1(248.2\ J/K) + 1/2(205.2\ J/K)]\end{aligned}$$

$= [256.8 \text{ J/K}] - [350.8 \text{ J/K}]$
$= -94.0 \text{ J/K}$

The number of moles of gas is decreasing.

Check: The units (J/K) are correct. The answer is negative, which is consistent with a decrease in the number of moles of gas.

(d) **Given:** $N_2O_4(g) + 4\,H_2(g) \rightarrow N_2(g) + 4\,H_2O(g)$ **Find:** ΔS°_{rxn}

Conceptual Plan: $\Delta S^\circ_{rxn} = \sum n_p S^\circ(\text{products}) - \sum n_r S^\circ(\text{reactants})$

Solution:

Reactant/Product	S°(J/mol K from Appendix IVB)
$N_2O_4(g)$	304.4
$H_2(g)$	130.7
$N_2(g)$	191.6
$H_2O(g)$	188.8

Be sure to pull data for the correct formula and phase:

$$\begin{aligned}\Delta S^\circ_{rxn} &= \sum n_p S^\circ(\text{products}) - \sum n_r S^\circ(\text{reactants}) \\ &= [1(S^\circ(N_2(g))) + 4(S^\circ(H_2O(g)))] - [1(S^\circ(N_2O_4(g))) + 4(S^\circ(H_2(g)))] \\ &= [1(191.6 \text{ J/K}) + 4(188.8 \text{ J/K})] - [1(304.4 \text{ J/K}) + 4(130.7 \text{ J/K})] \\ &= [946.8 \text{ J/K}] - [827.2 \text{ J/K}] \\ &= +119.6 \text{ J/K}\end{aligned}$$

The change is small because the number of moles of gas is constant.

Check: The units (J/K) are correct. The answer is positive, which is consistent with 1 mole of a complex gas and 4 moles of a simple gas going to 1 mole of a simple gas and 4 moles of a complex gas, respectively.

19.39 **Given:** $CH_2Cl_2(g)$ formed from elements in standard states **Find:** ΔS° and rationalize sign

Conceptual Plan: Write a balanced reaction, then $\Delta S^\circ_{rxn} = \sum n_p S^\circ(\text{products}) - \sum n_r S^\circ(\text{reactants})$.

Solution: $C(s) + H_2(g) + Cl_2(g) \rightarrow CH_2Cl_2(g)$

Reactant/Product	S°(J/mol K from Appendix IVB)
$C(s)$	5.7
$H_2(g)$	130.7
$Cl_2(g)$	223.1
$CH_2Cl_2(g)$	270.2

Be sure to pull data for the correct formula and phase.

$$\begin{aligned}\Delta S^\circ_{rxn} &= \sum n_p S^\circ(\text{products}) - \sum n_r S^\circ(\text{reactants}) \\ &= [1(S^\circ(CH_2Cl_2(g)))] - [1(S^\circ(C(s))) + 1(S^\circ(H_2(g))) + 1(S^\circ(Cl_2(g)))] \\ &= [1(270.2 \text{ J/K})] - [1(5.7 \text{ J/K}) + 1(130.7 \text{ J/K}) + 1(223.1 \text{ J/K})] \\ &= [270.2 \text{ J/K}] - [359.5 \text{ J/K}] \\ &= -89.3 \text{ J/K}\end{aligned}$$

The moles of gas are decreasing.

Check: The units (J/K) are correct. The answer is negative, which is consistent with 2 moles of gas going to 1 mole of gas.

19.40 **Given:** $NF_3(g)$ formed from elements in standard states **Find:** ΔS° and rationalize sign

Conceptual Plan: Write balanced reaction, then $\Delta S^\circ_{rxn} = \sum n_p S^\circ(\text{products}) - \sum n_r S^\circ(\text{reactants})$.

Solution: $\frac{1}{2}\,N_2(g) + \frac{3}{2}\,F_2(g) \rightarrow NF_3(g)$

Reactant/Product	S°(J/mol K from Appendix IVB)
$N_2(g)$	191.6
$F_2(g)$	202.79
$NF_3(g)$	260.8

Be sure to pull data for the correct formula and phase.

$$\Delta S^\circ_{rxn} = \sum n_p S^\circ(\text{products}) - \sum n_r S^\circ(\text{reactants})$$
$$= [1(S^\circ(NF_3(g)))] - [1/2(S^\circ(N_2(g))) + 3/2(S^\circ(F_2(g)))]$$
$$= [1(260.8\ J/K)] - [1/2(191.6\ J/K) + 3/2(202.79\ J/K)]$$
$$= [260.8\ J/K] - [399.\underline{9}85\ J/K]$$
$$= -139.2\ J/K$$

The moles of gas are decreasing.

Check: The units (J/K) are correct. The answer is negative, which is consistent with 2 moles of simple gases going to 1 mole of a complex gas.

Entropy Changes in the Surroundings

19.41 (a) $\Delta S_{sys} > 0$ because 6 moles of gas are being converted to 7 moles of gas. Because $\Delta H < 0$ and $\Delta S_{surr} > 0$, the reaction is spontaneous at all temperatures.

(b) $\Delta S_{sys} < 0$ because 2 moles of different gases are being converted to 2 moles of one gas. Because $\Delta H > 0$ and $\Delta S_{surr} < 0$, the reaction is nonspontaneous at all temperatures.

(c) $\Delta S_{sys} < 0$ because 3 moles of gas are being converted to 2 moles of gas. Because $\Delta H > 0$ and $\Delta S_{surr} < 0$, the reaction is nonspontaneous at all temperatures.

(d) $\Delta S_{sys} > 0$ because 9 moles of gas are being converted to 10 moles of gas. Because $\Delta H < 0$ and $\Delta S_{surr} > 0$, the reaction is spontaneous at all temperatures.

19.42 (a) $\Delta S_{sys} < 0$ because 3 moles of gas are being converted to 2 moles of gas. Because $\Delta H < 0$ and $\Delta S_{surr} > 0$, the reaction is spontaneous at low temperatures.

(b) $\Delta S_{sys} > 0$ because 2 moles of gas are being converted to 3 moles of gas. Because $\Delta H > 0$ and $\Delta S_{surr} < 0$, the reaction is spontaneous at high temperatures.

(c) $\Delta S_{sys} < 0$ because 3 moles of gas are being converted to 2 moles of gas. Because $\Delta H < 0$ and $\Delta S_{surr} > 0$, the reaction is spontaneous at low temperatures.

(d) $\Delta S_{sys} < 0$ because 1 mole of a complicated gas is being converted to 1 mole of gas and a solid. Because $\Delta H > 0$ and $\Delta S_{surr} < 0$, the reaction is nonspontaneous at all temperatures.

19.43 (a) **Given:** $\Delta H^\circ_{rxn} = -385$ kJ, $T = 298$ K **Find:** ΔS_{surr}

Conceptual Plan: kJ → J then $\Delta H^\circ_{rxn}, T \rightarrow \Delta S_{surr}$

$$\frac{1000\ J}{1\ kJ} \qquad \Delta S_{surr} = \frac{-\Delta H_{sys}}{T}$$

Solution: $-385\ \cancel{kJ} \times \dfrac{1000\ J}{1\ \cancel{kJ}} = -385{,}000$ J then

$$\Delta S_{surr} = \frac{-\Delta H_{sys}}{T} = \frac{-(-385{,}000\ J)}{298\ K} = 1290\ J/K = 1.29 \times 10^3\ J/K$$

Check: The units (J/K) are correct. The magnitude of the answer (10^3 J/K) makes sense because the kJ and the temperature started with very similar values and then a factor of 10^3 was applied.

(b) **Given:** $\Delta H^\circ_{rxn} = -385$ kJ, $T = 77$ K **Find:** ΔS_{surr}

Conceptual Plan: kJ → J then $\Delta H^\circ_{rxn}, T \rightarrow \Delta S_{surr}$

$$\frac{1000\ J}{1\ kJ} \qquad \Delta S_{surr} = \frac{-\Delta H_{sys}}{T}$$

Solution: $-385\ \cancel{kJ} \times \dfrac{1000\ J}{1\ \cancel{kJ}} = -385{,}000$ J then $\Delta S_{surr} = \dfrac{-\Delta H_{sys}}{T} = \dfrac{-(-385{,}000\ J)}{77\ K} = 5.00 \times 10^3\ J/K$

Check: The units (J/K) are correct. The magnitude of the answer (5×10^3 J/K) makes sense because the temperature is much lower than in part (a); so the answer should increase.

(c) **Given:** $\Delta H^\circ_{rxn} = +114$ kJ, $T = 298$ K **Find:** ΔS_{surr}

Conceptual Plan: kJ → J then $\Delta H^\circ_{rxn}, T \rightarrow \Delta S_{surr}$

$$\frac{1000\ J}{1\ kJ} \qquad \Delta S_{surr} = \frac{-\Delta H_{sys}}{T}$$

Solution: $+114\ \text{kJ} \times \dfrac{1000\ \text{J}}{1\ \text{kJ}} = +114{,}000\ \text{J}$ then $\Delta S_{\text{surr}} = \dfrac{-\Delta H_{\text{sys}}}{T} = \dfrac{-114{,}000\ \text{J}}{298\ \text{K}} = -383\ \text{J/K}$

Check: The units (J/K) are correct. The magnitude of the answer (−400 J/K) makes sense because the kJ are less and of the opposite sign than part (a); so the answer should decrease.

(d) **Given:** $\Delta H^\circ_{\text{rxn}} = +114\ \text{kJ}$, $T = 77\ \text{K}$ **Find:** ΔS_{surr}

Conceptual Plan: kJ → J then $\Delta H^\circ_{\text{rxn}}, T \rightarrow \Delta S_{\text{surr}}$

$\dfrac{1000\ \text{J}}{1\ \text{kJ}}$ $\qquad \Delta S_{\text{surr}} = \dfrac{-\Delta H_{\text{sys}}}{T}$

Solution: $+114\ \text{kJ} \times \dfrac{1000\ \text{J}}{1\ \text{kJ}} = +114{,}000\ \text{J}$ then

$$\Delta S_{\text{surr}} = \frac{-\Delta H_{\text{sys}}}{T} = \frac{-114{,}000\ \text{J}}{77\ \text{K}} = -1480\frac{\text{J}}{\text{K}} = -1.48 \times 10^3\ \text{J/K}$$

Check: The units (J/K) are correct. The magnitude of the answer (-2×10^3 J/K) makes sense because the temperature is much lower than in part (c); so the answer should increase.

19.44 **Given:** $\Delta H^\circ_{\text{rxn}} = -112\ \text{kJ}$, $\Delta S_{\text{rxn}} = 354\ \text{J/K}$ **Find:** T when $\Delta S^\circ_{\text{rxn}} = \Delta S_{\text{surr}}$

Conceptual Plan: kJ → J then set $\Delta S^\circ_{\text{rxn}} = \Delta S_{\text{surr}}$ then $\Delta H^\circ_{\text{rxn}}, \Delta S_{\text{surr}} \rightarrow T$

$\dfrac{1000\ \text{J}}{1\ \text{kJ}}$ $\qquad \Delta S_{\text{surr}} = \dfrac{-\Delta H_{\text{sys}}}{T}$

Solution: $-112\ \text{kJ} \times \dfrac{1000\ \text{J}}{1\ \text{kJ}} = -112{,}000\ \text{J}$ then set $\Delta S^\circ_{\text{rxn}} = 354\ \text{J/K} = \Delta S_{\text{surr}}$ then

$$T = \frac{-\Delta H_{\text{sys}}}{\Delta S_{\text{surr}}} = \frac{-(-112{,}000\ \text{J})}{354\ \dfrac{\text{J}}{\text{K}}} = +316\ \text{K}$$

Check: The units (K) are correct. The magnitude of the answer (316 K) makes sense because there is almost a factor of 300 between the enthalpy and the entropy.

19.45 (a) **Given:** $\Delta H^\circ_{\text{rxn}} = +115\ \text{kJ}$, $\Delta S_{\text{rxn}} = -263\ \text{J/K}$, $T = 298\ \text{K}$ **Find:** ΔS_{univ} and spontaneity

Conceptual Plan: kJ → J then $\Delta H^\circ_{\text{rxn}}, T \rightarrow \Delta S_{\text{surr}}$ then $\Delta S_{\text{rxn}}, \Delta S_{\text{surr}} \rightarrow \Delta S_{\text{univ}}$

$\dfrac{1000\ \text{J}}{1\ \text{kJ}}$ $\qquad \Delta S_{\text{surr}} = \dfrac{-\Delta H_{\text{sys}}}{T}$ $\qquad \Delta S_{\text{univ}} = \Delta S_{\text{sys}} + \Delta S_{\text{surr}}$

Solution: $115\ \text{kJ} \times \dfrac{1000\ \text{J}}{1\ \text{kJ}} = 115{,}000\ \text{J}$ then $\Delta S_{\text{surr}} = \dfrac{-\Delta H_{\text{sys}}}{T} = \dfrac{-(115{,}000\ \text{J})}{298\ \text{K}} = -38\underline{5}.906\ \text{J/K}$ then

$\Delta S_{\text{univ}} = \Delta S_{\text{sys}} + \Delta S_{\text{surr}} = -263\ \text{J/K} - 38\underline{5}.906\ \text{J/K} = -649\ \text{J/K}$; so the reaction is nonspontaneous.

Check: The units (J/K) are correct. The magnitude of the answer (−650 J/K) makes sense because both terms are negative; so the reaction is nonspontaneous.

(b) **Given:** $\Delta H^\circ_{\text{rxn}} = -115\ \text{kJ}$, $\Delta S_{\text{rxn}} = +263\ \text{J/K}$, $T = 298\ \text{K}$ **Find:** ΔS_{univ} and spontaneity

Conceptual Plan: kJ → J then $\Delta H^\circ_{\text{rxn}}, T \rightarrow \Delta S_{\text{surr}}$ then $\Delta S_{\text{rxn}}, \Delta S_{\text{surr}} \rightarrow \Delta S_{\text{univ}}$

$\dfrac{1000\ \text{J}}{1\ \text{kJ}}$ $\qquad \Delta S_{\text{surr}} = \dfrac{-\Delta H_{\text{sys}}}{T}$ $\qquad \Delta S_{\text{univ}} = \Delta S_{\text{sys}} + \Delta S_{\text{surr}}$

Solution: $-115\ \text{kJ} \times \dfrac{1000\ \text{J}}{1\ \text{kJ}} = -115{,}000\ \text{J}$ then $\Delta S_{\text{surr}} = \dfrac{-\Delta H_{\text{sys}}}{T} = \dfrac{-(-115{,}000\ \text{J})}{298\ \text{K}} = 38\underline{5}.906\ \text{J/K}$

then $\Delta S_{\text{univ}} = \Delta S_{\text{sys}} + \Delta S_{\text{surr}} = +263\ \text{J/K} + 38\underline{5}.906\ \text{J/K} = +649\ \text{J/K}$; so the reaction is spontaneous.

Check: The units (J/K) are correct. The magnitude of the answer (+650 J/K) makes sense because both terms were positive; so the reaction is spontaneous.

(c) **Given:** $\Delta H^\circ_{\text{rxn}} = -115\ \text{kJ}$, $\Delta S_{\text{rxn}} = -263\ \text{J/K}$, $T = 298\ \text{K}$ **Find:** ΔS_{univ} and spontaneity

Conceptual Plan: kJ → J then $\Delta H^\circ_{\text{rxn}}, T \rightarrow \Delta S_{\text{surr}}$ then $\Delta S_{\text{rxn}}, \Delta S_{\text{surr}} \rightarrow \Delta S_{\text{univ}}$

$\dfrac{1000\ \text{J}}{1\ \text{kJ}}$ $\qquad \Delta S_{\text{surr}} = \dfrac{-\Delta H_{\text{sys}}}{T}$ $\qquad \Delta S_{\text{univ}} = \Delta S_{\text{sys}} + \Delta S_{\text{surr}}$

Solution: $-115\ \text{kJ} \times \frac{1000\ \text{J}}{1\ \text{kJ}} = -115{,}000\ \text{J}$ then $\Delta S_{surr} = \frac{-\Delta H_{sys}}{T} = \frac{-(-115{,}000\ \text{J})}{298\ \text{K}} = +385.906\ \text{J/K}$
then $\Delta S_{univ} = \Delta S_{sys} + \Delta S_{surr} = -263\ \text{J/K} + 385.906\ \text{J/K} = +123\ \text{J/K}$; so the reaction is spontaneous.

Check: The units (J/K) are correct. The magnitude of the answer (120 J/K) makes sense because the larger term was positive; so the reaction is spontaneous.

(d) **Given:** $\Delta H^\circ_{rxn} = -115$ kJ, $\Delta S_{rxn} = -263$ J/K, $T = 615$ K **Find:** ΔS_{univ} and spontaneity
Conceptual Plan: kJ → J then $\Delta H^\circ_{rxn}, T \rightarrow \Delta S_{surr}$ then $\Delta S_{rxn}, \Delta S_{surr} \rightarrow \Delta S_{univ}$

$\frac{1000\ \text{J}}{1\ \text{kJ}}$ $\quad \Delta S_{surr} = \frac{-\Delta H_{sys}}{T}$ $\quad \Delta S_{univ} = \Delta S_{sys} + \Delta S_{surr}$

Solution: $-115\ \text{kJ} \times \frac{1000\ \text{J}}{1\ \text{kJ}} = -115{,}000\ \text{J}$ then $\Delta S_{surr} = \frac{-\Delta H_{sys}}{T} = \frac{-(-115{,}000\ \text{J})}{615\ \text{K}} = +186.992\ \text{J/K}$
then $\Delta S_{univ} = \Delta S_{sys} + \Delta S_{surr} = -263\ \text{J/K} + 186.992\ \text{J/K} = -76\ \text{J/K}$; so the reaction is nonspontaneous.

Check: The units (J/K) are correct. The magnitude of the answer (-80 J/K) makes sense because the larger term was negative, so the reaction is nonspontaneous.

19.46 (a) **Given:** $\Delta H^\circ_{rxn} = -95$ kJ, $\Delta S_{rxn} = -157$ J/K, $T = 298$ K **Find:** ΔS_{univ} and spontaneity
Conceptual Plan: kJ → J then $\Delta H^\circ_{rxn}, T \rightarrow \Delta S_{surr}$ then $\Delta S_{rxn}, \Delta S_{surr} \rightarrow \Delta S_{univ}$

$\frac{1000\ \text{J}}{1\ \text{kJ}}$ $\quad \Delta S_{surr} = \frac{-\Delta H_{sys}}{T}$ $\quad \Delta S_{univ} = \Delta S_{sys} + \Delta S_{surr}$

Solution: $-95\ \text{kJ} \times \frac{1000\ \text{J}}{1\ \text{kJ}} = +95{,}000\ \text{J}$ then $\Delta S_{surr} = \frac{-\Delta H_{sys}}{T} = \frac{-(-95{,}000\ \text{J})}{298\ \text{K}} = +318.79\ \text{J/K}$ then
$\Delta S_{univ} = \Delta S_{sys} + \Delta S_{surr} = -157\ \text{J/K} + 318.79\ \text{J/K} = +162\ \text{J/K} = +1.6 \times 10^2\ \text{J/K}$; so the reaction is spontaneous.

Check: The units (J/K) are correct. The magnitude of the answer ($+160$ J/K) makes sense because the surroundings gain more entropy than the system loses; so the reaction is spontaneous.

(b) **Given:** $\Delta H^\circ_{rxn} = -95$ kJ, $\Delta S_{rxn} = -157$ J/K, $T = 855$ K **Find:** ΔS_{univ} and spontaneity
Conceptual Plan: kJ → J then $\Delta H^\circ_{rxn}, T \rightarrow \Delta S_{surr}$ then $\Delta S_{rxn}, \Delta S_{surr} \rightarrow \Delta S_{univ}$

$\frac{1000\ \text{J}}{1\ \text{kJ}}$ $\quad \Delta S_{surr} = \frac{-\Delta H_{sys}}{T}$ $\quad \Delta S_{univ} = \Delta S_{sys} + \Delta S_{surr}$

Solution: $-95\ \text{kJ} \times \frac{1000\ \text{J}}{1\ \text{kJ}} = +95{,}000\ \text{J}$ then $\Delta S_{surr} = \frac{-\Delta H_{sys}}{T} = \frac{-(-95{,}000\ \text{J})}{855\ \text{K}} = +111.11\ \text{J/K}$ then
$\Delta S_{univ} = \Delta S_{sys} + \Delta S_{surr} = -157\ \text{J/K} + 111.11\ \text{J/K} = -46\ \text{J/K} = -5 \times 10^1\ \text{J/K}$; so the reaction is nonspontaneous.

Check: The units (J/K) are correct. The magnitude of the answer (-50 J/K) makes sense because at a higher temperature, the entropy of the surroundings is reduced so that the entropy reduction of the system dominates and the reaction is nonspontaneous.

(c) **Given:** $\Delta H^\circ_{rxn} = +95$ kJ, $\Delta S_{rxn} = -157$ J/K, $T = 298$ K **Find:** ΔS_{univ} and spontaneity
Conceptual Plan: kJ → J then $\Delta H^\circ_{rxn}, T \rightarrow \Delta S_{surr}$ then $\Delta S_{rxn}, \Delta S_{surr} \rightarrow \Delta S_{univ}$

$\frac{1000\ \text{J}}{1\ \text{kJ}}$ $\quad \Delta S_{surr} = \frac{-\Delta H_{sys}}{T}$ $\quad \Delta S_{univ} = \Delta S_{sys} + \Delta S_{surr}$

Solution: $+95\ \text{kJ} \times \frac{1000\ \text{J}}{1\ \text{kJ}} = +95{,}000\ \text{J}$ then $\Delta S_{surr} = \frac{-\Delta H_{sys}}{T} = \frac{-95{,}000\ \text{J}}{298\ \text{K}} = -318.79\ \text{J/K}$ then

$\Delta S_{univ} = \Delta S_{sys} + \Delta S_{surr} = -157\ \text{J/K} - 318.79\ \text{J/K} = -476\ \text{J/K} = -4.8 \times 10^2\ \text{J/K}$; so the reaction is nonspontaneous.

Check: The units (J/K) are correct. The magnitude of the answer (-480 J/K) makes sense because both terms are negative; so the sum is negative and the reaction is nonspontaneous.

(d) **Given:** $\Delta H^\circ_{rxn} = -95$ kJ, $\Delta S_{rxn} = +157$ J/K, $T = 398$ K **Find:** ΔS_{univ} and spontaneity
Conceptual Plan: kJ → J then $\Delta H^\circ_{rxn}, T \rightarrow \Delta S_{surr}$ then $\Delta S_{rxn}, \Delta S_{surr} \rightarrow \Delta S_{univ}$

$\frac{1000\ \text{J}}{1\ \text{kJ}}$ $\quad \Delta S_{surr} = \frac{-\Delta H_{sys}}{T}$ $\quad \Delta S_{univ} = \Delta S_{sys} + \Delta S_{surr}$

Solution: $-95\ \cancel{kJ} \times \frac{1000\ J}{1\ \cancel{kJ}} = -95{,}000\ J$ then $\Delta S_{surr} = \frac{-\Delta H_{sys}}{T} = \frac{-(-95{,}000\ J)}{398\ K} = +2\underline{3}8.69\ J/K$

then $\Delta S_{univ} = \Delta S_{sys} + \Delta S_{surr} = +157\ J/K + 2\underline{3}8.69\ J/K = +3\underline{9}6\ J/K = +4.0 \times 10^2\ J/K$; so the reaction is spontaneous.

Check: The units (J/K) are correct. The magnitude of the answer (400 J/K) makes sense because both terms are positive; so the sum is positive and the reaction is spontaneous.

Gibbs Free Energy

19.47 (a) **Given:** $\Delta H^\circ_{rxn} = +115$ kJ, $\Delta S_{rxn} = -263$ J/K, $T = 298$ K **Find:** ΔG and spontaneity

Conceptual Plan: J/K → kJ/K then $\Delta H^\circ_{rxn}, \Delta S_{rxn}, T \rightarrow \Delta G$

$\frac{1000\ J}{1\ kJ}$ $\Delta G = \Delta H_{rxn} - T\Delta S_{rxn}$

Solution: $-263\ \frac{\cancel{J}}{K} \times \frac{1\ kJ}{1000\ \cancel{J}} = -0.263\ kJ/K$ then

$\Delta G = \Delta H_{rxn} - T\Delta S_{rxn} = +115\ kJ - \left((298\ \cancel{K})\left(-0.263\ \frac{kJ}{\cancel{K}}\right)\right) = +1.93 \times 10^2\ kJ = +1.93 \times 10^5\ J$; so the reaction is nonspontaneous.

Check: The units (kJ) are correct. The magnitude of the answer (+190 kJ) makes sense because both terms are positive; so the reaction is nonspontaneous.

(b) **Given:** $\Delta H^\circ_{rxn} = -115$ kJ, $\Delta S_{rxn} = +263$ J/K, $T = 298$ K **Find:** ΔG and spontaneity

Conceptual Plan: J/K → kJ/K then $\Delta H^\circ_{rxn}, \Delta S_{rxn}, T \rightarrow \Delta G$

$\frac{1000\ J}{1\ kJ}$ $\Delta G = \Delta H_{rxn} - T\Delta S_{rxn}$

Solution: $+263\ \frac{\cancel{J}}{K} \times \frac{1\ kJ}{1000\ \cancel{J}} = +0.263\ kJ/K$ then

$\Delta G = \Delta H_{rxn} - T\Delta S_{rxn} = -115\ kJ - \left((298\ \cancel{K})\left(0.263\ \frac{kJ}{\cancel{K}}\right)\right) = -193\ kJ = -1.93 \times 10^2\ kJ = -1.93 \times 10^5\ J$; so the reaction is spontaneous.

Check: The units (kJ) are correct. The magnitude of the answer (−190 kJ) makes sense because both terms are negative; so the reaction is spontaneous.

(c) **Given:** $\Delta H^\circ_{rxn} = -115$ kJ, $\Delta S_{rxn} = -263$ J/K, $T = 298$ K **Find:** ΔG and spontaneity

Conceptual Plan: J/K → kJ/K then $\Delta H^\circ_{rxn}, \Delta S_{rxn}, T \rightarrow \Delta G$

$\frac{1000\ J}{1\ kJ}$ $\Delta G = \Delta H_{rxn} - T\Delta S_{rxn}$

Solution: $-263\ \frac{\cancel{J}}{K} \times \frac{1\ kJ}{1000\ \cancel{J}} = -0.263\ kJ/K$ then $\Delta G = \Delta H_{rxn} - T\Delta S_{rxn} =$

$-115\ kJ - \left((298\ \cancel{K})\left(-0.263\ \frac{kJ}{\cancel{K}}\right)\right) = -3\underline{6}.626\ kJ = -3.7 \times 10^1\ kJ = -3.7 \times 10^4\ J$; so the reaction is spontaneous.

Check: The units (kJ) are correct. The magnitude of the answer (−40 kJ) makes sense because the larger term was negative; so the reaction is spontaneous.

(d) **Given:** $\Delta H^\circ_{rxn} = -115$ kJ, $\Delta S_{rxn} = -263$ J/K, $T = 615$ K **Find:** ΔG and spontaneity

Conceptual Plan: J/K → kJ/K then $\Delta H^\circ_{rxn}, \Delta S_{rxn}, T \rightarrow \Delta G$

$\frac{1000\ J}{1\ kJ}$ $\Delta G = \Delta H_{rxn} - T\Delta S_{rxn}$

Solution: $-263\ \frac{\cancel{J}}{K} \times \frac{1\ kJ}{1000\ \cancel{J}} = -0.263\ kJ/K$ then

$\Delta G = \Delta H_{rxn} - T\Delta S_{rxn} = -115\ kJ - \left((615\ \cancel{K})\left(-0.263\ \frac{kJ}{\cancel{K}}\right)\right) = +47\ kJ = +4.7 \times 10^4\ J$; so the reaction is nonspontaneous.

Check: The units (J/K) are correct. The magnitude of the answer $(+47\text{ kJ})$ makes sense because the larger term was positive; so the reaction is nonspontaneous.

19.48 (a) **Given:** $\Delta H^\circ_{\text{rxn}} = -95\text{ kJ}$, $\Delta S_{\text{rxn}} = -157\text{ J/K}$, $T = 298\text{ K}$ **Find:** ΔG and spontaneity
Conceptual Plan: J/K → kJ/K then $\Delta H^\circ_{\text{rxn}}, \Delta S_{\text{rxn}}, T \rightarrow \Delta G$

$\frac{1\text{ kJ}}{1000\text{ J}}$ $\qquad \Delta G = \Delta H_{\text{rxn}} - T\Delta S_{\text{rxn}}$

Solution: $-157\frac{\text{J}}{\text{K}} \times \frac{1\text{ kJ}}{1000\text{ J}} = -0.157\text{ kJ/K}$ then

$\Delta G = \Delta H_{\text{rxn}} - T\Delta S_{\text{rxn}} = -95\text{ kJ} - \left((298\text{ K})\left(-0.157\frac{\text{kJ}}{\text{K}}\right)\right) = -48\text{ kJ}$; so the reaction is spontaneous.

Check: The units (kJ) are correct. The magnitude of the answer (-50 kJ) makes sense because the negative enthalpy dominates over the entropy term; so the reaction is spontaneous.

(b) **Given:** $\Delta H^\circ_{\text{rxn}} = -95\text{ kJ}$, $\Delta S_{\text{rxn}} = -157\text{ J/K}$, $T = 855\text{ K}$ **Find:** ΔG and spontaneity
Conceptual Plan: J/K → kJ/K then $\Delta H^\circ_{\text{rxn}}, \Delta S_{\text{rxn}}, T \rightarrow \Delta G$

$\frac{1\text{ kJ}}{1000\text{ J}}$ $\qquad \Delta G = \Delta H_{\text{rxn}} - T\Delta S_{\text{rxn}}$

Solution: $-157\frac{\text{J}}{\text{K}} \times \frac{1\text{ kJ}}{1000\text{ J}} = -0.157\text{ kJ/K}$ then

$\Delta G = \Delta H_{\text{rxn}} - T\Delta S_{\text{rxn}} = -95\text{ kJ} - \left((855\text{ K})\left(-0.157\frac{\text{kJ}}{\text{K}}\right)\right) = +39\text{ kJ}$; so the reaction is nonspontaneous.

Check: The units (kJ) are correct. The magnitude of the answer $(+40\text{ kJ})$ makes sense because at a higher temperature, the entropy term now dominates and the reaction is nonspontaneous.

(c) **Given:** $\Delta H^\circ_{\text{rxn}} = +95\text{ kJ}$, $\Delta S_{\text{rxn}} = -157\text{ J/K}$, $T = 298\text{ K}$ **Find:** ΔG and spontaneity
Conceptual Plan: J/K → kJ/K then $\Delta H^\circ_{\text{rxn}}, \Delta S_{\text{rxn}}, T \rightarrow \Delta G$

$\frac{1\text{ kJ}}{1000\text{ J}}$ $\qquad \Delta G = \Delta H_{\text{rxn}} - T\Delta S_{\text{rxn}}$

Solution: $-157\frac{\text{J}}{\text{K}} \times \frac{1\text{ kJ}}{1000\text{ J}} = -0.157\text{ kJ/K}$ then

$\Delta G = \Delta H_{\text{rxn}} - T\Delta S_{\text{rxn}} = +95\text{ kJ} - \left((298\text{ K})\left(-0.157\frac{\text{kJ}}{\text{K}}\right)\right) = +142\text{ kJ}$; so the reaction is nonspontaneous.

Check: The units (kJ) are correct. The magnitude of the answer $(+140\text{ kJ})$ makes sense because both terms are positive; so the sum is positive and the reaction is nonspontaneous.

(d) **Given:** $\Delta H^\circ_{\text{rxn}} = -95\text{ kJ}$, $\Delta S_{\text{rxn}} = +157\text{ J/K}$, $T = 398\text{ K}$ **Find:** ΔG and spontaneity
Conceptual Plan: J/K → kJ/K then $\Delta H^\circ_{\text{rxn}}, \Delta S_{\text{rxn}}, T \rightarrow \Delta G$

$\frac{1\text{ kJ}}{1000\text{ J}}$ $\qquad \Delta G = \Delta H_{\text{rxn}} - T\Delta S_{\text{rxn}}$

Solution: $+157\frac{\text{J}}{\text{K}} \times \frac{1\text{ kJ}}{1000\text{ J}} = +0.157\frac{\text{kJ}}{\text{K}}$ then

$\Delta G = \Delta H_{\text{rxn}} - T\Delta S_{\text{rxn}} = -95\text{ kJ} - \left((398\text{ K})\left(0.157\frac{\text{kJ}}{\text{K}}\right)\right) = -157\text{ kJ}$ so the reaction is spontaneous.

Check: The units (kJ) are correct. The magnitude of the answer (-160 kJ) makes sense because both terms are negative; so the sum is negative and the reaction is spontaneous.

19.49 **Given:** $\Delta H^\circ_{\text{rxn}} = -2217\text{ kJ}$, $\Delta S_{\text{rxn}} = +101.1\text{ J/K}$, $T = 25\text{ °C}$ **Find:** ΔG and spontaneity
Conceptual Plan: °C → K then J/K → kJ/K then $\Delta H^\circ_{\text{rxn}}, \Delta S_{\text{rxn}}, T \rightarrow \Delta G$

$\text{K} = 273.15 + \text{°C}$ $\qquad \frac{1\text{ kJ}}{1000\text{ J}}$ $\qquad \Delta G = \Delta H_{\text{rxn}} - T\Delta S_{\text{rxn}}$

Solution: $T = 273.15 + 25\text{ °C} = 298\text{ K}$ then $+101.1\frac{\text{J}}{\text{K}} \times \frac{1\text{ kJ}}{1000\text{ J}} = +0.1011\frac{\text{kJ}}{\text{K}}$ then

$\Delta G = \Delta H_{rxn} - T\Delta S_{rxn} = -2217 \text{ kJ} - \left((298 \text{ K})\left(0.1011 \frac{\text{kJ}}{\text{K}}\right)\right) = -2247 \text{ kJ} = -2.247 \times 10^6 \text{ J}$; so the reaction is spontaneous.

Check: The units (kJ) are correct. The magnitude of the answer (-2250 kJ) makes sense because both terms are negative; so the reaction is spontaneous.

19.50 **Given:** $\Delta H^\circ_{rxn} = -1269.8$ kJ, $\Delta S_{rxn} = -364.6$ J/K, $T = 25$ °C **Find:** ΔG and spontaneity
Conceptual Plan: °C → K then J/K → kJ/K then ΔH°_{rxn}, ΔS_{rxn}, $T \rightarrow \Delta G$

$$\text{K} = 273.15 + °\text{C} \qquad \frac{1 \text{ kJ}}{1000 \text{ J}} \qquad \Delta G = \Delta H_{rxn} - T\Delta S_{rxn}$$

Solution: $T = 273.15 + 25 \text{ °C} = 298$ K then $-364.6 \frac{\text{J}}{\text{K}} \times \frac{1 \text{ kJ}}{1000 \text{ J}} = -0.3646 \frac{\text{kJ}}{\text{K}}$ then

$\Delta G = \Delta H_{rxn} - T\Delta S_{rxn} = -1269.8 \text{ kJ} - \left((298 \text{ K})\left(-0.3646 \frac{\text{kJ}}{\text{K}}\right)\right) = -1161.1 \text{ kJ} = -1.1611 \times 10^6 \text{ J}$; so the reaction is spontaneous.

Check: The units (kJ) are correct. The magnitude of the answer (-1200 kJ) makes sense because the negative enthalpy term dominates over the positive entropy term; so the reaction is spontaneous.

19.51

ΔH	ΔS	ΔG	Low Temp.	High Temp.
−	+	−	Spontaneous	Spontaneous
−	−	Temp.-dependent	Spontaneous	Nonspontaneous
+	+	Temp.-dependent	Nonspontaneous	Spontaneous
+	−	+	Nonspontaneous	Nonspontaneous

19.52 (a) ΔH°_{rxn} for a condensation is negative, and ΔS_{rxn} is negative; so the reaction will be spontaneous at low temperatures (<100 °C).
(b) ΔH°_{rxn} for a sublimation is positive, and ΔS_{rxn} is positive; so the reaction will be spontaneous at high temperatures (>−78.5 °C).
(c) ΔH°_{rxn} for a bond breaking is positive, and ΔS_{rxn} is positive; so the reaction will be spontaneous at high temperatures.
(d) ΔH°_{rxn} is positive, and ΔS_{rxn} is positive; so the reaction will be spontaneous at high temperatures.

19.53 **Given:** methanol (CH_3OH) combustion at 25 °C **Find:** ΔH°_{rxn}, ΔS°_{rxn}, ΔG°_{rxn}, and spontaneity
Conceptual Plan: Write a balanced reaction then $\Delta H^\circ_{rxn} = \sum n_p H^\circ_f(\text{products}) - \sum n_r H^\circ_f(\text{reactants})$ then $\Delta S^\circ_{rxn} = \sum n_p S^\circ(\text{products}) - \sum n_r S^\circ(\text{reactants})$ then °C → K then J/K → kJ/K then

$$\text{K} = 273.15 + °\text{C} \qquad \frac{1 \text{ kJ}}{1000 \text{ J}}$$

ΔH°_{rxn}, ΔS°_{rxn}, $T \rightarrow \Delta G^\circ$.

$$\Delta G = \Delta H_{rxn} - T\Delta S_{rxn}$$

Solution: Combustion is combining with oxygen to form carbon dioxide and water.
$2\,CH_3OH(l) + 3\,O_2(g) \rightarrow 2\,CO_2(g) + 4\,H_2O(g)$

Reactant/Product	ΔH°_f(kJ/mol from Appendix IVB)
$CH_3OH(l)$	−238.6
$O_2(g)$	0.0
$CO_2(g)$	−393.5
$H_2O(g)$	−241.8

Be sure to pull data for the correct formula and phase.

$$\begin{aligned}\Delta H^\circ_{rxn} &= \sum n_p \Delta H^\circ_f(\text{products}) - \sum n_r \Delta H^\circ_f(\text{reactants}) \\ &= [2(\Delta H^\circ_f(CO_2(g))) + 4(\Delta H^\circ_f(H_2O(g)))] - [2(\Delta H^\circ_f(CH_3OH(l))) + 3(\Delta H^\circ_f(O_2(g)))] \\ &= [2(-393.5 \text{ kJ}) + 4(-241.8 \text{ kJ})] - [2(-238.6 \text{ kJ}) + 3(0.0 \text{ kJ})] \\ &= [-1754.2 \text{ kJ}] - [-477.2 \text{ kJ}] \\ &= -1277 \text{ kJ then}\end{aligned}$$

Reactant/Product	$S°$(J/mol K from Appendix IVB)
$CH_3OH(l)$	126.8
$O_2(g)$	205.2
$CO_2(g)$	213.8
$H_2O(g)$	188.8

Be sure to pull data for the correct formula and phase.

$$\begin{aligned}\Delta S°_{rxn} &= \sum n_p S°(\text{products}) - \sum n_r S°(\text{reactants}) \\ &= [2(S°(CO_2(g))) + 4(S°(H_2O(g)))] - [2(S°(CH_3OH(l))) + 3(S°(O_2(g)))] \\ &= [2(213.8\text{ J/K}) + 4(188.8\text{ J/K})] - [2(126.8\text{ J/K}) + 3(205.2\text{ J/K})] \\ &= [1182.8\text{ J/K}] - [869.2\text{ J/K}] \\ &= 313.6\text{ J/K}\end{aligned}$$

then $T = 273.15 + 25\text{ °C} = 298\text{ K}$ then $+313.6\frac{\cancel{J}}{\text{K}} \times \frac{1\text{ kJ}}{1000\cancel{J}} = +0.3136\text{ kJ/K}$ then

$\Delta G = \Delta H_{rxn} - T\Delta S_{rxn} = -1277\text{ kJ} - \left((298\cancel{K})\left(+0.3136\frac{\text{kJ}}{\cancel{K}}\right)\right) = -1370.\text{ kJ} = -1.370 \times 10^6\text{ J}$; so the reaction is spontaneous.

Check: The units (kJ, J/K, and kJ) are correct. Combustion reactions are exothermic, and we see a large negative enthalpy. We expect a large positive entropy because we have an increase in the number of moles of gas. The free energy is the sum of two negative terms; so we expect a large negative free energy, and the reaction is spontaneous.

19.54 **Given:** form glucose ($C_6H_{12}O_6$) and oxygen from sunlight, carbon dioxide, and water at 25 °C
Find: $\Delta H°_{rxn}$, $\Delta S°_{rxn}$, $\Delta G°_{rxn}$, and spontaneity
Conceptual Plan: Write a balanced reaction then $\Delta H°_{rxn} = \sum n_p H°_f(\text{products}) - \sum n_r H°_f(\text{reactants})$ then $\Delta S°_{rxn} = \sum n_p S°(\text{products}) - \sum n_r S°(\text{reactants})$ then °C → K then J/K → kJ/K then $\Delta H°_{rxn}, \Delta S°_{rxn}, T \rightarrow \Delta G°$.

$K = 273.15 + °C$ $\quad \frac{1\text{ kJ}}{1000\text{ J}}$ $\quad \Delta G = \Delta H_{rxn} - T\Delta S_{rxn}$

Solution: $6\,CO_2(g) + 6\,H_2O(l) \rightarrow C_6H_{12}O_6(s) + 6\,O_2(g)$

Reactant/Product	$\Delta H°_f$(kJ/mol from Appendix IVB)
$CO_2(g)$	−393.5
$H_2O(l)$	−285.8
$C_6H_{12}O_6(s)$	−1273.3
$O_2(g)$	0.0

Be sure to pull data for the correct formula and phase.

$$\begin{aligned}\Delta H°_{rxn} &= \sum n_p \Delta H°_f(\text{products}) - \sum n_r \Delta H°_f(\text{reactants}) \\ &= [1(\Delta H°_f(C_6H_{12}O_6(s))) + 6(\Delta H°_f(O_2(g)))] - [6(\Delta H°_f(CO_2(g))) + 6(\Delta H°_f(H_2O(l)))] \\ &= [1(-1273.3\text{ kJ}) + 6(0.0\text{ kJ})] - [6(-393.5\text{ kJ}) + 6(-285.8\text{ kJ})] \\ &= [-1273.3\text{ kJ}] - [-4075.8\text{ kJ}] \\ &= +2802.5\text{ kJ then}\end{aligned}$$

Reactant/Product	$S°$(J/mol K from Appendix IVB)
$CO_2(g)$	213.8
$H_2O(l)$	70.0
$C_6H_{12}O_6(s)$	212.1
$O_2(g)$	205.2

Be sure to pull data for the correct formula and phase.

$$\begin{aligned}\Delta S°_{rxn} &= \sum n_p S°(\text{products}) - \sum n_r S°(\text{reactants}) \\ &= [1(S°(C_6H_{12}O_6(s))) + 6(S°(O_2(g)))] - [6(S°(CO_2(g))) + 6(S°(H_2O(l)))] \\ &= [1(212.1\text{ J/K}) + 6(205.2\text{ J/K})] - [6(213.8\text{ J/K}) + 6(70.0\text{ J/K})] \\ &= [1443.3\text{ J/K}] - [1702.8\text{ J/K}] \\ &= -259.5\text{ J/K}\end{aligned}$$

then $T = 273.15 + 25\ °C = 298\ K$ then $-259.5\ \frac{\cancel{J}}{K} \times \frac{1\ kJ}{1000\ \cancel{J}} = -0.2595\ kJ/K$ then

$\Delta G = \Delta H_{rxn} - T\Delta S_{rxn} = +2802.5\ kJ - \left((298\ \cancel{K})\left(-0.2595\frac{kJ}{\cancel{K}}\right)\right) = +2879.8\ kJ = +2.8798 \times 10^6\ J$; so the reaction is nonspontaneous.

Check: The units (kJ, J/K, and kJ) are correct. The reaction requires the input of light energy, so we expect that this will be an endothermic reaction. We expect a negative entropy change because we are going from 6 moles of a gas and 6 moles of a liquid to 6 moles of a gas and 1 mole of a solid. The free energy is the sum of two positive terms; so we expect a large positive free energy, and the reaction is nonspontaneous. Photosynthesis does not happen on its own; light energy must be added to make the process move forward.

19.55 (a) **Given:** $N_2O_4(g) \rightarrow 2\ NO_2(g)$ at 25 °C

Find: $\Delta H°_{rxn}$, $\Delta S°_{rxn}$, $\Delta G°_{rxn}$, and spontaneity. Can temperature be changed to make it spontaneous?

Conceptual Plan: $\boldsymbol{\Delta H°_{rxn} = \sum n_p H°_f(\text{products}) - \sum n_r H°_f(\text{reactants})}$ **then**

$\boldsymbol{\Delta S°_{rxn} = \sum n_p S°(\text{products}) - \sum n_r S°(\text{reactants})}$ **then °C → K then J/K → kJ/K then**

$K = 273.15 + °C$ $\qquad \frac{1\ kJ}{1000\ J}$

$\boldsymbol{\Delta H°_{rxn}, \Delta S°_{rxn}, T \rightarrow \Delta G°}$

$\Delta G = \Delta H_{rxn} - T\Delta S_{rxn}$

Solution:

Reactant/Product	$\Delta H°_f$ (kJ/mol from Appendix IVB)
$N_2O_4(g)$	9.16
$NO_2(g)$	33.2

Be sure to pull data for the correct formula and phase.

$$\begin{aligned}\Delta H°_{rxn} &= \sum n_p \Delta H°_f(\text{products}) - \sum n_r \Delta H°_f(\text{reactants})\\ &= [2(\Delta H°_f(NO_2(g)))] - [1(\Delta H°_f(N_2O_4(g)))]\\ &= [2(33.2\ kJ)] - [1(9.16\ kJ)]\\ &= [66.4\ kJ] - [9.16\ kJ)]\\ &= +57.\underline{2}4\ kJ \text{ then}\end{aligned}$$

Reactant/Product	$S°$(J/mol K from Appendix IVB)
$N_2O_4(g)$	304.4
$NO_2(g)$	240.1

Be sure to pull data for the correct formula and phase.

$$\begin{aligned}\Delta S°_{rxn} &= \sum n_p S°(\text{products}) - \sum n_r S°(\text{reactants})\\ &= [2(S°(NO_2(g)))] - [1(S°(N_2O_4(g)))]\\ &= [2(240.1\ J/K)] - [1(304.4\ J/K)]\\ &= [480.2\ J/K] - [304.4\ J/K]\\ &= +175.8\ J/K \text{ then } T = 273.15 + 25\ °C = 298\ K \text{ then}\end{aligned}$$

$+175.8\ \frac{\cancel{J}}{K} \times \frac{1\ kJ}{1000\ \cancel{J}} = +0.1758\ kJ$ then

$\Delta G° = \Delta H°_{rxn} - T\Delta S°_{rxn} = 57.24 - \left((298\ \cancel{K})\left(+0.1758\frac{kJ}{\cancel{K}}\right)\right) = +4.\underline{8}516\ kJ = +4.9\ kJ = +4.9 \times 10^3\ J$;

so the reaction is nonspontaneous. It can be made spontaneous by raising the temperature.

Check: The units (kJ, J/K, and kJ) are correct. The reaction requires the breaking of a bond, so we expect that this will be an endothermic reaction. We expect a positive entropy change because we are increasing the number of moles of gas. Because the positive enthalpy term dominates at room temperature, the reaction is nonspontaneous. The second term can dominate if we raise the temperature high enough.

(b) **Given:** $NH_4Cl(s) \rightarrow HCl(g) + NH_3(g)$ at 25 °C

Find: $\Delta H°_{rxn}$, $\Delta S°_{rxn}$, $\Delta G°_{rxn}$, and spontaneity. Can temperature be changed to make it spontaneous?

Conceptual Plan: $\boldsymbol{\Delta H°_{rxn} = \sum n_p H°_f(\text{products}) - \sum n_r H°_f(\text{reactants})}$ **then**

$\Delta S^\circ_{rxn} = \sum n_p S^\circ(\text{products}) - \sum n_r S^\circ(\text{reactants})$ **then °C → K then J/K → kJ/K then**

$K = 273.15 + °C$ $\quad \frac{1\ kJ}{1000\ J}$

$\Delta H^\circ_{rxn}, \Delta S^\circ_{rxn}, T \rightarrow \Delta G^\circ$

$\Delta G = \Delta H_{rxn} - T\Delta S_{rxn}$

Solution:

Reactant/Product	ΔH°_f (kJ/mol from Appendix IVB)
$NH_4Cl(s)$	−314.4
$HCl(g)$	−92.3
$NH_3(g)$	−45.9

Be sure to pull data for the correct formula and phase.

$$\begin{aligned}\Delta H^\circ_{rxn} &= \sum n_p \Delta H^\circ_f(\text{products}) - \sum n_r \Delta H^\circ_f(\text{reactants})\\ &= [1(\Delta H^\circ_f(HCl(g))) + 1(\Delta H^\circ_f(NH_3(g)))] - [1(\Delta H^\circ_f(NH_4Cl(g)))]\\ &= [1(-92.3\text{ kJ}) + 1(-45.9\text{ kJ})] - [1(-314.4\text{ kJ})]\\ &= [-138.2\text{ kJ}] - [-314.4\text{ kJ}]\\ &= +176.2\text{ kJ then}\end{aligned}$$

Reactant/Product	S°(J/mol K from Appendix IVB)
$NH_4Cl(s)$	94.6
$HCl(g)$	186.9
$NH_3(g)$	192.8

Be sure to pull data for the correct formula and phase.

$$\begin{aligned}\Delta S^\circ_{rxn} &= \sum n_p S^\circ(\text{products}) - \sum n_r S^\circ(\text{reactants})\\ &= [1(S^\circ(HCl(g))) + 1(S^\circ(NH_3(g)))] - [1(S^\circ(NH_4Cl(g)))]\\ &= [1(186.9\text{ J/K}) + 1(192.8\text{ J/K})] - [1(94.6\text{ J/K})]\\ &= [379.7\text{ J/K}] - [94.6\text{ J/K}]\\ &= +285.1\text{ J/K}\end{aligned}$$

then $T = 273.15 + 25\ °C = 298\ K$ then $+285.1\frac{J}{K} \times \frac{1\ kJ}{1000\ J} = +0.2851\ kJ/K$ then

$\Delta G^\circ = \Delta H^\circ_{rxn} - T\Delta S^\circ_{rxn} = +176.2\text{ kJ} - \left((298\text{ K})\left(+0.2851\frac{kJ}{K}\right)\right) = +91.2\text{ kJ} = +9.12 \times 10^4\text{ J}$; so the reaction is nonspontaneous. It can be made spontaneous by raising the temperature.

Check: The units (kJ, J/K, and kJ) are correct. The reaction requires the breaking of a bond, so we expect that this will be an endothermic reaction. We expect a positive entropy change because we are increasing the number of moles of gas. Because the positive enthalpy term dominates at room temperature, the reaction is nonspontaneous. The second term can dominate if we raise the temperature high enough.

(c) **Given:** $3\ H_2(g) + Fe_2O_3(s) \rightarrow 2\ Fe(s) + 3\ H_2O(g)$ at 25 °C **Find:** $\Delta H^\circ_{rxn}, \Delta S^\circ_{rxn}, \Delta G^\circ_{rxn}$, and spontaneity. Can temperature be changed to make it spontaneous?

Conceptual Plan: $\Delta H^\circ_{rxn} = \sum n_p H^\circ_f(\text{products}) - \sum n_r H^\circ_f(\text{reactants})$ **then**

$\Delta S^\circ_{rxn} = \sum n_p S^\circ(\text{products}) - \sum n_r S^\circ(\text{reactants})$ **then °C → K then J/K → kJ/K then**

$K = 273.15 + °C$ $\quad \frac{1\ kJ}{1000\ J}$

$\Delta H^\circ_{rxn}, \Delta S^\circ_{rxn}, T \rightarrow \Delta G^\circ$

$\Delta G = \Delta H_{rxn} - T\Delta S_{rxn}$

Solution:

Reactant/Product	ΔH°_f (kJ/mol from Appendix IVB)
$H_2(g)$	0.0
$Fe_2O_3(s)$	−824.2
$Fe(s)$	0.0
$H_2O(g)$	−241.8

Be sure to pull data for the correct formula and phase.

$$\begin{aligned}\Delta H^\circ_{rxn} &= \sum n_p \Delta H^\circ_f(\text{products}) - \sum n_r \Delta H^\circ_f(\text{reactants})\\ &= [2(\Delta H^\circ_f(\text{Fe}(s))) + 3(\Delta H^\circ_f(\text{H}_2\text{O}(g)))] - [3(\Delta H^\circ_f(\text{H}_2(g))) + 1(\Delta H^\circ_f(\text{Fe}_2\text{O}_3(s)))]\\ &= [2(0.0\text{ kJ}) + 3(-241.8\text{ kJ})] - [3(0.0\text{ kJ}) + 1(-824.2\text{ kJ})]\\ &= [-725.4\text{ kJ}] - [-824.2\text{ kJ}]\\ &= +98.8\text{ kJ then}\end{aligned}$$

Reactant/Product	S°(J/mol K from Appendix IVB)
$H_2(g)$	130.7
$Fe_2O_3(s)$	87.4
$Fe(s)$	27.3
$H_2O(g)$	188.8

Be sure to pull data for the correct formula and phase.

$$\begin{aligned}\Delta S^\circ_{rxn} &= \sum n_p S^\circ(\text{products}) - \sum n_r S^\circ(\text{reactants})\\ &= [2(S^\circ(\text{Fe}(s))) + 3(S^\circ(\text{H}_2\text{O}(g)))] - [3(S^\circ(\text{H}_2(g))) + 1(S^\circ(\text{Fe}_2\text{O}_3(s)))]\\ &= [2(27.3\text{ J/K}) + 3(188.8\text{ J/K})] - [3(130.7\text{ J/K}) + 1(87.4\text{ J/K})]\\ &= [621.0\text{ J/K}] - [479.5\text{ J/K}]\\ &= +141.5\text{ J/K}\end{aligned}$$

then $T = 273.15 + 25\ ^\circ\text{C} = 298\text{ K}$ then $+141.5\frac{\cancel{\text{J}}}{\text{K}} \times \frac{1\text{ kJ}}{1000\ \cancel{\text{J}}} = +0.1415\text{ kJ/K}$ then

$\Delta G^\circ = \Delta H^\circ_{rxn} - T\Delta S^\circ_{rxn} = +98.8\text{ kJ} - \left((298\ \cancel{\text{K}})\left(+0.1415\frac{\text{kJ}}{\cancel{\text{K}}}\right)\right) = +56.6\text{ kJ} = +5.66 \times 10^4\text{ J}$; so the reaction is nonspontaneous. It can be made spontaneous by raising the temperature.

Check: The units (kJ, J/K, and kJ) are correct. The reaction requires the breaking of a bond, so we expect that this will be an endothermic reaction. We expect a positive entropy change because there is no change in the number of moles of gas, but the product gas is more complex. Because the positive enthalpy term dominates at room temperature, the reaction is nonspontaneous. The second term can dominate if we raise the temperature high enough. This process is the opposite of rusting, so we are not surprised that it is nonspontaneous.

(d) **Given:** $N_2(g) + 3\,H_2(g) \rightarrow 2\,NH_3(g)$ at 25 °C

Find: ΔH°_{rxn}, ΔS°_{rxn}, ΔG°_{rxn}, and spontaneity. Can temperature be changed to make it spontaneous?

Conceptual Plan: $\boldsymbol{\Delta H^\circ_{rxn} = \sum n_p H^\circ_f(\text{products}) - \sum n_r H^\circ_f(\text{reactants})}$ **then** $\boldsymbol{\Delta S^\circ_{rxn} = \sum n_p S^\circ(\text{products}) - \sum n_r S^\circ(\text{reactants})}$ **then °C → K then J/K → kJ/K then**

$K = 273.15 + ^\circ C$ $\qquad \frac{1\text{ kJ}}{1000\text{ J}}$

$\boldsymbol{\Delta H^\circ_{rxn}, \Delta S^\circ_{rxn}, T \rightarrow \Delta G^\circ}$

$\Delta G = \Delta H_{rxn} - T\Delta S_{rxn}$

Solution:

Reactant/Product	ΔH°_f(kJ/mol from Appendix IVB)
$N_2(g)$	0.0
$H_2(g)$	0.0
$NH_3(g)$	−45.9

Be sure to pull data for the correct formula and phase.

$$\begin{aligned}\Delta H^\circ_{rxn} &= \sum n_p \Delta H^\circ_f(\text{products}) - \sum n_r \Delta H^\circ_f(\text{reactants})\\ &= [2(\Delta H^\circ_f(\text{NH}_3(g)))] - [1(\Delta H^\circ_f(\text{N}_2(g))) + 3(\Delta H^\circ_f(\text{H}_2(g)))]\\ &= [2(-45.9\text{ kJ})] - [1(0.0\text{ kJ}) + 3(0.0\text{ kJ})]\\ &= [-91.8\text{ kJ}] - [0.0\text{ kJ}]\\ &= -91.8\text{ kJ then}\end{aligned}$$

Reactant/Product	S°(J/mol K from Appendix IVB)
$N_2(g)$	191.6
$H_2(g)$	130.7
$NH_3(g)$	192.8

Be sure to pull data for the correct formula and phase.

$$\begin{aligned}\Delta S^\circ_{rxn} &= \sum n_p S^\circ(\text{products}) - \sum n_r S^\circ(\text{reactants}) \\ &= [2(S^\circ(NH_3(g)))] - [1(S^\circ(N_2(g))) + 3(S^\circ(H_2(g)))] \\ &= [2(192.8\ J/K)] - [1(191.6\ J/K) + 3(130.7\ J/K)] \\ &= [385.6\ J/K] - [583.7\ J/K] \\ &= -198.1\ J/K\end{aligned}$$

then $T = 273.15 + 25\ °C = 298\ K$ then $-198.1\frac{\cancel{J}}{K} \times \frac{1\ kJ}{1000\ \cancel{J}} = -0.1981\ kJ/K$ then

$\Delta G^\circ = \Delta H^\circ_{rxn} - T\Delta S^\circ_{rxn} = -91.8\ kJ - \left((298\ \cancel{K})\left(-0.1981\frac{kJ}{\cancel{K}}\right)\right) = -32.8\ kJ = -3.28 \times 10^4\ J$; so the reaction is spontaneous.

Check: The units (kJ, J/k, and kJ) are correct. The reaction requires the breaking of a bond, so we expect that this will be an endothermic reaction. We expect a positive entropy change because we are increasing the number of moles of gas. Because the negative enthalpy term dominates at room temperature, the reaction is spontaneous. The second term can dominate if we raise the temperature high enough.

19.56 (a) **Given:** $2\ CH_4(g) \rightarrow C_2H_6(g) + H_2(g)$ at 25 °C **Find:** ΔH°_{rxn}, ΔS°_{rxn}, ΔG°_{rxn}, and spontaneity. Can temperature be changed to make it spontaneous?

Conceptual Plan: $\boldsymbol{\Delta H^\circ_{rxn} = \sum n_p H^\circ_f(\text{products}) - \sum n_r H^\circ_f(\text{reactants})}$ **then**

$\boldsymbol{\Delta S^\circ_{rxn} = \sum n_p S^\circ(\text{products}) - \sum n_r S^\circ(\text{reactants})}$ **then °C → K then J/K → kJ/K then**

$K = 273.15 + °C$ $\quad \frac{1\ kJ}{1000\ J}$

$\boldsymbol{\Delta H^\circ_{rxn}, \Delta S^\circ_{rxn}, T \rightarrow \Delta G^\circ}$

$\Delta G = \Delta H_{rxn} - T\Delta S_{rxn}$

Solution:

Reactant/Product	ΔH°_f(kJ/mol from Appendix IVB)
$CH_4(g)$	−74.6
$C_2H_6(g)$	−84.68
$H_2(g)$	0.0

Be sure to pull data for the correct formula and phase.

$$\begin{aligned}\Delta H^\circ_{rxn} &= \sum n_p \Delta H^\circ_f(\text{products}) - \sum n_r \Delta H^\circ_f(\text{reactants}) \\ &= [1(\Delta H^\circ_f(C_2H_6(g))) + 1(\Delta H^\circ_f(H_2(g)))] - [2(\Delta H^\circ_f(CH_4(g)))] \\ &= [1(-84.6\ kJ) + 1(0.0\ kJ)] - [2(-74.6\ kJ)] \\ &= [-84.6\ kJ] - [-149.2\ kJ] \\ &= +64.\underline{5}2\ kJ = 64.5\ kJ \text{ then}\end{aligned}$$

Reactant/Product	S°(J/mol K from Appendix IVB)
$CH_4(g)$	186.3
$C_2H_6(g)$	229.2
$H_2(g)$	130.7

Be sure to pull data for the correct formula and phase.

$$\begin{aligned}\Delta S^\circ_{rxn} &= \sum n_p S^\circ(\text{products}) - \sum n_r S^\circ(\text{reactants}) \\ &= [1(S^\circ(C_2H_6(g))) + 1(S^\circ(H_2(g)))] - [2(S^\circ(CH_4(g)))] \\ &= [1(229.2\ J/K) + 1(130.7\ J/K)] - [2(186.3\ J/K)] \\ &= [359.9\ J/K] - [372.6\ J/K] \\ &= -12.7\ J/K\end{aligned}$$

then $T = 273.15 + 25\ °C = 298\ K$ then $-12.7\frac{\cancel{J}}{K} \times \frac{1\ kJ}{1000\ \cancel{J}} = -0.0127\ kJ/K$ then

$\Delta G^\circ = \Delta H^\circ_{rxn} - T\Delta S^\circ_{rxn} = +64.\underline{5}2\ kJ - \left((298\ \cancel{K})\left(-0.0127\frac{kJ}{\cancel{K}}\right)\right) = +68.3\ kJ = +6.83 \times 10^4\ J$; so the reaction is nonspontaneous. Because both terms are positive, this reaction cannot be spontaneous at any temperature.

Check: The units (kJ, J/K, and kJ) are correct. The 2 moles of methane have a lower enthalpy than does 1 mole of ethane, so we expect that this will be an endothermic reaction. We expect a very small entropy change because the number of moles of gas is unchanged. Because both terms are positive, the reaction is nonspontaneous. Because both terms are positive, this reaction cannot be spontaneous at any temperature.

(b) **Given:** $2\ NH_3(g) \rightarrow N_2H_4(g) + H_2(g)$ at 25 °C

Find: ΔH°_{rxn}, ΔS°_{rxn}, ΔG°_{rxn}, and spontaneity. Can temperature be changed to make it spontaneous?

Conceptual Plan: $\Delta H^\circ_{rxn} = \sum n_p H^\circ_f(\text{products}) - \sum n_r H^\circ_f(\text{reactants})$ **then**

$\Delta S^\circ_{rxn} = \sum n_p S^\circ(\text{products}) - \sum n_r S^\circ(\text{reactants})$ **then °C → K then J/K → kJ/K then**

$K = 273.15 + °C$ $\qquad \frac{1\ kJ}{1000\ J}$

$\Delta H^\circ_{rxn}, \Delta S^\circ_{rxn}, T \rightarrow \Delta G^\circ$

$\Delta G = \Delta H_{rxn} - T\Delta S_{rxn}$

Solution:

Reactant/Product	ΔH°_f(kJ/mol from Appendix IVB)
$NH_3(g)$	−45.9
$N_2H_4(g)$	95.4
$H_2(g)$	0.0

Be sure to pull data for the correct formula and phase.

$$\begin{aligned}\Delta H^\circ_{rxn} &= \sum n_p \Delta H^\circ_f(\text{products}) - \sum n_r \Delta H^\circ_f(\text{reactants}) \\ &= [1(\Delta H^\circ_f(N_2H_4(g))) + 1(\Delta H^\circ_f(H_2(g)))] - [2(\Delta H^\circ_f(NH_3(g)))] \\ &= [1(95.4\ kJ) + 1(0.0\ kJ)] - [2(-45.9\ kJ)] \\ &= [95.4\ kJ] - [-91.8\ kJ] \\ &= +187.2\ kJ\ \text{then}\end{aligned}$$

Reactant/Product	S°(J/mol K from Appendix IVB)
$NH_3(g)$	192.8
$N_2H_4(g)$	238.5
$H_2(g)$	130.7

Be sure to pull data for the correct formula and phase.

$$\begin{aligned}\Delta S^\circ_{rxn} &= \sum n_p S^\circ(\text{products}) + \sum n_r S^\circ(\text{reactants}) \\ &= [1(S^\circ(N_2H_4(g))) + 1(S^\circ(H_2(g)))] - [2(S^\circ(NH_3(g)))] \\ &= [1(238.5\ J/K) + 1(130.7\ J/K)] - [2(192.8\ J/K)] \\ &= [369.2\ J/K] - [385.6\ J/K] \\ &= -16.4\ J/K\ \text{then}\end{aligned}$$

$$T = 273.15 + 25\ °C = 298\ K \text{ then } -16.4\frac{J}{K} \times \frac{1\ kJ}{1000\ J} = -0.0164\ kJ/K \text{ then}$$

$\Delta G^\circ = \Delta H^\circ_{rxn} - T\Delta S^\circ_{rxn} = +187.2\ kJ - \left((298\ K)\left(-0.0164\frac{kJ}{K}\right)\right) = +192.1\ kJ = +1.921 \times 10^5\ J$; so the reaction is nonspontaneous. Both terms are positive, so the reaction is not spontaneous at any temperature.

Check: The units (kJ, J/K, and kJ) are correct. N_2H_4 has such a high enthalpy of formation compared to ammonia that we expect this to be an endothermic reaction. We expect a very small entropy change because the number of moles of gas is unchanged. Because both terms are positive, the reaction is nonspontaneous. Because both terms are positive, this reaction cannot be spontaneous at any temperature.

(c) **Given:** $N_2(g) + O_2(g) \rightarrow 2\ NO(g)$ at 25 °C

Find: ΔH°_{rxn}, ΔS°_{rxn}, ΔG°_{rxn}, and spontaneity. Can temperature be changed to make it spontaneous?

Conceptual Plan: $\Delta H^\circ_{rxn} = \sum n_p H^\circ_f(\text{products}) - \sum n_r H^\circ_f(\text{reactants})$ **then**

$\Delta S^\circ_{rxn} = \sum n_p S^\circ(\text{products}) - \sum n_r S^\circ(\text{reactants})$ **then °C → K then J/K → kJ/K then**

$K = 273.15 + °C$ $\qquad \frac{1\ kJ}{1000\ J}$

$\Delta H^\circ_{rxn}, \Delta S^\circ_{rxn}, T \rightarrow \Delta G^\circ$

$\Delta G = \Delta H_{rxn} - T\Delta S_{rxn}$

Solution:

Reactant/Product	ΔH_f°(kJ/mol from Appendix IVB)
$N_2(g)$	0.0
$O_2(g)$	0.0
$NO(g)$	91.3

Be sure to pull data for the correct formula and phase.

$$\begin{aligned}\Delta H^\circ_{rxn} &= \sum n_p \Delta H_f^\circ(\text{products}) - \sum n_r \Delta H_f^\circ(\text{reactants})\\ &= [2(\Delta H_f^\circ(NO(g)))] - [1(\Delta H_f^\circ(N_2(g))) + 1(\Delta H_f^\circ(O_2(g)))]\\ &= [2(91.3\text{ kJ})] - [1(0.0\text{ kJ}) + 1(0.0\text{ kJ})]\\ &= [182.6\text{ kJ}] - [0.0\text{ kJ}]\\ &= +182.6\text{ kJ then}\end{aligned}$$

Reactant/Product	S°(J/mol K from Appendix IVB)
$N_2(g)$	191.6
$O_2(g)$	205.2
$NO(g)$	210.8

Be sure to pull data for the correct formula and phase.

$$\begin{aligned}\Delta S^\circ_{rxn} &= \sum n_p S^\circ(\text{products}) - \sum n_r S^\circ(\text{reactants})\\ &= [2(S^\circ(NO(g)))] - [1(S^\circ(N_2(g))) + 1(S^\circ(O_2(g)))]\\ &= [2(210.8\text{ J/K})] - [1(191.6\text{ J/K}) + 1(205.2\text{ J/K})]\\ &= [421.6\text{ J/K}] - [396.8\text{ J/K}]\\ &= +24.8\text{ J/K}\end{aligned}$$

then $T = 273.15 + 25\ ^\circ\text{C} = 298\text{ K}$ then $+24.8\frac{\cancel{J}}{\text{K}} \times \frac{1\text{ kJ}}{1000\ \cancel{J}} = +0.0248\text{ kJ/K}$ then

$\Delta G^\circ = \Delta H^\circ_{rxn} - T\Delta S^\circ_{rxn} = +182.6\text{ kJ} - \left((298\ \cancel{K})\left(0.0248\frac{\text{kJ}}{\cancel{K}}\right)\right) = +175.2\text{ kJ} = +1.752 \times 10^5\text{ J}$; so the reaction is nonspontaneous. It can be spontaneous at high temperatures.

Check: The units (kJ, J/K, and kJ) are correct. The enthalpy is twice the enthalpy of formation of NO. We expect a very small entropy change because the number of moles of gas is unchanged. Because the positive enthalpy term dominates at room temperature, the reaction is nonspontaneous. The second term can dominate if we raise the temperature high enough.

(d) **Given:** $2\ KClO_3(s) \rightarrow 2\ KCl(s) + 3\ O_2(g)$ at 25 °C

Find: ΔH°_{rxn}, ΔS°_{rxn}, ΔG°_{rxn}, and spontaneity. Can temperature be changed to make it spontaneous?

Conceptual Plan: $\boldsymbol{\Delta H^\circ_{rxn} = \sum n_p H_f^\circ(\text{products}) - \sum n_r H_f^\circ(\text{reactants})}$ **then**

$\boldsymbol{\Delta S^\circ_{rxn} = \sum n_p S^\circ(\text{products}) - \sum n_r S^\circ(\text{reactants})}$ **then °C → K then J/K → kJ/K then**

$K = 273.15 + ^\circ C$ $\quad \frac{1\text{ kJ}}{1000\text{ J}}$

$\boldsymbol{\Delta H^\circ_{rxn}, \Delta S^\circ_{rxn}, T \rightarrow \Delta G^\circ}$

$\Delta G = \Delta H_{rxn} - T\Delta S_{rxn}$

Solution:

Reactant/Product	ΔH_f°(kJ/mol from Appendix IVB)
$KClO_3(s)$	−397.7
$KCl(s)$	−436.5
$O_2(g)$	0.0

Be sure to pull data for the correct formula and phase.

$$\begin{aligned}\Delta H^\circ_{rxn} &= \sum n_p \Delta H_f^\circ(\text{products}) - \sum n_r \Delta H_f^\circ(\text{reactants})\\ &= [2(\Delta H_f^\circ(KCl(s))) + 3(\Delta H_f^\circ(O_2(g)))] - [2(\Delta H_f^\circ(KClO_3(s)))]\\ &= [2(-436.5\text{ kJ}) + 3(0.0\text{ kJ})] - [2(-397.7\text{ kJ})]\\ &= [-873.0\text{ kJ}] - [-795.4\text{ kJ}]\\ &= -77.6\text{ kJ then}\end{aligned}$$

Reactant/Product	S°(J/mol K from Appendix IVB)
$KClO_3(s)$	143.1
$KCl(s)$	82.6
$O_2(g)$	205.2

Be sure to pull data for the correct formula and phase.

$$\begin{aligned}\Delta S^\circ_{rxn} &= \sum n_p S^\circ(\text{products}) - \sum n_r S^\circ(\text{reactants}) \\ &= [2(S^\circ(KCl(s))) + 3(S^\circ(O_2(g)))] - [2(S^\circ(KClO_3(s)))] \\ &= [2(82.6\text{ J/K}) + 3(205.2\text{ J/K})] - [2(143.1\text{ J/K})] \\ &= [780.8\text{ J/K}] - [286.2\text{ J/K}] \\ &= +494.6\text{ J/K then}\end{aligned}$$

$T = 273.15 + 25\,°\text{C} = 298\text{ K}$ then $+494.6\frac{J}{\text{K}} \times \frac{1\text{ kJ}}{1000\text{ J}} = +0.4946\text{ kJ/K}$ then

$\Delta G^\circ = \Delta H^\circ_{rxn} - T\Delta S^\circ_{rxn} = -77.6\text{ kJ} - \left((298\text{ K})\left(0.4946\frac{\text{kJ}}{\text{K}}\right)\right) = -225.0\text{ kJ} = -2.250 \times 10^6\text{ J}$; so the reaction is spontaneous. Because both terms are negative, the reaction is spontaneous at all temperatures.

Check: The units (kJ, J/K, and kJ) are correct. The reaction is exothermic because the enthalpy of formation of KCl is less than that for $KClO_3$. We expect a positive entropy change because the number of moles of gas is increasing. Because both terms are negative, the reaction is spontaneous at all temperatures.

19.57 (a) **Given:** $N_2O_4(g) \rightarrow 2\,NO_2(g)$ at 25 °C **Find:** ΔG°_{rxn} and spontaneity and compare to Problem 19.55

Determine which method would show how free energy changes with temperature.

Conceptual Plan: $\Delta G^\circ_{rxn} = \sum n_p \Delta G^\circ_f(\text{products}) - \sum n_r \Delta G^\circ_f(\text{reactants})$ then compare to Problem 19.55

Solution:

Reactant/Product	ΔG°_f(kJ/mol from Appendix IVB)
$N_2O_4(g)$	99.8
$NO_2(g)$	51.3

Be sure to pull data for the correct formula and phase.

$$\begin{aligned}\Delta G^\circ_{rxn} &= \sum n_p \Delta G^\circ_f(\text{products}) - \sum n_r \Delta G^\circ_f(\text{reactants}) \\ &= [2(\Delta G^\circ_f(NO_2(g)))] - [1(\Delta G^\circ_f(N_2O_4(g)))] \\ &= [2(51.3\text{ kJ})] - [1(99.8\text{ kJ})] \\ &= [102.6\text{ kJ}] - [99.8\text{ kJ}] \\ &= +2.8\text{ kJ}\end{aligned}$$

So the reaction is nonspontaneous. The value is similar to that in Problem 19.55.

Check: The units (kJ) are correct. The free energy of the products is greater than that of the reactants; so the answer is positive, and the reaction is nonspontaneous. The answer is the same as that in Problem 19.55 within the error of the calculation.

(b) **Given:** $NH_4Cl(s) \rightarrow HCl(g) + NH_3(g)$ at 25 °C

Find: ΔG°_{rxn} and spontaneity and compare to Problem 19.55

Conceptual Plan: $\Delta G^\circ_{rxn} = \sum n_p \Delta G^\circ_f(\text{products}) - \sum n_r \Delta G^\circ_f(\text{reactants})$ then compare to Problem 19.55

Solution:

Reactant/Product	ΔG°_f(kJ/mol from Appendix IVB)
$NH_4Cl(s)$	−202.9
$HCl(g)$	−95.3
$NH_3(g)$	−16.4

Be sure to pull data for the correct formula and phase.

$$\begin{aligned}\Delta G^\circ_{rxn} &= \sum n_p \Delta G^\circ_f(\text{products}) - \sum n_r \Delta G^\circ_f(\text{reactants}) \\ &= [1(\Delta G^\circ_f(HCl(g))) + 1(\Delta G^\circ_f(NH_3(g)))] - [1(\Delta G^\circ_f(NH_4Cl(g)))] \\ &= [1(-95.3\text{ kJ}) + 1(-16.4\text{ kJ})] - [1(-202.9\text{ kJ})] \\ &= [-111.7\text{ kJ}] - [-202.9\text{ kJ}] \\ &= +91.2\text{ kJ}\end{aligned}$$

So the reaction is nonspontaneous. The result is the same as that in Problem 19.55.

Check: The units (kJ) are correct. The answer matches the one in Problem 19.55.

(c) **Given:** $3\,H_2(g) + Fe_2O_3(s) \rightarrow 2\,Fe(s) + 3\,H_2O(g)$ at 25 °C
Find: ΔG°_{rxn} and spontaneity and compare to Problem 19.55
Conceptual Plan: $\boldsymbol{\Delta G^\circ_{rxn} = \sum n_p \Delta G^\circ_f(\text{products}) - \sum n_r \Delta G^\circ_f(\text{reactants})}$ **then compare to Problem 19.55**
Solution:

Reactant/Product	ΔG°_f(kJ/mol from Appendix IVB)
$H_2(g)$	0.0
$Fe_2O_3(s)$	−742.2
$Fe(s)$	0.0
$H_2O(g)$	−228.6

Be sure to pull data for the correct formula and phase.

$$\begin{aligned}\Delta G^\circ_{rxn} &= \sum n_p \Delta G^\circ_f(\text{products}) - \sum n_r \Delta G^\circ_f(\text{reactants})\\ &= [2(\Delta G^\circ_f(Fe(s))) + 3(\Delta G^\circ_f(H_2O(g)))] - [3(\Delta G^\circ_f(H_2(g))) + 1(\Delta G^\circ_f(Fe_2O_3(s)))]\\ &= [2(0.0\text{ kJ}) + 3(-228.6\text{ kJ})] - [3(0.0\text{ kJ}) + 1(-742.2\text{ kJ})]\\ &= [-685.8\text{ kJ}] - [-742.2\text{ kJ}]\\ &= +56.4\text{ kJ}\end{aligned}$$

So the reaction is nonspontaneous. The value is similar to that in Problem 19.55.

Check: The units (kJ) are correct. The answer is the same as that in Problem 19.55 within the error of the calculation.

(d) **Given:** $N_2(g) + 3\,H_2(g) \rightarrow 2\,NH_3(g)$ at 25 °C
Find: ΔG°_{rxn} and spontaneity and compare to Problem 19.55
Conceptual Plan: $\boldsymbol{\Delta G^\circ_{rxn} = \sum n_p \Delta G^\circ_f(\text{products}) - \sum n_r \Delta G^\circ_f(\text{reactants})}$ **then compare to Problem 19.55**
Solution:

Reactant/Product	ΔG°_f(kJ/mol from Appendix IVB)
$N_2(g)$	0.0
$H_2(g)$	0.0
$NH_3(g)$	−16.4

Be sure to pull data for the correct formula and phase.

$$\begin{aligned}\Delta G^\circ_{rxn} &= \sum n_p \Delta G^\circ_f(\text{products}) - \sum n_r \Delta G^\circ_f(\text{reactants})\\ &= [2(\Delta G^\circ_f(NH_3(g)))] - [1(\Delta G^\circ_f(N_2(g))) + 3(\Delta G^\circ_f(H_2(g)))]\\ &= [2(-16.4\text{ kJ})] - [1(0.0\text{ kJ}) + 3(0.0\text{ kJ})]\\ &= [-32.8\text{ kJ}] - [0.0\text{ kJ}]\\ &= -32.8\text{ kJ}\end{aligned}$$

So the reaction is spontaneous. The result is the same as that in Problem 19.55.

Check: The units (kJ) are correct. The answer matches the one in Problem 19.55.

Values calculated by the two methods are comparable. The method using ΔH° and ΔS° is longer, but it can be used to determine how ΔG° changes with temperature.

19.58 (a) **Given:** $2\,CH_4(g) \rightarrow C_2H_6(g) + H_2(g)$ at 25 °C **Find:** ΔG°_{rxn} and spontaneity and compare to Problem 19.56
Determine which method would show how free energy changes with temperature.
Conceptual Plan: $\boldsymbol{\Delta G^\circ_{rxn} = \sum n_p \Delta G^\circ_f(\text{products}) - \sum n_r \Delta G^\circ_f(\text{reactants})}$ **then compare to Problem 19.56**
Solution:

Reactant/Product	ΔG°_f(kJ/mol from Appendix IVB)
$CH_4(g)$	−50.5
$C_2H_6(g)$	−32.0
$H_2(g)$	0.0

Be sure to pull data for the correct formula and phase.

$$\Delta G^\circ_{rxn} = \sum n_p \Delta G^\circ_f(\text{products}) - \sum n_r \Delta G^\circ_f(\text{reactants})$$

$= [1(\Delta G^\circ_f(C_2H_6(g))) + 1(\Delta G^\circ_f(H_2(g)))] - [2(\Delta G^\circ_f(CH_4(g)))]$
$= [1(-32.0\text{ kJ}) + 1(0.0\text{ kJ})] - [2(-50.5\text{ kJ})]$
$= [-32.0\text{ kJ}] - [-101.0\text{ kJ}]$
$= +69.0\text{ kJ}$

So the reaction is spontaneous. The value is similar to that in Problem 19.56.

Check: The units (kJ) are correct. The answer is the same as that in Problem 19.56 within the error of the calculation.

(b) **Given:** $2\ NH_3(g) \rightarrow N_2H_4(g) + H_2(g)$ at 25 °C
Find: ΔG°_{rxn} and spontaneity and compare to Problem 19.56
Conceptual Plan: $\boldsymbol{\Delta G^\circ_{rxn} = \sum n_p \Delta G^\circ_f(\text{products}) - \sum n_r \Delta G^\circ_f(\text{reactants})}$ **then compare to Problem 19.56**
Solution:

Reactant/Product	ΔG°_f(kJ/mol from Appendix IVB)
$NH_3(g)$	−16.4
$N_2H_4(g)$	159.4
$H_2(g)$	0.0

Be sure to pull data for the correct formula and phase.

$\Delta G^\circ_{rxn} = \sum n_p \Delta G^\circ_f(\text{products}) - \sum n_r \Delta G^\circ_f(\text{reactants})$
$= [1(\Delta G^\circ_f(N_2H_4(g))) + 1(\Delta G^\circ_f(H_2(g)))] - [2(\Delta G^\circ_f(NH_3(g)))]$
$= [1(159.4\text{ kJ}) + 1(0.0\text{ kJ})] - [2(-16.4\text{ kJ})]$
$= [159.4\text{ kJ}] - [-32.8\text{ kJ}]$
$= +192.2\text{ kJ}$

So the reaction is nonspontaneous. The value is similar to that in Problem 19.56.

Check: The units (kJ) are correct. The answer is the same as that in Problem 19.56 within the error of the calculation.

(c) **Given:** $N_2(g) + O_2(g) \rightarrow 2\ NO(g)$ at 25 °C
Find: ΔG°_{rxn} and spontaneity and compare to Problem 19.56
Conceptual Plan: $\boldsymbol{\Delta G^\circ_{rxn} = \sum n_p \Delta G^\circ_f(\text{products}) - \sum n_r \Delta G^\circ_f(\text{reactants})}$ **then compare to Problem 19.56**
Solution:

Reactant/Product	ΔG°_f(kJ/mol from Appendix IVB)
$N_2(g)$	0.0
$O_2(g)$	0.0
$NO(g)$	87.6

Be sure to pull data for the correct formula and phase.

$\Delta G^\circ_{rxn} = \sum n_p \Delta G^\circ_f(\text{products}) - \sum n_r \Delta G^\circ_f(\text{reactants})$
$= [2(\Delta G^\circ_f(NO(g)))] - [1(\Delta G^\circ_f(N_2(g))) + 1(\Delta G^\circ_f(O_2(g)))]$
$= [2(87.6\text{ kJ})] - [1(0.0\text{ kJ}) + 1(0.0\text{ kJ})]$
$= [175.2\text{ kJ}] - [0.0\text{ kJ}]$
$= +175.2\text{ kJ}$

So the reaction is nonspontaneous. The result is the same as that in Problem 19.56.

Check: The units (kJ) are correct. The answer matches the one in Problem 19.56.

(d) **Given:** $2\ KClO_3(s) \rightarrow 2\ KCl(s) + 3\ O_2(g)$ at 25 °C
Find: ΔG°_{rxn} and spontaneity and compare to Problem 19.56
Conceptual Plan: $\boldsymbol{\Delta G^\circ_{rxn} = \sum n_p \Delta G^\circ_f(\text{products}) - \sum n_r \Delta G^\circ_f(\text{reactants})}$ **then compare to Problem 19.56**
Solution:

Reactant/Product	ΔG°_f(kJ/mol from Appendix IVB)
$KClO_3(s)$	−296.3
$KCl(s)$	−408.5
$O_2(g)$	0.0

Be sure to pull data for the correct formula and phase.

$$\begin{aligned}\Delta G^\circ_{rxn} &= \sum n_p \Delta G^\circ_f(\text{products}) - \sum n_r \Delta G^\circ_f(\text{reactants})\\ &= [2(\Delta G^\circ_f(KCl(s))) + 1(\Delta G^\circ_f(O_2(g)))] - [2(\Delta G^\circ_f(KClO_3(s)))]\\ &= [2(-408.5\text{ kJ}) + 3(0.0\text{ kJ})] - [2(-296.3\text{ kJ})]\\ &= [-817.0\text{ kJ}] - [-592.6\text{ kJ}]\\ &= -224.4\text{ kJ}\end{aligned}$$

So the reaction is spontaneous. The value is similar to that in Problem 19.56.

Check: The units (kJ) are correct. The answer is the same as that in Problem 19.56 within the error of the calculation.

Values calculated by the two methods are comparable. The method using ΔH° and ΔS° is longer, but it can be used to determine how ΔG° changes with temperature.

19.59 **Given:** $2\,NO(g) + O_2(g) \rightarrow 2\,NO_2(g)$ **Find:** ΔG°_{rxn} and spontaneity at (a) 298 K, (b) 715 K, and (c) 855 K

Conceptual Plan: $\Delta H^\circ_{rxn} = \sum n_p H^\circ_f(\text{products}) - \sum n_r H^\circ_f(\text{reactants})$ **then**
$\Delta S^\circ_{rxn} = \sum n_p S^\circ(\text{products}) - \sum n_r S^\circ(\text{reactants})$ **then J/K → kJ/K then** $\Delta H^\circ_{rxn}, \Delta S^\circ_{rxn}, T \rightarrow \Delta G^\circ$

$\frac{1\text{ kJ}}{1000\text{ J}}$ $\qquad \Delta G = \Delta H_{rxn} - T\Delta S_{rxn}$

Solution:

Reactant/Product	ΔH°_f(kJ/mol from Appendix IVB)
$NO(g)$	91.3
$O_2(g)$	0.0
$NO_2(g)$	33.2

Be sure to pull data for the correct formula and phase.

$$\begin{aligned}\Delta H^\circ_{rxn} &= \sum n_p \Delta H^\circ_f(\text{products}) - \sum n_r \Delta H^\circ_f(\text{reactants})\\ &= [2(\Delta H^\circ_f(NO_2(g)))] - [2(\Delta H^\circ_f(NO(g))) + 1(\Delta H^\circ_f(O_2(g)))]\\ &= [2(33.2\text{ kJ})] - [2(91.3\text{ kJ}) + 1(0.0\text{ kJ})]\\ &= [66.4\text{ kJ}] - [182.6\text{ kJ}]\\ &= -116.2\text{ kJ then}\end{aligned}$$

Reactant/Product	S°(J/mol K from Appendix IVB)
$NO(g)$	210.8
$O_2(g)$	205.2
$NO_2(g)$	240.1

Be sure to pull data for the correct formula and phase.

$$\begin{aligned}\Delta S^\circ_{rxn} &= \sum n_p S^\circ(\text{products}) - \sum n_r S^\circ(\text{reactants})\\ &= [2(S^\circ(NO_2(g)))] - [2(S^\circ(NO(g))) + 1(S^\circ(O_2(g)))]\\ &= [2(240.1\text{ J/K})] - [2(210.8\text{ J/K}) + 1(205.2\text{ J/K})] \quad \text{then } -146.6\frac{\cancel{J}}{K} \times \frac{1\text{ kJ}}{1000\cancel{J}} = -0.1466\text{ kJ/K}\\ &= [480.2\text{ J/K}] - [626.8\text{ J/K}]\\ &= -146.6\text{ J/K}\end{aligned}$$

(a) $\Delta G^\circ = \Delta H^\circ_{rxn} - T\Delta S^\circ_{rxn} = -116.2\text{ kJ} - (298\cancel{K})\left(-0.1466\frac{kJ}{\cancel{K}}\right) = -72.5\text{ kJ} = -7.25 \times 10^4\text{ J}$; so the reaction is spontaneous.

(b) $\Delta G^\circ = \Delta H^\circ_{rxn} - T\Delta S^\circ_{rxn} = -116.2\text{ kJ} - (715\cancel{K})\left(-0.1466\frac{kJ}{\cancel{K}}\right) = -11.4\text{ kJ} = -1.14 \times 10^4\text{ J}$; so the reaction is spontaneous.

(c) $\Delta G^\circ = \Delta H^\circ_{rxn} - T\Delta S^\circ_{rxn} = -116.2\text{ kJ} - (855\cancel{K})\left(-0.1466\frac{kJ}{\cancel{K}}\right) = +9.1\text{ kJ} = +9.1 \times 10^3\text{ J}$; so the reaction is nonspontaneous.

Check: The units (kJ) are correct. The enthalpy term dominates at low temperatures, making the reaction spontaneous. As the temperature increases, the decrease in entropy starts to dominate and in the last case the reaction is nonspontaneous.

19.60 **Given:** $CaCO_3(s) \rightarrow CaO(s) + CO_2(g)$ **Find:** ΔG°_{rxn} and spontaneity at (a) 298 K, (b) 1055 K, and (c) 1455 K

Conceptual Plan: $\Delta H^\circ_{rxn} = \sum n_p H^\circ_f(\text{products}) - \sum n_r H^\circ_f(\text{reactants})$ **then**

$\Delta S^\circ_{rxn} = \sum n_p S^\circ(\text{products}) - \sum n_r S^\circ(\text{reactants})$ **then J/K → kJ/K then** $\Delta H^\circ_{rxn}, \Delta S^\circ_{rxn}, T \rightarrow \Delta G^\circ$

$\frac{1\text{ kJ}}{1000\text{ J}}$ $\Delta G = \Delta H_{rxn} - T\Delta S_{rxn}$

Solution:

Reactant/Product	ΔH°_f(kJ/mol from Appendix IVB)
$CaCO_3(s)$	−1207.6
$CaO(s)$	−634.9
$CO_2(g)$	−393.5

Be sure to pull data for the correct formula and phase.

$$\begin{aligned}\Delta H^\circ_{rxn} &= \sum n_p \Delta H^\circ_f(\text{products}) - \sum n_r \Delta H^\circ_f(\text{reactants})\\ &= [1(\Delta H^\circ_f(CaO(g))) + 1(\Delta H^\circ_f(CO_2(g)))] - [1(\Delta H^\circ_f(CaCO_2(g)))]\\ &= [1(-634.9\text{ kJ}) + 1(-393.5\text{ kJ})] - [1(-1207.6\text{ kJ})]\\ &= [-1028.4\text{ kJ}] - [-1207.6\text{ kJ}]\\ &= +179.2\text{ kJ then}\end{aligned}$$

Reactant/Product	S°(J/mol K from Appendix IVB)
$CaCO_3(s)$	91.7
$CaO(s)$	38.1
$CO_2(g)$	213.8

Be sure to pull data for the correct formula and phase.

$$\begin{aligned}\Delta S^\circ_{rxn} &= \sum n_p S^\circ(\text{products}) - \sum n_r S^\circ(\text{reactants})\\ &= [1(S^\circ(CaO(g))) + 1(S^\circ(CO_2(g)))] - [1(S^\circ(CaCO_2(g)))]\\ &= [1(38.1\text{ J/K}) + 1(213.8\text{ J/K})] - [1(91.7\text{ J/K})]\\ &= [251.9\text{ J/K}] - [91.7\text{ J/K}]\\ &= +160.2\text{ J/K then } 160.2\frac{\cancel{J}}{K} \times \frac{1\text{ kJ}}{1000\cancel{J}} = +0.1602\text{ kJ/K}\end{aligned}$$

(a) $\Delta G^\circ = \Delta H^\circ_{rxn} - T\Delta S^\circ_{rxn} = +179.2\text{ kJ} - (298\cancel{K})\left(+0.1602\frac{\text{kJ}}{\cancel{K}}\right) = +131.5\text{ kJ} = +1.315 \times 10^5\text{ J}$; so the reaction is nonspontaneous.

(b) $\Delta G^\circ = \Delta H^\circ_{rxn} - T\Delta S^\circ_{rxn} = +179.2\text{ kJ} - (1055\cancel{K})\left(+0.1602\frac{\text{kJ}}{\cancel{K}}\right) = +10.2\text{ kJ} = +1.02 \times 10^4\text{ J}$; so the reaction is nonspontaneous.

(c) $\Delta G^\circ = \Delta H^\circ_{rxn} - T\Delta S^\circ_{rxn} = +179.2\text{ kJ} - (1455\cancel{K})\left(+0.1602\frac{\text{kJ}}{\cancel{K}}\right) = -53.9\text{ kJ} = -5.39 \times 10^4\text{ J}$; so the reaction is spontaneous.

Check: The units (kJ) are correct. The enthalpy term dominates at low temperatures, making the reaction nonspontaneous. As the temperature increases, the increase in entropy starts to dominate and in the last case the reaction is spontaneous.

19.61 Because the first reaction has Fe_2O_3 as a product and the reaction of interest has it as a reactant, we need to reverse the first reaction. When the reaction direction is reversed, ΔG changes.

$Fe_2O_3(s) \rightarrow 2\,Fe(s) + 3/2\,O_2(g)$ $\qquad \Delta G^\circ = +742.2\text{ kJ}$

Because the second reaction has 1 mole of CO as a reactant and the reaction of interest has 3 moles of CO as a reactant, we need to multiply the second reaction and the ΔG by 3.

$3[CO(g) + 1/2\,O_2(g) \rightarrow CO_2(g)]$ $\qquad \Delta G^\circ = 3(-257.2\text{ kJ}) = -771.6\text{ kJ}$

Hess's law states that the ΔG of the net reaction is the sum of the ΔG of the steps.

The rewritten reactions are:

$Fe_2O_3(s) \rightarrow 2\,Fe(s) + \cancel{3/2\,O_2(g)}$	$\Delta G^\circ = +742.2\text{ kJ}$
$3\,CO(g) + \cancel{3/2\,O_2(g)} \rightarrow 3\,CO_2(g)$	$\Delta G^\circ = -771.6\text{ kJ}$
$Fe_2O_3(s) \rightarrow 3\,CO(g) \rightarrow 2\,Fe(s) + 3\,CO_2(g)$	$\Delta G^\circ_{rxn} = -29.4\text{ kJ}$

19.62 Because the first reaction has $CaCO_3$ as a product and the reaction of interest has it as a reactant, we need to reverse the first reaction. When the reaction direction is reversed, ΔG changes.

$CaCO_3(s) \rightarrow Ca(s) + CO_2(g) + 1/2\, O_2(g)$ $\quad \Delta G° = +734.4\text{ kJ}$

Because the second reaction has 2 moles of CaO as a product and the reaction of interest has 1 mole of CaO as a product, we need to multiply it by $\frac{1}{2}$. The ΔG of the second reaction is multiplied by $\frac{1}{2}$.

$1/2[2\,Ca(s) + O_2(g) \rightarrow 2\,CaO(s)]$ $\quad \Delta G° = 1/2(-1206.6\text{ kJ}) = -603.3\text{ kJ}$

Hess's law states that the ΔG of the net reaction is the sum of the ΔG of the steps.

The rewritten reactions are:

$CaCO_3(s) \rightarrow \cancel{Ca(s)} + CO_2(g) + \cancel{1/2\, O_2(g)}$ $\quad \Delta G° = +734.4\text{ kJ}$

$1/2[\cancel{2\,Ca(s)} + \cancel{O_2(g)} \rightarrow 2\,CaO(s)]$ $\quad \Delta G° = -603.3\text{ kJ}$

$CaCO_3(s) \rightarrow CaO(s) + CO_2(g)$ $\quad \Delta G°_{rxn} = +131.1\text{ kJ}$

Free Energy Changes, Nonstandard Conditions, and the Equilibrium Constant

19.63 (a) **Given:** $I_2(s) \rightarrow I_2(g)$ at 25.0 °C **Find:** $\Delta G°_{rxn}$

Conceptual Plan: $\Delta G°_{rxn} = \sum n_p \Delta G°_f(\text{products}) - \sum n_r \Delta G°_f(\text{reactants})$

Solution:

Reactant/Product	$\Delta G°_f$(kJ/mol from Appendix IVB)
$I_2(s)$	0.0
$I_2(g)$	19.3

Be sure to pull data for the correct formula and phase.

$$\begin{aligned}\Delta G°_{rxn} &= \sum n_p \Delta G°_f(\text{products}) - \sum n_r \Delta G°_f(\text{products}) \\ &= [1(\Delta G°_f(I_2(g)))] - [1(\Delta G°_f(I_2(s)))] \\ &= [1(19.3\text{ kJ})] - [1(0.0\text{ kJ})] \\ &= +19.3\text{ kJ; so the reaction is nonspontaneous.}\end{aligned}$$

Check: The units (kJ) are correct. The answer is positive because gases have higher free energy than do solids and the free energy change of the reaction is the same as the free energy of formation of gaseous iodine.

(b) **Given:** $I_2(s) \rightarrow I_2(g)$ at 25.0 °C (i) $P_{I_2} = 1.00$ mmHg; (ii) $P_{I_2} = 0.100$ mmHg **Find:** ΔG_{rxn}

Conceptual Plan: °C → K and mmHg → atm then $\Delta G°_{rxn}$, P_{I_2}, T → ΔG_{rxn}

$K = 273.15 + °C$ $\quad \frac{1\text{ atm}}{760\text{ mmHg}}$ $\quad \Delta G_{rxn} = \Delta G°_{rxn} + RT\ln Q$ where $Q = P_{I_2}$

Solution: $T = 273.15 + 25.0\text{ °C} = 298.2\text{ K}$ and (i) $1.00\,\cancel{\text{mmHg}} \times \frac{1\text{ atm}}{760\,\cancel{\text{mmHg}}} = 0.00131579\text{ atm}$

then $\Delta G_{rxn} = \Delta G°_{rxn} + RT\ln Q = \Delta G°_{rxn} + RT\ln P_{I_2} =$

$+19.3\text{ kJ} + \left(\left(8.314\frac{\cancel{\text{J}}}{\cancel{\text{K}}\cdot\text{mol}}\right)\left(\frac{1\text{ kJ}}{1000\,\cancel{\text{J}}}\right)(298.2\,\cancel{\text{K}})\ln(0.00131579)\right) = +2.9\text{ kJ}$; so the reaction is nonspontaneous.

(ii) $0.100\,\cancel{\text{mmHg}} \times \frac{1\text{ atm}}{760\,\cancel{\text{mmHg}}} = 0.000131579\text{ atm}$ then

$\Delta G_{rxn} = \Delta G°_{rxn} + RT\ln Q = \Delta G°_{rxn} + RT\ln P_{I_2} =$

$+19.3\text{ kJ} + \left(\left(8.314\frac{\cancel{\text{J}}}{\cancel{\text{K}}\cdot\text{mol}}\right)\left(\frac{1\text{ kJ}}{1000\,\cancel{\text{J}}}\right)(298.2\,\cancel{\text{K}})\ln(0.000131579)\right) = -2.9\text{ kJ}$; so the reaction is spontaneous.

Check: The units (kJ) are correct. The answer is positive at higher pressure because the pressure is higher than the vapor pressure of iodine. Once the desired pressure is below the vapor pressure (0.31 mmHg at 25.0 °C), the reaction becomes spontaneous.

(c) Iodine sublimes at room temperature because there is an equilibrium between the solid and the gas phases. The vapor pressure is low (0.31 mmHg at 25.0 °C), so a small amount of iodine can remain in the gas phase, which is consistent with the free energy values.

19.64 (a) **Given:** $CH_3OH(l) \rightarrow CH_3OH(g)$ at 25.0 °C **Find:** ΔG°_{rxn}

Conceptual Plan: $\Delta G^\circ_{rxn} = \sum n_p \Delta G^\circ_f(\text{products}) - \sum n_r \Delta G^\circ_f(\text{reactants})$

Solution:

Reactant/Product	ΔG°_f(kJ/mol from Appendix IVB)
$CH_3OH(l)$	−166.6
$CH_3OH(g)$	−162.3

Be sure to pull data for the correct formula and phase.

$$\begin{aligned}\Delta G^\circ_{rxn} &= \sum n_p \Delta G^\circ_f(\text{products}) - \sum n_r \Delta G^\circ_f(\text{reactants})\\ &= [1(\Delta G^\circ_f(CH_3OH\,(g)))] - [1(\Delta G^\circ_f(CH_3OH\,(l)))]\\ &= [1(-162.3\text{ kJ})] - [1(-166.6\text{ kJ})]\\ &= +4.3\text{ kJ; so the reaction is nonspontaneous.}\end{aligned}$$

Check: The units (kJ) are correct. The answer is positive because gases have higher free energy than do liquids.

(b) **Given:** $CH_3OH(l) \rightarrow CH_3OH(g)$ at 25.0 °C (i) $P_{CH_3OH} = 150.0$ mmHg; (ii) $P_{CH_3OH} = 100.0$ mmHg; (iii) $P_{CH_3OH} = 10.0$ mmHg **Find:** ΔG_{rxn}

Conceptual Plan: **°C → K and mmHg → atm then $\Delta G^\circ_{rxn}, P_{CH_3OH}, T \rightarrow \Delta G_{rxn}$**

$K = 273.15 + °C$ $\quad \frac{1\text{ atm}}{760\text{ mmHg}}$ $\quad \Delta G_{rxn} = \Delta G^\circ_{rxn} + RT\ln Q$ where $Q = P_{CH_3OH}$

Solution: $T = 273.15 + 25.0\text{ °C} = 298.2\text{ K}$ and (i) $150.0\text{ mmHg} \times \frac{1\text{ atm}}{760\text{ mmHg}} = 0.1973684\text{ atm}$ then

$$\Delta G_{rxn} = \Delta G^\circ_{rxn} + RT\ln Q = \Delta G^\circ_{rxn} + RT\ln P_{CH_3OH} =$$

$$+4.3\text{ kJ} + \left(\left(8.314\frac{\text{J}}{\text{K}\cdot\text{mol}}\right)\left(\frac{1\text{ kJ}}{1000\text{ J}}\right)(298.2\text{ K})\ln(0.1973684)\right) = +0.3\text{ kJ; so the reaction is nonspontaneous.}$$

(ii) $100.0\text{ mmHg} \times \frac{1\text{ atm}}{760\text{ mmHg}} = 0.131579\text{ atm}$ then

$$\Delta G_{rxn} = \Delta G^\circ_{rxn} + RT\ln Q = \Delta G^\circ_{rxn} + RT\ln P_{CH_3OH} =$$

$$+4.3\text{ kJ} + \left(\left(8.314\frac{\text{J}}{\text{K}\cdot\text{mol}}\right)\left(\frac{1\text{ kJ}}{1000\text{ J}}\right)(298.2\text{ K})\ln(0.131579)\right) = -0.7\text{ kJ; so the reaction is spontaneous.}$$

(iii) $10.0\text{ mmHg} \times \frac{1\text{ atm}}{760\text{ mmHg}} = 0.0131579\text{ atm}$ then

$$\Delta G_{rxn} = \Delta G^\circ_{rxn} + RT\ln Q = \Delta G^\circ_{rxn} + RT\ln P_{CH_3OH} =$$

$$+4.3\text{ kJ} + \left(\left(8.314\frac{\text{J}}{\text{K}\cdot\text{mol}}\right)\left(\frac{1\text{ kJ}}{1000\text{ J}}\right)(298.2\text{ K})\ln(0.0131579)\right) = -6.4\text{ kJ; so the reaction is spontaneous.}$$

Check: The units (kJ) are correct. The answer is positive at high pressures because the pressure is higher than the vapor pressure of methanol. Once the desired pressure is below the vapor pressure (143 mmHg at 25.0 °C), the reaction becomes spontaneous.

(c) Methanol evaporates at room temperature because there is an equilibrium between the liquid and the gas phases. The vapor pressure is moderate (143 mmHg at 25.0 °C), so a moderate amount of methanol can remain in the gas phase, which is consistent with the free energy values.

19.65 **Given:** $CH_3OH(g) \rightleftharpoons CO(g) + 2\,H_2(g)$ at 25 °C, $P_{CH_3OH} = 0.855$ atm, $P_{CO} = 0.125$ atm, $P_{H_2} = 0.183$ atm

Find: ΔG

Conceptual Plan:

$\Delta G^\circ_{rxn} = \sum n_p \Delta G^\circ_f(\text{products}) - \sum n_r \Delta G^\circ_f(\text{reactants})$ then °C → K then $\Delta G^\circ_{rxn}, P_{CH_3OH}, P_{CO}, P_{H_2}, T \rightarrow \Delta G$

$K = 273.15 + °C$ $\quad \Delta G_{rxn} = \Delta G^\circ_{rxn} + RT\ln Q$ $\quad$ where $Q = \frac{P_{CO}P^2_{H_2}}{P_{CH_3OH}}$

Solution:

Reactant/Product	ΔG°_f(kJ/mol from Appendix IVB)
$CH_3OH(g)$	−162.3
$CO(g)$	−137.2
$H_2(g)$	0.0

Be sure to pull data for the correct formula and phase.

$$\begin{aligned}\Delta G^\circ_{rxn} &= \sum n_p \Delta G^\circ_f(\text{products}) - \sum n_r \Delta G^\circ_f(\text{reactants})\\ &= [1(\Delta G^\circ_f(CO(g))) + 2(\Delta G^\circ_f(H_2(g)))] - [1(\Delta G^\circ_f(CH_3OH(g)))]\\ &= [1(-137.2\text{ kJ}) + 2(0.0\text{ kJ})] - [1(-162.3\text{ kJ})]\\ &= [-137.2\text{ kJ}] - [-162.3\text{ kJ}]\\ &= +25.1\text{ kJ}\end{aligned}$$

$T = 273.15 + 25\ ^\circ C = 298\text{ K}$ then $Q = \dfrac{P_{CO}P^2_{H_2}}{P_{CH_3OH}} = \dfrac{(0.125)(0.183)^2}{0.855} = 0.00489605$ then

$\Delta G_{rxn} = \Delta G^\circ_{rxn} + RT\ln Q = +25.1\text{ kJ} + \left(\left(8.314\dfrac{\text{J}}{\text{K}\cdot\text{mol}}\right)\left(\dfrac{1\text{ kJ}}{1000\text{ J}}\right)(298\text{ K})\ln(0.00489605)\right) = +11.9\text{ kJ};$

so the reaction is nonspontaneous.

Check: The units (kJ) are correct. The standard free energy for the reaction was positive, and the fact that Q was less than 1 made the free energy smaller. But the reaction at these conditions is still not spontaneous.

19.66 **Given:** $CO_2(g) + CCl_4(g) \rightleftharpoons 2\,COCl_2(g)$ at 25 °C, $P_{CO_2} = 0.112$ atm, $P_{CCl_4} = 0.174$ atm, $P_{COCl_2} = 0.744$ atm
Find: ΔG

Conceptual Plan:

$\boldsymbol{\Delta G^\circ_{rxn} = \sum n_p \Delta G^\circ_f(\text{products}) - \sum n_r \Delta G^\circ_f(\text{reactants})}$ **then °C → K then** $\boldsymbol{\Delta G^\circ_{rxn}, P_{CO_2}, P_{CCl_4}, P_{COCl_2}, T \rightarrow \Delta G}$

$K = 273.15 + ^\circ C \qquad \Delta G_{rxn} = \Delta G^\circ_{rxn} + RT\ln Q \quad \text{where } Q = \dfrac{P^2_{COCl_2}}{P_{CO_2}P_{CCl_4}}$

Solution:

Reactant/Product	ΔG°_f(kJ/mol from Appendix IVB)
$CO_2(g)$	−394.4
$CCl_4(g)$	−62.3
$COCl_2(g)$	−204.9

Be sure to pull data for the correct formula and phase.

$$\begin{aligned}\Delta G^\circ_{rxn} &= \sum n_p \Delta G^\circ_f(\text{products}) - \sum n_r \Delta G^\circ_f(\text{reactants})\\ &= [2(\Delta G^\circ_f(COCl_2(g)))] - [1(\Delta G^\circ_f(CO_2(g))) + 1(\Delta G^\circ_f(CCl_4(g)))]\\ &= [2(-204.9\text{ kJ})] - [1(-394.4\text{ kJ}) + 1(-62.3\text{ kJ})]\\ &= [-409.8\text{ kJ}] - [-456.7\text{ kJ}]\\ &= +46.9\text{ kJ}\end{aligned}$$

$T = 273.15 + 25\ ^\circ C = 298\text{ K}$ then $Q = \dfrac{P^2_{COCl_2}}{P_{CO_2}P_{CCl_4}} = \dfrac{(0.744)^2}{(0.112)(0.174)} = 28.4039$ then

$\Delta G_{rxn} = \Delta G^\circ_{rxn} + RT\ln Q = +46.9\text{ kJ} + \left(\left(8.314\dfrac{\text{J}}{\text{K}\cdot\text{mol}}\right)\left(\dfrac{1\text{ kJ}}{1000\text{ J}}\right)(298\text{ K})\ln(28.4039)\right) = +55.2\text{ kJ};$

so the reaction is nonspontaneous.

Check: The units (kJ) are correct. The standard free energy for the reaction was positive, and the fact that Q was greater than 1 made the free energy larger; so the reaction is less spontaneous in the forward direction at these conditions than at standard conditions.

19.67 (a) **Given:** $2\,CO(g) + O_2(g) \rightleftharpoons 2\,CO_2(g)$ at 25 °C **Find:** K

Conceptual Plan:

$\boldsymbol{\Delta G^\circ_{rxn} = \sum n_p \Delta G^\circ_f(\text{products}) - \sum n_r \Delta G^\circ_f(\text{reactants})}$ **then °C → K then** $\boldsymbol{\Delta G^\circ_{rxn}, T \rightarrow K}$

$K = 273.15 + ^\circ C \qquad \Delta G^\circ_{rxn} = -RT\ln K$

Solution:

Reactant/Product	ΔG°_f(kJ/mol from Appendix IVB)
$CO(g)$	−137.2
$O_2(g)$	0.0
$CO_2(g)$	−394.4

Be sure to pull data for the correct formula and phase.

$$\begin{aligned}\Delta G^\circ_{rxn} &= \sum n_p \Delta G^\circ_f(\text{products}) - \sum n_r \Delta G^\circ_f(\text{reactants}) \\ &= [2(\Delta G^\circ_f(CO_2(g)))] - [2(\Delta G^\circ_f(CO(g))) + 1(\Delta G^\circ_f(O_2(g)))] \\ &= [2(-394.4\text{ kJ})] - [2(-137.2\text{ kJ}) + 1(0.0\text{ kJ})] \\ &= [-788.8\text{ kJ}] - [-274.4\text{ kJ}] \\ &= -514.4\text{ kJ}\end{aligned}$$

$T = 273.15 + 25\ °C = 298\text{ K}$ then

$\Delta G^\circ_{rxn} = -RT \ln K$ Rearrange to solve for K.

$$K = e^{\frac{-\Delta G^\circ_{rxn}}{RT}} = e^{\frac{-(-514.4\text{ kJ}) \times \frac{1000\text{ J}}{1\text{ kJ}}}{\left(8.314\frac{\text{J}}{\text{K}\cdot\text{mol}}\right)(298\text{ K})}} = e^{207.623} = 1.48 \times 10^{90}$$

Check: The units (none) are correct. The standard free energy for the reaction was very negative, so we expect a very large K. The reaction is spontaneous, so mostly products are present at equilibrium.

(b) **Given:** $2\,H_2S(g) \rightleftharpoons 2\,H_2(g) + S_2(g)$ at 25 °C **Find:** K

Conceptual Plan:

$\boldsymbol{\Delta G^\circ_{rxn} = \sum n_p \Delta G^\circ_f(\text{products}) - \sum n_r \Delta G^\circ_f(\text{reactants})}$ **then °C → K then** $\boldsymbol{\Delta G^\circ_{rxn}, T \rightarrow K}$

$K = 273.15 + °C$ $\qquad \Delta G^\circ_{rxn} = -RT \ln K$

Solution:

Reactant/Product	ΔG°_f(kJ/mol from Appendix IVB)
$H_2S(g)$	−33.4
$H_2(g)$	0.0
$S_2(g)$	79.7

Be sure to pull data for the correct formula and phase.

$$\begin{aligned}\Delta G^\circ_{rxn} &= \sum n_p \Delta G^\circ_f(\text{products}) - \sum n_r \Delta G^\circ_f(\text{reactants}) \\ &= [2(\Delta G^\circ_f(H_2(g))) + 1(\Delta G^\circ_f(S_2(g)))] - [2(\Delta G^\circ_f(H_2S(g)))] \\ &= [2(0.0\text{ kJ})] + 1(79.7\text{ kJ})] - [2(-33.4\text{ kJ})] \\ &= [79.7\text{ kJ}] - [-66.8\text{ kJ}] \\ &= +146.5\text{ kJ}\end{aligned}$$

$T = 273.15 + 25\ °C = 298\text{ K}$ then

$\Delta G^\circ_{rxn} = -RT \ln K$ Rearrange to solve for K.

$$K = e^{\frac{-\Delta G^\circ_{rxn}}{RT}} = e^{\frac{-146.5\text{ kJ} \times \frac{1000\text{ J}}{1\text{ kJ}}}{\left(8.314\frac{\text{J}}{\text{K}\cdot\text{mol}}\right)(298\text{ K})}} = e^{-59.1305} = 2.09 \times 10^{-26}$$

Check: The units (none) are correct. The standard free energy for the reaction was positive, so we expect a small K. The reaction is nonspontaneous, so mostly reactants are present at equilibrium.

19.68 (a) **Given:** $2\,NO_2(g) \rightleftharpoons N_2O_4(g)$ at 25 °C **Find:** K

Conceptual Plan:

$\boldsymbol{\Delta G^\circ_{rxn} = \sum n_p \Delta G^\circ_f(\text{products}) - \sum n_r \Delta G^\circ_f(\text{reactants})}$ **then °C → K then** $\boldsymbol{\Delta G^\circ_{rxn}, T \rightarrow K}$

$K = 273.15 + °C$ $\qquad \Delta G^\circ_{rxn} = -RT \ln K$

Solution:

Reactant/Product	ΔG°_f(kJ/mol from Appendix IVB)
$NO_2(g)$	51.3
$N_2O_4(g)$	99.8

Be sure to pull data for the correct formula and phase.

$$\begin{aligned}\Delta G^\circ_{rxn} &= \sum n_p \Delta G^\circ_f(\text{products}) - \sum n_r \Delta G^\circ_f(\text{reactants})\\ &= [1(\Delta G^\circ_f(N_2O_4(g)))] - [2(\Delta G^\circ_f(NO_2(g)))]\\ &= [1(99.8\text{ kJ})] - [2(51.3\text{ kJ})]\\ &= [99.8\text{ kJ}] - [102.6\text{ kJ}]\\ &= -2.8\text{ kJ}\end{aligned}$$

$T = 273.15 + 25\ ^\circ\text{C} = 298\text{ K}$ then

$\Delta G^\circ_{rxn} = -RT\ln K$ Rearrange to solve for K.

$$K = e^{\frac{-\Delta G^\circ_{rxn}}{RT}} = e^{\frac{-(-2.8\text{ kJ})\times\frac{1000\text{ J}}{1\text{ kJ}}}{\left(8.314\frac{\text{J}}{\text{K}\cdot\text{mol}}\right)(298\text{ K})}} = e^{1.1301} = 3.1$$

Check: The units (none) are correct. The standard free energy for the reaction was very slightly negative, so we expect a K just over 1. The reaction is spontaneous, so mostly products are present at equilibrium.

(b) **Given:** $Br_2(g) + Cl_2(g) \rightleftharpoons 2\,BrCl(g)$ at 25 °C **Find:** K

Conceptual Plan:

$\Delta G^\circ_{rxn} = \sum n_p \Delta G^\circ_f(\text{products}) - \sum n_r \Delta G^\circ_f(\text{reactants})$ **then °C → K then** $\Delta G^\circ_{rxn}, T \rightarrow K$

$K = 273.15 + {}^\circ C$ $\qquad \Delta G^\circ_{rxn} = -RT\ln K$

Solution:

Reactant/Product	ΔG°_f(kJ/mol from Appendix IVB)
$Br_2(g)$	3.1
$Cl_2(g)$	0.0
$BrCl(g)$	−1.0

Be sure to pull data for the correct formula and phase.

$$\begin{aligned}\Delta G^\circ_{rxn} &= \sum n_p \Delta G^\circ_f(\text{products}) - \sum n_r \Delta G^\circ_f(\text{reactants})\\ &= [2(\Delta G^\circ_f(BrCl(g)))] - [1(\Delta G^\circ_f(Br_2(g))) + 1(\Delta G^\circ_f(Cl_2(g)))]\\ &= [2(-1.0\text{ kJ})] - [1(3.1\text{ kJ}) + 1(0.0\text{ kJ})]\\ &= [-2.0\text{ kJ}] - [3.1\text{ kJ}]\\ &= -5.1\text{ kJ}\end{aligned}$$

$T = 273.15 + 25\ ^\circ\text{C} = 298\text{ K}$ then $\Delta G^\circ_{rxn} = -RT\ln K$

Rearrange to solve for K. $K = e^{\frac{-\Delta G^\circ_{rxn}}{RT}} = e^{\frac{-(-5.1\text{ kJ})\times\frac{1000\text{ J}}{1\text{ kJ}}}{\left(8.314\frac{\text{J}}{\text{K}\cdot\text{mol}}\right)(298\text{ K})}} = e^{2.0585} = 7.8$

Check: The units (none) are correct. The standard free energy for the reaction was very slightly negative, so we expect a K just over 1. The reaction is spontaneous.

19.69 **Given:** $CO(g) + 2\,H_2(g) \rightleftharpoons CH_3OH(g)$ $K_p = 2.26 \times 10^4$ at 25 °C

Find: ΔG_{rxn} at (a) standard conditions; (b) at equilibrium; and (c) $P_{CH_3OH} = 1.0$ atm, $P_{CO} = P_{H_2} = 0.010$ atm

Conceptual Plan: °C → K then (a) $K, T \rightarrow \Delta G^\circ_{rxn}$ **then (b) at equilibrium,** $\Delta G_{rxn} = 0$ **then**

$K = 273.15 + {}^\circ C$ $\qquad \Delta G^\circ_{rxn} = -RT\ln K$

(c) $\Delta G^\circ_{rxn}, P_{CH_3OH}, P_{CO}, P_{H_2}, T \rightarrow \Delta G$

$\Delta G_{rxn} = \Delta G^\circ_{rxn} + RT\ln Q$ $\qquad$ where $Q = \frac{P_{CH_3OH}}{P_{CO}P^2_{H_2}}$

Solution: $T = 273.15 + 25\ ^\circ\text{C} = 298\text{ K}$ then

(a) $$\Delta G^\circ_{rxn} = -RT\ln K = -\left(8.314\frac{\text{J}}{\text{K}\cdot\text{mol}}\right)\left(\frac{1\text{ kJ}}{1000\text{ J}}\right)(298\text{ K})\ln(2.26\times10^4) = -24.8\text{ kJ}$$

(b) at equilibrium, $\Delta G_{rxn} = 0$

(c) $$Q = \frac{P_{CH_3OH}}{P_{CO}P^2_{H_2}} = \frac{1.0}{(0.010)(0.010)^2} = 1.0\times10^6 \text{ then}$$

$$\Delta G_{rxn} = \Delta G^\circ_{rxn} + RT \ln Q = -24.8 \text{ kJ} + \left(8.314 \frac{\cancel{J}}{\cancel{K} \cdot \text{mol}}\right)\left(\frac{1 \text{ kJ}}{1000 \cancel{J}}\right)(298 \cancel{K}) \ln(1.0 \times 10^6) = +9.4 \text{ kJ}$$

Check: The units (kJ) are correct. The K was greater than 1, so we expect a negative standard free energy for the reaction. At equilibrium, by definition, the free energy change is zero. Because the conditions give a $Q > K$, the reaction needs to proceed in the reverse direction, which means that the reaction is spontaneous in the reverse direction.

19.70 **Given:** $I_2(g) + Cl_2(g) \rightleftharpoons 2\,ICl(g) K_p = 81.9$ at 25 °C **Find:** ΔG_{rxn} at (a) standard conditions; (b) at equilibrium; and (c) $P_{ICl} = 2.55$ atm, $P_{I_2} = 0.325$ atm, $P_{Cl_2} = 0.221$ atm

Conceptual Plan: °C → K then (a) $K, T \rightarrow \Delta G^\circ_{rxn}$ then (b) at equilibrium, $\Delta G_{rxn} = 0$ then

$K = 273.15 + °C$ $\qquad \Delta G^\circ_{rxn} = -RT \ln K$

(c) $\Delta G^\circ_{rxn}, P_{ICl}, P_{I_2}, P_{Cl_2}, T \rightarrow \Delta G$

$\Delta G_{rxn} = \Delta G^\circ_{rxn} + RT \ln Q \quad$ where $Q = \dfrac{P^2_{ICl}}{P_{I_2} P_{Cl_2}}$

Solution: $T = 273.15 + 25\ °C = 298$ K then

(a) $$\Delta G^\circ_{rxn} = -RT \ln K = -\left(8.314 \frac{\cancel{J}}{\cancel{K} \cdot \text{mol}}\right)\left(\frac{1 \text{ kJ}}{1000 \cancel{J}}\right)(298 \cancel{K}) \ln(81.9) = -10.9 \text{ kJ}$$

(b) at equilibrium, $\Delta G_{rxn} = 0$

(c) $$Q = \frac{P^2_{ICl}}{P_{I_2} P_{Cl_2}} = \frac{(2.55)^2}{(0.325)(0.221)} = 90.\underline{5}325 \text{ then}$$

$$\Delta G_{rxn} = \Delta G^\circ_{rxn} + RT \ln Q = -10.9 \text{ kJ} + \left(8.314 \frac{\cancel{J}}{\cancel{K} \cdot \text{mol}}\right)\left(\frac{1 \text{ kJ}}{1000 \cancel{J}}\right)(298 \cancel{K}) \ln(90.\underline{5}325) = +0.3 \text{ kJ}$$

Check: The units (kJ) are correct. The K was greater than 1, so we expect a negative standard free energy for the reaction. At equilibrium, by definition, the free energy change is zero. Because the conditions give a Q just greater than K, the reaction needs to proceed in the reverse direction, which means that the reverse reaction is slightly spontaneous.

19.71 (a) **Given:** $2\,CO(g) + O_2(g) \rightleftharpoons 2\,CO_2(g)$ at 25 °C **Find:** K at 525 K

Conceptual Plan: $\Delta H^\circ_{rxn} = \sum n_p H^\circ_f(\text{products}) - \sum n_r H^\circ_f(\text{reactants})$ then $\Delta S^\circ_{rxn} = \sum n_p S^\circ(\text{products}) - \sum n_r S^\circ(\text{reactants})$ then J/K → kJ/K then $\Delta H^\circ_{rxn}, \Delta S_{rxn}, T \rightarrow \Delta G$

$\dfrac{1 \text{ kJ}}{1000 \text{ J}} \qquad \Delta G = \Delta H_{rxn} - T\Delta S_{rxn}$

then $\Delta G^\circ_{rxn}, T \rightarrow K$

$\Delta G^\circ_{rxn} = -RT \ln K$

Solution:

Reactant/Product	ΔH°_f(kJ/mol from Appendix IVB)
$CO(g)$	−110.5
$O_2(g)$	0.0
$CO_2(g)$	−393.5

Be sure to pull data for the correct formula and phase.

$$\begin{aligned}\Delta H^\circ_{rxn} &= \sum n_p \Delta H^\circ_f(\text{products}) - \sum n_r \Delta H^\circ_f(\text{reactants}) \\ &= [2(\Delta H^\circ_f(CO_2(g)))] - [2(\Delta H^\circ_f(CO(g))) + 1(\Delta H^\circ_f(O_2(g)))] \\ &= [2(-393.5 \text{ kJ})] - [2(-110.5 \text{ kJ}) + 1(0.0 \text{ kJ})] \text{ then} \\ &= [-787.0 \text{ kJ}] - [-221.0 \text{ kJ}] \\ &= -566.0 \text{ kJ}\end{aligned}$$

Reactant/Product	S°(J/mol K from Appendix IVB)
$CO(g)$	197.7
$O_2(g)$	205.2
$CO_2(g)$	213.8

Be sure to pull data for the correct formula and phase.

$$\begin{aligned}\Delta S^\circ_{rxn} &= \sum n_p S^\circ(\text{products}) - \sum n_r S^\circ(\text{reactants}) \\ &= [2(S^\circ(CO_2(g)))] - [2(S^\circ(CO(g))) + 1(S^\circ(O_2(g)))]\end{aligned}$$

$= [2(213.8\text{ J/K})] - [2(197.7\text{ J/K}) + 1(205.2\text{ J/K})]$
$= [427.6\text{ J/K}] - [600.6\text{ J/K}]$ then
$= -173.0\text{ K/K}$

$-173.0\frac{\cancel{J}}{K} \times \frac{1\text{ kJ}}{1000\cancel{J}} = -0.1730\text{ kJ/K}$ then

$\Delta G^\circ = \Delta H^\circ_{rxn} - T\Delta S^\circ_{rxn} = -566.0\text{ kJ} - (525\cancel{K})\left(-0.1730\frac{kJ}{\cancel{K}}\right) = -475.2\text{ kJ} = -4.752 \times 10^5\text{ J}$ then

$\Delta G^\circ_{rxn} = -RT\ln K$ Rearrange to solve for K.

$$K = e^{\frac{-\Delta G^\circ_{rxn}}{RT}} = e^{\frac{-(-4.752\times10^5\cancel{J})}{\left(8.314\frac{\cancel{J}}{\cancel{K}\cdot mol}\right)(525\cancel{K})}} = e^{108.870} = 1.91 \times 10^{47}$$

Check: The units (none) are correct. The free energy change was very negative, indicating a spontaneous reaction. This results in a very large K.

(b) **Given:** $2\,H_2S(g) \rightleftharpoons 2\,H_2(g) + S_2(g)$ at 25 °C **Find:** K at 525 K

Conceptual Plan: $\Delta H^\circ_{rxn} = \sum n_p H^\circ_f(\text{products}) - \sum n_r H^\circ_f(\text{reactants})$ **then**

$\Delta S^\circ_{rxn} = \sum n_p S^\circ(\text{products}) - \sum n_r S^\circ(\text{reactants})$ **then J/K → kJ/K then** $\Delta H^\circ_{rxn}, \Delta S_{rxn}, T \rightarrow \Delta G$

$\frac{1\text{ kJ}}{1000\text{ J}}$ $\Delta G = \Delta H_{rxn} - T\Delta S_{rxn}$

then $\Delta G^\circ_{rxn}, T \rightarrow K$

$\Delta G^\circ_{rxn} = -RT\ln K$

Solution:

Reactant/Product	ΔH°_f(kJ/mol from Appendix IVB)
$H_2S(g)$	−20.6
$H_2(g)$	0.0
$S_2(g)$	128.6

Be sure to pull data for the correct formula and phase.

$\Delta H^\circ_{rxn} = \sum n_p \Delta H^\circ_f(\text{products}) - \sum n_r \Delta H^\circ_f(\text{reactants})$
$= [2(\Delta H^\circ_f(H_2(g))) + 1(\Delta H^\circ_f(S_2(g)))] - [2(\Delta H^\circ_f(H_2S(g)))]$
$= [2(0.0\text{ kJ}) + 1(128.6\text{ kJ})] - [2(-20.6\text{ kJ})]$
$= [128.6\text{ kJ}] - [-41.2\text{ kJ}]$
$= +169.8\text{ kJ}$

Reactant/Product	S°(J/mol K from Appendix IVB)
$H_2S(g)$	205.8
$H_2(g)$	130.7
$S_2(g)$	228.2

Be sure to pull data for the correct formula and phase.

$\Delta S^\circ_{rxn} = \sum n_p S^\circ(\text{products}) - \sum n_r S^\circ(\text{reactants})$
$= [2(S^\circ(H_2(g))) + 1(S^\circ(S_2(g)))] - [2(S^\circ(H_2S(g)))]$
$= [2(130.7\text{ J/K}) + 1(228.2\text{ J/K})] - [2(205.8\text{ J/K})]$
$= [489.6\text{ J/K}] - [411.6\text{ J/K}]$
$= +78.0\text{ J/K}$

then $+78.0\frac{\cancel{J}}{K} \times \frac{1\text{ kJ}}{1000\cancel{J}} = +0.0780\text{ kJ/K}$ then

$\Delta G^\circ = \Delta H^\circ_{rxn} - T\Delta S^\circ_{rxn} = +169.8\text{ kJ} - (525\cancel{K})\left(+0.0780\frac{kJ}{\cancel{K}}\right) = +128.\underline{8}5\text{ kJ} = +1.28\underline{8}5 \times 10^5\text{ J}$ then

$\Delta G^\circ_{rxn} = -RT\ln K$ Rearrange to solve for K.

$$K = e^{\frac{-\Delta G^\circ_{rxn}}{RT}} = e^{\frac{-1.2885 \times 10^5 \,\text{J}}{\left(8.314 \frac{\text{J}}{\text{K} \cdot \text{mol}}\right)(525 \,\text{K})}} = e^{-29.5199} = 1.51 \times 10^{-13}$$

Check: The units (none) are correct. The free energy change is positive, indicating a nonspontaneous reaction. This results in a very small K.

19.72 (a) **Given:** $2\, NO_2(g) \rightleftharpoons N_2O_4(g)$ at 25 °C **Find:** K at 655 K

Conceptual Plan: $\Delta H^\circ_{rxn} = \sum n_p H^\circ_f(\text{products}) - \sum n_r H^\circ_f(\text{reactants})$ **then**

$\Delta S^\circ_{rxn} = \sum n_p S^\circ(\text{products}) - \sum n_r S^\circ(\text{reactants})$ **then J/K → kJ/K then** $\Delta H^\circ_{rxn}, \Delta S_{rxn}, T \rightarrow \Delta G$

$\frac{1\text{ kJ}}{1000\text{ J}}$ $\quad \Delta G = \Delta H_{rxn} - T\Delta S_{rxn}$

then $\Delta G^\circ_{rxn}, T \rightarrow K$

$\Delta G^\circ_{rxn} = -RT \ln K$

Solution:

Reactant/Product	ΔH°_f(kJ/mol from Appendix IVB)
$NO_2(g)$	33.2
$N_2O_4(g)$	9.16

Be sure to pull data for the correct formula and phase.

$$\begin{aligned}\Delta H^\circ_{rxn} &= \sum n_p \Delta H^\circ_f(\text{products}) - \sum n_r \Delta H^\circ_f(\text{reactants}) \\ &= [1(\Delta H^\circ_f(N_2O_4(g)))] - [2(\Delta H^\circ_f(NO_2(g)))] \\ &= [1(9.16\text{ kJ})] - [2(33.2\text{ kJ})] \\ &= [9.16\text{ kJ}] - [66.4\text{ kJ}] \\ &= -57.24\text{ kJ then}\end{aligned}$$

Reactant/Product	S°(J/mol K from Appendix IVB)
$NO_2(g)$	240.1
$N_2O_4(g)$	304.4

Be sure to pull data for the correct formula and phase.

$$\begin{aligned}\Delta S^\circ_{rxn} &= \sum n_p S^\circ(\text{products}) - \sum n_r S^\circ(\text{reactants}) \\ &= [1(S^\circ(N_2O_4(g)))] - [2(S^\circ(NO_2(g)))] \\ &= [1(304.4\text{ J/K})] - [2(240.1\text{ J/K})] \\ &= [304.4\text{ K/J}] - [480.2\text{ J/K}] \\ &= -175.8\text{ J/K}\end{aligned}$$

then $-175.8 \frac{\text{J}}{\text{K}} \times \frac{1\text{ kJ}}{1000\text{ J}} = -0.1758\text{ kJ/K}$ then

$$\Delta G^\circ = \Delta H^\circ_{rxn} - T\Delta S^\circ_{rxn} = -57.24\text{ kJ} - (655\text{ K})\left(-0.1758 \frac{\text{kJ}}{\text{K}}\right) = +57.909\text{ kJ} = +5.7909 \times 10^4\text{ J then}$$

$\Delta G^\circ_{rxn} = -RT \ln K$ Rearrange to solve for K.

$$K = e^{\frac{-\Delta G^\circ_{rxn}}{RT}} = e^{\frac{-5.7909 \times 10^4 \,\text{J}}{\left(8.314 \frac{\text{J}}{\text{K} \cdot \text{mol}}\right)(655 \,\text{K})}} = e^{-10.63395} = 2.41 \times 10^{-5}$$

Check: The units (none) are correct. The free energy change is positive, indicating a nonspontaneous reaction. This results in a very small K.

(b) **Given:** $Br_2(g) + Cl_2(g) \rightleftharpoons 2\, BrCl(g)$ at 25 °C **Find:** K at 655 K

Conceptual Plan: $\Delta H^\circ_{rxn} = \sum n_p H^\circ_f(\text{products}) - \sum n_r H^\circ_f(\text{reactants})$ **then**

$\Delta S^\circ_{rxn} = \sum n_p S^\circ(\text{products}) - \sum n_r S^\circ(\text{reactants})$ **then J/K → kJ/K then** $\Delta H^\circ_{rxn}, \Delta S_{rxn}, T \rightarrow \Delta G$

$\frac{1\text{ kJ}}{1000\text{ J}}$ $\quad \Delta G = \Delta H_{rxn} - T\Delta S_{rxn}$

then $\Delta G^\circ_{rxn}, T \rightarrow K$

$\Delta G^\circ_{rxn} = -RT \ln K$

Solution:

Reactant/Product	ΔH_f°(kJ/mol from Appendix IVB)
$Br_2(g)$	30.9
$Cl_2(g)$	0.0
$BrCl(g)$	14.6

Be sure to pull data for the correct formula and phase.

$$\begin{aligned}\Delta H^\circ_{rxn} &= \sum n_p \Delta H_f^\circ(\text{products}) - \sum n_r \Delta H_f^\circ(\text{reactants})\\ &= [2(\Delta H_f^\circ(BrCl(g)))] - [1(\Delta H_f^\circ(Br_2(g))) + 1(\Delta H_f^\circ(Cl_2(g)))]\\ &= [2(14.6\text{ kJ})] - [1(30.9\text{ kJ}) + 1(0.0\text{ kJ})]\\ &= [29.2\text{ kJ}] - [30.9\text{ kJ}]\\ &= -1.7\text{ kJ then}\end{aligned}$$

Reactant/Product	S°(J/mol K from Appendix IVB)
$Br_2(g)$	245.5
$Cl_2(g)$	223.1
$BrCl(g)$	240.0

Be sure to pull data for the correct formula and phase.

$$\begin{aligned}\Delta S^\circ_{rxn} &= \sum n_p S^\circ(\text{products}) - \sum n_r S^\circ(\text{reactants})\\ &= [2(S^\circ(BrCl(g)))] - [1(S^\circ(Br_2(g))) + 1(S^\circ(Cl_2(g)))]\\ &= [2(240.0\text{ J/K})] - [1(245.5\text{ J/K}) + 1(223.1\text{ J/K})]\\ &= [480.0\text{ J/K}] - [468.6\text{ J/K}]\\ &= +11.4\text{ J/K}\end{aligned}$$

then $+11.4\frac{\text{J}}{\text{K}} \times \frac{1\text{ kJ}}{1000\text{ J}} = +0.0114\text{ kJ/K}$ then

$$\Delta G^\circ = \Delta H^\circ_{rxn} - T\Delta S^\circ_{rxn} = -1.7\text{ kJ} - (655\text{ K})\left(+0.0114\frac{\text{kJ}}{\text{K}}\right) = -9.\underline{1}67\text{ kJ} = -9.\underline{1}67 \times 10^3\text{ J then}$$

$\Delta G^\circ_{rxn} = -RT\ln K$ Rearrange to solve for K.

$$K = e^{\frac{-\Delta G^\circ_{rxn}}{RT}} = e^{\frac{-(-9.\underline{1}67 \times 10^3\text{ J})}{\left(8.314\frac{\text{J}}{\text{K}\cdot\text{mol}}\right)(655\text{ K})}} = e^{1.\underline{6}834} = 5.38$$

Check: The units (none) are correct. The free energy change is positive, indicating a nonspontaneous reaction. This results in a small K.

19.73 **Given:** table of K_p versus temperature **Find:** ΔH°_{rxn}, ΔS°_{rxn}

Conceptual Plan: Plot ln K versus 1/T. The slope will be $-\Delta H^\circ_{rxn}/R$, and the intercept will be $\Delta S^\circ_{rxn}/R$.

Solution: Plot ln K versus 1/T. Because $\ln K = -\frac{\Delta H^\circ_{rxn}}{R}\frac{1}{T} + \frac{\Delta S^\circ_{rxn}}{R}$, the negative of the slope will be $-\frac{\Delta H^\circ_{rxn}}{R}$ and the intercept will be $\frac{\Delta S^\circ_{rxn}}{R}$. The slope can be determined by measuring $\Delta y/\Delta x$ on the plot or by using functions such as "add trendline" in Excel. Because the slope is $-60\underline{9}2.2$ K,

$$\Delta H^\circ_{rxn} = -slope\,R = -(-60\underline{9}2.2\text{ K})\left(8.314\frac{\text{J}}{\text{K}\cdot\text{mol}}\right)\left(\frac{1\text{ kJ}}{1000\text{ J}}\right) = 50.\underline{6}1\text{ kJ/mol} = 50.6\text{ kJ/mol}.$$ Because the intercept is $27.\underline{1}36\text{ K}^{-1}$, $\Delta S^\circ_{rxn} = intercept\,R = \left(\frac{27.\underline{1}36}{\text{K}}\right)\left(8.314\frac{\text{J}}{\text{K}\cdot\text{mol}}\right) = 22\underline{5}.609\text{ J/K} = 226\text{ J/K}.$

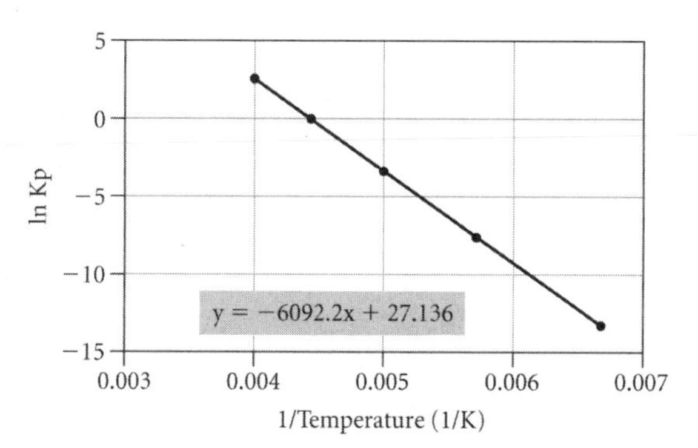

Check: The units are correct (kJ/mole and J/K). The plot is very linear. The numbers are typical for reactions. Because the slope is negative, the enthalpy change must be positive.

19.74 **Given:** table of K_p versus temperature **Find:** ΔH°_{rxn}, ΔS°_{rxn}

Conceptual Plan: Plot ln K versus $1/T$. The slope will be $-\Delta H^\circ_{rxn}/R$ and the intercept will be $\Delta S^\circ_{rxn}/R$.

Solution: Plot ln K versus $1/T$. Because $\ln K = -\frac{\Delta H^\circ_{rxn}}{R}\frac{1}{T} + \frac{\Delta S^\circ_{rxn}}{R}$, the negative of the slope will be $-\frac{\Delta H^\circ_{rxn}}{R}$ and the intercept will be $\frac{\Delta S^\circ_{rxn}}{R}$. The slope can be determined by measuring $\Delta y/\Delta x$ on the plot or by using functions such as "add trendline" in Excel. Because the slope is $-13\underline{7}04$ K,

$$\Delta H^\circ_{rxn} = -slope\,R = -(-13\underline{7}04\ \text{K})\left(8.314\frac{\text{J}}{\text{K}\cdot\text{mol}}\right)\left(\frac{1\ \text{kJ}}{1000\ \text{J}}\right) = 11\underline{3}.935\ \text{kJ/mol} = 114\ \text{kJ/mol}.$$

Because the intercept is $75.\underline{0}42\ \text{K}^{-1}$, $\Delta S^\circ_{rxn} = intercept\,R = \left(\frac{75.\underline{0}42}{\text{K}}\right)\left(8.314\frac{\text{J}}{\text{K}\cdot\text{mol}}\right) = 62\underline{3}.899\ \text{J/K} = 624\ \text{J/K}.$

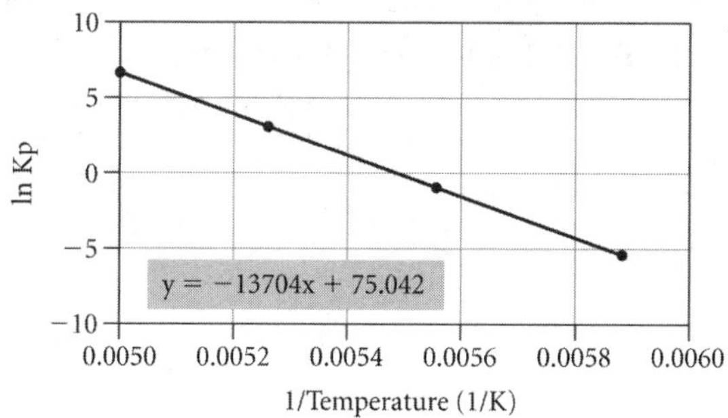

Check: The units are correct (kJ/mole and J/K). The plot is very linear. The numbers are typical for reactions. Because the slope is negative, the enthalpy change must be positive.

19.75 **Given:** $\Delta H^\circ_{rxn} = -25.8$ kJ/mol, $K = 1.4 \times 10^3$ at 298 K **Find:** K at 655 K

Conceptual Plan: $\Delta H^\circ_{rxn}, K_1, T_1, T_2 \rightarrow K_2$

$$\ln K = -\frac{\Delta H^\circ_{rxn}}{R}\frac{1}{T} + \frac{\Delta S^\circ_{rxn}}{R}$$

Solution: Because $\ln K = -\frac{\Delta H^\circ_{rxn}}{R}\frac{1}{T} + \frac{\Delta S^\circ_{rxn}}{R}$, $\ln K_1 + \frac{\Delta H^\circ_{rxn}}{R}\frac{1}{T_1} = \frac{\Delta S^\circ_{rxn}}{R} = \ln K_2 + \frac{\Delta H^\circ_{rxn}}{R}\frac{1}{T_2}$.

Rearrange to solve for K_2.

$$\ln K_2 = \ln K_1 + \frac{\Delta H^\circ_{rxn}}{R}\left(\frac{1}{T_1} - \frac{1}{T_2}\right) = \ln(1.4 \times 10^3) + \frac{\frac{-25.8\ \text{kJ}}{\text{mol}} \times \frac{1000\ \text{J}}{1\ \text{kJ}}}{8.314\frac{\text{J}}{\text{K}\cdot\text{mol}}}\left(\frac{1}{298\ \text{K}} - \frac{1}{655\ \text{K}}\right) = 1.\underline{5}6852 \text{ so}$$

$$K_2 = e^{1.\underline{5}6852} = 4.\underline{7}9954 = 4.8$$

Check: The units are correct (none). Because the reaction is exothermic, we expect the K to decrease with increasing temperature.

19.76 **Given:** $K_1 = 8.5 \times 10^3$ at 298 K, $K_2 = 0.65$ at 755 K **Find:** ΔH°_{rxn}

Conceptual Plan: $K_1, K_2, T_1, T_2 \rightarrow \Delta H^\circ_{rxn}$

$$\ln K = -\frac{\Delta H^\circ_{rxn}}{R}\frac{1}{T} + \frac{\Delta S^\circ_{rxn}}{R}$$

Solution: Because $\ln K = -\frac{\Delta H^\circ_{rxn}}{R}\frac{1}{T} + \frac{\Delta S^\circ_{rxn}}{R}$, $\ln K_1 + \frac{\Delta H^\circ_{rxn}}{R}\frac{1}{T_1} = \frac{\Delta S^\circ_{rxn}}{R} = \ln K_2 + \frac{\Delta H^\circ_{rxn}}{R}\frac{1}{T_2}$.

Rearrange to solve for ΔH°_{rxn}.

$$\Delta H^\circ_{rxn} = \frac{\ln\frac{K_2}{K_1}}{\left(\frac{1}{T_1} - \frac{1}{T_2}\right)} R = \frac{\ln\left(\frac{0.65}{8.5 \times 10^3}\right)}{\left(\frac{1}{298\ \text{K}} - \frac{1}{755\ \text{K}}\right)} \times 8.314\frac{\text{J}}{\text{K}\cdot\text{mol}} \times \frac{1\ \text{kJ}}{1000\ \text{J}} = -38.\underline{7}973\ \text{kJ/mol} = -38.8\ \text{kJ/mol}$$

Check: The units are correct (kJ/mol). Because K decreases with increasing temperature, the reaction is expected to be exothermic.

Cumulative Problems

19.77 (a) + because vapors have higher entropy than do liquids
(b) − because solids have less entropy than do liquids
(c) − because there is only one microstate for the final macrostate and there are six microstates for the initial macrostate

19.78 (a) + because vapors have higher entropy than do solids
(b) − because liquids have less entropy than do vapors
(c) + because there are 20 microstates for the final macrostate and there are only six microstates for the initial macrostate

19.79 (a) **Given:** $N_2(g) + O_2(g) \rightarrow 2\,NO(g)$ **Find:** ΔG°_{rxn} and K_p at 25 °C

Conceptual Plan: $\Delta H^\circ_{rxn} = \sum n_p H^\circ_f(\text{products}) - \sum n_r H^\circ_f(\text{reactants})$ **then**
$\Delta S^\circ_{rxn} = \sum n_p S^\circ(\text{products}) - \sum n_r S^\circ(\text{reactants})$ **then °C → K then J/K → kJ/K then**

$K = 273.15 + °C$ $\quad \frac{1\ \text{kJ}}{1000\ \text{J}}$

$\Delta H^\circ_{rxn}, \Delta S^\circ_{rxn}, T \rightarrow \Delta G$ **then** $\Delta G^\circ_{rxn}, T \rightarrow K$

$\Delta G = \Delta H_{rxn} - T\Delta S_{rxn}$ $\quad \Delta G^\circ_{rxn} = -RT \ln K$

Solution:

Reactant/Product	ΔH°_f(kJ/mol from Appendix IVB)
$N_2(g)$	0.0
$O_2(g)$	0.0
$NO(g)$	91.3

Be sure to pull data for the correct formula and phase.

$$\begin{aligned}\Delta H^\circ_{rxn} &= \sum n_p \Delta H^\circ_f(\text{products}) - \sum n_r \Delta H^\circ_f(\text{reactants}) \\ &= [2(\Delta H^\circ_f(NO(g)))] - [1(\Delta H^\circ_f(N_2(g))) + 1(\Delta H^\circ_f(O_2(g)))] \\ &= [2(91.3\ \text{kJ})] - [1(0.0\ \text{kJ}) + 1(0.0\ \text{kJ})] \\ &= [182.6\ \text{kJ}] - [0.0\ \text{kJ}] \\ &= +182.6\ \text{kJ then}\end{aligned}$$

Reactant/Product	S°(J/mol K from Appendix IVB)
$N_2(g)$	191.6
$O_2(g)$	205.2
$NO(g)$	210.8

Be sure to pull data for the correct formula and phase.

$$\begin{aligned}\Delta S^\circ_{rxn} &= \sum n_p S^\circ(\text{products}) - \sum n_r S^\circ(\text{reactants}) \\ &= [2(S^\circ(NO(g)))] - [1(S^\circ(N_2(g))) + 1(S^\circ(O_2(g)))] \\ &= [2(210.8\ \text{J/K})] - [1(191.6\ \text{J/K}) + 1(205.2\ \text{J/K})] \\ &= [421.6\ \text{J/K}] - [396.8\ \text{JK}] \\ &= +24.8\ \text{J/K}\end{aligned}$$

then $T = 273.15 + 25\ °\text{C} = 298\ \text{K}$ then $+24.8\frac{\cancel{\text{J}}}{\text{K}} \times \frac{1\ \text{kJ}}{1000\ \cancel{\text{J}}} = +0.0248\ \text{kJ/K}$ then

$$\Delta G° = \Delta H°_{rxn} - T\Delta S°_{rxn} = +182.6\ \text{kJ} - (298\ \cancel{\text{K}})\left(0.0248\frac{\text{kJ}}{\cancel{\text{K}}}\right) = +175.2\ \text{kJ} = +1.752 \times 10^5\ \text{J}$$

then $\Delta G°_{rxn} = -RT \ln K$ Rearrange to solve for K.

$$K = e^{\frac{-\Delta G°_{rxn}}{RT}} = e^{\frac{-1.752 \times 10^5\ \cancel{\text{J}}}{\left(8.314\frac{\cancel{\text{J}}}{\cancel{\text{K}} \cdot \text{mol}}\right)(298\ \cancel{\text{K}})}} = e^{-70.7144} = 1.95 \times 10^{-31}$$; so the reaction is nonspontaneous, and at equilibrium, mostly reactants are present.

Check: The units (kJ and none) are correct. The enthalpy is twice the enthalpy of formation of NO. We expect a very small entropy change because the number of moles of gas is unchanged. Because the positive enthalpy term dominates at room temperature, the free energy change is very positive and the reaction in the forward direction is nonspontaneous. This results in a very small K.

(b) **Given:** $N_2(g) + O_2(g) \rightarrow 2\ NO(g)$ **Find:** $\Delta G°_{rxn}$ at 2000 K

Conceptual Plan: Use results from part (a) $\Delta H°_{rxn}, \Delta S_{rxn}, T \rightarrow \Delta G$ **then** $\Delta G°_{rxn}, T \rightarrow K$

$\Delta G = \Delta H_{rxn} - T\Delta S_{rxn}$ $\quad$ $\Delta G°_{rxn} = -RT \ln K$

Solution: $\Delta G = \Delta H_{rxn} - T\Delta S_{rxn} = +182.6\ \text{kJ} - (2000\ \cancel{\text{K}})\left(0.0248\frac{\text{kJ}}{\cancel{\text{K}}}\right) = +133.0\ \text{kJ} = +1.330 \times 10^5\ \text{J}$

then $\Delta G°_{rxn} = -RT \ln K$ Rearrange to solve for K.

$$K = e^{\frac{-\Delta G°_{rxn}}{RT}} = e^{\frac{-1.330 \times 10^5\ \cancel{\text{J}}}{\left(8.314\frac{\cancel{\text{J}}}{\cancel{\text{K}} \cdot \text{mol}}\right)(2000\ \cancel{\text{K}})}} = e^{-7.998557} = 3.36 \times 10^{-4}$$; so the forward reaction is becoming more spontaneous.

Check: The units (kJ and none) are correct. As the temperature rises, the entropy term becomes more significant. The free energy change is reduced, and the K increases. The reaction is still nonspontaneous.

19.80 **Given:** $3\ NO_2(g) + H_2O(l) \rightarrow 2\ HNO_3(aq) + NO(g)$ **Find:** $\Delta G°_{rxn}$ and K_p at 25 °C

Conceptual Plan: $\Delta G°_{rxn} = \sum n_p \Delta G°_f(\text{products}) - \sum n_r \Delta G°_f(\text{reactants})$ **then** $\Delta G°_{rxn}, T \rightarrow K$

$\Delta G°_{rxn} = -RT \ln K$

Solution:

Reactant/Product	$\Delta G°_f$(kJ/mol from Appendix IVB)
$NO_2(g)$	51.3
$H_2O(l)$	−237.1
$HNO_3(aq)$	−110.9
$NO(g)$	87.6

Be sure to pull data for the correct formula and phase.

$$\begin{aligned}\Delta G°_{rxn} &= \sum n_p \Delta G°_f(\text{products}) - \sum n_r \Delta G°_f(\text{reactants}) \\ &= [2(\Delta G°_f(HNO_3(g))) + 1(\Delta G°_f(NO(g)))] - [3(\Delta G°_f(NO_2(g))) + 1(\Delta G°_f(H_2O(l)))] \\ &= [2(-1109.9\ \text{kJ}) + 1(87.6\ \text{kJ})] - [3(51.3\ \text{kJ}) + 1(-237.1\ \text{kJ})] \\ &= [-134.2\ \text{kJ}] - [-83.2\ \text{kJ}] \\ &= -51.0\ \text{kJ} = -5.10 \times 10^4\ \text{J}\end{aligned}$$

then $\Delta G°_{rxn} = -RT \ln K$

Rearrange to solve for K. $K = e^{\frac{-\Delta G°_{rxn}}{RT}} = e^{\frac{-(-5.10 \times 10^4\ \cancel{\text{J}})}{\left(8.314\frac{\cancel{\text{J}}}{\cancel{\text{K}} \cdot \text{mol}}\right)(298\ \cancel{\text{K}})}} = e^{20.5847} = 8.71 \times 10^8$; so the reaction is spontaneous.

Check: The units (kJ and none) are correct. The free energy change is negative, and the reaction is spontaneous. This results in a large K.

19.81 **Given:** $C_2H_4(g) + X_2(g) \rightarrow C_2H_4X_2(g)$ where X = Cl, Br, and I

Find: $\Delta H°_{rxn}$, $\Delta S°_{rxn}$, $\Delta G°_{rxn}$, and K at 25 °C and spontaneity trends with X and temperature

Conceptual Plan: $\Delta H°_{rxn} = \sum n_p H°_f(\text{products}) - \sum n_r H°_f(\text{reactants})$ **then**

$\Delta S^\circ_{rxn} = \sum n_p S^\circ(\text{products}) - \sum n_r S^\circ(\text{reactants})$ **then °C → K then J/K → kJ/K then**

$K = 273.15 + °C$ $\frac{1\ kJ}{1000\ J}$

$\Delta H^\circ_{rxn}, \Delta S^\circ_{rxn}, T \rightarrow \Delta G$ **then** $\Delta G^\circ_{rxn}, T \rightarrow K$

$\Delta G = \Delta H_{rxn} - T\Delta S_{rxn}$ $\Delta G^\circ_{rxn} = -RT \ln K$

Solution:

Reactant/Product	ΔH°_f(kJ/mol from Appendix IVB)
$C_2H_4(g)$	52.4
$Cl_2(g)$	0.0
$C_2H_4Cl_2(g)$	−129.7

Be sure to pull data for the correct formula and phase.

$$\begin{aligned}\Delta H^\circ_{rxn} &= \sum n_p \Delta H^\circ_f(\text{products}) - \sum n_r \Delta H^\circ_f(\text{reactants}) \\ &= [1(\Delta H^\circ_f(C_2H_4Cl_2(g)))] - [1(\Delta H^\circ_f(C_2H_4(g))) + 1(\Delta H^\circ_f(Cl_2(g)))] \\ &= [1(-129.7\ kJ)] - [1(52.4\ kJ) + 1(0.0\ kJ)] \\ &= [-129.7\ kJ] - [52.4\ kJ] \\ &= -182.1\ kJ \text{ then}\end{aligned}$$

Reactant/Product	S°(J/mol K from Appendix IVB)
$C_2H_4(g)$	219.3
$Cl_2(g)$	223.1
$C_2H_4Cl_2(g)$	308.0

Be sure to pull data for the correct formula and phase.

$$\begin{aligned}\Delta S^\circ_{rxn} &= \sum n_p S^\circ(\text{products}) - \sum n_r S^\circ(\text{reactants}) \\ &= [1(S^\circ(C_2H_4Cl_2(g)))] - [1(S^\circ(C_2H_4(g))) + 1(S^\circ(Cl_2(g)))] \\ &= [1(308.0\ J/K)] - [1(219.3\ J/K) + 1(223.1\ J/K)] \\ &= [308.0\ J/K] - [442.4\ J/K] \\ &= -134.4\ J/K\end{aligned}$$

then $T = 273.15 + 25\ °C = 298\ K$ then $-134.4 \frac{J}{K} \times \frac{1\ kJ}{1000\ J} = -0.1344\ kJ/K$ then

$$\Delta G^\circ = \Delta H^\circ_{rxn} - T\Delta S^\circ_{rxn} = -182.1\ kJ - (298\ K)\left(-0.1344 \frac{kJ}{K}\right) = -142.0\ kJ = -1.420 \times 10^5\ J \text{ then}$$

$\Delta G^\circ_{rxn} = -RT \ln K$ Rearrange to solve for K.

$$K = e^{\frac{-\Delta G^\circ_{rxn}}{RT}} = e^{\frac{-(-1.420 \times 10^5\ J)}{\left(8.314 \frac{J}{K \cdot mol}\right)(298\ K)}} = e^{57.314} = 7.78 \times 10^{24}; \text{ so the reaction is spontaneous.}$$

Reactant/Product	ΔH°_f(kJ/mol from Appendix IVB)
$C_2H_4(g)$	52.4
$Br_2(g)$	30.9
$C_2H_4Br_2(g)$	38.3

Be sure to pull data for the correct formula and phase.

$$\begin{aligned}\Delta H^\circ_{rxn} &= \sum n_p \Delta H^\circ_f(\text{products}) - \sum n_r \Delta H^\circ_f(\text{reactants}) \\ &= [1(\Delta H^\circ_f(C_2H_4Br_2(g)))] - [1(\Delta H^\circ_f(C_2H_4(g))) + 1(\Delta H^\circ_f(Br_2(g)))] \\ &= [1(38.3\ kJ)] - [1(52.4\ kJ) + 1(30.9\ kJ)] \\ &= [38.3\ kJ] - [83.3\ kJ] \\ &= -45.0\ kJ \text{ then}\end{aligned}$$

Reactant/Product	S°(J/mol K from Appendix IVB)
$C_2H_4(g)$	219.3
$Br_2(g)$	245.5
$C_2H_4Br_2(g)$	330.6

Be sure to pull data for the correct formula and phase.

$$\Delta S^\circ_{rxn} = \sum n_p S^\circ(\text{products}) - \sum n_r S^\circ(\text{reactants})$$
$$= [1(S^\circ(C_2H_4Br_2(g)))] - [1(S^\circ(C_2H_4(g))) + 1(S^\circ(Br_2(g)))]$$
$$= [1(330.6\ \text{J/K})] - [1(219.3\ \text{J/K})] + 1(245.5\ \text{J/K})]$$
$$= [330.6\ \text{J/K}] - [466.8\ \text{J/K}]$$
$$= -134.2\ \text{J/K}$$

then $-134.2\frac{\text{J}}{\text{K}} \times \frac{1\ \text{kJ}}{1000\ \text{J}} = -0.1342\ \text{kJ/K}$

then $\Delta G^\circ = \Delta H^\circ_{rxn} - T\Delta S^\circ_{rxn} = -45.0\ \text{kJ} - (298\ \text{K})\left(-0.1342\frac{\text{kJ}}{\text{K}}\right) = -5.0084\ \text{kJ} = -5.0084 \times 10^3\ \text{J}$

then $\Delta G^\circ_{rxn} = -RT \ln K$

Rearrange to solve for K. $K = e^{\frac{-\Delta G^\circ_{rxn}}{RT}} = e^{\frac{-(-5.0084 \times 10^3\ \text{J})}{\left(8.314\frac{\text{J}}{\text{K}\cdot\text{mol}}\right)(298\ \text{K})}} = e^{2.021495} = 7.5$; so the reaction is spontaneous.

Reactant/Product	ΔH°_f(kJ/mol from Appendix IVB)
$C_2H_4(g)$	52.4
$I_2(g)$	62.42
$C_2H_4I_2(g)$	66.5

Be sure to pull data for the correct formula and phase.

$$\Delta H^\circ_{rxn} = \sum n_p \Delta H^\circ_f(\text{products}) - \sum n_r \Delta H^\circ_f(\text{reactants})$$
$$= [1(\Delta H^\circ_f(C_2H_4I_2(g))) - [1(\Delta H^\circ_f(C_2H_4(g))) + 1(\Delta H^\circ_f(I_2(g)))]$$
$$= [1(66.5\ \text{kJ})] - [1(52.4\ \text{kJ}) + 1(62.42\ \text{kJ})]$$
$$= [66.5\ \text{kJ}] - [114.82\ \text{kJ}]$$
$$= -48.32\ \text{kJ then}$$

Reactant/Product	S°(J/mol K from Appendix IVB)
$C_2H_4(g)$	219.3
$I_2(g)$	260.69
$C_2H_4I_2(g)$	347.8

Be sure to pull data for the correct formula and phase.

$$\Delta S^\circ_{rxn} = \sum n_p S^\circ(\text{products}) - \sum n_r S^\circ(\text{reactants})$$
$$= [1(S^\circ(C_2H_4I_2(g)))] - [1(S^\circ(C_2H_4(g))) + 1(S^\circ(I_2(g)))]$$
$$= [1(347.8\ \text{J/K})] - [1(219.3\ \text{J/K}) + 1(260.69\ \text{J/K})]$$
$$= [347.8\ \text{J/K}] - [479.99\ \text{J/K}]$$
$$= -132.2\ \text{J/K}$$

then $-132.2\frac{\text{J}}{\text{K}} \times \frac{1\ \text{kJ}}{1000\ \text{J}} = -0.1322\ \text{kJ/K}$

then $\Delta G^\circ = \Delta H^\circ_{rxn} - T\Delta S^\circ_{rxn} = -48.32\ \text{kJ} - (298\ \text{K})\left(-0.1322\frac{\text{kJ}}{\text{K}}\right) = -8.9244\ \text{kJ} = -8.9244 \times 10^3\ \text{J}$

then $\Delta G^\circ_{rxn} = -RT \ln K$ Rearrange to solve for K.

$K = e^{\frac{-\Delta G^\circ_{rxn}}{RT}} = e^{\frac{-(-8.9244 \times 10^3\ \text{J})}{\left(8.314\frac{\text{J}}{\text{K}\cdot\text{mol}}\right)(298\ \text{K})}} = e^{3.6021} = 37$; so the reaction is spontaneous.

Cl_2 is the most spontaneous in the forward direction; K for Cl_2 is 7.78×10^{24}, K for Br_2 is 7.5, K for I_2 is 37. This means Cl_2 is the most spontaneous and Br_2 is the least. The entropy change in the reactions is very constant. The spontaneity is determined by the standard enthalpy of formation of the dihalogenated ethane. Higher temperatures make the forward reactions less spontaneous.

Check: The units (kJ and none) are correct. In general, the enthalpy change becomes less negative as we move to larger halogens (−182, −45, −48). The enthalpy term dominates at room temperature, and the free energy change is the same sign as the enthalpy change. The more negative the free energy change, the larger the K.

19.82 **Given:** $H_2(g) + X_2(g) \rightarrow 2\,HX(g)$ where X = Cl, Br, and I

Find: ΔH°_{rxn}, ΔS°_{rxn}, ΔG°_{rxn}, and K at 25 °C and spontaneity trends with X and temperature

Conceptual Plan: $\Delta H^\circ_{rxn} = \sum n_p H^\circ_f(\text{products}) - \sum n_r H^\circ_f(\text{reactants})$ **then**

$\Delta S^\circ_{rxn} = \sum n_p S^\circ(\text{products}) - \sum n_r S^\circ(\text{reactants})$ **then °C → K then J/K → kJ/K then**

$K = 273.15 + °C$ $\quad \frac{1\text{ kJ}}{1000\text{ J}}$

$\Delta H^\circ_{rxn}, \Delta S^\circ_{rxn}, T \rightarrow \Delta G$

$\Delta G = \Delta H_{rxn} - T\Delta S_{rxn}$

Solution:

Reactant/Product	ΔH°_f(kJ/mol from Appendix IVB)
$H_2(g)$	0.0
$Cl_2(g)$	0.0
$HCl(g)$	−92.3

Be sure to pull data for the correct formula and phase.

$$\begin{aligned}\Delta H^\circ_{rxn} &= \sum n_p \Delta H^\circ_f(\text{products}) - \sum n_r \Delta H^\circ_f(\text{reactants})\\ &= [2(\Delta H^\circ_f(HCl(g)))] - [1(\Delta H^\circ_f(H_2(g))) + 1(\Delta H^\circ_f(Cl_2(g)))]\\ &= [2(-92.3\text{ kJ})] - [1(0.0\text{ kJ}) + 1(0.0\text{ kJ})]\\ &= [-184.6\text{ kJ}] - [0.0]\\ &= -184.6\text{ kJ then}\end{aligned}$$

Reactant/Product	S°(J/mol K from Appendix IVB)
$H_2(g)$	130.7
$Cl_2(g)$	223.1
$HCl(g)$	186.9

Be sure to pull data for the correct formula and phase.

$$\begin{aligned}\Delta S^\circ_{rxn} &= \sum n_p S^\circ(\text{products}) - \sum n_r S^\circ(\text{reactants})\\ &= [2(S^\circ(HCl(g)))] - [1(S^\circ(H_2(g))) + 1(S^\circ(Cl_2(g)))]\\ &= [2(186.9\text{ J/K})] - [1(130.7\text{ J/K}) + 1(223.1\text{ J/K})]\\ &= [373.8\text{ J/K}] - [353.8\text{ J/K}]\\ &= +20.0\text{ J/K}\end{aligned}$$

then $T = 273.15 + 25\text{ °C} = 298\text{ K}$ then $+ 20.0\frac{\text{J}}{\text{K}} \times \frac{1\text{ kJ}}{1000\text{ J}} = +0.0200\text{ kJ/K}$ then

$$\Delta G^\circ = \Delta H^\circ_{rxn} - T\Delta S^\circ_{rxn} = -184.6\text{ kJ} - (298\text{ K})\left(+0.0200\frac{\text{kJ}}{\text{K}}\right) = -190.6\text{ kJ} = -1.906 \times 10^5\text{ J then}$$

$\Delta G^\circ_{rxn} = -RT\ln K$ Rearrange to solve for K.

$$K = e^{\frac{-\Delta G^\circ_{rxn}}{RT}} = e^{\left(\frac{-(-1.906 \times 10^5\text{ J})}{\left(8.314\frac{\text{J}}{\text{K}\cdot\text{mol}}\right)(298\text{ K})}\right)} = e^{76.9302} = 2.57 \times 10^{33}\text{; so the reaction is spontaneous.}$$

Reactant/Product	ΔH°_f(kJ/mol from Appendix IVB)
$H_2(g)$	0.0
$Br_2(g)$	30.9
$HBr(g)$	−36.3

Be sure to pull data for the correct formula and phase.

$$\begin{aligned}\Delta H^\circ_{rxn} &= \sum n_p \Delta H^\circ_f(\text{products}) - \sum n_r \Delta H^\circ_f(\text{reactants})\\ &= [2(\Delta H^\circ_f(HBr(g)))] - [1(\Delta H^\circ_f(H_2(g))) + 1(\Delta H^\circ_f(Br_2(g)))]\\ &= [2(-36.3\text{ kJ})] - [1(0.0\text{ kJ}) + 1(30.9\text{ kJ})]\\ &= [-72.6\text{ kJ}] - [30.9\text{ kJ}]\\ &= -103.5\text{ kJ then}\end{aligned}$$

Reactant/Product	S°(J/mol K from Appendix IVB)
$H_2(g)$	130.7
$Br_2(g)$	245.5
$HBr(g)$	198.7

Be sure to pull data for the correct formula and phase.

$$\begin{aligned}\Delta S^\circ_{rxn} &= \sum n_p S^\circ(\text{products}) - \sum n_r S^\circ(\text{reactants}) \\ &= [2(S^\circ(HBr(g)))] - [1(S^\circ(H_2(g))) + 1(S^\circ(Br_2(g)))] \\ &= [2(198.7\ J/K)] - [1(130.7\ J/K) + 1(245.5\ J/K)] \\ &= [397.4\ J/K] - [376.2\ J/K] \\ &= +21.2\ J/K\end{aligned}$$

then $+21.2\frac{\cancel{J}}{K} \times \frac{1\ kJ}{1000\ \cancel{J}} = +0.0212\ kJ/K$

then $\Delta G^\circ = \Delta H^\circ_{rxn} - T\Delta S^\circ_{rxn} = -103.5\ kJ - (298\ \cancel{K})\left(+0.0212\frac{kJ}{\cancel{K}}\right) = -109.\underline{8}176\ kJ = -1.09\underline{8}176 \times 10^5\ J$

then $\Delta G^\circ_{rxn} = -RT \ln K$ Rearrange to solve for K.

$$K = e^{\frac{-\Delta G^\circ_{rxn}}{RT}} = e^{\frac{-(-1.09\underline{8}176 \times 10^5\ \cancel{J})}{\left(8.314\frac{\cancel{J}}{\cancel{K}\cdot mol}\right)(298\ \cancel{K})}} = e^{44.\underline{3}247} = 1.78 \times 10^{19}\text{; so the reaction is spontaneous.}$$

Reactant/Product	ΔH°_f(kJ/mol from Appendix IVB)
$H_2(g)$	0.0
$I_2(g)$	62.42
$HI(g)$	26.5

Be sure to pull data for the correct formula and phase.

$$\begin{aligned}\Delta H^\circ_{rxn} &= \sum n_p \Delta H^\circ_f(\text{products}) - \sum n_r \Delta H^\circ_f(\text{reactants}) \\ &= [2(\Delta H^\circ_f(HI(g)))] - [1(\Delta H^\circ_f(H_2(g))) + 1(\Delta H^\circ_f(I_2(g)))] \\ &= [2(26.5\ kJ)] - [1(0.0\ kJ) + 1(62.42\ kJ)] \\ &= [53.0\ kJ] - [62.42\ kJ] \\ &= -9.\underline{4}2\ kJ\ \text{then}\end{aligned}$$

Reactant/Product	S°(J/mol K from Appendix IVB)
$H_2(g)$	130.7
$I_2(g)$	260.69
$HI(g)$	206.6

Be sure to pull data for the correct formula and phase.

$$\begin{aligned}\Delta S^\circ_{rxn} &= \sum n_p S^\circ(\text{products}) - \sum n_r S^\circ(\text{reactants}) \\ &= [2(S^\circ(HI(g)))] - [1(S^\circ(H_2(g))) + 1(S^\circ(I_2(g)))] \\ &= [2(206.6\ J/K)] - [1(130.7\ J/K) + 1(260.69\ J/K)] \\ &= [413.2\ J/K] - [391.39\ J/K] \\ &= +21.8\ J/K\end{aligned}$$

then $21.8\frac{\cancel{J}}{K} \times \frac{1\ kJ}{1000\ \cancel{J}} = +0.0218\ kJ/K$

then $\Delta G^\circ = \Delta H^\circ_{rxn} - T\Delta S^\circ_{rxn} = -9.\underline{4}2\ kJ - (298\ \cancel{K})\left(+0.0218\frac{kJ}{\cancel{K}}\right) = -15.\underline{9}164\ kJ = -1.5\underline{9}164 \times 10^4\ J$

then $\Delta G^\circ_{rxn} = -RT \ln K$ Rearrange to solve for K.

$$K = e^{\frac{-\Delta G^\circ_{rxn}}{RT}} = e^{\frac{-(-1.5\underline{9}164 \times 10^4\ \cancel{J})}{\left(8.314\frac{\cancel{J}}{\cancel{K}\cdot mol}\right)(298\ \cancel{K})}} = e^{6.4\underline{2}419} = 6.17 \times 10^2 = 617\text{; so the reaction is spontaneous.}$$

Cl_2 is the most spontaneous; I_2 is the least. The entropy change in the reactions is very constant. The spontaneity is determined by the standard enthalpy of formation of the acid. Higher temperatures make the reactions more spontaneous.

Check: The units (kJ and none) are correct. The enthalpy of the reaction is twice the enthalpy of formation of the acid. We expect a very small entropy change because the number of moles of gas is unchanged. Because both terms are negative, the free energy change is negative and the reaction is spontaneous. The more negative the free energy change, the larger the K.

19.83 (a) **Given:** $N_2O(g) + NO_2(g) \rightleftharpoons 3\ NO(g)$ at 298 K **Find:** ΔG°_{rxn}
Conceptual Plan: $\Delta G^\circ_{rxn} = \sum n_p \Delta G^\circ_f(\text{products}) - \sum n_r \Delta G^\circ_f(\text{reactants})$
Solution:

Reactant/Product	ΔG°_f(kJ/mol from Appendix IVB)
$N_2O(g)$	103.7
$NO_2(g)$	51.3
$NO(g)$	87.6

Be sure to pull data for the correct formula and phase.

$$\begin{aligned}\Delta G^\circ_{rxn} &= \sum n_p \Delta G^\circ_f(\text{products}) - \sum n_r \Delta G^\circ_f(\text{reactants})\\ &= [3(\Delta G^\circ_f(NO(g)))] - [1(\Delta G^\circ_f(N_2O(g))) + 1(\Delta G^\circ_f(NO_2(g)))]\\ &= [3(87.6\text{ kJ})] - [1(103.7\text{ kJ}) + 1(51.3\text{ kJ})]\\ &= [262.8\text{ kJ}] - [155.0\text{ kJ}]\\ &= +107.8\text{ kJ}\end{aligned}$$

The reaction is nonspontaneous.

Check: The units (kJ) are correct. The standard free energy for the reaction was positive, so the reaction is nonspontaneous.

(b) **Given:** $P_{N_2O} = P_{NO_2} = 1.0$ atm initially **Find:** P_{N_2O} when reaction ceases to be spontaneous
Conceptual Plan: Reaction will no longer be spontaneous when $Q = K$, so $\Delta G^\circ_{rxn}, T \rightarrow K$.

$$\Delta G^\circ_{rxn} = -RT \ln K$$

Then solve the equilibrium problem to get gas pressures. Because $K << 1$, the amount of NO generated will be very, very small compared to 1.0 atm; within experimental error, $P_{N_2O} = P_{NO_2} = 1.0$ atm. Simply solve for P_{NO}.

$$K = \frac{P^3_{NO}}{P_{N_2O}P_{NO_2}}$$

Solution: $\Delta G^\circ_{rxn} = -RT \ln K$ Rearrange to solve for K.

$$K = e^{\frac{-\Delta G^\circ_{rxn}}{RT}} = e^{\frac{-(+107.8\ \text{kJ}) \times \frac{1000\ \text{J}}{1\ \text{kJ}}}{\left(8.314 \frac{\text{J}}{\text{K}\cdot\text{mol}}\right)(298\ \text{K})}} = e^{-43.5103} = 1.27 \times 10^{-19}$$ Because $K = \frac{P^3_{NO}}{P_{N_2O}P_{NO_2}}$, rearrange to solve for P_{NO}. $P^3_{NO} = \sqrt[3]{K\,P_{N_2O}P_{NO_2}} = \sqrt[3]{(1.27 \times 10^{-19})(1.0)(1.0)} = 5.0 \times 10^{-7}$ atm

Note that the assumption that P_{N_2O} was very, very small was valid.

Check: The units (atm) are correct. Because the free energy change was positive, the K was very small. This leads us to expect that very little NO will be formed.

(c) **Given:** $N_2O(g) + NO_2(g) \rightleftharpoons 3\ NO(g)$ **Find:** temperature for spontaneity
Conceptual Plan: $\Delta H^\circ_{rxn} = \sum n_p H^\circ_f(\text{products}) - \sum n_r H^\circ_f(\text{reactants})$ **then**
$\Delta S^\circ_{rxn} = \sum n_p S^\circ(\text{products}) - \sum n_r S^\circ(\text{reactants})$ **then J/K → kJ/K then** $\Delta H^\circ_{rxn}, \Delta S_{rxn} \rightarrow T$

$$\frac{1\text{ kJ}}{1000\text{ J}} \qquad \Delta G = \Delta H_{rxn} - T\Delta S_{rxn}$$

Solution:

Reactant/Product	ΔH°_f(kJ/mol from Appendix IVB)
$N_2O(g)$	81.6
$NO_2(g)$	33.2
$NO(g)$	91.3

Be sure to pull data for the correct formula and phase.

$$\begin{aligned}\Delta H^\circ_{rxn} &= \sum n_p \Delta H^\circ_f(\text{products}) - \sum n_r \Delta H^\circ_f(\text{reactants})\\ &= [3(\Delta H^\circ_f(NO(g)))] - [1(\Delta H^\circ_f(N_2O(g))) + 1(\Delta H^\circ_f(NO_2(g)))]\\ &= [3(91.3\text{ kJ})] - [1(81.6\text{ kJ}) + 1(33.2\text{ kJ})]\\ &= [273.9\text{ kJ}] - [114.8\text{ kJ}]\\ &= +159.1\text{ kJ then}\end{aligned}$$

Reactant/Product	S°(J/mol K from Appendix IVB)
$N_2O(g)$	220.0
$NO_2(g)$	240.1
$NO(g)$	210.8

Be sure to pull data for the correct formula and phase.

$$\begin{aligned}\Delta S^\circ_{rxn} &= \sum n_p S^\circ(\text{products}) - \sum n_r S^\circ(\text{reactants})\\ &= [3(S^\circ(NO(g)))] - [1(S^\circ(N_2O(g))) + 1(S^\circ(NO_2(g)))]\\ &= [3(210.8\ \text{J/K})] - [1(220.0\ \text{J/K}) + 1(240.1\ \text{J/K})]\\ &= [632.4\ \text{J/K}] - [460.1\ \text{J/K}]\\ &= +172.3\ \text{J/K}\end{aligned}$$

then $+172.3 \frac{\text{J}}{\text{K}} \times \frac{1\ \text{kJ}}{1000\ \text{J}} = +0.1723\ \text{kJ/K}$. Because $\Delta G = \Delta H_{rxn} - T\Delta S_{rxn}$, set $\Delta G = 0$ and rearrange to solve for T. $T = \frac{\Delta H_{rxn}}{\Delta S_{rxn}} = \frac{+159.1\ \text{kJ}}{0.1723\ \frac{\text{kJ}}{\text{K}}} = +923.4\ \text{K}$

Check: The units (K) are correct. The reaction can be made more spontaneous by raising the temperature because the entropy change is positive (increase in the number of moles of gas).

19.84 (a) **Given:** $BaCO_3(s) \rightleftharpoons BaO(s) + CO_2(g)$ at 298 K **Find:** ΔG°_{rxn}

Conceptual Plan: $\Delta G^\circ_{rxn} = \sum n_p \Delta G^\circ_f(\text{products}) - \sum n_r \Delta G^\circ_f(\text{reactants})$

Solution:

Reactant/Product	ΔG°_f(kJ/mol from Appendix IVB)
$BaCO_3(s)$	−1134.4
$BaO(s)$	−520.3
$CO_2(g)$	−394.4

Be sure to pull data for the correct formula and phase.

$$\begin{aligned}\Delta G^\circ_{rxn} &= \sum n_p \Delta G^\circ_f(\text{products}) - \sum n_r \Delta G^\circ_f(\text{reactants})\\ &= [1(\Delta G^\circ_f(BaO(s))) + 1(\Delta G^\circ_f(CO_2(g)))] - [1(\Delta G^\circ_f(BaCO_3(s)))]\\ &= [1(-520.3\ \text{kJ}) + 1(-394.4\ \text{kJ})] - [1(-1134.4\ \text{kJ})]\\ &= [-914.7\ \text{kJ}] - [-1134.4\ \text{kJ}]\\ &= +219.7\ \text{kJ}\end{aligned}$$

The reaction is nonspontaneous.

Check: The units (kJ) are correct. The standard free energy for the reaction was positive, so the reaction is nonspontaneous.

(b) **Given:** $BaCO_3(s)$ initially in container **Find:** P_{CO_2} at equilibrium

Conceptual Plan: Reaction will be at equilibrium when $Q = K$, so $\Delta G^\circ_{rxn}, T \rightarrow K \rightarrow P_{CO_2}$.

$\Delta G^\circ_{rxn} = -RT \ln K$ $\quad K = P_{CO_2}$

Solution: $\Delta G^\circ_{rxn} = -RT \ln K$ Rearrange to solve for K.

$$K = e^{\frac{-\Delta G^\circ_{rxn}}{RT}} = e^{\frac{-(+219.7\ \text{kJ}) \times \frac{1000\ \text{J}}{1\ \text{kJ}}}{\left(8.314 \frac{\text{J}}{\text{K} \cdot \text{mol}}\right)(298\ \text{K})}} = e^{-88.6755} = 3.08 \times 10^{-39}$$ So $P_{CO_2} = 3.08 \times 10^{-39}$ atm.

Check: The units (atm) are correct. Because the free energy change was very positive, the K was very, very small. This leads us to expect that very little carbon dioxide will be formed.

(c) **Given:** $BaCO_3(s) \rightleftharpoons BaO(s) + CO_2(g)$ **Find:** temperature for $P_{CO_2} = 1.0$ atm

Conceptual Plan: $\Delta H^\circ_{rxn} = \sum n_p H^\circ_f(\text{products}) - \sum n_r H^\circ_f(\text{reactants})$ then $\Delta S^\circ_{rxn} = \sum n_p S^\circ(\text{products}) - \sum n_r S^\circ(\text{reactants})$ then J/K → kJ/K then $\Delta H^\circ_{rxn}, \Delta S_{rxn} \rightarrow T$

$\frac{1\ \text{kJ}}{1000\ \text{J}}$ $\quad \Delta G = \Delta H_{rxn} - T\Delta S_{rxn}$

Solution:

Reactant/Product	ΔH_f°(kJ/mol from Appendix IVB)
$BaCO_3(s)$	−1213.0
$BaO(s)$	−548.0
$CO_2(g)$	−393.5

Be sure to pull data for the correct formula and phase.

$$\begin{aligned}\Delta H^\circ_{rxn} &= \sum n_p \Delta H_f^\circ(\text{products}) - \sum n_r \Delta H_f^\circ(\text{reactants})\\ &= [1(\Delta H_f^\circ(BaO(s))) + 1(\Delta H_f^\circ(CO_2(g)))] - [1(\Delta H_f^\circ(BaCO_3(s)))]\\ &= [1(-548.0\text{ kJ}) + 1(-393.5\text{ kJ})] - [1(-1213.0\text{ kJ})]\\ &= [-941.5\text{ kJ}] - [-1213.0\text{ kJ}]\\ &= +271.5\text{ kJ then}\end{aligned}$$

Reactant/Product	S°(J/mol K from Appendix IVB)
$BaCO_3(s)$	112.1
$BaO(s)$	72.1
$CO_2(g)$	213.8

Be sure to pull data for the correct formula and phase.

$$\begin{aligned}\Delta S^\circ_{rxn} &= \sum n_p S^\circ(\text{products}) - \sum n_r S^\circ(\text{reactants})\\ &= [1(S^\circ(BaO(s))) + 1(S^\circ(CO_2(g)))] - [1(S^\circ(BaCO_3(s)))]\\ &= [1(72.1\text{ J/K}) + 1(213.8\text{ J/K})] - [1(112.1\text{ J/K})]\\ &= [285.9\text{ J/K}] - [112.1\text{ J/K}]\\ &= +173.8\text{ J/K then}\end{aligned}$$

$+173.8\dfrac{\cancel{J}}{K} \times \dfrac{1\text{ kJ}}{1000\,\cancel{J}} = +0.1738\text{ kJ/K}$ Because $\Delta G = \Delta H_{rxn} - T\Delta S_{rxn}$, set $\Delta G = 0$ and rearrange to solve for T. $T = \dfrac{\Delta H_{rxn}}{\Delta S_{rxn}} = \dfrac{+271.5\,\cancel{kJ}}{0.1738\dfrac{\cancel{kJ}}{K}} = +1562\text{ K}$ When $\Delta G = 0$ and $K = 1$, at 1562 K, $P_{CO_2} = 1.0$ atm.

Check: The units (K) are correct. The reaction can be made more spontaneous by raising the temperature because the entropy change is positive (increase in the number of moles of gas). We expect a high temperature because the enthalpy change is so positive.

19.85 (a) **Given:** $ATP(aq) + H_2O(l) \rightarrow ADP(aq) + P_i(aq)$ $\Delta G^\circ_{rxn} = -30.5$ kJ at 298 K **Find:** K

Conceptual Plan: $\Delta G^\circ_{rxn}, T \rightarrow K$

$\Delta G^\circ_{rxn} = -RT \ln K$

Solution: $\Delta G^\circ_{rxn} = -RT\ln K$ Rearrange to solve for K.

$$K = e^{\frac{-\Delta G^\circ_{rxn}}{RT}} = e^{\frac{-(-30.5\,\cancel{kJ}) \times \frac{1000\,\cancel{J}}{1\,\cancel{kJ}}}{\left(8.314\frac{\cancel{J}}{\cancel{K}\cdot mol}\right)(298\,\cancel{K})}} = e^{12.3104} = 2.22 \times 10^5$$

Check: The units (none) are correct. The free energy change is negative, and the reaction is spontaneous. This results in a large K.

(b) **Given:** oxidation of glucose drives reforming of ATP

Find: ΔG°_{rxn} of oxidation of glucose and moles ATP formed per mole of glucose

Conceptual Plan: Write a balanced reaction for glucose oxidation then $\Delta G^\circ_{rxn} = \sum n_p \Delta G_f^\circ(\text{products}) - \sum n_r \Delta G_f^\circ(\text{reactants})$ then ΔG°_{rxn}s → moles ATP/mole glucose.

$$\frac{\Delta G^\circ_{rxn}\text{ glucose oxidation}}{\Delta G^\circ_{rxn}\text{ ATP hydrolysis}}$$

Solution: $C_6H_{12}O_6(s) + 6\,O_2(g) \rightarrow 6\,CO_2(g) + 6\,H_2O(l)$

Reactant/Product	ΔG_f°(kJ/mol from Appendix IVB)
$C_6H_{12}O_6(s)$	−910.4
$O_2(g)$	0.0
$CO_2(g)$	−394.4
$H_2O(l)$	−237.1

Be sure to pull data for the correct formula and phase.

$$\begin{aligned}\Delta G^\circ_{rxn} &= \sum n_p \Delta G_f^\circ(\text{products}) - \sum n_r \Delta G_f^\circ(\text{reactants})\\ &= [6(\Delta G_f^\circ(CO_2(g))) + 6(\Delta G_f^\circ(H_2O(l)))] - [1(\Delta G_f^\circ(C_6H_{12}O_6(s))) + 6(\Delta G_f^\circ(O_2(g)))]\\ &= [6(-394.4\text{ kJ}) + 6(-237.1\text{ kJ})] - [1(-910.4\text{ kJ}) + 6(0.0\text{ kJ})]\\ &= [-3789.0\text{ kJ}] - [-910.4\text{ kJ}]\\ &= -2878.6\text{ kJ}\end{aligned}$$

So the reaction is very spontaneous. $\dfrac{2878.6\dfrac{\cancel{\text{kJ generated}}}{\text{mole glucose oxidized}}}{30.5\dfrac{\cancel{\text{kJ needed}}}{\text{mole ATP reformed}}} = 94.4\dfrac{\text{mole ATP reformed}}{\text{mole glucose oxidized}}$

Check: The units (mol) are correct. The free energy change for the glucose oxidation is large compared to the ATP hydrolysis, so we expect to reform many moles of ATP.

19.86 **Given:** $ATP(aq) + H_2O(l) \rightarrow ADP(aq) + P_i(aq)$ $\Delta G^\circ_{rxn} = -30.5$ kJ at 298 K
Find: ΔG_{rxn} when $[ATP] = 0.0031$ M, $[ADP] = 0.0014$ M and $[P_i] = 0.0048$ M
Conceptual Plan: $\mathbf{\Delta G^\circ_{rxn}, [ATP], [ADP], [P_i], T \rightarrow \Delta G}$

$$\Delta G_{rxn} = \Delta G^\circ_{rxn} + RT\ln Q \quad \text{where } Q = \frac{[ADP][P_i]}{[ATP]}$$

Solution: $Q = \dfrac{[ADP][P_i]}{[ATP]} = \dfrac{(0.0014)(0.0048)}{0.0031} = 0.00216774$ then

$$\Delta G_{rxn} = \Delta G^\circ_{rxn} + RT\ln Q = -30.5\text{ kJ} + \left(8.314\frac{\cancel{J}}{\cancel{K}\cdot\text{mol}}\right)\left(\frac{1\text{ kJ}}{1000\cancel{J}}\right)(298\cancel{K})\ln(0.00216774) = -45.7\text{ kJ}$$

Check: The units (kJ) are correct. The Q is less than 1, so we expect the free energy to be more negative than at standard conditions.

19.87 (a) **Given:** $2\,CO(g) + 2\,NO(g) \rightarrow N_2(g) + 2\,CO_2(g)$ **Find:** ΔG°_{rxn} and effect of increasing T on ΔG
Conceptual Plan: $\mathbf{\Delta G^\circ_{rxn} = \sum n_p \Delta G_f^\circ(products) - \sum n_r \Delta G_f^\circ(reactants)}$
Solution:

Reactant/Product	ΔG_f°(kJ/mol from Appendix IVB)
$CO(g)$	−137.2
$NO(g)$	87.6
$N_2(g)$	0.0
$CO_2(g)$	−394.4

Be sure to pull data for the correct formula and phase.

$$\begin{aligned}\Delta G^\circ_{rxn} &= \sum n_p \Delta G_f^\circ(\text{products}) - \sum n_r \Delta G_f^\circ(\text{reactants})\\ &= [1(\Delta G_f^\circ(N_2(g))) + 2(\Delta G_f^\circ(CO_2(g)))] - [2(\Delta G_f^\circ(CO(g))) + 2(\Delta G_f^\circ(NO(g)))]\\ &= [1(0.0\text{ kJ}) + 2(-394.4\text{ kJ})] - [2(-137.2\text{ kJ}) + 2(87.6\text{ kJ})]\\ &= [-788.8\text{ kJ}] - [-99.2\text{ kJ}]\\ &= -689.6\text{ kJ}\end{aligned}$$

Because the number of moles of gas is decreasing, the entropy change is negative; ΔG will become less negative with increasing temperature.

Check: The units (kJ) are correct. The free energy change is negative because the carbon dioxide has such a low free energy of formation.

(b) **Given:** $5\,H_2(g) + 2\,NO(g) \rightarrow 2\,NH_3(g) + 2\,H_2O(g)$ **Find:** ΔG°_{rxn} and effect of increasing T on ΔG

Conceptual Plan: $\Delta G^\circ_{rxn} = \sum n_p \Delta G^\circ_f(\text{products}) - \sum n_r \Delta G^\circ_f(\text{reactants})$
Solution:

Reactant/Product	ΔG°_f(kJ/mol from Appendix IVB)
$H_2(g)$	0.0
$NO(g)$	87.6
$NH_3(g)$	−16.4
$H_2O(g)$	−228.6

Be sure to pull data for the correct formula and phase.

$$\begin{aligned}\Delta G^\circ_{rxn} &= \sum n_p \Delta G^\circ_f(\text{products}) - \sum n_r \Delta G^\circ_f(\text{reactants}) \\ &= [2(\Delta G^\circ_f(NH_3(g))) + 2(\Delta G^\circ_f(H_2O(g)))] - [5(\Delta G^\circ_f(H_2(g))) + 2(\Delta G^\circ_f(NO(g)))] \\ &= [2(-16.4\text{ kJ}) + 2(-228.6\text{ kJ})] - [5(0.0\text{ kJ}) + 2(87.6\text{ kJ})] \\ &= [-490.0\text{ kJ}] - [175.2\text{ kJ}] \\ &= -665.2\text{ kJ}\end{aligned}$$

Because the number of moles of gas is decreasing, the entropy change is negative; so ΔG will become less negative with increasing temperature.

Check: The units (kJ) are correct. The free energy change is negative because ammonia and water have such a low free energy of formation.

(c) **Given:** $2\,H_2(g) + 2\,NO(g) \rightarrow N_2(g) + 2\,H_2O(g)$ **Find:** ΔG°_{rxn} and effect of increasing T on ΔG
Conceptual Plan: $\Delta G^\circ_{rxn} = \sum n_p \Delta G^\circ_f(\text{products}) - \sum n_r \Delta G^\circ_f(\text{reactants})$
Solution:

Reactant/Product	ΔG°_f(kJ/mol from Appendix IVB)
$H_2(g)$	0.0
$NO(g)$	87.6
$N_2(g)$	0.0
$H_2O(g)$	−228.6

Be sure to pull data for the correct formula and phase.

$$\begin{aligned}\Delta G^\circ_{rxn} &= \sum n_p \Delta G^\circ_f(\text{products}) - \sum n_r \Delta G^\circ_f(\text{reactants}) \\ &= [1(\Delta G^\circ_f(N_2(g))) + 2(\Delta G^\circ_f(H_2O(g)))] - [2(\Delta G^\circ_f(H_2(g))) + 2(\Delta G^\circ_f(NO(g)))] \\ &= [1(0.0\text{ kJ}) + 2(-228.6\text{ kJ})] - [2(0.0\text{ kJ}) + 2(87.6\text{ kJ})] \\ &= [-457.2\text{ kJ}] - [175.2\text{ kJ}] \\ &= -632.4\text{ kJ}\end{aligned}$$

Because the number of moles of gas is decreasing, the entropy change is negative; so ΔG will become less negative with increasing temperature.

Check: The units (kJ) are correct. The free energy change is negative because water has such a low free energy of formation.

(d) **Given:** $2\,NH_3(g) + 2\,O_2(g) \rightarrow N_2O(g) + 3\,H_2O(g)$ **Find:** ΔG°_{rxn} and effect of increasing T on ΔG
Conceptual Plan: $\Delta G^\circ_{rxn} = \sum n_p \Delta G^\circ_f(\text{products}) - \sum n_r \Delta G^\circ_f(\text{reactants})$
Solution:

Reactant/Product	ΔG°_f(kJ/mol from Appendix IVB)
$NH_3(g)$	−16.4
$O_2(g)$	0.0
$N_2O(g)$	103.7
$H_2O(g)$	−228.6

Be sure to pull data for the correct formula and phase.

$$\begin{aligned}\Delta G^\circ_{rxn} &= \sum n_p \Delta G^\circ_f(\text{products}) - \sum n_r \Delta G^\circ_f(\text{reactants}) \\ &= [1(\Delta G^\circ_f(N_2O(g))) + 3(\Delta G^\circ_f(H_2O(g)))] - [2(\Delta G^\circ_f(NH_3(g))) + 2(\Delta G^\circ_f(O_2(g)))] \\ &= [1(103.7\text{ kJ}) + 3(-228.6\text{ kJ})] - [2(-16.4\text{ kJ}) + 2(0.0\text{ kJ})] \\ &= [-582.1\text{ kJ}] - [-32.8\text{ kJ}] \\ &= -549.3\text{ kJ}\end{aligned}$$

Because the number of moles of gas is constant, the entropy change will be small and slightly negative; so ΔG will become less negative with increasing temperature.

Check: The units (kJ) are correct. The free energy change is negative because water has such a low free energy of formation. The entropy change is negative which is known once the $S°$ values are reviewed ($\Delta S_{rxn} = -9.6\ J/K$).

19.88 (a) **Given:** $NH_3(g) + HBr(g) \rightarrow NH_4Br(s)$ **Find:** $\Delta G°_{rxn}$ effect of decreasing T on ΔG
Conceptual Plan: $\boldsymbol{\Delta G°_{rxn} = \sum n_p \Delta G°_f(\text{products}) - \sum n_r \Delta G°_f(\text{reactants})}$
Solution:

Reactant/Product	$\Delta G°_f$(kJ/mol from Appendix IVB)
$NH_3(g)$	−16.4
$HBr(g)$	−53.4
$NH_4Br(s)$	−175.2

Be sure to pull data for the correct formula and phase.

$$\begin{aligned}\Delta G°_{rxn} &= \sum n_p \Delta G°_f(\text{products}) - \sum n_p \Delta G°_f(\text{reactants}) \\ &= [1(\Delta G°_f(NH_4Br(s)))] - [1(\Delta G°_f(NH_3(g))) + 1(\Delta G°_f(HBr(g)))] \\ &= [1(-175.2\ kJ)] - [1(-16.4\ kJ) + 1(-53.4\ kJ)] \\ &= [-175.2\ kJ] - [-69.8\ kJ] \\ &= -105.4\ kJ\end{aligned}$$

Because the number of moles of gas decreases, the entropy change will be negative and ΔG will become more negative with decreasing temperature.

Check: The units (kJ) are correct. The free energy change is negative because ammonium bromide has such a very negative energy of formation.

(b) **Given:** $CaCO_3(s) \rightarrow CaO(s) + CO_2(g)$ **Find:** $\Delta G°_{rxn}$ effect of decreasing T on ΔG
Conceptual Plan: $\boldsymbol{\Delta G°_{rxn} = \sum n_p \Delta G°_f(\text{products}) - \sum n_r \Delta G°_f(\text{reactants})}$
Solution:

Reactant/Product	$\Delta G°_f$(kJ/mol from Appendix IVB)
$CaCO_3(s)$	−1129.1
$CaO(s)$	−603.3
$CO_2(g)$	−394.4

Be sure to pull data for the correct formula and phase.

$$\begin{aligned}\Delta G°_{rxn} &= \sum n_p \Delta G°_f(\text{products}) - \sum n_r \Delta G°_f(\text{reactants}) \\ &= [1(\Delta G°_f(CaO(s))) + 1(\Delta G°_f(CO_2(g)))] - [1(\Delta G°_f(CaCO_3(s)))] \\ &= [1(-603.3\ kJ) + 1(-394.4\ kJ)] - [1(-1129.1\ kJ)] \\ &= [-997.7\ kJ] - [1129.1\ kJ] \\ &= +131.4\ kJ\end{aligned}$$

Because the number of moles of gas increases, the entropy change will be positive and ΔG will become more positive with decreasing temperature.

Check: The units (kJ) are correct. The free energy change is positive because calcium carbonate has such a large, positive free energy of formation.

(c) **Given:** $CH_4(g) + 3\ Cl_2(g) \rightarrow CHCl_3(g) + 3\ HCl(g)$ **Find:** $\Delta G°_{rxn}$ effect of decreasing T on ΔG
Conceptual Plan: $\boldsymbol{\Delta G°_{rxn} = \sum n_p \Delta G°_f(\text{products}) - \sum n_r \Delta G°_f(\text{reactants})}$
Solution:

Reactant/Product	$\Delta G°_f$(kJ/mol from Appendix IVB)
$CH_4(g)$	−50.5
$Cl_2(g)$	0.0
$CHCl_3(g)$	−70.4
$HCl(g)$	−95.3

Be sure to pull data for the correct formula and phase.

$$\begin{aligned}\Delta G^\circ_{rxn} &= \sum n_p \Delta G^\circ_f(\text{products}) - \sum n_r G^\circ_f(\text{reactants}) \\ &= [1(\Delta G^\circ_f(CHCl_3(g))) + 3(\Delta G^\circ_f(HCl(g)))] - [1(\Delta G^\circ_f(CH_4(g))) + 3(\Delta G^\circ_f(Cl_2(g)))] \\ &= [1(-70.4\text{ kJ}) + 3(-95.3\text{ kJ})] - [1(-50.5\text{ kJ}) + 3(0.0\text{ kJ})] \\ &= [-356.3\text{ kJ}] - [-50.5\text{ kJ}] \\ &= -305.8\text{ kJ}\end{aligned}$$

Because the number of moles of gas is constant, the entropy change will be small. The entropy change is so small that the magnitude of ΔG will remain almost constant with decreasing temperature.

Check: The units (kJ) are correct. The free energy change is negative because chloroform and hydrogen chloride have such low free energies of formation. The entropy change is slightly positive which is known once the S° values are reviewed ($\Delta S_{rxn} = +0.7\text{ J/K}$).

19.89 With one exception, the formation of any oxide of nitrogen at 298 K requires more moles of gas as reactants than are formed as products. For example, 1 mole of N_2O requires 0.5 mole of O_2 and 1 mole of N_2. One mole of N_2O_3 requires 1 mole of N_2 and 1.5 moles of O_2, and so on. The exception is NO, where 1 mole of NO requires 0.5 mole of O_2 and 0.5 mole of N_2: $\frac{1}{2}N_2(g) + \frac{1}{2}O_2(g) \rightarrow NO(g)$. This reaction has a positive ΔS because the molecular complexity increases when the product forms.

19.90 ΔG°_f becomes less negative as the atomic number increases because the bond length increases and the bond strength decreases. Less chemical energy is stored in the longer bonds. The ΔS°_f increases as the atomic number increases because of the halides in the hydrogen halides. This is because the hydrogen halide has a low-entropy component (hydrogen) and a high-entropy component (the halide, which increases as the atomic number of the halide increases).

19.91 **Given:** $X_2(g) \rightarrow 2\,X(g)$; $P_{\text{initial } X_2} = 755$ torr, $P_{\text{final X}} = 103$ torr at 298 K; $P_{\text{initial } X_2} = 748$ torr, $P_{\text{final X}} = 532$ torr at 755 K **Find:** ΔH°_{rxn}

Conceptual Plan: At each temperature $P_{\text{initial } X_2}, P_{\text{final X}} \rightarrow K$ then $K_1, K_2, T_1, T_2 \rightarrow \Delta H^\circ_{rxn}$,

Use stoichiometry to calculate $[X_2]_{\text{final}}$ and $K = \dfrac{P_X^2}{P_{X_2}}$ $\qquad$ $\ln K = -\dfrac{\Delta H^\circ_{rxn}}{R}\dfrac{1}{T} + \dfrac{\Delta S^\circ_{rxn}}{R}$

Solution: Use stoichiometry of $X_2(g) \rightarrow 2\,X(g)$; $P_{\text{final } X_2} = P_{\text{initial } X_2} - \frac{1}{2}P_{\text{final X}}$ so that at $T_1 = 298$ K, $P_{\text{final } X_2} = P_{\text{initial } X_2} - \frac{1}{2}P_{\text{final X}} = 755\text{ torr} - \frac{1}{2}(103\text{ torr}) = 703.5$ torr and at $T_2 = 755$ K, $P_{\text{final } X_2} = P_{\text{initial } X_2} - \frac{1}{2}P_{\text{final X}} = 748\text{ torr} - \frac{1}{2}(532\text{ torr}) = 482$ torr. Then $K = \dfrac{P_X^2}{P_{X_2}}$ so that at $T_1 = 298$ K,

$$K_1 = \frac{P_X^2}{P_{X_2}} = \frac{(103)^2}{703.5} = 15.08031 \text{ and at } T_2 = 755\text{ K}, K_2 = \frac{P_X^2}{P_{X_2}} = \frac{(532)^2}{482} = 587.1867.$$

Because $\ln K = -\dfrac{\Delta H^\circ_{rxn}}{R}\dfrac{1}{T} + \dfrac{\Delta S^\circ_{rxn}}{R}$, $\ln K_1 + \dfrac{\Delta H^\circ_{rxn}}{R}\dfrac{1}{T_1} = \dfrac{\Delta S^\circ_{rxn}}{R} = \ln K_2 + \dfrac{\Delta H^\circ_{rxn}}{R}\dfrac{1}{T_2}$.

Rearrange to solve for ΔH°_{rxn}.

$$\Delta H^\circ_{rxn} = \frac{\ln\dfrac{K_2}{K_1}}{\left(\dfrac{1}{T_1} - \dfrac{1}{T_2}\right)}R = \frac{\ln\left(\dfrac{587.1867}{15.08031}\right)}{\left(\dfrac{1}{298\text{ K}} - \dfrac{1}{755\text{ K}}\right)} \times 8.314\frac{\text{J}}{\text{K}\cdot\text{mol}} \times \frac{1\text{ kJ}}{1000\text{ J}} = 14.9889\text{ kJ/mol} = 15.0\text{ kJ/mol}$$

Note that an alternate way to generate the expressions for the two equilibrium constants is to use an ICE table.

Check: The units are correct (kJ/mol). Because K increases with increasing temperature, the reaction is expected to be endothermic.

19.92 **Given:** $N_2O_4(g) \rightarrow 2\,N_2O(g)$; $P_{\text{initial } N_2O_4} = 0.100$ atm, at equilibrium 58% decomposed at 298 K; $\Delta H^\circ_{rxn} = 55.3$ kJ/mol **Find:** percent decomposed at 388 K

Conceptual Plan: A 298 K, $P_{\text{initial } N_2O_4}$, percent decomposed $\rightarrow K_1$ then $\Delta H^\circ_{rxn}, K_1, T_1, T_2 \rightarrow K_2$

Use stoichiometry to calculate $[N_2O_4]_{\text{final}}$ *and* $[NO_2]_{\text{final}}$ *and* $K = \dfrac{P_{NO_2}^2}{P_{N_2O_4}}$ $\qquad$ $\ln K = -\dfrac{\Delta H^\circ_{rxn}}{R}\dfrac{1}{T} + \dfrac{\Delta S^\circ_{rxn}}{R}$

then $K_2, P_{\text{initial } N_2O_4} \rightarrow$ percent decomposed.

Use stoichiometry to calculate $[N_2O_4]_{\text{final}}$ and $[NO_2]_{\text{final}}$ and $K = \dfrac{P_{NO_2}^2}{P_{N_2O_4}}$

Solution: $K_1 = \frac{P^2_{NO_2}}{P_{N_2O_4}} = \frac{(2 \times 0.100 \times (0.58))^2}{0.100 \times (1 - 0.58)} = 0.3\underline{2}0381$. Because $\ln K = -\frac{\Delta H^\circ_{rxn}}{R}\frac{1}{T} + \frac{\Delta S^\circ_{rxn}}{R}$,

$\ln K_1 + \frac{\Delta H^\circ_{rxn}}{R}\frac{1}{T_1} = \frac{\Delta S^\circ_{rxn}}{R} = \ln K_2 + \frac{\Delta H^\circ_{rxn}}{R}\frac{1}{T_2}$. Rearrange to solve for K_2.

$$\ln K_2 = \ln K_1 + \frac{\Delta H^\circ_{rxn}}{R}\left(\frac{1}{T_1} - \frac{1}{T_2}\right) = \ln(0.3\underline{2}0381) + \frac{\frac{55.3\ \cancel{kJ}}{\cancel{mol}} \times \frac{1000\ \cancel{J}}{1\ \cancel{kJ}}}{8.314\frac{\cancel{J}}{\cancel{K}\cdot\cancel{mol}}}\left(\frac{1}{298\ \cancel{K}} - \frac{1}{388\ \cancel{K}}\right) = 4.\underline{0}3913 \text{ so}$$

$K_2 = e^{4.\underline{0}3913} = 5\underline{6}.77693$ then

Let $x = \left(\frac{\%\ \text{decomposed}}{100}\right)$ then $K_2 = \frac{P^2_{NO_2}}{P_{N_2O_4}} = \frac{(2x)^2}{0.100 - x} = 5\underline{6}.77693$ Rearrange to solve for x.

$(2x)^2 = 5\underline{6}.77693\,(0.100 - x) \rightarrow x^2 = 5.\underline{6}77693 - 5\underline{6}.77693x \rightarrow$

$x^2 + 5\underline{6}.77693\,x - 5\underline{6}.77693 = 0$ Using quadratic equation, $x = 0.099$ or $\frac{0.099}{0.100} \times 100\% = 99\%$ decomposed.

Note that an alternate way to generate the expressions for the two equilibrium constants is to use an ICE table.

Check: The units are correct (% decomposed). Because the reaction is endothermic, we expect the K to increase with increasing temperature. If we plug the % decomposed into the equation for K_2, we get the same value for K_2.

19.93 (a) $\Delta S_{univ} > 0$. The process is spontaneous. It is slow unless a spark is applied.

(b) $\Delta S_{univ} > 0$. Although the change in the system is not spontaneous, the overall change, which includes such processes as combustion or water flow to generate electricity, is spontaneous.

(c) $\Delta S_{univ} > 0$. The acorn oak/tree system is becoming more ordered, so the processes associated with growth are not spontaneous. But they are driven by spontaneous processes such as the generation of heat by the sun and the reactions that produce energy in the cell.

19.94 The Haber process is carried out at high temperatures because the reaction rate is so much larger at higher temperatures. In the Haber process, the ammonia product is continually removed via condensation; so the decreasing value of K is not very important to the overall process.

19.95 At equilibrium ($P_{H_2O} = 18.3$ mmHg), $\Delta G_{rxn} = 0$.

$$Q = P^6_{H_2O} = \left(\frac{18.3\ \cancel{mmHg}}{\frac{760\ \cancel{mmHg}}{1\ \text{atm}}}\right)^6 = (0.024\underline{0}7894\ \text{atm})^6 = 1.9\underline{4}9059 \times 10^{-10} \text{ and}$$

$\Delta G_{rxn} = \Delta G^\circ_{rxn} + RT\ln Q = 0$ Rearrange to solve for ΔG°_{rxn} (at $P_{H_2O} = 760$ mmHg = standard conditions).

$$\Delta G^\circ_{rxn} = \Delta G_{rxn} - RT\ln Q = 0 - \left(8.314\frac{\cancel{J}}{\cancel{K}\cdot\text{mol}}\right)\left(\frac{1\ \text{kJ}}{1000\ \cancel{J}}\right)(298\ \cancel{K})\ln(1.9\underline{4}9059 \times 10^{-10}) = 55.\underline{3}948\frac{\text{kJ}}{\text{mol}}$$

$$= 55.4\frac{\text{kJ}}{\text{mol}}$$

19.96 **Given:** AgCl, $\Delta H^\circ_{rxn} = 65.7$ kJ/mol, $S_1 = 1.33 \times 10^{-5}$ M at 25 °C (298 K) **Find:** S at 50.0 °C

Conceptual Plan: °C → K then $S_1 \rightarrow K_1$ then $\Delta H^\circ_{rxn}, K_1, T_1, T_2 \rightarrow K_2 \rightarrow S_2$

$K = 273.15 + °C$ $\quad K_{sp} = [Ag^+][Cl^-] = S^2 \quad \ln K = -\frac{\Delta H^\circ_{rxn}}{R}\frac{1}{T} + \frac{\Delta S^\circ_{rxn}}{R} \quad K_{sp} = S^2$

Solution: $T_1 = 273.15\,°C + 25 = 298$ K, and $T_2 = 273.15 + 50.0\,°C = 323.2$ K then

$K_{sp} = [Ag^+][Cl^-] = S^2 = (1.33 \times 10^{-5})^2 = 1.7\underline{6}89 \times 10^{-10}$. Because $\ln K = -\frac{\Delta H^\circ_{rxn}}{R}\frac{1}{T} + \frac{\Delta S^\circ_{rxn}}{R}$,

$\ln K_1 + \frac{\Delta H^\circ_{rxn}}{R}\frac{1}{T_1} = \frac{\Delta S^\circ_{rxn}}{R} = \ln K_2 + \frac{\Delta H^\circ_{rxn}}{R}\frac{1}{T_2}$ Rearrange to solve for K_2.

$$\ln K_2 = \ln K_1 + \frac{\Delta H^\circ_{rxn}}{R}\left(\frac{1}{T_1} - \frac{1}{T_2}\right) = \ln(1.7\underline{6}89 \times 10^{-10}) + \frac{\frac{65.7\ \cancel{kJ}}{\cancel{mol}} \times \frac{1000\ \cancel{J}}{1\ \cancel{kJ}}}{8.314\frac{\cancel{J}}{\cancel{K}\cdot\cancel{mol}}}\left(\frac{1}{298\ \cancel{K}} - \frac{1}{323.2\ \cancel{K}}\right) =$$

$-2.\underline{0}38032$ and so $K_2 = e^{-2.\underline{0}38032} = 1.\underline{4}09089 \times 10^{-9}$. Because $K_{sp} = [Ag^+][Cl^-] = S^2 = 1.\underline{4}09089 \times 10^{-9}$,

$S = 3.\underline{7}5378 \times 10^{-5}\text{ M} = 3.8 \times 10^{-5}\text{ M}$,

Check: The units are correct (M). Because the reaction is endothermic, we expect the K (and S) to increase with increasing temperature.

Challenge Problems

19.97 (a) **Given:** glutamate $(aq) + NH_3(aq) \rightarrow$ glutamine$(aq) + H_2O(l)$ $\Delta G^\circ_{rxn} = 14.2$ kJ at 298 K **Find:** K

Conceptual Plan: $\Delta G^\circ_{rxn}, T \rightarrow K$

$$\Delta G^\circ_{rxn} = -RT \ln K$$

Solution: $\Delta G^\circ_{rxn} = -RT \ln K$ Rearrange to solve for K.

$$K = e^{\frac{-\Delta G^\circ_{rxn}}{RT}} = e^{\frac{-(+14.2\text{ kJ}) \times \frac{1000\text{ J}}{1\text{ kJ}}}{\left(8.314\frac{\text{J}}{\text{K}\cdot\text{mol}}\right)(298\text{ K})}} = e^{-5.7\underline{3}142} = 3.24 \times 10^{-3}$$

Check: The units (none) are correct. The free energy change is positive, and the reaction is nonspontaneous. This results in a small K.

(b) **Given:** pair ATP hydrolysis with glutamate/NH_3 reaction **Find:** show coupled reactions, ΔG°_{rxn} and K

Conceptual Plan: Use the reaction mechanism shown, where A = NH_3 and B = glutamate ($C_5H_8O_4N^-$). Then calculate ΔG°_{rxn} by adding free energies of reactions then $\Delta G^\circ_{rxn}, T \rightarrow K$.

$$\Delta G^\circ_{rxn} = -RT \ln K$$

Solution:

$$NH_3(aq) + ATP(aq) + \cancel{H_2O(l)} \rightarrow \cancel{NH_3-P_i(aq)} + ADP(aq) \qquad \Delta G^\circ_{rxn} = -30.5\text{ kJ}$$
$$\cancel{NH_3-P_i(aq)} + C_5H_8O_4N^-(aq) \rightarrow C_5H_9O_3N_2^-(aq) + \cancel{H_2O(l)} + P_i(aq) \qquad \Delta G^\circ_{rxn} = +14.2\text{ kJ}$$

$$NH_3(aq) + C_5H_8O_4N^-(aq) + ATP(aq) \rightarrow C_5H_9O_3N_2^-(aq) + ADP(aq) + P_i(aq) \qquad \Delta G^\circ_{rxn} = -16.3\text{ kJ}$$

then $\Delta G^\circ_{rxn} = -RT \ln K$ Rearrange to solve for K.

$$K = e^{\frac{-\Delta G^\circ_{rxn}}{RT}} = e^{\frac{-(-16.3\text{ kJ}) \times \frac{1000\text{ J}}{1\text{ kJ}}}{\left(8.314\frac{\text{J}}{\text{K}\cdot\text{mol}}\right)(298\text{ K})}} = e^{6.5\underline{7}902} = 7.20 \times 10^2$$

Check: The units (none) are correct. The free energy change is negative, and the reaction is spontaneous. This results in a large K.

19.98 **Given:** flask configurations **Find:** entropy and rank as increasing entropy

Conceptual Plan: Calculate the number of possible states then $W \rightarrow S$ then rank.

$$W = \frac{n!}{(n-r!)\,r!} \text{ where } n = \#\text{ particles and } r = \#\text{ particle in one flask } S = k \ln W$$

Solution:

(a) $W = \dfrac{n!}{(n-r)!r!} = \dfrac{5!}{(0)\,!5!} = 1$ then $S = k \ln W = \left(1.38 \times 10^{-23}\dfrac{\text{J}}{\text{K}}\right) \ln 1 = 0$

(b) $W = \dfrac{n!}{(n-r)\,!r!} = \dfrac{5!}{(2)!3!} = 10$ then $S = k \ln W = \left(1.38 \times 10^{-23}\dfrac{\text{J}}{\text{K}}\right) \ln 10 = 3.18 \times 10^{-23}$ J/K

(c) $W = \dfrac{n!}{(n-r)!r!} = \dfrac{5!}{(1)\,!4!} = 5$ then $S = k \ln W = \left(1.38 \times 10^{-23}\dfrac{\text{J}}{\text{K}}\right) \ln 5 = 2.22 \times 10^{-23}$ J/K

So (a) < (c) < (b).

Check: The units (J/K) are correct. The more possibilities for rearranging particles, the higher the entropy.

19.99 (a) **Given:** $\frac{1}{2}H_2(g) + \frac{1}{2}Cl_2(g) \rightarrow HCl(g)$, define standard state as 2 atm **Find:** ΔG°_f

Conceptual Plan: $\Delta G^\circ_f, P_{H_2}, P_{Cl_2}, P_{HCl}, T \rightarrow$ **new** ΔG°_f

$$\Delta G_{rxn} = \Delta G^\circ_{rxn} + RT \ln Q \text{ where } Q = \frac{P_{HCl}}{P_{H_2}^{1/2}\,P_{Cl_2}^{1/2}}$$

Solution: $\Delta G^\circ_{rxn} = -95.3\text{ kJ/mol}$ and $Q = \frac{P_{HCl}}{P_{H_2}^{1/2} P_{Cl_2}^{1/2}} = \frac{2}{2^{1/2}\,2^{1/2}} = 1$ then

$\Delta G_{rxn} = \Delta G^\circ_{rxn} + RT\ln Q = -95.3\frac{\text{kJ}}{\text{mol}} + \left(8.314\frac{\text{J}}{\text{K}\cdot\text{mol}}\right)\left(\frac{1\text{ kJ}}{1000\text{ J}}\right)(298\text{ K})\ln(1) =$

$-95.3\text{ kJ/mol} = -95{,}300\text{ J/mol}$

Because the number of moles of reactants and products are the same, the decrease in volume affects the entropy of both equally; so there is no change in ΔG°_f.

Check: The units (kJ) are correct. The Q is 1, so ΔG°_f is unchanged under the new standard conditions.

(b) **Given:** $N_2(g) + \frac{1}{2}O_2(g) \rightarrow N_2O(g)$, define standard state as 2 atm **Find:** ΔG°_f

Conceptual Plan: $\Delta G^\circ_f, P_{N_2}, P_{O_2}, P_{N_2O}, T \rightarrow$ **new** ΔG°_f

$$\Delta G_{rxn} = \Delta G^\circ_{rxn} + RT\ln Q \text{ where } Q = \frac{P_{N_2O}}{P_{N_2} P_{O_2}^{1/2}}$$

Solution: $\Delta G^\circ_f = +103.7\text{ kJ/mol}$ and $Q = \frac{P_{N_2O}}{P_{N_2} P_{O_2}^{1/2}} = \frac{2}{2\,2^{1/2}} = \frac{1}{\sqrt{2}}$ then

$\Delta G_{rxn} = \Delta G^\circ_{rxn} + RT\ln Q = 103.7\frac{\text{kJ}}{\text{mol}} + \left(8.314\frac{\text{J}}{\text{K}\cdot\text{mol}}\right)\left(\frac{1\text{ kJ}}{1000\text{ J}}\right)(298\text{ K})\ln\left(\frac{1}{\sqrt{2}}\right) =$

$+102.8\text{ kJ/mol} = +102{,}800\text{ J/mol}$

The entropy of the reactants (1.5 mol) is decreased more than the entropy of the product (1 mol). Because the product is relatively more favored at lower volume, ΔG°_f is less positive.

Check: The units (kJ) are correct. The Q is less than 1, so ΔG°_f is reduced under the new standard conditions.

(c) **Given:** $1/2\,H_2(g) \rightarrow H(g)$, define standard state as 2 atm **Find:** ΔG°_f

Conceptual Plan: $\Delta G^\circ_f, P_{H_2}, P_H, T \rightarrow$ **new** ΔG°_f

$$\Delta G_{rxn} = \Delta G^\circ_{rxn} + RT\ln Q \text{ where } Q = \frac{P_H}{P_{H_2}^{1/2}}$$

Solution: $\Delta G^\circ_f = +203.3\text{ kJ/mol}$ and $Q = \frac{P_H}{P_{H_2}^{1/2}} = \frac{2}{2^{1/2}} = \sqrt{2}$ then

$\Delta G_{rxn} = \Delta G^\circ_{rxn} + RT\ln Q = +203.3\frac{\text{kJ}}{\text{mol}} + \left(8.314\frac{\text{J}}{\text{K}\cdot\text{mol}}\right)\left(\frac{1\text{ kJ}}{1000\text{ J}}\right)(298\text{ K})\ln(\sqrt{2}) =$

$+204.2\text{ kJ/mol} = +204{,}200\text{ J/mol}$

The entropy of the product (1 mol) is decreased more than the entropy of the reactant (1/2 mol). Because the product is relatively less favored, ΔG°_f is more positive.

Check: The units (kJ) are correct. The Q is greater than 1, so ΔG°_f is increased under the new standard conditions.

19.100 **Given:** $H_2O(l) \rightarrow H_2O(s)$ at $-10\,^\circ\text{C}$ $\Delta G_{freezing} = -210\text{ J/mol}$, $\Delta H_{fusion} = +5610\text{ J/mol}$ **Find:** ΔS_{univ} at $-10\,^\circ\text{C}$

Conceptual Plan: **°C → K then** $\Delta G_{freezing}, \Delta H_{fus}, T \rightarrow \Delta S_{univ}$

$$\text{K} = 273.15 + {}^\circ\text{C} \qquad \Delta G = \Delta H_{rxn} - T\Delta S_{rxn} \qquad \Delta S_{univ} = \Delta S_{sys} + \Delta S_{surr}$$

Solution: $T = -10\,^\circ\text{C} + 273.15 = 263\text{ K}$, $\Delta H_{freezing} = -\Delta H_{fusion} = -5610\text{ J/mol}$ then $\Delta G = \Delta H_{rxn} - T\Delta S_{rxn}$

Rearrange to solve for ΔS. $\Delta S_{freezing} = \frac{\Delta H_{freezing} - \Delta G_{freezing}}{T} = \Delta S_{sys}$ then

$\Delta S_{univ} = \Delta S_{sys} + \Delta S_{surr} = \frac{\Delta H_{freezing} - \Delta G_{freezing}}{T} - \frac{\Delta H_{freezing}}{T} = -\frac{\Delta G_{freezing}}{T} = \frac{-(-210\text{ J})}{263\text{ K}} =$

$+0.798\text{ J/K}$

Check: The units (J/K) are correct. The entropy for freezing should be negative because the solid has a more ordered structure.

19.101 **Given:** $K = 3.9 \times 10^5$ at 300 K and $K = 1.2 \times 10^{-1}$ at 500 K **Find:** ΔH°_{rxn}, ΔS°_{rxn}

Conceptual Plan: Plot ln K versus $1/T$. The slope will be $-\Delta H^\circ_{rxn}/R$, and the intercept will be $\Delta S^\circ_{rxn}/R$.

Solution: Plot ln K versus $1/T$. Because $\ln K = -\frac{\Delta H^\circ_{\text{rxn}}}{R}\frac{1}{T} + \frac{\Delta S^\circ_{\text{rxn}}}{R}$, the negative of the slope will be $-\frac{\Delta H^\circ_{\text{rxn}}}{R}$ and the intercept will be $\frac{\Delta S^\circ_{\text{rxn}}}{R}$. The slope can be determined by measuring $\Delta y/\Delta x$ on the plot or by using functions such as "add trendline" in Excel. Because the slope is $+11246$ K,

$$\Delta H^\circ_{\text{rxn}} = -slope\ R = -(+1\underline{1}246\ \text{K})\left(8.314\frac{\text{J}}{\text{K}\cdot\text{mol}}\right)\left(\frac{1\ \text{kJ}}{1000\ \text{J}}\right) = -9\underline{3}.499\ \text{kJ/mol} = -93\ \text{kJ/mol}$$

Because the intercept is $-2\underline{4}.612\ \text{K}^{-1}$,

$$\Delta S^\circ_{\text{rxn}} = intercept\ R = \left(\frac{-2\underline{4}.612}{\text{K}}\right)\left(8.314\frac{\text{J}}{\text{K}\cdot\text{mol}}\right) = -2\underline{0}4.62\ \text{J/K} = -2.0\times10^2\ \text{J/K}.$$

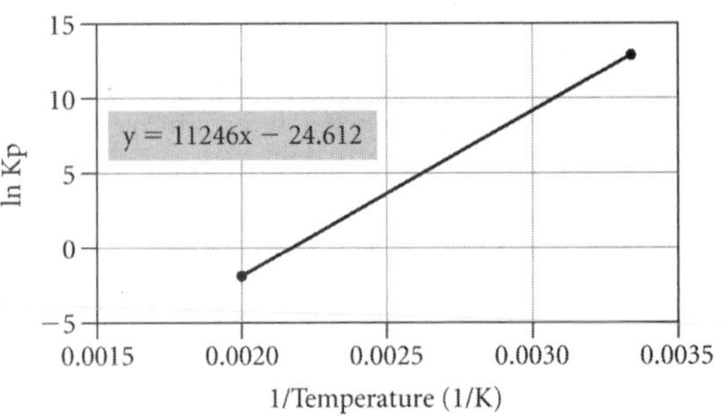

Note: This problem can also be solved using the mathematically equivalent equation in Problems 19.91 and 19.92.

Check: The units are correct (kJ/mol and J/K mol). The numbers are typical for reactions. Because the slope is positive, the enthalpy change must be negative. The entropy change is expected to be negative because the number of moles of gas is decreasing.

19.102 (a) **Given:** $NH_4NO_3(s) \rightarrow HNO_3(g) + NH_3(g)$ **Find:** $\Delta G^\circ_{\text{rxn}}$

Conceptual Plan: $\Delta G^\circ_{\text{rxn}} = \sum n_p \Delta G^\circ_f(\text{products}) - \sum n_r \Delta G^\circ_f(\text{reactants})$

Solution:

Reactant/Product	ΔG°_f(kJ/mol from Appendix IVB)
$NH_4NO_3(s)$	−183.9
$HNO_3(g)$	−73.5
$NH_3(g)$	−16.4

Be sure to pull data for the correct formula and phase.

$$\begin{aligned}\Delta G^\circ_{\text{rxn}} &= \sum n_p \Delta G^\circ_f(\text{products}) - \sum n_r \Delta G^\circ_f(\text{reactants})\\ &= [1(\Delta G^\circ_f(HNO_3(g))) + 1(\Delta G^\circ_f(NH_3(g)))] - [1(\Delta G^\circ_f(NH_4NO_3(s)))]\\ &= [1(-73.5\ \text{kJ}) + 1(-16.4\ \text{kJ})] - [1(-183.9\ \text{kJ})]\\ &= [-89.9\ \text{kJ}] - [-183.9\ \text{kJ}]\\ &= +94.0\ \text{kJ}\end{aligned}$$

Check: The units (kJ) are correct. The free energy change is positive because ammonium nitrate has such a low free energy of formation.

(b) **Given:** $NH_4NO_3(s) \rightarrow N_2O(g) + 2\ H_2O(g)$ **Find:** $\Delta G^\circ_{\text{rxn}}$

Conceptual Plan: $\Delta G^\circ_{\text{rxn}} = \sum n_p \Delta G^\circ_f(\text{products}) - \sum n_r \Delta G^\circ_f(\text{reactants})$

Solution:

Reactant/Product	ΔG°_f(kJ/mol from Appendix IVB)
$NH_4NO_3(s)$	−183.9
$N_2O(g)$	103.7
$H_2O(g)$	−228.6

Be sure to pull data for the correct formula and phase.

$$\begin{aligned}\Delta G^\circ_{\text{rxn}} &= \sum n_p \Delta G^\circ_f(\text{products}) - \sum n_r \Delta G^\circ_f(\text{reactants})\\ &= [1(\Delta G^\circ_f(N_2O(g))) + 2(\Delta G^\circ_f(H_2O(g)))] - [1(\Delta G^\circ_f(NH_4NO_3(s)))]\\ &= [1(103.7\ \text{kJ}) + 2(-228.6\ \text{kJ})] - [1(-183.9\ \text{kJ})]\\ &= [-353.5\ \text{kJ}] - [-183.9\ \text{kJ}]\\ &= -169.6\ \text{kJ}\end{aligned}$$

Check: The units (kJ) are correct. The free energy change is negative because water has such a low free energy of formation.

(c) **Given:** $NH_4NO_3(s) \rightarrow N_2(g) + \frac{1}{2}O_2(g) + 2\,H_2O(g)$ **Find:** ΔG°_{rxn}

Conceptual Plan: $\Delta G^\circ_{rxn} = \sum n_p \Delta G^\circ_f(\text{products}) - \sum n_r \Delta G^\circ_f(\text{reactants})$

Solution:

Reactant/Product	ΔG°_f(kJ/mol from Appendix IVB)
$NH_4NO_3(s)$	−183.9
$N_2(g)$	0.0
$O_2(g)$	0.0
$H_2O(g)$	−228.6

Be sure to pull data for the correct formula and phase.

$$\begin{aligned}\Delta G^\circ_{rxn} &= \sum n_p \Delta G^\circ_f(\text{products}) - \sum n_p \Delta G^\circ_f(\text{reactants})\\ &= [1(\Delta G^\circ_f(N_2(g))) + 1/2(\Delta G^\circ_f(O_2(g))) + 2(\Delta G^\circ_f(H_2O(g)))] - [1(\Delta G^\circ_f(NH_4NO_3(s)))]\\ &= [1(0.0\text{ kJ}) + 1/2(0.0\text{ kJ}) + 2(-228.6\text{ kJ})] - [1(-183.9\text{ kJ})]\\ &= [-457.2\text{ kJ}] - [-183.9\text{ kJ}]\\ &= -273.3\text{ kJ}\end{aligned}$$

Check: The units (kJ) are correct. The free energy change is negative because water has such a low free energy of formation.

The second and third reactions are spontaneous, so we would expect decomposition products of N_2O, N_2, O_2, and H_2O in the gas phase. It is still possible for ammonium nitrate to remain as a solid because the thermodynamics of the reaction say nothing about the kinetics of the reaction (reaction can be extremely slow). Because all of the products are gases, the decomposition of ammonium nitrate will result in a large increase in volume (explosion). Some of the products aid in combustion, which could facilitate the combustion of materials near the ammonium nitrate. Also, N_2O is known as laughing gas, which has anesthetic and toxic effects on humans. The solid should not be kept in tightly sealed containers.

19.103 **Given:** ΔH°_{vap} table **Find:** ΔS_{vap}; then compare values

Conceptual Plan: **°C → K then $\Delta H^\circ_{vap}, T \rightarrow \Delta S_{vap}$**

$K = 273.15 + °C$ $\qquad \Delta S_{vap} = \frac{-\Delta H_{vap}}{T}$

Solution:

Diethyl ether: $T = 273.15 + 34.6 = 307.8\text{ K}$ then $\Delta S_{vap} = \frac{\Delta H_{vap}}{T} = \frac{26.5\text{ kJ}}{307.8\text{ K}} = 0.0861\text{ kJ/K} = 86.1\text{ J/K}$

Acetone: $T = 273.15 + 56.1 = 329.3\text{ K}$ then $\Delta S_{vap} = \frac{\Delta H_{vap}}{T} = \frac{29.1\text{ kJ}}{329.3\text{ K}} = 0.0884\text{ kJ/K} = 88.4\text{ J/K}$

Benzene: $T = 273.15 + 79.8 = 353.0\text{ K}$ then $\Delta S_{vap} = \frac{\Delta H_{vap}}{T} = \frac{30.8\text{ kJ}}{353.0\text{ K}} = 0.0873\text{ kJ/K} = 87.3\text{ J/K}$

Chloroform: $T = 273.15 + 60.8 = 334.0\text{ K}$ then $\Delta S_{vap} = \frac{\Delta H_{vap}}{T} = \frac{29.4\text{ kJ}}{334.0\text{ K}} = 0.0880\text{ kJ/K} = 88.0\text{ J/K}$

Ethanol: $T = 273.15 + 77.8 = 351.0\text{ K}$ then $\Delta S_{vap} = \frac{\Delta H_{vap}}{T} = \frac{38.6\text{ kJ}}{351.0\text{ K}} = 0.110\text{ kJ/K} = 110.\text{ J/K}$

Water: $T = 273.15 + 100 = 373.15\text{ K}$ then $\Delta S_{vap} = \frac{\Delta H_{vap}}{T} = \frac{40.7\text{ kJ}}{373.15\text{ K}} = 0.109\text{ kJ/K} = 109\text{ J/K}$

The first four values are very similar because they have similar intermolecular forces between molecules (dispersion forces and/or dipole–dipole interactions). The values for ethanol and water are higher because the intermolecular forces between molecules are stronger due to hydrogen bonding. As a result, more energy is dispersed when these interactions are broken.

Conceptual Problems

19.104 (c) The spontaneity of a reaction says nothing about the speed of a reaction. It only states which direction the reaction will go as it approaches equilibrium.

19.105 Both (a) and (c) increase the entropy of the surroundings because they are exothermic reactions (adding thermal energy to the surroundings).

19.106 (b) has the largest decrease in the number of microstates from the initial to the final state. In (a), there are initially $\frac{9!}{4!\,4!\,1!} = 90$ microstates and $\frac{9!}{3!\,3!\,3!} = 1680$ microstates at the end; so $\Delta S > 0$. In (b), there are initially $\frac{9!}{4!\,2!\,3!} = 1260$ microstates and $\frac{9!}{6!\,3!\,0!} = 84$ microstates at the end; so $\Delta S < 0$. In (c), there are initially $\frac{9!}{3!\,4!\,2!} = 1260$ microstates and $\frac{9!}{3!\,4!\,2!} = 1260$ microstates at the end; so $\Delta S = 0$. Also, the final state in (b) has the least entropy.

19.107 (c) If the entropy of a system is increasing, the enthalpy of a reaction can be overcome (if necessary) by the entropy change as long as the temperature is high enough. If the entropy change of the system is decreasing, the reaction must be exothermic to be spontaneous because the entropy is working against spontaneity.

19.108 (c) Because the vapor pressure of water at 298 K is 23.78 mmHg or 0.03129 atm. As long as the desired pressure (0.010 atm) is less than the equilibrium vapor pressure of water, the reaction will be spontaneous.

19.109 Both (a) and (b) are true. Because $\Delta G_{rxn} = \Delta G^\circ_{rxn} + RT \ln Q$ and $\Delta G^\circ_{rxn} = -42.5$ kJ, for $\Delta G_{rxn} = 0$, the second term must be positive. This necessitates that $Q > 1$ or that we have more product than reactant. Any reaction at equilibrium has $\Delta G_{rxn} = 0$.

19.110 The relationship between ΔG°_{rxn} and K is $\Delta G^\circ_{rxn} = -RT \ln K$. When $K > 1$, the natural log of K is positive and $\Delta G^\circ_{rxn} < 0$. When $Q = 336$, the second term in $\Delta G_{rxn} = \Delta G^\circ_{rxn} + RT \ln Q$ is positive; so at high temperature, this term can dominate and make $\Delta G_{rxn} > 0$.

20 Electrochemistry

Review Questions

20.1 Oxidation is the loss of electrons that corresponds to an increase in oxidation state. Reduction is the gain of electrons that corresponds to a decrease in oxidation state. Balancing redox reactions can be more complicated than balancing other types of reactions because both the mass (or number of each type of atom) and the charge must be balanced. Redox reactions occurring in aqueous solutions can be balanced by using a special procedure called the half-reaction method of balancing. In this procedure, the overall equation is broken down into two half-reactions: one for oxidation and one for reduction. The half-reactions are balanced individually and then added together so that the number of electrons generated in the oxidation half-reaction is the same as the number of electrons consumed in the reduction half-reaction.

20.2 A voltaic (or galvanic) cell produces electric current from a spontaneous chemical reaction. An electrochemical cell, called an electrolytic cell, uses electric current to drive a nonspontaneous chemical reaction.

20.3 In all electrochemical cells, the electrode where oxidation occurs is called the anode. In a voltaic cell, the anode is labeled with a negative $(-)$ sign. The anode is negative because the oxidation reaction that occurs at the anode releases electrons. Electrons flow from the anode to the cathode (from negative to positive) through the wires connecting the electrodes.

20.4 In all electrochemical cells, the electrode where reduction occurs is called the cathode. The cathode of a voltaic cell is labeled with a $(+)$ sign. The cathode is positive because the reduction reaction that occurs at the cathode takes up electrons. Electrons flow from the anode to the cathode (from negative to positive) through the wires connecting the electrodes.

20.5 A salt bridge is a pathway by which ions can flow between the half-cells and complete an electrical circuit and maintain electroneutrality without the solutions in the half-cells totally mixing.

20.6 Electric current is measured in units of amperes (A). One ampere, or amp, represents the flow of 1 coulomb (a measure of electrical charge) per second: $1\ A = 1\ C/s$. Potential difference is a measure of the difference in potential energy (usually in joules) per unit of charge (coulombs). The SI unit of potential difference is the volt (V), which is equal to 1 joule per coulomb: $1\ V = 1\ J/C$. The potential difference is the drop in potential energy, and the current is a measure of how many electrons travel across this potential difference.

20.7 The standard cell potential E°_{cell} or standard emf is the cell potential under standard conditions (1 M concentration for reactants in solution and 1 atm pressure for gaseous reactants). The cell potential is a measure of the overall tendency of the redox reaction to occur spontaneously. The more positive the cell potential, the more spontaneous the reaction. A negative cell potential indicates a nonspontaneous reaction.

20.8 Electrochemical cells are often represented with a compact notation called a cell diagram or line notation. In this representation, the oxidation half-reaction is always written on the left and the reduction on the right. A double vertical line ($||$), indicating the salt bridge, separates the two half-reactions. Substances in different phases are separated by a single vertical line ($|$), which represents the boundary between the phases. For some redox reactions, the reactants and products of one or both of the half-reactions may be in the same phase. In these cases, the reactants and products are simply separated from each other with a comma in the line diagram. Such cells use an inert electrode, such as platinum (Pt) or graphite, as the anode or cathode (or both).

20.9 Inert electrodes, such as platinum (Pt) or graphite, are used as the anode or cathode (or both) when the reactants and products of one or both of the half-reactions are in the same phase.

20.10 The standard hydrogen electrode (SHE) half-cell is the half-cell that is normally chosen to have a potential of zero. This cell consists of an inert platinum electrode immersed in 1 M HCl with hydrogen gas at 1 atm bubbling through the solution. When the SHE acts as the cathode, the following half-reaction occurs: $2\,H^+(aq) + 2\,e^- \rightarrow H_2(g)\ E^\circ_{red} = 0.00\ V$. If we combine the SHE half-cell with another half-cell of interest, we can measure the voltage and assign it to the half-cell of interest.

20.11 The overall cell potential for any electrochemical cell will always be the sum of the half-cell potential for the oxidation reaction and the half-cell potential for the reduction reaction or $E^\circ_{cell} = E^\circ_{cathode} - E^\circ_{anode}$.

20.12 Half-reactions with large positive reduction half-cell potentials have a strong tendency to have the reduction reaction occur in the forward direction; therefore, the reactants are excellent oxidizing agents. Because half-reactions with large negative reduction half-cell potentials have a strong tendency for the reduction reaction to occur in the reverse direction, they have a strong tendency to be oxidized. These are excellent reducing agents.

20.13 In general, any reduction half-reaction will be spontaneous when paired with the reverse of a half-reaction below it in the table.

20.14 In general, metals whose reduction half-reactions lie below the reduction of H^+ to H_2 in Table 20.1 will dissolve in acids, while metals above it will not. An important exception to this rule is nitric acid (HNO_3), which can oxidize metals through the following reduction half-reaction: $NO_3^-(aq) + 4\,H^+(aq) + 3\,e^- \rightarrow NO(g) + 2\,H_2O(l)\ E^\circ_{red} = 0.96\ V$. Because this half-reaction is above the reduction of H^+ in Table 20.1, HNO_3 can oxidize metals (copper, for example) that can't be oxidized by HCl.

20.15 We have seen that a positive standard cell potential (E°_{cell}) corresponds to a spontaneous oxidation–reduction reaction. We also know (from Chapter 19) that the spontaneity of a reaction is determined by the sign of ΔG°. Therefore, E°_{cell} and ΔG° must be related. We also know from Section 19.9 that ΔG° for a reaction is related to the equilibrium constant (K) for the reaction. Because E°_{cell} and ΔG° are related, E°_{cell} and K must also be related. The equation that relates the three quantities is $\Delta G^\circ_{rxn} = -RT \ln K = -n F E^\circ_{cell}$.

20.16 A redox reaction with a small equilibrium constant ($K < 1$) will not be spontaneous; so it will have a negative E°_{cell} and a positive ΔG°_{rxn}.

20.17 The Nernst equation relates concentration and cell potential as follows: $E_{cell} = E^\circ_{cell} - \dfrac{0.0592\ V}{n} \log Q$, where E_{cell} is the cell potential in V, E°_{cell} is the standard cell potential in V, n is the number of moles of electrons transferred in the redox reaction, and Q is the reaction quotient. Increasing the concentration of the reactants decreases the value of Q, so it increases the cell potential. Increasing the concentration of the products increases the value of Q, so it decreases the cell potential.

20.18 The Nernst equation states that $E_{cell} = E^\circ_{cell} - \dfrac{0.0592\ V}{n} \log Q$. Under standard conditions, $Q = 1$ and (because $\log 1 = 0$) the second term drops out and $E_{cell} = E^\circ_{cell}$.

20.19 A concentration cell is a voltaic cell in which both half-reactions are the same but in which a difference in concentration drives the current flow (because the cell potential depends not only on the half-reactions occurring in the cell, but also on the concentrations of the reactants and products in those half-reactions).

Therefore, $E_{cell} = -\dfrac{0.0592\ V}{n} \log Q$.

20.20 In an inexpensive dry-cell battery, zinc is oxidized at the anode according to the following reaction: $Zn(s) \rightarrow Zn^{2+}(aq) + 2\,e^-$. The cathode is a carbon rod immersed in a moist paste of MnO_2 that also contains NH_4Cl. The MnO_2 is reduced to Mn_2O_3 according to the following reaction: $2\,MnO_2(s) + 2\,NH_4^+(aq) + 2\,e^- \rightarrow Mn_2O_3(s) + 2\,NH_3(aq) + H_2O(l)$. These two half-reactions produce a potential of about 1.5 volts.

The more common alkaline batteries employ slightly different half-reactions that use a base (therefore the name alkaline). The anode reaction is $Zn(s) + 2\,OH^-(aq) \rightarrow Zn(OH)_2(s) + 2\,e^-$. The cathode reaction is: $2\,MnO_2(s) + 2\,H_2O(l) + 2\,e^- \rightarrow 2\,MnO(OH)(s) + 2\,OH^-(aq)$.

20.21 Lead–acid storage batteries consist of six electrochemical cells wired in series in which each cell produces 2 V for a total of 12 V. Each cell contains a porous lead anode where oxidation occurs according to $Pb(s) + HSO_4^-(aq) \rightarrow PbSO_4(s) + H^+(aq) + 2\,e^-$ and a lead(IV) oxide cathode where reduction occurs according to $PbO_2(s) + HSO_4^-(aq) + 3\,H^+(aq) + 2\,e^- \rightarrow PbSO_4(s) + 2\,H_2O(l)$. As electric current is drawn from the battery, both the anode and the cathode become coated with $PbSO_4(s)$ and the solution becomes depleted of $HSO_4^-(aq)$. If the battery is run for a long time without being recharged, too much $PbSO_4(s)$ develops on the surface of the electrodes and the battery goes dead. The lead–acid storage battery can be recharged, however, by running electric current through it in reverse. The electric current, which must come from an external source such as an alternator in a car, causes the preceding reaction to occur in reverse, converting the $PbSO_4(s)$ back to $Pb(s)$ and $PbO_2(s)$, recharging the battery.

20.22 The most common types of portable rechargeable batteries include the nickel–cadmium (NiCad) battery, the nickel metal hydride (NiMH) battery, and the lithium ion battery.

The nickel–cadmium (NiCad) battery consists of an anode composed of solid cadmium and a cathode composed of $NiO(OH)(s)$. The electrolyte is usually $KOH(aq)$. During operation, the cadmium is oxidized and the NiO(OH) is reduced according to the following equations:

Anode reaction: $Cd(s) + 2\,OH^-(aq) \rightarrow Cd(OH)_2(s) + 2\,e^-$

Cathode reaction: $2\,NiO(OH)(s) + 2\,H_2O(l) + 2e^- \rightarrow 2\,Ni(OH)_2(s) + 2\,OH^-(aq)$

The overall reaction produces about 1.30 V. As current is drawn from the NiCad battery, solid cadmium hydroxide accumulates on the anode and solid nickel(II) hydroxide accumulates on the cathode. By running current in the opposite direction, the reactants can be regenerated from the products.

The NiMH battery uses the same cathode reaction as the NiCad battery but a different anode reaction. In the anode of a NiMH battery, hydrogen atoms held in a metal alloy are oxidized. If we let M represent the metal alloy, we can write the half-reactions as follows:

Anode reaction: $M \cdot H(s) + OH^-(aq) \rightarrow M(s) + H_2O(l) + e^-$

Cathode reaction: $NiO(OH)(s) + H_2O(l) + e^- \rightarrow Ni(OH)_2(s) + OH^-(aq)$

The lithium–ion battery is the newest and most expensive type of rechargeable battery. The lithium battery works differently than the other batteries we have examined so far, and the details of its operation are beyond our current scope. Briefly, we can think of the operation of the lithium battery as being due primarily to the motion of lithium ions from the anode to the cathode. The anode is composed of graphite into which lithium ions are incorporated between layers of carbon atoms. Upon discharge, the lithium ions spontaneously migrate to the cathode, which consists of a lithium transition metal oxide such as $LiCoO_2$ or $LiMn_2O_4$. The transition metal is reduced during this process. Upon recharging, the transition metal is oxidized, forcing the lithium to migrate back into the graphite. The flow of lithium ions from the anode to the cathode causes a corresponding flow of electrons in the external circuit.

20.23 Fuel cells are like batteries, but the reactants must be constantly replenished. Normal batteries lose their ability to generate voltage with use because the reactants become depleted as electric current is drawn from the battery. In a fuel cell, the reactants—the fuel—constantly flow through the battery, generating electric current as they undergo a redox reaction.

The most common fuel cell is the hydrogen–oxygen fuel cell. In the cell, hydrogen gas flows past the anode (a screen coated with a platinum catalyst) and undergoes oxidation: $2\,H_2(g) + 4\,OH^-(aq) \rightarrow 4\,H_2O(l) + 4\,e^-$. Oxygen gas flows past the cathode (a similar screen) and undergoes reduction: $O_2(g) + 2\,H_2O(l) + 4\,e^- \rightarrow 4\,OH^-(aq)$. The half-reactions sum to the following overall reaction: $2\,H_2(g) + O_2(g) \rightarrow 2\,H_2O(l)$. Notice that the only product is water.

20.24 In an electrolytic cell, the source of the electrons is the external power source. The external power source must draw electrons away from the anode; thus, the anode must be connected to the positive terminal of the battery (as shown in Figure 20.19). Similarly, the power source drives electrons toward the cathode (where they will be used in reduction), so the cathode must be connected to the negative terminal of the battery. Therefore, the charge labels (+ and −) on an electrolytic cell are opposite of what they are in a voltaic cell.

20.25 The anion is oxidized. The cation is reduced to the metal.

20.26 In the electrolysis of a mixture of ions, the cation that is most easily reduced (the one with the least negative, or most positive, reduction half-cell potential) is reduced first and the anion that is most easily oxidized (the one that has the least negative, or most positive, oxidation half-cell potential) is oxidized first.

20.27 In the electrolysis of aqueous NaCl solutions, two different reduction half-reactions are possible at the cathode, the reduction of Na^+ and the reduction of water.

$$2\,Na^+(l) + 2\,e^- \rightarrow 2\,Na(s) \qquad E^\circ_{red} = 2.71\text{ V}$$
$$2\,H_2O(l) + 2\,e^- \rightarrow H_2(g) + 2\,OH^-(aq) \qquad E_{red} = -0.41\text{ V }([OH^-] = 10^{-7}\text{ M})$$

The half-reaction that occurs most easily (the one with the least negative, or most positive, half-cell potential) is the one that actually occurs.

20.28 An overvoltage is an additional voltage that must be applied to get some nonspontaneous reactions to occur. Overvoltages are important because the desired reaction may not occur at all and the overvoltage can control what products are produced.

20.29 We can determine the number of moles of electrons that have flowed in a given electrolysis cell by measuring the total charge that has flowed through the cell, which in turn depends on the magnitude of current and the time the current has run. Because 1 ampere $= 1\text{ C/s}$, if we multiply the amount of current (in A) flowing through the cell by the time (in s) the current flowed, we can find the total charge that passed through the cell in that time: current $(\text{C/s}) \times$ time $(\text{s}) =$ charge (C). The relationship between charge and the number of moles of electrons is given by Faraday's constant, which corresponds to the charge in coulombs of 1 mol of electrons; so $F = 96{,}485\text{ C/mole e}^-$. These relationships can be used to solve problems involving the stoichiometry of electrolytic cells.

20.30 Corrosion is the (usually) gradual, nearly always undesired oxidation of metals that occurs when they are exposed to oxidizing agents in the environment. This metal oxidation is spontaneous when the half-reaction for the reduction of the metal ion lies below the half-reactions for the reduction of oxygen ($+1.23$ V) in Table 20.1. Many metals, such as aluminum, form oxides that coat the surface of the metal and prevent further corrosion.

20.31 Moisture must be present for many corrosion reactions to occur. The presence of water is necessary because water is a reactant in many oxidation reactions [such as $4\,Fe^{2+}(aq) + O_2(g) + (4 + 2n)\,H_2O(l) \rightarrow 2\,Fe_2O_3 \cdot n\,H_2O(s) + 8\,H^+(aq)$] and because charge (either electrons or ions) must be free to flow between the anodic and cathodic regions.

Additional electrolytes promote more corrosion. The presence of an electrolyte (such as sodium chloride) on the surface of iron promotes rusting because it enhances current flow. This is why cars rust so quickly in cold climates where roads are salted and in areas directly adjacent to beaches where saltwater mist is present.

The presence of acids promotes corrosion. Because H^+ ions are involved as a reactant in the reduction of oxygen, lower pH enhances the cathodic reaction and leads to faster corrosion.

20.32 The most obvious way to prevent rust formation is to keep iron dry. Without water, the redox reaction cannot occur. Another way to prevent rust formation is to coat the iron with a substance that is impervious to water. Cars, for example, are painted and sealed to prevent rust. A scratch in the paint, however, can lead to rusting of the underlying iron. Rust can also be prevented by placing a sacrificial electrode in electrical contact with the iron. The sacrificial electrode must be composed of a metal that oxidizes more easily than iron (i.e., it must be below iron in Table 20.1). The sacrificial electrode then oxidizes in place of the iron (just as the more easily oxidizable species in a mixture is the one to oxidize), protecting the iron from oxidation. Another way to protect iron from rusting is to coat it with a metal that oxidizes more easily than iron. Galvanized nails, for example, are coated with a thin layer of zinc. Because zinc has a more positive half-cell potential for oxidation, it oxidizes in place of the underlying iron (just as a sacrificial electrode does). The oxide of zinc is not crumbly and remains on the nail as a protective coating.

Balancing Redox Reactions

20.33 **Conceptual Plan: Separate the overall reaction into two half-reactions: one for oxidation and one for reduction. → Balance each half-reaction with respect to mass in the following order: (1) Balance all elements other than**

H and O, (2) balance O by adding H_2O, and (3) balance H by adding H^+. → Balance each half-reaction with respect to charge by adding electrons. (The sum of the charges on both sides of the equation should be made equal by adding electrons as necessary.) → Make the number of electrons in both half-reactions equal by multiplying one or both half-reactions by a small whole number. → Add the two half-reactions, canceling electrons and other species as necessary. → Verify that the reaction is balanced with respect to both mass and charge.

Solution:

(a) Separate: $K(s) \rightarrow K^+(aq)$ and $Cr^{3+}(aq) \rightarrow Cr(s)$

Balance elements: $K(s) \rightarrow K^+(aq)$ and $Cr^{3+}(aq) \rightarrow Cr(s)$

Add electrons: $K(s) \rightarrow K^+(aq) + e^-$ and $Cr^{3+}(aq) + 3\,e^- \rightarrow Cr(s)$

Equalize electrons: $3\,K(s) \rightarrow 3\,K^+(aq) + 3\,e^-$ and $Cr^{3+}(aq) + 3\,e^- \rightarrow Cr(s)$

Add half-reactions: $3\,K(s) + Cr^{3+}(aq) + \cancel{3e^-} \rightarrow 3\,K^+(aq) + \cancel{3e^-} + Cr(s)$

Cancel electrons: $3\,K(s) + Cr^{3+}(aq) \rightarrow 3\,K^+(aq) + Cr(s)$

Check:

Reactants	Products
3 K atoms	3 K atoms
1 Cr atom	1 Cr atom
+3 charge	+3 charge

(b) Separate: $Al(s) \rightarrow Al^{3+}(aq)$ and $Fe^{2+}(aq) \rightarrow Fe(s)$

Balance elements: $Al(s) \rightarrow Al^{3+}(aq)$ and $Fe^{2+}(aq) \rightarrow Fe(s)$

Add electrons: $Al(s) \rightarrow Al^{3+}(aq) + 3e^-$ and $Fe^{2+}(aq) + 2\,e^- \rightarrow Fe(s)$

Equalize electrons: $2\,Al(s) \rightarrow 2\,Al^{3+}(aq) + 6\,e^-$ and $3\,Fe^{2+}(aq) + 6\,e^- \rightarrow 3\,Fe(s)$

Add half-reactions: $2\,Al(s) + 3\,Fe^{2+}(aq) + \cancel{6e^-} \rightarrow 2\,Al^{3+}(aq) + \cancel{6e^-} + 3\,Fe(s)$

Cancel electrons: $2\,Al(s) + 3\,Fe^{2+}(aq) \rightarrow 2\,Al^{3+}(aq) + 3\,Fe(s)$

Check:

Reactants	Products
2 Al atoms	2 Al atoms
3 Fe atoms	3 Fe atoms
+6 charge	+6 charge

(c) Separate: $BrO_3^-(aq) \rightarrow Br^-(aq)$ and $N_2H_4(g) \rightarrow N_2(g)$

Balance non-H & O elements: $BrO_3^-(aq) \rightarrow Br^-(aq)$ and $N_2H_4(g) \rightarrow N_2(g)$

Balance O with H_2O: $BrO_3^-(aq) \rightarrow Br^-(aq) + 3\,H_2O(l)$ and $N_2H_4(g) \rightarrow N_2(g)$

Balance H with H^+: $BrO_3^-(aq) + 6\,H^+(aq) \rightarrow Br^-(aq) + 3\,H_2O(l)$ and $N_2H_4(g) \rightarrow N_2(g) + 4\,H^+(aq)$

Add electrons:

$BrO_3^-(aq) + 6\,H^+(aq) + 6\,e^- \rightarrow Br^-(aq) + 3\,H_2O(l)$ and $N_2H_4(g) \rightarrow N_2(g) + 4\,H^+(aq) + 4\,e^-$

Equalize electrons:

$2\,BrO_3^-(aq) + 12\,H^+(aq) + 12\,e^- \rightarrow 2\,Br^-(aq) + 6\,H_2O(l)$ and $3\,N_2H_4(g) \rightarrow 3\,N_2(g) + 12\,H^+(aq) + 12\,e^-$

Add half-reactions: $2\,BrO_3^-(aq) + \cancel{12\,H^+(aq)} + 3\,N_2H_4(g) + \cancel{12e^-} \rightarrow 2\,Br^-(aq) + 6\,H_2O\,(l) + 3\,N_2(g) + \cancel{12\,H^+(aq)} + \cancel{12e^-}$

Cancel electrons & others: $2\,BrO_3^-(aq) + 3\,N_2H_4(g) \rightarrow 2\,Br^-(aq) + 6\,H_2O(l) + 3\,N_2(g)$

Check:

Reactants	Products
2 Br atoms	2 Br atoms
6 O atoms	6 O atoms
12 H atoms	12 H atoms
6 N atoms	6 N atoms
−2 charge	−2 charge

20.34 **Conceptual Plan: Separate the overall reaction into two half-reactions: one for oxidation and one for reduction. → Balance each half-reaction with respect to mass in the following order: (1) Balance all elements other than H and O, (2) balance O by adding H_2O, and (3) balance H by adding H^+. → Balance each half-reaction with respect to charge by adding electrons. (The sum of the charges on both sides of the equation should be made equal by adding electrons as necessary.) → Make the number of electrons in both half-reactions equal by multiplying one or both half-reactions by a small whole number. → Add the two half-reactions, canceling electrons and other species as necessary. → Verify that the reaction is balanced with respect to both mass and charge.**

Solution:

(a)

Separate: $Zn(s) \rightarrow Zn^{2+}(aq)$ and $Sn^{2+}(aq) \rightarrow Sn(s)$

Balance elements: $Zn(s) \rightarrow Zn^{2+}(aq)$ and $Sn^{2+}(aq) \rightarrow Sn(s)$

Add electrons: $Zn(s) \rightarrow Zn^{2+}(aq) + 2\,e^-$ and $Sn^{2+}(aq) + 2\,e^- \rightarrow Sn(s)$

Equalize electrons: $Zn(s) \rightarrow Zn^{2+}(aq) + 2\,e^-$ and $Sn^{2+}(aq) + 2\,e^- \rightarrow Sn(s)$

Add half-reactions: $Zn(s) + Sn^{2+}(aq) + \cancel{2\,e^-} \rightarrow Zn^{2+}(aq) + \cancel{2\,e^-} + Sn(s)$

Cancel electrons: $Zn(s) + Sn^{2+}(aq) \rightarrow Zn^{2+}(aq) + Sn(s)$

Check:

Reactants	Products
1 Zn atom	1 Zn atom
1 Sn atom	1 Sn atom
+2 charge	+2 charge

(b)

Separate: $Mg(s) \rightarrow Mg^{2+}(aq)$ and $Cr^{3+}(aq) \rightarrow Cr(s)$

Balance elements: $Mg(s) \rightarrow Mg^{2+}(aq)$ and $Cr^{3+}(aq) \rightarrow Cr(s)$

Add electrons: $Mg(s) \rightarrow Mg^{2+}(aq) + 2\,e^-$ and $Cr^{3+}(aq) + 3\,e^- \rightarrow Cr(s)$

Equalize electrons: $3\,Mg(s) \rightarrow 3\,Mg^{2+}(aq) + 6\,e^-$ and $2\,Cr^{3+}(aq) + 6\,e^- \rightarrow 2\,Cr(s)$

Add half-reactions: $3\,Mg(s) \rightarrow 2\,Cr^{3+}(aq) + \cancel{6\,e^-} \rightarrow 3\,Mg^{2+}(aq) + \cancel{6\,e^-} + 2\,Cr(s)$

Cancel electrons: $3\,Mg(s) + 2\,Cr^{3+}(aq) \rightarrow 3\,Mg^{2+}(aq) + 2\,Cr(s)$

Check:

Reactants	Products
3 Mg atoms	3 Mg atoms
2 Cr atoms	2 Cr atoms
+6 charge	+6 charge

(c)

Separate: $MnO_4^-(aq) \rightarrow Mn^{2+}(aq)$ and $Al(s) \rightarrow Al^{3+}(aq)$

Balance non-H & O elements: $MnO_4^-(aq) \rightarrow Mn^{2+}(aq)$ and $Al(s) \rightarrow Al^{3+}(aq)$

Balance O with H_2O: $MnO_4^-(aq) \rightarrow Mn^{2+}(aq) + 4\,H_2O(l)$ and $Al(s) \rightarrow Al^{3+}(aq)$

Balance H with H^+: $MnO_4^-(aq) + 8\,H^+(aq) \rightarrow Mn^{2+}(aq) + 4\,H_2O(l)$ and $Al(s) \rightarrow Al^{3+}(aq)$

Add electrons: $MnO_4^-(aq) + 8\,H^+(aq) + 5\,e^- \rightarrow Mn^{2+}(aq) + 4\,H_2O(l)$ and $Al(s) \rightarrow Al^{3+}(aq) + 3\,e^-$

Equalize electrons:

$3\,MnO_4^-(aq) + 24\,H^+(aq) + 15\,e^- \rightarrow 3\,Mn^{2+}(aq) + 12\,H_2O(l)$ and $5\,Al(s) \rightarrow 5\,Al^{3+}(aq) + 15\,e^-$

Add half-reactions:

$3\,MnO_4^-(aq) + 24\,H^+(aq) + \cancel{15\,e^-} + 5\,Al(s) \rightarrow 3\,Mn^{2+}(aq) + 12\,H_2O(l) + 5\,Al^{3+}(aq) + \cancel{15\,e^-}$

Cancel electrons: $3\,MnO_4^-(aq) + 24\,H^+(aq) + 5\,Al(s) \rightarrow 3\,Mn^{2+}(aq) + 12\,H_2O(l) + 5\,Al^{3+}(aq)$

Check:

Reactants	Products
3 Mn atoms	3 Mn atoms
12 O atoms	12 O atoms
24 H atoms	24 H atoms
5 Al atoms	5 Al atoms
+21 charge	+21 charge

20.35 **Conceptual Plan: Separate the overall reaction into two half-reactions: one for oxidation and one for reduction. → Balance each half-reaction with respect to mass in the following order: (1) Balance all elements other than H and O, (2) balance O by adding H_2O, and (3) balance H by adding H^+. → Balance each half-reaction with respect to charge by adding electrons. (The sum of the charges on both sides of the equation should be made equal by adding electrons as necessary.) → Make the number of electrons in both half-reactions equal by multiplying one or both half-reactions by a small whole number. → Add the two half-reactions, canceling electrons and other species as necessary. → Verify that the reaction is balanced with respect to both mass and charge.**

Solution:

(a)

Separate: $PbO_2(s) \rightarrow Pb^{2+}(aq)$ and $I^-(aq) \rightarrow I_2(s)$

Balance non-H & O elements: $PbO_2(s) \rightarrow Pb^{2+}(aq)$ and $2\,I^-(aq) \rightarrow I_2(s)$

Balance O with H_2O: $PbO_2(s) \rightarrow Pb^{2+}(aq) + 2\,H_2O(l)$ and $2\,I^-(aq) \rightarrow I_2(s)$

Balance H with H^+: $PbO_2(s) + 4\,H^+(aq) \rightarrow Pb^{2+}(aq) + 2\,H_2O(l)$ and $2\,I^-(aq) \rightarrow I_2(s)$

Add electrons: $PbO_2(s) + 4\,H^+(aq) + 2\,e^- \rightarrow Pb^{2+}(aq) + 2\,H_2O(l)$ and $2\,I^-(aq) \rightarrow I_2(s) + 2\,e^-$

Equalize electrons: $PbO_2(s) + 4\,H^+(aq) + 2\,e^- \rightarrow Pb^{2+}(aq) + 2\,H_2O(l)$ and $2\,I^-(aq) \rightarrow I_2(s) + 2\,e^-$
Add half-reactions: $PbO_2(s) + 4\,H^+(aq) + \cancel{2\,e^-} + 2\,I^-(aq) \rightarrow Pb^{2+}(aq) + 2\,H_2O(l) + I_2(s) + \cancel{2\,e^-}$
Cancel electrons & others: $PbO_2(s) + 4\,H^+(aq) + 2\,I^-(aq) \rightarrow Pb^{2+}(aq) + 2\,H_2O(l) + I_2(s)$

Check:

Reactants	Products
1 Pb atom	1 Pb atom
2 O atoms	2 O atoms
4 H atoms	4 H atoms
2 I atoms	2 I atoms
+2 charge	+2 charge

(b) Separate: $MnO_4^-(aq) \rightarrow Mn^{2+}(aq)$ and $SO_3^{2-}(aq) \rightarrow SO_4^{2-}(aq)$
Balance non-H & O elements: $MnO_4^-(aq) \rightarrow Mn^{2+}(aq)$ and $SO_3^{2-}(aq) \rightarrow SO_4^{2-}(aq)$
Balance O with H_2O: $MnO_4^-(aq) \rightarrow Mn^{2+}(aq) + 4\,H_2O(l)$ and $SO_3^{2-}(aq) + H_2O(l) \rightarrow SO_4^{2-}(aq)$
Balance H with H^+:
$MnO_4^-(aq) + 8\,H^+(aq) \rightarrow Mn^{2+}(aq) + 4\,H_2O(l)$ and $SO_3^{2-}(aq) + H_2O(l) \rightarrow SO_4^{2-}(aq) + 2\,H^+(aq)$
Add electrons: $MnO_4^-(aq) + 8\,H^+(aq) + 5\,e^- \rightarrow Mn^{2+}(aq) + 4\,H_2O(l)$ and
$SO_3^{2-}(aq) + H_2O\,(l) \rightarrow SO_4^{2-}(aq) + 2\,H^+(aq) + 2\,e^-$
Equalize electrons: $2\,MnO_4^-(aq) + 16\,H^+(aq) + 10\,e^- \rightarrow 2\,Mn^{2+}(aq) + 8\,H_2O(l)$ and
$5\,SO_3^{2-}(aq) + 5\,H_2O(l) \rightarrow 5\,SO_4^{2-}(aq) + 10\,H^+(aq) + 10\,e^-$
Add half-reactions: $2\,MnO_4^-(aq) + 6\,\cancel{16}\,H^+(aq) + \cancel{10\,e^-} + 5\,SO_3^{2-}(aq) + \cancel{5\,H_2O(l)} \rightarrow$
$2\,Mn^{2+}(aq) + 3\,\cancel{8}\,H_2O(l) + 5\,SO_4^{2-}(aq) + \cancel{10\,H^+}(aq) + \cancel{10\,e^-}$
Cancel electrons: $2\,MnO_4^-(aq) + 6\,H^+(aq) + 5\,SO_3^{2-}(aq) \rightarrow 2\,Mn^{2+}(aq) + 3\,H_2O\,(l) + 5\,SO_4^{2-}(aq)$

Check:

Reactants	Products
2 Mn atoms	2 Mn atoms
23 O atoms	23 O atoms
6 H atoms	6 H atoms
5 S atoms	5 S atoms
−6 charge	−6 charge

(c) Separate: $S_2O_3^{2-}(aq) \rightarrow SO_4^{2-}(aq)$ and $Cl_2(g) \rightarrow Cl^-(aq)$
Balance non-H & O elements: $S_2O_3^{2-}(aq) \rightarrow 2\,SO_4^{2-}(aq)$ and $Cl_2(g) \rightarrow 2\,Cl^-(aq)$
Balance O with H_2O: $S_2O_3^{2-}(aq) + 5\,H_2O(l) \rightarrow 2\,SO_4^{2-}(aq)$ and $Cl_2(g) \rightarrow 2\,Cl^-(aq)$
Balance H with H^+: $S_2O_3^{2-}(aq) + 5\,H_2O(l) \rightarrow 2\,SO_4^{2-}(aq) + 10\,H^+(aq)$ and $Cl_2(g) \rightarrow 2\,Cl^-(aq)$
Add electrons:
$S_2O_3^{2-}(aq) + 5\,H_2O(l) \rightarrow 2\,SO_4^{2-}(aq) + 10\,H^+(aq) + 8\,e^-$ and $Cl_2(g) + 2\,e^- \rightarrow 2\,Cl^-(aq)$
Equalize electrons:
$S_2O_3^{2-}(aq) + 5\,H_2O(l) \rightarrow 2\,SO_4^{2-}(aq) + 10\,H^+(aq) + 8\,e^-$ and $4\,Cl_2(g) + 8\,e^- \rightarrow 8\,Cl^-(aq)$
Add half-reactions:
$S_2O_3^{2-}(aq) + 5\,H_2O(l) + 4\,Cl_2(g) + \cancel{8\,e^-} \rightarrow 2\,SO_4^{2-}(aq) + 10\,H^+(aq) + \cancel{8\,e^-} + 8\,Cl^-(aq)$
Cancel electrons: $S_2O_3^{2-}(aq) + 5\,H_2O(l) + 4\,Cl_2(g) \rightarrow 2\,SO_4^{2-}(aq) + 10\,H^+(aq) + 8\,Cl^-(aq)$

Check:

Reactants	Products
2 S atoms	2 S atoms
8 O atoms	8 O atoms
10 H atoms	10 H atoms
8 Cl atoms	8 Cl atoms
−2 charge	−2 charge

20.36 **Conceptual Plan: Separate the overall reaction into two half-reactions: one for oxidation and one for reduction. → Balance each half-reaction with respect to mass in the following order: (1) Balance all elements other than H and O, (2) balance O by adding H_2O, and (3) balance H by adding H^+. → Balance each half-reaction with respect to charge by adding electrons. (The sum of the charges on both sides of the equation should be made equal by adding electrons as necessary.) → Make the number of electrons in both half-reactions equal by multiplying one or both half-reactions by a small whole number. → Add the two half-reactions, canceling electrons and other species as necessary. → Verify that the reaction is balanced with respect to both mass and charge.**

Solution:

(a) Separate: $NO_2^-(aq) \rightarrow NO(g)$ and $I^-(aq) \rightarrow I_2(s)$

Balance non-H & O elements: $NO_2^-(aq) \rightarrow NO(g)$ and $2\,I^-(aq) \rightarrow I_2(s)$

Balance O with H_2O: $NO_2^-(aq) \rightarrow NO(g) + H_2O(l)$ and $2\,I^-(aq) \rightarrow I_2(s)$

Balance H with H^+: $NO_2^-(aq) + 2\,H^+(aq) \rightarrow NO(g) + H_2O(l)$ and $2\,I^-(aq) \rightarrow I_2(s)$

Add electrons: $NO_2^-(aq) + 2\,H^+(aq) + e^- \rightarrow NO(g) + H_2O(l)$ and $2\,I^-(aq) \rightarrow I_2(s) + 2\,e^-$

Equalize electrons: $2\,NO_2^-(aq) + 4\,H^+(aq) + 2\,e^- \rightarrow 2\,NO(g) + 2\,H_2O(l)$ and $2\,I^-(aq) \rightarrow I_2(s) + 2\,e^-$

Add half-reactions: $2\,NO_2^-(aq) + 4\,H^+(aq) + \cancel{2e^-} + 2\,I^-(aq) \rightarrow 2\,NO(g) + 2\,H_2O(l) + I_2(s) + \cancel{2e^-}$

Cancel electrons: $2\,NO_2^-(aq) + 4\,H^+(aq) + 2\,I^-(aq) \rightarrow 2\,NO(g) + 2\,H_2O(l) + I_2(s)$

Check:

Reactants	Products
2 N atoms	2 N atoms
4 O atoms	4 O atoms
4 H atoms	4 H atoms
2 I atoms	2 I atoms
0 charge	0 charge

(b) Separate: $ClO_4^-(aq) \rightarrow ClO_3^-(aq)$ and $Cl^-(aq) \rightarrow Cl_2(g)$

Balance non-H & O elements: $ClO_4^-(aq) \rightarrow ClO_3^-(aq)$ and $2\,Cl^-(aq) \rightarrow Cl_2(g)$

Balance O with H_2O: $ClO_4^-(aq) \rightarrow ClO_3^-(aq) + H_2O(l)$ and $2\,Cl^-(aq) \rightarrow Cl_2(g)$

Balance H with H^+: $ClO_4^-(aq) + 2\,H^+(aq) \rightarrow ClO_3^-(aq) + H_2O(l)$ and $2\,Cl^-(aq) \rightarrow Cl_2(g)$

Add electrons: $ClO_4^-(aq) + 2\,H^+(aq) + 2\,e^- \rightarrow ClO_3^-(aq) + H_2O(l)$ and $2\,Cl^-(aq) \rightarrow Cl_2(g) + 2\,e^-$

Equalize electrons: $ClO_4^-(aq) + 2\,H^+(aq) + 2\,e^- \rightarrow ClO_3^-(aq) + H_2O(l)$ and $2\,Cl^-(aq) \rightarrow Cl_2(g) + 2\,e^-$

Add half-reactions: $ClO_4^-(aq) + 2\,H^+(aq) + \cancel{2e^-} + 2\,Cl^-(aq) \rightarrow ClO_3^-(aq) + H_2O(l) + Cl_2(g) + \cancel{2e^-}$

Cancel electrons: $ClO_4^-(aq) + 2\,H^+(aq) + 2\,Cl^-(aq) \rightarrow ClO_3^-(aq) + H_2O(l) + Cl_2(g)$

Check:

Reactants	Products
3 Cl atoms	3 Cl atoms
4 O atoms	4 O atoms
2 H atoms	2 H atoms
−1 charge	−1 charge

(c) Separate: $NO_3^-(aq) \rightarrow NO(g)$ and $Sn^{2+}(aq) \rightarrow Sn^{4+}(aq)$

Balance non-H & O elements: $NO_3^-(aq) \rightarrow NO(g)$ and $Sn^{2+}(aq) \rightarrow Sn^{4+}(aq)$

Balance O with H_2O: $NO_3^-(aq) \rightarrow NO(g) + 2\,H_2O(l)$ and $Sn^{2+}(aq) \rightarrow Sn^{4+}(aq)$

Balance H with H^+: $NO_3^-(aq) + 4\,H^+(aq) \rightarrow NO(g) + 2\,H_2O(l)$ and $Sn^{2+}(aq) \rightarrow Sn^{4+}(aq)$

Add electrons: $NO_3^-(aq) + 4\,H^+(aq) + 3\,e^- \rightarrow NO(g) + 2\,H_2O(l)$ and $Sn^{2+}(aq) \rightarrow Sn^{4+}(aq) + 2\,e^-$

Equalize electrons:

$$2\,NO_3^-(aq) + 8\,H^+(aq) + 6\,e^- \rightarrow 2\,NO(g) + 4\,H_2O(l) \text{ and } 3\,Sn^{2+}(aq) \rightarrow 3\,Sn^{4+}(aq) + 6\,e^-$$

Add half-reactions:

$$2\,NO_3^-(aq) + 8\,H^+(aq) + \cancel{6e^-} + 3\,Sn^{2+}(aq) \rightarrow 2\,NO(g) + 4\,H_2O(l) + 3\,Sn^{4+}(aq) + \cancel{6e^-}$$

Cancel electrons: $2\,NO_3^-(aq) + 8\,H^+(aq) + 3\,Sn^{2+}(aq) \rightarrow 2\,NO(g) + 4\,H_2O(l) + 3\,Sn^{4+}(aq)$

Check:

Reactants	Products
2 N atoms	2 N atoms
6 O atoms	6 O atoms
8 H atoms	8 H atoms
3 Sn atoms	3 Sn atoms
+12 charge	+12 charge

20.37 **Conceptual Plan: Separate the overall reaction into two half-reactions: one for oxidation and one for reduction. → Balance each half-reaction with respect to mass in the following order: (1) Balance all elements other than H and O, (2) balance O by adding H_2O, (3) balance H by adding H^+, and (4) neutralize H^+ by adding enough OH^- to neutralize each H^+. Add the same number of OH^- ions to each side of the equation. → Balance each half-reaction with respect to charge by adding electrons. (The sum of the charges on both sides of the equation should be made equal by adding electrons as necessary.) → Make the number of electrons in both half-reactions equal by multiplying one or both half-reactions by a small whole number. → Add the two half-reactions, canceling electrons and other species as necessary. → Verify that the reaction is balanced with respect to both mass and charge.**

Solution:

(a) Separate: $ClO_2(aq) \rightarrow ClO_2^-(aq)$ and $H_2O_2(aq) \rightarrow O_2(g)$

Balance non-H & O elements: $ClO_2(aq) \rightarrow ClO_2^-(aq)$ and $H_2O_2(aq) \rightarrow O_2(g)$

Balance O with H_2O: $ClO_2(aq) \rightarrow ClO_2^-(aq)$ and $H_2O_2(aq) \rightarrow O_2(g)$

Balance H with H^+: $ClO_2(aq) \rightarrow ClO_2^-(aq)$ and $H_2O_2(aq) \rightarrow O_2(g) + 2\,H^+(aq)$

Neutralize H^+ with OH^-:

$$ClO_2(aq) \rightarrow ClO_2^-(aq) \text{ and } H_2O_2(aq) + 2\,OH^-(aq) \rightarrow O_2(g) + \underbrace{2\,H^+(aq) + 2\,OH^-(aq)}_{2\,H_2O(l)}$$

Add electrons: $ClO_2(aq) + e^- \rightarrow ClO_2^-(aq)$ and $H_2O_2(aq) + 2\,OH^-(aq) \rightarrow O_2(g) + 2\,H_2O(l) + 2\,e^-$

Equalize electrons:

$$2\,ClO_2(aq) + 2\,e^- \rightarrow 2\,ClO_2^-(aq) \text{ and } H_2O_2(aq) + 2\,OH^-(aq) \rightarrow O_2(g) + 2\,H_2O(l) + 2\,e^-$$

Add half-reactions:

$$2\,ClO_2(aq) + \cancel{2\,e^-} + H_2O_2(aq) + 2\,OH^-(aq) \rightarrow 2\,ClO_2^-(aq) + O_2(g) + 2\,H_2O(l) + \cancel{2\,e^-}$$

Cancel electrons: $2\,ClO_2(aq) + H_2O_2(aq) + 2\,OH^-(aq) \rightarrow 2\,ClO_2^-(aq) + O_2(g) + 2\,H_2O(l)$

Check:

Reactants	Products
2 Cl atoms	2 Cl atoms
8 O atoms	8 O atoms
4 H atoms	4 H atoms
−2 charge	−2 charge

(b) Separate: $MnO_4^-(aq) \rightarrow MnO_2(s)$ and $Al(s) \rightarrow Al(OH)_4^-(aq)$

Balance non-H & O elements: $MnO_4^-(aq) \rightarrow MnO_2(s)$ and $Al(s) \rightarrow Al(OH)_4^-(aq)$

Balance O with H_2O: $MnO_4^-(aq) \rightarrow MnO_2(s) + 2\,H_2O(l)$ and $Al(s) + 4\,H_2O(l) \rightarrow Al(OH)_4^-(aq)$

Balance H with H^+:

$MnO_4^-(aq) + 4\,H^+(aq) \rightarrow MnO_2(s) + 2\,H_2O(l)$ and $Al(s) + 4\,H_2O(l) \rightarrow Al(OH)_4^-(aq) + 4\,H^+(aq)$

Neutralize H^+ with OH^-: $MnO_4^- + \underbrace{4\,H^+(aq) + 4\,OH^-(aq)}_{2\,\cancel{4}\,H_2O(l)} \rightarrow MnO_2(s) + \cancel{2\,H_2O(l)} + 4\,OH^-(aq)$

and $Al(s) + \cancel{4\,H_2O(l)} + 4\,OH^-(aq) \rightarrow Al(OH)_4^-(aq) + \underbrace{4\,H^+(aq) + 4\,OH^-(aq)}_{\cancel{4\,H_2O(l)}}$

Add electrons: $MnO_4^-(aq) + 2\,H_2O(l) + 3\,e^- \rightarrow MnO_2(s) + 4\,OH^-(aq)$ and

$Al(s) + 4\,OH^-(aq) \rightarrow Al(OH)_4^-(aq) + 3\,e^-$

Equalize electrons: $MnO_4^-(aq) + 2\,H_2O(l) + 3\,e^- \rightarrow MnO_2(s) + 4\,OH^-(aq)$ and

$Al(s) + 4\,OH^-(aq) \rightarrow Al(OH)_4^-(aq) + 3\,e^-$

Add half-reactions:

$MnO_4^-(aq) + 2\,H_2O(l) + \cancel{3\,e^-} + Al(s) + \cancel{4\,OH^-(aq)} \rightarrow MnO_2(s) + \cancel{4\,OH^-(aq)} + Al(OH)_4^-(aq) + \cancel{3\,e^-}$

Cancel electrons: $MnO_4^-(aq) + 2\,H_2O(l) + Al(s) \rightarrow MnO_2(s) + Al(OH)_4^-(aq)$

Check:

Reactants	Products
1 Mn atom	1 Mn atom
6 O atoms	6 O atoms
4 H atoms	4 H atoms
1 Al atom	1 Al atom
−1 charge	−1 charge

(c) Separate: $Cl_2(g) \rightarrow Cl^-(aq)$ and $Cl_2(g) \rightarrow ClO^-(aq)$

Balance non-H & O elements: $Cl_2(g) \rightarrow 2\,Cl^-(aq)$ and $Cl_2(g) \rightarrow 2ClO^-(aq)$

Balance O with H_2O: $Cl_2(g) \rightarrow 2\,Cl^-(aq)$ and $Cl_2(g) + 2\,H_2O(l) \rightarrow 2\,ClO^-(aq)$

Balance H with H^+: $Cl_2(g) \rightarrow 2\,Cl^-(aq)$ and $Cl_2(g) + 2\,H_2O(l) \rightarrow 2\,ClO^-(aq) + 4\,H^+(aq)$

Neutralize H^+ with OH^-:

$Cl_2(g) \rightarrow 2\,Cl^-(aq)$ and $Cl_2(g) + \cancel{2\,H_2O(l)} + 4\,OH^-(aq) \rightarrow 2\,ClO^-(aq) + \underbrace{4\,H^+(aq) + 4\,OH^-(aq)}_{2\,\cancel{4}\,H_2O(l)}$

Add electrons: $Cl_2(g) + 2\,e^- \rightarrow 2\,Cl^-(aq)$ and $Cl_2(g) + 4\,OH^-(aq) \rightarrow 2\,ClO^-(aq) + 2\,H_2O(l) + 2\,e^-$

Equalize electrons: $Cl_2(g) + 2\,e^- \rightarrow 2\,Cl^-(aq)$ and $Cl_2(g) + 4\,OH^-(aq) \rightarrow 2\,ClO^-(aq) + 2\,H_2O(l) + 2\,e^-$

Add half-reactions: $Cl_2(g) + \cancel{2\,e^-} + Cl_2(g) + 4\,OH^-(aq) \rightarrow 2\,Cl^-(aq) + 2\,ClO^-(aq) + 2\,H_2O(l) + \cancel{2\,e^-}$

Cancel electrons: $2\,Cl_2(g) + 4\,OH^-(aq) \rightarrow 2\,Cl^-(aq) + 2\,ClO^-(aq) + 2\,H_2O(l)$

Simplify: $Cl_2(g) + 2\,OH^-(aq) \rightarrow Cl^-(aq) + ClO^-(aq) + H_2O(l)$

Check:

Reactants	Products
2 Cl atoms	2 Cl atoms
2 O atoms	2 O atoms
2 H atoms	2 H atoms
−2 charge	−2 charge

20.38 **Conceptual Plan: Separate the overall reaction into two half-reactions: one for oxidation and one for reduction. → Balance each half-reaction with respect to mass in the following order: (1) balance all elements other than H and O, (2) balance O by adding H_2O, (3) balance H by adding H^+, and (4) neutralize H^+ by adding enough OH^- to neutralize each H^+. Add the same number of OH^- ions to each side of the equation. → Balance each half-reaction with respect to charge by adding electrons. (The sum of the charges on both sides of the equation should be made equal by adding electrons as necessary.) → Make the number of electrons in both half-reactions equal by multiplying one or both half-reactions by a small whole number. → Add the two half-reactions, canceling electrons and other species as necessary. → Verify that the reaction is balanced with respect to both mass and charge.**

Solution:

(a) Separate: $MnO_4^-(aq) \rightarrow MnO_2(s)$ and $Br^-(aq) \rightarrow BrO_3^-(aq)$

Balance non-H & O elements: $MnO_4^-(aq) \rightarrow MnO_2(s)$ and $Br^-(aq) \rightarrow BrO_3^-(aq)$

Balance O with H_2O: $MnO_4^-(aq) \rightarrow MnO_2(s) + 2\,H_2O(l)$ and $Br^-(aq) + 3\,H_2O(l) \rightarrow BrO_3^-(aq)$

Balance H with H^+:

$MnO_4^-(aq) + 4\,H^+(aq) \rightarrow MnO_2(s) + 2\,H_2O(l)$ and $Br^-(aq) + 3\,H_2O(l) \rightarrow BrO_3^-(aq) + 6\,H^+(aq)$

Neutralize H^+ with OH^-: $MnO_4^-(aq) + \underbrace{4H^+(aq) + 4\,OH^-(aq)}_{2\,\cancel{4}\,H_2O(l)} \rightarrow MnO_2(s) + \cancel{2\,H_2O(l)} + 4\,OH^-(aq)$

and $Br^-(aq) + \cancel{3\,H_2O(l)} + 6\,OH^-(aq) \rightarrow BrO_3^-(aq) + \underbrace{6H^+(aq) + 6OH^-(aq)}_{3\,\cancel{6}\,H_2O(l)}$

Add electrons: $MnO_4^-(aq) + 2\,H_2O(l) + 3\,e^- \rightarrow MnO_2(s) + 4\,OH^-(aq)$ and

$Br^-(aq) + 6\,OH^-(aq) \rightarrow BrO_3^-(aq) + 3\,H_2O(l) + 6\,e^-$

Equalize electrons: $2\,MnO_4^-(aq) + 4\,H_2O(l) + 6\,e^- \rightarrow 2\,MnO_2(s) + 8\,OH^-(aq)$ and

$Br^-(aq) + 6\,OH^-(aq) \rightarrow BrO_3^-(aq) + 3\,H_2O(l) + 6\,e^-$

Add half-reactions: $2\,MnO_4^-(aq) + 1\,\cancel{4}\,H_2O(l) + \cancel{6e^-} + Br^-(aq) + \cancel{6\,OH^-(aq)} \rightarrow$

$2\,MnO_2(s) + 2\,\cancel{8}\,OH^-(aq) + BrO_3^-(aq) + \cancel{3\,H_2O(l)} + \cancel{6e^-}$

Cancel electrons & others: $2\,MnO_4^-(aq) + H_2O(l) + Br^-(aq) \rightarrow 2\,MnO_2(s) + 2\,OH^-(aq) + BrO_3^-(aq)$

Check:

Reactants	Products
2 Mn atoms	2 Mn atoms
9 O atoms	9 O atoms
2 H atoms	2 H atoms
1 Br atom	1 Br atom
−3 charge	−3 charge

(b) Separate: $Ag(s) + CN^-(aq) \rightarrow Ag(CN)_2^-(aq)$ and $O_2(g) \rightarrow$

Balance non-H & O elements: $Ag(s) + 2\,CN^-(aq) \rightarrow Ag(CN)_2^-(aq)$ and $O_2(g) \rightarrow$

Balance O with H_2O: $Ag(s) + 2\,CN^-(aq) \rightarrow Ag(CN)_2^-(aq)$ and $O_2(g) \rightarrow 2\,H_2O(l)$

Balance H with H^+: $Ag(s) + 2\,CN^-(aq) \rightarrow Ag(CN)_2^-(aq)$ and $O_2(g) + 4\,H^+(aq) \rightarrow 2\,H_2O(l)$

Neutralize H^+ with OH^-:

$Ag(s) + 2\,CN^-(aq) \rightarrow Ag(CN)_2^-(aq)$ and $O_2(g) + \underbrace{4H^+(aq) + 4OH^-(aq)}_{2\,\cancel{4}\,H_2O(l)} \rightarrow \cancel{2\,H_2O(l)} + 4\,OH^-(aq)$

Add electrons: $Ag(s) + 2\,CN^-(aq) \rightarrow Ag(CN)_2^-(aq) + e^-$ and $O_2(g) + 2\,H_2O\,(l) + 4\,e^- \rightarrow 4\,OH^-(aq)$

Equalize electrons:

$4\,Ag(s) + 8\,CN^-(aq) \rightarrow 4\,Ag(CN)_2^-(aq) + 4\,e^-$ and $O_2(g) + 2\,H_2O(l) + 4\,e^- \rightarrow 4\,OH^-(aq)$

Add half-reactions:

$4\,Ag(s) + 8\,CN^-(aq) + O_2(g) + 2\,H_2O(l) + \cancel{4e^-} \rightarrow 4\,Ag(CN)_2^-(aq) + \cancel{4e^-} + 4\,OH^-(aq)$

Cancel electrons: $4\,Ag(s) + 8\,CN^-(aq) + O_2(g) + 2\,H_2O(l) \rightarrow 4\,Ag(CN)_2^-(aq) + 4\,OH^-(aq)$

Check:

Reactants	Products
4 Ag atoms	4 Ag atoms
8 C atoms	8 C atoms
8 N atoms	8 N atoms
4 O atoms	4 O atoms
4 H atoms	4 H atoms
−8 charge	−8 charge

(c) Separate: $NO_2^-(aq) \rightarrow NH_3(g)$ and $Al(s) \rightarrow AlO_2^-(aq)$

Balance non-H & O elements: $NO_2^-(aq) \rightarrow NH_3(g)$ and $Al(s) \rightarrow AlO_2^-(aq)$

Balance O with H_2O: $NO_2^-(aq) \rightarrow NH_3(g) + 2\,H_2O(l)$ and $Al(s) + 2\,H_2O(l) \rightarrow AlO_2^-(aq)$

Balance H with H^+: $NO_2^-(aq) + 7\,H^+(aq) \rightarrow NH_3(g) + 2\,H_2O(l)$ and

$$Al(s) + 2\,H_2O(l) \rightarrow AlO_2^-(aq) + 4\,H^+(aq)$$

Neutralize H^+ with OH^-: $NO_2^-(aq) + \underbrace{7H^+(aq) + 7OH^-(aq)}_{5\,\cancel{7}\,H_2O(l)} \rightarrow NH_3(g) + \cancel{2\,H_2O(l)} + 7\,OH^-(aq)$ and

$$Al(s) + \cancel{2\,H_2O(l)} + 4\,OH^-(aq) \rightarrow AlO_2^-(aq) + \underbrace{4H^+(aq) + 4OH^-(aq)}_{2\,\cancel{4}\,H_2O(l)}$$

Add electrons: $NO_2^-(aq) + 5\,H_2O(l) + 6\,e^- \rightarrow NH_3(g) + 7\,OH^-(aq)$ and

$$Al(s) + 4\,OH^-(aq) \rightarrow AlO_2^-(aq) + 2\,H_2O(l) + 3\,e^-$$

Equalize electrons: $NO_2^-(aq) + 5\,H_2O(l) + 6\,e^- \rightarrow NH_3(g) + 7\,OH^-(aq)$ and

$$2\,Al(s) + 8\,OH^-(aq) \rightarrow 2\,AlO_2^-(aq) + 4\,H_2O(l) + 6\,e^-$$

Add half-reactions: $NO_2^-(aq) + {}^{1}\cancel{5}\,H_2O(l) + \cancel{6e^-} + 2\,Al(s) + {}^{1}\cancel{8}\,OH^-(aq) \rightarrow$

$$NH_3(g) + \cancel{7\,OH^-(aq)} + 2\,AlO_2^-(aq) + \cancel{4\,H_2O(l)} + \cancel{6e^-}$$

Cancel electrons & others: $NO_2^-(aq) + H_2O(l) + 2\,Al(s) + OH^-(aq) \rightarrow NH_3(g) + 2\,AlO_2^-(aq)$

Check:

Reactants	Products
1 N atom	1 N atom
4 O atoms	4 O atoms
3 H atoms	3 H atoms
2 Al atoms	2 Al atoms
−2 charge	−2 charge

Voltaic Cells, Standard Cell Potentials, and Direction of Spontaneity

20.39 **Given:** voltaic cell overall redox reaction

Find: Sketch voltaic cell, labeling anode, cathode, all species, and direction of electron flow

Conceptual Plan: Separate the overall reaction into two half-cell reactions and add electrons as needed to balance reactions. Put the anode reaction on the left (oxidation = electrons as product) and the cathode reaction on the right (reduction = electrons as reactant). Electrons flow from anode to cathode.

Solution:

(a) $2\,Ag^+(aq) + Pb(s) \rightarrow 2\,Ag(s) + Pb^{2+}(aq)$ separates to $2\,Ag^+(aq) \rightarrow 2\,Ag(s)$ and $Pb(s) \rightarrow Pb^{2+}(aq)$; then add electrons to balance to get the cathode reaction—$2\,Ag^+(aq) + 2\,e^- \rightarrow 2\,Ag(s)$— and the anode reaction—$Pb(s) \rightarrow Pb^{2+}(aq) + 2\,e^-$.

Because we have $Pb(s)$ as the reactant for the oxidation, it will be our anode. Because we have $Ag(s)$ as the product for the reduction, it will be our cathode. Simplify the cathode reaction, dividing all terms by 2.

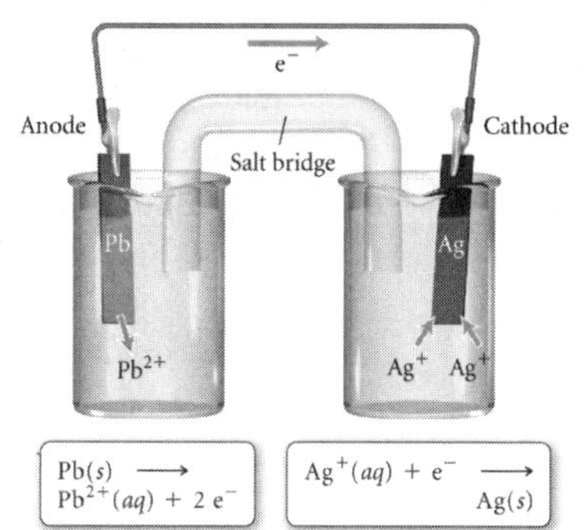

(b) $2\,ClO_2(g) + 2\,I^-(aq) \rightarrow 2\,ClO_2^-(aq) + I_2(s)$ separates to $2\,ClO_2(g) \rightarrow 2\,ClO_2^-(aq)$ and $2\,I^-(aq) \rightarrow I_2(s)$; then add electrons to balance to get the cathode reaction—$2\,ClO_2(g) + 2\,e^- \rightarrow 2\,ClO_2^-(aq)$—and the anode reaction—$2\,I^-(aq) \rightarrow I_2(s) + 2\,e^-$.
Because we have $I^-(aq)$ as the reactant for the oxidation, we will need to use Pt as our anode. Because we have $ClO_2^-(aq)$ as the product for the reduction, we will need to use Pt as our cathode. Because $ClO_2(g)$ is our reactant for the reduction, we need to use an electrode assembly like that used for a SHE. Simplify the cathode reaction, dividing all terms by 2.

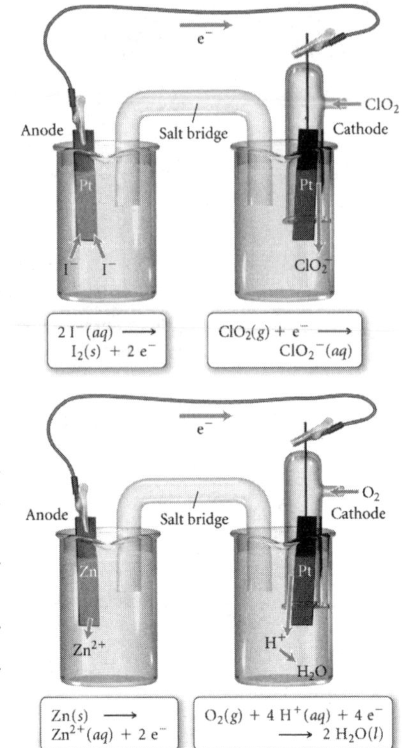

(c) $O_2(g) + 4\,H^+(aq) + 2\,Zn(s) \rightarrow 2\,H_2O(l) + 2\,Zn^{2+}(aq)$ separates to $O_2(g) + 4\,H^+(aq) \rightarrow 2\,H_2O(l)$ and $2\,Zn(s) \rightarrow 2\,Zn^{2+}(aq)$; then add electrons to balance to get the cathode reaction—$O_2(g) + 4\,H^+(aq) + 4\,e^- \rightarrow 2\,H_2O(l)$—and the anode reaction—$2\,Zn(s) \rightarrow 2\,Zn^{2+}(aq) + 4\,e^-$.
Because we have $Zn(s)$ as the reactant for the oxidation, it will be our anode. Because we have $H_2O(l)$ as the product for the reduction, we will need to use Pt as our cathode. Because $O_2(g)$ is our reactant for the reduction, we need to use an electrode assembly like that used for a SHE. Simplify the anode reaction, dividing all terms by 2.

20.40 **Given:** voltaic cell overall redox reaction
Find: Sketch voltaic cell, labeling anode, cathode, all species, and direction of electron flow
Conceptual Plan: Separate the overall reaction into two half-cell reactions and add electrons as needed to balance reactions. Put the anode reaction on the left (oxiden = electrons as product) and the cathode reaction on the right (reduction = electrons as reactant). Electrons flow from anode to cathode.
Solution:

(a) $Ni^{2+}(aq) + Mg(s) \rightarrow Ni(s) + Mg^{2+}(aq)$ separates to $Ni^{2+}(aq) \rightarrow Ni(s)$ and $Mg(s) \rightarrow Mg^{2+}(aq)$; then add electrons to balance to get the cathode reaction—$Ni^{2+}(aq) + 2\,e^- \rightarrow Ni(s)$—and the anode reaction—$Mg(s) \rightarrow Mg^{2+}(aq) + 2\,e^-$.
Because we have $Mg(s)$ as the reactant for the oxidation, it will be our anode. Because we have $Ni(s)$ as the product for the reduction, it will be our cathode.

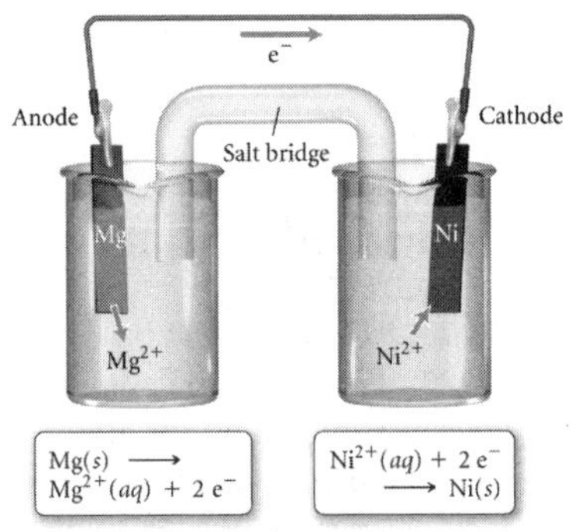

(b) $2\,H^+(aq) + Fe(s) \rightarrow H_2(g) + Fe^{2+}(aq)$ separates to $2\,H^+(aq) \rightarrow H_2(g)$ and $Fe(s) \rightarrow Fe^{2+}(aq)$; then add electrons to balance to get the cathode reaction—$2\,H^+(aq) + 2\,e^- \rightarrow H_2(g)$—and the anode reaction—$Fe(s) \rightarrow Fe^{2+}(aq) + 2\,e^-$.
Because we have $Fe(s)$ as the reactant for the oxidation, it will be our anode. Because we have $H_2(g)$ as the product for the reduction, we will need to use Pt as our cathode and the product can leave using an electrode assembly like that used for a SHE.

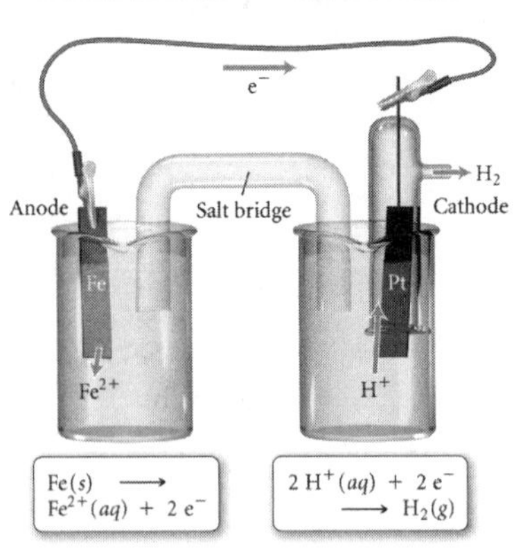

(c) $2\,NO_3^-(aq) + 8\,H^+(aq) + 3\,Cu(s) \rightarrow 2\,NO(g) + 4\,H_2O(l) + 3\,Cu^{2+}(aq)$ separates to $2\,NO_3^-(aq) + 8\,H^+(aq) \rightarrow 2\,NO(g) + 4\,H_2O(l)$ and $3\,Cu(s) \rightarrow 3\,Cu^{2+}(aq)$; then add electrons to balance to get the cathode reaction—$2\,NO_3^-(aq) + 8\,H^+(aq) + 6\,e^- \rightarrow 2\,NO(g) + 4\,H_2O(l)$—and the anode reaction—$3\,Cu(s) \rightarrow 3\,Cu^{2+}(aq) + 6\,e^-$.
Because we have $Cu(s)$ as the reactant for the oxidation, it will be our anode. Because we have $H_2O(l)$ and $NO(g)$ as the products for the reduction, we will need to use Pt as our cathode and the gaseous product can leave using an electrode assembly like that used for a SHE.

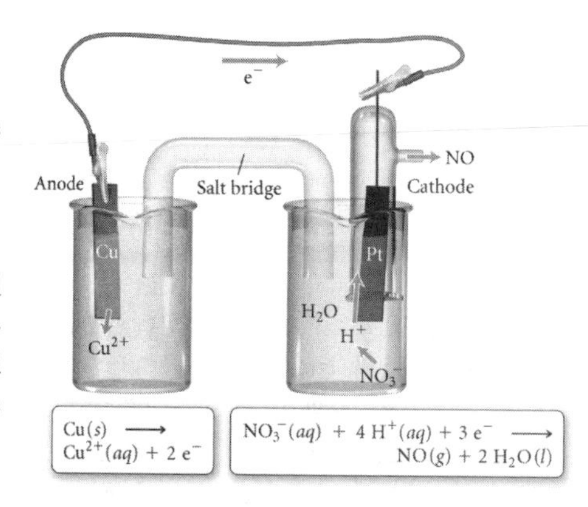

20.41 **Given:** overall reactions from Problem 20.39 **Find:** E°_{cell}
Conceptual Plan: Use Table 20.1 to look up half-reactions from the solution of Problem 20.39. Calculate the standard cell potential by subtracting the electrode potential of the anode from the electrode potential of the cathode: $E^\circ_{cell} = E^\circ_{cathode} - E^\circ_{anode}$.
Solution:
(a) $Ag^+(aq) + e^- \rightarrow Ag(s)$ $E^\circ_{red} = 0.80\text{ V} = E^\circ_{cathode}$ and $Pb(s) \rightarrow Pb^{2+}(aq) + 2\,e^-$ $E^\circ_{red} = -0.13\text{ V} = E^\circ_{anode}$. Then $E^\circ_{cell} = E^\circ_{cathode} - E^\circ_{anode} = 0.80\text{ V} - (-0.13\text{ V}) = 0.93\text{ V}$.
(b) $ClO_2(g) + e^- \rightarrow ClO_2^-(aq)$ $E^\circ_{red} = 0.95\text{ V} = E^\circ_{cathode}$ and $2\,I^-(aq) \rightarrow I_2(s) + 2\,e^-$ $E^\circ_{red} = +0.54\text{ V} = E^\circ_{anode}$. Then $E^\circ_{cell} = E^\circ_{cathode} - E^\circ_{anode} = 0.95\text{ V} - 0.54\text{ V} = 0.41\text{ V}$.
(c) $O_2(g) + 4\,H^+(aq) + 4\,e^- \rightarrow 2\,H_2O(l)$ $E^\circ_{red} = 1.23\text{ V}$ and $Zn(s) \rightarrow Zn^{2+}(aq) + 2\,e^-$ $E^\circ_{red} = -0.76\text{ V} = E^\circ_{anode}$. Then $E^\circ_{cell} = E^\circ_{cathode} - E^\circ_{anode} = 1.23\text{ V} - (-0.76\text{ V}) = 1.99\text{ V}$.
Check: The units (V) are correct. All of the voltages are positive, which is consistent with a voltaic cell.

20.42 **Given:** overall reactions from Problem 20.40 **Find:** E°_{cell}
Conceptual Plan: Use Table 20.1 to look up half-reactions from the solution of Problem 20.40. Calculate the standard cell potential by subtracting the electrode potential of the anode from the electrode potential of the cathode: $E^\circ_{cell} = E^\circ_{cathode} - E^\circ_{anode}$.
Solution:
(a) $Ni^{2+}(aq) + 2\,e^- \rightarrow Ni(s)$ $E^\circ_{red} = -0.23\text{ V} = E^\circ_{cathode}$ and $Mg(s) \rightarrow Mg^{2+}(aq) + 2\,e^-$ $E^\circ_{red} = -2.37\text{ V} = E^\circ_{anode}$. Then $E^\circ_{cell} = E^\circ_{cathode} - E^\circ_{anode} = -0.23\text{ V} - (-2.37\text{ V}) = 2.14\text{ V}$.
(b) $2\,H^+(aq) + 2\,e^- \rightarrow H_2(g)$ $E^\circ_{red} = 0.00\text{ V} = E^\circ_{cathode}$ and $Fe(s) \rightarrow Fe^{2+}(aq) + 2\,e^-$ $E^\circ_{red} = -0.45\text{ V} = E^\circ_{anode}$. Then $E^\circ_{cell} = E^\circ_{cathode} - E^\circ_{anode} = 0.00\text{ V} - (-0.45\text{ V}) = 0.45\text{ V}$.
(c) $NO_3^-(aq) + 4\,H^+(aq) + 3\,e^- \rightarrow NO(g) + 2\,H_2O(l)$ $E^\circ_{red} = 0.96\text{ V} = E^\circ_{cathode}$ and $Cu(s) \rightarrow Cu^{2+}(aq) + 2e^-$ $E^\circ_{red} = -0.34\text{ V} = E^\circ_{anode}$. Then $E^\circ_{cell} = E^\circ_{cathode} - E^\circ_{anode} = 0.96\text{ V} - 0.34\text{ V} = 0.62\text{ V}$.
Check: The units (V) are correct. All of the voltages are positive, which is consistent with a voltaic cell.

20.43 **Given:** voltaic cell drawing
Find: (a) Determine electron flow direction, anode, and cathode; (b) write balanced overall reaction and calculate E°_{cell}; (c) label electrodes as + and −; and (d) find directions of anions and cations from salt bridge.
Conceptual Plan: Look at each half-cell and write a reduction reaction by using electrode and solution composition and adding electrons to balance. Look up half-reactions and standard reduction potentials in Table 20.1. Because this is a voltaic cell, the cell potentials must be assigned to give a positive E°_{cell}. Calculate the standard cell potential by subtracting the electrode potential of the anode from the electrode potential of the cathode, $E^\circ_{cell} = E^\circ_{cathode} - E^\circ_{anode}$, choosing the electrode assignments to give a positive E°_{cell}.
(a) Label the electrode where the oxidation occurs as the anode. Label the electrode where the reduction occurs as the cathode. Electrons flow from anode to cathode.
(b) Take two half-cell reactions and multiply the reactions as necessary to equalize the number of electrons transferred. Add the two half-cell reactions and cancel electrons and any other species.
(c) Label anode as (−) and cathode as (+).
(d) Cations will flow from the salt bridge toward the cathode, and the anions will flow from the salt bridge toward the anode.

Solution:

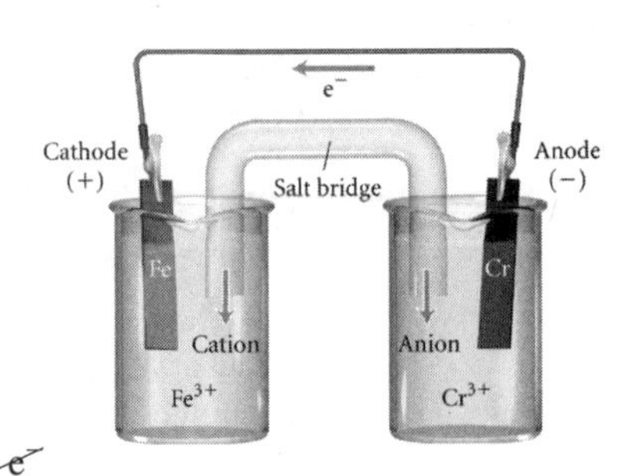

left side: $Fe^{3+}(aq) \rightarrow Fe(s)$; right side: $Cr^{3+}(aq) \rightarrow Cr(s)$.
Add electrons to balance $Fe^{3+}(aq) + 3\,e^- \rightarrow Fe(s)$ and right side—$Cr^{3+}(aq) + 3\,e^- \rightarrow Cr(s)$. Look up cell standard reduction potentials: $Fe^{3+}(aq) + 3\,e^- \rightarrow Fe(s)\ E^\circ_{red} = -0.036\ V$ and $Cr^{3+}(aq) + 3\,e^- \rightarrow Cr(s)$ $E^\circ_{red} = -0.73\ V$. To get a positive cell potential, the second reaction is the oxidation reaction (anode). $E^\circ_{cell} = E^\circ_{cathode} - E^\circ_{anode} = -0.036\ V - (-0.73V) = +0.69\ V$ ((a), (c), and (d)). (b) Add two half-reactions with the second reaction reversed. $Fe^{3+}(aq) + \cancel{3e^-} + Cr(s) \rightarrow Fe(s) + Cr^{3+}(aq) + \cancel{3e^-}$
Cancel electrons to get $Fe^{3+}(aq) + Cr(s) \rightarrow Fe(s) + Cr^{3+}(aq)$.

Check: All atoms and charge are balanced. The units (V) are correct. The cell potential is positive, which is consistent with a voltaic cell.

20.44 **Given:** voltaic cell drawing
Find: (a) Determine electron flow direction, anode, and cathode; (b) write balanced overall reaction and calculate E°_{cell}; (c) label electrodes as + and −; and (d) find directions of anions and cations from salt bridge.
Conceptual Plan: Look at each half-cell and write a reduction reaction by using electrode and solution composition and adding electrons to balance. Look up half-reactions and standard reduction potentials in Table 20.1. Because this is a voltaic cell, the cell potentials must be assigned to give a positive E°_{cell}. Calculate the standard cell potential by subtracting the electrode potential of the anode from the electrode potential of the cathode, $E^\circ_{cell} = E^\circ_{cathode} - E^\circ_{anode}$, choosing the electrode assignments to give a positive E°_{cell}.

(a) **Label the electrode where the oxidation occurs as the anode. Label the electrode where the reduction occurs as the cathode. Electrons flow from anode to cathode.**
(b) **Take two half-cell reactions and multiply the reactions as necessary to equalize the number of electrons transferred. Add the two half-cell reactions and cancel electrons and any other species.**
(c) **Label anode as (−) and cathode as (+).**
(d) **Cations will flow from the salt bridge toward the cathode, and the anions will flow from the salt bridge toward the anode.**

Solution: left side: $Pb^{2+}(aq) \rightarrow Pb(s)$; right side: $Cl_2(g) \rightarrow 2\,Cl^-(aq)$. Add electrons to balance $Pb^{2+}(aq) + 2\,e^- \rightarrow Pb(s)$ and right side—$Cl_2(g) + 2\,e^- \rightarrow 2\,Cl^-(aq)$. Look up cell standard reduction potentials: $Pb^{2+}(aq) + 2\,e^- \rightarrow Pb(s)\ E^\circ_{red} = -0.13\ V$ and $Cl_2(g) + 2\,e^- \rightarrow 2\,Cl^-(aq)\ E^\circ_{red} = 1.36\ V$. To get a positive cell potential, the first reaction is the oxidation reaction (anode).
$E^\circ_{cell} = E^\circ_{cathode} - E^\circ_{anode} = 1.36\ V - (-0.13\ V) = +1.49\ V$ ((a), (c), and (d)).
(b) Add two half-reactions with the first reaction reversed.
$Pb(s) + Cl_2(g) + \cancel{2e^-} \rightarrow Pb^{2+}(aq) + \cancel{2e^-} + 2\,Cl^-(aq)$
Cancel electrons to get $Pb(s) + Cl_2(g) \rightarrow Pb^2 + (aq) + 2\,Cl^-(aq)$.

Check: All atoms and charge are balanced. The units (V) are correct. The cell potential is positive, which is consistent with a voltaic cell.

Anode (−)
e$^-$
Cathode (+)
Pb(s)
$Cl_2(g)$
Salt bridge containing $NaNO_3(aq)$
Anion
1 M Pb^{2+}
Cation
1 M Cl^-

20.45 **Given:** overall reactions from Problem 20.39 **Find:** line notation
Conceptual Plan: Use the solution from Problem 20.39. Write the oxidation half-reaction components on the left and the reduction on the right. A double vertical line (| |), indicating the salt bridge, separates the two half-reactions. Substances in different phases are separated by a single vertical line (|), which represents the boundary between the phases. For some redox reactions, the reactants and products of one or both of the half-reactions may be in the same phase. In these cases, the reactants and products are separated from each other with a comma in the line diagram. Such cells use an inert electrode, such as platinum (Pt) or graphite, as the anode or cathode (or both).

Solution:

(a) Reduction reaction: $Ag^+(aq) + e^- \rightarrow Ag(s)$; oxidation reaction: $Pb(s) \rightarrow Pb^2 + (aq) + 2\,e^-$
so $Pb(s)\ |\ Pb^{2+}(aq)\ ||\ Ag^+(aq)\ |\ Ag(s)$
(b) Reduction reaction: $ClO_2(g) + e^- \rightarrow ClO_2^-(aq)$; oxidation reaction: $2\,I^-(aq) \rightarrow I_2(s) + 2\,e^-$
so $Pt(s)\ |\ I^-(aq)\ |\ I_2(s)\ ||\ ClO_2(g)\ |\ ClO_2^-(aq)\ |\ Pt(s)$

(c) Reduction reaction: $O_2(g) + 4\,H^+(aq) + 4\,e^- \rightarrow 2\,H_2O(l)$; oxidation reaction: $Zn(s) \rightarrow Zn^{2+}(aq) + 2\,e^-$ so $Zn(s) \mid Zn^{2+}(aq) \mid\mid O_2(g) \mid H^+(aq), H_2O(l) \mid Pt(s)$

20.46 **Given:** overall reactions from Problem 20.40 **Find:** line notation

Conceptual Plan: Use the solution from Problem 20.40. Write the oxidation half-reaction components on the left and the reduction on the right. A double vertical line (||), indicating the salt bridge, separates the two half-reactions. Substances in different phases are separated by a single vertical line (|), which represents the boundary between the phases. For some redox reactions, the reactants and products of one or both of the half-reactions may be in the same phase. In these cases, the reactants and products are separated from each other with a comma in the line diagram. Such cells use an inert electrode, such as platinum (Pt) or graphite, as the anode or cathode (or both).

Solution:

(a) Reduction reaction: $Ni^{2+}(aq) + 2\,e^- \rightarrow Ni(s)$; oxidation reaction: $Mg(s) \rightarrow Mg^{2+}(aq) + 2\,e^-$ so $Mg(s) \mid Mg^{2+}(aq) \mid\mid Ni^{2+}(aq) \mid Ni(s)$

(b) Reduction reaction: $2\,H^+(aq) + 2\,e^- \rightarrow H_2(g)$; oxidation reaction: $Fe(s) \rightarrow Fe^{2+}(aq) + 2\,e^-$ so $Fe(s) \mid Fe^{2+}(aq) \mid\mid H^+(aq) \mid H_2(g) \mid Pt(s)$

(c) Reduction reaction: $NO_3^-(aq) + 4\,H^+(aq) + 3\,e^- \rightarrow NO(g) + 2\,H_2O(l)$; oxidation reaction: $Cu(s) \rightarrow Cu^{2+}(aq) + 2\,e^-$ so $Cu(s) \mid Cu^{2+}(aq) \mid\mid NO_3^-(aq), H^+(aq), H_2O(l) \mid NO(g) \mid Pt(s)$

20.47 **Given:** $Sn(s) \mid Sn^{2+}(aq) \mid\mid NO_3^-(aq), H^+(aq), H_2O(l) \mid NO(g) \mid Pt(s)$

Find: Sketch voltaic cell, labeling anode, cathode, all species, direction of electron flow, and E°_{cell}

Conceptual Plan: Separate overall reaction into two half-cell reactions, knowing that the oxidation half-reaction components are on the left and the reduction half-reaction components are on the right. Add electrons as needed to balance reactions. Multiply the half-reactions by the appropriate factors so that an equal number of electrons are transferred. Add the half-cell reactions and cancel electrons. Put anode reaction on the left (oxidation = electrons as product) and cathode reaction on the right (reduction = electrons as reactant). Electrons flow from anode to cathode. Look up half-reactions in Table 20.1. Calculate the standard cell potential by subtracting the electrode potential of the anode from the electrode potential of the cathode: $E^\circ_{cell} = E^\circ_{cathode} - E^\circ_{anode}$.

Solution: Oxidation reaction (anode): $Sn(s) \rightarrow Sn^{2+}(aq) + 2\,e^-$ $E^\circ_{red} = -0.14$ V; reduction reaction (cathode): $NO_3^-(aq) + 4\,H^+(aq) + 3\,e^- \rightarrow NO(g) + 2\,H_2O(l)$ $E^\circ_{red} = 0.96$ V.

$E^\circ_{cell} = E^\circ_{cathode} - E^\circ_{anode} = 0.96\text{ V} - (-0.14\text{ V}) = 1.10\text{ V}$. Multiply the first reaction by 3 and the second reaction by 2 so that 6 electrons are transferred. $3\,Sn(s) \rightarrow 3\,Sn^{2+}(aq) + 6\,e^-$ and $2\,NO_3^-(aq) + 8\,H^+(aq) + 6\,e^- \rightarrow 2\,NO(g) + 4\,H_2O(l)$. Add the two half-reactions and cancel electrons. $3\,Sn(s) + 2\,NO_3^-(aq) + 8\,H^+(aq) + \cancel{6e^-} \rightarrow 3\,Sn^{2+}(aq) + \cancel{6e^-} + 2\,NO(g) + 4\,H_2O(l)$. So balanced reaction is $3\,Sn(s) + 2\,NO_3^-(aq) + 8\,H^+(aq) \rightarrow 3\,Sn^{2+}(aq) + 2\,NO(g) + 4\,H_2O(l)$.

Check: All atoms and charge are balanced. The units (V) are correct. The cell potential is positive, which is consistent with a voltaic cell.

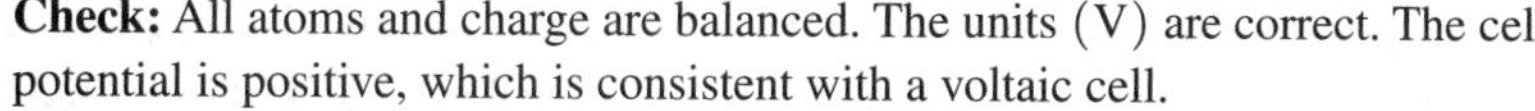

20.48 **Given:** $Mn(s) \mid Mn^{2+}(aq) \mid\mid ClO_2(g) \mid ClO_2^-(aq) \mid Pt(s)$

Find: Sketch voltaic cell, labeling anode, cathode, all species, direction of electron flow, and E°_{cell}

Conceptual Plan: Separate the overall reaction into two half-cell reactions, knowing that the oxidation half-reaction components are on the left and the reduction half-reaction components are on the right. Add electrons as needed to balance reactions. Multiply the half-reactions by the appropriate factors so that an equal number of electrons are transferred. Add the half-cell reactions and cancel electrons. Put the anode reaction on the left (oxidation = electrons as product) and the cathode reaction on the right (oxidation = electrons as reactant). Electrons flow from anode to cathode. Look up half-reactions in Table 20.1. Calculate the standard cell potential by subtracting the electrode potential of the anode from the electrode potential of the cathode: $E^\circ_{cell} = E^\circ_{cathode} - E^\circ_{anode}$.

Solution: Oxidation reaction (anode): $Mn(s) \rightarrow Mn^{2+}(aq) + 2\,e^- \; E^\circ_{red} = -1.18$ V; reduction reaction (cathode): $ClO_2(g) + e^- \rightarrow ClO_2^-(aq) \; E^\circ_{red} = 0.95$ V. $E^\circ_{cell} = E^\circ_{cathode} - E^\circ_{anode} = 0.95\text{ V} - (-1.18\text{ V}) = 2.13$ V. Multiply the second reaction by 2 so that 2 electrons are transferred. $Mn(s) \rightarrow Mn^{2+}(aq) + 2\,e^-$ and $2\,ClO_2(g) + 2\,e^- \rightarrow 2\,ClO_2^-(aq)$. Add the two half-reactions and cancel electrons. $Mn(s) + 2\,ClO_2(g) + \cancel{2e^-} \rightarrow Mn^{2+}(aq) + \cancel{2e^-} + 2\,ClO_2^-(aq)$. So balanced reaction is $Mn(s) + 2\,ClO_2(g) \rightarrow Mn^{2+}(aq) + 2\,ClO_2^-(aq)$.

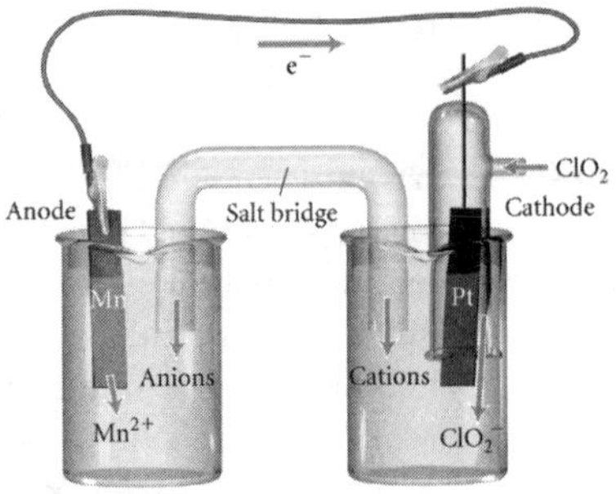

Check: All atoms and charge are balanced. The units (V) are correct. The cell potential is positive, which is consistent with a voltaic cell.

20.49 **Given:** overall reactions **Find:** spontaneity in forward direction

Conceptual Plan: Separate the overall reaction into two half-cell reactions and add electrons as needed to balance reactions. Look up half-reactions in Table 20.1. Calculate the standard cell potential by subtracting the electrode potential of the anode from the electrode potential of the cathode: $E^\circ_{cell} = E^\circ_{cathode} - E^\circ_{anode}$. If $E^\circ_{cell} = 0$, the reaction is spontaneous in the forward direction.

Solution:

(a) $Ni(s) + Zn^{2+}(aq) \rightarrow Ni^{2+}(aq) + Zn(s)$ separates to $Ni(s) \rightarrow Ni^{2+}(aq)$ and $Zn^{2+}(aq) \rightarrow Zn(s)$. Add electrons. $Ni(s) \rightarrow Ni^{2+}(aq) + 2\,e^-$ and $Zn^{2+}(aq) + 2\,e^- \rightarrow Zn(s)$. Look up cell potentials. Ni is oxidized, so $E^\circ_{red} = -0.23\text{ V} = E^\circ_{anode}$. Zn^{2+} is reduced, so $E^\circ_{cathode} = -0.76$ V. Then $E^\circ_{cell} = E^\circ_{cathode} - E^\circ_{anode} = -0.76\text{ V} - (-0.23\text{ V}) = -0.53$ V, so the reaction is nonspontaneous.

(b) $Ni(s) + Pb^{2+}(aq) \rightarrow Ni^{2+}(aq) + Pb(s)$ separates to $Ni(s) \rightarrow Ni^{2+}(aq)$ and $Pb^{2+}(aq) \rightarrow Pb(s)$. Add electrons. $Ni(s) \rightarrow Ni^{2+}(aq) + 2\,e^-$ and $Pb^{2+}(aq) + 2\,e^- \rightarrow Pb(s)$. Look up cell potentials. Ni is oxidized, so $E^\circ_{red} = -0.23\text{ V} = E^\circ_{anode}$. Pb^{2+} is reduced, so $E^\circ_{red} = -0.13\text{ V} = E^\circ_{cathode}$. Then $E^\circ_{cell} = E^\circ_{cathode} - E^\circ_{anode} = -0.13\text{ V} - (-0.23\text{ V}) = +0.10$ V, so the reaction is spontaneous.

(c) $Al(s) + 3\,Ag^+(aq) \rightarrow Al^{3+}(aq) + 3\,Ag(s)$ separates to $Al(s) \rightarrow Al^{3+}(aq)$ and $3\,Ag^+(aq) \rightarrow 3\,Ag(s)$. Add electrons. $Al(s) \rightarrow Al^{3+}(aq) + 3\,e^-$ and $3\,Ag^+(aq) + 3\,e^- \rightarrow 3\,Ag(s)$. Simplify the Ag reaction to $Ag^+(aq) + e^- \rightarrow Ag(s)$. Look up cell potentials. Al is oxidized, so $E^\circ_{red} = -1.66\text{ V} = E^\circ_{anode}$. Ag^+ is reduced, so $E^\circ_{red} = 0.80\text{ V} = E^\circ_{cathode}$. Then $E^\circ_{cell} = E^\circ_{cathode} - E^\circ_{anode} = 0.80\text{ V} - (-1.66\text{ V}) = +2.46$ V, so the reaction is spontaneous.

(d) $Pb(s) + Mn^{2+}(aq) \rightarrow Pb^{2+}(aq) + Mn(s)$ separates to $Pb(s) \rightarrow Pb^{2+}(aq)$ and $Mn^{2+}(aq) \rightarrow Mn(s)$. Add electrons. $Pb(s) \rightarrow Pb^{2+}(aq) + 2\,e^-$ and $Mn^{2+}(aq) + 2\,e^- \rightarrow Mn(s)$. Look up cell potentials. Pb is oxidized, so $E^\circ_{red} = -0.13\text{ V} = E^\circ_{anode}$. Mn^{2+} is reduced, so $E^\circ_{red} = -1.18\text{ V} = E^\circ_{cathode}$. Then $E^\circ_{cell} = E^\circ_{cathode} - E^\circ_{anode} = -1.18\text{ V} - (-0.13\text{ V}) = -1.05$ V, so the reaction is nonspontaneous.

Check: The units (V) are correct. If the voltage is positive, the reaction is spontaneous. If the voltage is negative, the reaction is nonspontaneous.

20.50 **Given:** overall reactions **Find:** spontaneity in reverse direction

Conceptual Plan: Separate the overall reaction into two half-cell reactions and add electrons as needed to balance reactions. Look up half-reactions in Table 20.1. Calculate the standard cell potential by subtracting the electrode potential of the anode from the electrode potential of the cathode: $E^\circ_{cell} = E^\circ_{cathode} - E^\circ_{anode}$. If $E^\circ_{cell} < 0$, the reaction is spontaneous in the reverse direction.

Solution:

(a) $Ca^{2+}(aq) + Zn(s) \rightarrow Ca(s) + Zn^{2+}(aq)$ separates to $Ca^{2+}(aq) \rightarrow Ca(s)$ and $Zn(s) \rightarrow Zn^{2+}(aq)$. Add electrons. $Ca^{2+}(aq) + 2e^- \rightarrow Ca(s)$ and $Zn(s) \rightarrow Zn^{2+}(aq) + 2\,e^-$. Look up cell potentials. Zn is oxidized, so $E^\circ_{red} = -0.76\text{ V} = E^\circ_{anode}$. Ca^{2+} is reduced, so $E^\circ_{red} = -2.76\text{ V} = E^\circ_{cathode}$. Then $E^\circ_{cell} = E^\circ_{cathode} - E^\circ_{anode} = -2.76\text{ V} - (-0.76\text{ V}) = -2.00$ V, so the reaction is spontaneous in the reverse direction.

(b) $2\,Ag^+(aq) + Ni(s) \rightarrow 2\,Ag(s) + Ni^{2+}(aq)$ separates to $2\,Ag^+(aq) \rightarrow 2\,Ag(s)$ and $Ni(s) \rightarrow Ni^{2+}(aq)$. Add electrons. $2\,Ag^+(aq) + 2\,e^- \rightarrow 2\,Ag(s)$ and $Ni(s) \rightarrow Ni^{2+}(aq) + 2\,e^-$. Simplify the Ag reaction to $Ag^+(aq) + e^- \rightarrow Ag(s)$. Look up cell potentials. Ni is oxidized, so $E^\circ_{red} = -0.23\text{ V} = E^\circ_{anode}$. Ag^+ is reduced, so $E^\circ_{red} = 0.80\text{ V} = E^\circ_{cathode}$. Then $E^\circ_{cell} = E^\circ_{cathode} - E^\circ_{anode} = 0.80\text{ V} - (-0.23\text{ V}) = +1.03$ V, so the reaction is nonspontaneous in the reverse direction.

(c) $Fe(s) + Mn^{2+}(aq) \rightarrow Fe^{2+}(aq) + Mn(s)$ separates to $Fe(s) \rightarrow Fe^{2+}(aq)$ and $Mn^{2+}(aq) \rightarrow Mn(s)$. Add electrons. $Fe(s) \rightarrow Fe^{2+}(aq) + 2\,e^-$ and $Mn^{2+}(aq) + 2\,e^- \rightarrow Mn(s)$. Look up cell potentials.

Fe is oxidized, so $E^\circ_{red} = -0.45$ V $= E^\circ_{anode}$. Mn^{2+} is reduced, so $E^\circ_{red} = -1.18$ V $= E^\circ_{cathode}$. Then $E^\circ_{cell} = E^\circ_{cathode} - E^\circ_{anode} = -1.18\text{ V} - (-0.45\text{ V}) = -0.73$ V, so the reaction is spontaneous in the reverse direction.

(d) $2\,Al(s) + 3\,Pb^{2+}(aq) \rightarrow 2\,Al^{3+}(aq) + 3\,Pb(s)$ separates to $2\,Al(s) \rightarrow 2\,Al^{3+}(aq)$ and $3\,Pb^{2+}(aq) \rightarrow 3\,Pb(s)$. Add electrons. $2\,Al(s) \rightarrow 2\,Al^{3+}(aq) + 6\,e^-$ and $3\,Pb^{2+}(aq) + 6\,e^- \rightarrow 3\,Pb(s)$. Simplify the reactions to $Al(s) \rightarrow Al^{3+}(aq) + 3\,e^-$ and $Pb^{2+}(aq) + 2\,e^- \rightarrow Pb(s)$. Look up cell potentials. Al is oxidized, so $E^\circ_{red} = -1.66$ V $= E^\circ_{anode}$. Pb^{2+} is reduced, so $E^\circ_{red} = -0.13$ V $= E^\circ_{cathode}$. Then $E^\circ_{cell} = E^\circ_{cathode} - E^\circ_{anode} = -0.13\text{ V} - (-1.66\text{ V}) = +1.53$ V, so the reaction is nonspontaneous in the reverse direction.

Check: The units (V) are correct. If the voltage is negative, the reaction is spontaneous in the reverse direction.

20.51 For a metal to be able to reduce an ion, it must be below it in Table 20.1 (need positive $E^\circ_{cell} = E^\circ_{cathode} - E^\circ_{anode}$). So we need a metal that is below Mn^{2+} but above Mg^{2+}. Aluminum is the only one in the table that meets those criteria.

20.52 For a metal to be oxidized into an ion by another species, it must be below it in Table 20.1 (need positive $E^\circ_{cell} = E^\circ_{cathode} - E^\circ_{anode}$). So we need a metal that is above Fe^{2+} but below Sn^{2+}. Nickel and cadmium meet those criteria.

20.53 In general, metals whose reduction half-reactions lie below the reduction of H^+ to H_2 in Table 20.1 will dissolve in acids, while metals above it will not. (a) Al and (c) Pb meet that criterion. To write the balanced redox reactions, pair the oxidation of the metal with the reduction of H^+ to $H_2(2\,H^+(aq) + 2\,e^- \rightarrow H_2(g))$. For Al, $Al(s) \rightarrow Al^{3+}(aq) + 3\,e^-$. To balance the number of electrons transferred, we need to multiply the Al reaction by 2 and the H^+ reaction by 3. So $2\,Al(s) \rightarrow 2\,Al^{3+}(aq) + 6\,e^-$ and $6\,H^+(aq) + 6\,e^- \rightarrow 3\,H_2(g)$. Adding the two reactions: $2\,Al(s) + 6\,H^+(aq) + \cancel{6\,e^-} \rightarrow 2\,Al^{3+}(aq) + \cancel{6\,e^-} + 3\,H_2(g)$. Simplify to: $2\,Al(s) + 6\,H^+(aq) \rightarrow 2\,Al^{3+}(aq) + 3\,H_2(g)$. For Pb, $Pb(s) \rightarrow Pb^{2+}(aq) + 2\,e^-$. Because each reaction involves two electrons, we can add the two reactions. $Pb(s) + 2\,H^+(aq) + \cancel{2\,e^-} \rightarrow Pb^{2+}(aq) + \cancel{2\,e^-} + H_2(g)$. Simplify to: $Pb(s) + 2\,H^+(aq) \rightarrow Pb^{2+}(aq) + H_2(g)$.

20.54 In general, metals whose reduction half-reactions lie below the reduction of H^+ to H_2 in Table 20.1 will dissolve in acids, while metals above it will not. Only (b) Fe meets that criterion. To write the balanced redox reactions, pair the oxidation of the metal with the reduction of H^+ to $H_2(2\,H^+(aq) + 2\,e^- \rightarrow H_2(g))$. For Fe, there are two possible reactions, $Fe(s) \rightarrow Fe^{3+}(aq) + 3\,e^-$ and $Fe(s) \rightarrow Fe^{2+}(aq) + 2\,e^-$. Because the second reaction is lower in Table 20.1, the cell potential will be more positive. This means that this reaction will be more spontaneous and thus preferred. Because each reaction involves two electrons, we can add the two reactions. $Fe(s) + 2\,H^+(aq) + \cancel{2\,e^-} \rightarrow Fe^{2+}(aq) + \cancel{2\,e^-} + H_2(g)$. Simplify to: $Fe(s) + 2\,H^+(aq) \rightarrow Fe^{2+}(aq) + H_2(g)$.

20.55 Nitric acid (HNO_3) oxidizes metals through the following reduction half-reaction: $NO_3^-(aq) + 4\,H^+(aq) + 3\,e^- \rightarrow NO(g) + 2\,H_2O(l)$ $E^\circ_{red} = 0.96$ V. Because this half-reaction is above the reduction of H^+ in Table 20.1, HNO_3 can oxidize metals (copper, for example) that cannot be oxidized by HCl. (a) Cu, which is below nitric acid in the table, will be oxidized, but (b) Au, which is above nitric acid in the table (and has a reduction potential of 1.50 V), will not be oxidized. To write the balanced redox reactions, pair the oxidation of the metal with the reduction of nitric acid: $(NO_3^-(aq) + 4\,H^+(aq) + 3\,e^- \rightarrow NO(g) + 2\,H_2O(l))$. For Cu, $Cu(s) \rightarrow Cu^{2+}(aq) + 2\,e^-$. To balance the number of electrons transferred, we need to multiply the Cu reaction by 3 and the nitric acid reaction by 2. So $3\,Cu(s) \rightarrow 3\,Cu^{2+}(aq) + 6\,e^-$ and $2\,NO_3^-(aq) + 8\,H^+(aq) + 6\,e^- \rightarrow 2\,NO(g) + 4\,H_2O(l)$. Adding the two reactions: $3\,Cu(s) + 2\,NO_3^-(aq) + 8\,H^+(aq) + \cancel{6\,e^-} \rightarrow 3\,Cu^{2+}(aq) + \cancel{6\,e^-} + 2\,NO(g) + 4\,H_2O(l)$. Simplify to: $3\,Cu(s) + 2\,NO_3^-(aq) + 8\,H^+(aq) \rightarrow 3\,Cu^{2+}(aq) + 2\,NO(g) + 4\,H_2O(l)$.

20.56 Iodic acid (HIO_3) oxidizes metals through the following reduction half-reaction: $IO_3^-(aq) + 6\,H^+(aq) + 5\,e^- \rightarrow \frac{1}{2}I_2(aq) + 3\,H_2O(l)$ $E^\circ_{red} = 1.20$ V. Because this half-reaction is above the reduction of H^+ in Table 20.1, HIO_3 can oxidize metals (copper, for example) that cannot be oxidized by HCl. (a) Au (which has a reduction potential of 1.50 V) will not be oxidized, but (b) Cr (which has a reduction potential of −0.73 V) will be oxidized by both HCl and HIO_3. To write the balanced redox reactions, pair the oxidation of the metal with the reduction of iodic acid $(IO_3^-(aq) + 6\,H^+(aq) + 5\,e^- \rightarrow \frac{1}{2}I_2(aq) + 3\,H_2O(l))$. For Cr, $Cr(s) \rightarrow Cr^{3+}(aq) + 3\,e^-$. To balance the number of electrons transferred, we need to multiply the Cr reaction by 5 and the iodic acid reaction by 3. So $5\,Cr(s) \rightarrow 5\,Cr^{3+}(aq) + 15\,e^-$ and $3\,IO_3^-(aq) + 18\,H^+(aq) + 15\,e^- \rightarrow 3/2\,I_2(aq) + 9\,H_2O(l)$. Adding the two reactions: $5\,Cr(s) + 3\,IO_3^-(aq) + 18\,H^+(aq) + \cancel{15\,e^-} \rightarrow 5\,Cr^{3+}(aq) + \cancel{15\,e^-} + 3/2\,I_2(aq) + 9\,H_2O(l)$. Simplify to: $5\,Cr(s) + 3\,IO_3^-(aq) + 18\,H^+(aq) \rightarrow 5\,Cr^{3+}(aq) + 3/2\,I_2(aq) + 9\,H_2O(l)$.

20.57 **Given:** overall reactions **Find:** E°_{cell} and spontaneity in forward direction

Conceptual Plan: Separate the overall reaction into two half-cell reactions and add electrons as needed to balance reactions. Look up half-reactions in Table 20.1. Calculate the standard cell potential by subtracting the electrode potential of the anode from the electrode potential of the cathode: $E^\circ_{cell} = E^\circ_{cathode} - E^\circ_{anode}$. If $E^\circ_{cell} > 0$, the reaction is spontaneous in the forward direction.

Solution:

(a) $2\,Cu(s) + Mn^{2+}(aq) \rightarrow 2\,Cu^+(aq) + Mn(s)$ separates to $2\,Cu(s) \rightarrow 2\,Cu^+(aq)$ and $Mn^{2+}(aq) \rightarrow Mn(s)$. Add electrons. $2\,Cu(s) \rightarrow 2\,Cu^+(aq) + 2\,e^-$ and $Mn^{2+}(aq) + 2\,e^- \rightarrow Mn(s)$. Simplify the Cu reaction to $Cu(s) \rightarrow Cu^+(aq) + e^-$. Look up cell potentials. Cu is oxidized, so $E^\circ_{red} = +0.52\text{ V} = E^\circ_{anode}$. Mn^{2+} is reduced, so $E^\circ_{red} = -1.18\text{ V} = E^\circ_{cathode}$. Then $E^\circ_{cell} = E^\circ_{cathode} - E^\circ_{anode} = -1.18\text{ V} - 0.52\text{ V} = -1.70\text{ V}$, so the reaction is nonspontaneous.

(b) $MnO_2(s) + 4\,H^+(aq) + Zn(s) \rightarrow Mn^{2+}(aq) + 2\,H_2O(l) + Zn^{2+}(aq)$ separates to $MnO_2(s) + 4\,H^+(aq) \rightarrow Mn^{2+}(aq) + 2\,H_2O(l)$ and $Zn(s) \rightarrow Zn^{2+}(aq)$. Add electrons. $MnO_2(s) + 4\,H^+(aq) + 2\,e^- \rightarrow Mn^{2+}(aq) + 2\,H_2O(l)$ and $Zn(s) \rightarrow Zn^{2+}(aq) + 2\,e^-$. Look up cell potentials. Zn is oxidized, so $E^\circ_{red} = -0.76\text{ V} = E^\circ_{anode}$. Mn is reduced, so $E^\circ_{red} = 1.21\text{ V} = E^\circ_{cathode}$. Then $E^\circ_{cell} = E^\circ_{cathode} - E^\circ_{anode} = 1.21\text{ V} - (-0.76\text{ V}) = +1.97\text{ V}$, so the reaction is spontaneous.

(c) $Cl_2(g) + 2\,F^-(aq) \rightarrow 2\,Cl^-(aq) + F_2(g)$ separates to $Cl_2(g) \rightarrow 2\,Cl^-(aq)$ and $2\,F^-(aq) \rightarrow F_2(g)$. Add electrons. $Cl_2(g) + 2\,e^- \rightarrow 2\,Cl^-(aq)$ and $2\,F^-(aq) \rightarrow F_2(g) + 2\,e^-$. Look up cell potentials. F^- is oxidized, so $E^\circ_{red} = 2.87\text{ V} = E^\circ_{anode}$. Cl is reduced, so $E^\circ_{red} = 1.36\text{ V} = E^\circ_{cathode}$. Then $E^\circ_{cell} = E^\circ_{cathode} - E^\circ_{anode} = 1.36\text{ V} - 2.87\text{ V} = -1.51\text{ V}$, so the reaction is nonspontaneous.

Check: The units (V) are correct. If the voltage is positive, the reaction is spontaneous.

20.58 **Given:** overall reactions **Find:** E°_{cell} and spontaneity in forward direction

Conceptual Plan: Separate the overall reaction into two half-cell reactions and add electrons as needed to balance reactions. Look up half-reactions in Table 20.1. Calculate the standard cell potential by subtracting the electrode potential of the anode from the electrode potential of the cathode: $E^\circ_{cell} = E^\circ_{cathode} - E^\circ_{anode}$. If $E^\circ_{cell} > 0$, the reaction is spontaneous in the forward direction.

Solution:

(a) $O_2(g) + 2\,H_2O(l) + 4\,Ag(s) \rightarrow 4\,OH^-(aq) + 4\,Ag^+(aq)$ separates to $O_2(g) + 2\,H_2O(l) \rightarrow 4\,OH^-(aq)$ and $4\,Ag(s) \rightarrow 4\,Ag^+(aq)$. Add electrons. $O_2(g) + 2\,H_2O(l) + 4\,e^- \rightarrow 4\,OH^-(aq)$ and $4\,Ag(s) \rightarrow 4\,Ag^+(aq) + 2\,e^-$. Simplify the Ag reaction to $Ag(s) \rightarrow Ag^+(aq) + e^-$. Look up cell potentials. Ag is oxidized, so $E^\circ_{red} = +0.80\text{ V} = E^\circ_{anode}$. O is reduced, so $E^\circ_{red} = 0.40\text{ V} = E^\circ_{cathode}$. Then $E^\circ_{cell} = E^\circ_{cathode} - E^\circ_{anode} = 0.40\text{ V} - 0.80\text{ V} = -0.40\text{ V}$, so the reaction is nonspontaneous.

(b) $Br_2(l) + 2\,I^-(aq) \rightarrow 2\,Br^-(aq) + I_2(g)$ separates to $Br_2(g) \rightarrow 2\,Br^-(aq)$ and $2\,I^-(aq) \rightarrow I_2(g)$. Add electrons. $Br_2(g) + 2\,e^- \rightarrow 2\,Br^-(aq)$ and $2\,I^-(aq) \rightarrow I_2(g) + 2\,e^-$. Look up cell potentials. I is oxidized, so $E^\circ_{red} = 0.54\text{ V} = E^\circ_{anode}$. Br is reduced, so $E^\circ_{red} = 1.09\text{ V} = E^\circ_{cathode}$. Then $E^\circ_{cell} = E^\circ_{cathode} - E^\circ_{anode} = 1.09\text{ V} - 0.54\text{ V} = +0.55\text{ V}$, so the reaction is spontaneous.

(c) $PbO_2(s) + 4\,H^+(aq) + Sn(s) \rightarrow Pb^{2+}(aq) + 2\,H_2O(l) + Sn^{2+}(aq)$ separates to $PbO_2(s) + 4\,H^+(aq) \rightarrow Pb^{2+}(aq) + 2\,H_2O(l)$ and $Sn(s) \rightarrow Sn^{2+}(aq)$. Add electrons. $PbO_2(s) + 4\,H^+(aq) + 2\,e^- \rightarrow Pb^{2+}(aq) + 2\,H_2O(l)$ and $Sn(s) \rightarrow Sn^{2+}(aq) + 2\,e^-$. Look up cell potentials. Sn is oxidized, so $E^\circ_{red} = -0.14\text{ V} = E^\circ_{anode}$. Pb is reduced, so $E^\circ_{red} = 1.46\text{ V} = E^\circ_{cathode}$. Then $E^\circ_{cell} = E^\circ_{cathode} - E^\circ_{anode} = 1.46\text{ V} - (-0.14\text{ V}) = +1.60\text{ V}$, so the reaction is spontaneous.

Check: The units (V) are correct. If the voltage is positive, the reaction is spontaneous.

20.59 (a) Pb^{2+}. The strongest oxidizing agent is the one with the reduction reaction that is closest to the top of Table 20.1 (most positive, least negative reduction potential).

20.60 (b) Al. The strongest reducing agent is the one with the reduction reaction that yields the metal that is closest to the bottom of Table 20.1 (most negative, least positive reduction potential).

Cell Potential, Free Energy, and the Equilibrium Constant

20.61 **Given:** overall reactions **Find:** ΔG°_{rxn} and spontaneity in forward direction

Conceptual Plan: Separate the overall reaction into two half-cell reactions and add electrons as needed to balance reactions. Look up half-reactions in Table 20.1. Calculate the standard cell potential by subtracting the

electrode potential of the anode from the electrode potential of the cathode: $E^\circ_{cell} = E^\circ_{cathode} - E^\circ_{anode}$. Then calculate ΔG°_{rxn} using $\Delta G^\circ_{rxn} = -nF\,E^\circ_{cell}$.

Solution:

(a) $Pb^{2+}(aq) + Mg(s) \rightarrow Pb(s) + Mg^{2+}(aq)$ separates to $Pb^{2+}(aq) \rightarrow Pb(s)$ and $Mg(s) \rightarrow Mg^{2+}(aq)$. Add electrons. $Pb^{2+}(aq) + 2\,e^- \rightarrow Pb(s)$ and $Mg(s) \rightarrow Mg^{2+}(aq) + 2\,e^-$. Look up cell potentials. Mg is oxidized, so $E^\circ_{red} = -2.37\text{ V} = E^\circ_{anode}$. Pb^{2+} is reduced, so $E^\circ_{red} = -0.13\text{ V} = E^\circ_{cathode}$. Then $E^\circ_{cell} = E^\circ_{cathode} - E^\circ_{anode} = -0.13\text{ V} - (-2.37\text{ V}) = +2.24\text{ V}$. $n = 2$, so $\Delta G^\circ_{rxn} = -nF\,E^\circ_{cell} =$ $-2\,\cancel{\text{mol e}^-} \times \dfrac{96{,}485\text{ C}}{\cancel{\text{mol e}^-}} \times 2.24\text{ V} = -2 \times 96{,}485\cancel{\text{C}} \times 2.24\dfrac{\text{J}}{\cancel{\text{C}}} = -4.32 \times 10^5\text{ J} = -432\text{ kJ}.$

(b) $Br_2(l) + 2\,Cl^-(aq) \rightarrow 2\,Br^-(aq) + Cl_2(g)$ separates to $Br_2(g) \rightarrow 2\,Br^-(aq)$ and $2\,Cl^-(aq) \rightarrow Cl_2(g)$. Add electrons. $Br_2(g) + 2\,e^- \rightarrow 2\,Br^-(aq)$ and $2\,Cl^-(aq) \rightarrow Cl_2(g) + 2\,e^-$. Look up cell potentials. Cl is oxidized, so $E^\circ_{red} = 1.36\text{ V} = E^\circ_{anode}$. Br is reduced, so $E^\circ_{red} = 1.09\text{ V} = E^\circ_{cathode}$. Then $E^\circ_{cell} = E^\circ_{cathode} - E^\circ_{anode} = 1.09\text{ V} - 1.36\text{ V} = -0.27\text{ V}$. $n = 2$, so $\Delta G^\circ_{rxn} = -nF\,E^\circ_{cell} = -2\,\cancel{\text{mol e}^-}$ $\times \dfrac{96{,}485\text{ C}}{\cancel{\text{mol e}^-}} \times -0.27\text{ V} = -2 \times 96{,}485\,\cancel{\text{C}} \times -0.27\dfrac{\text{J}}{\cancel{\text{C}}} = 5.2 \times 10^4\text{ J} = 52\text{ kJ}.$

(c) $MnO_2(s) + 4\,H^+(aq) + Cu(s) \rightarrow Mn^{2+}(aq) + 2\,H_2O(l) + Cu^{2+}(aq)$ separates to $MnO_2(s) + 4\,H^+(aq) \rightarrow Mn^{2+}(aq) + 2\,H_2O(l)$ and $Cu(s) \rightarrow Cu^{2+}(aq)$. Add electrons. $MnO_2(s) + 4\,H^+(aq) + 2\,e^- \rightarrow Mn^{2+}(aq) + 2\,H_2O(l)$ and $Cu(s) \rightarrow Cu^{2+}(aq) + 2\,e^-$. Look up cell potentials. Cu is oxidized, so $E^\circ_{red} = 0.34\text{ V} = E^\circ_{anode}$. Mn is reduced, so $E^\circ_{red} = 1.21\text{ V} = E^\circ_{cathode}$. Then $E^\circ_{cell} = E^\circ_{cathode} - E^\circ_{anode} = 1.21\text{ V} - 0.34\text{ V} = +0.87\text{ V}$. $n = 2$, so $\Delta G^\circ_{rxn} = -nF\,E^\circ_{cell} = -2\,\cancel{\text{mol e}^-} \times \dfrac{96{,}485\text{ C}}{\cancel{\text{mol e}^-}} \times 0.87\text{ V} = -2 \times 96{,}485\,\cancel{\text{C}} \times 0.87\dfrac{\text{J}}{\cancel{\text{C}}} = -1.7 \times 10^5\text{ J} =$ $-1.7 \times 10^2\text{ kJ}.$

Check: The units (kJ) are correct. If the voltage is positive, the reaction is spontaneous and the free energy change is negative.

20.62 **Given:** overall reactions **Find:** ΔG°_{rxn} and spontaneity in forward direction

Conceptual Plan: Separate the overall reaction into two half-cell reactions and add electrons as needed to balance reactions. Look up half-reactions in Table 20.1. Calculate the standard cell potential by subtracting the electrode potential of the anode from the electrode potential of the cathode: $E^\circ_{cell} = E^\circ_{cathode} - E^\circ_{anode}$. Then calculate ΔG°_{rxn} using $\Delta G^\circ_{rxn} = -nF\,E^\circ_{cell}$.

Solution:

(a) $2\,Fe^{3+}(aq) + 3\,Sn(s) \rightarrow 2\,Fe(s) + 3\,Sn^{2+}(aq)$ separates to $2\,Fe^{3+}(aq) \rightarrow 2\,Fe(s)$ and $3\,Sn(s) \rightarrow 3\,Sn^{2+}(aq)$. Add electrons. $2\,Fe^{3+}(aq) + 6\,e^- \rightarrow 2\,Fe(s)$ and $3\,Sn(s) \rightarrow 3\,Sn^{2+}(aq) + 6\,e^-$. Simplify reactions to $Fe^{3+}(aq) + 3\,e^- \rightarrow Fe(s)$ and $Sn(s) \rightarrow Sn^{2+}(aq) + 2\,e^-$. Look up cell potentials. Sn is oxidized, so $E^\circ_{red} = -0.14\text{ V} = E^\circ_{anode}$. Fe^{3+} is reduced, so $E^\circ_{red} = -0.036\text{ V} = E^\circ_{cathode}$. Then $E^\circ_{cell} = E^\circ_{cathode} - E^\circ_{anode} = -0.036\text{ V} - (-0.14\text{ V}) = +0.104\text{ V}$. $n = 6$, so $\Delta G^\circ_{rxn} = -nF\,E^\circ_{cell} =$ $-6\,\cancel{\text{mol e}^-} \times \dfrac{96{,}485\text{ C}}{\cancel{\text{mol e}^-}} \times 0.1\underline{0}4\text{ V} = -6 \times 96,485\,\cancel{\text{C}} \times 0.1\underline{0}4\dfrac{\text{J}}{\cancel{\text{C}}} = -6.0 \times 10^4\text{ J} = -6.0 \times 10^1\text{ kJ}.$

(b) $O_2(g) + 2\,H_2O(l) + 2\,Cu(s) \rightarrow 4\,OH^-(aq) + 2\,Cu^{2+}(aq)$ separates to $O_2(g) + 2\,H_2O(l) \rightarrow 4\,OH^-(aq)$ and $2\,Cu(s) \rightarrow 2\,Cu^{2+}(aq)$. Add electrons. $O_2(g) + 2\,H_2O(l) + 4\,e^- \rightarrow 4\,OH^-(aq)$ and $2\,Cu(s) \rightarrow 2Cu^{2+}(aq) + 4e^-$. Simplify the Cu reaction to $Cu(s) \rightarrow Cu^{2+}(aq) + 2\,e^-$. Look up cell potentials. Cu is oxidized, so $E^\circ_{red} = 0.34\text{ V} = E^\circ_{anode}$. O is reduced, so $E^\circ_{red} = 0.40\text{ V} = E^\circ_{cathode}$. Then $E^\circ_{cell} = E^\circ_{cathode} - E^\circ_{anode} = 0.40\text{ V} - 0.34\text{ V} = +0.06\text{ V}$. $n = 4$, so $\Delta G^\circ_{rxn} = -nF\,E^\circ_{cell} = -4\,\cancel{\text{mol e}^-} \times \dfrac{96,485\text{ C}}{\cancel{\text{mol e}^-}} \times 0.06\text{ V} = -4 \times 96,485\,\cancel{\text{C}} \times 0.06\dfrac{\text{J}}{\cancel{\text{C}}} = -2 \times 10^4\text{ J} =$ $-2 \times 10^1\text{ kJ}.$

(c) $Br_2(l) + 2\,I^-(aq) \rightarrow 2\,Br^-(aq) + I_2(g)$ separates to $Br_2(g) \rightarrow 2\,Br^-(aq)$ and $2\,I^-(aq) \rightarrow I_2(g)$. Add electrons. $Br_2(g) + 2\,e^- \rightarrow 2\,Br^-(aq)$ and $2\,I^-(aq) \rightarrow I_2(g) + 2\,e^-$. Look up cell potentials. I is oxidized, so $E^\circ_{red} = +0.54\text{ V} = E^\circ_{anode}$. Br is reduced, so $E^\circ_{red} = 1.09\text{ V} = E^\circ_{cathode}$. Then $E^\circ_{cell} = E^\circ_{cathode} - E^\circ_{anode} = 1.09\text{ V} - 0.54\text{ V} = +0.55\text{ V}$. $n = 2$, so $\Delta G^\circ_{rxn} = -nF\,E^\circ_{cell} =$ $-2\,\cancel{\text{mol e}^-} \times \dfrac{96,485\text{ C}}{\cancel{\text{mol e}^-}} \times 0.55\text{ V} = -2 \times 96,485\,\cancel{\text{C}} \times 0.55\dfrac{\text{J}}{\cancel{\text{C}}} = -1.1 \times 10^5\text{ J} = -1.1 \times 10^2\text{ kJ}.$

Check: The units (kJ) are correct. If the voltage is positive, the reaction is spontaneous and the free energy change is negative.

20.63 **Given:** overall reactions from Problem 20.61 **Find:** K

Conceptual Plan: °C → K then ΔG°_{rxn}, T = K

$K = 273.15 + °C \qquad \Delta G^\circ_{rxn} = -RT \ln K$

Solution: T = 273.15 + 25 °C = 298 K then

(a) $\Delta G^\circ_{rxn} = -RT \ln K$. Rearrange to solve for K.

$$K = e^{\frac{-\Delta G^\circ_{rxn}}{RT}} = e^{\frac{-(-432\ \text{kJ}) \times \frac{1000\ \text{J}}{1\ \text{kJ}}}{\left(8.314 \frac{\text{J}}{\text{K}\cdot\text{mol}}\right)(298\ \text{K})}} = e^{174.364} = 5.31 \times 10^{75}$$

(b) $\Delta G^\circ_{rxn} = -RT \ln K$. Rearrange to solve for K.

$$K = e^{\frac{-\Delta G^\circ_{rxn}}{RT}} = e^{\frac{-(-52\ \text{kJ}) \times \frac{1000\ \text{J}}{1\ \text{kJ}}}{\left(8.314 \frac{\text{J}}{\text{K}\cdot\text{mol}}\right)(298\ \text{K})}} = e^{-20.998} = 7.77 \times 10^{-10}$$

(c) $\Delta G^\circ_{rxn} = -RT \ln K$. Rearrange to solve for K.

$$K = e^{\frac{-\Delta G^\circ_{rxn}}{RT}} = e^{\frac{-(-170\ \text{kJ}) \times \frac{1000\ \text{J}}{1\ \text{kJ}}}{\left(8.314 \frac{\text{J}}{\text{K}\cdot\text{mol}}\right)(298\ \text{K})}} = e^{68.616} = 6.3 \times 10^{29}$$

Check: The units (none) are correct. If the voltage is positive, the reaction is spontaneous and the free energy change is negative and the equilibrium constant is large.

20.64 **Given:** overall reactions from Problem 20.62 **Find:** K

Conceptual Plan: °C → K then ΔG°_{rxn}, T = K

$K = 273.15 + °C \qquad \Delta G^\circ_{rxn} = -RT \ln K$

Solution: T = 273.15 + 25°C = 298 K then

(a) $\Delta G^\circ_{rxn} = -RT \ln K$. Rearrange to solve for K.

$$K = e^{\frac{-\Delta G^\circ_{rxn}}{RT}} = e^{\frac{-(-60\ \text{kJ}) \times \frac{1000\ \text{J}}{1\ \text{kJ}}}{\left(8.314 \frac{\text{J}}{\text{K}\cdot\text{mol}}\right)(298\ \text{K})}} = e^{24.217} = 3.3 \times 10^{10}$$

(b) $\Delta G^\circ_{rxn} = -RT \ln K$. Rearrange to solve for K.

$$K = e^{\frac{-\Delta G^\circ_{rxn}}{RT}} = e^{\frac{-(-20\ \text{kJ}) \times \frac{1000\ \text{J}}{1\ \text{kJ}}}{\left(8.314 \frac{\text{J}}{\text{K}\cdot\text{mol}}\right)(298\ \text{K})}} = e^{8.072} = 3 \times 10^{3}$$

(c) $\Delta G^\circ_{rxn} = -RT \ln K$. Rearrange to solve for K.

$$K = e^{\frac{-\Delta G^\circ_{rxn}}{RT}} = e^{\frac{-(-110\ \text{kJ}) \times \frac{1000\ \text{J}}{1\ \text{kJ}}}{\left(8.314 \frac{\text{J}}{\text{K}\cdot\text{mol}}\right)(298\ \text{K})}} = e^{44.398} = 1.9 \times 10^{19}$$

Check: The units (none) are correct. If the voltage is positive, the reaction is spontaneous and the free energy change is negative and the equilibrium constant is large.

20.65 **Given:** $Ni^{2+}(aq) + Cd(s) \rightarrow$ **Find:** K

Conceptual Plan: Write two half-cell reactions and add electrons as needed to balance the reactions. Look up half-reactions in Table 20.1. Calculate the standard cell potential by subtracting the electrode potential of the anode from the electrode potential of the cathode: $E^\circ_{cell} = E^\circ_{cathode} - E^\circ_{anode}$, then °C → K then E°_{cell}, n, T → K.

$K = 273.15 + °C \qquad \Delta G^\circ_{rxn} = -RT \ln K = -nFE^\circ_{cell}$

Solution: $Ni^{2+}(aq) + 2\,e^- \rightarrow Ni(s)$ and $Cd(s) \rightarrow Cd^{2+}(aq) + 2\,e^-$. Look up cell potentials. Cd is oxidized, so $E^\circ_{red} = -0.40\text{ V} = E^\circ_{anode}$. Ni^{2+} is reduced, so $E^\circ_{red} = -0.23\text{ V} = E^\circ_{cathode}$. Then $E^\circ_{cell} = E^\circ_{cathode} - E^\circ_{anode} = -0.23\text{ V} - (-0.40\text{ V}) = +0.17\text{ V}$. The overall reaction is $Ni^{2+}(aq) + Cd(s) \rightarrow Ni(s) + Cd^{2+}(aq)$. $n = 2$ and $T = 273.15 + 25\,°C = 298\text{ K}$ then $\Delta G^\circ_{rxn} = -RT\ln K = -nFE^\circ_{cell}$. Rearrange to solve for K.

$$K = e^{\frac{nFE^\circ_{cell}}{RT}} = e^{\frac{2\,\text{mol e}^- \times \frac{96{,}485\,\text{C}}{\text{mol e}^-} \times 0.17\frac{\text{J}}{\text{C}}}{\left(8.314\frac{\text{J}}{\text{K}\cdot\text{mol}}\right)(298\,\text{K})}} = e^{13.\underline{2}41} = 5.6 \times 10^5$$

Check: The units (none) are correct. If the voltage is positive, the reaction is spontaneous and the equilibrium constant is large.

20.66 **Given:** $Fe^{2+}(aq) + Zn(s) \rightarrow$ **Find:** K

Conceptual Plan: Write two half-cell reactions and add electrons as needed to balance the reactions. Look up half-reactions in Table 20.1. Calculate the standard cell potential by subtracting the electrode potential of the anode from the electrode potential of the cathode: $E^\circ_{cell} = E^\circ_{cathode} - E^\circ_{anode}$, then °C → K then $E^\circ_{cell}, n, T \rightarrow K$.

$K = 273.15 + °C$ $\quad \Delta G^\circ_{rxn} = -RT\ln K = -nFE^\circ_{cell}$

Solution: $Fe^{2+}(aq) + 2\,e^- \rightarrow Fe(s)$ and $Zn(s) \rightarrow Zn^{2+}(aq) + 2\,e^-$. Look up cell potentials. Zn is oxidized, so $E^\circ_{red} = -0.76\text{ V} = E^\circ_{anode}$. Fe^{2+} is reduced, so $E^\circ_{red} = -0.45\text{ V} = E^\circ_{cathode}$. Then $E^\circ_{cell} = E^\circ_{cathode} - E^\circ_{anode} = -0.45\text{ V} - (-0.76\text{ V}) = +0.31\text{ V}$. The overall reaction is $Fe^{2+}(aq) + Zn(s) \rightarrow Fe(s) + Zn^{2+}(aq)$. $n = 2$ and $T = 273.15 + 25°C = 298\text{ K}$ then $\Delta G^\circ_{rxn} = -RT\ln K = -nFE^\circ_{cell}$. Rearrange to solve for K.

$$K = e^{\frac{nFE^\circ_{cell}}{RT}} = e^{\frac{2\,\text{mol e}^- \times \frac{96{,}485\,\text{C}}{\text{mol e}^-} \times 0.31\frac{\text{J}}{\text{C}}}{\left(8.314\frac{\text{J}}{\text{K}\cdot\text{mol}}\right)(298\,\text{K})}} = e^{24.\underline{1}45} = 3.1 \times 10^{10}$$

Check: The units (none) are correct. If the voltage is positive, the reaction is spontaneous and the equilibrium constant is large.

20.67 **Given:** $n = 2$ and $K = 25$ **Find:** ΔG°_{rxn} and E°_{cell}

Conceptual Plan: $K, T \rightarrow \Delta G^\circ_{rxn}$ and $\Delta G^\circ_{rxn}, n \rightarrow E^\circ_{cell}$

$\Delta G^\circ_{rxn} = -RT\ln K$ $\quad \Delta G^\circ_{rxn} = -nFE^\circ_{cell}$

Solution: $\Delta G^\circ_{rxn} = -RT\ln K = -\left(8.314\dfrac{\text{J}}{\text{K}\cdot\text{mol}}\right)(298\,\text{K})\ln 25 = -7.\underline{9}7500 \times 10^3\text{ J} = -8.0\text{ kJ}$ and

$\Delta G^\circ_{rxn} = -nFE^\circ_{cell}$. Rearrange to solve for E°_{cell}.

$$E^\circ_{cell} = \frac{\Delta G^\circ_{rxn}}{-nF} = \frac{-7.\underline{9}7500 \times 10^3\text{ J}}{-2\,\text{mol e}^- \times \frac{96{,}485\text{ C}}{\text{mol e}^-}} = 0.041\frac{\text{V}\cdot\text{C}}{\text{C}} = 0.041\text{ V}$$

Check: The units (kJ and V) are correct. If $K > 1$, the voltage is positive and the free energy change is negative.

20.68 **Given:** $n = 3$ and $K = 0.050$ **Find:** ΔG°_{rxn} and E°_{cell}

Conceptual Plan: $K, T \rightarrow \Delta G^\circ_{rxn}$ and $\Delta G^\circ_{rxn}, n \rightarrow E^\circ_{cell}$

$\Delta G^\circ_{rxn} = -RT\ln K$ $\quad \Delta G^\circ_{rxn} = -nFE^\circ_{cell}$

Solution: $\Delta G^\circ_{rxn} = -RT\ln K = -\left(8.314\dfrac{\text{J}}{\text{K}\cdot\text{mol}}\right)(298\,\text{K})\ln 0.050 = 7.\underline{4}221 \times 10^3\text{ J} = 7.4\text{ kJ}$ and

$\Delta G^\circ_{rxn} = -nFE^\circ_{cell}$. Rearrange to solve for E°_{cell}.

$$E^\circ_{cell} = \frac{\Delta G^\circ_{rxn}}{-nF} = \frac{7.\underline{4}221 \times 10^3\text{ J}}{-3\,\text{mol e}^- \times \frac{96{,}485\text{ C}}{\text{mol e}^-}} = -0.026\frac{\text{V}\cdot\text{C}}{\text{C}} = -0.026\text{ V}$$

Check: The units (kJ and V) are correct. If $K < 1$, the voltage is negative and the free energy change is positive.

Nonstandard Conditions and the Nernst Equation

20.69 **Given:** $Sn^{2+}(aq) + Mn(s) \rightarrow Sn(s) + Mn^{2+}(aq)$
Find: (a) E°_{cell}; (b) E_{cell} when $[Sn^{2+}] = 0.0100$ M; $[Mn^{2+}] = 2.00$ M; and (c) E_{cell} when $[Sn^{2+}] = 2.00$ M; $[Mn^{2+}] = 0.0100$ M
Conceptual Plan: (a) Separate the overall reaction into two half-cell reactions and add electrons as needed to balance the reactions. Look up half-reactions in Table 20.1. Calculate the standard cell potential by subtracting the electrode potential of the anode from the electrode potential of the cathode: $E^\circ_{cell} = E^\circ_{cathode} - E^\circ_{anode}$.
(b) and (c) E°_{cell}, $[Sn^{2+}]$, $[Mn^{2+}]$, $n \rightarrow E_{cell}$

$$E_{cell} = E^\circ_{cell} - \frac{0.0592\text{ V}}{n}\log Q \quad \text{where } Q = \frac{[Mn^{2+}]}{[Sn^{2+}]}$$

Solution:

(a) Separate the overall reaction to $Sn^{2+}(aq) \rightarrow Sn(s)$ and $Mn(s) \rightarrow Mn^{2+}(aq)$. Add electrons. $Sn^{2+}(aq) + 2\,e^- \rightarrow Sn(s)$ and $Mn(s) \rightarrow Mn^{2+}(aq) + 2\,e^-$. Look up cell potentials. Mn is oxidized, so $E^\circ_{red} = -1.18\text{ V} = E^\circ_{anode}$. Sn^{2+} is reduced, so $E^\circ_{red} = -0.14\text{ V} = E^\circ_{cathode}$. Then $E^\circ_{cell} = E^\circ_{cathode} - E^\circ_{anode} = -0.14\text{ V} - (-1.18\text{ V}) = +1.04\text{ V}$.

(b) $$Q = \frac{[Mn^{2+}]}{[Sn^{2+}]} = \frac{2.00\text{ M}}{0.0100\text{ M}} = 200. \text{ and } n = 2 \text{ then}$$

$$E_{cell} = E^\circ_{cell} - \frac{0.0592\text{ V}}{n}\log Q = 1.04\text{ V} - \frac{0.0592\text{ V}}{2}\log 200. = +0.97\text{ V}$$

(c) $$Q = \frac{[Mn^{2+}]}{[Sn^{2+}]} = \frac{0.0100\text{ M}}{2.00\text{ M}} = 0.00500 \text{ and } n = 2 \text{ then}$$

$$E_{cell} = E^\circ_{cell} - \frac{0.0592\text{ V}}{n}\log Q = 1.04\text{ V} - \frac{0.0592\text{ V}}{2}\log 0.00500 = +1.11\text{ V}$$

Check: The units (V, V, and V) are correct. The Sn^{2+} reduction reaction is above the Mn^{2+} reduction reaction, so the standard cell potential will be positive. Having more products than reactants reduces the cell potential. Having more reactants than products raises the cell potential.

20.70 **Given:** $2\,Fe^{3+}(aq) + 3\,Mg(s) \rightarrow 2\,Fe(s) + 3\,Mg^{2+}(aq)$ **Find:** (a) E°_{cell}; (b) E_{cell} when $[Fe^{3+}] = 1.0 \times 10^{-3}$ M; $[Mg^{2+}] = 2.50$ M; and (c) E_{cell} when $[Fe^{3+}] = 2.00$ M; $[Mg^{2+}] = 1.5 \times 10^{-3}$ M
Conceptual Plan: (a) Separate the overall reaction into two half-cell reactions and add electrons as needed to balance reactions. Look up half-reactions in Table 20.1. Calculate the standard cell potential by subtracting the electrode potential of the anode from the electrode potential of the cathode: $E^\circ_{cell} = E^\circ_{cathode} - E^\circ_{anode}$. (b) and (c) E°_{cell}, $[Fe^{3+}]$, $[Mg^{2+}]$, $n \rightarrow E_{cell}$

$$E_{cell} = E^\circ_{cell} - \frac{0.0592\text{ V}}{n}\log Q \quad \text{where } Q = \frac{[Mg^{2+}]^3}{[Fe^{3+}]^2}$$

Solution:

(a) Separate the overall reaction to $2\,Fe^{3+}(aq) \rightarrow 2\,Fe(s)$ and $3\,Mg(s) \rightarrow 3\,Mg^{2+}(aq)$. Add electrons. $2\,Fe^{3+} + 6\,e^-(aq) \rightarrow 2\,Fe(s)$ and $3\,Mg(s) \rightarrow 3\,Mg^{2+}(aq) + 6\,e^-$. Look up cell potentials. Mg is oxidized, so $E^\circ_{red} = -2.37\text{ V} = E^\circ_{anode}$. Fe^{3+} is reduced, so $E^\circ_{red} = -0.036\text{ V} = E^\circ_{cathode}$. Then $E^\circ_{cell} = E^\circ_{cathode} - E^\circ_{anode} = -0.036\text{ V} - (-2.37\text{ V}) = +2.3\underline{3}4\text{ V} = +2.33\text{ V}$.

(b) $$Q = \frac{[Mg^{2+}]^3}{[Fe^{3+}]^2} = \frac{(2.50)^3}{(1.0 \times 10^{-3})^2} = 1.\underline{5}625 \times 10^7 \text{ and } n = 6 \text{ then}$$

$$E_{cell} = E^\circ_{cell} - \frac{0.0592\text{ V}}{n}\log Q = 2.3\underline{3}4\text{ V} - \frac{0.0592\text{ V}}{6}\log 1.\underline{5}625 \times 10^7 = +2.26\text{ V}$$

(c) $$Q = \frac{[Mg^{2+}]^3}{[Fe^{3+}]^2} = \frac{(1.5 \times 10^{-3})^3}{(2.00)^2} = 8.\underline{4}375 \times 10^{-10} \text{ and } n = 6 \text{ then}$$

$$E_{cell} = E^\circ_{cell} - \frac{0.0592\text{ V}}{n}\log Q = 2.3\underline{3}4\text{ V} - \frac{0.0592\text{ V}}{6}\log 8.\underline{4}375 \times 10^{-10} = +2.42\text{ V}$$

Check: The units (V, V, and V) are correct. The Fe^{3+} reduction reaction is above the Mg^{2+} reduction reaction, so the standard cell potential will be positive. Having more products than reactants reduces the cell potential. Having more reactants than products raises the cell potential.

20.71 **Given:** $Pb(s) \rightarrow Pb^{2+}(aq, 0.10\text{ M}) + 2\,e^-$ and $MnO_4^-(aq, 1.50\text{ M}) + 4\,H^+(aq, 2.0\text{ M}) + 3\,e^- \rightarrow MnO_2(s) + 2\,H_2O(l)$ **Find:** E_{cell}

Conceptual Plan: Look up half-reactions in Table 20.1. Calculate the standard cell potential by subtracting the electrode potential of the anode from the electrode potential of the cathode: $E^\circ_{cell} = E^\circ_{cathode} - E^\circ_{anode}$. Equalize the number of electrons transferred by multiplying the first reaction by 3 and the second reaction by 2. Add the two half-cell reactions and cancel the electrons.

Then E°_{cell}, $[Pb^{2+}]$, $[MnO_4^-]$, $[H^+]$, $n \rightarrow E_{cell}$.

$$E_{cell} = E^\circ_{cell} - \frac{0.0592\text{ V}}{n}\log Q \quad \text{where } Q = \frac{[Pb^{2+}]^3}{[MnO_4^-]^2[H^+]^8}$$

Solution: Pb is oxidized, so $E^\circ_{red} = -0.13\text{ V} = E^\circ_{anode}$. Mn is reduced, so $E^\circ_{red} = 1.68\text{ V} = E^\circ_{cathode}$. Then $E^\circ_{cell} = E^\circ_{cathode} - E^\circ_{anode} = 1.68\text{ V} - (-0.13\text{ V}) = +1.81\text{ V}$. Equalizing the electrons: $3\,Pb(s) \rightarrow 3\,Pb^{2+}(aq) + 6\,e^-$ and $2\,MnO_4^-(aq) + 8\,H^+(aq) + 6\,e^- \rightarrow 2\,MnO_2(s) + 4\,H_2O(l)$. Adding the two reactions: $3\,Pb(s) + 2\,MnO_4^-(aq) + 8\,H^+(aq) + \cancel{6e^-} \rightarrow 3\,Pb^{2+}(aq) + \cancel{6e^-} + 2\,MnO_2(s) + 4\,H_2O(l)$. Cancel the electrons: $3\,Pb(s) + 2\,MnO_4^-(aq) + 8\,H^+(aq) \rightarrow 3\,Pb^{2+}(aq) + 2\,MnO_2(s) + 4\,H_2O(l)$. So $n = 6$ and

$$Q = \frac{[Pb^{2+}]^3}{[MnO_4^-]^2[H^+]^8} = \frac{(0.10)^3}{(1.50)^2(2.0)^8} = 1.\underline{7}361 \times 10^{-6}. \text{ Then}$$

$$E_{cell} = E^\circ_{cell} - \frac{0.0592\text{ V}}{n}\log Q = 1.81\text{ V} - \frac{0.0592\text{ V}}{6}\log 1.\underline{7}361 \times 10^{-6} = +1.87\text{ V}.$$

Check: The units (V) are correct. The MnO_4^- reduction reaction is above the Pb^{2+} reduction reaction, so the standard cell potential will be positive. Having more reactants than products raises the cell potential.

20.72 **Given:** $Sn(s) \rightarrow Sn^{2+}(aq, 2.00\text{ M}) + 2\,e^-$ and $ClO_2(g, 0.100\text{ atm}) + e^- \rightarrow ClO_2^-(aq, 2.00\text{ M})$ **Find:** E_{cell}

Conceptual Plan: Look up half-reactions in Table 20.1. Calculate the standard cell potential by subtracting the electrode potential of the anode from the electrode potential of the cathode: $E^\circ_{cell} = E^\circ_{cathode} - E^\circ_{anode}$. Equalize the number of electrons transferred by multiplying the second reaction by 2. Add the two half-cell reactions and cancel the electrons. Then E°_{cell}, $[Sn^{2+}]$, P_{ClO_2}, $[ClO_2^-]$, $n \rightarrow E_{cell}$.

$$E_{cell} = E^\circ_{cell} - \frac{0.0592\text{ V}}{n}\log Q \quad \text{where } Q = \frac{[Sn^{2+}][ClO_2^-]^2}{P^2_{ClO_2}}$$

Solution: Sn is oxidized, so $E^\circ_{red} = -0.14\text{ V} = E^\circ_{anode}$. ClO_2 is reduced, so $E^\circ_{red} = 0.95\text{ V} = E^\circ_{cathode}$. Then $E^\circ_{cell} = E^\circ_{cathode} - E^\circ_{anode} = 0.95\text{ V} - (-0.14\text{ V}) = +1.09\text{ V}$. Equalizing the electrons: $Sn(s) \rightarrow Sn^{2+}(aq) + 2\,e^-$ and $2\,ClO_2(g) + 2\,e^- \rightarrow 2\,ClO_2^-(aq)$. Adding the two reactions: $Sn(s) + 2\,ClO_2(g) + \cancel{2e^-} \rightarrow Sn^{2+}(aq) + \cancel{2e^-} + 2\,ClO_2^-(aq)$. Cancel the electrons:

$$Sn(s) + 2\,ClO_2(g) \rightarrow Sn^{2+}(aq) + 2\,ClO_2^-(aq). \text{ So } n = 2 \text{ and } Q = \frac{[Sn^{2+}][ClO_2^-]^2}{P_{ClO_2}} = \frac{(2.00)(2.00)^2}{(0.100)^2} = 800.$$

$$\text{Then } E_{cell} = E^\circ_{cell} - \frac{0.0592\text{ V}}{n}\log Q = 1.09\text{ V} - \frac{0.0592\text{ V}}{2}\log 800. = +1.00\text{ V}.$$

Check: The units (V) are correct. The ClO_2 reduction reaction is above the Sn^{2+} reduction reaction, so the standard cell potential will be positive. Having more products than reactants lowers the cell potential.

20.73 **Given:** Zn/Zn^{2+} and Ni/Ni^{2+} half-cells in voltaic cell; initially, $[Ni^{2+}] = 1.50\text{ M}$ and $[Zn^{2+}] = 0.100\text{ M}$

Find: (a) initial E_{cell}, (b) E_{cell} when $[Ni^{2+}] = 0.500\text{ M}$, and (c) $[Ni^{2+}]$ and $[Zn^{2+}]$ when $E_{cell} = 0.45\text{ V}$

Conceptual Plan:

(a) **Write two half-cell reactions and add electrons as needed to balance reactions. Look up half-reactions in Table 20.1. Calculate the standard cell potential by subtracting the electrode potential of the anode from the electrode potential of the cathode: $E^\circ_{cell} = E^\circ_{cathode} - E^\circ_{anode}$. Choose the direction of the half-cell reactions so that $E^\circ_{cell} > 0$. Add two half-cell reactions and cancel electrons to generate overall reaction. Define Q based on overall reaction. Then E°_{cell}, $[Ni^{2+}]$, $[Zn^{2+}]$, $n \rightarrow E_{cell}$.**

$$E_{cell} = E^\circ_{cell} - \frac{0.0592\text{ V}}{n}\log Q$$

(b) **When $[Ni^{2+}] = 0.500$ M, $[Zn^{2+}] = 1.100$ M. (Because the stoichiometric coefficients for $Ni^{2+}: Zn^{2+}$ are 1:1 and the $[Ni^{2+}]$ drops by 1.00 M, the other concentration must rise by 1.00 M.) Then $E^\circ_{cell}, [Ni^{2+}], [Zn^{2+}], n \rightarrow E_{cell}$.**

$$E_{cell} = E^\circ_{cell} - \frac{0.0592\ V}{n} \log Q$$

(c) **$E^\circ_{cell}, E_{cell}, n \rightarrow [Zn^{2+}]/[Ni^{2+}] \rightarrow [Ni^{2+}], [Zn^{2+}]$**

$$E_{cell} = E^\circ_{cell} - \frac{0.0592\ V}{n} \log Q \quad [Ni^{2+}] + [Zn^{2+}] = 1.50\ M + 0.100\ M = 1.60\ M$$

Solution:

(a) $Zn^{2+}(aq) + 2\,e^- \rightarrow Zn(s)$ and $Ni^{2+}(aq) + 2\,e^- \rightarrow Ni(s)$. Look up cell potentials. For Zn, $E^\circ_{red} = -0.76$ V. For Ni, $E^\circ_{red} = -0.23$ V. To get a positive E°_{cell}, Zn is oxidized; so $E^\circ_{red} = -0.76\ V = E^\circ_{anode}$. Ni^{2+} is reduced, so $E^\circ_{red} = -0.23\ V = E^\circ_{cathode}$. Then $E^\circ_{cell} = E^\circ_{cathode} - E^\circ_{anode} = -0.23\ V - (-0.76\ V) = +0.53$ V. Adding the two half-cell reactions: $Zn(s) + Ni^{2+}(aq) + \cancel{2e^-} \rightarrow Zn^{2+}(aq) + \cancel{2e^-} + Ni(s)$. The overall reaction is

$Zn(s) + Ni^{2+}(aq) \rightarrow Zn^{2+}(aq) + Ni(s)$. Then $Q = \frac{[Zn^{2+}]}{[Ni^{2+}]} = \frac{0.100}{1.50} = 0.066\underline{6}667$ and $n = 2$. Then

$$E_{cell} = E^\circ_{cell} - \frac{0.0592\ V}{n}\log Q = 0.53\ V - \frac{0.0592\ V}{2}\log 0.066\underline{6}667 = +0.56\ V.$$

(b) $$Q = \frac{[Zn^{2+}]}{[Ni^{2+}]} = \frac{1.100}{0.500} = 2.20 \text{ then}$$

$$E_{cell} = E^\circ_{cell} - \frac{0.0592\ V}{n}\log Q = 0.53\ V - \frac{0.0592\ V}{2}\log 2.20 = +0.52\ V$$

(c) $$E_{cell} = E^\circ_{cell} - \frac{0.0592\ V}{n}\log Q \text{ so } 0.45\ V = 0.53\ V - \frac{0.0592\ V}{2}\log Q \rightarrow 0.08\ \cancel{V} = \frac{0.0592\ \cancel{V}}{2}\log Q \rightarrow$$

$\log Q = 2.\underline{7}0270 \rightarrow Q = 10^{2.\underline{7}0270} = 5\underline{0}4.31$. Then $Q = 5\underline{0}4.31 = \frac{[Zn^{2+}]}{1.60\ M - [Zn^{2+}]}$. Solving for

$[Zn^{2+}]: (5\underline{0}4.31)(1.60\ M - [Zn^{2+}]) = [Zn^{2+}] \rightarrow [Zn^{2+}] = \frac{8\underline{0}6.896\ M}{5\underline{0}5.31} = 1.\underline{5}9683\ M = 1.60$ M then

$[Ni^{2+}] = 1.60\ M - 1.\underline{5}9683\ M = 0.003$ M

Check: The units (V, V, and M) are correct. The standard cell potential is positive, and because there are more reactants than products, this raises the cell potential. As the reaction proceeds, reactants are converted to products; so the cell potential drops for parts (b) and (c).

20.74 **Given:** Pb/Pb^{2+} and Cu/Cu^{2+} half-cells in voltaic cell; initially, $[Pb^{2+}] = 0.0500$ M and $[Cu^{2+}] = 1.50$ M
Find: (a) initial E_{cell}, (b) E_{cell} when $[Cu^{2+}] = 0.200$ M, and (c) $[Pb^{2+}]$ and $[Cu^{2+}]$ when $E_{cell} = 0.35$ V
Conceptual Plan:

(a) **Write two half-cell reactions and add electrons as needed to balance reactions. Look up half-reactions in Table 20.1. Calculate the standard cell potential by subtracting the electrode potential of the anode from the electrode potential of the cathode: $E^\circ_{cell} = E^\circ_{cathode} - E^\circ_{anode}$. Choose the direction of the half-cell reactions so that $E^\circ_{cell} > 0$. Add two half-cell reactions and cancel electrons to generate overall reaction. Define Q based on overall reaction. Then $E^\circ_{cell}, [Pb^{2+}], [Cu^{2+}], n \rightarrow E_{cell}$.**

$$E_{cell} = E^\circ_{cell} - \frac{0.0592\ V}{n} \log Q$$

(b) **When $[Cu^{2+}] = 0.200$ M, $[Pb^{2+}] = 1.35$ M. (Because the stoichiometric coefficients for $Pb^{2+}: Cu^{2+}$ are 1:1 and the $[Cu^{2+}]$ drops by 1.30 M, the other concentration must rise by 1.30 M.) Then $E^\circ_{cell}, [Pb^{2+}], [Cu^{2+}], n \rightarrow E_{cell}$.**

$$E_{cell} = E^\circ_{cell} - \frac{0.0592\ V}{n} \log Q$$

(c) **$E^\circ_{cell}, E_{cell}, n \rightarrow [Pb^{2+}]/[Cu^{2+}] \rightarrow [Pb^{2+}], [Cu^{2+}]$**

$$E_{cell} = E^\circ_{cell} - \frac{0.0592\ V}{n} \log Q \quad [Pb^{2+}] + [Cu^{2+}] = 0.0500\ M + 1.50\ M = 1.55\ M$$

Solution:

(a) $Pb^{2+}(aq) + 2\,e^- \rightarrow Pb(s)$ and $Cu^{2+}(aq) + 2\,e^- \rightarrow Cu(s)$. Look up cell potentials. For Pb, $E^\circ_{red} = -0.13$ V. For Cu, $E^\circ_{red} = +0.34$ V. To get a positive E°_{cell}, Pb is oxidized; so $E^\circ_{red} = -0.13\ V = E^\circ_{anode}$. Cu^{2+} is reduced,

so $E^\circ_{red} = 0.34\text{ V} = E^\circ_{cathode}$. Then $E^\circ_{cell} = E^\circ_{cathode} - E^\circ_{anode} = 0.34\text{ V} - (-0.13\text{ V}) = +0.47\text{ V}$.
Adding the two half-cell reactions: $Pb(s) + Cu^{2+}(aq) + \cancel{2e^-} \rightarrow Pb^{2+}(aq) + \cancel{2e^-} + Cu(s)$. The overall reaction is $Pb(s) + Cu^{2+}(aq) \rightarrow Pb^{2+}(aq) + Cu(s)$. Then $Q = \frac{[Pb^{2+}]}{[Cu^{2+}]} = \frac{0.050}{1.50} = 0.03\underline{3}333$ and $n = 2$.

Then $E_{cell} = E^\circ_{cell} - \frac{0.0592\text{ V}}{n}\log Q = 0.47\text{ V} - \frac{0.0592\text{ V}}{2}\log 0.03\underline{3}333 = +0.51\text{ V}$.

(b) $Q = \frac{[Pb^{2+}]}{[Cu^{2+}]} = \frac{1.35}{0.200} = 6.75$ then

$E_{cell} = E^\circ_{cell} - \frac{0.0592\text{ V}}{n}\log Q = 0.47\text{ V} - \frac{0.0592\text{ V}}{2}\log 6.75 = +0.45\text{ V}$

(c) $E_{cell} = E^\circ_{cell} - \frac{0.0592\text{ V}}{n}\log Q$, so $0.35\text{ V} = 0.47\text{ V} - \frac{0.0592\text{ V}}{2}\log Q \rightarrow 0.12\cancel{\text{V}} = \frac{0.0592\cancel{\text{V}}}{2}\log Q \rightarrow$

$\log Q = 4.\underline{0}541 \rightarrow Q = 10^{4.\underline{0}541} = 1.\underline{1}327 \times 10^4$. Then $Q = 1.\underline{1}327 \times 10^4 = \frac{[Pb^{2+}]}{1.55\text{ M} - [Pb^{2+}]}$. Solving for $[Pb^{2+}]$: $(1.\underline{1}327 \times 10^4)(1.55\text{ M} - [Pb^{2+}]) = [Pb^{2+}] \rightarrow [Pb^{2+}] = \frac{1.\underline{7}557 \times 10^4\text{ M}}{1.\underline{1}328 \times 10^4} =$

$1.\underline{5}499\text{ M} = 1.5\text{ M}$ then $[Cu^{2+}] = 1.55\text{ M} - 1.\underline{5}499\text{ M} = 0.0\underline{0}01\text{ M} = 0.0\text{ M}$

Check: The units (V, V, and M) are correct. The standard cell potential is positive, and because there are more reactants than products, this raises the cell potential. As the reaction proceeds, reactants are converted to products; so the cell potential drops for parts (b) and (c).

20.75 **Given:** Zn/Zn^{2+} concentration cell, with $[Zn^{2+}] = 2.0\text{ M}$ in one half-cell and $[Zn^{2+}] = 1.0 \times 10^{-3}\text{ M}$ in other half-cell

Find: Sketch a voltaic cell, labeling the anode, the cathode, the reactions at electrodes, all species, and the direction of electron flow.

Conceptual Plan: In a concentration cell, the half-cell with the higher concentration is always the half-cell where the reduction takes place (contains the cathode). The two half-cell reactions are the same but reversed. Put anode reaction on the left (oxidation = electrons as product) and cathode reaction on the right (reduction = electrons as reactant). Electrons flow from anode to cathode.

Solution:

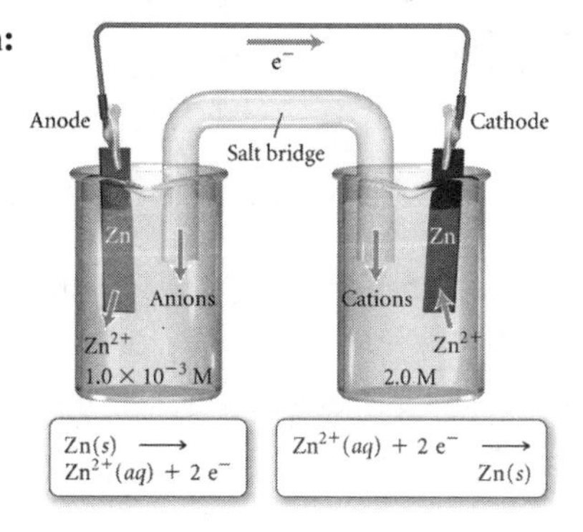

Check: The figure looks similar to the right side of Figure 20.12.

20.76 **Given:** Pb/Pb^{2+} concentration cell sketch

Find: (a) Label the anode and cathode, (b) indicate direction of electron flow, and (c) indicate what happens to $[Pb^{2+}]$ in each half-cell with time.

Conceptual Plan: (a) In a concentration cell, the half-cell with the higher concentration is always the half-cell where the reduction takes place (contains the cathode). (b) Electrons flow from anode to cathode. (c) Each half-cell reaction moves forward, so the direction of the concentration changes in each half-cell can be determined.

Solution:

(c) As the reaction proceeds, the left half-cell (cathode = reduction reaction) will decrease in concentration and the right half-cell (anode = oxidation reaction) will increase in concentration. Eventually, the two concentrations will be the same and the flow of electrons will stop.

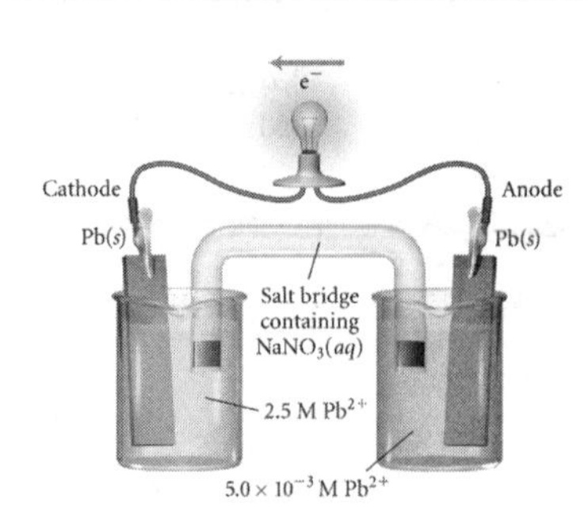

Check: The figure looks similar to the right side of Figure 20.12.

20.77 **Given:** Sn/Sn^{2+} concentration cell with $E_{cell} = 0.10$ V
Find: ratio of $[Sn^{2+}]$ in two half-cells
Conceptual Plan: Determine n, then $E^\circ_{cell}, E_{cell}, n \rightarrow Q$ = ratio of $[Sn^{2+}]$ in two half-cells.

$$E_{cell} = E^\circ_{cell} - \frac{0.0592\ V}{n}\log Q$$

Solution: Because $Sn^{2+}(aq) + 2\,e^- \rightarrow Sn(s)$, $n = 2$. In a concentration cell, $E^\circ_{cell} = 0$ V. So

$$E_{cell} = E^\circ_{cell} - \frac{0.0592\ V}{n}\log Q, \text{ so } 0.10\ V = 0.00V - \frac{0.0592\ V}{2}\log Q \rightarrow 0.10\ \cancel{V} = -\frac{0.0592\ \cancel{V}}{2}\log Q \rightarrow$$

$$\log Q = -3.3784 \rightarrow Q = 10^{-3.3784} = 4.2 \times 10^{-4} = \frac{[Sn^{2+}](ox)}{[Sn^{2+}](red)}.$$

Check: The units (none) are correct. Because the concentration in the reduction reaction half-cell is always greater than the concentration in the oxidation half-cell in a voltaic concentration cell, the Q or ratio of two cells is less than 1.

20.78 **Given:** Cu/Cu^{2+} concentration cell with $E_{cell} = 0.22$ V and $[Cu^{2+}] = 1.5 \times 10^{-3}$ M **Find:** $[Cu^{2+}]$ in other half-cell
Conceptual Plan: Determine n, then $E^\circ_{cell}, E_{cell}, n \rightarrow Q$ = ratio of $[Cu^{2+}]$ in two half-cells then

$$E_{cell} = E^\circ_{cell} - \frac{0.0592\ V}{n}\log Q$$

$[Cu^{2+}]$ one side, $Q \rightarrow [Cu^{2+}]$ other side.

$$Q = \frac{[Cu^{2+}](ox)}{[Cu^{2+}](red)}$$

Solution: Because $Sn^{2+}(aq) + 2e^- \rightarrow Sn(s)$, $n = 2$. In a concentration cell, $E^\circ_{cell} = 0$ V. So

$$E_{cell} = E^\circ_{cell} - \frac{0.0592\ V}{n}\log Q, \text{ so } 0.22\ V = 0.00\ V - \frac{0.0592\ V}{2}\log Q \rightarrow 0.22\ \cancel{V} = -\frac{0.0592\ \cancel{V}}{2}\log Q \rightarrow$$

$$\log Q = -7.4324 \rightarrow Q = 10^{-7.4324} = 3.6949 \times 10^{-8} = \frac{[Cu^{2+}](ox)}{[Cu^{2+}](red)}.$$

$$[Cu^{2+}](red) = \frac{[Cu^{2+}](ox)}{Q} = \frac{1.5 \times 10^{-3}\ M}{3.6949 \times 10^{-8}} = 4.1 \times 10^{4}\ M$$ The problem states that the given concentration is the higher concentration, so $[Cu^{2+}](ox) = [Cu^{2+}](red) \times Q = (1.5 \times 10^{-3}\ M)(3.6949 \times 10^{-8}) = 5.5 \times 10^{-11}$ M, which is low but possible.

Check: The units (M) are correct. Because the voltage is fairly high for a concentration cell, there must be a very small Q, which leads to an extremely low concentration on the oxidation side.

Batteries, Fuel Cells, and Corrosion

20.79 **Given:** alkaline battery **Find:** optimum mass ratio of Zn to MnO_2
Conceptual Plan: Look up alkaline battery reactions. Use stoichiometry to get mole ratio. Then

$$Zn(s) + 2OH^-(aq) \rightarrow Zn(OH)_2(s) + 2e^-$$
$$2\,MnO_2(s) + 2\,H_2O(l) + 2\,e^- \rightarrow 2\,MnO(OH)(s) + 2\,OH^-(aq)$$
$$\frac{1\ mol\ Zn}{2\ mol\ MnO_2}$$

mol Zn → g Zn Zn then mol MnO_2 → g MnO_2.

$$\frac{65.38\ g\ Zn}{1\ mol\ Zn} \qquad \frac{1\ mol\ MnO_2}{86.94\ g\ MnO_2}$$

Solution: $$\frac{1\ \cancel{mol\ Zn}}{2\ \cancel{mol\ MnO_2}} \times \frac{65.38\ g\ Zn}{1\ \cancel{mol\ Zn}} \times \frac{1\ \cancel{mol\ MnO_2}}{86.94\ g\ MnO_2} = 0.3760\ \frac{g\ Zn}{g\ MnO_2}$$

Check: The units (mass ratio) are correct. Because more moles of MnO_2 than zinc are needed and the molar mass is larger, the ratio is less than 1.

20.80 **Given:** lead storage battery, 1.00 g Pb oxidizes **Find:** mass $PbSO_4$

Conceptual Plan: Look up lead–acid battery reactions. Then g Pb → mol Pb → mol $PbSO_4$ → g $PbSO_4$.

$$\frac{1 \text{ mol Pb}}{207.2 \text{ g Pb}} \quad \frac{2 \text{ mol PbSO}_4}{1 \text{ mol Pb}} \quad \frac{303.27 \text{ g PbSO}_4}{1 \text{ mol PbSO}_4}$$

$$Pb(s) + HSO_4^-(aq) \rightarrow PbSO_4(s) + H^+(aq) + 2\,e^-$$

$$PbO_2(s) + HSO_4^-(aq) + 3\,H^+(aq) + 2\,e^- \rightarrow PbSO_4(s) + 2\,H_2O(l)$$

Solution: $1.00 \text{ g Pb} \times \dfrac{1 \text{ mol Pb}}{207.2 \text{ g Pb}} \times \dfrac{2 \text{ mol PbSO}_4}{1 \text{ mol Pb}} \times \dfrac{303.27 \text{ g PbSO}_4}{1 \text{ mol PbSO}_4} = 2.93 \text{ g PbSO}_4$

Check: The units (g) are correct. Because more moles of $PbSO_4$ are generated than moles of Pb consumed and the molar mass is larger, the mass is larger.

20.81 **Given:** $CH_4(g) + 2\,O_2(g) \rightarrow CO_2(g) + 2\,H_2O(g)$ **Find:** E°_{cell}

Conceptual Plan: $\Delta G^\circ_{rxn} = \sum n_p \Delta G^\circ_f(\text{products}) - \sum n_r \Delta G^\circ_f(\text{reactants})$ **and determine** n **then** $\Delta G^\circ_{rxn}, n \rightarrow E^\circ_{cell}$

$$\Delta G^\circ_{rxn} = -nFE^\circ_{cell}$$

Solution:

Reactant/Product	ΔG°_f(kJ/mol from Appendix IVB)
$CH_4(g)$	−50.5
$O_2(g)$	0.0
$CO_2(g)$	−394.4
$H_2O(g)$	−228.6

Be sure to pull data for the correct formula and phase.

$$\begin{aligned}\Delta G^\circ_{rxn} &= \sum n_p \Delta G^\circ_f(\text{products}) - \sum n_r \Delta G^\circ_f(\text{reactants})\\ &= [1(\Delta G^\circ_f(CO_2(g))) + 2(\Delta G^\circ_f(H_2O(g)))] - [1(\Delta G^\circ_f(CH_4(g))) + 2(\Delta G^\circ_f(O_2(g)))]\\ &= [1(-394.4 \text{ kJ}) + 2(-228.6 \text{ kJ})] - [1(-50.5 \text{ kJ}) + 2(0.0 \text{ kJ})]\\ &= [-851.6 \text{ kJ}] - [-50.5 \text{ kJ}]\\ &= -801.1 \text{ kJ} = -8.011 \times 10^5 \text{ J}\end{aligned}$$

Also, because one C atom goes from an oxidation state of −4 to +4 and four O atoms are going from 0 to −2, $n = 8$ and $\Delta G^\circ_{rxn} = -nFE^\circ_{cell}$. Rearrange to solve for E°_{cell}.

$$E^\circ_{cell} = \frac{\Delta G^\circ_{rxn}}{-nF} = \frac{-8.011 \times 10^5 \text{ J}}{-8 \text{ mol e}^- \times \dfrac{96{,}485 \text{ C}}{\text{mol e}^-}} = 1.038 \frac{\text{V} \cdot \text{C}}{\text{C}} = 1.038 \text{ V}$$

Check: The units (V) are correct. The cell voltage is positive, which is consistent with a spontaneous reaction.

20.82 **Given:** $CH_3CH_2OH(g) + O_2(g) \rightarrow HC_2H_3O_2(g) + H_2O(g)$ **Find:** E°_{cell}

Conceptual Plan: $\Delta G^\circ_{rxn} = \sum n_p \Delta G^\circ_f(\text{products}) - \sum n_r \Delta G^\circ_f(\text{reactants})$ **and determine** n **then** $\Delta G^\circ_{rxn}, n \rightarrow E^\circ_{cell}$

$$\Delta G^\circ_{rxn} = -nFE^\circ_{cell}$$

Solution:

Reactant/Product	ΔG°_f(kJ/mol from Appendix IVB)
$CH_3CH_2OH(g)$	−167.9
$O_2(g)$	0.0
$HC_2H_3O_2(g)$	−374.2
$H_2O(g)$	−228.6

Be sure to pull data for the correct formula and phase.

$$\begin{aligned}\Delta G^\circ_{rxn} &= \sum n_p \Delta G^\circ_f(\text{products}) - \sum n_r \Delta G^\circ_f(\text{reactants})\\ &= [1(\Delta G^\circ_f(HC_2H_3O_2(g))) + 1(\Delta G^\circ_f(H_2O(g)))] - [1(\Delta G^\circ_f(CH_3CH_2OH(g))) + 1(\Delta G^\circ_f(O_2(g)))]\\ &= [1(-374.2 \text{ kJ}) + 1(-228.6 \text{ kJ})] - [1(-167.9 \text{ kJ}) + 1(0.0 \text{ kJ})]\end{aligned}$$

$$= [-602.8 \text{ kJ}] - [-167.9 \text{ kJ}]$$
$$= -434.9 \text{ kJ} = -4.349 \times 10^5 \text{ J}$$

Also, because two O atoms are going from 0 to -2, $n = 4$; then $\Delta G^\circ_{rxn} = -n F E^\circ_{cell}$. Rearrange to solve for E°_{cell}.

$$E^\circ_{cell} = \frac{\Delta G^\circ_{rxn}}{-n F} = \frac{-4.349 \times 10^5 \text{ J}}{-4 \cancel{\text{mol e}^-} \times \frac{96,485 \text{ C}}{\cancel{\text{mol e}^-}}} = 1.127 \frac{\text{V} \cdot \cancel{\text{C}}}{\cancel{\text{C}}} = 1.127 \text{ V}$$

Check: The units (V) are correct. The cell voltage is positive, which is consistent with a spontaneous reaction.

20.83 When iron corrodes or rusts, it oxidizes to Fe^{2+}. For a metal to be able to protect iron, it must be more easily oxidized than iron or be below it in Table 20.1. (a) Zn and (c) Mn meet that criterion.

20.84 When iron corrodes or rusts, it oxidizes to Fe^{2+}. For a metal to be able to protect iron, it must be more easily oxidized than iron or be below it in Table 20.1. (a) Mg and (b) Cr meet that criterion.

Electrolytic Cells and Electrolysis

20.85 **Given:** electrolytic cell sketch

Find: (a) Label the anode and cathode and indicate half-reactions, (b) indicate direction of electron flow, and (c) label battery terminals and calculate minimum voltage to drive reaction.

Conceptual Plan: (a) Write two half-cell reactions and add electrons as needed to balance reactions. Look up half-reactions in Table 20.1. Calculate the standard cell potential by subtracting the electrode potential of the anode from the electrode potential of the cathode: $E^\circ_{cell} = E^\circ_{cathode} - E^\circ_{anode}$. Choose the direction of the half-cell reactions so that $E^\circ_{cell} < 0$. (b) Electrons flow from anode to cathode. (c) Each half-cell reaction moves forward, so direction of the concentration changes can be determined.

Solution:

(a) $Ni^{2+}(aq) + 2e^- \rightarrow Ni(s)$ and $Cd^{2+}(aq) + 2e^- \rightarrow Cd(s)$. Look up cell potentials. For Ni, $E^\circ_{red} = -0.23$ V. For Cd, $E^\circ_{red} = -0.40$ V. To get a negative cell potential, Ni is oxidized; so $E^\circ_{red} = -0.23 \text{ V} = E^\circ_{anode}$. Cd^{2+} is reduced, so $E^\circ_{red} = -0.23 \text{ V} = E^\circ_{cathode}$. Then $E^\circ_{cell} = E^\circ_{cathode} - E^\circ_{anode} = -0.40 \text{ V} - (-0.23 \text{ V}) = -0.17$ V. Because oxidation occurs at the anode, Ni is the anode and the reaction is $Ni(s) \rightarrow Ni^{2+}(aq) + 2e^-$. Because reduction takes place at the cathode, Cd is the cathode and the reaction is $Cd^{2+}(aq) + 2e^- \rightarrow Cd(s)$.

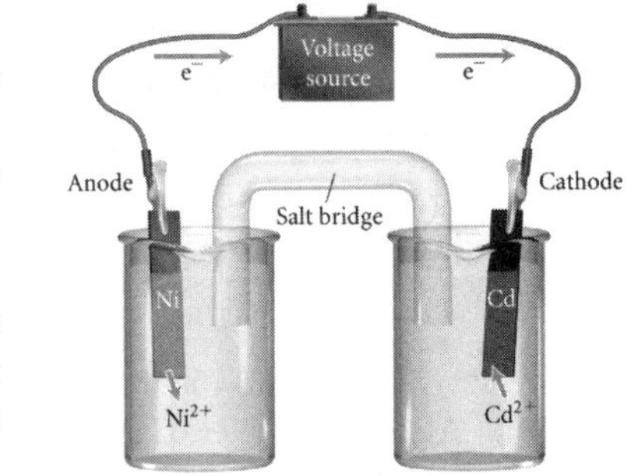

(c) Because reduction is occurring at the cathode, the battery terminal closest to the cathode is the negative terminal. Because the cell potential from part (a) is $= -0.17$ V, a minimum of 0.17 V must be applied by the battery.

Check: The reaction is nonspontaneous because the reduction of Ni^{2+} is above Cd^{2+}. Electrons still flow from the anode to the cathode. The reaction can proceed with the application of electrical energy. Applying a voltage drives the reaction but it's still not spontaneous.

20.86 **Given:** electrolytic cell Mn^{2+} reduced to Mn and Sn oxidized to Sn^{2+}

Find: Draw a cell and label the anode and cathode, write half-reactions, indicate the direction of electron flow, and calculate the minimum voltage to drive reaction.

Conceptual Plan: Write two half-cell reactions and add electrons as needed to balance reactions. Look up half-reactions in Table 20.1. Calculate the standard cell potential by subtracting the electrode potential of the anode from the electrode potential of the cathode: $E^\circ_{cell} = E^\circ_{cathode} - E^\circ_{anode}$. Anode is where oxidation occurs, so Sn is anode and Mn is cathode. Electrons flow from anode to cathode. The minimum potential needed is E°_{cell}.

Solution: $Mn^{2+}(aq) + 2\,e^- \rightarrow Mn(s)$ and $Sn^{2+}(aq) + 2\,e^- \rightarrow Sn(s)$. Look up cell potentials. For Mn^{2+}, $E^\circ_{red} = -1.18$ V. For Sn, $E^\circ_{red} = -0.14$ V. Because Sn is oxidized, $E^\circ_{red} = -0.14 \text{ V} = E^\circ_{anode}$. Then $E^\circ_{cell} = E^\circ_{cathode} - E^\circ_{anode} = -1.18 \text{ V} - (-0.14 \text{ V}) = -1.04$ V. Because oxidation occurs at the anode, the Sn is the anode and the reaction is $Sn(s) \rightarrow Sn^{2+}(aq) + 2\,e^-$. Because reduction takes place at the cathode, Mn is the cathode and the reaction is $Mn^{2+}(aq) + 2\,e^- \rightarrow Mn(s)$. Electrons flow from the anode to the cathode. Because the cell potential is -1.04 V, a minimum of 1.04 V must be applied by the battery.

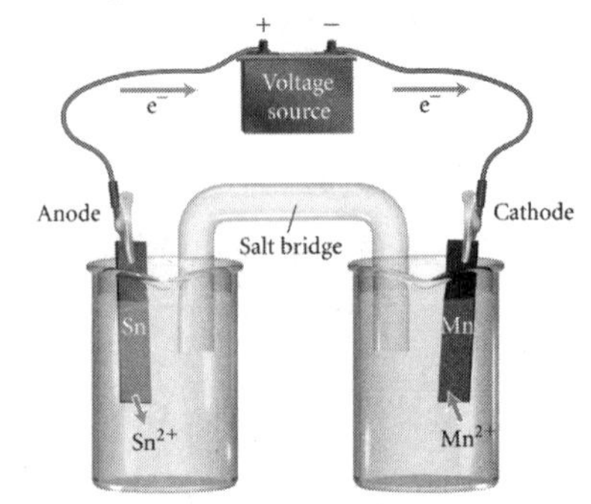

Check: The reaction is nonspontaneous because the reduction of Sn^{2+} is above Mn^{2+}. Electrons still flow from the anode to the cathode. The reaction can proceed with the application of electrical energy.

20.87 **Given:** electrolysis of molten KBr **Find:** write half-reactions

Conceptual Plan: Write two half-cell reactions, taking the cation and the anion to their elemental forms at high temperatures and adding electrons as needed to balance reactions.

Solution: KBr breaks apart to K^+ and Br^-. $K^+(l) + e^- \rightarrow K(l)$ and $2\,Br^-(l) \rightarrow Br_2(g) + 2\,e^-$.

Check: The cation is reduced, and the anion is oxidized. Mass and charge are balanced in the reactions.

20.88 **Given:** electrolysis of molten NaI **Find:** products

Conceptual Plan: The products are the cation and the anion in their elemental forms at high temperatures.

Solution: NaI breaks apart to Na^+ and I^-, so the elemental forms are $Na(l)$ and $I_2(g)$.

Check: The cation is reduced, and the anion is oxidized.

20.89 **Given:** electrolysis of mixture of molten KBr and molten LiBr **Find:** write half-reactions

Conceptual Plan: Write two half-cell reactions, taking the cation and the anion to their elemental forms at high temperatures and adding electrons as needed to balance reactions. Look up cation cell potentials to see which one generates the more spontaneous reaction.

Solution: KBr breaks apart to K^+ and Br^-. $K^+(l) + e^- \rightarrow K(l)$ and $2\,Br^-(l) \rightarrow Br_2(g) + 2\,e^-$. LiBr breaks apart to Li^+ and Br^-. $Li^+(l) + e^- \rightarrow Li(l)$ and $2\,Br^-(l) \rightarrow Br_2(g) + 2\,e^-$. The anions are the same, so the anode reaction is $2\,Br^-(l) \rightarrow Br_2(g) + 2\,e^-$. Looking up the cation reduction potentials, for K^+ $E^\circ_{red} = -2.92$ V and for Li^+ $E^\circ_{red} = -3.04$ V. Because the reduction potential is more positive for K^+, the reaction at the cathode is $K^+(l) + e^- \rightarrow K(l)$.

Check: The cation is reduced, and the anion is oxidized. Mass and charge are balanced in the reactions. The cation that is higher in Table 20.1 will be the reaction at the cathode.

20.90 **Given:** electrolysis of mixture of molten KI and molten KBr **Find:** products

Conceptual Plan: Write two half-cell reactions, taking the cation and the anion to their elemental forms at high temperatures and adding electrons as needed to balance reactions. Look up anion cell potentials to see which one generates the more spontaneous reaction.

Solution: KI breaks apart to K^+ and I^-. $K^+(l) + e^- \rightarrow K(l)$ and $2\,I^-(l) \rightarrow I_2(g) + 2\,e^-$. KBr breaks apart to K^+ and Br^-. $K^+(l) + e^- \rightarrow K(l)$ and $2\,Br^-(l) \rightarrow Br_2(g) + 2\,e^-$. The cations are the same, so the cathode reaction is $K^+(l) + e^- \rightarrow K(l)$ and the product will be $K(l)$. Looking up the anion reduction potentials, for Br^- $E^\circ_{anode} = E^\circ_{red} = +1.09$ V and for I^- $E^\circ_{anode} = E^\circ_{red} = +0.54$ V. Because the anode (oxidation) potential is more negative the reaction at the anode is $2\,I^-(l) \rightarrow I_2(g) + 2\,e^-$ and the product will be $I_2(g)$.

Check: The cation is reduced, and the anion is oxidized. The anion that is lower in Table 20.1 will be the reaction at the anode.

20.91 **Given:** electrolysis of aqueous solutions **Find:** write half-reactions

Conceptual Plan: Write two half-cell reactions, taking the cation and the anion to their elemental forms at standard conditions and adding electrons as needed to balance reactions. Look up cell potentials and compare to the cell potentials for the electrolysis of water to see which one generates the more spontaneous reaction.

Solution: The hydrolysis of water reactions are as follows: at a neutral pH, $2\,H_2O(l) \rightarrow O_2(g) + 4\,H^+(aq) + 4\,e^-$ where $E = -0.82$ V, and when $[OH^-] = 10^{-7}$ M. $2\,H_2O(l) + 2\,e^- \rightarrow H_2(g) + 2\,OH^-(aq)$ where $E = -0.41$ V.

(a) NaBr breaks apart to Na^+ and Br^-. $Na^+(aq) + e^- \rightarrow Na(s)$ and $2\,Br^-(aq) \rightarrow Br_2(l) + 2\,e^-$. Looking up the half-cell potentials, for Na^+ $E^\circ_{red} = -2.71$ V and for Br^- $E^\circ_{anode} = E^\circ_{red} = +1.09$ V. Because -0.82 V is more positive than $E^\circ_{anode} = -1.09$ V, the oxidation reaction will be $2\,H_2O(l) \rightarrow O_2(g) + 4\,H^+(aq) + 4\,e^-$ at the anode. Because -0.41 V is more positive than -2.71 V, the reduction reaction will be $2\,H_2O(l) + 2\,e^- \rightarrow H_2(g) + 2\,OH^-(aq)$ at the cathode.

(b) PbI_2 breaks apart to Pb^{2+} and I^-. $Pb^{2+}(aq) + 2\,e^- \rightarrow Pb(s)$ and $2\,I^-(aq) \rightarrow I_2(s) + 2\,e^-$. Looking up the half-cell potentials, for Pb^{2+} $E^\circ_{red} = -0.13$ V and for I^- $E^\circ_{anode} = E^\circ_{red} = -0.54$ V. Because $E^\circ_{anode} = -0.54$ V is more positive than -0.82 V, the oxidation reaction will be $2\,I^-(aq) \rightarrow I_2(s) + 2\,e^-$ at the anode. Because -0.13 V is more positive than -0.41 V, the reduction reaction will be $Pb^{2+}(aq) + 2\,e^- \rightarrow Pb\,(s)$ at the cathode.

(c) Na_2SO_4 breaks apart to Na^+ and SO_4^{2-}. $Na^+(aq) + e^- \rightarrow Na(s)$ and $SO_4^{2-}(aq) + 4\,H^+(aq) + 2\,e^- \rightarrow H_2SO_3(aq) + H_2O(l)$. Notice that both of these are reductions. Because S is in such a high oxidation state $(+6)$, it cannot be oxidized. Looking up the half-cell potentials, for Na^+ $E^\circ_{red} = -2.71$ V and for SO_4^{2-} $E^\circ_{red} = 0.20$

V. Because sodium sulfate solutions are neutral and not acidic, even though 0.20 V is more positive than -0.41 V and -2.71 V, the reduction reaction will not be: $SO_4^{2-}(aq) + 4\,H^+(aq) + 2\,e^- \rightarrow H_2SO_3(aq) + H_2O(l)$ at the cathode, which only occurs in acidic solutions. Thus, the reduction reaction will be $Na^+(aq) + e^- \rightarrow Na(s)$. Because only one oxidation reaction is possible, the oxidation reaction will be $2\,H_2O(l) \rightarrow O_2(g) + 4\,H^+(aq) + 4\,e^-$ at the anode.

Check: The most positive reactions are the reactions that will occur.

20.92 **Given:** electrolysis of aqueous solutions **Find:** write half-reactions

Conceptual Plan: Write two half-cell reactions, taking the cation and the anion to their elemental forms at standard conditions and adding electrons as needed to balance reactions. Look up cell potentials and compare to the cell potentials for the electrolysis of water to see which one generates the more spontaneous reaction.

Solution: The hydrolysis of water reactions are as follows: at a neutral pH, $2\,H_2O(l) \rightarrow O_2(g) + 4\,H^+(aq) + 4\,e^-$ where $E = -0.82$ V, and when $[OH^-] = 10^{-7}$ M. $2\,H_2O(l) + 2\,e^- \rightarrow H_2(g) + 2\,OH^-(aq)$ where $E = -0.41$ V.

(a) $Ni(NO_3)_2$ breaks apart to Ni^{2+} and NO_3^-. $Ni^{2+}(aq) + 2\,e^- \rightarrow Ni(s)$ and $NO_3^-(aq) + 4\,H^+(aq) + 3\,e^- \rightarrow NO(g) + 2\,H_2O(l)$. Notice that both of these are reductions. Because N is in such a high oxidation state (+5) it cannot be oxidized. Looking up the half-cell potentials, for Ni^{2+} $E^\circ_{red} = -0.23$ V and for NO_3^- $E^\circ_{red} = +0.96$ V. Because 0.96 V is more positive than -0.23 V and -0.41 V, the reduction reaction will be $NO_3^-(aq) + 4\,H^+(aq) + 3\,e^- \rightarrow NO(g) + 2\,H_2O(l)$ at the cathode. Because only one oxidation reaction is possible, the oxidation reaction will be $2\,H_2O(l) \rightarrow O_2(g) + 4\,H^+(aq) + 4\,e^-$ at the anode.

(b) KCl breaks apart to K^+ and Cl^-. $K^+(aq) + e^- \rightarrow K(s)$ and $2\,Cl^-(aq) \rightarrow Cl_2(s) + 2\,e^-$. Looking up the half-cell potentials, for K^+ $E^\circ_{cathode} = E^\circ_{red} = -2.92$ V and for Cl^- $E^\circ_{anode} = E^\circ_{red} = +1.36$ V. Because -0.82 V is more positive than -1.36 V, the oxidation reaction will be $2\,H_2O(l) \rightarrow O_2(g) + 4\,H^+(aq) + 4\,e^-$ at the anode. Because $E^\circ_{anode} = -0.41$ V is more positive than $E^\circ_{cathode} = -2.92$ V, the reduction reaction will be $2\,H_2O(l) + 2\,e^- \rightarrow H_2(g) + OH^-(aq)$ at the cathode.

(c) $CuBr_2$ breaks apart to Cu^{2+} and Br^-. $Cu^{2+}(aq) + 2\,e^- \rightarrow Cu(s)$ and $2\,Br^-(aq) \rightarrow Br_2(l) + 2\,e^-$. Looking up the half-cell potentials, for Cu^{2+} $E^\circ_{cathode} = E^\circ_{red} = +0.34$ V and for Br^- $E^\circ_{anode} = E^\circ_{red} = -1.09$ V. Because -0.82 V is more positive than $= -1.09$ V, the oxidation reaction will be $2\,H_2O(l) \rightarrow O_2(g) + 4\,H^+(aq) + 4\,e^-$ at the anode. Because $+0.34$ V is more positive than $= -0.41$ V, the reduction reaction will be $Cu^{2+}(aq) + 2\,e^- \rightarrow Cu(s)$ at the cathode.

Check: The most positive reactions are the reactions that will occur.

20.93 **Given:** electrolysis cell to electroplate Cu onto a metal surface

Find: Draw a cell and label the anode and cathode and write half-reactions.

Conceptual Plan: Write two half-cell reactions and add electrons as needed to balance reactions. The cathode reaction will be the reduction of Cu^{2+} to the metal. The anode will be the reverse reaction.

Solution:

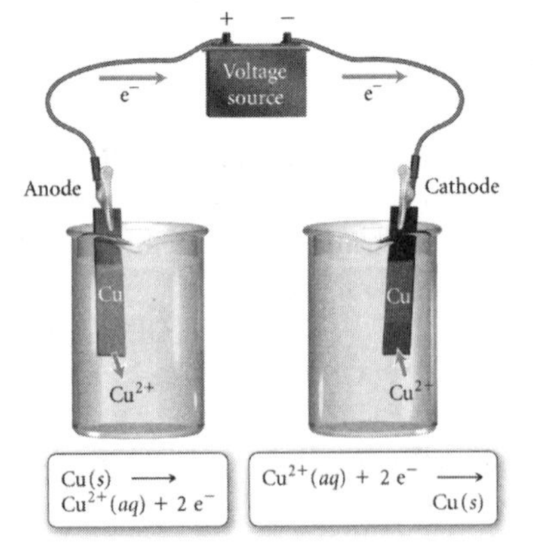

Check: The metal to be plated is the cathode because metal ions are converted to $Cu(s)$ on the surface of the metal.

20.94 **Given:** electrolysis cell to electroplate Ni onto a metal surface

Find: Draw a cell and label the anode and cathode and write half-reactions.

Conceptual Plan: Write two half-cell reactions and add electrons as needed to balance reactions. The cathode reaction will be the reduction of Ni^{2+} to the metal. The anode will be the reverse reaction.

Solution: Label the electrode where the oxidation occurs as the anode. Label the electrode where the reduction occurs as the cathode. Electrons flow from anode to cathode. Label anode as (−) and cathode as (+).

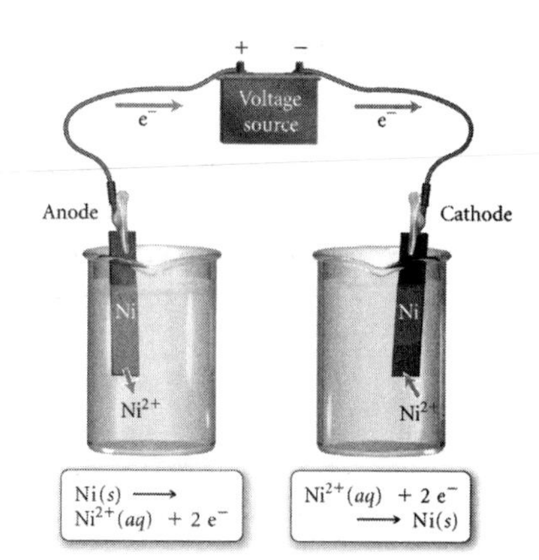

Check: The metal to be plated is the cathode because metal ions are converted to Ni(*s*) on the surface of the metal.

20.95 **Given:** Cu electroplating of 325 mg Cu at a current of 5.6 A; $Cu^{2+}(aq) + 2\,e^- \rightarrow Cu(s)$ **Find:** time

Conceptual Plan: mg Cu → g Cu → mol Cu → mol e⁻ → C → s

$$\frac{1\text{ g}}{1000\text{ mg}} \quad \frac{1\text{ mol Cu}}{63.55\text{ g Cu}} \quad \frac{2\text{ mol e}^-}{1\text{ mol Cu}} \quad \frac{96{,}485\text{ C}}{1\text{ mol e}^-} \quad \frac{1\text{ s}}{5.6\text{ C}}$$

Solution: $325\text{ mg Cu} \times \frac{1\text{ g Cu}}{1000\text{ mg Cu}} \times \frac{1\text{ mol Cu}}{63.55\text{ g Cu}} \times \frac{2\text{ mol e}^-}{1\text{ mol Cu}} \times \frac{96{,}485\text{ C}}{1\text{ mol e}^-} \times \frac{1\text{ s}}{5.6\text{ C}} = 180\text{ s}.$

Check: The units (*s*) are correct. Because far less than a mole of Cu is electroplated, the time is short.

20.96 **Given:** Ag electroplating at a current of 6.8 A for 72 min; $Ag^+(aq) + e^- \rightarrow Ag(s)$ **Find:** mass of Ag

Conceptual Plan: min → s → C → mol e⁻ → mol Ag → g Ag

$$\frac{60\text{ s}}{1\text{ min}} \quad \frac{6.8\text{ C}}{1\text{ s}} \quad \frac{1\text{ mol e}^-}{96{,}485\text{ C}} \quad \frac{1\text{ mol Ag}}{1\text{ mol e}^-} \quad \frac{107.87\text{ g Ag}}{1\text{ mol Ag}}$$

Solution: $72\text{ min} \times \frac{60\text{ s}}{1\text{ min}} \times \frac{6.8\text{ C}}{1\text{ s}} \times \frac{1\text{ mol e}^-}{96{,}485\text{ C}} \times \frac{1\text{ mol Ag}}{1\text{ mol e}^-} \times \frac{107.87\text{ g Ag}}{1\text{ mol Ag}} = 33\text{ g Ag}$

Check: The units (g) are correct. Because less than a mole of electrons is used, the mass is less than the molar mass of Ag.

20.97 **Given:** Na electrolysis, 1.0 kg in one hour **Find:** current

Conceptual Plan: $Na^+(l) + e^- \rightarrow Na(l)$ $\frac{\text{kg Na}}{\text{h}} \rightarrow \frac{\text{g Na}}{\text{h}} \rightarrow \frac{\text{mol Na}}{\text{h}} \rightarrow \frac{\text{mol e}^-}{\text{h}} \rightarrow \frac{\text{C}}{\text{h}} \rightarrow \frac{\text{C}}{\text{min}} \rightarrow \frac{\text{C}}{\text{s}}$

$$\frac{1000\text{ g}}{1\text{ kg}} \quad \frac{1\text{ mol Na}}{22.99\text{ g Na}} \quad \frac{1\text{ mol e}^-}{1\text{ mol Na}} \quad \frac{96{,}485\text{ C}}{1\text{ mol e}^-} \quad \frac{1\text{ h}}{60\text{ min}} \quad \frac{1\text{ min}}{60\text{ s}}$$

Solution:

$$\frac{1.0\text{ kg Na}}{1\text{ h}} \times \frac{1000\text{ g Na}}{1\text{ kg Na}} \times \frac{1\text{ mol Na}}{22.99\text{ g Na}} \times \frac{1\text{ mol e}^-}{1\text{ mol Na}} \times \frac{96{,}485\text{ C}}{1\text{ mol e}^-} \times \frac{1\text{ h}}{60\text{ min}} \times \frac{1\text{ min}}{60\text{ s}} = 1.2 \times 10^3 \frac{\text{C}}{\text{s}} = 1.2 \times 10^3\text{ A}$$

Check: The units (A) are correct. Because the amount per hour is so large, we expect a very large current.

20.98 **Given:** Al electrolysis at a current of 25 A for 1 hour; $Al^{3+}(aq) + 3e^- \rightarrow Al(s)$ **Find:** mass of Al

Conceptual Plan: h → min → s → C → mol e⁻ → mol Al → g Al

$$\frac{60\text{ min}}{1\text{ h}} \quad \frac{60\text{ s}}{1\text{ min}} \quad \frac{25\text{ C}}{1\text{ s}} \quad \frac{1\text{ mol e}^-}{96{,}485\text{ C}} \quad \frac{1\text{ mol Al}}{3\text{ mol e}^-} \quad \frac{26.98\text{ g Al}}{1\text{ mol Al}}$$

Solution: $1\text{ h} \times \frac{60\text{ min}}{1\text{ h}} \times \frac{60\text{ s}}{1\text{ min}} \times \frac{25\text{ C}}{1\text{ s}} \times \frac{1\text{ mol e}^-}{96{,}485\text{ C}} \times \frac{1\text{ mol Al}}{3\text{ mol e}^-} \times \frac{26.98\text{ g Al}}{1\text{ mol Al}} = 8.4\text{ g Al}$

Check: The units (g) are correct. Because three moles of electrons are used per mole of Al^{3+}, the mass is less than the molar mass of Al.

Cumulative Problems

20.99 **Given:** $MnO_4^-(aq) + Zn(s) \rightarrow Mn^{2+}(aq) + Zn^{2+}(aq)$, 0.500 M $KMnO_4$, and 2.85 g Zn

Find: balance equation and volume $KMnO_4$ solution

Conceptual Plan: Separate the overall reaction into two half-reactions: one for oxidation and one for reduction. → Balance each half-reaction with respect to mass in the following order: (1) Balance all elements other than H and O, (2) balance O by adding H_2O, and (3) balance H by adding H^+. → Balance each half-reaction with respect to charge by adding electrons. (The sum of the charges on both sides of the equation should be made equal by adding electrons as necessary.) → Make the number of electrons in both half-reactions equal by multiplying one or both half-reactions by a small whole number. → Add the two half-reactions, canceling electrons and other species as necessary. → Verify that the reaction is balanced with respect to both mass and charge.

Then g Zn → mol Zn → mol MnO_4^- → mol $KMnO_4$ → L $KMnO_4$ → mL $KMnO_4$.

$$\frac{1 \text{ mol Zn}}{65.38 \text{ g Zn}} \quad \frac{2 \text{ mol } MnO_4^-}{5 \text{ mol Zn}} \quad \frac{1 \text{ mol } KMnO_4}{1 \text{ mol } MnO_4^-} \quad \frac{1 \text{ L } KMnO_4}{0.500 \text{ mol } KMnO_4} \quad \frac{1000 \text{ mL } KMnO_4}{1 \text{ L } KMnO_4}$$

Solution:

Separate: $MnO_4^-(aq) \rightarrow Mn^{2+}(aq)$ and $Zn(s) \rightarrow Zn^{2+}(aq)$

Balance non-H & O elements: $MnO_4^-(aq) \rightarrow Mn^{2+}(aq)$ and $Zn(s) \rightarrow Zn^{2+}(aq)$

Balance O with H_2O: $MnO_4^-(aq) \rightarrow Mn^{2+}(aq) + 4\,H_2O(l)$ and $Zn(s) \rightarrow Zn^{2+}(aq)$

Balance H with H^+: $MnO_4^-(aq) + 8\,H^+(aq) \rightarrow Mn^{2+}(aq) + 4\,H_2O(l)$ and $Zn(s) \rightarrow Zn^{2+}(aq)$

Add electrons: $MnO_4^-(aq) + 8\,H^+(aq) + 5\,e^- \rightarrow Mn^{2+}(aq) + 4\,H_2O(l)$ and $Zn(s) \rightarrow Zn^{2+}(aq) + 2\,e^-$

Equalize electrons:

$2\,MnO_4^-(aq) + 16\,H^+(aq) + 10\,e^- \rightarrow 2\,Mn^{2+}(aq) + 8\,H_2O(l)$ and $5\,Zn(s) \rightarrow 5\,Zn^{2+}(aq) + 10\,e^-$

Add half-reactions:

$2\,MnO_4^-(aq) + 16\,H^+(aq) + \cancel{10\,e^-} + 5\,Zn(s) \rightarrow 2\,Mn^{2+}(aq) + 8\,H_2O(l) + 5\,Zn^{2+}(aq) + \cancel{10\,e^-}$

Cancel electrons: $2\,MnO_4^-(aq) + 16\,H^+(aq) + 5\,Zn(s) \rightarrow 2\,Mn^{2+}(aq) + 8\,H_2O(l) + 5\,Zn^{2+}(aq)$

$$2.85 \text{ g Zn} \times \frac{1 \text{ mol Zn}}{65.38 \text{ g Zn}} \times \frac{2 \text{ mol } MnO_4^-}{5 \text{ mol Zn}} \times \frac{1 \text{ mol } KMnO_4}{1 \text{ mol } MnO_4^-} \times \frac{1 \text{ L } KMnO_4}{0.500 \text{ mol } KMnO_4} \times \frac{1000 \text{ mL } KMnO_4}{1 \text{ L } KMnO_4^-}$$

$= 34.9$ mL $KMnO_4$

Check:

Reactants	Products
2 Mn atoms	2 Mn atoms
8 O atoms	8 O atoms
16 H atoms	16 H atoms
5 Zn atoms	5 Zn atoms
+14 charge	+14 charge

The units (mL) are correct. Because far less than a mole of zinc is used, less than a mole of permanganate is consumed; so the volume is less than a liter.

20.100 **Given:** $Cr_2O_7^{2-}(aq) + Cu(s) \rightarrow Cr^{3+}(aq) + Cu^{2+}(aq)$ 0.850 M $K_2Cr_2O_7$ and 5.25 g Zn

Find: balance equation and volume $K_2Cr_2O_7$ solution

Conceptual Plan: Separate the overall reaction into two half-reactions: one for oxidation and one for reduction. → Balance each half-reaction with respect to mass in the following order: (1) Balance all elements other than H and O, (2) balance O by adding H_2O, and (3) balance H by adding H^+. → Balance each half-reaction with respect to charge by adding electrons. (The sum of the charges on both sides of the equation should be made equal by adding electrons as necessary.) → Make the number of electrons in both half-reactions equal by multiplying one or both half-reactions by a small whole number. → Add the two half-reactions, canceling electrons and other species as necessary. → Verify that the reaction is balanced with respect to both mass and charge.

Then g Cu → mol Cu → mol $Cr_2O_7^{2-}$ → mol $K_2Cr_2O_7$ → L $K_2Cr_2O_7$ → mL $K_2Cr_2O_7$.

$$\frac{1 \text{ mol Cu}}{63.55 \text{ g Cu}} \quad \frac{1 \text{ mol } Cr_2O_7^{2-}}{3 \text{ mol Cu}} \quad \frac{1 \text{ mol } K_2Cr_2O_7}{1 \text{ mol } Cr_2O_7^{2-}} \quad \frac{1 \text{ L } K_2Cr_2O_7}{0.850 \text{ mol } K_2Cr_2O_7} \quad \frac{1000 \text{ mL } K_2Cr_2O_7}{1 \text{ L } K_2Cr_2O_7}$$

Solution:

Separate: $Cr_2O_7^{2-}(aq) \rightarrow Cr^{3+}(aq)$ and $Cu(s) \rightarrow Cu^{2+}(aq)$

Balance non-H & O elements: $Cr_2O_7^{2-}(aq) \rightarrow 2\,Cr^{3+}(aq)$ and $Cu(s) \rightarrow Cu^{2+}(aq)$

Balance O with H_2O: $Cr_2O_7^{2-}(aq) \rightarrow 2\,Cr^{3+}(aq) + 7\,H_2O(l)$ and $Cu(s) \rightarrow Cu^{2+}(aq)$

Balance H with H^+: $Cr_2O_7^{2-}(aq) + 14\,H^+(aq) \rightarrow 2\,Cr^{3+}(aq) + 7\,H_2O(l)$ and $Cu(s) \rightarrow Cu^{2+}(aq)$

Add electrons: $Cr_2O_7^{2-}(aq) + 14\,H^+(aq) + 6\,e^- \rightarrow 2\,Cr^{3+}(aq) + 7\,H_2O(l)$ and $Cu(s) \rightarrow Cu^{2+}(aq) + 2\,e^-$

Equalize electrons:

$Cr_2O_7^{2-}(aq) + 14\,H^+(aq) + 6\,e^- \rightarrow 2\,Cr^{3+}(aq) + 7\,H_2O(l)$ and $3\,Cu(s) \rightarrow 3\,Cu^{2+}(aq) + 6\,e^-$

Add half-reactions:

$Cr_2O_7^{2-}(aq) + 14\,H^+(aq) + \cancel{6\,e^-} + 3\,Cu(s) \rightarrow 2\,Cr^{3+}(aq) + 7\,H_2O(l) + 3\,Cu^{2+}(aq) + \cancel{6\,e^-}$

Cancel electrons: $Cr_2O_7^{2-}(aq) + 14\,H^+(aq) + 3\,Cu(s) \rightarrow 2\,Cr^{3+}(aq) + 7\,H_2O(l) + 3\,Cu^{2+}(aq)$

$$5.25\,\cancel{\text{g Cu}} \times \frac{1\,\cancel{\text{mol Cu}}}{63.55\,\cancel{\text{g Cu}}} \times \frac{1\,\cancel{\text{mol } Cr_2O_7^{2-}}}{3\,\cancel{\text{mol Cu}}} \times \frac{1\,\cancel{\text{mol } K_2Cr_2O_7}}{1\,\cancel{\text{mol } Cr_2O_7^{2-}}} \times \frac{1\,\cancel{\text{L } K_2Cr_2O_7}}{0.850\,\cancel{\text{mol } K_2Cr_2O_7}} \times \frac{1000\text{ mL } K_2Cr_2O_7}{1\,\cancel{\text{L } K_2Cr_2O_7}}$$

$= 32.4\text{ mL } K_2Cr_2O_7$

Check:

Reactants	Products
2 Cr atoms	2 Cr atoms
7 O atoms	7 O atoms
14 H atoms	14 H atoms
3 Cu atoms	3 Cu atoms
+12 charge	+12 charge

The units (mL) are correct. Because far less than a mole of copper is used, less than a mole of dichromate is consumed; so the volume is less than a liter.

20.101 **Given:** beaker with Al strip and Cu^{2+} ions **Find:** Draw sketch after Al is submerged for a few minutes.

Conceptual Plan: Write two half-cell reactions and add electrons as needed to balance reactions. Look up half-reactions in Table 20.1. Calculate the standard cell potential by subtracting the electrode potential of the anode from the electrode potential of the cathode: $E^\circ_{cell} = E^\circ_{cathode} - E^\circ_{anode}$. If $E^\circ_{cell} > 0$, the reaction is spontaneous in the forward direction and Al will dissolve and Cu will deposit.

Solution: $Al(s) \rightarrow Al^{3+}(aq)$ and $Cu^{2+}(aq) \rightarrow Cu(s)$. Add electrons.

$Al(s) \rightarrow Al^{3+}(aq) + 3\,e^-$ and $Cu^{2+}(aq) + 2\,e^- \rightarrow Cu(s)$. Look up cell potentials.

Al is oxidized, so $E^\circ_{anode} = E^\circ_{red} = -1.66$ V. Cu^{2+} is reduced, so $E^\circ_{cathode} = E^\circ_{red} = 0.34$ V.

Then $E^\circ_{cell} = E^\circ_{cathode} - E^\circ_{anode} = 0.34\text{ V} - (-1.66\text{ V}) + 2.00$ V, so the reaction is spontaneous. Al will dissolve to generate $Al^{3+}(aq)$, and $Cu(s)$ will deposit.

Check: The units (V) are correct. If the voltage is positive, the reaction is spontaneous; so Al will dissolve and Cu will deposit.

20.102 **Given:** Zn/Zn^{2+} and Ni/Ni^{2+} half-cells in voltaic cell

Find: Draw sketch after substantial amount of current is generated.

Conceptual Plan: Write two half-cell reactions and add electrons as needed to balance reactions. Look up half-reactions in Table 20.1. Calculate the standard cell potential by subtracting the electrode potential of the anode from the electrode potential of the cathode: $E^\circ_{cell} = E^\circ_{cathode} - E^\circ_{anode}$. Choose the direction of the half-cell reactions so that $E^\circ_{cell} > 0$. Add two half-cell reactions and cancel electrons to generate overall reaction. Because reaction is spontaneous, it will move forward.

Solution: $Zn^{2+}(aq) + 2\,e^- \rightarrow Zn(s)$ and $Ni^{2+}(aq) + 2\,e^- \rightarrow Ni(s)$. Look up cell potentials. For Zn, $E^\circ_{red} = -0.76$ V. For Ni, $E^\circ_{red} = -0.23$ V. To get a positive E°_{cell}, the sign of the Zn potential must be reversed. Zn is oxidized, so $E^\circ_{anode} = E^\circ_{red} = -0.76$ V. Ni^{2+} is reduced, so $E^\circ_{cathode} = E^\circ_{red} = -0.23$ V.

Then $E^\circ_{cell} = E^\circ_{cathode} - E^\circ_{anode} = -0.23\text{ V} - (-0.76\text{ V}) = +0.53$ V.

Adding the two half-cell reactions:

$Zn(s) + Ni^{2+}(aq) + \cancel{2\,e^-} \rightarrow Zn^{2+}(aq) + \cancel{2\,e^-} + Ni(s)$. The overall reaction is $Zn(s) + Ni^{2+}(aq) \rightarrow Zn^{2+}(aq) + Ni(s)$. Zn will dissolve to generate $Zn^{2+}(aq)$, and Ni (*s*) will deposit.

Check: The units (V) are correct. Because Ni is above Zn in Table 20.1, Ni is reduced. If the voltage is positive, the reaction is spontaneous; so Zn will dissolve, and Ni will deposit.

20.103 **Given:** (a) 2.15 g Al, (b) 4.85 g Cu, and (c) 2.42 g Ag in 3.5 M HI

Find: If metal dissolves, write a balanced reaction and the minimum amount of HI needed to dissolve the metal.

Conceptual Plan: In general, metals whose reduction half-reactions lie below the reduction of H^+ to H_2 in Table 20.1 will dissolve in acids, while metals above it will not. Stop here if metal does not dissolve. To write the balanced redox reactions, pair the oxidation of the metal with the reduction of H^+ to H_2 ($2\,H^+(aq) + 2\,e^- \rightarrow H_2(g)$). Balance the number of electrons transferred. Add the two reactions. Cancel electrons. Then g metal → mol metal → mol H^+ → L HI → mL HI.

$$\mathcal{M} \qquad \frac{x \text{ mol H}^+}{y \text{ mol metal}} \qquad \frac{1 \text{ L HI}}{3.5 \text{ mol HI}} \qquad \frac{1000 \text{ mL HI}}{1 \text{ L HI}}$$

Solution:

(a) Al meets this criterion. For Al, $Al(s) \rightarrow Al^{3+}(aq) + 3\,e^-$. We need to multiply the Al reaction by 2 and the H^+ reaction by 3. So $2\,Al(s) \rightarrow 2\,Al^{3+}(aq) + 6\,e^-$ and $6\,H^+(aq) + 6\,e^- \rightarrow 3\,H_2(g)$. Adding the half-reactions: $2\,Al(s) + 6\,H^+(aq) + \cancel{6\,e^-} \rightarrow 2\,Al^{3+}(aq) + \cancel{6\,e^-} + 3\,H_2(g)$. Simplify to $2\,Al(s) + 6\,H^+(aq) \rightarrow 2\,Al^{3+}(aq) + 3\,H_2(g)$. Then

$$2.15\cancel{\text{g Al}} \times \frac{1\,\cancel{\text{mol Al}}}{26.98\,\cancel{\text{g Al}}} \times \frac{6\,\cancel{\text{mol H}^+}}{2\,\cancel{\text{mol Al}}} \times \frac{1\,\cancel{\text{L HI}}}{3.5\,\cancel{\text{mol HI}}} \times \frac{1000 \text{ mL HI}}{1\,\cancel{\text{L HI}}} = 68.3 \text{ mL HI.}$$

(b) Cu does not meet this criterion, so it will not dissolve in HI.

(c) Ag does not meet this criterion, so it will not dissolve in HI.

Check: Only metals with negative reduction potentials will dissolve. The volume of acid needed is fairly small because the amount of metal is much less than 1 mole and the concentration of acid is high.

20.104 **Given:** (a) 5.90 g Au, (b) 2.55 g Cu, and (c) 4.83 g Sn in 6.0 M HNO_3

Find: If the metal dissolves, write the balanced reaction and minimum amount of HNO_3 needed to dissolve the metal.

Conceptual Plan: Nitric acid (HNO_3) oxidizes metals through the following reduction half-reaction: $NO_3^-(aq) + 4\,H^+(aq) + 3\,e^- \rightarrow NO(g) + 2\,H_2O(l)$ $E^\circ_{red} = 0.96$ V. Because this half-reaction is above the reduction of H^+ in Table 20.1, HNO_3 can oxidize metals that can't be oxidized by HCl.

Solution:

(a) Au (which has a reduction potential of 1.50 V) will not be oxidized, so it will not dissolve in HNO_3.

(b) Cu (which has a reduction potential of 0.34 V) will be oxidized. To write the balanced redox reactions, pair the oxidation of the copper ($Cu(s) \rightarrow Cu^{2+}(aq) + 2\,e^-$) with the reduction of nitric acid ($NO_3^-(aq) + 4\,H^+(aq) + 3\,e^- \rightarrow NO(g) + 2\,H_2O(l)$). To balance the number of electrons transferred, we need to multiply the Cu reaction by 3 and the nitric acid reaction by 2. So $3\,Cu(s) \rightarrow 3\,Cu^{2+}(aq) + 6\,e^-$ and $2\,NO_3^-(aq) + 8\,H^+(aq) + 6\,e^- \rightarrow 2\,NO(g) + 4\,H_2O(l)$. Adding the two reactions: $3\,Cu(s) + 2\,NO_3^-(aq) + 8\,H^+(aq) + \cancel{6\,e^-} \rightarrow 3\,Cu^{2+}(aq) + \cancel{6\,e^-} + 2\,NO(g) + 4\,H_2O(l)$. Simplify to $3\,Cu(s) + 2\,NO_3^-(aq) + 8\,H^+(aq) \rightarrow 3\,Cu^{2+}(aq) + 2\,NO(g) + 4\,H_2O(l)$. Then

$$2.55\,\cancel{\text{g Cu}} \times \frac{1\,\cancel{\text{mol Cu}}}{63.55\,\cancel{\text{g Cu}}} \times \frac{8\,\cancel{\text{mol H}^+}}{3\,\cancel{\text{mol Cu}}} \times \frac{1\,\cancel{\text{L HNO}_3}}{6.0\,\cancel{\text{mol HNO}_3}} \times \frac{1000 \text{ mL HNO}_3}{1\,\cancel{\text{L HNO}_3}} = 18 \text{ mL HNO}_3.$$

Use stoichiometric coefficient of H^+ because it is larger than the NO_3^- stoichiometric coefficient.

(c) Ni (which has a reduction potential of −0.23 V) will be oxidized. To write the balanced redox reactions, pair the oxidation of the zinc ($Ni(s) \rightarrow Ni^{2+}(aq) + 2\,e^-$) with the reduction of nitric acid ($NO_3^-(aq) + 4\,H^+(aq) + 3\,e^- \rightarrow NO(g) + 2\,H_2O(l)$). To balance the number of electrons transferred, we need to multiply the Ni reaction by 3 and the nitric acid reaction by 2. So $3\,Ni(s) \rightarrow 3\,Ni^{2+}(aq) + 6\,e^-$ and $2\,NO_3^-(aq) + 8\,H^+(aq) + 6\,e^- \rightarrow 2\,NO(g) + 4\,H_2O(l)$. Adding the two reactions: $3\,Ni(s) + 2\,NO_3^-(aq) + 8\,H^+(aq) + \cancel{6\,e^-} \rightarrow 3\,Ni^{2+}(aq) + \cancel{6\,e^-} + 2\,NO(g) + 4\,H_2O(l)$. Simplify to $3\,Ni(s) + 2\,NO_3^-(aq) + 8\,H^+(aq) \rightarrow 3\,Ni^{2+}(aq) + 2\,NO(g) + 4\,H_2O(l)$. Then

$$4.83\,\cancel{\text{g Ni}} \times \frac{1\,\cancel{\text{mol Ni}}}{65.38\,\cancel{\text{g Ni}}} \times \frac{8\,\cancel{\text{mol H}^+}}{3\,\cancel{\text{mol Ni}}} \times \frac{1\,\cancel{\text{L HNO}_3}}{6.0\,\cancel{\text{mol HNO}_3}} \times \frac{1000 \text{ mL HNO}_3}{1\,\cancel{\text{L HNO}_3}} = 33 \text{ mL HNO}_3.$$

Use the stoichiometric coefficient of H^+ because it is larger than the NO_3^- stoichiometric coefficient.

Check: Only metals with reduction potentials less than 0.96 V will dissolve. The volume of acid needed is fairly small because the amount of metal is much less than 1 mole and the concentration of acid is high.

20.105 **Given:** $Pt(s) \mid H_2(g, 1\text{ atm}) \mid H^+(aq, ?\text{ M}) \mid\mid Cu^{2+}(aq, 1.0\text{ M}) \mid Cu(s)$, $E_{cell} = 355$ mV **Find:** pH

Conceptual Plan: Write half-reactions from line notation. Look up half-reactions in Table 20.1. Calculate the standard cell potential by subtracting the electrode potential of the anode from the electrode

potential of the cathode: $E^\circ_{cell} = E^\circ_{cathode} - E^\circ_{anode}$. Add the two half-cell reactions and cancel the electrons. Then mV $\rightarrow$ V then $E^\circ_{cell}, E_{cell}, P_{H_2}, [Cu^{2+}], n \rightarrow [H^+] \rightarrow$ pH.

$$\frac{1\text{ V}}{1000\text{ mV}} \qquad E_{cell} = E^\circ_{cell} - \frac{0.0592\text{ V}}{n}\log Q \qquad \text{pH} = -\log[H^+]$$

Solution: The half-reactions are $H_2(g) \rightarrow 2\,H^+(aq) + 2\,e^-$ and $Cu^{2+}(aq) + 2\,e^- \rightarrow Cu(s)$. H is oxidized, so $E^\circ_{red} = 0.00\text{ V} = E^\circ_{anode}$. Cu is reduced, so $E^\circ_{red} = 0.34\text{ V} = E^\circ_{cathode}$. Then $E^\circ_{cell} = E^\circ_{cathode} - E^\circ_{anode}$ $= 0.34\text{ V} - 0.00\text{ V} = +0.34\text{ V}$. Adding the two reactions: $H_2(g) + Cu^{2+}(aq) + \cancel{2e^-} \rightarrow 2\,H^+(aq) + \cancel{2e^-} + Cu(s)$. Cancel the electrons: $H_2(g) + Cu^{2+}(aq) \rightarrow 2\,H^+(aq) + Cu(s)$. Then $355\,\cancel{\text{mV}} \times \frac{1\text{ V}}{1000\,\cancel{\text{mV}}} = 0.355\text{ V}$. So $n = 2$ and $Q = \frac{[H^+]^2}{P_{H_2}[Cu^{2+}]} = \frac{(x)^2}{(1)(1.0)} = x^2$. Then $E_{cell} = E^\circ_{cell} - \frac{0.0592\text{ V}}{n}\log Q$. Substitute values and solve for x. $0.355\text{ V} = 0.34\text{ V} - \frac{0.0592\text{ V}}{2}\log x^2 \rightarrow 0.0\underline{1}5\,\cancel{\text{V}} = -\frac{0.0592\,\cancel{\text{V}}}{2}\log x^2 \rightarrow -0.\underline{5}0676 = \log x^2 \rightarrow$ $x^2 = 10^{-0.\underline{5}0676} = 0.\underline{3}1134 \rightarrow x = 0.\underline{5}5798$ then $\text{pH} = -\log[H^+] = -\log[0.\underline{5}5798] = 0.\underline{2}5338 = 0.3$

Check: The units (none) are correct. The pH is acidic, which is consistent with dissolving a metal in acid.

20.106 **Given:** $Pt(s)\,|\,H_2(g, 1\text{ atm})\,|\,H^+(aq, 1.0\text{ M})\,||\,Au^{3+}(aq, ?\text{ M})\,|\,Au(s)$, $E_{cell} = 1.22\text{ V}$ **Find:** $[Au^{3+}]$
Conceptual Plan: Write half-reactions from line notation. Look up half-reactions in Table 20.1. Calculate the standard cell potential by subtracting the electrode potential of the anode from the electrode potential of the cathode: $E^\circ_{cell} = E^\circ_{cathode} - E^\circ_{anode}$. Equalize the number of electrons transferred; then add the two half-cell reactions and cancel the electrons.
Then $E^\circ_{cell}, E_{cell}, P_{H_2}, [H^+], n \rightarrow [Au^{3+}]$.

$$E_{cell} = E^\circ_{cell} - \frac{0.0592\text{ V}}{n}\log Q$$

Solution: The half-reactions are $H_2(g) \rightarrow 2\,H^+(aq) + 2\,e^-$ and $Au^{3+}(aq) + 3\,e^- \rightarrow Au(s)$. H is oxidized, so $E^\circ_{red} = 0.00\text{ V} = E^\circ_{anode}$. Au is reduced, so $E^\circ_{red} = 1.50\text{ V} = E^\circ_{cathode}$. Then $E^\circ_{cell} = E^\circ_{cathode} - E^\circ_{anode} =$ $1.50\text{ V} - 0.00\text{ V} = +1.50\text{ V}$. Equalize the number of electrons transferred by multiplying the first reaction by 3 and the second reaction by 2; so $3\,H_2(g) \rightarrow 6\,H^+(aq) + 6\,e^-$ and $2\,Au^{3+}(aq) + 6\,e^- \rightarrow 2\,Au(s)$. Adding the two reactions: $3\,H_2(g) + 2\,Au^{3+}(aq) + \cancel{6e^-} \rightarrow 6\,H^+(aq) + \cancel{6e^-} + 2\,Au(s)$. Cancel the electrons: $3\,H_2(g) + 2\,Au^{3+}(aq) \rightarrow 6\,H^+(aq) + 2\,Au(s)$. So $n = 6$ and $Q = \frac{[H^+]^6}{P^3_{H_2}[Au^{3+}]^2} = \frac{(1.0)^6}{(1)^3(x)^2} = x^{-2}$. Then $E_{cell} = E^\circ_{cell} - \frac{0.0592\text{ V}}{n}\log Q$. Substitute values and solve for x. $1.22\text{ V} = 1.50\text{ V} - \frac{0.0592\text{ V}}{6}\log x^{-2} \rightarrow 0.28\,\cancel{\text{V}} =$ $\frac{0.0592\,\cancel{\text{V}}}{6}\log x^{-2} \rightarrow 2\underline{8}.3784 = \log x^{-2} \rightarrow x^{-2} = 10^{2\underline{8}.3784} = 2.\underline{3}900 \times 10^{28} \rightarrow x = 6.\underline{4}685 \times 10^{-15}\text{ M}$

Check: The units (M) are correct. The concentration is expected to be low because the cell potential must be reduced.

20.107 You should be wary of the battery because the most a pairing of half-cell reactions can generate is 5 to 6 V, not 24 V.

20.108 **Given:** lithium oxidizing and fluorine gas reducing **Find:** E°_{cell} and why the battery is hard to produce
Conceptual Plan: Write two half-cell reactions, taking the elemental forms to cations and anions and adding electrons as needed to balance reactions. Look up half-reactions in Table 20.1. Calculate the standard cell potential by subtracting the electrode potential of the anode from the electrode potential of the cathode: $E^\circ_{cell} = E^\circ_{cathode} - E^\circ_{anode}$.
Solution: $Li(l) \rightarrow Li^+(l) + e^-$ and $F_2(g) + 2\,e^- \rightarrow 2\,F^-(l)$. Looking up the reduction potentials, for Li $E^\circ_{anode} = E^\circ_{red} = -3.04\text{ V}$ and for F_2 $E^\circ_{cathode} = E^\circ_{red} = +2.87\text{ V}$. So $E^\circ_{cell} = E^\circ_{cathode} - E^\circ_{anode} = 2.87\text{ V} - (-3.04\text{ V})$ $= 5.92\text{ V}$. One of the problems with using this system is getting a high concentration of F_2 because it is a gas and gases have low densities.

Check: The units (V) are correct. The metal is oxidized and the nonmetal is reduced, so we expect a high cell potential.

20.109 **Given:** Mg oxidation and Cu^{2+} reduction; initially, $[Mg^{2+}] = 1.0 \times 10^{-4}\text{ M}$ and $[Cu^{2+}] = 1.5\text{ M}$ in 1.0 L half-cells
Find: (a) initial E_{cell}, (b) E_{cell} after 5.0 A for 8.0 h, and (c) how long battery can deliver 5.0 A

Conceptual Plan:

(a) **Write the two half-cell reactions and add electrons as needed to balance reactions. Look up half-reactions in Table 20.1. Calculate the standard cell potential by subtracting the electrode potential of the anode from the electrode potential of the cathode: $E^\circ_{cell} = E^\circ_{cathode} - E^\circ_{anode}$. Add the two half-cell reactions, cancel electrons, and determine n. Then E°_{cell}, $[Mg^{2+}]$, $[Cu^{2+}]$, $n \rightarrow E_{cell}$.**

$$E_{cell} = E^\circ_{cell} - \frac{0.0592\ V}{n}\log Q$$

(b) **hr → min → s → C → mol e⁻ → mol Cu reduced → $[Cu^{2+}]$ and**

$$\frac{60\ min}{1\ hr}\quad \frac{60\ s}{1\ min}\quad \frac{5.0\ C}{1\ s}\quad \frac{1\ mol\ e^-}{96{,}485\ C}\quad \frac{1\ mol\ Cu^{2+}}{2\ mol\ e^-}\quad \text{because } V = 1.0\ L\quad [Cu^{2+}]_f = [Cu^{2+}]_i - \frac{mol\ Cu^{2+}\ reduced}{1.0\ L}$$

mol Cu reduced → mol Mg oxidized → $[Mg^{2+}]$

$$\frac{1\ mol\ Mg\ oxidized}{1\ mol\ Cu^{2+}\ reduced}\quad \text{because } V = 1.0\ L\quad [Mg^{2+}]_f = [Mg^{2+}]_i + \frac{mol\ Mg\ oxidized}{1.0\ L}$$

(c) **$[Cu^{2+}]$ → mol e⁻ → C → s → min → hr**

$$\frac{2\ mol\ e^-}{1\ mol\ Cu^{2+}}\quad \frac{96{,}485\ C}{1\ mol\ e^-}\quad \frac{1\ s}{5.0\ C}\quad \frac{1\ min}{60\ s}\quad \frac{1\ hr}{60\ min}$$

Solution:

(a) Write half-reactions and add electrons. $Cu^{2+}(aq) + 2\ e^- \rightarrow Cu(s)$ and $Mg(s) \rightarrow Mg^{2+}(aq) + 2\ e^-$. Look up cell potentials. Mg is oxidized, so $E^\circ_{red} = -2.37\ V = E^\circ_{anode}$. Cu^{2+} is reduced, so $E^\circ_{red} = 0.34\ V = E^\circ_{cathode}$. Then $E^\circ_{cell} = E^\circ_{cathode} - E^\circ_{anode} = 0.34\ V - (-2.37\ V) = +2.71\ V$. Add the two half-cell reactions: $Cu^{2+}(aq) + \cancel{2e^-} + Mg(s) \rightarrow Cu(s) + Mg^{2+}(aq) + \cancel{2e^-}$. Simplify to $Cu^{2+}(aq) + Mg(s) \rightarrow Cu(s) + Mg^{2+}(aq)$. So $Q = \frac{[Mg^{2+}]}{[Cu^{2+}]} = \frac{1.0 \times 10^{-4}}{1.5} = 6.\underline{6}667 \times 10^{-5}$ and $n = 2$.

Then $E_{cell} = E^\circ_{cell} - \frac{0.0592\ V}{n}\log Q = 2.71\ V - \frac{0.0592\ V}{2}\log(6.\underline{6}667 \times 10^{-5}) = +2.8\underline{3}361\ V = +2.83\ V.$

(b) $8.0\ \cancel{hr} \times \frac{60\ \cancel{min}}{1\ \cancel{hr}} \times \frac{60\ \cancel{s}}{1\ \cancel{min}} \times \frac{5.0\ \cancel{C}}{1\ \cancel{s}} \times \frac{1\ \cancel{mol\ e^-}}{96{,}485\ \cancel{C}} \times \frac{1\ mol\ Cu^{2+}}{2\ \cancel{mol\ e^-}} = 0.7\underline{4}623\ mol\ Cu^{2+}$ and

$[Cu^{2+}]_f = [Cu^{2+}]_i - \frac{mol\ Cu^{2+}\ reduced}{1.0\ L} = 1.5\ M - \frac{0.7\underline{4}623\ mol\ Cu^{2+}}{1.0\ L} = 0.7\underline{5}377\ M\ Cu^{2+}$ and

$0.7\underline{4}623\ \cancel{mol\ Cu^{2+}} \times \frac{1\ mol\ Mg\ oxidized}{1\ \cancel{mol\ Cu^{2+}\ reduced}} = 0.7\underline{4}623\ mol\ Mg$ oxidized and

$[Mg^{2+}]_f = [Mg^{2+}]_i + \frac{mol\ Mg\ oxidized}{1.0\ L} = 1.0 \times 10^{-4}\ M + \frac{0.7\underline{4}623\ mol\ Mg\ oxidized}{1.0\ L}$

$= 0.7\underline{4}633\ M\ Mg^{2+}$

$Q = \frac{[Mg^{2+}]}{[Cu^{2+}]} = \frac{0.7\underline{4}633}{0.7\underline{5}377} = 0.9\underline{9}013$ and $n = 2$ then

$E_{cell} = E^\circ_{cell} - \frac{0.0592\ V}{n}\log Q = 2.71\ V - \frac{0.0592\ V}{2}\log 0.9\underline{9}013 = +2.7\underline{1}013\ V = +2.71\ V$

(c) In 1.0 L, there are initially 1.5 moles of Cu^{2+}. So

$1.5\ \cancel{mol\ Cu^{2+}} \times \frac{2\ \cancel{mol\ e^-}}{1\ \cancel{mol\ Cu^{2+}}} \times \frac{96{,}485\ \cancel{C}}{1\ \cancel{mol\ e^-}} \times \frac{1\ \cancel{s}}{5.0\ \cancel{C}} \times \frac{1\ \cancel{min}}{60\ \cancel{s}} \times \frac{1\ hr}{60\ \cancel{min}} = 16\ hr.$

Check: The units (V, V, and hr) are correct. The Cu^{2+} reduction reaction is above the Mg^{2+} reduction reaction, so the standard cell potential will be positive. Having more reactants than products increases the cell potential. As the reaction proceeds, the potential drops. The concentrations drop by $\frac{1}{2}$ in 8 hours [part (b)], so all of it is consumed in 16 hours.

20.110 **Given:** Ag/Ag^+ concentration cell; initially, $[Ag^+] = 1.25\ M$ and $1.0 \times 10^{-3}\ M$ in 2.0 L half-cells
Find: (a) how long battery can deliver 2.5 A, (b) mass of Ag plated after 3.5 A for 5.5 h, and (c) how long battery can deliver 5.0 A to redissolve 1.00×10^2 g Ag with 10.0 A

Conceptual Plan:

(a) **Write the two half-cell reactions and add electrons as needed to balance reactions. Then $[Ag^+]$, V → mol Ag^+ → mol e^- → C → s → min → hr**

$$M = \frac{\text{mol } Ag^+}{\text{L}} \quad \frac{1 \text{ mol } e^-}{1 \text{ mol } Ag^+} \quad \frac{96{,}485 \text{ C}}{1 \text{ mol } e^-} \quad \frac{1 \text{ s}}{2.5 \text{ C}} \quad \frac{1 \text{ min}}{60 \text{ s}} \quad \frac{1 \text{ hr}}{60 \text{ min}}$$

(b) **hr → min → s → C → mol e^- → mol Ag → g Ag**

$$\frac{60 \text{ min}}{1 \text{ hr}} \quad \frac{60 \text{ s}}{1 \text{ min}} \quad \frac{3.5 \text{ C}}{1 \text{ s}} \quad \frac{1 \text{ mol } e^-}{96{,}485 \text{ C}} \quad \frac{1 \text{ mol Ag}}{1 \text{ mol } e^-} \quad \frac{107.87 \text{ g Ag}}{1 \text{ mol Ag}}$$

(c) **g Ag → mol Ag → mol e^- → C → s → min → hr**

$$\frac{1 \text{ mol Ag}}{107.87 \text{ g Ag}} \quad \frac{1 \text{ mol } e^-}{1 \text{ mol Ag}} \quad \frac{96{,}485 \text{ C}}{1 \text{ mol } e^-} \quad \frac{1 \text{ s}}{10.0 \text{ C}} \quad \frac{1 \text{ min}}{60 \text{ s}} \quad \frac{1 \text{ hr}}{60 \text{ min}}$$

Solution:

(a) Write half-reactions and add electrons. $Ag(aq) + e^- \rightarrow Ag(s)$. $M = \frac{\text{mol } Ag^+}{\text{L}}$, so

$$\text{mol } Ag^+ = M \times L = 1.25 \frac{\text{mol } Ag^+}{\text{L}} \times 2.0 \text{ L} = 2.50 \text{ mol } Ag^+ \text{ then}$$

$$2.50 \text{ mol } Ag^+ \times \frac{1 \text{ mol } e^-}{1 \text{ mol } Ag^+} \times \frac{96{,}485 \text{ C}}{1 \text{ mol } e^-} \times \frac{1 \text{ s}}{2.5 \text{ C}} \times \frac{1 \text{ min}}{60 \text{ s}} \times \frac{1 \text{ hr}}{60 \text{ min}} = 27 \text{ hr}$$

(b)
$$5.5 \text{ hr} \times \frac{60 \text{ min}}{1 \text{ hr}} \times \frac{60 \text{ s}}{1 \text{ min}} \times \frac{3.5 \text{ C}}{1 \text{ s}} \times \frac{1 \text{ mol } e^-}{96{,}485 \text{ C}} \times \frac{1 \text{ mol Ag}}{1 \text{ mol } e^-} \times \frac{107.87 \text{ g Ag}}{1 \text{ mol Ag}} = 77 \text{ g Ag}$$

(c)
$$1.00 \times 10^2 \text{ g Ag} \times \frac{1 \text{ mol Ag}}{107.87 \text{ g Ag}} \times \frac{1 \text{ mol } e^-}{1 \text{ mol Ag}} \times \frac{96{,}485 \text{ C}}{1 \text{ mol } e^-} \times \frac{1 \text{ s}}{10.0 \text{ C}} \times \frac{1 \text{ min}}{60 \text{ s}} \times \frac{1 \text{ hr}}{60 \text{ min}} = 2.48 \text{ hr}$$

Check: The units (hr, g, and hr) are correct. Because 2.5 moles need to be plated, we expect it to take a long time. The time is shorter by a factor of ~30, and the current is larger by about 1.5; so we expect less Ag to be plated in part (b) as compared to part (a). We expect a shorter time in part (c) than in part (b) because the current is so much larger. In parts (b) and (c), we are not exhausting the 2.5 moles initially in the half-cell.

20.111 **Given:** water electrolysis at 7.8 A; $H_2(g)$: $V = 25.0$ L, $P = 25.0$ atm; $T = 25\ °C$ **Find:** time

Conceptual Plan: °C → K then $V, P, T \rightarrow n$ then write half-reactions

$$K = °C + 273.15 \qquad PV = nRT$$

mol H_2 → mol e^- → C → s → min → hr

$$\frac{2 \text{ mol } e^-}{1 \text{ mol } H_2} \quad \frac{96{,}485 \text{ C}}{1 \text{ mol } e^-} \quad \frac{1 \text{ s}}{7.8 \text{ C}} \quad \frac{1 \text{ min}}{60 \text{ s}} \quad \frac{1 \text{ hr}}{60 \text{ min}}$$

Solution: $T = 25\ °C + 273.15 = 298$ K, then $PV = nRT$ Rearrange to solve for n.

$$n = \frac{PV}{RT} = \frac{25.0 \text{ atm} \times 25.0 \text{ L}}{0.08206 \frac{\text{L} \cdot \text{atm}}{\text{mol} \cdot \text{K}} \times 298 \text{ K}} = 25.\underline{5}583 \text{ mol } H_2.$$ The hydrolysis of water reactions are as follows:

$$2\,H_2O(l) \rightarrow O_2(g) + 4\,H^+(aq) + 4\,e^- \text{ and } 2\,H_2O(l) + 2\,e^- \rightarrow H_2(g) + 2\,OH^-(aq)$$

$$25.\underline{5}583 \text{ mol } H_2 \times \frac{2 \text{ mol } e^-}{1 \text{ mol } H_2} \times \frac{96{,}485 \text{ C}}{1 \text{ mol } e^-} \times \frac{1 \text{ s}}{7.8 \text{ C}} \times \frac{1 \text{ min}}{60 \text{ s}} \times \frac{1 \text{ hr}}{60 \text{ min}} = 176 \text{ hr}$$

Check: The units (hr) are correct. Because we have 25 L of gas at 25 atm and 25 °C, we expect ~25 moles of gas (remember that 1 mole of gas at STP = 22.4 L). A very long time is expected because we have so many moles of gas to generate.

20.112 **Given:** A fuel-cell breathalyzer test; 208 mL yields 324 mA for 10 s; $P = 1.0$ atm, $T = 25\ °C$

Find: percent by volume ethanol in breath

Conceptual Plan: Write the reaction and determine n, and mA → A then

$$\frac{1 \text{ A}}{1000 \text{ mA}}$$

s → C → mol e^- → mol CH_3CH_2OH and °C → K and mL → L then $V, n, T \rightarrow P_{CH_3CH_2OH}$

$$\frac{0.324 \text{ C}}{1 \text{ s}} \quad \frac{1 \text{ mol } e^-}{96{,}485 \text{ C}} \quad \frac{1 \text{ mol } CH_3CH_2OH}{4 \text{ mol } e^-} \qquad K = °C + 273.15 \qquad \frac{1 \text{ L}}{1000 \text{ mL}} \qquad PV = nRT$$

then $P_{CH_3CH_2OH}$, P_{Total}, → percent by volume CH_3CH_2OH.

$$\text{percent by volume } CH_3CH_2OH = \frac{P_{CH_3CH_2OH}}{P_{Total}} \times 100\%$$

Solution: $CH_3CH_2OH(g) + O_2(g) \rightarrow HC_2H_3O_2(g) + H_2O(g)$ because 2 O atoms are going from 0 to -2 so $n = 4$ and $324\ \cancel{mA} \times \frac{1\text{ A}}{1000\ \cancel{mA}} = 0.324$ A then

$$10\ \cancel{s} \times \frac{0.324\ \cancel{C}}{1\ \cancel{s}} \times \frac{1\ \cancel{mol\ e^-}}{96{,}485\ \cancel{C}} \times \frac{1\text{ mol } CH_3CH_2OH}{4\ \cancel{mol\ e^-}} = \underline{8}.3951 \times 10^{-6}\text{ mol } CH_3CH_2OH \text{ and}$$

$T = 25\ °C + 273.15 = 298$ K and $188\ \cancel{mL} \times \frac{1\text{ L}}{1000\ \cancel{mL}} = 0.188$ L then $PV = nRT$. Rearrange to solve for P.

$$P = \frac{nRT}{V} = \frac{\underline{8}.3951 \times 10^{-6}\ \cancel{mol} \times 0.08206 \frac{\cancel{L} \cdot \text{atm}}{\cancel{mol} \cdot \cancel{K}} \times 298\ \cancel{K}}{0.188\ \cancel{L}} = \underline{1}.09198 \times 10^{-3}\text{ atm } CH_3CH_2OH \text{ then}$$

$$\text{percent by volume } CH_3CH_2OH = \frac{P_{CH_3CH_2OH}}{P_{Total}} \times 100\% = \frac{1.09198 \times 10^{-3}\ \cancel{atm}}{1.0\ \cancel{atm}} \times 100\% =$$

$0.\underline{1}09198\% = 0.1\%$

Check: The units (%) are correct. Because the volume and current are so low, we expect a small percent by volume. This number is typical for a failing breathalyzer test.

20.113 **Given:** $Cu(s)\ |\ CuI(s)\ |\ I^-(aq, 1.0\text{ M})\ ||\ Cu^+(aq, 1.0\text{ M})\ |\ Cu(s)$, $K_{sp}(CuI) = 1.1 \times 10^{-12}$ **Find:** E_{cell}
Conceptual Plan: Write half-reactions from line notation. Because this is a concentration cell, $E°_{cell} = 0.00$ V. Then $K_{sp}, [I^-] \rightarrow [Cu^+]_{ox}$ then $E°_{cell}, [Cu^+]_{ox}, [Cu^+]_{red}, n \rightarrow E_{cell}$.

$K_{sp} = [Cu^+][I^-]$ $\qquad$ $E_{cell} = E°_{cell} - \frac{0.0592\text{ V}}{n}\log Q$

Solution: The half-reactions are $Cu(s) \rightarrow Cu^+(aq) + e^-$ and $Cu^+(aq) + e^- \rightarrow Cu(s)$. Because this is a concentration cell, $E°_{cell} = 0.00$ V and $n = 1$. Because $K_{sp} = [Cu^+][I^-]$, rearrange to solve for $[Cu^+]_{(ox)}$.

$$[Cu^+]_{ox} = \frac{K_{sp}}{[I^-]} = \frac{1.1 \times 10^{-12}}{1.0} = 1.1 \times 10^{-12}\text{ M then}$$

$$Q = \frac{[Cu^+]_{ox}}{[Cu^+]_{red}} = \frac{1.1 \times 10^{-12}}{1.0} = 1.1 \times 10^{-12}\text{ then}$$

$$E_{cell} = E°_{cell} - \frac{0.0592\text{ V}}{n}\log Q = 0.00\ V - \frac{0.0592\text{ V}}{1}\log(1.1 \times 10^{-12}) = 0.71\text{ V}$$

Check: The units (V) are correct. Because $[Cu^+]_{ox}$ is so low and $[Cu^+]_{red}$ is high, the Q is very small; so the voltage increase compared to the standard value is significant.

20.114 **Given:** $Zn(OH)_2(s) + 2\ e^- \rightarrow Zn(s) + 2\ OH^-(aq)$, $K_{sp}(Zn(OH)_2) = 1.8 \times 10^{-14}$ **Find:** E for half-cell
Conceptual Plan: In Table 20.1, look up the half-reaction for the Zn reduction ($E°_{red}$), determine n, then $K_{sp} \rightarrow [OH^-]$ then $E°_{cell}, [OH^-], n \rightarrow E_{cell}$.

$K_{sp} = [Zn^{2+}][OH^-]^2$ $\qquad$ $E_{cell} = E°_{cell} - \frac{0.0592\text{ V}}{n}\log Q$

Solution: The reduction half-reaction is $Zn^{2+}(aq) + 2\ e^- \rightarrow Zn(s)$ and $E°_{red} = -0.76$ V and $n = 2$. Because $K_{sp} = [Zn^{2+}][OH^-]^2 = S(2S)^2 = 4\ S^3$, rearrange to solve for $S = [OH^-]$.

$$S = [OH^-] = \sqrt[3]{\frac{K_{sp}}{4}} = \sqrt[3]{\frac{1.8 \times 10^{-14}}{4}} = 1.\underline{6}5096 \times 10^{-5}\text{ M then}$$

$$Q = [OH^-]^2 = (1.\underline{6}5096 \times 10^{-5})^2 = 2.\underline{7}257 \times 10^{-10}\text{ then}$$

$$E_{cell} = E°_{cell} - \frac{0.0592\text{ V}}{n}\log Q = -0.76\text{ V} - \frac{0.0592\text{ V}}{2}\log(2.\underline{7}257 \times 10^{-10}) = -0.48\text{ V}$$

Check: The units (V) are correct. Because $[OH^-]$ is so low, the Q is very small; so the voltage increase is significant, but the half-reaction is nonspontaneous.

20.115 **Given:** (a) disproportionation of $Mn^{2+}(aq)$ to $Mn(s)$ and $MnO_2(s)$ and (b) disproportionation of $MnO_2(s)$ to $Mn^{2+}(aq)$ and $MnO_4^-(s)$ in acidic solution **Find:** $\Delta G°_{rxn}$ and K
Conceptual Plan: Separate the overall reaction into two half-reactions: one for oxidation and one for reduction. → Balance each half-reaction with respect to mass in the following order: (1) Balance all elements other than H and O, (2) balance O by adding H_2O, and (3) balance H by adding H^+. → Balance each half-reaction with respect to charge by adding electrons. (The sum of the charges on both sides of the equation should be made

equal by adding electrons as necessary.) → Make the number of electrons in both half-reactions equal by multiplying one or both half-reactions by a small whole number. → Add the two half-reactions, canceling electrons and other species as necessary. → Verify that the reaction is balanced with respect to both mass and charge. Look up half-reactions in Table 20.1. Calculate the standard cell potential by subtracting the electrode potential of the anode from the electrode potential of the cathode: $E^\circ_{cell} = E^\circ_{cathode} - E^\circ_{anode}$. Then calculate ΔG°_{rxn} using $\Delta G^\circ_{rxn} = -n F E^\circ_{cell}$. Finally, °C → K then ΔG°_{rxn}, $T \to K$.

$K = 273.15 + °C$ $\quad \Delta G^\circ_{rxn} = -RT \ln K$

Solution:

(a) Separate: $Mn^{2+}(aq) \rightarrow MnO_2(s)$ and $Mn^{2+}(aq) \rightarrow Mn(s)$

Balance non-H & O elements: $Mn^{2+}(aq) \rightarrow MnO_2(s)$ and $Mn^{2+}(aq) \rightarrow Mn(s)$

Balance O with H_2O: $Mn^{2+}(aq) + 2\,H_2O(l) \rightarrow MnO_2(s)$ and $Mn^{2+}(aq) \rightarrow Mn(s)$

Balance H with H^+: $Mn^{2+}(aq) + 2\,H_2O(l) \rightarrow MnO_2(s) + 4\,H^+(aq)$ and $Mn^{2+}(aq) \rightarrow Mn(s)$

Add electrons: $Mn^{2+}(aq) + 2\,H_2O(l) \rightarrow MnO_2(s) + 4\,H^+(aq) + 2\,e^-$ and $Mn^{2+}(aq) + 2\,e^- \rightarrow Mn(s)$

Equalize electrons: $Mn^{2+}(aq) + 2\,H_2O(l) \rightarrow MnO_2(s) + 4\,H^+(aq) + 2\,e^-$ and $Mn^{2+}(aq) + 2\,e^- \rightarrow Mn(s)$

Add half-reactions: $Mn^{2+}(aq) + 2\,H_2O(l) + Mn^{2+}(aq) + \cancel{2e^-} \rightarrow MnO_2(s) + 4\,H^+(aq) + \cancel{2e^-} + Mn(s)$

Cancel electrons: $2\,Mn^{2+}(aq) + 2\,H_2O(l) \rightarrow MnO_2(s) + 4\,H^+(aq) + Mn(s)$

Look up cell potentials. Mn is oxidized in the first half-cell reaction, so $E^\circ_{anode} = E^\circ_{red} = +1.21$ V. Mn is reduced in the second half-cell reaction, so $E^\circ_{cathode} = E^\circ_{red} = -1.18$ V. Then $E^\circ_{cell} = E^\circ_{cathode} - E^\circ_{anode} = -1.18\text{ V} - 1.21\text{ V} = -2.39$ V and $n = 2$ so

$$\Delta G^\circ_{rxn} = -n F E^\circ_{cell} = -2\,\cancel{\text{mol e}^-} \times \frac{96{,}485\text{ C}}{\cancel{\text{mol e}^-}} \times -2.39\text{ V} = -2 \times 96{,}485\,\cancel{\text{C}} \times -2.39\frac{\text{J}}{\cancel{\text{C}}} =$$

$4.6\underline{1}198 \times 10^5$ J = 461 kJ and $T = 273.15 + 25\,°C = 298$ K then $\Delta G^\circ_{rxn} = -RT \ln K$. Rearrange to solve for K.

$$K = e^{\frac{-\Delta G^\circ_{rxn}}{RT}} = e^{\frac{-4.6\underline{1}198 \times 10^5\,\cancel{\text{J}}}{\left(8.314\frac{\cancel{\text{J}}}{\cancel{\text{K}}\cdot\text{mol}}\right)(298\,\cancel{\text{K}})}} = e^{-1\underline{8}6.149} = 1.43 \times 10^{-81}$$

Check:

Reactants	Products
2 Mn atoms	2 Mn atoms
2 O atoms	2 O atoms
4 H atoms	4 H atoms
+4 charge	+4 charge

The units (kJ and none) are correct. If the voltage is negative, the reaction is nonspontaneous, the free energy change is very positive, and the equilibrium constant is extremely small.

(b) Separate: $MnO_2(s) \rightarrow Mn^{2+}(aq)$ and $MnO_2(s) \rightarrow MnO_4^-(aq)$

Balance non-H & O elements: $MnO_2(s) \rightarrow Mn^{2+}(aq)$ and $MnO_2(s) \rightarrow MnO_4^-(aq)$

Balance O with H_2O: $MnO_2(s) \rightarrow Mn^{2+}(aq) + 2\,H_2O(l)$ and $MnO_2(s) + 2\,H_2O(l) \rightarrow MnO_4^-(aq)$

Balance H with H^+:

$MnO_2(s) + 4\,H^+(aq) \rightarrow Mn^{2+}(aq) + 2\,H_2O(l)$ and $MnO_2(s) + 2\,H_2O(l) \rightarrow MnO_4^-(aq) + 4\,H^+(aq)$

Add electrons: $MnO_2(s) + 4\,H^+(aq) + 2\,e^- \rightarrow Mn^{2+}(aq) + 2\,H_2O(l)$ and $MnO_2(s) + 2\,H_2O(l) \rightarrow MnO_4^-(aq) + 4\,H^+(aq) + 3\,e^-$

Equalize electrons: $3\,MnO_2(s) + 12\,H^+(aq) + 6\,e^- \rightarrow 3\,Mn^{2+}(aq) + 6\,H_2O(l)$ and $2\,MnO_2(s) + 4\,H_2O(l) \rightarrow 2\,MnO_4^-(aq) + 8\,H^+(aq) + 6\,e^-$

Add half-reactions: $3\,MnO_2(s) + 4\,\cancel{12\,H^+(aq)} + \cancel{6e^-} + 2\,MnO_2(s) + \cancel{4\,H_2O(l)} \rightarrow 3\,Mn^{2+}(aq) + 2\,\cancel{6}\,H_2O(l) + 2\,MnO_4^-(aq) + \cancel{8\,H^+(aq)} + \cancel{6e^-}$

Cancel electrons & species: $5\,MnO_2(s) + 4\,H^+(aq) \rightarrow 3\,Mn^{2+}(aq) + 2\,H_2O(l) + 2\,MnO_4^-(aq)$ Look up cell potentials. Mn is reduced in the first half-cell reaction, so $E^\circ_{cathode} = E^\circ_{red} = 1.21$ V. Mn is oxidized in the second half-cell reaction, so $E^\circ_{anode} = E^\circ_{red} = +1.68$ V. Then $E^\circ_{cell} = E^\circ_{cathode} - E^\circ_{anode} = 1.21\text{ V} - 1.68\text{ V} = -0.47$ V and $n = 6$ so

$$\Delta G^\circ_{rxn} = -nF E^\circ_{cell} = -6\,\cancel{\text{mol e}^-} \times \frac{96{,}485\text{ C}}{\cancel{\text{mol e}^-}} \times -0.47\text{ V} = -6 \times 96{,}485\,\cancel{\text{C}} \times -0.47\frac{\text{J}}{\cancel{\text{C}}}$$

$= 2.\underline{7}209 \times 10^5$ J = 270 kJ = 2.7×10^2 kJ and $T = 273.15 + 25\,°C = 298$ K then $\Delta G^\circ_{rxn} = -RT \ln K$

Rearrange to solve for K. $K = e^{\frac{-\Delta G^\circ_{rxn}}{RT}} = e^{\frac{-2.7209 \times 10^5 \text{ J}}{\left(8.314 \frac{\text{J}}{\text{K} \cdot \text{mol}}\right)(298 \text{ K})}} = e^{-109.82} = 2.0 \times 10^{-48}$

Check:

Reactants	Products
5 Mn atoms	5 Mn atoms
10 O atoms	10 O atoms
4 H atoms	4 H atoms
+4 charge	+4 charge

The units (kJ and none) are correct. If the voltage is negative, the reaction is nonspontaneous, the free energy change is very positive, and the equilibrium constant is extremely small. The voltage is less than in part (a), so the free energy change is not as large and the equilibrium constant is not as small.

20.116 **Given:** (a) reaction of $Cr^{2+}(aq)$ with $Cr_2O_7^{2-}(aq)$ in acidic solution to form $Cr^{3+}(aq)$ and (b) reaction of $Cr^{3+}(aq)$ with $Cr(s)$ to form $Cr^{2+}(aq)$ **Find:** ΔG°_{rxn} and K

Conceptual Plan: Separate the overall reaction into two half-reactions: one for oxidation and one for reduction. → Balance each half-reaction with respect to mass in the following order: (1) Balance all elements other than H and O, (2) balance O by adding H_2O, and (3) balance H by adding H^+. → Balance each half-reaction with respect to charge by adding electrons. (The sum of the charges on both sides of the equation should be made equal by adding electrons as necessary.) → Make the number of electrons in both half-reactions equal by multiplying one or both half-reactions by a small whole number. → Add the two half-reactions, canceling electrons and other species as necessary. → Verify that the reaction is balanced with respect to both mass and charge. Look up half-reactions in Table 20.1. Calculate the standard cell potential by subtracting the electrode potential of the anode from the electrode potential of the cathode: $E^\circ_{cell} = E^\circ_{cathode} - E^\circ_{anode}$. Then calculate ΔG°_{rxn} using $\Delta G^\circ_{rxn} = -nF\,E^\circ_{cell}$. Finally, °C → K then $\Delta G^\circ_{rxn}, T \rightarrow K$.

$\text{K} = 273.15 + {}^\circ\text{C}$ $\quad\quad$ $\Delta G^\circ_{rxn} = -RT \ln K$

Solution:

(a) Separate: $Cr^{2+}(aq) \rightarrow Cr^{3+}(aq)$ and $Cr_2O_7^{2-}(aq) \rightarrow Cr^{3+}(aq)$

Balance non-H & O elements: $Cr^{2+}(aq) \rightarrow Cr^{3+}(aq)$ and $Cr_2O_7^{2-}(aq) \rightarrow 2\,Cr^{3+}(aq)$

Balance O with H_2O: $Cr^{2+}(aq) \rightarrow Cr^{3+}(aq)$ and $Cr_2O_7^{2-}(aq) \rightarrow 2\,Cr^{3+}(aq) + 7\,H_2O(l)$

Balance H with H^+: $Cr^{2+}(aq) \rightarrow Cr^{3+}(aq)$ and $Cr_2O_7^{2-}(aq) + 14\,H^+(aq) \rightarrow 2\,Cr^{3+}(aq) + 7\,H_2O(l)$

Add electrons: $Cr^{2+}(aq) \rightarrow Cr^{3+}(aq) + e^-$ and $Cr_2O_7^{2-}(aq) + 14\,H^+(aq) + 6\,e^- \rightarrow 2\,Cr^{3+}(aq) + 7\,H_2O(l)$

Equalize electrons:

$6\,Cr^{2+}(aq) \rightarrow 6\,Cr^{3+}(aq) + 6\,e^-$ and $Cr_2O_7^{2-}(aq) + 14\,H^+(aq) + 6\,e^- \rightarrow 2\,Cr^{3+}(aq) + 7\,H_2O(l)$

Add half-reactions:

$6\,Cr^{2+}(aq) + Cr_2O_7^{2-}(aq) + 14\,H^+(aq) + \cancel{6e^-} \rightarrow 6\,Cr^{3+}(aq) + \cancel{6e^-}(aq) + 2\,Cr^{3+}(aq) + 7\,H_2O(l)$

Cancel electrons: $6\,Cr^{2+}(aq) + Cr_2O_7^{2-}(aq) + 14\,H^+(aq) \rightarrow 8\,Cr^{3+}(aq) + 7\,H_2O(l)$

Look up cell potentials. Cr is oxidized in the first half-cell reaction, so $E^\circ_{red} = -0.50\text{ V} = E^\circ_{anode}$. Cr is reduced in the second half-cell reaction, so $E^\circ_{red} = 1.33\text{ V} = E^\circ_{cathode}$. Then $E^\circ_{cell} = E^\circ_{cathode} - E^\circ_{anode} = 1.33\text{ V} - (-0.50\text{ V}) = +1.83\text{ V}$ and $n = 6$ so $\Delta G^\circ_{rxn} = -n\,F\,E^\circ_{cell} = -6\,\cancel{\text{mol e}^-} \times \frac{96{,}485\text{ C}}{\cancel{\text{mol e}^-}} \times 1.83\text{ V} = -6 \times 96{,}485\,\cancel{\text{C}} \times 1.83\frac{\text{J}}{\cancel{\text{C}}} = -1.05941 \times 10^6\text{ J} = -1.06 \times 10^3\text{ kJ}$ and $T = 273.15 + 25\,^\circ\text{C} = 298\text{ K}$ then $\Delta G^\circ_{rxn} = -RT \ln K$

Rearrange to solve for K.

$$K = e^{\frac{-\Delta G^\circ_{rxn}}{RT}} = e^{\frac{-(-1.05941 \times 10^6 \text{ J})}{\left(8.314 \frac{\text{J}}{\text{K} \cdot \text{mol}}\right)(298 \text{ K})}} = e^{427.600} = 5.06 \times 10^{185}$$

Check:

Reactants	Products
8 Cr atoms	8 Cr atoms
7 O atoms	7 O atoms
14 H atoms	14 H atoms
+24 charge	+24 charge

The units (kJ and none) are correct. If the voltage is positive, the reaction is spontaneous, the free energy change is very negative, and the equilibrium constant is extremely large.

(b) Separate: $Cr^{3+}(aq) \rightarrow Cr^{2+}(aq)$ and $Cr(s) \rightarrow Cr^{2+}(aq)$
Balance: $Cr^{3+}(aq) \rightarrow Cr^{2+}(aq)$ and $Cr(s) \rightarrow Cr^{2+}(aq)$
Add electrons: $Cr^{3+}(aq) + e^- \rightarrow Cr^{2+}(aq)$ and $Cr(s) \rightarrow Cr^{2+}(aq) + 2\,e^-$
Equalize electrons: $2\,Cr^{3+}(aq) + 2\,e^- \rightarrow 2\,Cr^{2+}(aq)$ and $Cr(s) \rightarrow Cr^{2+}(aq) + 2\,e^-$
Add half-reactions: $2\,Cr^{3+}(aq) + \cancel{2e^-} + Cr(s) \rightarrow 2\,Cr^{2+}(aq) + Cr^{2+}(aq) + \cancel{2e^-}$
Cancel electrons & species: $2\,Cr^{3+}(aq) + Cr(s) \rightarrow 3\,Cr^{2+}(aq)$
Look up cell potentials. Cr is reduced in the first half-cell reaction, so $E^\circ_{red} = E^\circ = -0.50\text{ V} = E^\circ_{cathode}$.
Cr is oxidized in the second half-cell reaction, so $E^\circ_{red} = -0.91\text{ V} = E^\circ_{anode}$. Then $E^\circ_{cell} = E^\circ_{cathode} - E^\circ_{anode} =$

$-0.50\text{ V} - (-0.91\text{ V}) = +0.41\text{ V}$ and $n = 2$ so $\Delta G^\circ_{rxn} = -n\,F\,E^\circ_{cell} = -2\,\cancel{\text{mol e}^-} \times \frac{96{,}485\text{ C}}{\cancel{\text{mol e}^-}} \times 0.41\text{ V} =$

$-2 \times 96{,}485\,\cancel{\text{C}} \times 0.41\frac{\text{J}}{\cancel{\text{C}}} = -7.\underline{9}118 \times 10^4\text{ J} = -79\text{ kJ}$ and $T = 273.15 + 25\text{ °C} = 298\text{ K}$

then $\Delta G^\circ_{rxn} = -RT\ln K$
Rearrange to solve for K.

$$K = e^{\frac{-\Delta G^\circ_{rxn}}{RT}} = e^{\frac{-(-7.\underline{9}118 \times 10^4\,\cancel{\text{J}})}{\left(8.314\frac{\cancel{\text{J}}}{\cancel{\text{K}}\cdot\text{mol}}\right)(298\,\cancel{\text{K}})}} = e^{3\underline{1}.934} = 7.4 \times 10^{13}$$

Check:

Reactants	Products
3 Cr atoms	3 Cr atoms
+6 charge	+6 charge

The units (kJ and none) are correct. If the voltage is positive, the reaction is spontaneous, the free energy change is negative, and the equilibrium constant is large. The voltage and n are less than in part (a), so the free energy change is not as negative and the equilibrium constant is not as large.

20.117 **Given:** metal, M, 50.9 g/mol, 1.20 g of metal reduced in 23.6 minutes at 6.42 A from molten chloride
Find: empirical formula of chloride
Conceptual Plan: min $\rightarrow$ s $\rightarrow$ C $\rightarrow$ mol e$^-$ and g M $\rightarrow$ mol M then mol e$^-$, mol M $\rightarrow$ charge $\rightarrow$ MCl_x

$\frac{60\text{ s}}{1\text{ min}}$ $\frac{6.42\text{ C}}{1\text{ s}}$ $\frac{1\text{ mol e}^-}{96{,}485\text{ C}}$ $\frac{1\text{ mol M}}{50.9\text{ g M}}$ $\frac{1\text{ mol e}^-}{1\text{ mol M}}$

Solution: $23.6\,\cancel{\text{min}} \times \frac{60\,\cancel{\text{s}}}{1\,\cancel{\text{min}}} \times \frac{6.42\,\cancel{\text{C}}}{1\,\cancel{\text{s}}} \times \frac{1\text{ mol e}^-}{96{,}485\,\cancel{\text{C}}} = 0.094\underline{2}190\text{ mol e}^-$ and

$1.20\,\cancel{\text{g M}} \times \frac{1\text{ mol M}}{50.9\,\cancel{\text{g M}}} = 0.023\underline{5}756\text{ mol M}$ then $\frac{0.094\underline{2}190\text{ mol e}^-}{0.023\underline{5}756\text{ mol M}} = 3.9\underline{9}646\frac{\text{e}^-}{\text{M}}$

So the empirical formula is MCl_4.

Check: The units (none) are correct. The result was an integer within the error of the measurements. The formula is typical for a metal salt. It could be vanadium, which has a +4 oxidation state.

20.118 **Given:** molten MF_3 electrolysis, 1.25 g of metal reduced in 16.2 minutes at 3.86 A **Find:** molar mass of metal
Conceptual Plan: min $\rightarrow$ s $\rightarrow$ C $\rightarrow$ mol e$^-$ $\rightarrow$ mol M then g M, mol M $\rightarrow$ molar mass

$\frac{60\text{ s}}{1\text{ min}}$ $\frac{3.86\text{ C}}{1\text{ s}}$ $\frac{1\text{ mol e}^-}{96{,}485\text{ C}}$ $\frac{1\text{ mol M}}{3\text{ mol e}^-}$ $\mathcal{M} = \frac{\text{g M}}{\text{mol M}}$

Solution: $16.2\,\cancel{\text{min}} \times \frac{60\,\cancel{\text{s}}}{1\,\cancel{\text{min}}} \times \frac{3.86\,\cancel{\text{C}}}{1\,\cancel{\text{s}}} \times \frac{1\,\cancel{\text{mol e}^-}}{96{,}485\,\cancel{\text{C}}} \times \frac{1\text{ mol M}}{3\,\cancel{\text{mol e}^-}} = 0.012\underline{9}620\text{ mol M}$ then

$$\mathcal{M} = \frac{1.25\text{ g M}}{0.012\underline{9}620\text{ mol M}} = 96.4\frac{\text{g}}{\text{mol}}$$

Check: The units (g/mol) are correct. The result was a number typical for metals, and it could be niobium, which is known to have a +3 oxidation state.

20.119 **Given:** 0.535g impure Sn; dissolve to form Sn^{2+} and titrate with 0.0344 L of 0.0448 M NO_3^- to generate NO
Find: percent by mass Sn

Conceptual Plan: Use balanced reaction from Problem 20.36(c) then
L → mol NO_3^- → mol Sn → g Sn → percent by mass Sn.

$M = \frac{mol}{L}$ $\frac{3 \text{ mol Sn}}{2 \text{ mol } NO_3^-}$ $\frac{118.71 \text{ g Sn}}{1 \text{ mol Sn}}$ percent by mass Sn $= \frac{\text{g Sn}}{\text{g sample}} \times 100\%$

Solution: $2\,NO_3^-(aq) + 8\,H^+(aq) + 3\,Sn^{2+}(aq) \rightarrow 2\,NO(g) + 4\,H_2O(l) + 3\,Sn^{4+}(aq)$

$$0.0344\,\text{L} \times \frac{0.0448 \text{ mol } NO_3^-}{1\,\text{L}} \times \frac{3 \text{ mol Sn}}{2 \text{ mol } NO_3^-} \times \frac{118.71 \text{ g Sn}}{1 \text{ mol Sn}} = 0.2744195 \text{ g Sn then}$$

$$\text{percent by mass Sn} = \frac{\text{g Sn}}{\text{g sample}} \times 100\% = \frac{0.2744195 \text{ g Sn}}{0.535 \text{ g sample}} \times 100\% = 51.3\% \text{ by mass Sn}$$

Check: The units (% by mass) are correct. The result was a number less than 100%.

20.120 **Given:** 0.0251 L Cu^+ solution titrated with 0.0322 L of 0.129 M $KMnO_4$ to generate Cu^{2+} and Mn^{2+}
Find: concentration of Cu solution
Conceptual Plan: Use the balanced reaction from Problem 20.99, but change metal and stoichiometry by a factor of 2 then L MnO_4^- → mol MnO_4^- → mol Cu^+ → [Cu^+].

$M = \frac{mol}{L}$ $\frac{5 \text{ mol } Cu^+}{2 \text{ mol } MnO_4^-}$ $M = \frac{mol}{L}$

Solution: $MnO_4^-(aq) + 8\,H^+(aq) + 5\,Cu^+(aq) \rightarrow Mn^{2+}(aq) + 4\,H_2O(l) + 5\,Cu^{2+}(aq)$

$$0.0322 \text{ L } MnO_4^- \times \frac{0.129 \text{ mol } MnO_4^-}{1 \text{ L } MnO_4^-} \times \frac{5 \text{ mol } Cu^+}{1 \text{ mol } MnO_4^-} = 0.020769 \text{ mol } Cu^+ \text{ then}$$

$$[Cu^{2+}] = \frac{0.020769 \text{ mol } Cu^{2+}}{0.0251 \text{ L}} = 0.827 \text{ M } Cu^{2+}$$

Check: The units (M) are correct. The Cu concentration is much higher than the $KMnO_4$ concentration because of the reaction stoichiometry.

20.121 **Given:** 1.25 L of a 0.552 M HBr solution, convert H^+ to $H_2(g)$ for 73 minutes at 11.32 A **Find:** pH
Conceptual Plan: min → s → C → mol e^- → mol H^+ consumed and L, M → mol H^+ initially then

$\frac{60 \text{ s}}{1 \text{ min}}$ $\frac{11.32 \text{ C}}{1 \text{ s}}$ $\frac{1 \text{ mol } e^-}{96{,}485 \text{ C}}$ $\frac{2 \text{ mol } H^+}{2 \text{ mol } e^-}$ $M = \frac{mol}{L}$

mol H^+ initially, mol H^+ consumed → mol H^+ remaining then mol H^+ remaining, L → [H^+] → pH

mol H^+ initially − mol H^+ consumed = mol H^+ remaining $M = \frac{mol}{L}$ $pH = -\log[H^+]$

Solution: $2\,H^+(aq) + 2\,e^- \rightarrow H_2(g)$

$$73 \text{ min} \times \frac{60 \text{ s}}{1 \text{ min}} \times \frac{11.32 \text{ C}}{1 \text{ s}} \times \frac{1 \text{ mol } e^-}{96{,}485 \text{ C}} \times \frac{2 \text{ mol } H^+}{2 \text{ mol } e^-} = 0.513879 \text{ mol } H^+ \text{ consumed and}$$

$$1.25 \text{ L } H^+ \times \frac{0.552 \text{ mol } H^+}{1 \text{ L } H^+} = 0.690 \text{ mol } H^+ \text{ initially then mol } H^+ \text{ initially} - \text{mol } H^+ \text{ consumed} =$$

mol H^+ remaining = 0.690 mol H^+ initially − 0.513879 mol H^+ consumed = 0.17612 mol H^+ remaining

$$\text{then } [H^+] = \frac{0.17612 \text{ mol } H^+}{1.25 \text{ L}} = 0.140897 \text{ M } H^+ \text{ then } pH = -\log[H^+] = = -\log 0.140897 = 0.85$$

Check: The units (none) are correct. The result is a pH higher than the initial pH [−log (0.552) = 0.258], as is expected.

20.122 **Given:** 215 mL of a 0.500 M NaCl solution, initially at pH = 7.00; after 15 minutes, a 10.0 mL aliquot is titrated with 22.8 mL 0.100 M HCl **Find:** current (A)
Conceptual Plan: Titration is neutralizing base generated in the hydrolysis, so
M HCl, mL HCl, → mol OH^- in aliquot mL solution, mL aliquot → mol OH^- in solution then min → s then

$\frac{1 \text{ min}}{60 \text{ s}}$

mol OH^- in solution, s → mol e^- → C/s.

$\frac{2 \text{ mol } e^-}{2 \text{ mol } OH^-}$ $\frac{96{,}485 \text{ C}}{1 \text{ mol } e^-}$

Solution: $2\,NaCl(aq) + 2\,H_2O(l) \rightarrow H_2(g) + Cl_2(g) + 2\,Na^+(aq) + 2\,OH^-(aq)$

$$22.8\ \cancel{mL\ HCl} \times \frac{0.100\ \cancel{mol\ HCl}}{1000\ \cancel{mL\ HCl}} \times \frac{1\ mol\ OH^-}{1\ \cancel{mol\ HCl}} = 0.00228\ mol\ OH^-$$

$$0.00228\ mol\ OH^-\ \text{in aliquot} \times \frac{215\ \cancel{mL}}{10.0\ \cancel{mL}} = 0.04\underline{9}02\ mol\ OH^-\ \text{in solution}$$

$$\frac{0.04\underline{9}02\ \cancel{mol\ OH^-\ in\ solution}}{15\ \cancel{min}} \times \frac{1\ \cancel{min}}{60\ s} \times \frac{2\ \cancel{mol\ e^-}}{2\ \cancel{mol\ OH^-}} \times \frac{96{,}485\ C}{1\ \cancel{mol\ e^-}} = 5.\underline{2}552\ \frac{C}{s} = 5.3\ A$$

Check: The units (A) are correct. The current is reasonable for an electrolysis process.

20.123 **Given:** MnO_2/Mn^{2+} electrode at pH 10.24 **Find:** $[Mn^{2+}]$ to get half-cell potential = 0.00 V

Conceptual Plan: pH → [H⁺] then

$$pH = -\log[H^+]$$

Write half-cell reactions and look up half-reactions in Table 20.1. Define Q based on half-cell reaction. Then $E^\circ_{half\text{-}cell}$, n, $[H^+] \rightarrow [Mn^{2+}]$.

$$E_{half\text{-}cell} = E^\circ_{half\text{-}cell} - \frac{0.0592\ V}{n}\log Q$$

Solution: Because $pH = -\log[H^+]$ so $[H^+] = 10^{-pH} = 10^{-10.24} = 5.\underline{7}5440 \times 10^{-11}\ M$

$MnO_2(s) + 4\,H^+(aq) + 2\,e^- \rightarrow Mn^{2+}(aq) + 2\,H_2O(l)$ $E^\circ_{half\text{-}cell} = 1.21\ V$, $n = 2$, and $Q = \dfrac{[Mn^{2+}]}{[H^+]^4}$

$$E_{half\text{-}cell} = E^\circ_{half\text{-}cell} - \frac{0.0592\ V}{n}\log Q \text{ so } 0.00\ V = 1.21\ V - \frac{0.0592\ V}{2}\log\frac{[Mn^{2+}]}{(5.\underline{7}5440 \times 10^{-11})^4} \rightarrow$$

$$1.21\ V = \frac{0.0592\ V}{2}\log\frac{[Mn^{2+}]}{(5.\underline{7}5440 \times 10^{-11})^4} \rightarrow 40.8\underline{7}84 = \log\frac{[Mn^{2+}]}{(5.\underline{7}5440 \times 10^{-11})^4} \rightarrow$$

$$\frac{[Mn^{2+}]}{(5.\underline{7}5440 \times 10^{-11})^4} = 10^{40.8\underline{7}84} = 7.\underline{5}5750 \times 10^{40} \rightarrow [Mn^{2+}] = 0.8\underline{2}8664\ M = 0.83\ M\ Mn^{2+}$$

Check: The units (M) are correct. The standard half-cell potential is very large and positive. Most of the shift toward 0.00 V is due to the fourth-order dependence in $[H^+]$ at a high pH, so the $[Mn^{2+}]$ is close to 1 M. So the concentration is reasonable.

20.124 **Given:** SHE **Find:** pH to get half-cell potential = −0.122 V

Conceptual Plan: Write the half-cell reaction. The standard half-cell potential is 0.00 V. Define Q based on half-cell reaction. Then $E^\circ_{half\text{-}cell}$, n → [H⁺] → pH.

$$E_{half\text{-}cell} = E^\circ_{half\text{-}cell} - \frac{0.0592\ V}{n}\log Q \qquad pH = -\log[H^+]$$

Solution: $2\,H^+(aq) + 2\,e^- \rightarrow H_2(g)$ $E^\circ_{half\text{-}cell} = 0.00\ V$, $n = 2$, and $Q = \dfrac{1}{[H^+]^2}$

$$E_{half\text{-}cell} = E^\circ_{half\text{-}cell} - \frac{0.0592\ V}{n}\log Q \text{ so } -0.122\ V = 0.00\ V - \frac{0.0592\ V}{2}\log[H^+]^{-2} \rightarrow$$

$4.1\underline{2}162 = \log[H^+]^{-2} \rightarrow [H^+]^{-2} = 10^{4.1\underline{2}162} = 1.\underline{3}2319 \times 10^4 \rightarrow [H^+] = 0.008\underline{6}9339\ M$

Because $pH = -\log[H^+] = -\log(0.008\underline{6}9339) = 2.0\underline{6}081 = 2.06$

Check: The units (none) are correct. There is an inverse second-order dependence in $[H^+]$, so we expect an acidic pH.

Challenge Problems

20.125 **Given:** hydrogen–oxygen fuel cell; 1.2×10^3 kWh of electricity/month **Find:** V of $H_2(g)$ at STP/month

Conceptual Plan: Write half-reactions. Look up half-reactions at pH 7. The reaction on the left is the oxidation. Calculate the standard cell potential by subtracting the electrode potential of the anode from the electrode potential of the cathode: $E^\circ_{cell} = E^\circ_{cathode} - E^\circ_{anode}$. Add the two half-cell reactions and cancel the electrons. Then kWh → J → C → mol e⁻ → mol H_2 → V.

$$\frac{3.60 \times 10^6\ J}{1\ kWh} \quad \frac{1\ C}{1.23\ J} \quad \frac{1\ mol\ e^-}{96{,}485\ C} \quad \frac{2\ mol\ H_2}{4\ mol\ e^-} \quad \text{at STP}\ \frac{22.414\ L}{1\ mol\ H_2}$$

Solution: $2\,H_2(g) + 4\,OH^-(aq) \rightarrow 4\,H_2O(l) + 4\,e^-$ where

$E^\circ_{red} = -0.83\ V = E^\circ_{anode}$ and $O_2(g) + 2\,H_2O(l) + 4\,e^- \rightarrow 4\,OH^-(aq)$ where

$E^\circ_{red} = 0.40\text{ V} = E^\circ_{cathode}$. $E^\circ_{cell} = E^\circ_{cathode} - E^\circ_{anode} = 0.40\text{ V} - (-0.83\text{ V}) = 1.23\text{ V} = 1.23\text{ J/C}$ and $n = 4$. Net reaction is $2\,H_2(g) + O_2(g) \rightarrow 2\,H_2O(l)$. Then

$$1.2 \times 10^3\,\cancel{\text{kWh}} \times \frac{3.60 \times 10^6\,\cancel{\text{J}}}{1\,\cancel{\text{kWh}}} \times \frac{1\,\cancel{\text{C}}}{1.23\,\cancel{\text{J}}} \times \frac{1\,\cancel{\text{mol e}^-}}{96,485\,\cancel{\text{C}}} \times \frac{2\,\cancel{\text{mol H}_2}}{4\,\cancel{\text{mol e}^-}} \times \frac{22.414\text{ L}}{1\,\cancel{\text{mol H}_2}} = 4.1 \times 10^5\text{ L}$$

Check: The units (L) are correct. A large volume is expected because we are trying to generate a large amount of electricity.

20.126 **Given:** voltaic cell to measure $[Cu^{2+}]$; SHE electrode paired with Cu^{2+}/Cu cell
Find: parameters to plot for a calibrations curve and the slope of the curve
Conceptual Plan: Write the two half-cell reactions and add electrons as needed to balance reactions. Look up half-reactions in Table 20.1. Calculate the standard cell potential by subtracting the electrode potential of the anode from the electrode potential of the cathode: $E^\circ_{cell} = E^\circ_{cathode} - E^\circ_{anode}$. Add the two half-cell reactions and cancel electrons and determine n. Then $E^\circ_{cell}, P_{H_2}, [H^+], [Cu^{2+}], n \rightarrow E_{cell}$.

$$E_{cell} = E^\circ_{cell} - \frac{0.0592\text{ V}}{n}\log Q$$

Solution: Write half-reactions and add electrons. $H_2(g) \rightarrow 2\,H^+(aq) + 2\,e^-$ and $Cu^{2+}(aq) + 2\,e^- \rightarrow Cu(s)$. Look up cell potentials. H is oxidized, so $E^\circ_{red} = 0.00\text{ V} = E^\circ_{anode}$. Cu^{2+} is reduced, so $E^\circ_{red} = 0.34\text{ V} = E^\circ_{cathode}$. Then $E^\circ_{cell} = E^\circ_{cathode} - E^\circ_{anode} = 0.34\text{ V} - 0.00\text{ V} = +0.34\text{ V}$. Add the two half-cell reactions:
$H_2(g) + Cu^{2+}(aq) + \cancel{2e^-} \rightarrow 2\,H^+(aq) + \cancel{2e^-} + Cu(s)$. Simplify to $H_2(g) + Cu^{2+}(aq) \rightarrow 2\,H^+(aq) + Cu(s)$.
So $Q = \dfrac{[H^+]^2}{P_{H_2}[Cu^{2+}]}$ and $n = 2$. Then $E_{cell} = E^\circ_{cell} - \dfrac{0.0592\text{ V}}{n}\log Q = 0.34\text{ V} - \dfrac{0.0592\text{ V}}{2}\log\dfrac{[H^+]^2}{P_{H_2}[Cu^{2+}]}$. If the anode half-cell is buffered at a constant pH and a constant P_{H_2} is used, two of the terms in Q are constant and can be pulled out of the expression so that

$$E_{cell} = 0.34\text{ V} - \frac{0.0592\text{ V}}{2}\left(\log\frac{[H^+]^2}{P_{H_2}} - \log[Cu^{2+}]\right)$$

$$= 0.34\text{ V} - (0.0296\text{ V})\log\frac{[H^+]^2}{P_{H_2}} + (0.0296\text{ V})\log[Cu^{2+}].$$

If we plot $\log[Cu^{2+}]$ versus E_{cell}, the slope will be 0.0296 V. Note: The value of both the slope and y-intercept are needed to calculate the copper concentration.

20.127 **Given:** Au^{3+}/Au electroplating; surface area $= 49.8\text{ cm}^2$, Au thickness $= 1.00 \times 10^{-3}$ cm, density $= 19.3\text{ g/cm}^3$; at 3.25 A **Find:** time
Conceptual Plan: Write the half-cell reaction and add electrons as needed to balance reactions. Then surface area, thickness $\rightarrow$ V $\rightarrow$ g Au $\rightarrow$ mol Au $\rightarrow$ mol e$^-$ $\rightarrow$ C $\rightarrow$ s

$V = surface\ area \times thickness$ $\quad \dfrac{19.3\text{ g Au}}{1\text{ cm}^3\text{ Au}}$ $\quad \dfrac{1\text{ mol Au}}{196.97\text{ g Au}}$ $\quad \dfrac{3\text{ mol e}^-}{1\text{ mol Au}}$ $\quad \dfrac{96,485\text{ C}}{1\text{ mol e}^-}$ $\quad \dfrac{1\text{ s}}{3.25\text{ C}}$

Solution: Write the half-reaction and add electrons. $Au^{3+}(aq) + 3\,e^- \rightarrow Au(s)$.
$V = surface\ area \times thickness = (49.8\text{ cm}^2)(1.00 \times 10^{-3}\text{ cm}) = 0.0498\text{ cm}^3$ then

$$0.0498\,\cancel{\text{cm}^3\text{ Au}} \times \frac{19.3\,\cancel{\text{g Au}}}{1\,\cancel{\text{cm}^3\text{ Au}}} \times \frac{1\,\cancel{\text{mol Au}}}{196.97\,\cancel{\text{g Au}}} \times \frac{3\,\cancel{\text{mol e}^-}}{1\,\cancel{\text{mol Au}}} \times \frac{96,485\,\cancel{\text{C}}}{1\,\cancel{\text{mol e}^-}} \times \frac{1\text{ s}}{3.25\,\cancel{\text{C}}} = 435\text{ s}$$

Check: The units (s) are correct. Because the layer is so thin, there is far less than a mole of gold; so the time is not very long. To be an economical process, it must be fairly quick.

20.128 **Given:** electrodeposit mixture Cu and Cd with 1.20 F (1 F = 1 mol e$^-$); total mass = 50.36 g
Find: mass of $CuSO_4$
Conceptual Plan: Write the half-cell reaction and add electrons as needed to balance reactions. Then F $\rightarrow$ mol e$^-$ $\rightarrow$ mol (Cu + Cd) then let x = g Cu so that (50.36 g $-$ x g) = g Cd then g Cu $\rightarrow$ mol Cu

$\dfrac{1\text{ mol e}^-}{1\text{ F}}$ $\quad \dfrac{1\text{ mol (Cu + Cd)}}{2\text{ mol e}^-}$ $\quad \dfrac{1\text{ mol Cu}}{63.55\text{ g Cu}}$

and g Cd $\rightarrow$ mol Cd then solve for x = g Cu $\rightarrow$ g $CuSO_4$.

$\dfrac{1\text{ mol Cd}}{112.41\text{ g Cd}}$ $\quad \dfrac{159.62\text{ g CuSO}_4}{63.55\text{ g Cu}}$

Solution: $Cu^{2+}(aq) + 2\,e^- \rightarrow Cu(s)$ and $Cd^{2+}(aq) + 2\,e^- \rightarrow Cd(s)$, so $n = 2$ for both metals.

$$1.20\,\cancel{F} \times \frac{1\,\cancel{mol\,e^-}}{1\,\cancel{F}} \times \frac{1\,mol(Cu + Cd)}{2\,\cancel{mol\,e^-}} = 0.600\,mol(Cu + Cd)$$ then let $x =$ g Cu so that $(50.36\,g - x\,g) =$ g Cd

then $x\,\cancel{g\,Cu} \times \frac{1\,mol\,Cu}{63.55\,\cancel{g\,Cu}} = \frac{x}{63.55}\,mol\,Cu$ and $(50.36 - x)\cancel{g\,Cd} \times \frac{1\,mol\,Cd}{112.41\,\cancel{g\,Cd}} = \frac{(50.36 - x)}{112.41}\,mol\,Cd$

then $0.600\,mol(Cu + Cd) = \frac{x}{63.55}\,mol\,Cu + \frac{(50.36 - x)}{112.41}\,mol\,Cd$. Solve for x.

$0.600 = 0.01573564x + 0.4480028 - 0.008896006x \rightarrow x = 22.\underline{2}2300\,g\,Cu \rightarrow$

$$22.\underline{2}2300\,\cancel{g\,Cu} \times \frac{159.62\,g\,CuSO_4}{63.55\,\cancel{g\,Cu}} = 55.8\,g\,CuSO_4$$

Check: The units (g) are correct. The result is reasonable because 0.600 mol Cu = 38.1 g and 0.600 mol Cd = 67.4 g and the amount deposited is between the two values.

20.129 **Given:** $C_2O_4^{2-} \rightarrow CO_2$ and $MnO_4^-(aq) \rightarrow Mn^{2+}(aq)$, 50.1 mL of MnO_4^- to titrate 0.339 g $Na_2C_2O_4$; 4.62 g U sample titrated by 32.5 mL MnO_4^-; and $UO^{2+} \rightarrow UO_2^{2+}$ **Find:** percent U in sample

Conceptual Plan: Separate the overall reaction into two half-reactions: one for oxidation and one for reduction. → Balance each half-reaction with respect to mass in the following order: (1) Balance all elements other than H and O, (2) balance O by adding H_2O, and (3) balance H by adding H^+. → Balance each half-reaction with respect to charge by adding electrons. (The sum of the charges on both sides of the equation should be made equal by adding electrons as necessary.) → Make the number of electrons in both half-reactions equal by multiplying one or both half-reactions by a small whole number. → Add the two half-reactions, canceling electrons and other species as necessary. → Verify that the reaction is balanced with respect to both mass and charge. Then

mL $MnO_4^- \rightarrow$ LMnO_4^- and g $Na_2C_2O_4 \rightarrow$ mol $Na_2C_2O_4 \rightarrow$ mol MnO_4^- then

$$\frac{1\,L\,MnO_4^-}{1000\,mL\,MnO_4^-} \qquad \frac{1\,mol\,Na_2C_2O_4}{134.00\,g\,Na_2C_2O_4} \quad \frac{2\,mol\,MnO_4^-}{5\,mol\,Na_2C_2O_4}$$

L MnO_4^-, mol $MnO_4^- \rightarrow$ M MnO_4^- then write U half-reactions and balance as above. →

$$M = \frac{mol\,MnO_4^-}{L}$$

Make the number of electrons in both half-reactions equal by multiplying one or both half-reactions by a small whole number. → Add the two half-reactions, canceling electrons and other species as necessary. → Verify that the reaction is balanced with respect to both mass and charge.

Then mL MnO_4^-, M $MnO_4^- \rightarrow$ mol $MnO_4^- \rightarrow$ mol U → g U then g U, g sample → % U.

$$M = \frac{mol\,MnO_4^-}{L} \qquad \frac{5\,mol\,U}{2\,mol\,MnO_4^-} \quad \frac{238.03\,g\,U}{1\,mol\,U} \qquad \text{percent U} = \frac{g\,U}{g\,sample} \times 100\%$$

Solution:

Separate: $MnO_4^-(aq) \rightarrow Mn^{2+}(aq)$ and $C_2O_4^{2-}(aq) \rightarrow CO_2(g)$

Balance non-H & O elements: $MnO_4^-(aq) \rightarrow Mn^{2+}(aq)$ and $C_2O_4^{2-}(aq) \rightarrow 2\,CO_2(g)$

Balance O with H_2O: $MnO_4^-(aq) \rightarrow Mn^{2+}(aq) + 4\,H_2O(l)$ and $C_2O_4^{2-}(aq) \rightarrow 2\,CO_2(g)$

Balance H with H^+: $MnO_4^-(aq) + 8\,H^+(aq) \rightarrow Mn^{2+}(aq) + 4\,H_2O(l)$ and $C_2O_4^{2-}(aq) \rightarrow 2\,CO_2(g)$

Add electrons: $MnO_4^-(aq) + 8\,H^+(aq) + 5\,e^- \rightarrow Mn^{2+}(aq) + 4\,H_2O(l)$ and $C_2O_4^{2-}(aq) \rightarrow 2\,CO_2(g) + 2\,e^-$

Equalize electrons:

$2\,MnO_4^-(aq) + 16\,H^+(aq) + 10\,e^- \rightarrow 2\,Mn^{2+}(aq) + 8\,H_2O(l)$ and $5\,C_2O_4^{2-}(aq) \rightarrow 10\,CO_2(g) + 10\,e^-$

Add half-reactions:

$2\,MnO_4^-(aq) + 16\,H^+(aq) + \cancel{10\,e^-} + 5\,C_2O_4^{2-}(aq) \rightarrow 2\,Mn^{2+}(aq) + 8\,H_2O(l) + 10\,CO_2(g) + \cancel{10\,e^-}$

Cancel electrons: $2\,MnO_4^-(aq) + 16\,H^+(aq) + 5\,C_2O_4^{2-}(aq) \rightarrow 2\,Mn^{2+}(aq) + 8\,H_2O(l) + 10\,CO_2(g)$

then $50.1\,\cancel{mL\,MnO_4^-} \times \frac{1\,L\,MnO_4^-}{1000\,\cancel{mL\,MnO_4^-}} = 0.0501\,L\,MnO_4^-$

$$0.399\,\cancel{g\,Na_2C_2O_4} \times \frac{1\,\cancel{mol\,Na_2C_2O_4}}{134.00\,\cancel{g\,Na_2C_2O_4}} \times \frac{2\,mol\,MnO_4^-}{5\,\cancel{mol\,Na_2C_2O_4}} = 0.0011\underline{9}10448\,mol\,MnO_4^-$$

$$M = \frac{0.0011\underline{9}10448\,mol\,MnO_4^-}{0.0501\,L} = 0.023\underline{7}73345\,M\,MnO_4^-$$

Separate: $MnO_4^-(aq) \rightarrow Mn^{2+}(aq)$ and $UO^{2+}(aq) \rightarrow UO_2^{2+}(aq)$

Balance non-H & O elements: $MnO_4^-(aq) \rightarrow Mn^{2+}(aq)$ and $UO^{2+}(aq) \rightarrow UO_2^{2+}(aq)$

Balance O with H_2O: $MnO_4^-(aq) \rightarrow Mn^{2+}(aq) + 4\,H_2O(l)$ and $UO^{2+}(aq) + H_2O(l) \rightarrow UO_2^{2+}(aq)$

Balance H with H^+:

$MnO_4^-(aq) + 8\,H^+(aq) \rightarrow Mn^{2+}(aq) + 4\,H_2O(l)$ and $UO^{2+}(aq) + H_2O(l) \rightarrow UO_2^{2+}(aq) + 2\,H^+(aq)$

Add electrons: $MnO_4^-(aq) + 8\,H^+(aq) + 5\,e^- \rightarrow Mn^{2+}(aq) + 4\,H_2O(l)$ and
$UO^{2+}(aq) + H_2O(l) \rightarrow UO_2^{2+}(aq) + 2\,H^+(aq) + 2\,e^-$

Equalize electrons: $2\,MnO_4^-(aq) + 16\,H^+(aq) + 10\,e^- \rightarrow 2\,Mn^{2+}(aq) + 8\,H_2O(l)$ and
$5\,UO^{2+}(aq) + 5\,H_2O(l) \rightarrow 5\,UO_2^{2+}(aq) + 10\,H^+(aq) + 10\,e^-$

Add half-reactions: $2\,MnO_4^-(aq) + 6\,\cancel{16}\,H^+(aq) + \cancel{10\,e^-} + 5\,UO^{2+}(aq) + \cancel{5\,H_2O(l)} \rightarrow$
$2\,Mn^{2+}(aq) + 3\,\cancel{8}\,H_2O(l) + 5\,UO_2^{2+}(aq) + \cancel{10\,H^+(aq)} + \cancel{10\,e^-}$

Cancel electrons & species: $2\,MnO_4^-(aq) + 6\,H^+(aq) + 5\,UO^{2+}(aq) \rightarrow 2\,Mn^{2+}(aq) + 3\,H_2O(l) + 5\,UO_2^{2+}(aq)$

$$32.5\,\cancel{\text{mL } MnO_4^-} \times \frac{0.02\underline{3}773345\,\cancel{\text{mol } MnO_4^-}}{1000\,\cancel{\text{mL } MnO_4^-}} \times \frac{5\,\cancel{\text{mol U}}}{2\,\cancel{\text{mol } MnO_4^-}} \times \frac{238.03\text{ g U}}{1\,\cancel{\text{mol U}}} = 0.45\underline{9}7750\text{ g U then}$$

$$\text{percent U} = \frac{\text{g U}}{\text{g sample}} \times 100\% = \frac{0.45\underline{9}7750\text{ g U}}{4.63\text{ g sample}} \times 100\% = 9.93\%$$

Check: first reaction

Reactants	Products
2 Mn atoms	2 Mn atoms
28 O atoms	28 O atoms
16 H atoms	16 H atoms
10 C atoms	10 C atoms
+4 charge	+4 charge

second reaction

Reactants	Products
2 Mn atoms	2 Mn atoms
13 O atoms	13 O atoms
6 H atoms	6 H atoms
5 U atoms	5 U atoms
+14 charge	+14 charge

The reactions are balanced. The units (%) are correct. The percentage is between 0 and 100%.

20.130 **Given:** 2.33 A applied to Cu electroplating of 1.74 g; also electroplate Au and Ag

Find: time, g Au, and g Ag

Conceptual Plan: g Cu → mol Cu → mol e^- → C → s then

$$\frac{1\text{ mol Cu}}{63.55\text{ g Cu}} \quad \frac{2\text{ mol } e^-}{1\text{ mol Cu}} \quad \frac{96{,}485\text{ C}}{1\text{ mol } e^-} \quad \frac{1\text{ s}}{2.33\text{ C}}$$

s → C → mol e^- → mol Au → g Au and s → C → mol e^- → mol Ag → g Ag

$$\frac{2.33\text{ C}}{1\text{ s}} \quad \frac{1\text{ mol } e^-}{96{,}485\text{ C}} \quad \frac{1\text{ mol Au}}{3\text{ mol } e^-} \quad \frac{196.97\text{ g Au}}{1\text{ mol Au}} \qquad \frac{2.33\text{ C}}{1\text{ s}} \quad \frac{1\text{ mol } e^-}{96{,}485\text{ C}} \quad \frac{1\text{ mol Ag}}{1\text{ mol } e^-} \quad \frac{107.87\text{ g Ag}}{1\text{ mol Ag}}$$

Solution: $Cu^{2+}(aq) + 2\,e^- \rightarrow Cu(s)$

$$1.74\,\cancel{\text{g Cu}} \times \frac{1\,\cancel{\text{mol Cu}}}{63.55\,\cancel{\text{g Cu}}} \times \frac{2\,\cancel{\text{mol } e^-}}{1\,\cancel{\text{mol Cu}}} \times \frac{96{,}485\,\cancel{\text{C}}}{1\,\cancel{\text{mol } e^-}} \times \frac{1\text{ s}}{2.33\,\cancel{\text{C}}} = 22\underline{6}7.61\text{ s} = 2270\text{ s or } 37.8\text{ min}$$

$Au^{3+}(aq) + 3\,e^- \rightarrow Au(s)$

$$22\underline{6}7.61\,\cancel{\text{s}} \times \frac{2.33\,\cancel{\text{C}}}{1\,\cancel{\text{s}}} \times \frac{1\,\cancel{\text{mol } e^-}}{96{,}485\,\cancel{\text{C}}} \times \frac{1\,\cancel{\text{mol Au}}}{3\,\cancel{\text{mol } e^-}} \times \frac{196.97\text{ g Au}}{1\,\cancel{\text{mol Au}}} = 3.5\underline{9}536\text{ g Au} = 3.60\text{ g Au}$$

$Ag^+(aq) + e^- \rightarrow Ag(s)$

$$22\underline{6}7.61\,\cancel{\text{s}} \times \frac{2.33\,\cancel{\text{C}}}{1\,\cancel{\text{s}}} \times \frac{1\,\cancel{\text{mol } e^-}}{96{,}485\,\cancel{\text{C}}} \times \frac{1\,\cancel{\text{mol Ag}}}{1\,\cancel{\text{mol } e^-}} \times \frac{107.87\text{ g Ag}}{1\,\cancel{\text{mol Ag}}} = 5.9\underline{0}6975\text{ g Ag} = 5.91\text{ g Ag}$$

Check: The units (s, g, and g) are correct. Because far less than a mole of Cu is electroplated, the time is short. The mass of Au is higher because of the relatively higher molar mass of Au (even taking into account the moles of electrons needed). The mass of Ag is highest because the molar mass is less than twice as high as Cu but takes only half the electrons per mole of metal.

20.131 The overall cell reaction for the first cell is $2\ Cu^{+}(aq) \rightarrow Cu^{2+}(aq) + Cu(s)$. The overall cell reaction for the second cell is $Cu^{+}(aq) \rightarrow Cu^{2+}(aq)$. The biggest difference in $E°$ is because $n = 1$ for the first cell and $n = 2$ for the second cell. Because $\Delta G°_{rxn} = -n\,F\,E°_{cell}$ for the first cell,

$$\Delta G°_{rxn} = -1\ \cancel{mol\ e^-} \times \frac{96{,}485\ C}{\cancel{mol\ e^-}} \times 0.364\ V = -1 \times 96{,}485\ \cancel{C} \times 0.364\frac{J}{\cancel{C}} = -35.1\ kJ$$, and for the

second cell,

$$\Delta G°_{rxn} = -2\ \cancel{mol\ e^-} \times \frac{96{,}485\ C}{\cancel{mol\ e^-}} \times 0.182\ V = -2 \times 96{,}485\ \cancel{C} \times 0.182\frac{J}{\cancel{C}} = -35.1\ kJ$$. Thus, $\Delta G°_{rxn}$ is the same.

Conceptual Problems

20.132 (b) If $E°_{cell} > 0$ and $E_{cell} = E°_{cell} - \frac{0.0592\ V}{n}\log Q < 0$, the second term dominates and is negative. This means that $Q > 1$. If $E_{cell} < 0$, then $K < 1$; thus, the reaction is nonspontaneous or $Q > K$.

20.133 (a) Looking for anion reductions that are between the reduction potentials of Cl_2 and Br_2. The only one that meets that criterion is the dichromate ion.

20.134 (b) If the free energy change is negative, this is a spontaneous reaction. This translates to a positive cell potential and a large equilibrium constant.

20.135 Because $K < 1$, the reaction must be nonspontaneous under standard conditions. Therefore, $E°_{cell}$ is negative and $\Delta G°_{rxn}$ is positive.

21 Radioactivity and Nuclear Chemistry

Review Questions

21.1 Radioactivity is the emission of subatomic particles or high-energy electromagnetic radiation by the nuclei of certain atoms. Radioactivity was discovered in 1896 by the French scientist Antoine-Henri Becquerel (1852–1908). Becquerel placed crystals—composed of potassium uranyl sulfate, a compound known to phosphoresce—on top of a photographic plate wrapped in black cloth. The photographic plate showed a bright exposure spot where the crystals had been. Becquerel, Marie Curie, and Pierre Curie received the Nobel Prize for the discovery of radioactivity.

21.2 Marie Curie discovered that elements besides uranium emitted radiation, such as polonium and radium. She changed the name uranic rays to radiation.

21.3 *A* is the mass number (number of protons + neutrons), *Z* is the atomic number (number of protons), and *X* is the chemical symbol of the element.

21.4 A proton is described as ${}^{1}_{1}\text{p}$, a neutron as ${}^{1}_{0}\text{n}$, and an electron as ${}^{0}_{-1}\text{e}$.

21.5 An alpha particle has the same symbol as a helium nucleus, ${}^{4}_{2}\text{He}$. When an element emits an alpha particle, the number of protons in its nucleus decreases by two and the mass number decreases by four, transforming it into a different element.

21.6 A beta particle has the same symbol as an electron, ${}^{0}_{-1}\text{e}$. When an atom emits a beta particle, its atomic number increases by one because it now has an additional proton. The mass number is unchanged.

21.7 Gamma rays are high-energy (short-wavelength) photons that have the symbol ${}^{0}_{0}\gamma$. A gamma ray has no charge and no mass. When a gamma-ray photon is emitted from a radioactive atom, it does not change the mass number or the atomic number of the element. Gamma rays, however, are usually emitted in conjunction with other types of radiation.

21.8 A positron is the antiparticle of the electron; that is, it has the same mass but opposite charge. It has a symbol of ${}^{0}_{+1}\text{e}$. When an atom emits a positron, its atomic number decreases by one because it has one fewer proton, but the mass number remains unchanged.

21.9 Electron capture occurs when a nucleus assimilates an electron from an inner orbital of its electron cloud. Like positron emission, the net effect of electron capture is the conversion of a proton into a neutron: ${}^{1}_{1}\text{p} + {}^{0}_{-1}\text{e} \rightarrow {}^{1}_{0}\text{n}$. When an atom undergoes electron capture, its atomic number decreases by one because it has one less proton and its mass number is unchanged.

21.10 (a) Ionizing power: gamma rays < beta particles ~ positrons < alpha particles
(b) Penetrating power: alpha particles < beta particles ~ positrons < gamma rays

21.11 For the lighter elements, the *N*/*Z* ratio of stable isotopes is about one (equal numbers of neutrons and protons). However, beyond about $Z = 20$, the *N*/*Z* ratio of stable nuclei begins to get larger (reaching about 1.5). Above $Z = 83$, stable nuclei do not exist. If the *N*/*Z* ratio is too high, it indicates the presence of too many neutrons, and so the neutrons will be converted to protons via beta decay. If the *N*/*Z* ratio is too low, it indicates the presence of too many protons, and therefore position emission or electron capture will be used to convert protons to neutrons.

21.12 Magic numbers are certain numbers of nucleons (N or $Z = 2, 8, 20, 28, 50, 82$ and $N = 126$) that have unique stability. Nuclei containing a magic number of protons or neutrons are particularly stable.

21.13 (a) Film-badge dosimeters consist of photographic film held in a small case that is pinned to clothing and are standard for most people working with or near radioactive substances. These badges are collected and processed (or developed) regularly to monitor a person's exposure. The more exposed the film has become in a given period of time, the more radioactivity to which the person has been exposed.

(b) A Geiger-Müller counter (commonly referred to as a Geiger counter) is an instrument that can detect radioactivity instantaneously. Particles emitted by radioactive nuclei pass through an argon-filled chamber. The energetic particles create a trail of ionized argon atoms. An applied high voltage between a wire within the chamber and the chamber itself causes these newly formed ions to produce an electrical signal that can be detected on a meter or turned into an audible click. Each click corresponds to a radioactive particle passing through the argon gas chamber. This clicking is the stereotypical sound most people associate with a radiation detector.

(c) A scintillation counter is another instrument that can detect radioactivity instantaneously. In this device, the radioactive emissions pass through a material (such as NaI or CsI) that emits ultraviolet or visible light in response to excitation by energetic particles. The radioactivity excites the atoms to a higher energy state. The atoms release this energy as light, which is then detected and turned into an electrical signal that can be read on a meter.

21.14 The half-life is the time it takes for one-half the parent nuclides in a radioactive sample to decay to the daughter nuclides and is identical to the concept of half-life for chemical reactions covered in Chapter 15. Thus, the relationship between the half-life of a nuclide and its rate constant is given by the same expression (Equation 15.19) that we derived for a first-order reaction in Section 15.4: $t_{1/2} = \frac{0.693}{k}$. Nuclides that decay quickly have short half-lives and large rate constants—they are considered very active (many decay events per unit time). Nuclides that decay slowly have long half-lives and are less active (fewer decay events per unit time).

21.15 Radiocarbon dating, a technique devised in 1949 by Willard Libby at the University of Chicago, is used by archeologists, geologists, anthropologists, and other scientists to estimate the ages of fossils and artifacts. Carbon-14 is constantly formed in the upper atmosphere by the neutron bombardment of nitrogen and then decays back to nitrogen by beta emission with a half-life of 5730 years. The continuous formation of carbon-14 in the atmosphere and its continuous decay back to nitrogen-14 produces a nearly constant equilibrium amount of atmospheric carbon-14, which is oxidized to carbon dioxide and incorporated into plants by photosynthesis. The C-14 then makes its way up the food chain and ultimately into all living organisms. As a result, all living plants, animals, and humans contain the same ratio of carbon-14 to carbon-12 (^{14}C:^{12}C) as is found in the atmosphere. When a living organism dies, however, it stops incorporating new carbon-14 into its tissues. The ^{14}C:^{12}C ratio then decreases with a half-life of 5730 years. The accuracy of carbon-14 dating can be checked against objects whose ages are known from historical sources.

To make C-14 dating more accurate, scientists have studied the carbon-14 content of western bristlecone pine trees, which can live up to 5000 years. The tree trunk contains growth rings corresponding to each year of the tree's life, and the wood laid down in each ring incorporates carbon derived from the carbon dioxide in the atmosphere at that time. The rings thus provide a record of the historical atmospheric carbon-14 content and allow for corrections to carbon-14 concentrations due to atmospheric changes.

The maximum age that can be estimated from carbon-14 dating is about 50,000 years—beyond that, the amount of carbon-14 becomes too low to measure accurately.

21.16 The ratio of uranium-238 to lead-206 within igneous rocks (rocks of volcanic origin) measures the time that has passed since the rock solidified (at which point the "radiometric clock" was reset). Because U-238 decays into Pb-206 with a half-life of 4.5×10^9 years, the relative amounts of U-238 and Pb-206 in a uranium-containing rock reveal its age. The method uses similar half-life calculations as the carbon dating method; only the half-life is much longer, so much older materials can be dated. The oldest rocks have an age of approximately 4.0 billion years, establishing a lower boundary for the age of the Earth (the Earth must be at least as old as its oldest rocks). The ages of about 70 meteorites that have struck the Earth have also been extensively studied and have been found to be about 4.5 billion years old. Because the meteorites were formed at the same time as our solar system (which includes the Earth), the best estimate for the age of the Earth is about 4.5 billion years.

21.17 Nuclear fission is the splitting of an atom into smaller products. The process emits enormous amounts of energy. Three researchers in Germany—Lise Meitner (1878–1968), Fritz Strassmann (1902–1980), and Otto Hahn (1879–1968)—repeating uranium bombardment experiments by Fermi, performed careful chemical analysis of the products. The nucleus of the neutron-bombarded uranium atom is split into barium, krypton, and other smaller products. A nuclear equation for a fission reaction, showing how uranium breaks apart into the daughter nuclides, is ${}^{235}_{92}U + {}^{1}_{0}n \rightarrow {}^{140}_{56}Ba + {}^{93}_{36}Kr + 3\,{}^{1}_{0}n$ + energy. The process produces three neutrons, which have the potential to initiate fission in three other U-235 atoms. Scientists quickly realized that a sample rich in U-235 could undergo a chain reaction in which neutrons produced by the fission of one uranium nucleus would induce fission in other uranium nuclei. The result would be a self-amplifying reaction capable of producing an enormous amount of energy—an atomic bomb. However, to make a bomb, a critical mass of U-235—enough U-235 to produce a self-sustaining reaction—would be necessary. Because an enormous amount of energy is produced, it can be used to produce steam to drive turbines (as shown in Figure 21.11).

21.18 The Manhattan Project was a top-secret endeavor with the main goal of building an atomic bomb before the Germans did. The project was led by physicist J. R. Oppenheimer (1904–1967) at a high-security research facility in Los Alamos, New Mexico. Four years later, on July 16, 1945, the world's first nuclear weapon was successfully detonated at a test site in New Mexico. The first atomic bomb exploded with a force equivalent to 18,000 tons of dynamite. Ironically, the Germans—who had not made a successful nuclear bomb—had already been defeated by this time. Instead, the atomic bomb was used on Japan. One bomb was dropped on Hiroshima, and a second bomb was dropped on Nagasaki. Together, the bombs killed approximately 200,000 people and forced Japan to surrender.

21.19 The advantages of using fission to generate electricity are that (1) a typical nuclear power plant generates enough electricity for a city of about 1 million people and uses about 50 kg of fuel per day (as opposed to a coal-burning power plant that uses about 2,000,000 kg of fuel to generate the same amount of electricity) and that (2) a nuclear power plant generates no air pollution and no greenhouse gases. (Coal-burning power plants also emit carbon dioxide, a greenhouse gas.) The disadvantages are as follows: (1) the danger of nuclear accidents, such as overheating and the release of radiation, and (2) waste disposal because the products of the reaction are radioactive and have long half-lives.

21.20 The mass of the products is usually different from the mass of the reactants because in nuclear reactions, matter can be converted to energy. The relationship between the amount of matter lost and the amount of energy formed is given by Einstein's famous equation relating the two quantities, $E = mc^2$, where E is the energy produced, m is the mass lost, and c is the speed of light.

21.21 This difference in mass between the products and the reactants is known as the mass defect. The energy corresponding to the mass defect, obtained by substituting the mass defect into the equation $E = mc^2$, is known as the nuclear binding energy, the amount of energy that would be required to break apart the nucleus into its component nucleons. The nuclear binding energy per nucleon peaks at a mass number of 60. The significance of this is that the nuclides with mass numbers of about 60 are among the most stable.

21.22 Nuclear fusion is the combination of two light nuclei to form a heavier one. Both fusion and fission emit large amounts of energy because they form daughter nuclides with greater binding energies per nucleon than do the parent nuclides.

21.23 An extremely high temperature is required for fusion to occur. To date, no material can withstand these temperatures. The U.S. Congress has reduced funding for these projects so that it is less likely that a viable process will be developed.

21.24 Transmutation is a process that results in the transformation of one element into another. Examples are ${}^{14}_{7}N + {}^{4}_{2}He \rightarrow {}^{17}_{8}O + {}^{1}_{1}H$ and ${}^{27}_{13}Al + {}^{4}_{2}He \rightarrow {}^{30}_{15}P + {}^{1}_{0}n$.

21.25 In a single-stage linear accelerator, a charged particle such as a proton is accelerated in an evacuated tube. The accelerating force is provided by a potential difference between the ends of the tube. In multistage linear accelerators such as the Stanford Linear Accelerator (SLAC) at Stanford University (Figure 21.14), a series of tubes of increasing length are connected to a source of alternating voltage, as shown in Figure 21.15. The voltage alternates in such a way that as a positively charged particle leaves a particular tube, that tube becomes positively charged, repelling the particle to the next tube. At the same time, the tube the particle is now approaching becomes negatively charged, pulling the particle toward it. This continues throughout the linear accelerator, allowing the particle to be accelerated to velocities up to 90% of the speed of light. Linear accelerators can be used to conduct nuclear transmutations, making nuclides that don't normally exist in nature.

21.26 In a cyclotron, a similarly alternating voltage as that used in linear accelerators is used to accelerate a charged particle, only this time the alternating voltage is applied between the two semicircular halves of the cyclotron (Figure 21.16). A charged particle originally in the middle of the two semicircles is accelerated back and forth between them. Additional magnets cause the particle to move in a spiral path. As the charged particle spirals out from the center, it gains speed and eventually exits the cyclotron aimed at the target.

21.27 The energy associated with radioactivity can ionize molecules. When radiation ionizes important molecules in living cells, problems can develop. The ingestion of radioactive materials, especially alpha and beta emitters, is particularly dangerous because the radioactivity is then inside the body and can do even more damage. The effects of radiation are divided into three different types: acute radiation damage, increased cancer risk, and genetic effects.

21.28 Different types of radiation have different ionizing power (gamma rays $<$ beta particles $\sim$ positrons $<$ alpha particles) and different penetrating power (alpha particles $<$ beta particles $\sim$ positrons $<$ gamma rays).

21.29 The biological effectiveness factor, or RBE (relative biological effectiveness), corrects the dosage (in rads) for the type of radiation. It is a correction factor that is usually multiplied by the dose in rads to obtain the dose in a unit called the rem (roentgen equivalent man). So dose in rads $\times$ biological effectiveness factor $=$ dose in rem. The biological effectiveness factor for alpha radiation, for example, is much higher than for gamma radiation.

21.30 The use of radioactivity in diagnosis usually involves a radiotracer, a radioactive nuclide that has been attached to a compound or introduced into a mixture to track the movement of the compound or mixture within the body. Tracers are useful in the diagnosis of disease because of two main factors: (1) the sensitivity with which radioactivity can be detected and (2) the identical chemical behavior of a radioactive nucleus and its nonradioactive counterpart. Because different elements are taken up preferentially by different organs or tissues, various radiotracers can be used to monitor metabolic activity and image a variety of organs and structures, including the kidneys, heart, brain, gallbladder, bones, and arteries. Radiotracers can also be used to locate infections or cancer within the body. Cancerous tumors can be detected because they naturally concentrate phosphorus. When a patient is given phosphorus-32 (a radioactive isotope of phosphorus), the tumors concentrate the radioactive phosphorous and become sources of radioactivity that can be detected.

Because radiation kills cells and because it is particularly effective at killing rapidly dividing cells, it is often used as a therapy for cancer (cancer cells reproduce much faster than normal cells). Gamma rays are focused on internal tumors to kill them.

Radioactive Decay and Nuclide Stability

21.31 **Conceptual Plan: Begin with the symbol for a parent nuclide on the left side of the equation and the symbol for a particle on the right side (except for electron capture). → Equalize the sum of the mass numbers and the sum of the atomic numbers on both sides of the equation by writing the appropriate mass number and atomic number for the unknown daughter nuclide. → Using the periodic table, deduce the identity of the unknown daughter nuclide from the atomic number and write its symbol.**

Solution:

(a) U-234 (alpha decay) ${}^{234}_{92}\text{U} \rightarrow {}^{?}_{?}? + {}^{4}_{2}\text{He}$ then ${}^{234}_{92}\text{U} \rightarrow {}^{230}_{90}? + {}^{4}_{2}\text{He}$ then ${}^{234}_{92}\text{U} \rightarrow {}^{230}_{90}\text{Th} + {}^{4}_{2}\text{He}$

(b) Th-230 (alpha decay) ${}^{230}_{90}\text{Th} \rightarrow {}^{?}_{?}? + {}^{4}_{2}\text{He}$ then ${}^{230}_{90}\text{Th} \rightarrow {}^{226}_{88}? + {}^{4}_{2}\text{He}$ then ${}^{230}_{90}\text{Th} \rightarrow {}^{226}_{88}\text{Ra} + {}^{4}_{2}\text{He}$

(c) Pb-214 (beta decay) ${}^{214}_{82}\text{Pb} \rightarrow {}^{?}_{?}? + {}^{0}_{-1}\text{e}$ then ${}^{214}_{82}\text{Pb} \rightarrow {}^{214}_{83}? + {}^{0}_{-1}\text{e}$ then ${}^{214}_{82}\text{Pb} \rightarrow {}^{214}_{83}\text{Bi} + {}^{0}_{-1}\text{e}$

(d) N-13 (positron emission) ${}^{13}_{7}\text{N} \rightarrow {}^{?}_{?}? + {}^{0}_{+1}\text{e}$ then ${}^{13}_{7}\text{N} \rightarrow {}^{13}_{6}? + {}^{0}_{+1}\text{e}$ then ${}^{13}_{7}\text{N} \rightarrow {}^{13}_{6}\text{C} + {}^{0}_{+1}\text{e}$

(e) Cr-51 (electron capture) ${}^{51}_{24}\text{Cr} + {}^{0}_{-1}\text{e} \rightarrow {}^{?}_{?}?$ then ${}^{51}_{24}\text{Cr} + {}^{0}_{-1}\text{e} \rightarrow {}^{51}_{23}?$ then ${}^{51}_{24}\text{Cr} + {}^{0}_{-1}\text{e} \rightarrow {}^{51}_{23}\text{V}$

Check: (a) $234 = 230 + 4, 92 = 90 + 2$, and thorium is atomic number 90. (b) $230 = 226 + 4, 90 = 88 + 2$, and radium is atomic number 88. (c) $214 = 214 + 0, 82 = 83 - 1$, and bismuth is atomic number 83. (d) $13 = 13 + 0, 7 = 6 + 1$, and carbon is atomic number 6. (e) $51 + 0 = 51, 24 - 1 = 23$, and vanadium is atomic number 23.

21.32 **Conceptual Plan: Begin with the symbol for a parent nuclide on the left side of the equation and the symbol for a particle on the right side (except for electron capture). → Equalize the sum of the mass numbers and the sum of the atomic numbers on both sides of the equation by writing the appropriate mass number and atomic**

number for the unknown daughter nuclide. → Using the periodic table, deduce the identity of the unknown daughter nuclide from the atomic number and write its symbol.

Solution:

(a) Po-210 (alpha decay) $^{210}_{84}Po \rightarrow ^{?}_{?}? + ^{4}_{2}He$ then $^{210}_{84}Po \rightarrow ^{206}_{82}? + ^{4}_{2}He$ then $^{210}_{84}Po \rightarrow ^{206}_{82}Pb + ^{4}_{2}He$

(b) Ac-227 (beta decay) $^{227}_{89}Ac \rightarrow ^{?}_{?}? + ^{0}_{-1}e$ then $^{227}_{89}Ac \rightarrow ^{227}_{90}? + ^{0}_{-1}e$ then $^{227}_{89}Ac \rightarrow ^{227}_{90}Th + ^{0}_{-1}e$

(c) Tl-207 (beta decay) $^{207}_{81}Tl \rightarrow ^{?}_{?}? + ^{0}_{-1}e$ then $^{207}_{81}Tl \rightarrow ^{207}_{82}? + ^{0}_{-1}e$ then $^{207}_{81}Tl \rightarrow ^{207}_{82}Pb + ^{0}_{-1}e$

(d) O-15 (positron emission) $^{15}_{8}O \rightarrow ^{?}_{?}? + ^{0}_{+1}e$ then $^{15}_{8}O \rightarrow ^{15}_{7}? + ^{0}_{+1}e$ then $^{15}_{8}O \rightarrow ^{15}_{7}N + ^{0}_{+1}e$

(e) Pd-103 (electron capture) $^{103}_{46}Pd + ^{0}_{-1}e \rightarrow ^{?}_{?}?$ then $^{103}_{46}Pd + ^{0}_{-1}e \rightarrow ^{103}_{45}?$ then $^{103}_{46}Pd + ^{0}_{-1}e \rightarrow ^{103}_{45}Rh$

Check: (a) $210 = 206 + 4, 84 = 82 + 2$, and lead is atomic number 82. (b) $227 = 227 + 0, 89 = 90 - 1$, and thorium is atomic number 90. (c) $207 = 207 + 0, 81 = 82 - 1$, and lead is atomic number 82. (d) $15 = 15 + 0, 8 = 7 + 1$, and nitrogen is atomic number 7. (e) $103 + 0 = 103, 46 - 1 = 45$, and rhodium is atomic number 45.

21.33 **Given:** Th-232 decay series: $\alpha, \beta, \beta, \alpha$ **Find:** balanced decay reactions

Conceptual Plan: Begin with the symbol for a parent nuclide on the left side of the equation and the symbol for a particle on the right side (except for electron capture). → Equalize the sum of the mass numbers and the sum of the atomic numbers on both sides of the equation by writing the appropriate mass number and atomic number for the unknown daughter nuclide. → Using the periodic table, deduce the identity of the unknown daughter nuclide from the atomic number and write its symbol. → Use the product of this reaction to write the next reaction.

Solution:

Th-232 (alpha decay) $^{232}_{90}Th \rightarrow ^{?}_{?}? + ^{4}_{2}He$ then $^{232}_{90}Th \rightarrow ^{228}_{88}? + ^{4}_{2}He$ then $^{232}_{90}Th \rightarrow ^{228}_{88}Ra + ^{4}_{2}He$

Ra-228 (beta decay) $^{228}_{88}Ra \rightarrow ^{?}_{?}? + ^{0}_{-1}e$ then $^{228}_{88}Ra \rightarrow ^{228}_{89}? + ^{0}_{-1}e$ then $^{228}_{88}Ra \rightarrow ^{228}_{89}Ac + ^{0}_{-1}e$

Ac-228 (beta decay) $^{228}_{89}Ac \rightarrow ^{?}_{?}? + ^{0}_{-1}e$ then $^{228}_{89}Ac \rightarrow ^{228}_{90}? + ^{0}_{-1}e$ then $^{228}_{89}Ac \rightarrow ^{228}_{90}Th + ^{0}_{-1}e$

Th-228 (alpha decay) $^{228}_{90}Th \rightarrow ^{?}_{?}? + ^{4}_{2}He$ then $^{228}_{90}Th \rightarrow ^{224}_{88}? + ^{4}_{2}He$ then $^{228}_{90}Th \rightarrow ^{224}_{88}Ra + ^{4}_{2}He$

Thus, the decay series is $^{232}_{90}Th \rightarrow ^{228}_{88}Ra + ^{4}_{2}He$, $^{228}_{88}Ra \rightarrow ^{228}_{89}Ac + ^{0}_{-1}e$, $^{228}_{89}Ac \rightarrow ^{228}_{90}Th + ^{0}_{-1}e$, $^{228}_{90}Th \rightarrow ^{224}_{88}Ra + ^{4}_{2}He$.

Check: $232 = 228 + 4, 90 = 88 + 2$, and radium is atomic number 88. $228 = 228 + 0, 88 = 89 - 1$, and actinium is atomic number 89. $228 = 228 + 0, 89 = 90 - 1$, and thorium is atomic number 90. $228 = 224 + 4, 90 = 88 + 2$, and radium is atomic number 88.

21.34 **Given:** Rn-220 decay series: $\alpha, \beta, \beta, \alpha$ **Find:** balanced decay reactions

Conceptual Plan: Begin with the symbol for a parent nuclide on the left side of the equation and the symbol for a particle on the right side (except for electron capture). → Equalize the sum of the mass numbers and the sum of the atomic numbers on both sides of the equation by writing the appropriate mass number and atomic number for the unknown daughter nuclide. → Using the periodic table, deduce the identity of the unknown daughter nuclide from the atomic number and write its symbol. → Use the product of this reaction to write the next reaction.

Solution:

Rn-220 (alpha decay) $^{220}_{86}Rn \rightarrow ^{?}_{?}? + ^{4}_{2}He$ then $^{220}_{86}Rn \rightarrow ^{216}_{84}? + ^{4}_{2}He$ then $^{220}_{86}Rn \rightarrow ^{216}_{84}Po + ^{4}_{2}He$

Po-216 (beta decay) $^{216}_{84}Po \rightarrow ^{?}_{?}? + ^{0}_{-1}e$ then $^{216}_{84}Po \rightarrow ^{216}_{85}? + ^{0}_{-1}e$ then $^{216}_{84}Po \rightarrow ^{216}_{85}At + ^{0}_{-1}e$

At-216 (beta decay) $^{216}_{85}At \rightarrow ^{?}_{?}? + ^{0}_{-1}e$ then $^{216}_{85}At \rightarrow ^{216}_{86}? + ^{0}_{-1}e$ then $^{216}_{85}At \rightarrow ^{216}_{86}Rn + ^{0}_{-1}e$

Rn-216 (alpha decay) $^{216}_{86}Rn \rightarrow ^{?}_{?}? + ^{4}_{2}He$ then $^{216}_{86}Rn \rightarrow ^{212}_{84}? + ^{4}_{2}He$ then $^{216}_{86}Rn \rightarrow ^{212}_{84}Po + ^{4}_{2}He$

Check: $220 = 216 + 4, 86 = 84 + 2$, and polonium is atomic number 84. $216 = 216 + 0, 84 = 85 - 1$, and astatine has atomic number 85. $216 = 216 + 0, 85 = 86 - 1$, and radon has atomic number 86. $216 = 212 + 4, 86 = 84 + 2$, and polonium has an atomic number of 84.

21.35 **Conceptual Plan: Equalize the sum of the mass numbers and the sum of the atomic numbers on both sides of the equation by writing the appropriate mass number and atomic number for the unknown species. → Using the periodic table and the list of particles, deduce the identity of the unknown species from the atomic number and write its symbol.**

Solution:

(a) $^{?}_{?}? \rightarrow ^{217}_{85}At + ^{4}_{2}He$ becomes $^{221}_{87}? \rightarrow ^{217}_{85}At + ^{4}_{2}He$ then $^{221}_{87}Fr \rightarrow ^{217}_{85}At + ^{4}_{2}He$

(b) $^{241}_{94}Pu \rightarrow ^{241}_{95}Am + ^{?}_{?}?$ becomes $^{241}_{94}Pu \rightarrow ^{241}_{95}Am + ^{0}_{-1}?$ then $^{241}_{94}Pu \rightarrow ^{241}_{95}Am + ^{0}_{-1}e$

(c) $^{19}_{11}Na \rightarrow ^{19}_{10}Ne + ^{?}_{?}?$ becomes $^{19}_{11}Na \rightarrow ^{19}_{10}Ne + ^{0}_{1}?$ then $^{19}_{11}Na \rightarrow ^{19}_{10}Ne + ^{0}_{+1}e$

(d) $^{75}_{34}Se + ^{?}_{?}? \rightarrow ^{75}_{33}As$ becomes $^{75}_{34}Se + ^{0}_{-1}? \rightarrow ^{75}_{33}As$ then $^{75}_{34}Se + ^{0}_{-1}e \rightarrow ^{75}_{33}As$

Check: (a) $221 = 217 + 4, 87 = 85 + 2$, and francium is atomic number 87. (b) $241 = 241 + 0, 94 = 95 - 1$, and the particle is a beta particle. (c) $19 = 19 + 0, 11 = 10 + 1$, and the particle is a positron. (d) $75 = 75 + 0, 34 - 1 = 33$, and the particle is an electron.

21.36 **Conceptual Plan: Equalize the sum of the mass numbers and the sum of the atomic numbers on both sides of the equation by writing the appropriate mass number and atomic number for the unknown species. → Using the periodic table and the list of particles, deduce the identity of the unknown species from the atomic number and write its symbol.**

Solution:

(a) ${}^{241}_{95}\text{Am} \rightarrow {}^{237}_{93}\text{Np} + {}^{?}_{?}?$ becomes ${}^{241}_{95}\text{Am} \rightarrow {}^{237}_{93}\text{Np} + {}^{4}_{2}?$ then ${}^{241}_{95}\text{Am} \rightarrow {}^{237}_{93}\text{Np} + {}^{4}_{2}\text{He}$

(b) ${}^{?}_{?}? \rightarrow {}^{233}_{92}\text{U} + {}^{0}_{-1}\text{e}$ becomes ${}^{233}_{91}? \rightarrow {}^{233}_{92}\text{U} + {}^{0}_{-1}\text{e}$ then ${}^{233}_{91}\text{Pa} \rightarrow {}^{233}_{92}\text{U} + {}^{0}_{-1}\text{e}$

(c) ${}^{237}_{93}\text{Np} \rightarrow {}^{?}_{?}? + {}^{4}_{2}\text{He}$ becomes ${}^{237}_{93}\text{Np} \rightarrow {}^{233}_{91}? + {}^{4}_{2}\text{He}$ then ${}^{237}_{93}\text{Np} \rightarrow {}^{233}_{91}\text{Pa} + {}^{4}_{2}\text{He}$

(d) ${}^{75}_{35}\text{Br} \rightarrow {}^{?}_{?}? + {}^{0}_{+1}\text{e}$ becomes ${}^{75}_{35}\text{Br} \rightarrow {}^{75}_{34}? + {}^{0}_{+1}\text{e}$ then ${}^{75}_{35}\text{Br} \rightarrow {}^{75}_{34}\text{Se} + {}^{0}_{+1}\text{e}$

Check: (a) $241 = 237 + 4, 95 = 93 + 2$, and the particle is an alpha particle. (b) $233 = 233 + 0, 91 = 92 - 1$, and protactinium is atomic number 91. (c) $237 = 233 + 4, 93 = 91 + 2$, and protactinium is atomic number 91. (d) $75 = 75 + 0, 35 = 34 + 1$, and selenium is atomic number 34.

21.37
(a) Mg-26: stable, N/Z ratio is close to 1, acceptable for low Z atoms
(b) Ne-25: not stable, N/Z ratio is much too high for low Z atom
(c) Co-51: not stable, N/Z ratio is less than 1, much too low
(d) Te-124: stable, N/Z ratio is acceptable for this Z

21.38
(a) Ti-48: stable, N/Z ratio is acceptable for this Z
(b) Cr-63: not stable, N/Z ratio is much too high for this Z
(c) Sn-102: not stable, N/Z ratio is close to 1, much too low for this Z
(d) Y-88: stable, N/Z ratio is acceptable for this Z

21.39 Sc, V, and Mn each have an odd number of protons. Atoms with an odd number of protons typically have fewer stable isotopes than those with an even number of protons.

21.40 Both aluminum and sodium, which have fewer stable isotopes, have an odd Z. Both of these atoms are small Z atoms, so the N/Z ratio should be close to 1. Only one option meets both criteria for aluminum and sodium. Neon and magnesium have an even Z, so they have more options for stable isotopes.

21.41
(a) Mo-109, $N = 67, Z = 42, N/Z = 1.6$, beta decay, because N/Z is too high
(b) Ru-90, $N = 46, Z = 44, N/Z = 1.0$, positron emission, because N/Z is too low
(c) P-27, $N = 15, Z = 12, N/Z = 0.8$, positron emission, because N/Z is too low
(d) Rn-196, $N = 110, Z = 86, N/Z = 1.3$, positron emission, because N/Z is too low

21.42
(a) Sb-132, $N = 81, Z = 51, N/Z = 1.6$, beta decay, because N/Z is too high
(b) Te-139, $N = 87, Z = 52, N/Z = 1.7$, beta decay, because N/Z is too high
(c) Fr-202, $N = 115, Z = 87, N/Z = 1.3$, positron emission, because N/Z is too low
(d) Ba-123, $N = 67, Z = 56, N/Z = 1.2$, positron emission, because N/Z is too low

21.43
(a) Cs-125, $N/Z = 70/55 = 1.3$; Cs-113, $N/Z = 58/55 = 1.1$; Cs-125 will have the longer half-life because it is closer to the proper N/Z ratio
(b) Fe-62, $N/Z = 36/26 = 1.4$; Fe-70, $N/Z = 44/26 = 1.7$; Fe-62 will have the longer half-life because it is closer to the proper N/Z ratio

21.44
(a) Cs-149, $N/Z = 94/55 = 1.7$; Cs-139, $N/Z = 84/55 = 1.5$; Cs-139 will have the longer half-life because it is closer to the proper N/Z ratio
(b) Fe-52, $N/Z = 26/26 = 1.0$; Fe-45, $N/Z = 19/26 = 0.7$; Fe-52 will have the longer half-life because it is closer to the proper N/Z ratio

The Kinetics of Radioactive Decay and Radiometric Dating

21.45 **Given:** U-235, $t_{1/2}$ for radioactive decay = 703 million years **Find:** t to 10.0% of initial amount
Conceptual Plan: Radioactive decay implies first-order kinetics, $t_{1/2} \rightarrow k$ then

$$t_{1/2} = \frac{0.693}{k}$$

$m_{\text{U-235 0}}, m_{\text{U-235 }t}, k \rightarrow t$

$$\ln N_t = -kt + \ln N_0$$

Solution: $t_{1/2} = \dfrac{0.693}{k}$ Rearrange to solve for k. $k = \dfrac{0.693}{t_{1/2}} = \dfrac{0.693}{703 \times 10^6 \text{ yr}} = 9.8\underline{5}7752 \times 10^{-10} \text{ yr}^{-1}$

Because $\ln m_{\text{U-235 }t} = -kt + \ln m_{\text{U-235 0}}$, rearrange to solve for t.

$$t = -\frac{1}{k} \ln \frac{m_{\text{U-235 }t}}{m_{\text{U-235 0}}} = -\frac{1}{9.8\underline{5}7752 \times 10^{-10} \text{ yr}^{-1}} \ln \frac{10.0\%}{100.0\%} = 2.34 \times 10^9 \text{ yr}$$

Check: The units (yr) are correct. The time is just over three half-lives, when 1/8 of the original amount will be left.

21.46 **Given:** initially, 0.050 mg Tc-99m, $t_{1/2}$ for radioactive decay = 6.0 h **Find:** t to 6.3×10^{-3} mg
Conceptual Plan: Radioactive decay implies first-order kinetics, $t_{1/2} \rightarrow k$ then

$$t_{1/2} = \frac{0.693}{k}$$

$m_{\text{Tc-99 m 0}}, m_{\text{Tc-99 m }t}, k \rightarrow t.$

$$\ln N_t = -kt + \ln N_0$$

Solution: $t_{1/2} = \dfrac{0.693}{k}$ Rearrange to solve for k. $k = \dfrac{0.693}{t_{1/2}} = \dfrac{0.693}{6.0 \text{ h}} = 0.1\underline{1}55 \text{ h}^{-1}$

Because $\ln m_{\text{Tc-99 m }t} = -kt + \ln m_{\text{Tc-99 m 0}}$, rearrange to solve for t.

$$t = -\frac{1}{k} \ln \frac{m_{\text{Tc-99 m }t}}{m_{\text{Tc-99 m 0}}} = -\frac{1}{0.1\underline{1}55 \text{ h}^{-1}} \ln \frac{6.3 \times 10^{-3} \cancel{\text{mg}}}{0.050 \cancel{\text{mg}}} = 18 \text{ h}$$

Check: The units (h) are correct. The time is about three half-lives, and the amount is about $1/8$, or $1/2^3$, of the original amount.

21.47 **Given:** $t_{1/2}$ for isotope decay = 3.8 days; 1.55 g isotope initially **Find:** mass of isotope after 5.5 days
Conceptual Plan: Radioactive decay implies first-order kinetics, $t_{1/2} \rightarrow k$ then

$$t_{1/2} = \frac{0.693}{k}$$

$m_{\text{isotope 0}}, t, k \rightarrow m_{\text{isotope }t}.$

$$\ln N_t = -kt + \ln N_0$$

Solution: $t_{1/2} = \dfrac{0.693}{k}$ Rearrange to solve for k. $k = \dfrac{0.693}{t_{1/2}} = \dfrac{0.693}{3.8 \text{ days}} = 0.1\underline{8}237 \text{ day}^{-1}$ Because
$\ln N_t = -kt + \ln N_0 = -(0.1\underline{8}237 \cancel{\text{day}^{-1}})(5.5 \cancel{\text{day}}) + \ln (1.55 \text{ g}) = -0.5\underline{6}478 \rightarrow N_t = e^{-0.5\underline{6}478} = 0.57 \text{ g}.$

Check: The units (g) are correct. The amount is consistent with a time between one and two half-lives.

21.48 **Given:** $t_{1/2}$ for I-131 = 8 days; 58 mg dose at 8 a.m. **Find:** mass of I-131 at 5 p.m. next day
Conceptual Plan: Radioactive decay implies first-order kinetics, $t_{1/2} \rightarrow k$ and determine days since dose

$$t_{1/2} = \frac{0.693}{k}$$

then $m_{\text{I-131 0}}, t, k \rightarrow m_{\text{I-131 }t}.$

$$\ln N_t = -kt + \ln N_0$$

Solution: $t_{1/2} = \dfrac{0.693}{k}$ Rearrange to solve for k. $k = \dfrac{0.693}{t_{1/2}} = \dfrac{0.693}{8 \text{ days}} = 0.0\underline{8}6625 \text{ day}^{-1}$ The time since the dose is one day plus 9 hours, or $(1 + 9/24)$ days = 1.375 days. Because $\ln m_{\text{I-131 }t} = -kt + \ln m_{\text{I-131 0}} =$
$-(0.0\underline{8}6625 \cancel{\text{day}^{-1}})(1.375 \cancel{\text{day}}) + \ln (58 \text{ mg}) = 3.\underline{9}413 \rightarrow m = e^{3.\underline{9}413} = 51 \text{ mg}$

Check: The units (mg) are correct. The amount is consistent with a time less than one half-life.

21.49 **Given:** F-18 initial decay rate $= 1.5 \times 10^5/s$, $t_{1/2}$ for F-18 $= 1.83$ h **Find:** t to decay rate of 2.5×10^3/s
Conceptual Plan: Radioactive decay implies first-order kinetics, $t_{1/2} \rightarrow k$ then Rate_0, Rate_t, $k \rightarrow t$.

$$t_{1/2} = \frac{0.693}{k} \qquad \ln\frac{\text{Rate}_t}{\text{Rate}_0} = -kt$$

Solution: $t_{1/2} = \dfrac{0.693}{k}$ Rearrange to solve for k. $k = \dfrac{0.693}{t_{1/2}} = \dfrac{0.693}{1.83 \text{ h}} = 0.378689 \text{ h}^{-1}$ Because $\ln\dfrac{\text{Rate}_t}{\text{Rate}_0} = -kt$ rearrange to solve for t.

$$t = -\frac{1}{k}\ln\frac{\text{Rate}_t}{\text{Rate}_0} = -\frac{1}{0.378689 \text{ h}^{-1}}\ln\frac{2.5 \times 10^3/\cancel{s}}{1.5 \times 10^5/\cancel{s}} = 10.8 \text{ h}$$

Check: The units (h) are correct. The time is between five and six half-lives, and the rate is just over $1/2^6$ of the original amount.

21.50 **Given:** Tl-201 initial decay rate $= 5.88 \times 10^4$/s, $t_{1/2}$ for Tl-201 $= 3.042$ days **Find:** t to decay rate of 287/s
Conceptual Plan: Radioactive decay implies first-order kinetics, $t_{1/2} \rightarrow k$ then Rate_0, Rate_t, $k \rightarrow t$.

$$t_{1/2} = \frac{0.693}{k} \qquad \ln\frac{\text{Rate}_t}{\text{Rate}_0} = -kt$$

Solution: $t_{1/2} = \dfrac{0.693}{k}$ Rearrange to solve for k. $k = \dfrac{0.693}{t_{1/2}} = \dfrac{0.69315}{3.042 \text{ days}} = 0.2278590 \text{ day}^{-1}$ Because $\ln\dfrac{\text{Rate}_t}{\text{Rate}_0} = -kt$, rearrange to solve for t.

$$t = -\frac{1}{k}\ln\frac{\text{Rate}_t}{\text{Rate}_0} = -\frac{1}{0.2278590 \text{ day}^{-1}}\ln\frac{287/\cancel{s}}{5.88 \times 10^4/\cancel{s}} = 23.358 \text{ days} = 23 \text{ days}$$

Check: The units (days) are correct. The time is between seven and eight half-lives, and the rate is just over $1/2^8$ of the original amount.

21.51 **Given:** boat analysis, C-14/C-12 $= 72.5\%$ of living organism **Find:** t
Other: $t_{1/2}$ for decay of C-14 $= 5730$ years
Conceptual Plan: Radioactive decay implies first-order kinetics, $t_{1/2} \rightarrow k$ then 72.5% of $m_{\text{C-14}\,0}$, $k \rightarrow t$.

$$t_{1/2} = \frac{0.693}{k} \qquad \ln N_t = -kt + \ln N_0$$

Solution: $t_{1/2} = \dfrac{0.693}{k}$ Rearrange to solve for k. $k = \dfrac{0.693}{t_{1/2}} = \dfrac{0.693}{5730 \text{ yr}} = 1.20942 \times 10^{-4} \text{ yr}^{-1}$ then $[\text{C-14}]_t = 0.725[\text{C-14}]_0$ Because $\ln m_{\text{C-14}\,t} = -kt + \ln m_{\text{C-14}\,0}$, rearrange to solve for t.

$$t = -\frac{1}{k}\ln\frac{m_{\text{C-14}\,t}}{m_{\text{C-14}\,0}} = -\frac{1}{1.20942 \times 10^{-4} \text{ yr}^{-1}}\ln\frac{0.725\,\cancel{m_{\text{C-14}\,0}}}{\cancel{m_{\text{C-14}\,0}}} = 2.66 \times 10^3 \text{ yr}$$

Check: The units (yr) are correct. The time to 72.5% decay is consistent with a time less than one half-life.

21.52 **Given:** peat analysis, C-14/C-12 $= 22.8\%$ of living organism **Find:** t
Other: $t_{1/2}$ for decay of C-14 $= 5730$ years
Conceptual Plan: Radioactive decay implies first-order kinetics, $t_{1/2} \rightarrow k$ then 72.5% of $[\text{C-14}]_0$, $k \rightarrow t$.

$$t_{1/2} = \frac{0.693}{k} \qquad \ln N_t = -kt + \ln N_0$$

Solution: $t_{1/2} = \dfrac{0.693}{k}$ Rearrange to solve for k. $k = \dfrac{0.693}{t_{1/2}} = \dfrac{0.693}{5730 \text{ yr}} = 1.20942 \times 10^{-4} \text{ yr}^{-1}$ then $[\text{C-14}]_t = 0.228[\text{C-14}]_0$. Because $\ln m_{\text{C-14}\,t} = -kt + \ln m_{\text{C-14}\,0}$, rearrange to solve for t.

$$t = -\frac{1}{k}\ln\frac{m_{\text{C-14}\,t}}{m_{\text{C-14}\,0}} = -\frac{1}{1.20942 \times 10^{-4} \text{ yr}^{-1}}\ln\frac{0.228\,\cancel{m_{\text{C-14}\,0}}}{\cancel{m_{\text{C-14}\,0}}} = 1.22 \times 10^4 \text{ yr}$$

Check: The units (yr) are correct. The time to 22.8% decay is consistent with a time just more than two half-lives.

21.53 **Given:** skull analysis, C-14 decay rate = 15.3 dis/min · gC in living organisms and 0.85 dis/min · gC in skull
Find: t **Other:** $t_{1/2}$ for decay of C-14 = 5730 years
Conceptual Plan: Radioactive decay implies first-order kinetics, $t_{1/2} \rightarrow k$ then $\text{Rate}_0, \text{Rate}_t, k \rightarrow t$.

$$t_{1/2} = \frac{0.693}{k} \qquad \ln\frac{\text{Rate}_t}{\text{Rate}_0} = -kt$$

Solution: $t_{1/2} = \frac{0.693}{k}$ Rearrange to solve for k. $k = \frac{0.693}{t_{1/2}} = \frac{0.693}{5730\text{ yr}} = 1.2\underline{0}942 \times 10^{-4}\text{ yr}^{-1}$

Because $\ln\frac{\text{Rate}_t}{\text{Rate}_0} = -kt$, rearrange to solve for t.

$$t = -\frac{1}{k}\ln\frac{\text{Rate}_t}{\text{Rate}_0} = -\frac{1}{1.2\underline{0}942 \times 10^{-4}\text{ yr}^{-1}}\ln\frac{0.85\ \cancel{\text{dis/min}\cdot\text{gC}}}{15.3\ \cancel{\text{dis/min}\cdot\text{gC}}} = 2.39 \times 10^4\text{ yr}$$

Check: The units (yr) are correct. The rate is 6% of initial value, and the time is consistent with a time just more than four half-lives.

21.54 **Given:** mammoth analysis C-14 decay rate = 15.3 dis/min · gC in living organisms and 0.48 dis/min · gC in mammoth
Find: when the mammoth lived **Other:** $t_{1/2}$ for decay of C-14 = 5730 years
Conceptual Plan: Radioactive decay implies first-order kinetics, $t_{1/2} \rightarrow k$ then $\text{Rate}_0, \text{Rate}_t, k \rightarrow t$.

$$t_{1/2} = \frac{0.693}{k} \qquad \ln\frac{\text{Rate}_t}{\text{Rate}_0} = -kt$$

Solution: $t_{1/2} = \frac{0.693}{k}$ Rearrange to solve for k. $k = \frac{0.693}{t_{1/2}} = \frac{0.693}{5730\text{ yr}} = 1.2\underline{0}942 \times 10^{-4}\text{ yr}^{-1}$

Because $\ln\frac{\text{Rate}_t}{\text{Rate}_0} = -kt$, rearrange to solve for t.

$$t = -\frac{1}{k}\ln\frac{\text{Rate}_t}{\text{Rate}_0} = -\frac{1}{1.2\underline{0}942 \times 10^{-4}\text{ yr}^{-1}}\ln\frac{0.48\ \cancel{\text{dis/min}\cdot\text{gC}}}{15.3\ \cancel{\text{dis/min}\cdot\text{gC}}} = 2.9 \times 10^4\text{ yr ago}$$

Check: The units (yr) are correct. The rate is 3% of initial value, and the time is consistent with a time just more than five half-lives.

21.55 **Given:** rock analysis, 0.438 g Pb-206 to every 1.00 g U-238, no Pb-206 initially **Find:** age of rock
Other: $t_{1/2}$ for decay of U-238 to Pb-206 = 4.5×10^9 years
Conceptual Plan: Radioactive decay implies first-order kinetics, $t_{1/2} \rightarrow k$ then

$$t_{1/2} = \frac{0.693}{k}$$

g Pb-206 → mol Pb-206 → mol U-238 → g U-238 then $m_{\text{U-238}\,t}, k \rightarrow t$.

$$\frac{1\text{ mol Pb-206}}{206\text{ g Pb-206}} \qquad \frac{1\text{ mol U-238}}{1\text{ mol Pb-206}} \qquad \frac{238\text{ g U-238}}{1\text{ mol U-238}} \qquad \ln N_t = -kt + \ln N_0$$

Solution: $t_{1/2} = \frac{0.693}{k}$ Rearrange to solve for k. $k = \frac{0.693}{t_{1/2}} = \frac{0.693}{4.5 \times 10^9\text{ yr}} = 1.\underline{5}4 \times 10^{-10}\text{ yr}^{-1}$ then

$0.438\ \cancel{\text{g Pb-206}} \times \frac{1\ \cancel{\text{mol Pb-206}}}{206\ \cancel{\text{g Pb-206}}} \times \frac{1\ \cancel{\text{mol U-238}}}{1\ \cancel{\text{mol Pb-206}}} \times \frac{238\text{ g U-238}}{1\ \cancel{\text{mol U-238}}} = 0.50\underline{6}039\text{ g U-238}$ Because

$\ln\frac{m_{\text{U-238}\,t}}{m_{\text{U-238}\,0}} = -kt$, rearrange to solve for t.

$$t = -\frac{1}{k}\ln\frac{m_{\text{U-238}\,t}}{m_{\text{U-238}\,0}} = -\frac{1}{1.\underline{5}4 \times 10^{-10}\text{ yr}^{-1}}\ln\frac{1.00\ \cancel{\text{g U-238}}}{(1.00 + 0.50\underline{6}039)\ \cancel{\text{g U-238}}} = 2.7 \times 10^9\text{ yr}$$

Check: The units (yr) are correct. The amount of Pb-206 is less than half the initial U-238 amount, and time is less than one half-life.

21.56 **Given:** meteor analysis, 0.855 g Pb-206 to every 1.00 g U-238, no Pb-206 initially **Find:** age of meteor
Other: $t_{1/2}$ for decay of U-238 to Pb-206 = 4.5×10^9 years
Conceptual Plan: Radioactive decay implies first-order kinetics, $t_{1/2} \rightarrow k$ then

$$t_{1/2} = \frac{0.693}{k}$$

g Pb-206 → mol Pb-206 → mol U-238 → g U-238 then $m_{U\text{-}238\,0}$, $m_{U\text{-}238\,t}$, $k \rightarrow t$.

$$\frac{1 \text{ mol Pb-206}}{206 \text{ g Pb-206}} \quad \frac{1 \text{ mol U-238}}{1 \text{ mol Pb-206}} \quad \frac{238 \text{ g U-238}}{1 \text{ mol U-238}} \qquad \ln N_t = -kt + \ln N_0$$

Solution: $t_{1/2} = \dfrac{0.693}{k}$ Rearrange to solve for k. $k = \dfrac{0.693}{t_{1/2}} = \dfrac{0.693}{4.5 \times 10^9 \text{ yr}} = 1.\underline{5}4 \times 10^{-10} \text{ yr}^1$ then

$0.855 \text{ g Pb-206} \times \dfrac{1 \text{ mol Pb-206}}{206 \text{ g Pb-206}} \times \dfrac{1 \text{ mol U-238}}{1 \text{ mol Pb-206}} \times \dfrac{238 \text{ g U-238}}{1 \text{ mol U-238}} = 0.98\underline{7}816 \text{ g U-238}$ Because

$\ln \dfrac{m_{U\text{-}238\,t}}{m_{U\text{-}238\,0}} = -kt$, rearrange to solve for t.

$$t = -\frac{1}{k} \ln \frac{m_{U\text{-}238\,t}}{m_{U\text{-}238\,0}} = -\frac{1}{1.\underline{5}4 \times 10^{-10} \text{ yr}^{-1}} \ln \frac{1.00 \text{ g U-238}}{(1.00 + 0.98\underline{7}816) \text{ g U-238}} = 4.5 \times 10^9 \text{ yr}$$

Check: The units (yr) are correct. The amount of Pb-206 is just less than the initial U-238 amount, and time is just under one half-life.

Fission and Fusion

21.57 **Given:** U-235 fission induced by neutrons to Xe-144 and Sr-90 **Find:** number of neutrons produced
Conceptual Plan: Write the species given on the appropriate side of the equation. → Equalize the sum of the mass numbers and the sum of the atomic numbers on both sides of the equation by writing the stoichiometric coefficient in front of the desired species.
Solution: ${}^{235}_{92}\text{U} + {}^{1}_{0}\text{n} \rightarrow {}^{144}_{54}\text{Xe} + {}^{90}_{38}\text{Sr} + ?{}^{1}_{0}\text{n}$ becomes ${}^{235}_{92}\text{U} + {}^{1}_{0}\text{n} \rightarrow {}^{144}_{54}\text{Xe} + {}^{90}_{38}\text{Sr} + 2{}^{1}_{0}\text{n}$, so two neutrons are produced.

Check: $235 + 1 = 144 + 90 + 2$, $92 + 0 = 54 + 38 + 0$, and no other particle is necessary to balance the equation.

21.58 **Given:** U-235 fission to Te-137 and Zr-97 **Find:** number of neutrons produced
Conceptual Plan: Write the species given on the appropriate side of the equation. → Equalize the sum of the mass numbers and the sum of the atomic numbers on both sides of the equation by writing the stoichiometric coefficient in front of the desired species.
Solution: ${}^{235}_{92}\text{U} + {}^{1}_{0}\text{n} \rightarrow {}^{137}_{52}\text{Te} + {}^{97}_{40}\text{Zr} + ?{}^{1}_{0}\text{n}$ becomes ${}^{235}_{92}\text{U} + {}^{1}_{0}\text{n} \rightarrow {}^{137}_{52}\text{Te} + {}^{97}_{40}\text{Zr} + 2{}^{1}_{0}\text{n}$, so two neutrons are produced.

Check: $235 + 1 = 137 + 97 + 2$, $92 + 0 = 52 + 40 + 0$, and no other particle is necessary to balance the equation.

21.59 **Given:** fusion of two H-2 atoms to form He-3 and one neutron **Find:** balanced equation
Conceptual Plan: Write the species given on the appropriate side of the equation. → Equalize the sum of the mass numbers and the sum of the atomic numbers on both sides of the equation by writing the stoichiometric coefficient in front of the desired species.
Solution: $2\,{}^{2}_{1}\text{H} \rightarrow {}^{3}_{2}\text{He} + {}^{1}_{0}\text{n}$

Check: $2(2) = 3 + 1$, $2(1) = 2 + 0$, and no other particle is necessary to balance the equation.

21.60 **Given:** fusion of H-3 and H-1 atoms to form He-4 **Find:** balanced equation
Conceptual Plan: Write the species given on the appropriate side of the equation. → Equalize the sum of the mass numbers and the sum of the atomic numbers on both sides of the equation by writing the stoichiometric coefficient in front of the desired species.
Solution: ${}^{3}_{1}\text{H} + {}^{1}_{1}\text{H} \rightarrow {}^{4}_{2}\text{He}$

Check: $3 + 1 = 4$, $1 + 1 = 2$, and no other particle is necessary to balance the equation.

21.61 **Given:** U-238 bombarded by neutrons to form U-239, which undergoes two beta decays to form Pu-239
Find: balanced equations
Conceptual Plan: Write the species given on the appropriate side of the equation. → Equalize the sum of the mass numbers and the sum of the atomic numbers on both sides of the equation by writing the stoichiometric coefficient in front of the desired species. → Use the product of this reaction to write the next reaction until the process is complete.
Solution: ${}^{238}_{92}\text{U} + ?{}^{1}_{0}\text{n} \rightarrow {}^{239}_{92}\text{U}$ becomes ${}^{238}_{92}\text{U} + {}^{1}_{0}\text{n} \rightarrow {}^{239}_{92}\text{U}$ then

beta decay ${}^{239}_{92}\text{U} \rightarrow {}^{?}_{?}? + {}^{0}_{-1}\text{e}$ becomes ${}^{239}_{92}\text{U} \rightarrow {}^{239}_{93}? + {}^{0}_{-1}\text{e}$ then ${}^{239}_{92}\text{U} \rightarrow {}^{239}_{93}\text{Np} + {}^{0}_{-1}\text{e}$ then
beta decay ${}^{239}_{93}\text{Np} \rightarrow {}^{?}_{?}? + {}^{0}_{-1}\text{e}$ becomes ${}^{239}_{93}\text{Np} \rightarrow {}^{239}_{94}? + {}^{0}_{-1}\text{e}$ then ${}^{239}_{93}\text{Np} \rightarrow {}^{239}_{94}\text{Pu} + {}^{0}_{-1}\text{e}$
The entire process is ${}^{238}_{92}\text{U} + {}^{1}_{0}\text{n} \rightarrow {}^{239}_{92}\text{U}$, ${}^{239}_{92}\text{U} \rightarrow {}^{239}_{93}\text{Np} + {}^{0}_{-1}\text{e}$, ${}^{239}_{93}\text{Np} \rightarrow {}^{239}_{94}\text{Pu} + {}^{0}_{-1}\text{e}$.

Check: $238 + 1 = 239, 92 + 0 = 92$, and no other particle is necessary to balance the equation. $239 = 239 + 0, 92 = 93 - 1$, and neptunium is atomic number 93. $239 = 239 + 0, 93 = 94 - 1$, and plutonium is atomic number 94.

21.62 **Given:** Al-27 is bombarded by a neutron and then undergoes an alpha decay and a beta decay
Find: balanced equations
Conceptual Plan: Write the species given on the appropriate side of the equation. → Equalize the sum of the mass numbers and the sum of the atomic numbers on both sides of the equation by writing the stoichiometric coefficient in front of the desired species. → Use the product of this reaction to write the next reaction until the process is complete.
Solution: ${}^{27}_{13}\text{Al} + {}^{1}_{0}\text{n} \rightarrow {}^{?}_{?}?$ becomes ${}^{27}_{13}\text{Al} + {}^{1}_{0}\text{n} \rightarrow {}^{28}_{13}?$ then ${}^{27}_{13}\text{Al} + {}^{1}_{0}\text{n} \rightarrow {}^{28}_{13}\text{Al}$ then
alpha decay ${}^{28}_{13}\text{Al} \rightarrow {}^{?}_{?}? + {}^{4}_{2}\text{He}$ becomes ${}^{28}_{13}\text{Al} \rightarrow {}^{24}_{11}? + {}^{4}_{2}\text{He}$ then ${}^{28}_{13}\text{Al} \rightarrow {}^{24}_{11}\text{Na} + {}^{4}_{2}\text{He}$ then
beta decay ${}^{24}_{11}\text{Na} \rightarrow {}^{?}_{?}? + {}^{0}_{-1}\text{e}$ becomes ${}^{24}_{11}\text{Na} \rightarrow {}^{24}_{12}? + {}^{0}_{-1}\text{e}$ then ${}^{24}_{11}\text{Na} \rightarrow {}^{24}_{12}\text{Mg} + {}^{0}_{-1}\text{e}$
The entire process is ${}^{27}_{13}\text{Al} + {}^{1}_{0}\text{n} \rightarrow {}^{28}_{13}\text{Al}$, ${}^{28}_{13}\text{Al} \rightarrow {}^{24}_{11}\text{Na} + {}^{4}_{2}\text{He}$, ${}^{24}_{11}\text{Na} \rightarrow {}^{24}_{12}\text{Mg} + {}^{0}_{-1}\text{e}$.

Check: $27 + 1 = 28, 13 + 0 = 13$, and aluminum is atomic number 13. $28 = 24 + 4, 13 = 11 + 2$, and sodium is atomic number 11. $24 = 24 + 0, 11 = 12 - 1$, and magnesium is atomic number 12.

21.63 **Given:** Rf-257 synthesized by bombarding Cf-249 with C-12 **Find:** balanced equation
Conceptual Plan: Write the species given on the appropriate side of the equation. → Equalize the sum of the mass numbers and the sum of the atomic numbers on both sides of the equation by writing the stoichiometric coefficient in front of the desired species. → Use the product of this reaction to write the next reaction until the process is complete.
Solution: ${}^{249}_{98}\text{Cf} + {}^{12}_{6}\text{C} \rightarrow {}^{257}_{?}\text{Rf} + ?\,{}^{?}_{?}?$ becomes ${}^{249}_{98}\text{Cf} + {}^{12}_{6}\text{C} \rightarrow {}^{257}_{104}\text{Rf} + 4\,{}^{1}_{0}\text{n}$.

Check: $249 + 12 = 257 - 4, 98 + 6 = 104 + 0$, and rutherfordium is atomic number 104, and four neutrons are needed to balance the equation.

21.64 **Given:** Bi-209 bombarded by Cr-54 to form Bh and one neutron **Find:** balanced equations
Conceptual Plan: Write the species given on the appropriate side of the equation. → Equalize the sum of the mass numbers and the sum of the atomic numbers on both sides of the equation by writing the stoichiometric coefficient in front of the desired species. → Use the product of this reaction to write the next reaction until the process is complete.
Solution: ${}^{209}_{83}\text{Bi} + {}^{54}_{24}\text{Cr} \rightarrow {}^{?}_{107}\text{Bh} + {}^{1}_{0}\text{n}$ becomes ${}^{209}_{83}\text{Bi} + {}^{54}_{24}\text{Cr} \rightarrow {}^{262}_{107}\text{Bh} + {}^{1}_{0}\text{n}$.

Check: $209 + 54 = 262 + 1, 83 + 24 = 107 + 0$, bohrium is atomic number 107, and no other particles are needed to balance the equation.

Energetics of Nuclear Reactions, Mass Defect, and Nuclear Binding Energy

21.65 **Given:** 1.0 g of matter converted to energy **Find:** energy
Conceptual Plan: g → kg → E

$$\frac{1\text{ kg}}{1000\text{ g}} \qquad E = mc^2$$

Solution: $1.0\,\cancel{\text{g}} \times \dfrac{1\text{ kg}}{1000\,\cancel{\text{g}}} = 0.0010\text{ kg}$ then $E = mc^2 = (0.0010\text{ kg})\left(2.9979 \times 10^8\,\dfrac{\text{m}}{\text{s}}\right)^2 = 9.0 \times 10^{13}\text{ J}$

Check: The units (J) are correct. The magnitude of the answer makes physical sense because we are converting a large quantity of amus to energy.

21.66 **Given:** 1.0×10^3 kWh of electricity/month from nuclear reaction **Find:** mass converted to energy/year
Conceptual Plan: kWh → J → kg → g then g/month → g/year

$$\frac{3.60 \times 10^6\text{ J}}{1\text{ kWh}} \qquad E = mc^2 \qquad \frac{1000\text{ g}}{1\text{ kg}} \qquad \frac{12\text{ months}}{1\text{ year}}$$

Solution: $1.0 \times 10^3\,\cancel{\text{kWh}} \times \dfrac{3.60 \times 10^6\text{ J}}{1\,\cancel{\text{kWh}}} = 3.6 \times 10^9\text{ J}$ Because $E = mc^2$, rearrange to solve for m.

$$m = \frac{E}{c^2} = \frac{3.6 \times 10^9\ \text{kg}\frac{\cancel{\text{m}^2}}{\cancel{\text{s}^2}}}{\left(2.9979 \times 10^8\ \text{kg}\frac{\cancel{\text{m}}}{\cancel{\text{s}}}\right)^2} = 4.0 \times 10^{-8}\ \cancel{\text{kg}} \times \frac{1000\ \text{g}}{1\ \cancel{\text{kg}}} = 4.0 \times 10^{-5}\ \text{g then}$$

$$\frac{4.0 \times 10^{-5}\ \text{g}}{1\ \cancel{\text{month}}} \times \frac{12\ \cancel{\text{month}}}{1\ \text{year}} = \frac{4.8 \times 10^{-4}\ \text{g}}{1\ \text{year}}$$

Check: The units (g) are correct. A small mass is expected because nuclear reactions generate a large amount of energy.

21.67 **Given:** (a) O-16 = 15.9949145 amu, (b) Ni-58 = 57.935346 amu, and (c) Xe-129 = 128.904780 amu
Find: mass defect and nuclear binding energy per nucleon
Conceptual Plan: ^A_ZX**, isotope mass → mass defect → nuclear binding energy per nucleon**

$$\text{mass defect} = Z(\text{mass } ^1_1\text{H}) + (A - Z)(\text{mass } ^1_0\text{n}) - \text{mass of isotope} \qquad \frac{931.5\ \text{MeV}}{(1\ \text{amu})(A\ \text{nucleons})}$$

Solution: mass defect $= Z(\text{mass } ^1_1\text{H}) + (A - Z)(\text{mass } ^1_0\text{n}) - \text{mass of isotope}$

(a) O-16 mass defect $= 8(1.00783\ \text{amu}) + (16 - 8)(1.00866\ \text{amu}) - 15.9949145\ \text{amu} =$
$0.137\underline{0}055\ \text{amu} = 0.13701\ \text{amu}$ and $0.137\underline{0}055\ \cancel{\text{amu}} \times \frac{931.5\ \text{MeV}}{(1\ \cancel{\text{amu}})(16\ \text{nucleons})} = 7.976\frac{\text{MeV}}{\text{nucleon}}$

(b) Ni-58 mass defect $= 28(1.00783\ \text{amu}) + (58 - 28)(1.00866\ \text{amu}) - 57.935346\ \text{amu} =$
$0.5436\underline{9}4\ \text{amu} = 0.54369\ \text{amu}$ and $0.5436\underline{9}4\ \cancel{\text{amu}} \times \frac{931.5\ \text{MeV}}{(1\ \cancel{\text{amu}})(58\ \text{nucleons})} = 8.732\frac{\text{MeV}}{\text{nucleon}}$

(c) Xe-129 mass defect $= 54(1.00783\ \text{amu}) + (129 - 54)(1.00866\ \text{amu}) - 128.904780\ \text{amu} =$
$1.16754\ \text{amu}$ and $1.16754\ \cancel{\text{amu}} \times \frac{931.5\ \text{MeV}}{(1\ \cancel{\text{amu}})(129\ \text{nucleons})} = 8.431\frac{\text{MeV}}{\text{nucleon}}$

Check: The units (amu and MeV/nucleon) are correct. The mass defect increases with an increasing number of nucleons, but the MeV/nucleon does not change by as much (on a relative basis).

21.68 **Given:** (a) Li-7 = 7.016003 amu, (b) Ti-48 = 47.947947 amu, and (c) Ag-107 = 106.905092 amu
Find: mass defect and nuclear binding energy per nucleon
Conceptual Plan: ^A_ZX**, isotope mass → mass defect → nuclear binding energy per nucleon**

$$\text{mass defect} = Z(\text{mass } ^1_1\text{H}) + (A - Z)(\text{mass } ^1_0\text{n}) - \text{mass of isotope} \qquad \frac{931.5\ \text{MeV}}{(1\ \text{amu})(A\ \text{nucleons})}$$

Solution: mass defect $= Z(\text{mass } ^1_1\text{H}) + (A - Z)(\text{mass } ^1_0\text{n}) - \text{mass of isotope}$

(a) Li-7 mass defect $= 3(1.00783\ \text{amu}) + (7 - 3)(1.00866\ \text{amu}) - 7.016003\ \text{amu} =$
$0.0421\underline{2}7\ \text{amu} = 0.04213\ \text{amu}$ and $0.0421\underline{2}7\ \cancel{\text{amu}} \times \frac{931.5\ \text{MeV}}{(1\ \cancel{\text{amu}})(7\ \text{nucleons})} = 5.606\frac{\text{MeV}}{\text{nucleon}}$

(b) Ti-48 mass defect $= 22(1.00783\ \text{amu}) + (48 - 22)(1.00866\ \text{amu}) - 47.947947\ \text{amu} =$
$0.4494\underline{7}3\ \text{amu} = 0.44947\ \text{amu}$ and $0.4494\underline{7}3\ \cancel{\text{amu}} \times \frac{931.5\ \text{MeV}}{(1\ \cancel{\text{amu}})(48\ \text{nucleons})} = 8.723\frac{\text{MeV}}{\text{nucleon}}$

(c) Ag-107 mass defect $= 47(1.00783\ \text{amu}) + (107 - 47)(1.00866\ \text{amu}) - 106.905092\ \text{amu} =$
$0.9825\underline{1}8\ \text{amu} = 0.98252\ \text{amu}$ and $0.9825\underline{1}8\ \cancel{\text{amu}} \times \frac{931.5\ \text{MeV}}{(1\ \cancel{\text{amu}})(107\ \text{nucleons})} = 8.553\frac{\text{MeV}}{\text{nucleon}}$

Check: The units (amu and MeV/nucleon) are correct. The mass defect increases with an increasing number of nucleons, but the MeV/nucleon does not change by as much (on a relative basis).

21.69 **Given:** $^{235}_{92}\text{U} + ^1_0\text{n} \rightarrow ^{144}_{54}\text{Xe} + ^{90}_{38}\text{Sr} + 2^1_0\text{n}$, U-235 = 235.043922 amu, Xe-144 = 143.9385 amu, and Sr-90 = 89.907738 amu **Find:** energy per g of U-235
Conceptual Plan: mass of products and reactants → mass defect → mass defect/g of U-235 then

$$\text{mass defect} = \Sigma\text{mass of reactants} - \Sigma\text{mass of products} \qquad \frac{\text{mass defect}}{235.043922\ \text{g U-235}}$$

g → kg → *E*

$\frac{1\text{ kg}}{1000\text{ g}}$ $E = mc^2$

Solution: mass defect = Σmass of reactants − Σmass of products; notice that we can cancel a neutron from each side to get ${}^{235}_{92}\text{U} \rightarrow {}^{144}_{54}\text{Xe} + {}^{90}_{38}\text{Sr} + {}^{1}_{0}\text{n}$ and

mass defect = 235.043922 g − (143.9385 g + 89.907738 g + 1.00866 g) = 0.189024 g

then $\frac{0.189024\text{ g}}{235.043922\text{ g U-235}} \times \frac{1\text{ kg}}{1000\text{ g}} = 8.04207 \times 10^{-7}\frac{\text{kg}}{\text{g U-235}}$ then

$$E = mc^2 = \left(8.04207 \times 10^{-7}\frac{\text{kg}}{\text{g U-235}}\right)\left(2.9979 \times 10^{8}\frac{\text{m}}{\text{s}}\right)^2 = 7.228 \times 10^{10}\frac{\text{J}}{\text{g U-235}}$$

Check: The units (J) are correct. A large amount of energy is expected per gram of fuel in a nuclear reactor.

21.70 **Given:** ${}^{235}_{92}\text{U} + {}^{1}_{0}\text{n} \rightarrow {}^{137}_{52}\text{Te} + {}^{97}_{40}\text{Zr} + 2{}^{1}_{0}\text{n}$, U-235 = 235.043922 amu, Te-137 = 136.9253 amu, and Zr-97 = 96.910950 amu **Find:** energy per mol of U-235

Conceptual Plan: mass of products and reactants → mass defect → mass defect/mol of U-235 then

mass defect = Σmass of reactants − Σmass of products $\frac{\text{mass defect}}{235.043922\text{ g U-235}}$

g → kg → *E*

$\frac{1\text{ kg}}{1000\text{ g}}$ $E = mc^2$

Solution: mass defect = Σmass of reactants − Σmass of products; notice that we can cancel a neutron from each side to get ${}^{235}_{92}\text{U} \rightarrow {}^{137}_{52}\text{Te} + {}^{97}_{40}\text{Zr} + {}^{1}_{0}\text{n}$ and

mass defect = 235.043922 g − (136.9253 g + 96.910950 g + 1.00866 g) = 0.199012 g/mol U-235

then $\frac{0.199012\text{ g}}{\text{mol U-235}} \times \frac{1\text{ kg}}{1000\text{ g}} = 1.99012 \times 10^{-4}\frac{\text{kg}}{\text{mol U-235}}$ then

$$E = mc^2 = \left(1.99012 \times 10^{-4}\frac{\text{kg}}{\text{mol U-235}}\right)\left(2.9979 \times 10^{8}\frac{\text{m}}{\text{s}}\right)^2 = 1.789 \times 10^{13}\frac{\text{J}}{\text{mol U-235}}$$

Check: The units (J) are correct. A large amount of energy is expected per gram of fuel in a nuclear reactor.

21.71 **Given:** $2{}^{2}_{1}\text{H} \rightarrow {}^{3}_{2}\text{He} + {}^{1}_{0}\text{n}$, H-2 = 2.014102 amu, and He-3 = 3.016029 amu **Find:** energy per g reactant

Conceptual Plan: mass of products and reactants → mass defect → mass defect/g of H-2 then

mass defect = Σmass of reactants − Σmass of products $\frac{\text{mass defect}}{2(2.014102\text{ g H-2})}$

g → kg → *E*

$\frac{1\text{ kg}}{1000\text{ g}}$ $E = mc^2$

Solution: mass defect = Σmass of reactants − Σmass of products and mass defect = 2(2.014102 g) − (3.016029 g + 1.00866 g) = 0.003515 g

then $\frac{0.003515\text{ g}}{2(2.014102\text{ g H-2})} \times \frac{1\text{ kg}}{1000\text{ g}} = 8.72597 \times 10^{-7}\frac{\text{kg}}{\text{g H-2}}$ then

$$E = mc^2 = \left(8.72597 \times 10^{-7}\frac{\text{kg}}{\text{g H-2}}\right)\left(2.9979 \times 10^{8}\frac{\text{m}}{\text{s}}\right)^2 = 7.84 \times 10^{10}\frac{\text{J}}{\text{g H-2}}$$

Check: The units (J) are correct. A large amount of energy is expected per gram of fuel in a fusion reaction.

21.72 **Given:** ${}^{3}_{1}\text{H} + {}^{1}_{1}\text{H} \rightarrow {}^{4}_{2}\text{He}$, H-3 = 3.016049 amu, H-1 = 1.007825 amu, and He-4 = 4.002603 amu

Find: energy per g reactant

Conceptual Plan: mass of products and reactants → mass defect → mass defect/g reactant then

mass defect = Σmass of reactants − Σmass of products $\frac{\text{mass defect}}{\text{g reactant}}$

g → kg → *E*

$\frac{1\ \text{kg}}{1000\ \text{g}}$ $E = mc^2$

Solution: mass defect = $\sum$mass of reactants − $\sum$mass of products and
mass defect = (3.016049 g + 1.007825 g) − 4.002603 g = 4.023874 g − 4.002603 g = 0.021271 g

then $\frac{0.021271\ \text{g}}{4.023874\ \text{g reactants}} \times \frac{1\ \text{kg}}{1000\ \text{g}} = 5.28\underline{6}1993 \times 10^{-6} \frac{\text{kg}}{\text{g reactants}}$ then

$$E = mc^2 = \left(5.28\underline{6}1993 \times 10^{-6} \frac{\text{kg}}{\text{g reactants}}\right)\left(2.9979 \times 10^{8} \frac{\text{m}}{\text{s}}\right)^2 = 4.7509 \times 10^{11} \frac{\text{J}}{\text{g reactants}}$$

Check: The units (J) are correct. A large amount of energy is expected per gram of fuel in a fusion reaction.

Effects and Applications of Radioactivity

21.73 **Given:** 75 kg human exposed to 32.8 rad and falling from chair **Find:** energy absorbed in each case
Conceptual Plan: rad, kg → J and assume d = 0.50 m chair height then mass, d → J

$1\ \text{rad} = \frac{0.01\ \text{J}}{1\ \text{kg body tissue}}$ $E = F \cdot d = mgd$

Solution: 32.8 rad $= 32.8 \frac{0.01\ \text{J}}{1\ \text{kg body tissue}} \times 75\ \text{kg} = 25\ \text{J}$ and

$$E = F \cdot d = mgd = 75\ \text{kg} \times 9.8 \frac{\text{m}}{\text{s}^2} \times 0.50\ \text{m} = 370\ \text{kg}\frac{\text{m}^2}{\text{s}^2} = 370\ \text{J}$$

Check: The units (J and J) are correct. Allowable radiation exposures are low because the radiation is very ionizing and thus damaging to tissue. Falling may have more energy, but it is not ionizing.

21.74 **Given:** 55 g mouse exposed to 20.5 rad **Find:** energy absorbed
Conceptual Plan: g → kg then rad, kg → J

$\frac{1\ \text{kg}}{1000\ \text{g}}$ $\frac{0.01\ \text{J}}{1\ \text{kg body tissue}}$

Solution: $55\ \text{g} \times \frac{1\ \text{kg}}{1000\ \text{g}} = 0.055\ \text{kg}$ then 20.5 rad $= 20.5 \frac{0.01\ \text{J}}{1\ \text{kg body tissue}} \times 0.055\ \text{kg} = 0.011\ \text{J}$

Check: The units (J) are correct. Allowable radiation exposures are low because the radiation is very ionizing and thus damaging to tissue.

21.75 **Given:** $t_{1/2}$ for F-18 = 1.83 h, 65% of F-18 makes it to the hospital, traveling at 60.0 miles/hour
Find: distance between hospital and cyclotron
Conceptual Plan: $t_{1/2} \to k$ then $m_{\text{F-18}\,0}, m_{\text{F-18}\,t}, k \to t$ then h → mi

$t_{1/2} = \frac{0.693}{k}$ $\ln\frac{m_{\text{F-18}t}}{m_{\text{F-18}0}} = -kt$ $\frac{60.0\ \text{mi}}{1\ \text{h}}$

Solution: $t_{1/2} = \frac{0.693}{k}$ Rearrange to solve for k. $k = \frac{0.693}{t_{1/2}} = \frac{0.693}{1.83\ \text{h}} = 0.37\underline{8}689\ \text{h}^{-1}$ Because

$\ln\frac{m_{\text{F-18}t}}{m_{\text{F-18}0}} = -kt$, rearrange to solve for t.

$$t = -\frac{1}{k}\ln\frac{m_{\text{F-18}\,t}}{m_{\text{F-18}\,0}} = -\frac{1}{0.37\underline{8}689\ \text{h}^{-1}}\ln\frac{0.65\ m_{\text{F-18}\,0}}{m_{\text{F-18}\,0}} = 1.\underline{1}376\ \text{h then}$$

$$1.\underline{1}376\ \text{h} \times \frac{60.0\ \text{mi}}{1\ \text{h}} = 68\ \text{mi}$$

Check: The units (mi) are correct. The time is less than one half-life, so the distance is less than 1.83 times the speed of travel.

21.76 **Given:** I-131, 155 mg, $t_{1/2}$ = 8.0 days **Find:** exposure (in Ci) after 4.0 h
Conceptual Plan: h → day and $t_{1/2} \to k$ then $m_{\text{I-131}0}, t, k \to m_{\text{I-131}\,t}$

$\frac{1\ \text{day}}{24\ \text{h}}$ $t_{1/2} = \frac{0.693}{k}$ $\ln N_t = -kt + \ln N_0$

then mg$_0$, mg$_t$ → mg decayed → g decayed → mol decayed → beta decays then h → min → s

$$\text{mg}_0 - \text{mg}_t = \text{mg decayed} \quad \frac{1\text{ g}}{1000\text{ mg}} \quad \frac{1\text{ mol I-131}}{131\text{ g I-131}} \quad \frac{6.022 \times 10^{23}\text{ beta decays}}{1\text{ mol I-131}} \quad \frac{60\text{ min}}{1\text{ h}} \quad \frac{60\text{ s}}{1\text{ min}}$$

then beta decays, s → beta decays/s → Ci

take ratio $\dfrac{1\text{ Ci}}{\dfrac{3.7 \times 10^{10}\text{ decays}}{\text{s}}}$

Solution: $4.0\text{ h} \times \dfrac{1\text{ day}}{24\text{ h}} = 0.1\underline{6}667\text{ day}$ then $t_{1/2} = \dfrac{0.693}{k}$ Rearrange to solve for k.

$k = \dfrac{0.693}{t_{1/2}} = \dfrac{0.693}{8.0\text{ days}} = 0.08\underline{6}625\text{ day}^{-1}$ Because

$\ln m_{\text{I-131}\,t} = -kt + \ln m_{\text{I-131}\,0} = -(0.08\underline{6}625\text{ day}^{-1})(0.1\underline{6}667\text{ day}) + \ln(155\text{ mg}) = 5.02\underline{8}99 \rightarrow$

$m_{\text{I-131}\,t} = e^{5.02\underline{8}99} = 152.\underline{7}79\text{ mg}$ then

$\text{mg}_0 - \text{mg}_t = \text{mg decayed} = 155\text{ mg} - 152.\underline{7}79\text{ mg} = \underline{2}.221\text{ mg I-131}$ then

$$\underline{2}.221\text{ mg I-131} \times \frac{1\text{ g I-131}}{1000\text{ mg I-131}} \times \frac{1\text{ mol I-131}}{131\text{ g I-131}} \times \frac{6.022 \times 10^{23}\text{ beta decays}}{1\text{ mol I-131}} = \underline{1}.0210 \times 10^{19}\text{ beta decays}$$

then $4.0\text{ h} \times \dfrac{60\text{ min}}{1\text{ h}} \times \dfrac{60\text{ s}}{1\text{ min}} = 1.44 \times 10^4\text{ s}$ then

$$\frac{\underline{1}.0210 \times 10^{19}\text{ beta decays}}{1.44 \times 10^4\text{ s}} \times \frac{1\text{ Ci}}{\dfrac{3.7 \times 10^{10}\text{ decays}}{\text{s}}} = \underline{1}.9163 \times 10^4\text{ Ci} = 2 \times 10^4\text{ Ci}$$

Check: The units (Ci) are correct. The amount that decays is large because the half-life is fairly short; so the dose is high.

Cumulative Problems

21.77 **Given:** incomplete reactions **Find:** balanced reaction and energy (in J/mol reactant)

Conceptual Plan: Equalize the sum of the mass numbers and the sum of the atomic numbers on both sides of the equation by writing the appropriate mass number and atomic number for the unknown species. → Using the periodic table and the list of particles, deduce the identity of the unknown species from the atomic number and write its symbol. Then mass of products & reactants → mass defect in g → mass defect in kg → E

$$\text{mass defect} = \Sigma\text{mass of reactants} - \Sigma\text{mass of products} \quad \frac{1\text{ kg}}{1000\text{ g}} \quad E = mc^2$$

Solution:

(a) ${}^{?}_{?}? + {}^{9}_{4}\text{Be} \rightarrow {}^{6}_{3}\text{Li} + {}^{4}_{2}\text{He}$ becomes ${}^{1}_{1}? + {}^{9}_{4}\text{Be} \rightarrow {}^{6}_{3}\text{Li} + {}^{4}_{2}\text{He}$ then ${}^{1}_{1}\text{H} + {}^{9}_{4}\text{Be} \rightarrow {}^{6}_{3}\text{Li} + {}^{4}_{2}\text{He}$

mass defect $= \Sigma$mass of reactants $- \Sigma$mass of products and

mass defect $= (1.00783\text{ g} + 9.012182\text{ g}) - (6.015122\text{ g} + 4.002603\text{ g}) = 0.0022\underline{8}7\text{ g}$

then $\dfrac{0.0022\underline{8}7\text{ g}}{2\text{ mol reactants}} \times \dfrac{1\text{ kg}}{1000\text{ g}} = 1.14\underline{3}5 \times 10^{-6}\dfrac{\text{kg}}{\text{mol reactants}}$ then

$$E = mc^2 = \left(1.14\underline{3}5 \times 10^{-6}\frac{\text{kg}}{\text{mol reactants}}\right)\left(2.9979 \times 10^8\frac{\text{m}}{\text{s}}\right)^2 = 1.03 \times 10^{11}\frac{\text{J}}{\text{mol reactants}}$$

Check: 1 + 9 = 6 + 4, 1 + 4 = 3 + 2. The units (J) are correct.

(b) ${}^{209}_{83}\text{Bi} + {}^{64}_{28}\text{Ni} \rightarrow {}^{272}_{111}\text{Rg} + {}^{?}_{?}?$ becomes ${}^{209}_{83}\text{Bi} + {}^{64}_{28}\text{Ni} \rightarrow {}^{272}_{111}\text{Rg} + {}^{1}_{0}?$ then ${}^{209}_{83}\text{Bi} + {}^{64}_{28}\text{Ni} \rightarrow {}^{272}_{111}\text{Rg} + {}^{1}_{0}\text{n}$

mass defect $= (208.980384\text{ g} + 63.927969\text{ g}) - (272.1535\text{ g} + 1.00866\text{ g}) = -0.253\underline{8}07\text{ g}$

(Note: Because this is negative, energy must be added.)

then $\dfrac{0.253\underline{8}07\text{ g}}{2\text{ mol reactants}} \times \dfrac{1\text{ kg}}{1000\text{ g}} = 1.26\underline{9}035 \times 10^{-4}\dfrac{\text{kg}}{\text{mol reactants}}$ then

$$E = mc^2 = \left(1.26\underline{9}035 \times 10^{-4}\frac{\text{kg}}{\text{mol reactants}}\right)\left(2.9979 \times 10^8\frac{\text{m}}{\text{s}}\right)^2 = 1.141 \times 10^{13}\frac{\text{J}}{\text{mol reactants}}$$

Check: 209 + 64 = 272 + 1, 83 + 28 = 111 + 0. The units (J) are correct.

(c) $^{179}_{74}\text{W} + ^{?}_{?}? \rightarrow ^{179}_{73}\text{Ta}$ becomes $^{179}_{74}\text{W} + ^{0}_{-1}? \rightarrow ^{179}_{73}\text{Ta}$ then $^{179}_{74}\text{W} + ^{0}_{-1}\text{e} \rightarrow ^{179}_{73}\text{Ta}$
mass defect = (178.94707 g + 0.00055 g) − 178.94593 g = 0.00169 g then

$$\frac{0.00169\ \cancel{g}}{2\ \text{mol reactants}} \times \frac{1\ \text{kg}}{1000\ \cancel{g}} = 8.45 \times 10^{-7}\ \frac{\text{kg}}{\text{mol reactants}}\ \text{then}$$

$$E = mc^2 = \left(8.45 \times 10^{-7}\ \frac{\text{kg}}{\text{mol reactants}}\right)\left(2.9979 \times 10^8\ \frac{\text{m}}{\text{s}}\right)^2 = 7.59 \times 10^{10}\ \frac{\text{J}}{\text{mol reactants}}$$

Check: 179 + 0 = 179, 74 − 1 = 73. The units (J) are correct.

21.78 **Given:** incomplete reactions **Find:** balanced reaction and energy (in J/mol reactant)
Conceptual Plan: Equalize the sum of the mass numbers and the sum of the atomic numbers on both sides of the equation by writing the appropriate mass number and atomic number for the unknown species. → Using the periodic table and the list of particles, deduce the identity of the unknown species from the atomic number and write its symbol. Then mass of products & reactants → mass defect in g → mass defect in kg → *E*

mass defect = Σmass of reactants − Σmass of products $\qquad \frac{1\ \text{kg}}{1000\ \text{g}} \qquad E = mc^2$

Solution:

(a) $^{27}_{13}\text{Al} + ^{4}_{2}\text{He} \rightarrow ^{30}_{15}\text{P} + ^{?}_{?}?$ becomes $^{27}_{13}\text{Al} + ^{4}_{2}\text{He} \rightarrow ^{30}_{15}\text{P} + ^{1}_{0}?$ then $^{27}_{13}\text{Al} + ^{4}_{2}\text{He} \rightarrow ^{30}_{15}\text{P} + ^{1}_{0}\text{n}$
mass defect = Σmass of reactants − Σmass of products and
mass defect = (26.981538 g + 4.002603 g) − (29.981801g + 1.00866 g) = −0.00632 g
(Note: Because this is negative, energy must be added.) then

$$\frac{0.00632\ \cancel{g}}{2\ \text{mol reactants}} \times \frac{1\ \text{kg}}{1000\ \cancel{g}} = 3.16 \times 10^{-6}\ \frac{\text{kg}}{\text{mol reactants}}\ \text{then}$$

$$E = mc^2 = \left(3.16 \times 10^{-6}\ \frac{\text{kg}}{\text{mol reactants}}\right)\left(2.9979 \times 10^8 \frac{\text{m}}{\text{s}}\right)^2 = 2.84 \times 10^{11}\ \frac{\text{J}}{\text{mol reactants}}$$

Check: 27 + 4 = 30 + 1, 13 + 2 = 15 + 0. The units (J) are correct.

(b) $^{32}_{16}\text{S} + ^{?}_{?}? \rightarrow ^{29}_{14}\text{Si} + ^{4}_{2}\text{He}$ becomes $^{32}_{16}\text{S} + ^{1}_{0}? \rightarrow ^{29}_{14}\text{Si} + ^{4}_{2}\text{He}$ then $^{32}_{16}\text{S} + ^{1}_{0}\text{n} \rightarrow ^{29}_{14}\text{Si} + ^{4}_{2}\text{He}$
mass defect = (31.972071 g + 1.00866 g) − (28.976495 g + 4.002603 g) = $0.0016\underline{3}3$ g

$$\text{then}\ \frac{0.0016\underline{3}3\ \cancel{g}}{2\ \text{mol reactants}} \times \frac{1\ \text{kg}}{1000\ \cancel{g}} = 8.1\underline{6}5 \times 10^{-7}\ \frac{\text{kg}}{\text{mol reactants}}\ \text{then}$$

$$E = mc^2 = \left(8.1\underline{6}5 \times 10^{-7}\ \frac{\text{kg}}{\text{mol reactants}}\right)\left(2.9979 \times 10^8\ \frac{\text{m}}{\text{s}}\right)^2 = 7.34 \times 10^{10}\ \frac{\text{J}}{\text{mol reactants}}$$

Check: 32 + 1 = 29 + 4, 16 + 0 = 14 + 2. The units (J) are correct.

(c) $^{241}_{95}\text{Am} \rightarrow ^{237}_{93}\text{Np} + ^{?}_{?}?$ becomes $^{241}_{95}\text{Am} \rightarrow ^{237}_{93}\text{Np} + ^{4}_{2}?$ then $^{241}_{95}\text{Am} \rightarrow ^{237}_{93}\text{Np} + ^{4}_{2}\text{He}$
mass defect = 241.056822 g − (237.048166 g + 4.002603 g) = 0.006053 g then

$$\frac{0.006053\ \cancel{g}}{1\ \text{mol reactants}} \times \frac{1\ \text{kg}}{1000\ \cancel{g}} = 6.053 \times 10^{-6}\ \frac{\text{kg}}{\text{mol reactants}}\ \text{then}$$

$$E = mc^2 = \left(6.053 \times 10^{-6}\ \frac{\text{kg}}{\text{mol reactants}}\right)\left(2.9979 \times 10^8 \frac{\text{m}}{\text{s}}\right)^2 = 5.440 \times 10^{11}\ \frac{\text{J}}{\text{mol reactants}}$$

Check: 241 = 237 + 4, 95 = 93 + 2. The units (J) are correct.

21.79 **Given:** (a) Ru-114, (b) Ra-216, (c) Zn-58, and (d) Ne-31 **Find:** Write a nuclear equation for the most likely decay.
Conceptual Plan: Referring to the Valley of Stability graph in Figure 21.5, decide on the most likely decay mode depending on N/Z (too large = beta decay, too low = positron emission). → Write the symbol for the parent nuclide on the left side of the equation and the symbol for a particle on the right side. → Equalize the sum of the mass numbers and the sum of the atomic numbers on both sides of the equation by writing the appropriate mass number and atomic number for the unknown daughter nuclide. → Using the periodic table, deduce the identity of the unknown daughter nuclide from the atomic number and write its symbol.

Solution:

(a) Ru-114 ($N/Z = 1.6$) will undergo beta decay $^{114}_{44}Ru \rightarrow {}^{?}_{?}? + {}^{0}_{-1}e$ then $^{114}_{44}Ru \rightarrow {}^{114}_{45}? + {}^{0}_{-1}e$ then $^{114}_{44}Ru \rightarrow {}^{114}_{45}Rh + {}^{0}_{-1}e$

(b) Ra-216 ($N/Z = 1.4$) will undergo positron emission $^{216}_{88}Ra \rightarrow {}^{?}_{?}? + {}^{0}_{+1}e$ then $^{216}_{88}Ra \rightarrow {}^{216}_{87}? + {}^{0}_{+1}e$ then $^{216}_{88}Ra \rightarrow {}^{216}_{87}Fr + {}^{0}_{+1}e$

(c) Zn-58 ($N/Z = 0.9$) will undergo positron emission $^{58}_{30}Zn \rightarrow {}^{?}_{?}? + {}^{0}_{+1}e$ then $^{58}_{30}Zn \rightarrow {}^{58}_{29}? + {}^{0}_{+1}e$ then $^{58}_{30}Zn \rightarrow {}^{58}_{29}Cu + {}^{0}_{+1}e$

(d) Ne-31 ($N/Z = 2$) will undergo beta decay $^{31}_{10}Ne \rightarrow {}^{?}_{?}? + {}^{0}_{-1}e$ then $^{31}_{10}Ne \rightarrow {}^{31}_{11}? + {}^{0}_{-1}e$ then $^{31}_{10}Ne \rightarrow {}^{31}_{11}Na + {}^{0}_{-1}e$

Check: (a) $114 = 114 + 0, 44 = 45 - 1$, and rhodium is atomic number 45. (b) $216 = 216 + 0, 88 = 87 + 1$, and francium is atomic number 87. (c) $58 = 58 + 0, 30 = 29 + 1$, and copper is atomic number 29. (d) $31 = 31 + 0, 10 = 11 - 1$, and sodium is atomic number 11.

21.80 **Given:** (a) Kr-74, (b) Th-221, (c) Ar-44, and (d) Nb-85 **Find:** Write a nuclear equation for the most likely decay.
Conceptual Plan: Referring to the Valley of Stability graph in Figure 21.5, decide on the most likely decay mode depending on N/Z (too large = beta decay, too low = positron emission). → Write the symbol for the parent nuclide on the left side of the equation and the symbol for a particle on the right side. → Equalize the sum of the mass numbers and the sum of the atomic numbers on both sides of the equation by writing the appropriate mass number and atomic number for the unknown daughter nuclide. → Using the periodic table, deduce the identity of the unknown daughter nuclide from the atomic number and write its symbol.
Solution:

(a) Kr-74 ($N/Z = 1$) will undergo positron emission $^{74}_{36}Kr \rightarrow {}^{?}_{?}? + {}^{0}_{+1}e$ then $^{74}_{36}Kr \rightarrow {}^{74}_{35}? + {}^{0}_{+1}e$ then $^{74}_{36}Kr \rightarrow {}^{74}_{35}Br + {}^{0}_{+1}e$

(b) Th-221 ($N/Z = 1.5$) will undergo positron emission $^{221}_{90}Th \rightarrow {}^{?}_{?}? + {}^{0}_{+1}e$ then $^{221}_{90}Th \rightarrow {}^{221}_{89}? + {}^{0}_{+1}e$ then $^{221}_{90}Th \rightarrow {}^{221}_{89}Ac + {}^{0}_{+1}e$

(c) Ar-44 ($N/Z = 1.4$) will undergo beta decay $^{44}_{18}Ar \rightarrow {}^{?}_{?}? + {}^{0}_{-1}e$ then $^{44}_{18}Ar \rightarrow {}^{44}_{19}? + {}^{0}_{-1}e$ then $^{44}_{18}Ar \rightarrow {}^{44}_{19}K + {}^{0}_{-1}e$

(d) Nb-85 ($N/Z = 1.1$) will undergo positron emission $^{85}_{41}Nb \rightarrow {}^{?}_{?}? + {}^{0}_{+1}e$ then $^{85}_{41}Nb \rightarrow {}^{85}_{40}? + {}^{0}_{+1}e$ then $^{85}_{41}Nb \rightarrow {}^{85}_{40}Zr + {}^{0}_{+1}e$

Check: (a) $74 = 74 + 0, 36 = 35 + 1$, and bromine is atomic number 35. (b) $221 = 221 + 0, 90 = 89 + 1$, and actinium is atomic number 89. (c) $44 = 44 + 0, 18 = 19 - 1$, and potassium is atomic number 19. (d) $85 = 85 + 0, 41 = 40 + 1$, and zirconium is atomic number 40.

21.81 **Given:** Bi-210, $t_{1/2} = 5.0$ days, 1.2 g Bi-210, 209.984105 amu, 5.5% absorbed
Find: beta emissions in 13.5 days and dose (in Ci)
Conceptual Plan: $t_{1/2} \rightarrow k$ **then** $m_{\text{Bi-210}\,0}, t, k \rightarrow m_{\text{Bi-210}\,t}$ **then**

$$t_{1/2} = \frac{0.693}{k} \qquad \ln N_t = -kt + \ln N_0$$

$g_0, g_t \rightarrow$ **g decayed → mol decayed → beta decays then day → h → min → s then**

$$g_0 - g_t = \text{g decayed} \qquad \frac{1 \text{ mol Bi-210}}{209.984105 \text{ g Bi-210}} \qquad \frac{6.022 \times 10^{23} \text{ beta decays}}{1 \text{ mol Bi-210}} \qquad \frac{24 \text{ h}}{1 \text{ day}} \qquad \frac{60 \text{ min}}{1 \text{ h}} \qquad \frac{60 \text{ s}}{1 \text{ min}}$$

beta decays, s → beta decays/s → Ci available → Ci absorbed

$$\text{take ratio} \qquad \frac{1 \text{ Ci}}{\frac{3.7 \times 10^{10} \text{ decays}}{\text{s}}} \qquad \frac{5.5 \text{ Ci absorbed}}{100 \text{ Ci emitted}}$$

Solution: $t_{1/2} = \frac{0.693}{k}$ Rearrange to solve for k. $k = \frac{0.693}{t_{1/2}} = \frac{0.693}{5.0 \text{ days}} = 0.1\underline{3}86 \text{ day}^{-1}$ Because

$\ln m_{\text{Bi-210}\,t} = -kt + \ln m_{\text{Bi-210}\,0} = -(0.1\underline{3}86 \cancel{\text{day}^{-1}})(13.5 \cancel{\text{day}}) + \ln(1.2 \text{ g}) = -1.\underline{6}888 \rightarrow$
$m_{\text{Bi-210}\,t} = e^{-1.\underline{6}888} = 0.1\underline{8}474 \text{ g}$ then $g_0 - g_t = \text{g decayed} = 1.2 \text{ g} - 0.1\underline{8}474 \text{ g} = 1.\underline{0}153 \text{ g Bi-210}$

then $1.\underline{0}153 \cancel{\text{g Bi-210}} \times \frac{1 \cancel{\text{mol Bi-210}}}{209.984105 \cancel{\text{g Bi-210}}} \times \frac{6.022 \times 10^{23} \text{ beta decays}}{1 \cancel{\text{mol Bi-210}}} = 2.\underline{9}117 \times 10^{21} \text{ beta decays} =$

2.9×10^{21} beta decays then $13.5 \cancel{\text{day}} \times \frac{24 \cancel{\text{h}}}{1 \cancel{\text{day}}} \times \frac{60 \cancel{\text{min}}}{1 \cancel{\text{h}}} \times \frac{60 \text{ s}}{1 \cancel{\text{min}}} = 1.1\underline{6}64 \times 10^{6} \text{ s}$ then

$$\frac{2.9117 \times 10^{21} \text{ beta decays}}{1.1664 \times 10^{6} \text{ s}} \times \frac{1 \text{ Ci}}{\frac{3.7 \times 10^{10} \text{ decays}}{\text{s}}} = 6.7468 \times 10^{4} \text{ Ci emitted} \times \frac{5.5 \text{ Ci absorbed}}{100 \text{ Ci emitted}} = 3700 \text{ Ci}$$

Check: The units (decays and Ci) are correct. The amount that decays is large because the time is over three half-lives and we have a relatively large amount of the isotope. Because the decay is large, the dosage is large.

21.82 **Given:** Po-218, $t_{1/2} = 3.0$ minutes, 55 mg Po-218, 218.008965 amu
Find: alpha emissions in 25.0 min and dose (in Ci)
Conceptual Plan: $t_{1/2} \rightarrow k$ **then** $m_{\text{Po-218 }0}, t, k \rightarrow m_{\text{Po-218 }t}$ **then**

$$t_{1/2} = \frac{0.693}{k} \qquad \ln N_t = -kt + \ln N_0$$

$mg_0, mg_t \rightarrow$ **mg decayed** $\rightarrow$ **g decayed** $\rightarrow$ **mol decayed** $\rightarrow$ **alpha decays then min** $\rightarrow$ **s**

$$mg_0 - mg_t = \text{mg decayed} \qquad \frac{1 \text{ g}}{1000 \text{ mg}} \qquad \frac{1 \text{ mol Po-218}}{218.008965 \text{ g Po-218}} \qquad \frac{6.022 \times 10^{23} \text{ alpha decays}}{1 \text{ mol Po-218}} \qquad \frac{60 \text{ s}}{1 \text{ min}}$$

then beta decays, s $\rightarrow$ **beta decay/s** $\rightarrow$ **Ci**

$$\text{take ratio} \qquad \frac{1 \text{ Ci}}{\frac{3.7 \times 10^{10} \text{ decays}}{\text{s}}}$$

Solution: $t_{1/2} = \frac{0.693}{k}$ Rearrange to solve for k. $k = \frac{0.693}{t_{1/2}} = \frac{0.693}{3.0 \text{ min}} = 0.231 \text{ min}^{-1}$ Because

$\ln m_{\text{Po-218 }t} = -kt + \ln m_{\text{Po-218 }0} = -(0.231 \text{ min}^{-1})(25.0 \text{ min}) + \ln(55 \text{ mg}) = -1.7677 \rightarrow$
$m_{\text{Po-218 }t} = e^{-1.7677} = 0.17073$ mg then
$mg_0 - mg_t = \text{mg decayed} = 55 \text{ mg} - 0.17073 \text{ mg} = 54.829$ mg Po-218 then 54.829 mg Po-218

$$54.829 \text{ mg Po-218} \times \frac{1 \text{ g Po-218}}{1000 \text{ mg Po-218}} \times \frac{1 \text{ mol Po-218}}{218.008965 \text{ g Po-218}} \times \frac{6.022 \times 10^{23} \text{ alpha decays}}{1 \text{ mol Po-218}} =$$

1.5145×10^{20} alpha decays $= 1.5 \times 10^{20}$ alpha decays then $25.0 \text{ min} \times \frac{60 \text{ s}}{1 \text{ min}} = 1.5 \times 10^{3}$ s then

$$\frac{1.5145 \times 10^{20} \text{ alpha decays}}{1.5 \times 10^{3} \text{ s}} \times \frac{1 \text{ Ci}}{\frac{3.7 \times 10^{10} \text{ decays}}{\text{s}}} = 2.7 \times 10^{6} \text{ Ci}$$

Check: The units (decays and Ci) are correct. The amount that decays is large because the time is over eight half-lives; so almost the entire isotope has decayed. Because the decay rate is large (small half-life), the dosage is large.

21.83 **Given:** Ra-226 (226.05402 amu) decays to Rn-224, $t_{1/2} = 1.6 \times 10^{3}$ yr, 25.0 g Ra-226, $T = 25.0$ °C, $P = 1.0$ atm
Find: V of Rn-224 gas produced in 5.0 day
Conceptual Plan: day $\rightarrow$ **yr then** $t_{1/2} \rightarrow k$ **then** $m_{\text{Ra-226 }0}\ t, k \rightarrow m_{\text{Ra-226 }t}$

$$\frac{1 \text{ yr}}{365.24 \text{ day}} \qquad t_{1/2} = \frac{0.693}{k} \qquad \ln N_t = -kt + \ln N_0$$

then $g_0, g_t \rightarrow$ **g decayed** $\rightarrow$ **mol decayed** $\rightarrow$ **mol Rn-224 formed then °C** $\rightarrow$ **K then** $P, n, T \rightarrow V$

$$g_0 - g_t = \text{g decayed} \qquad \frac{1 \text{ mol Ra-226}}{226.05402 \text{ g Ra-226}} \qquad \frac{1 \text{ mol Rn-224}}{1 \text{ mol Ra-226}} \qquad K = °C + 273.15 \qquad PV = nRT$$

Solution: $5.0 \text{ day} \times \frac{1 \text{ yr}}{365.24 \text{ day}} = 0.013690$ yr then $t_{1/2} = \frac{0.693}{k}$ Rearrange to solve for k.

$k = \frac{0.693}{t_{1/2}} = \frac{0.693}{1.6 \times 10^{3} \text{ yr}} = 4.33125 \times 10^{-4} \text{ yr}^{-1}$ Because

$\ln m_{\text{Ra-226 }t} = -kt + \ln m_{\text{Ra-226 }0} = -(4.33125 \times 10^{-4} \text{ yr}^{-1})(0.013690 \text{ yr}) + \ln(25.0 \text{ g}) = 3.21887 \rightarrow$
$m_{\text{Ra-226 }t} = e^{3.21887} = 24.999854$ g then
$g_0 - g_t = \text{g decayed} = 25.0 \text{ g} - 24.999854 \text{ g} = 0.000146$ g Ra-226 then

$$0.000146 \text{ g Ra-226} \times \frac{1 \text{mol Ra-226}}{226.05402 \text{ g Ra-226}} \times \frac{1 \text{ mol Rn-224}}{1 \text{ mol Ra-226}} = 6.4586 \times 10^{-7} \text{ mol Rn-224 then}$$

$T = 25.0 \text{ °C} + 273.15 = 298.2$ K then $PV = nRT$. Rearrange to solve for V.

$$V = \frac{nRT}{P} = \frac{6.4586 \times 10^{-7}\ \cancel{\text{mol}} \times 0.08206 \frac{\text{L} \cdot \cancel{\text{atm}}}{\cancel{\text{mol}} \cdot \cancel{\text{K}}} \times 298.2\ \cancel{\text{K}}}{1.0\ \cancel{\text{atm}}} = \underline{1}.5804 \times 10^{-5}\ \text{L} = 1.6 \times 10^{-5}\ \text{L}$$

Two significant figures are reported as requested in the problem.

Check: The units (L) are correct. The amount of gas is small because the time is so small compared to the half-life.

21.84 **Given:** U-235 (235.043922 amu) neutron-induced fission to Ba-140 and Kr-93, 1.00 g U-235, $T = 25.0\ °\text{C}$, $P = 1.0$ atm **Find:** V of Kr-93 gas produced

Conceptual Plan: Write the species given on the appropriate side of the equation. → Equalize the sum of the mass numbers and the sum of the atomic numbers on both sides of the equation by writing the stoichiometric coefficient in front of the desired species. Then

g U-235 → mol U-235 → mol Kr-93 formed then °C → K then $P, n, T \rightarrow V$.

$$\frac{1 \text{ mol U-235}}{235.043922 \text{ g U-235}} \quad \frac{1 \text{ mol Kr-93}}{1 \text{ mol U-235}} \quad \text{K} = °\text{C} + 273.15 \quad PV = nRT$$

Solution: ${}^{235}_{92}\text{U} + {}^{1}_{0}\text{n} \rightarrow {}^{140}_{56}\text{Ba} + {}^{93}_{36}\text{Kr} + ?{}^{1}_{0}\text{n}$ becomes ${}^{235}_{92}\text{U} + {}^{1}_{0}\text{n} \rightarrow {}^{140}_{56}\text{Ba} + {}^{93}_{36}\text{Kr} + 3{}^{1}_{0}\text{n}$ then

$$1.00\ \cancel{\text{g U-235}} \times \frac{1\ \cancel{\text{mol U-235}}}{235.043922\ \cancel{\text{g U-235}}} \times \frac{1 \text{ mol Kr-93}}{1\ \cancel{\text{mol U-235}}} = 4.2\underline{5}452 \times 10^{-3} \text{ mol Kr-93 then}$$

$T = 25.0\ °\text{C} + 273.15 = 298.2$ K then $PV = nRT$. Rearrange to solve for V.

$$V = \frac{nRT}{P} = \frac{4.2\underline{5}452 \times 10^{-3}\ \cancel{\text{mol}} \times 0.08206 \frac{\text{L} \cdot \cancel{\text{atm}}}{\cancel{\text{mol}} \cdot \cancel{\text{K}}} \times 298.2\ \cancel{\text{K}}}{1.0\ \cancel{\text{atm}}} = 0.1\underline{0}4109\ \text{L} = 0.10\ \text{L}$$

Check: $235 + 1 = 140 + 93 + 3(1), 92 + 0 = 56 + 36 + 3(0)$, and no other particle is necessary to balance the equation. The units (L) are correct. About 1/200 mole of gas is generated, so we expect the volume to be about 22/200 L.

21.85 **Given:** ${}^{0}_{1}\text{e} + {}^{0}_{-1}\text{e} \rightarrow 2{}^{0}_{0}\gamma$ **Find:** energy (in kJ/mol)

Conceptual Plan:

mass of products and reactants → mass defect (g) → kg → kg/mol → E (J/mol) → E (kJ/mol)

$$\text{mass defect} = \Sigma\text{mass of reactants} - \Sigma\text{mass of products} \quad \frac{1 \text{ kg}}{1000 \text{ g}}\ 2 \text{ mol} \quad E = mc^2 \quad \frac{1 \text{ kJ}}{1000 \text{ J}}$$

Solution: mass defect $= \Sigma$mass of reactants $- \Sigma$mass of products $= (0.00055 \text{ g} + 0.00055 \text{ g}) - 0 \text{ g} =$

$$0.00110 \text{ g then } \frac{0.00110\ \cancel{\text{g}}}{2 \text{ mol}} \times \frac{1 \text{ kg}}{1000\ \cancel{\text{g}}} = 5.50 \times 10^{-7} \frac{\text{kg}}{\text{mol}} \text{ then}$$

$$E = mc^2 = \left(5.50 \times 10^{-7} \frac{\text{kg}}{\text{mol}}\right)\left(2.9979 \times 10^{8} \frac{\text{m}}{\text{s}}\right)^2 = 4.9\underline{4}307 \times 10^{10} \frac{\cancel{\text{J}}}{\text{mol}} \times \frac{1 \text{ kJ}}{1000\ \cancel{\text{J}}} = 4.94 \times 10^{7} \frac{\text{kJ}}{\text{mol}}$$

Check: The units (kJ/mol) are correct. A large amount of energy is expected per mole of substance lost. The photon is in the gamma ray region of the electromagnetic spectrum.

21.86 **Given:** 1.0 MW power/day **Find:** minimum rate of mass loss required

Conceptual Plan: MW → MWh → kWh → J → kg → g

$$\frac{24 \text{ h}}{1 \text{ day}} \quad \frac{1000 \text{ kWh}}{1 \text{ MWh}} \quad \frac{3.60 \times 10^6 \text{ J}}{1 \text{ kWh}} \quad E = mc^2 \quad \frac{1000 \text{ g}}{1 \text{ kg}}$$

Solution: $1.0\ \cancel{\text{MW}} \times \frac{24\ \cancel{\text{h}}}{1 \text{ day}} \times \frac{1000\ \cancel{\text{kWh}}}{1\ \cancel{\text{MWh}}} \times \frac{3.60 \times 10^6 \text{ J}}{1\ \cancel{\text{kWh}}} = 8.64 \times 10^{10} \frac{\text{J}}{\text{day}}$ Because $E = mc^2$, rearrange to solve

$$\text{for } m.\ m = \frac{E}{c^2} = \frac{8.64 \times 10^{10} \frac{\text{kg}\ \cancel{\text{m}^2}}{\text{day}\ \cancel{\text{s}^2}}}{\left(2.9979 \times 10^{8} \frac{\cancel{\text{m}}}{\cancel{\text{s}}}\right)^2} = 9.6 \times 10^{-7} \frac{\cancel{\text{kg}}}{\text{day}} \times \frac{1000 \text{ g}}{1\ \cancel{\text{kg}}} = 9.6 \times 10^{-4} \frac{\text{g}}{\text{day}}$$

Check: The units (g/day) are correct. A large amount of energy is expected per gram of mass lost.

21.87 **Given:** ${}^{3}\text{He} = 3.016030$ amu **Find:** nuclear binding energy per atom

Conceptual Plan: ${}^{A}_{Z}X$**, isotope mass → mass defect → nuclear binding energy per nucleon**

$$\text{mass defect} = Z(\text{mass } {}^{1}_{1}\text{H}) + (A - Z)(\text{mass } {}^{1}_{0}\text{n}) - \text{mass of isotope} \quad \frac{931.5 \text{ MeV}}{1 \text{ amu}}$$

Solution: mass defect $= Z(\text{mass } {}^1_1\text{H}) + (A - Z)(\text{mass } {}^1_0\text{n}) -$ mass of isotope
He-3 mass defect $= 2(1.00783 \text{ amu}) + (3 - 2)(1.00866 \text{ amu}) - 3.016030 \text{ amu} = 0.00829 \text{ amu}$
and $0.00829 \text{ amu} \times \frac{931.5 \text{ MeV}}{1 \text{ amu}} = 7.72 \text{ MeV}$

Check: The units (MeV) are correct. The number of nucleons is small, so the MeV is not that large.

21.88 **Given:** $4\,{}^1_1\text{H} \rightarrow {}^4_2\text{He}$ **Find:** energy (in J/mol reactant)
Conceptual Plan: mass of products and reactants → mass defect in g → mass defect in kg → E

mass defect $= \sum$mass of reactants $- \sum$mass of products $\frac{1 \text{ kg}}{1000 \text{ g}}$ $E = mc^2$

Solution:
mass defect $= \sum$mass of reactants $- \sum$mass of products $= 4(1.00783 \text{ g}) - 4.002603 \text{ g} = 0.028717 \text{ g}$

then $\frac{0.028717 \text{ g}}{4 \text{ mol reactants}} \times \frac{1 \text{ kg}}{1000 \text{ g}} = 7.17925 \times 10^{-6} \frac{\text{kg}}{\text{mol reactants}}$ then

$$E = mc^2 = \left(7.17925 \times 10^{-6} \frac{\text{kg}}{\text{mol reactants}}\right)\left(2.9979 \times 10^8 \frac{\text{m}}{\text{s}}\right)^2 = 6.4523 \times 10^{11} \frac{\text{J}}{\text{mol reactants}}$$

Check: The units (J/mol) are correct.

21.89 **Given:** ${}^{247}\text{Es}$ and five neutrons made by bombarding ${}^{238}\text{U}$ **Find:** identity of bombarding particle
Conceptual Plan: Begin with the symbols for the nuclides given. → Equalize the sum of the mass numbers and the sum of the atomic numbers on both sides of the equation by writing the appropriate mass number and atomic number for the unknown daughter nuclide. → Using the periodic table, deduce the identity of the unknown nuclide from the atomic number and write its symbol.
Solution:
${}^{238}_{92}\text{U} + {}^{?}_{?}? \rightarrow {}^{247}_{99}\text{Es} + 5\,{}^1_0\text{n}$ then ${}^{238}_{92}\text{U} + {}^{14}_{7}? \rightarrow {}^{247}_{99}\text{Es} + 5\,{}^1_0\text{n}$ then ${}^{238}_{92}\text{U} + {}^{14}_{7}\text{N} \rightarrow {}^{247}_{99}\text{Es} + 5\,{}^1_0\text{n}$

Check: $238 + 14 = 247 + 5(1), 92 + 7 = 99 + 5(0)$, and nitrogen is atomic number 7.

21.90 **Given:** ${}^6\text{Li}$ reacts with ${}^2\text{H}$ to form two identical particles **Find:** identity of product particles
Conceptual Plan: Begin with the symbols for the nuclides given. → Equalize the sum of the mass numbers and the sum of the atomic numbers on both sides of the equation by writing the appropriate mass number and atomic number for the unknown daughter nuclide. → Using the periodic table, deduce the identity of the unknown nuclide from the atomic number and write its symbol.
Solution:
${}^6_3\text{Li} + {}^2_1\text{H} \rightarrow 2\,{}^?_??$ then ${}^6_3\text{Li} + {}^2_1\text{H} \rightarrow 2\,{}^4_2?$ then ${}^6_3\text{Li} + {}^2_1\text{H} \rightarrow 2\,{}^4_2\text{He}$ or two alpha particles

Check: $6 + 2 = 2(4), 3 + 1 = 2(2)$, and helium is atomic number 2.

21.91 **Given:** $t_{1/2}$ for decay of ${}^{238}\text{U} = 4.5 \times 10^9$ years, 1.6 g rock, 29 dis/s all radioactivity from U-238
Find: percent by mass ${}^{238}\text{U}$ in rock
Conceptual Plan: $t_{1/2} \rightarrow k$ and s → min → h → day → yr then Rate, $k \rightarrow N \rightarrow$ mol ${}^{238}\text{U} \rightarrow$ g ${}^{238}\text{U}$

$t_{1/2} = \frac{0.693}{k}$ $\frac{1 \text{ min}}{60 \text{ s}}$ $\frac{1 \text{ h}}{60 \text{ min}}$ $\frac{1 \text{ day}}{24 \text{ h}}$ $\frac{1 \text{ yr}}{365.24 \text{ day}}$ $\text{Rate} = kN \frac{1 \text{ mol dis}}{6.022 \times 10^{23} \text{ dis}}$ $\frac{238 \text{ g } {}^{238}\text{U}}{1 \text{ mol } {}^{238}\text{U}}$

then g ${}^{238}\text{U}$, g rock → percent by mass ${}^{238}\text{U}$

percent by mass ${}^{238}\text{U} = \frac{\text{g } {}^{238}\text{U}}{\text{g rock}} \times 100\%$

Solution: $t_{1/2} = \frac{0.693}{k}$ Rearrange to solve for k. $k = \frac{0.693}{t_{1/2}} = \frac{0.693}{4.5 \times 10^9 \text{ yr}} = 1.54 \times 10^{-10} \text{ yr}^{-1}$ and

$1 \text{ s} \times \frac{1 \text{ min}}{60 \text{ s}} \times \frac{1 \text{ h}}{60 \text{ min}} \times \frac{1 \text{ day}}{24 \text{ h}} \times \frac{1 \text{ yr}}{365.24 \text{ day}} = 3.16889554 \times 10^{-8} \text{ yr}$. Rate $= kN$. Rearrange to solve for N.

$$N = \frac{\text{Rate}}{k} = \frac{\dfrac{29 \text{ dis}}{3.16889554 \times 10^{-8} \text{ yr}}}{1.54 \times 10^{-10} \text{ yr}^{-1}} = 5.9425 \times 10^{18} \text{ dis then}$$

$$5.9425 \times 10^{18} \text{ dis} \times \frac{1 \text{ mol dis}}{6.022 \times 10^{23} \text{ dis}} \times \frac{238 \text{ g } ^{238}\text{U}}{1 \text{ mol } ^{238}\text{U}} = 2.3486 \times 10^{-3} \text{g } ^{238}\text{U then}$$

$$\text{percent by mass } ^{238}\text{U} = \frac{\text{g } ^{238}\text{U}}{\text{g rock}} \times 100\% = \frac{2.3486 \times 10^{-3} \text{ g } ^{238}\text{U}}{1.6 \text{ g rock}} \times 100\% = 0.15\%$$

Check: The units (%) are correct. The mass percent is low because the dis/s is low.

21.92 **Given:** $t_{1/2}$ for decay of ^{232}Th $= 1.4 \times 10^{10}$ years **Find:** number of dis emitted by 1.0 mol ^{232}Th in 1 min
Conceptual Plan: $t_{1/2} \rightarrow k$ **then** $N, k \rightarrow$ **Rate (dis/yr)** $\rightarrow$ **dis/day** $\rightarrow$ **dis/h** $\rightarrow$ **dis/min**

$t_{1/2} = \frac{0.693}{k}$ Rate $= kN$ $\frac{1 \text{ yr}}{365.24 \text{ day}}$ $\frac{1 \text{ day}}{24 \text{ h}}$ $\frac{1 \text{ h}}{60 \text{ min}}$

Solution: $t_{1/2} = \frac{0.693}{k}$ Rearrange to solve for k. $k = \frac{0.693}{t_{1/2}} = \frac{0.693}{1.4 \times 10^{10} \text{ yr}} = 4.95 \times 10^{-11} \text{ yr}^{-1}$ then

$$\text{Rate} = kN = (4.95 \times 10^{-11} \text{ yr}^{-1})(6.022 \times 10^{23} \text{ dis}) \times \frac{1 \text{ yr}}{365.24 \text{ day}} \times \frac{1 \text{ day}}{24 \text{ h}} = 3.4 \times 10^{9} \text{ dis/h}$$

$$3.4 \times 10^{9} \text{ dis/h} \times \frac{1 \text{ h}}{60 \text{ min}} = 5.7 \times 10^{7} \text{ dis/min}$$

Check: The units (dis/h and dis/min) are correct. The rate is high because the amount of ^{232}Th is high.

21.93 **Given:** $V = 1.50$ L, $P = 745$ mmHg, $T = 25.0$ °C, 3.55% Ra-220 by volume, $t_{1/2} = 55.6$ s
Find: number of alpha particles emitted in 5.00 min
Conceptual Plan: mmg $\rightarrow$ **atm and °C** $\rightarrow$ **K then** $P, V, T \rightarrow n_{\text{Total}} \rightarrow n_{\text{Ra-220}}$ **and min** $\rightarrow$ **s**

$\frac{1 \text{ atm}}{760 \text{ mmHg}}$ K = °C + 273.15 $PV = nRT$ $\frac{3.55 \text{ mol Ra-220 particles}}{100 \text{ mol gas particles}}$ $\frac{60 \text{ s}}{1 \text{ min}}$

then $t_{1/2} \rightarrow k$ **then** $n_{\text{Ra-220 0}}, t, k \rightarrow n_{\text{Ra-220 } t} \rightarrow$ **number of particles remaining** $\rightarrow$ **particles emitted**

$t_{1/2} = \frac{0.693}{k}$ $\ln N_t = -kt + \ln N_0$ $\frac{6.022 \times 10^{23} \text{ particles}}{1 \text{ mol}}$

Solution: $745 \text{ mmHg} \times \frac{1 \text{ atm}}{760 \text{ mmHg}} = 0.9802632$ atm and $T = 25.0 \text{ °C} + 273.15 = 298.2$ K then

$PV = nRT$. Rearrange to solve for n.

$$n = \frac{PV}{RT} = \frac{0.9802632 \text{ atm} \times 1.50 \text{ L}}{0.08206 \frac{\text{L} \cdot \text{atm}}{\text{mol} \cdot \text{K}} \times 298.2 \text{ K}} = 0.06008898 \text{ mol gas particles}$$

$$\text{then } 0.06008898 \text{ mol gas particles} \times \frac{3.55 \text{ mol Ra-220 particles}}{100 \text{ mol gas particles}} = 0.002133159 \text{ mol Ra-220 particles}$$

$5.00 \text{ min} \times \frac{60 \text{ s}}{1 \text{ min}} = 300. \text{ s then } t_{1/2} = \frac{0.693}{k}$. Rearrange to solve for k.

$$k = \frac{0.693}{t_{1/2}} = \frac{0.693}{55.6 \text{ s}} = 0.01246403 \text{ s}^{-1} \text{ Because}$$

$\ln m_{\text{Ra-220 } t} = -kt + \ln m_{\text{Ra-220 } 0} = -(0.01246403 \text{ s}^{-1})(300. \text{ s}) + \ln(0.002133159 \text{ mol}) = -9.889360 \rightarrow$
$m_{\text{Ra-220 } t} = e^{-9.889360} = 5.071139 \times 10^{-5}$ mol alpha particles remaining
The number of alpha particles emitted would be the difference between this and the initial number of moles.
0.002133158 mol − 0.00005071139 mol = 0.002082447 mol

$$0.002082447 \text{ mol} \times \frac{6.022 \times 10^{23} \text{ particles}}{1 \text{ mol}} = 1.254050 \times 10^{21} \text{ particles} = 1.25 \times 10^{21} \text{ particles}$$

Check: The units (particles) are correct. The number of particles is far less than a mole because we have far less than a mole of gas.

21.94 **Given:** 228 mL of 2.35 by mass $MgCl_2$; exactly $\frac{1}{2}$ of Mg is Mg-28, $t_{1/2} = 21$ h, $d = 1.02$ g/mL
Find: decay rate after 4.00 days
Conceptual Plan: mL → g solution → g $MgCl_2$ → mol Mg → atoms Mg → atoms Mg-28 and

$$\frac{1.02\text{ g}}{1\text{ mL}} \quad \frac{2.35\text{ g MgCl}_2}{100\text{ g solution}} \quad \frac{1\text{ mol Mg}}{95.21\text{ g MgCl}_2} \quad \frac{6.022\times10^{23}\text{ Mg atoms}}{1\text{ mol Mg}} \quad \frac{1\text{ Mg-28 atom}}{2\text{ Mg atoms}}$$

days → h then $t_{1/2} \rightarrow k$ then $N_{\text{Mg-28 }0}$, $t, k \rightarrow N_{\text{Mg-28 }t} \rightarrow$ Rate

$$\frac{1\text{ day}}{24\text{ h}} \quad t_{1/2} = \frac{0.693}{k} \quad \ln N_t = -kt + \ln N_0 \quad \text{Rate} = kN$$

Solution: $228\text{ mL solution} \times \frac{1.02\text{ g solution}}{1\text{ mL solution}} \times \frac{2.35\text{ g MgCl}_2}{100\text{ g solution}} \times \frac{1\text{ mol Mg}}{95.21\text{ g MgCl}_2} \times$

$\frac{6.022\times10^{23}\text{ Mg atoms}}{1\text{ mol Mg}} \times \frac{1\text{ Mg-28 atom}}{2\text{ Mg atoms}} = 1.728348\times10^{22}$ Mg-28 atoms

and $21\text{ h} \times \frac{1\text{ day}}{24\text{ h}} = 0.875$ day then $t_{1/2} = \frac{0.693}{k}$ Rearrange to solve for k.

$k = \frac{0.693}{t_{1/2}} = \frac{0.693}{0.875\text{ day}} = 0.792\text{ day}^{-1}$ Because

$\ln N_{\text{Mg-28 }t} = -kt + \ln N_{\text{Mg-28 }0} = -(0.792\text{ day}^{-1})(4.00\text{ day}) + \ln(1.728348\times10^{22}) = 48.03604 \rightarrow$

$N_{\text{Mg-28 }t} = e^{48.03604} = 7.2742\times10^{20}$ Mg-28 atoms. Finally, Rate $= kN$, so

Rate $= 0.792/\text{day} \times 7.2742\times10^{20} = 5.8\times10^{20}$ atoms/day.

Check: The units (atoms) are correct. The number of particles is far less than a mole because we have a dilute solution.

21.95 **Given:** ${}^{0}_{+1}\text{e} + {}^{0}_{-1}\text{e} \rightarrow 2\,{}^{0}_{0}\gamma$ **Find:** wavelength of gamma ray photons
Conceptual Plan:
mass of products and reactants → mass defect (g) → kg → kg/mol → E (J/mol) → E (kJ/mol)

$$\text{mass defect} = \Sigma\text{mass of reactants} - \Sigma\text{mass of products} \quad \frac{1\text{ kg}}{1000\text{ g}} \quad 2\text{ mol} \quad E = mc^2 \quad \frac{1\text{ kJ}}{1000\text{ J}}$$

This energy is for 2 moles of γ, so $E(\text{J}/2\text{ mol }\gamma) \rightarrow E(\text{J}/\gamma\text{ photon}) \rightarrow \lambda$.

$$\frac{1\text{ mol }\gamma}{6.022\times10^{23}\ \gamma\text{ photons}} \quad E = \frac{hc}{\lambda}$$

Solution: mass defect $= \Sigma$mass of reactants $- \Sigma$mass of products $= (0.00055\text{ g} + 0.00055\text{ g}) - 0\text{ g} = 0.00110$ g

then $\frac{0.00110\text{ g}}{2\text{ mol}} \times \frac{1\text{ kg}}{1000\text{ g}} = 5.50\times10^{-7}\frac{\text{kg}}{\text{mol}}$ then

$$E = mc^2 = \left(5.50\times10^{-7}\frac{\text{kg}}{\text{mol}}\right)\left(2.9979\times10^{8}\frac{\text{m}}{\text{s}}\right)^2 = 4.94307\times10^{10}\frac{\text{J}}{\text{mol}} \times \frac{1\text{ kJ}}{1000\text{ J}} = 4.94\times10^{7}\frac{\text{kJ}}{\text{mol}}$$

$E = 4.94307\times10^{10}\frac{\text{J}}{\text{mol }\gamma} \times \frac{1\text{ mol }\gamma}{6.022\times10^{23}\ \gamma\text{ photon}} = 8.20835\times10^{-14}\frac{\text{J}}{\gamma\text{ photon}}$ then $E = \frac{hc}{\lambda}$ Rearrange to solve

for λ. $\lambda = \frac{hc}{E} = \frac{(6.626\times10^{-34}\text{ J}\cdot\text{s})\left(2.9979\times10^{8}\frac{\text{m}}{\text{s}}\right)}{8.20835\times10^{-14}\frac{\text{J}}{\gamma\text{ photon}}} = 2.42\times10^{-12}\text{ m} = 2.42\text{ pm}$

Check: The units (m or pm) are correct. A large amount of energy is expected per mole of substance lost. The photon is in the gamma ray region of the electromagnetic spectrum.

21.96 **Given:** 1.0 mg U-235, $t_{1/2} = 7.1\times10^{8}$ yr **Find:** alpha particles emitted in 1.0 min
Conceptual Plan: Because the time is so small compared to the half-life, use Rate $= kN_0$ and calculate the number of particles emitted based on the initial rate.

mg → g → mol → atoms (N_0) and

$\frac{1\text{ g}}{1000\text{ mg}}$ $\frac{1\text{ mol}}{235\text{ g}}$ $\frac{6.022 \times 10^{23}\text{ atoms}}{1\text{ mol}}$

min → hr → days → yr then $t_{1/2}$ → k then $N_{\text{U-235 0}}$, t, k → alpha particles emitted

$\frac{1\text{ hr}}{60\text{ min}}$ $\frac{1\text{ day}}{24\text{ h}}$ $\frac{1\text{ yr}}{365.24\text{ days}}$ $t_{1/2} = \frac{0.693}{k}$ Rate $= kN_0$

Solution: $1.0\text{ mg} \times \frac{1\text{ g}}{1000\text{ mg}} \times \frac{1\text{ mol}}{235\text{ g}} \times \frac{6.022 \times 10^{23}\text{ atoms}}{1\text{ mol}} = 2.\underline{5}6255 \times 10^{18}\text{ atoms}$ and

$1.0\text{ min} \times \frac{1\text{ hr}}{60\text{ min}} \times \frac{1\text{ day}}{24\text{ hr}} \times \frac{1\text{ yr}}{365.24\text{ day}} = 1.\underline{9}01337 \times 10^{-6}\text{ yr}$ then $t_{1/2} = \frac{0.693}{k}$ Rearrange to solve for k.

$k = \frac{0.693}{t_{1/2}} = \frac{0.693}{7.1 \times 10^8\text{ yr}} = 9.\underline{7}6056 \times 10^{-10}\text{ yr}^{-1}$ then

Rate $= -kN_{\text{U-235 0}} = (9.\underline{7}6056 \times 10^{-10}\text{ yr}^{-1})(2.\underline{5}6255 \times 10^{18}\text{ particles}) = 2.\underline{5}01192 \times 10^9\text{ particles/yr}$ then →

$\frac{2.\underline{5}01192 \times 10^9\text{ particles}}{1\text{ yr}} \times 1.\underline{9}01337 \times 10^{-6}\text{ yr} = 4\underline{7}55.6\text{ particles} = 4800\text{ particles emitted}$

Check: The units (particles) are correct. The number of particles is small because the half-life is long and the observation time is small.

21.97 **Given:** ${}^2_1\text{H} + {}^2_1\text{H} \rightarrow {}^3_2\text{He} + {}^1_0\text{n}$ releases 3.3 MeV; ${}^2_1\text{H} + {}^2_1\text{H} \rightarrow {}^3_1\text{H} + {}^1_1\text{p}$ releases 4.0 MeV

Find: the energy change for ${}^3_2\text{He} + {}^1_0\text{n} \rightarrow {}^3_1\text{H} + {}^1_1\text{p}$ and why this can happen at a much lower temperature

Conceptual Plan: Use Hess's law to calculate the energy change and give the two reactions.

Solution:

${}^3_2\text{He} + {}^1_0\text{n} \rightarrow {}^2_1\text{H} + {}^2_1\text{H}$ $\Delta E = 3.3\text{ MeV}$

${}^2_1\text{H} + {}^2_1\text{H} \rightarrow {}^3_1\text{H} + {}^1_1\text{p}$ $\Delta E = -4.0\text{ MeV}$

${}^3_2\text{He} + {}^1_0\text{n} \rightarrow {}^3_1\text{H} + {}^1_1\text{p}$ $\Delta E = -0.7\text{ MeV}$

The energy change is much less, and there is no coulombic barrier for collision with a neutron; so the process can occur at lower temperatures.

Check: The units (MeV) are correct. Because one reaction releases energy and one requires energy, the resulting energy change is much smaller in magnitude.

21.98 **Given:** ${}^{18}_9\text{F} + {}^0_{-1}\text{e} \rightarrow {}^{18}_8\text{O}$ and ${}^{18}_9\text{F} \rightarrow {}^{18}_8\text{O} + {}^0_{+1}\text{e}$ **Find:** difference in energy released

Other: ${}^{18}_9\text{F} = 18.000950\text{ g}$ and ${}^{18}_8\text{O} = 17.9991598\text{ g}$

Conceptual Plan: For each reaction, calculate mass of products and reactants → mass defect in amu then

mass defect = Σmass of reactants − Σmass of products

calculate the difference between the two mass defects in amu → E.

$\frac{931.5\text{ MeV}}{1\text{ amu}}$

Solution: Because mass defect = Σmass of reactants − Σmass of products, for ${}^{18}_9\text{F} + {}^0_{-1}\text{e} \rightarrow {}^{18}_8\text{O}$,
mass defect $= (18.000950\text{ amu} + 0.0005486\text{ amu}) - (17.9991598\text{ amu}) = +0.00233\underline{8}8\text{ amu}$ and for
${}^{18}_9\text{F} \rightarrow {}^{18}_8\text{O} + {}^0_{+1}\text{e}$, mass defect $= (18.000950\text{ amu}) - (17.9991598\text{ amu} + 0.0005486\text{ amu}) = +0.00124\underline{1}6\text{ amu}$.
The mass defect difference between the two reactions is $0.00233\underline{8}8 - 0.00124\underline{1}6 = 0.00109\underline{7}2\text{ amu}$ = mass of two

electrons then $0.00109\underline{7}2\text{ amu} \times \frac{931.5\text{ MeV}}{1\text{ amu}} = 1.022\text{ MeV}$

Check: The units (MeV) are correct. The energy is smaller than that of most nuclear reactions because the mass difference is small.

Challenge Problems

21.99 (a) **Given:** 72,500 kg $Al(s)$ and $10\,Al(s) + 6\,NH_4ClO_4(s) \rightarrow 4\,Al_2O_3(s) + 2\,AlCl_3(s) + 12\,H_2O(g) + 3\,N_2(g)$ and 608,000 kg $O_2(g)$ that reacts with hydrogen to form gaseous water

Find: energy generated (ΔH°_{rxn})
Conceptual Plan: Write a balanced reaction for $O_2(g)$ then

$\Delta H^\circ_{rxn} = \Sigma n_p \Delta H^\circ_f(\text{products}) - \Sigma n_r \Delta H^\circ_f(\text{reactants})$ then

kg → g → mol → energy then add the results from the two reactions.

$\frac{1000\ g}{1\ kg}$ $\mathcal{M}$ ΔH°_{rxn}

Solution:

Reactant/Product	ΔH°_f(kJ/mol from Appendix IVB)
$Al(s)$	0.0
$NH_4ClO_4(s)$	−295
$Al_2O_3(s)$	−1675.7
$AlCl_3(s)$	−704.2
$H_2O(g)$	−241.8
$N_2(g)$	0.0

Be sure to pull data for the correct formula and phase.

$$\begin{aligned}\Delta H^\circ_{rxn} &= \Sigma n_p \Delta H^\circ_f(\text{products}) - \Sigma n_r \Delta H^\circ_f(\text{reactants})\\ &= [4(\Delta H^\circ_f(Al_2O_3(s))) + 2(\Delta H^\circ_f(AlCl_3(s))) + 12(\Delta H^\circ_f(H_2O(g))) + 3(\Delta H^\circ_f(N_2(g)))] +\\ &\quad -[10(\Delta H^\circ_f(Al(s))) + 6(\Delta H^\circ_f(NH_4ClO_4(s)))]\\ &= [4(-1675.7\text{ kJ}) + 2(-704.2\text{ kJ}) + 12(-241.8\text{ kJ}) + 3(0.0\text{ kJ})] - [10(0.0\text{ kJ}) + 6(-295\text{ kJ})]\\ &= [-11012.8\text{ kJ}] - [-1770.\text{ kJ}]\\ &= -924\underline{2}.8\text{ kJ}\end{aligned}$$

then $72{,}500\ \cancel{kg\ Al} \times \frac{1000\ \cancel{g\ Al}}{1\ \cancel{kg\ Al}} \times \frac{1\ \cancel{mol\ Al}}{26.98\ \cancel{g\ Al}} \times \frac{924\underline{2}.8\text{ kJ}}{10\ \cancel{mol\ Al}} = 2.4\underline{8}3703 \times 10^9\text{ kJ}$

balanced reaction: $H_2(g) + \frac{1}{2}O_2(g) \rightarrow H_2O(g)\ \Delta H^\circ_{rxn} = \Delta H^\circ_f(H_2O(g)) = -241.8\text{ KJ/mol}$ then

$608{,}000\ \cancel{kg\ O_2} \times \frac{1000\ \cancel{g\ O_2}}{1\ \cancel{kg\ O_2}} \times \frac{1\ \cancel{mol\ O_2}}{32.00\ \cancel{g\ O_2}} \times \frac{241.8\text{ kJ}}{0.5\ \cancel{mol\ O_2}} = 9.1\underline{8}84 \times 10^9\text{ kJ}$ So the total is

$2.4\underline{8}3703 \times 10^9\text{ kJ} + 9.1\underline{8}84 \times 10^9\text{ kJ} = 1.16\underline{7}2103 \times 10^{10}\text{ kJ} = 1.167 \times 10^{10}\text{ kJ}.$

Check: The units (kJ) are correct. The answer is very large because the reactions are very exothermic and the weight of reactants is so large.

(b) **Given:** ${}^1_1H + {}^{-1}_{-1}p + {}^0_{+1}e \rightarrow {}^0_0\gamma$ **Find:** mass of antimatter to give same energy as in part (a)
Conceptual Plan: Because the reaction is an annihilation reaction, no matter will be left; so the mass of antimatter is the same as the mass of the hydrogen. So kJ → J → kg → g

$\frac{1000\ J}{1\ kJ}$ $E = mc^2$ $\frac{1000\ g}{1\ kg}$

Solution: $1.16\underline{7}2103 \times 10^{10}\ \cancel{kJ} \times \frac{1000\ J}{1\ \cancel{kJ}} = 1.16\underline{7}2103 \times 10^{13}\ J$. Because $E = mc^2$, rearrange to solve for m.

$$m = \frac{E}{c^2} = \frac{1.16\underline{7}2103 \times 10^{13}\ kg\frac{\cancel{m^2}}{\cancel{s^2}}}{\left(2.9979 \times 10^8 \frac{\cancel{m}}{\cancel{s}}\right)^2} = 1.299 \times 10^{-4}\ \cancel{kg} \times \frac{1000\ g}{1\ \cancel{kg}} = 0.1299\text{ g total matter, 0.0649 g each of}$$

matter and antimatter

Check: The units (g) are correct. A small mass is expected because nuclear reactions generate a large amount of energy.

21.100 **Given:** 85.0 g animal, ingests 10.0 mg of substance with 2.55% by mass Pu-239, alpha emitter, $t_{1/2} = 24{,}110$ years **Find:** (a) initial exposure in Ci and (b) all radiation absorbed and 7.77×10^{-12} J/emission, RBE = 20, dose in rads in the first 4.0 hours and dose in rems in the first 4.0 hours

Conceptual Plan:

(a) $t_{1/2} \rightarrow k$ **and mg → g → g Pu-239 → mol Pu-239 → atoms Pu-239**

$t_{1/2} = \frac{0.693}{k}$ $\frac{1\text{ g}}{1000\text{ mg}}$ $\frac{2.55\text{ g Pu-239}}{100\text{ g substance}}$ $\frac{1\text{ mol Pu-239}}{239\text{ g Pu-239}}$ $\frac{6.022 \times 10^{23}\text{ Pu-239 atoms}}{1\text{ mol Pu-239}}$

then $N, k \rightarrow$ **Rate (dis/yr) → dis/day → dis/h → dis/min → dis/s → Ci**

Rate $= kN$ $\frac{1\text{ yr}}{365.24\text{ day}}$ $\frac{1\text{ day}}{24\text{ h}}$ $\frac{1\text{ h}}{60\text{ min}}$ $\frac{1\text{ min}}{60\text{ s}}$ $\frac{1\text{ Ci}}{\frac{3.7 \times 10^{10}\text{ decays}}{\text{s}}}$

(b) **h → min → s then dis/s, s → alpha decays → J and g → kg then J, animal mass → rad → rem**

$\frac{60\text{ min}}{1\text{ h}}$ $\frac{60\text{ s}}{1\text{ min}}$ multiply terms $\frac{7.77 \times 10^{-12}\text{ J}}{\text{decay}}$ $\frac{1\text{ kg}}{1000\text{ g}}$ $\frac{1\text{ rad}}{\frac{0.01\text{ J}}{\text{kg animal}}}$ rem = RBE × rad

Solution:

(a) $t_{1/2} = \frac{0.693}{k}$ Rearrange to solve for k. $k = \frac{0.693}{t_{1/2}} = \frac{0.693}{24{,}110\text{ yr}} = 2.874326 \times 10^{-5}\text{ yr}^{-1}$ then

$$10.0\text{ mg} \times \frac{1\text{ g}}{1000\text{ mg}} \times \frac{2.55\text{ g Pu-239}}{100\text{ g substance}} \times \frac{1\text{ mol Pu-239}}{239\text{ g Pu-239}} \times \frac{6.022 \times 10^{23}\text{ Pu-239 atoms}}{1\text{ mol Pu-239}}$$

$= 6.425146 \times 10^{17}$ Pu-239 atoms

Rate $= kN = (2.874326 \times 10^{-5}\text{ yr}^{-1})(6.425146 \times 10^{17}\text{ Pu-239 atoms}) \times$

$$\frac{1\text{ yr}}{365.24\text{ day}} \times \frac{1\text{ day}}{24\text{ h}} \times \frac{1\text{ h}}{60\text{ min}} \times \frac{1\text{ min}}{60\text{ s}} = 5.85230 \times 10^{5}\,\frac{\text{dis}}{\text{s}} \times \frac{1\text{ Ci}}{\frac{3.7 \times 10^{10}\text{ decays}}{\text{s}}} =$$

1.58170×10^{-5} Ci

(b) $4.0\text{ h} \times \frac{1\text{ day}}{24\text{ h}} \times \frac{1\text{ yr}}{365.24\text{ day}} = 4.5632 \times 10^{-4}$ yr. Because the time is so much less than the

$t_{1/2}(10^{-6}\%)$, the concentration is essentially constant. Use dis/s and time to get dose,

so $4.0\text{ h} \times \frac{60\text{ min}}{1\text{ h}} \times \frac{60\text{ s}}{1\text{ min}} = 1.44 \times 10^{4}\text{ s}$

$$5.85230 \times 10^{5}\frac{\text{dis}}{\text{s}} \times 1.44 \times 10^{4}\text{ s} = 8.42731 \times 10^{9}\text{ decays} \times \frac{7.77 \times 10^{-12}\text{ J}}{\text{decays}} = 6.54802 \times 10^{-2}\text{ J}$$

and $85.0\text{ g} \times \frac{1\text{ kg}}{1000\text{ g}} = 0.0850\text{ kg}$ then $\frac{6.54802 \times 10^{-2}\text{ J}}{0.0850\text{ kg}} \times \frac{1\text{ rad}}{\frac{0.01\text{ J}}{\text{kg animal}}} = 77$ rad and

rem = RBE × rad = 20 × 77 rad = 1.5×10^{3} rem and the animal will die.

Check: The units (Ci, rem, and rad) are correct. The number of Curies is small because of the conversion factor. The doses in rems and rad are high because it is an alpha emitter and the isotope was ingested.

21.101 **Given:** ${}^{235}_{92}\text{U} \rightarrow {}^{206}_{82}\text{Pb}$ and ${}^{232}_{90}\text{Th} \rightarrow {}^{206}_{82}\text{Pb}$ **Find:** decay series

Conceptual Plan: Write the species given on the appropriate side of the equation. → Equalize the sum of the mass numbers and the sum of the atomic numbers on both sides of the equation by writing the stoichiometric coefficient in front of the desired species.

Solution: ${}^{235}_{92}\text{U} \rightarrow {}^{?}_{82}\text{Pb} + ?{}^{4}_{2}\text{He} + ?{}^{0}_{-1}\text{e}$ becomes ${}^{235}_{92}\text{U} \rightarrow {}^{207}_{82}\text{Pb} + 7{}^{4}_{2}\text{He} + 4{}^{0}_{-1}\text{e}$.

${}^{232}_{90}\text{Th} \rightarrow {}^{?}_{82}\text{Pb} + ?{}^{4}_{2}\text{He} + ?{}^{0}_{-1}\text{e}$ becomes ${}^{232}_{90}\text{Th} \rightarrow {}^{208}_{82}\text{Pb} + 6{}^{4}_{2}\text{He} + 4{}^{0}_{-1}\text{e}$.

U-235 forms Pb-207 in 7 α-decays and 4 β-decays, and Th-232 forms Pb-208 in 6 α-decays and 4 β-decays.

Check: $235 = 207 + 7(4) + 4(0)$, and $92 = 82 + 7(2) + 4(-1)$. $232 = 208 + 6(4) + 4(0)$, and $90 = 82 + 6(2) + 4(-1)$. The mass of the Pb can be determined because alpha particles are large and need to be included as integer values. To make the masses balance requires more alpha particles than can be supported by the number of protons in the total equation. To account for this, an appropriate number of beta decays are added.

21.102 **Given:** $MH_2 \rightarrow He(g) + X(\text{noble gas})(g) + H_2(g)$, 0.025 mol MH_2; $V = 2.0$ L, $P = 0.55$ atm, $T = 298$ K, 82 min
Find: $t_{1/2}$
Conceptual Plan: $P, V, T \rightarrow n_{gas} \rightarrow n_{M\,t}$ **then** $n_{M\,0}, n_{M\,t}, t \rightarrow k \rightarrow t_{1/2}$

$$PV = nRT \quad n_{Mt} = 0.025 \text{ mol} - 1/3\, n_{H_2} \quad \ln N_t = -kt + \ln N_0 \quad t_{1/2} = \frac{0.693}{k}$$

Solution: $PV = nRT$. Rearrange to solve for n. $n = \dfrac{PV}{RT} = \dfrac{0.55 \text{ atm} \times 2.0 \text{ L}}{0.08206 \dfrac{\text{L} \cdot \text{atm}}{\text{mol} \cdot \text{K}} \times 298 \text{ K}} = 0.04\underline{4}9826$ mol gas

then $n_{M\,t} = 0.025 \text{ mol} - 1/3\, n_{gas} = 0.025 \text{ mol} - 1/3(0.04\underline{4}9826 \text{ mol}) = 0.01\underline{0}00579$ mol then

$\ln N_{M\,t} = -kt + \ln N_{M\,0}$ Rearrange to solve for k. $k = \dfrac{\ln \dfrac{N_{M\,0}}{N_{M\,t}}}{t} = \dfrac{\ln \dfrac{0.025 \text{ mol}}{0.01\underline{0}00579 \text{ mol}}}{82 \text{ min}} = 0.01\underline{1}672 \text{ min}^{-1}$

then $t_{1/2} = \dfrac{0.693}{k} = \dfrac{0.693}{0.01\underline{1}672 \text{ min}^{-1}} = 6\underline{2}.05664 \text{ min} = 62 \text{ min}$

Check: The units (min) are correct. The half-life is reasonable because most of the hydride decomposes in the 82 minutes.

21.103 **Given:** $H^{38}_{17}Cl(g) \rightarrow {}^{38}_{18}Ar(g) + {}^{0}_{-1}e + \frac{1}{2}H_2(g)$, 0.40 mol $H^{38}_{17}Cl(g)$; $V = 6.24$ L, $P = 1650$ mmHg, $t_{1/2} = 80.0$ min
Find: T
Conceptual Plan: Because $t = 2\,t_{1/2}$, **then** $N_{Cl\text{-}38\,t} = \frac{1}{4} N_{Cl\text{-}38\,0} \rightarrow n_{gas\,t}$ **then mmHg → atm and**

$$n_{gas\,t} = 3/2\, n_{HCl\,0} - 1/2\, n_{HCl\,t} \qquad \frac{1 \text{ atm}}{760 \text{ mmHg}}$$

$P, V, n_{gas\,t} \rightarrow T,$

$$PV = nRT$$

Solution: Because $t = 2\,t_{1/2}$, then $N_{Cl\text{-}38t} = \frac{1}{4} N_{Cl\text{-}38\,0} = \frac{1}{4}(0.40 \text{ mol}) = 0.10$ mol. As the HCl disintegrates, it produces argon gas, beta particles, and hydrogen gas, with a ratio of three particles produced (2 Ar, 1 H_2) for every two HCl molecules that decay. There was initially 0.40 mole of undisintegrated gas; now 0.10 mole HCl remains. So the total number of gas particles now in the container is

$n_{gas\,t} = 3/2\, n_{HCl\,0} - 1/2\, n_{HCl\,t} = 3/2(0.40 \text{ mol}) - 1/2\,(0.10 \text{ mol}) = 0.55$ mol gas then $PV = nRT$. Rearrange to

solve for T. $T = \dfrac{PV}{nR} = \dfrac{1650 \text{ mmHg} \times \dfrac{1 \text{ atm}}{760 \text{ mmHg}} \times 6.24 \text{ L}}{0.55 \text{ mol} \times 0.08206 \dfrac{\text{L} \cdot \text{atm}}{\text{mol} \cdot \text{K}}} = 3\underline{0}0.17 \text{ K} = 3.0 \times 10^2 \text{ K}$

Check: The units (K) are correct. The temperature is reasonable considering the volume of a gas at STP and the fact that most of the initial 0.40 mol has decomposed.

21.104 **Given:** ${}^{x}_{5}BF_3(g) + {}^{1}_{0}n \rightarrow {}^{x-3}_{3}LiF(s) + {}^{4}_{2}He(g) + F_2(g)$, 0.20 mol BF_3, react half of BF_3, $V = 3.0$ L, $T = 298$ K
Find: P
Conceptual Plan: Because one-half of BF_3 is reacted, $N_{B\,t} = \frac{1}{2} N_{B\,0} \rightarrow n_{gas\,t}$ **then** $V, n_{gas\,t}, T \rightarrow P.$

$$n_{gas\,t} = 1/2\, n_{BF_3\,0} + 2(n_{BF_3\,0} - n_{BF_3\,t}) \qquad PV = nRT$$

Solution: Because one-half of BF_3 is reacted, $N_{B\,t} = \frac{1}{2} N_{B\,0} = \frac{1}{2}(0.20 \text{ mol}) = 0.10$ mol BF_3 then

$n_{gas\,t} = 1/2\, n_{BF_3\,0} + 2\,(n_{BF_3\,0} - n_{BF_3\,t}) = \frac{1}{2}(0.20 \text{ mol}) + 2(0.20 \text{ mol} - 0.10 \text{ mol}) = 0.30$ mol then $PV = nRT$

Rearrange to solve for P. $P = \dfrac{nRT}{V} = \dfrac{0.30 \text{ mol} \times 0.08206 \dfrac{\text{L} \cdot \text{atm}}{\text{mol} \cdot \text{K}} \times 298 \text{ K}}{3.0 \text{ L}} = 2.\underline{4}4539 \text{ atm} = 2.4 \text{ atm}$

Check: The units (atm) are correct. The pressure is reasonable considering the conditions of a gas at STP and the fact that half of the initial 0.20 mol has decomposed.

Conceptual Problems

21.105 **Given:** $^{21}_{9}F \rightarrow ^{?}_{?}? + ^{0}_{-1}e$ **Find:** missing nucleus

Conceptual Plan: Write the species given on the appropriate side of the equation. → Equalize the sum of the mass numbers and the sum of the atomic numbers on both sides of the equation by writing the stoichiometric coefficient in front of the desired species.

Solution: $^{21}_{9}F \rightarrow ^{?}_{?}? + ^{0}_{-1}e$ becomes $^{21}_{9}F \rightarrow ^{21}_{10}Ne + ^{0}_{-1}e$.

Check: $21 = 21 + 0$, and $9 = 10 - 1$. Neon is atomic number 10, and no other species are needed to balance the equation.

21.106 7. Because $1/2^6 = 1.6\%$ and $1/2^7 = 0.8\%$.

21.107 Nuclide A is more dangerous because the half-life is shorter (18.5 days); so it decays faster.

21.108 The gamma emitter is a greater threat while you sleep because it can penetrate more tissue. The alpha particles will not penetrate the wall to enter your bedroom. The alpha emitter is a greater threat if you ingest it because it is more ionizing within the body and therefore its RBE is higher than gamma radiation when ingested.

21.109 Iodine is used by the thyroid gland to make hormones. Normally, we ingest iodine in foods, especially iodized salt. The thyroid gland cannot tell the difference between stable and radioactive iodine and will absorb both. KI tablets work by blocking radioactive iodine from entering the thyroid. When a person takes KI, the stable iodine in the tablet gets absorbed by the thyroid. Because KI contains so much stable iodine, the thyroid gland becomes "full" and cannot absorb any more iodine—either stable or radioactive—for the next 24 hours.

22 Organic Chemistry

Review Questions

22.1 Most common smells are caused by organic molecules; these molecules contain carbon combined with several other elements, including hydrogen, nitrogen, oxygen, and sulfur.

22.2 Organic chemistry is the study of compounds containing carbon combined with one or more of the elements hydrogen, nitrogen, oxygen, and sulfur, including their properties and their reactions.

22.3 Carbon is unique in the vast number of compounds it can form. Life needs diversity to exist, and carbon has the ability to form more compounds than any other element.

22.4 Carbon is unique in its ability to form four covalent bonds, its ability to form double and triple bonds, and its tendency to catenate (i.e., to form chains).

22.5 Silicon can form chains with itself. However, silicon's affinity for oxygen—the Si—O bond is 142 kJ/mol stronger than the Si—Si bond—coupled with the prevalence of oxygen in our atmosphere means that silicon–silicon chains are readily oxidized to form silicates (the silicon–oxygen compounds that compose a significant proportion of minerals). By contrast, the C—C bond (347 kJ/mol) and the C—O bond (359 kJ/mole) are nearly the same strength, allowing carbon chains to exist relatively peacefully in an oxygen-rich environment. In other words, silicon's affinity for oxygen robs it of the rich diversity that catenation provides to carbon.

22.6
- (a) When carbon forms four single bonds, the hybridization is sp^3 and the geometry is tetrahedral.
- (b) When carbon forms two single bonds and one double bond, the hybridization is sp^2 and the geometry is trigonal planar.
- (c) When carbon forms one single bond and one triple bond, the hybridization is *sp* and the geometry is linear.

22.7 Hydrocarbons—compounds that contain only carbon and hydrogen—are the simplest organic compounds. However, because of the uniqueness of carbon, many different kinds of hydrocarbons exist. Hydrocarbons are commonly used as fuels. Candle wax, oil, gasoline, liquid propane (LP) gas, and natural gas are all composed of hydrocarbons. Hydrocarbons are also the starting materials in the synthesis of many different consumer products, including fabrics, soaps, dyes, cosmetics, drugs, plastic, and rubber.

22.8 Hydrocarbons can be classified into four different types: alkanes, alkenes, alkynes, and aromatic hydrocarbons. Their generic formulas are C_nH_{2n+2}, C_nH_{2n}, C_nH_{2n-2}, and formulas containing a benzene ring, respectively, where $n =$ the number of carbon atoms.

22.9 A structural formula shows not only the numbers of each kind of atoms, but also how the atoms are bonded together. The condensed structural formula groups the hydrogen atoms together with the carbon atom to which they are bonded. Condensed structural formulas may show some of the bonds or none at all. The carbon skeleton formula shows the carbon–carbon bonds only as lines. Each end or bend of a line represents a carbon atom bonded to as many hydrogen atoms as necessary to form a total of four bonds. Space-filling or ball-and-stick models are three-dimensional representations that show how atoms are bonded together. The space-filling models show the relative size of the atoms that are bonded together.

22.10 Structural isomers are molecules with the same molecular formula but different structures. The different structures can lead to different boiling points, melting points, and chemical properties based on the types of intermolecular interactions that are possible (as discussed in Chapter 12).

22.11 Optical isomers are two molecules that are nonsuperimposable mirror images of each other. Optical isomers contain a carbon atom with four different substituent groups. Most properties of optical isomers are the same. The differences appear when they interact with polarized light and when they are placed in environments that can interact with one isomer and not the other.

22.12 Enantiomers are molecules that are nonsuperimposable mirror images and are therefore optical isomers of one other.

Any molecule that exhibits optical isomerism is said to be chiral. Every chiral molecule must have a chiral center, a carbon atom with four different substituents in a tetrahedral arrangement.

A dextrorotatory isomer (or the *d* isomer) is an optical isomer that rotates the polarization of light clockwise.

A levorotatory isomer (or the *l* isomer) is an optical isomer that rotates the polarization of light counterclockwise.

A racemic mixture is an equimolar mixture of both optical isomers that does not rotate the polarization of light at all.

22.13 Alkanes are often called saturated hydrocarbons because they are saturated (loaded to capacity) with hydrogen. Unsaturated hydrocarbons contain multiple bonds and therefore contain fewer hydrogen atoms.

22.14 The main difference in the way alkanes, alkenes, and alkynes are named is in the ending of the name. Alkane names end in *–ane*, alkene names end in *–ene*, and alkyne names end in *–yne*. The multiple bonds must also have their location in the molecule designated. If there are multiple bonds in a molecule, the multiple bond must be included in the longest continuous carbon chain.

22.15 Double bonds are composed of a sigma bond and a pi bond. Sigma bonds allow for free rotation about the bond. The pi bond restricts the rotational motion about the bond, resulting in isomers that are referred to as geometric isomers (same connectivity but different geometric positions). The two isomers are designated as *cis* (meaning "same side") and *trans* (meaning "opposite sides"). Cis–trans isomerism is common in alkenes and results in different boiling points and melting points and in different abilities to interact with rigid molecules.

22.16 Hydrocarbon combustion reactions are the burning of hydrocarbons in the presence of oxygen to form carbon dioxide and water. An example is $C_8H_{18}(l) + 25/2\, O_2(g) \rightarrow 8\, CO_2(g) + 9\, H_2O(g)$.

22.17 The most common types of reactions of alkanes are as follows:

- Hydrocarbon combustion in the presence of oxygen to form carbon dioxide and water. An example is $C_8H_{18}(l) + 25/2\, O_2(g) \rightarrow 8\, CO_2(g) + 9\, H_2O(g)$.
- Halogen substitution in the presence of a halogen gas to form a halogenated alkane and a hydrogen halide. An example is $CH_4(g) + Cl_2(g) \rightarrow CH_3Cl(g) + HCl(g)$.

22.18 (a) Substitution reactions involve replacing a hydrogen atom in a molecule with another functional group, such as a halogen.

(b) Addition reactions involve converting a pi bond to two sigma bonds. Two functional groups or hydrogen atoms are added to a molecule where the multiple bond once existed.

(c) Elimination reactions involve using concentrated acids to convert an alcohol to an alkene, eliminating water. They are also referred to as dehydration reactions where an OH from one carbon combines with an H from an adjacent carbon to form water. The removal of these atoms from the carbons allows the formation of a pi bond between the carbons.

22.19 The most common types of reactions of alkenes are as follows:

- Hydrocarbon combustion in the presence of oxygen to form carbon dioxide and water. An example is $C_2H_4(g) + 3\, O_2(g) \rightarrow 2\, CO_2(g) + 2\, H_2O(g)$.

- Hydrogenation in the presence of hydrogen to form an alkane. An example is $C_2H_4(g) + H_2(g) \rightarrow C_2H_6(g)$.
- Halogen addition in the presence of a halogen gas or hydrogen halide to form a halogenated alkane. An example is $CH_2CHCH_2CH_3(g) + HCl(g) \rightarrow CH_3CClHCH_2CH_3(g)$.

22.20 Markovnikov's rule states that when a polar reagent is added to an unsymmetrical alkene, the positive end (the least electronegative part) of the reagent adds to the carbon atom that has the most hydrogen atoms. In most reactions of this type, the positive end of the reagent is hydrogen; therefore, the hydrogen atom goes to the carbon atom that already contains the most hydrogen atoms. An example is $CH_2CHCH_2CH_3(g) + HCl(g) \rightarrow CH_3CHClCH_2CH_3(g)$.

22.21 The structure of benzene, C_6H_6, is a six-member ring where three pi bonds are delocalized on all six of the C—C bonds that form the ring. Benzene rings are represented as one or both of the Kekulé structures,

```
   H   H              H   H
    C—C                C=C
  //   \\             /   \
H—C     C—H  ⟷  H—C     C—H
   \   /            \\   //
    C=C                C—C
   H   H              H   H
```

, in a shorthand notation,

or as a ball-and-stick diagram with molecular orbitals, 

22.22 The most common reactions for aromatic compounds are substitution reactions. An example is $C_6H_6 + Cl_2 \rightarrow C_6ClH_5 + HCl$.

22.23 A functional group is a characteristic atom or group of atoms that is inserted into a hydrocarbon. Examples of functional groups include alcohols (—OH), halogens (—X, where X = F, Cl, Br, and I), and carboxylic acids (—COOH).

22.24 The generic structure for alcohols is R–OH, where R represents a hydrocarbon group. Methyl alcohol is CH_3OH, and ethyl alcohol is CH_3CH_2OH.

22.25 In organic chemistry, we think of oxidation and reduction from the point of view of the carbon atoms in the organic molecule. Thus, oxidation is the gaining of oxygen or the losing of hydrogen by a carbon atom. Reduction, then, is the loss of oxygen or the gaining of hydrogen by a carbon atom.

22.26 Alcohols undergo a number of reactions, including substitution, elimination (or dehydration), and oxidation. Alcohols also react with active metals to form strong bases. In a substitution reaction, an alcohol reacts with an acid, such as HBr, to form halogenated hydrocarbons as follows: $CH_3OH + HBr \rightarrow CH_3Br + H_2O$. In an elimination (or dehydration) reaction, concentrated acids, such as H_2SO_4, react with alcohols to eliminate water, forming an alkene. For example, ethanol eliminates water to form ethene according to the following reaction: $CH_3CH_2OH \xrightarrow{H_2SO_4} CH_2CH_2 + H_2O$. In an oxidation reaction, carbon atoms gain oxygen atoms and/or lose hydrogen atoms. For example, ethanol becomes acetic acid according to the following reaction: $CH_3CH_2OH \xrightarrow[H_2SO_4]{Na_2Cr_2O_7} CH_3COOH$. Alcohols react with active metals, such as alkali metals, forming an organic base and hydrogen gas. For example, $CH_3OH + Na \rightarrow CH_3O^-Na^+ + H_2$.

22.27 Aldehydes and ketones have the general structural formulas of RCHO and RCOR′, respectively, where R and R′ represent hydrocarbon groups. Both aldehydes and ketones contain a carbonyl group, an oxygen double-bonded to a carbon atom (—C(=O)—). Ketones have an R group attached to both sides of the carbonyl, while aldehydes have one R group and a hydrogen atom. An exception is formaldehyde, which is an aldehyde with two H atoms attached to the carbonyl group. Acetaldehyde is $CH_3—C(=O)—H$ and propanone is $CH_3—C(=O)—CH_3$.

22.28 Because oxidation reactions are used to generate aldehydes and ketones, reduction reactions can reverse these reactions and are common. Thus, butanone can be reduced to butanol $CH_3CH_2COCH_2 \xrightarrow[\text{reduction}]{} CH_3CH_2COHCH_2$. Addition reactions are also common, where the carbon–oxygen pi bond is broken and functional groups are added to the carbon and oxygen. An example is $CH_2O + HCN \xrightarrow{\text{NaCN}} NCC(OH)H_2$. Aldehydes can be further oxidized to carboxylic acids $CH_3CH_2CHO \xrightarrow[\text{oxidation}]{} CH_3CH_2COOH$.

22.29 Carboxylic acids and esters have the following general structural formulas of RCOOH and RCOOR', respectively, where R and R′ represent hydrocarbon groups. Both carboxylic acids and esters contain a carbonyl group, an oxygen double-bonded to a carbon atom $\left(-\overset{\overset{\displaystyle O}{\|}}{C}-\right)$. In carboxylic acids, the molecule ends in $-\overset{\overset{\displaystyle O}{\|}}{C}-O-H$ and esters have $-\overset{\overset{\displaystyle O}{\|}}{C}-O-$ inserted into a hydrocarbon chain. Acetic acid is $CH_3-\overset{\overset{\displaystyle O}{\|}}{C}-O-H$ and ethyl ethanoate is $CH_3-\overset{\overset{\displaystyle O}{\|}}{C}-O-CH_2-CH_3$.

22.30 Because carboxylic acids are acids, they can dissociate in water ($CH_3COOH \rightleftharpoons CH_3COO^- + H^+$) and participate in acid–base reactions, such as $CH_3COOH(aq) + NaOH(aq) \rightarrow CH_3COO^-Na^+(aq) + H_2O(l)$. An important reaction between carboxylic acids and alcohols is to form as an ester and water through a condensation reaction. An example is $CH_3COOH(aq) + CH_3CH_2OH(aq) \rightarrow CH_3COOCH_2CH_3(aq) + H_2O(l)$.

22.31 Ethers have the general structural formula of ROR', where R and R' represent hydrocarbon groups. The two hydrocarbon groups are linked through an oxygen atom. Diethyl ether is $CH_3-CH_2-O-CH_2-CH_3$, and ethyl propyl ether is $CH_3-CH_2-O-CH_2-CH_2-CH_3$.

22.32 Amines are organic compounds that contain nitrogen. One or more of the hydrogen atoms on an ammonia molecule are replaced by a hydrocarbon group. The simplest amines have the following general structural formula of RNH_2, where R represents a hydrocarbon group. All three hydrogen atoms can be replaced by different alkyl groups to give a formula of NRR'R". Methyl amine is CH_3NH_2, and diethylamine is $(CH_3CH_2)_2NH$.

22.33 Polymers are long chainlike molecules composed of repeating units called monomers. Copolymers are long chainlike molecules composed of two different monomers.

22.34 Addition polymers are polymers in which the monomers simply link together without the elimination of any atoms. Condensation polymers are polymers that eliminate an atom or a small group of atoms during the polymerization.

Hydrocarbons

22.35
(a) C_5H_{12} is an alkane because it follows the general formula C_nH_{2n+2}, where $n = 5$.
(b) C_3H_6 is an alkene because it follows the general formula C_nH_{2n}, where $n = 3$.
(c) C_7H_{12} is an alkyne because it follows the general formula C_nH_{2n-2}, where $n = 7$.
(d) $C_{11}H_{22}$ is an alkene because it follows the general formula C_nH_{2n}, where $n = 11$.

22.36
(a) C_8H_{16} is an alkene because it follows the general formula C_nH_{2n}, where $n = 8$.
(b) C_4H_6 is an alkyne because it follows the general formula C_nH_{2n-2}, where $n = 4$.
(c) C_7H_{16} is an alkane because it follows the general formula C_nH_{2n+2}, where $n = 7$.
(d) C_2H_2 is an alkyne because it follows the general formula C_nH_{2n-2}, where $n = 2$.

22.37 $CH_3-CH_2-CH_2-CH_2-CH_2-CH_2-CH_3$, $CH_3-CH(CH_3)-CH_2-CH_2-CH_2-CH_3$,

$CH_3-CH_2-CH(CH_3)-CH_2-CH_2-CH_3$, $CH_3-C(CH_3)_2-CH_2-CH_2-CH_3$,

$CH_3-CH_2-CH(CH_3)-CH(CH_3)-CH_3$, $CH_3-CH_2-C(CH_3)_2-CH_2-CH_3$,

$CH_3-CH(CH_3)-CH_2-CH(CH_3)-CH_3$, $CH_3-CH_2-CH(CH_2-CH_3)-CH_2-CH_3$, and $CH_3-C(CH_3)_2-CH(CH_3)-CH_3$

22.38 The 18 isomers are $CH_3-CH_2-CH_2-CH_2-CH_2-CH_2-CH_2-CH_3$,

$CH_3-CH(CH_3)-CH_2-CH_2-CH_2-CH_2-CH_3$, $CH_3-CH_2-CH(CH_3)-CH_2-CH_2-CH_2-CH_3$,

$CH_3-CH_2-CH_2-CH(CH_3)-CH_2-CH_2-CH_3$, $CH_3-C(CH_3)_2-CH_2-CH_2-CH_2-CH_3$,

$CH_3-CH(CH_3)-CH(CH_3)-CH_2-CH_2-CH_3$, $CH_3-CH(CH_3)-CH_2-CH(CH_3)-CH_2-CH_3$,

$CH_3-CH_2-C(CH_3)_2-CH_2-CH_2-CH_3$, $CH_3-CH_2-CH(CH_3)-CH(CH_3)-CH_2-CH_3$,

$CH_3-CH(CH_3)-CH_2-CH_2-CH(CH_3)-CH_3$, $CH_3-CH(CH_3)-CH(CH_3)-CH(CH_3)-CH_3$,

$CH_3-CH_2-CH(CH_2-CH_3)-CH_2-CH_2-CH_3$, $CH_3-C(CH_3)_2-C(CH_3)_2-CH_3$, $CH_3-CH(CH_3)-C(CH_3)_2-CH_2-CH_3$,

$CH_3-CH_2-C(CH_3)(CH_2-CH_3)-CH_2-CH_3$, $CH_3-CH_2-CH(CH_2-CH_3)-CH(CH_3)-CH_3$,

$CH_3-C(CH_3)_2-CH_2-CH(CH_3)-CH_3$, and $CH_3-C(CH_3)_2-CH(CH_3)-CH_2-CH_3$.

22.39 (a) No, this molecule will not because all four of the substituents are Cl atoms.
(b) Yes, this molecule will because the third carbon has four different substituent groups.
(c) Yes, this molecule will because the second carbon has four different substituent groups.
(d) No, each carbon has, at most, three different substituent groups.

22.40 (a) Yes, this molecule will because the third carbon from the left has four different substituent groups.
(b) No, each carbon has, at most, two different substituent groups.
(c) Yes, this molecule will because two carbons (the one with the amino and the bromine groups) have four different substituent groups.
(d) Yes, this molecule will because the middle carbon has four different substituent groups.

22.41 (a) They are enantiomers because they are non-superimposable mirror images of each other.
(b) They are the same because you can get the second molecule by rotating the first molecule counterclockwise about the C—H bond.
(c) They are enantiomers because they are non-superimposable mirror images of each other.

22.42 (a) They are the same because two of the substituent groups on the central carbon are the same.
(b) They are enantiomers because they are non-superimposable mirror images of each other.
(c) They are the same because you can get the second molecule by rotating the first molecule counterclockwise about the C—H bond of the optically active carbon (the one with the $—CCl_3$ group attached).

Alkanes

22.43 **Given:** alkane structures **Find:** name
Conceptual Plan: Count the number of carbon atoms in the longest continuous carbon chain to determine the base name of the compound. Find the prefix corresponding to this number of atoms in Table 22.5 and add the ending *-ane* to form the base name. → Consider every branch from the base chain to be a substituent. Name each substituent according to Table 22.6. → Beginning with the end closest to the branching, number the base chain and assign a number to each substituent. (If two substituents occur at equal distances from each end, go to the next substituent to determine from which end to start numbering.) → Write the name of the compound in the following format: (subst. #)-(subst. name)(base name). → If there are two or more substituents, give each one a number and list them alphabetically with hyphens between words and numbers. → If a compound has two or more identical substituents, designate the number of identical substituents with the prefix *di-* (2), *tri-* (3), or *tetra-* (4) before the substituent's name. Separate the numbers with a comma, indicating the positions of the substituents relative to each other. The prefixes are not taken into account when alphabetizing.
Solution:

(a) $CH_3—CH_2—CH_2—CH_2—CH_3$ has five carbons as the longest continuous chain. The prefix for 5 is penta-. There are no substituent groups on any of the carbons, so the name is pentane. Because this is a straight-chain molecule, it can be more specifically named *n*-pentane.

(b) $\boxed{CH_3—CH_2—CH—CH_3}$ (with CH_3 on the third carbon) has four carbons as the longest continuous chain. The prefix for 4 is but-, and the base name is butane. The only substituent group is a methyl group. $\boxed{C^4H_3—C^3H_2—C^2H—C^1H_3}$ (with $\boxed{CH_3}$ on C^2) If we start numbering the chain at the end closest to the methyl group, the methyl substituent is assigned the number 2. The name of the compound is 2-methylbutane.

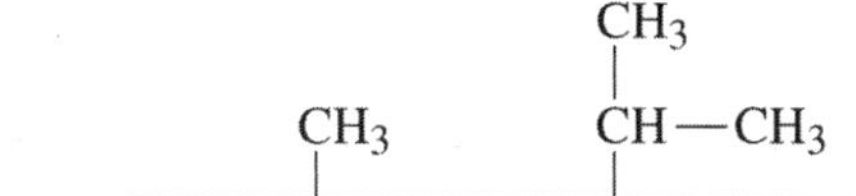

(c) $CH_3—CH(CH_3)—CH_2—CH(CH(CH_3)—CH_3)—CH_2—CH_2—CH_3$ has seven carbons as the longest continuous chain. The prefix for 7 is hept-, and the base name is heptane. The substituent groups are methyl and isopropyl groups.

$C^1H_3—C^2H(CH_3)—C^3H_2—C^4H(CH(CH_3)—CH_3)—C^5H_2—C^6H_2—C^7H_3$ If we start numbering the chain at the end closest to the methyl group, the methyl substituent is assigned the number 2 and the isopropyl group is assigned the number 4. Because *i* comes before *m*, the name of the compound is 4-isopropyl-2-methylheptane.

(d) $CH_3—CH(CH_3)—CH_2—CH(CH_2—CH_3)—CH_2—CH_3$ has six carbons as the longest continuous chain. The prefix for 6 is hex-, and the base name is hexane. The only substituent groups are methyl and ethyl groups.

$C^1H_3—C^2H(CH_3)—C^3H_2—C^4H(CH_2—CH_3)—C^5H_2—C^6H_3$ If we start numbering the chain at the end closest to the methyl group, the methyl substituent is assigned the number 2 and the ethyl group is assigned the number 4. Because *e* comes before *m*, the name of the compound is 4-ethyl-2-methylhexane.

22.44 **Given:** alkane structures **Find:** name

Conceptual Plan: Count the number of carbon atoms in the longest continuous carbon chain to determine the base name of the compound. Find the prefix corresponding to this number of atoms in Table 22.5 and add the ending *-ane* to form the base name. → Consider every branch from the base chain to be a substituent. Name each substituent according to Table 22.6. → Beginning with the end closest to the branching, number the base chain and assign a number to each substituent. (If two substituents occur at equal distances from each end, go to the next substituent to determine from which end to start numbering.) → Write the name of the compound in the following format: (subst. #)-(subst. name)(base name). → If there are two or more substituents, give each one a number and list them alphabetically with hyphens between words and numbers. → If a compound has two or more identical substituents, designate the number of identical substituents with the prefix *di-* (2), *tri-* (3), or *tetra-* (4) before the substituent's name. Separate the numbers with a comma, indicating the positions of the substituents relative to each other. The prefixes are not taken into account when alphabetizing.

Solution:

(a) $CH_3—CH(CH_3)—CH_3$ has three carbons as the longest continuous chain. The prefix for 3 is prop-, and the base name is propane. The only substituent group is a methyl group. $C^1H_3—C^2H(CH_3)—C^3H_3$ Because the branching is in the middle, it does not matter which end we start numbering. The methyl substituent is assigned the number 2. The name of the compound is 2-methylpropane.

(b) $CH_3—CH(CH_3)—CH_2—CH(CH_3)—CH_2—CH_3$ has six carbons as the longest continuous chain. The prefix for 6 is hex-, and the base name is hexane. The substituent groups are two methyl groups.

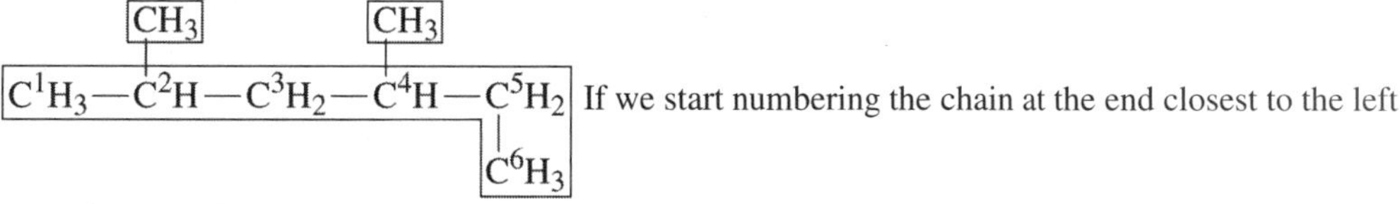

$C^1H_3—C^2H(CH_3)—C^3H_2—C^4H(CH_3)—C^5H_2—C^6H_3$ If we start numbering the chain at the end closest to the left methyl group, the methyl substituents are assigned the numbers 2 and 4. The name of the compound is 2, 4-dimethylhexane.

(c)

$$\begin{array}{c} \quad CH_3\,CH_3 \\ \boxed{CH_3-\overset{|}{\underset{|}{C}}-\overset{|}{\underset{|}{C}}-CH_3} \\ \quad CH_3\,CH_3 \end{array}$$

has four carbons as the longest continuous chain. The prefix for 4 is but-, and the base name is butane. The substituent groups are four methyl groups.

$$\begin{array}{c} \boxed{CH_3}\boxed{CH_3} \\ \boxed{C^1H_3-\overset{|}{\underset{|}{C^2}}-\overset{|}{\underset{|}{C^3}}-C^4H_3} \\ \boxed{CH_3}\boxed{CH_3} \end{array}$$

Because the branching is in the middle and symmetric in the molecule, it does not matter which end we start numbering. The methyl substituents are assigned the numbers 2, 2, 3, and 3. The name of the compound is 2,2,3,3-tetramethylbutane.

(d)

$$\begin{array}{l} \qquad\qquad\qquad\qquad\qquad CH_3 \\ \qquad\qquad\qquad\qquad\qquad | \\ \quad CH_3 \qquad\qquad\qquad\quad CH_2 \\ \quad | \qquad\qquad\qquad\qquad\quad | \\ \boxed{CH_3-CH-CH_2-CH-CH-CH_2-CH_2-CH_3} \\ \qquad\qquad\qquad\quad | \\ \qquad\qquad\qquad\quad CH_3 \end{array}$$

has eight carbons as the longest continuous chain.

The prefix for 8 is oct-, and the base name is octane. The only substituent groups are two methyl and one ethyl groups.

$$\begin{array}{l} \qquad\qquad\qquad\qquad\qquad \boxed{\begin{array}{c} CH_3 \\ | \\ CH_2 \end{array}} \\ \quad \boxed{CH_3} \\ \quad | \qquad\qquad\qquad\qquad\quad | \\ \boxed{C^1H_3-C^2H-C^3H_2-C^4H-C^5H-C^6H_2-C^7H_2-C^8H_3} \\ \qquad\qquad\qquad\quad | \\ \qquad\qquad\qquad\quad \boxed{CH_3} \end{array}$$

If we start numbering the chain at the end closest to the left methyl group, the methyl substituents are assigned the numbers 2 and 4 and the ethyl group is assigned the number 5. Because *e* comes before *m*, the name of the compound is 5-ethyl-2,4-dimethyloctane.

22.45 **Given:** alkane names **Find:** structure

Conceptual Plan: Find the number of carbon atoms corresponding to the prefix of the base name in Table 22.5. → Draw the base chain and number the carbons from left to right. → Using Table 22.6 and the prefix *di-* (2), *tri-* (3), or *tetra-* (4) before the substituent's name, determine each substituent. → Add the substituent to the proper carbon position in the chain. → Add hydrogen atoms to the base chain so that each carbon has four bonds.

Solution:

(a) 3-ethylhexane. The base name hexane indicates that there are six carbon atoms in the base chain. $C^1-C^2-C^3-C^4-C^5-C^6$. 3-ethyl designates that a $-CH_2CH_3$ group is in the third position.

$$\begin{array}{l} C^1-C^2-C^3-C^4-C^5-C^6 \\ \qquad\qquad | \\ \qquad\quad \boxed{CH_2-CH_3} \end{array}$$

Add hydrogens to the base chain to get the final molecule

$$\begin{array}{l} CH_3-CH_2-CH-CH_2-CH_2-CH_3. \\ \qquad\qquad\quad | \\ \qquad\qquad\quad CH_2-CH_3 \end{array}$$

(b) 3-ethyl-3-methylpentane. The base name pentane indicates that there are five carbon atoms in the base chain. $C^1-C^2-C^3-C^4-C^5$. 3-ethyl designates that a $-CH_2CH_3$ group is in the third position, and 3-methyl designates that a $-CH_3$ group is in the third position.

$$\begin{array}{l} \qquad\qquad \boxed{CH_3} \\ \qquad\qquad | \\ C^1-C^2-C^3-C^4-C^5 \\ \qquad\qquad | \\ \qquad\qquad \boxed{CH_2-CH_3} \end{array}$$

Add hydrogens to the base chain to get the final molecule

$$\begin{array}{l} \qquad\qquad\quad CH_3 \\ \qquad\qquad\quad | \\ CH_3-CH_2-C-CH_2-CH_3. \\ \qquad\qquad\quad | \\ \qquad\qquad\quad CH_2-CH_3 \end{array}$$

(c) 2,3-dimethylbutane. The base name butane indicates that there are four carbon atoms in the base chain. $C^1—C^2—C^3—C^4$. 2,3-dimethyl designates $—CH_3$ groups in the second and third positions.

$$\begin{array}{cccc} C^1— & C^2— & C^3— & C^4 \\ & | & | & \\ & \boxed{CH_3} & \boxed{CH_3} & \end{array}$$

. Add hydrogens to the base chain to get the final molecule

$$\begin{array}{cccc} CH_3— & CH— & CH— & CH_3. \\ & | & | & \\ & CH_3 & CH_3 & \end{array}$$

(d) 4,7-diethyl-2,2-dimethylnonane. The base name nonane indicates that there are nine carbon atoms in the base chain. $C^1—C^2—C^3—C^4—C^5—C^6—C^7—C^8—C^9$. 4,7-diethyl designates $—CH_2CH_3$ groups in the fourth and seventh positions, and 2,2-dimethyl designates two $—CH_3$ groups in the second

$$\begin{array}{ccccccccc} & \boxed{CH_3} & & & & & & & \\ & | & & & & & & & \\ C^1— & C^2— & C^3— & C^4— & C^5— & C^6— & C^7— & C^8— & C^9 \\ & | & & | & & & | & & \\ & \boxed{CH_3} & & \boxed{CH_2—CH_3} & & & \boxed{CH_2—CH_3} & & \end{array}$$

position. (above) Add hydrogens to the base chain to get the final molecule

$$\begin{array}{cccccccc} & CH_3 & & & & & & \\ & | & & & & & & \\ CH_3— & C— & CH_2— & CH— & CH_2— & CH_2— & CH— & CH_2—CH_3. \\ & | & & | & & & | & \\ & CH_3 & & CH_2—CH_3 & & & CH_2—CH_3 & \end{array}$$

22.46 **Given:** alkane names **Find:** structure

Conceptual Plan: Find the number of carbon atoms corresponding to prefix of the base name in Table 22.5. → Draw the base chain and number the carbons from left to right. → Using Table 22.6 and the prefix *di-* (2), *tri-* (3), or *tetra-* (4) before the substituent's name, determine each substituent. → Add the substituent to the proper carbon position in the chain. → Add hydrogen atoms to the base chain so that each carbon has four bonds.

Solution:

(a) 2,2-dimethylpentane. The base name pentane indicates that there are five carbon atoms in the base chain. $C^1—C^2—C^3—C^4—C^5$. 2, 2-dimethyl designates two $—CH_3$ groups in the second position.

$$\begin{array}{ccccc} & \boxed{CH_3} & & & \\ & | & & & \\ C^1— & C^2— & C^3— & C^4— & C^5 \\ & | & & & \\ & \boxed{CH_3} & & & \end{array}$$

. Add hydrogens to the base chain to get the final molecule

$$\begin{array}{ccccc} & CH_3 & & & \\ & | & & & \\ CH_3— & C— & CH_2— & CH_2— & CH_3. \\ & | & & & \\ & CH_3 & & & \end{array}$$

(b) 3-isopropylheptane. The base name heptane indicates that there are seven carbon atoms in the base chain. $C^1—C^2—C^3—C^4—C^5—C^6—C^7$. 3-isopropyl designates a $—CH(CH_3)_2$ group in the third position.

$$\begin{array}{ccccccc} C^1— & C^2— & C^3— & C^4— & C^5— & C^6— & C^7 \\ & & | & & & & \\ & & \boxed{\begin{array}{l} CH—CH_3 \\ | \\ CH_3 \end{array}} & & & & \end{array}$$

. Add hydrogens to the base chain to get the final molecule

$$\begin{array}{ccccccc} CH_3— & CH_2— & CH— & CH_2— & CH_2— & CH_2— & CH_3. \\ & & | & & & & \\ & & CH—CH_3 & & & & \\ & & | & & & & \\ & & CH_3 & & & & \end{array}$$

(c) 4-ethyl-2,2-dimethylhexane. The base name hexane indicates that there are six carbon atoms in the base chain. $C^1—C^2—C^3—C^4—C^5—C^6$. 4-ethyl designates an $—CH_2CH_3$ group in the fourth position, and 2,2-dimethyl designates two $—CH_3$ groups in the second position.

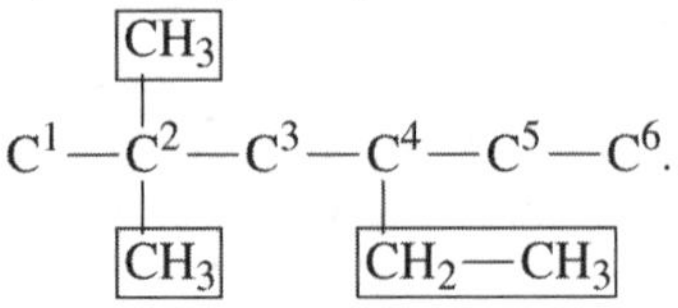

$$\begin{array}{l} \boxed{CH_3} \\ \;| \\ C^1—C^2—C^3—C^4—C^5—C^6 \\ \;\;\;\,| \qquad\quad\;\; | \\ \boxed{CH_3} \quad \boxed{CH_2—CH_3} \end{array}$$

Add hydrogens to the base chain to get the final molecule

$$\begin{array}{l} \qquad CH_3 \\ \qquad\;| \\ CH_3—C—CH_2—CH—CH_2—CH_3. \\ \qquad\;| \qquad\qquad | \\ \qquad CH_3 \qquad CH_2—CH_3 \end{array}$$

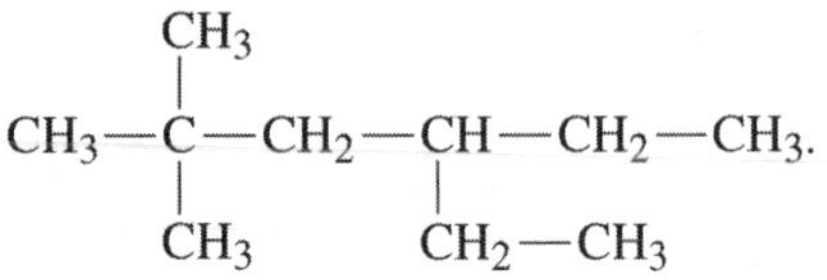

(d) 4,4-diethyloctane. The base name octane indicates that there are eight carbon atoms in the base chain. $C^1—C^2—C^3—C^4—C^5—C^6—C^7—C^8$. 4,4-diethyl designates two $—CH_2CH_3$ groups in the fourth position.

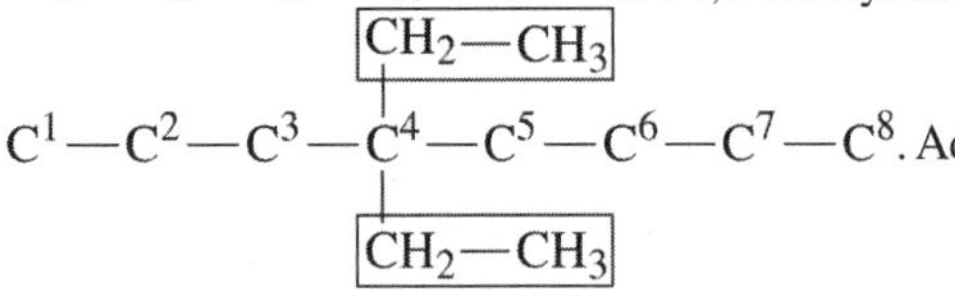

$$\begin{array}{l} \qquad\qquad\quad \boxed{CH_2—CH_3} \\ \qquad\qquad\quad | \\ C^1—C^2—C^3—C^4—C^5—C^6—C^7—C^8 \\ \qquad\qquad\quad | \\ \qquad\qquad\quad \boxed{CH_2—CH_3} \end{array}$$

Add hydrogens to the base chain to get the final molecule

$$\begin{array}{l} \qquad\qquad\qquad\quad CH_2—CH_3 \\ \qquad\qquad\qquad\quad | \\ CH_3—CH_2—CH_2—C—CH_2—CH_2—CH_2—CH_3. \\ \qquad\qquad\qquad\quad | \\ \qquad\qquad\qquad\quad CH_2—CH_3 \end{array}$$

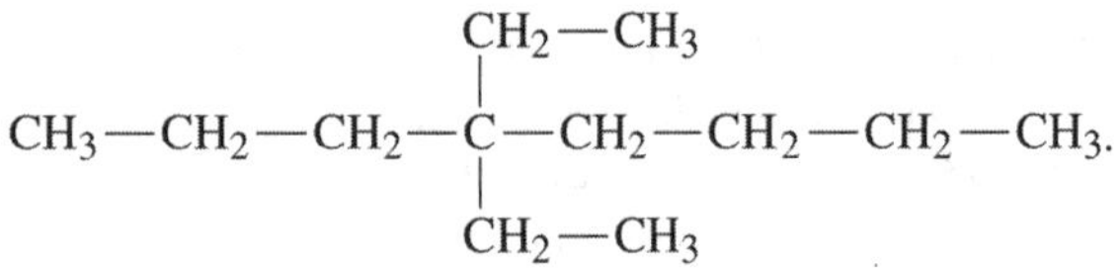

22.47 Hydrocarbon combustion in the presence of oxygen forms carbon dioxide and water. Balance the reaction.

(a) $CH_3CH_2CH_3(g) + 5\,O_2(g) \rightarrow 3\,CO_2(g) + 4\,H_2O(g)$

(b) $CH_3CH_2CH{=}CH_2(g) + 6\,O_2(g) \rightarrow 4\,CO_2(g) + 4\,H_2O(g)$

(c) $2\,CH{\equiv}CH(g) + 5\,O_2(g) \rightarrow 4\,CO_2(g) + 2\,H_2O(g)$

22.48 Hydrocarbon combustion in the presence of oxygen forms carbon dioxide and water. Balance the reaction.

(a) $2\,CH_3CH_2CH_2CH_3(g) + 13\,O_2(g) \rightarrow 8\,CO_2(g) + 10\,H_2O(g)$

(b) $2\,CH_2{=}CHCH_3(g) + 9\,O_2(g) \rightarrow 6\,CO_2(g) + 6\,H_2O(g)$

(c) $2\,CH{\equiv}CCH_2CH_3(g) + 11\,O_2(g) \rightarrow 8\,CO_2(g) + 6\,H_2O(g)$

22.49 Halogen substitution reactions remove a hydrogen atom from the alkane, replace it with a halogen atom, and generate a hydrogen halide. Assume one substitution on the hydrocarbon.

(a) $CH_3CH_3 + Br_2 \rightarrow CH_3CH_2Br + HBr$. Only one carbon-containing product is possible because the C—C bond freely rotates.

(b) $CH_3CH_2CH_3 + Cl_2 \rightarrow [CH_3CH_2CH_2Cl \text{ and } CH_3CHClCH_3] + HCl$. Two carbon-containing products are possible (on either the end carbon or the middle carbon) because the C—C bond freely rotates and the end carbons are equivalent before reaction.

(c) $CH_2Cl_2 + Br_2 \rightarrow CHBrCl_2 + HBr$. Only one carbon-containing product is possible because halogen substitution reactions only remove hydrogen atoms.

(d)

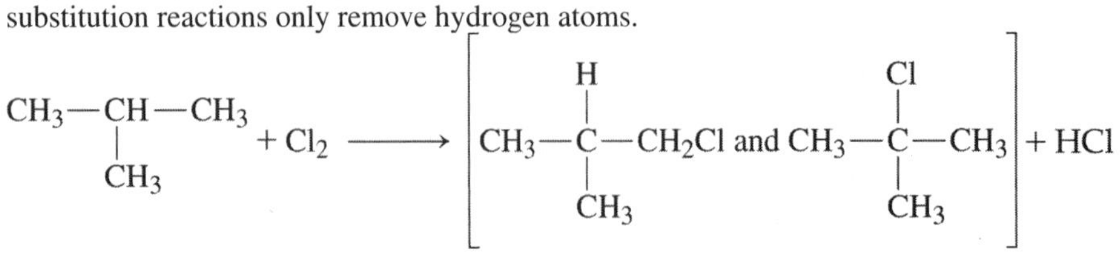

$$\begin{array}{l} CH_3—CH—CH_3 \\ \qquad\;| \\ \qquad CH_3 \end{array} + Cl_2 \longrightarrow \left[\begin{array}{l} \qquad\; H \\ \qquad\; | \\ CH_3—C—CH_2Cl \\ \qquad\; | \\ \qquad CH_3 \end{array} \text{ and } \begin{array}{l} \qquad\; Cl \\ \qquad\; | \\ CH_3—C—CH_3 \\ \qquad\; | \\ \qquad CH_3 \end{array} \right] + HCl$$

Two carbon-containing products are possible (on either an end carbon or the middle carbon) because the C—C bond freely rotates and the end carbons are all equivalent before reaction.

22.50 Halogen substitution reactions remove a hydrogen atom from the alkane, replace it with a halogen atom, and generate a hydrogen halide. Assume one substitution on the hydrocarbon.

(a) $CH_4 + Cl_2 \rightarrow CH_3Cl + HCl$. Only one carbon-containing product is possible because all of the hydrogens are equivalent.

(b) $CH_3CH_2Br + Br_2 \rightarrow [CH_2BrCH_2Br$ and $CH_3CHBr_2] + HBr$. Two carbon-containing products are possible (on either the left carbon or the right carbon) because the C—C bond freely rotates.

(c) $CH_3CH_2CH_2CH_3 + Cl_2 \rightarrow [CH_3CH_2CH_2CH_2Cl$ and $CH_3CHClCH_2CH_3] + HCl$. Two carbon-containing products are possible (on either the end carbons or the middle carbons) because the C—C bond freely rotates and the end and the middle carbons are equivalent before reaction.

(d) $CH_3CHBr_2 + Br_2 \rightarrow [CH_2BrCHBr_2$ and $CH_3CBr_3] + HBr$. Two carbon-containing products are possible (on either the left carbon or the right carbon) because the C—C bond freely rotates.

Alkenes and Alkynes

22.51 $CH_2{=}CH{-}CH_2{-}CH_2{-}CH_2{-}CH_3$, $CH_3{-}CH{=}CH{-}CH_2{-}CH_2{-}CH_3$, and $CH_3{-}CH_2{-}CH{=}CH{-}CH_2{-}CH_3$ are the only structural isomers. Remember that cis–trans isomerism generates geometric isomers, not structural isomers.

22.52 $HC{\equiv}C{-}CH_2{-}CH_2{-}CH_3$ and $CH_3{-}C{\equiv}C{-}CH_2{-}CH_3$ are the only structural isomers.

22.53 **Given:** alkene structures **Find:** name

Conceptual Plan: Count the number of carbon atoms in the longest continuous carbon chain that contains the multiple bond to determine the base name of the compound. Find the prefix corresponding to this number of atoms in Table 22.5 and add the ending *-ene* to form the base name. → Consider every branch from the base chain to be a substituent. Name each substituent according to Table 22.6. → Beginning with the end closest to the multiple bond, number the base chain and assign a number to each substituent. → Write the name of the compound in the following format: (subst. #)-(subst. name)(base name). → If there are two or more substituents, give each one a number and list them alphabetically with hyphens between words and numbers. → If a compound has two or more identical substituents, designate the number of identical substituents with the prefix *di-* (2), *tri-* (3), or *tetra-* (4) before the substituent's name. Separate the numbers with a comma, indicating the positions of the substituents relative to each other. The prefixes are not taken into account when alphabetizing.

Solution:

(a) $\boxed{CH_2{=}CH{-}CH_2{-}CH_3}$ has four carbons as the longest continuous chain. The prefix for 4 is but-, and the base name is butene. There are no substituent groups. $\boxed{C^1H_2{=}C^2H{-}C^3H_2{-}C^4H_3}$ Start numbering on the left because it is closer to the double bond. Because the double bond is between positions 1 and 2, the name of the compound is 1-butene.

(b)
```
     CH3  CH3
      |    |
 CH3—CH—C=CH—CH3
```
(longest chain boxed: $CH_3{-}CH{-}C{=}CH{-}CH_3$) has five carbons as the longest continuous chain. The prefix for 5 is pent-, and the base name is pentene. The substituent groups are two methyl groups.
```
     [CH3] [CH3]
       |     |
 C5H3—C4H—C3=C2H—C1H3
```
Start numbering the chain at the end closest to the double bond, so the double bond is between positions numbered 2 and 3 and the methyl substituents are assigned the numbers 3 and 4. The name of the compound is 3,4-dimethyl-2-pentene.

(c)
```
 CH2=CH—CH—CH2—CH2—CH3
         |
    CH3—CH
         |
         CH3
```
(longest chain boxed: $CH_2{=}CH{-}CH{-}CH_2{-}CH_2{-}CH_3$) has six carbons as the longest continuous chain. The prefix for 6 is hex-, and the base name is hexene. The only substituent group is an isopropyl group.
```
 C1H2=C2H—C3H—C4H2—C5H2—C6H3
           |
     [CH3—CH
           |
           CH3]
```
Start numbering on the left because it is closer to the double

bond. Because the double bond is between positions 1 and 2 and the isopropyl group is at position 3, the name of the compound is 3-isopropyl-1-hexene.

(d) 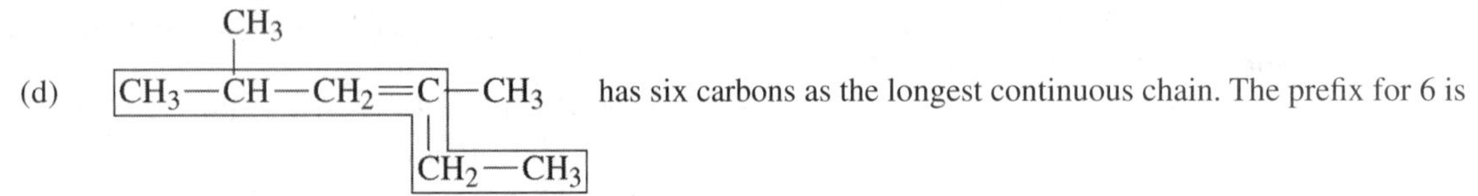 has six carbons as the longest continuous chain. The prefix for 6 is hex-, and the base name is hexene. The only substituent groups are two methyl groups.

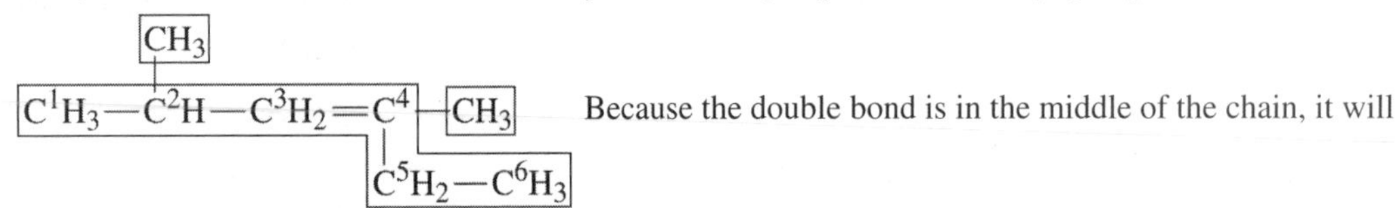 Because the double bond is in the middle of the chain, it will be at position 3 in both numbering schemes. If we start numbering the chain at the end closest to the left methyl group (closest to an end), the methyl substituents are assigned the numbers 2 and 4. The name of the compound is 2,4-dimethyl-3-hexene.

22.54 **Given:** alkene structures **Find:** name

Conceptual Plan: Count the number of carbon atoms in the longest continuous carbon chain that contains the multiple bond to determine the base name of the compound. Find the prefix corresponding to this number of atoms in Table 22.5 and add the ending *-ene* to form the base name. → Consider every branch from the base chain to be a substituent. Name each substituent according to Table 22.6. → Beginning with the end closest to the multiple bond, number the base chain and assign a number to each substituent. → Write the name of the compound in the following format: (subst. #)-(subst. name)(base name). → If there are two or more substituents, give each one a number and list them alphabetically with hyphens between words and numbers. → If a compound has two or more identical substituents, designate the number of identical substituents with the prefix *di-* (2), *tri-* (3), or *tetra-* (4) before the substituent's name. Separate the numbers with a comma, indicating the positions of the substituents relative to each other. The prefixes are not taken into account when alphabetizing.

Solution:

(a) $CH_2{-}CH_2{-}CH{=}CH{-}CH_2{-}CH_3$ has six carbons as the longest continuous chain. The prefix for 6 is hex-, and the base name is hexene. There are no substituent groups.

$C^1H_2{-}C^2H_2{-}C^3H{=}C^4H{-}C^5H_2{-}C^6H_3$ Because the double bond is in the middle of the molecule, it does not matter at which end the numbering is started. Because the double bond is between positions 3 and 4, the name of the compound is 3-hexene.

(b) $CH_3{-}CH(CH_3){-}CH{=}CH{-}CH_3$ has five carbons as the longest continuous chain. The prefix for 5 is pent-, and the base name is pentene. The only substituent group is a methyl group.

$C^5H_3{-}C^4H(CH_3){-}C^3H{=}C^2H{-}C^1H_3$ Start numbering the chain at the end closest to the double bond, so the double bond is between positions 2 and 3 and the methyl substituent is assigned the number 4. The name of the compound is 4-methyl-2-pentene.

(c) $CH_3{-}CH(CH_3){-}CH{=}C(CH_2{-}CH_3){-}CH(CH_3){-}CH_3$ has six carbons as the longest continuous chain. The prefix for 6 is hex-, and the base name is hexene. The substituent groups are two methyl groups and one ethyl group.

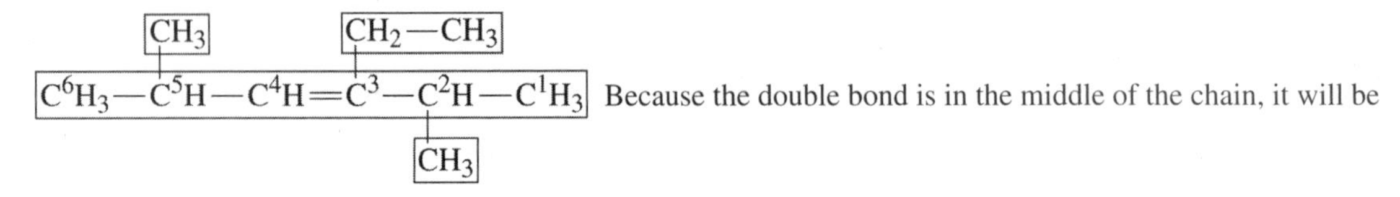 Because the double bond is in the middle of the chain, it will be

at position 3 in both numbering schemes. If we start numbering the chain at the end closest to the right methyl group (because there are two substituents on this end), the methyl substituents are assigned the numbers 2 and 5 and the ethyl substituent is assigned number 3. Because *e* comes before *m*, the name of the compound is 3-ethyl-2,5-dimethyl-3-hexene.

An alternative name is 4-isopropyl-2-methyl-3-hexene. $C^1H_3—C^2H(CH_3)—C^3H=C^4(C^5H_2—C^6H_3)—CH_2—CH_3$ (with CH_3 on CH_2)

(d) $CH_3—C(CH_3)_2—CH=C(CH_3)—CH_2—CH_3$ has six carbons as the longest continuous chain. The prefix for 6 is hex-, and the base name is hexene. The substituent groups are three methyl groups.

$C^1H_3—C^2(CH_3)_2—C^3H=C^4(CH_3)—C^5H—C^6H_3$ Because the double bond is in the middle of the chain, it will be at position 3 in both numbering schemes. If we start numbering the chain at the end closest to the left methyl groups (closest to an end), the methyl substituents are assigned the numbers 2, 2, and 4. The name of the compound is 2,2,4-trimethyl-3-hexene.

22.55 **Given:** alkyne structures **Find:** name

Conceptual Plan: Count the number of carbon atoms in the longest continuous carbon chain that contains the multiple bond to determine the base name of the compound. Find the prefix corresponding to this number of atoms in Table 22.5 and add the ending *-yne* to form the base name. → Consider every branch from the base chain to be a substituent. Name each substituent according to Table 22.6. → Beginning with the end closest to the multiple bond, number the base chain and assign a number to each substituent. → Write the name of the compound in the following format: (subst. #)-(subst. name)(base name). → If there are two or more substituents, give each one a number and list them alphabetically with hyphens between words and numbers. → If a compound has two or more identical substituents, designate the number of identical substituents with the prefix *di-* (2), *tri-* (3), or *tetra-* (4) before the substituent's name. Separate the numbers with a comma, indicating the positions of the substituents relative to each other. The prefixes are not taken into account when alphabetizing.

Solution:

(a) $CH_3—C\equiv C—CH_3$ has four carbons as the longest continuous chain. The prefix for 4 is but-, and the base name is butyne. There are no substituent groups. $C^1H_3—C^2\equiv C^3—C^4H_3$ Because the triple bond is in the middle of the molecule, it does not matter at which end the numbering is started. Because the triple bond is between positions 2 and 3, the name of the compound is 2-butyne.

(b) $CH_3—C\equiv C—C(CH_3)_2—CH_2—CH_3$ has six carbons as the longest continuous chain. The prefix for 6 is hex-, and the base name is hexyne. The only substituent groups are two methyl groups.

$C^1H_3—C^2\equiv C^3—C^4(CH_3)_2—C^5H_2—C^6H_3$ If we start numbering the chain at the end closest to the triple bond, the triple bond is between positions 2 and 3 and the methyl substituents are assigned the numbers 4 and 4. The name of the compound is 4,4-dimethyl-2-hexyne.

(c) 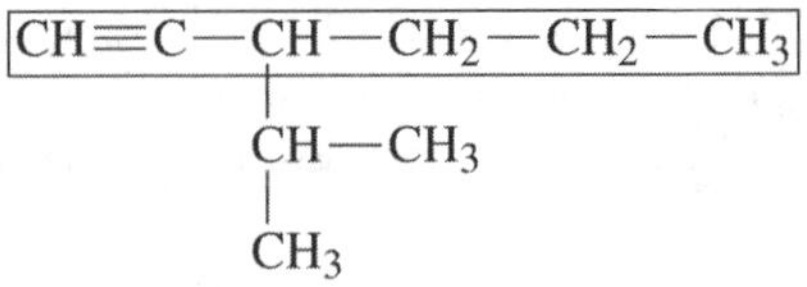 $CH{\equiv}C{-}CH(CH(CH_3){-}CH_3){-}CH_2{-}CH_2{-}CH_3$ has six carbons as the longest continuous chain. The prefix for 6 is hex-, and the base name is hexyne. The only substituent group is an isopropyl group.

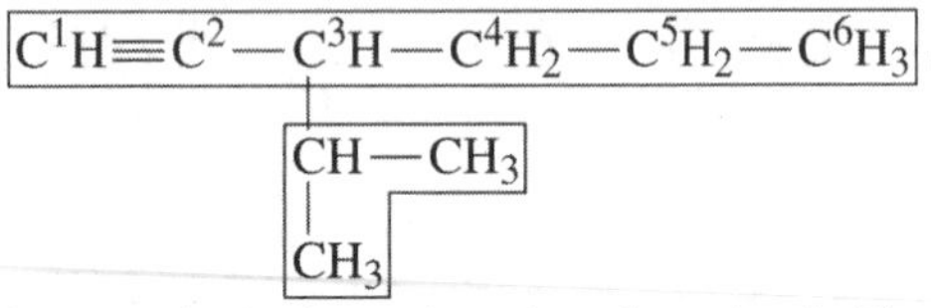

$C^1H{\equiv}C^2{-}C^3H(CH{-}CH_3\,|\,CH_3){-}C^4H_2{-}C^5H_2{-}C^6H_3$ Start numbering at the end closest to the triple bond. The isopropyl substituent is assigned number 3. The name of the compound is 3-isopropyl-1-hexyne.

(d) 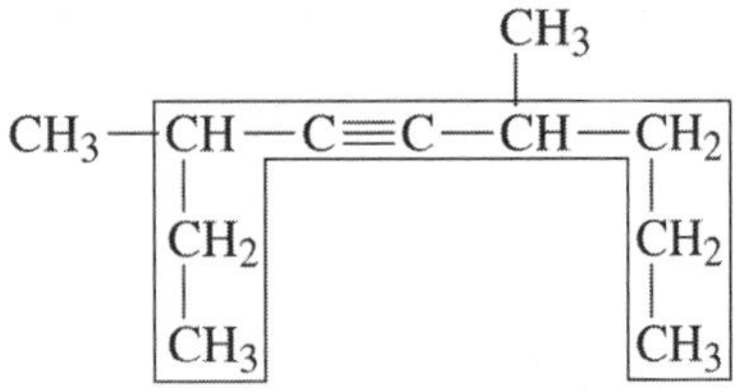 $CH_3{-}CH(CH_2{-}CH_3){-}C{\equiv}C{-}CH(CH_3){-}CH_2{-}CH_2{-}CH_3$ has nine carbons as the longest continuous chain. The prefix for 9 is non-, and the base name is nonyne. The substituent groups are two methyl groups.

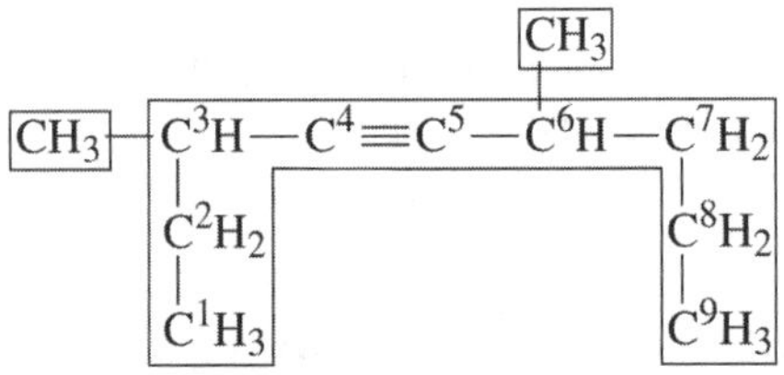

$CH_3{-}C^3H(C^2H_2{-}C^1H_3){-}C^4{\equiv}C^5{-}C^6H(CH_3){-}C^7H_2{-}C^8H_2{-}C^9H_3$ Start numbering at the bottom left and count clockwise along the chain. The triple bond is between positions 4 and 5, and the methyl substituents are assigned the numbers 3 and 6. The name of the compound is 3,6-dimethyl-4-nonyne.

22.56 **Given:** alkyne structures **Find:** name

Conceptual Plan: Count the number of carbon atoms in the longest continuous carbon chain that contains the multiple bond to determine the base name of the compound. Find the prefix corresponding to this number of atoms in Table 22.5 and add the ending *-yne* to form the base name. → Consider every branch from the base chain to be a substituent. Name each substituent according to Table 22.6. → Beginning with the end closest to the multiple bond, number the base chain and assign a number to each substituent. → Write the name of the compound in the following format: (subst. #)-(subst. name)(base name). → If there are two or more substituents, give each one a number and list them alphabetically with hyphens between words and numbers. → If a compound has two or more identical substituents, designate the number of identical substituents with the prefix *di-* (2), *tri-* (3), or *tetra-* (4) before the substituent's name. Separate the numbers with a comma, indicating the positions of the substituents relative to each other. The prefixes are not taken into account when alphabetizing.

Solution:

(a) 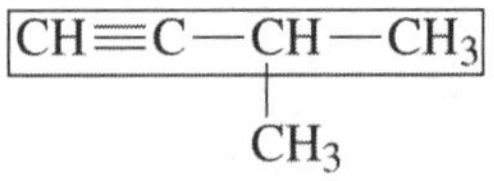 $CH{\equiv}C{-}CH(CH_3){-}CH_3$ has four carbons as the longest continuous chain. The prefix for 4 is but-, and the base name is butyne. The only substituent group is a methyl group. $C^1H{\equiv}C^2{-}C^3H(CH_3){-}C^4H_3$ Start numbering at the end closest to the triple bond. The triple bond is between positions 1 and 2, and the methyl substituent is assigned number 3. The name of the compound is 3-methyl-1-butyne.

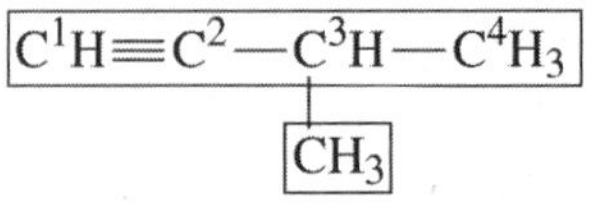

(b) 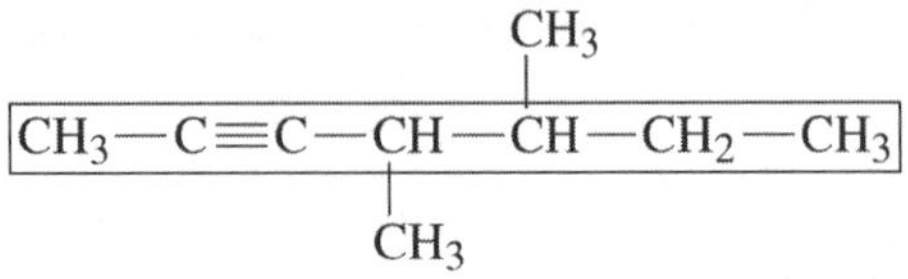 $CH_3{-}C{\equiv}C{-}CH(CH_3){-}CH(CH_3){-}CH_2{-}CH_3$ has seven carbons as the longest continuous chain. The prefix for 7 is hept-, and the base name is heptyne. The only substituent groups are two methyl groups.

$$\underset{\boxed{CH_3}}{\boxed{C^1H_3-C^2\equiv C^3-\overset{|}{C^4}H-\overset{\boxed{CH_3}}{\overset{|}{C^5}}H-C^6H_2-C^7H_3}}$$ If we start numbering the chain at the end closest to the triple bond, the triple bond is between positions 2 and 3 and the methyl substituents are assigned the numbers 4 and 5. The name of the compound is 4,5-dimethyl-2-heptyne.

(c) $\boxed{CH\equiv C-\underset{\underset{CH_3}{|}\,CH_2}{\overset{CH_3}{C}}-CH_2-CH_3}$ has five carbons as the longest continuous chain. The prefix for 5 is pent-, and the base name is pentyne. The substituent groups are a methyl group and an ethyl group.

$$\boxed{C^1H\equiv C^2-\overset{\boxed{CH_3}}{\underset{\boxed{CH_2-CH_3}}{C^3}}-C^4H_2-C^5H_3}$$ Start numbering at the end closest to the triple bond. The triple bond is between positions 1 and 2, the methyl and ethyl groups are assigned number 3. Because *e* comes before *m*, the name of the compound is 3-ethyl-3-methyl-1-pentyne.

(d) $\boxed{CH_3-C\equiv C-\underset{CH_2-CH_3}{CH}}-\underset{CH_3}{\overset{CH_3}{C}}-CH_3$ has six carbons as the longest continuous chain. The prefix for 6 is hex-, and the base name is hexyne. The only substituent group is a *tert*-butyl group.

$$\boxed{C^1H_3-C^2\equiv C^3-\underset{C^5H_2-C^6H_3}{C^4H}}-\boxed{\underset{CH_3}{\overset{CH_3}{C}}-CH_3}$$

Start numbering at the end nearest the triple bond and count clockwise along the chain. The triple bond is between positions 2 and 3, and the *tert*-butyl substituent is assigned the number 4. The name of the compound is 4-*tert*-butyl-2-hexyne. An alternative name is 4-ethyl-5,5-dimethyl-2-hexyne.

22.57 **Given:** hydrocarbon names **Find:** structure

Conceptual Plan: Find the number of carbon atoms corresponding to the prefix of the base name in Table 22.5. → Draw the base chain and number the carbons from left to right. → Determine the multiple bond type from the ending of the base name (*-ene* = double bond; *-yne* = triple bond) and place it in the appropriate position in the chain. → Using Table 22.6 and the prefix *di-* (2), *tri-* (3), or *tetra-* (4) before the substituent's name, determine each substituent. → Add the substituent to the proper carbon position in the chain. → Add hydrogen atoms to the base chain so that each carbon has four bonds.

Solution:

(a) 4-octyne. The base name octyne indicates that there are eight carbon atoms in the base chain. $C^1-C^2-C^3-C^4-C^5-C^6-C^7-C^8$. The *-yne* ending denotes a triple bond, and the 4 prefix places it between the fourth and fifth positions. $C^1-C^2-C^3-C^4\equiv C^5-C^6-C^7-C^8$. Add hydrogens to the base chain to get the final molecule $CH_3-CH_2-CH_2-C\equiv C-CH_2-CH_2-CH_3$.

(b) 3-nonene. The base name nonene indicates that there are nine carbon atoms in the base chain. $C^1-C^2-C^3-C^4-C^5-C^6-C^7-C^8-C^9$. The -ene ending denotes a double bond, and the 3 prefix places it between the third and fourth positions. $C^1-C^2-C^3=C^4-C^5-C^6-C^7-C^8-C^9$. Add hydrogens to the base chain to get the final molecule $CH_3-CH_2-CH=CH-CH_2-CH_2-CH_2-CH_2-CH_3$.

(c) 3,3-dimethyl-1-pentyne. The base name pentyne indicates that there are five carbon atoms in the base chain. $C^1—C^2—C^3—C^4—C^5$. The -yne ending denotes a triple bond, and the 1 prefix places it between the first and second positions. $C^1≡C^2—C^3—C^4—C^5$. 3,3-dimethyl indicates that there are two $—CH_3$ groups in the third position. $C^1≡C^2—C^3(\boxed{CH_3})(\boxed{CH_3})—C^4—C^5$. Add hydrogens to the base chain to get the final molecule

$$CH≡C—C(CH_3)(CH_3)—CH_2—CH_3.$$

(d) 5-ethyl-3,6-dimethyl-2-heptene. The base name heptene indicates that there are seven carbon atoms in the base chain. $C^1—C^2—C^3—C^4—C^5—C^6—C^7$. The -ene ending denotes a double bond, and the 2 prefix places it between the second and third positions. $C^1—C^2=C^3—C^4—C^5—C^6—C^7$. 5-ethyl designates that there is a $—CH_2CH_3$ group in the fifth position, and 3,6-dimethyl designates that there are $—CH_3$ groups in the third and sixth positions. $C^1—C^2=C^3(\boxed{CH_3})—C^4—C^5(\boxed{CH_2—CH_3})—C^6(\boxed{CH_3})—C^7$. Add hydrogens to the base chain to get the final molecule

$$CH_3—CH=C(CH_3)—CH_2—CH(CH_2—CH_3)—CH(CH_3)—CH_3.$$

22.58 **Given:** hydrocarbon names **Find:** structure

Conceptual Plan: Find the number of carbon atoms corresponding to the prefix of the base name in Table 22.5. → Draw the base chain and number the carbons from left to right. → Determine the multiple bond type from the ending of the base name (*-ene* = double bond; *-yne* = triple bond) and place it in the appropriate position in the chain. → Using Table 22.6 and the prefix *di-* (2), *tri-* (3), or *tetra-* (4) before the substituent's name, determine each substituent. → Add the substituent to the proper carbon position in the chain. → Add hydrogen atoms to the base chain so that each carbon has four bonds.

Solution:

(a) 2-hexene. The base name hexene indicates that there are six carbon atoms in the base chain. $C^1—C^2—C^3—C^4—C^5—C^6$. The -ene ending denotes a double bond, and the 2 prefix places it between the second and third positions. $C^1—C^2=C^3—C^4—C^5—C^6$. Add hydrogens to the base chain to get the final molecule $CH_3—CH=CH—CH_2—CH_2—CH_3$.

(b) 1-heptyne. The base name heptyne indicates that there are seven carbon atoms in the base chain. $C^1—C^2—C^3—C^4—C^5—C^6—C^7$. The -yne ending denotes a triple bond, and the 1 prefix places it between the first and second positions. $C^1≡C^2—C^3—C^4—C^5—C^6—C^7$. Add hydrogens to the base chain to get the final molecule $CH≡C—CH_2—CH_2—CH_2—CH_2—CH_3$.

(c) 4,4-dimethyl-2-hexene. The base name hexene indicates that there are six carbon atoms in the base chain. $C^1—C^2—C^3—C^4—C^5—C^6$. The -ene ending denotes a double bond, and the 2 prefix places it between the second and third positions. $C^1—C^2=C^3—C^4—C^5—C^6$. 4,4-dimethyl designates that there are two $—CH_3$ groups in the fourth position. $C^1—C^2=C^3—C^4(\boxed{CH_3})(\boxed{CH_3})—C^5—C^6$. Add hydrogens to the base chain to get the final molecule

$$CH_3—CH=CH—C(CH_3)(CH_3)—CH_2—CH_3.$$

(d) 3-ethyl-4-methyl-2-pentene. The base name pentene indicates that there are five carbon atoms in the base chain. $C^1—C^2—C^3—C^4—C^5$. The -ene ending denotes a double bond, and the 2 prefix places it between the second and third positions. $C^1—C^2{=}C^3—C^4—C^5$. 3-ethyl designates that there is a $—CH_2CH_3$ group in the third position, and 4-methyl designates that there is a $—CH_3$ group in the fourth position. $C^1—C^2{=}C^3(CH_2—CH_3)—C^4(CH_3)—C^5$. Add hydrogens to the base chain to get the final molecule

$CH_3—CH{=}C(CH_2—CH_3)—CH(CH_3)—CH_3$. An alternative name is 3-isopropyl-2-pentene.

22.59 Alkene addition reactions convert a double bond to a single bond and place the two halves of the other reactant on the two carbons that were in the double bond. Markovnikov's rule states that when a polar reagent is added to an unsymmetrical alkene, the positive end (the least electronegative part) of the reagent adds to the carbon atom that has the most hydrogen atoms. In most reactions of this type, the positive end of the reagent is hydrogen; therefore, the hydrogen atom goes to the carbon atom that already contains the most hydrogen atoms.

(a) $CH_3—CH{=}CH—CH_3 + Cl_2 \rightarrow CH_3—CH(Cl)—CH(Cl)—CH_3$

(b) $CH_3—CH(CH_3)—CH{=}CH—CH_3 + HBr \rightarrow CH_3—CH(CH_3)—CH_2—CH(Br)—CH_3$ or $CH_3—CH(CH_3)—CH(Br)—CH_2—CH_3$

(c) $CH_3—CH_2—CH{=}CH—CH_3 + Br_2 \rightarrow CH_3—CH_2—CH(Br)—CH(Br)—CH_3$

(d) $CH_3—CH(CH_3)—CH{=}C(CH_3)—CH_3 + HCl \rightarrow CH_3—CH(CH_3)—CH_2—C(CH_3)(Cl)—CH_3$

Only one product is formed according to Markovnikov's rule.

22.60 Alkene addition reactions convert a double bond to a single bond and place the two halves of the other reactant on the two carbons that were in the double bond. Markovnikov's rule states that when a polar reagent is added to an unsymmetrical alkene, the positive end (the least electronegative part) of the reagent adds to the carbon atom that has the most hydrogen atoms. In most reactions of this type, the positive end of the reagent is hydrogen; therefore, the hydrogen atom goes to the carbon atom that already contains the most hydrogen atoms.

(a) $CH_3—CH(CH_3)—CH{=}CH_2 + Br_2 \rightarrow CH_3—CH(CH_3)—CH(Br)—CH_2(Br)$

(b) $CH_2{=}CH—CH_3 + Cl_2 \rightarrow CH_2(Cl)—CH(Cl)—CH_3$

(c) $CH_3—C(CH_3)_2—CH{=}CH_2 + HCl \rightarrow CH_3—C(CH_3)_2—CH(Cl)—CH_3$

Only one product is formed according to Markovnikov's rule.

(d) 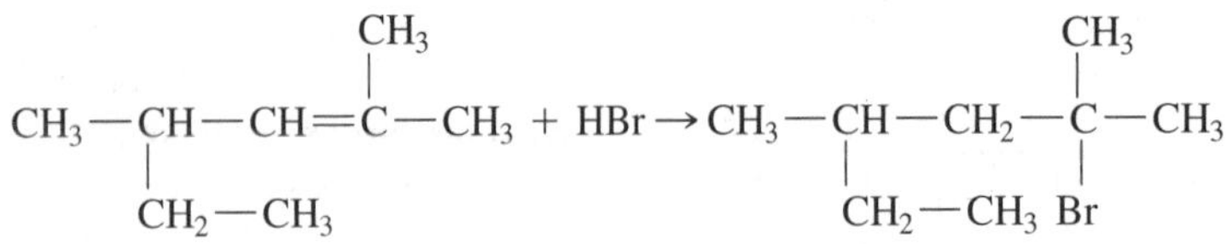
$$CH_3{-}\underset{\underset{CH_2-CH_3}{|}}{CH}{-}CH{=}\overset{\overset{CH_3}{|}}{C}{-}CH_3 + HBr \rightarrow CH_3{-}\underset{\underset{CH_2-CH_3}{|}}{CH}{-}CH_2{-}\overset{\overset{CH_3}{|}}{\underset{\underset{Br}{|}}{C}}{-}CH_3$$

Only one product is formed according to Markovnikov's rule.

22.61 Hydrogenation reactions convert a double bond to a single bond and place the hydrogen atoms on each of the two carbons that were in the double bond.

(a) $CH_2{=}CH{-}CH_3 + H_2 \rightarrow CH_3{-}CH_2{-}CH_3$

(b) 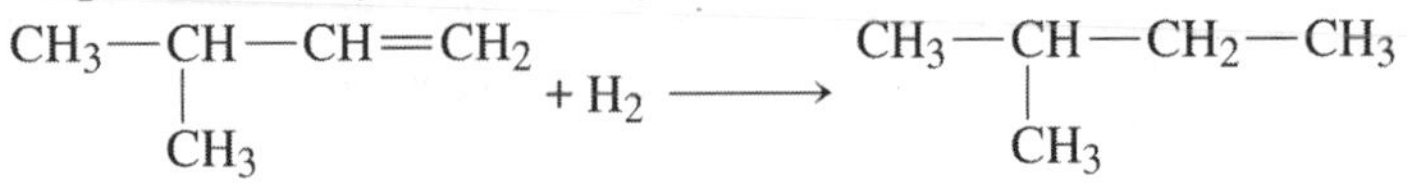
$$CH_3{-}\underset{\underset{CH_3}{|}}{CH}{-}CH{=}CH_2 + H_2 \longrightarrow CH_3{-}\underset{\underset{CH_3}{|}}{CH}{-}CH_2{-}CH_3$$

(c) 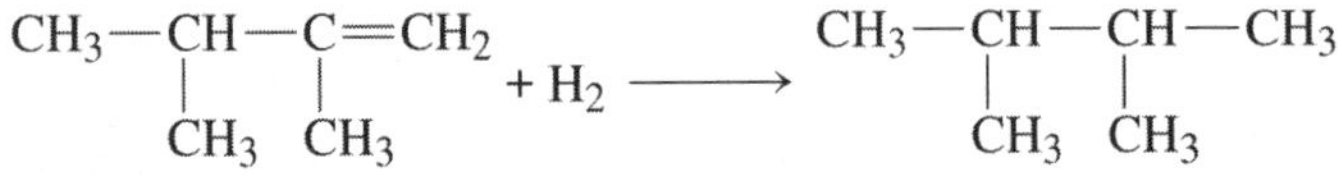
$$CH_3{-}\underset{\underset{CH_3}{|}}{CH}{-}\underset{\underset{CH_3}{|}}{C}{=}CH_2 + H_2 \longrightarrow CH_3{-}\underset{\underset{CH_3}{|}}{CH}{-}\underset{\underset{CH_3}{|}}{CH}{-}CH_3$$

22.62 Hydrogenation reactions convert a double bond to a single bond and place the hydrogen atoms on each of the two carbons that were in the double bond.

(a) $CH_3{-}CH_2{-}CH{=}CH_2 + H_2 \rightarrow CH_3{-}CH_2{-}CH_2{-}CH_3$

(b) 
$$CH_3{-}CH_2{-}\overset{\overset{CH_3}{|}}{C}{=}\overset{\overset{CH_3}{|}}{C}{-}CH_3 + H_2 \longrightarrow CH_3{-}CH_2{-}\overset{\overset{CH_3}{|}}{CH}{-}\overset{\overset{CH_3}{|}}{CH}{-}CH_3$$

(c)
$$CH_3{-}CH_2{-}\underset{\underset{CH_3}{|}}{C}{=}CH_2 + H_2 \longrightarrow CH_3{-}CH_2{-}\underset{\underset{CH_3}{|}}{CH}{-}CH_3$$

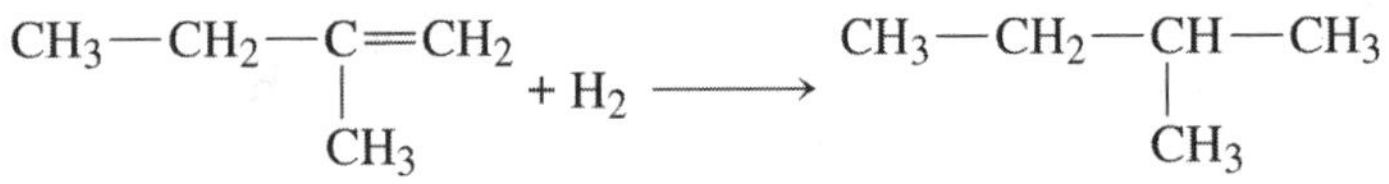

Aromatic Hydrocarbons

22.63 **Given:** monosubstituted benzene structures **Find:** name

Conceptual Plan: Consider the branch from the benzene ring to be a substituent. Name the substituent according to Table 22.6 or using the base name of a halogen with an *o* added at the end. → Write the name of the compound in the following format: (name of substituent)benzene.

Solution:

(a) $-CH_3$ or methyl is the substituent group, so the name is methylbenzene or toluene (common name).

(b) —Br or bromo is the substituent, so the name is bromobenzene.

(c) —Cl or chloro is the substituent, so the name is chlorobenzene.

22.64 **Given:** monosubstituted benzene structures **Find:** name

Conceptual Plan: Consider the branch from the benzene ring to be a substituent. Name the substituent according to Table 22.6 or using the base name of a halogen with an *o* added at the end. → Write the name of the compound in the following format: (name of substituent)benzene.

Solution:

(a) $-CH_2CH_3$ or ethyl is the substituent group, so the name is ethylbenzene.

(b) —F or fluoro is the substituent, so the name is fluorobenzene.

(c) $-C(CH_3)_3$ or *tert*-butyl is the substituent, so the name is *tert*-butylbenzene.

22.65 **Given:** hydrocarbon structures **Find:** name

Conceptual Plan: Count the number of carbon atoms in the longest continuous carbon chain (if a multiple bond is present, make sure this chain contains the multiple bond) to determine the base name of the compound. Find the prefix corresponding to this number of atoms in Table 22.5 and add the ending *-ane* for an alkane, *-ene* for an alkene, and *-yne* for an alkyne to form the base name. → Consider every branch from the base chain to be a substituent. Name each substituent according to Table 22.6. If a benzene ring is a substituent, use the phenyl group name. → Beginning with the end closest to the multiple bond, number the base chain and assign a num-

ber to each substituent. → Write the name of the compound in the following format: (subst. #)-(subst. name)(base name). → If there are two or more substituents, give each one a number and list them alphabetically with hyphens between words and numbers. → If a compound has two or more identical substituents, designate the number of identical substituents with the prefix *di-* (2), *tri-* (3), or *tetra-* (4) before the substituent's name. Separate the numbers with a comma, indicating the positions of the substituents relative to each other. The prefixes are not taken into account when alphabetizing.

Solution: Note: For simplicity, ∅ = (benzene ring)

(a)

$$\begin{array}{l} \qquad\qquad\quad CH_3 \qquad\qquad\quad \varnothing \\ CH_3-\boxed{CH-CH_2-CH-CH_2-CH-CH_2-CH_3} \\ \qquad\;\; | \\ \qquad\;\; \boxed{CH_2-CH_3} \end{array}$$

has nine carbons as the longest continuous chain.

The prefix for 9 is non- and the base name is nonane. The substituent groups are two methyl groups and a phenyl group.

$$\begin{array}{l} \qquad\qquad\qquad\quad \boxed{CH_3} \qquad\qquad\quad \boxed{\varnothing} \\ CH_3-\boxed{C^3H-C^4H_2-C^5H-C^6H_2-C^7H-C^8H_2-C^9H_3} \\ \qquad\;\; | \\ \qquad\;\; \boxed{C^2H_2-C^1H_3} \end{array}$$

Start numbering at the end closest to the methyl group. The substituent groups are two methyl groups at positions 3 and 5 and a phenyl group at position 7. Because *m* comes before *p,* the name of the compound is 3,5-dimethyl-7-phenylnonane. If the numbering is in the opposite direction, the name will be 5,7-dimethyl-3-phenylnonane, but this is not the approved name because IUPAC rules indicate that the first group cited receives the lowest number.

(b)

$$\begin{array}{l} \qquad\qquad\quad \varnothing \\ \boxed{CH_3-CH-CH=CH-CH_2-CH_2-CH_2-CH_3} \end{array}$$

has eight carbons as the longest continuous chain. The prefix for 8 is oct-, and because there is a double bond, the base name is octene. The only substituent group is a phenyl group.

$$\begin{array}{l} \qquad\qquad\quad \boxed{\varnothing} \\ \boxed{C^1H_3-C^2H-C^3H=C^4H-C^5H_2-C^6H_2-C^7H_2-C^8H_3} \end{array}$$

If we start numbering the chain at the end closest to the double bond, the double bond is between positions 3 and 4 and the phenyl substituent is assigned the number 2. The name of the compound is 2-phenyl-3-octene.

(c)

$$\begin{array}{l} \qquad\qquad\qquad\qquad\qquad\qquad \varnothing \\ \boxed{CH_3-C\equiv C-CH-CH-CH-CH_2-CH_3} \\ \qquad\qquad\qquad\; CH_3 \;\; CH_3 \end{array}$$

has eight carbons as the longest continuous chain. The prefix for 8 is oct-, and because there is a triple bond, the base name is octyne. The substituent groups are two methyl groups and a phenyl group.

$$\begin{array}{l} \qquad\qquad\qquad\qquad\qquad\qquad\qquad\quad \boxed{\varnothing} \\ \boxed{C^1H_3-C^2\equiv C^3-C^4H-C^5H-C^6H-C^7H_2-C^8H_3} \\ \qquad\qquad\qquad\quad \boxed{CH_3} \;\; \boxed{CH_3} \end{array}$$

Start numbering at the end closest to the triple bond. The triple bond is between positions 2 and 3, the methyl groups are assigned numbers 4 and 5, and the phenyl group is assigned number 6. Because *m* is before *p,* the name of the compound is 4,5-dimethyl-6-phenyl-2-octyne.

22.66 **Given:** hydrocarbon structures **Find:** name

Conceptual Plan: Count the number of carbon atoms in the longest continuous carbon chain (if a multiple bond is present, make sure that this chain contains the multiple bond) to determine the base name of the compound. Find the prefix corresponding to this number of atoms in Table 22.5 and add the ending *-ane* for an alkane, *-ene* for an alkene and *-yne* for an alkyne to form the base name. → Consider every branch from the base chain to be a substituent. Name each substituent according to Table 22.6. If a benzene ring is a substituent, use the phenyl group name. → Beginning with the end closest to the multiple bond, number the base chain and assign a number to each substituent. → Write the name of the compound in the following format: (subst. #)-(subst. name)(base name). → If there are two or more substituents, give each one a number and list them alphabetically with hyphens between words and numbers. → If a compound has two or more

identical substituents, designate the number of identical substituents with the prefix *di-* (2), *tri-* (3), or *tetra-* (4) before the substituent's name. Separate the numbers with a comma, indicating the positions of the substituents relative to each other. The prefixes are not taken into account when alphabetizing.

Solution: Note: For simplicity, ∅ = (benzene ring)

(a) $CH_3—CH_2—CH(∅)—CH(H_3C)—CH_2—CH_3$ has six carbons as the longest continuous chain. The prefix for 6 is hex-, and the base name is hexane. The substituent groups are a methyl group and a phenyl group.

$C^6H_3—C^5H_2—C^4H(∅)—C^3H(H_3C)—C^2H_2—C^1H_3$ Start numbering at the end closest to the methyl group. The substituent groups are a methyl group at position 3 and a phenyl group at position 4. Because *m* comes before *p*, the name of the compound is 3-methyl-4-phenylhexane. If the numbering is in the opposite direction, the name will be 4-methyl-3-phenylhexane, but this is not the approved name because IUPAC rules indicate that the first group cited receives the lowest number.

(b) $(∅)CH_2—CH_2—CH_2—CH_2—C{\equiv}C—CH_3$ has seven carbons as the longest continuous chain. The prefix for 7 is hept-, and because there is a triple bond, the base name is heptyne. The only substituent group is a phenyl group. $(∅)C^7H_2—C^6H_2—C^5H_2—C^4H_2—C^3{\equiv}C^2—C^1H_3$ If we start numbering the chain at the end closest to the triple bond, the triple bond is between positions 2 and 3 and the phenyl substituent is assigned the position 7. The name of the compound is 7-phenyl-2-heptyne.

(c) $CH_3—CH(CH_3)—CH(CH_3)—C(CH_3){=}CH—CH(∅)—CH_2—CH_3$ has eight carbons as the longest continuous chain. The prefix for 8 is oct-, and because there is a double bond, the base name is octene. The substituent groups are three methyl groups and a phenyl group. $C^1H_3—C^2H(CH_3)—C^3H(CH_3)—C^4(CH_3){=}C^5H—C^6H(∅)—C^7H_2—C^8H_3$

Because the double bond is in the middle of the molecule, start numbering at the end closest to the methyl group. The double bond is between positions 4 and 5; the methyl groups are assigned numbers 2, 3, and 4; and the phenyl group is assigned number 6. Because *m* is before *p*, the name of the compound is 2,3,4-trimethyl-6-phenyl-4-octene.

22.67 **Given:** disubstituted benzene structures **Find:** name

Conceptual Plan: Consider the branches from the benzene ring to be substituents. Name the substituent according to Table 22.6 or using the base name of a halogen with an *o* added at the end. → Number the benzene ring starting with the substituent that is first alphabetically and count in the direction that gets to the second substituent with the lower position number. → Write the name of the compound in the following format: (name of substituent)benzene. List them alphabetically with hyphens between words and numbers. → If a compound has two identical substituents, designate the number of identical substituents with the prefix *di-* (2) before the substituent's name. Separate the numbers with a comma, indicating the positions of the substituents relative to each other. → Alternative names are 1,2 = *ortho* or *o*; 1,3 = *meta* or *m*; and 1,4 = *para* or *p* replacing the numbers.

Solution:

(a) Both of the substituent groups are —Br or bromo groups. Start by giving one bromo an assignment of 1 and count in either direction to give the second bromo group an assignment of 4. The name is 1, 4-dibromobenzene or *p*-dibromobenzene.

(b) Both of the substituent groups are $—CH_2CH_3$ or ethyl groups. Start by giving the top ethyl an assignment of 1 and count in a clockwise direction to give the second ethyl group an assignment of 3. The name is 1,3-diethylbenzene or *m*-diethylbenzene.

(c) One substituent group is —Cl or chloro, and the other is —F or fluoro. Start by giving the chloro group an assignment of 1 and count in a counterclockwise direction to give the fluoro group an assignment of 2. The name is 1-chloro-2-fluorobenzene or *o*-chlorofluorobenzene.

22.68 **Given:** disubstituted benzene structures **Find:** name

Conceptual Plan: Consider the branches from the benzene ring to be substituents. Name the substituent according to Table 22.6 or using the base name of a halogen with an *o* added at the end. → Number the benzene ring starting with the substituent that is first alphabetically and count in the direction that gets to the second substituent with the lower position number. → Write the name of the compound in the following format: (name of substituent)benzene. List them alphabetically with hyphens between words and numbers. → If a compound has two identical substituents, designate the number of identical substituents with the prefix *di-* (2) before the substituent's name. Separate the numbers with a comma, indicating the positions of the substituents relative to each other. → Alternative names are 1,2 = *ortho* or *o*; 1,3 = *meta* or *m*; and 1,4 = *para* or *p* replacing the numbers.

Solution:

(a) One substituent group is —Br or bromo, and the other is —Cl or chloro. Start by giving the bromo group an assignment of 1 and count in a clockwise direction to give the chloro group an assignment of 2. The name is 1-bromo-2-chlorobenzene or *o*-bromochlorobenzene.

(b) One substituent group is —Cl or chloro, and the other is $—CH_2CH_3$ or ethyl. Start by giving the chloro group an assignment of 1 and count in either direction to give the ethyl group an assignment of 4. The name is 1-chloro-4-ethylbenzene or *p*-chloroethylbenzene.

(c) Both of the substituent groups are —I or iodo groups. Start by giving the top iodo an assignment of 1 and count in a clockwise direction to give the second iodo group an assignment of 3. The name is 1, 3-diiodobenzene or *m*-diiodobenzene.

22.69 **Given:** substituted benzene names **Find:** structures

Conceptual Plan: Start with a benzene ring. Identify the structure of the substituent(s) according to Table 22.6 or using the base name of a halogen with an *o* added at the end. → If only one substituent is present, simply attach it to the benzene ring. → If a compound has two identical substituents, they are designated with the prefix *di-* (2) before the substituent's name. → If there are two substituents, place the first one on the benzene ring and count clockwise around the ring to determine where to attach the second substituent. Note: 1,2 = *ortho* or *o*; 1,3 = *meta* or *m*; and 1,4 = *para* or *p* replacing the numbers.

Solution:

(a) isopropylbenzene. The isopropyl group is $—CH(CH_3)—CH_3$. Attach this to a benzene ring to make

CH3
|
HC—CH3.
|
(benzene ring)

(b) *meta*-dibromobenzene. There are two bromo or —Br groups. The positions are 1 and 3 because the designation is *meta*. Attach the —Br groups at the first and third positions to make

Br
(benzene ring)
Br.

(c) 1-chloro-4-methylbenzene. One substituent group is —Cl or chloro, and the other is $—CH_3$ or methyl. Start by giving the chloro group an assignment of 1 and count in a clockwise direction to give the methyl group an assignment of 4. Attach the two groups to make 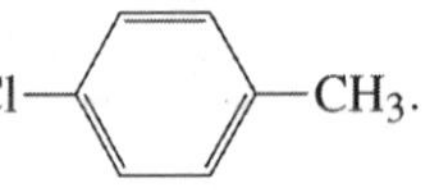.

22.70 **Given:** substituted benzene names **Find:** structures
Conceptual Plan: Start with a benzene ring. Identify the structure of the substituent(s) according to Table 22.6 or using the base name of a halogen with an *o* added at the end. → If only one substituent is present, simply attach it to the benzene ring. → If a compound has two identical substituents, they are designated with the prefix *di-* (2) before the substituent's name. → If there are two substituents, place the first one on the benzene ring and count clockwise around the ring to determine where to attach the second substituent. Note: 1,2 = *ortho* or *o*; 1,3 = *meta* or *m*; and 1,4 = *para* or *p* replacing the numbers.
Solution:

(a) ethylbenzene. The ethyl group is $—CH_2CH_3$. Attach this to a benzene ring to make

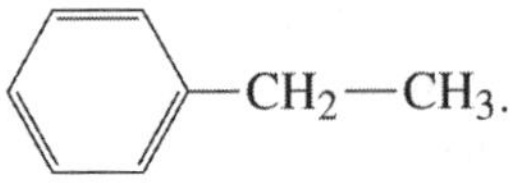

(b) 1-iodo-2-methylbenzene. The iodo group is —I, and the methyl group is $—CH_3$. Attach the —I to the benzene ring at the first position; then attach the $—CH_3$ to the adjacent carbon (second position) to make

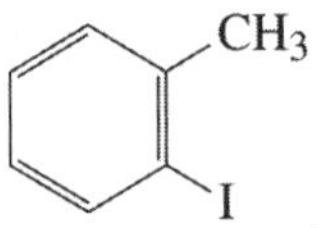

(c) *para*-diethylbenzene. There are two ethyl or $—CH_2CH_3$ groups. The positions are 1 and 4 because the designation is *para-*. Attach the $—CH_2CH_3$ at the first and fourth positions to make

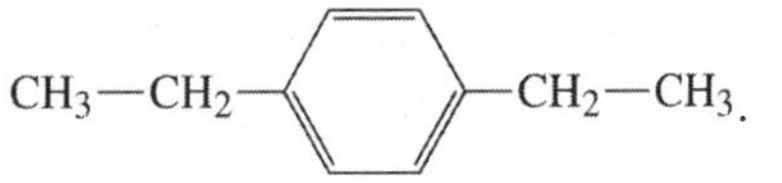

22.71 Aromatic substitution reactions substitute a halogen or alkyl group for a hydrogen (with a preference for the alkyl group) to generate a monosubstituted benzene and a hydrogen halide.

(a) 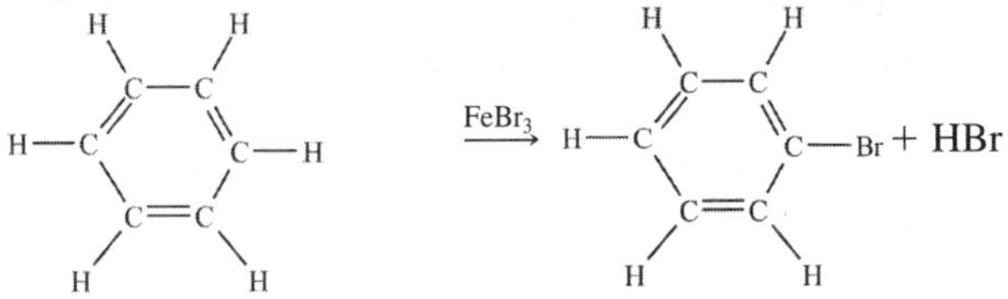

(b) 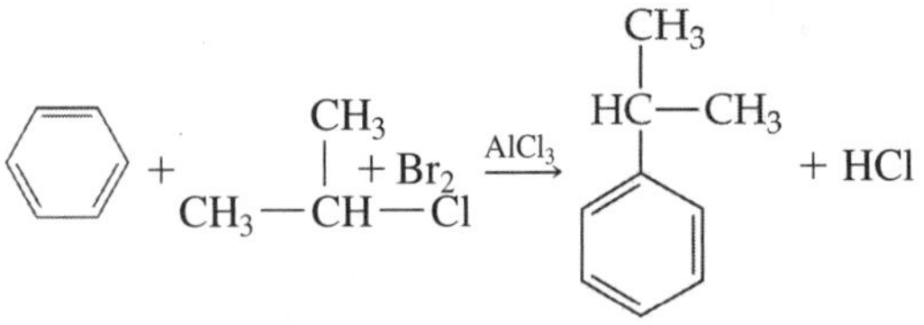

22.72 Aromatic substitution reactions substitute a halogen or alkyl group for a hydrogen (with a preference for the alkyl group) to generate a monosubstituted benzene and a hydrohalic acid.

(a) 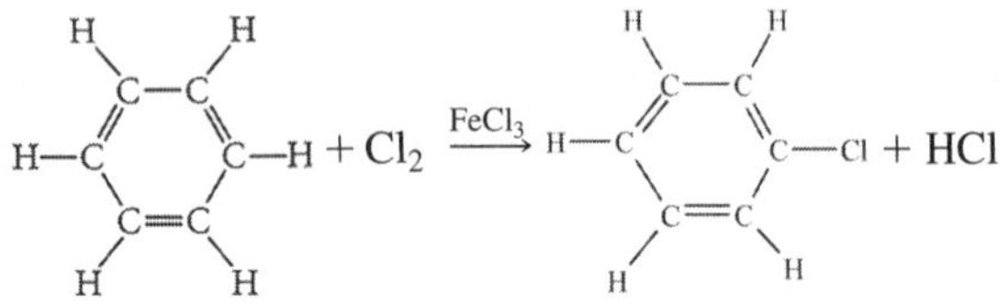

(b) $C_6H_6 + CH_3-C(CH_3)_2-Cl \xrightarrow{AlCl_3} H_3C-C(CH_3)_2-C_6H_5 + HCl$

Alcohols

22.73 **Given:** alcohol structures **Find:** name

Conceptual Plan: Alcohols are named like alkanes with the following differences: (1) The base chain is the longest continuous carbon chain that contains the —OH functional group. (2) The base name has the ending *-ol.* (3) The base chain is numbered to give the —OH group the lowest possible number. (4) A number indicating the position of the —OH group is inserted just before the base name.

Solution:

(a) $\boxed{CH_3-CH_2-CH_2}-OH$ The longest carbon chain has three carbons. The prefix for 3 is prop-, so the base name is propanol. The alcohol group is on the end, or first carbon, so the name is 1-propanol.

(b) $CH_3-CH(CH_2-CH_3)-CH_2-CH(OH)-CH_3$ The longest carbon chain has six carbons. The prefix for 6 is hex-, so the base name is hexanol. There is a methyl substituent group.

$\boxed{CH_3}-C^4H(C^5H_2-C^6H_3)-C^3H_2-C^2H(OH)-C^1H_3$ Start numbering at the end closest to the alcohol group. The alcohol group is assigned number 2, and the methyl group is assigned number 4. The name is 4-methyl-2-hexanol.

(c) $CH_3-CH(CH_3)-CH_2-CH(OH)-CH_2-CH(CH_3)-CH_3$ The longest carbon chain has seven carbons. The prefix for 7 is hept-, so the base name is heptanol. There are two methyl substituent groups.

$C^1H_3-C^2H(\boxed{CH_3})-C^3H_2-C^4H(OH)-C^5H_2-C^6H(\boxed{CH_3})-C^7H_3$ It does not matter at which end the numbering is started because substitutions are symmetrically attached. The alcohol group is assigned number 4, and the methyl groups are assigned numbers 2 and 6. The name is 2,6-dimethyl-4-heptanol.

(d) $CH_3-CH_2-C(HO)(H_3C)-CH_2-CH_3$ The longest carbon chain has five carbons. The prefix for 5 is pent-, so the base name is pentanol. There is a methyl substituent group. $C^1H_3-C^2H_2-C^3(HO)(\boxed{H_3C})-C^4H_2-C^5H_3$ It does not matter at which end the numbering is started because substitutions are symmetrically attached. The alcohol group is assigned number 3, and the methyl group is assigned number 3. The name is 3-methyl-3-pentanol.

22.74 **Given:** alcohol names **Find:** structure

Conceptual Plan: Find the number of carbon atoms corresponding to the prefix of the base name in Table 22.5. → Draw the base chain and number the carbons from left to right. → Add an —OH group to the specified position. → Using Table 22.6 and the prefix *di-* (2), *tri-* (3), or *tetra-* (4) before the substituent's name, determine each

substituent. → Add the substituent to the proper carbon position in the chain. → Add hydrogen atoms to the base chain so that each carbon has four bonds.

Solution:

(a) 2-butanol. The base name butanol indicates that there are four carbon atoms in the base chain. $C^1—C^2—C^3—C^4$. The 2- prefix denotes that the —OH group is in the second position

$$C^1—\underset{\displaystyle OH}{\underset{|}{C^2}}—C^3—C^4.$$

Add hydrogens to the base chain to get the final molecule $CH_3—\underset{\displaystyle OH}{\underset{|}{CH}}—CH_2—CH_3$.

(b) 2-methyl-1-propanol. The base name propanol indicates that there are three carbon atoms in the base chain. $C^1—C^2—C^3$. 2-methyl denotes that a $—CH_3$ group is in the second position, and the 1- prefix of the base name shows that the —OH group is in the first position $HO—C^1—\underset{\boxed{CH_3}}{\underset{|}{C^2}}—C^3$. Add hydrogens to the base chain to get the final molecule $HO—CH_2—\underset{\boxed{CH_3}}{\underset{|}{CH}}—CH_3$.

(c) 3-ethyl-1-hexanol. The base name hexanol indicates that there are six carbon atoms in the base chain. $C^1—C^2—C^3—C^4—C^5—C^6$. 3-ethyl denotes that a $—CH_2CH_3$ group is in the third position, and the 1- prefix of the base name shows that the —OH group is in the first position

$$HO—C^1—C^2—\underset{\boxed{CH_2CH_3}}{\underset{|}{C^3}}—C^4—C^5—C^6.$$

Add hydrogens to the base chain to get the final molecule $HO—CH_2—CH_2—\underset{\displaystyle CH_2CH_3}{\underset{|}{CH}}—CH_2—CH_2—CH_3$.

(d) 2-methyl-3-pentanol. The base name pentanol indicates that there are five carbon atoms in the base chain. $C^1—C^2—C^3—C^4—C^5$. 2-methyl denotes that a $—CH_3$ group is in the second position, and the 3- prefix of the base name shows that the —OH group is in the third position $C^1—\underset{\boxed{CH_3}}{\underset{|}{C^2}}—\underset{\displaystyle OH}{\underset{|}{C^3}}—C^4—C^5$.

Add hydrogens to the base chain to get the final molecule

$$CH_3—\underset{\displaystyle CH_3}{\underset{|}{CH}}—\underset{\displaystyle OH}{\underset{|}{CH}}—CH_2—CH_3.$$

22.75 In a substitution reaction, an alcohol reacts with an acid, such as HBr, to form halogenated hydrocarbons and water. In an elimination (or dehydration) reaction, concentrated acids, such as H_2SO_4, react with alcohols to eliminate water, forming an alkene. In an oxidation reaction, carbon atoms gain oxygen atoms and/or lose hydrogen atoms. In these reactions, the alcohol becomes a carboxylic acid. Active metals, such as Na, react with alcohols to produce a strong base and eliminate hydrogen.

(a) This is a substitution reaction, so $CH_3—CH_2—CH_2—OH + HBr \rightarrow CH_3—CH_2—CH_2—Br + H_2O$.

(b) This is an elimination reaction, so $CH_3—\underset{\displaystyle CH_3}{\underset{|}{CH}}—CH_2—OH \xrightarrow{H_2SO_4} CH_3—\underset{\displaystyle CH_3}{\underset{|}{C}}=CH_2 + H_2O$.

(c) Sodium is an active metal, so $CH_3—CH_2—OH + Na \rightarrow CH_3—CH_2—O^-Na^+ + 1/2\,H_2$.

(d) This is an oxidation reaction, so

$$CH_3—\underset{\displaystyle CH_3}{\overset{\displaystyle CH_3}{\overset{|}{\underset{|}{C}}}}—CH_2—CH_2—OH \xrightarrow[H_2SO_4]{Na_2Cr_2O_7} CH_3—\underset{\displaystyle CH_3}{\overset{\displaystyle CH_3}{\overset{|}{\underset{|}{C}}}}—CH_2—\overset{\displaystyle O}{\overset{\|}{C}}—OH.$$

22.76 In a substitution reaction, an alcohol reacts with an acid, such as HBr, to form halogenated hydrocarbons and water. In an elimination (or dehydration) reaction, concentrated acids, such as H_2SO_4, react with alcohols to eliminate water, forming an alkene. In an oxidation reaction, carbon atoms gain oxygen atoms and/or lose hydrogen atoms. In these reactions, the alcohol becomes a carboxylic acid. Active metals, such as Na, react with alcohols to produce a strong base and eliminate hydrogen.

(a) This is an elimination reaction, so

$$CH_3-\underset{\displaystyle CH_3}{\overset{\displaystyle CH_3}{\underset{|}{\overset{|}{C}}}}-OH \xrightarrow{H_2SO_4} CH_3-\underset{\displaystyle CH_3}{\underset{|}{C}}=CH_2 + H_2O.$$

(b) This is an oxidation reaction, so

$$CH_3-\overset{\displaystyle CH_3}{\overset{|}{CH}}-CH_2-CH_2-OH \xrightarrow[H_2SO_4]{Na_2Cr_2O_7} CH_3-\overset{\displaystyle CH_3}{\overset{|}{CH}}-CH_2-\overset{\displaystyle O}{\overset{||}{C}}-OH.$$

(c) This is a substitution reaction, so $CH_3-CH_2-OH + HCl \rightarrow CH_3-CH_2-Cl + H_2O$.

(d) Sodium is an active metal, so

$$CH_3-\overset{\displaystyle CH_3}{\overset{|}{CH}}-CH_2-OH + Na \rightarrow CH_3-\overset{\displaystyle CH_3}{\overset{|}{CH}}-CH_2-O^-Na^+ + 1/2\ H_2.$$

Aldehydes and Ketones

22.77 **Given:** aldehyde or ketone structures **Find:** name

Conceptual Plan: Simple aldehydes are systematically named according to the number of carbon atoms in the longest continuous carbon chain that contains the carbonyl group. Form the base name from the name of the corresponding alkane by dropping the *-e* and add the ending *-al*. Simple ketones are systematically named according to the longest continuous carbon chain containing the carbonyl group. Form the base name from the name of the corresponding alkane by dropping the letter *-e* and adding the ending *-one*. For ketones, number the chain to give the carbonyl group the lowest possible number. Because an aldehyde always has a carbonyl terminal group, the aldehyde is always at the 1 position even though it is not specified with a number in the name.

Solution:

(a) $\boxed{CH_3-\overset{\displaystyle O}{\overset{||}{C}}-CH_2-CH_3}$ The longest carbon chain has four carbons. The prefix for 4 is but-, and because this is a ketone, the base name is butanone. The position of the carbonyl carbon does not need to be specified because there is only one place it can be in the molecule (if it were on the end carbon, it would be an aldehyde, not a ketone). Because there are no other substituent groups, the name is butanone.

(b) $\boxed{CH_3-CH_2-CH_2-CH_2-\overset{\displaystyle O}{\overset{||}{CH}}}$ The longest carbon chain has five carbons. The prefix for 5 is pent-, and because this is an aldehyde, the base name is pentanal. Because there are no other substituent groups, the name is pentanal.

(c) $\boxed{CH_3-\underset{\displaystyle CH_3}{\overset{\displaystyle CH_3}{\underset{|}{\overset{|}{C}}}}-CH_2-\overset{\displaystyle CH_3}{\overset{|}{CH}}-CH_2-\overset{\displaystyle O}{\overset{||}{C}}}-H$ The longest carbon chain has six carbons. The prefix for 6 is hex-, and because this is an aldehyde, the base name is hexanal. There are three methyl substituent groups.

$\boxed{C^6H_3-\underset{\boxed{CH_3}}{\overset{\boxed{CH_3}}{\underset{|}{\overset{|}{C^5}}}}-C^4H_2-\overset{\boxed{CH_3}}{\overset{|}{C^3H}}-C^2H_2-\overset{\displaystyle O}{\overset{||}{C^1}}}-H$ Start numbering at the carbonyl carbon. The methyl groups are at positions 3, 5, and 5. The name of the molecule is 3,5,5-trimethylhexanal.

(d) 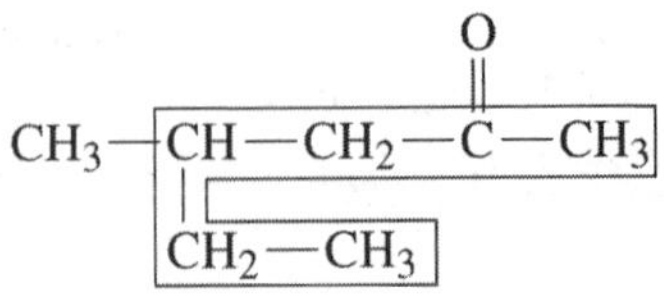 $CH_3-CH-CH_2-\overset{O}{\overset{\|}{C}}-CH_3$ with CH_2-CH_3 attached below CH. The longest carbon chain has six carbons. The prefix for 6 is hex-, and because this is a ketone, the base name is hexanone. There is one methyl substituent group.

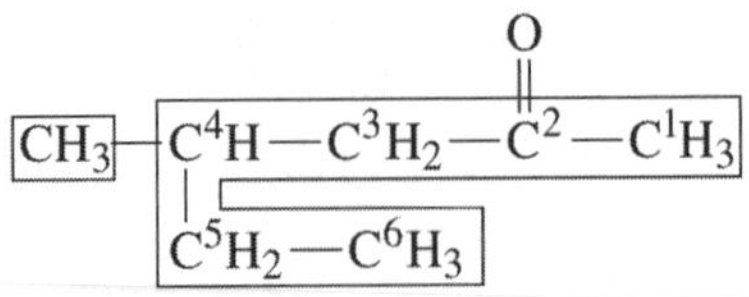

$\boxed{CH_3}-C^4H-C^3H_2-\overset{O}{\overset{\|}{C^2}}-C^1H_3$ with $C^5H_2-C^6H_3$ attached below C^4. Start numbering at the end closest to the carbonyl carbon. The carbonyl carbon is in position 2, and the methyl group is at position 4. The name of the molecule is 4-methyl-2-hexanone.

22.78 **Given:** aldehyde and ketone names **Find:** structure

Conceptual Plan: Find the number of carbon atoms corresponding to prefix of the base name in Table 22.5. → Draw the base chain and number the carbons from left to right. → Add an =O to the specified position. (For aldehydes, put it on the first carbon.) → Using Table 22.6 and the prefix *di-* (2), *tri-* (3), or *tetra-* (4) before the substituent's name, determine each substituent. → Add the substituent to the proper carbon position in the chain. → Add hydrogen atoms to the base chain so that each carbon has four bonds.

Solution:

(a) hexanal. The base name hexanal indicates that there are six carbon atoms in the base chain. $C^1-C^2-C^3-C^4-C^5-C^6$. The *-al* ending specifies that it is an aldehyde, and the carbonyl carbon is on the end. There are no substituent groups, so add hydrogens to the base chain to get the final molecule

$$CH_3-CH_2-CH_2-CH_2-CH_2-\overset{O}{\overset{\|}{C}}H.$$

(b) 2-pentanone. The base name pentanone indicates that there are five carbon atoms in the base chain. $C^1-C^2-C^3-C^4-C^5$. The *-one* ending specifies that it is a ketone. The 2- prefix of the base name designates that the =O group is in the second position. There are no substituent groups, so add hydrogens to the base chain to get the final molecule $CH_3-\overset{O}{\overset{\|}{C}}-CH_2-CH_2-CH_3$.

(c) 2-methylbutanal. The base name butanal indicates that there are four carbon atoms in the base chain. $C^1-C^2-C^3-C^4$. 2-methyl designates that a $-CH_3$ group is in the second position. The *-al* ending specifies that it is an aldehyde, so the =O group is in the first position. $\overset{O}{\overset{\|}{C^1}}-\overset{CH_3}{\overset{|}{C^2}}-C^3-C^4$. Add hydrogens to the base chain to get the final molecule $H-\overset{O}{\overset{\|}{C}}-\overset{CH_3}{\overset{|}{C}H}-CH_2-CH_3$.

(d) 4-heptanone. The base name heptanone indicates that there are seven carbon atoms in the base chain. $C^1-C^2-C^3-C^4-C^5-C^6-C^7$. The *-one* ending specifies that it is a ketone. The 4- prefix of the base name designates that the =O group is in the fourth position. There are no substituent groups, so add hydrogens to the base chain to get the final molecule $CH_3-CH_2-CH_2-\overset{O}{\overset{\|}{C}}-CH_2-CH_2-CH_3$.

22.79 This is an addition reaction,

$$CH_3-CH_2-CH_2-\overset{O}{\overset{\|}{C}}-H + H-C\equiv N \xrightarrow{NaCN} CH_3-CH_2-CH_2-\underset{H}{\underset{|}{\overset{OH}{\overset{|}{C}}}}-C\equiv N.$$

22.80 This is an addition reaction, $CH_3{-}\overset{\overset{\displaystyle O}{\|}}{C}{-}CH_2{-}CH_3 + H{-}C{\equiv}N \xrightarrow{NaCN} CH_3{-}\underset{\underset{\displaystyle CH_2-CH_3}{|}}{\overset{\overset{\displaystyle OH}{|}}{C}}{-}C{\equiv}N$.

Carboxylic Acids and Esters

22.81 **Given:** carboxylic acid or ester structures **Find:** name

Conceptual Plan: Carboxylic acids are systematically named according to the number of carbon atoms in the longest chain containing the —COOH functional group. Form the base name by dropping the *-e* from the name of the corresponding alkane and adding the ending *-oic acid.* Esters are systematically named as if they were derived from a carboxylic acid by replacing the H on the OH with an alkyl group. The R group from the parent acid forms the base name of the compound. Change the *-ic* on the name of the corresponding carboxylic acid to *-ate* and drop *acid.* The R group that replaced the H on the carboxylic acid is named as an alkyl group with the ending *-yl.* The —COOH terminal group is always at the 1 position even though it is not specified with a number in the name.

Solution:

(a) $\boxed{CH_3{-}CH_2{-}CH_2{-}\overset{\overset{\displaystyle O}{\|}}{C}}{-}O{-}\boxed{CH_3}$ The carbon chain that has the carbonyl carbon has four carbons. The prefix for 4 is but-, so the base name is butanoate. The R group that replaces the H of the carboxylic acid is a methyl group. The name is methylbutanoate.

(b) $\boxed{CH_3{-}CH_2{-}\overset{\overset{\displaystyle O}{\|}}{C}}{-}OH$ The carbon chain has three carbons. The prefix for 3 is prop- and there are no other substituents. The name is propanoic acid.

(c) $\boxed{CH_3{-}\underset{\underset{\displaystyle CH_3}{|}}{CH}{-}CH_2{-}CH_2{-}CH_2{-}\overset{\overset{\displaystyle O}{\|}}{C}}{-}OH$ The carbon chain has six carbons. The prefix for 6 is hex- so the base name is hexanoic acid. There is one methyl substituent group.

$\boxed{C^6H_3{-}\underset{\underset{\displaystyle \boxed{CH_3}}{|}}{C^5H}{-}C^4H_2{-}C^3H_2{-}C^2H_2{-}\overset{\overset{\displaystyle O}{\|}}{C^1}}{-}OH$ Number the chain starting with the carbonyl carbon.

The methyl is in the fifth position. The name of the molecule is 5-methylhexanoic acid.

(d) $\boxed{CH_3{-}CH_2{-}CH_2{-}CH_2{-}\overset{\overset{\displaystyle O}{\|}}{C}}{-}O{-}\boxed{CH_2{-}CH_3}$ The carbon chain that has the carbonyl carbon has five carbons. The prefix for 5 is pent-, so the base name is pentanoate. The R group that replaces the H of the carboxylic acid is an ethyl group. The name is ethylpentanoate.

22.82 **Given:** carboxylic acid and ester names **Find:** structure

Conceptual Plan: Find the number of carbon atoms corresponding to the prefix of the base name in Table 22.5. → Draw the base chain and number the carbons from right to left. → Add an =O and an —O to the last carbon on the right. → If the molecule is a carboxylic acid, add a hydrogen atom to the noncarbonyl oxygen. If the molecule is an ester, determine the length of the carbon chain using Table 22.5 and add this to the noncarbonyl oxygen. → Using Table 22.6 and the prefix *di-* (2), *tri-* (3), or *tetra-* (4) before the substituent's name, determine each substituent. → Add the substituent to the proper carbon position in the chain. → Add hydrogen atoms to the base chain so that each carbon has four bonds.

Solution:

(a) pentanoic acid. The base name pentanoic acid indicates that there are five carbon atoms in the base chain. $C^5—C^4—C^3—C^2—C^1$. The -oic acid ending specifies that it is a carboxylic acid. There are no substituent groups, so add hydrogens to the base chain to get the final molecule

$$CH_3—CH_2—CH_2—CH_2—\overset{\overset{\large O}{\|}}{C}—OH.$$

(b) methyl hexanoate. The base name hexanoate indicates that there are six carbon atoms in the base chain. $C^6—C^5—C^4—C^3—C^2—C^1$. Adding the two oxygens to the first carbon, $C^6—C^5—C^4—C^3—C^2—\overset{\overset{\large O}{\|}}{C^1}—O$. The methyl prefix indicates that there is a methyl group on the noncarbonyl oxygen. $C^6—C^5—C^4—C^3—C^2—\overset{\overset{\large O}{\|}}{C^1}—O—\boxed{CH3}$ There are no substituent groups, so add hydrogens to the base chain to get the final molecule $CH_3—CH_2—CH_2—CH_2—CH_2—\overset{\overset{\large O}{\|}}{C}—O—CH_3$.

(c) 3-ethylheptanoic acid. The base name heptanoic acid designates that there are seven carbon atoms in the base chain. $C^7—C^6—C^5—C^4—C^3—C^2—C^1$. The -oic acid ending designates that this is a carboxylic acid with a —COOH ending. 3-ethyl designates that a $—CH_2CH_3$ group is in the third position.

$$C^7—C^6—C^5—C^4—\underset{}{\overset{\boxed{CH_3—CH_2}}{\overset{|}{C^3}}}—C^2—\overset{\overset{\large O}{\|}}{C^1}—OH.$$

Add hydrogens to the base chain to get the final molecule

$$CH_3—CH_2—CH_2—CH_2—\overset{CH_3—CH_2}{\overset{|}{CH}}—CH_2—\overset{\overset{\large O}{\|}}{C}—OH.$$

(d) butyl ethanoate. The base name ethanoate indicates that there are two carbon atoms in the base chain. $C^2—C^1$. Adding the two oxygens to the first carbon, $C^2—\overset{\overset{\large O}{\|}}{C^1}—O$. The butyl prefix indicates that there is a butyl (four carbon) group on the noncarbonyl oxygen. $C^2—\overset{\overset{\large O}{\|}}{C^1}—O—\boxed{CH_2—CH_2—CH_2—CH_3}$. There are no substituent groups, so add hydrogens to the base chain to get the final molecule

$$CH_3—\overset{\overset{\large O}{\|}}{C}—O—CH_2—CH_2—CH_2—CH_3.$$

22.83 The reaction is a condensation reaction. The two noncarbonyl oxygen groups react linking the two molecules (or parts of a molecule) with the concomitant generation of a water molecule.

(a) $$CH_3—CH_2—CH_2—\overset{\overset{\large O}{\|}}{C}—OH + CH_3—CH_2—OH \xrightarrow{H_2SO_4}$$

$$CH_3—CH_2—CH_2—\overset{\overset{\large O}{\|}}{C}—O—CH_2—CH_3 + H_2O$$

(b)

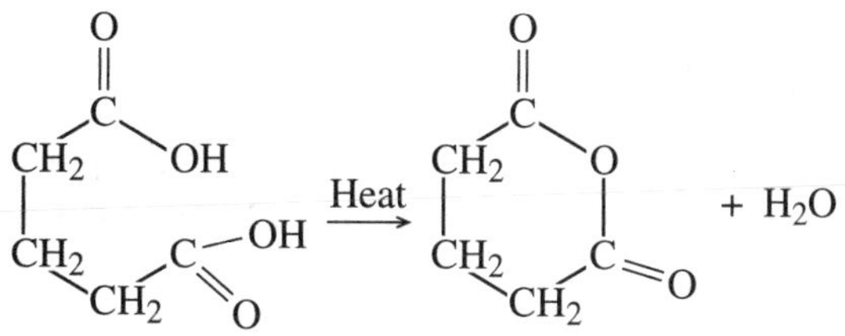

$$\begin{array}{c} \text{O} \\ \| \\ \text{CH}_2\text{—C—OH} \\ | \\ \text{CH}_2\text{—CH}_2\text{—C(=O)—OH} \end{array} \xrightarrow{\text{Heat}} \begin{array}{c} \text{O} \\ \| \\ \text{CH}_2\text{—C—O} \\ | \qquad\quad | \\ \text{CH}_2\text{—CH}_2\text{—C=O} \end{array} + H_2O$$

22.84 (a) This is an acid–base reaction, $CH_3—CH_2—\overset{\overset{\displaystyle O}{\|}}{C}—OH + NaOH \longrightarrow CH_3—CH_2—\overset{\overset{\displaystyle O}{\|}}{C}—O^-Na^+ + H_2O$.

(b) The reaction is a condensation reaction that links the two molecules where the noncarbonyl oxygens are in the reactants and generates water. $CH_3—CH_2—CH_2—\overset{\overset{\displaystyle O}{\|}}{C}—OH + CH_3—CH_2—CH_2—OH \xrightarrow{H_2SO_4}$

$$CH_3—CH_2—CH_2—\overset{\overset{\displaystyle O}{\|}}{C}—O—CH_2—CH_2—CH_3 + H_2O$$

Ethers

22.85 **Given:** ether structures **Find:** name

Conceptual Plan: Common names for ethers have the format (R group 1)(R group 2) ether, where the alkyl groups are those found in Table 22.5. List the names alphabetically. If the two R groups are different, use each of their names. If the two R groups are the same, use the prefix *di-*.

Solution:

(a) $\boxed{CH_3—CH_2—CH_2}—O—\boxed{CH_2—CH_3}$ The left carbon chain has three carbons, so it is a propyl group. The right carbon chain has two carbons, so it is an ethyl group. Listing them alphabetically, the name is ethyl propyl ether.

(b) $\boxed{CH_3—CH_2—CH_2—CH_2—CH_2}—O—\boxed{CH_2—CH_3}$ The left carbon chain has five carbons, so it is a pentyl group. The right carbon chain has two carbons, so it is an ethyl group. Listing them alphabetically, the name is ethyl pentyl ether.

(c) $\boxed{CH_3—CH_2—CH_2}—O—\boxed{CH_2—CH_2—CH_3}$ Both carbon chains have three carbons, so they are propyl groups. The name is dipropyl ether.

(d) $\boxed{CH_3—CH_2}—O—\boxed{CH_2—CH_2—CH_2—CH_3}$ The left carbon chain has two carbons, so it is an ethyl group. The right carbon chain has four carbons, so it is a butyl group. Listing them alphabetically, the name is butyl ethyl ether.

22.86 **Given:** ether names **Find:** structure

Conceptual Plan: Find the number of carbon atoms corresponding to the prefix of the two alkyl groups using Table 22.5. → Draw the two alkyl groups connecting them with an —O— linkage.

Solution:

(a) ethyl propyl ether. The ethyl group has two carbons, and the propyl group has three carbons. Linking the alkyl groups with an —O— linkage, the structure is $CH_3—CH_2—O—CH_2—CH_2—CH_3$.

(b) dibutyl ether. Each alkyl group is a butyl group with four carbons each. Linking the alkyl groups with an —O— linkage, the structure is $CH_3—CH_2—CH_2—CH_2—O—CH_2—CH_2—CH_2—CH_3$.

(c) methyl hexyl ether. The methyl group has one carbon, and the hexyl group has six carbons. Linking the alkyl groups with an —O— linkage, the structure is $CH_3—O—CH_2—CH_2—CH_2—CH_2—CH_2—CH_3$.

(d) dipentyl ether. Each alkyl group is a pentyl group with five carbons each. Linking the alkyl groups with an —O— linkage, the structure is $CH_3—CH_2—CH_2—CH_2—CH_2—O—CH_2—CH_2—CH_2—CH_2—CH_3$.

Amines

22.87 **Given:** amine structures **Find:** name

Conceptual Plan: Amines are systematically named by alphabetically listing the alkyl groups that are attached to the nitrogen atom and adding an *-amine* ending. The alkyl groups are found in Table 22.5. Use the prefix *di-* or *tri-* if the alkyl groups are the same.

Solution:

(a) $\boxed{CH_3-CH_2}-\underset{\displaystyle H}{\underset{|}{N}}-\boxed{CH_2-CH_3}$ Both carbon chains have two carbons, so they are ethyl groups. The name is diethylamine.

(b) $\boxed{CH_3-CH_2-CH_2}-\underset{\displaystyle H}{\underset{|}{N}}-\boxed{CH_3}$ The left carbon chain has three carbons, so it is a propyl group. The right carbon chain has one carbon, so it is a methyl group. Listing them alphabetically, the name is methylpropylamine.

(c) $\boxed{CH_3-CH_2-CH_2}-\overset{\displaystyle \boxed{CH_3}}{\overset{|}{N}}-\boxed{CH_2-CH_2-CH_2-CH_3}$ The left carbon chain has three carbons, so it is a propyl group. The top carbon chain has one carbon, so it is a methyl group. The right carbon chain has four carbons, so it is a butyl group. Listing them alphabetically, the name is butylmethylpropylamine.

22.88 **Given:** amine names **Find:** structure

Conceptual Plan: Find the number of carbon atoms corresponding to the prefix of the two alkyl groups using Table 22.5. → Draw the alkyl groups around a central nitrogen atom and add any necessary hydrogens to give three bonds to the nitrogen.

Solution:

(a) isopropylamine. The isopropyl group is $-\underset{\displaystyle CH_3}{\underset{|}{CH}}-CH_3$. Attach it to a nitrogen and add two hydrogen atoms to the nitrogen to get the final structure $H-\underset{\displaystyle H}{\underset{|}{N}}-\underset{\displaystyle CH_3}{\underset{|}{CH}}-CH_3$.

(b) triethylamine. All three alkyl groups are ethyl groups with two carbons each. Linking the alkyl groups around a nitrogen gives the structure $CH_3-CH_2-\underset{\displaystyle CH_2-CH_3}{\underset{|}{N}}-CH_2-CH_3$.

(c) butylethylamine. The first alkyl group is a butyl group with four carbons, and the other alkyl group is an ethyl group with two carbons. Attach the alkyl groups to a nitrogen and add a hydrogen atom to the nitrogen. The final structure is $CH_3-CH_2-CH_2-CH_2-\underset{\displaystyle H}{\underset{|}{N}}-CH_2-CH_3$.

22.89 (a) This reaction is an acid–base reaction, $CH_3NHCH_3(aq) + HCl(aq) \rightarrow (CH_3)_2NH_2^+(aq) + Cl^-(aq)$.

(b) Because the acid is a carboxylic acid, this reaction is a condensation reaction, $CH_3CH_2NH_2(aq) + CH_3CH_2COOH(aq) \rightarrow CH_3CH_2CONHCH_2CH_3(aq) + H_2O(l)$.

(c) This reaction is an acid–base reaction, $CH_3NH_2(aq) + H_2SO_4(aq) \rightarrow CH_3NH_3^+(aq) + HSO_4^-(aq)$.

22.90 (a) This reaction is an acid–base reaction, $N(CH_2CH_3)_3(aq) + HNO_3(aq) \rightarrow (CH_3CH_2)_3NH^+(aq) + NO_3^-(aq)$.

(b) This reaction is an addition reaction,

$$CH_3-\underset{\displaystyle H}{\underset{|}{N}}-\underset{\displaystyle CH_3}{\underset{|}{CH}}-CH_3\,(aq) + HCN\,(aq) \rightarrow CH_3-\underset{\displaystyle H-N{=}CH}{\underset{|}{N}}-\underset{\displaystyle CH_3}{\underset{|}{CH}}-CH_3\,(aq).$$

(c) Because the acid is a carboxylic acid, this reaction is a condensation reaction, $CH_3{-}NH{-}CH(CH_3){-}CH_3(aq)$ +

$$CH_3{-}CH(CH_3){-}CH_2{-}C({=}O){-}OH(aq) \rightarrow CH_3{-}CH(CH_3){-}CH_2{-}C({=}O){-}N(CH_3){-}CH(CH_3){-}CH_3(aq) + H_2O(l).$$

Polymers

22.91 Because the monomer is $F_2C{=}CF_2$, as the polymer forms, the double bond breaks to link with another monomer.

The structure is $\sim\!\!(CF_2{-}CF_2{-}CF_2{-}CF_2{-}CF_2{-}CF_2{-}CF_2{-}CF_2{-}CF_2{-}CF_2)\!\!\sim$.

22.92 The monomers $HClC{=}CCl_2$ and $H_2C{=}CHCl$ alternate, and as the polymer forms, the double bond breaks to link with an other monomer. The structure is

$$\sim\!\!(CH_2{-}CCl_2{-}CH_2{-}CHCl{-}CH_2{-}CCl_2{-}CH_2{-}CHCl{-}CH_2{-}CCl_2{-}CH_2{-}CHCl{-}CH_2{-}CCl_2{-}CH_2{-}CHCl)\!\!\sim.$$

22.93 In a condensation polymer, atoms are eliminated. Here, water is eliminated. The structure is

$$HO{-}C({=}O){-}C_6H_4{-}C({=}O){-}O{-}CH_2{-}CH_2{-}OH.$$

22.94 In a condensation polymer, atoms are eliminated. Here water is eliminated. The structure is

$$H{-}C({=}O){-}C_6H_4{-}C({=}O){-}N(N){-}C_6H_4{-}NH_2.$$

Cumulative Problems

22.95 **Given:** structures **Find:** identify molecule type and name

Conceptual Plan: Look for functional groups to identify molecule type. Review the appropriate set of naming rules to name the molecule.

Solution:

(a) $CH_3-CH(CH_3)-CH_2-C(=O)-O-CH_3$ The compound has the formula RCOOR′, so it is an ester.

$\boxed{CH_3-CH(CH_3)-CH_2-C(=O)}-O-\boxed{CH_3}$ The left carbon chain has four carbons, which translates to but- and a base name of butanoate. There is a methyl substituent on this chain. The right carbon chain has one carbon, so it is a methyl group. $\boxed{C^4H_3-C^3H(CH_3)-C^2H_2-C^1(=O)}-O-\boxed{CH_3}$ Number the left chain starting with the carbonyl carbon. The methyl group is at position 3. The name of the compound is methyl-3-methylbutanoate.

(b) $\boxed{CH_3-CH_2-CH(CH_3)-CH_2}-O-\boxed{CH_2-CH_3}$ The compound has the formula ROR′, so it is an ether. The left carbon chain has four carbons, so it is a butyl group. There is a methyl substituent on this chain. The right carbon chain has two carbons, so it is an ethyl group.

$\boxed{C^4H_3-C^3H_2-C^2H(\boxed{CH_3})-C^1H_2}-O-\boxed{CH_2-CH_3}$ Number the left chain starting with the carbon next to the oxygen. The methyl substituent is in position 2, so the alkyl group is 2-methylbutyl. Listing the alkyl group alphabetically, the name is ethyl 2-methylbutyl ether.

(c) The molecule is a disubstituted benzene, so it is an aromatic hydrocarbon. The substituents are an ethyl group and a methyl group. If the ethyl position is assigned the first position, the methyl group is in the third position. Listing the functional groups alphabetically, the name is 1-ethyl-3-methylbenzene or *m*-ethylmethylbenzene.

(benzene ring with CH_2-CH_3 and CH_3 substituents)

(d) $\boxed{CH_3-C\equiv C-CH(CH_3)-CH(CH_2-CH_3)-CH_2-CH_3}$ The molecule contains a triple bond, so it is an alkyne. The longest carbon chain is 7. The prefix for 7 is hept-, and the base name is heptyne. The substituent groups are a methyl and an ethyl group. $\boxed{C^1H_3-C^2\equiv C^3-C^4H(\boxed{CH_3})-C^5H(\boxed{CH_2-CH_3})-C^6H_2-C^7H_3}$ If we start numbering the chain at the end closest to the triple bond, the triple bond is between positions 2 and 3, the methyl substituent is assigned number 4, and the ethyl substituent is assigned number 5. Because *e* is before *m*, the name of the compound is 5-ethyl-4-methyl-2-heptyne.

(e) $\boxed{CH_3-CH_2-CH_2-CH(=O)}$ The molecule has the formula RCHO, so it is an aldehyde. The longest carbon chain has four carbons. The prefix for 4 is but-, and because this is an aldehyde, the base name is butanal. Because there are no other substituent groups, the name is butanal.

(f) 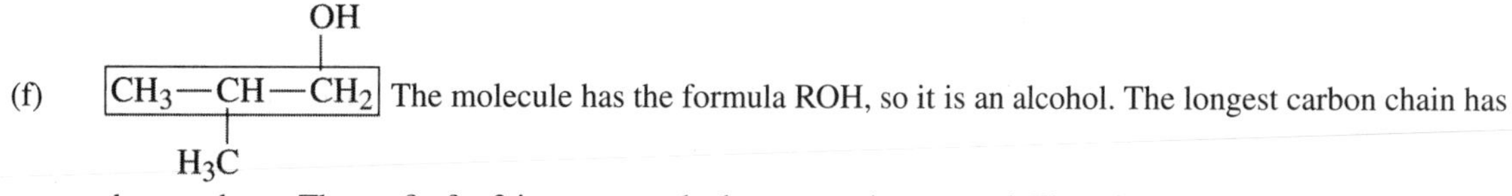 The molecule has the formula ROH, so it is an alcohol. The longest carbon chain has three carbons. The prefix for 3 is prop-, so the base name is propanol. There is a methyl substituent group.

$C^3H_3—C^2H(CH_3)—C^1H_2(OH)$ Start numbering the carbon with the —OH group. The alcohol group is assigned number 1, and the methyl group is assigned number 2. The name is 2-methyl-1-propanol.

22.96 **Given:** structures **Find:** identify molecule type and name

Conceptual Plan: Look for functional groups to identify molecule type. Review the appropriate set of naming rules to name the molecule.

Solution:

(a) $CH_3—HC(CH_3)—C(CH_3)=C(CH_3)—CH_3$ The molecule contains a double bond, so it is an alkene. The longest continuous chain has five carbons. The prefix for 5 is pent-, and the base name is pentene. The substituent groups are three methyl groups. $C^5H_3—HC^4(CH_3)—C^3(CH_3)=C^2(CH_3)—C^1H_3$ Start numbering at the end closest to the double bond. The double bond is between positions 2 and 3. The methyl substituents are assigned the numbers 2, 3, and 4. The name of the compound is 2,3,4-trimethyl-2-pentene.

(b) $CH_3—C(CH_3)_2—CH_2—CH(CH_3)—CH_2—CH_3$ The molecule contains only single bonds, so it is an alkane. The longest continuous chain has six carbons. The prefix for 6 is hex-, and the base name is hexane. The only substituent groups are three methyl groups. $C^1H_3—C^2(CH_3)_2—C^3H_2—C^4H(CH_3)—C^5H_2—C^6H_3$ If we start numbering the chain at the end closest to the left methyl groups, the methyl substituents are assigned the numbers 2, 2, and 4. The name of the compound is 2,2,4-trimethylhexane.

(c) $CH_3—CH_2—CH(CH_3)—CH_2—C(=O)—OH$ The compound has the formula RCOOH, so it is a carboxylic acid. The carbon chain has five carbons. The prefix for 5 is pent-, and because this is a carboxylic acid, the base name is pentanoic acid. There is one methyl substituent group. $C^5H_3—C^4H_2—C^3H(CH_3)—C^2H_2—C^1(=O)—OH$

Number the chain starting with the carbonyl carbon. The methyl is in the third position. The name of the molecule is 3-methylpentanoic acid.

(d) $CH_3—CH(CH_3)—N(H)—CH_2—CH_2—CH_2—CH_3$ The compound has the formula RR′NH, so it is an amine.

The left carbon chain is an isopropyl group. The right carbon chain has four carbons, so it is a butyl group. Listing them alphabetically, the name is butyl isopropylamine.

(e)

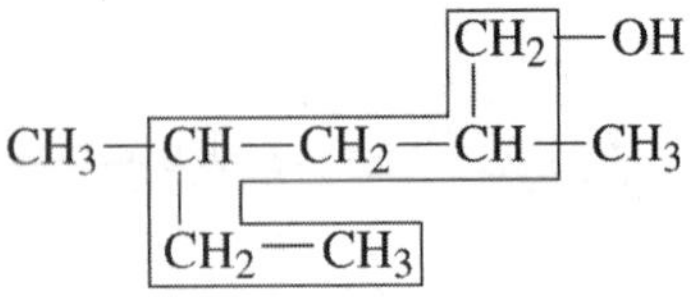

$$\begin{array}{l} \quad\quad\quad\quad\quad\quad\quad\quad\boxed{CH_2}-OH \\ CH_3-\boxed{CH-CH_2-CH}-CH_3 \\ \quad\quad\ \ \boxed{CH_2-CH_3} \end{array}$$

The molecule has the formula ROH, so it is an alcohol. The longest carbon chain has six carbons. The prefix for 6 is hex-, so the base name is hexanol. There are two methyl substituent groups.

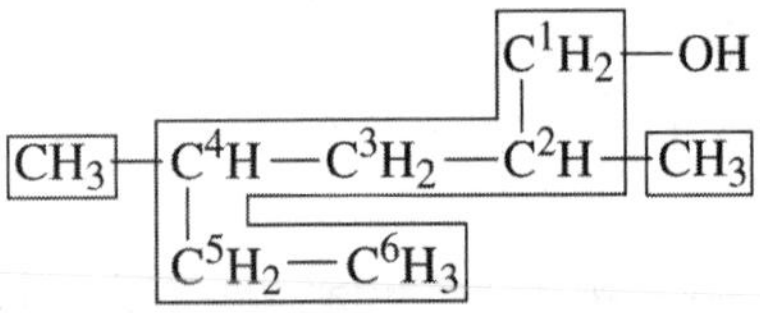

$$\begin{array}{l} \quad\quad\quad\quad\quad\quad\quad\quad\quad\boxed{C^1H_2}-OH \\ \boxed{CH_3}-\boxed{C^4H-C^3H_2-C^2H}-\boxed{CH_3} \\ \quad\quad\quad\ \ \boxed{C^5H_2-C^6H_3} \end{array}$$

Start numbering the carbon with the —OH group. The alcohol group is assigned number 1, and the methyl groups are assigned numbers 2 and 4. The name is 2,4-dimethyl-1-hexanol.

(f)

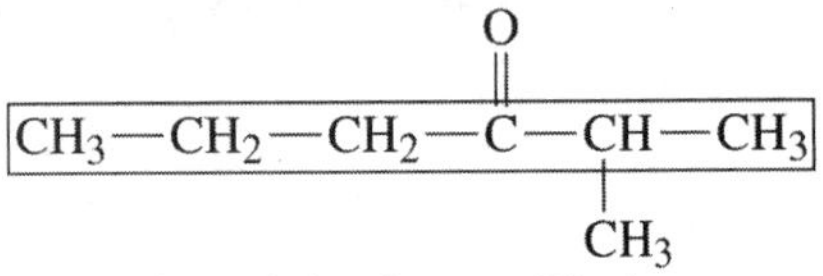

$$\boxed{CH_3-CH_2-CH_2-\overset{\overset{\displaystyle O}{\|}}{C}-\underset{\underset{\displaystyle CH_3}{|}}{CH}-CH_3}$$

The molecule contains a carbonyl carbon in the interior of the molecule, so it is a ketone. The longest carbon chain has six carbons. The prefix for 6 is hex-, and because this is a ketone, the base name is hexanone. There is one methyl substituent group.

$$\boxed{C^6H_3-C^5H_2-C^4H_2-\overset{\overset{\displaystyle O}{\|}}{C^3}-\underset{\underset{\displaystyle CH_3}{|}}{C^2H}-C^1H_3}$$

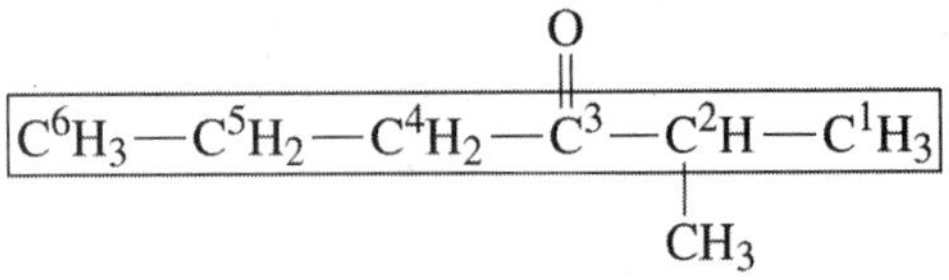

Start numbering at the end closest to the carbonyl. The carbonyl carbon is at position 3. The methyl group is at position 2. The name of the molecule is 2-methyl-3-hexanone.

22.97 **Given:** structures **Find:** name

Conceptual Plan: Look for functional groups to identify molecule type. Review the appropriate set of naming rules to name the molecule.

Solution:

(a)

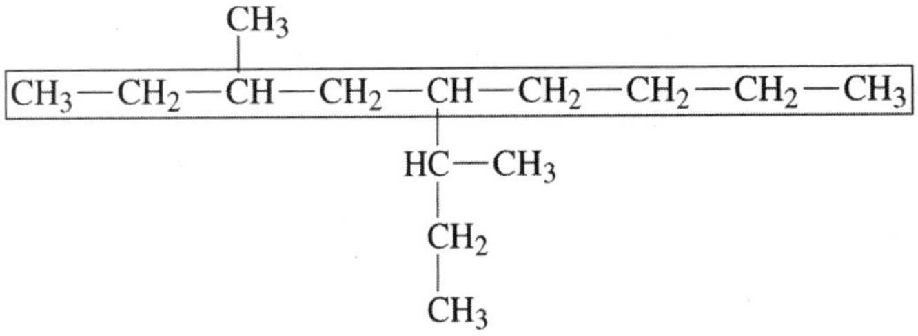

$$\boxed{CH_3-CH_2-\overset{\overset{\displaystyle CH_3}{|}}{CH}-CH_2-\underset{\underset{\underset{\underset{\underset{\displaystyle CH_3}{|}}{CH_2}}{|}}{HC-CH_3}}{|}}{CH}-CH_2-CH_2-CH_2-CH_3}$$

The molecule contains only single bonds, so it is an alkane. The longest continuous chain has nine carbons. The prefix for 9 is non-, and the base name is nonane. The substituent groups are a methyl group and an isobutyl group.

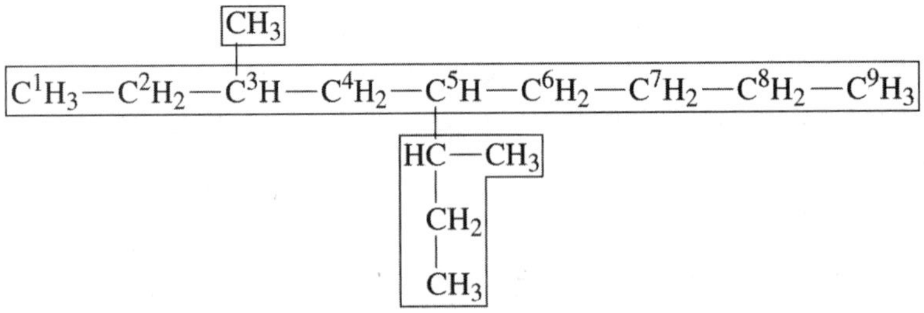

$$\boxed{C^1H_3-C^2H_2-\overset{\overset{\displaystyle \boxed{CH_3}}{|}}{C^3H}-C^4H_2-\underset{\underset{\displaystyle \boxed{\begin{array}{l}HC-CH_3\\ \ \ \ |\\ \ \ CH_2\\ \ \ \ |\\ \ \ CH_3\end{array}}}{|}}{C^5H}-C^6H_2-C^7H_2-C^8H_2-C^9H_3}$$

If we start numbering the chain at the end closest to the methyl group, the methyl substituent is assigned the number 3 and the isobutyl group is assigned the number 5. List the groups alphabetically to get the name of the compound, which is 5-isobutyl-3-methylnonane.

(b)

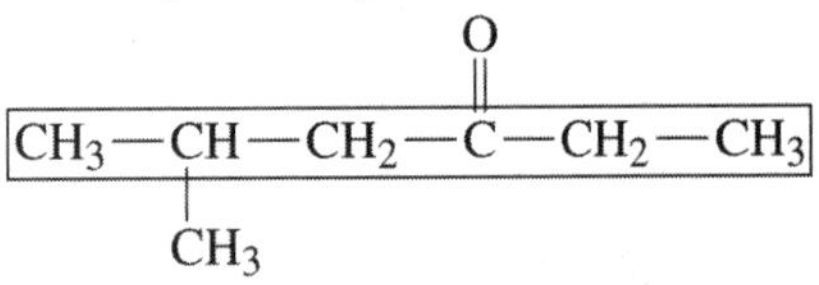

$$\boxed{CH_3-\underset{\underset{\displaystyle CH_3}{|}}{CH}-CH_2-\overset{\overset{\displaystyle O}{\|}}{C}-CH_2-CH_3}$$

The molecule contains a carbonyl carbon in the interior of the molecule, so it is a ketone. The longest carbon chain has six carbons. The prefix for 6 is hex-. Because this is a ketone, the base name is hexanone. There is one methyl substituent group.

$$C^6H_3{-}C^5H(CH_3){-}C^4H_2{-}C^3(=O){-}C^2H_2{-}C^1H_3$$ Start numbering at the end closest to the carbonyl.

The carbonyl carbon is at position 3. The methyl group is at position 5. The name of the molecule is 5-methyl-3-hexanone.

(c) $$CH_3{-}CH(CH_3){-}CH(OH){-}CH_3$$ The molecule has the formula ROH, so it is an alcohol. The longest carbon chain has four carbons. The prefix for 4 is but-, so the base name is butanol. The only substituent group is a methyl group. $$C^4H_3{-}C^3H(CH_3){-}C^2H(OH){-}C^1H_3$$ Start numbering at the end closest to the —OH group. The alcohol group is assigned number 2, and the methyl group is assigned number 3. The name is 3-methyl-2-butanol.

(d) $$CH_3{-}CH(CH_3){-}CH(CH_2CH_3){-}CH(CH_3){-}C{\equiv}C{-}H$$ The molecule contains a triple bond, so it is an alkyne. The longest carbon chain is 6. The prefix for 6 is hex-, and the base name is hexyne. The substituent groups are two methyl groups and an ethyl group. $$C^6H_3{-}C^5H(CH_3){-}C^4H(CH_2CH_3){-}C^3H(CH_3){-}C^2{\equiv}C^1{-}H$$ Start numbering the chain at the end closest to the triple bond, so the triple bond is between position numbers 1 and 2, the methyl substituents are assigned numbers 3 and 5, and the ethyl substituent is assigned number 4. Because *e* comes before *m*, the name of the compound is 4-ethyl-3,5-dimethyl-1-hexyne.

22.98 **Given:** structures **Find:** name

Conceptual Plan: Look for functional groups to identify molecule type. Review the appropriate set of naming rules to name the molecule.

Solution:

(a) $$CH_3{-}CH{=}CH{-}C(CH_3)(CH_2CH_3){-}CH(CH_3){-}CH_2{-}CH_3$$ The molecule contains a double bond, so it is an alkene.

The longest continuous chain has seven carbons. The prefix for 7 is hept-, and the base name is heptene. The substituent groups are two methyl groups and an ethyl group.

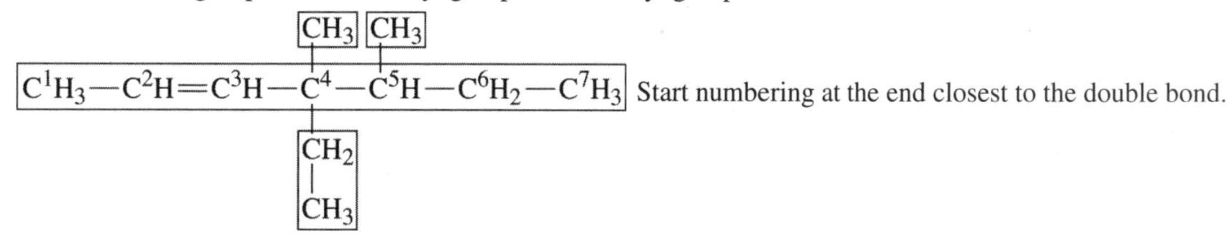

The double bond is between positions 2 and 3. The methyl substituents are assigned numbers 4 and 5, and the ethyl substituent is assigned number 4. Listing the groups alphabetically, the name of the compound is 4-ethyl-4,5-dimethyl-2-heptene.

(b) 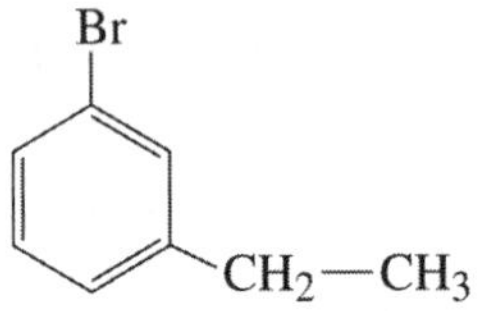 Br (benzene ring with $CH_2—CH_3$ substituent) The molecule is a disubstituted benzene, so it is an aromatic hydrocarbon. The substituents are an ethyl group and a bromo group. The bromo position is assigned the first position, and $—CH_2—CH_3$, the ethyl group, is in the third position. Listing the functional groups alphabetically, the name is 1-bromo-3-ethylbenzene or *m*-bromoethylbenzene.

(c) 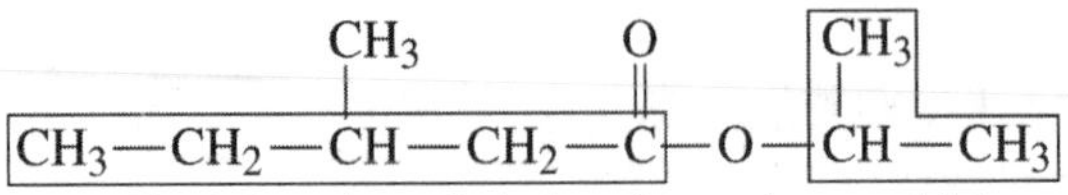 $CH_3—CH_2—CH(CH_3)—CH_2—C(=O)—O—CH(CH_3)—CH_3$ The compound has the formula RCOOR′, so it is an ester. The left carbon chain has five carbons, which translates to pent- and a base name of pentanoate. There is a methyl substituent on this chain. The right carbon chain has three carbons and is an isopropyl group.

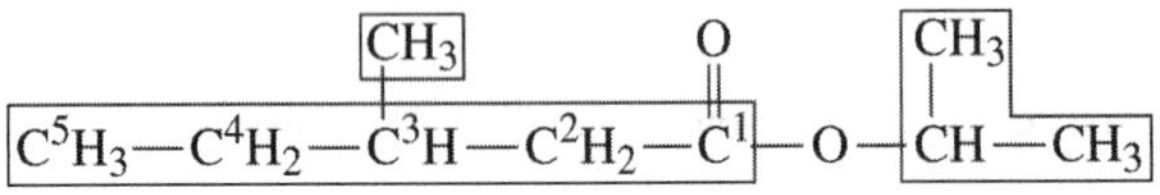

$C^5H_3—C^4H_2—C^3H(CH_3)—C^2H_2—C^1(=O)—O—CH(CH_3)—CH_3$ Number the left chain starting with the carbonyl carbon. The methyl group is at position 3. The name of the compound is isopropyl-3-methylpentanoate.

(d) 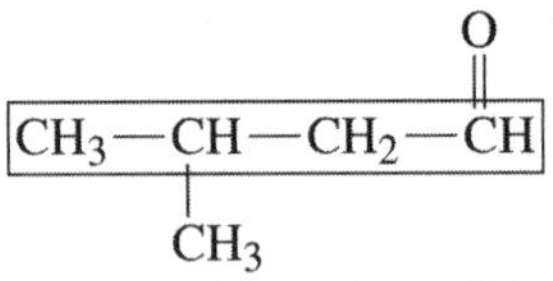 $CH_3—CH(CH_3)—CH_2—CH(=O)$ The molecule has the formula RCHO, so it is an aldehyde. The longest carbon chain has four carbons. The prefix for 4 is but-. Because this is an aldehyde, the base name is butanal. There is a methyl substituent. $C^4H_3—C^3H(CH_3)—C^2H_2—C^1H(=O)$ Start numbering with the carbonyl carbon. The methyl group is assigned number 3. The name is 3-methylbutanal.

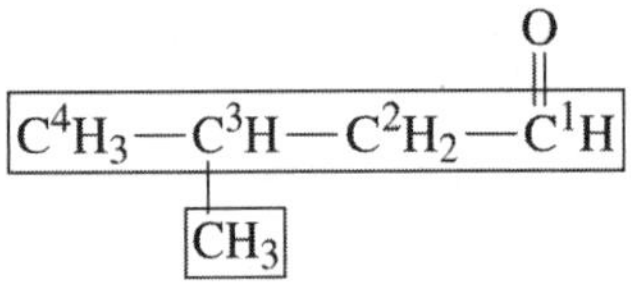

22.99 (a) These structures are structural isomers. The methyl and ethyl groups have been swapped.
(b) These structures are isomers. The second iodo group has been moved.
(c) These structures are the same molecule. The first methyl group is simply drawn in a different orientation.

22.100 (a) These structures are the same molecule. The alkyl groups are drawn in different orientations.
(b) These structures are isomers. In the first structure, there is a methyl substituent on the second carbon from the left end of the molecule. There is no methyl substituent on the second structure.
(c) These structures are the same molecule. The methyl group is drawn in a different orientation, and the orientation of the right carbon is changed.

22.101 **Given:** 15.5 kg 2-butene hydrogenation **Find:** minimum g of H_2 gas
Conceptual Plan: Write a balanced equation kg $C_4H_8 \rightarrow$ g $C_4H_8 \rightarrow$ mol $C_4H_8 \rightarrow$ mol $H_2 \rightarrow$ g H_2.

$$\frac{1000\ g}{1\ kg} \quad \frac{1\ mol\ C_4H_8}{56.10\ g\ C_4H_8} \quad \frac{1\ mol\ H_2}{1\ mol\ C_4H_8} \quad \frac{2.02\ g\ H_2}{1\ mol\ H_2}$$

Solution: $C_4H_8(g) + H_2(g) \rightarrow C_4H_{10}(g)$ then

$$15.5\ \cancel{kg\ C_4H_8} \times \frac{1000\ \cancel{g\ C_4H_8}}{1\ \cancel{kg\ C_4H_8}} \times \frac{1\ \cancel{mol\ C_4H_8}}{56.10\ \cancel{g\ C_4H_8}} \times \frac{1\ \cancel{mol\ H_2}}{1\ \cancel{mol\ C_4H_8}} \times \frac{2.02\ g\ H_2}{1\ \cancel{mol\ H_2}} = 558\ g\ H_2$$

Check: The units (g) are correct. The magnitude of the answer (600) makes physical sense because we are reacting 15,500 g of butane. Hydrogen is lighter, and the molar ratio is 1:1; so the amount of hydrogen will be significant but less than the mass of butene.

22.102 **Given:** 3.8 kg *n*-octane combustion **Find:** kg of CO_2 gas
Conceptual Plan: Write a balanced equation
kg $C_8H_{18} \rightarrow$ g $C_8H_{18} \rightarrow$ mol $C_8H_{18} \rightarrow$ mol $CO_2 \rightarrow$ g $CO_2 \rightarrow$ kg CO_2.

$$\frac{1000\ g}{1\ kg} \quad \frac{1\ mol\ C_8H_{18}}{114.22\ g\ C_8H_{18}} \quad \frac{8\ mol\ CO_2}{1\ mol\ C_8H_{18}} \quad \frac{44.01\ g\ CO_2}{1\ mol\ CO_2} \quad \frac{1\ kg}{1000\ g}$$

Solution: $C_8H_{18}(g) + 25/2\,O_2(g) \rightarrow 8\,CO_2(g) + 9\,H_2O(g)$ then

$$3.8\,\cancel{kg\,C_8H_{18}} \times \frac{1000\,\cancel{g\,C_8H_{18}}}{1\,\cancel{kg\,C_8H_{18}}} \times \frac{1\,\cancel{mol\,C_8H_{18}}}{114.22\,\cancel{g\,C_8H_{18}}} \times \frac{8\,\cancel{mol\,CO_2}}{1\,\cancel{mol\,C_8H_{18}}} \times \frac{44.01\,\cancel{g\,CO_2}}{1\,\cancel{mol\,CO_2}} \times \frac{1\,kg\,CO_2}{1000\,\cancel{g\,CO_2}} = 12\,kg\,CO_2$$

Check: The units (g) are correct. The magnitude of the answer (12) makes physical sense because we are reacting 4 kg of octane. Because all of the carbon is now bound to oxygen, which is much heavier than hydrogen, the weight increases dramatically.

22.103 (a) Combustion reaction—reaction with oxygen to form CO_2 and H_2O
(b) Alkane (halogen) substitution—reaction with halogen gas to substitute a hydrogen with a halogen
(c) Alcohol elimination—ROH reaction with concentrated acid to eliminate water from the alcohol to produce an alkene
(d) Aromatic (halogen) substitution—benzene reaction with halogen gas to substitute a hydrogen with a halogen

22.104 (a) Hydrogenation reaction— $CH_3—CH_2—C(CH_3)=CH_2 + H_2 \rightarrow CH_3—CH_2—CH(CH_3)—CH_3$

(b) Halogen substitution— $CH_3—CH_2—CH_2—CH_2—OH + HCl \rightarrow CH_3—CH_2—CH_2—CH_2—Cl + H_2O$

(c) Condensation reaction— $CH_3—CH_2—CH(CH_3)—CH_2—C(=O)—OH + CH_3—CH_2—OH \rightarrow$

$CH_3—CH_2—CH(CH_3)—CH_2—C(=O)—O—CH_2—CH_3 + H_2O$

(d) Amine acid–base reaction— $CH_3—CH_2—CH_2—CH_2—NH—CH_2—CH_3 + HCl \rightarrow$

$CH_3—CH_2—CH_2—CH_2—N^+H_2—CH_2—CH_3\ Cl^-$

22.105 **Given:** alkene names **Find:** structure

Conceptual Plan: Find the number of carbon atoms corresponding to the prefix of the base name in Table 22.5. → Draw the base chain and number the carbons from left to right. → Place the double bond in the appropriate position in the chain. → Using Table 22.6 and the prefix *di-* (2), *tri-* (3), or *tetra-* (4) before the substituent's name, determine each substituent. → Add the substituent to the proper carbon position in the chain. → Add hydrogen atoms to the base chain so that each carbon has four bonds. → Look for substitutions around double bonds to determine whether cis–trans isomerism is applicable and look for carbons with four different alkyl groups attached.

Solution:

(a) 3-methyl-1-pentene. The base name pentene designates that there are five carbon atoms in the base chain. $C^1—C^2—C^3—C^4—C^5$. The -ene ending designates that there is a double bond, and the 1 prefix designates that it is between the first and second positions. $C^1=C^2—C^3—C^4—C^5$. 3-methyl designates that there is a $—CH_3$ group in the third position. $C^1=C^2—C^3(\boxed{CH_3})—C^4—C^5$. Add hydrogens to the base chain to get the final molecule $CH_2=CH—CH(CH_3)—CH_2—CH_3$. There is optical stereoisomerism because the third carbon has four different groups attached.

(b) 3,5-dimethyl-2-hexene. The base name hexene designates that there are six carbon atoms in the base chain. $C^1—C^2—C^3—C^4—C^5—C^6$. The -ene ending designates that there is a double bond, and the 2 prefix designates that it is between the second and third positions. $C^1—C^2=C^3—C^4—C^5—C^6$. 3,5-dimethyl designates that there are $—CH_3$ groups in the third and fifth positions. $C^1—C^2=C^3(\boxed{CH_3})—C^4—C^5(\boxed{CH_3})—C^6$.

Add hydrogens to the base chain to get the final molecule $CH_3-CH=C(CH_3)-CH_2-CH(CH_3)-CH_3$. Cis–trans geometric isomerism is possible around the double bond because each carbon in the double bond has two different groups attached.

(c) 3-propyl-2-hexene. The base name hexene designates that there are six carbon atoms in the base chain. $C^1-C^2-C^3-C^4-C^5-C^6$. The -ene ending designates that there is a double bond, and the 2 prefix designates that it is between the second and third positions. $C^1-C^2=C^3-C^4-C^5-C^6$. 3-propyl designates that there is a $-CH_2CH_2CH_3$ group in the third position. $C^1-C^2=C^3(CH_2-CH_2-CH_3)-C^4-C^5-C^6$.

Add hydrogens to the base chain to get the final molecule $CH_3-CH=C(CH_2-CH_2-CH_3)-CH_2-CH_2-CH_3$. Cis–trans geometric isomerism is not possible around the double bond because the two groups on the third carbon are the same.

22.106 (c) and (d) exhibit optical stereoisomerism because there are carbons that have four different groups attached. The structures are $CH_3-CH(OH)-CH(CH_3)-CH_2-CH_3$ for 3-methyl-2-pentanol (carbons 2 and 3 are both optically active) and $CH_3-CH(CH_3)-CH(OH)-CH_2-CH_3$ for 2-methyl-3-pentanol (carbon 3 is optically active).

22.107 The 11 isomers are $HC(=O)-CH_2-CH_2-CH_3$ = aldehyde; $CH_3-C(=O)-CH_2-CH_3$ = ketone;

$CH_2=C-CH_2-CH_2-OH$ = alcohol and alkene; $CH_3-CH=CH-CH_2-OH$ = alcohol and alkene;

$CH_3-CH_2-CH=CH-OH$ = alcohol and alkene; $CH_3-CH_2-O-CH=CH_2$ = ether and alkene;

$CH_3-O-CH_2-CH=CH_2$ = ether and alkene; $CH_3-O-CH=CH-CH_3$ = ether and alkene;

$CH_3-CH_2-C(OH)=CH_2$ = alcohol and alkene; $CH_3-CH(OH)-CH=CH_2$ = alcohol and alkene; and

$CH_3-C(OH)=CH-CH_3$ = alcohol and alkene. There is also a twelfth isomer that is cyclic: H_2C-CH_2 / H_2C CH_2 joined through O (ring of four CH_2 and O) = ether.

22.108 There are seven isomers: five are amines and two are amides (the amine and carbonyl groups are on adjacent carbons), which are $HC(=O)-CH_2-CH_2-NH_2$ = aldehyde, $CH_3-C(=O)-CH_2-NH_2$ = ketone,

$CH_3-CH_2-C(=O)-NH_2$ = ketone, $HC(=O)-CH_2-N(H)-CH_3$ = aldehyde, $HC(=O)-N(CH_3)-CH_3$ = amide,

$HC(=O)-N(H)-CH_2-CH_3$ = amide, and $CH_3-C(=O)-N(H)-CH_3$ = amide.

22.109 In the acid form of the carboxylic acid, electron withdrawal by the C=O enhances acidity. The conjugate base, the carboxylate anion, is stabilized by resonance; so the two O atoms are equivalent and bear the negative charge equally. This resonance stability is not possible in an alcohol, where the carbonyl group is absent.

22.110 The more *s*-like an orbital, the more easily a negative charge can be stabilized due to its closer proximity to the nucleus. The alkyne has an *sp* hybrid orbital (50% *s*-character), whereas the alkene is an sp^2 hybrid orbital (33% *s*-character).

Challenge Problems

22.111 This is an internal condensation reaction. It requires only heat to cause it to happen.

(a)

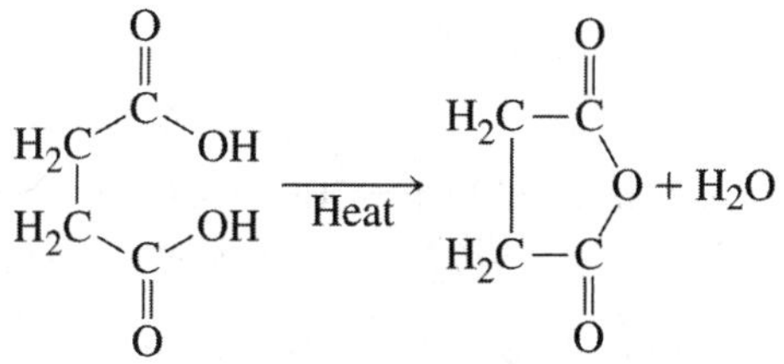

$$H_2C(COOH)-H_2C(COOH) \xrightarrow{\text{Heat}} \text{cyclic anhydride } (H_2C-C(=O)-O-C(=O)-CH_2) + H_2O$$

(b) This is a dehydration reaction,

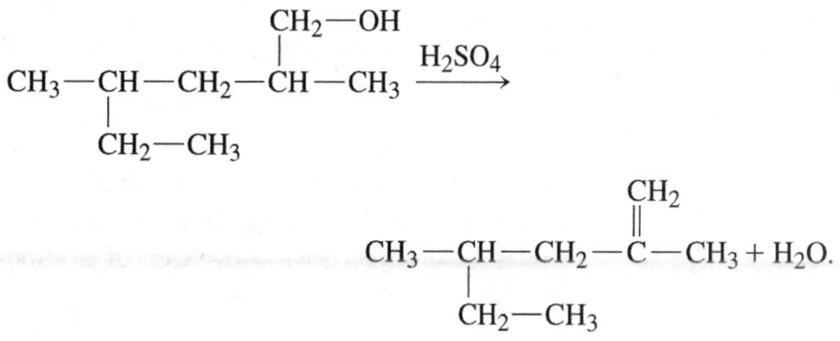

$$CH_3-CH(CH_2-CH_3)-CH_2-CH(CH_2-OH)-CH_3 \xrightarrow{H_2SO_4} CH_3-CH(CH_2-CH_3)-CH_2-C(=CH_2)-CH_3 + H_2O.$$

(c) This is an addition reaction that follows Markovnikov's rule,

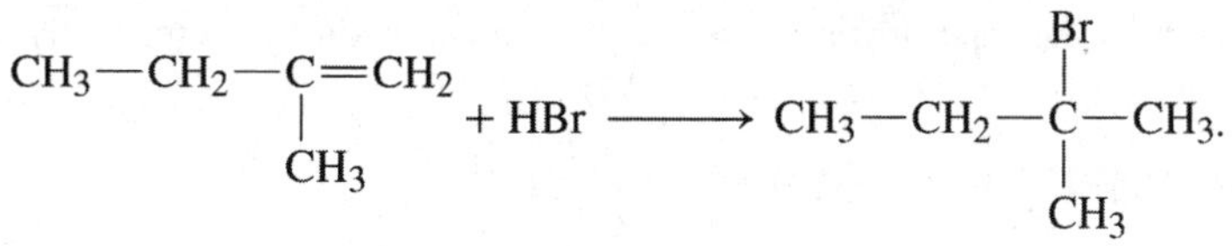

$$CH_3-CH_2-C(CH_3)=CH_2 + HBr \longrightarrow CH_3-CH_2-CBr(CH_3)-CH_3.$$

22.112

$$CH_3-CH(CH_3)-CH_2-CH_2-OH \xrightarrow[H_2SO_4]{\text{(a) } Na_2Cr_2O_7} CH_3-CH(CH_3)-CH_2-\overset{O}{\overset{\|}{C}}-OH \text{ (b)} \xrightarrow{\text{(c) } CH_3CH_2OH}$$

$$CH_3-CH(CH_3)-CH_2-\overset{O}{\overset{\|}{C}}-O-CH_2-CH_3 \xrightarrow{\text{(d) } Cl_2} CH_3-CCl(CH_3)-CH_2-\overset{O}{\overset{\|}{C}}-O-CH_2-CH_3 + HCl$$

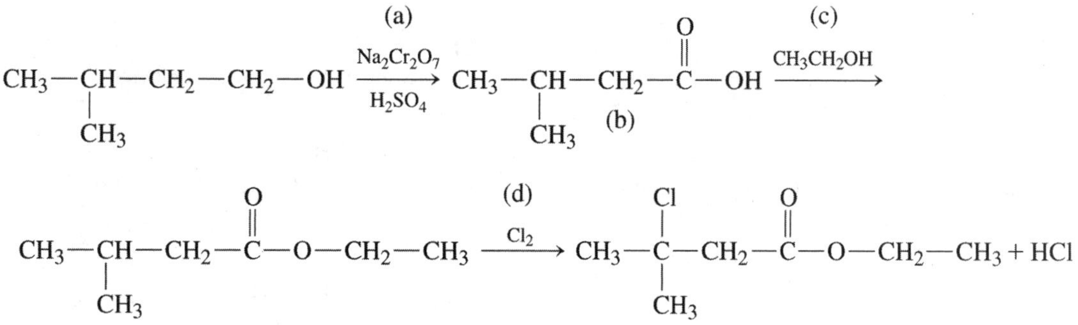

22.113 (a) Because there are six primary hydrogen atoms and two secondary hydrogen atoms, if they are equally reactive, we expect a ratio of 6:2 or 3:1.

(b) Assume that we generate 100 product molecules. The yield = (# hydrogen atoms)(reactivity). So 1° = 45 = (3)(reactivity 1°) and 2° = 55 = (1)(reactivity 2°). Taking the ratio, 2°:1° = 55:(45/3) = 55:15 = 11:3. The 2° hydrogens are much more reactive.

22.114 The two isomers are $CH_3—CH_2—CH_2—CH_3$ and $CH_3—CH(CH_3)—CH_3$. The products are

$CHCl_2—CH_2—CH_2—CH_3$, $CH_3—CCl_2—CH_2—CH_3$, $CH_2Cl—CHCl—CH_2—CH_3$,

$CH_2Cl—CH_2—CHCl—CH_3$, $CH_2Cl—CH_2—CH_2—CH_2Cl$, and $CH_3—CHCl—CHCl—CH_3$.

There are six products from the first isomer. The other products are $CHCl_2—CH(CH_3)—CH_3$,

$CH_2Cl—CH(CH_2Cl)—CH_3$, and $CH_2Cl—CCl(CH_3)—CH_3$. There are three products from the second isomer.

There are a total of 6 + 3 = 9 products.

22.115 The chiral products (that have four different groups on a carbon) are $CH_2Cl—CHCl—CH_2—CH_3$ (second carbon), $CH_2Cl—CH_2—CHCl—CH_3$ (third carbon), and $CH_3—CHCl—CHCl—CH_3$ (second and third carbons).

22.116 Nitromethane's two most important structures are $H—CH_2—N^{+}(=O)—O^{-}$ and $H—CH_2—N^{+}(—O^{-})=O$.

(a) The hybridization of the C atom is sp^3. The other hybrid orbitals are three sp^2 orbitals on the N atom, three sp^2 orbitals on the O atom with the double bond, and four sp^3 orbitals on the O atom with the single bond.
(b) The N═O bond is the shortest because multiple bonds are shorter than single bonds.
(c) The N═O bond is the strongest because two pairs of electrons are shared.
(d) The bond angles are less than 109.5° because the $—NO_2$ group is bulkier than the —H.

22.117 The first propagation step for F is very rapid and exothermic because of the strength of the H—F bond that forms. For I, the first propagation step is endothermic and slow because the H—I bond that forms is relatively weak. In addition, the C—F bond is much stronger than C—I (477 kJ versus 241 kJ).

22.118 The structure with no multiple bonds is cyclopropane: $(CH_2)_3$ (three-membered ring of C atoms, each carrying two H atoms).

The structure is much less stable because of the strain due to bond angles in the three-membered ring.

22.119 The three structures that have no dipole moment are $Cl—CH_2—CH_2—Cl$, trans-$HClC=CClH$ (H and Cl on opposite sides), and Cl—C≡C—Cl.

The structures with dipole moments are $H—CH_2—CHCl—Cl$, cis-$ClHC=CHCl$ (Cl atoms on the same side), and $H_2C=CCl_2$.

Conceptual Problems

22.120 (a) $CH_3—CH(=O)$ is more oxidized because the oxygen is double-bonded to the carbon.
(b) $CH_3—CH_2—OH$ is more oxidized because it has an oxygen.
(c) $CH_3—CH_2—C(=O)—OH$ is more oxidized because it has more oxygens.

22.121 For the structure to have only one product after a single bromination, all of the hydrogens must be equivalent.

The structure is

$$\begin{array}{ccccccc} & & CH_3 & & CH_3 & & \\ & & | & & | & & \\ CH_3 & — & C & — & C & — & CH_3 \\ & & | & & | & & \\ & & CH_3 & & CH_3 & & \end{array}$$

To name it, find the longest carbon chain:

$$\begin{array}{ccccccc} & & CH_3 & & CH_3 & & \\ & & | & & | & & \\ \hline CH_3 & — & C & — & C & — & CH_3 \\ \hline & & | & & | & & \\ & & CH_3 & & CH_3 & & \end{array}$$

. Because there are four carbons in this chain and it is an alkane, the base name is butane.

There are four methyl substituent groups.

$$\begin{array}{ccccccc} & & \boxed{CH_3} & & \boxed{CH_3} & & \\ & & | & & | & & \\ \hline C^1H_3 & — & C^2 & — & C^3 & — & C^4H_3 \\ \hline & & | & & | & & \\ & & \boxed{CH_3} & & \boxed{CH_3} & & \end{array}$$

It does not matter how the chain is numbered because it is symmetrically substituted. The methyl groups are assigned the numbers 2, 2, 3, and 3. The name is 2,2,3,3-tetramethylbutane.

22.122 (b) and (d) are chiral because each has one carbon atom with four different groups attached.

23 Transition Metals and Coordination Compounds

Review Questions

23.1 A transition metal atom forms an ion by losing the ns^2 (valence shell) electrons first.

23.2 Transition metals can have multiple oxidation states because they lose the ns^2 (valence shell) electrons and then can lose $(n - 1)d^x$ electrons.

23.3 The +2 oxidation state is common because most of the transition metals have two electrons occupying the *ns* orbitals. These electrons are lost first by the metal.

23.4 The third transition row elements are not larger than the second row transition elements because 14 of the 32 electrons added have gone into a $(n - 2)f$ sublevel, and although electrons in *f* orbitals are in lower principal quantum levels, they are not very effective at shielding the outer electrons from nuclear charge. Consequently, the outer electrons are held more tightly by the nucleus, offsetting the typical increase in size between the periods—an effect called the lanthanide contraction.

23.5 The electronegativity of the transition elements generally increases across a row, following the main-group trend. However, in contrast to the main-group trend, electronegativity increases from the first transition row to the second. There is little electronegativity difference between the second and third transition row. There is a slight increase from silver to gold. Therefore, Au is the most electronegative of the transition metals.

23.6 (a) The coordination number is the number of molecules or ions directly bound to the metal atom. It is the secondary valence of the metal atom.
(b) A ligand is a Lewis base (or electron donor) that forms a bond with the metal.
(c) Bidentate ligands can donate two pairs of electrons from two different atoms to the metal. Polydentate ligands can donate even more than two electron pairs from more than two atoms to the metal.
(d) A complex ion contains the metal ion in the center and the ligands—which can be neutral molecules or ions—arranged around it.
(e) A chelating agent is a ligand that is either a bidentate or polydentate ligand.

23.7 A ligand can be considered a Lewis base because it donates a pair of electrons. The transition metal ion would be a Lewis acid because it accepts the pair of electrons.

23.8 (a) In structural isomers, atoms are connected to one another in different ways, whereas in stereoisomers, atoms are connected the same way but the ligands have a different spatial arrangement about the metal atom.
(b) Linkage isomers occur when the ligands can coordinate to the metal in different orientations. Coordination isomers occur when a coordinated ligand exchanges places with the uncoordinated counterion.
(c) Geometric isomers result when the ligands bonded to the metal have a different spatial arrangement. Optical isomers are nonsuperimposable mirror images of one another.

23.9 Cis–trans isomerism occurs in square planar complexes of the general formula MA_2B_2 or octahedral complexes of the general formula MA_4B_2.

23.10 A complex ion will be optically active if the mirror image is nonsuperimposable on the original molecule.

23.11 Because of the spatial arrangement of the ligands, the normally degenerate *d* orbitals are split in energy. The difference between these split *d* orbitals is the crystal field splitting energy. The magnitude of the splitting depends on the particular complex. In strong-field complexes, the splitting is large; in weak-field complexes, the splitting is small. The magnitude of the crystal field splitting depends in large part on the ligands attached to the central metal ion.

23.12 Sc^{3+} has an electron configuration of [Ar], and Ti^{3+} has an electron configuration of $[\mathrm{Ar}]3d^1$. It is the presence of *d* electrons that gives Ti^{3+} compounds their color.

23.13 Zn^{2+} has a filled *d* subshell, and Cu^{2+} has nine *d* electrons. Because Zn^{2+} has this filled *d* subshell, the color of the compounds will be white. Cu^{2+} has an incomplete *d* subshell; so different ligands will cause a different crystal field splitting, and the compounds will have color.

23.14 Strong-field ligands cause a large split in the *d* orbitals, and it will take more energy to occupy the higher energy level than it does to pair the electrons in the lower energy level. Weak-field ligands cause a small split in the *d* orbital energy, so the electron pairing energy is greater than the energy of the split. Thus, complexes with strong-field ligands have fewer unpaired electrons relative to the free metal ion and are called low-spin complexes. Complexes with weak-field ligands have the same number of unpaired electrons as the free metal ion and are called high-spin complexes.

23.15 Almost all tetrahedral complexes are high-spin because of reduced ligand–metal interactions. The *d* orbitals in a tetrahedral complex are interacting with only four ligands, as opposed to six in the octahedral complex; so the value of Δ is generally smaller.

23.16 According to crystal field theory, the ligands cause the normally degenerate *d* orbitals to split in energy. The energy difference, the crystal field splitting energy Δ, is the energy absorbed by the complex that corresponds to a specific wavelength (color) of visible light. The color seen is the complementary color to the color absorbed.

Problems by Topic

Properties of Transition Metals

23.17 Identify the noble gas that precedes the element and put it in square brackets.
Determine the outer principal quantum level for the *s* orbital. Subtract 1 to obtain the quantum level for the *d* orbital. If the element is in the third or fourth transition series, include $(n - 2)f$ electrons in the configuration.
Count across the row to see how many electrons are in the neutral atom.
For an ion, remove the required number of electrons first from the *s* orbitals and then from the *d* orbitals.

(a) Ni; Ni^{2+}

The noble gas that precedes Ni is Ar. Ni is in the fourth period; so the orbitals we use are 4*s* and 3*d*, and Ni has 10 more electrons than Ar does.

Ni $[\mathrm{Ar}]4s^23d^8$

Ni will lose electrons from the 4*s* and then from the 3*d*.

Ni^{2+} $[\mathrm{Ar}]4s^03d^8$

(b) Mn; Mn^{4+}

The noble gas that precedes Mn is Ar. Mn is in the fourth period; so the orbitals we use are 4*s* and 3*d*, and Mn has seven more electrons than Ar does.

Mn $[\mathrm{Ar}]4s^23d^5$

Mn will lose electrons from the 4s and then from the 3*d*.

Mn^{4+} $[\mathrm{Ar}]4s^03d^3$

(c) Y; Y^+

The noble gas that precedes Y is Kr. Y is in the fifth period; so the orbitals we use are 5*s* and 4*d*, and Y has three more electrons than Kr does.

Y $[Kr]5s^2 4d^1$

Y will lose electrons from the 5*s* and then from the 4*d*.

Y^+ $[Kr]5s^1 4d^1$

(d) Ta; Ta^{2+}

The noble gas that precedes Ta is Xe. Ta is in the sixth period; so the orbitals we use are 6*s*, 5*d*, and 4*f*, and Ta has 19 more electrons than Xe does.

Ta $[Xe]6s^2 4f^{14} 5d^3$

Ta will lose electrons from the 6*s* and then from the 5*d*.

Ta^{2+} $[Xe]6s^0 4f^{14} 5d^3$

23.18 Identify the noble gas that precedes the element and put it in square brackets.
Determine the outer principal quantum level for the *s* orbital. Subtract 1 to obtain the quantum level for the *d* orbital. If the element is in the third or fourth transition series, include $(n - 2)f$ electrons in the configuration.
Count across the row to see how many electrons are in the neutral atom.
For an ion, remove the required number of electrons first from the *s* orbitals and then from the *d* orbitals.

(a) Zr; Zr^{2+}

The noble gas that precedes Zr is Kr. Zr is in the fifth period; so the orbitals we use are 5*s* and 4*d*, and Zr has four more electrons than Kr does.

Zr $[Kr]5s^2 4d^2$

Zr will lose electrons from the 5*s* and then from the 4*d*.

Zr^{2+} $[Kr]5s^0 4d^2$

(b) Co; Co^{2+}

The noble gas that precedes Co is Ar. Co is in the fourth period; so the orbitals we use are 4*s* and 3*d*, and Co has nine more electrons than Ar does.

Co $[Ar]4s^2 3d^7$

Co will lose electrons from the 5*s* and then from the 4*d*.

Co^{2+} $[Ar]4s^0 3d^7$

(c) Tc; Tc^{3+}

The noble gas that precedes Tc is Kr. Tc is in the fifth period; so the orbitals we use are 5*s* and 4*d*, and Tc has seven more electrons than Kr does.

Tc $[Kr]5s^2 4d^5$

Tc will lose electrons from the 5*s* and then from the 4*d*.

Tc^{3+} $[Kr]5s^0 4d^4$

(d) Os; Os^{4+}

The noble gas that precedes Os is Xe. Os is in the sixth period; so the orbitals we use are 6*s*, 5*d*, and 4*f*, and Os has 22 more electrons than Xe does.

Os $[Xe]6s^2 4f^{14} 5d^6$

Os will lose electrons from the 6*s* and then from the 5*d*.

Os^{4+} $[Xe]6s^0 4f^{14} 5d^4$

23.19 (a) V Highest oxidation state $= +5$. V $= [Ar]4s^2 3d^3$. Because V is to the left of Mn, it can lose all of the 4*s* and 3*d* electrons; so the highest oxidation state is +5.

(b) Re Highest oxidation state $= +7$. Re $= [Xe]6s^2 4f^{14} 5d^5$. Re can lose all of the 6*s* and 5*d* electrons, so the highest oxidation state is +7.

(c) Pd Highest oxidation state $= +4$. Pd $= [Kr]4d^{10}$. Metal ions with a d^8 electron configuration exhibit a square planar geometry. Pd can lose 4 electrons from the 5*s* and 4*d* orbitals, so the highest oxidation state is +4.

23.20 Reference Figure 23.4.

(a) +3 In the first row transition metals, the elements that can have a highest oxidation state of +3 are scandium, nickel, and copper.

(b) +7 In the first row transition metals, the element that can have a highest oxidation state of +7 is manganese.

(c) +4 In the first row transition metals, the elements that can have a highest oxidation state of +4 are titanium and cobalt.

Coordination Compounds

23.21 (a) $[Cr(H_2O)_6]^{3+}$ H_2O is neutral, so Cr has an oxidation state of +3. Six H_2O molecules are attached to each Cr, so the coordination number is 6.

(b) $[Co(NH_3)_3Cl_3]^-$ NH_3 is neutral, and Cl has charge of 1−. The sum of the oxidation state of Co and the charge of chloride ion = −1. $x + (3(-1)) = 1-$, $x = +2$; therefore, the oxidation state of Co is +2. The three NH_3 molecules and the three Cl^- ions are bound directly to the Co atom; therefore, the coordination number is 6.

(c) $[Cu(CN)_4]^{2-}$ CN has a charge of 1−. The sum of the oxidation state of Cu and the charge of the cyanide ion = 2−. $x + (4(-1)) = 2-$, $x = +2$; therefore, the oxidation state of Cu is +2. The four cyanide ions are directly bound to the Cu atom; therefore, the coordination number is 4.

(d) $[Ag(NH_3)_2]^+$ NH_3 is neutral, so Ag has an oxidation number of +1. Two NH_3 molecules are attached to each Ag atom, so the coordination number is 2.

23.22 (a) $[Co(NH_3)_5Br]^{2+}$ NH_3 is neutral, and Br has a charge of 1−. The sum of the oxidation state of Co and the charge of the bromide ion = 2+. $x + (1(-1)) = 2+$, $x = +3$; therefore, the oxidation state of Co is +3. The five NH_3 molecules and the bromide ion are directly bound to the Co atom; therefore, the coordination number is 6.

(b) $[Fe(CN)_6]^{4-}$ CN has a charge of 1−. The sum of the oxidation state of Fe and the charge of the cyanide ions = 4−. $x + (6(-1)) = 4-$, $x = +2$; therefore, the oxidation state of Fe is +2. The six cyanide ions are directly bound to the Fe atom; therefore, the coordination number is 6.

(c) $[Co(ox)_3]^{4-}$ ox (oxalate ion) has a charge of 2−. The sum of the oxidation state of Co and the charge of the oxalate ion = 4−, $x + (3(-2)) = 4-$, $x = +2$; therefore, the oxidation state of Co is +2. Oxalate ion is a bidentate ligand, and the oxalate ions are directly bound to the Co atom; therefore, the coordination number is 6.

(d) $[PdCl_4]^{2-}$ Cl has a charge of 1−. The sum of the oxidation state of Pd and the charge on the chloride ions = 2−. $x + (4(-1)) = 2-$, $x = +2$; therefore, the oxidation state of Pd is +2. The four chloride ions are directly bound to the Pd atom; therefore, the coordination number is 4.

23.23 (a) $[Cr(H_2O)_6]^{3+}$ is hexaaquachromium(III) ion.
$[Cr(H_2O)_6]^{3+}$ is a complex cation.
Name the ligand: H_2O is aqua.
Name the metal ion: Cr^{3+} is chromium(III).
Name the complex ion by adding the prefixes to indicate the number of each ligand, followed by the name of each ligand and the name of the metal ion: hexaaquachromium(III) ion.

(b) $[Cu(CN)_4]^{2-}$ is tetracyanocuprate(II) ion.
$[Cu(CN)_4]^{2-}$ is a complex anion.
Name the ligand: CN^- is cyano.
Name the metal ion: Cu^{2+} is cuprate(II) because the complex is an anion.
Name the complex ion by adding the prefix to indicate the number of each ligand, followed by the name of each ligand and the name of the metal ion: tetracyanocuprate(II) ion.

(c) $[Fe(NH_3)_5Br]SO_4$ is pentaamminebromoiron(III) sulfate.
$[Fe(NH_3)_5Br]^{2+}$ is a complex cation, SO_4^{2-} is sulfate.
Name the ligands in alphabetical order: NH_3 is ammine; Br^- is bromo.
Name the metal cation: Fe^{3+} is iron(III).
Name the complex ion by adding prefixes to indicate the number of each ligand, followed by the name of the ligand and the name of the metal ion: pentaamminebromoiron(III).
Name the compound by writing the name of the cation before the anion. The only space is between the ion names: pentaamminebromoiron(III) sulfate.

(d) $[Co(H_2O)_4(NH_3)(OH)]Cl_2$ is amminetetraaquahydroxocobalt(III) chloride.
$[Co(H_2O)_4(NH_3)(OH)]^{2+}$ is a complex cation; Cl^- is chloride.
Name the ligands in alphabetical order: NH_3 is ammine; H_2O is aqua; OH^- is hydroxo.
Name the metal cation: Co^{3+} is cobalt(III).
Name the complex ion by adding prefixes to indicate the number of each ligand, followed by the name of the ligand and the metal ion: amminetetraaquahydroxocobalt(III).
Name the compound by writing the name of the cation before the anion. The only space is between the ion names: amminetetraaquahydroxocobalt(III) chloride.

23.24 (a) $[Cu(en)_2]^{2+}$ is bis(ethylenediamine)cuprate(II) ion.
$[Cu(en)_2]^{2+}$ is a complex cation.
Name the ligand: en is ethylenediamine.
Name the metal cation: Cu^{2+} is cuprate(II).
Name the complex ion by adding a prefix to indicate the number of ligands, followed by the name of the ligand and the metal ion: bis(ethylenediamine)cuprate(II) ion.

(b) $[Mn(CO)_3(NO_2)_3]^{2+}$ is tricarbonyltrinitromanganese(V) ion.
$[Mn(CO)_3(NO_2)_3]^{2+}$ is a complex cation.
Name the ligands in alphabetical order: CO is carbonyl; NO_2^- is nitro.
Name the metal cation: Mn^{5+} is manganese(V).
Name the complex ion by adding the prefixes to indicate the number of each ligand, followed by the name of the ligand and the metal ion: tricarbonyltrinitromanganese(V) ion.

(c) $Na[Cr(H_2O)_2(ox)_2]$ is sodium diaquadioxalatochromate(III) ion.
Na^+ is sodium ion; $[Cr(H_2O)_2(ox)_2]^-$ is a complex anion.
Name the ligands in alphabetical order: H_2O is aqua; ox (oxalate) is oxalato.
Name the metal ion: Cr^{3+} is chromate(III) because the complex is an anion.
Name the complex ion by adding the prefixes to indicate the number of each ligands, followed by the name of the ligand and the metal ion: diaquadioxalatochromate(III). For bidentate or polydentate ligands the prefixes are bis (2), tris (3), and tetrakis (4).
Name the compound by writing the name of the cation before the anion. The only space is between the ion names: sodium diaquadioxalatochromate(III).

(d) $[Co(en)_3][Fe(CN)_6]$ is tris(ethylenediamine)cobalt(III) hexacyanoferrate(III).
$[Co(en)_3]^{3+}$ is a complex cation; $[Fe(CN)_6]^{3-}$ is a complex anion.
First, name the complex cation:
Name the ligand: en is ethylenediamine.
Name the metal ion: Co^{3+} is cobalt(III).
Name the complex ion by adding the prefix to indicate the number of ligand, followed by the name of the ligand and the metal ion: tris(ethylenediamine)cobalt(III). For bidentate or polydentate ligands so the prefixes are bis (2), tris (3), and tetrakis (4).
Second, name the complex anion:
Name the ligand: CN^- is cyano.
Name the metal ion: Fe^{3+} is ferrate(III) because the complex is an anion.
Name the complex ion by adding the prefix to indicate the number of ligand, followed by the name of the ligand and the metal ion: hexacyanoferrate(III) ion.
Name the compound by writing the name of the cation before the anion. The only space is between the ion names: tris(ethylenediamine)cobalt(III) hexacyanoferrate(III).

23.25 (a) Hexaamminechromium(III) is a complex ion with Cr^{3+} metal ion and six NH_3 ligands. $[Cr(NH_3)_6]^{3+}$

(b) Potassium hexacyanoferrate(III) is a compound with 3 K^+ cations and a complex anion with Fe^{3+} metal ion and six CN^- ligands. $K_3[Fe(CN)_6]$

(c) Ethylenediaminedithiocyanatocopper(II) is a compound with a Cu^{2+} metal ion, an ethylenediamine ligand, and two SCN^- ligands. $[Cu(en)(SCN)_2]$

(d) Tetraaquaplatinum(II) hexachloroplatinate(IV) is a complex compound with a complex cation that contains a Pt^{2+} metal ion, four H_2O ligands, and a complex anion that contains a Pt^{4+} metal ion and six Cl^- ligands. $[Pt(H_2O)_4][PtCl_6]$

23.26 (a) Hexaaquanickel(II) chloride is a compound with a complex cation that contains a Ni^{2+} metal ion, six H_2O ligands, and two Cl^- counterions. $[Ni(H_2O)_6]Cl_2$

(b) Pentacarbonylchloromanganese(I) is a coordination compound that contains a Mn^+ metal ion, five CO ligands, and one Cl^- ligand. $[Mn(CO)_5Cl]$

(c) Ammonium diaquatetrabromovanadate(III) is a coordination compound with an ammonium cation and a complex anion that contains a V^{3+} metal ion, two H_2O ligands, and four Br^- ligands. $NH_4[V(H_2O)_2Br_4]$

(d) Tris(ethylenediamine)cobalt(III) trioxalatoferrate(III) is a coordination compound with a complex cation that contains a Co^{3+} metal ion, three ethylenediamine ligands, and a complex anion that contains an Fe^{3+} metal ion and three oxalate ligands. $[Co(en)_3][Fe(ox)_3]$

23.27 (a) $[Co(NH_3)_3(CN)_3]$ is triamminetricyanocobalt(III).

(b) Because ethylenediamine is a bidentate ligand, you need three to have a coordination number of 6. $[Cr(en)_3]^{3+}$ is tris(ethylenediamine)chromium(III) ion.

23.28 (a) $[Fe(H_2O)_4(ONO)_2]^+$ is the tetraaquadinitritoiron(III) ion.

(b) The complex cation is $[V(en)_2Cl_2]^+$, the complex anion is $[NiCl_4]^{2-}$, and the coordination compound is $[V(en)_2Cl_2]_2[NiCl_4]$ and is named dichlorobis(ethylenediamine)vanadium(III) tetrachloronickelate(II).

Structure and Isomerism

23.29 In linkage isomers, the ligand coordinates to the metal in different ways.

pentaamminenitromanganese(III) ion

pentaamminenitritomanganese(III) ion

23.30 In linkage isomers, the ligand coordinates to the metal in different ways.

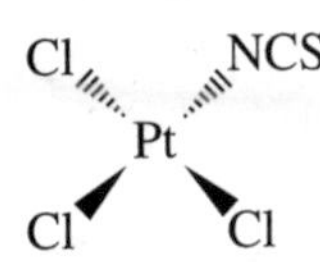

trichlorothiocyanatoplatinum(II) ion

trichloroisothiocyanatoplatinum(II) ion

23.31 Coordination isomers occur when a coordinated ligand exchanges places with the uncoordinated counterion.
$[Fe(H_2O)_5Cl]Cl \cdot H_2O$ pentaaquachloroiron(II) chloride monohydrate
$[Fe(H_2O)_4Cl_2] \cdot 2H_2O$ tetraaquadichloroiron(II) dihydrate

23.32 Coordination isomers occur when a coordinated ligand exchanges places with the uncoordinated counterion.
$[Co(en)_2(ox)][Cr(en)(ox)_2]$
bis(ethylenediamine)oxalatocobalt(III) ethylenediaminedioxalatochromate(III)
$[Cr(en)_2(ox)][Co(en)(ox)_2]$
bis(ethylenediamine)oxalatochromium(III) ethylenediamindioxalatocobaltate(III)
$[Cr(en)_3][Co(ox)_3]$
tris(ethylenediamine)chromium(III) trioxalatocobaltate(III)

23.33 Geometric isomers result when the ligands bonded to the metal have a different spatial arrangement.

(a) No, an octahedral complex must have at least two different ligands to have geometric isomers.
(b) Yes, there will be cis–trans isomers.
(c) Yes, there will be fac–mer isomers.
(d) No, a square planar complex must have at least two different ligands to have geometric isomers.
(e) Yes, there will be cis–trans isomers.

23.34 Geometric isomers result when the ligands bonded to the metal have a different spatial arrangement.

(a) Yes, there will be cis–trans isomers.
(b) No, an octahedral complex must have at least two different ligands to have geometric isomers.
(c) Yes, there will be cis–trans isomers.
(d) Yes, there will be cis–trans isomers.
(e) Yes, there will be cis–trans isomers.

23.35 (a) Square planar $[NiWXYZ]^{2+}$ would have three geometric isomers.

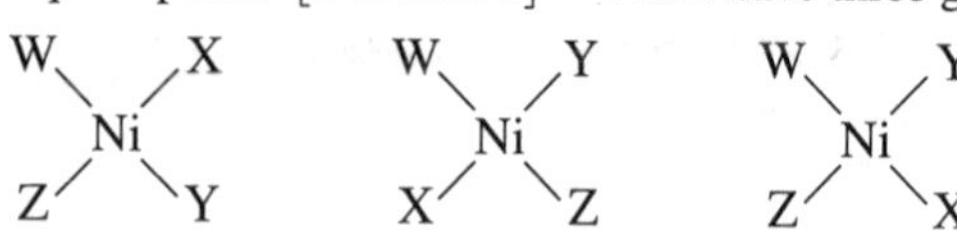

(b) Tetrahedral $[ZnWXYZ]^{2+}$ would have no geometric isomers but would have stereoisomers.

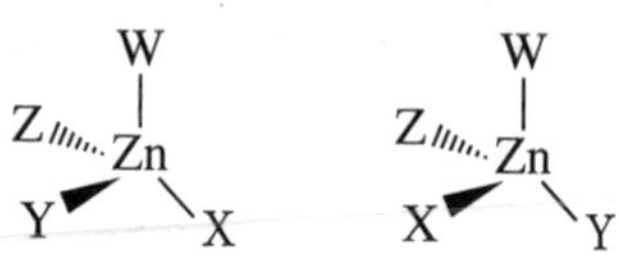

23.36 (a) $[Fe(CO)_3(Cl)_3]$ would have two geometric isomers (fac and mer).

fac mer

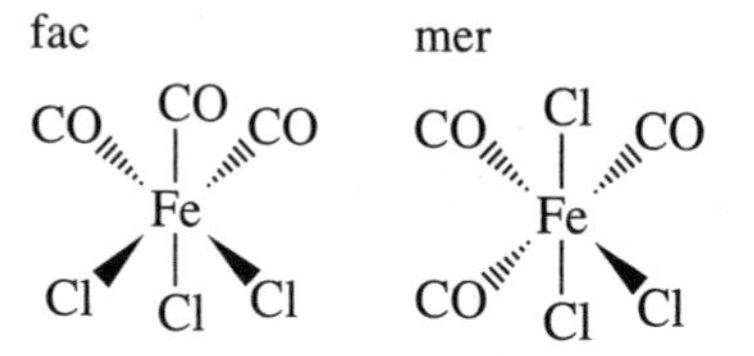

(b) $[Mn(CO)_2Cl_2Br_2]^+$ would have five isomers.

all trans CO trans; Cl, Br cis Cl trans; CO, Br cis

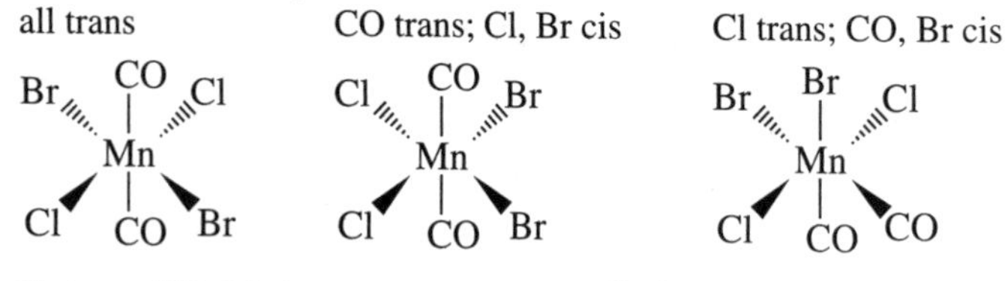

Br trans; CO, Cl cis all cis

23.37 (a) $[Cr(CO)_3(NH_3)_3]^{3+}$ has a coordination number of 6 and is octahedral. There will be fac and mer isomers and no optical isomers because after rotation, the mirror images are superimposable upon each other.

fac mer

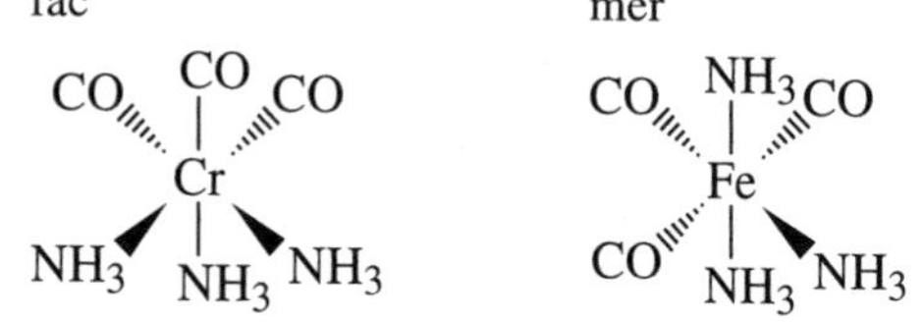

(b) $[Pd(CO)_2(H_2O)Cl]^+$ has a coordination number of 4 and is a d^8 complex, so it is square planar. There will be cis and trans isomers. There will be no optical isomers because after rotation, the mirror images are superimposable upon each other.

cis trans

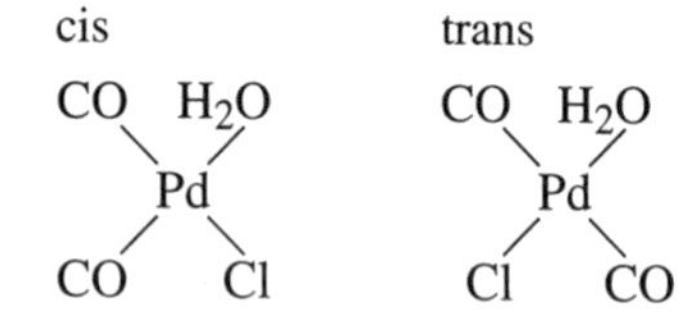

23.38 (a) $[Fe(CO)_4Cl_2]^+$ has a coordination number of 6 and is octahedral. There will be cis and trans isomers.

trans cis

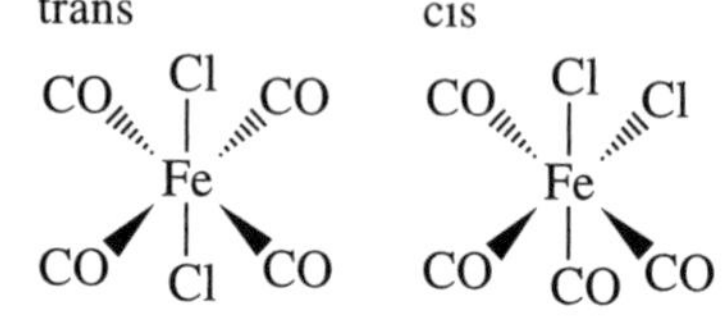

(b) $[Pt(en)Cl_2]$ has a coordination number of 4, and Pt is d^8; so it is square planar.

Cl
Pt en
Cl

23.39 $[Cr(CO)_2(ox)_2]^-$ has a coordination number of 6 and is octahedral. There will be cis and trans isomers. The cis isomer has a mirror image that is nonsuperimposable.

trans cis

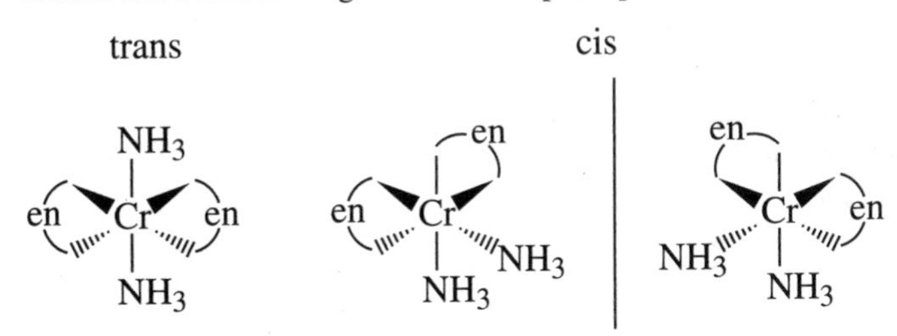

23.40 $[Fe(CO)_3Cl_3]$ has a coordination number of 6 and is octahedral. There will be fac and mer isomers. There are no optical isomers.

fac mer

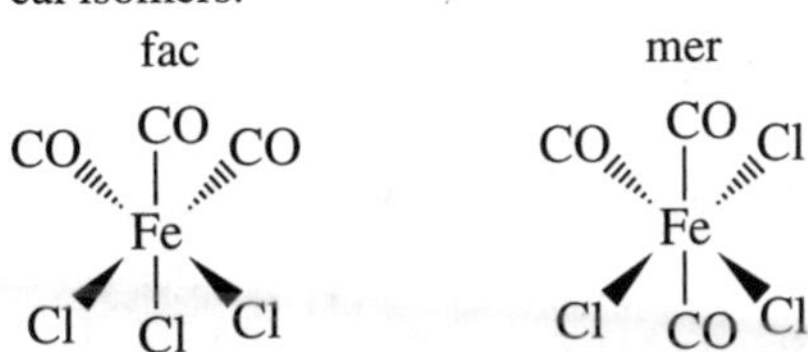

Bonding in Coordination Compounds

23.41 (a) Zn^{2+} d^{10}

Energy: upper: 1l 1l; lower: 1l 1l 1l

(b) Fe^{3+} d^5 high spin

Energy: upper: 1 1; lower: 1 1 1

low spin

Energy: upper: ___ ___; lower: 1l 1l 1

(c) V^{3+} d^2

Energy: upper: ___ ___; lower: 1 1 ___

(d) Co^{2+} d^7 high spin

Energy: upper: 1 1; lower: 1l 1l 1

23.42 (a) Cr^{3+} d^3

Energy: upper: ___ ___; lower: 1 1 1

(b) Cu^{2+} d^9

Energy: upper: 1l 1; lower: 1l 1l 1l

(c) Mn^{3+} d^4 high spin

Energy: upper: 1 ___; lower: 1 1 1

low spin

Energy: upper: ___ ___; lower: 1l 1 1

(d) Fe^{2+} d^6 low spin

Energy: upper: ___ ___; lower: 1l 1l 1l

23.43 **Given:** $[CrCl_6]^{3-}$ $\lambda = 735$ nm absorbance maximum **Find:** crystal field energy

Conceptual Plan: $\boldsymbol{\lambda \rightarrow \Delta}$ **(J)** $\rightarrow$ **kJ/ion** $\rightarrow$ **kJ/mol**

$$\Delta = \frac{hc}{\lambda} \qquad \frac{1\text{ kJ}}{1000\text{ J}} \qquad \frac{6.02 \times 10^{23}\text{ ions}}{\text{mol}}$$

Solution: $\dfrac{(6.626 \times 10^{-34}\ \text{J}\cdot\cancel{\text{s}})\left(3.00 \times 10^{8}\dfrac{\cancel{\text{m}}}{\cancel{\text{s}}}\right)}{(735\ \cancel{\text{nm}})\left(\dfrac{1 \times 10^{-9}\ \cancel{\text{m}}}{\cancel{\text{nm}}}\right)} = 2.7\underline{0}4 \times 10^{-19}\ \text{J/ion}$

$$2.7\underline{0}4 \times 10^{-19}\frac{\cancel{\text{J}}}{\cancel{\text{ion}}} \times \frac{1\ \text{kJ}}{1000\ \cancel{\text{J}}} \times \frac{6.02 \times 10^{23}\ \cancel{\text{ions}}}{\text{mol}} = 163\ \text{kJ/mol}$$

Check: Cl is a weak-field ligand and would be expected to have a relatively small Δ, which is consistent with a value of 163 kJ.

23.44 **Given:** $[Rh(NH_3)_6]^{3+}$ $\lambda = 295$ nm absorbance maximum **Find:** crystal field energy, Δ

Conceptual Plan: $\boldsymbol{\lambda \rightarrow \Delta\ (J) \rightarrow}$ **kJ/ion → kJ/mol**

$$\Delta = \frac{hc}{\lambda} \qquad \frac{1\ \text{kJ}}{1000\ \text{J}} \qquad \frac{6.02 \times 10^{23}\ \text{ions}}{\text{mol}}$$

Solution: $\dfrac{(6.626 \times 10^{-34}\ \text{J}\cdot\cancel{\text{s}})\left(3.00 \times 10^{8}\dfrac{\cancel{\text{m}}}{\cancel{\text{s}}}\right)}{(295\ \cancel{\text{nm}})\left(\dfrac{1 \times 10^{-9}\ \cancel{\text{m}}}{\cancel{\text{nm}}}\right)} = 6.7\underline{3}8 \times 10^{-19}\ \text{J/ion}$

$$6.7\underline{3}8 \times 10^{-19}\frac{\cancel{\text{J}}}{\cancel{\text{ion}}} \times \frac{1\ \text{kJ}}{1000\ \cancel{\text{J}}} \times \frac{6.02 \times 10^{23}\ \cancel{\text{ions}}}{\text{mol}} = 405\ \text{kJ/mol}$$

Check: NH_3 is a strong-field ligand and would be expected to have a relatively large Δ, which is consistent with a value of 405 kJ.

23.45 The crystal field ligand strength would be $CN^- > NH_3 > F^-$. The smaller the wavelength, the larger the energy and the greater the crystal field splitting observed. So $[Co(CN)_6]^{3-}$ would have a smaller wavelength than $[Co(NH_3)_6]^{3+}$, which would have a smaller wavelength than $[CoF_6]^{3-}$.
So $[Co(CN)_6]^{3-} = 290$ nm, which absorbs in the UV and would have a colorless solution.
$[Co(NH_3)_6]^{3+} = 440$ nm, which absorbs in the blue and would have an orange solution.
$[CoF_6]^{3-} = 770$ nm, which absorbs in the red and would have a green solution.

23.46 The green solution absorbs in the red, the yellow solution absorbs in the violet, and the purple solution absorbs in the yellow. So the green solution has the smallest Δ; the next would be the purple solution, and the largest Δ would be the yellow solution. The crystal field ligand strength would be $NH_3 > H_2O > Cl^-$.
So the green solution, which absorbs in the red, would have the smallest Δ and would be $[Cr(H_2O)_4Cl_2]^+$.
The purple solution, which absorbs in the yellow, would have the next largest Δ and would be $[Cr(H_2O)_6]^{3+}$.
The yellow solution, which absorbs in the violet, would have the largest Δ and would be $[Cr(NH_3)_6]^{3+}$.

23.47 Mn^{2+} is d^5, and there are five unpaired electrons; so the crystal field splitting energy, Δ, is small compared to the energy to pair the electrons. Therefore, NH_3 induces a weak field with Mn^{2+}.

23.48 Fe^{2+} is d^6; to be paramagnetic, there must be unpaired electrons. For all of the electrons to be unpaired, the crystal field splitting energy must be smaller than the energy to pair the electrons. Therefore, H_2O induces a weak field with Fe^{2+}.

23.49 (a) $[RhCl_6]^{3-}$ Rh^{3+} d^6 Cl^- is a weak-field ligand, so the value of Δ will be small. This gives four unpaired electrons.

Energy
↿ ↿
↿⇂ ↿ ↿

(b) $[Co(OH)_6]^{4-}$ Co^{2+} d^7 OH^- is a weak-field ligand, so the value of Δ will be small. This gives three unpaired electrons.

Energy
↿ ↿
↿⇂ ↿⇂ ↿

(c) cis-$[Fe(en)(NO_2)_2]^+$ Fe^{3+} d^5 en and NO_2^- are strong-field ligands, so the value of Δ will be large. This gives one unpaired electron.

Energy
___ ___
↑↓ ↑↓ ↑

23.50 (a) $[Cr(CN)_6]^{4-}$ Cr^{2+} d^4 CN^- is a strong-field ligand, so the value of Δ will be large. This gives two unpaired electrons.

Energy
___ ___
↑↓ ↑ ↑

(b) $[MnF_6]^{4-}$ Mn^{2+} d^5 F^- is a weak-field ligand, so the value of Δ will be small. This gives five unpaired electrons.

Energy
↑ ↑
↑ ↑ ↑

(c) $[Ru(en)_3]^{2+}$ Ru^{2+} d^6 en is a strong-field ligand, so the value of Δ will be large. This gives zero unpaired electrons.

Energy
___ ___
↑↓ ↑↓ ↑↓

23.51 $[CoCl_4]^{2-}$ Co^{2+} is d^7, and Cl^- is a weak-field ligand; so the value of Δ will be small, and there will be three unpaired electrons. The crystal field splitting for a tetrahedral structure is as follows:

Energy
↑ ↑ ↑
↑↓ ↑↓

23.52 $[PdCl_4]^{2-}$ is known to be diamagnetic. Pd complexes are low-spin complexes. Pd^{2+} is d^8. Use the crystal field splitting for a tetrahedral structure and a square planar structure and determine which would be diamagnetic.
tetrahedral (two unpaired electrons) square planar (zero unpaired electrons)

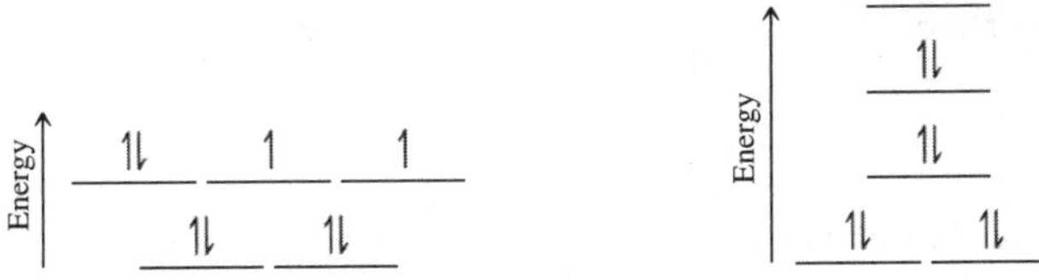

The tetrahedral structure would be paramagnetic, and the square planar would be diamagnetic; because the compound is diamagnetic, it must be square planar.

Applications of Coordination Compounds

23.53 Hemoglobin, cytochrome c, and chlorophyll all contain a porphyrin ligand.

23.54 (a) Hemoglobin contains Fe metal.
(b) Carbonic anhydrase contains Zn^{2+}.
(c) Chlorophyll contains Mg metal.
(d) Iron blue contains Fe^{2+} and Fe^{3+}.

23.55 Oxyhemoglobin is low-spin and a red color. Because it is low-spin, the crystal field splitting energy must be large. The red color means that the complex absorbs in the green region (~500 nm), which also indicates a large crystal field splitting energy; therefore, O_2 must be a strong-field ligand.
Deoxyhemoglobin is high-spin and a blue color. Because it is high-spin, the crystal field splitting energy must be small. The blue color means that the complex absorbs in the orange region (~600 nm), which also indicates a small crystal field splitting energy. Both of these are consistent with H_2O as a weak-field ligand.

23.56 To calculate the equilibrium constant for the new reaction:

Reverse reaction 1:	$HbO_2 \rightleftharpoons \cancel{Hb} + O_2$	$K = 1/K_1 = 1/(2 \times 10^{12})$
Add reaction 2:	$\cancel{Hb} + CO \rightleftharpoons HbCO$	$K_2 = 1 \times 10^{14}$
	$HbO_2 + CO \rightleftharpoons HbCO + O_2$	$K_3 = 5 \times 10^{1}$

Because $K > 1$, the reaction favors the products.

Cumulative Problems

23.57

(a)	Cr	[Ar]$4s^1 3d^5$	(b)	Cu	[Ar]$4s^1 3d^{10}$
	Cr^+	[Ar]$4s^0 3d^5$		Cu^+	[Ar]$4s^0 3d^{10}$
	Cr^{2+}	[Ar]$4s^0 3d^4$		Cu^{2+}	[Ar]$4s^0 3d^9$
	Cr^{3+}	[Ar]$4s^0 3d^3$			

23.58

(a)	Mo	[Kr]$5s^1 4d^5$	Ag	[Ar]$4s^1 3d^{10}$
	Mo^+	[Kr]$5s^0 4d^5$	Ag^+	[Ar]$4s^0 3d^{10}$
(b)	Ru	[Kr]$5s^1 4d^7$		
	Ru^{3+}	[Kr]$5s^0 4d^5$		
(c)	Rh	[Kr]$5s^1 4d^8$		
	Rh^{2+}	[Kr]$5s^0 4d^7$		
(d)	Pd	[Kr]$4d^{10}$		
	Pd^+	[Kr]$4d^9$		
	Pd^{2+}	[Kr]$4d^8$		

23.59 (a) 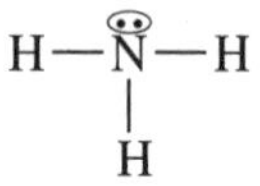

(b) 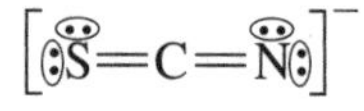 ligand can bond from either end.

(c) 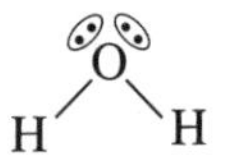

23.60 (a) 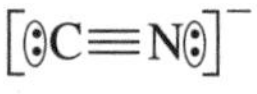 ligand can bond from either end.

(b) 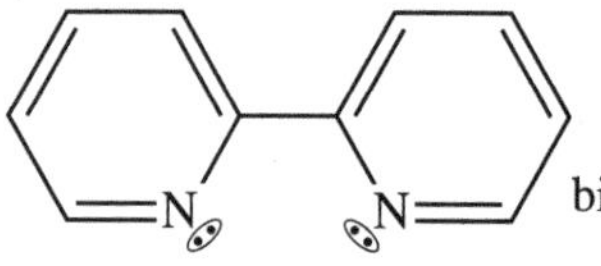 bipyridyl is a bidentate ligand.

(c) 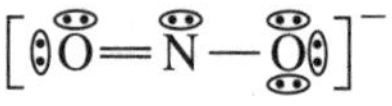 ligand can bond from the N or the O.

23.61 An octahedral complex has six ligands.

$MA_2B_2C_2$	will have cis–trans isomers: all cis; A trans, B,C cis; B trans, A,C cis; C trans, A,B cis; all trans.
MAB_2C_3	will have fac–mer isomers.
MA_2B_3C	will have fac–mer isomers.
MAB_3C_2	will have fac–mer isomers.
MA_3B_2C	will have fac–mer isomers.
MA_2BC_3	will have fac–mer isomers.
MA_3BC_2	will have fac–mer isomers.
$MABC_4$	will have AB cis and trans.
MAB_4C	will have AC cis and trans.
MA_4BC	will have BC cis and trans.

23.62 (a) $[Ni(gly)_2]$ Ni^{2+} d^8 square planar cis/trans isomers, no optical isomers

cis–diglycinenickel(II) *trans*–diglycinenickel(II)

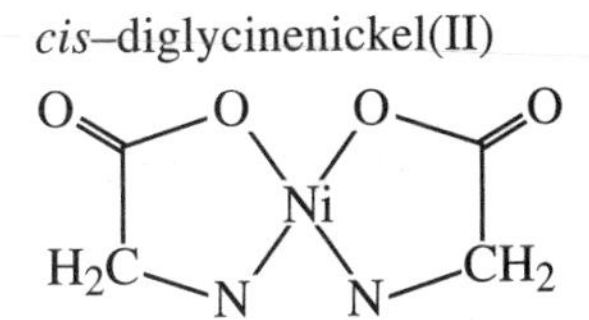

(b) $[Zn(gly)_2]$ Zn^{2+} d^{10} tetrahedral, optical isomers

diglycinezinc

(c) $[Fe(gly)_3]$ Fe^{3+} d^5 octahedral, fac isomer, optical isomers

fac–triglycineiron(III)

mer–triglycineiron(III)

mer isomer, optical isomers

23.63 $[Fe(ox)_3]^{3-}$ Fe^{3+} d^5 coordination number = 6, octahedral, structure has a nonsuperimposable mirror image.

23.64 (a) Square planar $[NiWXYZ]^{2+}$ would have three geometric isomers.

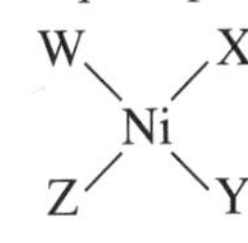

None of the isomers will be optically active. The mirror image of each can be rotated to superimpose

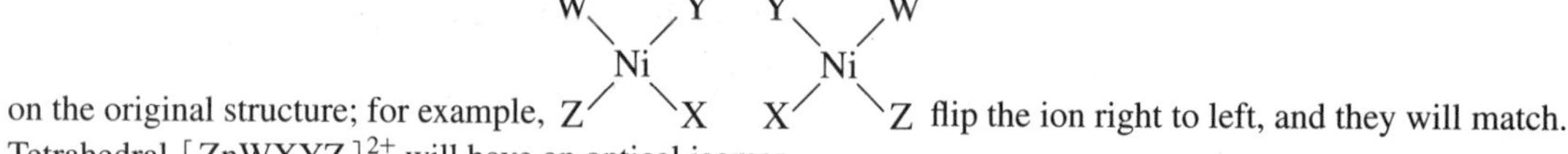

on the original structure; for example, flip the ion right to left, and they will match.

(b) Tetrahedral $[ZnWXYZ]^{2+}$ will have an optical isomer.

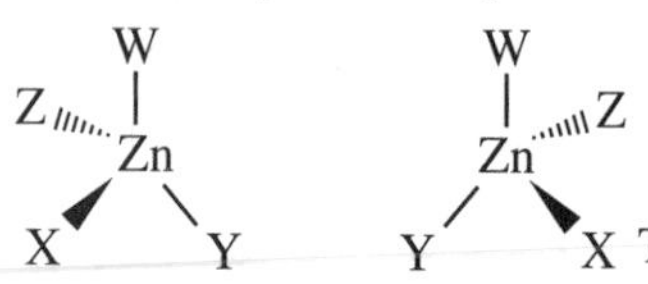

W Zn Z Y X The structures are nonsuperimposable; the ion is optically active.

23.65 $[Mn(CN)_6]^{3-}$ Mn^{3+} d^4 has a coordination number of 6 and is octahedral. CN^- is a strong-field ligand and will cause a large crystal field splitting energy, so the ion is low-spin.

Energy: ⇅ ↑ ↑ (lower level); ___ ___ (upper level). The ion will be paramagnetic with two unpaired electrons.

23.66 (a) A blue solution will absorb most strongly in the orange region, so the wavelength will be between 580 and 650 nm.
(b) A red solution will absorb most strongly in the green region, so the wavelength will be between 490 and 560 nm.
(c) A yellow solution will absorb most strongly in the violet region, so the wavelength will be between 400 and 430 nm.

23.67 There are five geometric isomers, one of which is chiral.

1. $[Ru(H_2O)_2(NH_3)_2Cl_2]^+$ (H₂O trans; NH₃, Cl, Cl, NH₃ in plane)
2. $[Ru(H_2O)_2(NH_3)_2Cl_2]^+$ (H₂O trans; NH₃, NH₃, Cl, Cl in plane)
3. $[Ru(H_2O)_2(NH_3)_2Cl_2]^+$ and its mirror image — chiral
4. $[Ru(H_2O)_2(NH_3)_2Cl_2]^+$
5. $[Ru(H_2O)_2(NH_3)_2Cl_2]^+$

23.68 **Given:** 0.32 mol NH_3; 0.47 L, 0.38M Ag^+, $K_f = 1.7 \times 10^7$ **Find:** equilibrium concentrations
Conceptual Plan: mol NH_3, V → $[NH_3]$. Set up an ICE table and determine equilibrium concentration.

$$[\,] = \frac{\text{mol}}{\text{vol}}$$

Solution: $[NH_3] = \dfrac{0.32 \text{ mol } NH_3}{0.47 \text{ L}} = 0.68 \text{ M}$

	$Ag^+(aq)$	+ $2NH_3(aq)$	⇌ $Ag(NH_3)_2^+(aq)$
Initial	0.38	0.68	0
Change	$-x$	$-2x$	$+x$
Equil	$0.38 - x$	$0.68 - 2x$	x

$$K_f = \frac{[Ag(NH_3)_2^+]}{[Ag^+][NH_3]^2} \qquad 1.7 \times 10^7 = \frac{x}{(0.38 - x)(0.68 - 2x)^2}$$

$-6.8 \times 10^7 x^3 + 7.208 \times 10^7 x^2 - 2.5432 \times 10^7 x + 0.2987104 \times 10^7 = 0$

Solve the cubic equation using a graphing calculator or an online calculator: $x = 0.3800, 0.3396, 0.3404$. The only possible solution is 0.3396.

So $[Ag(NH_3)_2^+] = x = 0.3396$ M; $[NH_3] = 0.68 - 2x = 0.0008$ M; $[Ag^+] = 0.38 - x = 0.0404$ M.

When the values are plugged into the K_f equation, the value is 1.3×10^7. The values need to be carried out further than normal significant figures because essentially all of the NH_3 is reacted.

Challenge Problems

23.69 **Given:** 46.7% Pt, 17.0% Cl, 14.8% P, 17.2% C, 4.34% H **Find:** formula, structures, and names for both compounds

Conceptual Plan: % composition → pseudoformula → formula

$n = \frac{g}{\text{molar mass}}$ divide by smallest

Solution: $46.7 \text{ g Pt} \times \frac{1 \text{ mol Pt}}{195.1 \text{ g}} = 0.23\underline{9}36 \text{ mol Pt}$ $17.0 \text{ g Cl} \times \frac{1 \text{ mol Cl}}{35.45 \text{ g Cl}} = 0.47\underline{9}55 \text{ mol Cl}$

$14.8 \text{ g P} \times \frac{1 \text{ mol P}}{30.97 \text{ g}} = 0.47\underline{7}88 \text{ mol P}$ $17.2 \text{ g C} \times \frac{1 \text{ mol C}}{12.01 \text{ g C}} = 1.4\underline{3}21 \text{ mol C}$

$4.34 \text{ g H} \times \frac{1 \text{ mol H}}{1.008 \text{ g}} = 4.3\underline{0}56 \text{ mol H}$

$Pt_{0.23936}Cl_{0.47955}P_{0.47788}C_{1.4321}H_{4.3056}$

$Pt_{\frac{0.23936}{0.23936}}Cl_{\frac{0.47955}{0.23936}}P_{\frac{0.47788}{0.23936}}C_{\frac{1.4321}{0.23936}}H_{\frac{4.3056}{0.23936}}$

$PtCl_2P_2C_6H_{18}$

$[Pt(P(CH_3)_3)_2Cl_2]$

cis–dichlorobis(trimethylphosphine)platinum(II)

Cl and $P(CH_3)_3$ (top), Pt, Cl and $P(CH_3)_3$ (bottom)

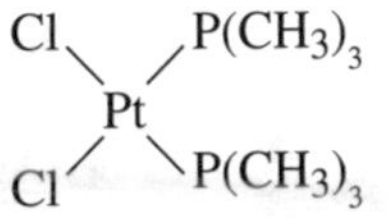

trans–dichlorobis(trimethylphosphine)platinum(II)

Cl and $P(CH_3)_3$ (top), Pt, $P(CH_3)_3$ and Cl (bottom)

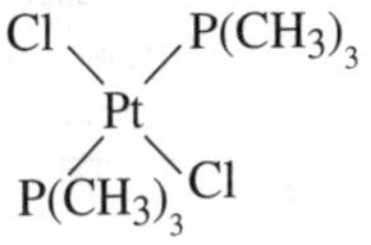

23.70

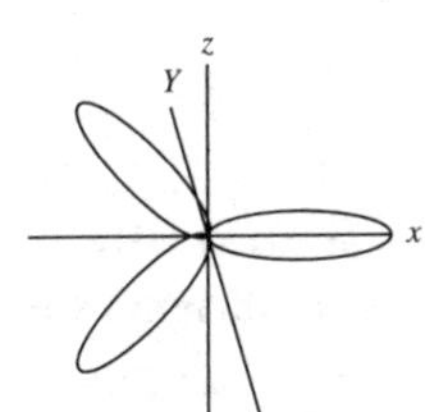

The trigonal planar complex ion has lobes along the x-axis and between the x- and y-axes. So the ligands will interact most strongly with the $x^2 - y^2$ and the xy orbitals and will not interact with the z^2, the xz, or the yz orbitals. So the crystal field splitting would look like the following:

Energy ↑
_____ $x^2 - y^2$ _____ xy
_____ z^2 _____ xz _____ yz

23.71

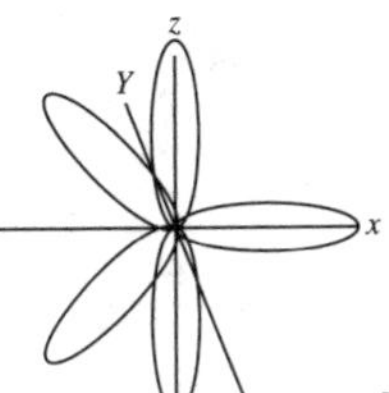

The trigonal bipyramidal complex ion has lobes along the x-axis and between the x- and y-axes and along the z-axis. So the ligands will interact most strongly with z^2 orbital and then with the $x^2 - y^2$ and the xy orbitals and will not interact with the xz or the yz orbitals. So the crystal field splitting would look like the following:

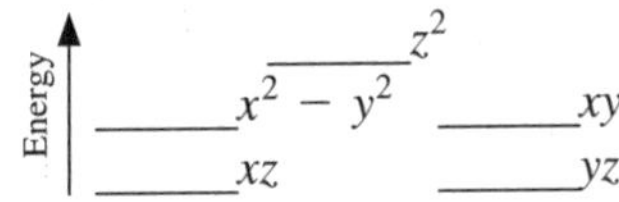

23.72 $[Ni(NH_3)_4]^{2+}$ Ni^{2+} is d^8 and square planar and paramagnetic.

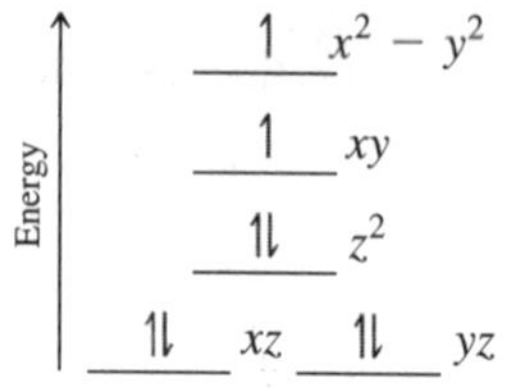

Energy

1 $x^2 - y^2$

1 xy

1l z^2

1l xz 1l yz

$[Ni(CN)_4]^{2-}$ Ni^{2+} is d^8 and square planar and diamagnetic.

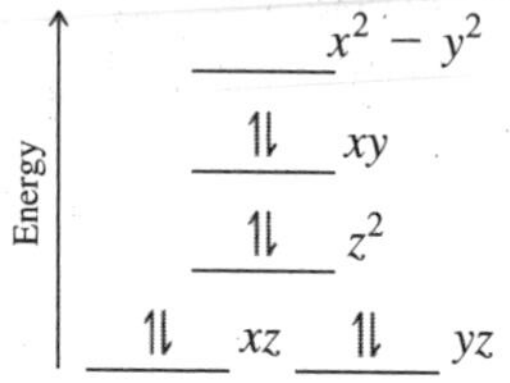

Energy

$x^2 - y^2$

1l xy

1l z^2

1l xz 1l yz

The CN^- ligand is a stronger field ligand than the NH_3 ligand; so there is a greater splitting between the $x^2 - y^2$ orbital and the xy orbital in the tetracyanonickel(II) complex ion than in the tetraamminenickel(II) complex ion.

23.73 (a) **Given:** $K_{sp}(NiS) = 3 \times 10^{-16}$ **Find:** solubility in water

Conceptual Plan: Write the reaction, prepare an ICE table, substitute into the equilibrium expression, and solve for S(molar solubility).

Solution: $NiS(s) \rightleftharpoons Ni^{2+}(aq) + S^{2-}(aq)$

	Ni^{2+}	S^{2-}
Initial	0.0	0.0
Change	S	S
Equil	S	S

$K_{sp} = [Ni^{2+}][S^{2-}]$ $3 \times 10^{-16} = S^2$

$S = 1.7 \times 10^{-8} = 2 \times 10^{-8}$

(b) **Given:** $K_{sp}(NiS) = 3 \times 10^{-16}$ $K_f\,[Ni(NH_3)_6]^{2+} = 2.0 \times 10^8$ **Find:** solubility in 3.0 M NH_3

Conceptual Plan: Sum the reaction, prepare an ICE table, substitute into the equilibrium expression, and solve for S(molar solubility).

Solution:

Reaction 1: $NiS(s) \rightleftharpoons Ni^{2+}(aq) + S^{2-}(aq)$ $K_{sp}(NiS) = 3 \times 10^{-16}$

Reaction 2: $Ni^{2} + (aq) + 6\,NH_3(aq) \rightleftharpoons Ni(NH_3)_6{}^{2+}(aq)$ $K_f = 2.0 \times 10^8$

Reaction 3: $NiS(s) + 6\,NH_3(aq) \rightleftharpoons Ni(NH_3)_6{}^{2+}(aq) + S^{2-}(aq)$ $K = K_{sp}\,K_f = 6.0 \times 10^{-8}$

	NH_3	$Ni(NH_3)_6{}^{2+}$	S^{2-}
Initial	3.0 M	0.0	0.0
Change	$-6S$	$+S$	$+S$
Equil	$3.0 - 6S$	S	S

$$K = \frac{[Ni(NH_3)_6{}^{2+}][S^{2-}]}{[NH_3]^6} \quad 6.0 \times 10^{-8} = \frac{(S)(S)}{(3.0 - S)^6}$$

Assume that $S << 3.0$.

$S^2 = 4.37 \times 10^{-5}$

$S = 6.6 \times 10^{-3}$

(c) NiS is more soluble in ammonia because the formation of the $[Ni(NH_3)_6]^{2+}$ complex ion is highly favorable. The formation removes Ni^{2+} ion from the solution, causing more of the NiS to dissolve.

23.74 **Given:** 2.0 M NaOH; $Zn(OH)_2(s)$ $K_{sp} = 3 \times 10^{-17}$; $Zn(OH)_4{}^{2-}(aq)$ $K_f = 2 \times 10^{15}$ **Find:** solubility

Conceptual Plan: Combine the solubility product reaction of $Zn(OH)_2(s)$ and the formation reaction of $Zn(OH)_4{}^{2-}$ and determine the K for the new reaction. Prepare an ICE table and solve for S (the molar solubility).

Solution:

$Zn(OH)_2(s) \rightleftharpoons Zn^{2+}(aq) + 2\,OH^-(aq)$ $\quad K_{sp} = 3 \times 10^{-17}$

$Zn^{2+}(aq) + 4\,OH^-(aq) \rightleftharpoons Zn(OH)_4^{2-}(aq)$ $\quad K_f = 2 \times 10^{15}$

$Zn(OH)_2(s) + 2\,OH^-(aq) \rightleftharpoons Zn(OH)_4^{2-}(aq)$ $\quad K = K_{sp}K_f = 0.06$

	$[OH^-]$	$[Zn(OH)_4^{2-}]$
Initial	2.0 M	0
Change	$-2S$	$+S$
Equil	$2.0 - 2S$	S

$$K = 0.06 = \frac{[Zn(OH)_4^{2-}]}{[OH^-]^2} = \frac{S}{(2.0 - 2S)^2}$$

Solve using the quadratic equation.

$S = 6$ or 0.1667; so $S = 0.17$ M, which is the solubility.

23.75 If the complexes exhibit lability, there will be an exchange of H_2O with the halide and some of the halide will be in solution. So prepare complexes with two different halides (e.g., Cl^- and Br^-) so that you have $[MCl_6]^{3-}$ and $[MBr_6]^{3-}$. Place both complexes in the same aqueous solution. If the complexes exhibit lability, a mixed complex containing both Cl^- and Br^- ligands, $[MCl_xBr_y]^{3-}$, will be formed because Cl^- and Br^- would be in solution when the ligands were exchanged with H_2O. If this mixed complex does not form, the $[MX_6]^{3-}$ complex does not exhibit lability.

23.76 $[Cu(H_2O)_4]^{2+}(aq) + 2\,en(aq) \rightarrow [Cu(en)_2]^{2+}(aq) + 4\,H_2O(l)$ $\quad K = 2 \times 10^{20}$

$[Cu(H_2O)_4]^{2+}(aq) + 4\,NH_3(aq) \rightarrow [Cu(NH_3)_4]^{2+}(aq) + 4\,H_2O(l)$ $\quad K = 1.7 \times 10^{13}$

The large difference in the K values for the two formation constants can be attributed to the effect of entropy. In the first reaction with the bidentate ligand, three molecules of reactants produce five molecules of product. While in the second reaction with the monodentate ligand, five molecules of reactants produce five molecules of products. Because there is a greater increase in the entropy of the first reaction, it would be more favorable.

$\Delta G^\circ = -RT \ln K$ and $\Delta G^\circ = \Delta H^\circ - T\Delta S^\circ$

Assume that ΔH° is approximately the same for each reaction; then

$\Delta S^\circ(en) = R \ln K(en)$ and $\Delta S^\circ(NH_3) = R \ln K(NH_3)$

$$= 8.314 \frac{J}{mol\,K} \ln 2 \times 10^{20} \qquad = 8.314 \frac{J}{mol\,K} \ln 1.7 \times 10^{13}$$

$$= 388.6 \frac{J}{mol\,K} \qquad = 253.3 \frac{J}{mol\,K}$$

The difference in entropy between the two reactions is as follows:

$$388.6 \frac{J}{mol\,K} - 253.3 \frac{J}{mol\,K} = 135.3 \frac{J}{mol\,K} = 1 \times 10^2 \frac{J}{mol\,K}$$

23.77 **Given:** 0.10 M NaI, $Cd(OH)_2(s)$ $K_{sp} = 7.2 \times 10^{-15}$; $Cd(I)_4^{2-}(aq)$ $K_f = 2 \times 10^6$ **Find:** pH

Conceptual Plan: Combine the solubility product reaction of $Cd(OH)_2(s)$ and the formation reaction of $Cd(I)_4^{2-}$ and determine the K for the new reaction. Prepare an ICE table and solve for S (the molar solubility). Then determine the $[OH^-]$, pOH, and pH.

Solution: $Cd(OH)_2(s) \rightleftharpoons Cd^{2+}(aq) + 2\,OH^-(aq)$ $\quad K_{sp} = 7.2 \times 10^{-15}$

$Cd^{2+}(aq) + 4\,I^-(aq) \rightleftharpoons Cd(I)_4^{2-}(aq)$ $\quad K_f = 2 \times 10^6$

$Cd(OH)_2(s) + 4\,I^-(aq) \rightleftharpoons Cd(I)_4^{2-}(aq) + 2\,OH^-(aq)$ $\quad K = 1.44 \times 10^{-8}$

	$[I^-]$	$[Cd(I)_4^{2-}]$	$[OH^-]$
Initial	0.10 M	0	0
Change	$-4S$	$+S$	$+2S$
Equil	$0.10 - 4S$	S	$2S$

$$K = 1.44 \times 10^{-8} = \frac{[Cd(I)_4^{2-}][OH^-]^2}{[I^-]^4} = \frac{(S)(2S)^2}{(0.10 - 4S)^4}$$ Assume that S << 0.10.

$$1.44 \times 10^{-8} = \frac{4S^3}{(0.10)^4}$$

$S = 7.11 \times 10^{-5}$; $[OH^-] = 2S = 2(7.01 \times 10^{-5}) = 1.42 \times 10^{-4}$

$pOH = -\log(1.42 \times 10^{-4}) = 3.85 = 3.9$ $pH = 14.0 - 3.9 = 10.1$

Conceptual Problems

23.78 M with ligand A forms a green solution. This means that the solution is absorbing red, which has a long wavelength, lower energy, and small Δ.
M with ligand B forms a violet solution. This means that the solution is absorbing yellow, which has a shorter wavelength and higher energy. This would have a larger Δ.
So ligand B results in the larger Δ.

23.79 Au would have the higher ionization energy. Ionization energy increases as you go down a group in the transition metals. Because there is a large increase in the number of protons and not a large increase in size, the ionization energy increases.

23.80 Fe^{3+} has a coordination number of 6 and is octahedral.
Fe $[Ar]4s^2 3d^6$
The octahedral splitting would be

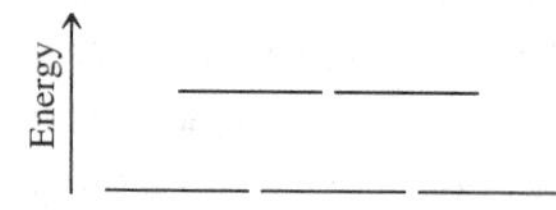

If the electrons were lost from 3*d* orbitals first, the electron configuration would be
Fe^{3+} $[Ar]4s^2 3d^3$ and there would be three *d* electrons.

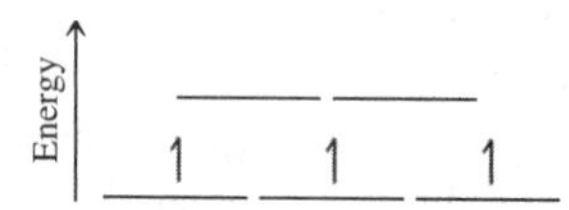

Because there are only three *d* electrons, it does not matter whether the complex is high-spin or low-spin; there will always be three unpaired electrons.
If the electrons were lost from the 4*s* orbitals first, the electron configuration would be
Fe^{3+} $[Ar]4s^0 3d^5$ and there would be five *d* electrons.
high-spin (all electrons unpaired) low-spin (electrons will pair)

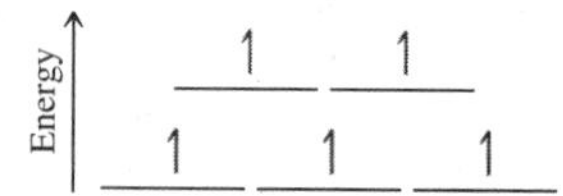

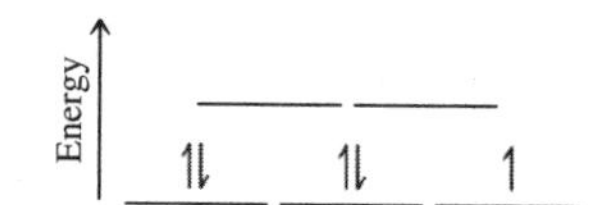

There are five unpaired electrons when the compound is high-spin and only one unpaired electron when the compound is low-spin. This influences the magnetic properties of the various complexes. This supports the idea that the electrons are lost from the 4*s* orbitals first.